MyMathLab Online Course
for *College Algebra with Intermediate Algebra: A Blended Course,* by Beecher/Penna/Johnson/Bittinger

Active Learning Figures

Active Learning Figures are interactive animations that allow students to examine visual representations of concepts through both guided and open-ended exploration. These are linked to chapter openers andvother locations throughout the text and MyMathLab. Accompanying exercises provide additional practice and reinforcement.

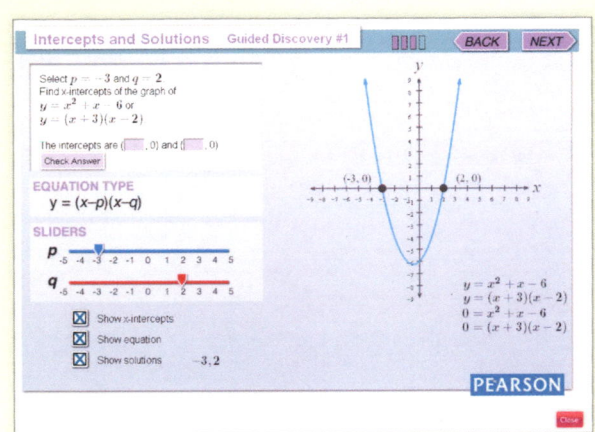

NEW! Guided Visualizations

These engaging interactive figures bring mathematical concepts to life, helping students visualize the concepts through directed explorations and purposeful manipulation. Excellent to use during lecture, *Guided Visualizations* are also assignable in MyMathLab with accompanying assessment to encourage active learning, critical thinking, and conceptual learning.

NEW! Enhanced Sample Assignments

These newly redesigned assignments put a powerful combination of author expertise and dynamic MyMathLab content at your fingertips. Comprised of author-selected exercises and our newest question types, including video assessment and interactive figures with assessment, these prebuilt assignments give you the best of the best for each section in a snap!

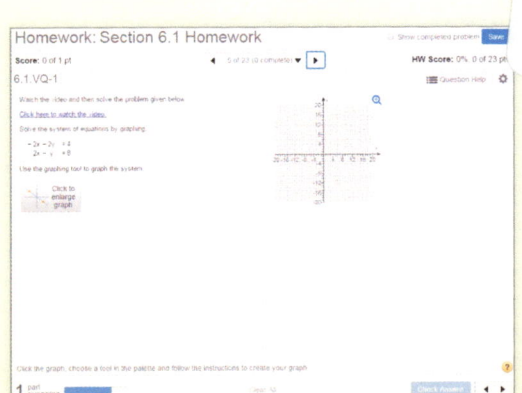

www.mymathlab.com

Connect the Concepts and Relate the Math

Learning Catalytics

Learning Catalytics is a "bring your own device" student engagement, assessment, and classroom intelligence system that can generate classroom discussion, guide your lecture, and promote peer-to-peer learning with real-time analytics. This resource is built into your MyMathLab course and will help boost student engagement, while motivating students to actively participate in their learning experience.

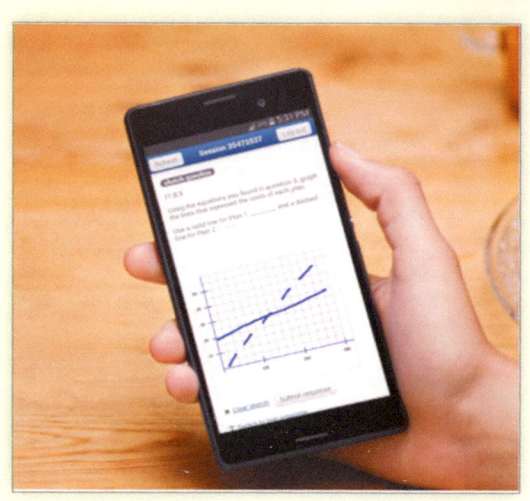

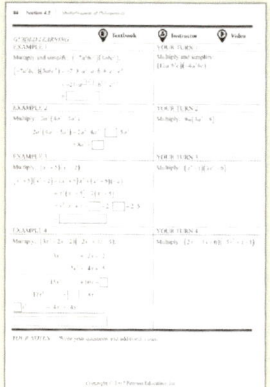

Video Notebook

The *Video Notebook* is an author-created notebook that offers fill-in-the-blank worksheets to accompany the section videos. Key vocabulary and definitions, theorems, and procedures are also included. After filling in the worksheet while watching the video, the student has an excellent study guide for review and test preparation. This is available in print and as a PDF in MyMathLab.

Skills for Success Modules

Get advice on how to succeed in college and prepare for future professions.
Activities and assignments are available for topics such as "Time Management," "Stress Management," and "Financial Literacy."

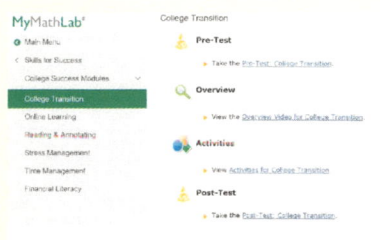

College Algebra with Intermediate Algebra: A Blended Course

Judith A. Beecher

Judith A. Penna

Barbara L. Johnson
Ivy Tech Community College of Indiana

Marvin L. Bittinger
Indiana University Purdue University Indianapolis

Pearson

Editorial Directors:	Chris Hoag, Marcia Horton
Editors in Chief:	Michael Hirsch, Anne Kelly
Aquisitions Editors:	Cathy Cantin, Chelsea Kharakozova
Editorial Assistants:	Alison Oehman, Ashley Gordon
Content Producer:	Ron Hampton
Media Producer:	Jonathan Wooding
TestGen Content Manager:	Marty Wright
MXL Content Manager:	Rebecca Williams
Marketing Manager:	Jennifer Edwards
Marketing Assistant:	Alexandra Habashi
Senior Author Support/ Technology Specialist:	Joe Vetere
Procurement Specialist:	Carol Melville
Cover Design:	Barbara Atkinson
Text Design, Art Editing, and Photo Research:	The Davis Group, Inc.
Composition:	Lumina Datamatics, Inc.
Illustrations:	Network Graphics, William Melvin, and Lumina Datamatics, Inc.
Cover Image:	Hailey P. Bell

Library of Congress Cataloging-in-Publication Data

Names: Beecher, Judith A.; Penna, Judith A.; Johnson, Barbara L.; Bittinger, Marvin L.
Title: College algebra with intermediate algebra : a blended course
Other titles: Intermediate algebra
   Description: 1st edition.; Boston : Pearson, [2017]
Identifiers: LCCN 2016020514; ISBN 9780134555263 (hardcover); ISBN 0134555260 (hardcover : student ed.)
   ISBN 9780134556505 (hardcover) |
   Subjects: LCSH: Algebra—Textbooks.
Classification: LCC QA154.3 .B44 2017 | DDC 512.9—dc23
   LC record available at https://lccn.loc.gov/2016020514

Contents

2   Graphs, Functions, and Applications   73

5 Rational Expressions, Equations, and Functions   289

6 Radical Expressions, Equations, and Functions 367

7 Quadratic Functions and Equations 437

8 — Polynomial Functions and Rational Functions   505

9 **Exponential Functions and Logarithmic Functions** **591**

10 **Matrices** **681**

Preface

▶ Algebra: A Streamlined Experience

We believe that helping students "see the math" is the key to success across the mathematics curriculum. With this in mind, we have developed an innovative new program, *College Algebra with Intermediate Algebra: A Blended Course,* that

- Focuses on visualization,
- Introduces functions and graphing early, and
- Makes connections between math concepts and the real world.

In intermediate algebra and college algebra courses today, the needs of instructors and students are changing:

- Many students enroll in the traditional two-course sequence,
- Others take co-requisite and accelerated courses, and
- Some enter college algebra without the firm grasp of prerequisite skills required for success in the course.

We designed *College Algebra with Intermediate Algebra: A Blended Course* to meet the needs of both instructors and students.

By eliminating the repetition of topics across the two-course sequence, we have created a streamlined course experience that makes better use of students' time and resources.

- Chapter R (Review of Basic Algebra) covers the prerequisite skills and concepts needed for the intermediate and college algebra material. Instructors can assign the chapter in an individualized instruction format (since students enter the course at various levels of math background), cover some or all of the chapter with the entire class at the beginning of the course, or refer students to it on a just-in-time basis as each topic is needed.
- With Chapter 2 (Graphs, Functions, and Applications), graphs and functions are introduced early and continue as a thread that runs through the course. This gives the instructor the opportunity to use the visual element of graphing to show students how solutions of equations, zeros of functions, and *x*-intercepts of graphs are related.
- Chapters 1 (Solving Linear Equations and Inequalities), 3 (Systems of Equations), 9 (Exponential Functions and Logarithmic Functions), and 11 (Conic Sections) deal with topics that are traditionally covered in both intermediate algebra and college algebra. Here, this repetition is eliminated.
- Chapters 4 (Polynomials and Polynomial Functions) and 5 (Rational Expressions, Equations, and Functions) provide thorough coverage of these topics, thus laying a solid foundation for the more advanced college algebra coverage of these topics in Chapter 8.
- Chapters 6 (Radical Expressions, Equations, and Functions) and 7 (Quadratic Functions and Equations) blend topics from intermediate algebra with those traditionally taught in college algebra.
- Chapters 10 (Matrices) and 12 (Sequences, Series, and Combinatorics) end the course with topics traditionally covered in college algebra.

Integrated review and reinforcement throughout helps students build the solid foundation of algebra skills and concepts that lead to success in math.

▶ Visualizing Concepts and Making Connections

Because the concept of a function can be challenging for students, we consistently use the language and notation of functions, visually relate functions to equations and graphs, and show how we use functions to model real data throughout the course. By introducing functions and graphs early (in Chapter 2), we are able to show both algebraic and graphical solutions to examples. Adding this element of visualization helps students quickly develop an understanding of the concepts.

- **Visualizing the Graph** asks students to match an equation with its graph by focusing on the characteristics of the equation and the corresponding attributes of the graph. Similarly, in the section exercise sets, students are often asked to match an equation with its graph or to determine the equation of a function by looking at its graph.
- **Translating for Success,** in the early chapters, gives students practice with translating applied problems to equations.
- **Active Learning Figures** and **Guided Visualizations** are interactive figures in MyMathLab that bring mathematical concepts to life, helping students visualize the concepts through directed explorations and purposeful manipulation.
- **Algebraic-Graphical Connections** and **Side-by-Side Examples** help students understand the connection between algebraic manipulation and the graphical interpretation of concepts.
- **Classify the Function** exercises ask students to identify a number of functions by their type (linear, quadratic, rational, etc.). Throughout the text, the variety of functions increases and these exercises become more challenging.
- **Technology Connections** demonstrate how a graphing calculator can be used to solve problems. Even if the student is not using graphing technology, the graphing calculator windows support visualization and reinforce conceptual understanding.
- **Connecting the Concepts,** for selected topics, invites students to stop and check their understanding of how concepts work together in one section or several sections. Concepts are summarized visually–using graphs, outlines, or charts–to help students make connections and reinforce their understanding of the concepts.

▶ Integrated Practice, Review, and Reinforcement

In the Text

Frequent opportunities for practice and review help students develop a solid understanding of algebra skills and concepts.

- **Now Try Exercises,** following most examples, direct students to work a similar problem in the section exercises for immediate reinforcement of the material just covered.
- **Mid-Chapter Reviews** offer mixed review exercise sets to help students reinforce their understanding of the concepts. Also included are **Collaborative Discussion and Writing** exercises for small group or class discussion of concepts.
- **Section Exercise Sets** cover concepts just presented and offer ongoing review of topics presented earlier in the course. **Vocabulary Exercises** check students' understanding of the language of mathematics and can serve as reading quizzes. **Skill Maintenance Exercises** allow students to review and reinforce previously learned material. **Synthesis Exercises** encourage critical thinking by asking students to apply multiple skills or concepts within a single exercise. Selected exercise sets start with a set of **Reading Check Exercises** designed to check the students' grasp of the concepts and skills in the section.
- **Chapter Summary and Reviews** offers comprehensive in-text practice and review. The **Study Guide,** with key concepts, terms, and examples, provides students with a concise and effective review of the chapter for test prep.
- **Chapter Tests** give students the opportunity to test themselves and target areas for further study before the class test.

In MyMathLab

With this blended course, we focused on MyMathLab features that reinforce visualization and conceptual understanding. **MyMathLab is closely integrated with the text** and offers a variety of question types for a robust online experience that mirrors our approach.

- **Videos**   The comprehensive video program offers topic coverage at the section, objective, and example levels, and students can use it hand-in-hand with the Video Notebook. Additionally, Chapter Test Prep videos let students watch an instructor work through step-by-step solutions to all the chapter test exercises in the textbook.
- **Video Notebook**   This author-created notebook contains fill-in-the-blank worksheets that accompany the example videos. Key vocabulary and definitions, theorems, and procedures are also included. After filling in the worksheet while watching the video, the student has an excellent study guide for review and test preparation. This is available in print to accompany the text or as a download in MyMathLab.
- **Active Learning Figures** and **Guided Visualizations** are interactive figures in MyMathLab that bring mathematical concepts to life, helping students visualize the concepts through directed explorations and purposeful manipulation. Guided Visualization figures are assignable in MyMathLab with accompanying assessment questions to guage conceptual understanding.
- **Activities** guide students to examine and connect key concepts while analyzing real-world data. These are available as PDFs in MyMathLab.
- **Adaptive Study Plan** for MyMathLab offers adaptive learning functionality that continuously analyzes students work and points them toward resources that will maximize their potential for understanding and success.
- **Skills for Success Modules** help students develop the good habits needed in the transition to college and then to professional life.
- **Enhanced Graphing Functionality** within the graphing utility allows graphing of 3-point quadratic functions, 4-point cubic functions, and transformations in exercises.
- **Suggested Homework Exercises** are preselected by the authors for each section. These are indicated in the Annotated Instructor's Edition by a blue underline within each end-of-section exercise set. These suggested exercises form the basis of the Sample Assignments in MyMathLab.
- **Sample Homework Assignments** are ready to assign in MyMathLab. Composed of author-selected exercises *and* our newest question types, including video assessment and interactive figures with assessment, these prebuilt assignments give you the best of the best for each section in a snap.
- **Learning Catalytics** is a "bring your own device" student engagement, assessment, and classroom-intelligence system that can generate classroom discussion, guide your lecture, and promote peer-to-peer learning with real-time analytics. This resource is built into your MyMathLab course and will help boost student engagement, while motivating students to actively participate in their learning experience.

▶ Acknowledgments

We wish to express our heartfelt thanks to a number of people who have contributed in vital ways to the development of this textbook. Our editors, Cathy Cantin and Chelsea Kharakozova, encouraged and supported our vision in introducing this new, blended approach to college algebra and its foundation in intermediate algebra. We appreciate the marketing insight provided by Jennifer Edwards, our marketing manager, and the support provided by the entire Pearson team, including Ron Hampton, content producer; Barbara Atkinson, cover designer; Alison Oehman and Ashley Gordon, editorial assistants; Alexandra Habashi, marketing assistant; Rebecca Williams, MathXL content manager; and Jon Wooding, media producer. We are also immensely grateful to Martha Morong for her editorial services and to Geri Davis for her text design and art editing.

The following reviewers made invaluable contributions to this text and we thank them for that:

Leslie Wenzel, *Garden City Community College*
Angela Everett, *Chattanooga State Community College*
Sandy Derry, *Butler Community College*
Ladorian Latin, *Franklin University*
Alvina Atkinson, *Georgia Gwinnett College*
Timothy Lucas, *Pepperdine University*
Elaine Clark, *University of New Mexico, Valencia*
Carol Abbott, *Ohio University, Lancaster*

Judith A. Beecher
Judith A. Penna
Barbara L. Johnson
Marvin L. Bittinger

Get the Most Out of MyMathLab®

MyMathLab is the leading online homework, tutorial, and assessment program for teaching and learning mathematics, built around Pearson's best-selling content. MyMathLab helps students and instructors improve results; it provides engaging experiences and personalized learning for each student so learning can happen in any environment. Plus, it offers flexible and time-saving course management features to allow instructors to easily manage their classes, regardless of course format.

Preparedness

One of the biggest challenges in many mathematics courses is making sure students are adequately prepared with the prerequisite skills needed to successfully complete their course work. MyMathLab offers a variety of content and course options to support students with just-in-time remediation and key-concept review.

In intermediate algebra and college algebra courses today, the needs of instructors and students are changing:

- Many students enroll in the traditional two-course sequence,
- Others take co-requisite accelerated courses, and
- Some enter college algebra without the firm grasp of prerequisite skills required for success in the course.

By eliminating the repetition of topics across the two-course sequence, we have created a streamlined course experience that makes better use of students' time and resources.

Adaptive Study Plan

The Study Plan makes studying more efficient and effective for every student. Performance and activity are assessed continually in real time. The data and analytics are used to provide personalized content—reinforcing concepts that target each student's strengths and weaknesses.

Used by more than 37 million students worldwide, MyMathLab delivers consistent, measurable gains in student learning outcomes, retention, and subsequent course success.

www.mymathlab.com

Resources for Success

MyMathLab® Online Course for
*College Algebra with Intermediate Algebra:
A Blended Course,* by Beecher/Penna/Johnson/Bittinger

Active Learning Figures

Active Learning Figures are interactive animations that allow students to examine visual representations of concepts through both guided and open-ended exploration. These are linked to chapter openers and other locations throughout the text and MyMathLab. Accompanying exercises provide additional practice and reinforcement.

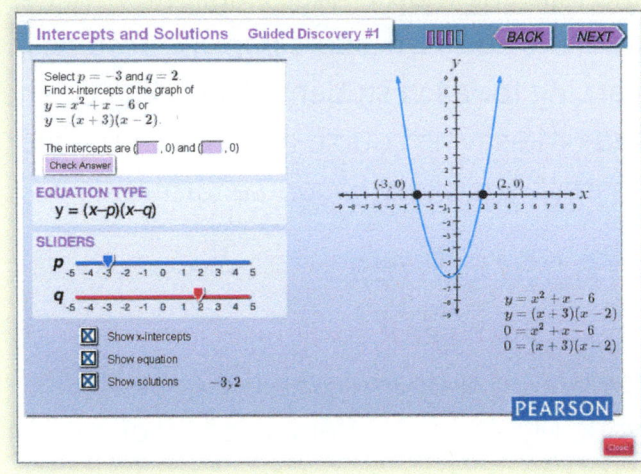

NEW! Guided Visualizations

These engaging interactive figures bring mathematical concepts to life, helping students visualize the concepts through directed explorations and purposeful manipulation. Excellent to use during lecture, *Guided Visualizations* are also assignable in MyMathLab with accompanying assessment to encourage active learning, critical thinking, and conceptual learning.

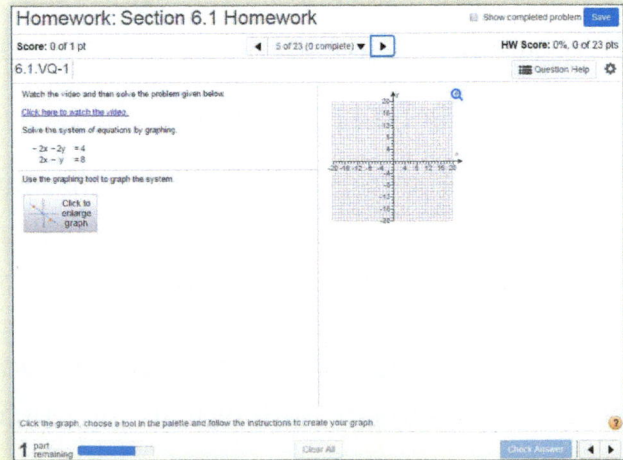

NEW! Enhanced Sample Assignments

These newly redesigned assignments put a powerful combination of author expertise and dynamic MyMathLab content at your fingertips. Comprising author-selected exercises and our newest question types, including video assessment and interactive figures with assessment, these prebuilt assignments give you the best of the best for each section in a snap!

www.mymathlab.com

Resources for Success

Instructor Resources

These resources can be downloaded from within MyMathLab and at www.pearsonhighered.com.

Annotated Instructor's Edition

The instructor's edition includes all answers to the exercise sets. Shorter answers are presented on the same page as the exercise; longer answers are in the back of the text. Sample homework assignments are indicated by a blue underline and may be assigned in MyMathLab. Available upon request from your Pearson sales representative.

Instructor's Solutions Manual (download only)

Written by Judy Penna, this resource contains solutions to even-numbered exercises in the exercise sets, and all solutions to Mid-Chapter Reviews and end-of-chapter exercises, including Chapter Tests and Chapter Reviews.

PowerPoint® Lecture Slides

Written and designed specifically for this text, these lectures slide provide an outline for presenting definitions, figures, and key examples from the text.

Instructor's Online Test Bank (download only)

For each chapter, contains four free-response and two multiple-choice test forms following the same format and having the same level of difficulty as the test in the text. It also provides four free-response and two multiple-choice forms of the final examination.

TestGen®

TestGen® (www.pearsoned.com/testgen) enables instructors to build, edit, print, and administer tests using a computerized bank of questions developed to cover all the objectives of the text.

Student Resources

These additional resources are designed to promote student success.

Video Notebook

This notebook can accompany the text and/or MyMathLab course. It contains fill-in-the-blank worksheets to accompany the video examples presented by the authors. Key definitions, theorems, and procedures are also included. After filling in the worksheet while watching the video, the student has an excellent study guide for review and test preparation. This is available for download within MyMathLab or as a printed resource that can be bundled with the text.

Comprehensive Video Program

The comprehensive video program offers topic coverage at the section, objective, and example levels, and students can use it hand-in-hand with the Video Notebook. Additionally, Chapter Test Prep videos let students watch an instructor work through step-by-step solutions to all the chapter test exercises in the textbook.

Student Solutions Manual

This resource, written by Judy Penna, contains worked-out solutions with step-by-step annotations for all odd-numbered exercises in the text exercise sets, as well as solutions for all the Mid-Chapter Review, Chapter Review, and Chapter Test exercises.

www.mymathlab.com

To the Student

GUIDE TO SUCCESS

Success can be planned. Combine goals and good study habits to create a plan for success that works for you. The following list contains study tips that we consider most helpful.

Skills for Success

- ▶ **Set goals and expect success.** Approach this class experience with a positive attitude.
- ▶ **Communicate with your instructor** when you need extra help.
- ▶ **Take your text with you to class and lab.** Each section in the text is designed with headings and boxed information that provide an outline for easy reference.
- ▶ **Ask questions in class, lab, and tutoring sessions.** Instructors encourage them, and other students probably have the same questions.
- ▶ **Carefully read the instructions** before working homework exercises **and include all steps.**
- ▶ **Form a study group** with fellow students. Verbalizing questions about topics that you do not understand can clarify the material for you.
- ▶ After each quiz or test, **write out corrected step-by-step solutions** to all missed questions. They will provide a valuable study guide for the midterm exam and the final exam.
- ▶ **MyMathLab has numerous tools to help you succeed.** Use MyMathLab to create a personalized study plan and practice skills with sample quizzes and tests. See pages xvii–xix for more information about additional resources available to you in MyMathLab.
- ▶ **Knowing math vocabulary is an important step toward success.** Review Vocabulary Exercises in the text and in MyMathLab.
- ▶ If you miss a lecture, **watch the video in the Multimedia Library** of MyMathLab that explains the concepts you missed.

In writing this textbook, we challenged ourselves to do everything possible to help you learn the concepts and skills contained between its covers so that you will be successful in this course and in the mathematics courses you take in the future. We realize that your time is both valuable and limited, so we communicate in a highly visual way that allows you to learn quickly and efficiently. We are confident that, if you invest an adequate amount of time in the learning process, this text will be of great value to you. We wish you a positive learning experience.

Judy Beecher
Judy Penna
Barbara Johnson
Marv Bittinger

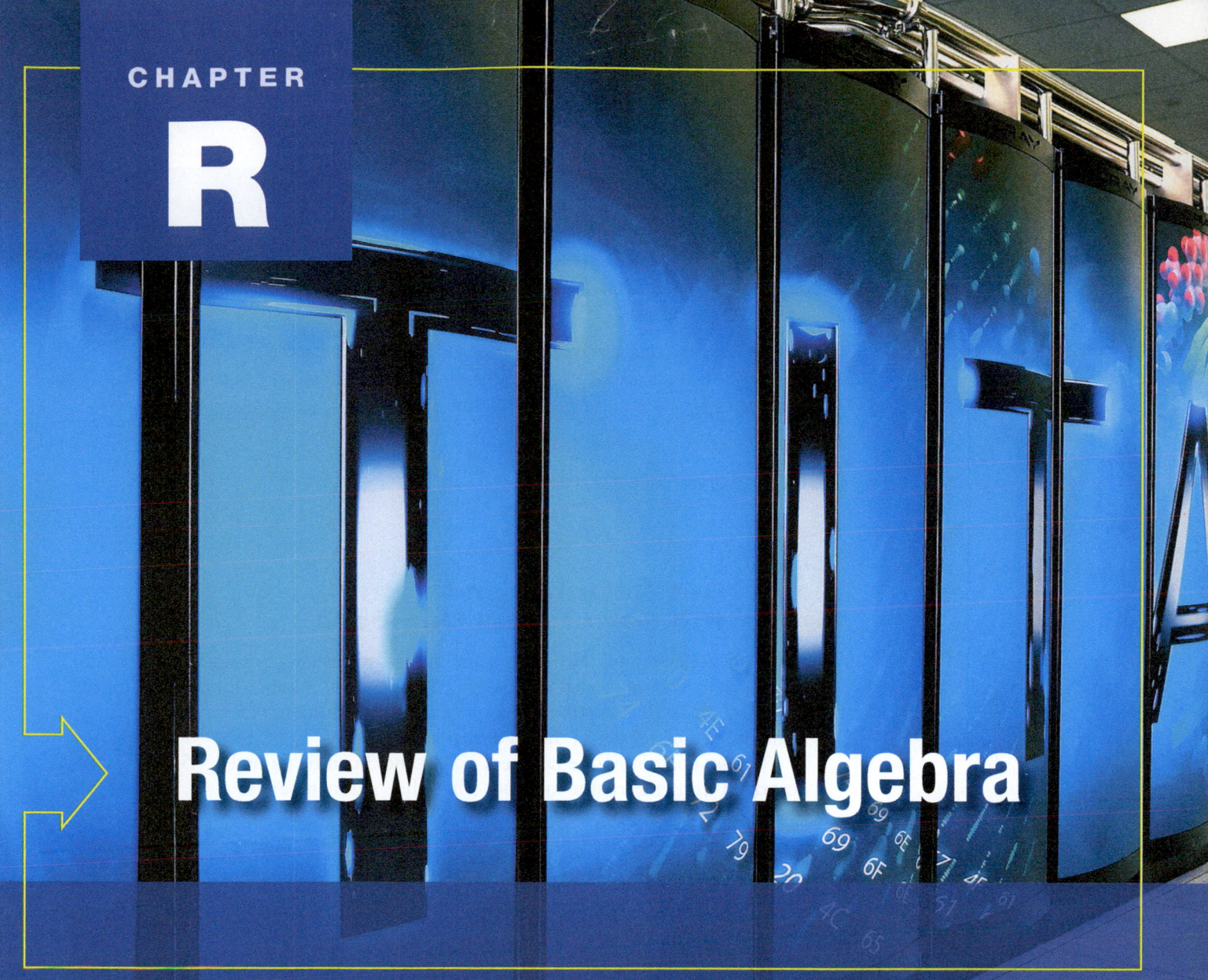

APPLICATION
This problem appears as Exercise 107 in Section R.7.

The Titan supercomputer can perform 20,000 trillion calculations per second. (*Source*: "Supercomputer Titan; 20,000 trillion calculations per second," by Evelyn Laeschke. October 29, 2012, on ip-192.com) How many calculations can be performed in one minute? in one hour?

R.1

PART 1 OPERATIONS
The Set of Real Numbers

▶ **a** Use roster notation and set-builder notation to name sets, and distinguish among various kinds of real numbers.

▶ **b** Determine which of two real numbers is greater and indicate which, using $<$ and $>$; given an inequality like $a < b$, write another inequality with the same meaning; and determine whether an inequality like $-2 \leq 3$ or $4 > 5$ is true.

▶ **c** Graph inequalities on the number line.

▶ **d** Find the absolute value of a real number.

▶ a Set Notation and the Set of Real Numbers

A **set** is a collection of objects. In mathematics, we usually consider sets of numbers, such as the set of **real numbers**. There is a real number for every point on the real-number line. A **subset** is a set contained within another set. We begin by examining some subsets of the set of real numbers.

The set containing the numbers $-5, 0,$ and 3 can be named $\{-5, 0, 3\}$. It is described using the **roster method**, which lists all members of a set. We use the roster method to describe three frequently used subsets of real numbers. Note that three dots are used to indicate that the pattern continues without end.

NATURAL NUMBERS, WHOLE NUMBERS, AND INTEGERS

Natural numbers are those numbers used for counting: $\{1, 2, 3, \ldots\}$.

Whole numbers are the set of natural numbers with 0 included: $\{0, 1, 2, 3, \ldots\}$.

Integers are the set of whole numbers and their opposites:

$$\{\ldots, -4, -3, -2, -1, 0, 1, 2, 3, 4, \ldots\}.$$

Natural numbers are also called **counting numbers**.

The integers can be illustrated on the number line as follows.

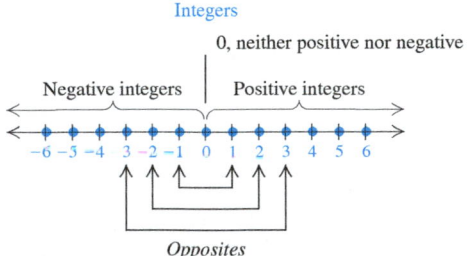

Opposites

The set of integers extends infinitely to the left and to the right of 0. The **opposite** of a number is found by reflecting it across the number 0. Thus the opposite of 3 is -3. The opposite of -4 is 4. The opposite of 0 is 0. We read a symbol like -3 as either "the opposite of 3" or "negative 3."

The natural numbers are called **positive integers**. The opposites of the natural numbers (those to the left of 0) are called **negative integers**. Zero is neither positive nor negative.

Other subsets of real numbers are described using **set-builder notation**. With this notation, instead of listing all members of a set, we specify conditions under which a number is in a set. For example, the set of all odd natural numbers less than 9 can be described and read as follows:

$$\{x \,|\, x \text{ is an odd number less than } 9\}.$$

The set of all x such that x is an odd number less than 9

Using roster notation, we can write this set as $\{1, 3, 5, 7\}$.

EXAMPLE 1   Name the set consisting of the first six even whole numbers using both roster notation and set-builder notation.

Roster notation: $\{0, 2, 4, 6, 8, 10\}$

Set-builder notation: $\{x \,|\, x \text{ is one of the first six even whole numbers}\}$

Now Try Exercises 13 and 19.

We can now describe the set of **rational numbers**.

RATIONAL NUMBERS

A **rational number** can be expressed as an integer divided by a nonzero integer. The set of rational numbers is

$$\left\{\frac{p}{q} \,\middle|\, p \text{ is an integer}, q \text{ is an integer, and } q \neq 0\right\}.$$

The decimal representation of rational numbers either terminates or has a repeating block of digits.

The following are examples of rational numbers:

$$\frac{5}{8}, \quad \frac{12}{-7}, \quad \frac{-17}{15}, \quad -\frac{9}{7}, \quad \frac{39}{1}, \quad \frac{0}{6}.$$

Note that $\frac{39}{1} = 39$. Thus the set of rational numbers contains the integers. Using long division, we can write a fraction in decimal notation:

$$\frac{5}{8} = 0.625 \quad \text{and} \quad \frac{6}{11} = 0.545454\ldots = 0.\overline{54}.$$

Terminating   **Repeating**

The bar in $0.\overline{54}$ indicates the repeating block of digits in decimal notation.

The number line has a point for every rational number.

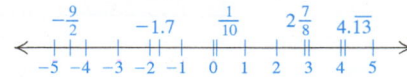

However, there are many points on the line for which there is no rational number. These points correspond to what are called **irrational numbers**.

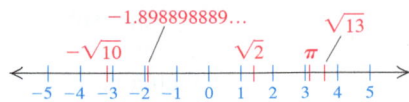

Numbers like π, $\sqrt{2}$, $-\sqrt{10}$, $\sqrt{13}$, and $-1.898898889\ldots$ are examples of irrational numbers. The decimal notation for an irrational number *neither* terminates *nor* repeats. Recall that decimal notation for rational numbers either terminates or has a repeating block of digits.

IRRATIONAL NUMBERS

Irrational numbers are numbers whose decimal representation neither terminates nor has a repeating block of digits. They cannot be represented as the quotient of two integers.

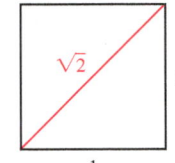

The irrational number $\sqrt{2}$ (read "the square root of 2") is the length of the diagonal of a square with sides of length 1. It is also the number that, when multiplied by itself, gives 2. No rational number can be multiplied by itself to get 2, although some approximations come close:

1.4 is an *approximation* of $\sqrt{2}$ because
$(1.4)^2 = (1.4)(1.4) = 1.96$;

1.41 is a better approximation because
$(1.41)^2 = (1.41)(1.41) = 1.9881$;

1.4142 is an even better approximation because
$(1.4142)^2 = (1.4142)(1.4142) = 1.99996164$.

We say that 1.4142 is a rational approximation of $\sqrt{2}$ because

$$(1.4142)^2 = 1.99996164 \approx 2.$$

The symbol $\approx$ means "is approximately equal to." We can find rational approximations for square roots and other irrational numbers using a calculator.

The set of all rational numbers, combined with the set of all irrational numbers, gives us the set of **real numbers**.

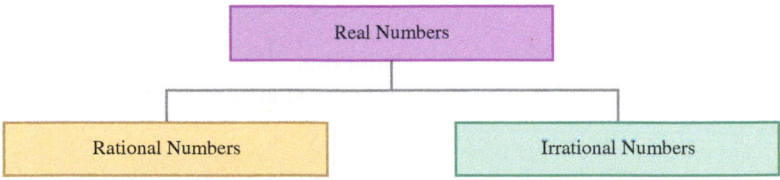

REAL NUMBERS

The set of **real numbers** is

$\{x \mid x$ is a rational number *or* x is an irrational number$\}$.

Every point on the number line represents some real number and every real number is represented by some point on the number line.

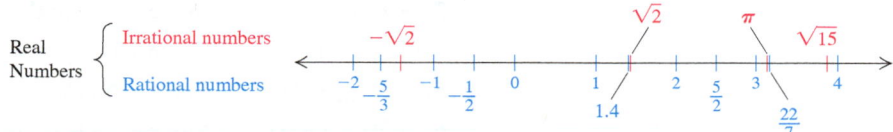

The following figure shows the relationships among various sets of real numbers.

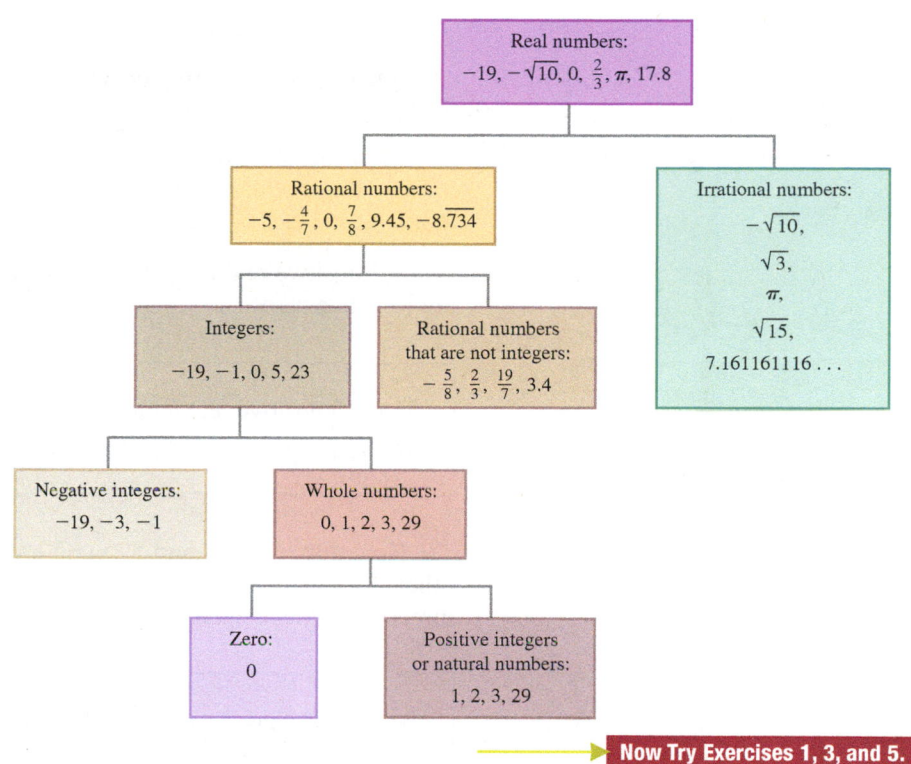

Now Try Exercises 1, 3, and 5.

▶ b   Order for the Real Numbers

Real numbers are named in order on the number line. For any two numbers on the line, the one to the left is less than the one to the right.

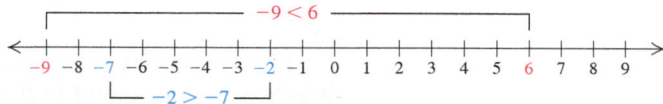

We use the symbol $<$ to mean "**is less than.**" The sentence $-9 < 6$ means "-9 is less than 6." The symbol $>$ means "**is greater than.**" The sentence $-2 > -7$ means "-2 is greater than -7." Sentences containing $<$ or $>$ are called **inequalities**.

EXAMPLES Use either $<$ or $>$ for $\square$ to write a true sentence.

2. $4 \; \square \; 9$ Since 4 is to the left of 9, 4 is less than 9, so $4 \; < \; 9$.

3. $-8 \; \square \; 3$ Since -8 is to the left of 3, we have $-8 \; < \; 3$.

4. $7 \; \square \; -12$ Since 7 is to the right of -12, then $7 \; > \; -12$.

5. $-5 \; \square \; -21$ Since -5 is to the right of -21, we have $-5 \; > \; -21$.

6. $4.79 \; \square \; 4.97$ Since 4.79 is to the left of 4.97, we have $4.79 \; < \; 4.97$.

7. $-2.7 \; \square \; -\dfrac{3}{2}$ Since $-\dfrac{3}{2} = -1.5$ and -2.7 is to the left of -1.5, we have

$$-2.7 \; < \; -\frac{3}{2}.$$

8. $\dfrac{5}{8} \; \square \; \dfrac{7}{11}$ We convert to decimal notation $\left(\dfrac{5}{8} = 0.625 \text{ and } \right.$

$\left. \dfrac{7}{11} = 0.6363 \ldots \right)$ and compare. Thus, $\dfrac{5}{8} \; < \; \dfrac{7}{11}$.

> **Now Try Exercises 31 and 37.**

All positive real numbers are greater than zero and all negative real numbers are less than zero.

> If x is a positive real number, then $x > 0$.
>
> If x is a negative real number, then $x < 0$.

Note that $-8 < 5$ and $5 > -8$ are both true. Every true inequality yields another true inequality if we interchange the numbers or variables and reverse the direction of the inequality sign.

> $a < b$ also has the meaning $b > a$.

EXAMPLES Write a different inequality with the same meaning.

9. $a < -5$ The inequality $-5 > a$ has the same meaning.

10. $-3 > -8$ The inequality $-8 < -3$ has the same meaning.

> **Now Try Exercise 41.**

Expressions like $a \leq b$ and $b \geq a$ are also **inequalities**. We read $\boldsymbol{a \leq b}$ as "**$\boldsymbol{a}$ is less than or equal to $\boldsymbol{b}$**"; $a \leq b$ is true if $a < b$ or if $a = b$. We read $\boldsymbol{a \geq b}$ as "**$\boldsymbol{a}$ is greater than or equal to $\boldsymbol{b}$**." If a is nonnegative, then $a \geq 0$.

EXAMPLES Determine whether each of the following is true or false.

11. $-8 \leq 5.7$ True since $-8 < 5.7$ is true.

12. $-8 \leq -8$ True since $-8 = -8$ is true.

13. $-7 \geq 4\frac{1}{3}$ False since neither $-7 > 4\frac{1}{3}$ nor $-7 = 4\frac{1}{3}$ is true.

14. $-\frac{2}{3} \geq -\frac{5}{4}$ True since $-\frac{2}{3} = -0.666\ldots$ and $-\frac{5}{4} = -1.25$ and $-0.666\ldots > -1.25$.

> **Now Try Exercises 45 and 47.**

▶ c Graphing Inequalities on the Number Line

A replacement for a variable that makes an inequality true is called a **solution** of the inequality. The set of all solutions is called the **solution set**. A **graph** of an inequality is a drawing that represents its solution set.

EXAMPLE 15 Graph the inequality $x > -3$ on the number line.

The solutions consist of all real numbers greater than -3, so we shade all numbers greater than -3. Since -3 is not a solution, we use a parenthesis at -3. The graph represents the solution set $\{x \mid x > -3\}$.

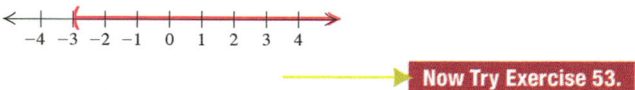

Now Try Exercise 53.

EXAMPLE 16 Graph the inequality $x \leq 2$ on the number line.

We make a drawing that represents the solution set $\{x \mid x \leq 2\}$. The graph consists of 2 as well as the numbers less than 2. We shade all numbers to the left of 2 and use a bracket at 2 to indicate that it is also a solution.

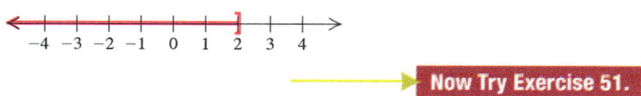

Now Try Exercise 51.

▶ d Absolute Value

We call the distance of a number from 0 on the number line the **absolute value** of the number. Since distance is always a nonnegative number, the absolute value of a number is always greater than or equal to 0.

The distance from -6 to 0 is 6.
The absolute value of -6 is 6.

The distance from 6 to 0 is 6.
The absolute value of 6 is 6.

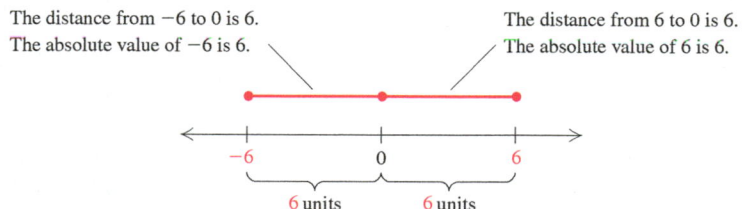

ABSOLUTE VALUE

The **absolute value** of a number is its distance from 0 on the number line. We use the symbol $|x|$ to represent the absolute value of a number x.

EXAMPLES Find the absolute value.

17. $|-7|$ The distance of -7 from 0 is 7, so $|-7|$ is 7.

18. $|12|$ The distance of 12 from 0 is 12, so $|12|$ is 12.

19. $|0|$ The distance of 0 from 0 is 0, so $|0|$ is 0.

20. $\left|\dfrac{4}{5}\right| = \dfrac{4}{5}$

21. $|-3.86| = 3.86$

Now Try Exercises 59 and 63.

R.1   Exercise Set

▶ **Reading Check**

Choose from the column on the right the set of numbers that matches the description.

RC1. ___ Natural numbers

RC2. ___ Whole numbers

RC3. ___ Integers

RC4. ___ Rational numbers

RC5. ___ Irrational numbers

RC6. ___ Real numbers

a) $\{\ldots, -3, -2, -1, 0, 1, 2, 3, \ldots\}$

b) $\{0, 1, 2, 3, \ldots\}$

c) $\{1, 2, 3, 4, \ldots\}$

d) $\{x \,|\, x \text{ is a rational number } or \text{ } x \text{ is an irrational number}\}$

e) $\{x \,|\, x \text{ cannot be represented as the quotient of two integers}\}$

f) $\left\{\dfrac{p}{q} \,\middle|\, p \text{ is an integer, } q \text{ is an integer, and } q \neq 0\right\}$

a *Given the numbers* $-6, 0, 1, -\frac{1}{2}, -4, \frac{7}{9}, 12, -\frac{6}{5},$ *$3.45, 5\frac{1}{2}, \sqrt{3}, \sqrt{25}, -\frac{12}{3}, 0.131331333133331 \ldots :$*

1. Name the natural numbers.

2. Name the whole numbers.

3. Name the rational numbers.

4. Name the integers.

5. Name the real numbers.

6. Name the irrational numbers.

Given the numbers $-\sqrt{5}, -3.43, -11, 12, 0,$ *$\frac{11}{34}, -\frac{7}{13}, \pi, -3.565665666566665 \ldots :$*

7. Name the whole numbers.

8. Name the natural numbers.

9. Name the integers.

10. Name the rational numbers.

11. Name the irrational numbers.

12. Name the real numbers.

Use roster notation to name each set.

13. The set of all letters in the word "math"

14. The set of all letters in the word "solve"

15. The set of all positive integers less than 13

16. The set of all odd whole numbers less than 13

17. The set of all even natural numbers

18. The set of all negative integers greater than -4

Use set-builder notation to name each set.

19. $\{0, 1, 2, 3, 4, 5\}$

20. $\{4, 5, 6, 7, 8, 9, 10\}$

21. The set of all rational numbers

22. The set of all real numbers

23. The set of all real numbers greater than -3

24. The set of all real numbers less than or equal to 21

b *Use either $<$ or $>$ for $\square$ to write a true sentence.*

25. $13 \;\square\; 0$

26. $18 \;\square\; 0$

27. $-8 \;\square\; 2$

28. $7 \;\square\; -7$

29. $-8 \;\square\; 8$

30. $0 \;\square\; -11$

31. $-8 \;\square\; -3$

32. $-6 \;\square\; -3$

33. $-2 \;\square\; -12$

34. $-7 \;\square\; -10$

35. $-9.9 \;\square\; -2.2$

36. $-13\frac{1}{5} \;\square\; \frac{11}{250}$

37. $37\frac{1}{5} \;\square\; -1\frac{67}{100}$

38. $-13.99 \;\square\; -8.45$

39. $\frac{6}{13} \;\square\; \frac{13}{25}$

40. $-\frac{14}{15} \;\square\; -\frac{27}{53}$

Write a different inequality with the same meaning.

41. $-8 > x$

42. $x < 7$

43. $-12.7 \leq y$

44. $10\frac{2}{3} \geq t$

Write true or false.

45. $6 \leq -6$

46. $-7 \leq -7$

47. $5 \geq -8.4$

48. $-11 \geq -13\frac{1}{2}$

c *Graph each inequality on the number line.*

49. $x < -2$

50. $x < -1$

51. $x \leq -2$

52. $x \geq -1$

53. $x > -3.3$

54. $x < 0$

55. $x \geq 2$

56. $x \leq 0$

d *Find the absolute value.*

57. $|-6|$

58. $|-3|$

59. $|28|$

60. $|16|$

61. $|-35|$

62. $|-127|$

63. $\left|-\frac{2}{3}\right|$

64. $\left|-\frac{13}{8}\right|$

65. $|42.8|$

66. $|16.4|$

67. $|986|$

68. $|465|$

69. $\left|\frac{0}{-7}\right|$

70. $\left|\frac{0}{-15}\right|$

▶ Synthesis

To the student and the instructor: *The Synthesis exercises found at the end of every exercise set challenge students to combine concepts or skills studied in that section or in preceding parts of the text.*

Use either $\leq$ *or* $\geq$ *for* $\square$ *to write a true sentence.*

71. $|-3| \,\square\, 5$

72. $|-5| \,\square\, |-2|$

73. $|4| \,\square\, |-7|$

74. $|-8| \,\square\, |8|$

75. List the following numbers in order from least to greatest.

$$\frac{1}{11}, \ 1.1\%, \ \frac{2}{7}, \ 0.3\%, \ 0.11, \ \frac{1}{8}\%, \ 0.009,$$

$$\frac{99}{1000}, \ 0.286, \ \frac{1}{8}, \ 1\%, \ \frac{9}{100}$$

R.2   Operations with Real Numbers

▶ **a**   Add real numbers.

▶ **b**   Find the opposite, or additive inverse, of a number.

▶ **c**   Subtract real numbers.

▶ **d**   Multiply real numbers.

▶ **e**   Divide real numbers.

We now review addition, subtraction, multiplication, and division of real numbers.

▶ a   Addition

To gain an understanding of addition of real numbers, we first add using the number line. To find $a + b$ using the number line, we start at 0, move to a, and then move according to b.

- If b is positive, move to the right.
- If b is negative, move to the left.
- If b is 0, stay at a.

EXAMPLES

1. $6 + (-8) = -2$: We begin at 0 and move 6 units right since 6 is positive. Then we move 8 units left since -8 is negative. The answer is -2.

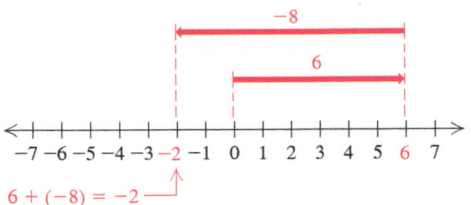

2. $-3 + 7 = 4$: We begin at 0 and move 3 units left since -3 is negative. Then we move 7 units right since 7 is positive. The answer is 4.

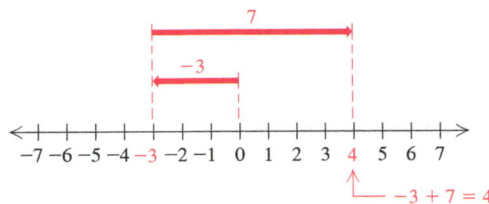

3. $-2 + (-5) = -7$: We begin at 0 and move 2 units left since -2 is negative. Then we move 5 units further left since -5 is negative. The answer is -7.

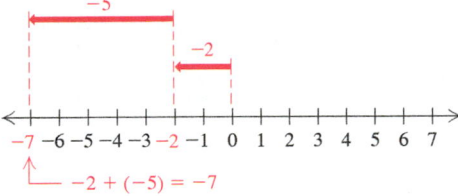

You may have noticed some patterns in the preceding examples. These lead us to rules for adding without using the number line.

Rules for Addition of Real Numbers

1. *Positive numbers*: Add the numbers. The result is positive.
2. *Negative numbers*: Add absolute values. Make the answer negative.
3. *A positive number and a negative number*:
   - If the numbers have the same absolute value, the answer is 0.
   - If the numbers have different absolute values, subtract the smaller absolute value from the larger. Then:
     a) If the positive number has the greater absolute value, make the answer positive.
     b) If the negative number has the greater absolute value, make the answer negative.
4. *One number is zero*: The sum is the other number.

Rule 4 is known as the **identity property of 0**. It says that for any real number a, $a + 0 = a$.

EXAMPLES   Add without using the number line.

4. $-13 + (-8) = -21$ Two negatives. Add the absolute values: $|-13| + |-8| = 13 + 8 = 21$. Make the answer *negative*: -21.

5. $-2.1 + 8.5 = 6.4$ One negative, one positive. Subtract the smaller absolute value from the larger: $8.5 - 2.1 = 6.4$. The *positive* number, 8.5, has the larger absolute value, so the answer is *positive*, 6.4.

6. $-48 + 31 = -17$ One negative, one positive. Subtract the smaller absolute value from the larger: $48 - 31 = 17$. The *negative* number, -48, has the larger absolute value, so the answer is *negative*, -17.

7. $2.6 + (-2.6) = 0$ One positive, one negative. The numbers have the same absolute value. The sum is 0.

8. $-\dfrac{5}{9} + 0 = -\dfrac{5}{9}$ One number is zero. The sum is $-\frac{5}{9}$.

9. $-\dfrac{2}{3} + \dfrac{5}{8} = -\dfrac{16}{24} + \dfrac{15}{24} = -\dfrac{1}{24}$ **Now Try Exercises 1, 9, and 21.**

▶ b   Opposites, or Additive Inverses

Suppose we add two numbers that are **opposites**, such as 4 and -4. The result is 0. When opposites are added, the result is always 0. Opposites are also called **additive inverses**. Every real number has an opposite, or additive inverse.

OPPOSITES, OR ADDITIVE INVERSES

Two numbers whose sum is 0 are called **opposites**, or **additive inverses**, of each other.

EXAMPLES   Find the opposite, or additive inverse, of each number.

10. 8.6   The opposite of 8.6 is -8.6 because $8.6 + (-8.6) = 0$.

11. 0   The opposite of 0 is 0 because $0 + 0 = 0$.

12. $-\frac{7}{9}$ The opposite of $-\frac{7}{9}$ is $\frac{7}{9}$ because $-\frac{7}{9} + \frac{7}{9} = 0$.

Now Try Exercise 29.

To name the opposite, or additive inverse, we use the symbol $-$, and read the symbolism $-a$ as "the opposite of a" or "the additive inverse of a."

CAUTION! A symbol such as -8 is usually read "negative 8." It could be read "the opposite of 8," because the opposite of 8 is -8. It could also be read "the additive inverse of 8," because the additive inverse of 8 is -8. When a variable is involved, as in a symbol like $-x$, it can be read "the opposite of x" or "the additive inverse of x" but *not* "negative x," because we do not know whether the symbol represents a positive number, a negative number, or 0. It is never correct to read -8 as "minus 8."

OPPOSITES, OR ADDITIVE INVERSES

For any real number a, the **opposite**, or **additive inverse**, of a, which is $-a$, is such that

$$a + (-a) = (-a) + a = 0.$$

EXAMPLE 13 Evaluate $-x$ and $-(-x)$ **(a)** when $x = 23$ and **(b)** when $x = -5$.

a) If $x = 23$, then $-x = -23 = -23$.   The opposite of 23 is -23.
   If $x = 23$, then $-(-x) = -(-23) = 23$.   The opposite of the opposite of 23 is 23.

b) If $x = -5$, then $-x = -(-5) = 5$.
   If $x = -5$, then $-(-x) = -(-(-5)) = -(5) = -5$.

                                            Now Try Exercise 25.

Note in Example 13(b) that an extra set of parentheses is used to show that we are substituting the negative number -5 for x. Symbolism like $--x$ is not considered meaningful.

Signs of Numbers

A negative number is sometimes said to have a "negative sign." A positive number is said to have a "positive sign." When we replace a number with its opposite, or additive inverse, we can say that we have "changed its sign."

EXAMPLES Change the sign. (Find the opposite, or additive inverse.)

14. -3 $-(-3) = 3$
15. $-\frac{3}{8}$ $-\left(-\frac{3}{8}\right) = \frac{3}{8}$
16. 0 $-0 = 0$
17. 14 $-(14) = -14$

We can now give a more formal definition of absolute value.

ABSOLUTE VALUE

For any real number a, the **absolute value** of a, denoted $|a|$, is given by

$$|a| = \begin{cases} a, & \text{if } a \geq 0, \\ -a, & \text{if } a < 0. \end{cases}$$
      For example, $|8| = 8$ and $|0| = 0$.
      For example, $|-5| = -(-5) = 5$.

(The absolute value of a is a if a is nonnegative. The absolute value of a is the opposite of a if a is negative.)

▶ c Subtraction

SUBTRACTION

The difference $a - b$ is the unique number c for which $a = b + c$.
That is, $a - b = c$ if c is the number such that $a = b + c$.

For example, $3 - 5 = -2$ because $3 = 5 + (-2)$. That is, -2 is the number that when added to 5 gives 3. Although this illustrates the formal definition of subtraction, we generally use the following when we subtract.

SUBTRACTING BY ADDING THE OPPOSITE

For any real numbers a and b,

$$a - b = a + (-b).$$

(We can subtract by adding the opposite (additive inverse) of the number being subtracted.)

EXAMPLES Subtract.

18. $3 - 5 = 3 + (-5) = -2$ **Changing the sign of 5 and adding**

19. $7 - (-3) = 7 + (3) = 10$ **Changing the sign of -3 and adding**

20. $-19.4 - 5.6 = -19.4 + (-5.6) = -25$

21. $-\dfrac{4}{3} - \left(-\dfrac{2}{5}\right) = -\dfrac{4}{3} + \dfrac{2}{5} = -\dfrac{20}{15} + \dfrac{6}{15} = -\dfrac{14}{15}$

> **Now Try Exercises 39, 41, and 51.**

▶ d Multiplication

We already know how to multiply two nonnegative numbers. To see how to multiply a positive number and a negative number, consider the following pattern in which multiplication is regarded as repeated addition:

This $4(-5) = (-5) + (-5) + (-5) + (-5) = -20$ This
number → $3(-5) = \qquad (-5) + (-5) + (-5) = -15$ ← number
decreases $2(-5) = \qquad\qquad (-5) + (-5) = -10$ increases
by 1 each $1(-5) = \qquad\qquad\qquad (-5) = -5$ by 5 each
time. $0(-5) = \qquad\qquad\qquad\qquad 0 = 0$ time.

This pattern illustrates that the product of a negative number and a positive number is negative.

THE PRODUCT OF A POSITIVE NUMBER AND A NEGATIVE NUMBER

To multiply a positive number and a negative number, multiply their absolute values. Then make the answer negative.

EXAMPLES Multiply.

22. $-3 \cdot 5 = -15$

23. $6 \cdot (-7) = -42$

24. $(-1.2)(4.5) = -5.4$

25. $3 \cdot \left(-\frac{1}{2}\right) = \frac{3}{1} \cdot \left(-\frac{1}{2}\right) = -\frac{3}{2}$

Note in Example 24 that the parentheses indicate multiplication.

> **Now Try Exercises 57 and 67.**

We can extend the above pattern still further to examine the product of two negative numbers.

This number decreases by →
1 each time.

$$2(-5) = (-5) + (-5) = -10$$
$$1(-5) = (-5) = -5$$
$$0(-5) = 0 = 0$$
$$-1(-5) = -(-5) = 5$$
$$-2(-5) = -(-5)-(-5) = 10$$

This number
← increases by
5 each time.

According to the pattern, the product of two negative numbers is positive.

THE PRODUCT OF TWO NEGATIVE NUMBERS

To multiply two negative numbers, multiply their absolute values. The answer is positive.

EXAMPLES Multiply.

26. $-3 \cdot (-5) = 15$

27. $-5.2(-10) = 52$

28. $(-8.8)(-3.5) = 30.8$

29. $\left(-\frac{3}{4}\right) \cdot \left(-\frac{5}{2}\right) = \frac{15}{8}$

> **Now Try Exercises 61 and 73.**

▶ **e Division**

DIVISION

The quotient $a \div b$, or $\dfrac{a}{b}$, where $b \neq 0$, is that unique real number c for which $a = b \cdot c$.

Using this definition and the rules for multiplying, we can see how to handle signs when dividing.

EXAMPLES Divide.

30. $\dfrac{10}{-2} = -5$, because $-5 \cdot (-2) = 10$

31. $\dfrac{-32}{4} = -8$, because $-8 \cdot (4) = -32$

32. $-25 \div (-5) = 5$, because $5 \cdot (-5) = -25$

33. $\dfrac{-10}{-40} = \dfrac{1}{4}$, or 0.25

34. $\dfrac{-10}{-3} = \dfrac{10}{3}$, or $3.\overline{3}$

The sign rules for division and multiplication are the same.

> To multiply or divide two real numbers:
>
> **1.** Multiply or divide the absolute values.
> **2.** If the signs are the same, then the answer is positive.
> **3.** If the signs are different, then the answer is negative.

Now Try Exercises 79 and 83.

Excluding Division by Zero

We cannot divide a nonzero number n by zero. By the definition of division, $n/0$ would be some number that when multiplied by 0 gives n. But when any number is multiplied by 0, the result is 0. Thus the only possibility for n would be 0.

Consider $0/0$. Using the definition of division, we might say that it is 5 because $5 \cdot 0 = 0$. We might also say that it is -8 because $-8 \cdot 0 = 0$. In fact, $0/0$ could be any number at all. So, division by 0 does not make sense. Division by 0 is not defined and is not possible.

EXAMPLES Divide, if possible.

35. $\dfrac{7}{0}$ Not defined: Division by 0.

36. $\dfrac{0}{7} = 0$ The quotient is 0 because $0 \cdot 7 = 0$.

37. $\dfrac{4}{x - x}$ Not defined: $x - x = 0$ for any x.

Now Try Exercise 87.

Division and Reciprocals

Two numbers whose product is 1 are called **reciprocals** (or **multiplicative inverses**) of each other.

> **PROPERTIES OF RECIPROCALS**
>
> Every nonzero real number a has a **reciprocal** (or **multiplicative inverse**) $1/a$. The reciprocal of a positive number is positive. The reciprocal of a negative number is negative.

EXAMPLES Find the reciprocal of each number.

38. $\dfrac{4}{5}$ The reciprocal is $\dfrac{5}{4}$, because $\dfrac{4}{5} \cdot \dfrac{5}{4} = 1$.

39. 8 The reciprocal is $\dfrac{1}{8}$, because $8 \cdot \dfrac{1}{8} = 1$.

40. $-\dfrac{2}{3}$ The reciprocal is $-\dfrac{3}{2}$, because $-\dfrac{2}{3} \cdot \left(-\dfrac{3}{2}\right) = 1$.

41. 0.25 The reciprocal is $\dfrac{1}{0.25}$, or 4, because $0.25 \cdot 4 = 1$.

> **Now Try Exercises 95 and 97.**

Remember that a number and its reciprocal (multiplicative inverse) have the same sign. Do *not* change the sign when taking the reciprocal of a number. On the other hand, when finding an opposite (additive inverse), be sure to change the sign.

We know that we can subtract by adding an opposite, or additive inverse. Similarly, we can divide by multiplying by a reciprocal.

RECIPROCALS AND DIVISION

For any real numbers a and b, $b \neq 0$,

$$a \div b = \frac{a}{b} = a \cdot \frac{1}{b}.$$

(To divide, we can multiply by the reciprocal of the divisor.)

We sometimes say that we "invert the divisor and multiply."

EXAMPLES Divide by multiplying by the reciprocal of the divisor.

42. $\dfrac{1}{4} \div \dfrac{3}{5} = \dfrac{1}{4} \cdot \dfrac{5}{3} = \dfrac{5}{12}$ *"Inverting" the divisor, $\dfrac{3}{5}$, and multiplying*

43. $\dfrac{2}{3} \div \left(-\dfrac{4}{9}\right) = \dfrac{2}{3} \cdot \left(-\dfrac{9}{4}\right) = -\dfrac{18}{12}$, or $-\dfrac{3}{2}$

44. $-\dfrac{5}{7} \div 3 = -\dfrac{5}{7} \cdot \dfrac{1}{3} = -\dfrac{5}{21}$

> **Now Try Exercises 103 and 105.**

The following properties can be used to make sign changes.

SIGN CHANGES IN FRACTION NOTATION

For any numbers a and b, $b \neq 0$,

$$\frac{-a}{b} = \frac{a}{-b} = -\frac{a}{b} \quad \text{and} \quad \frac{-a}{-b} = \frac{a}{b}.$$

R.2   Exercise Set

▶ **Reading Check**

Complete each statement with either "negative" or "positive."

RC1. A negative number has a _____ sign.

RC2. The opposite of a negative number is _____.

RC3. The reciprocal of a negative number is _____.

RC4. The absolute value of a negative number is _____.

RC5. When two negative numbers are multiplied, the result is _____.

RC6. The sum of 0 and a negative number is _____.

RC7. The sum of two negative numbers is _____.

RC8. The quotient of a negative number and a positive number is _____.

a *Add.*

1. $-10 + (-18)$

2. $-13 + (-12)$

3. $7 + (-2)$

4. $7 + (-5)$

5. $-8 + (-8)$

6. $-6 + (-6)$

7. $7 + (-11)$

8. $9 + (-12)$

9. $-16 + 6$

10. $-21 + 11$

11. $-26 + 0$

12. $0 + (-32)$

13. $-8.4 + 9.6$

14. $-6.3 + 8.2$

15. $-2.62 + (-6.24)$

16. $-2.73 + (-8.46)$

17. $-\dfrac{5}{9} + \dfrac{2}{9}$

18. $-\dfrac{3}{7} + \dfrac{1}{7}$

19. $-\dfrac{11}{12} + \left(-\dfrac{5}{12}\right)$

20. $-\dfrac{3}{8} + \left(-\dfrac{7}{8}\right)$

21. $\dfrac{2}{5} + \left(-\dfrac{3}{10}\right)$

22. $-\dfrac{3}{4} + \dfrac{1}{8}$

23. $-\dfrac{2}{5} + \dfrac{3}{4}$

24. $-\dfrac{5}{6} + \left(-\dfrac{7}{8}\right)$

b *Evaluate $-a$ for each of the following.*

25. $a = -4$

26. $a = -9$

27. $a = 3.7$

28. $a = 0$

Find the opposite (additive inverse).

29. 10

30. $-\dfrac{2}{3}$

31. 0

32. $-2x$

c *Subtract.*

33. $3 - 7$

34. $8 - 13$

35. $-5 - 9$

36. $-6 - 14$

37. $23 - 23$

38. $23 - (-23)$

39. $-23 - 23$

40. $-23 - (-23)$

41. $-6 - (-11)$

42. $-7 - (-12)$

43. $10 - (-5)$

44. $28 - (-16)$

45. $15.8 - 27.4$

46. $17.2 - 34.9$

47. $-18.01 - 11.24$

48. $-19.04 - 15.76$

49. $-\dfrac{21}{4} - \left(-\dfrac{7}{4}\right)$

50. $-\dfrac{16}{5} - \left(-\dfrac{3}{5}\right)$

51. $-\dfrac{1}{3} - \left(-\dfrac{1}{12}\right)$

52. $-\dfrac{7}{8} - \left(-\dfrac{5}{2}\right)$

53. $-\dfrac{3}{4} - \dfrac{5}{6}$

54. $-\dfrac{2}{3} - \dfrac{4}{5}$

55. $\dfrac{1}{3} - \dfrac{4}{5}$

56. $-\dfrac{4}{7} - \left(-\dfrac{5}{9}\right)$

d *Multiply.*

57. $3(-7)$

58. $5(-8)$

59. $-2 \cdot 4$

60. $-5 \cdot 9$

61. $-8(-3)$

62. $-5(-7)$

63. $-7 \cdot 16$

64. $-8 \cdot 19$

65. $-6(-5.7)$

66. $-7(-6.1)$

67. $-\dfrac{3}{5} \cdot \dfrac{4}{7}$

68. $-\dfrac{5}{4} \cdot \dfrac{11}{3}$

69. $-3\left(-\dfrac{2}{3}\right)$

70. $-5\left(-\dfrac{3}{5}\right)$

71. $-3(-4)(5)$

72. $-6(-8)(9)$

73. $(-4.2)(-6.3)$

74. $(-7.4)(-9.6)$

75. $-\dfrac{9}{11}\cdot\left(-\dfrac{11}{9}\right)$

76. $-\dfrac{13}{7}\cdot\left(-\dfrac{5}{2}\right)$

77. $-\dfrac{2}{3}\cdot\left(-\dfrac{2}{3}\right)\cdot\left(-\dfrac{2}{3}\right)$

78. $-\dfrac{4}{5}\cdot\left(-\dfrac{4}{5}\right)\cdot\left(-\dfrac{4}{5}\right)$

e *Divide, if possible.*

79. $\dfrac{-8}{4}$

80. $\dfrac{-16}{2}$

81. $\dfrac{56}{-8}$

82. $\dfrac{63}{-7}$

83. $-77 \div (-11)$

84. $-48 \div (-6)$

85. $\dfrac{-5.4}{-18}$

86. $\dfrac{-8.4}{-12}$

87. $\dfrac{5}{0}$

88. $\dfrac{92}{0}$

89. $\dfrac{0}{32}$

90. $\dfrac{0}{17}$

91. $\dfrac{9}{y - y}$

92. $\dfrac{2x - 2x}{2x - 2x}$

Find the reciprocal of each number.

93. $\dfrac{3}{4}$

94. $\dfrac{9}{10}$

95. $-\dfrac{7}{8}$

96. $-\dfrac{5}{6}$

97. 25

98. -65

99. 0.2

100. 0.8

101. $-\dfrac{a}{b}$

102. $\dfrac{1}{8x}$

Divide.

103. $\dfrac{2}{7} \div \left(-\dfrac{11}{3}\right)$

104. $\dfrac{3}{5} \div \left(-\dfrac{6}{7}\right)$

105. $-\dfrac{10}{3} \div \left(-\dfrac{2}{15}\right)$

106. $-\dfrac{12}{5} \div \left(-\dfrac{3}{10}\right)$

107. $18.6 \div (-3.1)$

108. $39.9 \div (-13.3)$

109. $(-75.5) \div (-15.1)$

110. $(-12.1) \div (-0.11)$

111. $-48 \div 0.4$

112. $520 \div (-0.13)$

113. $\dfrac{3}{4} \div \left(-\dfrac{2}{3}\right)$

114. $\dfrac{5}{8} \div \left(-\dfrac{1}{2}\right)$

115. $-\dfrac{5}{4} \div \left(-\dfrac{3}{4}\right)$

116. $-\dfrac{5}{9} \div \left(-\dfrac{5}{6}\right)$

117. $-\dfrac{2}{3} \div \left(-\dfrac{4}{9}\right)$

118. $-\dfrac{3}{5} \div \left(-\dfrac{5}{8}\right)$

119. $-\dfrac{3}{8} \div \left(-\dfrac{8}{3}\right)$

120. $-\dfrac{5}{8} \div \left(-\dfrac{5}{6}\right)$

121. $-6.6 \div 3.3$

122. $-44.1 \div (-6.3)$

123. $\dfrac{-12}{-13}$

124. $\dfrac{-1.9}{20}$

125. $\dfrac{48.6}{-30}$

126. $\dfrac{-17.8}{3.2}$

127. $\dfrac{-9}{17 - 17}$

128. $\dfrac{-8}{-6 + 6}$

129. Complete the following table.

Number	Opposite (Additive Inverse)	Reciprocal (Multiplicative Inverse)
$\dfrac{2}{3}$		
$-\dfrac{5}{4}$		
0		
1		
-4.5		
$x, x \neq 0$		

130. Complete the following table.

Number	Opposite (Additive Inverse)	Reciprocal (Multiplicative Inverse)
$-\dfrac{3}{8}$		
$\dfrac{7}{10}$		
-1		
0		
-6.4		
$a, a \neq 0$		

▶ ## Skill Maintenance

This heading indicates that the exercises that follow are Skill Maintenance exercises, which review any skill previously studied in the text. You can expect such exercises in every exercise set. Answers to all skill maintenance exercises are found at the back of the book. If you miss an exercise, restudy the objective shown in red.

Given the numbers $\sqrt{3}, -12.47, -13, 26, \pi, 0, -\dfrac{23}{32},$

$\dfrac{7}{11}, 4.57557555755557\ldots$: **[R.1a]**

131. Name the whole numbers.

132. Name the natural numbers.

133. Name the integers.

134. Name the irrational numbers.

135. Name the rational numbers.

136. Name the real numbers.

Use either $<$ *or* $>$ *for* ☐ *to write a true sentence.* **[R.1b]**

137. -7 ☐ 8

138. 5 ☐ $\frac{3}{8}$

139. -45.6 ☐ -23.8

140. 123 ☐ -10

▶ ## Synthesis

141. The reciprocal of an electric resistance is called *conductance*. When two resistors are connected in parallel, the conductance is the sum of the conductances,

$$\frac{1}{r_1} + \frac{1}{r_2}.$$

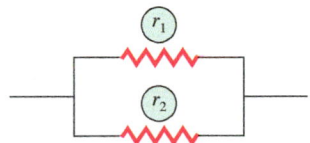

Find the conductance of two resistors of 12 ohms and 6 ohms when connected in parallel.

142. What number can be added to 11.7 to obtain $-7\frac{3}{4}$?

143. What number can be multiplied by -0.02 to obtain -625?

R.3   Exponential Notation and Order of Operations

▶ **a**   Rewrite expressions with whole-number exponents, and evaluate exponential expressions.

▶ **b**   Rewrite expressions with or without negative integers as exponents.

▶ **c**   Simplify expressions using the rules for order of operations.

▶ **a**   **Exponential Notation**

Exponential notation is a shorthand device. For $3 \cdot 3 \cdot 3 \cdot 3$, we write 3^4. In the **exponential notation** 3^4, the number 3 is called the **base** and the number 4 is called the **exponent**.

> **EXPONENTIAL NOTATION**
>
> Exponential notation a^n, where n is an integer greater than 1, means
>
> $$\underbrace{a \cdot a \cdot a \cdots a \cdot a.}_{n \text{ factors}}$$
>
> We read a^n as "a to the nth power," or simply "a to the nth."
> We can read a^2 as "a-squared" and a^3 as "a-cubed."

> **CAUTION!** a^n does *not* mean to multiply n times a. For example, 3^2 means $3 \cdot 3$, or 9, not $3 \cdot 2$, or 6.

EXAMPLES   Write exponential notation.

1. $7 \cdot 7 \cdot 7 = 7^3$ **2.** $xxxxx = x^5$ **3.** $\dfrac{2}{3} \cdot \dfrac{2}{3} \cdot \dfrac{2}{3} \cdot \dfrac{2}{3} = \left(\dfrac{2}{3}\right)^4$

> **Now Try Exercise 3.**

EXAMPLES   Evaluate.

4. $9^2 = 9 \cdot 9 = 81$

5. $\left(\dfrac{1}{2}\right)^3 = \dfrac{1}{2} \cdot \dfrac{1}{2} \cdot \dfrac{1}{2} = \dfrac{1}{8}$

6. $\left(\dfrac{7}{8}\right)^2 = \dfrac{7}{8} \cdot \dfrac{7}{8} = \dfrac{49}{64}$

7. $(0.1)^4 = (0.1)(0.1)(0.1)(0.1) = 0.0001$

8. $(-5)^3 = (-5)(-5)(-5) = -125$

9. $-(5^3) = -(5 \cdot 5 \cdot 5) = -125$

10. $-(10)^4 = -(10 \cdot 10 \cdot 10 \cdot 10) = -10,000$

11. $(-10)^4 = (-10)(-10)(-10)(-10) = 10,000$

Note that $-(10)^4 \neq (-10)^4$, as shown in Examples 10 and 11. In $-(10)^4$, the base is 10; in $(-10)^4$, the base is -10.

> **Now Try Exercises 15 and 17.**

When an exponent is an integer greater than 1, it tells how many times the base occurs as a factor. What happens when the exponent is 1 or 0? Look for a pattern below. Think of dividing by 10 on the right.

On this side, the exponents decrease by 1 at each step.

$$10^4 = 10 \cdot 10 \cdot 10 \cdot 10 = 10,000$$
$$10^3 = 10 \cdot 10 \cdot 10 = 1000$$
$$10^2 = 10 \cdot 10 = 100$$
$$10^1 = ?$$
$$10^0 = ?$$

On this side, we divide by 10 at each step.

In order for the pattern to continue, 10^1 would have to be 10 and 10^0 would have to be 1. We will *agree* that exponents of 1 and 0 have that meaning.

> **EXPONENTS OF 0 AND 1**
>
> For any number a, we agree that a^1 means a.
>
> For any nonzero number a, we agree that a^0 means 1.

EXAMPLES Rewrite without an exponent.

12. $4^1 = 4$ **13.** $(-97)^1 = -97$

14. $6^0 = 1$ **15.** $(-37.4)^0 = 1$

> **Now Try Exercises 21 and 23.**

Let's consider a justification for not defining 0^0. By examining the pattern $3^0 = 1$, $2^0 = 1$, and $1^0 = 1$, we might think that 0^0 should be 1. However, by examining the pattern $0^3 = 0$, $0^2 = 0$, and $0^1 = 0$, we might think that 0^0 should be 0. To avoid this confusion, mathematicians agree *not* to define 0^0.

▶ **b Negative Integers as Exponents**

How shall we define negative integers as exponents? Look for a pattern below. Again, think of dividing by 10 on the right.

> On this side, the exponents decrease by 1 at each step. $\begin{aligned} 10^2 &= 100 \\ 10^1 &= 10 \\ 10^0 &= 1 \\ 10^{-1} &= \ ? \\ 10^{-2} &= \ ? \end{aligned}$ On this side, we divide by 10 at each step.

In order for the pattern to continue, 10^{-1} would have to be $\frac{1}{10}$ and 10^{-2} would have to be $\frac{1}{100}$. This leads to the following agreement.

NEGATIVE EXPONENTS

For any nonzero real number a and any integer n,

$$a^{-n} = \frac{1}{a^n}.$$

EXAMPLES Rewrite using a positive exponent. Evaluate, if possible.

16. $y^{-5} = \dfrac{1}{y^5}$ **17.** $\dfrac{1}{t^{-4}} = t^4$

18. $(-2)^{-3} = \dfrac{1}{(-2)^3} = \dfrac{1}{(-2)(-2)(-2)} = \dfrac{1}{-8} = -\dfrac{1}{8}$

19. $\left(\dfrac{1}{2}\right)^{-3} = \dfrac{1}{\left(\frac{1}{2}\right)^3} = \dfrac{1}{\frac{1}{8}} = 1 \cdot \dfrac{8}{1} = 8$

20. $\left(\dfrac{2}{5}\right)^{-2} = \dfrac{1}{\left(\frac{2}{5}\right)^2} = \dfrac{1}{\frac{4}{25}} = 1 \cdot \dfrac{25}{4} = \dfrac{25}{4}$

> **Now Try Exercises 31 and 37.**

The numbers a^n and a^{-n} are reciprocals because

$$a^n \cdot a^{-n} = a^n \cdot \frac{1}{a^n} = \frac{a^n}{a^n} = 1.$$

> **CAUTION!** **A negative exponent does *not* necessarily indicate that a number is negative!** For example, 3^{-2} means $1/3^2$, or $1/9$, not -9.

For example, 7^3 and 7^{-3} are reciprocals:

$$7^3 \cdot 7^{-3} = 7^3 \cdot \frac{1}{7^3} = \frac{7^3}{7^3} = 1.$$

EXAMPLES Rewrite using a negative exponent.

21. $\dfrac{1}{x^2} = x^{-2}$

22. $\dfrac{1}{(-7)^4} = (-7)^{-4}$

> **Now Try Exercises 41 and 43.**

▶ c Order of Operations

What does $8 + 2 \cdot 5^3$ mean? If we add 8 and 2 and multiply by 5^3, or 125, we get 1250. If we multiply 2 times 125 and add 8, we get 258. Both results cannot be correct. To avoid such difficulties, we make agreements about which operations should be done first.

> **RULES FOR ORDER OF OPERATIONS**
> 1. Do all the calculations within grouping symbols, like parentheses, before operations outside.
> 2. Evaluate all exponential expressions.
> 3. Do all multiplications and divisions in order from left to right.
> 4. Do all additions and subtractions in order from left to right.

Most computers and calculators are programmed using these rules.

EXAMPLE 23 Simplify: $-43 \cdot 56 - 17$.

There are no parentheses or exponents so we begin with the multiplication.

$$-43 \cdot 56 - 17 = -2408 - 17 \qquad \text{Carrying out all multiplications and divisions in order from left to right}$$

$$= -2425 \qquad \text{Carrying out all additions and subtractions in order from left to right}$$

> **Now Try Exercise 53.**

EXAMPLE 24 Simplify: $8 + 2 \cdot 5^3$.

$$8 + 2 \cdot 5^3 = 8 + 2 \cdot 125 \qquad \text{Evaluating the exponential expression}$$
$$= 8 + 250 \qquad \text{Doing the multiplication}$$
$$= 258 \qquad \text{Adding} \qquad \text{Now Try Exercise 57.}$$

EXAMPLE 25 Simplify and compare: $(8 - 10)^2$ and $8^2 - 10^2$.

$$(8 - 10)^2 = (-2)^2 = 4;$$
$$8^2 - 10^2 = 64 - 100 = -36$$

We see that $(8 - 10)^2$ and $8^2 - 10^2$ are *not* the same.

> **Now Try Exercise 55.**

Order of Operations

Computations are usually entered on a graphing calculator in the same way in which we would write them. When an expression contains grouping symbols, we enter them using the **(** and **)** keys. Since a fraction bar acts as a grouping symbol, we often must supply parentheses when entering fraction expressions.

To calculate $\dfrac{45 + 135}{2 - 17}$, for example, we enter it as $(45 + 135) \div (2 - 17)$. The result is -12.

```
(45+135)/(2−17)
                    −12
```

Exercises

Calculate.

1. $48 \div 2 \cdot 3 - 4 \cdot 4$
2. $48 \div (2 \cdot 3 - 4) \cdot 4$
3. $\{(25 \cdot 30) \div [(2 \cdot 16) \div (4 \cdot 2)]\} + 15(45 \div 9)$
4. $\dfrac{17^2 - 311}{16 - 7}$

EXAMPLE 26 Simplify: $3^4 + 62 \cdot 8 - 2(29 + 33 \cdot 4)$.

$$3^4 + 62 \cdot 8 - 2(29 + 33 \cdot 4)$$
$$= 3^4 + 62 \cdot 8 - 2(29 + 132) \quad \text{Carrying out operations inside parentheses first; doing the multiplication}$$
$$= 3^4 + 62 \cdot 8 - 2(161) \quad \text{Adding inside parentheses}$$
$$= 81 + 62 \cdot 8 - 2(161) \quad \text{Evaluating the exponential expression}$$
$$= 81 + 496 - 2(161) \quad \text{Doing the multiplications in order from left to right}$$
$$= 81 + 496 - 322$$
$$= 577 - 322 \quad \text{Doing all additions and subtractions in order from left to right}$$
$$= 255$$

> **Now Try Exercise 59.**

When parentheses occur within parentheses, they can be different shapes, such as [ ] (also called "brackets") and { } (usually called "braces"). Parentheses, brackets, and braces all have the same meaning. When parentheses occur within parentheses, **computations in the *innermost* ones are to be done first**.

EXAMPLE 27 Simplify: $5 - \{6 - [3 - (7 + 3)]\}$.

$$5 - \{6 - [3 - (7 + 3)]\} = 5 - \{6 - [3 - 10]\} \quad \text{Adding } 7 + 3$$
$$= 5 - \{6 - [-7]\} \quad \text{Subtracting } 3 - 10$$
$$= 5 - 13 \quad \text{Subtracting } 6 - [-7]$$
$$= -8 \quad \text{Now Try Exercise 47.}$$

EXAMPLE 28 Simplify: $7 - [3(2 - 5) - 4(2 + 3)]$.

$$7 - [3(2 - 5) - 4(2 + 3)] = 7 - [3(-3) - 4(5)] \quad \text{Doing the calculations in the innermost grouping symbols first}$$
$$= 7 - [-9 - 20]$$
$$= 7 - [-29]$$
$$= 36 \quad \text{Now Try Exercise 49.}$$

In addition to parentheses, brackets, and braces, a fraction bar and absolute-value signs can act as grouping symbols.

EXAMPLE 29 Calculate: $\dfrac{12|7 - 9| + 8 \cdot 5}{3^2 + 2^3}$.

An equivalent expression with brackets as grouping symbols is

$$[12|7 - 9| + 8 \cdot 5] \div [3^2 + 2^3].$$

What this shows, in effect, is that we do the calculations in the numerator and in the denominator separately, and then divide the results:

$$\frac{12|7-9|+8\cdot5}{3^2+2^3} = \left.\begin{array}{c} \dfrac{12|-2|+8\cdot5}{9+8} \\[2ex] = \dfrac{12(2)+8\cdot5}{17} \end{array}\right\}$$

Subtracting inside the absolute-value signs before taking the absolute value

$$= \frac{24+40}{17} = \frac{64}{17}.$$

Now Try Exercise 105.

R.3   Exercise Set

▶ # Reading Check

Determine whether each statement is true or false.

RC1. If an expression contains a negative exponent, the entire expression is negative.

RC2. If an expression contains parentheses within parentheses, we simplify in the innermost set first.

RC3. Absolute-value bars can act as grouping symbols.

RC4. If we are using the rules for order of operations, subtractions are done before divisions.

RC5. We can read n^2 as "n-cubed."

RC6. For any nonzero number n, $n^0 = n$.

RC7. The reciprocal of 3 is 3^{-1}.

RC8. Using the rules for order of operations, we evaluate exponential expressions before we add.

a *Write exponential notation.*

1. $4\cdot4\cdot4\cdot4\cdot4$ **2.** $6\cdot6\cdot6$

3. $5\cdot5\cdot5\cdot5\cdot5\cdot5$ **4.** $x\cdot x\cdot x\cdot x$

5. mmm **6.** $ttttt$

7. $\dfrac{7}{12}\cdot\dfrac{7}{12}\cdot\dfrac{7}{12}\cdot\dfrac{7}{12}$

8. $(3.8)(3.8)(3.8)(3.8)(3.8)$

9. $(123.7)(123.7)$

10. $\left(-\dfrac{4}{5}\right)\left(-\dfrac{4}{5}\right)\left(-\dfrac{4}{5}\right)$

Evaluate.

11. 2^7 **12.** 9^3

13. $(-2)^5$ **14.** $(-7)^2$

15. $\left(\dfrac{1}{3}\right)^4$ **16.** $(0.1)^6$

17. $(-4)^3$ **18.** $(-3)^4$

19. $(-5.6)^2$ **20.** $\left(\dfrac{2}{3}\right)^4$

21. 5^1 **22.** $(\sqrt{6})^1$

23. 34^0 **24.** $\left(\dfrac{5}{2}\right)^1$

25. $(\sqrt{6})^0$ **26.** $(-4)^0$

27. $\left(\dfrac{7}{8}\right)^1$ **28.** $(-87)^0$

b *Rewrite using a positive exponent. Evaluate, if possible.*

29. $\left(\dfrac{1}{4}\right)^{-2}$

30. $\left(\dfrac{1}{5}\right)^{-3}$

31. $\left(\dfrac{2}{3}\right)^{-3}$

32. $\left(\dfrac{5}{2}\right)^{-4}$

33. y^{-5}

34. x^{-6}

35. $\dfrac{1}{a^{-2}}$

36. $\dfrac{1}{y^{-7}}$

37. $(-11)^{-1}$

38. $(-4)^{-3}$

Rewrite using a negative exponent.

39. $\dfrac{1}{3^4}$

40. $\dfrac{1}{9^2}$

41. $\dfrac{1}{b^3}$

42. $\dfrac{1}{n^5}$

43. $\dfrac{1}{(-16)^2}$

44. $\dfrac{1}{(-8)^6}$

c *Simplify.*

45. $12 - 4(5 - 1)$

46. $6 - 4(8 - 5)$

47. $9[8 - 7(5 - 2)]$

48. $10[7 - 4(8 - 5)]$

49. $[5(8 - 6) + 12] - [24 - (8 - 4)]$

50. $[9(7 - 4) + 19] - [25 - (7 + 3)]$

51. $[64 \div (-4)] \div (-2)$

52. $[48 \div (-3)] \div \left(-\dfrac{1}{4}\right)$

53. $19(-22) + 60$

54. $30 \cdot 10 - 18 \cdot 25$

55. $(5 + 7)^2; \quad 5^2 + 7^2$

56. $(9 - 12)^2; \quad 9^2 - 12^2$

57. $2^3 + 2^4 - 20 \cdot 30$

58. $7 \cdot 8 - 3^2 - 2^3$

59. $5^3 + 36 \cdot 72 - (18 + 25 \cdot 4)$

60. $4^3 + 20 \cdot 10 + 7^2 - 23$

61. $(13 \cdot 2 - 8 \cdot 4)^2$

62. $(9 \cdot 8 + 3 \cdot 3)^2$

63. $4000 \cdot (1 + 0.12)^3$

64. $5000 \cdot (4 + 1.16)^2$

65. $(20 \cdot 4 + 13 \cdot 8)^2 - (39 \cdot 15)^3$

66. $(43 \cdot 6 - 14 \cdot 7)^3 + (33 \cdot 34)^2$

67. $18 - 2 \cdot 3 - 9$

68. $18 - (2 \cdot 3 - 9)$

69. $(18 - 2 \cdot 3) - 9$

70. $(18 - 2)(3 - 9)$

71. $[24 \div (-3)] \div \left(-\dfrac{1}{2}\right)$

72. $[(-32) \div (-2)] \div (-2)$

73. $15 \cdot (-24) + 50$

74. $30 \cdot 20 - 15 \cdot 24$

75. $4 \div (8 - 10)^2 + 1$

76. $16 \div (19 - 15)^2 - 7$

77. $6^3 + 25 \cdot 71 - (16 + 25 \cdot 4)$

78. $5^3 + 20 \cdot 40 + 8^2 - 29$

79. $5000 \cdot (1 + 0.16)^3$

80. $4000 \cdot (3 + 1.14)^2$

81. $4 \cdot 5 - 2 \cdot 6 + 4$

82. $8(7 - 3)/4$

83. $4 \cdot (6 + 8)/(4 + 3)$

84. $4^3/8$

85. $[2 \cdot (5 - 3)]^2$

86. $5^3 - 7^2$

87. $8(-7) + 6(-5)$

88. $10(-5) + 1(-1)$

89. $19 - 5(-3) + 3$

90. $14 - 2(-6) + 7$

91. $9 \div (-3) + 16 \div 8$

92. $-32 - 8 \div 4 - (-2)$

93. $7 + 10 - (-10 \div 2)$

94. $(3 - 8)^2$

95. $5^2 - 8^2$

96. $28 - 10^3$

97. $20 + 4^3 \div (-8)$

98. $2 \times 10^3 - 5000$

99. $-7(3^4) + 18$

100. $6[9 - (3 - 4)]$

101. $9[(8 - 11) - 13]$

102. $1000 \div (-100) \div 10$

103. $256 \div (-32) \div (-4)$

104. $\dfrac{20 - 6^2}{9^2 + 3^2}$

105. $\dfrac{5^2 - |4^3 - 8|}{9^2 - 2^2 - 1^5}$

106. $\dfrac{4|6 - 7| - 5 \cdot 4}{6 \cdot 7 - 8|4 - 1|}$

107. $\dfrac{30(8 - 3) - 4(10 - 3)}{10|2 - 6| - 2(5 + 2)}$

108. $\dfrac{5^3 - 3^2 + 12 \cdot 5}{-32 \div (-16) \div (-4)}$

▶ Skill Maintenance

Find the absolute value. **[R.1d]**

109. $\left| -\dfrac{9}{7} \right|$

110. $|2.3|$

111. $|0|$

112. $|-900|$

Compute. **[R.2a, c, d]**

113. $23 - 56$

114. $-23 - 56$

115. $-23 - (-56)$

116. $-23 + (-56)$

117. $(-10)(2.3)$

118. $(-10)(-2.3)$

119. $10(-2.3)$

120. $\left(-\dfrac{2}{3} \right)\left(-\dfrac{15}{16} \right)$

▶ Synthesis

Simplify.

121. $(-2)^0 - (-2)^3 - (-2)^{-1} + (-2)^4 - (-2)^{-2}$

122. $2(6^1 \cdot 6^{-1} - 6^{-1} \cdot 6^0)$

123. Place parentheses in this statement to make it true:
$9 \cdot 5 + 2 - 8 \cdot 3 + 1 = 22$.

The symbol ⚟ means to use your calculator to work a particular exercise.

124. ⚟ Find each of the following.

$12345679 \cdot 9 = ?$

$12345679 \cdot 18 = ?$

$12345679 \cdot 27 = ?$

Then look for a pattern and find $12345679 \cdot 36$ without the use of a calculator.

125. ⚟ Find $(0.2)^{(-0.2)^{-1}}$.

126. ⚟ Determine which is larger: $(\pi)^{\sqrt{2}}$ or $(\sqrt{2})^{\pi}$.

127. Find $(2 + 3)^{-1}$ and $2^{-1} + 3^{-1}$ and determine whether they are equivalent.

R.4

PART 2 MANIPULATIONS
Introduction to Algebraic Expressions

▶ **a** Translate a phrase to an algebraic expression.

▶ **b** Evaluate an algebraic expression by substitution.

The study of algebra involves the use of equations to solve problems. Equations are constructed from algebraic expressions. The purpose of Part 2 of this chapter is to provide a review of the types of expressions encountered in algebra and ways in which we can manipulate them.

Algebraic Expressions and Their Use

In arithmetic, you worked with expressions such as

$$91 + 76, \quad 26 - 17, \quad 14 \cdot 35, \quad 7 \div 8, \quad \dfrac{7}{8}, \quad \text{and} \quad 5^2 - 3^2.$$

In algebra, we use these as well as expressions like

$$x + 76, \quad 26 - q, \quad 14 \cdot x, \quad d \div t, \quad \dfrac{d}{t}, \quad \text{and} \quad x^2 - y^2.$$

When a letter is used to represent various numbers, it is called a **variable**. Let $t =$ the number of hours that a passenger jet has been flying. Then t is a variable, because t changes as the flight continues. If a letter represents one particular number, it is called a **constant**. Let $d =$ the number of hours in a day. Then d is a constant.

An **algebraic expression** consists of variables, numbers, and operation signs, such as $+, -, \cdot, \div$. When an equals sign, $=$, is placed between two expressions, an **equation** is formed. The table at left lists examples of expressions and equations. Note that none of the expressions has an equals sign ($=$).

Equations can be used to solve applied problems. To illustrate this, consider the following bar graph, which shows the median pay for several occupations.

Algebraic Expressions	Equations
10	$t = 10$
$x - 5$	$x - 5 = 10$
$11x$	$x - 5 = 11x$
$y^2 + 2y$	$y^2 + 2y = 1 + y$

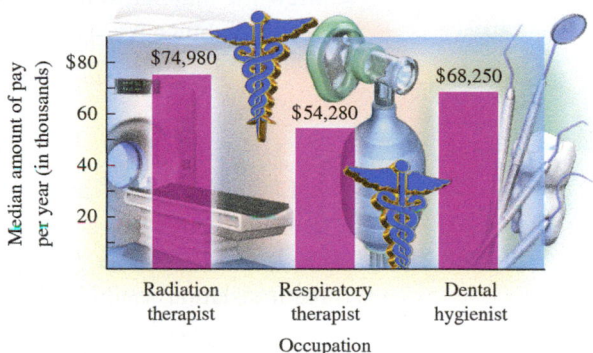

SOURCE: U.S. Bureau of Labor Statistics

Suppose we want to determine how much more a radiation therapist earns than a dental hygienist does. We can translate this problem to an equation:

Dental hygienist pay	plus	How much more	is	Radiation therapist pay
68,250	+	x	=	74,980.

We can then *solve* for x:

$$68{,}250 + x = 74{,}980$$
$$68{,}250 + x - 68{,}250 = 74{,}980 - 68{,}250 \qquad \text{\textcolor{red}{Subtracting 68,250}}$$
$$x = 6730.$$

We see that a radiation therapist earns $6730 more per year than a dental hygienist does.

▶ a   Translating to Algebraic Expressions

To translate problems to equations, we need to know that certain words correspond to certain symbols, as shown in the following table.

Key Words

Addition	Subtraction	Multiplication	Division
add	subtract	multiply	divide
sum	difference	product	quotient
plus	minus	times	divided by
total	decreased by	twice	ratio
increased by	less than	of	per
more than			

Expressions like *rs* represent products and can also be written as $r \cdot s$, $r \times s$, $(r)(s)$, or $r(s)$. The multipliers *r* and *s* are also called **factors**. A quotient $m \div 5$ can also be represented as $m/5$ or $\dfrac{m}{5}$.

EXAMPLE 1   Translate to an algebraic expression: Eight less than some number.

We can use any variable we wish to represent "some number," such as *x*, *y*, *t*, *m*, *n*, and so on. Here we let *t* represent the number. If we knew the number to be 23, then the translation of "eight less than 23" would be $23 - 8$. If we knew the number to be 345, then the translation of "eight less than 345" would be $345 - 8$. Since we are using a variable for the number, the translation is

$t - 8$.   *Caution!* $8 - t$ would be incorrect.

Now Try Exercise 3.

EXAMPLE 2   Translate to an algebraic expression: Twenty-two more than some number.

This time we let *y* represent the number. If we knew the number to be 47, then the translation would be $47 + 22$. Since we are using a variable, the translation is

$y + 22$.

Because addition is commutative, $22 + y$ is also a correct translation.

Now Try Exercise 1.

EXAMPLE 3   Translate to an algebraic expression: Five less than forty-three percent of the quotient of two numbers.

We let *r* and *s* represent the two numbers.

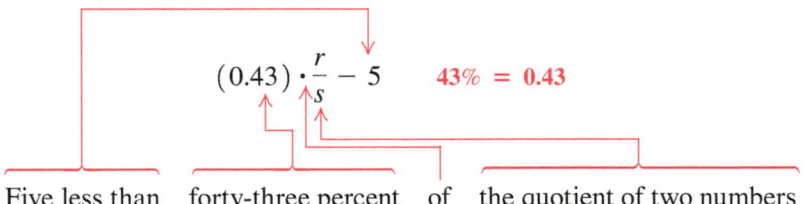

$$(0.43) \cdot \frac{r}{s} - 5 \qquad 43\% = 0.43$$

Five less than   forty-three percent   of   the quotient of two numbers

EXAMPLE 4   Translate each of the following to an algebraic expression.

Phrase	Algebraic Expression
Five *more than* some number	$n + 5$, or $5 + n$
Half *of* a number	$\dfrac{1}{2}t$, or $\dfrac{t}{2}$
Five *more than* three *times* some number	$3p + 5$, or $5 + 3p$
The *difference* of two numbers	$x - y$
Six *less than* the *product* of two numbers	$rs - 6$
Seventy-six percent *of* some number	$76\% \, z$, or $0.76z$
Eight *less than twice* some number	$2x - 8$

Now Try Exercise 21.

▶ b   Evaluating Algebraic Expressions

When we replace a variable with a number, we say that we are **substituting** for the variable. Carrying out the resulting calculation is called **evaluating the expression**. The result is called the **value** of the expression.

EXAMPLE 5   Evaluate $x - y$ when $x = 83$ and $y = 49$.

We substitute 83 for x and 49 for y and carry out the subtraction:

$$x - y = 83 - 49 = 34.$$

Now Try Exercise 31.

EXAMPLE 6   Evaluate a/b when $a = -63$ and $b = 7$.

We substitute -63 for a and 7 for b and carry out the division:

$$\frac{a}{b} = \frac{-63}{7} = -9.$$

Now Try Exercise 29.

EXAMPLE 7   Evaluate the expression $3xy + z$ when $x = 2$, $y = -5$, and $z = 7$.

We substitute and carry out the calculations according to the rules for order of operations:

$$3xy + z = 3(2)(-5) + 7 = -30 + 7 = -23.$$

Now Try Exercise 37.

In the next example, we use the formula for the area A of a triangle with a base of length b and a height of length h:

$$A = \tfrac{1}{2}bh.$$

EXAMPLE 8   *Area of a Triangular Sail.*   The base of a triangular sail is 6.4 m and the height is 8 m. Find the area of the sail.

We substitute 6.4 for b and 8 for h and multiply:

$$A = \tfrac{1}{2}bh = \tfrac{1}{2} \cdot 6.4 \cdot 8$$
$$= 25.6 \text{ m}^2.$$

Now Try Exercise 43.

EXAMPLE 9   Evaluate $5 + 2(a - 1)^2$ when $a = 4$.

$$
\begin{aligned}
5 + 2(a - 1)^2 &= 5 + 2(4 - 1)^2 &&\text{Substituting} \\
&= 5 + 2(3)^2 &&\text{Working within parentheses first} \\
&= 5 + 2(9) &&\text{Simplifying } 3^2 \\
&= 5 + 18 &&\text{Multiplying} \\
&= 23 &&\text{Adding}
\end{aligned}
$$

Now Try Exercise 39.

EXAMPLE 10   Evaluate $9 - x^3 + 6 \div 2y^2$ when $x = 2$ and $y = 5$.

$$
\begin{aligned}
9 - x^3 + 6 \div 2y^2 &= 9 - 2^3 + 6 \div 2(5)^2 &&\text{Substituting} \\
&= 9 - 8 + 6 \div 2 \cdot 25 &&\text{Simplifying } 2^3 \text{ and } 5^2 \\
&= 9 - 8 + 3 \cdot 25 &&\text{Dividing} \\
&= 9 - 8 + 75 &&\text{Multiplying} \\
&= 1 + 75 &&\text{Subtracting} \\
&= 76 &&\text{Adding}
\end{aligned}
$$

Now Try Exercise 35.

R.4   Exercise Set

► Reading Check

Choose from the column on the right an appropriate translation for each phrase. Choices may be used more than once or not at all.

RC1. _____ 5 less than some number

RC2. _____ A number decreased by 5

RC3. _____ A number increased by 5

RC4. _____ The product of a number and 5

RC5. _____ The sum of a number and 5

RC6. _____ The ratio of a number and 5

RC7. _____ 5 more than twice a number

RC8. _____ Twice the sum of a number and 5

a) $x + 5$

b) $2x + 5$

c) $x - 5$

d) $5 - x$

e) $5x$

f) $2(x + 5)$

g) $\dfrac{x}{5}$

---

a *Translate each phrase to an algebraic expression.*

1. 8 more than b

2. 11 more than t

3. 13.4 less than c

4. 0.203 less than d

5. 5 increased by q

6. 18 increased by z

7. b more than a

8. c more than d

9. x divided by y

10. c divided by h

11. x plus w

12. s added to t

13. m subtracted from n

14. p subtracted from q

15. The sum of p and q

16. The sum of a and b

17. Three times q

18. Twice z

19. -18 multiplied by m

20. The product of -6 and t

21. The product of 17% and your salary

22. 48% of the women attending

23. Megan drove at a speed of 75 mph for t hours on an interstate highway in Arizona. How far did Megan travel?

24. Joe had d dollars before spending \$19.95 on a movie. How much did Joe have after the purchase?

25. Jennifer had \$40 before spending x dollars on a pizza. How much remains?

26. Lance drove his pickup truck at a speed of 65 mph for t hours. How far did he travel?

b *Evaluate.*

27. $23z$, when $z = -4$

28. $57y$, when $y = -8$

29. $\dfrac{a}{b}$, when $a = -24$ and $b = -8$

30. $\dfrac{x}{y}$, when $x = 30$ and $y = -6$

31. $\dfrac{m - n}{8}$, when $m = 36$ and $n = 4$

32. $\dfrac{5}{p + q}$, when $p = 20$ and $q = 30$

33. $\dfrac{5z}{y}$, when $z = 9$ and $y = 2$

34. $\dfrac{18m}{n}$, when $m = 7$ and $n = 18$

35. $2c \div 3b$, when $b = 4$ and $c = 6$

36. $4x - y$, when $x = 3$ and $y = -2$

37. $25 - r^2 + s \div r^2$, when $r = 3$ and $s = 27$

38. $n^3 - 2 + p \div n^2$, when $n = 2$ and $p = 12$

39. $m + n(5 + n^2)$, when $m = 15$ and $n = 3$

40. $a^2 - 3(a - b)$, when $a = 10$ and $b = -8$

Simple Interest. The **simple interest** I on a principal of P dollars at interest rate r for t years is given by $I = Prt$.

41. Find the simple interest on a principal of $7345 at 6% for 1 year.

42. Find the simple interest on a principal of $18,000 at 4.6% for 2 years. (*Hint*: 4.6% = 0.046.)

43. *Area and Circumference of a Dining Table.* The area A of a circle with radius r is given by $A = \pi r^2$. The circumference C of the circle is given by $C = 2\pi r$. The radius of Ray and Mary's round oak dining table is 27 in. Find the area and the circumference of the table. Use 3.14 for π.

r = 27 in.

44. *Area of a Parallelogram.* The area A of a parallelogram with base b and height h is given by $A = bh$. Find the area of a flower garden that is shaped like a parallelogram with a height of 1.9 m and a base of 3.6 m.

h

b

▶ **Skill Maintenance**

Evaluate. **[R.3a]**

45. 3^5

46. $(-3)^5$

47. $(-10)^4$

48. $(-5.3)^2$

49. $\left(\dfrac{3}{5}\right)^2$

50. $(4.5)^0$

51. $(4.5)^1$

52. $(3x)^1$

Graph on the number line. **[R.1d]**

53. $y < -1$

−1 0

54. $0 \le x$

0

▶ **Synthesis**

Translate to an equation.

55. The distance d that a rapid transit train in the Denver airport travels in time t at a speed r is given by speed times time. Write an equation for d.

56. Marlana invests P dollars at 2.7% simple interest. Write an equation for the number of dollars N in the account 1 year from now.

Evaluate.

57. $\dfrac{x + y}{2} + \dfrac{3y}{2}$, when $x = 2$ and $y = 4$

58. $\dfrac{2.56y}{3.2x}$, when $y = 3$ and $x = 4$

R.5 Equivalent Algebraic Expressions

▶ **a** Determine whether two expressions are equivalent by completing a table of values.

▶ **b** Find equivalent fraction expressions by multiplying by 1, and simplify fraction expressions.

▶ **c** Use the commutative laws and the associative laws to find equivalent expressions.

▶ **d** Use the distributive laws to find equivalent expressions by multiplying and factoring.

▶ a Equivalent Expressions

It is often convenient to replace an expression with another expression that represents the same number. For example, instead of $x + 2x$, we might write $3x$, knowing that the two expressions represent the same number for any allowable replacement of x. In that sense, the expressions $x + 2x$ and $3x$ are **equivalent**, as are $5/x$ and $5x/x^2$, even though 0 is not an allowable replacement because division by 0 is not defined.

> **EQUIVALENT EXPRESSIONS**
>
> Two expressions that have the same value for all *allowable* replacements of the variables are called **equivalent expressions**.

EXAMPLE 1 Complete the following table by evaluating each of the expressions $x + 2x$, $3x$, and $8x - x$ for the given values. Then look for expressions that appear to be equivalent.

Value	$x + 2x$	$3x$	$8x - x$
$x = -2$			
$x = 5$			
$x = 0$			

We substitute and find the value of each expression. For example, for $x = -2$,

$$x + 2x = -2 + 2(-2) = -2 - 4 = -6,$$
$$3x = 3(-2) = -6, \quad \text{and}$$
$$8x - x = 8(-2) - (-2) = -16 + 2 = -14.$$

Value	$x + 2x$	$3x$	$8x - x$
$x = -2$	-6	-6	-14
$x = 5$	15	15	35
$x = 0$	0	0	0

Note that the values of $x + 2x$ and $3x$ are the same for the given values of x. Indeed, they are the same for any allowable real-number replacement of x, though we cannot substitute them all to find out. The expressions $x + 2x$ and $3x$ are **equivalent**. But the expressions $x + 2x$ and $8x - x$ are not equivalent, and the expressions $3x$ and $8x - x$ are not equivalent. Although $x + 2x$ and $8x - x$ have the same value for $x = 0$, they are not equivalent since values are not the same for *all x*. A similar statement is true for $3x$ and $8x - x$.

> **Now Try Exercise 1.**

▶ **b Equivalent Fraction Expressions**

Some properties of real numbers allow us to find equivalent expressions.

THE IDENTITY PROPERTY OF 1

For any real number a,

$$a \cdot 1 = 1 \cdot a = a.$$

(The number 1 is the **multiplicative identity**.)

We will often refer to the use of the identity property of 1 as "multiplying by 1." We can use multiplying by 1 to change from one fraction expression to an equivalent one with a different denominator.

EXAMPLE 2   Use multiplying by 1 to find an expression equivalent to $\frac{3}{5}$ with a denominator of $10x$.

Because $10x = 5 \cdot 2x$, we multiply by 1, using $2x/(2x)$ as a name for 1:

$$\frac{3}{5} = \frac{3}{5} \cdot 1 = \frac{3}{5} \cdot \frac{2x}{2x} = \frac{3 \cdot 2x}{5 \cdot 2x} = \frac{6x}{10x}.$$

Note that the expressions $3/5$ and $6x/(10x)$ are equivalent. They have the same value for any allowable replacement. Note too that 0 is not an allowable replacement in $6x/(10x)$, but for all nonzero real numbers, the expressions $3/5$ and $6x/(10x)$ have the same value.

> **Now Try Exercise 5.**

In algebra, we consider an expression like $3/5$ to be a "simplified" form of $6x/(10x)$. To find such simplified expressions, we reverse the identity property of 1 in order to "remove a factor of 1."

EXAMPLE 3   Simplify: $\dfrac{7x}{9x}$.

We do the reverse of what we did in Example 2:

$$\frac{7x}{9x} = \frac{7 \cdot x}{9 \cdot x}$$ **We factor the numerator and the denominator and then look for the largest common factor of both.**

$$= \frac{7}{9} \cdot \frac{x}{x}$$ **Factoring the expression**

$$= \frac{7}{9} \cdot 1 \qquad \frac{x}{x} = 1$$

$$= \frac{7}{9}.$$ **Removing a factor of 1 using the identity property of 1 in reverse**

> **Now Try Exercise 9.**

EXAMPLE 4   Simplify: $-\dfrac{24y}{16y}$.

$$-\frac{24y}{16y} = -\frac{3 \cdot 8y}{2 \cdot 8y} = -\frac{3}{2} \cdot \frac{8y}{8y} = -\frac{3}{2} \cdot 1 = -\frac{3}{2}$$

> **Now Try Exercise 11.**

▶ **c   The Commutative Laws and the Associative Laws**

Let's examine the expressions $x + y$ and $y + x$, as well as xy and yx.

EXAMPLE 5   Evaluate $x + y$ and $y + x$ when $x = 5$ and $y = 8$.

We substitute 5 for x and 8 for y in both expressions:

$$x + y = 5 + 8 = 13; \qquad y + x = 8 + 5 = 13.$$

EXAMPLE 6   Evaluate xy and yx when $x = 4$ and $y = 3$.

We substitute 4 for x and 3 for y in both expressions:

$$xy = 4 \cdot 3 = 12; \qquad yx = 3 \cdot 4 = 12.$$

Note that the expressions $x + y$ and $y + x$ have the same values no matter what the variables represent. Thus they are equivalent. When we add two numbers, the order in which we add does not matter. Similarly, when we multiply two numbers, the order in which we multiply does not matter. Thus the expressions xy and yx are equivalent. We say that addition and multiplication are *commutative*.

THE COMMUTATIVE LAWS

Addition.   For any numbers a and b,

$$a + b = b + a.$$

(We can change the order when adding without affecting the answer.)

Multiplication.   For any numbers a and b,

$$ab = ba.$$

(We can change the order when multiplying without affecting the answer.)

> **Now Try Exercises 13 and 15.**

Using a commutative law, we know that $x + 4$ and $4 + x$ are equivalent. Similarly, $5x$ and $x \cdot 5$ are equivalent. Thus, in an algebraic expression, we can replace one with the other and the result will be equivalent to the original expression.

Now let's examine the expressions $a + (b + c)$ and $(a + b) + c$. Note that these expressions use parentheses as grouping symbols, and they also involve three numbers. Calculations within grouping symbols are to be done first.

EXAMPLE 7 Evaluate $a + (b + c)$ and $(a + b) + c$ when $a = 4, b = 8,$ and $c = 5.$

$$a + (b + c) = 4 + (8 + 5)$$ Substituting
$$= 4 + 13$$ Calculating within parentheses first: adding 8 and 5
$$= 17;$$

$$(a + b) + c = (4 + 8) + 5$$ Substituting
$$= 12 + 5$$ Calculating within parentheses first: adding 4 and 8
$$= 17$$

EXAMPLE 8 Evaluate $a \cdot (b \cdot c)$ and $(a \cdot b) \cdot c$ when $a = 7, b = 4,$ and $c = 2.$

$$a \cdot (b \cdot c) = 7 \cdot (4 \cdot 2) = 7 \cdot 8 = 56;$$
$$(a \cdot b) \cdot c = (7 \cdot 4) \cdot 2 = 28 \cdot 2 = 56$$

When only addition is involved, changing the grouping does not change the answer. Likewise, when only multiplication is involved, changing the grouping does not change the answer.

> **THE ASSOCIATIVE LAWS**
>
> *Addition.* For any numbers a, b, and c,
> $$a + (b + c) = (a + b) + c.$$
> (Numbers can be grouped in any manner for addition.)
>
> *Multiplication.* For any numbers a, b, and c,
> $$a \cdot (b \cdot c) = (a \cdot b) \cdot c.$$
> (Numbers can be grouped in any manner for multiplication.)

Now Try Exercises 21 and 23.

EXAMPLE 9 Use the commutative laws and the associative laws to write at least three expressions equivalent to $(x + 8) + y.$

a) $(x + 8) + y = x + (8 + y)$ Using an associative law first and then the commutative law
$$= x + (y + 8)$$

b) $(x + 8) + y = y + (x + 8)$ Using a commutative law and then the commutative law again
$$= y + (8 + x)$$

c) $(x + 8) + y = (8 + x) + y$ Using a commutative law first and then the associative law
$$= 8 + (x + y)$$

Now Try Exercises 25 and 27.

Since grouping symbols can be placed any way we please when only additions or only multiplications are involved, we often omit them. For example, we write lwh instead of $(lw)h$ or $l(wh)$.

▶ **d The Distributive Laws**

Let's now examine two laws, each of which involves two operations. The first involves multiplication and addition.

EXAMPLE 10   Evaluate $8(x + y)$ and $8x + 8y$ when $x = 4$ and $y = 5$.

$$8(x + y) = 8(4 + 5) \qquad 8x + 8y = 8 \cdot 4 + 8 \cdot 5$$
$$= 8(9) \qquad\qquad\qquad\quad = 32 + 40$$
$$= 72; \qquad \longleftrightarrow \qquad = 72$$

The expressions $8(x + y)$ and $8x + 8y$ in Example 10 are equivalent. This fact is the result of a law called *the distributive law of multiplication over addition*. The other distributive law involves multiplication and subtraction.

THE DISTRIBUTIVE LAWS

The Distributive Law of Multiplication Over Addition
For any numbers a, b, and c,

$$a(b + c) = ab + ac, \quad \text{or} \quad (b + c)a = ba + ca.$$

(We can add and then multiply, or we can multiply and then add.)

The Distributive Law of Multiplication Over Subtraction
For any real numbers a, b, and c,

$$a(b - c) = ab - ac, \quad \text{or} \quad (b - c)a = ba - ca.$$

(We can subtract and then multiply, or we can multiply and then subtract.)

We often refer to "*the* distributive law" when we mean *either* or *both* of these laws.

Multiplying Expressions with Variables

The distributive laws are the basis of multiplication in algebra as well as in arithmetic. In the following examples, note that we multiply each number or letter inside the parentheses by the factor outside.

EXAMPLES   Multiply.

11. $4(x - 2) = 4 \cdot x - 4 \cdot 2 = 4x - 8$
12. $b(s - t + f) = bs - bt + bf$
13. $-3(y + 4) = -3 \cdot y + (-3) \cdot 4 = -3y - 12$
14. $-2x(y - 1) = -2x \cdot y - (-2x) \cdot 1 = -2xy + 2x$

> **Now Try Exercises 33 and 35.**

Factoring Expressions with Variables

The reverse of multiplying is called **factoring**. Factoring an expression involves factoring its *terms*. **Terms** of algebraic expressions are the parts separated by addition signs.

EXAMPLE 15   List the terms of $3x - 4y - 2z$.

We first find an equivalent expression that uses addition signs:

$$3x - 4y - 2z = 3x + (-4y) + (-2z). \qquad \text{Using the property } a - b = a + (-b)$$

Thus the terms are $3x$, $-4y$, and $-2z$.

> **Now Try Exercise 41.**

Factors are parts of products.

FACTORS

To **factor** an expression is to find an equivalent expression that is a product. If $N = a \cdot b$, then a and b are **factors** of N.

EXAMPLES Factor.

16. $8x + 8y = 8(x + y)$ 8 and $x + y$ are factors.
17. $cx - cy = c(x - y)$ c and $x - y$ are factors.

Now Try Exercise 45.

The distributive laws tell us that $8(x + y)$ and $8x + 8y$ are equivalent. We consider $8(x + y)$ to be **factored**. The factors are 8 and $x + y$. Whenever the terms of an expression have a factor in common, we can "remove" that factor, or "factor it out," using the distributive laws.

Generally, we try to factor out the largest factor common to all the terms. In the following example, we might factor out 3, but there is a larger factor common to the terms, 9. So we factor out the 9.

EXAMPLE 18 Factor: $9x + 27y$.

$$9x + 27y = 9 \cdot x + 9 \cdot (3y) = 9(x + 3y)$$

Now Try Exercise 55.

We often must supply a factor of 1 when factoring out a common factor, as in the next example, which is a formula involving simple interest.

EXAMPLE 19 Factor: $P + Prt$.

$$P + Prt = P \cdot 1 + P \cdot rt \quad \text{Writing } P \text{ as a product of } P \text{ and 1}$$
$$= P(1 + rt) \quad \text{Using the distributive law}$$

Now Try Exercise 51.

R.5   Exercise Set

▶ **Reading Check**

Choose the word or expression beneath each blank that best completes the statement.

RC1. The statement $3 + 7 = 7 + 3$ illustrates a(n) _____ law.
associative/commutative

RC2. The statement $4 \cdot 2 + 4 \cdot 8 = 4(2 + 8)$ illustrates a(n) _____ law.
commutative/distributive

RC3. The statement $5 \cdot (6 \cdot 7) = (5 \cdot 6) \cdot 7$ illustrates a(n) _____ law.
associative/distributive

RC4. The multiplicative identity is the number _____.
0/1

RC5. In the expression $7(5 + x)$, 7 and $(5 + x)$ are _____.
factors/terms

RC6. In the expression $7(5 + x)$, 5 and x are _____.
factors/terms

a *Complete each table by evaluating each expression for the given values. Then look for expressions that are equivalent.*

1.

Value	$2x + 3x$	$5x$	$2x - 3x$
$x = -2$			
$x = 5$			
$x = 0$			

2.

Value	$7x + 2x$	$5x$	$7x - 2x$
$x = -2$			
$x = 5$			
$x = 0$			

3.

Value	$4x + 8x$	$4(x + 3x)$	$4(x + 2x)$
$x = -1$			
$x = 3.2$			
$x = 0$			

4.

Value	$5(x - 2)$	$5x - 2$	$5x - 10$
$x = -1$			
$x = 4.6$			
$x = 0$			

b *Use multiplying by 1 to find an equivalent expression with the given denominator.*

5. $\dfrac{7}{8}$; $8x$

6. $\dfrac{4}{3}$; $3a$

7. $\dfrac{3}{4}$; $8a$

8. $\dfrac{3}{10}$; $50y$

Simplify.

9. $\dfrac{25x}{15x}$

10. $\dfrac{36y}{18y}$

11. $-\dfrac{100a}{25a}$

12. $\dfrac{-625t}{15t}$

c *Use a commutative law to find an equivalent expression.*

13. $w + 3$

14. $y + 5$

15. rt

16. cd

17. $4 + cd$

18. $pq + 14$

19. $yz + x$

20. $s + qt$

Use an associative law to find an equivalent expression.

21. $m + (n + 2)$

22. $5 \cdot (p \cdot q)$

23. $(7 \cdot x) \cdot y$

24. $(7 + p) + q$

Use the commutative laws and the associative laws to find three equivalent expressions.

25. $(a + b) + 8$

26. $(4 + x) + y$

27. $7 \cdot (a \cdot b)$

28. $(8 \cdot m) \cdot n$

d *Multiply.*

29. $4(a + 1)$

30. $3(c + 1)$

31. $8(x - y)$

32. $7(b - c)$

33. $-5(2a + 3b)$

34. $-2(3c + 5d)$

35. $2a(b - c + d)$

36. $5x(y - z + w)$

37. $2\pi r(h + 1)$

38. $P(1 + rt)$

39. $\dfrac{1}{2}h(a + b)$

40. $\dfrac{1}{4}\pi r(1 + s)$

List the terms of each of the following.

41. $4a - 5b + 6$

42. $5x - 9y + 12$

43. $2x - 3y - 2z$

44. $5a - 7b - 9c$

Factor.

45. $24x + 24y$

46. $9a + 9b$

47. $7p - 7$

48. $22x - 22$

49. $7x - 21$

50. $6y - 36$

51. $xy + x$

52. $ab + a$

53. $2x - 2y + 2z$

54. $3x + 3y - 3z$

55. $3x + 6y - 3$

56. $4a + 8b - 4$

57. $4w - 12z + 8$

58. $8m + 4n - 24$

59. $20x - 36y - 12$

60. $18a - 24b - 48$

61. $ab + ac - ad$

62. $xy - xz + xw$

63. $\dfrac{1}{4}\pi rr + \dfrac{1}{4}\pi rs$

64. $\dfrac{1}{2}ah + \dfrac{1}{2}bh$

▶ **Skill Maintenance**

Translate to an algebraic expression. **[R.4a]**

65. The square of the sum of two numbers

66. The sum of the squares of two numbers

Rewrite using a positive exponent. **[R.3b]**

67. x^{-4}

68. $\dfrac{1}{n^{-5}}$

Simplify. **[R.3c]**

69. $4 \cdot 2 - 5^2 - 3^2$

70. $10 \cdot 25 - 3 \cdot 2^3$

▶ **Synthesis**

Make substitutions to determine whether each pair of expressions is equivalent.

71. $x^2 + y^2$; $(x + y)^2$

72. $(a - b)(a + b)$; $a^2 - b^2$

73. $x^2 \cdot x^3$; x^5

74. $\dfrac{x^8}{x^4}$; x^2

| **R.6** | **Simplifying Algebraic Expressions** |

▶ **a**   Simplify an expression by collecting like terms.

▶ **b**   Simplify an expression by removing parentheses and collecting like terms.

We often wish to find a simpler expression equivalent to a given one.

▶ **a**   **Collecting Like Terms**

If two terms have the same variable factors, we say that they are **like terms**, or **similar terms**.

If two terms are simply numbers, they are also similar terms. We can simplify by **collecting**, or **combining**, **like terms**, using the distributive laws.

EXAMPLES   Collect like terms.

1. $3x + 5x = (3 + 5)x = 8x$ **Factoring out x using the distributive law**

2. $x - 3x = 1 \cdot x - 3 \cdot x = (1 - 3)x = -2x$ $x = 1 \cdot x$

3. $2x + 3y - 5x - 2y$
$= 2x + 3y + (-5x) + (-2y)$ **Subtracting by adding an opposite**
$= 2x + (-5x) + 3y + (-2y)$ **Using a commutative law**
$= (2 - 5)x + (3 - 2)y$ **Using a distributive law**
$= -3x + y$ **Simplifying**

4. $3x + 2x + 5 + 7 = (3 + 2)x + (5 + 7) = 5x + 12$

5. $4.2x - 6.7y - 5.8x + 23y = (4.2 - 5.8)x + (-6.7 + 23)y$
$= -1.6x + 16.3y$

6. $-\dfrac{1}{4}a + \dfrac{1}{2}b - \dfrac{3}{5}a - \dfrac{2}{5}b = \left(-\dfrac{1}{4} - \dfrac{3}{5}\right)a + \left(\dfrac{1}{2} - \dfrac{2}{5}\right)b$

$$= \left(-\dfrac{5}{20} - \dfrac{12}{20}\right)a + \left(\dfrac{5}{10} - \dfrac{4}{10}\right)b$$

$$= -\dfrac{17}{20}a + \dfrac{1}{10}b$$

> **Now Try Exercises 7 and 17.**

▶ **b Multiplying by −1 and Removing Parentheses**

What happens when we multiply a number by -1?

EXAMPLES

7. $-1 \cdot 9 = -9$ **8.** $-1 \cdot \left(-\dfrac{3}{5}\right) = \dfrac{3}{5}$ **9.** $-1 \cdot 0 = 0$

THE PROPERTY OF − 1

For any number a,

$$-1 \cdot a = -a.$$

(Negative 1 times a is the opposite of a. In other words, changing the sign is the same as multiplying by -1.)

By replacing $-$ with -1, we can find an equivalent expression for an opposite.

EXAMPLES Find an equivalent expression without parentheses.

10. $-(3x) = -1(3x)$ Replacing $-$ with -1 using the property of -1

$\qquad\quad = (-1 \cdot 3)x$ Using an associative law

$\qquad\quad = -3x$ Multiplying

11. $-(-9y) = -1(-9y)$ Replacing $-$ with -1

$\qquad\quad\; = [-1(-9)]y$ Using an associative law

$\qquad\quad\; = 9y$ Multiplying   ▶ **Now Try Exercise 25.**

EXAMPLES Find an equivalent expression without parentheses.

12. $-(4 + x) = -1(4 + x)$ Replacing $-$ with -1

$\qquad\quad\;\; = -1 \cdot 4 + (-1) \cdot x$ Multiplying using the distributive law

$\qquad\quad\;\; = -4 + (-x)$ Replacing $-1 \cdot x$ with $-x$

$\qquad\quad\;\; = -4 - x$ Adding an opposite is the same as subtracting.

13. $-(3x - 2y + 4) = -1(3x - 2y + 4)$

$\qquad\qquad\qquad\;\; = -1 \cdot 3x - (-1)2y + (-1)4$ Using the distributive law

$\qquad\qquad\qquad\;\; = -3x - (-2y) + (-4)$ Multiplying

$\qquad\qquad\qquad\;\; = -3x + 2y - 4$

14. $-(a - b) = -1(a - b) = -1 \cdot a - (-1) \cdot b$
$$= -a + b = b - a$$

Example 14 shows that the opposite of $a - b$ is $b - a$.

> **Now Try Exercises 31 and 35.**

THE OPPOSITE OF A DIFFERENCE

For any real numbers a and b,

$$-(a - b) = b - a.$$

(The opposite of $a - b$ is $b - a$.)

We can find an equivalent expression for an opposite by multiplying every term by -1. We could also say that we change the sign of every term inside the parentheses.

EXAMPLE 15   Find an equivalent expression without parentheses:

$$-(-9t + 7z - \tfrac{1}{4}w).$$

We have

$$-(-9t + 7z - \tfrac{1}{4}w) = 9t - 7z + \tfrac{1}{4}w. \qquad \text{Changing the sign of every term}$$

> **Now Try Exercise 39.**

Some expressions contain parentheses preceded by subtraction signs. These parentheses can be removed by changing the sign of *every* term inside.

EXAMPLES   Remove parentheses and simplify.

16. $6x - (4x + 2) = 6x - 4x - 2 \qquad \text{Changing the sign of every term inside}$
$$= 2x - 2 \qquad \text{Collecting like terms}$$

17. $3y - 4 - (9y - 7) = 3y - 4 - 9y + 7$
$$= -6y + 3, \text{ or } 3 - 6y$$

> **Now Try Exercise 45.**

In Examples 16 and 17, we see the reason for the word "simplify." The expression $2x - 2$ is equivalent to $6x - (4x + 2)$ but it is shorter.

If parentheses are preceded by an addition sign, *no* signs are changed when they are removed.

EXAMPLE 18   Remove parentheses and simplify.

$$3y + (3x - 8) - (5 - 12y) = 3y + 3x - 8 - 5 + 12y$$
$$= 15y + 3x - 13$$

> **Now Try Exercise 51.**

We can also simplify when an expression is multiplied by a number other than 1.

EXAMPLES   Remove parentheses and simplify.

19. $\frac{1}{3}(15x - 4) - (5x + 2y) + 1 = \frac{1}{3} \cdot 15x - \frac{1}{3} \cdot 4 - 5x - 2y + 1$

$$= 5x - \frac{4}{3} - 5x - 2y + 1$$

$$= -2y - \frac{1}{3}$$

20. $x - 3(x + y) = x - 3x - 3y$ Removing parentheses by multiplying $x + y$ by -3

$$= -2x - 3y$$ Collecting like terms

> **CAUTION!**   A common error is to forget to change this sign. *Remember*: When multiplying by a negative number, change the sign of *every* term inside the parentheses.

21. $3y - 2(4y - 5) = 3y - 8y + 10$ Removing parentheses by multiplying $4y - 5$ by -2

$$= -5y + 10$$ Collecting like terms

Now Try Exercises 49 and 53.

When expressions with grouping symbols contain variables, we still work from the inside out when simplifying, using the rules for order of operations.

EXAMPLE 22   Simplify: $[2(x + 7) - 4^2] - (2 - x)$.

$[2(x + 7) - 4^2] - (2 - x)$

$= [2x + 14 - 4^2] - (2 - x)$ Multiplying to remove the innermost grouping symbols using the distributive law

$= [2x + 14 - 16] - (2 - x)$ Evaluating the exponential expression

$= [2x - 2] - (2 - x)$ Collecting like terms inside the brackets

$= 2x - 2 - 2 + x$ Multiplying by -1 to remove the parentheses

$= 3x - 4$ Collecting like terms

Now Try Exercise 55.

EXAMPLE 23   Simplify: $6y - \{4[3(y - 2) - 4(y + 2)] - 3\}$.

$6y - \{4[3(y - 2) - 4(y + 2)] - 3\}$

$= 6y - \{4[3y - 6 - 4y - 8] - 3\}$ Multiplying to remove the innermost grouping symbols using the distributive law

$= 6y - \{4[-y - 14] - 3\}$ Collecting like terms inside the brackets

$= 6y - \{-4y - 56 - 3\}$ Multiplying to remove the brackets using the distributive law

$= 6y - \{-4y - 59\}$ Collecting like terms in the braces

$= 6y + 4y + 59$ Removing braces

$= 10y + 59$ Collecting like terms

Now Try Exercise 63.

R.6   Exercise Set

▶ ## Reading Check

Determine whether each statement is true or false.

RC1. The terms $6x$ and $-7x$ are like terms.

RC2. The terms $9y$ and $9c$ are similar terms.

RC3. Multiplying by -1 is the same as changing the sign.

RC4. The opposite of $5 - x$ is $x - 5$.

RC5. The expression $6x + (-7x)$ is equivalent to the expression $-x$.

RC6. The expression $-(5c + 6d - w)$ is equivalent to the expression $-5c - 6d - w$.

a *Collect like terms.*

1. $7x + 5x$

2. $6a + 9a$

3. $8b - 11b$

4. $9c - 12c$

5. $14y + y$

6. $13x + x$

7. $12a - a$

8. $15x - x$

9. $t - 9t$

10. $x - 6x$

11. $5x - 3x + 8x$

12. $3x - 11x + 2x$

13. $3x - 5y + 8x$

14. $4a - 9b + 10a$

15. $3c + 8d - 7c + 4d$

16. $12a + 3b - 5a + 6b$

17. $4x - 7 + 18x + 25$

18. $13p + 5 - 4p + 7$

19. $1.3x + 1.4y - 0.11x - 0.47y$

20. $0.17a + 1.7b - 12a - 38b$

21. $\dfrac{2}{3}a + \dfrac{5}{6}b - 27 - \dfrac{4}{5}a - \dfrac{7}{6}b$

22. $-\dfrac{1}{4}x - \dfrac{1}{2}x + \dfrac{1}{4}y + \dfrac{1}{2}y - 34$

b *The **perimeter** of a rectangle is the distance around it. The perimeter P is given by $P = 2(l + w)$.*

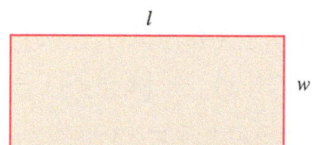

l

w

23. Find an equivalent expression for the perimeter formula $P = 2(l + w)$ by multiplying.

24. *Perimeter of a Football Field.* The standard football field has $l = 360$ ft and $w = 160$ ft. Evaluate both expressions in Exercise 23 to find the perimeter.

Find an equivalent expression without parentheses.

25. $-(-2c)$

26. $-(-5y)$

27. $-(b + 4)$

28. $-(a + 9)$

29. $-(b - 3)$

30. $-(x - 8)$

31. $-(t - y)$

32. $-(r - s)$

33. $-(x + y + z)$

34. $-(r + s + t)$

35. $-(8x - 6y + 13)$

36. $-(9a - 7b + 24)$

37. $-(-2c + 5d - 3e + 4f)$

38. $-(-4x + 8y - 5w + 9z)$

39. $-\left(-1.2x + 56.7y - 34z - \dfrac{1}{4}\right)$

40. $-\left(-x + 2y - \dfrac{2}{3}z - 56.3w\right)$

Simplify by removing parentheses and collecting like terms.

41. $a + (2a + 5)$

42. $x + (5x + 9)$

43. $4m - (3m - 1)$

44. $5a - (4a - 3)$

45. $5d - 9 - (7 - 4d)$

46. $6x - 7 - (9 - 3x)$

47. $-2(x + 3) - 5(x - 4)$

48. $-9(y + 7) - 6(y - 3)$

49. $5x - 7(2x - 3) - 4$

50. $8y - 4(5y - 6) + 9$

51. $8x - (-3y + 7) + (9x - 11)$

52. $-5t + (4t - 12) - 2(3t + 7)$

53. $\dfrac{1}{4}(24x - 8) - \dfrac{1}{2}(-8x + 6) - 14$

54. $-\dfrac{1}{2}(10t - w) + \dfrac{1}{4}(-28t + 4) + 1$

Simplify.

55. $7a - [9 - 3(5a - 2)]$

56. $14b - [7 - 3(9b - 4)]$

57. $5\{-2 + 3[4 - 2(3 + 5)]\}$

58. $7\{-7 + 8[5 - 3(4 + 6)]\}$

59. $[10(x + 3) - 4] + [2(x - 1) + 6]$

60. $[9(x + 5) - 7] + [4(x - 12) + 9]$

61. $[7(x + 5) - 19] - [4(x - 6) + 10]$

62. $[6(x + 4) - 12] - [5(x - 8) + 11]$

63. $3\{[7(x - 2) + 4] - [2(2x - 5) + 6]\}$

64. $4\{[8(x - 3) + 9] - [4(3x - 7) + 2]\}$

65. $4\{[5(x - 3) + 2^2] - 3[2(x + 5) - 9^2]\}$

66. $3\{[6(x - 4) + 5^2] - 2[5(x + 8) - 10^2]\}$

67. $2y + \{8[3(2y - 5) - (8y + 9)] + 6\}$

68. $7b - \{5[4(3b - 8) - (9b + 10)] + 14\}$

► **Skill Maintenance**

Evaluate. **[R.4b]**

69. $10 - x$, when $x = -3$

70. $3d \div 2c$, when $c = 5$ and $d = 10$

71. $a + n(n^2 - 1)$, when $a = 100$ and $n = -2$

72. $x^2 \div 3(y - z)$, when $x = 6, y = 8$, and $z = 10$

Divide. **[R.2e]**

73. $-256 \div 16$

74. $-256 \div (-16)$

75. $256 \div (-16)$

76. $-\dfrac{3}{8} \div \dfrac{9}{4}$

Multiply. **[R.5d]**

77. $8(a - b)$

78. $-8(2a - 3b + 4)$

79. $6x(a - b + 2c)$

80. $\dfrac{2}{3}(24x - 12y + 15)$

Factor. **[R.5d]**

81. $24a - 24$

82. $24a - 16b$

83. $ab - ac + a$

84. $15p + 45q - 10$

► **Synthesis**

Insert one pair of parentheses to convert the false statement into a true statement.

85. $3 - 8^2 + 9 = 34$

86. $2 \cdot 7 + 3^2 \cdot 5 = 104$

87. $5 \cdot 2^3 \div 3 - 4^4 = 40$

88. $2 - 7 \cdot 2^2 + 9 = -11$

Simplify.

89. $[11(a - 3) + 12a] - \{6[4(3b - 7) - (9b + 10)] + 11\}$

90. $-3[9(x - 4) + 5x] - 8\{3[5(3y + 4)] - 12\}$

91. $z - \{2z + [3z - (4z + 5x) - 6z] + 7z\} - 8z$

92. $\{x + [f - (f + x)] + [x - f]\} + 3x$

93. $x - \{x + 1 - [x + 2 - (x - 3 - \{x + 4 - [x - 5 + (x - 6)]\})]\}$

R.7 Properties of Exponents and Scientific Notation

▶ **a** Use exponential notation in multiplication and division.

▶ **b** Use exponential notation in raising a power to a power, and in raising a product or a quotient to a power.

▶ **c** Convert between decimal notation and scientific notation, and use scientific notation with multiplication and division.

We often need to find ways to determine *equivalent exponential expressions*. We do this with several rules or properties regarding exponents.

▶ a Multiplication and Division

To see how to multiply, or simplify, in an expression such as $a^3 \cdot a^2$, we use the definition of exponential notation:

$$a^3 \cdot a^2 = \underbrace{a \cdot a \cdot a}_{3 \text{ factors}} \cdot \underbrace{a \cdot a}_{2 \text{ factors}} = a^5.$$

The exponent in a^5 is the *sum* of those in $a^3 \cdot a^2$. In general, the exponents are added when we multiply if the base is the same in all factors.

THE PRODUCT RULE

For any number a and any integers m and n,

$$a^m \cdot a^n = a^{m+n}.$$

(When multiplying with exponential notation, add the exponents if the bases are the same.)

EXAMPLES Multiply and simplify.

1. $x^4 \cdot x^3 = x^{4+3} = x^7$

2. $4^5 \cdot 4^{-3} = 4^{5+(-3)} = 4^2 = 16$

3. $(-2)^{-3}(-2)^7 = (-2)^{-3+7}$
$= (-2)^4 = 16$

4. $(8x^n)(6x^{2n}) = 8 \cdot 6 \cdot x^n \cdot x^{2n}$
$= 48 \cdot x^{n+2n}$
$= 48x^{3n}$

5. $(8x^4y^{-2})(-3x^{-3}y) = 8 \cdot (-3) \cdot x^4 \cdot x^{-3} \cdot y^{-2} \cdot y^1$
$= -24x^{4-3}y^{-2+1}$
$= -24xy^{-1} = -\dfrac{24x}{y}$ Using $a^{-n} = \dfrac{1}{a^n}$

Note that we give answers using positive exponents. In some situations, this may not be appropriate, but we do so here.

Now Try Exercises 1 and 15.

Consider this division:

$$\frac{8^5}{8^3} = \frac{8 \cdot 8 \cdot 8 \cdot 8 \cdot 8}{8 \cdot 8 \cdot 8} = \frac{8 \cdot 8 \cdot 8}{8 \cdot 8 \cdot 8} \cdot 8 \cdot 8 = 8 \cdot 8 = 8^2.$$

We can obtain the result by subtracting exponents.

THE QUOTIENT RULE

For any nonzero number a and any integers m and n,

$$\frac{a^m}{a^n} = a^{m-n}.$$

(When dividing with exponential notation, subtract the exponent of the denominator from the exponent of the numerator, if the bases are the same.)

EXAMPLES   Divide and simplify.

6. $\dfrac{5^7}{5^3} = 5^{7-3} = 5^4$ **Subtracting exponents using the quotient rule**

7. $\dfrac{5^7}{5^{-3}} = 5^{7-(-3)} = 5^{7+3} = 5^{10}$ **Subtracting exponents (adding an opposite)**

8. $\dfrac{9^{-2}}{9^5} = 9^{-2-5} = 9^{-7} = \dfrac{1}{9^7}$

9. $\dfrac{7^{-4}}{7^{-5}} = 7^{-4-(-5)} = 7^{-4+5} = 7^1 = 7$

10. $\dfrac{16x^4y^7}{-8x^3y^9} = \dfrac{16}{-8} \cdot \dfrac{x^4}{x^3} \cdot \dfrac{y^7}{y^9} = -2x^{4-3}y^{7-9} = -2x^1y^{-2} = -\dfrac{2x}{y^2}$

The answers $\dfrac{-2x}{y^2}$ or $\dfrac{2x}{-y^2}$ would also be correct here.

11. $\dfrac{40x^{-2n}}{4x^{5n}} = \dfrac{40}{4} \cdot \dfrac{x^{-2n}}{x^{5n}} = 10x^{-2n-5n} = 10x^{-7n} = \dfrac{10}{x^{7n}}$

12. $\dfrac{14x^5y^{-3}}{4x^9y^{-5}} = \dfrac{14}{4} \cdot \dfrac{x^5}{x^9} \cdot \dfrac{y^{-3}}{y^{-5}} = \dfrac{7}{2} x^{5-9}y^{-3-(-5)}$

$\qquad\qquad = \dfrac{7}{2} x^{-4}y^2 = \dfrac{7}{2} \cdot \dfrac{1}{x^4} \cdot \dfrac{y^2}{1}$

$\qquad\qquad = \dfrac{7y^2}{2x^4}$

In exercises such as Examples 6–12 above, it may help to think as follows: After writing the base, write the top exponent. Then write a subtraction sign. Then write the bottom exponent. For example,

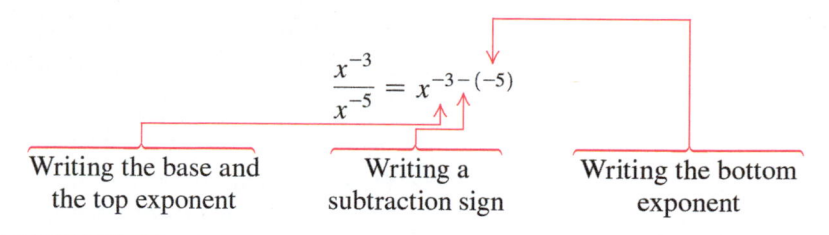

$$\frac{x^{-3}}{x^{-5}} = x^{-3-(-5)}$$

Writing the base and the top exponent Writing a subtraction sign Writing the bottom exponent

> **Now Try Exercises 27 and 35.**

▶ b Raising Powers to Powers and Products and Quotients to Powers

When an expression inside parentheses is raised to a power, the inside expression is the base. Thus, for $(5^2)^4$, we are raising 5^2 to the fourth power:

$$
\begin{aligned}
(5^2)^4 &= (5^2)(5^2)(5^2)(5^2) \\
&= (5 \cdot 5)(5 \cdot 5)(5 \cdot 5)(5 \cdot 5) \\
&= 5 \cdot 5 \cdot 5 \cdot 5 \cdot 5 \cdot 5 \cdot 5 \cdot 5 \qquad \text{\color{red}{Using an associative law}} \\
&= 5^8.
\end{aligned}
$$

Note that here we could have multiplied the exponents:

$$(5^2)^4 = 5^{2 \cdot 4} = 5^8.$$

THE POWER RULE

For any real number a and any integers m and n,

$$(a^m)^n = a^{mn}.$$

(To raise a power to a power, multiply the exponents.)

EXAMPLES Simplify.

13. $(x^5)^7 = x^{5 \cdot 7}$ **Multiply exponents.**   **14.** $(y^{-2})^{-2} = y^{(-2)(-2)}$
$\qquad\qquad = x^{35}$ $\qquad\qquad\qquad\qquad\qquad\qquad = y^4$

15. $(x^{-5})^4 = x^{-5 \cdot 4}$ **16.** $(x^4)^{-2t} = x^{4(-2t)}$
$\qquad\qquad = x^{-20} = \dfrac{1}{x^{20}}$ $\qquad\qquad\qquad\qquad = x^{-8t} = \dfrac{1}{x^{8t}}$

> **Now Try Exercise 43.**

Let's compare $2a^3$ and $(2a)^3$:

$$2a^3 = 2 \cdot a \cdot a \cdot a \qquad \text{\color{red}{The base is } } a.$$

and

$$(2a)^3 = (2a)(2a)(2a) \qquad \text{\color{red}{The base is } } 2a.$$
$$= (2 \cdot 2 \cdot 2)(a \cdot a \cdot a) \qquad \text{\color{red}{Using the commutative law and the}}$$
$$\text{\color{red}{associative law of multiplication}}$$
$$= 2^3 a^3 = 8a^3.$$

We see that $2a^3$ and $(2a)^3$ are *not* equivalent. We also see that we can evaluate the power $(2a)^3$ by raising each factor to the power 3. This leads us to the following rule for raising a product to a power.

RAISING A PRODUCT TO A POWER

For any real numbers a and b and any integer n,

$$(ab)^n = a^n b^n.$$

(To raise a product to the nth power, raise each factor to the nth power.)

EXAMPLES Simplify.

17. $(3x^2 y^{-2})^3 = 3^3 (x^2)^3 (y^{-2})^3 = 3^3 x^6 y^{-6} = 27 x^6 y^{-6} = \dfrac{27 x^6}{y^6}$

18. $(5x^3 y^{-5} z^2)^4 = 5^4 (x^3)^4 (y^{-5})^4 (z^2)^4 = 625 x^{12} y^{-20} z^8 = \dfrac{625 x^{12} z^8}{y^{20}}$

> **Now Try Exercise 47.**

There is a similar rule for raising a quotient to a power.

RAISING A QUOTIENT TO A POWER

For real numbers a and b, and any integer n,

$$\left(\frac{a}{b}\right)^n = \frac{a^n}{b^n}, b \neq 0; \quad \text{and} \quad \left(\frac{a}{b}\right)^{-n} = \left(\frac{b}{a}\right)^n = \frac{b^n}{a^n}, a \neq 0, b \neq 0.$$

(To raise a quotient to the nth power, raise the numerator to the nth power and divide by the denominator to the nth power.)

EXAMPLES Simplify. Write the answer using positive exponents.

19. $\left(\dfrac{x^2}{y^{-3}}\right)^{-5} = \dfrac{x^{2 \cdot (-5)}}{y^{-3 \cdot (-5)}} = \dfrac{x^{-10}}{y^{15}} = \dfrac{1}{x^{10} y^{15}}$

20. $\left(\dfrac{2x^3 y^{-2}}{3y^4}\right)^5 = \dfrac{(2x^3 y^{-2})^5}{(3y^4)^5} = \dfrac{2^5 (x^3)^5 (y^{-2})^5}{3^5 (y^4)^5}$

$$= \dfrac{32 x^{15} y^{-10}}{243 y^{20}} = \dfrac{32}{243} \cdot x^{15} \cdot y^{-10-20}$$

$$= \dfrac{32}{243} \cdot x^{15} y^{-30} = \dfrac{32}{243} \cdot \dfrac{x^{15}}{1} \cdot \dfrac{1}{y^{30}} = \dfrac{32 x^{15}}{243 y^{30}}$$

21. $\left[\dfrac{-3a^{-5}b^3}{2a^{-2}b^{-4}}\right]^{-2} = \dfrac{(-3a^{-5}b^3)^{-2}}{(2a^{-2}b^{-4})^{-2}} = \dfrac{(-3)^{-2}(a^{-5})^{-2}(b^3)^{-2}}{2^{-2}(a^{-2})^{-2}(b^{-4})^{-2}}$

$= \dfrac{(-3)^{-2}a^{10}b^{-6}}{2^{-2}a^4b^8} = (-3)^{-2} \cdot \dfrac{1}{2^{-2}} \cdot a^{10-4} \cdot b^{-6-8}$

$= \dfrac{1}{(-3)^2} \cdot \dfrac{2^2}{1} \cdot a^6 \cdot b^{-14}$

$= \dfrac{1}{9} \cdot \dfrac{4}{1} \cdot \dfrac{a^6}{1} \cdot \dfrac{1}{b^{14}} = \dfrac{4a^6}{9b^{14}}$

An alternative way to carry out Example 21 is to first write the expression with a positive exponent, as follows:

$\left[\dfrac{-3a^{-5}b^3}{2a^{-2}b^{-4}}\right]^{-2} = \left[\dfrac{2a^{-2}b^{-4}}{-3a^{-5}b^3}\right]^2 = \dfrac{(2a^{-2}b^{-4})^2}{(-3a^{-5}b^3)^2} = \dfrac{2^2(a^{-2})^2(b^{-4})^2}{(-3)^2(a^{-5})^2(b^3)^2}$

$= \dfrac{4a^{-4}b^{-8}}{9a^{-10}b^6} = \dfrac{4}{9}a^{-4-(-10)}b^{-8-6} = \dfrac{4}{9}a^6b^{-14} = \dfrac{4a^6}{9b^{14}}.$

▶ **Now Try Exercises 55 and 57.**

▶ c Scientific Notation

There are many kinds of symbolism, or *notation*, for numbers. You are already familiar with fraction notation, decimal notation, and percent notation. Now we study another, **scientific notation**, which is especially useful when representing very large or very small numbers and when estimating.

The following are examples of scientific notation:

- The planet Saturn is about 890,800,000 mi from the sun.

  890,800,000 mi $= 8.908 \times 10^8$ mi

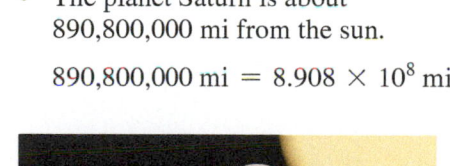

- Great Britain spent about $15 billion to stage the Summer 2012 Olympic Games.

  $15 billion $= \$15{,}000{,}000{,}000$
  $= \$1.5 \times 10^{10}$

  (*Source*: cnn.com)

- The diameter of a helium atom is about 0.000000022 cm.

  0.000000022 cm $= 2.2 \times 10^{-8}$ cm

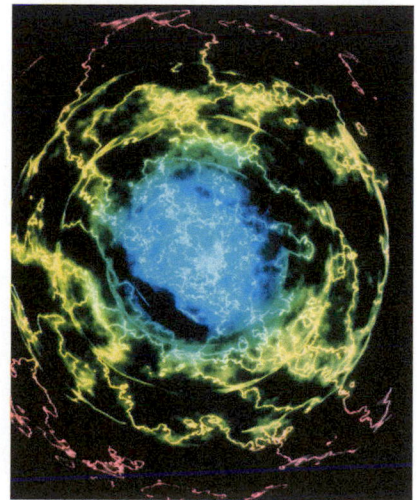

SCIENTIFIC NOTATION

Scientific notation for a number is an expression of the type

$$M \times 10^n,$$

where n is an integer, M is greater than or equal to 1 and less than 10 ($1 \le M < 10$), and M is expressed in decimal notation. 10^n is also considered to be scientific notation when $M = 1$.

You should try to make conversions to scientific notation mentally as much as possible. Here is a handy mental device.

A positive exponent in scientific notation indicates a large number (greater than or equal to 10) and a negative exponent indicates a small number (between 0 and 1).

EXAMPLES   Convert mentally to scientific notation.

22. Light travels 9,460,000,000,000 km in one year.

Think: Large number, so the exponent is positive.

$$9{,}460{,}000{,}000{,}000 = 9.46 \times 10^{12} \qquad 9.460{,}000{,}000{,}000.$$

12 places

23. The mass of a grain of sand is 0.0648 g (grams).

Think: Small number, so the exponent is negative.

$$0.0648 = 6.48 \times 10^{-2} \qquad 0.06.48$$

2 places

Now Try Exercises 79 and 81.

EXAMPLES   Convert mentally to decimal notation.

24. $4.893 \times 10^5 = 489{,}300 \qquad 4.89300.$

5 places

Positive exponent, indicating a large number.

25. $8.7 \times 10^{-8} = 0.000000087 \qquad 0.00000008.7$

8 places

Negative exponent, indicating a small number.

Now Try Exercises 87 and 89.

Each of the following is *not* scientific notation.

$$13.95 \times 10^{13}, \qquad\qquad 0.468 \times 10^{-8}$$

This number is greater than 10.   This number is less than 1.

We can use the properties of exponents when we multiply and divide in scientific notation.

EXAMPLE 26   Multiply and write scientific notation for the answer:

$$(3.1 \times 10^5)(4.5 \times 10^{-3}).$$

We apply the commutative laws and the associative laws to get

$$(3.1 \times 10^5)(4.5 \times 10^{-3}) = (3.1 \times 4.5)(10^5 \times 10^{-3}) = 13.95 \times 10^2.$$

To find scientific notation for the result, we convert 13.95 to scientific notation and then simplify:

$$13.95 \times 10^2 = (1.395 \times 10^1) \times 10^2 = 1.395 \times 10^3.$$

> **Now Try Exercise 95.**

EXAMPLE 27   Divide and write scientific notation for the answer:

$$\frac{6.4 \times 10^{-7}}{8.0 \times 10^6}.$$

$$\frac{6.4 \times 10^{-7}}{8.0 \times 10^6} = \frac{6.4}{8.0} \times \frac{10^{-7}}{10^6} \qquad \text{\color{red}{\textbf{Factoring, showing two divisions}}}$$

$$= 0.8 \times 10^{-13} \qquad \text{\color{red}{\textbf{Dividing; this result is not in scientific notation.}}}$$

$$= (8.0 \times 10^{-1}) \times 10^{-13} \qquad \text{\color{red}{\textbf{Converting 0.8 to scientific notation}}}$$

$$= 8.0 \times (10^{-1} \times 10^{-13}) \qquad \text{\color{red}{\textbf{Using the associative law of multiplication}}}$$

$$= 8.0 \times 10^{-14} \qquad \qquad \text{\color{red}{\textbf{Now Try Exercise 99.}}}$$

EXAMPLE 28   *Distance from Earth to Mars.*   When the Curiosity Mars Rover reached Mars on August 6, 2012, it took 14 min for a signal to reach Earth from Mars. If the signal travels at light speed, or 1.86×10^5 mi/sec, how far was Earth from Mars on that date? (*Source*: time.com)

We first convert minutes to seconds:

$$14 \text{ min} = 14 \times 60 \text{ sec} = 840 \text{ sec} = 8.4 \times 10^2 \text{ sec}.$$

Then we multiply rate and time to find distance:

$$\left(1.86 \times 10^5 \frac{\text{mi}}{\text{sec}}\right) \cdot (8.4 \times 10^2 \text{ sec}) = (1.86 \times 8.4)(10^5 \times 10^2) \text{ mi}$$
$$= 15.624 \times 10^7 \text{ mi}$$
$$= (1.5624 \times 10) \times 10^7 \text{ mi}$$
$$= 1.5624 \times 10^8 \text{ mi}.$$

On that date, Earth was 1.5624×10^8 mi, or 156,240,000 mi, from Mars.

> **Now Try Exercise 101.**

The following table lists the meanings of prefixes commonly used with units of measure.

Prefix	Meaning	Prefix	Meaning
exa-	10^{18}	atto-	10^{-18}
peta-	10^{15}	femto-	10^{-15}
tera-	10^{12}	pico-	10^{-12}
giga-	10^{9}	nano-	10^{-9}
mega-	10^{6}	micro-	10^{-6}
kilo-	10^{3}	milli-	10^{-3}
hecto-	10^{2}	centi-	10^{-2}

EXAMPLE 29 *Relative Size.* A molecule of hemoglobin is about 6.5 nanometers in diameter. A human egg is about 130 micrometers in diameter. How many times larger is a human egg than a hemoglobin molecule? (*Source:* University of Utah, Genetic Science Learning Center)

To determine how many times larger a human egg is than a hemoglobin molecule, we divide:

$$\frac{130 \text{ micrometers}}{6.5 \text{ nanometers}} = \frac{130 \times 10^{-6} \text{ meters}}{6.5 \times 10^{-9} \text{ meters}} = \frac{1.3 \times 10^{-4}}{6.5 \times 10^{-9}}$$

$$= \frac{1.3}{6.5} \times \frac{10^{-4}}{10^{-9}} = 0.2 \times 10^{5} = 2 \times 10^{4}.$$

A human egg is 2×10^{4}, or 20,000, times larger than a hemoglobin molecule.

Now Try Exercise 103.

Technology Connection

Scientific Notation

To enter a number in scientific notation on a graphing calculator, we first type the decimal portion of the number. Then we press **2ND** **EE**. (EE is the second operation associated with the **,** key.) Finally, we type the exponent, which can be at most two digits. The graphing calculator can be used to perform computations using scientific notation. To find the product in Example 26 and express the result in scientific notation, we first set the calculator in Scientific mode using the **MODE** key. The decimal portion of the number appears before a small E and the exponent follows the E, so we read the result shown at left as 1.395×10^{3}.

```
3.1E5*4.5E−3
              1.395E3
```

Exercises

Multiply or divide and express the answer in scientific notation.

1. $(5.13 \times 10^{8})(2.4 \times 10^{-13})$

2. $(7 \times 10^{9})(4 \times 10^{-5})$

3. $\dfrac{4.8 \times 10^{6}}{1.6 \times 10^{12}}$

4. $\dfrac{6 \times 10^{-10}}{5 \times 10^{4}}$

R.7 Exercise Set

▶ **Reading Check**

Match each expression with an equivalent expression from the column on the right.

RC1. _____ $4^5 \cdot 4^3$

RC2. _____ $(4^5)^3$

RC3. _____ $\dfrac{4^5}{4^3}$

RC4. _____ $\left(\dfrac{4}{5}\right)^3$

RC5. _____ $\left(\dfrac{4}{5}\right)^{-3}$

RC6. _____ $(4 \cdot 5)^3$

a) 4^{5-3}

b) $4^{5 \cdot 3}$

c) 4^{5+3}

d) $4^3 \cdot 5^3$

e) $\dfrac{4^3}{5^3}$

f) $\dfrac{5^3}{4^3}$

....

a *Multiply and simplify.*

1. $3^6 \cdot 3^3$

2. $8^2 \cdot 8^6$

3. $6^{-6} \cdot 6^2$

4. $9^{-5} \cdot 9^3$

5. $8^{-2} \cdot 8^{-4}$

6. $9^{-1} \cdot 9^{-6}$

7. $b^2 \cdot b^{-5}$

8. $a^4 \cdot a^{-3}$

9. $a^{-3} \cdot a^4 \cdot a^2$

10. $x^{-8} \cdot x^5 \cdot x^3$

11. $(2x)^3 \cdot (3x)^2$

12. $(9y)^2 \cdot (2y)^3$

13. $(14m^2n^3)(-2m^3n^2)$

14. $(6x^5y^{-2})(-3x^2y^3)$

15. $(-2x^{-3})(7x^{-8})$

16. $(6x^{-4}y^3)(-4x^{-8}y^{-2})$

17. $(15x^{4t})(7x^{-6t})$

18. $(9x^{-4n})(-4x^{-8n})$

19. $(2y^{3m})(-4y^{-9m})$

20. $(-3t^{-4a})(-5t^{-a})$

Divide and simplify.

21. $\dfrac{8^9}{8^2}$

22. $\dfrac{7^8}{7^2}$

23. $\dfrac{6^3}{6^{-2}}$

24. $\dfrac{5^{10}}{5^{-3}}$

25. $\dfrac{10^{-3}}{10^6}$

26. $\dfrac{12^{-4}}{12^8}$

27. $\dfrac{9^{-4}}{9^{-6}}$

28. $\dfrac{2^{-7}}{2^{-5}}$,

29. $\dfrac{x^{-4n}}{x^{6n}}$

30. $\dfrac{y^{-3t}}{y^{8t}}$

31. $\dfrac{w^{-11q}}{w^{-6q}}$

32. $\dfrac{m^{-7t}}{m^{-5t}}$

33. $\dfrac{a^3}{a^{-2}}$

34. $\dfrac{y^4}{y^{-5}}$

35. $\dfrac{27x^7z^5}{-9x^2z}$

36. $\dfrac{24a^5b^3}{-8a^4b}$

37. $\dfrac{-24x^6y^7}{18x^{-3}y^9}$

38. $\dfrac{14a^4b^{-3}}{-8a^8b^{-5}}$

39. $\dfrac{-18x^{-2}y^3}{-12x^{-5}y^5}$

40. $\dfrac{-14a^{14}b^{-5}}{-18a^{-2}b^{-10}}$

b *Simplify.*

41. $(4^3)^2$

42. $(5^4)^5$

43. $(8^4)^{-3}$

44. $(9^3)^{-4}$

45. $(6^{-4})^{-3}$

46. $(7^{-8})^{-5}$

47. $(5a^2b^2)^3$

48. $(2x^3y^4)^5$

49. $(-3x^3y^{-6})^{-2}$

50. $(-3a^2b^{-5})^{-3}$

51. $(-6a^{-2}b^3c)^{-2}$

52. $(-8x^{-4}y^5z^2)^{-4}$

53. $\left(\dfrac{4^{-3}}{3^4}\right)^3$

54. $\left(\dfrac{5^2}{4^{-3}}\right)^{-3}$

55. $\left(\dfrac{2x^3y^{-2}}{3y^{-3}}\right)^3$

56. $\left(\dfrac{-4x^4y^{-2}}{5x^{-1}y^4}\right)^{-4}$

57. $\left(\dfrac{125a^2b^{-3}}{5a^4b^{-2}}\right)^{-5}$

58. $\left(\dfrac{-200x^3y^{-5}}{8x^5y^{-7}}\right)^{-4}$

59. $\left(\dfrac{-6^5y^4z^{-5}}{2^{-2}y^{-2}z^3}\right)^6$

60. $\left(\dfrac{9^{-2}x^{-4}y}{3^{-3}x^{-3}y^2}\right)^8$

61. $[(-2x^{-4}y^{-2})^{-3}]^{-2}$

62. $[(-4a^{-4}b^{-5})^{-3}]^4$

63. $\left(\dfrac{3a^{-2}b}{5a^{-7}b^5}\right)^{-7}$ **64.** $\left(\dfrac{2x^2y^{-2}}{3x^8y^7}\right)^9$

65. $\dfrac{10^{2a+1}}{10^{a+1}}$ **66.** $\dfrac{11^{b+2}}{11^{3b-3}}$

67. $\dfrac{9a^{x-2}}{3a^{2x+2}}$ **68.** $\dfrac{-12x^{a+1}}{4x^{2-a}}$

69. $\dfrac{45x^{2a+4}y^{b+1}}{-9x^{a+3}y^{2+b}}$ **70.** $\dfrac{-28x^{b+5}y^{4+c}}{7x^{b-5}y^{c-4}}$

71. $(8^x)^{4y}$ **72.** $(7^{2p})^{3q}$

73. $(12^{3-a})^{2b}$ **74.** $(x^{a-1})^{3b}$

75. $(5x^{a-1}y^{b+1})^{2c}$

76. $(4x^{3a}y^{2b})^{5c}$

77. $\dfrac{4x^{2a+3}y^{2b-1}}{2x^{a+1}y^{b+1}}$ **78.** $\dfrac{25x^{a+b}y^{b-a}}{-5x^{a-b}y^{b+a}}$

C *Convert each number to scientific notation.*

79. 47,000,000,000

80. 2,600,000,000,000

81. 0.000000016

82. 0.000000263

83. *Coupon Redemptions.*   Shoppers redeemed 2,600,000,000 manufacturers' grocery coupons in a recent year. Write scientific notation for the number of coupons redeemed. (*Source*: CMS)

84. *Cell-Phone Subscribers.*   In 1985, there were 340 thousand cell-phone subscribers in the United States. By 2012, this number had increased to 322 million. Write the number of cell-phone subscribers in 1985 and in 2012 in scientific notation. (*Sources*: mobithinking.com; www.infoplease.com)

85. *Photoreceptor Rod.*   A photoreceptor rod is about 100 millionths of a meter long. Write 100 millionths in scientific notation. (*Source*: learn.genetics.utah. edu/content/begin/cells/scale)

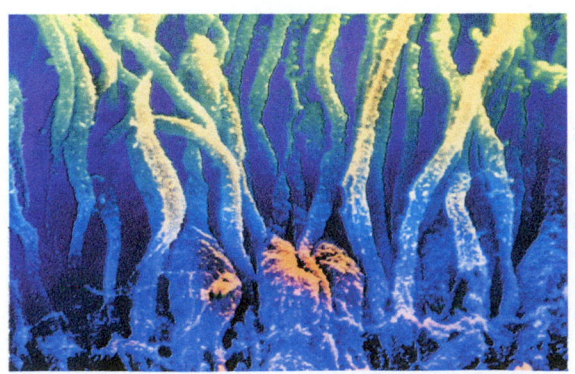

86. *Insect-Eating Lizard.*   A gecko is an insect-eating lizard. Its feet will adhere to virtually any surface because they contain millions of miniscule hairs, or setae, that are 200 billionths of a meter wide. Write 200 billionths in scientific notation. (*Source: The Proceedings of the National Academy of Sciences*, Dr. Kellar Autumn and Wendy Hansen of Lewis and Clark College, Portland, Oregon)

Convert each number to decimal notation.

87. 6.73×10^8

88. 9.24×10^7

89. The wavelength of a certain red light is 6.6×10^{-5} cm.

90. The mass of an electron is 9.11×10^{-28} g.

91. There were about 1.007 billion Facebook users worldwide in October 2012. (*Source*: Statista)

92. About 3.24 million tracks of rock music were sold in 2012. (*Source*: Nielsen)

Multiply and write the answer in scientific notation.

93. $(2.3 \times 10^6)(4.2 \times 10^{-11})$

94. $(6.5 \times 10^3)(5.2 \times 10^{-8})$

95. $(2.34 \times 10^{-8})(5.7 \times 10^{-4})$

96. $(3.26 \times 10^{-6})(8.2 \times 10^9)$

Divide and write the answer in scientific notation.

97. $\dfrac{8.5 \times 10^8}{3.4 \times 10^5}$

98. $\dfrac{5.1 \times 10^6}{3.4 \times 10^3}$

99. $\dfrac{4.0 \times 10^{-6}}{8.0 \times 10^{-3}}$

100. $\dfrac{7.5 \times 10^{-9}}{2.5 \times 10^{-4}}$

Write the answers to Exercises 101–110 in scientific notation.

101. *Seconds in 2000 Years.* About how many seconds are there in 2000 years? Assume that there are 365 days in one year.

102. *Hot Dog Consumption.* Americans consume 818 hot dogs per second in the summer. How many hot dogs are consumed in July? (July has 31 days.) (*Source*: National Hot Dog & Sausage Council; American Meat Institute)

103. *Word Knowledge.* There are 300,000 words in the English language. The average person knows about 10,000 of them. What part of the total number of words does the average person know?

104. *Astronomy.* The brightest star in the night sky, Sirius, is about 4.704×10^{13} mi from Earth. One light-year is 5.88×10^{12} mi. How many light-years is it from Earth to Sirius? (*Source*: *The Handy Science Answer Book*)

105. *Volume of a Plastic Sheet.* The volume of a rectangular solid is given by the length l times the width w times the height h: $V = lwh$. A sheet of plastic has a thickness of 150 micrometers. The sheet is 1.2 m by 79 m. Find the volume of the sheet.

106. *Orbit of Venus.* The circumference C of a circle is given by the formula $C = 2\pi r$, where r is the radius of the circle. Venus has a nearly circular orbit of the sun. The average distance from the sun to Venus is about 6.71×10^7 mi. How far does Venus travel in one orbit?

107. *Computer Calculations.* The Titan supercomputer can perform 20,000 trillion calculations per second. How many calculations can be performed in one minute? in one hour? (*Source*: "Supercomputer Titan; 20,000 trillion calculations per second," by Evelyn Laeschke. October 29, 2012, on ip-192.com)

108. *Amazon River Water Flow.* The average discharge at the mouth of the Amazon River is 4,200,000 cubic feet per second. How much water is discharged from the Amazon River in one hour? in one year?

109. *Printing and Engraving.* A ton of five-dollar bills is worth $4,540,000. How many pounds does a five-dollar bill weigh?

110. *Atoms in the Human Body.* A typical human body contains about 10^{26} atoms per kilogram of body mass. Of these, one-fourth are oxygen atoms. How many oxygen atoms are there in a 70-kg human body? (*Source*: Thomas Jefferson National Accelerator Facility, Office of Science Education, from Questions and Answers Archive, Brian Kross)

▶ ## Skill Maintenance

Simplify. **[R.3c], [R.6b]**

111. $9x - (-4y + 8) + (10x - 12)$

112. $-6t - (5t - 13) + 2(4 - 6t)$

113. $4^2 + 30 \cdot 10 - 7^3 + 16$

114. $5^4 - 38 \cdot 24 - (16 - 4 \cdot 18)$

115. $20 - 5 \cdot 4 - 8$ **116.** $20 - (5 \cdot 4 - 8)$

▶ ## Synthesis

Simplify.

117. $\dfrac{(2^{-2})^{-4} \cdot (2^3)^{-2}}{(2^{-2})^2 \cdot (2^5)^{-3}}$ **118.** $\left[\dfrac{(-3x^{-2}y^5)^{-3}}{(2x^4y^{-8})^{-2}} \right]^2$

119. $\left[\left(\dfrac{a^{-2}}{b^7} \right)^{-3} \cdot \left(\dfrac{a^4}{b^{-3}} \right)^2 \right]^{-1}$

Simplify. Assume that variables in exponents represent integers.

120. $(m^{x-b}n^{x+b})^x (m^b n^{-b})^x$ **121.** $\left[\dfrac{(2x^a y^b)^3}{(-2x^a y^b)^2} \right]^2$

122. $(x^b y^a \cdot x^a y^b)^c$

Chapter R Summary and Review

VOCABULARY REINFORCEMENT

Complete each statement with the correct term from the column on the right. Some of the choices may not be used.

associative
base
commutative
equation
exponent
factors
inequality
opposites
reciprocals
scientific
terms
variable

1. The sentence $x > -4$ is an example of a(n) _____ . **[R.1b]**

2. In the notation 7^4, the number 7 is called the _____ . **[R.3a]**

3. A(n) _____ is a letter that can represent various numbers. **[R.4a]**

4. In the expression $6x$, the multipliers 6 and x are _____ . **[R.4a]**

5. The number 5.93×10^7 is written in _____ notation. **[R.7c]**

6. The product of _____ is 1. **[R.2e]**

7. The sum of _____ is 0. **[R.2b]**

8. The _____ law for addition states that $a + b = b + a$. **[R.5c]**

CONCEPT REINFORCEMENT

Determine whether each statement is true or false.

1. For any numbers a and b, $a - b = b - a$. **[R.6b]**

2. Each member of the set of natural numbers is a member of the set of whole numbers. **[R.1a]**

3. The opposite of $-a$ when $a < 0$ is negative. **[R.2b]**

4. Zero is both positive and negative. **[R.1a]**

5. The absolute value of any real number is positive. **[R.1d]**

6. The reciprocal of a negative number is negative. **[R.2e]**

7. If c and d are real numbers and $c + d = 0$, then c and d are additive inverses. **[R.2b]**

8. The number 4.6×10^n, where n is an integer, is greater than 0 and less than 1 when $n < 0$. **[R.7c]**

REVIEW EXERCISES

▶ Part 1 Operations

1. Which of the following numbers are rational? **[R.1a]**

   $2, \ \sqrt{3}, \ -\dfrac{2}{3}, \ 0.45\overline{45}, \ -23.788$

2. Use set-builder notation to name the set of all real numbers less than or equal to 46. **[R.1a]**

3. Use $<$ or $>$ for $\square$ to write a true sentence:
   $-3.9 \ \square \ 2.9$. **[R.1b]**

4. Write a different inequality with the same meaning as $19 > x$. **[R.1b]**

Determine whether each of the following is true or false. **[R.1b]**

5. $-13 \geq 5$ **6.** $7.01 \leq 7.01$

Graph each inequality on the number line. **[R.1c]**

7. $x > -4$ **8.** $x \leq 1$

Find the absolute value. **[R.1d]**

9. $|-7.23|$ **10.** $|9 - 9|$

Add, subtract, multiply, or divide, if possible. **[R.2a, c, d, e]**

11. $6 + (-8)$ **12.** $-3.8 + (-4.1)$

13. $\dfrac{3}{4} + \left(-\dfrac{13}{7}\right)$ **14.** $-8 - (-3)$

15. $-17.3 - 9.4$ **16.** $\dfrac{3}{2} - \left(-\dfrac{13}{4}\right)$

17. $(-3.8)(-2.7)$ **18.** $-\dfrac{2}{3}\left(\dfrac{9}{14}\right)$

19. $-6(-7)(4)$ **20.** $-12 \div 3$

21. $\dfrac{-84}{-4}$ **22.** $\dfrac{49}{-7}$

23. $\dfrac{5}{6} \div \left(-\dfrac{10}{7}\right)$ **24.** $-\dfrac{5}{2} \div \left(-\dfrac{15}{16}\right)$

25. $\dfrac{21}{0}$ **26.** $-108 \div 4.5$

Evaluate $-a$ for each of the following. **[R.2b]**

27. $a = -7$ **28.** $a = 2.3$

29. $a = 0$

Write using exponential notation. **[R.3a]**

30. $a \cdot a \cdot a \cdot a \cdot a$

31. $\left(-\dfrac{7}{8}\right)\left(-\dfrac{7}{8}\right)\left(-\dfrac{7}{8}\right)$

32. Rewrite using a positive exponent: a^{-4}. **[R.3b]**

33. Rewrite using a negative exponent: $\dfrac{1}{x^8}$. **[R.3b]**

Simplify. **[R.3c]**

34. $2^3 - 3^4 + (13 \cdot 5 + 67)$

35. $64 \div (-4) + (-5)(20)$

▶ Part 2 Manipulations

Translate to an algebraic expression. **[R.4a]**

36. Five times some number

37. Twenty-eight percent of some number

38. Nine less than t

39. Eight less than the quotient of two numbers

Evaluate. **[R.4b]**

40. $5x - 7$, when $x = -2$

41. $\dfrac{x - y}{2}$, when $x = 4$ and $y = 20$

42. *Area of a Rug.*   The area A of a rectangle is given by the length l times the width w: $A = lw$. Find the area of a rectangular rug that measures 7 ft by 12 ft. **[R.4b]**

Complete each table by evaluating each expression for the given values. Then look for expressions that are equivalent. **[R.5a]**

43.

	$x^2 - 5$	$(x + 5)^2$	$(x - 5)^2$	$x^2 + 5$
$x = -1$				
$x = 10$				
$x = 0$				

44.

	$2x - 14$	$2x - 7$	$2(x - 7)$	$2x + 14$
$x = -1$				
$x = 10$				
$x = 0$				

45. Use multiplying by 1 to find an equivalent expression with the given denominator: **[R.5b]**

$\dfrac{7}{3}$; $9x$.

46. Simplify: $\dfrac{-84x}{7x}$. **[R.5b]**

Use a commutative law to find an equivalent expression. **[R.5c]**

47. $11 + a$ **48.** $8y$

Use an associative law to find an equivalent expression. **[R.5c]**

49. $(9 + a) + b$ **50.** $8(xy)$

Multiply. [R.5d]

51. $-3(2x - y)$ **52.** $4ab(2c + 1)$

Factor. [R.5d]

53. $5x + 10y - 5z$ **54.** $ptr + pts$

Collect like terms. [R.6a]

55. $2x + 6y - 5x - y$

56. $7c - 6 + 9c + 2 - 4c$

57. Find an equivalent expression without parentheses:
[R.6b]

$$-(-9c + 4d - 3).$$

Simplify. [R.6b]

58. $4(x - 3) - 3(x - 5)$

59. $12x - 3(2x - 5)$

60. $7x - [4 - 5(3x - 2)]$

61. $4m - 3[3(4m - 2) - (5m + 2) + 12]$

Multiply or divide, and simplify. [R.7a]

62. $(2x^4y^{-3})(-5x^3y^{-2})$

63. $\dfrac{-15x^2y^{-5}}{10x^6y^{-8}}$

Simplify. [R.7b]

64. $(-3a^{-4}bc^3)^{-2}$ **65.** $\left[\dfrac{-2x^4y^{-4}}{3x^{-2}y^6}\right]^{-4}$

Multiply or divide, and write scientific notation for the answer. [R.7c]

66. $\dfrac{2.2 \times 10^7}{3.2 \times 10^{-3}}$

67. $(3.2 \times 10^4)(4.1 \times 10^{-6})$

68. *Alpha Centauri.* Other than the sun, the star closest to Earth is Alpha Centauri. Its distance from Earth is about 2.4×10^{13} mi. One light-year = the distance that light travels in one year = 5.88×10^{12} mi. How many light-years is it from Earth to Alpha Centauri? (*Source: The Handy Science Answer Book*)

1 light-year = 5.88×10^{12} mi

Earth   Alpha Centauri

2.4×10^{13} mi

69. *Finance.* A **mil** is one thousandth of a dollar. The taxation rate in a certain school district is 5.0 mils for every dollar of assessed valuation. The assessed valuation for the district is 13.4 million dollars. How much tax revenue will be raised? [R.7c]

70. Evaluate $\dfrac{x - 4y}{3}$ when $x = 5$ and $y = -4$.

[R.4b]

A. -8 **B.** -7

C. $-\dfrac{11}{3}$ **D.** 7

71. Use the commutative laws and the associative laws to determine which expression is *not* equivalent to $2x + y$. [R.5c]

A. $2y + x$ **B.** $x \cdot 2 + y$
C. $y + 2x$ **D.** $y + x \cdot 2$

▶ **Synthesis**

72. Simplify: $(x^y \cdot x^{3y})^3$. [R.7b]

73. If $a = 2^x$ and $b = 2^{x+5}$, find $a^{-1}b$. [R.7a]

74. Which of the following expressions are equivalent?
[R.5d], [R.7b]

a) $3x - 3y$ **b)** $3x - y$
c) $x^{-2}x^5$ **d)** x^{-10}
e) x^{-3} **f)** $(x^{-2})^5$
g) $x(yz)$ **h)** $x(y + z)$
i) $3(x - y)$ **j)** $xy + xz$

▶ **Collaborative Discussion and Writing**

To the student and the instructor: *The Collaborative Discussion and Writing exercises are meant to be answered with one or more sentences. They can be discussed and answered collaboratively by the entire class or by small groups.*

75. List five examples of rational numbers that are not integers and explain why they are not. [R.1a]

76. Explain in your own words why $\frac{7}{0}$ is not defined. [R.2e]

77. If the base and the height of a triangle are each doubled, does its area double? Explain. (See Example 8 in Section R.4.) [R.4b]

78. If the base and the height of a parallelogram are each doubled, does its area double? Explain. (See Exercise 44 in Exercise Set R.4.) [R.4b]

79. A $20 bill weighs about 2.2×10^{-3} lb. A criminal claims to be carrying $5 million in $20 bills in his suitcase. Is this possible? Why or why not? [R.7c]

80. When a calculator indicates that $5^{17} = 7.629394531 \times 10^{11}$, you know that an approximation is being made. How can you tell? (*Hint:* What should the ones digit be?) [R.7c]

R   Chapter Test

▶ Part 1 Operations

1. Which of the following numbers are irrational?

$$-43, \quad \sqrt{7}, \quad -\frac{2}{3}, \quad 2.3\overline{76}, \quad \pi$$

2. Use set-builder notation to name the set of real numbers greater than 20.

3. Use $<$ or $>$ for $\square$ to write a true sentence:
$$-4.5 \ \square \ -8.7.$$

4. Write a different inequality with the same meaning as $a \le 5$.

Determine whether each of the following is true or false.

5. $-6 \ge -6$

6. $-8 \le -6$

7. Graph $x > -2$ on the number line.

Find the absolute value.

8. $|0|$

9. $\left| -\dfrac{7}{8} \right|$

Add, subtract, multiply, or divide, if possible.

10. $7 + (-9)$

11. $-5.3 + (-7.8)$

12. $-\dfrac{5}{2} + \left(-\dfrac{7}{2} \right)$

13. $-6 - (-5)$

14. $-18.2 - 11.5$

15. $\dfrac{19}{4} - \left(-\dfrac{3}{2} \right)$

16. $(-4.1)(8.2)$

17. $-\dfrac{4}{5} \left(-\dfrac{15}{16} \right)$

18. $-6(-4)(-11)2$

19. $-75 \div (-5)$

20. $\dfrac{-10}{2}$

21. $-\dfrac{5}{2} \div \left(-\dfrac{15}{16} \right)$

22. $-459.2 \div 5.6$

23. $\dfrac{-3}{0}$

Evaluate $-a$ for each of the following.

24. $a = -13$

25. $a = 0$

26. Write exponential notation: $q \cdot q \cdot q \cdot q$.

27. Rewrite using a negative exponent: $\dfrac{1}{a^9}$.

Simplify.

28. $1 - (2 - 5)^2 + 5 \div 10 \cdot 4^2$

29. $\dfrac{7(5 - 2 \cdot 3) - 3^2}{4^2 - 3^2}$

▶ Part 2 Manipulations

Translate to an algebraic expression.

30. Nine more than t

31. Twelve less than the quotient of two numbers

32. Evaluate $3x - 3y$ when $x = 2$ and $y = -4$.

33. *Area of a Triangular Stamp.* The area A of a triangle is given by $A = \frac{1}{2} bh$. Find the area of a triangular stamp whose base measures 3 cm and whose height measures 2.5 cm.

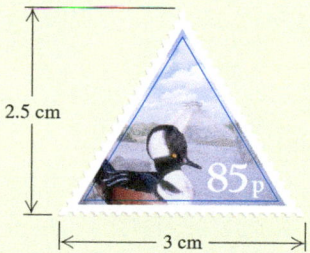

Complete a table by evaluating each expression for $x = -1, x = 10,$ and $x = 0$. Then determine whether the expressions are equivalent. Answer yes or no.

34. $x(x - 3); \ x^2 - 3x$

35. $3x + 5x^2; \ 8x^2$

36. Use multiplying by 1 to find an equivalent expression with the given denominator.
$$\frac{3}{4}; \ 36x$$

37. Simplify:
$$\frac{-54x}{-36x}.$$

Use a commutative law to find an equivalent expression.

38. pq **39.** $t + 4$

Use an associative law to find an equivalent expression.

40. $3 + (t + w)$

41. $(4a)b$

Multiply.

42. $-2(3a - 4b)$

43. $3\pi r(s + 1)$

Factor.

44. $ab - ac + 2ad$

45. $2ah + h$

Collect like terms.

46. $6y - 8x + 4y + 3x$

47. $4a - 7 + 17a + 21$

48. Find an equivalent expression without parentheses: $-(-9x + 7y - 22)$.

Simplify.

49. $-3(x + 2) - 4(x - 5)$

50. $4x - [6 - 3(2x - 5)]$

Multiply or divide, and simplify.

51. $\dfrac{-12x^3y^{-4}}{8x^7y^{-6}}$

52. $(3a^4b^{-2})(-2a^5b^{-3})$

53. $(5a^{4n})(-10a^{5n})$

54. $\dfrac{-60x^{3t}}{12x^{7t}}$

Simplify.

55. $(-3a^{-3}b^2c)^{-4}$

56. $\left[\dfrac{-5a^{-2}b^8}{10a^{10}b^{-4}}\right]^{-4}$

57. Convert to scientific notation: 0.0000437.

Multiply or divide, and write scientific notation for the answer.

58. $(8.7 \times 10^{-9})(4.3 \times 10^{15})$

59. $\dfrac{1.2 \times 10^{-12}}{6.4 \times 10^{-7}}$

60. *Mass of Pluto.* The mass of Earth is 5.98×10^{24} kg. The mass of the dwarf planet Pluto is about 0.002 times the mass of Earth. Find the mass of Pluto and express the answer in scientific notation.

  A. 29.9×10^{26} kg   **B.** 2.99×10^{27} kg

  C. 1.196×10^{22} kg   **D.** 11.96×10^{21} kg

▶ **Synthesis**

61. Which of the following expressions are equivalent?

  a) $x^{-3}x^{-4}$ **b)** x^{12}

  c) x^{-12} **d)** $5x + 5$

  e) $(x^{-3})^{-4}$ **f)** $5(x + 1)$

  g) $5x$ **h)** $5 + 5x$

  i) $5(xy)$ **j)** $(5x)y$

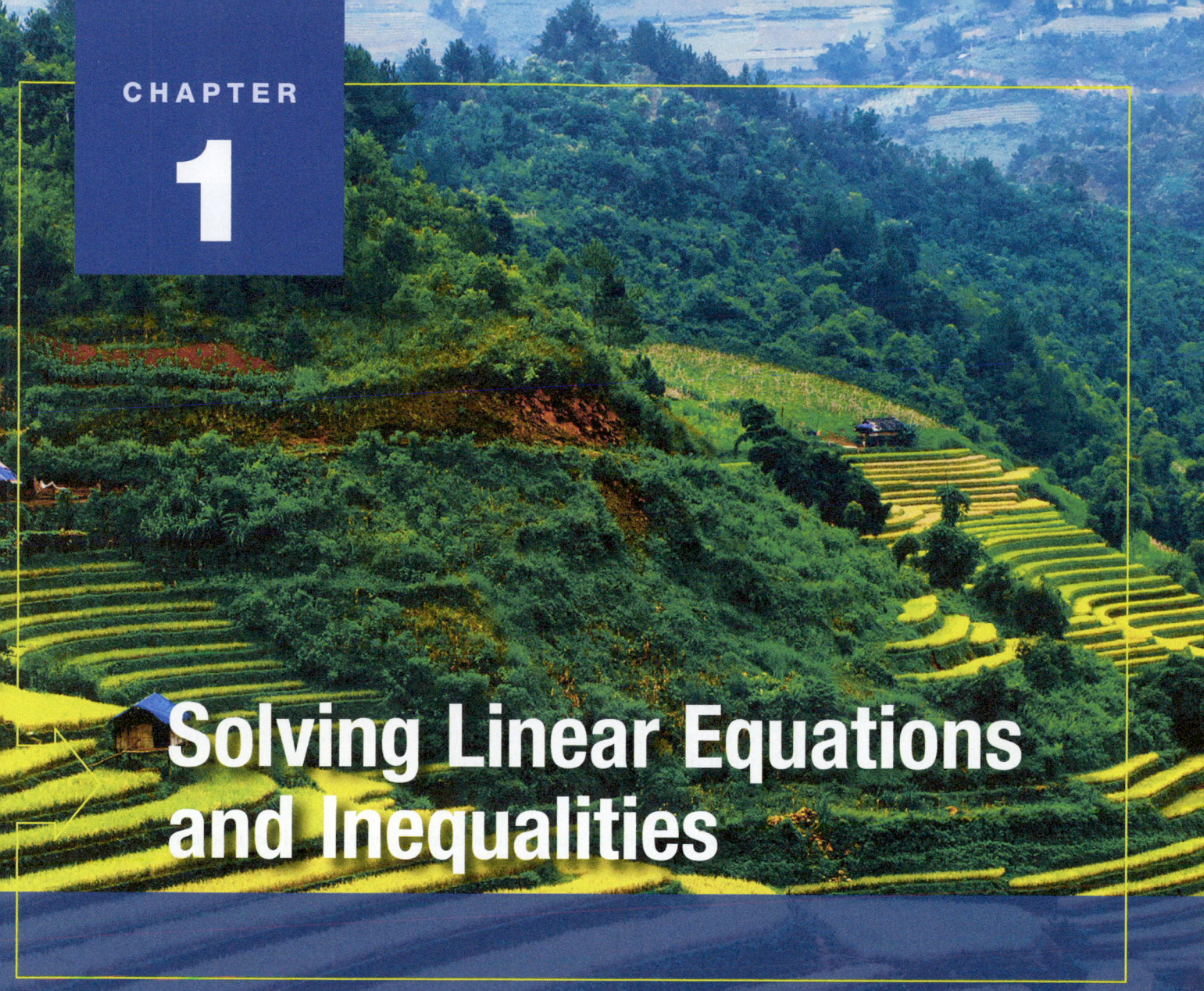

CHAPTER

1

Solving Linear Equations and Inequalities

APPLICATION   This problem appears as Example 1 in Section 1.3.

Rice is the grain with the second-highest worldwide production, after maize. Rice provides more than 20% of the calories consumed by humans. In 2012, China produced 143,000 metric tons of rice. This was 7500 metric tons more than five times the amount of rice produced by Vietnam. Find the amount of rice produced in Vietnam. (*Source*: United Nations' Food and Agriculture Organization)

1.1

Solving Equations

▶ **a**   Determine whether a given number is a solution of a given equation.

▶ **b**   Solve equations using the addition principle.

▶ **c**   Solve equations using the multiplication principle.

▶ **d**   Solve equations using the addition principle and the multiplication principle together, removing parentheses where appropriate.

▶ a   Equations and Solutions

In order to solve many kinds of problems, we must be able to solve *equations*. Some examples of equations are

$$15 - 10 = 2 + 3, \qquad x + 8 = 23, \qquad 5x - 2 = 9 - x.$$

EQUATION

An **equation** is a number sentence that says that the expressions on either side of the equals sign, =, represent the same number.

Some equations are true. Some are false. Some are neither true nor false.

EXAMPLES   Determine whether the equation is true, false, or neither.

1. $1 + 10 = 11$ Both expressions represent 11. The equation is *true*.

2. $7 - 8 = 9 - 13$ $7 - 8$ represents -1 and $9 - 13$ represents -4. The equation is *false*.

3. $x - 9 = 3$ The equation is *neither* true nor false, because we do not know what number x represents.   ▶■

If an equation contains a variable, then some replacements or values of the variable may make it true and some may make it false.

SOLUTION OF AN EQUATION

The replacements for the variable that make an equation true are called the **solutions** of the equation. The set of all solutions is called the **solution set** of the equation. When we find all the solutions, we say that we have **solved** the equation.

To determine whether a number is a solution of an equation, we evaluate the algebraic expression on each side of the equals sign by substitution. If the values are the same, then the number is a solution of the equation. If they are not, then the number is not a solution.

EXAMPLE 4 Determine whether 5 is a solution of $x + 6 = 11$.

$$x + 6 = 11 \qquad \text{Writing the equation}$$

$$5 + 6 \; \overset{?}{\mid} \; 11 \qquad \text{Substituting 5 for } x$$

$$11 \mid \qquad \text{TRUE}$$

Since the left-hand side and the right-hand side are the same, 5 is a solution of the equation.

Now Try Exercise 1.

EXAMPLE 5 Determine whether 18 is a solution of $2x - 3 = 5$.

$$2x - 3 = 5 \qquad \text{Writing the equation}$$

$$2 \cdot 18 - 3 \; \overset{?}{\mid} \; 5 \qquad \text{Substituting 18 for } x$$

$$36 - 3 \mid$$

$$33 \mid \qquad \text{FALSE}$$

Since the left-hand side and the right-hand side are not the same, 18 is not a solution of the equation.

Now Try Exercise 7.

To the Student: Now Try Exercises are found after nearly every example. This feature encourages active learning by asking you to do an exercise in the exercise set that is similar to the example you have just read.

Equivalent Equations

Consider the equation

$$x = 5.$$

The solution of this equation is easily "seen" to be 5. If we replace x with 5, we get

$$5 = 5, \quad \text{which is true.}$$

In Example 4, we saw that the solution of the equation $x + 6 = 11$ is also 5, but the fact that 5 is the solution is not so readily apparent. We now consider principles that allow us to start with one equation and end up with an *equivalent equation*, like $x = 5$, in which the variable is alone on one side, and for which the solution is read directly from the equation.

> **EQUIVALENT EQUATIONS**
>
> Equations with the same solutions are called **equivalent equations**.

▶ **b** **The Addition Principle**

One of the principles we use in solving equations involves addition. The equation $a = b$ says that a and b represent the same number. Suppose that $a = b$ is true and we then add a number c to a. We will get the same result if we add c to b, because a and b are the same number.

> **THE ADDITION PRINCIPLE**
>
> For any real numbers a, b, and c,
>
> $$a = b \quad \text{is equivalent to} \quad a + c = b + c.$$

When we use the addition principle, we sometimes say that we "add the same number on both sides of an equation." We can also "subtract the same number on both sides of an equation," because we can express subtraction as the addition of an opposite. That is,

$$a - c = b - c \quad \text{is equivalent to} \quad a + (-c) = b + (-c).$$

EXAMPLE 6   Solve: $x + 6 = 11$.

$$x + 6 = 11$$

$$x + 6 + (-6) = 11 + (-6)$$

$$x + 6 - 6 = 11 - 6$$

> Using the addition principle: adding -6 on both sides or subtracting 6 on both sides. Note that 6 and -6 are opposites.

$$x + 0 = 5$$ **Simplifying**

$$x = 5$$ **Using the identity property of 0:** $x + 0 = x$

Check: $x + 6 = 11$

$$\overline{5 + 6 \; ? \; 11}$$ **Substituting 5 for x**

$$11 \;\Big|\quad \text{TRUE}$$

The solution is 11.

Now Try Exercise 13.

In Example 6, we wanted to get x alone so that we could readily see the solution, so we added the opposite of 6. This eliminated the 6 on the left, giving us the *additive identity* 0, which when added to x is x. We began with $x + 6 = 11$. Using the addition principle, we derived a simpler equation, $x = 5$. The equations $x + 6 = 11$ and $x = 5$ are *equivalent*. The solution of $x = 5$ is the solution of $x + 6 = 11$.

EXAMPLE 7   Solve: $y - 4.7 = 13.9$.

$$y - 4.7 = 13.9$$

$$y - 4.7 + 4.7 = 13.9 + 4.7$$

> Using the addition principle: adding 4.7 on both sides. Note that -4.7 and 4.7 are opposites.

$$y + 0 = 18.6$$ **Simplifying**

$$y = 18.6$$ **Using the identity property of 0:** $y + 0 = y$

Check: $y - 4.7 = 13.9$

$$\overline{18.6 - 4.7 \; ? \; 13.9}$$ **Substituting 18.6 for y**

$$13.9 \;\Big|\quad \text{TRUE}$$

The solution is 18.6.

Now Try Exercise 21.

EXAMPLE 8   Solve: $-\frac{3}{8} + x = -\frac{5}{7}$.

$$-\frac{3}{8} + x = -\frac{5}{7}$$

$$\frac{3}{8} + \left(-\frac{3}{8}\right) + x = \frac{3}{8} + \left(-\frac{5}{7}\right)$$ **Using the addition principle: adding $\frac{3}{8}$**

$$0 + x = \frac{3}{8} - \frac{5}{7}$$

$$x = \frac{3}{8} \cdot \frac{7}{7} - \frac{5}{7} \cdot \frac{8}{8}$$ **Multiplying by 1 to obtain the least common denominator**

$$= \frac{21}{56} - \frac{40}{56}$$

$$= -\frac{19}{56}$$

Check:
$$-\frac{3}{8} + x = -\frac{5}{7}$$

$$
\begin{array}{c|c}
-\frac{3}{8} + \left(-\frac{19}{56}\right) \ ? \ -\frac{5}{7} & \\
-\frac{3}{8} \cdot \frac{7}{7} + \left(-\frac{19}{56}\right) & \\
-\frac{21}{56} + \left(-\frac{19}{56}\right) & \\
-\frac{40}{56} & \\
-\frac{5}{7} & \text{TRUE}
\end{array}
$$

Substituting $-\frac{19}{56}$ **for** x

The solution is $-\frac{19}{56}$.

Now Try Exercise 23.

▶ c   The Multiplication Principle

A second principle for solving equations involves multiplication. Suppose that $a = b$ is true and we multiply a by a nonzero number c. We get the same result if we multiply b by c, because a and b are the same number.

THE MULTIPLICATION PRINCIPLE

For any real numbers a, b, and c, $c \neq 0$,

$$a = b \quad \text{is equivalent to} \quad a \cdot c = b \cdot c.$$

EXAMPLE 9   Solve: $\frac{4}{5}x = 22$.

$$\frac{4}{5}x = 22$$

$$\frac{5}{4} \cdot \frac{4}{5}x = \frac{5}{4} \cdot 22 \qquad \textbf{Multiplying by } \tfrac{5}{4}\textbf{, the reciprocal of } \tfrac{4}{5}$$

$$1 \cdot x = \frac{55}{2} \qquad\qquad \textbf{Multiplying and simplifying}$$

$$x = \frac{55}{2} \qquad\qquad \textbf{Using the identity property of 1: } 1 \cdot x = x$$

Check:
$$\frac{4}{5}x = 22$$

$$
\begin{array}{c|c}
\frac{4}{5} \cdot \frac{55}{2} \ ? \ 22 & \\
22 & \text{TRUE}
\end{array}
$$

The solution is $\frac{55}{2}$.

Now Try Exercise 35.

In Example 9, in order to get x alone, we multiplied by the *multiplicative inverse*, or *reciprocal*, of $\frac{4}{5}$. When we multiplied, we got the *multiplicative identity* 1 times x, or $1 \cdot x$, which simplified to x. This enabled us to eliminate the $\frac{4}{5}$ on the left.

The multiplication principle also tells us that we can "divide by a nonzero number on both sides" because division is the same as multiplying by a reciprocal. That is,

$$\frac{a}{c} = \frac{b}{c} \quad \text{is equivalent to} \quad a \cdot \frac{1}{c} = b \cdot \frac{1}{c}, \quad \text{when } c \neq 0.$$

In a product like $\frac{4}{5}x$, the number in front of the variable is called the **coefficient**. When this number is in fraction notation, it is usually most convenient to multiply both sides by its reciprocal. If the coefficient is an integer or is in decimal notation, it is usually more convenient to divide by the coefficient.

EXAMPLE 10 Solve: $4x = 9$.

$$4x = 9$$

$$\frac{4x}{4} = \frac{9}{4} \qquad \text{Using the multiplication principle: multiplying on both sides by } \tfrac{1}{4} \text{ or dividing on both sides by the coefficient, 4}$$

$$1 \cdot x = \frac{9}{4} \qquad \text{Simplifying}$$

$$x = \frac{9}{4} \qquad \text{Using the identity property of 1: } 1 \cdot x = x$$

Check:

$$\begin{array}{c|c} 4x = 9 \\ \hline 4 \cdot \tfrac{9}{4} \;?\; 9 \\ 9 \;\bigg|\; \quad \text{TRUE} \end{array}$$

The solution is $\frac{9}{4}$.

Now Try Exercise 25.

EXAMPLE 11 Solve: $5.5 = -0.05y$.

$$5.5 = -0.05y$$

$$\frac{5.5}{-0.05} = \frac{-0.05y}{-0.05} \qquad \text{Dividing by } -0.05 \text{ on both sides}$$

$$-110 = 1 \cdot y$$

$$-110 = y$$

The check is left to the student. The solution is -110.

Now Try Exercise 33.

Note that equations are reversible. That is, $a = b$ is equivalent to $b = a$. Thus, in Example 11, $-110 = y$ and $y = -110$ are equivalent, and the solution of both equations is -110.

EXAMPLE 12 Solve: $-\dfrac{x}{4} = 10$.

$$-\frac{x}{4} = 10$$

$$-\frac{1}{4}x = 10 \qquad\qquad -\frac{x}{4} = -\frac{1}{4} \cdot x$$

$$-4 \cdot \left(-\frac{1}{4}\right)x = -4 \cdot 10 \qquad \begin{array}{l} -4 \text{ is the reciprocal of } -\tfrac{1}{4}; \text{ multiplying by } -4 \\ \text{on both sides} \end{array}$$

$$1 \cdot x = -40 \qquad\qquad \text{Simplifying}$$

$$x = -40$$

The check is left to the student. The solution is -40.

Now Try Exercise 29.

▶ d Using the Principles Together

Let's see how we can use the addition and multiplication principles together.

EXAMPLE 13 Solve: $3x - 4 = 13$.

$$3x - 4 = 13$$
$$3x - 4 + 4 = 13 + 4 \qquad \text{Using the addition principle: adding 4}$$
$$3x = 17 \qquad \text{Simplifying}$$
$$\frac{3x}{3} = \frac{17}{3} \qquad \text{Dividing by 3}$$
$$x = \frac{17}{3} \qquad \text{Simplifying}$$

Check:

$$\begin{array}{c|c} 3x - 4 = 13 \\ \hline 3 \cdot \frac{17}{3} - 4 \; ? \; 13 \\ 17 - 4 \\ 13 & \text{TRUE} \end{array}$$

The solution is $\frac{17}{3}$, or $5\frac{2}{3}$.

> In algebra, "improper" fraction notation, such as $\frac{17}{3}$, is quite "proper." We will generally use such notation rather than $5\frac{2}{3}$.

Now Try Exercise 39.

In a situation such as Example 13, it is easier to first use the addition principle. In a situation in which fractions or decimals are involved, it may be easier to use the multiplication principle first to clear them, but it is not mandatory.

EXAMPLE 14 Clear the fractions and solve: $\frac{3}{16}x + \frac{1}{2} = \frac{11}{8}$.

We multiply on both sides by the least common multiple of the denominators—in this case, 16:

$$\frac{3}{16}x + \frac{1}{2} = \frac{11}{8} \qquad \text{The LCM of the denominators is 16.}$$
$$16\left(\frac{3}{16}x + \frac{1}{2}\right) = 16\left(\frac{11}{8}\right) \qquad \text{Multiplying by 16}$$
$$16 \cdot \frac{3}{16}x + 16 \cdot \frac{1}{2} = 22 \qquad \text{Carrying out the multiplication. We use the distributive law on the left, being careful to multiply } both \text{ terms by 16.}$$
$$3x + 8 = 22 \qquad \text{Simplifying. The fractions are cleared.}$$
$$3x + 8 - 8 = 22 - 8 \qquad \text{Subtracting 8}$$
$$3x = 14$$
$$\frac{3x}{3} = \frac{14}{3} \qquad \text{Dividing by 3}$$
$$x = \frac{14}{3}.$$

The number $\frac{14}{3}$ checks and is the solution.

Now Try Exercise 45.

EXAMPLE 15 Clear the decimals and solve: $12.4 - 5.12x = 3.14x$.

We multiply on both sides by a power of ten—10, 100, 1000, and so on—to clear the equation of decimals. In this case, we use 10^2, or 100, because the greatest number of decimal places is 2.

$$12.4 - 5.12x = 3.14x$$

$$100(12.4 - 5.12x) = 100(3.14x) \qquad \text{\textbf{Multiplying by 100}}$$

$$100(12.4) - 100(5.12x) = 314x \qquad \text{\textbf{Carrying out the multiplication. We use the distributive law on the left.}}$$

$$1240 - 512x = 314x \qquad \text{\textbf{Simplifying}}$$

$$1240 - 512x + 512x = 314x + 512x \qquad \text{\textbf{Adding 512x}}$$

$$1240 = 826x$$

$$\frac{1240}{826} = \frac{826x}{826} \qquad \text{\textbf{Dividing by 826}}$$

$$x = \frac{1240}{826}, \text{ or } \frac{620}{413}$$

The solution is $\frac{620}{413}$.

> **Now Try Exercise 47.**

When there are like terms on the same side of an equation, we collect them. If there are like terms on opposite sides of an equation, we use the addition principle to get them on the same side of the equation.

EXAMPLE 16 Solve: $8x + 6 - 2x = -4x - 14$.

$$8x + 6 - 2x = -4x - 14$$

$$6x + 6 = -4x - 14 \qquad \text{\textbf{Collecting like terms on the left}}$$

$$4x + 6x + 6 = 4x - 4x - 14 \qquad \text{\textbf{Adding 4x}}$$

$$10x + 6 = -14 \qquad \text{\textbf{Collecting like terms}}$$

$$10x + 6 - 6 = -14 - 6 \qquad \text{\textbf{Subtracting 6}}$$

$$10x = -20$$

$$\frac{10x}{10} = \frac{-20}{10} \qquad \text{\textbf{Dividing by 10}}$$

$$x = -2$$

Check:

$$\begin{array}{c|c} \multicolumn{2}{c}{8x + 6 - 2x = -4x - 14} \\ \hline 8(-2) + 6 - 2(-2) & \stackrel{?}{=} -4(-2) - 14 \\ -16 + 6 + 4 & 8 - 14 \\ -6 & -6 \qquad \text{\textbf{TRUE}} \end{array}$$

The solution is -2.

> **Now Try Exercise 59.**

Special Cases

Some equations have no solution.

EXAMPLE 17   Solve: $-8x + 5 = 14 - 8x$.

We have

$$-8x + 5 = 14 - 8x$$
$$8x - 8x + 5 = 8x + 14 - 8x \qquad \text{Adding } 8x$$
$$5 = 14. \qquad \text{We get a false equation.}$$

No matter what number we use for x, we get a false sentence. Thus the equation has *no* solution.   **Now Try Exercise 51.**

There are some equations for which any real number is a solution.

EXAMPLE 18   Solve: $-8x + 5 = 5 - 8x$.

We have

$$-8x + 5 = 5 - 8x$$
$$8x - 8x + 5 = 8x + 5 - 8x \qquad \text{Adding } 8x$$
$$5 = 5. \qquad \text{We get a true equation.}$$

Replacing x with any real number gives a true sentence. Thus any real number is a solution. The equation has *infinitely* many solutions.   **Now Try Exercise 61.**

Equations Containing Parentheses

Equations containing parentheses can often be solved by first multiplying to remove parentheses and then proceeding as before.

EXAMPLE 19   Solve: $30 + 5(x + 3) = -3 + 5x + 48$.

We have

$$30 + 5(x + 3) = -3 + 5x + 48$$
$$30 + 5x + 15 = -3 + 5x + 48 \qquad \text{Multiplying, using the distributive law, to remove parentheses}$$
$$45 + 5x = 45 + 5x \qquad \text{Collecting like terms on each side}$$
$$45 + 5x - 5x = 45 + 5x - 5x \qquad \text{Subtracting } 5x$$
$$45 = 45. \qquad \text{Simplifying. We get a true equation.}$$

All real numbers are solutions.   **Now Try Exercise 71.**

EXAMPLE 20 Solve: $3(7 - 2x) = 14 - 8(x - 1)$.

$$3(7 - 2x) = 14 - 8(x - 1)$$
$$21 - 6x = 14 - 8x + 8 \qquad \text{Multiplying, using the distributive law, to remove parentheses}$$
$$21 - 6x = 22 - 8x \qquad \text{Collecting like terms}$$
$$21 - 6x + 8x = 22 - 8x + 8x \qquad \text{Adding } 8x$$
$$21 + 2x = 22 \qquad \text{Collecting like terms}$$
$$21 + 2x - 21 = 22 - 21 \qquad \text{Subtracting 21}$$
$$2x = 1$$
$$\frac{2x}{2} = \frac{1}{2} \qquad \text{Dividing by 2}$$
$$x = \frac{1}{2}$$

Check:

$$\begin{array}{c|c} 3(7 - 2x) = 14 - 8(x - 1) \\ \hline 3\left(7 - 2 \cdot \frac{1}{2}\right) \; ? \; 14 - 8\left(\frac{1}{2} - 1\right) \\ 3(7 - 1) \quad\Big|\quad 14 - 8\left(-\frac{1}{2}\right) \\ 3 \cdot 6 \quad\Big|\quad 14 + 4 \\ 18 \quad\Big|\quad 18 \qquad \textsf{TRUE} \end{array}$$

The solution is $\frac{1}{2}$.

> **Now Try Exercise 73.**

An Equation-Solving Procedure

1. Clear the equation of fractions or decimals if that is needed.
2. If parentheses occur, multiply to remove them using the distributive law.
3. Collect like terms on each side of the equation, if necessary.
4. Use the addition principle to get all terms with variables on one side and all other terms on the other side.
5. Collect like terms on each side again, if necessary.
6. Use the multiplication principle to solve for the variable.

Technology Connection

Checking Possible Solutions

Although a calculator is *not* required for this textbook, the book contains a series of *optional* discussions on using a graphing calculator. The keystrokes for the TI-84 Plus graphing calculator will be shown throughout. For keystrokes for other models of calculators, consult the user's manual for your particular model.

To check possible solutions of an equation on a calculator, we can substitute and carry out the calculations on each side of the equation. If the left-hand and the right-hand sides of the equation have the same value, then the number

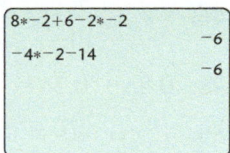

that was substituted is a solution of the equation. To check the possible solution -2 in the equation $8x + 6 - 2x = -4x - 14$ in Example 16, for instance, we first substitute -2 for x in the expression on the left side of the equation and get -6. Then we substitute -2 for x in the expression on the right side of the equation. Again, we get -6. Since the two sides of the equation have the same value when x is -2, we know that -2 is the solution of the equation.

A table can also be used to check possible solutions of equations. First, we press ⟨ Y= ⟩ to display the equation-editor screen. If an expression for Y_1 is currently entered, we place the cursor on it and press **CLEAR** to delete it. We do the same for any other entries that are present. Next, we position the cursor to the right of $Y_1 =$ and enter the left side of the equation. Then we position the cursor beside $Y_2 =$ and enter the right side of the equation. Now we press **2ND** ⟨ TBLSET ⟩ to display the Table Setup screen. (TBLSET is the second operation associated with the ⟨WINDOW⟩ key.) On the INDPNT line, we position the cursor on "ASK" and press **ENTER** to set up a table in ASK mode. (The settings for TblStart and ΔTbl are irrelevant in ASK mode.)

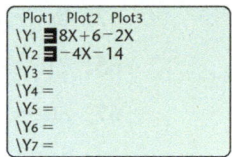 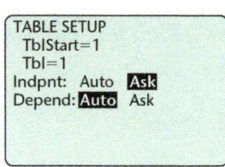 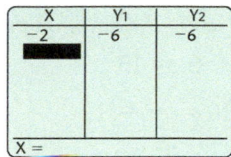

We press **2ND** ⟨ TABLE ⟩ to display the table. (TABLE is the second operation associated with the ⟨GRAPH⟩ key.) We enter the possible solution, -2, and see that $Y_1 = -6 = Y_2$ for this value of x. This confirms that the left and the right sides of the equation have the same value for $x = -2$, so -2 is the solution of the equation.

Exercises

1. Use substitution to check the solutions found in Examples 9, 13, and 15.

2. Use a table set in ASK mode to check the solutions found in Examples 16 and 20.

1.1 Exercise Set

▶ Reading Check

Choose from the column on the right the most appropriate first step in solving each equation.

RC1. $5 + x = -12$

RC2. $16 = x - 3$

RC3. $3 = -\dfrac{1}{9}x$

RC4. $5x = -16$

a) Divide by 3 on both sides.

b) Multiply by -9 on both sides.

c) Divide by 5 on both sides.

d) Subtract 5 on both sides.

e) Add $\dfrac{1}{9}$ on both sides.

f) Add 3 on both sides.

Remember to review the objectives before doing the exercises.

a Determine whether the given number is a solution of the given equation.

1. $17; \ x + 23 = 40$

2. $24; \ 47 - x = 23$

3. $-8; \ 2x - 3 = -18$

4. $-10; \ 3x + 14 = -27$

5. $45; \ \dfrac{-x}{9} = -2$

6. $32; \ \dfrac{-x}{8} = -3$

7. $10; \ 2 - 3x = 21$

8. $-11; \ 4 - 5x = 59$

9. $19; \ 5x + 7 = 102$

10. $9; \ 9y + 5 = 86$

11. $-11; \ 7(y - 1) = 84$

12. $-13; \ x + 5 = 5 + x$

b Solve using the addition principle. Don't forget to check.

13. $y + 6 = 13$

14. $x + 7 = 14$

15. $-20 = x - 12$

16. $-27 = y - 17$

17. $-8 + x = 19$

18. $-8 + r = 17$

19. $-12 + z = -51$

20. $-37 + x = -89$

21. $p - 2.96 = 83.9$

22. $z - 14.9 = -5.73$

23. $-\dfrac{3}{8} + x = -\dfrac{5}{24}$

24. $x + \dfrac{1}{12} = -\dfrac{5}{6}$

c Solve using the multiplication principle. Don't forget to check.

25. $3x = 18$

26. $5x = 30$

27. $-11y = 44$

28. $-4x = 124$

29. $-\dfrac{x}{7} = 21$

30. $-\dfrac{x}{3} = -25$

31. $-96 = -3z$

32. $-120 = -8y$

33. $4.8y = -28.8$

34. $0.39t = -2.73$

35. $\dfrac{3}{2}t = -\dfrac{1}{4}$

36. $-\dfrac{7}{6}y = -\dfrac{7}{8}$

d Solve using the principles together. Don't forget to check.

37. $6x - 15 = 45$

38. $4x - 7 = 81$

39. $5x - 10 = 45$

40. $6z - 7 = 11$

41. $9t + 4 = -104$

42. $5x + 7 = -108$

43. $-\dfrac{7}{3}x + \dfrac{2}{3} = -18$

44. $-\dfrac{9}{2}y + 4 = -\dfrac{91}{2}$

45. $\dfrac{6}{5}x + \dfrac{4}{10}x = \dfrac{32}{10}$

46. $\dfrac{9}{5}y + \dfrac{4}{10}y = \dfrac{66}{10}$

47. $0.9y - 0.7y = 4.2$

48. $0.8t - 0.3t = 6.5$

49. $8x + 48 = 3x - 12$

50. $15x + 40 = 8x - 9$

51. $7y - 1 = 27 + 7y$

52. $3x - 15 = 15 + 3x$

53. $3x - 4 = 5 + 12x$

54. $9t - 4 = 14 + 15t$

55. $5 - 4a = a - 13$

56. $6 - 7x = x - 14$

57. $3m - 7 = -7 - 4m - m$

58. $5x - 8 = -8 + 3x - x$

59. $5x + 3 = 11 - 4x + x$

60. $6y + 20 = 10 + 3y + y$

61. $-7 + 9x = 9x - 7$

62. $-3t + 4 = 5 - 3t$

63. $6y - 8 = 9 + 6y$

64. $5 - 2y = -2y + 5$

65. $2(x + 7) = 4x$

66. $3(y + 6) = 9y$

67. $80 = 10(3t + 2)$

68. $27 = 9(5y - 2)$

69. $180(n - 2) = 900$

70. $210(x - 3) = 840$

71. $5y - (2y - 10) = 25$

72. $8x - (3x - 5) = 40$

73. $7(3x + 6) = 11 - (x + 2)$

74. $3(4 - 2x) = 4 - (6x - 8)$

75. $2[9 - 3(-2x - 4)] = 12x + 42$

76. $-40x + 45 = 3[7 - 2(7x - 4)]$

77. $\dfrac{1}{8}(16y + 8) - 17 = -\dfrac{1}{4}(8y - 16)$

78. $\dfrac{1}{6}(12t + 48) - 20 = -\dfrac{1}{8}(24t - 144)$

79. $3[5 - 3(4 - t)] - 2$
$= 5[3(5t - 4) + 8] - 26$

80. $6[4(8 - y) - 5(9 + 3y)] - 21$
$= -7[3(7 + 4y) - 4]$

81. $\dfrac{2}{3}\left(\dfrac{7}{8} + 4x\right) - \dfrac{5}{8} = \dfrac{3}{8}$

82. $\dfrac{3}{4}\left(3x - \dfrac{1}{2}\right) + \dfrac{2}{3} = \dfrac{1}{3}$

83. $5(4x - 3) - 2(6 - 8x) + 10(-2x + 7)$
$= -4(9 - 12x)$

84. $9(4x + 7) - 3(5x - 8)$
$= 6\left(\dfrac{2}{3} - x\right) - 5\left(\dfrac{3}{5} + 2x\right)$

▶ Skill Maintenance

This heading indicates that the exercises that follow are Skill Maintenance exercises, which review any skill previously studied in the text. You will see them in virtually every exercise set. Answers to all skill maintenance exercises are found at the back of the book. If you miss an exercise, restudy the objective shown in red.

Multiply or divide, and simplify. **[R.7a]**

85. $a^{-9} \cdot a^{23}$

86. $\dfrac{a^{-9}}{a^{23}}$

87. $(6x^5 y^{-4})(-3x^{-3} y^{-7})$

88. $\dfrac{6x^5 y^{-4}}{-3x^{-3} y^{-7}}$

Multiply. **[R.5d]**

89. $2(6 - 10x)$

90. $-1(5 - 6x)$

91. $-4(3x - 2y + z)$

92. $5(-2x + 7y - 4)$

Factor. **[R.5d]**

93. $2x - 6y$

94. $-4x - 24y$

95. $4x - 10y + 2$

96. $-10x + 35y - 20$

97. Name the set consisting of the positive integers less than 10, using both roster notation and set-builder notation. **[R.1a]**

98. Name the set consisting of the negative integers greater than -9 using both roster notation and set-builder notation. **[R.1a]**

▶ Synthesis

To the student and the instructor: The Synthesis exercises found at the end of every exercise set challenge students to combine concepts or skills studied in that section or in preceding parts of the text.

Solve. (The symbol ▰ *indicates an exercise designed to be done using a calculator.)*

99. ▰ $4.23x - 17.898 = -1.65x - 42.454$

100. ▰ $-0.00458y + 1.7787 = 13.002y - 1.005$

101. $\dfrac{3x}{2} + \dfrac{5x}{3} - \dfrac{13x}{6} - \dfrac{2}{3} = \dfrac{5}{6}$

102. $\dfrac{2x - 5}{6} + \dfrac{4 - 7x}{8} = \dfrac{10 + 6x}{3}$

103. $x - \{3x - [2x - (5x - (7x - 1))]\} = x + 7$

104. $23 - 2\{4 + 3(x - 1)\} + 5\{x - 2(x + 3)\}$
$\quad = 7\{x - 2[5 - (2x + 3)]\}$

1.2   Formulas and Applications

▶ **a**   Evaluate formulas and solve a formula for a specified letter.

▶ a   Evaluating and Solving Formulas

A **formula** is an equation that represents or models a relationship between two or more quantities. For example, the relationship between the perimeter P of a square and the length s of its sides is given by the formula $P = 4s$. The formula $A = s^2$ represents the relationship between the area A of a square and the length s of its sides.

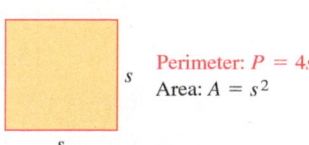

Perimeter: $P = 4s$
Area: $A = s^2$

Other important geometric formulas are $A = \pi r^2$ (for the area A of a circle of radius r), $C = \pi d$ (for the circumference C of a circle of diameter d), and $A = b \cdot h$ (for the area A of a parallelogram of height h and base b). A more complete list of geometric formulas appears on the last page of this text.

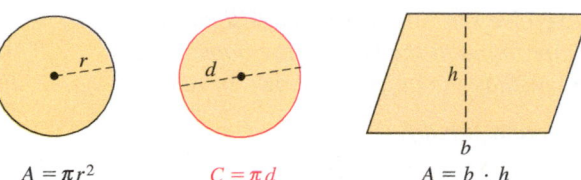

$A = \pi r^2$ $C = \pi d$ $A = b \cdot h$

EXAMPLE 1 *Body Mass Index.* **Body mass index** I can be used to determine whether an individual has a healthy weight for his or her height. An index in the range 18.5–24.9 indicates a normal weight. Body mass index is given by the formula, or model,

$$I = \frac{703W}{H^2},$$

where W is weight, in pounds, and H is height, in inches. (*Source:* Data from Centers for Disease Control and Prevention)

a) Gabby Douglas of the gold-medal winning 2012 U.S. Olympic gymnastics team is 4 ft 11 in. tall and weighs 90 lb. What is her body mass index?

b) NASCAR driver Jimmie Johnson has a body mass index of 23.0 and a height of 5 ft 11 in. What is his weight?

a) We substitute 90 lb for W and 4 ft 11 in., or $4 \cdot 12 + 11 = 59$ in., for H. Then we have

$$I = \frac{703W}{H^2} = \frac{703(90)}{59^2} \approx 18.2.$$

Thus Gabby Douglas's body mass index is 18.2.

b) We substitute 23.0 for I and 5 ft 11 in., or $5 \cdot 12 + 11 = 71$ in., for H and solve for W using the equation-solving principles introduced in Section 1.1:

$$I = \frac{703W}{H^2}$$

$$23.0 = \frac{703W}{71^2} \qquad \text{\color{red}\textbf{Substituting}}$$

$$23.0 = \frac{703W}{5041}$$

$$5041 \cdot 23.0 = 5041 \cdot \frac{703W}{5041} \qquad \text{\color{red}\textbf{Multiplying by 5041}}$$

$$115{,}943 = 703W \qquad \text{\color{red}\textbf{Simplifying}}$$

$$\frac{115{,}943}{703} = \frac{703W}{703} \qquad \text{\color{red}\textbf{Dividing by 703}}$$

$$165 \approx W$$

Jimmie Johnson weighs about 165 lb.   **Now Try Exercise 31(a).**

If we want to make repeated calculations of W, as in Example 1(b), it might be easier to first solve for W, getting it alone on one side of the equation. We "solve" for W as we did above, using the equation-solving principles of Section 1.1.

EXAMPLE 2   Solve for W: $I = \dfrac{703W}{H^2}$.

$$I = \frac{703W}{H^2}$$ **We want this letter alone.**

$$I \cdot H^2 = \frac{703W}{H^2} \cdot H^2$$ **Multiplying by H^2 on both sides to clear the fraction**

$$IH^2 = 703W$$ **Simplifying**

$$\frac{IH^2}{703} = \frac{703W}{703}$$ **Dividing by 703**

$$\frac{IH^2}{703} = W$$

> **Now Try Exercise 21.**

EXAMPLE 3   Solve for r: $H = 2r + 3m$.

$$H = 2r + 3m$$ **We want this letter alone.**

$$H - 3m = 2r$$ **Subtracting $3m$**

$$\frac{H - 3m}{2} = r$$ **Dividing by 2**

> **Now Try Exercise 5.**

EXAMPLE 4   Solve for b: $A = \frac{5}{2}(b - 20)$.

$$A = \tfrac{5}{2}(b - 20)$$ **We want this letter alone.**

$$2A = 5(b - 20)$$ **Multiplying by 2 to clear the fraction**

$$2A = 5b - 100$$ **Removing parentheses**

$$2A + 100 = 5b$$ **Adding 100**

$$\frac{2A + 100}{5} = b, \quad \text{or} \quad b = \frac{2A}{5} + \frac{100}{5} = \frac{2A}{5} + 20$$ **Dividing by 5**

> **Now Try Exercise 23.**

EXAMPLE 5   *Area of a Trapezoid.*   Solve for a: $A = \frac{1}{2}h(a + b)$. (To find the area of a trapezoid, take half the product of the height, h, and the sum of the lengths of the parallel sides, a and b.)

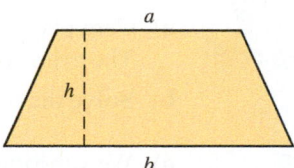

$$A = \tfrac{1}{2}h(a + b)$$ **We want this letter alone.**

$$2A = h(a + b)$$ **Multiplying by 2 to clear the fraction**

$$2A = ha + hb$$ **Using the distributive law**

$$2A - hb = ha$$ **Subtracting hb**

$$\frac{2A - hb}{h} = a, \quad \text{or} \quad a = \frac{2A}{h} - \frac{hb}{h} = \frac{2A}{h} - b$$ **Dividing by h**

> **Now Try Exercise 25.**

Note that there is more than one correct form of the answers in Examples 4 and 5. This is a common occurrence when we solve formulas.

We used the addition principle and the multiplication principle to solve equations in Section 1.1. In a similar manner, we use the same principles in this section to solve a formula for a given letter.

> To solve a formula for a given letter, identify the letter, and:
>
> **1.** Multiply on both sides to clear the fractions or decimals, if necessary.
> **2.** If parentheses occur, multiply to remove them using the distributive law.
> **3.** Collect like terms on each side, if necessary. This may require factoring if a variable is in more than one term.
> **4.** Using the addition principle, get all terms with the letter to be solved for on one side of the equation and all other terms on the other side.
> **5.** Collect like terms again, if necessary.
> **6.** Solve for the letter in question using the multiplication principle.

As indicated in step (3) above, sometimes we must factor to isolate a letter.

EXAMPLE 6 *Simple Interest.* Solve for P: $A = P + Prt$. (To find the amount A to which principal P, in dollars, will grow at simple interest rate r, in t years, add the principal P to the interest, Prt.)

$$A = P + Prt \qquad \textcolor{red}{\text{We want this letter alone.}}$$
$$A = P(1 + rt) \qquad \textcolor{red}{\text{Factoring (or collecting like terms)}}$$
$$\frac{A}{1 + rt} = P \qquad \textcolor{red}{\text{Dividing by } 1 + rt \text{ on both sides}}$$

Now Try Exercise 29.

EXAMPLE 7 *Chess Ratings.* The formula

$$R = r + \frac{400(W - L)}{N}$$

is used to establish a chess player's rating R, after he or she has played N games, where W is the number of wins, L is the number of losses, and r is the average rating of the opponents. (*Source*: Data from U.S. Chess Federation)

a) Cara plays 8 games in a chess tournament, winning 5 games and losing 3 games. The average rating of her opponents is 1205. Find Cara's chess rating.

b) Solve the formula for L.

a) We substitute 8 for N, 5 for W, 3 for L, and 1205 for r in the formula. Then we calculate R:

$$R = r + \frac{400(W - L)}{N} = 1205 + \frac{400(5 - 3)}{8} = 1305.$$

b) We solve as follows:

$$R = r + \frac{400(W - L)}{N}$$
We want this letter alone.

$$NR = N\left[r + \frac{400(W - L)}{N}\right]$$
Multiplying by N to clear the fraction

$$NR = N \cdot r + N \cdot \frac{400(W - L)}{N}$$
Multiplying using the distributive law

$$NR = Nr + 400(W - L)$$
Simplifying

$$NR - Nr = 400(W - L)$$
Subtracting Nr

$$NR - Nr = 400W - 400L$$
Using the distributive law

$$NR - Nr - 400W = -400L$$
Subtracting $400W$

$$\frac{NR - Nr - 400W}{-400} = L.$$
Dividing by -400

Other correct forms of the answer are

$$L = \frac{Nr + 400W - NR}{400} \quad \text{and} \quad L = W - \frac{NR - Nr}{400}.$$

Now Try Exercise 37.

| **1.2** | **Exercise Set** |

▶ # Reading Check

Choose from the column on the right the correct solution of each equation.

RC1. Solve $s = t + 4$ for t.

RC2. Solve $qs + 4r = t$ for q.

RC3. Solve $r = \dfrac{1}{4}q - t$ for t.

RC4. Solve $4q = 7r$ for q.

RC5. Solve $\dfrac{1}{4}s = t - q$ for s.

RC6. Solve $7r - t = 4s$ for r.

a) $q = \dfrac{7r}{4}$, or $\dfrac{7}{4}r$

b) $q = \dfrac{t - 4r}{s}$

c) $s = 4(t - q)$

d) $t = s - 4$

e) $r = \dfrac{4s + t}{7}$

f) $t = \dfrac{q - 4r}{4}$, or $\dfrac{1}{4}q - r$

a *Solve for the given letter.*

1. *Motion Formula:*

$$d = rt, \text{ for } r$$

(Distance d, speed r, time t)

Speed r Time t

Distance d

2. $d = rt$, for t

3. *Area of a Parallelogram:*

$$A = bh, \text{ for } h$$

(Area A, base b, height h)

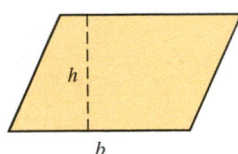

4. *Volume of a Sphere:*

$$V = \frac{4}{3}\pi r^3, \text{ for } r^3$$

(Volume V, radius r)

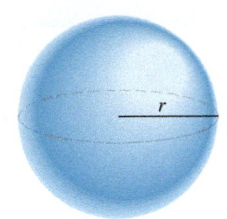

5. *Perimeter of a Rectangle:*

$$P = 2l + 2w, \text{ for } w$$

(Perimeter P, length l, and width w)

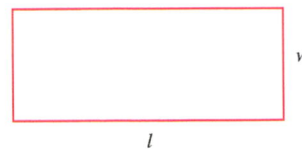

6. $P = 2l + 2w$, for l

7. *Area of a Triangle:*

$$A = \frac{1}{2}bh, \text{ for } b$$

(Area A, base b, height h)

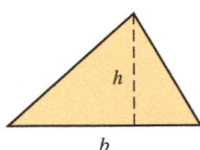

8. $A = \frac{1}{2}bh$, for h

9. *Average of Two Numbers:*

$$A = \frac{a + b}{2}, \text{ for } a$$

a $A = \frac{a+b}{2}$ b

10. $A = \dfrac{a + b}{2}$, for b

11. *Force:*

$$F = ma, \text{ for } m$$

(Force F, mass m, acceleration a)

12. $F = ma$, for a

13. *Simple Interest:*

$$I = Prt, \text{ for } t$$

(Interest I, principal P, interest rate r, time t)

14. $I = Prt$, for P

15. *Relativity:*

$$E = mc^2, \text{ for } c^2$$

(Energy E, mass m, speed of light c)

16. $E = mc^2$, for m **17.** $Q = \dfrac{p - q}{2}$, for p

18. $Q = \dfrac{p - q}{2}$, for q **19.** $Ax + By = c$, for y

20. $Ax + By = c$, for x **21.** $I = 1.08\,\dfrac{T}{N}$, for N

22. $F = \dfrac{mv^2}{r}$, for v^2

23. $C = \dfrac{3}{4}(m + 5)$, for m

24. $N = \dfrac{1}{3}M(t + w)$, for w

25. $n = \dfrac{1}{3}(a + b - c)$, for b

26. $t = \dfrac{1}{6}(x - y + z)$, for z

27. $d = R - Rst$, for R

28. $g = m + mnp$, for m

29. $T = B + Bqt$, for B

30. $Z = Q - Qab$, for Q

Basal Metabolic Rate. An individual's basal metabolic rate is the minimum number of calories required to sustain life when the individual is at rest. It can be thought of as the number of calories burned by an individual who sleeps all day. The Harris–Benedict formula for basal metabolic rate for a man is $R = 66 + 6.23w + 12.7h - 6.8a$. The formula for a woman is $R = 655 + 4.35w + 4.7h - 4.7a$. In each formula, R is in calories, w is weight, in pounds, h is height, in inches, and a is age, in years. (*Source*: Data from *Shapefit*)

31. a) Gary weighs 185 lb, is 5 ft 11 in. tall, and is 28 years old. Use the formula for the basal metabolic rate for a man to find Gary's basal metabolic rate.
   b) Solve the formula for w.

32. a) Alyssa weighs 145 lb, is 5 ft 6 in. tall, and is 32 years old. Use the formula for the basal metabolic rate for a woman to find Alyssa's basal metabolic rate.
   b) Solve the formula for h.

33. *Caloric Requirement.* The number of calories K required each day by a moderately active female who wants to maintain her weight is estimated by the formula

$$K = 1015.25 + 6.74w + 7.29h - 7.29a,$$

where w is weight, in pounds, h is height, in inches, and a is age, in years. (*Source*: Shapefit)

a) Serena is a moderately active 25-year-old woman who weighs 150 lb and is 5 ft 8 in. tall. Find the number of calories she requires each day in order to maintain her weight.
b) Solve the formula for a.

34. *Caloric Requirement.* The number of calories K required each day by a moderately active male who wants to maintain his weight is estimated by the formula

$$K = 102.3 + 9.66w + 19.69h - 10.54a,$$

where w is weight, in pounds, h is height, in inches, and a is age, in years. (*Source*: Shapefit)

a) Dan is a moderately active man who weighs 210 lb, is 6 ft 2 in. tall, and is 34 years old. Find the number of calories he requires each day in order to maintain his weight.
b) Solve the formula for a.

Projecting Birth Weight. Ultrasonic images of 29-week-old fetuses can be used to predict birth weight. One model, or formula, developed by Thurnau, is $P = 9.337da - 299$; a second model, developed by Weiner, is $P = 94.593c + 34.227a - 2134.616$. For the formulas, P is the estimated birth weight, in grams, d is the diameter of the fetal head, in centimeters, c is the circumference of the fetal head, in centimeters, and a is the circumference of the fetal abdomen, in centimeters. (*Sources*: Data from G. R. Thurnau, R. K. Tamura, R. E. Sabbagha, et al. *Am. J. Obstet Gynecol* 1983; **145**:557; C. P. Weiner, R. E. Sabbagha, N. Vaisrub, et al. *Obstet Gynecol* 1985; **65**:812.)

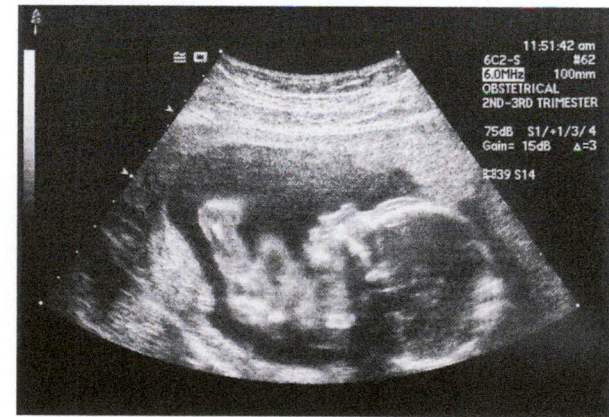

35. a) Use Thurnau's model to estimate the birth weight of a 29-week-old fetus when the diameter of the fetal head is 8.5 cm and the circumference of the fetal abdomen is 24.1 cm.
   b) Solve the formula for a.

36. a) Use Weiner's model to estimate the birth weight of a 29-week-old fetus when the circumference of the fetal head is 26.7 cm and the circumference of the fetal abdomen is 24.1 cm.
   b) Solve the formula for c.

37. *Young's Rule in Medicine.* Young's rule for determining the amount of a medicine dosage for a child is given by

$$c = \frac{ad}{a + 12},$$

where a is the child's age, in years, and d is the usual adult dosage, in milligrams. (*Warning!* Do not apply this formula without checking with a physician!) (*Source*: Data from June Looby Olsen, et al., *Medical Dosage Calculations*, 6th ed. Reading, MA: Addison Wesley Longman, p. A-31)

a) The usual adult dosage of a particular medication is 250 mg. Find the dosage for a child of age 3.

b) Solve the formula for d.

38. *Full-Time-Equivalent Students.* Colleges accommodate students who need to take different total-credit-hour loads. They determine the number of "full-time-equivalent" students, F, using the formula

$$F = \frac{n}{15},$$

where n is the total number of credits students enroll in for a given semester.

a) Determine the number of full-time-equivalent students on a campus in which students register for 42,690 credits.

b) Solve the formula for n.

▶ **Skill Maintenance**

Divide. **[R.2e]**

39. $\dfrac{80}{-16}$

40. $-2000 \div (-8)$

41. $-\dfrac{1}{2} \div \dfrac{1}{4}$

42. $120 \div (-4.8)$

43. $-\dfrac{2}{3} \div \left(-\dfrac{5}{6}\right)$

44. $\dfrac{-90}{-15}$

45. $\dfrac{-90}{15}$

46. $\dfrac{-80}{16}$

▶ **Synthesis**

Solve.

47. $A = \pi rs + \pi r^2$, for s

48. $s = v_1 t + \frac{1}{2} at^2$, for a; for v_1

49. $\dfrac{P_1 V_1}{T_1} = \dfrac{P_2 V_2}{T_2}$, for V_1; for P_2

50. $\dfrac{P_1 V_1}{T_1} = \dfrac{P_2 V_2}{T_2}$, for T_2; for P_1

51. In Exercise 13, you solved the formula $I = Prt$ for t. Now use the formula for t to determine how long it will take a deposit of \$75 to earn \$3 interest when invested at 5% simple interest.

52. The area of the shaded triangle ABE is 20 cm². Find the area of the trapezoid. (See Example 5.)

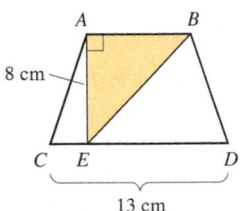

53. *Horsepower of an Engine.* The horsepower of an engine can be calculated by the formula

$$H = W\left(\frac{v}{234}\right)^3,$$

where W is the weight, in pounds, of the car, including the driver, fluids, and fuel, and v is the maximum velocity, or speed, in miles per hour, of the car attained a quarter mile after beginning acceleration.

a) Find the horsepower of a V-6, 2.8-liter engine if $W = 2700$ lb and $v = 83$ mph.

b) Find the horsepower of a 4-cylinder, 2.0-liter engine if $W = 3100$ lb and $v = 73$ mph.

1.3 Applications and Problem Solving

▶ **a** Solve applied problems by translating to equations.

▶ **b** Solve basic motion problems.

▶ a Five Steps for Problem Solving

One very important use of algebra is as a tool for problem solving. The following five-step strategy for solving problems will be used throughout this text.

FIVE STEPS FOR PROBLEM SOLVING

1. *Familiarize* yourself with the problem situation.
2. *Translate* the problem to an equation.
3. *Solve* the equation.
4. *Check* the answer in the original problem.
5. *State* the answer to the problem clearly.

Of the five steps, probably the most important is the first one: becoming familiar with the problem situation. Here are some hints for familiarization.

To familiarize yourself with the problem:

- If a problem is given in words, read it carefully.
- List the information given and the question to be answered. Choose a variable (or variables) to represent the unknown(s) and clearly state what the variable represents. Be descriptive! For example, let L = length (in meters), d = distance (in miles), and so on.
- Make a drawing and label it with known information. Also, indicate unknown information, using specific units if given.
- Find further information if necessary. Look up a formula at the back of this book or in a reference book. Look up the topic using an Internet search engine.
- Make a table that lists all the information you have collected. Look for patterns that may help in the translation to an equation.
- Think of a possible answer and check the guess. Note the manner in which the guess is checked. This will help you translate the problem to an equation.

Top Five Rice-Producing Countries

Country	Rice Production (in metric tons)
1. China	143,000
2. India	99,000
3. Indonesia	36,900
4. Bangladesh	33,800
5. Vietnam	?

Source: www.mapsofworld.com

EXAMPLE 1 *Rice Production.* Rice is the grain with the second-highest worldwide production, after maize. Rice provides more than 20% of the calories consumed by humans. In 2012, China produced 143,000 metric tons of rice. This was 7500 metric tons more than five times the amount of rice produced by Vietnam. (See the table at left.) Find the amount of rice produced in Vietnam. (*Source*: United Nations' Food and Agriculture Organization)

1. **Familiarize.** Let's say that Vietnam produced 30,000 metric tons. Then the amount produced in China would be

$$5(30,000) + 7500 = 150,000 + 7500 = 157,500 \text{ metric tons,}$$

which is more than 143,000 metric tons, the known amount for China. This tells us that our guess of 30,000 is too high. We let $t =$ the amount of rice, in metric tons, produced by Vietnam.

2. **Translate.** We translate as follows:

Five times the amount produced by Vietnam	plus	7500 metric tons	is	143,000 metric tons
↓	↓	↓	↓	↓
$5 \cdot t$	$+$	7500	$=$	$143,000.$

3. **Solve.** We solve the equation as follows:

$$5t + 7500 = 143,000$$
$$5t + 7500 - 7500 = 143,000 - 7500 \qquad \text{\color{red}{Subtracting 7500}}$$
$$5t = 135,500 \qquad \text{\color{red}{Simplifying}}$$
$$\frac{5t}{5} = \frac{135,500}{5} \qquad \text{\color{red}{Dividing by 5}}$$
$$t = 27,100.$$

4. **Check.** If Vietnam produced 27,100 metric tons of rice, then China produced $5(27,100) + 7500$, or 143,000, metric tons of rice. The amount checks.

5. **State.** In 2012, Vietnam produced 27,100 metric tons of rice.

Now Try Exercise 5.

EXAMPLE 2 *Solar Panel Support.* The cross section of a support for a solar energy panel is triangular. The second angle of the triangle is five times as large as the first angle. The third angle is 2° less than the first angle. Find the measures of the angles.

1. **Familiarize.** The second and third angles are described in terms of the first angle so we begin by assigning a variable to the first angle. Then we use that variable to describe the other two angles.

   We let $x =$ the measure of the first angle. Then $5x =$ the measure of the second angle and $x - 2 =$ the measure of the third angle. Recall that the sum of the measures of the angles of a triangle is 180°.

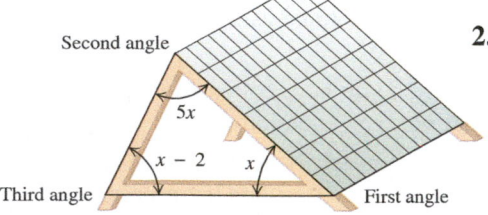

2. **Translate.** We have the following translation:

Measure of first angle	plus	Measure of second angle	plus	Measure of third angle	is	180°
↓	↓	↓	↓	↓	↓	↓
x	$+$	$5x$	$+$	$(x - 2)$	$=$	$180.$

3. **Solve.** We solve the equation as follows:

$$x + 5x + (x - 2) = 180$$
$$7x - 2 = 180 \qquad \text{\color{red}{Collecting like terms}}$$
$$7x - 2 + 2 = 180 + 2 \qquad \text{\color{red}{Adding 2}}$$
$$7x = 182 \qquad \text{\color{red}{Simplifying}}$$
$$\frac{7x}{7} = \frac{182}{7} \qquad \text{\color{red}{Dividing by 7}}$$
$$x = 26.$$

Thus the possible measures of the angles are

First angle: $x = 26°$;
Second angle: $5x = 5 \cdot 26 = 130°$;
Third angle: $x - 2 = 26 - 2 = 24°$.

4. **Check.** Do we have a solution of the *original* problem? The sum of the measures of the angles is

$$26° + 130° + 24° = 180°.$$

The measure of the second angle is five times the measure of the first angle: $130° = 5 \cdot 26°$. The measure of the third angle is 2° less than the measure of the first angle: $24° = 26° - 2°$. The answer checks.

5. **State.** The measure of the first angle is 26°, the measure of the second angle is 130°, and the measure of the third angle is 24°.

> **Now Try Exercise 3.**

EXAMPLE 3 *Price of Auto Detailing.* Using a coupon from an online coupon service, Edward paid $152.75 for full-service, interior and exterior, detailing of the cab of his truck. He paid 35% less than the original price for this service. What was the original price?

1. **Familiarize.** We let $x =$ the original price.

2. **Translate.**

Original price	minus	35%	of	Original price	is	New price
x	$-$	35%	$\cdot$	x	$=$	$152.75.$

    Translation

3. **Solve.** We solve the equation:

$$x - 35\% \cdot x = 152.75$$
$$1x - 0.35x = 152.75 \qquad \text{Replacing 35\% with 0.35}$$
$$(1 - 0.35)x = 152.75$$
$$0.65x = 152.75 \qquad \text{Collecting like terms}$$
$$x = 235. \qquad \text{Dividing by 0.65}$$

4. **Check.** If the original price were $235, we would have:

Price reduction: $35\% \cdot \$235 = 0.35 \cdot 235 = \$82.25$;
Reduced price: $\$235 - \$82.25 = \$152.75$.

We get the price that Edward paid, so the answer checks.

5. **State.** The original price of the full-service detailing of the cab of the truck was $235.

> **Now Try Exercise 7.**

EXAMPLE 4 *Real Estate Commission.* The Mendozas negotiated to pay the realtor who sold their house the following commission:

6% for the first $100,000 of the selling price,   and

5.5% for the amount that exceeded $100,000.

The realtor received a commission of $10,565 for selling the house. What was the selling price?

1. Familiarize. Let's make a guess or an estimate to become familiar with the problem. Suppose the house sold for $225,000. The realtor's commission would be

$$6\% \text{ of } \$100,000 = 0.06(\$100,000) = \$6000$$

plus

$$5.5\% \text{ times } (\$225,000 - \$100,000) = 0.055(\$125,000) = \$6875.$$

The total commission would be $6000 + $6875, or $12,875. Although our guess is not correct, the calculation we performed familiarizes us with the problem, and it also tells us that the house sold for less than $225,000, since $10,565 is less than $12,875. We let S = the selling price of the house.

2. Translate. We translate as follows:

Commission on the first $100,000	plus	Commission on the amount that exceeds $100,000	is	Total commission
$6\% \cdot 100,000$	$+$	$5.5\%(S - 100,000)$	$=$	$10,565.$

3. Solve. We solve the equation:

$$6\% \cdot 100,000 + 5.5\%(S - 100,000) = 10,565$$

$$0.06 \cdot 100,000 + 0.055(S - 100,000) = 10,565 \qquad \text{\color{red}{Converting to decimal notation}}$$

$$6000 + 0.055S - 0.055 \cdot 100,000 = 10,565. \qquad \text{\color{red}{Simplifying and using the distributive law}}$$

$$6000 + 0.055S - 5500 = 10,565 \qquad \text{\color{red}{Simplifying}}$$

$$0.055S + 500 = 10,565 \qquad \text{\color{red}{Collecting like terms}}$$

$$0.055S + 500 - 500 = 10,565 - 500 \qquad \text{\color{red}{Subtracting 500}}$$

$$0.055S = 10,065$$

$$\frac{0.055S}{0.055} = \frac{10,065}{0.055} \qquad \text{\color{red}{Dividing by 0.055}}$$

$$S = \$183,000.$$

4. Check. Performing the check is similar to the sample calculation in the *Familiarize* step. The check is left to the student.

5. State. The selling price of the house was $183,000.

Now Try Exercise 13.

EXAMPLE 5 *Insurance Premiums.* The mortality rate for smokers is much higher than it is for those who do not smoke. With all other factors being equal, the smoker can expect to pay a higher health-insurance premium. The equation

$$y = 8.33x + 32.81$$

can be used to estimate the monthly premium for a $250,000 term life insurance policy for a female smoker age 40 or older, where x is the issue age—that is, $x = 0$ corresponds to issue age 40, $x = 3$ corresponds to issue age 43, and so on. (*Source:* American General Life Insurance Company)

a) Estimate the monthly insurance premium for a female smoker who is 48 years old.

b) At what issue age would the monthly premium be approximately $167?

Since an equation is given, we know that we have a correct translation and thus we will not use the five-step problem-solving strategy in this case.

a) To estimate the monthly premium for a female smoker whose age is 48, we first note that 48 is 8 years beyond 40. We substitute 8 for x in the equation:

$$y = 8.33x + 32.81 = 8.33(8) + 32.81 = 99.45.$$

The estimated monthly insurance premium for a female smoker who is 48 years old when the policy is issued is $99.45.

b) To determine the issue age for a monthly premium of $167, we substitute 167 for y in the equation and solve for x:

$$y = 8.33x + 32.81$$
$$167 = 8.33x + 32.81 \qquad \textcolor{red}{\textbf{Substituting 167 for } \boldsymbol{y}}$$
$$167 - 32.81 = 8.33x + 32.81 - 32.81 \qquad \textcolor{red}{\textbf{Subtracting 32.81}}$$
$$134.19 = 8.33x$$
$$\frac{134.19}{8.33} = \frac{8.33x}{8.33} \qquad \textcolor{red}{\textbf{Dividing by 8.33}}$$
$$16 \approx x.$$

It is estimated that about 16 years after age 40, or at issue age 56, the monthly premium for a $250,000 term life insurance policy for a female smoker is $167.

▶ **Now Try Exercise 25.**

EXAMPLE 6 *Installing Seamless Guttering.* Seamless guttering is delivered on a continuous roll and sections are cut from the roll as needed. The Jordans know that they need 127 ft of guttering for six separate sections of their home and that the four shortest sections will be the same size. The longest piece of guttering is 13 ft more than twice the length of the midsize piece. The shortest piece of guttering is 10 ft less than the midsize piece. How long is each piece of guttering?

Longest piece = $2x + 13$

Midsize piece = x

Shortest piece = $x - 10$

1. **Familiarize.** All the pieces are described in terms of the midsize piece, so we begin by assigning a variable to that piece and using that variable to describe the lengths of the other pieces.

   We let $x =$ the length of the midsize piece, in feet, $2x + 13 =$ the length of the longest piece, and $x - 10 =$ the length of the shortest piece.

2. **Translate.** The sum of the length of the longest piece, plus the length of the midsize piece, plus four times the length of the shortest piece is 127 ft. This gives us the following translation:

Longest piece	plus	Midsize piece	plus	Four times the shortest piece	is	Total length
$(2x + 13)$	$+$	x	$+$	$4(x - 10)$	$=$	$127.$

3. **Solve.** We solve the equation, as follows:

$$(2x + 13) + x + 4(x - 10) = 127$$
$$2x + 13 + x + 4x - 40 = 127 \qquad \textcolor{red}{\textbf{Using the distributive law}}$$
$$7x - 27 = 127 \qquad \textcolor{red}{\textbf{Collecting like terms}}$$
$$7x - 27 + 27 = 127 + 27 \qquad \textcolor{red}{\textbf{Adding 27}}$$
$$7x = 154 \qquad \textcolor{red}{\textbf{Collecting like terms}}$$
$$\frac{7x}{7} = \frac{154}{7} \qquad \textcolor{red}{\textbf{Dividing by 7}}$$
$$x = 22.$$

4. **Check.** Do we have an answer to the *problem*? If the length of the midsize piece is 22 ft, then the length of the longest piece is

$$2 \cdot 22 + 13, \text{ or } 57 \text{ ft,}$$

and the length of the shortest piece is

$$22 - 10, \text{ or } 12 \text{ ft.}$$

The sum of the lengths of the longest piece, the midsize piece, and four times the shortest piece must be 127 ft:

$$57 \text{ ft} + 22 \text{ ft} + 4(12 \text{ ft}) = 127 \text{ ft.}$$

These lengths check.

5. **State.** The length of the longest piece is 57 ft, the length of the midsize piece is 22 ft, and the length of the shortest piece is 12 ft.

 Now Try Exercise 11.

Sometimes applied problems involve **consecutive integers** like 19, 20, 21, 22 or −34, −33, −32, −31. Consecutive integers can be represented in the form $x, x + 1, x + 2, x + 3$, and so on.

Some examples of **consecutive even integers** are 20, 22, 24, 26 and −34, −32, −30, −28. Consecutive even integers can be represented in the form $x, x + 2, x + 4, x + 6$, and so on, as can **consecutive odd integers** like 19, 21, 23, 25 and −33, −31, −29, −27.

EXAMPLE 7 *Artist's Prints.* Often artists will number in sequence a limited number of prints in order to increase their value. An artist creates 500 prints and saves three for his children. The numbers of those prints are consecutive integers whose sum is 189. Find the number of each of those prints.

1. **Familiarize.** The numbers of the prints are consecutive integers. Thus we let $x =$ the first integer, $x + 1 =$ the second, and $x + 2 =$ the third.

2. **Translate.** We translate as follows:

$$\underbrace{\text{First integer}}_{x} + \underbrace{\text{Second integer}}_{(x + 1)} + \underbrace{\text{Third integer}}_{(x + 2)} = \underset{189.}{189}$$

3. **Solve.** We solve the equation:

$$\begin{aligned}
x + (x + 1) + (x + 2) &= 189 \\
3x + 3 &= 189 && \text{Collecting like terms} \\
3x + 3 - 3 &= 189 - 3 && \text{Subtracting 3} \\
3x &= 186 \\
\frac{3x}{3} &= \frac{186}{3} && \text{Dividing by 3} \\
x &= 62.
\end{aligned}$$

Then $x + 1 = 62 + 1 = 63$ and $x + 2 = 62 + 2 = 64$.

4. **Check.** The numbers 62, 63, and 64 are consecutive integers, and their sum is 189. The numbers check.

5. **State.** The numbers of the prints are 62, 63, and 64.

 Now Try Exercise 15.

► b Basic Motion Problems

When a problem deals with speed, distance, and time, we can expect to use the following **motion formula**.

THE MOTION FORMULA

Distance = Rate (or speed) · Time

$$d = rt$$

EXAMPLE 8 *Moving Walkways.* A moving walkway in O'Hare Airport is 300 ft long and moves at a speed of 5 ft/sec. If Kate walks at a speed of 4 ft/sec, how long will it take her to travel the 300 ft using the moving walkway?

1. **Familiarize.** First read the problem very carefully. You might want to talk about it with a classmate or reword it in your mind. Organizing the information in a table can be very helpful.

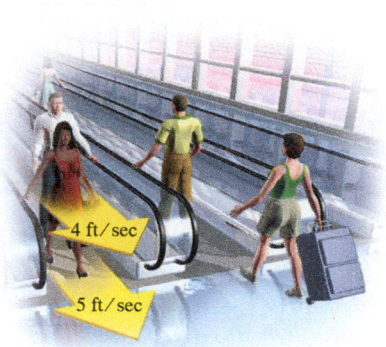

Distance to be traveled	300 ft
Kate's walking speed	4 ft/sec
Speed of the moving walkway	5 ft/sec
Kate's total speed on the walkway	?
Time required	?

Since Kate is walking on the walkway in the same direction in which it is moving, the two speeds can be added to determine Kate's total speed on the walkway. We can then complete the table, letting t = the time, in seconds, required to travel 300 ft on the moving walkway.

Distance to be traveled	300 ft
Kate's walking speed	4 ft/sec
Speed of the moving walkway	5 ft/sec
Kate's total speed on the walkway	9 ft/sec
Time required	t

2. **Translate.** To translate, we use the motion formula $d = rt$, where d = distance, r = speed, or rate, and t = time. We substitute 300 ft for d and 9 ft/sec for r:

$$d = rt$$
$$300 = 9 \cdot t.$$

3. **Solve.** We solve the equation:

$$300 = 9t$$

$$\frac{300}{9} = \frac{9t}{9} \qquad \textcolor{red}{\textbf{Dividing by 9}}$$

$$\frac{100}{3} = t.$$

4. **Check.** At a speed of 9 ft/sec and in a time of 100/3, or $33\frac{1}{3}$ sec, Kate would travel $d = 9 \cdot \frac{100}{3} = 300$ ft. This answer checks.

5. **State.** Kate will travel the distance of 300 ft in 100/3, or $33\frac{1}{3}$ sec.

> **Now Try Exercise 29.**

1.3   Exercise Set

▶ Reading Check

Choose from the column on the right the word that completes each step in the five steps for problem solving.

RC1. _____ yourself with the problem situation.

RC2. _____ the problem to an equation.

RC3. _____ the equation.

RC4. _____ the answer in the original problem.

RC5. _____ the answer to the problem clearly.

Solve

Familiarize

State

Translate

Check

---

a *Solve.*

1. *Diana Nyad.*   On September 2, 2013, Diana Nyad became the first person to swim from Havana, Cuba, to Key West, Florida, across the Florida Straits. She swam the 110-mi distance in approximately 53 hr. At 11.00 P.M. on Sunday, September 1, she was approximately three times as far from Cuba as she was from Florida. How far was she from the Florida coast? (*Source*: Data from ChicagoTribune.com, September 3, 2013)

2. *Ironman Triathlon.*   Held annually in Hawaii since 1978, the Ironman Triathlon championship is a series of long-distance races consisting of a 2.4-mi swim, a 112-mi bicycle ride, and a 26.2-mi marathon. At one point, a participant had completed twice as many miles as the number of miles left to complete. How many miles had he completed at that mark? (*Source*: Data from ironman.com)

3. *City Park.*   The residents of a downtown neighborhood designed a triangular-shaped park as part of a city beautification program. The park is bound by streets on all sides. The second angle of the triangle is 7° more than the first. The third angle is

7° less than twice the first. Find the measures of the angles.

4. *Angles of a Triangle.* The second angle of a triangle is three times as large as the first. The measure of the third angle is 25° greater than that of the first angle. How large are the angles?

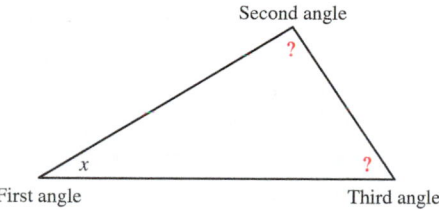

Second angle
?
First angle x
Third angle
?

5. *Climber Deaths.* Deaths of Mt. Everest climbers were first recorded in 1921. From that time through March 21, 2013, there have been 240 deaths. The total number of deaths from falling is 65. This number is 13 more than twice the number who died of exposure/frostbite. How many climbers died of exposure/frostbite? (*Source*: Data from *National Geographic*)

6. *Clearing Customs.* In 14 foreign airports, U.S. Customs and Border Protection provides "preclearance" for U.S.-bound passengers. From October 1, 2011, through September 30, 2012, 14.8 million people took advantage of this program and avoided long delays in Customs lines when returning to the United States. During this time period, 4,808,984 passengers cleared U.S. Customs in Toronto before entering the United States. This number of passengers is 103,264 more than four times the number of passengers who cleared U.S. Customs in Nassau, Bahamas, before entering the United States. How many passengers cleared U.S. Customs in Nassau? (*Source*: Data from U.S. Customs and Border Protection)

7. *Purchasing a Pencil Set.* Abby sees online a premium drawing-pencil set of 144 pieces that will be on sale at 30% off for a 24-hr period. She purchases the set for $176.40. What was the original price?

8. *Purchasing a Headphone.* Max purchased a professional headphone online. He paid $149.75. This amount included a 7% sales tax. What was the price of the headphone itself?

9. *Perimeter of an NBA Court.* The perimeter of an NBA-sized basketball court is 288 ft. The length is 44 ft longer than the width. Find the dimensions of the court. (*Source*: Data from the National Basketball Association)

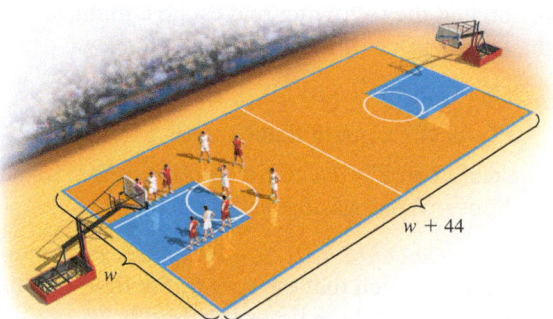

10. *Perimeter of a Tennis Court.* The width of a standard tennis court used for playing doubles is 42 ft less than the length. The perimeter of the court is 228 ft. Find the dimensions of the court. (*Source*: Data from *Dunlop Illustrated Encyclopedia of Facts*)

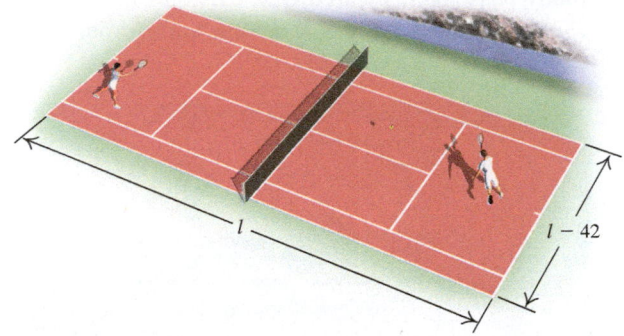

11. *Rope Cutting.* A rope that is 168 ft long is to be cut into three pieces such that the second piece is 6 ft less than three times the first, and the third is 2 ft more than two-thirds of the second. Find the length of the longest piece.

12. *Wire Cutting.* A piece of wire that is 100 cm long is to be cut into two pieces, each to be bent to make a square. The length of a side of one square is to be twice the length of a side of the other. How should the wire be cut?

13. *Real Estate Commission.* The Carlsons negotiated the following real estate commission on the selling price of their house:

7% for the first $100,000, and

5% for the amount that exceeds $100,000.

The realtor received a commission of $15,250 for selling the house. What was the selling price?

14. *Real Estate Commission.* The Hernandez family negotiated the following real estate commission on the selling price of their house:

8% for the first $100,000, and

3% for the amount that exceeds $100,000.

The realtor received a commission of $9200 for selling the house. What was the selling price?

15. *Consecutive Odd Integers.* Find three consecutive odd integers such that the sum of the first, two times the second, and three times the third is 70.

16. *Consecutive Even Integers.* Find three consecutive even integers such that the sum of the first, five times the second, and four times the third is 1226.

17. *Interstate Mile Markers.* U.S. interstate highways post numbered markers every mile to indicate location in case of an accident or break-down. In many states, the numbers increase from west to east. The sum of two consecutive mile markers on I-80 in Iowa is 459. Find the numbers on the markers. (*Source*: Data from Federal Highway Administration)

18. *Post-Office Box Numbers.* The sum of the numbers on two adjacent post-office boxes is 697. What are the numbers?

19. *School Photos.* Memory Makers prices its school photos as shown here.

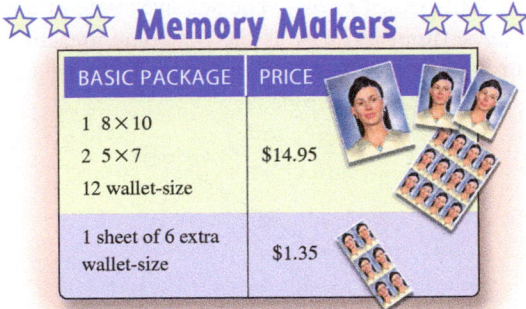

The Morris family purchases the basic package for each of its three children, along with extra wallet-size photos. How many wallet-size photos did they buy in all if their total bill for the photos is $57?

20. *Carpet Cleaning.* A1 Carpet Cleaners charges $75 to clean the first 200 sq ft of carpet. There is an additional charge of 25¢ per square foot for any footage that exceeds 200 sq ft and $1.40 per step for any carpeting on a staircase. A customer's cleaning bill was $253.95. This included the cleaning of a staircase with 13 steps. In addition to the staircase, how many square feet of carpet did the customer have cleaned?

21. *Original Salary.* An editorial assistant receives an 8% raise, bringing her salary to $42,066. What was her salary before the raise?

22. *Original Salary.* After a salesman receives a 5% raise, his new salary is $40,530. What was his old salary?

23. *Diabetes.* In 2010, there were 21.1 million Americans diagnosed with diabetes. This number of cases represents an increase of about 402% over the number of cases in 1973. How many cases of diabetes were there in 1973? (*Source:* Data from *National Geographic*, August 2013, Rich Cohen)

24. *Food Stamp Program.* Enrollment in the federal Supplemental Nutrition Assistance Program (SNAP) has greatly increased since 2008. In April 2013, 47.5 million people were recipients of assistance through the food stamp program. This was an increase of about 68.4% over the number receiving food stamps in 2008. How many people were receiving assistance in 2008? (*Source:* Data from U.S. Department of Agriculture)

25. *Internet Search Ads.* The number of Internet paid search ads is increasing as advertisers move away from traditional marketing methods. The equation

$$y = 6.5x + 41.6$$

can be used to project spending on Internet search ads, in billions of dollars, x years after 2012. (*Source:* Data from Zenith Optimedia)

a) Estimate spending on Internet search ads in 2014.

b) In what year will spending on Internet search ads reach $75 billion?

26. *Insurance Premiums.* The mortality rate for nonsmokers is much lower than it is for those who do smoke. With all other factors being equal, the nonsmoker can expect to pay a lower insurance premium. The equation

$$y = 2.06x + 10.08$$

can be used to estimate the monthly premium for a $250,000 term life insurance policy for a female nonsmoker age 40 or older, where x is the issue age—that is, $x = 0$ corresponds to issue age 40, $x = 3$ corresponds to issue age 43, and so on. (*Source:* Data from American General Life Insurance Company)

a) Estimate the monthly insurance premium for a female nonsmoker who is 50 years old.

b) At what issue age would the monthly premium be approximately $52?

b *Solve.*

27. *Cruising Altitude.* A Boeing 767 has been instructed to climb from its present altitude of 8000 ft to a cruising altitude of 29,000 ft. The plane ascends at a rate of 3500 ft/min. How long will it take the plane to reach the cruising altitude?

28. *Air Travel.* A pilot has been instructed to descend from an altitude of 26,000 ft to 11,000 ft. If the pilot descends at a rate of 2500 ft/min, how long will it take the plane to reach the new altitude?

29. *Boating.* Jen's motorboat travels at a speed of 10 mph in still water. Booth River flows at a speed of 2 mph. How long will it take Jen to travel 15 mi downstream? 15 mi upstream?

30. *Flight into a Headwind.* An airplane traveling 390 mph in still air encounters a 65-mph headwind. How long will it take the plane to travel 725 mi into the wind?

31. *Swimming.* Fran swims at a speed of 5 mph in still water. The Lazy River flows at a speed of 2.3 mph. How long will it take Fran to swim 1.8 mi upstream? 1.8 mi downstream?

32. *River Cruising.*   Now being used as a floating hotel and restaurant in Chattanooga, Tennessee, the *Delta Queen* is a sternwheel steamboat that once cruised the Mississippi River system. It was not uncommon for the *Delta Queen* to travel at a speed of 7 mph in still water and for the Mississippi to flow at a speed of 3 mph. At these rates, how long did it take the boat to cruise 2 mi upstream? (*Sources:* Data from *Delta Queen* information; *The Natchez Democrat,* February 12, 2009)

▶ **Skill Maintenance**

Simplify. [R.3c]

33. $5^2 - 2 \cdot 5 \cdot 12 + 12^2$

34. $16 \cdot 8 + 200 \div 25 \cdot 10$

35. $\dfrac{12|8 - 10| + 9 \cdot 6}{5^4 + 4^5}$

36. $\dfrac{(9 - 4)^2 + (8 - 11)^2}{4^2 + 2^2}$

▶ **Synthesis**

37. 📈 *Real Estate Prices.*   Home prices in Panduski increased 1% from 2013 to 2014. Prices dropped 3% from 2014 to 2015 and dropped another 7% from 2015 to 2016. If a house sold for $105,000 in 2016, what was it worth in 2013? (Round to the nearest dollar.)

38. *Adjusted Wages.*   Christina's salary is reduced $n\%$ during a period of financial difficulty. By what number should her salary be multiplied in order to bring it back to where it was before the reduction?

39. *Population Change.*   The yearly changes in the population census of Poplarville for three consecutive years are, respectively, a 20% increase, a 30% increase, and a 20% decrease. What is the total percent change from the beginning of the first year to the end of the third year, to the nearest percent?

40. *Watch Time.*   Your watch loses $1\frac{1}{2}$ sec every hour. You have a friend whose watch gains 1 sec every hour. The watches show the same time now. After how many more seconds will they show the same time again?

41. *Geometry.*   Consider the geometric figure below. Suppose that $L \parallel M$, $m\angle 8 = 5x + 25$, and $m\angle 4 = 8x + 4$. Find $m\angle 2$ and $m\angle 1$.

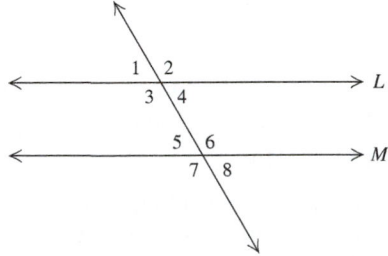

42. *Geometry.*   Suppose the figure $ABCD$ below is a square. Point A is folded onto the midpoint of $\overline{AB}$ and point D is folded onto the midpoint of $\overline{DC}$. The perimeter of the smaller figure formed is 25 in. Find the area of the square $ABCD$.

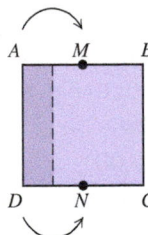

Mid-Chapter Review

Determine whether each statement is true or false.

1. $2x + 3 = 7$ and $x = 2$ are equivalent equations. [1.1a]

2. It is possible for an equation to be false. [1.1a]

3. Every equation has at least one solution. [1.1d]

4. When we solve an applied problem, we check the possible solution in the equation to which the problem was translated. [1.3a]

Determine whether the given number is a solution of the given equation. **[1.1a]**

5. $7; x + 5 = 12$

6. $\frac{1}{3}; 3x - 4 = 5$

7. $-24; \dfrac{-x}{8} = -3$

8. $9; 6(x - 3) = 36$

Solve. **[1.1b, c, d]**

9. $x - 7 = -10$

10. $-7x = 56$

11. $8x - 9 = 23$

12. $1 - x = 3x - 7$

13. $2 - 4y = -4y + 2$

14. $\frac{3}{4}y + 2 = \frac{7}{2}$

15. $5t - 9 = 7t - 4$

16. $4x - 11 = 11 + 4x$

17. $2(y - 4) = 8y$

18. $4y - (y - 1) = 16$

19. $t - 3(t - 4) = 9$

20. $6(2x + 3) = 10 - (4x - 5)$

Solve for the given letter. **[1.2a]**

21. $P = mn$, for n

22. $z = 3t + 3w$, for t

23. $N = \dfrac{r + s}{4}$, for s

24. $T = 1.5\dfrac{A}{B}$, for B

25. $H = \dfrac{2}{3}(t - 5)$, for t

26. $f = g + ghm$, for g

Solve. **[1.3a, b]**

27. *Female Medical School Graduates.*   The number of female medical school graduates in the United States totaled 8396 in 2011. This number was an increase of 21.3% from the total in 2002. What was the number of female medical school graduates in 2002? (*Source*: Data from the Henry J. Kaiser Family Foundation)

28. *Calories Burned While Walking.*   A person weighing 154 lb burns approximately 230 calories while walking 4.5 mph for 30 min. This number of calories is 50 calories less than twice the number burned by a 154-lb person while walking 3.5 mph for 30 min. How many calories would a 154-lb person burn walking 3.5 mph for 30 min? (*Source*: Data from the Centers for Disease Control and Prevention)

29. *Carpet Dimensions.*   The width of an Oriental carpet is 2 ft less than the length. The perimeter of the carpet is 24 ft. Find the dimensions of the carpet.

30. *Boating.*   Frederick's boat travels at a speed of 9 mph in still water. The Bailey River flows at a speed of 3 mph. How long will it take Frederick to travel 18 mi downstream? 18 mi upstream?

Collaborative Discussion and Writing

To the student and the instructor: The Discussion and Writing exercises are meant to be answered with one or more sentences. They can be discussed and answered collaboratively by the entire class or by small groups.

31. Explain the difference between equivalent expressions and equivalent equations.   **[1.1a]**

32. Devise an application in which it would be useful to solve the motion formula $d = rt$ for r.   **[1.2a]**

33. The equations

$$P = 2l + 2w \quad \text{and} \quad w = \frac{P}{2} - l$$

are equivalent formulas involving the perimeter P, length l, and width w of a rectangle. Devise a problem for which the second of the two formulas would be more useful.   **[1.2a]**

34. Explain why we can use the addition principle to subtract the same number on both sides of an equation and why we can use the multiplication principle to divide by the same nonzero number on both sides of an equation.   **[1.1b, c]**

35. How can a guess or an estimate help prepare you for the *Translate* step when solving problems?   **[1.3a]**

36. Why is it important to label clearly what a variable represents in an applied problem?   **[1.3a]**

1.4 Sets, Inequalities, and Interval Notation

▶ **a** Determine whether a given number is a solution of an inequality.

▶ **b** Write interval notation for the solution set or the graph of an inequality.

▶ **c** Solve an inequality using the addition principle and the multiplication principle and then graph the inequality.

▶ **d** Solve applied problems by translating to inequalities.

▶ a Inequalities

We can extend our equation-solving skills to the solving of inequalities.

INEQUALITY

An **inequality** is a sentence containing $<, >, \leq, \geq,$ or $\neq$.

Some examples of inequalities are

$$-2 < a, \qquad x > 4, \qquad x + 3 \leq 6,$$
$$6 - 7y \geq 10y - 4, \quad \text{and} \quad 5x \neq 10.$$

SOLUTION OF AN INEQUALITY

Any replacement or value for the variable that makes an inequality true is called a **solution** of the inequality. The set of all solutions is called the **solution set**. When all the solutions of an inequality have been found, we say that we have **solved** the inequality.

EXAMPLES Determine whether the given number is a solution of the inequality.

1. $x + 3 < 6$; 5

We substitute 5 for x and get $5 + 3 < 6$, or $8 < 6$, a *false* sentence. Therefore, 5 is not a solution.

2. $2x - 3 > -3$; 1

We substitute 1 for x and get $2(1) - 3 > -3$, or $-1 > -3$, a *true* sentence. Therefore, 1 is a solution.

3. $4x - 1 \leq 3x + 2$; -3

We substitute -3 for x and get $4(-3) - 1 \leq 3(-3) + 2$, or $-13 \leq -7$, a *true* sentence. Therefore, -3 is a solution. ▶ **Now Try Exercise 1.**

▶ b Inequalities and Interval Notation

The **graph** of an inequality is a drawing that represents its solutions. An inequality in one variable can be graphed on the number line.

EXAMPLE 4   Graph $x < 4$ on the number line.

The solutions are all real numbers less than 4, so we shade all numbers less than 4 on the number line. To indicate that 4 is not a solution, we use a right parenthesis ") " at 4.

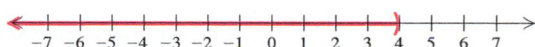

We can write the solution set for $x < 4$ using **set-builder notation**: $\{x \mid x < 4\}$. This is read "The set of all x such that x is less than 4."

Another way to write solutions of an inequality in one variable is to use **interval notation**. Interval notation uses parentheses ( ) and brackets [ ].

If a and b are real numbers such that $a < b$, we define the interval (a, b) as the set of all numbers between but not including a and b—that is, the set of all x for which $a < x < b$. Thus,

$$(a, b) = \{x \mid a < x < b\}.$$

The points a and b are the **endpoints** of the interval. The parentheses indicate that the endpoints are *not* included in the graph.

The interval $[a, b]$ is defined as the set of all numbers x for which $a \leq x \leq b$. Thus,

$$[a, b] = \{x \mid a \leq x \leq b\}.$$

The brackets indicate that the endpoints *are* included in the graph.*

The following intervals include one endpoint and exclude the other:

$$(a, b] = \{x \mid a < x \leq b\}. \quad \text{The graph excludes } a \text{ and includes } b.$$

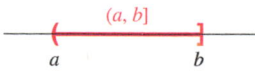

$$[a, b) = \{x \mid a \leq x < b\}. \quad \text{The graph includes } a \text{ and excludes } b.$$

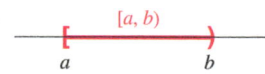

Some intervals extend without bound in one or both directions. We use the symbols ∞, read "infinity," and $-\infty$, read "negative infinity," to name these intervals. The notation (a, ∞) represents the set of all numbers greater than a—that is,

$$(a, \infty) = \{x \mid x > a\}.$$

> **CAUTION!**   Do not confuse the *interval* (a, b) with the *ordered pair* (a, b), which denotes a point in the plane, as we will see in Chapter 2. The context in which the notation appears usually makes the meaning clear.

*Some books use the representations $_{a\ b}$ and $_{a\ b}$ instead of, respectively, —) $_{a\ b}$ and $_{a\ b}$.

Similarly, the notation $(-\infty, a)$ represents the set of all numbers less than a—that is,

$$(-\infty, a) = \{x \mid x < a\}.$$

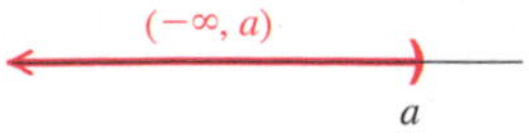

$(-\infty, a)$

The notations $[a, \infty)$ and $(-\infty, a]$ are used when we want to include the endpoint a. The interval $(-\infty, \infty)$ names the set of all real numbers.

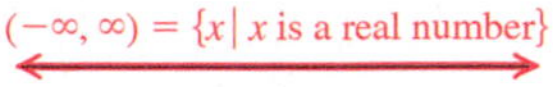

$$(-\infty, \infty) = \{x \mid x \text{ is a real number}\}$$

Interval notation is summarized in the following table.

Intervals: Notation and Graphs

Interval Notation	Set Notation	Graph
(a, b)	$\{x \mid a < x < b\}$	
$[a, b]$	$\{x \mid a \leq x \leq b\}$	
$[a, b)$	$\{x \mid a \leq x < b\}$	
$(a, b]$	$\{x \mid a < x \leq b\}$	
(a, ∞)	$\{x \mid x > a\}$	
$[a, \infty)$	$\{x \mid x \geq a\}$	
$(-\infty, b)$	$\{x \mid x < b\}$	
$(-\infty, b]$	$\{x \mid x \leq b\}$	
$(-\infty, \infty)$	$\{x \mid x \text{ is a real number}\}$	

> **CAUTION!** Whenever the symbol ∞ is included in interval notation, a right parenthesis ") " is used. Similarly, when $-\infty$ is included, a left parenthesis " (" is used.

EXAMPLES Write interval notation for the given set.

5. $\{x \mid -4 < x < 5\} = (-4, 5)$

6. $\{x \mid x \geq -2\} = [-2, \infty)$

7. $\{x \mid 7 > x \geq 1\} = \{x \mid 1 \leq x < 7\} = [1, 7)$

Now Try Exercise 7.

EXAMPLES Write interval notation for the given graph.

8.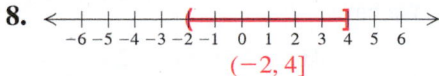

$(-2, 4]$

9.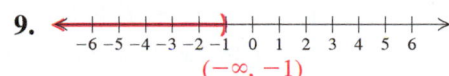

$(-\infty, -1)$

> **Now Try Exercise 13.**

▶ **c Solving Inequalities**

Two inequalities are **equivalent** if they have the same solution set. For example, the inequalities $x > 4$ and $4 < x$ are equivalent. Just as the addition principle for equations gives us equivalent equations, the addition principle for inequalities gives us equivalent inequalities.

THE ADDITION PRINCIPLE FOR INEQUALITIES

For any real numbers a, b, and c:

$$a < b \quad \text{is equivalent to} \quad a + c < b + c;$$
$$a > b \quad \text{is equivalent to} \quad a + c > b + c.$$

Similar statements hold for $\le$ and $\ge$.

Since subtracting c is the same as adding $-c$, there is no need for a separate subtraction principle.

EXAMPLE 10 Solve and graph: $x + 5 > 1$.

We have

$$x + 5 > 1$$
$$x + 5 - 5 > 1 - 5 \qquad \text{\color{red}{Using the addition principle:}}$$
$$\text{\color{red}{adding } -5 \text{ or subtracting } 5}$$
$$x > -4.$$

We used the addition principle to show that the inequalities $x + 5 > 1$ and $x > -4$ are equivalent. The solution set is $\{x \mid x > -4\}$ and consists of an infinite number of solutions. We cannot possibly check them all. Instead, we can perform a partial check by substituting one member of the solution set (here we use -1) into the original inequality:

$$\frac{x + 5 > 1}{-1 + 5 \overset{?}{} 1}$$
$$4 \mid \qquad \text{TRUE}$$

Since $4 > 1$ is true, we have a partial check. The solution set is $\{x \mid x > -4\}$, or $(-4, \infty)$. The graph is as follows:

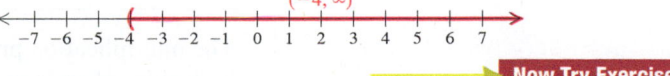

$(-4, \infty)$

> **Now Try Exercise 15.**

EXAMPLE 11   Solve and graph: $4x - 1 \geq 5x - 2$.

We have

$$4x - 1 \geq 5x - 2$$

$$4x - 1 + 2 \geq 5x - 2 + 2 \qquad \text{Adding 2}$$

$$4x + 1 \geq 5x \qquad \text{Simplifying}$$

$$4x + 1 - 4x \geq 5x - 4x \qquad \text{Subtracting } 4x$$

$$1 \geq x. \qquad \text{Simplifying}$$

The inequalities $1 \geq x$ and $x \leq 1$ have the same meaning and the same solutions. The solution set is $\{x \mid 1 \geq x\}$ or, more commonly, $\{x \mid x \leq 1\}$. Using interval notation, we write that the solution set is $(-\infty, 1]$. The graph is as follows:

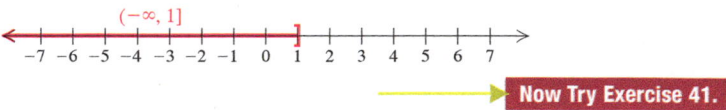

> **Now Try Exercise 41.**

The multiplication principle for inequalities differs from the multiplication principle for equations. Consider the true inequality

$$-4 < 9.$$

If we multiply both numbers by 2, we get another true inequality:

$$-4(2) < 9(2), \quad \text{or} \quad -8 < 18. \qquad \text{True}$$

If we multiply both numbers by -3, we get a false inequality:

$$-4(-3) < 9(-3), \quad \text{or} \quad 12 < -27. \qquad \text{False}$$

However, if we now *reverse* the inequality symbol above, we get a true inequality:

$$12 > -27. \qquad \text{True}$$

THE MULTIPLICATION PRINCIPLE FOR INEQUALITIES

For any real numbers a and b, and any *positive* number c:

$a < b$ is equivalent to $ac < bc$;

$a > b$ is equivalent to $ac > bc$.

For any real numbers a and b, and any *negative* number c:

$a < b$ is equivalent to $ac > bc$;

$a > b$ is equivalent to $ac < bc$.

Similar statements hold for $\leq$ and $\geq$.

Since division by c is the same as multiplication by $1/c$, there is no need for a separate division principle.

The multiplication principle tells us that when we multiply or divide on both sides of an inequality by a negative number, we must reverse the inequality symbol to obtain an equivalent inequality.

EXAMPLE 12 Solve and graph: $3y < \frac{3}{4}$.

We have

$$3y < \frac{3}{4}$$

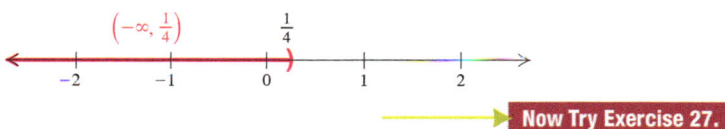

$$\frac{1}{3} \cdot 3y < \frac{1}{3} \cdot \frac{3}{4} \qquad \text{Multiplying by } \tfrac{1}{3}. \text{ Since } \tfrac{1}{3} > 0,$$
$$\qquad\qquad\qquad \text{the symbol stays the same.}$$

$$y < \frac{1}{4}. \qquad \text{Simplifying}$$

Any number less than $\frac{1}{4}$ is a solution. The solution set is $\left\{y \middle| y < \frac{1}{4}\right\}$, or $\left(-\infty, \frac{1}{4}\right)$. The graph is as follows:

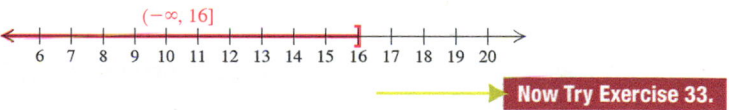

> **Now Try Exercise 27.**

EXAMPLE 13 Solve and graph: $-5x \geq -80$.

We have

$$-5x \geq -80$$

$$\frac{-5x}{-5} \leq \frac{-80}{-5} \qquad \text{Dividing by } -5. \text{ Since } -5 < 0, \text{ the}$$
$$\qquad\qquad\qquad \text{inequality symbol must be reversed.}$$

$$x \leq 16.$$

The solution set is $\{x \mid x \leq 16\}$, or $(-\infty, 16]$. The graph is as follows:

> **Now Try Exercise 33.**

We use the addition and multiplication principles together in solving inequalities in much the same way as in solving equations.

EXAMPLE 14 Solve: $16 - 7y \geq 10y - 4$.

We have

$$16 - 7y \geq 10y - 4$$
$$-16 + 16 - 7y \geq -16 + 10y - 4 \qquad \text{Adding } -16$$
$$-7y \geq 10y - 20 \qquad \text{Collecting like terms}$$
$$-10y + (-7y) \geq -10y + 10y - 20 \qquad \text{Adding } -10y$$
$$-17y \geq -20 \qquad \text{Collecting like terms}$$

$$\frac{-17y}{-17} \leq \frac{-20}{-17} \qquad \text{Dividing by } -17. \text{ The symbol}$$
$$\qquad\qquad\qquad \text{must be reversed.}$$

$$y \leq \frac{20}{17}. \qquad \text{Simplifying}$$

The solution set is $\left\{y \middle| y \leq \frac{20}{17},\right\}$ or $\left(-\infty, \frac{20}{17}\right]$.

> **Now Try Exercise 43.**

We can avoid multiplying or dividing by a negative number by using the addition principle in a different way. Let's rework Example 14 by adding $7y$ instead of $-10y$.

$$16 - 7y \geq 10y - 4$$

$$16 - 7y + 7y \geq 10y - 4 + 7y \qquad \text{Adding } 7y. \text{ This makes the coefficient of the } y\text{-term positive.}$$

$$16 \geq 17y - 4 \qquad \text{Collecting like terms}$$

$$16 + 4 \geq 17y - 4 + 4 \qquad \text{Adding 4}$$

$$20 \geq 17y \qquad \text{Collecting like terms}$$

$$\frac{20}{17} \geq \frac{17y}{17} \qquad \text{Dividing by 17. The symbol stays the same.}$$

$$\frac{20}{17} \geq y, \text{ or } y \leq \frac{20}{17}$$

EXAMPLE 15 Solve: $-3(x + 8) - 5x > 4x - 9$.

$$-3(x + 8) - 5x > 4x - 9$$

$$-3x - 24 - 5x > 4x - 9 \qquad \text{Using the distributive law}$$

$$-24 - 8x > 4x - 9 \qquad \text{Collecting like terms}$$

$$-24 - 8x + 8x > 4x - 9 + 8x \qquad \text{Adding } 8x$$

$$-24 > 12x - 9 \qquad \text{Collecting like terms}$$

$$-24 + 9 > 12x - 9 + 9 \qquad \text{Adding 9}$$

$$-15 > 12x$$

$$\qquad \text{Dividing by 12. The symbol stays the same.}$$

$$\frac{-15}{12} > \frac{12x}{12}$$

$$-\frac{5}{4} > x.$$

The solution set is $\left\{ x \mid -\frac{5}{4} > x \right\}$, or $\left\{ x \mid x < -\frac{5}{4} \right\}$, or $\left(-\infty, -\frac{5}{4} \right)$.

Now Try Exercise 49.

▶ **d Applications and Problem Solving**

Many problem-solving and applied situations translate to inequalities.

Important Words	Sample Sentence	Translation
is at least	Max is at least 5 years old.	$m \geq 5$
is at most	At most 6 people could fit in the elevator.	$n \leq 6$
cannot exceed	Total weight in the elevator cannot exceed 2000 pounds.	$w \leq 2000$
must exceed	The speed must exceed 15 mph.	$s > 15$
is between	Heather's income is between $23,000 and $35,000.	$23{,}000 < h < 35{,}000$
no more than	Bing weighs no more than 90 pounds.	$w \leq 90$
no less than	Saul would accept no less than $4000 for the piano.	$t \geq 4000$

The following phrases deserve special attention.

TRANSLATING "AT LEAST" AND "AT MOST"

A quantity x is **at least** some amount q: $x \geq q$.

  (If x is at least q, it cannot be less than q.)

A quantity x is **at most** some amount q: $x \leq q$.

  (If x is at most q, it cannot be more than q.)

EXAMPLE 16   *Physical Therapists.*   As a result of the aging population staying active longer than previous generations, the employment demand for physical therapists is expected to increase 39% from 2010 to 2020. The equation

$$P = 7745t + 198{,}600$$

can be used to estimate the number of licensed physical therapists in the work force, where t is the number of years since 2010. Determine the years for which the number of physical therapists will be more than 252,000. (*Source:* Data from U.S. Department of Labor)

1. **Familiarize.** We already have an equation. To become more familiar with it, we might make a substitution for t. Suppose that we want to know the number of physical therapists 8 years after 2010, or in 2018. We substitute 8 for t:

$$P = 7745(8) + 198{,}600 = 260{,}560.$$

We see that in 2018 the number of physical therapists will be more than 252,000. To find all the years in which the number of physical therapists exceeds 252,000, we could make other guesses less than 8, but it is more efficient to proceed to the next step.

2. **Translate.** The number of physical therapists is to be more than 252,000. Thus we have

$$P > 252{,}000.$$

We replace P with $7745t + 198{,}600$:

$$7745t + 198{,}600 > 252{,}000.$$

3. **Solve.** We solve the inequality:

$$7745t + 198{,}600 > 252{,}000$$
$$7745t > 53{,}400 \qquad \text{Subtracting 198,600}$$
$$t > 6.89. \qquad \text{Dividing by 7745 and rounding}$$

4. **Check.** As a partial check, we can substitute a value for t that is greater than 6.89. We did that in the *Familiarize* step and found that the number of physical therapists was more than 252,000.

5. **State.** The number of physical therapists will be more than 252,000 for years more than 6.89 years after 2010, so we have $\{t \mid t > 6.89\}$.

Now Try Exercise 83.

EXAMPLE 17   *Salary Plans.*   On her new job, Rose can be paid in one of two ways: *Plan A* is a salary of $600 per month, plus a commission of 4% of sales; and *Plan B* is a salary of $800 per month, plus a commission of 6% of sales in excess of $10,000. For what amount of monthly sales is plan A better than plan B, if we assume that sales are always more than $10,000?

1. **Familiarize.** Listing the given information in a table will be helpful.

Plan A: Monthly Income	Plan B: Monthly Income
$600 salary 4% of sales *Total*: $600 + 4% of sales	$800 salary 6% of sales over $10,000 *Total*: $800 + 6% of sales over $10,000

Next, suppose that Rose had sales of $12,000 in one month. Which plan would be better? Under plan A, she would earn $600 plus 4% of $12,000, or

$$600 + 0.04(12{,}000) = \$1080.$$

Since with plan B commissions are paid only on sales in excess of $10,000, Rose would earn $800 plus 6% of ($12,000 − $10,000), or

$$800 + 0.06(12{,}000 - 10{,}000) = 800 + 0.06(2000) = \$920.$$

This shows that for monthly sales of $12,000, plan A is better. Similar calculations will show that for sales of $30,000 per month, plan B is better. To determine *all* values for which plan A pays more money, we must solve an inequality that is based on the calculations above.

2. **Translate.** We let $S = $ the amount of monthly sales. If we examine the calculations in the *Familiarize* step, we see that the monthly income from plan A is $600 + 0.04S$ and from plan B is $800 + 0.06(S - 10{,}000)$. Thus we want to find all values of S for which

Income from plan A	is greater than	Income from plan B
↓	↓	↓
$600 + 0.04S$	$>$	$800 + 0.06(S - 10{,}000).$

3. **Solve.** We solve the inequality:

$$600 + 0.04S > 800 + 0.06(S - 10{,}000)$$

$$600 + 0.04S > 800 + 0.06S - 600 \qquad \textcolor{red}{\textbf{Using the distributive law}}$$

$$600 + 0.04S > 200 + 0.06S \qquad \textcolor{red}{\textbf{Collecting like terms}}$$

$$400 > 0.02S \qquad \textcolor{red}{\textbf{Subtracting 200 and 0.04S}}$$

$$20{,}000 > S, \text{ or } S < 20{,}000. \qquad \textcolor{red}{\textbf{Dividing by 0.02}}$$

4. **Check.** For $S = 20{,}000$, the income from plan A is

$$600 + 4\% \cdot 20{,}000, \text{ or } \$1400.$$

The income from plan B is

$$800 + 6\% \cdot (20{,}000 - 10{,}000), \text{ or } \$1400.$$

> **This confirms that for sales of $20,000, Rose's pay is the same under either plan.**

In the *Familiarize* step, we saw that for sales of $12,000, plan A pays more. Since $12{,}000 < 20{,}000$, this is a partial check. Since we cannot check all possible values of S, we will stop here.

5. **State.** For monthly sales of less than $20,000, plan A is better.

Now Try Exercise 77.

Translating for Success

The goal of these matching questions is to practice step (2), *Translate*, of the five-step problem-solving process. Translate each word problem to an equation or an inequality and select a correct translation from A–O.

A. $0.05(25{,}750) = x$

B. $x + 2x = 102$

C. $2x + 2(x + 6) = 102$

D. $150 - x \leq 102$

E. $x - 0.05x = 25{,}750$

F. $x + (x + 2) = 102$

G. $x + (x + 6) > 102$

H. $x + 5x = 150$

I. $x + 0.05x = 25{,}750$

J. $x + (2x + 6) = 102$

K. $x + (x + 1) = 102$

L. $102 + x > 150$

M. $0.05x = 25{,}750$

N. $102 + 5x > 150$

O. $x + (x + 6) = 102$

1. *Consecutive Integers.* The sum of two consecutive even integers is 102. Find the integers.

2. *Salary Increase.* After Susanna earned a 5% raise, her new salary was $25,750. What was her former salary?

3. *Dimensions of a Rectangle.* The length of a rectangle is 6 in. more than the width. The perimeter of the rectangle is 102 in. Find the length and the width.

4. *Population.* The population of Doddville is decreasing at a rate of 5% per year. The current population is 25,750. What was the population the previous year?

5. *Reading Assignment.* Quinn has 6 days to complete a 150-page reading assignment. How many pages must he read the first day so that he has no more than 102 pages left to read on the 5 remaining days?

6. *Numerical Relationship.* One number is 6 more than twice another. The sum of the numbers is 102. Find the numbers.

7. *DVD Collections.* Together Ella and Ken have 102 DVDs. If Ken has 6 more DVDs than Ella, how many does each have?

8. *Sales Commissions.* Will earns a commission of 5% on his sales. One year he earned commissions totaling $25,750. What were his total sales for the year?

9. *Fencing.* Jess has 102 ft of fencing that he plans to use to enclose two dog runs. The perimeter of one run is to be twice the perimeter of the other. Into what lengths should the fencing be cut?

10. *Quiz Scores.* Lupe has a total of 102 points on the first 6 quizzes in her sociology class. How many total points must she earn on the 5 remaining quizzes in order to have more than 150 points for the semester?

1.4   Exercise Set

▶ Reading Check

For each solution set expressed in set-builder notation, select from the column on the right the equivalent interval notation.

RC1. $\{x \mid a \le x < b\}$

RC2. $\{x \mid x < b\}$

RC3. $\{x \mid x \text{ is a real number}\}$

RC4. $\{x \mid a < x < b\}$

RC5. $\{x \mid a \le x \le b\}$

RC6. $\{x \mid x \ge a\}$

a) (a, b)

b) $[a, b)$

c) $(-\infty, \infty)$

d) $[a, \infty)$

e) $(-\infty, b]$

f) $(a, b]$

g) $[a, b]$

h) $(-\infty, b)$

i) (a, ∞)

a *Determine whether the given numbers are solutions of the inequality.*

1. $x - 2 \ge 6$; $-4, 0, 4, 8$

2. $3x + 5 \le -10$; $-5, -10, 0, 27$

3. $t - 8 > 2t - 3$; $0, -8, -9, -3, -\frac{7}{8}$

4. $5y - 7 < 8 - y$; $2, -3, 0, 3, \frac{2}{3}$

b *Write interval notation for the given set or graph.*

5. $\{x \mid x < 5\}$

6. $\{t \mid t \ge -5\}$

7. $\{x \mid -3 \le x \le 3\}$

8. $\{t \mid -10 < t \le 10\}$

9. $\{x \mid -4 > x > -8\}$

10. $\{x \mid 13 > x \ge 5\}$

11.

12.

13.

14.

c *Solve and graph on the number line.*

15. $x + 2 > 1$

16. $x + 8 > 4$

17. $y + 3 < 9$

18. $y + 4 < 10$

19. $a - 9 \le -31$

20. $a + 6 \le -14$

21. $t + 13 \ge 9$

22. $x - 8 \le 17$

23. $y - 8 > -14$

24. $y - 9 > -18$

25. $x - 11 \le -2$

26. $y - 18 \le -4$

27. $8x \ge 24$

28. $8t < -56$

29. $0.3x < -18$

30. $0.6x < 30$

31. $\frac{2}{3}x > 2$

32. $\frac{3}{5}x > -3$

Solve.

33. $-9x \ge -8.1$

34. $-5y \le 3.5$

35. $-\frac{3}{4}x \ge -\frac{5}{8}$

36. $-\frac{1}{8}y \le -\frac{9}{8}$

37. $2x + 7 < 19$

38. $5y + 13 > 28$

39. $5y + 2y \leq -21$

40. $-9x + 3x \geq -24$

41. $2y - 7 < 5y - 9$

42. $8x - 9 < 3x - 11$

43. $0.4x + 5 \leq 1.2x - 4$

44. $0.2y + 1 > 2.4y - 10$

45. $5x - \frac{1}{12} \leq \frac{5}{12} + 4x$

46. $2x - 3 < \frac{13}{4}x + 10 - 1.25x$

47. $4(4y - 3) \geq 9(2y + 7)$

48. $2m + 5 \geq 16(m - 4)$

49. $3(2 - 5x) + 2x < 2(4 + 2x)$

50. $2(0.5 - 3y) + y > (4y - 0.2)8$

51. $5[3m - (m + 4)] > -2(m - 4)$

52. $[8x - 3(3x + 2)] - 5 \geq 3(x + 4) - 2x$

53. $3(r - 6) + 2 > 4(r + 2) - 21$

54. $5(t + 3) + 9 < 3(t - 2) + 6$

55. $19 - (2x + 3) \leq 2(x + 3) + x$

56. $13 - (2c + 2) \geq 2(c + 2) + 3c$

57. $\frac{1}{4}(8y + 4) - 17 < -\frac{1}{2}(4y - 8)$

58. $\frac{1}{3}(6x + 24) - 20 > -\frac{1}{4}(12x - 72)$

59. $2[4 - 2(3 - x)] - 1 \geq 4[2(4x - 3) + 7] - 25$

60. $5[3(7 - t) - 4(8 + 2t)] - 20 \leq -6[2(6 + 3t) - 4]$

61. $\frac{4}{5}(7x - 6) < 40$

62. $\frac{2}{3}(4x - 3) > 30$

63. $\frac{3}{4}(3 + 2x) + 1 \geq 13$

64. $\frac{7}{8}(5 - 4x) - 17 \geq 38$

65. $\frac{3}{4}\left(3x - \frac{1}{2}\right) - \frac{2}{3} < \frac{1}{3}$

66. $\frac{2}{3}\left(\frac{7}{8} - 4x\right) - \frac{5}{8} < \frac{3}{8}$

67. $0.7(3x + 6) \geq 1.1 - (x + 2)$

68. $0.9(2x + 8) < 20 - (x + 5)$

69. $a + (a - 3) \leq (a + 2) - (a + 1)$

70. $0.8 - 4(b - 1) > 0.2 + 3(4 - b)$

d *Solve.*

Body Mass Index. Body mass index *I* can be used to determine whether an individual has a healthy weight for his or her height. An index in the range 18.5–24.9 indicates a normal weight. Body mass index is given by the formula, or model,

$$I = \frac{703W}{H^2},$$

where *W* is weight, in pounds, and *H* is height, in inches. Use this formula for Exercises 71 and 72. (*Source:* Data from Centers for Disease Control and Prevention)

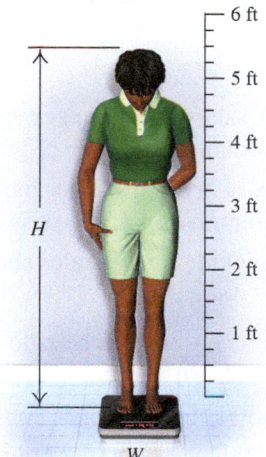

71. *Body Mass Index.* Alexandra's height is 62 in. Determine, in terms of an inequality, those weights *W* that will keep her body mass index below 25.

72. *Body Mass Index.* Josiah's height is 77 in. Determine, in terms of an inequality, those weights *W* that will keep his body mass index below 25.

73. *Grades.* David is taking an economics course in which there will be 4 tests, each worth 100 points. He has scores of 89, 92, and 95 on the first three tests. He must make a total of at least 360 in order to get an A. What scores on the last test will give David an A?

74. *Grades.* Elizabeth is taking a mathematics course in which there will be 5 tests, each worth 100 points. She has scores of 94, 90, and 89 on the first three tests. She must make a total of at least 450 in order to get an A. What scores on the fourth test will keep Elizabeth eligible for an A?

75. *Insurance Claims.* After a serious automobile accident, most insurance companies will replace the damaged car with a new one if repair costs exceed 80% of the N.A.D.A., or "blue-book," value of the car. Miguel's car recently sustained $9200 worth of damage but was not replaced. What was the blue-book value of his car?

76. *Delivery Service.* Jay's Express prices cross-town deliveries at $15 for the first 10 miles plus $1.25 for each additional mile. PDQ, Inc., prices its cross-town deliveries at $25 for the first 10 miles plus $0.75 for each additional mile. For what number of miles is PDQ less expensive?

77. *Salary Plans.* Imani can be paid in one of two ways:

Plan A: A salary of $400 per month plus a commission of 8% of gross sales;

Plan B: A salary of $610 per month, plus a commission of 5% of gross sales.

For what amount of gross sales should Imani select plan A?

78. *Salary Plans.* Aiden can be paid for his masonry work in one of two ways:

Plan A: $300 plus $9.00 per hour;

Plan B: Straight $12.50 per hour.

Suppose that the job takes n hours. For what values of n is plan B better for Aiden?

79. *Prescription Coverage.* Low Med offers two prescription-drug insurance plans. With plan 1, James would pay the first $150 of his prescription costs and 30% of all costs after that. With plan 2, James would pay the first $280 of costs, but only 10% of the rest. For what amount of prescription costs will plan 2 save James money? (Assume that his prescription costs exceed $280.)

80. *Insurance Benefits.* Bayside Insurance offers two plans. Under plan A, Giselle would pay the first $50 of her medical bills and 20% of all bills after that. Under plan B, Giselle would pay the first $250 of bills, but only 10% of the rest. For what amount of medical bills will plan B save Giselle money? (Assume that her bills will exceed $250.)

81. *Wedding Costs.* The Arnold Inn offers two plans for wedding parties. Under plan A, the inn charges $30 for each person in attendance. Under plan B, the inn charges $1300 plus $20 for each person in

excess of the first 25 who attend. For what size parties will plan B cost less? (Assume that more than 25 guests will attend.)

82. *Investing.* Matthew is about to invest $20,000, part at 3% and the rest at 4%. What is the most that he can invest at 3% and still be guaranteed at least $650 in interest per year?

83. *Renting Office Space.* An investment group is renovating a commercial building and will rent offices to small businesses. The formula

$$R = 2(s + 70)$$

can be used to determine the monthly rent for an office with s square feet. All utilities are included in the monthly payment. For what square footage will the rent be less than $2100?

84. *Temperatures of Solids.* The formula

$$C = \frac{5}{9}(F - 32)$$

can be used to convert Fahrenheit temperatures F to Celsius temperatures C.

a) Gold is a solid at Celsius temperatures less than 1063°C. Find the Fahrenheit temperatures for which gold is a solid.

b) Silver is a solid at Celsius temperatures less than 960.8°C. Find the Fahrenheit temperatures for which silver is a solid.

85. *Tuition and Fees at Two-Year Colleges.* The equation

$$C = 82t + 1923$$

can be used to estimate the average cost of tuition and fees at two-year public institutions of higher education, where t is the number of years since 2005. (*Source*: Data from National Center for Education Statistics, U.S. Department of Education)

a) What was the average cost of tuition and fees in 2010? in 2014?

b) For what years will the cost of tuition and fees be more than $3000?

86. *Dewpoint Spread.*   Pilots use the **dewpoint spread**, or the difference between the current temperature and the dewpoint (the temperature at which dew occurs), to estimate the height of the cloud cover. Each 3° of dewpoint spread corresponds to an increased height of cloud cover of 1000 ft. A plane, flying with limited instruments, must have a cloud cover higher than 3500 ft. What dewpoint spreads will allow the plane to fly?

Decrease of 3° per 1000 ft

3500 ft

▶ **Skill Maintenance**

Simplify. **[R.6b]**

87. $3a - 6(2a - 5b)$

88. $2(x - y) + 10(3x - 7y)$

89. $4(a - 2b) - 6(2a - 5b)$

90. $-3(2a - 3b) + 8b$

Factor. **[R.5d]**

91. $30x - 70y - 40$

92. $-12a + 30ab$

93. $-8x + 24y - 4$

94. $10n - 45mn + 100m$

Add or subtract. **[R.2a, c]**

95. $-2.3 - 8.9$

96. $-2.3 + 8.9$

97. $-2.3 + (-8.9)$

98. $-2.3 - (-8.9)$

▶ **Synthesis**

99. *Supply and Demand.*   The supply S and demand D for a certain product are given by

$$S = 460 + 94p \quad \text{and} \quad D = 2000 - 60p.$$

a) Find those values of p for which supply exceeds demand.

b) Find those values of p for which supply is less than demand.

Determine whether each statement is true or false. If false, give a counterexample.

100. For any real numbers x and y, if $x < y$, then $x^2 < y^2$.

101. For any real numbers a, b, c, and d, if $a < b$ and $c < d$, then $a + c < b + d$.

102. Determine whether the inequalities

$$x < 3 \quad \text{and} \quad 0 \cdot x < 0 \cdot 3$$

are equivalent. Give reasons to support your answer.

Solve.

103. $x + 5 \leq 5 + x$

104. $x + 8 < 3 + x$

105. $x^2 + 1 > 0$

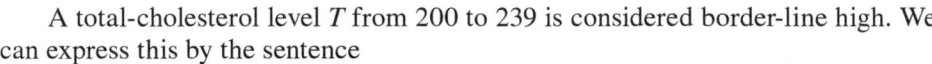

1.5 Intersections, Unions, and Compound Inequalities

▶ **a** Find the intersection of two sets. Solve and graph conjunctions of inequalities.

▶ **b** Find the union of two sets. Solve and graph disjunctions of inequalities.

▶ **c** Solve applied problems involving conjunctions and disjunctions of inequalities.

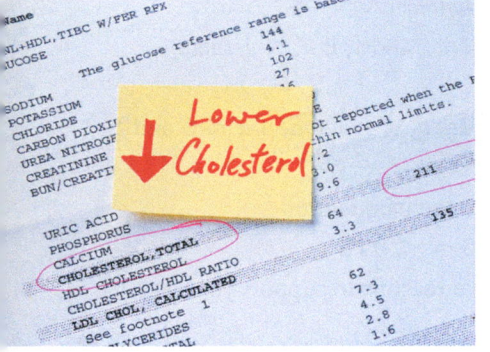

Cholesterol is a substance that is found in every cell of the human body. High levels of cholesterol can cause fatty deposits in the blood vessels that increase the risk of heart attack or stroke. A blood test can be used to measure *total cholesterol*. The following table shows the health risks associated with various cholesterol levels.

Total Cholesterol	Risk Level
Less than 200	Normal
From 200 to 239	Borderline high
240 or higher	High

A total-cholesterol level T from 200 to 239 is considered border-line high. We can express this by the sentence

$$200 \leq T \quad and \quad T \leq 239$$

or more simply by

$$200 \leq T \leq 239.$$

This is an example of a *compound inequality*. **Compound inequalities** consist of two or more inequalities joined by the word *and* or the word *or*. We now "solve" such sentences—that is, we find the set of all solutions.

▶ **a** Intersections of Sets and Conjunctions of Inequalities

> **INTERSECTION**
>
> The **intersection** of two sets A and B is the set of all members that are common to A and B. We denote the intersection of sets A and B as
>
> $A \cap B.$

The intersection of two sets is often illustrated as shown below.

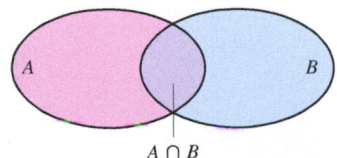

$A \cap B$

EXAMPLE 1 Find the intersection: $\{1, 2, 3, 4, 5\} \cap \{-2, -1, 0, 1, 2, 3\}$.

Only the numbers 1, 2, and 3 are common to the two sets, so the intersection is $\{1, 2, 3\}$.

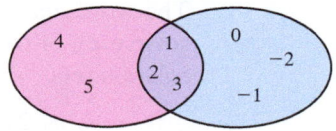

Now Try Exercise 1.

CONJUNCTION

When two or more sentences are joined by the word *and* to make a compound sentence, the new sentence is called a **conjunction** of the sentences.

The following is a conjunction of inequalities:

$-2 < x$ *and* $x < 1$.

A number is a solution of a conjunction if it is a solution of *both* inequalities. For example, 0 is a solution of $-2 < x$ *and* $x < 1$ because $-2 < 0$ *and* $0 < 1$. Shown below is the graph of $-2 < x$, followed by the graph of $x < 1$, and then by the graph of the conjunction $-2 < x$ *and* $x < 1$. As the graphs demonstrate, *the solution set of a conjunction is the intersection of the solution sets of the individual inequalities.*

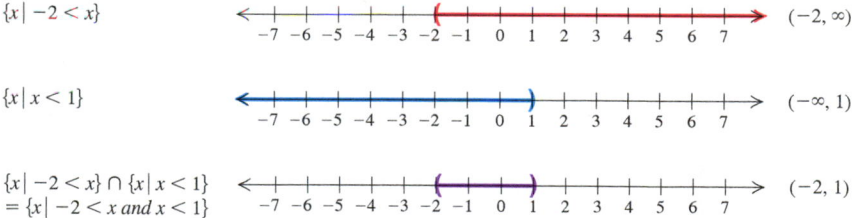

Because there are numbers that are both greater than -2 and less than 1, the conjunction $-2 < x$ *and* $x < 1$ can be abbreviated by $-2 < x < 1$. Thus the interval $(-2, 1)$ can be represented as $\{x \mid -2 < x < 1\}$, the set of all numbers that are *simultaneously* greater than -2 *and* less than 1. Note that, in general, for $a < b$,

$$a < x \quad and \quad x < b \quad \textbf{can be abbreviated} \quad a < x < b;$$
$$and \quad b > x \quad and \quad x > a \quad \textbf{can be abbreviated} \quad b > x > a.$$

CAUTION! "$a > x$ *and* $x < b$" cannot be abbreviated as "$a > x < b$".

"AND"; "INTERSECTION"

The word **"and"** corresponds to **"intersection"** and to the symbol "$\cap$". In order for a number to be a solution of a conjunction, it must make each part of the conjunction true.

Now Try Exercise 13.

EXAMPLE 2   Solve and graph: $-1 \leq 2x + 5 < 13$.

This inequality is an abbreviation for the conjunction

$$-1 \leq 2x + 5 \quad and \quad 2x + 5 < 13.$$

The word *and* corresponds to set *intersection*, $\cap$. To solve the conjunction, we solve each of the two inequalities separately and then find the intersection of the solution sets:

$$
\begin{array}{lll}
-1 \leq 2x + 5 & and & 2x + 5 < 13 \\
-6 \leq 2x & and & 2x < 8 \qquad \text{\color{red}Subtracting 5} \\
-3 \leq x & and & x < 4. \qquad \text{\color{red}Dividing by 2}
\end{array}
$$

We now abbreviate the result:

$$-3 \leq x < 4.$$

The solution set is $\{x \mid -3 \leq x < 4\}$, or, in interval notation, $[-3, 4)$. The graph is the intersection of the two separate solution sets.

$\{x \mid -3 \leq x\}$ $[-3, \infty)$

$\{x \mid x < 4\}$ $(-\infty, 4)$

$\{x \mid -3 \leq x\} \cap \{x \mid x < 4\}$
$= \{x \mid -3 \leq x < 4\}$ $[-3, 4)$

The steps above are generally combined as follows:

$$
\begin{array}{ll}
-1 \leq 2x + 5 < 13 & \text{\color{red}$2x + 5$ appears in both inequalities.} \\
-6 \leq 2x < 8 & \text{\color{red}Subtracting 5} \\
-3 \leq x < 4. & \text{\color{red}Dividing by 2}
\end{array}
$$

Such an approach saves some writing and will prove useful in Section 1.6.

Now Try Exercise 17.

EXAMPLE 3   Solve and graph: $2x - 5 \geq -3$ *and* $5x + 2 \geq 17$.

We first solve each inequality separately:

$$
\begin{array}{lll}
2x - 5 \geq -3 & and & 5x + 2 \geq 17 \\
2x \geq 2 & and & 5x \geq 15 \\
x \geq 1 & and & x \geq 3.
\end{array}
$$

Next, we find the intersection of the two separate solution sets:

$\{x \mid x \geq 1\}$ $[1, \infty)$

$\{x \mid x \geq 3\}$ $[3, \infty)$

$\{x \mid x \geq 1\} \cap \{x \mid x \geq 3\}$
$= \{x \mid x \geq 3\}$ $[3, \infty)$

The numbers common to both sets are those that are greater than or equal to 3. Thus the solution set is $\{x \mid x \geq 3\}$, or, in interval notation, $[3, \infty)$. You should check that any number in $[3, \infty)$ satisfies the conjunction whereas numbers outside $[3, \infty)$ do not.

———▶ **Now Try Exercise 19.**

EMPTY SET; DISJOINT SETS

Sometimes two sets have no elements in common. In such a case, we say that the intersection of the two sets is the **empty set**, denoted $\{\ \}$ or $\varnothing$. Two sets with an empty intersection are said to be **disjoint**.

$A \cap B = \varnothing$

EXAMPLE 4 Solve and graph: $2x - 3 > 1$ *and* $3x - 1 < 2$.

We solve each inequality separately:

$$2x - 3 > 1 \quad \text{and} \quad 3x - 1 < 2$$
$$2x > 4 \quad \text{and} \qquad 3x < 3$$
$$x > 2 \quad \text{and} \qquad x < 1.$$

The solution set is the intersection of the solution sets of the individual inequalities.

$\{x \mid x > 2\}$ 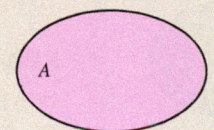 $(2, \infty)$

$\{x \mid x < 1\}$  $(-\infty, 1)$

$\{x \mid x > 2\} \cap \{x \mid x < 1\}$
$= \{x \mid x > 2 \text{ and } x < 1\}$
$= \varnothing$ $\qquad \varnothing$

Since no number is both greater than 2 and less than 1, the solution set is the empty set, $\varnothing$.

———▶ **Now Try Exercise 21.**

EXAMPLE 5 Solve: $3 \leq 5 - 2x < 7$.

We have

$$3 \leq 5 - 2x < 7$$
$$3 - 5 \leq 5 - 2x - 5 < 7 - 5 \qquad \textcolor{red}{\text{Subtracting 5}}$$
$$-2 \leq \quad -2x \quad < 2 \qquad \textcolor{red}{\text{Simplifying}}$$
$$\frac{-2}{-2} \geq \frac{-2x}{-2} > \frac{2}{-2} \qquad \textcolor{red}{\text{Dividing by } -2. \text{ The symbols must be reversed.}}$$
$$1 \geq x > -1. \qquad \textcolor{red}{\text{Simplifying}}$$

The solution set is $\{x \mid 1 \geq x > -1\}$, or $\{x \mid -1 < x \leq 1\}$, since the inequalities $1 \geq x > -1$ and $-1 < x \leq 1$ are equivalent. The solution, in interval notation, is $(-1, 1]$.

———▶ **Now Try Exercise 33.**

▶ **b** **Unions of Sets and Disjunctions of Inequalities**

> **UNION**
>
> The **union** of two sets A and B is the collection of elements belonging to A and/or B. We denote the union of A and B by
>
> $$A \cup B.$$

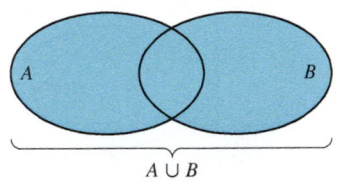

$A \cup B$

The union of two sets is often illustrated as shown at left.

EXAMPLE 6　Find the union: $\{2, 3, 4\} \cup \{3, 5, 7\}$.

The numbers in either or both sets are 2, 3, 4, 5, and 7, so the union is $\{2, 3, 4, 5, 7\}$. We don't list the number 3 twice.

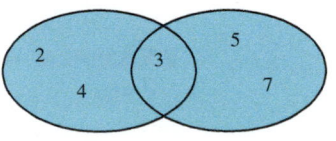

Now Try Exercise 5.

> **DISJUNCTION**
>
> When two or more sentences are joined by the word *or* to make a compound sentence, the new sentence is called a **disjunction** of the sentences.

The following is an example of a disjunction:

$$x < -3 \quad or \quad x > 3.$$

A number is a solution of a disjunction if it is a solution of at least one of the individual inequalities. For example, -7 is a solution of $x < -3 \ or \ x > 3$ because $-7 < -3$. Similarly, 5 is also a solution because $5 > 3$.

Shown below is the graph of $x < -3$, followed by the graph of $x > 3$, and then by the graph of the disjunction $x < -3 \ or \ x > 3$. As the graphs demonstrate, *the solution set of a disjunction is the union of the solution sets of the individual sentences.*

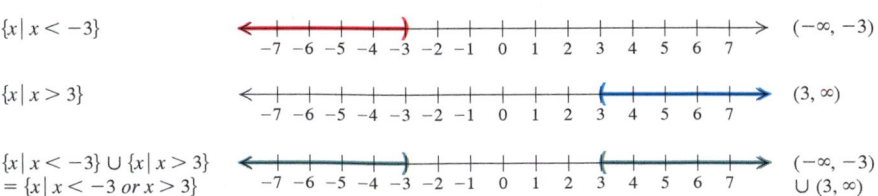

The solution set of

$$x < -3 \quad or \quad x > 3$$

is written $\{x | x < -3 \text{ or } x > 3\}$, or, in interval notation, $(-\infty, -3) \cup (3, \infty)$. This cannot be written in a more condensed form.

> **"OR"; "UNION"**
>
> The word "**or**" corresponds to "**union**" and the symbol "∪". In order for a number to be in the solution set of a disjunction, it must be in *at least one* of the solution sets of the individual sentences.

Now Try Exercise 41.

CAUTION! A compound inequality like

$$x < -4 \quad or \quad x \geq 2,$$

as in Example 7, *cannot* be expressed as $2 \leq x < -4$ because to do so would be to say that x is *simultaneously* less than -4 and greater than or equal to 2. No number is both less than -4 *and* greater than or equal to 2, but many are less than -4 *or* greater than or equal to 2.

EXAMPLE 7   Solve and graph: $7 + 2x < -1 \text{ or } 13 - 5x \leq 3$.

We solve each inequality separately, retaining the word *or*:

$$7 + 2x < -1 \quad or \quad 13 - 5x \leq 3$$
$$2x < -8 \quad or \quad -5x \leq -10$$

Dividing by -5. The symbol must be reversed.

$$x < -4 \quad or \quad x \geq 2.$$

To find the solution set of the disjunction, we consider the individual graphs. We graph $x < -4$ and then $x \geq 2$. Then we take the union of the graphs.

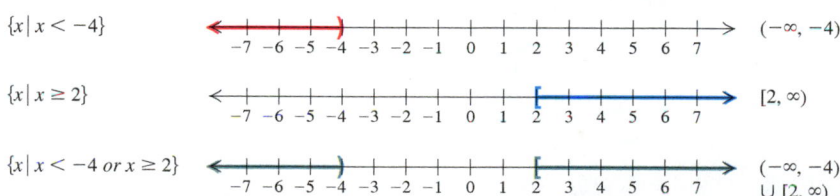

The solution set is written $\{x | x < -4 \text{ or } x \geq 2\}$, or, in interval notation, $(-\infty, -4) \cup [2, \infty)$.

Now Try Exercise 43.

EXAMPLE 8   Solve: $-2x - 5 < -2 \text{ or } x - 3 < -10$.

We solve the individual inequalities separately, retaining the word *or*:

$$-2x - 5 < -2 \quad or \quad x - 3 < -10$$
$$-2x < 3 \quad or \quad x < -7$$

Reversing the symbol

$$x > -\tfrac{3}{2} \quad or \quad x < -7.$$

Keep the word "or."

The solution set is written $\left\{x | x < -7 \text{ or } x > -\tfrac{3}{2}\right\}$, or, in interval notation, $(-\infty, -7) \cup \left(-\tfrac{3}{2}, \infty\right)$.

Now Try Exercise 49.

EXAMPLE 9   Solve: $3x - 11 < 4$ *or* $4x + 9 \geq 1$.

We solve the individual inequalities separately, retaining the word *or*:

$$3x - 11 < 4 \quad or \quad 4x + 9 \geq 1$$
$$3x < 15 \quad or \quad 4x \geq -8$$
$$x < 5 \quad or \quad x \geq -2.$$

To find the solution set, we first look at the individual graphs.

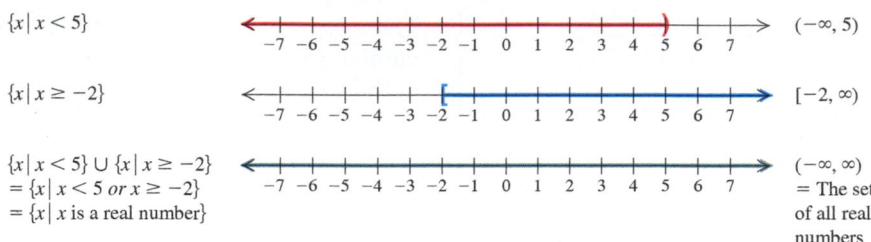

$\{x \mid x < 5\}$ $(-\infty, 5)$

$\{x \mid x \geq -2\}$ $[-2, \infty)$

$\{x \mid x < 5\} \cup \{x \mid x \geq -2\}$ $(-\infty, \infty)$
$= \{x \mid x < 5 \text{ or } x \geq -2\}$ = The set
$= \{x \mid x \text{ is a real number}\}$ of all real
numbers

Since any number is either less than 5 or greater than or equal to -2, the two sets fill the entire number line. Thus the solution set is the set of all real numbers, $(-\infty, \infty)$.   **Now Try Exercise 51.**

▶ c   Applications and Problem Solving

EXAMPLE 10   *Renting Office Space.*   The equation

$$R = 2(s + 70)$$

can be used to determine the monthly rent R for an office with s square footage in a renovated commercial building. All utilities are included in the monthly payment. A florist shop has a monthly rental budget between \$1720 and \$2560. What square footage can be rented and remain within budget?

1. **Familiarize.** We have an equation for calculating the monthly rent. Thus we can substitute a value into the formula. For 720 ft^2, the rent is found as follows:

$$R = 2(720 + 70) = 2 \cdot 790 = \$1580.$$

This familiarizes us with the equation and also tells us that the number of square feet that we are looking for must be larger than 720 since \$1580 is less than \$1720.

2. **Translate.** We want the monthly rent to be between \$1720 and \$2560, so we need to find those values of s for which $1720 < R < 2560$. Substituting $2(s + 70)$ for R, we have

$$1720 < 2(s + 70) < 2560.$$

3. **Solve.** We solve the inequality:

$$1720 < 2(s + 70) < 2560$$
$$\frac{1720}{2} < \frac{2(s + 70)}{2} < \frac{2560}{2} \qquad \text{\color{red}{Dividing by 2}}$$
$$860 < s + 70 < 1280$$
$$790 < s < 1210. \qquad \text{\color{red}{Subtracting 70}}$$

4. **Check.** We substitute some values as we did in the *Familiarize* step.

5. **State.** Square footage between 790 ft^2 and 1210 ft^2 can be rented for a budget between \$1720 and \$2560 per month.   **Now Try Exercise 59.**

1.5 Exercise Set

▶ **Reading Check**

Determine whether each statement is true or false.

RC1. A compound inequality like $x < 5$ *and* $x > -2$ can be expressed as $-2 < x < 5$.

RC2. A compound inequality like $x \geq 5$ *and* $x < -2$ can be expressed as $5 \leq x < -2$.

RC3. The solution set of $x < -4$ *and* $x > 4$ can be written as $\varnothing$.

RC4. The solution set of $x \leq 3$ *and* $x \geq 0$ can be written as $[0, 3]$.

a, b *Find the intersection or union.*

1. $\{9, 10, 11\} \cap \{9, 11, 13\}$

2. $\{1, 5, 10, 15\} \cap \{5, 15, 20\}$

3. $\{a, b, c, d\} \cap \{b, f, g\}$

4. $\{m, n, o, p\} \cap \{m, o, p\}$

5. $\{9, 10, 11\} \cup \{9, 11, 13\}$

6. $\{1, 5, 10, 15\} \cup \{5, 15, 20\}$

7. $\{a, b, c, d\} \cup \{b, f, g\}$

8. $\{m, n, o, p\} \cup \{m, o, p\}$

9. $\{2, 5, 7, 9\} \cap \{1, 3, 4\}$

10. $\{a, e, i, o, u\} \cap \{m, q, w, s, t\}$

11. $\{3, 5, 7\} \cup \varnothing$

12. $\{3, 5, 7\} \cap \varnothing$

a *Graph on the number line and write interval notation.*

13. $-4 < a$ *and* $a \leq 1$

14. $-\frac{5}{2} \leq m$ *and* $m < \frac{3}{2}$

15. $1 < x < 6$

16. $-3 \leq y \leq 4$

Solve and graph on the number line.

17. $-10 \leq 3x + 2$ *and* $3x + 2 < 17$

18. $-11 < 4x - 3$ *and* $4x - 3 \leq 13$

19. $3x + 7 \geq 4$ *and* $2x - 5 \geq -1$

20. $4x - 7 < 1$ *and* $7 - 3x > -8$

21. $4 - 3x \geq 10$ *and* $5x - 2 > 13$

22. $5 - 7x > 19$ *and* $2 - 3x < -4$

Solve.

23. $-4 < x + 4 < 10$

24. $-6 < x + 6 \leq 8$

25. $6 > -x \geq -2$

26. $3 > -x \geq -5$

27. $2 < x + 3 \leq 9$

28. $-6 \leq x + 1 < 9$

29. $1 < 3y + 4 \leq 19$

30. $5 \leq 8x + 5 \leq 21$

31. $-10 \leq 3x - 5 \leq -1$

32. $-6 \leq 2x - 3 < 6$

33. $-18 \leq -2x - 7 < 0$

34. $4 > -3m - 7 \geq 2$

35. $-\frac{1}{2} < \frac{1}{4}x - 3 \leq \frac{1}{2}$

36. $-\frac{2}{3} \leq 4 - \frac{1}{4}x < \frac{2}{3}$

37. $-4 \leq \dfrac{7 - 3x}{5} \leq 4$

38. $-3 < \dfrac{2x - 5}{4} < 8$

b *Graph on the number line and write interval notation.*

39. $x < -2$ *or* $x > 1$ **40.** $x < -4$ *or* $x > 0$

41. $x \leq -3$ *or* $x > 1$ **42.** $x \leq -1$ *or* $x > 3$

Solve and graph on the number line.

43. $x + 3 < -2$ *or* $x + 3 > 2$

44. $x - 2 < -1$ *or* $x - 2 > 3$

45. $2x - 8 \le -3 \text{ or } x - 1 \ge 3$

46. $x - 5 \le -4 \text{ or } 2x - 7 \ge 3$

47. $7x + 4 \ge -17 \text{ or } 6x + 5 \ge -7$

48. $4x - 4 < -8 \text{ or } 4x - 4 < 12$

Solve.

49. $7 > -4x + 5 \text{ or } 10 \le -4x + 5$

50. $6 > 2x - 1 \text{ or } -4 \le 2x - 1$

51. $3x - 7 > -10 \text{ or } 5x + 2 \le 22$

52. $3x + 2 < 2 \text{ or } 4 - 2x < 14$

53. $-2x - 2 < -6 \text{ or } -2x - 2 > 6$

54. $-3m - 7 < -5 \text{ or } -3m - 7 > 5$

55. $\dfrac{2}{3}x - 14 < -\dfrac{5}{6} \text{ or } \dfrac{2}{3}x - 14 > \dfrac{5}{6}$

56. $\dfrac{1}{4} - 3x \le -3.7 \text{ or } \dfrac{1}{4} - 5x \ge 4.8$

57. $\dfrac{2x - 5}{6} \le -3 \text{ or } \dfrac{2x - 5}{6} \ge 4$

58. $\dfrac{7 - 3x}{5} < -4 \text{ or } \dfrac{7 - 3x}{5} > 4$

C *Solve.*

59. *Pressure at Sea Depth.* The equation

$$P = 1 + \frac{d}{33}$$

gives the pressure P, in atmospheres (atm), at a depth of d feet in the sea. For what depths d is the pressure at least 1 atm and at most 7 atm?

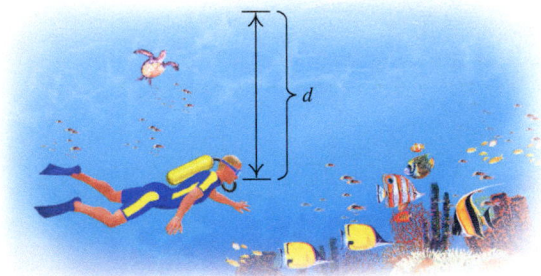

60. *Temperatures of Liquids.* The formula
$$C = \tfrac{5}{9}(F - 32)$$
can be used to convert Fahrenheit temperatures F to Celsius temperatures C.

a) Gold is a liquid for Celsius temperatures C such that $1063° \le C < 2660°$. Find such an inequality for the corresponding Fahrenheit temperatures.

b) Silver is a liquid for Celsius temperatures C such that $960.8° \le C < 2180°$. Find such an inequality for the corresponding Fahrenheit temperatures.

61. *Aerobic Exercise.* In order to achieve maximum results from aerobic exercise, one should maintain one's heart rate at a certain level. A 30-year-old woman with a resting heart rate of 60 beats per minute should keep her heart rate between 138 and 162 beats per minute while exercising. She checks her pulse for 10 sec while exercising. What should the number of beats be?

62. *Minimizing Tolls.* A $6.00 toll is charged to cross the bridge from mainland Florida to Sanibel Island. A six-month pass, costing $50.00, reduces the toll to $2.00. A one-year pass, costing $400, allows for free crossings. How many crossings per year does it take, on average, for the two six-month passes to be the most economical choice? Assume a constant number of trips per month. (*Source:* Data from leewayinfo.com)

63. *Body Mass Index.* Refer to Exercises 71 and 72 in Exercise Set 1.4. Alexandra's height is 62 in. What weights W will allow Alexandra to keep her body mass index I in the 18.5–24.9 range?

64. *Body Mass Index.* Refer to Exercises 71 and 72 in Exercise Set 1.4. Josiah's height is 77 in. What weights W will allow Josiah to keep his body mass index in the 18.5–24.9 range?

65. *Young's Rule in Medicine.* Refer to Exercise 37 in Exercise Set 1.2. The dosage of a medication for an 8-year-old child must stay between 100 mg and 200 mg. Find the equivalent adult dosage.

66. *Young's Rule in Medicine.* Refer to Exercise 65. The dosage of a medication for a 5-year-old child must stay between 50 mg and 100 mg. Find the equivalent adult dosage.

▶ ## Skill Maintenance

Solve. [1.1d]

67. $8y - 3 = 3 + 8y$

68. $-\dfrac{1}{2}t + 5 = -\dfrac{7}{2}t$

69. $20 = 4(3y - 7)$

70. $3x - (x - 1) = 19$

71. $-3 + 2x = 2x - 3$

72. $6(x - 5) = 2(x + 3)$

▶ ## Synthesis

Solve.

73. $x - 10 < 5x + 6 \leq x + 10$

74. $4m - 8 > 6m + 5 \text{ or } 5m - 8 < -2$

75. $-\dfrac{2}{15} \leq \dfrac{2}{3}x - \dfrac{2}{5} \leq \dfrac{2}{15}$

76. $2[5(3 - y) - 2(y - 2)] > y + 4$

77. $3x < 4 - 5x < 5 + 3x$

78. $2x - \dfrac{3}{4} < -\dfrac{1}{10} \text{ or } 2x - \dfrac{3}{4} > \dfrac{1}{10}$

79. $x + 4 < 2x - 6 \leq x + 12$

80. $2x + 3 \leq x - 6 \text{ or } 3x - 2 \leq 4x + 5$

Determine whether each sentence is true or false for all real numbers a, b, and c.

81. If $-b < -a$, then $a < b$.

82. If $a \leq c$ and $c \leq b$, then $b \geq a$.

83. If $a < c$ and $b < c$, then $a < b$.

84. If $-a < c$ and $-c > b$, then $a > b$.

85. What is the union of the set of all rational numbers with the set of all irrational numbers? the intersection?

1.6   Absolute-Value Equations and Inequalities

▶ **a**   Simplify expressions containing absolute-value symbols.

▶ **b**   Find the distance between two points on the number line.

▶ **c**   Solve equations with absolute-value expressions.

▶ **d**   Solve equations with two absolute-value expressions.

▶ **e**   Solve inequalities with absolute-value expressions.

▶ **a**   ## Properties of Absolute Value

We can think of the **absolute value** of a number as the number's distance from zero on the number line.

ABSOLUTE VALUE

The **absolute value** of x, denoted $|x|$, is defined as follows:

$$x \geq 0 \longrightarrow |x| = x; \qquad x < 0 \longrightarrow |x| = -x.$$

This definition tells us that, when x is nonnegative, the absolute value of x is x and, when x is negative, the absolute value of x is the opposite of x. For example, $|3| = 3$ and $|-3| = -(-3) = 3$. We see that absolute value is never negative.

Some simple properties of absolute value allow us to manipulate or simplify algebraic expressions.

PROPERTIES OF ABSOLUTE VALUE

a) $|ab| = |a| \cdot |b|$, for any real numbers a and b.
(The absolute value of a product is the product of the absolute values.)

b) $\left|\dfrac{a}{b}\right| = \dfrac{|a|}{|b|}$, for any real numbers a and b and $b \neq 0$.
(The absolute value of a quotient is the quotient of the absolute values.)

c) $|-a| = |a|$, for any real number a.
(The absolute value of the opposite of a number is the same as the absolute value of the number.)

EXAMPLES   Simplify, leaving as little as possible inside the absolute-value signs.

1. $|5x| = |5| \cdot |x| = 5|x|$

2. $|-3y| = |-3| \cdot |y| = 3|y|$

3. $|7x^2| = |7| \cdot |x^2| = 7|x^2| = 7x^2$ Since x^2 is never negative for any number x

4. $\left|\dfrac{6x}{-3x^2}\right| = \left|\dfrac{-2}{x}\right| = \dfrac{|-2|}{|x|} = \dfrac{2}{|x|}$ **Now Try Exercise 7.**

▶ b   Distance on the Number Line

The number line below shows that the distance between -3 and 2 is 5.

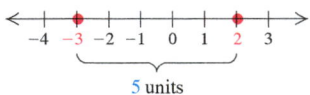

5 units

Another way to find the distance between two numbers on the number line is to determine the absolute value of the difference, as follows:

$$|-3 - 2| = |-5| = 5, \quad \text{or} \quad |2 - (-3)| = |5| = 5.$$

Note that the order in which we subtract does not matter because we are taking the absolute value after we have subtracted.

DISTANCE AND ABSOLUTE VALUE

For any real numbers a and b, the **distance** between them is $|a - b|$.

We should note that the distance is also $|b - a|$, because $a - b$ and $b - a$ are opposites and hence have the same absolute value.

EXAMPLE 5   Find the distance between -8 and -92 on the number line.

$$|-8 - (-92)| = |84| = 84, \quad \text{or} \quad |-92 - (-8)| = |-84| = 84$$

Now Try Exercise 17.

EXAMPLE 6 Find the distance between x and 0 on the number line.

$$|x - 0| = |x|$$

▶ c Equations with Absolute Value

EXAMPLE 7 Solve: $|x| = 4$. Then graph on the number line.

Note that $|x| = |x - 0|$, so that $|x - 0|$ is the distance from x to 0. Thus solutions of the equation $|x| = 4$, or $|x - 0| = 4$ are those numbers x whose distance from 0 is 4. Those numbers are -4 and 4. The solution set is $\{-4, 4\}$. The graph consists of just two points, as shown.

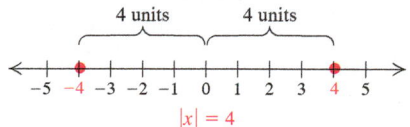

Now Try Exercise 25.

EXAMPLE 8 Solve: $|x| = 0$.

The only number whose absolute value is 0 is 0 itself. Thus the solution is 0. The solution set is $\{0\}$.

Now Try Exercise 29.

EXAMPLE 9 Solve: $|x| = -7$.

The absolute value of a number is always nonnegative. Thus there is no number whose absolute value is -7; consequently, the equation has no solution. The solution set is $\varnothing$.

Now Try Exercise 27.

Examples 7–9 lead us to the following principle for solving linear equations with absolute value.

THE ABSOLUTE VALUE PRINCIPLE

For any positive number p and any algebraic expression X:

a) The solution of $|X| = p$ is those numbers that satisfy $X = -p$ *or* $X = p$.

b) The equation $|X| = 0$ is equivalent to the equation $X = 0$.

c) The equation $|X| = -p$ has no solution.

We can use the absolute-value principle with the addition and multiplication principles to solve equations with absolute value.

EXAMPLE 10 Solve: $2|x| + 5 = 9$.

We first use the addition and multiplication principles to get $|x|$ by itself. Then we use the absolute-value principle.

$$2|x| + 5 = 9$$
$$2|x| = 4 \qquad \text{\textcolor{red}{Subtracting 5}}$$
$$|x| = 2 \qquad \text{\textcolor{red}{Dividing by 2}}$$
$$x = -2 \;\; or \;\; x = 2 \qquad \text{\textcolor{red}{Using the absolute-value principle}}$$

The solutions are -2 and 2. The solution set is $\{-2, 2\}$.

Now Try Exercise 45.

EXAMPLE 11   Solve: $|x - 2| = 3$.

We can consider solving this equation in two different ways.

Method 1:   This allows us to see the meaning of the solutions graphically. The solution set consists of those numbers that are 3 units from 2 on the number line.

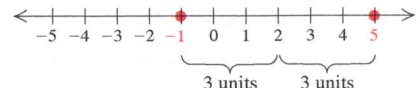

The solutions of $|x - 2| = 3$ are -1 and 5. The solution set is $\{-1, 5\}$.

Method 2:   This method is more efficient. We use the absolute-value principle, replacing X with $x - 2$ and p with 3. Then we solve each equation separately.

$$|X| = p$$
$$|x - 2| = 3$$
$$x - 2 = -3 \quad or \quad x - 2 = 3 \qquad \text{\color{red}{Absolute-value principle}}$$
$$x = -1 \quad or \qquad x = 5$$

The solutions are -1 and 5. The solution set is $\{-1, 5\}$.

> **Now Try Exercise 31.**

EXAMPLE 12   Solve: $|2x + 5| = 13$.

We use the absolute-value principle, replacing X with $2x + 5$ and p with 13:

$$|X| = p$$
$$|2x + 5| = 13$$
$$2x + 5 = -13 \quad or \quad 2x + 5 = 13 \qquad \text{\color{red}{Absolute-value principle}}$$
$$2x = -18 \quad or \qquad 2x = 8$$
$$x = -9 \quad or \qquad x = 4.$$

The solutions are -9 and 4. The solution set is $\{-9, 4\}$.

> **Now Try Exercise 33.**

EXAMPLE 13   Solve: $|4 - 7x| = -8$.

Since absolute value is always nonnegative, this equation has no solution. The solution set is $\varnothing$.

> **Now Try Exercise 53.**

▶ d   Equations with Two Absolute-Value Expressions

Sometimes equations have two absolute-value expressions. Consider $|a| = |b|$. This means that a and b are the same distance from 0. If a and b are the same distance from 0, then either they are the same number or they are opposites.

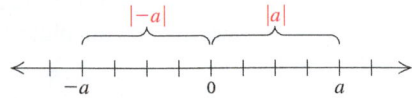

EXAMPLE 14 Solve: $|2x - 3| = |x + 5|$.

Either $2x - 3 = x + 5$ *or* $2x - 3 = -(x + 5)$. We solve each equation:

$$
\begin{array}{lcl}
2x - 3 = x + 5 & or & 2x - 3 = -(x + 5) \\
x - 3 = 5 & or & 2x - 3 = -x - 5 \\
x = 8 & or & 3x - 3 = -5 \\
x = 8 & or & 3x = -2 \\
x = 8 & or & x = -\frac{2}{3}.
\end{array}
$$

The solutions are 8 and $-\frac{2}{3}$. The solution set is $\left\{8, -\frac{2}{3}\right\}$.

Now Try Exercise 57.

EXAMPLE 15 Solve: $|x + 8| = |x - 5|$.

$$
\begin{array}{lcl}
x + 8 = x - 5 & or & x + 8 = -(x - 5) \\
8 = -5 & or & x + 8 = -x + 5 \\
8 = -5 & or & 2x = -3 \\
8 = -5 & or & x = -\frac{3}{2}
\end{array}
$$

The first equation has no solution. The solution of the second equation is $-\frac{3}{2}$. The solution set is $\left\{-\frac{3}{2}\right\}$.

Now Try Exercise 59.

▶ **e Inequalities with Absolute Value**

We can extend our methods for solving equations with absolute value to those for solving inequalities with absolute value.

EXAMPLE 16 Solve: $|x| = 4$. Then graph on the number line.

From Example 7, we know that the solutions are -4 and 4. The solution set is $\{-4, 4\}$. The graph consists of just two points, as shown here.

EXAMPLE 17 Solve: $|x| < 4$. Then graph.

The solutions of $|x| < 4$ are the solutions of $|x - 0| < 4$ and are those numbers x whose distance from 0 is less than 4. We can check by substituting or by looking at the number line that numbers like $-3, -2, -1, -\frac{1}{2}, -\frac{1}{4}, 0, \frac{1}{4}, \frac{1}{2}, 1, 2,$ and 3 are all solutions. In fact, the solutions are all the real numbers x between -4 and 4. The solution set is $\{x | -4 < x < 4\}$ or, in interval notation, $(-4, 4)$. The graph is as follows.

Now Try Exercise 71.

EXAMPLE 18 Solve: $|x| \geq 4$. Then graph.

The solutions of $|x| \geq 4$ are solutions of $|x - 0| \geq 4$ and are those numbers whose distance from 0 is greater than or equal to 4—in other words, those numbers x such that $x \leq -4$ *or* $x \geq 4$. The solution set is $\{x | x \leq -4 \text{ } or \text{ } x \geq 4\}$, or $(-\infty, -4] \cup [4, \infty)$. The graph is as follows.

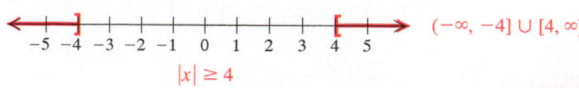

Now Try Exercise 73.

Examples 16–18 illustrate three cases of solving equations and inequalities with absolute value. The following is a general principle for solving equations and inequalities with absolute value.

SOLUTIONS OF ABSOLUTE-VALUE EQUATIONS AND INEQUALITIES

For any positive number p and any algebraic expression X:

a) The solutions of $|X| = p$ are those numbers that satisfy
$X = -p \text{ or } X = p.$

As an example, replacing X with $5x - 1$ and p with 8, we see that the solutions of $|5x - 1| = 8$ are those numbers x for which

$$5x - 1 = -8 \quad or \quad 5x - 1 = 8$$
$$5x = -7 \quad or \quad 5x = 9$$
$$x = -\tfrac{7}{5} \quad or \quad x = \tfrac{9}{5}.$$

The solution set is $\left\{-\tfrac{7}{5}, \tfrac{9}{5}\right\}$.

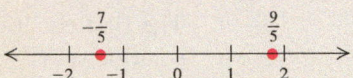

b) The solutions of $|X| < p$ are those numbers that satisfy
$-p < X < p.$

As an example, replacing X with $6x + 7$ and p with 5, we see that the solutions of $|6x + 7| < 5$ are those numbers x for which

$$-5 < 6x + 7 < 5$$
$$-12 < 6x < -2$$
$$-2 < x < -\tfrac{1}{3}.$$

The solution set is $\left\{x \mid -2 < x < -\tfrac{1}{3}\right\}$, or $\left(-2, -\tfrac{1}{3}\right)$.

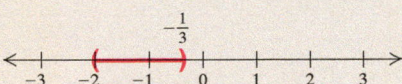

c) The solutions of $|X| > p$ are those numbers that satisfy
$X < -p \text{ or } X > p.$

As an example, replacing X with $2x - 9$ and p with 4, we see that the solutions of $|2x - 9| > 4$ are those numbers x for which

$$2x - 9 < -4 \quad or \quad 2x - 9 > 4$$
$$2x < 5 \quad or \quad 2x > 13$$
$$x < \tfrac{5}{2} \quad or \quad x > \tfrac{13}{2}.$$

The solution set is $\left\{x \mid x < \tfrac{5}{2} \text{ or } x > \tfrac{13}{2}\right\}$, or $\left(-\infty, \tfrac{5}{2}\right) \cup \left(\tfrac{13}{2}, \infty\right)$.

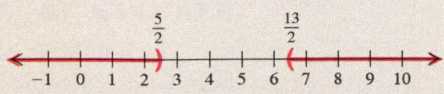

EXAMPLE 19 Solve: $|3x - 2| < 4$. Then graph.

We use part (b). In this case, X is $3x - 2$ and p is 4:

$$|X| < p$$
$$|3x - 2| < 4 \qquad \textcolor{red}{\textbf{Replacing } X \textbf{ with } 3x - 2 \textbf{ and } p \textbf{ with } 4}$$
$$-4 < 3x - 2 < 4$$
$$-2 < 3x < 6$$
$$-\tfrac{2}{3} < x < 2.$$

The solution set is $\left\{ x \mid -\tfrac{2}{3} < x < 2 \right\}$, or $\left(-\tfrac{2}{3}, 2 \right)$. The graph is as follows.

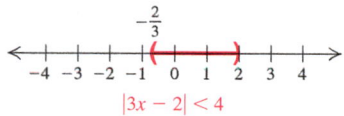

$|3x - 2| < 4$

Now Try Exercise 79.

EXAMPLE 20 Solve: $|8 - 4x| \le 5$. Then graph.

We use part (b). In this case, X is $8 - 4x$ and p is 5:

$$|X| \le p$$
$$|8 - 4x| \le 5 \qquad \textcolor{red}{\textbf{Replacing } X \textbf{ with } 8 - 4x \textbf{ and } p \textbf{ with } 5}$$
$$-5 \le 8 - 4x \le 5$$
$$-13 \le -4x \le -3$$
$$\tfrac{13}{4} \ge x \ge \tfrac{3}{4}. \qquad \textcolor{red}{\textbf{Dividing by } -4 \textbf{ and reversing the inequality symbols}}$$

The solution set is $\left\{ x \mid \tfrac{13}{4} \ge x \ge \tfrac{3}{4} \right\}$, or $\left\{ x \mid \tfrac{3}{4} \le x \le \tfrac{13}{4} \right\}$, or $\left[\tfrac{3}{4}, \tfrac{13}{4} \right]$.

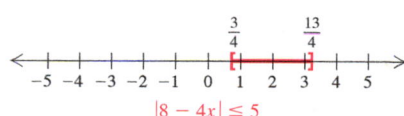

$|8 - 4x| \le 5$

Now Try Exercise 93.

EXAMPLE 21 Solve: $|4x + 2| \ge 6$. Then graph.

We use part (c). In this case, X is $4x + 2$ and p is 6:

$$|X| \ge p$$
$$|4x + 2| \ge 6 \qquad \textcolor{red}{\textbf{Replacing } X \textbf{ with } 4x + 2 \textbf{ and } p \textbf{ with } 6}$$
$$4x + 2 \le -6 \quad or \quad 4x + 2 \ge 6$$
$$4x \le -8 \quad or \qquad 4x \ge 4$$
$$x \le -2 \quad or \qquad x \ge 1.$$

The solution set is $\{ x \mid x \le -2 \ or \ x \ge 1 \}$, or $(-\infty, -2] \cup [1, \infty)$.

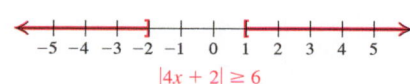

$|4x + 2| \ge 6$

Now Try Exercise 83.

1.6   Exercise Set

▶ **Reading Check**

Solve the inequality and then select the correct graph of the solution from the column on the right.

RC1. $|x| > 3$

RC2. $|x| \geq 3$

RC3. $|x| < 3$

RC4. $|x| = 3$

RC5. $|x| \leq 3$

RC6. $|x| > -3$

a) $\xleftarrow{\hspace{0.5em}}\underset{-5\;-4\;-3\;-2\;-1\;\;0\;\;1\;\;2\;\;3\;\;4\;\;5}{\rule{5em}{0.4pt}}\xrightarrow{\hspace{0.5em}}$

b) $\xleftarrow{\hspace{0.5em}}\underset{-5\;-4\;-3\;-2\;-1\;\;0\;\;1\;\;2\;\;3\;\;4\;\;5}{\rule{5em}{0.4pt}}\xrightarrow{\hspace{0.5em}}$

c) $\xleftarrow{\hspace{0.5em}}\underset{-5\;-4\;-3\;-2\;-1\;\;0\;\;1\;\;2\;\;3\;\;4\;\;5}{\rule{5em}{0.4pt}}\xrightarrow{\hspace{0.5em}}$

d) $\xleftarrow{\hspace{0.5em}}\underset{-5\;-4\;-3\;-2\;-1\;\;0\;\;1\;\;2\;\;3\;\;4\;\;5}{\rule{5em}{0.4pt}}\xrightarrow{\hspace{0.5em}}$

e) $\xleftarrow{\hspace{0.5em}}\underset{-5\;-4\;-3\;-2\;-1\;\;0\;\;1\;\;2\;\;3\;\;4\;\;5}{\rule{5em}{0.4pt}}\xrightarrow{\hspace{0.5em}}$

f) $\xleftarrow{\hspace{0.5em}}\underset{-5\;-4\;-3\;-2\;-1\;\;0\;\;1\;\;2\;\;3\;\;4\;\;5}{\rule{5em}{0.4pt}}\xrightarrow{\hspace{0.5em}}$

a   *Simplify, leaving as little as possible inside absolute-value signs.*

1. $|9x|$

2. $|26x|$

3. $|2x^2|$

4. $|8x^2|$

5. $|-2x^2|$

6. $|-20x^2|$

7. $|-6y|$

8. $|-17y|$

9. $\left|\dfrac{-2}{x}\right|$

10. $\left|\dfrac{y}{3}\right|$

11. $\left|\dfrac{x^2}{-y}\right|$

12. $\left|\dfrac{x^4}{-y}\right|$

13. $\left|\dfrac{-8x^2}{2x}\right|$

14. $\left|\dfrac{-9y^2}{3y}\right|$

15. $\left|\dfrac{4y^3}{-12y}\right|$

16. $\left|\dfrac{5x^3}{-25x}\right|$

b   *Find the distance between the points on the number line.*

17. $-8,\ -46$

18. $-7,\ -32$

19. $36, 17$

20. $52, 18$

21. $-3.9,\ 2.4$

22. $-1.8,\ -3.7$

23. $-5,\ 0$

24. $\dfrac{2}{3},\ -\dfrac{5}{6}$

c   *Solve.*

25. $|x| = 3$

26. $|x| = 5$

27. $|x| = -3$

28. $|x| = -9$

29. $|q| = 0$

30. $|y| = 7.4$

31. $|x - 3| = 12$

32. $|3x - 2| = 6$

33. $|2x - 3| = 4$

34. $|5x + 2| = 3$

35. $|4x - 9| = 14$

36. $|9y - 2| = 17$

37. $|x| + 7 = 18$

38. $|x| - 2 = 6.3$

39. $574 = 283 + |t|$

40. $-562 = -2000 + |x|$

41. $|5x| = 40$

42. $|2y| = 18$

43. $|3x| - 4 = 17$

44. $|6x| + 8 = 32$

45. $7|w| - 3 = 11$

46. $5|x| + 10 = 26$

47. $\left|\dfrac{2x - 1}{3}\right| = 5$

48. $\left|\dfrac{4 - 5x}{6}\right| = 7$

49. $|m + 5| + 9 = 16$

50. $|t - 7| - 5 = 4$

51. $10 - |2x - 1| = 4$

52. $2|2x - 7| + 11 = 25$

53. $|3x - 4| = -2$

54. $|x - 6| = -8$

55. $\left|\dfrac{5}{9} + 3x\right| = \dfrac{1}{6}$

56. $\left|\dfrac{2}{3} - 4x\right| = \dfrac{4}{5}$

d *Solve.*

57. $|3x + 4| = |x - 7|$

58. $|2x - 8| = |x + 3|$

59. $|x + 3| = |x - 6|$

60. $|x - 15| = |x + 8|$

61. $|2a + 4| = |3a - 1|$

62. $|5p + 7| = |4p + 3|$

63. $|y - 3| = |3 - y|$

64. $|m - 7| = |7 - m|$

65. $|5 - p| = |p + 8|$

66. $|8 - q| = |q + 19|$

67. $\left|\dfrac{2x - 3}{6}\right| = \left|\dfrac{4 - 5x}{8}\right|$

68. $\left|\dfrac{6 - 8x}{5}\right| = \left|\dfrac{7 + 3x}{2}\right|$

69. $\left|\frac{1}{2}x - 5\right| = \left|\frac{1}{4}x + 3\right|$

70. $\left|2 - \frac{2}{3}x\right| = \left|4 + \frac{7}{8}x\right|$

e *Solve.*

71. $|x| < 3$

72. $|x| \leq 5$

73. $|x| \geq 2$

74. $|y| > 12$

75. $|x - 1| < 1$

76. $|x + 4| \leq 9$

77. $5|x + 4| \leq 10$

78. $2|x - 2| > 6$

79. $|2x - 3| \leq 4$

80. $|5x + 2| \leq 3$

81. $|2y - 7| > 10$

82. $|3y - 4| > 8$

83. $|4x - 9| \geq 14$

84. $|9y - 2| \geq 17$

85. $|y - 3| < 12$

86. $|p - 2| < 6$

87. $|2x + 3| \leq 4$

88. $|5x + 2| \leq 13$

89. $|4 - 3y| > 8$

90. $|7 - 2y| > 5$

91. $|9 - 4x| \geq 14$

92. $|2 - 9p| \geq 17$

93. $|3 - 4x| < 21$

94. $|-5 - 7x| \leq 30$

95. $\left|\dfrac{1}{2} + 3x\right| \geq 12$

96. $\left|\dfrac{1}{4}y - 6\right| > 24$

97. $\left|\dfrac{x - 7}{3}\right| < 4$

98. $\left|\dfrac{x + 5}{4}\right| \leq 2$

99. $\left|\dfrac{2 - 5x}{4}\right| \geq \dfrac{2}{3}$

100. $\left|\dfrac{1 + 3x}{5}\right| > \dfrac{7}{8}$

101. $|m + 5| + 9 \leq 16$

102. $|t - 7| + 3 \geq 4$

103. $7 - |3 - 2x| \geq 5$

104. $16 \leq |2x - 3| + 9$

105. $\left|\dfrac{2x - 1}{3}\right| \leq 1$

106. $\left|\dfrac{3x - 2}{5}\right| \geq 1$

▶ **Skill Maintenance**

Solve.   **[1.4c]**

107. $-11x + 2x \geq -36$

108. $\dfrac{7}{9}y < -\dfrac{7}{10}$

109. $2(r - 1) + 4 < 3(r - 2) - 8$

Solve.

110. $8 > -x \geq 4$ **[1.5a]**

111. $-3 \leq 2x + 5 \text{ or } 10 > 2x - 1$ **[1.5b]**

112. $-2 \leq 6x - 4 < 20$ **[1.5a]**

▶ **Synthesis**

113. *Motion of a Spring.*   A weighted spring is bouncing up and down so that its distance d above the ground satisfies the inequality $|d - 6\text{ ft}| \leq \frac{1}{2}$ ft. Find all possible distances d.

114. *Container Sizes.*   A container company is manufacturing rectangular boxes of various sizes. The length of any box must exceed the width by at least 3 in., but the perimeter cannot exceed 24 in. What widths are possible?

$$l \geq w + 3,$$
$$2l + 2w \leq 24$$

Solve.

115. $|x + 5| > x$

116. $1 - \left|\frac{1}{4}x + 8\right| = \frac{3}{4}$

117. $|7x - 2| = x + 4$

118. $|x - 1| = x - 1$

119. $|x - 6| \leq -8$

120. $|3x - 4| > -2$

Find an equivalent inequality with absolute value.

121. $-3 < x < 3$

122. $-5 \leq y \leq 5$

123. $x \leq -6 \text{ or } x \geq 6$

124. $-5 < x < 1$

125. $x < -8 \text{ or } x > 2$

Chapter 1 Summary and Review

VOCABULARY REINFORCEMENT

Complete each statement with the correct term from the column on the right. Some of the choices may not be used and some may be used more than once.

1. A(n) _____ is a sentence containing $<, \leq, >, \geq$, or $\neq$. **[1.4a]**

2. Using _____ notation, we write the solution set for $x < 7$ as $\{x | x < 7\}$. **[1.4b]**

3. Using _____ notation, we write the solution set of $-5 \leq y < 16$ as $[-5, 16)$. **[1.4b]**

4. The _____ of two sets A and B is the set of all members that are common to A and B. **[1.5a]**

5. When two or more sentences are joined by the word *and* to make a compound sentence, the new sentence is called a(n) _____ of the sentences. **[1.5a]**

6. When two sets have no elements in common, the intersection of the two sets is the _____. **[1.5a]**

7. Two sets with an empty intersection are said to be _____. **[1.5a]**

8. The _____ of two sets A and B is the collection of elements belonging to A and/or B. **[1.5b]**

9. When two or more sentences are joined by the word *or* to make a compound sentence, the new sentence is called a(n) _____ of the sentences. **[1.5b]**

10. The _____ for equations states that for any real numbers a, b, and c, $a = b$ is equivalent to $a + c = b + c$. **[1.1b]**

11. The _____ for equations states that for any real numbers a, b, and c, $a = b$ is equivalent to $a \cdot c = b \cdot c$. **[1.1c]**

12. For any real numbers a and b, the _____ between them is $|a - b|$. **[1.6b]**

addition principle

multiplication principle

union

set-builder

empty set

absolute value

disjunction

inequality

intersection

distance

interval

disjoint sets

compound

conjunction

CONCEPT REINFORCEMENT

Determine whether each statement is true or false.

1. For any real numbers a, b, and c, $c \neq 0$, $a = b$ is equivalent to $a \cdot c = b \cdot c$. **[1.1c]**

2. When we solve $3B = mt + nt$ for t, we get $t = \dfrac{3B - mt}{n}$. **[1.2a]**

3. For any real numbers a, b, and c, $c \neq 0$, $a \leq b$ is equivalent to $ac \leq bc$. **[1.4c]**

4. The inequalities $x < 2$ and $x \leq 1$ are equivalent. **[1.4c]**

5. If x is negative, $|x| = -x$. **[1.6a]**

6. $|x|$ is always positive. **[1.6a]**

7. $|a - b| = |b - a|$. **[1.6b]**

STUDY GUIDE

Objective 1.1a   Determine whether a given number is a solution of a given equation.

Example   Determine whether 10 is a solution of $5x - 6 = 44$.

$$5x - 6 = 44$$
$$\overline{5 \cdot 10 - 6 \;\; ? \;\; 44}$$
$$50 - 6$$
$$44 \;\;\Big|\;\; \text{TRUE}$$

The number 10 is a solution of the equation.

Practice Exercise

1. Determine whether -3 is a solution of $28 - 7x = 7$.

Objective 1.1d   Solve equations using the addition principle and the multiplication principle together, removing parentheses where appropriate.

Example   Solve: $10y - 2(3y + 1) = 6$.

$$10y - 2(3y + 1) = 6$$
$$10y - 6y - 2 = 6 \qquad \textcolor{red}{\textbf{Removing parentheses}}$$
$$4y - 2 = 6 \qquad \textcolor{red}{\textbf{Collecting like terms}}$$
$$4y = 8 \qquad \textcolor{red}{\textbf{Adding 2}}$$
$$y = 2 \qquad \textcolor{red}{\textbf{Dividing by 4}}$$

The solution is 2.

Practice Exercise

2. Solve: $2(x + 2) = 5(x - 4)$.

Objective 1.2a   Evaluate formulas and solve a formula for a specified letter.

Example   Solve for z: $T = \dfrac{w + z}{3}$.

$$T = \frac{w + z}{3}$$
$$3 \cdot T = 3\left(\frac{w + z}{3}\right) \qquad \textcolor{red}{\textbf{Multiplying by 3 to}}$$
$$\textcolor{red}{\textbf{clear the fraction}}$$
$$3T = w + z \qquad \textcolor{red}{\textbf{Simplifying}}$$
$$3T - w = z \qquad \textcolor{red}{\textbf{Subtracting } w}$$

Practice Exercise

3. Solve for h: $F = \dfrac{1}{4}gh$.

Objective 1.4a   Determine whether a given number is a solution of an inequality.

Example   Determine whether -3 and 1 are solutions of the inequality $4 - x \geq 2 - 5x$.

We substitute -3 for x and get

$$4 - (-3) \geq 2 - 5(-3), \text{ or } 7 \geq 17,$$

a *false* sentence. Therefore, -3 is not a solution.
We substitute 1 for x and get

$$4 - 1 \geq 2 - 5 \cdot 1, \text{ or } 3 \geq -3,$$

a *true* sentence. Therefore, 1 is a solution.

Practice Exercise

4. Determine whether -2 and 5 are solutions of the inequality $8 - 3x \leq 3x + 6$.

Objective 1.4b   Write interval notation for the solution set of an inequality.

Example   Write interval notation for the solution set.

a) $\{x \mid x \leq -12\} = (-\infty, -12]$
b) $\{r \mid r > -1\} = (-1, \infty)$
c) $\{y \mid -8 \leq y < 9\} = [-8, 9)$
d) $\{x \mid 0 \geq x \geq -6\} = [-6, 0]$
e) $\{c \mid -25 < c \leq 25\} = (-25, 25]$

Practice Exercise

5. Write interval notation for the solution set.

   a) $\{t \mid t < -8\}$
   b) $\{x \mid -7 \leq x < 10\}$
   c) $\{b \mid b \geq 3\}$

Objective 1.4c   Solve an inequality using the addition principle and the multiplication principle and then graph the inequality.

Example   Solve and graph: $6x - 7 \leq 3x + 2$.

$$6x - 7 \leq 3x + 2$$
$$3x - 7 \leq 2 \qquad \text{Subtracting } 3x$$
$$3x \leq 9 \qquad \text{Adding } 7$$
$$x \leq 3 \qquad \text{Dividing by } 3$$

The solution set is $\{x \mid x \leq 3\}$, or $(-\infty, 3]$. We graph the solution set.

Practice Exercise

6. Solve and graph on the number line:
$5y + 5 < 2y - 1$.

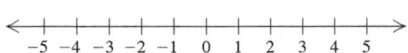

Objective 1.5a   Find the intersection of two sets. Solve and graph conjunctions of inequalities.

Example   Solve and graph: $-5 < 2x - 3 \leq 3$.

$$-5 < 2x - 3 \leq 3$$
$$-2 < 2x \leq 6 \qquad \text{Adding } 3$$
$$-1 < x \leq 3 \qquad \text{Dividing by } 2$$

The solution set is $\{x \mid -1 < x \leq 3\}$, or $(-1, 3]$. We graph the solution set.

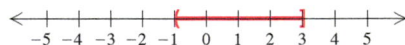

Practice Exercise

7. Solve and graph on the number line:
$-4 \leq 5z + 6 < 11$.

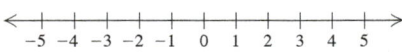

Objective 1.5b   Find the union of two sets. Solve and graph disjunctions of inequalities.

Example   Solve and graph:

$$2x + 1 \leq -5 \quad or \quad 3x + 1 > 7.$$
$$2x + 1 \leq -5 \quad or \quad 3x + 1 > 7$$
$$2x \leq -6 \quad or \quad 3x > 6$$
$$x \leq -3 \quad or \quad x > 2$$

The solution set is $\{x \mid x \leq -3 \text{ or } x > 2\}$, or $(-\infty, -3] \cup (2, \infty)$. We graph the solution set.

Practice Exercise

8. Solve and graph on the number line:
$z + 4 < 3 \text{ or } 4z + 1 \geq 5$.

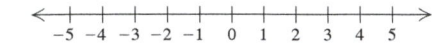

Objective 1.6a Simplify expressions containing absolute-value symbols.

Example Simplify: $\lvert -6c \rvert$. $$\lvert -6c \rvert = \lvert -6 \rvert \cdot \lvert c \rvert = 6\lvert c \rvert$$	**Practice Exercise** **9.** Simplify: $\lvert 8y^2 \rvert$.

Objective 1.6b Find the distance between two points on the number line.

Example Find the distance between -10 and 3 on the number line. $$\lvert -10 - 3 \rvert = \lvert -13 \rvert = 13$$	**Practice Exercise** **10.** Find the distance between 8 and -20 on the number line.

Objective 1.6c Solve equations with absolute-value expressions.

Example Solve: $\lvert y - 2 \rvert = 1$. $$\begin{aligned} y - 2 = -1 \quad &or \quad y - 2 = 1 \\ y = 1 \quad &or \quad y = 3 \end{aligned}$$ The solution set is $\{1, 3\}$.	**Practice Exercise** **11.** Solve: $\lvert 5x - 1 \rvert = 9$.

Objective 1.6d Solve equations with two absolute-value expressions.

Example Solve: $\lvert 4x - 4 \rvert = \lvert 2x + 8 \rvert$. $$\begin{aligned} 4x - 4 = 2x + 8 \quad &or \quad 4x - 4 = -(2x + 8) \\ 2x - 4 = 8 \quad &or \quad 4x - 4 = -2x - 8 \\ 2x = 12 \quad &or \quad 6x - 4 = -8 \\ x = 6 \quad &or \quad 6x = -4 \\ x = 6 \quad &or \quad x = -\frac{2}{3} \end{aligned}$$ The solution set is $\left\{ 6, -\frac{2}{3} \right\}$.	**Practice Exercise** **12.** Solve: $\lvert z + 4 \rvert = \lvert 3z - 2 \rvert$.

Objective 1.6e Solve inequalities with absolute-value expressions.

Example Solve: **(a)** $\lvert 5x + 3 \rvert < 2$; **(b)** $\lvert x + 3 \rvert \geq 1$. **(a)** $\lvert 5x + 3 \rvert < 2$ $$\begin{aligned} -2 &< 5x + 3 < 2 \\ -5 &< 5x < -1 \\ -1 &< x < -\frac{1}{5} \end{aligned}$$ The solution set is $\left\{ x \mid -1 < x < -\frac{1}{5} \right\}$, or $\left(-1, -\frac{1}{5} \right)$. **(b)** $\lvert x + 3 \rvert \geq 1$ $$\begin{aligned} x + 3 \leq -1 \quad &or \quad x + 3 \geq 1 \\ x \leq -4 \quad &or \quad x \geq -2 \end{aligned}$$ The solution set is $\{ x \mid x \leq -4 \ or \ x \geq -2 \}$, or $(-\infty, -4] \cup [-2, \infty)$.	**Practice Exercise** **13.** Solve: **(a)** $\lvert 2x + 3 \rvert < 5$; **(b)** $\lvert 3x + 2 \rvert \geq 8$.

REVIEW EXERCISES

Solve. **[1.1b, c, d]**

1. $-11 + y = -3$

2. $-7x = -3$

3. $-\frac{5}{3}x + \frac{7}{3} = -5$

4. $6(2x - 1) = 3 - (x + 10)$

5. $2.4x + 1.5 = 1.02$

6. $2(3 - x) - 4(x + 1) = 7(1 - x)$

Solve for the indicated letter. **[1.2a]**

7. $C = \frac{4}{11}d + 3$, for d

8. $A = 2a - 3b$, for b

9. *Interstate Mile Markers.* If you are traveling on a U.S. interstate highway, you will notice numbered markers every mile to tell your location in case of an accident or other emergency. In many states, the numbers on the markers increase from west to east. The sum of two consecutive mile markers on I-70 in Utah is 371. Find the numbers on the markers. (*Source*: Data from Federal Highway Administration) **[1.3a]**

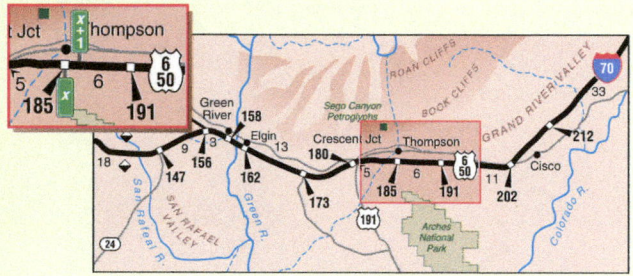

10. *Rope Cutting.* A piece of rope 27 m long is cut into two pieces so that one piece is four-fifths as long as the other. Find the length of each piece. **[1.3a]**

11. *Population Growth.* The population of Newcastle grew 12% from one year to the next to a total of 179,200. What was the former population? **[1.3a]**

12. *Moving Walkway.* A moving walkway in an airport is 360 ft long and moves at a speed of 6 ft/sec. If Arnie walks at a speed of 3 ft/sec, how long will it take him to walk the length of the moving walkway? **[1.3b]**

Write interval notation for the given set or graph. **[1.4b]**

13. $\{x \mid -8 \le x < 9\}$

14.

```
<-+++++++|++++|+++++|+++|+++->
  -80 -60 -40 -20  0  20  40  60  80
```

Solve and graph on the number line. Write interval notation for the solution set. **[1.4c]**

15. $x - 2 \le -4$　　　**16.** $x + 5 > 6$

Solve. **[1.4c]**

17. $a + 7 \le -14$　　**18.** $y - 5 \ge -12$

19. $4y > -16$　　　**20.** $-0.3y < 9$

21. $-6x - 5 < 13$　　**22.** $4y + 3 \le -6y - 9$

23. $-\frac{1}{2}x - \frac{1}{4} > \frac{1}{2} - \frac{1}{4}x$　　**24.** $0.3y - 8 < 2.6y + 15$

25. $-2(x - 5) \ge 6(x + 7) - 12$

26. *Moving Costs.* Metro Movers charges $85 plus $40 per hour to move households across town. Champion Moving charges $60 per hour for cross-town moves. For what lengths of time is Champion more expensive? **[1.4d]**

27. *Investments.* Joe plans to invest $30,000, part at 3% and part at 4%, for one year. What is the most that can be invested at 3% in order to make at least $1100 interest in one year? **[1.4d]**

Graph on the number line and write interval notation. **[1.5a, b]**

28. $-2 \le x < 5$　　　**29.** $x \le -2 \text{ or } x > 5$

30. Find the intersection: **[1.5a]**

$$\{1, 2, 5, 6, 9\} \cap \{1, 3, 5, 9\}.$$

31. Find the union: **[1.5b]**

$$\{1, 2, 5, 6, 9\} \cup \{1, 3, 5, 9\}.$$

Solve. **[1.5a, b]**

32. $2x - 5 < -7 \text{ and } 3x + 8 \ge 14$

33. $-4 < x + 3 \le 5$

34. $-15 < -4x - 5 < 0$

35. $3x < -9 \text{ or } -5x < -5$

36. $2x + 5 < -17 \text{ or } -4x + 10 \le 34$

37. $2x + 7 \le -5 \text{ or } x + 7 \ge 15$

Simplify. **[1.6a]**

38. $\left| -\dfrac{3}{x} \right|$　　**39.** $\left| \dfrac{2x}{y^2} \right|$　　**40.** $\left| \dfrac{12y}{-3y^2} \right|$

41. Find the distance between -23 and 39. **[1.6b]**

Solve. **[1.6c, d]**

42. $|x| = 6$　　　**43.** $|x - 2| = 7$

44. $|2x + 5| = |x - 9|$　　**45.** $|5x + 6| = -8$

Solve. **[1.6e]**

46. $|2x + 5| < 12$ **47.** $|x| \geq 3.5$

48. $|3x - 4| \geq 15$ **49.** $|x| < 0$

Greenhouse Gases. *The equation*

$$G = 0.506t + 18.3$$

is used to estimate global carbon dioxide emissions, in billions of metric tons, t years after 1980—that is, $t = 0$ corresponds to 1980, $t = 20$ corresponds to 2000, and so on. Use this equation in Exercises 50 and 51. (Source: Data from U.S. Department of Energy)

50. Estimate global carbon dioxide emissions in 2010. **[1.2a], [1.3a]**

   A. 23.36 billion metric tons
   B. 33.48 billion metric tons
   C. 38.54 billion metric tons
   D. 1035.4 billion metric tons

51. For what years are global carbon dioxide emissions predicted to be between 35 and 40 billion metric tons? **[1.5c]**

   A. Between 2013 and 2023
   B. Between 2011 and 2025
   C. Between 2020 and 2025
   D. Years after 2025

▶ ## Synthesis

52. Solve: $|2x + 5| \leq |x + 3|$. **[1.6d, e]**

▶ ## Collaborative Discussion and Writing

53. Explain in your own words why the inequality symbol must be reversed when both sides of an inequality are multiplied or divided by a negative number. **[1.4c]**

54. Explain in your own words why the solutions of the inequality $|x + 5| \leq 2$ can be interpreted as "all those numbers x whose distance from -5 is at most 2 units." **[1.6e]**

55. Describe the circumstances under which $[a, b] \cup [c, d] = [a, d]$. **[1.5b]**

56. Explain in your own words why the interval $[6, \infty)$ is only part of the solution set of $|x| \geq 6$. **[1.6e]**

57. Find the error or errors in each of the following steps: **[1.4c]**

$$7 - 9x + 6x < -9(x + 2) + 10x$$
$$7 - 9x + 6x < -9x + 2 + 10x \quad (1)$$
$$7 + 6x > 2 + 10x \quad (2)$$
$$-4x > 8 \quad (3)$$
$$x > -2. \quad (4)$$

58. Explain why the conjunction $3 < x$ and $x < 5$ is equivalent to $3 < x < 5$, but the disjunction $3 < x$ or $x < 5$ is not. **[1.5a, b]**

| **1** | **Chapter Test** |

Solve.

1. $x + 7 = 5$ **2.** $-12x = -8$

3. $x - \frac{3}{5} = \frac{2}{3}$ **4.** $3y - 4 = 8$

5. $1.7y - 0.1 = 2.1 - 0.3y$

6. $5(3x + 6) = 6 - (x + 8)$

7. Solve $A = 3B - C$ for B.

8. Solve $m = n - nt$ for n.

Solve.

9. *Room Dimensions.* A rectangular room has a perimeter of 48 ft. The width is two-thirds of the length. What are the dimensions of the room?

10. *Copy Budget.* Copy Solutions rents a copier for $240 per month plus 1.5¢ per copy. A law firm needs to lease a copy machine for use during a special case that they anticipate will take 3 months. If they allot a budget of $1500 for copying costs, how many copies can they make?

11. *Population Decrease.* The population of Baytown dropped 12% from one year to the next to a total of 158,400. What was the former population?

12. *Angles in a Triangle.* The measures of the angles of a triangle are three consecutive integers. Find the measures of the angles.

13. *Boating.* A paddleboat moves at a rate of 12 mph in still water. If the river's current moves at a rate of 3 mph, how long will it take the boat to travel 36 mi downstream? 36 mi upstream?

Write interval notation for the given set or graph.

14. $\{x \mid -3 < x \le 2\}$

15. ← + + (+ + + + + + + + + + + + →
    −6 −5 −4 −3 −2 −1 0 1 2 3 4 5 6

Solve and graph on the number line. Write interval notation for the solution set.

16. $x - 2 \le 4$

17. $-4y - 3 \ge 5$

Solve.

18. $x - 4 \ge 6$

19. $-0.6y < 30$

20. $3a - 5 \le -2a + 6$

21. $-5y - 1 > -9y + 3$

22. $4(5 - x) < 2x + 5$

23. $-8(2x + 3) + 6(4 - 5x)$
    $\ge 2(1 - 7x) - 4(4 + 6x)$

Solve.

24. *Moving Costs.* Mitchell Moving Company charges $105 plus $30 per hour to move households across town. Quick-Pak Moving charges $80 per hour for cross-town moves. For what lengths of time is Quick-Pak more expensive?

25. *Pressure at Sea Depth.* The equation

$$P = 1 + \frac{d}{33}$$

gives the pressure P, in atmospheres (atm), at a depth of d feet in the sea. For what depths d is the pressure at least 2 atm and at most 8 atm?

Graph on the number line and write interval notation.

26. $-3 \le x \le 4$

27. $x < -3 \text{ or } x > 4$

Solve.

28. $5 - 2x \le 1 \text{ and } 3x + 2 \ge 14$

29. $-3 < x - 2 < 4$

30. $-11 \le -5x - 2 < 0$

31. $-3x > 12 \text{ or } 4x > -10$

32. $x - 7 \le -5 \text{ or } x - 7 \ge -10$

33. $3x - 2 < 7 \text{ or } x - 2 > 4$

Simplify.

34. $\left| \dfrac{7}{x} \right|$

35. $\left| \dfrac{-6x^2}{3x} \right|$

36. Find the distance between 4.8 and −3.6.

37. Find the intersection:

$$\{1, 3, 5, 7, 9\} \cap \{3, 5, 11, 13\}.$$

38. Find the union:

$$\{1, 3, 5, 7, 9\} \cup \{3, 5, 11, 13\}.$$

Solve.

39. $|x| = 9$

40. $|x - 3| = 9$

41. $|x + 10| = |x - 12|$

42. $|2 - 5x| = -10$

43. $|4x - 1| < 4.5$

44. $|x| > 3$

45. $\left| \dfrac{6 - x}{7} \right| \le 15$

46. $|-5x - 3| \ge 10$

47. The solution of $2(3x - 6) + 5 = 1 - (x - 6)$ is which of the following?

  A. Less than 0
  B. Between 0 and 1
  C. Between 1 and 3
  D. Greater than 3

▶ **Synthesis**

Solve.

48. $|3x - 4| \le -3$

49. $7x < 8 - 3x < 6 + 7x$

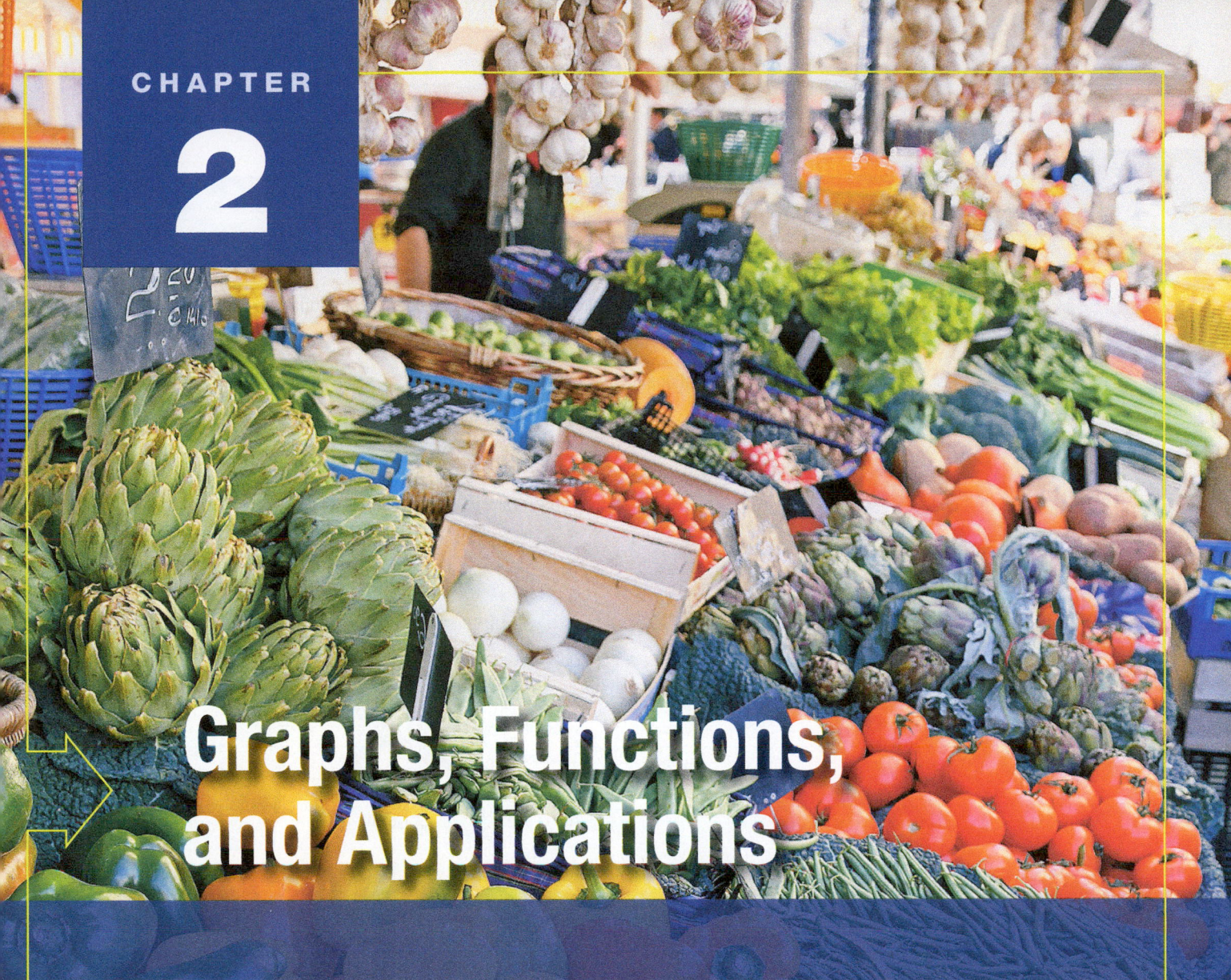

CHAPTER 2

Graphs, Functions, and Applications

APPLICATION   This problem appears as Example 7 in Section 2.7.

The number of farmers' markets has increased steadily in recent years. In 2002, there were 3137 farmers' markets. This number increased to 7864 in 2012. (*Source*: Data from U.S. Department of Agriculture) Assuming a constant rate of change, use these two data points to find a linear function that fits the data. Then use the function to determine the number of farmers' markets in 2010. In which year will the number of farmers' markets reach 12,000?

2.1 Graphs of Equations

▶ **a** Plot points associated with ordered pairs of numbers.

▶ **b** Determine whether an ordered pair of numbers is a solution of an equation.

▶ **c** Graph linear equations using tables.

▶ **d** Graph nonlinear equations using tables.

Graphs display information and can provide a visual approach to problem solving. We often see graphs in newspapers and magazines.

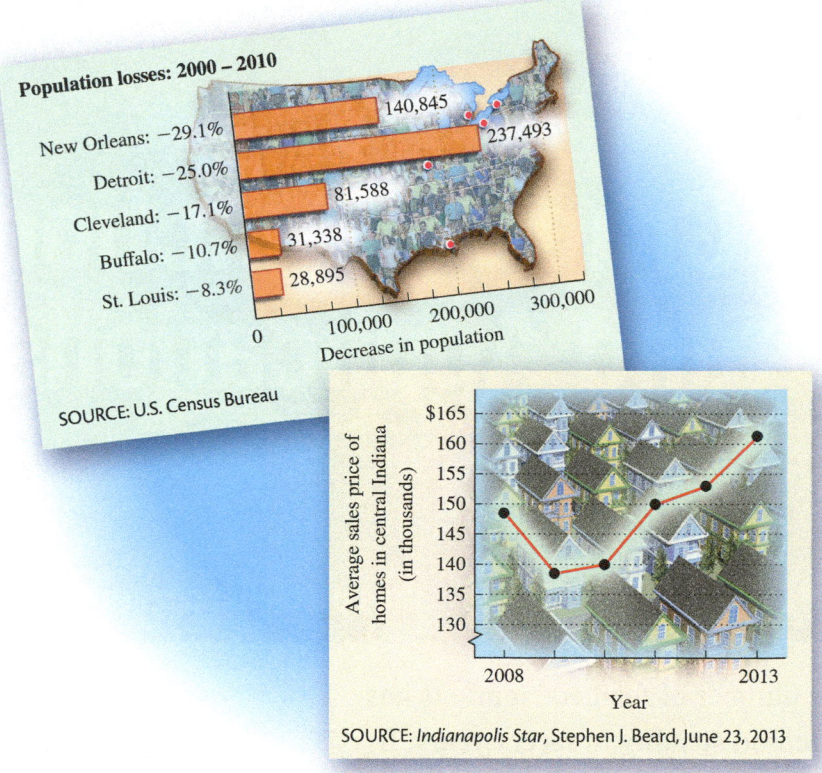

Population losses: 2000–2010

New Orleans: −29.1%   140,845
Detroit: −25.0%   237,493
Cleveland: −17.1%   81,588
Buffalo: −10.7%   31,338
St. Louis: −8.3%   28,895

Decrease in population

0   100,000   200,000   300,000

SOURCE: U.S. Census Bureau

Average sales price of homes in central Indiana (in thousands)

$165, 160, 155, 150, 145, 140, 135, 130

2008   2013

Year

SOURCE: *Indianapolis Star*, Stephen J. Beard, June 23, 2013

▶ a Plotting Ordered Pairs

We have already learned to graph numbers and inequalities in one variable on a line. To graph an equation that contains two variables, we graph pairs of numbers on a plane.

On the number line, each point is the graph of a number. On a plane, each point is the graph of a number pair. To locate points on a plane, we use two perpendicular number lines called **axes**. They cross at a point called the **origin**. The arrows show the positive directions on the axes. Consider the **ordered pair** $(2, 3)$. The numbers in an ordered pair are called **coordinates**. In $(2, 3)$, the **first coordinate** is 2 and

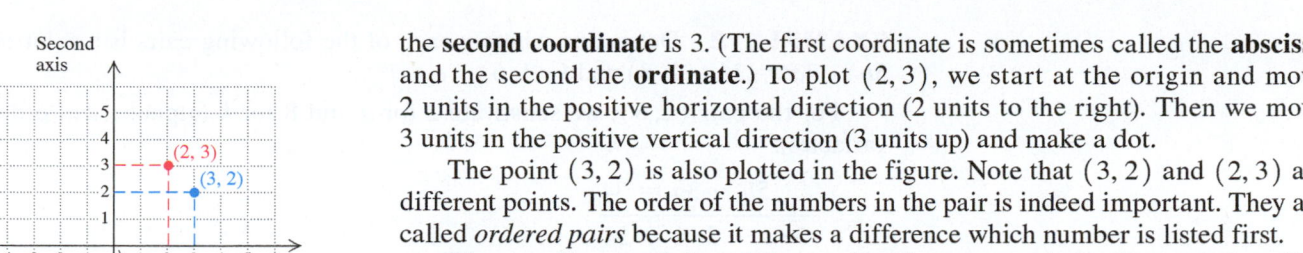

the **second coordinate** is 3. (The first coordinate is sometimes called the **abscissa** and the second the **ordinate**.) To plot $(2, 3)$, we start at the origin and move 2 units in the positive horizontal direction (2 units to the right). Then we move 3 units in the positive vertical direction (3 units up) and make a dot.

The point $(3, 2)$ is also plotted in the figure. Note that $(3, 2)$ and $(2, 3)$ are different points. The order of the numbers in the pair is indeed important. They are called *ordered pairs* because it makes a difference which number is listed first.

The coordinates of the origin are $(0, 0)$. In general, the first axis is called the *x*-axis and the second axis is called the *y*-axis. We call this the **Cartesian coordinate system** in honor of the great French mathematician and philosopher René Descartes (1596–1650).

EXAMPLE 1 Plot the points $(-4, 3)$, $(-5, -3)$, $(0, 4)$, and $(2.5, 0)$.

To plot $(-4, 3)$, we note that the first number, -4, tells us the distance in the first, or horizontal, direction. We move 4 units in the negative direction, *left*. The second number tells us the distance in the second, or vertical, direction. We move 3 units in the positive direction, *up*. The point $(-4, 3)$ is then marked, or plotted.

The points $(-5, -3)$, $(0, 4)$, and $(2.5, 0)$ are plotted in the same manner.

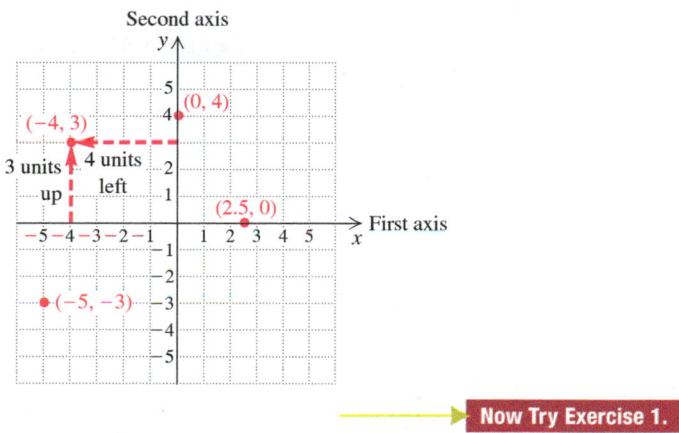

Now Try Exercise 1.

Quadrants

The axes divide the plane into four regions called **quadrants**, denoted by Roman numerals and numbered counterclockwise starting at the upper right. In region I (the *first* quadrant), both coordinates of a point are positive. In region II (the *second* quadrant), the first coordinate is negative and the second coordinate is positive. In the *third* quadrant, both coordinates are negative, and in the *fourth* quadrant, the first coordinate is positive and the second coordinate is negative.

Points with one or more 0's as coordinates, such as $(0, -5)$, $(4, 0)$, and $(0, 0)$ are on axes and *not* in quadrants.

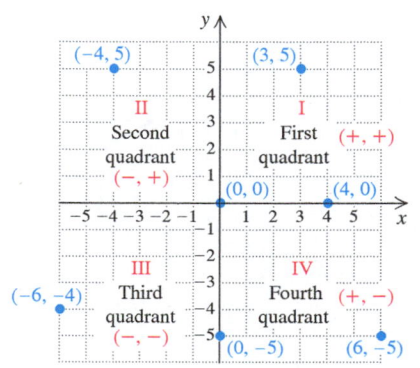

▶ b Solutions of Equations

If an equation has two variables, its solutions are pairs of numbers. When such a solution is written as an ordered pair, the first number listed in the pair generally replaces the variable that occurs first alphabetically.

EXAMPLE 2 Determine whether each of the following pairs is a solution of $5b - 3a = 34$: $(2, 8)$ and $(-1, 6)$.

For the pair $(2, 8)$, we substitute 2 for a and 8 for b (alphabetical order of variables):

$$\frac{5b - 3a = 34}{5 \cdot 8 - 3 \cdot 2 \overset{?}{} 34}$$
$$40 - 6$$
$$34 \qquad \text{TRUE}$$

Thus, $(2, 8)$ is a solution of the equation.

For $(-1, 6)$, we substitute -1 for a and 6 for b:

$$\frac{5b - 3a = 34}{5 \cdot 6 - 3 \cdot (-1) \overset{?}{} 34}$$
$$30 + 3$$
$$33 \qquad \text{FALSE}$$

Thus, $(-1, 6)$ is *not* a solution of the equation. ⟶ **Now Try Exercise 7.**

EXAMPLE 3 Show that the pairs $(-4, 3)$, $(0, 1)$, and $(4, -1)$ are solutions of $y = 1 - \frac{1}{2}x$. Then plot the three points and use them to help determine another pair that is a solution.

We replace x with the first coordinate and y with the second coordinate of each pair:

$$\frac{y = 1 - \frac{1}{2}x}{3 \overset{?}{} 1 - \frac{1}{2} \cdot (-4)}$$
$$1 + 2$$
$$3 \qquad \text{TRUE}$$

$$\frac{y = 1 - \frac{1}{2}x}{1 \overset{?}{} 1 - \frac{1}{2} \cdot (0)}$$
$$1 - 0$$
$$1 \qquad \text{TRUE}$$

$$\frac{y = 1 - \frac{1}{2}x}{-1 \overset{?}{} 1 - \frac{1}{2} \cdot (4)}$$
$$1 - 2$$
$$-1 \qquad \text{TRUE}$$

In each case, the substitution results in a true equation. Thus all the pairs are solutions of the equation.

We plot the points as shown at right. Note that the three points appear to "line up." That is, they appear to be on a straight line. We use a ruler and draw a line passing through $(-4, 3)$, $(0, 1)$, and $(4, -1)$.

The line appears to pass through $(2, 0)$ as well. Let's see if this pair is a solution of $y = 1 - \frac{1}{2}x$:

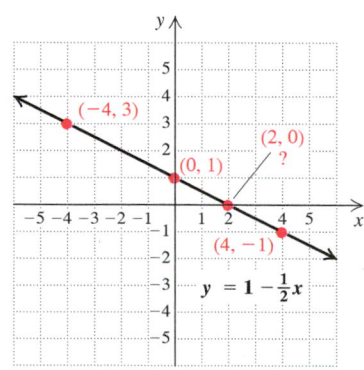

$$\frac{y = 1 - \frac{1}{2}x}{0 \overset{?}{} 1 - \frac{1}{2} \cdot (2)}$$
$$1 - 1$$
$$0 \qquad \text{TRUE}$$

We see that $(2, 0)$ is another solution of the equation.

Example 3 leads us to believe that any point on the line that passes through $(-4, 3)$, $(0, 1)$, and $(4, -1)$ represents a solution of $y = 1 - \frac{1}{2}x$. In fact, every solution of $y = 1 - \frac{1}{2}x$ is represented by a point on that line and every point on that line represents a solution. The line is said to be the *graph* of the equation.

> **Now Try Exercise 13.**

Technology Connection

Finding Solutions of Equations

A table of values representing ordered pairs that are solutions of an equation can be displayed on a graphing calculator. To do this for the equation in Example 4, $y = 2x$, we first access the equation-editor screen. Then we clear any equations that are present. (See the Technology Connection on p. 10 for the procedure for doing this.) Next, we enter the equation, display the table set-up screen, and set both INDPNT and DEPEND to AUTO.

We will display a table of values that starts with $x = -2$ (TBLSTART) and add 1 (ΔTBL) to the preceding x-value.

X	Y₁
−2	−4
−1	−2
0	0
1	2
2	4
3	6
4	8
X = −2	

Exercises

Create a table of ordered pairs that are solutions of the equation.

1. Example 5
2. Example 7

GRAPH OF AN EQUATION

The **graph** of an equation is a drawing that represents all its solutions.

▶ **c Graphs of Linear Equations**

Equations like $y = 1 - \frac{1}{2}x$ and $2x + 3y = 6$ are said to be **linear** because the graph of their solutions is a line. In general, a linear equation is any equation equivalent to one of the form $y = mx + b$ or $Ax + By = C$, where m, b, A, B, and C are constants (that is, they are numbers, not variables) and A and B are not both 0.

EXAMPLE 4 Graph: $y = 2x$.

We find some ordered pairs that are solutions. We list the pairs in a table. To find an ordered pair, we can choose *any* number for x and then determine y. For example, if we choose 3 for x, then $y = 2 \cdot 3 = 6$ (substituting into the equation $y = 2x$). We choose some negative values for x, as well as some positive ones. If a number takes us off the graph paper, we generally do not use it. Next, we plot these points. If we plotted *many* such points, they would appear to make a solid line. We draw the line with a ruler and label it $y = 2x$.

x	y	(x, y)
0	0	$(0, 0)$
1	2	$(1, 2)$
3	6	$(3, 6)$
−2	−4	$(-2, -4)$
−3	−6	$(-3, -6)$

Choose any x.
Compute y.
Form the pair.
Plot the points.

> **Now Try Exercise 19.**

> To graph a linear equation:
>
> 1. Select a value for one variable and calculate the corresponding value of the other variable. Form an ordered pair using alphabetical order as indicated by the variables.
> 2. Repeat step (1) to obtain at least two other ordered pairs. Two ordered pairs are essential. A third serves as a check.
> 3. Plot the ordered pairs and draw a straight line passing through the points.

EXAMPLE 5   Graph: $y = -\frac{1}{2}x + 3$.

By choosing even integers for x, we can avoid fraction values when calculating y. For example, if we choose 4 for x, we get

$$y = -\tfrac{1}{2}x + 3 = -\tfrac{1}{2}(4) + 3 = -2 + 3 = 1.$$

When x is -6, we get

$$y = -\tfrac{1}{2}x + 3 = -\tfrac{1}{2}(-6) + 3 = 3 + 3 = 6,$$

and when x is 0, we get

$$y = -\tfrac{1}{2}x + 3 = -\tfrac{1}{2}(0) + 3 = 0 + 3 = 3.$$

We list the results in a table. Then we plot the points corresponding to each pair.

x	y	(x, y)
4	1	$(4, 1)$
-6	6	$(-6, 6)$
0	3	$(0, 3)$

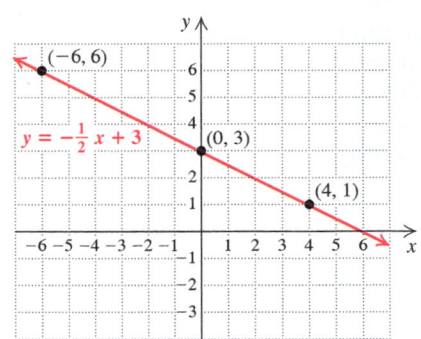

Note that the three points line up. If they did not, we would know that we had made a mistake. When only two points are plotted, an error is harder to detect. We use a ruler or other straightedge to draw a line through the points and then label the graph. Every point on the line represents a solution of $y = -\frac{1}{2}x + 3$.

> **Now Try Exercise 31.**

Calculating ordered pairs is usually easiest when y is isolated on one side of the equation, as in $y = 2x$ and $y = -\frac{1}{2}x + 3$. To graph an equation in which y is not isolated, we can use the addition principle and the multiplication principle to first solve for y. (See Sections 1.1 and 1.2.)

EXAMPLE 6 Graph: $3x + 5y = 10$.

We first solve for y:

$$3x + 5y = 10$$
$$3x + 5y - 3x = 10 - 3x \qquad \text{Subtracting } 3x$$
$$5y = 10 - 3x \qquad \text{Simplifying}$$
$$\tfrac{1}{5} \cdot 5y = \tfrac{1}{5} \cdot (10 - 3x) \qquad \text{Multiplying by } \tfrac{1}{5}, \text{ or dividing by 5}$$
$$y = \tfrac{1}{5} \cdot (10) - \tfrac{1}{5} \cdot (3x) \qquad \text{Using the distributive law}$$
$$y = 2 - \tfrac{3}{5}x, \text{ or } y = -\tfrac{3}{5}x + 2.$$

Thus the equation $3x + 5y = 10$ is equivalent to $y = -\tfrac{3}{5}x + 2$. We now find three ordered pairs, using multiples of 5 for x to avoid fractions.

x	y	(x, y)
0	2	$(0, 2)$
5	−1	$(5, -1)$
−5	5	$(-5, 5)$

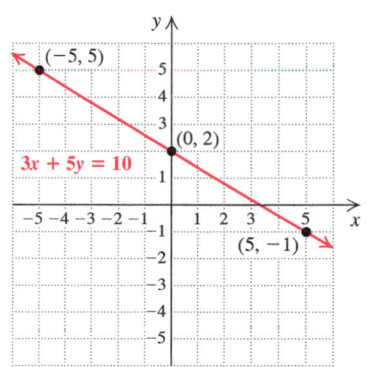

We plot the points, draw the line, and label the graph as shown.

Now Try Exercise 37.

Technology Connection

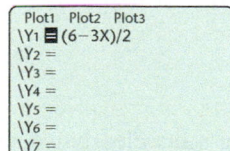

Graphing Equations

Equations must be solved for y before they can be graphed on the TI-84 Plus. Consider the equation $3x + 2y = 6$. Solving for y, we enter $y_1 = (6 - 3x)/2$ as described on p. 10. Then we select a window and press GRAPH to see the graph of the equation. (Press ZOOM 6 to select the standard window and graph as shown at left.)

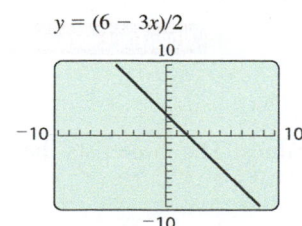

$y = (6 - 3x)/2$

Exercises

Graph each equation in the standard viewing window $[-10, 10, -10, 10]$, with Xscl = 1 and Yscl = 1.

1. $y = 2x - 1$
2. $3x + y = 2$
3. $y = 5x - 3$
4. $y = -4x + 5$
5. $y = \tfrac{2}{3}x - 3$
6. $y = -\tfrac{3}{4}x + 4$
7. $y = 3.104x - 6.21$
8. $2.98x + y = -1.75$

▶ d Graphing Nonlinear Equations

We have seen that equations whose graphs are straight lines are called **linear**. There are many equations whose graphs are not straight lines. Here are some examples.

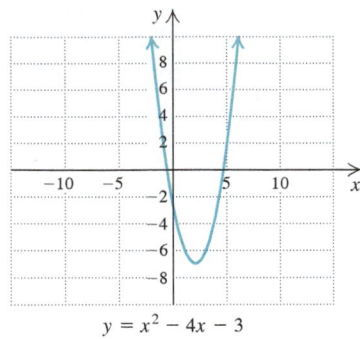

$$y = x^2 - 4x - 3$$

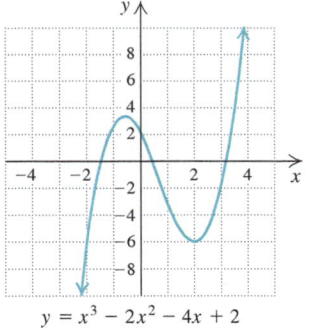

$$y = x^3 - 2x^2 - 4x + 2$$

Let's graph some of these **nonlinear equations**. We usually need to plot more than three points in order to get a good idea of the shape of the graph.

EXAMPLE 7 Graph: $y = x^2 - 5$.

We select numbers for x and find the corresponding values for y. For example, if we choose -2 for x, we get $y = (-2)^2 - 5 = 4 - 5 = -1$. The table lists several ordered pairs.

x	y
0	-5
-1	-4
1	-4
-2	-1
2	-1
-3	4
3	4

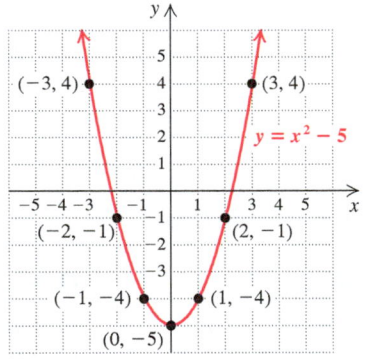

Next, we plot the points. The more points we plot, the more clearly we see the shape of the graph. Since the value of $x^2 - 5$ grows rapidly as x moves away from the origin, the graph rises steeply on either side of the y-axis.

Now Try Exercise 45.

EXAMPLE 8 Graph: $y = 1/x$.

We select x-values and find the corresponding y-values. The table lists the ordered pairs $\left(3, \frac{1}{3}\right)$, $\left(2, \frac{1}{2}\right)$, $(1, 1)$, and so on.

x	y
3	$\frac{1}{3}$
2	$\frac{1}{2}$
1	1
$\frac{1}{2}$	2
$-\frac{1}{2}$	-2
-1	-1
-2	$-\frac{1}{2}$
-3	$-\frac{1}{3}$

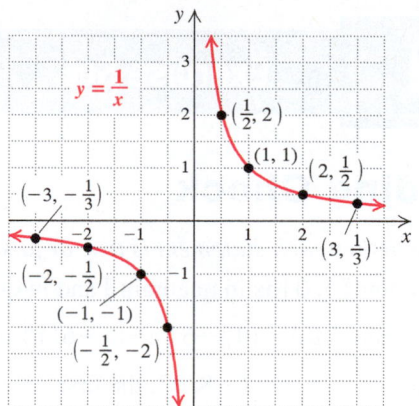

We plot these points, noting that each first coordinate is paired with its reciprocal. Since $1/0$ is undefined, we cannot use 0 as a first coordinate. Thus there are two "branches" to this graph—one on each side of the y-axis. Note that for x-values far to the right or far to the left of 0, the graph approaches, but does not touch, the x-axis; and for x-values close to 0, the graph approaches, but does not touch, the y-axis.

Now Try Exercise 47.

EXAMPLE 9 Graph: $y = |x|$.

We select numbers for x and find the corresponding values for y. For example, if we choose -1 for x, we get $y = |-1| = 1$. Several ordered pairs are listed in the table below.

x	y
-3	3
-2	2
-1	1
0	0
1	1
2	2
3	3

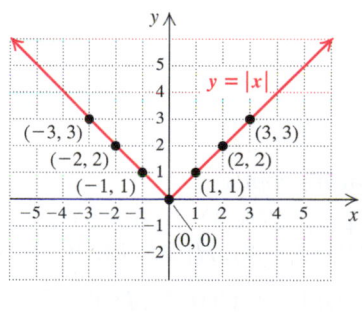

We plot these points, noting that the absolute value of a positive number is the same as the absolute value of its opposite. Thus the x-values 3 and -3 both are paired with the y-value 3. Note that the graph is V-shaped and centered at the origin.

Now Try Exercise 49.

With equations like $y = -\frac{1}{2}x + 3$, $y = x^2 - 5$, and $y = |x|$, which we have graphed in this section, it is understood that y is the **dependent variable** and x is the **independent variable**, since y is expressed in terms of x and consequently y is calculated after first choosing x.

2.1   Exercise Set

▶ **Reading Check**

Determine whether each statement is true or false.

RC1. The point $(5, 0)$ is in quadrant I and in quadrant IV.

RC2. The ordered pairs $(1, -6)$ and $(-6, 1)$ name the same point.

RC3. In the ordered pair $(-8, 3)$, the first coordinate, -8, is also called the abscissa.

RC4. To plot the point $(-2, 7)$, we start at the origin and move horizontally to -2. Then we move up vertically 7 units and make a "dot."

RC5. In the ordered pair $(4, -10)$, the second coordinate, -10, is also called the ordinate.

RC6. The point $(0, -3)$ is on the *x*-axis.

For each of the following equations, choose from the column on the right an equivalent equation.

RC7. $3x + 4y = 0$

RC8. $4y - 3x = 0$

RC9. $4x - 3y = -4$

RC10. $3y + 4x = -12$

a) $y = \dfrac{4}{3}x + \dfrac{4}{3}$

b) $y = \dfrac{3}{4}x$

c) $y = -\dfrac{4}{3}x - 4$

d) $y = -\dfrac{3}{4}x$

....

a *Plot the following points.*

1. $A(4, 1), B(2, 5), C(0, 3), D(0, -5), E(6, 0),$
$F(-3, 0), G(-2, -4), H(-5, 1), J(-6, 6)$

2. $A(-3, -5), B(1, 3), C(0, 7), D(0, -2),$
$E(5, 0), F(-4, 0), G(1, -7), H(-6, 4),$
$J(-3, 3)$

3. Plot the points $M(2, 3), N(5, -3)$, and $P(-2, -3)$. Draw $\overline{MN}, \overline{NP}$, and $\overline{MP}$. ($\overline{MN}$ means the line segment from *M* to *N*.) What kind of geometric figure is formed? What is its area?

4. Plot the points $Q(-4, 3), R(5, 3), S(2, -1)$, and $T(-7, -1)$. Draw $\overline{QR}, \overline{RS}, \overline{ST}$, and $\overline{TQ}$. What kind of figure is formed? What is its area?

b *Determine whether the given point is a solution of the equation.*

5. $(1, -1); \ y = 2x - 3$

6. $(3, 4); \ t = 4 - 3s$

7. $(3, 5); \ 4x - y = 7$

8. $(2, -1); \ 4r + 3s = 5$

9. $\left(0, \dfrac{3}{5}\right); \ 2a + 5b = 7$

10. $(-5, 1); \ 2p - 3q = -13$

In Exercises 11–16, an equation and two ordered pairs are given. Show that each pair is a solution of the equation. Then graph the equation and use the graph to determine another solution. Answers for solutions may vary, but the graphs do not.

11. $y = 4 - x; \ (-1, 5), (3, 1)$

12. $y = x - 3; \ (5, 2), (-1, -4)$

13. $3x + y = 7; \ (2, 1), (4, -5)$

14. $y = \dfrac{1}{2}x + 3; \ (4, 5), (-2, 2)$

15. $6x - 3y = 3; \ (1, 1), (-1, -3)$

16. $4x - 2y = 10; \ (0, -5), (4, 3)$

c *Graph.*

17. $y = x - 1$

18. $y = x + 1$

19. $y = x$

20. $y = -3x$

21. $y = \frac{1}{4}x$

22. $y = \frac{1}{3}x$

23. $y = 3 - x$

24. $y = x + 3$

25. $y = 5x - 2$

26. $y = \frac{1}{4}x + 2$

27. $y = \frac{1}{2}x + 1$

28. $y = \frac{1}{3}x - 4$

29. $x + y = 5$

30. $x + y = -4$

31. $y = -\frac{5}{3}x - 2$

32. $y = -\frac{5}{2}x + 3$

33. $x + 2y = 8$

34. $x + 2y = -6$

35. $y = \frac{3}{2}x + 1$

36. $y = -\frac{1}{2}x - 3$

37. $8y + 2x = 4$

38. $6x - 3y = -9$

39. $8y + 2x = -4$

40. $6y + 2x = 8$

d *Graph.*

41. $y = x^2$

42. $y = -x^2$
(*Hint:* $-x^2 = -1 \cdot x^2$.)

43. $y = x^2 + 2$

44. $y = 3 - x^2$

45. $y = x^2 - 3$

46. $y = x^2 - 3x$

47. $y = -\frac{1}{x}$

48. $y = \frac{3}{x}$

49. $y = |x - 2|$

50. $y = |x| + 2$

51. $y = x^3$

52. $y = x^3 - 2$

▶ Skill Maintenance

Solve. [1.5a, b]

53. $-3 < 2x - 5 \le 10$

54. $2x - 5 \ge -10 \ or \ -4x - 2 < 10$

55. $3x - 5 \le -12 \ or \ 3x - 5 \ge 12$

56. $-13 < 3x + 5 < 23$

Solve. [1.3a]

57. *Waiting Lists for Organ Transplants.* In August of 2013, there were more than 119,000 people on waiting lists for organ transplants. There were 113,162 people waiting for a kidney or a liver, and 81,528 fewer were waiting for a liver than for a kidney. How many were on the waiting list for a kidney? for a liver? (*Source:* Data from Organ Procurement and Transplantation Network)

58. *Landscaping.* Grass seed is being spread on a triangular traffic island. If the grass seed can cover an area of 200 ft² and the island's base is 16 ft long, how tall a triangle can the seed fill?

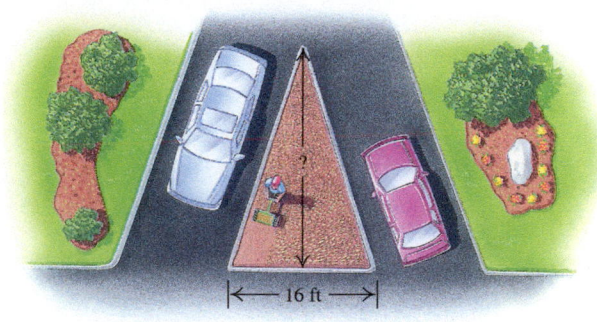

59. *Taxi Fare.* The fare for a taxi ride from Jen's office to the South Bay Health Center is $19.85. The driver charges $2.00 for the first $\frac{1}{2}$ mi and $1.05 for each additional $\frac{1}{4}$ mi. How far is it from Jen's office to the South Bay Health Center?

60. *Real Estate Commission.* The Clines negotiated the following real estate commission on the selling price of their house:

    7% for the first $100,000   and

    4% for the amount that exceeds $100,000.

The realtor received a commission of $16,200 for selling the house. What was the selling price?

▶ Synthesis

📈 *Use a graphing calculator to graph each of the equations in Exercises 61–64. Use a standard viewing window of* $[-10, 10, -10, 10]$, *with* Xscl $= 1$ *and* Yscl $= 1$.

61. $y = x^3 - 3x + 2$

62. $y = x - |x|$

63. $y = \frac{1}{x - 2}$

64. $y = \frac{1}{x^2}$

In Exercises 65–68, find an equation for the given graph.

65.

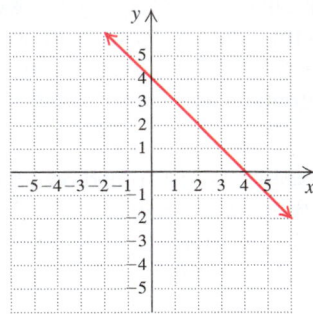

66.

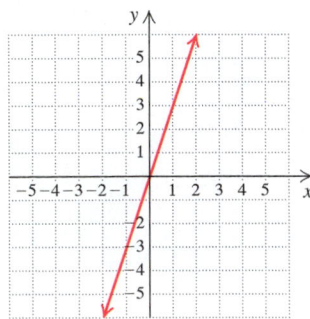

67.

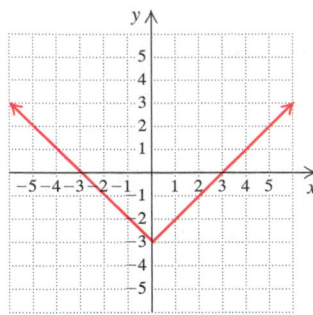

68.

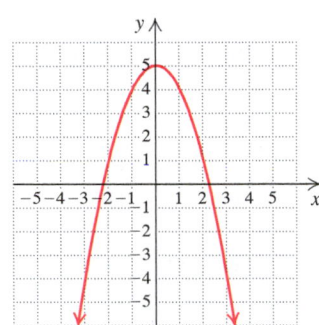

2.2 Functions and Graphs

▶ **a** Determine whether a correspondence is a function.

▶ **b** Given a function described by an equation, find function values (outputs) for specified values (inputs).

▶ **c** Draw the graph of a function.

▶ **d** Determine whether a graph is that of a function using the vertical-line test.

▶ **e** Solve applied problems involving functions and their graphs.

▶ a Identifying Functions

Consider the equation $y = 2x - 3$. If we substitute a value for x—say, 5—we get a value for y, 7:

$$y = 2x - 3 = 2(5) - 3 = 10 - 3 = 7.$$

The equation $y = 2x - 3$ is an example of a *function*, one of the most important concepts in mathematics.

In much the same way that ordered pairs form correspondences between first and second coordinates, a *function* is a correspondence from one set to another. For example:

To each student in a college, there corresponds his or her student ID.

To each item in a store, there corresponds its price.

To each real number, there corresponds the cube of that number.

In each case, the first set is called the **domain** and the second set is called the **range**. Each of these correspondences is a **function**, because given a member of the domain, there is *just one* member of the range to which it corresponds. Given a student, there is *just one* ID. Given an item, there is *just one* price. Given a real number, there is *just one* cube.

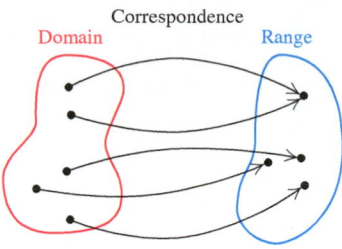

EXAMPLE 1 Determine whether the correspondence is a function.

$f:$

Domain	Range
1 ⟶	$107.40
2 ⟶	$ 34.10
3 ⟶	$ 29.60
4 ⟶	$ 19.60

$g:$

Domain	Range
3 ⟶	5
4 ⟶	9
5	
6	−7

$h:$

Domain	Range
Chicago	Cubs
	White Sox
Baltimore ⟶	Orioles
San Diego ⟶	Padres

$p:$

Domain	Range
Cubs	Chicago
White Sox	
Orioles ⟶	Baltimore
Padres ⟶	San Diego

The correspondence f *is* a function because each member of the domain is matched to *only one* member of the range.

The correspondence g *is* a function because each member of the domain is matched to *only one* member of the range. Note that a function allows two or more members of the domain to correspond to the same member of the range.

The correspondence h *is not* a function because one member of the domain, Chicago, is matched to *more than one* member of the range.

The correspondence p *is* a function because each member of the domain is matched to *only one* member of the range.

Now Try Exercises 1 and 5.

FUNCTION; DOMAIN; RANGE

A **function** is a correspondence between a first set, called the **domain**, and a second set, called the **range**, such that each member of the domain corresponds to **exactly one** member of the range.

EXAMPLE 2 Determine whether each correspondence is a function.

Domain	Correspondence	Range
a) The integers	Each number's square	A set of nonnegative integers
b) A set of presidents (listed below)	Each president's appointees to the Supreme Court	A set of Supreme Court Justices (listed below)

Appointing President	Supreme Court Justice
George H. W. Bush	Samuel A. Alito, Jr.
William Jefferson Clinton	Stephen G. Breyer
	Ruth Bader Ginsburg
George W. Bush	Elena Kagan
	John G. Roberts, Jr.
Barack H. Obama	Sonia M. Sotomayor
	Clarence Thomas

a) The correspondence *is* a function because each integer has *only one* square.

b) This correspondence *is not* a function because there is at least one member of the domain who is paired with more than one member of the range (William Jefferson Clinton with Stephen G. Breyer and Ruth Bader Ginsburg; George W. Bush with Samuel A. Alito, Jr., and John G. Roberts, Jr.; Barack H. Obama with Elena Kagan and Sonia M. Sotomayor).

> **Now Try Exercises 9 and 11.**

When a correspondence between two sets is not a function, it may still be an example of a **relation**.

RELATION

A **relation** is a correspondence between a first set, called the **domain**, and a second set, called the **range**, such that each member of the domain corresponds to **at least one** member of the range.

Thus, although the correspondences of Examples 1 and 2 are not all functions, they *are* all relations. A function is a special type of relation—one in which each member of the domain is paired with *exactly one* member of the range.

▶ b Finding Function Values

Most functions considered in mathematics are described by equations like $y = 2x + 3$ or $y = 4 - x^2$. We graph the function $y = 2x + 3$ by first performing calculations like the following:

for $x = 4, y = 2x + 3 = 2 \cdot 4 + 3 = 8 + 3 = 11$;

for $x = -5, y = 2x + 3 = 2 \cdot (-5) + 3 = -10 + 3 = -7$;

for $x = 0, y = 2x + 3 = 2 \cdot 0 + 3 = 0 + 3 = 3$;   and so on.

For $y = 2x + 3$, the **inputs** (members of the domain) are values of x substituted into the equation. The **outputs** (members of the range) are the resulting values of y. If we call the function f, we can use x to represent an arbitrary *input* and $f(x)$—read "f of x," or "f at x," or "the value of f at x"—to represent the corresponding *output*. In this notation, the function given by $y = 2x + 3$ is written as $f(x) = 2x + 3$ and the calculations above can be written more concisely as follows:

$$f(4) = 2 \cdot 4 + 3 = 8 + 3 = 11;$$
$$f(-5) = 2 \cdot (-5) + 3 = -10 + 3 = -7;$$
$$f(0) = 2 \cdot 0 + 3 = 0 + 3 = 3;$$ and so on.

CAUTION!   The notation $f(x)$ *does not mean* "f times x" and should not be read that way.

Thus instead of writing "when $x = 4$, the value of y is 11," we can simply write "$f(4) = 11$," which can also be read as "f of 4 is 11" or "for the input 4, the output of f is 11."

We can think of a function as a machine. Think of $f(4) = 11$ as putting 4, a member of the domain (an input), into the machine. The machine knows the correspondence $f(x) = 2x + 3$, multiplies 4 by 2 and adds 3, and produces 11, a member of the range (the output).

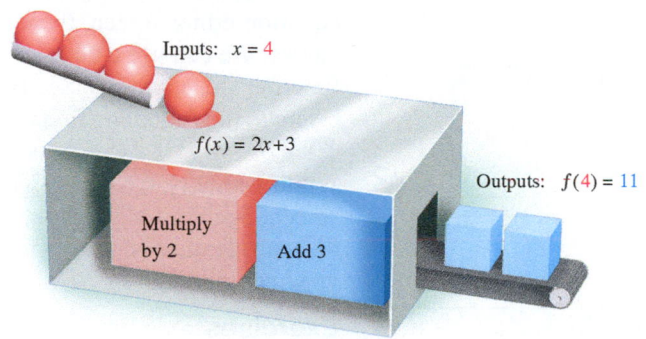

EXAMPLE 3   A function f is given by $f(x) = 3x^2 - 2x + 8$. Find each of the indicated function values.

a) $f(0)$ **b)** $f(-5)$ **c)** $f(7a)$

One way to find function values when a formula is given is to think of the formula with blanks, or placeholders, replacing the variable as follows:

$$f(\square) = 3\square^2 - 2\square + 8.$$

To find an output for a given input, we think: "Whatever goes in the blank on the left goes in the blank(s) on the right." With this in mind, let's complete the example.

a) $f(0) = 3 \cdot 0^2 - 2 \cdot 0 + 8 = 8$

b) $f(-5) = 3(-5)^2 - 2 \cdot (-5) + 8 = 3 \cdot 25 + 10 + 8 = 75 + 10 + 8 = 93$

c) $f(7a) = 3(7a)^2 - 2(7a) + 8 = 3 \cdot 49a^2 - 14a + 8 = 147a^2 - 14a + 8$

> **Now Try Exercise 19.**

EXAMPLE 4   Find the indicated function value.

a) $f(5)$, for $f(x) = 3x + 2$ **b)** $g(-2)$, for $g(x) = 7$

c) $F(a + 1)$, for $F(x) = 5x - 8$ **d)** $f(a + h)$, for $f(x) = -2x + 1$

a) $f(5) = 3 \cdot 5 + 2 = 15 + 2 = 17$

b) For the function given by $g(x) = 7$, all inputs share the same output, 7. Thus, $g(-2) = 7$. The function g is an example of a **constant function**.

c) $F(a + 1) = 5(a + 1) - 8 = 5a + 5 - 8 = 5a - 3$

d) $f(a + h) = -2(a + h) + 1 = -2a - 2h + 1$

> **Now Try Exercise 17.**

Technology Connection

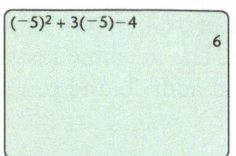

Figure 1.

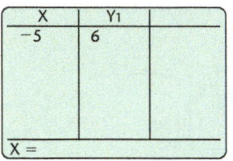

Figure 2.

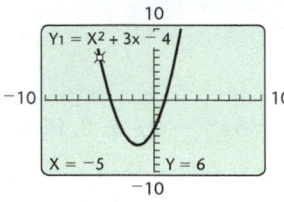

Figure 3.

Finding Function Values

We can find function values using a graphing calculator. One method is to substitute inputs directly into the formula. Consider the function $f(x) = x^2 + 3x - 4$. We find that $f(-5) = 6$. See Figure 1.

After we have entered the function as $y_1 = x^2 + 3x - 4$ on the equation-editor screen, there are other methods that we can use to find function values. We can use a table set in ASK mode and enter $x = -5$. We see that the function value, y_1, is 6. See Figure 2. We can also use the VALUE feature to evaluate the function. To do this, we first graph the function. Then we press **2ND** **CALC** **1** to access the VALUE feature. Next, we supply the desired x-value. Finally, we press **ENTER** to see X = −5, Y = 6 at the bottom of the screen. See Figure 3. Again, we see that the function value is 6. Note that when the VALUE feature is used to find a function value, the x-value must be in the viewing window.

Exercises

Find each function value.

1. $f(-5.1)$, for $f(x) = -3x + 2$

2. $f(3)$, for $f(x) = 4x^2 + x - 5$

▶ **c** **Graphs of Functions**

To graph a function, we find ordered pairs (x, y) or $(x, f(x))$, plot them, and connect the points. Note that y and $f(x)$ are used interchangeably—that is, $y = f(x)$—when we are working with functions and their graphs.

EXAMPLE 5 Graph: $f(x) = x + 2$.

A list of some function values is shown in the following table. We plot the points and connect them. The graph is a straight line. The vertical axis could also be labeled "$f(x)$" rather than "y".

x	$f(x)$
−4	−2
−3	−1
−2	0
−1	1
0	2
1	3
2	4
3	5
4	6

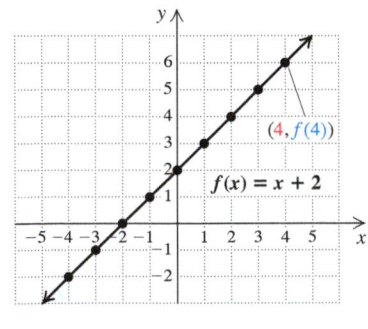

Now Try Exercise 33.

EXAMPLE 6 Graph: $g(x) = 4 - x^2$.

We calculate some function values, plot the corresponding points, and draw the curve.

$$g(0) = 4 - 0^2 = 4 - 0 = 4,$$
$$g(-1) = 4 - (-1)^2 = 4 - 1 = 3,$$
$$g(2) = 4 - 2^2 = 4 - 4 = 0,$$
$$g(-3) = 4 - (-3)^2 = 4 - 9 = -5$$

x	$g(x)$
-3	-5
-2	0
-1	3
0	4
1	3
2	0
3	-5

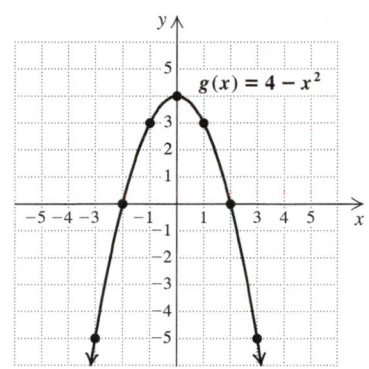

Now Try Exercise 47.

EXAMPLE 7 Graph: $h(x) = |x|$.

A list of some function values is shown in the following table. We plot the points and connect them. The graph is a V-shaped "curve" that rises on either side of the vertical axis.

x	$h(x)$
-3	3
-2	2
-1	1
0	0
1	1
2	2
3	3

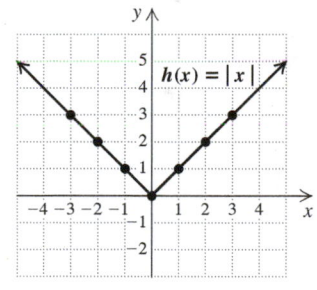

Now Try Exercise 41.

► **d The Vertical-Line Test**

Consider the graph of the function f described by $f(x) = x^2 - 5$ shown at right. It is also the graph of the equation $y = x^2 - 5$.

To find a function value, like $f(3)$, from a graph, we locate the input on the horizontal axis, move directly up or down to the graph of the function, and then move left or right to find the output on the vertical axis. Thus, $f(3) = 4$. Keep in mind that members of the domain are found on the horizontal axis, members of the range are found on the vertical axis, and the y on the vertical axis could also be labeled $f(x)$.

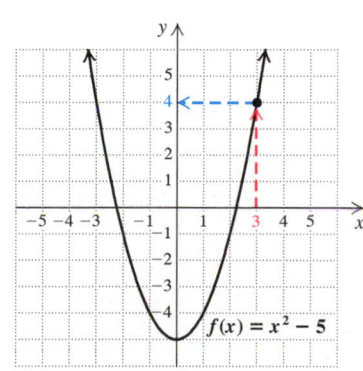

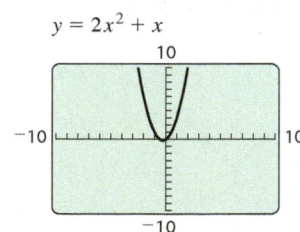

When one member of the domain is paired with two or more different members of the range, the correspondence is not a function. Thus, when a graph contains two or more different points with the same first coordinate, the graph cannot represent a function. Points sharing a common first coordinate are vertically above or below each other. (See the following graph.) This observation leads to the *vertical-line test*.

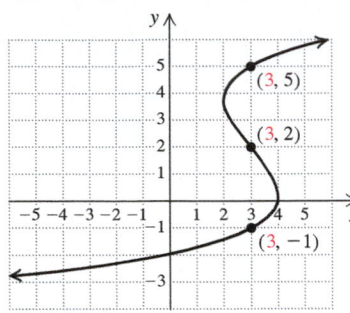

Since 3 is paired with more than one member of the range, the graph does not represent a function.

THE VERTICAL-LINE TEST

If it is possible for a vertical line to cross a graph more than once, then the graph is *not* the graph of a function.

EXAMPLE 8 Determine whether each of the following is the graph of a function.

a) 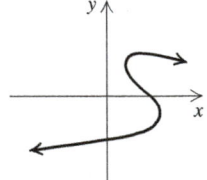 **b)**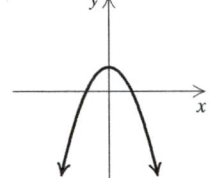

a) The graph *is not* that of a function because a vertical line can cross the graph at more than one point. The following graph shows one such vertical line.

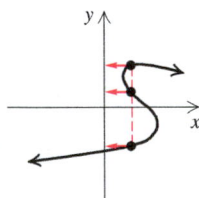

b) The graph *is* that of a function because no vertical line can cross the graph more than once.

Now Try Exercises 53 and 57.

▶ **e** **Applications of Functions and Their Graphs**

Functions are often described by graphs, whether or not an equation is given. To use a graph in an application, we note that each point on the graph represents a pair of values.

EXAMPLE 9 *IRS Instruction Booklet.* The following graph represents the number of pages in the IRS 1040 instruction booklet for years from 1965 through 2012. The number of pages is a function of the year. Note that no equation is given for the function.

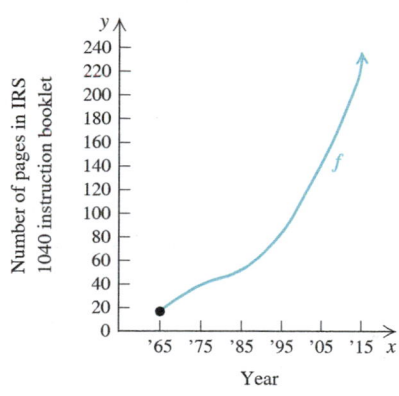

SOURCES: National Taxpayers Union;
Statista.com; Internal Revenue Service

a) How many pages are in the 1975 IRS 1040 instruction booklet? That is, find $f(1975)$.

b) How many pages are in the 2010 IRS 1040 instruction booklet? That is, find $f(2010)$.

a) To estimate the number of pages in the 1975 booklet, we locate 1975 on the horizontal axis and move directly up until we reach the graph. Then we move across to the vertical axis. We come to a point that is about 40, so we estimate that the number of pages in the 1975 booklet is 40.

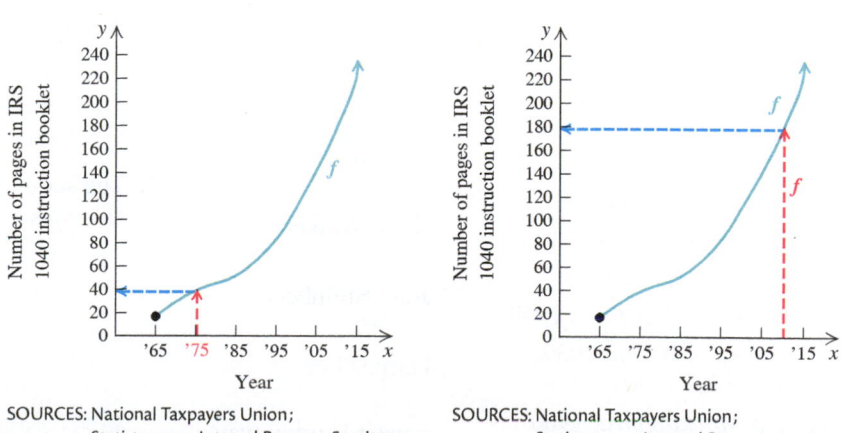

SOURCES: National Taxpayers Union;   SOURCES: National Taxpayers Union;
Statista.com; Internal Revenue Service   Statista.com; Internal Revenue Service

b) To estimate the number of pages in the 2010 booklet, we locate 2010 on the horizontal axis and move directly up until we reach the graph. Then we move across to the vertical axis. We come to a point that is about 180, so we estimate that the number of pages in the 2010 booklet is 180.

Now Try Exercise 61.

2.2 Exercise Set

▶ Reading Check

Use the graph at right to find the given function value by locating the input on the horizontal axis, moving directly up or down to the graph of the function, and then moving left or right to find the output on the vertical axis. As an example, finding $f(4) = -5$ is illustrated.

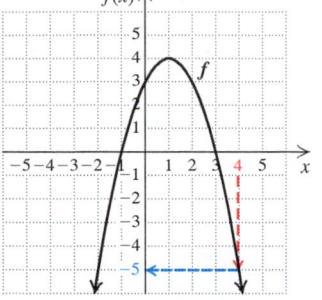

RC1. $f(2)$ **RC2.** $f(0)$

RC3. $f(-2)$ **RC4.** $f(3)$

---

a *Determine whether each correspondence is a function.*

1. Domain   Range

$2 \longrightarrow 9$
$5 \longrightarrow 8$
19

2. Domain   Range

$5 \longrightarrow 3$
$-3 \longrightarrow 7$
7
-7

3. Domain   Range

$-5 \longrightarrow 1$
5
8

4. Domain   Range

$6 \longrightarrow -6$
$7 \longrightarrow -7$
$3 \longrightarrow -3$

5. Domain   Range

$9 \longrightarrow 3$
$ \longrightarrow -3$
$16 \longrightarrow 4$
$ \longrightarrow -4$
$25 \longrightarrow 5$
$ \longrightarrow -5$

6. Domain

The Color Purple, 1982
(Pulitzer Prize 1983)

East of Eden, 1952

Fahrenheit 451, 1953

The Good Earth, 1931
(Pulitzer Prize 1932)

For Whom the Bell Tolls, 1940

The Grapes of Wrath, 1939
(Pulitzer Prize 1940)

To Kill a Mockingbird, 1960
(Pulitzer Prize 1961)

The Old Man and the Sea, 1952
(Pulitzer Prize 1953)

Range

Pearl Buck

Ray Bradbury

Alice Walker

John Steinbeck

Harper Lee

Ernest Hemingway

7. Domain   Range

Florida ⟷ Florida State University
University of Florida
University of Miami

Kansas ⟷ Baker University
Kansas State University
University of Kansas

8. Domain   Range

Colorado State University
University of Colorado ⟶ Colorado
University of Denver
Gonzaga University
University of Washington ⟶ Washington
Washington State University

Domain	**Correspondence**	**Range**
9. A set of numbers	The area of a triangle	A set of triangles
10. A family	Each person's height, in inches	A set of positive numbers
11. The set of U.S. Senators	The state that a Senator represents	The set of all states
12. The set of all states	Each state's members of the U.S. Senate	The set of U.S. Senators

UNITED STATES

Missouri

Claire McCaskill

Roy Blunt

b *Find the function values.*

13. $f(x) = x + 5$

  a) $f(4)$ **b)** $f(7)$
  c) $f(-3)$ **d)** $f(0)$
  e) $f(2.4)$ **f)** $f\left(\frac{2}{3}\right)$

14. $g(t) = t - 6$

  a) $g(0)$ **b)** $g(6)$
  c) $g(13)$ **d)** $g(-1)$
  e) $g(-1.08)$ **f)** $g\left(\frac{7}{8}\right)$

15. $h(p) = 3p$

  a) $h(-7)$ **b)** $h(5)$
  c) $h\left(\frac{2}{3}\right)$ **d)** $h(0)$
  e) $h(6a)$ **f)** $h(a + 1)$

16. $f(x) = -4x$

  a) $f(6)$ **b)** $f\left(-\frac{1}{2}\right)$
  c) $f(0)$ **d)** $f(-1)$
  e) $f(3a)$ **f)** $f(a - 1)$

17. $g(s) = 3s + 4$

  a) $g(1)$ **b)** $g(-7)$
  c) $g\left(\frac{2}{3}\right)$ **d)** $g(0)$
  e) $g(a - 2)$ **f)** $g(a + h)$

18. $h(x) = 19$, a constant function

  a) $h(4)$ **b)** $h(-6)$
  c) $h(12.5)$ **d)** $h(0)$
  e) $h\left(\frac{2}{3}\right)$ **f)** $h(a + 3)$

19. $f(x) = 2x^2 - 3x$

  a) $f(0)$ **b)** $f(-1)$
  c) $f(2)$ **d)** $f(10)$
  e) $f(-5)$ **f)** $f(4a)$

20. $f(x) = 3x^2 - 2x + 1$

  a) $f(0)$ **b)** $f(1)$
  c) $f(-1)$ **d)** $f(10)$
  e) $f(-3)$ **f)** $f(2a)$

21. $f(x) = |x| + 1$

  a) $f(0)$ **b)** $f(-2)$
  c) $f(2)$ **d)** $f(-10)$
  e) $f(a - 1)$ **f)** $f(a + h)$

22. $g(t) = |t - 1|$

  a) $g(4)$ **b)** $g(-2)$
  c) $g(-1)$ **d)** $g(100)$
  e) $g(5a)$ **f)** $g(a + 1)$

23. $f(x) = x^3$

  a) $f(0)$ **b)** $f(-1)$
  c) $f(2)$ **d)** $f(10)$
  e) $f(-5)$ **f)** $f(-3a)$

24. $f(x) = x^4 - 3$

  a) $f(1)$ **b)** $f(-1)$
  c) $f(0)$ **d)** $f(2)$
  e) $f(-2)$ **f)** $f(-a)$

25. *Average Age of Senators.* The function $A(s)$ given by

$$A(s) = 0.044s + 59$$

can be used to estimate the average age of senators in the U.S. Senate in the years 1945 to 2013. Let $A(s) =$ the average age of the senators and $s =$ the number of years since 1945—that is, $s = 0$ for 1945, $s = 20$ for 1965, and so on. What was the average age of U.S. Senators in 1980? in 2013? (*Sources:* Data from www.slate.com/; "Democracy or Gerontocracy," Brian Palmer, January 2, 2013; Congressional Research Service)

26. *Average Age of House Members.* The function $A(h)$ given by

$$A(h) = 0.059h + 53$$

can be used to estimate the average age of House members in the U.S. House of Representatives in the years 1945 to 2013. Let $A(h) =$ the average age of the House members and $h =$ the number of years since 1945. What is the average age of U.S.

House members in 1980? in 2013? (*Sources*: Data from www.slate.com/; "Democracy or Gerontocracy," Brian Palmer, January 2, 2013; Congressional Research Service)

27. *Pressure at Sea Depth.*   The function $P(d) = 1 + (d/33)$ gives the pressure, in *atmospheres* (atm), at a depth of d feet in the sea. Note that $P(0) = 1$ atm, $P(33) = 2$ atm, and so on. Find the pressure at 20 ft, 30 ft, and 100 ft.

28. *Temperature as a Function of Depth.*   The function $T(d) = 10d + 20$ gives the temperature, in degrees Celsius, inside the earth as a function of the depth d, in kilometers. Find the temperature at 5 km, 20 km, and 1000 km.

29. *Melting Snow.*   The function $W(d) = 0.112d$ approximates the amount of water, in centimeters, that results from d centimeters of snow melting. Find the amount of water that results from snow melting from depths of 16 cm, 25 cm, and 100 cm.

30. *Temperature Conversions.*   The function $C(F) = \frac{5}{9}(F - 32)$ determines the Celsius temperature that corresponds to F degrees Fahrenheit. Find

the Celsius temperature that corresponds to 62°F, 77°F, and 23°F.

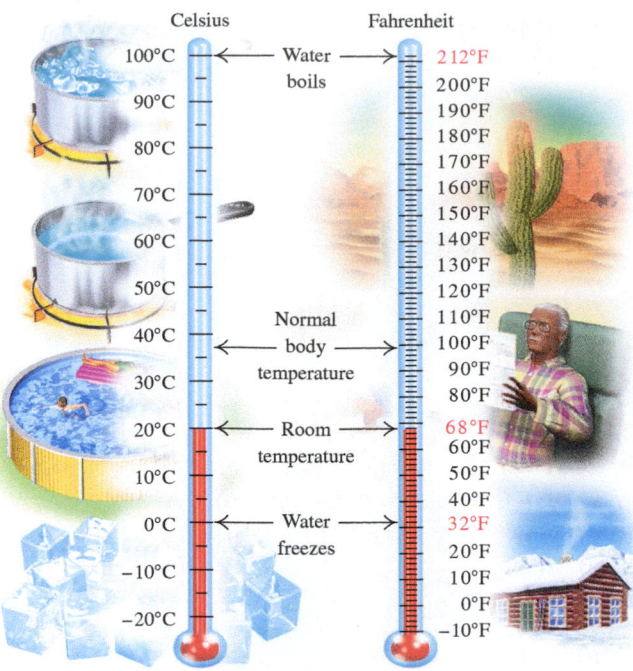

c   *Graph each function.*

31. $f(x) = -2x$

32. $g(x) = 3x$

33. $g(x) = 3x - 1$

34. $f(x) = 2x + 5$

35. $g(x) = -2x + 3$

36. $f(x) = -\frac{1}{2}x + 2$

37. $f(x) = \frac{1}{2}x + 1$

38. $f(x) = -\frac{3}{4}x - 2$

39. $f(x) = 2 - |x|$

40. $f(x) = |x| - 4$

41. $g(x) = |x - 1|$

42. $g(x) = |x + 3|$

43. $g(x) = x^2 + 2$

44. $f(x) = x^2 + 1$

45. $f(x) = x^2 - 2x - 3$

46. $g(x) = x^2 + 6x + 5$

47. $f(x) = -x^2 + 1$

48. $f(x) = -x^2 + 2$

49. $f(x) = x^3 + 1$

50. $f(x) = x^3 - 2$

d   *Determine whether each of the following is the graph of a function.*

51.
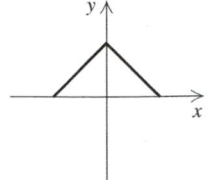

52.
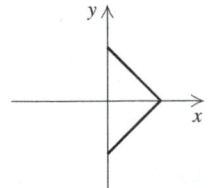

53.
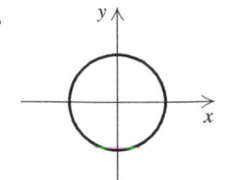

54.
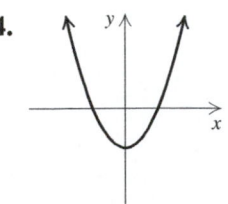

55.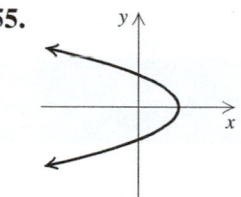

56.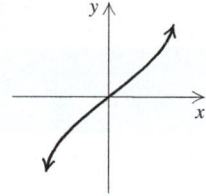

57.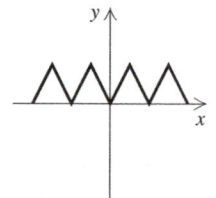

58.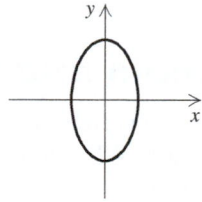

e *Solve.*

Living with Grandparents. *The following graph approximates the number of children in the United States who lived with only their grandparents in the years from 1991 through 2009. The number of children is a function f of the year x.*

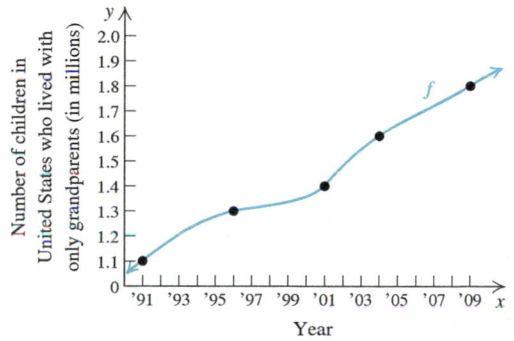

SOURCE: U.S. Census 2010

59. Approximate the number of children living with only grandparents in 2009. That is, find $f(2009)$.

60. Approximate the number of children living with only grandparents in 1996. That is, find $f(1996)$.

Pharmacists. *The following graph approximates the number of pharmacists in the United States in the years from 2002 through 2012. The number of pharmacists is a function g of the year x.*

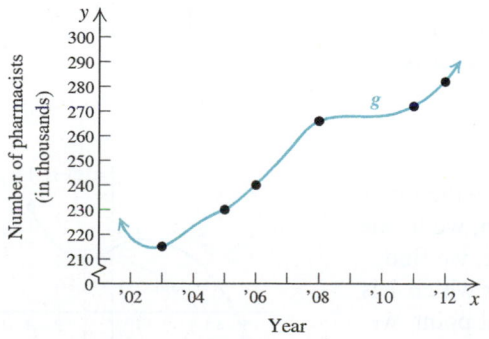

SOURCE: IDC; statista.com

61. Approximate the number of pharmacists in 2005.

62. Approximate the number of pharmacists in 2012.

▶ Skill Maintenance

Solve.

63. $-\dfrac{5}{3} + y = -\dfrac{1}{12} - \dfrac{5}{6}$ [1.1d]

64. $6x - 31 = 11 + 6(x - 7)$ [1.1d]

65. $4 - 7y > 2y - 32$ [1.4c]

66. $\dfrac{2}{3}(4x - 2) > 60$ [1.4c]

67. $7y - 2 = 3 + 7y$ [1.1d]

68. $4(x - 5) = 3(x + 2)$ [1.1d]

69. $-9w \geq -99.9$ [1.4c]

70. $\dfrac{1}{2}x + 10 < 8x - 5$ [1.4c]

71. $13x - 5 - x = 2(x + 5)$ [1.1d]

72. $\dfrac{1}{16}x + 4 = \dfrac{5}{8}x - 1$ [1.1d]

▶ Synthesis

73. Suppose that for some function g, $g(x - 6) = 10x - 1$. Find $g(-2)$.

74. Suppose that for some function h, $h(x + 5) = x^2 - 4$. Find $h(3)$.

For Exercises 75 and 76, let $f(x) = 3x^2 - 1$ and $g(x) = 2x + 5$.

75. Find $f(g(-4))$ and $g(f(-4))$.

76. Find $f(g(-1))$ and $g(f(-1))$.

77. Suppose that a function g is such that $g(-1) = -7$ and $g(3) = 8$. Find a formula for g if $g(x)$ is of the form $g(x) = mx + b$, where m and b are constants.

▶ **a**   Find the domain and the range of a function.

▶ a   Finding Domain and Range

The solutions of an equation in two variables consist of a set of ordered pairs. A set of ordered pairs is called a **relation**. When a set of ordered pairs is such that no two different pairs share a common first coordinate, we have a **function**. The **domain** is the set of all first coordinates, and the **range** is the set of all second coordinates.

EXAMPLE 1   Find the domain and the range of the function f whose graph is shown below.

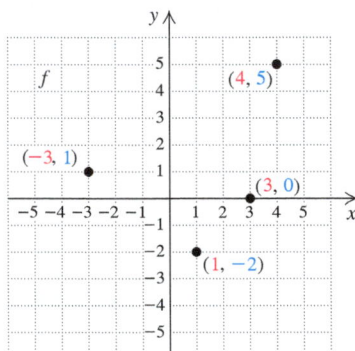

This function contains just four ordered pairs and it can be written as

$$\{(-3, 1), (1, -2), (3, 0), (4, 5)\}.$$

We can determine the domain and the range by reading the x- and y-values directly from the graph.

The domain is the set of all first coordinates, or x-values, $\{-3, 1, 3, 4\}$. The range is the set of all second coordinates, or y-values, $\{1, -2, 0, 5\}$.

EXAMPLE 2   For the function f whose graph is shown at right, determine each of the following.

a)   The number in the range that is paired with 1 from the domain. That is, find $f(1)$.

b)   The domain of f

c)   The numbers in the domain that are paired with 1 from the range. That is, find all x such that $f(x) = 1$.

d)   The range of f

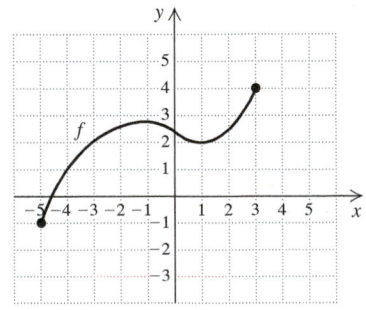

a)   To determine which number in the range is paired with 1 in the domain, we locate 1 on the horizontal axis. Next, we find the point on the graph of f for which 1 is the first coordinate. From that point, we can look to the vertical axis to find the corresponding y-coordinate, 2. The input 1 has the output 2—that is, $f(1) = 2$.

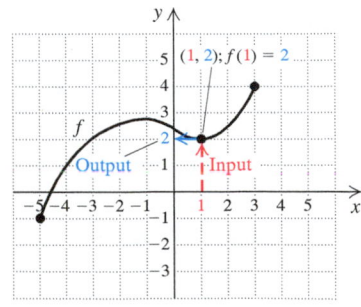

b) The domain of the function is the set of all *x*-values, or inputs, of the points on the graph. These extend from −5 to 3 and can be viewed as the curve's shadow, or projection, onto the *x*-axis. Thus the domain is $\{x | -5 \le x \le 3\}$, or, in interval notation, $[-5, 3]$.

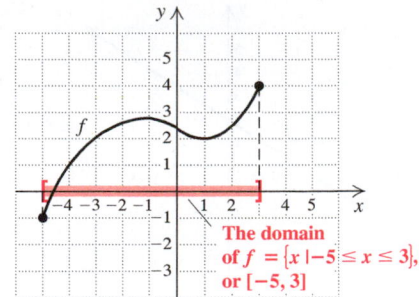

c) To determine which numbers in the domain are paired with 1 in the range, we locate 1 on the vertical axis. From there, we look left and right to the graph of *f* to find any points for which 1 is the second coordinate (output). One such point exists, $(-4, 1)$. For this function, we note that −4 is the only member of the domain paired with 1. For other functions, there might be more than one member of the domain paired with a member of the range.

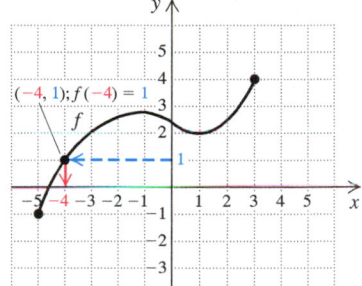

d) The range of the function is the set of all *y*-values, or outputs, of the points on the graph. These extend from −1 to 4 and can be viewed as the curve's shadow, or projection, onto the *y*-axis. Thus the range is $\{y | -1 \le y \le 4\}$, or, in interval notation, $[-1, 4]$.

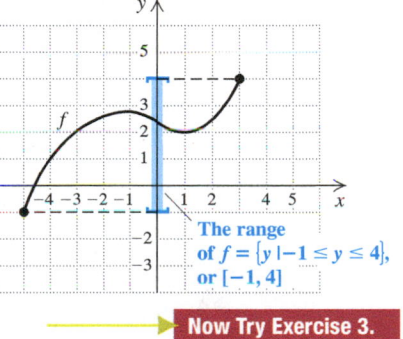

Now Try Exercise 3.

EXAMPLE 3 Find the domain and the range of the function *h* whose graph is shown below.

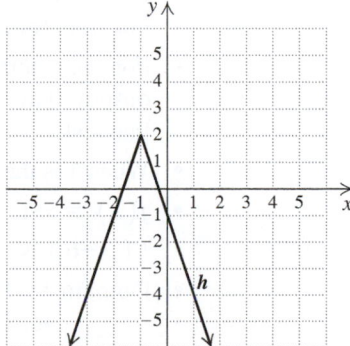

Since no endpoints are indicated, the graph extends indefinitely horizontally. Thus the domain, or the set of inputs, is the set of all real numbers. The range, or the set of outputs, is the set of all *y*-values of the points on the graph. Thus the range is $\{y | y \le 2\}$, or $(-\infty, 2]$.

Now Try Exercise 7.

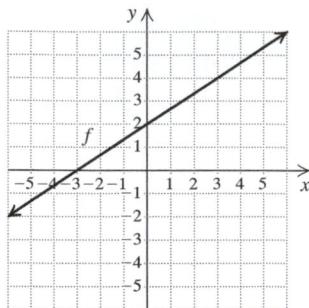

EXAMPLE 4   Find the domain and the range of the function f whose graph is shown at left.

Since no endpoints are indicated, the graph extends indefinitely both horizontally and vertically. Thus the domain is the set of all real numbers. Likewise, the range is the set of all real numbers.

> **Now Try Exercise 5.**

When a function is given by an equation or a formula, the domain is understood to be the largest set of real numbers (inputs) for which function values (outputs) can be calculated. That is, the domain is the set of all possible allowable inputs into the formula. To find the domain, think, "What can we substitute?"

EXAMPLE 5   Find the domain: $f(x) = |x|$.

We ask, "What can we substitute?" Is there any number x for which we cannot calculate $|x|$? The answer is no. Thus the domain of f is the set of all real numbers.

> **Now Try Exercise 17.**

EXAMPLE 6   Find the domain: $f(x) = \dfrac{3}{2x - 5}$.

We ask, "What can we substitute?" Is there any number x for which we cannot calculate $3/(2x - 5)$? Since $3/(2x - 5)$ cannot be calculated when the denominator $2x - 5$ is 0, we solve the following equation to find those real numbers that must be excluded from the domain of f:

$$2x - 5 = 0 \qquad \text{\textcolor{red}{Setting the denominator equal to 0}}$$
$$2x = 5 \qquad \text{\textcolor{red}{Adding 5}}$$
$$x = \tfrac{5}{2}. \qquad \text{\textcolor{red}{Dividing by 2}}$$

Thus, $\frac{5}{2}$ is not in the domain, whereas all other real numbers are.

The domain of f is $\left\{ x \mid x \text{ is a real number } and\ x \neq \frac{5}{2} \right\}$. In interval notation, the domain is $\left(-\infty, \frac{5}{2} \right) \cup \left(\frac{5}{2}, \infty \right)$.

> **Now Try Exercise 15.**

Functions: A Review

The following is a review of the function concepts. Use the graph below to visualize the concepts.

Function Concepts

- Formula for f: $f(x) = x^2 - 7$

- For every input of f, there is exactly one output. The graph passas the vertical line test.

- When 1 is the input, -6 is the output.

- $f(1) = -6$

- $(1, -6)$ is on the graph.

- Domain = The set of all inputs
  = The set of all real numbers

- Range = The set of all outputs
  = $\{ y \mid y \geq -7 \}$
  = $[-7, \infty)$

Graph

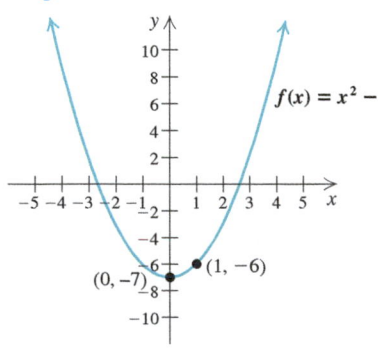

2.3   Exercise Set

► Reading Check

Choose from the column on the right the domain of the function. Some choices may be used more than once; others not at all.

RC1. $f(x) = 5 - x$

RC2. $f(x) = \dfrac{-5}{5 - x}$

RC3. $f(x) = |5 - x|$

RC4. $f(x) = \dfrac{5}{|x - 5|}$

RC5. $f(x) = 5 - |x|$

RC6. $f(x) = \dfrac{x - 5}{x + 5}$

a) All real numbers

b) $\{x \,|\, x \text{ is a real number } and \ x \neq 5\}$

c) $\{x \,|\, x \text{ is a real number } and \ x \neq -5 \ and \ x \neq 5\}$

d) $\{x \,|\, x \text{ is a real number } and \ x \neq -5\}$

---

a *In Exercises 1–8, the graph is that of a function. Determine for each one* **(a)** $f(1)$; **(b)** *the domain;* **(c)** *all x-values such that* $f(x) = 2$; *and* **(d)** *the range. An open dot indicates that the point does not belong to the graph.*

1.

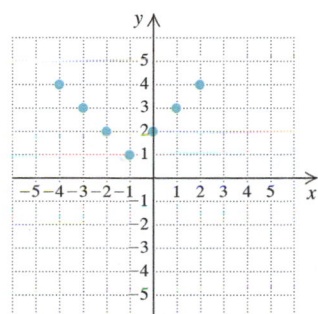

2.

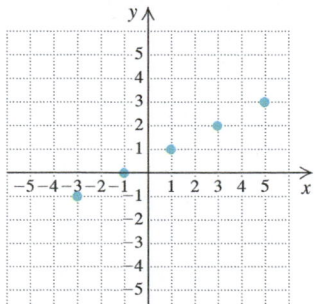

3.

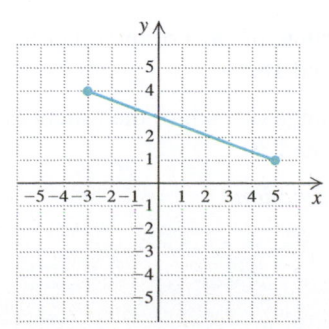

4.

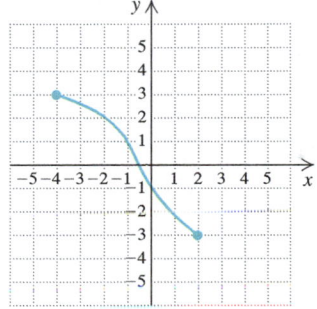

5.

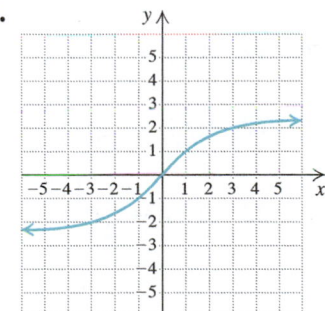

6.

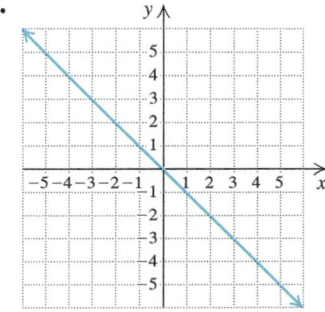

7.

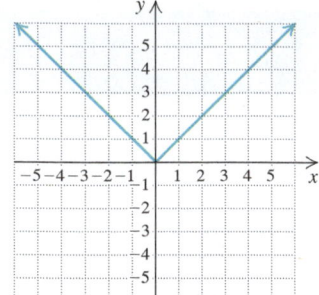

8.

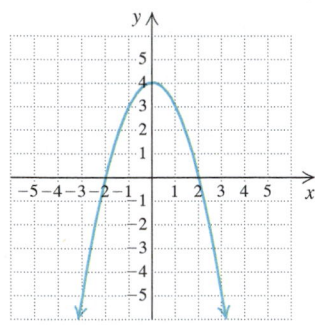

Find the domain.

9. $f(x) = \dfrac{2}{x + 3}$

10. $f(x) = \dfrac{7}{5 - x}$

11. $f(x) = 2x + 1$

12. $f(x) = 4 - 5x$

13. $f(x) = x^2 + 3$

14. $f(x) = x^2 - 2x + 3$

15. $f(x) = \dfrac{8}{5x - 14}$

16. $f(x) = \dfrac{x - 2}{3x + 4}$

17. $f(x) = |x| - 4$

18. $f(x) = |x - 4|$

19. $f(x) = \dfrac{x^2 - 3x}{|4x - 7|}$

20. $f(x) = \dfrac{4}{|2x - 3|}$

21. $g(x) = \dfrac{1}{x - 1}$

22. $g(x) = \dfrac{-11}{4 + x}$

23. $g(x) = x^2 - 2x + 1$

24. $g(x) = 8 - x^2$

25. $g(x) = x^3 - 1$

26. $g(x) = 4x^3 + 5x^2 - 2x$

27. $g(x) = \dfrac{7}{20 - 8x}$

28. $g(x) = \dfrac{2x - 3}{6x - 12}$

29. $g(x) = |x + 7|$

30. $g(x) = |x| + 1$

31. $g(x) = \dfrac{-2}{|4x + 5|}$

32. $g(x) = \dfrac{x^2 + 2x}{|10x - 20|}$

33. For the function f whose graph is shown below, find $f(-1)$, $f(0)$, and $f(1)$.

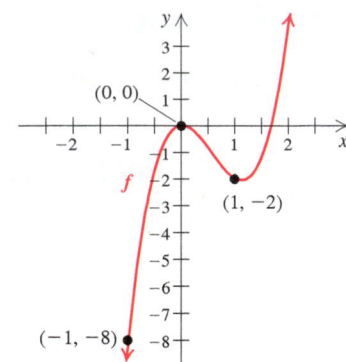

34. For the function g whose graph is shown below, find all the x-values for which $g(x) = 1$.

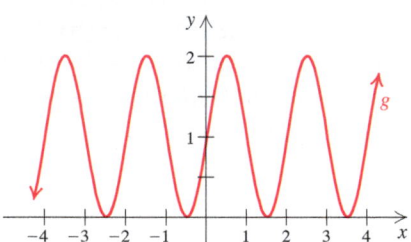

▶ Skill Maintenance

Solve. **[1.6c, d]**

35. $|x| = 8$

36. $|x| = -8$

37. $|x - 7| = 11$

38. $|2x + 3| = 13$

39. $|3x - 4| = |x + 2|$

40. $|5x - 6| = |3 - 8x|$

41. $|3x - 8| = -11$

42. $|3x - 8| = 0$

▶ Synthesis

43. ◥◤ Determine the range of each of the functions in Exercises 9, 14, 17, and 18.

44. ◥◤ Determine the range of each of the functions in Exercises 22, 23, 24, and 30.

Find the domain of each function.

45. $f(x) = \sqrt[3]{x - 1}$

46. $g(x) = \sqrt{2 - x}$

2.4   The Algebra of Functions

▶ **a**   Given two functions f and g, find their sum, difference, product, and quotient.

▶ **b**   For a pair of functions f and g, determine the domain of $f + g$, $f - g$, $f \cdot g$, and f/g.

We now examine four ways in which functions can be combined.

▶ a   The Sum, Difference, Product, or Quotient of Two Functions

Suppose that a is in the domain of two functions, f and g. The input a is paired with $f(a)$ by f and with $g(a)$ by g. The outputs can then be added to get $f(a) + g(a)$.

EXAMPLE 1   Let $f(x) = x + 4$ and $g(x) = x^2 + 1$. Find $f(2) + g(2)$.

Solution   Because 2 is in the domain of each function, we can compute $f(2)$ and $g(2)$.
  Since

$$f(2) = 2 + 4 = 6 \quad \text{and} \quad g(2) = 2^2 + 1 = 5,$$

we have

$$f(2) + g(2) = 6 + 5 = 11. \qquad\qquad \blacktriangleright \boxed{\textbf{Now Try Exercise 1.}}$$

  In Example 1, suppose that we were to write $f(x) + g(x)$ as $(x + 4) + (x^2 + 1)$, or $f(x) + g(x) = x^2 + x + 5$. This can be regarded as a "new" function. The notation $(f + g)(x)$ is generally used to indicate the output of a function formed in this manner. Similar notations exist for subtraction, multiplication, and division of functions.

THE ALGEBRA OF FUNCTIONS

If f and g are functions and x is in the domain of both functions, then:

1. $(f + g)(x) = f(x) + g(x)$;
2. $(f - g)(x) = f(x) - g(x)$;
3. $(f \cdot g)(x) = f(x) \cdot g(x)$;
4. $(f/g)(x) = f(x)/g(x)$, provided $g(x) \neq 0$.

EXAMPLE 2   For $f(x) = x^2 - x$ and $g(x) = x + 2$, find the following.

a) $(f + g)(4)$
b) $(f - g)(x)$ and $(f - g)(-1)$
c) $(f/g)(x)$ and $(f/g)(-3)$
d) $(f \cdot g)(4)$

Solution

a) Since $f(4) = 4^2 - 4 = 12$ and $g(4) = 4 + 2 = 6$, we have

$$(f + g)(4) = f(4) + g(4)$$
$$= 12 + 6 \qquad \text{Substituting}$$
$$= 18.$$

Alternatively, we could first find $(f + g)(x)$:

$$(f + g)(x) = f(x) + g(x)$$
$$= x^2 - x + x + 2$$
$$= x^2 + 2. \qquad \text{Collecting like terms}$$

Thus,

$$(f + g)(4) = 4^2 + 2 = 18. \qquad \text{Our results match.}$$

b) We have

$$(f - g)(x) = f(x) - g(x)$$
$$= x^2 - x - (x + 2) \qquad \text{Substituting}$$
$$= x^2 - 2x - 2. \qquad \text{Removing parentheses and collecting like terms}$$

Then,

$$(f - g)(-1) = (-1)^2 - 2(-1) - 2 \qquad \text{Using } (f - g)(x) \text{ is faster than using } f(x) - g(x).$$
$$= 1. \qquad \text{Simplifying}$$

> **CAUTION!** Although $(f + g)(3) = f(3) + g(3)$, it is *not true* that $f(2 + 3)$ is the same as $f(2) + f(3)$. In the expression $f(2 + 3)$, the *inputs* are added; for $(f + g)(3)$, the functions themselves are added.

c) We have

$$(f/g)(x) = f(x)/g(x)$$
$$= \frac{x^2 - x}{x + 2}. \qquad \text{We assume that } x \neq -2.$$

Then,

$$(f/g)(-3) = \frac{(-3)^2 - (-3)}{-3 + 2} \qquad \text{Substituting}$$
$$= \frac{12}{-1} = -12.$$

d) Using our work in part (a), we have

$$(f \cdot g)(4) = f(4) \cdot g(4)$$
$$= 12 \cdot 6$$
$$= 72.$$

Now Try Exercises 15, 17, and 23.

▶ b Determining Domain

To find $(f + g)(a)$, $(f - g)(a)$, or $(f \cdot g)(a)$, we must know that $f(a)$ and $g(a)$ exist. This means that a must be in the domain of both f and g.

EXAMPLE 3 Let

$$f(x) = \frac{5}{x} \quad \text{and} \quad g(x) = \frac{2x - 6}{x + 1}.$$

Find the domain of $f + g$, the domain of $f - g$, and the domain of $f \cdot g$.

Solution Note that because division by 0 is undefined, we have

$$\text{Domain of } f = \{x \mid x \text{ is a real number } and \; x \neq 0\}$$

and

$$\text{Domain of } g = \{x \mid x \text{ is a real number } and \; x \neq -1\}.$$

In order to find $(f + g)(x)$, $(f - g)(x)$, or $(f \cdot g)(x)$, we must know that x is in *both* of the above domains. Thus,

$$\text{Domain of } f + g = \text{Domain of } f - g = \text{Domain of } f \cdot g$$
$$= \{x \mid x \text{ is a real number } and \; x \neq 0 \; and \; x \neq -1\}.$$

> **Now Try Exercise 45.**

Suppose that for $f(x) = x^2 - x$ and $g(x) = x + 2$, we want to find $(f/g)(-2)$. Finding $f(-2)$ and $g(-2)$ poses no problem:

$$f(-2) = 6 \quad \text{and} \quad g(-2) = 0;$$

but then

$$(f/g)(-2) = f(-2)/g(-2)$$
$$= 6/0. \qquad \textcolor{red}{\textbf{Division by 0 is undefined.}}$$

Thus, although -2 is in the domain of both f and g, it is not in the domain of f/g. That is, since $g(x) = 0$ when $x = -2$, the domain of f/g must exclude -2.

To find the domain of the sum, the difference, the product, or the quotient of two functions f and g:

1. Find the domain of f and the domain of g.
2. The functions $f + g$, $f - g$, and $f \cdot g$ have the same domain. It is the intersection of the domains of f and g, or, in other words, the set of all values common to the domains of f and g.
3. Find any values of x for which $g(x) = 0$.
4. The domain of f/g is the set found in step (2) (the set of all values common to the domains of f and g) *excluding* any values of x found in step (3).

EXAMPLE 4 Given

$$f(x) = \frac{1}{x - 3} \quad \text{and} \quad g(x) = 2x - 7,$$

find the domains of $f + g$, $f - g$, $f \cdot g$, and f/g.

Solution We first find the domain of f and the domain of g:

The domain of f is $\{x \mid x \text{ is a real number } and \; x \neq 3\}$.

The domain of g is $\{x \mid x \text{ is a real number}\}$.

The domain of $f + g$, $f - g$, and $f \cdot g$ is the set of all elements common to the domains of f and g. This consists of all real numbers except 3.

$$\text{The domain of } f + g = \text{the domain of } f - g = \text{the domain of } f \cdot g$$
$$= \{x \mid x \text{ is a real number } and \; x \neq 3\}.$$

Because we cannot divide by 0, the domain of f/g must also exclude any values of x for which $g(x)$ is 0. We determine those values by solving $g(x) = 0$:

$$g(x) = 0$$
$$2x - 7 = 0 \quad \text{Replacing } g(x) \text{ with } 2x - 7$$
$$2x = 7$$
$$x = \tfrac{7}{2}.$$

The domain of f/g is the domain of the sum, the difference, and the product of f and g, found above, excluding $\tfrac{7}{2}$.

The domain of $f/g = \{x \mid x \text{ is a real number } and \ x \neq 3 \ and \ x \neq \tfrac{7}{2}\}$.

Now Try Exercise 51.

2.4   Exercise Set

▶ Reading Check

Given that $f(x) = x^2 - 1$ *and* $g(x) = x + 3$, *match each expression with an equivalent expression from the column on the right.*

RC1. $(f + g)(x)$

RC2. $(f - g)(x)$

RC3. $(f \cdot g)(x)$

RC4. $(g - f)(x)$

RC5. $(g \cdot g)(x)$

a) $x^2 + 6x + 9$

b) $x^3 + 3x^2 - x - 3$

c) $x^2 - x - 4$

d) $x^2 + x + 2$

e) $-x^2 + x + 4$

---

a *Let* $f(x) = -2x + 3$ *and* $g(x) = x^2 - 5$. *Find each of the following.*

1. $f(3) + g(3)$

2. $f(4) + g(4)$

3. $f(1) - g(1)$

4. $f(2) - g(2)$

5. $f(-2) \cdot g(-2)$

6. $f(-1) \cdot g(-1)$

7. $f(-4)/g(-4)$

8. $f(3)/g(3)$

9. $g(1) - f(1)$

10. $g(-3)/f(-3)$

11. $(f + g)(x)$

12. $(f - g)(x)$

13. $(g - f)(x)$

14. $(g/f)(x)$

Let $F(x) = x^2 - 2$ *and* $G(x) = 5 - x$. *Find each of the following.*

15. $(F + G)(x)$

16. $(F + G)(a)$

17. $(F - G)(3)$

18. $(F - G)(2)$

19. $(F \cdot G)(a)$

20. $(G \cdot F)(x)$

21. $(F/G)(x)$

22. $(G - F)(x)$

23. $(G/F)(-2)$

24. $(F/G)(-1)$

25. $(F + F)(1)$

26. $(G \cdot G)(6)$

Let

$$r(x) = \frac{5}{x^2} \quad \text{and} \quad t(x) = \frac{3}{2x}.$$

Find each of the following.

27. $(r \cdot t)(x)$

28. $(r/t)(x)$

29. $(r - t)(x)$

30. $(r + t)(x)$

31. $(t/r)(x)$

32. $(t - r)(x)$

The following graph shows the number of births in the United States, in millions, from 1970–2010. Here, $C(t)$ represents the number of Caesarean section births, $B(t)$ the number of non-Caesarean section births, and $N(t)$ the total number of births in year t.

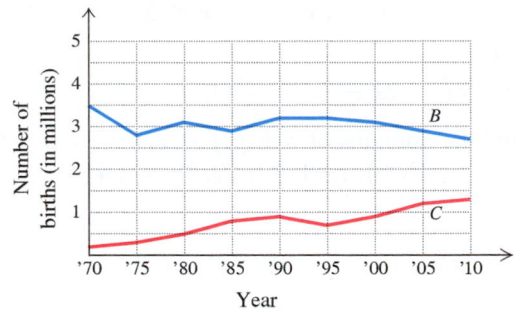

Source: National Center for Health Statistics

33. Use estimates of $C(2005)$ and $B(2005)$ to estimate $N(2005)$.

34. Use estimates of $C(1985)$ and $B(1985)$ to estimate $N(1985)$.

In the following graph, $S(t)$ represents the number of gallons of carbonated soft drinks, $M(t)$ the number of gallons of milk, $J(t)$ the number of gallons of fruit juice, and $W(t)$ the number of gallons of bottled water consumed by the average American in year t.

Beverage consumption

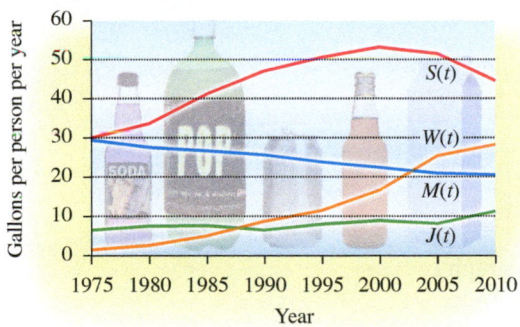

Source: Economic Research Service, U.S. Department of Agriculture

35. Use estimates of $S(2010)$ and $W(2010)$ to estimate $(S - W)(2010)$.

36. Use estimates of $M(2010)$ and $J(2010)$ to estimate $(M - J)(2010)$.

b *For each pair of functions f and g, determine the domain of the sum, the difference, and the product of the two functions.*

37. $f(x) = x^2$,
$g(x) = 7x - 4$

38. $f(x) = 5x - 1$,
$g(x) = 2x^2$

39. $f(x) = \dfrac{1}{x + 5}$,
$g(x) = 4x^3$

40. $f(x) = 3x^2$,
$g(x) = \dfrac{1}{x - 9}$

41. $f(x) = \dfrac{2}{x}$,
$g(x) = x^2 - 4$

42. $f(x) = x^3 + 1$,
$g(x) = \dfrac{5}{x}$

43. $f(x) = x + \dfrac{2}{x - 1}$,
$g(x) = 3x^3$

44. $f(x) = 9 - x^2$,
$g(x) = \dfrac{3}{x + 6} + 2x$

45. $f(x) = \dfrac{3}{2x + 9}$,
$g(x) = \dfrac{5}{1 - x}$

46. $f(x) = \dfrac{5}{3 - x}$,
$g(x) = \dfrac{1}{4x - 1}$

For each pair of functions f and g, determine the domain of f/g.

47. $f(x) = x^4$,
$g(x) = x - 3$

48. $f(x) = 2x^3$,
$g(x) = 5 - x$

49. $f(x) = 3x - 2$,
$g(x) = 2x + 8$

50. $f(x) = 5 + x$,
$g(x) = 6 - 2x$

51. $f(x) = \dfrac{3}{x - 4}$,
$g(x) = 5 - x$

52. $f(x) = \dfrac{1}{2 - x}$,
$g(x) = 7 + x$

53. $f(x) = \dfrac{2x}{x + 1}$,
$g(x) = 2x + 5$

54. $f(x) = \dfrac{7x}{x - 2}$,
$g(x) = 3x + 7$

For Exercises 55–60, consider the functions F and G as shown.

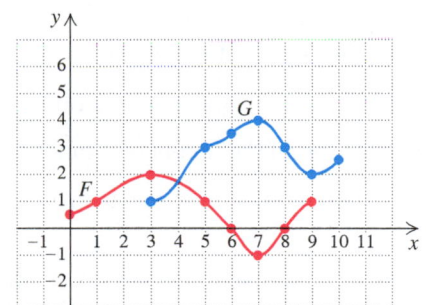

55. Determine $(F + G)(5)$ and $(F + G)(7)$.

56. Determine $(F \cdot G)(6)$ and $(F \cdot G)(9)$.

57. Determine $(G - F)(7)$ and $(G - F)(3)$.

58. Determine $(F/G)(3)$ and $(F/G)(7)$.

59. Find the domains of F, G, $F + G$, and F/G.

60. Find the domains of $F - G$, $F \cdot G$, and G/F.

▶ # Skill Maintenance

Solve. [1.3a]

61. One angle of a triangle measures twice the second angle. The third angle measures three times the second angle. Find the measures of the angles.

62. In one basketball game, Terrence scored 5 fewer points than Isaiah. Together, they scored 27 points. How many points did Terrence score?

Solve. [1.1d]

63. $7x + 16 = 5x - 20$

64. $2y - 11 = -11 - 3y - y$

▶ # Synthesis

65. Find the domain of F/G, if

$$F(x) = \frac{1}{x-4} \quad \text{and} \quad G(x) = \frac{x^2 - 4}{x - 3}.$$

66. Find the domain of f/g, if

$$f(x) = \frac{3x}{2x + 5} \quad \text{and} \quad g(x) = \frac{x^4 - 1}{3x + 9}.$$

67. Find the domain of m/n, if

$$m(x) = 3x \quad \text{for } -1 < x < 5$$

and

$$n(x) = 2x - 3.$$

68. Find the domains of $f + g, f - g, f \cdot g$, and f/g, if

$$f = \{(-2, 1), (-1, 2), (0, 3), (1, 4), (2, 5)\}$$

and

$$g = \{(-4, 4), (-3, 3), (-2, 4), (-1, 0), (0, 5), (1, 6)\}.$$

Mid-Chapter Review

Determine whether each statement is true or false.

1. Every function is a relation. [2.2a]

2. It is possible for one input of a function to have two or more outputs. [2.2a]

3. It is possible for all the inputs of a function to have the same output. [2.2a]

4. If it is possible for a vertical line to cross a graph more than once, the graph is not the graph of a function. [2.2d]

5. If the domain of a function is the set of real numbers, then the range is the set of real numbers. [2.3a]

Determine whether the given point is a solution of the equation. [2.1b]

6. $(-2, -1); \ 5y + 6 = 4x$

7. $\left(\frac{1}{2}, 0\right); \ 8a = 4 - b$

Determine whether the correspondence is a function. [2.2a]

8. *Domain*   *Range*

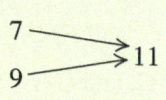

9. *Domain*   *Range*

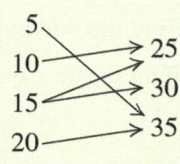

10. Find the domain and the range. [2.3a]

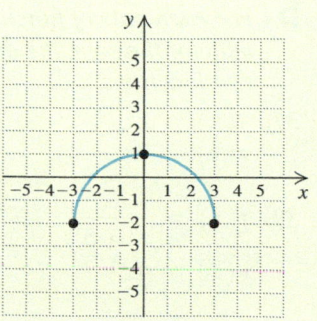

Find the function value. **[2.2b]**

11. $g(x) = 2 + x;\quad g(-5)$

12. $f(x) = x - 7;\quad f(0)$

13. $h(x) = 8;\quad h\left(\frac{1}{2}\right)$

14. $f(x) = 3x^2 - x + 5;\quad f(-1)$

15. $g(p) = p^4 - p^3;\quad g(10)$

16. $f(t) = \frac{1}{2}t + 3;\quad f(-6)$

Determine whether each of the following is the graph of a function. **[2.2d]**

17.

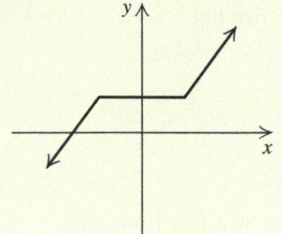

18.

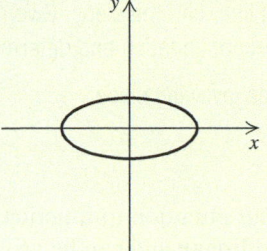

19.

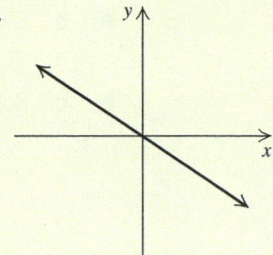

Find the domain. **[2.3a]**

20. $g(x) = \dfrac{3}{12 - 3x}$

21. $f(x) = x^2 - 10x + 3$

22. $h(x) = \dfrac{x - 2}{x + 2}$

23. $f(x) = |x - 4|$

Graph. **[2.1c], [2.2c]**

24. $g(x) = -\dfrac{2}{3}x - 2$

25. $f(x) = x - 1$

26. $h(x) = 2x + \dfrac{1}{2}$

27. $g(x) = |x| - 3$

28. $f(x) = 1 + x^2$

29. $f(x) = -\dfrac{1}{4}x$

Let $f(x) = 3x - 1$ and $g(x) = x^2 + 2$. Find each of the following. **[2.4a]**

30. $(f - g)(x)$

31. $f(-2) \cdot g(-2)$

32. $(g/f)(a)$

For each pair of functions f and g, determine the domains of $f + g, f - g, f \cdot g$, and f/g. **[2.4b]**

33. $f(x) = 5x^2,$
$g(x) = x + 4$

34. $f(x) = \dfrac{7}{x - 9},$
$g(x) = 6 - x$

Collaborative Discussion and Writing

35. Is it possible for a function to have more numbers as outputs than as inputs? Why or why not? **[2.2a]**

36. Without making a drawing, how can you tell that the graph of $y = x - 30$ passes through three quadrants? **[2.1c]**

37. For a given function f, it is known that $f(2) = -3$. Give as many interpretations of this fact as you can. **[2.2b], [2.3a]**

38. Explain the difference between the domain and the range of a function. **[2.3a]**

2.5 Linear Functions: Graphs and Slope

▶ **a** Find the *y*-intercept of a line from the equation $y = mx + b$ or $f(x) = mx + b$.

▶ **b** Given two points on a line, find the slope. Given a linear equation, derive the equivalent slope–intercept equation and determine the slope and the *y*-intercept.

▶ **c** Solve applied problems involving slope.

We now turn our attention to functions whose graphs are straight lines. Such functions are called **linear** and can be written in the form $f(x) = mx + b$.

LINEAR FUNCTION

A **linear function** f is any function that can be described by $f(x) = mx + b$.

Compare the two equations $7y + 2x = 11$ and $y = 3x + 5$. Both are linear equations because their graphs are straight lines. Each can be expressed in the form $f(x) = mx + b$.

The equation $y = 3x + 5$ can be expressed as $f(x) = mx + b$, where $m = 3$ and $b = 5$.

The equation $7y + 2x = 11$ also has an equivalent form $f(x) = mx + b$. To see this, we solve for y:

$$7y + 2x = 11$$
$$7y + 2x - 2x = -2x + 11 \qquad \text{Subtracting } 2x$$
$$7y = -2x + 11$$
$$\frac{7y}{7} = \frac{-2x + 11}{7} \qquad \text{Dividing by } 7$$
$$y = -\frac{2}{7}x + \frac{11}{7}. \qquad \text{Simplifying}$$

We now have an equivalent function in the form $f(x) = mx + b$:

$$f(x) = -\frac{2}{7}x + \frac{11}{7}, \qquad \text{where} \quad m = -\frac{2}{7} \quad \text{and} \quad b = \frac{11}{7}.$$

In this section, we consider the effects of the constants m and b on the graphs of linear functions.

▶ a The Constant *b*: The *y*-Intercept

Let's first explore the effect of the constant b.

EXAMPLE 1 Graph $y = 2x$ and $y = 2x + 3$ using the same set of axes. Then compare the graphs.

We first make a table of solutions of both equations. Next, we plot these points. Drawing a red line for $y = 2x$ and a blue line for $y = 2x + 3$, we note that the graph of $y = 2x + 3$ is simply the graph of $y = 2x$ shifted, or *translated*, up 3 units. The lines are parallel.

x	y $y = 2x$	y $y = 2x + 3$
0	0	3
1	2	5
−1	−2	1
2	4	7
−2	−4	−1

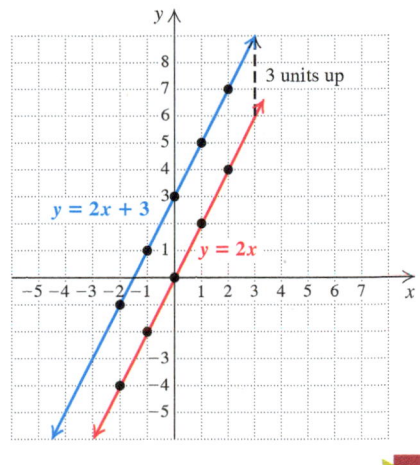

Technology Connection

Exploring b

We can use a graphing calculator to explore the effect of the constant b on the graph of a function of the form $f(x) = mx + b$. Graph $y_1 = x$ in the standard $[-10, 10, -10, 10]$ viewing window. Then graph $y_2 = x + 4$, followed by $y_3 = x - 3$, in the same viewing window.

Exercises

1. Compare the graph of y_2 with the graph of y_1.

2. Compare the graph of y_3 with the graph of y_1.

EXAMPLE 2 Graph $f(x) = \frac{1}{3}x$ and $g(x) = \frac{1}{3}x - 2$ using the same set of axes. Then compare the graphs.

We first make a table of solutions of both equations. By choosing multiples of 3, we can avoid fractions.

x	f(x) $f(x) = \frac{1}{3}x$	g(x) $g(x) = \frac{1}{3}x - 2$
0	0	−2
3	1	−1
−3	−1	−3
6	2	0

We then plot these points. Drawing a red line for $f(x) = \frac{1}{3}x$ and a blue line for $g(x) = \frac{1}{3}x - 2$, we see that the graph of $g(x) = \frac{1}{3}x - 2$ is simply the graph of $f(x) = \frac{1}{3}x$ shifted, or translated, down 2 units. The lines are parallel.

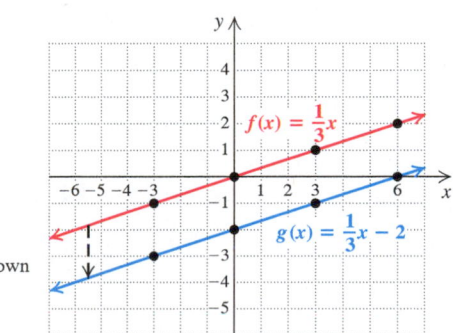

In Example 1, we saw that the graph of $y = 2x + 3$ is parallel to the graph of $y = 2x$ and that it passes through the point $(0, 3)$. Similarly, in Example 2, we saw that the graph of $y = \frac{1}{3}x - 2$ is parallel to the graph of $y = \frac{1}{3}x$ and that it passes through the point $(0, -2)$. In general, the graph of $y = mx + b$ is a line parallel to $y = mx$, passing through the point $(0, b)$. The point $(0, b)$ is called the **y-intercept** because it is the point at which the graph crosses the y-axis. Often it is convenient to refer simply to the number b as the y-intercept. The constant b has the effect of shifting the graph of $y = mx$ up or down $|b|$ units to obtain the graph of $y = mx + b$.

---

y-INTERCEPT

The y-intercept of the graph of
$f(x) = mx + b$ is the point $(0, b)$
or, simply, b.

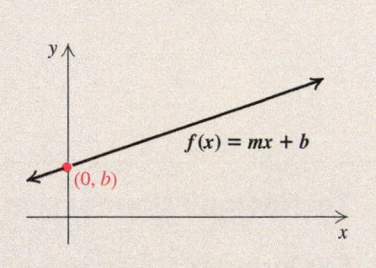

EXAMPLE 3   Find the y-intercept: $y = -5x + 4$.

$$y = -5x + 4 \qquad (0, 4) \text{ is the } y\text{-intercept.}$$

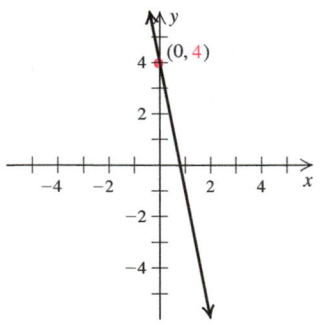

EXAMPLE 4   Find the y-intercept: $f(x) = 6.3x - 7.8$.

$$f(x) = 6.3x - 7.8 \qquad (0, -7.8) \text{ is the } y\text{-intercept.}$$

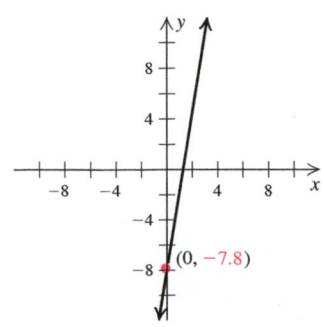

▶ **b   The Constant *m*: Slope**

Look again at the graphs in Examples 1 and 2. Note that the slant of each red line seems to match the slant of each blue line. This leads us to believe that the number *m* in the equation $y = mx + b$ is related to the slant of the line. Let's consider some examples.

Graphs with *m* < 0:

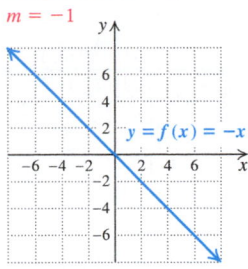

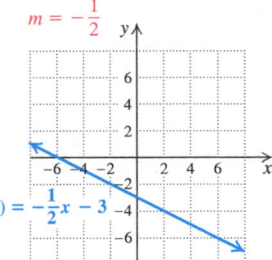

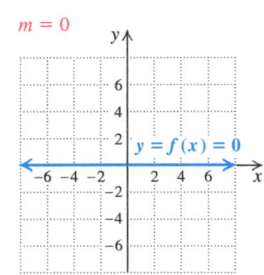

Graphs with *m* = 0:

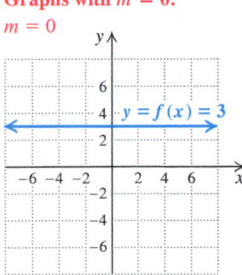

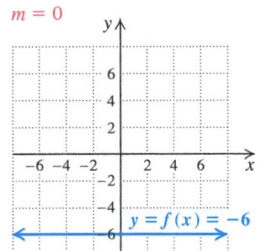

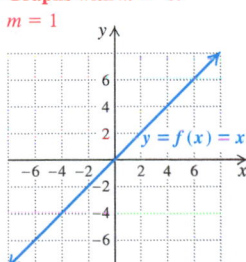

Graphs with *m* > 0:

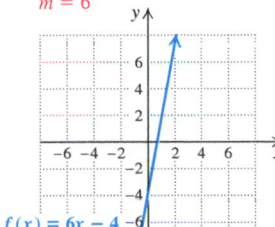

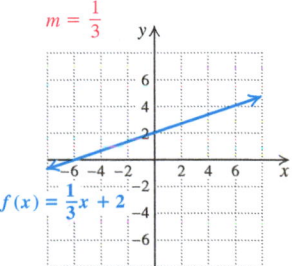

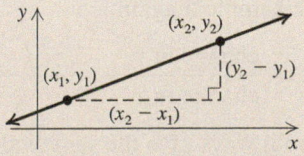

Note that

$m < 0 \longrightarrow$ The graph slants down from left to right;

$m = 0 \longrightarrow$ the graph is horizontal; and

$m > 0 \longrightarrow$ the graph slants up from left to right.

The following definition enables us to visualize the slant and attach a number, a geometric ratio, or *slope*, to the line.

SLOPE

The **slope** of a line containing points (x_1, y_1) and (x_2, y_2) is given by

$$m = \frac{\text{rise}}{\text{run}}$$

$$= \frac{\text{change in } y}{\text{change in } x} = \frac{y_2 - y_1}{x_2 - x_1} = \frac{y_1 - y_2}{x_1 - x_2}.$$

Consider a line with two points marked P_1 and P_2, as follows. As we move from P_1 to P_2, the y-coordinate changes from 1 to 3 and the x-coordinate changes from 2 to 7. The change in y is $3 - 1$, or 2. The change in x is $7 - 2$, or 5.

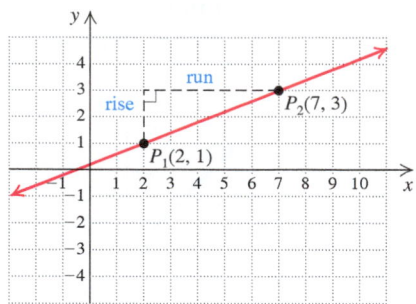

We call the change in y the **rise** and the change in x the **run**. The ratio rise/run is the same for any two points on a line. We call this ratio the **slope**. Slope describes the slant of a line. The slope of the line in the graph above is given by

$$\frac{\text{rise}}{\text{run}}, \quad \text{or} \quad \frac{\text{change in } y}{\text{change in } x}, \quad \text{or} \quad \frac{2}{5}.$$

Whenever x increases by 5 units, y increases by 2 units. Equivalently, whenever x increases by 1 unit, y increases by $\frac{2}{5}$ unit.

EXAMPLE 5 Graph the line containing the points $(-4, 3)$ and $(2, -5)$ and find the slope.

The graph is shown below. Going from $(-4, 3)$ to $(2, -5)$, we see that the change in y, or the rise, is $-5 - 3$, or -8. The change in x, or the run, is $2 - (-4)$, or 6.

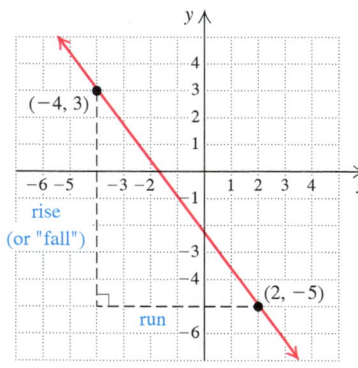

$$\text{Slope} = \frac{\text{rise}}{\text{run}} = \frac{\text{change in } y}{\text{change in } x}$$

$$= \frac{-5 - 3}{2 - (-4)}$$

$$= \frac{-8}{6} = -\frac{8}{6}, \text{ or } -\frac{4}{3}$$

▶ **Now Try Exercise 21.**

The formula

$$m = \frac{y_2 - y_1}{x_2 - x_1} = \frac{y_1 - y_2}{x_1 - x_2}$$

tells us that we can subtract in two ways. We must remember, however, to subtract the x-coordinates in the same order that we subtract the y-coordinates.

Let's do Example 5 again:

$$\text{Slope} = \frac{\text{change in } y}{\text{change in } x} = \frac{3 - (-5)}{-4 - 2} = \frac{8}{-6} = -\frac{8}{6} = -\frac{4}{3}.$$

We see that both ways give the same value, $-\frac{4}{3}$, for the slope.

The slope of a line tells how it slants. A line with positive slope slants up from left to right. The larger the positive number, the steeper the slant. A line with negative slope slants downward from left to right. The smaller the negative number, the steeper the line.

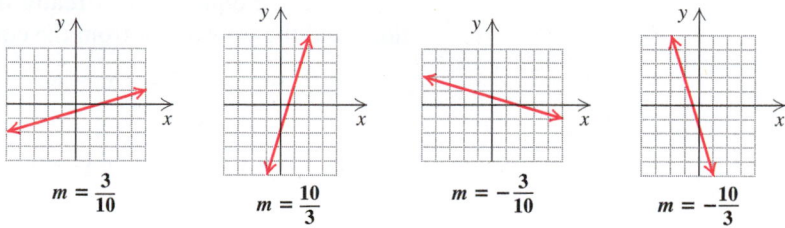

$$m = \frac{3}{10} \qquad m = \frac{10}{3} \qquad m = -\frac{3}{10} \qquad m = -\frac{10}{3}$$

How can we find the slope from a given equation? Let's consider the equation $y = 2x + 3$, which is in the form $y = mx + b$. We can find two points by choosing convenient values for x—say, 0 and 1—and substituting to find the corresponding y-values.

If $x = 0, y = 2 \cdot 0 + 3 = 3$.

If $x = 1, y = 2 \cdot 1 + 3 = 5$.

We find two points on the line to be

$$(0, 3) \quad \text{and} \quad (1, 5).$$

The slope of the line is found as follows, using the definition of slope:

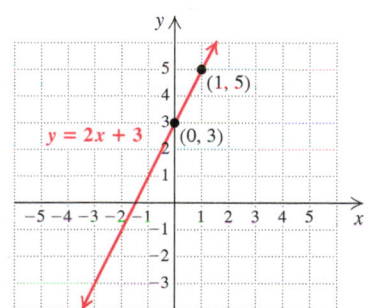

$$m = \frac{\text{change in } y}{\text{change in } x}$$

$$= \frac{5 - 3}{1 - 0} = \frac{2}{1} = 2.$$

The slope is 2. Note that this is the coefficient of the x-term in the equation $y = 2x + 3$.

If we had chosen different points on the line—say, $(-2, -1)$ and $(4, 11)$—the slope would still be 2, as we see in the following calculation:

$$m = \frac{11 - (-1)}{4 - (-2)} = \frac{11 + 1}{4 + 2} = \frac{12}{6} = 2.$$

We see that the slope of the line $y = mx + b$ is indeed the constant m, the coefficient of x.

SLOPE

The **slope** of the line $y = mx + b$ is m.

From a linear equation in the form $y = mx + b$, we can read the slope and the y-intercept of the graph directly.

SLOPE–INTERCEPT EQUATION

The equation $y = mx + b$ is called the **slope–intercept equation**. The slope is m and the y-intercept is $(0, b)$.

Note that any graph of an equation $y = mx + b$ passes the vertical-line test and thus represents a function.

EXAMPLE 6   Find the slope and the y-intercept of $y = 5x - 4$.

Since the equation is already in the form $y = mx + b$, we simply read the slope and the y-intercept from the equation:

$$y = 5x - 4.$$

The slope is 5.   The y-intercept is $(0, -4)$.

> **Now Try Exercise 1.**

EXAMPLE 7   Find the slope and the y-intercept of $2x + 3y = 8$.

We first solve for y so we can easily read the slope and the y-intercept:

$$2x + 3y = 8$$
$$3y = -2x + 8 \qquad \text{Subtracting } 2x$$
$$\frac{3y}{3} = \frac{-2x + 8}{3} \qquad \text{Dividing by 3}$$
$$y = -\frac{2}{3}x + \frac{8}{3}. \qquad \text{Finding the form } y = mx + b$$

The slope is $-\frac{2}{3}$.   The y-intercept is $(0, \frac{8}{3})$. ────▶ **Now Try Exercise 9.**

▶ c   Applications

Slope has many real-world applications. For example, numbers like 2%, 3%, and 6% are often used to represent the *grade* of a road, a measure of how steep a road on a hill or a mountain is. A 3% grade $\left(3\% = \frac{3}{100}\right)$ means that for every horizontal distance of 100 ft that the road runs, the road rises 3 ft, and a -3% grade means that for every horizontal distance of 100 ft, the road drops 3 ft. (Normally, the road-grade signs do not include negative signs, since it is obvious whether you are climbing or descending.)

Road grade = $\frac{a}{b}$
(expressed as a percent)

An athlete might change the grade of a treadmill during a workout. An escape ramp on an airliner might have a slope of about −0.6.

Architects and carpenters use slope when designing and building stairs, ramps, or roof pitches. Another application occurs in hydrology. The strength or force of a river depends on how far the river falls vertically compared to how far it flows horizontally. Slope can also be considered as a **rate of change**.

EXAMPLE 8   *Student Debt.*   The average educational debt per college student at his or her graduation has steadily increased. In 1993, the average debt was $14,500 (in 2011 dollars). By 2011, this amount had increased to $26,600. Find the rate of change in the average student debt with respect to time, in years. (*Sources*: Data from Higher Education Research Institute, UCLA; Sallie Mae; NCES; FinAid; the College Board; McKinsey Global Institute; *Time*, October 29, 2012)

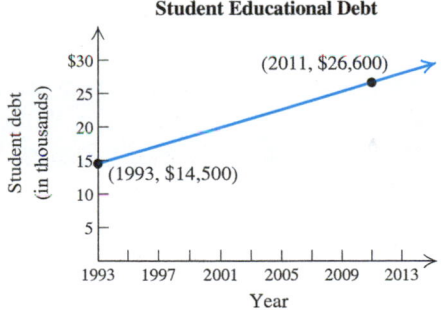

The rate of change with respect to time, in years, is given by

$$\text{Rate of change} = \frac{\$26{,}600 - \$14{,}500}{2011 - 1993}$$

$$= \frac{\$12{,}100}{18 \text{ years}}$$

$$\approx \$672 \text{ per year.}$$

The average student debt at graduation is increasing at a rate of about $672 per year.

Now Try Exercise 31.

EXAMPLE 9 *Volume of Mail.* The volume of first-class mail through the U.S. Postal Service has been decreasing since 2005. Find the rate of change of the volume of first-class mail with respect to time, in years.

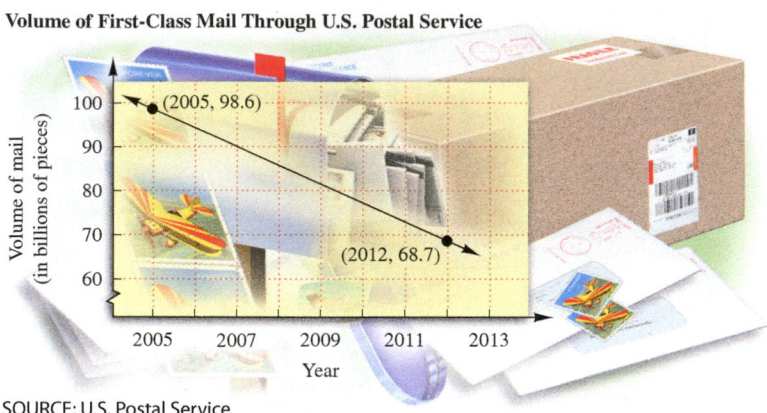

Volume of First-Class Mail Through U.S. Postal Service

SOURCE: U.S. Postal Service

Since the graph is linear, we can use any pair of points to determine the rate of change:

$$\text{Rate of change} = \frac{68.7 \text{ billion} - 98.6 \text{ billion}}{2012 - 2005} = \frac{-29.9 \text{ billion}}{7 \text{ years}} \approx -4.27 \text{ billion per year.}$$

The volume of first-class mail through the U.S. Postal Service is decreasing at a rate of about 4.27 billion pieces per year.

Now Try Exercise 33.

2.5 Exercise Set

▶ Reading Check

Choose from the column on the right the slope of each line.

RC1.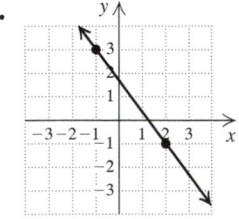

RC2.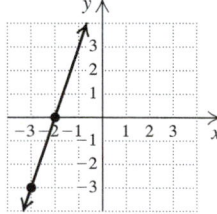

RC3.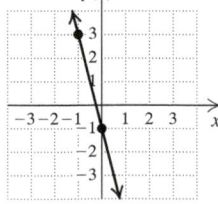

RC4.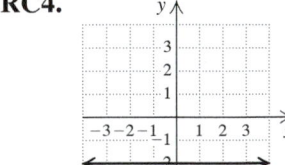

RC5.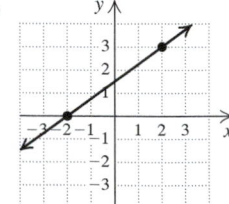

RC6.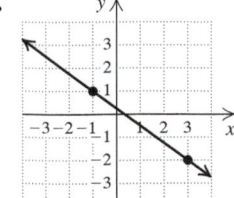

a) $-\dfrac{3}{4}$

b) 3

c) 0

d) -4

e) $\dfrac{3}{4}$

f) $-\dfrac{4}{3}$

a, b *Find the slope and the y-intercept of each equation.*

1. $y = 4x + 5$

2. $y = -5x + 10$

3. $f(x) = -2x - 6$

4. $g(x) = -5x + 7$

5. $y = -\frac{3}{8}x - \frac{1}{5}$

6. $y = \frac{15}{7}x + \frac{16}{5}$

7. $g(x) = 0.5x - 9$

8. $f(x) = -3.1x + 5$

9. $2x - 3y = 8$

10. $-8x - 7y = 24$

11. $9x = 3y + 6$

12. $9y + 36 - 4x = 0$

13. $3 - \frac{1}{4}y = 2x$

14. $5x = \frac{2}{3}y - 10$

15. $17y + 4x + 3 = 7 + 4x$

16. $3y - 2x = 5 + 9y - 2x$

b *Find the slope of each line.*

17.

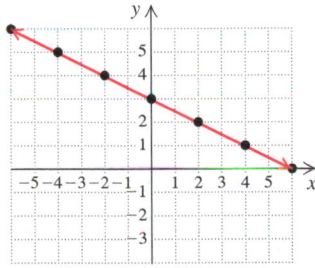

18.

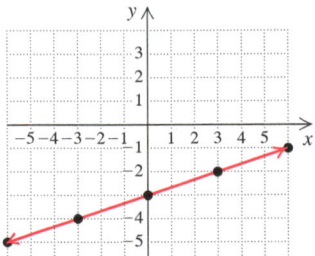

19.

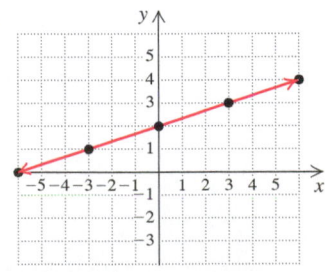

20.

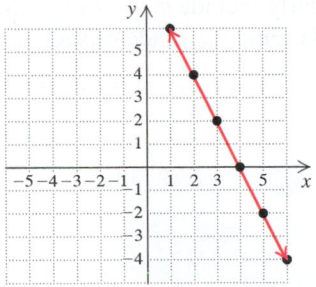

Find the slope of the line containing the given pair of points.

21. $(6, 9)$ and $(4, 5)$

22. $(8, 7)$ and $(2, -1)$

23. $(9, -4)$ and $(3, -8)$

24. $(17, -12)$ and $(-9, -15)$

25. $(-16.3, 12.4)$ and $(-5.2, 8.7)$

26. $(14.4, -7.8)$ and $(-12.5, -17.6)$

C *Find the slope (or rate of change).*

27. Find the slope (or grade) of the treadmill.

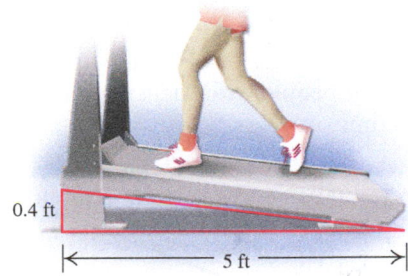

0.4 ft

5 ft

28. Find the slope (or head) of the river.

43.33 ft

1238 ft

29. Find the slope (or pitch) of the roof.

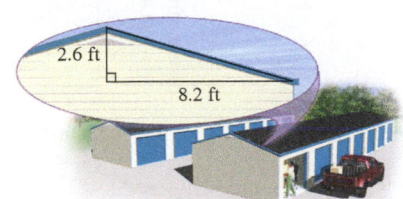

2.6 ft

8.2 ft

30. Public buildings regularly include steps with 7-in. risers and 11-in. treads. Find the grade of such a stairway.

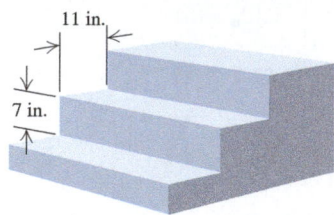

31. *Luxury Purchases.* Find the rate of change in luxury purchases in China with respect to time, in years.

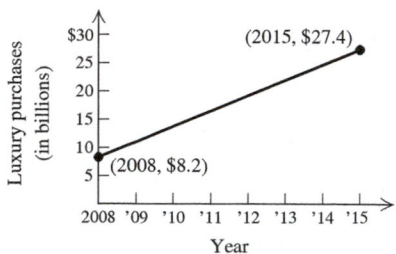

*Estimated values for 2010–2015
SOURCE: McKinsey Insights, China-Wealthy Consumer Studies (2008, 2010)

32. *People with Alzheimer's.* Find the rate of change in the number of people with Alzheimer's disease with respect to time, in years.

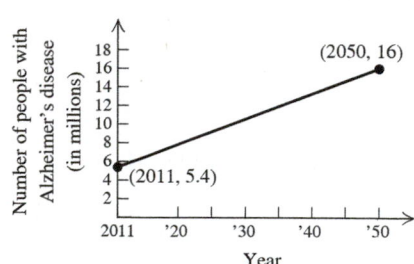

*Estimated values for 2012–2050
SOURCE: Alzheimer's Association

Find the rate of change.

33.

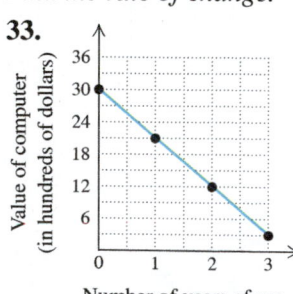

34.

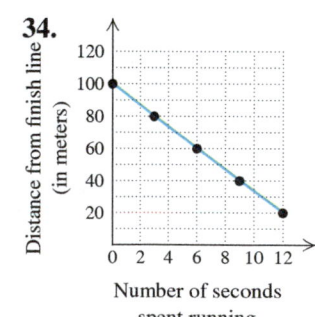

35.

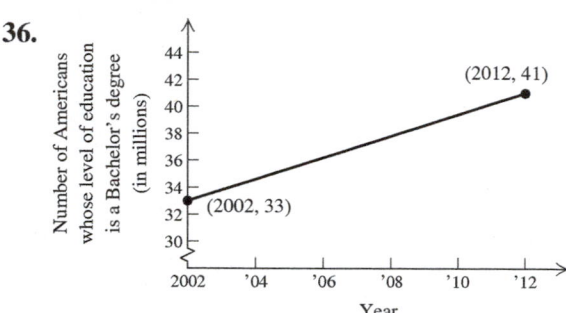

SOURCES: FDIC Consumer News Winter 2004/2005; *USA TODAY*, April 13, 2012; www.greenbaypressgazette.com, January 4, 2013

36.

Number of Americans whose level of education is a Bachelor's degree (in millions)

(2012, 41)

(2002, 33)

2002   '04   '06   '08   '10   '12
Year

SOURCE: U.S. Census Bureau

▶ **Skill Maintenance**

Simplify. **[R.3c], [R.6b]**

37. $3^2 - 24 \cdot 56 + 144 \div 12$

38. $9\{2x - 3[5x + 2(-3x + y^0 - 2)]\}$

39. $10\{2x + 3[5x - 2(-3x + y^1 - 2)]\}$

40. $5^4 \div 625 \div 5^2 \cdot 5^7 \div 5^3$

Solve. **[1.3a]**

41. One side of a square is 5 yd less than a side of an equilateral triangle. If the perimeter of the square is the same as the perimeter of the triangle, what is the length of a side of the square? of the triangle?

Solve. **[1.6c, e]**

42. $|5x - 8| \geq 32$

43. $|5x - 8| < 32$

44. $|5x - 8| = 32$

45. $|5x - 8| = -32$

2.6 More on Graphing Linear Equations

▶ **a** Graph linear equations using intercepts.

▶ **b** Given a linear equation in slope–intercept form, use the slope and the *y*-intercept to graph the line.

▶ **c** Graph linear equations of the form $x = a$ or $y = b$.

▶ **d** Given the equations of two lines, determine whether their graphs are parallel or whether they are perpendicular.

▶ a Graphing Using Intercepts

The **x-intercept** of the graph of a linear equation or function is the point at which the graph crosses the *x*-axis. The **y-intercept** is the point at which the graph crosses the *y*-axis. We know from geometry that only one line can be drawn through two given points. Thus, if we know the intercepts, we can graph the line. To ensure that a computation error has not been made, it is a good idea to calculate a third point as a check.

Many equations of the type $Ax + By = C$ can be graphed conveniently using intercepts.

---

x- AND y-INTERCEPTS

A **y-intercept** is a point $(0, b)$. To find b, let $x = 0$ and solve for y.

An **x-intercept** is a point $(a, 0)$. To find a, let $y = 0$ and solve for x.

EXAMPLE 1 Find the intercepts of $3x + 2y = 12$ and then graph the line.

y-intercept: To find the *y*-intercept, we let $x = 0$ and solve for y:

$$3x + 2y = 12$$
$$3 \cdot 0 + 2y = 12 \qquad \text{Substituting 0 for } x$$
$$2y = 12$$
$$y = 6.$$

The *y*-intercept is $(0, 6)$.

x-intercept: To find the *x*-intercept, we let $y = 0$ and solve for x:

$$3x + 2y = 12$$
$$3x + 2 \cdot 0 = 12 \qquad \text{Substituting 0 for } y$$
$$3x = 12$$
$$x = 4.$$

The *x*-intercept is $(4, 0)$.

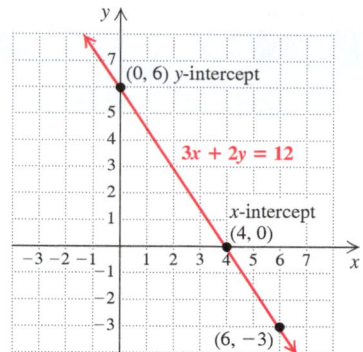

We plot these points and draw the line, using a third point as a check. We choose $x = 6$ and solve for y:

$$3(6) + 2y = 12$$
$$18 + 2y = 12$$
$$2y = -6$$
$$y = -3.$$

We plot $(6, -3)$ and note that it is on the line so the graph is probably correct.

➤ **Now Try Exercise 5.**

When both the x-intercept and the y-intercept are $(0, 0)$, as is the case with an equation such as $y = 2x$, whose graph passes through the origin, another point will have to be calculated and a third point used as a check.

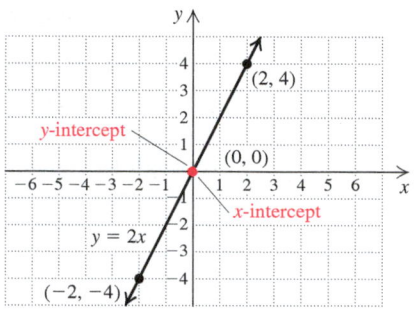

Technology Connection

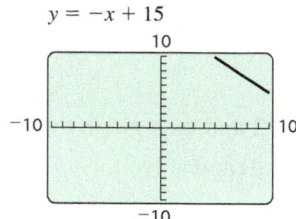

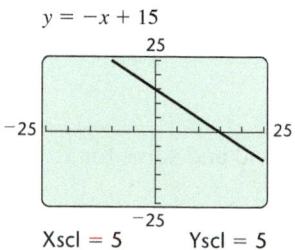

Xscl = 5   Yscl = 5

Viewing the Intercepts

Knowing the intercepts of a linear equation helps us determine a good viewing window for the graph of the equation. For example, when we graph the equation $y = -x + 15$ in the standard window, we see only a small portion of the graph in the upper right-hand corner of the screen, as shown at left above.

Using algebra, as we did in Example 1, we can find that the intercepts of the graph of this equation are $(0, 15)$ and $(15, 0)$. This tells us that, if we are to see a portion of the graph that includes the intercepts, both Xmax and Ymax should be greater than 15. We can try different window settings until we find one that suits us. One good choice, shown at left below, is $[-25, 25, -25, 25]$, with Xscl = 5 and Yscl = 5.

Exercises

Find the intercepts of the equation algebraically. Then graph the equation on a graphing calculator, choosing window settings that allow the intercepts to be seen clearly. (Settings may vary.)

1. $y = -3.2x - 16$
2. $y - 4.25x = 85$
3. $6x + 5y = 90$
4. $5x - 6y = 30$
5. $8x + 3y = 9$
6. $y = 0.4x - 5$
7. $y = 1.2x - 12$
8. $4x - 5y = 2$

▶ **b   Graphing Using the Slope and the y-Intercept**

We can also graph a line using its slope and *y*-intercept.

EXAMPLE 2   Graph: $y = -\frac{2}{3}x + 1$.

This equation is in slope–intercept form, $y = mx + b$. The *y*-intercept is $(0, 1)$. We plot $(0, 1)$. We can think of the slope $\left(m = -\frac{2}{3}\right)$ as $\frac{-2}{3}$.

$$m = \frac{\text{Rise}}{\text{Run}} = \frac{-2}{3} \qquad \begin{array}{l} \textcolor{red}{\textbf{Move 2 units down.}} \\ \textcolor{red}{\textbf{Move 3 units right.}} \end{array}$$

Starting at the *y*-intercept and using the slope, we find another point by moving 2 units down (since the numerator is *negative* and corresponds to the change in *y*) and 3 units to the right (since the denominator is *positive* and corresponds to the change in *x*). We get to a new point, $(3, -1)$. In a similar manner, we can move from the point $(3, -1)$ to find another point, $(6, -3)$.

We could also think of the slope $\left(m = -\frac{2}{3}\right)$ as $\frac{2}{-3}$.

$$m = \frac{\text{Rise}}{\text{Run}} = \frac{2}{-3} \qquad \begin{array}{l} \textcolor{red}{\textbf{Move 2 units up.}} \\ \textcolor{red}{\textbf{Move 3 units left.}} \end{array}$$

Then we can start again at $(0, 1)$, but this time we move 2 units up (since the numerator is *positive* and corresponds to the change in *y*) and 3 units to the left (since the denominator is *negative* and corresponds to the change in *x*). We get another point on the graph, $(-3, 3)$, and from it we can obtain $(-6, 5)$ and others in a similar manner. We plot the points and draw the line.

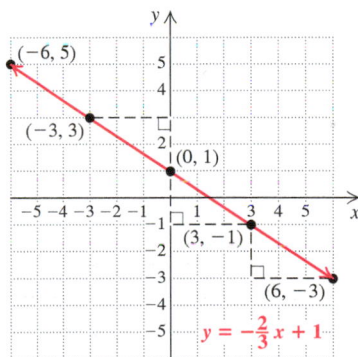

▶ **Now Try Exercise 19.**

EXAMPLE 3   Graph: $f(x) = \frac{2}{5}x + 4$.

First, we plot the *y*-intercept, $(0, 4)$. We then consider the slope $\frac{2}{5}$. A slope of $\frac{2}{5}$ tells us that, for every 2 units that the graph rises, it runs 5 units horizontally in the positive direction, or to the right. Thus, starting at the *y*-intercept and using the slope, we find another point by moving 2 units up (since the numerator is *positive* and corresponds to the change in *y*) and 5 units to the right (since the denominator is *positive* and corresponds to the change in *x*). We get to a new point, $(5, 6)$.

We can also think of the slope $\frac{2}{5}$ as $\frac{-2}{-5}$. A slope of $\frac{-2}{-5}$ tells us that, for every 2 units that the graph drops, it runs 5 units horizontally in the negative direction, or to the left. We again start at the *y*-intercept, $(0, 4)$. We move 2 units down (since the numerator is *negative* and corresponds to the change in *y*) and 5 units to the left (since the denominator is *negative* and corresponds to the change in *x*). We get to another new point, $(-5, 2)$. We plot the points and draw the line.

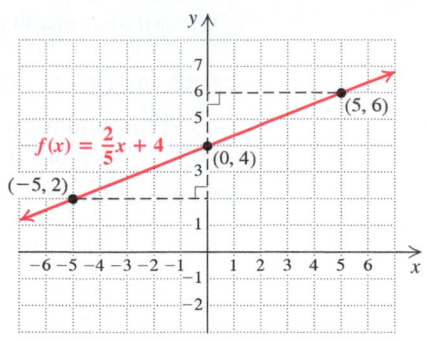

Now Try Exercise 17.

▶ c Horizontal Lines and Vertical Lines

Some equations have graphs that are parallel to one of the axes. This happens when either *A* or *B* is 0 in $Ax + By = C$. These equations have a missing variable; that is, there is only one variable in the equation. In the following example, *x* is missing.

EXAMPLE 4   Graph: $y = 3$.

Since *x* is missing, any number for *x* will do. Thus all ordered pairs $(x, 3)$ are solutions. The graph is a **horizontal line** parallel to the *x*-axis.

x	y	
−1	3	
0	3	← *y*-intercept
2	3	

↑ Choose *any* number for *x*.   ↑ Regardless of *x*, *y* must be 3.

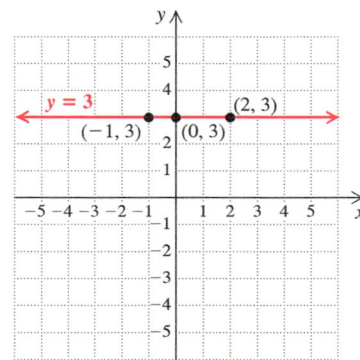

What about the slope of a horizontal line? In Example 4, consider the points $(-1, 3)$ and $(2, 3)$, which are on the line $y = 3$. The change in *y* is $3 - 3$, or 0. The change in *x* is $-1 - 2$, or -3. Thus,

$$m = \frac{3 - 3}{-1 - 2} = \frac{0}{-3} = 0.$$

Any two points on a horizontal line have the same *y*-coordinate. Thus the change in *y* is always 0, so the slope is 0.

Now Try Exercise 31.

We can also determine the slope by noting that $y = 3$ can be written in slope–intercept form as $y = 0x + 3$, or $f(x) = 0x + 3$. From this equation, we read that the slope is 0. A function of this type is called a **constant function**. We can express it in the form $y = b$, or $f(x) = b$. Its graph is a horizontal line that crosses the y-axis at $(0, b)$.

In the following example, y is missing and the graph is parallel to the y-axis.

EXAMPLE 5 Graph: $x = -2$.

Since y is missing, any number for y will do. Thus all ordered pairs $(-2, y)$ are solutions. The graph is a **vertical line** parallel to the y-axis.

x	y
−2	0	← *x*-intercept
−2	3
−2	−4

↑ ↑ —— Choose *any* number for *y*.
Regardless of *y*,
x must be −2.

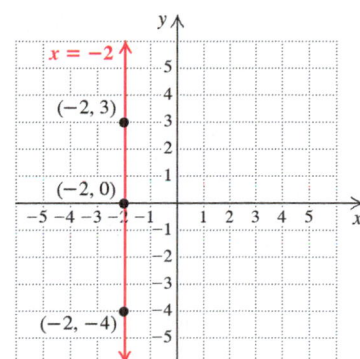

This graph is not the graph of a function because it fails the vertical-line test. The vertical line itself crosses the graph more than once.

Now Try Exercise 29.

What about the slope of a vertical line? In Example 5, consider the points $(-2, 3)$ and $(-2, -4)$, which are on the line $x = -2$. The change in y is $3 - (-4)$, or 7. The change in x is $-2 - (-2)$, or 0. Thus,

$$m = \frac{3 - (-4)}{-2 - (-2)} = \frac{7}{0}. \quad \text{\color{red}{Not defined}}$$

Since division by 0 is not defined, the slope of this line is not defined. Any two points on a vertical line have the same x-coordinate. Thus the change in x is always 0, so the slope of any vertical line is not defined.

The following summarizes the characteristics of horizontal lines and vertical lines and their equations.

HORIZONTAL LINE; VERTICAL LINE

The graph of $y = b$, or $f(x) = b$, is a **horizontal line** with y-intercept $(0, b)$. It is the graph of a constant function with slope 0.

The graph of $x = a$ is a **vertical line** with x-intercept $(a, 0)$. The slope is not defined. It is not the graph of a function.

We have graphed linear equations in several ways in this chapter. Although, in general, you can use any method that works best for you, we list some guidelines in the margin at left.

To graph a linear equation:

1. Is the equation of the type $x = a$ or $y = b$? If so, the graph will be a line parallel to an axis; $x = a$ is vertical and $y = b$ is horizontal.
2. If the line is of the type $y = mx$, both intercepts are the origin, $(0, 0)$. Plot $(0, 0)$ and one other point.
3. If the line is of the type $y = mx + b$, plot the y-intercept and one other point.
4. If the equation is of the form $Ax + By = C$, graph using intercepts. If the intercepts are too close together, choose another point farther from the origin.
5. In all cases, use a third point as a check.

▶ **d Parallel Lines and Perpendicular Lines**

Parallel Lines

Parallel lines extend indefinitely without intersecting. If two lines are vertical, they are parallel. How can we tell whether nonvertical lines are parallel? We examine their slopes and y-intercepts.

PARALLEL LINES

Vertical lines are parallel. Two nonvertical lines are **parallel** if they have the *same* slope and *different* y-intercepts.

EXAMPLE 6 Determine if the graphs of $y - 3x = 1$ and $3x + 2y = -2$ are parallel.

To determine if lines are parallel, we first find their slopes. To do this, we find the slope–intercept form of each equation by solving for y:

$$y - 3x = 1 \qquad\qquad 3x + 2y = -2$$
$$y = 3x + 1; \qquad\qquad 2y = -3x - 2$$
$$y = \tfrac{1}{2}(-3x - 2)$$
$$y = -\tfrac{3}{2}x - 1.$$

The slopes, 3 and $-\tfrac{3}{2}$, are different. Thus the lines are not parallel, as the graphs at right confirm.

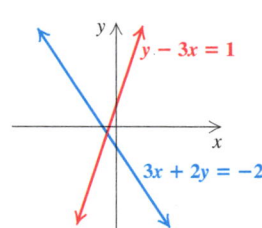

Now Try Exercise 43.

EXAMPLE 7 Determine if the graphs of $3x - y = -5$ and $y - 3x = -2$ are parallel.

We first find the slope–intercept form of each equation by solving for y:

$$3x - y = -5 \qquad\qquad y - 3x = -2$$
$$-y = -3x - 5 \qquad\qquad y = 3x - 2.$$
$$-1(-y) = -1(-3x - 5)$$
$$y = 3x + 5;$$

The slopes, 3, are the same. The y-intercepts, $(0, 5)$ and $(0, -2)$, are different. Thus the lines are parallel, as the graphs illustrate.

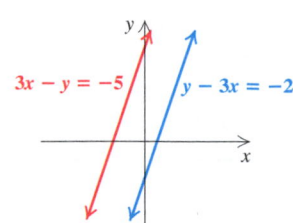

Now Try Exercise 45.

Perpendicular Lines

If one line is vertical and another is horizontal, they are perpendicular. For example, the lines $x = 5$ and $y = -3$ are perpendicular. Otherwise, how can we tell whether two lines are perpendicular?

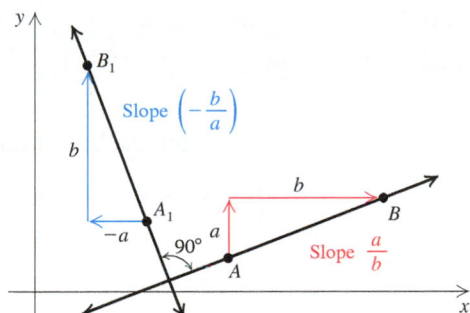

Consider a line $\overleftrightarrow{AB}$, as shown in the figure above, with slope a/b. Then think of rotating the line $90°$ to get a line $\overleftrightarrow{A_1B_1}$ perpendicular to $\overleftrightarrow{AB}$. For the new line, the rise and the run are interchanged, but the run is now negative. Thus the slope of the new line is $-b/a$, which is the opposite of the reciprocal of the slope of the first line. Also note that when we multiply the slopes, we get

$$\frac{a}{b}\left(-\frac{b}{a}\right) = -1.$$

This is the condition under which lines will be perpendicular.

PERPENDICULAR LINES

Two lines are **perpendicular** if the product of their slopes is -1. (If one line has slope m, then the slope of a line perpendicular to it is $-1/m$. That is, to find the slope of a line perpendicular to a given line, we take the reciprocal of the given slope and change the sign.)

Lines are also perpendicular if one of them is vertical $(x = a)$ and one of them is horizontal $(y = b)$.

EXAMPLE 8   Determine whether the graphs of $5y = 4x + 10$ and $4y = -5x + 4$ are perpendicular.

To determine whether the lines are perpendicular, we determine whether the product of their slopes is -1. We first find the slope–intercept form of each equation by solving for y.

We have

$$5y = 4x + 10 \qquad\qquad 4y = -5x + 4$$
$$y = \tfrac{1}{5}(4x + 10) \qquad y = \tfrac{1}{4}(-5x + 4)$$
$$y = \tfrac{4}{5}x + 2; \qquad\qquad y = -\tfrac{5}{4}x + 1.$$

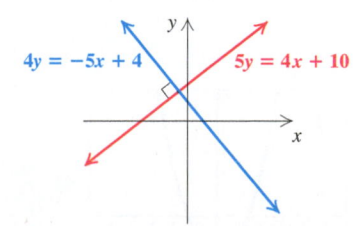

The slope of the first line is $\tfrac{4}{5}$, and the slope of the second line is $-\tfrac{5}{4}$. The product of the slopes is $\tfrac{4}{5} \cdot \left(-\tfrac{5}{4}\right) = -1$. Thus the lines are perpendicular.

Now Try Exercise 49.

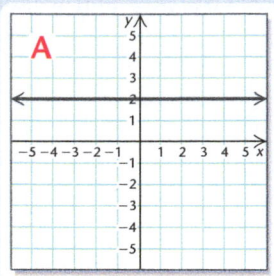

Visualizing the Graph

Match each equation with its graph.

1. $y = 2 - x$

2. $x - y = 2$

3. $x + 2y = 2$

4. $2x - 3y = 6$

5. $x = 2$

6. $y = 2$

7. $y = |x + 2|$

8. $y = |x| + 2$

9. $y = x^2 - 2$

10. $y = 2 - x^2$

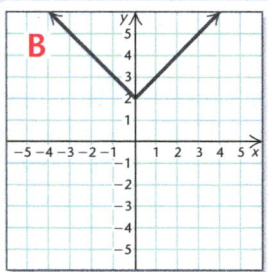

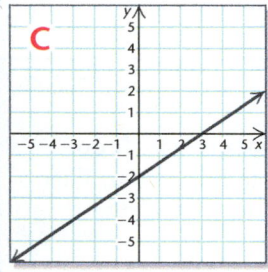

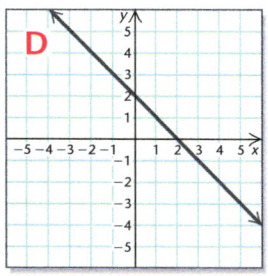

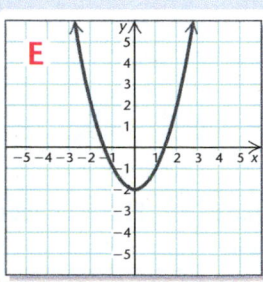

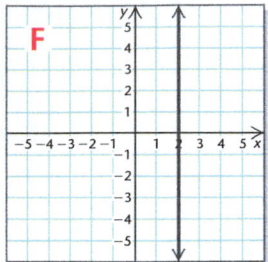

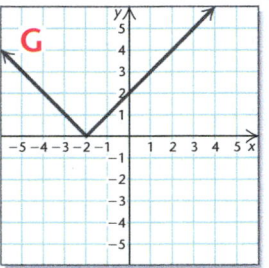

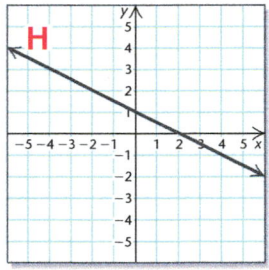

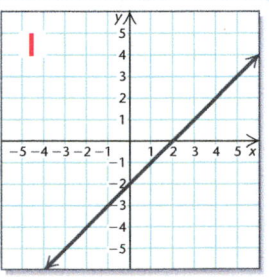

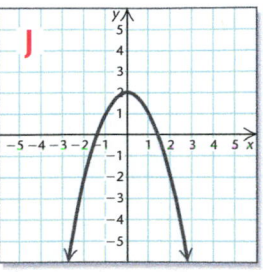

2.6 Exercise Set

▶ **Reading Check**

Determine whether each statement is true or false.

RC1. The graphs of the lines $x = -4$ and $y = 5$ are perpendicular.

RC2. The y-intercept of $y = -2x + 7$ is $(0, -2)$.

RC3. Two lines are perpendicular if the product of their slopes is 1.

RC4. The x-intercept of $x = -\frac{2}{7}$ is $\left(-\frac{2}{7}, 0\right)$.

RC5. The slope of a horizontal line is 0.

RC6. Two nonvertical lines are parallel if they have the same slope and the same y-intercepts.

a *Find the intercepts and then graph the line.*

1. $x - 2 = y$

2. $x + 3 = y$

3. $x + 3y = 6$

4. $x - 2y = 4$

5. $2x + 3y = 6$

6. $5x - 2y = 10$

7. $f(x) = -2 - 2x$

8. $g(x) = 5x - 5$

9. $5y = -15 + 3x$

10. $5x - 10 = 5y$

11. $2x - 3y = 6$

12. $4x + 5y = 20$

13. $2.8y - 3.5x = -9.8$

14. $10.8x - 22.68 = 4.2y$

15. $5x + 2y = 7$

16. $3x - 4y = 10$

b *Graph using the slope and the y-intercept.*

17. $y = \frac{5}{2}x + 1$

18. $y = \frac{2}{5}x - 4$

19. $f(x) = -\frac{5}{2}x - 4$

20. $f(x) = \frac{2}{5}x + 3$

21. $x + 2y = 4$

22. $x - 3y = 6$

23. $4x - 3y = 12$

24. $2x + 6y = 12$

25. $f(x) = \frac{1}{3}x - 4$

26. $g(x) = -0.25x + 2$

27. $5x + 4 \cdot f(x) = 4$
(*Hint*: Solve for $f(x)$.)

28. $3 \cdot f(x) = 4x + 6$

c *Graph and, if possible, determine the slope.*

29. $x = 1$

30. $x = -4$

31. $y = -1$

32. $y = \frac{3}{2}$

33. $f(x) = -6$

34. $f(x) = 2$

35. $y = 0$

36. $x = 0$

37. $2 \cdot f(x) + 5 = 0$

38. $4 \cdot g(x) + 3x = 12 + 3x$

39. $7 - 3x = 4 + 2x$

40. $3 - f(x) = 2$

d *Determine whether the graphs of the given pair of lines are parallel.*

41. $x + 6 = y,$
$y - x = -2$

42. $2x - 7 = y,$
$y - 2x = 8$

43. $y + 3 = 5x$,
$3x - y = -2$

44. $y + 8 = -6x$,
$-2x + y = 5$

45. $y = 3x + 9$,
$2y = 6x - 2$

46. $y + 7x = -9$,
$-3y = 21x + 7$

47. $12x = 3$,
$-7x = 10$

48. $5y = -2$,
$\frac{3}{4}x = 16$

Determine whether the graphs of the given pair of lines are perpendicular.

49. $y = 4x - 5$,
$4y = 8 - x$

50. $2x - 5y = -3$,
$2x + 5y = 4$

51. $x + 2y = 5$,
$2x + 4y = 8$

52. $y = -x + 7$,
$y = x + 3$

53. $2x - 3y = 7$,
$2y - 3x = 10$

54. $x = y$,
$y = -x$

55. $2x = 3$,
$-3y = 6$

56. $-5y = 10$,
$y = -\frac{4}{9}$

▶ Skill Maintenance

Write in scientific notation. **[R.7c]**

57. 53,000,000,000

58. 0.000047

59. 0.018

60. 99,902,000

Write in decimal notation. **[R.7c]**

61. 2.13×10^{-5}

62. 9.01×10^{8}

63. 2×10^{4}

64. 8.5677×10^{-2}

Factor. **[R.5d]**

65. $9x - 15y$

66. $12a + 21ab$

67. $21p - 7pq + 14p$

68. $64x - 128y + 256$

69. *Heaviest Pumpkin.*   In September 2012, Ron Wallace of Greene, Rhode Island, set a world record for the heaviest pumpkin. The previous record, set in 2010, was 1810.5 lb. The new record is 706.75 lb less than 1.5 times the record set in 2010. What was the record weight set in 2012? (*Source*: Data from www.huffingtonpost.com) **[1.3a]**

70.   Graph: $f(x) = -x^2 + 3x - 1$. **[2.2c]**

▶ Synthesis

71. Find the value of a such that the graphs of $5y = ax + 5$ and $\frac{1}{4}y = \frac{1}{10}x - 1$ are parallel.

72. Find the value of k such that the graphs of $x + 7y = 70$ and $y + 3 = kx$ are perpendicular.

73. Write an equation of the line that has x-intercept $(-3, 0)$ and y-intercept $\left(0, \frac{2}{5}\right)$.

74. Find the coordinates of the point of intersection of the graphs of the equations $x = -4$ and $y = 5$.

75. Write an equation for the x-axis. Is this a function?

76. Write an equation for the y-axis. Is this a function?

77. Find the value of m in $y = mx + 3$ so that the x-intercept of its graph will be $(4, 0)$.

78. Find the value of b in $2y = -7x + 3b$ so that the y-intercept of its graph will be $(0, -13)$.

2.7 Finding Equations of Lines; Applications

▶ **a** Find an equation of a line when the slope and the *y*-intercept are given.

▶ **b** Find an equation of a line when the slope and a point are given.

▶ **c** Find an equation of a line when two points are given.

▶ **d** Given a line and a point not on the given line, find an equation of the line parallel to the line and containing the point, and find an equation of the line perpendicular to the line and containing the point.

▶ **e** Solve applied problems involving linear functions.

In this section, we will learn to find an equation of a line for which we have been given two pieces of information.

▶ a Finding an Equation of a Line When the Slope and the *y*-Intercept Are Given

If we know the slope and the *y*-intercept of a line, we can find an equation of the line using the slope–intercept equation $y = mx + b$.

EXAMPLE 1   A line has slope -0.7 and *y*-intercept $(0, 13)$. Find an equation of the line.

We use the slope–intercept equation and substitute -0.7 for m and 13 for b:

$$y = mx + b$$
$$y = -0.7x + 13.$$

▶ **Now Try Exercise 1.**

▶ b Finding an Equation of a Line When the Slope and a Point Are Given

Suppose we know the slope of a line and the coordinates of one point on the line. We can use the slope–intercept equation to find an equation of the line. Or, we can use the **point–slope equation**. We first develop a formula for such a line.

Suppose that a line of slope m passes through the point (x_1, y_1). For any other point (x, y) on this line, we must have

$$\frac{y - y_1}{x - x_1} = m.$$

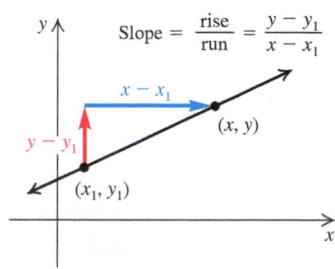

$$\text{Slope} = \frac{\text{rise}}{\text{run}} = \frac{y - y_1}{x - x_1}$$

It is tempting to use this last equation as an equation of the line of slope m that passes through (x_1, y_1). The only problem with this form is that when x and y are replaced with x_1 and y_1, we have $\frac{0}{0} = m$, a false equation. To avoid this difficulty, we multiply by $x - x_1$ on both sides and simplify:

$$\frac{y - y_1}{x - x_1}(x - x_1) = m(x - x_1) \qquad \text{\color{red}{Multiplying by } } x - x_1 \text{ \color{red}{on both sides}}$$

$$y - y_1 = m(x - x_1). \qquad \text{\color{red}{Removing a factor of 1: }} \frac{x - x_1}{x - x_1} = 1$$

This is the *point–slope* form of a linear equation.

> ### POINT-SLOPE EQUATION
> The **point–slope equation** of a line with slope m, passing through (x_1, y_1), is
> $$y - y_1 = m(x - x_1).$$

If we know the slope of a line and a point on the line, we can find an equation of the line using either the point–slope equation,

$$y - y_1 = m(x - x_1),$$

or the slope–intercept equation,

$$y = mx + b.$$

EXAMPLE 2 Find an equation of the line with slope 5 and containing the point $\left(\frac{1}{2}, -1\right)$.

Using the Point–Slope Equation: We consider $\left(\frac{1}{2}, -1\right)$ to be (x_1, y_1) and 5 to be the slope m, and substitute:

$$y - y_1 = m(x - x_1) \qquad \text{\color{red}{Point–slope equation}}$$
$$y - (-1) = 5\left(x - \tfrac{1}{2}\right) \qquad \text{\color{red}{Substituting}}$$
$$y + 1 = 5x - \tfrac{5}{2} \qquad \text{\color{red}{Simplifying}}$$
$$y = 5x - \tfrac{5}{2} - 1$$
$$y = 5x - \tfrac{5}{2} - \tfrac{2}{2}$$
$$y = 5x - \tfrac{7}{2}.$$

Using the Slope–Intercept Equation: The point $\left(\frac{1}{2}, -1\right)$ is on the line, so it is a solution of the equation. Thus we can substitute $\frac{1}{2}$ for x, -1 for y, and 5 for m in $y = mx + b$. Then we solve for b:

$$y = mx + b \qquad \text{\color{red}{Slope–intercept equation}}$$
$$-1 = 5 \cdot \left(\tfrac{1}{2}\right) + b \qquad \text{\color{red}{Substituting}}$$
$$-1 = \tfrac{5}{2} + b$$
$$-1 - \tfrac{5}{2} = b$$
$$-\tfrac{2}{2} - \tfrac{5}{2} = b$$
$$-\tfrac{7}{2} = b. \qquad \text{\color{red}{Solving for } } b$$

We then use the slope–intercept equation $y = mx + b$ again and substitute 5 for m and $-\frac{7}{2}$ for b:

$$y = 5x - \tfrac{7}{2}.$$

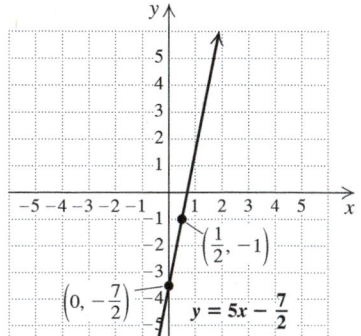

 **Now Try Exercise 11.**

▶ **c** **Finding an Equation of a Line When Two Points Are Given**

We can also use the slope–intercept equation or the point–slope equation to find an equation of a line when two points are given.

EXAMPLE 3 Find an equation of the line containing the points $(2, 3)$ and $(-6, 1)$.

First, we find the slope:

$$m = \frac{3-1}{2-(-6)} = \frac{2}{8}, \text{ or } \frac{1}{4}.$$

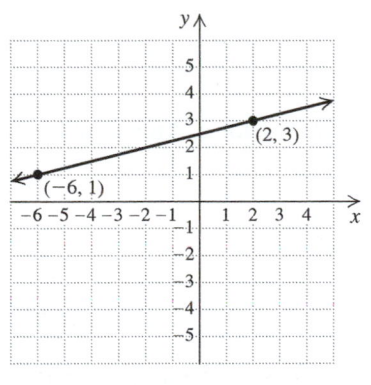

Now we have the slope and two points. We then proceed as we did in Example 2, using either point, and either the point–slope equation or the slope–intercept equation.

Using the Point–Slope Equation: We choose $(2, 3)$ and substitute 2 for x_1, 3 for y_1, and $\frac{1}{4}$ for m:

$$y - y_1 = m(x - x_1) \qquad \text{Point–slope equation}$$
$$y - 3 = \tfrac{1}{4}(x - 2) \qquad \text{Substituting}$$
$$y - 3 = \tfrac{1}{4}x - \tfrac{1}{2}$$
$$y = \tfrac{1}{4}x - \tfrac{1}{2} + 3$$
$$y = \tfrac{1}{4}x - \tfrac{1}{2} + \tfrac{6}{2}$$
$$y = \tfrac{1}{4}x + \tfrac{5}{2}.$$

Using the Slope–Intercept Equation: We choose $(2, 3)$ and substitute 2 for x, 3 for y, and $\frac{1}{4}$ for m:

$$y = mx + b \qquad \text{Slope–intercept equation}$$
$$3 = \tfrac{1}{4} \cdot 2 + b \qquad \text{Substituting}$$
$$3 = \tfrac{1}{2} + b$$
$$3 - \tfrac{1}{2} = \tfrac{1}{2} + b - \tfrac{1}{2}$$
$$\tfrac{6}{2} - \tfrac{1}{2} = b$$
$$\tfrac{5}{2} = b. \qquad \text{Solving for } b$$

Finally, we use the slope–intercept equation $y = mx + b$ again and substitute $\frac{1}{4}$ for m and $\frac{5}{2}$ for b:

$$y = \tfrac{1}{4}x + \tfrac{5}{2}.$$

▶ **Now Try Exercise 21.**

▶ **d** **Finding an Equation of a Line Parallel or Perpendicular to a Given Line Through a Point Not on the Line**

We can also use the methods of Example 2 to find an equation of a line parallel or perpendicular to a given line and containing a point not on the line.

EXAMPLE 4 Find an equation of the line containing the point $(-1, 3)$ and parallel to the line $2x + y = 10$.

A line parallel to the given line $2x + y = 10$ must have the same slope as the given line. To find that slope, we first find the slope–intercept equation by solving for y:

$$2x + y = 10$$
$$y = -2x + 10.$$

Thus the line we want to find through $(-1, 3)$ must also have slope -2.

Using the Point–Slope Equation: We use the point $(-1, 3)$ and the slope -2, substituting -1 for x_1, 3 for y_1, and -2 for m:

$$y - y_1 = m(x - x_1)$$
$$y - 3 = -2(x - (-1)) \quad \text{Substituting}$$
$$y - 3 = -2(x + 1) \quad \text{Simplifying}$$
$$y - 3 = -2x - 2$$
$$y = -2x + 1.$$

Using the Slope–Intercept Equation: We substitute -1 for x, 3 for y, and -2 for m in $y = mx + b$. Then we solve for b:

$$y = mx + b$$
$$3 = -2(-1) + b \quad \text{Substituting}$$
$$3 = 2 + b$$
$$1 = b. \quad \text{Solving for } b$$

We then use the equation $y = mx + b$ again and substitute -2 for m and 1 for b:

$$y = -2x + 1.$$

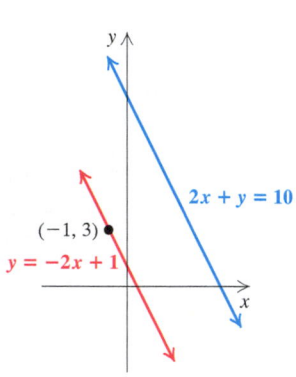

The given line $2x + y = 10$, or $y = -2x + 10$, and the line $y = -2x + 1$ have the same slope but different y-intercepts. Thus their graphs are parallel.

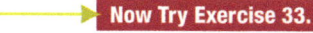

Now Try Exercise 33.

EXAMPLE 5 Find an equation of the line containing the point $(2, -3)$ and perpendicular to the line $4y - x = 20$.

To find the slope of the given line, we first find its slope–intercept form by solving for y:

$$4y - x = 20$$
$$4y = x + 20$$
$$\frac{4y}{4} = \frac{x + 20}{4} \quad \text{Dividing by 4}$$
$$y = \tfrac{1}{4}x + 5.$$

We know that the slope of the perpendicular line must be the opposite of the reciprocal of $\frac{1}{4}$. Thus the new line through $(2, -3)$ must have slope -4.

Using the Point–Slope Equation: We use the point $(2, -3)$ and the slope -4, substituting 2 for x_1, -3 for y_1, and -4 for m:

$$y - y_1 = m(x - x_1)$$
$$y - (-3) = -4(x - 2) \quad \text{Substituting}$$
$$y + 3 = -4x + 8$$
$$y = -4x + 5.$$

Using the Slope–Intercept Equation: We substitute 2 for x, -3 for y, and -4 for m in $y = mx + b$. Then we solve for b:

$$y = mx + b$$
$$-3 = -4(2) + b \quad \text{Substituting}$$
$$-3 = -8 + b$$
$$5 = b. \quad \text{Solving for } b$$

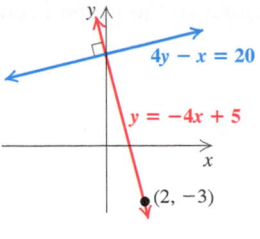

Finally, we use the equation $y = mx + b$ again and substitute -4 for m and 5 for b:

$$y = -4x + 5.$$

The product of the slopes of the lines $4y - x = 20$ and $y = -4x + 5$ is $\frac{1}{4} \cdot (-4) = -1$. Thus their graphs are perpendicular.

> **Now Try Exercise 39.**

▶ e Applications of Linear Functions

When the essential parts of a problem are described in mathematical language, we say that we have a **mathematical model**. We have already studied many kinds of mathematical models in this text—for example, the formulas in Section 1.2 and the functions in Section 2.2. Here we study linear functions as models.

EXAMPLE 6 *Cost of a Necklace.* Amelia's Beads offers a class in designing necklaces. For a necklace made of 6-mm beads, 4.23 beads per inch are needed. The cost of a necklace of 6-mm gemstone beads is $7 for the clasp and the crimps and approximately $1.70 per inch.

a) Formulate a linear function that models the total cost of a necklace $C(n)$, where n is the length of the necklace, in inches.

b) Graph the model.

c) Use the model to determine the cost of a 30-in. necklace.

a) The problem describes a situation in which cost per inch is charged in addition to the fixed cost of the clasp and the crimps. The total cost of a 16-in. necklace is

$$\$7 + \$1.70 \cdot 16 = \$34.20.$$

For a 17-in. necklace, the total cost is

$$\$7 + \$1.70 \cdot 17 = \$35.90.$$

These calculations lead us to generalize that for a necklace that is n inches long, the total cost is given by $C(n) = 7 + 1.7n$, where $n > 0$ since the length of the necklace cannot be negative or zero. (Actually most necklaces are at least 14 in. long.) The notation $C(n)$ indicates that the cost C is a function of the length n.

b) Before we draw the graph, we rewrite the model in slope–intercept form:

$$C(n) = 1.7n + 7.$$

The y-intercept is $(0, 7)$ and the slope, or rate of change, is 1.70, or $\frac{\$17}{10}$, per inch. We first plot $(0, 7)$; from that point, we move 17 units up and 10 units to the right to the point $(10, 24)$. We then draw a line through these points. We also calculate a third value as a check:

$$C(20) = 1.7 \cdot 20 + 7 = 41.$$

The point $(20, 41)$ lines up with the other two points so the graph is correct.

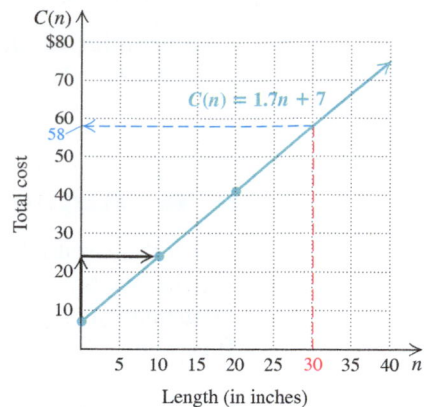

c) To determine the total cost of a 30-in. necklace, we find $C(30)$:

$$C(30) = 1.7 \cdot 30 + 7 = 58.$$

From the graph, we see that the input 30 corresponds to the output 58. Thus we see that a 30-in. necklace costs $58.

Now Try Exercise 45.

In the following example, we use two points and find an equation for the linear function through these points. Then we use the equation to estimate.

EXAMPLE 7 *Farmers' Markets.* The number of farmers' markets has increased steadily in recent years. The following table lists data regarding the correspondence between the year and the number of farmers' markets.

Year, x (in number of years since 2002)	Number of Farmers' Markets, n
2002, 0	3137
2012, 10	7864

Source: Data from U.S. Department of Agriculture

a) Assuming a constant rate of change, use the two data points to find a linear function that fits the data.

b) Use the function to determine the number of farmers' markets in 2010.

c) In which year will the number of farmers' markets reach 12,000?

a) We let x = the number of years since 2002 and N = the number of farmers' markets. The table gives us two ordered pairs, $(0, 3137)$ and $(10, 7864)$. We use them to find a linear function that fits the data. First, we find the slope:

$$m = \frac{7864 - 3137}{10 - 0} = \frac{4727}{10} = 472.7.$$

Next, we find an equation $N = mx + b$ that fits the data. One of the data points, $(0, 3137)$, is the y-intercept. Thus we know b in the slope–intercept equation, $y = mx + b$. We use the equation $N = mx + b$ and substitute 472.7 for m and 3137 for b:

$$N = 472.7x + 3137.$$

Using function notation, we have

$$N(x) = 472.7x + 3137.$$

b) To determine the number of farmers' markets in 2010, we substitute 8 for x (2010 is 8 years since 2002) in the function $N(x) = 472.7x + 3137$:

$$N(x) = 472.7x + 3137$$
$$N(8) = 472.7(8) + 3137 \qquad \text{Substituting}$$
$$= 3781.6 + 3137$$
$$= 6918.6 \approx 6919.$$

There were about 6919 farmers' markets in 2010.

c) To find the year in which the number of farmers' markets will reach 12,000, we substitute 12,000 for $N(x)$ and solve for x:

$$N(x) = 472.7x + 3137$$
$$12{,}000 = 472.7x + 3137 \qquad \text{Substituting}$$
$$8863 = 472.7x \qquad \text{Subtracting 3137}$$
$$19 \approx x. \qquad \text{Dividing by 472.7}$$

The number of farmers' markets will reach 12,000 about 19 years after 2002, or in 2021.

> **Now Try Exercise 49.**

2.7   Exercise Set

► Reading Check

For the given equation, determine the slope of the line (a) parallel to the given line and (b) perpendicular to the given line.

RC1. $y = \dfrac{4}{11}x - 2$

RC2. $y = -5$

RC3. $2x - y = -4$

RC4. $y - \dfrac{4}{3} = -\dfrac{5}{6}x$

RC5. $x = 3$

RC6. $10x + 5y = 14$

a   *Find an equation of the line having the given slope and y-intercept.*

1. Slope: -8; y-intercept: $(0, 4)$

2. Slope: 5; y-intercept: $(0, -3)$

3. Slope: 2.3; y-intercept: $(0, -1)$

4. Slope: -9.1; y-intercept: $(0, 2)$

Find a linear function $f(x) = mx + b$ whose graph has the given slope and y-intercept.

5. Slope: $-\frac{7}{3}$; y-intercept: $(0, -5)$

6. Slope: $\frac{4}{5}$; y-intercept: $(0, 28)$

7. Slope: $\frac{2}{3}$; y-intercept: $\left(0, \frac{5}{8}\right)$

8. Slope: $-\frac{7}{8}$; y-intercept: $\left(0, -\frac{7}{11}\right)$

b   *Find an equation of the line having the given slope and containing the given point.*

9. $m = 5$, $(4, 3)$

10. $m = 4$, $(5, 2)$

11. $m = -3$, $(9, 6)$

12. $m = -2$, $(2, 8)$

13. $m = 1$, $(-1, -7)$

14. $m = 3$, $(-2, -2)$

15. $m = -2$, $(8, 0)$

16. $m = -3$, $(-2, 0)$

17. $m = 0$, $(0, -7)$

18. $m = 0, (0, 4)$

19. $m = \frac{2}{3}, (1, -2)$

20. $m = -\frac{4}{5}, (2, 3)$

c *Find an equation of the line containing the given pair of points.*

21. $(1, 4)$ and $(5, 6)$

22. $(2, 5)$ and $(4, 7)$

23. $(-3, -3)$ and $(2, 2)$

24. $(-1, -1)$ and $(9, 9)$

25. $(-4, 0)$ and $(0, 7)$

26. $(0, -5)$ and $(3, 0)$

27. $(-2, -3)$ and $(-4, -6)$

28. $(-4, -7)$ and $(-2, -1)$

29. $(0, 0)$ and $(6, 1)$

30. $(0, 0)$ and $(-4, 7)$

31. $\left(\frac{1}{4}, -\frac{1}{2}\right)$ and $\left(\frac{3}{4}, 6\right)$

32. $\left(\frac{2}{3}, \frac{3}{2}\right)$ and $\left(-3, \frac{5}{6}\right)$

d *Write an equation of the line containing the given point and parallel to the given line.*

33. $(3, 7)$; $x + 2y = 6$

34. $(0, 3)$; $2x - y = 7$

35. $(2, -1)$; $5x - 7y = 8$

36. $(-4, -5)$; $2x + y = -3$

37. $(-6, 2)$; $3x = 9y + 2$

38. $(-7, 0)$; $2y + 5x = 6$

Write an equation of the line containing the given point and perpendicular to the given line.

39. $(2, 5)$; $2x + y = 3$

40. $(4, 1)$; $x - 3y = 9$

41. $(3, -2)$; $3x + 4y = 5$

42. $(-3, -5)$; $5x - 2y = 4$

43. $(0, 9)$; $2x + 5y = 7$

44. $(-3, -4)$; $6y - 3x = 2$

e *Solve.*

45. *School Fund-Raiser.* A school club is raising funds by having a "Shred It Day," when residents of the community can bring in their sensitive documents to be shredded. The club is charging $10 for the first three paper bags full of documents and $5 for each additional bag.

a) Formulate a linear function that models the total cost $C(x)$ of shredding the first three bags plus x additional bags of documents.

b) Graph the model.

c) Use the model to determine the total cost of shredding 7 bags of documents.

46. *Fitness Club Costs.* A fitness club charges an initiation fee of $165 plus $24.95 per month.

a) Formulate a linear function that models the total cost $C(t)$ of a club membership for t months.

b) Graph the model.

c) Use the model to determine the total cost of a 14-month membership.

47. *Value of a Lawn Mower.* A landscaping business purchased a ZTR commercial lawn mower for $9400. The value $V(t)$ of the mower depreciates (declines) at a rate of $85 per month.

a) Formulate a linear function that models the value $V(t)$ of the mower after t months.

b) Graph the model.

c) Use the model to determine the value of the mower after 18 months.

48. *Value of a Computer.* True Tone Graphics bought a computer for $3800. The value $V(t)$ of the computer depreciates at a rate of $50 per month.

a) Formulate a linear function that models the value $V(t)$ of the computer after t months.

b) Graph the model.

c) Use the model to determine the value of the computer after $10\frac{1}{2}$ months.

In Exercises 49–54, assume that a constant rate of change exists for each model formed.

49. *Organic-Food Sales.* The following table lists data regarding sales, in billions of dollars, of organic food in 2004 and in 2012.

Year, x (in number of years since 2004)	Organic Food Sales (in billions)
2004, 0	$11
2012, 8	27

Source: Data from Nutrition Business Journal

a) Use the two data points to find a linear function that fits the data. Let $x =$ the number of years since 2004 and $S(x) =$ the total sales, in billions of dollars, of organic food.

b) Use the function of part (a) to estimate and predict the sales of organic food in 2008 and in 2017.

50. *Cost of Diabetes.*   The following table lists data regarding the health-care and work-related costs of diabetes in 2007 and in 2012.

Year, x (in number of years since 2007)	Costs of Diabetes (in billions)
2007, 0	$174
2012, 5	245

Source: Data from American Diabetes Association

a) Use the two data points to find a function that fits the data. Let $x =$ the number of years since 2007 and $D(x) =$ the costs, in billions of dollars, of diabetes.

b) Use the function of part (a) to estimate the costs of diabetes in 2010 and in 2015.

51. *Auto Dealers.*   At the close of 1995, there were 22,800 new-auto dealers in the United States. By the end of 2012, this number had dropped to 17,540. Let $D(x) =$ the number of new-auto dealerships and $x =$ the number of years since 1995. (*Source*: Data from Urban Science Automotive Dealer Census)

a) Find a linear function that fits the data.

b) Use the function of part (a) to estimate the number of new-auto dealerships in 2000.

c) At this rate of decrease, when will the number of new-auto dealerships be 15,500?

52. *Records in the 400-Meter Run.*   In 1930, the record for the 400-m run was 46.8 sec. In 1970, it was 43.8 sec. Let $R(t) =$ the record in the 400-m run and $t =$ the number of years since 1930.

a) Find a linear function that fits the data.

b) Use the function of part (a) to estimate the record in 2003 and in 2006.

c) When will the record be 40 sec?

53. *Life Expectancy in South Africa.*   In 2003, the life expectancy in South Africa was 46.56 years. In 2011, it was 49.33 years. Let $E(t) =$ life expectancy and $t =$ the number of years since 2003. (*Source*: Data from *CIA World Factbook* 2003–2012)

a) Find a linear function that fits the data.

b) Use the function of part (a) to estimate life expectancy in 2016.

54. *Life Expectancy in Monaco.*   In 2003, the life expectancy in Monaco was 79.27 years. In 2011, it was 89.73 years. Let $E(t) =$ life expectancy and $t =$ the number of years since 2003. (*Source*: Data from CIA World Factbook 2003–2012)

a) Find a linear function that fits the data.

b) Use the function of part (a) to estimate life expectancy in 2016.

▶ **Skill Maintenance**

Solve.   **[1.4c], [1.5a], [1.6c, d, e]**

55. $2x + 3 > 51$

56. $|2x + 3| = 51$

57. $2x + 3 \leq 51$

58. $2x + 3 \leq 5x - 4$

59. $|2x + 3| \leq 13$

60. $|2x + 3| = |x - 4|$

61. $|5x - 4| = -8$

62. $-12 \leq 2x + 3 < 51$

▶ **Synthesis**

63. Find k such that the line containing the points $(-3, k)$ and $(4, 8)$ is parallel to the line containing the points $(5, 3)$ and $(1, -6)$.

64. Find an equation of the line passing through the point $(4, 5)$ and perpendicular to the line passing through the points $(-1, 3)$ and $(2, 9)$.

Chapter 2 Summary and Review

VOCABULARY REINFORCEMENT

Complete each statement with the correct term from the column on the right. Some of the choices may be used more than once, and some may not be used at all.

1. The graph of $x = a$ is a(n) _____ line with x-intercept $(a, 0)$. **[2.6c]**

2. The _____ equation of a line with slope m and passing through (x_1, y_1) is $y - y_1 = m(x - x_1)$. **[2.7b]**

3. A(n) _____ is a correspondence between a first set, called the _____, and a second set called the _____, such that each member of the _____ corresponds to _____ member of the _____. **[2.2a]**

4. The _____ of a line containing points (x_1, y_1) and (x_2, y_2) is given by m = the change in y/the change in x, also described as rise/run. **[2.5b]**

5. Two lines are _____ if the product of their slopes is -1. **[2.6d]**

6. The equation $y = mx + b$ is called the _____ equation of a line with slope m and y-intercept $(0, b)$. **[2.5b]**

7. Lines are _____ if they have the same slope and different y-intercepts. **[2.6d]**

x-intercept

y-intercept

at least one

exactly one

slope–intercept

point–slope

slope

function

relation

parallel

perpendicular

vertical

horizontal

domain

range

CONCEPT REINFORCEMENT

Determine whether each statement is true or false.

1. The slope of a vertical line is 0. **[2.6c]**

2. A line with slope 1 slants less steeply than a line with slope -5. **[2.5b]**

3. Parallel lines have the same slope and y-intercept. **[2.6d]**

STUDY GUIDE

Objective 2.2a   Determine whether a correspondence is a function.

Example   Determine whether each correspondence is a function.

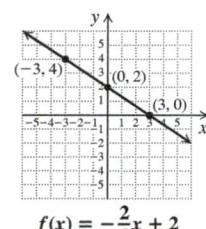

Domain   *Range*   *Domain*   *Range*

The correspondence f is a function because each member of the domain is matched to *only one* member of the range. The correspondence g *is not* a function because one member of the domain, Q, is matched to more than one member of the range.

Practice Exercise

1. Determine whether the correspondence is a function.

Domain   *Range*

h: 11   20
15   31
19

Objective 2.2b   Given a function described by an equation, find function values (outputs) for specified values (inputs).

Example   Find the indicated function value.

a) $f(0)$, for $f(x) = -x + 6$

b) $g(5)$, for $g(x) = -10$

c) $h(-1)$, for $h(x) = 4x^2 + x$

a) $f(x) = -x + 6$: $f(0) = -0 + 6 = 6$

b) $g(x) = -10$: $g(5) = -10$

c) $h(x) = 4x^2 + x$: $h(-1) = 4(-1)^2 + (-1) = 3$

Practice Exercise

2. Find $g(0), g(-2)$, and $g(6)$ for $g(x) = \frac{1}{2}x - 2$.

Objective 2.2c   Draw the graph of a function.

Example   Graph: $f(x) = -\frac{2}{3}x + 2$.

By choosing multiples of 3 for x, we can avoid fraction values for y. If $x = -3$, then $y = -\frac{2}{3} \cdot (-3) + 2 = 2 + 2 = 4$. We list three ordered pairs in a table, plot the points, draw the line, and label the graph.

x	$f(x)$
3	0
0	2
-3	4

$$f(x) = -\frac{2}{3}x + 2$$

Practice Exercise

3. Graph: $f(x) = \frac{2}{5}x - 3$.

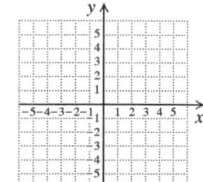

x	$f(x)$

Objective 2.2d   Determine whether a graph is that of a function using the vertical-line test.

Example   Determine whether each of the following is the graph of a function.

a) 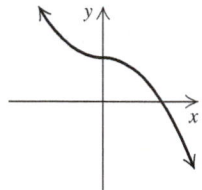   **b)**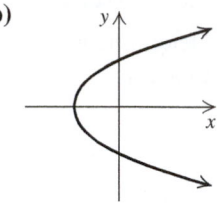

a) The graph is that of a function because no vertical line can cross the graph at more than one point.

b) The graph is not that of a function because a vertical line can cross the graph more than once.

Practice Exercise

4. Determine whether the graph is the graph of a function.

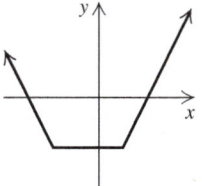

Objective 2.3a   Find the domain and the range of a function.

Example   For the function *f* whose graph is shown below, determine the domain and the range.

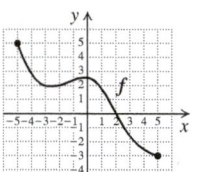   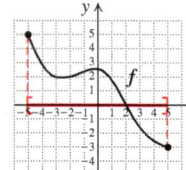   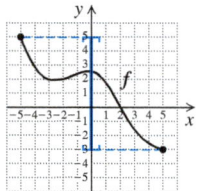

Domain: $[-5, 5]$; range: $[-3, 5]$

Example   Find the domain of $g(x) = \dfrac{x + 1}{2x - 6}$.

Since $(x + 1)/(2x - 6)$ cannot be calculated when the denominator $2x - 6$ is 0, we solve $2x - 6 = 0$ to find the real numbers that must be excluded from the domain of *g*:

$$2x - 6 = 0$$
$$2x = 6$$
$$x = 3.$$

Thus, 3 is not in the domain. The domain of *g* is $\{x \mid x \text{ is a real number } and \ x \neq 3\}$, or $(-\infty, 3) \cup (3, \infty)$.

Practice Exercises

5. For the function *g* whose graph is shown below, determine the domain and the range.

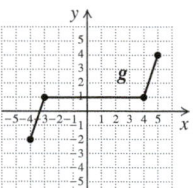

6. Find the domain of

$$h(x) = \frac{x - 3}{3x + 9}.$$

Objective 2.4a Given two functions f and g, find their sum, difference, product, and quotient.

Example For $f(x) = x + 4$ and $g(x) = 1 - x^2$, find:

a) $(f + g)(3)$

$$(f + g)(x) = f(x) + g(x)$$
$$= x + 4 + 1 - x^2$$
$$= -x^2 + x + 5$$
$$(f + g)(3) = -3^2 + 3 + 5 = -1$$

b) $(f - g)(-2)$

$$(f - g)(x) = f(x) - g(x)$$
$$= x + 4 - (1 - x^2)$$
$$= x + 4 - 1 + x^2$$
$$= x^2 + x + 3$$
$$(f - g)(-2) = (-2)^2 + (-2) + 3 = 5$$

c) $(f/g)(x) = \dfrac{f(x)}{g(x)} = \dfrac{x + 4}{1 - x^2}$

d) $(f \cdot g)(3)$

$$f(x) = x + 4; f(3) = 3 + 4 = 7$$
$$g(x) = 1 - x^2; g(3) = 1 - 3^2 = -8$$
$$(f \cdot g)(3) = f(3) \cdot g(3) = 7(-8) = -56$$

Practice Exercise

7. For $f(x) = 2x^2 + 3$ and $g(x) = 1 - x$, find:

a) $(f + g)(x)$
b) $(f - g)(-4)$
c) $(f/g)(6)$
d) $(f \cdot g)(1)$

Objective 2.4b For a pair of functions f and g, determine the domain of $f + g, f - g, f \cdot g$, and f/g.

Example For $f(x) = \dfrac{1}{x + 8}$ and $g(x) = 3x + 5$, find the domains of $f + g, f - g, f \cdot g$, and f/g.

The domain of f
$= \{x | x \text{ is a real number } and \; x \neq -8\}$.

The domain of g
$= \{x | x \text{ is a real number}\}$.

The domain of $f + g$
$=$ the domain of $f - g$
$=$ the domain of $f \cdot g$
$= \{x | x \text{ is a real number } and \; x \neq -8\}$.

Because we cannot divide by 0, we determine the values of x for which $g(x) = 0$.

$$g(x) = 0$$
$$3x + 5 = 0$$
$$3x = -5$$
$$x = -\frac{5}{3}$$

The domain of f/g
$= \{x | x \text{ is a real number } and \; x \neq -8 \; and \; x \neq -\frac{5}{3}\}$.

Practice Exercise

8. For $f(x) = \dfrac{6}{4x + 3}$ and $g(x) = x - 2$, find the domains of $f + g, f - g, f \cdot g$, and f/g.

Objective 2.5b   Given two points on a line, find the slope. Given a linear equation, derive the equivalent slope–intercept equation and determine the slope and the *y*-intercept.

Example Find the slope of the line containing $(-5, 6)$ and $(-1, -4)$.

$$m = \frac{\text{change in } y}{\text{change in } x} = \frac{6 - (-4)}{-5 - (-1)} = \frac{6 + 4}{-5 + 1} = \frac{10}{-4} = -\frac{5}{2}$$

Example Find the slope and the *y*-intercept of

$$4x - 2y = 20.$$

We first solve for *y*:

$$4x - 2y = 20$$
$$-2y = -4x + 20 \qquad \text{Subtracting } 4x$$
$$y = 2x - 10. \qquad \text{Dividing by } -2$$

The slope is 2, and the *y*-intercept is $(0, -10)$.

Practice Exercises

9. Find the slope of the line containing $(2, -8)$ and $(-3, 2)$.

10. Find the slope and the *y*-intercept of

$$3x = -6y + 12.$$

Objective 2.6a   Graph linear equations using intercepts.

Example Find the intercepts of $x - 2y = 6$ and then graph the line.

To find the *y*-intercept, we let $x = 0$ and solve for *y*:

$$0 - 2y = 6 \qquad \text{Substituting 0 for } x$$
$$-2y = 6$$
$$y = -3.$$

The *y*-intercept is $(0, -3)$.

To find the *x*-intercept, we let $y = 0$ and solve for *x*:

$$x - 2 \cdot 0 = 6 \qquad \text{Substituting 0 for } y$$
$$x = 6.$$

The *x*-intercept is $(6, 0)$.

We plot these points and draw the line, using a third point as a check. We let $x = -2$ and solve for *y*:

$$-2 - 2y = 6$$
$$-2y = 8$$
$$y = -4.$$

We plot $(-2, -4)$ and note that it is on the line. Thus the graph is correct.

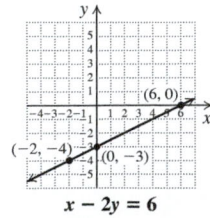

$x - 2y = 6$

Practice Exercise

11. Find the intercepts of $3y - 3 = x$ and then graph the line.

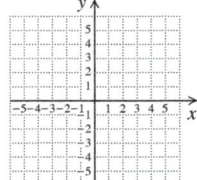

Objective 2.6b Given a linear equation in slope–intercept form, use the slope and the y-intercept to graph the line.

Example Graph using the slope and the y-intercept:

$$y = -\frac{3}{2}x + 5.$$

This equation is in slope–intercept form, $y = mx + b$. The y-intercept is $(0, 5)$. We plot $(0, 5)$. We can think of the slope $\left(m = -\frac{3}{2}\right)$ as $\frac{-3}{2}$.

Starting at the y-intercept, we use the slope to find another point on the graph. We move 3 units down and 2 units to the right. We get a new point: $(2, 2)$.

To get a third point for a check, we start at $(2, 2)$ and move 3 units down and 2 units to the right to the point $(4, -1)$. We plot the points and draw the line.

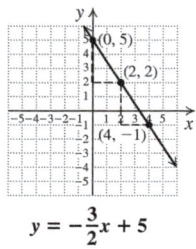

$$y = -\frac{3}{2}x + 5$$

Practice Exercise

12. Graph using the slope and the y-intercept:

$$y = \frac{1}{4}x - 3.$$

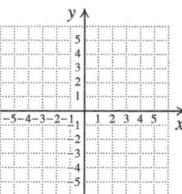

Objective 2.6c Graph linear equations of the form $x = a$ or $y = b$.

Example Graph: $y = -1$.

All ordered pairs $(x, -1)$ are solutions; y is -1 at each point. The graph is a horizontal line that intersects the y-axis at $(0, -1)$.

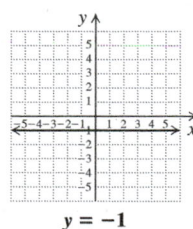

$$y = -1$$

Example Graph: $x = 2$.

All ordered pairs $(2, y)$ are solutions; x is 2 at each point. The graph is a vertical line that intersects the x-axis at $(2, 0)$.

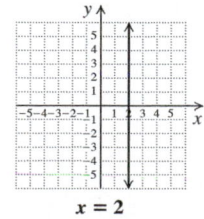

$$x = 2$$

Practice Exercises

13. Graph: $y = 3$.

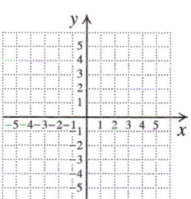

14. Graph: $x = -\frac{5}{2}$.

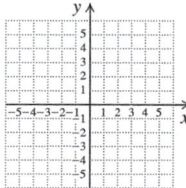

Objective 2.6d   Given the equations of two lines, determine whether their graphs are parallel or whether they are perpendicular.

Example   Determine whether the graphs of the given pair of lines are parallel, perpendicular, or neither.

a) $2y - x = 16,$
   $x + \frac{1}{2}y = 4$

b) $5x - 3 = 2y,$
   $2y + 12 = 5x$

a) Writing each equation in slope–intercept form, we have $y = \frac{1}{2}x + 8$ and $y = -2x + 8$. The slopes are $\frac{1}{2}$ and -2. The product of the slopes is -1: $\frac{1}{2} \cdot (-2) = -1$. The graphs are perpendicular.

b) Writing each equation in slope–intercept form, we have $y = \frac{5}{2}x - \frac{3}{2}$ and $y = \frac{5}{2}x - 6$. The slopes are the same, $\frac{5}{2}$, and the y-intercepts are different. The graphs are parallel.

Practice Exercises

Determine whether the graphs of the given pair of lines are parallel, perpendicular, or neither.

15. $-3x + 8y = -8,$
    $8y = 3x + 40$

16. $5x - 2y = -8,$
    $2x + 5y = 15$

Objective 2.7a   Find an equation of a line when the slope and the y-intercept are given.

Example   A line has slope 0.8 and y-intercept $(0, -17)$. Find an equation of the line.

We use the slope–intercept equation and substitute 0.8 for m and -17 for b:

$$y = mx + b \qquad \textcolor{red}{\textbf{Slope–intercept equation}}$$
$$y = 0.8x - 17.$$

Practice Exercise

17. A line has slope -8 and y-intercept $(0, 0.3)$. Find an equation of the line.

Objective 2.7b   Find an equation of a line when the slope and a point are given.

Example   Find an equation of the line with slope -2 and containing the point $\left(\frac{1}{3}, -1\right)$.

Using the *point–slope equation*, we substitute -2 for m, $\frac{1}{3}$ for x_1, and -1 for y_1:

$$y - (-1) = -2\left(x - \frac{1}{3}\right) \qquad \textcolor{red}{\textbf{Using } y - y_1 = m(x - x_1)}$$

$$y + 1 = -2x + \frac{2}{3}$$

$$y = -2x - \frac{1}{3}.$$

Using the *slope–intercept equation*, we substitute -2 for m, $\frac{1}{3}$ for x, and -1 for y, and then solve for b:

$$\textcolor{red}{-1 = -2 \cdot \frac{1}{3} + b} \qquad \textcolor{red}{\textbf{Using } y = mx + b}$$

$$-1 = -\frac{2}{3} + b$$

$$-\frac{1}{3} = b.$$

Then, substituting -2 for m and $-\frac{1}{3}$ for b in the slope–intercept equation $y = mx + b$, we have $y = -2x - \frac{1}{3}$.

Practice Exercise

18. Find an equation of the line with slope -4 and containing the point $\left(\frac{1}{2}, -3\right)$.

Objective 2.7c Find an equation of a line when two points are given.

Example Find an equation of the line containing the points $(-3, 9)$ and $(1, -2)$.

We first find the slope:

$$\frac{9 - (-2)}{-3 - 1} = \frac{11}{-4} = -\frac{11}{4}.$$

Using the slope–intercept equation and the point $(1, -2)$, we substitute $-\frac{11}{4}$ for m, 1 for x, and -2 for y, and then solve for b. We could also have used the point $(-3, 9)$.

$$y = mx + b$$
$$-2 = -\frac{11}{4} \cdot 1 + b$$
$$-\frac{8}{4} = -\frac{11}{4} + b$$
$$\frac{3}{4} = b$$

Then substituting $-\frac{11}{4}$ for m and $\frac{3}{4}$ for b in $y = mx + b$, we have $y = -\frac{11}{4}x + \frac{3}{4}$.

Practice Exercise

19. Find an equation of the line containing the points $(-2, 7)$ and $(4, -3)$.

Objective 2.7d Given a line and a point not on the given line, find an equation of the line parallel to the line and containing the point, and find an equation of the line perpendicular to the line and containing the point.

Example Write an equation of the line containing $(-1, 1)$ and parallel to $3y - 6x = 5$.

Solving $3y - 6x = 5$ for y, we get $y = 2x + \frac{5}{3}$. The slope of the given line is 2.

A line parallel to the given line must have the same slope, 2. We substitute 2 for m, -1 for x_1, and 1 for y_1 in the point–slope equation:

$$y - 1 = 2[x - (-1)] \qquad \text{Using } y - y_1 = m(x - x_1)$$
$$y - 1 = 2(x + 1)$$
$$y - 1 = 2x + 2$$
$$y = 2x + 3. \qquad \text{Line parallel to the given line and passing through } (-1, 1)$$

Example Write an equation of the line containing the point $(2, -4)$ and perpendicular to $6x + 2y = 13$.

Solving $6x + 2y = 13$ for y, we get $y = -3x + \frac{13}{2}$. The slope of the given line is -3.

The slope of a line perpendicular to the given line is the opposite of the reciprocal of -3, or $\frac{1}{3}$. We substitute $\frac{1}{3}$ for m, 2 for x_1, and -4 for y_1 in the point–slope equation:

$$y - (-4) = \frac{1}{3}(x - 2) \qquad \text{Using } y - y_1 = m(x - x_1)$$
$$y + 4 = \frac{1}{3}x - \frac{2}{3}$$
$$y = \frac{1}{3}x - \frac{14}{3}. \qquad \text{Line perpendicular to the given line and passing through } (2, -4)$$

Practice Exercises

20. Write an equation of the line containing the point $(2, -5)$ and parallel to $4x - 3y = 6$.

21. Write an equation of the line containing $(2, -5)$ and perpendicular to $4x - 3y = 6$.

REVIEW EXERCISES

Determine whether each correspondence is a function. [2.2a]

1. 3 ⟶ a
5 ⟶ b
7 ⟶ c
9 ⟶ d
⟶ e

2. 1 ⟶ a
2 ⟶ b
3 ⟶ c
4 ⟶ d
5

Find the function values. [2.2b]

3. $g(x) = -2x + 5$; $g(0)$ and $g(-1)$

4. $f(x) = 3x^2 - 2x + 7$; $f(0)$ and $f(-1)$

5. *Tuition Cost.* The function $C(t) = 309.2t + 3717.7$ can be used to approximate the average cost of tuition and fees for in-state students at public four-year colleges, where t is the number of years after 2000. Estimate the average cost of tuition and fees in 2010. That is, find $C(10)$. (*Source*: Data from U.S. National Center for Education Statistics) [2.2b]

Graph. [2.1c, d], [2.2c]

6. $y = -3x + 2$ **7.** $g(x) = \frac{5}{2}x - 3$

8. $f(x) = |x - 3|$ **9.** $y = 3 - x^2$

Determine whether each of the following is the graph of a function. [2.2d]

10.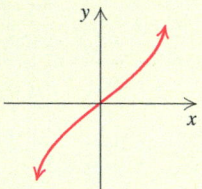

11.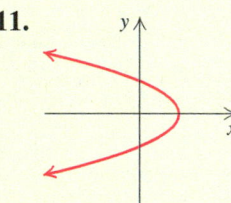

12. For the following graph of a function f, determine
(a) $f(2)$; **(b)** the domain; **(c)** all x-values such that $f(x) = 2$; and **(d)** the range. [2.3a]

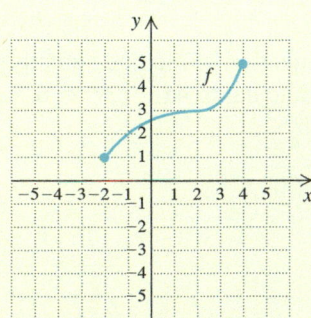

Find the domain. [2.3a]

13. $f(x) = \dfrac{5}{x - 4}$ **14.** $g(x) = x - x^2$

For $f(x) = x^2 - 3x$ and $g(x) = x + 10$, find each of the following. [2.4a]

15. $(f/g)(x)$ **16.** $(f - g)(-2)$

For each pair of functions f and g, determine the domains of $f + g$, $f - g$, $f \cdot g$, and f/g. [2.4b]

17. $f(x) = -2x^2$,
$g(x) = x + 6$

18. $f(x) = \dfrac{3}{7 - x}$,
$g(x) = 5 - x$

Find the slope and the y-intercept. [2.5a, b]

19. $y = -3x + 2$ **20.** $4y + 2x = 8$

21. Find the slope, if it exists, of the line containing the points $(13, 7)$ and $(10, -4)$. [2.5b]

Find the intercepts. Then graph the equation. [2.6a]

22. $2y + x = 4$ **23.** $2y = 6 - 3x$

Graph using the slope and the y-intercept. [2.6b]

24. $g(x) = -\frac{2}{3}x - 4$ **25.** $f(x) = \frac{5}{2}x + 3$

Graph. [2.6c]

26. $x = -3$ **27.** $f(x) = 4$

Determine whether the graphs of the given pair of lines are parallel or perpendicular. [2.6d]

28. $y + 5 = -x$,
$x - y = 2$

29. $3x - 5 = 7y$,
$7y - 3x = 7$

30. $4y + x = 3$,
$2x + 8y = 5$

31. $x = 4$,
$y = -3$

32. Find a linear function $f(x) = mx + b$ whose graph has slope 4.7 and y-intercept $(0, -23)$. [2.7a]

33. Find an equation of the line having slope -3 and containing the point $(3, -5)$. [2.7b]

34. Find an equation of the line containing the points $(-2, 3)$ and $(-4, 6)$. [2.7c]

35. Find an equation of the line containing the given point and parallel to the given line:
$(14, -1)$; $5x + 7y = 8$.   [2.7d]

36. Find an equation of the line containing the given point and perpendicular to the given line:

$(5, 2); \quad 3x + y = 5.$ **[2.7d]**

37. *Records in the 400-Meter Run.* The following table shows data regarding the Summer Olympics winning times in the men's 400-m run. **[2.7e]**

Year	Summer Olympics Winning Time in Men's 400-m Run (in seconds)
1972, 0	44.66
2012, 40	43.94

a) Use the two data points to find a linear function that fits the data. Let x = the number of years since 1972 and $R(x)$ = the Summer Olympics winning time x years from 1972.

b) Use the function to estimate the winning time in the men's 400-m run in 2000 and in 2010.

38. What is the domain of $f(x) = \dfrac{x + 3}{x - 2}$? **[2.3a]**

A. $\{x \mid x \geq -3\}$
B. $\{x \mid x \text{ is a real number } and \ x \neq -3 \ and \ x \neq 2\}$
C. $\{x \mid x \text{ is a real number } and \ x \neq 2\}$
D. $\{x \mid x > -3\}$

39. Find an equation of the line containing the point $(-2, 1)$ and perpendicular to $3y - \frac{1}{2}x = 0$. **[2.7d]**

A. $6x + y = -11$
B. $y = -\dfrac{1}{6}x - 11$
C. $y = -2x - 3$
D. $2x + \dfrac{1}{3} = 0$

► Synthesis

40. Homespun Jellies charges $2.49 for each jar of preserves. Shipping charges are $3.75 per order for handling, plus $0.60 per jar. Find a linear function for determining the cost of buying and shipping x jars of preserves. **[2.7e]**

► Collaborative Discussion and Writing

41. Under what conditions will the x-intercept and the y-intercept of a line be the same? What would the equation for such a line look like? **[2.6a]**

42. Explain the usefulness of the concept of slope when describing a line. **[2.5b, c], [2.6b], [2.7a, b, c, d]**

43. A student makes a mistake when using a graphing calculator to draw $4x + 5y = 12$ and the following screen appears. Use algebra to show that a mistake has been made. What do you think the mistake was? **[2.5b]**

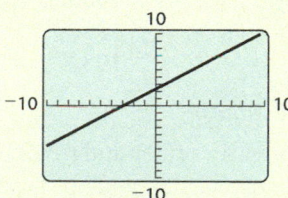

44. *Computer Repair.* The cost $R(t)$, in dollars, of computer repair at PC Pros is given by

$R(t) = 50t + 35,$

where t is the number of hours that the repair requires. Determine m and b in this application and explain their meaning. **[2.7e]**

45. Explain why the slope of a vertical line is not defined but the slope of a horizontal line is 0. **[2.6c]**

46. A student makes a mistake when using a graphing calculator to draw $5x - 2y = 3$ and the following screen appears. Use algebra to show that a mistake has been made. What do you think the mistake was? **[2.5b]**

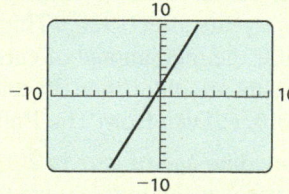

2   Chapter Test

Determine whether each correspondence is a function.

1.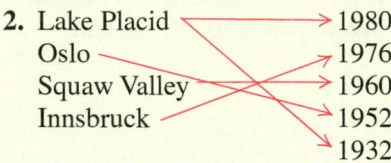
cat   dog
fish   worm
dog   cat
tiger   fish
teacher ⟶ student

2. Lake Placid   ⟶ 1980
Oslo   ⟶ 1976
Squaw Valley   ⟶ 1960
Innsbruck   ⟶ 1952
           1932

Find the function values.

3. $f(x) = -3x - 4$; $f(0)$ and $f(-2)$

4. $g(x) = x^2 + 7$; $g(0)$ and $g(-1)$

5. $h(x) = -6$; $h(-4)$ and $h(-6)$

6. $f(x) = |x + 7|$; $f(-10)$ and $f(-7)$

Graph.

7. $y = -2x - 5$

8. $f(x) = -\dfrac{3}{5}x$

9. $g(x) = 2 - |x|$

10. $f(x) = x^2 + 2x - 3$

11. $y = f(x) = -3$

12. $2x = -4$

13. *Median Age of Cars.*   The function
$$A(t) = 0.233t + 5.87$$
can be used to estimate the median age of cars in the United States t years after 1990. (This means, for example, that if the median age of cars is 3 years, then half the cars are older than 3 years and half are younger.) (*Source*: Data from The Polk Co.)

a) Find the median age of cars in 2005.

b) In what year was the median age of cars 7.734 years?

Determine whether each of the following is the graph of a function.

14.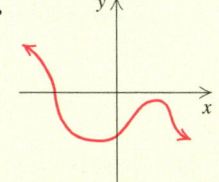

15.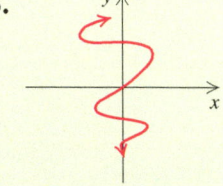

Find the domain.

16. $f(x) = \dfrac{8}{2x + 3}$ **17.** $g(x) = 5 - x^2$

18. For the following graph of function f, determine
(a) $f(1)$; **(b)** the domain; **(c)** all x-values such that $f(x) = 2$; and **(d)** the range.

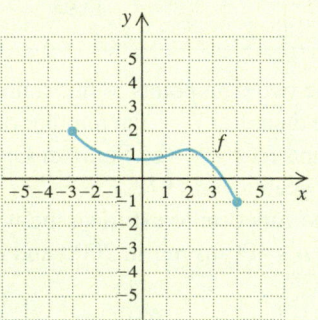

19. For $f(x) = -4x + 3$ and $g(x) = x^2 - 1$, find $(f - g)(-2)$ and $(f/g)(x)$.

20. For $f(x) = \dfrac{4}{3 - x}$ and $g(x) = 2x + 1$, determine the domain of $f + g, f - g, f \cdot g$, and f/g.

Find the slope and the y-intercept.

21. $f(x) = -\frac{3}{5}x + 12$

22. $-5y - 2x = 7$

Find the slope, if it exists, of the line containing the following points.

23. $(-2, -2)$ and $(6, 3)$

24. $(-3.1, 5.2)$ and $(-4.4, 5.2)$

25. Find the slope, or rate of change, of the graph below.

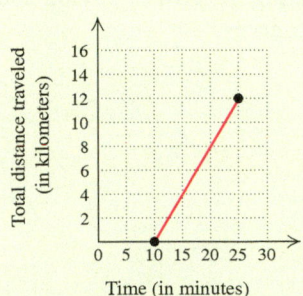

26. Find the intercepts. Then graph the equation.

$$2x + 3y = 6$$

27. Graph using the slope and the y-intercept:

$$f(x) = -\frac{2}{3}x - 1.$$

Determine whether the graphs of the given pair of lines are parallel or perpendicular.

28. $4y + 2 = 3x,$
$-3x + 4y = -12$

29. $y = -2x + 5,$
$2y - x = 6$

30. Find an equation of the line that has the given characteristics:

slope: -3; y-intercept: $(0, 4.8)$.

31. Find a linear function $f(x) = mx + b$ whose graph has the given slope and y-intercept:

slope: 5.2; y-intercept: $\left(0, -\frac{5}{8}\right)$.

32. Find an equation of the line having the given slope and containing the given point:

$m = -4$; $(1, -2)$.

33. Find an equation of the line containing the given pair of points:

$(4, -6)$ and $(-10, 15)$.

34. Find an equation of the line containing the given point and parallel to the given line:

$(4, -1)$; $x - 2y = 5$.

35. Find an equation of the line containing the given point and perpendicular to the given line:

$(2, 5)$; $x + 3y = 2$.

36. *Median Age of Men at First Marriage.*   The following table lists data regarding the median age of men at first marriage in 1970 and in 2010.

Year	Median Age of Men at First Marriage
1970, 0	23.2
2010, 40	28.2

Source: Data from U.S. Census Bureau

a) Use the two data points to find a linear function that fits the data. Let $x =$ the number of years since 1970 and $A(x) =$ the median age at first marriage x years from 1970.

b) Use the function to estimate the median age of men at first marriage in 2008 and in 2015.

37. Find an equation of the line having slope -2 and containing the point $(3, 1)$.

A. $y - 1 = 2(x - 3)$
B. $y - 1 = -2(x - 3)$
C. $x - 1 = -2(y - 3)$
D. $x - 1 = 2(y - 3)$

▶ **Synthesis**

38. Find k such that the line $3x + ky = 17$ is perpendicular to the line $8x - 5y = 26$.

39. Find a formula for a function f for which
$f(-2) = 3$.

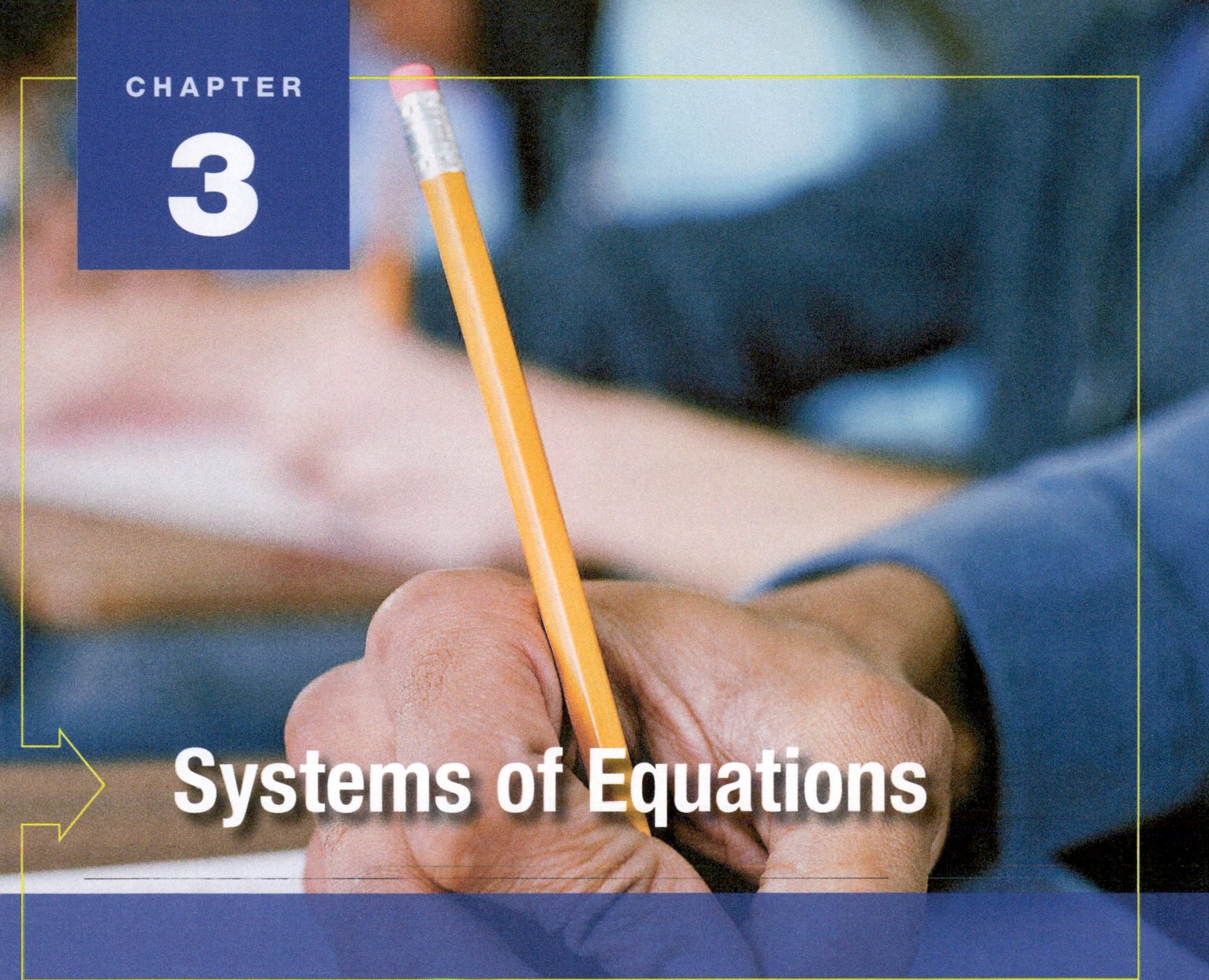

Systems of Equations

APPLICATION   This problem appears as Exercise 1 in Section 3.6.

More than 1.66 million members of the class of 2012 took the Scholastic Aptitude Test. Students taking the SAT receive a critical reading score, a mathematics score, and a writing score. The average total score of the students from the class of 2012 was 1498. The average math score exceeded the average reading score by 18 points. The average math score was 470 points less than the sum of the average reading and writing scores. (*Source*: College Board) Find the average score on each part of the test.

3.1   Systems of Equations in Two Variables

▶ **a**   Solve a system of two linear equations or two functions by graphing and determine whether a system is consistent or inconsistent and whether the equations in a system are dependent or independent.

We can solve many applied problems more easily by translating them to two or more equations in two or more variables than by translating to a single equation. Let's look at such a problem.

School Enrollment.   In 2012, approximately 50 million children were enrolled in public elementary and secondary schools in the United States. There were 20 million more students enrolled in prekindergarten–grade 8 than there were in grades 9–12. How many were enrolled at each level? (*Source*: National Center for Education Statistics)

To solve, we first let

x = the number enrolled in prekindergarten–grade 8, and

y = the number enrolled in grades 9–12,

where x and y are in millions of students. The problem gives us two statements that can be translated to equations.

First, we consider the total number enrolled:

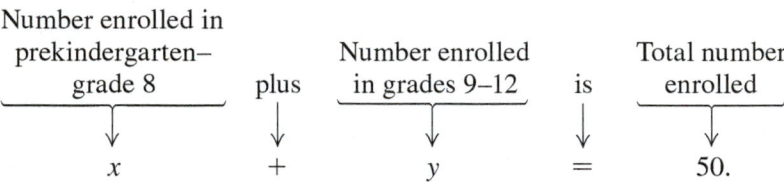

The second statement of the problem compares the enrollment at the two levels:

$$\underbrace{\text{Number enrolled in prekindergarten–grade 8}}_{x} \quad \underbrace{\text{is}}_{=} \quad \underbrace{\substack{\text{20 million more than} \\ \text{the number enrolled} \\ \text{in grades 9–12}}}_{y + 20.}$$

We have now translated the problem to a pair of equations, or a **system of equations**:

$$x + y = 50,$$
$$x = y + 20.$$

A **solution** of a system of two equations in two variables is an ordered pair that makes *both* equations true. If we graph a system of equations, the point at which the graphs intersect will be a solution of *both* equations. To find the solution of the system above, we graph both equations, as shown here.

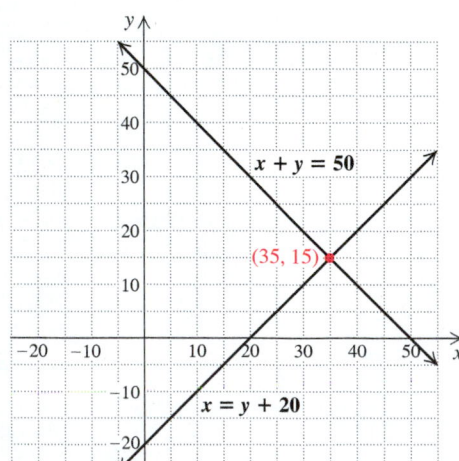

We see that the graphs intersect at the point $(35, 15)$—that is, $x = 35$ and $y = 15$. These numbers check in the statement of the original problem. This tells us that 35 million students were enrolled in prekindergarten–grade 8, and 15 million students were enrolled in grades 9–12.

▶ **a Solving Systems of Equations Graphically**

One Solution

EXAMPLE 1 Solve this system graphically:

$$y - x = 1,$$
$$y + x = 3.$$

We draw the graph of each equation and find the coordinates of the point of intersection.

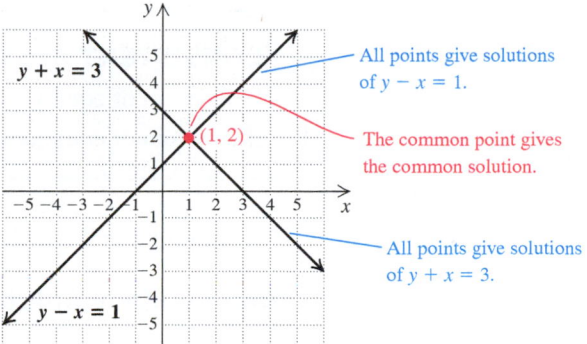

The point of intersection has coordinates that make *both* equations true. The solution seems to be the point $(1, 2)$. However, solving by graphing may give only approximate answers. Thus we check the pair $(1, 2)$ in both equations.

Check:

$y - x = 1$		$y + x = 3$	
$2 - 1 \ ? \ 1$		$2 + 1 \ ? \ 3$	
1	TRUE	3	TRUE

The solution is $(1, 2)$.

No Solution

Sometimes the equations in a system have graphs that are parallel lines.

EXAMPLE 2 Solve graphically:

$$f(x) = -3x + 5,$$
$$g(x) = -3x - 2.$$

Note that this system is written using function notation. We graph the functions. The graphs have the same slope, -3, and different y-intercepts, so they are parallel.

There is no point at which they cross, so the system has no solution. No matter what point we try, it will *not* check in *both* equations. The solution set is thus the empty set, denoted $\varnothing$, or $\{\ \}$.

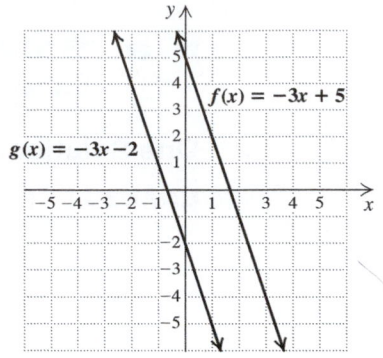

CONSISTENT SYSTEMS AND INCONSISTENT SYSTEMS

If a system of equations has at least one solution, then it is **consistent**.

If a system of equations has no solution, then it is **inconsistent**.

The system in Example 1 is consistent. The system in Example 2 is inconsistent.

Infinitely Many Solutions

Sometimes the equations in a system have the same graph. In such a case, the equations have an *infinite* number of solutions in common.

EXAMPLE 3   Solve graphically:

$$3y - 2x = 6,$$
$$-12y + 8x = -24.$$

We graph the equations and see that the graphs are the same. Thus any solution of one of the equations is a solution of the other. Each equation has an infinite number of solutions, two of which are shown on the graph.

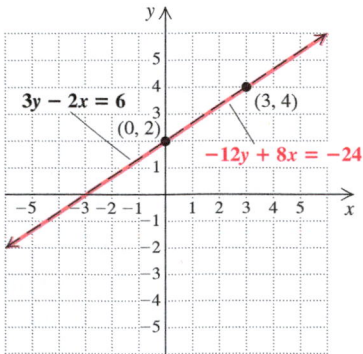

We check one such solution, $(0, 2)$, which is the y-intercept of each equation.

Check:

$$
\begin{array}{c|c}
3y - 2x = 6 & -12y + 8x = -24 \\
\hline
3(2) - 2(0) \;?\; 6 & -12(2) + 8(0) \;?\; -24 \\
6 - 0 & -24 + 0 \\
6 \quad \text{TRUE} & -24 \quad \text{TRUE}
\end{array}
$$

We leave it to the student to check that $(3, 4)$ is a solution of both equations. If $(0, 2)$ and $(3, 4)$ are solutions, then all points on the line containing them will be solutions. The system has an infinite number of solutions.

DEPENDENT EQUATIONS AND INDEPENDENT EQUATIONS

If for a system of two equations in two variables:

the graphs of the equations are the same line, then the equations are **dependent**.

the graphs of the equations are different lines, then the equations are **independent**.

When we graph a system of two equations, one of the following three things can happen.

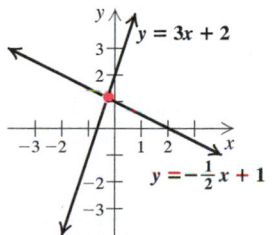

One solution.
Graphs intersect.
The system is consistent and *the equations are independent*.

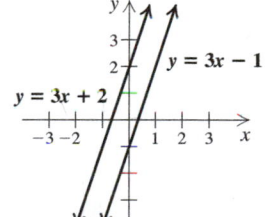

No solution.
Graphs are parallel.
The system is inconsistent and *the equations are independent*.

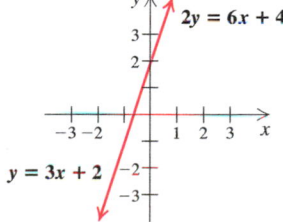

Infinitely many solutions.
Equations have the same graph. *The system is consistent* and *the equations are dependent*.

Let's summarize what we know about the systems of equations shown in Examples 1–3.

	Number of Solutions	**Graphs of Equations**
Example 1	**1** System is consistent.	**Different** Equations are independent.
Example 2	**0** System is inconsistent.	**Different** Equations are independent.
Example 3	**Infinitely many** System is consistent.	**Same** Equations are dependent.

Now Try Exercises 5, 13, and 15.

Algebraic–Graphical Connection

Consider the equation $-2x + 13 = 4x - 17$. Let's solve it algebraically:

$$-2x + 13 = 4x - 17$$

$13 = 6x - 17$	**Adding 2x**
$30 = 6x$	**Adding 17**
$5 = x.$	**Dividing by 6**

We can also solve the equation graphically, as we see in the following two methods. Using method 1, we graph two functions. The solution of the original equation is the x-coordinate of the point of intersection. Using method 2, we graph one function. The solution of the original equation is the x-coordinate of the x-intercept of the graph.

Method 1: Solve $-2x + 13 = 4x - 17$ graphically.

We let $f(x) = -2x + 13$ and $g(x) = 4x - 17$. Graphing the system of equations, we get the graph shown below.

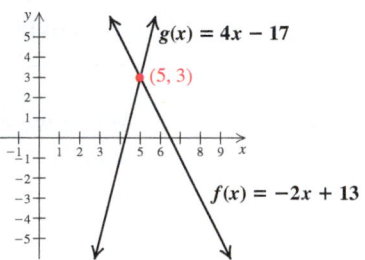

The point of intersection of the two graphs is $(5, 3)$. Note that the x-coordinate of this point is 5. This is the value of x for which $-2x + 13 = 4x - 17$, so it is the solution of the equation.

Method 2: Solve $-2x + 13 = 4x - 17$ graphically.

Adding $-4x$ and 17 on both sides, we obtain an equation with 0 on one side: $-6x + 30 = 0$. This time we let $f(x) = -6x + 30$ and $g(x) = 0$. Since the graph of $g(x) = 0$, or $y = 0$, is the x-axis, we need only graph $f(x) = -6x + 30$ and see where it crosses the x-axis.

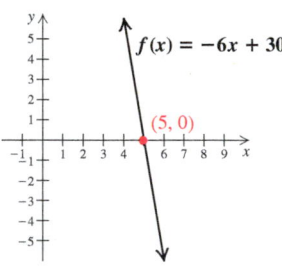

Note that the x-intercept of $f(x) = -6x + 30$ is $(5, 0)$, or just 5. This x-value is the solution of the equation $-2x + 13 = 4x - 17$.

Technology Connection

Solving Systems of Equations

We can solve a system of two equations in two variables using a graphing calculator. Consider the system of equations in Example 1:

$$y - x = 1,$$
$$y + x = 3.$$

First, we solve the equations for y, obtaining $y = x + 1$ and $y = -x + 3$. Next, we enter $y_1 = x + 1$ and $y_2 = -x + 3$ on the equation-editor screen and graph the equations. We can use the standard viewing window, $[-10, 10, -10, 10]$.

We will use the INTERSECT feature to find the coordinates of the point of intersection of the lines. To access this feature, we press **2ND** **CALC** **5**. (CALC is the second operation associated with the **TRACE** key.) The query "First curve?" appears on the graph screen. The blinking cursor is positioned on the graph of y_1. We press **ENTER** to indicate that this is the first curve involved in the intersection. Next, the query "Second curve?" appears and the blinking cursor is positioned on the graph of y_2. We press **ENTER** to indicate that this is the second curve. Now the query "Guess?" appears. We use the ▷ and ◁ keys to move the cursor close to the point of intersection or we enter an x-value close to the first coordinate of the point of intersection. Then we press **ENTER**. The coordinates of the point of intersection of the graphs, $x = 1$, $y = 2$, appear at the bottom of the screen. Thus the solution of the system of equations is $(1, 2)$.

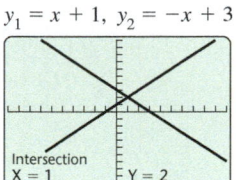

$y_1 = x + 1,\ y_2 = -x + 3$

Intersection
X = 1   Y = 2

Exercises

Use a graphing calculator to solve each system of equations.

1. $x + y = 5,$
   $y = x + 1$

2. $y = x + 3,$
   $2x - y = -7$

3. $x - y = -6,$
   $y = 2x + 7$

4. $x + 4y = -1,$
   $x - y = 4$

3.1   Exercise Set

▶ Reading Check

Determine whether each statement is true or false.

RC1. Every system of equations has one solution.

RC2. A solution of a system of equations in two variables is an ordered pair.

RC3. Graphs of two lines may have one point, no points, or an infinite number of points in common.

RC4. If a system of two equations has only one solution, the system is consistent and the equations in the system are independent.

a *Solve each system of equations graphically. Then classify the system as consistent or inconsistent and the equations as dependent or independent.*

1. $x + y = 4,$
    $x - y = 2$

2. $x - y = 3,$
    $x + y = 5$

3. $2x - y = 4,$
    $2x + 3y = -4$

4. $3x + y = 5,$
    $x - 2y = 4$

5. $2x + y = 6,$
    $3x + 4y = 4$

6. $2y = 6 - x,$
    $3x - 2y = 6$

7. $f(x) = x - 1,$
    $g(x) = -2x + 5$

8. $f(x) = x + 1,$
    $g(x) = \frac{2}{3}x$

9. $2u + v = 3,$
    $2u = v + 7$

10. $2b + a = 11,$
    $a - b = 5$

11. $f(x) = -\frac{1}{3}x - 1,$
    $g(x) = \frac{4}{3}x - 6$

12. $f(x) = -\frac{1}{4}x + 1,$
    $g(x) = \frac{1}{2}x - 2$

13. $6x - 2y = 2,$
    $9x - 3y = 1$

14. $y - x = 5,$
    $2x - 2y = 10$

15. $2x - 3y = 6,$
    $3y - 2x = -6$

16. $y = 3 - x,$
    $2x + 2y = 6$

17. $x = 4,$
    $y = -5$

18. $x = -3,$
    $y = 2$

19. $y = -x - 1,$
    $4x - 3y = 17$

20. $a + 2b = -3,$
    $b - a = 6$

Matching. *Each of Exercises 21–26 shows the graph of a system of equations and its solution. First, classify the system as consistent or inconsistent and the equations as dependent or independent. Then match it with one of the appropriate systems of equations (A)–(F), which follow.*

21. Solution: $(3, 3)$

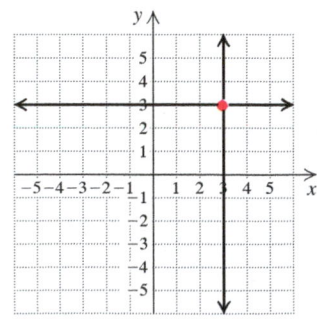

22. Solution: $(1, 1)$

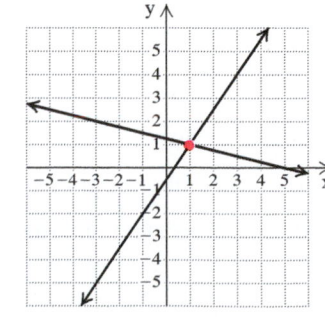

23. Solutions: Infinitely many

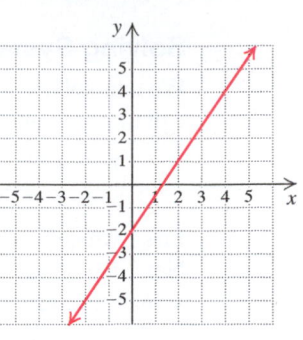

24. Solution: $(4, -3)$

25. Solution: No solution

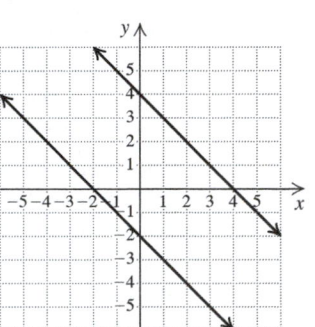

26. Solution: $(-1, 3)$

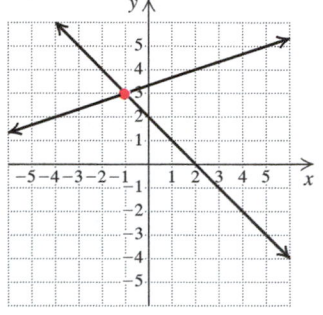

A. $3y - x = 10,$
    $x = -y + 2$

B. $9x - 6y = 12,$
    $y = \frac{3}{2}x - 2$

C. $2y - 3x = -1,$
    $x + 4y = 5$

D. $x + y = 4,$
    $y = -x - 2$

E. $\frac{1}{2}x + y = -1,$
    $y = -3$

F. $x = 3,$
    $y = 3$

▶ Skill Maintenance

Solve. **[1.1d]**

27. $3x + 4 = x - 2$

28. $\frac{3}{4}x + 2 = \frac{2}{5}x - 5$

29. $4x - 5x = 8x - 9 + 11x$

30. $5(10 - 4x) = -3(7x - 4)$

▶ Synthesis

 *Use a graphing calculator to solve each system of equations. Round all answers to the nearest hundredth. You may need to solve for y first.*

31. $2.18x + 7.81y = 13.78,$
    $5.79x - 3.45y = 8.94$

32. $f(x) = 123.52x + 89.32,$
    $g(x) = -89.22x + 33.76$

Solve graphically.

33. $y = |x|,$
    $x + 4y = 15$

34. $x - y = 0,$
    $y = x^2$

Algebra *(continued)*

The Distance Formula

The distance from (x_1, y_1) to (x_2, y_2) is given by

$$d = \sqrt{(x_2 - x_1)^2 + (y_2 - y_1)^2}.$$

The Midpoint Formula

The midpoint of the line segment from (x_1, y_1) to (x_2, y_2) is given by

$$\left(\frac{x_1 + x_2}{2}, \frac{y_1 + y_2}{2} \right).$$

Formulas Involving Lines

The slope of the line containing points (x_1, y_1) and (x_2, y_2) is given by

$$m = \frac{y_2 - y_1}{x_2 - x_1}.$$

Slope–intercept equation: $y = f(x) = mx + b$
Horizontal line: $y = b$ or $f(x) = b$
Vertical line: $x = a$
Point–slope equation: $y - y_1 = m(x - x_1)$

The Quadratic Formula

The solutions of $ax^2 + bx + c = 0, a \neq 0$, are given by

$$x = \frac{-b \pm \sqrt{b^2 - 4ac}}{2a}.$$

Compound Interest Formulas

Compounded n times per year: $A = P\left(1 + \dfrac{r}{n}\right)^{nt}$

Compounded continuously: $P(t) = P_0 e^{kt}$

Properties of Exponential and Logarithmic Functions

$$\log_a x = y \leftrightarrow x = a^y \qquad\qquad a^x = a^y \leftrightarrow x = y$$

$$\log_a MN = \log_a M + \log_a N \qquad \log_a M^p = p \log_a M$$

$$\log_a \frac{M}{N} = \log_a M - \log_a N$$

$$\log_b M = \frac{\log_a M}{\log_a b}$$

$$\log_a a = 1 \qquad\qquad\qquad \log_a 1 = 0$$

$$\log_a a^x = x \qquad\qquad\qquad a^{\log_a x} = x$$

Conic Sections

Circle: $(x - h)^2 + (y - k)^2 = r^2$

Ellipse: $\dfrac{(x - h)^2}{a^2} + \dfrac{(y - k)^2}{b^2} = 1,$

$\dfrac{(x - h)^2}{b^2} + \dfrac{(y - k)^2}{a^2} = 1$

Parabola: $(x - h)^2 = 4p(y - k),$

$(y - k)^2 = 4p(x - h)$

Hyperbola: $\dfrac{(x - h)^2}{a^2} - \dfrac{(y - k)^2}{b^2} = 1,$

$\dfrac{(y - k)^2}{a^2} - \dfrac{(x - h)^2}{b^2} = 1$

Arithmetic Sequences and Series

$$a_1, \quad a_1 + d, \quad a_1 + 2d, \quad a_1 + 3d, \ldots$$

$$a_{n+1} = a_n + d \qquad\qquad a_n = a_1 + (n - 1)d$$

$$S_n = \frac{n}{2}(a_1 + a_n)$$

Geometric Sequences and Series

$$a_1, \quad a_1 r, \quad a_1 r^2, \quad a_1 r^3, \ldots$$

$$a_{n+1} = a_n r \qquad\qquad a_n = a_1 r^{n-1}$$

$$S_n = \frac{a_1(1 - r^n)}{1 - r} \qquad\qquad S_\infty = \frac{a_1}{1 - r}, |r| < 1$$

Algebra

Properties of Real Numbers

Commutative:
$$a + b = b + a; \quad ab = ba$$

Associative:
$$a + (b + c) = (a + b) + c;$$
$$a(bc) = (ab)c$$

Additive Identity:
$$a + 0 = 0 + a = a$$

Additive Inverse:
$$-a + a = a + (-a) = 0$$

Multiplicative Identity:
$$a \cdot 1 = 1 \cdot a = a$$

Multiplicative Inverse:
$$a \cdot \frac{1}{a} = \frac{1}{a} \cdot a = 1, a \neq 0$$

Distributive:
$$a(b + c) = ab + ac$$

Exponents and Radicals

$$a^m \cdot a^n = a^{m+n} \qquad \frac{a^m}{a^n} = a^{m-n}$$

$$(a^m)^n = a^{mn} \qquad (ab)^m = a^m b^m$$

$$\left(\frac{a}{b}\right)^m = \frac{a^m}{b^m} \qquad a^{-n} = \frac{1}{a^n}$$

If n is even, $\sqrt[n]{a^n} = |a|$.

If n is odd, $\sqrt[n]{a^n} = a$.

$$\sqrt[n]{a} \cdot \sqrt[n]{b} = \sqrt[n]{ab}, \quad a, b \geq 0$$

$$\sqrt[n]{\frac{a}{b}} = \frac{\sqrt[n]{a}}{\sqrt[n]{b}}$$

$$\sqrt[n]{a^m} = (\sqrt[n]{a})^m = a^{m/n}$$

Special-Product Formulas

$$(a + b)(a - b) = a^2 - b^2$$
$$(a + b)^2 = a^2 + 2ab + b^2$$
$$(a - b)^2 = a^2 - 2ab + b^2$$
$$(a + b)^3 = a^3 + 3a^2b + 3ab^2 + b^3$$
$$(a - b)^3 = a^3 - 3a^2b + 3ab^2 - b^3$$
$$(a + b)^n = \sum_{k=0}^{n} \binom{n}{k} a^{n-k}b^k, \text{ where}$$
$$\binom{n}{k} = \frac{n!}{k!(n-k)!}$$
$$= \frac{n(n-1)(n-2)\cdots[n-(k-1)]}{k!}$$

Factoring Formulas

$$a^2 - b^2 = (a + b)(a - b)$$
$$a^2 + 2ab + b^2 = (a + b)^2$$
$$a^2 - 2ab + b^2 = (a - b)^2$$
$$a^3 + b^3 = (a + b)(a^2 - ab + b^2)$$
$$a^3 - b^3 = (a - b)(a^2 + ab + b^2)$$

Interval Notation

$$(a, b) = \{x \mid a < x < b\}$$
$$[a, b] = \{x \mid a \leq x \leq b\}$$
$$(a, b] = \{x \mid a < x \leq b\}$$
$$[a, b) = \{x \mid a \leq x < b\}$$
$$(-\infty, a) = \{x \mid x < a\}$$
$$(a, \infty) = \{x \mid x > a\}$$
$$(-\infty, a] = \{x \mid x \leq a\}$$
$$[a, \infty) = \{x \mid x \geq a\}$$

Absolute Value

$$|a| \geq 0$$

For $a > 0$,
$$|X| = a \rightarrow X = -a \quad or \quad X = a,$$
$$|X| < a \rightarrow -a < X < a,$$
$$|X| > a \rightarrow X < -a \quad or \quad X > a.$$

Equation-Solving Principles

$$a = b \rightarrow a + c = b + c$$
$$a = b \rightarrow ac = bc$$
$$a = b \rightarrow a^n = b^n$$
$$ab = 0 \leftrightarrow a = 0 \quad or \quad b = 0$$
$$x^2 = k \rightarrow x = \sqrt{k} \quad or \quad x = -\sqrt{k}$$

Inequality-Solving Principles

$$a < b \rightarrow a + c < b + c$$
$$a < b \text{ and } c > 0 \rightarrow ac < bc$$
$$a < b \text{ and } c < 0 \rightarrow ac > bc$$

(Algebra continued)

Geometry

Plane Geometry

Rectangle
Area: $A = lw$
Perimeter: $P = 2l + 2w$

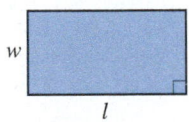

Square
Area: $A = s^2$
Perimeter: $P = 4s$

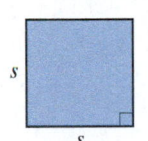

Triangle
Area: $A = \frac{1}{2}bh$

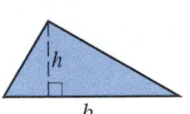

Sum of Angle Measures
$A + B + C = 180°$

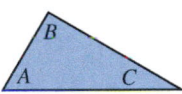

Right Triangle
Pythagorean theorem (equation):
$$a^2 + b^2 = c^2$$

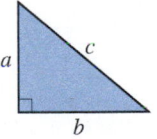

Parallelogram
Area: $A = bh$

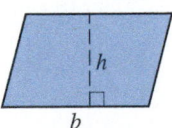

Trapezoid
Area: $A = \frac{1}{2}h(a + b)$

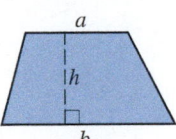

Circle
Area: $A = \pi r^2$
Circumference:
$C = \pi d = 2\pi r$

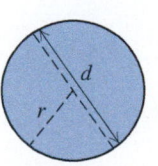

Solid Geometry

Rectangular Solid
Volume: $V = lwh$

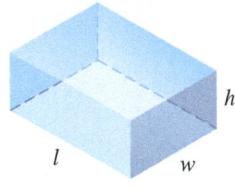

Cube
Volume: $V = s^3$

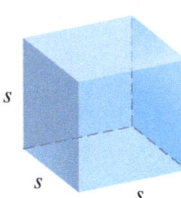

Right Circular Cylinder
Volume: $V = \pi r^2 h$
Lateral surface area:
$L = 2\pi rh$
Total surface area:
$S = 2\pi rh + 2\pi r^2$

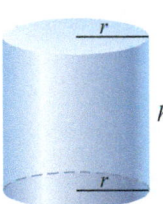

Right Circular Cone
Volume: $V = \frac{1}{3}\pi r^2 h$
Lateral surface area:
$L = \pi rs$
Total surface area:
$S = \pi r^2 + \pi rs$
Slant height:
$s = \sqrt{r^2 + h^2}$

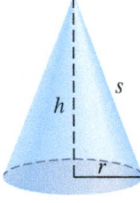

Sphere
Volume: $V = \frac{4}{3}\pi r^3$
Surface area: $S = 4\pi r^2$

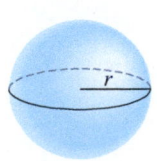

Chapter 9

Page 591: Igor Mojzes/Fotolia **Page 592:** Pkstock/Fotolia **Page 598:** Africa Studio/Fotolia **Page 599:** Blend Images/Glow Images **Page 609:** David Madison/Digital Vision/Getty Images **Page 610:** Yanik Chauvin/Fotolia **Page 619:** Kolett/Fotolia; Viappy/Fotolia **Page 620:** Amble Design/Shutterstock; Igor Mojzes/Fotolia; Gladskikh Tatiana/Shutterstock **Page 621:** Jean-Marie Guyon/123rf **Page 634:** Huaji/Shutterstock **Page 635:** Stephen Rees/Shutterstock **Page 660:** Wiktor Bubniak/Fotolia **Page 663:** Zev Radovan/BibleLandPicturesAlamy **Page 664:** Andrey_Popov/Shutterstock **Page 665:** STEVE HEALEY; Nyiragongo/Fotolia **Page 666:** Cozyta/Fotolia; Jonathan Little/Alamy; Petoo/Fotolia **Page 667:** AP Images; Rjack/Fotolia

Chapter 10

Page 681: Epa european pressphoto agency b.v/Alamy **Page 689:** Roman Milert/123rd **Page 697:** GWImages/Shutterstock **Page 698:** Aleph Studio/Shutterstock **Page 706:** Epa european pressphoto agency b.v/Alamy

Chapter 11

Page 721: Wiskerke/Alamy **Page 729:** Information Unlimited **Page 736:** Jason Tench/Shutterstock **Page 738:** Wiskerke/Alamy **Page 747:** Frank Oppermann/Shutterstock **Page 759:** Fotosearch LBRF/Age footstock; Denis Nata/Fotolia; Pavel Losevsky/Shutterstock

Chapter 12

Page 769: Jeff Lewis/Icon SMI 910/Newscom **Page 776:** Carola Schubbel/Fotolia **Page 783:** Epitavi/Shutterstock **Page 784:** Shariff Che'Lah/123rf **Page 790:** Jeff Lewis/Icon SMI 910/Newscom **Page 823:** Hero Images/Getty Images **Page 829:** Vojtech Vlk/Fotolia **Page 835:** John Greim/Age Fotostock

Credits

3.2 Solving by Substitution

▶ **a** Solve systems of equations in two variables by the substitution method.

▶ **b** Solve applied problems by solving systems of two equations using substitution.

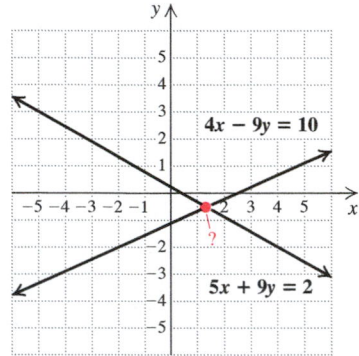

Consider this system of equations:

$$5x + 9y = 2,$$
$$4x - 9y = 10.$$

What is the solution? It is rather difficult to tell exactly by graphing. It would appear that fractions are involved. It turns out that the solution is

$$\left(\frac{4}{3}, -\frac{14}{27}\right).$$

Solving by graphing, though useful in many applied situations, is not always fast or accurate in cases where solutions are not integers. We need techniques involving algebra to determine the solution exactly. Because these techniques use algebra, they are called **algebraic methods**.

▶ a The Substitution Method

One nongraphical method for solving systems is known as the **substitution method**.

EXAMPLE 1 Solve this system:

$$x + y = 4, \qquad \textbf{(1)}$$
$$x = y + 1. \qquad \textbf{(2)}$$

Equation (2) says that x and $y + 1$ name the same number. Thus we can substitute $y + 1$ for x in equation (1):

$$x + y = 4 \qquad \text{\color{red}{Equation (1)}}$$
$$(y + 1) + y = 4. \qquad \text{\color{red}{Substituting } y + 1 \text{ for } x}$$

Since this equation has only one variable, we can solve for y using methods learned earlier:

$$(y + 1) + y = 4$$
$$2y + 1 = 4 \qquad \text{\color{red}{Removing parentheses and collecting like terms}}$$
$$2y = 3 \qquad \text{\color{red}{Subtracting 1}}$$
$$y = \tfrac{3}{2}. \qquad \text{\color{red}{Dividing by 2}}$$

We return to the original pair of equations and substitute $\frac{3}{2}$ for y in *either* equation so that we can solve for x. Calculation will be easier if we choose equation (2) since it is already solved for x:

$$x = y + 1 \qquad \text{\color{red}{Equation (2)}}$$
$$= \tfrac{3}{2} + 1 \qquad \text{\color{red}{Substituting } \tfrac{3}{2} \text{ for } y}$$
$$= \tfrac{3}{2} + \tfrac{2}{2} = \tfrac{5}{2}.$$

We obtain the ordered pair $\left(\frac{5}{2}, \frac{3}{2}\right)$. Even though we solved for *y first*, it is still the *second* coordinate since *x* is before *y* alphabetically. We check to be sure that the ordered pair is a solution.

Check:

$$x + y = 4$$
$$\frac{5}{2} + \frac{3}{2} \;?\; 4$$
$$\frac{8}{2}$$
$$4 \qquad \text{TRUE}$$

$$x = y + 1$$
$$\frac{5}{2} \;?\; \frac{3}{2} + 1$$
$$\frac{3}{2} + \frac{2}{2}$$
$$\frac{5}{2} \qquad \text{TRUE}$$

Since $\left(\frac{5}{2}, \frac{3}{2}\right)$ checks, it is the solution. Even though exact fraction solutions are difficult to determine graphically, a graph can help us to visualize whether the solution is reasonable.

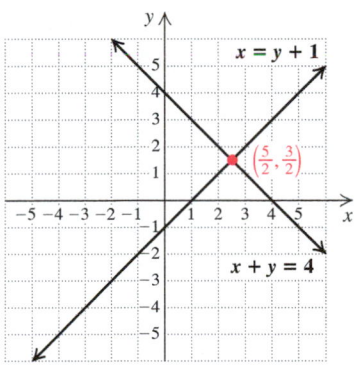

Now Try Exercise 3.

Suppose neither equation of a pair has a variable alone on one side. We then solve one equation for one of the variables.

EXAMPLE 2   Solve this system:

$$2x + y = 6, \qquad \text{(1)}$$
$$3x + 4y = 4. \qquad \text{(2)}$$

First, we solve one equation for one variable. Since the coefficient of *y* is 1 in equation (1), it is the easier one to solve for *y*:

$$y = 6 - 2x. \qquad \text{(3)}$$

Next, we substitute $6 - 2x$ for *y* in equation (2) and solve for *x*:

$$3x + 4(6 - 2x) = 4 \qquad \text{**Substituting $6 - 2x$ for y**}$$

> **CAUTION!**   Remember to use parentheses when you substitute. Then remove them properly.

$$3x + 24 - 8x = 4 \qquad \text{**Multiplying to remove parentheses**}$$
$$24 - 5x = 4 \qquad \text{**Collecting like terms**}$$
$$-5x = -20 \qquad \text{**Subtracting 24**}$$
$$x = 4. \qquad \text{**Dividing by -5**}$$

In order to find *y*, we return to either of the original equations, (1) or (2), or equation (3), which we solved for *y*. It is generally easier to use an equation like (3), where we have solved for the specific variable. We substitute 4 for *x* in equation (3) and solve for *y*:

$$y = 6 - 2x = 6 - 2(4) = 6 - 8 = -2.$$

We obtain the ordered pair $(4, -2)$.

Check:

$2x + y = 6$		$3x + 4y = 4$	
$2(4) + (-2)$? 6		$3(4) + 4(-2)$? 4	
$8 - 2$		$12 - 8$	
6	TRUE	4	TRUE

Since $(4, -2)$ checks, it is the solution.

→ **Now Try Exercise 5.**

EXAMPLE 3 Solve this system of equations:

$$y = -3x + 5, \quad \text{(1)}$$
$$y = -3x - 2. \quad \text{(2)}$$

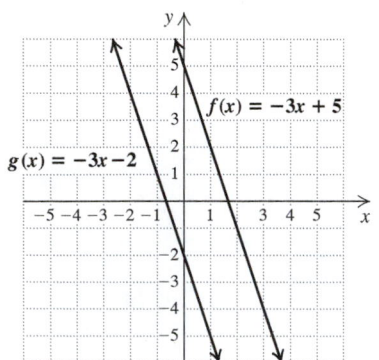

$g(x) = -3x - 2$
$f(x) = -3x + 5$

The graphs of the equations in the system are shown at left. Since the graphs are parallel, there is no solution. Let's try to solve this system algebraically using substitution. We substitute $-3x - 2$ for y in equation (1):

$$-3x - 2 = -3x + 5 \qquad \text{Substituting } -3x - 2 \text{ for } y$$
$$-2 = 5. \qquad \text{Adding } 3x$$

We have a false equation. The equation has no solution. This means that the system has **no solution**.

→ **Now Try Exercise 11.**

EXAMPLE 4 Solve this system of equations:

$$x = 2y - 1, \quad \text{(1)}$$
$$4y - 2x = 2. \quad \text{(2)}$$

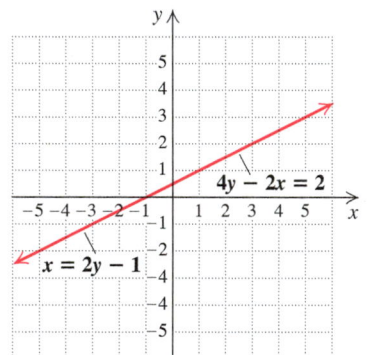

$4y - 2x = 2$
$x = 2y - 1$

The graphs of the equations in the system are shown at left. Since the graphs are the same, there is an infinite number of solutions.

Let's try to solve this system algebraically using substitution. We substitute $2y - 1$ for x in equation (2):

$$4y - 2(2y - 1) = 2 \qquad \text{Substituting } 2y - 1 \text{ for } x$$
$$4y - 4y + 2 = 2 \qquad \text{Removing parentheses}$$
$$2 = 2. \qquad \text{Simplifying; } 4y - 4y = 0$$

We have a true equation. Any value of y will make this equation true. This means that the system has **infinitely many solutions**.

→ **Now Try Exercise 15.**

SPECIAL CASES

When solving a system of two linear equations in two variables:

1. If a false equation is obtained, then the system has no solution.
2. If a true equation is obtained, then the system has an infinite number of solutions.

▶ **b Solving Applied Problems Involving Two Equations**

Many applied problems are easier to solve if we first translate to a system of two equations rather than to a single equation.

EXAMPLE 5 *Architecture.* The architects who designed the John Hancock Building in Chicago created a visually appealing building that slants on the sides. The ground floor is in the shape of a rectangle that is larger than the rectangle formed by the top floor. The ground floor has a perimeter of 860 ft. The length is 100 ft more than the width. Find the length and the width.

1. **Familiarize.** We first make a drawing and label it, using l for length and w for width. We recall or look up the formula for perimeter: $P = 2l + 2w$. This formula can be found at the back of this book.

2. **Translate.** We translate as follows:

$$\underbrace{\text{The perimeter}}_{\downarrow} \quad \underset{\downarrow}{\text{is}} \quad \underset{\downarrow}{860 \text{ ft.}}$$
$$2l + 2w \quad = \quad 860.$$

We can also write a second equation:

$$\underbrace{\text{The length}}_{\downarrow} \quad \underset{\downarrow}{\text{is}} \quad \underbrace{\begin{array}{c}100 \text{ ft more than} \\ \text{the width.}\end{array}}_{\downarrow}$$
$$l \quad = \quad w + 100.$$

We now have a system of equations:

$$2l + 2w = 860, \qquad (1)$$
$$l = w + 100. \qquad (2)$$

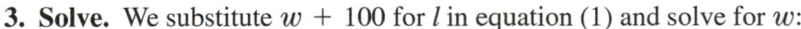

$l = w + 100$ w

3. **Solve.** We substitute $w + 100$ for l in equation (1) and solve for w:

$2(w + 100) + 2w = 860$	**Substituting in equation (1)**
$2w + 200 + 2w = 860$	**Multiplying to remove parentheses**
$4w + 200 = 860$	**Collecting like terms**
$\left.\begin{array}{r} 4w = 660 \\ w = 165. \end{array}\right\}$	**Solving for w**

Next, we substitute 165 for w in equation (2) and solve for l:

$$l = 165 + 100 = 265.$$

4. **Check.** Consider the dimensions 265 ft and 165 ft. The length is 100 ft more than the width. The perimeter is $2(265 \text{ ft}) + 2(165 \text{ ft})$, or 860 ft. The dimensions 265 ft and 165 ft check in the original problem.

5. **State.** The length is 265 ft, and the width is 165 ft.

Now Try Exercise 19.

3.2 Exercise Set

▶ Reading Check

Determine whether each statement is true or false.

RC1. The substitution method is an algebraic method for solving systems of equations.

RC2. We can find solutions of systems of equations involving fractions using the substitution method.

RC3. When we are writing the solution of a system, the value that we found first is always the first number in the ordered pair.

RC4. When solving using substitution, if we obtain a false equation, then the system has infinitely many solutions.

a *Solve each system of equations by the substitution method.*

1. $y = 5 - 4x,$
$2x - 3y = 13$

2. $x = 8 - 4y,$
$3x + 5y = 3$

3. $2y + x = 9,$
$x = 3y - 3$

4. $9x - 2y = 3,$
$3x - 6 = y$

5. $3s - 4t = 14,$
$5s + t = 8$

6. $m - 2n = 3,$
$4m + n = 1$

7. $9x - 2y = -6,$
$7x + 8 = y$

8. $t = 4 - 2s,$
$t + 2s = 6$

9. $-5s + t = 11,$
$4s + 12t = 4$

10. $5x + 6y = 14,$
$-3y + x = 7$

11. $2x - 3 = y,$
$y - 2x = 1$

12. $4p - 2w = 16,$
$5p + 7w = 1$

13. $3a - b = 7,$
$2a + 2b = 5$

14. $3x + y = 4,$
$12 - 3y = 9x$

15. $2x - 6y = 4,$
$3y + 2 = x$

16. $5x + 3y = 4,$
$x - 4y = 3$

17. $2x + 2y = 2,$
$3x - y = 1$

18. $4x + 13y = 5,$
$-6x + y = 13$

b *Solve.*

19. *Archaeology.* The remains of an ancient ball court in Monte Alban, Mexico, include a rectangular playing alley with a perimeter of about 60 m. The length of the alley is five times the width. Find the length and the width of the playing alley.

20. *Soccer Field.* The perimeter of a soccer field is 340 m. The length exceeds the width by 50 m. Find the length and the width.

21. *Supplementary Angles.*   **Supplementary angles** are angles whose sum is 180°. Two supplementary angles are such that one angle is 12° less than three times the other. Find the measures of the angles.

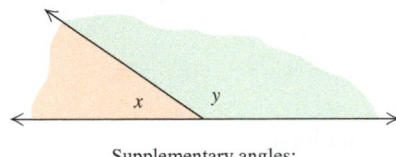

Supplementary angles:
$x + y = 180°$

22. *Complementary Angles.*   **Complementary angles** are angles whose sum is 90°. Two complementary angles are such that one angle is 6° more than five times the other. Find the measures of the angles.

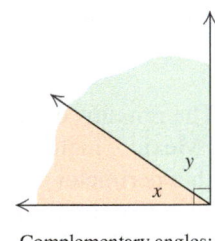

Complementary angles:
$x + y = 90°$

23. *Hockey Points.*   At one time, hockey teams received two points when they won a game and one point when they tied. One season, a team won a championship with 60 points. They won 9 more games than they tied. How many wins and how many ties did the team have?

24. *Airplane Seating.*   An airplane has a total of 152 seats. The number of coach-class seats is 5 more than six times the number of first-class seats. How many of each type of seat are there on the plane?

▶ **Skill Maintenance**

25. Find the slope of the line $y = 1.3x - 7$. **[2.4b]**

26. Simplify: $-9(y + 7) - 6(y - 4)$. **[R.6b]**

27. Solve $A = \dfrac{pq}{7}$ for p. **[1.2a]**

28. Find the slope of the line containing the points $(-2, 3)$ and $(-5, -4)$. **[2.4b]**

Solve. **[1.1d]**

29. $-4x + 5(x - 7) = 8x - 6(x + 2)$

30. $-12(2x - 3) = 16(4x - 5)$

▶ **Synthesis**

31. Two solutions of $y = mx + b$ are $(1, 2)$ and $(-3, 4)$. Find m and b.

32. Solve for x and y in terms of a and b:
$$5x + 2y = a,$$
$$x - y = b.$$

33. *Design.*   A piece of posterboard has a perimeter of 156 in. If you cut 6 in. off the width, the length becomes four times the width. What are the dimensions of the original piece of posterboard?

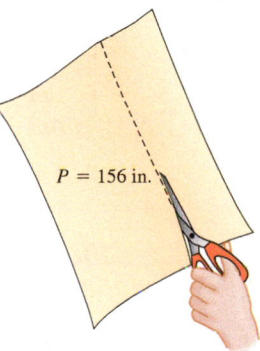

$P = 156$ in.

34. *Nontoxic Scouring Powder.*   A nontoxic scouring powder is made up of 4 parts baking soda and 1 part vinegar. How much of each ingredient is needed for a 16-oz mixture?

3.3   Solving by Elimination

▶ **a**   Solve systems of equations in two variables by the elimination method.

▶ **b**   Solve applied problems by solving systems of two equations using elimination.

▶ **a   The Elimination Method**

The **elimination method** for solving systems of equations makes use of the *addition principle* for equations. Some systems are much easier to solve using the elimination method rather than the substitution method.

EXAMPLE 1 Solve this system:

$$2x - 3y = 0, \qquad \textbf{(1)}$$
$$-4x + 3y = -1. \qquad \textbf{(2)}$$

The key to the advantage of the elimination method in this case is the $-3y$ in one equation and the $3y$ in the other. These terms are opposites. If we add them, these terms will add to 0, and in effect, the variable y will have been "eliminated."

We will use the addition principle for equations, adding the same number on both sides of the equation. According to equation (2), $-4x + 3y$ and -1 are the same number. Thus we can use a vertical form and add $-4x + 3y$ on the left side of equation (1) and -1 on the right side:

$$
\begin{array}{ll}
2x - 3y = 0 & \textbf{(1)} \\
\underline{-4x + 3y = -1} & \textbf{(2)} \\
-2x + 0y = -1 & \text{\color{red}Adding} \\
-2x + 0 = -1 & \\
-2x = -1. &
\end{array}
$$

We have eliminated the variable y, which is why we call this the *elimination method.** We now have an equation with just one variable, which we solve for x:

$$-2x = -1$$
$$x = \tfrac{1}{2}.$$

Next, we substitute $\tfrac{1}{2}$ for x in either equation and solve for y:

$$
\begin{array}{ll}
2 \cdot \tfrac{1}{2} - 3y = 0 & \text{\color{red}Substituting in equation (1)} \\
1 - 3y = 0 & \\
-3y = -1 & \text{\color{red}Subtracting 1} \\
y = \tfrac{1}{3}. & \text{\color{red}Dividing by } -3
\end{array}
$$

We obtain the ordered pair $\left(\tfrac{1}{2}, \tfrac{1}{3}\right)$.

Check:

$$
\begin{array}{c|c}
2x - 3y = 0 & -4x + 3y = -1 \\
\hline
2\left(\tfrac{1}{2}\right) - 3\left(\tfrac{1}{3}\right) \;?\; 0 & -4\left(\tfrac{1}{2}\right) + 3\left(\tfrac{1}{3}\right) \;?\; -1 \\
1 - 1 & -2 + 1 \\
0 \;\big|\; \text{\small TRUE} & -1 \;\big|\; \text{\small TRUE}
\end{array}
$$

Since $\left(\tfrac{1}{2}, \tfrac{1}{3}\right)$ checks, it is the solution. We can also see this in the graph shown at left.

▶ Now Try Exercise 3.

In order to eliminate a variable, we sometimes use the multiplication principle to multiply one or both of the equations by a particular number before adding.

EXAMPLE 2 Solve this system:

$$3x + 3y = 15, \qquad \textbf{(1)}$$
$$2x + 6y = 22. \qquad \textbf{(2)}$$

---

* This method is also called the *addition method.*

If we add directly, we will not eliminate a variable. However, note that if the $3y$ in equation (1) were $-6y$, we could eliminate y. Thus we multiply by -2 on both sides of equation (1) and add:

$$
\begin{array}{ll}
-6x - 6y = -30 & \text{\color{red}\textbf{Multiplying by} -2 \textbf{on both sides of equation (1)}} \\
\underline{2x + 6y = 22} & \text{\color{red}\textbf{Equation (2)}} \\
-4x + 0 = -8 & \text{\color{red}\textbf{Adding}} \\
-4x = -8 & \\
x = 2. & \text{\color{red}\textbf{Solving for} x}
\end{array}
$$

Then

$$
\begin{array}{ll}
2 \cdot 2 + 6y = 22 & \text{\color{red}\textbf{Substituting 2 for} x \textbf{in equation (2)}} \\
4 + 6y = 22 & \\
6y = 18 & \text{\color{red}\textbf{Solving for} y} \\
y = 3. &
\end{array}
$$

We obtain $(2, 3)$, or $x = 2$, $y = 3$.

Check:

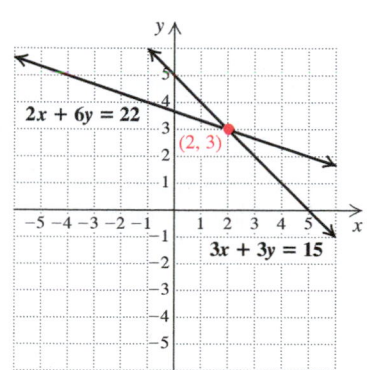

$$
\begin{array}{c|c}
3x + 3y = 15 & 2x + 6y = 22 \\
\hline
3(2) + 3(3) \overset{?}{} 15 & 2(2) + 6(3) \overset{?}{} 22 \\
6 + 9 & 4 + 18 \\
15 \quad \text{\small TRUE} & 22 \quad \text{\small TRUE}
\end{array}
$$

Since $(2, 3)$ checks, it is the solution. We can also see this in the graph at left.

➔ **Now Try Exercise 7.**

Sometimes we must multiply twice in order to make two terms opposites.

EXAMPLE 3 Solve this system:

$$
\begin{array}{ll}
2x + 3y = 17, & \text{\color{red}(1)} \\
5x + 7y = 29. & \text{\color{red}(2)}
\end{array}
$$

We must first multiply in order to make one pair of terms with the same variable opposites. We decide to do this with the x-terms in each equation. We multiply equation (1) by 5 and equation (2) by -2. Then we get $10x$ and $-10x$, which are opposites.

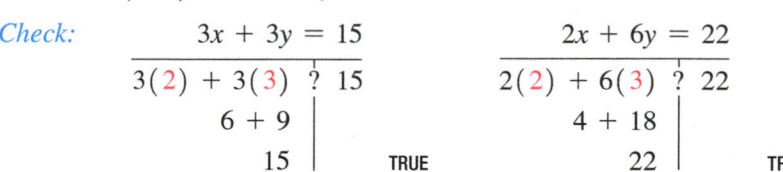

$$
\begin{array}{lll}
\textit{From equation (1):} & 10x + 15y = 85 & \text{\color{red}\textbf{Multiplying by 5}} \\
\textit{From equation (2):} & \underline{-10x - 14y = -58} & \text{\color{red}\textbf{Multiplying by} -2} \\
& 0 + y = 27 & \text{\color{red}\textbf{Adding}} \\
& y = 27 & \text{\color{red}\textbf{Solving for} y}
\end{array}
$$

Then

$$
\begin{array}{ll}
2x + 3 \cdot 27 = 17 & \text{\color{red}\textbf{Substituting 27 for} y \textbf{in equation (1)}} \\
2x + 81 = 17 & \\
2x = -64 & \text{\color{red}\textbf{Solving for} x} \\
x = -32. &
\end{array}
$$

We check the ordered pair $(-32, 27)$.

Check:

$$2x + 3y = 17$$

$$\overline{2(-32) + 3(27) \overset{?}{\;} 17}$$

$$-64 + 81$$

$$17 \quad \bigg| \quad \text{TRUE}$$

$$5x + 7y = 29$$

$$\overline{5(-32) + 7(27) \overset{?}{\;} 29}$$

$$-160 + 189$$

$$29 \quad \bigg| \quad \text{TRUE}$$

We obtain $(-32, 27)$, or $x = -32$, $y = 27$, as the solution.

➤ **Now Try Exercise 13.**

When solving a system of equations using the elimination method, it helps to first write the equations in the form $Ax + By = C$. When decimals or fractions occur, it also helps to *clear* before solving.

EXAMPLE 4   Solve this system:

$$0.2x + 0.3y = 1.7,$$

$$\tfrac{1}{7}x + \tfrac{1}{5}y = \tfrac{29}{35}.$$

We have

$$0.2x + 0.3y = 1.7, \;\longrightarrow \overset{\textstyle\color{red}{\textbf{Multiplying by 10}}}{\color{red}{\textbf{to clear decimals}}} \longrightarrow 2x + 3y = 17,$$

$$\tfrac{1}{7}x + \tfrac{1}{5}y = \tfrac{29}{35} \;\longrightarrow \overset{\textstyle\color{red}{\textbf{Multiplying by 35}}}{\color{red}{\textbf{to clear fractions}}} \longrightarrow 5x + 7y = 29.$$

We multiplied by 10 to clear the decimals. Multiplication by 35, the least common denominator, clears the fractions. The problem is now identical to Example 3. The solution is $(-32, 27)$, or $x = -32$, $y = 27$.

➤ **Now Try Exercise 23.**

> To use the elimination method to solve systems of two equations:
>
> 1. Write both equations in the form $Ax + By = C$.
> 2. Clear any decimals or fractions.
> 3. Choose a variable to eliminate.
> 4. Make the chosen variable's terms opposites by multiplying one or both equations by appropriate numbers if necessary.
> 5. Eliminate a variable by adding the respective sides of the equations and then solve for the remaining variable.
> 6. Substitute in either of the original equations to find the value of the other variable.

Some systems have no solution. How do we recognize such systems if we are solving using elimination?

EXAMPLE 5   Solve this system:

$$y + 3x = 5, \qquad \text{(1)}$$

$$y + 3x = -2. \qquad \text{(2)}$$

If we find the slope–intercept equations for this system, we get

$$y = -3x + 5,$$

$$y = -3x - 2.$$

The graphs are parallel lines. The system has no solution.

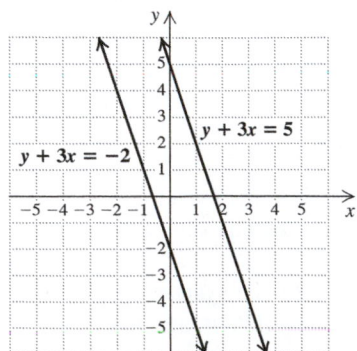

Let's attempt to solve the system by the elimination method:

$$y + 3x = 5 \qquad \text{Equation (1)}$$
$$\underline{-y - 3x = 2} \qquad \text{Multiplying equation (2) by } -1$$
$$0 = 7. \qquad \text{Adding, we obtain a false equation.}$$

The *x*-terms and the *y*-terms are eliminated and we have a *false* equation. If we obtain a false equation, such as $0 = 7$, when solving algebraically, we know that the system has **no solution**. The system is inconsistent, and the equations are independent.

 **Now Try Exercise 15.**

Some systems have infinitely many solutions. How can we recognize such a situation when we are solving systems using an algebraic method?

EXAMPLE 6 Solve this system:

$$3y - 2x = 6, \qquad \textbf{(1)}$$
$$-12y + 8x = -24. \qquad \textbf{(2)}$$

The graphs are the same line. The system has an infinite number of solutions.

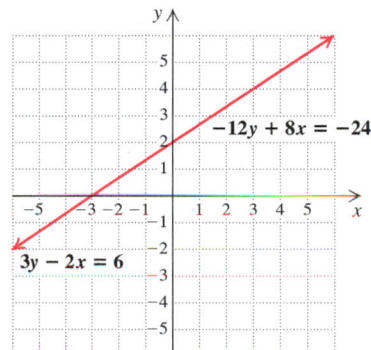

Suppose we try to solve this system by the elimination method:

$$12y - 8x = 24 \qquad \text{Multiplying equation (1) by 4}$$
$$\underline{-12y + 8x = -24} \qquad \text{Equation (2)}$$
$$0 = 0. \qquad \text{Adding, we obtain a true equation.}$$

We have eliminated both variables, and what remains is a true equation, $0 = 0$. It can be expressed as $0 \cdot x + 0 \cdot y = 0$, and is true for all numbers *x* and *y*. If an ordered pair is a solution of one of the original equations, then it will be a solution of the other. The system has an **infinite number of solutions**. The system is consistent, and the equations are dependent.

Now Try Exercise 9.

SPECIAL CASES

When solving a system of two linear equations in two variables:

1. If a false equation is obtained, such as $0 = 7$, then the system has no solution. The system is *inconsistent*, and the equations are *independent*.
2. If a true equation is obtained, such as $0 = 0$, then the system has an infinite number of solutions. The system is *consistent*, and the equations are *dependent*.

Comparing Methods

We can solve systems of equations graphically, or we can solve them algebraically using substitution or elimination. When deciding which method to use, consider the information in this table as well as directions from your instructor.

Method	Strengths	Weaknesses
Graphical	Can "see" solutions.	Inexact when solutions involve numbers that are not integers. Solutions may not appear on the part of the graph drawn.
Substitution	Yields exact solutions. Convenient to use when a variable has a coefficient of 1.	Can introduce extensive computations with fractions. Cannot "see" solutions quickly.
Elimination	Yields exact solutions. Convenient to use when no variable has a coefficient of 1. The preferred method for systems of three or more equations in three or more variables. (See Section 3.5.)	Cannot "see" solutions quickly.

▶ b Solving Applied Problems Using Elimination

Let's now solve an applied problem using the elimination method.

EXAMPLE 7 *Stimulating the Hometown Economy.* To stimulate the economy in his town of Brewton, Alabama, in 2009, Danny Cottrell, co-owner of The Medical Center Pharmacy, gave each of his full-time employees $700 and each part-time employee $300. He asked that each person donate 15% to a charity of his or her choice and spend the rest locally. The money was paid in $2 bills, a rarely used currency, so that the business community could easily see how the money circulated. Cottrell gave away a total of $16,000 to his 24 employees. How many full-time employees and how many part-time employees were there? (*Source: The Press-Register,* March 4, 2009)

1. **Familiarize.** We let $f =$ the number of full-time employees and $p =$ the number of part-time employees. Each full-time employee received $700, so a total of $700f$ was paid to them. Similarly, the part-time employees received a total of $300p$. Thus a total of $700f + 300p$ was given away.

2. **Translate.** We translate to two equations.

$$\underbrace{\text{Total amount given away}}_{700f + 300p} \quad \underset{=}{\text{is}} \quad \underset{16{,}000}{\$16{,}000.}$$

$$\underbrace{\text{Total number of employees}}_{f + p} \quad \underset{=}{\text{is}} \quad \underset{24}{24.}$$

We now have a system of equations:

$$700f + 300p = 16{,}000, \qquad (1)$$
$$f + p = 24. \qquad (2)$$

3. **Solve.** First, we multiply by -300 on both sides of equation (2) and add:

$$
\begin{array}{rll}
700f + 300p & = 16{,}000 & \text{Equation (1)} \\
-300f - 300p & = -7200 & \text{Multiplying by } -300 \text{ on both sides of equation (2)} \\
\hline
400f & = 8800 & \text{Adding} \\
f & = 22. & \text{Solving for } f
\end{array}
$$

Next, we substitute 22 for f in equation (2) and solve for p:

$$22 + p = 24$$
$$p = 2.$$

4. **Check.** If there are 22 full-time employees and 2 part-time employees, there is a total of $22 + 2$, or 24, employees. The 22 full-time employees received a total of $\$700 \cdot 22$, or $\$15{,}400$, and the 2 part-time employees received a total of $\$300 \cdot 2$, or $\$600$. Then a total of $\$15{,}400 + \$600$, or $\$16{,}000$, was given away. The numbers check in the original problem.

5. **State.** There were 22 full-time employees and 2 part-time employees.

> **Now Try Exercise 35.**

3.3   Exercise Set

▶ Reading Check

Choose from the column on the right the word that best completes each sentence. Words may be used more than once.

RC1. If a system of equations has a solution, then it is _____ .

RC2. If a system of equations has no solution, then it is _____ .

RC3. If a system of equations has infinitely many solutions, then it is _____ .

RC4. If the graphs of the equations in a system of two equations in two variables are the same line, then the equations are _____ .

RC5. If the graphs of the equations in a system of two equations in two variables are parallel, then the system is _____ .

RC6. If the graphs of the equations in a system of two equations in two variables intersect at one point, then the equations are _____ .

consistent

inconsistent

dependent

independent

a *Solve each system of equations using the elimination method.*

1. $x + 3y = 7,$
   $-x + 4y = 7$

2. $x + y = 9,$
   $2x - y = -3$

3. $9x + 5y = 6,$
   $2x - 5y = -17$

4. $2x - 3y = 18,$
   $2x + 3y = -6$

5. $5x + 3y = -11,$
   $3x - y = -1$

6. $2x + 3y = -9,$
   $5x - 6y = -9$

7. $5r - 3s = 19,$
   $2r - 6s = -2$

8. $2a + 3b = 11,$
   $4a - 5b = -11$

9. $2x + 3y = 1,$
   $4x + 6y = 2$

10. $3x - 2y = 1,$
    $-6x + 4y = -2$

11. $5x - 9y = 7,$
    $7y - 3x = -5$

12. $5x + 4y = 2,$
    $2x - 8y = 4$

13. $3x + 2y = 24,$
    $2x + 3y = 26$

14. $5x + 3y = 25,$
    $3x + 4y = 26$

15. $2x - 4y = 5,$
    $2x - 4y = 6$

16. $3x - 5y = -2,$
    $5y - 3x = 7$

17. $2a + b = 12,$
    $a + 2b = -6$

18. $10x + y = 306,$
    $10y + x = 90$

19. $\frac{1}{3}x + \frac{1}{5}y = 7,$
    $\frac{1}{6}x - \frac{2}{5}y = -4$

20. $\frac{2}{3}x + \frac{1}{7}y = -11,$
    $\frac{1}{7}x - \frac{1}{3}y = -10$

21. $\frac{1}{5}x + \frac{1}{2}y = 6,$
    $\frac{2}{5}x - \frac{3}{2}y = -8$

22. $\frac{2}{3}x + \frac{3}{5}y = -17,$
    $\frac{1}{2}x - \frac{1}{3}y = -1$

23. $\frac{1}{2}x - \frac{1}{3}y = -4,$
    $\frac{1}{4}x + \frac{5}{6}y = 4$

24. $\frac{4}{3}x + \frac{3}{2}y = 4,$
    $\frac{5}{6}x - \frac{1}{8}y = -6$

25. $0.3x - 0.2y = 4,$
    $0.2x + 0.3y = 0.5$

26. $0.7x - 0.3y = 0.5,$
    $-0.4x + 0.7y = 1.3$

27. $0.05x + 0.25y = 22,$
    $0.15x + 0.05y = 24$

28. $1.3x - 0.2y = 12,$
    $0.4x + 17y = 89$

b *Solve. Use the elimination method when solving the translated system.*

29. *Finding Numbers.* The sum of two numbers is 63. The larger number minus the smaller number is 9. Find the numbers.

30. *Finding Numbers.* The sum of two numbers is 2. The larger number minus the smaller number is 20. Find the numbers.

31. *Finding Numbers.* The sum of two numbers is 3. Three times the larger number plus two times the smaller number is 24. Find the numbers.

32. *Finding Numbers.* The sum of two numbers is 9. Two times the larger number plus three times the smaller number is 2. Find the numbers.

33. *Complementary Angles.* Two angles are complementary. (**Complementary angles** are angles whose sum is 90°.) Their difference is 6°. Find the angles.

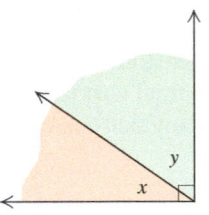

Complementary angles:
$x + y = 90°$

34. *Supplementary Angles.* Two angles are supplementary. (**Supplementary angles** are angles whose sum is 180°.) Their difference is 22°. Find the angles.

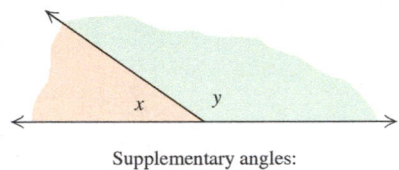

Supplementary angles:
$x + y = 180°$

35. *Basketball Scoring.* In their championship game, the Eastside Golden Eagles scored 60 points on a combination of two-point shots and three-point shots. If they made a total of 27 shots, how many of each kind of shot was made?

36. *Basketball Scoring.* Wilt Chamberlain once scored 100 points, setting a record for points scored in an NBA game. Chamberlain took only two-point shots and (one-point) foul shots and made a total of 64 shots. How many shots of each type did he make?

37. Each course offered during the winter session at New Heights Community College is worth either 3 credits or 4 credits. The members of the Touring Concert Chorale took a total of 33 courses during the winter session, worth a total of 107 credits. How many of each type of class did the chorale members take?

38. Daphne's Lawn and Garden Center offered customers who bought a custom lawn-care package a free ornamental tree, either an Eastern Redbud or a Kousa Dogwood. The center's cost for each Eastern Redbud was $37, and its cost for each Kousa Dogwood was $45. A total of 18 customers took advantage of the offer. The center's total cost for the promotional items was $754. How many patrons chose each type of ornamental tree?

▶ **Skill Maintenance**

Given the function $f(x) = 3x^2 - x + 1$, find each of the following function values. **[2.2b]**

39. $f(0)$ **40.** $f(-1)$

41. $f(-2)$ **42.** $f(2a)$

43. Find the domain of the function
$$f(x) = \frac{x-5}{x+7}. \text{ [2.3a]}$$

44. Find the domain and the range of the function $g(x) = 5 - x^2$. **[2.3a]**

45. Find an equation of the line with slope $-\frac{3}{5}$ and y-intercept $(0, -7)$. **[2.6a]**

46. Find an equation of the line containing the points $(-10, 2)$ and $(-2, 10)$. **[2.6c]**

▶ **Synthesis**

47. Use the INTERSECT feature to solve the following system of equations. You may need to first solve for y. Round answers to the nearest hundredth.
$$3.5x - 2.1y = 106.2,$$
$$4.1x + 16.7y = -106.28$$

48. Solve:
$$\frac{x+y}{2} - \frac{x-y}{5} = 1,$$
$$\frac{x-y}{2} + \frac{x+y}{6} = -2.$$

49. The solution of this system is $(-5, -1)$. Find A and B.
$$Ax - 7y = -3,$$
$$x - By = -1$$

50. Find an equation to pair with $6x + 7y = -4$ such that $(-3, 2)$ is a solution of the system.

51. The points $(0, -3)$ and $\left(-\frac{3}{2}, 6\right)$ are two of the solutions of the equation $px - qy = -1$. Find p and q.

52. Determine a and b for which $(-4, -3)$ will be a solution of the system
$$ax + by = -26,$$
$$bx - ay = 7.$$

3.4   Solving Applied Problems: Two Equations

▶ **a** Solve applied problems involving total value and mixture using systems of two equations.

▶ **b** Solve applied problems involving motion using systems of two equations.

▶ **a** Total-Value Problems and Mixture Problems

Systems of equations can be a useful tool in solving applied problems. Using systems often makes the *Translate* step easier than using a single equation. The first kind of problem we consider involves quantities of items purchased and the total value, or cost, of the items. We refer to this type of problem as a **total-value problem**.

EXAMPLE 1 *Lunch Orders.* In order to pick up lunch, Cathy collected $181.50 from her co-workers for a total of 21 salads and sandwiches. When she got to the deli, she forgot how many of each were ordered. If salads cost $7.50 and sandwiches cost $9.50, how many of each should she buy?

1. **Familiarize.** Let's begin by guessing that 5 salads were ordered. Since there was a total of 21 orders, this means that 16 sandwiches were ordered. The total cost of the order would then be

$$\underbrace{\$7.50(5)}_{\text{Cost of salads}} + \underbrace{\$9.50(16)}_{\text{Cost of sandwiches}} = \$37.50 + \$152 = \$189.50.$$

The guess is incorrect, but we can use the same process to translate the problem to a system of equations. We also note that our guess resulted in a total that was too high, so there were more salads and fewer sandwiches ordered than we guessed.

We let d = the number of salads and w = the number of sandwiches ordered. The cost of d salads is $\$7.50d$, and the cost of w sandwiches is $\$9.50w$. Organizing the information in a table can help us translate the information to a system of equations.

	Salads	Sandwiches	Total	
Number of Orders	d	w	21	$\longrightarrow d + w = 21$
Cost Per Order	$7.50	$9.50		
Total Cost	$7.50d	$9.50w	$181.50	$\longrightarrow 7.50d + 9.50w = 181.50$

2. **Translate.** The first row of the table gives us one equation:

$$d + w = 21.$$

The last row of the table gives us a second equation:

$$7.50d + 9.50w = 181.50.$$

Clearing decimals in the second equation gives us the following system of equations:

$$d + w = 21, \qquad \text{(1)}$$
$$75d + 95w = 1815. \qquad \text{(2)}$$

3. **Solve.** We use the elimination method to solve the system of equations. We eliminate d by multiplying by -75 on both sides of equation (1) and then adding the result to equation (2):

$$\begin{array}{ll} -75d - 75w = -1575 & \text{Multiplying equation (1) by } -75 \\ \underline{75d + 95w = 1815} & \text{Equation (2)} \\ 20w = 240 & \text{Adding} \\ w = 12. & \text{Dividing by 20} \end{array}$$

Next, we substitute 12 for w in equation (1) and solve for d:

$$\begin{array}{ll} d + w = 21 & \text{Equation (1)} \\ d + 12 = 21 & \text{Substituting 12 for } w \\ d = 9. & \text{Solving for } d \end{array}$$

We obtain $(9, 12)$, or $d = 9, w = 12$.

4. **Check.** We check in the original problem.

Total number of orders: $d + w = 9 + 12 = 21$
Cost of salads: $\$7.50d = \$7.50(9) = \$67.50$
Cost of sandwiches: $\$9.50w = \$9.50(12) = \underline{\$114.00}$
                    Total $= \$181.50$

The numbers check.

5. **State.** Cathy should buy 9 salads and 12 sandwiches.

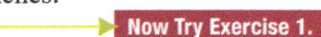

The following problem, similar to Example 1, is called a **mixture problem**.

EXAMPLE 2   *Blending Spices.*   Spice It Up sells ground sumac for $1.85 per ounce and ground turmeric for $1.35 per ounce. Ethan wants to make a 20-oz seasoning blend of the two spices that sells for $1.65 per ounce. How much of each should he use?

1. **Familiarize.** Suppose that Ethan uses 4 oz of sumac. Since he wants a total of 20 oz, he will need 16 oz of turmeric. We compare the value of the spices separately with the desired value of the blend:

Spices purchased separately: $\$1.85(4) + \$1.35(16)$, or $29.
Blend: $\$1.65(20) = \$33$

Since these amounts are not the same, our guess is not correct, but these calculations help us to translate the problem.

We let $s =$ the number of ounces of sumac and $t =$ the number of ounces of turmeric. Then we organize the information in a table as follows.

	Sumac	Turmeric	Blend	
Number of Ounces	s	t	20	$\rightarrow s + t = 20$
Price Per Ounce	$1.85	$1.35	$1.65	
Value of Spices	$1.85s	$1.35t	$1.65(20), or $33	$\rightarrow 1.85s + 1.35t = 33$

2. **Translate.** The total number of ounces in the blend is 20, so we have one equation:

$s + t = 20.$

The value of the sumac is $1.85s$, and the value of the turmeric is $1.35t$. These amounts are in dollars. Since the total value is to be $1.65(20)$, or 33, we have

$1.85s + 1.35t = 33.$

We can multiply by 100 on both sides of the second equation in order to clear the decimals. Thus we have translated to a system of equations:

$s + t = 20,$ **(1)**
$185s + 135t = 3300.$ **(2)**

3. Solve. We will solve this system using substitution, but elimination is also an appropriate method to use. When equation (1) is solved for t, we get $t = 20 - s$. We substitute $20 - s$ for t in equation (2) and solve for s:

$$185s + 135(20 - s) = 3300 \qquad \text{\color{red}Substituting}$$
$$185s + 2700 - 135s = 3300 \qquad \text{\color{red}Using the distributive law}$$
$$50s = 600 \qquad \text{\color{red}Subtracting 2700 and collecting like terms}$$
$$s = 12.$$

We have $s = 12$. Substituting 12 for s in the equation $t = 20 - s$, we obtain $t = 20 - 12$, or 8.

4. Check. We check in a manner similar to our guess in the *Familiarize* step.

Total number of ounces: $\quad 12 + 8 = 20$

Value of the blend: $\qquad 1.85(12) + 1.35(8) = 33$

Thus the number of ounces of each spice checks.

5. State. Ethan should use 12 oz of sumac and 8 oz of turmeric.

Now Try Exercise 11.

EXAMPLE 3 *Student Loans.* Jeron's student loans totaled $16,200. Part was a Perkins loan made at 5% interest and the rest was a Stafford loan made at 4% interest. After one year, Jeron's loans accumulated $715 in interest. What was the amount of each loan?

1. Familiarize. Listing the given information in a table will help. The columns in the table come from the formula for simple interest: $I = Prt$. We let $x =$ the number of dollars in the Perkins loan and $y =$ the number of dollars in the Stafford loan.

	Perkins Loan	Stafford Loan	Total	
Principal	x	y	$16,200	$\rightarrow x + y = 16{,}200$
Rate of Interest	5%	4%		
Time	1 year	1 year		
Interest	$0.05x$	$0.04y$	$715	$\rightarrow 0.05x + 0.04y = 715$

2. Translate. The total of the amounts of the loans is found in the first row of the table. This gives us one equation:

$$x + y = 16{,}200.$$

Look at the last row of the table. The interest totals $715. This gives us a second equation:

$$5\%x + 4\%y = 715, \quad \text{or} \quad 0.05x + 0.04y = 715.$$

After we multiply on both sides to clear the decimals, we have

$$5x + 4y = 71{,}500.$$

3. **Solve.** Using either elimination or substitution, we solve the resulting system:

$$x + y = 16{,}200,$$
$$5x + 4y = 71{,}500.$$

We find that $x = 6700$ and $y = 9500$.

4. **Check.** The sum is $6700 + $9500, or $16,200. The interest from $6700 at 5% for one year is 5%($6700), or $335. The interest from $9500 at 4% for one year is 4%($9500), or $380. The total amount of interest is $335 + $380, or $715. The numbers check in the problem.

5. **State.** The Perkins loan was for $6700, and the Stafford loan was for $9500.

Now Try Exercise 13.

EXAMPLE 4 *Mixing Fertilizers.* Nature's Landscapes carries two kinds of fertilizer containing nitrogen and water. "Gently Green" is 5% nitrogen and "Sun Saver" is 15% nitrogen. Nature's Landscapes needs to combine the two types of solution in order to make 90 L of a solution that is 12% nitrogen. How much of each brand should be used?

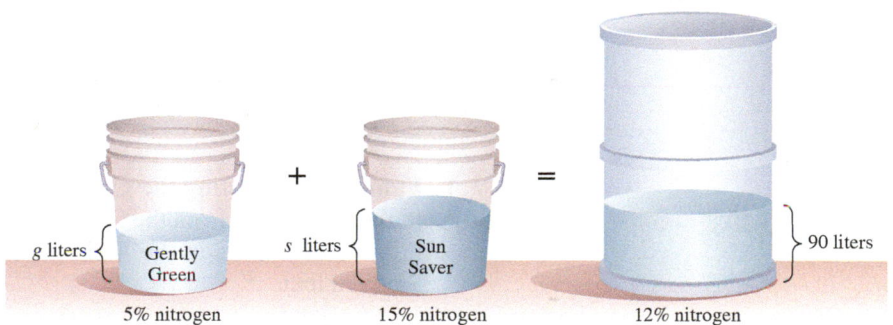

1. **Familiarize.** We first guess to become familiar with the problem.

We choose two numbers that total 90 L—say, 40 L of Gently Green and 50 L of Sun Saver—for the amounts of each fertilizer. Will the resulting mixture have the correct percentage of nitrogen? To find out, we multiply as follows:

$$5\%(40\,\text{L}) = 2\,\text{L of nitrogen} \quad \text{and} \quad 15\%(50\,\text{L}) = 7.5\,\text{L of nitrogen}.$$

Thus the total amount of nitrogen in the mixture is 2 L + 7.5 L, or 9.5 L. The final mixture of 90 L is supposed to be 12% nitrogen. Now

$$12\%(90\,\text{L}) = 10.8\,\text{L}.$$

Since 9.5 L and 10.8 L are not the same, our guess is incorrect. But these calculations help us to make the translation.

We let $g = $ the number of liters of Gently Green and $s = $ the number of liters of Sun Saver in the mixture.

	Gently Green	Sun Saver	Mixture	
Number of Liters	g	s	90	→ $g + s = 90$
Percent of Nitrogen	5%	15%	12%	
Amount of Nitrogen	$0.05g$	$0.15s$	0.12 × 90, or 10.8 liters	→ $0.05g + 0.15s = 10.8$

2. Translate. If we add g and s in the first row, we get 90, and this gives us one equation:

$$g + s = 90.$$

If we add the amounts of nitrogen listed in the third row, we get 10.8, and this gives us another equation:

$$5\%g + 15\%s = 10.8, \quad \text{or}$$
$$0.05g + 0.15s = 10.8.$$

After clearing the decimals, we have the following system:

$$g + s = 90, \qquad \text{(1)}$$
$$5g + 15s = 1080. \qquad \text{(2)}$$

3. Solve. We solve the system using elimination. We multiply equation (1) by -5 and add the result to equation (2):

$$
\begin{array}{rl}
-5g - 5s = -450 & \textbf{Multiplying equation (1) by } -5 \\
\underline{5g + 15s = 1080} & \textbf{Equation (2)} \\
10s = 630 & \textbf{Adding} \\
s = 63. & \textbf{Dividing by 10}
\end{array}
$$

Next, we substitute 63 for s in equation (1) and solve for g:

$$
\begin{array}{ll}
g + 63 = 90 & \textbf{Substituting in equation (1)} \\
g = 27. & \textbf{Solving for } g
\end{array}
$$

We obtain $(27, 63)$, or $g = 27, s = 63$.

4. Check. Remember that g is the number of liters of Gently Green, with 5% nitrogen, and s is the number of liters of Sun Saver, with 15% nitrogen.

Total number of liters of mixture: $g + s = 27 + 63 = 90$ L

Amount of nitrogen: $5\%(27) + 15\%(63) = 1.35 + 9.45 = 10.8$ L

Percentage of nitrogen in mixture: $\dfrac{10.8}{90} = 0.12 = 12\%$

The numbers check in the original problem.

5. State. Nature's Landscapes should mix 27 L of Gently Green and 63 L of Sun Saver.

▶ **Now Try Exercise 7.**

Tips for Solving Motion Problems

1. Make a drawing using an arrow or arrows to represent distance and the direction of each object in motion.
2. Organize the information in a table or a chart.
3. Look for as many things as you can that are the same, so you can write equations.

▶ **b Motion Problems**

When a problem deals with speed, distance, and time, we can expect to use the following *motion formula*.

THE MOTION FORMULA

$$\text{Distance} = \text{Rate (or speed)} \cdot \text{Time}$$
$$d = rt$$

EXAMPLE 5 *Auto Travel.* Keri left Monday morning to drive to a seminar that began Monday evening. An hour after she had left the office, her assistant, Matt, realized that she had forgotten to take a large portfolio needed for a presentation. Knowing Keri would not answer her cell phone when driving, Matt left immediately with the portfolio to try to catch up with her. If Keri drove at a speed of 55 mph and Matt drove at a speed of 65 mph, how long did it take Matt to catch up with her? Assume that neither driver stopped to take a break.

1. **Familiarize.** We first make a drawing. From the drawing, we see that when Matt catches up with Keri, the distances from the office are the same. We let d = the distance, in miles. If we let t = the time, in hours, for Matt to catch Keri, then $t + 1$ = the time traveled by Keri at a slower speed.

Matt's car   Keri's car
65 mph   55 mph
t hours, d miles $t + 1$ hours, d miles

We organize the information in a table as follows.

$$d = r \cdot t$$

	Distance	Rate	Time	
Keri	d	55	$t+1$	$\longrightarrow d = 55(t+1)$
Matt	d	65	t	$\longrightarrow d = 65t$

2. **Translate.** Using $d = rt$ in each row of the table, we get an equation. Thus we have a system of equations:

$$d = 55(t+1), \qquad \text{(1)}$$
$$d = 65t. \qquad \text{(2)}$$

3. **Solve.** We solve the system using the substitution method:

$$65t = 55(t+1) \qquad \text{Substituting } 65t \text{ for } d \text{ in equation (1)}$$
$$65t = 55t + 55 \qquad \text{Multiplying to remove parentheses on the right}$$
$$10t = 55 \qquad$$
$$t = 5.5. \qquad \text{Solving for } t$$

Matt's time is 5.5 hr, which means that Keri's time is 5.5 + 1, or 6.5 hr.

4. **Check.** At 65 mph, Matt will travel $65 \cdot 5.5$, or 357.5 mi, in 5.5 hr. At 55 mph, Keri will travel $55 \cdot 6.5$, or the same 357.5 mi, in 6.5 hr. The distances are the same, so the numbers check.

5. **State.** Matt caught up with Keri in 5.5 hr.   **Now Try Exercise 21.**

EXAMPLE 6 *Marine Travel.* A Coast-Guard patrol boat travels 4 hr on a trip downstream with a 6-mph current. The return trip against the same current takes 5 hr. Find the speed of the boat in still water.

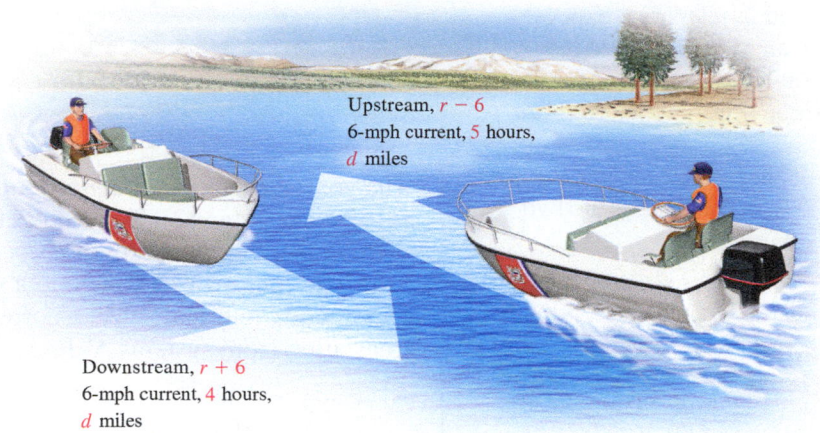

Upstream, $r - 6$
6-mph current, 5 hours,
d miles

Downstream, $r + 6$
6-mph current, 4 hours,
d miles

1. **Familiarize.** We first make a drawing. From the drawing, we see that the distances are the same. We let $d =$ the distance, in miles, and $r =$ the speed of the boat in still water, in miles per hour. Then, when the boat is traveling downstream, its speed is $r + 6$. (The current helps the boat along.) When it is traveling upstream, its speed is $r - 6$. (The current holds the boat back.) We can organize the information in a table. We use the formula $d = rt$.

$$d \quad = \quad r \quad \cdot \quad t$$

	Distance	Rate	Time	
Downstream	d	$r + 6$	4	$\rightarrow d = (r + 6)4$
Upstream	d	$r - 6$	5	$\rightarrow d = (r - 6)5$

2. **Translate.** From each row of the table, we get an equation, $d = rt$:

$$d = 4r + 24, \quad (1)$$
$$d = 5r - 30. \quad (2)$$

3. **Solve.** We solve the system using the substitution method:

$$4r + 24 = 5r - 30 \qquad \text{Substituting } 4r + 24 \text{ for } d \text{ in equation (2)}$$
$$\left. \begin{array}{l} 24 = r - 30 \\ 54 = r. \end{array} \right\} \quad \text{Solving for } r$$

4. **Check.** If $r = 54$, then $r + 6 = 60$; and $60 \cdot 4 = 240$ mi, the distance traveled downstream. If $r = 54$, then $r - 6 = 48$; and $48 \cdot 5 = 240$ mi, the distance traveled upstream. The distances are the same.

   When checking your answer, always ask, "Have I found what the problem asked for?" We could solve for a certain variable but still have not answered the question of the original problem. For example, we might have found speed when the problem wanted distance. In this problem, we want the speed of the boat in still water, and that is r.

5. **State.** The speed in still water is 54 mph. ▶ **Now Try Exercise 23.**

Translating for Success

The goal of these matching questions is to practice step (2), *Translate*, of the five-step problem-solving process. Translate each word problem to a system of equations and select a correct translation from systems A–J.

A. $x = y + 248,$
$x + y = 1094$

B. $5x = 2y - 3,$
$y = \frac{2}{3}x + 5$

C. $y = \frac{1}{2}x,$
$2x + 2y = 192$

D. $2x = 7 + y,$
$x + y = 180$

E. $x + y = 192,$
$x = 2y$

F. $x + y = 180,$
$x = 2y + 7$

G. $x - 1094 = y,$
$3x - 4y = 248$

H. $3\%x + 2.5\%y = 97.50,$
$x + y = 2500$

I. $2x = 5 + \frac{2}{3}y,$
$3y = 15x - 4$

J. $x = (y + 2.5) \cdot 3,$
$3.5(y - 2.5) = x$

1. *Office Expense.*   The monthly telephone expense for an office is $1094 less than the janitorial expense. Three times the janitorial expense minus four times the telephone expense is $248. What is the total of the two expenses?

2. *Dimensions of a Triangle.* The sum of the base and the height of a triangle is 192 in. The height is twice the base. Find the base and the height.

3. *Supplementary Angles.* Two supplementary angles are such that twice one angle is 7° more than the other. Find the measures of the angles.

4. *SAT Scores.*   The total of Megan's writing and math scores on the SAT was 1094. Her math score was 248 points higher than her writing score. What were her math and writing SAT scores?

5. *Sightseeing Boat.*   A sightseeing boat travels 3 hr on a trip downstream with a 2.5-mph current. The return trip against the same current takes 3.5 hr. Find the speed of the boat in still water.

6. *Running Distances.*   Each day Tricia runs 5 mi more than two-thirds the distance that Chris runs. Five times the distance that Chris runs is 3 mi less than twice the distance that Tricia runs. How far does Tricia run daily?

7. *Dimensions of a Rectangle.* The perimeter of a rectangle is 192 in. The width is half the length. Find the length and the width.

8. *Mystery Numbers.*   Teka asked her students to determine the two numbers that she placed in a sealed envelope. Twice the smaller number is 5 more than two-thirds the larger number. Three times the larger number is 4 less than fifteen times the smaller. Find the numbers.

9. *Supplementary Angles.*   Two supplementary angles are such that one angle is 7° more than twice the other. Find the measures of the angles.

10. *Student Loans.*   Brandt's student loans totaled $2500. Part was borrowed at 3% interest and the rest at 2.5%. After one year, Brandt had accumulated $97.50 in interest. What was the amount of each loan?

3.4 | Exercise Set

▶ Reading Check

Consider the following mixture problem and the table used to translate the problem.

Cherry Breeze is 30% fruit juice and Berry Choice is 15% fruit juice. How much of each should be used in order to make 10 gal of a drink that is 20% fruit juice?

Choose from the options below the expression that best fits each numbered space in the table.

2   10   15   0.15y

	Cherry Breeze	Berry Choice	Mixture
Gallons of Drink	x	y	RC1. __
Percent of Fruit Juice	30%	RC2. __ %	20%
Gallons of Fruit Juice in Mixture	0.3x	RC3. ____	RC4. __

a *Solve.*

1. *Entertainment.* For her personal-finance class, Laura was required to estimate her annual entertainment expenditures. She discovered that during the previous year, she spent $225.32 on a total of 68 e-books and game applications. If each book cost $3.99 and each game cost $1.99, how many books and how many games did she purchase?

2. *Flowers.* Kevin's Floral Emporium offers two types of sunflowers for sale by the stem. When in season, the small ones sell for $2.50 per stem, and the large ones sell for $3.95 per stem. One late summer weekend, Kevin sold a total of 118 stems for $376.20. How many of each size did he sell?

3. *Furniture Polish.* A nontoxic furniture polish can be made by combining vinegar and olive oil. The amount of oil should be three times the amount of vinegar. How much of each ingredient is needed in order to make 30 oz of furniture polish?

4. *Nontoxic Floor Wax.* A nontoxic floor wax can be made by combining lemon juice and food-grade linseed oil. The amount of oil should be twice the amount of lemon juice. How much of each ingredient is needed in order to make 32 oz of floor wax? (The mix should be spread with a rag and buffed when dry.)

5. *Balloon Bouquets.* When the Southeast Cougars women's soccer team won the state championship, the parent boosters welcomed the team back to school with a balloon bouquet for each of the 18 players. The parents spent a total of $86.76 (excluding tax) on foil balloons that cost $1.99 each and latex school-color balloons that cost $0.12 each. Each player received 9 balloons, and all the balloon bouquets were identical. How many of each type of balloon did each bouquet include?

6. *Chocolate Assortments.* For a fundraiser, the Greenfield Merchants Association spent a total of $1872 on an assortment of chocolate truffles at $2.95 each and chocolate cream mints at $1.79 each. They then packaged 75 boxes to sell, each containing 12 pieces of candy. If the contents of the boxes were

identical, how many of each kind of candy did each box contain?

7. *Catering.*   Stella's Catering is planning a wedding reception. The bride and groom would like to serve a nut mixture containing 25% peanuts. Stella has available mixtures that are either 40% or 10% peanuts. How much of each type should be mixed in order to get a 10-lb mixture that is 25% peanuts?

8. *Blending Granola.*   Deep Thought Granola is 25% nuts and dried fruit. Oat Dream Granola is 10% nuts and dried fruit. How much of Deep Thought and how much of Oat Dream should be mixed in order to form a 20-lb batch of granola that is 19% nuts and dried fruit?

9. *Ink Remover.*   Etch Clean Graphics uses one cleanser that is 25% acid and a second that is 50% acid. How many liters of each should be mixed in order to get 10 L of a solution that is 40% acid?

10. *Livestock Feed.*   Soybean meal is 16% protein and corn meal is 9% protein. How many pounds of each should be mixed in order to get a 350-lb mixture that is 12% protein?

11. *Vegetable Seeds.*   Tara's website, verdantveggies.com, specializes in the sale of rare or unusual vegetable seeds. Tara sells packets of sweet-pepper seeds for $2.85 each and packets of hot-pepper seeds for $4.29 each. She also offers a 16-packet mixed-pepper assortment combining packets of both types of seeds at $3.30 per packet. How many packets of each type of seed are in the assortment?

Sweet peppers   Hot peppers   Assorted

12. *Flower Bulbs.*   Heritage Bulbs sells heirloom flower bulbs. Acuminata tulip bulbs cost $4.85 each, and Cafe Brun tulip bulbs cost $9.50 each. An assortment of 12 of these bulbs is priced at $7.95 per bulb. How many of each type of bulb are in the assortment?

13. *Food Science.*   The following bar graph shows the milk fat percentages in three dairy products. How many pounds each of whole milk and cream should be mixed in order to form 200 lb of milk for cream cheese?

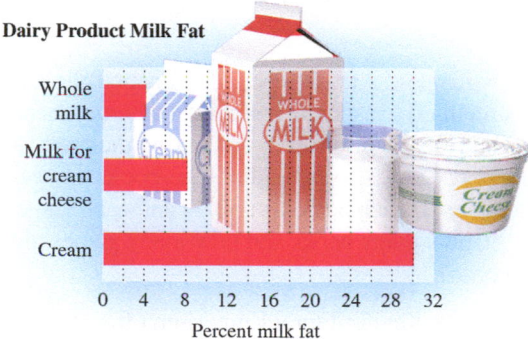

14. *Automotive Maintenance.*   Arctic Antifreeze is 18% alcohol and Frost No-More is 10% alcohol. How many liters of Arctic Antifreeze should be mixed with 7.5 L of Frost No-More in order to get a mixture that is 15% alcohol?

15. *Student Loans.*   Sarah's two student loans totaled $12,000. One of her loans was at 6% simple interest and the other at 3%. After one year, Sarah owed $585 in interest. What was the amount of each loan?

16. *Investments.*   Ana and Johnny made two investments totaling $45,000. In one year, these investments yielded $2430 in simple interest. Part of the money was invested at 4% and the rest at 6%. How much was invested at each rate?

17. *Investments.*   William opened two investment accounts for his daughter's college fund. The first year, these investments, which totaled $3200, yielded $155 in simple interest. Part of the money was invested at 5.5% and the rest at 4%. How much was invested at each rate?

18. *Student Loans.*   Cole's two student loans totaled $31,000. One of his loans was at 2.8% simple interest and the other at 4.5%. After one year, Cole owed $1024.40 in interest. What was the amount of each loan?

19. *Making Change.*   Juan goes to a bank and gets change for a $50 bill consisting of all $5 bills and $1 bills. There are 22 bills in all. How many of each kind are there?

20. *Making Change.*   Christina makes a $9.25 purchase at a bookstore with a $20 bill. The store has no bills and gives her the change in quarters and dollar coins. There are 19 coins in all. How many of each kind are there?

b *Solve.*

21. *Train Travel.*   A train leaves Danville Junction and travels north at a speed of 75 mph. Two hours later, a second train leaves on a parallel track and travels north at 125 mph. How far from the station will they meet?

Trains meet here

t − 2 hours   *t* hours

Danville Junction

75 mph, *d* miles

125 mph, *d* miles

22. *Car Travel.*   Max leaves Kansas City and drives east at a speed of 80 km/h. One hour later, Olivia leaves Kansas City traveling in the same direction as Max but at 96 km/h. Assuming neither driver stops for a break, how far from Kansas City will they be when Olivia catches up with Max?

23. *Canoeing.*   Darren paddled for 4 hr with a 6-km/h current to reach a campsite. The return trip against the same current took 10 hr. Find the speed of Darren's canoe in still water.

24. *Boating.*   Mia's motorboat took 3 hr to make a trip downstream with a 6-mph current. The return trip against the same current took 5 hr. Find the speed of the boat in still water.

25. *Air Travel.*   Christie pilots her Cessna 150 plane for 270 mi against a headwind in 3 hr. The flight would take 1 hr and 48 min with a tailwind of the same speed. Find the headwind and the speed of the plane in still air.

26. *Air Travel.*   Rod is a pilot for Crossland Airways. He computes his flight time against a headwind for a trip of 2900 mi at 5 hr. The flight would take 4 hr and 50 min if the headwind were half as great. Find the headwind and the plane's air speed in still air.

27. *Air Travel.*   Two airplanes start at the same time and fly toward each other from points 1000 km apart at rates of 420 km/h and 330 km/h. After how many hours will they meet?

28. *Air Travel.*   Two planes start at the same time and travel toward each other from cities that are 780 km apart at rates of 190 km/h and 200 km/h. In how many hours will they meet?

29. 🖩 *Point of No Return.*   A plane flying the 3458-mi trip from New York City to London has a 50-mph tailwind. The flight's *point of no return* is the point at which the flight time required to return to New York is the same as the time required to continue to London. If the speed of the plane in still air is 360 mph, how far is New York from the point of no return?

30. 🖩 *Point of No Return.*   A plane is flying the 2553-mi trip from Los Angeles to Honolulu into a 60-mph headwind. If the speed of the plane in still air is 310 mph, how far from Los Angeles is the plane's point of no return? (See Exercise 29.)

▶ **Skill Maintenance**

31. Find the intersection:

$$\{2, 4, 6, 8, 10\} \cap \{6, 7, 8, 9, 10\}. \quad \textbf{[1.5a]}$$

32. Find the union:

$$\{2, 4, 6, 8, 10\} \cup \{6, 7, 8, 9, 10\}. \quad \textbf{[1.5b]}$$

Simplify. **[1.6a]**

33. $|3a|$

34. $|-7x^2|$

35. $\left|\dfrac{-3}{y}\right|$

36. $\left|\dfrac{a^4}{c}\right|$

▶ **Synthesis**

37. *Automotive Maintenance.* The radiator in Michelle's car contains 16 L of antifreeze and water. This mixture is 30% antifreeze. How much of this mixture should she drain and replace with pure antifreeze so that there will be a mixture of 50% antifreeze?

38. *Physical Exercise.* Natalie jogs and walks to school each day. She averages 4 km/h walking and 8 km/h jogging. The distance from home to school is 6 km and Natalie makes the trip in 1 hr. How far does she jog in a trip?

39. *Fuel Economy.* Ashlee's SUV gets 18 miles per gallon (mpg) in city driving and 24 mpg in highway driving. The SUV is driven 465 mi on 23 gal of gasoline. How many miles were driven in the city and how many were driven on the highway?

40. *Siblings.* Phil and Maria are siblings. Maria has twice as many brothers as she has sisters. Phil has the same number of brothers as he has sisters. How many girls and how many boys are in the family?

Mid-Chapter Review

Determine whether each statement is true or false.

1. If, when solving a system of two linear equations in two variables, a false equation is obtained, the system has infinitely many solutions. **[3.2a], [3.3a]**

2. Every system of equations has at least one solution. **[3.1a]**

3. If the graphs of two linear equations intersect, then the system is consistent. **[3.1a]**

4. The intersection of the graphs of the lines $x = a$ and $y = b$ is (a, b). **[3.1a]**

Solve each system of equations graphically. Then classify the system as consistent or inconsistent and the equations as dependent or independent. **[3.1a]**

5. $y = x - 6,$
    $y = 4 - x$

6. $x + y = 3,$
    $3x + y = 3$

7. $y = 2x - 3,$
    $4x - 2y = 6$

8. $x - y = 3,$
    $2y - 2x = 6$

Solve using the substitution method. **[3.2a]**

9. $x = y + 2,$
    $2x - 3y = -2$

10. $y = x - 5,$
    $x - 2y = 8$

11. $4x + 3y = 3,$
    $y = x + 8$

12. $3x - 2y = 1,$
    $x = y + 1$

Solve using the elimination method. **[3.3a]**

13. $2x + y = 2,$
    $x - y = 4$

14. $x - 2y = 13,$
    $x + 2y = -3$

15. $3x - 4y = 5,$
    $5x - 2y = -1$

16. $3x + 2y = 11,$
    $2x + 3y = 9$

17. $x - 2y = 5,$
    $3x - 6y = 10$

18. $4x - 6y = 2,$
    $-2x + 3y = -1$

19. $\dfrac{1}{2}x + \dfrac{1}{3}y = 1,$
    $\dfrac{1}{5}x - \dfrac{3}{4}y = 11$

20. $0.2x + 0.3y = 0.6,$
    $0.1x - 0.2y = -2.5$

Solve.

21. *Garden Dimensions.* A landscape architect designs a garden with a perimeter of 44 ft. The width is 2 ft less than the length. Find the length and the width. **[3.2b]**

22. *Investments.* Sandy made two investments totaling $5000. Part of the money was invested at 2% and the rest at 3%. In one year, these investments earned $129 in simple interest. How much was invested at each rate? **[3.4a]**

23. *Mixing Solutions.*   A lab technician wants to mix a solution that is 20% acid with a second solution that is 50% acid in order to get 84 L of a solution that is 30% acid. How many liters of each solution should be used? **[3.4a]**

24. *Boating.*   Monica's motorboat took 5 hr to make a trip downstream with a 6-mph current. The return trip against the same current took 8 hr. Find the speed of the boat in still water. **[3.4b]**

Collaborative Discussion and Writing

25. Explain how to find the solution of $\frac{3}{4}x + 2 = \frac{2}{5}x - 5$ in two ways graphically and in two ways algebraically. **[3.1a], [3.2a], [3.3a]**

26. Write a system of equations with the given solution. Answers may vary. **[3.1a], [3.2a], [3.3a]**

    a) $(4, -3)$
    b) No solution
    c) Infinitely many solutions

27. Describe a method that could be used to create an inconsistent system of equations. **[3.1a], [3.2a], [3.3a]**

28. Describe a method that could be used to create a system of equations with dependent equations. **[3.1a], [3.2a], [3.3a]**

3.5   Systems of Equations in Three Variables

▶ **a**   Solve systems of three equations in three variables.

▶ a Solving Systems in Three Variables

A **linear equation in three variables** is an equation equivalent to one of the type $Ax + By + Cz = D$. A **solution** of a system of three equations in three variables is an ordered triple (x, y, z) that makes *all three* equations true.

The substitution method can be used to solve systems of three equations, but it is not efficient unless a variable has already been eliminated from one or more of the equations. Therefore, we will use only the elimination method.* The first step is to eliminate a variable and obtain a system of two equations in two variables.

EXAMPLE 1   Solve the following system of equations:

$$
\begin{array}{rl}
x + y + z = 4, & \quad\text{(1)} \\
x - 2y - z = 1, & \quad\text{(2)} \\
2x - y - 2z = -1. & \quad\text{(3)}
\end{array}
$$

a) We first use *any* two of the three equations to get an equation in two variables. In this case, let's use equations (1) and (2) and add to eliminate z:

$$
\begin{array}{ll}
x + y + z = 4 & \quad\text{(1)} \\
\underline{x - 2y - z = 1} & \quad\text{(2)} \\
2x - y \phantom{{}- z}\; = 5. & \quad\text{(4)} \qquad \textbf{Adding to eliminate } z
\end{array}
$$

b) We use a *different* pair of equations and eliminate the **same variable** that we did in part (a). Let's use equations (1) and (3) and again eliminate z.

---

*Other methods for solving systems of equations are considered in Sections 6.3 and 6.5.

> **CAUTION!**   A common error is to eliminate a different variable the second time.

$$x + y + z = 4, \quad (1)$$
$$2x - y - 2z = -1; \quad (3)$$

$$2x + 2y + 2z = 8 \qquad \text{Multiplying equation (1) by 2}$$
$$\underline{2x - y - 2z = -1} \quad (3)$$
$$4x + y = 7 \quad (5) \qquad \text{Adding to eliminate } z$$

c) Now we solve the resulting system of equations, (4) and (5). That solution will give us two of the numbers. Note that we now have two equations in two variables. Had we eliminated two *different* variables in parts (a) and (b), this would not be the case.

$$2x - y = 5 \quad (4)$$
$$\underline{4x + y = 7} \quad (5)$$
$$6x = 12 \qquad \text{Adding}$$
$$x = 2$$

We can use either equation (4) or (5) to find y. We choose equation (5):

$$4x + y = 7 \quad (5)$$
$$4(2) + y = 7 \qquad \text{Substituting 2 for } x$$
$$8 + y = 7$$
$$y = -1.$$

d) We now have $x = 2$ and $y = -1$. To find the value for z, we use any of the original three equations, substitute, and solve for z. Let's use equation (1) and substitute our two numbers in it:

$$x + y + z = 4 \quad (1)$$
$$2 + (-1) + z = 4 \qquad \text{Substituting 2 for } x \text{ and } -1 \text{ for } y$$
$$\left. \begin{array}{l} 1 + z = 4 \\ z = 3. \end{array} \right\} \quad \text{Solving for } z$$

We have obtained the ordered triple $(2, -1, 3)$. To check, we substitute $(2, -1, 3)$ into each of the three equations using alphabetical order of the variables.

Check:

$$\frac{x + y + z = 4}{2 + (-1) + 3 \ ? \ 4}$$
$$4 \ \bigg| \quad \text{TRUE}$$

$$\frac{x - 2y - z = 1}{2 - 2(-1) - 3 \ ? \ 1}$$
$$2 + 2 - 3 \ \bigg|$$
$$1 \ \bigg| \quad \text{TRUE}$$

$$\frac{2x - y - 2z = -1}{2(2) - (-1) - 2 \cdot 3 \ ? \ -1}$$
$$4 + 1 - 6 \ \bigg|$$
$$-1 \ \bigg| \quad \text{TRUE}$$

The triple $(2, -1, 3)$ checks and is the solution.

To use the elimination method to solve systems of three equations:

1. Write all equations in the standard form, $Ax + By + Cz = D$.
2. Clear any decimals or fractions.
3. Choose a variable to eliminate. Then use *any* two of the three equations to eliminate that variable, getting an equation in two variables.
4. Next, use a different pair of equations and get another equation in *the same two variables*. That is, eliminate the same variable that you did in step (3).
5. Solve the resulting system (pair) of equations. That will give two of the numbers.
6. Then use any of the original three equations to find the third number.

Now Try Exercise 1.

EXAMPLE 2 Solve this system:

$$4x - 2y - 3z = 5, \quad \text{(1)}$$
$$-8x - y + z = -5, \quad \text{(2)}$$
$$2x + y + 2z = 5. \quad \text{(3)}$$

a) The equations are in standard form and do not contain decimals or fractions.

b) We decide to eliminate the variable y since the y-terms are opposites in equations (2) and (3). We add:

$$\begin{array}{ll} -8x - y + z = -5 & \text{(2)} \\ \underline{2x + y + 2z = 5} & \text{(3)} \\ -6x + 3z = 0. & \text{(4)} \quad \textbf{Adding} \end{array}$$

c) We use another pair of equations to get an equation in the same two variables, x and z. We use equations (1) and (3) and eliminate y:

$$\begin{array}{ll} 4x - 2y - 3z = 5, & \text{(1)} \\ 2x + y + 2z = 5; & \text{(3)} \end{array}$$

$$\begin{array}{lll} 4x - 2y - 3z = 5 & \text{(1)} \\ \underline{4x + 2y + 4z = 10} & & \textbf{Multiplying equation (3) by 2} \\ 8x + z = 15. & \text{(5)} & \textbf{Adding} \end{array}$$

d) Next, we solve the resulting system of equations (4) and (5). That will give us two of the numbers:

$$\begin{array}{ll} -6x + 3z = 0, & \text{(4)} \\ 8x + z = 15. & \text{(5)} \end{array}$$

We multiply equation (5) by -3 and then add:

$$\begin{array}{ll} -6x + 3z = 0 & \text{(4)} \\ \underline{-24x - 3z = -45} & \textbf{Multiplying equation (5) by } -3 \\ -30x = -45 & \textbf{Adding} \\ x = \frac{-45}{-30} = \frac{3}{2}. \end{array}$$

We now use equation (5) to find z:

$$\begin{array}{ll} 8x + z = 15 & \text{(5)} \\ 8\left(\frac{3}{2}\right) + z = 15 & \textbf{Substituting } \frac{3}{2} \textbf{ for } x \\ 12 + z = 15 \\ z = 3. & \textbf{Solving for } z \end{array}$$

e) Next, we use any of the original equations and substitute to find the third number, y. We choose equation (3) since the coefficient of y there is 1:

$$2x + y + 2z = 5 \qquad \textbf{(3)}$$
$$2\left(\tfrac{3}{2}\right) + y + 2(3) = 5 \qquad \color{red}\textbf{Substituting } \tfrac{3}{2} \textbf{ for } x \textbf{ and } 3 \textbf{ for } z$$
$$\left.\begin{aligned}3 + y + 6 &= 5 \\ y + 9 &= 5 \\ y &= -4.\end{aligned}\right\} \qquad \color{red}\textbf{Solving for } y$$

The solution is $\left(\tfrac{3}{2}, -4, 3\right)$. The check is as follows.

Check:

$$\frac{4x - 2y - 3z = 5}{4 \cdot \tfrac{3}{2} - 2(-4) - 3(3) \;?\; 5}$$
$$6 + 8 - 9 \quad\Big|$$
$$5 \quad\Big| \qquad \textsf{TRUE}$$

$$\frac{-8x - y + z = -5}{-8 \cdot \tfrac{3}{2} - (-4) + 3 \;?\; -5}$$
$$-12 + 4 + 3 \quad\Big|$$
$$-5 \quad\Big| \qquad \textsf{TRUE}$$

$$\frac{2x + y + 2z = 5}{2 \cdot \tfrac{3}{2} + (-4) + 2(3) \;?\; 5}$$
$$3 - 4 + 6 \quad\Big|$$
$$5 \quad\Big| \qquad \textsf{TRUE}$$

$\longrightarrow$ **Now Try Exercise 17.**

In Example 3, two of the equations have a missing variable.

EXAMPLE 3   Solve this system:

$$\begin{aligned}x + y + z &= 180, &\textbf{(1)}\\ x \qquad\;\; - z &= -70, &\textbf{(2)}\\ 2y - z &= 0. &\textbf{(3)}\end{aligned}$$

We note that there is no y in equation (2). In order to have a system of two equations in the variables x and z, we need to find another equation without a y. We use equations (1) and (3) to eliminate y:

$$\begin{aligned}x + y + z &= 180, &\textbf{(1)}\\ 2y - z &= 0; &\textbf{(3)}\end{aligned}$$

$$\begin{aligned}-2x - 2y - 2z &= -360 &\color{red}\textbf{Multiplying equation (1) by } -2\\ 2y - \;\; z &= \quad 0 &\textbf{(3)}\\ \hline -2x \qquad\; - 3z &= -360. &\textbf{(4)} \qquad \color{red}\textbf{Adding}\end{aligned}$$

Next, we solve the resulting system of equations (2) and (4):

$$\begin{aligned}x - \;\; z &= -70, &\textbf{(2)}\\ -2x - 3z &= -360; &\textbf{(4)}\end{aligned}$$

$$\begin{aligned}2x - 2z &= -140 &\color{red}\textbf{Multiplying equation (2) by 2}\\ -2x - 3z &= -360 &\textbf{(4)}\\ \hline -5z &= -500 &\color{red}\textbf{Adding}\\ z &= 100.\end{aligned}$$

To find x, we substitute 100 for z in equation (2) and solve for x:

$$x - z = -70$$
$$x - 100 = -70$$
$$x = 30.$$

To find y, we substitute 100 for z in equation (3) and solve for y:

$$2y - z = 0$$
$$2y - 100 = 0$$
$$2y = 100$$
$$y = 50.$$

The triple $(30, 50, 100)$ is the solution. The check is left to the student.

Now Try Exercise 23.

It is possible for a system of three equations to have no solution, that is, to be inconsistent. An example is the system

$$x + y + z = 14,$$
$$x + y + z = 11,$$
$$2x - 3y + 4z = -3.$$

Note the first two equations. It is not possible for a sum of three numbers to be both 14 and 11. Thus the system has no solution. We will not consider such systems here, nor will we consider systems with infinitely many solutions, which also exist.

3.5 Exercise Set

▶ Reading Check

Choose from the column on the right the option that is an example of each term. Choices may be used more than once.

RC1. A linear equation in three variables

RC2. A system of equations in three variables

RC3. A solution of a linear equation in three variables

RC4. A solution of a system of equations in three variables

a) $(4, -3, 0)$

b) $a + b - c = 1$

c) $a + 3b - c = 1,$
$2a + 3b - c = -1,$
$a - 2b + 3c = 10$

a Solve.

1. $x + y + z = 2,$
$2x - y + 5z = -5,$
$-x + 2y + 2z = 1$

2. $2x - y - 4z = -12,$
$2x + y + z = 1,$
$x + 2y + 4z = 10$

3. $2x - y + z = 5,$
$6x + 3y - 2z = 10,$
$x - 2y + 3z = 5$

4. $x - y + z = 4,$
$3x + 2y + 3z = 7,$
$2x + 9y + 6z = 5$

5. $2x - 3y + z = 5,$
$x + 3y + 8z = 22,$
$3x - y + 2z = 12$

6. $6x - 4y + 5z = 31,$
$5x + 2y + 2z = 13,$
$x + y + z = 2$

7. $3a - 2b + 7c = 13,$
$a + 8b - 6c = -47,$
$7a - 9b - 9c = -3$

8. $x + y + z = 0,$
$2x + 3y + 2z = -3,$
$-x + 2y - 3z = -1$

9. $2x + 3y + z = 17,$
$x - 3y + 2z = -8,$
$5x - 2y + 3z = 5$

10. $2x + y - 3z = -4,$
$4x - 2y + z = 9,$
$3x + 5y - 2z = 5$

11. $2x + y + z = -2,$
$2x - y + 3z = 6,$
$3x - 5y + 4z = 7$

12. $2x + y + 2z = 11,$
$3x + 2y + 2z = 8,$
$x + 4y + 3z = 0$

13. $x - y + z = 4,$
$5x + 2y - 3z = 2,$
$3x - 7y + 4z = 8$

14. $2x + y + 2z = 3,$
$x + 6y + 3z = 4,$
$3x - 2y + z = 0$

15. $4x - y - z = 4,$
$2x + y + z = -1,$
$6x - 3y - 2z = 3$

16. $2r + s + t = 6,$
$3r - 2s - 5t = 7,$
$r + s - 3t = -10$

17. $a - 2b - 5c = -3,$
$3a + b - 2c = -1,$
$2a + 3b + c = 4$

18. $x + 4y - z = 5,$
$2x - y + 3z = -5,$
$4x + 3y + z = 5$

19. $2r + 3s + 12t = 4,$
$4r - 6s + 6t = 1,$
$r + s + t = 1$

20. $10x + 6y + z = 7,$
$5x - 9y - 2z = 3,$
$15x - 12y + 2z = -5$

21. $a + 2b + c = 1,$
$7a + 3b - c = -2,$
$a + 5b + 3c = 2$

22. $3p + 2r = 11,$
$q - 7r = 4,$
$p - 6q = 1$

23. $x + y + z = 57,$
$-2x + y = 3,$
$x - z = 6$

24. $4a + 9b = 8,$
$8a + 6c = -1,$
$6b + 6c = -1$

25. $r + s = 5,$
$3s + 2t = -1,$
$4r + t = 14$

26. $a - 5c = 17,$
$b + 2c = -1,$
$4a - b - 3c = 12$

27. $x + y + z = 105,$
$10y - z = 11,$
$2x - 3y = 7$

▶ Skill Maintenance

Solve for the indicated letter. [1.2a]

28. $F = 3ab,$ for a

29. $Q = 4(a + b),$ for a

30. $F = \frac{1}{2}t(c - d),$ for d

31. $F = \frac{1}{2}t(c - d),$ for c

32. $Ax + By = c,$ for y

33. $Ax - By = c,$ for y

Find the slope and the y-intercept. [2.4b]

34. $y = -\frac{2}{3}x - \frac{5}{4}$

35. $y = 5 - 4x$

36. $2x - 5y = 10$

37. $7x - 6.4y = 20$

▶ Synthesis

Solve.

38. $w + x - y + z = 0,$
$w - 2x - 2y - z = -5,$
$w - 3x - y + z = 4,$
$2w - x - y + 3z = 7$

39. $w + x + y + z = 2,$
$w + 2x + 2y + 4z = 1,$
$w - x + y + z = 6,$
$w - 3x - y + z = 2$

3.6 Solving Applied Problems: Three Equations

▶ **a** Solve applied problems using systems of three equations.

▶ a Using Systems of Three Equations

Solving systems of three or more equations is important in many applications occurring in the natural and social sciences, business, and engineering.

EXAMPLE 1 *Jewelry Design.* Kim is designing a triangular-shaped pendant for a client of her custom jewelry business. The largest angle of the triangle is 70° greater than the smallest angle. The largest angle is twice as large as the remaining angle. Find the measure of each angle.

1. **Familiarize.** We first make a drawing. We let x = the smallest angle, z = the largest angle, and y = the remaining angle.

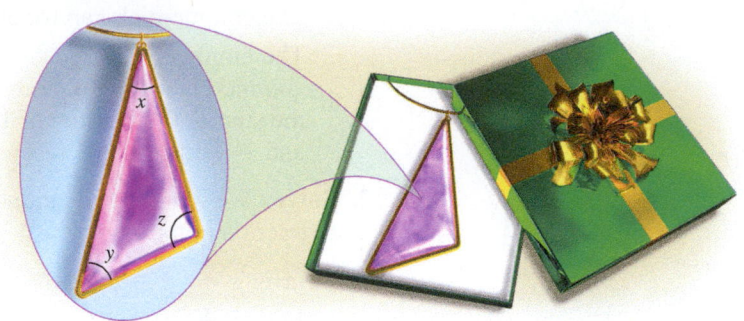

2. **Translate.** In order to translate the problem, we use the fact that the sum of the measures of the angles of a triangle is 180°:

   $$x + y + z = 180.$$

   There are two statements in the problem that we can translate directly.

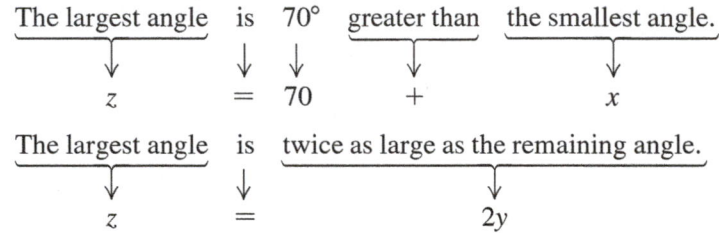

   We now have a system of three equations:

   $$
   \begin{array}{lll}
   x + y + z = 180, & & \color{red}{x + y + z = 180,} \\
   x + 70 = z, & \text{or} & \color{red}{x \quad - z = -70,} \\
   2y = z; & & \color{red}{2y - z = 0.}
   \end{array}
   $$

3. **Solve.** Solving the system, we find that the solution is $(30, 50, 100)$.

4. **Check.** The sum of the numbers is 180. The largest angle measures 100° and the smallest measures 30°, so the largest angle is 70° greater than the smallest. The largest angle is twice as large as 50°, the remaining angle. We have an answer to the problem.

5. **State.** The measures of the angles of the triangle are 30°, 50°, and 100°.

> **Now Try Exercise 3.**

EXAMPLE 2 *Super Bowl Commercials.* For commercials aired during Super Bowl XLVII, advertisers paid an average of \$3.8 million to air a 30-sec commercial. Even at this rate, a number of commercials were longer than 30 sec. A total of 42 commercials ran for either 30 sec, 1 min, or $1\frac{1}{2}$ min. Together, these 42 commercials ran for 30 min. The number of 30-sec commercials was 10 more than the sum of the number of 1-min and $1\frac{1}{2}$-min commercials. How many of each length commercial aired during the Super Bowl? (*Sources*: businessinsider.com, kantarmediana.com)

1. **Familiarize.** As we read the problem, we note that the price paid to air the commercials is not needed to solve the problem. We also note that the units of time are not all the same, so we convert 30 sec to $\frac{1}{2}$ min. We let x = the number of $\frac{1}{2}$-min commercials, y = the number of 1-min commercials, and z = the number of $1\frac{1}{2}$-min commercials.

2. Translate. We can now translate three statements to equations.

A total of 42 commercials ran. $\rightarrow$ $x + y + z = 42$

The commercials ran for 30 min. $\rightarrow$ $\frac{1}{2}x + y + 1\frac{1}{2}z = 30$

The number of 30-sec commercials was 10 more than the sum of the number of 1-min and $1\frac{1}{2}$-min commercials. $\rightarrow$ $x = y + z + 10$

3. Solve. We write the equations in standard form and convert the mixed numeral to fraction notation:

$$x + y + z = 42,$$
$$\frac{1}{2}x + y + \frac{3}{2}z = 30,$$
$$x - y - z = 10.$$

After clearing fractions, we have the system

$$x + y + z = 42, \quad \text{(1)}$$
$$x + 2y + 3z = 60, \quad \text{(2)}$$
$$x - y - z = 10. \quad \text{(3)}$$

This system is unusual, because we can eliminate *both* y and z by adding equations (1) and (3):

$$
\begin{array}{ll}
x + y + z = 42 & \text{(1)} \\
\underline{x - y - z = 10} & \text{(3)} \\
2x \qquad\quad = 52 & \\
\quad\; x = 26. &
\end{array}
$$

We can now substitute 26 for x in equations (1) and (2) and solve for y and z.
Equation (1) becomes

$$26 + y + z = 42 \qquad \text{Substituting 26 for } x$$
$$y + z = 16. \qquad \text{Simplifying}$$

Equation (2) becomes

$$26 + 2y + 3z = 60 \qquad \text{Substituting 26 for } x$$
$$2y + 3z = 34. \qquad \text{Simplifying}$$

We then solve the following system for y and z:

$$y + z = 16, \quad \text{(4)}$$
$$2y + 3z = 34. \quad \text{(5)}$$

Multiplying equation (4) by -2 and adding, we have

$$
\begin{array}{l}
-2y - 2z = -32 \\
\underline{2y + 3z = 34} \\
\; z = 2.
\end{array}
$$

Finally, we find y by substituting 2 for z in equation (4):

$$y + 2 = 16 \qquad \text{Substituting 2 for } z$$
$$y = 14.$$

We have $x = 26$, $y = 14$, and $z = 2$.

4. Check. We check our answers in each statement of the problem.

- The total number of commercials is $26 + 14 + 2 = 42$.
- The total time for the commercials is

$$\tfrac{1}{2}(26) + (14) + 1\tfrac{1}{2}(2) = 13 + 14 + 3 = 30.$$

- Ten more than the sum of the number of 1-min and $1\tfrac{1}{2}$-min commercials is $14 + 2 + 10 = 26$, which is the number of 30-sec commercials. The answer checks.

5. State. There were 26 30-sec commercials, 14 1-min commercials, and 2 $1\tfrac{1}{2}$-min commercials.

▶ **Now Try Exercise 13.**

3.6 Exercise Set

▶ # Reading Check

Match each statement with an appropriate translation from the column on the right.

RC1. The sum of three numbers is 60.

RC2. The first number minus the second number plus the third number is 60.

RC3. The first number is 60 more than the sum of the other two numbers.

RC4. The first number is 60 less than the sum of the other two numbers.

a) $x = y + z + 60$

b) $x = y + z - 60$

c) $x + y + z = 60$

d) $x - y + z = 60$

---

a *Solve.*

1. *Scholastic Aptitude Test.* More than 1.66 million members of the class of 2012 took the Scholastic Aptitude Test. Students taking the SAT receive a critical reading score, a mathematics score, and a writing score. The average total score of the students from the class of 2012 was 1498. The average math score exceeded the average reading score by 18 points. The average math score was 470 points less than the sum of the average reading and writing scores. Find the average score on each part of the test. (*Source*: College Board)

2. *Fat Content of Fast Food.* A meal at McDonald's consisting of a Big Mac, a medium order of fries, and a 12-oz vanilla milkshake contains 59 g of fat. The Big Mac has 13 more grams of fat than the milkshake. The total fat content of the fries and the shake exceeds that of the Big Mac by 3 g. Find the fat content of each food item. (*Source*: McDonald's)

3. *Triangle Measures.* In triangle *ABC,* the measure of angle *B* is three times that of angle *A*. The measure of angle *C* is 20° more than that of angle *A*. Find the measure of each angle.

4. *Triangle Measures.* In triangle *ABC,* the measure of angle *B* is twice the measure of angle *A*. The measure of angle *C* is 80° more than that of angle *A*. Find the measure of each angle.

5. *Finding Numbers.* The sum of three numbers is 55. The difference of the largest and the smallest is 49, and the sum of the two smaller numbers is 13. Find the numbers.

6. *History.* Find the year in which the first U.S. transcontinental railroad was completed. The following are some facts about the number. The sum of the digits in the year is 24. The ones digit is 1 more than the hundreds digit. Both the tens and the ones digits are multiples of 3.

7. *Smoothies.* Jamba Juice sells fruit and veggie smoothies in three sizes: a 16-oz "Sixteen," a 22-oz "Original," and a 30-oz "Power." A Sixteen smoothie sells for $3.90, an Original smoothie for $4.90, and a Power smoothie for $5.70. One hot summer afternoon, Elliot sold 34 smoothies for a total of $163. In all, he sold 752 oz of smoothies. How many of each size did he sell? (*Source*: Jamba Juice)

8. *Coffee.* A Starbucks® store on campus sells coffee in three sizes: a 12-oz tall, a 16-oz grande, and a 20-oz venti. A tall coffee sells for $1.75, a grande coffee for $1.95, and a venti coffee for $2.25. One morning, Brandie served 50 coffees for a total of $98.70. She made the coffee in 80-oz batches, and used exactly 10 of the batches during the morning. How many of each size did she sell? (*Source*: Starbucks®)

9. *Cholesterol Levels.* Recent studies indicate that a child's intake of cholesterol should be no more than 300 mg per day. By eating 1 egg, 1 cupcake, and 1 slice of pizza, a child consumes 302 mg of cholesterol. If the child eats 2 cupcakes and 3 slices of pizza, he or she takes in 65 mg of cholesterol. By eating 2 eggs and 1 cupcake, a child consumes 567 mg of cholesterol. How much cholesterol is in each item?

10. *Book Sale.* Katie, Rachel, and Logan went together to a library book sale. Katie bought 22 children's books, 10 paperbacks, and 5 hardbacks for a total of $63.50. Rachel bought 12 paperbacks and 15 hardbacks for a total of $52.50. Logan bought 8 children's books and 6 hardbacks for a total of $29.00. How much did each type of book cost?

11. *Automobile Pricing.* A recent basic model of a particular automobile had a price of $14,685. The basic model with the added features of automatic transmission and power door locks was $16,070. The basic model with air conditioning (AC) and power door locks was $15,580. The basic model with AC and automatic transmission was $15,925. What was the individual cost of each of the three options?

12. *Computer Pricing.* Lindsay plans to buy a new desktop computer for gaming. The base price of the computer is $480. If she upgrades the processor and the memory, the price of the computer is $745. If she upgrades the memory and the graphics card, the price of the computer is $690. If she upgrades the processor and the graphics card, the price of the computer is $805. What is the price of each upgrade?

13. *Veterinary Expenditure.* The sum of the average amounts Americans spent, per animal, for veterinary expenses for dogs, cats, and birds in a recent year was $290. The average expenditure per dog exceeded the sum of the averages for cats and birds by $110. The amount spent per cat was nine times the amount spent per bird. Find the average amount spent on each type of animal. (*Source*: American Veterinary Medical Association)

14. *Nutrition Facts.* A meal at Subway consisting of a 6-in. turkey breast sandwich, a bowl of minestrone soup, and a chocolate chip cookie contains 580 calories. The number of calories in the sandwich is 20 less than in the soup and the cookie together. The cookie has 120 calories more than the soup. Find the number of calories in each item. (*Source*: Subway)

15. *Nutrition.* A dietician in a hospital prepares meals under the guidance of a physician. Suppose that for a particular patient a physician prescribes a meal to have 800 calories, 55 g of protein, and 220 mg of vitamin C. The dietician prepares a meal of roast beef, baked potato, and broccoli according to the data in the following table. How many servings of each food are needed in order to satisfy the doctor's orders?

Food	Calories	Protein (in grams)	Vitamin C (in milligrams)
Roast beef, 3oz	300	20	0
Baked potato	100	5	20
Broccoli, 156 g	50	5	100

16. *Nutrition.* Repeat Exercise 15 but replace the broccoli with asparagus, for which one 180-g serving contains 50 calories, 5 g of protein, and 44 mg of vitamin C. Which meal would you prefer eating?

17. *Investments.* A business class divided an imaginary investment of $80,000 among three mutual funds. The first fund grew by 2%, the second by 6%, and the third by 3%. Total earnings were $2250. The earnings from the first fund were $150 more than the earnings from the third. How much was invested in each fund?

18. *Student Loans.* Terrence owes $32,000 in student loans. The interest rate on his Perkins loan is 5%, the rate on his Stafford loan is 4%, and the rate on his bank loan is 7%. Interest for one year totaled $1500. The interest for one year from the Perkins loan is $220 more than the interest from the bank loan. What is the amount of each loan?

19. *Golf.* On an 18-hole golf course, there are par-3 holes, par-4 holes, and par-5 holes. A golfer who shoots par on every hole has a total of 70. There are twice as many par-4 holes as there are par-5 holes. How many of each type of hole are there on the golf course?

20. *Basketball Scoring.* The New York Knicks once scored a total of 92 points on a combination of 2-point field goals, 3-point field goals, and 1-point foul shots. Altogether, the Knicks made 50 baskets and 19 more 2-pointers than foul shots. How many shots of each kind were made?

21. *Lens Production.* When Sight-Rite's three polishing machines, A, B, and C, are all working, 5700 lenses can be polished in one week. When only A and B are working, 3400 lenses can be polished in one week. When only B and C are working, 4200 lenses can be polished in one week. How many lenses can be polished in a week by each machine alone?

22. *Telemarketing.* Steve, Teri, and Isaiah can process 740 telephone orders per day. Steve and Teri together can process 470 orders, while Teri and Isaiah together can process 520 orders per day. How many orders can each person process alone?

▶ Skill Maintenance

Graph each function. **[2.2c]**

23. $f(x) = 2x - 3$

24. $g(x) = |x + 1|$

25. $h(x) = x^2 - 2$

Determine whether each of the following is the graph of a function. **[2.2d]**

26.

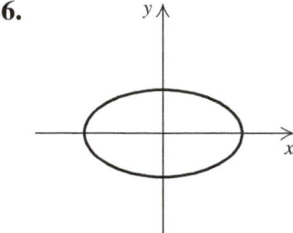

27. 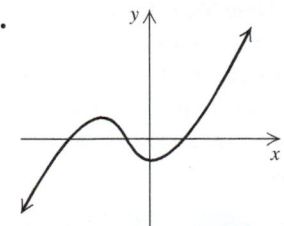   **28.**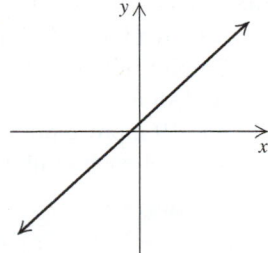

▶ **Synthesis**

29. Find the sum of the angle measures at the tips of the star in this figure.

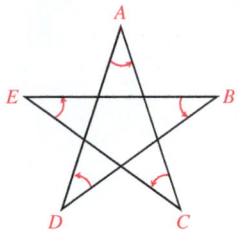

30. *Sharing Raffle Tickets.*   Hal gives Tom as many raffle tickets as Tom has and Gary as many as Gary has. In like manner, Tom then gives Hal and Gary as many tickets as each then has. Similarly, Gary gives Hal and Tom as many tickets as each then has. If each finally has 40 tickets, with how many tickets does Tom begin?

31. *Digits.*   Find a three-digit positive integer such that the sum of all three digits is 14, the tens digit is 2 more than the ones digit, and if the digits are reversed, the number is unchanged.

32. *Ages.*   Tammy's age is the sum of the ages of Carmen and Dennis. Carmen's age is 2 more than the sum of the ages of Dennis and Mark. Dennis's age is four times Mark's age. The sum of all four ages is 42. How old is Tammy?

3.7   Systems of Inequalities and Linear Programming

▶ **a**   Graph linear inequalities.

▶ **b**   Graph systems of linear inequalities.

▶ **c**   Solve linear programming problems.

A graph of an inequality is a drawing that represents its solutions. We have already seen that an inequality in one variable can be graphed on the number line. An inequality in two variables can be graphed on a coordinate plane.

SOLVE LINEAR INEQUALITIES

REVIEW SECTION 1.4

▶ **a**   **Graphs of Linear Inequalities**

A statement like $5x - 4y < 20$ is a linear inequality in two variables.

> **LINEAR INEQUALITY IN TWO VARIABLES**
>
> A **linear inequality in two variables** is an inequality that can be written in the form
>
> $$Ax + By < C,$$
>
> where A, B, and C are real numbers and A and B are not both zero. The symbol $<$ may be replaced with $\leq$, $>$, or $\geq$.

A solution of a linear inequality in two variables is an ordered pair (x, y) for which the inequality is true. For example, $(1, 3)$ is a solution of $5x - 4y < 20$ because $5 \cdot 1 - 4 \cdot 3 < 20$, or $-7 < 20$, is true. On the other hand, $(2, -6)$ is not a solution of $5x - 4y < 20$ because $5 \cdot 2 - 4 \cdot (-6) \not< 20$, or $34 \not< 20$.

The **solution set** of an inequality is the set of all ordered pairs that make it true. The **graph of an inequality** represents its solution set.

EXAMPLE 1 Graph: $y < x + 3$.

Solution We begin by graphing the **related equation** $y = x + 3$. We use a dashed line because the inequality symbol is $<$. This indicates that the line itself is not in the solution set of the inequality.

Note that the line divides the coordinate plane into two regions called **half-planes**. One of these half-planes satisfies the inequality. Either *all* points in a half-plane are in the solution set of the inequality or *none* is.

To determine which half-plane satisfies the inequality, we try a **test point** in either region. The point $(0, 0)$ is usually a convenient choice so long as it does not lie on the line.

$$y < x + 3$$
$$\overline{0 \;?\; 0 + 3}$$
$$0 \;\Big|\; 3 \qquad \text{TRUE} \qquad \mathbf{0 < 3 \text{ is true.}}$$

Since $(0, 0)$ satisfies the inequality, so do all points in the half-plane that contains $(0, 0)$. We shade this region to show the solution set of the inequality.

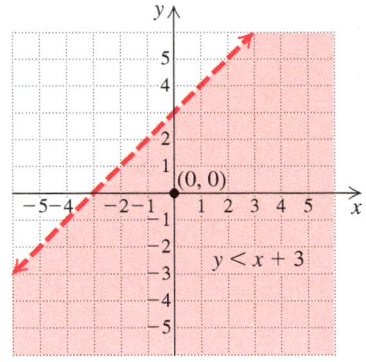

> **Now Try Exercise 13.**

Technology Connection

One way to graph the inequality in Example 1 on a graphing calculator is to first enter the related equation, $y = x + 3$. Then select the "shade below" graph style. Note that we must keep in mind that the line $y = x + 3$ is not included in the solution set.

Some calculators have an application Inequalz on the APPS menu that can be used to graph an inequality. When this application is used, the inequality $y < x + 3$ is entered directly and the graph of the related equation appears as a dashed line.

$y = x + 3$

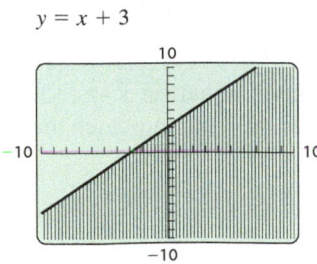

$y < x + 3$

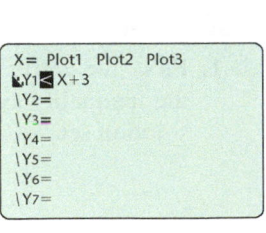

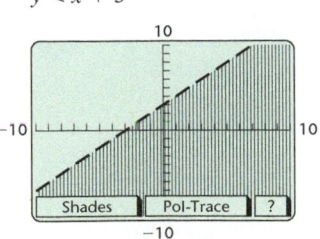

In general, we use the following procedure to graph linear inequalities in two variables by hand.

> **To graph a linear inequality in two variables:**
>
> 1. Replace the inequality symbol with an equals sign and graph this related equation. If the inequality symbol is $<$ or $>$, draw the line dashed. If the inequality symbol is $\leq$ or $\geq$, draw the line solid.
> 2. The graph consists of a half-plane on one side of the line and, if the line is solid, the line as well. To determine which half-plane to shade, test a point not on the line in the original inequality. If that point is a solution, shade the half-plane containing that point. If not, shade the opposite half-plane.

EXAMPLE 2 Graph: $3x + 4y \geq 12$.

Solution

1. First, we graph the related equation $3x + 4y = 12$. We use a solid line because the inequality symbol is $\geq$. This indicates that the line is included in the solution set.

2. To determine which half-plane to shade, we test a point in either region. We choose $(0, 0)$.

$$3x + 4y \geq 12$$
$$\overline{3 \cdot 0 + 4 \cdot 0 \ ? \ 12}$$
$$0 \ \bigm| \ 12 \quad \text{FALSE} \qquad 0 \geq 12 \text{ is false.}$$

Because $(0, 0)$ is *not* a solution, all the points in the half-plane that does *not* contain $(0, 0)$ are solutions. We shade that region, as shown in the following figure.

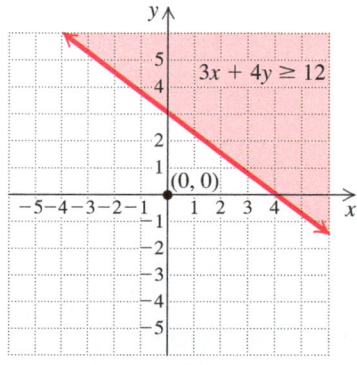

Now Try Exercise 17.

EXAMPLE 3 Graph $x > -3$ on a plane.

Solution

1. First, we graph the related equation $x = -3$. We use a dashed line because the inequality symbol is $>$. This indicates that the line is not included in the solution set.

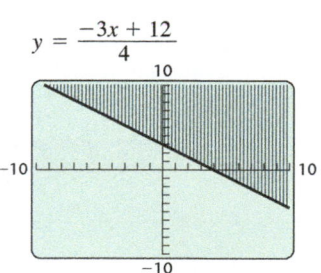

2. The inequality tells us that all points (x, y) for which $x > -3$ are solutions. These are the points to the right of the line. We can also use a test point to determine the solutions. We choose $(5, 1)$.

$$x > -3$$
$$\overline{5 \ ? \ -3} \quad \text{TRUE} \qquad 5 > -3 \text{ is true.}$$

Because $(5, 1)$ is a solution, we shade the region containing that point—that is, the region to the right of the dashed line.

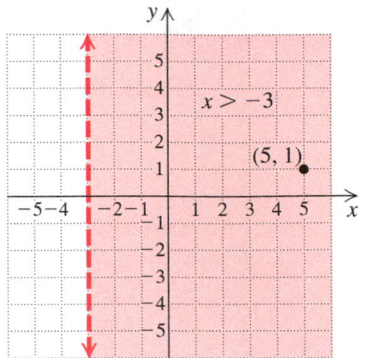

> **Now Try Exercise 23.**

EXAMPLE 4 Graph $y \le 4$ on a plane.

Solution

1. First, we graph the related equation $y = 4$. We use a solid line because the inequality symbol is $\le$.

2. The inequality tells us that all points (x, y) for which $y \le 4$ are solutions of the inequality. These are the points on or below the line. We can also use a test point to determine the solutions. We choose $(-2, 5)$.

$$y \le 4$$
$$\overline{5 \ ? \ 4} \quad \text{FALSE} \qquad 5 \le 4 \text{ is false.}$$

Because $(-2, 5)$ is not a solution, we shade the half-plane that does not contain that point.

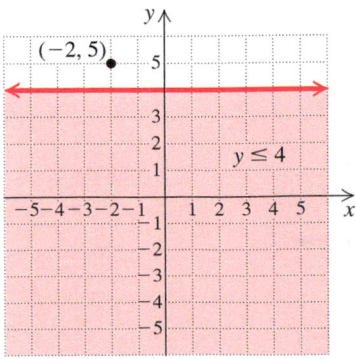

> **Now Try Exercise 25.**

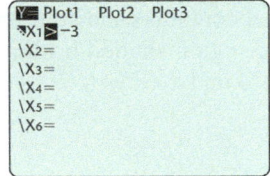
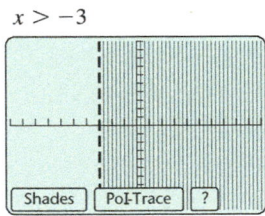
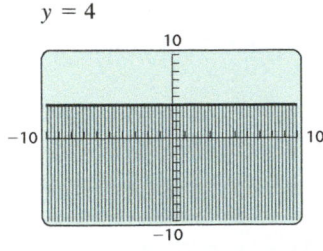

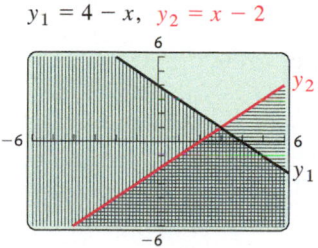
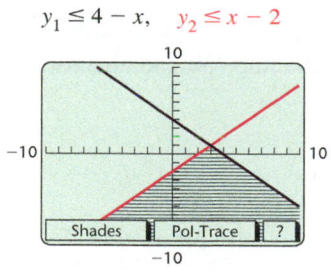
▶ **b Systems of Linear Inequalities**

A system of inequalities consists of two or more inequalities considered simultaneously. For example,

$$x + y \leq 4,$$
$$x - y \geq 2$$

is a system of *two linear inequalities in two variables.*

   A solution of a system of inequalities is an ordered pair that is a solution of each inequality in the system. To graph a system of linear inequalities, we graph each inequality and determine the region that is common to *all* the solution sets.

EXAMPLE 5   Graph the solution set of the system

$$x + y \leq 4,$$
$$x - y \geq 2.$$

Solution   We graph $x + y \leq 4$ by first graphing the equation $x + y = 4$ using a solid line. Next, we choose $(0, 0)$ as a test point and find that it is a solution of $x + y \leq 4$, so we shade the half-plane containing $(0, 0)$ using red. Next, we graph $x - y = 2$ using a solid line. We find that $(0, 0)$ is not a solution of $x - y \geq 2$, so we shade the half-plane that does not contain $(0, 0)$ using green. The arrows near the ends of each line help to indicate the half-plane that contains each solution set.

   The solution set of the system of equations is the region shaded both red and green, or brown, including parts of the lines $x + y = 4$ and $x - y = 2$.

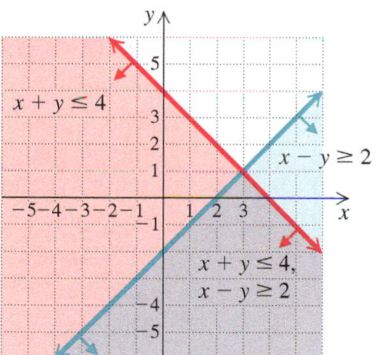

   A system of inequalities may have a graph that consists of a polygon and its interior. As we will see later in this section, in many applications we will need to know the vertices of such a polygon.

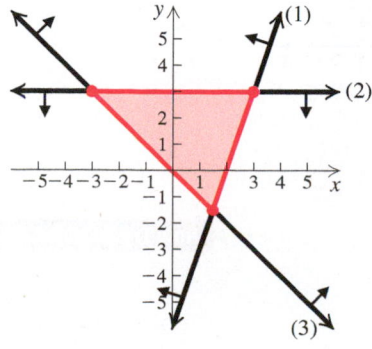

EXAMPLE 6   Graph the following system of inequalities and find the coordinates of any vertices formed:

$$3x - y \leq 6, \quad \textbf{(1)}$$
$$y - 3 \leq 0, \quad \textbf{(2)}$$
$$x + y \geq 0. \quad \textbf{(3)}$$

Solution   We graph the related equations $3x - y = 6$, $y - 3 = 0$, and $x + y = 0$ using solid lines shown at left. The half-plane containing the solution set for each inequality is indicated by the arrows near the ends of each line. We shade the region common to all three solution sets.

To find the vertices, we solve three systems of equations. The system of equations from inequalities (1) and (2) is

$$3x - y = 6,$$
$$y - 3 = 0.$$

Solving, we obtain the vertex $(3, 3)$.

The system of equations from inequalities (1) and (3) is

$$3x - y = 6,$$
$$x + y = 0.$$

Solving, we obtain the vertex $\left(\frac{3}{2}, -\frac{3}{2}\right)$.

The system of equations from inequalities (2) and (3) is

$$y - 3 = 0,$$
$$x + y = 0.$$

Solving, we obtain the vertex $(-3, 3)$.

 **Now Try Exercise 55.**

▶ c Applications: Linear Programming

In many applications, we want to find a maximum value or a minimum value. In business, for example, we might want to maximize profit and minimize cost. **Linear programming** can tell us how to do this.

In our study of linear programming, we will consider linear functions of two variables that are to be maximized or minimized subject to several conditions, or **constraints**. These constraints are expressed as inequalities. The solution set of the system of inequalities made up of the constraints contains all the **feasible solutions** of a linear programming problem. The function that we want to maximize or minimize is called the **objective function**.

It can be shown that the maximum and minimum values of the objective function occur at a vertex of the region of feasible solutions. Thus we have the following procedure.

Linear Programming Procedure

To find the maximum or minimum value of a linear objective function subject to a set of constraints:

1. Graph the region of feasible solutions.
2. Determine the coordinates of the vertices of the region.
3. Evaluate the objective function at each vertex. The largest and smallest of those values are the maximum and minimum values of the function, respectively.

EXAMPLE 7 *Maximizing Profit.* Dovetail Carpentry Shop makes bookcases and desks. Each bookcase requires 5 hr of woodworking and 4 hr of finishing. Each desk requires 10 hr of woodworking and 3 hr of finishing. Each month the shop has 600 hr of labor available for woodworking and 240 hr for finishing. The profit on each bookcase is $40 and on each desk is $75. How many of each product should be made each month in order to maximize profit? What is the maximum profit?

Solution We let $x =$ the number of bookcases to be produced and $y =$ the number of desks. Then the profit P is given by the function

$$P = 40x + 75y.$$ *To emphasize that P is a function of two variables, we sometimes write $P(x, y) = 40x + 75y$.*

We know that x bookcases require $5x$ hr of woodworking and y desks require $10y$ hr of woodworking. Since there is no more than 600 hr of labor available for woodworking, we have one constraint:

$$5x + 10y \le 600.$$

Similarly, the bookcases and the desks require $4x$ hr and $3y$ hr of finishing, respectively. There is no more than 240 hr of labor available for finishing, so we have a second constraint:

$$4x + 3y \le 240.$$

We also know that $x \ge 0$ and $y \ge 0$ because the carpentry shop cannot make a negative number of either product.

Thus we want to maximize the objective function

$$P = 40x + 75y$$

subject to the constraints

$$5x + 10y \le 600,$$
$$4x + 3y \le 240,$$
$$x \ge 0,$$
$$y \ge 0.$$

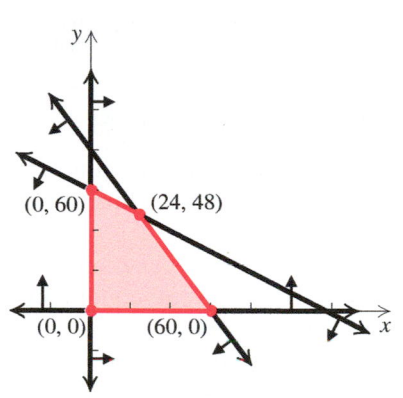

We graph the system of inequalities and determine the vertices, as shown in the figure at left.

Next, we evaluate the objective function P at each vertex.

Vertices (x, y)	Profit $P = 40x + 75y$	
$(0, 0)$	$P = 40 \cdot 0 + 75 \cdot 0 = 0$	
$(60, 0)$	$P = 40 \cdot 60 + 75 \cdot 0 = 2400$	
$(24, 48)$	$P = 40 \cdot 24 + 75 \cdot 48 = 4560$	← Maximum
$(0, 60)$	$P = 40 \cdot 0 + 75 \cdot 60 = 4500$	

Technology Connection

We can create a table in which an objective function is evaluated at each vertex of a system of inequalities if the system has been graphed using the Inequalz application in the APPS menu.

The carpentry shop will make a maximum profit of $4560 when 24 bookcases and 48 desks are produced and sold.

Now Try Exercise 65.

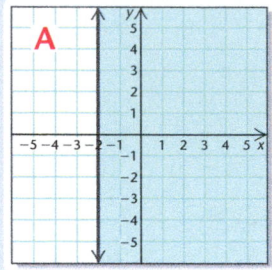

A

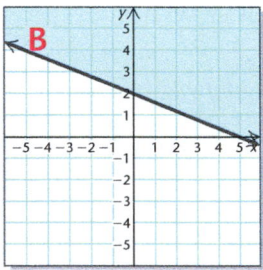

B

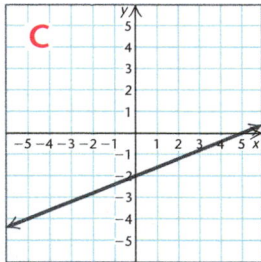

C

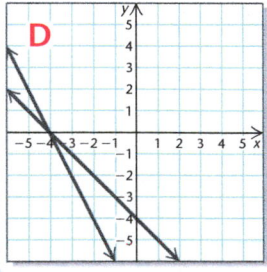

D

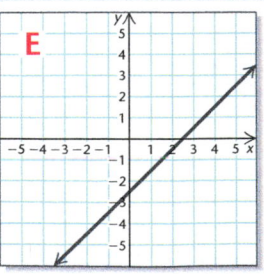

E

Visualizing the Graph

Match the equation, inequality, system of equations, or system of inequalities with its graph.

1. $x + y = -4,$
$2x + y = -8$

2. $2x + 5y \geq 10$

3. $2x - 2y = 5$

4. $2x - 5y = 10$

5. $-2y < 8$

6. $5x - 2y = 10$

7. $2x = 10$

8. $5x + 2y < 10,$
$2x - 5y > 10$

9. $5x \geq -10$

10. $y - 2x < 8$

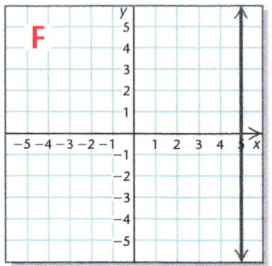

F

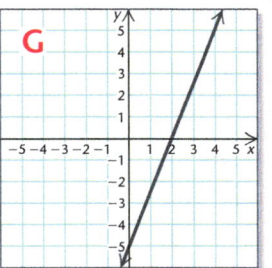

G

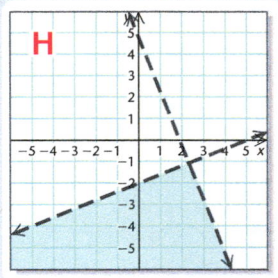

H

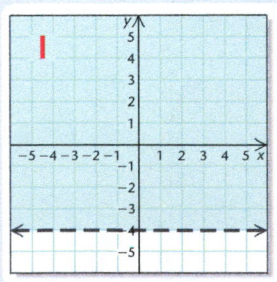

I

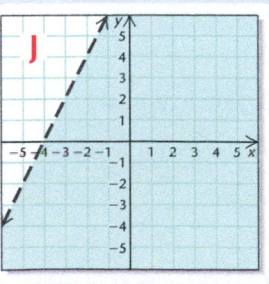

J

3.7 Exercise Set

a *In Exercises 1–8, match the inequality with one of the graphs (a)–(h) that follow.*

a)
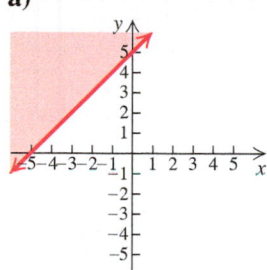

b)
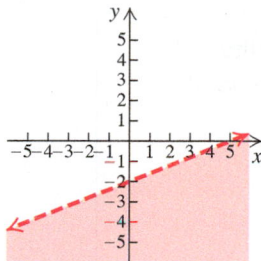

c)
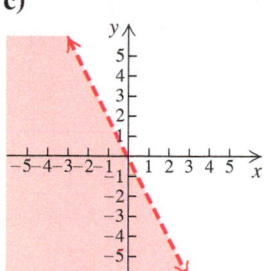

d)
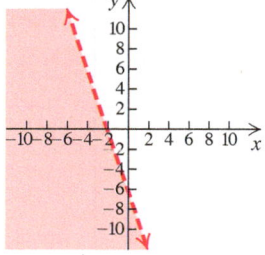

e)
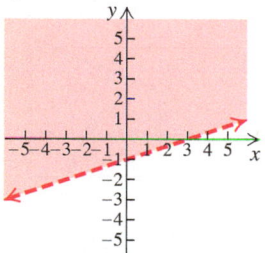

f)
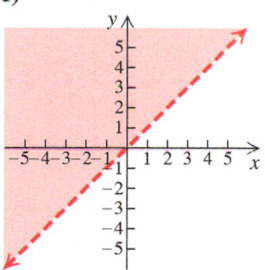

g)
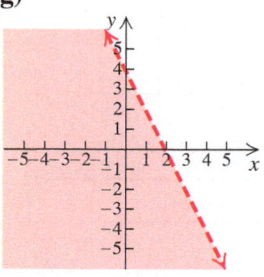

h)
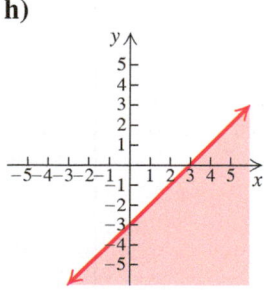

1. $y > x$

2. $y < -2x$

3. $y \le x - 3$

4. $y \ge x + 5$

5. $2x + y < 4$

6. $3x + y < -6$

7. $2x - 5y > 10$

8. $3x - 9y < 9$

Graph.

9. $y > 2x$

10. $2y < x$

11. $y + x \ge 0$

12. $y - x < 0$

13. $y > x - 3$

14. $y \le x + 4$

15. $x + y < 4$

16. $x - y \ge 5$

17. $3x - 2y \le 6$

18. $2x - 5y < 10$

19. $3y + 2x \ge 6$

20. $2y + x \le 4$

21. $3x - 2 \le 5x + y$

22. $2x - 6y \ge 8 + 2y$

23. $x < -4$

24. $y > -3$

25. $y \ge 5$

26. $x \le 5$

27. $-4 < y < -1$
(*Hint*: Think of this as $-4 < y$ and $y < -1$.)

28. $-3 \le x \le 3$

29. $y \ge |x|$

30. $y \le |x + 2|$

b *In Exercises 31–36, match the system of inequalities with one of the graphs (a)–(f) that follow.*

a)
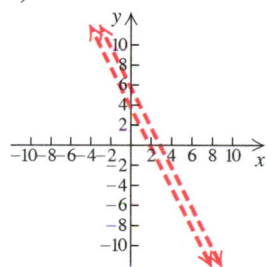

b)
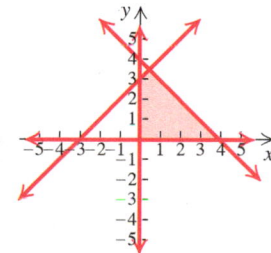

c)
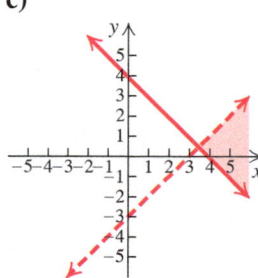

d)
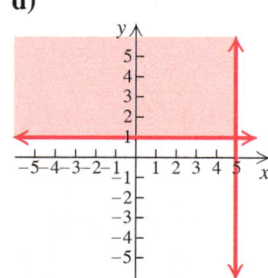

e)
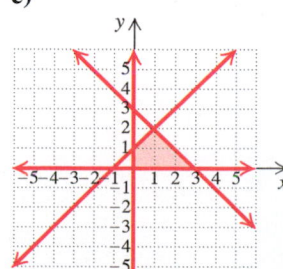

f)
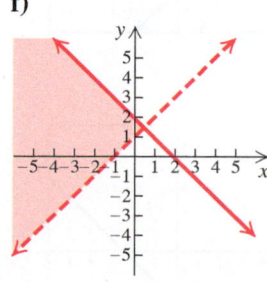

31. $y > x + 1,$
$y \le 2 - x$

32. $y < x - 3,$
$y \ge 4 - x$

33. $2x + y < 4,$
$4x + 2y > 12$

34. $x \le 5,$
$y \ge 1$

35. $x + y \le 4,$
$x - y \ge -3,$
$x \ge 0,$
$y \ge 0$

36. $x - y \ge -2,$
$x + y \le 6,$
$x \ge 0,$
$y \ge 0$

Find a system of inequalities with the given graph. Answers may vary.

37.

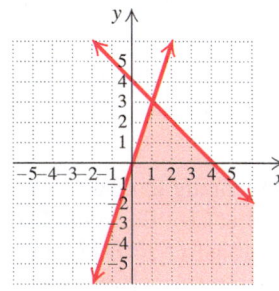

38.

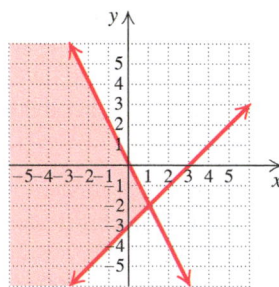

39.

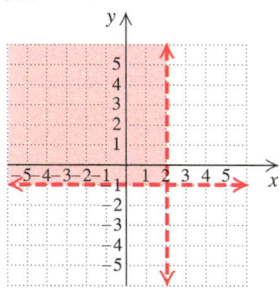

40.

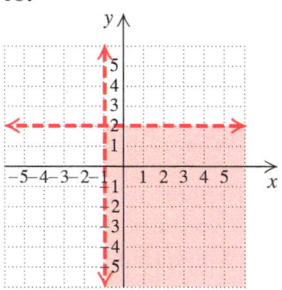

41.

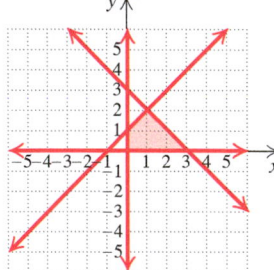

42.

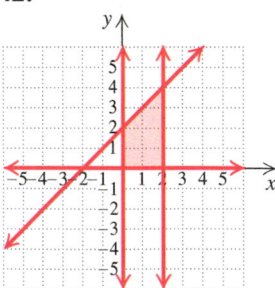

Graph the system of inequalities. Then find the coordinates of the vertices.

43. $y \le x,$
$y \ge 3 - x$

44. $y \le x,$
$y \ge 5 - x$

45. $y \ge x,$
$y \le 4 - x$

46. $y \ge x,$
$y \le 2 - x$

47. $y \ge -3,$
$x \ge 1$

48. $y \le -2,$
$x \ge 2$

49. $x \le 3,$
$y \ge 2 - 3x$

50. $x \ge -2,$
$y \le 3 - 2x$

51. $x + y \le 1,$
$x - y \le 2$

52. $y + 3x \ge 0,$
$y + 3x \le 2$

53. $2y - x \le 2,$
$y + 3x \ge -1$

54. $y \le 2x + 1,$
$y \ge -2x + 1,$
$x - 2 \le 0$

55. $x - y \le 2,$
$x + 2y \ge 8,$
$y - 4 \le 0$

56. $x + 2y \le 12,$
$2x + y \le 12,$
$x \ge 0,$
$y \ge 0$

57. $4y - 3x \ge -12,$
$4y + 3x \ge -36,$
$y \le 0,$
$x \le 0$

58. $8x + 5y \le 40,$
$x + 2y \le 8,$
$x \ge 0,$
$y \ge 0$

59. $3x + 4y \ge 12,$
$5x + 6y \le 30,$
$1 \le x \le 3$

60. $y - x \ge 1,$
$y - x \le 3,$
$2 \le x \le 5$

C *Find the maximum value and the minimum value of the function and the values of x and y for which they occur.*

61. $P = 17x - 3y + 60,$ subject to
$6x + 8y \le 48,$
$0 \le y \le 4,$
$0 \le x \le 7.$

62. $Q = 28x - 4y + 72,$ subject to
$5x + 4y \ge 20,$
$0 \le y \le 4,$
$0 \le x \le 3.$

63. $F = 5x + 36y,$ subject to
$5x + 3y \le 34,$
$3x + 5y \le 30,$
$x \ge 0,$
$y \ge 0.$

64. $G = 16x + 14y,$ subject to
$3x + 2y \le 12,$
$7x + 5y \le 29,$
$x \ge 0,$
$y \ge 0.$

65. *Maximizing Mileage.* Jazmin owns a pickup truck and a moped. She can afford 12 gal of gasoline to be split between the truck and the moped. Jazmin's truck gets 20 mpg and, with the fuel currently in the tank, can hold at most an additional 10 gal of gas. Her moped gets 100 mpg and can hold at most 3 gal of gas. How many gallons of gasoline should each vehicle use if Jazmin wants to travel as far as possible on the 12 gal of gas? What is the maximum number of miles that she can travel?

66. *Maximizing Income.* Golden Harvest Foods makes jumbo biscuits and regular biscuits. The oven can cook at most 200 biscuits per day. Each jumbo biscuit requires 2 oz of flour, each regular biscuit requires 1 oz of flour, and there are 300 oz of flour available. The income from each jumbo biscuit is $0.45 and from each regular biscuit is $0.30. How many of each size biscuit should be made in order to maximize income? What is the maximum income?

67. *Maximizing Profit.* Waterbrook Farm includes 240 acres of cropland. The farm owner wishes to plant this acreage in both corn and soybeans. The profit per acre in corn production is $325 and in soybeans is $180. A total of 320 hr of labor is available. Each acre of corn requires 2 hr of labor, whereas each acre of soybean requires 1 hr of labor. How should the land be divided between corn and soybeans in order to yield the maximum profit? What is the maximum profit?

68. *Maximizing Profit.* Norris Mill can convert logs into lumber and plywood. In a given week, the mill can turn out 400 units of production, of which at least 100 units of lumber and at least 150 units of plywood are required by regular customers. The profit is $25 per unit of lumber and $38 per unit of plywood. How many units of each should the mill produce in order to maximize the profit? What is the maximum profit?

69. *Minimizing Cost.* An animal feed to be mixed from soybean meal and oats must contain at least 120 lb of protein, 24 lb of fat, and 10 lb of mineral ash. Each 100-lb sack of soybean meal costs $20 and contains 50 lb of protein, 8 lb of fat, and 5 lb of mineral ash. Each 100-lb sack of oats costs $8 and contains 15 lb of protein, 5 lb of fat, and 1 lb of mineral ash. How many sacks of each should be used in order to satisfy the minimum requirements at minimum cost? What is the minimum cost?

70. *Minimizing Cost.* Suppose that in the preceding exercise the oats were replaced by alfalfa, which costs $10 per 100-lb sack and contains 20 lb of protein, 6 lb of fat, and 8 lb of mineral ash. How much of each would now be required in order to minimize the cost? What is the minimum cost?

71. *Maximizing Income.* Francisco is planning to invest up to $40,000 in corporate and municipal bonds. The least he is allowed to invest in corporate bonds is $6000, and he does not want to invest more than $22,000 in corporate bonds. He also does not want to invest more than $30,000 in municipal bonds. The interest is 3% on corporate bonds and $4\frac{1}{4}\%$ on municipal bonds. This is simple interest for one year. How much should he invest in each type of bond in order to maximize his income? What is the maximum income?

72. *Maximizing Income.* Mila is planning to invest up to $22,000 in certificates of deposit at City Bank and People's Bank. She wants to invest at least $2000 but no more than $14,000 at City Bank. People's Bank does not insure more than a $15,000 investment, so she will invest no more than that in People's Bank. The interest is $2\frac{1}{2}\%$ at City Bank and $1\frac{3}{4}\%$ at People's Bank. This is simple interest for one year. How much should she invest in each bank in order to maximize her income? What is the maximum income?

73. *Minimizing Transportation Cost.* An airline with two types of airplanes, P_1 and P_2, has contracted with a tour group to provide transportation for a minimum of 2000 first-class, 1500 business-class, and 2400 economy-class passengers. For a certain trip, airplane P_1 costs $12 thousand to operate and can accommodate 40 first-class, 40 business-class, and 120 economy-class passengers, whereas airplane P_2 costs $10 thousand to operate and can accommodate 80 first-class, 30 business-class, and 40 economy-class passengers. How many of each type of airplane should be used in order to minimize the operating cost? What is the minimum operating cost?

74. *Minimizing Transportation Cost.* Suppose that in the preceding exercise a new airplane P_3 becomes available, having an operating cost for the same trip of $15 thousand and accommodating 40 first-class, 40 tourist-class, and 80 economy-class passengers. If airplane P_1 were replaced by airplane P_3, how many of P_2 and P_3 should be used in order to minimize the operating cost? What is the minimum operating cost?

75. *Maximizing Profit.* It takes Fena Tailoring 3 hr of cutting and 6 hr of sewing to make a tiered silk organza bridal dress. It takes 6 hr of cutting and 3 hr

of sewing to make a lace sheath bridal dress. The shop has at most 27 hr per week available for cutting and at most 36 hr per week for sewing. The profit is $320 on an organza dress and $305 on a lace dress. How many of each kind of bridal dress should be made each week in order to maximize profit? What is the maximum profit?

76. *Maximizing Profit.* Cambridge Metal Works manufactures two sizes of gears. The smaller gear requires 4 hr of machining and 1 hr of polishing and yields a profit of $45. The larger gear requires 1 hr of machining and 1 hr of polishing and yields a profit of $30. The firm has available at most 24 hr per day for machining and 9 hr per day for polishing. How many of each type of gear should be produced each day in order to maximize profit? What is the maximum profit?

77. *Minimizing Nutrition Cost.* Suppose that it takes 12 units of carbohydrates and 6 units of protein to satisfy Jacob's minimum weekly requirements. A particular type of meat contains 2 units of carbohydrates and 2 units of protein per pound. A particular cheese contains 3 units of carbohydrates and 1 unit of protein per pound. The meat costs $3.50 per pound and the cheese costs $4.60 per pound. How many pounds of each are needed in order to minimize the cost and still meet the minimum requirements? What is the minimum cost?

78. *Minimizing Salary Cost.* The Spring Hill school board is analyzing education costs for Hill Top School. It wants to hire teachers and teacher's aides to make up a faculty that satisfies its needs at minimum cost. The average annual salary for a teacher is $53,000 and for a teacher's aide is $23,600. The school building can accommodate a faculty of no more than 50 but needs at least 20 faculty members to function properly. The school must have at least

12 aides, but the number of teachers must be at least twice the number of aides in order to accommodate the expectations of the community. How many teachers and teacher's aides should be hired in order to minimize salary costs? What is the minimum salary cost?

79. *Maximizing Animal Support in a Forest.* A certain area of forest is populated by two species of animal, which scientists refer to as A and B for simplicity. The forest supplies two kinds of food, referred to as F_1 and F_2. For one year, each member of species A requires 1 unit of F_1 and 0.5 unit of F_2. Each member of species B requires 0.2 unit of F_1 and 1 unit of F_2. The forest can normally supply at most 600 units of F_1 and 525 units of F_2 per year. What is the maximum total number of these animals that the forest can support?

80. *Maximizing Animal Support in a Forest.* Refer to Exercise 79. If there is a wet spring, then supplies of food increase to 1080 units of F_1 and 810 units of F_2. In this case, what is the maximum total number of these animals that the forest can support?

▶ **Skill Maintenance**

Solve.

81. $-5 \leq x + 2 < 4$ **[1.6]**

82. $|x - 3| \geq 2$ **[3.5]**

83. $x^2 - 2x \leq 3$ **[4.6]**

84. $\dfrac{x - 1}{x + 2} > 4$ **[4.6]**

▶ **Synthesis**

Graph the system of inequalities.

85. $y \geq x^2 - 2,$
$y \leq 2 - x^2$

86. $y < x + 1,$
$y \geq x^2$

Graph the inequality.

87. $|x + y| \leq 1$

88. $|x| + |y| \leq 1$

89. $|x| > |y|$

90. $|x - y| > 0$

91. *Allocation of Resources.* Comfort-by-Design Furniture produces chairs and sofas. Each chair requires 20 ft of wood, 1 lb of foam rubber, and 2 yd^2 of fabric. Each sofa requires 100 ft of wood, 50 lb of foam rubber, and 20 yd^2 of fabric. The manufacturer has in stock 1900 ft of wood, 500 lb of foam rubber, and 240 yd^2 of fabric. The chairs can be sold for $200 each and the sofas for $750 each. How many of each should be produced in order to maximize income? What is the maximum income?

Chapter 3 Summary and Review

VOCABULARY REINFORCEMENT

Complete each statement with the correct term from the column on the right. Some of the choices may be used more than once and some may not be used at all.

line

half-plane

independent

dependent

consistent

inconsistent

pair

triple

1. A solution of a system of two equations in two variables is an ordered _____ that makes both equations true. **[3.1a]**

2. A(n) _____ system of equations has at least one solution. **[3.1a]**

3. A solution of a system of three equations in three variables is an ordered _____ that makes all three equations true. **[3.5a]**

4. If, for a system of two equations in two variables, the graphs of the equations are different lines, then the equations are _____. **[3.1a]**

5. The graph of an inequality like $x > 2y$ is a(n) _____. **[3.7b]**

CONCEPT REINFORCEMENT

Determine whether each statement is true or false.

1. A system of equations with infinitely many solutions is inconsistent. **[3.1a]**

2. It is not possible for the equations in an inconsistent system of two equations to be dependent. **[3.1a]**

3. When $(0, b)$ is a solution of each equation in a system of two equations, the graphs of the two equations have the same y-intercept. **[3.1a]**

4. The system of equations $x = 4$ and $y = -4$ is inconsistent. **[3.1a]**

STUDY GUIDE

Objective 3.1a   Solve a system of two linear equations or two functions by graphing and determine whether a system is consistent or inconsistent and whether the equations in a system are dependent or independent.

Example   Solve this system of equations graphically. Then classify the system as consistent or inconsistent and the equations as dependent or independent.

$$x - y = 3,$$
$$y = 2x - 4$$

Practice Exercise

1. Solve this system of equations graphically. Then classify the system as consistent or inconsistent and the equations as dependent or independent.

$$x + 3y = 1,$$
$$x + y = 3$$

(continued)

We graph the equations. The point of intersection appears to be $(1, -2)$. This checks in both equations, so it is the solution. The system has one solution, so it is consistent and the equations are independent.

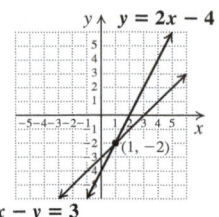

Objective 3.2a   Solve systems of equations in two variables by the substitution method.

Example   Solve the system

$$x - 2y = 1, \qquad \textbf{(1)}$$
$$2x - 3y = 3. \qquad \textbf{(2)}$$

We solve equation (1) for x, since the coefficient of x is 1 in that equation:

$$x - 2y = 1$$
$$x = 2y + 1. \qquad \textbf{(3)}$$

Next, we substitute for x in equation (2) and solve for y:

$$2x - 3y = 3$$
$$2(2y + 1) - 3y = 3$$
$$4y + 2 - 3y = 3$$
$$y + 2 = 3$$
$$y = 1.$$

Then we substitute 1 for y in equation (1), (2), or (3) and find x. We choose equation (3) since it is already solved for x:

$$x = 2y + 1 = 2 \cdot 1 + 1 = 2 + 1 = 3$$

Check:

$$\begin{array}{c|c} x - 2y = 1 \\ \hline 3 - 2 \cdot 1 \;?\; 1 \\ 3 - 2 \\ 1 \;\bigg|\; \text{TRUE} \end{array} \qquad \begin{array}{c|c} 2x - 3y = 3 \\ \hline 2 \cdot 3 - 3 \cdot 1 \;?\; 3 \\ 6 - 3 \\ 3 \;\bigg|\; \text{TRUE} \end{array}$$

The ordered pair $(3, 1)$ checks in both equations, so it is the solution of the system of equations.

Practice Exercise

2. Solve the system

$$2x + y = 2,$$
$$3x + 2y = 5$$

using the substitution method.

Objective 3.3a   Solve systems of equations in two variables by the elimination method.

Example   Solve the system

$$2a + 3b = -1, \qquad \textbf{(1)}$$
$$3a + 2b = 6. \qquad \textbf{(2)}$$

We could eliminate either a or b. In this case, we decide to eliminate the a-terms. We multiply equation (1) by 3 and equation (2) by -2 and then add and solve for b:

$$\begin{array}{r} 6a + 9b = -3 \\ -6a - 4b = -12 \\ \hline 5b = -15 \\ b = -3. \end{array}$$

Practice Exercise

3. Solve the system

$$2x + 3y = 5,$$
$$3x + 4y = 6$$

using the elimination method.

(continued)

Next, we substitute -3 for b in either of the original equations:

$$2a + 3b = -1 \quad \textbf{(1)}$$
$$2a + 3(-3) = -1$$
$$2a - 9 = -1$$
$$2a = 8$$
$$a = 4.$$

The ordered pair $(4, -3)$ checks in both equations, so it is a solution of the system of equations.

Objective 3.4a Solve applied problems involving total value and mixture using systems of two equations.

Example To start a small business, Michael took two loans totaling $18,000. One of the loans was at 7% interest and the other at 8%. After one year, Michael owed $1365 in interest. What was the amount of each loan?

1. **Familiarize.** We let x and y represent the amounts of the two loans. Next, we organize the information in a table and use the simple interest formula, $I = Prt$.

	Loan 1	Loan 2	Total
Principal	x	y	$18,000
Rate of Interest	7%	8%	
Time	1 year	1 year	
Interest	7%x, or 0.07x	8%y, or 0.08y	$1365

2. **Translate.** The total amount of the loans is found in the first row of the table. This gives us one equation:

$$x + y = 18,000.$$

From the last row of the table, we see that the interest totals $1365. This gives us a second equation:

$$0.07x + 0.08y = 1365.$$

3. **Solve.** We solve the resulting system of equations:

$$x + y = 18,000, \quad \textbf{(1)}$$
$$0.07x + 0.08y = 1365. \quad \textbf{(2)}$$

We multiply by -0.07 on both sides of equation (1) and add:

$$-0.07x - 0.07y = -1260$$
$$\underline{0.07x + 0.08y = \quad 1365} \quad \textbf{(2)}$$
$$0.01y = \quad 105 \quad \text{Adding}$$
$$y = 10,500. \quad \text{Solving for } y$$

Practice Exercise

4. Jaretta made two investments totaling $23,000. In one year, these investments yielded $1237 in simple interest. Part of the money was invested at 6% and the rest at 5%. How much was invested at each rate?

Then

$$x + 10{,}500 = 18{,}000 \qquad \text{Substituting 10,500 for } y \text{ in equation (1)}$$

$$x = 7500. \qquad \text{Solving for } x$$

We find that $x = 7500$ and $y = 10{,}500$.

4. **Check.** The sum is $7500 + \$10{,}500$, or $18{,}000$. The interest from $7500 at 7% for one year is $7\%(\$7500)$, or 525. The interest from $10{,}500 at 8% for one year is $8\%(\$10{,}500)$, or 840. The total amount of interest is $525 + \$840$, or 1365. The numbers check in the problem.

5. **State.** Michael took loans of $7500 at 7% interest and $10{,}500 at 8% interest.

Objective 3.5a Solve systems of three equations in three variables.

Example Solve:

$$\begin{aligned} x - y - z &= -2, & \textbf{(1)} \\ 2x + 3y + z &= 3, & \textbf{(2)} \\ 5x - 2y - 2z &= -1. & \textbf{(3)} \end{aligned}$$

The equations are in standard form and do not contain decimals or fractions. We choose to eliminate z since the z-terms in equations (1) and (2) are opposites.

First, we add these two equations:

$$\begin{aligned} x - y - z &= -2 \\ 2x + 3y + z &= 3 \\ \hline 3x + 2y \phantom{{}+ z} &= 1. \quad \textbf{(4)} \end{aligned}$$

Next, we multiply equation (2) by 2 and add it to equation (3) to eliminate z from another pair of equations:

$$\begin{aligned} 4x + 6y + 2z &= 6 \\ 5x - 2y - 2z &= -1 \\ \hline 9x + 4y \phantom{{}+ 2z} &= 5. \quad \textbf{(5)} \end{aligned}$$

Now we solve the system consisting of equations (4) and (5). We multiply equation (4) by -2 and add:

$$\begin{aligned} -6x - 4y &= -2 \\ 9x + 4y &= 5 \\ \hline 3x \phantom{{}+ 4y} &= 3 \\ x &= 1. \end{aligned}$$

Then we use either equation (4) or (5) to find y:

$$\begin{aligned} 3x + 2y &= 1 \qquad \textbf{(4)} \\ 3 \cdot 1 + 2y &= 1 \\ 3 + 2y &= 1 \\ 2y &= -2 \\ y &= -1. \end{aligned}$$

Practice Exercise

5. Solve:

$$\begin{aligned} x - y + z &= 9, \\ 2x + y + 2z &= 3, \\ 4x + 2y - 3z &= -1. \end{aligned}$$

(continued)

Finally, we use one of the original equations to find z:

$$2x + 3y + z = 3 \qquad \textbf{(2)}$$
$$2 \cdot 1 + 3(-1) + z = 3$$
$$-1 + z = 3$$
$$z = 4.$$

Check:

$$\frac{x - y - z = -2}{1 - (-1) - 4 \;?\; -2}$$
$$1 + 1 - 4$$
$$-2 \quad | \quad \text{TRUE}$$

$$\frac{2x + 3y + z = 3}{2 \cdot 1 + 3(-1) + 4 \;?\; 3}$$
$$2 - 3 + 4$$
$$3 \quad | \quad \text{TRUE}$$

$$\frac{5x - 2y - 2z = -1}{5 \cdot 1 - 2(-1) - 2 \cdot 4 \;?\; -1}$$
$$5 + 2 - 8$$
$$-1 \quad | \quad \text{TRUE}$$

The ordered triple $(1, -1, 4)$ checks in all three equations, so it is the solution of the system of equations.

Objective 3.7a   Graph linear inequalities in two variables.

Example   Graph: $2x + y \leq 4$.

First, we graph the line $2x + y = 4$. The intercepts are $(0, 4)$ and $(2, 0)$. We draw the line solid because the inequality symbol is $\leq$. Next, we choose a test point not on the line and determine whether it is a solution of the inequality. We choose $(0, 0)$, since it is usually an easy point to use.

$$\frac{2x + y \leq 4}{2 \cdot 0 + 0 \;?\; 4}$$
$$0 \quad | \quad \text{TRUE}$$

Since $(0, 0)$ is a solution, we shade the half-plane that contains $(0, 0)$.

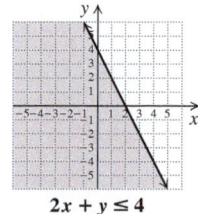

$2x + y \leq 4$

Practice Exercise

6. Graph: $3x - 2y > 6$.

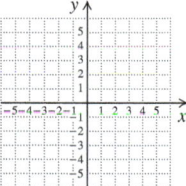

Objective 3.7b Graph systems of linear inequalities and find coordinates of any vertices.

Example Graph this system of inequalities and find the coordinates of any vertices formed:

$$x - 2y \geq -2, \quad \textbf{(1)}$$
$$3x - y \leq 4, \quad \textbf{(2)}$$
$$y \geq -1. \quad \textbf{(3)}$$

We graph the related equations using solid lines. Then we indicate the region for each inequality by arrows at the ends of the line. Next, we shade the region of overlap.

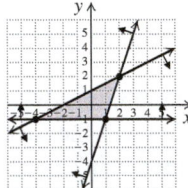

To find the vertices, we solve three different systems of related equations. From (1) and (2), we solve

$$x - 2y = -2,$$
$$3x - y = 4$$

to find the vertex $(2, 2)$. From (1) and (3), we solve

$$x - 2y = -2,$$
$$y = -1$$

to find the vertex $(-4, -1)$. From (2) and (3), we solve

$$3x - y = 4,$$
$$y = -1$$

to find the vertex $(1, -1)$.

Practice Exercise

7. Graph this system of inequalities and find the coordinates of any vertices found:

$$x - 2y \leq 4,$$
$$x + y \leq 4,$$
$$x - 1 \geq 0.$$

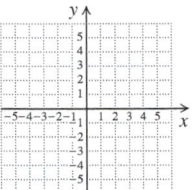

Objective 3.7c Solve linear programming problems.

Example Maximize $G = 8x - 5y$ subject to

$$x + y \leq 3,$$
$$x \geq 0,$$
$$y \geq 1.$$

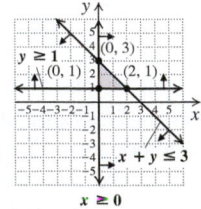

Practice Exercise

8. Maximize $P = 4x - 5y$ subject to

$$x - y \leq 2,$$
$$x + 2y \geq 8,$$
$$y \leq 4.$$

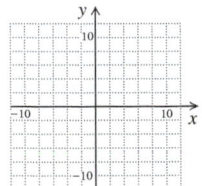

(continued)

Vertex	$G = 8x - 5y$
$(0, 1)$	$G = 8 \cdot 0 - 5 \cdot 1 = -5$
$(0, 3)$	$G = 8 \cdot 0 - 5 \cdot 3 = -15$
$(2, 1)$	$G = 8 \cdot 2 - 5 \cdot 1 = 11$

The maximum is 11 when $x = 2$ and $y = 1$.

REVIEW EXERCISES

Solve graphically. Then classify the system as consistent or inconsistent and the equations as dependent or independent. [3.1a]

1. $4x - y = -9,$
$x - y = -3$

2. $15x + 10y = -20,$
$3x + 2y = -4$

3. $y - 2x = 4,$
$y - 2x = 5$

Solve by the substitution method. [3.2a]

4. $2x - 3y = 5,$
$x = 4y + 5$

5. $y = x + 2,$
$y - x = 8$

6. $7x - 4y = 6,$
$y - 3x = -2$

Solve by the elimination method. [3.3a]

7. $x + 3y = -3,$
$2x - 3y = 21$

8. $3x - 5y = -4,$
$5x - 3y = 4$

9. $\frac{1}{3}x + \frac{2}{9}y = 1,$
$\frac{3}{2}x + \frac{1}{2}y = 6$

10. $1.5x - 3 = -2y,$
$3x + 4y = 6$

Solve.

11. *Retail Sales.* Paint Town sold 45 paintbrushes, one kind at $8.50 each and another at $9.75 each. In all, $398.75 was taken in for the brushes. How many of each kind were sold? [3.4a]

12. *Orange Drink Mixtures.* "Orange Thirst" is 15% orange juice and "Quencho" is 5% orange juice. How many liters of each should be combined in order to get 10 L of a mixture that is 10% orange juice? [3.4a]

13. *Train Travel.* A train leaves Watsonville at noon traveling north at 44 mph. One hour later, another

train, going 52 mph, travels north on a parallel track. How many hours will the second train travel before it overtakes the first train? [3.4b]

Solve. [3.5a]

14. $x + 2y + z = 10,$
$2x - y + z = 8,$
$3x + y + 4z = 2$

15. $3x + 2y + z = 1,$
$2x - y - 3z = 1,$
$-x + 3y + 2z = 6$

16. $2x - 5y - 2z = -4,$
$7x + 2y - 5z = -6,$
$-2x + 3y + 2z = 4$

17. $x + y + 2z = 1,$
$x - y + z = 1,$
$x + 2y + z = 2$

18. *Triangle Measure.* In triangle *ABC*, the measure of angle *A* is four times the measure of angle *C*, and the measure of angle *B* is 45° more than the measure of angle *C*. What are the measures of the angles of the triangle? [3.6a]

19. *Popcorn.* Paul paid a total of $49 for 1 bag of caramel nut crunch popcorn, 1 bag of plain popcorn, and 1 bag of mocha choco latte popcorn. The price of the caramel nut crunch popcorn was six times the price of the plain popcorn and $16 more than the mocha choco latte popcorn. What was the price of each type of popcorn? [3.6a]

Graph. [3.7a]

20. $2x + 3y < 12$ **21.** $y \le 0$

22. $x + y \ge 1$

Graph. Find the coordinates of any vertices formed. [3.7b]

23. $y \ge -3,$
$\quad x \ge 2$

24. $x + 3y \ge -1,$
$\quad x + 3y \le 4$

25. $x - y \le 3,$
$\quad x + y \ge -1,$
$\quad\quad y \le 2$

26. *Maximizing a Test Score.* Jackson is taking a test that contains questions in group A worth 7 points each and questions in group B worth 12 points each. The total number of questions answered must be at least 8. If Jackson knows that group A questions take 8 min each and group B questions take 10 min each and the maximum time for the test is 80 min, how many questions from each group must he answer correctly in order to maximize his score? What is the maximum score? [3.7c]

27. The sum of two numbers is -2. The sum of twice one number and the other is 4. One number is which of the following? [3.3b]

A. -6 B. 2   C. 6   D. 8

28. *Motorcycle Travel.* Sally and Elliot travel on motorcycles toward each other from Chicago and Indianapolis, which are about 350 km apart, and they are biking at rates of 110 km/h and 90 km/h. They started at the same time. In how many hours will they meet? [3.4b]

A. 1.75 hr   B. 3.9 hr
C. 3.2 hr   D. 17.5 hr

► **Synthesis**

29. Solve graphically: [2.1d], [3.1a]
$$y = x + 2,$$
$$y = x^2 + 2.$$

► **Collaborative Discussion and Writing**

30. Write a problem for a classmate to solve. Design the problem so the answer is "The florist sold 14 hanging baskets and 9 flats of petunias." [3.4a]

31. Exercise 21 in Exercise Set 3.6 can be solved mentally after a careful reading of the problem. Explain how this can be done. [3.6a]

32. *Ticket Revenue.* A pops-concert audience of 100 people consists of adults, senior citizens, and children. The ticket prices are $10 each for adults, $3 each for senior citizens, and $0.50 each for children. The total amount of money taken in is $100. How many adults, senior citizens, and children are in attendance? Does there seem to be some information missing? Do some careful reasoning and explain. [3.6a]

33. When graphing linear inequalities, Ron always shades above the line when he sees a $\ge$ symbol. Is this wise? Why or why not? [3.7a]

3 **Chapter Test**

Solve graphically. Then classify the system as consistent or inconsistent and the equations as dependent or independent.

1. $y = 3x + 7,$
$\quad 3x + 2y = -4$

2. $y = 3x + 4,$
$\quad y = 3x - 2$

3. $y - 3x = 6,$
$\quad 6x - 2y = -12$

Solve by the substitution method.

4. $4x + 3y = -1,$
$\quad y = 2x - 7$

5. $x = 3y + 2,$
$\quad 2x - 6y = 4$

6. $x + 2y = 6,$
$\quad 2x + 3y = 7$

Solve by the elimination method.

7. $2x + 5y = 3,$
$\quad -2x + 3y = 5$

8. $x + y = -2,$
$\quad 4x - 6y = -3$

9. $\dfrac{2}{3}x - \dfrac{4}{5}y = 1,$
$\quad \dfrac{1}{3}x - \dfrac{2}{5}y = 2$

Solve.

10. *Tennis Court.* The perimeter of a standard tennis court used for playing doubles is 288 ft. The width of the court is 42 ft less than the length. Find the length and the width.

11. *Air Travel.* An airplane flew for 5 hr with a 20-km/h tailwind and returned in 7 hr against the same wind. Find the speed of the plane in still air.

12. *Chicken Dinners.* High Flyin' Wings charges $12 for a bucket of chicken wings and $7 for a chicken dinner. After filling 28 orders for buckets and dinners during a football game, the waiters had collected $281. How many buckets and how many dinners did they sell?

13. *Mixing Solutions.* A chemist has one solution that is 20% salt and a second solution that is 45% salt. How many liters of each should be used in order to get 20 L of a solution that is 30% salt?

14. Solve:
$$6x + 2y - 4z = 15,$$
$$-3x - 4y + 2z = -6,$$
$$4x - 6y + 3z = 8.$$

15. *Repair Rates.* An electrician, a carpenter, and a plumber are hired to work on a house. The electrician earns $21 per hour, the carpenter $19.50 per hour, and the plumber $24 per hour. The first day on the job, they worked a total of 21.5 hr and earned a total of $469.50. If the plumber worked 2 hr more than the carpenter did, how many hours did the electrician work?

Graph. Find the coordinates of any vertices formed.

16. $y \geq x - 2$

17. $x - 6y < -6$

Graph. Find the coordinates of any vertices formed.

18. $x + y \geq 3,$
$x - y \geq 5$

19. $2y - x \geq -4,$
$2y + 3x \leq -6,$
$y \leq 0,$
$x \leq 0$

20. *Maximizing Profit.* Jane's Cakes prepares pound cakes and carrot cakes. In a given week, at most 100 cakes can be prepared, of which 25 pound cakes and 15 carrot cakes are required by regular customers. The profit from each pound cake is $6, and the profit from each carrot cake is $8. How many of each type of cake should be prepared in order to maximize the profit? What is the maximum profit?

21. A business class divided an imaginary $30,000 investment among three funds. The first fund grew 2%, the second grew 3%, and the third grew 5%. Total earnings were $990. The earnings from the third fund were $280 more than the earnings from the first. How much was invested at 5%?

A. $9000

B. $10,000

C. $11,000

D. $12,000

▶ **Synthesis**

22. The graph of the function $f(x) = mx + b$ contains the points $(-1, 3)$ and $(-2, -4)$. Find m and b.

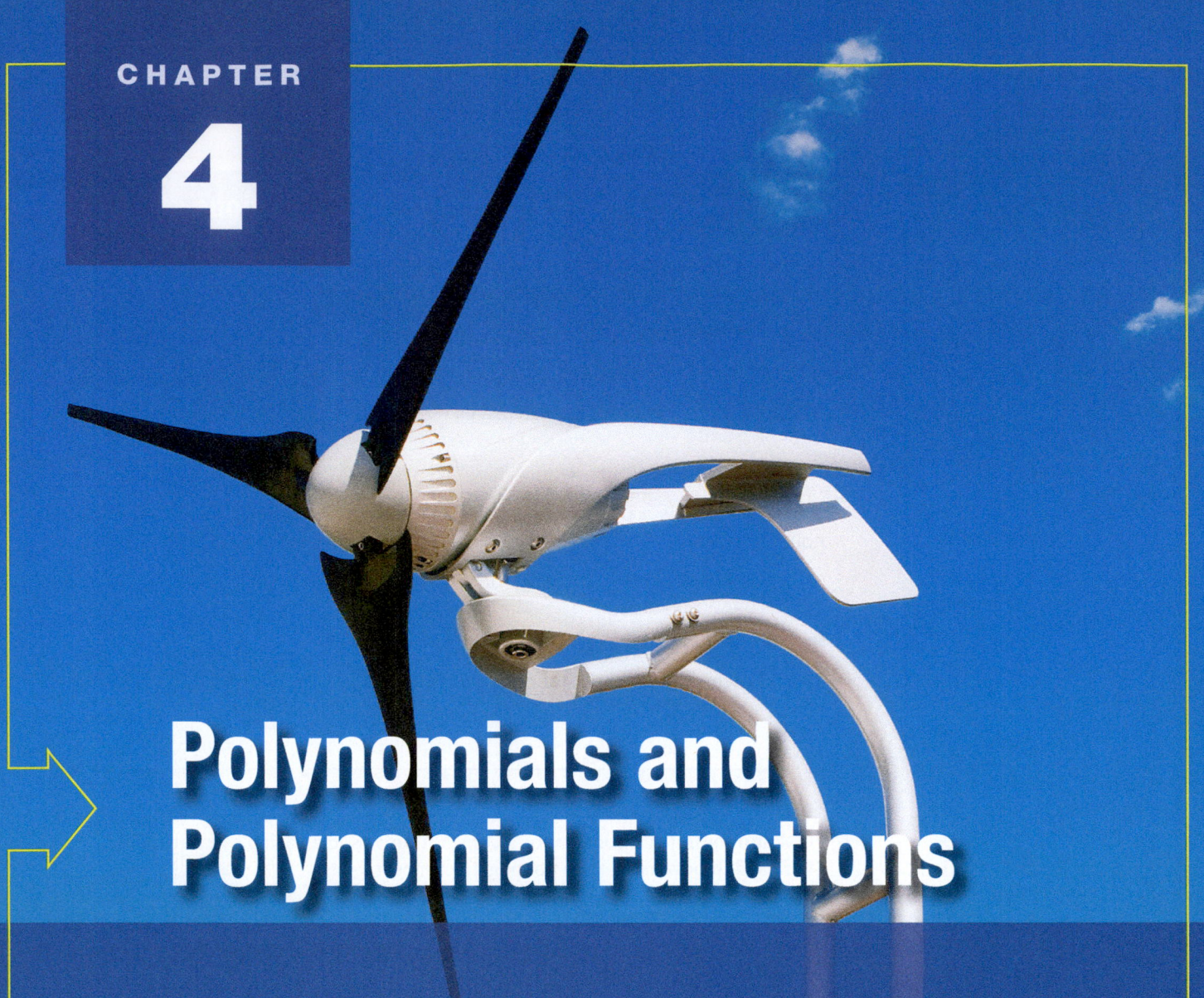

Polynomials and Polynomial Functions

APPLICATION   This problem appears as Exercise 23 in Exercise Set 4.1.

The number P of watts of power generated by a particular home-sized turbine at a wind speed of x miles per hour can be approximated by the polynomial function $P(x) = 0.0157x^3 + 0.1163x^2 - 1.3396x + 3.7063$. Estimate the power, in watts, generated by a 25-mph wind.

4.1   Introduction to Polynomials and Polynomial Functions

▶ **a**   Identify terms, degrees, and coefficients in polynomials; identify types of polynomials; and arrange polynomials in ascending order or descending order.

▶ **b**   Evaluate a polynomial function for given inputs.

▶ **c**   Collect like terms in a polynomial and add polynomials.

▶ **d**   Find the opposite of a polynomial and subtract polynomials.

A **polynomial** is a particular type of algebraic expression. Some examples of polynomials are

$$x + 7, \qquad abc, \qquad 5t^2 - 6t + 1, \quad \text{and} \quad 7.$$

▶ a   Polynomial Expressions

The following are examples of *monomials*:

$$0, \qquad -3, \qquad z, \qquad 8x, \qquad -7y^2, \qquad 4a^2b^3, \qquad 1.3p^4q^5r^7.$$

> **MONOMIAL**
>
> A **monomial** is a constant or a constant times some variable or variables raised to powers that are nonnegative integers.

Expressions like these are called **polynomials in one variable**:

$$5x^2, \qquad 2y^2 + 5y - 3, \qquad 5a^4 - 3a^2 + \tfrac{1}{4}a - 8.$$

Expressions like these are called **polynomials in several variables**:

$$15x^3y^2, \qquad a - b, \qquad \tfrac{1}{2}xy^2z - 4x^3z + y^3 + 9.$$

> **POLYNOMIAL**
>
> A **polynomial** is a monomial or a combination of sums and/or differences of monomials.

The following are algebraic expressions that are not polynomials:

$$(1)\ \frac{y^2 - 3}{y^2 + 4}, \qquad (2)\ 8x^4 - 2x^3 + \frac{1}{x}, \qquad (3)\ \frac{2xy}{x^3 - y^3}.$$

Expressions (1) and (3) are not polynomials because they represent quotients. In expression (2), although we can write $1/x$ as x^{-1}, this is not a monomial because the exponent is negative.

The **terms** of a polynomial are separated by $+$ signs. The polynomial $5x^3y - 7xy^2 - y^3 + 2$ has four terms:

$$5x^3y, \qquad -7xy^2, \qquad -y^3, \quad \text{and} \quad 2.$$

The **coefficients** of the terms are 5, -7, -1, and 2. The term 2 is called a **constant term**.

The **degree of a term** is the sum of the exponents of the variables, if there are variables. For example,

the degree of the term $9x^5$ is 5   and

the degree of the term $0.6a^2b^7$ is $2 + 7$, or 9.

The degree of a nonzero constant term, such as 2, is 0. We can express 2 as $2x^0$.

Because we can express 0 as $0 = 0x^5 = 0x^{12}$, and so on, using any exponent we wish, the term 0 has *no* degree.

The **degree of a polynomial** is the same as the degree of its term of highest degree. For example,

the degree of the polynomial $4 - x^3 + 5x^2 - x^6$ is 6.

The **leading term** of a polynomial is the term of highest degree. Its coefficient is called the **leading coefficient**. For example,

the leading term of $9x^2 - 5x^3 + x - 10$ is $-5x^3$ and

the leading coefficient is -5.

EXAMPLE 1   Identify the terms, the degree of each term, and the degree of the polynomial $2x^3 + 8x^2 - 17x - 3$. Then identify the leading term, the leading coefficient, and the constant term.

Term	$2x^3$	$8x^2$	$-17x$	-3
Degree of Term	3	2	1	0
Degree of Polynomial	3			
Leading Term	$2x^3$			
Leading Coefficient	2			
Constant Term	-3			

> **Now Try Exercise 1.**

EXAMPLE 2   Identify the terms, the degree of each term, and the degree of the polynomial $6x^2 + 8x^2y^3 - 17xy - 24xy^2z^4 + 2y + 3$. Then identify the leading term, the leading coefficient, and the constant term.

Term	$6x^2$	$8x^2y^3$	$-17xy$	$-24xy^2z^4$	$2y$	3
Degree of Term	2	5	2	7	1	0
Degree of Polynomial	7					
Leading Term	$-24xy^2z^4$					
Leading Coefficient	-24					
Constant Term	3					

> **Now Try Exercise 5.**

The following are some names for certain types of polynomials.

Type	Definition: Polynomial of	Examples
Monomial	One term	$4, \quad -3p, \quad -7a^2b^3, \quad 0, \quad xyz$
Binomial	Two terms	$2x + 7, \quad a^2 - 3b, \quad 5x^3 + 8x$
Trinomial	Three terms	$x^2 - 7x + 12, \quad 4a^2 + 2ab + b^2$

We generally arrange polynomials in one variable in **descending order** so that the exponents *decrease* from left to right. Sometimes they may be written so that the exponents *increase* from left to right, which is **ascending order**. In general, if an exercise is written in a particular order, we write the answer in that same order.

EXAMPLE 3 Consider $12 + x^2 - 7x$. Arrange in descending order and then in ascending order.

Descending order: $x^2 - 7x + 12$

Ascending order: $12 - 7x + x^2$ ⟶ **Now Try Exercises 7 and 13.**

EXAMPLE 4 Consider $x^4 + 2 - 5x^2 + 3x^3y + 7xy^2$. Arrange in descending powers of x and then in ascending powers of x.

Descending powers of x: $x^4 + 3x^3y - 5x^2 + 7xy^2 + 2$

Ascending powers of x: $2 + 7xy^2 - 5x^2 + 3x^3y + x^4$ ⟶ **Now Try Exercises 9 and 15.**

▶ **b   Evaluating Polynomial Functions**

A polynomial function is one like $P(x) = 5x^7 + 3x^5 - 4x^2 - 5$, in which the algebraic expression used to describe the function is a polynomial. To find the outputs of a polynomial function for a given input, we substitute the input for each occurrence of the variable.

EXAMPLE 5 For the polynomial function $P(x) = -x^2 + 4x - 1$, find $P(2), P(10)$, and $P(-10)$.

$P(2) = -2^2 + 4(2) - 1 = -4 + 8 - 1 = 3;$

$P(10) = -10^2 + 4(10) - 1 = -100 + 40 - 1 = -61;$

$P(-10) = -(-10)^2 + 4(-10) - 1 = -100 - 40 - 1 = -141$ ⟶ **Now Try Exercise 19.**

EXAMPLE 6 *Veterinary Medicine.*   Gentamicin is an antibiotic frequently used by veterinarians. The concentration C, in micrograms per milliliter (mcg/mL), of Gentamicin in a horse's bloodstream t hours after injection can be approximated by the polynomial function

$C(t) = -0.005t^4 + 0.003t^3 + 0.35t^2 + 0.5t.$

a) Evaluate $C(2)$ to find the concentration 2 hr after injection.

b) Use only the graph below to estimate $C(4)$.

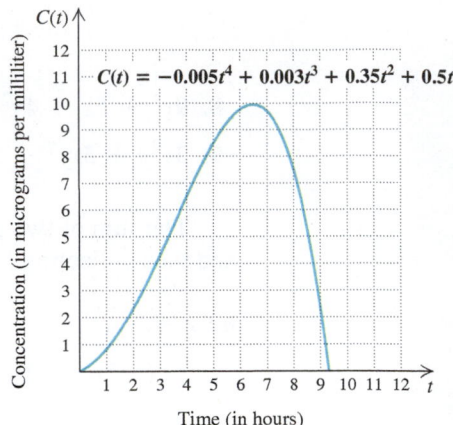

a) We evaluate the function when $t = 2$:

$$C(t) = -0.005t^4 + 0.003t^3 + 0.35t^2 + 0.5t$$
$$C(2) = -0.005(2)^4 + 0.003(2)^3 + 0.35(2)^2 + 0.5(2)$$
$$= -0.005(16) + 0.003(8) + 0.35(4) + 0.5(2)$$
$$= -0.08 + 0.024 + 1.4 + 1$$
$$= 2.344.$$

We carry out the calculation using the rules for order of operations.

The concentration after 2 hr is about 2.344 mcg/mL.

b) To estimate $C(4)$, the concentration after 4 hr, we locate 4 on the horizontal axis. From there we move vertically to the graph of the function and then horizontally to the $C(t)$-axis. This locates a value of about 6.5. Thus,

$$C(4) \approx 6.5.$$

The concentration after 4 hr is about 6.5 mcg/mL.

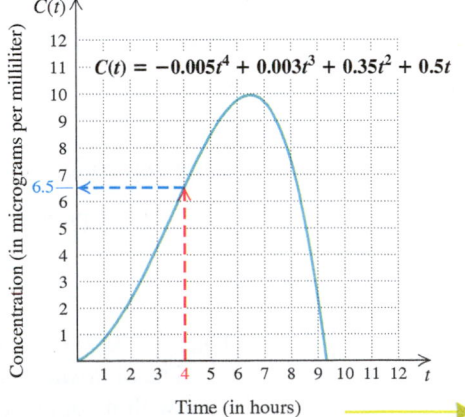

Now Try Exercise 25.

▶ c Adding Polynomials

When two terms have the same variable(s) raised to the same power(s), they are called **like terms**, or **similar terms**, and they can be "collected," or "combined," using the distributive laws.

EXAMPLES   Collect like terms.

7. $3x^2 - 4y + 2x^2 = 3x^2 + 2x^2 - 4y$ *Rearranging using the commutative law for addition*

$\qquad\qquad\qquad = (3 + 2)x^2 - 4y$ *Using the distributive law*

$\qquad\qquad\qquad = 5x^2 - 4y$ *Adding the coefficients of x^2*

8. $9x^3 + 5x - 4x^2 - 2x^3 + 5x^2 = 7x^3 + x^2 + 5x$

9. $3x^2y + 5xy^2 - 3x^2y - xy^2 = 4xy^2$ → **Now Try Exercise 41.**

The sum of two polynomials can be found by writing a plus sign between them and then collecting like terms to simplify the expression.

EXAMPLE 10   Add: $(-3x^3 + 2x - 4) + (4x^3 + 3x^2 + 2)$.

$(-3x^3 + 2x - 4) + (4x^3 + 3x^2 + 2) = x^3 + 3x^2 + 2x - 2$

→ **Now Try Exercise 47.**

EXAMPLE 11   Add: $13x^3y + 3x^2y - 5y$ and $x^3y + 4x^2y - 3xy$.

$(13x^3y + 3x^2y - 5y) + (x^3y + 4x^2y - 3xy) = 14x^3y + 7x^2y - 3xy - 5y$

→ **Now Try Exercise 55.**

In order to use columns to add, we write the polynomials one under the other, listing like terms under one another and leaving spaces for missing terms.

EXAMPLE 12   Add: $4ax^2 + 4bx - 5$ and $-6ax^2 + 8$.

$$\begin{array}{r} 4ax^2 + 4bx - 5 \\ \underline{-6ax^2 \qquad\ + 8} \\ -2ax^2 + 4bx + 3 \end{array}$$

→ **Now Try Exercise 53.**

▶ d Subtracting Polynomials

If the sum of two polynomials is 0, the polynomials are **opposites**, or **additive inverses**, of each other. For example,

$$(3x^2 - 5x + 2) + (-3x^2 + 5x - 2) = 0,$$

so the opposite of $3x^2 - 5x + 2$ is $-3x^2 + 5x - 2$. We can say the same thing using algebraic symbolism, as follows:

The opposite of $(3x^2 - 5x + 2)$ is $(-3x^2 + 5x - 2)$.

$\qquad - \qquad\qquad (3x^2 - 5x + 2) \quad = \quad -3x^2 + 5x - 2$

Thus, $-(3x^2 - 5x + 2)$ and $-3x^2 + 5x - 2$ are equivalent.

The *opposite* of a polynomial P can be symbolized by $-P$ or by replacing each term with its opposite. The two expressions for the opposite are equivalent.

EXAMPLE 13 Write two equivalent expressions for the opposite of

$$7xy^2 - 6xy - 4y + 3.$$

First expression: $-(7xy^2 - 6xy - 4y + 3)$ **Writing an inverse sign in front**

Second expression: $-7xy^2 + 6xy + 4y - 3$ **Writing the opposite of each term**

> **Now Try Exercise 61.**

To subtract a polynomial, we add its opposite.

EXAMPLE 14 Subtract: $(-5x^2 + 4) - (2x^2 + 3x - 1)$.

We have

$$(-5x^2 + 4) - (2x^2 + 3x - 1)$$
$$= (-5x^2 + 4) + [-(2x^2 + 3x - 1)] \quad \text{\color{red}{\textbf{Adding the opposite}}}$$
$$= (-5x^2 + 4) + (-2x^2 - 3x + 1) \quad \color{red}{\substack{\textbf{$-2x^2 - 3x + 1$}\\\textbf{is equivalent to}\\\textbf{$-(2x^2 + 3x - 1)$.}}}$$

$$= -7x^2 - 3x + 5. \quad \text{\color{red}{\textbf{Adding}}}$$

With practice, you may find that you can skip some steps, by mentally taking the opposite of each term and then combining like terms. Eventually, all you will write is the answer.

$(-5x^2 + 4) - (2x^2 + 3x - 1)$ *Think:*
$= -7x^2 - 3x + 5$ $-5x^2 - 2x^2 = -5x^2 + (-2x^2) = -7x^2,$
                    $0x - 3x = 0x + (-3x) = -3x,$
                    $4 - (-1) = 4 + 1 = 5.$

> **Now Try Exercise 67.**

To use columns for subtraction, we mentally change the signs of the terms being subtracted.

EXAMPLE 15 Subtract:

$$(4x^2y - 6x^3y^2 + x^2y^2) - (4x^2y + x^3y^2 + 3x^2y^3 - 8x^2y^2).$$

Write: (Subtract)   *Think:* (Add)

$$\begin{array}{ll} 4x^2y - 6x^3y^2 \qquad\quad + x^2y^2 \\ \underline{-(4x^2y + \ \ x^3y^2 + 3x^2y^3 - 8x^2y^2)} \end{array} \longleftrightarrow \begin{array}{l} 4x^2y - 6x^3y^2 \qquad\quad + x^2y^2 \\ \underline{-4x^2y - \ \ x^3y^2 - 3x^2y^3 + 8x^2y^2} \\ \qquad\qquad -7x^3y^2 - 3x^2y^3 + 9x^2y^2 \end{array}$$

Take the opposite of each term mentally and add.

> **Now Try Exercise 75.**

4.1 Exercise Set

▶ Reading Check

Choose from the column on the right the expression that best fits each description.

RC1. A binomial

RC2. A trinomial

RC3. A polynomial with more than one term written in ascending order

RC4. A polynomial in several variables

RC5. The coefficient of the term $7x^5$

RC6. The degree of the term $6xy^2z$

RC7. The constant term in the polynomial $3x^9 - 7x + 5$

RC8. The leading coefficient in the polynomial $5x^3 - 6x + x^4 + 7$

a) $8x - 9x^2 + x^4 + 2x^5$

b) $7x^2yz^3$

c) $3a^4 - 9$

d) $4x^2 + 8x + 2$

e) 1

f) 4

g) 5

h) 7

---

a *Identify the terms, the degree of each term, and the degree of the polynomial. Then identify the leading term, the leading coefficient, and the constant term.*

1. $-9x^4 - x^3 + 7x^2 + 6x - 8$

2. $y^3 - 5y^2 + y + 1$

3. $t^3 + 4t^7 + s^2t^4 - 2$

4. $a^2 + 9b^5 - a^4b^3 - 11$

5. $u^7 + 8u^2v^6 + 3uv + 4u - 1$

6. $2p^6 + 5p^4w^4 - 13p^3w + 7p^2 - 10$

Arrange in descending powers of y.

7. $23 - 4y^3 + 7y - 6y^2$

8. $5 - 8y + 6y^2 + 11y^3 - 18y^4$

9. $x^2y^2 + x^3y - xy^3 + 1$

10. $x^3y - x^2y^2 + xy^3 + 6$

11. $2by - 9b^5y^5 - 8b^2y^3$

12. $dy^6 - 2d^7y^2 + 3cy^5 - 7y - 2d$

Arrange in ascending powers of x.

13. $12x + 5 + 8x^5 - 4x^3$

14. $-3x^2 + 8x + 2$

15. $-9x^3y + 3xy^3 + x^2y^2 + 2x^4$

16. $5x^2y^2 - 9xy + 8x^3y^2 - 5x^4$

17. $4ax - 7ab + 4x^6 - 7ax^2$

18. $5xy^8 - 3ax^5 + 4ax^3 - 12a + 5x^5$

b *Evaluate each polynomial function for the given values of the variable.*

19. $P(x) = 3x^2 - 2x + 5; P(4), P(-2), P(0)$

20. $f(x) = -7x^3 + 10x^2 - 13; f(4), f(-1), f(0)$

21. $p(x) = 9x^3 + 8x^2 - 4x - 9;$
$p(-3), p(0), p(1), p\left(\frac{1}{2}\right)$

22. $Q(x) = 6x^3 - 11x - 4;$
$Q(-2), Q\left(\frac{1}{3}\right), Q(0), Q(10)$

23. *Wind Energy.* The number P of watts of power generated by a particular home-sized turbine at a wind speed of x miles per hour can be approximated by the polynomial function

$P(x) = 0.0157x^3 + 0.1163x^2 - 1.3396x + 3.7063.$

Estimate the power, in watts, generated by a 25-mph wind.

24. *Golf Ball Stacks.* The stack of golf balls pictured below is formed by square layers of golf balls. The number N of balls in the stack is given by the polynomial function

$$N(x) = \tfrac{1}{3}x^3 + \tfrac{1}{2}x^2 + \tfrac{1}{6}x,$$

where x is the number of layers. How many golf balls are in the stacks?

25. *Medicine.* Ibuprofen is a medication used to relieve pain. The polynomial function

$$M(t) = 0.5t^4 + 3.45t^3 - 96.65t^2 + 347.7t,$$
$$0 \le t \le 6,$$

can be used to estimate the number of milligrams of ibuprofen in the bloodstream t hours after 400 mg of the medication has been swallowed. (*Source*: Based on data from Dr. P. Carey, Burlington, VT)

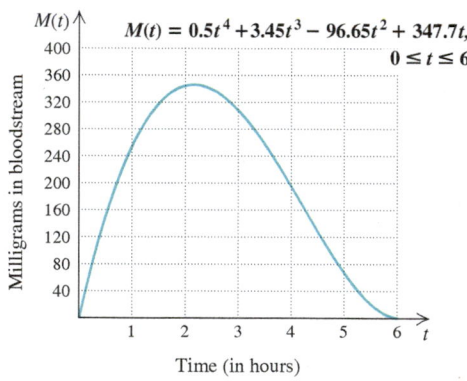

a) Use the graph above to estimate the number of milligrams of ibuprofen in the bloodstream 2 hr after 400 mg has been swallowed.

b) Use the graph above to estimate the number of milligrams of ibuprofen in the bloodstream 4 hr after 400 mg has been swallowed.

c) Approximate $M(5)$.

d) Approximate $M(3)$.

26. *Median Income by Age.* The polynomial function

$$I(x) = -0.0560x^4 + 7.9980x^3 - 436.1840x^2$$
$$+ 11{,}627.8376x - 90{,}625.0001,$$
$$13 \le x \le 65,$$

can be used to approximate the median income I by age x of a person living in the United States. The graph is shown below.

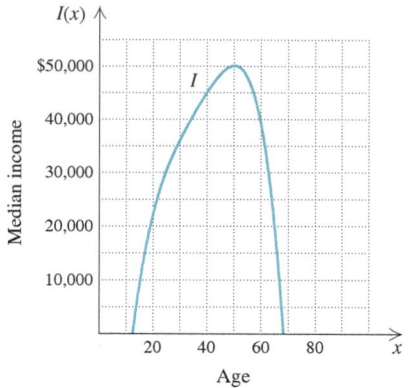

SOURCES: U.S. Census Bureau; The Conference Board: Simmons Bureau of Labor Statistics

a) Evaluate $I(22)$ to estimate the median income of a 22-year-old.

b) Use only the graph to estimate $I(40)$.

27. *Total Revenue.* A firm is marketing a new style of sunglasses. The firm determines that when it sells x pairs of sunglasses, its total revenue is

$$R(x) = 240x - 0.5x^2 \text{ dollars.}$$

a) What is the total revenue from the sale of 50 pairs of sunglasses?

b) What is the total revenue from the sale of 95 pairs of sunglasses?

28. *Total Cost.* A firm determines that the total cost, in dollars, of producing x pairs of sunglasses is given by

$$C(x) = 5000 + 0.4x^2.$$

a) What is the total cost of producing 50 pairs of sunglasses?

b) What is the total cost of producing 95 pairs of sunglasses?

Total Profit. **Total profit P** *is defined as total revenue R minus total cost C, and is given by the function*

$$P(x) = R(x) - C(x).$$

For each of the following, find the total profit $P(x)$.

29. $R(x) = 280x - 0.4x^2, C(x) = 7000 + 0.6x^2$

30. $R(x) = 280x - 0.7x^2, C(x) = 8000 + 0.5x^2$

Magic Number. *In a recent baseball season, the Arizona Diamondbacks were leading the San Francisco Giants for the Western Division championship of the National League. In the following table, the number in parentheses, 18, was the* **magic number** *for the Diamondbacks. It means that any combination of Diamondbacks wins and Giants losses that totals 18 would ensure the championship*

WEST	W	L	Pct.	GB
Arizona (18)	81	62	.566	–
San Francisco	80	64	.556	$1\frac{1}{2}$
Los Angeles	78	65	.545	3
San Diego	70	73	.490	11
Colorado	62	80	.437	$18\frac{1}{2}$

Magic number in parentheses

for the Diamondbacks. *The magic number M is given by the polynomial*

$$M = G - W_1 - L_2 + 1,$$

where W_1 is the number of wins for the first-place team, L_2 is the number of losses for the second-place team, and G is the total number of games in the season, which is 162 in the major leagues. When the magic number reaches 1, a tie for the championship is clinched. When the magic number reaches 0, the championship is clinched. For the situation shown below, $G = 162$, $W_1 = 81$, and $L_2 = 64$. Then the magic number is

$$M = G - W_1 - L_2 + 1$$
$$= 162 - 81 - 64 + 1$$
$$= 18.$$

31. Compute the magic number for Atlanta.

East	W	L	Pct.	GB
Atlanta (?)	78	64	.549	—
Philadelphia	75	68	.524	$3\frac{1}{2}$
New York	71	73	.493	8
Florida	66	77	.462	$12\frac{1}{2}$
Montreal	61	82	.427	$17\frac{1}{2}$

32. Compute the magic number for Houston.

Central	W	L	Pct.	GB
Houston (?)	84	59	.587	—
St. Louis	78	64	.549	$5\frac{1}{2}$
Chicago	78	65	.545	6
Milwaukee	63	80	.441	21
Cincinnati	58	86	.403	$26\frac{1}{2}$
Pittsburgh	55	88	.385	29

33. Compute the magic number for New York.

East	W	L	Pct.	GB
New York (?)	86	57	.601	—
Boston	72	69	.511	13
Toronto	70	73	.490	16
Baltimore	55	87	.387	$30\frac{1}{2}$
Tampa Bay	50	93	.350	36

34. Compute the magic number for Cleveland.

Central	W	L	Pct.	GB
Cleveland (?)	82	62	.569	—
Minnesota	76	68	.528	6
Chicago	74	70	.514	8
Detroit	57	86	.399	$24\frac{1}{2}$
Kansas City	57	86	.399	$24\frac{1}{2}$

C *Collect like terms.*

35. $6x^2 - 7x^2 + 3x^2$

36. $-2y^2 - 7y^2 + 5y^2$

37. $7x - 2y - 4x + 6y$

38. $a - 8b - 5a + 7b$

39. $3a + 9 - 2 + 8a - 4a + 7$

40. $13x + 14 - 6 - 7x + 3x + 5$

41. $3a^2b + 4b^2 - 9a^2b - 6b^2$

42. $5x^2y^2 + 4x^3 - 8x^2y^2 - 12x^3$

43. $8x^2 - 3xy + 12y^2 + x^2 - y^2 + 5xy + 4y^2$

44. $a^2 - 2ab + b^2 + 9a^2 + 5ab - 4b^2 + a^2$

45. $4x^2y - 3y + 2xy^2 - 5x^2y + 7y + 7xy^2$

46. $3xy^2 + 4xy - 7xy^2 + 7xy + x^2y$

Add.

47. $(3x^2 + 5y^2 + 6) + (2x^2 - 3y^2 - 1)$

48. $(11y^2 + 6y - 3) + (9y^2 - 2y + 9)$

49. $(2a - c + 3b) + (4a - 2b + 2c)$

50. $(8x + z - 7y) + (5x + 10y - 4z)$

51. $(a^2 - 3b^2 + 4c^2) + (-5a^2 + 2b^2 - c^2)$

52. $(x^2 - 5y^2 - 9z^2) + (-6x^2 + 9y^2 - 2z^2)$

53. $(x^2 + 3x - 2xy - 3) + (-4x^2 - x + 3xy + 2)$

54. $(5a^2 - 3b + ab + 6) + (-a^2 + 8b - 8ab - 4)$

55. $(7x^2y - 3xy^2 + 4xy) + (-2x^2y - xy^2 + xy)$

56. $(7ab - 3ac + 5bc) + (13ab - 15ac - 8bc)$

57. $(2r^2 + 12r - 11) + (6r^2 - 2r + 4) +$
$(r^2 - r - 2)$

58. $(5x^2 + 19x - 23) + (-7x^2 - 11x + 12) +$
$(-x^2 - 9x + 8)$

59. $\left(\frac{2}{3}xy + \frac{5}{6}xy^2 + 5.1x^2y\right) +$
$\left(-\frac{4}{5}xy + \frac{3}{4}xy^2 - 3.4x^2y\right)$

60. $\left(\frac{1}{8}xy - \frac{3}{5}x^3y^2 + 4.3y^3\right) + \left(-\frac{1}{3}xy - \frac{3}{4}x^3y^2 - 2.9y^3\right)$

d *Write two equivalent expressions for the opposite of the polynomial.*

61. $5x^3 - 7x^2 + 3x - 6$

62. $-8y^4 - 18y^3 + 4y - 9$

63. $-13y^2 + 6ay^4 - 5by^2$

64. $9ax^5y^3 - 8by^5 - abx - 16ay$

Subtract.

65. $(7x - 2) - (-4x + 5)$

66. $(8y + 1) - (-5y - 2)$

67. $(-3x^2 + 2x + 9) - (x^2 + 5x - 4)$

68. $(-9y^2 + 4y + 8) - (4y^2 + 2y - 3)$

69. $(5a + c - 2b) - (3a + 2b - 2c)$

70. $(z + 8x - 4y) - (4x + 6y - 3z)$

71. $(3x^2 - 2x - x^3) - (5x^2 - x^3 - 8x)$

72. $(8y^2 - 4y^3 - 3y) - (3y^2 - 9y - 7y^3)$

73. $(5a^2 + 4ab - 3b^2) - (9a^2 - 4ab + 2b^2)$

74. $(9y^2 - 14yz - 8z^2) - (12y^2 - 8yz + 4z^2)$

75. $(6ab - 4a^2b + 6ab^2) - (3ab^2 - 10ab - 12a^2b)$

76. $(10xy - 4x^2y^2 - 3y^3) - (-9x^2y^2 + 4y^3 - 7xy)$

77. $(0.09y^4 - 0.052y^3 + 0.93) -$
$(0.03y^4 - 0.084y^3 + 0.94y^2)$

78. $(1.23x^4 - 3.122x^3 + 1.11x) -$
$(0.79x^4 - 8.734x^3 + 0.04x^2 + 6.71x)$

79. $\left(\frac{5}{8}x^4 - \frac{1}{4}x^2 - \frac{1}{2}\right) - \left(-\frac{3}{8}x^4 + \frac{3}{4}x^2 + \frac{1}{2}\right)$

80. $\left(\frac{5}{6}y^4 - \frac{1}{2}y^2 - 7.8y + \frac{1}{3}\right) -$
$\left(-\frac{3}{8}y^4 + \frac{3}{4}y^2 + 3.4y - \frac{1}{5}\right)$

▶ **Skill Maintenance**

Graph. **[2.1c, d], [2.2c]**

81. $f(x) = \frac{2}{3}x - 1$

82. $g(x) = |x| - 1$

83. $g(x) = \dfrac{4}{x - 3}$

84. $f(x) = 1 - x^2$

Solve.

85. $-3x - 7 = x - 5$ **[1.1d]**

86. $\frac{1}{3}t - \frac{1}{2} = \frac{1}{6}t$ **[1.1d]**

87. $x - (7 - x) = 2(x + 3)$ **[1.1d]**

88. $-9y \leq -18$ **[1.4c]**

Graph using the slope and the y-intercept. **[2.6b]**

89. $y = \frac{4}{3}x + 2$

90. $y = -0.4x + 1$

91. $y = 0.4x - 3$

92. $y = -\frac{2}{3}x - 4$

▶ **Synthesis**

93. *Triangular Layers.* The number of spheres in a triangular pyramid with x triangular layers is given by the function
$$N(x) = \tfrac{1}{6}x^3 + \tfrac{1}{2}x^2 + \tfrac{1}{3}x.$$
The volume of a sphere of radius r is given by the function
$$V(r) = \tfrac{4}{3}\pi r^3,$$
where π can be approximated as 3.14.

Chocolate Heaven has a window display of truffles piled in triangular pyramid formations, each 5 layers deep. If the diameter of each truffle is 3 cm, find the volume of chocolate in each triangular pyramid in the display.

94. *Surface Area.* Find a polynomial function that gives the outside surface area of a box like this one, with dimensions as shown.

95. 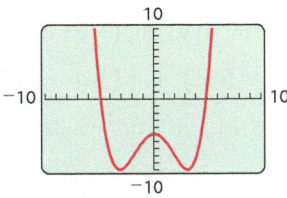 A student who is trying to graph
$f(x) = 0.05x^4 - x^2 + 5$ gets the following screen.

How can the student tell at a glance that a mistake has been made?

Perform the indicated operations. Assume that the exponents are natural numbers.

96. $(3x^{6a} - 5x^{5a} + 4x^{3a} + 8) - (2x^{6a} + 4x^{4a} + 3x^{3a} + 2x^{2a})$

97. $(47x^{4a} + 3x^{3a} + 22x^{2a} + x^a + 1) + (37x^{3a} + 8x^{2a} + 3)$

4.2 Multiplication of Polynomials

▶ **a** Multiply any two polynomials.

▶ **b** Use the FOIL method to multiply two binomials.

▶ **c** Use a rule to square a binomial.

▶ **d** Use a rule to multiply a sum and a difference of the same two terms.

▶ **e** For functions f described by second-degree polynomials, find and simplify notation like $f(a + h)$ and $f(a + h) - f(a)$.

▶ **a** Multiplication of any Two Polynomials

Multiplying Monomials

Monomials are expressions like $10x^2$, $8x^5$, and $-7a^2b^3$. To multiply monomials, we first multiply their coefficients. Then we multiply the variables using the commutative and associative laws and the rules for exponents.

EXAMPLES Multiply and simplify.

1. $(10x^2)(8x^5) = (10 \cdot 8)(x^2 \cdot x^5)$
$\qquad\qquad\quad = 80x^{2+5}$ **Adding exponents**
$\qquad\qquad\quad = 80x^7$

2. $(-8x^4y^7)(5x^3y^2) = (-8 \cdot 5)(x^4 \cdot x^3)(y^7 \cdot y^2)$
$\qquad\qquad\qquad\quad = -40x^{4+3}y^{7+2}$ **Adding exponents**
$\qquad\qquad\qquad\quad = -40x^7y^9$

> **Now Try Exercise 5.**

Multiplying Monomials and Binomials

THE DISTRIBUTIVE LAW

REVIEW SECTION R.5

The distributive law is the basis for multiplying polynomials other than monomials. We first multiply a monomial and a binomial.

EXAMPLE 3 Multiply: $2x(3x - 5)$.

$$2x \cdot (3x - 5) = 2x \cdot 3x - 2x \cdot 5 \qquad \text{Using the distributive law}$$
$$= 6x^2 - 10x \qquad \text{Multiplying monomials}$$

Now Try Exercise 7.

EXAMPLE 4 Multiply: $3a^2b(a^2 - b^2)$.

$$3a^2b \cdot (a^2 - b^2) = 3a^2b \cdot a^2 - 3a^2b \cdot b^2 \qquad \text{Using the distributive law}$$
$$= 3a^4b - 3a^2b^3$$

Now Try Exercise 9.

Multiplying Binomials

Next, we multiply two binomials. To do so, we use the distributive law twice, first considering one of the binomials as a single expression and multiplying it by each term of the other binomial.

EXAMPLE 5 Multiply: $(3y^2 + 4)(y^2 - 2)$.

$$(3y^2 + 4)(y^2 - 2) = (3y^2 + 4) \cdot y^2 - (3y^2 + 4) \cdot 2 \qquad \text{Using the distributive law}$$
$$= [3y^2 \cdot y^2 + 4 \cdot y^2] - [3y^2 \cdot 2 + 4 \cdot 2] \qquad \text{Using the distributive law}$$
$$= 3y^2 \cdot y^2 + 4 \cdot y^2 - 3y^2 \cdot 2 - 4 \cdot 2 \qquad \text{Removing parentheses}$$
$$= 3y^4 + 4y^2 - 6y^2 - 8 \qquad \text{Multiplying the monomials}$$
$$= 3y^4 - 2y^2 - 8 \qquad \text{Collecting like terms}$$

Now Try Exercise 13.

Multiplying Any Two Polynomials

To find a quick way to multiply any two polynomials, let's consider another example.

EXAMPLE 6 Multiply: $(p + 2)(p^4 - 2p^3 + 3)$.

By the distributive law, we have

$$(p + 2)(p^4 - 2p^3 + 3)$$
$$= (p + 2)(p^4) - (p + 2)(2p^3) + (p + 2)(3)$$
$$= p(p^4) + 2(p^4) - p(2p^3) - 2(2p^3) + p(3) + 2(3)$$
$$= p^5 + 2p^4 - 2p^4 - 4p^3 + 3p + 6$$
$$= p^5 - 4p^3 + 3p + 6. \qquad \text{Collecting like terms}$$

Now Try Exercise 23.

From the preceding examples, we can see how to multiply any two polynomials.

PRODUCT OF TWO POLYNOMIALS

To multiply two polynomials P and Q, select one of the polynomials, say P. Then multiply each term of P by every term of Q and collect like terms.

We can use columns when doing long multiplications. We multiply each term at the top by every term at the bottom, keeping like terms in columns and *adding spaces for missing terms*. Then we add.

EXAMPLE 7   Multiply: $(5x^3 + 3x^2 + x - 4)(-2x^2 + 3x + 6)$.

$$
\begin{array}{r}
5x^3 + 3x^2 + x - 4 \\
-2x^2 + 3x + 6 \\
\hline
30x^3 + 18x^2 + 6x - 24 \\
15x^4 + 9x^3 + 3x^2 - 12x \\
-10x^5 - 6x^4 - 2x^3 + 8x^2 \\
\hline
-10x^5 + 9x^4 + 37x^3 + 29x^2 - 6x - 24
\end{array}
$$

   Multiplying by 6
   Multiplying by 3x
   Multiplying by $-2x^2$

▶ **Now Try Exercise 27.**

EXAMPLE 8   Multiply: $(5x^3 + x - 4)(-2x^2 + 3x + 6)$.

$$
\begin{array}{r}
5x^3 + x - 4 \\
-2x^2 + 3x + 6 \\
\hline
30x^3 + 6x - 24 \\
15x^4 + 3x^2 - 12x \\
-10x^5 - 2x^3 + 8x^2 \\
\hline
-10x^5 + 15x^4 + 28x^3 + 11x^2 - 6x - 24
\end{array}
$$

   Multiplying by 6
   Multiplying by 3x
   Multiplying by $-2x^2$

▶ **Now Try Exercise 29.**

▶ b   Product of Two Binomials Using the FOIL Method

We now consider some **special products**. Let's find a faster special-product rule for the product of two binomials. Consider $(x + 7)(x + 4)$. We multiply each term of $(x + 7)$ by each term of $(x + 4)$:

$$(x + 7)(x + 4) = x \cdot x + x \cdot 4 + 7 \cdot x + 7 \cdot 4.$$

This multiplication illustrates a pattern that occurs whenever two binomials are multiplied:

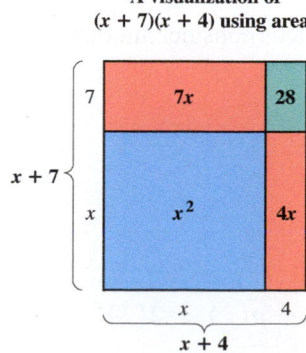

A visualization of $(x + 7)(x + 4)$ using areas

$$
\begin{array}{cccc}
\textbf{First} & \textbf{Outside} & \textbf{Inside} & \textbf{Last} \\
\text{terms} & \text{terms} & \text{terms} & \text{terms}
\end{array}
$$

$$(x + 7)(x + 4) = x \cdot x \;+\; 4x \;+\; 7x \;+\; 7(4) = x^2 + 11x + 28.$$

This special method of multiplying is called the **FOIL method**. Keep in mind that this method is based on the distributive law.

THE FOIL METHOD

To multiply two binomials, $A + B$ and $C + D$, multiply the **F**irst terms AC, the **O**utside terms AD, the **I**nside terms BC, and then the **L**ast terms BD. Then collect like terms, if possible.

$$(A + B)(C + D) = AC + AD + BC + BD$$

1. Multiply **F**irst terms: AC.
2. Multiply **O**utside terms: AD.
3. Multiply **I**nside terms: BC.
4. Multiply **L**ast terms: BD.

    ↓

    FOIL

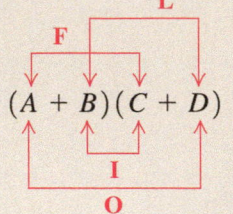

EXAMPLE 9 Multiply: $(x + 5)(x - 8)$.

$$
\begin{aligned}
&\qquad\quad\ \ \overset{\text{F}}{}\ \ \overset{\text{O}}{}\ \ \overset{\text{I}}{}\ \ \overset{\text{L}}{}\\
(x + 5)(x - 8) &= x^2 - 8x + 5x - 40\\
&= x^2 - 3x - 40 \qquad \text{\color{red}Collecting like terms}
\end{aligned}
$$

We write the result in descending order since the original binomials are in descending order.

▶ **Now Try Exercise 37.**

EXAMPLES

$$
\begin{aligned}
&\qquad\qquad\qquad\qquad\ \ \ \overset{\text{F}}{}\qquad\ \overset{\text{O}}{}\qquad\ \overset{\text{I}}{}\qquad\ \ \overset{\text{L}}{}\\
\mathbf{10.}\ &(3xy + 2x)(x^2 + 2xy^2) = 3x^3y + 6x^2y^3 + 2x^3 + 4x^2y^2
\end{aligned}
$$

11. $(2x - 3)(y + 2) = 2xy + 4x - 3y - 6$

12. $(2x + 3y)(x - 4y) = 2x^2 - 8xy + 3xy - 12y^2$
$$\qquad\qquad\qquad\quad = 2x^2 - 5xy - 12y^2 \qquad \text{\color{red}Collecting like terms}$$

▶ **Now Try Exercise 51.**

▶ **c Squares of Binomials**

We can use the FOIL method to develop special products for the square of a binomial:

$$
\begin{aligned}
(A + B)^2 &= (A + B)(A + B) & (A - B)^2 &= (A - B)(A - B)\\
&= A^2 + AB + AB + B^2 & &= A^2 - AB - AB + B^2\\
&= A^2 + 2AB + B^2; & &= A^2 - 2AB + B^2.
\end{aligned}
$$

A visualization of $(A + B)^2$ using areas

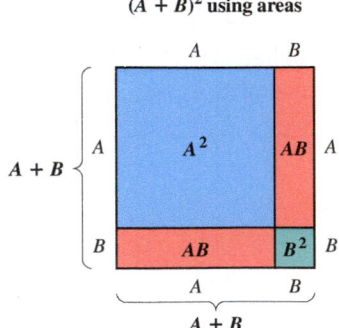

SQUARE OF A BINOMIAL

The **square of a binomial** is the square of the first term, plus twice the product of the two terms, plus the square of the last term.

$$(A + B)^2 = A^2 + 2AB + B^2;$$
$$(A - B)^2 = A^2 - 2AB + B^2$$

CAUTION! In general,

$$(AB)^2 = A^2B^2, \quad \text{but} \quad (A+B)^2 \neq A^2 + B^2.$$

EXAMPLES Multiply.

$$(A - B)^2 = A^2 - 2 \ A \ B + B^2$$

13. $(y - 5)^2 = y^2 - 2(y)(5) + 5^2$
$$= y^2 - 10y + 25$$

$$(A + B)^2 = A^2 + 2 \ A \ B + B^2$$

14. $(2x + 3y)^2 = (2x)^2 + 2(2x)(3y) + (3y)^2$
$$= 4x^2 + 12xy + 9y^2$$

15. $(3x^2 + 5xy^2)^2 = (3x^2)^2 + 2(3x^2)(5xy^2) + (5xy^2)^2$
$$= 9x^4 + 30x^3y^2 + 25x^2y^4$$

16. $\left(\frac{1}{2}a^2 - b^3\right)^2 = \left(\frac{1}{2}a^2\right)^2 - 2\left(\frac{1}{2}a^2\right)(b^3) + (b^3)^2$
$$= \frac{1}{4}a^4 - a^2b^3 + b^6$$ **Now Try Exercises 43 and 53.**

▶ **d Products of Sums and Differences**

Another special case of a product of two binomials is the product of the sum and the difference of the same two terms. Note the following:

$$\begin{array}{cccc} F & O & I & L \\ \downarrow & \downarrow & \downarrow & \downarrow \end{array}$$

$$(A + B)(A - B) = A^2 - AB + AB - B^2 = A^2 - B^2.$$

PRODUCT OF A SUM AND A DIFFERENCE

The product of the sum and the difference of the same two terms is the square of the first term minus the square of the second term (the difference of their squares).

$$(A + B)(A - B) = A^2 - B^2 \qquad \text{This is called a}$$
difference of squares.

EXAMPLES Multiply. (Say the rule as you work.)

$$(A + B)(A - B) = A^2 - B^2$$

17. $(y + 5)(y - 5) = y^2 - 5^2 = y^2 - 25$

18. $(2xy^2 + 3x)(2xy^2 - 3x) = (2xy^2)^2 - (3x)^2 = 4x^2y^4 - 9x^2$

19. $(0.2t - 1.4m)(0.2t + 1.4m) = (0.2t)^2 - (1.4m)^2 = 0.04t^2 - 1.96m^2$

20. $\left(\frac{2}{3}n - m^2\right)\left(\frac{2}{3}n + m^2\right) = \left(\frac{2}{3}n\right)^2 - (m^2)^2 = \frac{4}{9}n^2 - m^4$

Now Try Exercises 63 and 69.

EXAMPLES Multiply.

21. $(5y + 4 + 3x)(5y + 4 - 3x) = (5y + 4)^2 - (3x)^2$

$$= 25y^2 + 40y + 16 - 9x^2$$

Here we treat the binomial $5y + 4$ as the first expression, A, and $3x$ as the second, B.

22. $(3xy^2 + 4y)(-3xy^2 + 4y) = (4y + 3xy^2)(4y - 3xy^2)$

$$= (4y)^2 - (3xy^2)^2$$

$$= 16y^2 - 9x^2y^4$$

> **Now Try Exercises 73 and 81.**

Try to multiply polynomials mentally. When several types are mixed, first check to see what types of polynomials are to be multiplied. Then use the quickest method. Sometimes we might use more than one method to find a product. Remember that FOIL *always* works for multiplying binomials!

EXAMPLE 23 Multiply: $(s - 5t)(s + 5t)(s^2 - 25t^2)$.

We first note that $s - 5t$ and $s + 5t$ can be multiplied using the rule $(A - B)(A + B) = A^2 - B^2$. Then we have the product of two identical binomials, so we square, using $(A - B)^2 = A^2 - 2AB + B^2$.

$(s - 5t)(s + 5t)(s^2 - 25t^2)$

$= (s^2 - 25t^2)(s^2 - 25t^2)$ Using $(A - B)(A + B) = A^2 - B^2$

$= (s^2 - 25t^2)^2$

$= (s^2)^2 - 2(s^2)(25t^2) + (25t^2)^2$ Using $(A - B)^2 = A^2 - 2AB + B^2$

$= s^4 - 50s^2t^2 + 625t^4$

> **Now Try Exercise 79.**

Technology Connection

X	Y₁	Y₂
-2	-2	-2
-1	-2	-2
0	-2	-2
1	4	4
2	22	22
3	58	58
4	118	118

X=-2

Figure 1.

NORMAL SCI ENG
FLOAT 0123456789
RADIAN DEGREE
FUNC PAR POL SEQ
CONNECTED DOT
SEQUENTIAL SIMUL
REAL a+bi re^θi
FULL HORIZ G-T

Figure 2.

Checking Using Tables and Graphs

A partial check of operations with polynomials in one variable can be done using tables or graphs. For example, a table set in AUTO mode can be used to check the addition $(-3x^3 + 2x - 4) + (4x^3 + 3x^2 + 2) = x^3 + 3x^2 + 2x - 2$. To do so, we enter $y_1 = (-3x^3 + 2x - 4) + (4x^3 + 3x^2 + 2)$ and $y_2 = x^3 + 3x^2 + 2x - 2$. If the addition has been done correctly, the values of y_1 and y_2 will be the same regardless of the table settings used. (See Figure 1.)

To check an operation graphically, we compare two graphs. This is often easier to do when two graph styles are used. To check the product $(x + 3)(x - 2) = x^2 + x - 6$, we first use **MODE** to select the SEQUENTIAL mode. (See Figure 2.)

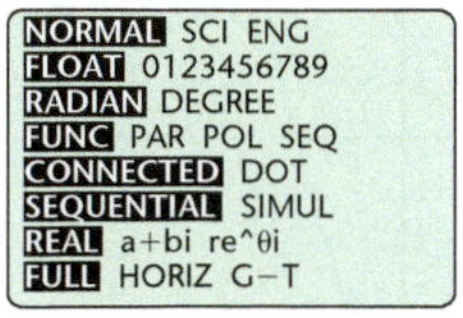

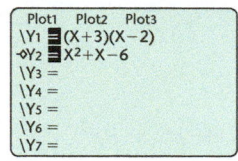

Figure 3.

$y_1 = (x + 3)(x - 2),$
$y_2 = x^2 + x - 6$

Figure 4.

Next, on the Y= screen, we enter $y_1 = (x + 3)(x - 2)$ and $y_2 = x^2 + x - 6$. We will select the line-graph style for y_1 and the path style for y_2. To select these graph styles, we use ⟨ to position the cursor over the icon to the left of the equation and press **ENTER** repeatedly until the desired style of icon appears, as shown in Figure 3.

The graphing calculator will graph y_1 first as a solid curve. Then it will graph y_2 as the circular cursor traces the leading edge of the graph, allowing us to determine visually whether the graphs coincide. In this case, the graphs appear to coincide, so the multiplication is probably correct. (See Figure 4.)

Exercises Use a table or graphs to determine whether each sum, difference, or product is correct.

1. $(x + 4)(x + 3) = x^2 + 7x + 12$
2. $(3x + 2)(x - 1) = 3x^2 + x - 2$
3. $(x - 1)(x - 1) = x^2 + 1$
4. $(x - 2)(x + 2) = x^2 - 4$
5. $(2x^2 + 3x - 6) + (5x^2 - 7x + 4) = 7x^2 + 4x - 2$
6. $(7x^5 + 2x^4 - 5x) - (-x^5 - 2x^4 + 3) = 8x^5 + 4x^4 - 5x - 3$

▶ **e   Using Function Notation**

Algebraic/Graphical Connection

Since $(x - 2)(x + 2) = x^2 - 4$, we know that $x^2 - 4$ and $(x - 2)(x + 2)$ are equivalent expressions.

From the viewpoint of functions, if

$$f(x) = (x - 2)(x + 2)$$

and

$$g(x) = x^2 - 4,$$

then for any given input x, the outputs $f(x)$ and $g(x)$ are identical. Thus the graphs of these functions are identical and we say that f and g represent the same function.

x	$f(x)$	$g(x)$
3	5	5
2	0	0
1	-3	-3
0	-4	-4
-1	-3	-3
-2	0	0
-3	5	5

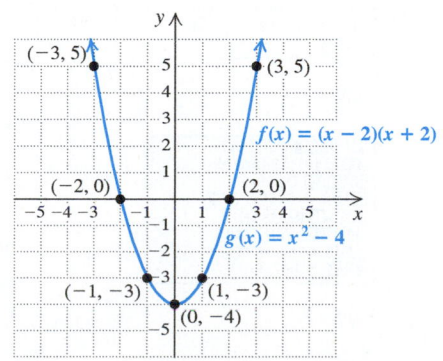

Our work with multiplying can be used when manipulating functions.

EXAMPLE 24 Given $f(x) = x^2 - 4x + 5$, find and simplify $f(a + 3)$ and $f(a + h) - f(a)$.

To find $f(a + 3)$, we replace x with $a + 3$. Then we simplify:

$$f(a + 3) = (a + 3)^2 - 4(a + 3) + 5$$
$$= a^2 + 6a + 9 - 4a - 12 + 5 = a^2 + 2a + 2.$$

To find $f(a + h) - f(a)$, we replace x with $a + h$ for $f(a + h)$ and x with a for $f(a)$. Then we simplify:

$$f(a + h) - f(a) = [(a + h)^2 - 4(a + h) + 5] - [a^2 - 4a + 5]$$
$$= a^2 + 2ah + h^2 - 4a - 4h + 5 - a^2 + 4a - 5$$
$$= 2ah + h^2 - 4h.$$

Now Try Exercise 85.

4.2   Exercise Set

► Reading Check

Determine whether each statement is true or false.

RC1. We use the distributive law to multiply polynomials.

RC2. The square of a binomial is a binomial.

RC3. We can use FOIL to multiply any two binomials.

RC4. The product of two monomials is always a monomial.

a *Multiply.*

1. $8y^2 \cdot 3y$

2. $-5x^2 \cdot 6xy$

3. $2x(-10x^2y)$

4. $-7ab^2(4a^2b^2)$

5. $(5x^5y^4)(-2xy^3)$

6. $(2a^2bc^2)(-3ab^5c^4)$

7. $2z(7 - x)$

8. $4a(a^2 - 3a)$

9. $6ab(a + b)$

10. $2xy(2x - 3y)$

11. $5cd(3c^2d - 5cd^2)$

12. $a^2(2a^2 - 5a^3)$

13. $(5x + 2)(3x - 1)$

14. $(2a - 3b)(4a - b)$

15. $(s + 3t)(s - 3t)$

16. $(y + 4)(y - 4)$

17. $(x - y)(x - y)$

18. $(a + 2b)(a + 2b)$

19. $(x^3 + 8)(x^3 - 5)$

20. $(2x^4 - 7)(3x^3 + 5)$

21. $(a^2 - 2b^2)(a^2 - 3b^2)$

22. $(2m^2 - n^2)(3m^2 - 5n^2)$

23. $(x - 4)(x^2 + 4x + 16)$

24. $(y + 3)(y^2 - 3y + 9)$

25. $(x + y)(x^2 - xy + y^2)$

26. $(a - b)(a^2 + ab + b^2)$

27. $(a^2 + a - 1)(a^2 + 4a - 5)$

28. $(x^2 - 2x + 1)(x^2 + x + 2)$

29. $(4a^2b - 2ab + 3b^2)(ab - 2b + a)$

30. $(2x^2 + y^2 - 2xy)(x^2 - 2y^2 - xy)$

31. $\left(x + \frac{1}{4}\right)\left(x + \frac{1}{4}\right)$

32. $\left(b - \frac{1}{3}\right)\left(b - \frac{1}{3}\right)$

33. $\left(\frac{1}{2}x - \frac{2}{3}\right)\left(\frac{1}{4}x + \frac{1}{3}\right)$

34. $\left(\frac{2}{3}a + \frac{1}{6}b\right)\left(\frac{1}{3}a - \frac{5}{6}b\right)$

35. $(1.3x - 4y)(2.5x + 7y)$

36. $(40a - 0.24b)(0.3a + 10b)$

b, c *Multiply.*

37. $(a + 8)(a + 5)$ **38.** $(x + 2)(x + 3)$

39. $(y + 7)(y - 4)$ **40.** $(y - 2)(y + 3)$

41. $\left(3a + \frac{1}{2}\right)^2$ **42.** $\left(2x - \frac{1}{3}\right)^2$

43. $(x - 2y)^2$ **44.** $(2s + 3t)^2$

45. $\left(b - \frac{1}{3}\right)\left(b - \frac{1}{2}\right)$ **46.** $\left(x - \frac{1}{2}\right)\left(x - \frac{1}{4}\right)$

47. $(2x + 9)(x + 2)$ **48.** $(3b + 2)(2b - 5)$

49. $(20a - 0.16b)^2$ **50.** $(10p^2 + 2.3y)^2$

51. $(2x - 3y)(2x + y)$

52. $(2a - 3b)(2a - b)$

53. $(x^3 + 2)^2$ **54.** $(y^4 - 7)^2$

55. $(2x^2 - 3y^2)^2$ **56.** $(3s^2 + 4t^2)^2$

57. $(a^3b^2 + 1)^2$ **58.** $(x^2y - xy^3)^2$

59. $(0.1a^2 - 5b)^2$ **60.** $(6m + 0.45p^2)^2$

61. *Compound Interest.* Suppose that P dollars is invested in a savings account at interest rate i, compounded annually, for 2 years. The amount A in the account after 2 years is given by

$$A = P(1 + i)^2.$$

Find an equivalent expression for A without parentheses.

62. *Compound Interest.* Suppose that P dollars is invested in a savings account at interest rate i, compounded semiannually, for 1 year. The amount A in the account after 1 year is given by

$$A = P\left(1 + \frac{i}{2}\right)^2.$$

Find an equivalent expression for A without parentheses.

d *Multiply.*

63. $(d + 8)(d - 8)$ **64.** $(y - 3)(y + 3)$

65. $(2c + 3)(2c - 3)$ **66.** $(1 - 2x)(1 + 2x)$

67. $(6m - 5n)(6m + 5n)$

68. $(3x + 7y)(3x - 7y)$

69. $(x^2 + yz)(x^2 - yz)$

70. $(2a^2 + 5ab)(2a^2 - 5ab)$

71. $(-mn + m^2)(mn + m^2)$

72. $(1.6 + cw)(-1.6 + cw)$

73. $(-3pt + 4p^2)(4p^2 + 3pt)$

74. $(-10xy + 5x^2)(5x^2 + 10xy)$

75. $\left(\frac{1}{2}p - \frac{2}{3}n\right)\left(\frac{1}{2}p + \frac{2}{3}n\right)$

76. $\left(\frac{3}{5}ab + 4c\right)\left(\frac{3}{5}ab - 4c\right)$

77. $(x + 1)(x - 1)(x^2 + 1)$

78. $(y - 2)(y + 2)(y^2 + 4)$

79. $(a - b)(a + b)(a^2 - b^2)$

80. $(2x - y)(2x + y)(4x^2 - y^2)$

81. $(a + b + 1)(a + b - 1)$

82. $(m + n + 2)(m + n - 2)$

83. $(2x + 3y + 4)(2x + 3y - 4)$

84. $(3a - 2b + c)(3a - 2b - c)$

e *For each of the following functions, find* $f(t - 1)$, $f(p + 1)$, $f(a + h) - f(a)$, $f(t - 2) + c$, *and* $f(a) + 5$.

85. $f(x) = 5x + x^2$

86. $f(x) = 4x + 2x^2$

87. $f(x) = 3x^2 - 7x + 8$

88. $f(x) = 3x^2 - 4x + 7$

89. $f(x) = 5x - x^2$

90. $f(x) = 4x - 2x^2$

91. $f(x) = 4 + 3x - x^2$

92. $f(x) = 2 - 4x - 3x^2$

▶ **Skill Maintenance**

Solve. **[3.4b]**

93. *Auto Travel.* Rachel leaves on a business trip, forgetting her laptop computer. Her sister discovers Rachel's laptop 2 hr later, and knowing that Rachel needs it for her sales presentation and that Rachel normally travels at a speed of 55 mph, she decides to follow her at a speed of 75 mph. After how long will Rachel's sister catch up with her?

94. *Air Travel.* An airplane flew for 5 hr against a 20-mph headwind. The return trip with the wind took 4 hr. Find the speed of the plane in still air.

Solve. **[3.2a], [3.3a]**

95. $5x + 9y = 2,$ **96.** $x + 4y = 13,$
     $4x - 9y = 10$ $5x - 7y = -16$

97. $2x - 3y = 1,$ **98.** $9x - 8y = -2,$
     $4x - 6y = 2$ $3x + 2y = 3$

▶ **Synthesis**

99. 〽 Use the TABLE feature or the GRAPH feature of a graphing calculator to check your answers to Exercises 28, 40, and 77.

100. 〽 Use the TABLE feature or the GRAPH feature of a graphing calculator to determine whether each of the following is correct.

a) $(x - 1)^2 = x^2 - 1$
b) $(x - 2)(x + 3) = x^2 + x - 6$
c) $(x - 1)^3 = x^3 - 3x^2 + 3x - 1$
d) $(x + 1)^4 = x^4 + 1$

Multiply. Assume that variables in exponents represent natural numbers.

101. $(z^{n^2})^{n^3}(z^{4n^3})^{n^2}$

102. $y^3 z^n (y^{3n} z^3 - 4yz^{2n})$

103. $(r^2 + s^2)^2 (r^2 + 2rs + s^2)(r^2 - 2rs + s^2)$

104. $(y - 1)^6 (y + 1)^6$

105. $\left(3x^5 - \frac{5}{11}\right)^2$

106. $(4x^2 + 2xy + y^2)(4x^2 - 2xy + y^2)$

107. $(x^a + y^b)(x^a - y^b)(x^{2a} + y^{2b})$

108. $\left(x - \frac{1}{7}\right)\left(x^2 + \frac{1}{7}x + \frac{1}{49}\right)$

109. $(x - 1)(x^2 + x + 1)(x^3 + 1)$

110. $(x^{a-b})^{a+b}$

4.3   Introduction to Factoring

▶ **a**   Factor polynomials whose terms have a common factor.

▶ **b**   Factor certain polynomials with four terms by grouping.

Factoring is the reverse of multiplication. To **factor** an expression is to find an equivalent expression that is a product. For example, we can *factor* $x^2 - 9$: $x^2 - 9 = (x + 3)(x - 3)$. We say that $x + 3$ and $x - 3$ are **factors** of $x^2 - 9$ and that $(x + 3)(x - 3)$ is a **factorization**.

> **FACTOR AND FACTORIZATION**
>
> To **factor** a polynomial is to express it as a product.
>
> A **factor** of a polynomial P is a polynomial that can be used to express P as a product.
>
> A **factorization** of a polynomial P is an expression that names P as a product of factors.

▶ **a   Terms with Common Factors**

To multiply a monomial and a polynomial with more than one term, we multiply each term by the monomial using the distributive laws. To factor, we do the reverse. We express a polynomial as a product using the distributive laws in reverse. Compare.

Multiply

$5x(x^2 - 3x + 1)$
$= 5x \cdot x^2 - 5x \cdot 3x + 5x \cdot 1$
$= 5x^3 - 15x^2 + 5x$

Factor

$5x^3 - 15x^2 + 5x$
$= 5x \cdot x^2 - 5x \cdot 3x + 5x \cdot 1$
$= 5x(x^2 - 3x + 1)$

EXAMPLE 1 Factor: $4y^2 - 8$.

$$4y^2 - 8 = 4 \cdot y^2 - 4 \cdot 2 \qquad \text{\color{red}{4 is the largest common factor.}}$$
$$= 4(y^2 - 2) \qquad \text{\color{red}{Factoring out the common factor 4}}$$

▶ **Now Try Exercise 9.**

> **CAUTION!** Be careful not to confuse terms with factors! The terms of $x^2 - 9$ are x^2 and -9. Terms are used to form sums. Factors of $x^2 - 9$ are $x - 3$ and $x + 3$. Factors are used to form products.

In some cases, there is more than one common factor. In Example 2 below, for instance, 5 is a common factor, x^3 is a common factor, and $5x^3$ is a common factor. If there is more than one common factor, we generally choose the one with the largest coefficient and the largest exponent.

EXAMPLES Factor.

2. $5x^4 - 20x^3 = 5x^3 \cdot x - 5x^3 \cdot 4$
$$= 5x^3(x - 4) \qquad \text{\color{red}{Multiply mentally to check your answer.}}$$

3. $12x^2y - 20x^3y = 4x^2y(3 - 5x)$

▶ **Now Try Exercise 5.**

The polynomials in Examples 1–3 have been **factored completely**. They cannot be factored further. The factors in the resulting factorization are said to be **prime polynomials**.

EXAMPLE 4 Factor: $10a^6b^2 - 4a^5b^3 + 2a^4b^4$.

First, we look for the greatest positive common factor in the coefficients:

$$10, -4, 2 \longrightarrow \text{Greatest common factor} = 2.$$

Second, we look for the greatest common factor in the powers of a:

$$a^6, a^5, a^4 \longrightarrow \text{Greatest common factor} = a^4.$$

Third, we look for the greatest common factor in the powers of b:

$$b^2, b^3, b^4 \longrightarrow \text{Greatest common factor} = b^2.$$

Thus, $2a^4b^2$ is the greatest common factor of the given polynomial. Then

$$10a^6b^2 - 4a^5b^3 + 2a^4b^4 = 2a^4b^2 \cdot 5a^2 - 2a^4b^2 \cdot 2ab + 2a^4b^2 \cdot b^2$$
$$= 2a^4b^2(5a^2 - 2ab + b^2).$$

▶ **Now Try Exercise 17.**

When the leading coefficient is a negative number, we generally factor out a negative coefficient.

EXAMPLES Factor out a common factor with a negative coefficient.

5. $-4x - 24 = -4(x + 6)$

6. $-2x^2 + 6x - 10 = -2(x^2 - 3x + 5)$

▶ **Now Try Exercise 23.**

EXAMPLE 7 *Height of a Rocket.* A water rocket is launched upward with an initial velocity of 96 ft/sec. Its height h, in feet, after t seconds is given by the function $h(t) = -16t^2 + 96t$.

a) Find an equivalent expression for $h(t)$ by factoring out a common factor with a negative coefficient.

b) Check your factoring by evaluating both expressions for $h(t)$ at $t = 2$.

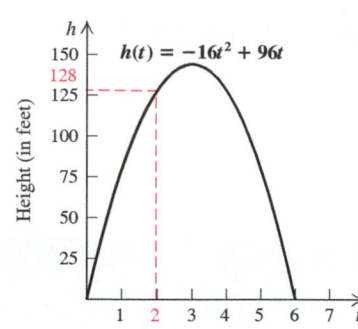

a) We factor out $-16t$ as follows:

$$h(t) = -16t^2 + 96t = -16t(t - 6).$$

b) We check as follows:

$$h(2) = -16 \cdot 2^2 + 96 \cdot 2 = 128;$$
$$h(2) = -16 \cdot 2(2 - 6) = 128. \quad \text{Using the factorization}$$

Now Try Exercise 33.

▶ b Factoring by Grouping

In expressions of four or more terms, there may be a *common binomial factor*. We proceed as in the following examples.

EXAMPLE 8 Factor: $(a - b)(x + 5) + (a - b)(x - y^2)$.

$$(a - b)(x + 5) + (a - b)(x - y^2) = (a - b)[(x + 5) + (x - y^2)]$$
$$= (a - b)(2x + 5 - y^2)$$

Now Try Exercise 39.

In Example 9, we group before we factor.

EXAMPLE 9 Factor: $y^3 + 3y^2 + 4y + 12$.

$$y^3 + 3y^2 + 4y + 12 = (y^3 + 3y^2) + (4y + 12) \qquad \text{Grouping}$$
$$= y^2(y + 3) + 4(y + 3) \qquad \text{Factoring each binomial}$$
$$= (y + 3)(y^2 + 4) \qquad \text{Factoring out the common factor } y + 3$$

Now Try Exercise 45.

EXAMPLE 10 Factor: $3x^3 - 6x^2 - x + 2$.

First, we factor out the greatest common factor in the first two terms:

$$3x^3 - 6x^2 = 3x^2(x - 2).$$

Next, we look at the third and fourth terms to see if we can factor them in order to have $x - 2$ as a factor. We see that if we factor out -1, we get $x - 2$:

$$-x + 2 = -1 \cdot (x - 2).$$

Finally, we factor out the common factor $x - 2$:

$$3x^3 - 6x^2 - x + 2 = (3x^3 - 6x^2) + (-x + 2)$$
$$= 3x^2(x - 2) + (-x + 2)$$
$$= 3x^2(x - 2) - 1(x - 2) \qquad \textit{Check:}$$
$$\qquad\qquad\qquad -1(x - 2) = -x + 2$$
$$= (x - 2)(3x^2 - 1). \qquad \text{Factoring out the common factor } x - 2$$

Now Try Exercise 47.

EXAMPLE 11 Factor: $4x^3 - 15 + 20x^2 - 3x$.

$$4x^3 - 15 + 20x^2 - 3x = 4x^3 + 20x^2 - 3x - 15 \qquad \text{Rearranging}$$
$$= 4x^2(x + 5) - 3(x + 5) \qquad \textit{Check:}$$
$$\qquad\qquad\qquad -3(x + 5) = -3x - 15$$
$$= (x + 5)(4x^2 - 3) \qquad \text{Factoring out } x + 5$$

Now Try Exercise 53.

Not all polynomials with four terms can be factored by grouping. An example is $x^3 + x^2 + 3x - 3$. Note that in a grouping like $x^2(x + 1) + 3(x - 1)$, the expressions $x + 1$ and $x - 1$ are not the same. No grouping allows us to factor out a common binomial.

4.3 Exercise Set

▶ Reading Check

Choose from the column on the right the word that best completes each statement.

RC1. To factor a polynomial is to express it as a(n) _____.

RC2. In the expression $x(x - 2)$, x and $x - 2$ are _____.

RC3. The expression $x(x - 2)$ is a(n) _____ of $x^2 - 2x$.

RC4. A polynomial that cannot be factored is said to be _____.

RC5. The expression $4x$ is a(n) _____ factor of the terms of the polynomial $8x + 12x^3$.

RC6. When we factor by grouping, we look for a common _____ factor.

binomial

common

factorization

factors

prime

product

⋯⋯⋯

a *Factor.*

1. $6a^2 + 3a$

2. $4x^2 + 2x$

3. $x^3 + 9x^2$

4. $y^3 + 8y^2$

5. $8x^2 - 4x^4$

6. $6x^2 + 3x^4$

7. $4x^2y - 12xy^2$

8. $5x^2y^3 + 15x^3y^2$

9. $3y^2 - 3y - 9$

10. $5x^2 - 5x + 15$

11. $4ab - 6ac + 12ad$

12. $8xy + 10xz - 14xw$

13. $10a^4 + 15a^2 - 25a - 30$

14. $12t^5 - 20t^4 + 8t^2 - 16$

15. $15x^2y^5z^3 - 12x^4y^4z^7$

16. $21a^3b^5c^7 - 14a^7b^6c^2$

17. $14a^4b^3c^5 + 21a^3b^5c^4 - 35a^4b^4c^3$

18. $9x^3y^6z^2 - 12x^4y^4z^4 + 15x^2y^5z^3$

Factor out a common factor with a negative coefficient.

19. $-5x - 45$

20. $-3t + 18$

21. $-6a - 84$

22. $-8t + 40$

23. $-2x^2 + 2x - 24$

24. $-2x^2 + 16x - 20$

25. $-3y^2 + 24y$

26. $-7x^2 - 56y$

27. $-a^4 + 2a^3 - 13a^2 - 1$

28. $-m^3 - m^2 + m - 2$

29. $-3y^3 + 12y^2 - 15y + 24$

30. $-4m^4 - 32m^3 + 64m - 12$

31. *Volume of a Propane Gas Tank.* A propane gas tank is shaped like a circular cylinder with half of a sphere at each end. The volume of the tank with length h and radius r of the cylindrical section is given by the polynomial

$$\pi r^2 h + \tfrac{4}{3}\pi r^3.$$

Find an equivalent expression by factoring out a common factor.

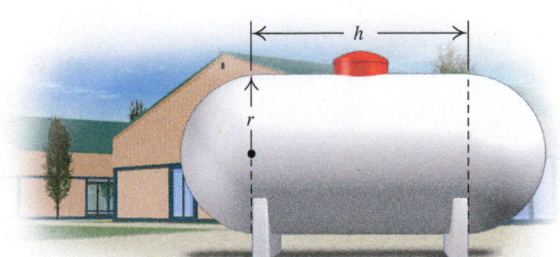

32. *Triangular Layers.* The stack of truffles shown below is formed by triangular layers of truffles. The number N of truffles in the stack is given by the polynomial function

$$N(x) = \tfrac{1}{6}x^3 + \tfrac{1}{2}x^2 + \tfrac{1}{3}x,$$

where x is the number of layers. Find an equivalent expression for $N(x)$ by factoring out a common factor.

33. *Height of a Baseball.* A baseball is popped up with an upward velocity of 72 ft/sec. Its height h, in feet, after t seconds is given by

$$h(t) = -16t^2 + 72t.$$

  a) Find an equivalent expression for $h(t)$ by factoring out a common factor with a negative coefficient.

  b) Perform a partial check of part (a) by evaluating both expressions for $h(t)$ at $t = 2$.

34. *Number of Diagonals.* The number of diagonals of a polygon having n sides is given by the polynomial function

$$P(n) = \tfrac{1}{2}n^2 - \tfrac{3}{2}n.$$

Find an equivalent expression for $P(n)$ by factoring out a common factor.

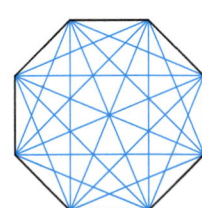

35. *Total Revenue.* Perfect Sound is marketing a new kind of home theater chair. The firm determines that when it sells x chairs, the total revenue R is given by the polynomial function

$$R(x) = 280x + 0.4x^2 \text{ dollars.}$$

Find an equivalent expression for $R(x)$ by factoring out $0.4x$.

36. *Total Cost.* Perfect Sound determines that the total cost C of producing x home theater chairs is given by the polynomial function

$$C(x) = 0.18x + 0.6x^2.$$

Find an equivalent expression for $C(x)$ by factoring out $0.6x$.

b *Factor.*

37. $a(b - 2) + c(b - 2)$

38. $a(x^2 - 3) - 2(x^2 - 3)$

39. $(x - 2)(x + 5) + (x - 2)(x + 8)$

40. $(m - 4)(m + 3) + (m - 4)(m - 3)$

41. $y^8 - 7y^7 + y - 7$ **42.** $b^5 - 3b^4 + b - 3$

43. $ac + ad + bc + bd$ **44.** $xy + xz + wy + wz$

45. $b^3 - b^2 + 2b - 2$

46. $y^3 - y^2 + 3y - 3$

47. $y^3 + 8y^2 - 5y - 40$

48. $t^3 + 6t^2 - 2t - 12$

49. $24x^3 + 72x - 36x^2 - 108$

50. $10a^3 + 50a - 15a^2 - 75$

51. $a^4 - a^3 + a^2 + a$

52. $p^6 + p^5 - p^3 + p^2$

53. $2y^4 + 6y^2 - 5y^2 - 15$

54. $2xy + x^2y - 6 - 3x$

▶ Skill Maintenance

Solve.

55. $|x - 3| = 10$ **[1.6c]**

56. $|2a - 3| = |3a + 5|$ **[1.6d]**

57. $|2 - x| \le 12$ **[1.6e]**

58. $|3y - 7| + 2 > 8$ **[1.6e]**

59. $8 \le x - 7 \le 10$ **[1.5a]**

60. $-2 < -3x + 1 < 0$ **[1.5a]**

61. $2x - 7 > 6 \text{ or } 3x + 1 < 2$ **[1.5b]**

62. $-m + 3 \le 2 \text{ or } m + 5 > 5m - 1$ **[1.5b]**

▶ Synthesis

Complete each of the following.

63. $x^5y^4 + \underline{\hspace{1cm}} = x^3y(\underline{\hspace{1cm}} + xy^5)$

64. $a^3b^7 - \underline{\hspace{1cm}} = \underline{\hspace{1cm}}(ab^4 - c^2)$

Factor.

65. $rx^2 - rx + 5r + sx^2 - sx + 5s$

66. $3a^2 + 6a + 30 + 7a^2b + 14ab + 70b$

67. $a^4x^4 + a^4x^2 + 5a^4 + a^2x^4 + a^2x^2 + 5a^2 + 5x^4 + 5x^2 + 25$
  (*Hint*: Use three groups of three.)

Factor out the smallest power of x in each of the following.

68. $x^{1/2} + 5x^{3/2}$ **69.** $x^{1/3} - 7x^{4/3}$

70. $x^{3/4} + x^{1/2} - x^{1/4}$ **71.** $x^{1/3} - 5x^{1/2} + 3x^{3/4}$

Factor. Assume that all exponents are natural numbers.

72. $2x^{3a} + 8x^a + 4x^{2a}$

73. $3a^{n+1} + 6a^n - 15a^{n+2}$

74. $4x^{a+b} + 7x^{a-b}$

75. $7y^{2a+b} - 5y^{a+b} + 3y^{a+2b}$

4.4   Factoring Trinomials: $x^2 + bx + c$

▶ **a**   Factor trinomials of the type $x^2 + bx + c$.

▶ a   Factoring Trinomials: $x^2 + bx + c$

We now consider factoring trinomials of the type $x^2 + bx + c$. We use a refined trial-and-error process that is based on the FOIL method.

Constant Term Positive

Recall the FOIL method of multiplying two binomials:

$$\overset{\text{F} \quad\;\; \text{O} \quad\;\; \text{I} \quad\;\; \text{L}}{(x + 3)(x + 5) = x^2 + 5x + 3x + 15}$$
$$= x^2 \quad\;\; + 8x \quad\;\; + 15.$$

The product is a trinomial. To factor $x^2 + 8x + 15$, we think of FOIL in reverse. Since the first term of the trinomial is x^2, the first term of each binomial factor is x. We want to find numbers p and q such that

$$x^2 + 8x + 15 = (x + p)(x + q).$$

We now look for two numbers whose product is 15 and whose sum is 8. Those numbers are 3 and 5. Thus the factorization is

$$(x + 3)(x + 5), \quad \text{or} \quad (x + 5)(x + 3).$$

Thus we can factor using the following general form in reverse:

$$(x + p)(x + q) = x^2 + (p + q)x + pq.$$

EXAMPLE 1   Factor: $x^2 + 9x + 8$.

Think of FOIL in reverse. The first term of each factor is x. We are looking for numbers p and q such that

$$x^2 + 9x + 8 = (x + p)(x + q) = x^2 + (p + q)x + pq.$$

We look for two numbers p and q whose product is 8 and whose sum is 9. Since both 8 and 9 are positive, we need consider only positive factors.

Pairs of Factors	Sums of Factors
2, 4	6
1, 8	9

The numbers we need are 1 and 8.

The factorization is $(x + 1)(x + 8)$. We can check by multiplying:

$$(x + 1)(x + 8) = x^2 + 9x + 8.$$

Now Try Exercise 1.

> When the constant term of a trinomial is positive, we look for two factors with the same sign (both positive or both negative). The sign is that of the middle term.

EXAMPLE 2 Factor: $20 - 9y + y^2$.

We begin by writing the trinomial in descending order:

$$y^2 - 9y + 20.$$

Since the constant term, 20, is positive and the coefficient of the middle term, -9, is negative, we look for a factorization of 20 in which both factors are negative. Their sum must be -9.

Pairs of Factors	Sums of Factors
$-1, -20$	-21
$-2, -10$	-12
$-4, \ -5$	-9 ←

The numbers we need are -4 and -5.

The factorization is $(y - 4)(y - 5)$.

Now Try Exercise 3.

Constant Term Negative

> When the constant term of a trinomial is negative, we look for two factors whose product is negative. One of them must be positive and the other negative. Their sum must be the coefficient of the middle term.

EXAMPLE 3 Factor: $x^3 - x^2 - 30x$.

Always look first for the largest common factor. This time x is the common factor. We first factor it out:

$$x^3 - x^2 - 30x = x(x^2 - x - 30).$$

Now consider $x^2 - x - 30$. We look for a factorization of -30 the constant term. One factor will be positive and one factor will be negative. The sum of the factors must be -1, the coefficient of the middle term, so the negative factor must have the larger absolute value. Thus we consider only pairs of factors in which the negative factor has the larger absolute value.

Pairs of Factors	Sums of Factors
$1, -30$	-29
$2, -15$	-13
$3, -10$	-7
$5, \ -6$	-1 ←

The numbers we need are 5 and -6.

The factorization of $x^2 - x - 30$ is $(x + 5)(x - 6)$. But do not forget the common factor! The factorization of the original trinomial is

$$x(x + 5)(x - 6).$$

▶ **Now Try Exercise 31.**

EXAMPLE 4 Factor: $x^2 + 17x - 110$.

Since the constant term, -110, is negative, factorizations of -110 will have one positive factor and one negative factor. The sum of the factors must be 17, so the positive factor must have the larger absolute value.

Pairs of Factors	Sums of Factors
-1, 110	109
-2, 55	53
-5, 22	17 ←
-10, 11	1

We consider only pairs of factors in which the positive term has the larger absolute value.
The numbers we need are -5 and 22.

The factorization is $(x - 5)(x + 22)$.

▶ **Now Try Exercise 9.**

Some trinomials are not factorable.

EXAMPLE 5 Factor: $x^2 - x - 7$.

There are no factors of -7 whose sum is -1. This trinomial is *not* factorable into binomials.

▶ **Now Try Exercise 27.**

> To factor $x^2 + bx + c$:
>
> **1.** First arrange in descending order.
> **2.** Use a trial-and-error procedure that looks for factors of c whose sum is b.
>   - If c is positive, then the signs of the factors are the same as the sign of b.
>   - If c is negative, then one factor is positive and the other is negative. (If the sum of the two factors is the opposite of b, changing the signs of each factor will give the desired factors whose sum is b.)
> **3.** Check your result by multiplying.

The procedure considered here can also be applied to a trinomial with more than one variable.

EXAMPLE 6 Factor: $x^2 - 2xy - 48y^2$.

We look for numbers p and q such that

$$x^2 - 2xy - 48y^2 = (x + py)(x + qy).$$

Our thinking is much the same as if we were factoring $x^2 - 2x - 48$. We look for factors of -48 whose sum is -2. Those factors are 6 and -8. Then

$$x^2 - 2xy - 48y^2 = (x + 6y)(x - 8y).$$

We can check by multiplying.

▶ **Now Try Exercise 29.**

EXAMPLE 7 Factor: $x^4 + 2x^2 - 15$.

We look for numbers p and q such that

$$x^4 + 2x^2 - 15 = (x^2 + p)(x^2 + q).$$

The constant term is negative and the middle term is positive. Thus we look for pairs of factors of -15, such that the positive factor has the larger absolute value and the sum of the factors is 2. Those factors are -3 and 5. The desired factorization is

$$(x^2 - 3)(x^2 + 5).$$

Now Try Exercise 25.

Leading Coefficient of -1

EXAMPLE 8 Factor: $-x^2 + 5x + 14$.

Note that this trinomial has a leading coefficient of -1. Before factoring in such a case, we can factor out a -1:

$$\begin{aligned} -x^2 + 5x + 14 &= -1(x^2 - 5x - 14) \\ &= -1(x - 7)(x + 2). \quad \text{Factoring } x^2 - 5x - 14 \end{aligned}$$

We can also express this answer two other ways by multiplying through either binomial by -1. Thus each of the following is a correct answer:

$$\begin{aligned} -x^2 + 5x + 14 &= -1(x - 7)(x + 2); \\ &= (-x + 7)(x + 2); \quad \text{Multiplying } x - 7 \text{ by } -1 \\ &= (x - 7)(-x - 2). \quad \text{Multiplying } x + 2 \text{ by } -1 \end{aligned}$$

Now Try Exercise 21.

4.4 Exercise Set

▶ # Reading Check

Choose from the column on the right the phrase that best completes each statement.

RC1. To factor $x^2 + 19x - 20$, we look for a factorization of -20 in which _____.

RC2. To factor $x^2 - 11x - 12$, we look for a factorization of -12 in which _____.

RC3. To factor $x^2 + 7x + 12$, we look for a factorization of 12 in which _____.

RC4. To factor $x^2 - 12x + 20$, we look for a factorization of 20 in which _____.

a) both factors are positive

b) both factors are negative

c) the positive factor has the greater absolute value

d) the negative factor has the greater absolute value

a *Factor.*

1. $x^2 + 13x + 36$

2. $x^2 + 9x + 18$

3. $t^2 - 8t + 15$

4. $y^2 - 10y + 21$

5. $x^2 - 8x - 33$

6. $t^2 - 15 - 2t$

7. $2y^2 - 16y + 32$

8. $2a^2 - 20a + 50$

9. $p^2 + 3p - 54$

10. $m^2 + m - 72$

11. $12x + x^2 + 27$

12. $10y + y^2 + 24$

13. $y^2 - \dfrac{2}{3}y + \dfrac{1}{9}$

14. $p^2 + \dfrac{2}{5}p + \dfrac{1}{25}$

15. $t^2 - 4t + 3$

16. $y^2 - 14y + 45$

17. $5x + x^2 - 14$

18. $x + x^2 - 90$

19. $x^2 + 5x + 6$

20. $y^2 + 8y + 7$

21. $56 + x - x^2$

22. $32 + 4y - y^2$

23. $32y + 4y^2 - y^3$

24. $56x + x^2 - x^3$

25. $x^4 + 11x^2 - 80$

26. $y^4 + 5y^2 - 84$

27. $x^2 - 3x + 7$

28. $x^2 + 12x + 13$

29. $x^2 + 12xy + 27y^2$

30. $p^2 - 5pq - 24q^2$

31. $2x^2 - 8x - 90$

32. $3x^2 - 21x - 90$

33. $-z^2 + 36 - 9z$

34. $24 - a^2 - 10a$

35. $x^4 + 50x^2 + 49$

36. $p^4 + 80p^2 + 79$

37. $x^6 + 11x^3 + 18$

38. $x^6 - x^3 - 42$

39. $x^8 - 11x^4 + 24$

40. $x^8 - 7x^4 + 10$

41. $y^2 - 0.8y + 0.16$

42. $a^2 + 1.4a + 0.49$

43. $12 - b^{10} - b^{20}$

44. $8 - 7t^{15} - t^{30}$

▶ Skill Maintenance

Solve. [3.4a]

45. *Mixing Rice.* Countryside Rice is 90% white rice and 10% wild rice. Mystic Rice is 50% wild rice. How much of each type should be used to create a 25-lb batch of rice that is 35% wild rice?

46. *Wages.* Takako worked a total of 17 days last month at her father's restaurant. She earned $50 per day during the week and $60 per day during the weekend. Last month Takako earned $940. How many weekdays did she work?

Determine whether each of the following is the graph of a function. [2.2d]

47.

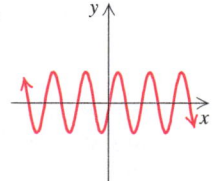

48.

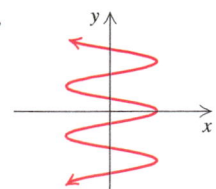

49.

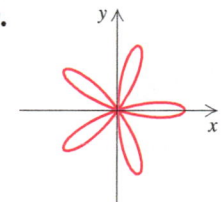

50.

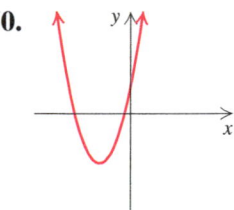

Find the domain of f. [2.3a]

51. $f(x) = x^2 - 2$

52. $f(x) = 3 - 2x$

53. $f(x) = \dfrac{3}{4x - 7}$

54. $f(x) = 3 - |x|$

▶ Synthesis

55. Find all integers m for which $x^2 + mx + 75$ can be factored.

56. Find all integers q for which $x^2 + qx - 32$ can be factored.

57. One of the factors of $x^2 - 345x - 7300$ is $x + 20$. Find the other factor.

58. 📈 Use the TABLE feature and the GRAPH feature of a graphing calculator to check your answers to Exercises 1–6.

Mid-Chapter Review

Determine whether each statement is true or false.

1. The polynomial $5x + 2x^2 - 4x^3$ can be factored. [4.3a]

2. The expression $17x^{-2}y^3$ is a monomial. [4.1a]

3. The degree of a polynomial is the same as the degree of the leading term. [4.1a]

4. The opposite of $-x^2 + x$ is $x - x^2$. [4.1d]

5. The binomial $144 - x^2$ is a difference of squares. [4.2d]

For each polynomial, identify the terms, the degree of each term, and the degree of the polynomial. Then identify the leading term, the leading coefficient, and the constant term. **[4.1a]**

6. $-a^7 + a^4 - a + 8$

7. $3x^4 + 2x^3w^5 - 12x^2w + 4x^2 - 1$

8. Arrange in ascending powers of y: $-2y + 5 - y^3 + y^9 - 2y^4$. **[4.1a]**

9. Arrange in descending powers of x: $2qx - 9qr + 2x^5 - 4qx^2$. **[4.1a]**

Evaluate each polynomial function for the given values of the variable. **[4.1b]**

10. $h(x) = -x^3 - 4x + 5$; $h(0), h(-2)$, and $h\left(\dfrac{1}{2}\right)$

11. $f(x) = \dfrac{1}{2}x^4 - x^3$; $f(-1), f(1)$, and $f(0)$

12. Given $f(x) = x^2 + 2x - 9$, find and simplify $f(a - 2)$ and $f(a + h) - f(a)$. **[4.2e]**

Add, subtract, or multiply.

13. $(3a^2 - 7b + ab + 2) + (-5a^2 + 4b - 5ab - 3)$ **[4.1c]**

14. $(x^2 + 10x - 4) + (9x^2 - 2x + 1) + (x^2 - x - 5)$ **[4.1c]**

15. $(b - 12)(b + 1)$ **[4.2b]**

16. $c^2(3c^2 - c^3)$ **[4.2a]**

17. $(y^4 - 6)(y^4 + 3)$ **[4.2b]**

18. $(7y^2 - 2y^3 - 5y) - (y^2 - 3y - 6y^3)$ **[4.1d]**

19. $(8x - 11) - (-x + 1)$ **[4.1d]**

20. $(4x - 5)^2$ **[4.2c]**

21. $(2x + 5)^2$ **[4.2c]**

22. $(0.01x - 0.5y) - (2.5y - 0.1x)$ **[4.1d]**

23. $-13x^2 \cdot 10xy$ **[4.2a]**

24. $(x + y)(x^2 - 2xy + 3y^2)$ **[4.2a]**

25. $(5x - 7)(2x + 9)$ **[4.2b]**

26. $(9x - 4)(9x + 4)$ **[4.2d]**

Factor.

27. $5h^2 + 7h$ **[4.3a]**

28. $x^2 + 8x - 20$ **[4.4a]**

29. $21 - 4b - b^2$ **[4.4a]**

30. $m^2 + \dfrac{2}{7}m + \dfrac{1}{49}$ **[4.4a]**

31. $2xy - x^2y - 5x + 10$ **[4.3b]**

32. $3w^2 - 6w + 3$ **[4.4a]**

33. $t^3 + 3t^2 + t + 3$ **[4.3b]**

34. $24xy^6z^4 - 16x^4y^3z$ **[4.3a]**

35. $x^2 + 8x + 6$ **[4.4a]**

Collaborative Discussion and Writing

36. Explain in your own words why $-(a - b) = b - a$. **[4.1d]**, **[4.3a]**

37. Is the sum of two binomials always a binomial? Why or why not? **[4.1c]**

38. Is it true that if a polynomial's coefficients and exponents are all prime numbers, then the polynomial itself is prime? Why or why not? **[4.3a]**

39. Under what conditions would it be easier to evaluate a polynomial function after it has been factored? **[4.1b]**, **[4.4a]**

40. Explain the error in each of the following.

 a) $(a + 3)^2 = a^2 + 9$ **[4.2c]**
 b) $(a - b)(a - b) = a^2 - b^2$ **[4.2c]**
 c) $(x + 3)(x - 4) = x^2 - 12$ **[4.2b]**
 d) $(p + 7)(p - 7) = p^2 + 49$ **[4.2d]**
 e) $(t - 3)^2 = t^2 - 9$ **[4.2c]**

41. Checking the factorization of a second-degree polynomial by making a single replacement is only a *partial* check. Write an *incorrect* factorization and explain how evaluating both the polynomial and the factorization might catch a possible error. **[4.4a]**

▶ **a**   Factor trinomials of the type $ax^2 + bx + c$, $a \neq 1$, by the FOIL method.

▶ **b**   Factor trinomials of the type $ax^2 + bx + c$, $a \neq 1$, by the ac–method.

Now we learn to factor trinomials of the type $ax^2 + bx + c, a \neq 1$. We use two methods: the FOIL method and the ac-method.

▶ a   The FOIL Method

MULTIPLYING USING FOIL

REVIEW SECTION 4.2

We first consider the **FOIL method** for factoring trinomials of the type $ax^2 + bx + c, a \neq 1$. Consider the following multiplication.

$$\begin{array}{c} \text{F} \qquad \text{O} \qquad \text{I} \qquad \text{L} \\ (3x + 2)(4x + 5) = 12x^2 + 15x + 8x + 10 \\ = 12x^2 + 23x + 10 \end{array}$$

To factor $12x^2 + 23x + 10$, we must reverse what we just did. We look for two binomials whose product is this trinomial. The product of the First terms must be $12x^2$. The product of the Outside terms plus the product of the Inside terms must be $23x$. The product of the Last terms must be 10. In general, finding such an answer involves trial and error. We use the following method.

The Foil Method for Factoring Trinomials

1. Factor out the largest common factor. The remaining trinomial is $ax^2 + bx + c$.

2. Find two First terms whose product is ax^2:

$$(\boxed{}x +)(\boxed{}x +) = ax^2 + bx + c.$$

FOIL

3. Find two Last terms whose product is c:

$$(\ x + \boxed{})(\ x + \boxed{}) = ax^2 + bx + c.$$

FOIL

4. Repeat steps (2) and (3), if necessary, until a combination is found for which the sum of the Outside and Inside products is bx:

$$(\boxed{}x + \boxed{})(\boxed{}x + \boxed{}) = ax^2 + bx + c.$$

I

O

FOIL

5. Check by multiplying.

To the instructor: Here we present two ways to factor general trinomials: the FOIL method and the ac-method. You can teach both methods and let the student use the one he or she prefers or you can select just one for the student.

EXAMPLE 1   Factor: $3x^2 + 10x - 8$.

1. First, we factor out the largest common factor, if any. There is none (other than 1 or -1).

2. Next, we factor the first term, $3x^2$. The only possibility is $3x \cdot x$. The desired factorization is then of the form $(3x + \ )(x + \ )$.

3. We then factor the last term, -8, which is negative. The possibilities are $(-8)(1)$, $8(-1)$, $2(-4)$, and $(-2)(4)$. They can be written in either order.

4. We look for combinations of factors from steps (2) and (3) such that the sum of the outside and the inside products is the middle term, $10x$:

$$(3x - 8)(x + 1) = 3x^2 - 5x - 8;$$
$3x$ ... $-8x$ Wrong middle term

$$(3x + 8)(x - 1) = 3x^2 + 5x - 8;$$
$-3x$ ... $8x$ Wrong middle term

$$(3x + 2)(x - 4) = 3x^2 - 10x - 8;$$
$-12x$ ... $2x$ Wrong middle term

$$(3x - 2)(x + 4) = 3x^2 + 10x - 8.$$
$12x$ ... $-2x$ Correct middle term!

5. *Check*: $(3x - 2)(x + 4) = 3x^2 + 10x - 8$.

> **Now Try Exercise 1.**

EXAMPLE 2   Factor: $18x^6 - 57x^5 + 30x^4$.

1. First, we factor out the largest common factor, if any. The expression $3x^4$ is common to all terms, so we factor it out: $3x^4(6x^2 - 19x + 10)$.

2. Next, we factor the trinomial $6x^2 - 19x + 10$. We factor the first term, $6x^2$, and get $6x \cdot x$, or $3x \cdot 2x$. We then have these as possibilities for factorizations: $(3x + \ )(2x + \ )$ or $(6x + \ )(x + \ )$.

3. We then factor the last term, 10, which is positive. The possibilities are $(10)(1)$, $(-10)(-1)$, $(5)(2)$, and $(-5)(-2)$. They can be written in either order. Since the middle term, $-19x$, is negative, we consider only $(-10)(-1)$ and $(-5)(-2)$.

4. We look for combinations of factors from steps (2) and (3) such that the sum of the outside and the inside products is the middle term, $-19x$. We begin by using these factors with $(3x + \ )(2x + \ )$. Should we not find the correct factorization, we will consider $(6x + \ )(x + \ )$.

$$(3x - 10)(2x - 1) = 6x^2 - 23x + 10;$$
$-3x$ ... $-20x$ Wrong middle term

$$(3x - 1)(2x - 10) = 6x^2 - 32x + 10;$$
$-30x$ ... $-2x$ Wrong middle term

$$(3x - 5)(2x - 2) = 6x^2 - 16x + 10;$$
$-6x$ ... $-10x$ Wrong middle term

$$(3x - 2)(2x - 5) = 6x^2 - 19x + 10$$
$-15x$ ... $-4x$ Correct middle term!

We have a correct answer. We need not consider $(6x + \square)(x + \square)$.

The factorization of $6x^2 - 19x + 10$ is $(3x - 2)(2x - 5)$. But do not forget the common factor! We must include it in order to get a complete factorization of the original trinomial:

$$18x^6 - 57x^5 + 30x^4 = 3x^4(3x - 2)(2x - 5).$$

5. *Check:* $3x^4(3x - 2)(2x - 5) = 3x^4(6x^2 - 19x + 10)$
$$= 18x^6 - 57x^5 + 30x^4.$$

> **Now Try Exercise 13.**

From Examples 1 and 2, we can make some observations that might speed up the factoring.

- Note in Example 1 that changing the signs in the binomial factors changed the signs of the middle terms.
- In Example 2, we examined the signs of the terms before forming possible products. If a and c are both positive, the signs of the factors of c will match the sign of b.
- In Example 2, look again at the possibility $(3x - 1)(2x - 10)$. Without multiplying, we can reject such a possibility, noting that the expression $2x - 10$ has a common factor, 2. But we removed the largest common factor before we began. If this expression were a factorization, then 2 would have to be a common factor along with $3x^4$.

Given that we factored out the largest common factor at the outset, we can now eliminate factorizations that have a common factor.

Tips for Factoring $ax^2 + bx + c$ Using the Foil Method

1. If the largest common factor has been factored out of the original trinomial, then no binomial factor can have a common factor (other than 1 or -1).

2. a) If the signs of all the terms are positive, then the signs of all the terms of the binomial factors are positive.
  b) If a and c are positive and b is negative, then the signs of the factors of c are negative.
  c) If a is positive and c is negative, then the factors of c will have opposite signs.

3. Be systematic about your trials. Keep track of those you have tried and those you have not.

4. Changing the signs of the factors of c will change the sign of the middle term.

Keep in mind that this method of factoring trinomials of the type $ax^2 + bx + c$ involves trial and error. As you practice, you will find that you will need fewer trials to arrive at the factorization.

The procedure considered here can also be applied to a trinomial with more than one variable.

EXAMPLE 3 Factor: $30m^2 + 23mn - 11n^2$.

1. First, we factor out the largest common factor, if any. In this polynomial, there is no common factor (other than 1 or -1).

2. Next, we factor the first term, $30m^2$, and get the following possibilities:

   $30m \cdot m$, $15m \cdot 2m$, $10m \cdot 3m$,   and $6m \cdot 5m$.

   We then have these as possibilities for factorizations:

   $(30m + \ \ )(m + \ \ )$, $(15m + \ \ )(2m + \ \ )$,

   $(10m + \ \ )(3m + \ \ )$, $(6m + \ \ )(5m + \ \ )$.

3. We then factor the last term, $-11n^2$, which is negative. The possibilities are $-11n \cdot n$ and $11n \cdot (-n)$.

4. We look for combinations of factors from steps (2) and (3) such that the sum of the outside and the inside products is the middle term, $23mn$. Since the coefficient of the middle term is positive, let's begin our search using $11n \cdot (-n)$. Should we not find the correct factorization, we will consider $-11n \cdot n$.

   $(30m + 11n)(m - n) = 30m^2 - 19mn - 11n^2$;   **Note that changing the order of $11n$ and $-n$ changes the middle term.**

   $(30m - n)(m + 11n) = 30m^2 + 329mn - 11n^2$;

   $(15m + 11n)(2m - n) = 30m^2 + 7mn - 11n^2$;

   $(15m - n)(2m + 11n) = 30m^2 + 163mn - 11n^2$;

   $(10m + 11n)(3m - n) = 30m^2 + 23mn - 11n^2$ ← **Correct middle term**

   We have a correct answer: $30m^2 + 23mn - 11n^2$. The factorization of $30m^2 + 23mn - 11n^2$ is $(10m + 11n)(3m - n)$.

5. *Check:* $(10m + 11n)(3m - n) = 30m^2 + 23mn - 11n^2$.

   Now Try Exercise 43.

▶ b The *ac*–Method

The second method of factoring trinomials of the type $ax^2 + bx + c$, $a \neq 1$, is known as the ***ac*-method**, or the **grouping method**.

We can factor $x^2 + 7x + 10$ by "splitting" the middle term, $7x$, and using factoring by grouping:

$$x^2 + 7x + 10 = x^2 + 2x + 5x + 10$$
$$= x(x + 2) + 5(x + 2)$$
$$= (x + 2)(x + 5).$$

If the leading coefficient is not 1, as in $6x^2 + 23x + 20$, we use a method for factoring similar to what we just did with $x^2 + 7x + 10$.

The *ac*–Method for Factoring Trinomials

1. Factor out the largest common factor. The remaining trinomial is $ax^2 + bx + c$.
2. Multiply the leading coefficient a and the constant c.
3. Try to factor the product ac so that the sum of the factors is b. That is, find integers p and q such that $pq = ac$ and $p + q = b$.
4. Split the middle term, writing it as a sum using the factors found in step (3).
5. Factor by grouping.
6. Check by multiplying.

EXAMPLE 4 Factor: $6x^2 + 23x + 20$.

1. First, factor out a common factor, if any. There is none (other than 1 or -1).

2. Multiply the leading coefficient, 6, and the constant, 20: $6 \cdot 20 = 120$.

3. Then look for a factorization of 120 in which the sum of the factors is the coefficient of the middle term, 23. Since both 120 and 23 are positive, we need consider only positive factors of 120.

Pairs of Factors	Sums of Factors	Pairs of Factors	Sums of Factors
1, 120	121	5, 24	29
2, 60	62	6, 20	26
3, 40	43	8, 15	23
4, 30	34	10, 12	22

4. Split the middle term: $23x = 8x + 15x$.

5. Factor by grouping:

$$6x^2 + 23x + 20 = 6x^2 + 8x + 15x + 20 \qquad \text{Substituting } 8x + 15x \text{ for } 23x$$
$$= 2x(3x + 4) + 5(3x + 4) \qquad \text{Factoring by grouping}$$
$$= (3x + 4)(2x + 5).$$

We could also split the middle term as $15x + 8x$. We still get the same factorization, although the factors are in a different order:

$$6x^2 + 23x + 20 = 6x^2 + 15x + 8x + 20$$
$$= 3x(2x + 5) + 4(2x + 5)$$
$$= (2x + 5)(3x + 4).$$

6. *Check*: $(3x + 4)(2x + 5) = 6x^2 + 23x + 20$.

Now Try Exercise 23.

EXAMPLE 5 Factor: $6x^4 - 116x^3 - 80x^2$.

1. First, factor out the largest common factor, if any. The expression $2x^2$ is common to all three terms: $2x^2(3x^2 - 58x - 40)$.

2. Now, factor the trinomial $3x^2 - 58x - 40$. Multiply the leading coefficient, 3, and the constant, -40: $3(-40) = -120$.

3. Next, try to factor -120 so that the sum of the factors is -58. Since the coefficient of the middle term, -58, is negative, the negative factor of -120 must have the larger absolute value.

Pairs of Factors	Sums of Factors	Pairs of Factors	Sums of Factors
1, -120	-119	5, -24	-19
2, -60	-58	6, -20	-14
3, -40	-37	8, -15	-7
4, -30	-26	10, -12	-2

4. Split the middle term, $-58x$, as follows: $-58x = 2x - 60x$.

5. Factor by grouping:

$$3x^2 - 58x - 40 = 3x^2 + 2x - 60x - 40 \qquad \text{Substituting } 2x - 60x \text{ for } -58x$$

$$= x(3x + 2) - 20(3x + 2) \qquad \text{Factoring by grouping}$$

$$= (3x + 2)(x - 20).$$

The factorization of $3x^2 - 58x - 40$ is $(3x + 2)(x - 20)$. But don't forget the common factor! The factorization of the original polynomial is

$$6x^4 - 116x^3 - 80x^2 = 2x^2(3x + 2)(x - 20).$$

6. *Check:* $2x^2(3x + 2)(x - 20) = 2x^2(3x^2 - 58x - 40)$
$$= 6x^4 - 116x^3 - 80x^2.$$

> **Now Try Exercise 19.**

4.5 Exercise Set

▶ Reading Check

Choose the word or expression shown under each blank that best completes the description of factoring $2x^2 - 3x + 1$ using the FOIL method.

RC1. The product of the First terms must be

_____.
$2x^2 / -3x / 1$

RC2. The product of the Last terms must be

_____.
$2x^2 / -3x / 1$

RC3. The sum of the Outside and the Inside products

must be _____.
$2x^2 / -3x / 1$

RC4. Both Last terms must be _____.
positive/negative.

Choose the word or expression shown under each blank that best completes the description of factoring $10x^2 + 21x + 2$ using the ac-method.

RC5. Multiply the _____ coefficient 10 and
leading/largest

the _____ 2. The product is 20.
constant/degree

RC6. Find two integers whose _____ is 20
sum/product

and whose _____ is 21. The integers
sum/product

are 20 and 1.

RC7. Split the middle term, _____, writing it
$10x^2 / 21x / 2$

as the sum of $20x$ and x.

RC8. _____ by grouping:
Factor/Add

$$10x^2 + 20x + x + 2 = 10x(x + 2) + 1(x + 2)$$
$$= (x + 2)(10x + 1).$$

---

a, b *Factor.*

1. $3x^2 - 14x - 5$

2. $8x^2 - 6x - 9$

3. $10y^3 + y^2 - 21y$

4. $6x^3 + x^2 - 12x$

5. $3c^2 - 20c + 32$

6. $12b^2 - 8b + 1$

7. $35y^2 + 34y + 8$

8. $9a^2 + 18a + 8$

9. $4t + 10t^2 - 6$

10. $8x + 30x^2 - 6$

11. $8x^2 - 16 - 28x$

12. $18x^2 - 24 - 6x$

13. $18a^2 - 51a + 15$

14. $30a^2 - 85a + 25$

15. $30t^2 + 85t + 25$

16. $18y^2 + 51y + 15$

17. $12x^3 - 31x^2 + 20x$

18. $15x^3 - 19x^2 - 10x$

19. $14x^4 - 19x^3 - 3x^2$

20. $70x^4 - 68x^3 + 16x^2$

21. $3a^2 - a - 4$

22. $6a^2 - 7a - 10$

23. $9x^2 + 15x + 4$

24. $6y^2 - y - 2$

25. $3 + 35z - 12z^2$

26. $8 - 6a - 9a^2$

27. $-4t^2 - 4t + 15$

28. $-12a^2 + 7a - 1$

29. $3x^3 - 5x^2 - 2x$

30. $18y^3 - 3y^2 - 10y$

31. $24x^2 - 2 - 47x$

32. $15y^2 - 10 - 15y$

33. $-8t^3 - 8t^2 + 30t$

34. $-36a^3 + 21a^2 - 3a$

35. $-24x^3 + 2x + 47x^2$

36. $-15y^3 + 10y + 47y^2$

37. $21x^2 + 37x + 12$

38. $10y^2 + 23y + 12$

39. $40x^4 + 16x^2 - 12$

40. $24y^4 + 2y^2 - 15$

41. $12a^2 - 17ab + 6b^2$

42. $20p^2 - 23pq + 6q^2$

43. $2x^2 + xy - 6y^2$

44. $8m^2 - 6mn - 9n^2$

45. $12x^2 - 58xy + 56y^2$

46. $30a^2 + 21ab - 36b^2$

47. $9x^2 - 30xy + 25y^2$

48. $4p^2 + 12pq + 9q^2$

49. $3x^6 + 4x^3 - 4$

50. $2p^8 + 11p^4 + 15$

51. *Height of a Thrown Baseball.*   Suppose that a baseball is thrown upward with an initial velocity of 80 ft/sec from a height of 224 ft. Its height h, in feet, after t seconds is given by the function

$$h(t) = -16t^2 + 80t + 224.$$

a) What is the height of the ball after 0 sec? 1 sec? 3 sec? 4 sec? 6 sec?

b) Find an equivalent expression for $h(t)$ by factoring.

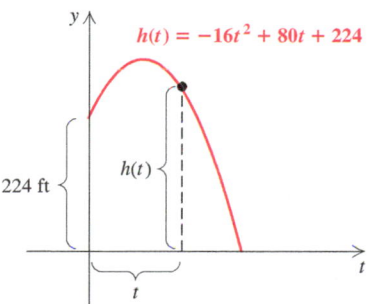

52. *Fireworks.*   Suppose that a bottle rocket is launched upward with an initial velocity of 96 ft/sec from a height of 880 ft. Its height h, in feet, after t seconds is given by the function

$$h(t) = -16t^2 + 96t + 880.$$

a) What is the height of the bottle rocket after 0 sec? 1 sec? 3 sec? 8 sec? 10 sec?

b) Find an equivalent expression for $h(t)$ by factoring.

▶ Skill Maintenance

Solve. **[3.5a]**

53. $x + 2y - z = 0,$
$4x + 2y + 5z = 6,$
$2x - y + z = 5$

54. $2x + y + 2z = 5,$
$4x - 2y - 3z = 5,$
$-8x - y + z = -5$

55. $2x + 9y + 6z = 5,$
$x - y + z = 4,$
$3x + 2y + 3z = 7$

56. $x - 3y + 2z = -8,$
$2x + 3y + z = 17,$
$5x - 2y + 3z = 5$

Determine whether the graphs of the given pairs of lines are parallel or perpendicular. **[2.6d]**

57. $y - 2x = 18,$
$2x - 7 = y$

58. $21x + 7 = -3y,$
$y + 7x = -9$

59. $2x + 5y = 4,$
$2x - 5y = -3$

60. $y + x = 7,$
$y - x = 3$

Find an equation of the line containing the given pair of points. **[2.7c]**

61. $(-2, -3)$ and $(5, -4)$

62. $(2, -3)$ and $(5, -4)$

63. $(-10, 3)$ and $(7, -4)$

64. $\left(-\frac{2}{3}, 1\right)$ and $\left(\frac{4}{3}, -4\right)$

▶ Synthesis

65. Use the TABLE feature and the GRAPH feature of a graphing calculator to check your answers to Exercises 2, 17, and 28.

66. Use the TABLE feature and the GRAPH feature of a graphing calculator to check your answers to Exercises 4, 11, and 32.

Factor. Assume that variables in exponents represent positive integers.

67. $7a^2b^2 + 6 + 13ab$

68. $2x^4y^6 - 3x^2y^3 - 20$

69. $9x^2y^2 - 4 + 5xy$

70. $\frac{1}{4}p^2 - \frac{2}{5}p + \frac{4}{25}$

71. $x^{2a} + 5x^a - 24$

72. $4x^{2a} - 4x^a - 3$

4.6 Special Factoring

▶ **a** Factor trinomial squares.

▶ **b** Factor differences of squares.

▶ **c** Factor certain polynomials with four terms by grouping and possibly using the factoring of a trinomial square or the difference of squares.

▶ **d** Factor sums and differences of cubes.

In this section, we consider some special factoring methods.

▶ a Trinomial Squares

Consider the trinomial $x^2 + 6x + 9$. To factor it, we can look for factors of 9 whose sum is 6. We see that these factors are 3 and 3 and the factorization is

$$x^2 + 6x + 9 = (x + 3)(x + 3) = (x + 3)^2.$$

Note that the result is the square of a binomial. We also call $x^2 + 6x + 9$ a **trinomial square**, or **perfect-square trinomial**. We now develop a faster procedure for factoring trinomial squares.

How can we recognize when an expression is a trinomial square? Look at $A^2 + 2AB + B^2$ and $A^2 - 2AB + B^2$.

How to recognize a **trinomial square**:

a) The two expressions A^2 and B^2 must be squares.

b) There must be no minus sign before either A^2 or B^2.

c) Multiplying A and B (expressions whose squares are A^2 and B^2) and doubling the result gives either the remaining term or its opposite.

EXAMPLES Determine whether the polynomial is a trinomial square.

1. $x^2 + 10x + 25$

a) Two terms are squares: x^2 and 25.

b) There is no minus sign before either x^2 or 25.

c) If we multiply the expressions whose squares are x^2 and 25, x and 5, and double the product, we get $10x$, the remaining term.

Thus this is a trinomial square.

2. $4x + 16 + 3x^2$

a) Only one term, 16, is a square ($3x^2$ is not a square because 3 is not a perfect square and $4x$ is not a square because x is not a square).

Thus this is not a trinomial square.

3. $100y^2 + 81 - 180y$

(It can help to first write this in descending order: $100y^2 - 180y + 81$.)

a) Two of the terms, $100y^2$ and 81, are squares.

b) There is no minus sign before either $100y^2$ or 81.

c) If we multiply the expressions whose squares are $100y^2$ and 81, $10y$ and 9, and double the product, we get $2(10y)(9) = 180y$. This is the opposite of the remaining term, $-180y$.

Thus this is a trinomial square.

The factors of a trinomial square are two identical binomials. We use the following equations.

TRINOMIAL SQUARES

$A^2 + 2AB + B^2 = (A + B)^2;$
$A^2 - 2AB + B^2 = (A - B)^2$

EXAMPLE 4 Factor: $x^2 - 10x + 25$.

$$x^2 - 10x + 25 = (x - 5)^2$$

Note the sign!

We find the square terms and write their square roots with a minus sign between them.

Now Try Exercise 1.

EXAMPLE 5 Factor: $16y^2 + 49 + 56y$.

$$16y^2 + 49 + 56y = 16y^2 + 56y + 49$$

Rewriting in descending order

$$= (4y + 7)^2$$

We find the square terms and write their square roots with a plus sign between them.

Now Try Exercise 7.

EXAMPLE 6 Factor: $-20xy + 4y^2 + 25x^2$.

We have

$$-20xy + 4y^2 + 25x^2 = 4y^2 - 20xy + 25x^2$$

Writing descending order in y

$$= (2y - 5x)^2.$$

This square can also be expressed as

$$25x^2 - 20xy + 4y^2 = (5x - 2y)^2.$$

Now Try Exercise 25.

In factoring, we must always remember to look *first* for the largest factor common to all the terms.

EXAMPLE 7 Factor: $-4y^2 - 144y^8 + 48y^5$.

$$-4y^2 - 144y^8 + 48y^5$$
$$= -4y^2(1 + 36y^6 - 12y^3)$$

Factoring out the common factor $-4y^2$

$$= -4y^2(1 - 12y^3 + 36y^6)$$

Changing order

$$= -4y^2(1 - 6y^3)^2$$

Factoring the trinomial square

Now Try Exercise 17.

▶ **b   Differences of Squares**

The following are *differences of squares:*

$$x^2 - 9, \qquad 49 - 4y^2, \qquad a^2 - 49b^2.$$

To factor a difference of squares such as $x^2 - 9$, we use the following special product:

$$(A + B)(A - B) = A^2 - B^2.$$

Equations are reversible, so we have the following.

> THE PRODUCT OF THE SUM AND
> THE DIFFERENCE OF THE SAME
> TWO TERMS
>
> REVIEW SECTION 4.2

FACTORING A DIFFERENCE OF SQUARES

$A^2 - B^2 = (A + B)(A - B)$

To factor a difference of squares $A^2 - B^2$, we find A and B, which are square roots of the expressions A^2 and B^2. We then use A and B to form two factors. One is the sum $A + B$, and the other is the difference $A - B$.

EXAMPLE 8   Factor: $x^2 - 9$.

$$
\begin{array}{c}
A^2 - B^2 = (A + B)(A - B) \\
x^2 - 9 = x^2 - 3^2 = (x + 3)(x - 3)
\end{array}
$$

> Now Try Exercise 33.

EXAMPLE 9   Factor: $25y^6 - 49x^2$.

$$
\begin{array}{c}
A^2 \quad - \quad B^2 \quad = \quad (A \ + \ B) \ (A \ - \ B) \\
25y^6 - 49x^2 = (5y^3)^2 - (7x)^2 = (5y^3 + 7x)(5y^3 - 7x)
\end{array}
$$

> Now Try Exercise 37.

EXAMPLE 10   Factor: $x^2 - \frac{1}{16}$.

$$x^2 - \tfrac{1}{16} = x^2 - \left(\tfrac{1}{4}\right)^2 = \left(x + \tfrac{1}{4}\right)\left(x - \tfrac{1}{4}\right)$$

> Now Try Exercise 49.

Common factors should always be factored out.

EXAMPLE 11   Factor: $5 - 5x^2y^6$.

There is a common factor, 5.

$$
\begin{aligned}
5 - 5x^2y^6 &= 5(1 - x^2y^6) \\
&= 5[1^2 - (xy^3)^2] \\
&= 5(1 + xy^3)(1 - xy^3)
\end{aligned}
$$

Factoring out the common factor 5
Recognizing the difference of squares; $x^2y^6 = (x^1y^3)^2 = (xy^3)^2$
Factoring the difference of squares

> Now Try Exercise 39.

EXAMPLE 12   Factor: $2x^4 - 8y^4$.

There is a common factor, 2.

$$
\begin{aligned}
2x^4 - 8y^4 &= 2(x^4 - 4y^4) \\
&= 2[(x^2)^2 - (2y^2)^2] \\
&= 2(x^2 + 2y^2)(x^2 - 2y^2)
\end{aligned}
$$

Factoring out the common factor 2
Recognizing the difference of squares
Factoring the difference of squares

> Now Try Exercise 47.

EXAMPLE 13   Factor: $16x^4y - 81y$.

There is a common factor, y.

$$
\begin{aligned}
16x^4y - 81y &= y(16x^4 - 81) && \text{Factoring out the common factor } y \\
&= y[(4x^2)^2 - 9^2] \\
&= y(4x^2 + 9)(4x^2 - 9) && \text{Factoring the difference of squares} \\
&= y(4x^2 + 9)(2x + 3)(2x - 3) && \text{Factoring } 4x^2 - 9, \text{ which is} \\
& && \text{also a difference of squares}
\end{aligned}
$$

> **Now Try Exercise 41.**

> **CAUTION!**   We cannot factor a sum of squares as the square of a binomial. In particular,
>
> $$A^2 + B^2 \neq (A + B)^2.$$

In Example 13, note that $4x^2 + 9$ is a sum of two squares, and it cannot be factored further. Also note that one of the factors, $4x^2 - 9$, could be factored further. Whenever that is possible, you should do so. That way you will be factoring *completely.*

► c   More Factoring by Grouping

Sometimes when factoring a polynomial with four terms, we get a factor that can be factored further using other methods we have learned.

EXAMPLE 14   Factor completely: $x^3 + 3x^2 - 4x - 12$.

$$
\begin{aligned}
x^3 + 3x^2 - 4x - 12 &= x^2(x + 3) - 4(x + 3) \\
&= (x + 3)(x^2 - 4) \\
&= (x + 3)(x + 2)(x - 2)
\end{aligned}
$$

> **Now Try Exercise 53.**

A difference of squares can have more than two terms. For example, one of the squares may be a trinomial. We can factor by a type of grouping.

EXAMPLE 15   Factor completely: $x^2 + 6x + 9 - y^2$.

$$
\begin{aligned}
x^2 + 6x + 9 - y^2 &= (x^2 + 6x + 9) - y^2 && \text{Grouping as a trinomial} \\
& && \text{minus } y^2 \text{ to show a difference} \\
& && \text{of squares} \\
&= (x + 3)^2 - y^2 \\
&= (x + 3 + y)(x + 3 - y)
\end{aligned}
$$

> **Now Try Exercise 61.**

► d   Sums or Differences of Cubes

We can factor the sum or the difference of two expressions that are cubes.

Consider the following products:

$$
\begin{aligned}
(A + B)(A^2 - AB + B^2) &= A(A^2 - AB + B^2) + B(A^2 - AB + B^2) \\
&= A^3 - A^2B + AB^2 + A^2B - AB^2 + B^3 \\
&= A^3 + B^3
\end{aligned}
$$

and

$$
\begin{aligned}
(A - B)(A^2 + AB + B^2) &= A(A^2 + AB + B^2) - B(A^2 + AB + B^2) \\
&= A^3 + A^2B + AB^2 - A^2B - AB^2 - B^3 \\
&= A^3 - B^3.
\end{aligned}
$$

The above equations (reversed) show how we can factor a sum or a difference of two cubes. Each factors as a product of a binomial and a trinomial.

N	N^3
0.2	0.008
0.1	0.001
0	0
1	1
2	8
3	27
4	64
5	125
6	216
7	343
8	512
9	729
10	1000

> **SUM OR DIFFERENCE OF CUBES**
>
> $A^3 + B^3 = (A + B)(A^2 - AB + B^2);$
> $A^3 - B^3 = (A - B)(A^2 + AB + B^2)$

Note that what we are considering here is a sum or a difference of cubes. We are not cubing a binomial. For example, $(A + B)^3$ is *not* the same as $A^3 + B^3$. The table of cubes in the margin is helpful.

EXAMPLE 16 Factor: $x^3 - 27$.

We have

$$\overset{A^3 - B^3}{\downarrow \quad \downarrow}$$
$$x^3 - 27 = x^3 - 3^3.$$

In one set of parentheses, we write the cube root of the first term, x. Then we write the cube root of the second term, -3. This gives us the expression $x - 3$:

$$(x - 3)(\qquad).$$

To get the next factor, we think of $x - 3$ and do the following:

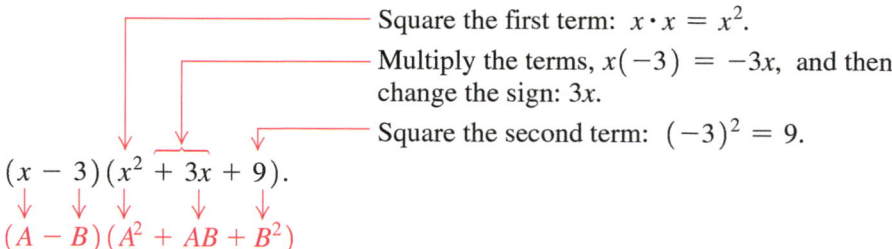

- Square the first term: $x \cdot x = x^2$.
- Multiply the terms, $x(-3) = -3x$, and then change the sign: $3x$.
- Square the second term: $(-3)^2 = 9$.

$$(x - 3)(x^2 + 3x + 9).$$
$$(A - B)(A^2 + AB + B^2)$$

Note that we cannot factor $x^2 + 3x + 9$. It is not a trinomial square nor can it be factored by trial and error. Check this on your own.

▶ **Now Try Exercise 71.**

EXAMPLE 17 Factor: $125x^3 + y^3$.

We have

$$125x^3 + y^3 = (5x)^3 + y^3.$$

In one set of parentheses, we write the cube root of the first term, $5x$. Then we write the cube root of the second term, y. This gives us the expression $5x + y$:

$$(5x + y)(\qquad).$$

To get the next factor, we think of $5x + y$ and do the following:

- Square the first term: $(5x)(5x) = 25x^2$.
- Multiply the terms, $5x \cdot y = 5xy$, and then change the sign: $-5xy$.
- Square the second term: $y \cdot y = y^2$.

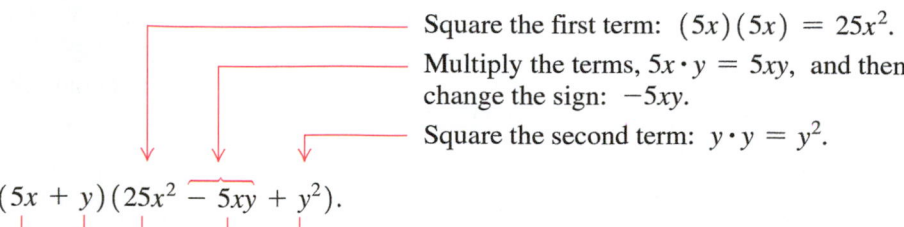

$$(5x + y)(25x^2 - 5xy + y^2).$$
$$(A + B)(A^2 - AB + B^2)$$

▶ **Now Try Exercise 77.**

EXAMPLE 18 Factor: $128y^7 - 250x^6y$.

We first look for the largest common factor:

$$128y^7 - 250x^6y = 2y(64y^6 - 125x^6)$$
$$= 2y[(4y^2)^3 - (5x^2)^3]$$
$$= 2y(4y^2 - 5x^2)(16y^4 + 20x^2y^2 + 25x^4).$$

> **Now Try Exercise 91.**

EXAMPLE 19 Factor: $a^6 - b^6$.

We factor a difference of squares:

$$a^6 - b^6 = (a^3)^2 - (b^3)^2 = (a^3 + b^3)(a^3 - b^3).$$

One factor is a sum of two cubes, and the other factor is a difference of two cubes. We factor them:

$$a^6 - b^6 = (a + b)(a^2 - ab + b^2)(a - b)(a^2 + ab + b^2).$$

We have now factored completely.

> **Now Try Exercise 95.**

In Example 19, had we thought of factoring first as a difference of two cubes, we would have had

$$(a^2)^3 - (b^2)^3 = (a^2 - b^2)(a^4 + a^2b^2 + b^4)$$
$$= (a + b)(a - b)(a^4 + a^2b^2 + b^4).$$

In this case, we might have missed some factors; $a^4 + a^2b^2 + b^4$ can be factored as $(a^2 - ab + b^2)(a^2 + ab + b^2)$, but we probably would not have known to do such factoring.

When you can factor as either a difference of squares or a difference of cubes, factor as a difference of squares first.

EXAMPLE 20 Factor: $64a^6 - 729b^6$.

We have

$$64a^6 - 729b^6 = (8a^3)^2 - (27b^3)^2$$
$$= (8a^3 - 27b^3)(8a^3 + 27b^3) \quad \text{Factoring a difference of squares}$$
$$= [(2a)^3 - (3b)^3][(2a)^3 + (3b)^3].$$

Each factor is a sum or a difference of cubes. We factor each:

$$= (2a - 3b)(4a^2 + 6ab + 9b^2)(2a + 3b)(4a^2 - 6ab + 9b^2).$$

> **Now Try Exercise 97.**

Factoring Summary

Sum of cubes:	$A^3 + B^3 = (A + B)(A^2 - AB + B^2)$;
Difference of cubes:	$A^3 - B^3 = (A - B)(A^2 + AB + B^2)$;
Difference of squares:	$A^2 - B^2 = (A + B)(A - B)$;
Sum of squares:	$A^2 + B^2$ cannot be factored as the square of a binomial: $A^2 + B^2 \neq (A + B)^2$.

1

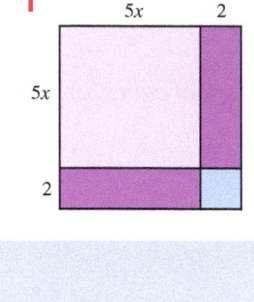

2

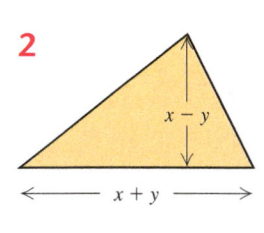

3

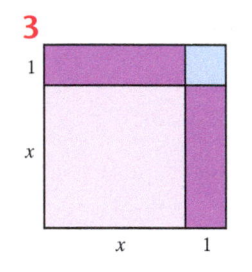

4

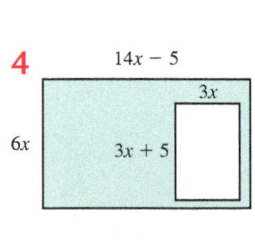

5

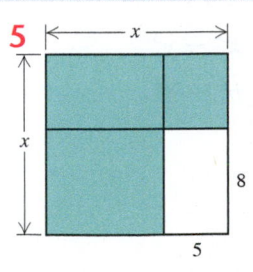

Visualizing for Success

In each of Exercises 1–10, find two algebraic expressions from the following list for the shaded area of the figure.

A. $(5x + 2)^2$

B. $13x$

C. $400 - 4x^2$

D. $x^2 - (x - 2y)^2$

E. $25x^2 + 20x + 4$

F. $\frac{1}{2}(x^2 - y^2)$

G. $(x + 1)^2$

H. $4y(x - y)$

I. $4(10 - x)(10 + x)$

J. $\frac{1}{2}(x - y)(x + y)$

K. $x^2 + 2x + 1$

L. $6x(14x - 5) - 3x(3x + 5)$

M. $x^2 + 9x + 20$

N. $(x - 2)^2$

O. $(x + 4)(x + 5)$

P.
$8(x - 5) + (x - 5)(x - 8) + 5(x - 8)$

Q. $x^2 - 40$

R. $5x + 8x$

S. $15x(5x - 3)$

T. $x^2 - 4x + 4$

6

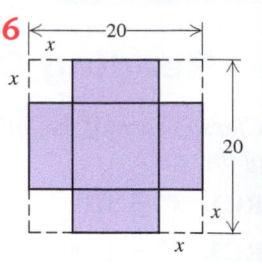

7

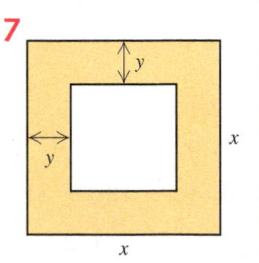

8

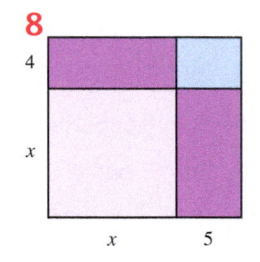

9

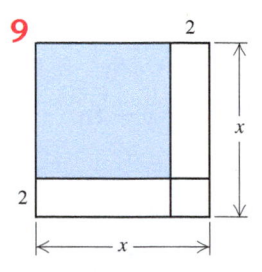

10

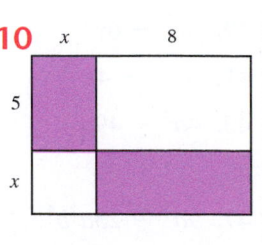

4.6 Exercise Set

▶ **Reading Check**

Classify each of the following as a trinomial square, a difference of squares, a difference of cubes, a sum of cubes, or none of these.

RC1. $x^2 - 100$ **RC2.** $x^2 - 20x + 100$

RC3. $x^3 - 1000$ **RC4.** $4x^2 + 49$

RC5. $16x^2 + 40x + 25$ **RC6.** $x^2 - 9x + 6$

RC7. $27x^3 + 1$ **RC8.** $36x^4 - 1$

a *Factor.*

1. $x^2 - 4x + 4$ **2.** $y^2 - 16y + 64$

3. $y^2 + 18y + 81$ **4.** $x^2 + 8x + 16$

5. $x^2 + 1 + 2x$ **6.** $x^2 + 1 - 2x$

7. $9y^2 + 12y + 4$ **8.** $25x^2 - 60x + 36$

9. $-18y^2 + y^3 + 81y$ **10.** $24a^2 + a^3 + 144a$

11. $12a^2 + 36a + 27$ **12.** $20y^2 + 100y + 125$

13. $2x^2 - 40x + 200$ **14.** $32x^2 + 48x + 18$

15. $1 - 8d + 16d^2$ **16.** $64 + 25y^2 - 80y$

17. $3a^3 - 6a^2 + 3a$ **18.** $5c^3 + 20c^2 + 20c$

19. $0.25x^2 + 0.30x + 0.09$ **20.** $0.04x^2 - 0.28x + 0.49$

21. $p^2 - 2pq + q^2$ **22.** $m^2 + 2mn + n^2$

23. $a^2 + 4ab + 4b^2$ **24.** $49x^2 - 14xy + y^2$

25. $25a^2 - 30ab + 9b^2$ **26.** $49p^2 - 84pw + 36w^2$

27. $y^6 + 26y^3 + 169$ **28.** $p^6 - 10p^3 + 25$

29. $16x^{10} - 8x^5 + 1$ **30.** $9x^{10} + 12x^5 + 4$

31. $x^4 + 2x^2y^2 + y^4$ **32.** $a^6 - 2a^3b^4 + b^8$

b *Factor.*

33. $p^2 - 49$ **34.** $m^2 - 64$

35. $y^4 - 8y^2 + 16$ **36.** $y^4 - 18y^2 + 81$

37. $p^2q^2 - 25$ **38.** $a^2b^2 - 81$

39. $6x^2 - 6y^2$ **40.** $8x^2 - 8y^2$

41. $4xy^4 - 4xz^4$ **42.** $25ab^4 - 25az^4$

43. $4a^3 - 49a$ **44.** $9x^3 - 25x$

45. $3x^8 - 3y^8$ **46.** $2a^9 - 32a$

47. $9a^4 - 25a^2b^4$ **48.** $16x^6 - 121x^2y^4$

49. $\frac{1}{36} - z^2$ **50.** $\frac{1}{100} - y^2$

51. $0.04x^2 - 0.09y^2$ **52.** $0.01x^2 - 0.04y^2$

c *Factor.*

53. $m^3 - 7m^2 - 4m + 28$ **54.** $x^3 + 8x^2 - x - 8$

55.   **56.** $p^2q - 25q + 3p^2 - 75$

57. $(a + b)^2 - 100$ **58.** $(p - 7)^2 - 144$

59. $144 - (p - 8)^2$ **60.** $100 - (x - 4)^2$

61. $a^2 + 2ab + b^2 - 9$

62. $x^2 - 2xy + y^2 - 25$

63. $r^2 - 2r + 1 - 4s^2$

64. $c^2 + 4cd + 4d^2 - 9p^2$

65. $2m^2 + 4mn + 2n^2 - 50b^2$

66. $12x^2 + 12x + 3 - 3y^2$

67. $9 - (a^2 + 2ab + b^2)$

68. $16 - (x^2 - 2xy + y^2)$

d *Factor.*

69. $z^3 + 27$ **70.** $a^3 + 8$

71. $x^3 - 1$ **72.** $c^3 - 64$

73. $8 - 27b^3$ **74.** $64 - 125x^3$

75. $8a^3 + 1$ **76.** $27x^3 + 1$

77. $8x^3 + 27$ **78.** $27y^3 + 64$

79. $a^3 - b^3$ **80.** $x^3 - y^3$

81. $a^3 + \frac{1}{8}$ **82.** $b^3 + \frac{1}{27}$

83. $x^3 + 0.001$

84. $y^3 + 0.125$

85. $2y^3 - 128$

86. $3z^3 - 3$

87. $24a^3 + 3$

88. $54x^3 + 2$

89. $rs^3 + 64r$

90. $ab^3 + 125a$

91. $5x^5 - 40x^2z^3$

92. $2y^3z^4 - 54z^7$

93. $64x^6 - 8t^6$

94. $125c^6 - 8d^6$

95. $z^6 - 1$ **96.** $t^6 + 1$

97. $t^6 + 64y^6$ **98.** $p^6 - q^6$

99. $8w^9 - z^9$ **100.** $a^9 + 64b^9$

101. $\frac{1}{8}c^3 + d^3$ **102.** $\frac{27}{125}x^3 - y^3$

103. $0.001x^3 - 0.008y^3$ **104.** $0.125r^3 - 0.216s^3$

▶ Skill Maintenance

Solve. [3.2a], [3.3a]

105. $7x - 2y = -11,$
$2x + 7y = 18$

106. $y = 3x - 8,$
$4x - 6y = 100$

107. $x - y = -12,$
$x + y = 14$

108. $7x - 2y = -11,$
$2y - 7x = -18$

Graph the given system of inequalities and determine coordinates of any vertices formed. [3.7c]

109. $x - y \le 5,$
$x + y \ge 3$

110. $x - y \le 5,$
$x + y \ge 3,$
$x \le 6$

111. $x - y \ge 5,$
$x + y \le 3,$
$x \ge 1$

112. $x - y \ge 5,$
$x + y \le 3$

Given the line and a point not on the line, find an equation through the point parallel to the given line, and find an equation through the point perpendicular to the given line. [2.7d]

113. $x - y = 5; (-2, -4)$

114. $2x - 3y = 6; (1, -7)$

115. $y = -\frac{1}{2}x + 3; (4, 5)$

116. $x - 4y = -10; (6, 0)$

▶ Synthesis

117. Given that $P(x) = x^3$, use factoring to simplify $P(a + h) - P(a)$.

118. Given that $P(x) = x^4$, use factoring to simplify $P(a + h) - P(a)$.

119. *Volume of Carpeting.* The volume of a carpet that is rolled up can be estimated by the polynomial $\pi R^2 h - \pi r^2 h$ where R and r are as shown in figure below.

a) Factor the polynomial.

b) Use both the original form and the factored form to find the volume of a roll for which $R = 50$ cm, $r = 10$ cm, and $h = 4$ m. Use 3.14 for π.

120. Show how the geometric model below can be used to verify the formula for factoring $a^3 - b^3$.

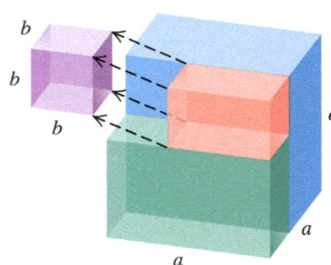

Factor. Assume that variables in exponents represent positive integers.

121. $5c^{100} - 80d^{100}$

122. $9x^{2n} - 6x^n + 1$

123. $x^{6a} + y^{3b}$

124. $a^3x^3 - b^3y^3$

125. $3x^{3a} + 24y^{3b}$

126. $\frac{8}{27}x^3 + \frac{1}{64}y^3$

127. $\frac{1}{24}x^3y^3 + \frac{1}{3}z^3$

128. $7x^3 - \frac{7}{8}$

129. $(x + y)^3 - x^3$

130. $(1 - x)^3 + (x - 1)^6$

131. $(a + 2)^3 - (a - 2)^3$

132. $y^4 - 8y^3 - y + 8$

| **4.7** | **Factoring: A General Strategy** |

▶ **a**   Factor polynomials completely using any of the methods considered in this chapter.

▶ a   A General Factoring Strategy

Factoring is an important algebraic skill, used for solving equations and many other manipulations of algebraic symbolism. We now consider polynomials of many types and learn to use a general strategy for factoring. The key is to recognize the type of polynomial to be factored.

A Strategy for Factoring

a) Always look for a *common factor* (other than 1 or -1). If there are any, factor out the largest one.

b) Then look at the number of terms.

Two terms: Try factoring as a difference of squares first. Next, try factoring as a sum or a difference of cubes. Do *not* try to factor a *sum* of squares: $A^2 + B^2$.

Three terms: Determine whether the expression is a trinomial square. If it is, you know how to factor. If not, try the trial-and-error method or the *ac*-method.

Four or more terms: Try factoring by grouping and removing a common binomial factor. Next, try grouping into a difference of squares, one of which is a trinomial.

c) Always *factor completely*. If a factor with more than one term can be factored, you should factor it.

d) Always *check* by multiplying.

EXAMPLE 1   Factor: $10a^2x - 40b^2x$.

a) We look first for a common factor:

$$10x(a^2 - 4b^2).   \text{Factoring out the largest common factor}$$

b) The factor $a^2 - 4b^2$ has only two terms. It is a difference of squares. We factor it, keeping the common factor: $10x(a + 2b)(a - 2b)$.

c) Have we factored completely? Yes, because none of the factors with more than one term can be factored further using polynomials of smaller degree.

d) *Check*: $10x(a + 2b)(a - 2b) = 10x(a^2 - 4b^2) = 10xa^2 - 40xb^2$, or $10a^2x - 40b^2x$.

> **Now Try Exercise 43.**

EXAMPLE 2   Factor: $10x^6 + 40y^2$.

a) We remove the largest common factor: $10(x^6 + 4y^2)$.

b) In the parentheses, there are two terms, a sum of squares, which cannot be factored.

c) We have factored $10x^6 + 40y^2$ completely as $10(x^6 + 4y^2)$.

d) *Check*: $10(x^6 + 4y^2) = 10x^6 + 40y^2$.

EXAMPLE 3 Factor: $2x^2 + 50a^2 - 20ax$.

a) We remove the largest common factor: $2(x^2 + 25a^2 - 10ax)$.

b) In the parentheses, there are three terms. The trinomial is a square. We factor it: $2(x - 5a)^2$.

c) None of the factors with more than one term can be factored further.

d) *Check*: $2(x - 5a)^2 = 2(x^2 - 10ax + 25a^2) = 2x^2 - 20ax + 50a^2$, or $2x^2 + 50a^2 - 20ax$.

> **Now Try Exercise 39.**

EXAMPLE 4 Factor: $3x + 12 + ax^2 + 4ax$.

a) There is no common factor (other than 1 or −1).

b) There are four terms. We try grouping to remove a common binomial factor:

$$3(x + 4) + ax(x + 4) \qquad \text{\color{red}{Factoring two grouped binomials}}$$
$$= (x + 4)(3 + ax). \qquad \text{\color{red}{Factoring out the common binomial factor}}$$

c) None of the factors with more than one term can be factored further.

d) *Check*: $(x + 4)(3 + ax) = 3x + ax^2 + 12 + 4ax$, or $3x + 12 + ax^2 + 4ax$.

> **Now Try Exercise 29.**

EXAMPLE 5 Factor: $x^6 - y^6$.

a) We look for a common factor. There isn't one (other than 1 or −1).

b) There are only two terms. The binomial is a difference of squares and a difference of cubes. We factor first as a difference of squares: $(x^3)^2 - (y^3)^2$. We factor it: $(x^3 + y^3)(x^3 - y^3)$.

c) One factor is a sum of two cubes, and the other factor is a difference of two cubes. We factor them:

$$x^6 - y^6 = (x^3 + y^3)(x^3 - y^3)$$

$$= (x + y)(x^2 - xy + y^2)(x - y)(x^2 + xy + y^2).$$

We have now factored completely because none of the factors can be factored further using polynomials of smaller degree.

d) *Check*: $(x + y)(x^2 - xy + y^2)(x - y)(x^2 + xy + y^2)$
$= (x^3 + y^3)(x^3 - y^3) = (x^3)^2 - (y^3)^2 = x^6 - y^6$.

> **Now Try Exercise 21.**

EXAMPLE 6 Factor: $x^3 - xy^2 + x^2y - y^3$.

a) There is no common factor (other than 1 or −1).

b) There are four terms. We factor by grouping:

$$x(x^2 - y^2) + y(x^2 - y^2) \qquad \text{\color{red}{Factoring two grouped binomials}}$$
$$= (x^2 - y^2)(x + y). \qquad \text{\color{red}{Factoring out the common binomial factor}}$$

c) The factor $x^2 - y^2$ can be factored further, giving

$$(x + y)(x - y)(x + y). \qquad \text{\color{red}{Factoring a difference of squares}}$$

None of the factors with more than one term can be factored further, so we have factored completely.

d) *Check:* $(x + y)(x - y)(x + y) = (x^2 - y^2)(x + y)$
$$= x^3 + x^2y - y^2x - y^3, \text{or}$$
$$x^3 - xy^2 + x^2y - y^3.$$

> **Now Try Exercise 35.**

EXAMPLE 7   Factor: $6x^2 - 20x - 16$.

a) We remove the largest common factor: $2(3x^2 - 10x - 8)$.

b) In the parentheses, there are three terms. The trinomial is not a square. We factor:
$2(x - 4)(3x + 2)$.

c) We cannot factor further.

d) *Check:* $2(x - 4)(3x + 2) = 2(3x^2 - 10x - 8) = 6x^2 - 20x - 16$.

> **Now Try Exercise 9.**

EXAMPLE 8   Factor: $y^2 - 9a^2 + 12y + 36$.

a) There is no common factor (other than 1 or -1).

b) There are four terms. We try grouping to remove a common binomial factor, but that is not possible. We try grouping as a difference of squares:

$$(y^2 + 12y + 36) - 9a^2 = (y + 6)^2 - (3a)^2$$
$$= (y + 6 + 3a)(y + 6 - 3a). \quad \text{\textcolor{red}{Factoring the difference of squares}}$$

c) No factor with more than one term can be factored further.

d) *Check:* $(y + 6 + 3a)(y + 6 - 3a) = [(y + 6) + 3a][(y + 6) - 3a]$
$$= (y + 6)^2 - (3a)^2$$
$$= y^2 + 12y + 36 - 9a^2, \text{or}$$
$$y^2 - 9a^2 + 12y + 36.$$

> **Now Try Exercise 23.**

4.7   Exercise Set

▶ Reading Check

Choose from the column on the right the word that best completes each step in the factoring strategy.

RC1. Always look first for a _____ factor.

RC2. If there are two terms, determine whether the binomial is a _____ of squares, a sum of cubes, or a difference of cubes.

RC3. If there are three terms, determine whether the trinomial is a _____.

RC4. If there are four terms, try factoring by _____.

RC5. Always factor _____.

RC6. Always _____ by multiplying.

check
common
completely
difference
grouping
square

a *Factor completely.*

1. $y^2 - 225$
2. $x^2 - 400$
3. $2x^2 + 11x + 12$
4. $8a^2 + 18a - 5$
5. $5x^4 - 20$
6. $3xy^2 - 75x$
7. $p^2 + 36 + 12p$
8. $a^2 + 49 + 14a$
9. $2x^2 - 10x - 132$
10. $3y^2 - 15y - 252$
11. $9x^2 - 25y^2$
12. $16a^2 - 81b^2$
13. $4m^4 - 100$
14. $2x^2 - 288$
15. $6w^2 + 12w - 18$
16. $8z^2 - 8z - 16$
17. $2xy^2 - 50x$
18. $3a^3b - 108ab$
19. $225 - (a - 3)^2$
20. $625 - (t - 10)^2$
21. $m^6 - 1$
22. $64t^6 - 1$
23. $x^2 + 6x - y^2 + 9$
24. $t^2 + 10t - p^2 + 25$
25. $250x^3 - 128y^3$
26. $27a^3 - 343b^3$
27. $8m^3 + m^6 - 20$
28. $-37x^2 + x^4 + 36$
29. $ac + cd - ab - bd$
30. $xw - yw + xz - yz$
31. $50b^2 - 5ab - a^2$
32. $9c^2 + 12cd - 5d^2$
33. $-7x^2 + 2x^3 + 4x - 14$
34. $9m^2 + 3m^3 + 8m + 24$
35. $2x^3 + 6x^2 - 8x - 24$
36. $3x^3 + 6x^2 - 27x - 54$
37. $16x^3 + 54y^3$
38. $250a^3 + 54b^3$
39. $6y - 60x^2y - 9xy$
40. $2b - 28a^2b + 10ab$
41. $a^8 - b^8$
42. $2x^4 - 32$
43. $a^3b - 16ab^3$
44. $x^3y - 25xy^3$
45. $\frac{1}{16}x^2 - \frac{1}{6}xy^2 + \frac{1}{9}y^4$
46. $36x^2 + 15x + \frac{25}{16}$
47. $5x^3 - 5x^2y - 5xy^2 + 5y^3$
48. $a^3 - ab^2 + a^2b - b^3$
49. $42ab + 27a^2b^2 + 8$
50. $-23xy + 20x^2y^2 + 6$
51. $8y^4 - 125y$
52. $64p^4 - p$
53. $a^2 - b^2 - 6b - 9$
54. $m^2 - n^2 - 8n - 16$

55. $q^2 - 10q + 25 - r^2$
56. $y^2 - 14y + 49 - z^2$

▶ Skill Maintenance

Solve. **[3.2b]**

57. *Exam Scores.* There are 75 questions on a college entrance examination. Two points are awarded for each correct answer, and one half point is deducted for each incorrect answer. A score of 100 indicates how many correct and how many incorrect answers, assuming that all questions are answered?

58. *Perimeter.* A pentagon with all five sides the same length has the same perimeter as an octagon with all eight sides the same length. One side of the pentagon is 2 less than three times the length of one side of the octagon. Find the perimeters.

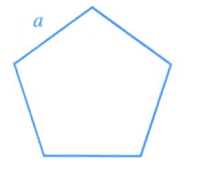 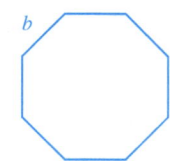

▶ Synthesis

Factor. Assume that variables in exponents represent natural numbers.

59. $30y^4 - 97xy^2 + 60x^2$
60. $3x^2y^2z + 25xyz^2 + 28z^3$
61. $5x^3 - \frac{5}{27}$
62. $9y^3 - \frac{9}{1000}$
63. $(x - p)^2 - p^2$
64. $s^6 - 729t^6$
65. $(y - 1)^4 - (y - 1)^2$
66. $27x^{6s} + 64y^{3t}$
67. $4x^2 + 4xy + y^2 - r^2 + 6rs - 9s^2$
68. $c^4d^4 - a^{16}$
69. $c^{2w+1} + 2c^{w+1} + c$
70. $24x^{2a} - 6$
71. $3(x + 1)^2 + 9(x + 1) - 12$
72. $8(a - 3)^2 - 64(a - 3) + 128$
73. $x^6 - 2x^5 + x^4 - x^2 + 2x - 1$
74. $1 - \dfrac{x^{27}}{1000}$
75. $y^9 - y$
76. $(m - 1)^3 - (m + 1)^3$

Applications of Polynomial Equations and Functions

▶ **a** Solve quadratic and other polynomial equations by first factoring and then using the principle of zero products.

▶ **b** Solve applied problems involving quadratic and other polynomial equations that can be solved by factoring.

Whenever two polynomials are set equal to each other, we have a **polynomial equation**. Some examples of polynomial equations are

$$4x^3 + x^2 + 5x = 6x - 3,$$
$$x^2 - x = 6,$$
and $\quad 3y^4 + 2y^2 + 2 = 0.$

A second-degree polynomial equation in one variable is often called a **quadratic equation**. Of the equations listed above, only $x^2 - x = 6$ is a quadratic equation.

Polynomial equations, and quadratic equations in particular, occur frequently in applications, so the ability to solve them is an important skill. One way of solving certain polynomial equations involves factoring.

▶ a The Principle of Zero Products

When we multiply two or more numbers, if either factor is 0, then the product is 0. Conversely, if a product is 0, then at least one of the factors must be 0. This property of 0 gives us a new principle for solving equations.

THE PRINCIPLE OF ZERO PRODUCTS

For any real numbers a and b:

If $ab = 0$, then $a = 0$ *or* $b = 0$ (or both).

If $a = 0$ *or* $b = 0$, then $ab = 0$.

To solve an equation using the principle of zero products, we first write it in *standard form*: with 0 on one side of the equation and the leading coefficient positive.

EXAMPLE 1 Solve: $x^2 - x = 6$.

In order to use the principle of zero products, we must have 0 on one side of the equation, so we subtract 6 on both sides:

$$x^2 - x - 6 = 0. \qquad \textcolor{red}{\text{Getting 0 on one side}}$$

We need a factorization on the other side, so we factor the polynomial:

$$(x - 3)(x + 2) = 0. \qquad \textcolor{red}{\text{Factoring}}$$

We now have two expressions, $x - 3$ and $x + 2$, whose product is 0. Using the principle of zero products, we set each expression or factor equal to 0:

$$x - 3 = 0 \quad or \quad x + 2 = 0. \qquad \textcolor{red}{\text{Using the principle of zero products}}$$

This gives us two simple linear equations. We solve them separately,

$$x = 3 \quad or \quad x = -2,$$

and check in the original equation as follows.

Check:

$$\frac{x^2 - x = 6}{3^2 - 3 \;?\; 6}$$
$$9 - 3$$
$$6 \qquad \text{TRUE}$$

$$\frac{x^2 - x = 6}{(-2)^2 - (-2) \;?\; 6}$$
$$4 + 2$$
$$6 \qquad \text{TRUE}$$

The numbers 3 and -2 are both solutions.

> To solve an equation using the principle of zero products:
> 1. Obtain a 0 on one side of the equation.
> 2. Factor the other side.
> 3. Set each factor equal to 0.
> 4. Solve the resulting equations.

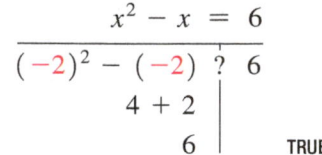 **Now Try Exercise 1.**

When you solve an equation using the principle of zero products, you can always check by substitution as we did in Example 1.

CAUTION! When we are using the principle of zero products, it is important to be sure that there is a 0 on one side of the equation. If neither side of the equation is 0, the procedure will not work.

For example, consider $x^2 - x = 6$ in Example 1 as

$$x(x - 1) = 6.$$

Suppose we reasoned as follows, setting factors equal to 6:

$$x = 6 \quad or \quad x - 1 = 6 \qquad \text{\textcolor{red}{This step is incorrect!}}$$
$$x = 7.$$

Neither 6 nor 7 checks, as shown below:

$$\frac{x(x - 1) = 6}{6(6 - 1) \;?\; 6}$$
$$6(5)$$
$$30 \qquad \text{\textcolor{red}{FALSE}}$$

$$\frac{x(x - 1) = 6}{7(7 - 1) \;?\; 6}$$
$$7(6)$$
$$42 \qquad \text{\textcolor{red}{FALSE}}$$

EXAMPLE 2 Solve: $7y + 3y^2 = -2$.

Since there must be a 0 on one side of the equation, we add 2 to get 0 on the right-hand side and arrange in descending order. Then we factor and use the principle of zero products.

$$7y + 3y^2 = -2$$
$$3y^2 + 7y + 2 = 0 \qquad \text{\textcolor{red}{Getting 0 on one side}}$$
$$(3y + 1)(y + 2) = 0 \qquad \text{\textcolor{red}{Factoring}}$$
$$3y + 1 = 0 \quad or \quad y + 2 = 0 \qquad \text{\textcolor{red}{Using the principle of zero products}}$$
$$y = -\tfrac{1}{3} \quad or \qquad y = -2$$

The solutions are $-\tfrac{1}{3}$ and -2.

 **Now Try Exercise 25.**

EXAMPLE 3   Solve: $5b^2 = 10b$.

$$5b^2 = 10b$$
$$5b^2 - 10b = 0 \qquad \text{Getting 0 on one side}$$
$$5b(b - 2) = 0 \qquad \text{Factoring}$$
$$5b = 0 \quad or \quad b - 2 = 0 \qquad \text{Using the principle of zero products}$$
$$b = 0 \quad or \qquad b = 2$$

The solutions are 0 and 2.   **Now Try Exercise 31.**

EXAMPLE 4   Solve: $x^2 - 6x + 9 = 0$.

$$x^2 - 6x + 9 = 0 \qquad \text{Getting 0 on one side}$$
$$(x - 3)(x - 3) = 0 \qquad \text{Factoring}$$
$$x - 3 = 0 \quad or \quad x - 3 = 0 \qquad \text{Using the principle of zero products}$$
$$x = 3 \quad or \qquad x = 3$$

There is only one solution, 3.   **Now Try Exercise 5.**

EXAMPLE 5   Solve: $3x^3 - 9x^2 = 30x$.

$$3x^3 - 9x^2 = 30x$$
$$3x^3 - 9x^2 - 30x = 0 \qquad \text{Getting 0 on one side}$$
$$3x(x^2 - 3x - 10) = 0 \qquad \text{Factoring out a common factor}$$
$$3x(x + 2)(x - 5) = 0 \qquad \text{Factoring the trinomial}$$
$$3x = 0 \quad or \quad x + 2 = 0 \quad or \quad x - 5 = 0 \qquad \text{Using the principle of zero products}$$
$$x = 0 \quad or \qquad x = -2 \quad or \qquad x = 5$$

The solutions are 0, -2, and 5.   **Now Try Exercise 35.**

EXAMPLE 6   Given that $f(x) = 3x^2 - 4x$, find all values of x for which $f(x) = 4$.

We want all numbers x for which $f(x) = 4$. Since $f(x) = 3x^2 - 4x$, we must have

$$3x^2 - 4x = 4 \qquad \text{Setting } f(x) \text{ equal to 4}$$
$$3x^2 - 4x - 4 = 0 \qquad \text{Getting 0 on one side}$$
$$(3x + 2)(x - 2) = 0 \qquad \text{Factoring}$$
$$3x + 2 = 0 \quad or \quad x - 2 = 0$$
$$x = -\tfrac{2}{3} \quad or \qquad x = 2.$$

We can check as follows.

$$f\left(-\tfrac{2}{3}\right) = 3\left(-\tfrac{2}{3}\right)^2 - 4\left(-\tfrac{2}{3}\right) = 3 \cdot \tfrac{4}{9} + \tfrac{8}{3} = \tfrac{4}{3} + \tfrac{8}{3} = \tfrac{12}{3} = 4;$$
$$f(2) = 3(2)^2 - 4(2) = 3 \cdot 4 - 8 = 12 - 8 = 4$$

To have $f(x) = 4$, we must have $x = -\tfrac{2}{3}$ or $x = 2$.

Now Try Exercise 47.

EXAMPLE 7 Find the domain of F if $F(x) = \dfrac{x - 2}{x^2 + 2x - 15}$.

The domain of F is the set of all values for which

$$\frac{x - 2}{x^2 + 2x - 15}$$

is a real number. Since division by 0 is undefined, $F(x)$ cannot be calculated for any x-value for which the denominator, $x^2 + 2x - 15$, is 0. To make sure that these values are *excluded*, we solve:

$$x^2 + 2x - 15 = 0 \qquad \text{\textcolor{red}{Setting the denominator equal to 0}}$$
$$(x - 3)(x + 5) = 0 \qquad \text{\textcolor{red}{Factoring}}$$
$$x - 3 = 0 \quad or \quad x + 5 = 0$$
$$x = 3 \quad or \quad x = -5. \qquad \text{\textcolor{red}{These are the values to \textit{exclude}.}}$$

The domain of F is $\{x \mid x$ is a real number *and* $x \neq -5$ *and* $x \neq 3\}$.

> **Now Try Exercise 51.**

EXAMPLE 8 Use the graph of $f(x) = x^2 + 6x + 8$ to find the x-intercepts of the graph and the solutions of the equation $x^2 + 6x + 8 = 0$.

The graph of the function $f(x) = x^2 + 6x + 8$ and its x-intercepts are shown below.

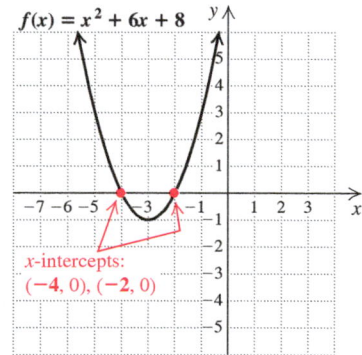

The x-intercepts are $(-4, 0)$ and $(-2, 0)$. These pairs are also the points of intersection of the graphs of $f(x) = x^2 + 6x + 8$ and $g(x) = 0$ (the x-axis). Thus, -4 and -2 are the solutions of the equation $x^2 + 6x + 8 = 0$.

To verify this, let's solve the equation $x^2 + 6x + 8 = 0$:

$$x^2 + 6x + 8 = 0$$
$$(x + 4)(x + 2) = 0 \qquad \qquad \text{\textcolor{red}{Factoring}}$$
$$x + 4 = 0 \quad or \quad x + 2 = 0 \qquad \text{\textcolor{red}{Principle of zero products}}$$
$$x = -4 \quad or \quad x = -2.$$

We see that the solutions of $0 = x^2 + 6x + 8$, -4 and -2, are the first coordinates of the x-intercepts, $(-4, 0)$ and $(-2, 0)$, of the graph of $f(x) = x^2 + 6x + 8$.

> **Now Try Exercise 59.**

Technology Connection

Solving Quadratic Equations

To solve the equation $x^2 - x = 6$ graphically, we can first write the equation with 0 on one side. We get $x^2 - x - 6 = 0$. Next, we graph $y = x^2 - x - 6$ in a window that shows the x-intercepts. The standard window works well in this case.

The solutions of the equation are the values of x for which $x^2 - x - 6 = 0$. These are also the first coordinates of the x-intercepts of the graph. To find the solution corresponding to the leftmost x-intercept, we first select the ZERO feature from the CALC menu. The prompt "Left Bound?" appears. We use the ⓒ key or the ⓓ key to move the cursor to the left of the intercept and press **ENTER**. Now, the prompt "Right Bound?" appears. We move the cursor to the right of the intercept and press **ENTER**. Next, the prompt "Guess?" appears. We move the cursor close to the intercept and press **ENTER** again. We now see the cursor positioned at the leftmost x-intercept and the coordinates of that point, $x = -2$, $y = 0$, are displayed. Thus, $x^2 - x - 6 = 0$ when $x = -2$. This is one solution of the equation.

We repeat this procedure to find the first coordinate of the other x-intercept. We see that $x = 3$ at that point. Thus the solutions of the equation are -2 and 3.

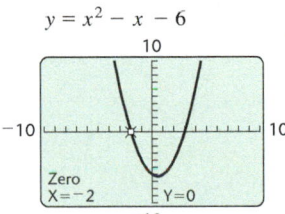

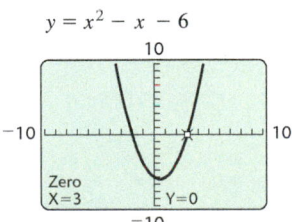

This equation could also be solved by entering $y_1 = x^2 - x$ and $y_2 = 6$ and finding the first coordinate of the points of intersection using the INTERSECT feature.

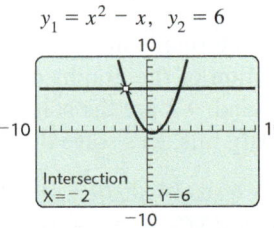

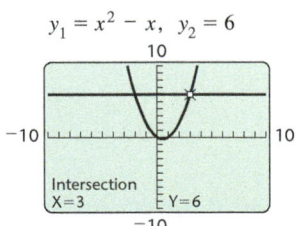

Exercise

1. Solve the equations in Examples 2–5 graphically. Note that, regardless of the variable used in an example, each equation should be entered on the equation-editor screen in terms of x.

▶ b Applications and Problem Solving

Some problems can be translated to quadratic equations. The problem-solving process is the same one we use for other kinds of applied problems.

EXAMPLE 9 *Prize Tee Shirts.* During intermission at sporting events, team mascots commonly use a powerful slingshot to launch tightly rolled tee shirts into the stands. The height $h(t)$, in feet, of an airborne tee shirt t seconds after having been launched can be approximated by

$$h(t) = -15t^2 + 75t + 10.$$

After peaking, a rolled-up tee shirt is caught by a fan 70 ft above ground level. For how long was the tee shirt in the air?

1. **Familiarize.** We make a drawing and label it, using the information provided. (See Figure 1.) We could evaluate $h(t)$ for a few values of t. Note that t cannot be negative, since it represents time from launch.

2. **Translate.** The function is given. Since we are asked to determine how long it will take for the shirt to reach someone 70 ft above ground level, we are interested in the value of t for which $h(t) = 70$:

$$-15t^2 + 75t + 10 = 70.$$

3. **Solve.** We solve by factoring:

$$-15t^2 + 75t + 10 = 70$$
$$-15t^2 + 75t - 60 = 0 \qquad \text{Subtracting 70}$$
$$\left. \begin{aligned} -15(t^2 - 5t + 4) &= 0 \\ -15(t - 4)(t - 1) &= 0 \end{aligned} \right\} \qquad \text{Factoring}$$
$$t - 4 = 0 \quad or \quad t - 1 = 0$$
$$t = 4 \quad or \qquad t = 1.$$

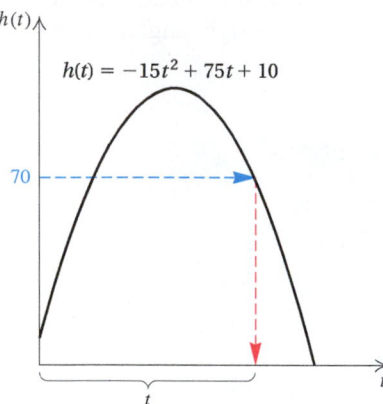

Figure 1.

The solutions appear to be 4 and 1.

4. **Check.** We have

$$h(4) = -15 \cdot 4^2 + 75 \cdot 4 + 10 = -240 + 300 + 10 = 70 \text{ ft};$$
$$h(1) = -15 \cdot 1^2 + 75 \cdot 1 + 10 = -15 + 75 + 10 = 70 \text{ ft}.$$

Both 1 and 4 check, as we can also see from the graph. (See Figure 2.)

However, the problem states that the tee shirt is caught on the way down from its peak height. Thus we reject the solution 1 since that would indicate when the height of the tee shirt was 70 ft on the way up.

5. **State.** The tee shirt was in the air for 4 sec.

Now Try Exercise 81.

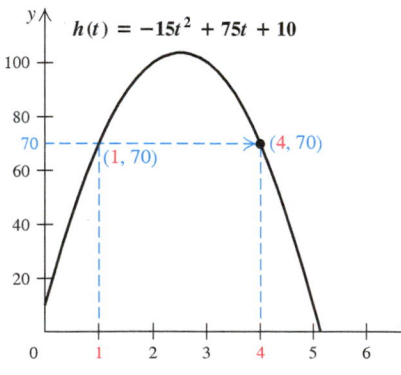

Figure 2.

The following example involves the **Pythagorean theorem**, which relates the lengths of the sides of a right triangle. A **right triangle** has a 90°, or right, angle, which is denoted by a symbol like ⌐. The longest side, opposite the 90° angle, is called the **hypotenuse**. The other sides, called **legs**, form the two sides of the right angle.

THE PYTHAGOREAN THEOREM

The sum of the squares of the lengths of the legs of a right triangle is equal to the square of the length of the hypotenuse:

$$a^2 + b^2 = c^2.$$

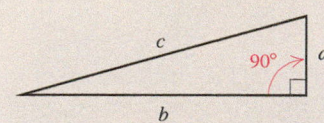

EXAMPLE 10 *Carpentry.* In order to build a deck at a right angle to her house, Geri places a stake in the ground a precise distance from the back wall of her house. This stake will combine with two marks on the house to form a right triangle. From a course in geometry, Geri remembers that there are three consecutive integers that can work as sides of a right triangle. Find the measurements of that triangle.

1. **Familiarize.** Recall that x, $x + 1$, and $x + 2$ can be used to represent three unknown consecutive integers. Since $x + 2$ is the largest number, it must represent the hypotenuse. The legs serve as the sides of the right angle, so one leg must be formed by the marks on the house.

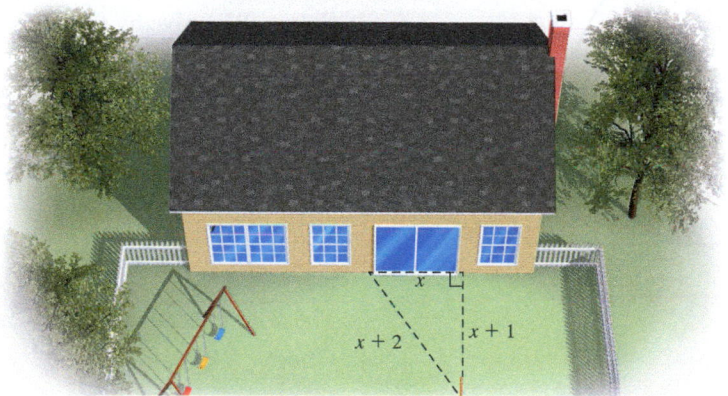

2. **Translate.** Applying the Pythagorean theorem, we translate as follows:

$$a^2 + b^2 = c^2$$
$$x^2 + (x + 1)^2 = (x + 2)^2.$$

3. **Solve.** We solve the equation as follows:

$$x^2 + (x^2 + 2x + 1) = x^2 + 4x + 4 \qquad \text{Squaring the binomials}$$
$$2x^2 + 2x + 1 = x^2 + 4x + 4 \qquad \text{Collecting like terms}$$
$$x^2 - 2x - 3 = 0 \qquad \text{Subtracting } x^2 + 4x + 4$$
$$(x - 3)(x + 1) = 0 \qquad \text{Factoring}$$
$$x - 3 = 0 \quad or \quad x + 1 = 0$$
$$x = 3 \quad or \qquad x = -1.$$

4. **Check.** The integer -1 cannot be a length of a side because it is negative. For $x = 3$, we have $x + 1 = 4$, and $x + 2 = 5$. Since $3^2 + 4^2 = 5^2$, the lengths 3, 4, and 5 determine a right triangle. Thus, 3, 4, and 5 check.

5. **State.** If the marks on the house are 3 yd apart, Geri should locate the stake at the point in the yard that is precisely 4 yd from one mark and 5 yd from the other mark.

Now Try Exercise 77.

1. *Car Travel.* Two cars leave town at the same time going in different directions. One travels 50 mph and the other travels 55 mph. In how many hours will they be 200 mi apart?

2. *Mixture of Solutions.* Solution A is 27% alcohol and solution B is 55% alcohol. How much of each should be used in order to make 10 L of a solution that is 48% alcohol?

3. *Triangle Dimensions.* The base of a triangle is 3 cm less than the height. The area is 27 cm². Find the height and the base.

4. *Three Numbers.* The sum of three numbers is 38. The first number is 3 less than twice the second number. The second number minus the third number is −7. What are the numbers?

5. *Supplementary Angles.* Two angles are supplementary. One angle measures 27° more than three times the measure of the other. Find the measure of each angle.

Translating for Success

Translate each word problem to an equation or a system of equations and select a correct translation from choices A–Q.

A. $x + y + z = 38,$
$x = 2y - 3,$
$y - z = -7$

B. $\frac{1}{2}x(x - 3) = 27$

C. $x + y = 180,$
$x = 3y - 27$

D. $x^2 + 36 = (x + 4)^2$

E. $x^2 + (x + 4)^2 = 36$

F. $x + y = 10,$
$0.27x + 0.55y = 4.8$

G. $x + y = 45,$
$10x - 7y = 402$

H. $x + y + z = 180,$
$y - 3x - 38 = 0,$
$x - z = 7$

I. $x + y = 90,$
$x = 3y + 10$

J. $2x + 2(x - 3) = 27$

K. $x + y + z = 38,$
$x - 2y = 3,$
$x - z = -7$

L. $x + y = 10,$
$27x + 55y = 4.8$

M. $55x - 50x = 200$

N. $x(x - 3) = 27$

O. $x + y = 45,$
$7x + 10y = 402$

P. $x + y = 180,$
$x = 3y + 27$

Q. $50x + 55x = 200$

6. *Triangle Dimensions.* The length of one leg of a right triangle is 6 m. The length of the hypotenuse is 4 m longer than the length of the other leg. Find the lengths of the hypotenuse and the other leg.

7. *Pizza Sales.* Todd's fraternity sold 45 pizzas over a football weekend. Small pizzas sold for $7 each and large pizzas for $10 each. The total amount of the sales was $402. How many of each size pizza were sold?

8. *Angle Measures.* The second angle of a triangle measures 38° more than three times the measure of the first. The measure of the third angle is 7° less than the first. Find the measures of each angle of the triangle.

9. *Complementary Angles.* Two angles are complementary. One angle measures 10° more than three times the measure of the other. Find the measure of each angle.

10. *Rectangle Dimensions.* The base of a rectangle is 3 cm less than the height. The area is 27 cm². Find the height and the base.

4.8   **Exercise Set**

▶ Reading Check

Determine whether each sentence is true or false.

RC1. If $(x + 2)(x + 3) = 24$, then $x + 2 = 24$ *or* $x + 3 = 24$.

RC2. A quadratic equation always has two different solutions.

RC3. The number 0 is never a solution of a quadratic equation.

RC4. The Pythagorean theorem states that the sum of the lengths of the legs of a right triangle is equal to the length of the hypotenuse.

a *Solve.*

1. $x^2 + 3x = 28$

2. $y^2 - 4y = 45$

3. $y^2 + 9 = 6y$

4. $r^2 + 4 = 4r$

5. $x^2 + 20x + 100 = 0$

6. $y^2 + 10y + 25 = 0$

7. $9x + x^2 + 20 = 0$

8. $8y + y^2 + 15 = 0$

9. $x^2 + 8x = 0$

10. $t^2 + 9t = 0$

11. $x^2 - 25 = 0$

12. $p^2 - 49 = 0$

13. $z^2 = 144$

14. $y^2 = 64$

15. $y^2 + 2y = 63$

16. $a^2 + 3a = 40$

17. $32 + 4x - x^2 = 0$

18. $27 + 6t - t^2 = 0$

19. $3b^2 + 8b + 4 = 0$

20. $9y^2 + 15y + 4 = 0$

21. $8y^2 - 10y + 3 = 0$

22. $4x^2 + 11x + 6 = 0$

23. $6z - z^2 = 0$

24. $8y - y^2 = 0$

25. $12z^2 + z = 6$

26. $6x^2 - 7x = 10$

27. $7x^2 - 7 = 0$

28. $4y^2 - 36 = 0$

29. $10 - r - 21r^2 = 0$

30. $28 + 5a - 12a^2 = 0$

31. $15y^2 = 3y$

32. $18x^2 = 9x$

33. $14 = x(x - 5)$

34. $x(x - 5) = 24$

35. $2x^3 - 2x^2 = 12x$

36. $50y + 5y^3 = 35y^2$

37. $2x^3 = 128x$

38. $147y = 3y^3$

39. $t^4 - 26t^2 + 25 = 0$

40. $x^4 - 13x^2 + 36 = 0$

41. $(a - 4)(a + 4) = 20$

42. $(t - 6)(t + 6) = 45$

43. $x(5 + 12x) = 28$

44. $a(1 + 21a) = 10$

45. Given that $f(x) = x^2 + 12x + 40$, find all values of x such that $f(x) = 8$.

46. Given that $f(x) = x^2 + 14x + 50$, find all values of x such that $f(x) = 5$.

47. Given that $g(x) = 2x^2 + 5x$, find all values of x such that $g(x) = 12$.

48. Given that $g(x) = 2x^2 - 15x$, find all values of x such that $g(x) = -7$.

49. Given that $h(x) = 12x + x^2$, find all values of x such that $h(x) = -27$.

50. Given that $h(x) = 4x - x^2$, find all values of x such that $h(x) = -32$.

Find the domain of the function f given by each of the following.

51. $f(x) = \dfrac{3}{x^2 - 4x - 5}$

52. $f(x) = \dfrac{2}{x^2 - 7x + 6}$

53. $f(x) = \dfrac{x}{6x^2 - 54}$

54. $f(x) = \dfrac{2x}{5x^2 - 20}$

55. $f(x) = \dfrac{x - 5}{25x^2 - 10x + 1}$

56. $f(x) = \dfrac{1 + x}{9x^2 + 30x + 25}$

57. $f(x) = \dfrac{7}{5x^3 - 35x^2 + 50x}$

58. $f(x) = \dfrac{3}{2x^3 - 2x^2 - 12x}$

In each of Exercises 59–62, an equation $ax^2 + bx + c = 0$ is given. Use only the graph of $f(x) = ax^2 + bx + c$ to find the x-intercepts of the graph and the solutions of the equation $ax^2 + bx + c = 0$.

59. $x^2 - 4x - 45 = 0$

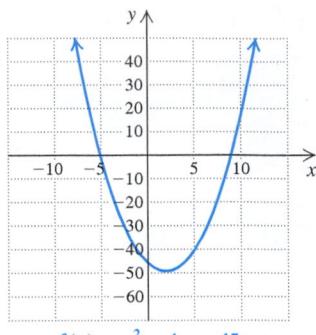

$f(x) = x^2 - 4x - 45$

60. $-x^2 - 3x + 40 = 0$

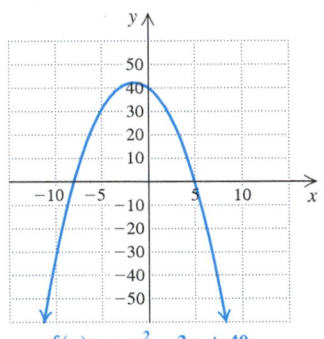

$f(x) = -x^2 - 3x + 40$

61. $32 + 4x - x^2 = 0$

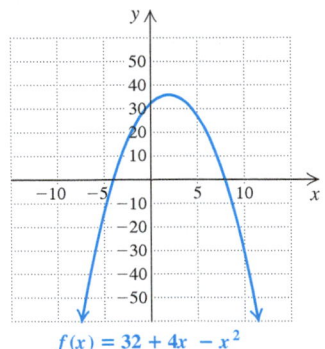

$f(x) = 32 + 4x - x^2$

62. $3x^2 - 12x = 0$

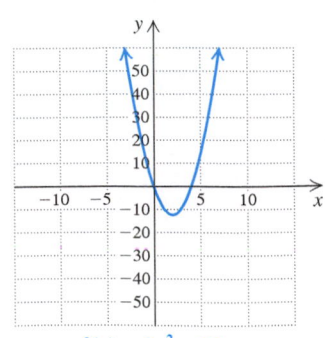

$f(x) = 3x^2 - 12x$

b *Solve.*

63. *Book Area.* A book is 5 cm longer than it is wide. The area is 84 cm². Find the length and the width.

64. *Area of an Envelope.* An envelope is 4 cm longer than it is wide. The area is 96 cm². Find the length and the width.

65. *Tent Design.* The triangular entrance to a tent is 2 ft taller than it is wide. The area of the entrance is 12 ft². Find the height and the base.

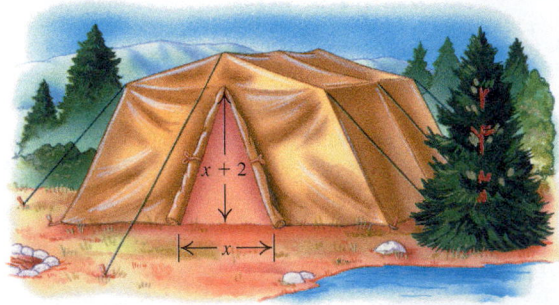

66. *Sailing.* A triangular sail is 9 m taller than it is wide. The area is 56 m². Find the height and the base of the sail.

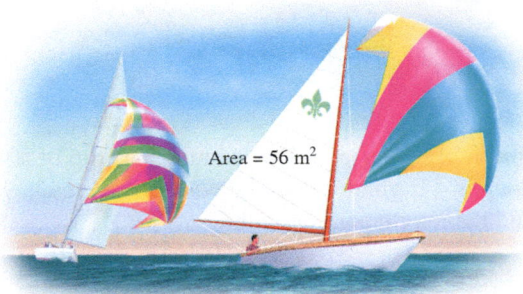

67. *Geometry.* If each of the sides of a square is lengthened by 6 cm, the area becomes 144 cm². Find the length of a side of the original square.

68. *Geometry.* If each of the sides of a square is lengthened by 4 m, the area becomes 49 m². Find the length of a side of the original square.

69. *Consecutive Even Integers.* Three consecutive even integers are such that the square of the third is 76 more than the square of the second. Find the three integers.

70. *Consecutive Even Integers.* Three consecutive even integers are such that the square of the first plus the square of the third is 136. Find the three integers.

71. *Framing a Picture.* A picture frame measures 12 cm by 20 cm, and 84 cm² of picture shows. Find the width of the frame.

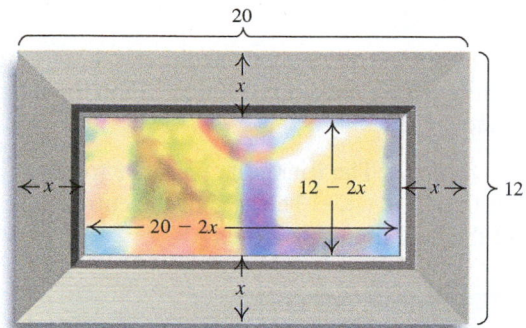

72. *Flower Bed Design.* A rectangular flower bed is to be 3 m longer than it is wide. The flower bed will have an area of 108 m². What will its dimensions be?

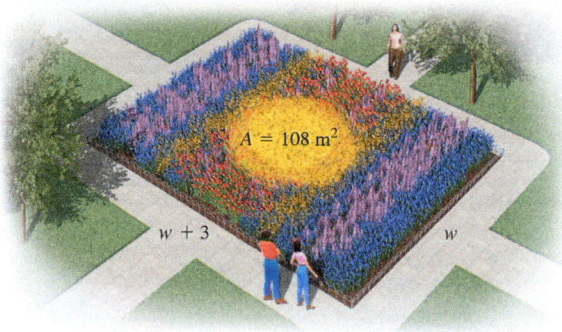

$A = 108 \text{ m}^2$

73. *Workbench Design.* The length of the top of a workbench is 4 ft greater than the width. The area is 96 ft². Find the length and the width.

74. *Framing a Picture.* A picture frame measures 14 cm by 20 cm, and 160 cm² of picture shows. Find the width of the frame.

75. *Antenna Wires.* A wire is stretched from the ground to the top of an antenna tower, as shown. The wire is 20 ft long. The height of the tower is 4 ft greater than the distance *d* from the tower's base to the end of the wire. Find the distance *d* and the height of the tower.

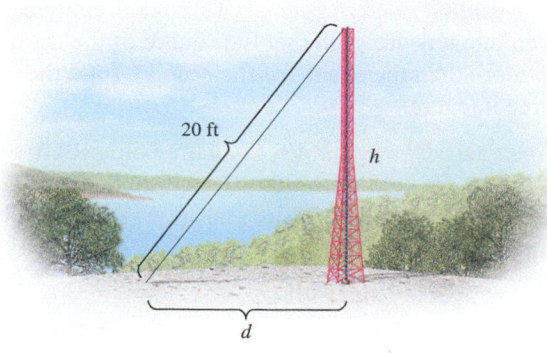

76. *Parking Lot Design.* A rectangular parking lot is 50 ft longer than it is wide. Determine the dimensions of the parking lot if it measures 250 ft diagonally.

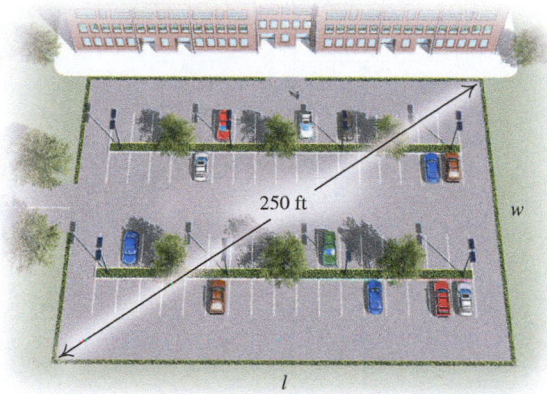

77. *Ladder Location.* The foot of an extension ladder is 9 ft from a wall. The height that the ladder reaches on the wall and the length of the ladder are consecutive integers. How long is the ladder?

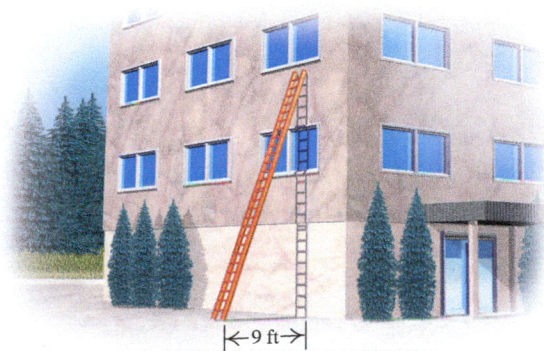

78. *Child's Block.* The lengths of the sides of a right triangle formed by a child's wooden block are such that one leg has length 5 cm. The lengths of the other sides are consecutive integers. Find the lengths of the other sides of the triangle.

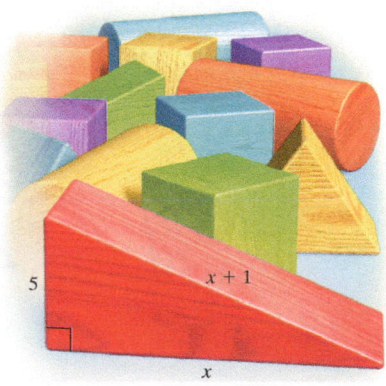

79. *Triangle Dimensions.* The lengths of the sides of a right triangle are consecutive even integers. Find the lengths of the sides.

80. *Triangle Dimensions.* The lengths of the hypotenuse and one leg of a right triangle are consecutive odd integers. The other leg is 9 ft shorter than the hypotenuse. Find the lengths of the sides

81. *Fireworks.* Suppose that a bottle rocket is launched upward with an initial velocity of 96 ft/sec and from a height of 880 ft. Its height h, in feet, after t seconds is given by

$$h(t) = -16t^2 + 96t + 880.$$

After how long will the rocket reach the ground?

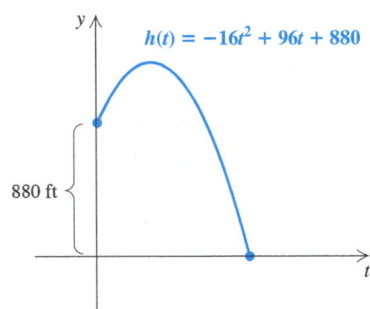

82. *Safety Flares.* Suppose that a flare is launched upward with an initial velocity of 80 ft/sec and from a height of 224 ft. Its height h, in feet, after t seconds is given by

$$h(t) = -16t^2 + 80t + 224.$$

After how long will the flare reach the ground?

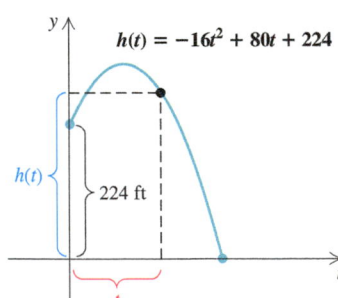

▶ **Skill Maintenance**

Find the distance between the given pair of points on the number line. [1.6b]

83. $-3, -4$

84. $3.6, 4.9$

85. $-\frac{3}{5}, \frac{2}{3}$

86. $0, -1023$

Find an equation of the line containing the given pair of points. [2.7c]

87. $(-2, 7)$ and $(-8, -4)$

88. $(-2, 7)$ and $(8, -4)$

89. $(-2, 7)$ and $(8, 4)$

90. $(-24, 10)$ and $(-86, -42)$

▶ **Synthesis**

91. Following is the graph of $f(x) = -x^2 - 2x + 3$. Use *only* the graph to solve $-x^2 - 2x + 3 = 0$ and $-x^2 - 2x + 3 \geq -5$.

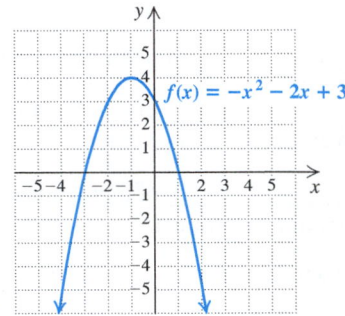

92. Following is the graph of $f(x) = x^4 - 3x^3$. Use *only* the graph to solve $x^4 - 3x^3 = 0$, $x^4 - 3x^3 \leq 0$, and $x^4 - 3x^3 > 0$.

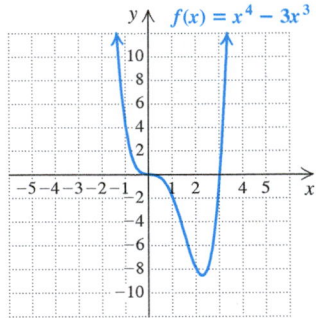

93. Use a graphing calculator to solve each equation.
   a) $x^4 - 3x^3 - x^2 + 5 = 0$
   b) $x^4 - 3x^3 - x^2 + 5 = 5$
   c) $x^4 - 3x^3 - x^2 + 5 = -8$
   d) $x^4 = 1 + 3x^3 + x^2$

94. Solve each of the following equations.
   a) $(8x + 11)(12x^2 - 5x - 2) = 0$
   b) $(3x^2 - 7x - 20)(x - 5) = 0$
   c) $3x^3 + 6x^2 - 27x - 54 = 0$
       (*Hint*: Factor by grouping.)
   d) $2x^3 + 6x^2 = 8x + 24$

Chapter 4 Summary and Review

VOCABULARY REINFORCEMENT

Complete each statement with the correct term from the column on the right. Some of the choices may be used more than once and some may not be used at all.

product

difference

factor

factorization

grouping

ascending

descending

monomial

binomial

trinomial

zero

1. When the terms of a polynomial are written such that the exponents increase from left to right, we say that the polynomial is written in _____ order. **[4.1a]**

2. To _____ a polynomial is to express it as a product. **[4.3a]**

3. A(n) _____ of a polynomial P is a polynomial that can be used to express P as a product. **[4.3a]**

4. A(n) _____ of a polynomial is an expression that names that polynomial as a product. **[4.3a]**

5. When factoring a polynomial with four terms, try factoring by _____ . **[4.7a]**

6. A trinomial square is the square of a(n) _____. **[4.6a]**

7. The principle of _____ products states that if $ab = 0$, then $a = 0$ or $b = 0$. **[4.8a]**

8. The factorization of a(n) _____ of squares is the product of the sum and the difference of two terms. **[4.6c]**

CONCEPT REINFORCEMENT

Determine whether each statement is true or false.

1. According to the principle of zero products, if $ab = 0$, then $a = 0$ and $b = 0$. **[4.8a]**

2. The binomial $27 - t^3$ is a difference of cubes. **[4.6d]**

3. The expression $5x^2 - 6y^{-1}$ is a binomial. **[4.1a]**

STUDY GUIDE

Objective 4.1a Identify terms, degrees, and coefficients in polynomials; identify types of polynomials; and arrange polynomials in ascending order or descending order.

Example Identify the terms, the degree of each term, and the degree of the polynomial. Then identify the leading term, the leading coefficient, and the constant term:

$$-x^5 + 3x^4 - 7x^3 - 2x^2 + x - 10.$$

Terms: $-x^5, 3x^4, -7x^3, -2x^2, x, -10$

Degree of each term: 5, 4, 3, 2, 1, 0

Degree of polynomial: 5

Leading term: $-x^5$

Leading coefficient: -1

Constant term: -10

Practice Exercise

1. Identify the terms, the degree of each term, and the degree of the polynomial. Then identify the leading term, the leading coefficient, and the constant term:

$$-6x^4 + 5x^3 - x^2 + 10x - 1.$$

Objective 4.1d Find the opposite of a polynomial and subtract polynomials.

Example Subtract: $(4t^2 - t - t^3) - (7t^2 - t^3 - 5t)$.

$$
\begin{aligned}
&(4t^2 - t - t^3) - (7t^2 - t^3 - 5t) \\
&= (4t^2 - t - t^3) + (-7t^2 + t^3 + 5t) \\
&= 4t^2 - t - t^3 - 7t^2 + t^3 + 5t \\
&= -3t^2 + 4t
\end{aligned}
$$

Practice Exercise

2. Subtract:

$$(3y^2 - 6y^3 + 7y) - (y^2 - 10y - 8y^3 + 8).$$

Objective 4.2b Use the FOIL method to multiply two binomials.

Example Multiply: $(7a - b)(4a + 9b)$.

$$
\begin{array}{cccc}
\text{F} & \text{O} & \text{I} & \text{L}
\end{array}
$$
$$
\begin{aligned}
(7a - b)(4a + 9b) &= 28a^2 + 63ab - 4ab - 9b^2 \\
&= 28a^2 + 59ab - 9b^2
\end{aligned}
$$

Practice Exercise

3. Multiply: $(3x - 5y)(x + 2y)$.

Objective 4.2c Use a rule to square a binomial.

Example Multiply: $(3q - 4)^2$.

$$
\begin{aligned}
(A - B)^2 &= A^2 - 2AB + B^2 \\
(3q - 4)^2 &= (3q)^2 - 2(3q)(4) + 4^2 \\
&= 9q^2 - 24q + 16
\end{aligned}
$$

Practice Exercise

4. Multiply: $(2y + 7)^2$.

Objective 4.2d   Use a rule to multiply a sum and a difference of the same two terms.

Example   Multiply: $(8x + 5)(8x - 5)$.

$$(A + B)(A - B) = A^2 - B^2$$
$$(8x + 5)(8x - 5) = (8x)^2 - 5^2$$
$$= 64x^2 - 25$$

Practice Exercise

5. Multiply: $(5d + 10)(5d - 10)$.

Objective 4.2e   For functions f described by second-degree polynomials, find and simplify notation like $f(a + h)$ and $f(a + h) - f(a)$.

Example   Given $f(x) = 2x - x^2$, find $f(x - 1)$ and $f(a + h) - f(a)$.

$$f(x - 1) = 2(x - 1) - (x - 1)^2$$
$$= 2(x - 1) - (x^2 - 2x + 1)$$
$$= 2x - 2 - x^2 + 2x - 1 = -x^2 + 4x - 3;$$

$$f(a + h) - f(a)$$
$$= [2(a + h) - (a + h)^2] - [2a - a^2]$$
$$= [2(a + h) - (a^2 + 2ah + h^2)] - [2a - a^2]$$
$$= 2a + 2h - a^2 - 2ah - h^2 - 2a + a^2$$
$$= -h^2 - 2ah + 2h$$

Practice Exercise

6. Given $f(x) = 3x^2 - x + 2$, find $f(x + 1)$ and $f(a + h) - f(a)$.

Objective 4.3b   Factor certain polynomials with four terms by grouping.

Example   Factor: $x^3 - 6x^2 + 3x - 18$.

$$x^3 - 6x^2 + 3x - 18 = (x^3 - 6x^2) + (3x - 18)$$
$$= x^2(x - 6) + 3(x - 6)$$
$$= (x - 6)(x^2 + 3)$$

Practice Exercise

7. Factor: $y^3 + 3y^2 - 8y - 24$.

Objective 4.5a   Factor trinomials of the type $ax^2 + bx + c$, $a \neq 1$, by the FOIL method.

Example   Factor $15x^2 - 4x - 3$ by the FOIL method.

The terms of $15x^2 - 4x - 3$ do not have a common factor. We factor the first term, $15x^2$, and get $15x \cdot x$ and $5x \cdot 3x$. We then have

$$(15x + \square)(x + \square) \text{ and } (5x + \square)(3x + \square)$$

as possible factorizations. We then factor the last term, -3. The possibilities are $(-3)(1)$ and $(3)(-1)$. We look for combinations of factors such that the sum of the outside product and the inside product is the middle term, $-4x$.

$(15x - 3)(x + 1)$; $(5x - 3)(3x + 1)$; → **Correct middle term, $-4x$**

$(15x + 3)(x - 1)$; $(5x + 3)(3x - 1)$;

$(15x + 1)(x - 3)$; $(5x + 1)(3x - 3)$;

$(15x - 1)(x + 3)$; $(5x - 1)(3x + 3)$

Thus, $15x^2 - 4x - 3 = (5x - 3)(3x + 1)$.

Practice Exercise

8. Factor $3x^2 + 19x - 72$ by the FOIL method.

Objective 4.5b Factor trinomials of the type $ax^2 + bx + c$, $a \neq 1$, by the ac-method.

Example Factor $6x^2 - 19x - 36$ by the ac-method.

Note that there are no common factors. We multiply the leading coefficient, 6, and the constant, -36: $6(-36) = -216$. Next, we try to factor -216 so that the sum of the factors is -19. Since -19 is negative, the negative factor of -216 must have the larger absolute value.

Pairs of Factors	Sum
1, -216	-215
2, -108	-106
3, -72	-69
4, -54	-50

Pairs of Factors	Sum
6, -36	-30
8, -27	-19
9, -24	-15
12, -18	-6

Next, we split the middle term using the factors 8 and -27:

$$6x^2 - 19x - 36 = 6x^2 + 8x - 27x - 36$$
$$= 2x(3x + 4) - 9(3x + 4)$$
$$= (3x + 4)(2x - 9).$$

Practice Exercise

9. Factor $10x^2 - 33x - 7$ by the ac-method.

Objective 4.6a Factor trinomial squares.

Example Factor: $4x^2 - 44x + 121$.

$$A^2 - 2AB + B^2 = (A - B)^2$$
$$4x^2 - 44x + 121 = (2x)^2 - 44x + 11^2 = (2x - 11)^2$$

Practice Exercise

10. Factor: $81x^2 - 72x + 16$.

Objective 4.6b Factor differences of squares.

Example Factor: $64y^2 - 9$.

$$A^2 - B^2 = (A + B)(A - B)$$
$$64y^2 - 9 = (8y)^2 - 3^2 = (8y + 3)(8y - 3)$$

Practice Exercise

11. Factor: $100t^2 - 1$.

Objective 4.6d Factor sums and differences of cubes.

Example Factor: $8w^3 + 125$.

$$A^3 + B^3 = (A + B)(A^2 - AB + B^2)$$
$$8w^3 + 125 = (2w)^3 + 5^3$$
$$= (2w + 5)(4w^2 - 10w + 25)$$

Example Factor: $125x^3 - 8$.

$$A^3 - B^3 = (A - B)(A^2 + AB + B^2)$$
$$125x^3 - 8 = (5x)^3 - 2^3 = (5x - 2)(25x^2 + 10x + 4)$$

Practice Exercises

12. Factor: $216x^3 + 1$.

13. Factor: $1000y^3 - 27$.

Objective 4.8a   Solve quadratic and other polynomial equations by first factoring and then using the principle of zero products.

Example Solve: $5x^2 + 11x = 12$.	**Practice Exercise**

Example   Solve: $5x^2 + 11x = 12$.

$5x^2 + 11x - 12 = 0$ **Getting 0 on one side**

$(5x - 4)(x + 3) = 0$ **Factoring**

$5x - 4 = 0 \ or \ x + 3 = 0$ **Using the principle of zero products**

$5x = 4 \ or \quad\quad x = -3$

$x = \frac{4}{5} \ or \quad\quad x = -3$

The solutions are -3 and $\frac{4}{5}$.

Practice Exercise

14. Solve: $3x^2 - x = 14$.

REVIEW EXERCISES

1. Given the polynomial **[4.1a]**

   $3x^6y - 7x^8y^3 + 2x^3 - 3y^2$:

   a) Identify the degree of each term and the degree of the polynomial.
   b) Identify the leading term and the leading coefficient.
   c) Arrange in ascending powers of x.
   d) Arrange in descending powers of y.

Evaluate the polynomial function for the given values. **[4.1b]**

2. $P(x) = x^3 - x^2 + 4x$; $P(0)$ and $P(-1)$

3. $P(x) = 4 - 2x - x^2$; $P(-2)$ and $P(5)$

Collect like terms. **[4.1c]**

4. $8x + 13y - 15x + 10y$

5. $3ab - 10 + 5ab^2 - 2ab + 7ab^2 + 14$

6. *Youth Football.*   The number of children ages 7 to 17 who played football in a given year can be estimated by

   $f(t) = 0.25t^2 - 0.81t + 5.54,$

   where $f(t)$ is the number of participants, in millions, t years after 2008. Use the following graph to estimate the number of children participating in

football in 2010. (*Source*: National Sporting Goods Association) **[4.1b]**

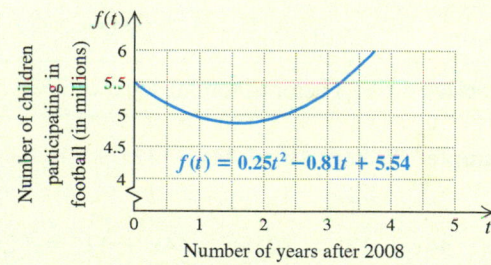

Number of years after 2008

Add, subtract, or multiply. **[4.1c, d], [4.2a, b, c, d]**

7. $(-6x^3 - 4x^2 + 3x + 1) + (5x^3 + 2x + 6x^2 + 1)$

8. $(4x^3 - 2x^2 - 7x + 5) + (8x^2 - 3x^3 - 9 + 6x)$

9. $(-9xy^2 - xy + 6x^2y) + (-5x^2y - xy + 4xy^2) + (12x^2y - 3xy^2 + 6xy)$

10. $(3x - 5) - (-6x + 2)$

11. $(4a - b + 3c) - (6a - 7b - 4c)$

12. $(9p^2 - 4p + 4) - (-7p^2 + 4p + 4)$

13. $(6x^2 - 4xy + y^2) - (2x^2 + 3xy - 2y^2)$

14. $(3x^2y)(-6xy^3)$

15. $(x^4 - 2x^2 + 3)(x^4 + x^2 - 1)$

16. $(4ab + 3c)(2ab - c)$

17. $(2x + 5y)(2x - 5y)$

18. $(2x - 5y)^2$

19. $(5x^2 - 7x + 3)(4x^2 + 2x - 9)$

20. $(x^2 + 4y^3)^2$

21. $(x - 5)(x^2 + 5x + 25)$

22. $\left(x - \frac{1}{3}\right)\left(x - \frac{1}{6}\right)$

23. Given that $f(x) = x^2 - 2x - 7$, find and simplify $f(a - 1)$ and $f(a + h) - f(a)$. **[4.2e]**

Factor. **[4.3a, b], [4.4a], [4.5a, b], [4.6a, b, c, d], [4.7a]**

24. $9y^4 - 3y^2$

25. $15x^4 - 18x^3 + 21x^2 - 9x$

26. $a^2 - 12a + 27$

27. $3m^2 + 14m + 8$

28. $25x^2 + 20x + 4$

29. $4y^2 - 16$

30. $ax + 2bx - ay - 2by$

31. $4x^4 + 4x^2 + 20$

32. $27x^3 - 8$

33. $0.064b^3 - 0.125c^3$

34. $y^5 - y$

35. $2z^8 - 16z^6$

36. $54x^6y - 2y$

37. $1 + a^3$

38. $36x^2 - 120x + 100$

39. $6t^2 + 17pt + 5p^2$

40. $x^3 + 2x^2 - 9x - 18$

41. $a^2 - 2ab + b^2 - 4t^2$

Solve. **[4.8a]**

42. $x^2 - 20x = -100$ **43.** $6b^2 - 13b + 6 = 0$

44. $8y^2 = 14y$ **45.** $r^2 = 16$

46. Given that $f(x) = x^2 - 7x - 40$, find all values of x such that $f(x) = 4$. **[4.8a]**

47. Find the domain of the function f given by

$$f(x) = \frac{x - 3}{3x^2 + 19x - 14}.\ \text{[4.8a]}$$

Solve. **[4.8b]**

48. *Photograph Dimensions.* A photograph is 3 in. longer than it is wide. When a 2-in. matte border is placed around the photograph, the total area of

the photograph and the border is 108 in². Find the dimensions of the photograph.

49. The sum of the squares of three consecutive odd integers is 83. Find the integers.

50. *Area.* The number of *square units* in the area of a square is 7 more than six times the number of *units* in the length of a side. What is the length of a side of the square?

51. Which of the following is a factor of $t^3 - 64$? **[4.6d]**

  A. $t - 4$ **B.** $t^2 - 4t + 16$
  C. $t^2 + 8t + 16$ **D.** $t + 4$

52. Which of the following is a factor of
$$hm + 5hn - gm - 5gn?\ \text{[4.3b]}$$

  A. $m - n$ **B.** $h + g$
  C. $m + 5n$ **D.** $m - 5n$

▶ Synthesis

Factor. **[4.6d]**

53. $128x^6 - 2y^6$

54. $(x + 1)^3 - (x - 1)^3$

55. Multiply:
$$[a - (b - 1)][(b - 1)^2 + a(b - 1) + a^2].$$
[4.6d]

56. Solve: $64x^3 = x$. **[4.8a]**

▶ Collaborative Discussion and Writing

57. Under what conditions, if any, can the sum of two squares be factored? Explain. **[4.3a], [4.6b]**

58. Explain how to use the *ac*-method to factor trinomials of the type $ax^2 + bx + c, a \neq 1$. **[4.5b]**

59. Annie claims that she can add any two polynomials but finds subtraction difficult. What advice would you offer her? **[4.1d]**

60. Suppose that you are given a detailed graph of $y = P(x)$, where $P(x)$ is a polynomial. How could you use the graph to solve the equation $P(x) = 0$? $P(x) = 4$? **[4.8a]**

61. 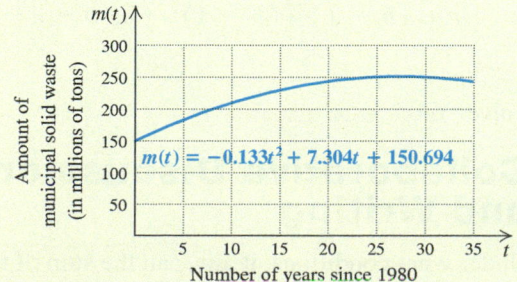 Explain how you could use factoring or graphing to explain why $x^3 - 8 \neq (x - 2)^3$. **[4.6d]**

62. Emily has factored a particular polynomial as $(a - b)(x - y)$. George factors the same polynomial and gets $(b - a)(y - x)$. Who is correct and why? **[4.3a], [4.7a]**

63. Explain how one could write a quadratic equation that has 5 and -3 as solutions. Can the number of solutions of a quadratic equation exceed two? Why or why not? **[4.8a]**

64. In this chapter, we learned to solve equations that we could not have solved before. Describe these new equations and the way we go about solving them. How is the procedure different from those we have used before now? **[4.8a]**

---

4   Chapter Test

1. Given the polynomial
$$3xy^3 - 4x^2y + 5x^5y^4 - 2x^4y:$$

  a) Identify the degree of each term and the degree of the polynomial.
  b) Identify the leading term and the leading coefficient.
  c) Arrange in ascending powers of x.
  d) Arrange in descending powers of y.

2. Given that $P(x) = 2x^3 + 3x^2 - x + 4$, find $P(0)$ and $P(-2)$.

3. *Solid-Waste Generation.* The amount of municipal solid waste $m(t)$ generated in the United States, in millions of tons, can be estimated by the polynomial function given by
$$m(t) = -0.133t^2 + 7.304t + 150.694,$$
where t is the number of years after 1980. Use the graph to estimate the amount of municipal solid waste generated in 2010. (*Source*: United States Environmental Protection Agency)

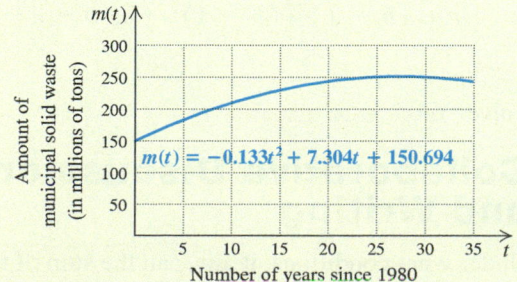

4. Collect like terms: $5xy - 2xy^2 - 2xy + 5xy^2$.

Add, subtract, or multiply.

5. $(-6x^3 + 3x^2 - 4y) + (3x^3 - 2y - 7y^2)$

6. $(4a^3 - 2a^2 + 6a - 5) + (3a^3 - 3a + 2 - 4a^2)$

7. $(5m^3 - 4m^2n - 6mn^2 - 3n^3) + (9mn^2 - 4n^3 + 2m^3 + 6m^2n)$

8. $(9a - 4b) - (3a + 4b)$

9. $(4x^2 - 3x + 7) - (-3x^2 + 4x - 6)$

10. $(6y^2 - 2y - 5y^3) - (4y^2 - 7y - 6y^3)$

11. $(-4x^2y)(-16xy^2)$

12. $(6a - 5b)(2a + b)$

13. $(x - y)(x^2 - xy - y^2)$

14. $(3m^2 + 4m - 2)(-m^2 - 3m + 5)$

15. $(4y - 9)^2$

16. $(x - 2y)(x + 2y)$

17. Given that $f(x) = x^2 - 5x$, find and simplify $f(a + 10)$ and $f(a + h) - f(a)$.

Factor.

18. $9x^2 + 7x$

19. $24y^3 + 16y^2$

20. $y^3 + 5y^2 - 4y - 20$

21. $p^2 - 12p - 28$

22. $12m^2 + 20m + 3$

23. $9y^2 - 25$

24. $3r^3 - 3$

25. $9x^2 + 25 - 30x$

26. $(z + 1)^2 - b^2$

27. $x^8 - y^8$

28. $y^2 + 8y + 16 - 100t^2$

29. $20a^2 - 5b^2$

30. $24x^2 - 46x + 10$

31. $16a^7b + 54ab^7$

Solve.

32. $x^2 - 18 = 3x$

33. $5y^2 - 125 = 0$

34. $2x^2 + 21 = -17x$

35. Given that $f(x) = 3x^2 - 15x + 11$, find all values of x such that $f(x) = 11$.

36. Find the domain of the function f given by

$$f(x) = \frac{3 - x}{x^2 + 2x + 1}.$$

Solve.

37. *Photograph Dimensions.* A photograph is 3 cm longer than it is wide. Its area is 40 cm^2. Find its length and its width.

38. *Ladder Location.* The foot of an extension ladder is 10 ft from a wall. The ladder is 2 ft longer than the distance that it reaches up the wall. How far up the wall does the ladder reach?

39. *Number of Games in a League.* If there are n teams in a league and each team plays every other team once, the total number of games played is given by the polynomial function $f(n) = \frac{1}{2}n^2 - \frac{1}{2}n$. Find an equivalent expression for $f(n)$ by factoring completely.

40. Factor: $8x^3 - 1$.
   A. $(2x - 1)(2x - 1)(2x - 1)$
   B. $(2x - 1)(2x + 1)$
   C. $(2x - 1)(4x^2 + 2x + 1)$
   D. $(2x + 1)(4x^2 - 2x + 1)$

▶ **Synthesis**

41. Factor: $6x^{2n} - 7x^n - 20$.

42. If $pq = 5$ and $(p + q)^2 = 29$, find the value of $p^2 + q^2$.

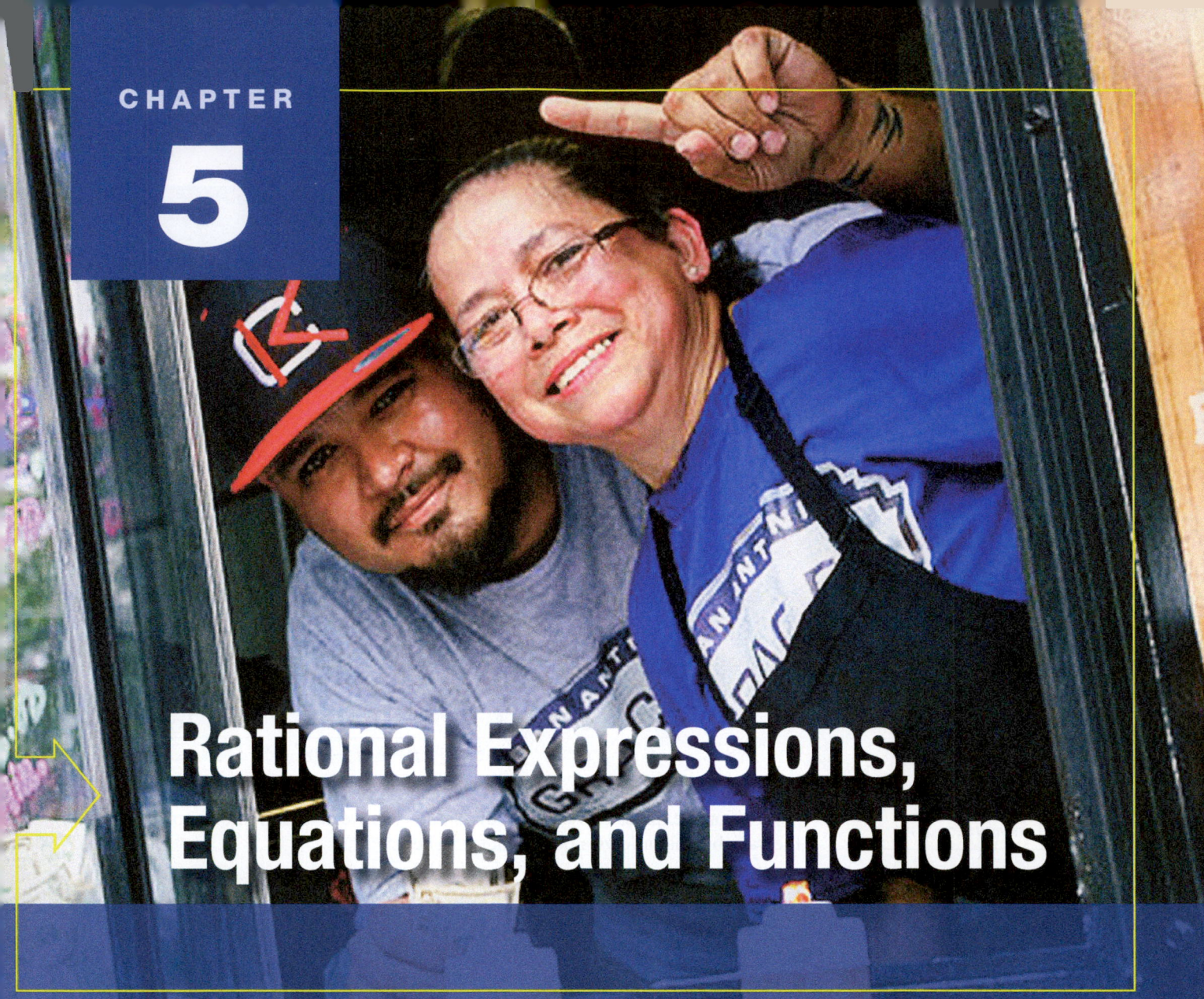

CHAPTER

5

Rational Expressions, Equations, and Functions

APPLICATION   This problem appears as Example 2 in Section 5.6.

A convention planner is organizing a business luncheon for 2300 guests in San Antonio, Texas. The chosen menu will include tacos and tamales. To complete the order, two food caterers are needed. Reading that Gracie's Kitchen won the 2013 Annual Twisted Taco Truck Throwdown, the organizer wants to hire Gracie as one of the caterers. Suppose Gracie and a second caterer, together, can complete the order in 12 hr. If the second caterer would take 10 hr longer than Gracie to complete the order alone, how long would it take each, working alone, to prepare the luncheon?

5.1   Rational Expressions and Functions: Multiplying, Dividing, and Simplifying

5.2   LCMs, LCDs, Addition, and Subtraction

5.3   Division of Polynomials

5.4   Complex Rational Expressions

Mid-Chapter Review

5.5   Solving Rational Equations

5.6   Applications and Proportions

Translating for Success

5.7   Formulas and Applications

5.8   Variation and Applications

Summary and Review

Chapter Test

289

5.1

Rational Expressions and Functions: Multiplying, Dividing, and Simplifying

▶ **a** Find all numbers for which a rational expression is not defined or that are not in the domain of a rational function, and state the domain of the function.

▶ **b** Multiply a rational expression by 1, using an expression like *A*/*A*.

▶ **c** Simplify rational expressions.

▶ **d** Multiply rational expressions and simplify.

▶ **e** Divide rational expressions and simplify.

▶ a Rational Expressions and Functions

An expression that consists of the quotient of two polynomials, where the polynomial in the denominator is nonzero, is called a **rational expression**. The following are examples of rational expressions:

$$\frac{7}{8}, \quad \frac{z}{-6}, \quad \frac{a}{b}, \quad \frac{8}{y+5}, \quad \frac{t^4 - 5t}{t^2 - 3t - 28}, \quad \frac{x^2 + 7xy - 4}{x^3 - y^3}.$$

> **EXCLUDING DIVISION BY 0**
>
> **REVIEW SECTION R.2**

Note that every rational number is a rational expression.

Rational expressions indicate division. Thus we cannot make a replacement of the variable that allows a denominator to be 0.

EXAMPLE 1 Find all numbers for which the rational expression

$$\frac{2x + 1}{x - 3}$$

is not defined.

When *x* is replaced with 3, the denominator is 0, and the rational expression is not defined:

$$\frac{2x + 1}{x - 3} = \frac{2 \cdot 3 + 1}{3 - 3} = \frac{7}{0}. \longleftarrow \text{Division by 0 is not defined.}$$

You can check some replacements other than 3 to see that it appears that 3 is the only replacement that is not allowable. Thus the rational expression is not defined for the number 3.

> **Now Try Exercise 1.**

You may have noticed that the procedure in Example 1 is similar to one that we have performed when finding the domain of a function.

> **DOMAIN OF A FUNCTION**
>
> **REVIEW SECTION 2.3**

EXAMPLE 2 Find the domain of *f* if $f(x) = \dfrac{2x + 1}{x - 3}$.

The domain is the set of all replacements for which the rational expression is defined. We must exclude the replacements that make the denominator 0. We find them by first setting the denominator equal to 0. Solving $x - 3 = 0$ for *x*, we get $x = 3$. The domain of *f* is $\{x \mid x \text{ is a real number } and\ x \neq 3\}$, or, in interval notation, $(-\infty, 3) \cup (3, \infty)$.

> **Now Try Exercise 5.**

Algebraic/Graphical Connection

Let's make a visual check of Example 2 by looking at the following graph.

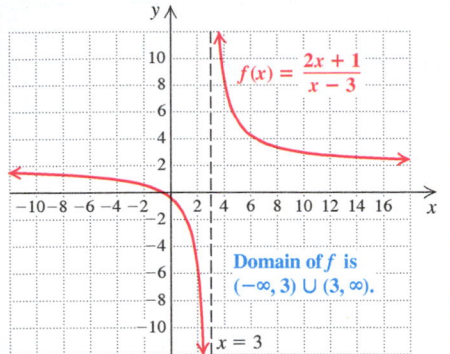

Domain of f is
$(-\infty, 3) \cup (3, \infty)$.

Note that the graph consists of two unconnected "branches." If a vertical line were drawn at $x = 3$, shown dashed here, it would not touch the graph of f. Thus 3 is not in the domain of f.

EXAMPLE 3   Find all numbers for which the rational expression

$$\frac{t^4 - 5t}{t^2 - 3t - 28}$$

is not defined.

The rational expression is not defined for a replacement that makes the denominator 0. To determine those replacements to exclude, we set the denominator equal to 0 and solve:

$$t^2 - 3t - 28 = 0 \qquad \text{\color{red}Setting the denominator equal to 0}$$
$$(t - 7)(t + 4) = 0 \qquad \text{\color{red}Factoring}$$
$$t - 7 = 0 \quad or \quad t + 4 = 0 \qquad \text{\color{red}Using the principle of zero products}$$
$$t = 7 \quad or \qquad t = -4.$$

Thus the expression is not defined for the replacements 7 and -4.

> **Now Try Exercise 3.**

EXAMPLE 4   Find the domain of g if

$$g(t) = \frac{t^4 - 5t}{t^2 - 3t - 28}.$$

We proceed as we did in Example 3. The expression is not defined for the replacements 7 and -4. Thus the domain is $\{t \mid t$ is a real number *and* $t \neq 7$ *and* $t \neq -4\}$, or, in interval notation, $(-\infty, -4) \cup (-4, 7) \cup (7, \infty)$.

> **Now Try Exercise 11.**

▶ **b   Finding Equivalent Rational Expressions**

Calculations with rational expressions are similar to those with rational numbers.

Multiplying Rational Expressions

To multiply rational expressions, multiply numerators and multiply denominators:

$$\frac{A}{B} \cdot \frac{C}{D} = \frac{AC}{BD}.$$

For example, we have the following:

$$\frac{3}{5} \cdot \frac{2}{7} = \frac{3 \cdot 2}{5 \cdot 7} = \frac{6}{35}, \qquad \frac{3x}{4} \cdot \frac{5x}{7} = \frac{(3x)(5x)}{4 \cdot 7} = \frac{15x^2}{28},$$

and $$\frac{x + 3}{y - 4} \cdot \frac{x^3}{y + 5} = \frac{(x + 3)x^3}{(y - 4)(y + 5)}. \qquad \text{\color{red}Multiplying numerators and multiplying denominators}$$

For purposes of our work in this chapter, it is better in the example above to leave the numerator $(x + 3)x^3$ and the denominator $(y - 4)(y + 5)$ in factored form because it is easier to simplify if we do not multiply.

Before discussing simplifying rational expressions, we first consider multiplying by 1.

> Any rational expression with the same numerator and denominator is a symbol for 1:
>
> $$\frac{73}{73} = 1, \qquad \frac{x-y}{x-y} = 1, \qquad \frac{4x^2 - 5}{4x^2 - 5} = 1, \qquad \frac{-1}{-1} = 1, \qquad \frac{x+5}{x+5} = 1.$$

We can multiply by 1 to get equivalent expressions—for example,

$$\frac{7}{9} = \frac{7}{9} \cdot \frac{4}{4} = \frac{7 \cdot 4}{9 \cdot 4} = \frac{28}{36} \quad \text{and} \quad \frac{5}{6} = \frac{5}{6} \cdot \frac{x}{x} = \frac{5 \cdot x}{6 \cdot x} = \frac{5x}{6x}.$$

As another example, let's multiply $(x + y)/5$ by 1, using the symbol $(x - y)/(x - y)$:

$$\frac{x+y}{5} \cdot \frac{x-y}{x-y} = \frac{(x+y)(x-y)}{5(x-y)}. \qquad \text{Multiplying by } \frac{x-y}{x-y}, \text{which is 1}$$

We know that the expressions

$$\frac{x+y}{5} \quad \text{and} \quad \frac{(x+y)(x-y)}{5(x-y)}$$

are equivalent. This means that they will name the same number for all replacements that do not make a denominator 0.

EXAMPLES   Multiply to obtain an equivalent expression.

5. $\dfrac{x^2 + 3}{x - 1} \cdot 1 = \dfrac{x^2 + 3}{x - 1} \cdot \dfrac{x + 1}{x + 1} = \dfrac{(x^2 + 3)(x + 1)}{(x - 1)(x + 1)}$ Using $\dfrac{x + 1}{x + 1}$ for 1

6. $1 \cdot \dfrac{x - 4}{x - y} = \dfrac{-1}{-1} \cdot \dfrac{x - 4}{x - y} = \dfrac{-1 \cdot (x - 4)}{-1 \cdot (x - y)}$ Using $\dfrac{-1}{-1}$ for 1

$$= \frac{-x + 4}{-x + y} = \frac{4 - x}{y - x}$$

> **Now Try Exercise 13.**

▶ c Simplifying Rational Expressions

We simplify rational expressions using the identity property of 1 in reverse. That is, we "remove" factors that are equal to 1. We first factor the numerator and the denominator and then factor the rational expression, so that a factor is equal to 1. We also say, accordingly, that we "remove a factor of 1."

EXAMPLE 7   Simplify: $\dfrac{120}{320}$.

$$\frac{120}{320} = \frac{40 \cdot 3}{40 \cdot 8}$$ Factoring the numerator and the denominator, looking for common factors

$$= \frac{40}{40} \cdot \frac{3}{8}$$ Factoring the rational expression; $\dfrac{40}{40}$ is a factor of 1

$$= 1 \cdot \frac{3}{8}$$ $\dfrac{40}{40} = 1$

$$= \frac{3}{8}$$ Removing a factor of 1

EXAMPLES Simplify.

8. $\dfrac{5x^2}{x} = \dfrac{5x \cdot x}{1 \cdot x}$ **Factoring the numerator and the denominator**

$= \dfrac{5x}{1} \cdot \dfrac{x}{x}$ **Factoring the rational expression; $\dfrac{x}{x}$ is a factor of 1**

$= 5x \cdot 1$ $\dfrac{x}{x} = 1$

$= 5x$ **Removing a factor of 1**

In this example, we supplied a 1 in the denominator. This can always be done, but it is not necessary.

9. $\dfrac{4a + 8}{2} = \dfrac{2(2a + 4)}{2 \cdot 1}$ **Factoring the numerator and the denominator**

$= \dfrac{2}{2} \cdot \dfrac{2a + 4}{1}$ **Factoring the rational expression; $\dfrac{2}{2}$ is a factor of 1**

$= \dfrac{2a + 4}{1}$ **Removing a factor of 1**

$= 2a + 4$ **Now Try Exercises 17 and 21.**

EXAMPLES Simplify.

10. $\dfrac{2x^2 + 4x}{6x^2 + 2x} = \dfrac{2x(x + 2)}{2x(3x + 1)}$ **Factoring the numerator and the denominator**

$= \dfrac{2x}{2x} \cdot \dfrac{x + 2}{3x + 1}$ **Factoring the rational expression**

$= \dfrac{x + 2}{3x + 1}$ **Removing a factor of 1**

11. $\dfrac{x^2 - 1}{2x^2 - x - 1} = \dfrac{(x - 1)(x + 1)}{(2x + 1)(x - 1)}$ **Factoring the numerator and the denominator**

$= \dfrac{x + 1}{2x + 1} \cdot \dfrac{x - 1}{x - 1}$ **Factoring the rational expression**

$= \dfrac{x + 1}{2x + 1}$ **Removing a factor of 1**

12. $\dfrac{9x^2 + 6xy - 3y^2}{12x^2 - 12y^2} = \dfrac{3(x + y)(3x - y)}{3(4)(x + y)(x - y)}$ **Factoring the numerator and the denominator**

$= \dfrac{3(x + y)}{3(x + y)} \cdot \dfrac{3x - y}{4(x - y)}$ **Factoring the rational expression**

$= \dfrac{3x - y}{4(x - y)}$ **Removing a factor of 1**

For purposes of later work, we generally do not multiply out the numerator and the denominator after simplifying rational expressions.

Now Try Exercise 27.

Technology Connection

Checking Multiplication and Simplification

We can use the TABLE feature as a partial check that rational expressions have been multiplied and/or simplified correctly. To check the simplification in Example 11,

$$\frac{x^2 - 1}{2x^2 - x - 1} = \frac{x + 1}{2x + 1},$$

we first enter

$$y_1 = (x^2 - 1)/(2x^2 - x - 1) \text{ and}$$
$$y_2 = (x + 1)/(2x + 1).$$

Then, using AUTO mode, we look at a table of values of y_1 and y_2. (See p. 77.) If the simplification is correct, the values should be the same for all replacements for which the rational expression is defined. The ERROR messages indicate that -0.5 and 1 are replacements in the first rational expression for which the expression is not defined and -0.5 is a replacement in the second rational expression for which the expression is not defined. For all other numbers, we see that y_1 and y_2 are the same, so the simplification appears to be correct. Remember, this is only a partial check since we cannot check all possible values of x.

X	Y₁	Y₂
-1.5	.25	.25
-1	0	0
-.5	ERROR	ERROR
0	1	1
.5	.75	.75
1	ERROR	.66667
1.5	.625	.625

X=-1.5

Exercises

Use the TABLE feature to determine whether each of the following is correct.

1. $\dfrac{5x^2}{x} = 5x$

2. $\dfrac{2x^2 + 4x}{6x^2 + 2x} = \dfrac{x + 2}{3x + 1}$

3. $\dfrac{x^2 - 3x + 2}{x^2 - 1} = \dfrac{x + 2}{x - 1}$

4. $\dfrac{x^2 - 16}{x^2 - 4} = 4$

5. $\dfrac{x^2 - 5x}{x^2} \cdot \dfrac{4}{x^2 - 25} = \dfrac{4}{x(x + 5)}$

Canceling

Canceling is a shortcut that you may have used for removing a factor of 1 when working with fraction notation or rational expressions. With great concern, we mention it here as a possible way to speed up your work. **Canceling can be done for removing factors of 1 only in products.** It *cannot* be done in sums or when adding expressions together. Our concern is that canceling be done with care and understanding. Example 12 might have been done faster as follows:

$$\frac{9x^2 + 6xy - 3y^2}{12x^2 - 12y^2} = \frac{\cancel{3}(\cancel{x + y})(3x - y)}{\cancel{3}(4)(\cancel{x + y})(x - y)}$$

$$= \frac{3x - y}{4(x - y)}.$$

When a factor of 1 is noted, it is "canceled" as shown.

Removing a factor of 1:
$\dfrac{3(x + y)}{3(x + y)} = 1$

CAUTION!   The difficulty with canceling is that it can be applied incorrectly in situations such as the following:

$$\frac{\cancel{2} + 3}{\cancel{2}} = 3, \qquad \frac{\cancel{4} + 1}{\cancel{4} + 2} = \frac{1}{2}, \qquad \frac{1\cancel{5}}{\cancel{5}4} = \frac{1}{4}.$$

Wrong!   Wrong!   Wrong!

In each of these situations, the expressions canceled are *not* factors of 1. Factors are parts of products. For example, in $2 \cdot 3$, 2 and 3 are factors, but in $2 + 3$, 2 and 3 are *not* factors. **If you can't factor, you can't cancel! If in doubt, don't cancel!**

Opposites in Rational Expressions

Expressions of the form $a - b$ and $b - a$ are opposites, or additive inverses, of each other. When either of these binomials is multiplied by -1, the result is the other binomial:

$$\left.\begin{array}{l} -1(a - b) = -a + b = b - a; \\ -1(b - a) = -b + a = a - b. \end{array}\right\}$$ **Multiplication by -1 reverses the order in which subtraction occurs.**

Consider

$$\frac{x - 8}{8 - x}.$$

At first glance, the numerator and the denominator do not appear to have any common factors other than 1. But $x - 8$ and $8 - x$ are opposites of each other. Therefore, we can rewrite one as the opposite of the other by factoring out a -1.

EXAMPLE 13   Simplify: $\dfrac{x - 8}{8 - x}$.

$$\begin{aligned} \frac{x - 8}{8 - x} &= \frac{x - 8}{-(x - 8)} \qquad \text{Rewriting } 8 - x \text{ as } -(x - 8). \text{ See Section R.6.}\\ &= \frac{1(x - 8)}{-1(x - 8)} \\ &= \frac{1}{-1} \cdot \frac{x - 8}{x - 8} \\ &= -1 \cdot 1 \qquad \text{Note that } \frac{1}{-1} = -1, \text{ not } 1. \\ &= -1 \end{aligned}$$

Now Try Exercise 33.

▶ d   Multiplying and Simplifying

After multiplying, we generally simplify, if possible. That is one reason why we leave the numerator and the denominator in factored form. Even so, we might need to factor them further in order to simplify.

EXAMPLES Multiply and simplify.

14. $\dfrac{x+2}{x-3} \cdot \dfrac{x^2-4}{x^2+x-2} = \dfrac{(x+2)(x^2-4)}{(x-3)(x^2+x-2)}$ Multiplying the numerators and the denominators

$$= \dfrac{(x+2)(x+2)(x-2)}{(x-3)(x+2)(x-1)}$$ Factoring the numerator and the denominator

$$= \dfrac{(x+2)\cancel{(x+2)}(x-2)}{(x-3)\cancel{(x+2)}(x-1)}$$ Removing a factor of 1: $\dfrac{x+2}{x+2} = 1$

$$= \dfrac{(x+2)(x-2)}{(x-3)(x-1)}$$ Simplifying

15. $\dfrac{a^3-b^3}{a^2-b^2} \cdot \dfrac{a^2+2ab+b^2}{a^2+ab+b^2}$

$$= \dfrac{(a^3-b^3)(a^2+2ab+b^2)}{(a^2-b^2)(a^2+ab+b^2)}$$

$$= \dfrac{(a-b)(a^2+ab+b^2)(a+b)(a+b)}{(a-b)(a+b)(a^2+ab+b^2)\cdot 1}$$ Factoring the numerator and the denominator

$$= \dfrac{\cancel{(a-b)}\cancel{(a^2+ab+b^2)}\cancel{(a+b)}(a+b)}{\cancel{(a-b)}\cancel{(a+b)}\cancel{(a^2+ab+b^2)}\cdot 1}$$

Removing a factor of 1: $\dfrac{(a-b)(a^2+ab+b^2)(a+b)}{(a-b)(a^2+ab+b^2)(a+b)} = 1$

$$= \dfrac{a+b}{1} = a+b$$ Simplifying   **Now Try Exercise 41.**

▶ **e** **Dividing and Simplifying**

Two expressions are *reciprocals* (or *multiplicative inverses*) of each other if their product is 1. To find the reciprocal of a rational expression, we interchange the numerator and the denominator.

The reciprocal of $\dfrac{3}{7}$ is $\dfrac{7}{3}$.   The reciprocal of $y-8$ is $\dfrac{1}{y-8}$.

The reciprocal of $\dfrac{x+2y}{x+y-1}$ is $\dfrac{x+y-1}{x+2y}$.

DIVIDING FRACTION NOTATION

REVIEW SECTION R.2

We divide rational expressions in the same way that we divide fraction notation in arithmetic.

> **Dividing Rational Expressions**
>
> To divide by a rational expression, multiply by its reciprocal:
> $$\dfrac{A}{B} \div \dfrac{C}{D} = \dfrac{A}{B} \cdot \dfrac{D}{C} = \dfrac{AD}{BC}.$$
> Then factor and simplify if possible.

For example,

$$\dfrac{2}{3} \div \dfrac{4}{5} = \dfrac{2}{3} \cdot \dfrac{5}{4} = \dfrac{2 \cdot 5}{3 \cdot 2 \cdot 2} = \dfrac{5}{3 \cdot 2} \cdot \dfrac{2}{2} = \dfrac{5}{6} \cdot 1 = \dfrac{5}{6}.$$

EXAMPLE 16 Divide and simplify.

$$\frac{x-2}{x+1} \div \frac{x+5}{x-3} = \frac{x-2}{x+1} \cdot \frac{x-3}{x+5} = \frac{(x-2)(x-3)}{(x+1)(x+5)}$$

EXAMPLE 17 Divide and simplify.

$$\frac{a^2-1}{a-1} \div \frac{a^2-2a+1}{a+1}$$

$$= \frac{a^2-1}{a-1} \cdot \frac{a+1}{a^2-2a+1} \qquad \text{Multiplying by the reciprocal of the divisor}$$

$$= \frac{(a^2-1)(a+1)}{(a-1)(a^2-2a+1)} \qquad \text{Multiplying the numerators and the denominators}$$

$$= \frac{(a+1)(a-1)(a+1)}{(a-1)(a-1)(a-1)} \qquad \text{Factoring the numerator and the denominator}$$

$$= \frac{(a+1)\cancel{(a-1)}(a+1)}{(a-1)\cancel{(a-1)}(a-1)} \qquad \text{Removing a factor of 1: } \frac{a-1}{a-1} = 1$$

$$= \frac{(a+1)(a+1)}{(a-1)(a-1)} \qquad \text{Simplifying}$$

Now Try Exercise 55.

EXAMPLE 18 Perform the indicated operations and simplify:

$$\frac{c^3-d^3}{(c+d)^2} \div (c-d) \cdot (c+d).$$

Using the rules for order of operations, we do the division first:

$$\frac{c^3-d^3}{(c+d)^2} \div (c-d) \cdot (c+d) = \frac{c^3-d^3}{(c+d)^2} \cdot \frac{1}{c-d} \cdot (c+d)$$

$$= \frac{(c-d)(c^2+cd+d^2)(c+d)}{(c+d)(c+d)(c-d)} = \frac{\cancel{(c-d)}(c^2+cd+d^2)\cancel{(c+d)}}{(c+d)\cancel{(c+d)}\cancel{(c-d)}}$$

$$= \frac{c^2+cd+d^2}{c+d}.$$

Now Try Exercise 65.

5.1 Exercise Set

▶ Reading Check

Choose from selections (a)–(h) below an expression that is equivalent to the given expression.

a) $\dfrac{1}{x} \cdot \dfrac{x}{8}$ **b)** $\dfrac{1}{8x}$ **c)** $\dfrac{8}{x}$ **d)** $\dfrac{x}{8} \div \dfrac{x+1}{8}$ **e)** $\dfrac{x-8}{8}$ **f)** 1   **g)** $8-x$ **h)** $\dfrac{1}{x-8}$

RC1. $\dfrac{1}{x} \div \dfrac{1}{8}$

RC2. The opposite of $x-8$

RC3. The reciprocal of $\dfrac{8}{x-8}$

RC4. $\dfrac{1}{x} \div \dfrac{8}{x}$

RC5. $\dfrac{1}{x} \div 8$

RC6. The reciprocal of $x-8$

RC7. $\dfrac{x}{8} \cdot \dfrac{8}{x}$

RC8. $\dfrac{x}{8} \cdot \dfrac{8}{x+1}$

a *Find all numbers for which the rational expression is not defined.*

1. $\dfrac{5t^2 - 64}{3t + 17}$

2. $\dfrac{x^2 + x + 105}{5x - 45}$

3. $\dfrac{x^3 - x^2 + x + 2}{x^2 + 12x + 35}$

4. $\dfrac{x^2 - 3x - 4}{x^2 - 18x + 77}$

Find the domain. Write both set-builder notation and interval notation for the answer.

5. $f(x) = \dfrac{4x - 5}{x + 7}$

6. $f(r) = \dfrac{5r + 3}{r - 6}$

7. $g(x) = \dfrac{7}{3x - x^2}$

8. $g(x) = \dfrac{9}{8x + x^2}$

9. $f(t) = \dfrac{5t^2 - 64}{3t + 17}$

10. $f(x) = \dfrac{x^2 + x + 105}{5x - 45}$

11. $f(x) = \dfrac{x^3 - x^2 + x + 2}{x^2 + 12x + 35}$

12. $f(x) = \dfrac{x^2 - 3x - 4}{x^2 - 18x + 77}$

b *Multiply to obtain an equivalent expression. Do not simplify.*

13. $\dfrac{7x}{7x} \cdot \dfrac{x + 2}{x + 8}$

14. $\dfrac{2 - y^2}{8 - y} \cdot \dfrac{-1}{-1}$

15. $\dfrac{q - 5}{q + 3} \cdot \dfrac{q + 5}{q + 5}$

16. $\dfrac{p + 1}{p + 4} \cdot \dfrac{p - 4}{p - 4}$

c *Simplify.*

17. $\dfrac{15y^5}{5y^4}$

18. $\dfrac{7w^3}{28w^2}$

19. $\dfrac{16p^3}{24p^7}$

20. $\dfrac{48t^5}{56t^{11}}$

21. $\dfrac{9a - 27}{9}$

22. $\dfrac{6a - 30}{6}$

23. $\dfrac{12x - 15}{21}$

24. $\dfrac{18a - 2}{22}$

25. $\dfrac{4y - 12}{4y + 12}$

26. $\dfrac{8x + 16}{8x - 16}$

27. $\dfrac{t^2 - 16}{t^2 - 8t + 16}$

28. $\dfrac{p^2 - 25}{p^2 + 10p + 25}$

29. $\dfrac{x^2 - 9x + 8}{x^2 + 3x - 4}$

30. $\dfrac{y^2 + 8y - 9}{y^2 - 5y + 4}$

31. $\dfrac{w^3 - z^3}{w^2 - z^2}$

32. $\dfrac{a^2 - b^2}{a^3 + b^3}$

33. $\dfrac{10 - x}{x - 10}$

34. $\dfrac{x - 8}{8 - x}$

35. $\dfrac{a - b}{4b - 4a}$

36. $\dfrac{2p - 2q}{8 - p}$

d *Multiply and simplify.*

37. $\dfrac{x^4}{3x + 6} \cdot \dfrac{5x + 10}{5x^7}$

38. $\dfrac{10t}{6t - 12} \cdot \dfrac{20t - 40}{30t^3}$

39. $\dfrac{x^2 - 16}{x^2} \cdot \dfrac{x^2 - 4x}{x^2 - x - 12}$

40. $\dfrac{y^2 + 10y + 25}{y^2 - 9} \cdot \dfrac{y^2 - 3y}{y + 5}$

41. $\dfrac{y^2 - 16}{2y + 6} \cdot \dfrac{y + 3}{y - 4}$

42. $\dfrac{m^2 - n^2}{4m + 4n} \cdot \dfrac{m + n}{m - n}$

43. $\dfrac{x^2 - 2x - 35}{2x^3 - 3x^2} \cdot \dfrac{4x^3 - 9x}{7x - 49}$

44. $\dfrac{y^2 - 10y + 9}{y^2 - 1} \cdot \dfrac{y + 4}{y^2 - 5y - 36}$

45. $\dfrac{c^3 + 8}{c^2 - 4} \cdot \dfrac{c^2 - 4c + 4}{c^2 - 2c + 4}$

46. $\dfrac{x^3 - 27}{x^2 - 9} \cdot \dfrac{x^2 - 6x + 9}{x^2 + 3x + 9}$

47. $\dfrac{x^2 - y^2}{x^3 - y^3} \cdot \dfrac{x^2 + xy + y^2}{x^2 + 2xy + y^2}$

48. $\dfrac{4x^2 - 9y^2}{8x^3 - 27y^3} \cdot \dfrac{4x^2 + 6xy + 9y^2}{4x^2 + 12xy + 9y^2}$

e *Divide and simplify.*

49. $\dfrac{12x^8}{3y^4} \div \dfrac{16x^3}{6y}$

50. $\dfrac{9a^7}{8b^2} \div \dfrac{12a^2}{24b^7}$

51. $\dfrac{3y + 15}{y} \div \dfrac{y + 5}{y}$

52. $\dfrac{6x + 12}{x} \div \dfrac{x + 2}{x^3}$

53. $\dfrac{y^2 - 9}{y} \div \dfrac{y + 3}{y + 2}$

54. $\dfrac{x^2 - 4}{x} \div \dfrac{x - 2}{x + 4}$

55. $\dfrac{4a^2 - 1}{a^2 - 4} \div \dfrac{2a - 1}{a - 2}$

56. $\dfrac{25x^2 - 4}{x^2 - 9} \div \dfrac{5x - 2}{x + 3}$

57. $\dfrac{x^2 - 16}{x^2 - 10x + 25} \div \dfrac{3x - 12}{x^2 - 3x - 10}$

58. $\dfrac{y^2 - 36}{y^2 - 8y + 16} \div \dfrac{3y - 18}{y^2 - y - 12}$

59. $\dfrac{y^3 + 3y}{y^2 - 9} \div \dfrac{y^2 + 5y - 14}{y^2 + 4y - 21}$

60. $\dfrac{a^3 + 4a}{a^2 - 16} \div \dfrac{a^2 + 8a + 15}{a^2 + a - 20}$

61. $\dfrac{x^3 - 64}{x^3 + 64} \div \dfrac{x^2 - 16}{x^2 - 4x + 16}$

62. $\dfrac{8y^3 + 27}{64y^3 - 1} \div \dfrac{4y^2 - 9}{16y^2 + 4y + 1}$

63. $\dfrac{8x^3y^3 + 27x^3}{64x^3y^3 - x^3} \div \dfrac{4x^2y^2 - 9x^2}{16x^2y^2 + 4x^2y + x^2}$

64. $\dfrac{x^3y - 64y}{x^3y + 64y} \div \dfrac{x^2y^2 - 16y^2}{x^2y^2 - 4xy^2 + 16y^2}$

Perform the indicated operations and simplify.

65. $\dfrac{r^2 - 4s^2}{r + 2s} \div (r + 2s) \cdot \dfrac{2s}{r - 2s}$

66. $\dfrac{d^2 - d}{d^2 - 6d + 8} \cdot \dfrac{d - 2}{d^2 + 5d} \div \dfrac{5d}{d^2 - 9d + 20}$

67. $\dfrac{y^2 - 2y}{y^2 + y - 2} \cdot \dfrac{y - 1}{y^2 + 4y + 4} \div \dfrac{y^2 + 2y - 8}{y^4}$

68. $\dfrac{9x^2}{x^2 - 16y^2} \div \dfrac{1}{x^2 + 4xy} \cdot \dfrac{x - 4y}{3x}$

▶ **Skill Maintenance**

Determine the domain and the range of each function. **[2.3a]**

69.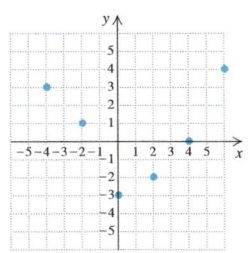

70.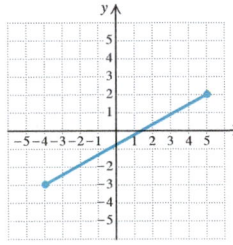

71.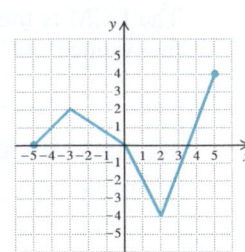

72.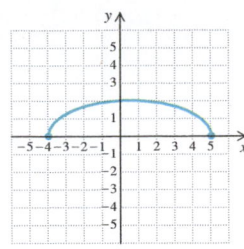

Factor. **[4.7a]**

73. $21p^2 + p - 10$

74. $12m^2 - 26m - 10$

75. $2x^3 - 16x^2 - 66x$

76. $10y^2 + 80y - 650$

77. Find an equation of the line with slope $-\frac{2}{3}$ and y-intercept $(0, -5)$. **[2.7a]**

78. Find an equation of the line having slope $-\frac{2}{7}$ and containing the point $(-4, 8)$. **[2.7b]**

▶ **Synthesis**

Simplify.

79. $\dfrac{x(x + 1) - 2(x + 3)}{(x + 1)(x + 2)(x + 3)}$

80. $\dfrac{a^3 - 2a^2 + 2a - 4}{a^3 - 2a^2 - 3a + 6}$

81. Let $g(x) = \dfrac{2x + 3}{4x - 1}$. Find $g(5)$, $g(0)$, $g\left(\frac{1}{4}\right)$, and $g(a + h)$.

5.2   LCMs, LCDs, Addition, and Subtraction

▶ **a**   Find the LCM of several algebraic expressions by factoring.

▶ **b**   Add and subtract rational expressions.

▶ **c**   Simplify combined additions and subtractions of rational expressions.

▶ a   Finding LCMs by Factoring

To add rational expressions when denominators are different, we first find a common denominator. Let's review the procedure used in arithmetic first. To do the addition

$$\frac{5}{42} + \frac{7}{12},$$

we find a common denominator. We look for the least common multiple (LCM) of 42 and 12. That number becomes the least common denominator (LCD).

To find the LCM, we factor both numbers completely (into primes).

$$42 = 2 \cdot 3 \cdot 7 \longleftarrow \text{ Any multiple of 42 has these factors.}$$

$$12 = 2 \cdot 2 \cdot 3 \longleftarrow \text{ Any multiple of 12 has these factors.}$$

The LCM is the number that has 2 as a factor twice, 3 as a factor once, and 7 as a factor once: LCM $= 2 \cdot 2 \cdot 3 \cdot 7$, or 84.

FINDING LCMs

To find the LCM, use each factor the greatest number of times that it occurs in any one prime factorization.

EXAMPLE 1 Find the LCM of 18 and 24.

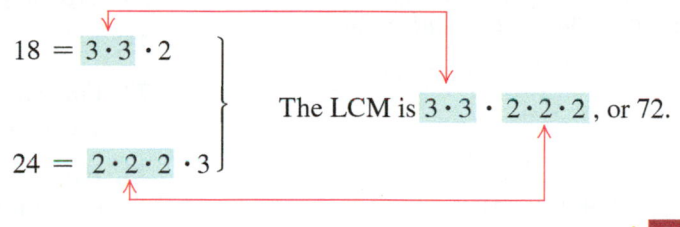

$$18 = 3 \cdot 3 \cdot 2$$

$$24 = 2 \cdot 2 \cdot 2 \cdot 3$$

The LCM is $3 \cdot 3 \cdot 2 \cdot 2 \cdot 2$, or 72.

Now Try Exercise 1.

Now let's return to adding $\frac{5}{42}$ and $\frac{7}{12}$:

$$\frac{5}{42} + \frac{7}{12} = \frac{5}{2 \cdot 3 \cdot 7} + \frac{7}{2 \cdot 2 \cdot 3}. \qquad \text{\color{red}{Factoring the denominators}}$$

The LCD is the LCM of the denominators, $2 \cdot 2 \cdot 3 \cdot 7$. To get this LCD in the first denominator, we need a factor of 2. In the second denominator, we need a factor of 7. We multiply by 1, as follows:

$$\frac{5}{2 \cdot 3 \cdot 7} \cdot \frac{2}{2} + \frac{7}{2 \cdot 2 \cdot 3} \cdot \frac{7}{7} = \frac{10}{2 \cdot 2 \cdot 3 \cdot 7} + \frac{49}{2 \cdot 2 \cdot 3 \cdot 7} = \frac{59}{2 \cdot 2 \cdot 3 \cdot 7} = \frac{59}{84}.$$

Multiplying the first fraction by $\frac{2}{2}$ gave us an equivalent fraction with a denominator that is the LCD. Multiplying the second fraction by $\frac{7}{7}$ also gave us an equivalent fraction with a denominator that is the LCD. Once we had a common denominator, we added the numerators and keep the common denominator.

We find the LCM of algebraic expressions in the same way that we find the LCM of natural numbers.

Our reasoning for learning how to find LCMs is so that we will be able to add rational expressions. For example, to do the addition

$$\frac{7}{12xy^2} + \frac{8}{15x^3y},$$

we first need to find the LCM of $12xy^2$ and $15x^3y$, which is $60x^3y^2$.

EXAMPLES

2. Find the LCM of $12xy^2$ and $15x^3y$.

We factor each expression completely. To find the LCM, we use each factor the greatest number of times that it occurs in any one prime factorization.

$$12xy^2 = 2 \cdot 2 \cdot 3 \cdot x \cdot y \cdot y;$$
$$15x^3y = 3 \cdot 5 \cdot x \cdot x \cdot x \cdot y \qquad \text{\color{red}{Factoring}}$$

$$\text{\color{red}{$12xy^2$ is a factor.}}$$
$$\text{LCM} = 2 \cdot 2 \cdot 3 \cdot 5 \cdot x \cdot x \cdot x \cdot y \cdot y = 60x^3y^2$$
$$\text{\color{red}{$15x^3y$ is a factor.}}$$

The LCM of $12xy^2$ and $15x^3y$ is $60x^3y^2$.

3. Find the LCM of $x^2 + 2x + 1$, $5x^2 - 5x$, and $x^2 - 1$.

$$x^2 + 2x + 1 = (x + 1)(x + 1);$$
$$5x^2 - 5x = 5x(x - 1);$$
$$x^2 - 1 = (x + 1)(x - 1)$$

Factoring

Both factors of $x^2 - 1$ are already present in the previous factorizations.

$$\text{LCM} = 5x(x + 1)(x + 1)(x - 1)$$

4. Find the LCM of $x^2 - y^2$, $x^3 + y^3$, and $x^2 + 2xy + y^2$.

$$x^2 - y^2 = (x - y)(x + y);$$
$$x^3 + y^3 = (x + y)(x^2 - xy + y^2);$$
$$x^2 + 2xy + y^2 = (x + y)(x + y) = (x + y)^2$$

Factoring

$$\text{LCM} = (x - y)(x + y)^2(x^2 - xy + y^2)$$

> **Now Try Exercises 15 and 21.**

Recall that $-(x - 3) = -1(x - 3) = 3 - x$. If $(x - 3)(x + 2)$ is an LCM, then $-1(x - 3)(x + 2) = (3 - x)(x + 2)$ is also an LCM.

If, when we are finding LCMs, factors that are opposites occur, we do not use both of them. For example, if $a - b$ occurs in one factorization and $b - a$ occurs in another, we do not use both, since they are opposites.

EXAMPLE 5 Find the LCM of $x^2 - y^2$ and $3y - 3x$.

$$x^2 - y^2 = (x + y)(x - y)$$
$$3y - 3x = 3(y - x), \text{ or } -3(x - y)$$

We can use $(x - y)$ or $(y - x)$, but we do not use both.

$$\text{LCM} = 3(x + y)(x - y), \text{ or } 3(x + y)(y - x), \text{ or } -3(x + y)(x - y)$$

In most cases, we would use the form $3(x + y)(x - y)$.

> **Now Try Exercise 23.**

▶ b Adding and Subtracting Rational Expressions

> **Addition and Subtraction with Like Denominators**
>
> To add or subtract when denominators are the same, add or subtract the numerators and keep the same denominator.
>
> $$\frac{A}{C} + \frac{B}{C} = \frac{A + B}{C} \quad \text{and} \quad \frac{A}{C} - \frac{B}{C} = \frac{A - B}{C}, \quad \text{where } C \neq 0.$$
>
> Then factor and simplify if possible.

EXAMPLE 6 Add: $\dfrac{3 + x}{x} + \dfrac{4}{x}$.

$$\frac{3 + x}{x} + \frac{4}{x} = \frac{3 + x + 4}{x}$$

Adding numerators and keeping the same denominator

$$= \frac{7 + x}{x}$$

CAUTION! This expression does *not* simplify to 7: $\dfrac{7 + x}{x} \neq 7$.

Example 6 shows that

$$\frac{3+x}{x} + \frac{4}{x} \quad \text{and} \quad \frac{7+x}{x}$$

are equivalent expressions. They name the same number for all replacements for which the rational expressions are defined.

EXAMPLE 7 Add: $\dfrac{4x^2 - 5xy}{x^2 - y^2} + \dfrac{2xy - y^2}{x^2 - y^2}$.

$$\frac{4x^2 - 5xy}{x^2 - y^2} + \frac{2xy - y^2}{x^2 - y^2} = \frac{4x^2 - 3xy - y^2}{x^2 - y^2} \quad \color{red}{\text{Adding the numerators}}$$

$$= \frac{(4x + y)(x - y)}{(x + y)(x - y)} \quad \color{red}{\text{Factoring the numerator and the denominator}}$$

$$= \frac{(4x + y)\cancel{(x - y)}}{(x + y)\cancel{(x - y)}} \quad \color{red}{\text{Removing a factor of 1: } \frac{x - y}{x - y} = 1}$$

$$= \frac{4x + y}{x + y}$$

▶ **Now Try Exercise 35.**

EXAMPLE 8 Subtract: $\dfrac{4x + 5}{x + 3} - \dfrac{x - 2}{x + 3}$.

$$\frac{4x + 5}{x + 3} - \frac{x - 2}{x + 3} = \frac{4x + 5 - (x - 2)}{x + 3} \quad \color{red}{\text{Subtracting numerators}}$$

$$= \frac{4x + 5 - x + 2}{x + 3}$$

$$= \frac{3x + 7}{x + 3}$$

> **A common error: forgetting these parentheses. If you forget them, you will be subtracting only *part* of the second numerator.**

▶ **Now Try Exercise 33.**

Addition and Subtraction with Different Denominators

To add or subtract rational expressions with different denominators:

1. Find the LCM of the denominators. This is the least common denominator (LCD).
2. For each rational expression, find an equivalent expression with the LCD. To do so, multiply by 1 using an expression for 1 made up of factors of the LCD that are missing from the original denominator.
3. Add or subtract the numerators. Write the result over the LCD.
4. Simplify, if possible.

When denominators are different, we find the least common denominator, LCD. The procedure we will use is as shown at left.

EXAMPLE 9 Add: $\dfrac{2a}{5} + \dfrac{3b}{2a}$.

We first find the LCD: $\left.\begin{array}{c} 5 \\ 2a \end{array}\right\}$ LCD $= 5 \cdot 2a$, or $10a$.

Now we multiply each expression by 1. We choose symbols for 1 that will give us the LCD in each denominator. In this case, we use $2a/(2a)$ and $5/5$:

$$\frac{2a}{5} \cdot \frac{2a}{2a} + \frac{3b}{2a} \cdot \frac{5}{5} = \frac{4a^2}{10a} + \frac{15b}{10a} = \frac{4a^2 + 15b}{10a}.$$

Multiplying the first term by $2a/(2a)$ gave us a denominator of $10a$. Multiplying the second term by $\frac{5}{5}$ also gave us a denominator of $10a$.

▶

EXAMPLE 10 Add: $\dfrac{3x^2 + 3xy}{x^2 - y^2} + \dfrac{2 - 3x}{x - y}$.

We first find the LCD of the denominators. To do so, we first factor:

$$\left.\begin{array}{l} x^2 - y^2 = (x + y)(x - y) \\ x - y = x - y \end{array}\right\} \quad \text{LCD} = (x + y)(x - y).$$

The first expression already has the LCD. We multiply by 1 to get the LCD in the second expression. Then we add and simplify if possible.

$$\frac{3x^2 + 3xy}{(x + y)(x - y)} + \frac{2 - 3x}{x - y} \cdot \frac{x + y}{x + y} \qquad \text{Multiplying by 1 to get the LCD}$$

$$= \frac{3x^2 + 3xy}{(x + y)(x - y)} + \frac{(2 - 3x)(x + y)}{(x - y)(x + y)}$$

$$= \frac{3x^2 + 3xy}{(x + y)(x - y)} + \frac{2x + 2y - 3x^2 - 3xy}{(x - y)(x + y)} \qquad \begin{array}{l}\text{Multiplying in the}\\ \text{second numerator}\end{array}$$

$$= \frac{3x^2 + 3xy + 2x + 2y - 3x^2 - 3xy}{(x + y)(x - y)} \qquad \text{Adding the numerators}$$

$$= \frac{2x + 2y}{(x + y)(x - y)} \qquad \text{Combining like terms}$$

$$= \frac{2(x + y)}{(x + y)(x - y)} \qquad \text{Factoring the numerator}$$

$$= \frac{2\cancel{(x + y)}}{\cancel{(x + y)}(x - y)} \qquad \text{Removing a factor of 1: } \frac{x + y}{x + y} = 1$$

$$= \frac{2}{x - y}$$

> **Now Try Exercise 45.**

EXAMPLE 11 Subtract: $\dfrac{2y + 1}{y^2 - 7y + 6} - \dfrac{y + 3}{y^2 - 5y - 6}$.

$$\frac{2y + 1}{y^2 - 7y + 6} - \frac{y + 3}{y^2 - 5y - 6}$$

$$= \frac{2y + 1}{(y - 6)(y - 1)} - \frac{y + 3}{(y - 6)(y + 1)} \qquad \text{LCD} = (y - 6)(y - 1)(y + 1)$$

$$= \frac{2y + 1}{(y - 6)(y - 1)} \cdot \frac{y + 1}{y + 1} - \frac{y + 3}{(y - 6)(y + 1)} \cdot \frac{y - 1}{y - 1} \qquad \begin{array}{l}\text{Multiplying by 1}\\ \text{to get the LCD}\end{array}$$

$$= \frac{(2y + 1)(y + 1) - (y + 3)(y - 1)}{(y - 6)(y - 1)(y + 1)} \qquad \text{Subtracting the numerators}$$

$$= \frac{(2y^2 + 3y + 1) - (y^2 + 2y - 3)}{(y - 6)(y - 1)(y + 1)} \qquad \begin{array}{l}\text{Multiplying. Note the use of}\\ \text{parentheses.}\end{array}$$

$$= \frac{2y^2 + 3y + 1 - y^2 - 2y + 3}{(y - 6)(y - 1)(y + 1)}$$

$$= \frac{y^2 + y + 4}{(y - 6)(y - 1)(y + 1)} \qquad \begin{array}{l}\text{The numerator cannot be factored.}\\ \text{The rational expression is simplified.}\end{array}$$

> **Now Try Exercise 55.**

> **SUBTRACTING POLYNOMIALS**
>
> REVIEW SECTION 4.1

We generally do not multiply out a numerator or a denominator if it has two or more factors (other than monomials). This will be helpful when we solve equations.

Denominators That Are Opposites

When one denominator is the opposite of the other, we can first multiply either expression by 1 using $-1/-1$.

EXAMPLE 12 Add: $\dfrac{a}{2a} + \dfrac{a^3}{-2a}$.

$$\dfrac{a}{2a} + \dfrac{a^3}{-2a} = \dfrac{a}{2a} + \dfrac{a^3}{-2a} \cdot \dfrac{-1}{-1} \qquad \text{\color{red}{Multiplying by 1, using }} \dfrac{-1}{-1}$$

> This is equal to **1** (not -1).

$$= \dfrac{a}{2a} + \dfrac{-a^3}{2a}$$

$$= \dfrac{a - a^3}{2a} \qquad \text{\color{red}{Adding numerators}}$$

$$= \dfrac{a(1 + a)(1 - a)}{2a} \qquad \text{\color{red}{Factoring}}$$

$$= \dfrac{\cancel{a}(1 + a)(1 - a)}{2\cancel{a}} \qquad \text{\color{red}{Removing a factor of 1: }} \dfrac{a}{a} = 1$$

$$= \dfrac{(1 + a)(1 - a)}{2}$$

> **Now Try Exercise 35.**

EXAMPLE 13 Subtract: $\dfrac{x^2}{5y} - \dfrac{x^3}{-5y}$.

$$\dfrac{x^2}{5y} - \dfrac{x^3}{-5y} = \dfrac{x^2}{5y} - \dfrac{x^3}{-5y} \cdot \dfrac{-1}{-1} \qquad \text{\color{red}{Multiplying by }} \dfrac{-1}{-1}$$

$$= \dfrac{x^2}{5y} - \dfrac{-x^3}{5y}$$

$$= \dfrac{x^2 - (-x^3)}{5y} \qquad \text{Don't forget these parentheses!}$$

$$= \dfrac{x^2 + x^3}{5y}, \text{ or } \dfrac{x^2(1 + x)}{5y}$$

EXAMPLE 14 Subtract: $\dfrac{5x}{x - 2y} - \dfrac{3y - 7}{2y - x}$.

$$\dfrac{5x}{x - 2y} - \dfrac{3y - 7}{2y - x} = \dfrac{5x}{x - 2y} - \dfrac{3y - 7}{2y - x} \cdot \dfrac{-1}{-1}$$

$$= \dfrac{5x}{x - 2y} - \dfrac{-3y + 7}{x - 2y} \qquad \text{Remember: } (2y - x)(-1) = -2y + x = x - 2y.$$

$$= \dfrac{5x - (-3y + 7)}{x - 2y} \qquad \text{\color{red}{Subtracting numerators}}$$

$$= \dfrac{5x + 3y - 7}{x - 2y}$$

> **Now Try Exercise 37.**

▶ c Combined Additions and Subtractions

EXAMPLE 15 Perform the indicated operations and simplify.

$$\frac{2x}{x^2 - 4} + \frac{5}{2 - x} - \frac{1}{2 + x}$$

$$= \frac{2x}{(x - 2)(x + 2)} + \frac{5}{2 - x} - \frac{1}{2 + x}$$

$$= \frac{2x}{(x - 2)(x + 2)} + \frac{5}{2 - x} \cdot \frac{-1}{-1} - \frac{1}{x + 2} \qquad \text{Multiplying by } \frac{-1}{-1}$$

$$= \frac{2x}{(x - 2)(x + 2)} + \frac{-5}{x - 2} - \frac{1}{x + 2} \qquad \text{LCD} = (x - 2)(x + 2)$$

$$= \frac{2x}{(x - 2)(x + 2)} + \frac{-5}{x - 2} \cdot \frac{x + 2}{x + 2} - \frac{1}{x + 2} \cdot \frac{x - 2}{x - 2} \qquad \begin{matrix}\text{Multiplying by 1}\\\text{to get the LCD}\end{matrix}$$

$$= \frac{2x - 5(x + 2) - (x - 2)}{(x - 2)(x + 2)} \qquad \text{Adding and subtracting the numerators}$$

$$= \frac{2x - 5x - 10 - x + 2}{(x - 2)(x + 2)} \qquad \text{Removing parentheses}$$

$$= \frac{-4x - 8}{(x - 2)(x + 2)} = \frac{-4(x + 2)}{(x - 2)(x + 2)} \qquad \text{Removing a factor of 1: } \frac{x + 2}{x + 2} = 1$$

$$= \frac{-4}{x - 2}, \text{ or } -\frac{4}{x - 2}$$

Another correct form of the answer is $\dfrac{4}{2 - x}$. It is found by multiplying by $-1/-1$.

Now Try Exercise 71.

5.2 Exercise Set

▶ Reading Check

When we are subtracting rational expressions, it is important to use parentheses to make sure that we subtract the entire numerator. In Exercises RC1–RC3, complete each numerator by (a) filling in the parentheses, (b) removing the parentheses, and (c) collecting like terms.

RC1. $\dfrac{10x}{x - 7} - \dfrac{3x + 5}{x - 7} = \overset{\text{(a)}}{\dfrac{10x - (\qquad)}{x - 7}} = \overset{\text{(b)}}{\dfrac{}{x - 7}} = \overset{\text{(c)}}{\dfrac{}{x - 7}}$

RC2. $\dfrac{7}{4 + a} - \dfrac{4 - 9a}{4 + a} = \overset{\text{(a)}}{\dfrac{7 - (\qquad)}{4 + a}} = \overset{\text{(b)}}{\dfrac{}{4 + a}} = \overset{\text{(c)}}{\dfrac{}{4 + a}}$

RC3. $\dfrac{9y - 2}{y^2 - 10} - \dfrac{y + 1}{y^2 - 10} = \overset{\text{(a)}}{\dfrac{9y - 2 - (\qquad)}{y^2 - 10}} = \overset{\text{(b)}}{\dfrac{}{y^2 - 10}} = \overset{\text{(c)}}{\dfrac{}{y^2 - 10}}$

a *Find the LCM by factoring.*

1. 15, 40   **2.** 12, 32   **3.** 18, 48

4. 45, 54   **5.** 30, 105   **6.** 24, 60

7. 9, 15, 5   **8.** 27, 35, 63

Add. Find the LCD first.

9. $\dfrac{5}{6} + \dfrac{4}{15}$ **10.** $\dfrac{5}{12} + \dfrac{13}{18}$

11. $\dfrac{7}{36} + \dfrac{1}{24}$ **12.** $\dfrac{11}{30} + \dfrac{19}{75}$

13. $\dfrac{3}{4} + \dfrac{7}{30} + \dfrac{1}{16}$ **14.** $\dfrac{5}{8} + \dfrac{7}{12} + \dfrac{11}{40}$

Find the LCM.

15. $21x^2y,\ 7xy$ **16.** $18a^2b,\ 50ab^3$

17. $y^2 - 100,\ 10y + 100$ **18.** $r^2 - s^2,\ rs + s^2$

19. $15ab^2,\ 3ab,\ 10a^3b$ **20.** $6x^2y^2,\ 9x^3y,\ 15y^3$

21. $5y - 15,\ y^2 - 6y + 9$

22. $x^2 + 10x + 25,\ x^2 + 2x - 15$

23. $y^2 - 25,\ 5 - y$ **24.** $x^2 - 36,\ 6 - x$

25. $2r^2 - 5r - 12,\ 3r^2 - 13r + 4,\ r^2 - 16$

26. $2x^2 - 5x - 3,\ 2x^2 - x - 1,\ x^2 - 6x + 9$

27. $x^5 + 4x^3,\ x^3 - 4x^2 + 4x$

28. $9x^3 + 9x^2 - 18x,\ 6x^5 + 24x^4 + 24x^3$

29. $x^5 - 2x^4 + x^3,\ 2x^3 + 2x,\ 5x + 5$

30. $x^5 - 4x^4 + 4x^3,\ 3x^2 - 12,\ 2x + 4$

b *Add or subtract. Then simplify. If a denominator has two or more factors (other than monomials), leave it in factored form.*

31. $\dfrac{x - 2y}{x + y} + \dfrac{x + 9y}{x + y}$ **32.** $\dfrac{a - 8b}{a + b} + \dfrac{a + 13b}{a + b}$

33. $\dfrac{4y + 3}{y - 2} - \dfrac{y - 2}{y - 2}$ **34.** $\dfrac{3t + 2}{t - 4} - \dfrac{t - 4}{t - 4}$

35. $\dfrac{a^2}{a - b} + \dfrac{b^2}{b - a}$ **36.** $\dfrac{r^2}{r - s} + \dfrac{s^2}{s - r}$

37. $\dfrac{6}{y} - \dfrac{7}{-y}$ **38.** $\dfrac{4}{x} - \dfrac{9}{-x}$

39. $\dfrac{4a - 2}{a^2 - 49} + \dfrac{5 + 3a}{49 - a^2}$ **40.** $\dfrac{2y - 3}{y^2 - 1} - \dfrac{4 - y}{1 - y^2}$

41. $\dfrac{a^3}{a - b} + \dfrac{b^3}{b - a}$ **42.** $\dfrac{x^3}{x^2 - y^2} + \dfrac{y^3}{y^2 - x^2}$

43. $\dfrac{y - 2}{y + 4} + \dfrac{y + 3}{y - 5}$

44. $\dfrac{x - 2}{x + 3} + \dfrac{x + 2}{x - 4}$

45. $\dfrac{4xy}{x^2 - y^2} + \dfrac{x - y}{x + y}$

46. $\dfrac{5ab}{a^2 - b^2} + \dfrac{a + b}{a - b}$

47. $\dfrac{9x + 2}{3x^2 - 2x - 8} + \dfrac{7}{3x^2 + x - 4}$

48. $\dfrac{3y + 2}{2y^2 - y - 10} + \dfrac{8}{2y^2 - 7y + 5}$

49. $\dfrac{4}{x + 1} + \dfrac{x + 2}{x^2 - 1} + \dfrac{3}{x - 1}$

50. $\dfrac{-2}{y + 2} + \dfrac{5}{y - 2} + \dfrac{y + 3}{y^2 - 4}$

51. $\dfrac{x - 1}{3x + 15} - \dfrac{x + 3}{5x + 25}$

52. $\dfrac{y - 2}{4y + 8} - \dfrac{y + 6}{5y + 10}$

53. $\dfrac{5ab}{a^2 - b^2} - \dfrac{a - b}{a + b}$

54. $\dfrac{6xy}{x^2 - y^2} - \dfrac{x + y}{x - y}$

55. $\dfrac{3y}{y^2 - 7y + 10} - \dfrac{2y}{y^2 - 8y + 15}$

56. $\dfrac{5x}{x^2 - 6x + 8} - \dfrac{3x}{x^2 - x - 12}$

57. $\dfrac{y}{y^2 - y - 20} + \dfrac{2}{y + 4}$

58. $\dfrac{6}{y^2 + 6y + 9} + \dfrac{5}{y^2 - 9}$

59. $\dfrac{3y + 2}{y^2 + 5y - 24} + \dfrac{7}{y^2 + 4y - 32}$

60. $\dfrac{3y + 2}{y^2 - 7y + 10} + \dfrac{2y}{y^2 - 8y + 15}$

61. $\dfrac{3x - 1}{x^2 + 2x - 3} - \dfrac{x + 4}{x^2 - 9}$

62. $\dfrac{3p - 2}{p^2 + 2p - 24} - \dfrac{p - 3}{p^2 - 16}$

c *Perform the indicated operations and simplify.*

63. $\dfrac{1}{x+1} - \dfrac{x}{x-2} + \dfrac{x^2+2}{x^2-x-2}$

64. $\dfrac{2}{y+3} - \dfrac{y}{y-1} + \dfrac{y^2+2}{y^2+2y-3}$

65. $\dfrac{y-3}{y-4} - \dfrac{y+2}{y+4} + \dfrac{y-7}{y^2-16}$

66. $\dfrac{x-1}{x-2} - \dfrac{x+1}{x+2} + \dfrac{x-6}{x^2-4}$

67. $\dfrac{y+2}{y+4} + \dfrac{y-7}{y^2-16} - \dfrac{y-3}{y-4}$

68. $\dfrac{x-6}{x^2-4} - \dfrac{x-1}{x-2} - \dfrac{x+1}{x+2}$

69. $\dfrac{4x}{x^2-1} + \dfrac{3x}{1-x} - \dfrac{4}{x-1}$

70. $\dfrac{5y}{1-2y} - \dfrac{2y}{2y+1} + \dfrac{3}{4y^2-1}$

71. $\dfrac{1}{x+y} + \dfrac{1}{y-x} - \dfrac{2x}{x^2-y^2}$

72. $\dfrac{1}{b-a} + \dfrac{1}{a+b} - \dfrac{2b}{a^2-b^2}$

73. $\dfrac{x+5}{x-3} - \dfrac{x+2}{x+1} - \dfrac{6x+10}{x^2-2x-3}$

74. $\dfrac{13x+2}{x^2+3x-10} - \dfrac{x+2}{x-2} + \dfrac{x-3}{x+5}$

▶ Skill Maintenance

Graph. **[3.7a]**

75. $2x - 3y > 6$

76. $y - x > 3$

77. $5x + 3y \le 15$

78. $5x - 3y \le 15$

Factor. **[4.6d]**

79. $t^3 - 8$

80. $q^3 + 125$

81. $23x^4 + 23x$

82. $64a^3 - 27b^3$

83. Find an equation of the line that passes through the point $(4, -6)$ and is parallel to the line $3y + 8x = 10$. **[2.7d]**

84. Find an equation of the line that passes through the point $(-2, 3)$ and is perpendicular to the line $5y + 4x = 7$. **[2.7d]**

▶ Synthesis

85. Determine the domain and the range of the function graphed below.

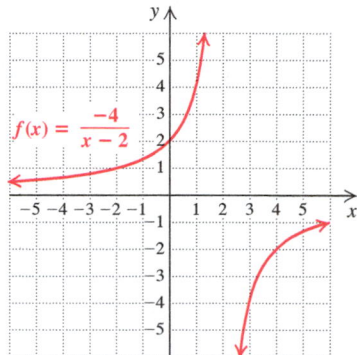

Find the LCM.

86. 18, 42, 82, 120, 300, 700

87. $x^8 - x^4,\ x^5 - x^2,\ x^5 - x^3,\ x^5 + x^2$

88. The LCM of two expressions is $8a^4b^7$. One of the expressions is $2a^3b^7$. List all possibilities for the other expression.

Perform the indicated operations and simplify.

89. $\dfrac{x+y+1}{y-(x+1)} + \dfrac{x+y-1}{x-(y-1)} - \dfrac{x-y-1}{1-(y-x)}$

90. $\dfrac{b-c}{a-(b-c)} - \dfrac{b-a}{(b-a)-c}$

91. $\dfrac{x}{x^4-y^4} - \dfrac{1}{x^2+2xy+y^2}$

92. $\dfrac{x^2}{3x^2-5x-2} - \dfrac{2x}{3x+1} \cdot \dfrac{1}{x-2}$

5.3   Division of Polynomials

▶ **a** Divide a polynomial by a monomial.

▶ **b** Divide a polynomial by a divisor that is not a monomial, and if there is a remainder, express the result in two ways.

▶ **c** Use synthetic division to divide a polynomial by a binomial of the type $x - a$.

A rational expression represents division. "Long" division of polynomials, like division of real numbers, relies on our multiplication and subtraction skills.

▶ **a**   Divisor a Monomial

RULES OF EXPONENTS

REVIEW SECTION R.7

We first consider division by a monomial (a term like $45x^{10}$ or $48a^2b^5$). When we are dividing a monomial by a monomial, we can use the rules of exponents and subtract exponents when the bases are the same. For example,

$$\frac{45x^{10}}{3x^4} = \frac{45}{3}x^{10-4} = 15x^6 \quad \text{and} \quad \frac{48a^2b^5}{-3ab^2} = \frac{48}{-3}a^{2-1}b^{5-2} = -16ab^3.$$

When we divide a polynomial by a monomial, we break up the division into a sum of quotients of monomials. To do so, we reverse the rule for adding fractions. That is, since

$$\frac{A}{C} + \frac{B}{C} = \frac{A+B}{C}, \quad \text{we know that} \quad \frac{A+B}{C} = \frac{A}{C} + \frac{B}{C}.$$

EXAMPLE 1 Divide $12x^3 + 8x^2 + x + 4$ by $4x$.

$$\frac{12x^3 + 8x^2 + x + 4}{4x} \qquad \text{Writing a fraction expression}$$

$$= \frac{12x^3}{4x} + \frac{8x^2}{4x} + \frac{x}{4x} + \frac{4}{4x} \qquad \text{Dividing each term of the numerator by the monomial}$$

$$= 3x^2 + 2x + \frac{1}{4} + \frac{1}{x} \qquad \text{Doing the four indicated divisions}$$

Now Try Exercise 1.

EXAMPLE 2 Divide: $(8x^4y^5 - 3x^3y^4 + 5x^2y^3) \div (x^2y^3)$.

$$\frac{8x^4y^5 - 3x^3y^4 + 5x^2y^3}{x^2y^3} = \frac{8x^4y^5}{x^2y^3} - \frac{3x^3y^4}{x^2y^3} + \frac{5x^2y^3}{x^2y^3}$$

$$= 8x^2y^2 - 3xy + 5$$

Now Try Exercise 5.

> **Dividing by a Monomial**
>
> To divide a polynomial by a monomial, divide each term by the monomial.

▶ **b Divisor Not a Monomial**

When the divisor is not a monomial, we use a procedure very much like long division in arithmetic.

EXAMPLE 3 Divide $x^2 + 5x + 8$ by $x + 3$.

We have

$$
\begin{array}{r}
x \\
x + 3 \overline{\smash{)}x^2 + 5x + 8} \\
x^2 + 3x \\
\hline
2x
\end{array}
$$

Divide the first term by the first term: $x^2/x = x$.

Multiply x above by the divisor, $x + 3$.

Subtract: $(x^2 + 5x) - (x^2 + 3x) = x^2 + 5x - x^2 - 3x$
$= 2x$.

We now "bring down" the next term of the dividend—in this case, 8.

$$
\begin{array}{r}
x + 2 \\
x + 3 \overline{\smash{)}x^2 + 5x + 8} \\
x^2 + 3x \\
\hline
2x + 8 \\
2x + 6 \\
\hline
2
\end{array}
$$

Divide the first term by the first term: $2x/x = 2$.

The 8 has been "brought down."

Multiply 2 above by the divisor, $x + 3$.

Subtract: $(2x + 8) - (2x + 6) = 2x + 8 - 2x - 6$
$= 2$.

The answer is $x + 2$, R 2; or

$$x + 2 + \frac{2}{x + 3}.$$

> **This expression is the remainder over the divisor.**

Note that the answer is not a polynomial unless the remainder is 0.

To check, we multiply the quotient by the divisor and add the remainder to see if we get the dividend:

$$
\underbrace{(x + 3)}_{\text{Divisor}} \cdot \underbrace{(x + 2)}_{\text{Quotient}} + \underbrace{2}_{\text{Remainder}} = \overbrace{(x^2 + 5x + 6) + 2}^{\text{Dividend}}
$$
$$
= x^2 + 5x + 8.
$$

The answer checks.

Now Try Exercise 7.

EXAMPLE 4   Divide: $(5x^4 + x^3 - 3x^2 - 6x - 8) \div (x - 1)$.

$$
\begin{array}{r}
5x^3 + 6x^2 + 3x - 3 \\
x - 1 \overline{) 5x^4 + x^3 - 3x^2 - 6x - 8}
\end{array}
$$

$\underline{5x^4 - 5x^3}$ ⟵ Subtract:
$\qquad 6x^3 - 3x^2$ $(5x^4 + x^3) - (5x^4 - 5x^3) = 6x^3.$

$\qquad \underline{6x^3 - 6x^2}$ ⟵ Subtract:
$\qquad\qquad 3x^2 - 6x$ $(6x^3 - 3x^2) - (6x^3 - 6x^2) = 3x^2.$

$\qquad\qquad \underline{3x^2 - 3x}$ ⟵ Subtract:
$\qquad\qquad\qquad -3x - 8$ $(3x^2 - 6x) - (3x^2 - 3x) = -3x.$

$\qquad\qquad\qquad \underline{-3x + 3}$ ⟵Subtract:
$\qquad\qquad\qquad\qquad -11$ $(-3x - 8) - (-3x + 3) = -11.$

The answer is $5x^3 + 6x^2 + 3x - 3$, R -11; or

$$5x^3 + 6x^2 + 3x - 3 + \frac{-11}{x - 1}.$$

Now Try Exercise 9.

When dividing polynomials, remember to always arrange the polynomials in descending order. In a polynomial division, if there are *missing* terms in the dividend, either write them with 0 coefficients or leave space for them. For example, in $125y^3 - 8$, we say that "the y^2-term and the y-term are **missing**." We could write them in as follows: $125y^3 + 0y^2 + 0y - 8$.

EXAMPLE 5   Divide: $(125y^3 - 8) \div (5y - 2)$.

$$
\begin{array}{r}
25y^2 + 10y + 4 \\
5y - 2 \overline{) 125y^3 + 0y^2 + 0y - 8}
\end{array}
$$

When there are missing terms, we can write them in.

$\underline{125y^3 - 50y^2}$ ⟵ Subtract:
$\qquad 50y^2 + 0y$ $125y^3 - (125y^3 - 50y^2) = 50y^2.$

$\qquad \underline{50y^2 - 20y}$ ⟵ Subtract: $50y^2 - (50y^2 - 20y) = 20y.$
$\qquad\qquad 20y - 8$

$\qquad\qquad \underline{20y - 8}$ ⟵ Subtract: $(20y - 8) - (20y - 8) = 0.$
$\qquad\qquad\qquad 0$

The answer is $25y^2 + 10y + 4$.

Another way to deal with missing terms is to leave space for them, as we see in Example 6.

EXAMPLE 6 Divide: $(x^4 - 9x^2 - 5) \div (x - 2)$.

Note that the x^3-term and the x-term are missing in the dividend.

$$
\begin{array}{r}
x^3 + 2x^2 - 5x - 10 \\
x - 2 \overline{\smash{)}x^4 \qquad\quad - 9x^2 \qquad\qquad - 5} \\
\underline{x^4 - 2x^3} \\
2x^3 - 9x^2 \\
\underline{2x^3 - 4x^2} \\
- 5x^2 \\
\underline{- 5x^2 + 10x} \\
- 10x - 5 \\
\underline{- 10x + 20} \\
- 25
\end{array}
$$

We leave spaces for missing terms.

Subtract: $x^4 - (x^4 - 2x^3) = 2x^3$.

Subtract: $(2x^3 - 9x^2) - (2x^3 - 4x^2) = -5x^2$.

Subtract: $-5x^2 - (-5x^2 + 10x) = -10x$.

Subtract: $(-10x - 5) - (-10x + 20) = -25$.

The answer is $x^3 + 2x^2 - 5x - 10$, R -25; or

$$x^3 + 2x^2 - 5x - 10 + \frac{-25}{x - 2}.$$

Now Try Exercise 13.

When dividing, we may "come out even" (have a remainder of 0) or we may not. If not, how long should we keep working? We continue until the degree of the remainder is less than the degree of the divisor, as in the next example.

EXAMPLE 7 Divide: $(6x^3 + 9x^2 - 5) \div (x^2 - 2x)$.

$$
\begin{array}{r}
6x + 21 \\
x^2 - 2x \overline{\smash{)}6x^3 + 9x^2 + 0x - 5} \\
\underline{6x^3 - 12x^2} \\
21x^2 + 0x \\
\underline{21x^2 - 42x} \\
42x - 5
\end{array}
$$

We have a missing term. We can write it in.

The degree of the remainder, 1, is less than the degree of the divisor, 2, so we are finished.

The answer is $6x + 21$, R $(42x - 5)$; or

$$6x + 21 + \frac{42x - 5}{x^2 - 2x}.$$

Now Try Exercise 19.

▶ **c Synthetic Division**

To divide a polynomial by a binomial of the type $x - a$, we can streamline the general procedure by a process called **synthetic division**.

Compare the following. In **A**, we perform a division. In **B**, we also divide but we do not write the variables. In **B**, we eliminate some duplication of writing.

A.

$$
\begin{array}{r}
4x^2 + 5x + 11 \\
x - 2\overline{)\,4x^3 - 3x^2 +\ \ x +\ \ 7} \\
\underline{4x^3 - 8x^2} \\
5x^2 +\ \ x \\
\underline{5x^2 - 10x} \\
11x +\ \ 7 \\
\underline{11x - 22} \\
29
\end{array}
$$

B.

$$
\begin{array}{r}
4 + 5 + 11 \\
1 - 2\overline{)\,4 - 3 +\ \ 1 +\ \ 7} \\
\underline{4 - 8} \\
5 +\ \ 1 \\
\underline{5 - 10} \\
11 +\ \ 7 \\
\underline{11 - 22} \\
29
\end{array}
$$

In **B**, there is still some duplication of writing. Also, since we can subtract by adding the opposite, we can use 2 instead of -2 and then add instead of subtracting.

C. Synthetic Division

a) $\underline{2}\big|4 \ \ -3 \ \ 1 \ \ 7$
Write 2, the number that is subtracted in the divisor $x - 2$, and the coefficients of the dividend.

<hr style="width:25%;margin-left:0">

4   Bring down the first coefficient.

b) 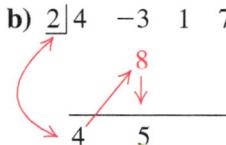
Multiply 4 by 2 to get 8. Add 8 and -3.

c) 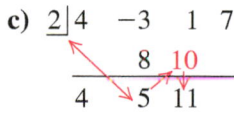
Multiply 5 by 2 to get 10. Add 10 and 1.

d) 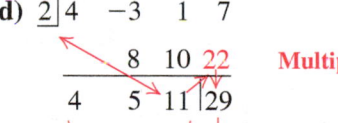
Multiply 11 by 2 to get 22. Add 22 and 7.

Quotient   Remainder

The last number, 29, is the remainder. The other numbers are the coefficients of the quotient with that of the term of highest degree first. Note that the degree of the term of highest degree is 1 less than the degree of the dividend.

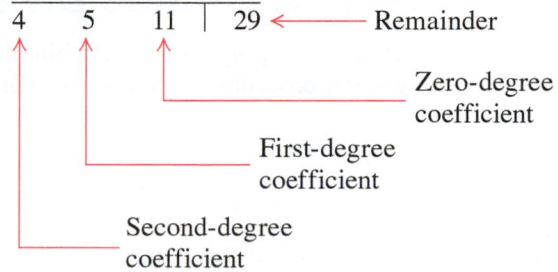

The answer is $4x^2 + 5x + 11$, R 29; or $4x^2 + 5x + 11 + \dfrac{29}{x - 2}$.

It is important to remember that in order for synthetic division to work, the divisor must be of the form $x - a$, that is, a variable minus a constant. The coefficient of the variable must be 1.

EXAMPLE 8 Use synthetic division to divide:

$$(x^3 + 6x^2 - x - 30) \div (x - 2).$$

We have

$$
\begin{array}{r|rrrr}
2 & 1 & 6 & -1 & -30 \\
  &   & 2 & 16 & 30 \\
\hline
  & 1 & 8 & 15 & 0
\end{array}
$$

The answer is $x^2 + 8x + 15$, R 0; or just $x^2 + 8x + 15$.

Now Try Exercise 21.

When there are missing terms, be sure to write 0's for their coefficients.

EXAMPLES Use synthetic division to divide.

9. $(2x^3 + 7x^2 - 5) \div (x + 3)$

There is no x-term, so we must write a 0 for its coefficient. Note that $x + 3 = x - (-3)$, so we write -3 at the left.

$$
\begin{array}{r|rrrr}
-3 & 2 & 7 & 0 & -5 \\
   &   & -6 & -3 & 9 \\
\hline
   & 2 & 1 & -3 & 4
\end{array}
$$

The answer is $2x^2 + x - 3$, R 4; or $2x^2 + x - 3 + \dfrac{4}{x + 3}$.

10. $(x^3 + 4x^2 - x - 4) \div (x + 4)$

Note that $x + 4 = x - (-4)$, so we write -4 at the left.

$$
\begin{array}{r|rrrr}
-4 & 1 & 4 & -1 & -4 \\
   &   & -4 & 0 & 4 \\
\hline
   & 1 & 0 & -1 & 0
\end{array}
$$

The answer is $x^2 - 1$.

11. $(x^4 - 1) \div (x - 1)$

The divisor is $x - 1$, so we write 1 at the left.

$$
\begin{array}{r|rrrrr}
1 & 1 & 0 & 0 & 0 & -1 \\
  &   & 1 & 1 & 1 & 1 \\
\hline
  & 1 & 1 & 1 & 1 & 0
\end{array}
$$
 There are 3 missing terms.

The answer is $x^3 + x^2 + x + 1$.

Now Try Exercise 37.

5.3 Exercise Set

▶ **Reading Check**

In order for synthetic division to work, the divisor must be of the form x − a; that is, a variable minus a constant. The coefficient of the variable must be 1. For each divisor in Exercises RC1–RC6, determine the constant a.

RC1. $(x^2 - x + 3) \div (x - 4)$

RC2. $(x^3 + 2x^2 + 5) \div (x + 6)$

RC3. $(2x^2 + 4x - 7) \div (x - 7)$

RC4. $(4x^4 - x^3 + x^2 - x) \div (x - 6)$

RC5. $(x^4 - 6x^2 - x + 4) \div (x + 4)$

RC6. $(10x^2 - 6) \div (x + 7)$

a *Divide and check.*

1. $\dfrac{24x^6 + 18x^5 - 36x^2}{6x^2}$

2. $\dfrac{30y^8 - 15y^6 + 40y^4}{5y^4}$

3. $\dfrac{45y^7 - 20y^4 + 15y^2}{5y^2}$

4. $\dfrac{60x^8 + 44x^5 - 28x^3}{4x^3}$

5. $(32a^4b^3 + 14a^3b^2 - 22a^2b) \div (2a^2b)$

6. $(7x^3y^4 - 21x^2y^3 + 28xy^2) \div (7xy)$

b *Divide.*

7. $(x^2 + 10x + 21) \div (x + 3)$

8. $(y^2 - 8y + 16) \div (y - 4)$

9. $(a^2 - 8a - 16) \div (a + 4)$

10. $(y^2 - 10y - 25) \div (y - 5)$

11. $(x^2 + 7x + 14) \div (x + 5)$

12. $(t^2 - 7t - 9) \div (t - 3)$

13. $(4y^3 + 6y^2 + 14) \div (2y + 4)$

14. $(6x^3 - x^2 - 10) \div (3x + 4)$

15. $(10y^3 + 6y^2 - 9y + 10) \div (5y - 2)$

16. $(6x^3 - 11x^2 + 11x - 2) \div (2x - 3)$

17. $(2x^4 - x^3 - 5x^2 + x - 6) \div (x^2 + 2)$

18. $(3x^4 + 2x^3 - 11x^2 - 2x + 5) \div (x^2 - 2)$

19. $(2x^5 - x^4 + 2x^3 - x) \div (x^2 - 3x)$

20. $(2x^5 + 3x^3 + x^2 - 4) \div (x^2 + x)$

c *Use synthetic division to divide.*

21. $(x^3 - 2x^2 + 2x - 5) \div (x - 1)$

22. $(x^3 - 2x^2 + 2x - 5) \div (x + 1)$

23. $(a^2 + 11a - 19) \div (a + 4)$

24. $(a^2 + 11a - 19) \div (a - 4)$

25. $(x^3 - 7x^2 - 13x + 3) \div (x - 2)$

26. $(x^3 - 7x^2 - 13x + 3) \div (x + 2)$

27. $(3x^3 + 7x^2 - 4x + 3) \div (x + 3)$

28. $(3x^3 + 7x^2 - 4x + 3) \div (x - 3)$

29. $(y^3 - 3y + 10) \div (y - 2)$

30. $(x^3 - 2x^2 + 8) \div (x + 2)$

31. $(3x^4 - 25x^2 - 18) \div (x - 3)$

32. $(6y^4 + 15y^3 + 28y + 6) \div (y + 3)$

33. $(x^3 - 8) \div (x - 2)$

34. $(y^3 + 125) \div (y + 5)$

35. $(y^4 - 16) \div (y - 2)$

36. $(x^5 - 32) \div (x - 2)$

37. $(y^8 - 1) \div (y + 1)$

38. $(y^6 - 2) \div (y - 1)$

▶ **Skill Maintenance**

Graph. **[3.7a]**

39. $2x - 3y < 6$

40. $5x + 3y \le 15$

41. $y > 4$

42. $x \le -2$

Graph. **[2.2c]**

43. $f(x) = x^2$

44. $g(x) = x^2 - 3$

45. $f(x) = 3 - x^2$

46. $f(x) = x^2 + 6x + 6$

Solve. **[4.8a]**

47. $x^2 - 5x = 0$ **48.** $25y^2 = 64$

49. $12x^2 = 17x + 5$

50. $12x^2 + 11x + 2 = 0$

▶ **Synthesis**

51. Let $f(x) = 4x^3 + 16x^2 - 3x - 45$. Find $f(-3)$ and then solve $f(x) = 0$.

52. Let $f(x) = 6x^3 - 13x^2 - 79x + 140$. Find $f(4)$ and then solve $f(x) = 0$.

53. When $x^2 - 3x + 2k$ is divided by $x + 2$, the remainder is 7. Find k.

54. Find k such that when $x^3 - kx^2 + 3x + 7k$ is divided by $x + 2$, the remainder is 0.

Divide.

55. $(4a^3b + 5a^2b^2 + a^4 + 2ab^3) \div (a^2 + 2b^2 + 3ab)$

56. $(a^7 + b^7) \div (a + b)$

5.4   Complex Rational Expressions

  ▶ **a**   Simplify complex rational expressions.

  ▶ **b**   Find the difference quotient for a function.

▶ a   Simplifying Complex Rational Expressions

A **complex rational expression** is a rational expression that contains rational expressions within its numerator and/or its denominator. Here are some examples:

$$\frac{\frac{2}{3}}{\frac{4}{5}}, \qquad \frac{1 + \frac{5}{x}}{4x}, \qquad \frac{\frac{x - y}{x + y}}{\frac{2x - y}{3x + y}}, \qquad \frac{\frac{3x}{5} - \frac{2}{x}}{\frac{4x}{3} + \frac{7}{6x}}.$$

The rational expressions within each complex rational expression are red.

There are two methods that can be used to simplify complex rational expressions. We will consider both of them.

Method 1: Multiplying by the LCM of All the Denominators

To the instructor and the student: Students can be instructed to try both methods and then choose the one that works best for them, or one method can be chosen by the instructor.

Method 1. To simplify a complex rational expression:

1. First, find the LCM of all the denominators of all the rational expressions occurring within both the numerator and the denominator of the (original) complex rational expression.
2. Multiply by 1 using LCM/LCM.
3. If possible, simplify.

EXAMPLE 1 Simplify: $\dfrac{x + \dfrac{1}{5}}{x - \dfrac{1}{3}}$.

We first find the LCM of all the denominators of all the rational expressions occurring in both the numerator and the denominator of the complex rational expression. The denominators are 3 and 5. The LCM of these denominators is $3 \cdot 5$, or 15. We multiply by $15/15$.

$$\dfrac{x + \dfrac{1}{5}}{x - \dfrac{1}{3}} = \left(\dfrac{x + \dfrac{1}{5}}{x - \dfrac{1}{3}} \right) \cdot \dfrac{15}{15} \qquad \text{\color{red}{Multiplying by 1}}$$

$$= \dfrac{\left(x + \dfrac{1}{5} \right) \cdot 15}{\left(x - \dfrac{1}{3} \right) \cdot 15} \qquad \text{\color{red}{Multiplying the numerators and the denominators}}$$

$$= \dfrac{15x + \dfrac{1}{5} \cdot 15}{15x - \dfrac{1}{3} \cdot 15} \qquad \text{\color{red}{Carrying out the multiplications using the distributive laws}}$$

$$= \dfrac{15x + 3}{15x - 5}, \text{ or } \dfrac{3(5x + 1)}{5(3x - 1)} \qquad \text{\color{red}{No further simplification is possible.}}$$

Now Try Exercise 1.

In Example 1, if you feel more comfortable doing so, you can always write denominators of 1 where there are no denominators written. In this case, you could start out by writing

$$\dfrac{\dfrac{x}{1} + \dfrac{1}{5}}{\dfrac{x}{1} - \dfrac{1}{3}}.$$

EXAMPLE 2 Simplify: $\dfrac{1 + \dfrac{1}{x}}{1 - \dfrac{1}{x^2}}$.

We first find the LCM of all the denominators of all the rational expressions occurring in both the numerator and the denominator of the complex rational expression. The denominators are x and x^2. The LCM of these denominators is x^2. We multiply by x^2/x^2.

$$\frac{1 + \dfrac{1}{x}}{1 - \dfrac{1}{x^2}} = \left(\frac{1 + \dfrac{1}{x}}{1 - \dfrac{1}{x^2}} \right) \cdot \frac{x^2}{x^2}$$

The LCM of the denominators is x^2.
We multiply by 1: $\dfrac{x^2}{x^2}$.

$$= \frac{\left(1 + \dfrac{1}{x} \right) \cdot x^2}{\left(1 - \dfrac{1}{x^2} \right) \cdot x^2}$$

Multiplying the numerators and the denominators

$$= \frac{x^2 + \dfrac{1}{x} \cdot x^2}{x^2 - \dfrac{1}{x^2} \cdot x^2}$$

Carrying out the multiplications using the distributive laws

$$= \frac{x^2 + x}{x^2 - 1}$$

$$= \frac{x(x + 1)}{(x + 1)(x - 1)}$$

Factoring

$$= \frac{x\cancel{(x + 1)}}{\cancel{(x + 1)}(x - 1)}$$

Removing a factor of 1: $\dfrac{x + 1}{x + 1} = 1$

$$= \frac{x}{x - 1}$$

→ **Now Try Exercise 9.**

EXAMPLE 3 Simplify: $\dfrac{\dfrac{1}{a} + \dfrac{1}{b}}{\dfrac{1}{a^3} + \dfrac{1}{b^3}}$.

The denominators are a, b, a^3, and b^3. The LCM of these denominators is a^3b^3. We multiply by $a^3b^3/(a^3b^3)$.

$$\frac{\dfrac{1}{a} + \dfrac{1}{b}}{\dfrac{1}{a^3} + \dfrac{1}{b^3}} = \left(\frac{\dfrac{1}{a} + \dfrac{1}{b}}{\dfrac{1}{a^3} + \dfrac{1}{b^3}} \right) \cdot \frac{a^3b^3}{a^3b^3}$$

The LCM of the denominators is a^3b^3.
We multiply by 1: $\dfrac{a^3b^3}{a^3b^3}$.

$$= \frac{\left(\dfrac{1}{a} + \dfrac{1}{b} \right) \cdot a^3b^3}{\left(\dfrac{1}{a^3} + \dfrac{1}{b^3} \right) \cdot a^3b^3}$$

Multiplying the numerators and the denominators

$$= \frac{\dfrac{1}{a} \cdot a^3b^3 + \dfrac{1}{b} \cdot a^3b^3}{\dfrac{1}{a^3} \cdot a^3b^3 + \dfrac{1}{b^3} \cdot a^3b^3}$$

Carrying out the multiplications using a distributive law

$$= \frac{a^2b^3 + a^3b^2}{b^3 + a^3} = \frac{a^2b^2(b + a)}{(b + a)(b^2 - ba + a^2)} \qquad \textcolor{red}{\text{Factoring}}$$

$$= \frac{a^2b^2\cancel{(b + a)}}{\cancel{(b + a)}(b^2 - ba + a^2)} \qquad \textcolor{red}{\text{Removing a factor of 1: } \frac{b + a}{b + a} = 1}$$

$$= \frac{a^2b^2}{b^2 - ba + a^2}.$$

Now Try Exercise 27.

Method 2: Adding or Subtracting in the Numerator and the Denominator

> *Method 2.* To simplify a complex rational expression:
> 1. Add or subtract, as necessary, to get a single rational expression in the numerator.
> 2. Add or subtract, as necessary, to get a single rational expression in the denominator.
> 3. Divide the numerator by the denominator.
> 4. If possible, simplify.

We will redo Examples 1–3 using this method.

EXAMPLE 4 Simplify: $\dfrac{x + \dfrac{1}{5}}{x - \dfrac{1}{3}}$.

$$\frac{x + \dfrac{1}{5}}{x - \dfrac{1}{3}} = \frac{x \cdot \dfrac{\textcolor{red}{5}}{\textcolor{red}{5}} + \dfrac{1}{5}}{x - \dfrac{1}{3}} = \frac{\dfrac{5x + 1}{5}}{x - \dfrac{1}{3}}$$

To get a single rational expression in the numerator, we note that the LCM in the numerator is 5. We multiply by 1 and add.

$$= \frac{\dfrac{5x + 1}{5}}{x \cdot \dfrac{3}{3} - \dfrac{1}{3}} = \frac{\dfrac{5x + 1}{5}}{\dfrac{3x - 1}{3}}$$

To get a single rational expression in the denominator, we note that the LCM in the denominator is 3. We multiply by 1 and subtract.

$$= \frac{5x + 1}{5} \cdot \frac{3}{3x - 1}$$

Multiplying by the reciprocal of the denominator

$$= \frac{3(5x + 1)}{5(3x - 1)}, \text{ or } \frac{15x + 3}{15x - 5}$$

No further simplification is possible.

Now Try Exercise 1.

EXAMPLE 5 Simplify: $\dfrac{1 + \dfrac{1}{x}}{1 - \dfrac{1}{x^2}}$.

$$\frac{1 + \dfrac{1}{x}}{1 - \dfrac{1}{x^2}} = \frac{1 \cdot \dfrac{x}{x} + \dfrac{1}{x}}{1 \cdot \dfrac{x^2}{x^2} - \dfrac{1}{x^2}}$$

Finding the LCM in the numerator and multiplying by 1

Finding the LCM in the denominator and multiplying by 1

$$= \frac{\dfrac{x}{x} + \dfrac{1}{x}}{\dfrac{x^2}{x^2} - \dfrac{1}{x^2}}$$

$$= \frac{\dfrac{x + 1}{x}}{\dfrac{x^2 - 1}{x^2}}$$

Adding in the numerator and subtracting in the denominator

$$= \frac{x + 1}{x} \cdot \frac{x^2}{x^2 - 1}$$

Multiplying by the reciprocal of the denominator

$$= \frac{(x + 1) \cdot x \cdot x}{x(x - 1)(x + 1)}$$

Factoring and removing a factor of 1: $\dfrac{x(x + 1)}{x(x + 1)} = 1$

$$= \frac{x}{x - 1}$$

Now Try Exercise 9.

EXAMPLE 6 Simplify: $\dfrac{\dfrac{1}{a} + \dfrac{1}{b}}{\dfrac{1}{a^3} + \dfrac{1}{b^3}}$.

The LCM in the numerator is ab, and the LCM in the denominator is a^3b^3.

$$\frac{\dfrac{1}{a} + \dfrac{1}{b}}{\dfrac{1}{a^3} + \dfrac{1}{b^3}} = \frac{\dfrac{1}{a} \cdot \dfrac{b}{b} + \dfrac{1}{b} \cdot \dfrac{a}{a}}{\dfrac{1}{a^3} \cdot \dfrac{b^3}{b^3} + \dfrac{1}{b^3} \cdot \dfrac{a^3}{a^3}} = \frac{\dfrac{b}{ab} + \dfrac{a}{ab}}{\dfrac{b^3}{a^3b^3} + \dfrac{a^3}{a^3b^3}}$$

$$= \frac{\dfrac{b + a}{ab}}{\dfrac{b^3 + a^3}{a^3b^3}}$$

Adding in the numerator and the denominator

$$= \frac{b + a}{ab} \cdot \frac{a^3b^3}{b^3 + a^3}$$

Multiplying by the reciprocal of the denominator

$$= \frac{(b + a)a^3b^3}{ab(b^3 + a^3)} = \frac{(b + a) \cdot ab \cdot a^2b^2}{ab(b + a)(b^2 - ba + a^2)}$$

$$= \frac{a^2b^2}{b^2 - ba + a^2}$$

Now Try Exercise 27.

▶ b   Difference Quotients

The slope of a line can be considered as an *average rate of change*. Here let's consider a nonlinear function f and draw a line through two points $(x, f(x))$ and $(x + h, f(x + h))$ as shown below.

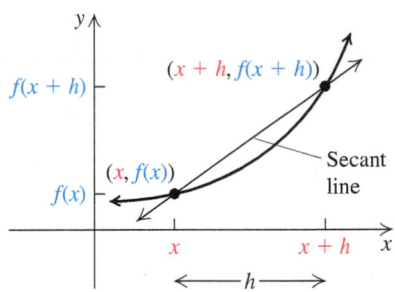

The slope of the line, called a **secant line**, is

$$m = \frac{y_2 - y_1}{x_2 - x_1} = \frac{f(x + h) - f(x)}{(x + h) - x},$$

which simplifies to

$$\frac{f(x + h) - f(x)}{h}. \qquad \textcolor{red}{\textbf{Difference quotient}}$$

This ratio is called the **difference quotient**, or the **average rate of change**. In calculus, it is important to be able to find and simplify difference quotients.

EXAMPLE 7   For the function f given by $f(x) = 2x - 3$, find and simplify the difference quotient

$$\frac{f(x + h) - f(x)}{h}.$$

$$
\begin{aligned}
\frac{f(x + h) - f(x)}{h} &= \frac{2(x + h) - 3 - (2x - 3)}{h} \qquad \textcolor{red}{\textbf{Substituting}} \\[2mm]
&= \frac{2x + 2h - 3 - 2x + 3}{h} \qquad \textcolor{red}{\textbf{Removing parentheses}} \\[2mm]
&= \frac{2\cancel{h}}{\cancel{h}} \\[2mm]
&= 2 \qquad \textcolor{red}{\textbf{Simplifying}}
\end{aligned}
$$

> **Now Try Exercise 31.**

EXAMPLE 8   For the function f given by $f(x) = \dfrac{1}{x}$, find and simplify the difference quotient

$$\frac{f(x + h) - f(x)}{h}.$$

$$
\begin{aligned}
\frac{f(x + h) - f(x)}{h} &= \frac{\dfrac{1}{x + h} - \dfrac{1}{x}}{h} \qquad \textcolor{red}{\textbf{Substituting}} \\[4mm]
&= \frac{\dfrac{1}{x + h} \cdot \dfrac{x}{x} - \dfrac{1}{x} \cdot \dfrac{x + h}{x + h}}{h}
\end{aligned}
$$

$\textcolor{red}{\text{The LCD of } \dfrac{1}{x + h} \text{ and } \dfrac{1}{x} \text{ is } x(x + h).}$

$$= \frac{\dfrac{x}{x(x+h)} - \dfrac{x+h}{x(x+h)}}{h}$$

$$= \frac{\dfrac{x - (x+h)}{x(x+h)}}{h} \qquad \text{Subtracting in the numerator}$$

$$= \frac{\dfrac{x - x - h}{x(x+h)}}{h} \qquad \text{Removing parentheses}$$

$$= \frac{\dfrac{-h}{x(x+h)}}{h} \qquad \text{Simplifying the numerator}$$

$$= \frac{-h}{x(x+h)} \cdot \frac{1}{h} \qquad \text{Multiplying by the reciprocal of the divisor}$$

$$= \frac{-h \cdot 1}{x \cdot (x+h) \cdot h}$$

$$= \frac{-1 \cdot \cancel{h}}{x \cdot (x+h) \cdot \cancel{h}} \qquad \text{Rewriting } -h \cdot 1 \text{ as } -1 \cdot h$$

$$= \frac{-1}{x(x+h)}, \text{ or } -\frac{1}{x(x+h)}$$

> **Now Try Exercise 37.**

EXAMPLE 9 For the function f given by $f(x) = 2x^2 - x - 3$, find and simplify the difference quotient

$$\frac{f(x+h) - f(x)}{h}.$$

We first find $f(x+h)$:

$$f(x+h) = 2(x+h)^2 - (x+h) - 3 \qquad \substack{\text{Substituting } x + h \text{ for } x \text{ in} \\ f(x) = 2x^2 - x - 3}$$

$$= 2[x^2 + 2xh + h^2] - (x+h) - 3$$

$$= 2x^2 + 4xh + 2h^2 - x - h - 3.$$

Then we have

$$\frac{f(x+h) - f(x)}{h} = \frac{[2x^2 + 4xh + 2h^2 - x - h - 3] - [2x^2 - x - 3]}{h}$$

$$= \frac{2x^2 + 4xh + 2h^2 - x - h - 3 - 2x^2 + x + 3}{h}$$

$$= \frac{4xh + 2h^2 - h}{h}$$

$$= \frac{\cancel{h}(4x + 2h - 1)}{\cancel{h} \cdot 1} = \frac{4x + 2h - 1}{1} = 4x + 2h - 1.$$

> **Now Try Exercise 45.**

5.4 Exercise Set

▶ Reading Check

Consider the expression $\dfrac{\dfrac{7}{x} - \dfrac{2}{3}}{\dfrac{4}{x}}$. *Choose from the column on the right the term that best completes the statement.*

numerator
denominator
opposite
reciprocal
complex
least common denominator

RC1. The expression given above is a(n) _____ rational expression.

RC2. The expression $\dfrac{7}{x} - \dfrac{2}{3}$ is the _____ of the above expression.

RC3. The _____ of the rational expressions $\dfrac{7}{x}, \dfrac{2}{3}$, and $\dfrac{4}{x}$ above is $3x$.

RC4. After subtracting in the numerator, we multiply the numerator by the _____ of the denominator, $\dfrac{4}{x}$.

a Simplify.

1. $\dfrac{2 + \dfrac{3}{5}}{4 - \dfrac{1}{2}}$

2. $\dfrac{\dfrac{3}{8} - 5}{\dfrac{2}{3} + 6}$

3. $\dfrac{\dfrac{2}{3} + \dfrac{4}{5}}{\dfrac{3}{4} - \dfrac{1}{2}}$

4. $\dfrac{\dfrac{5}{8} - \dfrac{2}{3}}{\dfrac{3}{4} + \dfrac{5}{6}}$

5. $\dfrac{\dfrac{x}{y^2}}{\dfrac{y^3}{x^2}}$

6. $\dfrac{\dfrac{a^3}{b^5}}{\dfrac{a^4}{b^2}}$

7. $\dfrac{\dfrac{9x^2 - y^2}{xy}}{\dfrac{3x - y}{y}}$

8. $\dfrac{\dfrac{a^2 - 16b^2}{ab}}{\dfrac{a + 4b}{b}}$

9. $\dfrac{\dfrac{1}{a} + 2}{\dfrac{1}{a} - 1}$

10. $\dfrac{\dfrac{1}{t} + 6}{\dfrac{1}{t} - 5}$

11. $\dfrac{x - \dfrac{1}{x}}{x + \dfrac{1}{x}}$

12. $\dfrac{y + \dfrac{1}{y}}{y - \dfrac{1}{y}}$

13. $\dfrac{\dfrac{3}{x} + \dfrac{4}{y}}{\dfrac{4}{x} - \dfrac{3}{y}}$

14. $\dfrac{\dfrac{2}{y} + \dfrac{5}{z}}{\dfrac{1}{y} - \dfrac{4}{z}}$

15. $\dfrac{a - \dfrac{3a}{b}}{b - \dfrac{b}{a}}$

16. $\dfrac{1 - \dfrac{2}{3x}}{x - \dfrac{4}{9x}}$

17. $\dfrac{\dfrac{1}{a} + \dfrac{1}{b}}{\dfrac{a^2 - b^2}{ab}}$

18. $\dfrac{\dfrac{1}{x} - \dfrac{1}{y}}{\dfrac{x^2 - y^2}{xy}}$

19. $\dfrac{\dfrac{1}{x + h} - \dfrac{1}{x}}{h}$

20. $\dfrac{\dfrac{1}{a - h} - \dfrac{1}{a}}{h}$

It may help you to write h as $\dfrac{h}{1}$.

21. $\dfrac{\dfrac{x^2 - x - 12}{x^2 - 2x - 15}}{\dfrac{x^2 + 8x + 12}{x^2 - 5x - 14}}$

22. $\dfrac{\dfrac{y^2 - y - 6}{y^2 - 5y - 14}}{\dfrac{y^2 + 6y + 5}{y^2 - 6y - 7}}$

23. $\dfrac{\dfrac{1}{x+2}+\dfrac{4}{x-3}}{\dfrac{2}{x-3}-\dfrac{7}{x+2}}$

24. $\dfrac{\dfrac{1}{y-4}+\dfrac{1}{y+5}}{\dfrac{6}{y+5}+\dfrac{2}{y-4}}$

25. $\dfrac{\dfrac{6}{x^2-4}-\dfrac{5}{x+2}}{\dfrac{7}{x^2-4}-\dfrac{4}{x-2}}$

26. $\dfrac{\dfrac{1}{x^2-1}+\dfrac{5}{x^2-5x+4}}{\dfrac{1}{x^2-1}+\dfrac{2}{x^2+3x+2}}$

27. $\dfrac{\dfrac{1}{z^2}-\dfrac{1}{w^2}}{\dfrac{1}{z^3}+\dfrac{1}{w^3}}$

28. $\dfrac{\dfrac{1}{b^2}-\dfrac{1}{c^2}}{\dfrac{1}{b^3}-\dfrac{1}{c^3}}$

29. $\dfrac{\dfrac{3}{x^2+2x-3}-\dfrac{1}{x^2-3x-10}}{\dfrac{3}{x^2-6x+5}-\dfrac{1}{x^2+5x+6}}$

30. $\dfrac{\dfrac{1}{a^2+7a+12}+\dfrac{1}{a^2+a-6}}{\dfrac{1}{a^2+2a-8}+\dfrac{1}{a^2+5a+4}}$

b *For each function f, construct and simplify the difference quotient*

$$\frac{f(x+h)-f(x)}{h}.$$

31. $f(x)=3x-5$

32. $f(x)=4x-1$

33. $f(x)=6x+2$

34. $f(x)=5x+3$

35. $f(x)=\dfrac{1}{3}x+1$

36. $f(x)=-\dfrac{1}{2}x+7$

37. $f(x)=\dfrac{1}{3x}$

38. $f(x)=\dfrac{1}{2x}$

39. $f(x)=-\dfrac{1}{4x}$

40. $f(x)=-\dfrac{1}{x}$

41. $f(x)=x^2+1$

42. $f(x)=x^2-3$

43. $f(x)=4-x^2$

44. $f(x)=2-x^2$

45. $f(x)=3x^2-2x+1$

46. $f(x)=5x^2+4x$

47. $f(x)=4+5|x|$

48. $f(x)=2|x|+3x$

49. $f(x)=x^3$

50. $f(x)=x^3-2x$

51. $f(x)=\dfrac{x-4}{x+3}$

52. $f(x)=\dfrac{x}{2-x}$

▶ Skill Maintenance

Solve. **[1.3a]**

53. *Tax Code.* The 1984 publication explaining the federal tax code contained 26,300 pages. The 2012 publication contained 5292 fewer pages than three times the number of pages for 1984. How long was the tax code for 2012? (*Source*: CCH Inc.)

54. *Moving Freight.* Most freight in the United States is moved by truck. The total percent of freight moved by truck and rail is 84%. If the percent of freight moved by truck is 9% more than four times the percent moved by rail, what percent is moved by truck? (*Source*: U.S. Freight Transportation Forecast to 2020)

Factor. **[4.3a], [4.4a], [4.6d]**

55. $4x^3+20x^2+6x$

56. y^3+8

57. y^3-8

58. $2x^3-32x^2+126x$

59. $1000x^3+1$

60. $1-1000a^3$

61. y^3-64x^3

62. $\dfrac{1}{8}a^3-343$

63. Solve for s: $T=\dfrac{r+s}{3}$. **[1.2a]**

64. Graph: $f(x)=-3x+2$. **[2.2c]**

65. Given that $f(x)=x^2-3$, find $f(-5)$. **[2.2b]**

66. Solve: $|2x-5|=7$. **[1.6c]**

▶ Synthesis

Simplify.

67. $\dfrac{5x^{-1}-5y^{-1}+10x^{-1}y^{-1}}{6x^{-1}-6y^{-1}+12x^{-1}y^{-1}}$

68. $\left[\dfrac{\dfrac{x+3}{x-3}+1}{\dfrac{x+3}{x-3}-1}\right]^8$

Find the reciprocal and simplify.

69. $x^2-\dfrac{1}{x}$

70. $\dfrac{1-\dfrac{1}{a}}{a-1}$

71. $\dfrac{a^3+b^3}{a+b}$

72. $x^2+x+1+\dfrac{1}{x}+\dfrac{1}{x^2}$

Mid-Chapter Review

Determine whether each statement is true or false.

1. For synthetic division, the divisor must be in the form $x - a$. **[5.3c]**

2. The sum of two rational expressions is the sum of the numerators over the sum of the denominators. **[5.2b]**

3. The domain of $f(x) = \dfrac{(x - 5)(x + 4)}{x - 4}$ is $\{x \mid x \neq 5 \text{ and } x \neq -4 \text{ and } x \neq 4\}$. **[5.1a]**

Find the domain of each function. **[5.1a]**

4. $f(x) = \dfrac{x + 5}{x^2 - 100}$

5. $g(x) = \dfrac{-3}{x - 7}$

6. $h(x) = \dfrac{x^2 - 9}{x^2 + 8x - 9}$

Simplify. **[5.1c]**

7. $\dfrac{24p^2}{36p^9}$

8. $\dfrac{42y - 3}{33}$

9. $\dfrac{x^2 - y^2}{x^3 + y^3}$

10. $\dfrac{x^2 - x - 30}{x^2 - 4x - 12}$

11. $\dfrac{9a - 18}{9a + 18}$

12. $\dfrac{3 - t}{t^2 - t - 6}$

Find the LCM. **[5.2a]**

13. $x^3,\ 14x^2y,\ 35x^4y^5$

14. $x^2 - 25,\ x^2 - 10x + 25,\ x^2 + 3x - 40$

Perform the indicated operations and simplify.

15. $\dfrac{45}{x^2 - 1} \div \dfrac{x + 1}{x - 1}$ **[5.1e]**

16. $\dfrac{3x - 1}{x + 6} + \dfrac{x}{x - 2}$ **[5.2b]**

17. $\dfrac{q}{q + 2} - \dfrac{q + 1}{q}$ **[5.2b]**

18. $\dfrac{2y}{y^2 + 2y - 3} - \dfrac{3y + 1}{y^2 + y - 2}$ **[5.2b]**

19. $\dfrac{\dfrac{1}{b} - 1}{\dfrac{1}{b^2} - 1}$ **[5.4a]**

20. $\dfrac{w^2 - z^2}{5w - 5z} \cdot \dfrac{w - z}{w + z}$ **[5.1d]**

21. $\dfrac{t^3 - 8}{2t + 3} \cdot \dfrac{2t^2 + t - 3}{t - 2}$ **[5.1d]**

22. $\dfrac{5c}{3} + \dfrac{2a}{5c}$ **[5.2b]**

23. $\dfrac{x^2 - 4x}{x^2 + 2x} \div \dfrac{x^2 - 8x + 16}{x^2 + 4x + 4}$ **[5.1e]**

Divide and if there is a remainder, express it in two ways. Use synthetic division in Exercises 26–28. **[5.3b, c]**

24. $(6x^2 - 5x + 11) \div (2x - 3)$

25. $(x^4 - 1) \div (x + 1)$

26. $(2x^3 - x^2 + 5x - 4) \div (x + 2)$

27. $(x^2 - 4x - 12) \div (x - 6)$

28. $(x^4 - 3x^2 + 2) \div (x + 3)$

29. $(15x^2 - 2x + 6) \div (5x + 1)$

For each function f, construct and simplify the difference quotient

$$\frac{f(x + h) - f(x)}{h}.$$ **[5.4b]**

30. $f(x) = 4x - 3$

31. $f(x) = 6 - x^2$

Collaborative Discussion and Writing

32. Explain how synthetic division can be useful when factoring a polynomial. **[5.3c]**

33. Do addition, subtraction, multiplication, and division of polynomials always result in a polynomial? Why or why not? **[5.1d], [5.2b], [5.3b]**

34. Is it possible to understand how to simplify rational expressions without first understanding how to multiply? Why or why not? **[5.1c]**

35. Janine found that the sum of two rational expressions was $(3 - x)/(x - 5)$, but the answer at the back of the book was $(x - 3)/(5 - x)$. Was Janine's answer correct? Why or why not? **[5.2b]**

36. Nancy *incorrectly* simplifies $(x + 2)/x$ as follows:

$$\frac{x + 2}{x} = \frac{\cancel{x} + 2}{\cancel{x}} = 1 + 2 = 3.$$

She insists that this is correct because when x is replaced with 1, her answer checks. Explain her error. **[5.1c]**

37. Explain why it is easier to use method 1 than method 2 to simplify the following expression. **[5.4a]**

$$\frac{\dfrac{a}{b} + \dfrac{c}{d}}{\dfrac{a}{b} - \dfrac{c}{d}}$$

5.5   Solving Rational Equations

▶ **a**   Solve rational equations.

▶ **a   Rational Equations**

In Sections 5.1–5.4, we studied operations with *rational expressions*. These expressions do not have equals signs. Although we can perform operations on and simplify rational expressions, we cannot solve them. Note the following examples:

$$\frac{x^2 - 6x + 9}{x^2 - 4} \cdot \frac{x - 2}{x - 3}, \qquad \frac{x + y}{x - y} \div \frac{x^2 + y}{x^2 - y^2}, \quad \text{and} \quad \frac{a + 7}{a^2 - 16} + \frac{5}{5a - 15}.$$

Operation signs occur. There are no equals signs!

Most often, the result of our manipulation is another rational expression that is not cleared of fractions.

Equations *do have* equals signs, and we can clear them of fractions as we did in Section 1.1. A **rational**, or **fraction, equation** is an equation containing one or more rational expressions. Here are some examples:

$$\frac{2}{3} - \frac{5}{6} = \frac{1}{x}, \qquad x + \frac{6}{x} = 5, \quad \text{and} \quad \frac{2x}{x-3} - \frac{6}{x} = \frac{18}{x^2 - 3x}.$$

There are equals signs as well as operation signs.

Solving Rational Equations

To solve a rational equation, the first step is to clear the equation of fractions. To do this, multiply all terms on both sides of the equation by the LCM of all the denominators. Then carry out the equation-solving process as discussed in Chapters 1 and 4.

EXAMPLE 1 Solve: $\dfrac{2}{3} - \dfrac{5}{6} = \dfrac{1}{x}$.

The LCM of all the denominators is $6x$, or $2 \cdot 3 \cdot x$. Using the multiplication principle of Chapter 1, we multiply all terms on both sides of the equation by the LCM.

Multiplying, we have

$$(2 \cdot 3 \cdot x) \cdot \left(\frac{2}{3} - \frac{5}{6} \right) = (2 \cdot 3 \cdot x) \cdot \frac{1}{x} \qquad \textbf{Multiplying both sides by the LCM}$$

$$2 \cdot 3 \cdot x \cdot \frac{2}{3} - 2 \cdot 3 \cdot x \cdot \frac{5}{6} = 2 \cdot 3 \cdot x \cdot \frac{1}{x} \qquad \textbf{Multiplying to remove parentheses}$$

> **When clearing fractions, be sure to multiply *every* term in the equation by the LCM.**

$$2 \cdot x \cdot 2 - x \cdot 5 = 2 \cdot 3$$
$$4x - 5x = 6$$
$$-x = 6$$
$$-1 \cdot x = 6$$
$$x = -6$$

Check:

$$\frac{2}{3} - \frac{5}{6} = \frac{1}{x}$$

$$\frac{2}{3} - \frac{5}{6} \; ? \; \frac{1}{-6}$$

$$\frac{4}{6} - \frac{5}{6} \;\Big|\; -\frac{1}{6}$$

$$-\frac{1}{6} \;\Big|\; \quad \text{TRUE}$$

The solution is -6.

Now Try Exercise 3.

CAUTION! *Clearing fractions* is a valid procedure only when solving equations, *not* when adding, subtracting, multiplying, or dividing rational expressions.

EXAMPLE 2 Solve: $\dfrac{x+1}{2} - \dfrac{x-3}{3} = 3$.

The LCM of all the denominators is $2 \cdot 3$, or 6. We multiply all terms on both sides of the equation by the LCM.

$$2 \cdot 3 \cdot \left(\frac{x+1}{2} - \frac{x-3}{3} \right) = 2 \cdot 3 \cdot 3 \qquad \text{Multiplying both sides by the LCM}$$

$$2 \cdot 3 \cdot \frac{x+1}{2} - 2 \cdot 3 \cdot \frac{x-3}{3} = 2 \cdot 3 \cdot 3 \qquad \text{Multiplying to remove parentheses}$$

$$3(x+1) - 2(x-3) = 18 \qquad \text{Simplifying}$$

$$\left. \begin{array}{r} 3x + 3 - 2x + 6 = 18 \\ x + 9 = 18 \end{array} \right\} \quad \begin{array}{l} \text{Multiplying and} \\ \text{collecting like terms} \end{array}$$

$$x = 9$$

Check:

$$\frac{x+1}{2} - \frac{x-3}{3} = 3$$

$$\frac{9+1}{2} - \frac{9-3}{3} \ ?\ 3$$

$$5 - 2$$

$$3 \quad \text{| \ TRUE}$$

The solution is 9.

→ **Now Try Exercise 5.**

Checking Possible Solutions

When we multiply all terms on both sides of an equation by the LCM, the resulting equation might yield numbers that are *not* solutions of the original equation. Thus we must *always* check possible solutions in the original equation.

1. If you have carried out all algebraic procedures correctly, you need only check to see whether a number makes a denominator 0 in the original equation. If it does, it is not a solution.
2. To be sure that no computational errors have been made and that you indeed have a solution, a complete check is necessary, as we did in Examples 1 and 2.

The next example illustrates the importance of checking all possible solutions.

EXAMPLE 3 Solve: $\dfrac{2x}{x-3} - \dfrac{6}{x} = \dfrac{18}{x^2 - 3x}$.

The LCM of the denominators is $x(x-3)$. We multiply all terms on both sides by $x(x-3)$.

$$x(x-3)\left(\frac{2x}{x-3} - \frac{6}{x}\right) = x(x-3)\left(\frac{18}{x^2-3x}\right) \quad \text{Multiplying both sides by the LCM}$$

$$x(x-3)\cdot\frac{2x}{x-3} - x(x-3)\cdot\frac{6}{x} = x(x-3)\cdot\left(\frac{18}{x^2-3x}\right) \quad \text{Multiplying to remove parentheses}$$

$$2x^2 - 6(x-3) = 18 \qquad \text{Simplifying}$$

$$2x^2 - 6x + 18 = 18$$

$$2x^2 - 6x = 0$$

$$2x(x-3) = 0 \qquad \text{Factoring}$$

$$2x = 0 \quad or \quad x - 3 = 0 \qquad \text{Using the principle of zero products}$$

$$x = 0 \quad or \qquad x = 3$$

> **THE PRINCIPLE OF ZERO PRODUCTS**
>
> REVIEW SECTION 4.8

The numbers 0 and 3 are possible solutions. We look at the original equation and see that each makes a denominator 0, so neither is a solution. We can carry out a check, as follows.

Check:

For 0:

$$\dfrac{2x}{x-3} - \dfrac{6}{x} = \dfrac{18}{x^2 - 3x}$$

$$\dfrac{2(0)}{0-3} - \dfrac{6}{0} \overset{?}{\vert} \dfrac{18}{0^2 - 3(0)}$$

$$0 - \dfrac{6}{0} \,\Big\vert\, \dfrac{18}{0} \qquad \text{NOT DEFINED}$$

For 3:

$$\dfrac{2x}{x-3} - \dfrac{6}{x} = \dfrac{18}{x^2 - 3x}$$

$$\dfrac{2(3)}{3-3} - \dfrac{6}{3} \overset{?}{\vert} \dfrac{18}{3^2 - 3(3)}$$

$$\dfrac{6}{0} - 2 \,\Big\vert\, \dfrac{18}{0} \qquad \text{NOT DEFINED}$$

The equation has *no solution*.   ▶ **Now Try Exercise 33.**

EXAMPLE 4 Solve: $\dfrac{x^2}{x-2} = \dfrac{4}{x-2}$.

The LCM of the denominators is $x - 2$. We multiply all terms on both sides by $x - 2$.

$$(x-2)\cdot\frac{x^2}{x-2} = (x-2)\cdot\frac{4}{x-2}$$

$$x^2 = 4 \qquad \text{Simplifying}$$

$$x^2 - 4 = 0$$

$$(x+2)(x-2) = 0$$

$$x + 2 = 0 \quad or \quad x - 2 = 0 \qquad \text{Using the principle of zero products}$$

$$x = -2 \quad or \qquad x = 2$$

Technology Connection

Checking Solutions of Rational Equations

We can use a table to check possible solutions of rational equations. Consider the equation in Example 4,

$$\frac{x^2}{x - 2} = \frac{4}{x - 2},$$

and the possible solutions that were found, -2 and 2. To check these solutions, we enter $y_1 = x^2/(x - 2)$ and $y_2 = 4/(x - 2)$ on the equation-editor screen. Then, with a table set in ASK mode, we enter $x = -2$. (See p. 10.) Since y_1 and y_2 have the same value, we know that the equation is true when $x = -2$, and thus -2 is a solution. Now we enter $x = 2$. The ERROR messages indicate that 2 is not a solution because it is not an allowable replacement for x in the equation.

X	Y₁	Y₂
-2	-1	-1
2	ERROR	ERROR

X=

Exercise

1. Use a graphing calculator to check the possible solutions found in Examples 1, 2, and 3.

Check: For 2:

$$\frac{x^2}{x - 2} = \frac{4}{x - 2}$$

$$\frac{2^2}{2 - 2} \;\overset{?}{\vert}\; \frac{4}{2 - 2}$$

$$\frac{4}{0} \;\Big\vert\; \frac{4}{0} \qquad \text{NOT DEFINED}$$

For -2:

$$\frac{x^2}{x - 2} = \frac{4}{x - 2}$$

$$\frac{(-2)^2}{-2 - 2} \;\overset{?}{\vert}\; \frac{4}{-2 - 2}$$

$$\frac{4}{-4} \;\Big\vert\; \frac{4}{-4}$$

$$-1 \;\Big\vert\; -1 \qquad \text{TRUE}$$

The number -2 is a solution, but 2 is not (it results in division by 0).

> **Now Try Exercise 15.**

EXAMPLE 5 Solve: $\dfrac{2}{x - 1} = \dfrac{3}{x + 1}$.

The LCM of the denominators is $(x - 1)(x + 1)$. We multiply all terms on both sides by $(x - 1)(x + 1)$.

$$(x - 1)(x + 1) \cdot \frac{2}{x - 1} = (x - 1)(x + 1) \cdot \frac{3}{x + 1} \qquad \text{Multiplying}$$

$$2(x + 1) = 3(x - 1) \qquad \text{Simplifying}$$

$$2x + 2 = 3x - 3$$

$$5 = x$$

The check is left to the student. The number 5 checks and is the solution.

> **Now Try Exercise 13.**

EXAMPLE 6 Solve: $\dfrac{2}{x + 5} + \dfrac{1}{x - 5} = \dfrac{16}{x^2 - 25}$.

The LCM of the denominators is $(x + 5)(x - 5)$. We multiply all terms on both sides by $(x + 5)(x - 5)$.

$$(x + 5)(x - 5) \cdot \left[\frac{2}{x + 5} + \frac{1}{x - 5} \right]$$

$$= (x + 5)(x - 5) \cdot \frac{16}{x^2 - 25}$$

$$(x + 5)(x - 5) \cdot \frac{2}{x + 5} + (x + 5)(x - 5) \cdot \frac{1}{x - 5}$$

$$= (x + 5)(x - 5) \cdot \frac{16}{x^2 - 25}$$

$$2(x - 5) + (x + 5) = 16$$

$$2x - 10 + x + 5 = 16$$

$$3x - 5 = 16$$

$$3x = 21$$

$$x = 7$$

Check:

$$\frac{2}{x+5} + \frac{1}{x-5} = \frac{16}{x^2 - 25}$$

$$\frac{2}{7+5} + \frac{1}{7-5} \; \overset{?}{\underset{|}{}} \; \frac{16}{7^2 - 25}$$

$$\frac{2}{12} + \frac{1}{2} \; \bigg| \; \frac{16}{49 - 25}$$

$$\frac{8}{12} \; \bigg| \; \frac{16}{24}$$

$$\frac{2}{3} \; \bigg| \; \frac{2}{3} \qquad \text{TRUE}$$

The solution is 7.

> **Now Try Exercise 25.**

EXAMPLE 7 Given that $f(x) = x + 6/x$, find all values of x for which $f(x) = 5$.

Since $f(x) = x + 6/x$, we want to find all values of x for which

$$x + \frac{6}{x} = 5.$$

The LCM of the denominators is x. We multiply all terms on both sides by x:

$$x\left(x + \frac{6}{x}\right) = x \cdot 5 \qquad \text{Multiplying by } x \text{ on both sides}$$

$$x \cdot x + x \cdot \frac{6}{x} = 5x$$

$$x^2 + 6 = 5x \qquad \text{Simplifying}$$

$$x^2 - 5x + 6 = 0 \qquad \text{Getting 0 on one side}$$

$$(x - 3)(x - 2) = 0 \qquad \text{Factoring}$$

$$x - 3 = 0 \quad or \quad x - 2 = 0 \qquad \text{Using the principle of zero products}$$

$$x = 3 \quad or \qquad x = 2.$$

Check: For $x = 3, f(3) = 3 + \dfrac{6}{3} = 3 + 2 = 5.$

For $x = 2, f(2) = 2 + \dfrac{6}{2} = 2 + 3 = 5.$

The solutions are 2 and 3.

> **Now Try Exercise 41.**

Algebraic/Graphical Connection

Let's make a visual check of Example 7 by looking at a graph. We can think of the equation

$$x + \frac{6}{x} = 5$$

as the intersection of the graphs of

$$f(x) = x + \frac{6}{x} \quad \text{and} \quad g(x) = 5.$$

We see in the graph that there are two points of intersection, at $x = 2$ and at $x = 3$.

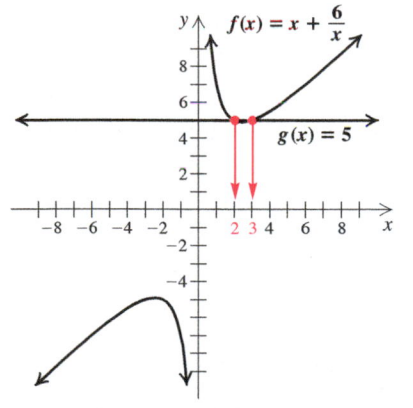

CAUTION! In this section, we have introduced a new use of the LCM. Before, you used the LCM in adding or subtracting rational expressions. Now we are working with equations. There are equals signs. We clear the fractions by multiplying all terms on both sides of the equation by the LCM. This eliminates the denominators. *Do not* make the mistake of trying to "clear the fractions" when you do not have an equation!

5.5 Exercise Set

▶ Reading Check

One of the common difficulties with studying the material in this chapter is being sure about the task at hand. Are you combining expressions using operations to get another rational expression, *or are you solving equations for which the results are numbers that are* solutions *of an equation? To practice making these decisions, determine for each of the following exercises the type of answer you should get: "Rational expression" or "Solutions." You need not complete the mathematical operations.*

RC1. Add: $\dfrac{6w}{w^2 - 1} + \dfrac{w}{w^2 - w}$.

RC2. Solve: $\dfrac{5}{y - 3} - \dfrac{30}{y^2 - 9} = 1$.

RC3. Subtract: $\dfrac{2}{a - 2} - \dfrac{1}{a + 2}$.

RC4. Divide: $\dfrac{x + 4}{x - 2} \div \dfrac{6x}{x^2 - 4}$.

RC5. Solve: $\dfrac{x^2}{x - 1} = \dfrac{1}{x - 1}$.

RC6. Solve: $\dfrac{10}{y} + y = -2$.

RC7. Multiply: $\dfrac{2t^2}{t^2 - 25} \cdot \dfrac{t^2 + 10t + 25}{t^8}$.

RC8. Solve: $\dfrac{7}{x - 4} - \dfrac{2}{x + 4} = \dfrac{1}{x^2 - 16}$.

a Solve. Don't forget to check!

1. $\dfrac{y}{10} = \dfrac{2}{5} + \dfrac{3}{8}$

2. $\dfrac{3}{8} + \dfrac{1}{3} = \dfrac{t}{12}$

3. $\dfrac{1}{4} - \dfrac{5}{6} = \dfrac{1}{a}$

4. $\dfrac{5}{8} - \dfrac{2}{5} = \dfrac{1}{y}$

5. $\dfrac{x}{3} - \dfrac{x}{4} = 12$

6. $\dfrac{y}{5} - \dfrac{y}{3} = 15$

7. $x + \dfrac{8}{x} = -9$

8. $y + \dfrac{22}{y} = -13$

9. $\dfrac{3}{y} + \dfrac{7}{y} = 5$

10. $\dfrac{4}{3y} - \dfrac{3}{y} = \dfrac{10}{3}$

11. $\dfrac{1}{2} = \dfrac{z - 5}{z + 1}$

12. $\dfrac{x - 6}{x + 9} = \dfrac{2}{7}$

13. $\dfrac{3}{y + 1} = \dfrac{2}{y - 3}$

14. $\dfrac{4}{x - 1} = \dfrac{3}{x + 2}$

15. $\dfrac{y - 1}{y - 3} = \dfrac{2}{y - 3}$

16. $\dfrac{x - 2}{x - 4} = \dfrac{2}{x - 4}$

17. $\dfrac{x + 1}{x} = \dfrac{3}{2}$

18. $\dfrac{y + 2}{y} = \dfrac{5}{3}$

19. $\dfrac{1}{2} - \dfrac{4}{9x} = \dfrac{4}{9} - \dfrac{1}{6x}$

20. $-\dfrac{1}{3} - \dfrac{5}{4y} = \dfrac{3}{4} - \dfrac{1}{6y}$

21. $\dfrac{60}{x} - \dfrac{60}{x - 5} = \dfrac{2}{x}$

22. $\dfrac{50}{y} - \dfrac{50}{y - 2} = \dfrac{4}{y}$

23. $\dfrac{7}{5x - 2} = \dfrac{5}{4x}$

24. $\dfrac{5}{y + 4} = \dfrac{3}{y - 2}$

25. $\dfrac{x}{x - 2} + \dfrac{x}{x^2 - 4} = \dfrac{x + 3}{x + 2}$

26. $\dfrac{3}{y - 2} + \dfrac{2y}{4 - y^2} = \dfrac{5}{y + 2}$

27. $\dfrac{6}{x^2 - 4x + 3} - \dfrac{1}{x - 3} = \dfrac{1}{4x - 4}$

28. $\dfrac{1}{2x + 10} = \dfrac{8}{x^2 - 25} - \dfrac{2}{x - 5}$

29. $\dfrac{5}{y + 3} = \dfrac{1}{4y^2 - 36} + \dfrac{2}{y - 3}$

30. $\dfrac{7}{x - 2} - \dfrac{8}{x + 5} = \dfrac{1}{2x^2 + 6x - 20}$

31. $\dfrac{a}{2a - 6} - \dfrac{3}{a^2 - 6a + 9} = \dfrac{a - 2}{3a - 9}$

32. $\dfrac{1}{x-2} = \dfrac{2}{x+4} + \dfrac{2x-1}{x^2+2x-8}$

33. $\dfrac{2x+3}{x-1} = \dfrac{10}{x^2-1} + \dfrac{2x-3}{x+1}$

34. $\dfrac{y}{y+1} + \dfrac{3y+5}{y^2+4y+3} = \dfrac{2}{y+3}$

35. $\dfrac{3x}{x+2} + \dfrac{72}{x^3+8} = \dfrac{24}{x^2-2x+4}$

36. $\dfrac{4}{x+3} + \dfrac{7}{x^2-3x+9} = \dfrac{108}{x^3+27}$

37. $\dfrac{5x}{x-7} - \dfrac{35}{x+7} = \dfrac{490}{x^2-49}$

38. $\dfrac{3x}{x+2} + \dfrac{6}{x} + 4 = \dfrac{12}{x^2+2x}$

39. $\dfrac{x^2}{x^2-4} = \dfrac{x}{x+2} - \dfrac{2x}{2-x}$

For the given rational function f, find all values of x for which $f(x)$ has the indicated value.

40. $f(x) = 2x - \dfrac{15}{x}; \; f(x) = 1$

41. $f(x) = 2x - \dfrac{6}{x}; \; f(x) = 1$

42. $f(x) = \dfrac{x-5}{x+1}; \; f(x) = \dfrac{3}{5}$

43. $f(x) = \dfrac{x-3}{x+2}; \; f(x) = \dfrac{1}{5}$

44. $f(x) = \dfrac{12}{x} - \dfrac{12}{2x}; \; f(x) = 8$

45. $f(x) = \dfrac{6}{x} - \dfrac{6}{2x}; \; f(x) = 5$

▶ Skill Maintenance

Factor. [4.6d]

46. $4t^3 + 500$

47. $1 - t^6$

48. $a^3 + 8b^3$

49. $a^3 - 8b^3$

Solve. [4.8a]

50. $x^2 - 6x + 9 = 0$

51. $(x-3)(x+4) = 0$

52. $x^2 - 49 = 0$

53. $12x^2 - 11x + 2 = 0$

Solve. [2.5c]

54. *Counterfeit Money.* During January 2003 in Indiana, the amount of counterfeit money removed from circulation was $11,492. This amount increased to $54,548 for the month of January 2013. Find the rate of change in the amount of counterfeit money removed with respect to time in years. Round the answer to the nearest dollar. (*Source*: U.S. Secret Service)

55. *New Housing Permits.* In the Indianapolis metropolitan area, 15,054 new housing permits were issued in 2001. By 2012, this number had dropped to approximately 4100 permits. Find the rate of change in the number of new housing permits issued with respect to time in years. (*Sources*: Builders Association of Greater Indianapolis; Market Graphics Research Group; Metropolitan Indianapolis Board of Realtors)

▶ Synthesis

56.
a) Use the INTERSECT feature of a graphing calculator to find the points of intersection of the graphs of

$$f(x) = \dfrac{1}{1+x} + \dfrac{x}{1-x} \quad \text{and} \quad g(x) = \dfrac{1}{1-x} - \dfrac{x}{1+x}.$$

b) Use the algebraic methods of this section to check your answers to part (a).
c) Explain which procedure you prefer.

57.
a) Use the INTERSECT feature of a graphing calculator to find the points of intersection of the graphs of

$$f(x) = \dfrac{x+3}{x+2} - \dfrac{x+4}{x+3} \quad \text{and} \quad g(x) = \dfrac{x+5}{x+4} - \dfrac{x+6}{x+5}.$$

b) Use the algebraic methods of this section to check your answers to part (a).
c) Explain which procedure you prefer.

5.6 Applications and Proportions

▶ **a** Solve work problems and certain basic problems using rational equations.

▶ **b** Solve applied problems involving proportions.

▶ **c** Solve motion problems using rational equations.

▶ **a** Work Problems

EXAMPLE 1 *Filling Sandbags.* The Sandbagger Corporation sells machines that fill sandbags at a job site. The Sandbagger™ can fill an order of 8000 sandbags in 5 hr. The MultiBagger™ can fill the same order in 8 hr. If both machines are used together, how long would it take to fill an order of 8000 sandbags?

1. **Familiarize.** We familiarize ourselves with the problem by considering two incorrect ways of translating the problem to mathematical language.

   a) A common *incorrect* way to translate the problem is to average the two times:

   $$\frac{5 + 8}{2} \text{ hr} = \frac{13}{2} \text{ hr, or } 6\frac{1}{2} \text{ hr.}$$

   Let's think about this. Using only the Sandbagger, the job is completed in 5 hr. If the two baggers are used together, the time it takes to complete the order should be less than 5 hr. Thus we reject $6\frac{1}{2}$ hr as a solution, but we do have a partial check on any answer we get. The answer should be less than 5 hr.

   b) Another *incorrect* way to translate the problem is as follows. Suppose the two machines are used in such a way that half of the job is done by the Sandbagger and the other half by the Multibagger. Then

   the Sandbagger fills $\frac{1}{2}$ of the bags in $\frac{1}{2}(5 \text{ hr})$, or $2\frac{1}{2}$ hr,

   and

   the Multibagger fills $\frac{1}{2}$ of the bags in $\frac{1}{2}(8 \text{ hr})$, or 4 hr.

   But time is wasted since the Sandbagger completed its part $1\frac{1}{2}$ hr earlier than the Multibagger. In effect, the machines were not used together to complete the job as fast as possible. If the Sandbagger is used in addition to the MultiBagger after completing its half, the entire job could be finished in a time somewhere between $2\frac{1}{2}$ hr and 4 hr.

   We proceed to a translation by considering how much of the job is finished in 1 hr, 2 hr, 3 hr, and so on. It takes the Sandbagger 5 hr to fill 8000 bags alone. Then in 1 hr, it can do $\frac{1}{5}$ of the job. It takes the MultiBagger 8 hr to complete the job alone. Then in 1 hr, it can do $\frac{1}{8}$ of the job. Both baggers together can complete

   $$\frac{1}{5} + \frac{1}{8}, \text{ or } \frac{13}{40} \text{ of the job in 1 hr.}$$

In 2 hr, the Sandbagger can do $2\left(\frac{1}{5}\right)$ of the job and the MultiBagger can do $2\left(\frac{1}{8}\right)$ of the job. Both baggers together can complete

$$2\left(\frac{1}{5}\right) + 2\left(\frac{1}{8}\right), \text{ or } \frac{13}{20} \text{ of the job in 2 hr.}$$

Continuing this reasoning, we can form a table like the one below. From the table, we see that if the baggers work together for 4 hr, the fraction of the job that will be completed is $1\frac{3}{10}$, which is more of the job than needs to be done. We also see that the answer is somewhere between 3 hr and 4 hr. What we are looking for is a number of hours t such that the fraction of the job that is completed is 1; that is, the job is just completed—not more $\left(1\frac{3}{10}\right)$ and not less $\left(\frac{39}{40}\right)$.

	Fraction of Job Completed		
Time	**Sandbagger**	**Multibagger**	**Together**
1 hr	$\frac{1}{5}$	$\frac{1}{8}$	$\frac{1}{5} + \frac{1}{8}$, or $\frac{13}{40}$
2 hr	$2\left(\frac{1}{5}\right)$	$2\left(\frac{1}{8}\right)$	$2\left(\frac{1}{5}\right) + 2\left(\frac{1}{8}\right)$, or $\frac{13}{20}$
3 hr	$3\left(\frac{1}{5}\right)$	$3\left(\frac{1}{8}\right)$	$3\left(\frac{1}{5}\right) + 3\left(\frac{1}{8}\right)$, or $\frac{39}{40}$
4 hr	$4\left(\frac{1}{5}\right)$	$4\left(\frac{1}{8}\right)$	$4\left(\frac{1}{5}\right) + 4\left(\frac{1}{8}\right)$, or $1\frac{3}{10}$
t hr	$t\left(\frac{1}{5}\right)$	$t\left(\frac{1}{8}\right)$	$t\left(\frac{1}{5}\right) + t\left(\frac{1}{8}\right)$

2. **Translate.** From the table, we see that the time we are looking for is some number t for which

$$t\left(\frac{1}{5}\right) + t\left(\frac{1}{8}\right) = 1, \quad \text{or} \quad \frac{t}{5} + \frac{t}{8} = 1,$$

where 1 represents the idea that the entire job is completed in time t.

3. **Solve.** We solve the equation:

$$\frac{t}{5} + \frac{t}{8} = 1$$

$$40\left(\frac{t}{5} + \frac{t}{8}\right) = 40 \cdot 1 \qquad \textcolor{red}{\textbf{The LCM is } 5 \cdot 2 \cdot 2 \cdot 2, \textbf{ or } 40.}$$
$$\textcolor{red}{\textbf{We multiply by 40.}}$$

$$40 \cdot \frac{t}{5} + 40 \cdot \frac{t}{8} = 40 \qquad \textcolor{red}{\textbf{Using the distributive law}}$$

$$8t + 5t = 40 \qquad \textcolor{red}{\textbf{Simplifying}}$$

$$13t = 40$$

$$t = \frac{40}{13}, \text{ or } 3\frac{1}{13} \text{ hr.}$$

4. **Check.** The check can be done by using $\frac{40}{13}$ for t and substituting into the original equation:

$$\frac{40}{13}\left(\frac{1}{5}\right) + \frac{40}{13}\left(\frac{1}{8}\right) = \frac{8}{13} + \frac{5}{13} = \frac{13}{13} = 1.$$

We also have a partial check in what we learned from the *Familiarize* step. The answer, $3\frac{1}{13}$ hr, is between 3 hr and 4 hr (see the table), and it is less than 5 hr, the time it takes the Sandbagger to do the job alone.

5. **State.** It takes $3\frac{1}{13}$ hr for the two baggers to complete the job working together.

▶ **Now Try Exercise 1.**

THE WORK PRINCIPLE

Suppose that a is the time it takes A to do a job, b is the time it takes B to do the same job, and t is the time it takes them to do the job working together. Then

$$\frac{t}{a} + \frac{t}{b} = 1.$$

Gracie and Phillip Ramos of San Antonio, Texas, own and operate Gracie's Kitchen food truck.

EXAMPLE 2 *Catering a Business Luncheon.*   A convention planner is organizing a business luncheon for 2300 guests in San Antonio, Texas. The chosen menu will include tacos and tamales. To complete the order, two food caterers are needed. Reading that Gracie's Kitchen won the 2013 Annual Twisted Taco Truck Throwdown, the organizer wants to hire Gracie as one of the caterers. Suppose Gracie and a second caterer together can complete the order in 12 hr. If the second caterer would take 10 hr longer than Gracie to complete the order alone, how long would it take each, working alone, to prepare the luncheon?

1. **Familiarize.** Comparing this problem to Example 1, we note that we do not know the times required by each caterer to complete the task had each worked alone. We let

   $h =$ the amount of time, in hours, that it would take Gracie working alone.

   Then

   $h + 10 =$ the amount of time, in hours, that it would take the second caterer working alone.

   We also know that $t = 12$ hr = total time. Thus,

   $\dfrac{12}{h} =$ the fraction of the job that Gracie could finish in 12 hr

   and

   $\dfrac{12}{h + 10} =$ the fraction of the job that the second caterer could finish in 12 hr.

2. **Translate.** Using the work principle, we know that

   $$\frac{t}{a} + \frac{t}{b} = 1 \qquad \text{\color{red}{Using the work principle}}$$

   $$\frac{12}{h} + \frac{12}{h + 10} = 1. \qquad \text{\color{red}{Substituting } \frac{12}{h} \text{ for } \frac{t}{a} \text{ and } \frac{12}{h + 10} \text{ for } \frac{t}{b}}$$

3. Solve. We solve the equation:

$$\frac{12}{h} + \frac{12}{h + 10} = 1$$

$$h(h + 10)\left(\frac{12}{h} + \frac{12}{h + 10}\right) = h(h + 10) \cdot 1 \qquad \text{We multiply by the LCM, which is } h(h + 10).$$

$$h(h + 10) \cdot \frac{12}{h} + h(h + 10) \cdot \frac{12}{h + 10} = h(h + 10) \qquad \text{Using the distributive law}$$

$$(h + 10) \cdot 12 + h \cdot 12 = h^2 + 10h \qquad \text{Simplifying}$$

$$12h + 120 + 12h = h^2 + 10h$$

$$0 = h^2 - 14h - 120 \qquad \text{Getting 0 on one side}$$

$$0 = (h - 20)(h + 6) \qquad \text{Factoring}$$

$$h - 20 = 0 \quad or \quad h + 6 = 0 \qquad \text{Using the principle of zero products}$$

$$h = 20 \quad or \qquad h = -6.$$

4. Check. Since negative time has no meaning in the problem, we reject -6 as a solution to the original problem. The number 20 checks since if Gracie takes 20 hr alone and the second caterer takes $20 + 10$, or 30 hr alone, in 12 hr, working together, they would have completed

$$\frac{12}{20} + \frac{12}{30} = \frac{3}{5} + \frac{2}{5} = \frac{5}{5}, \text{ or 1 job.}$$

5. State. It would take Gracie 20 hr and the second caterer 30 hr to complete the task alone.

▶ **Now Try Exercise 5.**

▶ **b** **Applications Involving Proportions**

Any rational expression a/b represents a **ratio**. Percent can be considered a ratio. For example, 67% is the ratio of 67 to 100, or $67/100$. The ratio of two different kinds of measure is called a **rate**. Speed is an example of a rate. For example, Usain Bolt, from Jamaica, completed the 100-m run in the 2012 Summer Olympics in London with a time of 9.63 sec. His speed, or rate, was

$$\frac{100 \text{ m}}{9.63 \text{ sec}}, \quad \text{or} \quad 10.4 \frac{\text{m}}{\text{sec}}. \qquad \text{Rounded to the nearest tenth}$$

He also holds the world record for this event with a time of 9.58 sec, which he raced on August 16, 2009, in Berlin, Germany.

PROPORTION

An equality of ratios, $A/B = C/D$, read "A is to B as C is to D," is called a **proportion**. The numbers named in a true proportion are said to be **proportional** to each other.

We can use proportions to solve applied problems by expressing a ratio in two ways, as shown below. For example, suppose that it takes 8 gal of gas to drive for 120 mi, and we want to determine how much gas will be required to drive for 550 mi. If we assume that the car uses gas at the same rate throughout the trip, the ratios are the same, and we can write a proportion. We let x represent the number of gallons it takes to drive 550 mi.

Miles $\longrightarrow \dfrac{120}{8} = \dfrac{550}{x} \longleftarrow$ **Miles**
Gallons $\longrightarrow$ $\qquad\qquad\qquad \longleftarrow$ **Gallons**

To solve this proportion, we note that the LCM is $8x$. Thus we multiply by $8x$.

$$8x \cdot \frac{120}{8} = 8x \cdot \frac{550}{x} \qquad \text{\color{red}{Multiplying by } } 8x$$

$$x \cdot 120 = 8 \cdot 550 \qquad \text{\color{red}{Simplifying}}$$

$$120x = 8 \cdot 550$$

$$x = \frac{8 \cdot 550}{120} \qquad \text{\color{red}{Dividing by } } 120$$

$$x \approx 36.67$$

Thus, 36.67 gal will be required to drive 550 mi.

It is common to use **cross products** to solve proportions, as follows:

$$\frac{120}{8} = \frac{550}{x}$$

If $\dfrac{A}{B} = \dfrac{C}{D}$, then $AD = BC$.

$$120 \cdot x = 8 \cdot 550 \qquad \text{\color{red}{$120 \cdot x$ and $8 \cdot 550$ are called \textit{cross products}. Note that this is the equation that results from clearing fractions above.}}$$

$$x = \frac{8 \cdot 550}{120}$$

$$x \approx 36.67.$$

EXAMPLE 3   *Calories Burned.*   Jayden, who weighs 170 lb, will burn 345 calories in 45 min while hiking. How many calories will he burn if he hikes for 2 hr? (*Source*: The American Dietetic Association's *Complete Food & Nutrition Guide*)

1. **Familiarize.** We let $c =$ the number of calories burned in 2 hr.

2. **Translate.** Next, we translate to a proportion. We make each side the ratio of number of minutes to number of calories, with number of minutes in the numerator and number of calories in the denominator. We substitute 120 min for 2 hr.

Minutes $\longrightarrow \dfrac{45}{345} = \dfrac{120}{c} \longleftarrow$ **Minutes**
Calories $\longrightarrow$ $\qquad\qquad\qquad \longleftarrow$ **Calories**

3. **Solve.** We solve the proportion:

$$\frac{45}{345} = \frac{120}{c}$$

$$45c = 345 \cdot 120 \qquad \text{\color{red}{Equating cross products}}$$

$$c = \frac{345 \cdot 120}{45} \qquad \text{\color{red}{Dividing by } } 45$$

$$c = 920. \qquad \text{\color{red}{Multiplying and dividing}}$$

4. Check. We substitute into the proportion and check cross products:

$$\frac{45}{345} = \frac{120}{920};$$

$$45 \cdot 920 = 41{,}400; \quad 345 \cdot 120 = 41{,}400.$$

Since the cross products are the same, the answer checks.

5. State. In 120 min, or 2 hr, Jayden will burn 920 calories.

 **Now Try Exercise 11.**

EXAMPLE 4 *Estimating Wild Burro Population.* Wild burros still exist in at least five U.S. states. To estimate the number in Arizona, a ranger catches 383 wild burros, tags them, and releases them. Later, 103 burros are caught, and it is found that 11 of them are tagged. Estimate how many wild burros there are in Arizona. (*Source*: U.S. Department of the Interior Bureau of Land Management, The Wild Horse and Burro Program)

1. Familiarize. We let $B =$ the number of wild burros in Arizona. For the purposes of this example, we assume that the tagged burros mix freely with others in the state. We also assume that when some burros have been captured, the ratio of those tagged to the total number captured is the same as the ratio of burros originally tagged to the total number of wild burros in the state. For example, if 1 of every 3 burros captured later is tagged, we would assume that 1 of every 3 burros in the state was originally tagged.

2. Translate. We translate to a proportion, as follows:

Burros tagged originally $\longrightarrow$ $\dfrac{383}{B} = \dfrac{11}{103}$. $\longleftarrow$ Tagged burros caught
Burros in Arizona $\longrightarrow$ $\qquad\qquad\qquad$ $\longleftarrow$ Burros caught

3. Solve. We solve the proportion:

$$383 \cdot 103 = B \cdot 11 \qquad \text{Equating cross products}$$

$$\frac{383 \cdot 103}{11} = B \qquad \text{Dividing by 11}$$

$$3586 \approx B. \qquad \text{Multiplying and dividing and rounding}$$

4. Check. We substitute into the proportion and check cross products:

$$\frac{383}{3586} = \frac{11}{103}; \quad 383 \cdot 103 = 39{,}449; \quad 3586 \cdot 11 = 39{,}446.$$

The cross products are close but not exact because we rounded the total.

5. State. We estimate that there are about 3586 wild burros in Arizona.

Now Try Exercise 21.

▶ c Motion Problems

We considered motion problems earlier in Sections 1.3 and 3.4. To translate them, we know that we can use either the basic motion formula, $d = rt$, or either of two formulas, $r = d/t$ or $t = d/r$, which can be derived from $d = rt$.

MOTION FORMULAS

The following are the formulas for motion problems:

$$d = rt \longrightarrow \text{Distance} = \text{Rate} \cdot \text{Time};$$

$$r = \frac{d}{t} \longrightarrow \text{Rate} = \frac{\text{Distance}}{\text{Time}}; \qquad t = \frac{d}{r} \longrightarrow \text{Time} = \frac{\text{Distance}}{\text{Rate}}.$$

EXAMPLE 5 *Bicycling.* A racer is bicycling 15 km/h faster than a person on a mountain bike. In the time it takes the racer to travel 80 km, the person on the mountain bike has gone 50 km. Find the speed of each bicyclist.

1. **Familiarize.** Let's guess that the person on the mountain bike is going 10 km/h. The racer would then be traveling $10 + 15$, or 25 km/h. At 25 km/h, the racer will travel 80 km in $\frac{80}{25} = 3.2$ hr. Going 10 km/h, the mountain bike will cover 50 km in $\frac{50}{10} = 5$ hr. Since $3.2 \neq 5$, our guess is wrong, but we can see that if r is the rate, in kilometers per hour, of the slower bike, then the rate of the racer, who is traveling 15 km/h faster, is $r + 15$.

   Making a drawing and organizing the facts in a chart can be helpful.

50 km

r km/h

80 km

$r + 15$ km/h

	Distance	Speed	Time
Mountain Bike	50	r	t
Racing Bike	80	$r + 15$	t

$\longrightarrow 50 = rt \longrightarrow t = \dfrac{50}{r}$

$\longrightarrow 80 = (r + 15)t \longrightarrow t = \dfrac{80}{r + 15}$

2. **Translate.** The time is the same for both bikes. Using the formula $d = rt$ and then $t = d/r$ across both rows of the table, we find two expressions for time and can equate them as

$$\frac{50}{r} = \frac{80}{r + 15}.$$

3. **Solve.** We solve the equation:

$$\frac{50}{r} = \frac{80}{r + 15}$$

$$r(r + 15) \cdot \frac{50}{r} = r(r + 15) \cdot \frac{80}{r + 15} \qquad \text{The LCM is } r(r + 15). \text{ We multiply by } r(r + 15).$$

$$(r + 15) \cdot 50 = r \cdot 80 \qquad \text{Simplifying. We can also obtain this by equating cross products.}$$

$$50r + 750 = 80r \qquad \text{Using the distributive law}$$

$$750 = 30r \qquad \text{Subtracting } 50r$$

$$\frac{750}{30} = r \qquad \text{Dividing by 30}$$

$$25 = r.$$

4. **Check.** If our answer checks, the mountain bike is going 25 km/h and the racing bike is going 25 + 15, or 40 km/h.

Traveling 80 km at 40 km/h, the racer is riding for $\frac{80}{40} = 2$ hr. Traveling 50 km at 25 km/h, the person on the mountain bike is riding for $\frac{50}{25} = 2$ hr. Our answer checks since the two times are the same.

5. **State.** The speed of the racer is 40 km/h, and the speed of the person on the mountain bike is 25 km/h.

Now Try Exercise 31.

EXAMPLE 6 *Transporting by Barge.* A river barge travels 98 mi downstream in the same time it takes to travel 52 mi upstream. The speed of the current in the river is 2.3 mph. Find the speed of the barge in still water.

1. **Familiarize.** We first make a drawing. We let s = the speed of the barge in still water and t = the time, and then organize the facts in a table.

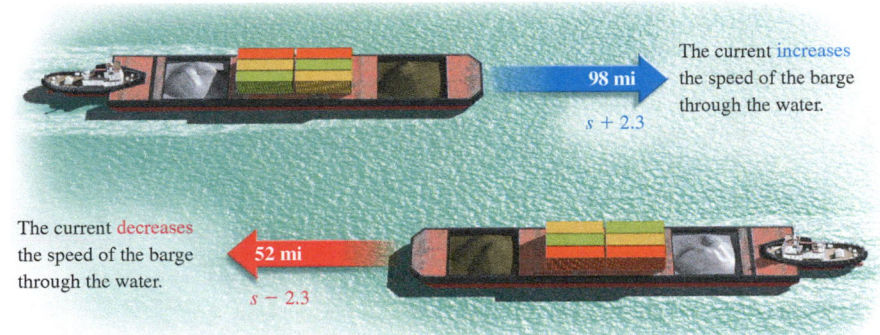

The current increases the speed of the barge through the water.

98 mi
$s + 2.3$

The current decreases the speed of the barge through the water.

52 mi
$s - 2.3$

	Distance	Speed	Time
Downstream	98	$s + 2.3$	t
Upstream	52	$s - 2.3$	t

$\longrightarrow 98 = (s + 2.3)t \longrightarrow t = \dfrac{98}{s + 2.3}$

$\longrightarrow 52 = (s - 2.3)t \longrightarrow t = \dfrac{52}{s - 2.3}$

2. **Translate.** Using the formula $t = d/r$ across both rows of the table, we find two expressions for time and equate them as

$$\frac{98}{s + 2.3} = \frac{52}{s - 2.3}.$$

3. **Solve.** We solve the equation:

$$\frac{98}{s + 2.3} = \frac{52}{s - 2.3}$$

$$(s + 2.3)(s - 2.3)\left(\frac{98}{s + 2.3}\right) = (s + 2.3)(s - 2.3)\left(\frac{52}{s - 2.3}\right)$$

$$(s - 2.3)98 = (s + 2.3)52$$

$$98s - 225.4 = 52s + 119.6$$

$$46s = 345$$

$$s = 7.5.$$

4. **Check.** Downstream, the speed of the barge is 7.5 + 2.3, or 9.8 mph. Dividing the distance, 98 mi, by the speed, 9.8 mph, we get 10 hr. Upstream, the speed of the barge is 7.5 − 2.3, or 5.2 mph. Dividing the distance, 52 mi, by the speed, 5.2 mph, we get 10 hr. Since the times are the same, the answer checks.

5. **State.** The speed of the barge in still water is 7.5 mph.

Now Try Exercise 29.

Translating for Success

1. Sums of Squares. The sum of the squares of two consecutive odd integers is 650. Find the integers.

2. Estimating Fish Population. To determine the number of fish in a lake, a conservationist catches 225 of them, tags them, and releases them back into the lake. Later, 108 fish are caught, and it is found that 15 of them are tagged. Estimate how many fish are in the lake.

3. Consecutive Integers. The sum of two consecutive even integers is 650. Find the integers.

4. Sums of Squares. The sum of the squares of two consecutive integers is 685. Find the integers.

5. Hockey Results. A hockey team played 81 games in a season. They won 1 fewer game than three times the number of ties and lost 8 fewer games than they won. How many games did they win? lose? tie?

Translate each word problem to an equation or a system of equations and select a correct translation from choices A–O.

A. $x + (x + 2) = 650$

B. $\dfrac{225}{x} = \dfrac{15}{108}$

C. $x^2 + (x + 1)^2 = 685$

D. $\dfrac{30}{x + 3} = \dfrac{40}{x}$

E. $x + y + z = 81,$
$x = 3y - 1,$
$z = x - 8$

F. $x + y + z = 81,$
$x - 1 = 3y,$
$z = x - 8$

G. $x^2 + (x + 5)^2 = 650$

H. $x + y + z = 650,$
$x + y = 480,$
$y + z = 685$

I. $\dfrac{40}{x + 3} = \dfrac{30}{x}$

J. $\dfrac{15}{x} = \dfrac{108}{225}$

K. $x + y + z = 685,$
$x + y = 480,$
$y + z = 650$

L. $x^2 + (x + 2)^2 = 685$

M. $\dfrac{x}{3} + \dfrac{x}{8} = 1$

N. $x^2 + (x + 2)^2 = 650$

O. $x = y + 3,$
$2x + 2y = 81$

6. Sides of a Square. If each side of a square is increased by 5 ft, the area of the original square plus the area of the enlarged square is 650 ft². Find the length of a side of the original square.

7. Bicycling. The speed of one mountain biker is 3 km/h faster than the speed of another biker. The first biker travels 40 km in the same amount of time that it takes the second to travel 30 km. Find the speed of each biker.

8. PDQ Shopping Network. Sarah, Claire, and Maggie can process 685 telephone orders per day for PDQ shopping network. Sarah and Claire together can process 480 orders per day, while Claire and Maggie can process 650 orders. How many orders can each process alone?

9. Filling Time. A spa can be filled in 3 hr by hose A alone and in 8 hr by hose B alone. How long would it take to fill the spa if both hoses are working?

10. Rectangle Dimensions. The length of a rectangle is 3 ft longer than its width. Find the dimensions of the rectangle such that the perimeter of the rectangle is 81 ft.

5.6 Exercise Set

► Reading Check

Choose from the column on the right the word that best completes the statement. Some words may be used more than once, and some words may not be used.

RC1. If two triangles are similar, then their _____ angles have the _____ measures and their corresponding sides are _____.

RC2. A ratio of two quantities is their _____.

RC3. An equality of ratios, $\frac{A}{B} = \frac{C}{D}$, is called a(n) _____.

RC4. Distance equals _____ times time.

RC5. Rate equals _____ divided by time.

RC6. Time equals _____ divided by rate.

RC7. The numbers named in a true proportion are said to be _____ to each other.

same

different

product

quotient

distance

rate

proportion

proportional

similar

corresponding

---

a *Solve.*

1. *Painting a House.* Jose can paint a house in 28 hr. His brother, Miguel, can paint the same house in 36 hr. Working together, how long will it take them to paint the house?

2. *Filling a Pool.* An in-ground backyard pool can be filled in 12 hr if water enters through a pipe alone, or in 30 hr if water enters through a hose alone. If water is entering through both the pipe and the hose, how long will it take to fill the pool?

3. *Printing Tee Shirts.* In 30 hr, one machine can print tee shirts honoring the winning team in a national championship sporting event. Another machine can complete the same order in only 24 hr. If both machines are used, how long will it take to print the order?

4. *Washing Elephants.* Leah can wash the zoo's elephants in 3 hr. Ian, who is less experienced, needs 4 hr to do the same job. Working together, how long will it take them to wash the elephants?

5. *Placing Wrappers on Canned Goods.* Machine A can place wrappers on a batch of canned goods in 4 fewer hours than machine B. Together, they can complete the job in 1.5 hr. How long would it take each machine working alone?

6. *Cutting Firewood.* Tom can cut and split a cord of firewood in 6 fewer hours than Henry can. When they work together, it takes them 4 hr. How long would it take each of them to do the job alone?

7. *Clearing a Lot.* A commercial contractor needs to clear a plot of land for a new bank. Ryan can clear the lot in 7.5 hr. Ethan can do the same job in 10.5 hr. How long will it take them, working together, to clear the land? (*Hint*: You may find that multiplying by $\frac{1}{10}$ on both sides of the equation will clear the decimals.)

8. *Sorting Donations.* At the neighborhood food pantry, Grace can sort a morning's donations in 4.5 hr. Caleb can do the same job in 3 hr. Working together, how long would it take them to sort the food donations?

9. *Skimming a Swimming Pool.* Cole can skim a swimming pool with a leaf net three times as fast as Jim can. Together they can skim the pool in 6 min. How long would it take each to skim the pool alone?

10. *Pressing Shirts.* Karla and William press shirts for Perfection Laundry. Each week an amusement park drops off 320 shirts to be laundered and pressed. Karla can press shirts twice as fast as William. Together they can press 320 shirts in 11 hr. How long would it take each to press this order alone?

b *Solve.*

11. *Three-Point Field Goals.* After 15 games of an NBA season, a player had scored 9 three-point field goals. Assuming that he would continue to score at the same rate, how many three-point field goals would he score in the entire 82-game season?

12. *Touchdown Pace.* After 5 games of a recent NFL season, a quarterback had passed for 12 touchdowns. Assuming that he would continue to pass for touchdowns at the same rate, how many touchdown passes would he throw in the entire 16-game season?

13. *Models of Indy Cars.* Mattel, Inc., recently added some made-to-scale models of Indy Cars to their Hot Wheels® product line. The length of an IRL car is 15 ft. Its width is 7 ft. The width of the die-cast replica is 3.5 in. Find the length of the model. (*Source*: Mattel, IndyCar.com)

14. *USS Constitution.* For a woodworking class, Alexis is building a scale model of the *USS Constitution*, known as "Old Ironsides." The length of the ship at the water line is 175 ft; the beam (or width) is 43.5 ft. The width of the model is 6.75 in. Find the length of the model.

15. *Weight on Moon.* The ratio of the weight of an object on the moon to the weight of an object on Earth is 0.16 to 1. How much will a 180-lb astronaut weigh on the moon?

16. *Weight on Mars.* The ratio of the weight of an object on Mars to the weight of an object on Earth is 0.378 to 1. How much will a 120-lb astronaut weigh on Mars?

17. *Coffee Consumption.* Coffee beans from 14 trees are required to produce 7.7 kg of coffee. (This is the average amount that each person in the United States drinks each year.) The beans from how many trees are required to produce 638 kg of coffee?

18. *Human Blood.* 10 cm^3 of a normal specimen of human blood contains 1.2 g of hemoglobin. How many grams does 32 cm^3 of the same blood contain?

19. *Wine Production.* In a recent year, a winery produced 4320 bottles of wine from 8 tons of grapes. They expect the demand to reach 7850 bottles next year. How many tons of grapes will they need? (*Source*: www.napanow.com)

20. *Wind Turbines.* As of 2013, Indiana had 923 wind turbines on 5 wind farms. Data show that when operating at full capacity, 15 300-ft tall wind turbines can power 5250 homes. How many homes can be powered by 923 wind turbines? (*Source:* Indiana Office of Energy Development, E.ON Climate & Renewables)

21. *Rope Cutting.* A rope is 28 ft long. How can the rope be cut in such a way that the ratio of the resulting two segments is 3 to 5?

22. Consider the numbers 1, 2, 3, and 5. If the same number is added to each of the numbers, it is found that the ratio of the first new number to the second is the same as the ratio of the third new number to the fourth. Find the number.

23. *Retaining Wall.* On average, a retaining wall requires approximately 1017 kg of stone for each 3.2 m². The area includes the face of the wall and its upper surface. How many kilograms of stone are needed for 65 m²? Round the answer to the nearest kilogram. (*Source:* Reed, David, *The Art and Craft of Stonescaping.* Sterling Publishing Co., Inc.: New York, 2000)

24. *Corona Arch.* The photograph below shows Corona Arch in Moab, Utah, one of the favorite hiking places of one of your authors, Marv Bittinger. He appears at the bottom of the photograph. Assume that an $8\frac{1}{2}$-in. by 11-in. photograph has been printed from a digital file and that in that photo Marv is $\frac{11}{32}$, or 0.34375 in. tall, and the height of the arch in the photo is $7\frac{5}{8}$, or 7.625 in. Given that Marv is 73 in. tall, find the actual height H of the arch.

25. *Estimating Wildlife Populations.* To determine the number of trout in a lake, a conservationist catches 112 trout, tags them, and releases them back into the lake. Later, 82 trout are caught; 32 of them are tagged. How many trout are in the lake?

26. *Estimating Wildlife Populations.* To determine the number of deer in a game preserve, a conservationist catches 318 deer, tags them, and lets them loose. Later, 168 deer are caught; 56 of them are tagged. How many deer are in the preserve?

C *Solve.*

27. *Jet Travel.* A Boeing 747 flies 2420 mi with the wind. In the same amount of time, it can fly 2140 mi against the wind. The cruising speed (in still air) is 570 mph. Find the speed of the wind. (*Source:* Boeing)

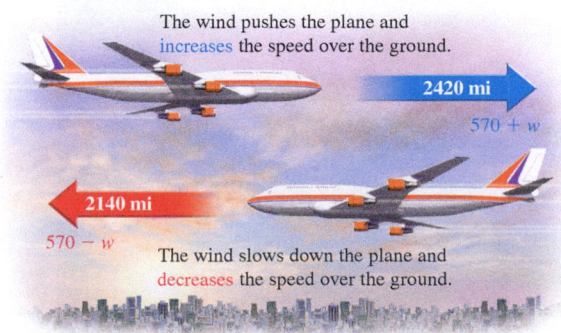

The wind pushes the plane and increases the speed over the ground.

2420 mi

570 + w

2140 mi

570 − w

The wind slows down the plane and decreases the speed over the ground.

28. *Transporting Cargo.* Boeing has a jumbo jet that is used to transport cargo. This jet flies 1430 mi with the wind. In the same amount of time, it can fly 1320 mi against the wind. The cruising speed (in still air) is 550 mph. Find the speed of the wind. (*Source:* Boeing)

29. *Kayaking.* The speed of the current in the Wabash River is 3 mph. Brooke's kayak can travel 4 mi upstream in the same time that it takes to travel 10 mi downstream. What is the speed of Brooke's kayak in still water?

30. *Boating.* The current in the Animas River moves at a rate of 4 mph. Sydney's dinghy motors 6 mi upstream in the same time that it takes to motor 12 mi downstream. What is the speed of the dinghy in still water?

31. *Tour Travel.* Adventure Tours has 6 leisure-tour trolleys that travel 15 mph slower than their 3 express-tour buses. The bus travels 132 mi in the time it takes the trolley to travel 99 mi. Find the speed of each mode of transportation.

32. *Hiking.* Vanessa hikes 2 mph slower than Xavier. In the time it takes Xavier to hike 8 mi, Vanessa hikes 5 mi. Find the speed of each person.

33. *Moving Sidewalks.* A moving sidewalk at an airport moves at a rate of 1.8 ft/sec. Walking on the moving sidewalk, Thomas travels 105 ft forward in the time it takes to travel 51 ft in the opposite direction. How fast would Thomas be walking on a nonmoving sidewalk?

34. *Moving Sidewalks.* A moving sidewalk moves at a rate of 1.7 ft/sec. Walking on the moving sidewalk, Hunter can travel 120 ft forward in the same time it takes to travel 52 ft in the opposite direction. How fast would Hunter be walking on a nonmoving sidewalk?

▶ **Skill Maintenance**

In Exercises 35–38, the graph is that of a function. Determine the domain and the range. **[2.3a]**

35.

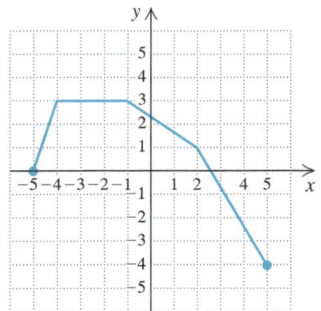

36.

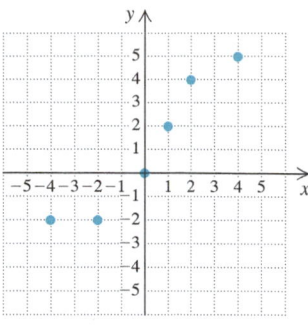

37.

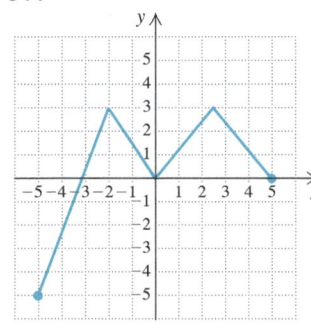

38.

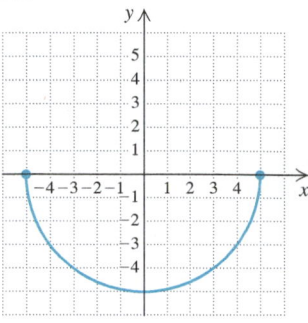

Graph on a plane. **[3.7a]**

39. $x - 4y \geq 4$

40. $x \geq 3$

Graph. **[2.2c]**

41. $f(x) = |x + 3|$

42. $f(x) = 5 - |x|$

▶ **Synthesis**

43. Three trucks, A, B, and C, working together, can move a load of mulch in t hours. When working alone, it takes A 1 extra hour to move the mulch, B 6 extra hours, and C t extra hours. Find t.

44. *Escalators.* Together, a 100-cm wide escalator and a 60-cm wide escalator can empty a 1575-person auditorium in 14 min. The wider escalator moves twice as many people as the narrower one does. How many people per hour does the 60-cm wide escalator move?

45. *Gas Mileage.* An automobile gets 22.5 miles per gallon (mpg) in city driving and 30 mpg in highway driving. The car is driven 465 mi on a full tank of 18.4 gal of gasoline. How many miles were driven in the city and how many were driven on the highway?

46. *Travel by Car.* Mackenzie drives to work at 50 mph and arrives 1 min late. She drives to work at 60 mph and arrives 5 min early. How far does Mackenzie live from work?

5.7 Formulas and Applications

▶ **a** Solve a formula for a letter.

▶ a Formulas

Formulas occur frequently as mathematical models.

> To solve a rational formula for a given letter, identify the letter, and:
>
> 1. Multiply on both sides to clear fractions or decimals, if that is needed.
> 2. Multiply if necessary to remove parentheses.
> 3. Get all terms with the letter to be solved for on one side of the equation and all other terms on the other side, using the addition principle.
> 4. Factor out the unknown if it appears in more than one term.
> 5. Solve for the letter in question, using the multiplication principle.

EXAMPLE 1 *Optics.* The formula $f = L/d$ defines a camera's "f-stop," where L is the *focal length* (the distance from the lens to the film) and d is the *aperture* (the diameter of the lens). Solve the formula for d.

We solve this equation as we did the rational equations in Section 5.5:

$$f = \frac{L}{d} \qquad \text{We want the letter } d \text{ alone.}$$

$$d \cdot f = d \cdot \frac{L}{d} \qquad \text{The LCM is } d. \text{ We multiply by } d.$$

$$df = L \qquad \text{Simplifying}$$

$$d = \frac{L}{f}. \qquad \text{Dividing by } f$$

The formula $d = L/f$ can now be used to find the aperture if we know the focal length and the f-stop.

▶ **Now Try Exercise 1.**

EXAMPLE 2 *Astronomy.* The formula $L = \dfrac{dR}{D - d}$, where D is the diameter of the sun, d is the diameter of the earth, R is the earth's distance from the sun, and L is some fixed distance, is used to calculate when lunar eclipses occur. Solve the formula for D.

$$L = \frac{dR}{D - d} \qquad \text{We want the letter } D \text{ alone.}$$

$$(D - d) \cdot L = (D - d) \cdot \frac{dR}{D - d} \qquad \text{The LCM is } D - d. \text{ We multiply by } D - d.$$

$$(D - d)L = dR \qquad \text{Simplifying}$$
$$DL - dL = dR$$
$$DL = dR + dL \qquad \text{Adding } dL$$
$$D = \frac{dR + dL}{L} \qquad \text{Dividing by } L$$

Since D appears by itself on one side and not on the other, we have solved for D.

→ **Now Try Exercise 5.**

EXAMPLE 3   Solve the formula $L = \dfrac{dR}{D - d}$ for d.

We proceed as we did in Example 2 until we reach the equation

$$DL - dL = dR. \qquad \text{We want } d \text{ alone.}$$

We must get all terms containing d alone on one side:

$$DL - dL = dR$$
$$DL = dR + dL \qquad \text{Adding } dL$$
$$DL = d(R + L) \qquad \text{Factoring out the letter } d$$
$$\frac{DL}{R + L} = d. \qquad \text{Dividing by } R + L$$

We now have d alone on one side, so we have solved the formula for d.

→ **Now Try Exercise 7.**

CAUTION!   If, when you are solving an equation for a letter, the letter appears on both sides of the equation, you know the answer is wrong. The letter must be alone on one side and *not* occur on the other.

EXAMPLE 4   *Resistance.*   The formula

$$\frac{1}{R} = \frac{1}{r_1} + \frac{1}{r_2}$$

involves the resistance R of two resistors r_1 and r_2 connected in parallel.* Solve the formula for r_1.

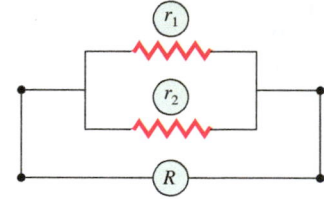

We multiply by the LCM, which is Rr_1r_2:

$$Rr_1r_2 \cdot \frac{1}{R} = Rr_1r_2 \cdot \left(\frac{1}{r_1} + \frac{1}{r_2} \right) \qquad \text{Multiplying by the LCM}$$
$$Rr_1r_2 \cdot \frac{1}{R} = Rr_1r_2 \cdot \frac{1}{r_1} + Rr_1r_2 \cdot \frac{1}{r_2} \qquad \text{Multiplying to remove parentheses}$$
$$r_1r_2 = Rr_2 + Rr_1. \qquad \text{Simplifying by removing factors of 1}$$

We might be tempted at this point to multiply by $1/r_2$ to get r_1 alone on the left, *but* note that there is an r_1 on the right. We must get all the terms involving r_1 on the *same side* of the equation.

$$r_1r_2 - Rr_1 = Rr_2 \qquad \text{Subtracting } Rr_1$$
$$r_1(r_2 - R) = Rr_2 \qquad \text{Factoring out } r_1$$
$$r_1 = \frac{Rr_2}{r_2 - R} \qquad \text{Dividing by } r_2 - R \text{ to get } r_1 \text{ alone}$$

→ **Now Try Exercise 3.**

*Note that R, r_1, and r_2 are all different variables. It is common to use subscripts, as in r_1 (read "r sub 1") and r_2, to distinguish variables.

5.7   Exercise Set

▶ Reading Check

Solve each equation for y. Choose from the column on the right an equivalent equation.

RC1. $\dfrac{x}{y} = \dfrac{w}{t}$

RC2. $w = \dfrac{(x + y)t}{4}$

RC3. $\dfrac{1}{w} = \dfrac{1}{x} + \dfrac{1}{y}$

RC4. $w = \dfrac{(x + t)y}{4}$

a) $y = \dfrac{4w - xt}{t}$ b) $y = \dfrac{xw}{t}$

c) $y = \dfrac{xt}{w}$ d) $y = \dfrac{4w}{x + t}$

e) $y = \dfrac{wx}{x - w}$ f) $y = \dfrac{xw}{y - x}$

· ·

a *Solve.*

1. $\dfrac{W_1}{W_2} = \dfrac{d_1}{d_2}$, for W_2 **2.** $\dfrac{W_1}{W_2} = \dfrac{d_1}{d_2}$, for d_1

3. $\dfrac{1}{R} = \dfrac{1}{r_1} + \dfrac{1}{r_2}$, for r_2 (Electricity formula)

4. $\dfrac{1}{R} = \dfrac{1}{r_1} + \dfrac{1}{r_2}$, for R (Electricity formula)

5. $s = \dfrac{(v_1 + v_2)t}{2}$, for t

6. $s = \dfrac{(v_1 + v_2)t}{2}$, for v_1

7. $R = \dfrac{gs}{g + s}$, for s

8. $I = \dfrac{2V}{V + 2r}$, for V

9. $\dfrac{1}{p} + \dfrac{1}{q} = \dfrac{1}{f}$, for p (Optics formula)

10. $\dfrac{1}{p} + \dfrac{1}{q} = \dfrac{1}{f}$, for f (Optics formula)

11. $\dfrac{t}{a} + \dfrac{t}{b} = 1$, for a (Work formula)

12. $\dfrac{t}{a} + \dfrac{t}{b} = 1$, for b (Work formula)

13. $I = \dfrac{nE}{E + nr}$, for E

14. $I = \dfrac{nE}{E + nr}$, for n

15. $I = \dfrac{704.5W}{H^2}$, for H^2

16. $S = \dfrac{H}{m(t_1 - t_2)}$, for t_1

17. $\dfrac{E}{e} = \dfrac{R + r}{r}$, for r

18. $\dfrac{E}{e} = \dfrac{R + r}{r}$, for e

19. $V = \dfrac{1}{3}\pi h^2(3R - h)$, for R

20. $A = P(1 + rt)$, for r (Interest formula)

21. $S = 2\pi rh + 2\pi r^2$, for h (Surface area of a cylinder)

22. *Escape Velocity.*   The formula
$$\dfrac{V^2}{R^2} = \dfrac{2g}{R + h}$$
is used to find a satellite's *escape velocity V*, where R is a planet's radius, h is the satellite's height above the planet, and g is the planet's acceleration due to gravity. Solve the formula for h.

23. *Average Speed.*   The formula
$$v = \dfrac{d_2 - d_1}{t_2 - t_1}$$
gives an object's average speed v when that object has traveled d_1 miles in t_1 hours and d_2 miles in t_2 hours. Solve the formula for t_2.

24. *Earned Run Average.* The formula

$$A = 9 \cdot \frac{R}{I}$$

gives a pitcher's *earned run average A*, where *R* is the number of earned runs, and *I* is the number of innings pitched. How many earned runs were given up if a pitcher's earned run average is 2.4 after 45 innings? Solve the formula for *I*.

▶ **Skill Maintenance**

Solve. [3.6a]

25. *Coin Value.* There are 50 dimes in a roll of dimes, 40 nickels in a roll of nickels, and 40 quarters in a roll of quarters. Rob has 12 rolls of coins with a total value of $70.00. He has 3 more rolls of nickels than dimes. How many of each roll of coins does he have?

Given that $f(x) = x^3 - x$, *find each of the following.* [2.2b]

26. $f(-2)$ **27.** $f(2)$

28. $f(0)$ **29.** $f(2a)$

30. Find the slope of the line containing the points $(-2, 5)$ and $(8, -3)$. [2.5b]

31. Find an equation of the line containing the points $(-2, 5)$ and $(8, -3)$. [2.7c]

▶ **Synthesis**

32. *Escape Velocity.* (Refer to Exercise 22.) A satellite's escape velocity is 6.5 mi/sec, the radius of the earth is 3960 mi, and the acceleration due to gravity is 32.2 ft/sec². How far is the satellite from the surface of the earth?

5.8   Variation and Applications

▶ **a**   Find an equation of direct variation given a pair of values of the variables.

▶ **b**   Solve applied problems involving direct variation.

▶ **c**   Find an equation of inverse variation given a pair of values of the variables.

▶ **d**   Solve applied problems involving inverse variation.

▶ **e**   Find equations of other kinds of variation given values of the variables.

▶ **f**   Solve applied problems involving other kinds of variation.

We now extend our study of formulas and functions by considering applications involving variation.

▶ **a**   **Equations of Direct Variation**

A substitute teacher earns $65 per day. For 1 day, $65 is earned; for 2 days, $130 is earned; for 3 days, $195 is earned; and so on. We plot this information on a graph,

using the number of hours as the first coordinate and the amount earned as the second coordinate to form a set of ordered pairs:

$$(1, 65), \qquad (2, 130),$$
$$(3, 195), \qquad (4, 260),$$

and so on.

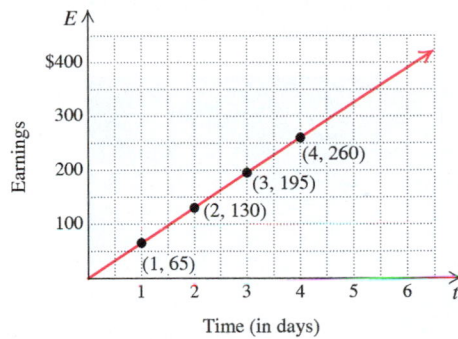

Note that the ratio of the second coordinate to the first coordinate is the same number for each point:

$$\frac{65}{1} = 65, \qquad \frac{130}{2} = 65, \qquad \frac{195}{3} = 65, \qquad \frac{260}{4} = 65, \quad \text{and so on.}$$

Whenever a situation produces pairs of numbers in which the *ratio is constant*, we say that there is **direct variation**. Here the amount earned varies directly as the time:

$$\frac{E}{t} = 65 \text{ (a constant),} \quad \text{or} \quad E = 65t,$$

or, using function notation, $E(t) = 65t$. The equation is an **equation of direct variation**. The coefficient, 65 in the situation above, is called the **variation constant**. In this case, it is the rate of change of earnings with respect to time.

The graph of $y = kx, k > 0$, always goes through the origin and rises from left to right. Note that as x increases, y increases. The constant k is also the slope of the line.

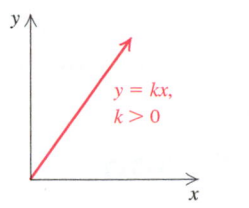

DIRECT VARIATION

If a situation gives rise to a linear function $f(x) = kx$, or $y = kx$, where k is a positive constant, we say that we have **direct variation**, or that **y varies directly as x**, or that **y is directly proportional to x**. The number k is called the **variation constant**, or the **constant of proportionality**.

EXAMPLE 1 Find the variation constant and an equation of variation in which y varies directly as x, and $y = 32$ when $x = 2$.

We know that $(2, 32)$ is a solution of $y = kx$. Thus,

$$y = kx$$
$$32 = k \cdot 2 \qquad \text{Substituting}$$
$$\frac{32}{2} = k, \text{ or } k = 16. \qquad \text{Solving for } k$$

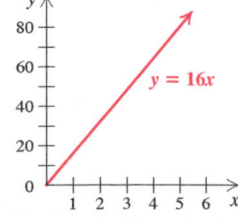

The variation constant, 16, is the rate of change of y with respect to x. The equation of variation is $y = 16x$.

Now Try Exercise 1.

▶ b Applications of Direct Variation

EXAMPLE 2 *Water from Melting Snow.* The number of centimeters W of water produced from melting snow varies directly as S, the number of centimeters of snow. Meteorologists have found that, under certain conditions, 150 cm of snow will melt to 16.8 cm of water. To how many centimeters of water will 200 cm of snow melt?

We first find the variation constant using the data and then find an equation of variation:

$$W = kS \qquad \text{\textcolor{red}{W varies directly as S.}}$$
$$16.8 = k \cdot 150 \qquad \text{\textcolor{red}{Substituting}}$$
$$\frac{16.8}{150} = k \qquad \text{\textcolor{red}{Solving for k}}$$
$$0.112 = k. \qquad \text{\textcolor{red}{This is the variation constant.}}$$

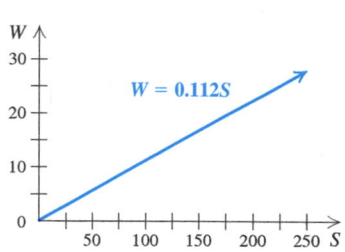

The equation of variation is $W = 0.112S$.

Next, we use the equation to find how many centimeters of water will result from melting 200 cm of snow:

$$W = 0.112S$$
$$ = 0.112(200) \qquad \text{\textcolor{red}{Substituting}}$$
$$ = 22.4.$$

Thus, 200 cm of snow will melt to 22.4 cm of water.

Now Try Exercise 7.

▶ c Equations of Inverse Variation

A bus is traveling a distance of 20 mi. At a speed of 5 mph, the trip will take 4 hr; at 20 mph, it will take 1 hr; at 40 mph, it will take $\frac{1}{2}$ hr; and so on. We plot this information on a graph, using speed as the first coordinate and time as the second coordinate to determine a set of ordered pairs:

$$(5, 4), \qquad (10, 2),$$
$$(20, 1), \qquad \left(40, \tfrac{1}{2}\right),$$

and so on.

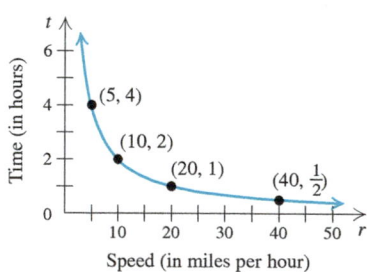

Note that the products of the coordinates are all the same number:

$$5 \cdot 4 = 20, \qquad 20 \cdot 1 = 20, \qquad 40 \cdot \tfrac{1}{2} = 20, \quad \text{and so on.}$$

Whenever a situation produces pairs of numbers in which the *product is constant*, we say that there is **inverse variation**. Here the time varies inversely as the speed:

$$rt = 20 \,(\text{a constant}), \quad \text{or} \quad t = \frac{20}{r}.$$

The equation is an **equation of inverse variation**. The constant, 20 in the situation above, is called the **variation constant**. Note that as the first number (speed) increases, the second number (time) decreases.

It is helpful to look at the graph of $y = k/x$, $k > 0$. The graph is like the one shown below for positive values of x. Note that as x increases, y decreases.

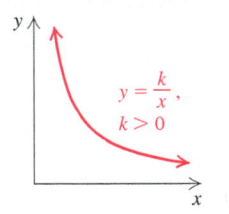

$$y = \frac{k}{x}, \quad k > 0$$

INVERSE VARIATION

If a situation gives rise to a function $f(x) = k/x$, or $y = k/x$, where k is a positive constant, we say that we have **inverse variation**, or that **y varies inversely as x**, or that **y is inversely proportional to x**. The number k is called the **variation constant**, or the **constant of proportionality**.

EXAMPLE 3   Find the variation constant and an equation of variation in which y varies inversely as x, and $y = 32$ when $x = 0.2$.

We know that $(0.2, 32)$ is a solution of $y = k/x$. We substitute:

$$y = \frac{k}{x}$$

$$32 = \frac{k}{0.2} \qquad \text{Substituting}$$

$$(0.2)32 = k \qquad \text{Solving for } k$$

$$6.4 = k.$$

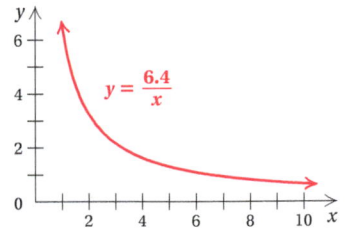

$$y = \frac{6.4}{x}$$

The variation constant is 6.4. The equation of variation is $y = \dfrac{6.4}{x}$.

▶ **Now Try Exercise 15.**

▶ **d   Applications of Inverse Variation**

EXAMPLE 4   *Musical Pitch.*   The pitch P of a musical tone varies inversely as its wavelength W. One tone has a pitch of 550 vibrations per second and a wavelength of 1.92 ft. Find the pitch of another tone that has a wavelength of 3.2 ft.

We first find the variation constant using the data given and then find an equation of variation:

$$P = \frac{k}{W} \qquad \textit{P varies inversely as W.}$$

$$550 = \frac{k}{1.92} \qquad \textbf{Substituting}$$

$$1056 = k. \qquad \textbf{Solving for } k, \textbf{ the variation constant}$$

The equation of variation is $P = \dfrac{1056}{W}$.

Next, we use the equation to find the pitch of a tone that has a wavelength of 3.2 ft:

$$P = \frac{1056}{W} \qquad \textbf{Equation of variation}$$

$$= \frac{1056}{3.2} \qquad \textbf{Substituting}$$

$$= 330.$$

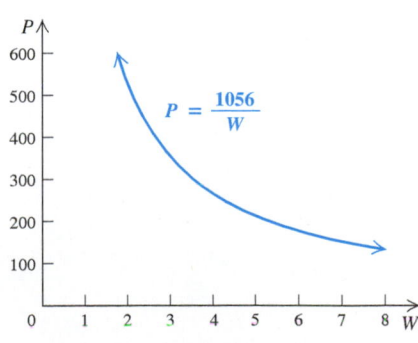

$$P = \frac{1056}{W}$$

The pitch of a musical tone that has a wavelength of 3.2 ft is 330 vibrations per second.

▶ **Now Try Exercise 21.**

▶ e Other Kinds of Variation

We now look at other kinds of variation. Consider the equation for the area of a circle, in which A and r are variables and π is a constant:

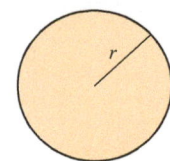

$$A = \pi r^2, \quad \text{or, as a function,} \quad A(r) = \pi r^2.$$

We say that the area *varies directly* as the square of the radius.

> y varies inversely as the *n*th power of x if there is some positive constant k such that
>
> $$y = kx^n.$$

EXAMPLE 5 Find an equation of variation in which y varies directly as the square of x, and $y = 12$ when $x = 2$.

We write an equation of variation and find k:

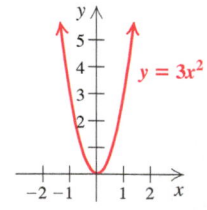

$$y = kx^2$$
$$12 = k \cdot 2^2$$
$$12 = k \cdot 4$$
$$3 = k.$$

Thus, $y = 3x^2$.

Now Try Exercise 29.

From the law of gravity, we know that the weight W of an object *varies inversely* as the square of its distance d from the center of the earth:

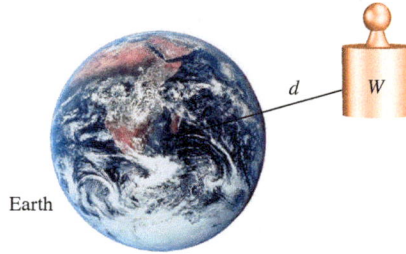

$$W = \frac{k}{d^2}.$$

Earth

> y varies inversely as the *n*th power of x if there is some positive constant k such that
>
> $$y = \frac{k}{x^n}.$$

EXAMPLE 6 Find an equation of variation in which W varies inversely as the square of d, and $W = 3$ when $d = 5$.

$$W = \frac{k}{d^2}$$

$$3 = \frac{k}{5^2} \qquad \text{Substituting}$$

$$3 = \frac{k}{25}$$

$$75 = k$$

Thus, $W = \dfrac{75}{d^2}$.

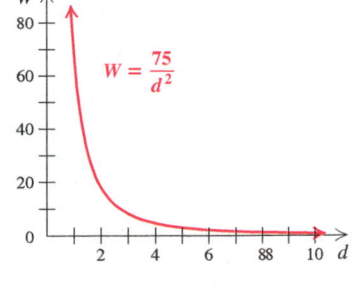

> **Now Try Exercise 31.**

Consider the equation for the area A of a triangle with height h and base b: $A = \frac{1}{2}bh$. We say that the area **varies jointly** as the height and the base.

---

y varies jointly as x and z if there is some positive constant k such that

$$y = kxz.$$

EXAMPLE 7 Find an equation of variation in which y varies jointly as x and z, and $y = 42$ when $x = 2$ and $z = 3$.

$$y = kxz$$
$$42 = k \cdot 2 \cdot 3 \qquad \text{Substituting}$$
$$42 = k \cdot 6$$
$$7 = k$$

Thus, $y = 7xz$.

> **Now Try Exercise 33.**

Different types of variation can be combined. For example, the equation

$$y = k \cdot \frac{xz^2}{w}$$

asserts that y varies jointly as x and the square of z, and inversely as w.

EXAMPLE 8 Find an equation of variation in which y varies jointly as x and z and inversely as the square of w, and $y = 105$ when $x = 3$, $z = 20$, and $w = 2$.

$$y = k \cdot \frac{xz}{w^2}$$

$$105 = k \cdot \frac{3 \cdot 20}{2^2} \qquad \text{Substituting}$$

$$105 = k \cdot 15$$

$$7 = k$$

Thus, $y = 7 \cdot \dfrac{xz}{w^2}$.

> **Now Try Exercise 37.**

▶ f   Other Applications of Variation

EXAMPLE 9 *Volume of a Tree.*   The volume of wood V in a tree varies jointly as the height h and the square of the girth g (girth is distance around). If the volume of a redwood tree is 216 m^3 when the height is 30 m and the girth is 1.5 m, what is the height of a tree whose volume is 960 m^3 and girth is 2 m?

We first find k using the first set of data. Then we solve for h using the second set of data.

$$V = khg^2$$
$$216 = k \cdot 30 \cdot 1.5^2$$
$$3.2 = k$$

Then the equation of variation is $V = 3.2hg^2$. We substitute the second set of data into the equation:

$$960 = 3.2 \cdot h \cdot 2^2$$
$$75 = h.$$

Therefore, the height of the tree is 75 m.

Now Try Exercise 43.

EXAMPLE 10 *TV Signal.*   The intensity I of a TV signal varies inversely as the square of the distance d from the transmitter. If the intensity is 23 watts per square meter (W/m^2) at a distance of 2 km, what is the intensity at a distance of 6 km?

We first find k using the first set of data. Then we solve for I using the second set of data.

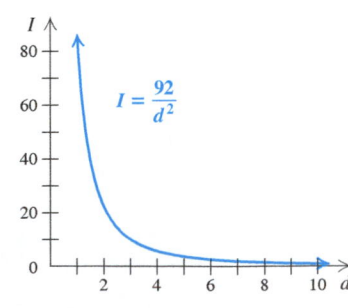

$$I = \frac{k}{d^2}$$
$$23 = \frac{k}{2^2}$$
$$92 = k$$

Then the equation of variation is $I = 92/d^2$. We substitute the second distance into the equation:

$$I = \frac{92}{d^2} = \frac{92}{6^2} \approx 2.56. \qquad \text{Rounded to the nearest hundredth}$$

Therefore, at 6 km, the intensity is about 2.56 W/m^2.

Now Try Exercise 39.

5.8   Exercise Set

▶ # Reading Check

Match each description of variation with the appropriate equation of variation listed on the right.

RC1. *a* varies directly as *z*.

RC2. *y* is inversely proportional to *b*.

RC3. *y* varies jointly as *z* and *b* and inversely as *c*.

RC4. *x* is directly proportional to *c*.

RC5. *b* varies directly as *z*.

RC6. *a* varies inversely as *z*.

RC7. *c* varies inversely as *x*.

RC8. *y* varies jointly as *c* and *x*.

a) $a = \dfrac{k}{z}$ **b)** $y = kcx$

c) $b = kz$ **d)** $y = \dfrac{k}{b}$

e) $x = ky$ **f)** $a = kz$

g) $c = \dfrac{k}{x}$ **h)** $y = \dfrac{kzb}{c}$

i) $x = kc$ **j)** $y = \dfrac{kbc}{z}$

---

a *Find the variation constant and an equation of variation in which y varies directly as x and the following are true.*

1. $y = 40$ when $x = 8$ **2.** $y = 54$ when $x = 12$

3. $y = 4$ when $x = 30$ **4.** $y = 3$ when $x = 33$

5. $y = 0.9$ when $x = 0.4$ **6.** $y = 0.8$ when $x = 0.2$

b *Solve.*

7. *Shipping by Semi Truck.*   The number of semi trucks *T* needed to ship metal varies directly as the weight *W* of the metal. It takes 75 semi trucks to ship 1500 tons of metal. How many trucks are needed for 3500 tons of metal? (*Source*: www.scrappy.com)

8. *Shipping by Rail Cars.*   The number of rail cars *R* needed to ship metal varies directly as the weight *W* of the metal. It takes approximately 21 rail cars to ship 1500 tons of metal. How many rail cars are needed for 3500 tons of metal? (*Source*: www.scrappy.com)

9. *Fat Intake.*   The maximum number of grams of fat that should be in a diet varies directly as a person's weight. A person weighing 120 lb should have no more than 60 g of fat per day. What is the maximum daily fat intake for a person weighing 180 lb?

10. *Relative Aperture.*   The relative aperture, or f-stop, of a 23.5-mm diameter lens is directly proportional to the focal length *F* of the lens. If a 150-mm focal length has an f-stop of 6.3, find the f-stop of a 23.5-mm diameter lens with a focal length of 80 mm.

11. *Aluminum Usage.*   The number *N* of aluminum cans used each year varies directly as the number of people using them. If 250 people use 60,000 cans in one year, how many cans are used each year in St. Louis, Missouri, which has a population of 318,172?

12. *Hooke's Law.*   Hooke's law states that the distance *d* that a spring is stretched by a hanging object varies directly as the weight *w* of the object. If a spring is stretched 40 cm by a 3-kg barbell, what is the distance stretched by a 5-kg barbell?

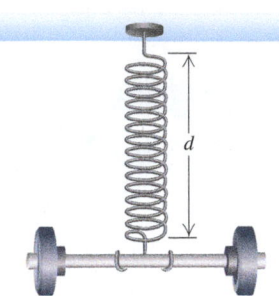

13. *Mass of Water in Human Body.*   The number of kilograms *W* of water in a human body varies directly as the mass of the body. A 96-kg person contains 64 kg of water. How many kilograms of water are in a 60-kg person?

14. *Weight on Mars.*   The weight *M* of an object on Mars varies directly as its weight *E* on Earth. A person who weighs 95 lb on Earth weighs 38 lb on Mars. How much would a 100-lb person weigh on Mars?

c *Find the variation constant and an equation of variation in which y varies inversely as x and the following are true.*

15. $y = 14$ when $x = 7$ **16.** $y = 1$ when $x = 8$

17. $y = 3$ when $x = 12$ **18.** $y = 12$ when $x = 5$

19. $y = 0.1$ when $x = 0.5$ **20.** $y = 1.8$ when $x = 0.3$

d *Solve.*

21. *Work Rate.*   The time *T* required to do a job varies inversely as the number of people *P* working. It takes 5 hr for 7 bricklayers to build a park wall. How long will it take 10 bricklayers to complete the job?

22. *Pumping Rate.*   The time *t* required to empty a tank varies inversely as the rate *r* of pumping. If a pump can empty a tank in 45 min at the rate of 600 kL/min, how long will it take the pump to empty the same tank at the rate of 1000 kL/min?

23. *Current and Resistance.*   The current *I* in an electrical conductor varies inversely as the resistance *R* of the conductor. If the current is $\frac{1}{2}$ ampere when the resistance is 240 ohms, what is the current when the resistance is 540 ohms?

24. *Wavelength and Frequency.*   The wavelength *W* of a radio wave varies inversely as its frequency *F*. A wave with a frequency of 1200 kilohertz has a length of 300 meters. What is the length of a wave with a frequency of 800 kilohertz?

25. *Beam Weight.*   The weight *W* that a horizontal beam can support varies inversely as the length *L* of the beam. Suppose that a 12-ft beam can support 1200 lb. How many pounds can a 15-ft beam support?

26. *Musical Pitch.*   The pitch *P* of a musical tone varies inversely as its wavelength *W*. One tone has a pitch of 440 vibrations per second and a wavelength of 2.4 ft. Find the wavelength of another tone that has a pitch of 275 vibrations per second.

27. *Rate of Travel.*   The time *t* required to drive a fixed distance varies inversely as the speed *r*. It takes 5 hr at a speed of 80 km/h to drive a fixed distance. How long will it take to drive the same distance at a speed of 70 km/h?

28. *Volume and Pressure.*   The volume *V* of a gas varies inversely as the pressure *P* upon it. The volume of a gas is 200 cm^3 under a pressure of 32 kg/cm^2. What will its volume be under a pressure of 40 kg/cm^2?

e *Find an equation of variation in which the following are true.*

29. *y* varies directly as the square of *x*, and $y = 0.15$ when $x = 0.1$

30. *y* varies directly as the square of *x*, and $y = 6$ when $x = 3$

31. *y* varies inversely as the square of *x*, and $y = 0.15$ when $x = 0.1$

32. *y* varies inversely as the square of *x*, and $y = 6$ when $x = 3$

33. *y* varies jointly as *x* and *z*, and $y = 56$ when $x = 7$ and $z = 8$

34. *y* varies directly as *x* and inversely as *z*, and $y = 4$ when $x = 12$ and $z = 15$

35. *y* varies jointly as *x* and the square of *z*, and $y = 105$ when $x = 14$ and $z = 5$

36. *y* varies jointly as *x* and *z* and inversely as *w*, and $y = \frac{3}{2}$ when $x = 2$, $z = 3$, and $w = 4$

37. y varies jointly as x and z and inversely as the product of w and p, and $y = \frac{3}{28}$ when $x = 3$, $z = 10$, $w = 7$, and $p = 8$

38. y varies jointly as x and z and inversely as the square of w, and $y = \frac{12}{5}$ when $x = 16$, $z = 3$, and $w = 5$

f *Solve.*

39. *Intensity of Light.* The intensity I of light from a light bulb varies inversely as the square of the distance d from the bulb. Suppose that I is 90 W/m^2 (watts per square meter) when the distance is 5 m. How much *further* would it be to a point where the intensity is 40 W/m^2?

40. *Stopping Distance of a Car.* The stopping distance d of a car after the brakes have been applied varies directly as the square of the speed r. If a car traveling 60 mph can stop in 200 ft, how fast can a car travel and still stop in 72 ft?

41. *Weight of an Astronaut.* The weight W of an object varies inversely as the square of the distance d from the center of the earth. At sea level (3978 mi from the center of the earth), an astronaut weighs 220 lb. Find his weight when he is 200 mi above the surface of the earth and the spacecraft is not in motion.

42. *Combined Gas Law.* The volume V of a given mass of a gas varies directly as the temperature T and inversely as the pressure P. If $V = 231$ cm^3 when $T = 42°$ and $P = 20$ kg/cm^2, what is the volume when $T = 30°$ and $P = 15$ kg/cm^2?

43. *Earned-Run Average.* A pitcher's earned-run average E varies directly as the number R of earned runs allowed and inversely as the number I of innings pitched. In 2013, Jon Lester of the Boston Red Sox had an earned-run average of 3.75. He gave up 89 earned runs in 213.1 innings. How many earned runs would he have given up had he pitched 235 innings with the same average? Round to the nearest whole number. (*Source:* Major League Baseball)

44. *Atmospheric Drag.* Wind resistance, or atmospheric drag, tends to slow down moving objects. Atmospheric drag varies jointly as an object's surface area A and velocity v. If a car traveling at a speed of 40 mph with a surface area of 37.8 ft^2 experiences a drag of 222 N (Newtons), how fast must a car with 51 ft^2 of surface area travel in order to experience a drag force of 430 N?

45. *Water Flow.* The amount Q of water emptied by a pipe varies directly as the square of the diameter d. A pipe 5 in. in diameter will empty 225 gal of water over a fixed time period. If we assume the same kind of flow, how many gallons of water are emptied in the same amount of time by a pipe that is 9 in. in diameter?

46. *Weight of a Sphere.* The weight W of a sphere of a given material varies directly as its volume V, and its volume V varies directly as the cube of its diameter.

   a) Find an equation of variation relating the weight W to the diameter d.

   b) An iron ball that is 5 in. in diameter is known to weigh 25 lb. Find the weight of an iron ball that is 8 in. in diameter.

▶ Skill Maintenance

Write interval notation for the given set.

47. $\{y \mid y \geq -8\}$ **[1.4b]**

48. $\{x \mid x \text{ is a real number}\}$ **[1.4b]**

49. $\{t \mid -5 \leq t < 15\}$ **[1.4b]**

50. $\{a \mid a < -4 \text{ or } a \geq 0\}$ **[1.5b]**

51. $\{q \mid q < -\frac{1}{2}\}$ **[1.4b]**

52. $\{x \mid x > 2 \text{ or } x \leq -\frac{5}{8}\}$ **[1.5b]**

Solve.

53. $3x - y = 7,$
$y = 1 - x$ **[3.2a]**

54. $6x - 2y = 12,$
$3x + 2y = 24$ **[3.3a]**

▶ Synthesis

55. In each of the following equations, state whether y varies directly as x, inversely as x, or neither directly nor inversely as x.

   a) $7xy = 14$ **b)** $x - 2y = 12$

   c) $-2x + 3y = 0$ **d)** $x = \frac{3}{4}y$

56. *Area of a Circle.* The area of a circle varies directly as the square of the length of a diameter. What is the variation constant?

Chapter 5 Summary and Review

VOCABULARY REINFORCEMENT

Complete each statement with the correct term from the column on the right. Some of the choices may be used more than once, and some may not be used.

1. An equality of ratios, $A/B = C/D$, is called a(n) _____. [5.6b]

2. An expression that consists of the quotient of two polynomials, where the polynomial in the denominator is nonzero, is called a(n) _____ expression. [5.1a]

3. A(n) _____ equation is an equation containing one or more rational expressions. [5.5a]

4. If a situation gives rise to a function $f(x) = k/x$, or $y = k/x$, where k is a(n) _____ constant, we say that we have _____ variation. [5.8c]

5. A(n) _____ rational expression is a rational expression that contains rational expressions within its numerator and/or its denominator. [5.4a]

6. If a situation gives rise to a linear function $f(x) = kx$, or $y = kx$, where k is a(n) _____ constant, we say that we have _____ variation. The number k is called the variation _____. [5.8a]

positive

negative

proportional

proportion

rational

complex

constant

inverse

direct

joint

CONCEPT REINFORCEMENT

Determine whether each statement is true or false.

1. If y is inversely proportional to x, then the rational function $f(x) = k/x$ can model the situation. [5.8c]

2. Clearing fractions is a valid procedure only when solving equations, not when adding, subtracting, multiplying, or dividing rational expressions. [5.5a]

STUDY GUIDE

Objective 5.1a Find all numbers for which a rational expression is not defined or that are not in the domain of a rational function, and state the domain of the function.

Example Find the domain of $f(x) = \dfrac{x^2 - 12x + 27}{x^2 + 6x - 16}$.

The rational expression is not defined for a replacement that makes the denominator 0. We set the denominator equal to 0 and solve for x.

Practice Exercise

1. Find the domain of

$$f(x) = \frac{x^2 + 3x - 28}{x^2 + 3x - 54}.$$

(continued)

$$x^2 + 6x - 16 = 0$$
$$(x + 8)(x - 2) = 0$$
$$x + 8 = 0 \quad or \quad x - 2 = 0$$
$$x = -8 \quad or \quad x = 2$$

The expression is not defined for replacements -8 and 2. Thus the domain is

$$\{x \mid x \text{ is a real number } and \ x \neq -8 \ and \ x \neq 2\},$$
or $(-\infty, -8) \cup (-8, 2) \cup (2, \infty).$

Objective 5.1c Simplify rational expressions.

Example Simplify: $\dfrac{a^2 - 1}{a^2 + 7a - 8}$.

$$\frac{a^2 - 1}{a^2 + 7a - 8} = \frac{(a + 1)(a - 1)}{(a + 8)(a - 1)} = \frac{a + 1}{a + 8} \cdot \frac{a - 1}{a - 1} = \frac{a + 1}{a + 8}$$

Practice Exercise

2. Simplify:

$$\frac{b^2 - 9}{b^2 - 5b - 24}.$$

Objective 5.1e Divide rational expressions and simplify.

Example Divide and simplify: $\dfrac{t^2 + 2t + 4}{3t^2 + 6t} \div \dfrac{t^3 - 8}{t^3 + 2t^2}$.

$$\frac{t^2 + 2t + 4}{3t^2 + 6t} \div \frac{t^3 - 8}{t^3 + 2t^2} = \frac{t^2 + 2t + 4}{3t^2 + 6t} \cdot \frac{t^3 + 2t^2}{t^3 - 8}$$

$$= \frac{(t^2 + 2t + 4)(t)(t)(t + 2)}{3t(t + 2)(t - 2)(t^2 + 2t + 4)}$$

$$= \frac{t}{3(t - 2)}$$

Practice Exercise

3. Divide and simplify:

$$\frac{w^3 - 125}{w^3 + 8w^2 + 15w} \div \frac{w - 5}{w^3 - 25w}.$$

Objective 5.2a Find the LCM of several algebraic expressions by factoring.

Example Find the LCM of x^2, $16x^2 - 25$, and $4x^3 - 15x^2 - 25x$.

We factor each expression completely:

$$x^2 = x \cdot x \, ;$$
$$16x^2 - 25 = (4x + 5)(4x - 5);$$
$$4x^3 - 15x^2 - 25x = x(4x + 5)(x - 5).$$
$$\text{LCM} = x \cdot x \cdot (4x + 5)(4x - 5)(x - 5)$$
$$= x^2(4x + 5)(4x - 5)(x - 5)$$

Practice Exercise

4. Find the LCM of x^4, $x^5 - 9x^3$, and $2x^2 + 11x + 15$.

Objective 5.2b Add and subtract rational expressions.

Example Subtract:

$$\frac{x - y}{x^2 + 3xy + 2y^2} - \frac{3y}{x^2 + 6xy + 5y^2}.$$

First, we factor the denominators.

$$\frac{x - y}{(x + 2y)(x + y)} - \frac{3y}{(x + 5y)(x + y)} \quad \text{The LCM is } (x + 2y)(x + y)(x + 5y).$$

$$= \frac{x - y}{(x + 2y)(x + y)} \cdot \frac{x + 5y}{x + 5y} - \frac{3y}{(x + 5y)(x + y)} \cdot \frac{x + 2y}{x + 2y}$$

$$= \frac{(x - y)(x + 5y)}{(x + 2y)(x + y)(x + 5y)} - \frac{3y(x + 2y)}{(x + 5y)(x + y)(x + 2y)}$$

$$= \frac{(x^2 + 4xy - 5y^2) - (3xy + 6y^2)}{(x + 2y)(x + y)(x + 5y)}$$

$$= \frac{x^2 + 4xy - 5y^2 - 3xy - 6y^2}{(x + 2y)(x + y)(x + 5y)} = \frac{x^2 + xy - 11y^2}{(x + 2y)(x + y)(x + 5y)}$$

Practice Exercise

5. Subtract:

$$\frac{r + s}{r^2 + rs - 2s^2} - \frac{5s}{r^2 - s^2}.$$

Objective 5.3b Divide a polynomial by a divisor that is not a monomial, and if there is a remainder, express the result in two ways.

Example Divide: $(y^2 - 2y + 13) \div (y + 2)$.

$$\begin{array}{r} y - 4 \\ y + 2 \overline{\smash{)}\, y^2 - 2y + 13} \\ \underline{y^2 + 2y } \\ -4y + 13 \\ \underline{-4y - 8} \\ 21 \end{array}$$

The answer is $y - 4$, R 21, or

$$y - 4 + \frac{21}{y + 2}.$$

Practice Exercise

6. Divide:

$$(y^2 - 5y + 9) \div (y - 1).$$

Objective 5.3c Use synthetic division to divide a polynomial by a binomial of the type $x - a$.

Example Use synthetic division to divide:

$$(x^3 - 2x^2 - 6) \div (x + 2).$$

There is no x-term, so we write 0 for its coefficient. Note that $x + 2 = x - (-2)$, so we write -2 on the left.

$$\begin{array}{r|rrrr} -2 & 1 & -2 & 0 & -6 \\ & & -2 & 8 & -16 \\ \hline & 1 & -4 & 8 & \,|\, -22 \end{array}$$

The answer is $x^2 - 4x + 8$, R -22, or

$$x^2 - 4x + 8 + \frac{-22}{x + 2}, \text{ or}$$

$$x^2 - 4x + 8 - \frac{22}{x + 2}$$

Practice Exercise

7. Use synthetic division to divide:

$$(x^3 - 5x^2 - 1) \div (x + 3).$$

Objective 5.4a Simplify complex rational expressions.

Example Simplify: $\dfrac{\dfrac{2}{x} - \dfrac{5}{y}}{\dfrac{5}{x} + \dfrac{2}{y}}$.

The LCM of all denominators is xy.

$$\dfrac{\dfrac{2}{x} - \dfrac{5}{y}}{\dfrac{5}{x} + \dfrac{2}{y}} = \dfrac{\dfrac{2}{x} - \dfrac{5}{y}}{\dfrac{5}{x} + \dfrac{2}{y}} \cdot \dfrac{xy}{xy} = \dfrac{\dfrac{2}{x} \cdot xy - \dfrac{5}{y} \cdot xy}{\dfrac{5}{x} \cdot xy + \dfrac{2}{y} \cdot xy} = \dfrac{2y - 5x}{5y + 2x}$$

Practice Exercise

8. Simplify:

$$\dfrac{\dfrac{2}{a} + \dfrac{8}{b}}{\dfrac{8}{a} - \dfrac{2}{b}}.$$

Objective 5.4b Find the difference quotient for a function.

Example For the function $f(x) = x^2 - 4$, construct and simplify the difference quotient.

$$\dfrac{f(x + h) - f(x)}{h} = \dfrac{[(x + h)^2 - 4] - (x^2 - 4)}{h}$$

$$= \dfrac{x^2 + 2xh + h^2 - 4 - x^2 + 4}{h}$$

$$= \dfrac{2xh + h^2}{h} = \dfrac{h(2x + h)}{h}$$

$$= 2x + h$$

Practice Exercise

9. For the function $f(x) = 6x - 3$, construct and simplify the difference quotient.

Objective 5.5a Solve rational equations.

Example Solve:

$$\dfrac{12}{x^2 - 6x - 7} - \dfrac{3}{x - 7} = \dfrac{1}{x + 1}.$$

The LCM of the denominators is $(x - 7)(x + 1)$. We multiply all terms on both sides by $(x - 7)(x + 1)$.

$$(x - 7)(x + 1)\left(\dfrac{12}{x^2 - 6x - 7} - \dfrac{3}{x - 7}\right) = (x - 7)(x + 1) \cdot \dfrac{1}{x + 1}$$

$$12 - 3(x + 1) = x - 7$$
$$12 - 3x - 3 = x - 7$$
$$9 - 3x = x - 7$$
$$-4x = -16$$
$$x = 4$$

We must always check possible solutions. The number 4 checks in the original equation. The solution is 4.

Practice Exercise

10. Solve:

$$\dfrac{5}{x - 4} - \dfrac{3}{x + 5} = \dfrac{4}{x^2 + x - 20}.$$

Objective 5.8a Find an equation of direct variation given a pair of values of the variables.

Example Find the variation constant and an equation of variation in which y varies directly as x, and $y = 44$ when $x = \frac{11}{5}$.

$$y = kx \qquad \text{Direct variation}$$
$$44 = k \cdot \frac{11}{5} \qquad \text{Substituting}$$
$$\frac{5}{11} \cdot 44 = k$$
$$20 = k$$

The variation constant is 20. The equation of variation is $y = 20x$.

Practice Exercise

11. Find the variation constant and an equation of variation in which y varies directly as x, and $y = 62$ when $x = \frac{2}{3}$.

Objective 5.8c Find an equation of inverse variation given a pair of values of the variables.

Example Find the variation constant and an equation of variation in which y varies inversely as x, and $y = \frac{5}{18}$ when $x = 2$.

$$y = \frac{k}{x} \qquad \text{Inverse variation}$$
$$\frac{5}{18} = \frac{k}{2} \qquad \text{Substituting}$$
$$2 \cdot \frac{5}{18} = k$$
$$\frac{5}{9} = k$$

The variation constant is $\frac{5}{9}$. The equation of variation is $y = \frac{\frac{5}{9}}{x}$, or $y = \frac{5}{9x}$.

Practice Exercise

12. Find the variation constant and an equation of variation in which y varies inversely as x, and $y = \frac{3}{10}$ when $x = 15$.

REVIEW EXERCISES

1. Find all numbers for which the rational expression
$$\frac{x^2 - 3x + 2}{x^2 - 9}$$
is not defined. **[5.1a]**

2. Find the domain of f where
$$f(x) = \frac{x^2 - 3x + 2}{x^2 - 9}. \quad \textbf{[5.1a]}$$

Simplify. **[5.1c]**

3. $\dfrac{4x^2 - 7x - 2}{12x^2 + 11x + 2}$

4. $\dfrac{a^2 + 2a + 4}{a^3 - 8}$

Find the LCM. **[5.2a]**

5. $6x^3,\ 16x^2$

6. $x^2 - 49,\ 3x + 1$

7. $x^2 + x - 20,\ x^2 + 3x - 10$

Perform the indicated operations and simplify. **[5.1d, e], [5.2b, c]**

8. $\dfrac{y^2 - 64}{2y + 10} \cdot \dfrac{y + 5}{y + 8}$

9. $\dfrac{x^3 - 8}{x^2 - 25} \cdot \dfrac{x^2 + 10x + 25}{x^2 + 2x + 4}$

10. $\dfrac{9a^2 - 1}{a^2 - 9} \div \dfrac{3a + 1}{a + 3}$

11. $\dfrac{x^3 - 64}{x^2 - 16} \div \dfrac{x^2 + 5x + 6}{x^2 - 3x - 18}$

12. $\dfrac{x}{x^2 + 5x + 6} - \dfrac{2}{x^2 + 3x + 2}$

13. $\dfrac{2x^2}{x - y} + \dfrac{2y^2}{x + y}$

14. $\dfrac{3}{y + 4} - \dfrac{y}{y - 1} + \dfrac{y^2 + 3}{y^2 + 3y - 4}$

Divide.

15. $(16ab^3c - 10ab^2c^2 + 12a^2b^2c) \div (4ab)$ **[5.3a]**

16. $(y^2 - 20y + 64) \div (y - 6)$ **[5.3b]**

17. $(6x^4 + 3x^2 + 5x + 4) \div (x^2 + 2)$ **[5.3b]**

Divide using synthetic division. Show your work. **[5.3c]**

18. $(x^3 + 5x^2 + 4x - 7) \div (x - 4)$

19. $(3x^4 - 5x^3 + 2x - 7) \div (x + 1)$

Simplify. **[5.4a]**

20. $\dfrac{3 + \dfrac{3}{y}}{4 + \dfrac{4}{y}}$

21. $\dfrac{\dfrac{2}{a} + \dfrac{2}{b}}{\dfrac{4}{a^3} + \dfrac{4}{b^3}}$

22. $\dfrac{\dfrac{x^2 - 5x - 36}{x^2 - 36}}{\dfrac{x^2 + x - 12}{x^2 - 12x + 36}}$

23. $\dfrac{\dfrac{4}{x + 3} - \dfrac{2}{x^2 - 3x + 2}}{\dfrac{3}{x - 2} + \dfrac{1}{x^2 + 2x - 3}}$

For each function f, construct and simplify the difference quotient. **[5.4b]**

24. $f(x) = 2x + 7$

25. $f(x) = 3 - x^2$

26. $f(x) = \dfrac{4}{x}$

Solve. **[5.5a]**

27. $\dfrac{x}{4} + \dfrac{x}{7} = 1$

28. $\dfrac{5}{3x + 2} = \dfrac{3}{2x}$

29. $\dfrac{4x}{x + 1} + \dfrac{4}{x} + 9 = \dfrac{4}{x^2 + x}$

30. $\dfrac{90}{x^2 - 3x + 9} - \dfrac{5x}{x + 3} = \dfrac{405}{x^3 + 27}$

31. $\dfrac{2}{x - 3} + \dfrac{1}{4x + 20} = \dfrac{1}{x^2 + 2x - 15}$

32. Given that
$$f(x) = \dfrac{6}{x} + \dfrac{4}{x},$$
find all x for which $f(x) = 5$.

33. *House Painting.* David can paint the outside of a house in 12 hr. Bill can paint the same house in 9 hr. How long would it take them working together to paint the house? **[5.6a]**

34. *Boat Travel.* The current of the Gold River is 6 mph. A boat travels 50 mi downstream in the same time that it takes to travel 30 mi upstream. Complete the table below and then find the speed of the boat in still water. **[5.6c]**

	Distance	Speed	Time
Downstream			
Upstream			

35. *Travel Distance.* Fred operates a potato-chip delivery route. He drives 800 mi in 3 days. How far will he travel in 15 days? **[5.6b]**

Solve for the indicated letter. **[5.7a]**

36. $W = \dfrac{cd}{c + d}$, for d; for c

37. $S = \dfrac{p}{a} + \dfrac{t}{b}$, for b; for t

38. Find an equation of variation in which y varies directly as x, and $y = 100$ when $x = 25$. **[5.8a]**

39. Find an equation of variation in which y varies inversely as x, and $y = 100$ when $x = 25$. **[5.8c]**

40. *Pumping Time.* The time t required to empty a tank varies inversely as the rate r of pumping. If a pump can empty a tank in 35 min at the rate of 800 kL per minute, how long will it take the pump to empty the same tank at the rate of 1400 kL per minute? **[5.8d]**

41. *Test Score.* The score N on a test varies directly as the number of correct responses a. Ellen answers 29 questions correctly and earns a score of 87. What would Ellen's score have been if she had answered 25 questions correctly? **[5.8b]**

42. *Power of Electric Current.* The power P expended by heat in an electric circuit of fixed resistance varies directly as the square of the current C in the circuit. A circuit expends 180 watts when a current of 6 amperes is flowing. What is the amount of heat expended when the current is 10 amperes? **[5.8f]**

43. Find the domain of $f(x) = \dfrac{x^2 - x}{x^2 - 2x - 35}$. **[5.1a]**

   A. $(0, 1)$
   B. $(-\infty, -5) \cup (-5, 7) \cup (7, \infty)$
   C. $(-5, 7)$
   D. $(-\infty, 0) \cup (0, 1) \cup (1, \infty)$

44. Find the LCM of x^5, $x - 4$, $x^2 - 4$, and $x^2 - 4x$. **[5.2a]**

   A. $x(x - 4)^2$
   B. $(x - 4)(x + 4)$
   C. $x^5(x - 4)(x - 2)(x + 2)$
   D. $x^5(x - 4)^2$

▶ Synthesis

45. Find the reciprocal and simplify: $\dfrac{a - b}{a^3 - b^3}$. **[5.1c, e]**

46. Solve: $\dfrac{5}{x - 13} - \dfrac{5}{x} = \dfrac{65}{x^2 - 13x}$. **[5.5a]**

▶ Collaborative Discussion and Writing

47. Discuss at least three different uses of the LCM studied in this chapter. **[5.2b], [5.4a], [5.5a]**

48. You have learned to solve a new kind of equation in this chapter. Explain how this type differs from those you have studied previously and how the equation-solving process differs. **[5.5a]**

49. Explain why it is sufficient, when checking a possible solution of a rational equation, to verify that the number in question does not make a denominator 0. **[5.5a]**

50. If y varies directly as x and x varies inversely as z, how does y vary with regard to z? Why? **[5.8a, c, e]**

51. Explain how you might easily create rational equations for which there is no solution. (See Example 4 of Section 5.5 for a hint.) **[5.5a]**

52. Which is easier to solve for x? Explain why. **[5.7a]**

$$\frac{1}{38} + \frac{1}{47} = \frac{1}{x} \quad \text{or} \quad \frac{1}{a} + \frac{1}{b} = \frac{1}{x}$$

---

5   Chapter Test

1. Find all numbers for which the rational expression
$$\frac{x^2 - 16}{x^2 - 3x + 2}$$
is not defined.

2. Find the domain of f where
$$f(x) = \frac{x^2 - 16}{x^2 - 3x + 2}.$$

Simplify.

3. $\dfrac{12x^2 + 11x + 2}{4x^2 - 7x - 2}$ **4.** $\dfrac{p^3 + 1}{p^2 - p - 2}$

5. Find the LCM of $x^2 + x - 6$ and $x^2 + 8x + 15$.

Perform the indicated operations and simplify.

6. $\dfrac{2x^2 + 20x + 50}{x^2 - 4} \cdot \dfrac{x + 2}{x + 5}$

7. $\dfrac{x}{x^2 + 11x + 30} - \dfrac{5}{x^2 + 9x + 20}$

8. $\dfrac{y^2 - 16}{2y + 6} \div \dfrac{y - 4}{y + 3}$

9. $\dfrac{x^2}{x - y} + \dfrac{y^2}{y - x}$

10. $\dfrac{1}{x + 1} - \dfrac{x + 2}{x^2 - 1} + \dfrac{3}{x - 1}$

11. $\dfrac{a}{a - b} + \dfrac{b}{a^2 + ab + b^2} - \dfrac{2}{a^3 - b^3}$

Divide.

12. $(20r^2s^3 + 15r^2s^2 - 10r^3s^3) \div (5r^2s)$

13. $(y^3 + 125) \div (y + 5)$

14. $(4x^4 + 3x^3 - 5x - 2) \div (x^2 + 1)$

Divide using synthetic division. Show your work.

15. $(x^3 + 3x^2 + 2x - 6) \div (x - 3)$

16. $(3x^3 + 22x^2 - 160) \div (x + 4)$

Simplify.

17. $\dfrac{1 - \dfrac{1}{x^2}}{1 - \dfrac{1}{x}}$

18. $\dfrac{\dfrac{1}{a^3} + \dfrac{1}{b^3}}{\dfrac{1}{a} + \dfrac{1}{b}}$

19. For the function $f(x) = \dfrac{1}{2}x + 4$, construct and simplify the difference quotient.

20. Given that
$$f(x) = \frac{2}{x - 1} + \frac{2}{x + 2},$$
find all x for which $f(x) = 1$.

Solve.

21. $\dfrac{2}{x - 1} = \dfrac{3}{x + 3}$

22. $\dfrac{7x}{x + 3} + \dfrac{21}{x - 3} = \dfrac{126}{x^2 - 9}$

23. $\dfrac{2x}{x + 7} = \dfrac{5}{x + 1}$

24. $\dfrac{1}{3x - 6} - \dfrac{1}{x^2 - 4} = \dfrac{3}{x + 2}$

25. *Completing a Puzzle.* Working together, Rachel and Jessie can complete a jigsaw puzzle in 1.5 hr. Rachel takes 4 hr longer than Jessie does when working alone. How long would it take Jessie to complete the puzzle?

26. *Bicycle Travel.* David can bicycle at a rate of 12 mph when there is no wind. Against the wind, David bikes 8 mi in the same time that it takes to bike 14 mi with the wind. What is the speed of the wind?

27. *Predicting Paint Needs.* Logan and Noah run a summer painting company to defray their college expenses. They need 4 gal of paint to paint 1700 ft^2 of clapboard. How much paint would they need for a building with 6000 ft^2 of clapboard?

Solve for the indicated letter.

28. $T = \dfrac{ab}{a - b}$, for a; for b

29. $Q = \dfrac{2}{a} - \dfrac{t}{b}$, for a

30. Find an equation of variation in which Q varies jointly as x and y, and $Q = 25$ when $x = 2$ and $y = 5$.

31. Find an equation of variation in which y varies inversely as x, and $y = 10$ when $x = 25$.

32. *Income vs. Time.* Kaylee's income I varies directly as the time t worked. She gets a job that pays $550 for 40 hr of work. What is she paid for working 72 hr, assuming that there is no change in pay scale for overtime?

33. *Time and Speed.* The time t required to drive a fixed distance varies inversely as the speed r. It takes 5 hr at 60 km/h to drive a fixed distance. How long would it take to drive that same distance at 40 km/h?

34. *Area of a Balloon.* The surface area of a balloon varies directly as the square of its radius. The area is 314 cm^2 when the radius is 5 cm. What is the area when the radius is 7 cm?

35. Find the LCM of $6x^2$, $3x^2 - 3y^2$, and $x^2 - 2xy - 3y^2$.
A. $3x^2(2x + y)(x - 3y)$
B. $6x(x + y)(x - y)(x - 3y)$
C. $3x^2(x + y)^2(x - y)$
D. $6x^2(x + y)(x - y)(x - 3y)$

▶ **Synthesis**

36. Solve: $\dfrac{6}{x - 15} - \dfrac{6}{x} = \dfrac{90}{x^2 - 15x}$.

37. Find the x- and y-intercepts of the function f given by
$$f(x) = \frac{\dfrac{5}{x + 4} - \dfrac{3}{x - 2}}{\dfrac{2}{x - 3} + \dfrac{1}{x + 4}}.$$

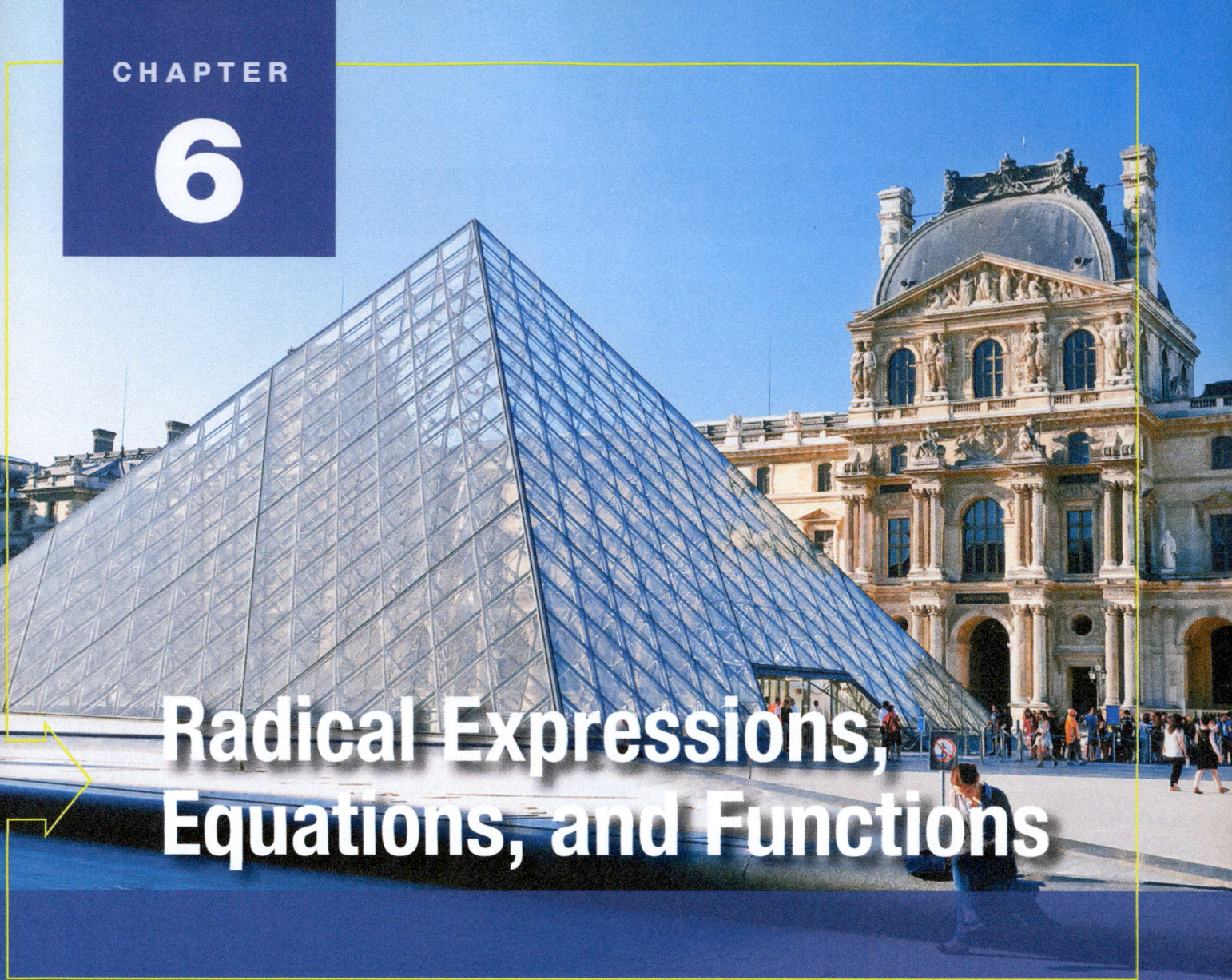

Radical Expressions, Equations, and Functions

APPLICATION   This problem appears as Exercise 17 in Exercise Set 6.7.

A large glass and metal pyramid designed by I. M. Pei attracts visitors to the entrance of the Louvre Museum in Paris, France. If a plumb line from the highest point of the pyramid to the center of the base is 71 ft long and each side of the four equilateral triangles measures 100 ft, how far is it from the center of the base to a corner of the base?

6.1 Radical Expressions and Functions

▶ **a** Find principal square roots and their opposites, approximate square roots, identify radicands, find outputs of square-root functions, graph square-root functions, and find the domains of square-root functions.

▶ **b** Simplify radical expressions with perfect-square radicands.

▶ **c** Find cube roots, simplifying certain expressions, and find outputs of cube-root functions.

▶ **d** Simplify expressions involving odd roots and even roots.

In this section, we consider roots, such as square roots and cube roots. We define the symbolism and consider methods of manipulating symbols to get equivalent expressions.

▶ a   Square Roots and Square-Root Functions

When we raise a number to the second power, we say that we have **squared** the number. Sometimes we may need to find the number that was squared. We call this process **finding a square root** of a number.

> **SQUARE ROOT**
>
> The number c is a **square root** of a if $c^2 = a$.

For example:

5 is a *square root* of 25 because $5^2 = 5 \cdot 5 = 25$;

-5 is a *square root* of 25 because $(-5)^2 = (-5)(-5) = 25$.

The number -4 does not have a real-number square root because there is no real number c such that $c^2 = -4$.

> **PROPERTIES OF SQUARE ROOTS**
>
> Every positive real number has two real-number square roots.
>
> The number 0 has just one square root, 0 itself.
>
> Negative numbers do not have real-number square roots.*

EXAMPLE 1   Find the two square roots of 64.

The square roots of 64 are 8 and -8 because $8^2 = 64$ and $(-8)^2 = 64$.

▶ **Now Try Exercise 1.**

---

* In Chapter 7, we will consider a number system in which negative numbers do have square roots.

PRINCIPAL SQUARE ROOT

The **principal square root** of a nonnegative number is its nonnegative square root. The symbol $\sqrt{a}$ represents the principal square root of a. To name the negative square root of a, we can write $-\sqrt{a}$.

EXAMPLES Simplify.

2. $\sqrt{25} = 5$

> **Remember:** $\sqrt{}$ indicates the principal (nonnegative) square root.

3. $-\sqrt{25} = -5$

4. $\sqrt{\dfrac{81}{64}} = \dfrac{9}{8}$ because $\left(\dfrac{9}{8}\right)^2 = \dfrac{9}{8} \cdot \dfrac{9}{8} = \dfrac{81}{64}$.

5. $\sqrt{0.0049} = 0.07$ because $(0.07)^2 = (0.07)(0.07) = 0.0049$.

6. $-\sqrt{0.000001} = -0.001$

7. $\sqrt{0} = 0$

8. $\sqrt{-25}$ Does not exist as a real number. Negative numbers do not have real-number square roots.

> **Now Try Exercises 9 and 13.**

Table of Square Roots	
$\sqrt{1} = 1$	$\sqrt{196} = 14$
$\sqrt{4} = 2$	$\sqrt{225} = 15$
$\sqrt{9} = 3$	$\sqrt{256} = 16$
$\sqrt{16} = 4$	$\sqrt{289} = 17$
$\sqrt{25} = 5$	$\sqrt{324} = 18$
$\sqrt{36} = 6$	$\sqrt{361} = 19$
$\sqrt{49} = 7$	$\sqrt{400} = 20$
$\sqrt{64} = 8$	$\sqrt{441} = 21$
$\sqrt{81} = 9$	$\sqrt{484} = 22$
$\sqrt{100} = 10$	$\sqrt{529} = 23$
$\sqrt{121} = 11$	$\sqrt{576} = 24$
$\sqrt{144} = 12$	$\sqrt{625} = 25$
$\sqrt{169} = 13$	

We found exact square roots in Examples 1–8. It would be helpful to memorize the table of exact square roots at left. We often need to use rational numbers to *approximate* square roots that are irrational. Such expressions can be found using a calculator with a square-root key.

EXAMPLES Use a calculator to approximate each of the following.

	Number	Using a calculator with a 10-digit readout	Rounded to three decimal places
9.	$\sqrt{11}$	3.316624790	3.317
10.	$\sqrt{487}$	22.06807649	22.068
11.	$-\sqrt{7297.8}$	-85.42716196	-85.427
12.	$\sqrt{\dfrac{463}{557}}$	.9117229728	0.912

> **Now Try Exercise 15.**

RADICAL; RADICAL EXPRESSION; RADICAND

The symbol $\sqrt{}$ is called a **radical**.

An expression written with a radical is called a **radical expression**.

The expression written under the radical is called the **radicand**.

These are radical expressions:

$$\sqrt{5}, \qquad \sqrt{a}, \qquad -\sqrt{5x}, \qquad \sqrt{y^2 + 7}.$$

The radicands in these expressions are 5, a, $5x$, and $y^2 + 7$, respectively.

EXAMPLE 13 Identify the radicand in $x\sqrt{x^2 - 9}$.

The radicand is the expression under the radical, $x^2 - 9$.

> **Now Try Exercise 19.**

Since each nonnegative real number x has exactly one principal square root, the symbol $\sqrt{x}$ represents exactly one real number and thus can be used to define a square-root function:

$$f(x) = \sqrt{x}.$$

The domain of this function is the set of nonnegative real numbers. In interval notation, the domain is $[0, \infty)$.

EXAMPLE 14 For the given function, find the indicated function values:

$$f(x) = \sqrt{3x - 2}; \quad f(1), f(5), \text{ and } f(0).$$

We have

$$
\begin{aligned}
f(1) &= \sqrt{3 \cdot 1 - 2} && \textbf{Substituting 1 for } x\\
&= \sqrt{3 - 2} = \sqrt{1} = 1; && \textbf{Simplifying and taking the square root}\\
f(5) &= \sqrt{3 \cdot 5 - 2} && \textbf{Substituting 5 for } x\\
&= \sqrt{13} \approx 3.606; && \textbf{Simplifying and approximating}\\
f(0) &= \sqrt{3 \cdot 0 - 2} && \textbf{Substituting 0 for } x\\
&= \sqrt{-2}. && \textbf{Negative radicand. No real-number function value exists; 0 is not in the domain of } f.
\end{aligned}
$$

> **Now Try Exercise 23.**

> **DOMAIN OF A FUNCTION**
>
> **REVIEW SECTION 2.3**

EXAMPLE 15 Find the domain of $g(x) = \sqrt{x + 2}$.

The expression $\sqrt{x + 2}$ is a real number only when $x + 2$ is nonnegative. Thus the domain of $g(x) = \sqrt{x + 2}$ is the set of all x-values for which $x + 2 \geq 0$. We solve as follows:

$$x + 2 \geq 0$$
$$x \geq -2. \quad \textbf{Adding } -2$$

The domain of $g = \{x \mid x \geq -2\} = [-2, \infty)$. —————▶ **Now Try Exercise 27.**

EXAMPLE 16 Graph: **(a)** $f(x) = \sqrt{x}$; **(b)** $g(x) = \sqrt{x + 2}$.

We first find outputs as we did in Example 14. We can either select inputs that have exact outputs or use a calculator to make approximations. Once ordered pairs have been calculated, a smooth curve can be drawn.

a)

x	$f(x) = \sqrt{x}$	$(x, f(x))$
0	0	$(0, 0)$
1	1	$(1, 1)$
3	1.7	$(3, 1.7)$
4	2	$(4, 2)$
7	2.6	$(7, 2.6)$
9	3	$(9, 3)$

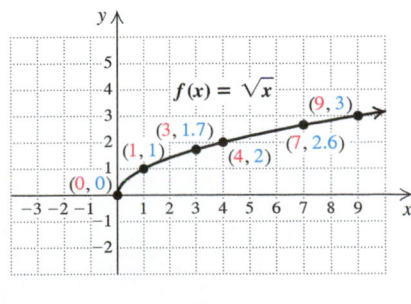

We can see from the table and the graph that the domain of f is $[0, \infty)$. The range is also the set of nonnegative real numbers $[0, \infty)$.

b)

x	$g(x) = \sqrt{x + 2}$	$(x, g(x))$
-2	0	$(-2, 0)$
-1	1	$(-1, 1)$
0	1.4	$(0, 1.4)$
3	2.2	$(3, 2.2)$
5	2.6	$(5, 2.6)$
10	3.5	$(10, 3.5)$

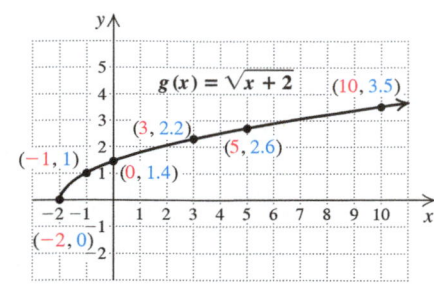

We can see from the table, the graph, and Example 15 that the domain of g is $[-2, \infty)$. The range is the set of nonnegative real numbers $[0, \infty)$.

Now Try Exercise 31.

▶ b Finding $\sqrt{a^2}$

In the expression $\sqrt{a^2}$, the radicand is a perfect square. It is tempting to think that $\sqrt{a^2} = a$, but we see below that this is not always the case.

Suppose $a = 5$. Then we have $\sqrt{5^2}$, which is $\sqrt{25}$, or 5.

Suppose $a = -5$. Then we have $\sqrt{(-5)^2}$, which is $\sqrt{25}$, or 5.

Suppose $a = 0$. Then we have $\sqrt{0^2}$, which is $\sqrt{0}$, or 0.

The symbol $\sqrt{a^2}$ never represents a negative number. It represents the principal square root of a^2. Note the following.

Simplifying $\sqrt{a^2}$

$a \geq 0 \longrightarrow \sqrt{a^2} = a$

If a is positive or 0, the principal square root of a^2 is a.

$a < 0 \longrightarrow \sqrt{a^2} = -a$

If a is negative, the principal square root of a^2 is the opposite of a.

In all cases, the radical expression $\sqrt{a^2}$ represents the absolute value of a.

PRINCIPAL SQUARE ROOT OF a^2

For any real number a, $\sqrt{a^2} = |a|$. The principal (nonnegative) square root of a^2 is the absolute value of a.

The absolute value is used to ensure that the principal square root is nonnegative, which is as it is defined.

EXAMPLES Find each of the following. Assume that letters can represent any real number.

17. $\sqrt{(-16)^2} = |-16|$, or 16

18. $\sqrt{(3b)^2} = |3b| = |3| \cdot |b| = 3|b|$

> $|3b|$ can be simplified to $3|b|$ because the absolute value of any product is the product of the absolute values. That is, $|a \cdot b| = |a| \cdot |b|$.

19. $\sqrt{(x-1)^2} = |x-1|$

20. $\sqrt{x^2 + 8x + 16} = \sqrt{(x+4)^2}$
$= |x+4|$

> **CAUTION!** $|x+4|$ is *not* the same as $|x| + 4$.

> **Now Try Exercises 43 and 47.**

▶ c Cube Roots

> **CUBE ROOT**
>
> The number c is the **cube root** of a, written $\sqrt[3]{a}$, if the third power of c is a—that is, if $c^3 = a$, then $\sqrt[3]{a} = c$.

For example:

> 2 is the *cube root* of 8 because $2^3 = 2 \cdot 2 \cdot 2 = 8$;
>
> -4 is the *cube root* of -64 because $(-4)^3 = (-4)(-4)(-4) = -64$.

We talk about *the* cube root of a number rather than *a* cube root because of the following.

> Every real number has exactly one cube root in the system of real numbers. The symbol $\sqrt[3]{a}$ represents *the* cube root of a.

EXAMPLES Find each of the following.

21. $\sqrt[3]{8} = 2$ because $2^3 = 8$.

22. $\sqrt[3]{-27} = -3$

23. $\sqrt[3]{-\dfrac{216}{125}} = -\dfrac{6}{5}$

24. $\sqrt[3]{0.001} = 0.1$

25. $\sqrt[3]{x^3} = x$

26. $\sqrt[3]{-8} = -2$

27. $\sqrt[3]{0} = 0$

28. $\sqrt[3]{-8y^3} = \sqrt[3]{(-2y)^3} = -2y$

> **Now Try Exercises 51 and 53.**

When we are determining a cube root, no absolute-value signs are needed because a real number has just one cube root. The real-number cube root of a positive number is positive. The real-number cube root of a negative number is negative. The cube root of 0 is 0. That is, $\sqrt[3]{a^3} = a$ whether $a > 0$, $a < 0$, or $a = 0$.

Since the symbol $\sqrt[3]{x}$ represents exactly one real number, it can be used to define a cube-root function: $f(x) = \sqrt[3]{x}$.

EXAMPLE 29   For the given function, find the indicated function values:
$$f(x) = \sqrt[3]{x}; \quad f(125), f(0), f(-8), \text{ and } f(-10).$$

We have

$$f(125) = \sqrt[3]{125} = 5;$$
$$f(0) = \sqrt[3]{0} = 0;$$
$$f(-8) = \sqrt[3]{-8} = -2;$$
$$f(-10) = \sqrt[3]{-10} \approx -2.154.$$

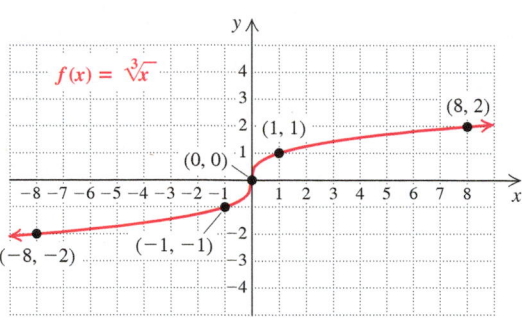

The graph of $f(x) = \sqrt[3]{x}$ is shown above for reference. Note that both the domain and the range consist of the entire set of real numbers, $(-\infty, \infty)$.

Now Try Exercise 59.

▶ d   Odd and Even *k*th Roots

In the expression $\sqrt[k]{a}$, we call k the **index** and assume $k \geq 2$.

Odd Roots

The 5th root of a number a is the number c for which $c^5 = a$. There are also 7th roots, 9th roots, and so on. Whenever the number k in $\sqrt[k]{}$ is an odd number, we say that we are taking an **odd root**.

Every number has just one real-number odd root. If the number is positive, then the root is positive. If the number is negative, then the root is negative. If the number is 0, then the root is 0. For example, $\sqrt[3]{8} = 2$, $\sqrt[3]{-8} = -2$, and $\sqrt[3]{0} = 0$. Absolute-value signs are *not* needed when we are finding odd roots.

> If k is an *odd* natural number, then for any real number a,
> $$\sqrt[k]{a^k} = a.$$

EXAMPLES   Find each of the following.

30. $\sqrt[5]{32} = 2$

31. $\sqrt[5]{-32} = -2$

32. $-\sqrt[5]{32} = -2$

33. $-\sqrt[5]{-32} = -(-2) = 2$

34. $\sqrt[7]{x^7} = x$

35. $\sqrt[7]{128} = 2$

36. $\sqrt[7]{-128} = -2$

37. $\sqrt[7]{0} = 0$

38. $\sqrt[5]{a^5} = a$

39. $\sqrt[9]{(x-1)^9} = x - 1$

Now Try Exercises 65 and 79.

Even Roots

When the index k in $\sqrt[k]{}$ is an even number, we say that we are taking an **even root**. When the index is 2, we do not write it. Every positive real number has two

real-number *k*th roots when *k* is even. One of those roots is positive and one is negative. Negative real numbers do not have real-number *k*th roots when *k* is even. When we are finding even *k*th roots, absolute-value signs are sometimes necessary, as we have seen with square roots. For example,

$$\sqrt{64} = 8, \quad \sqrt[6]{64} = 2, \quad -\sqrt[6]{64} = -2, \quad \sqrt[6]{64x^6} = \sqrt[6]{(2x)^6} = |2x| = 2|x|.$$

Note that in $\sqrt[6]{64x^6}$, we need absolute-value signs because a variable is involved.

EXAMPLES   Find each of the following. Assume that variables can represent any real number.

40. $\sqrt[4]{16} = 2$

41. $-\sqrt[4]{16} = -2$

42. $\sqrt[4]{-16}$
   Does not exist as a real number.

43. $\sqrt[4]{81x^4} = \sqrt[4]{(3x)^4} = |3x| = 3|x|$

44. $\sqrt[6]{(y+7)^6} = |y+7|$

45. $\sqrt{81y^2} = \sqrt{(9y)^2} = |9y| = 9|y|$

> **Now Try Exercises 63 and 69.**

The following is a summary of how absolute value is used when we are taking even roots or odd roots.

SIMPLIFYING

For any real number *a*:

a) $\sqrt[k]{a^k} = |a|$ when *k* is an *even* natural number. We use absolute value when *k* is even unless *a* is nonnegative.

b) $\sqrt[k]{a^k} = a$ when *k* is an *odd* natural number greater than 1. We do not use absolute value when *k* is odd.

6.1   Exercise Set

▶ Reading Check

Choose from the columns on the right the domain of the given function. Some of the choices may not be used, and some may be used more than once.

RC1. $f(x) = \sqrt{9-x}$

RC2. $f(x) = \sqrt{x+9} + 3$

RC3. $g(x) = \sqrt{x-3}$

RC4. $h(x) = x + 9$

RC5. $f(x) = 3 - x$

RC6. $g(x) = 3 - \sqrt{3-x}$

a) $[-3, \infty)$ b) $[-9, \infty)$
c) $(3, \infty)$ d) $(-\infty, -9)$
e) $(-\infty, -3]$ f) $[9, \infty)$
g) $(-\infty, 3]$ h) $[3, \infty)$
i) $(-\infty, \infty)$ j) $(-\infty, 9]$

....

a *Find the square roots.*

1. 16

2. 225

3. 144

4. 9

5. 400

6. 81

Simplify.

7. $-\sqrt{\dfrac{49}{36}}$

8. $-\sqrt{\dfrac{361}{9}}$

9. $\sqrt{196}$

10. $\sqrt{441}$

11. $\sqrt{0.0036}$ **12.** $\sqrt{0.04}$

13. $\sqrt{-225}$ **14.** $\sqrt{-64}$

Use a calculator to approximate to three decimal places.

15. $\sqrt{347}$ **16.** $-\sqrt{1839.2}$

17. $\sqrt{\dfrac{285}{74}}$ **18.** $\sqrt{\dfrac{839.4}{19.7}}$

Identify the radicand.

19. $9\sqrt{y^2 + 16}$ **20.** $-3\sqrt{p^2 - 10}$

21. $x^4 y^5 \sqrt{\dfrac{x}{y - 1}}$ **22.** $a^2 b^2 \sqrt{\dfrac{a^2 - b}{b}}$

For the given function, find the indicated function values.

23. $f(x) = \sqrt{5x - 10};\quad f(6), f(2), f(1),$ and $f(-1)$

24. $t(x) = -\sqrt{2x + 1};\quad t(4), t(0), t(-1),$ and $t\left(-\frac{1}{2}\right)$

25. $g(x) = \sqrt{x^2 - 25};\quad g(-6), g(3), g(6),$ and $g(13)$

26. $F(x) = \sqrt{x^2 + 1};\quad F(0), F(-1),$ and $F(-10)$

27. Find the domain of the function f in Exercise 23.

28. Find the domain of the function t in Exercise 24.

29. *Parking-Lot Arrival Spaces.*   The attendants at a parking lot park cars in temporary spaces before the cars are taken to long-term parking stalls. The number N of such spaces needed is approximated by the function

$$N(a) = 2.5\sqrt{a},$$

where a is the average number of arrivals in peak hours. What is the number of spaces needed when the average number of arrivals is 66? 100?

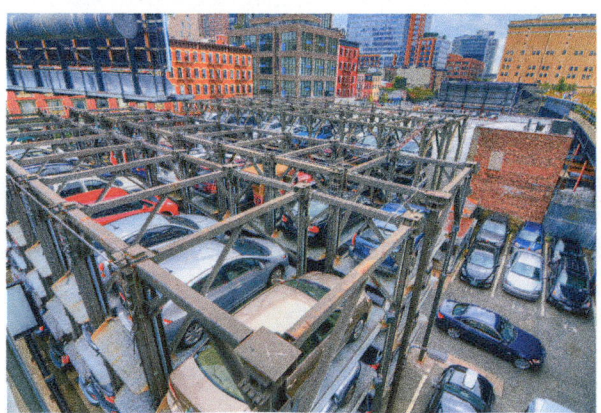

30. *Body Surface Area.*   Body surface area B can be estimated using the Mosteller formula

$$B = \sqrt{\dfrac{h \times w}{3600}},$$

where B is in square meters, h is height, in centimeters, and w is weight, in kilograms. Estimate the body surface area of **(a)** a woman whose height is

165 cm and whose weight is 63 kg and **(b)** a man whose height is 183 cm and whose weight is 100 kg. Round to the nearest tenth.

Graph.

31. $f(x) = 2\sqrt{x}$ **32.** $g(x) = 3 - \sqrt{x}$

33. $F(x) = -3\sqrt{x}$ **34.** $f(x) = 2 + \sqrt{x - 1}$

35. $f(x) = \sqrt{x}$ **36.** $g(x) = -\sqrt{x}$

37. $f(x) = \sqrt{x - 2}$ **38.** $g(x) = \sqrt{x + 3}$

39. $f(x) = \sqrt{12 - 3x}$ **40.** $g(x) = \sqrt{8 - 4x}$

41. $g(x) = \sqrt{3x + 9}$ **42.** $f(x) = \sqrt{3x - 6}$

b   *Find each of the following. Assume that letters can represent any real number.*

43. $\sqrt{16x^2}$ **44.** $\sqrt{25t^2}$

45. $\sqrt{(-12c)^2}$ **46.** $\sqrt{(-9d)^2}$

47. $\sqrt{(p + 3)^2}$ **48.** $\sqrt{(2 - x)^2}$

49. $\sqrt{x^2 - 4x + 4}$ **50.** $\sqrt{9t^2 - 30t + 25}$

c   *Simplify.*

51. $\sqrt[3]{27}$ **52.** $-\sqrt[3]{64}$

53. $\sqrt[3]{-64x^3}$ **54.** $\sqrt[3]{-125y^3}$

55. $\sqrt[3]{-216}$ **56.** $-\sqrt[3]{-1000}$

57. $\sqrt[3]{0.343(x + 1)^3}$ **58.** $\sqrt[3]{0.000008(y - 2)^3}$

For the given function, find the indicated function values.

59. $f(x) = \sqrt[3]{x + 1};\quad f(7), f(26), f(-9),$ and $f(-65)$

60. $g(x) = -\sqrt[3]{2x - 1};\quad g(-62), g(0), g(-13),$ and $g(63)$

61. $f(x) = -\sqrt[3]{3x + 1};\quad f(0), f(-7), f(21),$ and $f(333)$

62. $g(t) = \sqrt[3]{t - 3};\quad g(30), g(-5), g(1),$ and $g(67)$

d   *Find each of the following. Assume that letters can represent any real number.*

63. $-\sqrt[4]{625}$ **64.** $-\sqrt[4]{256}$

65. $\sqrt[5]{-1}$ **66.** $\sqrt[5]{-32}$

67. $\sqrt[5]{-\dfrac{32}{243}}$ **68.** $\sqrt[5]{-\dfrac{1}{32}}$

69. $\sqrt[6]{x^6}$ **70.** $\sqrt[8]{y^8}$

71. $\sqrt[4]{(5a)^4}$ **72.** $\sqrt[4]{(7b)^4}$

73. $\sqrt[10]{(-6)^{10}}$ **74.** $\sqrt[12]{(-10)^{12}}$

75. $\sqrt[414]{(a + b)^{414}}$ **76.** $\sqrt[1999]{(2a + b)^{1999}}$

77. $\sqrt[7]{y^7}$ **78.** $\sqrt[3]{(-6)^3}$

79. $\sqrt[5]{(x - 2)^5}$ **80.** $\sqrt[9]{(2xy)^9}$

▶ Skill Maintenance

Solve. **[4.8a]**

81. $x^2 + x - 2 = 0$

82. $x^2 + x = 0$

83. $4x^2 - 49 = 0$

84. $2x^2 - 26x + 72 = 0$

85. $3x^2 + x = 10$

86. $4x^2 - 20x + 25 = 0$

87. $4x^3 - 20x^2 + 25x = 0$

88. $x^3 - x^2 = 0$

Simplify.

89. $(a^3b^2c^5)^3$ **[R.7b]**

90. $(5a^7b^8)(2a^3b)$ **[R.7a]**

▶ Synthesis

91. Find the domain of
$$f(x) = \frac{\sqrt{x+3}}{\sqrt{2-x}}.$$

92. ⌁ Use a graphing calculator to check your answers to Exercises 35, 39, and 41.

93. Use only the graph of $f(x) = \sqrt{x}$, shown below, to approximate $\sqrt{3}$, $\sqrt{5}$, and $\sqrt{10}$. Answers may vary.

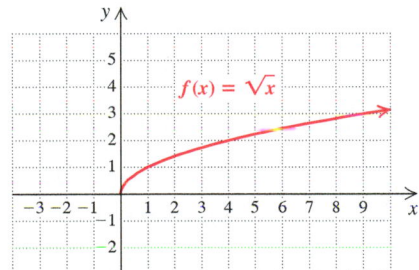

94. Use only the graph of $f(x) = \sqrt[3]{x}$, shown below, to approximate $\sqrt[3]{4}$, $\sqrt[3]{6}$, and $\sqrt[3]{-5}$. Answers may vary.

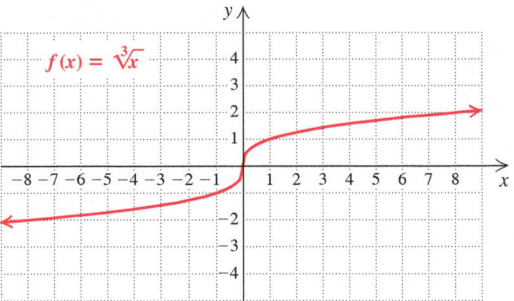

95. Use the TABLE, TRACE, and GRAPH features of a graphing calculator to find the domain and the range of each of the following functions.

a) $f(x) = \sqrt[3]{x}$

b) $g(x) = \sqrt[3]{4x-5}$

c) $q(x) = 2 - \sqrt{x+3}$

d) $h(x) = \sqrt[4]{x}$

e) $t(x) = \sqrt[4]{x-3}$

6.2 Rational Numbers as Exponents

▶ **a** Write expressions with or without rational exponents, and simplify, if possible.

▶ **b** Write expressions without negative exponents, and simplify, if possible.

▶ **c** Use the laws of exponents with rational exponents.

▶ **d** Use rational exponents to simplify radical expressions.

In this section, we give meaning to expressions such as $a^{1/3}$, $7^{-1/2}$, and $(3x)^{0.84}$, which have rational numbers as exponents. We will see that using such notation can help simplify certain radical expressions.

▶ **a   Rational Exponents**

Expressions like $a^{1/2}$, $5^{-1/4}$, and $(2y)^{4/5}$ have not yet been defined. We will define such expressions so that the general properties of exponents hold.

Consider $a^{1/2} \cdot a^{1/2}$. If we want to multiply by adding exponents, it must follow that $a^{1/2} \cdot a^{1/2} = a^{1/2+1/2}$, or a^1. Thus we should define $a^{1/2}$ to be a square root of a. Similarly, $a^{1/3} \cdot a^{1/3} \cdot a^{1/3} = a^{1/3+1/3+1/3}$, or a^1, so $a^{1/3}$ should be defined to mean $\sqrt[3]{a}$.

$a^{1/n}$

For any *nonnegative* real number a and any natural number index n $(n \neq 1)$,

$$a^{1/n} \quad \text{means} \quad \sqrt[n]{a} \quad (\text{the nonnegative } n\text{th root of } a).$$

With rational exponents, we assume that the bases are nonnegative.

EXAMPLES   Rewrite without rational exponents, and simplify, if possible.

1. $27^{1/3} = \sqrt[3]{27} = 3$ **2.** $(abc)^{1/5} = \sqrt[5]{abc}$

3. $x^{1/2} = \sqrt{x}$ **An index of 2 is not written.**

Now Try Exercises 3 and 5.

EXAMPLES   Rewrite with rational exponents.

4. $\sqrt[5]{7xy} = (7xy)^{1/5}$ **We need parentheses around the radicand.**

5. $8\sqrt[3]{xy} = 8(xy)^{1/3}$ **6.** $\sqrt[7]{\dfrac{x^3y}{9}} = \left(\dfrac{x^3y}{9}\right)^{1/7}$

Now Try Exercise 15.

How should we define $a^{2/3}$? If the general properties of exponents are to hold, we have $a^{2/3} = (a^{1/3})^2$, or $(a^2)^{1/3}$, or $(\sqrt[3]{a})^2$, or $\sqrt[3]{a^2}$. We define this accordingly.

$a^{m/n}$

For any natural numbers m and n $(n \neq 1)$ and any nonnegative real number a,

$$a^{m/n} \quad \text{means} \quad \sqrt[n]{a^m}, \quad \text{or} \quad (\sqrt[n]{a})^m.$$

EXAMPLES   Rewrite without rational exponents, and simplify, if possible.

7. $(27)^{2/3} = \sqrt[3]{27^2}$ **8.** $4^{3/2} = \sqrt[2]{4^3}$
              $= (\sqrt[3]{27})^2$ $= (\sqrt[2]{4})^3$
              $= 3^2$ $= 2^3$
              $= 9$ $= 8$

Now Try Exercise 7.

EXAMPLES   Rewrite with rational exponents.

The index becomes the denominator of the rational exponent.

9. $\sqrt[3]{9^4} = 9^{4/3}$ **10.** $\left(\sqrt[4]{7xy}\right)^5 = (7xy)^{5/4}$

> **Now Try Exercise 17.**

▶ b Negative Rational Exponents

NEGATIVE INTEGER EXPONENTS

REVIEW SECTION R.3

Negative rational exponents have a meaning similar to that of negative integer exponents.

$a^{-m/n}$

For any rational number m/n and any positive real number a,

$$a^{-m/n} \quad \text{means} \quad \frac{1}{a^{m/n}};$$

that is, $a^{m/n}$ and $a^{-m/n}$ are reciprocals.

EXAMPLES   Rewrite with positive exponents, and simplify, if possible.

11. $9^{-1/2} = \dfrac{1}{9^{1/2}} = \dfrac{1}{\sqrt{9}} = \dfrac{1}{3}$

12. $(5xy)^{-4/5} = \dfrac{1}{(5xy)^{4/5}}$

13. $64^{-2/3} = \dfrac{1}{64^{2/3}} = \dfrac{1}{\left(\sqrt[3]{64}\right)^2} = \dfrac{1}{4^2} = \dfrac{1}{16}$

14. $4x^{-2/3}y^{1/5} = 4 \cdot \dfrac{1}{x^{2/3}} \cdot y^{1/5} = \dfrac{4y^{1/5}}{x^{2/3}}$

15. $\left(\dfrac{3r}{7s}\right)^{-5/2} = \left(\dfrac{7s}{3r}\right)^{5/2}$ Since $\left(\dfrac{a}{b}\right)^{-n} = \left(\dfrac{b}{a}\right)^{n}$

> **Now Try Exercises 21 and 29.**

Technology Connection

```
7^(2/3)
           3.65930571
14^−1.9
           .006642885
```

Rational Exponents

We can use a graphing calculator to approximate rational roots of real numbers. To approximate $7^{2/3}$, note that the parentheses around the exponent are necessary. If they are not used, the calculator will read the expression as $7^2 \div 3$. To approximate $14^{-1.9}$, note that parentheses are not required when the exponent is expressed in a single decimal number. The display indicates that $7^{2/3} \approx 3.659$ and $14^{-1.9} \approx 0.007$.

Exercises

Approximate each of the following.

1. $5^{3/4}$ **2.** $8^{4/7}$ **3.** $29^{-3/8}$

4. $73^{0.56}$ **5.** $34^{-2.78}$ **6.** $32^{0.2}$

► **c Laws of Exponents**

The same laws hold for rational-number exponents as for integer exponents. We list them for review.

> For any real number a and any rational exponents m and n:
>
> **1.** $a^m \cdot a^n = a^{m+n}$ In multiplying, we add exponents if the bases are the same.
>
> **2.** $\dfrac{a^m}{a^n} = a^{m-n}$ In dividing, we subtract exponents if the bases are the same.
>
> **3.** $(a^m)^n = a^{m \cdot n}$ To raise a power to a power, we multiply the exponents.
>
> **4.** $(ab)^m = a^m b^m$ To raise a product to a power, we raise each factor to the power.
>
> **5.** $\left(\dfrac{a}{b}\right)^n = \dfrac{a^n}{b^n}$ To raise a quotient to a power, we raise both the numerator and the denominator to the power.

EXAMPLES   Use the laws of exponents to simplify.

16. $3^{1/5} \cdot 3^{3/5} = 3^{1/5+3/5} = 3^{4/5}$ **Adding exponents**

17. $\dfrac{7^{1/4}}{7^{1/2}} = 7^{1/4-1/2} = 7^{1/4-2/4} = 7^{-1/4} = \dfrac{1}{7^{1/4}}$ **Subtracting exponents**

18. $(7.2^{2/3})^{3/4} = 7.2^{2/3 \cdot 3/4} = 7.2^{6/12} = 7.2^{1/2}$ **Multiplying exponents**

19. $(a^{-1/3}b^{2/5})^{1/2} = a^{-1/3 \cdot 1/2} \cdot b^{2/5 \cdot 1/2}$ **Raising a product to a power and multiplying exponents**

$$= a^{-1/6}b^{1/5} = \dfrac{b^{1/5}}{a^{1/6}}$$ ⟶ **Now Try Exercises 41 and 47.**

► **d Simplifying Radical Expressions**

Rational exponents can be used to simplify some radical expressions. The procedure is as follows.

> **Simplifying Radical Expressions**
>
> **1.** Convert radical expressions to exponential expressions.
> **2.** Use arithmetic and the laws of exponents to simplify.
> **3.** Convert back to radical notation when appropriate.
>
> *Important:* This procedure works only when we assume that a negative number has not been raised to an even power in the radicand. With this assumption, no absolute-value signs will be needed.

EXAMPLES   Use rational exponents to simplify.

20. $\sqrt[6]{x^3} = x^{3/6}$ **Converting to an exponential expression**

$\qquad = x^{1/2}$ **Simplifying the exponent**

$\qquad = \sqrt{x}$ **Converting back to radical notation**

21. $\sqrt[6]{4} = 4^{1/6}$ **Converting to exponential notation**

$= (2^2)^{1/6}$ **Renaming 4 as 2^2**

$= 2^{2/6}$ **Using $(a^m)^n = a^{mn}$; multiplying exponents**

$= 2^{1/3}$ **Simplifying the exponent**

$= \sqrt[3]{2}$ **Converting back to radical notation**

22. $\sqrt[8]{a^2b^4} = (a^2b^4)^{1/8}$ **Converting to exponential notation**

$= a^{2/8} \cdot b^{4/8}$ **Using $(ab)^n = a^nb^n$**

$= a^{1/4} \cdot b^{1/2}$ **Simplifying the exponents**

$= a^{1/4} \cdot b^{2/4}$ **Rewriting $\frac{1}{2}$ with a denominator of 4**

$= (ab^2)^{1/4}$ **Using $a^nb^n = (ab)^n$**

$= \sqrt[4]{ab^2}$ **Converting back to radical notation**

> **Now Try Exercises 59 and 71.**

We can use properties of rational exponents to write a single radical expression for a product or a quotient.

EXAMPLE 23 Use rational exponents to write a single radical expression for $\sqrt[3]{5} \cdot \sqrt{2}$.

$\sqrt[3]{5} \cdot \sqrt{2} = 5^{1/3} \cdot 2^{1/2}$ **Converting to exponential notation**

$= 5^{2/6} \cdot 2^{3/6}$ **Rewriting so that exponents have a common denominator**

$= (5^2 \cdot 2^3)^{1/6}$ **Using $a^nb^n = (ab)^n$**

$= \sqrt[6]{5^2 \cdot 2^3}$ **Converting back to radical notation**

$= \sqrt[6]{200}$ **Multiplying under the radical**

> **Now Try Exercise 79.**

EXAMPLE 24 Write a single radical expression for $a^{1/2}b^{-1/2}c^{5/6}$.

$a^{1/2}b^{-1/2}c^{5/6} = a^{3/6}b^{-3/6}c^{5/6}$ **Rewriting so that exponents have a common denominator**

$= (a^3b^{-3}c^5)^{1/6}$ **Using $a^nb^n = (ab)^n$**

$= \sqrt[6]{a^3b^{-3}c^5}$ **Converting to radical notation**

> **Now Try Exercise 85.**

EXAMPLE 25 Write a single radical expression for $\dfrac{x^{5/6} \cdot y^{3/8}}{x^{4/9} \cdot y^{1/4}}$.

$\dfrac{x^{5/6} \cdot y^{3/8}}{x^{4/9} \cdot y^{1/4}} = x^{5/6-4/9} \cdot y^{3/8-1/4}$ **Subtracting exponents**

$= x^{15/18-8/18} \cdot y^{3/8-2/8}$ **Finding common denominators so that exponents can be subtracted**

$= x^{7/18} \cdot y^{1/8}$ **Carrying out the subtraction of exponents**

$= x^{28/72} \cdot y^{9/72}$ **Rewriting so that the exponents have a common denominator**

$= (x^{28}y^9)^{1/72}$ **Using $a^nb^n = (ab)^n$**

$= \sqrt[72]{x^{28}y^9}$ **Converting to radical notation**

> **Now Try Exercise 87.**

EXAMPLES Use rational exponents to simplify.

26. $\sqrt[6]{(5x)^3} = (5x)^{3/6}$ **Converting to exponential notation**

$= (5x)^{1/2}$ **Simplifying the exponent**

$= \sqrt{5x}$ **Converting back to radical notation**

27. $\sqrt[5]{t^{20}} = t^{20/5}$ **Converting to exponential notation**

$= t^4$ **Simplifying the exponent**

28. $\left(\sqrt[3]{pq^2c}\right)^{12} = (pq^2c)^{12/3}$ **Converting to exponential notation**

$= (pq^2c)^4$ **Simplifying the exponent**

$= p^4q^8c^4$ **Using $(ab)^n = a^nb^n$**

29. $\sqrt{\sqrt[3]{x}} = \sqrt{x^{1/3}}$ **Converting the radicand to exponential notation**

$= (x^{1/3})^{1/2}$ **Try to go directly to this step.**

$= x^{1/6}$ **Multiplying exponents**

$= \sqrt[6]{x}$ **Converting back to radical notation**

→ **Now Try Exercise 83.**

6.2 Exercise Set

▶ Reading Check

Match the expression with an equivalent expression from the columns on the right.

RC1. $\dfrac{c^2}{c^5}$ **RC2.** $c^{-2/5}$ **a)** c^{2+5} **b)** $\dfrac{1}{c^{2/5}}$

RC3. $\sqrt{c}$ **RC4.** $c^{2/5}$ **c)** $c^{1/2}$ **d)** $c^{2 \cdot 5}$

e) $c^{5/2}$ **f)** $-c^{5+2}$

RC5. $\sqrt{c^5}$ **RC6.** $(c^2)^5$ **g)** $\left(\sqrt[5]{c}\right)^2$ **h)** c^{2-5}

RC7. $-c^5 \cdot c^2$ **RC8.** $c^2 c^5$

a *Rewrite without rational exponents, and simplify, if possible.*

1. $y^{1/7}$ **2.** $x^{1/6}$

3. $8^{1/3}$ **4.** $16^{1/2}$

5. $(a^3b^3)^{1/5}$ **6.** $(x^2y^2)^{1/3}$

7. $16^{3/4}$ **8.** $4^{7/2}$

9. $49^{3/2}$ **10.** $27^{4/3}$

Rewrite with rational exponents.

11. $\sqrt{17}$ **12.** $\sqrt{x^3}$

13. $\sqrt[3]{18}$ **14.** $\sqrt[3]{23}$

15. $\sqrt[5]{xy^2z}$ **16.** $\sqrt[7]{x^3y^2z^2}$

17. $\left(\sqrt{3mn}\right)^3$ **18.** $\left(\sqrt[3]{7xy}\right)^4$

19. $\left(\sqrt[7]{8x^2y}\right)^5$ **20.** $\left(\sqrt[6]{2a^5b}\right)^7$

b *Rewrite with positive exponents, and simplify, if possible.*

21. $27^{-1/3}$ **22.** $100^{-1/2}$

23. $100^{-3/2}$ **24.** $16^{-3/4}$

25. $3x^{-1/4}$ **26.** $8y^{-1/7}$

27. $(2rs)^{-3/4}$ **28.** $(5xy)^{-5/6}$

29. $2a^{3/4}b^{-1/2}c^{2/3}$ **30.** $5x^{-2/3}y^{4/5}z$

31. $\left(\dfrac{7x}{8yz}\right)^{-3/5}$ **32.** $\left(\dfrac{2ab}{3c}\right)^{-5/6}$

33. $\dfrac{1}{x^{-2/3}}$ **34.** $\dfrac{1}{a^{-7/8}}$

35. $2^{-1/3}x^4y^{-2/7}$ **36.** $3^{-5/2}a^3b^{-7/3}$

37. $\dfrac{7x}{\sqrt[3]{z}}$ **38.** $\dfrac{6a}{\sqrt[4]{b}}$

39. $\dfrac{5a}{3c^{-1/2}}$ **40.** $\dfrac{2z}{5x^{-1/3}}$

c *Use the laws of exponents to simplify. Write the answers with positive exponents.*

41. $5^{3/4} \cdot 5^{1/8}$ **42.** $11^{2/3} \cdot 11^{1/2}$

43. $\dfrac{7^{5/8}}{7^{3/8}}$ **44.** $\dfrac{3^{5/8}}{3^{-1/8}}$

45. $\dfrac{4.9^{-1/6}}{4.9^{-2/3}}$ **46.** $\dfrac{2.3^{-3/10}}{2.3^{-1/5}}$

47. $(6^{3/8})^{2/7}$ **48.** $(3^{2/9})^{3/5}$

49. $a^{2/3} \cdot a^{5/4}$ **50.** $x^{3/4} \cdot x^{2/3}$

51. $(a^{2/3} \cdot b^{5/8})^4$ **52.** $(x^{-1/3} \cdot y^{-2/5})^{-15}$

53. $(x^{2/3})^{-3/7}$ **54.** $(a^{-3/2})^{2/9}$

55. $\left(\dfrac{x^{3/4}}{y^{1/2}}\right)^{-2/3}$ **56.** $\left(\dfrac{a^{-3/2}}{b^{-5/3}}\right)^{1/3}$

57. $(m^{-1/4} \cdot n^{-5/6})^{-12/5}$ **58.** $(x^{3/8} \cdot y^{5/2})^{4/3}$

d *Use rational exponents to simplify. Write the answer in radical notation if appropriate.*

59. $\sqrt[6]{a^2}$ **60.** $\sqrt[6]{t^4}$

61. $\sqrt[3]{x^{15}}$ **62.** $\sqrt[4]{a^{12}}$

63. $\sqrt[6]{x^{-18}}$ **64.** $\sqrt[5]{a^{-10}}$

65. $(\sqrt[3]{ab})^{15}$ **66.** $(\sqrt[7]{cd})^{14}$

67. $\sqrt[14]{128}$ **68.** $\sqrt[6]{81}$

69. $\sqrt[6]{4x^2}$ **70.** $\sqrt[3]{8y^6}$

71. $\sqrt{x^4 y^6}$ **72.** $\sqrt[4]{16x^4 y^2}$

73. $\sqrt[5]{32c^{10}d^{15}}$

Use rational exponents to write a single radical expression.

74. $\sqrt[3]{3}\sqrt{3}$ **75.** $\sqrt[3]{7} \cdot \sqrt[4]{5}$

76. $\sqrt[7]{11} \cdot \sqrt[6]{13}$ **77.** $\sqrt[4]{5} \cdot \sqrt[5]{7}$

78. $\sqrt[3]{y}\sqrt[5]{3y}$ **79.** $\sqrt{x}\sqrt[3]{2x}$

80. $(\sqrt[3]{x^2 y^5})^{12}$ **81.** $(\sqrt[5]{a^2 b^4})^{15}$

82. $\sqrt[4]{\sqrt{x}}$ **83.** $\sqrt[3]{\sqrt[6]{m}}$

84. $a^{2/3} \cdot b^{3/4}$ **85.** $x^{1/3} \cdot y^{1/4} \cdot z^{1/6}$

86. $\dfrac{x^{8/15} \cdot y^{7/5}}{x^{1/3} \cdot y^{-1/5}}$ **87.** $\left(\dfrac{c^{-4/5}d^{5/9}}{c^{3/10}d^{1/6}}\right)^3$

88. $\sqrt[3]{\sqrt[4]{xy}}$

► Skill Maintenance

Solve. **[1.6c]**

89. $|7x - 5| = 9$ **90.** $|3x| = 120$

91. $8 - |2x + 5| = -2$ **92.** $\left|\dfrac{1}{2} + x\right| = \dfrac{7}{8}$

► Synthesis

93. Use the SIMULTANEOUS mode to graph
$$y_1 = x^{1/2}, \quad y_2 = 3x^{2/5}, \quad y_3 = x^{4/7}, \quad y_4 = \tfrac{1}{5}x^{3/4}.$$
Then, looking only at coordinates, match each graph with its equation.

94. Simplify:
$$\left(\sqrt[10]{\sqrt[5]{x^{15}}}\right)^5 \left(\sqrt[5]{\sqrt[10]{x^{15}}}\right)^5.$$

6.3   Simplifying Radical Expressions

► **a**   Multiply and simplify radical expressions.

► **b**   Divide and simplify radical expressions.

► **a   Multiplying and Simplifying Radical Expressions**

Note that $\sqrt{4}\sqrt{25} = 2 \cdot 5 = 10$. Also $\sqrt{4 \cdot 25} = \sqrt{100} = 10$. Likewise,
$$\sqrt[3]{27}\sqrt[3]{8} = 3 \cdot 2 = 6 \quad \text{and} \quad \sqrt[3]{27 \cdot 8} = \sqrt[3]{216} = 6.$$

These examples suggest the following.

THE PRODUCT RULE FOR RADICALS

For any nonnegative real numbers a and b and any index k,

$$\sqrt[k]{a} \cdot \sqrt[k]{b} = \sqrt[k]{a \cdot b}, \quad \text{or} \quad a^{1/k} \cdot b^{1/k} = (ab)^{1/k}.$$

The index must be the same throughout.

(To multiply, multiply the radicands.)

EXAMPLES   Multiply.

1. $\sqrt{3} \cdot \sqrt{5} = \sqrt{3 \cdot 5} = \sqrt{15}$

2. $\sqrt{5a}\sqrt{2b} = \sqrt{5a \cdot 2b} = \sqrt{10ab}$

3. $\sqrt[3]{4}\sqrt[3]{5} = \sqrt[3]{4 \cdot 5} = \sqrt[3]{20}$

4. $\sqrt[4]{\dfrac{y}{5}} \, \sqrt[4]{\dfrac{7}{x}} = \sqrt[4]{\dfrac{y}{5} \cdot \dfrac{7}{x}} = \sqrt[4]{\dfrac{7y}{5x}}$

> **CAUTION!**   A common error is to omit the index in the answer.

Keep in mind that the product rule can be used only when the indexes are the same. When indexes differ, we can use rational exponents as we did in Examples 23 and 24 of Section 6.2.

EXAMPLE 5   Multiply: $\sqrt{5x} \cdot \sqrt[4]{3y}$.

$$\sqrt{5x} \cdot \sqrt[4]{3y} = (5x)^{1/2}(3y)^{1/4} \qquad \text{Converting to exponential notation}$$
$$= (5x)^{2/4}(3y)^{1/4} \qquad \text{Rewriting so that exponents have a common denominator}$$
$$= \left[(5x)^2(3y) \right]^{1/4} \qquad \text{Using } a^n b^n = (ab)^n$$
$$= \left[(25x^2)(3y) \right]^{1/4} \qquad \text{Squaring } 5x$$
$$= \sqrt[4]{(25x^2)(3y)} \qquad \text{Converting back to radical notation}$$
$$= \sqrt[4]{75x^2y} \qquad \text{Multiplying under the radical}$$

Now Try Exercise 41.

We can reverse the product rule to simplify a product. We simplify the root of a product by taking the root of each factor separately.

FACTORING RADICAL EXPRESSIONS

For any nonnegative real numbers a and b and any index k,

$$\sqrt[k]{ab} = \sqrt[k]{a} \cdot \sqrt[k]{b}, \quad \text{or} \quad (ab)^{1/k} = a^{1/k} \cdot b^{1/k}.$$

(Take the kth root of each factor separately.)

Compare the following:

$$\sqrt{50} = \sqrt{10 \cdot 5} = \sqrt{10}\sqrt{5};$$
$$\sqrt{50} = \sqrt{25 \cdot 2} = \sqrt{25}\sqrt{2} = 5\sqrt{2}.$$

In the second case, the radicand is written with the perfect-square factor 25. If you do not recognize any perfect-square factors, try factoring the radicand into its prime factors. For example,

$$\sqrt{50} = \sqrt{2 \cdot \underbrace{5 \cdot 5}} = 5\sqrt{2}.$$

Perfect square (a pair of the same numbers)

Square-root radical expressions in which the radicand has no perfect-square factors, such as $5\sqrt{2}$, are considered to be in simplest form. A procedure for simplifying kth roots follows.

Simplifying *k*th Roots

To simplify a radical expression by factoring:

1. Look for the largest factors of the radicand that are perfect kth powers (where k is the index).
2. Then take the kth root of the resulting factors.
3. A radical expression, with index k, is *simplified* when its radicand has no factors that are perfect kth powers.

EXAMPLES Simplify by factoring.

6. $\sqrt{50} = \sqrt{\underbrace{25} \cdot 2} = \sqrt{25} \cdot \sqrt{2} = \sqrt{5 \cdot 5} \cdot \sqrt{2} = 5\sqrt{2}$

This factor is a perfect square.

7. $\sqrt[3]{32} = \sqrt[3]{\underbrace{8} \cdot 4} = \sqrt[3]{8} \cdot \sqrt[3]{4} = \sqrt[3]{2 \cdot 2 \cdot 2} \cdot \sqrt[3]{2 \cdot 2} = 2\sqrt[3]{4}$

This factor is a perfect cube (third power).

8. $\sqrt[4]{48} = \sqrt[4]{\underbrace{16} \cdot 3} = \sqrt[4]{16} \cdot \sqrt[4]{3} = \sqrt[4]{2 \cdot 2 \cdot 2 \cdot 2} \cdot \sqrt[4]{3} = 2\sqrt[4]{3}$

This factor is a perfect fourth power.

Now Try Exercises 1 and 5.

Frequently, expressions under radicals do not contain negative numbers raised to even powers. In such cases, absolute-value notation is not necessary. **For this reason, we will no longer use absolute-value notation.**

EXAMPLES Simplify by factoring. Assume that no radicands were formed by raising negative numbers to even powers.

9. $\sqrt{5x^2} = \sqrt{5 \cdot x^2}$ Factoring the radicand

$\qquad\;\; = \sqrt{5} \cdot \sqrt{x^2}$ Factoring into two radicals

$\qquad\;\; = \sqrt{5} \cdot x$ Taking the square root of x^2

Absolute-value notation is not needed because we assume that x is not negative.

10. $\sqrt{18x^2y} = \sqrt{9 \cdot 2 \cdot x^2 \cdot y}$ **Looking for perfect-square factors and factoring the radicand**

$\qquad\qquad = \sqrt{9 \cdot x^2 \cdot 2 \cdot y}$

$\qquad\qquad = \sqrt{9} \cdot \sqrt{x^2} \cdot \sqrt{2 \cdot y}$ **Factoring into several radicals**

$\qquad\qquad = 3x\sqrt{2y}$ **Taking square roots**

11. $\sqrt{216x^5y^3} = \sqrt{36 \cdot 6 \cdot x^4 \cdot x \cdot y^2 \cdot y}$ **Looking for perfect-square factors and factoring the radicand**

$\qquad\qquad = \sqrt{36 \cdot x^4 \cdot y^2 \cdot 6 \cdot x \cdot y}$

$\qquad\qquad = \sqrt{36}\sqrt{x^4}\sqrt{y^2}\sqrt{6xy}$ **Factoring into several radicals**

$\qquad\qquad = 6x^2y\sqrt{6xy}$ **Taking square roots**

Let's look at this example another way. We do a complete factorization and look for pairs of factors. Each pair of factors makes a square:

$$\sqrt{216x^5y^3} = \sqrt{2 \cdot 2 \cdot 2 \cdot 3 \cdot 3 \cdot 3 \cdot x \cdot x \cdot x \cdot x \cdot x \cdot y \cdot y \cdot y}$$

Each pair of factors makes a perfect square.

$$= 2 \cdot 3 \cdot x \cdot x \cdot y \cdot \sqrt{2 \cdot 3 \cdot x \cdot y}$$
$$= 6x^2y\sqrt{6xy}.$$

12. $\sqrt[3]{16a^7b^{11}} = \sqrt[3]{8 \cdot 2 \cdot a^6 \cdot a \cdot b^9 \cdot b^2}$ **Factoring the radicand. The index is 3, so we look for the largest powers that are multiples of 3 because these are perfect cubes.**

$\qquad\qquad = \sqrt[3]{8} \cdot \sqrt[3]{a^6} \cdot \sqrt[3]{b^9} \cdot \sqrt[3]{2ab^2}$ **Factoring into radicals**

$\qquad\qquad = 2a^2b^3\sqrt[3]{2ab^2}$ **Taking cube roots**

Let's look at this example another way. We do a complete factorization and look for triples of factors. Each triple of factors makes a cube:

$\sqrt[3]{16a^7b^{11}}$

$= \sqrt[3]{2 \cdot 2 \cdot 2 \cdot 2 \cdot a \cdot a \cdot a \cdot a \cdot a \cdot a \cdot a \cdot b \cdot b \cdot b \cdot b \cdot b \cdot b \cdot b \cdot b \cdot b \cdot b \cdot b}$

Each triple of factors makes a cube.

$= 2 \cdot a \cdot a \cdot b \cdot b \cdot b \cdot \sqrt[3]{2 \cdot a \cdot b \cdot b}$
$= 2a^2b^3\sqrt[3]{2ab^2}.$ **Now Try Exercises 15 and 17.**

Sometimes after we have multiplied, we can simplify by factoring.

EXAMPLES Multiply and simplify. Assume that no radicands were formed by raising negative numbers to even powers.

13. $\sqrt{20}\sqrt{8} = \sqrt{20 \cdot 8} = \sqrt{4 \cdot 5 \cdot 4 \cdot 2} = 4\sqrt{10}$

14. $3\sqrt[3]{25} \cdot 2\sqrt[3]{5} = 3 \cdot 2 \cdot \sqrt[3]{25} \cdot \sqrt[3]{5} = 6 \cdot \sqrt[3]{25 \cdot 5}$

$\qquad\qquad\qquad\qquad\qquad\qquad = 6 \cdot \sqrt[3]{5 \cdot 5 \cdot 5}$

$\qquad\qquad\qquad\qquad\qquad\qquad = 6 \cdot 5 = 30$

15. $\sqrt[3]{18y^3}\sqrt[3]{4x^2} = \sqrt[3]{18y^3 \cdot 4x^2}$ **Multiplying radicands**

$\qquad\qquad = \sqrt[3]{2 \cdot 3 \cdot 3 \cdot y \cdot y \cdot y \cdot 2 \cdot 2 \cdot x \cdot x}$

$\qquad\qquad = 2 \cdot y \cdot \sqrt[3]{3 \cdot 3 \cdot x \cdot x}$

$\qquad\qquad = 2y\sqrt[3]{9x^2}$ **Now Try Exercises 23 and 33.**

▶ **b** **Dividing and Simplifying Radical Expressions**

Note that $\dfrac{\sqrt[3]{27}}{\sqrt[3]{8}} = \dfrac{3}{2}$ and that $\sqrt[3]{\dfrac{27}{8}} = \dfrac{3}{2}$. This example suggests the following.

THE QUOTIENT RULE FOR RADICALS

For any nonnegative number a, any positive number b, and any index k,

$$\frac{\sqrt[k]{a}}{\sqrt[k]{b}} = \sqrt[k]{\frac{a}{b}}, \quad \text{or} \quad \frac{a^{1/k}}{b^{1/k}} = \left(\frac{a}{b}\right)^{1/k}.$$

(To divide, divide the radicands. After doing this, you can sometimes simplify by taking roots.)

EXAMPLES Divide and simplify. Assume that no radicands were formed by raising negative numbers to even powers.

16. $\dfrac{\sqrt{80}}{\sqrt{5}} = \sqrt{\dfrac{80}{5}} = \sqrt{16} = 4$ | **We divide the radicands.** |

17. $\dfrac{5\sqrt[3]{32}}{\sqrt[3]{2}} = 5\sqrt[3]{\dfrac{32}{2}} = 5\sqrt[3]{16} = 5\sqrt[3]{8 \cdot 2} = 5\sqrt[3]{8}\sqrt[3]{2} = 5 \cdot 2\sqrt[3]{2} = 10\sqrt[3]{2}$

18. $\dfrac{\sqrt{72xy}}{2\sqrt{2}} = \dfrac{1}{2} \cdot \dfrac{\sqrt{72xy}}{\sqrt{2}} = \dfrac{1}{2}\sqrt{\dfrac{72xy}{2}} = \dfrac{1}{2}\sqrt{36xy} = \dfrac{1}{2}\sqrt{36}\sqrt{xy}$

$\qquad\qquad = \dfrac{1}{2} \cdot 6\sqrt{xy} = 3\sqrt{xy}$

       ➤ **Now Try Exercises 53 and 63.**

We can simplify the root of a quotient by taking the roots of the numerator and of the denominator separately.

---

kTH ROOTS OF QUOTIENTS

For any nonnegative number a, any positive number b, and any index k,

$$\sqrt[k]{\frac{a}{b}} = \frac{\sqrt[k]{a}}{\sqrt[k]{b}}, \quad \text{or} \quad \left(\frac{a}{b}\right)^{1/k} = \frac{a^{1/k}}{b^{1/k}}.$$

(Take the kth roots of the numerator and of the denominator separately.)

EXAMPLES Simplify by taking the roots of the numerator and the denominator. Assume that no radicands were formed by raising negative numbers to even powers.

19. $\sqrt[3]{\dfrac{27}{125}} = \dfrac{\sqrt[3]{27}}{\sqrt[3]{125}} = \dfrac{3}{5}$ | **We take the cube root of the numerator and of the denominator.** |

20. $\sqrt{\dfrac{25}{y^2}} = \dfrac{\sqrt{25}}{\sqrt{y^2}} = \dfrac{5}{y}$ | **We take the square root of the numerator and of the denominator.** |

21. $\sqrt{\dfrac{16x^3}{y^4}} = \dfrac{\sqrt{16x^3}}{\sqrt{y^4}} = \dfrac{\sqrt{16x^2 \cdot x}}{\sqrt{y^4}} = \dfrac{\sqrt{16x^2} \cdot \sqrt{x}}{\sqrt{y^4}} = \dfrac{4x\sqrt{x}}{y^2}$

22. $\sqrt[3]{\dfrac{27y^5}{343x^3}} = \dfrac{\sqrt[3]{27y^5}}{\sqrt[3]{343x^3}} = \dfrac{\sqrt[3]{27y^3 \cdot y^2}}{\sqrt[3]{343x^3}} = \dfrac{\sqrt[3]{27y^3} \cdot \sqrt[3]{y^2}}{\sqrt[3]{343x^3}} = \dfrac{3y\sqrt[3]{y^2}}{7x}$

> **Now Try Exercises 79 and 85.**

We are assuming here that no variable represents 0 or a negative number. Thus we need not be concerned about zero denominators or absolute value.

When indexes differ, we can use rational exponents.

EXAMPLE 23   Divide and simplify: $\dfrac{\sqrt[3]{a^2b^4}}{\sqrt{ab}}$.

$\dfrac{\sqrt[3]{a^2b^4}}{\sqrt{ab}} = \dfrac{(a^2b^4)^{1/3}}{(ab)^{1/2}}$ **Converting to exponential notation**

$= \dfrac{a^{2/3}b^{4/3}}{a^{1/2}b^{1/2}}$ **Using the product and power rules**

$= a^{2/3-1/2}b^{4/3-1/2}$ **Subtracting exponents**

$= a^{4/6-3/6}b^{8/6-3/6}$ **Finding common denominators so exponents can be subtracted**

$= a^{1/6}b^{5/6}$

$= (ab^5)^{1/6}$ **Using $a^n b^n = (ab)^n$**

$= \sqrt[6]{ab^5}$ **Converting back to radical notation**

> **Now Try Exercise 71.**

6.3   Exercise Set

▶ Reading Check

Determine whether each statement is true or false.

RC1. For any nonnegative real numbers a and b and any index k, $\sqrt[k]{a} \cdot \sqrt[k]{b} = \sqrt[k]{ab}$.

RC2. For $q > 0$, $\sqrt{q^2 - 100} = q + 10$.

RC3. The expression $\sqrt{Y}$ is not simplified if Y contains a factor that contains a perfect square.

RC4. For any nonnegative number a, any positive number b, and any index k, $\dfrac{\sqrt[k]{a}}{\sqrt[k]{b}} = \sqrt[k]{\dfrac{a}{b}}$.

a *Simplify by factoring. Assume that no radicands were formed by raising negative numbers to even powers.*

1. $\sqrt{24}$ **2.** $\sqrt{20}$

3. $\sqrt{90}$ **4.** $\sqrt{18}$

5. $\sqrt[3]{250}$ **6.** $\sqrt[3]{108}$

7. $\sqrt{180x^4}$ **8.** $\sqrt{175y^6}$

9. $\sqrt[3]{54x^8}$ **10.** $\sqrt[3]{40y^3}$

11. $\sqrt[3]{80t^8}$ **12.** $\sqrt[3]{108x^5}$

13. $\sqrt[4]{80}$ **14.** $\sqrt[4]{32}$

15. $\sqrt{32a^2b}$ **16.** $\sqrt{75p^3q^4}$

17. $\sqrt[4]{243x^8y^{10}}$ **18.** $\sqrt[4]{162c^4d^6}$

19. $\sqrt[5]{96x^7y^{15}}$ **20.** $\sqrt[5]{p^{14}q^9r^{23}}$

Multiply and simplify. Assume that no radicands were formed by raising negative numbers to even powers.

21. $\sqrt{10}\sqrt{5}$ **22.** $\sqrt{6}\sqrt{3}$

23. $\sqrt{15}\sqrt{6}$ **24.** $\sqrt{2}\sqrt{32}$

25. $\sqrt[3]{2}\sqrt[3]{4}$ **26.** $\sqrt[3]{9}\sqrt[3]{3}$

27. $\sqrt{45}\sqrt{60}$ **28.** $\sqrt{24}\sqrt{75}$

29. $\sqrt{3x^3}\sqrt{6x^5}$ **30.** $\sqrt{5a^7}\sqrt{15a^3}$

31. $\sqrt{5b^3}\sqrt{10c^4}$ **32.** $\sqrt{2x^3y}\sqrt{12xy}$

33. $\sqrt[3]{5a^2}\sqrt[3]{2a}$ **34.** $\sqrt[3]{7x}\sqrt[3]{3x^2}$

35. $\sqrt[3]{y^4}\sqrt[3]{16y^5}$ **36.** $\sqrt[3]{s^2t^4}\sqrt[3]{s^4t^6}$

37. $\sqrt[4]{16}\sqrt[4]{64}$ **38.** $\sqrt[5]{64}\sqrt[5]{16}$

39. $\sqrt{12a^3b}\sqrt{8a^4b^2}$ **40.** $\sqrt{30x^3y^4}\sqrt{18x^2y^5}$

41. $\sqrt{2}\sqrt[3]{5}$ **42.** $\sqrt{6}\sqrt[3]{5}$

43. $\sqrt[4]{3}\sqrt{2}$ **44.** $\sqrt[3]{5}\sqrt[4]{2}$

45. $\sqrt{a}\sqrt[4]{a^3}$ **46.** $\sqrt[3]{x^2}\sqrt[6]{x^5}$

47. $\sqrt[5]{b^2}\sqrt{b^3}$ **48.** $\sqrt[4]{a^3}\sqrt[3]{a^2}$

49. $\sqrt{xy^3}\sqrt[3]{x^2y}$ **50.** $\sqrt{y^5z}\sqrt[3]{yz^4}$

51. $\sqrt{2a^3b}\sqrt[4]{8ab^2}$ **52.** $\sqrt[4]{9ab^3}\sqrt{3a^4b}$

b *Divide and simplify. Assume that all expressions under radicals represent positive numbers.*

53. $\dfrac{\sqrt{90}}{\sqrt{5}}$ **54.** $\dfrac{\sqrt{98}}{\sqrt{2}}$

55. $\dfrac{\sqrt{35q}}{\sqrt{7q}}$ **56.** $\dfrac{\sqrt{30x}}{\sqrt{10x}}$

57. $\dfrac{\sqrt[3]{54}}{\sqrt[3]{2}}$ **58.** $\dfrac{\sqrt[3]{40}}{\sqrt[3]{5}}$

59. $\dfrac{\sqrt{56xy^3}}{\sqrt{8x}}$ **60.** $\dfrac{\sqrt{52ab^3}}{\sqrt{13a}}$

61. $\dfrac{\sqrt[3]{96a^4b^2}}{\sqrt[3]{12a^2b}}$ **62.** $\dfrac{\sqrt[3]{189x^5y^7}}{\sqrt[3]{7x^2y^2}}$

63. $\dfrac{\sqrt{128xy}}{2\sqrt{2}}$ **64.** $\dfrac{\sqrt{48ab}}{2\sqrt{3}}$

65. $\dfrac{\sqrt[4]{48x^9y^{13}}}{\sqrt[4]{3xy^5}}$ **66.** $\dfrac{\sqrt[5]{64a^{11}b^{28}}}{\sqrt[5]{2ab^2}}$

67. $\dfrac{\sqrt[3]{a}}{\sqrt{a}}$ **68.** $\dfrac{\sqrt{x}}{\sqrt[4]{x}}$

69. $\dfrac{\sqrt[3]{a^2}}{\sqrt[4]{a}}$ **70.** $\dfrac{\sqrt[3]{x^2}}{\sqrt[5]{x}}$

71. $\dfrac{\sqrt[4]{x^2y^3}}{\sqrt[3]{xy}}$ **72.** $\dfrac{\sqrt[5]{a^4b^2}}{\sqrt[3]{ab^2}}$

Simplify.

73. $\sqrt{\dfrac{25}{36}}$ **74.** $\sqrt{\dfrac{49}{64}}$

75. $\sqrt{\dfrac{16}{49}}$ **76.** $\sqrt{\dfrac{100}{81}}$

77. $\sqrt[3]{\dfrac{125}{27}}$ **78.** $\sqrt[3]{\dfrac{343}{1000}}$

79. $\sqrt{\dfrac{49}{y^2}}$ **80.** $\sqrt{\dfrac{121}{x^2}}$

81. $\sqrt{\dfrac{25y^3}{x^4}}$ **82.** $\sqrt{\dfrac{36a^5}{b^6}}$

83. $\sqrt[3]{\dfrac{81y^5}{64}}$ **84.** $\sqrt[3]{\dfrac{8z^7}{125}}$

85. $\sqrt[3]{\dfrac{27a^4}{8b^3}}$ **86.** $\sqrt[3]{\dfrac{64x^7}{216y^6}}$

87. $\sqrt[4]{\dfrac{81x^4}{16}}$ **88.** $\sqrt[4]{\dfrac{256}{81x^8}}$

89. $\sqrt[4]{\dfrac{16a^{12}}{b^4c^{16}}}$ **90.** $\sqrt[4]{\dfrac{81x^4}{y^8z^4}}$

91. $\sqrt[5]{\dfrac{32x^8}{y^{10}}}$ **92.** $\sqrt[5]{\dfrac{32b^{10}}{243a^{20}}}$

93. $\sqrt[5]{\dfrac{w^7}{z^{10}}}$ **94.** $\sqrt[5]{\dfrac{z^{11}}{w^{20}}}$

95. $\sqrt[6]{\dfrac{x^{13}}{y^6z^{12}}}$ **96.** $\sqrt[6]{\dfrac{p^9q^{24}}{r^{18}}}$

▶ # Skill Maintenance

Solve. **[4.8b]**

97. The sum of a number and its square is 90. Find the number.

98. *Triangle Dimensions.*   The base of a triangle is 2 in. longer than the height. The area is 12 in². Find the height and the base.

Solve. **[5.5a]**

99. $\dfrac{12x}{x-4} - \dfrac{3x^2}{x+4} = \dfrac{384}{x^2-16}$

100. $\dfrac{4x}{x+5} + \dfrac{20}{x} = \dfrac{100}{x^2+5x}$

▶ # Synthesis

101. *Pendulums.*   The **period** of a pendulum is the time it takes to complete one cycle, swinging to and fro. For a pendulum that is L centimeters long, the period T is given by the function

$$T(L) = 2\pi\sqrt{\dfrac{L}{980}},$$

where T is in seconds. Find, to the nearest hundredth of a second, the period of a pendulum of length **(a)** 65 cm; **(b)** 98 cm; **(c)** 120 cm. Use a calculator's π key if possible.

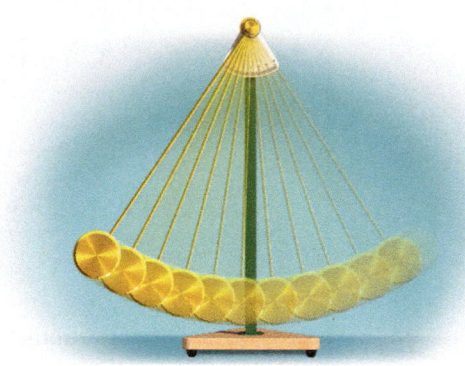

Simplify.

102. $\dfrac{\sqrt[3]{x^3 - y^3}}{\sqrt[3]{x - y}}$

103. $\dfrac{\sqrt{44x^2y^9z}\,\sqrt{22y^9z^6}}{(\sqrt{11xy^8z^2})^2}$

6.4   Addition, Subtraction, and More Multiplication

▶ **a**   Add or subtract with radical notation and simplify.

▶ **b**   Multiply expressions involving radicals in which some factors contain more than one term.

▶ ## a   Addition and Subtraction

Any two real numbers can be added. For example, the sum of 7 and $\sqrt{3}$ can be expressed as $7 + \sqrt{3}$. We cannot simplify this sum. However, when we have **like radicals** (radicals having the same index and radicand), we can use the distributive laws to simplify by collecting like radical terms. For example,

$$7\sqrt{3} + \sqrt{3} = 7\sqrt{3} + 1\sqrt{3} = (7+1)\sqrt{3} = 8\sqrt{3}.$$

EXAMPLES   Add or subtract. Simplify by collecting like radical terms, if possible.

1. $6\sqrt{7} + 4\sqrt{7} = (6+4)\sqrt{7}$ **Using a distributive law**
$\phantom{6\sqrt{7} + 4\sqrt{7}} = 10\sqrt{7}$

2. $8\sqrt[3]{2} - 7x\sqrt[3]{2} + 5\sqrt[3]{2} = (8 - 7x + 5)\sqrt[3]{2}$ **Factoring out $\sqrt[3]{2}$**

$= (13 - 7x)\sqrt[3]{2}$

> These parentheses are necessary!

3. $3\sqrt[5]{4x} + 7\sqrt[5]{4x} - \sqrt[3]{4x} = (3 + 7)\sqrt[5]{4x} - \sqrt[3]{4x}$

$= 10\sqrt[5]{4x} - \sqrt[3]{4x}$

> Note that these expressions have the same *radicand*, but they are not like radicals because they do not have the same *index*.

> **Now Try Exercises 1 and 3.**

Sometimes we need to simplify radicals by factoring.

EXAMPLES Add or subtract. Simplify by collecting like radical terms, if possible.

4. $3\sqrt{8} - 5\sqrt{2} = 3\sqrt{4 \cdot 2} - 5\sqrt{2}$ **Factoring 8**

$= 3\sqrt{4} \cdot \sqrt{2} - 5\sqrt{2}$ **Factoring $\sqrt{4 \cdot 2}$ into two radicals**

$= 3 \cdot 2\sqrt{2} - 5\sqrt{2}$ **Taking the square root of 4**

$= 6\sqrt{2} - 5\sqrt{2}$

$= (6 - 5)\sqrt{2}$ **Collecting like radical terms**

$= \sqrt{2}$

5. $5\sqrt{2} - 4\sqrt{3}$ No simplification is possible.

6. $5\sqrt[3]{16y^4} + 7\sqrt[3]{2y} = 5\sqrt[3]{8y^3 \cdot 2y} + 7\sqrt[3]{2y}$ **Factoring the first radical**

$= 5\sqrt[3]{8y^3} \cdot \sqrt[3]{2y} + 7\sqrt[3]{2y}$

$= 5 \cdot 2y \cdot \sqrt[3]{2y} + 7\sqrt[3]{2y}$ **Taking the cube root of $8y^3$**

$= 10y\sqrt[3]{2y} + 7\sqrt[3]{2y}$

$= (10y + 7)\sqrt[3]{2y}$ **Collecting like radical terms**

> **Now Try Exercises 13 and 21.**

▶ **b** **More Multiplication**

To multiply expressions in which some factors contain more than one term, we use the procedures for multiplying polynomials.

EXAMPLES Multiply.

7. $\sqrt{3}(x - \sqrt{5}) = \sqrt{3} \cdot x - \sqrt{3} \cdot \sqrt{5}$ **Using a distributive law**

$= x\sqrt{3} - \sqrt{15}$ **Multiplying radicals**

8. $\sqrt[3]{y}(\sqrt[3]{y^2} + \sqrt[3]{2}) = \sqrt[3]{y} \cdot \sqrt[3]{y^2} + \sqrt[3]{y} \cdot \sqrt[3]{2}$ **Using a distributive law**

$= \sqrt[3]{y^3} + \sqrt[3]{2y}$ **Multiplying radicals**

$= y + \sqrt[3]{2y}$ **Simplifying $\sqrt[3]{y^3}$**

> **Now Try Exercise 37.**

EXAMPLE 9 Multiply: $\left(4\sqrt{3} + \sqrt{2}\right)\left(\sqrt{3} - 5\sqrt{2}\right)$.

$$\left(4\sqrt{3} + \sqrt{2}\right)\left(\sqrt{3} - 5\sqrt{2}\right) = \overset{F}{4\left(\sqrt{3}\right)^2} - \overset{O}{20\sqrt{3}\cdot\sqrt{2}} + \overset{I}{\sqrt{2}\cdot\sqrt{3}} - \overset{L}{5\left(\sqrt{2}\right)^2}$$

$$= 4\cdot 3 - 20\sqrt{6} + \sqrt{6} - 5\cdot 2$$

$$= 12 - 20\sqrt{6} + \sqrt{6} - 10$$

$$= 2 - 19\sqrt{6} \qquad \text{Collecting like terms}$$

> **Now Try Exercise 65.**

EXAMPLE 10 Multiply: $\left(\sqrt{a} + \sqrt{3}\right)\left(\sqrt{b} + \sqrt{3}\right)$. Assume that all expressions under radicals represent nonnegative numbers.

$$\left(\sqrt{a} + \sqrt{3}\right)\left(\sqrt{b} + \sqrt{3}\right) = \sqrt{a}\sqrt{b} + \sqrt{a}\sqrt{3} + \sqrt{3}\sqrt{b} + \sqrt{3}\sqrt{3}$$

$$= \sqrt{ab} + \sqrt{3a} + \sqrt{3b} + 3$$

> **Now Try Exercise 71.**

> **THE PRODUCT OF THE SUM AND THE DIFFERENCE OF THE SAME TWO TERMS**
>
> **REVIEW SECTION 4.2**

EXAMPLE 11 Multiply: $\left(\sqrt{5} + \sqrt{7}\right)\left(\sqrt{5} - \sqrt{7}\right)$.

$$\left(\sqrt{5} + \sqrt{7}\right)\left(\sqrt{5} - \sqrt{7}\right) = \left(\sqrt{5}\right)^2 - \left(\sqrt{7}\right)^2 \qquad \text{This is now a difference of two squares:}$$

$$(A - B)(A + B) = A^2 - B^2.$$

$$= 5 - 7 = -2 \qquad \text{Now Try Exercise 51.}$$

EXAMPLE 12 Multiply: $\left(\sqrt{a} + \sqrt{b}\right)\left(\sqrt{a} - \sqrt{b}\right)$. Assume that no radicands were formed by raising negative numbers to even powers.

$$\left(\sqrt{a} + \sqrt{b}\right)\left(\sqrt{a} - \sqrt{b}\right) = \left(\sqrt{a}\right)^2 - \left(\sqrt{b}\right)^2$$

$$= a - b \qquad \boxed{\text{No radicals}}$$

> **Now Try Exercise 63.**

> Expressions of the form $\sqrt{a} + \sqrt{b}$ and $\sqrt{a} - \sqrt{b}$ are called **conjugates**. Their product is always an expression that has no radicals.

EXAMPLE 13 Multiply: $\left(\sqrt{3} + x\right)^2$.

$$\left(\sqrt{3} + x\right)^2 = \left(\sqrt{3}\right)^2 + 2\cdot\sqrt{3}\cdot x + x^2 \qquad \text{Squaring a binomial}$$

$$= 3 + 2x\sqrt{3} + x^2$$

> **Now Try Exercise 75.**

6.4 Exercise Set

▶ # Reading Check

Like radical terms have the same *index and the* same *radicand. Determine whether the given pair of terms are like radicals. Answer yes or no.*

RC1. $4\sqrt[3]{5y}, \ 2\sqrt[3]{5y}$

RC2. $5, \ 5\sqrt{2}$

RC3. $\sqrt[7]{x^2y^3}, \ \sqrt[7]{x^2y^2}$

RC4. $q\sqrt[4]{q^3}, \ 2\sqrt[4]{q^3}$

RC5. $-4\sqrt{3}, \ \sqrt{3}$

RC6. $x\sqrt[3]{y}, \ y\sqrt[3]{x}$

RC7. $3\sqrt[5]{a - b}, \ 3\sqrt[4]{a - b}$

RC8. $\dfrac{1}{4}\sqrt[3]{\dfrac{x^2}{y}}, \ 4\sqrt[3]{\dfrac{x^2}{y}}$

a *Add or subtract. Then simplify by collecting like radical terms, if possible. Assume that no radicands were formed by raising negative numbers to even powers.*

1. $7\sqrt{5} + 4\sqrt{5}$
2. $2\sqrt{3} + 9\sqrt{3}$
3. $6\sqrt[3]{7} - 5\sqrt[3]{7}$
4. $13\sqrt[5]{3} - 8\sqrt[5]{3}$
5. $4\sqrt[3]{y} + 9\sqrt[3]{y}$
6. $6\sqrt[4]{t} - 3\sqrt[4]{t}$
7. $5\sqrt{6} - 9\sqrt{6} - 4\sqrt{6}$
8. $3\sqrt{10} - 8\sqrt{10} + 7\sqrt{10}$
9. $4\sqrt[3]{3} - \sqrt{5} + 2\sqrt[3]{3} + \sqrt{5}$
10. $5\sqrt{7} - 8\sqrt[4]{11} + \sqrt{7} + 9\sqrt[4]{11}$
11. $8\sqrt{27} - 3\sqrt{3}$
12. $9\sqrt{50} - 4\sqrt{2}$
13. $8\sqrt{45} + 7\sqrt{20}$
14. $9\sqrt{12} + 16\sqrt{27}$
15. $18\sqrt{72} + 2\sqrt{98}$
16. $12\sqrt{45} - 8\sqrt{80}$
17. $3\sqrt[3]{16} + \sqrt[3]{54}$
18. $\sqrt[3]{27} - 5\sqrt[3]{8}$
19. $2\sqrt{128} - \sqrt{18} + 4\sqrt{32}$
20. $5\sqrt{50} - 2\sqrt{18} + 9\sqrt{32}$
21. $\sqrt{5a} + 2\sqrt{45a^3}$
22. $4\sqrt{3x^3} - \sqrt{12x}$
23. $\sqrt[3]{24x} - \sqrt[3]{3x^4}$
24. $\sqrt[3]{54x} - \sqrt[3]{2x^4}$
25. $7\sqrt{27x^3} + \sqrt{3x}$
26. $2\sqrt{45x^3} - \sqrt{5x}$
27. $\sqrt{4} + \sqrt{18}$
28. $\sqrt[3]{8} - \sqrt[3]{24}$
29. $5\sqrt[3]{32} - \sqrt[3]{108} + 2\sqrt[3]{256}$
30. $3\sqrt[3]{8x} - 4\sqrt[3]{27x} + 2\sqrt[3]{64x}$
31. $\sqrt[3]{6x^4} + \sqrt[3]{48x} - \sqrt[3]{6x}$
32. $\sqrt[4]{80x^5} - \sqrt[4]{405x^9} + \sqrt[4]{5x}$
33. $\sqrt{4a-4} + \sqrt{a-1}$
34. $\sqrt{9y+27} + \sqrt{y+3}$
35. $\sqrt{x^3 - x^2} + \sqrt{9x-9}$
36. $\sqrt{4x-4} + \sqrt{x^3 - x^2}$

b *Multiply. Assume that no radicands were formed by raising negative numbers to even powers.*

37. $\sqrt{5}(4 - 2\sqrt{5})$
38. $\sqrt{6}(2 + \sqrt{6})$
39. $\sqrt{3}(\sqrt{2} - \sqrt{7})$
40. $\sqrt{2}(\sqrt{5} - \sqrt{2})$
41. $\sqrt{3}(-4\sqrt{3} + 6)$
42. $\sqrt{2}(-5\sqrt{2} - 7)$

43. $\sqrt{3}(2\sqrt{5} - 3\sqrt{4})$
44. $\sqrt{2}(3\sqrt{10} - 2\sqrt{2})$
45. $\sqrt[3]{2}(\sqrt[3]{4} - 2\sqrt[3]{32})$
46. $\sqrt[3]{3}(\sqrt[3]{9} - 4\sqrt[3]{21})$
47. $3\sqrt[3]{y}(2\sqrt[3]{y^2} - 4\sqrt[3]{y})$
48. $2\sqrt[3]{y^2}(5\sqrt[3]{y} + 4\sqrt[3]{y^2})$
49. $\sqrt[3]{a}(\sqrt[3]{2a^2} + \sqrt[3]{16a^2})$
50. $\sqrt[3]{x}(\sqrt[3]{3x^2} - \sqrt[3]{81x^2})$
51. $(\sqrt{3} - \sqrt{2})(\sqrt{3} + \sqrt{2})$
52. $(\sqrt{5} + \sqrt{6})(\sqrt{5} - \sqrt{6})$
53. $(\sqrt{8} + 2\sqrt{5})(\sqrt{8} - 2\sqrt{5})$
54. $(\sqrt{18} + 3\sqrt{7})(\sqrt{18} - 3\sqrt{7})$
55. $(7 + \sqrt{5})(7 - \sqrt{5})$
56. $(4 - \sqrt{3})(4 + \sqrt{3})$
57. $(2 - \sqrt{3})(2 + \sqrt{3})$
58. $(11 - \sqrt{2})(11 + \sqrt{2})$
59. $(\sqrt{8} + \sqrt{5})(\sqrt{8} - \sqrt{5})$
60. $(\sqrt{6} - \sqrt{7})(\sqrt{6} + \sqrt{7})$
61. $(3 + 2\sqrt{7})(3 - 2\sqrt{7})$
62. $(6 - 3\sqrt{2})(6 + 3\sqrt{2})$
63. $(\sqrt{c} + \sqrt{d})(\sqrt{c} - \sqrt{d})$
64. $(\sqrt{x} - \sqrt{y})(\sqrt{x} + \sqrt{y})$
65. $(3 - \sqrt{5})(2 + \sqrt{5})$
66. $(2 + \sqrt{6})(4 - \sqrt{6})$
67. $(\sqrt{3} + 1)(2\sqrt{3} + 1)$
68. $(4\sqrt{3} + 5)(\sqrt{3} - 2)$
69. $(2\sqrt{7} - 4\sqrt{2})(3\sqrt{7} + 6\sqrt{2})$
70. $(4\sqrt{5} + 3\sqrt{3})(3\sqrt{5} - 4\sqrt{3})$
71. $(\sqrt{a} + \sqrt{2})(\sqrt{a} + \sqrt{3})$
72. $(2 - \sqrt{x})(1 - \sqrt{x})$
73. $(2\sqrt[3]{3} + \sqrt[3]{2})(\sqrt[3]{3} - 2\sqrt[3]{2})$
74. $(3\sqrt[3]{7} + \sqrt[3]{6})(2\sqrt[3]{7} - 3\sqrt[3]{6})$
75. $(2 + \sqrt{3})^2$
76. $(\sqrt{5} + 1)^2$
77. $(\sqrt[5]{9} - \sqrt[5]{3})(\sqrt[5]{8} + \sqrt[5]{27})$
78. $(\sqrt[3]{8x} - \sqrt[3]{5y})^2$

▶ ## Skill Maintenance

Multiply or divide and simplify.

79. $\dfrac{x^3 + 4x}{x^2 - 16} \div \dfrac{x^2 + 8x + 15}{x^2 + x - 20}$ **[5.1e]**

80. $\dfrac{a^2 - 4}{a} \div \dfrac{a - 2}{a + 4}$ **[5.1e]**

81. $\dfrac{a^3 + 8}{a^2 - 4} \cdot \dfrac{a^2 - 4a + 4}{a^2 - 2a + 4}$ **[5.1d]**

82. $\dfrac{y^3 - 27}{y^2 - 9} \cdot \dfrac{y^2 - 6y + 9}{y^2 + 3y + 9}$ **[5.1d]**

Simplify. **[5.4a]**

83. $\dfrac{x - \dfrac{1}{3}}{x + \dfrac{1}{4}}$

84. $\dfrac{1 - \dfrac{1}{x}}{1 - \dfrac{1}{x^2}}$

85. $\dfrac{\dfrac{1}{p} - \dfrac{1}{q}}{\dfrac{1}{p^2} - \dfrac{1}{q^2}}$

86. $\dfrac{\dfrac{1}{a} + \dfrac{1}{b}}{\dfrac{1}{a^3} + \dfrac{1}{b^3}}$

Solve. **[1.6c, d, e]**

87. $|3x + 7| = 22$

88. $|3x + 7| < 22$

89. $|3x + 7| \geq 22$

90. $|3x + 7| = |2x - 5|$

▶ ## Synthesis

91. Graph the function $f(x) = \sqrt{(x - 2)^2}$. What is the domain?

92. Use a graphing calculator to check your answers to Exercises 5, 22, and 72.

Multiply and simplify.

93. $\sqrt{9 + 3\sqrt{5}}\,\sqrt{9 - 3\sqrt{5}}$

94. $\left(\sqrt{x + 2} - \sqrt{x - 2}\right)^2$

95. $\left(\sqrt{3} + \sqrt{5} - \sqrt{6}\right)^2$

96. $\sqrt[3]{y}\left(1 - \sqrt[3]{y}\right)\left(1 + \sqrt[3]{y}\right)$

97. $\left(\sqrt[3]{9} - 2\right)\left(\sqrt[3]{9} + 4\right)$

98. $\left[\sqrt{3} + \sqrt{2} + \sqrt{1}\right]^4$

Mid-Chapter Review

Determine whether each statement is true or false.

1. Every real number has two real-number square roots. **[6.1a]**

2. If $\sqrt[3]{q}$ is negative, then q is negative. **[6.1c]**

3. $a^{m/n}$ and $a^{n/m}$ are reciprocals. **[6.2b]**

4. To multiply radicals with the same index, we multiply the radicands. **[6.3a]**

Simplify. **[6.1a]**

5. $\sqrt{81}$

6. $-\sqrt{144}$

7. $\sqrt{\dfrac{16}{25}}$

8. $\sqrt{-9}$

9. For $f(x) = \sqrt{2x + 3}$, find $f(3)$ and $f(-2)$. **[6.1a]**

10. Find the domain of $f(x) = \sqrt{4 - x}$. **[6.1a]**

Graph. **[6.1a]**

11. $f(x) = -2\sqrt{x}$

12. $g(x) = \sqrt{x + 1}$

Find each of the following. Assume that letters can represent any real number. **[6.1b, c, d]**

13. $\sqrt{36z^2}$ **14.** $\sqrt{x^2 - 8x + 16}$ **15.** $\sqrt[3]{-64}$

16. $-\sqrt[3]{27a^3}$ **17.** $\sqrt[5]{32}$ **18.** $\sqrt[10]{y^{10}}$

Rewrite without rational exponents and simplify, if possible. **[6.2a]**

19. $125^{1/3}$ **20.** $(a^3b)^{1/4}$

Rewrite with rational exponents. **[6.2a]**

21. $\sqrt[5]{16}$ **22.** $\sqrt[3]{6m^2n}$

Simplify. Write the answer with positive exponents. **[6.2b, c]**

23. $3^{1/4} \cdot 3^{-5/8}$ **24.** $\dfrac{7^{6/5}}{7^{2/5}}$

25. $(x^{3/4}y^{-2/3})^2$ **26.** $(n^{-3/5})^{5/4}$

Use rational exponents to simplify. Write the answer in radical notation. **[6.2d]**

27. $\sqrt[6]{16}$ **28.** $\left(\sqrt[10]{ab}\right)^5$

Use rational exponents to write a single radical expression. **[6.2d]**

29. $\sqrt{y}\,\sqrt[3]{y}$ **30.** $a^{2/3}b^{3/5}$

Perform the indicated operation and simplify. Assume that no radicands were formed by raising negative numbers to even powers. **[6.3a, b], [6.4a, b]**

31. $\sqrt{5}\sqrt{15}$ **32.** $\sqrt[3]{4x^2y}\,\sqrt[3]{6xy^4}$ **33.** $\dfrac{\sqrt[3]{80}}{\sqrt[3]{2}}$

34. $\sqrt{\dfrac{49a^5}{b^8}}$ **35.** $5\sqrt{7} + 6\sqrt{7}$ **36.** $3\sqrt{18x^3} - 6\sqrt{32x}$

37. $\sqrt{3}(2 - 5\sqrt{3})$ **38.** $(1 - \sqrt{x})(3 - \sqrt{x})$ **39.** $(\sqrt{m} - \sqrt{n})(\sqrt{m} + \sqrt{n})$

40. $(\sqrt{7} + 2)^2$ **41.** $(2\sqrt{3} + 3\sqrt{5})(3\sqrt{3} - 4\sqrt{5})$

Collaborative Discussion and Writing

42. Does the *n*th root of x^2 always exist? Why or why not? **[6.1a]**

43. Explain how to formulate a radical expression that can be used to define a function f with a domain of $\{x | x \le 5\}$. **[6.1a]**

44. Explain why $\sqrt[3]{x^6} = x^2$ for any value of x, but $\sqrt{x^6} = x^3$ only when $x \ge 0$. **[6.2d]**

45. Is the quotient of two irrational numbers always an irrational number? Why or why not? **[6.3b]**

6.5 More on Division of Radical Expressions

▶ **a** Rationalize the denominator of a radical expression having one term in the denominator.

▶ **b** Rationalize the denominator of a radical expression having two terms in the denominator.

▶ **a Rationalizing Denominators**

Sometimes in mathematics it is useful to find an equivalent expression without a radical in the denominator. This provides a standard notation for expressing results. The procedure for finding such an expression is called **rationalizing the denominator**. We carry this out by multiplying by 1.

EXAMPLE 1 Rationalize the denominator: $\sqrt{\dfrac{7}{3}}$.

We multiply by 1, using $\sqrt{3}/\sqrt{3}$. We do this so that the denominator of the radicand will be a perfect square.

$$\sqrt{\frac{7}{3}} = \frac{\sqrt{7}}{\sqrt{3}} \cdot \frac{\sqrt{3}}{\sqrt{3}} = \frac{\sqrt{7} \cdot \sqrt{3}}{\sqrt{3} \cdot \sqrt{3}}$$

$$= \frac{\sqrt{21}}{\sqrt{3^2}} = \frac{\sqrt{21}}{3}$$

The radicand is a perfect square.

Now Try Exercise 1.

EXAMPLE 2 Rationalize the denominator: $\sqrt[3]{\dfrac{7}{25}}$.

We first factor the denominator:

$$\sqrt[3]{\frac{7}{25}} = \sqrt[3]{\frac{7}{5 \cdot 5}}.$$

To get a perfect cube in the denominator, we consider the index 3 and the factors. We have 2 factors of 5, and we need 3 factors of 5. We achieve this by multiplying by 1, using $\sqrt[3]{5}/\sqrt[3]{5}$.

$$\sqrt[3]{\frac{7}{25}} = \frac{\sqrt[3]{7}}{\sqrt[3]{25}} = \frac{\sqrt[3]{7}}{\sqrt[3]{5 \cdot 5}} \cdot \frac{\sqrt[3]{5}}{\sqrt[3]{5}}$$

$$= \frac{\sqrt[3]{7} \cdot \sqrt[3]{5}}{\sqrt[3]{5 \cdot 5} \cdot \sqrt[3]{5}}$$

$$= \frac{\sqrt[3]{35}}{\sqrt[3]{5^3}} = \frac{\sqrt[3]{35}}{5}$$

The radicand is a perfect cube.

Now Try Exercise 7.

EXAMPLE 3 Rationalize the denominator: $\sqrt{\dfrac{2a}{5b}}$. Assume that no radicands were formed by raising negative numbers to even powers.

$$\sqrt{\frac{2a}{5b}} = \frac{\sqrt{2a}}{\sqrt{5b}} \qquad \text{\color{red}{Converting to a quotient of radicals}}$$

$$= \frac{\sqrt{2a}}{\sqrt{5b}} \cdot \frac{\sqrt{5b}}{\sqrt{5b}} \qquad \text{\color{red}{Multiplying by 1}}$$

$$= \frac{\sqrt{10ab}}{\sqrt{5^2 b^2}} \qquad \text{\color{red}{The radicand in the denominator is a perfect square.}}$$

$$= \frac{\sqrt{10ab}}{5b}$$

> **Now Try Exercise 15.**

EXAMPLE 4 Rationalize the denominator: $\dfrac{\sqrt[3]{a}}{\sqrt[3]{9x}}$.

We factor the denominator:

$$\frac{\sqrt[3]{a}}{\sqrt[3]{9x}} = \frac{\sqrt[3]{a}}{\sqrt[3]{3 \cdot 3 \cdot x}}.$$

To choose the symbol for 1, we look at $3 \cdot 3 \cdot x$. To make it a cube, we need another 3 and two more x's. Thus we multiply by 1, using $\sqrt[3]{3x^2} / \sqrt[3]{3x^2}$:

$$\frac{\sqrt[3]{a}}{\sqrt[3]{9x}} = \frac{\sqrt[3]{a}}{\sqrt[3]{3 \cdot 3 \cdot x}} \cdot \frac{\sqrt[3]{3x^2}}{\sqrt[3]{3x^2}} \qquad \text{\color{red}{Multiplying by 1}}$$

$$= \frac{\sqrt[3]{3ax^2}}{\sqrt[3]{3^3 x^3}} \qquad \text{\color{red}{The radicand in the denominator is a perfect cube.}}$$

$$= \frac{\sqrt[3]{3ax^2}}{3x}.$$

> **Now Try Exercise 11.**

EXAMPLE 5 Rationalize the denominator: $\dfrac{3x}{\sqrt[5]{2x^2 y^3}}$.

$$\frac{3x}{\sqrt[5]{2x^2 y^3}} = \frac{3x}{\sqrt[5]{2 \cdot x \cdot x \cdot y \cdot y \cdot y}}$$

$$= \frac{3x}{\sqrt[5]{2x^2 y^3}} \cdot \frac{\sqrt[5]{2^4 x^3 y^2}}{\sqrt[5]{2^4 x^3 y^2}}$$

$$= \frac{3x \sqrt[5]{16x^3 y^2}}{\sqrt[5]{2^5 x^5 y^5}} \qquad \text{\color{red}{The radicand in the denominator is a perfect fifth power.}}$$

$$= \frac{3x \sqrt[5]{16x^3 y^2}}{2xy}$$

$$= \frac{x}{x} \cdot \frac{3 \sqrt[5]{16x^3 y^2}}{2y}$$

$$= \frac{3 \sqrt[5]{16x^3 y^2}}{2y}$$

> **Now Try Exercise 19.**

▶ **b   Rationalizing When There Are Two Terms**

Certain pairs of expressions containing square roots, such as $c - \sqrt{b}$, $c + \sqrt{b}$ and $\sqrt{a} - \sqrt{b}$, $\sqrt{a} + \sqrt{b}$, are called **conjugates**. The product of such a pair of conjugates has no radicals in it. (See Example 12 of Section 6.4.) Thus when we wish to rationalize a denominator that has two terms and one or more of them involves a square-root radical, we multiply by 1 using the conjugate of the denominator to write a symbol for 1.

EXAMPLES   In each of the following, what symbol for 1 would you use to rationalize the denominator?

Expression	*Symbol for 1*

6. $\dfrac{3}{x + \sqrt{7}}$ $\dfrac{x - \sqrt{7}}{x - \sqrt{7}}$

> Change the operation sign in the denominator to obtain the conjugate. Use the conjugate for the numerator and denominator of the symbol for 1.

7. $\dfrac{\sqrt{7} + 4}{3 - 2\sqrt{5}}$ $\dfrac{3 + 2\sqrt{5}}{3 + 2\sqrt{5}}$

EXAMPLE 8   Rationalize the denominator: $\dfrac{4}{\sqrt{3} + x}$.

$$\frac{4}{\sqrt{3} + x} = \frac{4}{\sqrt{3} + x} \cdot \frac{\sqrt{3} - x}{\sqrt{3} - x}$$

$$= \frac{4\left(\sqrt{3} - x\right)}{\left(\sqrt{3} + x\right)\left(\sqrt{3} - x\right)}$$

$$= \frac{4\sqrt{3} - 4x}{3 - x^2}$$

> **Now Try Exercise 21.**

EXAMPLE 9   Rationalize the denominator: $\dfrac{4 + \sqrt{2}}{\sqrt{5} - \sqrt{2}}$.

$$\frac{4 + \sqrt{2}}{\sqrt{5} - \sqrt{2}} = \frac{4 + \sqrt{2}}{\sqrt{5} - \sqrt{2}} \cdot \frac{\sqrt{5} + \sqrt{2}}{\sqrt{5} + \sqrt{2}}$$

Multiplying by 1, using the conjugate of $\sqrt{5} - \sqrt{2}$, which is $\sqrt{5} + \sqrt{2}$

$$= \frac{\left(4 + \sqrt{2}\right)\left(\sqrt{5} + \sqrt{2}\right)}{\left(\sqrt{5} - \sqrt{2}\right)\left(\sqrt{5} + \sqrt{2}\right)}$$

Multiplying numerators and denominators

$$= \frac{4\sqrt{5} + 4\sqrt{2} + \sqrt{2}\sqrt{5} + \left(\sqrt{2}\right)^2}{\left(\sqrt{5}\right)^2 - \left(\sqrt{2}\right)^2}$$

Using $(A - B)(A + B) = A^2 - B^2$ in the denominator

$$= \frac{4\sqrt{5} + 4\sqrt{2} + \sqrt{10} + 2}{5 - 2}$$

$$= \frac{4\sqrt{5} + 4\sqrt{2} + \sqrt{10} + 2}{3}$$

> **Now Try Exercise 29.**

6.5   Exercise Set

▶ ## Reading Check

Choose from the column on the right the symbol for 1 that you would use to rationalize the denominator. Some choices may be used more than once and others may not be used.

RC1. $\dfrac{2}{x - \sqrt{3}}$ **RC2.** $\dfrac{\sqrt{5}}{\sqrt{3}}$

RC3. $\dfrac{\sqrt[4]{9x^2}}{\sqrt[4]{x^3}}$ **RC4.** $\dfrac{\sqrt[3]{3y}}{\sqrt[3]{9y^2}}$

RC5. $\dfrac{x + \sqrt{3}}{\sqrt{3}}$ **RC6.** $\dfrac{1}{3\sqrt{x}}$

a) $\dfrac{\sqrt{3x}}{\sqrt{3x}}$ b) $\dfrac{x - \sqrt{3}}{x - \sqrt{3}}$

c) $\dfrac{\sqrt{3}}{\sqrt{3}}$ d) $\dfrac{\sqrt{x}}{\sqrt{x}}$

e) $\dfrac{\sqrt[4]{x}}{\sqrt[4]{x}}$ f) $\dfrac{\sqrt{9}}{\sqrt{9}}$

g) $\dfrac{x + \sqrt{3}}{x + \sqrt{3}}$ h) $\dfrac{\sqrt[3]{3y}}{\sqrt[3]{3y}}$

---

a *Rationalize the denominator. Assume that no radicands were formed by raising negative numbers to even powers.*

1. $\sqrt{\dfrac{5}{3}}$ **2.** $\sqrt{\dfrac{8}{7}}$

3. $\sqrt{\dfrac{11}{2}}$ **4.** $\sqrt{\dfrac{17}{6}}$

5. $\dfrac{2\sqrt{3}}{7\sqrt{5}}$ **6.** $\dfrac{3\sqrt{5}}{8\sqrt{2}}$

7. $\sqrt[3]{\dfrac{16}{9}}$ **8.** $\sqrt[3]{\dfrac{1}{3}}$

9. $\dfrac{\sqrt[3]{3a}}{\sqrt[3]{5c}}$ **10.** $\dfrac{\sqrt[3]{7x}}{\sqrt[3]{3y}}$

11. $\dfrac{\sqrt[3]{2y^4}}{\sqrt[3]{6x^4}}$ **12.** $\dfrac{\sqrt[3]{3a^4}}{\sqrt[3]{7b^2}}$

13. $\dfrac{1}{\sqrt[4]{st}}$ **14.** $\dfrac{1}{\sqrt[3]{yz}}$

15. $\sqrt{\dfrac{3x}{20}}$ **16.** $\sqrt{\dfrac{7a}{32}}$

17. $\sqrt[3]{\dfrac{4}{5x^5y^2}}$ **18.** $\sqrt[3]{\dfrac{7c}{100ab^5}}$

19. $\sqrt[4]{\dfrac{1}{8x^7y^3}}$ **20.** $\dfrac{2x}{\sqrt[5]{18x^8y^6}}$

b *Rationalize the denominator. Assume that no radicands were formed by raising negative numbers to even powers.*

21. $\dfrac{9}{6 - \sqrt{10}}$ **22.** $\dfrac{3}{8 + \sqrt{5}}$

23. $\dfrac{-4\sqrt{7}}{\sqrt{5} + \sqrt{3}}$ **24.** $\dfrac{-5\sqrt{2}}{\sqrt{7} - \sqrt{5}}$

25. $\dfrac{6\sqrt{3}}{3\sqrt{2} - \sqrt{5}}$ **26.** $\dfrac{34\sqrt{5}}{2\sqrt{5} - \sqrt{3}}$

27. $\dfrac{3 + \sqrt{5}}{\sqrt{2} + \sqrt{5}}$ **28.** $\dfrac{2 + \sqrt{3}}{\sqrt{3} + \sqrt{5}}$

29. $\dfrac{\sqrt{3} - \sqrt{2}}{\sqrt{3} - \sqrt{7}}$ **30.** $\dfrac{\sqrt{5} - \sqrt{3}}{\sqrt{5} - \sqrt{2}}$

31. $\dfrac{\sqrt{5} - 2\sqrt{6}}{\sqrt{3} - 4\sqrt{5}}$ **32.** $\dfrac{\sqrt{6} - 3\sqrt{5}}{\sqrt{3} - 2\sqrt{7}}$

33. $\dfrac{2 - \sqrt{a}}{3 + \sqrt{a}}$ **34.** $\dfrac{5 + \sqrt{x}}{8 - \sqrt{x}}$

35. $\dfrac{2 + 3\sqrt{x}}{3 + 2\sqrt{x}}$ **36.** $\dfrac{5 + 2\sqrt{y}}{4 + 3\sqrt{y}}$

37. $\dfrac{5\sqrt{3} - 3\sqrt{2}}{3\sqrt{2} - 2\sqrt{3}}$ **38.** $\dfrac{7\sqrt{2} + 4\sqrt{3}}{4\sqrt{3} - 3\sqrt{2}}$

39. $\dfrac{\sqrt{x} - \sqrt{y}}{\sqrt{x} + \sqrt{y}}$ **40.** $\dfrac{\sqrt{a} + \sqrt{b}}{\sqrt{a} - \sqrt{b}}$

▶ # Skill Maintenance

Solve. [5.5a]

41. $\dfrac{1}{2} - \dfrac{1}{3} = \dfrac{5}{t}$

42. $\dfrac{5}{x - 1} + \dfrac{9}{x^2 + x + 1} = \dfrac{15}{x^3 - 1}$

Divide and simplify. [5.1e]

43. $\dfrac{1}{x^3 - y^3} \div \dfrac{1}{(x - y)(x^2 + xy + y^2)}$

44. $\dfrac{2x^2 - x - 6}{x^2 + 4x + 3} \div \dfrac{2x^2 + x - 3}{x^2 - 1}$

▶ # Synthesis

45. Use a graphing calculator to check your answers to Exercises 15 and 16.

46. Express each of the following as the product of two radical expressions.

   a) $x - 5$
   b) $x - a$

Simplify. (Hint: Rationalize the denominator.)

47. $\sqrt{a^2 - 3} - \dfrac{a^2}{\sqrt{a^2 - 3}}$

48. $\dfrac{1}{4 + \sqrt{3}} + \dfrac{1}{\sqrt{3}} + \dfrac{1}{\sqrt{3} - 4}$

6.6   Solving Radical Equations

▶ **a**   Solve radical equations with one radical term.

▶ **b**   Solve radical equations with two radical terms.

▶ **c**   Solve applied problems involving radical equations.

▶ **a**   **The Principle of Powers**

A **radical equation** has variables in one or more radicands. For example,

$$\sqrt[3]{2x} + 1 = 5 \quad \text{and} \quad \sqrt{x} + \sqrt{4x - 2} = 7$$

are radical equations. To solve such an equation, we need a new equation-solving principle. Suppose that an equation $a = b$ is true. If we square both sides, we get another true equation: $a^2 = b^2$. This can be generalized.

> **THE PRINCIPLE OF POWERS**
>
> For any natural number n, if an equation $a = b$ is true, then $a^n = b^n$ is true.

However, if an equation $a^n = b^n$ is true, it *may not* be true that $a = b$, if n is even. For example, $3^2 = (-3)^2$ is true, but $3 = -3$ is not true. Thus we *must check* the possible solutions when we solve an equation using the principle of powers.

To solve an equation with a radical term, we first isolate the radical term on one side of the equation. Then we use the principle of powers.

EXAMPLE 1   Solve: $\sqrt{x} - 3 = 4$.

We have

$$\sqrt{x} - 3 = 4$$
$$\sqrt{x} = 7 \qquad \text{\color{red}Adding to isolate the radical}$$
$$(\sqrt{x})^2 = 7^2 \qquad \text{\color{red}Using the principle of powers (squaring)}$$
$$x = 49. \qquad \color{red}\sqrt{x} \cdot \sqrt{x} = x$$

The number 49 is a possible solution. But we *must* check in order to be sure!

Check:
$$\sqrt{x} - 3 = 4$$
$$\overline{\sqrt{49} - 3 \;?\; 4}$$
$$7 - 3 \;\Big|$$
$$4 \;\Big|\quad \text{TRUE}$$

The solution is 49.

Now Try Exercise 3.

EXAMPLE 2 Solve: $\sqrt{x} = -3$.

We might note at the outset that this equation has no solution because the principal square root of a number is never negative. Let's continue as above for comparison.

$$\sqrt{x} = -3$$
$$(\sqrt{x})^2 = (-3)^2$$
$$x = 9$$

Check:
$$\sqrt{x} = -3$$
$$\overline{\sqrt{9} \;?\; -3}$$
$$3 \;\Big|\quad \text{FALSE}$$

> **CAUTION!** The principle of powers does not always give equivalent equations. For this reason, a check is a must!

The number 9 does *not* check. Thus the equation $\sqrt{x} = -3$ has no real-number solution. Note that the solution of the equation $x = 9$ is 9, but the equation $\sqrt{x} = -3$ has *no* solution. Thus the equations $x = 9$ and $\sqrt{x} = -3$ are *not* equivalent equations.

Now Try Exercise 11.

EXAMPLE 3 Solve: $x - 7 = 2\sqrt{x + 1}$.

The radical term is already isolated. We proceed with the principle of powers:

$$x - 7 = 2\sqrt{x + 1}$$
$$(x - 7)^2 = \left(2\sqrt{x + 1}\right)^2 \qquad \text{Using the principle of powers (squaring)}$$
$$(x - 7)(x - 7) = \left(2\sqrt{x + 1}\right)\left(2\sqrt{x + 1}\right)$$
$$x^2 - 14x + 49 = 2^2\left(\sqrt{x + 1}\right)^2$$
$$x^2 - 14x + 49 = 4(x + 1)$$
$$x^2 - 14x + 49 = 4x + 4$$
$$x^2 - 18x + 45 = 0$$
$$(x - 3)(x - 15) = 0 \qquad \text{Factoring}$$
$$x - 3 = 0 \quad \text{or} \quad x - 15 = 0 \qquad \text{Using the principle of zero products}$$
$$x = 3 \quad \text{or} \qquad\quad x = 15.$$

> **THE PRINCIPLE OF ZERO PRODUCTS**
>
> REVIEW SECTION 4.8

The possible solutions are 3 and 15.

Check:

For 3:
$$x - 7 = 2\sqrt{x + 1}$$
$$\overline{3 - 7 \;?\; 2\sqrt{3 + 1}}$$
$$-4 \;\Big|\; 2\sqrt{4}$$
$$\Big|\; 2(2)$$
$$\Big|\; 4 \qquad \text{FALSE}$$

For 15:
$$x - 7 = 2\sqrt{x + 1}$$
$$\overline{15 - 7 \;?\; 2\sqrt{15 + 1}}$$
$$8 \;\Big|\; 2\sqrt{16}$$
$$\Big|\; 2(4)$$
$$\Big|\; 8 \qquad \text{TRUE}$$

The number 3 does *not* check, but the number 15 does check. The solution is 15.

Now Try Exercise 21.

The number 3 in Example 3 is what is sometimes called an *extraneous solution*, but such terminology is risky to use at best because the number 3 is in *no way a solution* of the original equation.

Algebraic/Graphical Connection

We can visualize or check the solutions of a radical equation graphically. Consider the equation of Example 3: $x - 7 = 2\sqrt{x + 1}$. We can examine the solutions by graphing the equations

$$y = x - 7 \quad \text{and} \quad y = 2\sqrt{x + 1}$$

using the same set of axes. A hand-drawn graph of $y = 2\sqrt{x + 1}$ would involve approximating square roots on a calculator.

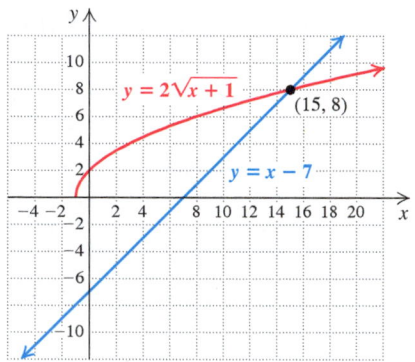

It appears from the graph that when $x = 15$, the values of $y = x - 7$ and $y = 2\sqrt{x + 1}$ are the same, 8. We can check this as we did in Example 3. Note too that the graphs *do not* intersect at $x = 3$.

Technology Connection

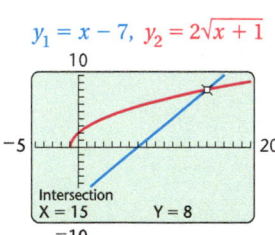

Solving Radical Equations

We can solve radical equations graphically. Consider the equation in Example 3,

$$x - 7 = 2\sqrt{x + 1}.$$

We first graph each side of the equation. We enter $y_1 = x - 7$ and $y_2 = 2\sqrt{x + 1}$ on the equation-editor screen and graph the equations using the window $[-5, 20, -10, 10]$. Note that there is one point of intersection. We use the INTERSECT feature to find its coordinates. The first coordinate, 15, is the value of x for which $y_1 = y_2$, or $x - 7 = 2\sqrt{x + 1}$. It is the solution of the equation. Note that the graph shows a single solution whereas the algebraic solution in Example 3 yields two possible solutions, 3 and 15, that must be checked. The algebraic check shows that 15 is the only solution.

Exercise

1. Solve the equations in Examples 1 and 4 graphically.

EXAMPLE 4 Solve: $x = \sqrt{x + 7} + 5$.

We have

$$x = \sqrt{x + 7} + 5$$

$$x - 5 = \sqrt{x + 7} \qquad \text{Subtracting 5 to isolate the radical term}$$

$$(x - 5)^2 = (\sqrt{x + 7})^2 \qquad \text{Using the principle of powers (squaring both sides)}$$

$$x^2 - 10x + 25 = x + 7$$

$$x^2 - 11x + 18 = 0$$

$$(x - 9)(x - 2) = 0 \qquad \text{Factoring}$$

$$x = 9 \quad or \quad x = 2. \qquad \text{Using the principle of zero products}$$

The possible solutions are 9 and 2.

Check: For 9:

$$x = \sqrt{x + 7} + 5$$

$$\overline{9 \; ? \; \sqrt{9 + 7} + 5}$$

$$\sqrt{16} + 5$$

$$4 + 5$$

$$9 \qquad \text{TRUE}$$

For 2:

$$x = \sqrt{x + 7} + 5$$

$$\overline{2 \; ? \; \sqrt{2 + 7} + 5}$$

$$\sqrt{9} + 5$$

$$3 + 5$$

$$8 \qquad \text{FALSE}$$

Since 9 checks but 2 does not, the solution is 9.   **Now Try Exercise 23.**

EXAMPLE 5 Solve: $\sqrt[3]{2x + 1} + 5 = 0$.

We have

$$\sqrt[3]{2x + 1} + 5 = 0$$

$$\sqrt[3]{2x + 1} = -5 \qquad \text{Subtracting 5. This isolates the radical term.}$$

$$(\sqrt[3]{2x + 1})^3 = (-5)^3 \qquad \text{Using the principle of powers (raising to the third power)}$$

$$2x + 1 = -125$$

$$2x = -126 \qquad \text{Subtracting 1}$$

$$x = -63.$$

Check:

$$\sqrt[3]{2x + 1} + 5 = 0$$

$$\overline{\sqrt[3]{2 \cdot (-63) + 1} + 5 \; ? \; 0}$$

$$\sqrt[3]{-126 + 1} + 5$$

$$\sqrt[3]{-125} + 5$$

$$-5 + 5$$

$$0 \qquad \text{TRUE}$$

The solution is -63.   **Now Try Exercise 17.**

▶ b Equations with Two Radical Terms

A general strategy for solving radical equations, including those with two radical terms, is as follows.

Solving Radical Equations

To solve radical equations:

1. Isolate one of the radical terms.
2. Use the principle of powers.
3. If a radical remains, perform steps (1) and (2) again.
4. Check possible solutions.

EXAMPLE 6 Solve: $\sqrt{x-3} + \sqrt{x+5} = 4$.

$$\sqrt{x-3} + \sqrt{x+5} = 4$$
$$\sqrt{x-3} = 4 - \sqrt{x+5}$$

Subtracting $\sqrt{x+5}$. This isolates one of the radical terms.

THE SQUARE OF A BINOMIAL

REVIEW SECTION 4.6

$$\left(\sqrt{x-3}\right)^2 = \left(4 - \sqrt{x+5}\right)^2$$

Using the principle of powers (squaring both sides)

$$x - 3 = 16 - 8\sqrt{x+5} + (x+5)$$

Using $(A - B)^2 = A^2 - 2AB + B^2$.

$$-3 = 21 - 8\sqrt{x+5}$$

Subtracting x and collecting like terms

$$-24 = -8\sqrt{x+5}$$

Isolating the remaining radical term

$$3 = \sqrt{x+5}$$

Dividing by -8

$$3^2 = \left(\sqrt{x+5}\right)^2$$

Squaring

$$9 = x + 5$$
$$4 = x$$

The number 4 checks and is the solution.

Now Try Exercise 31.

EXAMPLE 7 Solve: $\sqrt{2x-5} = 1 + \sqrt{x-3}$.

$$\sqrt{2x-5} = 1 + \sqrt{x-3}$$
$$\left(\sqrt{2x-5}\right)^2 = \left(1 + \sqrt{x-3}\right)^2$$

One radical is already isolated. We square both sides.

$$2x - 5 = 1 + 2\sqrt{x-3} + (x-3)$$
$$2x - 5 = 2\sqrt{x-3} + x - 2$$
$$x - 3 = 2\sqrt{x-3}$$

Isolating the remaining radical term

$$(x-3)^2 = \left(2\sqrt{x-3}\right)^2$$

Squaring both sides

$$x^2 - 6x + 9 = 4(x-3)$$
$$x^2 - 6x + 9 = 4x - 12$$
$$x^2 - 10x + 21 = 0$$
$$(x-7)(x-3) = 0$$

Factoring

$$x = 7 \quad or \quad x = 3$$

Using the principle of zero products

The possible solutions are 7 and 3. We check.

Check:

For 7:

$$\sqrt{2x - 5} = 1 + \sqrt{x - 3}$$

$$\begin{array}{c|c} \hline \sqrt{2(7) - 5} & \text{?} & 1 + \sqrt{7 - 3} \\ \sqrt{14 - 5} & & 1 + \sqrt{4} \\ \sqrt{9} & & 1 + 2 \\ 3 & & 3 \qquad \text{TRUE} \end{array}$$

For 3:

$$\sqrt{2x - 5} = 1 + \sqrt{x - 3}$$

$$\begin{array}{c|c} \hline \sqrt{2(3) - 5} & \text{?} & 1 + \sqrt{3 - 3} \\ \sqrt{6 - 5} & & 1 + \sqrt{0} \\ \sqrt{1} & & 1 + 0 \\ 1 & & 1 \qquad \text{TRUE} \end{array}$$

The numbers 7 and 3 check and are the solutions.

 **Now Try Exercise 37.**

EXAMPLE 8 Solve: $\sqrt{x + 2} - \sqrt{2x + 2} + 1 = 0$.

We first isolate one radical.

$$\sqrt{x + 2} - \sqrt{2x + 2} + 1 = 0$$

$$\sqrt{x + 2} + 1 = \sqrt{2x + 2} \qquad \text{\color{red}Adding } \sqrt{2x + 2} \text{ \color{red}to isolate a radical term}$$

$$\left(\sqrt{x + 2} + 1\right)^2 = \left(\sqrt{2x + 2}\right)^2 \qquad \text{\color{red}Squaring both sides}$$

$$x + 2 + 2\sqrt{x + 2} + 1 = 2x + 2$$

$$2\sqrt{x + 2} = x - 1$$

$$\left(2\sqrt{x + 2}\right)^2 = (x - 1)^2$$

$$4(x + 2) = x^2 - 2x + 1$$

$$4x + 8 = x^2 - 2x + 1$$

$$0 = x^2 - 6x - 7$$

$$0 = (x - 7)(x + 1) \qquad \text{\color{red}Factoring}$$

$$x - 7 = 0 \quad \text{or} \quad x + 1 = 0 \qquad \text{\color{red}Using the principle of zero products}$$

$$x = 7 \quad \text{or} \qquad x = -1$$

The possible solutions are 7 and -1.

Check: For 7:

$$\sqrt{x + 2} - \sqrt{2x + 2} + 1 = 0$$

$$\begin{array}{c|c} \hline \sqrt{7 + 2} - \sqrt{2 \cdot 7 + 2} + 1 & \text{?} & 0 \\ \sqrt{9} - \sqrt{16} + 1 & & \\ 3 - 4 + 1 & & \\ 0 & & \qquad \text{TRUE} \end{array}$$

For -1:

$$\sqrt{x + 2} - \sqrt{2x + 2} + 1 = 0$$

$$\begin{array}{c|c} \hline \sqrt{-1 + 2} - \sqrt{2 \cdot (-1) + 2} + 1 & \text{?} & 0 \\ \sqrt{1} - \sqrt{0} + 1 & & \\ 1 - 0 + 1 & & \\ 2 & & \qquad \text{FALSE} \end{array}$$

The number 7 checks, but -1 does not. The solution is 7.

 **Now Try Exercise 39.**

▶ **c Applications**

Speed of Sound.   Many applications translate to radical equations. For example, at a temperature of t degrees Fahrenheit, sound travels at a rate of S feet per second, where

$$S = 21.9\sqrt{5t + 2457}.$$

EXAMPLE 9   *Concert Series at the Capitol.*   During the annual summer concert series in Washington, D.C., military bands perform on the west steps of the Capitol. A scientific instrument at one of these concerts determined that the sound of the music was traveling at a rate of 1170 ft/sec. What was the air temperature at the concert?

We substitute 1170 for S in the formula $S = 21.9\sqrt{5t + 2457}$:

$$1170 = 21.9\sqrt{5t + 2457}.$$

Then we solve the equation for t:

$$
\begin{aligned}
1170 &= 21.9\sqrt{5t + 2457} \\
\frac{1170}{21.9} &= \sqrt{5t + 2457} && \text{\color{red}\textbf{Dividing by 21.9}} \\
\left(\frac{1170}{21.9}\right)^2 &= \left(\sqrt{5t + 2457}\right)^2 && \text{\color{red}\textbf{Squaring both sides}} \\
2854.2 &\approx 5t + 2457 && \text{\color{red}\textbf{Simplifying}} \\
397.2 &\approx 5t && \text{\color{red}\textbf{Subtracting 2457}} \\
79 &\approx t. && \text{\color{red}\textbf{Dividing by 5}}
\end{aligned}
$$

The temperature at the concert was about 79°F.

Now Try Exercise 53.

6.6 Exercise Set

▶ **Reading Check**

Choose from the column on the right the term that best completes each statement. Not every word will be used.

RC1. The equation $\sqrt{4 - 11x} = 3$ is a(n) _____ equation.

RC2. When we square both sides of an equation, we are using the principle of _____.

RC3. To solve an equation with a radical term, we first _____ the radical term on one side of the equation.

RC4. A radical equation has variables in one or more _____.

RC5. A check is essential when we raise both sides of an equation to a(n) _____ power.

even
radical
isolate
odd
radicands
square roots
powers
raise
rational
principle

a *Solve.*

1. $\sqrt{2x - 3} = 4$

2. $\sqrt{5x + 2} = 7$

3. $\sqrt{6x + 1} = 8$

4. $\sqrt{3x - 4} = 6$

5. $\sqrt{y + 7} - 4 = 4$

6. $\sqrt{x - 1} - 3 = 9$

7. $\sqrt{5y + 8} = 10$

8. $\sqrt{2y + 9} = 5$

9. $\sqrt[3]{x} = -1$

10. $\sqrt[3]{y} = -2$

11. $\sqrt{x + 2} = -4$

12. $\sqrt{y - 3} = -2$

13. $\sqrt[3]{x + 5} = 2$

14. $\sqrt[3]{x - 2} = 3$

15. $\sqrt[4]{y - 3} = 2$

16. $\sqrt[4]{x + 3} = 3$

17. $\sqrt[3]{6x + 9} + 8 = 5$

18. $\sqrt[3]{3y + 6} + 2 = 3$

19. $8 = \dfrac{1}{\sqrt{x}}$

20. $\dfrac{1}{\sqrt{y}} = 3$

21. $x - 7 = \sqrt{x - 5}$

22. $x - 5 = \sqrt{x + 7}$

23. $2\sqrt{x + 1} + 7 = x$

24. $\sqrt{2x + 7} - 2 = x$

25. $3\sqrt{x - 1} - 1 = x$

26. $x - 1 = \sqrt{x + 5}$

27. $x - 3 = \sqrt{27 - 3x}$

28. $x - 1 = \sqrt{1 - x}$

b *Solve.*

29. $\sqrt{3y + 1} = \sqrt{2y + 6}$

30. $\sqrt{5x - 3} = \sqrt{2x + 3}$

31. $\sqrt{y - 5} + \sqrt{y} = 5$

32. $\sqrt{x - 9} + \sqrt{x} = 1$

33. $3 + \sqrt{z - 6} = \sqrt{z + 9}$

34. $\sqrt{4x - 3} = 2 + \sqrt{2x - 5}$

35. $\sqrt{20 - x} + 8 = \sqrt{9 - x} + 11$

36. $4 + \sqrt{10 - x} = 6 + \sqrt{4 - x}$

37. $\sqrt{4y + 1} - \sqrt{y - 2} = 3$

38. $\sqrt{y + 15} - \sqrt{2y + 7} = 1$

39. $\sqrt{x + 2} + \sqrt{3x + 4} = 2$

40. $\sqrt{6x + 7} - \sqrt{3x + 3} = 1$

41. $\sqrt{3x - 5} + \sqrt{2x + 3} + 1 = 0$

42. $\sqrt{2m - 3} + 2 - \sqrt{m + 7} = 0$

43. $2\sqrt{t - 1} - \sqrt{3t - 1} = 0$

44. $3\sqrt{2y + 3} - \sqrt{y + 10} = 0$

c *Solve.*

Sighting to the Horizon. *How far can you see to the horizon from a given height? The function*

$$D = 1.2\sqrt{h}$$

can be used to approximate the distance D, in miles, that a person can see to the horizon from a height h, in feet.

45. An observation deck near the top of the Willis Tower in Chicago is 1353 ft high. How far can a tourist see to the horizon from this deck?

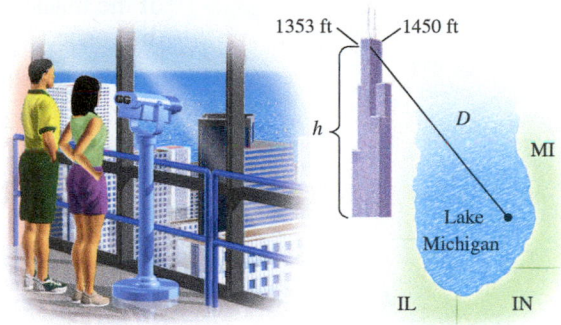

46. The roof of the Willis Tower is 1450 ft high. How far can a worker see to the horizon from the top of the Willis Tower?

47. Sarah can see 31.3 mi to the horizon from the top of a cliff. What is the height of Sarah's eyes?

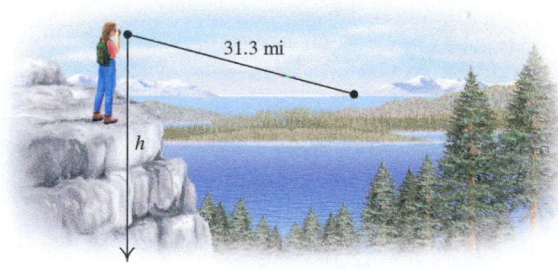

48. A steeplejack can see 13 mi to the horizon from the top of a building. What is the height of the steeplejack's eyes?

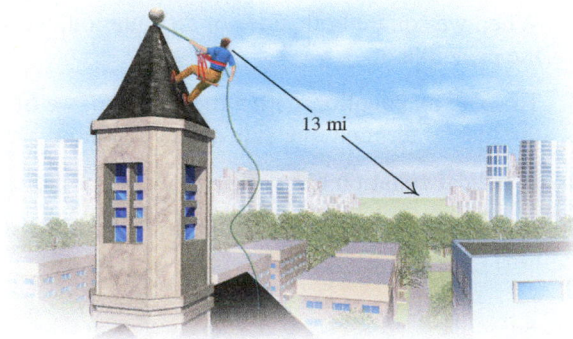

49. A technician can see 30.4 mi to the horizon from the top of a radio tower. How high is the tower?

50. A person can see 230 mi to the horizon from an airplane window. How high is the airplane?

Speed of a Skidding Car. *After an accident, how do police determine the speed at which the car had been traveling? The formula*

$$r = 2\sqrt{5L}$$

can be used to approximate the speed r, in miles per hour, of a car that has left a skid mark of length L, in feet. Use this formula for Exercises 51 and 52.

51. How far will a car skid at 55 mph? at 75 mph?

52. How far will a car skid at 65 mph? at 100 mph?

Temperature and the Speed of Sound. *Solve Exercises 53 and 54 using the formula* $S = 21.9\sqrt{5t + 2457}$ *from Example 9.*

53. At a recent concert by Carrie Underwood, sound traveled at a rate of 1176 ft/sec. What was the temperature at the time?

54. During blasting for avalanche control in Utah's Wasatch Mountains, sound traveled at a rate of 1113 ft/sec. What was the temperature at the time?

Period of a Swinging Pendulum. *The formula* $T = 2\pi\sqrt{L/32}$ *can be used to find the period T, in seconds, of a pendulum of length L, in feet.*

55. What is the length of a pendulum that has a period of 1.0 sec? Use 3.14 for π.

56. What is the length of a pendulum that has a period of 2.0 sec? Use 3.14 for π.

57. The pendulum in Jean's grandfather clock has a period of 2.2 sec. Find the length of the pendulum. Use 3.14 for π.

58. A playground swing has a period of 3.1 sec. Find the length of the swing's chain. Use 3.14 for π.

▶ **Skill Maintenance**

Solve. **[5.6a]**

59. *Painting a Room.* Julia can paint a room in 8 hr. George can paint the same room in 10 hr. How long will it take them, working together, to paint the same room?

60. *Delivering Leaflets.* Jeff can distribute leaflets to homes three times as fast as Grace can. If they work together, it takes them 1 hr to complete the job. How long would it take each to deliver the leaflets alone?

Solve. **[5.6b]**

61. *Bicycle Travel.* A cyclist traveled 702 mi in 14 days. At this same rate, how far would the cyclist have traveled in 56 days?

62. *Earnings.* Dharma earned $696.64 working for 56 hr at a fruit stand. How many hours must she work in order to earn $1044.96?

Solve. [4.8a]

63. $x^2 + 2.8x = 0$

64. $3x^2 - 5x = 0$

65. $x^2 - 64 = 0$

66. $2x^2 = x + 21$

For each of the following functions, find and simplify
$f(a + h) - f(a)$. [4.2e]

67. $f(x) = x^2$

68. $f(x) = x^2 - x$

69. $f(x) = 2x^2 - 3x$

70. $f(x) = 2x^2 + 3x - 7$

▶ Synthesis

71. 〰 Use a graphing calculator to check your answers to Exercises 4, 9, 33, and 38.

72. 〰 Use a graphing calculator to solve
$\sqrt{2x + 1} + \sqrt{5x - 4} = \sqrt{10x + 9}$.

Solve.

73. $\sqrt{\sqrt{y + 49} - \sqrt{y}} = \sqrt{7}$

74. $\sqrt[3]{x^2 + x + 15} - 3 = 0$

75. $\sqrt{\sqrt{x^2 + 9x + 34}} = 2$

76. $6\sqrt{y} + 6y^{-1/2} = 37$

77. $\sqrt{x - 2} - \sqrt{x + 2} + 2 = 0$

78. $\sqrt{\sqrt{x} + 4} = \sqrt{x} - 2$

79. $\sqrt{a^2 + 30a} = a + \sqrt{5a}$

80. $\sqrt{x + 1} - \dfrac{2}{\sqrt{x + 1}} = 1$

81. $\dfrac{x - 1}{\sqrt{x^2 + 3x + 6}} = \dfrac{1}{4}$

82. $2\sqrt{x - 1} - \sqrt{3x - 5} = \sqrt{x - 9}$

83. $\sqrt{y + 1} - \sqrt{2y - 5} = \sqrt{y - 2}$

84. Evaluate: $\sqrt{7 + 4\sqrt{3}} - \sqrt{7 - 4\sqrt{3}}$.

6.7　Applications Involving Powers and Roots

▶ **a**　Solve applied problems involving the Pythagorean theorem and powers and roots.

▶ a　Applications

There are many kinds of applied problems that involve powers and roots. Many also make use of right triangles and the Pythagorean theorem: $a^2 + b^2 = c^2$.

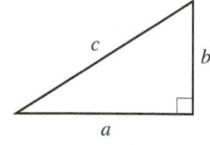

EXAMPLE 1　*Computer Screen Size.*　The viewable image size of a widescreen computer measures 21 in. diagonally and has a width of 18.3 in. What is its height?

Using the Pythagorean theorem, $a^2 + b^2 = c^2$, we substitute 18.3 for b and 21 for c and then solve for a:

$$a^2 + b^2 = c^2$$
$$a^2 + 18.3^2 = 21^2 \quad \textbf{Substituting}$$
$$a^2 + 334.89 = 441$$
$$a^2 = 106.11$$
$$a = \sqrt{106.11}$$
$$a \approx 10.3.$$

> We consider only the positive root since length cannot be negative.

The exact answer is $\sqrt{106.11}$. This is approximately equal to 10.3. Thus the height of the viewable image is about 10.3 in.

Now Try Exercise 23.

EXAMPLE 2   Find the length of the hypotenuse of this right triangle. Give an exact answer and an approximation to three decimal places.

$$7^2 + 4^2 = c^2 \quad \textbf{Substituting}$$
$$49 + 16 = c^2$$
$$65 = c^2$$

Exact answer: $c = \sqrt{65}$

Approximation: $c \approx 8.062$ **Using a calculator**

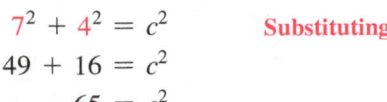

> **Now Try Exercise 1.**

EXAMPLE 3   Find the missing length b in this right triangle. Give an exact answer and an approximation to three decimal places.

$$1^2 + b^2 = (\sqrt{11})^2 \quad \textbf{Substituting}$$
$$1 + b^2 = 11$$
$$b^2 = 10$$

Exact answer: $b = \sqrt{10}$

Approximation: $b \approx 3.162$ **Using a calculator**

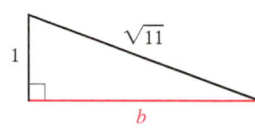

> **Now Try Exercise 7.**

EXAMPLE 4   *Construction.*   Darla is laying out the footing of a house. To see if the corner is square, she measures 16 ft from the corner along one wall and 12 ft from the corner along the other wall. How long should the diagonal be between those two points if the corner is a right angle?

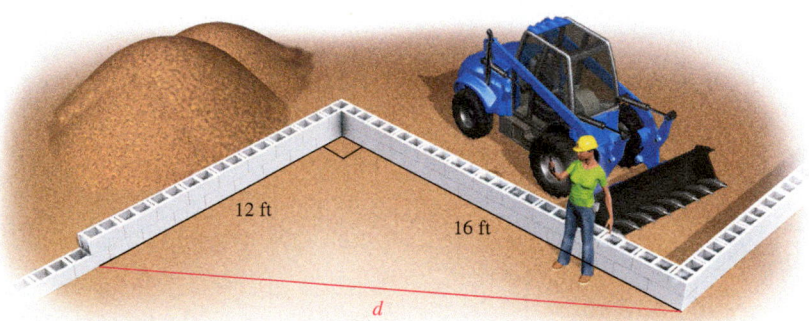

We make a drawing and let d = the length of the diagonal. It is the length of the hypotenuse of a right triangle whose legs are 12 ft and 16 ft. We substitute these values in the Pythagorean theorem to find d:

$$d^2 = 12^2 + 16^2$$
$$d^2 = 144 + 256$$
$$d^2 = 400$$
$$d = \sqrt{400} = 20.$$

The length of the diagonal should be 20 ft.

> **Now Try Exercise 15.**

EXAMPLE 5 *Road-Pavement Messages.* In a psychological study, it was determined that the ideal length L of the letters of a word painted on pavement is given by

$$L = \frac{0.000169 d^{2.27}}{h},$$

where d is the distance of a car from the lettering and h is the height of the eye above the road. All units are in feet. For a person h feet above the road, a message d feet away will be the most readable if the length of the letters is L. Find L, given that $h = 4$ ft and $d = 180$ ft.

We substitute 4 for h and 180 for d and calculate L using a calculator with an exponentiation key $\boxed{y^x}$, or ⬛:

$$L = \frac{0.000169(180)^{2.27}}{4} \approx 5.6 \text{ ft.}$$

Now Try Exercise 13.

Translating for Success

1. *Angles of a Triangle.* The second angle of a triangle is four times as large as the first. The third is 27° less than the sum of the other angles. Find the measures of the angles.

2. *Lengths of a Rectangle.* The area of a rectangle is 180 ft². The length is 26 ft greater than the width. Find the length and the width.

3. *Boat Travel.* The speed of a river is 3 mph. A boat can go 72 mi upstream and 24 mi downstream in a total time of 16 hr. Find the speed of the boat in still water.

4. *Coin Mixture.* A collection of nickels and quarters is worth $13.85. There are 85 coins in all. How many of each coin are there?

5. *Perimeter.* The perimeter of a rectangle is 180 ft. The length is 26 ft greater than the width. Find the length and the width.

Translate each word problem to an equation or a system of equations and select a correct translation from equations A–O.

A. $12^2 + 12^2 = x^2$

B. $x(x + 26) = 180$

C. $10{,}311 + 5\%x = x$

D. $x + y = 85,$
$5x + 25y = 13.85$

E. $x^2 + 4^2 = 12^2$

F. $\dfrac{240}{x - 18} = \dfrac{384}{x}$

G. $x + 5\%x = 10{,}311$

H. $\dfrac{x}{65} + 1 = \dfrac{x}{85}$

I. $\dfrac{x}{65} + \dfrac{x}{85} = 1$

J. $x + y + z = 180,$
$y = 4x,$
$z = x + y - 27$

K. $2x + 2(x + 26) = 180$

L. $\dfrac{384}{x - 18} = \dfrac{240}{x}$

M. $x + y = 85,$
$0.05x + 0.25y = 13.85$

N. $2x + 2(x + 24) = 240$

O. $\dfrac{72}{x - 3} + \dfrac{24}{x + 3} = 16$

6. *Shoveling Time.* It takes Marv 65 min to shovel 4 in. of snow from his driveway. It takes Elaine 85 min to do the same job. How long would it take if they worked together?

7. *Money Borrowed.* Claire borrows some money at 5% simple interest. After 1 year, $10,311 pays off her loan. How much did she originally borrow?

8. *Plank Height.* A 12-ft plank is leaning against a shed. The bottom of the plank is 4 ft from the building. How high up the side of the shed is the top of the plank?

9. *Train Speeds.* The speed of train A is 18 mph slower than the speed of train B. Train A travels 240 mi in the same time that it takes train B to travel 384 mi. Find the speed of train A.

10. *Diagonal of a Square.* Find the length of a diagonal of a square swimming pool whose sides are 12 ft long.

6.7	**Exercise Set**

▶ **Reading Check**

For each right triangle, choose from the column on the right the equation that can be used to find the missing side.

RC1.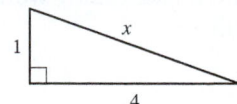

RC2.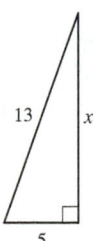

a) $12^2 + x^2 = 5^2$

b) $12^2 + 13^2 = x^2$

c) $5^2 + x^2 = 13^2$

d) $x^2 + 4^2 = \left(\sqrt{17}\right)^2$

e) $1^2 + 4^2 = x^2$

f) $5^2 + 13^2 = x^2$

g) $5^2 + 12^2 = x^2$

h) $\left(\sqrt{17}\right)^2 + x^2 = 4^2$

RC3.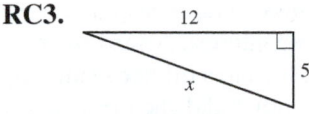

RC4.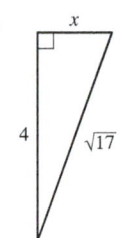

a *In a right triangle, find the length of the side not given. Give an exact answer and, where appropriate, an approximation to three decimal places.*

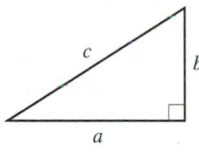

1. $a = 3$, $b = 5$ 2. $a = 8$, $b = 10$

3. $a = 15$, $b = 15$ 4. $a = 8$, $b = 8$

5. $b = 12$, $c = 13$ 6. $a = 5$, $c = 12$

7. $c = 7$, $a = \sqrt{6}$ 8. $c = 10$, $a = 4\sqrt{5}$

9. $b = 1$, $c = \sqrt{13}$ 10. $a = 1$, $c = \sqrt{12}$

11. $a = 1$, $c = \sqrt{n}$ 12. $c = 2$, $a = \sqrt{n}$

In the following problems, give an exact answer and, where appropriate, an approximation to three decimal places.

13. *Road-Pavement Messages.* Using the formula of Example 5, find the length L of a road-pavement message when $h = 4$ ft and $d = 200$ ft.

14. *Road-Pavement Messages.* Using the formula of Example 5, find the length L of a road-pavement message when $h = 8$ ft and $d = 300$ ft.

15. *Guy Wire.* How long is a guy wire reaching from the top of a 10-ft pole to a point on the ground 4 ft from the pole?

16. *Softball Diamond.* A slow-pitch softball diamond is actually a square 65 ft on a side. How far is it from home to second base?

17. *Pyramide du Louvre.* A large glass and metal pyramid designed by I. M. Pei attracts visitors to the entrance of the Louvre Museum in Paris, France. If

a plumb line from the highest point of the pyramid to the center of the base is 71 ft long and each side of the four equilateral triangles measures 100 ft, how far is it from the center of the base to a corner of the base? (*Source*: www.glassonweb.com)

18. *Central Park.*   New York City's rectangular Central Park in Manhattan runs 13,725 ft from 59th Street to 110th Street. A diagonal of the park is 13,977 ft. Find the width of the park.

19. *Bridge Expansion.*   During the summer heat, a 2-mi bridge expands 2 ft in length. If we assume that the bulge occurs straight up the middle, how high is the bulge? (The answer may surprise you. In reality, bridges are built with expansion spaces to avoid such buckling.)

20. *Triangle Areas.*   Triangle *ABC* has sides of lengths 25 ft, 25 ft, and 30 ft. Triangle *PQR* has sides of lengths 25 ft, 25 ft, and 40 ft. Which triangle has the greater area and by how much?

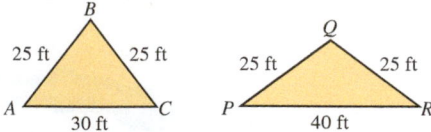

21. Find all ordered pairs on the *x*-axis of a Cartesian coordinate system that are 5 units from the point $(0, 4)$.

22. Find all ordered pairs on the *y*-axis of a Cartesian coordinate system that are 5 units from the point $(3, 0)$.

23. *Speaker Placement.*   A stereo receiver is in a corner of a 12-ft by 14-ft room. Speaker wire will run under the floor, diagonally, to a speaker in the far corner. If 4 ft of slack is required on each end, how long should the piece of wire be?

24. *Distance Over Water.*   To determine the distance between two points on opposite sides of a pond, a surveyor locates two stakes at either end of the pond and uses instrumentation to place a third stake so that the distance across the pond is the length of a hypotenuse. If the third stake is 90 m from one stake and 70 m from the other, how wide is the pond?

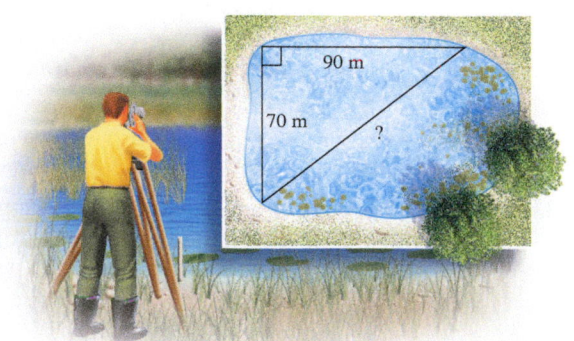

25. *Plumbing.* Plumbers use the Pythagorean theorem to calculate pipe length. If a pipe is to be offset, as shown in the figure, the *travel*, or length, of the pipe, is calculated using the lengths of the *advance* and the *offset*. Find the travel if the offset is 17.75 in. and the advance is 10.25 in.

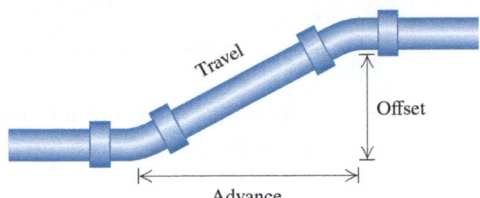

26. *Ramps for the Disabled.* Laws regarding access ramps for the disabled state that a ramp must be in the form of a right triangle, where every vertical length (leg) of 1 ft has a horizontal length (leg) of 12 ft. What is the length of a ramp with a 12-ft horizontal leg and a 1-ft vertical leg?

27. The length and the width of a rectangle are given by consecutive integers. The area of the rectangle is 90 cm^2. Find the length of a diagonal of the rectangle.

28. The diagonal of a square has length $8\sqrt{2}$ ft. Find the length of a side of the square.

29. Each side of a regular octagon has length s. Find a formula for the distance d between the parallel sides of the octagon.

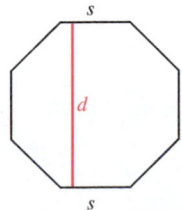

30. The two equal sides of an isosceles right triangle are of length s. Find a formula for the length of the hypotenuse.

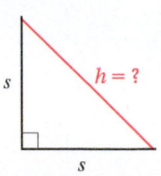

► **Skill Maintenance**

Solve. **[5.6c]**

31. *Commuter Travel.* The speed of the Zionsville Flash commuter train is 14 mph faster than that of the Carmel Crawler. The Flash travels 290 mi in the same time that it takes the Crawler to travel 230 mi. Find the speed of each train.

32. *Marine Travel.* A motor boat travels three times as fast as the current in the Saskatee River. A trip up the river and back takes 10 hr, and the total distance of the trip is 100 mi. Find the speed of the current.

Solve.

33. $2x^2 + 11x - 21 = 0$ **[4.8a]**

34. $x^2 + 24 = 11x$ **[4.8a]**

35. $\dfrac{x+2}{x+3} = \dfrac{x-4}{x-5}$ **[5.5a]**

36. $3x^2 - 12 = 0$ **[4.8a]**

37. $\dfrac{x-5}{x-7} = \dfrac{4}{3}$ **[5.5a]**

38. $\dfrac{x-1}{x-3} = \dfrac{6}{x-3}$ **[5.5a]**

► **Synthesis**

39. *Roofing.* Kit's cottage, which is 24 ft wide and 32 ft long, needs a new roof. By counting clapboards that are 4 in. apart, Kit determines that the peak of the roof is 6 ft higher than the sides. If one packet of shingles covers $33\frac{1}{3}$ sq ft, how many packets will the job require?

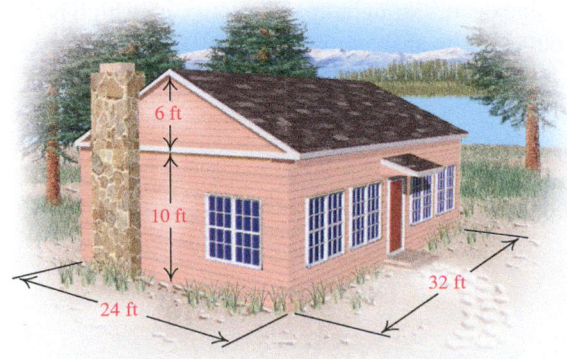

40. *Windchill Temperature.* Because wind enhances the loss of heat from the skin, we feel colder when there is wind than when there is not. The *windchill temperature* is what the temperature would have to be with no wind in order to give the same chilling effect as with the wind.

A formula for finding the windchill temperature, T_W, is

$$T_W = 35.74 + 0.6215T - 35.75V^{0.16} + 0.4275TV^{0.16},$$

where T is the actual temperature given by a thermometer, in degrees Fahrenheit, and V is the wind speed, in miles per hour. This formula can be used only when the wind speed is *above* 3 mph. Use a calculator to find the windchill temperature in each

case. Round to the nearest degree. (*Source*: National Weather Service)

a) $T = 40°F$, $V = 25$ mph
b) $T = 20°F$, $V = 25$ mph
c) $T = 10°F$, $V = 20$ mph
d) $T = 10°F$, $V = 40$ mph
e) $T = -5°F$, $V = 35$ mph
f) $T = -15°F$, $V = 35$ mph

6.8 Increasing, Decreasing, and Piecewise Functions; Applications

▶ **a** Determine the intervals on which a function is increasing, decreasing, or constant from its graph.

▶ **b** Determine relative maxima and minima of a function from its graph.

▶ **c** Given an application, find a function that models the application. Find the domain of the function and function values.

▶ **d** Graph functions defined piecewise.

Because functions occur in so many real-world situations, it is important to be able to analyze them carefully.

▶ a Increasing, Decreasing, and Constant Functions

On a given interval, if the graph of a function rises from left to right, it is said to be **increasing** on that interval. If the graph drops from left to right, it is said to be **decreasing**. If the function values stay the same on the interval, the function is said to be **constant**.

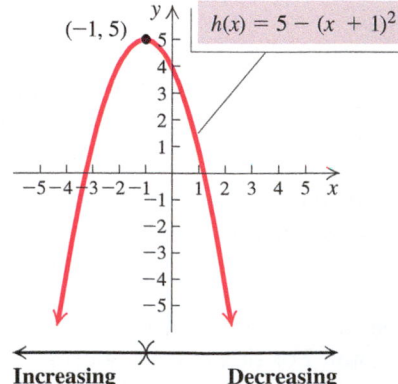

Increasing   Decreasing

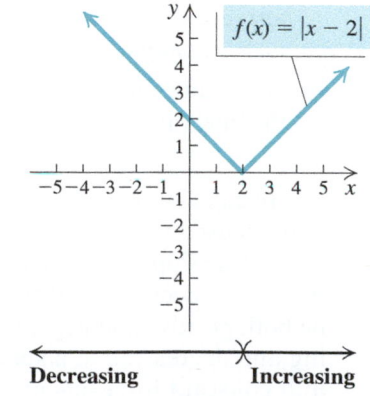

Decreasing   Increasing

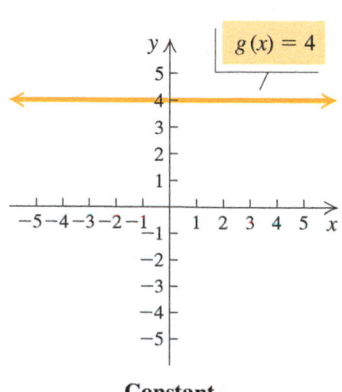

Constant

We are led to the following definitions.

INCREASING, DECREASING, AND CONSTANT FUNCTIONS

A function f is said to be **increasing** on an *open* interval I, if for all a and b in that interval, $a < b$ implies $f(a) < f(b)$. (See Fig. 1 below.)

A function f is said to be **decreasing** on an *open* interval I, if for all a and b in that interval, $a < b$ implies $f(a) > f(b)$. (See Fig. 2.)

A function f is said to be **constant** on an *open* interval I, if for all a and b in that interval, $f(a) = f(b)$. (See Fig. 3.)

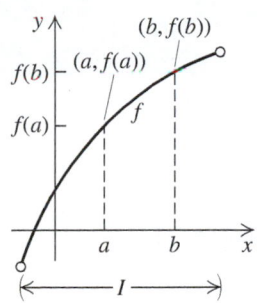

For $a < b$ in I, $f(a) < f(b)$; f is *increasing* on I.

Figure 1.

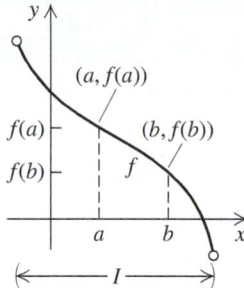

For $a < b$ in I, $f(a) > f(b)$; f is *decreasing* on I.

Figure 2.

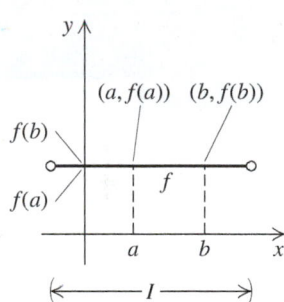

For all a and b in I, $f(a) = f(b)$; f is *constant* on I.

Figure 3.

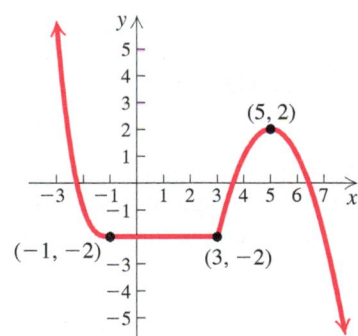

EXAMPLE 1 Determine the intervals on which the function in the figure at left is **(a)** increasing; **(b)** decreasing; **(c)** constant.

Solution When expressing interval(s) on which a function is increasing, decreasing, or constant, we consider only values in the *domain* of the function. Since the domain of this function is $(-\infty, \infty)$, we consider all real values of x.

a) As x-values (that is, values in the domain) increase from $x = 3$ to $x = 5$, the y-values (that is, values in the range) increase from -2 to 2. Thus the function is increasing on the interval $(3, 5)$.

b) As x-values increase from negative infinity to -1, y-values decrease; y-values also decrease as x-values increase from 5 to positive infinity. Thus the function is decreasing on the intervals $(-\infty, -1)$ and $(5, \infty)$.

c) As x-values increase from -1 to 3, y remains -2. The function is constant on the interval $(-1, 3)$.

> **Now Try Exercise 5.**

In calculus, the slope of a line tangent to the graph of a function at a particular point is used to determine whether the function is increasing, decreasing, or neither. If the slope is positive, the function is increasing; if the slope is negative, the function is decreasing; if the slope is 0, the function is constant. Since slope cannot be both positive and negative at the same point, a function cannot be both increasing and decreasing at a specific point. For this reason, **increasing, decreasing, and constant intervals are expressed in *open-interval* notation**. In Example 1, if $[3, 5]$ had been used for the increasing interval and $[5, \infty)$ for a decreasing interval, the function would be both increasing and decreasing at $x = 5$. This is not possible.

▶ b Relative Maximum and Minimum Values

Consider the graph shown below. Note the "peaks" and "valleys" at the x-values c_1, c_2, and c_3. The function value $f(c_2)$ is called a **relative maximum** (plural, **maxima**). Each of the function values $f(c_1)$ and $f(c_3)$ is called a **relative minimum** (plural, **minima**).

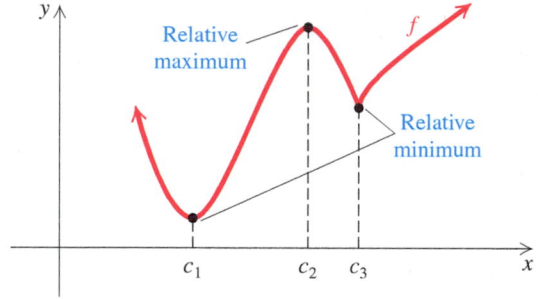

RELATIVE MAXIMA AND MINIMA

Suppose that f is a function for which $f(c)$ exists for some c in the domain of f. Then:

$f(c)$ is a **relative maximum** if there exists an *open* interval I containing c such that $f(c) > f(x)$, for all x in I where $x \neq c$; and

$f(c)$ is a **relative minimum** if there exists an *open* interval I containing c such that $f(c) < f(x)$, for all x in I where $x \neq c$.

Simply stated, $f(c)$ is a *relative maximum* if $(c, f(c))$ is the highest point in some *open* interval, and $f(c)$ is a *relative minimum* if $(c, f(c))$ is the lowest point in some *open* interval.

If you take a calculus course, you will learn a method for determining exact values of relative maxima and minima. In Section 7.5, we will find exact maximum and minimum values of quadratic functions algebraically.

EXAMPLE 2 Using the graph shown below, determine any relative maxima or minima of the function $f(x) = 0.1x^3 - 0.6x^2 - 0.1x + 2$ and the intervals on which the function is increasing or decreasing.

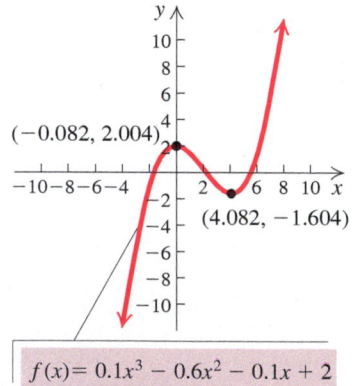

$f(x) = 0.1x^3 - 0.6x^2 - 0.1x + 2$

Solution We see that the *relative maximum* value of the function is 2.004. It occurs when $x = -0.082$. We also see the *relative minimum*: -1.604 at $x = 4.082$.

We can approximate relative maximum and minimum values with the MAXIMUM and MINIMUM features from the CALC menu on a graphing calculator.

$y = 0.1x^3 - 0.6x^2 - 0.1x + 2$

Maximum
X = −.0816657   Y = 2.0041105

$y = 0.1x^3 - 0.6x^2 - 0.1x + 2$

Minimum
X = 4.0816679   Y = −1.604111

We note that the graph starts rising, or increasing, from the left and stops increasing at the relative maximum. From this point, the graph decreases to the relative minimum and then begins to rise again. The function is *increasing* on the intervals

$$(-\infty, -0.082) \quad \text{and} \quad (4.082, \infty)$$

and *decreasing* on the interval

$$(-0.082, 4.082).$$

Let's summarize our results.

Relative Maximum	2.004 at $x = -0.082$
Relative Minimum	-1.604 at $x = 4.082$
Increasing	$(-\infty, -0.082), (4.082, \infty)$
Decreasing	$(-0.082, 4.082)$

▶ **Now Try Exercise 15.**

▶ c   Applications of Functions

Many real-world situations can be modeled by functions.

EXAMPLE 3   *Car Distance.*   Two nurses, Kiara and Matias, drive away from a hospital at right angles to each other. Kiara's speed is 35 mph and Matias's is 40 mph.

a) Express the distance between the cars as a function of time, $d(t)$.

b) Find the domain of the function.

Solution

a) Suppose 1 hr goes by. At that time, Kiara has traveled 35 mi and Matias has traveled 40 mi. We can use the Pythagorean theorem to find the distance between them. This distance would be the length of the hypotenuse of a right triangle with legs measuring 35 mi and 40 mi. After 2 hr, the triangle's legs would measure $2 \cdot 35$, or 70 mi, and $2 \cdot 40$, or 80 mi. Noting that the distances will always be changing, we make a drawing and let $t =$ the time, in hours, that Kiara and Matias have been driving since leaving the hospital.

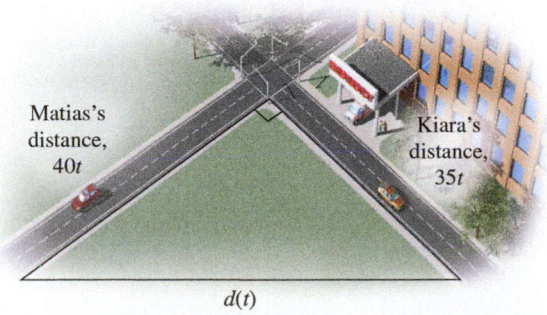

Matias's distance, $40t$

Kiara's distance, $35t$

$d(t)$

After t hours, Kiara has traveled $35t$ miles and Matias $40t$ miles. We now use the Pythagorean theorem:

$$[d(t)]^2 = (35t)^2 + (40t)^2.$$

Table 2.

x $(x \geq 3)$	$g(x) = -x$
3	-3
4	-4
6	-6

b) We graph $g(x) = -x$ *only* for inputs x greater than or equal to 3. That is, we use $g(x) = -x$ only for x-values in the interval $[3, \infty)$. Some ordered pairs that are solutions of this piece of the function are shown in Table 2.

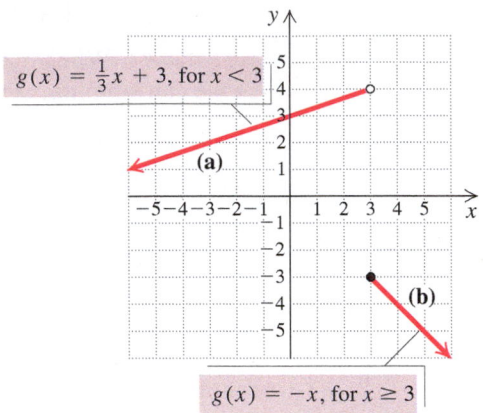

$g(x) = \frac{1}{3}x + 3$, for $x < 3$

(a)

(b)

$g(x) = -x$, for $x \geq 3$

> **Now Try Exercise 33.**

Table 3.

x $(x \leq 0)$	$f(x) = 4$
-5	4
-2	4
0	4

Table 4.

x $(0 < x \leq 2)$	$f(x) = 4 - x^2$
$\frac{1}{2}$	$3\frac{3}{4}$
1	3
2	0

Table 5.

x $(x > 2)$	$f(x) = 2x - 6$
$2\frac{1}{2}$	-1
3	0
5	4

EXAMPLE 7 Graph the function defined as

$$f(x) = \begin{cases} 4, & \text{for } x \leq 0, \\ 4 - x^2, & \text{for } 0 < x \leq 2, \\ 2x - 6, & \text{for } x > 2. \end{cases}$$

Solution We create the graph in three pieces, or parts.

a) We graph $f(x) = 4$ *only* for inputs x less than or equal to 0. That is, we use $f(x) = 4$ only for x-values in the interval $(-\infty, 0]$. Some ordered pairs that are solutions of this piece of the function are shown in Table 3.

b) We graph $f(x) = 4 - x^2$ *only* for inputs x greater than 0 and less than or equal to 2. That is, we use $f(x) = 4 - x^2$ only for x-values in the interval $(0, 2]$. Some ordered pairs that are solutions of this piece of the function are shown in Table 4.

c) We graph $f(x) = 2x - 6$ *only* for inputs x greater than 2. That is, we use $f(x) = 2x - 6$ only for x-values in the interval $(2, \infty)$. Some ordered pairs that are solutions of this piece of the function are shown in Table 5.

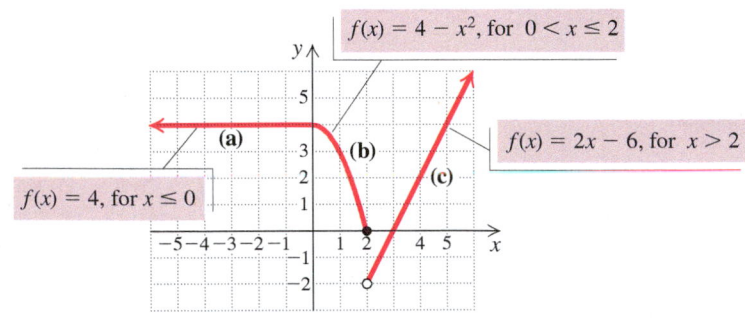

$f(x) = 4 - x^2$, for $0 < x \leq 2$

(a)

(b)

(c)

$f(x) = 2x - 6$, for $x > 2$

$f(x) = 4$, for $x \leq 0$

> **Now Try Exercise 39.**

EXAMPLE 8   Graph the function defined as

$$f(x) = \begin{cases} \dfrac{x^2 - 4}{x + 2}, & \text{for } x \neq -2, \\ 3, & \text{for } x = -2. \end{cases}$$

Solution   When $x \neq -2$, the denominator of $(x^2 - 4)/(x + 2)$ is nonzero, so we can simplify:

$$\frac{x^2 - 4}{x + 2} = \frac{(x + 2)(x - 2)}{x + 2} = x - 2.$$

Thus,

$$f(x) = x - 2, \quad \text{for } x \neq -2.$$

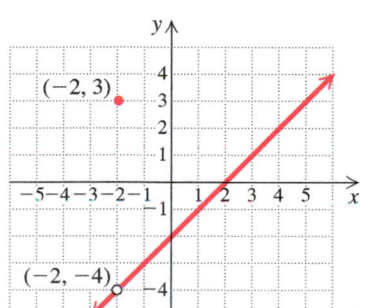

The graph of this part of the function consists of a line with a "hole" at the point $(-2, -4)$, indicated by the open circle. The hole occurs because the piece of the function represented by $(x^2 - 4)/(x + 2)$ is not defined for $x = -2$. By the definition of the function, we see that $f(-2) = 3$, so we plot the point $(-2, 3)$ above the open circle.

Now Try Exercise 41.

A piecewise function with importance in calculus and computer programming is the **greatest integer function**, f, denoted $f(x) = [\![x]\!]$, or $\text{int}(x)$.

Technology Connection

To graph the greatest integer function

$$f(x) = [\![x]\!]$$

on a graphing calculator, select the greatest integer function from the MATH NUM menu. Notice that the graph does not show the open dots at the endpoints of segments.

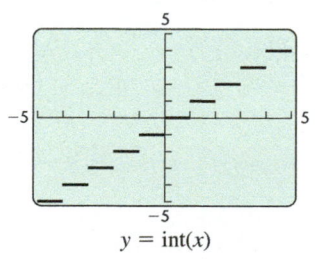

$y = \text{int}(x)$

GREATEST INTEGER FUNCTION

$f(x) = [\![x]\!] =$ the greatest integer *less than or equal to* x.

The greatest integer function pairs each input with the greatest integer *less than or equal to* that input. Thus x-values 1, $1\frac{1}{2}$, and 1.8 are all paired with the y-value 1. Other pairings are shown below.

$$\begin{array}{cccc}
-4 & -1 & 0 & 2 \\
-3.6 \rightarrow -4 & -\frac{7}{8} \rightarrow -1 & \frac{1}{10} \rightarrow 0 & 2.1 \rightarrow 2 \\
-3\frac{1}{4} & -0.25 & 0.99 & 2\frac{3}{4}
\end{array}$$

EXAMPLE 9   Graph $f(x) = [\![x]\!]$ and determine its domain and range.

Solution   The greatest integer function can also be defined as a piecewise function with an infinite number of statements.

$$f(x) = [\![x]\!] = \begin{cases} \vdots \\ -3, & \text{for } -3 \leq x < -2, \\ -2, & \text{for } -2 \leq x < -1, \\ -1, & \text{for } -1 \leq x < 0, \\ 0, & \text{for } 0 \leq x < 1, \\ 1, & \text{for } 1 \leq x < 2, \\ 2, & \text{for } 2 \leq x < 3, \\ 3, & \text{for } 3 \leq x < 4, \\ \vdots \end{cases}$$

$f(x) = [\![x]\!]$

We see that the domain of this function is the set of all real numbers, $(-\infty, \infty)$, and the range is the set of all integers, $\{\ldots, -3, -2, -1, 0, 1, 2, 3, \ldots\}$.

Now Try Exercise 45.

6.8	Exercise Set

▶ Reading Check

The piecewise function f is defined as

$$f(x) = \begin{cases} x - 3, & \text{for } x < -2, \\ x^2, & \text{for } -2 \le x < 5, \\ 7x, & \text{for } x > 5. \end{cases}$$

Choose the expression from the list at right that should be used to find each function value.

RC1. $f(-10)$ **a)** $x - 3$

RC2. $f(10)$ **b)** x^2

RC3. $f(0)$ **c)** $7x$

RC4. $f(-2)$

RC5. $f(5)$

---

a *Determine the intervals on which the function is*
(a) *increasing;* **(b)** *decreasing;* **(c)** *constant.*

1.
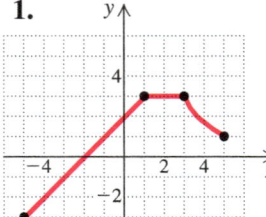

2.
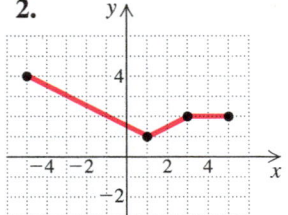

3.
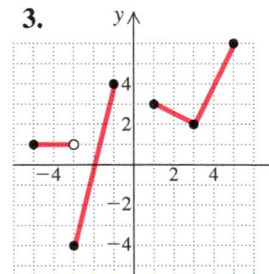

4.
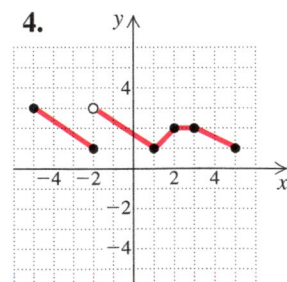

5.
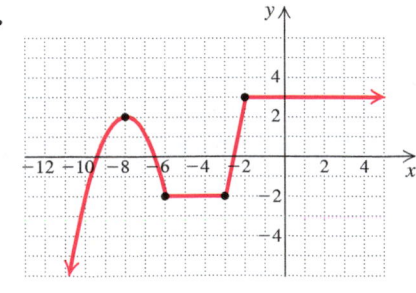

6.
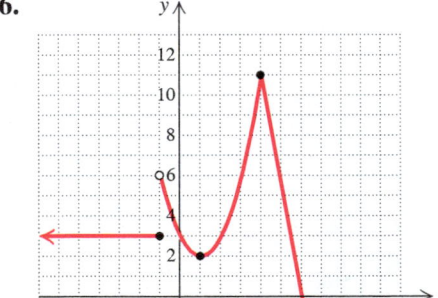

7.–12. Determine the domain and the range of each of the functions graphed in Exercises 1–6.

a, b *Using the graph, determine any relative maxima or minima of the function and the intervals on which the function is increasing or decreasing.*

13. $f(x) = -x^2 + 5x - 3$

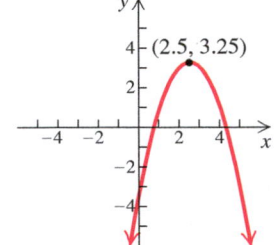

14. $f(x) = x^2 - 2x + 3$

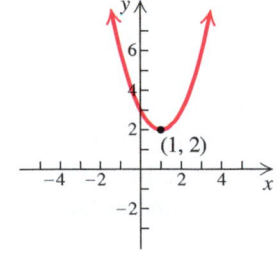

15. $f(x) = \frac{1}{4}x^3 - \frac{1}{2}x^2 - x + 2$

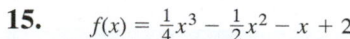

$(-0.667, 2.370)$

$(2, 0)$

16. $f(x) = -0.09x^3 + 0.5x^2 - 0.1x + 1$

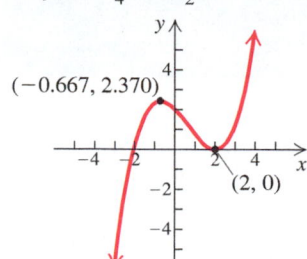

$(3.601, 2.921)$

$(0.103, 0.995)$

C

17. *Lumberyard.* Rick's lumberyard has 480 yd of fencing with which to enclose a rectangular area. If the enclosed area is x yards long, express its area as a function of its length.

18. *Triangular Flag.* A seamstress is designing a triangular flag so that the length of the base of the triangle, in inches, is 7 less than twice the height h. Express the area of the flag as a function of the height.

19. *Blimp Distance.* The Goodyear Blimp can be seen flying at an altitude of 3500 ft above the Motor Speedway during the Indianapolis 500 race. The slanted distance directly to the Pagoda at the start–

finish line is d feet. Express the horizontal distance h as a function of d.

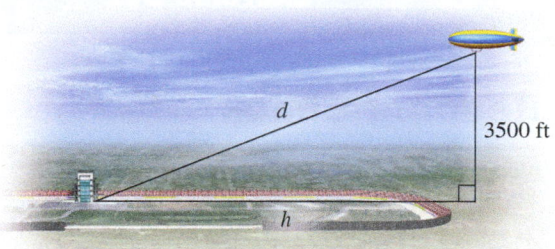

d

3500 ft

h

20. *Rising Balloon.* A hot-air balloon rises straight up from the ground at a rate of 120 ft/min. The balloon is tracked from a rangefinder on the ground at point P, which is 400 ft from the release point Q of the basket. Let $d =$ the distance from the balloon to the rangefinder and $t =$ the time, in minutes, since the balloon was released. Express d as a function of t.

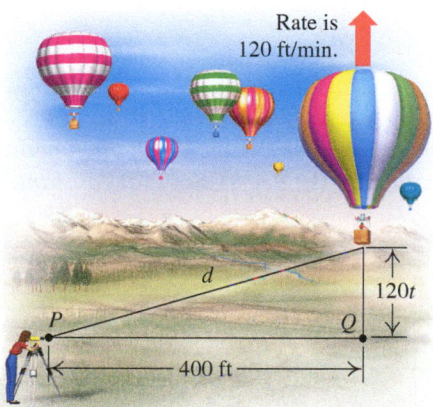

Rate is 120 ft/min.

d

$120t$

P

Q

400 ft

21. *Inscribed Rhombus.* A rhombus is inscribed in a rectangle that is w meters wide with a perimeter of 40 m. Each vertex of the rhombus is a midpoint of a side of the rectangle. Express the area of the rhombus as a function of the width of the rectangle.

w

22. *Carpet Area.* A carpet installer uses 46 ft of linen tape to bind the edges of a rectangular hall runner. If the runner is w feet wide, express its area as a function of the width.

23. *Golf Distance Finder.* A device used in golf to estimate the distance *d*, in yards, to a hole measures the size *s*, in inches, that the 7-ft pin appears to be in a viewfinder. Express the distance *d* as a function of *s*.

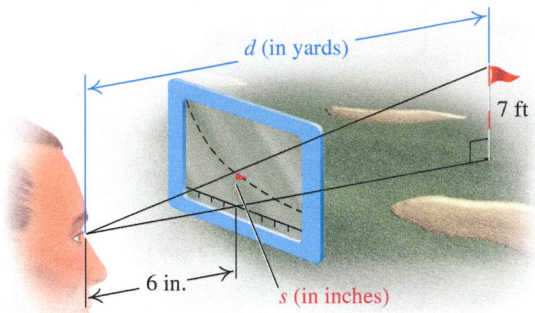

d (in yards)

7 ft

6 in.

s (in inches)

24. *Gas Tank Volume.* A gas tank has ends that are hemispheres of radius *r* feet. The cylindrical midsection is 6 ft long. Express the volume of the tank as a function of *r*.

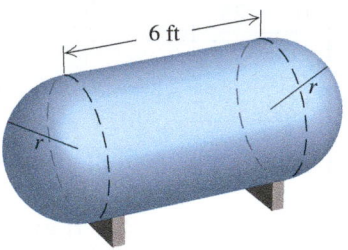

6 ft

r

r

25. *Swimming Areas.* A summer camp has 240 ft of float line with which to rope off three adjacent rectangular areas of a lake for swimming lessons, one for each of three levels of swimming ability. A beach forms one side of the swimming areas. Suppose the width of each area is *x* yards.

x   *x*   *x*   *x*

a) Express the total area of the three swimming areas as a function of *x*.
b) Find the domain of the function.
c) Using the graph of the function shown below, determine the dimensions that yield the maximum area.

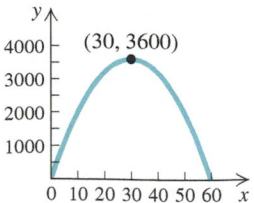

(30, 3600)

y

4000
3000
2000
1000

0 10 20 30 40 50 60 *x*

26. *Play Space.* A car dealership has 24 ft of dividers with which to enclose a rectangular play space in a corner of a customer lounge. The sides against the wall require no partition. Suppose the play space is *x* feet long.

a) Express the area of the play space as a function of *x*.
b) Find the domain of the function.
c) Using the graph shown below, determine the dimensions that yield the maximum area.

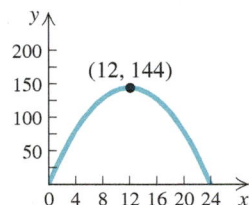

y

200
150
100
50

(12, 144)

0 4 8 12 16 20 24 *x*

27. *Volume of a Box.* From a 12-cm by 12-cm piece of cardboard, square corners are cut out so that the sides can be folded up to make a box.

x

x

a) Express the volume of the box as a function of the side *x*, in centimeters, of a cut-out square.
b) Find the domain of the function.
c) Using the graph of the function shown below, determine the dimensions that yield the maximum volume.

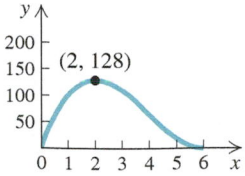

y

200
150
100
50

(2, 128)

0 1 2 3 4 5 6 *x*

28. *Office File.* Designs Unlimited plans to produce a one-component vertical file by bending the long side of an 8-in. by 14-in. sheet of plastic along two lines to form a ⊔ shape.

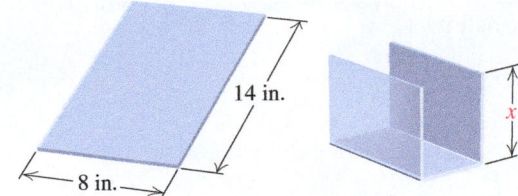

a) Express the volume of the file as a function of the height x, in inches, of the file.

b) Find the domain of the function.

c) Using the graph of the function shown below, determine how tall the file should be in order to maximize the volume that the file can hold.

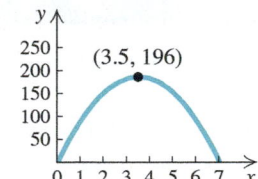

d *For each piecewise function, find the specified function values.*

29. $g(x) = \begin{cases} x + 4, & \text{for } x \le 1, \\ 8 - x, & \text{for } x > 1 \end{cases}$

$g(-4), g(0), g(1),$ and $g(3)$

30. $f(x) = \begin{cases} 3, & \text{for } x \le -2, \\ \frac{1}{2}x + 6, & \text{for } x > -2 \end{cases}$

$f(-5), f(-2), f(0),$ and $f(2)$

31. $h(x) = \begin{cases} -3x - 18, & \text{for } x < -5, \\ 1, & \text{for } -5 \le x < 1, \\ x + 2, & \text{for } x \ge 1 \end{cases}$

$h(-5), h(0), h(1),$ and $h(4)$

32. $f(x) = \begin{cases} -5x - 8, & \text{for } x < -2, \\ \frac{1}{2}x + 5, & \text{for } -2 \le x \le 4, \\ 10 - 2x, & \text{for } x > 4 \end{cases}$

$f(-4), f(-2), f(4),$ and $f(6)$

Graph each of the following.

33. $f(x) = \begin{cases} \frac{1}{2}x, & \text{for } x < 0, \\ x + 3, & \text{for } x \ge 0 \end{cases}$

34. $f(x) = \begin{cases} -\frac{1}{3}x + 2, & \text{for } x \le 0, \\ x - 5, & \text{for } x > 0 \end{cases}$

35. $f(x) = \begin{cases} -\frac{3}{4}x + 2, & \text{for } x < 4, \\ -1, & \text{for } x \ge 4 \end{cases}$

36. $h(x) = \begin{cases} 2x - 1, & \text{for } x < 2, \\ 2 - x, & \text{for } x \ge 2 \end{cases}$

37. $f(x) = \begin{cases} x + 1, & \text{for } x \le -3, \\ -1, & \text{for } -3 < x < 4, \\ \frac{1}{2}x, & \text{for } x \ge 4 \end{cases}$

38. $f(x) = \begin{cases} 4, & \text{for } x \le -2, \\ x + 1, & \text{for } -2 < x < 3, \\ -x, & \text{for } x \ge 3 \end{cases}$

39. $g(x) = \begin{cases} \frac{1}{2}x - 1, & \text{for } x < 0, \\ 3, & \text{for } 0 \le x \le 1, \\ -2x, & \text{for } x > 1 \end{cases}$

40. $f(x) = \begin{cases} \dfrac{x^2 - 9}{x + 3}, & \text{for } x \ne -3, \\ 5, & \text{for } x = -3 \end{cases}$

41. $f(x) = \begin{cases} 2, & \text{for } x = 5, \\ \dfrac{x^2 - 25}{x - 5}, & \text{for } x \ne 5 \end{cases}$

42. $f(x) = \begin{cases} \dfrac{x^2 + 3x + 2}{x + 1}, & \text{for } x \ne -1, \\ 7, & \text{for } x = -1 \end{cases}$

43. $f(x) = [x]$ **44.** $f(x) = 2[x]$

45. $g(x) = 1 + [x]$ **46.** $h(x) = \frac{1}{2}[x] - 2$

47.–52. Find the domain and the range of each of the functions defined in Exercises 33–38.

Determine the domain and the range of the piecewise function. Then write an equation for the function.

53.
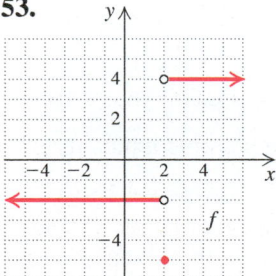

54.
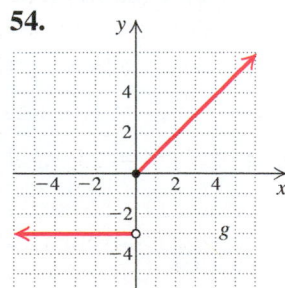

55.
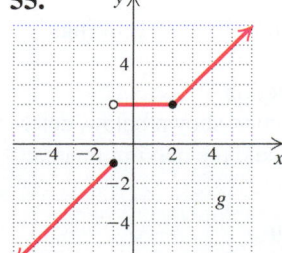

56.
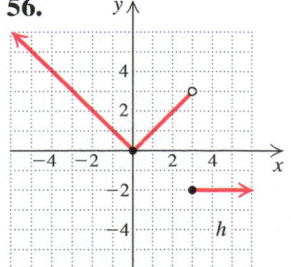

57.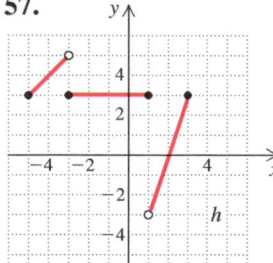

58.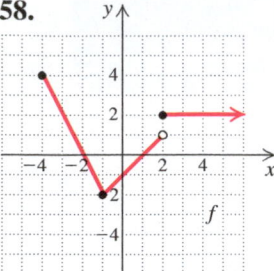

▶ Skill Maintenance

59. Given $f(x) = 5x^2 - 7$, find each of the following. **[2.2b]**

a) $f(-3)$ b) $f(3)$
c) $f(a)$ d) $f(-a)$

60. Given $f(x) = 4x^3 - 5x$, find each of the following. **[2.2b]**

a) $f(2)$ b) $f(-2)$
c) $f(a)$ d) $f(-a)$

61. Write an equation of the line perpendicular to the graph of the line $8x - y = 10$ and containing the point $(-1, 1)$. **[2.7d]**

62. Find the slope and the y-intercept of the line with equation $2x - 9y + 1 = 0$. **[2.5b]**

▶ Synthesis

63. *Parking Costs.* A parking garage charges $3 for up to (but not including) 1 hr of parking, $6 for up to 2 hr of parking, $9 for up to 3 hr of parking, and so on. Let $C(t) =$ the cost of parking for t hours.

a) Graph the function.
b) Write an equation for $C(t)$ using the greatest integer notation $[\![t]\!]$.

64. If $[\![x + 2]\!] = -3$, what are the possible inputs for x?

65. If $([\![x]\!])^2 = 25$, what are the possible inputs for x?

66. *Minimizing Power Line Costs.* A power line is constructed from a power station at point A to an island at point I, which is 1 mi directly out in the water from a point B on the shore. Point B is 4 mi downshore from the power station at A. It costs $5000 per mile to lay the power line under water and $3000 per mile to lay the power line under ground. The line comes to the shore at point S downshore from A. Let $x =$ the distance from B to S.

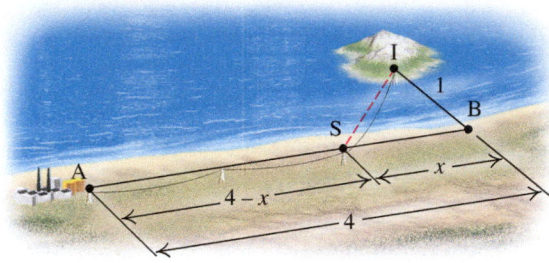

a) Express the cost C of laying the line as a function of x.
b) At what distance x from point B should the line come to shore in order to minimize cost?

67. *Volume of an Inscribed Cylinder.* A right circular cylinder of height h and radius r is inscribed in a right circular cone with a height of 10 ft and a base with radius 6 ft.

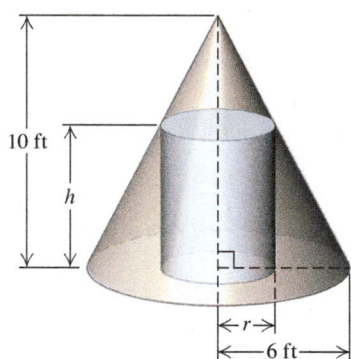

a) Express the height h of the cylinder as a function of r.
b) Express the volume V of the cylinder as a function of r.
c) Express the volume V of the cylinder as a function of h.

Chapter 6 Summary and Review

VOCABULARY REINFORCEMENT

Complete each statement with the correct term from the column on the right. Some of the choices may not be used and some may be used more than once.

radicand
index
square
open
closed
cube
conjugate
radical
principal
rationalizing
increasing
decreasing

1. The number c is the _____ root of a, written $\sqrt[3]{a}$, if the third power of c is a—that is, if $c^3 = a$, then $\sqrt[3]{a} = c$. **[6.1c]**

2. On a given interval, if the graph of a function rises from left to right, it is _____ on that interval. **[6.8a]**

3. For any real number a, $\sqrt{a^2} = |a|$. The _____ (nonnegative) square root of a^2 is the absolute value of a. **[6.1b]**

4. To find an equivalent expression without a radical in the denominator is called _____ the denominator. **[6.5a]**

5. The symbol $\sqrt{}$ is called a(n) _____ . **[6.1a]**

6. Intervals on which a function is increasing, decreasing, or constant are expressed using _____-interval notation. **[6.8a]**

7. The number c is a(n) _____ root of a if $c^2 = a$. **[6.1a]**

8. The expression written under the radical is called the _____. **[6.1a]**

9. The _____ of a radical expression $\sqrt{a} + \sqrt{b}$ is $\sqrt{a} - \sqrt{b}$. **[6.5b]**

10. In the expression $\sqrt[k]{a}$, we call the k the _____. **[6.1d]**

CONCEPT REINFORCEMENT

Determine whether each statement is true or false.

1. For any negative number a, we have $\sqrt{a^2} = -a$. **[6.1a]**

2. For any real numbers $\sqrt[m]{a}$ and $\sqrt[n]{b}$, $\sqrt[m]{a} \cdot \sqrt[n]{b} = \sqrt[mn]{ab}$. **[6.3a]**

3. For any real numbers $\sqrt[n]{a}$ and $\sqrt[n]{b}$, $\sqrt[n]{a} + \sqrt[n]{b} = \sqrt[n]{a + b}$. **[6.4a]**

4. If $x^2 = 4$, then $x = 2$. **[6.6a]**

5. Every real number has a cube root. **[6.1c]**

6. The greatest integer function pairs each input with the greatest integer less than or equal to that input. **[6.8d]**

STUDY GUIDE

Objective 6.1b Simplify radical expressions with perfect-square radicands.

Example Simplify: $\sqrt{16x^2}$.

$$\sqrt{16x^2} = \sqrt{(4x)^2} = |4x| = |4| \cdot |x| = 4|x|$$

Example Simplify: $\sqrt{x^2 - 6x + 9}$.

$$\sqrt{x^2 - 6x + 9} = \sqrt{(x - 3)^2} = |x - 3|$$

Practice Exercises

1. Simplify: $\sqrt{36y^2}$.

2. Simplify: $\sqrt{a^2 + 4a + 4}$.

Objective 6.2a Write expressions with or without rational exponents, and simplify, if possible.

Example Rewrite $x^{1/4}$ without a rational exponent.
Recall that $a^{1/n}$ means $\sqrt[n]{a}$. Then

$$x^{1/4} = \sqrt[4]{x}.$$

Example Rewrite $\left(\sqrt[3]{4xy^2}\right)^4$ with a rational exponent.
Recall that $\left(\sqrt[n]{a}\right)^m$ means $a^{m/n}$. Then

$$\left(\sqrt[3]{4xy^2}\right)^4 = (4xy^2)^{4/3}.$$

Practice Exercises

3. Rewrite $z^{3/5}$ without a rational exponent.

4. Rewrite $\left(\sqrt{6ab}\right)^5$ with a rational exponent.

Objective 6.2b Write expressions without negative exponents, and simplify, if possible.

Example Rewrite $8^{-2/3}$ with a positive exponent, and simplify, if possible.
Recall that $a^{-m/n}$ means $\dfrac{1}{a^{m/n}}$. Then

$$8^{-2/3} = \frac{1}{8^{2/3}} = \frac{1}{\left(\sqrt[3]{8}\right)^2} = \frac{1}{2^2} = \frac{1}{4}.$$

Practice Exercise

5. Rewrite $9^{-3/2}$ with a positive exponent, and simplify, if possible.

Objective 6.2d Use rational exponents to simplify radical expressions.

Example Use rational exponents to simplify: $\sqrt[6]{x^2y^4}$.

$$\begin{aligned}
\sqrt[6]{x^2y^4} &= (x^2y^4)^{1/6} \\
&= x^{2/6}y^{4/6} \\
&= x^{1/3}y^{2/3} \\
&= (xy^2)^{1/3} \\
&= \sqrt[3]{xy^2}
\end{aligned}$$

Practice Exercise

6. Use rational exponents to simplify: $\sqrt[8]{a^6b^2}$.

Objective 6.3a Multiply and simplify radical expressions.

Example Multiply and simplify: $\sqrt[3]{6xy^2}\,\sqrt[3]{9y}$.

$$\begin{aligned}
\sqrt[3]{6xy^2}\,\sqrt[3]{9y} &= \sqrt[3]{6xy^2 \cdot 9y} \\
&= \sqrt[3]{54xy^3} \\
&= \sqrt[3]{27y^3 \cdot 2x} \\
&= \sqrt[3]{27y^3}\,\sqrt[3]{2x} \\
&= 3y\sqrt[3]{2x}
\end{aligned}$$

Practice Exercise

7. Multiply and simplify. Assume that all expressions under radicals represent nonnegative numbers.

$$\sqrt{5y}\,\sqrt{30y}$$

Objective 6.3b Divide and simplify radical expressions.

Example Divide and simplify: $\dfrac{\sqrt{24x^5}}{\sqrt{6x}}$.

$$\frac{\sqrt{24x^5}}{\sqrt{6x}} = \sqrt{\frac{24x^5}{6x}} = \sqrt{4x^4} = 2x^2$$

Practice Exercise

8. Divide and simplify: $\dfrac{\sqrt{20a}}{\sqrt{5}}$.

Objective 6.4a Add or subtract with radical notation and simplify.

Example Subtract: $5\sqrt{2} - 4\sqrt{8}$.

$$
\begin{aligned}
5\sqrt{2} - 4\sqrt{8} &= 5\sqrt{2} - 4\sqrt{4 \cdot 2} \\
&= 5\sqrt{2} - 4\sqrt{4}\sqrt{2} \\
&= 5\sqrt{2} - 4 \cdot 2\sqrt{2} = 5\sqrt{2} - 8\sqrt{2} \\
&= (5 - 8)\sqrt{2} = -3\sqrt{2}
\end{aligned}
$$

Practice Exercise

9. Subtract: $\sqrt{48} - 2\sqrt{3}$.

Objective 6.4b Multiply expressions involving radicals in which some factors contain more than one term.

Example Multiply: $\left(3 - \sqrt{6}\right)\left(2 + 4\sqrt{6}\right)$.

We use FOIL:

$$
\begin{aligned}
\left(3 - \sqrt{6}\right)&\left(2 + 4\sqrt{6}\right) \\
&= 3 \cdot 2 + 3 \cdot 4\sqrt{6} - \sqrt{6} \cdot 2 - \sqrt{6} \cdot 4\sqrt{6} \\
&= 6 + 12\sqrt{6} - 2\sqrt{6} - 4 \cdot 6 \\
&= 6 + 12\sqrt{6} - 2\sqrt{6} - 24 \\
&= -18 + 10\sqrt{6}.
\end{aligned}
$$

Practice Exercise

10. Multiply: $\left(5 - \sqrt{x}\right)^2$.

Objective 6.6a Solve radical equations with one radical term.

Example Solve: $x = \sqrt{x - 2} + 4$.

First, we subtract 4 on both sides to isolate the radical. Then we square both sides of the equation.

$$
\begin{aligned}
x &= \sqrt{x - 2} + 4 \\
x - 4 &= \sqrt{x - 2} \\
(x - 4)^2 &= \left(\sqrt{x - 2}\right)^2 \\
x^2 - 8x + 16 &= x - 2 \\
x^2 - 9x + 18 &= 0 \\
(x - 3)(x - 6) &= 0 \\
x - 3 = 0 \quad &or \quad x - 6 = 0 \\
x = 3 \quad &or \qquad x = 6
\end{aligned}
$$

We must check both possible solutions. When we do, we find that 6 checks, but 3 does not. Thus the solution is 6.

Practice Exercise

11. Solve: $3 + \sqrt{x - 1} = x$.

Objective 6.6b Solve radical equations with two radical terms.

Example Solve: $1 = \sqrt{x + 9} - \sqrt{x}$.

$$1 = \sqrt{x + 9} - \sqrt{x}$$

$\sqrt{x} + 1 = \sqrt{x + 9}$ **Isolating one radical**

$\left(\sqrt{x} + 1\right)^2 = \left(\sqrt{x + 9}\right)^2$ **Squaring both sides**

$x + 2\sqrt{x} + 1 = x + 9$

$2\sqrt{x} = 8$ **Isolating the remaining radical**

$\sqrt{x} = 4$

$\left(\sqrt{x}\right)^2 = 4^2$

$x = 16$

The number 16 checks. It is the solution.

Practice Exercise

12. Solve: $\sqrt{x + 3} - \sqrt{x - 2} = 1$.

Objective 6.8a Determine the intervals on which a function is increasing, decreasing, or constant from its graph.

Example Determine the intervals on which the function is **(a)** increasing; **(b)** decreasing; **(c)** constant.

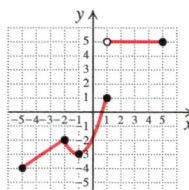

a) As x-values increase from -5 to -2, y-values increase from -4 to -2; y-values also increase as x-values increase from -1 to 1. Thus the function is increasing on the intervals $(-5, -2)$ and $(-1, 1)$.

b) As x-values increase from -2 to -1, y-values decrease from -2 to -3, so the function is decreasing on the interval $(-2, -1)$.

c) As x-values increase from 1 to 5, y remains 5, so the function is constant on the interval $(1, 5)$.

Practice Exercise

13. Determine the intervals on which the function is **(a)** increasing; **(b)** decreasing; **(c)** constant.

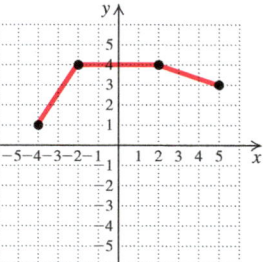

Objective 6.8b Determine relative maxima and minima of a function from its graph.

Example Determine any relative maxima or minima of the function.

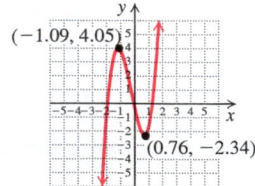

We see from the graph that the function has one relative maximum, 4.05. It occurs when $x = -1.09$. We also see that there is one relative minimum, -2.34. It occurs when $x = 0.76$.

Practice Exercise

14. Determine any relative maxima or minima of the function.

$$f(x) = 0.3x^4 + 2x^2 - x - 1$$

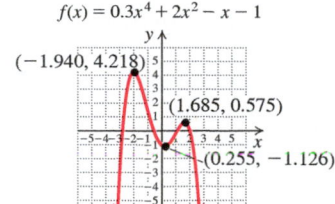

Objective 6.8c   Given an application, find a function that models the application. Find the domain of the function and function values.

See Example 3 and 4 on pp. 418–420.

Practice Exercise

15. Demetrius is designing a triangular garden. The base of the garden is 10 ft longer than twice the height h. Express the area of the garden as a function of the height.

Objective 6.8d   Graph functions defined piecewise.

Graph the function defined as

$$f(x) = \begin{cases} 2x - 3, & \text{for } x < 1, \\ x + 1, & \text{for } x \geq 1. \end{cases}$$

We create the graph in two parts. First, we graph $f(x) = 2x - 3$ for inputs x less than 1. Then we graph $f(x) = x + 1$ for inputs x greater than or equal to 1.

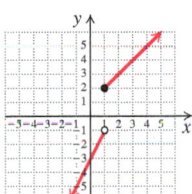

Practice Exercise

16. Graph the function defined as

$$f(x) = \begin{cases} -x, & \text{for } x \leq -4, \\ \frac{1}{2}x + 1, & \text{for } x > -4. \end{cases}$$

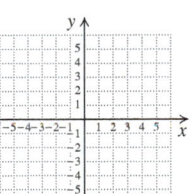

REVIEW EXERCISES

Use a calculator to approximate to three decimal places. [6.1a]

1. $\sqrt{778}$

2. $\sqrt{\dfrac{963.2}{23.68}}$

3. For the given function, find the indicated function values. [6.1a]

$$f(x) = \sqrt{3x - 16}; \ f(0), f(-1), f(1), \text{ and } f\left(\tfrac{41}{3}\right)$$

4. Find the domain of the function f in Exercise 3. [6.1a]

Simplify. Assume that letters represent any real number. [6.1b]

5. $\sqrt{81a^2}$

6. $\sqrt{(-7z)^2}$

7. $\sqrt{(6 - b)^2}$

8. $\sqrt{x^2 + 6x + 9}$

Simplify. [6.1c]

9. $\sqrt[3]{-1000}$

10. $\sqrt[3]{-\dfrac{1}{27}}$

11. For the given function, find the indicated function values. [6.1c]

$$f(x) = \sqrt[3]{x + 2}; \ f(6), f(-10), \text{ and } f(25)$$

Simplify. Assume that letters represent any real number. [6.1d]

12. $\sqrt[10]{x^{10}}$

13. $-\sqrt[13]{(-3)^{13}}$

Rewrite without rational exponents, and simplify, if possible. [6.2a]

14. $a^{1/5}$

15. $64^{3/2}$

Rewrite with rational exponents. [6.2a]

16. $\sqrt{31}$

17. $\sqrt[5]{a^2b^3}$

Rewrite with positive exponents, and simplify, if possible. [6.2b]

18. $49^{-1/2}$

19. $(8xy)^{-2/3}$

20. $5a^{-3/4}b^{1/2}c^{-2/3}$

21. $\dfrac{3a}{\sqrt[4]{t}}$

Use the laws of exponents to simplify. Write answers with positive exponents. **[6.2c]**

22. $(x^{-2/3})^{3/5}$

23. $\dfrac{7^{-1/3}}{7^{-1/2}}$

Use rational exponents to simplify. Write the answer in radical notation if appropriate. **[6.2d]**

24. $\sqrt[3]{x^{21}}$

25. $\sqrt[3]{27x^6}$

Use rational exponents to write a single radical expression. **[6.2d]**

26. $x^{1/3}y^{1/4}$

27. $\sqrt[4]{x}\sqrt[3]{x}$

Simplify by factoring. Assume that all expressions under radicals represent nonnegative numbers. **[6.3a]**

28. $\sqrt{245}$

29. $\sqrt[3]{-108}$

30. $\sqrt[3]{250a^2b^6}$

Simplify. Assume that no radicands were formed by raising negative numbers to even powers. **[6.3b]**

31. $\sqrt{\dfrac{49}{36}}$

32. $\sqrt[3]{\dfrac{64x^6}{27}}$

33. $\sqrt[4]{\dfrac{16x^8}{81y^{12}}}$

Perform the indicated operations and simplify. Assume that no radicands were formed by raising negative numbers to even powers. **[6.3a, b]**, **[6.4a]**

34. $\sqrt{5x}\sqrt{3y}$

35. $\sqrt[3]{a^5b}\sqrt[3]{27b}$

36. $\sqrt[3]{a}\sqrt[5]{b^3}$

37. $\dfrac{\sqrt[3]{60xy^3}}{\sqrt[3]{10x}}$

38. $\dfrac{\sqrt{75x}}{2\sqrt{3}}$

39. $\dfrac{\sqrt[3]{x^2}}{\sqrt[4]{x}}$

40. $5\sqrt[3]{x} + 2\sqrt[3]{x}$

41. $2\sqrt{75} - 7\sqrt{3}$

42. $\sqrt{50} + 2\sqrt{18} + \sqrt{32}$

43. $\sqrt[3]{8x^4} + \sqrt[3]{xy^6}$

Multiply. **[6.4b]**

44. $\left(\sqrt{5} - 3\sqrt{8}\right)\left(\sqrt{5} + 2\sqrt{8}\right)$

45. $\left(1 - \sqrt{7}\right)^2$

46. $\left(\sqrt[3]{27} - \sqrt[3]{2}\right)\left(\sqrt[3]{27} + \sqrt[3]{2}\right)$

Rationalize the denominator. **[6.5a, b]**

47. $\sqrt{\dfrac{8}{3}}$

48. $\dfrac{2}{\sqrt{a} + \sqrt{b}}$

Solve. **[6.6a, b]**

49. $x - 3 = \sqrt{5 - x}$

50. $\sqrt[4]{x + 3} = 2$

51. $\sqrt{x + 8} - \sqrt{3x + 1} = 1$

Automotive Repair. For an engine with a displacement of 2.8 L, the function given by

$$d(n) = 0.75\sqrt{2.8n}$$

can be used to determine the diameter of the carburetor's opening, $d(n)$, in millimeters, where n is the number of rpm's at which the engine achieves peak performance. (*Source: macdizzy.com*) **[6.6c]**

52. 🖩 If a carburetor's opening is 81 mm, for what number of rpm's will the engine produce peak power?

53. 🖩 If a carburetor's opening is 84 mm, for what number of rpm's will the engine produce peak power?

54. *Length of a Side of a Square.* The diagonal of a square has length $9\sqrt{2}$ cm. Find the length of a side of the square. **[6.7a]**

55. *Bookcase Width.* A bookcase is 5 ft tall and has a 7-ft diagonal brace, as shown. How wide is the bookcase? **[6.7a]**

In a right triangle, find the length of the side not given. Give an exact answer and, where appropriate, an answer to three decimal places. **[6.7a]**

56. $a = 7$, $b = 24$

57. $a = 2$, $c = 5\sqrt{2}$

Use the following graph of a function for Exercises 58 and 59.

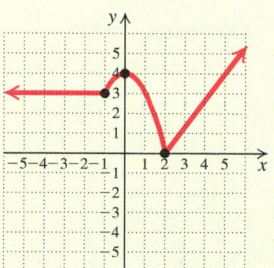

58. Determine the intervals on which the function is **(a)** increasing, **(b)** decreasing, and **(c)** constant. **[6.8a]**

59. Determine any relative maxima or minima of the function. **[6.8b]**

60. *Fenced Patio.* Syd has 48 ft of rolled bamboo fence to enclose a rectangular patio. The house forms one side of the patio. Suppose two sides of the patio are each x feet. Express the area of the patio as a function of x. **[6.8c]**

61. *Inscribed Rectangle.* A rectangle is inscribed in a semicircle of radius 2, as shown. The variable $x =$ half the length of the rectangle. Express the area of the rectangle as a function of x. **[6.8c]**

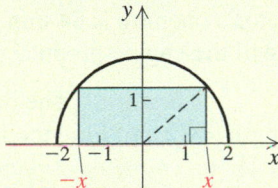

62. *Minimizing Surface Area.* A container firm is designing an open-top rectangular box, with a square base, that will hold 108 in^3. Let $x =$ the length of a side of the base.

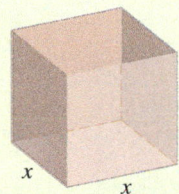

a) Express the surface area as a function of x. **[6.8c]**
b) Find the domain of the function. **[6.8c]**
c) Using the following graph, determine the dimensions that will minimize the surface area of the box. **[6.8c]**

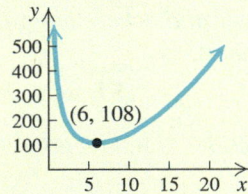

Graph each of the following. **[6.8d]**

63. $f(x) = \begin{cases} x^3, & \text{for } x < -2, \\ |x|, & \text{for } -2 \le x \le 2, \\ \sqrt{x-1}, & \text{for } x > 2 \end{cases}$

64. $f(x) = \begin{cases} \dfrac{x^2 - 1}{x + 1}, & \text{for } x \ne -1, \\ 3, & \text{for } x = -1 \end{cases}$

65. $f(x) = [\![x]\!]$

66. $f(x) = [\![x - 3]\!]$

67. For the function in Exercise 63, find $f(-1), f(5)$, $f(-2)$, and $f(-3)$. **[6.8d]**

68. For the function in Exercise 64, find $f(-2), f(-1)$, $f(0)$, and $f(4)$. **[6.8d]**

69. Graph: $f(x) = \sqrt{x}$. **[6.1a]**

70. Which of the following does *not* exist as a real number? **[6.1a]**

   A. $\sqrt{8}$ **B.** $\sqrt{-8}$ **C.** $\sqrt[3]{8}$ **D.** $\sqrt[3]{-8}$

▶ **Synthesis**

71. Solve: $\sqrt{11x} + \sqrt{6 + x} = 6$. **[6.6a]**

Graph each function. Estimate the intervals on which the function is increasing or decreasing and any relative maxima or minima. **[6.8a, b]**

72. $f(x) = |x + 3| - 5$

73. $f(x) = x^2 - 6x + 10$

▶ **Collaborative Discussion and Writing**

74. Find the domain of
$$f(x) = (x + 5)^{1/2}(x + 7)^{-1/2}$$
and explain how you found your answer. **[6.1a], [6.2b]**

75. 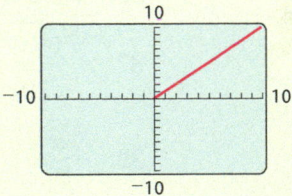 Ron is puzzled. When he uses a graphing calculator to graph $y = \sqrt{x} \cdot \sqrt{x}$, he gets the following screen. Explain why Ron did not get the complete line $y = x$. **[6.1a], [6.3a]**

76. In what way(s) is collecting like radical terms the same as collecting like monomial terms? **[6.4a]**

77. Is checking solutions of equations necessary when the principle of powers is used with an odd power n? Why or why not? **[6.1d], [6.6a, b]**

78. A student *incorrectly* claims that
$$\frac{5 + \sqrt{2}}{\sqrt{18}} = \frac{5 + \sqrt{1}}{\sqrt{9}} = \frac{5 + 1}{3} = 2.$$

How could you convince the student that a mistake has been made? How would you explain the correct way of rationalizing the denominator? **[6.5a]**

6   Chapter Test

1. Use a calculator to approximate $\sqrt{148}$ to three decimal places.

2. For the given function, find the indicated function values.
$$f(x) = \sqrt{8 - 4x}; \quad f(1) \text{ and } f(3)$$

3. Find the domain of the function f in Exercise 2.

Simplify. Assume that letters represent any real number.

4. $\sqrt{(-3q)^2}$

5. $\sqrt{x^2 + 10x + 25}$

6. $\sqrt[3]{-\dfrac{1}{1000}}$

7. $\sqrt[5]{x^5}$

8. $\sqrt[10]{(-4)^{10}}$

Rewrite without rational exponents, and simplify, if possible.

9. $a^{2/3}$

10. $32^{3/5}$

Rewrite with rational exponents.

11. $\sqrt{37}$

12. $\left(\sqrt{5xy^2}\right)^5$

Rewrite with positive exponents, and simplify, if possible.

13. $1000^{-1/3}$

14. $8a^{3/4}b^{-3/2}c^{-2/5}$

Use the laws of exponents to simplify. Write answers with positive exponents.

15. $(x^{2/3}y^{-3/4})^{12/5}$

16. $\dfrac{2.9^{-5/8}}{2.9^{2/3}}$

Use rational exponents to simplify. Write the answer in radical notation if appropriate. Assume that no radicands were formed by raising negative numbers to even powers.

17. $\sqrt[8]{x^2}$

18. $\sqrt[4]{16x^6}$

Use rational exponents to write a single radical expression.

19. $a^{2/5}b^{1/3}$

20. $\sqrt[4]{2y}\sqrt[3]{y}$

Simplify by factoring. Assume that no radicands were formed by raising negative numbers to even powers.

21. $\sqrt{148}$

22. $\sqrt[4]{80}$

23. $\sqrt[3]{24a^{11}b^{13}}$

Simplify. Assume that no radicands were formed by raising negative numbers to even powers.

24. $\sqrt[3]{\dfrac{16x^5}{y^6}}$

25. $\sqrt{\dfrac{25x^2}{36y^4}}$

Perform the indicated operations and simplify. Assume that no radicands were formed by raising negative numbers to even powers.

26. $\sqrt[3]{2x}\sqrt[3]{5y^2}$

27. $\sqrt[4]{x^3y^2}\sqrt{xy}$

28. $\dfrac{\sqrt[5]{x^3y^4}}{\sqrt[5]{xy^2}}$

29. $\dfrac{\sqrt{300a}}{5\sqrt{3}}$

30. Add: $3\sqrt{128} + 2\sqrt{18} + 2\sqrt{32}$.

Multiply.

31. $\left(\sqrt{20} + 2\sqrt{5}\right)\left(\sqrt{20} - 3\sqrt{5}\right)$

32. $\left(3 + \sqrt{x}\right)^2$

33. Rationalize the denominator: $\dfrac{1 + \sqrt{2}}{3 - 5\sqrt{2}}$.

Solve.

34. $\sqrt[5]{x - 3} = 2$

35. $\sqrt{x - 6} = \sqrt{x + 9} - 3$

36. $\sqrt{x - 1} + 3 = x$

37. *Length of a Side of a Square.* The diagonal of a square has length $7\sqrt{2}$ ft. Find the length of a side of the square.

38. *Sighting to the Horizon.* A person can see 72 mi to the horizon from an airplane window. How high is the airplane? Use the formula $D = 1.2\sqrt{h}$, where D is in miles and h is in feet.

In a right triangle, find the length of the side not given. Give an exact answer and an approximation to three decimal places.

39. $a = 7, \ b = 7$

40. $a = 1, \ c = \sqrt{5}$

41. Determine the intervals on which the function is (a) increasing; (b) decreasing; (c) constant.

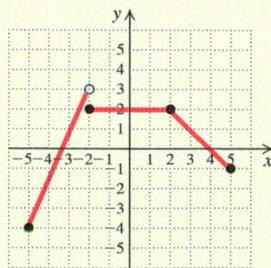

42. Determine any relative maxima or minima of the function.

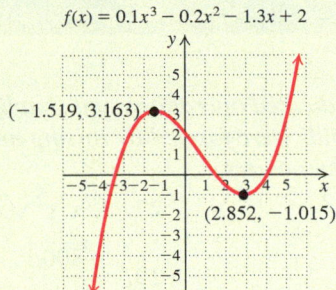

$f(x) = 0.1x^3 - 0.2x^2 - 1.3x + 2$

$(-1.519, 3.163)$

$(2.852, -1.015)$

43. *Triangular Pennant.*   A softball team is designing a triangular pennant such that the height is 6 in. less than four times the length of the base b. Express the area of the pennant as a function of b.

44. Graph:

$$f(x) = \begin{cases} x^2, & \text{for } x < -1, \\ |x|, & \text{for } -1 \le x \le 1, \\ \sqrt{x - 1}, & \text{for } x > 1. \end{cases}$$

45. For the function in Exercise 44, find $f\left(-\frac{7}{8}\right)$, $f(5)$, and $f(-4)$.

46. Which of the following describes the solution(s) of the equation $x - 4 = \sqrt{x - 2}$?

A. There is exactly one solution, and it is positive.

B. There are one positive solution and one negative solution.

C. There are two positive solutions.

D. There is no solution.

▶ **Synthesis**

47. Find the domain of

$$f(x) = \frac{\sqrt{1 - x}}{x + 4}.$$

48. Solve: $\sqrt{2x - 2} + \sqrt{7x + 4} = \sqrt{13x + 10}$.

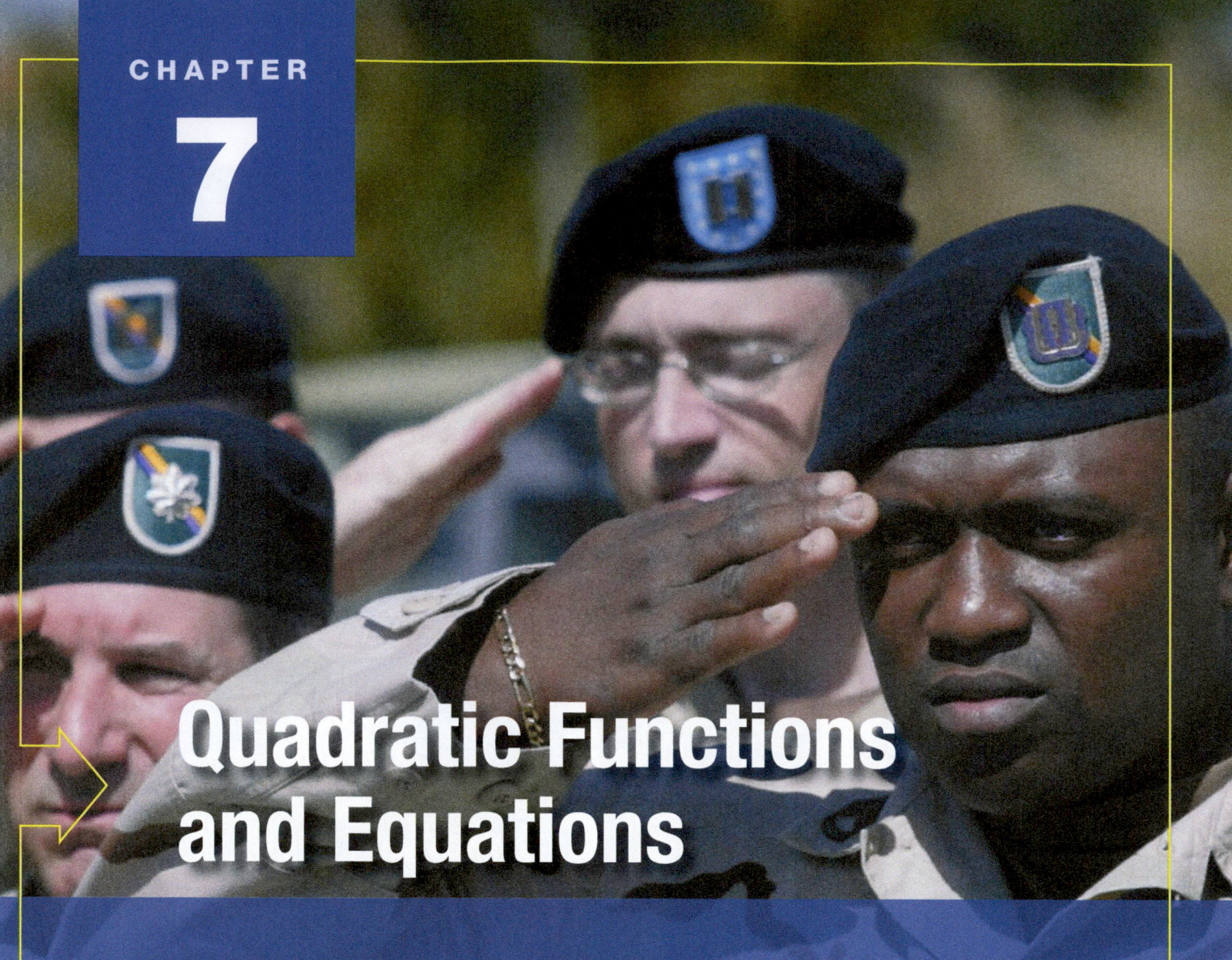

Quadratic Functions and Equations

APPLICATION   This problem appears as Exercise 95 in Section 7.4.

The number of U.S. forces in Afghanistan decreased to approximately 34,000 in 2014 from a high of about 100,000 in 2010. The amount of U.S. funding for Afghan security forces also decreased during this period. The function

$$f(x) = -1.321x^2 + 5.156x + 5.517$$

can be used to estimate the amount of U.S. funding for Afghan security forces, in billions of dollars, x years after 2009 (*Sources:* U.S. Department of Defense; Brookings Institution; International Security Assistance Force; ESRI). In what year was the amount of U.S. funding about $10.5 billion?

7.1 Symmetry

▶ Determine whether a graph is symmetric with respect to the *x*-axis, the *y*-axis, and the origin.

▶ Determine whether a function is even, odd, or neither even nor odd.

▶ Symmetry

Symmetry occurs often in nature and in art. For example, when viewed from the front, the bodies of most animals are at least approximately symmetric. This means that each eye is the same distance from the center of the bridge of the nose, each shoulder is the same distance from the center of the chest, and so on. Architects have used symmetry for thousands of years to enhance the beauty of buildings.

An understanding in mathematics helps us graph and analyze equations and functions.

Consider the points $(4, 2)$ and $(4, -2)$ that appear on the graph of $x = y^2$, as shown below. Points like these have the same *x*-value but opposite *y*-values and are **reflections** of each other across the *x*-axis. If, for any point (x, y) on a graph, the point $(x, -y)$ is also on the graph, then the graph is said to be **symmetric with respect to the x-axis**. If we fold the graph on the *x*-axis, the parts above and below the *x*-axis will coincide.

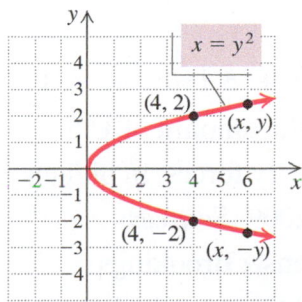

Consider the points $(3, 4)$ and $(-3, 4)$ that appear on the graph of $y = x^2 - 5$, as shown below. Points like these have the same y-value but opposite x-values and are **reflections** of each other across the y-axis. If, for any point (x, y) on a graph, the point $(-x, y)$ is also on the graph, then the graph is said to be **symmetric with respect to the y-axis**. If we fold the graph on the y-axis, the parts to the left and right of the y-axis will coincide.

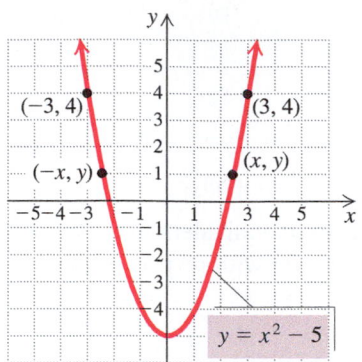

Consider the points $\left(-3, \sqrt{7}\right)$ and $\left(3, -\sqrt{7}\right)$ that appear on the graph of $x^2 = y^2 + 2$, as shown below. Note that if we take the opposites of the coordinates of one pair, we get the other pair. If, for any point (x, y) on a graph, the point $(-x, -y)$ is also on the graph, then the graph is said to be **symmetric with respect to the origin**. Visually, if we rotate the graph $180°$ about the origin, the resulting figure coincides with the original.

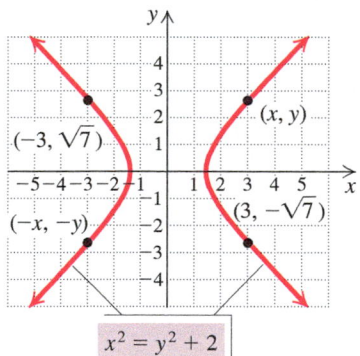

ALGEBRAIC TESTS OF SYMMETRY

x-axis: If replacing y with $-y$ produces an equivalent equation, then the graph is *symmetric with respect to the x-axis.*

y-axis: If replacing x with $-x$ produces an equivalent equation, then the graph is *symmetric with respect to the y-axis.*

Origin: If replacing x with $-x$ and y with $-y$ produces an equivalent equation, then the graph is *symmetric with respect to the origin.*

EXAMPLE 1 Test $y = x^2 + 2$ for symmetry with respect to the x-axis, the y-axis, and the origin.

Algebraic Solution

x-Axis:
We replace y with $-y$:

$$y = x^2 + 2$$
$$-y = x^2 + 2$$
$$y = -x^2 - 2. \qquad \text{\color{red}\textbf{Multiplying by} } -1 \text{ \color{red}\textbf{on both sides}}$$

The resulting equation *is not* equivalent to the original equation, so the graph *is not* symmetric with respect to the x-axis.

y-Axis:
We replace x with $-x$:

$$y = x^2 + 2$$
$$y = (-x)^2 + 2$$
$$y = x^2 + 2. \qquad \text{\color{red}\textbf{Simplifying}}$$

The resulting equation *is* equivalent to the original equation, so the graph *is* symmetric with respect to the y-axis.

Origin:
We replace x with $-x$ and y with $-y$:

$$y = x^2 + 2$$
$$-y = (-x)^2 + 2$$
$$-y = x^2 + 2 \qquad \text{\color{red}\textbf{Simplifying}}$$
$$y = -x^2 - 2.$$

The resulting equation *is not* equivalent to the original equation, so the graph *is not* symmetric with respect to the origin.

Visualizing the Solution

Let's look at the graph of $y = x^2 + 2$.

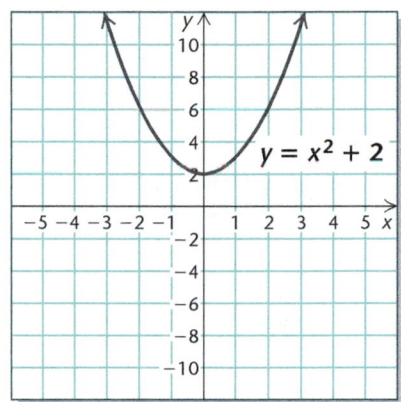

Note that if the graph were folded on the x-axis, the parts above and below the x-axis would not coincide. If it were folded on the y-axis, the parts to the left and right of the y-axis would coincide. If we rotated it 180° about the origin, the resulting graph would not coincide with the original graph.

Thus we see that the graph *is not* symmetric with respect to the x-axis or the origin. The graph *is* symmetric with respect to the y-axis.

Now Try Exercise 11.

EXAMPLE 2 Test $x^2 + y^4 = 5$ for symmetry with respect to the x-axis, the y-axis, and the origin.

Algebraic Solution

x-Axis:
We replace y with $-y$:

$$x^2 + y^4 = 5$$
$$x^2 + (-y)^4 = 5$$
$$x^2 + y^4 = 5.$$

The resulting equation *is* equivalent to the original equation. Thus the graph *is* symmetric with respect to the x-axis.

y-Axis:
We replace x with $-x$:

$$x^2 + y^4 = 5$$
$$(-x)^2 + y^4 = 5$$
$$x^2 + y^4 = 5.$$

The resulting equation *is* equivalent to the original equation, so the graph *is* symmetric with respect to the y-axis.

Origin:
We replace x with $-x$ and y with $-y$:

$$x^2 + y^4 = 5$$
$$(-x)^2 + (-y)^4 = 5$$
$$x^2 + y^4 = 5.$$

The resulting equation *is* equivalent to the original equation, so the graph *is* symmetric with respect to the origin.

Visualizing the Solution

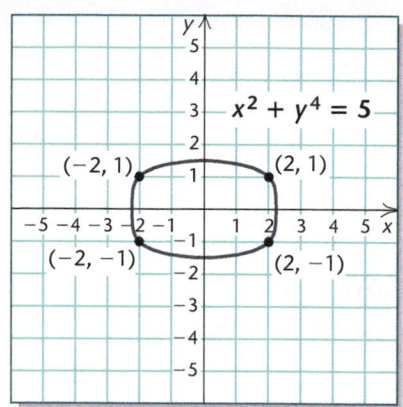

From the graph of the equation, we see symmetry with respect to both axes and with respect to the origin.

Now Try Exercise 21.

Algebraic Procedure for Determining Even Functions and Odd Functions

Given the function $f(x)$:

1. Find $f(-x)$ and simplify. If $f(x) = f(-x)$, then f is even.
2. Find $-f(x)$, simplify, and compare with $f(-x)$ from step (1). If $f(-x) = -f(x)$, then f is odd.

Except for the function $f(x) = 0$, a function cannot be *both* even and odd. Thus if $f(x) \neq 0$ and we see in step (1) that $f(x) = f(-x)$ (that is, f is even), we need not continue.

▶ Even Functions and Odd Functions

Now we relate symmetry to graphs of functions.

EVEN FUNCTIONS AND ODD FUNCTIONS

If the graph of a function f is symmetric with respect to the y-axis, we say that it is an **even function**. That is, for each x in the domain of f, $f(x) = f(-x)$.

If the graph of a function f is symmetric with respect to the origin, we say that it is an **odd function**. That is, for each x in the domain of f, $f(-x) = -f(x)$.

An algebraic procedure for determining even functions and odd functions is shown at left. Below we show an even function and an odd function. Many functions are neither even nor odd.

EXAMPLE 3 Determine whether each of the following functions is even, odd, or neither.

a) $f(x) = 5x^7 - 6x^3 - 2x$

b) $h(x) = 5x^6 - 3x^2 - 7$

a) Algebraic Solution

$f(x) = 5x^7 - 6x^3 - 2x$

1. $f(-x) = 5(-x)^7 - 6(-x)^3 - 2(-x)$
   $= 5(-x^7) - 6(-x^3) + 2x$

   $(-x)^7 = (-1 \cdot x)^7 = (-1)^7 x^7 = -x^7; \ (-x)^3 = -x^3$

   $= -5x^7 + 6x^3 + 2x$

   We see that $f(x) \neq f(-x)$. Thus, f is not even.

2. $-f(x) = -(5x^7 - 6x^3 - 2x)$
   $= -5x^7 + 6x^3 + 2x$

   We see that $f(-x) = -f(x)$. Thus, f is odd.

Visualizing the Solution

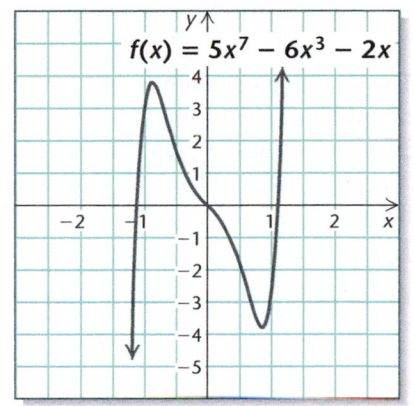

We see that the graph appears to be symmetric with respect to the origin. The function is odd.

b) Algebraic Solution

$h(x) = 5x^6 - 3x^2 - 7$

1. $h(-x) = 5(-x)^6 - 3(-x)^2 - 7$
   $= 5x^6 - 3x^2 - 7$

   We see that $h(x) = h(-x)$. Thus the function is even.

Visualizing the Solution

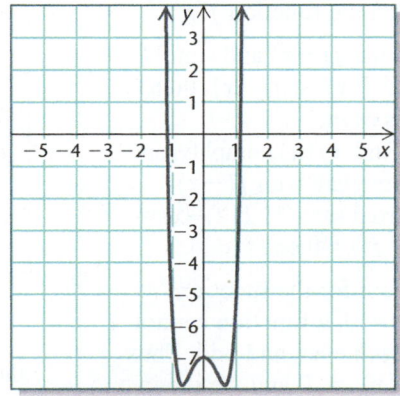

$h(x) = 5x^6 - 3x^2 - 7$

We see that the graph appears to be symmetric with respect to the y-axis. The function is even.

Now Try Exercises 39 and 41.

7.1   Exercise Set

Determine visually whether the graph is symmetric with respect to the x-axis, the y-axis, and the origin.

1.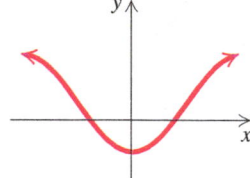

2.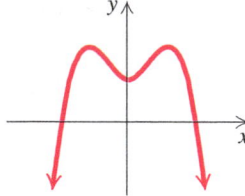

3.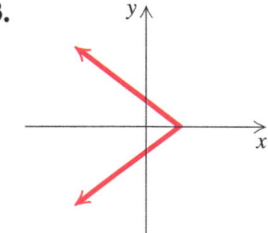

4.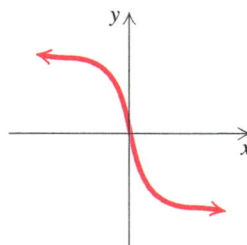

5.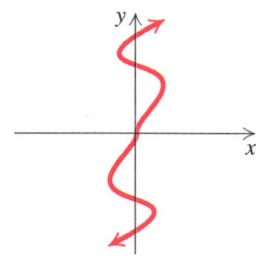

6.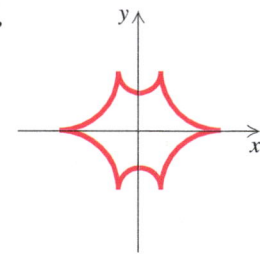

First, graph the equation and determine visually whether it is symmetric with respect to the x-axis, the y-axis, and the origin. Then verify your assertion algebraically.

7. $y = |x| - 2$

8. $y = |x + 5|$

9. $5y = 4x + 5$

10. $2x - 5 = 3y$

11. $5y = 2x^2 - 3$

12. $x^2 + 4 = 3y$

13. $y = \dfrac{1}{x}$

14. $y = -\dfrac{4}{x}$

Determine whether the graph is symmetric with respect to the x-axis, the y-axis, and the origin.

15. $5x - 5y = 0$

16. $6x + 7y = 0$

17. $3x^2 - 2y^2 = 3$

18. $5y = 7x^2 - 2x$

19. $y = |2x|$

20. $y^3 = 2x^2$

21. $2x^4 + 3 = y^2$

22. $2y^2 = 5x^2 + 12$

23. $3y^3 = 4x^3 + 2$

24. $3x = |y|$

25. $xy = 12$

26. $xy - x^2 = 3$

Find the point that is symmetric to the given point with respect to the x-axis, the y-axis, and the origin.

27. $(-5, 6)$

28. $\left(\frac{7}{2}, 0\right)$

29. $(-10, -7)$

30. $\left(1, \frac{3}{8}\right)$

31. $(0, -4)$

32. $(8, -3)$

Determine visually whether the function is even, odd, or neither even nor odd.

33.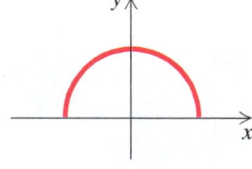

34.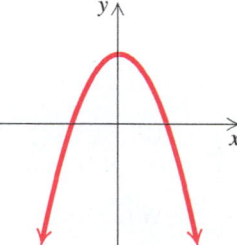

35.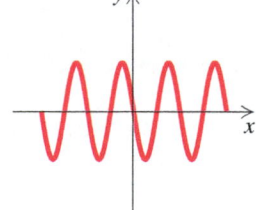

36.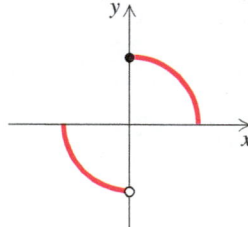

37.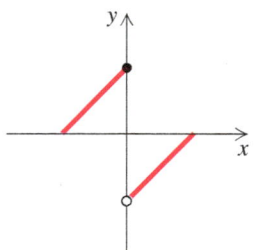

38.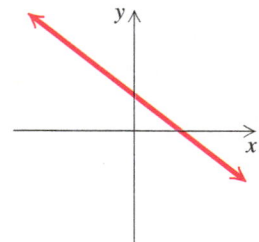

Determine whether the function is even, odd, or neither even nor odd.

39. $f(x) = -3x^3 + 2x$

40. $f(x) = 7x^3 + 4x - 2$

41. $f(x) = 5x^2 + 2x^4 - 1$

42. $f(x) = x + \dfrac{1}{x}$

43. $f(x) = x^{17}$

44. $f(x) = \sqrt[3]{x}$

45. $f(x) = x - |x|$

46. $f(x) = \dfrac{1}{x^2}$

47. $f(x) = 8$

48. $f(x) = \sqrt{x^2 + 1}$

▶ # Skill Maintenance

49. Graph: $f(x) = \begin{cases} x - 2, & \text{for } x \le -1, \\ 3, & \text{for } -1 < x \le 2, \\ x, & \text{for } x > 2. \end{cases}$ [6.8]

50. *Peace Corps Volunteers.*   Since 1961, there has been a total of 6688 Peace Corps volunteers from the University of California–Berkeley and the University of Wisconsin–Madison. The number of volunteers from the University of California–Berkeley is 464 more than the number of volunteers from the University of Wisconsin–Madison. (*Source:* Peace Corps 2014) Find the number of Peace Corps volunteers from each university. [1.3]

▶ # Synthesis

Determine whether the function is even, odd, or neither even nor odd.

51. $f(x) = x\sqrt{10 - x^2}$

52. $f(x) = \dfrac{x^2 + 1}{x^3 - 1}$

Determine whether the graph is symmetric with respect to the x-axis, the y-axis, and the origin.

53. $x^3 = y^2(2 - x)$

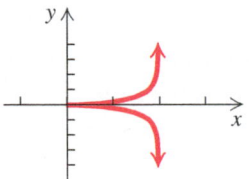

54. $(x^2 + y^2)^2 = 2xy$

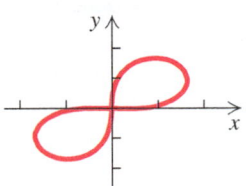

55. Show that if f is *any* function, then the function E defined by

$$E(x) = \frac{f(x) + f(-x)}{2}$$

is even.

56. Show that if f is *any* function, then the function O defined by

$$O(x) = \frac{f(x) - f(-x)}{2}$$

is odd.

57. Consider the functions E and O of Exercises 55 and 56.
a) Show that $f(x) = E(x) + O(x)$. This means that every function can be expressed as the sum of an even function and an odd function.
b) Let $f(x) = 4x^3 - 11x^2 + \sqrt{x} - 10$. Express f as a sum of an even function and an odd function.

Determine whether the statement is true or false.

58. The product of two odd functions is odd.

59. The sum of two even functions is even.

60. The product of an even function and an odd function is odd.

7.2   Transformations

▶ Given the graph of a function, graph its transformation under translations, reflections, stretchings, and shrinkings.

▶ ## Transformations of Functions

The graphs of some basic functions are shown on the following page. Others can be seen on the inside back cover.

Identity function:
$y = x$

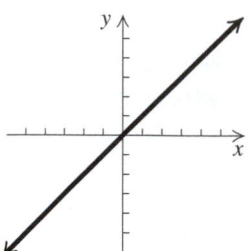

Squaring function:
$y = x^2$

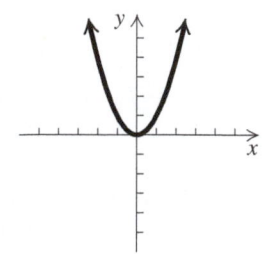

Square root function:
$y = \sqrt{x}$

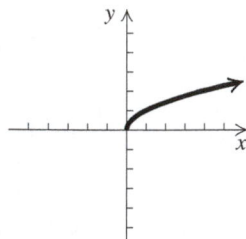

Cubing function:
$y = x^3$

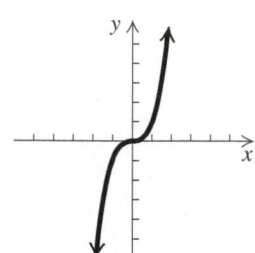

Cube root function:
$y = \sqrt[3]{x}$

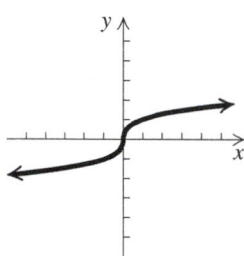

Reciprocal function:
$y = \dfrac{1}{x}$

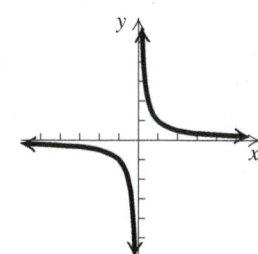

Absolute-value function:
$y = |x|$

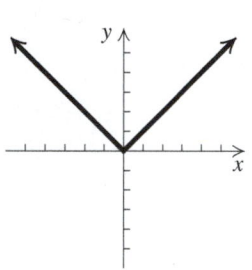

These functions can be considered building blocks for many other functions. We can create graphs of new functions by shifting them horizontally or vertically, stretching or shrinking them, and reflecting them across an axis. We now consider these **transformations**.

▶ Vertical Translations and Horizontal Translations

Suppose that we have a function given by $y = f(x)$. Let's explore the graphs of the new functions $y = f(x) + b$ and $y = f(x) - b$, for $b > 0$.

Consider the functions $y = \frac{1}{5}x^4$, $y = \frac{1}{5}x^4 + 5$, and $y = \frac{1}{5}x^4 - 3$ and compare their graphs. What pattern do you see? Test it with some other functions.

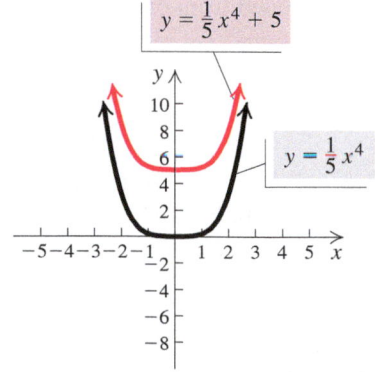

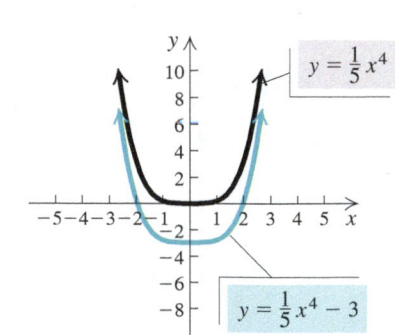

The effect of adding a constant to or subtracting a constant from $f(x)$ in $y = f(x)$ is a shift of the graph of $f(x)$ up or down. Such a shift is called a **vertical translation**.

VERTICAL TRANSLATION

For $b > 0$:

the graph of $y = f(x) + b$ is the graph of $y = f(x)$ shifted *up* b units;

the graph of $y = f(x) - b$ is the graph of $y = f(x)$ shifted *down* b units.

Suppose that we have a function given by $y = f(x)$. Let's explore the graphs of the new functions $y = f(x - d)$ and $y = f(x + d)$, for $d > 0$.

Consider the functions $y = \frac{1}{5}x^4$, $y = \frac{1}{5}(x - 3)^4$, and $y = \frac{1}{5}(x + 7)^4$ and compare their graphs. What pattern do you observe? Test it with some other functions.

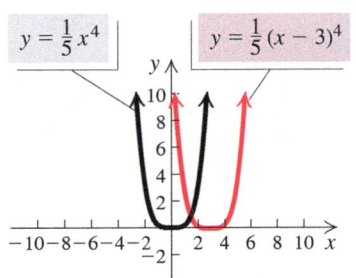
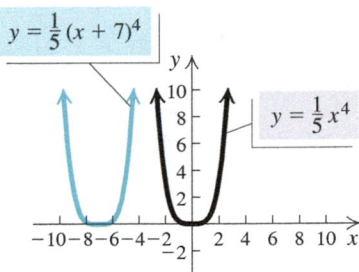

The effect of subtracting a constant from the x-value or adding a constant to the x-value in $y = f(x)$ is a shift of the graph of $f(x)$ to the right or to the left. Such a shift is called a **horizontal translation**.

HORIZONTAL TRANSLATION

For $d > 0$:

the graph of $y = f(x - d)$ is the graph of $y = f(x)$ shifted *right* d units;

the graph of $y = f(x + d)$ is the graph of $y = f(x)$ shifted *left* d units.

EXAMPLE 1 Graph each of the following. Before doing so, describe how each graph can be obtained from one of the basic graphs shown on the preceding page.

a) $g(x) = x^2 - 6$

b) $h(x) = |x - 4|$

c) $g(x) = \sqrt{x} + 2$

d) $h(x) = \sqrt{x + 2} - 3$

Solution

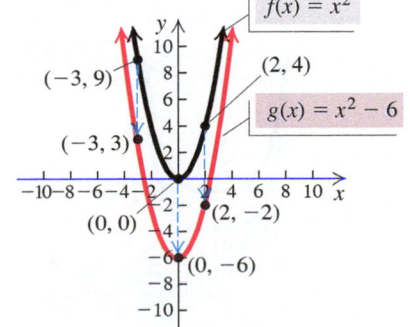

Figure 1.

a) To graph $g(x) = x^2 - 6$, think of the graph of $f(x) = x^2$. Since $g(x) = f(x) - 6$, the graph of $g(x) = x^2 - 6$ is the graph of $f(x) = x^2$, shifted, or translated, *down* 6 units. (See Fig. 1.)

Let's compare some points on the graphs of f and g.

Points on f: $\qquad (-3, 9), \qquad (0, 0), \qquad (2, 4)$

Corresponding points on g: $\quad (-3, 3), \qquad (0, -6), \qquad (2, -2)$

We note that the y-coordinate of a point on the graph of g is 6 less than the corresponding y-coordinate on the graph of f.

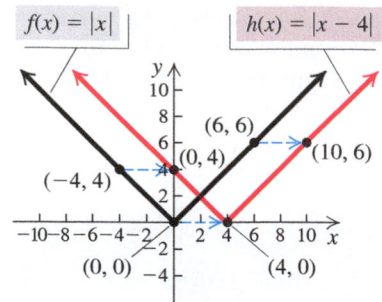

Figure 2.

b) To graph $h(x) = |x - 4|$, think of the graph of $f(x) = |x|$. Since $h(x) = f(x - 4)$, the graph of $h(x) = |x - 4|$ is the graph of $f(x) = |x|$ shifted *right* 4 units. (See Fig. 2.)

Let's again compare points on the two graphs.

Points on f: $\qquad (-4, 4), \qquad (0, 0), \qquad (6, 6)$

Corresponding $\qquad\qquad\qquad \downarrow \qquad\qquad \downarrow \qquad\qquad \downarrow$
points on h: $\qquad (0, 4), \qquad (4, 0), \qquad (10, 6)$

Noting points on f and h, we see that the *x*-coordinate of a point on the graph of h is 4 more than the *x*-coordinate of the corresponding point on f.

c) To graph $g(x) = \sqrt{x + 2}$, think of the graph of $f(x) = \sqrt{x}$. Since $g(x) = f(x + 2)$, the graph of $g(x) = \sqrt{x + 2}$ is the graph of $f(x) = \sqrt{x}$, shifted *left* 2 units. (See Fig. 3.)

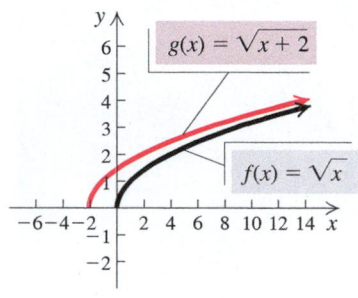

Figure 3.

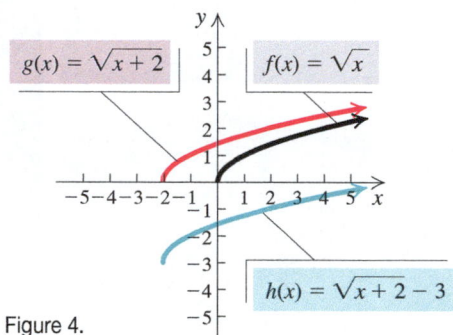

Figure 4.

d) To graph $h(x) = \sqrt{x + 2} - 3$, think of the graph of $f(x) = \sqrt{x}$. In part (c), we found that the graph of $g(x) = \sqrt{x + 2}$ is the graph of $f(x) = \sqrt{x}$ shifted *left* 2 units. Since $h(x) = g(x) - 3$, we shift the graph of $g(x) = \sqrt{x + 2}$ *down* 3 units. Together, the graph of $f(x) = \sqrt{x}$ is shifted *left* 2 units and *down* 3 units. (See Fig. 4.)

Now Try Exercises 3 and 15.

▶ **Reflections**

Suppose that we have a function given by $y = f(x)$. Let's explore the graphs of the new functions $y = -f(x)$ and $y = f(-x)$.

Compare the functions $y = f(x)$ and $y = -f(x)$ by looking at the graphs of $y = \frac{1}{5}x^4$ and $y = -\frac{1}{5}x^4$ shown on the left below. What do you see? Test your observation with some other functions y_1 and y_2 where $y_2 = -y_1$.

Compare the functions $y = f(x)$ and $y = f(-x)$ by looking at the graphs of $y = 2x^3 - x^4 + 5$ and $y = 2(-x)^3 - (-x)^4 + 5$ shown on the right below. What do you see? Test your observation with some other functions in which x is replaced with $-x$.

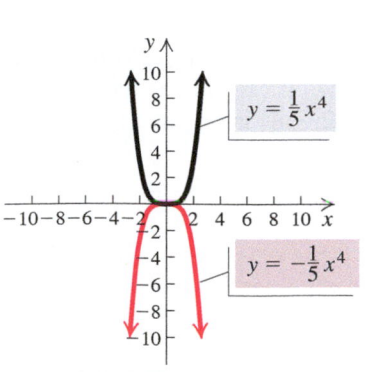

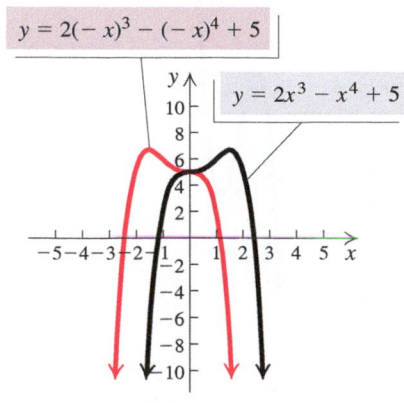

Given the graph of $y = f(x)$, we can reflect each point *across the x-axis* to obtain the graph of $y = -f(x)$. We can reflect each point of $y = f(x)$ *across the y-axis* to obtain the graph of $y = f(-x)$. The new graphs are called **reflections** of $y = f(x)$.

The following photographs illustrate reflection.

REFLECTIONS

The graph of $y = -f(x)$ is the **reflection** of the graph of $y = f(x)$ across the x-axis.

The graph of $y = f(-x)$ is the **reflection** of the graph of $y = f(x)$ across the y-axis.

If a point (x, y) is on the graph of $y = f(x)$, then $(x, -y)$ is on the graph of $y = -f(x)$, and $(-x, y)$ is on the graph of $y = f(-x)$.

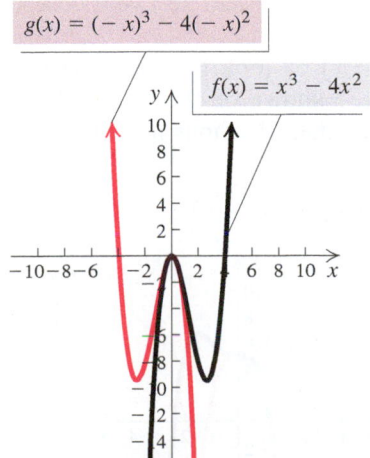

$g(x) = (-x)^3 - 4(-x)^2$

$f(x) = x^3 - 4x^2$

EXAMPLE 2 Graph each of the following. Before doing so, describe how each graph can be obtained from the graph of $f(x) = x^3 - 4x^2$.

a) $g(x) = (-x)^3 - 4(-x)^2$ **b)** $h(x) = 4x^2 - x^3$

Solution

a) We first note that

$$f(-x) = (-x)^3 - 4(-x)^2 = g(x).$$

Thus the graph of g is a *reflection* of the graph of f across the y-axis. (See the figure at left.) If (x, y) is on the graph of f, then $(-x, y)$ is on the graph of g. For example, $(2, -8)$ is on f and $(-2, -8)$ is on g.

b) We first note that

$$-f(x) = -(x^3 - 4x^2)$$
$$= -x^3 + 4x^2$$
$$= 4x^2 - x^3$$
$$= h(x).$$

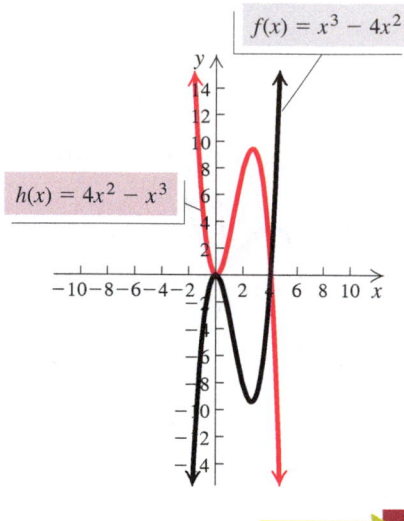

Thus the graph of h is a reflection of the graph of f across the x-axis. (See the figure at right.) If (x, y) is on the graph of f, then $(x, -y)$ is on the graph of h. For example, $(2, -8)$ is on f and $(2, 8)$ is on h.

▶ **Vertical and Horizontal Stretchings and Shrinkings**

Suppose that we have a function given by $y = f(x)$. Let's explore the graphs of the new functions $y = af(x)$ and $y = f(cx)$.

Consider the functions $y = f(x) = x^3 - x$, $y = \frac{1}{10}(x^3 - x) = \frac{1}{10}f(x)$, $y = 2(x^3 - x) = 2f(x)$, and $y = -2(x^3 - x) = -2f(x)$ and compare their graphs. What pattern do you observe? Test it with some other functions.

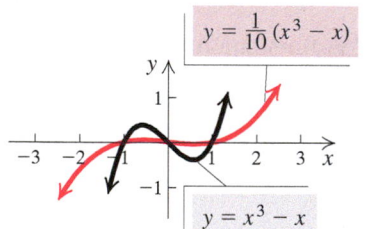

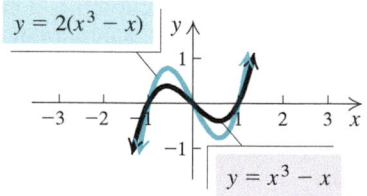

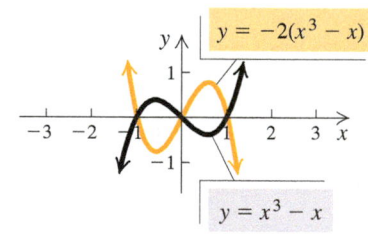

Consider any function f given by $y = f(x)$. Multiplying $f(x)$ by any constant a, where $|a| > 1$, to obtain $g(x) = af(x)$ will *stretch* the graph vertically away from the x-axis. If $0 < |a| < 1$, then the graph will be flattened or *shrunk* vertically toward the x-axis. If $a < 0$, the graph is also reflected across the x-axis.

VERTICAL STRETCHING AND SHRINKING

The graph of $y = af(x)$ can be obtained from the graph of $y = f(x)$ by

stretching vertically for $|a| > 1$, or
shrinking vertically for $0 < |a| < 1$.

For $a < 0$, the graph is also reflected across the x-axis. (The y-coordinates of the graph of $y = af(x)$ can be obtained by multiplying the y-coordinates of $y = f(x)$ by a.)

Consider the functions $y = f(x) = x^3 - x$, $y = (2x)^3 - (2x) = f(2x)$, $y = \left(\frac{1}{2}x\right)^3 - \left(\frac{1}{2}x\right) = f\left(\frac{1}{2}x\right)$, and $y = \left(-\frac{1}{2}x\right)^3 - \left(-\frac{1}{2}x\right) = f\left(-\frac{1}{2}x\right)$ and compare their graphs. What pattern do you observe? Test it with some other functions.

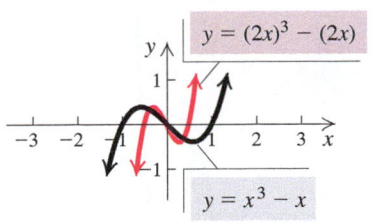

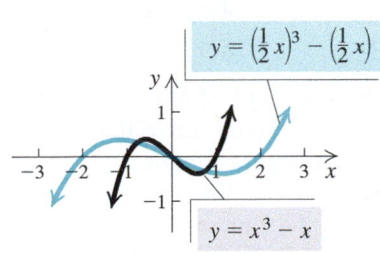

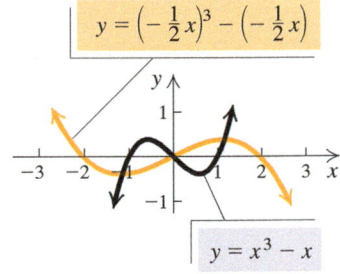

The constant c in the equation $g(x) = f(cx)$ will *shrink* the graph of $y = f(x)$ horizontally toward the y-axis if $|c| > 1$. If $0 < |c| < 1$, the graph will be *stretched* horizontally away from the y-axis. If $c < 0$, the graph is also reflected across the y-axis.

HORIZONTAL STRETCHING AND SHRINKING

The graph of $y = f(cx)$ can be obtained from the graph of $y = f(x)$ by

shrinking horizontally for $|c| > 1$, or

stretching horizontally for $0 < |c| < 1$.

For $c < 0$, the graph is also reflected across the y-axis. (The x-coordinates of the graph of $y = f(cx)$ can be obtained by dividing the x-coordinates of the graph of $y = f(x)$ by c.)

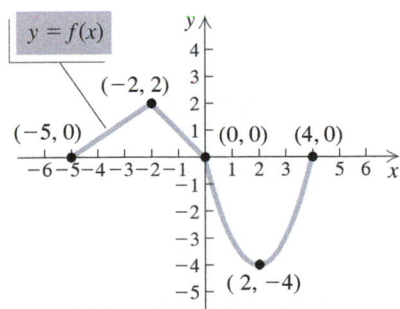

EXAMPLE 3 Shown at left is a graph of $y = f(x)$ for some function f. No formula for f is given. Graph each of the following.

a) $g(x) = 2f(x)$ **b)** $h(x) = \frac{1}{2}f(x)$ **c)** $r(x) = f(2x)$

d) $s(x) = f\left(\frac{1}{2}x\right)$ **e)** $t(x) = f\left(-\frac{1}{2}x\right)$

Solution

a) Since $|2| > 1$, the graph of $g(x) = 2f(x)$ is a vertical stretching of the graph of $y = f(x)$ by a factor of 2. We can consider the key points $(-5, 0)$, $(-2, 2)$, $(0, 0)$, $(2, -4)$, and $(4, 0)$ on the graph of $y = f(x)$. The transformation multiplies each y-coordinate by 2 to obtain the key points $(-5, 0)$, $(-2, 4)$, $(0, 0)$, $(2, -8)$, and $(4, 0)$ on the graph of $g(x) = 2f(x)$, as shown below.

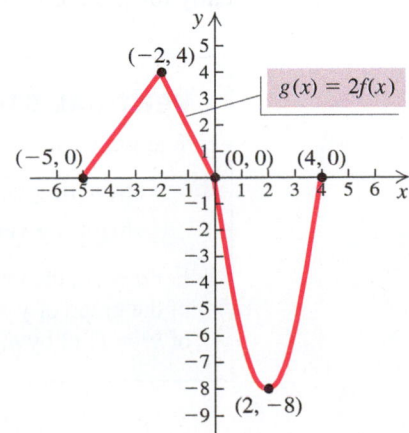

b) Since $\left|\frac{1}{2}\right| < 1$, the graph of $h(x) = \frac{1}{2}f(x)$ is a vertical shrinking of the graph of $y = f(x)$ by a factor of $\frac{1}{2}$. We again consider the key points $(-5, 0)$, $(-2, 2)$, $(0, 0)$, $(2, -4)$, and $(4, 0)$ on the graph of $y = f(x)$. The transformation multiplies each y-coordinate by $\frac{1}{2}$ to obtain the key points $(-5, 0)$, $(-2, 1)$, $(0, 0)$, $(2, -2)$, and $(4, 0)$ on the graph of $h(x) = \frac{1}{2}f(x)$. The graph is shown on the left below.

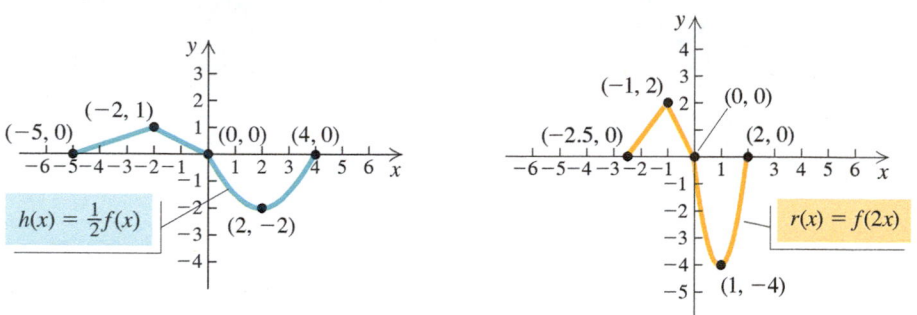

c) Since $|2| > 1$, the graph of $r(x) = f(2x)$ is a horizontal shrinking of the graph of $y = f(x)$. We consider the key points $(-5, 0)$, $(-2, 2)$, $(0, 0)$, $(2, -4)$, and $(4, 0)$ on the graph of $y = f(x)$. The transformation divides each x-coordinate by 2 to obtain the key points $(-2.5, 0)$, $(-1, 2)$, $(0, 0)$, $(1, -4)$, and $(2, 0)$ on the graph of $r(x) = f(2x)$. The graph is shown on the right above.

d) Since $\left|\frac{1}{2}\right| < 1$, the graph of $s(x) = f\left(\frac{1}{2}x\right)$ is a horizontal stretching of the graph of $y = f(x)$. We consider the key points $(-5, 0)$, $(-2, 2)$, $(0, 0)$, $(2, -4)$, and $(4, 0)$ on the graph of $y = f(x)$. The transformation divides each x-coordinate by $\frac{1}{2}$ (which is the same as multiplying by 2) to obtain the key points $(-10, 0)$, $(-4, 2)$, $(0, 0)$, $(4, -4)$, and $(8, 0)$ on the graph of $s(x) = f\left(\frac{1}{2}x\right)$. The graph is shown below.

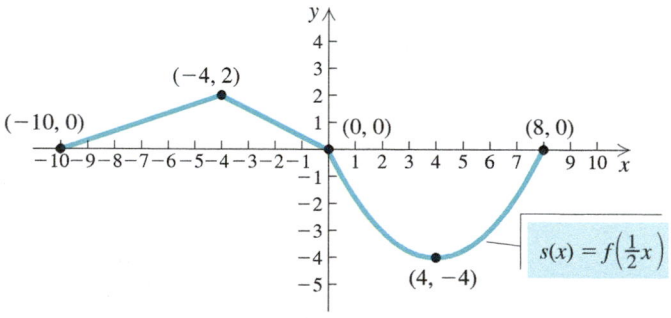

e) The graph of $t(x) = f\left(-\frac{1}{2}x\right)$ can be obtained by reflecting the graph in part (d) across the y-axis.

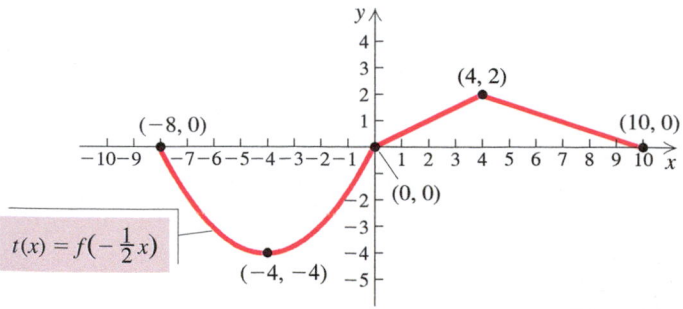

Now Try Exercises 59 and 61.

y = f(x)
(−2, 2)
(−5, 0)
(0, 0)
(4, 0)
(2, −4)

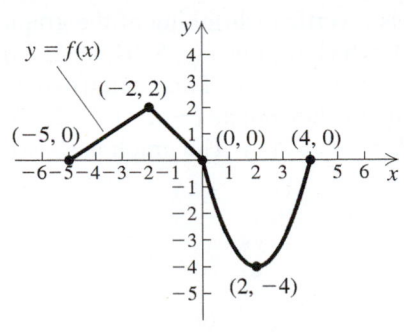

EXAMPLE 4 Use the graph of $y = f(x)$ shown at left to graph

$$y = -2f(x - 3) + 1.$$

Solution

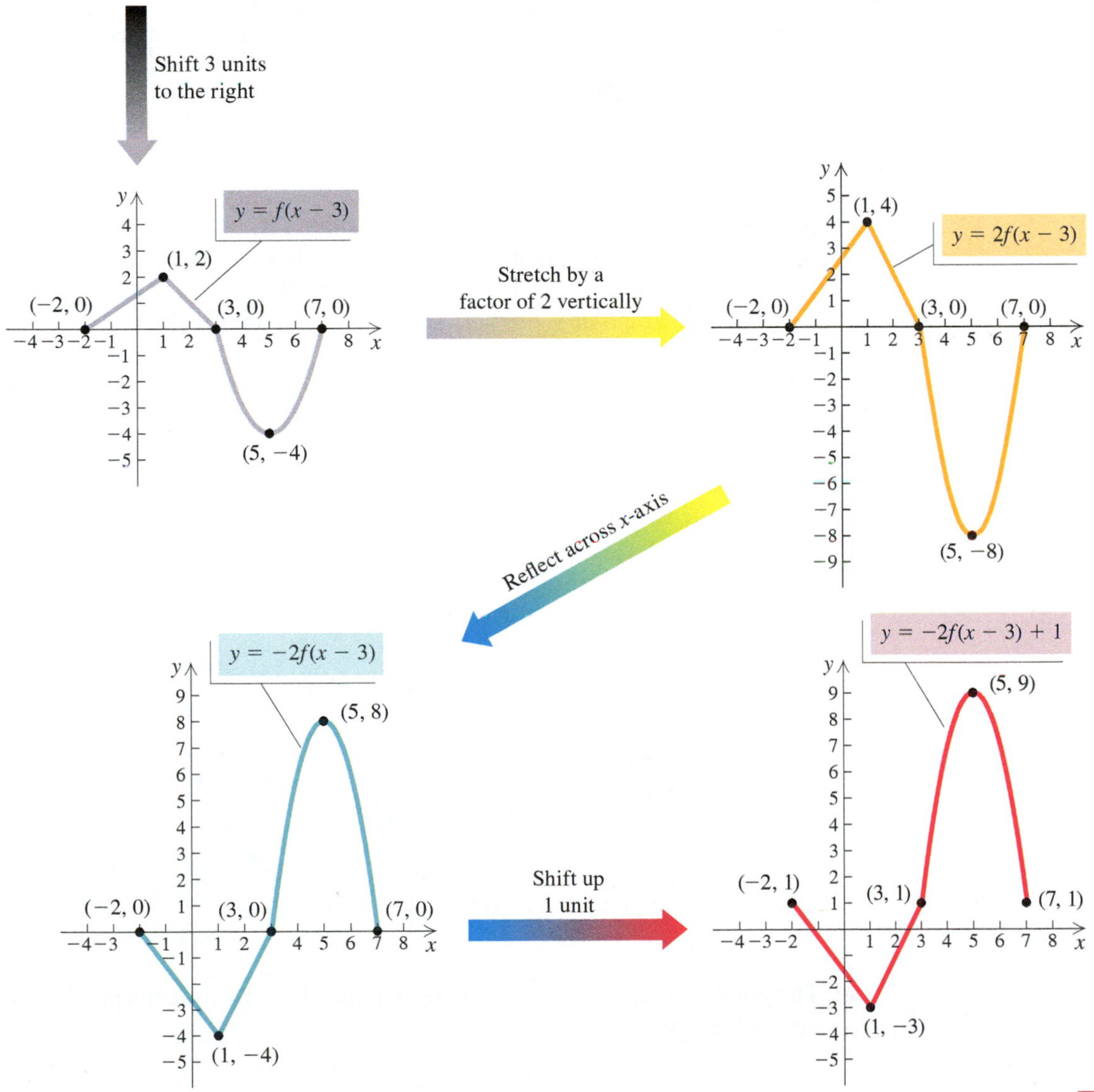

Shift 3 units
to the right
y = f(x − 3)
(1, 2)
(−2, 0)
(3, 0)
(7, 0)
(5, −4)
Stretch by a
factor of 2 vertically
y = 2f(x − 3)
(1, 4)
(−2, 0)
(3, 0)
(7, 0)
(5, −8)
Reflect across x-axis
y = −2f(x − 3)
(5, 8)
(−2, 0)
(3, 0)
(7, 0)
(1, −4)
Shift up
1 unit
y = −2f(x − 3) + 1
(5, 9)
(−2, 1)
(3, 1)
(7, 1)
(1, −3)

Now Try Exercise 63.

Summary of Transformations of $y = f(x)$

Vertical Translation: $y = f(x) \pm b$

For $b > 0$:

the graph of $y = f(x) + b$ is the graph of $y = f(x)$ shifted *up* b units;

the graph of $y = f(x) - b$ is the graph of $y = f(x)$ shifted *down* b units.

Horizontal Translation: $y = f(x \mp d)$

For $d > 0$:

the graph of $y = f(x - d)$ is the graph of $y = f(x)$ shifted *right* d units;

the graph of $y = f(x + d)$ is the graph of $y = f(x)$ shifted *left* d units.

Reflections

Across the x-axis:

The graph of $y = -f(x)$ is the reflection of the graph of $y = f(x)$ across the x-axis.

Across the y-axis:

The graph of $y = f(-x)$ is the reflection of the graph of $y = f(x)$ across the y-axis.

Vertical Stretching or Shrinking: $y = af(x)$

The graph of $y = af(x)$ can be obtained from the graph of $y = f(x)$ by

stretching vertically for $|a| > 1$, or

shrinking vertically for $0 < |a| < 1$.

For $a < 0$, the graph is also reflected across the x-axis.

Horizontal Stretching or Shrinking: $y = f(cx)$

The graph of $y = f(cx)$ can be obtained from the graph of $y = f(x)$ by

shrinking horizontally for $|c| > 1$, or

stretching horizontally for $0 < |c| < 1$.

For $c < 0$, the graph is also reflected across the y-axis.

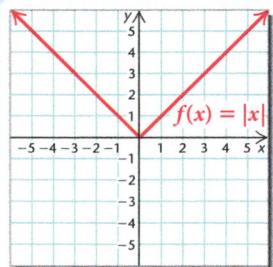

Visualizing the Graph

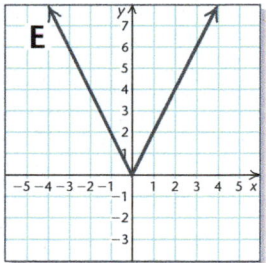

Match the function with its graph. Use transformation graphing techniques to obtain the graph of g from the basic function $f(x) = |x|$ shown at top left.

1. $g(x) = -2|x|$

2. $g(x) = |x - 1| + 1$

3. $g(x) = -\left|\dfrac{1}{3}x\right|$

4. $g(x) = |2x|$

5. $g(x) = |x + 2|$

6. $g(x) = |x| + 3$

7. $g(x) = -\dfrac{1}{2}|x - 4|$

8. $g(x) = \dfrac{1}{2}|x| - 3$

9. $g(x) = -|x| - 2$

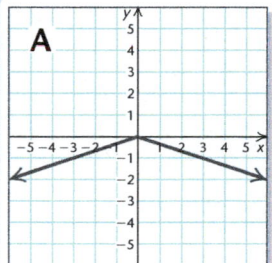

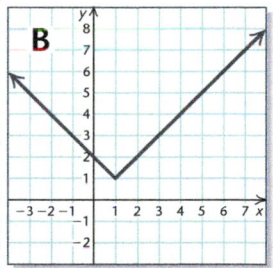

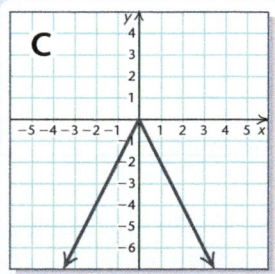

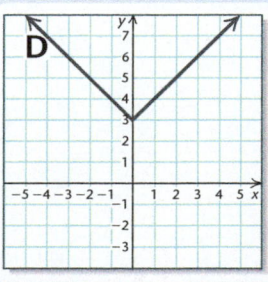

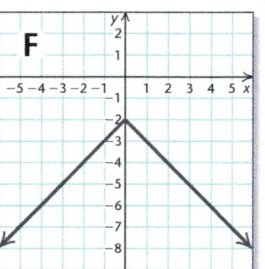

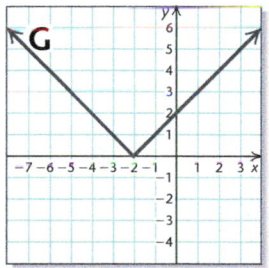

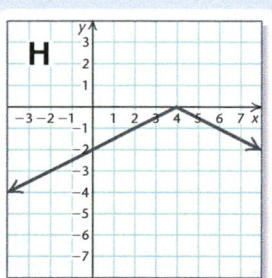

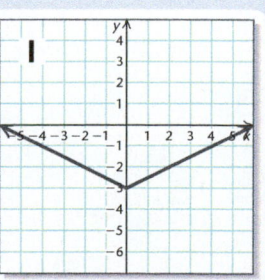

7.2 Exercise Set

Describe how the graph of the function can be obtained from one of the basic graphs on p. 445. Then graph the function.

1. $f(x) = (x - 3)^2$

2. $g(x) = x^2 + \frac{1}{2}$

3. $g(x) = x - 3$

4. $g(x) = -x - 2$

5. $h(x) = -\sqrt{x}$

6. $g(x) = \sqrt{x - 1}$

7. $h(x) = \dfrac{1}{x} + 4$

8. $g(x) = \dfrac{1}{x - 2}$

9. $h(x) = -3x + 3$

10. $f(x) = 2x + 1$

11. $h(x) = \frac{1}{2}|x| - 2$

12. $g(x) = -|x| + 2$

13. $g(x) = -(x - 2)^3$

14. $f(x) = (x + 1)^3$

15. $g(x) = (x + 1)^2 - 1$

16. $h(x) = -x^2 - 4$

17. $g(x) = \frac{1}{3}x^3 + 2$

18. $h(x) = (-x)^3$

19. $f(x) = \sqrt{x + 2}$

20. $f(x) = -\frac{1}{2}\sqrt{x} - 1$

21. $f(x) = \sqrt[3]{x} - 2$

22. $h(x) = \sqrt[3]{x + 1}$

Describe how the graph of the function can be obtained from one of the basic graphs on p. 445.

23. $g(x) = |3x|$

24. $f(x) = \frac{1}{2}\sqrt[3]{x}$

25. $h(x) = \dfrac{2}{x}$

26. $f(x) = |x - 3| - 4$

27. $f(x) = 3\sqrt{x} - 5$

28. $f(x) = 5 - \dfrac{1}{x}$

29. $g(x) = \left|\frac{1}{3}x\right| - 4$

30. $f(x) = \frac{2}{3}x^3 - 4$

31. $f(x) = -\frac{1}{4}(x - 5)^2$

32. $f(x) = (-x)^3 - 5$

33. $f(x) = \dfrac{1}{x + 3} + 2$

34. $g(x) = \sqrt{-x} + 5$

35. $h(x) = -(x - 3)^2 + 5$

36. $f(x) = 3(x + 4)^2 - 3$

The point $(-12, 4)$ is on the graph of $y = f(x)$. Find the corresponding point on the graph of $y = g(x)$.

37. $g(x) = \frac{1}{2}f(x)$

38. $g(x) = f(x - 2)$

39. $g(x) = f(-x)$

40. $g(x) = f(4x)$

41. $g(x) = f(x) - 2$

42. $g(x) = f\left(\frac{1}{2}x\right)$

43. $g(x) = 4f(x)$

44. $g(x) = -f(x)$

Given that $f(x) = x^2 + 3$, match the function g with a transformation of f from one of A–D.

45. $g(x) = x^2 + 4$ **A.** $f(x - 2)$

46. $g(x) = 9x^2 + 3$ **B.** $f(x) + 1$

47. $g(x) = (x - 2)^2 + 3$ **C.** $2f(x)$

48. $g(x) = 2x^2 + 6$ **D.** $f(3x)$

Write an equation for a function that has a graph with the given characteristics.

49. The shape of $y = x^2$, but upside-down and shifted right 8 units

50. The shape of $y = \sqrt{x}$, but shifted left 6 units and down 5 units

51. The shape of $y = |x|$, but shifted left 7 units and up 2 units

52. The shape of $y = x^3$, but upside-down and shifted right 5 units

53. The shape of $y = 1/x$, but shrunk horizontally by a factor of 2 and shifted down 3 units

54. The shape of $y = x^2$, but shifted right 6 units and up 2 units

55. The shape of $y = x^2$, but upside-down and shifted right 3 units and up 4 units

56. The shape of $y = |x|$, but stretched horizontally by a factor of 2 and shifted down 5 units

57. The shape of $y = \sqrt{x}$, but reflected across the y-axis and shifted left 2 units and down 1 unit

58. The shape of $y = 1/x$, but reflected across the x-axis and shifted up 1 unit

A graph of $y = f(x)$ follows. No formula for f is given. In Exercises 59–66, graph the given function.

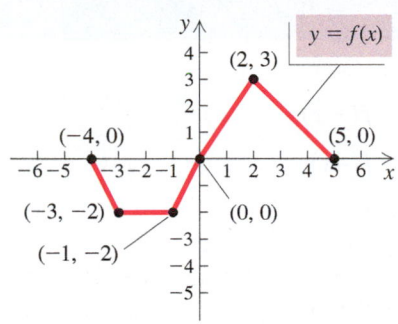

59. $g(x) = -2f(x)$

60. $g(x) = \frac{1}{2}f(x)$

61. $g(x) = f\left(-\frac{1}{2}x\right)$

62. $g(x) = f(2x)$

63. $g(x) = -\frac{1}{2}f(x - 1) + 3$

64. $g(x) = -3f(x + 1) - 4$

65. $g(x) = f(-x)$

66. $g(x) = -f(x)$

A graph of $y = g(x)$ follows. No formula for g is given. In Exercises 67–70, graph the given function.

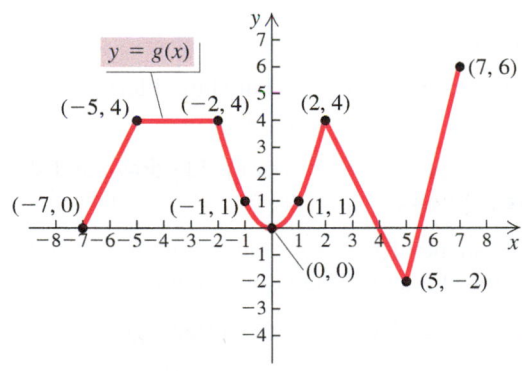

67. $h(x) = -g(x + 2) + 1$

68. $h(x) = \frac{1}{2}g(-x)$

69. $h(x) = g(2x)$

70. $h(x) = 2g(x - 1) - 3$

The graph of the function f is shown in figure (a) below. In each of Exercises 71–78, match the function g with one of the graphs (a)–(h) that follow. Some graphs may be used more than once and some may not be used at all.

a)

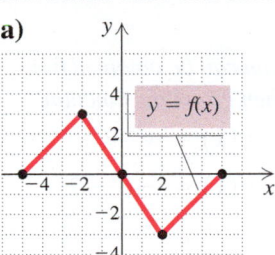

b)

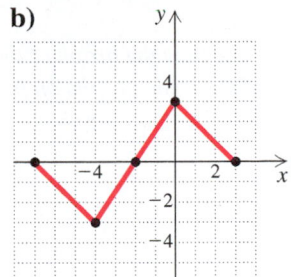

c)

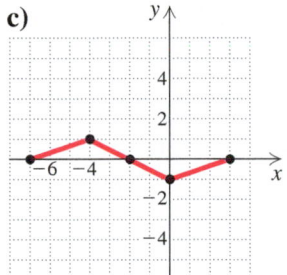

d)

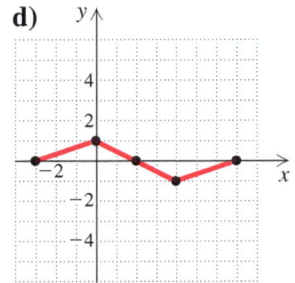

e)

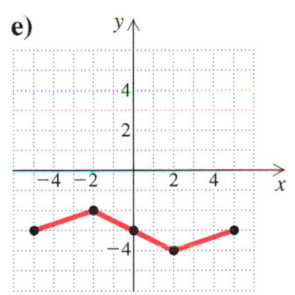

f)

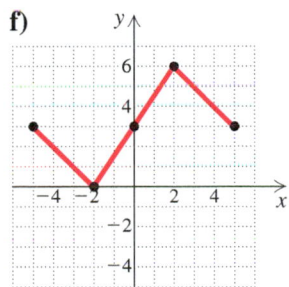

g)

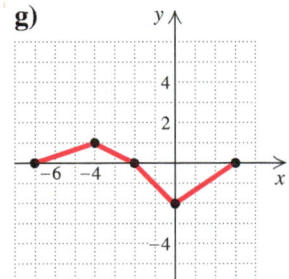

h)

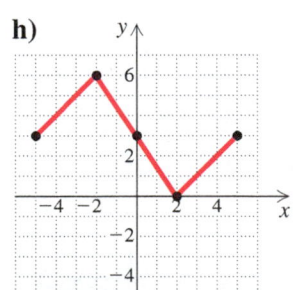

71. $g(x) = f(-x) + 3$

72. $g(x) = f(x) + 3$

73. $g(x) = -f(x) + 3$

74. $g(x) = -f(-x)$

75. $g(x) = \frac{1}{3}f(x - 2)$

76. $g(x) = \frac{1}{3}f(x) - 3$

77. $g(x) = \frac{1}{3}f(x + 2)$

78. $g(x) = -f(x + 2)$

For each pair of functions, determine if $g(x) = f(-x)$.

79. $f(x) = 2x^4 - 35x^3 + 3x - 5$,
$g(x) = 2x^4 + 35x^3 - 3x - 5$

80. $f(x) = \frac{1}{4}x^4 + \frac{1}{5}x^3 - 81x^2 - 17$,
$g(x) = \frac{1}{4}x^4 + \frac{1}{5}x^3 + 81x^2 - 17$

A graph of the function $f(x) = x^3 - 3x^2$ *is shown below. Exercises 81–84 show graphs of functions transformed from this one. Find a formula for each function.*

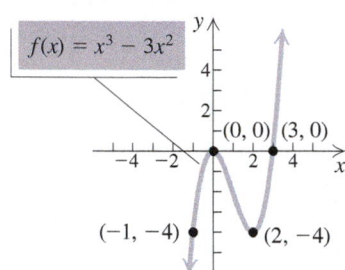

81.

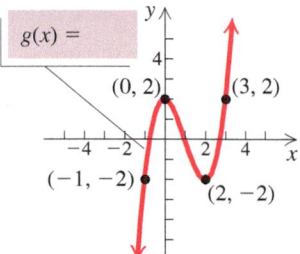

82.

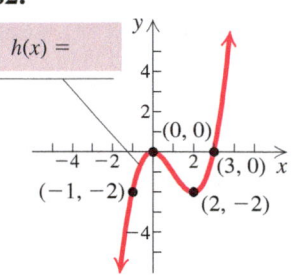

83.

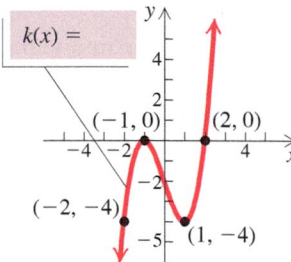

84.

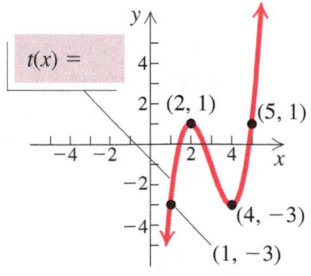

▶ Skill Maintenance

Determine algebraically whether the graph is symmetric with respect to the x-axis, the y-axis, and the origin. **[7.1]**

85. $y = 3x^4 - 3$

86. $y^2 = x$

87. $2x - 5y = 0$

Solve. **[1.3]**

88. *Federal Tax Rules.* The number of pages of U.S. federal tax rules that explain the tax code and regulations totaled 74,608 in 2014 (for tax year 2013). This number was an increase of 84.2% over the number of pages in 1995 (for tax year 1994). (*Source:* Wolters Kluwer, CCH: 2014) Find the number of pages of federal tax rules in 1995.

89. *Guns with Airline Passengers.* In 2013, the Transportation Security Administration found 1828 guns with travelers preparing to board an airplane. This number was 418 less than twice the number of guns discovered in 2010. (*Source:* Transportation Security Administration data by Northwestern University Medill National Security Journalism Initiative) How many guns were found with airline travelers in 2010?

90. *Acres of Pumpkins.* In 2012, 16,200 acres of pumpkins were harvested in Illinois. This amount was about 54.5% of the total number of acres of pumpkins harvested in Michigan, Ohio, and Illinois together. (*Source:* U.S. Department of Agriculture) Find the total number of acres of pumpkins harvested in Michigan, Ohio, and Illinois.

▶ Synthesis

Use the following graph of the function f for Exercises 91 and 92.

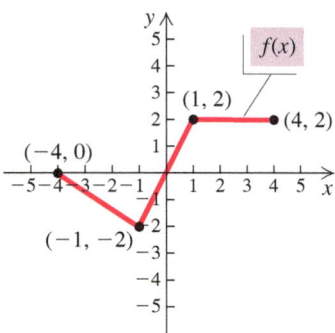

91. Graph: $y = |f(x)|$.   **92.** Graph: $y = f(|x|)$.

Use the following graph of the function g for Exercises 93 and 94.

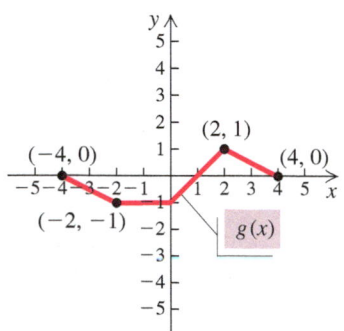

93. Graph: $y = g(|x|)$.   **94.** Graph: $y = |g(x)|$.

95. If $(-1, 5)$ is a point on the graph of $y = f(x)$, find b such that $(2, b)$ is on the graph of $y = f(x - 3)$.

96. The graph of $f(x) = |x|$ passes through the points $(-3, 3)$, $(0, 0)$, and $(3, 3)$. Transform this function to one whose graph passes through the points $(5, 1)$, $(8, 4)$, and $(11, 1)$.

7.3 The Complex Numbers

▶ Perform computations involving complex numbers.

For some functions $f(x)$, there are no real-number values of x for which $f(x) = 0$. In order to solve $f(x) = 0$ for these functions, we must consider the **complex-number system**.

▶ The Complex-Number System

We know that the square root of a negative number is not a real number. For example, $\sqrt{-1}$ is not a real number because there is no real number x such that $x^2 = -1$. This means that certain equations, like $x^2 = -1$ or $x^2 + 1 = 0$, do not have real-number solutions, and for certain functions, like $f(x) = x^2 + 1$, there are no real numbers for which $f(x) = 0$. Consider the graph of $f(x) = x^2 + 1$.

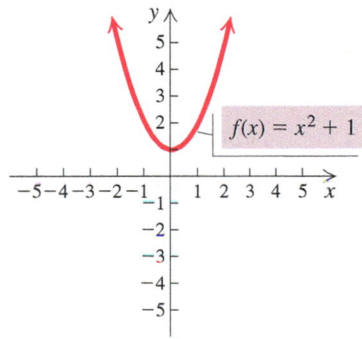

We see that the graph does not cross the x-axis and thus has no x-intercepts. This illustrates that there are no real numbers for which $f(x) = x^2 + 1$ is equal to zero. Thus there are no real-number solutions of the corresponding equation $x^2 + 1 = 0$.

We can define a nonreal number that is a solution of the equation $x^2 + 1 = 0$.

THE NUMBER *i*

The number i is defined such that

$$i = \sqrt{-1} \quad \text{and} \quad i^2 = -1.$$

To express roots of negative numbers in terms of i, we can use the fact that

$$\sqrt{-p} = \sqrt{-1 \cdot p} = \sqrt{-1} \cdot \sqrt{p} = i\sqrt{p}$$

when p is a positive real number.

EXAMPLE 1 Express each number in terms of i.

a) $\sqrt{-7}$ **b)** $\sqrt{-16}$ **c)** $-\sqrt{-13}$

d) $-\sqrt{-64}$ **e)** $\sqrt{-48}$

Solution

a) $\sqrt{-7} = \sqrt{-1 \cdot 7} = \sqrt{-1} \cdot \sqrt{7}$
$= i\sqrt{7}, \text{ or } \sqrt{7}i \leftarrow$

> *i* is *not* under the radical.

b) $\sqrt{-16} = \sqrt{-1 \cdot 16} = \sqrt{-1} \cdot \sqrt{16}$
$= i \cdot 4 = 4i$

c) $-\sqrt{-13} = -\sqrt{-1 \cdot 13} = -\sqrt{-1} \cdot \sqrt{13}$
$= -i\sqrt{13}, \text{ or } -\sqrt{13}i \leftarrow$

d) $-\sqrt{-64} = -\sqrt{-1 \cdot 64} = -\sqrt{-1} \cdot \sqrt{64}$
$= -i \cdot 8 = -8i$

e) $\sqrt{-48} = \sqrt{-1 \cdot 48} = \sqrt{-1} \cdot \sqrt{48}$
$= i\sqrt{16 \cdot 3}$
$= i \cdot 4\sqrt{3}$
$= 4i\sqrt{3}, \text{ or } 4\sqrt{3}i \leftarrow$

Now Try Exercises 1, 7, and 9.

The complex numbers are formed by adding real numbers and multiples of *i*.

COMPLEX NUMBERS

A **complex number** is a number of the form $a + bi$, where *a* and *b* are real numbers. The number *a* is said to be the **real part** of $a + bi$ and the number *b* is said to be the **imaginary part** of $a + bi$.*

Note that either *a* or *b* or both can be 0. When $b = 0$, $a + bi = a + 0i = a$, so every real number is a complex number. Complex numbers like $3 + 4i$ and $17i$, in which $b \neq 0$, are called **imaginary numbers**. Complex numbers like $17i$ and $-4i$, in which $a = 0$ and $b \neq 0$, are sometimes called **pure imaginary numbers**. The relationships among various types of complex numbers are shown in the following figure.

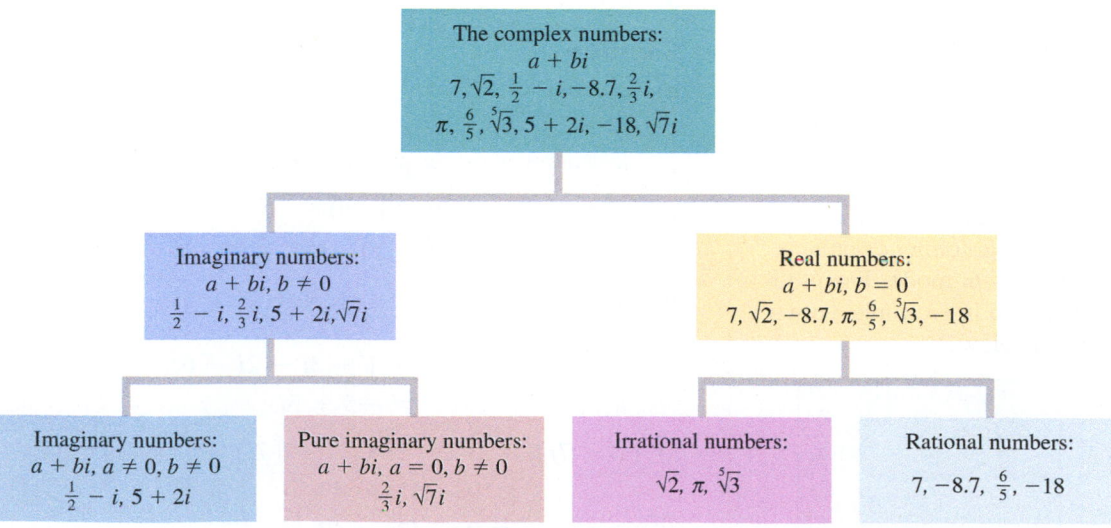

---

*Sometimes *bi* is considered to be the imaginary part.

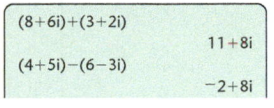
▶ **Addition and Subtraction**

The complex numbers obey the commutative, associative, and distributive laws. Thus we can add and subtract them as we do binomials. We collect the real parts and the imaginary parts of complex numbers just as we collect like terms in binomials.

EXAMPLE 2 Add or subtract and simplify each of the following.

a) $(8 + 6i) + (3 + 2i)$ **b)** $(4 + 5i) - (6 - 3i)$

Solution

a) $(8 + 6i) + (3 + 2i) = (8 + 3) + (6i + 2i)$

Collecting the real parts and the imaginary parts

$$= 11 + (6 + 2)i = 11 + 8i$$

b) $(4 + 5i) - (6 - 3i) = (4 - 6) + [5i - (-3i)]$

Note that 6 and $-3i$ are both being subtracted.

$$= -2 + 8i$$

> **Now Try Exercises 11 and 21.**

▶ **Multiplication**

When $\sqrt{a}$ and $\sqrt{b}$ are real numbers, $\sqrt{a} \cdot \sqrt{b} = \sqrt{ab}$, but this is not true when $\sqrt{a}$ and $\sqrt{b}$ are not real numbers. Thus,

$$\sqrt{-2} \cdot \sqrt{-5} = \sqrt{-1} \cdot \sqrt{2} \cdot \sqrt{-1} \cdot \sqrt{5}$$
$$= i\sqrt{2} \cdot i\sqrt{5}$$
$$= i^2\sqrt{10} = -1\sqrt{10} = -\sqrt{10} \quad \text{is correct!}$$

But

$$\sqrt{-2} \cdot \sqrt{-5} = \sqrt{(-2)(-5)} = \sqrt{10} \quad \text{is wrong!}$$

Keeping this and the fact that $i^2 = -1$ in mind, we multiply with imaginary numbers in much the same way that we do with real numbers.

EXAMPLE 3 Multiply and simplify each of the following.

a) $\sqrt{-16} \cdot \sqrt{-25}$ **b)** $(1 + 2i)(1 + 3i)$ **c)** $(3 - 7i)^2$

Solution

a) $\sqrt{-16} \cdot \sqrt{-25} = \sqrt{-1} \cdot \sqrt{16} \cdot \sqrt{-1} \cdot \sqrt{25}$

$$= i \cdot 4 \cdot i \cdot 5$$
$$= i^2 \cdot 20$$
$$= -1 \cdot 20 \quad i^2 = -1$$
$$= -20$$

b) $(1 + 2i)(1 + 3i) = 1 + 3i + 2i + 6i^2$ **Multiplying each term of one number by every term of the other (FOIL)**

$$= 1 + 3i + 2i - 6 \quad i^2 = -1$$
$$= -5 + 5i \quad \text{Collecting like terms}$$

c) $(3 - 7i)^2 = 3^2 - 2 \cdot 3 \cdot 7i + (7i)^2$ **Recall that $(A - B)^2 = A^2 - 2AB + B^2$.**

$$= 9 - 42i + 49i^2$$
$$= 9 - 42i - 49 \quad i^2 = -1$$
$$= -40 - 42i$$

> **Now Try Exercises 31, 39, and 55.**

Recall that -1 raised to an *even* power is 1, and -1 raised to an *odd* power is -1. Simplifying powers of i can then be done by using the fact that $i^2 = -1$ and expressing the given power of i in terms of i^2. Consider the following:

$$i = \sqrt{-1},$$
$$i^2 = -1,$$
$$i^3 = i^2 \cdot i = (-1)i = -i,$$
$$i^4 = (i^2)^2 = (-1)^2 = 1,$$
$$i^5 = i^4 \cdot i = (i^2)^2 \cdot i = (-1)^2 \cdot i = 1 \cdot i = i,$$
$$i^6 = (i^2)^3 = (-1)^3 = -1,$$
$$i^7 = i^6 \cdot i = (i^2)^3 \cdot i = (-1)^3 \cdot i = -1 \cdot i = -i,$$
$$i^8 = (i^2)^4 = (-1)^4 = 1.$$

Note that the powers of i cycle through the values i, -1, $-i$, and 1.

EXAMPLE 4   Simplify each of the following.

a) i^{37} **b)** i^{58}

c) i^{75} **d)** i^{80}

Solution

a) $i^{37} = i^{36} \cdot i = (i^2)^{18} \cdot i = (-1)^{18} \cdot i = 1 \cdot i = i$

b) $i^{58} = (i^2)^{29} = (-1)^{29} = -1$

c) $i^{75} = i^{74} \cdot i = (i^2)^{37} \cdot i = (-1)^{37} \cdot i = -1 \cdot i = -i$

d) $i^{80} = (i^2)^{40} = (-1)^{40} = 1$ **Now Try Exercises 79 and 83.**

These powers of i can also be simplified in terms of i^4 rather than i^2. Consider i^{37} in Example 4(a), for instance. When we divide 37 by 4, we get 9 with a remainder of 1. Then $37 = 4 \cdot 9 + 1$, so

$$i^{37} = (i^4)^9 \cdot i = 1^9 \cdot i = 1 \cdot i = i.$$

The other examples shown above can be done in a similar manner.

▶ Conjugates and Division

Conjugates of complex numbers are defined as follows.

CONJUGATE OF A COMPLEX NUMBER

The **conjugate** of a complex number $a + bi$ is $a - bi$. The numbers $a + bi$ and $a - bi$ are **complex conjugates**.

Each of the following pairs of numbers are complex conjugates:

$$-3 + 7i \text{ and } -3 - 7i; \quad 14 - 5i \text{ and } 14 + 5i; \quad \text{and} \quad 8i \text{ and } -8i.$$

The product of a complex number and its conjugate is a real number.

EXAMPLE 5 Multiply each of the following.

a) $(5 + 7i)(5 - 7i)$ **b)** $(8i)(-8i)$

Solution

a)
$$
\begin{aligned}
(5 + 7i)(5 - 7i) &= 5^2 - (7i)^2 \qquad \text{Using } (A + B)(A - B) = A^2 - B^2 \\
&= 25 - 49i^2 \\
&= 25 - 49(-1) \\
&= 25 + 49 \\
&= 74
\end{aligned}
$$

b)
$$
\begin{aligned}
(8i)(-8i) &= -64i^2 \\
&= -64(-1) \\
&= 64
\end{aligned}
$$

> **Now Try Exercise 49.**

Conjugates are used when we divide complex numbers.

EXAMPLE 6 Divide $2 - 5i$ by $1 - 6i$.

Solution We write fraction notation and then multiply by 1, using the conjugate of the denominator to form the symbol for 1.

$$
\begin{aligned}
\frac{2 - 5i}{1 - 6i} &= \frac{2 - 5i}{1 - 6i} \cdot \frac{1 + 6i}{1 + 6i} \qquad \text{Note that } 1 + 6i \text{ is the conjugate} \\
&\qquad\qquad\qquad\qquad \text{of the divisor, } 1 - 6i. \\
&= \frac{(2 - 5i)(1 + 6i)}{(1 - 6i)(1 + 6i)} \\
&= \frac{2 + 12i - 5i - 30i^2}{1 - 36i^2} \\
&= \frac{2 + 7i + 30}{1 + 36} \qquad i^2 = -1 \\
&= \frac{32 + 7i}{37} \\
&= \frac{32}{37} + \frac{7}{37}i. \qquad \text{Writing the quotient in the form } a + bi
\end{aligned}
$$

> **Now Try Exercise 69.**

Technology Connection

With a graphing calculator set in $a + bi$ mode, we can divide complex numbers and express the real and imaginary parts in fraction form, just as we did in Example 6.

```
(2−5i)/(1−6i) ▶Frac
            32   7
            ── + ── i
            37   37
```

7.3   Exercise Set

Express the number in terms of i.

1. $\sqrt{-3}$ **2.** $\sqrt{-21}$

3. $\sqrt{-25}$ **4.** $\sqrt{-100}$

5. $-\sqrt{-33}$ **6.** $-\sqrt{-59}$

7. $-\sqrt{-81}$ **8.** $-\sqrt{-9}$

9. $\sqrt{-98}$ **10.** $\sqrt{-28}$

Simplify. Write answers in the form $a + bi$, where a and b are real numbers.

11. $(-5 + 3i) + (7 + 8i)$

12. $(-6 - 5i) + (9 + 2i)$

13. $(4 - 9i) + (1 - 3i)$

14. $(7 - 2i) + (4 - 5i)$

15. $(12 + 3i) + (-8 + 5i)$

16. $(-11 + 4i) + (6 + 8i)$

17. $(-1 - i) + (-3 - i)$

18. $(-5 - i) + (6 + 2i)$

19. $\left(3 + \sqrt{-16}\right) + \left(2 + \sqrt{-25}\right)$

20. $\left(7 - \sqrt{-36}\right) + \left(2 + \sqrt{-9}\right)$

21. $(10 + 7i) - (5 + 3i)$

22. $(-3 - 4i) - (8 - i)$

23. $(13 + 9i) - (8 + 2i)$

24. $(-7 + 12i) - (3 - 6i)$

25. $(6 - 4i) - (-5 + i)$

26. $(8 - 3i) - (9 - i)$

27. $(-5 + 2i) - (-4 - 3i)$

28. $(-6 + 7i) - (-5 - 2i)$

29. $(4 - 9i) - (2 + 3i)$

30. $(10 - 4i) - (8 + 2i)$

31. $\sqrt{-4} \cdot \sqrt{-36}$

32. $\sqrt{-49} \cdot \sqrt{-9}$

33. $\sqrt{-81} \cdot \sqrt{-25}$

34. $\sqrt{-16} \cdot \sqrt{-100}$

35. $7i(2 - 5i)$

36. $3i(6 + 4i)$

37. $-2i(-8 + 3i)$

38. $-6i(-5 + i)$

39. $(1 + 3i)(1 - 4i)$

40. $(1 - 2i)(1 + 3i)$

41. $(2 + 3i)(2 + 5i)$

42. $(3 - 5i)(8 - 2i)$

43. $(-4 + i)(3 - 2i)$

44. $(5 - 2i)(-1 + i)$

45. $(8 - 3i)(-2 - 5i)$

46. $(7 - 4i)(-3 - 3i)$

47. $\left(3 + \sqrt{-16}\right)\left(2 + \sqrt{-25}\right)$

48. $\left(7 - \sqrt{-16}\right)\left(2 + \sqrt{-9}\right)$

49. $(5 - 4i)(5 + 4i)$

50. $(5 + 9i)(5 - 9i)$

51. $(3 + 2i)(3 - 2i)$

52. $(8 + i)(8 - i)$

53. $(7 - 5i)(7 + 5i)$

54. $(6 - 8i)(6 + 8i)$

55. $(4 + 2i)^2$

56. $(5 - 4i)^2$

57. $(-2 + 7i)^2$

58. $(-3 + 2i)^2$

59. $(1 - 3i)^2$

60. $(2 - 5i)^2$

61. $(-1 - i)^2$

62. $(-4 - 2i)^2$

63. $(3 + 4i)^2$

64. $(6 + 5i)^2$

65. $\dfrac{3}{5 - 11i}$

66. $\dfrac{i}{2 + i}$

67. $\dfrac{5}{2 + 3i}$

68. $\dfrac{-3}{4 - 5i}$

69. $\dfrac{4 + i}{-3 - 2i}$

70. $\dfrac{5 - i}{-7 + 2i}$

71. $\dfrac{5 - 3i}{4 + 3i}$

72. $\dfrac{6 + 5i}{3 - 4i}$

73. $\dfrac{2 + \sqrt{3}i}{5 - 4i}$

74. $\dfrac{\sqrt{5} + 3i}{1 - i}$

75. $\dfrac{1 + i}{(1 - i)^2}$

76. $\dfrac{1 - i}{(1 + i)^2}$

77. $\dfrac{4 - 2i}{1 + i} + \dfrac{2 - 5i}{1 + i}$

78. $\dfrac{3 + 2i}{1 - i} + \dfrac{6 + 2i}{1 - i}$

Simplify.

79. i^{11}

80. i^7

81. i^{35}

82. i^{24}

83. i^{64}

84. i^{42}

85. $(-i)^{71}$

86. $(-i)^6$

87. $(5i)^4$

88. $(2i)^5$

▶ Skill Maintenance

Vocabulary Reinforcement

In each of Exercises 89–96, fill in the blank with the correct term. Some of the given choices will not be used.

function
relation
x-intercept
y-intercept
perpendicular
parallel
horizontal lines
vertical lines
increasing

decreasing
constant
symmetric with respect to the x-axis
symmetric with respect to the y-axis
symmetric with respect to the origin

89. A(n) _____ is a point $(0, b)$. **[2.6]**

90. A(n) _____ is a point $(a, 0)$. **[2.6]**

91. A(n) _____ is a correspondence such that each member of the domain corresponds to at least one member of the range. **[2.2]**

92. A(n) _____ is a correspondence such that each member of the domain corresponds to exactly one member of the range. **[2.2]**

93. _____ are given by equations of the type $y = b$, or $f(x) = b$. **[2.5]**

94. Nonvertical lines are _____ if and only if they have the same slope and different y-intercepts. **[2.5]**

95. A function f is said to be _____ on an open interval I if, for all a and b in that interval, $a < b$ implies $f(a) > f(b)$. **[6.8]**

96. For an equation $y = f(x)$, if replacing x with $-x$ produces an equivalent equation, then the graph is _____. **[7.1]**

▶ **Synthesis**

Determine whether the statement is true or false.

97. The sum of two numbers that are conjugates of each other is always a real number.

98. The conjugate of a sum is the sum of the conjugates of the individual complex numbers.

99. The conjugate of a product is the product of the conjugates of the individual complex numbers.

Let $z = a + bi$ and $\bar{z} = a - bi$.

100. Find a general expression for $1/z$.

101. Find a general expression for $z\bar{z}$.

102. Solve $z + 6\bar{z} = 7$ for z.

103. Multiply and simplify:

$$[x - (3 + 4i)][x - (3 - 4i)]$$

Mid-Chapter Review

Determine whether the statement is true or false.

1. The product of a complex number and its conjugate is a real number. **[7.3]**

2. If replacing x with $-x$ produces an equivalent equation, then the graph is symmetric with respect to the x-axis. **[7.1]**

3. The effect of adding a constant to or subtracting a constant from $f(x)$ in $y = f(x)$ is a shift of the graph of $f(x)$ left or right. **[7.2]**

4. The graph of $y = f(-x)$ is the reflection of the graph of $y = f(x)$ across the y-axis. **[7.2]**

Determine whether the graph is symmetric with respect to the x-axis, the y-axis, and the origin. **[7.1]**

5. $y = \left|\frac{1}{2}x\right|$

6. $5x^2 - 3 = y^2$

7. $xy = 6$

8. $2y = 3x^2 + x$

Find the point that is symmetric to the given point with respect to the x-axis, the y-axis, and the origin. **[7.1]**

9. $(-7, 2)$

10. $\left(0, -\frac{1}{3}\right)$

Determine whether the function is even, odd, or neither even nor odd. **[7.1]**

11. $f(x) = x^3 - x + 5$

12. $f(x) = 3x + \dfrac{6}{x}$

13. $f(x) = x^2 - x^4 + 2$

14. $f(x) = -5$

The point $(9, -6)$ *is on the graph of* $y = f(x)$. *Find the corresponding point on the graph of* $y = g(x)$. **[7.2]**

15. $g(x) = \frac{1}{3}f(x)$

16. $g(x) = f(x - 3)$

17. $g(x) = f(-x)$

18. $g(x) = f(2x)$

Write an equation for a function that has a graph with the given characteristics. **[7.2]**

19. The shape of $y = x^2$, but shifted left 3 units and down 4 units.

20. The shape of $y = \sqrt{x}$, but reflected across the x-axis and shifted right 5 units.

A graph of $y = f(x)$ *follows. No formula for* f *is given. In Exercises 21–24, graph the given function.* **[7.2]**

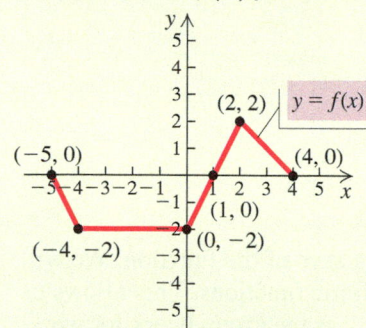

21. $g(x) = -\frac{1}{2}f(x)$

22. $g(x) = f(x - 1)$

23. $g(x) = f(-x)$

24. $g(x) = f(2x) - 1$

Express the number in terms of i. **[7.3]**

25. $\sqrt{-36}$

26. $\sqrt{-5}$

27. $-\sqrt{-16}$

28. $\sqrt{-32}$

Simplify. Write answers in the form $a + bi$, *where* a *and* b *are real numbers.* **[7.3]**

29. $(3 - 2i) + (-4 + 3i)$

30. $(-5 + i) - (2 - 4i)$

31. $(2 + 3i)(4 - 5i)$

32. $\dfrac{3 + i}{-2 + 5i}$

Simplify. **[3.1]**

33. i^{13}

34. i^{44}

35. $(-i)^5$

36. $(2i)^6$

Collaborative Discussion and Writing

37. Is the sum of two imaginary numbers always an imaginary number? Explain your answer. **[7.3]**

38. Given the graph of $y = f(x)$, explain and contrast the effect of the constant c on the graphs of $y = f(cx)$ and $y = cf(x)$. **[7.2]**

39. Consider the constant function $f(x) = 0$. Determine whether the graph of this function is symmetric with respect to the x-axis, the y-axis, and/or the origin. Determine whether this function is even or odd. **[7.1]**

40. Describe conditions under which you would know whether a polynomial function

$$f(x) = a_n x^n + a_{n-1} x^{n-1} + \cdots + a_2 x^2 + a_1 x + a_0$$

is even or odd without using an algebraic procedure. Explain. **[7.1]**

7.4 Quadratic Equations, Functions, Zeros, and Models

▶ Find zeros of quadratic functions and solve quadratic equations by using the principle of zero products, by using the principle of square roots, by completing the square, and by using the quadratic formula.

▶ Solve equations that are reducible to quadratic.

▶ Solve applied problems using quadratic equations.

▶ Quadratic Equations and Quadratic Functions

An input for which a function's output is 0 is called a **zero** of the function. We will restrict our attention in this section to zeros of quadratic functions. This allows us to become familiar with the concept of a zero, and it lays the groundwork for working with zeros of other types of functions in succeeding chapters.

> **ZEROS OF FUNCTIONS**
>
> An input c of a function f is called a **zero** of the function if the output for the function is 0 when the input is c. That is, c is a zero of f if $f(c) = 0$.

In this section, we will explore the relationship between the solutions of quadratic equations and the zeros of quadratic functions. We define quadratic equations and quadratic functions as follows.

> **QUADRATIC EQUATIONS**
>
> A **quadratic equation** is an equation that can be written in the form
>
> $$ax^2 + bx + c = 0, \quad a \neq 0,$$
>
> where a, b, and c are real numbers.
>
> **QUADRATIC FUNCTIONS**
>
> A **quadratic function** f is a function that can be written in the form
>
> $$f(x) = ax^2 + bx + c, \quad a \neq 0,$$
>
> where a, b, and c are real numbers.

A quadratic equation written in the form $ax^2 + bx + c = 0$ is said to be in **standard form**.

The *zeros* of a quadratic function $f(x) = ax^2 + bx + c$ are the *solutions* of the associated quadratic equation $ax^2 + bx + c = 0$. (These solutions are sometimes called *roots* of the equation.) Quadratic functions can have real-number or imaginary-number zeros and quadratic equations can have real-number or imaginary-number solutions. If the zeros or solutions are real numbers, they are also the first coordinates of the *x*-intercepts of the graph of the quadratic function.

The following principles allow us to solve many quadratic equations.

EQUATION-SOLVING PRINCIPLES

The Principle of Zero Products: If $ab = 0$ is true, then $a = 0$ *or* $b = 0$, and if $a = 0$ *or* $b = 0$, then $ab = 0$.

The Principle of Square Roots: If $x^2 = k$, then $x = \sqrt{k}$ *or* $x = -\sqrt{k}$.

EXAMPLE 1 Solve: $2x^2 - x = 3$.

Algebraic Solution

We have

$$2x^2 - x = 3$$
$$2x^2 - x - 3 = 0 \qquad \text{Subtracting 3 on both sides}$$
$$(x + 1)(2x - 3) = 0 \qquad \text{Factoring}$$
$$x + 1 = 0 \quad or \quad 2x - 3 = 0 \qquad \text{Using the principle of zero products}$$
$$x = -1 \quad or \quad 2x = 3$$
$$x = -1 \quad or \quad x = \tfrac{3}{2}.$$

Check: For $x = -1$:

$$\frac{2x^2 - x = 3}{2(-1)^2 - (-1) \;\overset{?}{\mid}\; 3}$$
$$2 \cdot 1 + 1$$
$$2 + 1$$
$$3 \;\mid\; 3 \quad \text{TRUE}$$

For $x = \tfrac{3}{2}$:

$$\frac{2x^2 - x = 3}{2\left(\tfrac{3}{2}\right)^2 - \tfrac{3}{2} \;\overset{?}{\mid}\; 3}$$
$$2 \cdot \tfrac{9}{4} - \tfrac{3}{2}$$
$$\tfrac{9}{2} - \tfrac{3}{2}$$
$$\tfrac{6}{2}$$
$$3 \;\mid\; 3 \quad \text{TRUE}$$

The solutions are -1 and $\tfrac{3}{2}$.

Visualizing the Solution

The solutions of the equation $2x^2 - x = 3$, or the equivalent equation $2x^2 - x - 3 = 0$, are the zeros of the function $f(x) = 2x^2 - x - 3$. They are also the first coordinates of the *x*-intercepts of the graph of $f(x) = 2x^2 - x - 3$.

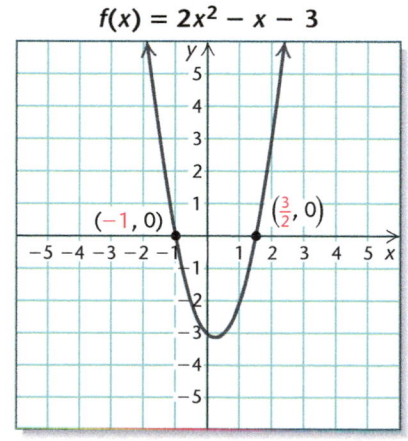

$$f(x) = 2x^2 - x - 3$$

The solutions are -1 and $\tfrac{3}{2}$.

Now Try Exercise 3.

EXAMPLE 2 Solve: $2x^2 - 10 = 0$.

Algebraic Solution

We have

$$2x^2 - 10 = 0$$
$$2x^2 = 10 \qquad \text{Adding 10 on both sides}$$
$$x^2 = 5 \qquad \text{Dividing by 2 on both sides}$$
$$x = \sqrt{5} \quad or \quad x = -\sqrt{5}. \qquad \text{Using the principle of square roots}$$

Check:

$$\begin{array}{c|c} 2x^2 - 10 = 0 & \\ \hline 2\left(\pm\sqrt{5}\right)^2 - 10 \ ? \ 0 & \text{We can check both solutions at once.} \\ 2 \cdot 5 - 10 & \\ 10 - 10 & \\ 0 \ \big| \ 0 \quad \text{TRUE} \end{array}$$

The solutions are $\sqrt{5}$ and $-\sqrt{5}$, or $\pm\sqrt{5}$.

Visualizing the Solution

The solutions of the equation $2x^2 - 10 = 0$ are the zeros of the function $f(x) = 2x^2 - 10$. Note that they are also the first coordinates of the x-intercepts of the graph of $f(x) = 2x^2 - 10$.

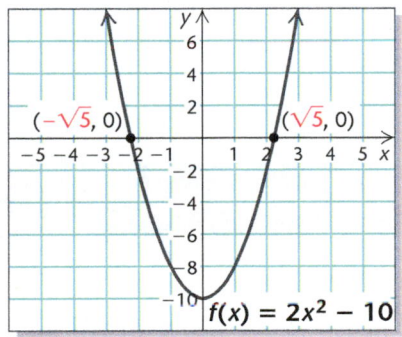

The solutions are $-\sqrt{5}$ and $\sqrt{5}$.

Now Try Exercise 7.

We have seen that some quadratic equations can be solved by factoring and using the principle of zero products. For example, consider the equation $x^2 - 3x - 4 = 0$:

$$x^2 - 3x - 4 = 0$$
$$(x + 1)(x - 4) = 0 \qquad \text{Factoring}$$
$$x + 1 = 0 \quad or \quad x - 4 = 0 \qquad \text{Using the principle of zero products}$$
$$x = -1 \quad or \qquad x = 4.$$

The equation $x^2 - 3x - 4 = 0$ has *two real-number* solutions, -1 and 4. These are the zeros of the associated quadratic function $f(x) = x^2 - 3x - 4$ and the first coordinates of the x-intercepts of the graph of this function. (See Fig. 1.)

Next, consider the equation $x^2 - 6x + 9 = 0$. Again, we factor and use the principle of zero products:

$$x^2 - 6x + 9 = 0$$
$$(x - 3)(x - 3) = 0 \qquad \text{Factoring}$$
$$x - 3 = 0 \quad or \quad x - 3 = 0 \qquad \text{Using the principle of zero products}$$
$$x = 3 \quad or \qquad x = 3.$$

The equation $x^2 - 6x + 9 = 0$ has *one real-number* solution, 3. It is the zero of the quadratic function $g(x) = x^2 - 6x + 9$ and the first coordinate of the x-intercept of the graph of this function. (See Fig. 2.)

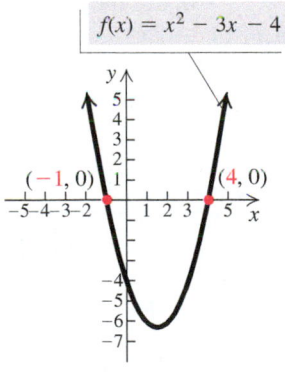

Two real-number zeros
Two x-intercepts

Figure 1.

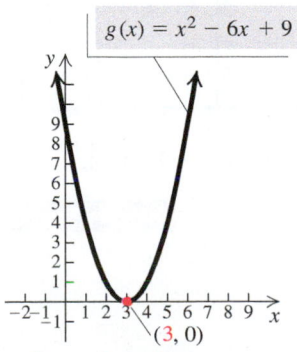

One real-number zero
One x-intercept

Figure 2.

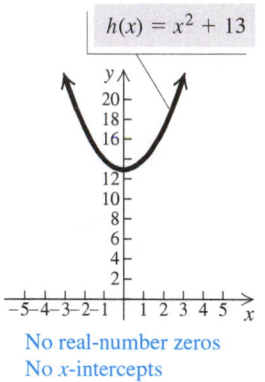

$h(x) = x^2 + 13$

No real-number zeros
No x-intercepts

Figure 3.

The principle of square roots can be used to solve quadratic equations like $x^2 + 13 = 0$:

$$x^2 + 13 = 0$$
$$x^2 = -13$$
$$x = \pm \sqrt{-13} \qquad \text{Using the principle of square roots}$$
$$x = \pm \sqrt{13}i. \qquad \sqrt{-13} = \sqrt{-1} \cdot \sqrt{13} = i \cdot \sqrt{13} = \sqrt{13}i$$

The equation has *two imaginary-number* solutions, $-\sqrt{13}i$ and $\sqrt{13}i$. These are the zeros of the associated quadratic function $h(x) = x^2 + 13$. Since the zeros are not real numbers, the graph of the function has no x-intercepts. (See Fig. 3.)

▶ Completing the Square

Consider the function $f(x) = x^2 - 6x - 10$. Since $x^2 - 6x - 10$ cannot be factored using rational numbers and since it is not in the form $x^2 = k$, we need a new procedure to find the zeros of $f(x) = x^2 - 6x - 10$ or the solutions of the associated equation $x^2 - 6x - 10 = 0$. If we wish to find zeros or solutions, we can use a procedure called **completing the square** and then use the principle of square roots.

EXAMPLE 3   Find the zeros of $f(x) = x^2 - 6x - 10$ by completing the square.

Solution   We find the values of x for which $f(x) = 0$; that is, we solve the associated equation $x^2 - 6x - 10 = 0$. Our goal is to find an equivalent equation of the form $x^2 + bx + c = d$ in which $x^2 + bx + c$ is a perfect square. Since

$$x^2 + bx + \left(\frac{b}{2}\right)^2 = \left(x + \frac{b}{2}\right)^2,$$

the number c is found by taking half the coefficient of the x-term and squaring it. Thus for the equation $x^2 - 6x - 10 = 0$, we have

$$x^2 - 6x - 10 = 0$$
$$x^2 - 6x \qquad = 10 \qquad \text{\textcolor{red}{Adding 10}}$$
$$x^2 - 6x + 9 = 10 + 9 \qquad \text{\textcolor{red}{Adding 9 to complete the square:}}$$
$$\text{\textcolor{red}{$\left(\dfrac{b}{2}\right)^2 = \left(\dfrac{-6}{2}\right)^2 = (-3)^2 = 9$}}$$
$$x^2 - 6x + 9 = 19.$$

Because $x^2 - 6x + 9$ is a perfect square, we are able to write it as $(x - 3)^2$, the square of a binomial. We can then use the principle of square roots to finish the solution:

$$(x - 3)^2 = 19 \qquad \text{\textcolor{red}{Factoring}}$$
$$x - 3 = \pm \sqrt{19} \qquad \text{\textcolor{red}{Using the principle of square roots}}$$
$$x = 3 \pm \sqrt{19}. \qquad \text{\textcolor{red}{Adding 3}}$$

Therefore, the solutions of the equation are $3 + \sqrt{19}$ and $3 - \sqrt{19}$, or simply $3 \pm \sqrt{19}$. The zeros of $f(x) = x^2 - 6x - 10$ are also $3 + \sqrt{19}$ and $3 - \sqrt{19}$, or $3 \pm \sqrt{19}$.

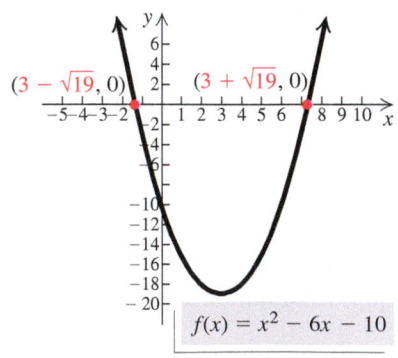

$(3 - \sqrt{19}, 0)$ $(3 + \sqrt{19}, 0)$

$f(x) = x^2 - 6x - 10$

We can find decimal approximations for $3 \pm \sqrt{19}$ using a calculator:

$$3 + \sqrt{19} \approx 7.359 \quad \text{and} \quad 3 - \sqrt{19} \approx -1.359.$$

The zeros are approximately 7.359 and -1.359.

▶ **Now Try Exercise 31.**

Technology Connection

Approximations for the zeros of the quadratic function $f(x) = x^2 - 6x - 10$ in Example 3 can be found using the Zero feature.

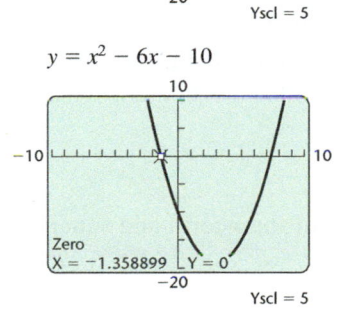

$y = x^2 - 6x - 10$

Zero
X = 7.3588989   Y = 0
Yscl = 5

$y = x^2 - 6x - 10$

Zero
X = −1.358899   Y = 0
Yscl = 5

Before we can complete the square, the coefficient of the x^2-term must be 1. When it is not, we divide by the x^2-coefficient on both sides of the equation.

EXAMPLE 4 Solve: $2x^2 - 1 = 3x$.

Solution We have

$$2x^2 - 1 = 3x$$

$$2x^2 - 3x - 1 = 0 \qquad \text{Subtracting } 3x. \text{ We are unable to factor } 2x^2 - 3x - 1.$$

$$2x^2 - 3x = 1 \qquad \text{Adding 1}$$

$$x^2 - \frac{3}{2}x = \frac{1}{2} \qquad \text{Dividing by 2 to make the } x^2\text{-coefficient 1}$$

$$x^2 - \frac{3}{2}x + \frac{9}{16} = \frac{1}{2} + \frac{9}{16} \qquad \text{Completing the square: } \frac{1}{2}\left(-\frac{3}{2}\right) = -\frac{3}{4} \text{ and } \left(-\frac{3}{4}\right)^2 = \frac{9}{16}; \text{ adding } \frac{9}{16}$$

$$\left(x - \frac{3}{4}\right)^2 = \frac{17}{16} \qquad \text{Factoring and simplifying}$$

$$x - \frac{3}{4} = \pm\frac{\sqrt{17}}{4} \qquad \text{Using the principle of square roots and the quotient rule for radicals}$$

$$x = \frac{3}{4} \pm \frac{\sqrt{17}}{4} \qquad \text{Adding } \frac{3}{4}$$

$$x = \frac{3 \pm \sqrt{17}}{4}.$$

The solutions are

$$\frac{3 + \sqrt{17}}{4} \quad \text{and} \quad \frac{3 - \sqrt{17}}{4}, \quad \text{or} \quad \frac{3 \pm \sqrt{17}}{4}.$$

▶ **Now Try Exercise 35.**

To solve a quadratic equation by completing the square:

1. Isolate the terms with variables on one side of the equation and arrange them in descending order.
2. Divide by the coefficient of the squared term if that coefficient is not 1.
3. Complete the square by taking half the coefficient of the first-degree term and adding its square on both sides of the equation.
4. Express one side of the equation as the square of a binomial.
5. Use the principle of square roots.
6. Solve for the variable.

▶ Using the Quadratic Formula

Because completing the square works for *any* quadratic equation, it can be used to solve the general quadratic equation $ax^2 + bx + c = 0$ for x. The result will be a formula that can be used to solve any quadratic equation quickly.

Consider any quadratic equation in standard form:

$$ax^2 + bx + c = 0, \quad a \neq 0.$$

For now, we assume that $a > 0$ and solve by completing the square. As the steps are carried out, compare them with those of Example 4.

$$ax^2 + bx + c = 0 \qquad \text{Standard form}$$

$$ax^2 + bx = -c \qquad \text{Adding } -c$$

$$x^2 + \frac{b}{a}x = -\frac{c}{a} \qquad \text{Dividing by } a$$

Half of $\dfrac{b}{a}$ is $\dfrac{b}{2a}$, and $\left(\dfrac{b}{2a}\right)^2 = \dfrac{b^2}{4a^2}$. Thus we add $\dfrac{b^2}{4a^2}$:

$$x^2 + \frac{b}{a}x + \frac{b^2}{4a^2} = -\frac{c}{a} + \frac{b^2}{4a^2} \qquad \text{Adding } \frac{b^2}{4a^2} \text{ to complete the square}$$

$$\left(x + \frac{b}{2a}\right)^2 = -\frac{4ac}{4a^2} + \frac{b^2}{4a^2} \qquad \begin{array}{l}\text{Factoring on the left; finding a} \\ \text{common denominator on the} \\ \text{right: } -\frac{c}{a} = -\frac{c}{a} \cdot \frac{4a}{4a} = -\frac{4ac}{4a^2}\end{array}$$

$$\left(x + \frac{b}{2a}\right)^2 = \frac{b^2 - 4ac}{4a^2}$$

$$x + \frac{b}{2a} = \pm\sqrt{\frac{b^2 - 4ac}{4a^2}}$$

$$x + \frac{b}{2a} = \pm\frac{\sqrt{b^2 - 4ac}}{2a} \qquad \begin{array}{l}\text{Using the principle of square} \\ \text{roots and the quotient} \\ \text{rule for radicals. Since} \\ a > 0, \sqrt{4a^2} = 2a.\end{array}$$

$$x = -\frac{b}{2a} \pm \frac{\sqrt{b^2 - 4ac}}{2a} \qquad \text{Adding } -\frac{b}{2a}$$

$$x = \frac{-b \pm \sqrt{b^2 - 4ac}}{2a}.$$

It can also be shown that this result holds if $a < 0$.

THE QUADRATIC FORMULA

The solutions of $ax^2 + bx + c = 0, a \neq 0$, are given by

$$x = \frac{-b \pm \sqrt{b^2 - 4ac}}{2a}.$$

EXAMPLE 5   Solve $3x^2 + 2x = 7$. Find exact solutions and approximate solutions rounded to three decimal places.

Solution   After writing the equation in standard form, we are unable to factor, so we identify a, b, and c in order to use the quadratic formula:

$$3x^2 + 2x = 7$$
$$3x^2 + 2x - 7 = 0;$$
$$a = 3, \quad b = 2, \quad c = -7.$$

We then use the quadratic formula:

$$x = \frac{-b \pm \sqrt{b^2 - 4ac}}{2a}$$

$$= \frac{-2 \pm \sqrt{2^2 - 4(3)(-7)}}{2(3)} \qquad \text{Substituting}$$

$$= \frac{-2 \pm \sqrt{4 + 84}}{6} = \frac{-2 \pm \sqrt{88}}{6}$$

$$= \frac{-2 \pm \sqrt{4 \cdot 22}}{6} = \frac{-2 \pm 2\sqrt{22}}{6} = \frac{2(-1 \pm \sqrt{22})}{2 \cdot 3}$$

$$= \frac{2}{2} \cdot \frac{-1 \pm \sqrt{22}}{3} = \frac{-1 \pm \sqrt{22}}{3}.$$

The exact solutions are

$$\frac{-1 - \sqrt{22}}{3} \quad \text{and} \quad \frac{-1 + \sqrt{22}}{3}.$$

Using a calculator, we approximate the solutions to be -1.897 and 1.230.

Now Try Exercise 41.

EXAMPLE 6   Solve: $x^2 + 5x + 8 = 0$.

Algebraic Solution

To find the solutions, we use the quadratic formula. For $x^2 + 5x + 8 = 0$, we have

$$a = 1, \quad b = 5, \quad c = 8;$$

$$x = \frac{-b \pm \sqrt{b^2 - 4ac}}{2a}$$

$$= \frac{-5 \pm \sqrt{5^2 - 4(1)(8)}}{2 \cdot 1} \qquad \text{Substituting}$$

$$= \frac{-5 \pm \sqrt{25 - 32}}{2}$$

$$= \frac{-5 \pm \sqrt{-7}}{2} = \frac{-5 \pm \sqrt{7}i}{2}.$$

The solutions are $-\dfrac{5}{2} - \dfrac{\sqrt{7}}{2}i$ and $-\dfrac{5}{2} + \dfrac{\sqrt{7}}{2}i$.

Visualizing the Solution

The graph of the function
$f(x) = x^2 + 5x + 8$ has no x-intercepts.

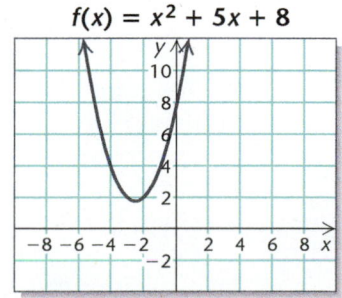

$f(x) = x^2 + 5x + 8$

Thus the function has no real-number zeros and there are no real-number solutions of the associated equation $x^2 + 5x + 8 = 0$.

Now Try Exercise 47.

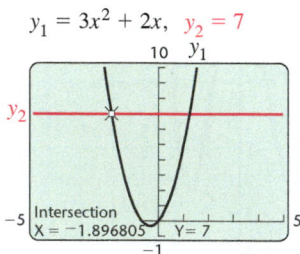
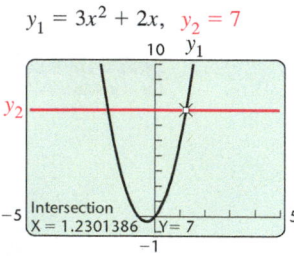
▶ The Discriminant

From the quadratic formula, we know that the solutions x_1 and x_2 of a quadratic equation are given by

$$x_1 = \frac{-b + \sqrt{b^2 - 4ac}}{2a} \quad \text{and} \quad x_2 = \frac{-b - \sqrt{b^2 - 4ac}}{2a}.$$

The expression $b^2 - 4ac$ shows the nature of the solutions. This expression is called the **discriminant**. If it is 0, then it makes no difference whether we choose the plus sign or the minus sign in the formula. That is, $x_1 = -\dfrac{b}{2a} = x_2$, so there is just one solution. In this case, we sometimes say that there is one repeated real solution. If the discriminant is positive, there will be two real solutions. If it is negative, we will be taking the square root of a negative number; hence there will be two imaginary-number solutions, and they will be complex conjugates.

DISCRIMINANT

For $ax^2 + bx + c = 0$, where a, b, and c are real numbers:

$b^2 - 4ac = 0 \rightarrow$ One real-number solution;

$b^2 - 4ac > 0 \rightarrow$ Two different real-number solutions;

$b^2 - 4ac < 0 \rightarrow$ Two different imaginary-number solutions, complex conjugates.

In Example 5, the discriminant, 88, is positive, indicating that there are two different real-number solutions. The negative discriminant, -7, in Example 6 indicates that there are two different imaginary-number solutions.

▶ Equations Reducible to Quadratic

Some equations can be treated as quadratic, provided we make a suitable substitution. For example, consider the following:

$$x^4 - 5x^2 + 4 = 0$$
$$(x^2)^2 - 5x^2 + 4 = 0 \qquad x^4 = (x^2)^2$$
$$u^2 - 5u + 4 = 0. \qquad \text{Substituting } u \text{ for } x^2$$

The equation $u^2 - 5u + 4 = 0$ can be solved for u by factoring or using the quadratic formula. Then we can reverse the substitution, replacing u with x^2, and solve for x. Equations like the one above are said to be **reducible to quadratic**, or **quadratic in form**.

EXAMPLE 7 Solve: $x^4 - 5x^2 + 4 = 0$.

Algebraic Solution

We let $u = x^2$ and substitute:

$$u^2 - 5u + 4 = 0 \qquad \text{Substituting } u \text{ for } x^2$$
$$(u - 1)(u - 4) = 0 \qquad \text{Factoring}$$
$$u - 1 = 0 \quad or \quad u - 4 = 0 \qquad \text{Using the principle of zero products}$$
$$u = 1 \quad or \quad u = 4.$$

Don't stop here! We must solve for the original variable. We substitute x^2 for u and solve for x:

$$x^2 = 1 \quad or \quad x^2 = 4$$
$$x = \pm 1 \quad or \quad x = \pm 2. \qquad \text{Using the principle of square roots}$$

The solutions are -1, 1, -2, and 2.

Visualizing the Solution

The solutions of the given equation are the zeros of $f(x) = x^4 - 5x^2 + 4$. Note that the zeros occur at the x-values -2, -1, 1, and 2.

$$f(x) = x^4 - 5x^2 + 4$$

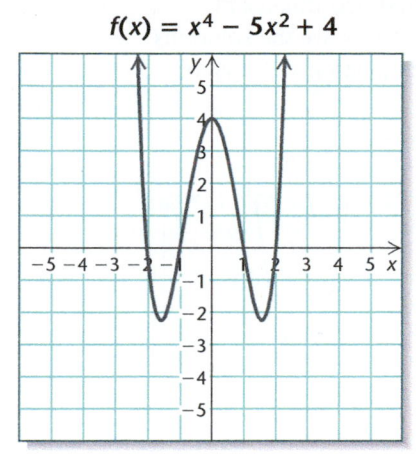

Now Try Exercise 79.

EXAMPLE 8 Solve: $t^{2/3} - 2t^{1/3} - 3 = 0$.

Solution We let $u = t^{1/3}$ and substitute:

$$t^{2/3} - 2t^{1/3} - 3 = 0$$
$$(t^{1/3})^2 - 2t^{1/3} - 3 = 0$$
$$u^2 - 2u - 3 = 0 \qquad \text{Substituting } u \text{ for } t^{1/3}$$
$$(u + 1)(u - 3) = 0 \qquad \text{Factoring}$$
$$u + 1 = 0 \quad or \quad u - 3 = 0 \qquad \text{Using the principle of zero products}$$
$$u = -1 \quad or \quad u = 3.$$

Now we must solve for the original variable, t. We substitute $t^{1/3}$ for u and solve for t:

$$t^{1/3} = -1 \quad or \quad t^{1/3} = 3$$
$$(t^{1/3})^3 = (-1)^3 \quad or \quad (t^{1/3})^3 = 3^3 \qquad \text{Cubing on both sides}$$
$$t = -1 \quad or \quad t = 27.$$

The solutions are -1 and 27.   **Now Try Exercise 87.**

▶ Applications

EXAMPLE 9 *Museums in China.* The number of museums in China increased from approximately 2000 in the year 2000 to over 3500 by the end of 2012. In 2012, a record 451 new museums opened. For comparison, in the United States, only 20–40 new museums were opened per year from 2000 to 2008. The function

$$h(x) = 30.992x^2 + 4.108x + 2294.594$$

can be used to estimate the number of museums in China, x years after 2005. (*Source:* The Economist/www.economist.com) Use this function to answer the following.

a) Estimate the number of museums that will be in China in 2017 if the number of new museums that open per year continues to grow in the same pattern.

b) In what year was the number of museums in China 2600?

Solution

a) For 2017, $x = 2017 - 2005 = 12$. We substitute 12 for x and find $h(12)$:

$$h(x) = 30.992x^2 + 4.108x + 2294.594$$
$$h(12) = 30.992(12)^2 + 4.108(12) + 2294.594$$
$$h(12) = 4462.848 + 49.296 + 2294.594 \approx 6807.$$

In 2017, there will be approximately 6807 museums in China.

b) We substitute 2600 for $h(x)$ and solve for x:

$$h(x) = 30.992x^2 + 4.108x + 2294.594$$
$$2600 = 30.992x^2 + 4.108x + 2294.594$$
$$0 = 30.992x^2 + 4.108x - 305.406.$$

We then use the quadratic formula, with $a = 30.992$, $b = 4.108$, and $c = -305.406$:

$$x = \frac{-b \pm \sqrt{b^2 - 4ac}}{2a}$$

$$x = \frac{-4.108 \pm \sqrt{(4.108)^2 - 4(30.992)(-305.406)}}{2(30.992)}$$

$$x = \frac{-4.108 \pm \sqrt{37,877.44667}}{61.984}$$

$$x = 3.074 \quad or \quad x = -3.206.$$

Because we are looking for a year after 2005, we use the positive solution. Thus there were 2600 museums in China about 3 years after 2005, or in 2008.

Now Try Exercise 97.

EXAMPLE 10   *Sales of New Homes.*   Sales of new homes have increased in recent years. The function

$$h(x) = 22.1x^2 - 72.2x + 371.9$$

can be used to estimate the sales of new homes, in thousands, in the United States, where x is the number of years after 2009 (*Source:* IHS Global Insight). In what year were the number of sales of new homes about 563,400, or 563.4 thousands?

Solution   We substitute 563.4 for $h(x)$ and solve for x:

$$563.4 = 22.1x^2 - 72.2x + 371.9$$
$$0 = 22.1x^2 - 72.2x - 191.5.$$

We then use the quadratic formula, with $a = 22.1$, $b = -72.2$, and $c = -191.5$:

$$x = \frac{-(-72.2) \pm \sqrt{(-72.2)^2 - 4(22.1)(-191.5)}}{2(22.1)} \qquad \text{Substituting}$$

$$x = \frac{72.2 \pm \sqrt{22,141.44}}{44.2}$$

$$x = 5 \quad or \quad x \approx -1.7.$$

Because we are looking for a year after 2009, we use the positive solution. Thus there were about 563,400 sales of new homes 5 years after 2009, or in 2014.

Now Try Exercise 95.

EXAMPLE 11 *Train Speeds.* Two trains leave a station at the same time. One train travels due west, and the other travels due south. The train traveling west travels 20 km/h faster than the train traveling south. After 2 hr, the trains are 200 km apart. Find the speed of each train.

Solution

1. **Familiarize.** First, we make a drawing. We let r = the speed of the train traveling south, in kilometers per hour. Then $r + 20$ = the speed of the train traveling west, in kilometers per hour. We use the motion formula $d = rt$, where d is the distance, r is the rate (or speed), and t is the time. After 2 hr, the train traveling south has traveled $2r$ kilometers, and the train traveling west has traveled $2(r + 20)$ kilometers. We add these distances to the drawing.

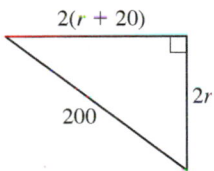

2. **Translate.** We use the Pythagorean theorem, $a^2 + b^2 = c^2$, where a and b are the lengths of the legs of a right triangle and c is the length of the hypotenuse:

$$[2(r + 20)]^2 + (2r)^2 = 200^2.$$

3. **Carry out.** We solve the equation:

$$[2(r + 20)]^2 + (2r)^2 = 200^2$$

$$4(r^2 + 40r + 400) + 4r^2 = 40{,}000$$

$$4r^2 + 160r + 1600 + 4r^2 = 40{,}000$$

$$8r^2 + 160r + 1600 = 40{,}000 \qquad \text{\textcolor{red}{Collecting like terms}}$$

$$8r^2 + 160r - 38{,}400 = 0 \qquad \text{\textcolor{red}{Subtracting 40,000}}$$

$$r^2 + 20r - 4800 = 0 \qquad \text{\textcolor{red}{Dividing by 8}}$$

$$(r + 80)(r - 60) = 0 \qquad \text{\textcolor{red}{Factoring}}$$

$$r + 80 = 0 \quad \text{or} \quad r - 60 = 0 \qquad \text{\textcolor{red}{Principle of zero products}}$$

$$r = -80 \quad \text{or} \qquad r = 60.$$

4. **Check.** Since speed cannot be negative, we need check only 60. If the speed of the train traveling south is 60 km/h, then the speed of the train traveling west is $60 + 20$, or 80 km/h. In 2 hr, the train heading south travels $60 \cdot 2$, or 120 km, and the train heading west travels $80 \cdot 2$, or 160 km. Then they are $\sqrt{120^2 + 160^2}$, or $\sqrt{40{,}000}$, or 200 km apart. The answer checks.

5. **State.** The speed of the train heading south is 60 km/h, and the speed of the train heading west is 80 km/h.

> **Now Try Exercise 101.**

CONNECTING THE CONCEPTS

Zeros, Solutions, and Intercepts

The zeros of a function $y = f(x)$ are also the solutions of the equation $f(x) = 0$, and the real-number zeros are the first coordinates of the x-intercepts of the graph of the function.

FUNCTION	ZEROS OF THE FUNCTION; SOLUTIONS OF THE EQUATION	x-INTERCEPTS OF THE GRAPH

Linear Function

$f(x) = 2x - 4$, or
$\quad y = 2x - 4$

To find the **zero** of $f(x)$, we solve $f(x) = 0$:

$$2x - 4 = 0$$
$$2x = 4$$
$$x = 2.$$

The **solution** of the equation $2x - 4 = 0$ is 2. This is the zero of the function $f(x) = 2x - 4$; that is, $f(2) = 0$.

The zero of $f(x)$ is the first coordinate of the **x-intercept** of the graph of $y = f(x)$.

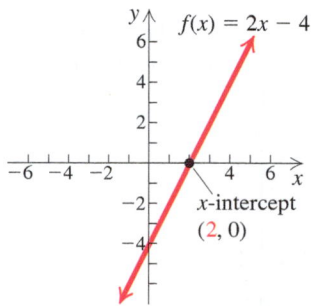

Quadratic Function

$g(x) = x^2 - 3x - 4$, or
$\quad y = x^2 - 3x - 4$

To find the **zeros** of $g(x)$, we solve $g(x) = 0$:

$$x^2 - 3x - 4 = 0$$
$$(x + 1)(x - 4) = 0$$
$$x + 1 = 0 \quad or \quad x - 4 = 0$$
$$x = -1 \quad or \quad x = 4.$$

The **solutions** of the equation $x^2 - 3x - 4 = 0$ are -1 and 4. They are the zeros of the function $g(x)$; that is, $g(-1) = 0$ and $g(4) = 0$.

The real-number zeros of $g(x)$ are the first coordinates of the **x-intercepts** of the graph of $y = g(x)$.

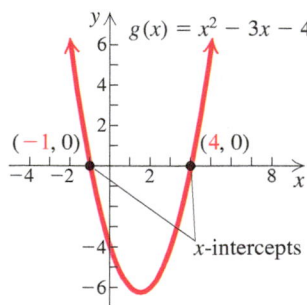

7.4 Exercise Set

Solve.

1. $(2x - 3)(3x - 2) = 0$

2. $(2x + 3)(5x - 2) = 0$

3. $x^2 - 8x - 20 = 0$

4. $x^2 + 6x + 8 = 0$

5. $3x^2 + x - 2 = 0$

6. $10x^2 - 16x + 6 = 0$

7. $4x^2 - 12 = 0$

8. $6x^2 = 36$

9. $3x^2 = 21$

10. $2x^2 - 20 = 0$

11. $5x^2 + 10 = 0$

12. $4x^2 + 12 = 0$

13. $x^2 + 16 = 0$

14. $x^2 + 25 = 0$

15. $2x^2 = 6x$

16. $18x + 9x^2 = 0$

17. $3y^3 - 5y^2 - 2y = 0$

18. $3t^3 + 2t = 5t^2$

19. $7x^3 + x^2 - 7x - 1 = 0$
(*Hint:* Factor by grouping.)

20. $3x^3 + x^2 - 12x - 4 = 0$
(*Hint:* Factor by grouping.)

In Exercises 21–28, use the given graph to find each of the following: (**a**) *the x-intercept(s) and* (**b**) *the zero(s) of the function.*

21.

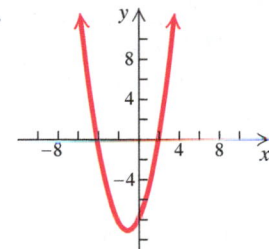

22.

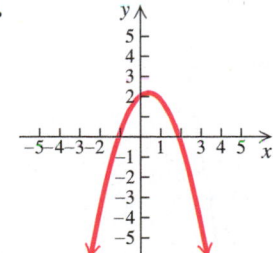

23.

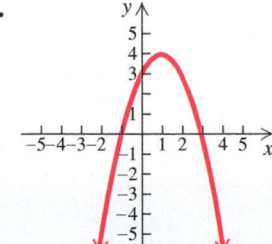

24.

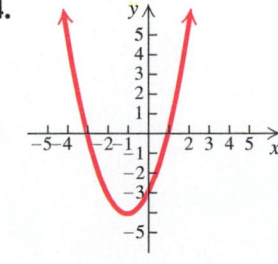

25.

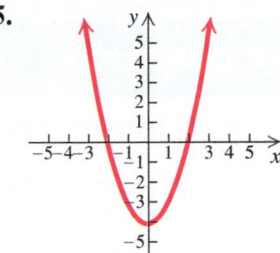

26.

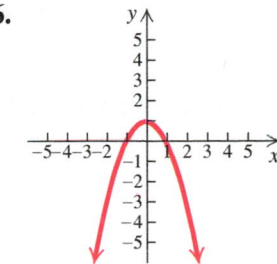

27.

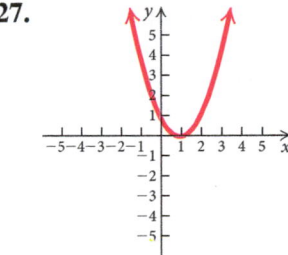

28.

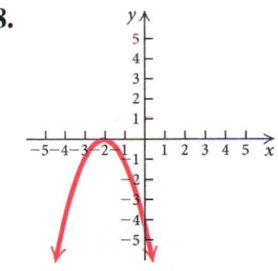

Solve by completing the square to obtain exact solutions.

29. $x^2 + 6x = 7$

30. $x^2 + 8x = -15$

31. $x^2 = 8x - 9$

32. $x^2 = 22 + 10x$

33. $x^2 + 8x + 25 = 0$

34. $x^2 + 6x + 13 = 0$

35. $3x^2 + 5x - 2 = 0$

36. $2x^2 - 5x - 3 = 0$

Use the quadratic formula to find exact solutions.

37. $x^2 - 2x = 15$

38. $x^2 + 4x = 5$

39. $5m^2 + 3m = 2$

40. $2y^2 - 3y - 2 = 0$

41. $3x^2 + 6 = 10x$

42. $3t^2 + 8t + 3 = 0$

43. $x^2 + x + 2 = 0$

44. $x^2 + 1 = x$

45. $5t^2 - 8t = 3$

46. $5x^2 + 2 = x$

47. $3x^2 + 4 = 5x$

48. $2t^2 - 5t = 1$

49. $x^2 - 8x + 5 = 0$

50. $x^2 - 6x + 3 = 0$

51. $3x^2 + x = 5$

52. $5x^2 + 3x = 1$

53. $2x^2 + 1 = 5x$

54. $4x^2 + 3 = x$

55. $5x^2 + 2x = -2$

56. $3x^2 + 3x = -4$

For each of the following, find the discriminant, $b^2 - 4ac$, and then determine whether one real-number solution, two different real-number solutions, or two different imaginary-number solutions exist.

57. $4x^2 = 8x + 5$

58. $4x^2 - 12x + 9 = 0$

59. $x^2 + 3x + 4 = 0$

60. $x^2 - 2x + 4 = 0$

61. $5t^2 - 7t = 0$

62. $5t^2 - 4t = 11$

Find the zeros of the function. Give exact answers and approximate solutions rounded to three decimal places when possible.

63. $f(x) = x^2 + 6x + 5$

64. $f(x) = x^2 - x - 2$

65. $f(x) = x^2 - 3x - 3$

66. $f(x) = 3x^2 + 8x + 2$

67. $f(x) = x^2 - 5x + 1$

68. $f(x) = x^2 - 3x - 7$

69. $f(x) = x^2 + 2x - 5$

70. $f(x) = x^2 - x - 4$

71. $f(x) = 2x^2 - x + 4$

72. $f(x) = 2x^2 + 3x + 2$

73. $f(x) = 3x^2 - x - 1$

74. $f(x) = 3x^2 + 5x + 1$

75. $f(x) = 5x^2 - 2x - 1$

76. $f(x) = 4x^2 - 4x - 5$

77. $f(x) = 4x^2 + 3x - 3$

78. $f(x) = x^2 + 6x - 3$

Solve.

79. $x^4 - 3x^2 + 2 = 0$

80. $x^4 + 3 = 4x^2$

81. $x^4 + 3x^2 = 10$

82. $x^4 - 8x^2 = 9$

83. $y^4 + 4y^2 - 5 = 0$

84. $y^4 - 15y^2 - 16 = 0$

85. $x - 3\sqrt{x} - 4 = 0$
$\left(\textit{Hint}: \text{Let } u = \sqrt{x}.\right)$

86. $2x - 9\sqrt{x} + 4 = 0$

87. $m^{2/3} - 2m^{1/3} - 8 = 0$
$(\textit{Hint}: \text{Let } u = m^{1/3}.)$

88. $t^{2/3} + t^{1/3} - 6 = 0$

89. $x^{1/2} - 3x^{1/4} + 2 = 0$

90. $x^{1/2} - 4x^{1/4} = -3$

91. $(2x - 3)^2 - 5(2x - 3) + 6 = 0$
(Hint: Let $u = 2x - 3$.)

92. $(3x + 2)^2 + 7(3x + 2) - 8 = 0$

93. $(2t^2 + t)^2 - 4(2t^2 + t) + 3 = 0$

94. $12 = (m^2 - 5m)^2 + (m^2 - 5m)$

Funding for Afghan Security. *The number of U.S. forces in Afghanistan decreased to approximately 34,000 in 2014 from a high of about 100,000 in 2010. The amount of U.S. funding for Afghan security forces also decreased during this period. The function*

$$f(x) = -1.321x^2 + 5.156x + 5.517$$

can be used to estimate the amount of U.S. funding for Afghan security forces, in billions of dollars, x years after 2009 (Source: U.S. Department of Defense; Brookings Institution; International Security Assistance Force; ESRI). Use this function for Exercises 95 and 96.

95. In what year was the amount of U.S. funding for Afghan security forces about \$10.5 billion?

96. In what year was the amount of U.S. funding for Afghan security forces about \$5.0 billion?

Multigenerational Households. *After declining between 1940 and 1980, the number of multigenerational American households has been increasing since 1980. The function*

$$h(x) = 0.012x^2 - 0.583x + 35.727$$

can be used to estimate the number of multigenerational households in the United States, in millions, x years after 1940 (Source: Pew Research Center). Use this function for Exercises 97 and 98.

97. In what year were there 40 million multigenerational households?

98. In what year were there 55 million multigenerational households?

Time of a Free Fall. *The formula* $s = 16t^2$ *is used to approximate the distance s, in feet, that an object falls freely from rest in t seconds. Use this formula for Exercises 99 and 100.*

99. The Taipei 101 Tower, also known as the Taipei Financial Center, in Taipei, Taiwan, is 1670 ft tall. How long would it take an object dropped from the top to reach the ground?

100. At 630 ft, the Gateway Arch in St. Louis is the tallest man-made monument in the United States. How long would it take an object dropped from the top to reach the ground?

101. The length of a rectangular poster is 1 ft more than the width, and a diagonal of the poster is 5 ft. Find the length and the width.

102. One leg of a right triangle is 7 cm less than the length of the other leg. The length of the hypotenuse is 13 cm. Find the lengths of the legs.

103. One number is 5 greater than another. The product of the numbers is 36. Find the numbers.

104. One number is 6 less than another. The product of the numbers is 72. Find the numbers.

105. *Box Construction.* An open box is made from a 10-cm by 20-cm piece of tin by cutting a square from each corner and folding up the edges. The area of the resulting base is 96 cm^2. What is the length of the sides of the squares?

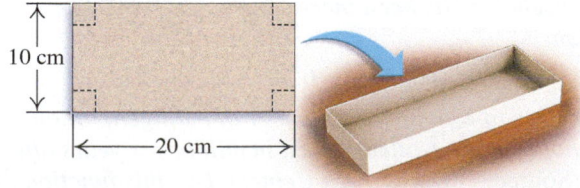

106. *Petting Zoo Dimensions.* At the Glen Island Zoo, 170 m of fencing was used to enclose a rectangular petting area of 1750 m^2. Find the dimensions of the petting area.

107. *Dimensions of a Rug.* Find the dimensions of a rectangular Persian rug whose perimeter is 28 ft and whose area is 48 ft^2.

108. *Picture Frame Dimensions.* The rectangular frame on a picture is 8 in. by 10 in. outside and is of uniform width. What is the width of the frame if 48 in^2 of the picture shows?

8 in.

10 in.

State whether the function is linear or quadratic.

109. $f(x) = 4 - 5x$

110. $f(x) = 4 - 5x^2$

111. $f(x) = 7x^2$

112. $f(x) = 23x + 6$

113. $f(x) = 1.2x - (3.6)^2$

114. $f(x) = 2 - x - x^2$

▶ Skill Maintenance

Cost of a Super Bowl Ad. *The cost of a 30-sec Super Bowl ad has increased more than 70% since 2004. The function*

$$C(x) = 0.17x + 2.25$$

can be used to estimate the cost of a 30-sec ad, in millions of dollars, x years after 2004 (Source: Katar Media). Use this function for Exercises 115 and 116. **[2.2]**

115. Estimate the cost of a 30-sec Super Bowl ad in 2014.

116. When will the cost of a 30-sec Super Bowl ad reach $5.0 million?

Determine whether the graph is symmetric with respect to the x-axis, the y-axis, and the origin. **[7.1]**

117. $3x^2 + 4y^2 = 5$

118. $y^3 = 6x^2$

Determine whether the function is even, odd, or neither even nor odd. **[7.1]**

119. $f(x) = 2x^3 - x$

120. $f(x) = 4x^2 + 2x - 3$

▶ ## Synthesis

For each equation in Exercises 121–124, under the given condition: **(a)** *find k and* **(b)** *find a second solution.*

121. $kx^2 - 17x + 33 = 0$; one solution is 3

122. $kx^2 - 2x + k = 0$; one solution is -3

123. $x^2 - kx + 2 = 0$; one solution is $1 + i$

124. $x^2 - (6 + 3i)x + k = 0$; one solution is 3

Solve.

125. $(x - 2)^3 = x^3 - 2$

126. $(x + 1)^3 = (x - 1)^3 + 26$

127. $(6x^3 + 7x^2 - 3x)(x^2 - 7) = 0$

128. $\left(x - \frac{1}{5}\right)\left(x^2 - \frac{1}{4}\right) + \left(x - \frac{1}{5}\right)\left(x^2 + \frac{1}{8}\right) = 0$

129. $x^2 + x - \sqrt{2} = 0$

130. $x^2 + \sqrt{5}x - \sqrt{3} = 0$

131. $2t^2 + (t - 4)^2 = 5t(t - 4) + 24$

132. $9t(t + 2) - 3t(t - 2) = 2(t + 4)(t + 6)$

133. $\sqrt{x - 3} - \sqrt[4]{x - 3} = 2$

134. $x^2 + 3x + 1 - \sqrt{x^2 + 3x + 1} = 8$

135. $\left(y + \frac{2}{y}\right)^2 + 3y + \frac{6}{y} = 4$

136. Solve $\frac{1}{2}at^2 + v_0t + x_0 = 0$ for t.

7.5

Analyzing Graphs of Quadratic Functions

▶ Find the vertex, the axis of symmetry, and the maximum or minimum value of a quadratic function using the method of completing the square.

▶ Graph quadratic functions.

▶ Solve applied problems involving maximum and minimum function values.

▶ ## Graphing Quadratic Functions of the Type $f(x) = a(x - h)^2 + k$

The graph of a quadratic function is called a **parabola**. The graph of every parabola evolves from the graph of the squaring function $f(x) = x^2$ using transformations.

TRANSFORMATIONS
REVIEW SECTION 7.2

We get the graph of $f(x) = a(x - h)^2 + k$ from the graph of $f(x) = x^2$ as follows:

$$f(x) = x^2$$

$$\downarrow$$

$$f(x) = ax^2 \qquad \text{Vertical stretching or shrinking with a reflection across the } x\text{-axis if } a < 0$$

$$\downarrow$$

$$f(x) = a(x - h)^2 \qquad \text{Horizontal translation}$$

$$\downarrow$$

$$f(x) = a(x - h)^2 + k. \qquad \text{Vertical translation}$$

Consider the following graphs of the form $f(x) = a(x - h)^2 + k$. The point (h, k) at which the graph turns is called the **vertex**. The maximum or minimum value of $f(x)$ occurs at the vertex. Each graph has a line $x = h$ that is called the **axis of symmetry**.

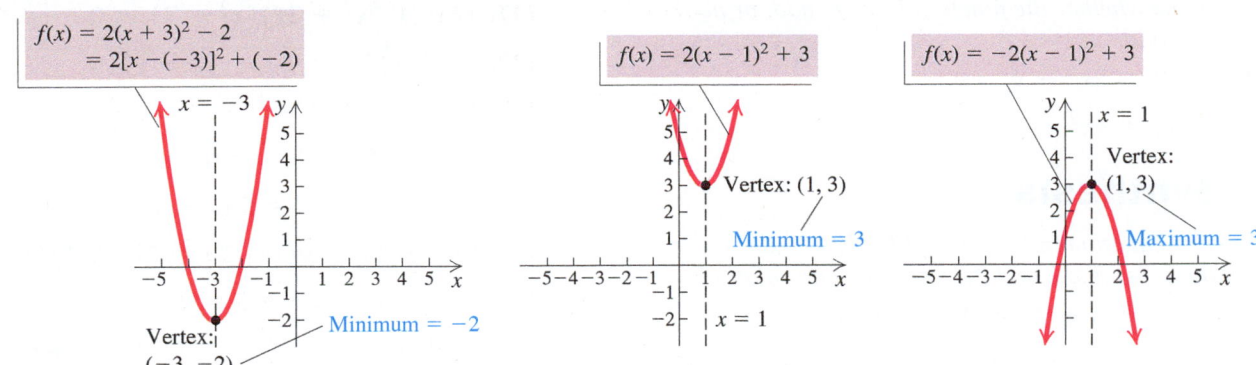

CONNECTING THE CONCEPTS

Graphing Quadratic Functions

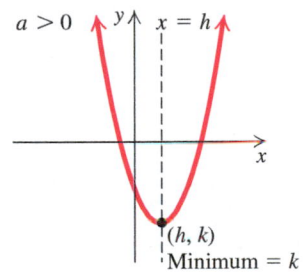

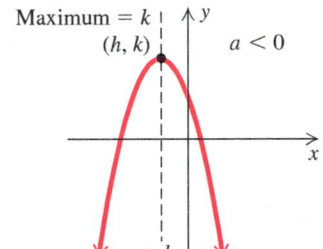

The graph of the function
$f(x) = a(x - h)^2 + k$ is a parabola that

- opens up if $a > 0$ and down if $a < 0$;
- has (h, k) as the vertex;
- has $x = h$ as the axis of symmetry;
- has k as a minimum value (output) if $a > 0$;
- has k as a maximum value if $a < 0$.

As we saw in Section 7.2, the constant a serves to stretch or shrink the graph vertically. As a parabola is stretched vertically, it becomes narrower, and as it is shrunk vertically, it becomes wider. That is, as $|a|$ increases, the graph becomes narrower, and as $|a|$ gets close to 0, the graph becomes wider.

If the equation is in the form $f(x) = a(x - h)^2 + k$, we can learn a great deal about the graph without actually graphing the function.

Function	$f(x) = 3\left(x - \frac{1}{4}\right)^2 - 2$ $= 3\left(x - \frac{1}{4}\right)^2 + (-2)$	$g(x) = -3(x + 5)^2 + 7$ $= -3[x - (-5)]^2 + 7$
Vertex	$\left(\frac{1}{4}, -2\right)$	$(-5, 7)$
Axis of Symmetry	$x = \frac{1}{4}$	$x = -5$
Maximum	None (3 > 0, so the graph opens up.)	7 (−3 < 0, so the graph opens down.)
Minimum	−2 (3 > 0, so the graph opens up.)	None (−3 < 0, so the graph opens down.)

Note that the vertex (h, k) is used to find the maximum or minimum value of the function. The maximum or minimum value is the number k, *not* the ordered pair (h, k).

▶ Graphing Quadratic Functions of the Type $f(x) = ax^2 + bx + c$, $a \neq 0$

We now use a modification of the method of completing the square as an aid in graphing and analyzing quadratic functions of the form $f(x) = ax^2 + bx + c$, $a \neq 0$.

EXAMPLE 1   Find the vertex, the axis of symmetry, and the maximum or minimum value of $f(x) = x^2 + 10x + 23$. Then graph the function.

Solution   To express $f(x) = x^2 + 10x + 23$ in the form $f(x) = a(x - h)^2 + k$, we complete the square on the terms involving x. To do so, we take half the coefficient of x and square it, obtaining $(10/2)^2$, or 25. We now add and subtract that number on the *right side*:

$$f(x) = x^2 + 10x + 23 = x^2 + 10x + 25 - 25 + 23.$$

Since $25 - 25 = 0$, the new expression for the function is equivalent to the original expression. Note that this process differs from the one we used to complete the square in order to solve a quadratic equation, where we added the same number on both sides of the equation to obtain an equivalent equation. Instead, when we complete the square to write a function in the form $f(x) = a(x - h)^2 + k$, we add and subtract the same number on the one side. The entire process is shown below:

$$f(x) = x^2 + 10x + 23 \qquad \text{Note that 25 completes the square for } x^2 + 10x.$$

$$= x^2 + 10x + 25 - 25 + 23 \qquad \text{Adding } 25 - 25, \text{ or 0, on the right side}$$

$$= (x^2 + 10x + 25) - 25 + 23 \qquad \text{Regrouping}$$

$$= (x + 5)^2 - 2 \qquad \text{Factoring and simplifying}$$

$$= [x - (-5)]^2 + (-2). \qquad \text{Writing in the form } f(x) = a(x - h)^2 + k$$

Keeping in mind that this function will have a minimum value since $a > 0$ ($a = 1$), from this form of the function we know the following:

Vertex: $(-5, -2)$;

Axis of symmetry: $x = -5$;

Minimum value of the function: -2.

To graph the function, we first plot the vertex and find several points on either side of it. Then we plot these points and connect them with a smooth curve. We see that the points $(-4, -1)$ and $(-3, 2)$ are reflections of the points $(-6, -1)$ and $(-7, 2)$, respectively, across the axis of symmetry, $x = -5$.

x	$f(x)$	
-5	-2	← **Vertex**
-4	-1	
-3	2	
-6	-1	
-7	2	

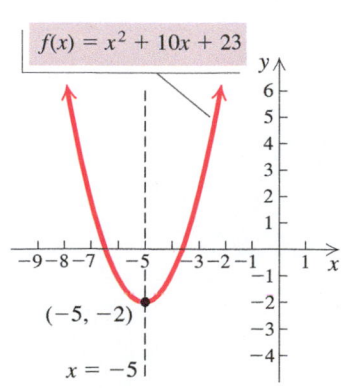

$f(x) = x^2 + 10x + 23$

$(-5, -2)$

$x = -5$

The graph of $f(x) = x^2 + 10x + 23$, or $f(x) = [x - (-5)]^2 + (-2)$, shown above, is a shift of the graph of $y = x^2$ left 5 units and down 2 units.

> **Now Try Exercise 3.**

Keep in mind that the axis of symmetry is not part of the graph; it is a characteristic of the graph. If you fold the graph on its axis of symmetry, the two halves of the graph will coincide.

EXAMPLE 2 Find the vertex, the axis of symmetry, and the maximum or minimum value of $g(x) = x^2/2 - 4x + 8$. Then graph the function.

Solution We complete the square in order to write the function in the form $g(x) = a(x - h)^2 + k$. First, we factor $\frac{1}{2}$ out of the first two terms. This makes the coefficient of x^2 within the parentheses 1:

$$g(x) = \frac{x^2}{2} - 4x + 8$$

$$= \frac{1}{2}(x^2 - 8x) + 8.$$

Factoring $\frac{1}{2}$ out of the first two terms:
$$\frac{x^2}{2} - 4x = \frac{1}{2} \cdot x^2 - \frac{1}{2} \cdot 8x$$

Next, we complete the square inside the parentheses: Half of -8 is -4, and $(-4)^2 = 16$. We add and subtract 16 inside the parentheses:

$$g(x) = \tfrac{1}{2}(x^2 - 8x + 16 - 16) + 8$$

$$= \tfrac{1}{2}(x^2 - 8x + 16) - \tfrac{1}{2} \cdot 16 + 8$$

Using the distributive law to remove -16 from within the parentheses

$$= \tfrac{1}{2}(x^2 - 8x + 16) - 8 + 8$$

$$= \tfrac{1}{2}(x - 4)^2 + 0, \text{ or } \tfrac{1}{2}(x - 4)^2.$$

Factoring and simplifying

We know the following:

Vertex: $(4, 0)$;

Axis of symmetry: $x = 4$;

Minimum value of the function: 0.

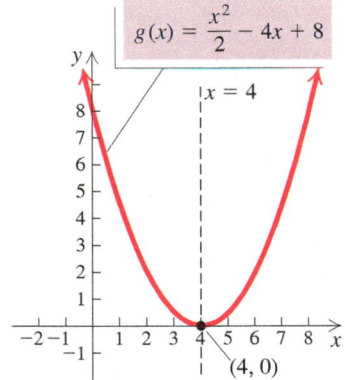

Finally, we plot the vertex and several points on either side of it and draw the graph of the function. The graph of g is a vertical shrinking of the graph of $y = x^2$ along with a shift right 4 units.

> **Now Try Exercise 9.**

EXAMPLE 3 Find the vertex, the axis of symmetry, and the maximum or minimum value of $f(x) = -2x^2 + 10x - \frac{23}{2}$. Then graph the function.

Solution We have

$$f(x) = -2x^2 + 10x - \tfrac{23}{2}$$

$$= -2(x^2 - 5x) - \tfrac{23}{2}$$

Factoring -2 out of the first two terms

$$= -2\left(x^2 - 5x + \tfrac{25}{4} - \tfrac{25}{4}\right) - \tfrac{23}{2}$$

Completing the square inside the parentheses

$$= -2\left(x^2 - 5x + \tfrac{25}{4}\right) - 2\left(-\tfrac{25}{4}\right) - \tfrac{23}{2}$$

Using the distributive law to remove $-\frac{25}{4}$ from within the parentheses

$$= -2\left(x^2 - 5x + \tfrac{25}{4}\right) + \tfrac{25}{2} - \tfrac{23}{2}$$

$$= -2\left(x - \tfrac{5}{2}\right)^2 + 1.$$

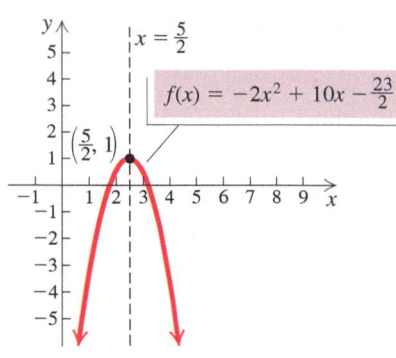

This form of the function yields the following:

Vertex: $\left(\frac{5}{2}, 1\right)$;

Axis of symmetry: $x = \frac{5}{2}$;

Maximum value of the function: 1.

The graph is found by shifting the graph of $f(x) = x^2$ right $\frac{5}{2}$ units, reflecting it across the x-axis, stretching it vertically, and shifting it up 1 unit.

In many situations, we want to use a formula to find the coordinates of the vertex directly from the equation $f(x) = ax^2 + bx + c$. One way to develop such a formula is to observe that the x-coordinate of the vertex is centered between the x-intercepts, or zeros, of the function. By averaging the two solutions of $ax^2 + bx + c = 0$, we find a formula for the x-coordinate of the vertex:

$$x\text{-coordinate of vertex} = \frac{\dfrac{-b - \sqrt{b^2 - 4ac}}{2a} + \dfrac{-b + \sqrt{b^2 - 4ac}}{2a}}{2}$$

$$= \frac{\dfrac{-2b}{2a}}{2} = \frac{-\dfrac{b}{a}}{2}$$

$$= -\frac{b}{a} \cdot \frac{1}{2} = -\frac{b}{2a}.$$

We use this value of x to find the y-coordinate of the vertex, $f\left(-\dfrac{b}{2a}\right)$.

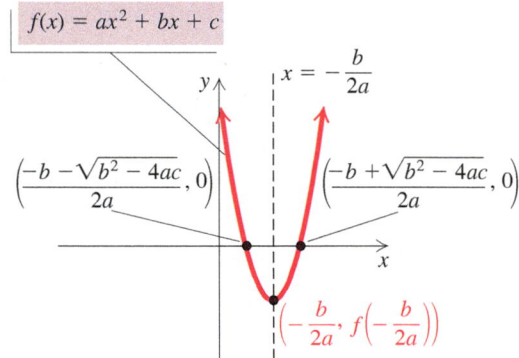

THE VERTEX OF A PARABOLA

The **vertex** of the graph of $f(x) = ax^2 + bx + c$ is

$$\left(-\frac{b}{2a}, f\left(-\frac{b}{2a}\right)\right).$$

We calculate the   We substitute to find
x-coordinate.   the y-coordinate.

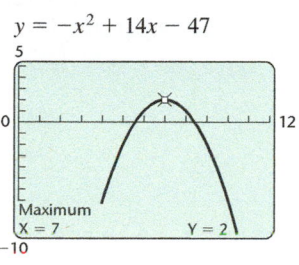
EXAMPLE 4   For the function $f(x) = -x^2 + 14x - 47$:

a) Find the vertex.

b) Determine whether there is a maximum or a minimum value and find that value.

c) Find the range.

d) On what intervals is the function increasing? decreasing?

Solution   There is no need to graph the function.

a) The x-coordinate of the vertex is

$$-\frac{b}{2a} = -\frac{14}{2(-1)} = -\frac{14}{-2} = 7.$$

Since

$$f(7) = -7^2 + 14 \cdot 7 - 47 = -49 + 98 - 47 = 2,$$

the vertex is $(7, 2)$.

b) Since a is negative ($a = -1$), the graph opens down so the second coordinate of the vertex, 2, is the maximum value of the function.

c) The range is $(-\infty, 2]$.

d) Since the graph opens down, function values increase as we approach the vertex from the left and decrease as we move to the right from the vertex. Thus the function is increasing on the interval $(-\infty, 7)$ and decreasing on $(7, \infty)$.

▶ **Now Try Exercise 31.**

▶ Applications

Many real-world situations involve finding the maximum or the minimum value of a quadratic function.

EXAMPLE 5 *Maximizing Area.*   A landscaper has enough stone to enclose a rectangular koi pond next to an existing garden wall of the Englemans' house with 24 ft of stone wall. If the garden wall forms one side of the rectangle, what is the maximum area that the landscaper can enclose? What dimensions of the koi pond will yield this area?

Solution   We will use the five-step problem-solving strategy.

1. **Familiarize.** We first make a drawing of the situation, using w to represent the width of the koi pond, in feet. Then $(24 - 2w)$ feet of stone is available for the length. Suppose the koi pond were 1 ft wide. Then its length would be $24 - 2 \cdot 1 = 22$ ft, and its area would be $(22\text{ ft})(1\text{ ft}) = 22\text{ ft}^2$. If the koi pond were 2 ft wide, its length would be $24 - 2 \cdot 2 = 20$ ft, and its area would be $(20\text{ ft})(2\text{ ft}) = 40\text{ ft}^2$. This is larger than the first area we found, but we do not know if it is the maximum possible area. To find the maximum area, we will find a function that represents the area and then determine its maximum value.

$24 - 2w$ w

2. **Translate.** We write a function for the area of the koi pond. We have

$$A(w) = (24 - 2w)w \qquad A = lw; l = 24 - 2w$$
$$= -2w^2 + 24w,$$

where $A(w)$ is the area of the koi pond, in square feet, as a function of the width w.

3. **Carry out.** To solve this problem, we need to determine the maximum value of $A(w)$ and find the dimensions for which that maximum occurs. Since A is a quadratic function and w^2 has a negative coefficient, we know that the function has a maximum value that occurs at the vertex of the graph of the function. The first coordinate of the vertex, $(w, A(w))$, is

$$w = -\frac{b}{2a} = -\frac{24}{2(-2)} = -\frac{24}{-4} = 6.$$

Thus, if $w = 6$ ft, then the length $l = 24 - 2 \cdot 6 = 12$ ft, and the area is $(12 \text{ ft})(6 \text{ ft}) = 72 \text{ ft}^2$.

4. **Check.** As a partial check, we note that $72 \text{ ft}^2 > 40 \text{ ft}^2$, the larger of the two areas that we found in the *Familiarize* step. We could also complete the square to write the function in the form $A(w) = a(w - h)^2 + k$, and then use this form of the function to determine the coordinates of the vertex. We get

$$A(w) = -2(w - 6)^2 + 72.$$

This confirms that the vertex is $(6, 72)$, so the answer checks.

5. **State.** The maximum possible area is 72 ft^2 when the koi pond is 6 ft wide and 12 ft long. ▸ **Now Try Exercise 45.**

EXAMPLE 6 *Height of a Rocket.* A model rocket is launched with an initial velocity of 100 ft/sec from the top of a hill that is 20 ft high. Its height, in feet, t seconds after it has been launched is given by the function $s(t) = -16t^2 + 100t + 20$. Determine the time at which the rocket reaches its maximum height and find the maximum height.

Solution

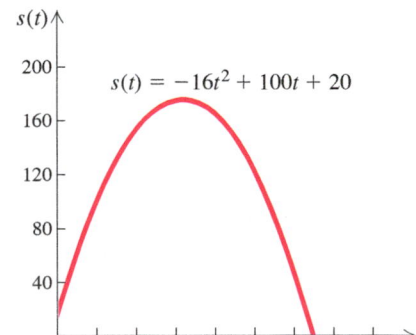

1., 2. **Familiarize** and **Translate.** We are given the function in the statement of the problem: $s(t) = -16t^2 + 100t + 20$.

3. **Carry out.** We need to find the maximum value of the function and the value of t for which it occurs. Since $s(t)$ is a quadratic function and t^2 has a negative coefficient, we know that the maximum value of the function occurs at the vertex of the graph of the function. The first coordinate of the vertex gives the time t at which the rocket reaches its maximum height. It is

$$t = -\frac{b}{2a} = -\frac{100}{2(-16)} = -\frac{100}{-32} = 3.125.$$

The second coordinate of the vertex gives the maximum height of the rocket. We substitute in the function to find it:

$$s(3.125) = -16(3.125)^2 + 100(3.125) + 20 = 176.25.$$

4. **Check.** As a check, we can complete the square to write the function in the form $s(t) = a(t - h)^2 + k$ and determine the coordinates of the vertex from this form of the function. We get

$$s(t) = -16(t - 3.125)^2 + 176.25.$$

This confirms that the vertex is $(3.125, 176.25)$, so the answer checks.

5. **State.** The rocket reaches a maximum height of 176.25 ft. This occurs 3.125 sec after it has been launched. ▸ **Now Try Exercise 41.**

EXAMPLE 7 *Determining the Height of an Elevator Shaft.* Jared drops a screwdriver from the top of an elevator shaft. Exactly 5 sec later, he hears the sound of the screwdriver hitting the bottom of the shaft. The speed of sound is 1100 ft/sec. How tall is the elevator shaft?

Solution

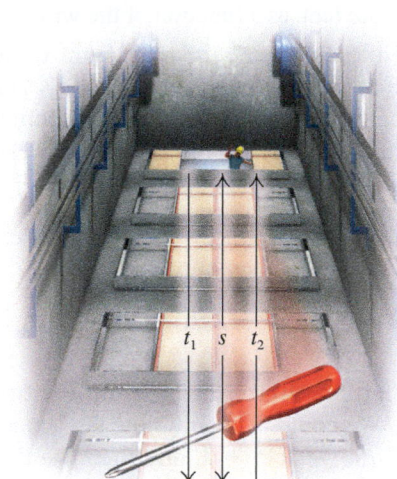

1. **Familiarize.** We first make a drawing and label it with known and unknown information. We let s = the height of the elevator shaft, in feet, t_1 = the time, in seconds, that it takes for the screwdriver to hit the bottom of the elevator shaft, and t_2 = the time, in seconds, that it takes for the sound to reach the top of the elevator shaft. This gives us the equation

$$t_1 + t_2 = 5. \tag{1}$$

2. **Translate.** Can we find any relationship between the two times and the distance s? Often in problem solving you may need to look up related formulas in a physics book, another mathematics book, or on the Internet. We find that the formula

$$s = 16t^2$$

gives the distance, in feet, that a dropped object falls in t seconds. The time t_1 that it takes the screwdriver to hit the bottom of the elevator shaft can be found as follows:

$$s = 16t_1^2, \quad \text{or} \quad \frac{s}{16} = t_1^2, \quad \text{so} \quad t_1 = \frac{\sqrt{s}}{4}. \quad \color{red}{\textbf{Taking the positive square root}} \tag{2}$$

To find an expression for t_2, the time that it takes the sound to travel to the top of the well, recall that *Distance = Rate · Time*. Thus,

$$s = 1100t_2, \quad \text{or} \quad t_2 = \frac{s}{1100}. \tag{3}$$

We now have expressions for t_1 and t_2, both in terms of s. Substituting into equation (1), we obtain

$$t_1 + t_2 = 5, \quad \text{or} \quad \frac{\sqrt{s}}{4} + \frac{s}{1100} = 5. \tag{4}$$

3. **Carry out.** We solve equation (4) for s. Multiplying by 1100, we get

$$275\sqrt{s} + s = 5500, \quad \text{or} \quad s + 275\sqrt{s} - 5500 = 0.$$

This equation is reducible to quadratic with $u = \sqrt{s}$. Substituting, we get

$$u^2 + 275u - 5500 = 0.$$

Using the quadratic formula, we can solve for u:

$$u = \frac{-b \pm \sqrt{b^2 - 4ac}}{2a}$$
$$= \frac{-275 + \sqrt{275^2 - 4 \cdot 1 \cdot (-5500)}}{2 \cdot 1} \quad \color{red}{\textbf{We want only the positive solution.}}$$
$$= \frac{-275 + \sqrt{97,625}}{2}$$
$$\approx 18.725.$$

Since $u \approx 18.725$, we have

$$\sqrt{s} \approx 18.725$$
$$s \approx 350.6. \quad \color{red}{\textbf{Squaring both sides and rounding to the nearest tenth}}$$

4. **Check.** To check, we can substitute 350.6 for s in equation (4) and see that $t_1 + t_2 \approx 5$. We leave the computation to the student.

5. **State.** The height of the elevator shaft is about 350.6 ft.

Now Try Exercise 55.

Technology Connection

We can solve the equation in Example 7 graphically using the Intersect feature. It will probably require some trial and error to determine an appropriate window.

$$y_1 = \frac{\sqrt{x}}{4} + \frac{x}{1100}, \quad y_2 = 5$$

Intersection
X = 350.62555   Y = 5

Visualizing the Graph

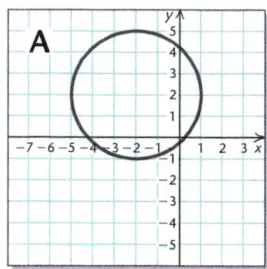

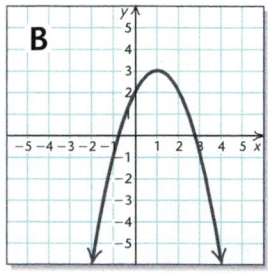

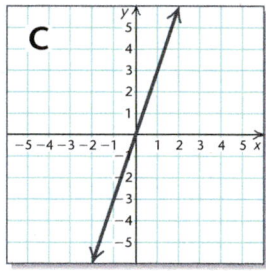

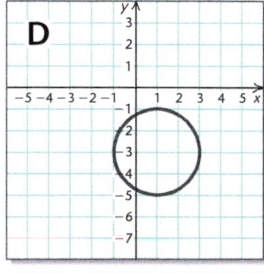

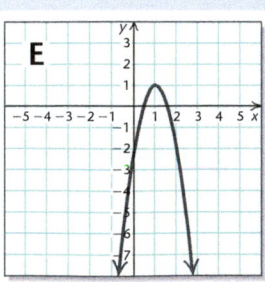

Match the equation with its graph.

1. $y = 3x$

2. $y = -(x - 1)^2 + 3$

3. $(x + 2)^2 + (y - 2)^2 = 9$

4. $y = 3$

5. $2x - 3y = 6$

6. $(x - 1)^2 + (y + 3)^2 = 4$

7. $y = -2x + 1$

8. $y = 2x^2 - x - 4$

9. $x = -2$

10. $y = -3x^2 + 6x - 2$

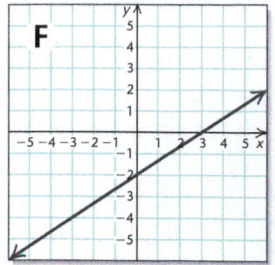

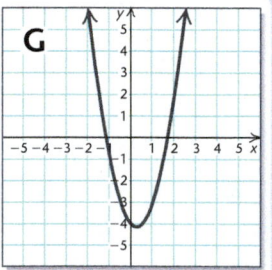

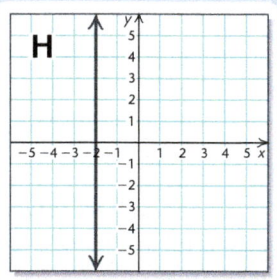

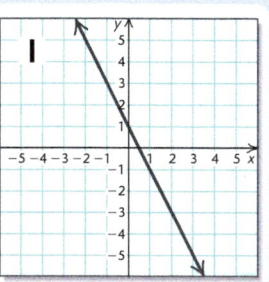

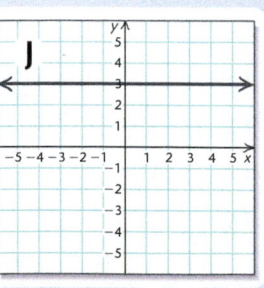

7.5 Exercise Set

In Exercises 1 and 2, use the given graph to find each of the following: **(a)** *the vertex;* **(b)** *the axis of symmetry; and* **(c)** *the maximum or the minimum value of the function.*

1.

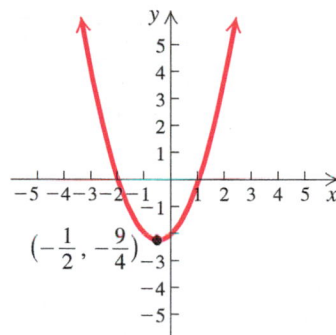

$\left(-\dfrac{1}{2}, -\dfrac{9}{4}\right)$

2.

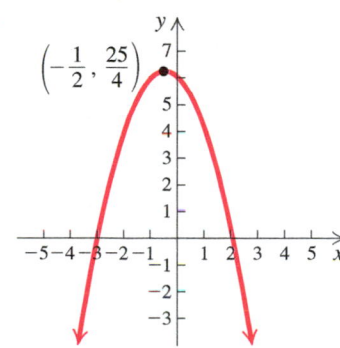

$\left(-\dfrac{1}{2}, \dfrac{25}{4}\right)$

In Exercises 3–16, **(a)** *find the vertex;* **(b)** *find the axis of symmetry;* **(c)** *determine whether there is a maximum or a minimum value, and find that value; and* **(d)** *graph the function.*

3. $f(x) = x^2 - 8x + 12$ **4.** $g(x) = x^2 + 7x - 8$

5. $f(x) = x^2 - 7x + 12$ **6.** $g(x) = x^2 - 5x + 6$

7. $f(x) = x^2 + 4x + 5$

8. $f(x) = x^2 + 2x + 6$

9. $g(x) = \dfrac{x^2}{2} + 4x + 6$

10. $g(x) = \dfrac{x^2}{3} - 2x + 1$

11. $g(x) = 2x^2 + 6x + 8$

12. $f(x) = 2x^2 - 10x + 14$

13. $f(x) = -x^2 - 6x + 3$

14. $f(x) = -x^2 - 8x + 5$

15. $g(x) = -2x^2 + 2x + 1$

16. $f(x) = -3x^2 - 3x + 1$

In Exercises 17–24, match the equation with one of the graphs (a)–(h) that follow.

a)
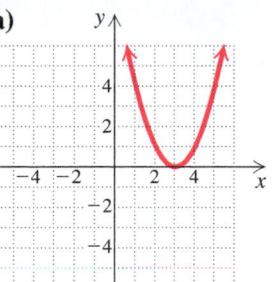

b)
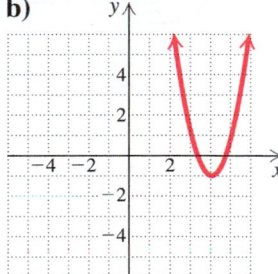

c)
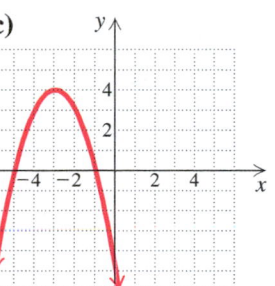

d)
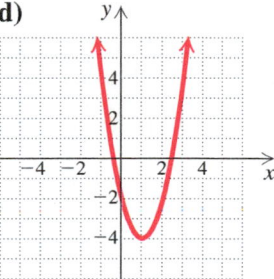

e)
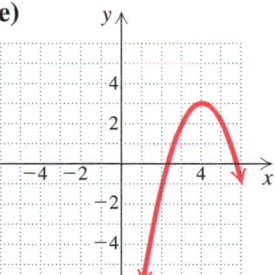

f)
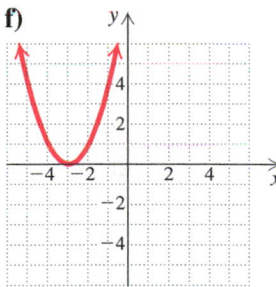

g)
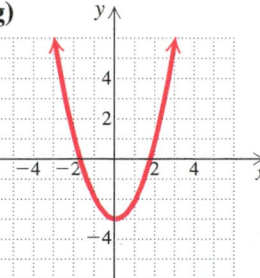

h)
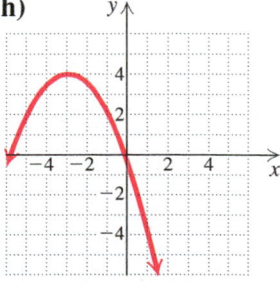

17. $y = (x + 3)^2$ **18.** $y = -(x - 4)^2 + 3$

19. $y = 2(x - 4)^2 - 1$ **20.** $y = x^2 - 3$

21. $y = -\frac{1}{2}(x + 3)^2 + 4$ **22.** $y = (x - 3)^2$

23. $y = -(x + 3)^2 + 4$ **24.** $y = 2(x - 1)^2 - 4$

In Exercises 25–30, determine whether the statement is true or false.

25. The function $f(x) = -3x^2 + 2x + 5$ has a maximum value.

26. The vertex of the graph of $f(x) = ax^2 + bx + c$ is $-\dfrac{b}{2a}$.

27. The graph of $h(x) = (x + 2)^2$ can be obtained by translating the graph of $h(x) = x^2$ right 2 units.

28. The vertex of the graph of the function $g(x) = 2(x - 4)^2 - 1$ is $(-4, -1)$.

29. The axis of symmetry of the function $f(x) = -(x + 2)^2 - 4$ is $x = -2$.

30. The minimum value of the function $f(x) = 3(x - 1)^2 + 5$ is 5.

In Exercises 31–40, **(a)** *find the vertex;* **(b)** *determine whether there is a maximum or a minimum value, and find that value;* **(c)** *find the range; and* **(d)** *find the intervals on which the function is increasing and the intervals on which the function is decreasing.*

31. $f(x) = x^2 - 6x + 5$

32. $f(x) = x^2 + 4x - 5$

33. $f(x) = 2x^2 + 4x - 16$

34. $f(x) = \frac{1}{2}x^2 - 3x + \frac{5}{2}$

35. $f(x) = -\frac{1}{2}x^2 + 5x - 8$

36. $f(x) = -2x^2 - 24x - 64$

37. $f(x) = 3x^2 + 6x + 5$

38. $f(x) = -3x^2 + 24x - 49$

39. $g(x) = -4x^2 - 12x + 9$

40. $g(x) = 2x^2 - 6x + 5$

41. *Height of a Ball.* A ball is thrown directly upward from a height of 6 ft with an initial velocity of 20 ft/sec. The function $s(t) = -16t^2 + 20t + 6$ gives the height of the ball, in feet, t seconds after it has been thrown. Determine the time at which the ball reaches its maximum height and find the maximum height.

42. *Height of a Projectile.* A stone is thrown directly upward from a height of 30 ft with an initial velocity of 60 ft/sec. The height of the stone, in feet, t seconds after it has been thrown is given by the function $s(t) = -16t^2 + 60t + 30$. Determine the time at which the stone reaches its maximum height and find the maximum height.

43. *Height of a Rocket.* A model rocket is launched with an initial velocity of 120 ft/sec from a height of 80 ft. The height of the rocket, in feet, t seconds after it has been launched is given by the function $s(t) = -16t^2 + 120t + 80$. Determine the time at which the rocket reaches its maximum height and find the maximum height.

44. *Height of a Rocket.* A model rocket is launched with an initial velocity of 150 ft/sec from a height of 40 ft. The function $s(t) = -16t^2 + 150t + 40$ gives the height of the rocket, in feet, t seconds after it has been launched. Determine the time at which the rocket reaches its maximum height and find the maximum height.

45. *Maximizing Volume.* Mendoza Manufacturing plans to produce a one-compartment vertical file by bending the long side of a 10-in. by 18-in. sheet of plastic along two lines to form a ⊔-shape. How tall should the file be in order to maximize the volume that it can hold?

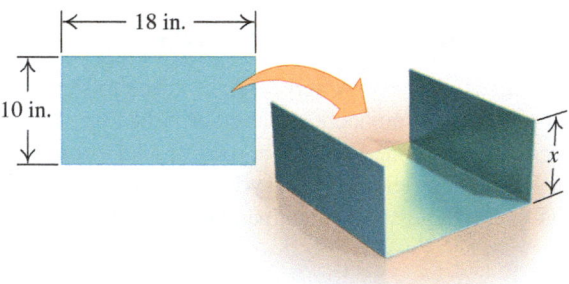

46. *Maximizing Area.* A fourth-grade class decides to enclose a rectangular garden, using the side of the school as one side of the rectangle. What is the maximum area that the class can enclose with 32 ft of fence? What should the dimensions of the garden be in order to yield this area?

47. *Maximizing Area.* The sum of the base and the height of a triangle is 20 cm. Find the dimensions for which the area is a maximum.

48. *Maximizing Area.* The sum of the base and the height of a parallelogram is 69 cm. Find the dimensions for which the area is a maximum.

49. *Minimizing Cost.* Designs for #1 Canines has determined that when x hundred wooden doghouses are built, the average cost per doghouse is given by
$$C(x) = 0.1x^2 - 0.7x + 1.625,$$

where $C(x)$ is in hundreds of dollars. How many doghouses should be built in order to minimize the average cost per doghouse?

Maximizing Profit.   In business, profit is the difference between revenue and cost; that is,

$$\text{Total profit} = \text{Total revenue} - \text{Total cost,}$$
$$P(x) = R(x) - C(x),$$

where x is the number of units sold. Find the maximum profit and the number of units that must be sold in order to yield the maximum profit for each of the following.

50. $R(x) = 5x$, $C(x) = 0.001x^2 + 1.2x + 60$

51. $R(x) = 50x - 0.5x^2$, $C(x) = 10x + 3$

52. $R(x) = 20x - 0.1x^2$, $C(x) = 4x + 2$

53. *Maximizing Area.*   A berry farmer needs to separate and enclose two adjacent rectangular fields, one for blueberries and one for strawberries. If a lake forms one side of the fields and 240 yd of fencing is available, what is the largest total area that can be enclosed?

54. *Norman Window.*   A Norman window is a rectangle with a semicircle on top. Sky Blue Windows is designing a Norman window that will require 24 ft of trim on the outer edges. What dimensions will allow the maximum amount of light to enter a house?

55. *Finding the Depth of a Well.*   Two seconds after a chlorine tablet has been dropped into a well, a splash is heard. The speed of sound is 1100 ft/sec. How far is the top of the well from the water? (*Hint*: See Example 7.)

56. *Finding the Height of a Cliff.*   A water balloon is dropped from a cliff. Exactly 3 sec later, the sound of the balloon hitting the ground reaches the top of the cliff. How high is the cliff? (*Hint*: See Example 7.)

▶ Skill Maintenance

For each function f, construct and simplify the difference quotient

$$\frac{f(x + h) - f(x)}{h}. \quad [5.4]$$

57. $f(x) = 3x - 7$ **58.** $f(x) = 2x^2 - x + 4$

A graph of $y = f(x)$ follows. No formula is given for f. Graph each of the following. **[7.2]**

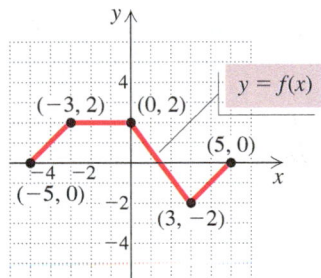

59. $g(x) = -2f(x)$ **60.** $g(x) = f(2x)$

▶ Synthesis

61. Find c such that
$$f(x) = -0.2x^2 - 3x + c$$
has a maximum value of -225.

62. Find b such that
$$f(x) = -4x^2 + bx + 3$$
has a maximum value of 50.

63. Graph: $f(x) = (|x| - 5)^2 - 3$.

64. Find a quadratic function with vertex $(4, -5)$ and containing the point $(-3, 1)$.

65. *Minimizing Area.*   A 24-in. piece of string is cut into two pieces. One piece is used to form a circle while the other is used to form a square. How should the string be cut so that the sum of the areas is a minimum?

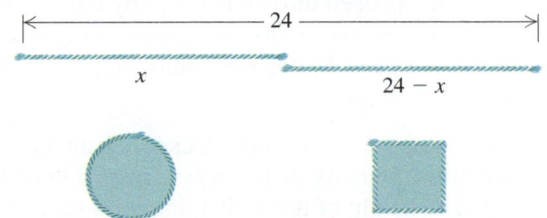

Chapter 7 Summary and Review

STUDY GUIDE

KEY TERMS AND CONCEPTS	EXAMPLE

SECTION 7.1: SYMMETRY

Algebraic Tests of Symmetry

x-axis: If replacing y with $-y$ produces an equivalent equation, then the graph is *symmetric with respect to the x-axis.*

y-axis: If replacing x with $-x$ produces an equivalent equation, then the graph is *symmetric with respect to the y-axis.*

Origin: If replacing x with $-x$ and y with $-y$ produces an equivalent equation, then the graph is *symmetric with respect to the origin.*

Test $y = 2x^3$ for symmetry with respect to the x-axis, the y-axis, and the origin.

x-axis: We replace y with $-y$:

$$-y = 2x^3$$
$$y = -2x^3. \quad \textcolor{red}{\textbf{Multiplying by } -1}$$

The resulting equation *is not* equivalent to the original equation, so the graph *is not* symmetric with respect to the x-axis.

y-axis: We replace x with $-x$:

$$y = 2(-x)^3$$
$$y = -2x^3.$$

The resulting equation *is not* equivalent to the original equation, so the graph *is not* symmetric with respect to the y-axis.

Origin: We replace x with $-x$ and y with $-y$:

$$-y = 2(-x)^3$$
$$-y = -2x^3$$
$$y = 2x^3.$$

The resulting equation *is* equivalent to the original equation, so the graph *is* symmetric with respect to the origin.

Even Functions and Odd Functions

If the graph of a function is symmetric with respect to the y-axis, we say that it is an **even function**. That is, for each x in the domain of f, $f(x) = f(-x)$.

If the graph of a function is symmetric with respect to the origin, we say that it is an **odd function**. That is, for each x in the domain of f, $f(-x) = -f(x)$.

Determine whether each function is even, odd, or neither.

a) $g(x) = 2x^2 - 4$ **b)** $h(x) = x^5 - 3x^3 - x$

a) We first find $g(-x)$ and simplify:

$$g(-x) = 2(-x)^2 - 4$$
$$= 2x^2 - 4.$$

$g(x) = g(-x)$, so g is even. Since a function other than $f(x) = 0$ cannot be *both* even and odd and g is even, we need not test to see if it is an odd function.

b) We first find $h(-x)$ and simplify:

$$h(-x) = (-x)^5 - 3(-x)^3 - (-x)$$
$$= -x^5 + 3x^3 + x.$$

$h(x) \neq h(-x)$, so h is *not* even.
   Next, we find $-h(x)$ and simplify:

$$-h(x) = -(x^5 - 3x^3 - x)$$
$$= -x^5 + 3x^3 + x.$$

$h(-x) = -h(x)$, so h is odd.

SECTION 7.2: TRANSFORMATIONS

Vertical Translation

For $b > 0$:

the graph of $y = f(x) + b$ is the graph of $y = f(x)$ shifted *up* b units;

the graph of $y = f(x) - b$ is the graph of $y = f(x)$ shifted *down* b units.

Horizontal Translation

For $d > 0$:

the graph of $y = f(x - d)$ is the graph of $y = f(x)$ shifted *right* d units;

the graph of $y = f(x + d)$ is the graph of $y = f(x)$ shifted *left* d units.

Reflections

The graph of $y = -f(x)$ is the **reflection** of $y = f(x)$ across the x-axis.

The graph of $y = f(-x)$ is the **reflection** of $y = f(x)$ across the y-axis.

If a point (x, y) is on the graph of $y = f(x)$, then $(x, -y)$ is on the graph of $y = -f(x)$, and $(-x, y)$ is on the graph of $y = f(-x)$.

Graph $g(x) = (x - 2)^2 + 1$. Before doing so, describe how the graph can be obtained from the graph of $f(x) = x^2$.

First, note that the graph of $h(x) = (x - 2)^2$ is the graph of $f(x) = x^2$ shifted right 2 units. Then the graph of $g(x) = (x - 2)^2 + 1$ is the graph of $h(x) = (x - 2)^2$ shifted up 1 unit. Thus the graph of g is obtained by shifting the graph of $f(x) = x^2$ right 2 units and up 1 unit.

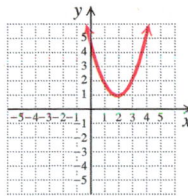

$$g(x) = (x - 2)^2 + 1$$

Graph each of the following. Before doing so, describe how each graph can be obtained from the graph of $f(x) = x^2 - x$.

a) $g(x) = x - x^2$ **b)** $h(x) = (-x)^2 - (-x)$

a) Note that

$$-f(x) = -(x^2 - x)$$
$$= -x^2 + x$$
$$= x - x^2$$
$$= g(x).$$

Thus the graph is a reflection of the graph of $f(x) = x^2 - x$ across the x-axis.

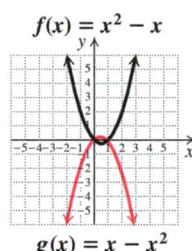

$$f(x) = x^2 - x$$
$$g(x) = x - x^2$$

b) Note that

$$f(-x) = (-x)^2 - (-x) = h(x).$$

Thus the graph of $h(x) = (-x)^2 - (-x)$ is a reflection of the graph of $f(x) = x^2 - x$ across the y-axis.

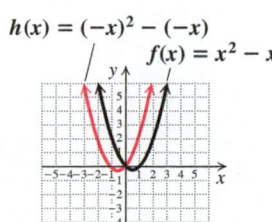

$$h(x) = (-x)^2 - (-x)$$
$$f(x) = x^2 - x$$

Vertical Stretching and Shrinking

The graph of $y = af(x)$ can be obtained from the graph of $y = f(x)$ by:

stretching vertically for $|a| > 1$, or

shrinking vertically for $0 < |a| < 1$.

For $a < 0$, the graph is also reflected across the x-axis.

(The y-coordinates of the graph of $y = af(x)$ can be obtained by multiplying the y-coordinates of $y = f(x)$ by a.)

Horizontal Stretching and Shrinking

The graph of $y = f(cx)$ can be obtained from the graph of $y = f(x)$ by:

shrinking horizontally for $|c| > 1$, or

stretching horizontally for $0 < |c| < 1$.

For $c < 0$, the graph is also reflected across the y-axis.

(The x-coordinates of the graph of $y = f(cx)$ can be obtained by dividing the x-coordinates of $y = f(x)$ by c.)

A graph of $y = g(x)$ is shown below. Use this graph to graph each of the given equations.

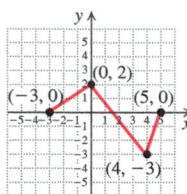

a) $f(x) = g(2x)$ **b)** $f(x) = -2g(x)$
c) $f(x) = \frac{1}{2}g(x)$ **d)** $f(x) = g\left(\frac{1}{2}x\right)$

a) Since $|2| > 1$, the graph of $f(x) = g(2x)$ is a horizontal shrinking of the graph of $y = g(x)$. The transformation divides each x-coordinate of g by 2.

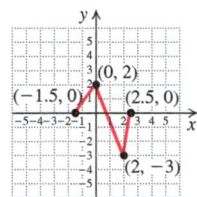

b) Since $|-2| > 1$, the graph of $f(x) = -2g(x)$ is a vertical stretching of the graph of $y = g(x)$. The transformation multiplies each y-coordinate of g by 2. Since $-2 < 0$, the graph is also reflected across the x-axis.

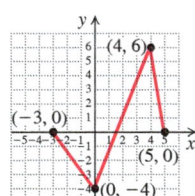

c) Since $\left|\frac{1}{2}\right| < 1$, the graph of $f(x) = \frac{1}{2}g(x)$ is a vertical shrinking of the graph of $y = g(x)$. The transformation multiplies each y-coordinate of g by $\frac{1}{2}$.

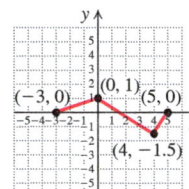

(continued)

d) Since $\left|\frac{1}{2}\right| < 1$, the graph of $f(x) = g\left(\frac{1}{2}x\right)$ is a horizontal stretching of the graph of $y = g(x)$. The transformation divides each x-coordinate of g by $\frac{1}{2}$ (which is the same as multiplying by 2).

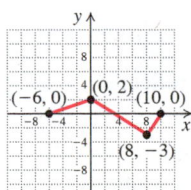

SECTION 7.3: THE COMPLEX NUMBERS

The number i is defined such that $i = \sqrt{-1}$ and $i^2 = -1$.

Express each number in terms of i.

$$\sqrt{-5} = \sqrt{-1 \cdot 5} = \sqrt{-1} \cdot \sqrt{5} = i\sqrt{5}, \text{ or } \sqrt{5}i;$$

$$-\sqrt{-36} = -\sqrt{-1 \cdot 36} = -\sqrt{-1} \cdot \sqrt{36} = -i \cdot 6 = -6i$$

A **complex number** is a number of the form $a + bi$, where a and b are real numbers. The number a is said to be the **real part** of $a + bi$, and the number b is said to be the **imaginary part** of $a + bi$.

To **add** or **subtract complex numbers**, we add or subtract the real parts, and we add or subtract the imaginary parts.

Add or subtract.

$$(-3 + 4i) + (5 - 8i) = (-3 + 5) + (4i - 8i)$$
$$= 2 - 4i;$$

$$(6 - 7i) - (10 + 3i) = (6 - 10) + (-7i - 3i)$$
$$= -4 - 10i$$

When we **multiply complex numbers**, we must keep in mind the fact that $i^2 = -1$.

Note that $\sqrt{a} \cdot \sqrt{b} \neq \sqrt{ab}$ when $\sqrt{a}$ and $\sqrt{b}$ are not real numbers.

Multiply.

$$\sqrt{-4} \cdot \sqrt{-100} = \sqrt{-1} \cdot \sqrt{4} \cdot \sqrt{-1} \cdot \sqrt{100}$$
$$= i \cdot 2 \cdot i \cdot 10 = i^2 \cdot 20$$
$$= -1 \cdot 20 \qquad i^2 = -1$$
$$= -20;$$

$$(2 - 5i)(3 + i) = 6 + 2i - 15i - 5i^2$$
$$= 6 - 13i - 5(-1)$$
$$= 6 - 13i + 5$$
$$= 11 - 13i$$

The **conjugate of a complex number** $a + bi$ is $a - bi$. The numbers $a + bi$ and $a - bi$ are **complex conjugates**.

Conjugates are used when we **divide complex numbers**.

Divide.

$$\frac{5 - 2i}{3 + i} = \frac{5 - 2i}{3 + i} \cdot \frac{3 - i}{3 - i} \qquad \begin{array}{l} 3 - i \text{ is the conjugate} \\ \text{of the divisor, } 3 + i. \end{array}$$

$$= \frac{15 - 5i - 6i + 2i^2}{9 - i^2}$$

$$= \frac{15 - 11i - 2}{9 + 1} \qquad i^2 = -1$$

$$= \frac{13 - 11i}{10}$$

$$= \frac{13}{10} - \frac{11}{10}i$$

SECTION 7.4: QUADRATIC EQUATIONS, FUNCTIONS, ZEROS, AND MODELS

A **quadratic equation** is an equation that can be written in the form

$$ax^2 + bx + c = 0, \quad a \neq 0,$$

where a, b, and c are real numbers.

A **quadratic function** f is a function that can be written in the form

$$f(x) = ax^2 + bx + c, \quad a \neq 0,$$

where a, b, and c are real numbers.

The **zeros** of a quadratic function $f(x) = ax^2 + bx + c$ are the *solutions* of the associated quadratic equation $ax^2 + bx + c = 0$ and the x-coordinates of the x-intercepts of the graph of the function.

$3x^2 - 2x + 4 = 0$ and $5 - 4x = x^2$ are examples of quadratic equations. The equation $3x^2 - 2x + 4 = 0$ is written in **standard form**.

The functions $f(x) = 2x^2 + x + 1$ and $f(x) = 5x^2 - 4$ are examples of quadratic functions.

The solutions of $x^2 - x - 6 = 0$ are -2 and 3. The zeros of $f(x) = x^2 - x - 6$ are -2 and 3. The x-intercepts of the graph of $f(x) = x^2 - x - 6$ are $(-2, 0)$ and $(3, 0)$.

The Principle of Zero Products

If $ab = 0$ is true, then $a = 0$ *or* $b = 0$, and if $a = 0$ *or* $b = 0$, then $ab = 0$.

Solve: $3x^2 - 4 = 11x$.

$$3x^2 - 4 = 11x$$
$$3x^2 - 11x - 4 = 0 \qquad \text{Subtracting } 11x \text{ on both sides to get } 0 \text{ on one side of the equation}$$

$$(3x + 1)(x - 4) = 0 \qquad \text{Factoring}$$
$$3x + 1 = 0 \quad or \quad x - 4 = 0 \qquad \text{Using the principle of zero products}$$

$$3x = -1 \quad or \qquad x = 4$$
$$x = -\frac{1}{3} \quad or \qquad x = 4$$

The solutions are $-\dfrac{1}{3}$ and 4.

The Principle of Square Roots

If $x^2 = k$, then $x = \sqrt{k}$ *or* $x = -\sqrt{k}$.

Solve: $3x^2 - 18 = 0$.

$$3x^2 - 18 = 0$$
$$3x^2 = 18 \qquad \text{Adding 18 on both sides}$$
$$x^2 = 6 \qquad \text{Dividing by 3 on both sides}$$
$$x = \sqrt{6} \quad or \quad x = -\sqrt{6} \qquad \text{Using the principle of square roots}$$

The solutions are $\sqrt{6}$ and $-\sqrt{6}$, or $\pm\sqrt{6}$.

To solve a quadratic equation by **completing the square**:

1. Isolate the terms with variables on one side of the equation and arrange them in descending order.

2. Divide by the coefficient of the squared term if that coefficient is not 1.

3. Complete the square by taking half the coefficient of the first-degree term and adding its square on both sides of the equation.

4. Express one side of the equation as the square of a binomial.

5. Use the principle of square roots.

6. Solve for the variable.

Solve: $2x^2 - 3 = 6x$.

$$2x^2 - 3 = 6x$$

$$2x^2 - 6x - 3 = 0 \qquad \textcolor{red}{\text{Subtracting } 6x}$$

$$2x^2 - 6x = 3 \qquad \textcolor{red}{\text{Adding } 3}$$

$$x^2 - 3x = \frac{3}{2} \qquad \textcolor{red}{\begin{array}{l}\text{Dividing by 2 to make the}\\ x^2\text{-coefficient 1}\end{array}}$$

$$x^2 - 3x + \frac{9}{4} = \frac{3}{2} + \frac{9}{4} \qquad \textcolor{red}{\begin{array}{l}\text{Completing the square:}\\ \frac{1}{2}(-3) = -\frac{3}{2} \text{ and}\\ (-\frac{3}{2})^2 = \frac{9}{4}; \text{ adding } \frac{9}{4}\end{array}}$$

$$\left(x - \frac{3}{2}\right)^2 = \frac{15}{4} \qquad \textcolor{red}{\begin{array}{l}\text{Factoring and}\\ \text{simplifying}\end{array}}$$

$$x - \frac{3}{2} = \pm\frac{\sqrt{15}}{2} \qquad \textcolor{red}{\begin{array}{l}\text{Using the principle of square}\\ \text{roots and the quotient rule}\\ \text{for radicals}\end{array}}$$

$$x = \frac{3}{2} \pm \frac{\sqrt{15}}{2} = \frac{3 \pm \sqrt{15}}{2}$$

The solutions are $\dfrac{3 + \sqrt{15}}{2}$ and $\dfrac{3 - \sqrt{15}}{2}$, or $\dfrac{3 \pm \sqrt{15}}{2}$.

The solutions of $ax^2 + bx + c = 0$, $a \neq 0$, can be found using the **quadratic formula**:

$$x = \frac{-b \pm \sqrt{b^2 - 4ac}}{2a}.$$

Solve: $x^2 - 6 = 3x$.

$$x^2 - 6 = 3x$$

$$x^2 - 3x - 6 = 0 \qquad \textcolor{red}{\text{Standard form}}$$

$$a = 1, b = -3, c = -6$$

$$x = \frac{-b \pm \sqrt{b^2 - 4ac}}{2a}$$

$$= \frac{-(-3) \pm \sqrt{(-3)^2 - 4(1)(-6)}}{2 \cdot 1}$$

$$= \frac{3 \pm \sqrt{9 + 24}}{2}$$

$$= \frac{3 \pm \sqrt{33}}{2} \qquad \textcolor{red}{\text{Exact solutions}}$$

Using a calculator, we approximate the solutions to be 4.372 and -1.372.

Discriminant

For $ax^2 + bx + c = 0$, where a, b, and c are real numbers:

$b^2 - 4ac = 0 \rightarrow$ One real-number solution;

$b^2 - 4ac > 0 \rightarrow$ Two different real-number solutions;

$b^2 - 4ac < 0 \rightarrow$ Two different imaginary-number solutions, complex conjugates.

For $x^2 - 6 = 3x$, we see that $b^2 - 4ac$ is 33. Since 33 is positive, there are two different real-number solutions.

For $2x^2 - x + 4 = 0$, with $a = 2$, $b = -1$, and $c = 4$, the discriminant, $(-1)^2 - 4 \cdot 2 \cdot 4 = 1 - 32 = -31$, is negative, so there are two different imaginary-number (or nonreal) solutions.

For $x^2 - 6x + 9 = 0$, with $a = 1$, $b = -6$, and $c = 9$, the discriminant, $(-6)^2 - 4 \cdot 1 \cdot 9 = 36 - 36 = 0$, is 0 so there is one real-number solution.

Equations **reducible to quadratic,** or **quadratic in form,** can be treated as quadratic equations if a suitable substitution is made.

Solve: $x^4 - x^2 - 12 = 0$.

$x^4 - x^2 - 12 = 0$ Let $u = x^2$. Then $u^2 = (x^2)^2 = x^4$.

$u^2 - u - 12 = 0$ **Substituting**

$(u - 4)(u + 3) = 0$

$u - 4 = 0$ or $u + 3 = 0$

$u = 4$ or $u = -3$ **Solving for u**

$x^2 = 4$ or $x^2 = -3$

$x = \pm 2$ or $x = \pm\sqrt{3}i$ **Solving for x**

The solutions are 2, -2, $\sqrt{3}i$, and $-\sqrt{3}i$.

SECTION 7.5: ANALYZING GRAPHS OF QUADRATIC FUNCTIONS

Graphing Quadratic Equations

The graph of the function $f(x) = a(x - h)^2 + k$ is a parabola that:

- opens up if $a > 0$ and down if $a < 0$;
- has (h, k) as the vertex;
- has $x = h$ as the axis of symmetry;
- has k as a minimum value (output) if $a > 0$;
- has k as a maximum value if $a < 0$.

We can use a modification of the technique of completing the square as an aid in analyzing and graphing quadratic functions.

Find the vertex, the axis of symmetry, and the maximum or minimum value of $f(x) = 2x^2 + 12x + 12$.

$f(x) = 2x^2 + 12x + 12$

$= 2(x^2 + 6x) + 12$ **Note that 9 completes the square for $x^2 + 6x$.**

$= 2(x^2 + 6x + 9 - 9) + 12$ **Adding 9 − 9, or 0, inside the parentheses**

$= 2(x^2 + 6x + 9) - 2 \cdot 9 + 12$ **Using the distributive law to remove −9 from within the parentheses**

$= 2(x + 3)^2 - 6$

$= 2[x - (-3)]^2 + (-6)$

The function is now written in the form $f(x) = a(x - h)^2 + k$ with $a = 2$, $h = -3$, and $k = -6$. Because $a > 0$, we know the graph opens up and thus the function has a minimum value. We also know the following:

Vertex (h, k): $(-3, -6)$;

Axis of symmetry $x = h$: $x = -3$;

Minimum value of the function k: -6.

To graph the function, we first plot the vertex and then find several points on either side of it. We plot these points and connect them with a smooth curve.

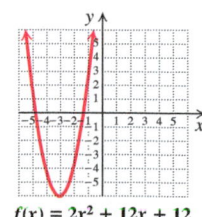

$f(x) = 2x^2 + 12x + 12$

The Vertex of a Parabola The **vertex** of the graph of $f(x) = ax^2 + bx + c$ is $$\left(-\frac{b}{2a}, f\left(-\frac{b}{2a}\right)\right).$$ ↑ ↑ We calculate the We substitute to x-coordinate. find the y-coordinate.	Find the vertex of the function $f(x) = -3x^2 + 6x + 1$. $$-\frac{b}{2a} = -\frac{6}{2(-3)} = 1$$ $$f(1) = -3 \cdot 1^2 + 6 \cdot 1 + 1 = 4$$ The vertex is $(1, 4)$.
Some applied problems can be solved by finding the maximum or minimum value of a quadratic function.	See Examples 5–7 on pp. 486–488.

REVIEW EXERCISES

Determine whether the statement is true or false.

1. We can use the quadratic formula to solve any quadratic equation. **[7.4]**

2. The function $f(x) = -3(x + 4)^2 - 1$ has a maximum value. **[7.5]**

3. The graph of $y = (x - 2)^2$ is the graph of $y = x^2$ shifted right 2 units. **[7.2]**

4. The graph of $y = -x^2$ is the reflection of the graph of $y = x^2$ across the x-axis. **[7.2]**

Graph the given equation and determine visually whether it is symmetric with respect to the x-axis, the y-axis, and the origin. Then verify your assertion algebraically. **[7.1]**

5. $x^2 + y^2 = 4$

6. $y^2 = x^2 + 3$

7. $x + y = 3$

8. $y = x^2$

9. $y = x^3$

10. $y = x^4 - x^2$

Determine visually whether the function is even, odd, or neither even nor odd. **[7.1]**

11.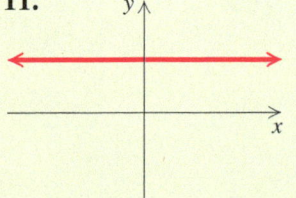

12.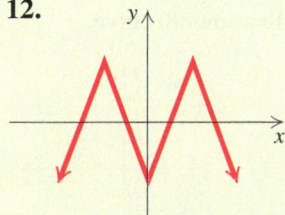

13.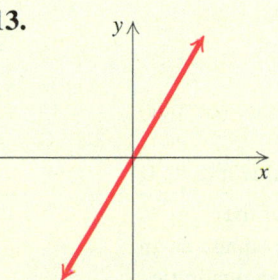

14.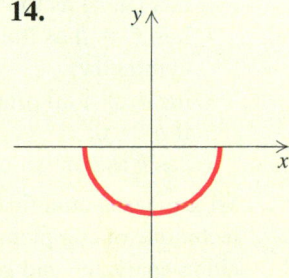

In Exercises 15–20, test whether the function is even, odd, or neither even nor odd. **[7.1]**

15. $f(x) = 9 - x^2$

16. $f(x) = x^3 - 2x + 4$

17. $f(x) = x^7 - x^5$

18. $f(x) = |x|$

19. $f(x) = \sqrt{16 - x^2}$

20. $f(x) = \dfrac{10x}{x^2 + 1}$

Write an equation for a function that has a graph with the given characteristics. **[7.2]**

21. The shape of $y = x^2$, but shifted left 3 units

22. The shape of $y = \sqrt{x}$, but upside down and shifted right 3 units and up 4 units

23. The shape of $y = |x|$, but stretched vertically by a factor of 2 and shifted right 3 units

A graph of $y = f(x)$ is shown below. No formula for f is given. Graph each of the following. **[7.2]**

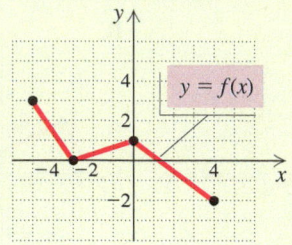

24. $y = f(x - 1)$

25. $y = f(2x)$

26. $y = -2f(x)$

27. $y = 3 + f(x)$

Express in terms of i. **[7.3]**

28. $-\sqrt{-40}$

29. $\sqrt{-12} \cdot \sqrt{-20}$

30. $\dfrac{\sqrt{-49}}{-\sqrt{-64}}$

Simplify each of the following. Write the answer in the form $a + bi$, where a and b are real numbers. **[7.3]**

31. $(6 + 2i) + (-4 - 3i)$

32. $(3 - 5i) - (2 - i)$

33. $(6 + 2i)(-4 - 3i)$

34. $\dfrac{2 - 3i}{1 - 3i}$

35. i^{23}

Solve. **[7.4]**

36. $(2y + 5)(3y - 1) = 0$

37. $x^2 + 4x - 5 = 0$

38. $3x^2 + 2x = 8$

39. $5x^2 = 15$

40. $x^2 + 10 = 0$

Find the zero(s) of the function. **[7.3]**

41. $f(x) = x^2 - 2x + 1$

42. $f(x) = x^2 + 2x - 15$

43. $f(x) = 2x^2 - x - 5$

44. $f(x) = 3x^2 + 2x + 3$

Solve by completing the square to obtain exact solutions. Show your work. **[7.4]**

45. $x^2 - 3x = 18$

46. $3x^2 - 12x - 6 = 0$

Solve. Give exact solutions. **[7.4]**

47. $3x^2 + 10x = 8$

48. $r^2 - 2r + 10 = 0$

49. $x^2 = 10 + 3x$

50. $x = 2\sqrt{x} - 1$

51. $y^4 - 3y^2 + 1 = 0$

52. $(x^2 - 1)^2 - (x^2 - 1) - 2 = 0$

53. $(p + 2)(3p + 2)(p - 3) = 0$

54. $x^3 + 5x^2 - 4x - 20 = 0$

*In Exercises 55 and 56, complete the square to **(a)** find the vertex; **(b)** find the axis of symmetry; **(c)** determine whether there is a maximum or minimum value and find that value; **(d)** find the range; and **(e)** graph the function.* **[7.5]**

55. $f(x) = -4x^2 + 3x - 1$

56. $f(x) = 5x^2 - 10x + 3$

In Exercises 57–60, match the equation with one of the figures (a)–(d) that follow. **[7.5]**

a)

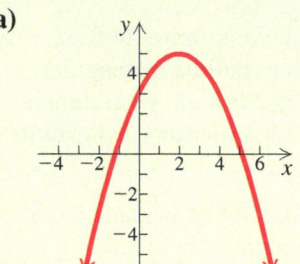

b)

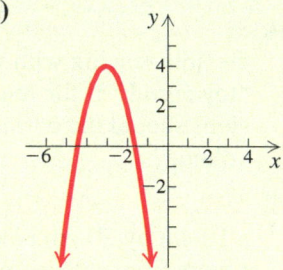

c)

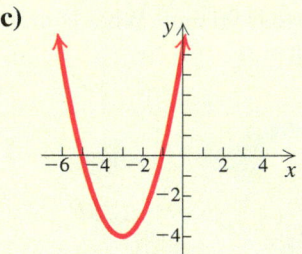

d)

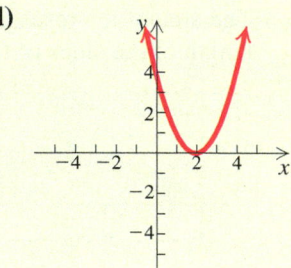

57. $y = (x - 2)^2$

58. $y = (x + 3)^2 - 4$

59. $y = -2(x + 3)^2 + 4$

60. $y = -\frac{1}{2}(x - 2)^2 + 5$

61. *Legs of a Right Triangle.*   The hypotenuse of a right triangle is 50 ft. One leg is 10 ft longer than the other. What are the lengths of the legs? **[7.4]**

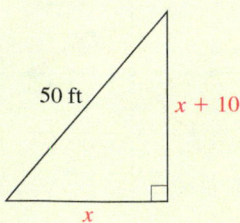

62. *Bicycling Speed.* Harry and Rebecca leave a camp-site, Harry biking due north and Rebecca biking due east. Harry bikes 7 km/h slower than Rebecca. After 4 hr, they are 68 km apart. Find the speed of each bicyclist. **[7.4]**

63. *Sidewalk Width.* A 60-ft by 80-ft parking lot is torn up to install a sidewalk of uniform width around its perimeter. The new area of the parking lot is two-thirds of the old area. How wide is the sidewalk? **[7.4]**

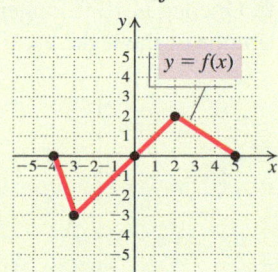

64. *Maximizing Volume.* The Garcias have 24 ft of flexible fencing with which to build a rectangular "toy corral." If the fencing is 2 ft high, what dimensions should the corral have in order to maximize its volume? **[7.5]**

65. *Dimensions of a Box.* An open box is made from a 10-cm by 20-cm piece of aluminum by cutting a square from each corner and folding up the edges. The area of the resulting base is 90 cm². What is the length of the sides of the squares? **[7.4]**

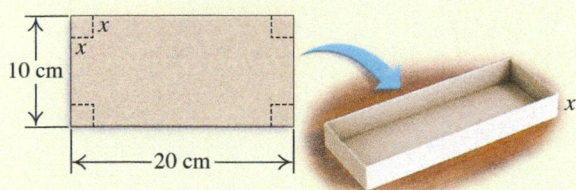

66. For $b > 0$, the graph of $y = f(x) + b$ is the graph of $y = f(x)$ shifted in which of the following ways? **[7.2]**

A. Right b units   **B.** Left b units
C. Up b units   **D.** Down b units

67. The graph of the function f is shown below.

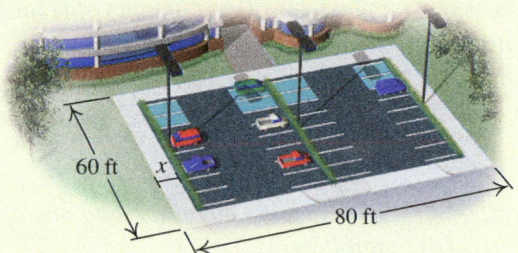

The graph of $g(x) = -\frac{1}{2}f(x) + 1$ is which of the following? **[7.2]**

A.
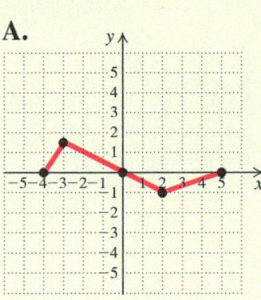

B.
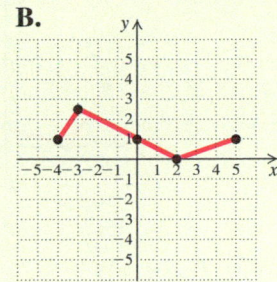

C.
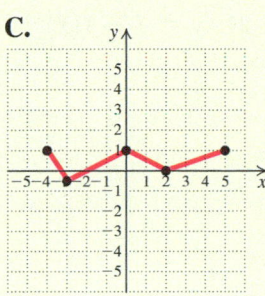

D.
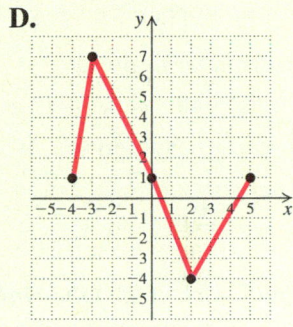

68. Find the zeros of $f(x) = 2x^2 - 5x + 1$. **[7.4]**

A. $\dfrac{5 \pm \sqrt{17}}{2}$ **B.** $\dfrac{5 \pm \sqrt{17}}{4}$

C. $\dfrac{5 \pm \sqrt{33}}{4}$ **D.** $\dfrac{-5 \pm \sqrt{17}}{4}$

69. The graph of $f(x) = (x - 2)^2 - 3$ is which of the following? **[7.5]**

A.
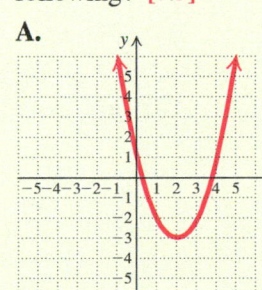

B.
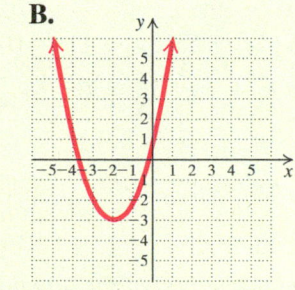

C.
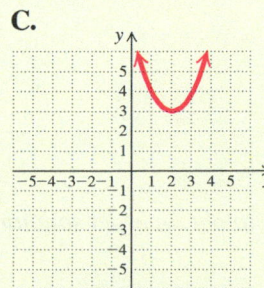

D.
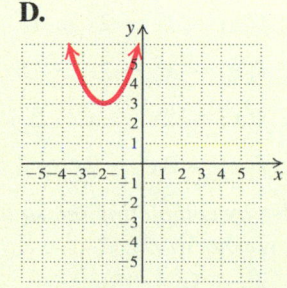

► Synthesis

70. Prove that the sum of two odd functions is odd.
[7.1]

71. Describe how the graph of $y = -f(-x)$ is obtained from the graph of $y = f(x)$. [7.2]

Solve.

72. $(2y - 2)^2 + y - 1 = 5$ [7.4]

73. $\sqrt{x + 2} + \sqrt[4]{x + 2} - 2 = 0$ [7.4]

74. At the beginning of the year, $3500 was deposited in a savings account. One year later, $4000 was deposited in another account. The interest rate was the same for both accounts. At the end of the second year, there was a total of $8518.35 in the accounts. What was the annual interest rate? [7.4]

75. Find b such that $f(x) = -3x^2 + bx - 1$ has a maximum value of 2. [7.5]

► Collaborative Discussion and Writing

76. Given that $f(x) = 4x^3 - 2x + 7$, find each of the following. Then discuss how each expression differs from the other. [7.2]

a) $f(x) + 2$
b) $f(x + 2)$
c) $f(x) + f(2)$

77. The graph of a quadratic function can have 0, 1, or 2 x-intercepts. How can you predict the number of x-intercepts without drawing the graph or (completely) solving an equation? [7.4]

78. Discuss two ways in which we used completing the square in this chapter. [7.4], [7.5]

79. Suppose that the graph of $f(x) = ax^2 + bx + c$ has x-intercepts $(x_1, 0)$ and $(x_2, 0)$. What are the x-intercepts of $g(x) = -ax^2 - bx - c$? Explain. [7.5]

80. Is the product of two imaginary numbers always an imaginary number? Explain your answer. [7.3]

81. Is it possible for a quadratic function to have one real zero and one imaginary zero? Why or why not? [7.4]

82. If the graphs of
$$f(x) = a_1(x - h_1)^2 + k_1$$
and
$$g(x) = a_2(x - h_2)^2 + k_2$$
have the same shape, what, if anything, can you conclude about the a's, the h's, and the k's? Explain your answer. [7.5]

7	**Chapter Test**

1. Determine whether the graph of $y = x^4 - 2x^2$ is symmetric with respect to the x-axis, the y-axis, and the origin.

2. Test whether the function
$$f(x) = \frac{2x}{x^2 + 1}$$
is even, odd, or neither even nor odd. Show your work.

3. Write an equation for a function that has the shape of $y = x^2$, but shifted right 2 units and down 1 unit.

4. Write an equation for a function that has the shape of $y = x^2$, but shifted left 2 units and down 3 units.

5. The graph of a function $y = f(x)$ is shown below. No formula for f is given. Graph $y = -\frac{1}{2}f(x)$.

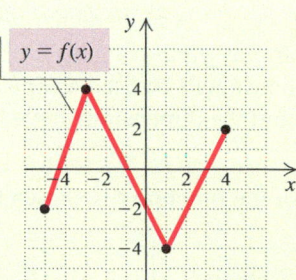

Express in terms of i.

6. $\sqrt{-43}$

7. $-\sqrt{-25}$

Simplify.

8. $(5 - 2i) - (2 + 3i)$

9. $(3 + 4i)(2 - i)$

10. $\dfrac{1 - i}{6 + 2i}$

11. i^{33}

Solve. Find exact solutions.

12. $(2x - 1)(x + 5) = 0$

13. $6x^2 - 36 = 0$

14. $x^2 + 4 = 0$

15. $x^2 - 2x - 3 = 0$

16. $x^2 - 5x + 3 = 0$

17. $2t^2 - 3t + 4 = 0$

18. $x + 5\sqrt{x} - 36 = 0$

19. Solve $x^2 + 4x = 1$ by completing the square. Find the exact solutions. Show your work.

Find the zeros of each function.

20. $f(x) = 4x^2 - 11x - 3$

21. $f(x) = 2x^2 - x - 7$

22. The tallest structure in the United States, at 2063 ft, is the KTHI-TV tower in Blanchard, North Dakota (*Source: The Cambridge Fact Finder*). How long would it take an object falling freely from the top to reach the ground? (Use the formula $s = 16t^2$, where s is the distance, in feet, that an object falls freely from rest in t seconds.)

23. For the graph of the function $f(x) = -x^2 + 2x + 8$, **(a)** find the vertex; **(b)** find the axis of symmetry; **(c)** state whether there is a maximum or a minimum value and find that value; **(d)** find the range; and **(e)** graph the function.

24. *Maximizing Area.* A homeowner wants to fence a rectangular play yard using 80 ft of fencing. The side of the house will be used as one side of the rectangle. Find the dimensions for which the area is a maximum.

25. The graph of the function f is shown below.

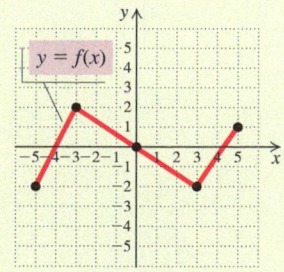

The graph of $g(x) = 2f(x) - 1$ is which of the following?

A.

B.

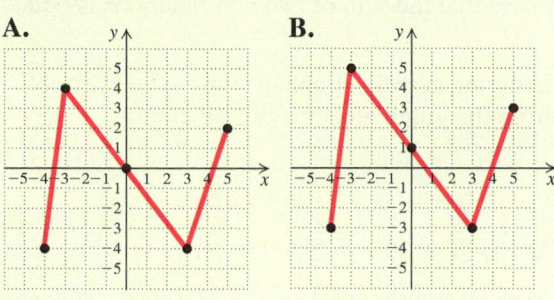

C.

D.

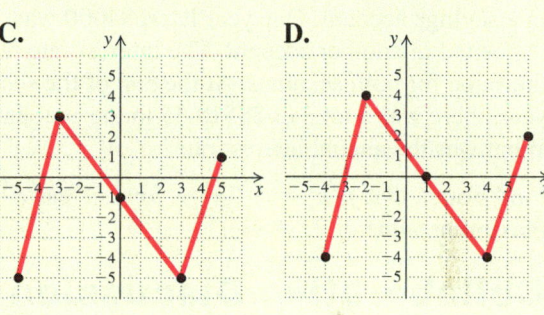

26. The graph of $f(x) = (x - 1)^2 - 2$ is which of the following?

A.   **B.**

C.   **D.**

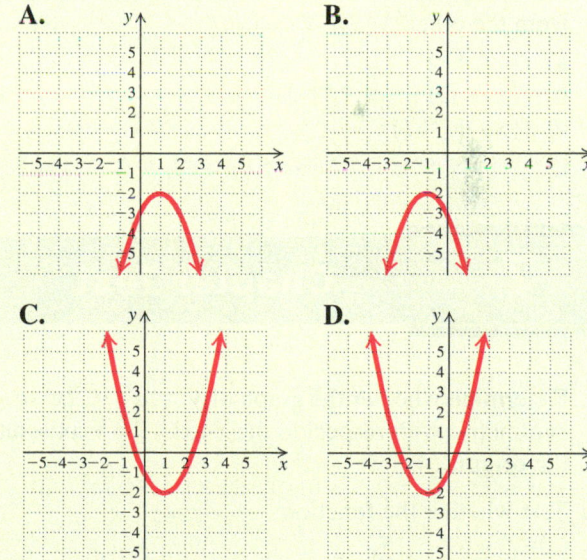

▶ Synthesis

27. If $(-3, 1)$ is a point on the graph of $y = f(x)$, what point do you know is on the graph of $y = f(3x)$?

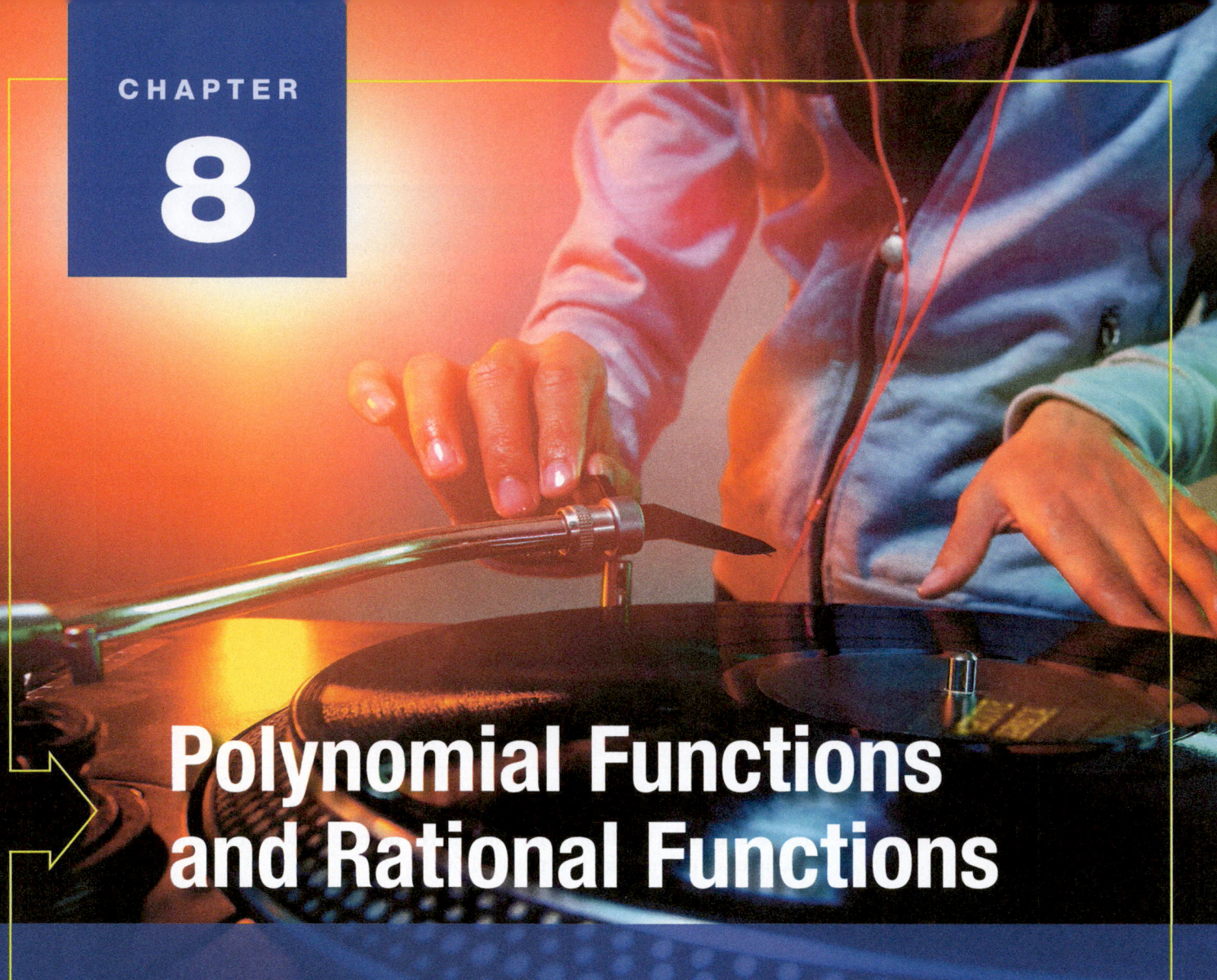

CHAPTER
8

Polynomial Functions and Rational Functions

APPLICATION This problem appears as Exercise 47 in Section 8.1.

Vinyl record albums are making a comeback. Sales of vinyl albums rose 32% from 2012 to 2013. The sales data over the years 2001 to 2013 are modeled by the quartic function

$$f(x) = -0.000913x^4 + 0.248x^3 - 0.1515x^2 + 0.2136x + 1.2779,$$

where x is the number of years after 2001 and $f(x)$ is the number of albums in millions (*Source:* Nielsen SoundScan). Find the number of vinyl albums sold in 2008, in 2012, and in 2016.

8.1

Polynomial Functions and Models

▶ Determine the behavior of the graph of a polynomial function using the leading-term test.

▶ Factor polynomial functions and find their zeros and their multiplicities.

▶ Solve applied problems using polynomial models.

There are many different kinds of functions. The constant, linear, and quadratic functions that we studied in Chapters 2 and 7 are part of a larger group of functions called *polynomial functions*.

POLYNOMIAL FUNCTION

A **polynomial function** P is given by

$$P(x) = a_n x^n + a_{n-1} x^{n-1} + a_{n-2} x^{n-2} + \cdots + a_1 x + a_0,$$

where the coefficients $a_n, a_{n-1}, \ldots, a_1, a_0$ are real numbers and the exponents are whole numbers.

The first nonzero coefficient, a_n, is called the **leading coefficient**. The term $a_n x^n$ is called the **leading term**. The **degree** of the polynomial function is n. Some examples of polynomial functions follow.

POLYNOMIAL FUNCTION	EXAMPLE	DEGREE	LEADING TERM	LEADING COEFFICIENT
Constant	$f(x) = 3 \quad (f(x) = 3 = 3x^0)$	0	3	3
Linear	$f(x) = \frac{2}{3}x + 5 \quad (f(x) = \frac{2}{3}x + 5 = \frac{2}{3}x^1 + 5)$	1	$\frac{2}{3}x$	$\frac{2}{3}$
Quadratic	$f(x) = 4x^2 - x + 3$	2	$4x^2$	4
Cubic	$f(x) = x^3 + 2x^2 + x - 5$	3	x^3	1
Quartic	$f(x) = -x^4 - 1.1x^3 + 0.3x^2 - 2.8x - 1.7$	4	$-x^4$	-1

The function $f(x) = 0$ can be described in many ways:

$$f(x) = 0 = 0x^2 = 0x^{15} = 0x^{48},$$

and so on. For this reason, we say that the constant function $f(x) = 0$ has no degree.

Functions such as

$$f(x) = \frac{2}{x} + 5, \quad \text{or } 2x^{-1} + 5, \quad \text{and} \quad g(x) = \sqrt{x} - 6, \quad \text{or } x^{1/2} - 6,$$

are *not* polynomial functions because the exponents -1 and $\frac{1}{2}$ are *not* whole numbers.

From our study of functions in Chapters 2 and 4–7, we know how to find or at least estimate many characteristics of a polynomial function. Let's consider two examples for review.

Quadratic Function

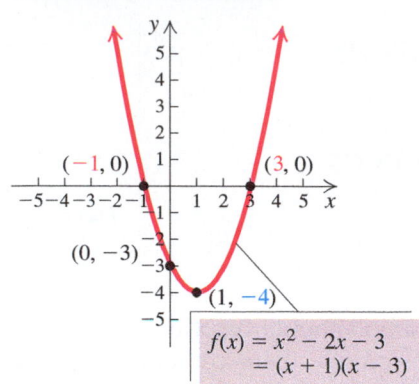

Function: $f(x) = x^2 - 2x - 3$
$= (x + 1)(x - 3)$

Zeros: -1, 3

x-intercepts: $(-1, 0)$, $(3, 0)$

y-intercept: $(0, -3)$

Minimum: -4 at $x = 1$

Maximum: None

Domain: All real numbers, $(-\infty, \infty)$

Range: $[-4, \infty)$

$f(x) = x^2 - 2x - 3$
$= (x + 1)(x - 3)$

Cubic Function

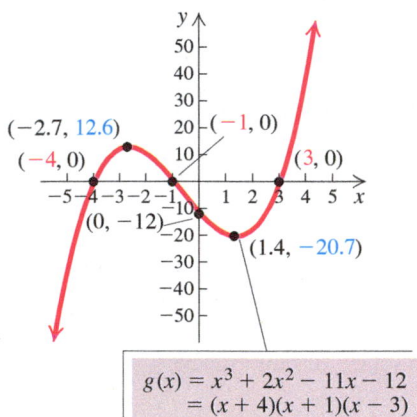

Function: $g(x) = x^3 + 2x^2 - 11x - 12$
$= (x + 4)(x + 1)(x - 3)$

Zeros: -4, -1, 3

x-intercepts: $(-4, 0)$, $(-1, 0)$, $(3, 0)$

y-intercept: $(0, -12)$

Relative minimum: -20.7 at $x = 1.4$

Relative maximum: 12.6 at $x = -2.7$

Domain: All real numbers, $(-\infty, \infty)$

Range: All real numbers, $(-\infty, \infty)$

$g(x) = x^3 + 2x^2 - 11x - 12$
$= (x + 4)(x + 1)(x - 3)$

All graphs of polynomial functions have some characteristics in common. Compare the following graphs on this page and the next. How do the graphs of polynomial functions differ from the graphs of nonpolynomial functions? Describe some characteristics of the graphs of polynomial functions that you observe.

Polynomial Functions

$f(x) = x^2 + 3x + 1$

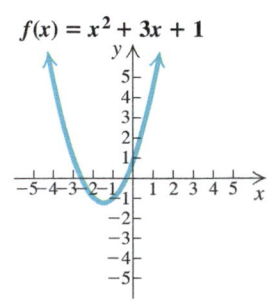

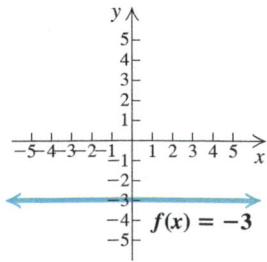

$f(x) = -3$

$f(x) = 2x^3 + x^2 + x - 1$

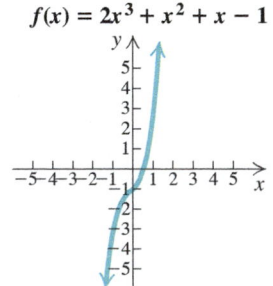

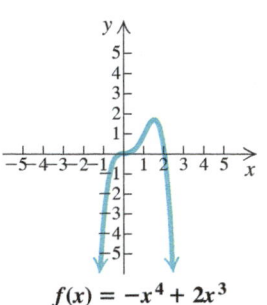

$f(x) = -x^4 + 2x^3$

Nonpolynomial Functions

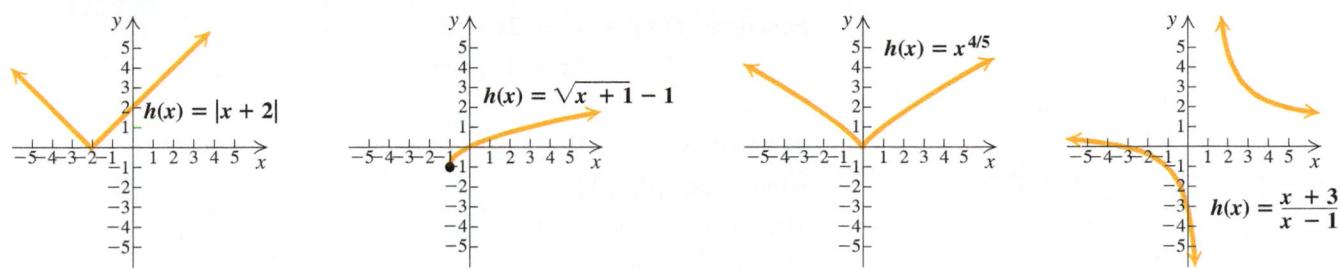

<table>
<tr><td>

DOMAIN OF A FUNCTION

REVIEW SECTION 2.3

</td></tr>
</table>

You probably noted that the graph of a polynomial function is *continuous*; that is, it has no holes or breaks. It is also smooth; there are no sharp corners. Furthermore, the *domain* of a polynomial function is the set of all real numbers, $(-\infty, \infty)$.

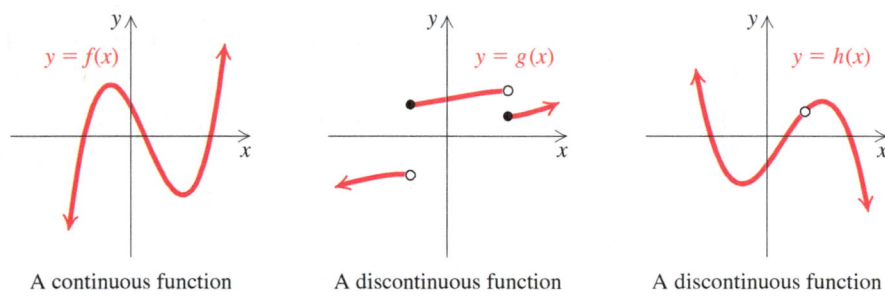

A continuous function   A discontinuous function   A discontinuous function

The *domain* of a polynomial function is the set of all real numbers, $(-\infty, \infty)$.

▶ The Leading-Term Test

The behavior of the graph of a polynomial function as x becomes very large $(x \to \infty)$ or very small $(x \to -\infty)$ is referred to as the end behavior of the graph. The leading term of a polynomial function determines its end behavior.

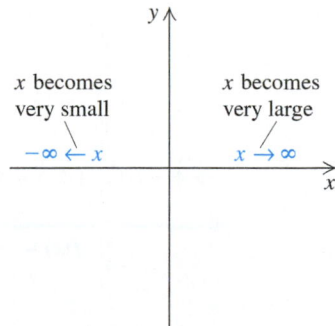

Using the graphs shown on the following page, let's see if we can discover some general patterns by comparing the end behavior of even- and odd-degree functions. We also note the effect of positive and negative leading coefficients.

Even Degree

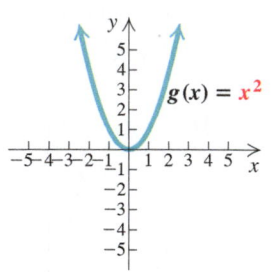

$g(x) = x^2$

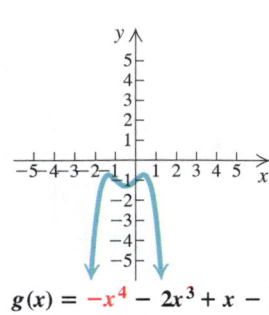

$g(x) = -x^4 - 2x^3 + x - 1$

$g(x) = \frac{1}{2}x^6 + 3$

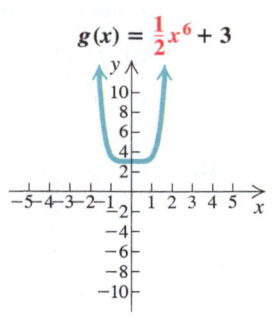

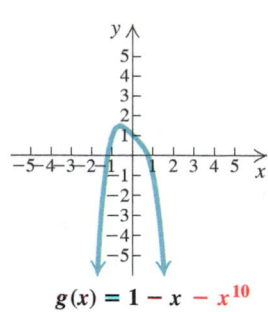

$g(x) = 1 - x - x^{10}$

Odd Degree

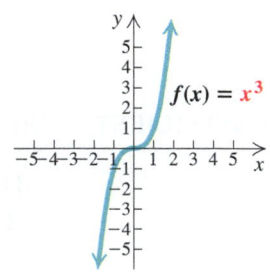

$f(x) = x^3$

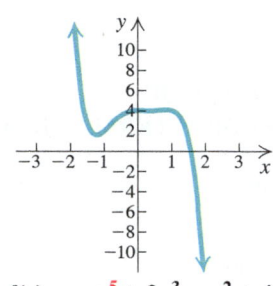

$f(x) = -x^5 + 2x^3 - x^2 + 4$

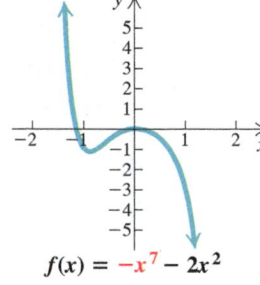

$f(x) = -x^7 - 2x^2$

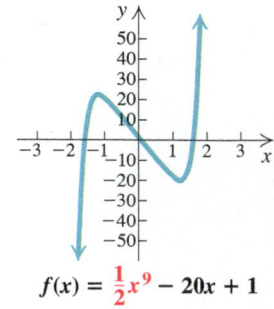

$f(x) = \frac{1}{2}x^9 - 20x + 1$

We can summarize our observations as follows.

THE LEADING-TERM TEST

If $a_n x^n$ is the leading term of a polynomial function, then the behavior of the graph as $x \to \infty$ or as $x \to -\infty$ can be described in one of the four following ways.

n	$a_n > 0$	$a_n < 0$
Even		
Odd		

The 〰 portion of the graph is not determined by this test.

EXAMPLE 1 Using the leading-term test, match each of the following functions with one of the graphs A–D that follow.

a) $f(x) = 3x^4 - 2x^3 + 3$ **b)** $f(x) = -5x^3 - x^2 + 4x + 2$

c) $f(x) = x^5 + \frac{1}{4}x + 1$ **d)** $f(x) = -x^6 + x^5 - 4x^3$

A. 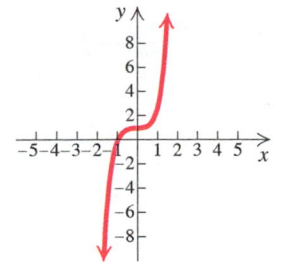   **B.** 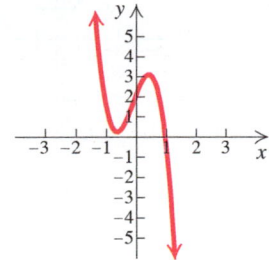   **C.** 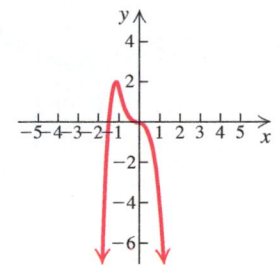   **D.**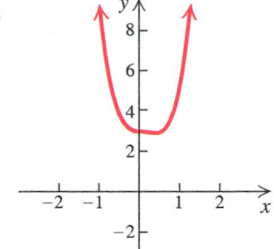

Solution

	LEADING TERM	DEGREE OF LEADING TERM	SIGN OF LEADING COEFFICIENT	GRAPH
a)	$3x^4$	4, even	3, positive	D
b)	$-5x^3$	3, odd	-5, negative	B
c)	x^5	5, odd	1, positive	A
d)	$-x^6$	6, even	-1, negative	C

> **Now Try Exercise 19.**

▶ **Finding Zeros of Polynomial Functions**

Let's review the meaning of the real zeros of a function and their connection to the *x*-intercepts of the function's graph.

CONNECTING THE CONCEPTS

Zeros, Solutions, and Intercepts

FUNCTION

ZEROS OF THE FUNCTION; SOLUTIONS OF THE EQUATION

ZEROS OF THE FUNCTION; *x*-INTERCEPTS OF THE GRAPH

Quadratic Polynomial

$g(x) = x^2 - 2x - 8$
    $= (x + 2)(x - 4),$

or

    $y = (x + 2)(x - 4)$

To find the **zeros** of $g(x)$, we solve $g(x) = 0$:

$$x^2 - 2x - 8 = 0$$
$$(x + 2)(x - 4) = 0$$
$$x + 2 = 0 \quad or \quad x - 4 = 0$$
$$x = -2 \quad or \quad x = 4.$$

The **solutions** of $x^2 - 2x - 8 = 0$ are −2 and 4. They are the zeros of the function $g(x)$; that is,

$$g(-2) = 0 \quad and \quad g(4) = 0.$$

The real-number zeros of $g(x)$ are the *x*-coordinates of the **x-intercepts** of the graph of $y = g(x)$.

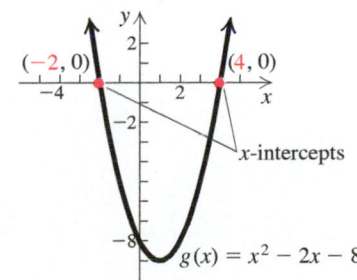

(continued)

Cubic Polynomial

$h(x)$
$= x^3 + 2x^2 - 5x - 6$
$= (x + 3)(x + 1)(x - 2),$
or
$y = (x + 3)(x + 1)(x - 2)$

To find the **zeros** of $h(x)$, we solve $h(x) = 0$:

$$x^3 + 2x^2 - 5x - 6 = 0$$
$$(x + 3)(x + 1)(x - 2) = 0$$
$$x + 3 = 0 \quad \text{or} \quad x + 1 = 0 \quad \text{or} \quad x - 2 = 0$$
$$x = -3 \quad \text{or} \quad\quad x = -1 \quad \text{or} \quad\quad x = 2.$$

The **solutions** of $x^3 + 2x^2 - 5x - 6 = 0$ are $-3, -1$, and 2. They are the zeros of the function $h(x)$; that is,

$$h(-3) = 0,$$
$$h(-1) = 0, \quad \text{and}$$
$$h(2) = 0.$$

The real-number zeros of $h(x)$ are the x-coordinates of the **x-intercepts** of the graph of $y = h(x)$.

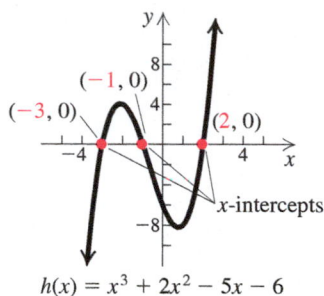

$h(x) = x^3 + 2x^2 - 5x - 6$

The connection between the real-number zeros of a function and the x-intercepts of the graph of the function is easily seen in the preceding examples. If c is a real zero of a function (that is, $f(c) = 0$), then $(c, 0)$ is an x-intercept of the graph of the function.

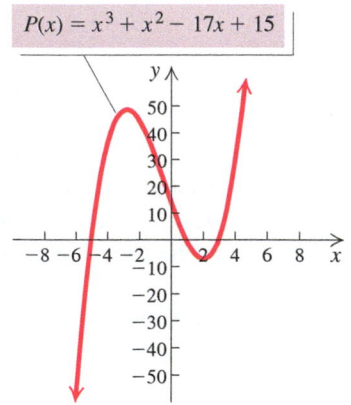

$P(x) = x^3 + x^2 - 17x + 15$

EXAMPLE 2   Consider $P(x) = x^3 + x^2 - 17x + 15$. Determine whether each of the numbers 2 and -5 is a zero of $P(x)$.

Solution   We first evaluate $P(2)$:

$$P(2) = (2)^3 + (2)^2 - 17(2) + 15 = -7. \quad \text{Substituting 2 into the polynomial}$$

Since $P(2) \neq 0$, we know that 2 is *not* a zero of the polynomial function.
   We then evaluate $P(-5)$:

$$P(-5) = (-5)^3 + (-5)^2 - 17(-5) + 15 = 0. \quad \text{Substituting } -5 \text{ into the polynomial}$$

Since $P(-5) = 0$, we know that -5 is a zero of $P(x)$.

Now Try Exercise 23.

Let's take a closer look at the polynomial function

$$h(x) = x^3 + 2x^2 - 5x - 6$$

(see Connecting the Concepts above). The factors of $h(x)$ are

$$x + 3, \quad x + 1, \quad \text{and} \quad x - 2,$$

and the zeros are

$$-3, \quad -1, \quad \text{and} \quad 2.$$

We note that when the polynomial is expressed as a product of linear factors, each factor determines a zero of the function. Thus if we know the linear factors of a polynomial function $f(x)$, we can easily find the zeros of $f(x)$ by solving the equation $f(x) = 0$ using the principle of zero products.

EXAMPLE 3   Find the zeros of

$$f(x) = 5(x - 2)(x - 2)(x - 2)(x + 1)$$
$$= 5(x - 2)^3(x + 1).$$

Solution   To solve the equation $f(x) = 0$, we use the principle of zero products, solving $x - 2 = 0$ and $x + 1 = 0$. The zeros of $f(x)$ are 2 and -1.

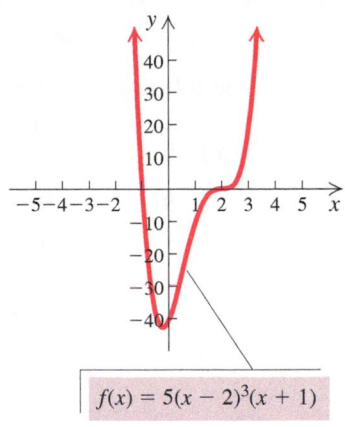

$$f(x) = 5(x - 2)^3(x + 1)$$

EXAMPLE 4   Find the zeros of

$$g(x) = -(x - 1)(x - 1)(x + 2)(x + 2)$$
$$= -(x - 1)^2(x + 2)^2.$$

Solution   To solve the equation $g(x) = 0$, we use the principle of zero products, solving $x - 1 = 0$ and $x + 2 = 0$. The zeros of $g(x)$ are 1 and -2.

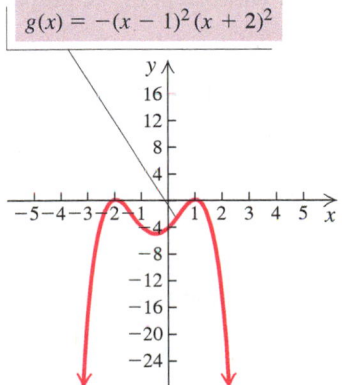

$$g(x) = -(x - 1)^2(x + 2)^2$$

Let's consider the occurrences of the zeros in the functions in Examples 3 and 4 and their relationship to the graphs of those functions. In Example 3, the factor $x - 2$ occurs three times. In a case like this, we say that the zero we obtain from this factor, 2, has a **multiplicity** of 3. The factor $x + 1$ occurs one time. The zero we obtain from this factor, -1, has a *multiplicity* of 1.

In Example 4, the factors $x - 1$ and $x + 2$ each occur two times. Thus both zeros, 1 and -2, have a *multiplicity* of 2.

Note, in Example 3, that the zeros have odd multiplicities and the graph crosses the x-axis at both -1 and 2. But in Example 4, the zeros have even multiplicities and the graph is tangent to (touches but does not cross) the x-axis at -2 and 1. This leads us to the following generalization.

EVEN MULTIPLICITY AND ODD MULTIPLICITY

If $(x - c)^k$, $k \geq 1$, is a factor of a polynomial function $P(x)$ and $(x - c)^{k+1}$ is not a factor and:

- if k is odd, then the graph crosses the x-axis at $(c, 0)$;
- if k is even, then the graph is tangent to the x-axis at $(c, 0)$.

Some polynomials can be factored by grouping. Then we use the principle of zero products to find their zeros.

EXAMPLE 5 Find the zeros of

$$f(x) = x^3 - 2x^2 - 9x + 18.$$

Solution We factor by grouping, as follows:

$$
\begin{aligned}
f(x) &= x^3 - 2x^2 - 9x + 18 \\
&= x^2(x - 2) - 9(x - 2) && \text{Grouping } x^3 \text{ with } -2x^2 \text{ and } -9x \\
&&& \text{with 18 and factoring each group} \\
&= (x - 2)(x^2 - 9) && \text{Factoring out } x - 2 \\
&= (x - 2)(x + 3)(x - 3). && \text{Factoring } x^2 - 9
\end{aligned}
$$

Then, by the principle of zero products, the solutions of the equation $f(x) = 0$ are 2, -3, and 3. These are the zeros of $f(x)$.

Now Try Exercise 39.

Technology Connection

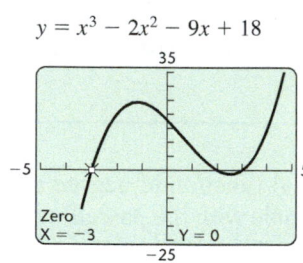

$y = x^3 - 2x^2 - 9x + 18$

Zero
X = −3   Y = 0

Using the Zero feature, we can determine the zeros of the function in Example 5. The window at left shows the calculator display when we find the leftmost zero. The other zeros, 2 and 3, can be found in the same manner.

Find the real zeros of the function f given by $f(x) = 0.1x^3 - 0.6x^2 - 0.1x + 2$. Approximate the zeros to three decimal places.

We use a graphing calculator to create a graph that clearly shows the curvature. It appears that there are three zeros, one near -2, one near 2, and one near 6. We use the ZERO feature to find them. The zeros are approximately -1.680, 2.154, and 5.526.

$y = 0.1x^3 - 0.6x^2 - 0.1x + 2$

X=−1.680	Y=0
X=2.154	Y=0
X=5.526	Y=0

Other factoring techniques can also be used.

EXAMPLE 6 Find the zeros of

$$f(x) = x^4 + 4x^2 - 45.$$

Solution We factor as follows:

$$f(x) = x^4 + 4x^2 - 45 = (x^2 - 5)(x^2 + 9).$$

We now solve the equation $f(x) = 0$ to determine the zeros. We use the principle of zero products:

$$
\begin{aligned}
(x^2 - 5)(x^2 + 9) &= 0 \\
x^2 - 5 = 0 \quad &or \quad x^2 + 9 = 0 \\
x^2 = 5 \quad &or \quad x^2 = -9 \\
x = \pm\sqrt{5} \quad &or \quad x = \pm\sqrt{-9} = \pm 3i.
\end{aligned}
$$

The solutions are $\pm\sqrt{5}$ and $\pm 3i$. These are the zeros of $f(x)$.

Now Try Exercise 37.

Only the real-number zeros of a function correspond to the *x*-intercepts of its graph. For instance, the real-number zeros of the function in Example 6, $-\sqrt{5}$ and $\sqrt{5}$, can be seen on the graph of the function below, but the nonreal zeros, $-3i$ and $3i$, cannot.

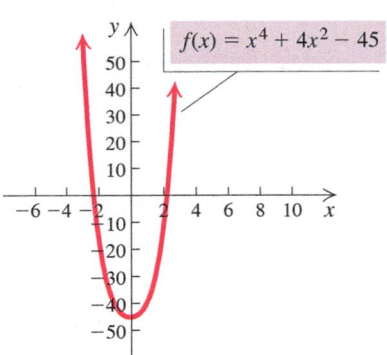

$f(x) = x^4 + 4x^2 - 45$

Every polynomial function of degree n, with $n \geq 1$, has at least one zero and at most n zeros.

This is often stated as follows: "Every polynomial function of degree n, with $n \geq 1$, has *exactly n* zeros." This statement is compatible with the preceding statement, if one takes multiplicities into account.

▶ Polynomial Models

Polynomial functions have many uses as models in science, engineering, and business. The simplest use of polynomial functions in applied problems occurs when we merely evaluate a polynomial function. In such cases, a model has already been developed.

EXAMPLE 7 *Ibuprofen in the Bloodstream.* The polynomial function

$$M(t) = 0.5t^4 + 3.45t^3 - 96.65t^2 + 347.7t$$

can be used to estimate the number of milligrams of the pain relief medication ibuprofen in the bloodstream t hours after 400 mg of the medication has been taken. Find the number of milligrams in the bloodstream at $t = 0, 0.5, 1, 1.5,$ and so on, up to 6 hr. Round the function values to the nearest tenth.

Solution Using a calculator, we compute function values:

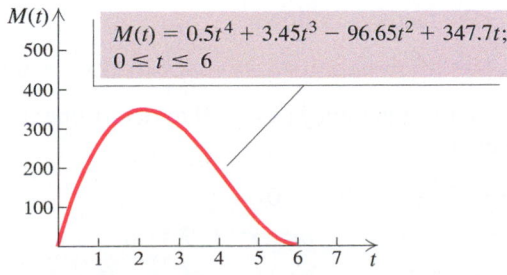

$M(t) = 0.5t^4 + 3.45t^3 - 96.65t^2 + 347.7t;$
$0 \leq t \leq 6$

$M(0) = 0,$	$M(3.5) = 255.9,$
$M(0.5) = 150.2,$	$M(4) = 193.2,$
$M(1) = 255,$	$M(4.5) = 126.9,$
$M(1.5) = 318.3,$	$M(5) = 66,$
$M(2) = 344.4,$	$M(5.5) = 20.2,$
$M(2.5) = 338.6,$	$M(6) = 0.$
$M(3) = 306.9,$	

Now Try Exercise 49.

Recall that the domain of a polynomial function, unless restricted by a statement of the function, is $(-\infty, \infty)$. The implications of the application in Example 7 restrict the domain of the function. If we assume that a patient had not taken any of the medication before, it seems reasonable that $M(0) = 0$; that is, at time 0, there is 0 mg of the medication in the bloodstream. After the medication has been taken, $M(t)$ will be positive for a period of time and eventually decrease back to 0 when $t = 6$ and not increase again (unless another dose is taken). Thus the restricted domain is $[0, 6]$.

Technology Connection

X	Y₁
0	0
.5	150.15
1	255
1.5	318.26
2	344.4
2.5	338.63
3	306.9

X =

We can evaluate the function in Example 7 with the TABLE feature of a graphing calculator set in AUTO mode. We start at 0 and use a step-value of 0.5.

As discussed above, the domain of $M(t)$ is $[0, 6]$. To determine the range, we find the relative maximum value of the function using the MAXIMUM feature.

The maximum is about 345.76 mg. It occurs approximately 2.15 hr, or 2 hr 9 min, after the initial dose has been taken. The range is about $[0, 345.76]$.

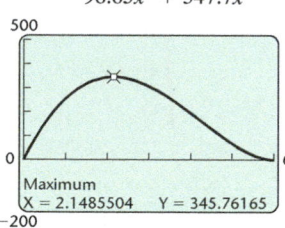

$$y = 0.5x^4 + 3.45x^3 - 96.65x^2 + 347.7x$$

Maximum
X = 2.1485504   Y = 345.76165

8.1   Exercise Set

Determine the leading term, the leading coefficient, and the degree of the polynomial. Then classify the polynomial function as constant, linear, quadratic, cubic, or quartic.

1. $g(x) = \frac{1}{2}x^3 - 10x + 8$

2. $f(x) = 15x^2 - 10 + 0.11x^4 - 7x^3$

3. $h(x) = 0.9x - 0.13$

4. $f(x) = -6$

5. $g(x) = 305x^4 + 4021$

6. $h(x) = 2.4x^3 + 5x^2 - x + \frac{7}{8}$

7. $h(x) = -5x^2 + 7x^3 + x^4$

8. $f(x) = 2 - x^2$

9. $g(x) = 4x^3 - \frac{1}{2}x^2 + 8$

10. $f(x) = 12 + x$

In Exercises 11–18, select one of the following four sketches to describe the end behavior of the graph of the function.

a)    b)

c) 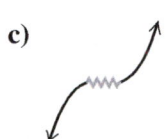   d)

11. $f(x) = -3x^3 - x + 4$

12. $f(x) = \frac{1}{4}x^4 + \frac{1}{2}x^3 - 6x^2 + x - 5$

13. $f(x) = -x^6 + \frac{3}{4}x^4$

14. $f(x) = \frac{2}{5}x^5 - 2x^4 + x^3 - \frac{1}{2}x + 3$

15. $f(x) = -3.5x^4 + x^6 + 0.1x^7$

16. $f(x) = -x^3 + x^5 - 0.5x^6$

17. $f(x) = 10 + \frac{1}{10}x^4 - \frac{2}{5}x^3$

18. $f(x) = 2x + x^3 - x^5$

In Exercises 19–22, use the leading-term test to match the function with one of the graphs (a)–(d) that follow.

a)

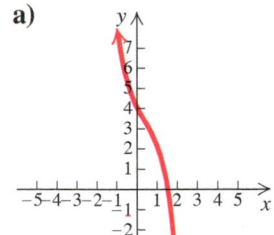

b)

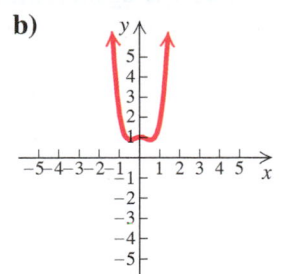

c)

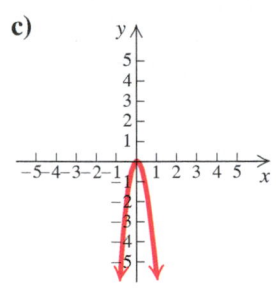

d)

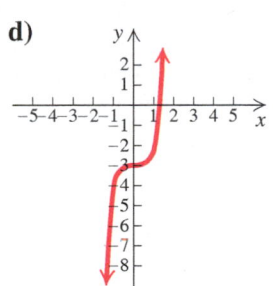

19. $f(x) = -x^6 + 2x^5 - 7x^2$

20. $f(x) = 2x^4 - x^2 + 1$

21. $f(x) = x^5 + \frac{1}{10}x - 3$

22. $f(x) = -x^3 + x^2 - 2x + 4$

23. Use substitution to determine whether 4, 5, and -2 are zeros of
$$f(x) = x^3 - 9x^2 + 14x + 24.$$

24. Use substitution to determine whether 2, 3, and -1 are zeros of
$$f(x) = 2x^3 - 3x^2 + x + 6.$$

25. Use substitution to determine whether 2, 3, and -1 are zeros of
$$g(x) = x^4 - 6x^3 + 8x^2 + 6x - 9.$$

26. Use substitution to determine whether 1, -2, and 3 are zeros of
$$g(x) = x^4 - x^3 - 3x^2 + 5x - 2.$$

Find the zeros of the polynomial function and state the multiplicity of each.

27. $f(x) = (x + 3)^2(x - 1)$

28. $f(x) = (x + 5)^3(x - 4)(x + 1)^2$

29. $f(x) = -2(x - 4)(x - 4)(x - 4)(x + 6)$

30. $f(x) = \left(x + \frac{1}{2}\right)(x + 7)(x + 7)(x + 5)$

31. $f(x) = (x^2 - 9)^3$

32. $f(x) = (x^2 - 4)^2$

33. $f(x) = x^3(x - 1)^2(x + 4)$

34. $f(x) = x^2(x + 3)^2(x - 4)(x + 1)^4$

35. $f(x) = -8(x - 3)^2(x + 4)^3x^4$

36. $f(x) = (x^2 - 5x + 6)^2$

37. $f(x) = x^4 - 4x^2 + 3$

38. $f(x) = x^4 - 10x^2 + 9$

39. $f(x) = x^3 + 3x^2 - x - 3$

40. $f(x) = x^3 - x^2 - 2x + 2$

41. $f(x) = 2x^3 - x^2 - 8x + 4$

42. $f(x) = 3x^3 + x^2 - 48x - 16$

Determine whether the statement is true or false.

43. If $P(x) = (x - 3)^4(x + 1)^3$, then the graph of the polynomial function $y = P(x)$ crosses the x-axis at $(3, 0)$.

44. If $P(x) = (x + 2)^2\left(x - \frac{1}{4}\right)^5$, then the graph of the polynomial function $y = P(x)$ crosses the x-axis at $\left(\frac{1}{4}, 0\right)$.

45. If $P(x) = (x - 2)^3(x + 5)^6$, then the graph of $y = P(x)$ is tangent to the x-axis at $(-5, 0)$.

46. If $P(x) = (x + 4)^2(x - 1)^2$, then the graph of $y = P(x)$ is tangent to the x-axis at $(4, 0)$.

47. *Vinyl Album Sales.* Vinyl record albums are making a comeback. Sales of vinyl albums rose 32% from 2012 to 2013. The sales data over the years 2001 to 2013 are modeled by the quartic function
$$f(x) = -0.000913x^4 + 0.0248x^3$$
$$- 0.1515x^2 + 0.2136x + 1.2779,$$
where x is the number of years after 2001 and $f(x)$ is the number of albums in millions (*Source:* Nielsen SoundScan). Find the number of vinyl albums sold in 2008, in 2012, and in 2016.

48. *Railroad Miles.* The greatest combined length of U.S.-owned operating railroad track existed in 1916, when industrial activity increased during World War I. The total length has decreased ever since. The data over the years 1900 to 2011 are modeled by the quartic function

$$f(x) = -0.002391x^4 + 0.949686x^3$$
$$- 123.648199x^2 + 4729.3635x$$
$$+ 198,846.4097,$$

where x is the number of years after 1900 and $f(x)$ is in miles (*Source:* Association of American Railroads). Find the number of miles of operating railroad track in the United States in 1916, in 1960, in 2000, and in 2016.

49. *Dog Years.* A dog's life span is typically much shorter than that of a human. The cubic function

$$d(x) = 0.010255x^3 - 0.340119x^2$$
$$+ 7.397499x + 6.618361,$$

where x is the dog's age, in years, approximates the equivalent human age in years. Estimate the equivalent human age for dogs that are 3, 12, and 16 years old.

50. *Threshold Weight.* In a study performed by Alvin Shemesh, it was found that the **threshold weight W**, defined as the weight above which the risk of death rises dramatically, is given by

$$W(h) = \left(\frac{h}{12.3}\right)^3,$$

where W is in pounds and h is a person's height, in inches. Find the threshold weight of a person who is 5 ft 7 in. tall.

51. *Projectile Motion.* A stone thrown downward with an initial velocity of 34.3 m/sec will travel a distance of s meters, where

$$s(t) = 4.9t^2 + 34.3t$$

and t is in seconds. If a stone is thrown downward at 34.3 m/sec from a height of 294 m, how long will it take the stone to hit the ground?

52. *Games in a Sports League.* If there are x teams in a sports league and all the teams play each other twice, a total of $N(x)$ games are played, where

$$N(x) = x^2 - x.$$

A softball league has 9 teams, each of which plays the others twice. If the league pays $110 per game for the field and the umpires, how much will it cost to play the entire schedule?

53. *Prison Admissions.* Since 2006, total admissions to state and federal prisons have been declining (*Source:* Bureau of Justice Statistics). The quartic function

$$p(x) = 6.213x^4 - 432.347x^3$$
$$+ 1922.987x^2 + 20,503.912x$$
$$+ 638,684.984,$$

where x is the number of years after 2001, can be used to estimate the number of admissions to state and federal prisons from 2001 to 2012. Estimate the number of prison admissions in 2003, in 2006, and in 2011.

54. *Obesity.* The percentage of adults who are obese is rising (*Source:* Gallup–Healthways Well-Being Index). The cubic function

$$f(x) = 0.102x^3 - 0.764x^2$$
$$+ 1.595x + 25.494,$$

where x is the number of years after 2008, can be used to estimate the percentage of adults who are obese. Using this function, estimate the percentage of adults who were obese in 2009 and in 2013.

55. *Interest Compounded Annually.* When P dollars is invested at interest rate i, compounded annually, for t years, the investment grows to A dollars, where

$$A = P(1 + i)^t.$$

Trevor's parents deposit $8000 in a savings account when Trevor is 16 years old. The principal plus interest is to be used for a truck when Trevor is 18 years old. Find the interest rate i if the $8000 grows to $9039.75 in 2 years.

56. *Interest Compounded Annually.* When *P* dollars is invested at interest rate *i*, compounded annually, for *t* years, the investment grows to *A* dollars, where
$$A = P(1 + i)^t.$$
When Sara enters the 11th grade, her grandparents deposit $10,000 in a college savings account. Find the interest rate *i* if the $10,000 grows to $11,193.64 in 2 years.

▶ # Skill Maintenance

Solve.

57. $2y - 3 \geq 1 - y + 5$ **[1.4]**

58. $-3 \leq 2x + 1 < 5$ **[1.5]**

59. $3x - 1 < -7 \text{ or } 3x - 1 > 5$ **[1.5]**

60. $|x + 6| \leq 7$ **[1.6]**

61. Find the domain of $f(x) = \dfrac{6}{3x + 1}$. **[2.3]**

62. Find the slope of the line containing the points $(4, -3)$ and $(-2, 5)$. **[2.4]**

Find the zeros of the function. **[7.4]**

63. $f(x) = x^2 - 5x + 4$

64. $f(x) = 2x^2 + 7x - 15$

▶ # Synthesis

Determine the degree and the leading term of the polynomial function.

65. $f(x) = (x^5 - 1)^2(x^2 + 2)^3$

66. $f(x) = (10 - 3x^5)^2(5 - x^4)^3(x + 4)$

8.2   Graphing Polynomial Functions

▶ Graph polynomial functions.

▶ Use the intermediate value theorem to determine whether a function has a real zero between two given real numbers.

▶ ## Graphing Polynomial Functions

In addition to using the leading-term test and finding the zeros of the function, it is helpful to consider the following facts when graphing a polynomial function.

> If $P(x)$ is a polynomial function of degree *n*, the graph of the function has:
>
> • at most *n* real zeros, and thus at most *n* *x*-intercepts;
>
> • at most $n - 1$ turning points.
>
> (Turning points on a graph, also called relative maxima and minima, occur when the function changes from decreasing to increasing or from increasing to decreasing.)

EXAMPLE 1   Graph the polynomial function $h(x) = -2x^4 + 3x^3$.

Solution

1. First, we use the leading-term test to determine the end behavior of the graph. The leading term is $-2x^4$. The degree, 4, is even, and the coefficient, -2, is negative. Thus the end behavior of the graph as $x \to \infty$ and as $x \to -\infty$ can be sketched as follows.

2. The zeros of the function are the first coordinates of the x-intercepts of the graph. To find the zeros, we solve $h(x) = 0$ by factoring and using the principle of zero products.

$$-2x^4 + 3x^3 = 0$$
$$-x^3(2x - 3) = 0 \qquad \text{\color{red}Factoring}$$
$$-x^3 = 0 \quad or \quad 2x - 3 = 0 \qquad \text{\color{red}Using the principle of zero products}$$
$$x = 0 \quad or \qquad x = \frac{3}{2}.$$

The zeros of the function are 0 and $\frac{3}{2}$. Note that the multiplicity of 0 is 3 and the multiplicity of $\frac{3}{2}$ is 1. The x-intercepts are $(0, 0)$ and $\left(\frac{3}{2}, 0\right)$.

3. The zeros divide the x-axis into three intervals:

$$(-\infty, 0), \qquad \left(0, \frac{3}{2}\right), \quad \text{and} \quad \left(\frac{3}{2}, \infty\right).$$

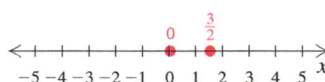

The sign of $h(x)$ is the same for all values of x in each of the three intervals. That is, **$h(x)$ is positive for all x-values in an interval or $h(x)$ is negative for all x-values in an interval**. To determine which, we choose a test value for x from each interval and find $h(x)$.

Interval	$(-\infty, 0)$	$\left(0, \frac{3}{2}\right)$	$\left(\frac{3}{2}, \infty\right)$
Test Value, x	-1	1	2
Function Value, $h(x)$	-5	1	-8
Sign of $h(x)$	$-$	$+$	$-$
Location of Points on Graph	Below x-axis	Above x-axis	Below x-axis

This test-point procedure also gives us three points to plot. In this case, we have $(-1, -5)$, $(1, 1)$, and $(2, -8)$.

4. To determine the y-intercept, we find $h(0)$:

$$h(x) = -2x^4 + 3x^3$$
$$h(0) = -2 \cdot 0^4 + 3 \cdot 0^3 = 0.$$

The y-intercept is $(0, 0)$.

5. A few additional points are helpful when completing the graph.

x	$h(x)$
-1.5	-20.25
-0.5	-0.5
0.5	0.25
2.5	-31.25

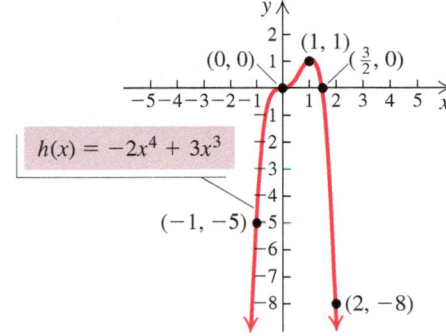

6. The degree of h is 4. The graph of h can have at most 4 x-intercepts and at most 3 turning points. In fact, it has 2 x-intercepts and 1 turning point. The zeros, 0 and $\frac{3}{2}$, each have odd multiplicities: 3 for 0 and 1 for $\frac{3}{2}$. Since the multiplicities are odd, the graph crosses the x-axis at 0 and $\frac{3}{2}$. The end behavior of the graph is what we described in step (1). As $x \to \infty$ and also as $x \to -\infty$, $h(x) \to -\infty$. The graph appears to be correct. ▶ **Now Try Exercise 23.**

The following is a procedure for graphing polynomial functions.

To graph a polynomial function:

1. Use the leading-term test to determine the end behavior.

2. Find the zeros of the function by solving $f(x) = 0$. Any real zeros are the first coordinates of the x-intercepts.

3. Use the x-intercepts (zeros) to divide the x-axis into intervals and choose a test point in each interval to determine the sign of all function values in that interval.

4. Find $f(0)$. This gives the y-intercept of the function.

5. If necessary, find additional function values to determine the general shape of the graph and then draw the graph.

6. As a partial check, use the facts that the graph has at most n x-intercepts and at most $n - 1$ turning points. Multiplicity of zeros can also be considered in order to check where the graph crosses or is tangent to the x-axis.

EXAMPLE 2   Graph the polynomial function

$$f(x) = 2x^3 + x^2 - 8x - 4.$$

Solution

1. The leading term is $2x^3$. The degree, 3, is odd, and the coefficient, 2, is positive. Thus the end behavior of the graph will appear as follows.

2. To find the zeros, we solve $f(x) = 0$. Here we can use factoring by grouping.

$$2x^3 + x^2 - 8x - 4 = 0$$
$$x^2(2x + 1) - 4(2x + 1) = 0 \qquad \text{Factoring by grouping}$$
$$(2x + 1)(x^2 - 4) = 0$$
$$(2x + 1)(x + 2)(x - 2) = 0 \qquad \text{Factoring a difference of squares}$$

The zeros are $-\frac{1}{2}$, -2, and 2. Each is of multiplicity 1. The x-intercepts are $(-2, 0)$, $\left(-\frac{1}{2}, 0\right)$, and $(2, 0)$.

3. The zeros divide the x-axis into four intervals:

$$(-\infty, -2), \qquad \left(-2, -\frac{1}{2}\right), \qquad \left(-\frac{1}{2}, 2\right), \quad \text{and} \quad (2, \infty).$$

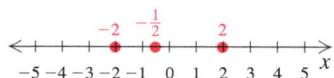

We choose a test value for x from each interval and find $f(x)$.

Interval	$(-\infty, -2)$	$\left(-2, -\frac{1}{2}\right)$	$\left(-\frac{1}{2}, 2\right)$	$(2, \infty)$
Test Value, x	-3	-1	1	3
Function Value, $f(x)$	-25	3	-9	35
Sign of $f(x)$	$-$	$+$	$-$	$+$
Location of Points on Graph	Below x-axis	Above x-axis	Below x-axis	Above x-axis

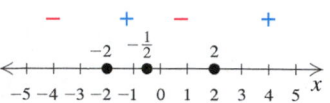

The test values and corresponding function values also give us four points on the graph: $(-3, -25)$, $(-1, 3)$, $(1, -9)$, and $(3, 35)$.

4. To determine the *y*-intercept, we find $f(0)$:

$$f(x) = 2x^3 + x^2 - 8x - 4$$
$$f(0) = 2 \cdot 0^3 + 0^2 - 8 \cdot 0 - 4 = -4.$$

The *y*-intercept is $(0, -4)$.

5. We find a few additional points and complete the graph.

x	$f(x)$
−2.5	−9
−1.5	3.5
0.5	−7.5
1.5	−7

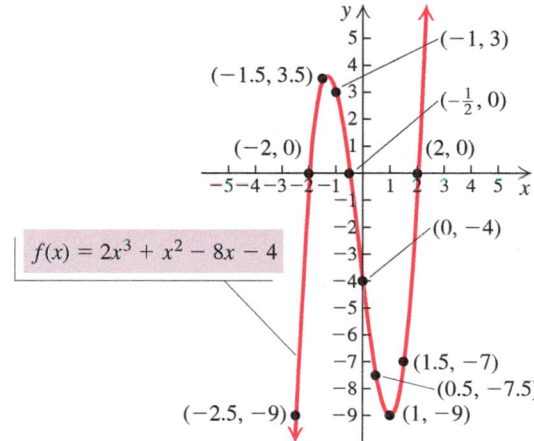

$f(x) = 2x^3 + x^2 - 8x - 4$

6. The degree of *f* is 3. The graph of *f* can have at most 3 *x*-intercepts and at most 2 turning points. It has 3 *x*-intercepts and 2 turning points. Each zero has a multiplicity of 1; thus the graph crosses the *x*-axis at -2, $-\frac{1}{2}$, and 2. The graph has the end behavior described in step (1). As $x \to -\infty$, $h(x) \to -\infty$, and as $x \to \infty$, $h(x) \to \infty$. The graph appears to be correct.

▶ **Now Try Exercise 33.**

Some polynomials are difficult to factor. In the next example, the polynomial is given in factored form. In Sections 8.3 and 8.4, we will learn methods that facilitate determining factors of such polynomials.

EXAMPLE 3   Graph the polynomial function

$$g(x) = x^4 - 7x^3 + 12x^2 + 4x - 16$$
$$= (x + 1)(x - 2)^2(x - 4).$$

Solution

1. The leading term is x^4. The degree, 4, is even, and the coefficient, 1, is positive. The sketch below shows the end behavior.

2. To find the zeros, we solve $g(x) = 0$:

$$(x + 1)(x - 2)^2(x - 4) = 0.$$

The zeros are -1, 2, and 4; 2 is of multiplicity 2; the others are of multiplicity 1. The *x*-intercepts are $(-1, 0)$, $(2, 0)$, and $(4, 0)$.

3. The zeros divide the *x*-axis into four intervals:

$$(-\infty, -1), \quad (-1, 2), \quad (2, 4), \quad \text{and} \quad (4, \infty).$$

We choose a test value for *x* from each interval and find $g(x)$.

Interval	$(-\infty, -1)$	$(-1, 2)$	$(2, 4)$	$(4, \infty)$
Test Value, *x*	-1.25	1	3	4.25
Function Value, $g(x)$	≈ 13.9	-6	-4	≈ 6.6
Sign of $g(x)$	$+$	$-$	$-$	$+$
Location of Points on Graph	Above *x*-axis	Below *x*-axis	Below *x*-axis	Above *x*-axis

The test values and the corresponding function values also give us four points on the graph: $(-1.25, 13.9)$, $(1, -6)$, $(3, -4)$, and $(4.25, 6.6)$.

4. To determine the *y*-intercept, we find $g(0)$:

$$g(x) = x^4 - 7x^3 + 12x^2 + 4x - 16$$
$$g(0) = 0^4 - 7 \cdot 0^3 + 12 \cdot 0^2 + 4 \cdot 0 - 16 = -16.$$

The *y*-intercept is $(0, -16)$.

5. We find a few additional points and draw the graph.

x	$g(x)$
-0.5	-14.1
0.5	-11.8
1.5	-1.6
2.5	-1.3
3.5	-5.1

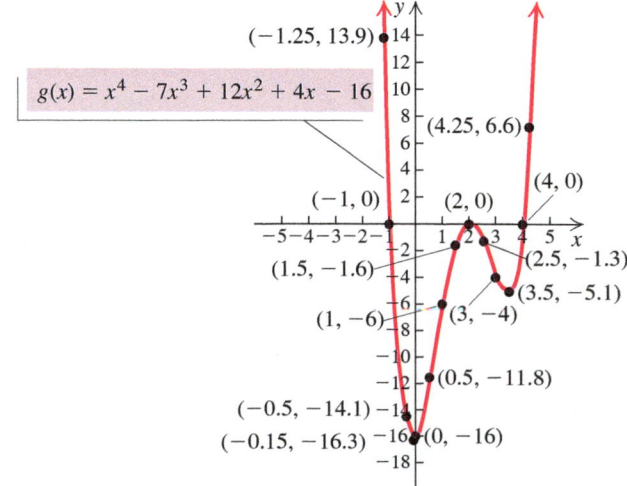

6. The degree of g is 4. The graph of g can have at most 4 x-intercepts and at most 3 turning points. It has 3 x-intercepts and 3 turning points. One of the zeros, 2, has a multiplicity of 2, so the graph is tangent to the x-axis at 2. The other zeros, -1 and 4, each have a multiplicity of 1 so the graph crosses the x-axis at -1 and 4. The graph has the end behavior described in step (1). As $x \to \infty$ and as $x \to -\infty$, $g(x) \to \infty$. The graph appears to be correct.

> **Now Try Exercise 19.**

▶ The Intermediate Value Theorem

Polynomial functions are continuous, hence their graphs are unbroken. The domain of a polynomial function, unless restricted by the statement of the function, is $(-\infty, \infty)$. Suppose two polynomial function values $P(a)$ and $P(b)$ have opposite signs. Since P is continuous, its graph must be a curve from $(a, P(a))$ to $(b, P(b))$ without a break. Then it follows that the curve must cross the x-axis at at least one point c between a and b; that is, the function has a zero at c between a and b.

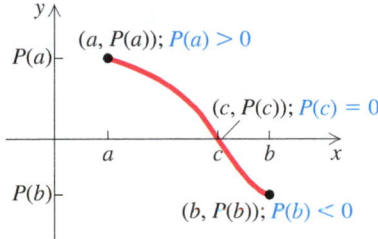

THE INTERMEDIATE VALUE THEOREM

For any polynomial function $P(x)$ with real coefficients, suppose that for $a \neq b$, $P(a)$ and $P(b)$ are of opposite signs. Then the function has at least one real zero between a and b.

    The intermediate value theorem *cannot* be used to determine whether there is a real zero between a and b when $P(a)$ and $P(b)$ have the *same* sign.

EXAMPLE 4   Using the intermediate value theorem, determine, if possible, whether the function has at least one real zero between a and b.

a) $f(x) = x^3 + x^2 - 6x$; $a = -4, b = -2$

b) $f(x) = x^3 + x^2 - 6x$; $a = -1, b = 3$

c) $g(x) = \frac{1}{3}x^4 - x^3$; $a = -\frac{1}{2}, b = \frac{1}{2}$

d) $g(x) = \frac{1}{3}x^4 - x^3$; $a = 1, b = 2$

Solution   We find $f(a)$ and $f(b)$ or $g(a)$ and $g(b)$ and determine whether they differ in sign. The graphs of $f(x)$ and $g(x)$ provide visual checks of the conclusions.

a) $f(-4) = (-4)^3 + (-4)^2 - 6(-4) = -24$,
    $f(-2) = (-2)^3 + (-2)^2 - 6(-2) = 8$

Note that $f(-4)$ is negative and $f(-2)$ is positive. By the intermediate value theorem, since $f(-4)$ and $f(-2)$ have opposite signs, then $f(x)$ has at least one zero between -4 and -2. The following graph confirms this.

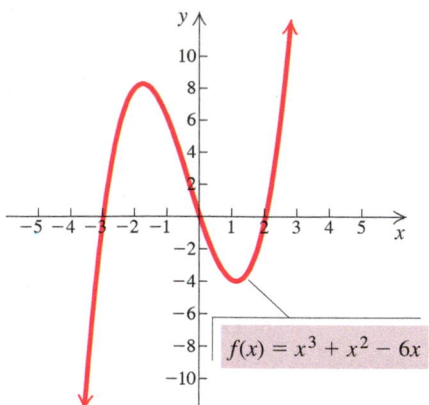

b) $f(-1) = (-1)^3 + (-1)^2 - 6(-1) = 6,$

$f(3) = 3^3 + 3^2 - 6(3) = 18$

Both $f(-1)$ and $f(3)$ are positive. Thus the intermediate value theorem *cannot be used* to determine whether there is a real zero between -1 and 3. Note that the graph of $f(x)$ above shows that there are two zeros between -1 and 3.

c) $g\left(-\frac{1}{2}\right) = \frac{1}{3}\left(-\frac{1}{2}\right)^4 - \left(-\frac{1}{2}\right)^3 = \frac{7}{48},$

$g\left(\frac{1}{2}\right) = \frac{1}{3}\left(\frac{1}{2}\right)^4 - \left(\frac{1}{2}\right)^3 = -\frac{5}{48}$

Since $g\left(-\frac{1}{2}\right)$ and $g\left(\frac{1}{2}\right)$ have opposite signs, $g(x)$ has at least one zero between $-\frac{1}{2}$ and $\frac{1}{2}$. The following graph confirms this.

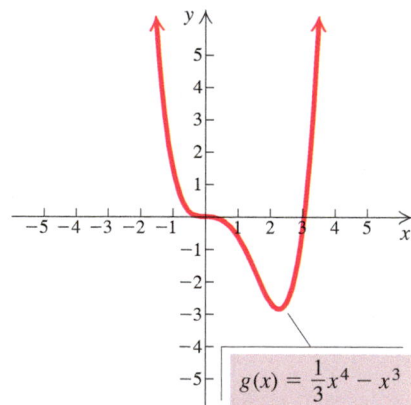

d) $g(1) = \frac{1}{3}(1)^4 - 1^3 = -\frac{2}{3},$

$g(2) = \frac{1}{3}(2)^4 - 2^3 = -\frac{8}{3}$

Both $g(1)$ and $g(2)$ are negative. Thus the intermediate value theorem *cannot be used* to determine whether there is a real zero between 1 and 2. Note that the graph of $g(x)$ above shows that there are no zeros between 1 and 2.

> **Now Try Exercises 39 and 43.**

A
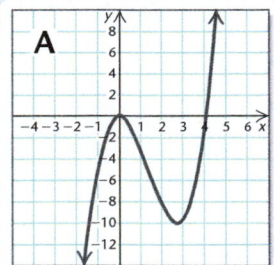

B
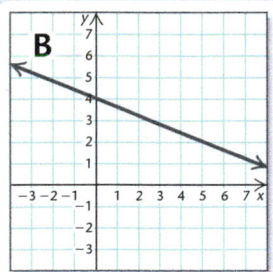

C
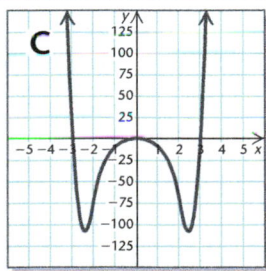

D
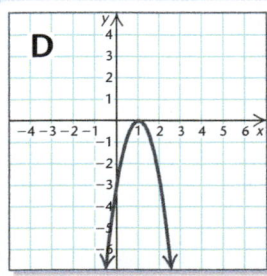

E
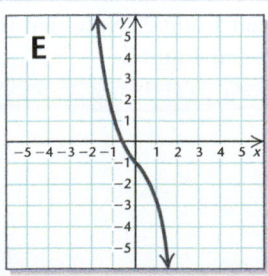

Visualizing the Graph

Match the function with its graph.

1. $f(x) = -x^4 - x + 5$

2. $f(x) = -3x^2 + 6x - 3$

3. $f(x) = x^4 - 4x^3 + 3x^2 + 4x - 4$

4. $f(x) = -\dfrac{2}{5}x + 4$

5. $f(x) = x^3 - 4x^2$

6. $f(x) = x^6 - 9x^4$

7. $f(x) = x^5 - 3x^3 + 2$

8. $f(x) = -x^3 - x - 1$

9. $f(x) = x^2 + 7x + 6$

10. $f(x) = \dfrac{7}{2}$

F
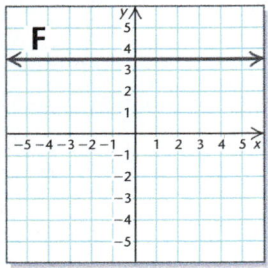

G
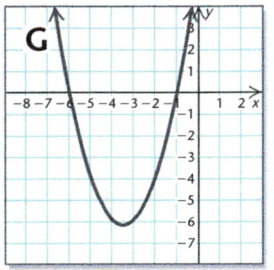

H
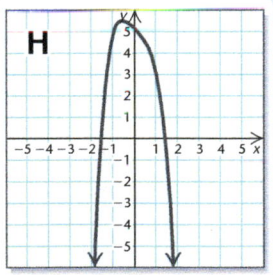

I
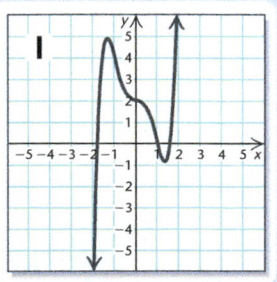

J
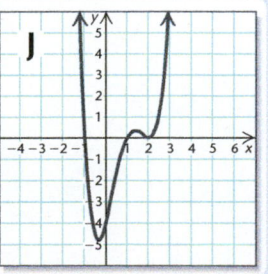

8.2 Exercise Set

For each function in Exercises 1–6, state:

a) *the maximum number of real zeros that the function can have;*

b) *the maximum number of x-intercepts that the graph of the function can have; and*

c) *the maximum number of turning points that the graph of the function can have.*

1. $f(x) = x^5 - x^2 + 6$

2. $f(x) = -x^2 + x^4 - x^6 + 3$

3. $f(x) = x^{10} - 2x^5 + 4x - 2$

4. $f(x) = \frac{1}{4}x^3 + 2x^2$ **5.** $f(x) = -x - x^3$

6. $f(x) = -3x^4 + 2x^3 - x - 4$

In Exercises 7–12, use the leading-term test and your knowledge of y-intercepts to match the function with one of the graphs (a)–(f) that follow.

a)
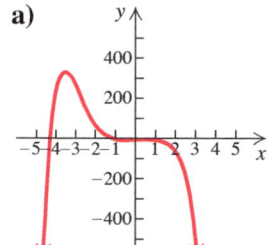

b)
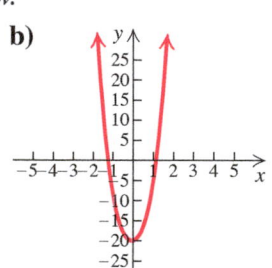

c)
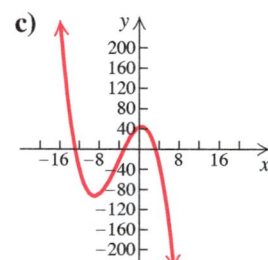

d)
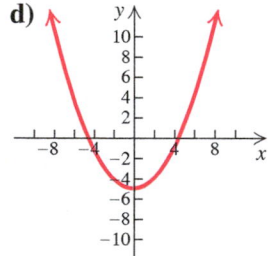

e)
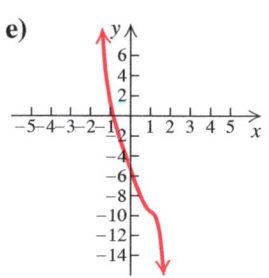

f)
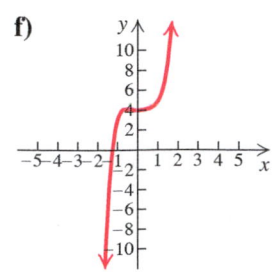

7. $f(x) = \frac{1}{4}x^2 - 5$

8. $f(x) = -0.5x^6 - x^5 + 4x^4 - 5x^3 - 7x^2 + x - 3$

9. $f(x) = x^5 - x^4 + x^2 + 4$

10. $f(x) = -\frac{1}{3}x^3 - 4x^2 + 6x + 42$

11. $f(x) = x^4 - 2x^3 + 12x^2 + x - 20$

12. $f(x) = -0.3x^7 + 0.11x^6 - 0.25x^5 + x^4 + x^3 - 6x - 5$

Graph the polynomial function. Follow the steps outlined in the procedure on p. 520.

13. $f(x) = -x^3 - 2x^2$

14. $g(x) = x^4 - 4x^3 + 3x^2$

15. $h(x) = x^2 + 2x - 3$

16. $f(x) = x^2 - 5x + 4$

17. $h(x) = x^5 - 4x^3$

18. $f(x) = x^3 - x$

19. $h(x) = x(x - 4)(x + 1)(x - 2)$

20. $f(x) = x(x - 1)(x + 3)(x + 5)$

21. $g(x) = -\frac{1}{4}x^3 - \frac{3}{4}x^2$ **22.** $f(x) = \frac{1}{2}x^3 + \frac{5}{2}x^2$

23. $g(x) = -x^4 - 2x^3$ **24.** $h(x) = x^3 - 3x^2$

25. $f(x) = -\frac{1}{2}(x - 2)(x + 1)^2(x - 1)$

26. $g(x) = (x - 2)^3(x + 3)$

27. $g(x) = -x(x - 1)^2(x + 4)^2$

28. $h(x) = -x(x - 3)(x - 3)(x + 2)$

29. $f(x) = (x - 2)^2(x + 1)^4$

30. $g(x) = x^4 - 9x^2$

31. $g(x) = -(x - 1)^4$

32. $h(x) = (x + 2)^3$

33. $h(x) = x^3 + 3x^2 - x - 3$

34. $g(x) = -x^3 + 2x^2 + 4x - 8$

35. $f(x) = 6x^3 - 8x^2 - 54x + 72$

36. $h(x) = x^5 - 5x^3 + 4x$

Graph each piecewise function.

37. $g(x) = \begin{cases} -x + 3, & \text{for } x \leq -2, \\ 4, & \text{for } -2 < x < 1, \\ \frac{1}{2}x^3, & \text{for } x \geq 1 \end{cases}$

38. $h(x) = \begin{cases} -x^2, & \text{for } x < -2, \\ x + 1, & \text{for } -2 \leq x < 0, \\ x^3 - 1, & \text{for } x \geq 0 \end{cases}$

Using the intermediate value theorem, determine, if possible, whether the function f has at least one real zero between a and b.

39. $f(x) = x^3 + 3x^2 - 9x - 13;\ a = -5, b = -4$

40. $f(x) = x^3 + 3x^2 - 9x - 13;\ a = 1, b = 2$

41. $f(x) = 3x^2 - 2x - 11;\ a = -3, b = -2$

42. $f(x) = 3x^2 - 2x - 11;\ a = 2, b = 3$

43. $f(x) = x^4 - 2x^2 - 6;\ a = 2, b = 3$

44. $f(x) = 2x^5 - 7x + 1;\ a = 1, b = 2$

45. $f(x) = x^3 - 5x^2 + 4;\ a = 4, b = 5$

46. $f(x) = x^4 - 3x^2 + x - 1;\ a = -3, b = -2$

▶ **Skill Maintenance**

In Exercises 47–52, match the equation with one of the graphs (a)–(f) that follow.

a)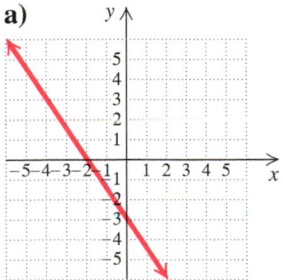

b)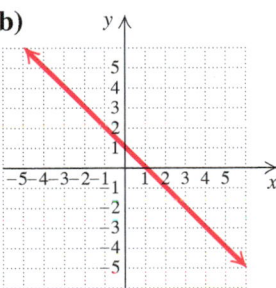

c)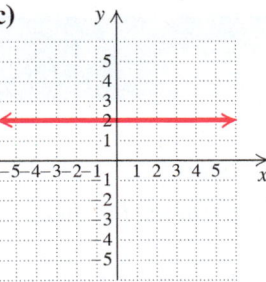

d)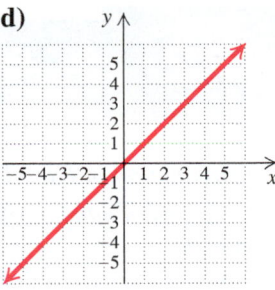

e)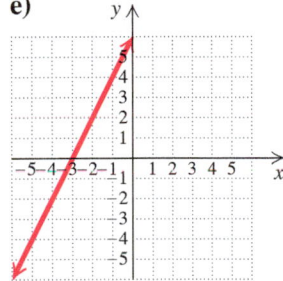

f)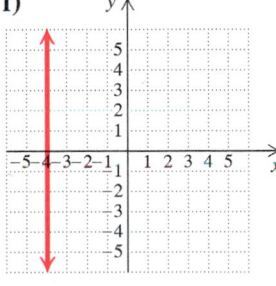

47. $y = x$ **[2.1]**

48. $x = -4$ **[2.6]**

49. $y - 2x = 6$ **[2.1]**

50. $3x + 2y = -6$ **[2.1]**

51. $y = 1 - x$ **[2.1]**

52. $y = 2$ **[2.6]**

Solve.

53. $2x - \frac{1}{2} = 4 - 3x$ **[1.1]**

54. $x^3 - x^2 - 12x = 0$ **[7.4]**

55. $6x^2 - 23x - 55 = 0$ **[7.4]**

56. $\frac{3}{4}x + 10 = \frac{1}{5} + 2x$ **[1.1]**

8.3   Polynomial Division; The Remainder Theorem and the Factor Theorem

▶ Perform long division with polynomials and determine whether one polynomial is a factor of another.

▶ Use synthetic division to divide a polynomial by $x - c$.

▶ Use the remainder theorem to find a function value $f(c)$.

▶ Use the factor theorem to determine whether $x - c$ is a factor of $f(x)$.

In general, finding exact zeros of many polynomial functions is neither easy nor straightforward. In this section and the one that follows, we develop concepts that help us find exact zeros of certain polynomial functions with degree 3 or greater.

    Consider the polynomial

$$h(x) = x^3 + 2x^2 - 5x - 6 = (x + 3)(x + 1)(x - 2).$$

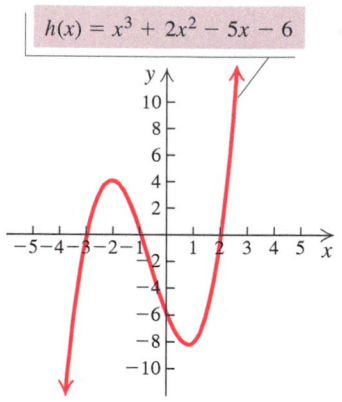

$h(x) = x^3 + 2x^2 - 5x - 6$

The factors are

$$x + 3, \quad x + 1, \quad \text{and} \quad x - 2,$$

and the zeros are

$$-3, \quad -1, \quad \text{and} \quad 2.$$

When a polynomial is expressed as a product of linear factors, each factor determines a zero of the function. Thus if we know the factors of a polynomial, we can easily find the zeros. The "reverse" is also true: If we know the zeros of a polynomial function, we can find the factors of the polynomial.

▶ **Division and Factors**

When we divide one polynomial by another, we obtain a quotient and a remainder. If the remainder is 0, then the divisor is a factor of the dividend.

EXAMPLE 1 Divide to determine whether $x + 1$ and $x - 3$ are factors of

$$x^3 + 2x^2 - 5x - 6.$$

Solution We divide $x^3 + 2x^2 - 5x - 6$ by $x + 1$.

$$
\begin{array}{r}
\overbrace{x^2 + x - 6}^{\text{Quotient}} \\
x + 1 \overline{\smash{\big)}\,x^3 + 2x^2 - 5x - 6} \quad \leftarrow \text{Dividend} \\
\underline{x^3 + x^2} \phantom{{}- 5x - 6} \\
x^2 - 5x \phantom{{}- 6} \\
\underline{x^2 + x} \phantom{{}- 6} \\
-6x - 6 \\
\underline{-6x - 6} \\
0 \quad \leftarrow \text{Remainder}
\end{array}
$$

Divisor

Since the remainder is 0, we know that $x + 1$ is a factor of $x^3 + 2x^2 - 5x - 6$. In fact, we know that

$$x^3 + 2x^2 - 5x - 6 = (x + 1)(x^2 + x - 6).$$

We divide $x^3 + 2x^2 - 5x - 6$ by $x - 3$.

$$
\begin{array}{r}
x^2 + 5x + 10 \\
x - 3 \overline{\smash{\big)}\,x^3 + 2x^2 - 5x - 6} \\
\underline{x^3 - 3x^2} \phantom{{}- 5x - 6} \\
5x^2 - 5x \phantom{{}- 6} \\
\underline{5x^2 - 15x} \phantom{{}- 6} \\
10x - 6 \\
\underline{10x - 30} \\
24 \quad \leftarrow \text{Remainder}
\end{array}
$$

Since the remainder, 24, is not 0, we know that $x - 3$ is *not* a factor of $x^3 + 2x^2 - 5x - 6$.

Now Try Exercise 3.

When we divide a polynomial $P(x)$ by a divisor $d(x)$, a polynomial $Q(x)$ is the quotient and a polynomial $R(x)$ is the remainder. The quotient $Q(x)$ must have degree less than that of the dividend $P(x)$. The remainder $R(x)$ must either be 0 or have degree less than that of the divisor $d(x)$.

As in arithmetic, to check a division, we multiply the quotient by the divisor and add the remainder, to see if we get the dividend. Thus these polynomials are related as follows:

$$P(x) = d(x) \cdot Q(x) + R(x).$$

Dividend   Divisor   Quotient   Remainder

For instance, if $P(x) = x^3 + 2x^2 - 5x - 6$ and $d(x) = x - 3$, as in Example 1, then $Q(x) = x^2 + 5x + 10$ and $R(x) = 24$, and

$$P(x) = d(x) \cdot Q(x) + R(x)$$
$$x^3 + 2x^2 - 5x - 6 = (x - 3) \cdot (x^2 + 5x + 10) + 24$$
$$= x^3 + 5x^2 + 10x - 3x^2 - 15x - 30 + 24$$
$$= x^3 + 2x^2 - 5x - 6.$$

▶ ## The Remainder Theorem and Synthetic Division

Consider the function

$$h(x) = x^3 + 2x^2 - 5x - 6.$$

When we divided $h(x)$ by $x + 1$ and $x - 3$ in Example 1, the remainders were 0 and 24, respectively. Let's now find the function values $h(-1)$ and $h(3)$:

$$h(-1) = (-1)^3 + 2(-1)^2 - 5(-1) - 6 = 0;$$
$$h(3) = (3)^3 + 2(3)^2 - 5(3) - 6 = 24.$$

Note that the function values are the same as the remainders. This suggests the following theorem.

THE REMAINDER THEOREM

If a number c is substituted for x in the polynomial $f(x)$, then the result $f(c)$ is the remainder that would be obtained by dividing $f(x)$ by $x - c$. That is, if $f(x) = (x - c) \cdot Q(x) + R$, then $f(c) = R$.

Proof (Optional).   The equation $f(x) = d(x) \cdot Q(x) + R(x)$, where $d(x) = x - c$, is the basis of this proof. If we divide $f(x)$ by $x - c$, we obtain a quotient $Q(x)$ and a remainder $R(x)$ related as follows:

$$f(x) = (x - c) \cdot Q(x) + R(x).$$

The remainder $R(x)$ must either be 0 or have degree less than $x - c$. Thus, $R(x)$ must be a constant. Let's call this constant R. The equation above is true for any replacement of x, so we replace x with c. We get

$$f(c) = (c - c) \cdot Q(c) + R$$
$$= 0 \cdot Q(c) + R$$
$$= R.$$

Thus the function value $f(c)$ is the remainder obtained when we divide $f(x)$ by $x - c$. ■

The remainder theorem motivates us to find a rapid way of dividing by $x - c$ in order to find function values. To streamline division, we can arrange the work so that duplicate and unnecessary writing is avoided. Consider the following:

$$(4x^3 - 3x^2 + x + 7) \div (x - 2).$$

A.
$$
\begin{array}{r}
4x^2 + 5x + 11 \\
x - 2 \overline{)4x^3 - 3x^2 + x + 7} \\
\underline{4x^3 - 8x^2} \\
5x^2 + x \\
\underline{5x^2 - 10x} \\
11x + 7 \\
\underline{11x - 22} \\
29
\end{array}
$$

B.
$$
\begin{array}{r}
4 \quad 5 \quad 11 \\
1 - 2 \overline{)4 - 3 + 1 + 7} \\
\underline{4 - 8} \\
5 + 1 \\
\underline{5 - 10} \\
11 + 7 \\
\underline{11 - 22} \\
29
\end{array}
$$

The division in (B) is the same as that in (A), but we wrote only the coefficients. The red numerals are duplicated, so we look for an arrangement in which they are not duplicated. In place of the divisor in the form $x - c$, we can simply use c and then add rather than subtract. When the procedure is "collapsed," we have the algorithm known as **synthetic division**.

C. *Synthetic Division*

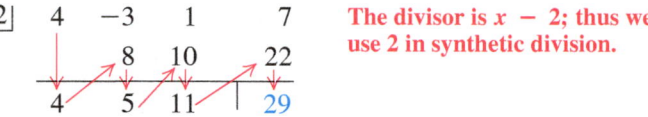

The divisor is $x - 2$; thus we use 2 in synthetic division.

We "bring down" the 4. Then we multiply it by the 2 to get 8 and add to get 5. We then multiply 5 by 2 to get 10, add, and so on. The last number, 29, is the remainder. The others, 4, 5, and 11, are the coefficients of the quotient, $4x^2 + 5x + 11$. (Note that the degree of the quotient is 1 less than the degree of the dividend when the degree of the divisor is 1.)

When using synthetic division, we write a 0 for a missing term in the dividend.

EXAMPLE 2 Use synthetic division to find the quotient and the remainder:

$$(2x^3 + 7x^2 - 5) \div (x + 3).$$

Solution First, we note that $x + 3 = x - (-3)$.

$$
\begin{array}{r|rrrr}
-3 & 2 & 7 & 0 & -5 \\
   &   & -6 & -3 & 9 \\
\hline
   & 2 & 1 & -3 & \,|\, 4
\end{array}
$$

Note: We must write a 0 for the missing x-term.

The quotient is $2x^2 + x - 3$. The remainder is 4. ⟶ **Now Try Exercise 13.**

We can now use synthetic division to find polynomial function values.

EXAMPLE 3 Given that $f(x) = 2x^5 - 3x^4 + x^3 - 2x^2 + x - 8$, find $f(10)$.

Solution By the remainder theorem, $f(10)$ is the remainder when $f(x)$ is divided by $x - 10$. We use synthetic division to find that remainder.

$$
\begin{array}{r|rrrrrr}
10 & 2 & -3 & 1 & -2 & 1 & -8 \\
   &   & 20 & 170 & 1710 & 17{,}080 & 170{,}810 \\
\hline
   & 2 & 17 & 171 & 1708 & 17{,}081 & \,|\, 170{,}802
\end{array}
$$

Thus, $f(10) = 170{,}802$. ⟶ **Now Try Exercise 25.**

Compare the computations in Example 3 with those in a direct substitution:

$$f(10) = 2(10)^5 - 3(10)^4 + (10)^3 - 2(10)^2 + 10 - 8$$
$$= 2 \cdot 100{,}000 - 3 \cdot 10{,}000 + 1000 - 2 \cdot 100 + 10 - 8$$
$$= 200{,}000 - 30{,}000 + 1000 - 200 + 10 - 8$$
$$= 170{,}802.$$

EXAMPLE 4 Determine whether 5 is a zero of $g(x)$, where

$$g(x) = x^4 - 26x^2 + 25.$$

Solution We use synthetic division and the remainder theorem to find $g(5)$.

$$
\begin{array}{r|rrrrr}
5 & 1 & 0 & -26 & 0 & 25 \\
  &   & 5 & 25 & -5 & -25 \\
\hline
  & 1 & 5 & -1 & -5\,| & 0
\end{array}
$$

Writing 0's for missing terms:
$x^4 + 0x^3 - 26x^2 + 0x + 25$

Since $g(5) = 0$, the number 5 is a zero of $g(x)$. ▶ **Now Try Exercise 31.**

EXAMPLE 5 Determine whether i is a zero of $f(x)$, where

$$f(x) = x^3 - 3x^2 + x - 3.$$

Solution We use synthetic division and the remainder theorem to find $f(i)$.

$$
\begin{array}{r|rrrr}
i & 1 & -3 & 1 & -3 \\
  &   & i & -3i-1 & 3 \\
\hline
  & 1 & -3+i & -3i\,| & 0
\end{array}
$$

$i(-3 + i) = -3i + i^2 = -3i - 1$

$i(-3i) = -3i^2 = 3$

Since $f(i) = 0$, the number i is a zero of $f(x)$. ▶ **Now Try Exercise 35.**

▶ Finding Factors of Polynomials

We now consider a useful result that follows from the remainder theorem.

> **THE FACTOR THEOREM**
>
> For a polynomial $f(x)$, if $f(c) = 0$, then $x - c$ is a factor of $f(x)$.

Proof (Optional). If we divide $f(x)$ by $x - c$, we obtain a quotient and a remainder, related as follows:

$$f(x) = (x - c) \cdot Q(x) + f(c).$$

Then if $f(c) = 0$, we have

$$f(x) = (x - c) \cdot Q(x),$$

so $x - c$ is a factor of $f(x)$. ∎

The factor theorem is very useful in factoring polynomials and hence in solving polynomial equations and finding zeros of polynomial functions. If we know a zero of a polynomial function, we know a factor.

EXAMPLE 6 Let $f(x) = x^3 - 3x^2 - 6x + 8$. Factor $f(x)$ and solve the equation $f(x) = 0$.

Solution We look for linear factors of the form $x - c$. Let's try $x + 1$, or $x - (-1)$. (In the next section, we will learn a method for choosing the numbers to try for c.) We use synthetic division to determine whether $f(-1) = 0$.

$$
\begin{array}{r|rrrr}
-1 & 1 & -3 & -6 & 8 \\
   &   & -1 & 4 & 2 \\
\hline
   & 1 & -4 & -2 & \,\big|\ 10
\end{array}
$$

Since $f(-1) \ne 0$, we know that $x + 1$ *is not a factor* of $f(x)$. We now try $x - 1$.

$$
\begin{array}{r|rrrr}
1 & 1 & -3 & -6 & 8 \\
  &   & 1 & -2 & -8 \\
\hline
  & 1 & -2 & -8 & \,\big|\ 0
\end{array}
$$

Since $f(1) = 0$, we know that $x - 1$ *is one factor* of $f(x)$ and the quotient, $x^2 - 2x - 8$, is another. Thus,

$$f(x) = (x - 1)(x^2 - 2x - 8).$$

The trinomial $x^2 - 2x - 8$ is easily factored, so we have

$$f(x) = (x - 1)(x - 4)(x + 2).$$

Our goal is to solve the equation $f(x) = 0$. To do so, we use the principle of zero products:

$$(x - 1)(x - 4)(x + 2) = 0$$
$$x - 1 = 0 \quad or \quad x - 4 = 0 \quad or \quad x + 2 = 0$$
$$x = 1 \quad or \quad\quad x = 4 \quad or \quad\quad x = -2.$$

The solutions of the equation $x^3 - 3x^2 - 6x + 8 = 0$ are -2, 1, and 4. They are also the zeros of the function $f(x) = x^3 - 3x^2 - 6x + 8$.

> **Now Try Exercise 41.**

Technology Connection

In Example 6, we can use a table set in ASK mode to check the solutions of the equation

$$x^3 - 3x^2 - 6x + 8 = 0.$$

We check the solutions, -2, 1, and 4, of $f(x) = 0$ by evaluating $f(-2), f(1)$, and $f(4)$.

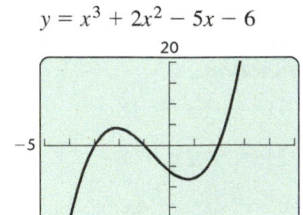

X	Y₁
-2	0
1	0
4	0

X =

CONNECTING THE CONCEPTS

Consider the function

$$f(x) = (x - 2)(x + 3)(x + 1), \quad or \quad f(x) = x^3 + 2x^2 - 5x - 6,$$

$y = x^3 + 2x^2 - 5x - 6$

and its graph.

We can make the following statements:

- -3 is a zero of f.
- $f(-3) = 0$.
- -3 is a solution of $f(x) = 0$.
- $(-3, 0)$ is an x-intercept of the graph of f.
- 0 is the remainder when $f(x)$ is divided by $x - (-3)$.
- $x - (-3)$ is a factor of f.

Similar statements are also true for -1 and 2.

8.3 Exercise Set

1. For the function
$$f(x) = x^4 - 6x^3 + x^2 + 24x - 20,$$
use long division to determine whether each of the following is a factor of $f(x)$.
   a) $x + 1$ b) $x - 2$ c) $x + 5$

2. For the function
$$h(x) = x^3 - x^2 - 17x - 15,$$
use long division to determine whether each of the following is a factor of $h(x)$.
   a) $x + 5$ b) $x + 1$ c) $x + 3$

3. For the function
$$g(x) = x^3 - 2x^2 - 11x + 12,$$
use long division to determine whether each of the following is a factor of $g(x)$.
   a) $x - 4$ b) $x - 3$ c) $x - 1$

4. For the function
$$f(x) = x^4 + 8x^3 + 5x^2 - 38x + 24,$$
use long division to determine whether each of the following is a factor of $f(x)$.
   a) $x + 6$ b) $x + 1$ c) $x - 4$

In each of the following, a polynomial $P(x)$ and a divisor $d(x)$ are given. Use long division to find the quotient $Q(x)$ and the remainder $R(x)$ when $P(x)$ is divided by $d(x)$. Express $P(x)$ in the form $d(x) \cdot Q(x) + R(x)$.

5. $P(x) = x^3 - 8$,
   $d(x) = x + 2$

6. $P(x) = 2x^3 - 3x^2 + x - 1$,
   $d(x) = x - 3$

7. $P(x) = x^3 + 6x^2 - 25x + 18$,
   $d(x) = x + 9$

8. $P(x) = x^3 - 9x^2 + 15x + 25$,
   $d(x) = x - 5$

9. $P(x) = x^4 - 2x^2 + 3$,
   $d(x) = x + 2$

10. $P(x) = x^4 + 6x^3$,
    $d(x) = x - 1$

Use synthetic division to find the quotient and the remainder.

11. $(2x^4 + 7x^3 + x - 12) \div (x + 3)$

12. $(x^3 - 7x^2 + 13x + 3) \div (x - 2)$

13. $(x^3 - 2x^2 - 8) \div (x + 2)$

14. $(x^3 - 3x + 10) \div (x - 2)$

15. $(3x^3 - x^2 + 4x - 10) \div (x + 1)$

16. $(4x^4 - 2x + 5) \div (x + 3)$

17. $(x^5 + x^3 - x) \div (x - 3)$

18. $(x^7 - x^6 + x^5 - x^4 + 2) \div (x + 1)$

19. $(x^4 - 1) \div (x - 1)$

20. $(x^5 + 32) \div (x + 2)$

21. $(2x^4 + 3x^2 - 1) \div \left(x - \frac{1}{2}\right)$

22. $(3x^4 - 2x^2 + 2) \div \left(x - \frac{1}{4}\right)$

Use synthetic division to find the function values. Then check your work using a graphing calculator.

23. $f(x) = x^3 - 6x^2 + 11x - 6$; find $f(1)$, $f(-2)$, and $f(3)$.

24. $f(x) = x^3 + 7x^2 - 12x - 3$; find $f(-3)$, $f(-2)$, and $f(1)$.

25. $f(x) = x^4 - 3x^3 + 2x + 8$; find $f(-1)$, $f(4)$, and $f(-5)$.

26. $f(x) = 2x^4 + x^2 - 10x + 1$; find $f(-10)$, $f(2)$, and $f(3)$.

27. $f(x) = 2x^5 - 3x^4 + 2x^3 - x + 8$; find $f(20)$ and $f(-3)$.

28. $f(x) = x^5 - 10x^4 + 20x^3 - 5x - 100$; find $f(-10)$ and $f(5)$.

29. $f(x) = x^4 - 16$; find $f(2)$, $f(-2)$, $f(3)$, and $f(1 - \sqrt{2})$.

30. $f(x) = x^5 + 32$; find $f(2)$, $f(-2)$, $f(3)$, and $f(2 + 3i)$.

Using synthetic division, determine whether the numbers are zeros of the polynomial function.

31. $-3, 2$; $f(x) = 3x^3 + 5x^2 - 6x + 18$

32. $-4, 2$; $f(x) = 3x^3 + 11x^2 - 2x + 8$

33. $-3, 1$; $h(x) = x^4 + 4x^3 + 2x^2 - 4x - 3$

34. $2, -1$; $g(x) = x^4 - 6x^3 + x^2 + 24x - 20$

35. $i, -2i$; $g(x) = x^3 - 4x^2 + 4x - 16$

36. $\frac{1}{3}, 2$; $h(x) = x^3 - x^2 - \frac{1}{9}x + \frac{1}{9}$

37. $-3, \frac{1}{2}$; $f(x) = x^3 - \frac{7}{2}x^2 + x - \frac{3}{2}$

38. $i, -i, -2$; $f(x) = x^3 + 2x^2 + x + 2$

Factor the polynomial function $f(x)$. Then solve the equation $f(x) = 0$.

39. $f(x) = x^3 + 4x^2 + x - 6$

40. $f(x) = x^3 + 5x^2 - 2x - 24$

41. $f(x) = x^3 - 6x^2 + 3x + 10$

42. $f(x) = x^3 + 2x^2 - 13x + 10$

43. $f(x) = x^3 - x^2 - 14x + 24$

44. $f(x) = x^3 - 3x^2 - 10x + 24$

45. $f(x) = x^4 - 7x^3 + 9x^2 + 27x - 54$

46. $f(x) = x^4 - 4x^3 - 7x^2 + 34x - 24$

47. $f(x) = x^4 - x^3 - 19x^2 + 49x - 30$

48. $f(x) = x^4 + 11x^3 + 41x^2 + 61x + 30$

Sketch the graph of the polynomial function. Follow the procedure outlined on p. 520. Use synthetic division and the remainder theorem to find the zeros.

49. $f(x) = x^4 - x^3 - 7x^2 + x + 6$

50. $f(x) = x^4 + x^3 - 3x^2 - 5x - 2$

51. $f(x) = x^3 - 7x + 6$

52. $f(x) = x^3 - 12x + 16$

53. $f(x) = -x^3 + 3x^2 + 6x - 8$

54. $f(x) = -x^4 + 2x^3 + 3x^2 - 4x - 4$

▶ Skill Maintenance

Solve. Find exact solutions. **[7.4]**

55. $2x^2 + 12 = 5x$

56. $7x^2 + 4x = 3$

Consider the function
$$g(x) = x^2 + 5x - 14$$
in Exercises 57–59.

57. What are the inputs if the output is -14? **[7.4]**

58. What is the output if the input is 3? **[2.2]**

59. Given an output of -20, find the corresponding inputs. **[7.4]**

60. *Movie Ticket Price.*   The average price of a movie ticket has increased linearly over the years, rising from \$2.69 in 1980 to \$8.38 in 2013 (*Source:* Motion Picture Association of America). Using these two data points, find a linear function, $f(x) = mx + b$, that models the data. Let x represent the number of years after 1980. Then use this function to estimate the average price of a movie ticket in 1995 and in 2018. **[2.7]**

61. The sum of the base and the height of a triangle is 30 in. Find the dimensions for which the area is a maximum. **[7.4]**

▶ Synthesis

In Exercises 62 and 63, a graph of a polynomial function is given. On the basis of the graph:

a) *Find as many factors of the polynomial as you can.*

b) *Construct a polynomial function with the zeros shown in the graph.*

c) *Can you find any other polynomial functions with the given zeros?*

d) *Can you find any other polynomial functions with the given zeros and the same graph?*

62.

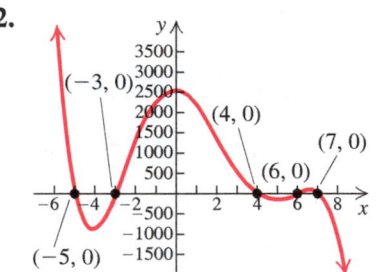

63.

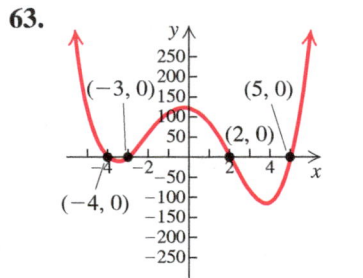

64. For what values of k will the remainder be the same when $x^2 + kx + 4$ is divided by $x - 1$ and by $x + 1$?

65. Find k such that $x + 2$ is a factor of $x^3 - kx^2 + 3x + 7k$.

66. *Beam Deflection.*   A beam rests at two points A and B and has a concentrated load applied to its center, as shown below. Let $y = $ the deflection, in feet, of the beam at a distance of x feet from A. Under certain conditions, this deflection is given by
$$y = \frac{1}{13}x^3 - \frac{1}{14}x.$$

Find the zeros of the polynomial in the interval $[0, 2]$.

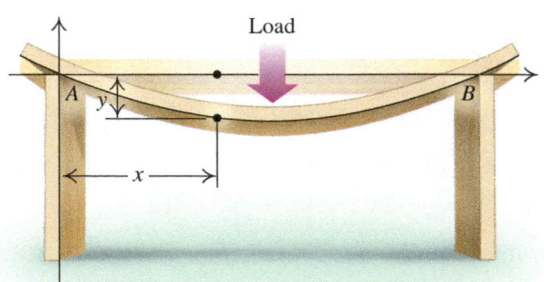

Solve.

67. $\dfrac{2x^2}{x^2 - 1} + \dfrac{4}{x + 3} = \dfrac{12x - 4}{x^3 + 3x^2 - x - 3}$

68. $\dfrac{6x^2}{x^2 + 11} + \dfrac{60}{x^3 - 7x^2 + 11x - 77} = \dfrac{1}{x - 7}$

69. Find a 15th-degree polynomial for which $x - 1$ is a factor. Answers may vary.

Use synthetic division to divide.

70. $(x^4 - y^4) \div (x - y)$

71. $(x^3 + 3ix^2 - 4ix - 2) \div (x + i)$

72. $(x^2 - 4x - 2) \div [x - (3 + 2i)]$

73. $(x^2 - 3x + 7) \div (x - i)$

Mid-Chapter Review

Determine whether the statement is true or false.

1. The y-intercept of the graph of the function $P(x) = 5 - 2x^3$ is $(5, 0)$. **[8.2]**

2. The degree of the polynomial $x - \frac{1}{2}x^4 - 3x^6 + x^5$ is 6. **[8.1]**

3. If $f(x) = (x + 7)(x - 8)$, then $f(8) = 0$. **[8.3]**

4. If $f(12) = 0$, then $x + 12$ is a factor of $f(x)$. **[8.3]**

Find the zeros of the polynomial function and state the multiplicity of each. **[8.1]**

5. $f(x) = (x^2 - 10x + 25)^3$

6. $h(x) = 2x^3 + x^2 - 50x - 25$

7. $g(x) = x^4 - 3x^2 + 2$

8. $f(x) = -6(x - 3)^2(x + 4)$

In Exercises 9–12, match the function with one of the graphs (a)–(d) that follow. **[8.2]**

a)
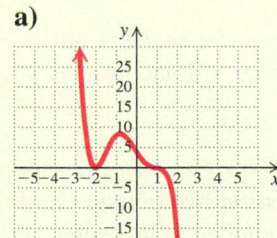

b)
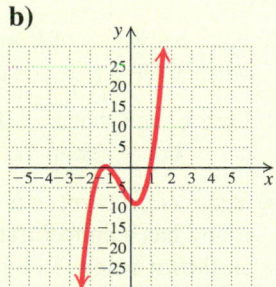

c)
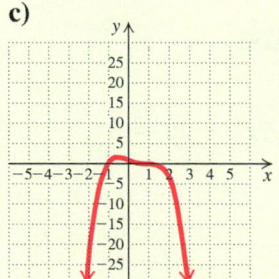

d)
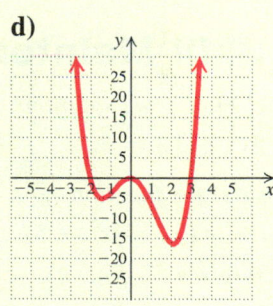

9. $f(x) = x^4 - x^3 - 6x^2$

10. $f(x) = -(x - 1)^3(x + 2)^2$

11. $f(x) = 6x^3 + 8x^2 - 6x - 8$

12. $f(x) = -(x - 1)^3(x + 1)$

Using the intermediate value theorem, determine, if possible, whether the function has at least one real zero between a and b. **[8.2]**

13. $f(x) = x^3 - 2x^2 + 3$; $a = -2, b = 0$

14. $f(x) = x^3 - 2x^2 + 3$; $a = -\frac{1}{2}, b = 1$

15. For the polynomial $P(x) = x^4 - 6x^3 + x - 2$ and the divisor $d(x) = x - 1$, use long division to find the quotient $Q(x)$ and the remainder $R(x)$ when $P(x)$ is divided by $d(x)$. Express $P(x)$ in the form $d(x) \cdot Q(x) + R(x)$. **[8.3]**

Use synthetic division to find the quotient and the remainder. **[8.3]**

16. $(3x^4 - x^3 + 2x^2 - 6x + 6) \div (x - 2)$

17. $(x^5 - 5) \div (x + 1)$

Use synthetic division to find the function values. **[8.3]**

18. For $g(x) = x^3 - 9x^2 + 4x - 10$, find $g(-5)$.

19. For $f(x) = 20x^2 - 40x$, find $f\left(\frac{1}{2}\right)$.

20. For $f(x) = 5x^4 + x^3 - x$, find $f\left(-\sqrt{2}\right)$.

Using synthetic division, determine whether the numbers are zeros of the polynomial function. [8.3]

21. $-3i, 3; \ f(x) = x^3 - 4x^2 + 9x - 36$

22. $-1, 5; \ f(x) = x^6 - 35x^4 + 259x^2 - 225$

Factor the polynomial function $f(x)$. *Then solve the equation* $f(x) = 0$. [8.3]

23. $h(x) = x^3 - 2x^2 - 55x + 56$

24. $g(x) = x^4 - 2x^3 - 13x^2 + 14x + 24$

Collaborative Discussion and Writing

25. How is the range of a polynomial function related to the degree of the polynomial? [8.1]

26. Is it possible for the graph of a polynomial function to have no y-intercept? no x-intercepts? Explain your answer. [8.2]

27. Explain why values of a function must be all positive or all negative between consecutive zeros. [8.2]

28. In synthetic division, why is the degree of the quotient 1 less than that of the dividend? [8.3]

8.4

Theorems about Zeros of Polynomial Functions

▶ Find a polynomial with specified zeros.

▶ For a polynomial function with integer coefficients, find the rational zeros and the other zeros, if possible.

▶ Use Descartes' rule of signs to find information about the number of real zeros of a polynomial function with real coefficients.

We will now allow the coefficients of a polynomial to be complex numbers. In certain cases, we will restrict the coefficients to be real numbers, rational numbers, or integers, as shown in the following examples.

Polynomial	Type of Coefficient
$5x^3 - 3x^2 + (2 + 4i)x + i$	Complex
$5x^3 - 3x^2 + \sqrt{2}x - \pi$	Real
$5x^3 - 3x^2 + \frac{2}{3}x - \frac{7}{4}$	Rational
$5x^3 - 3x^2 + 8x - 11$	Integer

▶ The Fundamental Theorem of Algebra

A linear, or first-degree, polynomial function $f(x) = mx + b$ (where $m \neq 0$) has just one zero, $-b/m$. It can be shown that any quadratic polynomial function $f(x) = ax^2 + bx + c$ with complex numbers for coefficients has at least one, and at most two, complex zeros. The following theorem is a generalization. No proof is given in this text.

THE FUNDAMENTAL THEOREM OF ALGEBRA

Every polynomial function of degree n, with $n \geq 1$, has at least one zero in the set of complex numbers.

Recall that the zeros of a polynomial function $f(x)$ are the solutions of the polynomial equation $f(x) = 0$. We now develop some concepts that can help in finding zeros. First, we consider one of the results of the fundamental theorem of algebra.

Every polynomial function f of degree n, with $n \geq 1$, can be factored into n linear factors (not necessarily unique); that is,

$$f(x) = a_n(x - c_1)(x - c_2) \cdots (x - c_n).$$

▶ Finding Polynomials with Given Zeros

Given several numbers, we can find a polynomial function with those numbers as its zeros.

EXAMPLE 1 Find a polynomial function of degree 3, having the zeros 1, $3i$, and $-3i$.

Solution Such a function has factors $x - 1$, $x - 3i$, and $x - (-3i)$, or $x + 3i$, so we have

$$f(x) = a_n(x - 1)(x - 3i)(x + 3i).$$

The number a_n can be any nonzero number. The simplest polynomial function will be obtained if we let it be 1. If we then multiply the factors, we obtain

$$
\begin{aligned}
f(x) &= (x - 1)(x^2 - 9i^2) &&\text{Multiplying } (x - 3i)(x + 3i)\\
&= (x - 1)(x^2 + 9) &&-9i^2 = -9(-1) = 9\\
&= x^3 - x^2 + 9x - 9.
\end{aligned}
$$

> **Now Try Exercise 3.**

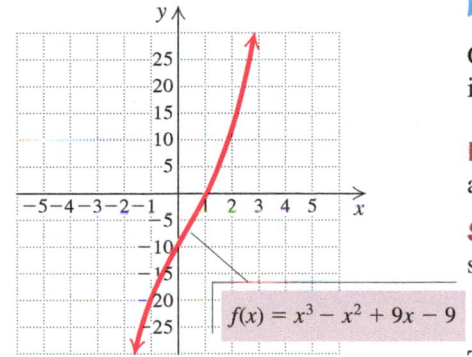

$f(x) = x^3 - x^2 + 9x - 9$

EXAMPLE 2 Find a polynomial function of degree 5 with -1 as a zero of multiplicity 3, 4 as a zero of multiplicity 1, and 0 as a zero of multiplicity 1.

Solution Proceeding as in Example 1, letting $a_n = 1$, we obtain

$$
\begin{aligned}
f(x) &= [x - (-1)]^3(x - 4)(x - 0)\\
&= (x + 1)^3(x - 4)x\\
&= (x^3 + 3x^2 + 3x + 1)(x^2 - 4x)\\
&= x^5 - x^4 - 9x^3 - 11x^2 - 4x.
\end{aligned}
$$

> **Now Try Exercise 13.**

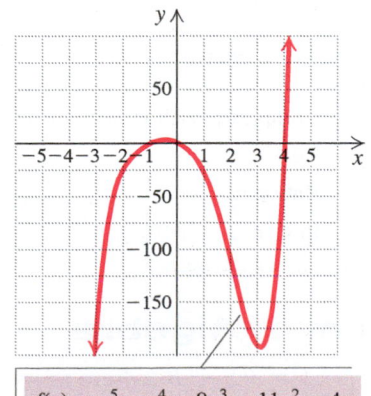

$f(x) = x^5 - x^4 - 9x^3 - 11x^2 - 4x$

▶ Zeros of Polynomial Functions with Real Coefficients

Consider the quadratic equation $x^2 - 2x + 2 = 0$, with real coefficients. Its solutions are $1 + i$ and $1 - i$. Note that they are complex conjugates. This generalizes to any polynomial equation with real coefficients.

NONREAL ZEROS: $a + bi$ and $a - bi$, $b \neq 0$

If a complex number $a + bi$, $b \neq 0$, is a zero of a polynomial function $f(x)$ with *real* coefficients, then its conjugate, $a - bi$, is also a zero. For example, if $2 + 7i$ is a zero of a polynomial function $f(x)$ with real coefficients, then its conjugate, $2 - 7i$, is also a zero. (Nonreal zeros occur in conjugate pairs.)

In order for the preceding to be true, it is essential that the coefficients be *real* numbers.

▶ Rational Coefficients

When a polynomial function has *rational* numbers for coefficients, certain irrational zeros also occur in pairs, as described in the following theorem.

IRRATIONAL ZEROS: $a + c\sqrt{b}$ and $a - c\sqrt{b}$, b IS NOT A PERFECT SQUARE

If $a + c\sqrt{b}$, where a, b, and c are rational and b is not a perfect square, is a zero of a polynomial function $f(x)$ with *rational* coefficients, then its conjugate, $a - c\sqrt{b}$, is also a zero. For example, if $-3 + 5\sqrt{2}$ is a zero of a polynomial function $f(x)$ with rational coefficients, then its conjugate, $-3 - 5\sqrt{2}$, is also a zero. (Irrational zeros occur in conjugate pairs.)

EXAMPLE 3   Suppose that a polynomial function of degree 6 with rational coefficients has

$$-2 + 5i, \qquad -2i, \quad \text{and} \quad 1 - \sqrt{3}$$

as three of its zeros. Find the other zeros.

Solution   Since the coefficients are rational (and thus real), the other zeros are the conjugates of the given zeros:

$$-2 - 5i, \qquad 2i, \quad \text{and} \quad 1 + \sqrt{3}.$$

There are no other zeros because a polynomial function of degree 6 can have at most 6 zeros.   ▶ **Now Try Exercise 19.**

EXAMPLE 4   Find a polynomial function of lowest degree with rational coefficients that has $-\sqrt{3}$ and $1 + i$ as two of its zeros.

Solution   The function must also have the zeros $\sqrt{3}$ and $1 - i$. Because we want to find the polynomial function of lowest degree with the known zeros, we will not include additional zeros; that is, we will write a polynomial function of degree 4. Thus if we let $a_n = 1$, the polynomial function is

$$
\begin{aligned}
f(x) &= \left[x - \left(-\sqrt{3}\right)\right]\left[x - \sqrt{3}\right]\left[x - (1 + i)\right]\left[x - (1 - i)\right] \\
&= \left(x + \sqrt{3}\right)\left(x - \sqrt{3}\right)\left[(x - 1) - i\right]\left[(x - 1) + i\right] \\
&= (x^2 - 3)\left[(x - 1)^2 - i^2\right] \\
&= (x^2 - 3)\left[x^2 - 2x + 1 + 1\right] \\
&= (x^2 - 3)(x^2 - 2x + 2) \\
&= x^4 - 2x^3 - x^2 + 6x - 6.
\end{aligned}
$$

▶ **Now Try Exercise 39.**

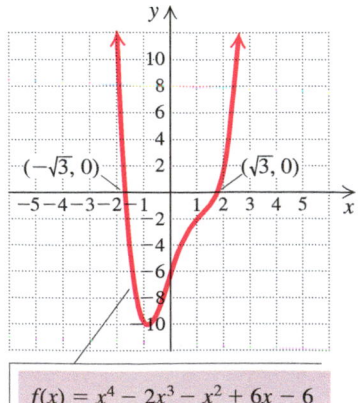

$(-\sqrt{3}, 0)$ $(\sqrt{3}, 0)$

$f(x) = x^4 - 2x^3 - x^2 + 6x - 6$

▶ Integer Coefficients and the Rational Zeros Theorem

It is not always easy to find the zeros of a polynomial function. However, if a polynomial function has integer coefficients, there is a procedure that will yield all the rational zeros.

THE RATIONAL ZEROS THEOREM

Let

$$P(x) = a_n x^n + a_{n-1} x^{n-1} + \cdots + a_1 x + a_0,$$

where all the coefficients are integers. Consider a rational number denoted by p/q, where p and q are relatively prime (having no common factor besides -1 and 1). If p/q is a zero of $P(x)$, then p is a factor of a_0 and q is a factor of a_n.

Technology Connection

We can narrow the list of possibilities for the rational zeros of a function by examining its graph.

$$y = 3x^4 - 11x^3 + 10x - 4$$

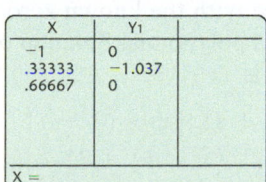

From the graph of the function in Example 5, we see that of the possibilities for rational zeros, only the numbers

$$-1, \ \tfrac{1}{3}, \ \text{and} \ \tfrac{2}{3}$$

might actually be rational zeros. We can check these with the TABLE feature. Note that the graphing calculator converts $\frac{1}{3}$ and $\frac{2}{3}$ to decimal notation.

X	Y₁
−1	0
.33333	−1.037
.66667	0

X =

Since $f(-1) = 0$ and $f\left(\frac{2}{3}\right) = 0$, -1 and $\frac{2}{3}$ are zeros. Since $f\left(\frac{1}{3}\right) \neq 0$, $\frac{1}{3}$ is not a zero.

EXAMPLE 5   Given $f(x) = 3x^4 - 11x^3 + 10x - 4$:

a) Find the rational zeros and then the other zeros; that is, solve $f(x) = 0$.

b) Factor $f(x)$ into linear factors.

Solution

a) Because the degree of $f(x)$ is 4, there are at most 4 distinct zeros. The rational zeros theorem says that if a rational number p/q is a zero of $f(x)$, then p must be a factor of -4 and q must be a factor of 3. Thus the possibilities for p/q are

$$\frac{\textit{Possibilities for } p}{\textit{Possibilities for } q} : \quad \frac{\pm 1, \ \pm 2, \ \pm 4}{\pm 1, \ \pm 3};$$

$$\textit{Possibilities for } p/q: \quad 1, -1, 2, -2, 4, -4, \tfrac{1}{3}, -\tfrac{1}{3}, \tfrac{2}{3}, -\tfrac{2}{3}, \tfrac{4}{3}, -\tfrac{4}{3}.$$

To find which are zeros, we could use substitution, but synthetic division is usually more efficient. It is easier to consider the integers first. Then we consider the fractions, if the integers do not produce all the zeros.

We try 1:

$$
\begin{array}{r|rrrrr}
1 & 3 & -11 & 0 & 10 & -4 \\
  &   & 3 & -8 & -8 & 2 \\
\hline
  & 3 & -8 & -8 & 2 & -2.
\end{array}
$$

Since $f(1) = -2$, 1 is not a zero.

We try -1:

$$
\begin{array}{r|rrrrr}
-1 & 3 & -11 & 0 & 10 & -4 \\
   &   & -3 & 14 & -14 & 4 \\
\hline
   & 3 & -14 & 14 & -4 & 0.
\end{array}
$$

We have $f(-1) = 0$, so -1 is a zero. Thus, $x + 1$ is a factor of $f(x)$. Using the results of the synthetic division, we can express $f(x)$ as

$$f(x) = (x + 1)(3x^3 - 14x^2 + 14x - 4).$$

We now consider the factor $3x^3 - 14x^2 + 14x - 4$ and check the other possible zeros. We use synthetic division again, to determine whether -1 is a zero of multiplicity 2 of $f(x)$:

$$
\begin{array}{r|rrrr}
-1 & 3 & -14 & 14 & -4 \\
 & & -3 & 17 & -31 \\
\hline
 & 3 & -17 & 31 & -35.
\end{array}
$$

We see that -1 is not a double zero. We leave it to the student to verify that 2, -2, 4, and -4 are not zeros. There are no other zeros that are integers, so we start checking the fractions. Let's try $\frac{2}{3}$.

$$
\begin{array}{r|rrrr}
2/3 & 3 & -14 & 14 & -4 \\
 & & 2 & -8 & 4 \\
\hline
 & 3 & -12 & 6 & 0
\end{array}
$$

Since the remainder is 0, we know that $x - \frac{2}{3}$ is a factor of $3x^3 - 14x^2 + 14x - 4$ and is also a factor of $f(x)$. Thus, $\frac{2}{3}$ is a zero of $f(x)$.

Using the results of the synthetic division, we can factor further:

$$f(x) = (x + 1)\left(x - \tfrac{2}{3}\right)(3x^2 - 12x + 6) \qquad \text{\color{red}\textbf{Using the results of the last synthetic division}}$$

$$= (x + 1)\left(x - \tfrac{2}{3}\right) \cdot 3 \cdot (x^2 - 4x + 2). \qquad \text{\color{red}\textbf{Removing a factor of 3}}$$

The quadratic formula can be used to find the values of x for which $x^2 - 4x + 2 = 0$. Those values are also zeros of $f(x)$:

$$x = \frac{-b \pm \sqrt{b^2 - 4ac}}{2a}$$

$$= \frac{-(-4) \pm \sqrt{(-4)^2 - 4 \cdot 1 \cdot 2}}{2 \cdot 1} \qquad \text{\color{red}$a = 1, b = -4, \text{ and } c = 2$}$$

$$= \frac{4 \pm \sqrt{8}}{2} = \frac{4 \pm 2\sqrt{2}}{2} = \frac{2(2 \pm \sqrt{2})}{2} = 2 \pm \sqrt{2}.$$

The rational zeros are -1 and $\frac{2}{3}$. The other zeros are $2 \pm \sqrt{2}$.

b) The complete factorization of $f(x)$ is

$$f(x) = 3(x + 1)\left(x - \tfrac{2}{3}\right)\left[x - \left(2 - \sqrt{2}\right)\right]\left[x - \left(2 + \sqrt{2}\right)\right], \quad \text{or}$$
$$(x + 1)(3x - 2)\left[x - \left(2 - \sqrt{2}\right)\right]\left[x - \left(2 + \sqrt{2}\right)\right].$$

\color{red}Replacing $3\left(x - \tfrac{2}{3}\right)$ with $(3x - 2)$

\color{red}Now Try Exercise 55.

EXAMPLE 6   Given $f(x) = 2x^5 - x^4 - 4x^3 + 2x^2 - 30x + 15$:

a) Find the rational zeros and then the other zeros; that is, solve $f(x) = 0$.

b) Factor $f(x)$ into linear factors.

Solution

a) Because the degree of $f(x)$ is 5, there are at most 5 distinct zeros. According to the rational zeros theorem, any rational zero of f must be of the form p/q, where p is a factor of 15 and q is a factor of 2. The possibilities are

$$\frac{\textit{Possibilities for } p}{\textit{Possibilities for } q}: \quad \frac{\pm 1, \pm 3, \pm 5, \pm 15}{\pm 1, \pm 2};$$

$$\textit{Possibilities for } p/q: \quad 1, -1, 3, -3, 5, -5, 15, -15, \tfrac{1}{2}, -\tfrac{1}{2}, \tfrac{3}{2}, -\tfrac{3}{2},$$
$$\tfrac{5}{2}, -\tfrac{5}{2}, \tfrac{15}{2}, -\tfrac{15}{2}.$$

We use synthetic division to check each of the possibilities. Let's try 1 and −1.

$$
\begin{array}{r|rrrrrr}
1 & 2 & -1 & -4 & 2 & -30 & 15 \\
  &   & 2 & 1 & -3 & -1 & -31 \\
\hline
  & 2 & 1 & -3 & -1 & -31 & \;-16
\end{array}
$$

$$
\begin{array}{r|rrrrrr}
-1 & 2 & -1 & -4 & 2 & -30 & 15 \\
   &   & -2 & 3 & 1 & -3 & 33 \\
\hline
   & 2 & -3 & -1 & 3 & -33 & \;48
\end{array}
$$

Since $f(1) = -16$ and $f(-1) = 48$, neither 1 nor −1 is a zero of the function. We leave it to the student to verify that the other integer possibilities are not zeros. We now check $\tfrac{1}{2}$.

$$
\begin{array}{r|rrrrrr}
1/2 & 2 & -1 & -4 & 2 & -30 & 15 \\
    &   & 1 & 0 & -2 & 0 & -15 \\
\hline
    & 2 & 0 & -4 & 0 & -30 & \;0
\end{array}
$$

This means that $x - \tfrac{1}{2}$ is a factor of $f(x)$. We write the factorization and try to factor further:

$$
\begin{aligned}
f(x) &= \left(x - \tfrac{1}{2}\right)\left(2x^4 - 4x^2 - 30\right) \\
&= \left(x - \tfrac{1}{2}\right) \cdot 2 \cdot (x^4 - 2x^2 - 15) \qquad \text{\color{red}Factoring out the 2} \\
&= \left(x - \tfrac{1}{2}\right) \cdot 2 \cdot (x^2 - 5)(x^2 + 3). \qquad \text{\color{red}Factoring the trinomial}
\end{aligned}
$$

We now solve the equation $f(x) = 0$ to determine the zeros. We use the principle of zero products:

$$
\left(x - \tfrac{1}{2}\right) \cdot 2 \cdot (x^2 - 5)(x^2 + 3) = 0
$$

$$
\begin{array}{ccccc}
x - \tfrac{1}{2} = 0 & \text{or} & x^2 - 5 = 0 & \text{or} & x^2 + 3 = 0 \\
x = \tfrac{1}{2} & \text{or} & x^2 = 5 & \text{or} & x^2 = -3 \\
x = \tfrac{1}{2} & \text{or} & x = \pm\sqrt{5} & \text{or} & x = \pm\sqrt{3}\,i.
\end{array}
$$

There is only one rational zero, $\tfrac{1}{2}$. The other zeros are $\pm\sqrt{5}$ and $\pm\sqrt{3}\,i$.

b) The factorization into linear factors is

$$
f(x) = 2\left(x - \tfrac{1}{2}\right)\left(x + \sqrt{5}\right)\left(x - \sqrt{5}\right)\left(x + \sqrt{3}\,i\right)\left(x - \sqrt{3}\,i\right), \quad \text{or}
$$

$$
(2x - 1)\left(x + \sqrt{5}\right)\left(x - \sqrt{5}\right)\left(x + \sqrt{3}\,i\right)\left(x - \sqrt{3}\,i\right).
$$

Replacing $2\left(x - \tfrac{1}{2}\right)$ **with** $(2x - 1)$

Now Try Exercise 61.

▶ Descartes' Rule of Signs

The development of a rule that helps determine the number of positive real zeros and the number of negative real zeros of a polynomial function is credited to the French mathematician René Descartes. To use the rule, we must have the polynomial arranged in descending order or ascending order, with no zero terms written in and the constant term not 0. Then we determine the number of *variations of sign*, that is, the number of times, in reading through the polynomial, that successive coefficients are of different signs.

EXAMPLE 7 Determine the number of variations of sign in the polynomial function $P(x) = 2x^5 - 3x^2 + x + 4$.

Solution This polynomial is written in descending order. We have

$$P(x) = \underbrace{2x^5 - 3x^2 + x + 4}$$

From positive to negative; a variation

Both positive; no variation

From negative to positive; a variation

The number of variations of sign is 2.

Note the following:

$$P(-x) = 2(-x)^5 - 3(-x)^2 + (-x) + 4$$
$$= -2x^5 - 3x^2 - x + 4.$$

We see that the number of variations of sign in $P(-x)$ is 1. It occurs as we go from $-x$ to 4.

We now state Descartes' rule, without proof.

DESCARTES' RULE OF SIGNS

Let $P(x)$, written in descending order or ascending order, be a polynomial function with real coefficients and a nonzero constant term. The number of positive real zeros of $P(x)$ is either:

1. The same as the number of variations of sign in $P(x)$, or
2. Less than the number of variations of sign in $P(x)$ by a positive even integer.

The number of negative real zeros of $P(x)$ is either:

3. The same as the number of variations of sign in $P(-x)$, or
4. Less than the number of variations of sign in $P(-x)$ by a positive even integer.

A zero of multiplicity m must be counted m times.

In each of Examples 8–10, what does Descartes' rule of signs tell you about the number of positive real zeros and the number of negative real zeros?

EXAMPLE 8 $P(x) = 2x^5 - 5x^2 - 3x + 6$

Solution The number of variations of sign in $P(x)$ is 2. Therefore, the number of positive real zeros is either 2 or less than 2 by 2, 4, 6, and so on. Thus the number of positive real zeros is either 2 or 0, since a negative number of zeros has no meaning.

$$P(-x) = -2x^5 - 5x^2 + 3x + 6$$

The number of variations of sign in $P(-x)$ is 1. Thus there is exactly 1 negative real zero. Since nonreal, complex conjugates occur in pairs, we also know the possible ways in which nonreal zeros might occur. The table shown at left summarizes all the possibilities for real zeros and nonreal zeros of $P(x)$.

Total Number of Zeros	5	
Positive Real	2	0
Negative Real	1	1
Nonreal	2	4

Now Try Exercise 93.

EXAMPLE 9 $P(x) = 5x^4 - 3x^3 + 7x^2 - 12x + 4$

Solution There are 4 variations of sign. Thus the number of positive real zeros is either

$$4 \quad \text{or} \quad 4 - 2 \quad \text{or} \quad 4 - 4.$$

That is, the number of positive real zeros is 4, 2, or 0.

$$P(-x) = 5x^4 + 3x^3 + 7x^2 + 12x + 4$$

There are 0 changes in sign, so there are no negative real zeros. The table at left summarizes all the possibilities for real zeros and nonreal zeros of $P(x)$.

▶ **Now Try Exercise 81.**

Total Number of Zeros	4		
Positive Real	4	2	0
Negative Real	0	0	0
Nonreal	0	2	4

EXAMPLE 10 $P(x) = 6x^6 - 2x^2 - 5x$

Solution As written, the polynomial does not satisfy the conditions of Descartes' rule of signs because the constant term is 0. But because x is a factor of every term, we know that the polynomial has 0 as a zero. We can then factor as follows:

$$P(x) = x(6x^5 - 2x - 5).$$

Now we analyze $Q(x) = 6x^5 - 2x - 5$ and $Q(-x) = -6x^5 + 2x - 5$. The number of variations of sign in $Q(x)$ is 1. Therefore, there is exactly 1 positive real zero. The number of variations of sign in $Q(-x)$ is 2. Thus the number of negative real zeros is 2 or 0. The same results apply to $P(x)$. Since nonreal, complex conjugates occur in pairs, we know the possible ways in which nonreal zeros might occur. The table at left summarizes all the possibilities for real zeros and nonreal zeros of $P(x)$.

▶ **Now Try Exercise 95.**

Total Number of Zeros	6	
0 as a Zero	1	1
Positive Real	1	1
Negative Real	2	0
Nonreal	2	4

8.4 Exercise Set

Find a polynomial function of degree 3 with the given numbers as zeros.

1. $-2, 3, 5$

2. $-1, 0, 4$

3. $-3, 2i, -2i$

4. $2, i, -i$

5. $\sqrt{2}, -\sqrt{2}, 3$

6. $-5, \sqrt{3}, -\sqrt{3}$

7. $1 - \sqrt{3}, 1 + \sqrt{3}, -2$

8. $-4, 1 - \sqrt{5}, 1 + \sqrt{5}$

9. $1 + 6i, 1 - 6i, -4$

10. $1 + 4i, 1 - 4i, -1$

11. $-\frac{1}{3}, 0, 2$

12. $-3, 0, \frac{1}{2}$

13. Find a polynomial function of degree 5 with -1 as a zero of multiplicity 3, 0 as a zero of multiplicity 1, and 1 as a zero of multiplicity 1.

14. Find a polynomial function of degree 4 with -2 as a zero of multiplicity 1, 3 as a zero of multiplicity 2, and -1 as a zero of multiplicity 1.

15. Find a polynomial function of degree 4 with $a_4 = 1$ and with -1 as a zero of multiplicity 3 and 0 as a zero of multiplicity 1.

16. Find a polynomial function of degree 5 with $a_5 = 1$ and with $-\frac{1}{2}$ as a zero of multiplicity 2, 0 as a zero of multiplicity 1, and 1 as a zero of multiplicity 2.

Suppose that a polynomial function of degree 4 with rational coefficients has the given numbers as zeros. Find the other zero(s).

17. $-1, \sqrt{3}, \frac{11}{3}$

18. $-\sqrt{2}, -1, \frac{4}{5}$

19. $-i, 2 - \sqrt{5}$

20. $i, -3 + \sqrt{3}$

21. $3i, 0, -5$

22. $3, 0, -2i$

23. $-4 - 3i, 2 - \sqrt{3}$

24. $6 - 5i, -1 + \sqrt{7}$

Suppose that a polynomial function of degree 5 with rational coefficients has the given numbers as zeros. Find the other zero(s).

25. $-\frac{1}{2}, \sqrt{5}, -4i$

26. $\frac{3}{4}, -\sqrt{3}, 2i$

27. $-5, 0, 2 - i, 4$

28. $-2, 3, 4, 1 - i$

29. $6, -3 + 4i, 4 - \sqrt{5}$

30. $-3 - 3i, 2 + \sqrt{13}, 6$

31. $-\frac{3}{4}, \frac{3}{4}, 0, 4 - i$

32. $-0.6, 0, 0.6, -3 + \sqrt{2}$

Find a polynomial function of lowest degree with rational coefficients that has the given numbers as some of its zeros.

33. $1 + i, 2$

34. $2 - i, -1$

35. $4i$

36. $-5i$

37. $-4i, 5$

38. $3, -i$

39. $1 - i, -\sqrt{5}$

40. $2 - \sqrt{3}, 1 + i$

41. $\sqrt{5}, -3i$

42. $-\sqrt{2}, 4i$

Given that the polynomial function has the given zero, find the other zeros.

43. $f(x) = x^3 + 5x^2 - 2x - 10; \ -5$

44. $f(x) = x^3 - x^2 + x - 1; \ 1$

45. $f(x) = x^4 - 5x^3 + 7x^2 - 5x + 6; \ -i$

46. $f(x) = x^4 - 16; \ 2i$

47. $f(x) = x^3 - 6x^2 + 13x - 20; \ 4$

48. $f(x) = x^3 - 8; \ 2$

List all possible rational zeros of the function.

49. $f(x) = x^5 - 3x^2 + 1$

50. $f(x) = x^7 + 37x^5 - 6x^2 + 12$

51. $f(x) = 2x^4 - 3x^3 - x + 8$

52. $f(x) = 3x^3 - x^2 + 6x - 9$

53. $f(x) = 15x^6 + 47x^2 + 2$

54. $f(x) = 10x^{25} + 3x^{17} - 35x + 6$

For each polynomial function, (a) find the rational zeros and then the other zeros; that is, solve $f(x) = 0$; and (b) factor $f(x)$ into linear factors.

55. $f(x) = x^3 + 3x^2 - 2x - 6$

56. $f(x) = x^3 - x^2 - 3x + 3$

57. $f(x) = 3x^3 - x^2 - 15x + 5$

58. $f(x) = 4x^3 - 4x^2 - 3x + 3$

59. $f(x) = x^3 - 3x + 2$

60. $f(x) = x^3 - 2x + 4$

61. $f(x) = 2x^3 + 3x^2 + 18x + 27$

62. $f(x) = 2x^3 + 7x^2 + 2x - 8$

63. $f(x) = 5x^4 - 4x^3 + 19x^2 - 16x - 4$

64. $f(x) = 3x^4 - 4x^3 + x^2 + 6x - 2$

65. $f(x) = x^4 - 3x^3 - 20x^2 - 24x - 8$

66. $f(x) = x^4 + 5x^3 - 27x^2 + 31x - 10$

67. $f(x) = x^3 - 4x^2 + 2x + 4$

68. $f(x) = x^3 - 8x^2 + 17x - 4$

69. $f(x) = x^3 + 8$

70. $f(x) = x^3 - 8$

71. $f(x) = \frac{1}{3}x^3 - \frac{1}{2}x^2 - \frac{1}{6}x + \frac{1}{6}$

72. $f(x) = \frac{2}{3}x^3 - \frac{1}{2}x^2 + \frac{2}{3}x - \frac{1}{2}$

Find only the rational zeros of the function.

73. $f(x) = x^4 + 2x^3 - 5x^2 - 4x + 6$

74. $f(x) = x^4 - 3x^3 - 9x^2 - 3x - 10$

75. $f(x) = x^3 - x^2 - 4x + 3$

76. $f(x) = 2x^3 + 3x^2 + 2x + 3$

77. $f(x) = x^4 + 2x^3 + 2x^2 - 4x - 8$

78. $f(x) = x^4 + 6x^3 + 17x^2 + 36x + 66$

79. $f(x) = x^5 - 5x^4 + 5x^3 + 15x^2 - 36x + 20$

80. $f(x) = x^5 - 3x^4 - 3x^3 + 9x^2 - 4x + 12$

What does Descartes' rule of signs tell you about the number of positive real zeros and the number of negative real zeros of the function?

81. $f(x) = 3x^5 - 2x^2 + x - 1$

82. $g(x) = 5x^6 - 3x^3 + x^2 - x$

83. $h(x) = 6x^7 + 2x^2 + 5x + 4$

84. $P(x) = -3x^5 - 7x^3 - 4x - 5$

85. $F(p) = 3p^{18} + 2p^4 - 5p^2 + p + 3$

86. $H(t) = 5t^{12} - 7t^4 + 3t^2 + t + 1$

87. $C(x) = 7x^6 + 3x^4 - x - 10$

88. $g(z) = -z^{10} + 8z^7 + z^3 + 6z - 1$

89. $h(t) = -4t^5 - t^3 + 2t^2 + 1$

90. $P(x) = x^6 + 2x^4 - 9x^3 - 4$

91. $f(y) = y^4 + 13y^3 - y + 5$

92. $Q(x) = x^4 - 2x^2 + 12x - 8$

93. $r(x) = x^4 - 6x^2 + 20x - 24$

94. $f(x) = x^5 - 2x^3 - 8x$

95. $R(x) = 3x^5 - 5x^3 - 4x$

96. $f(x) = x^4 - 9x^2 - 6x + 4$

Sketch the graph of the polynomial function. Follow the procedure outlined on p. 520. Use the rational zeros theorem when finding the zeros.

97. $f(x) = 4x^3 + x^2 - 8x - 2$

98. $f(x) = 3x^3 - 4x^2 - 5x + 2$

99. $f(x) = 2x^4 - 3x^3 - 2x^2 + 3x$

100. $f(x) = 4x^4 - 37x^2 + 9$

▶ Skill Maintenance

*For Exercises 101 and 102, complete the square to **(a)** find the vertex; **(b)** find the axis of symmetry; and **(c)** determine whether there is a maximum or a minimum function value and find that value.*

101. $f(x) = x^2 - 8x + 10$ **[7.5]**

102. $f(x) = 3x^2 - 6x - 1$ **[7.5]**

Find the zeros of the function. **[7.4]**

103. $f(x) = 3x^2 + 12x$

104. $g(x) = x^2 - 8x - 33$

Determine the leading term, the leading coefficient, and the degree of the polynomial. Then describe the end behavior of the function's graph and classify the polynomial function as constant, linear, quadratic, cubic, or quartic. **[4.1]**

105. $g(x) = -x^3 - 2x^2$

106. $f(x) = -x^2 - 3x + 6$

107. $f(x) = -\frac{4}{9}$

108. $h(x) = x - 2$

109. $g(x) = x^4 - 2x^3 + x^2 - x + 2$

110. $h(x) = x^3 + \frac{1}{2}x^2 - 4x - 3$

▶ Synthesis

111. Consider $f(x) = 2x^3 - 5x^2 - 4x + 3$. Find the solutions of each equation.

    a) $f(x) = 0$
    b) $f(x - 1) = 0$
    c) $f(x + 2) = 0$
    d) $f(2x) = 0$

112. Use the rational zeros theorem and the equation $x^4 - 12 = 0$ to show that $\sqrt[4]{12}$ is irrational.

Find the rational zeros of the function.

113. $P(x) = 2x^5 - 33x^4 - 84x^3 + 2203x^2 - 3348x - 10{,}080$

114. $P(x) = x^6 - 6x^5 - 72x^4 - 81x^2 + 486x + 5832$

8.5   Rational Functions

▶ For a rational function, find the domain and graph the function, identifying all of the asymptotes.

▶ Solve applied problems involving rational functions.

Now we turn our attention to functions that represent the quotient of two polynomials. Whereas the sum, the difference, or the product of two polynomials is a polynomial, in general the quotient of two polynomials is *not* itself a polynomial.

A *rational number* can be expressed as the quotient of two integers, p/q, where $q \neq 0$. A *rational function* is formed by the quotient of two polynomials, $p(x)/q(x)$, where $q(x) \neq 0$. Here are some examples of rational functions and their graphs.

$$f(x) = \frac{1}{x}$$

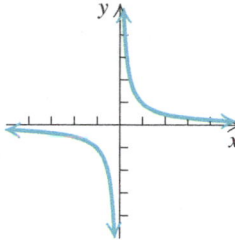

$$f(x) = \frac{1}{x^2}$$

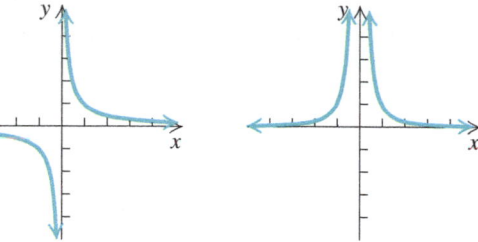

$$f(x) = \frac{x - 3}{x^2 + x - 2}$$

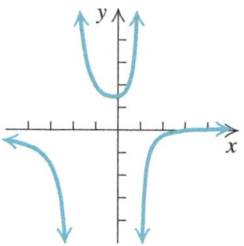

$$f(x) = \frac{2x + 5}{2x - 6}$$

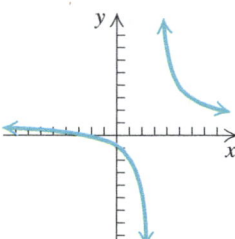

$$f(x) = \frac{x^2 + 2x - 3}{x^2 - x - 2}$$

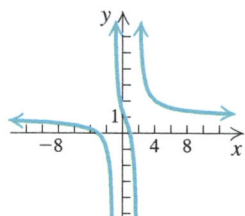

$$f(x) = \frac{-x^2}{x + 1}$$

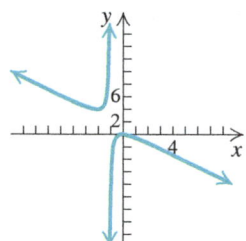

RATIONAL FUNCTION

A **rational function** is a function f that is a quotient of two polynomials. That is,

$$f(x) = \frac{p(x)}{q(x)},$$

where $p(x)$ and $q(x)$ are polynomials and where $q(x)$ is not the zero polynomial. The domain of f consists of all inputs x for which $q(x) \neq 0$.

▶ **The Domain of a Rational Function**

EXAMPLE 1 Consider

$$f(x) = \frac{1}{x - 3}.$$

Find the domain and graph f.

DOMAINS OF FUNCTIONS

REVIEW SECTION 2.3

Solution When the denominator $x - 3$ is 0, we have $x = 3$, so the only input that results in a denominator of 0 is 3. Thus the domain is

$$\{x \mid x \neq 3\}, \text{ or } (-\infty, 3) \cup (3, \infty).$$

The graph of this function is the graph of $y = 1/x$ translated right 3 units.

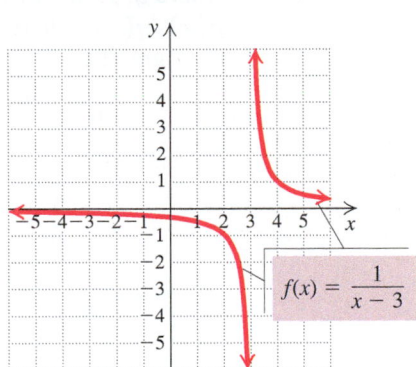

EXAMPLE 2   Determine the domain of each of the functions illustrated at the beginning of this section.

Solution   The domain of each rational function will be the set of all real numbers except those values that make the denominator 0. To determine those exceptions, we set the denominator equal to 0 and solve for x.

FUNCTION	DOMAIN
$f(x) = \dfrac{1}{x}$	$\{x \mid x \neq 0\}$, or $(-\infty, 0) \cup (0, \infty)$
$f(x) = \dfrac{1}{x^2}$	$\{x \mid x \neq 0\}$, or $(-\infty, 0) \cup (0, \infty)$
$f(x) = \dfrac{x-3}{x^2 + x - 2} = \dfrac{x-3}{(x+2)(x-1)}$	$\{x \mid x \neq -2 \text{ and } x \neq 1\}$, or $(-\infty, -2) \cup (-2, 1) \cup (1, \infty)$
$f(x) = \dfrac{2x+5}{2x-6} = \dfrac{2x+5}{2(x-3)}$	$\{x \mid x \neq 3\}$, or $(-\infty, 3) \cup (3, \infty)$
$f(x) = \dfrac{x^2 + 2x - 3}{x^2 - x - 2} = \dfrac{x^2 + 2x - 3}{(x+1)(x-2)}$	$\{x \mid x \neq -1 \text{ and } x \neq 2\}$, or $(-\infty, -1) \cup (-1, 2) \cup (2, \infty)$
$f(x) = \dfrac{-x^2}{x+1}$	$\{x \mid x \neq -1\}$, or $(-\infty, -1) \cup (-1, \infty)$

As a partial check of the domains, we can observe the discontinuities (breaks) in the graphs of these functions. (See p. 547.)

▶ **Asymptotes**

Vertical Asymptotes

Look at the graph of $f(x) = 1/(x - 3)$, shown at left. (Also see Example 1.) Let's explore what happens as x-values get closer and closer to 3 from the left. We then explore what happens as x-values get closer and closer to 3 from the right.

From the left:

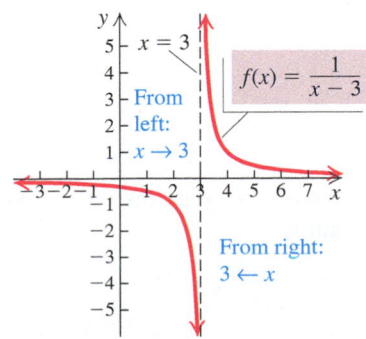

Vertical asymptote: $x = 3$

x	2	$2\frac{1}{2}$	$2\frac{99}{100}$	$2\frac{9999}{10,000}$	$2\frac{999,999}{1,000,000}$	$\longrightarrow 3$
$f(x)$	-1	-2	-100	$-10,000$	$-1,000,000$	$\longrightarrow -\infty$

From the right:

x	4	$3\frac{1}{2}$	$3\frac{1}{100}$	$3\frac{1}{10,000}$	$3\frac{1}{1,000,000}$	→ 3
$f(x)$	1	2	100	10,000	1,000,000	→ ∞

We see that as x-values get closer and closer to 3 from the left, the function values (y-values) decrease without bound (that is, they approach negative infinity, $-\infty$). Similarly, as the x-values approach 3 from the right, the function values increase without bound (that is, they approach positive infinity, ∞). We write this as

$$f(x) \to -\infty \text{ as } x \to 3^- \quad \text{and} \quad f(x) \to \infty \text{ as } x \to 3^+.$$

We read "$f(x) \to -\infty$ as $x \to 3^-$" as "$f(x)$ decreases without bound as x approaches 3 from the left." We read "$f(x) \to \infty$ as $x \to 3^+$" as "$f(x)$ increases without bound as x approaches 3 from the right." The notation $x \to 3$ means that x gets as close to 3 as possible without being equal to 3. The vertical line $x = 3$ is said to be a *vertical asymptote* for this curve.

In general, the line $x = a$ is a **vertical asymptote** for the graph of f if any of the following is true:

$$f(x) \to \infty \text{ as } x \to a^-, \quad \text{or} \quad f(x) \to -\infty \text{ as } x \to a^-, \quad \text{or}$$
$$f(x) \to \infty \text{ as } x \to a^+, \quad \text{or} \quad f(x) \to -\infty \text{ as } x \to a^+.$$

The following figures show the four ways in which a vertical asymptote can occur.

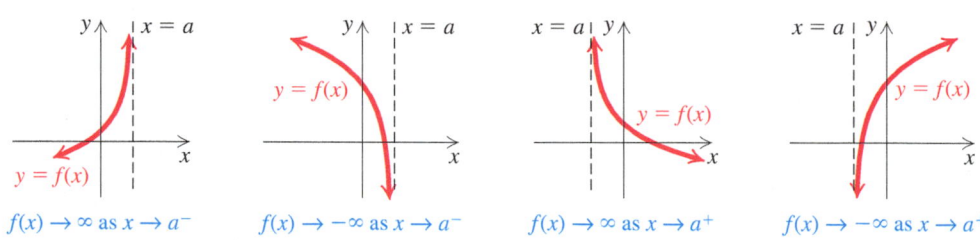

$f(x) \to \infty$ as $x \to a^-$ $f(x) \to -\infty$ as $x \to a^-$ $f(x) \to \infty$ as $x \to a^+$ $f(x) \to -\infty$ as $x \to a^+$

The vertical asymptotes of a rational function $f(x) = p(x)/q(x)$ are found by determining the zeros of $q(x)$ that are not also zeros of $p(x)$. If $p(x)$ and $q(x)$ are polynomials with no common factors other than constants, we need determine only the zeros of the denominator $q(x)$.

DETERMINING VERTICAL ASYMPTOTES

For a rational function $f(x) = p(x)/q(x)$, where $p(x)$ and $q(x)$ are polynomials with *no common factors* other than constants, if a is a zero of the denominator, then the line $x = a$ is a vertical asymptote for the graph of the function.

EXAMPLE 3   Determine the vertical asymptotes for the graph of each of the following functions.

a) $f(x) = \dfrac{2x - 11}{x^2 + 2x - 8}$ **b)** $h(x) = \dfrac{x^2 - 4x}{x^3 - x}$

c) $g(x) = \dfrac{x - 2}{x^3 - 5x}$

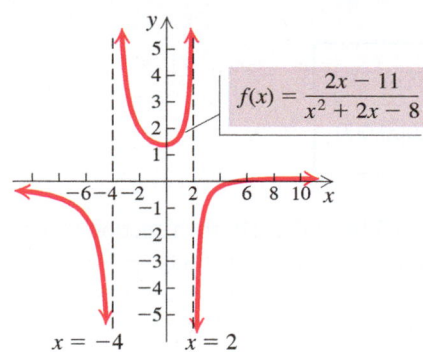

Figure 1.

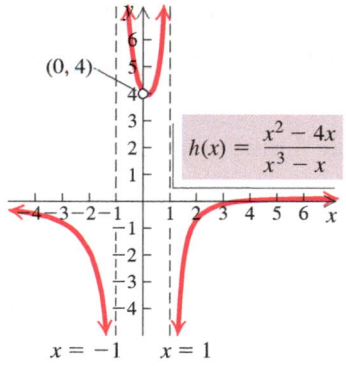

Figure 2.

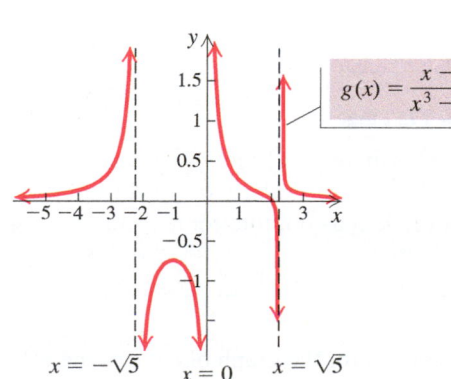

Figure 3.

Solution

a) First, we factor the denominator:

$$f(x) = \frac{2x - 11}{x^2 + 2x - 8} = \frac{2x - 11}{(x + 4)(x - 2)}.$$

The numerator and the denominator have no common factors. The zeros of the denominator are -4 and 2. Thus the vertical asymptotes for the graph of $f(x)$ are the lines $x = -4$ and $x = 2$. (See Fig. 1.)

b) We factor the numerator and the denominator:

$$h(x) = \frac{x^2 - 4x}{x^3 - x} = \frac{x(x - 4)}{x(x^2 - 1)} = \frac{x(x - 4)}{x(x + 1)(x - 1)}.$$

The domain of the function is $\{x \mid x \neq -1 \text{ and } x \neq 0 \text{ and } x \neq 1\}$, or $(-\infty, -1)$ $\cup$ $(-1, 0)$ $\cup$ $(0, 1)$ $\cup$ $(1, \infty)$. Note that the numerator and the denominator share a common factor, x. The vertical asymptotes of $h(x)$ are found by determining the zeros of the denominator, $x(x + 1)(x - 1)$, that are *not* also zeros of the numerator, $x(x - 4)$. The zeros of $x(x + 1)(x - 1)$ are 0, -1, and 1. The zeros of $x(x - 4)$ are 0 and 4. Thus, although the denominator has three zeros, the graph of $h(x)$ has only two vertical asymptotes, $x = -1$ and $x = 1$. (See Fig. 2.)

The rational expression $[x(x - 4)]/[x(x + 1)(x - 1)]$ can be simplified. Thus,

$$h(x) = \frac{x(x - 4)}{x(x + 1)(x - 1)} = \frac{x - 4}{(x + 1)(x - 1)},$$

where $x \neq 0$, $x \neq -1$, and $x \neq 1$. The graph of $h(x)$ is the graph of

$$h(x) = \frac{x - 4}{(x + 1)(x - 1)}$$

with the point where $x = 0$ missing. To determine the y-coordinate of the hole, we substitute 0 for x:

$$h(0) = \frac{0 - 4}{(0 + 1)(0 - 1)} = \frac{-4}{1 \cdot (-1)} = 4.$$

Thus the hole is located at $(0, 4)$.

c) We factor the denominator:

$$g(x) = \frac{x - 2}{x^3 - 5x} = \frac{x - 2}{x(x^2 - 5)}.$$

The numerator and the denominator have no common factors. We find the zeros of the denominator, $x(x^2 - 5)$. Solving $x(x^2 - 5) = 0$, we get

$$x = 0 \quad or \quad x^2 - 5 = 0$$
$$x = 0 \quad or \quad x^2 = 5$$
$$x = 0 \quad or \quad x = \pm\sqrt{5}.$$

The zeros of the denominator are 0, $\sqrt{5}$, and $-\sqrt{5}$. Thus the vertical asymptotes are the lines $x = 0$, $x = \sqrt{5}$, and $x = -\sqrt{5}$. (See Fig. 3.)

> **Now Try Exercises 15 and 19.**

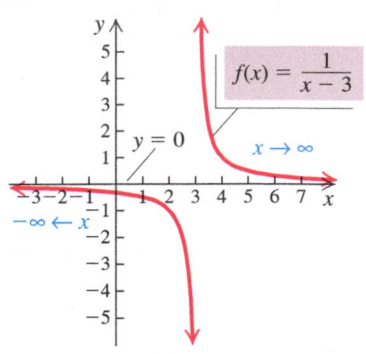

Horizontal asymptote: $y = 0$

Horizontal Asymptotes

Looking again at the graph of $f(x) = 1/(x - 3)$, shown at left (also see Example 1), let's explore what happens to $f(x) = 1/(x - 3)$ as x increases without bound (approaches positive infinity, ∞) and as x decreases without bound (approaches negative infinity, $-\infty$).

x increases without bound:

x	100	5000	1,000,000	$\longrightarrow \infty$
$f(x)$	≈ 0.0103	≈ 0.0002	≈ 0.000001	$\longrightarrow 0$

x decreases without bound:

x	-300	-8000	$-1,000,000$	$\longrightarrow -\infty$
$f(x)$	≈ -0.0033	≈ -0.0001	≈ -0.000001	$\longrightarrow 0$

We see that

$$\frac{1}{x - 3} \rightarrow 0 \text{ as } x \rightarrow \infty \quad \text{and} \quad \frac{1}{x - 3} \rightarrow 0 \text{ as } x \rightarrow -\infty.$$

Since $y = 0$ is the equation of the x-axis, we say that the curve approaches the x-axis asymptotically and that the x-axis is a *horizontal asymptote* for the curve.

In general, the line $y = b$ is a **horizontal asymptote** for the graph of f if either or both of the following are true:

$$f(x) \rightarrow b \text{ as } x \rightarrow \infty \quad \text{or} \quad f(x) \rightarrow b \text{ as } x \rightarrow -\infty.$$

The following figures illustrate four ways in which horizontal asymptotes can occur. In each case, the curve gets close to the line $y = b$ either as $x \rightarrow \infty$ or as $x \rightarrow -\infty$. Keep in mind that the symbols ∞ and $-\infty$ convey the idea of increasing without bound and decreasing without bound, respectively.

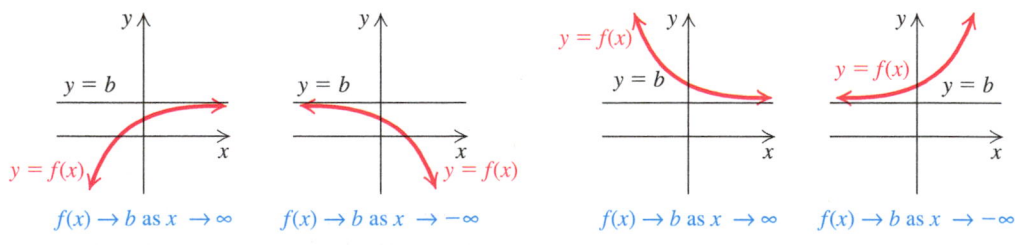

How can we determine a horizontal asymptote? As x gets very large or very small, the value of a polynomial function $p(x)$ is dominated by the function's leading term. Because of this, if $p(x)$ and $q(x)$ have the *same* degree, the value of $p(x)/q(x)$ as $x \rightarrow \infty$ or as $x \rightarrow -\infty$ is dominated by the ratio of the numerator's leading coefficient to the denominator's leading coefficient.

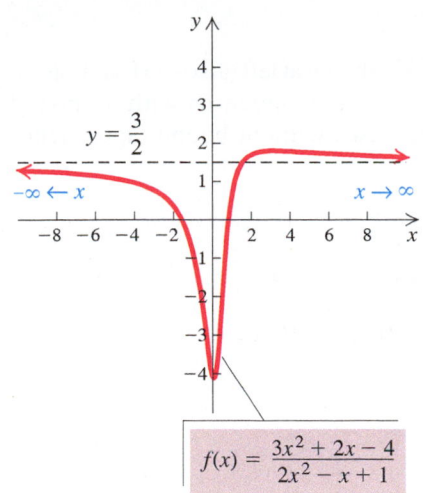

$$y = \frac{3}{2}$$

$$f(x) = \frac{3x^2 + 2x - 4}{2x^2 - x + 1}$$

For $f(x) = (3x^2 + 2x - 4)/(2x^2 - x + 1)$, we see that the numerator, $3x^2 + 2x - 4$, is dominated by $3x^2$ and the denominator, $2x^2 - x + 1$, is dominated by $2x^2$, so $f(x)$ approaches $3x^2/2x^2$, or $3/2$, as x gets very large or very small:

$$\frac{3x^2 + 2x - 4}{2x^2 - x + 1} \to \frac{3}{2}, \text{ or } 1.5, \text{ as } x \to \infty, \quad \text{and}$$

$$\frac{3x^2 + 2x - 4}{2x^2 - x + 1} \to \frac{3}{2}, \text{ or } 1.5, \text{ as } x \to -\infty.$$

We say that the curve approaches the horizontal line $y = \frac{3}{2}$ asymptotically and that $y = \frac{3}{2}$ is a *horizontal asymptote* for the curve.

It follows that when the numerator and the denominator of a rational function have the same degree, the line $y = a/b$ is the horizontal asymptote, where a and b are the leading coefficients of the numerator and the denominator, respectively.

EXAMPLE 4 Find the horizontal asymptote: $f(x) = \dfrac{-7x^4 - 10x^2 + 1}{11x^4 + x - 2}$.

Solution The numerator and the denominator have the same degree. The ratio of the leading coefficients is $-\frac{7}{11}$, so the line $y = -\frac{7}{11}$, or $-0.\overline{63}$, is the horizontal asymptote.

> **Now Try Exercise 21.**

To check Example 4, we could evaluate the function for a very large and a very small value of x. Another check, one that is useful in calculus, is to multiply by 1, using $(1/x^4)/(1/x^4)$:

$$f(x) = \frac{-7x^4 - 10x^2 + 1}{11x^4 + x - 2} \cdot \frac{\dfrac{1}{x^4}}{\dfrac{1}{x^4}} = \frac{\dfrac{-7x^4}{x^4} - \dfrac{10x^2}{x^4} + \dfrac{1}{x^4}}{\dfrac{11x^4}{x^4} + \dfrac{x}{x^4} - \dfrac{2}{x^4}}$$

$$= \frac{-7 - \dfrac{10}{x^2} + \dfrac{1}{x^4}}{11 + \dfrac{1}{x^3} - \dfrac{2}{x^4}}.$$

As $|x|$ becomes very large, each expression whose denominator is a power of x tends toward 0. Specifically, as $x \to \infty$ or as $x \to -\infty$, we have

$$f(x) \to \frac{-7 - 0 + 0}{11 + 0 - 0}, \quad \text{or} \quad f(x) \to -\frac{7}{11}.$$

The horizontal asymptote is $y = -\frac{7}{11}$, or $-0.\overline{63}$.

We now investigate the occurrence of a horizontal asymptote when the degree of the numerator is less than the degree of the denominator.

EXAMPLE 5 Find the horizontal asymptote: $f(x) = \dfrac{2x + 3}{x^3 - 2x^2 + 4}$.

Solution Let $p(x) = 2x + 3$, $q(x) = x^3 - 2x^2 + 4$, and $f(x) = p(x)/q(x)$. Note that as $x \to \infty$, the value of $q(x)$ grows much faster than the value of $p(x)$. Because of this, the ratio $p(x)/q(x)$ shrinks toward 0. As $x \to -\infty$, the ratio $p(x)/q(x)$ behaves in a similar manner. The horizontal asymptote is $y = 0$, the x-axis. This is the case for all rational functions for which the degree of the numerator is less than the degree of the denominator. Note in Example 1 that $y = 0$, the x-axis, is the horizontal asymptote of $f(x) = 1/(x - 3)$.

> **Now Try Exercise 23.**

The following statements describe the two ways in which a horizontal asymptote occurs.

DETERMINING A HORIZONTAL ASYMPTOTE

- When the numerator and the denominator of a rational function have the same degree, the line $y = a/b$ is the horizontal asymptote, where a and b are the leading coefficients of the numerator and the denominator, respectively.

- When the degree of the numerator of a rational function is less than the degree of the denominator, the x-axis, or $y = 0$, is the horizontal asymptote.

- When the degree of the numerator of a rational function is greater than the degree of the denominator, there is no horizontal asymptote.

EXAMPLE 6 Graph

$$g(x) = \frac{2x^2 + 1}{x^2}.$$

Include and label all asymptotes.

Solution Since 0 is the zero of the denominator and is not a zero of the numerator, the y-axis, $x = 0$, is the vertical asymptote. Note also that the degree of the numerator is the same as the degree of the denominator. Thus, $y = 2/1$, or 2, is the horizontal asymptote.

To draw the graph, we first draw the asymptotes with dashed lines. Then we compute and plot some ordered pairs and draw the two branches of the curve.

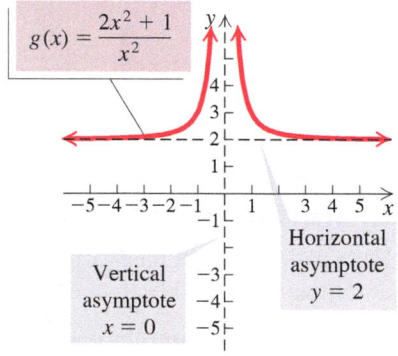

x	$g(x)$
-2	2.25
-1.5	$2.\overline{4}$
-1	3
-0.5	6
0.5	6
1	3
1.5	$2.\overline{4}$
2	2.25

> **Now Try Exercise 41.**

Oblique Asymptotes

Sometimes a line that is neither horizontal nor vertical is an asymptote. Such a line is called an **oblique asymptote**, or a **slant asymptote**.

EXAMPLE 7 Find all the asymptotes of

$$f(x) = \frac{2x^2 - 3x - 1}{x - 2}.$$

Solution The line $x = 2$ is the vertical asymptote because 2 is the zero of the denominator and is not a zero of the numerator. There is no horizontal asymptote because the degree of the numerator is greater than the degree of the denominator. When the degree of the numerator is 1 greater than the degree of the denominator, we divide to find an equivalent expression:

$$\frac{2x^2 - 3x - 1}{x - 2} = (2x + 1) + \frac{1}{x - 2}.$$

$$
\begin{array}{r}
2x + 1 \\
x - 2 \overline{)2x^2 - 3x - 1} \\
\underline{2x^2 - 4x} \\
x - 1 \\
\underline{x - 2} \\
1
\end{array}
$$

Now we see that when $x \to \infty$ or $x \to -\infty$, $1/(x - 2) \to 0$ and the value of $f(x) \to 2x + 1$. This means that as $|x|$ becomes very large, the graph of $f(x)$ gets very close to the graph of $y = 2x + 1$. Thus the line $y = 2x + 1$ is the oblique asymptote.

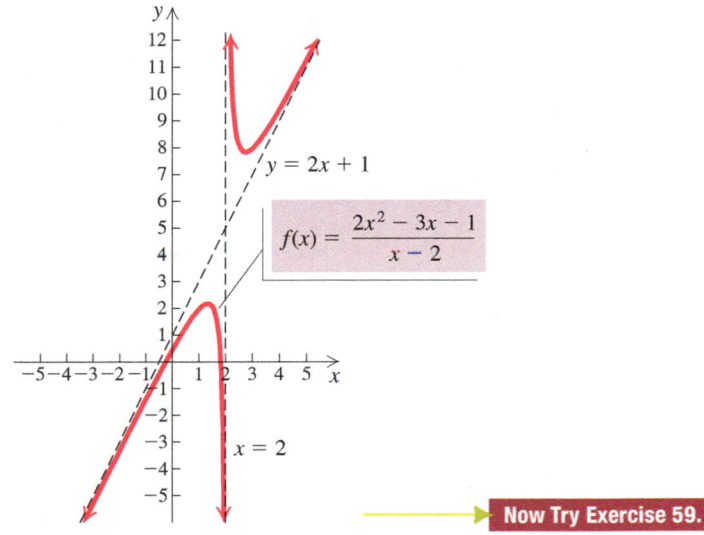

Now Try Exercise 59.

OCCURRENCE OF LINES AS ASYMPTOTES OF RATIONAL FUNCTIONS

For a rational function $f(x) = p(x)/q(x)$, where $p(x)$ and $q(x)$ have no common factors other than constants:

Vertical asymptotes occur at any x-values that make the denominator 0.

The x-axis is the horizontal asymptote when the degree of the numerator is less than the degree of the denominator.

A horizontal asymptote other than the x-axis occurs when the numerator and the denominator have the same degree.

An oblique asymptote occurs when the degree of the numerator is 1 greater than the degree of the denominator.

There can be only one horizontal asymptote or one oblique asymptote and never both.

An asymptote is *not* part of the graph of the function.

The following statements are also true.

Crossing an Asymptote

- The graph of a rational function *never crosses* a vertical asymptote.
- The graph of a rational function *might cross* a horizontal asymptote or an oblique asymptote but does not necessarily do so.

Shown below is an outline of a procedure that we can follow to create accurate graphs of rational functions.

To graph a rational function $f(x) = p(x)/q(x)$, where $p(x)$ and $q(x)$ have no common factor other than constants:

1. Find any real zeros of the denominator. Determine the domain of the function and sketch any vertical asymptotes.
2. Find the horizontal asymptote or the oblique asymptote, if there is one, and sketch it.
3. Find any zeros of the function. The zeros are found by determining the zeros of the numerator. These are the first coordinates of the x-intercepts of the graph.
4. Find $f(0)$. This gives the y-intercept, $(0, f(0))$, of the function.
5. Find other function values to determine the general shape. Then draw the graph.

EXAMPLE 8 Graph: $f(x) = \dfrac{2x + 3}{3x^2 + 7x - 6}$.

Solution

1. We find the zeros of the denominator by solving $3x^2 + 7x - 6 = 0$. Since

$$3x^2 + 7x - 6 = (3x - 2)(x + 3),$$

the zeros are $\frac{2}{3}$ and -3. Thus the domain excludes $\frac{2}{3}$ and -3 and is

$$\left(-\infty, -3\right) \cup \left(-3, \tfrac{2}{3}\right) \cup \left(\tfrac{2}{3}, \infty\right).$$

Since neither zero of the denominator is a zero of the numerator, the graph has vertical asymptotes $x = -3$ and $x = \frac{2}{3}$. We sketch these as dashed lines.

2. Because the degree of the numerator is less than the degree of the denominator, the x-axis, $y = 0$, is the horizontal asymptote.

3. To find the zeros of the numerator, we solve $2x + 3 = 0$ and get $x = -\frac{3}{2}$. Thus, $-\frac{3}{2}$ is the zero of the function, and the pair $\left(-\frac{3}{2}, 0\right)$ is the x-intercept.

4. We find $f(0)$:

$$f(0) = \frac{2 \cdot 0 + 3}{3 \cdot 0^2 + 7 \cdot 0 - 6}$$

$$= \frac{3}{-6} = -\frac{1}{2}.$$

The point $\left(0, -\frac{1}{2}\right)$ is the y-intercept.

5. We find other function values to determine the general shape. We choose values in each interval of the domain as shown in the table below and then draw the graph. Note that the graph of this function crosses its horizontal asymptote at $x = -\frac{3}{2}$.

x	y
-4.5	-0.26
-3.25	-1.19
-2.5	0.42
-0.5	-0.23
0.5	-2.29
0.75	4.8
1.5	0.53
3.5	0.18

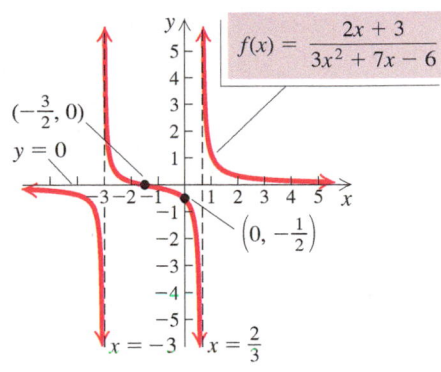

$$f(x) = \frac{2x + 3}{3x^2 + 7x - 6}$$

> **Now Try Exercise 65.**

EXAMPLE 9 Graph: $g(x) = \dfrac{x^2 - 1}{x^2 + x - 6}$.

Solution

1. We find the zeros of the denominator by solving $x^2 + x - 6 = 0$. Since

$$x^2 + x - 6 = (x + 3)(x - 2),$$

the zeros are -3 and 2. Thus the domain excludes the x-values -3 and 2 and is

$$(-\infty, -3) \cup (-3, 2) \cup (2, \infty).$$

Since neither zero of the denominator is a zero of the numerator, the graph has vertical asymptotes $x = -3$ and $x = 2$. We sketch these as dashed lines.

2. The numerator and the denominator have the same degree, so the horizontal asymptote is determined by the ratio of the leading coefficients: $1/1$, or 1. Thus, $y = 1$ is the horizontal asymptote. We sketch it with a dashed line.

3. To find the zeros of the numerator, we solve $x^2 - 1 = 0$. The solutions are -1 and 1. Thus, -1 and 1 are the zeros of the function and the pairs $(-1, 0)$ and $(1, 0)$ are the x-intercepts.

4. We find $g(0)$:

$$g(0) = \frac{0^2 - 1}{0^2 + 0 - 6} = \frac{-1}{-6} = \frac{1}{6}.$$

Thus, $\left(0, \frac{1}{6}\right)$ is the y-intercept.

5. We find other function values to determine the general shape and then draw the graph.

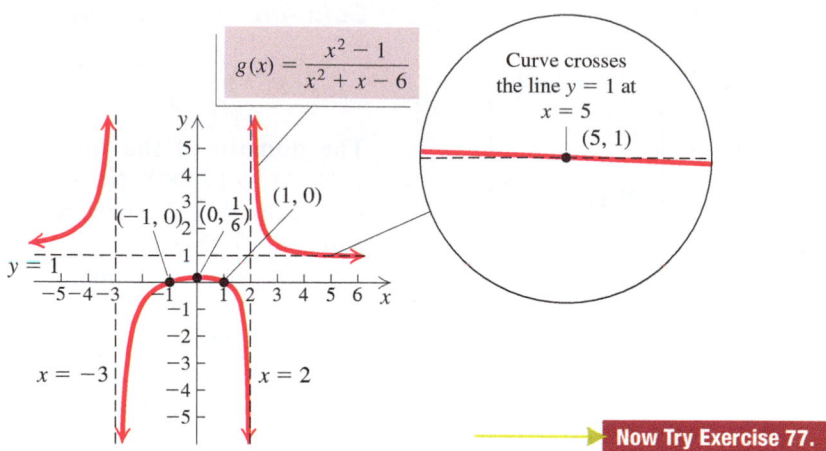

Now Try Exercise 77.

The magnified portion of the graph in Example 9 above shows another situation in which a graph can cross its horizontal asymptote. The point where $g(x)$ crosses $y = 1$ can be found by setting $g(x) = 1$ and solving for x:

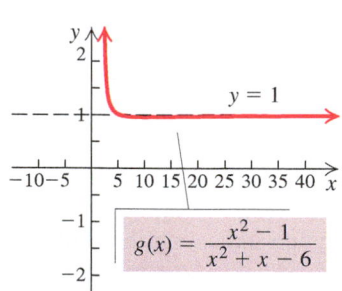

$$\frac{x^2 - 1}{x^2 + x - 6} = 1$$

$$x^2 - 1 = x^2 + x - 6$$

$$-1 = x - 6 \qquad \text{Subtracting } x^2$$

$$5 = x. \qquad \text{Adding 6}$$

The point of intersection is $(5, 1)$. Note the behavior of the curve after it crosses the horizontal asymptote at $x = 5$. (See the graph at left.) It continues to decrease for a short interval and then begins to increase, getting closer and closer to $y = 1$ as $x \to \infty$.

Graphs of rational functions can also cross an oblique asymptote. The graph of

$$f(x) = \frac{2x^3}{x^2 + 1}$$

shown below crosses its oblique asymptote $y = 2x$. **Remember, graphs can cross horizontal asymptotes or oblique asymptotes, but they cannot cross vertical asymptotes.**

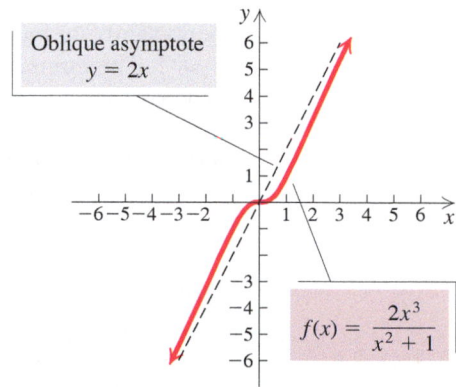

Let's now graph a rational function $f(x) = p(x)/q(x)$, where $p(x)$ and $q(x)$ have a common factor, $x - c$. The graph of such a function has a "hole" in it. We first saw this situation in Example 3(b), where the common factor was x.

x	y
−3	$-\frac{1}{2}$
−2	−1
−1	Not defined
0	1
1	$\frac{1}{2}$
2	Not defined
3	$\frac{1}{4}$

EXAMPLE 10   Graph: $g(x) = \dfrac{x-2}{x^2-x-2}$.

Solution   We first express the denominator in factored form:

$$g(x) = \frac{x-2}{x^2-x-2} = \frac{x-2}{(x+1)(x-2)}.$$

The domain of the function is $\{x \,|\, x \neq -1 \text{ and } x \neq 2\}$, or $(-\infty, -1) \cup (-1, 2) \cup (2, \infty)$. Note that the numerator and the denominator have the common factor $x-2$. The zeros of the denominator are -1 and 2, and the zero of the numerator is 2. Since -1 is the only zero of the denominator that is *not* a zero of the numerator, the graph of the function has $x = -1$ as its *only* vertical asymptote. The degree of the numerator is less than the degree of the denominator, so $y = 0$ is the horizontal asymptote. There are no zeros of the function and thus no x-intercepts, because 2 is the only zero of the numerator and 2 is not in the domain of the function. Since $g(0) = 1$, $(0, 1)$ is the y-intercept. The rational expression

$$\frac{x-2}{(x+1)(x-2)}$$

can be simplified. Thus,

$$g(x) = \frac{x-2}{(x+1)(x-2)} = \frac{1}{x+1}, \quad \text{where } x \neq -1 \text{ and } x \neq 2.$$

The graph of $g(x)$ is the graph of $y = 1/(x+1)$ with the point where $x = 2$ missing. To determine the coordinates of the "hole," we substitute 2 for x in $g(x) = 1/(x+1)$:

$$g(2) = \frac{1}{2+1} = \frac{1}{3}.$$

Thus the "hole" is located at $\left(2, \frac{1}{3}\right)$. We draw the graph indicating the "hole" when $x = 2$ with an open circle.

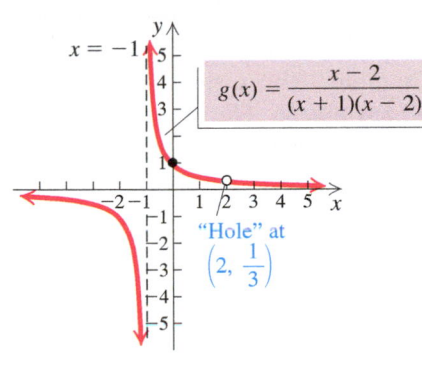

$x = -1$

$g(x) = \dfrac{x-2}{(x+1)(x-2)}$

"Hole" at $\left(2, \dfrac{1}{3}\right)$

Now Try Exercise 49.

EXAMPLE 11   Graph: $f(x) = \dfrac{-2x^2-x+15}{x^2-x-12}$.

Solution   We first express the numerator and the denominator in factored form:

$$f(x) = \frac{-2x^2-x+15}{x^2-x-12} = \frac{-(2x^2+x-15)}{x^2-x-12} = \frac{-(2x-5)(x+3)}{(x-4)(x+3)}.$$

The domain of the function is $\{x \,|\, x \neq -3 \text{ and } x \neq 4\}$, or $(-\infty, -3) \cup (-3, 4) \cup (4, \infty)$. The numerator and the denominator have the common factor $x+3$. The zeros of the denominator are -3 and 4, and the zeros of the numerator are -3 and $\frac{5}{2}$. Since 4 is the only zero of the denominator that is *not* a zero of the numerator, the graph of the function has $x = 4$ as its *only* vertical asymptote.

The degrees of the numerator and the denominator are the same, so the line $y = \frac{-2}{1} = -2$ is the horizontal asymptote. The zeros of the numerator are $\frac{5}{2}$ and -3. Because -3 is not in the domain of the function, the only x-intercept is $\left(\frac{5}{2}, 0\right)$. Since $f(0) = -\frac{15}{12} = -\frac{5}{4}$, then $\left(0, -\frac{5}{4}\right)$ is the y-intercept. The rational function

$$\frac{-(2x-5)(x+3)}{(x-4)(x+3)}$$

can be simplified. Thus,

$$f(x) = \frac{-(2x-5)(x+3)}{(x-4)(x+3)} = \frac{-(2x-5)}{x-4}, \quad \text{where } x \neq -3 \text{ and } x \neq 4.$$

The graph of $f(x)$ is the graph of $y = -(2x - 5)/(x - 4)$ with the point where $x = -3$ missing. To determine the coordinates of the hole, we substitute -3 for x in $f(x) = (-2x + 5)/(x - 4)$:

$$f(-3) = \frac{-[2(-3) - 5]}{-3 - 4} = \frac{-[-11]}{-7} = \frac{11}{-7} = -\frac{11}{7}.$$

Thus the hole is located at $\left(-3, -\frac{11}{7}\right)$. We draw the graph indicating the hole when $x = -3$ with an open circle.

x	y
-5	-1.67
-4	-1.63
-3	Not defined
-2	-1.5
-1	-1.4
0	-1.25
1	-1
2	-0.5

x	y
3	1
3.5	4
4	Not defined
5	-5
6	-3.5
7	-3
8	-2.75

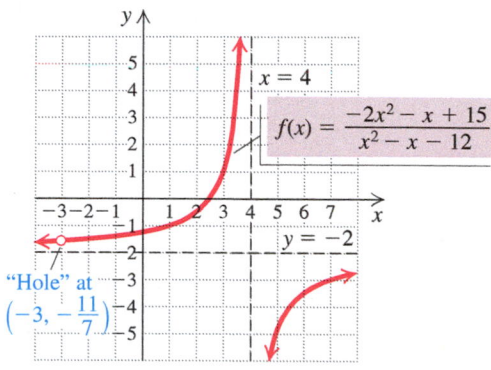

$$f(x) = \frac{-2x^2 - x + 15}{x^2 - x - 12}$$

"Hole" at $\left(-3, -\frac{11}{7}\right)$

▶ Now Try Exercise 67.

▶ **Applications**

EXAMPLE 12 *Temperature During an Illness.* A person's temperature T, in degrees Fahrenheit, during an illness is given by the function

$$T(t) = \frac{4t}{t^2 + 1} + 98.6,$$

where time t is given in hours since the onset of the illness. The graph of this function is shown at left.

a) Find the temperature at $t = 0, 1, 2, 5, 12,$ and 24.

b) Find the horizontal asymptote of the graph of $T(t)$. Complete:

$$T(t) \rightarrow \boxed{} \text{ as } t \rightarrow \infty.$$

c) Give the meaning of the answer to part (b) in terms of the application.

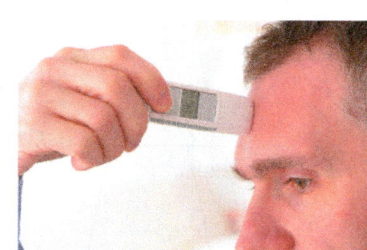

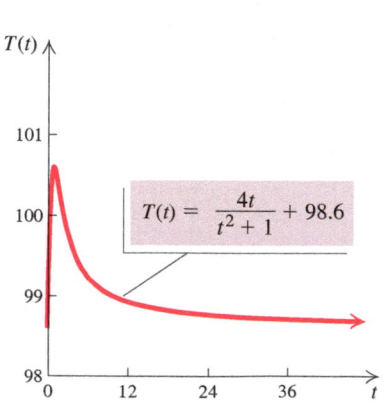

$$T(t) = \frac{4t}{t^2 + 1} + 98.6$$

Solution

a) We have

$$T(0) = 98.6, \qquad T(1) = 100.6, \qquad T(2) = 100.2,$$
$$T(5) \approx 99.369, \qquad T(12) \approx 98.931, \quad \text{and} \quad T(24) \approx 98.766.$$

b) Since

$$T(t) = \frac{4t}{t^2 + 1} + 98.6 = \frac{98.6t^2 + 4t + 98.6}{t^2 + 1},$$

the horizontal asymptote is $y = 98.6/1$, or 98.6. Then it follows that $T(t) \rightarrow 98.6$ as $t \rightarrow \infty$.

c) As time goes on, the temperature returns to "normal," which is $98.6°F$.

▶ Now Try Exercise 83.

A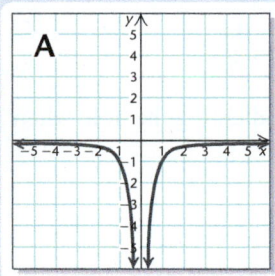

B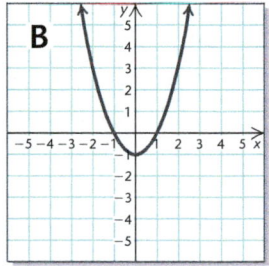

C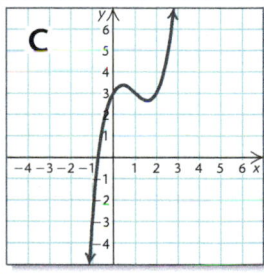

D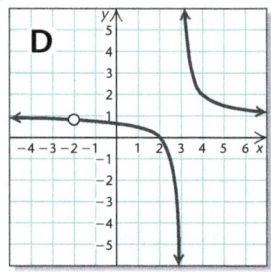

E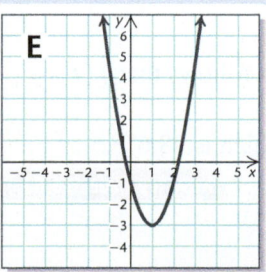

Visualizing the Graph

Match the function with its graph.

1. $f(x) = -\dfrac{1}{x^2}$

2. $f(x) = x^3 - 3x^2 + 2x + 3$

3. $f(x) = \dfrac{x^2 - 4}{x^2 - x - 6}$

4. $f(x) = -x^2 + 4x - 1$

5. $f(x) = \dfrac{x - 3}{x^2 + x - 6}$

6. $f(x) = \dfrac{3}{4}x + 2$

7. $f(x) = x^2 - 1$

8. $f(x) = x^4 - 2x^2 - 5$

9. $f(x) = \dfrac{8x - 4}{3x + 6}$

10. $f(x) = 2x^2 - 4x - 1$

F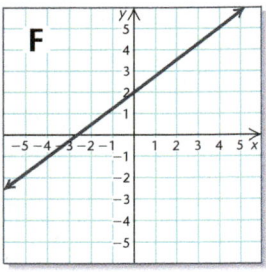

G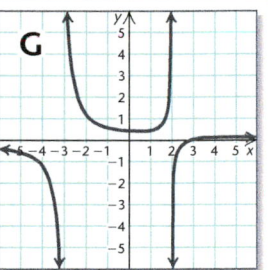

H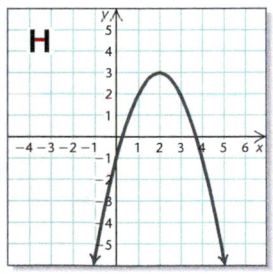

I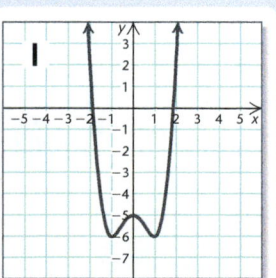

J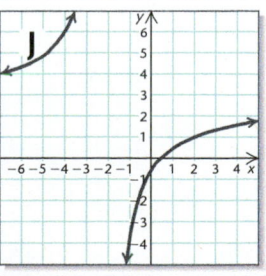

8.5 Exercise Set

Determine the domain of the function.

1. $f(x) = \dfrac{x^2}{2 - x}$

2. $f(x) = \dfrac{1}{x^3}$

3. $f(x) = \dfrac{x + 1}{x^2 - 6x + 5}$

4. $f(x) = \dfrac{(x + 4)^2}{4x - 3}$

5. $f(x) = \dfrac{3x - 4}{3x + 15}$

6. $f(x) = \dfrac{x^2 + 3x - 10}{x^2 + 2x}$

In Exercises 7–12, use your knowledge of asymptotes and intercepts to match the equation with one of the graphs (a)–(f) that follow. List all asymptotes.

a)

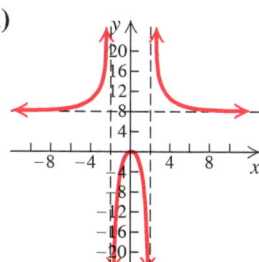

b)

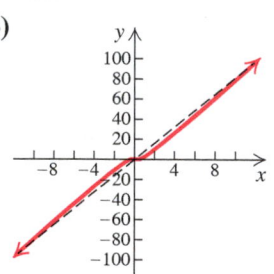

c)

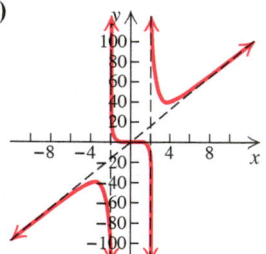

d)

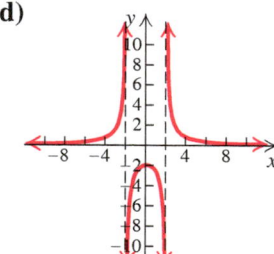

e)

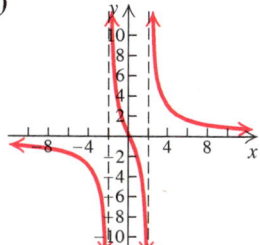

f)

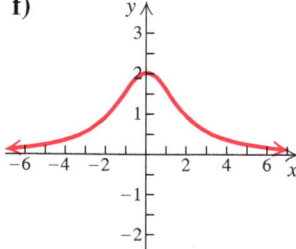

7. $f(x) = \dfrac{8}{x^2 - 4}$

8. $f(x) = \dfrac{8}{x^2 + 4}$

9. $f(x) = \dfrac{8x}{x^2 - 4}$

10. $f(x) = \dfrac{8x^2}{x^2 - 4}$

11. $f(x) = \dfrac{8x^3}{x^2 - 4}$

12. $f(x) = \dfrac{8x^3}{x^2 + 4}$

Determine the vertical asymptotes of the graph of the function.

13. $g(x) = \dfrac{1}{x^2}$

14. $f(x) = \dfrac{4x}{x^2 + 10x}$

15. $h(x) = \dfrac{x + 7}{2 - x}$

16. $g(x) = \dfrac{x^4 + 2}{x}$

17. $f(x) = \dfrac{3 - x}{(x - 4)(x + 6)}$

18. $h(x) = \dfrac{x^2 - 4}{x(x + 5)(x - 2)}$

19. $g(x) = \dfrac{x^3}{2x^3 - x^2 - 3x}$

20. $f(x) = \dfrac{x + 5}{x^2 + 4x - 32}$

Determine the horizontal asymptote of the graph of the function.

21. $f(x) = \dfrac{3x^2 + 5}{4x^2 - 3}$

22. $g(x) = \dfrac{x + 6}{x^3 + 2x^2}$

23. $h(x) = \dfrac{x^2 - 4}{2x^4 + 3}$

24. $f(x) = \dfrac{x^5}{x^5 + x}$

25. $g(x) = \dfrac{x^3 - 2x^2 + x - 1}{x^2 - 16}$

26. $h(x) = \dfrac{8x^4 + x - 2}{2x^4 - 10}$

Determine the oblique asymptote of the graph of the function.

27. $g(x) = \dfrac{x^2 + 4x - 1}{x + 3}$

28. $f(x) = \dfrac{x^2 - 6x}{x - 5}$

29. $h(x) = \dfrac{x^4 - 2}{x^3 + 1}$

30. $g(x) = \dfrac{12x^3 - x}{6x^2 + 4}$

31. $f(x) = \dfrac{x^3 - x^2 + x - 4}{x^2 + 2x - 1}$

32. $h(x) = \dfrac{5x^3 - x^2 + x - 1}{x^2 - x + 2}$

Graph the function. Be sure to label all the asymptotes.
List the domain and the x- and y-intercepts.

33. $f(x) = \dfrac{1}{x}$

34. $g(x) = \dfrac{1}{x^2}$

35. $h(x) = -\dfrac{4}{x^2}$

36. $f(x) = -\dfrac{6}{x}$

37. $g(x) = \dfrac{x^2 - 4x + 3}{x + 1}$

38. $h(x) = \dfrac{2x^2 - x - 3}{x - 1}$

39. $f(x) = \dfrac{-2}{x - 5}$

40. $f(x) = \dfrac{1}{x - 5}$

41. $f(x) = \dfrac{2x + 1}{x}$

42. $f(x) = \dfrac{3x - 1}{x}$

43. $f(x) = \dfrac{x + 3}{x^2 - 9}$

44. $f(x) = \dfrac{x - 1}{x^2 - 1}$

45. $f(x) = \dfrac{x}{x^2 + 3x}$

46. $f(x) = \dfrac{3x}{3x - x^2}$

47. $f(x) = \dfrac{1}{(x - 2)^2}$

48. $f(x) = \dfrac{-2}{(x - 3)^2}$

49. $f(x) = \dfrac{x^2 + 2x - 3}{x^2 + 4x + 3}$

50. $f(x) = \dfrac{x^2 - x - 2}{x^2 - 5x - 6}$

51. $f(x) = \dfrac{1}{x^2 + 3}$

52. $f(x) = \dfrac{-1}{x^2 + 2}$

53. $f(x) = \dfrac{x^2 - 4}{x - 2}$

54. $f(x) = \dfrac{x^2 - 9}{x + 3}$

55. $f(x) = \dfrac{x - 1}{x + 2}$

56. $f(x) = \dfrac{x - 2}{x + 1}$

57. $f(x) = \dfrac{x^2 + 3x}{2x^3 - 5x^2 - 3x}$

58. $f(x) = \dfrac{3x}{x^2 + 5x + 4}$

59. $f(x) = \dfrac{x^2 - 9}{x + 1}$

60. $f(x) = \dfrac{x^3 - 4x}{x^2 - x}$

61. $f(x) = \dfrac{x^2 + x - 2}{2x^2 + 1}$

62. $f(x) = \dfrac{x^2 - 2x - 3}{3x^2 + 2}$

63. $g(x) = \dfrac{3x^2 - x - 2}{x - 1}$

64. $f(x) = \dfrac{2x^2 - 5x - 3}{2x + 1}$

65. $f(x) = \dfrac{x - 1}{x^2 - 2x - 3}$

66. $f(x) = \dfrac{x + 2}{x^2 + 2x - 15}$

67. $f(x) = \dfrac{3x^2 + 11x - 4}{x^2 + 2x - 8}$

68. $f(x) = \dfrac{2x^2 - 3x - 9}{x^2 - 2x - 3}$

69. $f(x) = \dfrac{x - 3}{(x + 1)^3}$

70. $f(x) = \dfrac{x + 2}{(x - 1)^3}$

71. $f(x) = \dfrac{x^3 + 1}{x}$

72. $f(x) = \dfrac{x^3 - 1}{x}$

73. $f(x) = \dfrac{x^3 + 2x^2 - 15x}{x^2 - 5x - 14}$

74. $f(x) = \dfrac{x^3 + 2x^2 - 3x}{x^2 - 25}$

75. $f(x) = \dfrac{5x^4}{x^4 + 1}$

76. $f(x) = \dfrac{x + 1}{x^2 + x - 6}$

77. $f(x) = \dfrac{x^2}{x^2 - x - 2}$

78. $f(x) = \dfrac{x^2 - x - 2}{x + 2}$

Find a rational function that satisfies the given conditions.
Answers may vary, but try to give the simplest answer
possible.

79. Vertical asymptotes $x = -4, x = 5$

80. Vertical asymptotes $x = -4, x = 5$; x-intercept $(-2, 0)$

81. Vertical asymptotes $x = -4, x = 5$; horizontal asymptote $y = \frac{3}{2}$; x-intercept $(-2, 0)$

82. Oblique asymptote $y = x - 1$

83. *Medical Dosage.*   The function

$$N(t) = \dfrac{0.8t + 1000}{5t + 4}, \quad t \geq 15,$$

gives the body concentration $N(t)$, in parts per million, of a certain dosage of medication after time t, in hours.

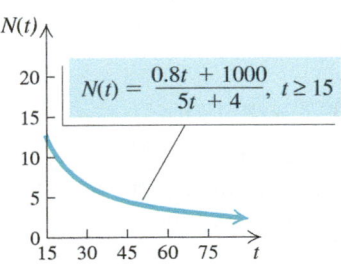

a) Find the horizontal asymptote of the graph and complete the following:

$$N(t) \to \boxed{} \text{ as } t \to \infty.$$

b) Explain the meaning of the answer to part (a) in terms of the application.

84. *Average Cost.*   The average cost per light, in dollars, for a company to produce x roadside emergency lights is given by the function

$$A(x) = \frac{2x + 100}{x}, \quad x > 0.$$

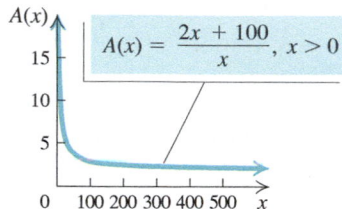

a) Find the horizontal asymptote of the graph and complete the following:

$A(x) \rightarrow \boxed{}$ as $x \rightarrow \infty$.

b) Explain the meaning of the answer to part (a) in terms of the application.

85. *Population Growth.*   The population P, in thousands, of a resort community is given by

$$P(t) = \frac{500t}{2t^2 + 9},$$

where t is the time, in months, since the city council raised the property taxes.

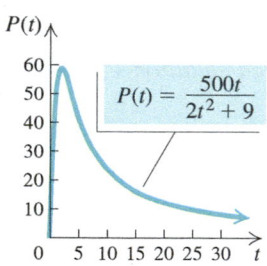

a) Find the population at $t = 0, 1, 3,$ and 8 months.

b) Find the horizontal asymptote of the graph and complete the following:

$P(t) \rightarrow \boxed{}$ as $t \rightarrow \infty$.

c) Explain the meaning of the answer to part (b) in terms of the application.

▶ Skill Maintenance

Vocabulary Reinforcement

In each of Exercises 86–94, fill in the blank with the correct term. Some of the given choices will not be used. Others will be used more than once.

x-intercept	even function
y-intercept	domain
odd function	range
slope	difference
horizontal lines	quotient
vertical lines	$f(x) = f(-x)$
point–slope equation	$f(-x) = -f(x)$
slope–intercept equation	

86. A function is a correspondence between a first set, called the _____, and a second set, called the _____, such that each member of the _____ corresponds to exactly one member of the _____. **[2.2]**

87. The _____ of a line containing (x_1, y_1) and (x_2, y_2) is given by $(y_2 - y_1)/(x_2 - x_1)$. **[2.5]**

88. The _____ of the line with slope m and y-intercept $(0, b)$ is $y = mx + b$. **[2.5]**

89. The _____ of the line with slope m passing through (x_1, y_1) is $y - y_1 = m(x - x_1)$. **[2.7]**

90. A(n) _____ is a point $(a, 0)$. **[2.6]**

91. For each x in the domain of an odd function f, _____. **[7.1]**

92. _____ are given by equations of the type $x = a$. **[2.6]**

93. For a function $f(x)$, the _____ is $\dfrac{f(x + h) - f(x)}{h}$. **[5.4]**

94. A(n) _____ is a point $(0, b)$. **[2.6]**

▶ Synthesis

Find the nonlinear asymptote of the function.

95. $f(x) = \dfrac{x^5 + 2x^3 + 4x^2}{x^2 + 2}$

96. $f(x) = \dfrac{x^4 + 3x^2}{x^2 + 1}$

Graph the function.

97. $f(x) = \dfrac{2x^3 + x^2 - 8x - 4}{x^3 + x^2 - 9x - 9}$

98. $f(x) = \dfrac{x^3 + 4x^2 + x - 6}{x^2 - x - 2}$

Polynomial Inequalities and Rational Inequalities

▶ Solve polynomial inequalities.

▶ Solve rational inequalities.

We will use a combination of algebraic methods and graphical methods to solve polynomial inequalities and rational inequalities.

▶ Polynomial Inequalities

Just as a quadratic equation can be written in the form $ax^2 + bx + c = 0$, a **quadratic inequality** can be written in the form $ax^2 + bx + c \;\square\; 0$, where $\square$ is $<$, $>$, $\leq$, or $\geq$. Here are some examples of quadratic inequalities:

$$x^2 - 4x - 5 < 0 \quad \text{and} \quad -\tfrac{1}{2}x^2 + 4x - 7 \geq 0.$$

When the inequality symbol in a polynomial inequality is replaced with an equals sign, a **related equation** is formed. Polynomial inequalities can be solved once the related equation has been solved.

EXAMPLE 1 Solve: $x^2 - 4x - 5 > 0$.

Solution We are asked to find all x-values for which $x^2 - 4x - 5 > 0$. To locate these values, we graph $f(x) = x^2 - 4x - 5$. Then we note that whenever its graph passes through an x-intercept, the function changes sign. Thus to solve $x^2 - 4x - 5 > 0$, we first solve the *related equation* $x^2 - 4x - 5 = 0$ to find all zeros of the function:

$$x^2 - 4x - 5 = 0$$
$$(x + 1)(x - 5) = 0.$$

The zeros are -1 and 5. Thus the x-intercepts of the graph are $(-1, 0)$ and $(5, 0)$, as shown below.

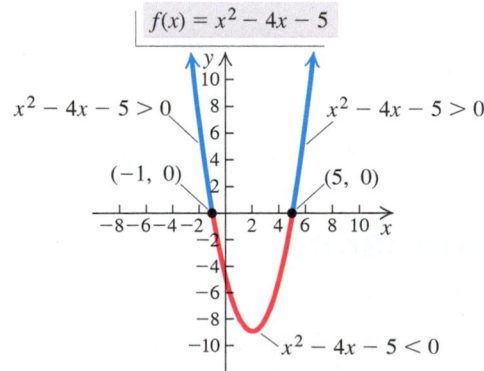

The zeros divide the x-axis into three intervals:

$$(-\infty, -1), \quad (-1, 5), \quad \text{and} \quad (5, \infty).$$

The sign of $x^2 - 4x - 5$ is the same for all values of x in a given interval. Thus we choose a test value for x from each interval and find $f(x)$. We can also determine the sign of $f(x)$ in each interval by simply looking at the graph of the function.

Interval	$(-\infty, -1)$	$(-1, 5)$	$(5, \infty)$
Test Value	$f(-2) = 7$	$f(0) = -5$	$f(7) = 16$
Sign of $f(x)$	Positive	Negative	Positive

Since we are solving $x^2 - 4x - 5 > 0$, the solution set consists of only two of the three intervals, those in which the sign of $f(x)$ is positive. Since the inequality sign is $>$, we do not include the endpoints of the intervals in the solution set. The solution set is $(-\infty, -1) \cup (5, \infty)$, or $\{x \mid x < -1 \text{ or } x > 5\}$.

> **Now Try Exercise 27.**

EXAMPLE 2   Solve: $x^2 + 3x - 5 \le x + 3$.

Solution   By subtracting $x + 3$, we form an equivalent inequality:

$$x^2 + 3x - 5 - x - 3 \le 0$$
$$x^2 + 2x - 8 \le 0.$$

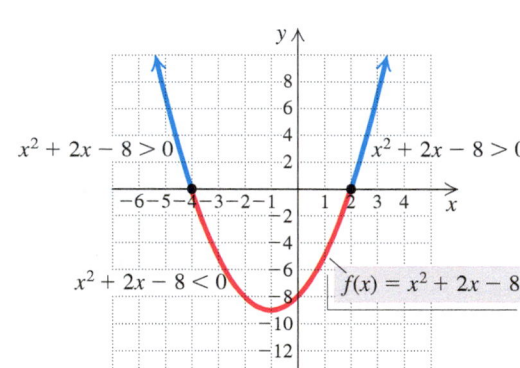

We need to find all x-values for which $x^2 + 2x - 8 \le 0$. To visualize these values, we first graph $f(x) = x^2 + 2x - 8$ and then determine the zeros of the function. To find the zeros, we solve the related equation $x^2 + 2x - 8 = 0$:

$$x^2 + 2x - 8 = 0$$
$$(x + 4)(x - 2) = 0.$$

The zeros are -4 and 2. Thus the x-intercepts of the graph are $(-4, 0)$ and $(2, 0)$, as shown in the figure at left.

The zeros divide the x-axis into three intervals:

$$(-\infty, -4), \quad (-4, 2), \quad \text{and} \quad (2, \infty).$$

We choose a test value for x from each interval and find $f(x)$. The sign of $x^2 + 2x - 8$ is the same for all values of x in a given interval.

Interval	$(-\infty, -4)$	$(-4, 2)$	$(2, \infty)$
Test Value	$f(-5) = 7$	$f(0) = -8$	$f(4) = 16$
Sign of $f(x)$	Positive	Negative	Positive

Function values are negative on the interval $(-4, 2)$. We can also see from the graph where the function values are negative. Since the inequality symbol is $\le$, we include the endpoints of the interval in the solution set. The solution set of $x^2 + 3x - 5 \le x + 3$ is $[-4, 2]$, or $\{x \mid -4 \le x \le 2\}$.

> **Now Try Exercise 29.**

Quadratic inequalities are one type of **polynomial inequality**. Other examples of polynomial inequalities are

$$-2x^4 + x^2 - 3 < 7, \qquad \tfrac{2}{3}x + 4 \geq 0, \quad \text{and} \quad 4x^3 - 2x^2 > 5x + 7.$$

EXAMPLE 3   Solve: $x^3 - x > 0$.

Solution   We are asked to find all x-values for which $x^3 - x > 0$. To locate these values, we graph $f(x) = x^3 - x$. Then we note that whenever the function changes sign, its graph passes through an x-intercept. Thus to solve $x^3 - x > 0$, we first solve the related equation $x^3 - x = 0$ to find all zeros of the function:

$$x^3 - x = 0$$
$$x(x^2 - 1) = 0$$
$$x(x + 1)(x - 1) = 0.$$

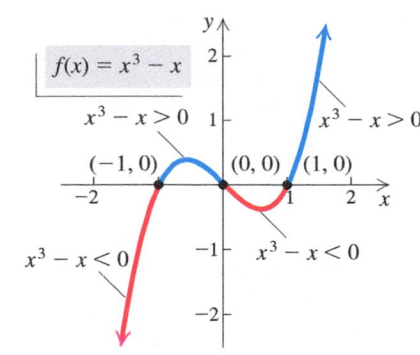

The zeros are -1, 0, and 1. Thus the x-intercepts of the graph are $(-1, 0)$, $(0, 0)$, and $(1, 0)$, as shown in the figure at left. The zeros divide the x-axis into four intervals:

$$(-\infty, -1), \qquad (-1, 0), \qquad (0, 1), \quad \text{and} \quad (1, \infty).$$

The sign of $x^3 - x$ is the same for all values of x in a given interval. Thus we choose a test value for x from each interval and find $f(x)$. We can also determine the sign of $f(x)$ in each interval by simply looking at the graph of the function.

Interval	$(-\infty, -1)$	$(-1, 0)$	$(0, 1)$	$(1, \infty)$
Test Value	$f(-2) = -6$	$f(-0.5) = 0.375$	$f(0.5) = -0.375$	$f(2) = 6$
Sign of $f(x)$	Negative	Positive	Negative	Positive

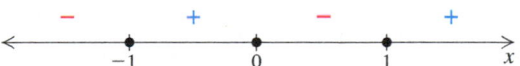

Since we are solving $x^3 - x > 0$, the solution set consists of only two of the four intervals, those in which the sign of $f(x)$ is *positive*. We see that the solution set is $(-1, 0) \cup (1, \infty)$, or $\{x \mid -1 < x < 0 \ or \ x > 1\}$.

Now Try Exercise 39.

To solve a polynomial inequality:

1. Find an equivalent inequality with $P(x)$ on one side and 0 on the other.
2. Change the inequality symbol to an equals sign and solve the related equation; that is, solve $P(x) = 0$.
3. Use the solutions to divide the x-axis into intervals. Then select a test value from each interval and determine the polynomial's sign on the interval.
4. Determine the intervals for which the inequality is satisfied and write interval notation or set-builder notation for the solution set. Include the endpoints of the intervals in the solution set if the inequality symbol is $\leq$ or $\geq$.

EXAMPLE 4 Solve: $3x^4 + 10x \leq 11x^3 + 4$.

Solution By subtracting $11x^3 + 4$, we form the equivalent inequality

$$3x^4 - 11x^3 + 10x - 4 \leq 0.$$

Algebraic Solution

To solve the related equation

$$3x^4 - 11x^3 + 10x - 4 = 0,$$

we need to use the theorems of Section 8.4. We solved this equation in Example 5 in Section 8.4. The solutions are

$$-1, \quad 2 - \sqrt{2}, \quad \tfrac{2}{3}, \quad \text{and} \quad 2 + \sqrt{2},$$

or approximately

$$-1, \quad 0.586, \quad 0.667, \quad \text{and} \quad 3.414.$$

These numbers divide the x-axis into five intervals:

$$(-\infty, -1), \left(-1, 2 - \sqrt{2}\right), \left(2 - \sqrt{2}, \tfrac{2}{3}\right),$$
$$\left(\tfrac{2}{3}, 2 + \sqrt{2}\right), \text{and} \left(2 + \sqrt{2}, \infty\right).$$

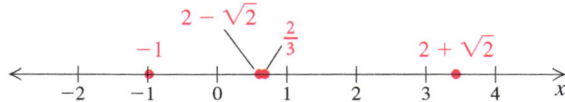

We then let $f(x) = 3x^4 - 11x^3 + 10x - 4$ and, using test values for $f(x)$, determine the sign of $f(x)$ in each interval.

INTERVAL	TEST VALUE	SIGN OF $f(x)$
$(-\infty, -1)$	$f(-2) = 112$	$+$
$\left(-1, 2 - \sqrt{2}\right)$	$f(0) = -4$	$-$
$\left(2 - \sqrt{2}, 2/3\right)$	$f(0.6) = 0.0128$	$+$
$\left(2/3, 2 + \sqrt{2}\right)$	$f(1) = -2$	$-$
$\left(2 + \sqrt{2}, \infty\right)$	$f(4) = 100$	$+$

Function values are negative in the intervals $\left(-1, 2 - \sqrt{2}\right)$ and $\left(\tfrac{2}{3}, 2 + \sqrt{2}\right)$. Since the inequality sign is $\leq$, we include the endpoints of the intervals in the solution set. The solution set is

$$\left[-1, 2 - \sqrt{2}\right] \cup \left[\tfrac{2}{3}, 2 + \sqrt{2}\right], \quad \text{or}$$
$$\left\{x \mid -1 \leq x \leq 2 - \sqrt{2} \text{ or } \tfrac{2}{3} \leq x \leq 2 + \sqrt{2}\right\}.$$

Visualizing the Solution

Observing the graph of the function

$$f(x) = 3x^4 - 11x^3 + 10x - 4$$

and a closeup view of the graph on the interval $(0, 1)$, we see the intervals on which $f(x) \leq 0$. The values of $f(x)$ are less than or equal to 0 in two intervals.

$f(x) = 3x^4 - 11x^3 + 10x - 4$

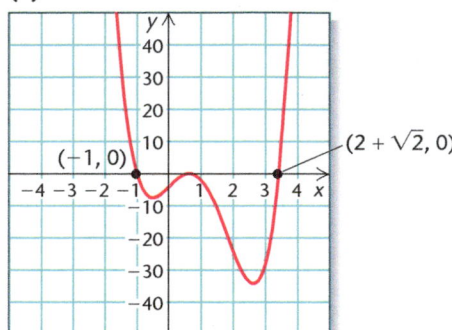

$f(x) = 3x^4 - 11x^3 + 10x - 4$

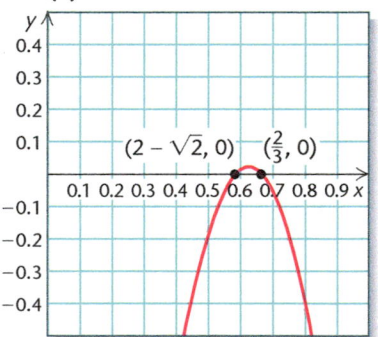

The solution set of the inequality

$$3x^4 - 11x^3 + 10x - 4 \leq 0$$

is

$$\left[-1, 2 - \sqrt{2}\right] \cup \left[\tfrac{2}{3}, 2 + \sqrt{2}\right].$$

Now Try Exercise 45.

Technology Connection

Polynomial inequalities can be solved quickly with a graphing calculator. Consider the inequality in Example 4:

$$3x^4 + 10x \le 11x^3 + 4, \quad \text{or}$$
$$3x^4 - 11x^3 + 10x - 4 \le 0.$$

We graph the related equation

$$y = 3x^4 - 11x^3 + 10x - 4$$

and use the ZERO feature. We see in the window on the left below that two of the zeros are -1 and approximately 3.414 $\left(2 + \sqrt{2} \approx 3.414\right)$. However, this window leaves us uncertain about the number of zeros of the function on the interval $[0, 1]$.

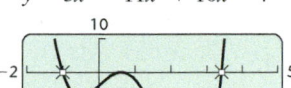

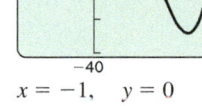

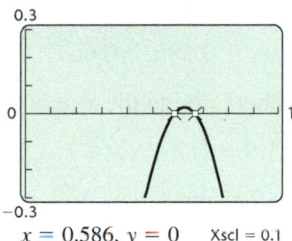

$x = -1, \quad y = 0$ $x = 0.586, y = 0$ Xscl = 0.1
$x = 3.414, y = 0$ $x = 0.667, y = 0$ Yscl = 0.1

The window on the right above shows another view of the zeros on the interval $[0, 1]$. Those zeros are about 0.586 and 0.667 $\left(2 - \sqrt{2} \approx 0.586; \frac{2}{3} \approx 0.667\right)$. The intervals to be considered are $(-\infty, -1)$, $(-1, 0.586)$, $(0.586, 0.667)$, $(0.667, 3.414)$, and $(3.414, \infty)$. We note on the graph where the function is negative. Then, including appropriate endpoints, we find that the solution set is approximately

$$[-1, 0.586] \cup [0.667, 3.414], \quad \text{or}$$
$$\{x \,|\, -1 \le x \le 0.586 \,\text{ or }\, 0.667 \le x \le 3.414\}.$$

▶ Rational Inequalities

Some inequalities involve rational expressions and functions. These are called **rational inequalities**. To solve rational inequalities, we must make some adjustments to the preceding method.

EXAMPLE 5   Solve: $\dfrac{3x}{x + 6} < 0$.

Solution   We look for all values of x for which the related function

$$f(x) = \frac{3x}{x + 6}$$

is not defined or is 0. These are called **critical values**.

The denominator tells us that $f(x)$ is not defined when $x = -6$. Next, we solve $f(x) = 0$:

$$\frac{3x}{x + 6} = 0$$

$$(x + 6) \cdot \frac{3x}{x + 6} = (x + 6) \cdot 0 \qquad \text{Multiplying by } x + 6$$

$$3x = 0$$

$$x = 0.$$

The critical values are -6 and 0. These values divide the x-axis into three intervals:

$$(-\infty, -6), \qquad (-6, 0), \quad \text{and} \quad (0, \infty).$$

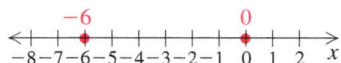

We then use a test value to determine the sign of $f(x)$ in each interval.

Interval	$(-\infty, -6)$	$(-6, 0)$	$(0, \infty)$
Test Value	$f(-8) = 12$	$f(-2) = -\dfrac{3}{2}$	$f(3) = 1$
Sign of $f(x)$	Positive	Negative	Positive

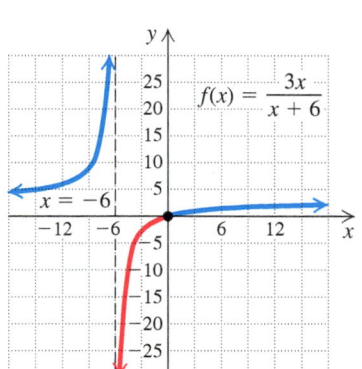

Function values are negative on only the interval $(-6, 0)$. Although $f(0) = 0$, the inequality symbol is $<$, so we know that 0 is not included in the solution set. Note that since -6 is not in the domain of f, -6 cannot be part of the solution set. The solution set is

$$(-6, 0), \quad \text{or} \quad \{x \mid -6 < x < 0\}.$$

The graph of $f(x)$ shows where $f(x)$ is positive and where it is negative.

Now Try Exercise 57.

EXAMPLE 6   Solve: $\dfrac{x + 1}{2x - 4} \le 1$.

Solution   We first subtract 1 on both sides in order to find an equivalent inequality with 0 on one side:

$$\frac{x + 1}{2x - 4} - 1 \le 0.$$

Algebraic Solution

We look for all values of x for which the related function

$$f(x) = \frac{x+1}{2x-4} - 1$$

is not defined or is 0. These are called **critical values**.

A look at the denominator shows that $f(x)$ is not defined for $x = 2$. Next, we solve $f(x) = 0$:

$$\frac{x+1}{2x-4} - 1 = 0$$

$$(2x-4)\left(\frac{x+1}{2x-4} - 1\right) = (2x-4)\cdot 0$$

$$x + 1 - (2x-4)\cdot 1 = 0$$

$$x + 1 - 2x + 4 = 0$$

$$-x = -5$$

$$x = 5.$$

The critical values are 2 and 5. These values divide the x-axis into three intervals:

$$(-\infty, 2) \quad (2, 5), \quad \text{and} \quad (5, \infty).$$

We then use a test value to determine the sign of $f(x)$ in each interval.

INTERVAL	TEST VALUE	SIGN OF $f(x)$
$(-\infty, 2)$	$f(0) = -\dfrac{5}{4}$	$-$
$(2, 5)$	$f(3) = 1$	$+$
$(5, \infty)$	$f(6) = -\dfrac{1}{8}$	$-$

Function values are negative on the intervals $(-\infty, 2)$ and $(5, \infty)$. Since $f(5) = 0$ and the inequality symbol is $\leq$, we know that 5 is in the solution set. Note that since 2 is not in the domain of f, it cannot be part of the solution set. The solution set is $(-\infty, 2) \cup [5, \infty)$.

Visualizing the Solution

The graph of the related function

$$f(x) = \frac{x+1}{2x-4} - 1$$

confirms the two critical values found algebraically: 2 where $f(x)$ is not defined and 5 where $f(x) = 0$.

$$f(x) = \frac{x+1}{2x-4} - 1$$

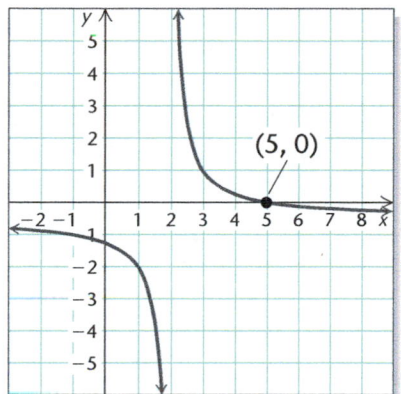

The graph shows where $f(x)$ is negative. Note that 2 cannot be in the solution set since $f(x)$ is not defined for this value. We do include 5, however, since the inequality symbol is $\leq$ and $f(5) = 0$.

The solution set is

$$(-\infty, 2) \cup [5, \infty).$$

Now Try Exercise 59.

EXAMPLE 7   Solve: $\dfrac{x-3}{x+4} \geq \dfrac{x+2}{x-5}$.

Solution   We first subtract $(x+2)/(x-5)$ on both sides in order to find an equivalent inequality with 0 on one side:

$$\frac{x-3}{x+4} - \frac{x+2}{x-5} \geq 0.$$

Algebraic Solution

We look for all values of x for which the related function

$$f(x) = \frac{x-3}{x+4} - \frac{x+2}{x-5}$$

is not defined or is 0. These are called **critical values**.

A look at the denominators shows that $f(x)$ is not defined for $x = -4$ and $x = 5$. Next, we solve $f(x) = 0$:

$$\frac{x-3}{x+4} - \frac{x+2}{x-5} = 0$$

$$(x+4)(x-5)\left(\frac{x-3}{x+4} - \frac{x+2}{x-5}\right) = (x+4)(x-5)\cdot 0$$

$$(x-5)(x-3) - (x+4)(x+2) = 0$$

$$(x^2 - 8x + 15) - (x^2 + 6x + 8) = 0$$

$$-14x + 7 = 0$$

$$x = \tfrac{1}{2}.$$

The critical values are -4, $\tfrac{1}{2}$, and 5. These values divide the x-axis into four intervals:

$$(-\infty, -4), \quad \left(-4, \tfrac{1}{2}\right), \quad \left(\tfrac{1}{2}, 5\right), \quad \text{and} \quad (5, \infty).$$

We then use a test value to determine the sign of $f(x)$ in each interval.

INTERVAL	TEST VALUE	SIGN OF $f(x)$
$(-\infty, -4)$	$f(-5) = 7.7$	$+$
$\left(-4, \tfrac{1}{2}\right)$	$f(-2) = -2.5$	$-$
$\left(\tfrac{1}{2}, 5\right)$	$f(3) = 2.5$	$+$
$(5, \infty)$	$f(6) = -7.7$	$-$

Function values are positive in the intervals $(-\infty, -4)$ and $\left(\tfrac{1}{2}, 5\right)$. Since $f\left(\tfrac{1}{2}\right) = 0$ and the inequality symbol is $\geq$, we know that $\tfrac{1}{2}$ must be in the solution set. Note that since neither -4 nor 5 is in the domain of f, they cannot be part of the solution set.

The solution set is $(-\infty, -4) \cup \left[\tfrac{1}{2}, 5\right)$.

Visualizing the Solution

The graph of the related function

$$f(x) = \frac{x-3}{x+4} - \frac{x+2}{x-5}$$

confirms the three critical values found algebraically: -4 and 5 where $f(x)$ is not defined and $\tfrac{1}{2}$ where $f(x) = 0$.

$$f(x) = \frac{x-3}{x+4} - \frac{x+2}{x-5}$$

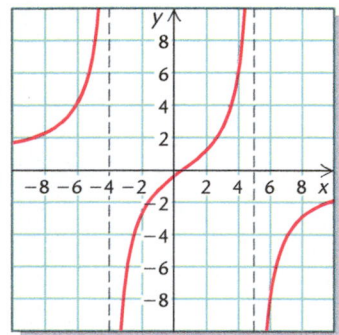

The graph shows where $f(x)$ is positive and where it is negative. Note that -4 and 5 cannot be in the solution set since $f(x)$ is not defined for these values. We do include $\tfrac{1}{2}$, however, since the inequality symbol is $\geq$ and $f\left(\tfrac{1}{2}\right) = 0$.

The solution set is

$$(-\infty, -4) \cup \left[\tfrac{1}{2}, 5\right).$$

Now Try Exercise 61.

The following is a method for solving rational inequalities.

To solve a rational inequality:

1. Find an equivalent inequality with 0 on one side.
2. Change the inequality symbol to an equals sign and solve the related equation.
3. Find values of the variable for which the related rational expression is not defined.
4. The numbers found in steps (2) and (3) are called critical values. Use the critical values to divide the x-axis into intervals. Then determine the function's sign in each interval using an x-value from the interval or using the graph of the equation.
5. Select the intervals for which the inequality is satisfied and write interval notation or set-builder notation for the solution set. If the inequality symbol is $\leq$ or $\geq$, then the solutions to step (2) should be included in the solution set. The x-values found in step (3) are never included in the solution set.

It works well to use a combination of algebraic methods and graphical methods to solve polynomial inequalities and rational inequalities. The algebraic methods give exact numbers for the critical values, and the graphical methods usually allow us to see easily what intervals satisfy the inequality.

8.6   Exercise Set

For the function $f(x) = x^2 + 2x - 15$, solve each of the following.

1. $f(x) = 0$ **2.** $f(x) < 0$

3. $f(x) \leq 0$ **4.** $f(x) > 0$

5. $f(x) \geq 0$

For the function $g(x) = \dfrac{x - 2}{x + 4}$, solve each of the following.

6. $g(x) = 0$ **7.** $g(x) > 0$

8. $g(x) \leq 0$ **9.** $g(x) \geq 0$

10. $g(x) < 0$

For the function

$$h(x) = \frac{7x}{(x - 1)(x + 5)},$$

solve each of the following.

11. $h(x) = 0$ **12.** $h(x) \leq 0$

13. $h(x) \geq 0$ **14.** $h(x) > 0$

15. $h(x) < 0$

For the function $g(x) = x^5 - 9x^3$, solve each of the following.

16. $g(x) = 0$

17. $g(x) < 0$

18. $g(x) \leq 0$

19. $g(x) > 0$

20. $g(x) \geq 0$

In Exercises 21–24, a related function is graphed. Solve the given inequality.

21. $x^3 + 6x^2 < x + 30$

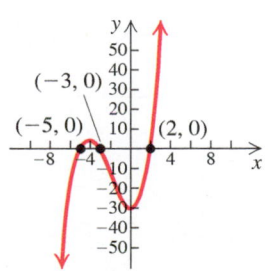

22. $x^4 - 27x^2 - 14x + 120 \geq 0$

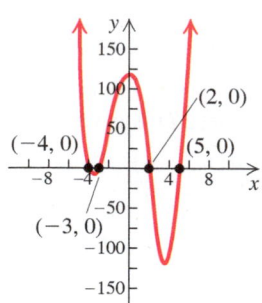

23. $\dfrac{8x}{x^2 - 4} \geq 0$

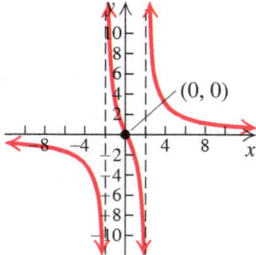

24. $\dfrac{8}{x^2 - 4} < 0$

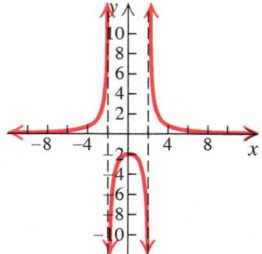

Solve.

25. $(x - 1)(x + 4) < 0$

26. $(x + 3)(x - 5) < 0$

27. $x^2 + x - 2 > 0$

28. $x^2 - x - 6 > 0$

29. $x^2 - x - 5 \geq x - 2$

30. $x^2 + 4x + 7 \geq 5x + 9$

31. $x^2 > 25$

32. $x^2 \leq 1$

33. $4 - x^2 \leq 0$

34. $11 - x^2 \geq 0$

35. $6x - 9 - x^2 < 0$

36. $x^2 + 2x + 1 \leq 0$

37. $x^2 + 12 < 4x$

38. $x^2 - 8 > 6x$

39. $4x^3 - 7x^2 \leq 15x$

40. $2x^3 - x^2 < 5x$

41. $x^3 + 3x^2 - x - 3 \geq 0$

42. $x^3 + x^2 - 4x - 4 \geq 0$

43. $x^3 - 2x^2 < 5x - 6$

44. $x^3 + x \leq 6 - 4x^2$

45. $x^5 + x^2 \geq 2x^3 + 2$

46. $x^5 + 24 > 3x^3 + 8x^2$

47. $2x^3 + 6 \leq 5x^2 + x$

48. $2x^3 + x^2 < 10 + 11x$

49. $x^3 + 5x^2 - 25x \leq 125$

50. $x^3 - 9x + 27 \geq 3x^2$

51. $0.1x^3 - 0.6x^2 - 0.1x + 2 < 0$

52. $19.2x^3 + 12.8x^2 + 144 \geq 172.8x + 3.2x^4$

List the critical values of the related function. Then solve the inequality.

53. $\dfrac{1}{x + 4} > 0$ **54.** $\dfrac{1}{x - 3} \leq 0$

55. $\dfrac{-4}{2x + 5} < 0$ **56.** $\dfrac{-2}{5 - x} \geq 0$

57. $\dfrac{2x}{x - 4} \geq 0$ **58.** $\dfrac{5x}{x + 1} < 0$

59. $\dfrac{x + 1}{x - 2} \geq 3$ **60.** $\dfrac{x}{x - 5} < 2$

61. $\dfrac{x - 4}{x + 3} - \dfrac{x + 2}{x - 1} \leq 0$ **62.** $\dfrac{x + 1}{x - 2} - \dfrac{x - 3}{x - 1} < 0$

63. $\dfrac{x + 6}{x - 2} > \dfrac{x - 8}{x - 5}$ **64.** $\dfrac{x - 7}{x + 2} \geq \dfrac{x - 9}{x + 3}$

65. $x - 2 > \dfrac{1}{x}$ **66.** $4 \geq \dfrac{4}{x} + x$

67. $\dfrac{2}{x^2 - 4x + 3} \leq \dfrac{5}{x^2 - 9}$

68. $\dfrac{3}{x^2 - 4} \leq \dfrac{5}{x^2 + 7x + 10}$

69. $\dfrac{3}{x^2 + 1} \geq \dfrac{6}{5x^2 + 2}$

70. $\dfrac{4}{x^2 - 9} < \dfrac{3}{x^2 - 25}$

71. $\dfrac{5}{x^2 + 3x} < \dfrac{3}{2x + 1}$

72. $\dfrac{2}{x^2 + 3} > \dfrac{3}{5 + 4x^2}$

73. $\dfrac{5x}{7x - 2} > \dfrac{x}{x + 1}$

74. $\dfrac{x^2 - x - 2}{x^2 + 5x + 6} < 0$

75. $\dfrac{x}{x^2 + 4x - 5} + \dfrac{3}{x^2 - 25} \le \dfrac{2x}{x^2 - 6x + 5}$

76. $\dfrac{2x}{x^2 - 9} + \dfrac{x}{x^2 + x - 12} \ge \dfrac{3x}{x^2 + 7x + 12}$

77. *Temperature During an Illness.* A person's temperature T, in degrees Fahrenheit, during an illness is given by the function

$$T(t) = \dfrac{4t}{t^2 + 1} + 98.6,$$

where t is the time since the onset of the illness, in hours. Find the interval on which the temperature was over 100°F. (See Example 12 in Section 8.5.)

78. *Population Growth.* The population P, in thousands, of a resort community is given by

$$P(t) = \dfrac{500t}{2t^2 + 9},$$

where t is the time, in months, since the city council raised the property taxes. Find the interval on which the population was 40,000 or greater. (See Exercise 85 in Exercise Set 8.5.)

79. *Total Profit.* Flexl, Inc., determines that its total profit is given by the function

$$P(x) = -3x^2 + 630x - 6000.$$

a) Flexl makes a profit for those nonnegative values of x for which $P(x) > 0$. Find the values of x for which Flexl makes a profit.

b) Flexl loses money for those nonnegative values of x for which $P(x) < 0$. Find the values of x for which Flexl loses money.

80. *Height of a Thrown Object.* The function

$$S(t) = -16t^2 + 32t + 1920$$

gives the height S, in feet, of an object thrown upward from a cliff that is 1920 ft high. Here t is the time, in seconds, that the object is in the air.

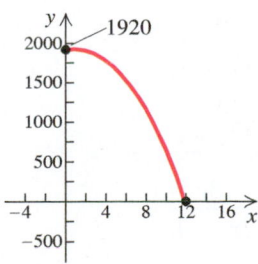

a) For what times is the height greater than 1920 ft?

b) For what times is the height less than 640 ft?

81. *Number of Diagonals.* A polygon with n sides has D diagonals, where D is given by the function

$$D(n) = \dfrac{n(n - 3)}{2}.$$

Find the number of sides n if

$$27 \le D \le 230.$$

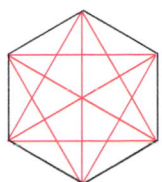

82. *Number of Handshakes.* If there are n people in a room, the number N of possible handshakes by all the people in the room is given by the function

$$N(n) = \dfrac{n(n - 1)}{2}.$$

For what number n of people is

$$66 \le N \le 300?$$

▶ **Skill Maintenance**

Solve each system of equations. [3.2], [3.3]

83. $2x + y = 2,$
    $x = y + 7$

84. $2x + 3y = 4,$
    $3x - 2y = -7$

In Exercises 85 and 86, **(a)** *find the vertex;* **(b)** *determine whether there is a maximum or a minimum value and find that value; and* **(c)** *find the range.*

85. $h(x) = -2x^2 + 3x - 8$ [3.3]

86. $g(x) = x^2 - 10x + 2$ [3.3]

► Synthesis

Solve.

87. $|x^2 - 5| = 5 - x^2$

88. $x^4 - 6x^2 + 5 > 0$

89. $2|x|^2 - |x| + 2 \le 5$

90. $(7 - x)^{-2} < 0$

91. $\left| 1 + \dfrac{1}{x} \right| < 3$

92. $\left| 2 - \dfrac{1}{x} \right| \le 2 + \left| \dfrac{1}{x} \right|$

93. Write a quadratic inequality for which the solution set is $(-4, 3)$.

94. Write a polynomial inequality for which the solution set is $[-4, 3] \cup [7, \infty)$.

Find the domain of the function.

95. $f(x) = \sqrt{\dfrac{72}{x^2 - 4x - 21}}$

96. $f(x) = \sqrt{x^2 - 4x - 21}$

Chapter 8 Summary and Review

STUDY GUIDE

KEY TERMS AND CONCEPTS

EXAMPLES

SECTION 8.1: POLYNOMIAL FUNCTIONS AND MODELS

Polynomial Function

$$P(x) = a_n x^n + a_{n-1}x^{n-1} + a_{n-2}x^{n-2} + \cdots + a_1 x + a_0,$$

where the coefficients $a_n, a_{n-1}, \ldots, a_1, a_0$ are real numbers and the exponents are whole numbers.

The first nonzero coefficient, a_n, is called the **leading coefficient**. The term $a_n x^n$ is called the **leading term**. The **degree** of the polynomial function is n.

Classifying polynomial functions by degree:

Type	Degree
Constant	0
Linear	1
Quadratic	2
Cubic	3
Quartic	4

Consider the polynomial

$$P(x) = \frac{1}{3}x^2 + x - 4x^5 + 2.$$

Leading term: $-4x^5$

Leading coefficient: -4

Degree of polynomial: 5

Classify the following polynomial functions:

Function	Type
$f(x) = -2$	Constant
$f(x) = 0.6x - 11$	Linear
$f(x) = 5x^2 + x - 4$	Quadratic
$f(x) = 5x^3 - x + 10$	Cubic
$f(x) = -x^4 + 8x^3 + x$	Quartic

The Leading-Term Test

If $a_n x^n$ is the leading term of a polynomial function, then the behavior of the graph as $x \to \infty$ and as $x \to -\infty$ can be described in one of the following four ways.

a) If n is even, and $a_n > 0$:

b) If n is even, and $a_n < 0$:

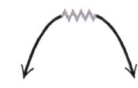

c) If n is odd, and $a_n > 0$:

d) If n is odd, and $a_n < 0$:

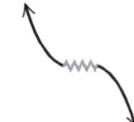

Using the leading-term test, describe the end behavior of the graph of each function by selecting one of (a)–(d) shown at left.

$$h(x) = -2x^6 + x^4 - 3x^2 + x$$

The leading term $a_n x^n$ is $-2x^6$. Since 6 is even and $-2 < 0$, the shape is shown in (b).

$$g(x) = 4x^3 - 8x + 1$$

The leading term, $a_n x^n$, is $4x^3$. Since 3 is odd and $4 > 0$, the shape is shown in (c).

Zeros of Functions

If c is a real zero of a function $f(x)$ (that is, $f(c) = 0$), then $x - c$ is a factor of $f(x)$ and $(c, 0)$ is an x-intercept of the graph of the function.

If we know the linear factors of a polynomial function $f(x)$, we can find the zeros of $f(x)$ by solving the equation $f(x) = 0$ using the principle of zero products.

Every function of degree n, with $n \geq 1$, has at least one zero and at most n zeros.

To find the zeros of

$$f(x) = -2(x - 3)(x + 8)^2,$$

solve $-2(x - 3)(x + 8)^2 = 0$ using the principle of zero products:

$$x - 3 = 0 \quad or \quad x + 8 = 0$$
$$x = 3 \quad or \quad x = -8.$$

The zeros of $f(x)$ are 3 and -8.

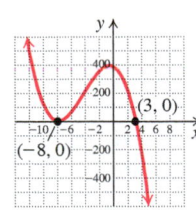

$f(x) = -2(x - 3)(x + 8)^2$

To find the zeros of

$$h(x) = x^4 - 12x^2 - 64,$$

solve $h(x) = 0$:

$$x^4 - 12x^2 - 64 = 0$$
$$(x^2 - 16)(x^2 + 4) = 0$$
$$(x + 4)(x - 4)(x^2 + 4) = 0$$
$$x + 4 = 0 \quad or \quad x - 4 = 0 \quad or \quad x^2 + 4 = 0$$
$$x = -4 \quad or \quad x = 4 \quad or \quad x^2 = -4$$
$$= \pm\sqrt{-4}$$
$$= \pm 2i.$$

The zeros of $h(x)$ are $-4, 4$, and $\pm 2i$.

Even and Odd Multiplicity

If $(x - c)^k, k \geq 1$, is a factor of a polynomial function $P(x)$ and $(x - c)^{k+1}$ is not a factor and:

- k is odd, then the graph crosses the x-axis at $(c, 0)$;
- k is even, then the graph is tangent to the x-axis at $(c, 0)$.

For $f(x) = -2(x - 3)(x + 8)^2$ graphed above, note that for the factor $x - 3$, or $(x - 3)^1$, the exponent 1 is odd and the graph crosses the x-axis at $(3, 0)$. For the factor $(x + 8)^2$, the exponent 2 is even and the graph is tangent to the x-axis at $(-8, 0)$.

SECTION 8.2: GRAPHING POLYNOMIAL FUNCTIONS

If $P(x)$ is a polynomial function of degree n, the graph of the function has:

- at most n real zeros, and thus at most n x-intercepts, and
- at most $n - 1$ turning points.

To Graph a Polynomial Function

1. Use the leading-term test to determine the end behavior.

2. Find the zeros of the function by solving $f(x) = 0$. Any real zeros are the first coordinates of the x-intercepts.

3. Use the x-intercepts (zeros) to divide the x-axis into intervals and choose a test point in each interval to determine the sign of all function values in that interval. For all x-values in an interval, $f(x)$ is either always positive for all values or always negative for all values.

4. Find $f(0)$. This gives the y-intercept of the function.

5. If necessary, find additional function values to determine the general shape of the graph and then draw the graph.

Graph: $h(x) = x^4 - 12x^2 - 16x = x(x - 4)(x + 2)^2$.

1. The leading term is x^4. Since 4 is even and $1 > 0$, the end behavior of the graph can be sketched as follows.

2. Solve $x(x - 4)(x + 2)^2 = 0$. The solutions are 0, 4, and -2. The zeros of $h(x)$ are 0, 4, and -2. The x-intercepts are $(0, 0)$, $(4, 0)$, and $(-2, 0)$. The multiplicity of 0 and 4 is 1. The graph will cross the x-axis at 0 and 4. The multiplicity of -2 is 2. The graph is tangent to the x-axis at -2.

3. The zeros divide the x-axis into four intervals.

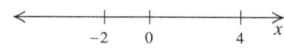

Interval	$(-\infty, -2)$	$(-2, 0)$	$(0, 4)$	$(4, \infty)$
Test Value	-3	-1	1	5
Function Value, $h(x)$	21	5	-27	245
Sign of $h(x)$	$+$	$+$	$-$	$+$
Location of Points on Graph	Above x-axis	Above x-axis	Below x-axis	Above x-axis

Four points on the graph are $(-3, 21)$, $(-1, 5)$, $(1, -27)$, and $(5, 245)$.

4. Find $h(0)$:

$$h(0) = 0(0 - 4)(0 + 2)^2 = 0.$$

The y-intercept is $(0, 0)$.

(continued)

5. Find additional points and draw the graph.

x	$h(x)$
-2.5	4.1
-1.5	2.1
-0.5	5.1
0.5	-10.9
2	-64
3	-75

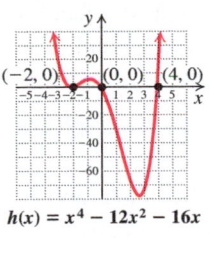

$h(x) = x^4 - 12x^2 - 16x$

The Intermediate Value Theorem

For any polynomial function $P(x)$ with real coefficients, suppose that for $a \neq b$, $P(a)$ and $P(b)$ are of opposite signs. Then the function has at least one real zero between a and b.

The intermediate value theorem *cannot* be used to determine whether there is a real zero between a and b when $P(a)$ and $P(b)$ have the *same* sign.

Use the intermediate value theorem to determine, if possible, whether the function has a zero between a and b.

$$f(x) = 2x^3 - 5x^2 + x - 2, \quad a = 2, b = 3;$$
$$f(2) = -4; \ f(3) = 10$$

Since $f(2)$ and $f(3)$ have opposite signs, $f(x)$ has at least one zero between 2 and 3.

$$f(x) = 2x^3 - 5x^2 + x - 2, \quad a = -2, b = -1;$$
$$f(-2) = -40; \ f(-1) = -10$$

Both $f(-2)$ and $f(-1)$ are negative. Thus the intermediate value theorem does not allow us to determine whether there is a real zero between -2 and -1.

SECTION 8.3: POLYNOMIAL DIVISION; THE REMAINDER THEOREM AND THE FACTOR THEOREM

Polynomial Division

$$P(x) = d(x) \cdot Q(x) + R(x)$$

Dividend   Divisor   Quotient   Remainder

When we divide a polynomial $P(x)$ by a divisor $d(x)$, a polynomial $Q(x)$ is the quotient and a polynomial $R(x)$ is the remainder. The quotient $Q(x)$ must have degree less than that of the dividend $P(x)$. The remainder $R(x)$ must either be 0 or have degree less than that of the divisor $d(x)$. If $R(x) = 0$, then the divisor $d(x)$ is a factor of the dividend.

Given $P(x) = x^4 - 6x^3 + 9x^2 + 4x - 12$ and $d(x) = x + 2$, use long division to find the quotient and the remainder when $P(x)$ is divided by $d(x)$. Express $P(x)$ in the form $d(x) \cdot Q(x) + R(x)$.

$$
\begin{array}{r}
x^3 - 8x^2 + 25x - 46 \\
x + 2 \overline{)\, x^4 - 6x^3 + 9x^2 + 4x - 12} \\
\underline{x^4 + 2x^3} \\
-8x^3 + 9x^2 \\
\underline{-8x^3 - 16x^2} \\
25x^2 + 4x \\
\underline{25x^2 + 50x} \\
-46x - 12 \\
\underline{-46x - 92} \\
80
\end{array}
$$

$Q(x) = x^3 - 8x^2 + 25x - 46$ and $R(x) = 80$. Thus, $P(x) = (x + 2)(x^3 - 8x^2 + 25x - 46) + 80$. Since $R(x) \neq 0$, $x + 2$ is not a factor of $P(x)$.

The Remainder Theorem

If a number c is substituted for x in the polynomial $f(x)$, then the result $f(c)$ is the remainder that would be obtained by dividing $f(x)$ by $x - c$. That is, if $f(x) = (x - c) \cdot Q(x) + R$, then $f(c) = R$.

The long-division process can be streamlined with synthetic division. Synthetic division can also be used to find polynomial function values.

Repeat the division shown above using synthetic division. Note that the divisor $x + 2 = x - (-2)$.

$$
\begin{array}{r|rrrrr}
-2 & 1 & -6 & 9 & 4 & -12 \\
   &   & -2 & 16 & -50 & 92 \\
\hline
   & 1 & -8 & 25 & -46 & \mid 80
\end{array}
$$

Again, note that $Q(x) = x^3 - 8x^2 + 25x - 46$ and $R(x) = 80$. Since $R(x) \neq 0$, $x - (-2)$, or $x + 2$, is not a factor of $P(x)$.

Now divide $P(x)$ by $x - 3$.

$$
\begin{array}{r|rrrrr}
3 & 1 & -6 & 9 & 4 & -12 \\
  &   & 3 & -9 & 0 & 12 \\
\hline
  & 1 & -3 & 0 & 4 & \mid 0
\end{array}
$$

$Q(x) = x^3 - 3x^2 + 4$ and $R(x) = 0$. Since $R(x) = 0$, $x - 3$ is a factor of $P(x)$.

For $f(x) = 2x^5 - x^3 - 3x^2 - 4x + 15$, find $f(-2)$.

$$
\begin{array}{r|rrrrrr}
-2 & 2 & 0 & -1 & -3 & -4 & 15 \\
   &   & -4 & 8 & -14 & 34 & -60 \\
\hline
   & 2 & -4 & 7 & -17 & 30 & \mid -45
\end{array}
$$

Thus, $f(-2) = -45$.

The Factor Theorem

For a polynomial $f(x)$, if $f(c) = 0$, then $x - c$ is a factor of $f(x)$.

Let $g(x) = x^4 + 8x^3 + 6x^2 - 40x + 25$. Factor $g(x)$ and solve $g(x) = 0$.

Use synthetic division to look for factors of the form $x - c$. Let's try $x + 5$.

$$
\begin{array}{r|rrrrr}
-5 & 1 & 8 & 6 & -40 & 25 \\
   &   & -5 & -15 & 45 & -25 \\
\hline
   & 1 & 3 & -9 & 5 & \mid 0
\end{array}
$$

Since $g(-5) = 0$, the number -5 is a zero of $g(x)$ and $x - (-5)$, or $x + 5$, is a factor of $g(x)$. This gives us

$$g(x) = (x + 5)(x^3 + 3x^2 - 9x + 5).$$

Let's try $x + 5$ again with the factor $x^3 + 3x^2 - 9x + 5$.

$$
\begin{array}{r|rrrr}
-5 & 1 & 3 & -9 & 5 \\
   &   & -5 & 10 & -5 \\
\hline
   & 1 & -2 & 1 & \mid 0
\end{array}
$$

Now we have

$$g(x) = (x + 5)^2(x^2 - 2x + 1).$$

The trinomial $x^2 - 2x + 1$ easily factors, so

$$g(x) = (x + 5)^2(x - 1)^2.$$

Solve $g(x) = 0$. The solutions of $(x + 5)^2(x - 1)^2 = 0$ are -5 and 1. They are also the zeros of $g(x)$.

SECTION 8.4: THEOREMS ABOUT ZEROS OF POLYNOMIAL FUNCTIONS

The Fundamental Theorem of Algebra

Every polynomial function of degree n, $n \geq 1$, with complex coefficients has at least one zero in the system of complex numbers.

Every polynomial function f of degree n, with $n \geq 1$, can be factored into n linear factors (not necessarily unique); that is,

$$f(x) = a_n(x - c_1)(x - c_2)\cdots(x - c_n).$$

Nonreal Zeros

$a + bi$ and **$a - bi$, $b \neq 0$**

If a complex number $a + bi$, $b \neq 0$, is a zero of a polynomial function $f(x)$ with *real* coefficients, then its conjugate, $a - bi$, is also a zero. (Nonreal zeros occur in conjugate pairs.)

Irrational Zeros

$a + c\sqrt{b}$, and **$a - c\sqrt{b}$,**
b not a perfect square

If $a + c\sqrt{b}$, where a, b, and c are rational and b is not a perfect square, is a zero of a polynomial function $f(x)$ with *rational* coefficients, then its conjugate, $a - c\sqrt{b}$, is also a zero. (Irrational zeros occur in conjugate pairs.)

Find a polynomial function of degree 5 with -4 and 2 as zeros of multiplicity 1, and -1 as a zero of multiplicity 3.

$$\begin{aligned} f(x) &= [x - (-4)][x - 2][x - (-1)]^3 \\ &= (x + 4)(x - 2)(x + 1)^3 \\ &= x^5 + 5x^4 + x^3 - 17x^2 - 22x - 8 \end{aligned}$$

Find a polynomial function with rational coefficients of lowest degree with $1 - i$ and $\sqrt{7}$ as two of its zeros.

If $1 - i$ is a zero, then $1 + i$ is also a zero. If $\sqrt{7}$ is a zero, then $-\sqrt{7}$ is also a zero.

$$\begin{aligned} f(x) &= [x - (1 - i)][x - (1 + i)][x - \sqrt{7}] \times \\ &\quad [x - (-\sqrt{7})] \\ &= [(x - 1) + i][(x - 1) - i](x - \sqrt{7})(x + \sqrt{7}) \\ &= [(x - 1)^2 - i^2](x^2 - 7) \\ &= (x^2 - 2x + 1 + 1)(x^2 - 7) \\ &= (x^2 - 2x + 2)(x^2 - 7) \\ &= x^4 - 2x^3 - 5x^2 + 14x - 14 \end{aligned}$$

The Rational Zeros Theorem

Consider the polynomial function

$$P(x) = a_n x^n + a_{n-1}x^{n-1} + a_{n-2}x^{n-2}$$
$$+ \cdots + a_1 x + a_0,$$

where all the coefficients are integers and $n \geq 1$. Also, consider a rational number p/q, where p and q have no common factor other than -1 and 1. If p/q is a zero of $P(x)$, then p is a factor of a_0 and q is a factor of a_n.

For $f(x) = 2x^4 - 9x^3 - 16x^2 - 9x - 18$, solve $f(x) = 0$ and factor $f(x)$ into linear factors.

There are at most 4 distinct zeros. Any rational zeros of f must be of the form p/q, where p is a factor of -18 and q is a factor of 2.

Possibilities for p: $\pm 1, \pm 2, \pm 3, \pm 6, \pm 9, \pm 18$
Possibilities for q: $\underline{\pm 1, \pm 2}$
Possibilities for p/q: $1, -1, 2, -2, 3, -3, 6, -6, 9, -9,$
$\qquad 18, -18, \dfrac{1}{2}, -\dfrac{1}{2}, \dfrac{3}{2}, -\dfrac{3}{2}, \dfrac{9}{2}, -\dfrac{9}{2}$

Use synthetic division to check the possibilities. We leave it to the student to verify that ± 1, ± 2, and ± 3 are not zeros. Let's try 6.

$$\begin{array}{r|rrrrr} 6 & 2 & -9 & -16 & -9 & -18 \\ & & 12 & 18 & 12 & 18 \\ \hline & 2 & 3 & 2 & 3 & 0 \end{array}$$

(continued)

Since $f(6) = 0$, 6 is a zero and $x - 6$ is a factor of $f(x)$. Now express $f(x)$ as

$$f(x) = (x - 6)(2x^3 + 3x^2 + 2x + 3).$$

Consider the factor $2x^3 + 3x^2 + 2x + 3$ and check the other possibilities. Let's try $-\frac{3}{2}$.

$$
\begin{array}{r|rrrr}
-\frac{3}{2} & 2 & 3 & 2 & 3 \\
 & & -3 & 0 & -3 \\
\hline
 & 2 & 0 & 2 & 0
\end{array}
$$

Since $f\left(-\frac{3}{2}\right) = 0$, $-\frac{3}{2}$ is also a zero and $x + \frac{3}{2}$ is a factor of $f(x)$. We express $f(x)$ as

$$
\begin{aligned}
f(x) &= (x - 6)\left(x + \tfrac{3}{2}\right)(2x^2 + 2) \\
&= 2(x - 6)\left(x + \tfrac{3}{2}\right)(x^2 + 1).
\end{aligned}
$$

Now solve the equation $f(x) = 0$ to determine the zeros. We see that the only rational zeros are 6 and $-\frac{3}{2}$. The other zeros are $\pm i$.

The factorization into linear factors is

$$
\begin{aligned}
f(x) &= 2(x - 6)\left(x + \tfrac{3}{2}\right)(x - i)(x + i), \quad \text{or} \\
&\quad (x - 6)(2x + 3)(x - i)(x + i).
\end{aligned}
$$

Descartes' Rule of Signs

Let $P(x)$, written in descending or ascending order, be a polynomial function with real coefficients and a nonzero constant term. The number of positive real zeros of $P(x)$ is either:

1. The same as the number of variations of sign in $P(x)$, or
2. Less than the number of variations of sign in $P(x)$ by a positive even integer.

The number of negative real zeros of $P(x)$ is either:

3. The same as the number of variations of sign in $P(-x)$, or
4. Less than the number of variations of sign in $P(-x)$ by a positive even integer.

A zero of multiplicity m must be counted m times.

Determine the number of positive real zeros and the number of negative real zeros of

$$P(x) = 4x^5 - x^4 - 2x^3 + 8x - 10.$$

There are 3 variations of sign in $P(x)$. Thus the number of positive real zeros is 3 or 1.

$$P(-x) = -4x^5 - x^4 + 2x^3 - 8x - 10$$

There are 2 variations of sign in $P(-x)$. Thus the number of negative real zeros is 2 or 0.

SECTION 8.5: RATIONAL FUNCTIONS

Rational Function

$$f(x) = \frac{p(x)}{q(x)},$$

where $p(x)$ and $q(x)$ are polynomials and $q(x)$ is not the zero polynomial. The domain of $f(x)$ consists of all x for which $q(x) \neq 0$.

Determine the domain of each function.

FUNCTION	DOMAIN
$f(x) = \dfrac{1}{x^5}$	$(-\infty, 0) \cup (0, \infty)$
$f(x) = \dfrac{x + 6}{x^2 + 2x - 8}$	
$\quad = \dfrac{x + 6}{(x - 2)(x + 4)}$	$(-\infty, -4) \cup (-4, 2) \cup (2, \infty)$

Vertical Asymptotes

For a rational function $f(x) = p(x)/q(x)$, where $p(x)$ and $q(x)$ are polynomials with *no common factors* other than constants, if a is a zero of the denominator, then the line $x = a$ is a vertical asymptote for the graph of the function.

Horizontal Asymptotes

When the numerator and the denominator have the same degree, the line $y = a/b$ is the horizontal asymptote, where a and b are the leading coefficients of the numerator and the denominator, respectively.

When the degree of the numerator is less than the degree of the denominator, the x-axis, or $y = 0$, is the horizontal asymptote.

When the degree of the numerator is greater than the degree of the denominator, there is *no* horizontal asymptote.

Oblique Asymptotes

When the degree of the numerator is 1 greater than the degree of the denominator, there is an oblique asymptote.

Determine the vertical, horizontal, and oblique asymptotes of the graph of the function.

FUNCTION	ASYMPTOTES
$f(x) = \dfrac{x^2 - 2}{x - 1}$	Vertical: $x = 1$ Horizontal: None Oblique: $y = x + 1$
$f(x) = \dfrac{3x - 4}{x^2 + 6x - 7}$	Vertical: $x = -7; x = 1$ Horizontal: $y = 0$ Oblique: None
$f(x) = \dfrac{2x^2 + 9x - 5}{3x^2 + 13x + 12}$	Vertical: $x = -\dfrac{4}{3}; x = -3$ Horizontal: $y = \dfrac{2}{3}$ Oblique: None

To Graph a Rational Function

$f(x) = p(x)/q(x)$, where $p(x)$ and $q(x)$ have no common factor other than constants:

1. Find any real zeros of the denominator. Determine the domain of the function and sketch any vertical asymptotes.

2. Find the horizontal asymptote or the oblique asymptote, if there is one, and sketch it.

Graph: $g(x) = \dfrac{x^2 - 4}{x^2 + 4x - 5}$.

Domain: The zeros of the denominator are -5 and 1. The domain is $(-\infty, -5) \cup (-5, 1) \cup (1, \infty)$.

Vertical asymptotes: Since neither zero of the denominator is a zero of the numerator, the graph has vertical asymptotes at $x = -5$ and $x = 1$.

Horizontal asymptote: The degree of the numerator is the same as the degree of the denominator, so the horizontal asymptote is determined by the ratio of the leading coefficients: $1/1$, or 1. The horizontal asymptote is $y = 1$.

(continued)

3. Find any zeros of the function. The zeros are found by determining the zeros of the numerator. These are the first coordinates of the *x*-intercepts of the graph.

4. Find $f(0)$. This gives the *y*-intercept, $(0, f(0))$, of the graph.

5. Find other function values to determine the general shape. Then draw the graph.

Crossing an Asymptote

The graph of a rational function never crosses a vertical asymptote.

The graph of a rational function might cross a horizontal asymptote but does not necessarily do so.

Oblique asymptote: None

Zeros of g: Solving $g(x) = 0$ gives us -2 and 2, so the zeros are -2 and 2.

x-intercepts: $(-2, 0)$ and $(2, 0)$

y-intercept: $\left(0, \dfrac{4}{5}\right)$, because $g(0) = \dfrac{4}{5}$

Other values:

x	y
-8	2.22
-6	4.57
-4	-2.4
-3	-0.63
-1	0.38
0.5	1.36
1.5	-0.54
3	0.31
4	0.44

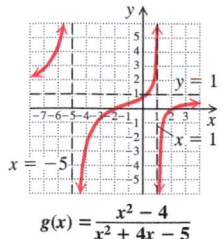

$$g(x) = \frac{x^2 - 4}{x^2 + 4x - 5}$$

Graph: $f(x) = \dfrac{x + 3}{x^2 - x - 12}$.

Domain: The zeros of the denominator are -3 and 4. The domain is $(-\infty, -3) \cup (-3, 4) \cup (4, \infty)$.

Vertical asymptote: Since 4 is the only zero of the denominator that is not a zero of the numerator, the only vertical asymptote is $x = 4$.

Horizontal asymptote: Because the degree of the numerator is less than the degree of the denominator, the *x*-axis, $y = 0$, is the horizontal asymptote.

Oblique asymptote: None

Zeros of f: The equation $f(x) = 0$ has no solutions that are in the domain of the function, so there are no zeros of f.

x-intercepts: None

y-intercept: $\left(0, -\dfrac{1}{4}\right)$ because $f(0) = -\dfrac{1}{4}$

(continued)

Hole in the graph:

$$f(x) = \frac{x + 3}{(x + 3)(x - 4)} = \frac{1}{x - 4},$$

where $x \neq -3$ and $x \neq 4$.

To determine the coordinates of the hole, substitute -3 for x in $f(x) = 1/(x - 4)$:

$$f(-3) = \frac{1}{-3 - 4} = -\frac{1}{7}.$$

The hole is located at $\left(-3, -\frac{1}{7}\right)$.

Other values:

x	y
-4	-0.13
-2	-0.17
1	-0.33
3	-1
5	1
7	0.33

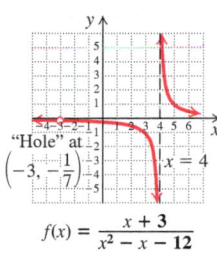

"Hole" at $\left(-3, -\frac{1}{7}\right)$

$x = 4$

$$f(x) = \frac{x + 3}{x^2 - x - 12}$$

SECTION 8.6: POLYNOMIAL INEQUALITIES AND RATIONAL INEQUALITIES

To Solve a Polynomial Inequality

1. Find an equivalent inequality with 0 on one side.

2. Change the inequality symbol to an equals sign and solve the related equation.

3. Use the solutions to divide the x-axis into intervals. Then select a test value from each interval and determine the polynomial's sign on the interval.

4. Determine the intervals for which the inequality is satisfied and write interval notation or set-builder notation for the solution set. Include the endpoints of the intervals in the solution set if the inequality symbol is $\leq$ or $\geq$.

Solve: $x^3 - 3x^2 \leq 6x - 8$.

Equivalent inequality: $x^3 - 3x^2 - 6x + 8 \leq 0$.

First, solve the related equation:

$$x^3 - 3x^2 - 6x + 8 = 0.$$

The solutions are -2, 1, and 4. The numbers divide the x-axis into 4 intervals. Next, let $f(x) = x^3 - 3x^2 - 6x + 8$ and, using test values for $f(x)$, determine the sign of $f(x)$ in each interval.

INTERVAL	TEST VALUE	SIGN OF $f(x)$
$(-\infty, -2)$	$f(-3) = -28$	$-$
$(-2, 1)$	$f(0) = 8$	$+$
$(1, 4)$	$f(2) = -8$	$-$
$(4, \infty)$	$f(6) = 80$	$+$

Test values are negative in the intervals $(-\infty, -2)$ and $(1, 4)$. Since the inequality sign is $\leq$, include the endpoints of the intervals in the solution set.

The solution set is

$$(-\infty, -2] \cup [1, 4].$$

To Solve a Rational Inequality

1. Find an equivalent inequality with 0 on one side.

2. Change the inequality symbol to an equals sign and solve the related equation.

3. Find values of the variable for which the related rational function is not defined.

4. The numbers found in steps (2) and (3) are called *critical values*. Use the critical values to divide the *x*-axis into intervals. Then determine the function's sign in each interval using an *x*-value from the interval or using the graph of the equation.

5. Select the intervals for which the inequality is satisfied and write interval notation or set-builder notation for the solution set. If the inequality symbol is $\leq$ or $\geq$, then the solutions to step (2) should be included in the solution set. The *x*-values found in step (3) are never included in the solution set.

Solve: $\dfrac{x-1}{x+5} > \dfrac{x+3}{x-2}$.

Equivalent inequality: $\dfrac{x-1}{x+5} - \dfrac{x+3}{x-2} > 0$

Related function: $f(x) = \dfrac{x-1}{x+5} - \dfrac{x+3}{x-2}$

The function is not defined for $x = -5$ and $x = 2$. Solving $f(x) = 0$, we get $x = -\frac{13}{11}$. The critical values are -5, $-\frac{13}{11}$, and 2. These divide the *x*-axis into four intervals.

INTERVAL	TEST VALUE	SIGN OF $f(x)$
$(-\infty, -5)$	$f(-6) = 6.63$	$+$
$\left(-5, -\frac{13}{11}\right)$	$f(-2) = -0.75$	$-$
$\left(-\frac{13}{11}, 2\right)$	$f(0) = 1.3$	$+$
$(2, \infty)$	$f(3) = -5.75$	$-$

Test values are positive in the intervals $(-\infty, -5)$ and $\left(-\frac{13}{11}, 2\right)$. Since $f\left(-\frac{13}{11}\right) = 0$ and -5 and 2 are not in the domain of f, -5, $-\frac{13}{11}$, and 2 cannot be part of the solution set. The solution set is

$$(-\infty, -5) \cup \left(-\tfrac{13}{11}, 2\right).$$

REVIEW EXERCISES

Determine whether the statement is true or false.

1. If $f(x) = (x + a)(x + b)(x - c)$, then $f(-b) = 0$. **[8.3]**

2. The graph of a rational function never crosses a vertical asymptote. **[8.5]**

3. For the function $g(x) = x^4 - 8x^2 - 9$, the only possible rational zeros are $1, -1, 3$, and -3. **[8.4]**

4. The graph of $P(x) = x^6 - x^8$ has at most 6 *x*-intercepts. **[8.2]**

5. The domain of the function

$$f(x) = \frac{x - 4}{(x + 2)(x - 3)}$$

is $(-\infty, -2) \cup (3, \infty)$. **[8.5]**

Determine the leading term, the leading coefficient, and the degree of the polynomial. Then classify the polynomial function as constant, linear, quadratic, cubic, or quartic. [8.1]

6. $f(x) = 7x^2 - 5 + 0.45x^4 - 3x^3$

7. $h(x) = -25$

8. $g(x) = 6 - 0.5x$

9. $f(x) = \frac{1}{3}x^3 - 2x + 3$

Use the leading-term test to describe the end behavior of the graph of the function. [8.1]

10. $f(x) = -\frac{1}{2}x^4 + 3x^2 + x - 6$

11. $f(x) = x^5 + 2x^3 - x^2 + 5x + 4$

Find the zeros of the polynomial function and state the multiplicity of each. [8.1]

12. $g(x) = \left(x - \frac{2}{3}\right)(x + 2)^3(x - 5)^2$

13. $f(x) = x^4 - 26x^2 + 25$

14. $h(x) = x^3 + 4x^2 - 9x - 36$

15. *Interest Compounded Annually.* When P dollars is invested at interest rate i, compounded annually, for t years, the investment grows to A dollars, where
$$A = P(1 + i)^t.$$

 a) Find the interest rate i if \$6250 grows to \$6760 in 2 years. [8.1]

 b) Find the interest rate i if \$1,000,000 grows to \$1,215,506.25 in 4 years. [8.1]

Sketch the graph of the polynomial function.

16. $f(x) = -x^4 + 2x^3$ [8.2]

17. $g(x) = (x - 1)^3(x + 2)^2$ [8.2]

18. $h(x) = x^3 + 3x^2 - x - 3$ [8.2]

19. $f(x) = x^4 - 5x^3 + 6x^2 + 4x - 8$ [8.2], [8.3], [8.4]

20. $g(x) = 2x^3 + 7x^2 - 14x + 5$ [8.2], [8.4]

Using the intermediate value theorem, determine, if possible, whether the function f has a zero between a and b. [8.2]

21. $f(x) = 4x^2 - 5x - 3$; $a = 1, b = 2$

22. $f(x) = x^3 - 4x^2 + \frac{1}{2}x + 2$; $a = -1, b = 1$

In each of the following, a polynomial $P(x)$ and a divisor $d(x)$ are given. Use long division to find the quotient $Q(x)$ and the remainder $R(x)$ when $P(x)$ is divided by $d(x)$. Express $P(x)$ in the form $d(x) \cdot Q(x) + R(x)$. [8.3]

23. $P(x) = 6x^3 - 2x^2 + 4x - 1$,
  $d(x) = x - 3$

24. $P(x) = x^4 - 2x^3 + x + 5$,
  $d(x) = x + 1$

Use synthetic division to find the quotient and the remainder. [8.3]

25. $(x^3 + 2x^2 - 13x + 10) \div (x - 5)$

26. $(x^4 + 3x^3 + 3x^2 + 3x + 2) \div (x + 2)$

27. $(x^5 - 2x) \div (x + 1)$

Use synthetic division to find the indicated function value. [8.3]

28. $f(x) = x^3 + 2x^2 - 13x + 10$; $f(-2)$

29. $f(x) = x^4 - 16$; $f(-2)$

30. $f(x) = x^5 - 4x^4 + x^3 - x^2 + 2x - 100$; $f(-10)$

Using synthetic division, determine whether the given numbers are zeros of the polynomial function. [8.3]

31. $-i, -5$; $f(x) = x^3 - 5x^2 + x - 5$

32. $-1, -2$; $f(x) = x^4 - 4x^3 - 3x^2 + 14x - 8$

33. $\frac{1}{3}, 1$; $f(x) = x^3 - \frac{4}{3}x^2 - \frac{5}{3}x + \frac{2}{3}$

34. $2, -\sqrt{3}$; $f(x) = x^4 - 5x^2 + 6$

Factor the polynomial $f(x)$. Then solve the equation $f(x) = 0$. [8.3], [8.4]

35. $f(x) = x^3 + 2x^2 - 7x + 4$

36. $f(x) = x^3 + 4x^2 - 3x - 18$

37. $f(x) = x^4 - 4x^3 - 21x^2 + 100x - 100$

38. $f(x) = x^4 - 3x^2 + 2$

Find a polynomial function of degree 3 with the given numbers as zeros. [8.4]

39. $-4, -1, 2$

40. $-3, 1 - i, 1 + i$

41. $\frac{1}{2}, 1 - \sqrt{2}, 1 + \sqrt{2}$

42. Find a polynomial function of degree 4 with -5 as a zero of multiplicity 3 and $\frac{1}{2}$ as a zero of multiplicity 1. [8.4]

43. Find a polynomial function of degree 5 with -3 as a zero of multiplicity 2, 2 as a zero of multiplicity 1, and 0 as a zero of multiplicity 2. [8.4]

Suppose that a polynomial function of degree 5 with rational coefficients has the given zeros. Find the other zero(s). [8.4]

44. $-\frac{2}{3}, \sqrt{5}, i$

45. $0, 1 + \sqrt{3}, -\sqrt{3}$

46. $-\sqrt{2}, \frac{1}{2}, 1, 2$

Find a polynomial function of lowest degree with rational coefficients and the following as some of its zeros. **[8.4]**

47. $\sqrt{11}$

48. $-i, 6$

49. $-1, 4, 1 + i$

50. $\sqrt{5}, -2i$

51. $\frac{1}{3}, 0, -3$

List all possible rational zeros. **[8.4]**

52. $h(x) = 4x^5 - 2x^3 + 6x - 12$

53. $g(x) = 3x^4 - x^3 + 5x^2 - x + 1$

54. $f(x) = x^3 - 2x^2 + x - 24$

For each polynomial function, **(a)** *find the rational zeros and then the other zeros; that is, solve* $f(x) = 0$; *and* **(b)** *factor* $f(x)$ *into linear factors.* **[8.4]**

55. $f(x) = 3x^5 + 2x^4 - 25x^3 - 28x^2 + 12x$

56. $f(x) = x^3 - 2x^2 - 3x + 6$

57. $f(x) = x^4 - 6x^3 + 9x^2 + 6x - 10$

58. $f(x) = x^3 + 3x^2 - 11x - 5$

59. $f(x) = 3x^3 - 8x^2 + 7x - 2$

60. $f(x) = x^5 - 8x^4 + 20x^3 - 8x^2 - 32x + 32$

61. $f(x) = x^6 + x^5 - 28x^4 - 16x^3 + 192x^2$

62. $f(x) = 2x^5 - 13x^4 + 32x^3 - 38x^2 + 22x - 5$

What does Descartes' rule of signs tell you about the number of positive real zeros and the number of negative real zeros of each of the following polynomial functions? **[8.4]**

63. $f(x) = 2x^6 - 7x^3 + x^2 - x$

64. $h(x) = -x^8 + 6x^5 - x^3 + 2x - 2$

65. $g(x) = 5x^5 - 4x^2 + x - 1$

Graph the function. Be sure to label all the asymptotes. List the domain and the x- and y-intercepts. **[8.5]**

66. $f(x) = \dfrac{x^2 - 5}{x + 2}$

67. $f(x) = \dfrac{5}{(x - 2)^2}$

68. $f(x) = \dfrac{x^2 + x - 6}{x^2 - x - 20}$

69. $f(x) = \dfrac{x - 2}{x^2 - 2x - 15}$

In Exercises 70 and 71, find a rational function that satisfies the given conditions. Answers may vary, but try to give the simplest answer possible. **[8.5]**

70. Vertical asymptotes $x = -2, x = 3$

71. Vertical asymptotes $x = -2, x = 3$; horizontal asymptote $y = 4$; x-intercept $(-3, 0)$

72. *Medical Dosage.*   The function

$$N(t) = \frac{0.7t + 2000}{8t + 9}, \quad t \geq 5,$$

gives the body concentration $N(t)$, in parts per million, of a certain dosage of medication after time t, in hours.

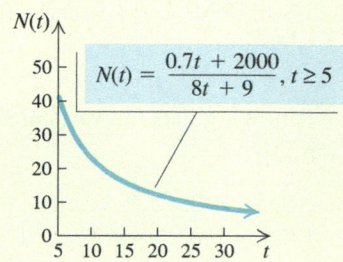

a) Find the horizontal asymptote of the graph and complete the following:

$$N(t) \rightarrow \boxed{} \text{ as } t \rightarrow \infty. \quad \textbf{[8.5]}$$

b) Explain the meaning of the answer to part (a) in terms of the application. **[8.5]**

Solve. **[8.6]**

73. $x^2 - 9 < 0$

74. $2x^2 > 3x + 2$

75. $(1 - x)(x + 4)(x - 2) \leq 0$

76. $\dfrac{x - 2}{x + 3} < 4$

77. *Height of a Rocket.*   The function

$$S(t) = -16t^2 + 80t + 224$$

gives the height S, in feet, of a model rocket launched with a velocity of 80 ft/sec from a hill that is 224 ft high, where t is the time, in seconds.

a) Determine when the rocket reaches the ground. **[8.1]**

b) On what interval is the height greater than 320 ft? **[8.1], [8.6]**

78. *Population Growth.*   The population P, in thousands, of Novi is given by

$$P(t) = \frac{8000t}{4t^2 + 10},$$

where t is the time, in months. Find the interval on which the population was 400,000 or greater. **[8.6]**

79. Which of the following is the domain of the function

$$g(x) = \frac{x^2 + 2x - 3}{x^2 - 5x + 6}? \text{ [8.5]}$$

A. $(-\infty, 2) \cup (2, 3) \cup (3, \infty)$
B. $(-\infty, -3) \cup (-3, 1) \cup (1, \infty)$
C. $(-\infty, 2) \cup (3, \infty)$
D. $(-\infty, -3) \cup (1, \infty)$

80. Which of the following lists the vertical asymptotes of the function

$$f(x) = \frac{x - 4}{(x + 1)(x - 2)(x + 4)}? \text{ [8.5]}$$

A. $x = 1$, $x = -2$, and $x = 4$
B. $x = -1$, $x = 2$, $x = -4$, and $x = 4$
C. $x = -1$, $x = 2$, and $x = -4$
D. $x = 4$

81. The graph of $f(x) = -\frac{1}{2}x^4 + x^3 + 1$ is which of the following? **[8.2]**

A.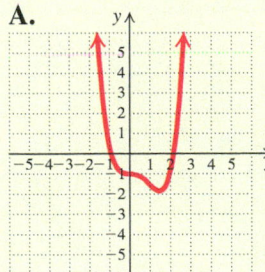

B.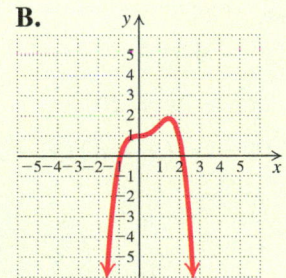

C.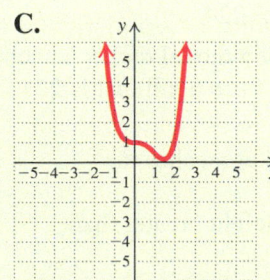

D.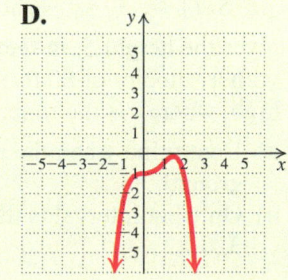

▶ **Synthesis**

Solve.

82. $x^2 \geq 5 - 2x$ **[8.6]**

83. $\left| 1 - \dfrac{1}{x^2} \right| < 3$ **[8.6]**

84. $x^4 - 2x^3 + 3x^2 - 2x + 2 = 0$ **[8.4]**

85. $(x - 2)^{-3} < 0$ **[8.6]**

86. Express $x^3 - 1$ as a product of linear factors. **[8.4]**

87. Find k such that $x + 3$ is a factor of $x^3 + kx^2 + kx - 15$. **[8.3]**

88. When $x^2 - 4x + 3k$ is divided by $x + 5$, the remainder is 33. Find the value of k. **[8.3]**

Find the domain of the function. **[8.5]**

89. $f(x) = \sqrt{x^2 + 3x - 10}$

90. $f(x) = \sqrt{x^2 - 3.1x + 2.2} + 1.75$

91. $f(x) = \dfrac{1}{\sqrt{5 - |7x + 2|}}$

▶ **Collaborative Discussion and Writing**

92. Explain the difference between a polynomial function and a rational function. **[8.1]**, **[8.5]**

93. Is it possible for a third-degree polynomial with rational coefficients to have no real zeros? Why or why not? **[8.4]**

94. Explain and contrast the three types of asymptotes considered for rational functions. **[8.5]**

95. If $P(x)$ is an even function, and by Descartes' rule of signs, $P(x)$ has one positive real zero, how many negative real zeros does $P(x)$ have? Explain. **[8.4]**

96. Explain why the graph of a rational function cannot have both a horizontal asymptote and an oblique asymptote. **[8.5]**

97. Under what circumstances would a quadratic inequality have a solution set that is a closed interval? **[8.6]**

8	**Chapter Test**

Determine the leading term, the leading coefficient, and the degree of the polynomial. Then classify the polynomial as constant, linear, quadratic, cubic, or quartic.

1. $f(x) = 2x^3 + 6x^2 - x^4 + 11$

2. $h(x) = -4.7x + 29$

3. Find the zeros of the polynomial function and state the multiplicity of each:

$$f(x) = x(3x - 5)(x - 3)^2(x + 1)^3.$$

4. *Hybrid Automobiles.* In 2004, only 84,199 hybrid automobiles were sold, while in 2012, 431,798 were sold (*Source:* U.S. Department of Transportation). The quartic function

$$f(x) = 897.690x^4 - 10,349.487x^3$$
$$+ 19,202.137x^2 + 91,597.838x$$
$$+ 88,209.580,$$

where x is the number of years after 2004, can be used to estimate the number of hybrid automobiles sold in years 2004 to 2012. Use this function to estimate the number of hybrid automobiles sold in 2008 and in 2011.

Sketch the graph of the polynomial function.

5. $f(x) = x^3 - 5x^2 + 2x + 8$

6. $f(x) = -2x^4 + x^3 + 11x^2 - 4x - 12$

Using the intermediate value theorem, determine, if possible, whether the function has a zero between a and b.

7. $f(x) = -5x^2 + 3;\ a = 0, b = 2$

8. $g(x) = 2x^3 + 6x^2 - 3;\ a = -2, b = -1$

9. Use long division to find the quotient $Q(x)$ and the remainder $R(x)$ when $P(x)$ is divided by $d(x)$. Express $P(x)$ in the form $d(x) \cdot Q(x) + R(x)$. Show your work.

$$P(x) = x^4 + 3x^3 + 2x - 5,$$
$$d(x) = x - 1$$

10. Use synthetic division to find the quotient and the remainder. Show your work.

$$(3x^3 - 12x + 7) \div (x - 5)$$

11. Use synthetic division to find $P(-3)$ for $P(x) = 2x^3 - 6x^2 + x - 4$. Show your work.

12. Use synthetic division to determine whether -2 is a zero of $f(x) = x^3 + 4x^2 + x - 6$. Answer yes or no. Show your work.

13. Find a polynomial function of degree 4 with -3 as a zero of multiplicity 2 and 0 and 6 as zeros of multiplicity 1.

14. Suppose that a polynomial function of degree 5 with rational coefficients has 1, $\sqrt{3}$, and $2 - i$ as zeros. Find the other zeros.

Find a polynomial function of lowest degree with rational coefficients and the following as some of its zeros.

15. $-10, 3i$

16. $0, -\sqrt{3}, 1 - i$

List all possible rational zeros.

17. $f(x) = 2x^3 + x^2 - 2x + 12$

18. $h(x) = 10x^4 - x^3 + 2x - 5$

For each polynomial function, (a) find the rational zeros and then the other zeros; that is, solve $f(x) = 0$; and (b) factor $f(x)$ into linear factors.

19. $f(x) = x^3 + x^2 - 5x - 5$

20. $f(x) = 2x^4 - 11x^3 + 16x^2 - x - 6$

21. $f(x) = x^3 + 4x^2 + 4x + 16$

22. $f(x) = 3x^4 - 11x^3 + 15x^2 - 9x + 2$

23. What does Descartes' rule of signs tell you about the number of positive real zeros and the number of negative real zeros of the following function?

$$g(x) = -x^8 + 2x^6 - 4x^3 - 1$$

Graph the function. Be sure to label all the asymptotes. List the domain and the x- and y-intercepts.

24. $f(x) = \dfrac{2}{(x - 3)^2}$

25. $f(x) = \dfrac{x + 3}{x^2 - 3x - 4}$

26. Find a rational function that has vertical asymptotes $x = -1$ and $x = 2$ and x-intercept $(-4, 0)$.

Solve.

27. $2x^2 > 5x + 3$

28. $\dfrac{x + 1}{x - 4} \le 3$

29. The function $S(t) = -16t^2 + 64t + 192$ gives the height S, in feet, of a model rocket launched with a velocity of 64 ft/sec from a hill that is 192 ft high.

  a) Determine how long it will take the rocket to reach the ground.

  b) Find the interval on which the height of the rocket is greater than 240 ft.

30. The graph of $f(x) = x^3 - x^2 - 2$ is which of the following?

A.

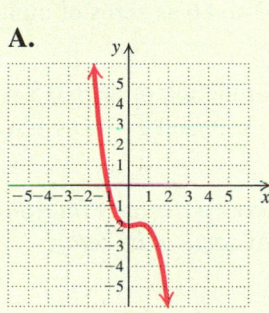

B.

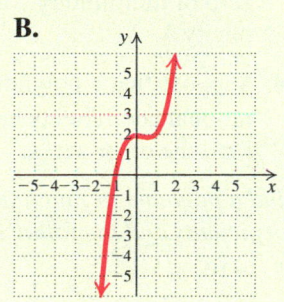

C.

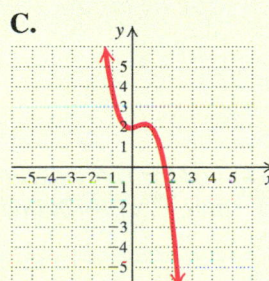

D.

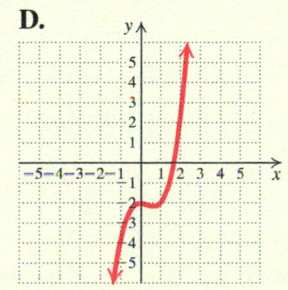

▶ **Synthesis**

31. Find the domain of $f(x) = \sqrt{x^2 + x - 12}$.

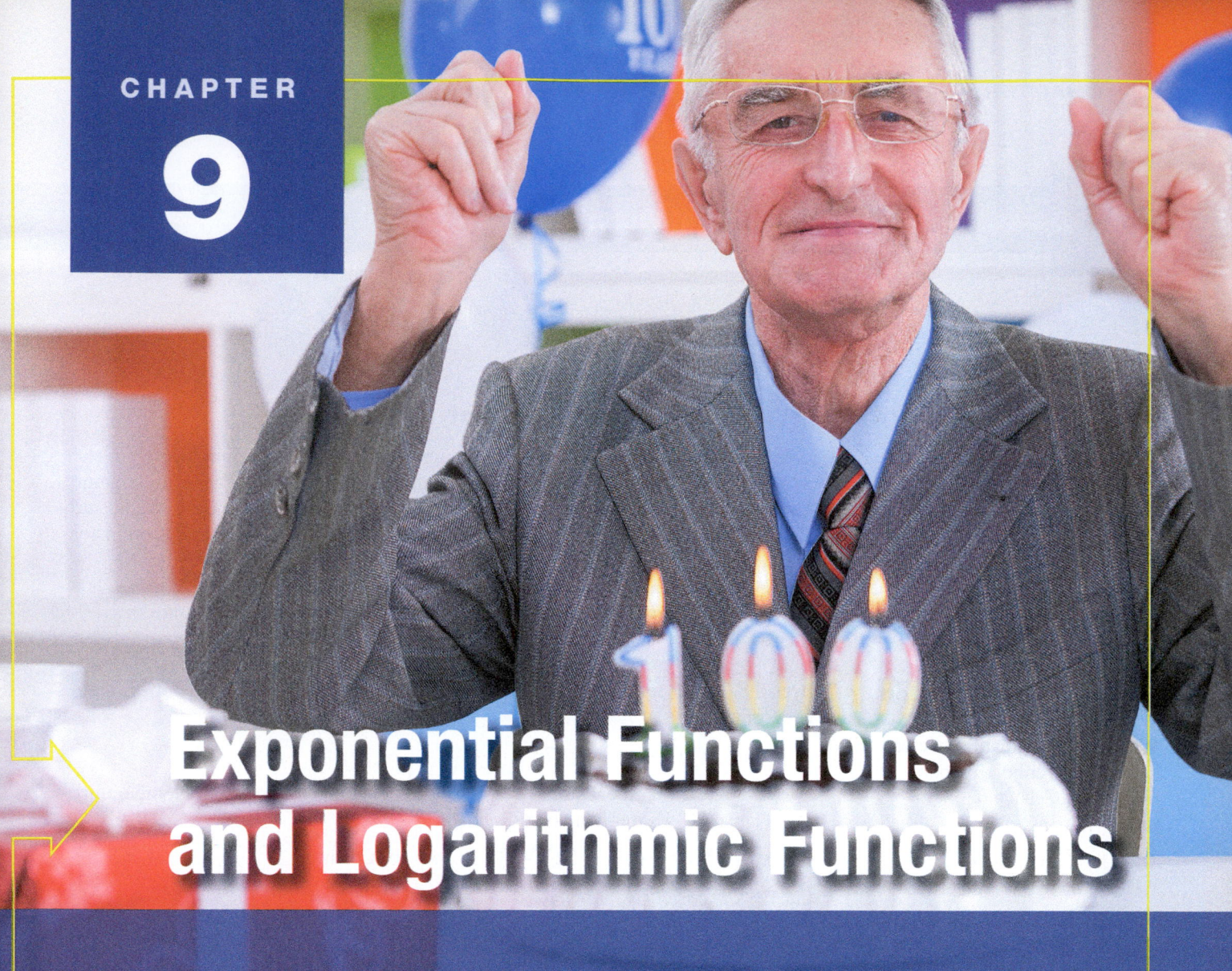

Exponential Functions and Logarithmic Functions

APPLICATION   This problem appears as Exercise 69 in Section 9.3.

The centenarian population in the United States has grown over 65% in the last 30 years. In 1980, there were only 32,194 residents ages 100 and over. This number had grown to 53,364 by 2010. (*Sources*: Population Projections Program; U.S. Census Bureau; U.S. Department of Commerce; "What People Who Live to 100 Have in Common," by Emily Brandon, *U.S. News and World Report*, January 7, 2013) The exponential function

$$H(t) = 80,040.68(1.0481)^t,$$

where t is the number of years after 2015, can be used to project the number of centenarians, in thousands. Use this function to project the centenarian population in 2020 and in 2050.

9.1 The Composition of Functions

▶ Find the composition of two functions and the domain of the composition.

▶ Decompose a function as a composition of two functions.

▶ ### The Composition of Functions

In real-world situations, it is not uncommon for the output of a function to depend on some input that is itself an output of another function. For instance, the amount that a person pays as state income tax usually depends on the amount of adjusted gross income on the person's federal tax return, which, in turn, depends on his or her annual earnings. Such functions are called **composite functions**.

To see how composite functions work, suppose a chemistry student needs a formula to convert Fahrenheit temperatures to Kelvin units. The formula

$$c(t) = \tfrac{5}{9}(t - 32)$$

gives the Celsius temperature $c(t)$ that corresponds to the Fahrenheit temperature t. The formula

$$k(c(t)) = c(t) + 273$$

gives the Kelvin temperature $k(c(t))$ that corresponds to the Celsius temperature $c(t)$. Thus, 50° Fahrenheit corresponds to

$$c(50) = \tfrac{5}{9}(50 - 32) = \tfrac{5}{9}(18) = 10° \text{ Celsius}$$

and 10° Celsius corresponds to

$$k(c(50)) = k(10) = 10 + 273 = 283 \text{ Kelvin units,}$$

which is usually written 283 K. We see that 50° Fahrenheit is the same as 283 K. This two-step procedure can be used to convert any Fahrenheit temperature to Kelvin units.

	°F Fahrenheit	°C Celsius	K Kelvin
Boiling point of water	212°	100°	373 K
	50° ➡	10° ➡	283 K
Freezing point of water	32°	0°	273 K
Absolute zero	−460°	−273°	0 K

A student making numerous conversions might look for a formula that converts directly from Fahrenheit to Kelvin. Such a formula can be found by substitution:

$$k(c(t)) = c(t) + 273$$

$$= \frac{5}{9}(t - 32) + 273 \qquad \text{Substituting } \tfrac{5}{9}(t - 32) \text{ for } c(t)$$

$$= \frac{5}{9}t - \frac{160}{9} + 273$$

$$= \frac{5}{9}t - \frac{160}{9} + \frac{2457}{9}$$

$$= \frac{5t + 2297}{9}. \qquad \text{Simplifying}$$

Since the formula found above expresses the Kelvin temperature as a new function K of the Fahrenheit temperature t, we can write

$$K(t) = \frac{5t + 2297}{9},$$

where $K(t)$ is the Kelvin temperature corresponding to the Fahrenheit temperature, t. Here we have $K(t) = k(c(t))$. The new function K is called the **composition** of k and c and can be denoted $k \circ c$ (read "k composed with c," "the composition of k and c," or "k circle c").

COMPOSITION OF FUNCTIONS

The **composite function** $f \circ g$, the **composition** of f and g, is defined as

$$(f \circ g)(x) = f(g(x)),$$

where x is in the domain of g and $g(x)$ is in the domain of f.

EXAMPLE 1   Given that $f(x) = 2x - 5$ and $g(x) = x^2 - 3x + 8$, find each of the following.

a) $(f \circ g)(x)$ and $(g \circ f)(x)$ 　　　　　 **b)** $(f \circ g)(7)$ and $(g \circ f)(7)$
c) $(g \circ g)(1)$ 　　　　　　　　　　　　 **d)** $(f \circ f)(x)$

Solution   Consider each function separately:

$$f(x) = 2x - 5 \qquad \text{This function multiplies each input by 2 and then subtracts 5.}$$

and

$$g(x) = x^2 - 3x + 8. \qquad \text{This function squares an input, subtracts three times the input from the result, and then adds 8.}$$

a) To find $(f \circ g)(x)$, we substitute $g(x)$ for x in the equation for $f(x)$:

$$(f \circ g)(x) = f(g(x)) = f(x^2 - 3x + 8) \qquad x^2 - 3x + 8 \text{ is the input for } f.$$

$$= 2(x^2 - 3x + 8) - 5 \qquad f \text{ multiplies the input by 2 and then subtracts 5.}$$

$$= 2x^2 - 6x + 16 - 5$$

$$= 2x^2 - 6x + 11.$$

To find $(g \circ f)(x)$, we substitute $f(x)$ for x in the equation for $g(x)$:

$$(g \circ f)(x) = g(f(x)) = g(2x - 5) \qquad \text{\color{red}{2x − 5 is the input for g.}}$$

$$= (2x - 5)^2 - 3(2x - 5) + 8 \qquad \text{\color{red}{g squares the input, subtracts three times the input, and then adds 8.}}$$

$$= 4x^2 - 20x + 25 - 6x + 15 + 8$$
$$= 4x^2 - 26x + 48.$$

b) To find $(f \circ g)(7)$, we first find $g(7)$. Then we use $g(7)$ as an input for f:

$$(f \circ g)(7) = f(g(7)) = f(7^2 - 3 \cdot 7 + 8)$$
$$= f(36) = 2 \cdot 36 - 5$$
$$= 72 - 5 = 67.$$

To find $(g \circ f)(7)$, we first find $f(7)$. Then we use $f(7)$ as an input for g:

$$(g \circ f)(7) = g(f(7)) = g(2 \cdot 7 - 5)$$
$$= g(9) = 9^2 - 3 \cdot 9 + 8$$
$$= 81 - 27 + 8 = 62.$$

We could also find $(f \circ g)(7)$ and $(g \circ f)(7)$ by substituting 7 for x in the equations that we found in part (a):

$$(f \circ g)(x) = 2x^2 - 6x + 11$$
$$(f \circ g)(7) = 2 \cdot 7^2 - 6 \cdot 7 + 11 = 67;$$

$$(g \circ f)(x) = 4x^2 - 26x + 48$$
$$(g \circ f)(7) = 4 \cdot 7^2 - 26 \cdot 7 + 48 = 62.$$

c) $(g \circ g)(1) = g(g(1)) = g(1^2 - 3 \cdot 1 + 8)$
$$= g(1 - 3 + 8) = g(6)$$
$$= 6^2 - 3 \cdot 6 + 8$$
$$= 36 - 18 + 8 = 26$$

d) $(f \circ f)(x) = f(f(x)) = f(2x - 5)$
$$= 2(2x - 5) - 5$$
$$= 4x - 10 - 5 = 4x - 15$$

> **Now Try Exercises 1 and 15.**

Example 1 illustrates that, as a rule, $(f \circ g)(x) \neq (g \circ f)(x)$. We can see this graphically, as shown in the graphs at left.

EXAMPLE 2 Given that $f(x) = \sqrt{x}$ and $g(x) = x - 3$:

a) Find $f \circ g$ and $g \circ f$.

b) Find the domain of $f \circ g$ and the domain of $g \circ f$.

Solution

a) $(f \circ g)(x) = f(g(x)) = f(x - 3) = \sqrt{x - 3}$
$(g \circ f)(x) = g(f(x)) = g(\sqrt{x}) = \sqrt{x} - 3$

Technology Connection

We can check our work in Example 1(b) using a graphing calculator. We enter the following on the equation-editor screen:

$$y_1 = 2x - 5$$

and

$$y_2 = x^2 - 3x + 8.$$

Then, on the home screen, we find $(f \circ g)(7)$ and $(g \circ f)(7)$ using the function notations Y1(Y2(7)) and Y2(Y1(7)), respectively.

$y_1 = 2x - 5, \quad y_2 = x^2 - 3x + 8$

Y1(Y2(7))	
	67
Y2(Y1(7))	
	62

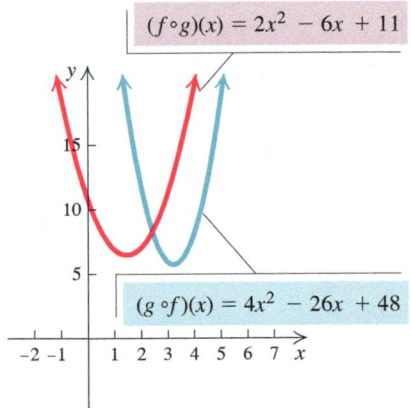

$(f \circ g)(x) = 2x^2 - 6x + 11$

$(g \circ f)(x) = 4x^2 - 26x + 48$

b) Since $f(x)$ is not defined for negative radicands, the domain of $f(x)$ is $\{x | x \geq 0\}$, or $[0, \infty)$. Any real number can be an input for $g(x)$, so the domain of $g(x)$ is $(-\infty, \infty)$.

Since the inputs of $f \circ g$ are outputs of g, the domain of $f \circ g$ consists of the values of x in the domain of g, $(-\infty, \infty)$, for which $g(x)$ is nonnegative. (Recall that the inputs of $f(x)$ must be nonnegative.) Thus we have

$$g(x) \geq 0$$
$$x - 3 \geq 0 \qquad \text{Substituting } x - 3 \text{ for } g(x)$$
$$x \geq 3.$$

We see that the domain of $f \circ g$ is $\{x | x \geq 3\}$, or $[3, \infty)$.

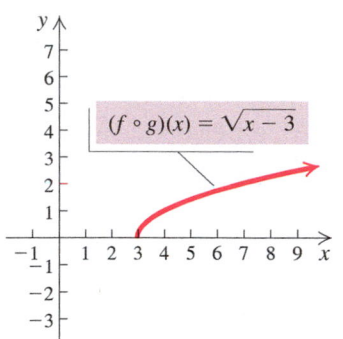

Figure 1.

We can also find the domain of $f \circ g$ by examining the composite function itself, $(f \circ g)(x) = \sqrt{x - 3}$. Since any real number can be an input for g, the only restriction on $f \circ g$ is that the radicand must be nonnegative. We have

$$x - 3 \geq 0$$
$$x \geq 3.$$

Again, we see that the domain of $f \circ g$ is $\{x | x \geq 3\}$, or $[3, \infty)$. The graph in Fig. 1 confirms this.

The inputs of $g \circ f$ are outputs of f, so the domain of $g \circ f$ consists of the values of x in the domain of f, $[0, \infty)$, for which $g(x)$ is defined. Since g can accept *any* real number as an input, any output from f is acceptable, so the entire domain of f is the domain of $g \circ f$. That is, the domain of $g \circ f$ is $\{x | x \geq 0\}$, or $[0, \infty)$.

We can also examine the composite function itself to find its domain. First, recall that the domain of f is $\{x | x \geq 0\}$, or $[0, \infty)$. Then consider $(g \circ f)(x) = \sqrt{x} - 3$. The radicand cannot be negative, so we have $x \geq 0$. As above, we see that the domain of $g \circ f$ is the domain of f, $\{x | x \geq 0\}$, or $[0, \infty)$. The graph in Fig. 2 confirms this.

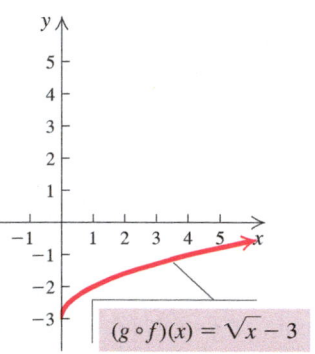

Figure 2.

Now Try Exercise 27.

EXAMPLE 3 Given that $f(x) = \dfrac{1}{x - 2}$ and $g(x) = \dfrac{5}{x}$, find $f \circ g$ and $g \circ f$ and the domain of each.

Solution We have

$$(f \circ g)(x) = f(g(x)) = f\left(\frac{5}{x}\right) = \frac{1}{\dfrac{5}{x} - 2} = \frac{1}{\dfrac{5 - 2x}{x}} = \frac{x}{5 - 2x};$$

$$(g \circ f)(x) = g(f(x)) = g\left(\frac{1}{x - 2}\right) = \frac{5}{\dfrac{1}{x - 2}} = 5(x - 2).$$

Values of x that make a denominator 0 are not in the domains of these functions. Since $x - 2 = 0$ when $x = 2$, the domain of f is $\{x | x \neq 2\}$. The denominator of g is x, so the domain of g is $\{x | x \neq 0\}$.

The inputs of $f \circ g$ are outputs of g, so the domain of $f \circ g$ consists of the values of x in the domain of g for which $g(x) \neq 2$. (Recall that 2 cannot be an input of f.)

Since the domain of g is $\{x \mid x \neq 0\}$, 0 is not in the domain of $f \circ g$. In addition, we must find the value(s) of x for which $g(x) = 2$. We have

$$g(x) = 2$$

$$\frac{5}{x} = 2 \qquad \text{Substituting } \frac{5}{x} \text{ for } g(x)$$

$$5 = 2x$$

$$\frac{5}{2} = x.$$

This tells us that $\frac{5}{2}$ is also *not* in the domain of $f \circ g$. Then the domain of $f \circ g$ is

$$\left\{x \mid x \neq 0 \text{ and } x \neq \tfrac{5}{2}\right\}, \text{ or } (-\infty, 0) \cup \left(0, \tfrac{5}{2}\right) \cup \left(\tfrac{5}{2}, \infty\right).$$

We can also examine the composite function $f \circ g$ to find its domain. First, recall that 0 is not in the domain of g, so it cannot be in the domain of $(f \circ g)(x) = x/(5 - 2x)$. We must also exclude the value(s) of x for which the denominator of $f \circ g$ is 0. We have

$$5 - 2x = 0$$

$$5 = 2x$$

$$\frac{5}{2} = x.$$

Again, we see that $\frac{5}{2}$ is also not in the domain, so the domain of $f \circ g$ is

$$\left\{x \mid x \neq 0 \text{ and } x \neq \tfrac{5}{2}\right\}, \text{ or } (-\infty, 0) \cup \left(0, \tfrac{5}{2}\right) \cup \left(\tfrac{5}{2}, \infty\right).$$

Since the inputs of $g \circ f$ are outputs of f, the domain of $g \circ f$ consists of the values of x in the domain of f for which $f(x) \neq 0$. (Recall that 0 cannot be an input of g.) The domain of f is $\{x \mid x \neq 2\}$, so 2 is not in the domain of $g \circ f$. Next, we determine whether there are values of x for which $f(x) = 0$:

$$f(x) = 0$$

$$\frac{1}{x - 2} = 0 \qquad \text{Substituting } \frac{1}{x - 2} \text{ for } f(x)$$

$$(x - 2) \cdot \frac{1}{x - 2} = (x - 2) \cdot 0 \qquad \text{Multiplying by } x - 2$$

$$1 = 0. \qquad \text{False equation}$$

We see that there are no values of x for which $f(x) = 0$, so there are no additional restrictions on the domain of $g \circ f$. Thus the domain of $g \circ f$ is

$$\{x \mid x \neq 2\}, \text{ or } (-\infty, 2) \cup (2, \infty).$$

We can also examine $g \circ f$ to find its domain. First, recall that 2 is not in the domain of f, so it cannot be in the domain of $(g \circ f)(x) = 5(x - 2)$. Since $5(x - 2)$ is defined for all real numbers, there are no additional restrictions on the domain of $g \circ f$. The domain is

$$\{x \mid x \neq 2\}, \text{ or } (-\infty, 2) \cup (2, \infty). \qquad \longrightarrow \boxed{\text{Now Try Exercise 23.}}$$

▶ Decomposing a Function as a Composition

In calculus, one often needs to recognize how a function can be expressed as the composition of two functions. In this way, we are "decomposing" the function.

EXAMPLE 4 If $h(x) = (2x - 3)^5$, find $f(x)$ and $g(x)$ such that $h(x) = (f \circ g)(x)$.

Solution The function $h(x)$ raises $(2x - 3)$ to the 5th power. Two functions that can be used for the composition are

$$f(x) = x^5 \quad \text{and} \quad g(x) = 2x - 3.$$

We can check by forming the composition:

$$h(x) = (f \circ g)(x) = f(g(x)) = f(2x - 3) = (2x - 3)^5.$$

This is the most "obvious" solution. There can be other less obvious solutions. For example, if

$$f(x) = (x + 7)^5 \quad \text{and} \quad g(x) = 2x - 10,$$

then

$$
\begin{aligned}
h(x) = (f \circ g)(x) &= f(g(x)) \\
&= f(2x - 10) \\
&= [2x - 10 + 7]^5 = (2x - 3)^5.
\end{aligned}
$$

> **Now Try Exercise 39.**

EXAMPLE 5 If $h(x) = \dfrac{1}{(x + 3)^3}$, find $f(x)$ and $g(x)$ such that $h(x) = (f \circ g)(x)$.

Solution Two functions that can be used are

$$f(x) = \frac{1}{x} \quad \text{and} \quad g(x) = (x + 3)^3.$$

We check by forming the composition:

$$h(x) = (f \circ g)(x) = f(g(x)) = f((x + 3)^3) = \frac{1}{(x + 3)^3}.$$

There are other functions that can be used as well. For example, if

$$f(x) = \frac{1}{x^3} \quad \text{and} \quad g(x) = x + 3,$$

then

$$h(x) = (f \circ g)(x) = f(g(x)) = f(x + 3) = \frac{1}{(x + 3)^3}.$$

> **Now Try Exercise 41.**

9.1 Exercise Set

Given that $f(x) = 3x + 1$, $g(x) = x^2 - 2x - 6$, *and* $h(x) = x^3$, *find each of the following.*

1. $(f \circ g)(-1)$

2. $(g \circ f)(-2)$

3. $(h \circ f)(1)$

4. $(g \circ h)\left(\frac{1}{2}\right)$

5. $(g \circ f)(5)$

6. $(f \circ g)\left(\frac{1}{3}\right)$

7. $(f \circ h)(-3)$

8. $(h \circ g)(3)$

9. $(g \circ g)(-2)$

10. $(g \circ g)(3)$

11. $(h \circ h)(2)$

12. $(h \circ h)(-1)$

13. $(f \circ f)(-4)$

14. $(f \circ f)(1)$

15. $(h \circ h)(x)$

16. $(f \circ f)(x)$

Find $(f \circ g)(x)$ and $(g \circ f)(x)$ and the domain of each.

17. $f(x) = x + 3$, $g(x) = x - 3$

18. $f(x) = \frac{4}{5}x$, $g(x) = \frac{5}{4}x$

19. $f(x) = x + 1$, $g(x) = 3x^2 - 2x - 1$

20. $f(x) = 3x - 2$, $g(x) = x^2 + 5$

21. $f(x) = x^2 - 3$, $g(x) = 4x - 3$

22. $f(x) = 4x^2 - x + 10$, $g(x) = 2x - 7$

23. $f(x) = \frac{4}{1 - 5x}$, $g(x) = \frac{1}{x}$

24. $f(x) = \frac{6}{x}$, $g(x) = \frac{1}{2x + 1}$

25. $f(x) = 3x - 7$, $g(x) = \frac{x + 7}{3}$

26. $f(x) = \frac{2}{3}x - \frac{4}{5}$, $g(x) = 1.5x + 1.2$

27. $f(x) = 2x + 1$, $g(x) = \sqrt{x}$

28. $f(x) = \sqrt{x}$, $g(x) = 2 - 3x$

29. $f(x) = 20$, $g(x) = 0.05$

30. $f(x) = x^4$, $g(x) = \sqrt[4]{x}$

31. $f(x) = \sqrt{x + 5}$, $g(x) = x^2 - 5$

32. $f(x) = x^5 - 2$, $g(x) = \sqrt[5]{x + 2}$

33. $f(x) = x^2 + 2$, $g(x) = \sqrt{3 - x}$

34. $f(x) = 1 - x^2$, $g(x) = \sqrt{x^2 - 25}$

35. $f(x) = \frac{1 - x}{x}$, $g(x) = \frac{1}{1 + x}$

36. $f(x) = \frac{1}{x - 2}$, $g(x) = \frac{x + 2}{x}$

37. $f(x) = x^3 - 5x^2 + 3x + 7$, $g(x) = x + 1$

38. $f(x) = x - 1$, $g(x) = x^3 + 2x^2 - 3x - 9$

Find $f(x)$ and $g(x)$ such that $h(x) = (f \circ g)(x)$.
Answers may vary.

39. $h(x) = (4 + 3x)^5$

40. $h(x) = \sqrt[3]{x^2 - 8}$

41. $h(x) = \frac{1}{(x - 2)^4}$

42. $h(x) = \frac{1}{\sqrt{3x + 7}}$

43. $h(x) = \frac{x^3 - 1}{x^3 + 1}$

44. $h(x) = |9x^2 - 4|$

45. $h(x) = \left(\frac{2 + x^3}{2 - x^3}\right)^6$

46. $h(x) = (\sqrt{x} - 3)^4$

47. $h(x) = \sqrt{\frac{x - 5}{x + 2}}$

48. $h(x) = \sqrt{1 + \sqrt{1 + x}}$

49. $h(x) = (x + 2)^3 - 5(x + 2)^2 + 3(x + 2) - 1$

50. $h(x) = 2(x - 1)^{5/3} + 5(x - 1)^{2/3}$

51. *Ripple Spread.* A stone is thrown into a pond, creating a circular ripple that spreads over the pond in such a way that the radius is increasing at a rate of 3 ft/sec.

   a) Find a function $r(t)$ for the radius in terms of t.
   b) Find a function $A(r)$ for the area of the ripple in terms of the radius r.
   c) Find $(A \circ r)(t)$. Explain the meaning of this function.

52. The surface area S of a right circular cylinder is given by the formula $S = 2\pi rh + 2\pi r^2$. If the height is twice the radius, find each of the following.

   a) A function $S(r)$ for the surface area as a function of r
   b) A function $S(h)$ for the surface area as a function of h

53. *Blouse Sizes.* A blouse that is size x in Japan is size $s(x)$ in the United States, where $s(x) = x - 3$. A blouse that is size x in the United States is size $t(x)$ in Australia, where $t(x) = x + 4$. (*Source:* www.onlineconversion.com) Find a function that will

convert blouse sizes in Japan to blouse sizes in Australia.

54. A manufacturer of tools, selling rechargeable drills to a chain of home improvement stores, charges $6 more per drill than its manufacturing cost m. The stores then sell each drill for 150% of the price that it paid the manufacturer. Find a function $P(m)$ for the price at the home improvement stores.

▶ Skill Maintenance

Consider the following linear equations. Without graphing them, answer the questions in Exercises 55–62. [2.5], [2.6]

a) $y = x$
b) $y = -5x + 4$
c) $y = \frac{2}{3}x + 1$
d) $y = -0.1x + 6$
e) $y = 3x - 5$
f) $y = -x - 1$
g) $2x - 3y = 6$
h) $6x + 3y = 9$

55. Which, if any, have y-intercept $(0, 1)$?

56. Which, if any, have the same y-intercept?

57. Which slope down from left to right?

58. Which has the steepest slope?

59. Which pass(es) through the origin?

60. Which, if any, have the same slope?

61. Which, if any, are parallel?

62. Which, if any, are perpendicular?

▶ Synthesis

63. Let $p(a)$ represent the number of pounds of grass seed required to seed a lawn with area a. Let $c(s)$ represent the cost of s pounds of grass seed. Which composition makes sense: $(c \circ p)(a)$ or $(p \circ c)(s)$? What does it represent?

64. Write equations of two functions f and g such that $f \circ g = g \circ f = x$. (In Section 9.2, we will study inverse functions. If $f \circ g = g \circ f = x$, functions f and g are *inverses* of each other.)

9.2

Inverse Functions

▶ Determine whether a function is one-to-one, and if it is, find a formula for its inverse.

▶ Simplify expressions of the type $(f \circ f^{-1})(x)$ and $(f^{-1} \circ f)(x)$.

▶ Inverses

When we go from an output of a function back to its input or inputs, we get an inverse relation. When that relation is a function, we have an inverse function.

Consider the relation h given as follows:

$$h = \{(-8, 5), (4, -2), (-7, 1), (3.8, 6.2)\}.$$

RELATIONS

REVIEW SECTION 2.2

Suppose we *interchange* the first and second coordinates. The relation we obtain is called the **inverse** of the relation h and is given as follows:

$$\text{Inverse of } h = \{(5, -8), (-2, 4), (1, -7), (6.2, 3.8)\}.$$

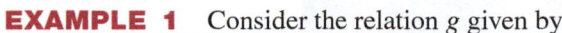

INVERSE RELATION: ORDERED PAIRS

Interchanging the first and second coordinates of each ordered pair in a relation produces the **inverse relation**.

EXAMPLE 1   Consider the relation g given by

$$g = \{(2,4), (-1,3), (-2,0)\}.$$

Graph the relation in blue. Find the inverse and graph it in red.

Solution   The relation g is shown in blue in the figure at left. The inverse of the relation is

$$\{(4,2), (3,-1), (0,-2)\}$$

and is shown in red. The pairs in the inverse are reflections of the pairs in g across the line $y = x$.

Now Try Exercise 1.

INVERSE RELATION: EQUATION

If a relation is defined by an equation, interchanging the variables produces an equation of the **inverse relation**.

EXAMPLE 2   Find an equation for the inverse of the relation

$$y = x^2 - 5x.$$

Solution   We interchange x and y and obtain an equation of the inverse:

$$x = y^2 - 5y.$$

Now Try Exercise 9.

If a relation is given by an equation, then the solutions of the inverse can be found from those of the original equation by interchanging the first and second coordinates of each ordered pair. Thus the graphs of a relation and its inverse are always reflections of each other across the line $y = x$. This is illustrated with the equations of Example 2 in the tables and graph below. We will explore inverses and their graphs later in this section.

$x = y^2 - 5y$	y
6	-1
0	0
-6	2
-4	4

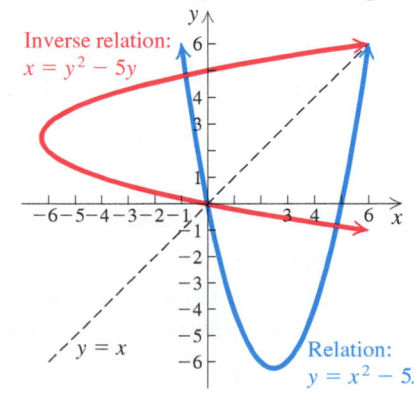

x	$y = x^2 - 5x$
-1	6
0	0
2	-6
4	-4

► **Inverses and One-to-One Functions**

Let's consider the following two functions.

Year (domain)	First-Class Postage Cost, in cents (range)
2006 ———→	39
2007 ———→	41
2008 ———→	42
2009	
2010 ———→	44
2011	
2012 ———→	45
2013 ———→	46
2014 ———→	49

Number (domain)	Cube (range)
−3 ———→	−27
−2 ———→	−8
−1 ———→	−1
0 ———→	0
1 ———→	1
2 ———→	8
3 ———→	27

Source: U.S. Postal Service

Suppose we reverse the arrows. Are these inverse relations functions?

Year (range)	First-Class Postage Cost, in cents (domain)
2006 ←———	39
2007 ←———	41
2008 ←———	42
2009	
2010 ←———	44
2011	
2012 ←———	45
2013 ←———	46
2014 ←———	49

Number (range)	Cube (domain)
−3 ←———	−27
−2 ←———	−8
−1 ←———	−1
0 ←———	0
1 ←———	1
2 ←———	8
3 ←———	27

Source: U.S. Postal Service

We see that the inverse of the postage function is not a function. Like all functions, each input in the postage function has exactly one output. However, the output for 2009, 2010, and 2011 is 44. Thus in the inverse of the postage function, the input 44 has three outputs, 2009, 2010, and 2011. When two or more inputs of a function have the same output, the inverse relation cannot be a function. In the cubing function, each output corresponds to exactly one input, so its inverse is also a function. The cubing function is an example of a **one-to-one function**.

ONE-TO-ONE FUNCTIONS

A function f is **one-to-one** if different inputs have different outputs—that is,

$$\text{if} \quad a \neq b, \quad \text{then} \quad f(a) \neq f(b).$$

Or a function f is **one-to-one** if when the outputs are the same, the inputs are the same—that is,

$$\text{if} \quad f(a) = f(b), \quad \text{then} \quad a = b.$$

If the inverse of a function f is also a function, it is named f^{-1} (read "f-inverse").

The -1 in f^{-1} is *not* an exponent!

Do *not* misinterpret the -1 in f^{-1} as a negative exponent: f^{-1} does *not* mean the reciprocal of f and $f^{-1}(x)$ is *not* equal to $\dfrac{1}{f(x)}$.

ONE-TO-ONE FUNCTIONS AND INVERSES

- If a function f is one-to-one, then its inverse f^{-1} is a function.
- The domain of a one-to-one function f is the range of the inverse f^{-1}.
- The range of a one-to-one function f is the domain of the inverse f^{-1}.

$$\begin{matrix} D_f & & D_{f^{-1}} \\ R_f & & R_{f^{-1}} \end{matrix}$$

- A function that is increasing over its entire domain or is decreasing over its entire domain is a one-to-one function.

EXAMPLE 3   Given the function f described by $f(x) = 2x - 3$, prove that f is one-to-one (that is, it has an inverse that is a function).

Solution   To show that f is one-to-one, we show that if $f(a) = f(b)$, then $a = b$. Assume that $f(a) = f(b)$ for a and b in the domain of f. Since $f(a) = 2a - 3$ and $f(b) = 2b - 3$, we have

$$2a - 3 = 2b - 3$$
$$2a = 2b \qquad \color{red}{\text{Adding 3}}$$
$$a = b. \qquad \color{red}{\text{Dividing by 2}}$$

Thus, if $f(a) = f(b)$, then $a = b$. This shows that f is one-to-one.

> **Now Try Exercise 17.**

EXAMPLE 4   Given the function g described by $g(x) = x^2$, prove that g is not one-to-one.

Solution   We can prove that g is not one-to-one by finding two numbers a and b for which $a \neq b$ and $g(a) = g(b)$. Two such numbers are -3 and 3, because $-3 \neq 3$ and $g(-3) = g(3) = 9$. Thus g is not one-to-one.

> **Now Try Exercise 21.**

The following graphs show a function, in blue, and its inverse, in red. To determine whether the inverse is a function, we can apply the vertical-line test to its graph. By reflecting each such vertical line across the line $y = x$, we obtain an equivalent **horizontal-line test** for the original function.

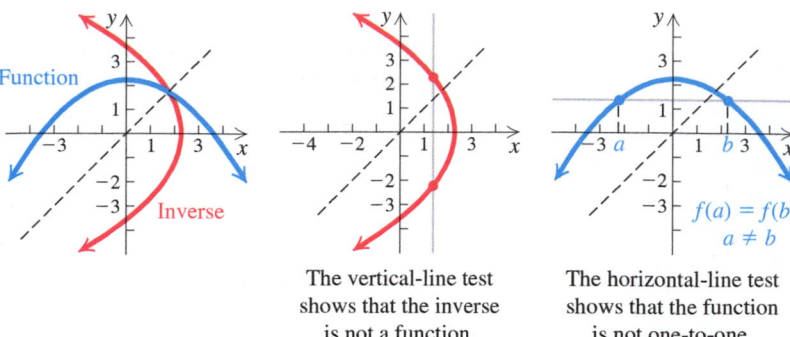

The vertical-line test
shows that the inverse
is not a function.

The horizontal-line test
shows that the function
is not one-to-one.

HORIZONTAL-LINE TEST

If it is possible for a horizontal line to intersect the graph of a function more than once, then the function is *not* one-to-one and its inverse is *not* a function.

EXAMPLE 5 From the graphs shown, determine whether each function is one-to-one and thus has an inverse that is a function.

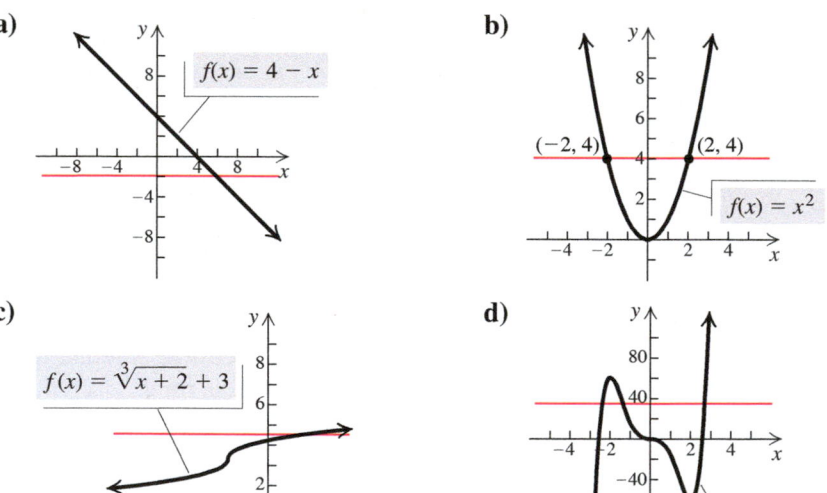

a) $f(x) = 4 - x$

b) $f(x) = x^2$, $(-2, 4)$, $(2, 4)$

c) $f(x) = \sqrt[3]{x + 2} + 3$

d) $f(x) = 3x^5 - 20x^3$

Solution   For each function, we apply the horizontal-line test.

RESULT	REASON
a) One-to-one; inverse is a function	No horizontal line intersects the graph more than once.
b) Not one-to-one; inverse is not a function	There are many horizontal lines that intersect the graph more than once. Note that where the line $y = 4$ intersects the graph, the first coordinates are -2 and 2. Although these are different inputs, they have the same output, 4.
c) One-to-one; inverse is a function	No horizontal line intersects the graph more than once.
d) Not one-to-one; inverse is not a function	There are many horizontal lines that intersect the graph more than once.

> **Now Try Exercises 25 and 27.**

▶ Finding Formulas for Inverses

Suppose that a function is described by a formula. If it has an inverse that is a function, we proceed as follows to find a formula for f^{-1}.

Obtaining a Formula for an Inverse

If a function f is one-to-one, a formula for its inverse can generally be found as follows:

1. Replace $f(x)$ with y.
2. Interchange x and y.
3. Solve for y.
4. Replace y with $f^{-1}(x)$.

EXAMPLE 6   Determine whether the function $f(x) = 2x - 3$ is one-to-one, and if it is, find a formula for $f^{-1}(x)$.

Solution   The graph of f is shown at left. It passes the horizontal-line test. Thus it is one-to-one and its inverse is a function. We also proved that f is one-to-one in Example 3. We find a formula for $f^{-1}(x)$.

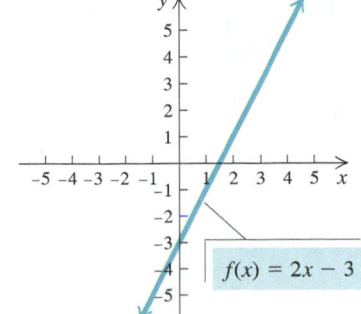

$f(x) = 2x - 3$

1. Replace $f(x)$ with y: $y = 2x - 3$
2. Interchange x and y: $x = 2y - 3$
3. Solve for y: $x + 3 = 2y$
$$\frac{x + 3}{2} = y$$

4. Replace y with $f^{-1}(x)$: $f^{-1}(x) = \dfrac{x + 3}{2}$.

> **Now Try Exercise 47.**

Consider

$$f(x) = 2x - 3 \quad \text{and} \quad f^{-1}(x) = \frac{x+3}{2}$$

from Example 6. For the input 5, we have

$$f(5) = 2 \cdot 5 - 3 = 10 - 3 = 7.$$

The output is 7. Now we use 7 for the input in the inverse:

$$f^{-1}(7) = \frac{7+3}{2} = \frac{10}{2} = 5.$$

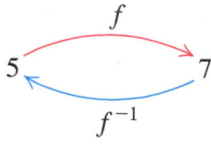

The function f takes the number 5 to 7. The inverse function f^{-1} takes the number 7 back to 5.

EXAMPLE 7 Graph

$$f(x) = 2x - 3 \quad \text{and} \quad f^{-1}(x) = \frac{x+3}{2}$$

using the same set of axes. Then compare the two graphs.

Solution The graphs of f and f^{-1} are shown at left. The solutions of the inverse function can be found from those of the original function by interchanging the first and second coordinates of each ordered pair.

x	$f(x) = 2x - 3$
-1	-5
0	-3 ←——— *y*-intercept
2	1
3	3

x	$f^{-1}(x) = \dfrac{x+3}{2}$
-5	-1
-3	0 ←——— *x*-intercept
1	2
3	3

When we interchange x and y in finding a formula for the inverse of $f(x) = 2x - 3$, we are in effect reflecting the graph of that function across the line $y = x$. For example, when the coordinates of the *y*-intercept, $(0, -3)$, of the graph of f are reversed, we get the *x*-intercept, $(-3, 0)$, of the graph of f^{-1}. If we were to graph $f(x) = 2x - 3$ in wet ink and fold along the line $y = x$, the graph of $f^{-1}(x) = (x + 3)/2$ would be formed by the ink transferred from f.

> The graph of f^{-1} is a reflection of the graph of f across the line $y = x$.

Technology Connection

On some graphing calculators, we can graph the inverse of a function after graphing the function itself by accessing a drawing feature.

EXAMPLE 8 Consider $g(x) = x^3 + 2$.

a) Determine whether the function is one-to-one.

b) If it is one-to-one, find a formula for its inverse.

c) Graph the function and its inverse.

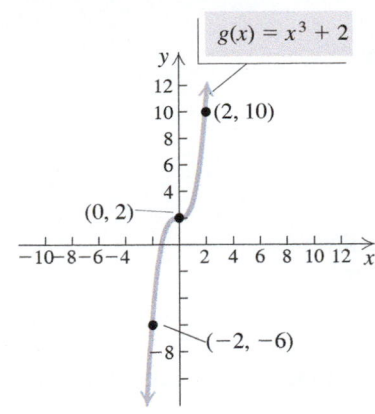

x	$g(x)$
-2	-6
-1	1
0	2
1	3
2	10

x	$g^{-1}(x)$
-6	-2
1	-1
2	0
3	1
10	2

Solution

a) The graph of $g(x) = x^3 + 2$ is shown at left. It passes the horizontal-line test and thus has an inverse that is a function. We also know that $g(x)$ is one-to-one because it is an increasing function over its entire domain.

b) We follow the procedure for finding an inverse.

1. Replace $g(x)$ with y: $y = x^3 + 2$
2. Interchange x and y: $x = y^3 + 2$
3. Solve for y: $x - 2 = y^3$
     $$\sqrt[3]{x - 2} = y$$
4. Replace y with $g^{-1}(x)$: $g^{-1}(x) = \sqrt[3]{x - 2}$.

We can test a point as a partial check:

$$g(x) = x^3 + 2$$
$$g(3) = 3^3 + 2 = 27 + 2 = 29.$$

Will $g^{-1}(29) = 3$? We have

$$g^{-1}(x) = \sqrt[3]{x - 2}$$
$$g^{-1}(29) = \sqrt[3]{29 - 2} = \sqrt[3]{27} = 3.$$

Since $g(3) = 29$ and $g^{-1}(29) = 3$, we can be reasonably certain that the formula for $g^{-1}(x)$ is correct.

c) To find the graph of the inverse function, we reflect the graph of $g(x) = x^3 + 2$ across the line $y = x$. This can be done by plotting points.

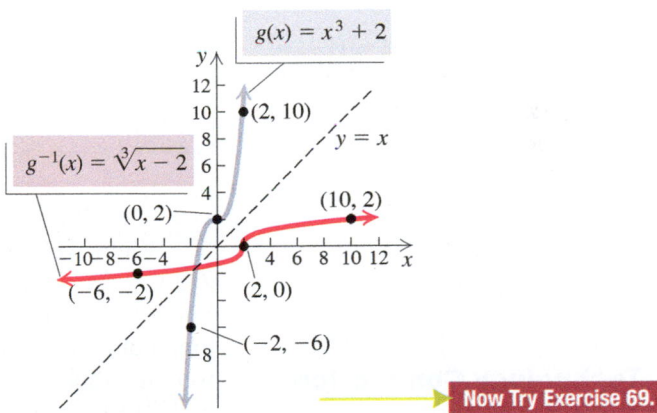

> **Now Try Exercise 69.**

▶ Inverse Functions and Composition

Suppose that we were to use some input a for a one-to-one function f and find its output, $f(a)$. The function f^{-1} would then take that output back to a. Similarly, if we began with an input b for the function f^{-1} and found its output, $f^{-1}(b)$, the original function f would then take that output back to b. This is summarized as follows.

> If a function f is one-to-one, then f^{-1} is the unique function such that each of the following holds:
>
> $$(f^{-1} \circ f)(x) = f^{-1}(f(x)) = x, \quad \text{for each } x \text{ in the domain of } f, \text{ and}$$
> $$(f \circ f^{-1})(x) = f(f^{-1}(x)) = x, \quad \text{for each } x \text{ in the domain of } f^{-1}.$$

EXAMPLE 9 Given that $f(x) = 5x + 8$, use composition of functions to show that

$$f^{-1}(x) = \frac{x - 8}{5}.$$

Solution We find $(f^{-1} \circ f)(x)$ and $(f \circ f^{-1})(x)$ and check to see that each is x:

$$(f^{-1} \circ f)(x) = f^{-1}(f(x))$$
$$= f^{-1}(5x + 8) = \frac{(5x + 8) - 8}{5} = \frac{5x}{5} = x;$$

$$(f \circ f^{-1})(x) = f(f^{-1}(x))$$
$$= f\left(\frac{x - 8}{5}\right) = 5\left(\frac{x - 8}{5}\right) + 8 = x - 8 + 8 = x.$$

> **Now Try Exercise 77.**

▶ Restricting a Domain

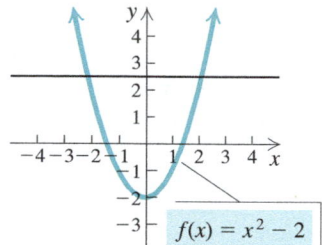

$f(x) = x^2 - 2$

$f^{-1}(x) = \sqrt{x + 2}$

$f(x) = x^2 - 2,$
for inputs $x \geq 0$

In the case in which the inverse of a function is not a function, the domain of the function can be restricted to allow the inverse to be a function. Let's consider the function $f(x) = x^2 - 2$. It is not one-to-one. The graph is shown at left.

Suppose that we had tried to find a formula for the inverse as follows:

$$y = x^2 - 2 \qquad \textcolor{red}{\text{Replacing } f(x) \text{ with } y}$$
$$x = y^2 - 2 \qquad \textcolor{red}{\text{Interchanging } x \text{ and } y}$$
$$x + 2 = y^2$$
$$\pm \sqrt{x + 2} = y. \qquad \textcolor{red}{\text{Solving for } y}$$

This is not the equation of a function. An input of, say, 2 would yield two outputs, -2 and 2. In such cases, it is convenient to consider "part" of the function by restricting the domain of $f(x)$. For example, if we restrict the domain of $f(x) = x^2 - 2$ to nonnegative numbers, then its inverse is a function, as shown at left by the graphs of $f(x) = x^2 - 2$, $x \geq 0$, and $f^{-1}(x) = \sqrt{x + 2}$.

9.2   Exercise Set

Find the inverse of the relation.

1. $\{(7, 8), (-2, 8), (3, -4), (8, -8)\}$

2. $\{(0, 1), (5, 6), (-2, -4)\}$

3. $\{(-1, -1), (-3, 4)\}$

4. $\{(-1, 3), (2, 5), (-3, 5), (2, 0)\}$

Find an equation of the inverse relation.

5. $y = 4x - 5$

6. $2x^2 + 5y^2 = 4$

7. $x^3y = -5$

8. $y = 3x^2 - 5x + 9$

9. $x = y^2 - 2y$

10. $x = \frac{1}{2}y + 4$

Graph the equation by substituting and plotting points. Then reflect the graph across the line $y = x$ to obtain the graph of its inverse.

11. $x = y^2 - 3$ **12.** $y = x^2 + 1$

13. $y = 3x - 2$ **14.** $x = -y + 4$

15. $y = |x|$ **16.** $x + 2 = |y|$

Given the function f, prove that f is one-to-one using the definition of a one-to-one function on p. 602.

17. $f(x) = \frac{1}{3}x - 6$ **18.** $f(x) = 4 - 2x$

19. $f(x) = x^3 + \frac{1}{2}$ **20.** $f(x) = \sqrt[3]{x}$

Given the function g, prove that g is not one-to-one using the definition of a one-to-one function on p. 602.

21. $g(x) = 1 - x^2$ **22.** $g(x) = 3x^2 + 1$

23. $g(x) = x^4 - x^2$ **24.** $g(x) = \frac{1}{x^6}$

Using the horizontal-line test, determine whether the function is one-to-one.

25. $f(x) = 2.7^x$ **26.** $f(x) = 2^{-x}$

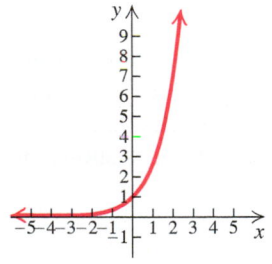 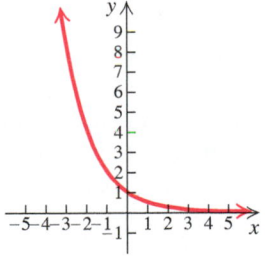

27. $f(x) = 4 - x^2$ **28.** $f(x) = x^3 - 3x + 1$

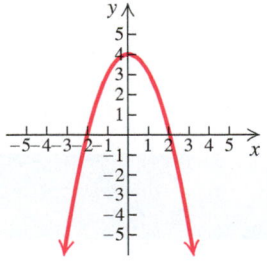 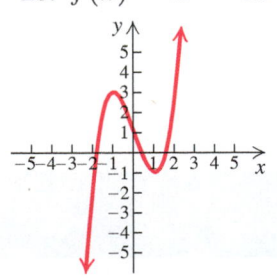

29. $f(x) = \dfrac{8}{x^2 - 4}$ **30.** $f(x) = \sqrt{\dfrac{10}{4 + x}}$

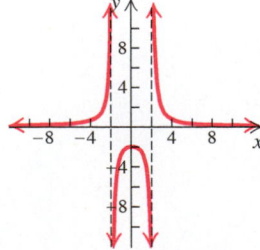 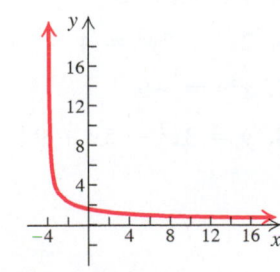

31. $f(x) = \sqrt[3]{x + 2} - 2$ **32.** $f(x) = \dfrac{8}{x}$

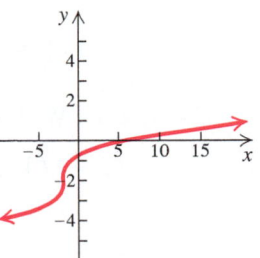 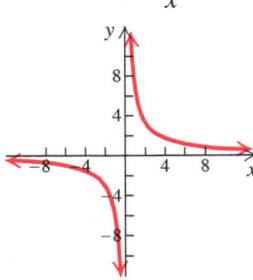

Graph the function and determine whether the function is one-to-one using the horizontal-line test.

33. $f(x) = 5x - 8$ **34.** $f(x) = 3 + 4x$

35. $f(x) = 1 - x^2$ **36.** $f(x) = |x| - 2$

37. $f(x) = |x + 2|$ **38.** $f(x) = -0.8$

39. $f(x) = -\dfrac{4}{x}$ **40.** $f(x) = \dfrac{2}{x + 3}$

41. $f(x) = \frac{2}{3}$ **42.** $f(x) = \frac{1}{2}x^2 + 3$

43. $f(x) = \sqrt{25 - x^2}$ **44.** $f(x) = -x^3 + 2$

In Exercises 45–60, for each function:

a) *Determine whether it is one-to-one.*

b) *If the function is one-to-one, find a formula for the inverse.*

45. $f(x) = x + 4$

46. $f(x) = 7 - x$

47. $f(x) = 2x - 1$

48. $f(x) = 5x + 8$

49. $f(x) = \dfrac{4}{x + 7}$

50. $f(x) = -\dfrac{3}{x}$

51. $f(x) = \dfrac{x + 4}{x - 3}$

52. $f(x) = \dfrac{5x - 3}{2x + 1}$

53. $f(x) = x^3 - 1$

54. $f(x) = (x + 5)^3$

55. $f(x) = x\sqrt{4 - x^2}$

56. $f(x) = 2x^2 - x - 1$

57. $f(x) = 5x^2 - 2,\ x \geq 0$

58. $f(x) = 4x^2 + 3,\ x \geq 0$

59. $f(x) = \sqrt{x + 1}$

60. $f(x) = \sqrt[3]{x - 8}$

Find the inverse by thinking about the operations of the function and then reversing, or undoing, them. Check your work algebraically.

FUNCTION	**INVERSE**
61. $f(x) = 3x$	$f^{-1}(x) =$
62. $f(x) = \frac{1}{4}x + 7$	$f^{-1}(x) =$
63. $f(x) = -x$	$f^{-1}(x) =$
64. $f(x) = \sqrt[3]{x} - 5$	$f^{-1}(x) =$
65. $f(x) = \sqrt[3]{x - 5}$	$f^{-1}(x) =$
66. $f(x) = x^{-1}$	$f^{-1}(x) =$

Each graph in Exercises 67–72 is the graph of a one-to-one function f. Sketch the graph of the inverse function f^{-1}.

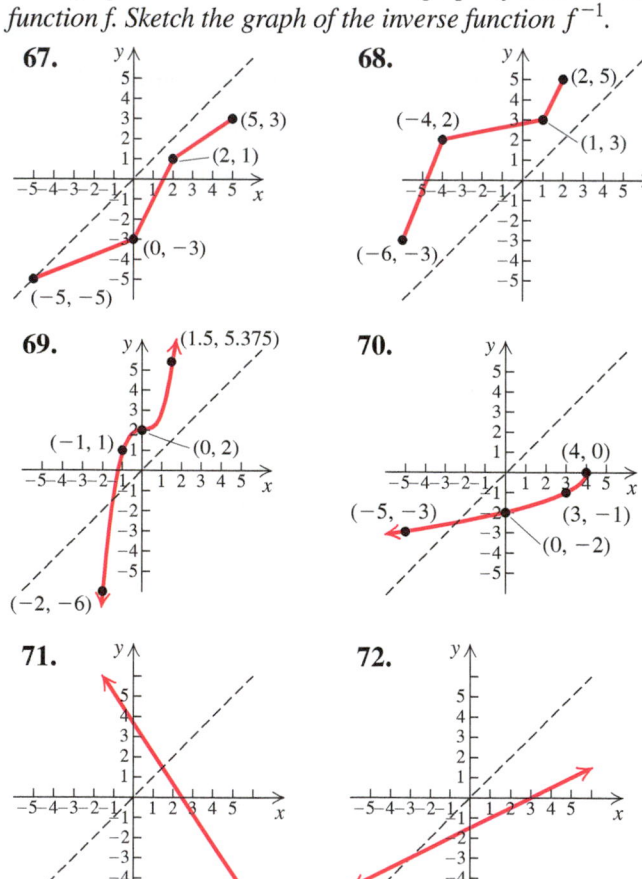

67.

68.

69.

70.

71.

72.

For the function f, use composition of functions to show that f^{-1} is as given.

73. $f(x) = \frac{7}{8}x, \; f^{-1}(x) = \frac{8}{7}x$

74. $f(x) = \frac{x + 5}{4}, \; f^{-1}(x) = 4x - 5$

75. $f(x) = \frac{1 - x}{x}, \; f^{-1}(x) = \frac{1}{x + 1}$

76. $f(x) = \sqrt[3]{x + 4}, \; f^{-1}(x) = x^3 - 4$

77. $f(x) = \frac{2}{5}x + 1, \; f^{-1}(x) = \frac{5x - 5}{2}$

78. $f(x) = \frac{x + 6}{3x - 4}, \; f^{-1}(x) = \frac{4x + 6}{3x - 1}$

Find the inverse of the given one-to-one function f. Give the domain and the range of f and of f^{-1}, and then graph both f and f^{-1} on the same set of axes.

79. $f(x) = 5x - 3$

80. $f(x) = 2 - x$

81. $f(x) = \frac{2}{x}$

82. $f(x) = -\frac{3}{x + 1}$

83. $f(x) = \frac{1}{3}x^3 - 2$

84. $f(x) = \sqrt[3]{x} - 1$

85. $f(x) = \frac{x + 1}{x - 3}$

86. $f(x) = \frac{x - 1}{x + 2}$

87. Find $f(f^{-1}(5))$ and $f^{-1}(f(a))$:
$$f(x) = x^3 - 4.$$

88. Find $(f^{-1}(f(p)))$ and $f(f^{-1}(1253))$:
$$f(x) = \sqrt[5]{\frac{2x - 7}{3x + 4}}.$$

89. *Hitting Lessons.* A summer little-league baseball team determines that the cost per player of a group hitting lesson is given by the formula
$$C(x) = \frac{72 + 2x}{x},$$
where x is the number of players in the group and $C(x)$ is in dollars.

a) Determine the cost per player of a group hitting lesson when there are 2, 5, and 8 players in the group.

b) Find a formula for the inverse of the function and explain what it represents.

c) Use the inverse function to determine the number of players in the group lesson when the cost per player is $74, $20, and $11.

90. *Women's Shoe Sizes.* A function that will convert women's shoe sizes in the United States to those in Australia is

$$s(x) = \frac{2x - 3}{2}$$

(*Source*: www.onlineconversion.com).

a) Determine the women's shoe sizes in Australia that correspond to sizes 5, $7\frac{1}{2}$, and 8 in the United States.

b) Find a formula for the inverse of the function and explain what it represents.

c) Use the inverse function to determine the women's shoe sizes in the United States that correspond to sizes 3, $5\frac{1}{2}$, and 7 in Australia.

91. *E-Commerce Holiday Sales.* Retail e-commerce holiday season sales (November and December), in billions of dollars, x years after 2008 is given by the function

$$H(x) = 6.58x + 27.7$$

(*Source:* statista.com).

a) Determine the total amount of e-commerce holiday sales in 2010 and in 2013.

b) Find $H^{-1}(x)$ and explain what it represents.

92. *Converting Temperatures.* The following formula can be used to convert Fahrenheit temperatures x to Celsius temperatures $T(x)$:

$$T(x) = \frac{5}{9}(x - 32).$$

a) Find $T(-13°)$ and $T(86°)$.

b) Find $T^{-1}(x)$ and explain what it represents.

▶ Skill Maintenance

Consider the quadratic functions (a)–(h) that follow. Without graphing them, answer the questions below. **[7.5]**

a) $f(x) = 2x^2$

b) $f(x) = -x^2$

c) $f(x) = \frac{1}{4}x^2$

d) $f(x) = -5x^2 + 3$

e) $f(x) = \frac{2}{3}(x - 1)^2 - 3$

f) $f(x) = -2(x + 3)^2 + 1$

g) $f(x) = (x - 3)^2 + 1$

h) $f(x) = -4(x + 1)^2 - 3$

93. Which functions have a maximum value?

94. Which graphs open up?

95. Consider (a) and (c). Which graph is narrower?

96. Consider (d) and (e). Which graph is narrower?

97. Which graph has vertex $(-3, 1)$?

98. For which is the line of symmetry $x = 0$?

▶ Synthesis

99. The function $f(x) = x^2 - 3$ is not one-to-one. Restrict the domain of f so that its inverse is a function. Find the inverse and state the restriction on the domain of the inverse.

100. Consider the function f given by

$$f(x) = \begin{cases} x^3 + 2, & \text{for } x \le -1, \\ x^2, & \text{for } -1 < x < 1, \\ x + 1, & \text{for } x \ge 1. \end{cases}$$

Does f have an inverse that is a function? Why or why not?

101. Find three examples of functions that are their own inverses; that is, $f = f^{-1}$.

102. Given the function $f(x) = ax + b, a \ne 0$, find the values of a and b for which $f^{-1}(x) = f(x)$.

9.3 Exponential Functions and Graphs

▶ Graph exponential equations and exponential functions.

▶ Solve applied problems involving exponential functions and their graphs.

We now turn our attention to the study of a set of functions that are very rich in application. Consider the following graphs. Each one illustrates an *exponential function*. In this section, we consider such functions and some important applications.

Skype Users Online at Same Time

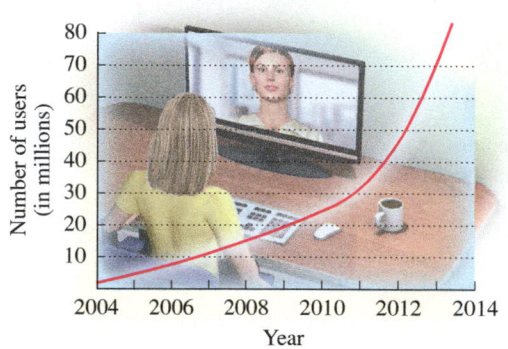

Source: Skype Numerology Blog

Postseason Bowl Games

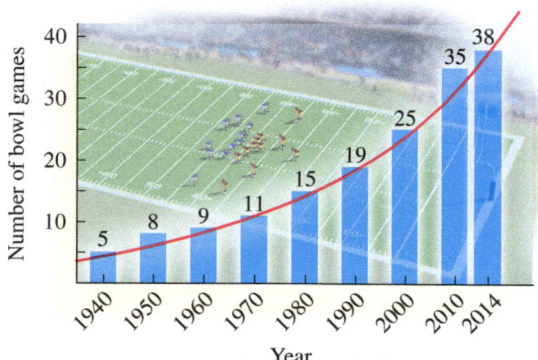

Source: USA TODAY research, College Football Data Warehouse

▶ Graphing Exponential Functions

We now define exponential functions. We assume that a^x has meaning for any real number x and any positive real number a and that the laws of exponents still hold, though we will not prove them here.

EXPONENTIAL FUNCTION

The function $f(x) = a^x$, where x is a real number, $a > 0$ and $a \neq 1$, is called the **exponential function, base a**.

We require the **base** to be positive in order to avoid the imaginary numbers that would occur by taking even roots of negative numbers—an example is $(-1)^{1/2}$, the square root of -1, which is not a real number. The restriction $a \neq 1$ is made to exclude the constant function $f(x) = 1^x = 1$, which does not have an inverse that is a function because it is not one-to-one.

The following are examples of exponential functions:

$$f(x) = 2^x, \qquad f(x) = \left(\frac{1}{2}\right)^x, \qquad f(x) = (3.57)^x.$$

Note that, in contrast to functions like $f(x) = x^5$ and $f(x) = x^{1/2}$ in which the variable is the base of an exponential expression, the variable in an exponential function is *in the exponent*.

Let's now consider graphs of exponential functions.

EXAMPLE 1 Graph the exponential function $y = f(x) = 2^x$.

Solution We compute some function values and list the results in a table.

$$f(0) = 2^0 = 1; \quad f(-1) = 2^{-1} = \frac{1}{2^1} = \frac{1}{2};$$
$$f(1) = 2^1 = 2;$$
$$f(2) = 2^2 = 4; \quad f(-2) = 2^{-2} = \frac{1}{2^2} = \frac{1}{4};$$
$$f(3) = 2^3 = 8;$$
$$f(-3) = 2^{-3} = \frac{1}{2^3} = \frac{1}{8}.$$

x	y $y = f(x) = 2^x$	(x, y)
0	1	(0, 1)
1	2	(1, 2)
2	4	(2, 4)
3	8	(3, 8)
−1	$\frac{1}{2}$	$\left(-1, \frac{1}{2}\right)$
−2	$\frac{1}{4}$	$\left(-2, \frac{1}{4}\right)$
−3	$\frac{1}{8}$	$\left(-3, \frac{1}{8}\right)$

Next, we plot these points and connect them with a smooth curve. Be sure to plot enough points to determine how steeply the curve rises.

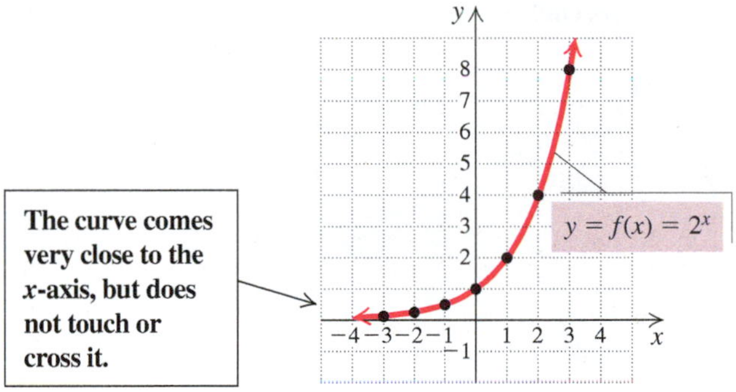

HORIZONTAL ASYMPTOTES
REVIEW SECTION 8.5

Note that as x increases, the function values increase without bound. As x decreases, the function values decrease, getting close to 0. That is, as $x \to -\infty$, $y \to 0$. Thus the x-axis, or the line $y = 0$, is a horizontal asymptote. As the x-inputs decrease, the curve gets closer and closer to this line, but does not cross it.

Now Try Exercise 11.

EXAMPLE 2 Graph the exponential function $y = f(x) = \left(\frac{1}{2}\right)^x$.

Solution Before we plot points and draw the curve, note that

$$y = f(x) = \left(\frac{1}{2}\right)^x = (2^{-1})^x = 2^{-x}.$$

This tells us that this graph is a reflection of the graph of $y = 2^x$ across the y-axis. For example, if $(3, 8)$ is a point of the graph of $g(x) = 2^x$, then $(-3, 8)$ is a point of the graph of $f(x) = 2^{-x}$. Selected points are listed in the table at left.

Points of $g(x) = 2^x$	Points of $f(x) = \left(\frac{1}{2}\right)^x = 2^{-x}$
(0, 1)	(0, 1)
(1, 2)	(−1, 2)
(2, 4)	(−2, 4)
(3, 8)	(−3, 8)
$\left(-1, \frac{1}{2}\right)$	$\left(1, \frac{1}{2}\right)$
$\left(-2, \frac{1}{4}\right)$	$\left(2, \frac{1}{4}\right)$
$\left(-3, \frac{1}{8}\right)$	$\left(3, \frac{1}{8}\right)$

Next, we plot these points and connect them with a smooth curve.

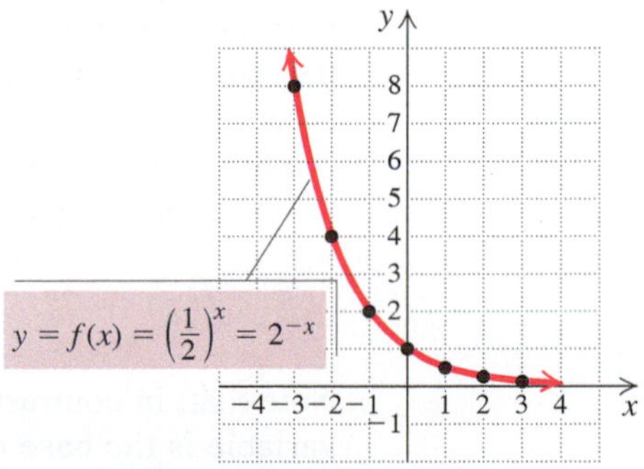

Note that as x increases, the function values decrease, getting close to 0. The x-axis, $y = 0$, is the horizontal asymptote. As x decreases, the function values increase without bound.

Now Try Exercise 15.

CONNECTING THE CONCEPTS

Properties of Exponential Functions

Let's list and compare some characteristics of exponential functions, keeping in mind that the definition of an exponential function, $f(x) = a^x$, requires that a be positive and different from 1.

$f(x) = a^x, a > 0, a \neq 1$

Continuous

One-to-one

Domain: $(-\infty, \infty)$

Range: $(0, \infty)$

Increasing if $a > 1$

Decreasing if $0 < a < 1$

Horizontal asymptote is x-axis

y-intercept: $(0, 1)$

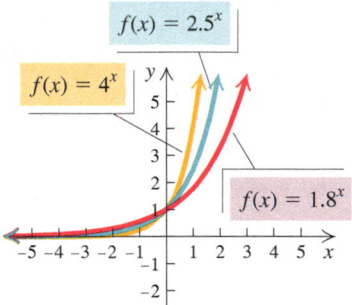

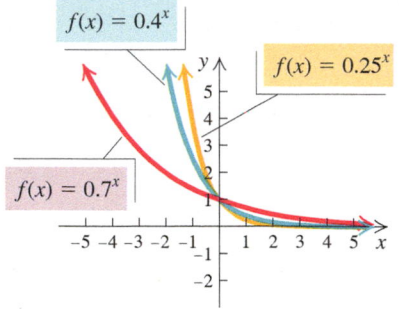

TRANSFORMATIONS

REVIEW SECTION 7.2

To graph other types of exponential functions, keep in mind the ideas of translation, stretching, and reflection. All these concepts allow us to visualize the graph before drawing it.

EXAMPLE 3 Graph each of the following. Before doing so, describe how each graph can be obtained from the graph of $f(x) = 2^x$.

a) $f(x) = 2^{x-2}$ **b)** $f(x) = 2^x - 4$ **c)** $f(x) = 5 - 0.5^x$

Solution

a) The graph of $f(x) = 2^{x-2}$ is the graph of $y = 2^x$ shifted *right* 2 units.

x	$f(x)$
-1	$\frac{1}{8}$
0	$\frac{1}{4}$
1	$\frac{1}{2}$
2	1
3	2
4	4
5	8

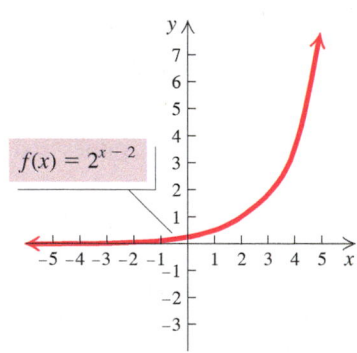

b) The graph of $f(x) = 2^x - 4$ is the graph of $y = 2^x$ shifted *down* 4 units.

x	$f(x)$
-2	$-3\frac{3}{4}$
-1	$-3\frac{1}{2}$
0	-3
1	-2
2	0
3	4

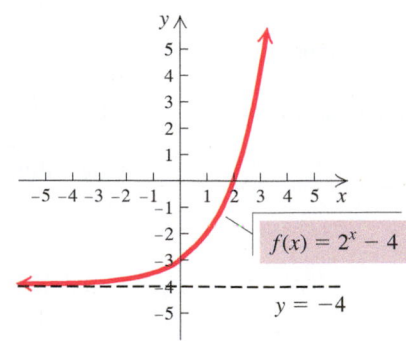

c) The graph of $f(x) = 5 - 0.5^x = 5 - \left(\frac{1}{2}\right)^x = 5 - 2^{-x}$ is a reflection of the graph of $y = 2^x$ across the y-axis, followed by a reflection across the x-axis and then a shift *up* 5 units.

x	$f(x)$
-3	-3
-2	1
-1	3
0	4
1	$4\frac{1}{2}$
2	$4\frac{3}{4}$

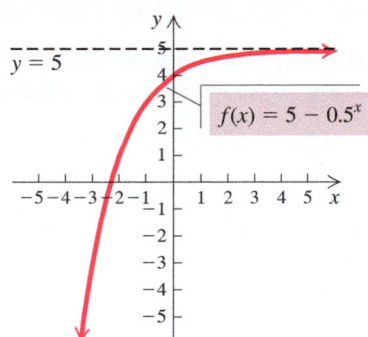

$f(x) = 5 - 0.5^x$

$y = 5$

> **Now Try Exercises 27 and 33.**

► Applications

One of the most frequent applications of exponential functions occurs with compound interest.

EXAMPLE 4 *Compound Interest.* The amount of money A to which a principal P will grow after t years at interest rate r (in decimal form), compounded n times per year, is given by the formula

$$A = P\left(1 + \frac{r}{n}\right)^{nt}.$$

Suppose that \$100,000 is invested at 6.5% interest, compounded semiannually.

a) Find a function for the amount to which the investment grows after t years.

b) Find the amount of money in the account at $t = 0, 4, 8,$ and 10 years.

c) Graph the function.

Solution

a) Since $P = \$100{,}000$, $r = 6.5\% = 0.065$, and $n = 2$, we can substitute these values and write the following function:

$$A(t) = 100{,}000\left(1 + \frac{0.065}{2}\right)^{2 \cdot t} = \$100{,}000(1.0325)^{2t}.$$

b) We can compute function values with a calculator:

$$A(0) = 100{,}000(1.0325)^{2 \cdot 0} = \$100{,}000;$$
$$A(4) = 100{,}000(1.0325)^{2 \cdot 4} \approx \$129{,}157.75;$$
$$A(8) = 100{,}000(1.0325)^{2 \cdot 8} \approx \$166{,}817.25;$$
$$A(10) = 100{,}000(1.0325)^{2 \cdot 10} \approx \$189{,}583.79.$$

c) We use the function values computed in part (b) and others if we wish, and draw the graph as follows. Note that the axes are scaled differently because of the large values of A and that t is restricted to nonnegative values, because negative time values have no meaning here.

Technology Connection

We can find the function values in Example 4(b) using the VALUE feature from the CALC menu.

$y = 100{,}000(1.0325)^{2x}$

500,000

0 ⎿_____⏌ 30
0

Xscl = 5, Yscl = 50,000

We could also use the TABLE feature.

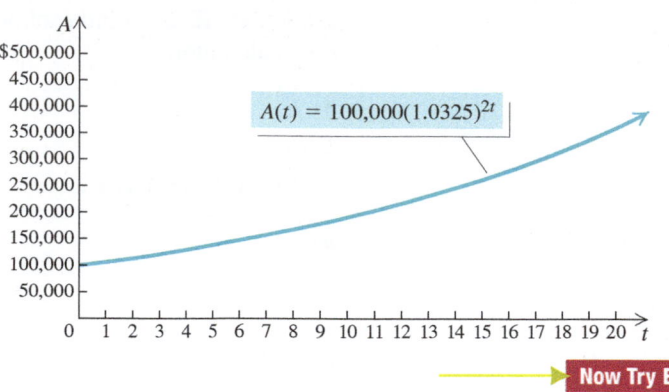

Now Try Exercise 51.

▶ The Number *e*

We now consider a very special number in mathematics. In 1741, Leonhard Euler named this number *e*. Though you may not have encountered it before, you will see here and in future mathematics courses that it has many important applications. To explain this number, we use the compound interest formula $A = P(1 + r/n)^{nt}$ discussed in Example 4. Suppose that \$1 is invested at 100% interest for 1 year. Since $P = 1, r = 100\% = 1$, and $t = 1$, the formula above becomes a function A defined in terms of the number of compounding periods n:

$$A = P\left(1 + \frac{r}{n}\right)^{nt} = 1\left(1 + \frac{1}{n}\right)^{n \cdot 1} = \left(1 + \frac{1}{n}\right)^{n}.$$

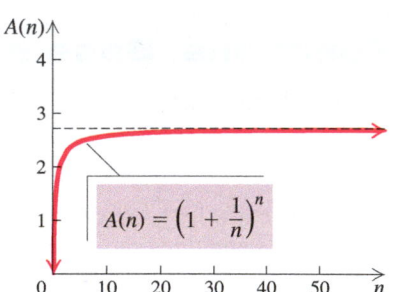

Let's visualize this function using its graph, shown at left, and explore the values of $A(n)$ as $n \to \infty$. Consider the graph for larger and larger values of n. Does this function have a horizontal asymptote?

Let's find some function values using a calculator.

n, Number of Compounding Periods	$A(n) = \left(1 + \dfrac{1}{n}\right)^{n}$
1 (compounded annually)	\$2.00
2 (compounded semiannually)	2.25
3	2.3704
4 (compounded quarterly)	2.4414
5	2.4883
100	2.7048
365 (compounded daily)	2.7146
8760 (compounded hourly)	2.7181

It appears from these values that the graph does have a horizontal asymptote, $y \approx 2.7$. As the values of n get larger and larger, the function values get closer and closer to the number Euler named *e*. Its decimal representation does not terminate or repeat; it is irrational.

$$e = 2.7182818284 \ldots$$

EXAMPLE 5 Find each value of e^x, to four decimal places, using the e^x key on a calculator.

a) e^3 **b)** $e^{-0.23}$ **c)** e^0

Solution

FUNCTION VALUE	READOUT	ROUNDED
a) e^3	e^3 20.08553692	20.0855
b) $e^{-0.23}$	$e^{-.23}$ .7945336025	0.7945
c) e^0	e^0 1	1

> **Now Try Exercises 1 and 3.**

▶ # Graphs of Exponential Functions, Base e

We now demonstrate ways in which to graph exponential functions.

EXAMPLE 6 Graph $f(x) = e^x$ and $g(x) = e^{-x}$.

Solution We can compute points for each equation using the e^x key on a calculator. (See the following table.) Then we plot these points and draw the graphs of the functions.

x	$f(x) = e^x$	$g(x) = e^{-x}$
−2	0.135	7.389
−1	0.368	2.718
0	1	1
1	2.718	0.368
2	7.389	0.135

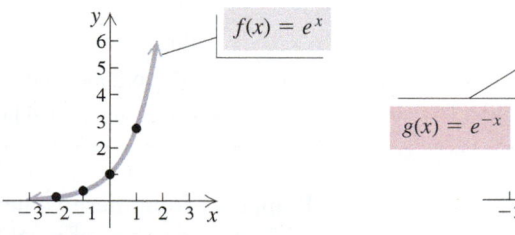

Note that the graphs are reflections of each other across the y-axis.

> **Now Try Exercise 23.**

EXAMPLE 7 Graph each of the following. Before doing so, describe how each graph can be obtained from the graph of $y = e^x$.

a) $f(x) = e^{x+3}$ **b)** $f(x) = e^{-0.5x}$ **c)** $f(x) = 1 - e^{-2x}$

Solution

a) The graph of $f(x) = e^{x+3}$ is a translation of the graph of $y = e^x$ left 3 units.

x	$f(x)$
-7	0.018
-5	0.135
-3	1
-1	7.389
0	20.086

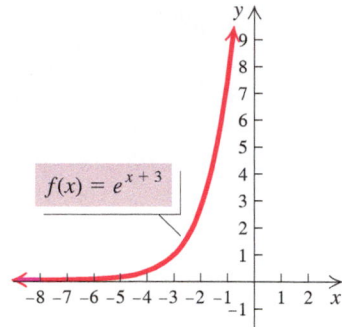

b) We note that the graph of $f(x) = e^{-0.5x}$ is a horizontal stretching of the graph of $y = e^x$ followed by a reflection across the y-axis.

x	$f(x)$
-2	2.718
-1	1.649
0	1
1	0.607
2	0.368

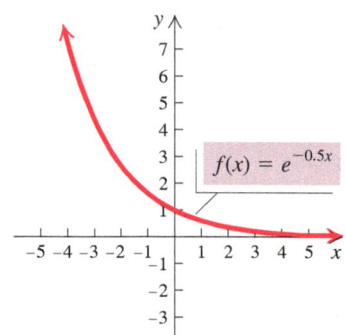

c) The graph of $f(x) = 1 - e^{-2x}$ is a horizontal shrinking of the graph of $y = e^x$, followed by a reflection across the y-axis, then across the x-axis, and followed by a translation up 1 unit.

x	$f(x)$
-1	-6.389
0	0
1	0.865
2	0.982
3	0.998

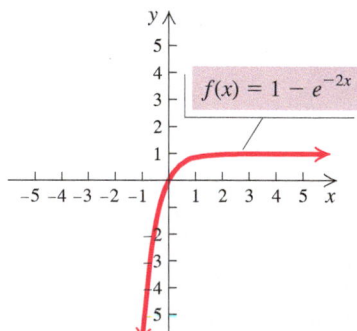

Now Try Exercises 41 and 47.

9.3   Exercise Set

Find each of the following, to four decimal places, using a calculator.

1. e^4

2. e^{10}

3. $e^{-2.458}$

4. $\left(\dfrac{1}{e^3}\right)^2$

In Exercises 5–10, match the function with one of the graphs (a)–(f) that follow.

a)
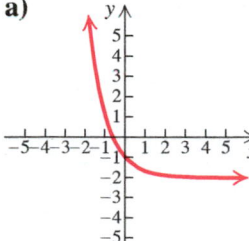

b)
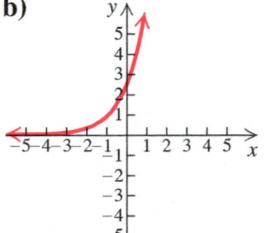

c)
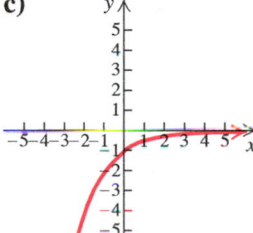

d)
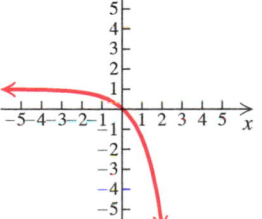

e)
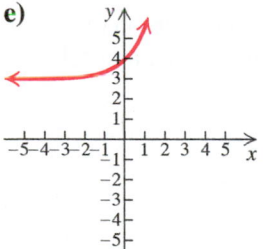

f)
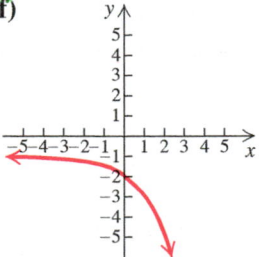

5. $f(x) = -2^x - 1$

6. $f(x) = -\left(\frac{1}{2}\right)^x$

7. $f(x) = e^x + 3$

8. $f(x) = e^{x+1}$

9. $f(x) = 3^{-x} - 2$

10. $f(x) = 1 - e^x$

Graph the function by substituting and plotting points.

11. $f(x) = 3^x$

12. $f(x) = 5^x$

13. $f(x) = 6^x$

14. $f(x) = 3^{-x}$

15. $f(x) = \left(\frac{1}{4}\right)^x$

16. $f(x) = \left(\frac{2}{3}\right)^x$

17. $y = -2^x$

18. $y = 3 - 3^x$

19. $f(x) = -0.25^x + 4$

20. $f(x) = 0.6^x - 3$

21. $f(x) = 1 + e^{-x}$

22. $f(x) = 2 - e^{-x}$

23. $y = \frac{1}{4}e^x$

24. $y = 2e^{-x}$

25. $f(x) = 1 - e^{-x}$

26. $f(x) = e^x - 2$

Sketch the graph of the function. Describe how each graph can be obtained from the graph of a basic exponential function.

27. $f(x) = 2^{x+1}$

28. $f(x) = 2^{x-1}$

29. $f(x) = 2^x - 3$

30. $f(x) = 2^x + 1$

31. $f(x) = 2^{1-x} + 2$

32. $f(x) = 5 - 2^{-x}$

33. $f(x) = 4 - 3^{-x}$

34. $f(x) = 2^{x-1} - 3$

35. $f(x) = \left(\frac{3}{2}\right)^{x-1}$

36. $f(x) = 3^{4-x}$

37. $f(x) = 2^{x+3} - 5$

38. $f(x) = -3^{x-2}$

39. $f(x) = 3 \cdot 2^{x-1} + 1$

40. $f(x) = 2 \cdot 3^{x+1} - 2$

41. $f(x) = e^{2x}$

42. $f(x) = e^{-0.2x}$

43. $f(x) = \dfrac{1}{2}(1 - e^x)$

44. $f(x) = 3(1 + e^x) - 2$

45. $y = e^{-x+1}$

46. $y = e^{2x} + 1$

47. $f(x) = 2(1 - e^{-x})$

48. $f(x) = 1 - e^{-0.01x}$

Graph the piecewise function.

49. $f(x) = \begin{cases} e^{-x} - 4, & \text{for } x < -2, \\ x + 3, & \text{for } -2 \le x < 1, \\ x^2, & \text{for } x \ge 1 \end{cases}$

50. $g(x) = \begin{cases} 4, & \text{for } x \le -3, \\ x^2 - 6, & \text{for } -3 < x < 0, \\ e^x, & \text{for } x \ge 0 \end{cases}$

51. *Compound Interest.*   Suppose that \$82,000 is invested at $4\frac{1}{2}\%$ interest, compounded quarterly.

a) Find the function for the amount to which the investment grows after t years.

b) Find the amount of money in the account at $t = 0, 2, 5,$ and 10 years.

52. *Compound Interest.*   Suppose that \$750 is invested at 7% interest, compounded semiannually.

a) Find the function for the amount to which the investment grows after t years.

b) Find the amount of money in the account at $t = 1, 6, 10, 15,$ and 25 years.

53. *Interest on a CD.*   On Elizabeth's sixth birthday, her grandparents present her with a \$3000 certificate of deposit (CD) that earns 5% interest, compounded quarterly. If the CD matures on her sixteenth birthday, what amount will be available then?

54. *Interest in a College Trust Fund.* Following the birth of his child, Benjamin deposits $10,000 in a college trust fund where interest is 3.9%, compounded semiannually.

a) Find a function for the amount in the account after t years.

b) Find the amount of money in the account at $t = 0$, 4, 8, 10, 18, and 21 years.

In Exercises 55–62, use the compound-interest formula to find the account balance A with the given conditions:

P = principal,
r = interest rate,
n = number of compounding periods per year,
t = time, in years,
A = account balance.

	P	r	Compounded	n	t	A
55.	$3,000	4%	Semiannually		2	
56.	$12,500	3%	Quarterly		3	
57.	$120,000	2.5%	Annually		10	
58.	$120,000	2.5%	Quarterly		10	
59.	$53,500	$5\frac{1}{2}$%	Quarterly		$6\frac{1}{2}$	
60.	$6,250	$6\frac{3}{4}$%	Semiannually		$4\frac{1}{2}$	
61.	$17,400	8.1%	Daily		5	
62.	$900	7.3%	Daily		$7\frac{1}{4}$	

63. *Alternative-Fuel Vehicles.* The sales of alternative-fuel vehicles have more than tripled since 1995 (*Source:* Energy Information Administration). The exponential function

$$A(x) = 246{,}855(1.0931)^x,$$

where x is the number of years after 1995, can be used to estimate the number of alternative-fuel vehicles in a given year. Find the number of alternative-fuel vehicles in 1998 and in 2010. Then project the number of alternative-fuel vehicles in 2018.

64. *Increasing CPU Power.* The central processing unit (CPU) power in computers has increased significantly over the years. The CPU power in Macintosh computers has grown exponentially from 8 MHz in 1984 to 3400 MHz in 2013 (*Source:* Apple). The exponential function

$$M(t) = 7.91477(1.26698)^t,$$

where t is the number of years after 1984, can be used to estimate the CPU power in a Macintosh computer in a given year. Find the CPU power of a Macintosh Performa 5320CD in 1995 and of an iMac G6 in 2009. Round to the nearest one MHz.

65. *E-Cigarette Sales.* The electronic cigarette was launched in 2007, and since then sales have increased from about $20 million in 2008 to about $500 million in 2012 (*Sources:* UBS; forbes.com). The exponential function

$$S(x) = 20.913(2.236)^x,$$

where x is the number of years after 2008, models the sales, in millions of dollars. Use this function to estimate the sales of e-cigarettes in 2011 and in 2015. Round to the nearest million dollars.

66. *Foreign High School Students.* The number of foreign students studying in high schools in the United States grew from 6541 in 2007 to 65,452 in 2012 (*Source:* Council on Standards for International Educational Travel). The increase can be modeled by the exponential function

$$E(x) = 6541(1.5851)^x,$$

where x is the number of years after 2007. Find the number of foreign students enrolled in U.S. high schools in 2009 and in 2011. Then use this function to project the number of foreign students enrolled in 2016.

67. *Skype Concurrent Users.* The number of concurrent users of Skype has increased dramatically since 2004 (*Source:* Skype Numerology Blog). By 2013, Skype could connect concurrently 70 million users online. The exponential function

$$P(t) = 2.307(1.483)^t,$$

where t is the number of years after 2004, models this increase in millions of users. Estimate the number of

Skype users that could be online concurrently in 2005, in 2009, and in 2012. Round to the nearest million users.

68. *U.S. Imports.* The amount of imports to the United States has increased exponentially since 1980 (*Sources*: U.S. Census Bureau; U.S. Bureau of Economic Analysis; U.S. Department of Commerce). The exponential function

$$P(x) = 307.368(1.072)^x,$$

where x is the number of years after 1980, can be used to estimate the total amount of U.S. imports, in billions of dollars. Find the total amount of imports to the United States in 1990, in 2000, and in 2012. Round to the nearest billion dollars.

69. *Centenarian Population.* The centenarian population in the United States has grown over 65% in the last 30 years.

In 1980, there were only 32,194 residents ages 100 and over. This number had grown to 53,364 by 2010.

(*Sources*: Population Projections Program; U.S. Census Bureau; U.S. Department of Commerce; "What People Who Live to 100 Have in Common," by Emily Brandon, *U.S. News and World Report*, January 7, 2013) The exponential function

$$H(t) = 80,040.68(1.0481)^t,$$

where t is the number of years after 2015, can be used to project the number of centenarians. Use this function to project the centenarian population in 2020 and in 2050.

70. *Bachelor's Degrees Earned.* The exponential function

$$D(t) = 347(1.024)^t$$

gives the number of bachelor's degrees, in thousands, earned in the United States t years after 1970 (*Sources*: National Center for Educational Statistics; U.S. Department of Education). Find the number of bachelor's degrees earned in 1985, in 2000, and in 2014. Then estimate the number of bachelor's degrees that will be earned in 2020. Round to the nearest thousand degrees.

71. *Charitable Giving.* Over the last four decades, the amount of charitable giving in the United States has grown exponentially from approximately $20.7 billion in 1969 to approximately $316.2 billion in 2012 (*Sources*: Giving USA Foundation; Volunteering in America by the Corporation for National & Community Service; National Philanthropic Trust; School of Philanthropy, Indiana University Purdue University Indianapolis). The exponential function

$$G(x) = 20.7(1.066)^x,$$

where x is the number of years after 1969, can be used to estimate the amount of charitable giving, in billions of dollars, in a given year. Find the amount of charitable giving in 1982, in 1995, and in 2010. Round to the nearest billion dollars.

72. *Tennis Participation.* In recent years, U.S. tennis participation has increased. The exponential function

$$T(x) = 23.7624(1.0752)^x,$$

where $T(x)$ is in millions and x is the number of years after 2006, models the number of people who had played tennis at least once in a given year (*Source*: USTA/Coyne Public Relations). Estimate the number who had played tennis at least once in 2010. Then use this function to project participation in tennis in 2018. Round to the nearest million participants.

73. *Salvage Value.* A restaurant purchased a 72-in. range with six burners for $6982. The value of the range each year is 85% of the value of the preceding year. After t years, its value, in dollars, is given by the exponential function

$$V(t) = 6982(0.85)^t.$$

Find the value of the range after 0, 1, 2, 5, and 8 years. Round to the nearest dollar.

74. *Salvage Value.* A landscape company purchased a backhoe for $56,395. The value of the backhoe each year is 90% of the value of the preceding year. After t years, its value, in dollars, is given by the exponential function

$$V(t) = 56,395(0.9)^t.$$

Find the value of the backhoe after 0, 1, 3, 6, and 10 years. Round to the nearest dollar.

75. *Advertising.* A company begins an Internet advertising campaign to market a new telephone. The percentage of the target market that buys a product is generally a function of the length of the advertising campaign. The estimated percentage is given by the exponential function

$$f(t) = 100(1 - e^{-0.04t}),$$

where t is the number of days of the campaign. Find $f(25)$, the percentage of the target market that has bought the product after a 25-day advertising campaign.

76. *Growth of a Stock.* The value of a stock is given by the function

$$V(t) = 58(1 - e^{-1.1t}) + 20,$$

where V is the value of the stock after time t, in months. Find $V(1)$, $V(2)$, $V(4)$, $V(6)$, and $V(12)$.

▶ **Skill Maintenance**

Simplify. [7.3]

77. $(1 - 4i)(7 + 6i)$

78. $\dfrac{2 - i}{3 + i}$

Find the x-intercepts and the zeros of the function.

79. $f(x) = 2x^2 - 13x - 7$ [7.4]

80. $h(x) = x^3 - 3x^2 + 3x - 1$ [8.4]

81. $h(x) = x^4 - x^2$ [8.1]

82. $g(x) = x^3 + x^2 - 12x$ [8.1]

Solve.

83. $x^3 + 6x^2 - 16x = 0$ [8.1]

84. $3x^2 - 6 = 5x$ [7.4]

▶ **Synthesis**

85. Which is larger, 7^π or π^7? 70^{80} or 80^{70}?

86. For the function f, construct and simplify the difference quotient.

$$f(x) = 2e^x - 3$$

9.4

Logarithmic Functions and Graphs

▶ Find common logarithms and natural logarithms with and without a calculator.

▶ Convert between exponential equations and logarithmic equations.

▶ Change logarithmic bases.

▶ Graph logarithmic functions.

▶ Solve applied problems involving logarithmic functions.

We now consider *logarithmic*, or *logarithm*, *functions*. These functions are inverses of exponential functions and have many applications.

▶ Logarithmic Functions

We have noted that every exponential function (with $a > 0$ and $a \neq 1$) is one-to-one. Thus such a function has an inverse that is a function. In this section, we will name these inverse functions logarithmic functions and use them in applications.

EXAMPLE 1 Graph: $x = 2^y$.

Solution Note that x is alone on one side of the equation. We can find ordered pairs that are solutions by choosing values for y and then computing the corresponding x-values.

For $y = 0, x = 2^0 = 1$.
For $y = 1, x = 2^1 = 2$.
For $y = 2, x = 2^2 = 4$.
For $y = 3, x = 2^3 = 8$.

For $y = -1, x = 2^{-1} = \dfrac{1}{2^1} = \dfrac{1}{2}$.

For $y = -2, x = 2^{-2} = \dfrac{1}{2^2} = \dfrac{1}{4}$.

For $y = -3, x = 2^{-3} = \dfrac{1}{2^3} = \dfrac{1}{8}$.

x		
$x = 2^y$	y	(x, y)
1	0	$(1, 0)$
2	1	$(2, 1)$
4	2	$(4, 2)$
8	3	$(8, 3)$
$\dfrac{1}{2}$	-1	$\left(\dfrac{1}{2}, -1\right)$
$\dfrac{1}{4}$	-2	$\left(\dfrac{1}{4}, -2\right)$
$\dfrac{1}{8}$	-3	$\left(\dfrac{1}{8}, -3\right)$

(1) Choose values for y.
(2) Compute values for x.

We plot the points and connect them with a smooth curve. Note that the curve does not touch or cross the y-axis. The y-axis is a vertical asymptote.

Note too that this curve is the graph of $y = 2^x$ reflected across the line $y = x$, as we would expect for an inverse. The inverse of $y = 2^x$ is $x = 2^y$.

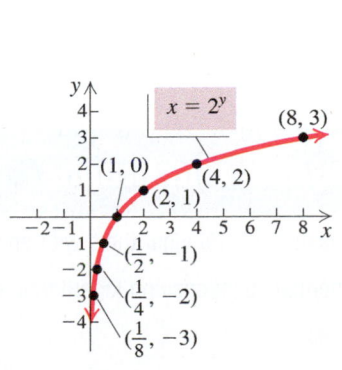

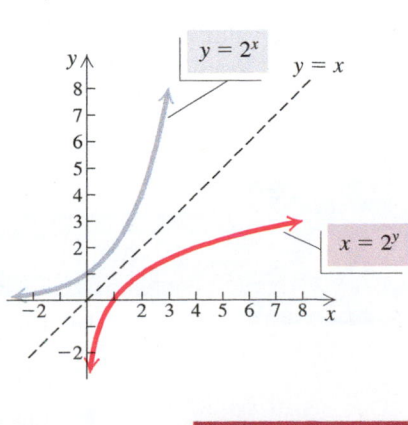

Now Try Exercise 1.

To find a formula for f^{-1} when $f(x) = 2^x$, we use the method discussed in Section 9.2:

1. Replace $f(x)$ with y: $y = 2^x$

2. Interchange x and y: $x = 2^y$

3. Solve for y: $y =$ the power to which we raise 2 to get x.

4. Replace y with $f^{-1}(x)$: $f^{-1}(x) =$ the power to which we raise 2 to get x.

Mathematicians have defined a new symbol to replace the words "the power to which we raise 2 to get x." That symbol is "$\log_2 x$," read "the logarithm, base 2, of x."

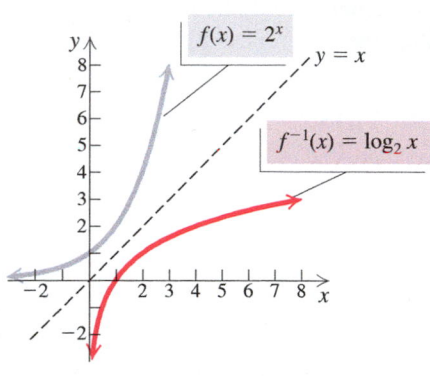

Logarithmic Function, Base 2

"$\log_2 x$," read "the logarithm, base 2, of x," means "the power to which we raise 2 to get x."

Thus if $f(x) = 2^x$, then $f^{-1}(x) = \log_2 x$. For example, $f^{-1}(8) = \log_2 8 = 3$, because *3 is the power to which we raise 2 to get 8*. Similarly, $\log_2 13$ is the power to which we raise 2 to get 13. As yet, we have no simpler way to say this other than

"$\log_2 13$ is the power to which we raise 2 to get 13."

Later, however, we will learn how to approximate this expression using a calculator.

For any exponential function $f(x) = a^x$, its inverse is called a **logarithmic function, base a**. The graph of the inverse can be obtained by reflecting the graph of $y = a^x$ across the line $y = x$, to obtain $x = a^y$. Then $x = a^y$ is equivalent to $y = \log_a x$. We read $\log_a x$ as "the logarithm, base a, of x."

The inverse of $f(x) = a^x$ is given by $f^{-1}(x) = \log_a x$.

LOGARITHMIC FUNCTION, BASE a

We define $y = \log_a x$ as that number y such that $x = a^y$, where $x > 0$ and a is a positive constant other than 1.

Let's look at the graphs of $f(x) = a^x$ and $f^{-1}(x) = \log_a x$ for $a > 1$ and for $0 < a < 1$.

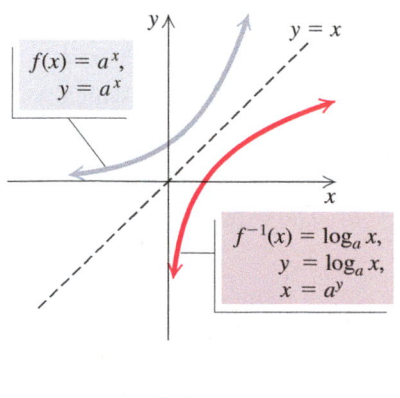

$a > 1$

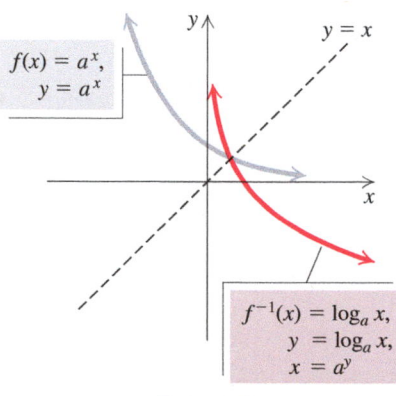

$0 < a < 1$

Note that the graphs of $f(x)$ and $f^{-1}(x)$ are reflections of each other across the line $y = x$.

CONNECTING THE CONCEPTS

Comparing Exponential Functions and Logarithmic Functions

In the following table, we compare exponential functions and logarithmic functions with bases a greater than 1. Similar statements could be made for a, where $0 < a < 1$. It is helpful to visualize the differences by carefully observing the graphs.

EXPONENTIAL FUNCTION

$y = a^x$
$f(x) = a^x$
$a > 1$
Continuous
One-to-one
Domain: All real
  numbers, $(-\infty, \infty)$
Range: All positive real
  numbers, $(0, \infty)$
Increasing
Horizontal asymptote is x-axis:
  $(a^x \to 0$ as $x \to -\infty)$
y-intercept: $(0, 1)$
There is no x-intercept.

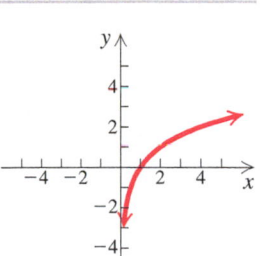

LOGARITHMIC FUNCTION

$x = a^y$
$f^{-1}(x) = \log_a x$
$a > 1$
Continuous
One-to-one
Domain: All positive real
  numbers, $(0, \infty)$
Range: All real
  numbers, $(-\infty, \infty)$
Increasing
Vertical asymptote is y-axis:
  $(\log_a x \to -\infty$ as $x \to 0^+)$
x-intercept: $(1, 0)$
There is no y-intercept.

▶ Finding Certain Logarithms

Let's use the definition of logarithms to find some logarithmic values.

EXAMPLE 2 Find each of the following logarithms.

a) $\log_{10} 10{,}000$ **b)** $\log_{10} 0.01$ **c)** $\log_2 8$

d) $\log_9 3$ **e)** $\log_6 1$ **f)** $\log_8 8$

Solution

a) The exponent to which we raise 10 to obtain 10,000 is 4; thus $\log_{10} 10{,}000 = 4$.

b) We have $0.01 = \dfrac{1}{100} = \dfrac{1}{10^2} = 10^{-2}$. The exponent to which we raise 10 to get 0.01 is -2, so $\log_{10} 0.01 = -2$.

c) $8 = 2^3$. The exponent to which we raise 2 to get 8 is 3, so $\log_2 8 = 3$.

d) $3 = \sqrt{9} = 9^{1/2}$. The exponent to which we raise 9 to get 3 is $\frac{1}{2}$, so $\log_9 3 = \frac{1}{2}$.

e) $1 = 6^0$. The exponent to which we raise 6 to get 1 is 0, so $\log_6 1 = 0$.

f) $5 = 5^1$. The exponent to which we raise 5 to get 5 is 1, so $\log_5 5 = 1$.

▶ Now Try Exercises 9 and 15.

Examples 2(e) and 2(f) illustrate two important properties of logarithms. The property $\log_a 1 = 0$ follows from the fact that $a^0 = 1$. Thus, $\log_5 1 = 0$, $\log_{10} 1 = 0$, and so on. The property $\log_a a = 1$ follows from the fact that $a^1 = a$. Thus, $\log_5 5 = 1$, $\log_{10} 10 = 1$, and so on.

$$\log_a 1 = 0 \quad \text{and} \quad \log_a a = 1, \quad \text{for any logarithmic base } a.$$

▶ Converting Between Exponential Equations and Logarithmic Equations

In dealing with logarithmic functions, it is helpful to remember that a logarithm of a number is an *exponent*. It is the exponent y in $x = a^y$. You might think to yourself, "the logarithm, base a, of a number x is the power to which a must be raised to get x."

We are led to the following. (The symbol $\longleftrightarrow$ means that the two statements are equivalent; that is, when one is true, the other is true. The words "if and only if" can be used in place of $\longleftrightarrow$.)

$$\log_a x = y \longleftrightarrow x = a^y \qquad \text{A logarithm is an exponent!}$$

EXAMPLE 3 Convert each of the following to a logarithmic equation.

a) $16 = 2^x$ **b)** $10^{-3} = 0.001$ **c)** $e^t = 70$

Solution

The exponent is the logarithm.

a) $16 = 2^x \rightarrow \log_2 16 = x$

The base remains the same.

b) $10^{-3} = 0.001 \rightarrow \log_{10} 0.001 = -3$

c) $e^t = 70 \rightarrow \log_e 70 = t$

> **Now Try Exercise 37.**

EXAMPLE 4 Convert each of the following to an exponential equation.

a) $\log_2 32 = 5$ **b)** $\log_a Q = 8$ **c)** $x = \log_t M$

Solution

The logarithm is the exponent.

a) $\log_2 32 = 5 \qquad 2^5 = 32$

The base remains the same.

b) $\log_a Q = 8 \rightarrow a^8 = Q$

c) $x = \log_t M \rightarrow t^x = M$

> **Now Try Exercise 45.**

▶ ## Finding Logarithms on a Calculator

Before calculators became so widely available, base-10 logarithms, or **common logarithms**, were used extensively to simplify complicated calculations. In fact, that is why logarithms were invented. The abbreviation **log**, with no base written, is used to represent common logarithms, or base-10 logarithms. Thus,

$$\log x \quad \text{means} \quad \log_{10} x.$$

For example, log 29 means $\log_{10} 29$. Let's compare log 29 with log 10 and log 100:

$$\left.\begin{array}{l} \log 10 = \log_{10} 10 = 1 \\ \log 29 = \; ? \\ \log 100 = \log_{10} 100 = 2 \end{array}\right\}$$ Since 29 is between 10 and 100, it seems reasonable that log 29 is between 1 and 2.

On a calculator, the key for common logarithms is generally marked **LOG**. Using that key, we find that

$$\log 29 \approx 1.462397998 \approx 1.4624$$

rounded to four decimal places. Since $1 < 1.4624 < 2$, our answer seems reasonable. This also tells us that $10^{1.4624} \approx 29$.

EXAMPLE 5 Find each of the following common logarithms on a calculator. If you are using a graphing calculator, set the calculator in REAL mode. Round to four decimal places.

a) log 645,778   **b)** log 0.0000239   **c)** log (-3)

Solution

FUNCTION VALUE	READOUT	ROUNDED
a) log 645,778	log(645778) 5.810083246	5.8101
b) log 0.0000239	log(0.0000239) −4.621602099	−4.6216
c) log (-3)	ERR:NONREAL ANS *	Does not exist as a real number

Since 5.810083246 is the power to which we raise 10 to get 645,778, we can check part (a) by finding $10^{5.810083246}$. We can check part (b) in a similar manner. In part (c), log (-3) does not exist as a real number because there is no real-number power to which we can raise 10 to get -3. The number 10 raised to any real-number power is positive. The common logarithm of a negative number does not exist as a real number. Recall that logarithmic functions are inverses of exponential functions, and since the range of an exponential function is $(0, \infty)$, the domain of $f(x) = \log_a x$ is $(0, \infty)$.

Now Try Exercises 57 and 61.

---

*If the graphing calculator is set in $a + bi$ mode, the readout is $.4771212547 + 1.364376354i$.

▶ **Natural Logarithms**

Logarithms, base e, are called **natural logarithms**. The abbreviation "ln" is generally used for natural logarithms. Thus,

ln x **means** $\log_e x$.

For example, ln 53 means $\log_e 53$. On a calculator, the key for natural logarithms is generally marked **LN**. Using that key, we find that

    $\ln 53 \approx 3.970291914 \approx 3.9703$

rounded to four decimal places. This also tells us that $e^{3.9703} \approx 53$.

EXAMPLE 6 Find each of the following natural logarithms on a calculator. If you are using a graphing calculator, set the calculator in REAL mode. Round to four decimal places.

a) ln 645,778   **b)** ln 0.0000239   **c)** ln (-5)

d) ln e **e)** ln 1

Solution

FUNCTION VALUE	READOUT		ROUNDED
a) ln 645,778	ln(645778) 13.37821107		13.3782
b) ln 0.0000239	ln(0.0000239) −10.6416321		−10.6416
c) ln (-5)	ERR:NONREAL ANS	*	Does not exist
d) ln e	ln(e) 1		1
e) ln 1	ln(1) 0		0

    Since 13.37821107 is the power to which we raise e to get 645,778, we can check part (a) by finding $e^{13.37821107}$. We can check parts (b), (d), and (e) in a similar manner. In parts (d) and (e), note that $\ln e = \log_e e = 1$ and $\ln 1 = \log_e 1 = 0$.

          Now Try Exercises 65 and 67.

    $\ln 1 = 0$ and $\ln e = 1$

---

*If the graphing calculator is set in $a + bi$ mode, the readout is $1.609437912 + 3.141592654i$.

▶ **Changing Logarithmic Bases**

Most calculators give the values of both common logarithms and natural logarithms. To find a logarithm with a base other than 10 or e, we can use the following conversion formula.

THE CHANGE-OF-BASE FORMULA

For any logarithmic bases a and b, and any positive number M,

$$\log_b M = \frac{\log_a M}{\log_a b}.$$

We will prove this result in the next section.

EXAMPLE 7   Find $\log_5 8$ using common logarithms.

Solution   First, we let $a = 10$, $b = 5$, and $M = 8$. Then we substitute into the change-of-base formula:

$$\log_5 8 = \frac{\log_{10} 8}{\log_{10} 5} \qquad \textcolor{red}{\textbf{Substituting}}$$

$$\approx 1.2920. \qquad \textcolor{red}{\textbf{Using a calculator}}$$

Since $\log_5 8$ is the power to which we raise 5 to get 8, we would expect this power to be greater than 1 $(5^1 = 5)$ and less than 2 $(5^2 = 25)$, so the result is reasonable.

Now Try Exercise 69.

We can also use base e for a conversion.

EXAMPLE 8   Find $\log_5 8$ using natural logarithms.

Solution   Substituting e for a, 5 for b, and 8 for M, we have

$$\log_5 8 = \frac{\log_e 8}{\log_e 5}$$

$$= \frac{\ln 8}{\ln 5} \approx 1.2920.$$

Note that we get the same value using base e for the conversion that we did using base 10 in Example 7.

Now Try Exercise 75.

▶ **Graphs of Logarithmic Functions**

Let's now consider graphs of logarithmic functions.

EXAMPLE 9   Graph: $y = f(x) = \log_5 x$.

Solution   The equation $y = \log_5 x$ is equivalent to $x = 5^y$. We can find ordered pairs that are solutions by choosing values for y and computing the corresponding x-values. We then plot points, remembering that x is still the first coordinate.

Technology Connection

To graph $y = \log_5 x$ in Example 9 with a graphing calculator, we can first change the base to 10 or e. Here we change from base 5 to base e:

$$y = \log_5 x = \frac{\ln x}{\ln 5}.$$

The graph is shown below.

$$y = \log_5 x = \frac{\ln x}{\ln 5}$$

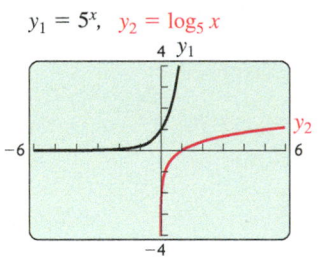

Some graphing calculators can graph inverses without the need to first find an equation of the inverse. If we begin with $y_1 = 5^x$, the graphs of both y_1 and its inverse, $y_2 = \log_5 x$, will be drawn as shown below.

$y_1 = 5^x$, $\quad y_2 = \log_5 x$

For $y = 0$, $x = 5^0 = 1$.

For $y = 1$, $x = 5^1 = 5$.

For $y = 2$, $x = 5^2 = 25$.

For $y = 3$, $x = 5^3 = 125$.

For $y = -1$, $x = 5^{-1} = \frac{1}{5}$.

For $y = -2$, $x = 5^{-2} = \frac{1}{25}$.

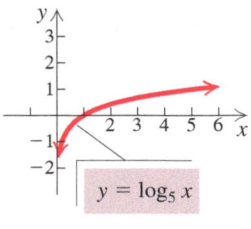

x, or 5^y	y
1	0
5	1
25	2
125	3
$\frac{1}{5}$	-1
$\frac{1}{25}$	-2

↑ (1) Select y.

↑ (2) Compute x.

→ **Now Try Exercise 5.**

EXAMPLE 10 Graph: $g(x) = \ln x$.

Solution To graph $y = g(x) = \ln x$, we select values for x and use the **LN** key on a calculator to find the corresponding values of $\ln x$. We then plot points and draw the curve.

x	$g(x) = \ln x$
0.5	-0.7
1	0
2	0.7
3	1.1
4	1.4
5	1.6

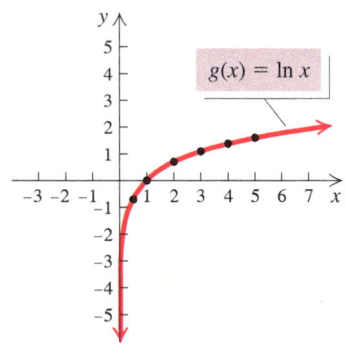

We could also write $g(x) = \ln x$, or $y = \ln x$, as $x = e^y$, select values for y, and use a calculator to find the corresponding values of x.

→ **Now Try Exercise 7.**

Recall that the graph of $f(x) = \log_a x$, for any base a, has the x-intercept $(1, 0)$. The domain is the set of positive real numbers, and the range is the set of all real numbers. The y-axis is the vertical asymptote.

EXAMPLE 11 Graph each of the following. Before doing so, describe how each graph can be obtained from the graph of $y = \ln x$. Give the domain and the vertical asymptote of each function.

a) $f(x) = \ln(x + 3)$

b) $f(x) = 3 - \frac{1}{2}\ln x$

c) $f(x) = |\ln(x - 1)|$

Solution

a) The graph of $f(x) = \ln(x + 3)$ is a shift of the graph of $y = \ln x$ left 3 units. The domain is the set of all real numbers greater than -3, $(-3, \infty)$. The line $x = -3$ is the vertical asymptote.

x	f(x)
−2.9	−2.303
−2	0
0	1.099
2	1.609
4	1.946

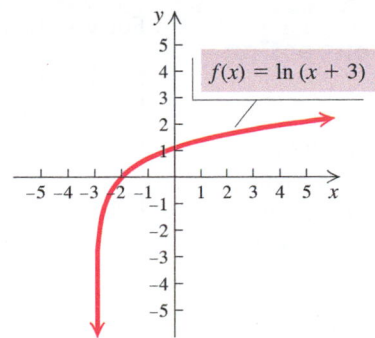

b) The graph of $f(x) = 3 - \frac{1}{2} \ln x$ is a vertical shrinking of the graph of $y = \ln x$, followed by a reflection across the x-axis, and then a translation up 3 units. The domain is the set of all positive real numbers, $(0, \infty)$. The y-axis is the vertical asymptote.

x	f(x)
0.1	4.151
1	3
3	2.451
6	2.104
9	1.901

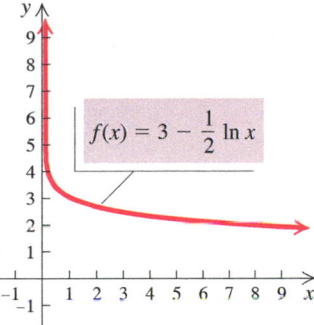

c) The graph of $f(x) = |\ln (x - 1)|$ is a translation of the graph of $y = \ln x$ right 1 unit. Then the absolute value has the effect of reflecting negative outputs across the x-axis. The domain is the set of all real numbers greater than 1, $(1, \infty)$. The line $x = 1$ is the vertical asymptote.

x	f(x)
1.1	2.303
2	0
4	1.099
6	1.609
8	1.946

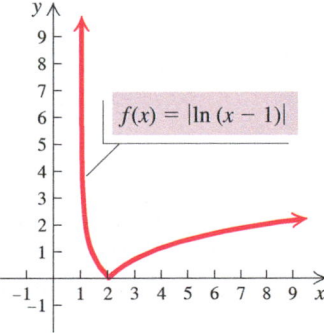

Now Try Exercise 89.

▶ **Applications**

EXAMPLE 12 *Walking Speed.* In a study by psychologists Bornstein and Bornstein, it was found that the average walking speed w, in feet per second, of a person living in a city of population P, in thousands, is given by the function

$$w(P) = 0.37 \ln P + 0.05$$

(*Source*: *International Journal of Psychology*).

a) The population of Billings, Montana, is 106,954. Find the average walking speed of people living in Billings.

b) The population of Chicago, Illinois, is 2,714,856. Find the average walking speed of people living in Chicago.

Solution

a) Since P is in thousands and $106,954 = 106.954$ thousand, we substitute 106.954 for P:

$$w(106.954) = 0.37 \ln 106.954 + 0.05 \qquad \text{Substituting}$$
$$\approx 1.8. \qquad \text{Finding the natural logarithm and simplifying}$$

The average walking speed of people living in Billings is about 1.8 ft/sec.

b) We substitute 2714.856 for P:

$$w(2714.856) = 0.37 \ln 2714.856 + 0.05 \qquad \text{Substituting}$$
$$\approx 3.0.$$

The average walking speed of people living in Chicago is about 3.0 ft/sec.

Now Try Exercise 95(d).

EXAMPLE 13 *Earthquake Magnitude.* Measured on the Richter scale, the magnitude R of an earthquake of intensity I is defined as

$$R = \log \frac{I}{I_0},$$

where I_0 is a minimum intensity used for comparison. We can think of I_0 as a threshold intensity that is the weakest earthquake that can be recorded on a seismograph. If one earthquake is 10 times as intense as another, its magnitude on the Richter scale is 1 greater than that of the other. If one earthquake is 100 times as intense as another, its magnitude on the Richter scale is 2 higher, and so on. Thus an earthquake whose magnitude is 7 on the Richter scale is 10 times as intense as an earthquake whose magnitude is 6. Earthquake intensities can be interpreted as multiples of the minimum intensity I_0.

The undersea Tohoku earthquake and tsunami, near the northeast coast of Honshu, Japan, on March 11, 2011, had an intensity of $10^{9.0} \cdot I_0$ (*Source:* earthquake.usgs.gov). They caused extensive loss of life and severe structural damage to buildings, railways, and roads. What was the magnitude on the Richter scale?

Solution We substitute into the formula:

$$R = \log \frac{I}{I_0} = \log \frac{10^{9.0} \cdot I_0}{I_0} = \log 10^{9.0} = 9.0.$$

The magnitude of the earthquake was 9.0 on the Richter scale.

Now Try Exercise 97(a).

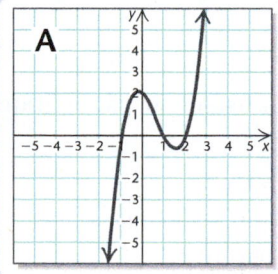

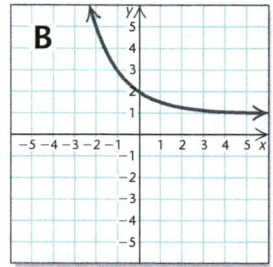

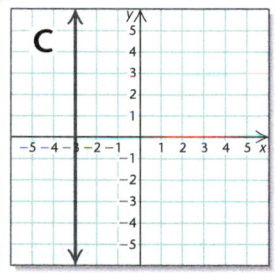

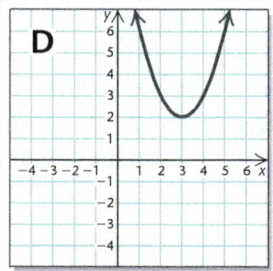

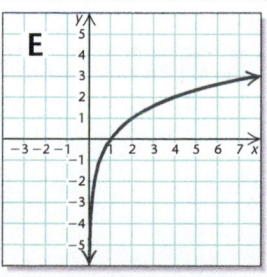

Visualizing the Graph

Match the equation or function with its graph.

1. $f(x) = 4^x$

2. $f(x) = \ln x - 3$

3. $(x + 3)^2 + y^2 = 9$

4. $f(x) = 2^{-x} + 1$

5. $f(x) = \log_2 x$

6. $f(x) = x^3 - 2x^2 - x + 2$

7. $x = -3$

8. $f(x) = e^x - 4$

9. $f(x) = (x - 3)^2 + 2$

10. $3x = 6 + y$

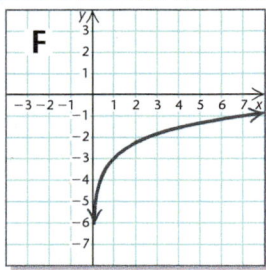

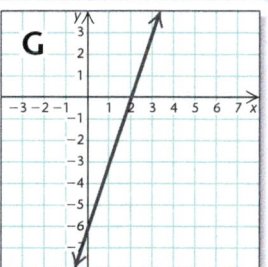

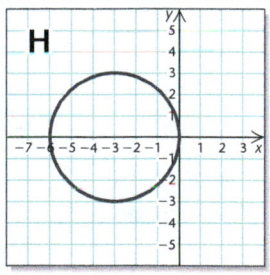

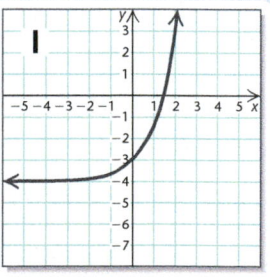

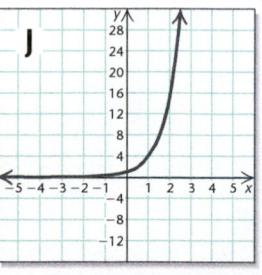

9.4 Exercise Set

Graph.

1. $x = 3^y$
2. $x = 4^y$
3. $x = \left(\frac{1}{2}\right)^y$
4. $x = \left(\frac{4}{3}\right)^y$
5. $y = \log_3 x$
6. $y = \log_4 x$
7. $f(x) = \log x$
8. $f(x) = \ln x$

Find each of the following. Do not use a calculator.

9. $\log_2 16$
10. $\log_3 9$
11. $\log_5 125$
12. $\log_2 64$
13. $\log 0.001$
14. $\log 100$
15. $\log_2 \frac{1}{4}$
16. $\log_8 2$
17. $\ln 1$
18. $\ln e$
19. $\log 10$
20. $\log 1$
21. $\log_5 5^4$
22. $\log \sqrt{10}$
23. $\log_3 \sqrt[4]{3}$
24. $\log 10^{8/5}$
25. $\log 10^{-7}$
26. $\log_5 1$
27. $\log_{49} 7$
28. $\log_3 3^{-2}$
29. $\ln e^{3/4}$
30. $\log_2 \sqrt{2}$
31. $\log_4 1$
32. $\ln e^{-5}$
33. $\ln \sqrt{e}$
34. $\log_{64} 4$

Convert to a logarithmic equation.

35. $10^3 = 1000$
36. $5^{-3} = \frac{1}{125}$
37. $8^{1/3} = 2$
38. $10^{0.3010} = 2$
39. $e^3 = t$
40. $Q^t = x$
41. $e^2 = 7.3891$
42. $e^{-1} = 0.3679$
43. $p^k = 3$
44. $e^{-t} = 4000$

Convert to an exponential equation.

45. $\log_5 5 = 1$
46. $t = \log_4 7$
47. $\log 0.01 = -2$
48. $\log 7 = 0.845$
49. $\ln 30 = 3.4012$
50. $\ln 0.38 = -0.9676$
51. $\log_a M = -x$
52. $\log_t Q = k$
53. $\log_a T^3 = x$
54. $\ln W^5 = t$

Find each of the following using a calculator. Round to four decimal places.

55. $\log 3$
56. $\log 8$
57. $\log 532$
58. $\log 93{,}100$
59. $\log 0.57$
60. $\log 0.082$
61. $\log (-2)$
62. $\ln 50$
63. $\ln 2$
64. $\ln (-4)$
65. $\ln 809.3$
66. $\ln 0.00037$
67. $\ln (-1.32)$
68. $\ln 0$

Find the logarithm using common logarithms and the change-of-base formula. Round to four decimal places.

69. $\log_4 100$
70. $\log_3 20$
71. $\log_{100} 0.3$
72. $\log_\pi 100$
73. $\log_{200} 50$
74. $\log_{5.3} 1700$

Find the logarithm using natural logarithms and the change-of-base formula. Round to four decimal places.

75. $\log_3 12$
76. $\log_4 25$
77. $\log_{100} 15$
78. $\log_9 100$

Graph the function and its inverse using the same set of axes. Use any method.

79. $f(x) = 3^x, \; f^{-1}(x) = \log_3 x$
80. $f(x) = \log_4 x, \; f^{-1}(x) = 4^x$
81. $f(x) = \log x, \; f^{-1}(x) = 10^x$
82. $f(x) = e^x, \; f^{-1}(x) = \ln x$

For each of the following functions, briefly describe how the graph can be obtained from the graph of a basic logarithmic function. Then graph the function. Give the domain and the vertical asymptote of each function.

83. $f(x) = \log_2 (x + 3)$
84. $f(x) = \log_3 (x - 2)$
85. $y = \log_3 x - 1$
86. $y = 3 + \log_2 x$
87. $f(x) = 4 \ln x$
88. $f(x) = \frac{1}{2} \ln x$
89. $y = 2 - \ln x$
90. $y = \ln (x + 1)$
91. $f(x) = \frac{1}{2} \log (x - 1) - 2$
92. $f(x) = 5 - 2 \log (x + 1)$

Graph the piecewise function.

93. $g(x) = \begin{cases} 5, & \text{for } x \le 0, \\ \log x + 1, & \text{for } x > 0 \end{cases}$

94. $f(x) = \begin{cases} 1 - x, & \text{for } x \le -1, \\ \ln (x + 1), & \text{for } x > -1 \end{cases}$

95. *Walking Speed.* Refer to Example 12. Various cities and their populations are given below. Find the average walking speed in each city. Round to the nearest tenth of a foot per second.

a) El Paso, Texas: 672,538
b) Phoenix, Arizona: 1,488,750
c) Birmingham, Alabama: 212,038
d) Milwaukee, Wisconsin: 598,916
e) Honolulu, Hawaii: 345,610
f) Charlotte, North Carolina: 775,202
g) Omaha, Nebraska: 421,570
h) Sydney, Australia: 3,908,643

96. *Forgetting.* Students in an accounting class took a final exam and then took equivalent forms of the exam at monthly intervals thereafter. The average score $S(t)$, as a percent, after t months was found to be given by the function

$$S(t) = 78 - 15 \log (t + 1), \quad t \geq 0.$$

a) What was the average score when the students initially took the test, $t = 0$?
b) What was the average score after 4 months? after 24 months?

97. *Earthquake Magnitude.* Refer to Example 13. Various locations of earthquakes and their intensities are given below. Find the magnitude of each earthquake on the Richter scale.

a) San Francisco, California, 1906: $10^{7.7} \cdot I_0$
b) Chile, 1960: $10^{9.5} \cdot I_0$
c) Iran, 2003: $10^{6.6} \cdot I_0$
d) Turkey, 1999: $10^{7.6} \cdot I_0$
e) Peru, 2007: $10^{8.0} \cdot I_0$
f) China, 2008: $10^{7.9} \cdot I_0$
g) Spain, 2011: $10^{5.1} \cdot I_0$
h) Sumatra, 2004: $10^{9.3} \cdot I_0$

98. *pH of Substances in Chemistry.* In chemistry, the pH of a substance is defined as

$$pH = -\log [H^+],$$

where H^+ is the hydrogen ion concentration, in moles per liter. Find the pH of each substance.

SUBSTANCE	HYDROGEN ION CONCENTRATION
a) Pineapple juice	1.6×10^{-4}
b) Hair conditioner	0.0013
c) Mouthwash	6.3×10^{-7}
d) Eggs	1.6×10^{-8}
e) Tomatoes	6.3×10^{-5}

99. Find the hydrogen ion concentration of each substance, given the pH. (See Exercise 98.) Express the answer in scientific notation.

SUBSTANCE	pH
a) Tap water	7
b) Rainwater	5.4
c) Orange juice	3.2
d) Wine	4.8

100. *Advertising.* A model for advertising response is given by the function

$$N(a) = 1000 + 200 \ln a, \quad a \geq 1,$$

where $N(a)$ is the number of units sold when a is the amount spent on advertising, in thousands of dollars.

a) How many units were sold after spending $1000 ($a = 1$) on advertising?
b) How many units were sold after spending $5000?

101. *Loudness of Sound.* The **loudness** L, in bels (after Alexander Graham Bell), of a sound of intensity I is defined to be

$$L = \log \frac{I}{I_0},$$

where I_0 is the minimum intensity detectable by the human ear (such as the tick of a watch at 20 ft under quiet conditions). If a sound is 10 times as intense as another, its loudness is 1 bel greater than that of the other. If a sound is 100 times as intense as another, its loudness is 2 bels greater, and so on. The

bel is a large unit, so a subunit, the **decibel**, is generally used. For L, in decibels, the formula is

$$L = 10 \log \frac{I}{I_0}.$$

Find the loudness, in decibels, of each sound with the given intensity.

SOUND	INTENSITY
a) Jet engine at 100 ft	$10^{14} \cdot I_0$
b) Loud rock concert	$10^{11.5} \cdot I_0$
c) Bird calls	$10^{4} \cdot I_0$
d) Normal conversation	$10^{6.5} \cdot I_0$
e) Thunder	$10^{12} \cdot I_0$
f) Loudest sound possible	$10^{19.4} \cdot I_0$

▶ **Skill Maintenance**

Find the slope and the y-intercept of the line. [2.5]

102. $3x - 10y = 14$

103. $y = 6$

104. $x = -4$

Use synthetic division to find the function values. [8.3]

105. $g(x) = x^3 - 6x^2 + 3x + 10$; find $g(-5)$

106. $f(x) = x^4 - 2x^3 + x - 6$; find $f(-1)$

Find a polynomial function of degree 3 with the given numbers as zeros. Answers may vary. [8.4]

107. $\sqrt{7}, -\sqrt{7}, 0$ **108.** $4i, -4i, 1$

▶ **Synthesis**

Simplify.

109. $\dfrac{\log_5 8}{\log_5 2}$ **110.** $\dfrac{\log_3 64}{\log_3 16}$

Find the domain of the function.

111. $f(x) = \log_5 x^3$ **112.** $f(x) = \log_4 x^2$

113. $f(x) = \ln |x|$ **114.** $f(x) = \log(3x - 4)$

Solve.

115. $\log_2 (2x + 5) < 0$ **116.** $\log_2 (x - 3) \geq 4$

In Exercises 117–120, match the equation with one of the figures (a)–(d) that follow.

a)

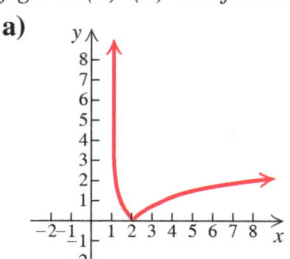

b)

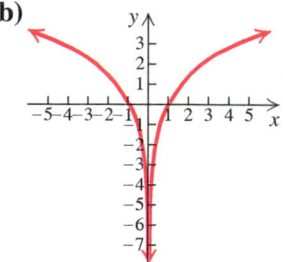

c)

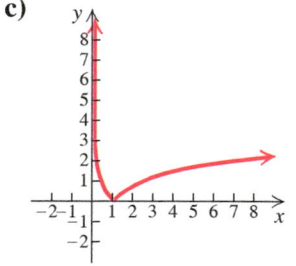

d)

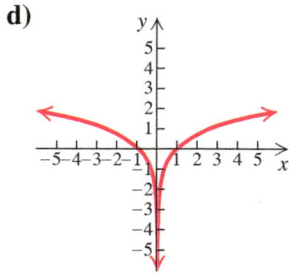

117. $f(x) = \ln |x|$ **118.** $f(x) = |\ln x|$

119. $f(x) = \ln x^2$ **120.** $g(x) = |\ln (x - 1)|$

Mid-Chapter Review

Determine whether the statement is true or false.

1. In general, $(f \circ g)(x) \neq (g \circ f)(x)$. [9.1]

2. The domain of all logarithmic functions is $[1, \infty)$. [9.4]

3. The range of a one-to-one function f is the domain of its inverse f^{-1}. [9.2]

4. The y-intercept of $f(x) = e^{-x}$ is $(0, -1)$. [9.3]

Given that $f(x) = 5x - 4$, $g(x) = x^3 + 1$, and $h(x) = x^2 - 2x + 3$, find each of the following. [9.1]

5. $(f \circ g)(1)$ **6.** $(g \circ h)(2)$

7. $(f \circ f)(0)$ **8.** $(h \circ f)(-1)$

Find $(f \circ g)(x)$ and $(g \circ f)(x)$ and the domain of each. **[9.1]**

9. $f(x) = \dfrac{1}{2}x, \; g(x) = 6x + 4$

10. $f(x) = 3x + 2, \; g(x) = \sqrt{x}$

For each function, determine whether it is one-to-one, and if the function is one-to-one, find a formula for its inverse. **[9.2]**

11. $f(x) = -\dfrac{2}{x}$

12. $f(x) = 3 + x^2$

13. $f(x) = \dfrac{5}{x - 2}$

14. Given the function $f(x) = \sqrt{x - 5}$, use composition of functions to show that $f^{-1}(x) = x^2 + 5$. **[9.2]**

15. Given the one-to-one function $f(x) = x^3 + 2$, find the inverse, give the domain and the range of f and f^{-1}, and graph both f and f^{-1} on the same set of axes. **[9.2]**

Match the function with one of the graphs (a)–(h) that follow. **[9.3], [9.4]**

a)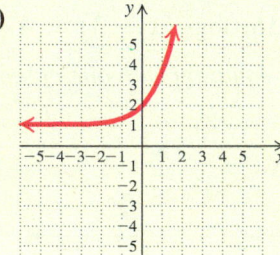

b)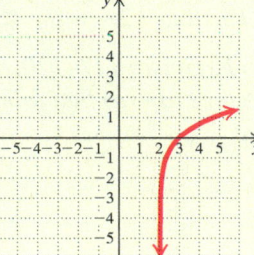

c)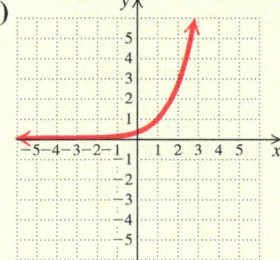

d)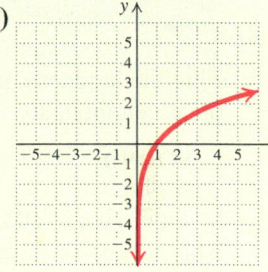

e)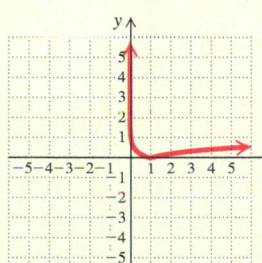

f)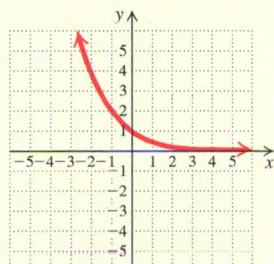

g)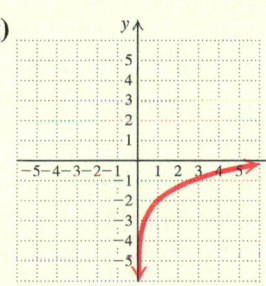

h)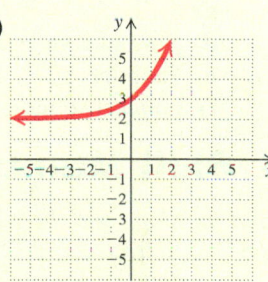

16. $y = \log_2 x$

17. $f(x) = 2^x + 2$

18. $f(x) = e^{x-1}$

19. $f(x) = \ln x - 2$

20. $f(x) = \ln (x - 2)$

21. $y = 2^{-x}$

22. $f(x) = |\log x|$

23. $f(x) = e^x + 1$

24. Suppose that \$3200 is invested at $4\frac{1}{2}\%$ interest, compounded quarterly. Find the amount of money in the account in 6 years. **[9.3]**

Find each of the following without a calculator. **[9.4]**

25. $\log_4 1$

26. $\ln e^{-4/5}$

27. $\log 0.01$

28. $\ln e^2$

29. $\ln 1$

30. $\log_2 \dfrac{1}{16}$

31. $\log 1$

32. $\log_3 27$

33. $\log \sqrt[4]{10}$

34. $\ln e$

35. Convert $e^{-6} = 0.0025$ to a logarithmic equation. **[9.4]**

36. Convert $\log T = r$ to an exponential equation. **[9.4]**

Find the logarithm using the change-of-base formula. **[9.4]**

37. $\log_3 20$

38. $\log_\pi 10$

Collaborative Discussion and Writing

39. Nora determines the domain of $f \circ g$ by examining only the formula for $(f \circ g)(x)$. Is her approach valid? Why or why not? **[9.1]**

40. Explain why an even function f does not have an inverse f^{-1} that is a function. **[9.2]**

41. Describe the differences between the graph of $f(x) = x^3$ and the graph of $g(x) = 3^x$. **[9.3]**

42. If $\log b < 0$, what can you say about b? **[9.4]**

9.5

Properties of Logarithmic Functions

▶ Convert from logarithms of products, powers, and quotients to expressions in terms of individual logarithms, and conversely.

▶ Simplify expressions of the type $\log_a a^x$ and $a^{\log_a x}$.

We now establish some properties of logarithmic functions. These properties are based on the corresponding rules for exponents.

▶ Logarithms of Products

The first property of logarithms corresponds to the product rule for exponents: $a^m \cdot a^n = a^{m+n}$.

THE PRODUCT RULE

For any positive numbers M and N and any logarithmic base a,

$$\log_a MN = \log_a M + \log_a N.$$

(The logarithm of a product is the sum of the logarithms of the factors.)

EXAMPLE 1   Express as a sum of logarithms: $\log_3 (9 \cdot 27)$.

Solution   We have

$$\log_3(9 \cdot 27) = \log_3 9 + \log_3 27.$$ Using the product rule

As a check, note that

$$\log_3(9 \cdot 27) = \log_3 243 = 5$$ $3^5 = 243$

and $\log_3 9 + \log_3 27 = 2 + 3 = 5.$ $3^2 = 9; 3^3 = 27$

Now Try Exercise 1.

EXAMPLE 2   Express as a single logarithm: $\log_2 p^3 + \log_2 q$.

Solution   We have

$$\log_2 p^3 + \log_2 q = \log_2 (p^3 q).$$

Now Try Exercise 35.

A Proof of the Product Rule.   Let $\log_a M = x$ and $\log_a N = y$. Converting to exponential equations, we have $a^x = M$ and $a^y = N$. Then

$$MN = a^x \cdot a^y = a^{x+y}.$$

Converting back to a logarithmic equation, we get

$$\log_a MN = x + y.$$

Remembering what x and y represent, we know it follows that

$$\log_a MN = \log_a M + \log_a N.$$

▶ ## Logarithms of Powers

The second property of logarithms corresponds to the power rule for exponents: $(a^m)^n = a^{mn}$.

THE POWER RULE

For any positive number M, any logarithmic base a, and any real number p,

$$\log_a M^p = p \log_a M.$$

(The logarithm of a power of M is the exponent times the logarithm of M.)

EXAMPLE 3 Express each of the following as a product.

a) $\log_a 11^{-3}$ **b)** $\log_a \sqrt[4]{7}$ **c)** $\ln x^6$

Solution

a) $\log_a 11^{-3} = -3 \log_a 11$ **Using the power rule**

b) $\log_a \sqrt[4]{7} = \log_a 7^{1/4}$ **Writing exponential notation**

            $= \frac{1}{4} \log_a 7$ **Using the power rule**

c) $\ln x^6 = 6 \ln x$ **Using the power rule**

▶ **Now Try Exercises 13 and 15.**

A Proof of the Power Rule. Let $x = \log_a M$. The equivalent exponential equation is $a^x = M$. Raising both sides to the power p, we obtain

$$(a^x)^p = M^p, \quad \text{or} \quad a^{xp} = M^p.$$

Converting back to a logarithmic equation, we get

$$\log_a M^p = xp.$$

But $x = \log_a M$, so substituting gives us

$$\log_a M^p = (\log_a M)p = p \log_a M.$$ ■

▶ ## Logarithms of Quotients

The third property of logarithms corresponds to the quotient rule for exponents: $a^m/a^n = a^{m-n}$.

THE QUOTIENT RULE

For any positive numbers M and N, and any logarithmic base a,

$$\log_a \frac{M}{N} = \log_a M - \log_a N.$$

(The logarithm of a quotient is the logarithm of the numerator minus the logarithm of the denominator.)

EXAMPLE 4   Express as a difference of logarithms: $\log_t \dfrac{8}{w}$.

Solution   We have

$$\log_t \frac{8}{w} = \log_t 8 - \log_t w. \qquad \text{\color{red}Using the quotient rule}$$

Now Try Exercise 17.

EXAMPLE 5   Express as a single logarithm: $\log_b 64 - \log_b 16$.

Solution   We have

$$\log_b 64 - \log_b 16 = \log_b \frac{64}{16} = \log_b 4.$$

Now Try Exercise 37.

A Proof of the Quotient Rule.   The proof follows from both the product rule and the power rule:

$$
\begin{aligned}
\log_a \frac{M}{N} &= \log_a MN^{-1} \\[4pt]
&= \log_a M + \log_a N^{-1} \qquad &&\text{\color{red}Using the product rule} \\
&= \log_a M + (-1)\log_a N \qquad &&\text{\color{red}Using the power rule} \\
&= \log_a M - \log_a N.
\end{aligned}
$$

Common Errors

$\log_a MN \neq (\log_a M)(\log_a N)$	The logarithm of a product is *not* the product of the logarithms.
$\log_a (M + N) \neq \log_a M + \log_a N$	The logarithm of a sum is *not* the sum of the logarithms.
$\log_a \dfrac{M}{N} \neq \dfrac{\log_a M}{\log_a N}$	The logarithm of a quotient is *not* the quotient of the logarithms.
$(\log_a M)^p \neq p \log_a M$	The power of a logarithm is *not* the exponent times the logarithm.

▶ **Applying the Properties**

EXAMPLE 6   Express each of the following in terms of sums and differences of logarithms.

a) $\log_a \dfrac{x^2 y^5}{z^4}$ b) $\log_a \sqrt[3]{\dfrac{a^2 b}{c^5}}$ c) $\log_b \dfrac{a y^5}{m^3 n^4}$

Solution

$$
\begin{aligned}
\textbf{a)}\quad \log_a \frac{x^2 y^5}{z^4} &= \log_a (x^2 y^5) - \log_a z^4 \qquad &&\text{\color{red}Using the quotient rule} \\[4pt]
&= \log_a x^2 + \log_a y^5 - \log_a z^4 \qquad &&\text{\color{red}Using the product rule} \\
&= 2\log_a x + 5\log_a y - 4\log_a z \qquad &&\text{\color{red}Using the power rule}
\end{aligned}
$$

b) $\log_a \sqrt[3]{\dfrac{a^2b}{c^5}} = \log_a \left(\dfrac{a^2b}{c^5}\right)^{1/3}$ *Writing exponential notation*

$= \dfrac{1}{3} \log_a \dfrac{a^2b}{c^5}$ *Using the power rule*

$= \dfrac{1}{3} \left(\log_a a^2b - \log_a c^5\right)$ *Using the quotient rule. The parentheses are necessary.*

$= \dfrac{1}{3} \left(2 \log_a a + \log_a b - 5 \log_a c\right)$ *Using the product rule and the power rule*

$= \dfrac{1}{3} \left(2 + \log_a b - 5 \log_a c\right)$ $\log_a a = 1$

$= \dfrac{2}{3} + \dfrac{1}{3} \log_a b - \dfrac{5}{3} \log_a c$ *Multiplying to remove parentheses*

c) $\log_b \dfrac{ay^5}{m^3n^4} = \log_b ay^5 - \log_b m^3n^4$ *Using the quotient rule*

$= \left(\log_b a + \log_b y^5\right) - \left(\log_b m^3 + \log_b n^4\right)$ *Using the product rule*

$= \log_b a + \log_b y^5 - \log_b m^3 - \log_b n^4$ *Removing parentheses*

$= \log_b a + 5 \log_b y - 3 \log_b m - 4 \log_b n$ *Using the power rule*

> **Now Try Exercises 25 and 31.**

EXAMPLE 7 Express as a single logarithm:

$$5 \log_b x - \log_b y + \dfrac{1}{4} \log_b z.$$

Solution We have

$5 \log_b x - \log_b y + \dfrac{1}{4} \log_b z = \log_b x^5 - \log_b y + \log_b z^{1/4}$ *Using the power rule*

$= \log_b \dfrac{x^5}{y} + \log_b z^{1/4}$ *Using the quotient rule*

$= \log_b \dfrac{x^5 z^{1/4}}{y},$ or $\log_b \dfrac{x^5 \sqrt[4]{z}}{y}.$ *Using the product rule*

> **Now Try Exercise 41.**

EXAMPLE 8 Express as a single logarithm:

$$\ln(3x + 1) - \ln(3x^2 - 5x - 2).$$

Solution We have

$\ln(3x + 1) - \ln(3x^2 - 5x - 2)$

$= \ln \dfrac{3x + 1}{3x^2 - 5x - 2}$ *Using the quotient rule*

$= \ln \dfrac{3x + 1}{(3x + 1)(x - 2)}$ *Factoring*

$= \ln \dfrac{1}{x - 2}.$ *Simplifying*

> **Now Try Exercise 45.**

EXAMPLE 9 Given that $\log_a 2 \approx 0.301$ and $\log_a 3 \approx 0.477$, find each of the following, if possible.

a) $\log_a 6$ **b)** $\log_a \dfrac{2}{3}$ **c)** $\log_a 81$

d) $\log_a \dfrac{1}{4}$ **e)** $\log_a 5$ **f)** $\dfrac{\log_a 3}{\log_a 2}$

Solution

a) $\log_a 6 = \log_a (2 \cdot 3) = \log_a 2 + \log_a 3$ **Using the product rule**
$$\approx 0.301 + 0.477$$
$$\approx 0.778$$

b) $\log_a \frac{2}{3} = \log_a 2 - \log_a 3$ **Using the quotient rule**
$$\approx 0.301 - 0.477 \approx -0.176$$

c) $\log_a 81 = \log_a 3^4 = 4 \log_a 3$ **Using the power rule**
$$\approx 4(0.477) \approx 1.908$$

d) $\log_a \frac{1}{4} = \log_a 1 - \log_a 4$ **Using the quotient rule**
$$= 0 - \log_a 2^2$$ $\log_a 1 = 0; 4 = 2^2$
$$= -2 \log_a 2$$ **Using the power rule**
$$\approx -2(0.301) \approx -0.602$$

e) $\log_a 5$ *cannot* be found using these properties and the given information.
$$\log_a 5 \neq \log_a 2 + \log_a 3$$ $\log_a 2 + \log_a 3 = \log_a (2 \cdot 3) = \log_a 6$

f) $\dfrac{\log_a 3}{\log_a 2} \approx \dfrac{0.477}{0.301} \approx 1.585$ **We simply divide, not using any of the properties.**

▶ **Now Try Exercises 53 and 55.**

▶ ## Simplifying Expressions of the Type $\log_a a^x$ and $a^{\log_a x}$

We have two final properties of logarithms to consider. The first follows from the product rule: Since $\log_a a^x = x \log_a a = x \cdot 1 = x$, we have $\log_a a^x = x$. This property also follows from the definition of a logarithm: x is the power to which we raise a in order to get a^x.

THE LOGARITHM OF A BASE TO A POWER

For any base a and any real number x,

$$\log_a a^x = x.$$

(The logarithm, base a, of a to a power is the power.)

EXAMPLE 10   Simplify each of the following.

a) $\log_a a^8$ **b)** $\ln e^{-t}$ **c)** $\log 10^{3k}$

Solution

a) $\log_a a^8 = 8$ 8 is the power to which we raise a in order to get a^8.

b) $\ln e^{-t} = \log_e e^{-t} = -t$ $\ln e^x = x$

c) $\log 10^{3k} = \log_{10} 10^{3k} = 3k$ ▶ **Now Try Exercises 65 and 73.**

Let $M = \log_a x$. Then $a^M = x$. Substituting $\log_a x$ for M, we obtain $a^{\log_a x} = x$. This also follows from the definition of a logarithm: $\log_a x$ is the power to which a is raised in order to get x.

A BASE TO A LOGARITHMIC POWER

For any base a and any positive real number x,

$$a^{\log_a x} = x.$$

(The number a raised to the power $\log_a x$ is x.)

EXAMPLE 11 Simplify each of the following.

a) $4^{\log_4 k}$ b) $e^{\ln 5}$ c) $10^{\log 7t}$

Solution

a) $4^{\log_4 k} = k$

b) $e^{\ln 5} = e^{\log_e 5} = 5$

c) $10^{\log 7t} = 10^{\log_{10} 7t} = 7t$ ➤ **Now Try Exercises 69 and 71.**

A Proof of the Change-of-Base Formula. We close this section by proving the change-of-base formula and summarizing the properties of logarithms considered thus far in this chapter. In Section 9.4, we used the change-of-base formula,

$$\log_b M = \frac{\log_a M}{\log_a b},$$

to make base conversions in order to find logarithmic values using a calculator. Let $x = \log_b M$. Then

$$b^x = M \qquad \text{Definition of logarithm}$$
$$\log_a b^x = \log_a M \qquad \text{Taking the logarithm on both sides}$$
$$x \log_a b = \log_a M \qquad \text{Using the power rule}$$
$$x = \frac{\log_a M}{\log_a b}, \qquad \text{Dividing by } \log_a b$$

so $\quad x = \log_b M = \dfrac{\log_a M}{\log_a b}.$ ■

CHANGE-OF-BASE FORMULA

REVIEW SECTION 9.4

Following is a summary of the properties of logarithms.

Summary of the Properties of Logarithms

The Product Rule: $\log_a MN = \log_a M + \log_a N$

The Power Rule: $\log_a M^p = p \log_a M$

The Quotient Rule: $\log_a \dfrac{M}{N} = \log_a M - \log_a N$

The Change-of-Base Formula: $\log_b M = \dfrac{\log_a M}{\log_a b}$

Other Properties: $\log_a a = 1, \qquad \log_a 1 = 0,$
                         $\log_a a^x = x, \qquad a^{\log_a x} = x$

9.5 Exercise Set

Express as a sum of logarithms.

1. $\log_3 (81 \cdot 27)$ **2.** $\log_2 (8 \cdot 64)$

3. $\log_5 (5 \cdot 125)$ **4.** $\log_4 (64 \cdot 4)$

5. $\log_t 8Y$ **6.** $\log 0.2x$

7. $\ln xy$ **8.** $\ln ab$

Express as a product.

9. $\log_b t^3$ **10.** $\log_a x^4$

11. $\log y^8$ **12.** $\ln y^5$

13. $\log_c K^{-6}$ **14.** $\log_b Q^{-8}$

15. $\ln \sqrt[3]{4}$ **16.** $\ln \sqrt{a}$

Express as a difference of logarithms.

17. $\log_t \dfrac{M}{8}$ **18.** $\log_a \dfrac{76}{13}$

19. $\log \dfrac{x}{y}$ **20.** $\ln \dfrac{a}{b}$

21. $\ln \dfrac{r}{s}$ **22.** $\log_b \dfrac{3}{w}$

Express in terms of sums and differences of logarithms.

23. $\log_a 6xy^5z^4$ **24.** $\log_a x^3y^2z$

25. $\log_b \dfrac{p^2q^5}{m^4b^9}$ **26.** $\log_b \dfrac{x^2y}{b^3}$

27. $\ln \dfrac{2}{3x^3y}$ **28.** $\log \dfrac{5a}{4b^2}$

29. $\log \sqrt{r^3t}$ **30.** $\ln \sqrt[3]{5x^5}$

31. $\log_a \sqrt{\dfrac{x^6}{p^5q^8}}$ **32.** $\log_c \sqrt[3]{\dfrac{y^3z^2}{x^4}}$

33. $\log_a \sqrt[4]{\dfrac{m^8n^{12}}{a^3b^5}}$ **34.** $\log_a \sqrt{\dfrac{a^6b^8}{a^2b^5}}$

Express as a single logarithm and, if possible, simplify.

35. $\log_a 75 + \log_a 2$ **36.** $\log 0.01 + \log 1000$

37. $\log 10{,}000 - \log 100$ **38.** $\ln 54 - \ln 6$

39. $\frac{1}{2}\log n + 3 \log m$ **40.** $\frac{1}{2}\log a - \log 2$

41. $\frac{1}{2}\log_a x + 4 \log_a y - 3 \log_a x$

42. $\frac{2}{5}\log_a x - \frac{1}{3}\log_a y$

43. $\ln x^2 - 2 \ln \sqrt{x}$

44. $\ln 2x + 3(\ln x - \ln y)$

45. $\ln (x^2 - 4) - \ln (x + 2)$

46. $\log (x^3 - 8) - \log (x - 2)$

47. $\log (x^2 - 5x - 14) - \log (x^2 - 4)$

48. $\log_a \dfrac{a}{\sqrt{x}} - \log_a \sqrt{ax}$

49. $\ln x - 3[\ln (x - 5) + \ln (x + 5)]$

50. $\frac{2}{3}[\ln (x^2 - 9) - \ln (x + 3)] + \ln (x + y)$

51. $\frac{3}{2}\ln 4x^6 - \frac{4}{5}\ln 2y^{10}$

52. $120(\ln \sqrt[5]{x^3} + \ln \sqrt[3]{y^2} - \ln \sqrt[4]{16z^5})$

Given that $\log_a 2 \approx 0.301$, $\log_a 7 \approx 0.845$, and $\log_a 11 \approx 1.041$, find each of the following, if possible. Round the answer to the nearest thousandth.

53. $\log_a \frac{2}{11}$ **54.** $\log_a 14$

55. $\log_a 98$ **56.** $\log_a \frac{1}{7}$

57. $\dfrac{\log_a 2}{\log_a 7}$ **58.** $\log_a 9$

Given that $\log_b 2 \approx 0.693$, $\log_b 3 \approx 1.099$, and $\log_b 5 \approx 1.609$, find each of the following, if possible. Round the answer to the nearest thousandth.

59. $\log_b 125$ **60.** $\log_b \frac{5}{3}$

61. $\log_b \frac{1}{6}$ **62.** $\log_b 30$

63. $\log_b \dfrac{3}{b}$ **64.** $\log_b 15b$

Simplify.

65. $\log_p p^3$ **66.** $\log_t t^{2713}$

67. $\log_e e^{|x-4|}$ **68.** $\log_q q^{\sqrt{3}}$

69. $3^{\log_3 4x}$ **70.** $5^{\log_5 (4x-3)}$

71. $10^{\log w}$ **72.** $e^{\ln x^3}$

73. $\ln e^{8t}$ **74.** $\log 10^{-k}$

75. $\log_b \sqrt{b}$ **76.** $\log_b \sqrt{b^3}$

▶ Skill Maintenance

In each of Exercises 77–86, classify the function as linear, quadratic, cubic, quartic, rational, exponential, or logarithmic.

77. $f(x) = 5 - x^2 + x^4$ [8.1]

78. $f(x) = 2^x$ [9.3]   **79.** $f(x) = -\frac{3}{4}$ [2.5]

80. $f(x) = 4^x - 8$ [9.3]   **81.** $f(x) = -\frac{3}{x}$ [8.5]

82. $f(x) = \log x + 6$ [9.4]

83. $f(x) = -\frac{1}{3}x^3 - 4x^2 + 6x + 42$ [8.1]

84. $f(x) = \dfrac{x^2 - 1}{x^2 + x - 6}$ [8.5]

85. $f(x) = \frac{1}{2}x + 3$ [2.5]

86. $f(x) = 2x^2 - 6x + 3$ [7.5]

▶ Synthesis

Solve for x.

87. $5^{\log_5 8} = 2x$ **88.** $\ln e^{3x-5} = -8$

Express as a single logarithm and, if possible, simplify.

89. $\log_a (x^2 + xy + y^2) + \log_a (x - y)$

90. $\log_a (a^{10} - b^{10}) - \log_a (a + b)$

Express as a sum or a difference of logarithms.

91. $\log_a \dfrac{x - y}{\sqrt{x^2 - y^2}}$ **92.** $\log_a \sqrt{9 - x^2}$

93. Given that $\log_a x = 2$, $\log_a y = 3$, and $\log_a z = 4$, find

$$\log_a \dfrac{\sqrt[4]{y^2 z^5}}{\sqrt[4]{x^3 z^{-2}}}.$$

Determine whether each of the following is true. Assume that a, x, M, and N are positive.

94. $\log_a M + \log_a N = \log_a (M + N)$

95. $\log_a M - \log_a N = \log_a \dfrac{M}{N}$

96. $\dfrac{\log_a M}{\log_a N} = \log_a M - \log_a N$

97. $\dfrac{\log_a M}{x} = \log_a M^{1/x}$

98. $\log_a x^3 = 3 \log_a x$

99. $\log_a 8x = \log_a x + \log_a 8$

100. $\log_N (MN)^x = x \log_N M + x$

Suppose that $\log_a x = 2$. Find each of the following.

101. $\log_a \left(\dfrac{1}{x}\right)$ **102.** $\log_{1/a} x$

103. Simplify:

$\log_{10} 11 \cdot \log_{11} 12 \cdot \log_{12} 13 \cdots \log_{998} 999 \cdot \log_{999} 1000$.

Write each of the following without using logarithms.

104. $\log_a x + \log_a y - mz = 0$

105. $\ln a - \ln b + xy = 0$

Prove each of the following for any base a and any positive number x.

106. $\log_a \left(\dfrac{1}{x}\right) = -\log_a x = \log_{1/a} x$

107. $\log_a \left(\dfrac{x + \sqrt{x^2 - 5}}{5}\right) = -\log_a (x - \sqrt{x^2 - 5})$

9.6 Solving Exponential Equations and Logarithmic Equations

▶ Solve exponential equations.

▶ Solve logarithmic equations.

▶ Solving Exponential Equations

Equations with variables in the exponents, such as

$$3^x = 20 \quad \text{and} \quad 2^{5x} = 64,$$

are called **exponential equations**.

Sometimes, as is the case with the equation $2^{5x} = 64$, we can write each side as a power of the same number:

$$2^{5x} = 2^6.$$

We can then set the exponents equal and solve:

$$5x = 6$$
$$x = \tfrac{6}{5}, \text{ or } 1.2.$$

We use the following property to solve exponential equations.

BASE–EXPONENT PROPERTY

For any $a > 0, a \neq 1$,

$$a^x = a^y \longleftrightarrow x = y.$$

ONE-TO-ONE FUNCTIONS

REVIEW SECTION 9.2

This property follows from the fact that for any $a > 0, a \neq 1$, $f(x) = a^x$ is a one-to-one function. If $a^x = a^y$, then $f(x) = f(y)$. Then since f is one-to-one, it follows that $x = y$. Conversely, if $x = y$, it follows that $a^x = a^y$, since we are raising a to the same power in each case.

EXAMPLE 1 Solve: $2^{3x-7} = 32$.

Algebraic Solution

Note that $32 = 2^5$. Thus we can write each side as a power of the same number:

$$2^{3x-7} = 2^5.$$

Since the bases are the same number, 2, we can use the base–exponent property and set the exponents equal:

$$3x - 7 = 5$$
$$3x = 12$$
$$x = 4.$$

Check:

$$\begin{array}{c|c} 2^{3x-7} = 32 \\ \hline 2^{3(4)-7} \stackrel{?}{=} 32 \\ 2^{12-7} \\ 2^5 \\ 32 & 32 \quad \text{TRUE} \end{array}$$

The solution is 4.

Visualizing the Solution

We graph $y = 2^{3x-7}$ and $y = 32$. The first coordinate of the point of intersection of the graphs is the value of x for which $2^{3x-7} = 32$ and is thus the solution of the equation.

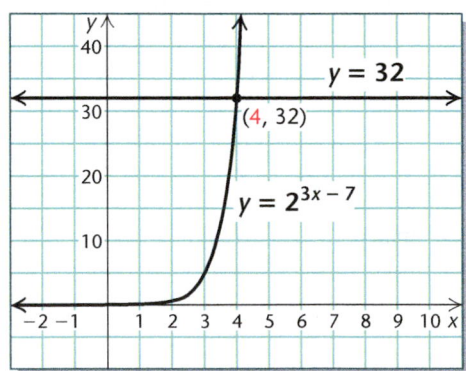

The solution of $2^{3x-7} = 32$ is 4.

Now Try Exercise 7.

Another property that is used when solving some exponential equations and logarithmic equations is as follows.

PROPERTY OF LOGARITHMIC EQUALITY

For any $M > 0, N > 0, a > 0$, and $a \neq 1$,

$$\log_a M = \log_a N \leftrightarrow M = N.$$

This property follows from the fact that for any $a > 0, a \neq 1, f(x) = \log_a x$ is a one-to-one function. If $\log_a x = \log_a y$, then $f(x) = f(y)$. Then since f is one-to-one, it follows that $x = y$. Conversely, if $x = y$, it follows that $\log_a x = \log_a y$, since we are taking the logarithm of the same number in each case.

When it does not seem possible to write each side as a power of the same base, we can use the property of logarithmic equality and take the logarithm with any base on each side and then use the power rule for logarithms.

EXAMPLE 2 Solve: $3^x = 20$.

Algebraic Solution

We have

$$3^x = 20$$

$$\log 3^x = \log 20 \qquad \text{Taking the common logarithm on both sides}$$

$$x \log 3 = \log 20 \qquad \text{Using the power rule}$$

$$x = \frac{\log 20}{\log 3}. \qquad \text{Dividing by log 3}$$

This is an exact answer. We cannot simplify further, but we can approximate using a calculator:

$$x = \frac{\log 20}{\log 3} \approx 2.7268.$$

We can check this by finding $3^{2.7268}$:

$$3^{2.7268} \approx 20.$$

The solution is about 2.7268.

Visualizing the Solution

We graph $y = 3^x$ and $y = 20$. The first coordinate of the point of intersection of the graphs is the value of x for which $3^x = 20$ and is thus the solution of the equation.

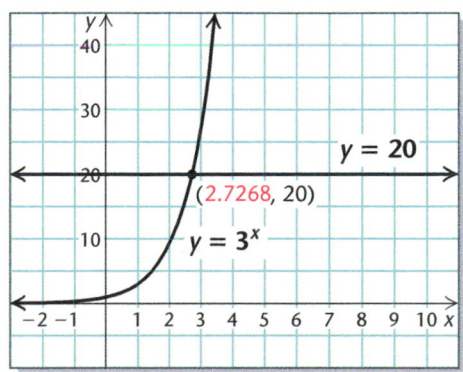

The solution is approximately 2.7268.

Now Try Exercise 11.

In Example 2, we took the common logarithm on both sides of the equation. Any base will give the same result. Let's try base 3. We have

$$3^x = 20$$
$$\log_3 3^x = \log_3 20$$
$$x = \log_3 20 \qquad \color{red}{\log_a a^x = x}$$
$$x = \frac{\log 20}{\log 3} \qquad \color{red}{\textbf{Using the change-of-base formula}}$$
$$x \approx 2.7268.$$

Note that we must change the base in order to do the final calculation.

EXAMPLE 3   Solve: $100e^{0.08t} = 2500$.

Algebraic Solution

It will make our work easier if we take the natural logarithm when working with equations that have e as a base.
   We have

$$100e^{0.08t} = 2500$$
$$e^{0.08t} = 25 \qquad \color{red}{\textbf{Dividing by 100}}$$
$$\ln e^{0.08t} = \ln 25 \qquad \color{red}{\substack{\textbf{Taking the natural} \\ \textbf{logarithm on both sides}}}$$
$$0.08t = \ln 25 \qquad \color{red}{\substack{\textbf{Finding the logarithm} \\ \textbf{of a base to a power:} \\ \log_a a^x = x}}$$
$$t = \frac{\ln 25}{0.08} \qquad \color{red}{\textbf{Dividing by 0.08}}$$
$$t \approx 40.2.$$

The solution is about 40.2.

Visualizing the Solution

The first coordinate of the point of intersection of the graphs of $y = 100e^{0.08t}$ and $y = 2500$ is about 40.2. This is the solution of the equation.

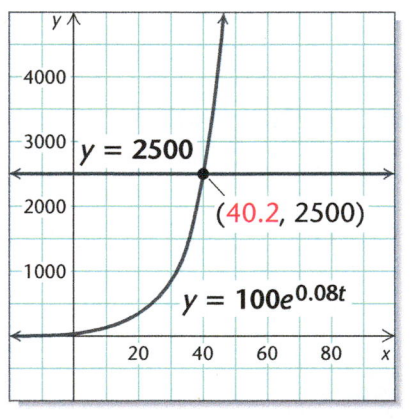

Now Try Exercise 19.

EXAMPLE 4 Solve: $4^{x+3} = 3^{-x}$.

Algebraic Solution

We have

$$4^{x+3} = 3^{-x}$$

$$\log 4^{x+3} = \log 3^{-x} \qquad \text{Taking the common logarithm on both sides}$$

$$(x+3)\log 4 = -x \log 3 \qquad \text{Using the power rule}$$

$$x \log 4 + 3 \log 4 = -x \log 3 \qquad \text{Removing parentheses}$$

$$x \log 4 + x \log 3 = -3 \log 4 \qquad \text{Adding } x \log 3 \text{ and subtracting } 3 \log 4$$

$$x(\log 4 + \log 3) = -3 \log 4 \qquad \text{Factoring on the left}$$

$$x = \frac{-3 \log 4}{\log 4 + \log 3} \qquad \text{Dividing by } \log 4 \ + \ \log 3$$

$$x \approx -1.6737.$$

The solution is about -1.6737.

Visualizing the Solution

We graph $y = 4^{x+3}$ and $y = 3^{-x}$. The first coordinate of the point of intersection of the graphs is the value of x for which $4^{x+3} = 3^{-x}$ and is thus the solution of the equation.

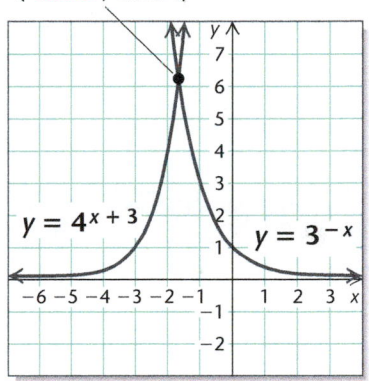

$(-1.6737, 6.2884)$

$y = 4^{x+3}$

$y = 3^{-x}$

The solution is approximately -1.6737.

Now Try Exercise 21.

Technology Connection

$y_1 = 100e^{0.08x}, \quad y_2 = 2500$

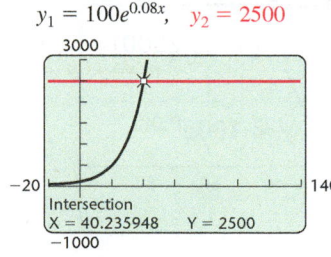

3000

−20   140

Intersection
X = 40.235948   Y = 2500
−1000

We can solve the equations in Examples 1–4 using the Intersect feature. In Example 3, for instance, we graph $y = 100e^{0.08x}$ and $y = 2500$ and use the INTERSECT feature to find the coordinates of the point of intersection.

The first coordinate of the point of intersection is the solution of the equation $100e^{0.08x} = 2500$. The solution is about 40.2. We could also write the equation in the form $100e^{0.08x} - 2500 = 0$ and use the Zero feature.

EQUATIONS REDUCIBLE TO QUADRATIC

REVIEW SECTION 7.4

EXAMPLE 5 Solve: $e^x + e^{-x} - 6 = 0$.

Algebraic Solution

In this case, we have more than one term with x in the exponent:

$$e^x + e^{-x} - 6 = 0$$

$$e^x + \frac{1}{e^x} - 6 = 0 \qquad \text{Rewriting } e^{-x} \text{ with a positive exponent}$$

$$e^{2x} + 1 - 6e^x = 0. \qquad \text{Multiplying by } e^x \text{ on both sides}$$

This equation is reducible to quadratic with $u = e^x$:

$$u^2 - 6u + 1 = 0.$$

We use the quadratic formula with $a = 1$, $b = -6$, and $c = 1$:

$$u = \frac{-b \pm \sqrt{b^2 - 4ac}}{2a}$$

$$u = \frac{-(-6) \pm \sqrt{(-6)^2 - 4 \cdot 1 \cdot 1}}{2 \cdot 1}$$

$$u = \frac{6 \pm \sqrt{32}}{2} = \frac{6 \pm 4\sqrt{2}}{2}$$

$$u = \frac{2(3 \pm 2\sqrt{2})}{2}$$

$$u = 3 \pm 2\sqrt{2}$$

$$e^x = 3 \pm 2\sqrt{2}. \qquad \text{Replacing } u \text{ with } e^x$$

We now take the natural logarithm on both sides:

$$\ln e^x = \ln\left(3 \pm 2\sqrt{2}\right)$$

$$x = \ln\left(3 \pm 2\sqrt{2}\right). \qquad \text{Using } \ln e^x = x$$

Approximating each of the solutions, we obtain 1.76 and -1.76.

Visualizing the Solution

The solutions of the equation

$$e^x + e^{-x} - 6 = 0$$

are the zeros of the function

$$f(x) = e^x + e^{-x} - 6.$$

Note that the solutions are also the first coordinates of the x-intercepts of the graph of the function.

$$f(x) = e^x + e^{-x} - 6$$

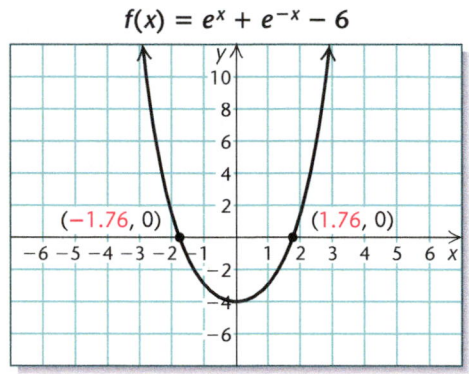

The leftmost zero is about -1.76. The zero on the right is about 1.76. The solutions of the equation are approximately -1.76 and 1.76.

Now Try Exercise 25.

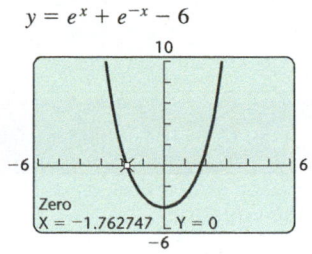

To use a graphing calculator to solve the equation in Example 5, we graph the function $y = e^x + e^{-x} - 6$ and use the ZERO feature to find the zeros.

The leftmost zero is about -1.76. Using the ZERO feature one more time, we find that the other zero is about 1.76.

▶ **Solving Logarithmic Equations**

Equations containing variables in logarithmic expressions, such as $\log_2 x = 4$ and $\log x + \log (x + 3) = 1$, are called **logarithmic equations**. To solve logarithmic equations algebraically, we first try to obtain a single logarithmic expression on one side and then write an equivalent exponential equation.

EXAMPLE 6 Solve: $\log_3 x = -2$.

Algebraic Solution

We have

$$\log_3 x = -2$$
$$3^{-2} = x \qquad \text{Converting to an exponential equation}$$
$$\frac{1}{3^2} = x$$
$$\frac{1}{9} = x.$$

Check:

$$\begin{array}{c|c} \log_3 x = -2 \\ \hline \log_3 \frac{1}{9} \;?\; -2 \\ \log_3 3^{-2} \\ \quad -2 \;\bigg|\; -2 \quad \text{TRUE} \end{array}$$

The solution is $\frac{1}{9}$.

Visualizing the Solution

When we graph $y = \log_3 x$ and $y = -2$, we find that the first coordinate of the point of intersection of the graphs is $\frac{1}{9}$.

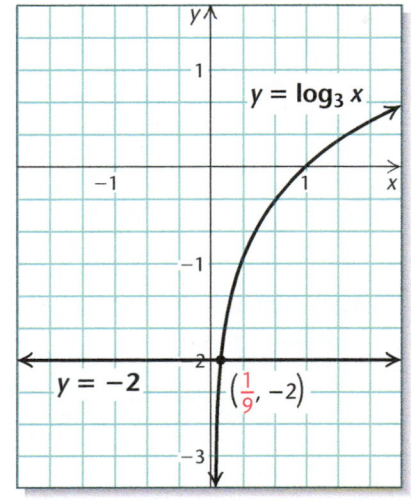

The solution of $\log_3 x = -2$ is $\frac{1}{9}$.

Now Try Exercise 33.

EXAMPLE 7 Solve: $\log x + \log(x + 3) = 1$.

Algebraic Solution

In this case, we have common logarithms. Writing the base of 10 will help us understand the problem:

$$\log_{10} x + \log_{10}(x + 3) = 1$$
$$\log_{10}[x(x + 3)] = 1 \quad \text{Using the product rule to obtain a single logarithm}$$
$$x(x + 3) = 10^1 \quad \text{Writing an equivalent exponential equation}$$
$$x^2 + 3x = 10$$
$$x^2 + 3x - 10 = 0$$
$$(x - 2)(x + 5) = 0 \quad \text{Factoring}$$
$$x - 2 = 0 \quad or \quad x + 5 = 0$$
$$x = 2 \quad or \quad x = -5.$$

Check: For 2:

$$\log x + \log(x + 3) = 1$$
$$\overline{\log 2 + \log(2 + 3) \; ? \; 1}$$
$$\log 2 + \log 5$$
$$\log(2 \cdot 5)$$
$$\log 10$$
$$1 \; | \; 1 \quad \text{TRUE}$$

For -5:

$$\log x + \log(x + 3) = 1$$
$$\overline{\log(-5) + \log(-5 + 3) \; ? \; 1} \quad \text{FALSE}$$

The number -5 is not a solution because negative numbers do not have real-number logarithms. The solution is 2.

Visualizing the Solution

The solution of the equation

$$\log x + \log(x + 3) = 1$$

is the zero of the function

$$f(x) = \log x + \log(x + 3) - 1.$$

The solution is also the first coordinate of the *x*-intercept of the graph of the function.

$$f(x) = \log x + \log(x + 3) - 1$$

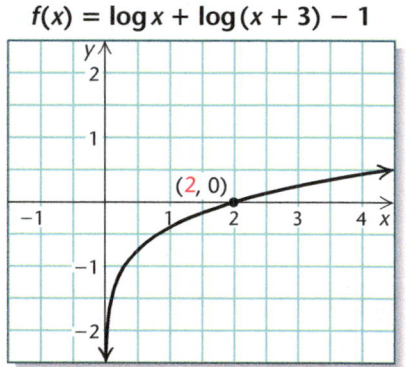

The solution of the equation is 2. From the graph, we can easily see that there is only one solution.

Now Try Exercise 41.

Technology Connection

$y_1 = \log x + \log (x + 3)$, $y_2 = 1$

Intersection
X = 2   Y = 1

$y_1 = \log x + \log (x + 3) - 1$

Zero
X = 2   Y = 0

In Example 7, we can graph the equations

$$y_1 = \log x + \log (x + 3)$$

and

$$y_2 = 1$$

and use the Intersect feature. The first coordinate of the point of intersection is the solution of the equation.

We could also graph the function

$$y = \log x + \log (x + 3) - 1$$

and use the Zero feature. The zero of the function is the solution of the equation.

With either method, we see that the solution is 2. Note that the graphical solution gives only the one *true* solution.

EXAMPLE 8 Solve: $\log_3 (2x - 1) - \log_3 (x - 4) = 2$.

Algebraic Solution

We have

$$\log_3 (2x - 1) - \log_3 (x - 4) = 2$$

$$\log_3 \frac{2x - 1}{x - 4} = 2 \qquad \text{Using the quotient rule}$$

$$\frac{2x - 1}{x - 4} = 3^2 \qquad \text{Writing an equivalent exponential equation}$$

$$\frac{2x - 1}{x - 4} = 9$$

$$(x - 4) \cdot \frac{2x - 1}{x - 4} = 9(x - 4) \qquad \text{Multiplying by the LCD, } x - 4$$

$$2x - 1 = 9x - 36$$

$$35 = 7x$$

$$5 = x.$$

Check:
$$\log_3 (2x - 1) - \log_3 (x - 4) = 2$$

$$\overline{\log_3 (2 \cdot 5 - 1) - \log_3 (5 - 4) \; ? \; 2}$$

$$\log_3 9 - \log_3 1 \;\Big|$$

$$2 - 0 \;\Big|$$

$$2 \;\Big|\; 2 \quad \text{TRUE}$$

The solution is 5.

Visualizing the Solution

We see that the first coordinate of the point of intersection of the graphs of

$$y = \log_3 (2x - 1) - \log_3 (x - 4)$$

and

$$y = 2$$

is 5.

$$y = \log_3 (2x - 1) - \log_3 (x - 4)$$

The solution is 5.

Now Try Exercise 45.

EXAMPLE 9 Solve: $\ln(4x + 6) - \ln(x + 5) = \ln x$.

Algebraic Solution

We have

$$\ln(4x + 6) - \ln(x + 5) = \ln x$$

$$\ln\frac{4x + 6}{x + 5} = \ln x \qquad \text{Using the quotient rule}$$

$$\frac{4x + 6}{x + 5} = x \qquad \text{Using the property of logarithmic equality}$$

$$(x + 5) \cdot \frac{4x + 6}{x + 5} = x(x + 5) \qquad \text{Multiplying by } x + 5$$

$$4x + 6 = x^2 + 5x$$

$$0 = x^2 + x - 6$$

$$0 = (x + 3)(x - 2) \qquad \text{Factoring}$$

$$x + 3 = 0 \quad \text{or} \quad x - 2 = 0$$

$$x = -3 \quad \text{or} \qquad x = 2.$$

The number -3 is not a solution because $4(-3) + 6 = -6$ and $\ln(-6)$ is not a real number. The value 2 checks and is the solution.

Visualizing the Solution

The solution of the equation

$$\ln(4x + 6) - \ln(x + 5) = \ln x$$

is the zero of the function

$$f(x) = \ln(4x + 6) - \ln(x + 5) - \ln x.$$

The solution is also the first coordinate of the x-intercept of the graph of the function.

$$f(x) = \ln(4x + 6) - \ln(x + 5) - \ln x$$

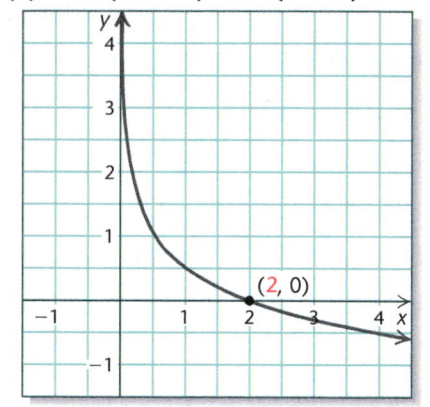

The solution of the equation is 2. From the graph, we can easily see that there is only one solution.

Now Try Exercise 43.

9.6 Exercise Set

Solve the exponential equation.

1. $3^x = 81$

2. $2^x = 32$

3. $2^{2x} = 8$

4. $3^{7x} = 27$

5. $2^x = 33$

6. $2^x = 40$

7. $5^{4x-7} = 125$

8. $4^{3x-5} = 16$

9. $27 = 3^{5x} \cdot 9^{x^2}$

10. $3^{x^2+4x} = \frac{1}{27}$

11. $84^x = 70$

12. $28^x = 10^{-3x}$

13. $10^{-x} = 5^{2x}$

14. $15^x = 30$

15. $e^{-c} = 5^{2c}$

16. $e^{4t} = 200$

17. $e^t = 1000$

18. $e^{-t} = 0.04$

19. $e^{-0.03t} = 0.08$

20. $1000e^{0.09t} = 5000$

21. $3^x = 2^{x-1}$

22. $5^{x+2} = 4^{1-x}$

23. $(3.9)^x = 48$

24. $250 - (1.87)^x = 0$

25. $e^x + e^{-x} = 5$

26. $e^x - 6e^{-x} = 1$

27. $3^{2x-1} = 5^x$

28. $2^{x+1} = 5^{2x}$

29. $2e^x = 5 - e^{-x}$

30. $e^x + e^{-x} = 4$

Solve the logarithmic equation.

31. $\log_5 x = 4$ **32.** $\log_2 x = -3$

33. $\log x = -4$ **34.** $\log x = 1$

35. $\ln x = 1$ **36.** $\ln x = -2$

37. $\log_{64} \frac{1}{4} = x$ **38.** $\log_{125} \frac{1}{25} = x$

39. $\log_2 (10 + 3x) = 5$

40. $\log_5 (8 - 7x) = 3$

41. $\log x + \log (x - 9) = 1$

42. $\log_2 (x + 1) + \log_2 (x - 1) = 3$

43. $\log_2 (x + 20) - \log_2 (x + 2) = \log_2 x$

44. $\log (x + 5) - \log (x - 3) = \log 2$

45. $\log_8 (x + 1) - \log_8 x = 2$

46. $\log x - \log (x + 3) = -1$

47. $\log x + \log (x + 4) = \log 12$

48. $\log_3 (x + 14) - \log_3 (x + 6) = \log_3 x$

49. $\log (x + 8) - \log (x + 1) = \log 6$

50. $\ln x - \ln (x - 4) = \ln 3$

51. $\log_4 (x + 3) + \log_4 (x - 3) = 2$

52. $\ln (x + 1) - \ln x = \ln 4$

53. $\log (2x + 1) - \log (x - 2) = 1$

54. $\log_5 (x + 4) + \log_5 (x - 4) = 2$

55. $\ln (x + 8) + \ln (x - 1) = 2 \ln x$

56. $\log_3 x + \log_3 (x + 1) = \log_3 2 + \log_3 (x + 3)$

Solve.

57. $\log_6 x = 1 - \log_6 (x - 5)$

58. $2^{x^2 - 9x} = \dfrac{1}{256}$

59. $9^{x-1} = 100(3^x)$

60. $2 \ln x - \ln 5 = \ln (x + 10)$

61. $e^x - 2 = -e^{-x}$

62. $2 \log 50 = 3 \log 25 + \log (x - 2)$

▶ Skill Maintenance

In Exercises 63–66:

a) *Find the vertex.*

b) *Find the axis of symmetry.*

c) *Determine whether there is a maximum or a minimum value and find that value.* [7.5]

63. $g(x) = x^2 - 6$

64. $f(x) = -x^2 + 6x - 8$

65. $G(x) = -2x^2 - 4x - 7$

66. $H(x) = 3x^2 - 12x + 16$

▶ Synthesis

Solve using any method.

67. $\dfrac{e^x + e^{-x}}{e^x - e^{-x}} = 3$

68. $\ln (\ln x) = 2$

69. $\sqrt{\ln x} = \ln \sqrt{x}$

70. $\ln \sqrt[4]{x} = \sqrt{\ln x}$

71. $(\log_3 x)^2 - \log_3 x^2 = 3$

72. $\log_3 (\log_4 x) = 0$

73. $\ln x^2 = (\ln x)^2$

74. $x \left(\ln \frac{1}{6} \right) = \ln 6$

75. $5^{2x} - 3 \cdot 5^x + 2 = 0$

76. $x^{\log x} = \dfrac{x^3}{100}$

77. $\ln x^{\ln x} = 4$

78. $\left| 2^{x^2} - 8 \right| = 3$

79. $\dfrac{\sqrt{(e^{2x} \cdot e^{-5x})^{-4}}}{e^x \div e^{-x}} = e^7$

80. Given that $a = (\log_{125} 5)^{\log_5 125}$, find the value of $\log_3 a$.

81. Given that $a = \log_8 225$ and $b = \log_2 15$, express a as a function of b.

82. Given that $f(x) = e^x - e^{-x}$, find $f^{-1}(x)$ if it exists.

9.7 Applications and Models: Growth and Decay; Compound Interest

▶ Solve applied problems involving exponential growth and decay.

▶ Solve applied problems involving compound interest.

Exponential functions and logarithmic functions with base e are rich in applications to many fields such as business, science, psychology, and sociology.

▶ Population Growth

The function

$$P(t) = P_0 e^{kt}, \quad k > 0,$$

is a model of many kinds of population growth, whether it be a population of people, bacteria, smartphones, or money. In this function, P_0 is the population at time 0, P is the population after time t, and k is called the **exponential growth rate**. The graph of such an equation is shown at right.

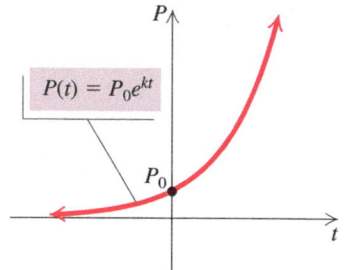

$$P(t) = P_0 e^{kt}$$

EXAMPLE 1   *Population Growth of Ghana.*   In 2013, the population of Ghana, located on the west coast of Africa, was about 25.2 million, and the exponential growth rate was 2.19% per year (*Source*: *CIA World Factbook*, 2014).

a) Find the exponential growth function.

b) Estimate the population in 2018.

c) After how long will the population be double what it was in 2013?

d) At this growth rate, when will the population be 40 million?

Technology Connection

We can find function values using a graphing calculator. Below, we find $P(5)$ from Example 1(b) with the VALUE feature from the CALC menu. We see that $P(5) \approx 28.1$.

$y_1 = 25.2e^{0.0219x}$

X = 5   Y = 28.116146

Solution

a) At $t = 0$ (2013), the population was 25.2 million, and the exponential growth rate was 2.19% per year. We substitute 25.2 for P_0 and 2.19%, or 0.0219, for k to obtain the exponential growth function

$$P(t) = 25.2e^{0.0219t},$$

where t is the number of years after 2013 and $P(t)$ is in millions.

b) In 2018, $t = 5$; that is, 5 years have passed since 2013. To find the population in 2018, we substitute 5 for t:

$$P(5) = 25.2e^{0.0219(5)} = 25.2e^{0.1095} \approx 28.1.$$

The population will be about 28.1 million, or 28,100,000, in 2018.

c) We are looking for the time T for which $P(T) = 2 \cdot 25.2$, or 50.4. The number T is called the **doubling time**. To find T, we solve the equation

$$50.4 = 25.2e^{0.0219T}.$$

Algebraic Solution

We have

$$50.4 = 25.2e^{0.0219T} \qquad \text{Substituting 50.4 for } P(T)$$

$$2 = e^{0.0219T} \qquad \text{Dividing by 25.2}$$

$$\ln 2 = \ln e^{0.0219T} \qquad \text{Taking the natural logarithm on both sides}$$

$$\ln 2 = 0.0219T \qquad \ln e^x = x$$

$$\frac{\ln 2}{0.0219} = T \qquad \text{Dividing by 0.0219}$$

$$31.7 \approx T.$$

The population of Ghana will be double what it was in 2013 about 31.7 years after 2013.

Visualizing the Solution

From the graphs of $y = 50.4$ and $y = 25.2e^{0.0219T}$, we see that the first coordinate of their point of intersection is about 31.7.

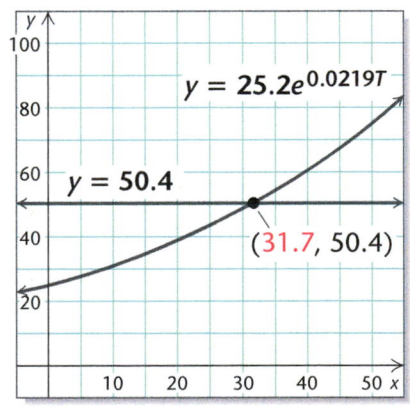

$$y = 25.2e^{0.0219T}$$

$$y = 50.4$$

$$(31.7, 50.4)$$

The solution of the equation is approximately 31.7.

d) To determine the time t for which $P(t) = 40$, we solve the following equation for t:

$$40 = 25.2e^{0.0219t} \qquad \text{\color{red}\textbf{Substituting 40 for } P(t)}$$

$$\frac{40}{25.2} = e^{0.0219t} \qquad \text{\color{red}\textbf{Dividing by 25.2}}$$

$$\ln\frac{40}{25.2} = \ln e^{0.0219t} \qquad \text{\color{red}\textbf{Taking the natural logarithm on both sides}}$$

$$\ln\frac{40}{25.2} = 0.0219t \qquad \text{\color{red}\textbf{ln } e^x = x}$$

$$\frac{\ln\dfrac{40}{25.2}}{0.0219} = t \qquad \text{\color{red}\textbf{Dividing by 0.0219}}$$

$$21 \approx t.$$

The population of Ghana will be 40 million about 21 years after 2013.

→ **Now Try Exercise 1.**

Technology Connection

$y_1 = 25.2e^{0.0219x}, y_2 = 50.4$

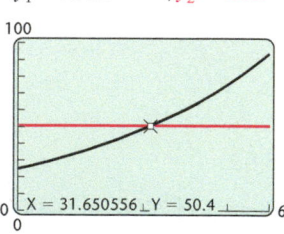

X = 31.650556 Y = 50.4

Using the Intersect feature in Example 1(c), we graph the equations

$$y_1 = 25.2e^{0.0219x} \quad \text{and} \quad y_2 = 50.4$$

and find the first coordinate of their point of intersection. It is about 31.7, so the population of Ghana will be double that of 2013 about 31.7 years after 2013.

▶ Interest Compounded Continuously

When interest is paid on interest, we call it **compound interest**. Suppose that an amount P_0 is invested in a savings account at interest rate k, **compounded continuously**. The amount $P(t)$ in the account after t years is given by the exponential function

$$P(t) = P_0e^{kt}.$$

EXAMPLE 2   *Interest Compounded Continuously.*   Suppose that $2000 is invested at interest rate k, compounded continuously, and grows to $2504.65 in 5 years.

a) What is the interest rate?

b) Find the exponential growth function.

c) What will the balance be after 10 years?

d) After how long will the $2000 have doubled?

Technology Connection

We can also find k in Example 2(a) by graphing the equations

$$y_1 = 2000e^{5x}$$

and

$$y_2 = 2504.65$$

and use the Intersect feature to approximate the first coordinate of the point of intersection.

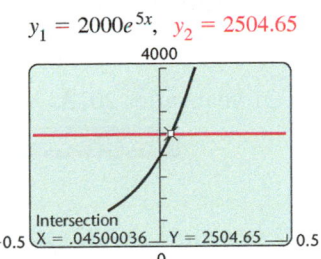

$y_1 = 2000e^{5x}, \quad y_2 = 2504.65$

The interest rate is about 0.045, or 4.5%.

Solution

a) At $t = 0$, $P(0) = P_0 = \$2000$. Thus the exponential growth function is of the form

$$P(t) = 2000e^{kt}.$$

We know that $P(5) = \$2504.65$. We substitute and solve for k:

$$2504.65 = 2000e^{k(5)} \qquad \text{Substituting 2504.65 for } P(t) \text{ and 5 for } t$$

$$2504.65 = 2000e^{5k}$$

$$\frac{2504.65}{2000} = e^{5k} \qquad \text{Dividing by 2000}$$

$$\ln\frac{2504.65}{2000} = \ln e^{5k} \qquad \text{Taking the natural logarithm}$$

$$\ln\frac{2504.65}{2000} = 5k \qquad \text{Using } \ln e^x = x$$

$$\frac{\ln\dfrac{2504.65}{2000}}{5} = k \qquad \text{Dividing by 5}$$

$$0.045 \approx k.$$

The interest rate is about 0.045, or 4.5%.

b) Substituting 0.045 for k in the function $P(t) = 2000e^{kt}$, we see that the exponential growth function is

$$P(t) = 2000e^{0.045t}.$$

c) The balance after 10 years is

$$P(10) = 2000e^{0.045(10)} = 2000e^{0.45} \approx \$3136.62.$$

d) To find the doubling time T, we set $P(T) = 2 \cdot P_0 = 2 \cdot \$2000 = \$4000$ and solve for T. We solve

$$4000 = 2000e^{0.045T}.$$

Algebraic Solution

We have

$$4000 = 2000e^{0.045T}$$
$$2 = e^{0.045T} \qquad \text{Dividing by 2000}$$
$$\ln 2 = \ln e^{0.045T} \qquad \text{Taking the natural logarithm}$$
$$\ln 2 = 0.045T \qquad \ln e^x = x$$
$$\frac{\ln 2}{0.045} = T \qquad \text{Dividing by 0.045}$$
$$15.4 \approx T.$$

Thus the original investment of $2000 will double in about 15.4 years.

Visualizing the Solution

The solution of the equation

$$4000 = 2000e^{0.045T}$$

or

$$2000e^{0.045T} - 4000 = 0,$$

is the zero of the function

$$y = 2000e^{0.045T} - 4000.$$

Note the zero from the graph shown here.

$$y = 2000e^{0.045T} - 4000$$

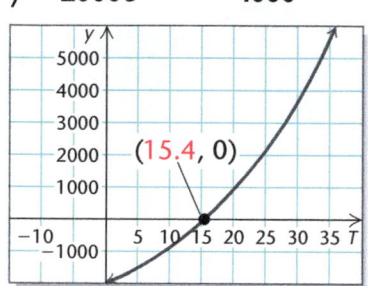

The zero is about 15.4. Thus the solution of the equation is approximately 15.4.

Now Try Exercise 7.

Technology Connection

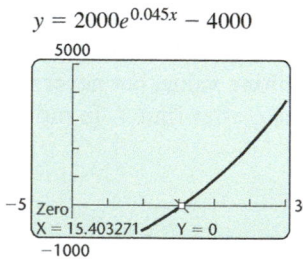

$$y = 2000e^{0.045x} - 4000$$

The amount of money in Example 2 will have doubled when $P(t) = 2 \cdot P_0 = 4000$, or when $2000e^{0.045t} = 4000$. We use the Zero feature. We graph the equation

$$y = 2000e^{0.045x} - 4000$$

and find the zero of the function. The zero of the function is the solution of the equation. The zero is about 15.4, so the original investment of $2000 will double in about 15.4 years.

We can find a general expression relating the growth rate k and the doubling time T by solving the following equation:

$$2P_0 = P_0e^{kT} \qquad \text{Substituting } 2P_0 \text{ for } P \text{ and } T \text{ for } t$$
$$2 = e^{kT} \qquad \text{Dividing by } P_0$$
$$\ln 2 = \ln e^{kT} \qquad \text{Taking the natural logarithm}$$
$$\ln 2 = kT \qquad \text{Using } \ln e^x = x$$
$$\frac{\ln 2}{k} = T.$$

Philippines

GROWTH RATE AND DOUBLING TIME

The **growth rate** k and the **doubling time** T are related by

$$kT = \ln 2, \quad \text{or} \quad k = \frac{\ln 2}{T}, \quad \text{or} \quad T = \frac{\ln 2}{k}.$$

Note that the relationship between k and T does not depend on P_0.

EXAMPLE 3   *Population Growth.*   The population of the Philippines is now doubling every 37.7 years (*Source: CIA World Factbook*, 2014). What is the exponential growth rate?

Solution   We have

$$k = \frac{\ln 2}{T} = \frac{\ln 2}{37.7} \approx 0.0184 \approx 1.84\%.$$

The growth rate of the population of the Philippines is about 1.84% per year.

Now Try Exercise 3(e).

▶ Models of Limited Growth

The model $P(t) = P_0 e^{kt}$, $k > 0$, has many applications involving unlimited population growth. However, in some populations, there can be factors that prevent a population from exceeding some limiting value—perhaps a limitation on food, living space, or other natural resources. One model of such growth is

$$P(t) = \frac{a}{1 + be^{-kt}}.$$

This is called a **logistic function**. This function increases toward a *limiting value a* as $t \to \infty$. Thus, $y = a$ is the horizontal asymptote of the graph of $P(t)$.

EXAMPLE 4   *Limited Population Growth in a Lake.*   A lake is stocked with 400 fish of a new variety. The size of the lake, the availability of food, and the number of other fish restrict the growth of that type of fish in the lake to a limiting value of 2500. The population gets closer and closer to this limiting value, but never reaches it. The population of the new variety of fish in the lake after time t, in months, is given by the function

$$P(t) = \frac{2500}{1 + 5.25e^{-0.32t}}.$$

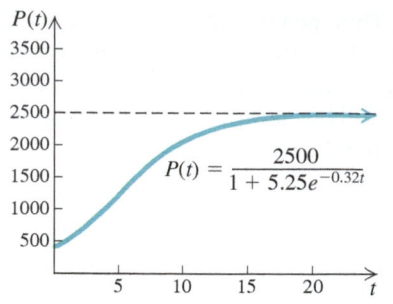

The graph of $P(t)$ is the curve shown at left. Note that this function increases toward a limiting value of 2500. The graph has $y = 2500$ as a horizontal asymptote. Find the population after 0, 1, 5, 10, 15, and 20 months.

Solution Using a calculator, we compute the function values. We find that

$$P(0) = 400, \qquad P(10) \approx 2059,$$
$$P(1) \approx 520, \qquad P(15) \approx 2396,$$
$$P(5) \approx 1214, \qquad P(20) \approx 2478.$$

Thus the population will be about 400 after 0 months, 520 after 1 month, 1214 after 5 months, 2059 after 10 months, 2396 after 15 months, and 2478 after 20 months.

Now Try Exercise 17(b).

Another model of limited growth is provided by the function

$$P(t) = L(1 - e^{-kt}), \quad k > 0,$$

which is shown graphed below. This function also increases toward a limiting value L, as $t \to \infty$, so $y = L$ is the horizontal asymptote of the graph of $P(t)$.

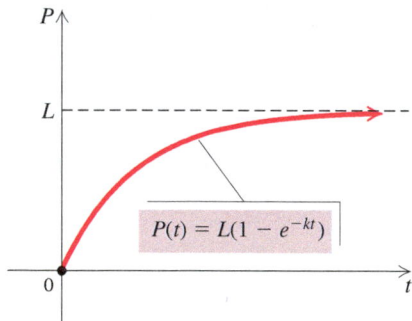

▶ Exponential Decay

The function

$$P(t) = P_0 e^{-kt}, \quad k > 0,$$

is an effective model of the decline, or decay, of a population. An example is the decay of a radioactive substance. In this case, P_0 is the amount of the substance at time $t = 0$, and $P(t)$ is the amount of the substance left after time t, where k is a positive constant that depends on the situation. The constant k is called the **decay rate**.

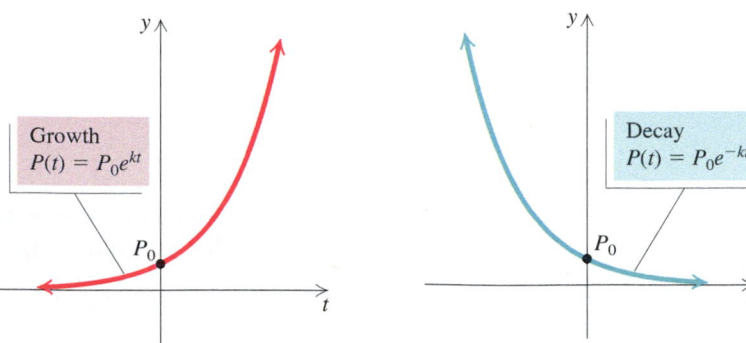

The **half-life** of bismuth (Bi-210) is 5 days. This means that half of an amount of bismuth will cease to be radioactive in 5 days. The effect of half-life T for non-negative inputs is shown in the following graph. The exponential function gets close to 0, but never reaches 0, as t gets very large. Thus, according to an exponential decay model, a radioactive substance never completely decays.

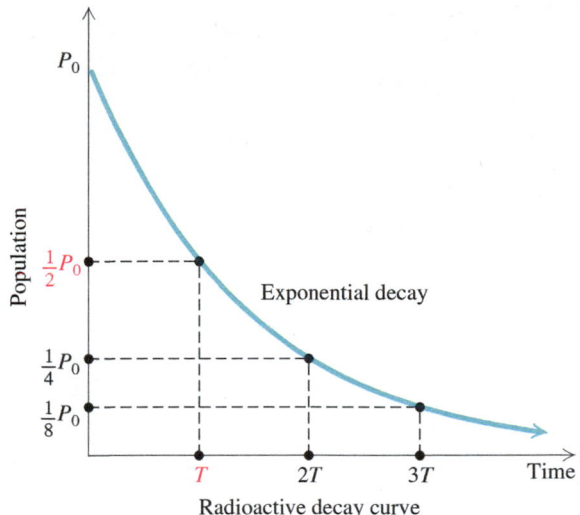

Radioactive decay curve

We can find a general expression relating the decay rate k and the half-life time T by solving the following equation:

$$\frac{1}{2}P_0 = P_0e^{-kT} \qquad \text{Substituting } \frac{1}{2}P_0 \text{ for } P \text{ and } T \text{ for } t$$

$$\frac{1}{2} = e^{-kT} \qquad \text{Dividing by } P_0$$

$$\ln\frac{1}{2} = \ln e^{-kT} \qquad \text{Taking the natural logarithm}$$

$$\ln 2^{-1} = -kT \qquad \frac{1}{2} = 2^{-1}; \ln e^x = x$$

$$-\ln 2 = -kT \qquad \text{Using the power rule}$$

$$\frac{\ln 2}{k} = T. \qquad \text{Dividing by } -k$$

DECAY RATE AND HALF-LIFE

The decay rate k and the half-life T are related by

$$kT = \ln 2, \quad \text{or} \quad k = \frac{\ln 2}{T}, \quad \text{or} \quad T = \frac{\ln 2}{k}.$$

Note that the relationship between decay rate and half-life is the same as that between growth rate and doubling time.

How can scientists determine that an animal bone has lost 30% of its carbon-14? The assumption is that the percentage of carbon-14 in the atmosphere is the same as that in living plants and animals. When a plant or an animal dies, the amount of carbon-14 that it contains decays exponentially. A scientist can burn an animal bone and use a Geiger counter to determine the percentage of the smoke that is carbon-14. The amount by which this varies from the percentage in the atmosphere tells how much carbon-14 has been lost.

The process of carbon-14 dating was developed by the American chemist Willard E. Libby in 1952. It is known that the radioactivity in a living plant is 16 disintegrations per gram per minute. Since the half-life of carbon-14 is 5750 years, an object with an activity of 8 disintegrations per gram per minute is 5750 years old, one with an activity of 4 disintegrations per gram per minute is 11,500 years old, and so on. Carbon-14 dating can be used to measure the age of objects up to 40,000 years old. Beyond such an age, it is too difficult to measure the radioactivity and some other method would have to be used.

Carbon-14 dating was used to find the age of the Dead Sea Scrolls. It was also used to refute the authenticity of the Shroud of Turin, presumed to have covered the body of Christ.

In 1947, a Bedouin youth looking for a stray goat climbed into a cave at Kirbet Qumran on the shores of the Dead Sea near Jericho and came upon earthenware jars containing an incalculable treasure of ancient manuscripts. Shown here are fragments of those Dead Sea Scrolls, a portion of some 600 or so texts found so far and which concern the Jewish books of the Bible. Officials date them before A.D. 70, making them the oldest Biblical manuscripts by 1000 years.

EXAMPLE 5 *Carbon Dating.* The radioactive element carbon-14 has a half-life of 5750 years. The percentage of carbon-14 present in the remains of organic matter can be used to determine the age of that organic matter. Archaeologists discovered that the linen wrapping from one of the Dead Sea Scrolls had lost 22.3% of its carbon-14 at the time it was found. How old was the linen wrapping?

Solution We first find k when the half-life T is 5750 years:

$$k = \frac{\ln 2}{T}$$

$$k = \frac{\ln 2}{5750} \qquad \text{Substituting 5750 for } T$$

$$k = 0.00012.$$

Now we have the function

$$P(t) = P_0 e^{-0.00012t}.$$

(This function can be used for any subsequent carbon-dating problem.) If the linen wrapping has lost 22.3% of its carbon-14 from an initial amount P_0, then 77.7% P_0 is the amount present. To find the age t of the wrapping, we solve the following equation for t:

$$77.7\% P_0 = P_0 e^{-0.00012t} \qquad \text{Substituting 77.7\% } P_0 \text{ for } P$$

$$0.777 = e^{-0.00012t} \qquad \text{Dividing by } P_0 \text{ and writing 77.7\% as 0.777}$$

$$\ln 0.777 = \ln e^{-0.00012t} \qquad \text{Taking the natural logarithm on both sides}$$

$$\ln 0.777 = -0.00012t \qquad \ln e^x = x$$

$$\frac{\ln 0.777}{-0.00012} = t \qquad \text{Dividing by } -0.00012$$

$$2103 \approx t.$$

Thus the linen wrapping on the Dead Sea Scrolls was about 2103 years old when it was found.

Now Try Exercise 9.

9.7   Exercise Set

1. *Population Growth of Houston.*   The Houston–Woodlands—Sugar Land metropolitan area is the fifth largest metropolitan area in the United States. In 2012, the population of this area was 6.18 million, and the exponential growth rate was 2.14% per year.

a) Find the exponential growth function.

b) Estimate the population of the Houston–Woodlands—Sugar Land metropolitan area in 2018.

c) When will the population of this metropolitan area be 8 million?

d) Find the doubling time.

2. *Population Growth of Rabbits.*   Under ideal conditions, a population of rabbits has an exponential growth rate of 11.7% per day. Consider an initial population of 100 rabbits.

a) Find the exponential growth function.

b) What will the population be after 7 days? after 2 weeks?

c) Find the doubling time.

3. *Population Growth.*   Complete the following table.

Country	Growth Rate, k	Doubling Time, T
a) United States		77.0 years
b) Bolivia		42.5 years
c) Uganda	3.32%	
d) Australia	1.11%	
e) Sweden		385 years
f) Laos	2.32%	
g) India	1.28%	
h) China		150.7 years
i) Guinea		26.3 years
j) Hong Kong	0.39%	

4. *E-Book Sales.*   The revenue from e-book sales (not including educational textbooks) accounted for only 0.5% of U.S. publishing sales in 2006. This percentage grew to 22.6% in 2012. (*Source:* Association of American Publishers) Assuming that the exponential growth model applies:

a) Find the value of k and write the function.

b) Estimate the percentage of U.S. publishing sales that were e-book sales in 2009 and in 2010. Round to the nearest tenth.

5. *Population Growth of Haiti.*   The population of Haiti has a growth rate of 0.99% per year. In 2013, the population was 9,893,934, and the land area of Haiti is 32,961,561,600 square yards. (*Source:* U.S. Census Bureau) Assuming that this growth rate continues and is exponential, after how long will there be one person for every square yard of land?

6. *Picasso Painting.*   In May 2010, a 1932 painting, *Nude, Green Leaves, and Bust,* by Pablo Picasso, sold at a New York City art auction for $106.5 million to an anonymous buyer. The painting had belonged to the estate of Sydney and Francis Brody, who bought it for $17,000 in 1952 from a New York art dealer, who had acquired it from Picasso in 1936. (*Sources:* Associated Press, "Picasso Painting Fetches World Record $106.5M at NYC Auction," by Ula Ilnytzky, May 4, 2010; Online Associated Newspapers, May 6, 2010)

Assuming that the value A_0 of the painting has grown exponentially:

a) Find the value of k, and determine the exponential growth function, assuming that $A_0 = 17,000$ and t is the number of years after 1952.
b) Estimate the value of the painting in 2020.
c) What is the doubling time for the value of the painting?
d) After how long will the value of the painting be $240 million, assuming that there is no change in the growth rate?

7. *Interest Compounded Continuously.*   Suppose that $10,000 is invested at an interest rate of 5.4% per year, compounded continuously.

a) Find the exponential function that describes the amount in the account after time t, in years.
b) What is the balance after 1 year? 2 years? 5 years? 10 years?
c) What is the doubling time?

8. *Interest Compounded Continuously.*   Complete the following table.

Initial Investment at $t = 0$, P_0	Interest Rate, k	Doubling Time, T	Amount After 5 Years
a) $35,000	3.2%		
b) $5000			$7,130.90
c)	5.6%		$9,923.47
d)		11 years	$17,539.32
e) $109,000			$136,503.18
f)		46.2 years	$19,552.82

9. *Carbon Dating.*   In 1970, Amos Flora of Flora, Indiana, discovered teeth and jawbones while dredging a creek. Scientists determined that the bones were from a mastodon and had lost 77.2% of their carbon-14. How old were the bones at the time they were discovered? (*Sources*: "Farm Yields Bones Thousands of Years Old," by Dan McFeely, *Indianapolis Star,* October 20, 2008; Field Museum of Chicago, Bill Turnbull, anthropologist)

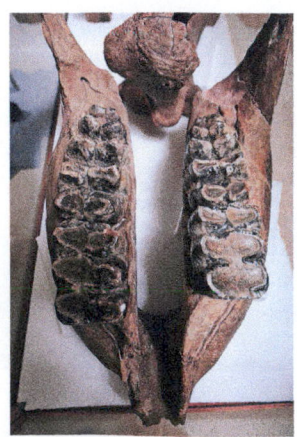

10. *Tomb in the Valley of the Kings.*   In February 2006, in the Valley of the Kings in Egypt, a team of archaeologists uncovered the first tomb since King Tut's tomb was found in 1922. The tomb contained five wooden sarcophagi that contained mummies. The archaeologists believe that the mummies are from the 18th Dynasty, about 3300 to 3500 years ago. Determine the amount of carbon-14 that the mummies have lost.

11. *Radioactive Decay.*   Complete the following table.

Radioactive Substance	Decay Rate, k	Half-Life, T
a) Polonium (Po-218)		3.1 min
b) Lead (Pb-210)		22.3 years
c) Iodine (I-125)	1.15% per day	
d) Krypton (Kr-85)	6.5% per year	
e) Strontium (Sr-90)		29.1 years
f) Uranium (U-232)		70.0 years
g) Plutonium (Pu-239)		24,100 years

12. *Advertising Revenue.*   The amount of advertising revenue in U.S. newspapers has declined continually since 2006. In 2006, the advertising revenue was $49.3 billion, and in 2013 that amount had decreased to $20.7 billion (*Source*: Newspaper Association of America). Assuming that the amount of newspaper advertising revenue decreased according to the exponential decay model:

a) Find the value of k, and write an exponential function that describes the advertising revenue after time t, in years, where t is the number of years after 2006.

b) Estimate the advertising revenue in 2008 and in 2012.
c) At this decay rate, when will the advertising revenue be $16 billion?

13. *Married Adults.* The data in the following table show that the percentage of adults in the United States who are currently married is declining.

Year	Percent of Adults Who Are Married
1960	72.2%
1980	62.3
2000	57.4
2010	51.4
2012	50.5

Sources: Pew Research Center; U.S. Census Bureau

Assuming that the percentage of adults who are married will continue to decrease according to the exponential decay model:

a) Use the data for 1960 and 2012 to find the value of k and to write an exponential function that describes the percent of adults married after time t, in years, where t is the number of years after 1960.
b) Estimate the percent of adults who are married in 2015 and in 2018.
c) At this decay rate, in which year will the percent of adults who are married be 40%?

14. *Lamborghini 350 GT.* The market value of the 1964–1965 Lamborghini 350 GT has had a recent upswing. In a decade, the car's value increased from $66,000 in 1999 to $220,000 in 2009 (*Source*: "1964–1965 Lamborghini 350 GT," by David LaChance, *Hemmings Motor News*, July, 2010, p. 28).

Assuming that the value V_0 of the car has grown exponentially:

a) Find the value of k, and determine the exponential growth function, assuming that $V_0 = 66,000$ and t is the number of years after 1999.
b) Estimate the value of the car in 2011.
c) After how long was the value of the car $300,000, assuming that there is no change in the growth rate?

15. *Oil Consumption.* In 1980, China consumed 1.85 million barrels of oil per day. By 2012, that consumption had grown to 10.28 million barrels per day. (*Sources*: U.S. Energy Information Administration; NextBigThingInvestor.com)

Assuming that the consumption of oil C_0 in China has grown exponentially:

a) Find the value of k, and determine the exponential growth function, assuming that $C_0 = 1.85$ and t is the number of years after 1980.
b) Estimate the consumption of oil in 2005.
c) What is the doubling time for the consumption of oil in China?
d) After how long will the consumption of oil in China be 13 million barrels per day, assuming that there is no change in the growth rate?

16. *T206 Wagner Baseball Card.* In 1909, the Pittsburgh Pirates shortstop Honus Wagner forced the American Tobacco Company to withdraw his baseball card because it was packaged with cigarettes. Fewer than 60 of the Wagner cards still exist. In 1971, a Wagner card sold for $1000; and in September 2007, a card in near-mint condition was purchased for a record $2.8 million (*Source*: *USA Today*, 9/6/07; Kathy Willens/AP).

Assuming that the value W_0 of the baseball card has grown exponentially:

a) Find the value of k, and determine the exponential growth function, assuming that $W_0 = 1000$ and t is the number of years since 1971.
b) Estimate the value of the Wagner card in 2011.
c) What is the doubling time for the value of the card?
d) After how long was the value of the Wagner card $3 million, assuming that there is no change in the growth rate?

17. *Spread of an Epidemic.* In a town whose population is 3500, a disease creates an epidemic. The number of people N infected t days after the disease has begun is given by the function

$$N(t) = \frac{3500}{1 + 19.9e^{-0.6t}}.$$

a) How many are initially infected with the disease $(t = 0)$?
b) Find the number infected after 2 days, 5 days, 8 days, 12 days, and 16 days.
c) Using this model, can you say whether all 3500 people will ever be infected? Explain.

18. *Limited Population Growth in a Lake.* A lake is stocked with 640 fish of a new variety. The size of the lake, the availability of food, and the number of other fish restrict the growth of that type of fish in

the lake to a limiting value of 3040. The population of the new variety of fish in the lake after time t, in months, is given by the function

$$P(t) = \frac{3040}{1 + 3.75e^{-0.32t}}.$$

Find the population after 0, 1, 5, 10, 15, and 20 months.

Newton's Law of Cooling. *Suppose that a body with temperature T_1 is placed in surroundings with temperature T_0 different from that of T_1. The body will either cool or warm to temperature $T(t)$ after time t, in minutes, where*

$$T(t) = T_0 + (T_1 - T_0)e^{-kt}.$$

Use this law in Exercises 19–22.

19. A cup of coffee with temperature 105°F is placed in a freezer with temperature 0°F. After 5 min, the temperature of the coffee is 70°F. What will its temperature be after 10 min?

20. A pan of lasagna baked at 375°F is taken out of the oven at 11:15 A.M. into a kitchen that is 72°F. After 3 min, the temperature of the lasagna is 365°F. What will the temperature of the lasagna be at 11:30 A.M.?

21. A chilled jello salad that has a temperature of 43°F is taken from the refrigerator and placed on the dining room table in a room that is 68°F. After 12 min, the temperature of the salad is 55°F. What will the temperature of the salad be after 20 min?

22. *When Was the Murder Committed?* The police discover the body of a murder victim. Critical to solving the crime is determining when the murder was committed. The coroner arrives at the murder scene at 12:00 P.M. She immediately takes the temperature of the body and finds it to be 94.6°F. She then takes the temperature 1 hr later and finds it to be 93.4°F. The temperature of the room is 70°F. When was the murder committed?

▶ Skill Maintenance

Vocabulary Reinforcement

In Exercises 23–28, choose the correct name of the principle or the rule from the given choices.

principle of zero products
multiplication principle for equations
product rule
addition principle for inequalities
power rule
multiplication principle for inequalities
principle of square roots
quotient rule

23. For any real numbers a, b, and c: If $a < b$ and $c > 0$ are true, then $ac < bc$ is true. If $a < b$ and $c < 0$ are true, then $ac > bc$ is true. **[1.4]**
_____

24. For any positive numbers M and N and any logarithmic base a, $\log_a MN = \log_a M + \log_a N$. **[9.5]**
_____

25. If $ab = 0$ is true, then $a = 0$ or $b = 0$, and if $a = 0$ or $b = 0$, then $ab = 0$. **[7.4]** _____

26. If $x^2 = k$, then $x = \sqrt{k}$ or $x = -\sqrt{k}$. **[7.4]**
_____

27. For any positive number M, any logarithmic base a, and any real number p, $\log_a M^p = p \log_a M$. **[9.5]**
_____

28. For any real numbers a, b, and c: If $a = b$ is true, then $ac = bc$ is true. **[1.1]** _____

▶ Synthesis

29. *Supply and Demand.* The supply function and the demand function for the sale of a certain type of DVD player are given by

$$S(p) = 150e^{0.004p} \quad \text{and} \quad D(p) = 480e^{-0.003p},$$

respectively, where $S(p)$ is the number of DVD players that the company is willing to sell at price p and $D(p)$ is the quantity that the public is willing to buy at price p. Find p such that $D(p) = S(p)$. This is called the **equilibrium price**.

30. *Carbon Dating.* Recently, while digging in Chaco Canyon, New Mexico, archaeologists found corn pollen that was 4000 years old (*Source: American Anthropologist*). This was evidence that Native Americans had been cultivating crops in the Southwest centuries earlier than scientists had thought. What percent of the carbon-14 had been lost from the pollen?

31. *Present Value.* Following the birth of a child, a grandparent wants to make an initial investment P_0 that will grow to $50,000 for the child's education at age 18. Interest is compounded continuously at 5.2%. What should the initial investment be? Such an amount is called the **present value** of $50,000 due 18 years from now.

32. *Present Value.*
a) Solve $P = P_0 e^{kt}$ for P_0.
b) Referring to Exercise 31, find the present value of $50,000 due 18 years from now at interest rate 6.4%, compounded continuously.

33. *Electricity.* The formula

$$i = \frac{V}{R}\left[1 - e^{-(R/L)t}\right]$$

occurs in the theory of electricity. Solve for t.

34. *The Beer–Lambert Law.* A beam of light enters a medium such as water or smog with initial intensity I_0. Its intensity decreases depending on the thickness (or concentration) of the medium. The intensity I at a depth (or concentration) of x units is given by

$$I = I_0 e^{-\mu x}.$$

The constant μ (the Greek letter "mu") is called the **coefficient of absorption**, and it varies with the medium. For sea water, $\mu = 1.4$.
a) What percentage of light intensity I_0 remains in sea water at a depth of 1 m? 3 m? 5 m? 50 m?
b) Plant life cannot exist below 10 m. What percentage of I_0 remains at 10 m?

35. Given that $y = ae^x$, take the natural logarithm on both sides. Let $Y = \ln y$. Consider Y as a function of x. What kind of function is Y?

36. Given that $y = ax^b$, take the natural logarithm on both sides. Let $Y = \ln y$ and $X = \ln x$. Consider Y as a function of X. What kind of function is Y?

Chapter 9 Summary and Review

STUDY GUIDE

KEY TERMS AND CONCEPTS	EXAMPLES

SECTION 9.1: THE COMPOSITION OF FUNCTIONS

The **composition of functions**, $f \circ g$, is defined as

$$(f \circ g)(x) = f(g(x)),$$

where x is in the domain of g and $g(x)$ is in the domain of f.

Given that $f(x) = 2x - 1$ and $g(x) = \sqrt{x}$, find each of the following.

a) $(f \circ g)(4)$ **b)** $(g \circ g)(625)$

c) $(f \circ g)(x)$ **d)** $(g \circ f)(x)$

e) The domain of $f \circ g$ and the domain of $g \circ f$

a) $(f \circ g)(4) = f(g(4)) = f(\sqrt{4}) = f(2) = 2 \cdot 2 - 1 = 4 - 1 = 3$

b) $(g \circ g)(625) = g(g(625)) = g(\sqrt{625}) = g(25) = \sqrt{25} = 5$

c) $(f \circ g)(x) = f(g(x)) = f(\sqrt{x}) = 2\sqrt{x} - 1$

d) $(g \circ f)(x) = g(f(x)) = g(2x - 1) = \sqrt{2x - 1}$

e) The domain and the range of $f(x)$ are both $(-\infty, \infty)$, and the domain and the range of $g(x)$ are both $[0, \infty)$. Since the inputs of $f \circ g$ are outputs of g and since f can accept any real number as an input, the domain of $f \circ g$ consists of all real numbers that are outputs of g, or $[0, \infty)$.

The inputs of $g \circ f$ consist of all real numbers that are in the domain of g. Thus we must have $2x - 1 \geq 0$, or $x \geq \frac{1}{2}$, so the domain of $g \circ f$ is $\left[\frac{1}{2}, \infty\right)$.

When we **decompose** a function, we write it as the composition of two functions.

If $h(x) = \sqrt{3x + 7}$, find $f(x)$ and $g(x)$ such that $h(x) = (f \circ g)(x)$.

    This function finds the square root of $3x + 7$, so one decomposition is $f(x) = \sqrt{x}$ and $g(x) = 3x + 7$.

    There are other correct answers, but this one is probably the most obvious.

SECTION 9.2: INVERSE FUNCTIONS

Inverse Relation

If a relation is defined by an equation, then interchanging the variables produces an equation of the inverse relation.

Given $y = -5x + 7$, find an equation of the inverse relation.

$$y = -5x + 7 \qquad \text{Relation}$$
$$\downarrow \qquad\quad \downarrow$$
$$x = -5y + 7 \qquad \text{Inverse relation}$$

One-to-One Functions

A function f is one-to-one if different inputs have different outputs—that is,

$$\text{if } a \neq b, \quad \text{then} \quad f(a) \neq f(b).$$

Or a function f is one-to-one if when the outputs are the same, the inputs are the same—that is,

$$\text{if } f(a) = f(b), \quad \text{then} \quad a = b.$$

Prove that $f(x) = 16 - 3x$ is one-to-one.

Show that if $f(a) = f(b)$, then $a = b$. Assume $f(a) = f(b)$. Since $f(a) = 16 - 3a$ and $f(b) = 16 - 3b$,

$$16 - 3a = 16 - 3b$$
$$-3a = -3b$$
$$a = b.$$

Thus, if $f(a) = f(b)$, then $a = b$ and f is one-to-one.

Horizontal-Line Test

If it is possible for a horizontal line to intersect the graph of a function more than once, then the function is *not* one-to-one and its inverse is *not* a function.

One-to-One Functions and Inverses

- If a function f is one-to-one, then its inverse f^{-1} is a function.
- The domain of a one-to-one function f is the range of the inverse f^{-1}.
- The range of a one-to-one function f is the domain of the inverse f^{-1}.
- A function that is increasing over its entire domain or is decreasing over its entire domain is a one-to-one function.

The -1 in f^{-1} is *not* an exponent.

Using its graph, determine whether each function is one-to-one.

a) 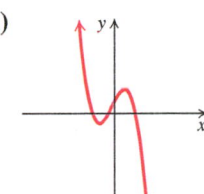   b)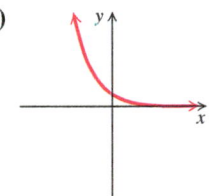

a) There are many horizontal lines that intersect the graph more than once. Thus the function is *not* one-to-one and its inverse is *not* a function.

b) No horizontal line intersects the graph more than once. Thus the function is one-to-one and its inverse is a function.

Obtaining a Formula for an Inverse

If a function f is one-to-one, a formula for its inverse can generally be found as follows:

1. Replace $f(x)$ with y.
2. Interchange x and y.
3. Solve for y.
4. Replace y with $f^{-1}(x)$.

The graph of f^{-1} is a reflection of the graph of f across the line $y = x$.

Given the one-to-one function $f(x) = 2 - x^3$, find a formula for its inverse. Then graph the function and its inverse on the same set of axes.

$$f(x) = 2 - x^3$$

1. $y = 2 - x^3$ **Replacing $f(x)$ with y**

2. $x = 2 - y^3$ **Interchanging x and y**

3. Solve for y:

$$y^3 = 2 - x$$ **Adding y^3 and subtracting x**
$$y = \sqrt[3]{2 - x}.$$

4. $f^{-1}(x) = \sqrt[3]{2 - x}$ **Replacing y with $f^{-1}(x)$**

$$f(x) = 2 - x^3 \text{ and } f^{-1}(x) = \sqrt[3]{2 - x}$$

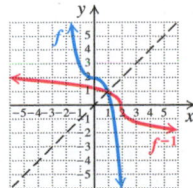

If a function f is one-to-one, then f^{-1} is the unique function such that each of the following holds:

$$(f^{-1} \circ f)(x) = f^{-1}(f(x)) = x,$$

for each x in the domain of f, and

$$(f \circ f^{-1})(x) = f(f^{-1}(x)) = x,$$

for each x in the domain of f^{-1}.

Given $f(x) = \dfrac{3 + x}{x}$, use composition of functions to show that $f^{-1}(x) = \dfrac{3}{x - 1}$.

$$(f^{-1} \circ f)(x) = f^{-1}(f(x))$$

$$= f^{-1}\left(\frac{3 + x}{x}\right) = \frac{3}{\dfrac{3 + x}{x} - 1}$$

$$= \frac{3}{\dfrac{3 + x - x}{x}} = \frac{3}{\dfrac{3}{x}} = 3 \cdot \frac{x}{3} = x;$$

$$(f \circ f^{-1})(x) = f(f^{-1}(x)) = f\left(\frac{3}{x - 1}\right)$$

$$= \frac{3 + \dfrac{3}{x - 1}}{\dfrac{3}{x - 1}} = \frac{3(x - 1) + 3}{x - 1} \cdot \frac{x - 1}{3}$$

$$= \frac{3x - 3 + 3}{3} = \frac{3x}{3} = x$$

SECTION 9.3: EXPONENTIAL FUNCTIONS AND GRAPHS

Exponential Function

$y = a^x$, or $f(x) = a^x$, $\quad a > 0, a \neq 1$

    Continuous

    One-to-one

    Domain: $(-\infty, \infty)$

    Range: $(0, \infty)$

    Increasing if $a > 1$

    Decreasing if $0 < a < 1$

    Horizontal asymptote is x-axis

    y-intercept: $(0, 1)$

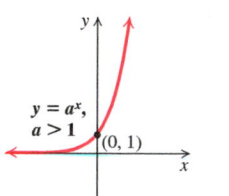

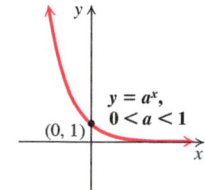

Graph: $f(x) = 2^x$, $g(x) = 2^{-x}$, $h(x) = 2^{x-1}$, and $t(x) = 2^x - 1$.

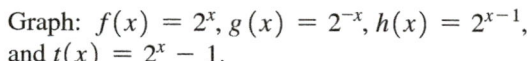

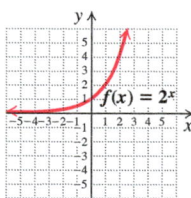

$f(x) = 2^x$

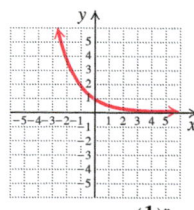

$g(x) = 2^{-x} = \left(\frac{1}{2}\right)^x$

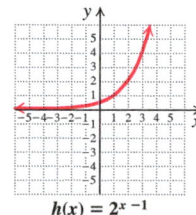

$h(x) = 2^{x-1}$

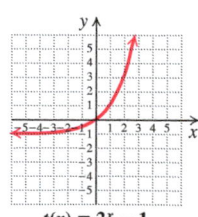

$t(x) = 2^x - 1$

Compound Interest

The amount of money A to which a principal P will grow after t years at interest rate r (in decimal form), compounded n times per year, is given by the formula

$$A = P\left(1 + \frac{r}{n}\right)^{nt}.$$

Suppose that \$5000 is invested at 3.5% interest, compounded quarterly. Find the money in the account after 3 years.

$$A = P\left(1 + \frac{r}{n}\right)^{nt} = 5000\left(1 + \frac{0.035}{4}\right)^{4 \cdot 3}$$

$$\approx \$5551.02$$

The Number e

$$e = 2.7182818284\ldots$$

Find each of the following, to four decimal places, using a calculator.

$$e^{-3} \approx 0.0498;$$
$$e^{4.5} \approx 90.0171$$

Graph: $f(x) = e^x$ and $g(x) = e^{-x+2} - 4$.

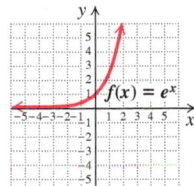

$f(x) = e^x$

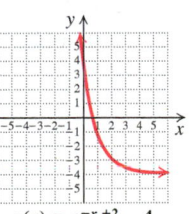

$g(x) = e^{-x+2} - 4$

SECTION 9.4: LOGARITHMIC FUNCTIONS AND GRAPHS

Logarithmic Function

$$y = \log_a x, \quad x > 0, a > 0, a \neq 1$$

Continuous

One-to-one

Domain: $(0, \infty)$

Range: $(-\infty, \infty)$

Increasing if $a > 1$

Vertical asymptote is y-axis

x-intercept: $(1, 0)$

The inverse of an exponential function $f(x) = a^x$ is given by $f^{-1}(x) = \log_a x$.

Graph: $f(x) = \log_2 x$ and $g(x) = \ln(x - 1) + 2$.

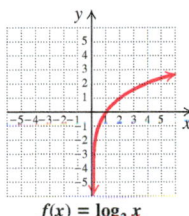

$f(x) = \log_2 x$

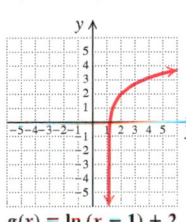

$g(x) = \ln(x - 1) + 2$

A logarithm is an exponent:

$$\log_a x = y \longleftrightarrow x = a^y.$$

Convert each logarithmic equation to an exponential equation.

$$\log_4 \frac{1}{16} = -2 \longleftrightarrow 4^{-2} = \frac{1}{16};$$
$$\ln R = 3 \longleftrightarrow e^3 = R$$

Convert each exponential equation to a logarithmic equation.

$$e^{-5} = 0.0067 \longleftrightarrow \ln 0.0067 = -5;$$
$$7^2 = 49 \longleftrightarrow \log_7 49 = 2$$

$\log x$ means $\log_{10} x$ **Common logarithms**

$\ln x$ means $\log_e x$ **Natural logarithms**

For any logarithm base a,

$$\log_a 1 = 0 \quad \text{and} \quad \log_a a = 1.$$

For the logarithm base e,

$$\ln 1 = 0 \quad \text{and} \quad \ln e = 1.$$

Find each of the following without using a calculator.

$$\log 100 = 2; \qquad \log 10^{-5} = -5;$$
$$\ln 1 = 0; \qquad \log_9 9 = 1;$$
$$\ln \sqrt[3]{e} = \frac{1}{3}; \qquad \log_2 64 = 6;$$
$$\log_8 1 = 0; \qquad \ln e = 1$$

Find each of the following using a calculator and rounding to four decimal places.

$$\ln 223 \approx 5.4072; \qquad \log \frac{2}{9} \approx -0.6532;$$

$$\log (-8) \quad \text{Does not exist;} \qquad \ln 0.06 \approx -2.8134$$

The Change-of-Base Formula

For any logarithmic bases a and b, and any positive number M,

$$\log_b M = \frac{\log_a M}{\log_a b}.$$

Find $\log_3 11$ using common logarithms:

$$\log_3 11 = \frac{\log 11}{\log 3} \approx 2.1827.$$

Find $\log_3 11$ using natural logarithms:

$$\log_3 11 = \frac{\ln 11}{\ln 3} \approx 2.1827.$$

Earthquake Magnitude

The magnitude R, measured on the Richter scale, of an earthquake of intensity I is defined as

$$R = \log \frac{I}{I_0},$$

where I_0 is a minimum intensity used for comparison.

What is the magnitude on the Richter scale of an earthquake of intensity $10^{6.8} \cdot I_0$?

$$R = \log \frac{I}{I_0} = \log \frac{10^{6.8} \cdot I_0}{I_0} = \log 10^{6.8} = 6.8$$

SECTION 9.5: PROPERTIES OF LOGARITHMIC FUNCTIONS

The Product Rule

For any positive numbers M and N, and any logarithmic base a,

$$\log_a MN = \log_a M + \log_a N.$$

The Power Rule

For any positive number M, any logarithmic base a, and any real number p,

$$\log_a M^p = p \log_a M.$$

The Quotient Rule

For any positive numbers M and N, and any logarithmic base a,

$$\log_a \frac{M}{N} = \log_a M - \log_a N.$$

Express $\log_c \sqrt{\dfrac{c^2 r}{b^3}}$ in terms of sums and differences of logarithms.

$$
\begin{aligned}
\log_c \sqrt{\frac{c^2 r}{b^3}} &= \log_c \left(\frac{c^2 r}{b^3} \right)^{1/2} \\
&= \tfrac{1}{2} \log_c \left(\frac{c^2 r}{b^3} \right) \\
&= \tfrac{1}{2} (\log_c c^2 r - \log_c b^3) \\
&= \tfrac{1}{2} (\log_c c^2 + \log_c r - \log_c b^3) \\
&= \tfrac{1}{2} (2 + \log_c r - 3 \log_c b) \\
&= 1 + \tfrac{1}{2} \log_c r - \tfrac{3}{2} \log_c b
\end{aligned}
$$

Express $\ln (3x^2 + 5x - 2) - \ln (x + 2)$ as a single logarithm.

$$
\begin{aligned}
\ln (3x^2 + 5x - 2) - \ln (x + 2) &= \ln \frac{3x^2 + 5x - 2}{x + 2} \\
&= \ln \frac{(3x - 1)(x + 2)}{x + 2} \\
&= \ln (3x - 1)
\end{aligned}
$$

Given $\log_a 7 \approx 0.8451$ and $\log_a 5 \approx 0.6990$, find $\log_a \frac{1}{7}$ and $\log_a 35$.

$$\log_a \frac{1}{7} = \log_a 1 - \log_a 7 \approx 0 - 0.8451 \approx -0.8451;$$

$$\log_a 35 = \log_a (7 \cdot 5) = \log_a 7 + \log_a 5$$
$$\approx 0.8451 + 0.6990$$
$$\approx 1.5441$$

For any base a and any real number x,

$$\log_a a^x = x.$$

For any base a and any positive real number x,

$$a^{\log_a x} = x.$$

Simplify each of the following.

$$8^{\log_8 k} = k; \qquad \log 10^{43} = 43;$$
$$\log_a a^4 = 4; \qquad e^{\ln 2} = 2$$

SECTION 9.6: SOLVING EXPONENTIAL EQUATIONS AND LOGARITHMIC EQUATIONS

The Base–Exponent Property

For any $a > 0, a \neq 1$,

$$a^x = a^y \longleftrightarrow x = y.$$

Solve: $3^{2x-3} = 81$.

$$3^{2x-3} = 3^4 \qquad \textcolor{red}{81 = 3^4}$$
$$2x - 3 = 4$$
$$2x = 7$$
$$x = \tfrac{7}{2}$$

The solution is $\frac{7}{2}$.

The Property of Logarithmic Equality

For any $M > 0, N > 0, a > 0$, and $a \neq 1$,

$$\log_a M = \log_a N \longleftrightarrow M = N.$$

Solve: $6^{x-2} = 2^{-3x}$.

$$\log 6^{x-2} = \log 2^{-3x}$$
$$(x - 2)\log 6 = -3x \log 2$$
$$x \log 6 - 2 \log 6 = -3x \log 2$$
$$x \log 6 + 3x \log 2 = 2 \log 6$$
$$x(\log 6 + 3 \log 2) = 2 \log 6$$
$$x = \frac{2 \log 6}{\log 6 + 3 \log 2}$$
$$x \approx 0.9257$$

The solution is about 0.9257.

Solve: $\log_3 (x - 2) + \log_3 x = 1$.

$$\log_3 [x(x - 2)] = 1$$
$$x(x - 2) = 3^1$$
$$x^2 - 2x - 3 = 0$$
$$(x - 3)(x + 1) = 0$$
$$x - 3 = 0 \quad or \quad x + 1 = 0$$
$$x = 3 \quad or \qquad x = -1$$

The number -1 is not a solution because negative numbers do not have real-number logarithms. The value 3 checks and is the solution.

Solve: $\ln(x+10) - \ln(x+4) = \ln x$.

$$\ln\frac{x+10}{x+4} = \ln x$$

$$\frac{x+10}{x+4} = x$$

$$x+10 = x(x+4)$$

$$x+10 = x^2 + 4x$$

$$0 = x^2 + 3x - 10$$

$$0 = (x+5)(x-2)$$

$$x+5 = 0 \quad \text{or} \quad x-2 = 0$$

$$x = -5 \quad \text{or} \quad x = 2$$

The number -5 is not a solution because $-5 + 4 = -1$ and $\ln(-1)$ is not a real number. The value 2 checks and is the solution.

SECTION 9.7: APPLICATIONS AND MODELS: GROWTH AND DECAY; COMPOUND INTEREST

Exponential Growth Model

$$P(t) = P_0 e^{kt}, \quad k > 0$$

Doubling Time

$$kT = \ln 2, \quad \text{or} \quad k = \frac{\ln 2}{T},$$

$$\text{or} \quad T = \frac{\ln 2}{k}$$

In July 2013, the population of the United States was 316.7 million, and the exponential growth rate was 0.9% per year (*Source*: CIA World Factbook 2014). After how long will the population be double what it was in 2013? Estimate the population in 2020.

With a population growth rate of 0.9%, or 0.009, the doubling time T is

$$T = \frac{\ln 2}{k} = \frac{\ln 2}{0.009} \approx 77.$$

The population of the United States will be double what it was in 2013 about 77 years after 2013.

The exponential growth function is

$$P(t) = 316.7 e^{0.009t},$$

where t is the number of years after 2013 and $P(t)$ is in millions. Since in 2020, $t = 7$, we substitute 7 for t:

$$P(7) = 316.7 e^{0.009 \cdot 7} = 316.7 e^{0.063} \approx 337.3.$$

The population will be about 337.3 million, or 337,300,000, in 2020.

Interest Compounded Continuously

$$P(t) = P_0 e^{kt}, \quad k > 0$$

Suppose that $20,000 is invested at interest rate k, compounded continuously, and grows to $23,236.68 in 3 years. What is the interest rate? What will the balance be in 8 years?

The exponential growth function is of the form $P(t) = 20,000 e^{kt}$. Given that $P(3) = \$23,236.68$, substituting 3 for t and 23,236.68 for $P(t)$ gives

$$23,236.68 = 20,000 e^{k(3)}$$

to get $k \approx 0.05$, or 5%.

We then substitute 0.05 for k and 8 for t and determine $P(8)$:

$$P(8) = 20,000 e^{0.05(8)} = 20,000 e^{0.4} \approx \$29,836.49.$$

Exponential Decay Model

$$P(t) = P_0 e^{-kt}, \quad k > 0$$

Half-Life

$$kT = \ln 2, \quad \text{or} \quad k = \frac{\ln 2}{T},$$

$$\text{or} \quad T = \frac{\ln 2}{k}$$

Archaeologists discovered an animal bone that had lost 65.2% of its carbon-14 at the time it was found. How old was the bone?

The decay rate for carbon-14 is 0.012%, or 0.00012. If the bone has lost 65.2% of its carbon-14 from an initial amount P_0, then 34.8% P_0 is the amount present. We substitute 34.8% P_0 for $P(t)$ and solve:

$$34.8\% \, P_0 = P_0 e^{-0.00012t}$$

$$0.348 = e^{-0.00012t}$$

$$\ln 0.348 = -0.00012t$$

$$\frac{\ln 0.348}{-0.00012} = t$$

$$8796 \approx t.$$

The bone was about 8796 years old when it was found.

REVIEW EXERCISES

Determine whether the statement is true or false.

1. In general, for functions f and g, the domain of $f \circ g$ = the domain of $g \circ f$. [9.1]

2. The domain of a one-to-one function f is the range of the inverse f^{-1}. [9.2]

3. The x-intercept of $f(x) = \log x$ is $(0, 1)$. [9.4]

4. The graph of f^{-1} is a reflection of the graph of f across $y = 0$. [9.2]

5. If it is not possible for a horizontal line to intersect the graph of a function more than once, then the function is one-to-one and its inverse is a function. [9.2]

6. The range of all exponential functions is $[0, \infty)$. [9.3]

7. The horizontal asymptote of $y = 2^x$ is $y = 0$. [9.3]

Given that $f(x) = 2x - 1$, $g(x) = x^2 + 4$, and $h(x) = 3 - x^3$, find each of the following. [9.1]

8. $(f \circ g)(1)$

9. $(g \circ f)(1)$

10. $(h \circ f)(-2)$

11. $(g \circ h)(3)$

12. $(f \circ h)(-1)$

13. $(h \circ g)(2)$

14. $(f \circ f)(x)$

15. $(h \circ h)(x)$

For each pair of functions in Exercises 16 and 17:

a) *Find $(f \circ g)(x)$ and $(g \circ f)(x)$.* [9.1]

b) *Find the domain of $f \circ g$ and the domain of $g \circ f$.* [9.1]

16. $f(x) = \frac{4}{x^2}$, $g(x) = 3 - 2x$

17. $f(x) = 3x^2 + 4x$, $g(x) = 2x - 1$

Find $f(x)$ and $g(x)$ such that $h(x) = (f \circ g)(x)$. [9.1]

18. $h(x) = \sqrt{5x + 2}$

19. $h(x) = 4(5x - 1)^2 + 9$

20. Find the inverse of the relation
$$\{(1.3, -2.7), (8, -3), (-5, 3), (6, -3), (7, -5)\}.$$
[9.2]

21. Find an equation of the inverse relation. [9.2]

a) $y = -2x + 3$

b) $y = 3x^2 + 2x - 1$

c) $0.8x^3 - 5.4y^2 = 3x$

Graph the function and determine whether the function is one-to-one using the horizontal-line test. [9.2]

22. $f(x) = -|x| + 3$

23. $f(x) = x^2 + 1$

24. $f(x) = 2x - \frac{3}{4}$

25. $f(x) = -\frac{6}{x + 1}$

In Exercises 26–31, given the function:

a) *Sketch the graph and determine whether the function is one-to-one.* [9.2], [9.4]

b) *If it is one-to-one, find a formula for the inverse.* [9.2], [9.4]

26. $f(x) = 2 - 3x$

27. $f(x) = \frac{x + 2}{x - 1}$

28. $f(x) = \sqrt{x - 6}$

29. $f(x) = x^3 - 8$

30. $f(x) = 3x^2 + 2x - 1$

31. $f(x) = e^x$

For the function f, use composition of functions to show that f^{-1} *is as given.* **[9.2]**

32. $f(x) = 6x - 5$, $f^{-1}(x) = \dfrac{x+5}{6}$

33. $f(x) = \dfrac{x+1}{x}$, $f^{-1}(x) = \dfrac{1}{x-1}$

Find the inverse of the given one-to-one function f. Give the domain and the range of f and of f^{-1} *and then graph both f and* f^{-1} *on the same set of axes.* **[9.2]**

34. $f(x) = 2 - 5x$

35. $f(x) = \dfrac{x-3}{x+2}$

36. Find $f(f^{-1}(657))$:

$$f(x) = \dfrac{4x^5 - 16x^{37}}{119x}, \quad x > 1. \ \textbf{[9.2]}$$

37. Find $f(f^{-1}(a))$: $f(x) = \sqrt[3]{3x - 4}$. **[9.2]**

Graph the function.

38. $f(x) = \left(\frac{1}{3}\right)^x$ **[9.3]**

39. $f(x) = 1 + e^x$ **[9.3]**

40. $f(x) = -e^{-x}$ **[9.3]**

41. $f(x) = \log_2 x$ **[9.4]**

42. $f(x) = \frac{1}{2}\ln x$ **[9.3]**

43. $f(x) = \log x - 2$ **[9.4]**

In Exercises 44–49, match the equation with one of the figures (a)–(f) that follow.

a)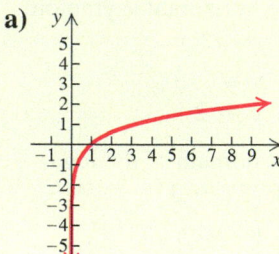

b)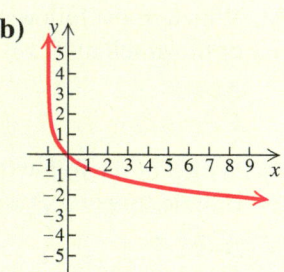

c)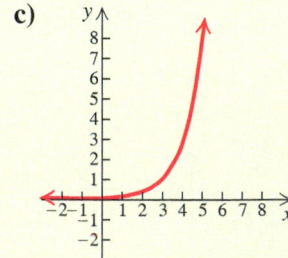

d)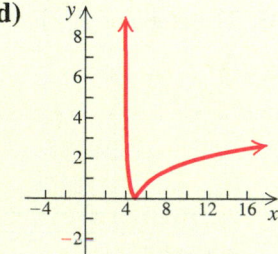

e)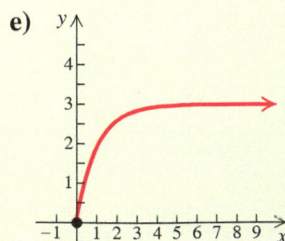

f)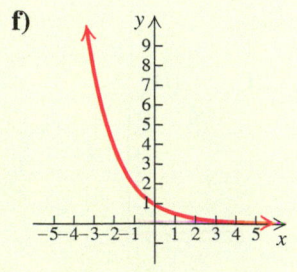

44. $f(x) = e^{x-3}$ **[9.3]**

45. $f(x) = \log_3 x$ **[9.4]**

46. $y = -\log_3(x + 1)$ **[9.4]**

47. $y = \left(\frac{1}{2}\right)^x$ **[9.3]**

48. $f(x) = 3(1 - e^{-x})$, $x \geq 0$ **[9.3]**

49. $f(x) = |\ln(x - 4)|$ **[9.4]**

Find each of the following. Do not use a calculator. **[9.4]**

50. $\log_5 125$

51. $\log 100{,}000$

52. $\ln e$

53. $\ln 1$

54. $\log 10^{1/4}$

55. $\log_3 \sqrt{3}$

56. $\log 1$

57. $\log 10$

58. $\log_2 \sqrt[3]{2}$

59. $\log 0.01$

Convert to an exponential equation. **[9.4]**

60. $\log_4 x = 2$

61. $\log_a Q = k$

Convert to a logarithmic equation. **[9.4]**

62. $4^{-3} = \frac{1}{64}$

63. $e^x = 80$

Find each of the following using a calculator. Round to four decimal places. **[9.4]**

64. $\log 11$

65. $\log 0.234$

66. $\ln 3$

67. $\ln 0.027$

68. $\log(-3)$

69. $\ln 0$

Find the logarithm using the change-of-base formula. **[9.4]**

70. $\log_5 24$

71. $\log_8 3$

Express as a single logarithm and, if possible, simplify. **[9.5]**

72. $3 \log_b x - 4 \log_b y + \frac{1}{2} \log_b z$

73. $\ln(x^3 - 8) - \ln(x^2 + 2x + 4) + \ln(x + 2)$

Express in terms of sums and differences of logarithms. **[9.5]**

74. $\ln \sqrt[4]{wr^2}$

75. $\log \sqrt[3]{\dfrac{M^2}{N}}$

Given that $\log_a 2 \approx 0.301$, $\log_a 5 \approx 0.699$, *and* $\log_a 6 = 0.778$, *find each of the following.* **[9.5]**

76. $\log_a 3$

77. $\log_a 50$

78. $\log_a \frac{1}{5}$

79. $\log_a \sqrt[3]{5}$

Simplify. **[9.5]**

80. $\ln e^{-5k}$

81. $\log_5 5^{-6t}$

Solve. **[9.6]**

82. $\log_4 x = 2$

83. $3^{1-x} = 9^{2x}$

84. $e^x = 80$

85. $4^{2x-1} - 3 = 61$

86. $\log_{16} 4 = x$

87. $\log_x 125 = 3$

88. $\log_2 x + \log_2(x - 2) = 3$

89. $\log(x^2 - 1) - \log(x - 1) = 1$

90. $\log x^2 = \log x$ **91.** $e^{-x} = 0.02$

92. *Saving for College.*   Following the birth of triplets, the grandparents deposit $30,000 in a college trust fund that earns 4.2% interest, compounded quarterly.

   a) Find a function for the amount in the account after t years. **[9.3]**

   b) Find the amount in the account at $t = 0, 6, 12,$ and 18 years. **[9.3]**

93. *Wind Power Capacity.*   Global wind power capacity is increasing exponentially. The total capacity, in gigawatts (GW), can be estimated with the exponential function

$$W(t) = 29.9(1.26)^t,$$

where t is the number of years after 2002 (*Source*: REN21). Find the global wind power capacity in 2005 and in 2010. Then use this function to estimate the capacity in 2016. **[9.3]**

94. How long will it take an investment to double if it is invested at 4.5%, compounded continuously? **[9.7]**

95. The population of a metropolitan area consisting of 8 counties doubled in 26 years. What was the exponential growth rate? **[9.7]**

96. How old is a skeleton that has lost 27% of its carbon-14? **[9.7]**

97. The hydrogen ion concentration of milk is 2.3×10^{-6}. What is the pH? (See Exercise 98 in Exercise Set 9.4.) **[9.4]**

98. *Earthquake Magnitude.*   The earthquake in Kashgar, China, on February 25, 2003, had an intensity of $10^{6.3} \cdot I_0$ (*Source*: U.S. Geological Survey). What is the magnitude on the Richter scale? **[9.4]**

99. What is the loudness, in decibels, of a sound whose intensity is $1000I_0$? (See Exercise 101 in Exercise Set 9.4.) **[9.4]**

100. *Walking Speed.*   The average walking speed w, in feet per second, of a person living in a city of population P, in thousands, is given by the function

$$w(P) = 0.37 \ln P + 0.05.$$

   a) The population of Wichita, Kansas, is 353,823. Find the average walking speed. **[9.4]**

   b) A city's population has an average walking speed of 3.4 ft/sec. Find the population. **[9.7]**

101. *Social Security Distributions.*   Cash Social Security distributions were $35 million, or $0.035 billion, in 1940. This amount has increased exponentially to $786 billion in 2012. (*Source*: Pew Research Center) Assuming that the exponential growth model applies:

   a) Find the exponential growth rate k. **[9.7]**

   b) Find the exponential growth function. **[9.7]**

   c) Estimate the total cash distributions in 1970, in 2000, and in 2015. **[9.7]**

   d) In what year will the cash benefits reach $2 trillion? **[9.7]**

102. *The Population of Cambodia.*   The population of Cambodia was 15.2 million in 2013, and the exponential growth rate was 1.67% per year (*Source*: U.S. Census Bureau, World Population Profile).

   a) Find the exponential growth function. **[9.7]**

   b) What will the population be in 2017? in 2020? **[9.7]**

   c) When will the population be 18 million? **[9.7]**

   d) What is the doubling time? **[9.7]**

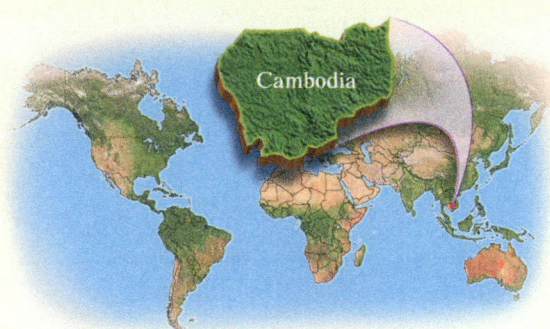

103. Which of the following is the horizontal asymptote of the graph of $f(x) = e^{x-3} + 2$? **[9.3]**

   A. $y = -2$ **B.** $y = -3$

   C. $y = 3$ **D.** $y = 2$

104. Which of the following is the domain of the logarithmic function $f(x) = \log(2x - 3)$? **[9.4]**

   A. $\left(\frac{3}{2}, \infty\right)$ **B.** $\left(-\infty, \frac{3}{2}\right)$

   C. $(3, \infty)$ **D.** $(-\infty, \infty)$

105. The graph of $f(x) = 2^{x-2}$ is which of the following? **[9.3]**

A.

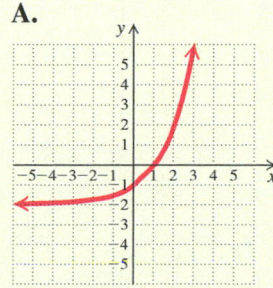

B.

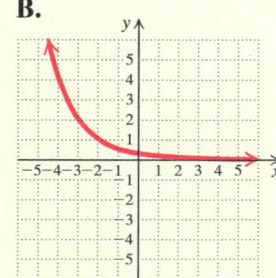

C.

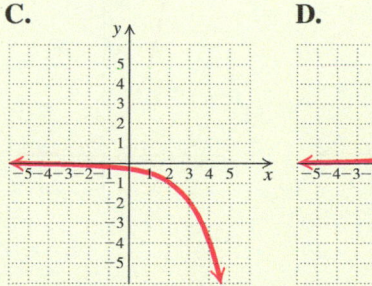

D.

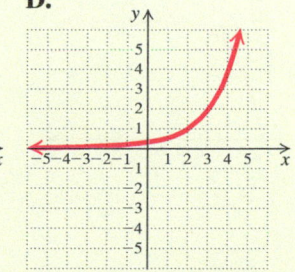

106. The graph of $f(x) = \log_2 x$ is which of the following? [9.4]

A.

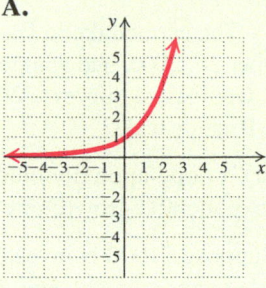

B.

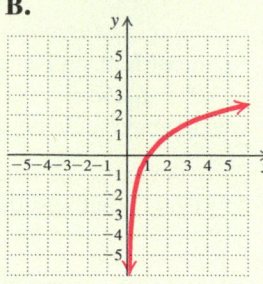

C.

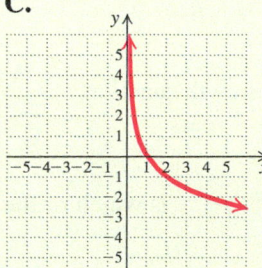

D.

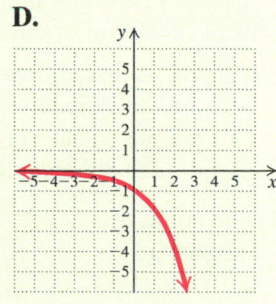

▶ **Synthesis**

Solve. [9.6]

107. $|\log_4 x| = 3$

108. $\log x = \ln x$

109. $5^{\sqrt{x}} = 625$

▶ **Collaborative Discussion and Writing**

110. *Atmospheric Pressure.* Atmospheric pressure P at an altitude a is given by

$$P = P_0 e^{-0.00005a},$$

where P_0 is the pressure at sea level, approximately 14.7 lb/in^2 (pounds per square inch). Explain how a barometer, or some device for measuring atmospheric pressure, can be used to find the height of a skyscraper. [9.7]

111. Explain how the graph of $f(x) = \ln x$ can be used to obtain the graph of $g(x) = e^{x-2}$. [9.4]

112. Describe the difference between $f^{-1}(x)$ and $[f(x)]^{-1}$. [9.2]

113. Explain the errors, if any, in the following:

$$\log_a ab^3 = (\log_a a)(\log_a b^3) = 3\log_a b.$$

[9.5]

---

9 **Chapter Test**

Given that $f(x) = x^2 - 1$, $g(x) = 4x + 3$, and $h(x) = 3x^2 + 2x + 4$, find each of the following.

1. $(g \circ h)(2)$ **2.** $(f \circ g)(-1)$

3. $(h \circ f)(1)$ **4.** $(g \circ g)(x)$

For $f(x) = \sqrt{x - 5}$ and $g(x) = x^2 + 1$:

5. Find $(f \circ g)(x)$ and $(g \circ f)(x)$.

6. Find the domain of $(f \circ g)(x)$ and the domain of $(g \circ f)(x)$.

7. Find $f(x)$ and $g(x)$ such that
$$h(x) = (f \circ g)(x) = (2x - 7)^4.$$

8. Find the inverse of the relation
$$\{(-2, 5), (4, 3), (0, -1), (-6, -3)\}.$$

Determine whether the function is one-to-one. Answer yes or no.

9.

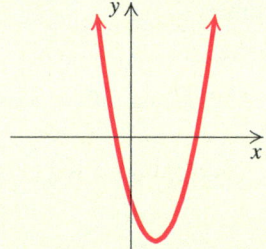

10.

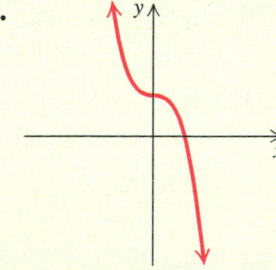

In Exercises 11–14, given the function:

a) *Sketch the graph and determine whether the function is one-to-one.*

b) *If it is one-to-one, find a formula for the inverse.*

11. $f(x) = x^3 + 1$ **12.** $f(x) = 1 - x$

13. $f(x) = \dfrac{x}{2 - x}$ **14.** $f(x) = x^2 + x - 3$

15. Use composition of functions to show that f^{-1} is as given:

$$f(x) = -4x + 3, \qquad f^{-1}(x) = \frac{3 - x}{4}.$$

16. Find the inverse of the one-to-one function

$$f(x) = \frac{1}{x - 4}.$$

Give the domain and the range of f and of f^{-1} and then graph both f and f^{-1} on the same set of axes.

Graph the function.

17. $f(x) = 4^{-x}$

18. $f(x) = \log x$

19. $f(x) = e^x - 3$

20. $f(x) = \ln(x + 2)$

Find each of the following. Do not use a calculator.

21. $\log 0.00001$

22. $\ln e$

23. $\ln 1$

24. $\log_4 \sqrt[5]{4}$

25. Convert to an exponential equation: $\ln x = 4$.

26. Convert to a logarithmic equation: $3^x = 5.4$.

Find each of the following using a calculator. Round to four decimal places.

27. $\ln 16$

28. $\log 0.293$

29. Find $\log_6 10$ using the change-of-base formula.

30. Express as a single logarithm:

$$2 \log_a x - \log_a y + \tfrac{1}{2} \log_a z.$$

31. Express $\ln \sqrt[5]{x^2 y}$ in terms of sums and differences of logarithms.

32. Given that $\log_a 2 \approx 0.328$ and $\log_a 8 \approx 0.984$, find $\log_a 4$.

33. Simplify: $\ln e^{-4t}$.

Solve.

34. $\log_{25} 5 = x$

35. $\log_3 x + \log_3 (x + 8) = 2$

36. $3^{4-x} = 27^x$

37. $e^x = 65$

38. *Earthquake Magnitude.*   The earthquake in Bam, in southeast Iran, on December 26, 2003, had an intensity of $10^{6.6} \cdot I_0$ (*Source*: U.S. Geological Survey). What was its magnitude on the Richter scale?

39. *Growth Rate.*   A country's population doubled in 45 years. What was the exponential growth rate?

40. *Compound Interest.*   Suppose \$1000 is invested at interest rate k, compounded continuously, and grows to \$1144.54 in 3 years.

   a) Find the interest rate.
   b) Find the exponential growth function.
   c) Find the balance after 8 years.
   d) Find the doubling time.

41. The graph of $f(x) = 2^{x-1} + 1$ is which of the following?

A.
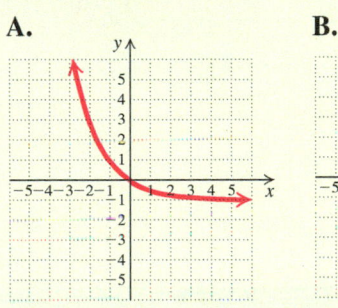

B.
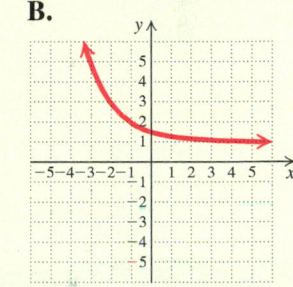

C.
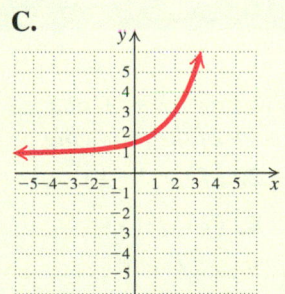

D.
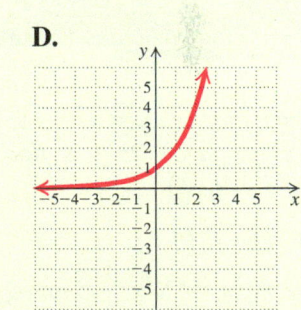

▶ **Synthesis**

42. Solve: $4^{\sqrt[3]{x}} = 8$.

Matrices

APPLICATION   This problem appears as Exercise 39 in Exercise Set 10.3.

At the 2014 Paralympic Games in Sochi, Russia, the top three countries—the Russian Federation, Ukraine, and the United States—won a total of 123 medals. Ukraine won 7 more medals than the United States. The Russian Federation won 37 more medals than the total amount won by Ukraine and the United States. (*Source*: International Paralympic Committee) How many medals did each of the top three countries win?

10.1 Matrices and Systems of Equations

▶ Solve systems of equations using matrices.

▶ Matrices and Row-Equivalent Operations

In this section, we consider additional techniques for solving systems of equations. You have probably observed that when we solve a system of equations, we perform computations with the coefficients and the constants and continually rewrite the variables. We can streamline the solution process by omitting the variables until a solution is found. For example, the system

$$2x - 3y = 7,$$
$$x + 4y = -2$$

can be written more simply as

$$\begin{bmatrix} 2 & -3 & | & 7 \\ 1 & 4 & | & -2 \end{bmatrix}.$$

The vertical line replaces the equals signs.

A rectangular array of numbers like the one above is called a **matrix** (pl., **matrices**). The matrix above is called an **augmented matrix** for the given system of equations, because it contains not only the coefficients but also the constant terms. The matrix

$$\begin{bmatrix} 2 & -3 \\ 1 & 4 \end{bmatrix}$$

is called the **coefficient matrix** of the system.

The **rows** of a matrix are horizontal, and the **columns** are vertical. The augmented matrix above has 2 rows and 3 columns, and the coefficient matrix has 2 rows and 2 columns. A matrix with m rows and n columns is said to be of **order** $m \times n$. Thus the order of the augmented matrix above is 2×3, and the order of the coefficient matrix is 2×2. When $m = n$, a matrix is said to be **square**. The coefficient matrix above is a square matrix. The numbers 2 and 4 lie on the **main diagonal** of the coefficient matrix. The numbers in a matrix are called **entries**, or **elements**.

▶ Gaussian Elimination with Matrices

One method for solving a system of equations in three variables is to transform the original system to an equivalent system (one with the same solution set) of the form

$$Ax + By + Cz = D,$$
$$Ey + Fz = G,$$
$$Hz = K.$$

Then we solve the third equation for z, **back-substitute** that value for z in the second equation and solve for y, and finally back-substitute for y and z in the first equation and solve for x. This method is called **Gaussian elimination**, named for the German mathematician Karl Friedrich Gauss (1777–1855).

We can also use this method when a system of equations is written as an augmented matrix. We use a series of *row-equivalent operations* to produce a *row-equivalent matrix* that corresponds to the system of equations in the form shown above. We can then reinsert the variables and complete the solution.

Although illustrated here for a system of three equations in three variables, Gaussian elimination can be used to solve any system of linear equations.

ROW-EQUIVALENT OPERATIONS

1. Interchange any two rows.
2. Multiply each entry in a row by the same nonzero constant.
3. Add a nonzero multiple of one row to another row.

EXAMPLE 1 Solve the following system:

$$2x - y + 4z = -3,$$
$$x - 2y - 10z = -6,$$
$$3x \quad\quad + 4z = 7.$$

Solution First, we write the augmented matrix, writing 0 for the missing y-term in the last equation:

$$\left[\begin{array}{ccc|c} 2 & -1 & 4 & -3 \\ 1 & -2 & -10 & -6 \\ 3 & 0 & 4 & 7 \end{array}\right].$$

Our goal is to find a row-equivalent matrix of the form

$$\left[\begin{array}{ccc|c} 1 & a & b & c \\ 0 & 1 & d & e \\ 0 & 0 & 1 & f \end{array}\right].$$

The variables can then be reinserted to form equations from which we can complete the solution. This is done by working from the bottom equation to the top and using back-substitution.

The first step is to multiply and/or interchange rows so that each number in the first column below the first number is a multiple of that number. In this case, we interchange the first and second rows to obtain a 1 in the upper left-hand corner.

$$\left[\begin{array}{ccc|c} 1 & -2 & -10 & -6 \\ 2 & -1 & 4 & -3 \\ 3 & 0 & 4 & 7 \end{array}\right] \quad \begin{array}{l} \color{red}{\textbf{New row 1} = \textbf{row 2}} \\ \color{red}{\textbf{New row 2} = \textbf{row 1}} \\ \end{array}$$

Next, we multiply the first row by -2 and add it to the second row. We also multiply the first row by -3 and add it to the third row.

$$\left[\begin{array}{ccc|c} 1 & -2 & -10 & -6 \\ 0 & 3 & 24 & 9 \\ 0 & 6 & 34 & 25 \end{array}\right] \quad \begin{array}{l} \color{red}{\textbf{Row 1 is unchanged.}} \\ \color{red}{\textbf{New row 2} = -2(\textbf{row 1}) + \textbf{row 2}} \\ \color{red}{\textbf{New row 3} = -3(\textbf{row 1}) + \textbf{row 3}} \end{array}$$

Now we multiply the second row by $\frac{1}{3}$ to get a 1 in the second row, second column.

$$\left[\begin{array}{ccc|c} 1 & -2 & -10 & -6 \\ 0 & 1 & 8 & 3 \\ 0 & 6 & 34 & 25 \end{array}\right] \quad \color{red}{\textbf{New row 2} = \tfrac{1}{3}(\textbf{row 2})}$$

Then we multiply the second row by -6 and add it to the third row.

$$\left[\begin{array}{ccc|c} 1 & -2 & -10 & -6 \\ 0 & 1 & 8 & 3 \\ 0 & 0 & -14 & 7 \end{array}\right] \quad \color{red}{\textbf{New row 3} = -6(\textbf{row 2}) + \textbf{row 3}}$$

Row-equivalent operations can be performed on a graphing calculator. For example, to interchange the first and second rows of the augmented matrix, as we did in the first step in Example 1, we enter the matrix as matrix **A** and select "rowSwap" from the MATRIX MATH menu. Some graphing calculators will not automatically store the matrix produced using a row-equivalent operation, so when several operations are to be performed in succession, it is helpful to store the result of each operation as it is produced. In the window below, we see both the matrix produced by the rowSwap operation and the indication that this matrix is stored as matrix **B**.

```
rowSwap([A],1,2)→[B]
[[1  −2  −10  −6]
 [2  −1   4   −3]
 [3   0   4    7]]
```

Finally, we multiply the third row by $-\frac{1}{14}$ to get a 1 in the third row, third column.

$$\left[\begin{array}{ccc|c} 1 & -2 & -10 & -6 \\ 0 & 1 & 8 & 3 \\ 0 & 0 & 1 & -\frac{1}{2} \end{array}\right] \quad \textcolor{red}{\textbf{New row 3} = -\frac{1}{14}(\textbf{row 3})}$$

Now we can write the system of equations that corresponds to the last matrix above:

$$\begin{aligned} x - 2y - 10z &= -6, &\quad \textbf{(1)} \\ y + 8z &= 3, &\quad \textbf{(2)} \\ z &= -\tfrac{1}{2}. &\quad \textbf{(3)} \end{aligned}$$

We back-substitute $-\frac{1}{2}$ for z in equation (2) and solve for y:

$$\begin{aligned} y + 8\left(-\tfrac{1}{2}\right) &= 3 \\ y - 4 &= 3 \\ y &= 7. \end{aligned}$$

Next, we back-substitute 7 for y and $-\frac{1}{2}$ for z in equation (1) and solve for x:

$$\begin{aligned} x - 2 \cdot 7 - 10\left(-\tfrac{1}{2}\right) &= -6 \\ x - 14 + 5 &= -6 \\ x - 9 &= -6 \\ x &= 3. \end{aligned}$$

The triple $\left(3, 7, -\frac{1}{2}\right)$ checks in the original system of equations, so it is the solution.

> **Now Try Exercise 27.**

The procedure followed in Example 1 is called **Gaussian elimination with matrices**. The last matrix in Example 1 is in **row-echelon form**. To be in this form, a matrix must have the following properties.

ROW-ECHELON FORM

1. If a row does not consist entirely of 0's, then the first nonzero element in the row is a 1 (called a **leading 1**).
2. For any two successive nonzero rows, the leading 1 in the lower row is farther to the right than the leading 1 in the higher row.
3. All the rows consisting entirely of 0's are at the bottom of the matrix.

If a fourth property is also satisfied, a matrix is said to be in **reduced row-echelon form**:

4. Each column that contains a leading 1 has 0's everywhere else.

EXAMPLE 2 Which of the following matrices are in row-echelon form? Which, if any, are in reduced row-echelon form?

a) $\left[\begin{array}{ccc|c} 1 & -3 & 5 & -2 \\ 0 & 1 & -4 & 3 \\ 0 & 0 & 1 & 10 \end{array}\right]$
b) $\left[\begin{array}{cc|c} 0 & -1 & 2 \\ 0 & 1 & 5 \end{array}\right]$
c) $\left[\begin{array}{cccc|c} 1 & -2 & -6 & 4 & 7 \\ 0 & 3 & 5 & -8 & -1 \\ 0 & 0 & 1 & 9 & 2 \end{array}\right]$

d) $\left[\begin{array}{ccc|c} 1 & 0 & 0 & -2.4 \\ 0 & 1 & 0 & 0.8 \\ 0 & 0 & 1 & 5.6 \end{array}\right]$
e) $\left[\begin{array}{cccc|c} 1 & 0 & 0 & 0 & \frac{2}{3} \\ 0 & 1 & 0 & 0 & -\frac{1}{4} \\ 0 & 0 & 1 & 0 & \frac{6}{7} \\ 0 & 0 & 0 & 0 & 0 \end{array}\right]$
f) $\left[\begin{array}{ccc|c} 1 & -4 & 2 & 5 \\ 0 & 0 & 0 & 0 \\ 0 & 1 & -3 & -8 \end{array}\right]$

Solution   The matrices in (a), (d), and (e) satisfy the row-echelon criteria and, thus, are in row-echelon form. In (b) and (c), the first nonzero elements of the first and second rows, respectively, are not 1. In (f), the row consisting entirely of 0's is not at the bottom of the matrix. Thus the matrices in (b), (c), and (f) are not in row-echelon form. In (d) and (e), not only are the row-echelon criteria met but each column that contains a leading 1 also has 0's elsewhere, so these matrices are in reduced row-echelon form.

▶ Gauss–Jordan Elimination

We have seen that with Gaussian elimination we perform row-equivalent operations on a matrix to obtain a row-equivalent matrix in row-echelon form. When we continue to apply these operations until we have a matrix in *reduced* row-echelon form, we are using **Gauss–Jordan elimination**. This method is named for Karl Friedrich Gauss and Wilhelm Jordan (1842–1899).

EXAMPLE 3   Use Gauss–Jordan elimination to solve the system of equations in Example 1.

Solution   Using Gaussian elimination in Example 1, we obtained the matrix

$$\begin{bmatrix} 1 & -2 & -10 & | & -6 \\ 0 & 1 & 8 & | & 3 \\ 0 & 0 & 1 & | & -\frac{1}{2} \end{bmatrix}.$$

We continue to perform row-equivalent operations until we have a matrix in reduced row-echelon form. We multiply the third row by 10 and add it to the first row. We also multiply the third row by -8 and add it to the second row.

$$\begin{bmatrix} 1 & -2 & 0 & | & -11 \\ 0 & 1 & 0 & | & 7 \\ 0 & 0 & 1 & | & -\frac{1}{2} \end{bmatrix}$$ **New row 1 = 10(row 3) + row 1**
New row 2 = −8(row 3) + row 2

Next, we multiply the second row by 2 and add it to the first row.

$$\begin{bmatrix} 1 & 0 & 0 & | & 3 \\ 0 & 1 & 0 & | & 7 \\ 0 & 0 & 1 & | & -\frac{1}{2} \end{bmatrix}$$ **New row 1 = 2(row 2) + row 1**

Writing the system of equations that corresponds to this matrix, we have

$$x \qquad\quad = 3,$$
$$\quad y \qquad = 7,$$
$$\qquad z = -\tfrac{1}{2}.$$

We can actually read the solution, $\left(3, 7, -\frac{1}{2}\right)$, directly from the last column of the reduced row-echelon matrix.

Now Try Exercise 27.

EXAMPLE 4   Solve the following system:

$$3x - 4y - z = 6,$$
$$2x - y + z = -1,$$
$$4x - 7y - 3z = 13.$$

Solution   First, we write the augmented matrix and use Gauss–Jordan elimination.

$$\begin{bmatrix} 3 & -4 & -1 & | & 6 \\ 2 & -1 & 1 & | & -1 \\ 4 & -7 & -3 & | & 13 \end{bmatrix}$$

Technology Connection

After an augmented matrix is entered in a graphing calculator, reduced row-echelon form can be found directly using the "rref" operation from the MATRIX MATH menu.

```
rref([A])▶Frac
  [[1 0 0 3    ]
   [0 1 0 7    ]
   [0 0 1 −1/2]]
```

The application PolySmlt from the APPS menu can also be used to solve a system of equations.

We then multiply the second and third rows by 3 so that each number in the first column below the first number, 3, is a multiple of that number.

$$\begin{bmatrix} 3 & -4 & -1 & | & 6 \\ 6 & -3 & 3 & | & -3 \\ 12 & -21 & -9 & | & 39 \end{bmatrix}$$ New row 2 = 3(row 2)
New row 3 = 3(row 3)

Next, we multiply the first row by -2 and add it to the second row. We also multiply the first row by -4 and add it to the third row.

$$\begin{bmatrix} 3 & -4 & -1 & | & 6 \\ 0 & 5 & 5 & | & -15 \\ 0 & -5 & -5 & | & 15 \end{bmatrix}$$ New row 2 = -2(row 1) + row 2
New row 3 = -4(row 1) + row 3

Now we add the second row to the third row.

$$\begin{bmatrix} 3 & -4 & -1 & | & 6 \\ 0 & 5 & 5 & | & -15 \\ 0 & 0 & 0 & | & 0 \end{bmatrix}$$ New row 3 = row 2 + row 3

We can stop at this stage because we have a row consisting entirely of 0's. The last row of the matrix corresponds to the equation $0 = 0$, which is true for all values of x, y, and z. Therefore, the equations are dependent and the system is equivalent to

$$3x - 4y - z = 6,$$
$$5y + 5z = -15.$$

This particular system has infinitely many solutions. (A system containing dependent equations could be inconsistent.)

Solving the second equation for y gives us

$$y = -z - 3.$$

Substituting $-z - 3$ for y in the first equation and solving for x, we get

$$3x - 4(-z - 3) - z = 6$$
$$3x + 4z + 12 - z = 6$$
$$3x + 3z + 12 = 6$$
$$3x = -3z - 6$$
$$x = -z - 2.$$

Then the solutions of this system are of the form

$$(-z - 2, -z - 3, z),$$

where z can be any real number.   ▶ **Now Try Exercise 33.**

Similarly, if we obtain a row whose only nonzero entry occurs in the last column, we have an inconsistent system of equations. For example, in the matrix

$$\begin{bmatrix} 1 & 0 & 3 & | & -2 \\ 0 & 1 & 5 & | & 4 \\ 0 & 0 & 0 & | & 6 \end{bmatrix},$$

the last row corresponds to the false equation $0 = 6$, so we know the original system of equations has no solution.

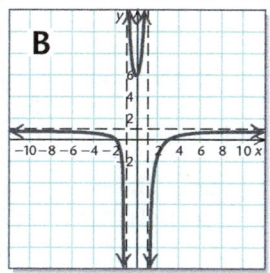

Visualizing the Graph

Match the equation or system of equations with its graph.

1. $2x - 3y = 6$

2. $f(x) = x^2 - 2x - 3$

3. $f(x) = -x^2 + 4$

4. $(x - 2)^2 + (y + 3)^2 = 9$

5. $f(x) = x^3 - 2$

6. $f(x) = -(x - 1)^2(x + 1)^2$

7. $f(x) = \dfrac{x - 1}{x^2 - 4}$

8. $f(x) = \dfrac{x^2 - x - 6}{x^2 - 1}$

9. $x - y = -1,$
$2x - y = 2$

10. $3x - y = 3,$
$2y = 6x - 6$

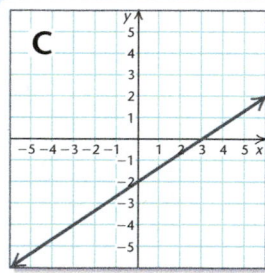

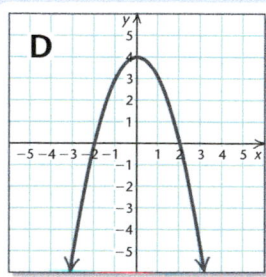

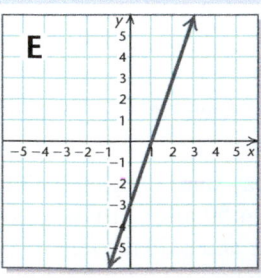

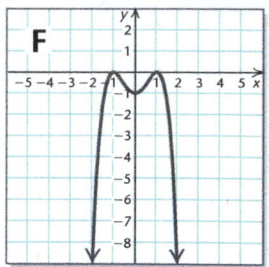

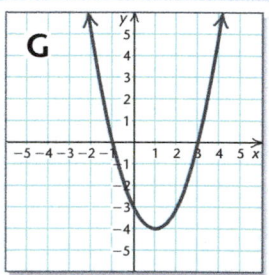

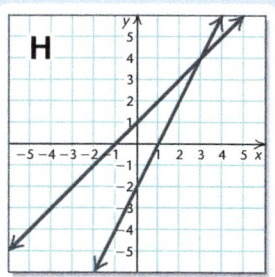

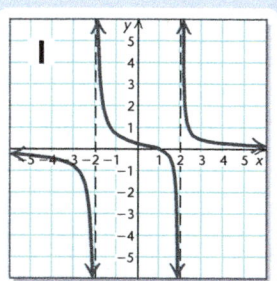

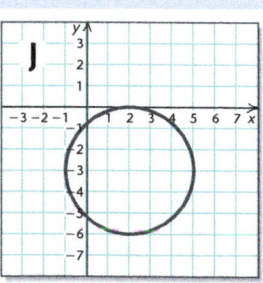

10.1 Exercise Set

Determine the order of the matrix.

1. $\begin{bmatrix} 1 & -6 \\ -3 & 2 \\ 0 & 5 \end{bmatrix}$

2. $\begin{bmatrix} 7 \\ -5 \\ -1 \\ 3 \end{bmatrix}$

3. $\begin{bmatrix} 2 & -4 & 0 & 9 \end{bmatrix}$

4. $\begin{bmatrix} -8 \end{bmatrix}$

5. $\begin{bmatrix} 1 & -5 & -8 \\ 6 & 4 & -2 \\ -3 & 0 & 7 \end{bmatrix}$

6. $\begin{bmatrix} 13 & 2 & -6 & 4 \\ -1 & 18 & 5 & -12 \end{bmatrix}$

Write the augmented matrix for the system of equations.

7. $2x - y = 7,$
   $x + 4y = -5$

8. $3x + 2y = 8,$
   $2x - 3y = 15$

9. $x - 2y + 3z = 12,$
   $2x \quad\quad - 4z = 8,$
   $\quad 3y + z = 7$

10. $x + y - z = 7,$
    $\quad\quad 3y + 2z = 1,$
    $-2x - 5y \quad = 6$

Write the system of equations that corresponds to the augmented matrix.

11. $\left[\begin{array}{cc|c} 3 & -5 & 1 \\ 1 & 4 & -2 \end{array}\right]$

12. $\left[\begin{array}{cc|c} 1 & 2 & -6 \\ 4 & 1 & -3 \end{array}\right]$

13. $\left[\begin{array}{ccc|c} 2 & 1 & -4 & 12 \\ 3 & 0 & 5 & -1 \\ 1 & -1 & 1 & 2 \end{array}\right]$

14. $\left[\begin{array}{ccc|c} -1 & -2 & 3 & 6 \\ 0 & 4 & 1 & 2 \\ 2 & -1 & 0 & 9 \end{array}\right]$

Solve the system of equations using Gaussian elimination or Gauss–Jordan elimination.

15. $4x + 2y = 11,$
    $3x - y = 2$

16. $2x + y = 1,$
    $3x + 2y = -2$

17. $5x - 2y = -3,$
    $2x + 5y = -24$

18. $2x + y = 1,$
    $3x - 6y = 4$

19. $3x + 4y = 7,$
    $-5x + 2y = 10$

20. $5x - 3y = -2,$
    $4x + 2y = 5$

21. $3x + 2y = 6,$
    $2x - 3y = -9$

22. $x - 4y = 9,$
    $2x + 5y = 5$

23. $x - 3y = 8,$
    $-2x + 6y = 3$

24. $4x - 8y = 12,$
    $-x + 2y = -3$

25. $-2x + 6y = 4,$
    $3x - 9y = -6$

26. $6x + 2y = -10,$
    $-3x - y = 6$

27. $x + 2y - 3z = 9,$
    $2x - y + 2z = -8,$
    $3x - y - 4z = 3$

28. $x - y + 2z = 0,$
    $x - 2y + 3z = -1,$
    $2x - 2y + z = -3$

29. $4x - y - 3z = 1,$
    $8x + y - z = 5,$
    $2x + y + 2z = 5$

30. $3x + 2y + 2z = 3,$
    $x + 2y - z = 5,$
    $2x - 4y + z = 0$

31. $x - 2y + 3z = -4,$
    $3x + y - z = 0,$
    $2x + 3y - 5z = 1$

32. $2x - 3y + 2z = 2,$
    $x + 4y - z = 9,$
    $-3x + y - 5z = 5$

33. $2x - 4y - 3z = 3,$
    $x + 3y + z = -1,$
    $5x + y - 2z = 2$

34. $x + y - 3z = 4,$
    $4x + 5y + z = 1,$
    $2x + 3y + 7z = -7$

35. $p + q + r = 1,$
    $p + 2q + 3r = 4,$
    $4p + 5q + 6r = 7$

36. $m + n + t = 9,$
    $m - n - t = -15,$
    $3m + n + t = 2$

37. $a + b - c = 7,$
    $a - b + c = 5,$
    $3a + b - c = -1$

38. $a - b + c = 3,$
    $2a + b - 3c = 5,$
    $4a + b - c = 11$

39. $-2w + 2x + 2y - 2z = -10,$
    $w + x + y + z = -5,$
    $3w + x - y + 4z = -2,$
    $w + 3x - 2y + 2z = -6$

40. $-w + 2x - 3y + z = -8,$
    $-w + x + y - z = -4,$
    $w + x + y + z = 22,$
    $-w + x - y - z = -14$

Use Gaussian elimination or Gauss–Jordan elimination in Exercises 41–44.

41. *Borrowing.* Greenfield Manufacturing borrowed $30,000 to buy a new piece of equipment. Part of the money was borrowed at 8%, part at 10%, and part at 12%. The annual interest was $3040, and the total amount borrowed at 8% and at 10% was twice the amount borrowed at 12%. How much was borrowed at each rate?

42. *Time of Return.* The Patels pay their babysitter $11 per hour before 11 P.M. and $14.50 after 11 P.M. One evening, they went out for 6 hr and paid the sitter $73. What time did they return?

43. *Stamp Purchase.* For her business, Olivia spent $86.80 on both 49¢ and 21¢ stamps. She bought a total of 200 stamps. How many of each type did she buy?

44. *Advertising Expense.* eAuction.com spent a total of $11 million on advertising in fiscal years 2010, 2011, and 2012. The amount spent in 2012 was three times the amount spent in 2010. The amount spent in 2011 was $3 million less than the amount spent in 2012. How much was spent on advertising each year?

▶ Skill Maintenance

In each of Exercises 45–52, fill in the blank with the correct term. Some of the given choices will not be used.

Descartes' rule of
  signs
the leading-term test
the intermediate value
  theorem
the fundamental
  theorem of algebra
polynomial function
rational function
quadratic formula

constant function
horizontal asymptote
vertical asymptote
oblique asymptote
direct variation
inverse variation
horizontal line
vertical line
parallel
perpendicular

45. Two lines with slopes m_1 and m_2 are _____ if and only if the product of their slopes is -1. **[2.6]**

46. We can use _____ to determine the behavior of the graph of a polynomial function as $x \longrightarrow \infty$ or as $x \longrightarrow -\infty$. **[8.1]**

47. If it is possible for a(n) _____ to cross a graph more than once, then the graph is not the graph of a function. **[2.2]**

48. The graph of $y = b$, or $f(x) = b$, is a _____. **[2.6]**

49. A(n) _____ is a function that is a quotient of two polynomials. **[8.5]**

50. If a situation gives rise to a function $f(x) = k/x$, or $y = k/x$, where k is a positive constant, we say that we have _____. **[5.8]**

51. A(n) _____ of a rational function $p(x)/q(x)$, where $p(x)$ and $q(x)$ have no common factors other than constants, occurs at an x-value that makes the denominator 0. **[8.5]**

52. When the numerator and the denominator of a rational function have the same degree, the graph of the function has a(n) _____. **[8.5]**

▶ Synthesis

In Exercises 53 and 54, three solutions of the equation $y = ax^2 + bx + c$ are given. Use a system of three equations in three variables and Gaussian elimination or Gauss–Jordan elimination to find the constants a, b, and c and write the equation.

53. $(-3, 12)$, $(-1, -7)$, and $(1, -2)$

54. $(-1, 0)$, $(1, -3)$, and $(3, -22)$

55. Find two different row-echelon forms of
$$\begin{bmatrix} 1 & 5 \\ 3 & 2 \end{bmatrix}.$$

56. Consider the system of equations
$$\begin{aligned} x - y + 3z &= -8, \\ 2x + 3y - z &= 5, \\ 3x + 2y + 2kz &= -3k. \end{aligned}$$
For what value(s) of k, if any, will the system have:

a) no solution?
b) exactly one solution?
c) infinitely many solutions?

Solve using matrices.

57. $y = x + z,$
   $3y + 5z = 4,$
   $x + 4 = y + 3z$

58. $x + y = 2z,$
   $2x - 5z = 4,$
   $x - z = y + 8$

59. $\quad x - 4y + 2z = 7,$
   $3x + y + 3z = -5$

60. $\quad x - y - 3z = 3,$
   $-x + 3y + z = -7$

61. $\quad 4x + 5y = 3,$
   $-2x + y = 9,$
   $3x - 2y = -15$

62. $2x - 3y = -1,$
   $-x + 2y = -2,$
   $3x - 5y = 1$

10.2   Matrix Operations

▶ Add, subtract, and multiply matrices when possible.

▶ Write a matrix equation equivalent to a system of equations.

In addition to solving systems of equations, matrices are useful in many other types of applications. In this section, we study matrices and some of their properties.

A capital letter is generally used to name a matrix, and lower-case letters with double subscripts generally denote its entries. For example, a_{47}, read "*a* sub four seven," indicates the entry in the fourth row and the seventh column. A general term is represented by a_{ij}. The notation a_{ij} indicates the entry in row i and column j. In general, we can write a matrix as

$$\mathbf{A} = [a_{ij}] = \begin{bmatrix} a_{11} & a_{12} & a_{13} & \cdots & a_{1n} \\ a_{21} & a_{22} & a_{23} & \cdots & a_{2n} \\ a_{31} & a_{32} & a_{33} & \cdots & a_{3n} \\ \vdots & \vdots & \vdots & & \vdots \\ a_{m1} & a_{m2} & a_{m3} & \cdots & a_{mn} \end{bmatrix}.$$

The matrix above has m rows and n columns; that is, its order is $m \times n$.

Two matrices are **equal** if they have the same order and corresponding entries are equal.

▶ Matrix Addition and Subtraction

To add or subtract matrices, we add or subtract their corresponding entries. The matrices must have the same order for this to be possible.

ADDITION AND SUBTRACTION OF MATRICES

Given two $m \times n$ matrices $\mathbf{A} = [a_{ij}]$ and $\mathbf{B} = [b_{ij}]$, their sum is

$$\mathbf{A} + \mathbf{B} = [a_{ij} + b_{ij}]$$

and their difference is

$$\mathbf{A} - \mathbf{B} = [a_{ij} - b_{ij}].$$

Addition of matrices is both commutative and associative.

EXAMPLE 1   Find $\mathbf{A} + \mathbf{B}$, if possible, for each of the following.

a) $\mathbf{A} = \begin{bmatrix} -5 & 0 \\ 4 & \frac{1}{2} \end{bmatrix}$, $\mathbf{B} = \begin{bmatrix} 6 & -3 \\ 2 & 3 \end{bmatrix}$

b) $\mathbf{A} = \begin{bmatrix} 1 & 3 \\ -1 & 5 \\ 6 & 0 \end{bmatrix}$, $\mathbf{B} = \begin{bmatrix} -1 & -2 \\ 1 & -2 \\ -3 & 1 \end{bmatrix}$

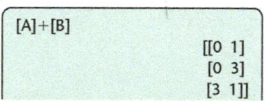
Solution We have a pair of 2 × 2 matrices in part (a) and a pair of 3 × 2 ma-trices in part (b). Since each pair of matrices has the same order, we can add the corresponding entries.

a) $\mathbf{A} + \mathbf{B} = \begin{bmatrix} -5 & 0 \\ 4 & \frac{1}{2} \end{bmatrix} + \begin{bmatrix} 6 & -3 \\ 2 & 3 \end{bmatrix}$

$= \begin{bmatrix} -5 + 6 & 0 + (-3) \\ 4 + 2 & \frac{1}{2} + 3 \end{bmatrix} = \begin{bmatrix} 1 & -3 \\ 6 & 3\frac{1}{2} \end{bmatrix}$

b) $\mathbf{A} + \mathbf{B} = \begin{bmatrix} 1 & 3 \\ -1 & 5 \\ 6 & 0 \end{bmatrix} + \begin{bmatrix} -1 & -2 \\ 1 & -2 \\ -3 & 1 \end{bmatrix}$

$= \begin{bmatrix} 1 + (-1) & 3 + (-2) \\ -1 + 1 & 5 + (-2) \\ 6 + (-3) & 0 + 1 \end{bmatrix} = \begin{bmatrix} 0 & 1 \\ 0 & 3 \\ 3 & 1 \end{bmatrix}$

> **Now Try Exercise 5.**

EXAMPLE 2 Find $\mathbf{C} - \mathbf{D}$, if possible, for each of the following.

a) $\mathbf{C} = \begin{bmatrix} 1 & 2 \\ -2 & 0 \\ -3 & -1 \end{bmatrix}, \quad \mathbf{D} = \begin{bmatrix} 1 & -1 \\ 1 & 3 \\ 2 & 3 \end{bmatrix}$

b) $\mathbf{C} = \begin{bmatrix} 5 & -6 \\ -3 & 4 \end{bmatrix}, \quad \mathbf{D} = \begin{bmatrix} -4 \\ 1 \end{bmatrix}$

Solution

a) Since the order of each matrix is 3 × 2, we can subtract corresponding entries:

$\mathbf{C} - \mathbf{D} = \begin{bmatrix} 1 & 2 \\ -2 & 0 \\ -3 & -1 \end{bmatrix} - \begin{bmatrix} 1 & -1 \\ 1 & 3 \\ 2 & 3 \end{bmatrix}$

$= \begin{bmatrix} 1 - 1 & 2 - (-1) \\ -2 - 1 & 0 - 3 \\ -3 - 2 & -1 - 3 \end{bmatrix} = \begin{bmatrix} 0 & 3 \\ -3 & -3 \\ -5 & -4 \end{bmatrix}.$

b) **C** is a 2 × 2 matrix and **D** is a 2 × 1 matrix. Since the matrices do not have the same order, we cannot subtract.

> **Now Try Exercise 13.**

The **opposite**, or **additive inverse**, of a matrix is obtained by replacing each entry with its opposite.

EXAMPLE 3 Find $-\mathbf{A}$ and $\mathbf{A} + (-\mathbf{A})$ for

$\mathbf{A} = \begin{bmatrix} 1 & 0 & 2 \\ 3 & -1 & 5 \end{bmatrix}.$

Solution   To find $-\mathbf{A}$, we replace each entry of $\mathbf{A}$ with its opposite.

$$-\mathbf{A} = \begin{bmatrix} -1 & 0 & -2 \\ -3 & 1 & -5 \end{bmatrix},$$

$$\mathbf{A} + (-\mathbf{A}) = \begin{bmatrix} 1 & 0 & 2 \\ 3 & -1 & 5 \end{bmatrix} + \begin{bmatrix} -1 & 0 & -2 \\ -3 & 1 & -5 \end{bmatrix}$$

$$= \begin{bmatrix} 0 & 0 & 0 \\ 0 & 0 & 0 \end{bmatrix}$$

A matrix having 0's for all its entries is called a **zero matrix**. When a zero matrix is added to a second matrix of the same order, the second matrix is unchanged. Thus a zero matrix is an **additive identity**. For example,

$$\begin{bmatrix} 2 & 3 & -4 \\ 0 & 6 & 5 \end{bmatrix} + \begin{bmatrix} 0 & 0 & 0 \\ 0 & 0 & 0 \end{bmatrix} = \begin{bmatrix} 2 & 3 & -4 \\ 0 & 6 & 5 \end{bmatrix}.$$

The matrix

$$\begin{bmatrix} 0 & 0 & 0 \\ 0 & 0 & 0 \end{bmatrix}$$

is the additive identity for any 2×3 matrix.

▶ Scalar Multiplication

When we find the product of a number and a matrix, we obtain a **scalar product**.

> **SCALAR PRODUCT**
>
> The **scalar product** of a number k and a matrix $\mathbf{A}$ is the matrix denoted $k\mathbf{A}$, obtained by multiplying each entry of $\mathbf{A}$ by the number k. The number k is called a **scalar**.

Technology Connection

Scalar products, like those in Example 4, can be found using a graphing calculator.

```
3[A]
            [[-9   0]
             [ 12  15]]
(-1)[A]
            [[3    0 ]
             [-4   -5]]
```

EXAMPLE 4   Find $3\mathbf{A}$ and $(-1)\mathbf{A}$ for

$$\mathbf{A} = \begin{bmatrix} -3 & 0 \\ 4 & 5 \end{bmatrix}.$$

Solution   We have

$$3\mathbf{A} = 3\begin{bmatrix} -3 & 0 \\ 4 & 5 \end{bmatrix} = \begin{bmatrix} 3(-3) & 3 \cdot 0 \\ 3 \cdot 4 & 3 \cdot 5 \end{bmatrix} = \begin{bmatrix} -9 & 0 \\ 12 & 15 \end{bmatrix},$$

$$(-1)\mathbf{A} = -1\begin{bmatrix} -3 & 0 \\ 4 & 5 \end{bmatrix} = \begin{bmatrix} -1(-3) & -1 \cdot 0 \\ -1 \cdot 4 & -1 \cdot 5 \end{bmatrix} = \begin{bmatrix} 3 & 0 \\ -4 & -5 \end{bmatrix}.$$

Now Try Exercise 9.

The properties of matrix addition and scalar multiplication are similar to the properties of addition and multiplication of real numbers.

**PROPERTIES OF MATRIX ADDITION
AND SCALAR MULTIPLICATION**

For any $m \times n$ matrices **A**, **B**, and **C** and any scalars k and l:

$\mathbf{A} + \mathbf{B} = \mathbf{B} + \mathbf{A}.$ **Commutative property of addition**

$\mathbf{A} + (\mathbf{B} + \mathbf{C}) = (\mathbf{A} + \mathbf{B}) + \mathbf{C}.$ **Associative property of addition**

$(kl)\mathbf{A} = k(l\mathbf{A}).$ **Associative property of scalar multiplication**

$k(\mathbf{A} + \mathbf{B}) = k\mathbf{A} + k\mathbf{B}.$ **Distributive property**

$(k + l)\mathbf{A} = k\mathbf{A} + l\mathbf{A}.$ **Distributive property**

There exists a unique matrix **0** such that:

$\mathbf{A} + \mathbf{0} = \mathbf{0} + \mathbf{A} = \mathbf{A}.$ **Additive identity property**

There exists a unique matrix $-\mathbf{A}$ such that:

$\mathbf{A} + (-\mathbf{A}) = -\mathbf{A} + \mathbf{A} = \mathbf{0}.$ **Additive inverse property**

EXAMPLE 5 *Production.* Waterworks, Inc., manufactures three types of kayaks in its two plants. The following table lists the number of each style produced at each plant in April.

	Whitewater Kayak	Ocean Kayak	Crossover Kayak
Madison Plant	150	120	100
Greensburg Plant	180	90	130

a) Write a 2×3 matrix **A** that represents the information in the table.

b) The manufacturer increased production by 20% in May. Find a matrix **M** that represents the increased production figures.

c) Find the matrix $\mathbf{A} + \mathbf{M}$ and tell what it represents.

Solution

a) Write the entries in the table in a 2×3 matrix **A**.

$$\mathbf{A} = \begin{bmatrix} 150 & 120 & 100 \\ 180 & 90 & 130 \end{bmatrix}$$

b) The production in May will be represented by $\mathbf{A} + 20\%\mathbf{A}$, or $\mathbf{A} + 0.2\mathbf{A}$, or $1.2\mathbf{A}$. Thus,

$$\mathbf{M} = (1.2)\begin{bmatrix} 150 & 120 & 100 \\ 180 & 90 & 130 \end{bmatrix} = \begin{bmatrix} 180 & 144 & 120 \\ 216 & 108 & 156 \end{bmatrix}.$$

c) $\mathbf{A} + \mathbf{M} = \begin{bmatrix} 150 & 120 & 100 \\ 180 & 90 & 130 \end{bmatrix} + \begin{bmatrix} 180 & 144 & 120 \\ 216 & 108 & 156 \end{bmatrix}$

$= \begin{bmatrix} 330 & 264 & 220 \\ 396 & 198 & 286 \end{bmatrix}$

The matrix $\mathbf{A} + \mathbf{M}$ represents the total production of each of the three types of kayaks at each plant in April and May.   **Now Try Exercise 29.**

▶ Products of Matrices

Matrix multiplication is defined in such a way that it can be used in solving systems of equations and in many applications.

MATRIX MULTIPLICATION

For an $m \times n$ matrix $\mathbf{A} = [a_{ij}]$ and an $n \times p$ matrix $\mathbf{B} = [b_{ij}]$, the **product $\mathbf{AB} = [c_{ij}]$** is an $m \times p$ matrix, where

$$c_{ij} = a_{i1} \cdot b_{1j} + a_{i2} \cdot b_{2j} + a_{i3} \cdot b_{3j} + \cdots + a_{in} \cdot b_{nj}.$$

In other words, the entry c_{ij} in $\mathbf{AB}$ is obtained by multiplying the entries in row i of $\mathbf{A}$ by the corresponding entries in column j of $\mathbf{B}$ and adding the results.

> Note that we can multiply two matrices only when the number of columns in the first matrix is equal to the number of rows in the second matrix.

EXAMPLE 6   For

$$\mathbf{A} = \begin{bmatrix} 3 & 1 & -1 \\ 2 & 0 & 3 \end{bmatrix}, \quad \mathbf{B} = \begin{bmatrix} 1 & 6 \\ 3 & -5 \\ -2 & 4 \end{bmatrix}, \quad \text{and} \quad \mathbf{C} = \begin{bmatrix} 4 & -6 \\ 1 & 2 \end{bmatrix},$$

find each of the following, if possible.

a) $\mathbf{AB}$ b) $\mathbf{BA}$

c) $\mathbf{BC}$ d) $\mathbf{AC}$

Solution

a) $\mathbf{A}$ is a 2×3 matrix and $\mathbf{B}$ is a 3×2 matrix, so $\mathbf{AB}$ will be a 2×2 matrix.

$$\mathbf{AB} = \begin{bmatrix} 3 & 1 & -1 \\ 2 & 0 & 3 \end{bmatrix} \begin{bmatrix} 1 & 6 \\ 3 & -5 \\ -2 & 4 \end{bmatrix}$$

$$= \begin{bmatrix} 3 \cdot 1 + 1 \cdot 3 + (-1)(-2) & 3 \cdot 6 + 1(-5) + (-1)(4) \\ 2 \cdot 1 + 0 \cdot 3 + 3(-2) & 2 \cdot 6 + 0(-5) + 3 \cdot 4 \end{bmatrix}$$

$$= \begin{bmatrix} 8 & 9 \\ -4 & 24 \end{bmatrix}$$

b) $\mathbf{B}$ is a 3×2 matrix and $\mathbf{A}$ is a 2×3 matrix, so $\mathbf{BA}$ will be a 3×3 matrix.

$$\mathbf{BA} = \begin{bmatrix} 1 & 6 \\ 3 & -5 \\ -2 & 4 \end{bmatrix} \begin{bmatrix} 3 & 1 & -1 \\ 2 & 0 & 3 \end{bmatrix}$$

$$= \begin{bmatrix} 1 \cdot 3 + 6 \cdot 2 & 1 \cdot 1 + 6 \cdot 0 & 1(-1) + 6 \cdot 3 \\ 3 \cdot 3 + (-5)(2) & 3 \cdot 1 + (-5)(0) & 3(-1) + (-5)(3) \\ -2 \cdot 3 + 4 \cdot 2 & -2 \cdot 1 + 4 \cdot 0 & -2(-1) + 4 \cdot 3 \end{bmatrix} = \begin{bmatrix} 15 & 1 & 17 \\ -1 & 3 & -18 \\ 2 & -2 & 14 \end{bmatrix}$$

> Note in parts (a) and (b) that $\mathbf{AB} \neq \mathbf{BA}$. Multiplication of matrices is generally not commutative.

Matrix multiplication can be performed on a graphing calculator. The products in Examples 6(a) and 6(b) are shown below.

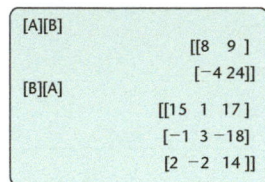

When the product **AC** in Example 6(d) is entered on a graphing calculator, an ERROR message is returned, indicating that the dimensions of the matrices are mismatched.

c) **B** is a 3×2 matrix and **C** is a 2×2 matrix, so **BC** will be a 3×2 matrix.

$$\mathbf{BC} = \begin{bmatrix} 1 & 6 \\ 3 & -5 \\ -2 & 4 \end{bmatrix} \begin{bmatrix} 4 & -6 \\ 1 & 2 \end{bmatrix}$$

$$= \begin{bmatrix} 1 \cdot 4 + 6 \cdot 1 & 1(-6) + 6 \cdot 2 \\ 3 \cdot 4 + (-5)(1) & 3(-6) + (-5)(2) \\ -2 \cdot 4 + 4 \cdot 1 & -2(-6) + 4 \cdot 2 \end{bmatrix} = \begin{bmatrix} 10 & 6 \\ 7 & -28 \\ -4 & 20 \end{bmatrix}$$

d) The product **AC** is not defined because the number of columns of **A**, 3, is not equal to the number of rows of **C**, 2. ⟶ **Now Try Exercises 23 and 25.**

EXAMPLE 7 *Bakery Profit.* Two of the items sold at Sweet Treats Bakery are gluten-free bagels and gluten-free doughnuts. The following table lists the number of dozens of each product that are sold at the bakery's three stores one week.

	Main Street Store	Avon Road Store	Dalton Avenue Store
Bagels (in dozens)	25	30	20
Doughnuts (in dozens)	40	35	15

The bakery's profit on one dozen bagels is $5, and its profit on one dozen doughnuts is $6. Use matrices to find the total profit on these items at each store for the given week.

Solution We can write the table showing the sales of the products as a 2×3 matrix:

$$\mathbf{S} = \begin{bmatrix} 25 & 30 & 20 \\ 40 & 35 & 15 \end{bmatrix}.$$

The profit per dozen for each product can also be written as a matrix:

$$\mathbf{P} = \begin{bmatrix} 5 & 6 \end{bmatrix}.$$

Then the total profit at each store is given by the matrix product **PS**:

$$\mathbf{PS} = \begin{bmatrix} 5 & 6 \end{bmatrix} \begin{bmatrix} 25 & 30 & 20 \\ 40 & 35 & 15 \end{bmatrix}$$

$$= \begin{bmatrix} 5 \cdot 25 + 6 \cdot 40 & 5 \cdot 30 + 6 \cdot 35 & 5 \cdot 20 + 6 \cdot 15 \end{bmatrix}$$

$$= \begin{bmatrix} 365 & 360 & 190 \end{bmatrix}.$$

The total profit on the sale of gluten-free bagels and gluten-free doughnuts for the given week was $365 at the Main Street store, $360 at the Avon Road store, and $190 at the Dalton Avenue store. ⟶ **Now Try Exercise 33.**

A matrix that consists of a single row, like **P** in Example 7, is called a **row matrix**. Similarly, a matrix that consists of a single column, like

$$\begin{bmatrix} 8 \\ -3 \\ 5 \end{bmatrix},$$

is called a **column matrix**.

We have already seen that matrix multiplication is generally not commutative. Nevertheless, matrix multiplication does have some properties that are similar to those for multiplication of real numbers.

> **PROPERTIES OF MATRIX MULTIPLICATION**
>
> For matrices **A**, **B**, and **C**, assuming that the indicated operations are possible:
>
> $\mathbf{A}(\mathbf{BC}) = (\mathbf{AB})\mathbf{C}.$ Associative property of multiplication
>
> $\mathbf{A}(\mathbf{B} + \mathbf{C}) = \mathbf{AB} + \mathbf{AC}.$ Distributive property
>
> $(\mathbf{B} + \mathbf{C})\mathbf{A} = \mathbf{BA} + \mathbf{CA}.$ Distributive property

▶ **Matrix Equations**

We can write a matrix equation equivalent to a system of equations.

EXAMPLE 8 Write a matrix equation equivalent to the following system of equations:

$$\begin{aligned} 4x + 2y - z &= 3, \\ 9x \quad\quad + z &= 5, \\ 4x + 5y - 2z &= 1. \end{aligned}$$

Solution We write the coefficients on the left in a matrix. We then write the product of that matrix and the column matrix containing the variables and set the result equal to the column matrix containing the constants on the right:

$$\begin{bmatrix} 4 & 2 & -1 \\ 9 & 0 & 1 \\ 4 & 5 & -2 \end{bmatrix} \begin{bmatrix} x \\ y \\ z \end{bmatrix} = \begin{bmatrix} 3 \\ 5 \\ 1 \end{bmatrix}.$$

If we let

$$\mathbf{A} = \begin{bmatrix} 4 & 2 & -1 \\ 9 & 0 & 1 \\ 4 & 5 & -2 \end{bmatrix}, \quad \mathbf{X} = \begin{bmatrix} x \\ y \\ z \end{bmatrix}, \quad \text{and} \quad \mathbf{B} = \begin{bmatrix} 3 \\ 5 \\ 1 \end{bmatrix},$$

we can write this matrix equation as $\mathbf{AX} = \mathbf{B}.$ ——▶ **Now Try Exercise 39.**

10.2 Exercise Set

Find x and y.

1. $\begin{bmatrix} 5 & x \end{bmatrix} = \begin{bmatrix} y & -3 \end{bmatrix}$

2. $\begin{bmatrix} 6x \\ 25 \end{bmatrix} = \begin{bmatrix} -9 \\ 5y \end{bmatrix}$

3. $\begin{bmatrix} 3 & 2x \\ y & -8 \end{bmatrix} = \begin{bmatrix} 3 & -2 \\ 1 & -8 \end{bmatrix}$

4. $\begin{bmatrix} x - 1 & 4 \\ y + 3 & -7 \end{bmatrix} = \begin{bmatrix} 0 & 4 \\ -2 & -7 \end{bmatrix}$

For Exercises 5–20, let

$$A = \begin{bmatrix} 1 & 2 \\ 4 & 3 \end{bmatrix}, \qquad B = \begin{bmatrix} -3 & 5 \\ 2 & -1 \end{bmatrix},$$

$$C = \begin{bmatrix} 1 & -1 \\ -1 & 1 \end{bmatrix}, \qquad D = \begin{bmatrix} 1 & 1 \\ 1 & 1 \end{bmatrix},$$

$$E = \begin{bmatrix} 1 & 3 \\ 2 & 6 \end{bmatrix}, \qquad F = \begin{bmatrix} 3 & 3 \\ -1 & -1 \end{bmatrix},$$

$$0 = \begin{bmatrix} 0 & 0 \\ 0 & 0 \end{bmatrix}, \qquad I = \begin{bmatrix} 1 & 0 \\ 0 & 1 \end{bmatrix}.$$

Find each of the following.

5. $A + B$ **6.** $B + A$ **7.** $E + 0$

8. $2A$ **9.** $3F$ **10.** $(-1)D$

11. $3F + 2A$ **12.** $A - B$

13. $B - A$ **14.** AB

15. BA **16.** $0F$

17. CD **18.** EF

19. AI **20.** IA

Find the product, if possible.

21. $\begin{bmatrix} -1 & 0 & 7 \\ 3 & -5 & 2 \end{bmatrix} \begin{bmatrix} 6 \\ -4 \\ 1 \end{bmatrix}$

22. $\begin{bmatrix} 6 & -1 & 2 \end{bmatrix} \begin{bmatrix} 1 & 4 \\ -2 & 0 \\ 5 & -3 \end{bmatrix}$

23. $\begin{bmatrix} -2 & 4 \\ 5 & 1 \\ -1 & -3 \end{bmatrix} \begin{bmatrix} 3 & -6 \\ -1 & 4 \end{bmatrix}$

24. $\begin{bmatrix} 2 & -1 & 0 \\ 0 & 5 & 4 \end{bmatrix} \begin{bmatrix} -3 & 1 & 0 \\ 0 & 2 & -1 \\ 5 & 0 & 4 \end{bmatrix}$

25. $\begin{bmatrix} 1 \\ -5 \\ 3 \end{bmatrix} \begin{bmatrix} -6 & 5 & 8 \\ 0 & 4 & -1 \end{bmatrix}$

26. $\begin{bmatrix} 2 & 0 & 0 \\ 0 & -1 & 0 \\ 0 & 0 & 3 \end{bmatrix} \begin{bmatrix} 0 & -4 & 3 \\ 2 & 1 & 0 \\ -1 & 0 & 6 \end{bmatrix}$

27. $\begin{bmatrix} 1 & -4 & 3 \\ 0 & 8 & 0 \\ -2 & -1 & 5 \end{bmatrix} \begin{bmatrix} 3 & 0 & 0 \\ 0 & -4 & 0 \\ 0 & 0 & 1 \end{bmatrix}$

28. $\begin{bmatrix} 4 \\ -5 \end{bmatrix} \begin{bmatrix} 2 & 0 \\ 6 & -7 \\ 0 & -3 \end{bmatrix}$

29. *Produce.* The produce manager at Dugan's Market orders 40 lb of tomatoes, 20 lb of zucchini, and 30 lb of onions from a local farmer one week.

  a) Write a 1×3 matrix **A** that represents the amount of each item ordered.

  b) The following week, the produce manager increases his order by 10%. Find a matrix **B** that represents this order.

  c) Find $A + B$ and tell what the entries represent.

30. *Budget.* For the month of June, Maggie budgets $320 for food, $140 for clothes, and $80 for entertainment.

  a) Write a 1×3 matrix **B** that represents the amount budgeted for each of these items.

  b) After receiving a raise, Maggie increases the amount budgeted for each item in July by 5%. Find a matrix **R** that represents the new amounts.

  c) Find $B + R$ and tell what the entries represent.

31. *Nutrition.* A 3-oz serving of roasted, skinless chicken breast contains 140 Cal, 27 g of protein, 3 g of fat, 13 mg of calcium, and 64 mg of sodium. One-half cup of potato salad contains 180 Cal, 4 g of protein, 11 g of fat, 24 mg of calcium, and 662 mg of sodium. One broccoli spear contains 50 Cal, 5 g of protein, 1 g of fat, 82 mg of calcium, and 20 mg of sodium. (*Source*: *Home and Garden Bulletin No. 72*, U.S. Government Printing Office, Washington, D.C. 20402)

  a) Write 1×5 matrices **C**, **P**, and **B** that represent the nutritional values of each food.

  b) Find $C + 2P + 3B$ and tell what the entries represent.

32. *Nutrition.* One slice of cheese pizza contains 290 Cal, 15 g of protein, 9 g of fat, and 39 g of carbohydrates. One-half cup of gelatin dessert contains 70 Cal, 2 g of protein, 0 g of fat, and 17 g of carbohydrates. One cup of whole milk contains 150 Cal, 8 g of protein, 8 g of fat, and 11 g of carbohydrates. (*Source*: *Home and Garden Bulletin No. 72*, U.S. Government Printing Office, Washington, D.C. 20402)

  a) Write 1×4 matrices **P**, **G**, and **M** that represent the nutritional values of each food.

  b) Find $3P + 2G + 2M$ and tell what the entries represent.

33. *Food Service Management.* The food service manager at a large hospital is concerned about maintaining reasonable food costs. The following table lists the cost per serving, in dollars, for items on four menus.

Menu	Meat	Potato	Vegetable	Salad	Dessert
1	1.50	0.30	0.36	0.45	0.64
2	1.55	0.28	0.48	0.57	0.75
3	1.62	0.52	0.65	0.38	0.53
4	1.70	0.43	0.40	0.42	0.68

On a particular day, a dietician orders 65 meals from menu 1, 48 from menu 2, 93 from menu 3, and 57 from menu 4.

a) Write the information in the table as a 4 × 5 matrix **M**.
b) Write a row matrix **N** that represents the number of each menu ordered.
c) Find the product **NM**.
d) State what the entries of **NM** represent.

34. *Food Service Management.* A college food service manager uses a table like the one below to list the number of units of ingredients, by weight, required for various menu items.

	White Cake	Bread	Coffee Cake	Sugar Cookies
Flour	1	2.5	0.75	0.5
Milk	0	0.5	0.25	0
Eggs	0.75	0.25	0.5	0.5
Butter	0.5	0	0.5	1

The cost per unit of each ingredient is 25 cents for flour, 34 cents for milk, 54 cents for eggs, and 83 cents for butter.

a) Write the information in the table as a 4 × 4 matrix **M**.
b) Write a row matrix **C** that represents the cost per unit of each ingredient.
c) Find the product **CM**.
d) State what the entries of **CM** represent.

35. *Production Cost.* Karin supplies two small campus coffee shops with homemade chocolate chip cookies, oatmeal cookies, and peanut butter cookies. The following table shows the number of each type of cookie, in dozens, that Karin sold in one week.

	Mugsy's Coffee Shop	The Coffee Club
Chocolate Chip	8	15
Oatmeal	6	10
Peanut Butter	4	3

Karin spends $4 for the ingredients for one dozen chocolate chip cookies, $2.50 for the ingredients for one dozen oatmeal cookies, and $3 for the ingredients for one dozen peanut butter cookies.

a) Write the information in the table as a 3 × 2 matrix **S**.
b) Write a row matrix **C** that represents the cost, per dozen, of the ingredients for each type of cookie.
c) Find the product **CS**.
d) State what the entries of **CS** represent.

36. *Profit.* A manufacturer produces exterior plywood, interior plywood, and fiberboard, which are shipped to two distributors. The following table shows the number of units of each type of product that are shipped to each warehouse.

	Distributor 1	Distributor 2
Exterior Plywood	900	500
Interior Plywood	450	1000
Fiberboard	600	700

The profits from each unit of exterior plywood, interior plywood, and fiberboard are $8, $10, and $7, respectively.

a) Write the information in the table as a 3 × 2 matrix **M**.
b) Write a row matrix **P** that represents the profit per unit of each type of product.
c) Find the product **PM**.
d) State what the entries of **PM** represent.

37. *Profit.* In Exercise 35, suppose that Karin's profits on one dozen chocolate chip, oatmeal, and peanut butter cookies are $7.50, $4.80, and $6.25, respectively.

a) Write a row matrix **P** that represents this information.

b) Use the matrices **S** and **P** to find Karin's total profit from each coffee shop.

38. *Production Cost.* In Exercise 36, suppose that the manufacturer's production costs for each unit of exterior plywood, interior plywood, and fiberboard are $20, $25, and $15, respectively.

a) Write a row matrix **C** that represents this information.

b) Use the matrices **M** and **C** to find the total production cost for the products shipped to each distributor.

Write a matrix equation equivalent to the system of equations.

39. $2x - 3y = 7,$
$\quad x + 5y = -6$

40. $-x + y = 3,$
$\quad 5x - 4y = 16$

41. $\quad x + y - 2z = 6,$
$\quad 3x - y + z = 7,$
$\quad 2x + 5y - 3z = 8$

42. $3x - y + z = 1,$
$\quad x + 2y - z = 3,$
$\quad 4x + 3y - 2z = 11$

43. $3x - 2y + 4z = 17,$
$\quad 2x + y - 5z = 13$

44. $3x + 2y + 5z = 9,$
$\quad 4x - 3y + 2z = 10$

45. $-4w + x - y + 2z = 12,$
$\quad w + 2x - y - z = 0,$
$\quad -w + x + 4y - 3z = 1,$
$\quad 2w + 3x + 5y - 7z = 9$

46. $12w + 2x + 4y - 5z = 2,$
$\quad -w + 4x - y + 12z = 5,$
$\quad 2w - x + 4y = 13,$
$\quad 2x + 10y + z = 5$

▶ Skill Maintenance

In Exercises 47–50:

a) *Find the vertex.*

b) *Find the axis of symmetry.*

c) *Determine whether there is a maximum or a minimum value and find that value.*

d) *Graph the function.* [7.5]

47. $f(x) = x^2 - x - 6$

48. $f(x) = 2x^2 - 5x - 3$

49. $f(x) = -x^2 - 3x + 2$

50. $f(x) = -3x^2 + 4x + 4$

▶ Synthesis

For Exercises 51–54, let

$$A = \begin{bmatrix} -1 & 0 \\ 2 & 1 \end{bmatrix} \quad and \quad B = \begin{bmatrix} 1 & -1 \\ 0 & 2 \end{bmatrix}.$$

51. Show that
$$(A + B)(A - B) \neq A^2 - B^2,$$
where
$$A^2 = AA \quad and \quad B^2 = BB.$$

52. Show that
$$(A + B)(A + B) \neq A^2 + 2AB + B^2.$$

53. Show that
$$(A + B)(A - B) = A^2 + BA - AB - B^2.$$

54. Show that
$$(A + B)(A + B) = A^2 + BA + AB + B^2.$$

Mid-Chapter Review

Determine whether the statement is true or false.

1. One of the properties of a matrix written in row-echelon form is that all the rows consisting entirely of 0's are at the bottom of the matrix. **[10.1]**

2. If Gaussian elimination produces a matrix with a row whose only nonzero entry occurs in the last column, the original system of equations has infinitely many solutions. **[10.1]**

3. We can multiply two matrices only when the number of columns in the first matrix is equal to the number of rows in the second matrix. **[10.2]**

4. Addition of matrices is not commutative. **[10.2]**

Solve the system of equations using Gaussian elimination or Gauss–Jordan elimination. **[10.1]**

5. $2x + y = 5,$
$3x + 2y = 6$

6. $2x - 3y = 8,$
$3x + 2y = -1$

7. $x - y = 4,$
$-3x + 3y = -5$

8. $x - 3y = 1,$
$-2x + 4y = -2$

9. $3x + 2y - 3z = -2,$
$2x + 3y + 2z = -2,$
$x + 4y + 4z = 1$

10. $x + 2y + 3z = 4,$
$x - 2y + z = 2,$
$2x - 6y + 4z = 7$

Solve using Gaussian elimination or Gauss–Jordan elimination.

11. *e-Commerce.* computerwarehouse.com charges $8 to ship orders up to 10 lb, $12 for orders from 10 lb up to 15 lb, and $15 for orders of 15 lb or more. One day, shipping charges for 150 orders totaled $1620. The number of orders under 10 lb was three times the number of orders weighing 15 lb or more. Find the number of packages shipped at each rate. **[10.1]**

For Exercises 12–19, let

$$A = \begin{bmatrix} 3 & -1 \\ 5 & 4 \end{bmatrix}, \quad B = \begin{bmatrix} -2 & 6 \\ 1 & -3 \end{bmatrix}, \quad C = \begin{bmatrix} -4 & 1 & -1 \\ 2 & 3 & -2 \end{bmatrix}, \quad and \quad D = \begin{bmatrix} -2 & 3 & 0 \\ 1 & -1 & 2 \\ -3 & 4 & 1 \end{bmatrix}.$$

Find each of the following. **[10.2]**

12. $A + B$

13. $B - A$

14. $4D$

15. $2A + 3B$

16. AB

17. BA

18. BC

19. DC

20. Write a matrix equation equivalent to the following system of equations: **[10.2]**
$$2x - y + 3z = 7,$$
$$x + 2y - z = 3,$$
$$3x - 4y + 2z = 5.$$

Collaborative Discussion and Writing

21. Explain how a matrix in reduced row-echelon form differs from a matrix in row-echelon form. **[10.1]**

22. Explain in your own words why the following augmented matrix represents a system of dependent equations. **[10.1]**

$$\begin{bmatrix} 1 & -3 & 2 & | & -5 \\ 0 & 1 & -4 & | & 8 \\ 0 & 0 & 0 & | & 0 \end{bmatrix}$$

23. Antonin says that two matrices can be multiplied only when they have the same number of rows and the same number of columns. How would you respond? **[10.2]**

24. Is it true that if $\mathbf{AB} = \mathbf{0}$, for matrices $\mathbf{A}$ and $\mathbf{B}$, then $\mathbf{A} = \mathbf{0}$ or $\mathbf{B} = \mathbf{0}$? Why or why not? **[10.2]**

10.3

Inverses of Matrices

▶ Find the inverse of a square matrix, if it exists.

▶ Use inverses of matrices to solve systems of equations.

In this section, we continue our study of matrix algebra, finding the **multiplicative inverse**, or simply **inverse**, of a square matrix, if it exists. Then we use such inverses to solve systems of equations.

▶ The Identity Matrix

Recall that, for real numbers, $a \cdot 1 = 1 \cdot a = a$; 1 is the multiplicative identity. A multiplicative identity matrix is very similar to the number 1.

IDENTITY MATRIX

For any positive integer n, the $n \times n$ **identity matrix** is an $n \times n$ matrix with 1's on the main diagonal and 0's elsewhere and is denoted by

$$\mathbf{I} = \begin{bmatrix} 1 & 0 & 0 & \cdots & 0 \\ 0 & 1 & 0 & \cdots & 0 \\ 0 & 0 & 1 & \cdots & 0 \\ \vdots & \vdots & \vdots & & \vdots \\ 0 & 0 & 0 & \cdots & 1 \end{bmatrix}$$

Then $\mathbf{AI} = \mathbf{IA} = \mathbf{A}$, for any $n \times n$ matrix $\mathbf{A}$.

EXAMPLE 1 For

$$\mathbf{A} = \begin{bmatrix} 4 & -7 \\ -3 & 2 \end{bmatrix} \quad \text{and} \quad \mathbf{I} = \begin{bmatrix} 1 & 0 \\ 0 & 1 \end{bmatrix},$$

find each of the following.

a) AI   **b) IA**

Solution

a) $\mathbf{AI} = \begin{bmatrix} 4 & -7 \\ -3 & 2 \end{bmatrix} \begin{bmatrix} 1 & 0 \\ 0 & 1 \end{bmatrix}$

$$= \begin{bmatrix} 4 \cdot 1 - 7 \cdot 0 & 4 \cdot 0 - 7 \cdot 1 \\ -3 \cdot 1 + 2 \cdot 0 & -3 \cdot 0 + 2 \cdot 1 \end{bmatrix} = \begin{bmatrix} 4 & -7 \\ -3 & 2 \end{bmatrix} = \mathbf{A}$$

b) $\mathbf{IA} = \begin{bmatrix} 1 & 0 \\ 0 & 1 \end{bmatrix} \begin{bmatrix} 4 & -7 \\ -3 & 2 \end{bmatrix}$

$$= \begin{bmatrix} 1 \cdot 4 + 0(-3) & 1(-7) + 0 \cdot 2 \\ 0 \cdot 4 + 1(-3) & 0(-7) + 1 \cdot 2 \end{bmatrix}$$

$$= \begin{bmatrix} 4 & -7 \\ -3 & 2 \end{bmatrix} = \mathbf{A}$$

▶ The Inverse of a Matrix

Recall that for every nonzero real number a, there is a multiplicative inverse $1/a$, or a^{-1}, such that $a \cdot a^{-1} = a^{-1} \cdot a = 1$. The multiplicative inverse of a matrix behaves in a similar manner.

INVERSE OF A MATRIX

For an $n \times n$ matrix $\mathbf{A}$, if there is a matrix $\mathbf{A}^{-1}$ for which $\mathbf{A}^{-1} \cdot \mathbf{A} = \mathbf{I} = \mathbf{A} \cdot \mathbf{A}^{-1}$, then $\mathbf{A}^{-1}$ is the **inverse** of $\mathbf{A}$.

We read $\mathbf{A}^{-1}$ as "$\mathbf{A}$ inverse." Note that not every matrix has an inverse.

EXAMPLE 2 Verify that

$$\mathbf{B} = \begin{bmatrix} 4 & -3 \\ 3 & -2 \end{bmatrix} \quad \text{is the inverse of} \quad \mathbf{A} = \begin{bmatrix} -2 & 3 \\ -3 & 4 \end{bmatrix}.$$

Solution   We show that $\mathbf{BA} = \mathbf{I} = \mathbf{AB}$.

$$\mathbf{BA} = \begin{bmatrix} 4 & -3 \\ 3 & -2 \end{bmatrix} \begin{bmatrix} -2 & 3 \\ -3 & 4 \end{bmatrix} = \begin{bmatrix} 1 & 0 \\ 0 & 1 \end{bmatrix} = \mathbf{I}.$$

$$\mathbf{AB} = \begin{bmatrix} -2 & 3 \\ -3 & 4 \end{bmatrix} \begin{bmatrix} 4 & -3 \\ 3 & -2 \end{bmatrix} = \begin{bmatrix} 1 & 0 \\ 0 & 1 \end{bmatrix} = \mathbf{I}.$$

Now Try Exercise 1.

We can find the inverse of a square matrix, if it exists, by using row-equivalent operations as in the Gauss–Jordan elimination method. For example, consider the matrix

$$\mathbf{A} = \begin{bmatrix} -2 & 3 \\ -3 & 4 \end{bmatrix}.$$

To find its inverse, we first form an **augmented matrix** consisting of **A** on the left side and the 2×2 identity matrix on the right side:

$$\begin{bmatrix} -2 & 3 & 1 & 0 \\ -3 & 4 & 0 & 1 \end{bmatrix}. \qquad \text{Augmented matrix}$$

The 2×2 matrix **A**   The 2×2 identity matrix

Then we attempt to transform the augmented matrix to one of the form

$$\begin{bmatrix} 1 & 0 & a & b \\ 0 & 1 & c & d \end{bmatrix}.$$

The 2×2 identity matrix   The matrix $\mathbf{A}^{-1}$

If we can do this, the matrix on the right, $\begin{bmatrix} a & b \\ c & d \end{bmatrix}$, is $\mathbf{A}^{-1}$.

EXAMPLE 3 Find $\mathbf{A}^{-1}$, where

$$\mathbf{A} = \begin{bmatrix} -2 & 3 \\ -3 & 4 \end{bmatrix}.$$

Solution First, we write the augmented matrix. Then we transform it to the desired form.

$$\begin{bmatrix} -2 & 3 & 1 & 0 \\ -3 & 4 & 0 & 1 \end{bmatrix}$$

$$\begin{bmatrix} 1 & -\frac{3}{2} & -\frac{1}{2} & 0 \\ -3 & 4 & 0 & 1 \end{bmatrix} \qquad \text{New row 1} = -\frac{1}{2}(\text{row 1})$$

$$\begin{bmatrix} 1 & -\frac{3}{2} & -\frac{1}{2} & 0 \\ 0 & -\frac{1}{2} & -\frac{3}{2} & 1 \end{bmatrix} \qquad \text{New row 2} = 3(\text{row 1}) + \text{row 2}$$

$$\begin{bmatrix} 1 & -\frac{3}{2} & -\frac{1}{2} & 0 \\ 0 & 1 & 3 & -2 \end{bmatrix} \qquad \text{New row 2} = -2(\text{row 2})$$

$$\begin{bmatrix} 1 & 0 & 4 & -3 \\ 0 & 1 & 3 & -2 \end{bmatrix} \qquad \text{New row 1} = \frac{3}{2}(\text{row 2}) + \text{row 1}$$

Thus,

$$\mathbf{A}^{-1} = \begin{bmatrix} 4 & -3 \\ 3 & -2 \end{bmatrix},$$

which we verified in Example 2.

Now Try Exercise 5.

Technology Connection

The $\boxed{x^{-1}}$ key on a graphing calculator can also be used to find the inverse of a matrix like the one in Example 3.

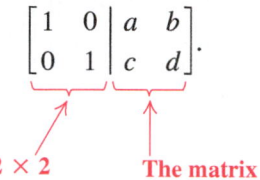

```
[A]⁻¹
          [[4 −3]
           [3 −2]]
```

EXAMPLE 4 Find $\mathbf{A}^{-1}$, where

$$\mathbf{A} = \begin{bmatrix} 1 & 2 & -1 \\ 3 & 5 & 3 \\ 2 & 4 & 3 \end{bmatrix}.$$

Solution First, we write the augmented matrix. Then we transform it to the desired form.

$$\begin{bmatrix} 1 & 2 & -1 & | & 1 & 0 & 0 \\ 3 & 5 & 3 & | & 0 & 1 & 0 \\ 2 & 4 & 3 & | & 0 & 0 & 1 \end{bmatrix}$$

$$\begin{bmatrix} 1 & 2 & -1 & | & 1 & 0 & 0 \\ 0 & -1 & 6 & | & -3 & 1 & 0 \\ 0 & 0 & 5 & | & -2 & 0 & 1 \end{bmatrix}$$ New row 2 $= -3(\text{row } 1) + \text{row } 2$
New row 3 $= -2(\text{row } 1) + \text{row } 3$

$$\begin{bmatrix} 1 & 2 & -1 & | & 1 & 0 & 0 \\ 0 & -1 & 6 & | & -3 & 1 & 0 \\ 0 & 0 & 1 & | & -\frac{2}{5} & 0 & \frac{1}{5} \end{bmatrix}$$ New row 3 $= \frac{1}{5}(\text{row } 3)$

$$\begin{bmatrix} 1 & 2 & 0 & | & \frac{3}{5} & 0 & \frac{1}{5} \\ 0 & -1 & 0 & | & -\frac{3}{5} & 1 & -\frac{6}{5} \\ 0 & 0 & 1 & | & -\frac{2}{5} & 0 & \frac{1}{5} \end{bmatrix}$$ New row 1 $= \text{row } 3 + \text{row } 1$
New row 2 $= -6(\text{row } 3) + \text{row } 2$

$$\begin{bmatrix} 1 & 0 & 0 & | & -\frac{3}{5} & 2 & -\frac{11}{5} \\ 0 & -1 & 0 & | & -\frac{3}{5} & 1 & -\frac{6}{5} \\ 0 & 0 & 1 & | & -\frac{2}{5} & 0 & \frac{1}{5} \end{bmatrix}$$ New row 1 $= 2(\text{row } 2) + \text{row } 1$

$$\begin{bmatrix} 1 & 0 & 0 & | & -\frac{3}{5} & 2 & -\frac{11}{5} \\ 0 & 1 & 0 & | & \frac{3}{5} & -1 & \frac{6}{5} \\ 0 & 0 & 1 & | & -\frac{2}{5} & 0 & \frac{1}{5} \end{bmatrix}$$ New row 2 $= -1(\text{row } 2)$

Thus,

$$\mathbf{A}^{-1} = \begin{bmatrix} -\frac{3}{5} & 2 & -\frac{11}{5} \\ \frac{3}{5} & -1 & \frac{6}{5} \\ -\frac{2}{5} & 0 & \frac{1}{5} \end{bmatrix}.$$

Now Try Exercise 11.

If a matrix has an inverse, we say that it is **invertible**, or **nonsingular**. When we cannot obtain the identity matrix on the left using the Gauss–Jordan method, then no inverse exists. This occurs when we obtain a row consisting entirely of 0's in either of the two matrices in the augmented matrix. In this case, we say that $\mathbf{A}$ is a **singular matrix**.

Technology Connection

When we try to find the inverse of a singular matrix using a graphing calculator, the calculator returns an error message similar to ERR: SINGULAR MATRIX.

MATRIX EQUATIONS

REVIEW SECTION 10.2

▶ **Solving Systems of Equations**

We can write a system of n linear equations in n variables as a matrix equation $\mathbf{AX} = \mathbf{B}$. If $\mathbf{A}$ has an inverse, then the system of equations has a unique solution that can be found by solving for $\mathbf{X}$, as follows:

$$\mathbf{AX} = \mathbf{B}$$
$$\mathbf{A}^{-1}(\mathbf{AX}) = \mathbf{A}^{-1}\mathbf{B} \quad \text{Multiplying by } \mathbf{A}^{-1} \text{ on the left on both sides}$$
$$(\mathbf{A}^{-1}\mathbf{A})\mathbf{X} = \mathbf{A}^{-1}\mathbf{B} \quad \text{Using the associative property of matrix multiplication}$$
$$\mathbf{IX} = \mathbf{A}^{-1}\mathbf{B} \quad \mathbf{A}^{-1}\mathbf{A} = \mathbf{I}$$
$$\mathbf{X} = \mathbf{A}^{-1}\mathbf{B}. \quad \mathbf{IX} = \mathbf{X}$$

> **MATRIX SOLUTIONS OF SYSTEMS OF EQUATIONS**
>
> For a system of n linear equations in n variables, $\mathbf{AX} = \mathbf{B}$, if $\mathbf{A}$ is an invertible matrix, then the unique solution of the system is given by
>
> $$\mathbf{X} = \mathbf{A}^{-1}\mathbf{B}.$$

> Since matrix multiplication is not commutative in general, care must be taken to multiply *on the left* by $\mathbf{A}^{-1}$.

EXAMPLE 5   Use an inverse matrix to solve the following system of equations:

$$-2x + 3y = 4,$$
$$-3x + 4y = 5.$$

Solution   We write an equivalent matrix equation, $\mathbf{AX} = \mathbf{B}$:

$$\underbrace{\begin{bmatrix} -2 & 3 \\ -3 & 4 \end{bmatrix}}_{\mathbf{A}} \cdot \underbrace{\begin{bmatrix} x \\ y \end{bmatrix}}_{\mathbf{X}} = \underbrace{\begin{bmatrix} 4 \\ 5 \end{bmatrix}}_{\mathbf{B}}$$

In Example 3, we found that

$$\mathbf{A}^{-1} = \begin{bmatrix} 4 & -3 \\ 3 & -2 \end{bmatrix}.$$

We also verified this in Example 2. Now we have

$$\mathbf{X} = \mathbf{A}^{-1}\mathbf{B}$$
$$\begin{bmatrix} x \\ y \end{bmatrix} = \begin{bmatrix} 4 & -3 \\ 3 & -2 \end{bmatrix}\begin{bmatrix} 4 \\ 5 \end{bmatrix} = \begin{bmatrix} 1 \\ 2 \end{bmatrix}.$$

The solution of the system of equations is $(1, 2)$. ▸ **Now Try Exercise 25.**

Technology Connection

To use a graphing calculator to solve the system of equations in Example 5, we enter $\mathbf{A}$ and $\mathbf{B}$ and then enter the notation $\mathbf{A}^{-1}\mathbf{B}$ on the home screen.

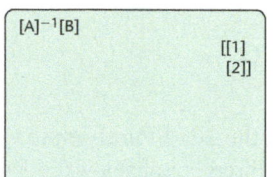

```
[A]⁻¹[B]
                    [[1]
                     [2]]
```

10.3   Exercise Set

Determine whether **B** *is the inverse of* **A**.

1. $\mathbf{A} = \begin{bmatrix} 1 & -3 \\ -2 & 7 \end{bmatrix}$, $\mathbf{B} = \begin{bmatrix} 7 & 3 \\ 2 & 1 \end{bmatrix}$

2. $\mathbf{A} = \begin{bmatrix} 3 & 2 \\ 4 & 3 \end{bmatrix}$, $\mathbf{B} = \begin{bmatrix} 3 & -2 \\ -4 & 3 \end{bmatrix}$

3. $\mathbf{A} = \begin{bmatrix} -1 & -1 & 6 \\ 1 & 0 & -2 \\ 1 & 0 & -3 \end{bmatrix}$, $\mathbf{B} = \begin{bmatrix} 2 & 3 & 2 \\ 3 & 3 & 4 \\ 1 & 1 & 1 \end{bmatrix}$

4. $\mathbf{A} = \begin{bmatrix} -2 & 0 & -3 \\ 5 & 1 & 7 \\ -3 & 0 & 4 \end{bmatrix}$, $\mathbf{B} = \begin{bmatrix} 4 & 0 & -3 \\ 1 & 1 & 1 \\ -3 & 0 & 2 \end{bmatrix}$

Use the Gauss–Jordan method to find $\mathbf{A}^{-1}$, *if it exists. Check your answers by finding* $\mathbf{A}^{-1}\mathbf{A}$ *and* $\mathbf{AA}^{-1}$.

5. $\mathbf{A} = \begin{bmatrix} 3 & 2 \\ 5 & 3 \end{bmatrix}$

6. $\mathbf{A} = \begin{bmatrix} 3 & 5 \\ 1 & 2 \end{bmatrix}$

7. $\mathbf{A} = \begin{bmatrix} 6 & 9 \\ 4 & 6 \end{bmatrix}$

8. $\mathbf{A} = \begin{bmatrix} -4 & -6 \\ 2 & 3 \end{bmatrix}$

9. $\mathbf{A} = \begin{bmatrix} 4 & -3 \\ 1 & -2 \end{bmatrix}$

10. $\mathbf{A} = \begin{bmatrix} 0 & -1 \\ 1 & 0 \end{bmatrix}$

11. $\mathbf{A} = \begin{bmatrix} 3 & 1 & 0 \\ 1 & 1 & 1 \\ 1 & -1 & 2 \end{bmatrix}$

12. $\mathbf{A} = \begin{bmatrix} 1 & 0 & 1 \\ 2 & 1 & 0 \\ 1 & -1 & 1 \end{bmatrix}$

13. $A = \begin{bmatrix} 1 & -4 & 8 \\ 1 & -3 & 2 \\ 2 & -7 & 10 \end{bmatrix}$ **14.** $A = \begin{bmatrix} -2 & 5 & 3 \\ 4 & -1 & 3 \\ 7 & -2 & 5 \end{bmatrix}$

15. $A = \begin{bmatrix} 2 & 3 & 2 \\ 3 & 3 & 4 \\ -1 & -1 & -1 \end{bmatrix}$

16. $A = \begin{bmatrix} 1 & 2 & 3 \\ 2 & -1 & -2 \\ -1 & 3 & 3 \end{bmatrix}$

17. $A = \begin{bmatrix} 1 & 2 & -1 \\ -2 & 0 & 1 \\ 1 & -1 & 0 \end{bmatrix}$

18. $A = \begin{bmatrix} 7 & -1 & -9 \\ 2 & 0 & -4 \\ -4 & 0 & 6 \end{bmatrix}$

19. $A = \begin{bmatrix} 1 & 3 & -1 \\ 0 & 2 & -1 \\ 1 & 1 & 0 \end{bmatrix}$

20. $A = \begin{bmatrix} -1 & 0 & -1 \\ -1 & 1 & 0 \\ 0 & 1 & 1 \end{bmatrix}$

21. $A = \begin{bmatrix} 1 & 2 & 3 & 4 \\ 0 & 1 & 3 & -5 \\ 0 & 0 & 1 & -2 \\ 0 & 0 & 0 & -1 \end{bmatrix}$

22. $A = \begin{bmatrix} -2 & -3 & 4 & 1 \\ 0 & 1 & 1 & 0 \\ 0 & 4 & -6 & 1 \\ -2 & -2 & 5 & 1 \end{bmatrix}$

23. $A = \begin{bmatrix} 1 & -14 & 7 & 38 \\ -1 & 2 & 1 & -2 \\ 1 & 2 & -1 & -6 \\ 1 & -2 & 3 & 6 \end{bmatrix}$

24. $A = \begin{bmatrix} 10 & 20 & -30 & 15 \\ 3 & -7 & 14 & -8 \\ -7 & -2 & -1 & 2 \\ 4 & 4 & -3 & 1 \end{bmatrix}$

In Exercises 25–28, a system of equations is given, together with the inverse of the coefficient matrix. Use the inverse of the coefficient matrix to solve the system of equations.

25. $\begin{array}{l} 11x + 3y = -4, \\ 7x + 2y = 5; \end{array}$ $A^{-1} = \begin{bmatrix} 2 & -3 \\ -7 & 11 \end{bmatrix}$

26. $\begin{array}{l} 8x + 5y = -6, \\ 5x + 3y = 2; \end{array}$ $A^{-1} = \begin{bmatrix} -3 & 5 \\ 5 & -8 \end{bmatrix}$

27. $\begin{array}{l} 3x + y \quad = 2, \\ 2x - y + 2z = -5, \\ x + y + z = 5; \end{array}$ $A^{-1} = \dfrac{1}{9}\begin{bmatrix} 3 & 1 & -2 \\ 0 & -3 & 6 \\ -3 & 2 & 5 \end{bmatrix}$

28. $\begin{array}{l} y - z = -4, \\ 4x + y \quad = -3, \\ 3x - y + 3z = 1; \end{array}$ $A^{-1} = \dfrac{1}{5}\begin{bmatrix} -3 & 2 & -1 \\ 12 & -3 & 4 \\ 7 & -3 & 4 \end{bmatrix}$

Solve the system of equations using the inverse of the coefficient matrix of the equivalent matrix equation.

29. $\begin{array}{l} 4x + 3y = 2, \\ x - 2y = 6 \end{array}$ **30.** $\begin{array}{l} 2x - 3y = 7, \\ 4x + y = -7 \end{array}$

31. $\begin{array}{l} 5x + y = 2, \\ 3x - 2y = -4 \end{array}$ **32.** $\begin{array}{l} x - 6y = 5, \\ -x + 4y = -5 \end{array}$

33. $\begin{array}{l} x \quad + z = 1, \\ 2x + y \quad = 3, \\ x - y + z = 4 \end{array}$ **34.** $\begin{array}{l} x + 2y + 3z = -1, \\ 2x - 3y + 4z = 2, \\ -3x + 5y - 6z = 4 \end{array}$

35. $\begin{array}{l} 2x + 3y + 4z = 2, \\ x - 4y + 3z = 2, \\ 5x + y + z = -4 \end{array}$ **36.** $\begin{array}{l} x + y \quad = 2, \\ 3x \quad + 2z = 5, \\ 2x + 3y - 3z = 9 \end{array}$

37. $\begin{array}{l} 2w - 3x + 4y - 5z = 0, \\ 3w - 2x + 7y - 3z = 2, \\ w + x - y + z = 1, \\ -w - 3x - 6y + 4z = 6 \end{array}$

38. $\begin{array}{l} 5w - 4x + 3y - 2z = -6, \\ w + 4x - 2y + 3z = -5, \\ 2w - 3x + 6y - 9z = 14, \\ 3w - 5x + 2y - 4z = -3 \end{array}$

39. *Paralympic Medals.* At the 2014 Paralympic Games in Sochi, Russia, the top three countries—the Russian Federation, Ukraine, and the United States—won a total of 123 medals. Ukraine won 7 more medals than the United States. The Russian Federation won 37 more medals than the total amount won by Ukraine and the United States. (*Source*: International Paralympic Committee) How many medals did each of the top three countries win?

40. *Lunch Cost.* Coworkers Jan and Richard purchase lunch from a food truck. Jan buys 1 beef taco and 2 fruit cups for $5.25. Richard buys 3 beef tacos and 1 fruit cup for $8.25. Find the price of each item.

41. *Landscaping Cost.* Green-Up Landscaping bought 4 tons of topsoil, 3 tons of mulch, and 6 tons of pea gravel for $2825. The next week, the firm bought 5 tons of topsoil, 2 tons of mulch, and 5 tons of pea gravel for $2663. Pea gravel costs $17 less per ton than topsoil. Find the price per ton for each item.

42. *Investment.* Trevor receives $230 per year in simple interest from three investments totaling $8500. Part is invested at 2.2%, part at 2.65%, and the rest at 3.05%. There is $1500 more invested at 3.05% than at 2.2%. Find the amount invested at each rate.

▶ **Skill Maintenance**

Use synthetic division to find the function values.

43. $f(x) = x^3 - 6x^2 + 4x - 8$; find $f(-2)$ [8.3]

44. $f(x) = 2x^4 - x^3 + 5x^2 + 6x - 4$; find $f(3)$ [8.3]

Solve.

45. $2x^2 + x = 7$ [7.4]

46. $\dfrac{1}{x+1} - \dfrac{6}{x-1} = 1$ [5.5], [7.4]

47. $\sqrt{2x+1} - 1 = \sqrt{2x - 4}$ [6.6], [7.4]

48. $x - \sqrt{x} - 6 = 0$ [7.4]

Factor the polynomial $f(x)$. [8.3]

49. $f(x) = x^3 - 3x^2 - 6x + 8$

50. $f(x) = x^4 + 2x^3 - 16x^2 - 2x + 15$

▶ **Synthesis**

State the conditions under which $\mathbf{A}^{-1}$ exists. Then find a formula for $\mathbf{A}^{-1}$.

51. $\mathbf{A} = \begin{bmatrix} x \end{bmatrix}$

52. $\mathbf{A} = \begin{bmatrix} x & 0 \\ 0 & y \end{bmatrix}$

53. $\mathbf{A} = \begin{bmatrix} 0 & 0 & x \\ 0 & y & 0 \\ z & 0 & 0 \end{bmatrix}$

54. $\mathbf{A} = \begin{bmatrix} x & 1 & 1 & 1 \\ 0 & y & 0 & 0 \\ 0 & 0 & z & 0 \\ 0 & 0 & 0 & w \end{bmatrix}$

10.4   Determinants and Cramer's Rule

▶ Evaluate determinants of square matrices.

▶ Use Cramer's rule to solve systems of equations.

▶ **Determinants of Square Matrices**

With every square matrix, we associate a number called its *determinant*.

DETERMINANT OF A 2 × 2 MATRIX

The **determinant** of the matrix $\begin{bmatrix} a & c \\ b & d \end{bmatrix}$ is denoted $\begin{vmatrix} a & c \\ b & d \end{vmatrix}$ and is defined as

$$\begin{vmatrix} a & c \\ b & d \end{vmatrix} = ad - bc.$$

EXAMPLE 1   Evaluate: $\begin{vmatrix} \sqrt{2} & -3 \\ -4 & -\sqrt{2} \end{vmatrix}$.

Solution

$\begin{vmatrix} \sqrt{2} & -3 \\ -4 & -\sqrt{2} \end{vmatrix}$ **The arrows indicate the products involved.**

$$= \sqrt{2}(-\sqrt{2}) - (-4)(-3)$$
$$= -2 - 12$$
$$= -14$$

> **Now Try Exercise 1.**

We now consider a way to evaluate determinants of square matrices of order 3×3 or higher.

▶ Evaluating Determinants Using Cofactors

Often we first find minors and cofactors of matrices in order to evaluate determinants.

MINOR

For a square matrix $\mathbf{A} = [a_{ij}]$, the **minor** M_{ij} of an entry a_{ij} is the determinant of the matrix formed by deleting the ith row and the jth column of $\mathbf{A}$.

EXAMPLE 2   For the matrix

$$\mathbf{A} = [a_{ij}] = \begin{bmatrix} -8 & 0 & 6 \\ 4 & -6 & 7 \\ -1 & -3 & 5 \end{bmatrix},$$

find each of the following.

a) M_{11} **b)** M_{23}

Solution

a) For M_{11}, we delete the first row and the first column and find the determinant of the 2×2 matrix formed by the remaining entries.

$$\begin{bmatrix} -8 & 0 & 6 \\ 4 & -6 & 7 \\ -1 & -3 & 5 \end{bmatrix} \qquad M_{11} = \begin{vmatrix} -6 & 7 \\ -3 & 5 \end{vmatrix}$$
$$= (-6) \cdot 5 - (-3) \cdot 7$$
$$= -30 - (-21)$$
$$= -30 + 21$$
$$= -9$$

b) For M_{23}, we delete the second row and the third column and find the determinant of the 2×2 matrix formed by the remaining entries.

$$\begin{bmatrix} -8 & 0 & 6 \\ 4 & -6 & 7 \\ -1 & -3 & 5 \end{bmatrix} \qquad M_{23} = \begin{vmatrix} -8 & 0 \\ -1 & -3 \end{vmatrix}$$
$$= -8(-3) - (-1)0$$
$$= 24$$

> **Now Try Exercise 9.**

COFACTOR

For a square matrix $\mathbf{A} = [a_{ij}]$, the **cofactor** A_{ij} of an entry a_{ij} is given by

$$A_{ij} = (-1)^{i+j}M_{ij},$$

where M_{ij} is the minor of a_{ij}.

EXAMPLE 3   For the matrix given in Example 2, find each of the following.

a) A_{11} **b)** A_{23}

Solution

a) In Example 2, we found that $M_{11} = -9$. Then

$$A_{11} = (-1)^{1+1}(-9) = (1)(-9) = -9.$$

b) In Example 2, we found that $M_{23} = 24$. Then

$$A_{23} = (-1)^{2+3}(24) = (-1)(24) = -24.$$

> **Now Try Exercise 11.**

Note that minors and cofactors are *numbers*. They are *not matrices*.

Consider the matrix $\mathbf{A}$ given by

$$\mathbf{A} = \begin{bmatrix} a_{11} & a_{12} & a_{13} \\ a_{21} & a_{22} & a_{23} \\ a_{31} & a_{32} & a_{33} \end{bmatrix}.$$

The determinant of the matrix, denoted $|\mathbf{A}|$, can be found by multiplying each element of the first column by its cofactor and adding:

$$|\mathbf{A}| = a_{11}A_{11} + a_{21}A_{21} + a_{31}A_{31}.$$

Because

$$A_{11} = (-1)^{1+1}M_{11} = M_{11},$$
$$A_{21} = (-1)^{2+1}M_{21} = -M_{21},$$
and $A_{31} = (-1)^{3+1}M_{31} = M_{31},$

we can write

$$|\mathbf{A}| = a_{11} \cdot \begin{vmatrix} a_{22} & a_{23} \\ a_{32} & a_{33} \end{vmatrix} - a_{21} \cdot \begin{vmatrix} a_{12} & a_{13} \\ a_{32} & a_{33} \end{vmatrix} + a_{31} \cdot \begin{vmatrix} a_{12} & a_{13} \\ a_{22} & a_{23} \end{vmatrix}.$$

It can be shown that we can determine $|\mathbf{A}|$ by choosing *any* row or column, multiplying each element in that row or column by its cofactor, and adding. This is called *expanding* across a row or down a column. We just expanded down the first column. We now define the determinant of a square matrix of any order.

> **DETERMINANT OF ANY SQUARE MATRIX**
>
> For any square matrix $\mathbf{A}$ of order $n \times n$ $(n > 1)$, we define the **determinant** of $\mathbf{A}$, denoted $|\mathbf{A}|$, as follows. Choose any row or column. Multiply each element in that row or column by its cofactor and add the results. The determinant of a 1×1 matrix is simply the element of the matrix. The value of a determinant will be the same no matter which row or column is chosen.

Technology Connection

Determinants can be evaluated on a graphing calculator. After entering a matrix, we select the determinant operation from the MATRIX MATH menu and enter the name of the matrix. The calculator will return the value of the determinant of the matrix. For example, for

$$\mathbf{A} = \begin{bmatrix} 1 & 6 & -1 \\ -3 & -5 & 3 \\ 0 & 4 & 2 \end{bmatrix},$$

we have

```
det ([A])
                    26
```

EXAMPLE 4 Evaluate $|\mathbf{A}|$ by expanding across the third row.

$$\mathbf{A} = \begin{bmatrix} -8 & 0 & 6 \\ 4 & -6 & 7 \\ -1 & -3 & 5 \end{bmatrix}$$

Solution We have

$$
\begin{aligned}
|\mathbf{A}| &= (-1)A_{31} + (-3)A_{32} + 5A_{33} \\
&= (-1)(-1)^{3+1} \cdot \begin{vmatrix} 0 & 6 \\ -6 & 7 \end{vmatrix} + (-3)(-1)^{3+2} \cdot \begin{vmatrix} -8 & 6 \\ 4 & 7 \end{vmatrix} \\
&\quad + 5(-1)^{3+3} \cdot \begin{vmatrix} -8 & 0 \\ 4 & -6 \end{vmatrix} \\
&= (-1) \cdot 1 \cdot [0 \cdot 7 - (-6)6] + (-3)(-1)[-8 \cdot 7 - 4 \cdot 6] \\
&\quad + 5 \cdot 1 \cdot [-8(-6) - 4 \cdot 0] \\
&= -[36] + 3[-80] + 5[48] \\
&= -36 - 240 + 240 = -36.
\end{aligned}
$$

The value of this determinant is -36 no matter which row or column we expand on.

> **Now Try Exercise 13.**

▶ Cramer's Rule

Determinants can be used to solve systems of linear equations. Consider a system of two linear equations:

$$
\begin{aligned}
a_1 x + b_1 y &= c_1, \\
a_2 x + b_2 y &= c_2.
\end{aligned}
$$

Solving this system using the elimination method, we obtain

$$x = \frac{c_1 b_2 - c_2 b_1}{a_1 b_2 - a_2 b_1} \quad \text{and} \quad y = \frac{a_1 c_2 - a_2 c_1}{a_1 b_2 - a_2 b_1}.$$

The numerators and the denominators of these expressions can be written as determinants:

$$x = \frac{\begin{vmatrix} c_1 & b_1 \\ c_2 & b_2 \end{vmatrix}}{\begin{vmatrix} a_1 & b_1 \\ a_2 & b_2 \end{vmatrix}} \quad \text{and} \quad y = \frac{\begin{vmatrix} a_1 & c_1 \\ a_2 & c_2 \end{vmatrix}}{\begin{vmatrix} a_1 & b_1 \\ a_2 & b_2 \end{vmatrix}}.$$

If we let

$$D = \begin{vmatrix} a_1 & b_1 \\ a_2 & b_2 \end{vmatrix}, \qquad D_x = \begin{vmatrix} c_1 & b_1 \\ c_2 & b_2 \end{vmatrix}, \quad \text{and} \quad D_y = \begin{vmatrix} a_1 & c_1 \\ a_2 & c_2 \end{vmatrix},$$

we have

$$x = \frac{D_x}{D} \quad \text{and} \quad y = \frac{D_y}{D}.$$

This procedure for solving systems of equations is known as *Cramer's rule.*

CRAMER'S RULE FOR 2 × 2 SYSTEMS

The solution of the system of equations

$$a_1 x + b_1 y = c_1,$$
$$a_2 x + b_2 y = c_2$$

is given by

$$x = \frac{D_x}{D}, \qquad y = \frac{D_y}{D},$$

where

$$D = \begin{vmatrix} a_1 & b_1 \\ a_2 & b_2 \end{vmatrix}, \qquad D_x = \begin{vmatrix} c_1 & b_1 \\ c_2 & b_2 \end{vmatrix}, \qquad D_y = \begin{vmatrix} a_1 & c_1 \\ a_2 & c_2 \end{vmatrix}, \quad \text{and} \quad D \neq 0.$$

Note that the denominator D contains the coefficients of x and y, in the same position as in the original equations. For x, the numerator is obtained by replacing the x-coefficients in D (the a's) with the c's. For y, the numerator is obtained by replacing the y-coefficients in D (the b's) with the c's.

EXAMPLE 5 Solve using Cramer's rule:

$$2x + 5y = 7,$$
$$5x - 2y = -3.$$

Solution We have

$$x = \frac{\begin{vmatrix} 7 & 5 \\ -3 & -2 \end{vmatrix}}{\begin{vmatrix} 2 & 5 \\ 5 & -2 \end{vmatrix}} = \frac{7(-2) - (-3)5}{2(-2) - 5 \cdot 5} = \frac{1}{-29} = -\frac{1}{29},$$

$$y = \frac{\begin{vmatrix} 2 & 7 \\ 5 & -3 \end{vmatrix}}{\begin{vmatrix} 2 & 5 \\ 5 & -2 \end{vmatrix}} = \frac{2(-3) - 5 \cdot 7}{-29} = \frac{-41}{-29} = \frac{41}{29}.$$

The solution is $\left(-\frac{1}{29}, \frac{41}{29}\right)$. **Now Try Exercise 29.**

Cramer's rule works only when a system of equations has a unique solution. This occurs when $D \neq 0$. If $D = 0$ and D_x and D_y are also 0, then the equations are dependent. If $D = 0$ and D_x and/or D_y is not 0, then the system is inconsistent.

Cramer's rule can be extended to a system of n linear equations in n variables. We consider a 3 × 3 system.

Technology Connection

To use Cramer's rule to solve the system of equations in Example 5 on a graphing calculator, we first enter the matrices corresponding to D, D_x, and D_y. We enter

$$\mathbf{A} = \begin{bmatrix} 2 & 5 \\ 5 & -2 \end{bmatrix},$$

$$\mathbf{B} = \begin{bmatrix} 7 & 5 \\ -3 & -2 \end{bmatrix},$$

and

$$\mathbf{C} = \begin{bmatrix} 2 & 7 \\ 5 & -3 \end{bmatrix}.$$

Then

$$x = \frac{\det(\mathbf{B})}{\det(\mathbf{A})}$$

and

$$y = \frac{\det(\mathbf{C})}{\det(\mathbf{A})}.$$

CRAMER'S RULE FOR 3 × 3 SYSTEMS

The solution of the system of equations

$$a_1 x + b_1 y + c_1 z = d_1,$$
$$a_2 x + b_2 y + c_2 z = d_2,$$
$$a_3 x + b_3 y + c_3 z = d_3$$

is given by

$$x = \frac{D_x}{D}, \qquad y = \frac{D_y}{D}, \qquad z = \frac{D_z}{D},$$

where

$$D = \begin{vmatrix} a_1 & b_1 & c_1 \\ a_2 & b_2 & c_2 \\ a_3 & b_3 & c_3 \end{vmatrix}, \qquad D_x = \begin{vmatrix} d_1 & b_1 & c_1 \\ d_2 & b_2 & c_2 \\ d_3 & b_3 & c_3 \end{vmatrix},$$

$$D_y = \begin{vmatrix} a_1 & d_1 & c_1 \\ a_2 & d_2 & c_2 \\ a_3 & d_3 & c_3 \end{vmatrix}, \qquad D_z = \begin{vmatrix} a_1 & b_1 & d_1 \\ a_2 & b_2 & d_2 \\ a_3 & b_3 & d_3 \end{vmatrix}, \quad \text{and} \quad D \neq 0.$$

Note that the determinant D_x is obtained from D by replacing the x-coefficients with d_1, d_2, and d_3. D_y and D_z are obtained in a similar manner. As with a system of two equations, Cramer's rule cannot be used if $D = 0$. If $D = 0$ and D_x, D_y, and D_z are 0, then the equations are dependent. If $D = 0$ and one of D_x, D_y, or D_z is not 0, then the system is inconsistent.

EXAMPLE 6   Solve using Cramer's rule:

$$x - 3y + 7z = 13,$$
$$x + y + z = 1,$$
$$x - 2y + 3z = 4.$$

Solution   We have

$$D = \begin{vmatrix} 1 & -3 & 7 \\ 1 & 1 & 1 \\ 1 & -2 & 3 \end{vmatrix} = -10, \qquad D_x = \begin{vmatrix} 13 & -3 & 7 \\ 1 & 1 & 1 \\ 4 & -2 & 3 \end{vmatrix} = 20,$$

$$D_y = \begin{vmatrix} 1 & 13 & 7 \\ 1 & 1 & 1 \\ 1 & 4 & 3 \end{vmatrix} = -6, \qquad D_z = \begin{vmatrix} 1 & -3 & 13 \\ 1 & 1 & 1 \\ 1 & -2 & 4 \end{vmatrix} = -24.$$

Then

$$x = \frac{D_x}{D} = \frac{20}{-10} = -2, \qquad y = \frac{D_y}{D} = \frac{-6}{-10} = \frac{3}{5}, \qquad z = \frac{D_z}{D} = \frac{-24}{-10} = \frac{12}{5}.$$

The solution is $\left(-2, \frac{3}{5}, \frac{12}{5}\right)$.

In practice, it is not necessary to evaluate D_z. When we have found values for x and y, we can substitute them into one of the equations to find z.

Now Try Exercise 37.

10.4 Exercise Set

Evaluate the determinant.

1. $\begin{vmatrix} 5 & 3 \\ -2 & -4 \end{vmatrix}$

2. $\begin{vmatrix} -8 & 6 \\ -1 & 2 \end{vmatrix}$

3. $\begin{vmatrix} 4 & -7 \\ -2 & 3 \end{vmatrix}$

4. $\begin{vmatrix} -9 & -6 \\ 5 & 4 \end{vmatrix}$

5. $\begin{vmatrix} -2 & -\sqrt{5} \\ -\sqrt{5} & 3 \end{vmatrix}$

6. $\begin{vmatrix} \sqrt{5} & -3 \\ 4 & 2 \end{vmatrix}$

7. $\begin{vmatrix} x & 4 \\ x & x^2 \end{vmatrix}$

8. $\begin{vmatrix} y^2 & -2 \\ y & 3 \end{vmatrix}$

Use the following matrix for Exercises 9–16:

$$\mathbf{A} = \begin{bmatrix} 7 & -4 & -6 \\ 2 & 0 & -3 \\ 1 & 2 & -5 \end{bmatrix}.$$

9. Find M_{11}, M_{32}, and M_{22}.

10. Find M_{13}, M_{31}, and M_{23}.

11. Find A_{11}, A_{32}, and A_{22}.

12. Find A_{13}, A_{31}, and A_{23}.

13. Evaluate $|\mathbf{A}|$ by expanding across the second row.

14. Evaluate $|\mathbf{A}|$ by expanding down the second column.

15. Evaluate $|\mathbf{A}|$ by expanding down the third column.

16. Evaluate $|\mathbf{A}|$ by expanding across the first row.

Use the following matrix for Exercises 17–22:

$$\mathbf{A} = \begin{bmatrix} 1 & 0 & 0 & -2 \\ 4 & 1 & 0 & 0 \\ 5 & 6 & 7 & 8 \\ -2 & -3 & -1 & 0 \end{bmatrix}.$$

17. Find M_{12} and M_{44}.

18. Find M_{41} and M_{33}.

19. Find A_{22} and A_{34}.

20. Find A_{24} and A_{43}.

21. Evaluate $|\mathbf{A}|$ by expanding across the first row.

22. Evaluate $|\mathbf{A}|$ by expanding down the third column.

Evaluate the determinant.

23. $\begin{vmatrix} 3 & 1 & 2 \\ -2 & 3 & 1 \\ 3 & 4 & -6 \end{vmatrix}$

24. $\begin{vmatrix} 3 & -2 & 1 \\ 2 & 4 & 3 \\ -1 & 5 & 1 \end{vmatrix}$

25. $\begin{vmatrix} x & 0 & -1 \\ 2 & x & x^2 \\ -3 & x & 1 \end{vmatrix}$

26. $\begin{vmatrix} x & 1 & -1 \\ x^2 & x & x \\ 0 & x & 1 \end{vmatrix}$

Solve using Cramer's rule.

27. $-2x + 4y = 3,$
$3x - 7y = 1$

28. $5x - 4y = -3,$
$7x + 2y = 6$

29. $2x - y = 5,$
$x - 2y = 1$

30. $3x + 4y = -2,$
$5x - 7y = 1$

31. $2x + 9y = -2,$
$4x - 3y = 3$

32. $2x + 3y = -1,$
$3x + 6y = -0.5$

33. $2x + 5y = 7,$
$3x - 2y = 1$

34. $3x + 2y = 7,$
$2x + 3y = -2$

35. $3x + 2y - z = 4,$
$3x - 2y + z = 5,$
$4x - 5y - z = -1$

36. $3x - y + 2z = 1,$
$x - y + 2z = 3,$
$-2x + 3y + z = 1$

37. $3x + 5y - z = -2,$
$x - 4y + 2z = 13,$
$2x + 4y + 3z = 1$

38. $3x + 2y + 2z = 1,$
$5x - y - 6z = 3,$
$2x + 3y + 3z = 4$

39. $x - 3y - 7z = 6,$
$2x + 3y + z = 9,$
$4x + y = 7$

40. $x - 2y - 3z = 4,$
$3x - 2z = 8,$
$2x + y + 4z = 13$

41.
$$6y + 6z = -1,$$
$$8x + 6z = -1,$$
$$4x + 9y = 8$$

42.
$$3x + 5y = 2,$$
$$2x - 3z = 7,$$
$$4y + 2z = -1$$

► Skill Maintenance

Determine whether the function is one-to-one, and if it is, find a formula for $f^{-1}(x)$. **[9.2]**

43. $f(x) = 3x + 2$

44. $f(x) = x^2 - 4$

45. $f(x) = |x| + 3$

46. $f(x) = \sqrt[3]{x} + 1$

Simplify. Write answers in the form $a + bi$, where a and b are real numbers. **[7.3]**

47. $(3 - 4i) - (-2 - i)$

48. $(5 + 2i) + (1 - 4i)$

49. $(1 - 2i)(6 + 2i)$

50. $\dfrac{3 + i}{4 - 3i}$

► Synthesis

Solve.

51. $\begin{vmatrix} y & 2 \\ 3 & y \end{vmatrix} = y$

52. $\begin{vmatrix} x & -3 \\ -1 & x \end{vmatrix} \geq 0$

53. $\begin{vmatrix} 2 & x & 1 \\ 1 & 2 & -1 \\ 3 & 4 & -2 \end{vmatrix} = -6$

54. $\begin{vmatrix} m + 2 & -3 \\ m + 5 & -4 \end{vmatrix} = 3m - 5$

Rewrite the expression using a determinant. Answers may vary.

55. $a^2 + b^2$

56. $\frac{1}{2}h(a + b)$

57. $2\pi r^2 + 2\pi rh$

58. $x^2y^2 - Q^2$

Chapter 10 Summary and Review

STUDY GUIDE

KEY TERMS AND CONCEPTS	EXAMPLES

SECTION 10.1: MATRICES AND SYSTEMS OF EQUATIONS

A **matrix** (pl., **matrices**) is a rectangular array of numbers called **entries**, or **elements**, of the matrix.

Row 1 → Row 2 →
$$\begin{bmatrix} 3 & -2 & 5 \\ -1 & 4 & -3 \end{bmatrix}$$
Column 1 Column 2 Column 3

This matrix has 2 rows and 3 columns. Its **order** is 2×3.

We can apply the **row-equivalent operations** on p. 683 to use **Gaussian elimination** with matrices to solve systems of equations.

Solve: $x - 2y = 8,$ $2x + y = 1.$

We write the augmented matrix and transform it to **row-echelon form** or **reduced row-echelon form**:

$$\begin{bmatrix} 1 & -2 & | & 8 \\ 2 & 1 & | & 1 \end{bmatrix} \rightarrow \begin{bmatrix} 1 & -2 & | & 8 \\ 0 & 1 & | & -3 \end{bmatrix} \rightarrow \begin{bmatrix} 1 & 0 & | & 2 \\ 0 & 1 & | & -3 \end{bmatrix}$$

Row-echelon form   Reduced row-echelon form

Thus we have $x = 2$, $y = -3$. The solution is $(2, -3)$.

SECTION 10.2: MATRIX OPERATIONS

Matrices of the same order can be added or subtracted by adding or subtracting their corresponding entries.	Find each of the following. $\begin{bmatrix} 3 & -4 \\ -1 & 2 \end{bmatrix} + \begin{bmatrix} -5 & -1 \\ 3 & 0 \end{bmatrix} = \begin{bmatrix} 3 + (-5) & -4 + (-1) \\ -1 + 3 & 2 + 0 \end{bmatrix}$ $\qquad\qquad\qquad\qquad = \begin{bmatrix} -2 & -5 \\ 2 & 2 \end{bmatrix}$ $\begin{bmatrix} 3 & -4 \\ -1 & 2 \end{bmatrix} - \begin{bmatrix} -5 & -1 \\ 3 & 0 \end{bmatrix} = \begin{bmatrix} 3 - (-5) & -4 - (-1) \\ -1 - 3 & 2 - 0 \end{bmatrix}$ $\qquad\qquad\qquad\qquad = \begin{bmatrix} 8 & -3 \\ -4 & 2 \end{bmatrix}$
The **scalar product** of a number k and a matrix $\mathbf{A}$ is the matrix $k\mathbf{A}$ obtained by multiplying each entry of $\mathbf{A}$ by k. The number k is called a **scalar**. The properties of matrix addition and scalar multiplication are given on p. 693.	For $\mathbf{A} = \begin{bmatrix} 2 & 3 & -1 \\ -4 & -2 & 5 \end{bmatrix}$, find $2\mathbf{A}$. $2\mathbf{A} = 2\begin{bmatrix} 2 & 3 & -1 \\ -4 & -2 & 5 \end{bmatrix} = \begin{bmatrix} 2 \cdot 2 & 2 \cdot 3 & 2 \cdot (-1) \\ 2 \cdot (-4) & 2 \cdot (-2) & 2 \cdot 5 \end{bmatrix}$ $\qquad = \begin{bmatrix} 4 & 6 & -2 \\ -8 & -4 & 10 \end{bmatrix}$
For an $m \times n$ matrix $\mathbf{A} = [a_{ij}]$ and an $n \times p$ matrix $\mathbf{B} = [b_{ij}]$, the **product** $\mathbf{AB} = [c_{ij}]$ is an $m \times p$ matrix, where $\quad c_{ij} = a_{i1} \cdot b_{1j} + a_{i2} \cdot b_{2j}$ $\qquad\quad + a_{i3} \cdot b_{3j} + \cdots + a_{in} \cdot b_{nj}.$ The properties of matrix multiplication are given on p. 696.	For $\mathbf{A} = \begin{bmatrix} 4 & -1 & 3 \\ 0 & -2 & 1 \end{bmatrix}$ and $\mathbf{B} = \begin{bmatrix} -3 & 1 \\ 3 & 4 \\ 2 & -1 \end{bmatrix}$, find $\mathbf{AB}$. $\mathbf{AB} = \begin{bmatrix} 4 & -1 & 3 \\ 0 & -2 & 1 \end{bmatrix} \begin{bmatrix} -3 & 1 \\ 3 & 4 \\ 2 & -1 \end{bmatrix}$ $\quad = \begin{bmatrix} 4 \cdot (-3) + (-1) \cdot 3 + 3 \cdot 2 & 4 \cdot 1 + (-1) \cdot 4 + 3 \cdot (-1) \\ 0 \cdot (-3) + (-2) \cdot 3 + 1 \cdot 2 & 0 \cdot 1 + (-2) \cdot 4 + 1 \cdot (-1) \end{bmatrix}$ $\quad = \begin{bmatrix} -9 & -3 \\ -4 & -9 \end{bmatrix}$
We can write a matrix equation equivalent to a system of equations.	Write a matrix equation equivalent to the system of equations: $\qquad 2x - 3y = 6,$ $\qquad x - 4y = 1.$ This system of equations can be written as $\qquad \begin{bmatrix} 2 & -3 \\ 1 & -4 \end{bmatrix} \begin{bmatrix} x \\ y \end{bmatrix} = \begin{bmatrix} 6 \\ 1 \end{bmatrix}.$

SECTION 10.3: INVERSES OF MATRICES

The $n \times n$ **identity matrix I** is an $n \times n$ matrix with 1's on the main diagonal and 0's elsewhere.

For any $n \times n$ matrix **A**,

$$\mathbf{AI} = \mathbf{IA} = \mathbf{A}.$$

For an $n \times n$ matrix **A**, if there is a matrix $\mathbf{A}^{-1}$ for which $\mathbf{A}^{-1} \cdot \mathbf{A} = \mathbf{I} = \mathbf{A} \cdot \mathbf{A}^{-1}$, then $\mathbf{A}^{-1}$ is the **inverse** of **A**.

For a system of n linear equations in n variables, $\mathbf{AX} = \mathbf{B}$, if **A** has an inverse, then the solution of the system of equations is given by

$$\mathbf{X} = \mathbf{A}^{-1}\mathbf{B}.$$

Since matrix multiplication is not commutative, in general, **B** *must* be multiplied *on the left* by $\mathbf{A}^{-1}$.

The inverse of an $n \times n$ matrix **A** can be found by first writing an augmented matrix consisting of **A** on the left side and the $n \times n$ identity matrix on the right side. Then row-equivalent operations are used to transform the augmented matrix to a matrix with the $n \times n$ identity matrix on the left side and the inverse on the right side.

See Examples 3 and 4 on pp. 703 and 704.

Use an inverse matrix to solve the following system of equations:

$$x - y = 1,$$
$$x - 2y = -1.$$

First, we write an equivalent matrix equation:

$$\underset{\mathbf{A}}{\begin{bmatrix} 1 & -1 \\ 1 & -2 \end{bmatrix}} \cdot \underset{\mathbf{X}}{\begin{bmatrix} x \\ y \end{bmatrix}} = \underset{\mathbf{B}}{\begin{bmatrix} 1 \\ -1 \end{bmatrix}}.$$

Then we find $\mathbf{A}^{-1}$ and multiply *on the left* by $\mathbf{A}^{-1}$:

$$\mathbf{X} = \mathbf{A}^{-1} \cdot \mathbf{B}$$

$$\begin{bmatrix} x \\ y \end{bmatrix} = \begin{bmatrix} 2 & -1 \\ 1 & -1 \end{bmatrix} \begin{bmatrix} 1 \\ -1 \end{bmatrix} = \begin{bmatrix} 3 \\ 2 \end{bmatrix}.$$

The solution is $(3, 2)$.

SECTION 10.4: DETERMINANTS AND CRAMER'S RULE

Determinant of a 2 × 2 Matrix

The determinant of the matrix $\begin{bmatrix} a & c \\ b & d \end{bmatrix}$ is denoted by $\begin{vmatrix} a & c \\ b & d \end{vmatrix}$ and is defined as

$$\begin{vmatrix} a & c \\ b & d \end{vmatrix} = ad - bc.$$

Evaluate: $\begin{vmatrix} 3 & -4 \\ 2 & 1 \end{vmatrix}$.

$$\begin{vmatrix} 3 & -4 \\ 2 & 1 \end{vmatrix} = 3 \cdot 1 - 2(-4) = 3 + 8 = 11$$

The **determinant** of any **square matrix** can be found by *expanding across a row* or *down a column*. See p. 710.

See Example 4 on p. 710.

We can use determinants to solve systems of linear equations.

Cramer's rule for a 2×2 system is given on p. 711. Cramer's rule for a 3×3 system is given on p. 712.

Solve: $2x - 3y = 2,$
$\quad\quad 6x + 6y = 1.$

$$x = \frac{\begin{vmatrix} 2 & -3 \\ 1 & 6 \end{vmatrix}}{\begin{vmatrix} 2 & -3 \\ 6 & 6 \end{vmatrix}}, \quad\quad y = \frac{\begin{vmatrix} 2 & 2 \\ 6 & 1 \end{vmatrix}}{\begin{vmatrix} 2 & -3 \\ 6 & 6 \end{vmatrix}}$$

$$x = \frac{15}{30} = \frac{1}{2}, \quad\quad y = \frac{-10}{30} = -\frac{1}{3}.$$

The solution is $\left(\frac{1}{2}, -\frac{1}{3} \right)$.

REVIEW EXERCISES

Determine whether the statement is true or false.

1. For any $m \times n$ matrices $\mathbf{A}$ and $\mathbf{B}$, $\mathbf{A} + \mathbf{B} = \mathbf{B} + \mathbf{A}$. [10.2]

2. In general, matrix multiplication is commutative. [10.2]

3. Every matrix has an inverse. [10.3]

4. Cramer's rule works only when a system of equations has a unique solution. [10.4]

Solve the system of equations using Gaussian elimination or Gauss–Jordan elimination. [10.1]

5. $x + 2y = 5,$
$\quad 2x - 5y = -8$

6. $3x + 4y + 2z = 3,$
$\quad 5x - 2y - 13z = 3,$
$\quad 4x + 3y - 3z = 6$

7. $3x + 5y + z = 0,$
$\quad 2x - 4y - 3z = 0,$
$\quad x + 3y + z = 0$

8. $w + x + y + z = -2,$
$\quad -3w - 2x + 3y + 2z = 10,$
$\quad 2w + 3x + 2y - z = -12,$
$\quad 2w + 4x - y + z = 1$

Solve using Gaussian elimination or Gauss–Jordan elimination. [10.1]

9. *Coins.* The value of 75 coins, consisting of nickels and dimes, is $5.95. How many of each kind of coin are there?

10. *Test Scores.* A student has a total of 226 on three tests. The sum of the scores on the first and second tests exceeds the score on the third test by 62. The first score exceeds the second by 6. Find the three scores.

For Exercises 11–18, let

$$\mathbf{A} = \begin{bmatrix} 1 & -1 & 0 \\ 2 & 3 & -2 \\ -2 & 0 & 1 \end{bmatrix}, \quad \mathbf{B} = \begin{bmatrix} -1 & 0 & 6 \\ 1 & -2 & 0 \\ 0 & 1 & -3 \end{bmatrix},$$

and

$$\mathbf{C} = \begin{bmatrix} -2 & 0 \\ 1 & 3 \end{bmatrix}.$$

Find each of the following, if possible. [10.2]

11. $\mathbf{A} + \mathbf{B}$

12. $-3\mathbf{A}$

13. $-\mathbf{A}$

14. $\mathbf{AB}$

15. $\mathbf{B} + \mathbf{C}$

16. $\mathbf{A} - \mathbf{B}$

17. $\mathbf{BA}$

18. $\mathbf{A} + 3\mathbf{B}$

19. *Food Service Management.* The following table lists the cost per serving, in dollars, for items on four menus that are served at an NFL football training camp.

Menu	Meat	Potato	Vegetable	Salad	Dessert
1	2.25	0.38	0.55	0.33	0.85
2	3.09	0.42	0.46	0.48	0.51
3	2.40	0.31	0.59	0.36	0.64
4	1.80	0.29	0.34	0.55	0.52

On a particular day, a dietician orders 41 meals from menu 1, 18 from menu 2, 39 from menu 3, and 36 from menu 4.

a) Write the information in the table as a 4×5 matrix $\mathbf{M}$. [10.2]

b) Write a row matrix $\mathbf{N}$ that represents the number of each menu ordered. [10.2]

c) Find the product $\mathbf{NM}$. [10.2]

d) State what the entries of $\mathbf{NM}$ represent. [10.2]

Find $\mathbf{A}^{-1}$, if it exists. [10.3]

20. $\mathbf{A} = \begin{bmatrix} -2 & 0 \\ 1 & 3 \end{bmatrix}$

21. $\mathbf{A} = \begin{bmatrix} 0 & 0 & 3 \\ 0 & -2 & 0 \\ 4 & 0 & 0 \end{bmatrix}$

22. $\mathbf{A} = \begin{bmatrix} 1 & 0 & 0 & 0 \\ 0 & 4 & -5 & 0 \\ 0 & 2 & 2 & 0 \\ 0 & 0 & 0 & 1 \end{bmatrix}$

23. Write a matrix equation equivalent to this system of equations:
$$3x - 2y + 4z = 13,$$
$$x + 5y - 3z = 7,$$
$$2x - 3y + 7z = -8. \quad [10.2]$$

Solve the system of equations using the inverse of the coefficient matrix of the equivalent matrix equation. [10.3]

24. $2x + 3y = 5,$
$\quad 3x + 5y = 11$

25. $5x - y + 2z = 17,$
$\quad 3x + 2y - 3z = -16,$
$\quad 4x - 3y - z = 5$

26. $w - x - y + z = -1,$
$2w + 3x - 2y - z = 2,$
$-w + 5x + 4y - 2z = 3,$
$3w - 2x + 5y + 3z = 4$

Evaluate the determinant. **[10.4]**

27. $\begin{vmatrix} 1 & -2 \\ 3 & 4 \end{vmatrix}$

28. $\begin{vmatrix} \sqrt{3} & -5 \\ -3 & -\sqrt{3} \end{vmatrix}$

29. $\begin{vmatrix} 2 & -1 & 1 \\ 1 & 2 & -1 \\ 3 & 4 & -3 \end{vmatrix}$

30. $\begin{vmatrix} 1 & -1 & 2 \\ -1 & 2 & 0 \\ -1 & 3 & 1 \end{vmatrix}$

Solve using Cramer's rule. **[10.4]**

31. $5x - 2y = 19,$
$7x + 3y = 15$

32. $x + y = 4,$
$4x + 3y = 11$

33. $3x - 2y + z = 5,$
$4x - 5y - z = -1,$
$3x + 2y - z = 4$

34. $2x - y - z = 2,$
$3x + 2y + 2z = 10,$
$x - 5y - 3z = -2$

35. Solve: $2x + y = 7,$
$x - 2y = 6.$ **[10.1]**

A. x and y are both positive numbers.
B. x and y are both negative numbers.
C. x is positive and y is negative.
D. x is negative and y is positive.

36. Which of the following is *not* a row-equivalent operation on a matrix? **[10.1]**

A. Interchange any two columns.
B. Interchange any two rows.
C. Add two rows.
D. Multiply each entry in a row by -3.

▶ **Synthesis**

37. One year, Lucia invested a total of $40,000, part at 4%, part at 5%, and the rest at $5\frac{1}{2}$%. The total

amount of interest received on the investments was $1990. The interest received on the $5\frac{1}{2}$% investment was $590 more than the interest received on the 4% investment. How much was invested at each rate? **[10.1]**

Solve.

38. $\dfrac{2}{3x} + \dfrac{4}{5y} = 8,$
$\dfrac{5}{4x} - \dfrac{3}{2y} = -6$ **[10.1]**

39. $\dfrac{3}{x} - \dfrac{4}{y} + \dfrac{1}{z} = -2,$
$\dfrac{5}{x} + \dfrac{1}{y} - \dfrac{2}{z} = 1,$
$\dfrac{7}{x} + \dfrac{3}{y} + \dfrac{2}{z} = 19$ **[10.2]**

▶ **Collaborative Discussion and Writing**

40. For square matrices **A** and **B**, is it true, in general, that $(\mathbf{AB})^2 = \mathbf{A}^2\mathbf{B}^2$? Explain. **[10.2]**

41. Given the system of equations
$a_1x + b_1y = c_1,$
$a_2x + b_2y = c_2,$

explain why the equations are dependent or the system is inconsistent when

$\begin{vmatrix} a_1 & b_1 \\ a_2 & b_2 \end{vmatrix} = 0.$ **[10.4]**

42. If the lines $a_1x + b_1y = c_1$ and $a_2x + b_2y = c_2$ are parallel, what can you say about the values of

$\begin{vmatrix} a_1 & b_1 \\ a_2 & b_2 \end{vmatrix},$ $\begin{vmatrix} c_1 & b_1 \\ c_2 & b_2 \end{vmatrix},$ and $\begin{vmatrix} a_1 & c_1 \\ a_2 & c_2 \end{vmatrix}$? **[10.4]**

---

10 Chapter Test

Solve Exercises 1–7 using Gaussian elimination or Gauss–Jordan elimination.

1. $3x + 2y = 1,$
$2x - y = -11$

2. $2x - y = 3,$
$2y = 4x - 6$

3. $x - y = 4,$
$3y = 3x - 8$

4. $2x - 3y = 8,$
$5x - 2y = 9$

5. $4x + 2y + z = 4,$
$3x - y + 5z = 4,$
$5x + 3y - 3z = -2$

6. *Ticket Sales.* One evening, 620 tickets were sold for Clearview Community College's talent show. Tickets cost $8 each for students and $12 each for nonstudents. Total receipts were $5592. How many of each type of ticket were sold?

7. Hui, Ashlyn, and Sheriann can process 352 telephone orders per day. Hui and Ashlyn together can process 224 orders per day while Hui and Sheriann together can process 248 orders per day. How many orders can each of them process alone?

For Exercises 8–13, let

$$A = \begin{bmatrix} 1 & -1 & 3 \\ -2 & 5 & 2 \end{bmatrix}, \qquad B = \begin{bmatrix} -5 & 1 \\ -2 & 4 \end{bmatrix},$$

and

$$C = \begin{bmatrix} 3 & -4 \\ -1 & 0 \end{bmatrix}.$$

Find each of the following, if possible.

8. $B + C$ **9.** $A - C$ **10.** CB

11. AB **12.** $2A$ **13.** C^{-1}

14. *Food Service Management.* The following table lists the cost per serving, in dollars, for items on three lunch menus served at a senior citizens' center.

Menu	Main Dish	Side Dish	Dessert
1	1.55	1.00	0.99
2	1.70	0.95	1.01
3	1.65	0.99	0.96

On a particular day, 26 Menu 1 meals, 18 Menu 2 meals, and 23 Menu 3 meals are served.

a) Write the information in the table as a 3×3 matrix **M**.

b) Write a row matrix **N** that represents the number of each menu served.

c) Find the product **NM**.

d) State what the entries of **NM** represent.

15. Write a matrix equation equivalent to the system of equations

$$3x - 4y + 2z = -8,$$
$$2x + 3y + z = 7,$$
$$x - 5y - 3z = 3.$$

16. Solve the system of equations using the inverse of the coefficient matrix of the equivalent matrix equation.

$$3x + 2y + 6z = 2,$$
$$x + y + 2z = 1,$$
$$2x + 2y + 5z = 3$$

Evaluate the determinant.

17. $\begin{vmatrix} 3 & -5 \\ 8 & 7 \end{vmatrix}$ **18.** $\begin{vmatrix} 2 & -1 & 4 \\ -3 & 1 & -2 \\ 5 & 3 & -1 \end{vmatrix}$

19. Solve using Cramer's rule. Show your work.

$$5x + 2y = -1,$$
$$7x + 6y = 1$$

▶ Synthesis

20. Three solutions of the equation $Ax - By = Cz - 8$ are $(2, -2, 2)$, $(-3, -1, 1)$, and $(4, 2, 9)$. Find A, B, and C.

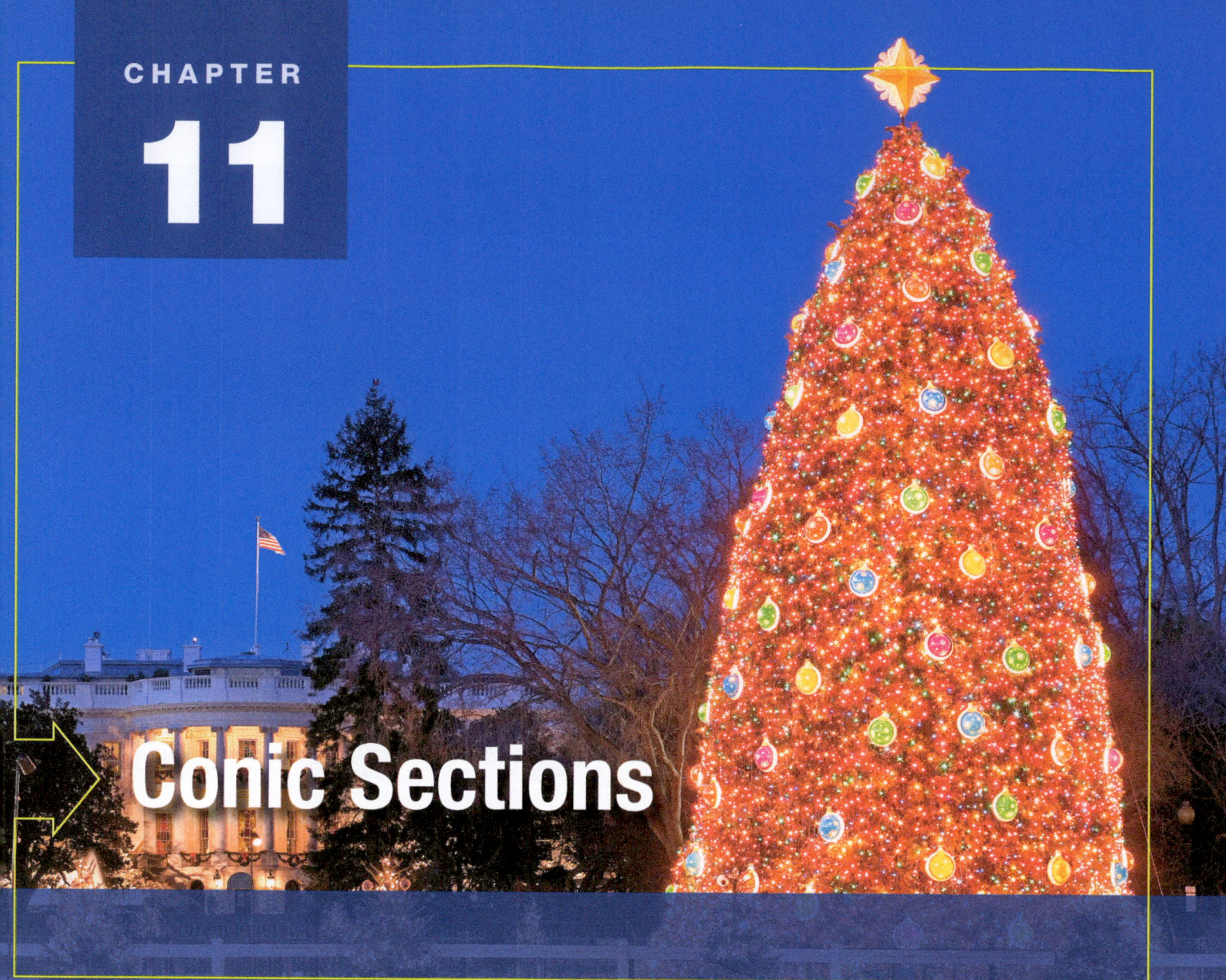

CHAPTER
11

Conic Sections

APPLICATION This problem appears as Exercise 54 in Section 11.2.

The lighting of the National Christmas Tree located on the Ellipse, a large grassy area south of the White House, marks the beginning of the holiday season in Washington, D.C. This area of the lawn is actually an ellipse with major axis of length 1048 ft and minor axis of length 898 ft. Assuming that a coordinate system is superimposed on the area in such a way that the center is at the origin and the major and minor axes are on the *x*- and *y*-axes of the coordinate system, respectively, find an equation of the ellipse.

11.1 The Parabola

▶ Given an equation of a parabola, complete the square, if necessary, and then find the vertex, the focus, and the directrix and graph the parabola.

A **conic section** is formed when a right circular cone with two parts, called *nappes*, is intersected by a plane. One of four types of curves can be formed: a parabola, a circle, an ellipse, or a hyperbola.

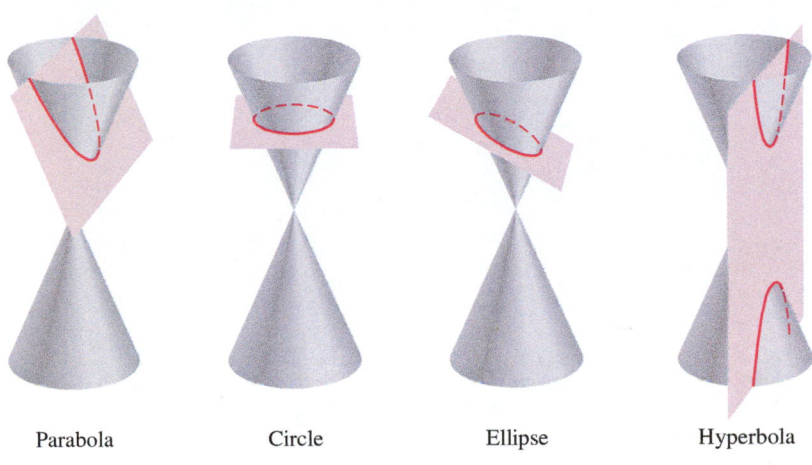

Parabola   Circle   Ellipse   Hyperbola

Conic Sections

Conic sections can be defined algebraically using second-degree equations of the form $Ax^2 + Bxy + Cy^2 + Dx + Ey + F = 0$. In addition, they can be defined geometrically as a set of points that satisfy certain conditions.

▶ Parabolas

The graph of the quadratic function $f(x) = ax^2 + bx + c, a \neq 0$, is a parabola. A parabola can be defined geometrically.

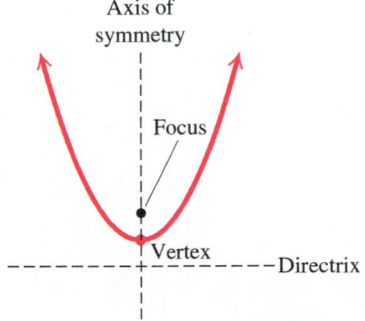

> **PARABOLA**
>
> A **parabola** is the set of all points in a plane equidistant from a fixed line (the **directrix**) and a fixed point not on the line (the **focus**).

The line that is perpendicular to the directrix and contains the focus is the **axis of symmetry**. The **vertex** is the midpoint of the segment between the focus and the directrix. (See the figure at left.)

Let's derive the standard equation of a parabola with vertex $(0, 0)$ and directrix $y = -p$, where $p > 0$. We place the coordinate axes as shown in Fig. 1. The y-axis is the axis of symmetry and contains the focus F. The distance from the focus to the vertex is the same as the distance from the vertex to the directrix. Thus the coordinates of F are $(0, p)$.

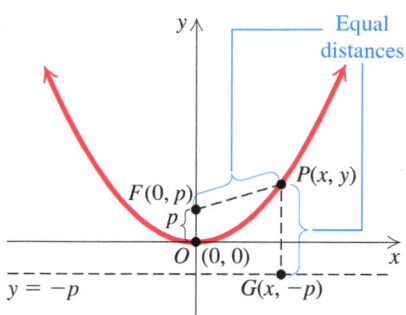

Figure 1.

Let $P(x, y)$ be any point on the parabola and consider $\overline{PG}$ perpendicular to the line $y = -p$. The coordinates of G are $(x, -p)$. By the definition of a parabola,

$$PF = PG.$$ The distance from P to the focus is the same as the distance from P to the directrix.

THE DISTANCE FORMULA

The **distance** d between any two points

$$(x_1, y_1) \text{ and } (x_2, y_2)$$

is given by

$$d = \sqrt{(x_2 - x_1)^2 + (y_2 - y_1)^2}.$$

Then using the distance formula, we have

$$\sqrt{(x - 0)^2 + (y - p)^2} = \sqrt{(x - x)^2 + [y - (-p)]^2}$$
$$x^2 + y^2 - 2py + p^2 = y^2 + 2py + p^2 \quad \text{Squaring both sides and squaring the binomials}$$
$$x^2 = 4py.$$

We have shown that if $P(x, y)$ is on the parabola shown in Fig. 1, then its coordinates satisfy this equation. The converse is also true, but we will not prove it here.

Note that if $p > 0$, as above, the graph opens up. If $p < 0$, the graph opens down.

The equation of a parabola with vertex $(0, 0)$ and directrix $x = -p$ is derived similarly. Such a parabola opens either to the right $(p > 0)$, as shown in Fig. 2, or to the left $(p < 0)$.

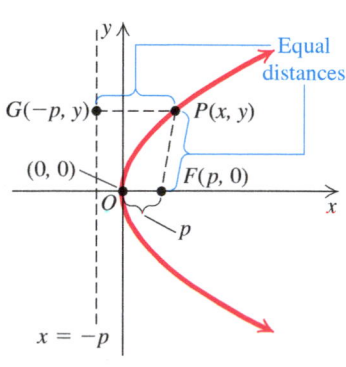

Figure 2.

STANDARD EQUATION OF A PARABOLA WITH VERTEX AT THE ORIGIN

The standard equation of a parabola with vertex $(0, 0)$ and directrix $y = -p$ is

$$x^2 = 4py.$$

The focus is $(0, p)$, and the y-axis is the axis of symmetry.

The standard equation of a parabola with vertex $(0, 0)$ and directrix $x = -p$ is

$$y^2 = 4px.$$

The focus is $(p, 0)$, and the x-axis is the axis of symmetry.

EXAMPLE 1 Find the vertex, the focus, and the directrix of the parabola $y = -\frac{1}{12}x^2$. Then graph the parabola.

Solution We write $y = -\frac{1}{12}x^2$ in the form $x^2 = 4py$:

$$-\frac{1}{12}x^2 = y \qquad \text{Given equation}$$

$$x^2 = -12y \qquad \text{Multiplying by } -12 \text{ on both sides}$$

$$x^2 = 4(-3)y. \qquad \text{Standard form}$$

Since the equation can be written in the form $x^2 = 4py$, we know that the vertex is $(0, 0)$.

We have $p = -3$, so the focus is $(0, p)$, or $(0, -3)$. The directrix is

$$y = -p = -(-3) = 3.$$

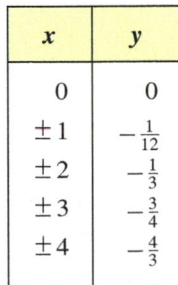

x	y
0	0
± 1	$-\frac{1}{12}$
± 2	$-\frac{1}{3}$
± 3	$-\frac{3}{4}$
± 4	$-\frac{4}{3}$

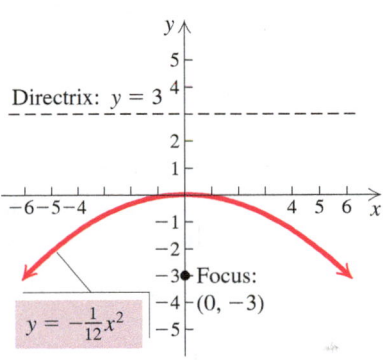

Now Try Exercise 7.

EXAMPLE 2 Find an equation of the parabola with vertex $(0, 0)$ and focus $(5, 0)$. Then graph the parabola.

Solution The focus is on the x-axis so the line of symmetry is the x-axis. Thus the equation is of the type

$$y^2 = 4px.$$

Since the focus $(5, 0)$ is 5 units to the right of the vertex, $p = 5$ and the equation is

$$y^2 = 4(5)x, \quad \text{or} \quad y^2 = 20x.$$

x	y^2	y	(x, y)
0	0	0	$(0, 0)$
1	20	$\pm\sqrt{20}$	$(1, 4.47)$
			$(1, -4.47)$
2	40	$\pm\sqrt{40}$	$(2, 6.32)$
			$(2, -6.32)$
3	60	$\pm\sqrt{60}$	$(3, 7.75)$
			$(3, -7.75)$

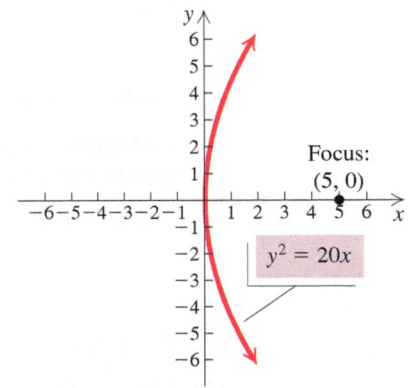

Now Try Exercise 17.

Technology Connection

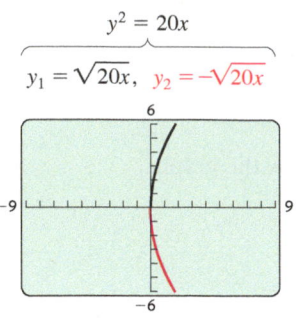

We can use a graphing calculator to graph parabolas. Consider the parabola in Example 2. It might be necessary to solve the equation for y before entering it in the calculator:

$$y^2 = 20x$$
$$y = \pm\sqrt{20x}.$$

We now graph $y_1 = \sqrt{20x}$ and $y_2 = -\sqrt{20x}$ or $y_1 = \sqrt{20x}$ and $y_2 = -y_1$ in a squared viewing window.

On some graphing calculators, the Conics application from the APPS menu can be used to graph parabolas. This method will be discussed following Example 4.

▶ ## Finding Standard Form by Completing the Square

If a parabola with vertex at the origin is translated horizontally $|h|$ units and vertically $|k|$ units, it has an equation as follows.

STANDARD EQUATION OF A PARABOLA WITH VERTEX (h, k) AND VERTICAL AXIS OF SYMMETRY

The standard equation of a parabola with vertex (h, k) and vertical axis of symmetry is

$$(x - h)^2 = 4p(y - k),$$

where the vertex is (h, k), the focus is $(h, k + p)$, and the directrix is $y = k - p$.

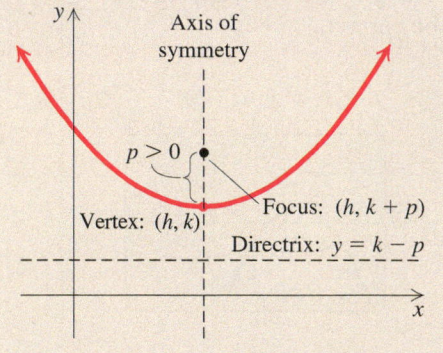

(When $p < 0$, the parabola opens down.)

STANDARD EQUATION OF A PARABOLA WITH VERTEX (h, k) AND HORIZONTAL AXIS OF SYMMETRY

The standard equation of a parabola with vertex (h, k) and horizontal axis of symmetry is

$$(y - k)^2 = 4p(x - h),$$

where the vertex is (h, k), the focus is $(h + p, k)$, and the directrix is $x = h - p.$

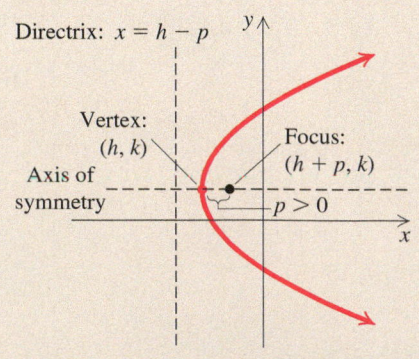

(When $p < 0$, the parabola opens to the left.)

> **COMPLETING THE SQUARE**
>
> **REVIEW SECTION 7.4**

We can complete the square on equations of the form

$$y = ax^2 + bx + c \quad \text{or} \quad x = ay^2 + by + c$$

in order to write them in standard form.

EXAMPLE 3   For the parabola

$$x^2 + 6x + 4y + 5 = 0,$$

find the vertex, the focus, and the directrix. Then draw the graph.

Solution   We first complete the square:

$$x^2 + 6x + 4y + 5 = 0$$

$$\begin{array}{ll} x^2 + 6x = -4y - 5 & \text{Subtracting } 4y \text{ and } 5 \text{ on both sides} \\ x^2 + 6x + 9 = -4y - 5 + 9 & \text{Adding 9 on both sides to complete the square on the left side} \end{array}$$

$$\begin{array}{ll} x^2 + 6x + 9 = -4y + 4 & \\ (x + 3)^2 = -4(y - 1) & \text{Factoring} \\ [x - (-3)]^2 = 4(-1)(y - 1). & \text{Writing standard form:} \\ & (x - h)^2 = 4p(y - k) \end{array}$$

We see that $h = -3$, $k = 1$, and $p = -1$, so we have the following:

Vertex (h, k):	$(-3, 1)$;
Focus $(h, k + p)$:	$(-3, 1 + (-1))$, or $(-3, 0)$;
Directrix $y = k - p$:	$y = 1 - (-1)$, or $y = 2$.

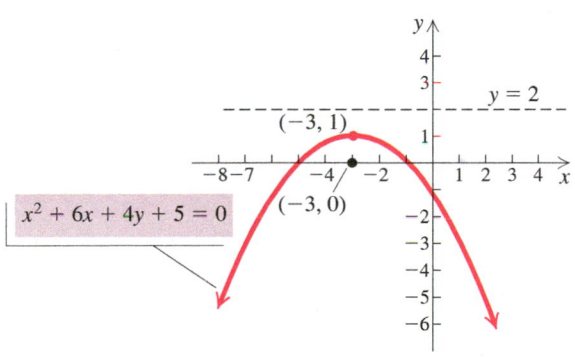

Now Try Exercise 25.

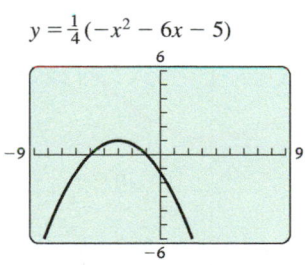
EXAMPLE 4   For the parabola

$$y^2 - 2y - 8x - 31 = 0,$$

find the vertex, the focus, and the directrix. Then draw the graph.

Solution   We first complete the square:

$$y^2 - 2y - 8x - 31 = 0$$

$$\begin{array}{ll} y^2 - 2y = 8x + 31 & \text{Adding } 8x \text{ and } 31 \text{ on both sides} \\ y^2 - 2y + 1 = 8x + 31 + 1 & \text{Adding 1 on both sides to complete the square on the left side} \end{array}$$

$$\begin{array}{ll} y^2 - 2y + 1 = 8x + 32 & \\ (y - 1)^2 = 8(x + 4) & \text{Factoring} \\ (y - 1)^2 = 4(2)[x - (-4)]. & \text{Writing standard form:} \\ & (y - k)^2 = 4p(x - h) \end{array}$$

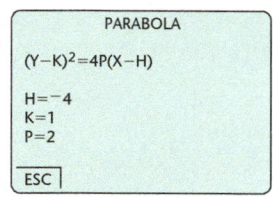
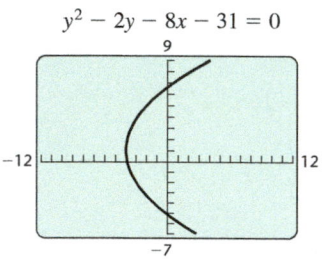
We see that $h = -4$, $k = 1$, and $p = 2$, so we have the following:

Vertex (h, k): $(-4, 1)$;

Focus $(h + p, k)$: $(-4 + 2, 1)$, or $(-2, 1)$;

Directrix $x = h - p$: $x = -4 - 2$, or $x = -6$.

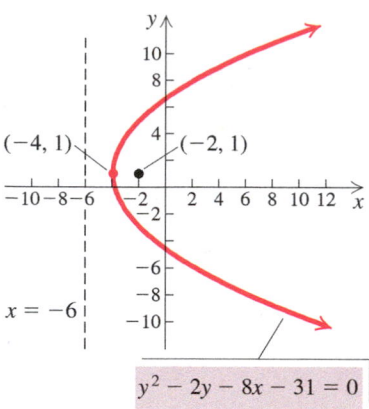

$$y^2 - 2y - 8x - 31 = 0$$

Now Try Exercise 31.

▶ Applications

Parabolas have many applications. For example, cross sections of car headlights, flashlights, and searchlights are parabolas. The bulb is located at the focus and light from that point is reflected outward parallel to the axis of symmetry. Satellite dishes and field microphones used at sporting events often have parabolic cross sections. Incoming radio waves or sound waves parallel to the axis are reflected into the focus.

Similarly, in solar cooking, a parabolic mirror is mounted on a rack with a cooking pot hung in the focal area. Incoming sun rays parallel to the axis are reflected into the focus, producing a temperature high enough for cooking.

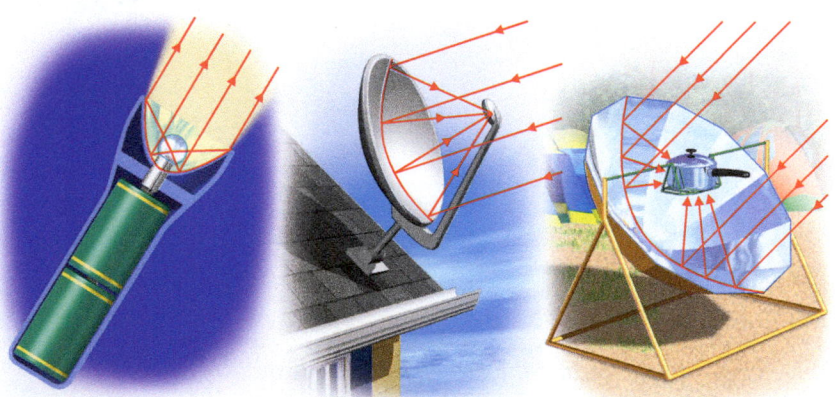

11.1   Exercise Set

In Exercises 1–6, match the equation with one of the graphs (a)–(f) that follow.

a)

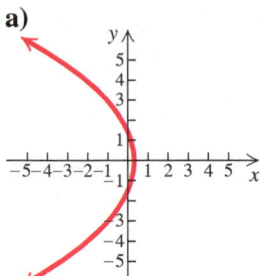

b)

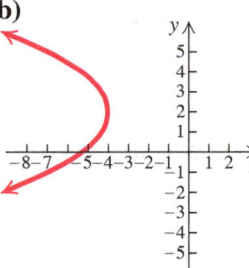

c)

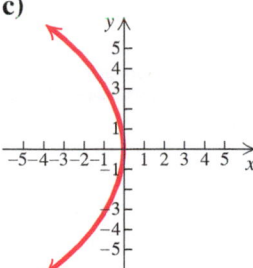

d)

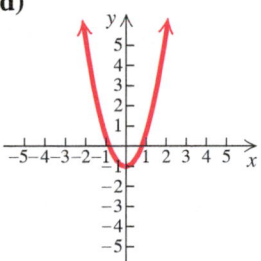

e)

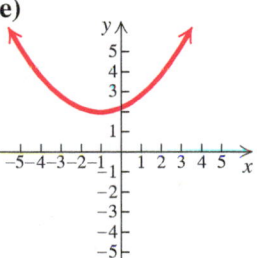

f)

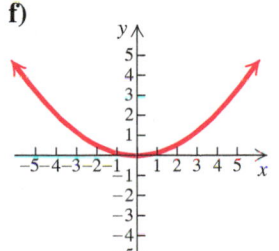

1. $x^2 = 8y$

2. $y^2 = -10x$

3. $(y - 2)^2 = -3(x + 4)$

4. $(x + 1)^2 = 5(y - 2)$

5. $13x^2 - 8y - 9 = 0$

6. $41x + 6y^2 = 12$

Find the vertex, the focus, and the directrix. Then draw the graph.

7. $x^2 = 20y$

8. $x^2 = 16y$

9. $y^2 = -6x$

10. $y^2 = -2x$

11. $x^2 - 4y = 0$

12. $y^2 + 4x = 0$

13. $x = 2y^2$

14. $y = \frac{1}{2}x^2$

Find an equation of a parabola satisfying the given conditions.

15. Vertex $(0, 0)$, focus $(-3, 0)$

16. Vertex $(0, 0)$, focus $(0, 10)$

17. Focus $(7, 0)$, directrix $x = -7$

18. Focus $\left(0, \frac{1}{4}\right)$, directrix $y = -\frac{1}{4}$

19. Focus $(0, -\pi)$, directrix $y = \pi$

20. Focus $\left(-\sqrt{2}, 0\right)$, directrix $x = \sqrt{2}$

21. Focus $(3, 2)$, directrix $x = -4$

22. Focus $(-2, 3)$, directrix $y = -3$

Find the vertex, the focus, and the directrix. Then draw the graph.

23. $(x + 2)^2 = -6(y - 1)$

24. $(y - 3)^2 = -20(x + 2)$

25. $x^2 + 2x + 2y + 7 = 0$

26. $y^2 + 6y - x + 16 = 0$

27. $x^2 - y - 2 = 0$

28. $x^2 - 4x - 2y = 0$

29. $y = x^2 + 4x + 3$

30. $y = x^2 + 6x + 10$

31. $y^2 - y - x + 6 = 0$

32. $y^2 + y - x - 4 = 0$

33. *Satellite Dish.*   An engineer designs a satellite dish with a parabolic cross section. The dish is 15 ft wide at the opening, and the focus is placed 4 ft from the vertex.

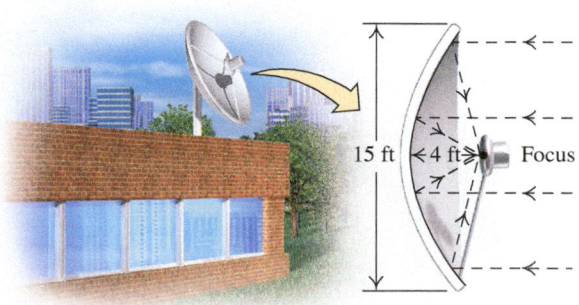

a) Position a coordinate system with the origin at the vertex and the *x*-axis on the parabola's axis of symmetry and find an equation of the parabola.

b) Find the depth of the satellite dish at the vertex.

34. *Flashlight Mirror.* A heavy-duty flashlight mirror has a parabolic cross section with diameter 6 in. and depth 1 in.

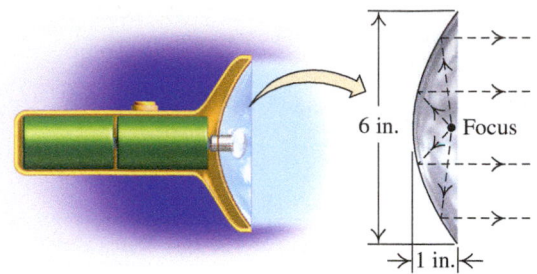

6 in.   • Focus

→|1 in.|←

a) Position a coordinate system with the origin at the vertex and the *x*-axis on the parabola's axis of symmetry and find an equation of the parabola.

b) How far from the vertex should the bulb be positioned if it is to be placed at the focus?

35. *Spotlight.* A spotlight has a parabolic cross section that is 4 ft wide at the opening and 1.5 ft deep at the vertex. How far from the vertex is the focus?

36. *Ultrasound Receiver.* Information Unlimited designed and sells the Ultrasonic Receiver, which detects sounds unable to be heard by the human ear. The HT90P can detect mechanical and electrical sounds such as leaking gases, air, corona, and motor friction noises. It can also be used to hear bats, insects, and even beading water. The receiver has a parabolic cross section and is 2.625 in. deep. The focus is 3.287 in. from the vertex. (*Source:* Information Unlimited, Amherst, NH, Robert Iannini, President) Find the diameter of the outside edge of the receiver.

► **Skill Maintenance**

Consider the following linear equations. Without graphing them, answer the questions below.

a) $y = 2x$ **b)** $y = \frac{1}{3}x + 5$
c) $y = -3x - 2$ **d)** $y = -0.9x + 7$
e) $y = -5x + 3$ **f)** $y = x + 4$
g) $8x - 4y = 7$ **h)** $3x + 6y = 2$

37. Which has/have *x*-intercept $\left(\frac{2}{3}, 0\right)$? **[2.6]**

38. Which has/have *y*-intercept $(0, 7)$? **[2.5], [2.6]**

39. Which slant up from left to right? **[2.5]**

40. Which has the least steep slant? **[2.5]**

41. Which has/have slope $\frac{1}{3}$? **[2.5]**

42. Which, if any, contain the point $(3, 7)$? **[2.1]**

43. Which, if any, are parallel? **[2.6]**

44. Which, if any, are perpendicular? **[2.6]**

► **Synthesis**

45. Find an equation of the parabola with a vertical axis of symmetry and vertex $(-1, 2)$ and containing the point $(-3, 1)$.

46. Find an equation of a parabola with a horizontal axis of symmetry and vertex $(-2, 1)$ and containing the point $(-3, 5)$.

47. *Suspension Bridge.* The parabolic cables of a 200-ft portion of the roadbed of a suspension bridge are positioned as shown below. Vertical cables are to be spaced every 20 ft along this portion of the roadbed. Calculate the lengths of these vertical cables.

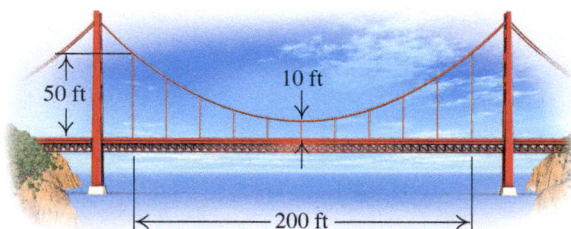

50 ft

10 ft

←——— 200 ft ———→

11.2   The Circle and the Ellipse

▶ Given an equation of a circle, complete the square, if necessary, and then find the center and the radius and graph the circle.

▶ Given an equation of an ellipse, complete the square, if necessary, and then find the center, the vertices, and the foci and graph the ellipse.

▶ Circles

We can define a circle geometrically.

> **CIRCLE**
>
> A **circle** is the set of all points in a plane that are at a fixed distance from a fixed point (the **center**) in the plane.

We state the standard equation of a circle with center (h, k) and radius r.

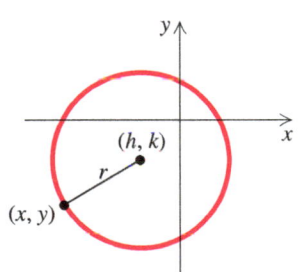

> **STANDARD EQUATION OF A CIRCLE**
>
> The standard equation of a circle with center (h, k) and radius r is
> $$(x - h)^2 + (y - k)^2 = r^2.$$

EXAMPLE 1   For the circle
$$x^2 + y^2 - 16x + 14y + 32 = 0,$$
find the center and the radius. Then graph the circle.

Solution   First, we complete the square twice:
$$x^2 + y^2 - 16x + 14y + 32 = 0$$
$$x^2 - 16x \quad + y^2 + 14y \qquad = -32$$
$$x^2 - 16x + 64 + y^2 + 14y + 49 = -32 + 64 + 49$$

$$[\tfrac{1}{2}(-16)]^2 = (-8)^2 = 64 \text{ and } (\tfrac{1}{2} \cdot 14)^2 = 7^2 = 49;$$
adding 64 and 49 on both sides to complete the square twice on the left side

$$(x - 8)^2 + (y + 7)^2 = 81$$
$$(x - 8)^2 + [y - (-7)]^2 = 9^2. \qquad \textbf{Writing standard form}$$

The center is $(8, -7)$ and the radius is 9. We graph the circle as shown below.

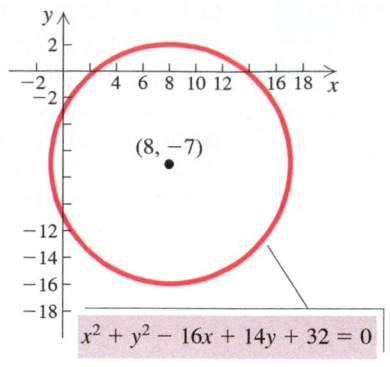

$$x^2 + y^2 - 16x + 14y + 32 = 0$$

Now Try Exercise 7.

Technology Connection

When we use the Conics CIRCLE APP to graph a circle, it is not necessary to write the equation in standard form or to solve it for y first. We enter the coefficients of x^2, y^2, x, and y and also the constant term when the equation is written in the form $ax^2 + ay^2 + bx + cy + d = 0$. For the circle in Example 1, we enter 1 for A, -16 for B, 14 for C, and 32 for D.

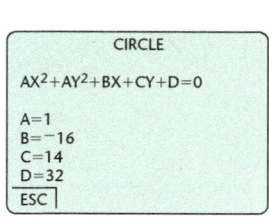

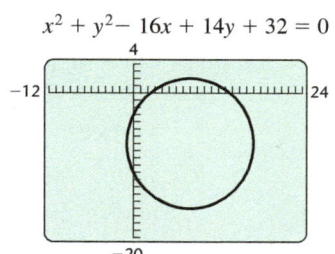

$$x^2 + y^2 - 16x + 14y + 32 = 0$$

Some graphing calculators have a DRAW feature that provides a quick way to graph a circle when the center and the radius are known.

► Ellipses

We have studied two conic sections, the parabola and the circle. Now we turn our attention to a third, the *ellipse*.

ELLIPSE

An **ellipse** is the set of all points in a plane, the sum of whose distances from two fixed points (the **foci**) is constant. The **center** of an ellipse is the midpoint of the segment between the foci.

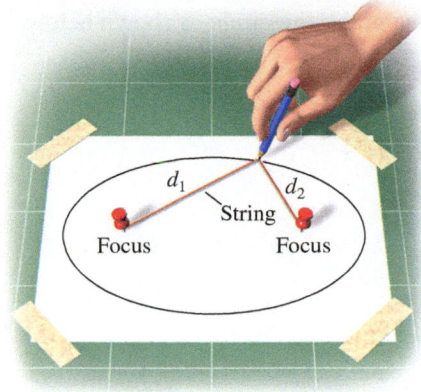

We can draw an ellipse by first placing two thumbtacks in a piece of cardboard. These are the foci (singular, *focus*). We then attach a piece of string to the tacks. Its length is the constant sum of the distances $d_1 + d_2$ from the foci to any point on the ellipse. Next, we trace a curve with a pencil held tight against the string. The figure traced is an ellipse.

Let's first consider the ellipse shown below with center at the origin. The points F_1 and F_2 are the foci. The segment $\overline{A'A}$ is the **major axis**, and the points A' and A are the **vertices**. The segment $\overline{B'B}$ is the **minor axis**, and the points B' and B are the **y-intercepts**. Note that the major axis of an ellipse is longer than the minor axis.

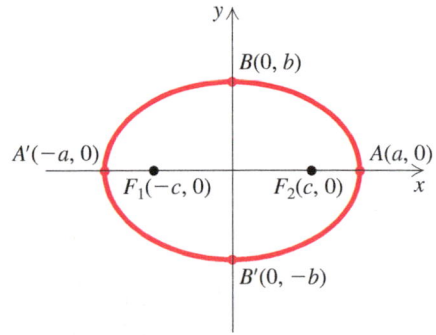

STANDARD EQUATION OF AN ELLIPSE WITH CENTER AT THE ORIGIN

Major Axis Horizontal

$$\frac{x^2}{a^2} + \frac{y^2}{b^2} = 1, \quad a > b > 0$$

Vertices: $(-a, 0), (a, 0)$

y-intercepts: $(0, -b), (0, b)$

Foci: $(-c, 0), (c, 0)$, where $c^2 = a^2 - b^2$

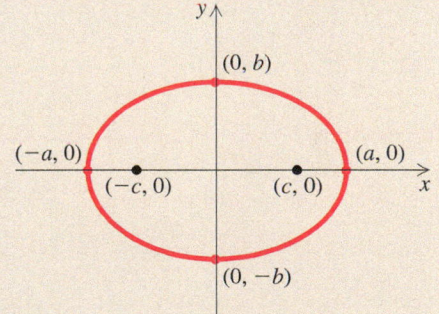

Major Axis Vertical

$$\frac{x^2}{b^2} + \frac{y^2}{a^2} = 1, \quad a > b > 0$$

Vertices: $(0, -a), (0, a)$

x-intercepts: $(-b, 0), (b, 0)$

Foci: $(0, -c), (0, c)$, where $c^2 = a^2 - b^2$

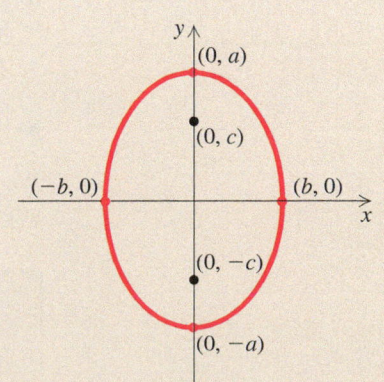

EXAMPLE 2 Find the standard equation of the ellipse with vertices $(-5, 0)$ and $(5, 0)$ and foci $(-3, 0)$ and $(3, 0)$. Then graph the ellipse.

Solution Since the foci are on the x-axis and the origin is the midpoint of the segment between them, the major axis is horizontal and $(0, 0)$ is the center of the ellipse. Thus the equation is of the form

$$\frac{x^2}{a^2} + \frac{y^2}{b^2} = 1.$$

Since the vertices are $(-5, 0)$ and $(5, 0)$ and the foci are $(-3, 0)$ and $(3, 0)$, we know that $a = 5$ and $c = 3$. These values can be used to find b^2:

$$c^2 = a^2 - b^2$$
$$3^2 = 5^2 - b^2$$
$$9 = 25 - b^2$$
$$b^2 = 16.$$

Thus the equation of the ellipse is

$$\frac{x^2}{5^2} + \frac{y^2}{4^2} = 1, \quad \text{or} \quad \frac{x^2}{25} + \frac{y^2}{16} = 1.$$

To graph the ellipse, we plot the vertices $(-5, 0)$ and $(5, 0)$. Since $b^2 = 16$, we know that $b = 4$ and the y-intercepts are $(0, -4)$ and $(0, 4)$. We plot these points as well and connect the four points we have plotted with a smooth curve.

> **Now Try Exercise 31.**

Technology Connection

When the equation of an ellipse is written in standard form, we can use the Conics ELLIPSE APP on a graphing calculator to graph it. The standard equation of the ellipse in Example 2 is

$$\frac{x^2}{5^2} + \frac{y^2}{4^2} = 1.$$

Note that the center is $(0, 0)$. We enter 5 for A, 4 for B, 0 for H, and 0 for K.

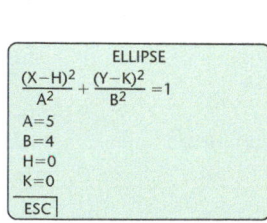

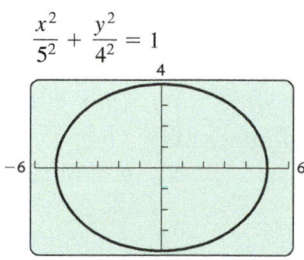

EXAMPLE 3 For the ellipse

$$9x^2 + 4y^2 = 36,$$

find the vertices and the foci. Then draw the graph.

Solution   We first find standard form:

$$9x^2 + 4y^2 = 36$$

$$\frac{9x^2}{36} + \frac{4y^2}{36} = \frac{36}{36}$$ **Dividing by 36 on both sides to get 1 on the right side**

$$\frac{x^2}{4} + \frac{y^2}{9} = 1$$

$$\frac{x^2}{2^2} + \frac{y^2}{3^2} = 1.$$ **Writing standard form**

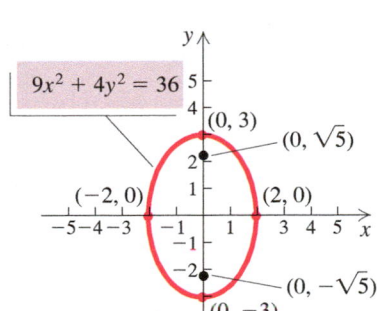

Thus, $a = 3$ and $b = 2$. The major axis is vertical, so the vertices are $(0, -3)$ and $(0, 3)$. Since we know that $c^2 = a^2 - b^2$, we have $c^2 = 3^2 - 2^2 = 9 - 4 = 5$, so $c = \sqrt{5}$ and the foci are $\left(0, -\sqrt{5}\right)$ and $\left(0, \sqrt{5}\right)$.

To graph the ellipse, we plot the vertices. Note also that since $b = 2$, the x-intercepts are $(-2, 0)$ and $(2, 0)$. We plot these points as well and connect the four points we have plotted with a smooth curve.

Now Try Exercise 25.

If the center of an ellipse is not at the origin but at some point (h, k), then we can think of an ellipse with center at the origin being translated horizontally $|h|$ units and vertically $|k|$ units.

STANDARD EQUATION OF AN ELLIPSE WITH CENTER AT (h, k)

Major Axis Horizontal

$$\frac{(x - h)^2}{a^2} + \frac{(y - k)^2}{b^2} = 1, \quad a > b > 0$$

Vertices: $(h - a, k), (h + a, k)$

Length of minor axis: $2b$

Foci: $(h - c, k), (h + c, k)$, where $c^2 = a^2 - b^2$

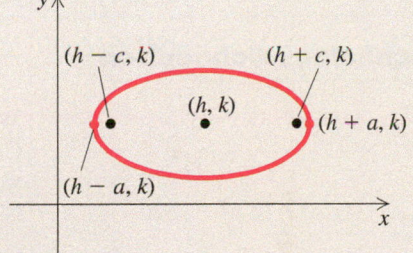

Major Axis Vertical

$$\frac{(x - h)^2}{b^2} + \frac{(y - k)^2}{a^2} = 1, \quad a > b > 0$$

Vertices: $(h, k - a), (h, k + a)$

Length of minor axis: $2b$

Foci: $(h, k - c), (h, k + c)$, where $c^2 = a^2 - b^2$

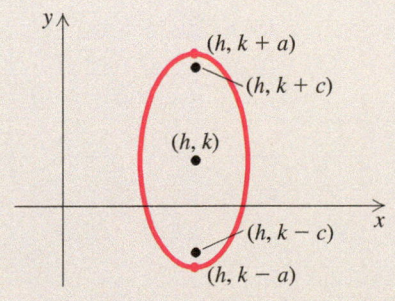

EXAMPLE 4   For the ellipse

$$4x^2 + y^2 + 24x - 2y + 21 = 0,$$

find the center, the vertices, and the foci. Then draw the graph.

Solution First, we complete the square twice to get standard form:

$$4x^2 + y^2 + 24x - 2y + 21 = 0$$
$$4(x^2 + 6x \qquad) + (y^2 - 2y \qquad) = -21$$
$$4(x^2 + 6x + 9 - 9) + (y^2 - 2y + 1 - 1) = -21 \qquad \text{Completing the square twice}$$
$$4(x^2 + 6x + 9) + 4(-9) + (y^2 - 2y + 1) + (-1) = -21$$
$$4(x + 3)^2 - 36 + (y - 1)^2 - 1 = -21$$
$$4(x + 3)^2 + (y - 1)^2 = 16 \qquad \text{Adding 37 on both sides}$$
$$\tfrac{1}{16}[4(x + 3)^2 + (y - 1)^2] = \tfrac{1}{16} \cdot 16$$
$$\frac{(x + 3)^2}{4} + \frac{(y - 1)^2}{16} = 1$$
$$\frac{[x - (-3)]^2}{2^2} + \frac{(y - 1)^2}{4^2} = 1. \qquad \text{Writing standard form}$$

The center is $(-3, 1)$. Note that $a = 4$ and $b = 2$. The major axis is vertical, so the vertices are 4 units above and below the center:

$$(-3, 1 + 4) \text{ and } (-3, 1 - 4), \quad \text{or} \quad (-3, 5) \text{ and } (-3, -3).$$

We know that $c^2 = a^2 - b^2$, so $c^2 = 4^2 - 2^2 = 16 - 4 = 12$ and $c = \sqrt{12}$, or $2\sqrt{3}$. Then the foci are $2\sqrt{3}$ units above and below the center:

$$\left(-3, 1 + 2\sqrt{3}\right) \quad \text{and} \quad \left(-3, 1 - 2\sqrt{3}\right).$$

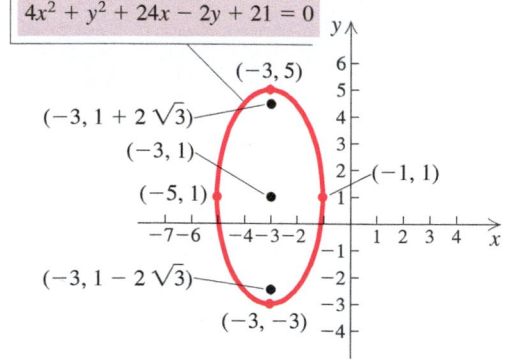

To graph the ellipse, we plot the vertices. Note also that since $b = 2$, two other points on the graph are the endpoints of the minor axis, 2 units right and left of the center:

$$(-3 + 2, 1) \quad \text{and} \quad (-3 - 2, 1),$$

or

$$(-1, 1) \quad \text{and} \quad (-5, 1).$$

We plot these points as well and connect the four points with a smooth curve, as shown at left.

> **Now Try Exercise 43.**

Technology Connection

When the equation of an ellipse is written in standard form, we can use the Conics ELLIPSE APP to graph it. The standard equation of the ellipse in Example 4 is

$$\frac{[x - (-3)]^2}{2^2} + \frac{(y - 1)^2}{4^2} = 1.$$

Note that the center is $(-3, 1)$. We enter 4 for A, 2 for B, -3 for H, and 1 for K.

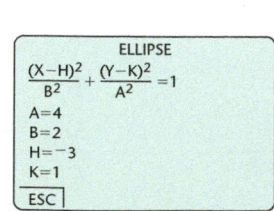

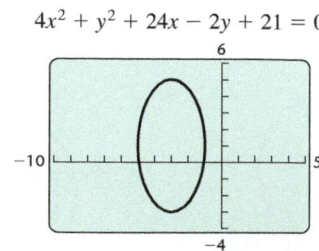
$4x^2 + y^2 + 24x - 2y + 21 = 0$

▶ Applications

An exciting medical application of an ellipse is a device called a *lithotripter*. One type of this device uses electromagnetic technology to generate a shock wave to pulverize kidney stones. The wave originates at one focus of an ellipse and is reflected to the kidney stone, which is positioned at the other focus. Recovery time following the use of this technique is much shorter than with conventional surgery.

Ellipses have many other applications. Planets travel around the sun in elliptical orbits with the sun at one focus, for example, and satellites travel around the earth in elliptical orbits as well.

A room with an ellipsoidal ceiling is known as a *whispering gallery*. In such a room, a word whispered at one focus can be clearly heard at the other. Whispering galleries are found in the rotunda of the Capitol Building in Washington, D.C., and in St. Paul's Cathedral in London.

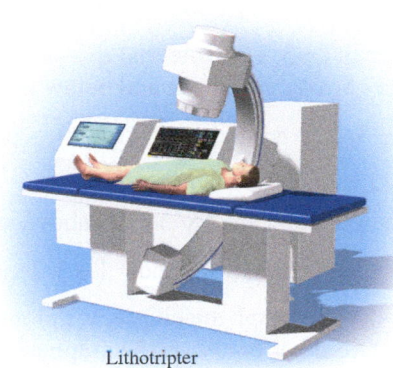

Lithotripter

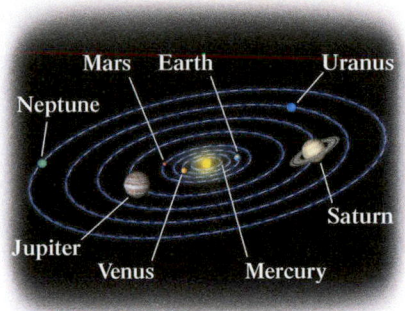

11.2   Exercise Set

In Exercises 1–6, match the equation with one of the graphs (a)–(f) that follow.

a)

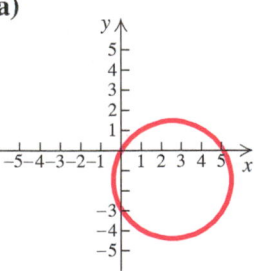

b)

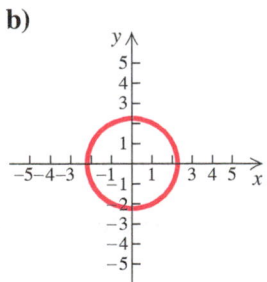

c)

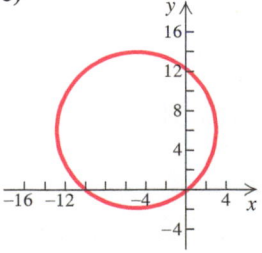

d)

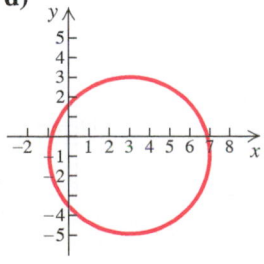

e)

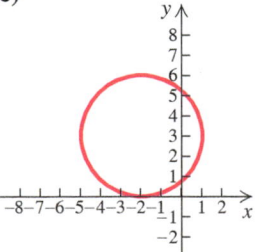

f)

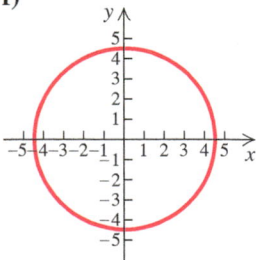

1. $x^2 + y^2 = 5$

2. $y^2 = 20 - x^2$

3. $x^2 + y^2 - 6x + 2y = 6$

4. $x^2 + y^2 + 10x - 12y = 3$

5. $x^2 + y^2 - 5x + 3y = 0$

6. $x^2 + 4x - 2 = 6y - y^2 - 6$

Find the center and the radius of the circle with the given equation. Then draw the graph.

7. $x^2 + y^2 - 14x + 4y = 11$

8. $x^2 + y^2 + 2x - 6y = -6$

9. $x^2 + y^2 + 6x - 2y = 6$

10. $x^2 + y^2 - 4x + 2y = 4$

11. $x^2 + y^2 + 4x - 6y - 12 = 0$

12. $x^2 + y^2 - 8x - 2y - 19 = 0$

13. $x^2 + y^2 - 6x - 8y + 16 = 0$

14. $x^2 + y^2 - 2x + 6y + 1 = 0$

15. $x^2 + y^2 + 6x - 10y = 0$

16. $x^2 + y^2 - 7x - 2y = 0$

17. $x^2 + y^2 - 9x = 7 - 4y$

18. $y^2 - 6y - 1 = 8x - x^2 + 3$

In Exercises 19–22, match the equation with one of the graphs (a)–(d) that follow.

a)

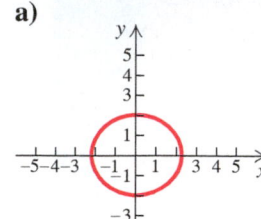

b)

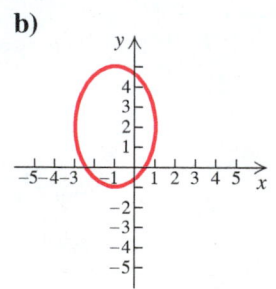

c)

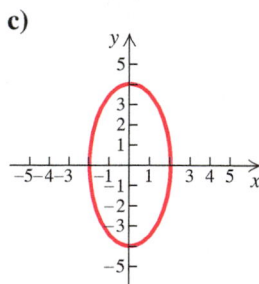

d)

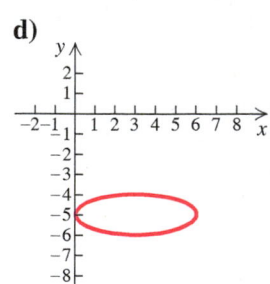

19. $16x^2 + 4y^2 = 64$

20. $4x^2 + 5y^2 = 20$

21. $x^2 + 9y^2 - 6x + 90y = -225$

22. $9x^2 + 4y^2 + 18x - 16y = 11$

Find the vertices and the foci of the ellipse with the given equation. Then draw the graph.

23. $\dfrac{x^2}{4} + \dfrac{y^2}{1} = 1$

24. $\dfrac{x^2}{25} + \dfrac{y^2}{36} = 1$

25. $16x^2 + 9y^2 = 144$

26. $9x^2 + 4y^2 = 36$

27. $2x^2 + 3y^2 = 6$

28. $5x^2 + 7y^2 = 35$

29. $4x^2 + 9y^2 = 1$

30. $25x^2 + 16y^2 = 1$

Find an equation of an ellipse satisfying the given conditions.

31. Vertices: $(-7, 0)$ and $(7, 0)$; foci: $(-3, 0)$ and $(3, 0)$

32. Vertices: $(0, -6)$ and $(0, 6)$; foci: $(0, -4)$ and $(0, 4)$

33. Vertices: $(0, -8)$ and $(0, 8)$; length of minor axis: 10

34. Vertices: $(-5, 0)$ and $(5, 0)$; length of minor axis: 6

35. Foci: $(-2, 0)$ and $(2, 0)$; length of major axis: 6

36. Foci: $(0, -3)$ and $(0, 3)$; length of major axis: 10

Find the center, the vertices, and the foci of the ellipse. Then draw the graph.

37. $\dfrac{(x - 1)^2}{9} + \dfrac{(y - 2)^2}{4} = 1$

38. $\dfrac{(x - 1)^2}{1} + \dfrac{(y - 2)^2}{4} = 1$

39. $\dfrac{(x + 3)^2}{25} + \dfrac{(y - 5)^2}{36} = 1$

40. $\dfrac{(x - 2)^2}{16} + \dfrac{(y + 3)^2}{25} = 1$

41. $3(x + 2)^2 + 4(y - 1)^2 = 192$

42. $4(x - 5)^2 + 3(y - 4)^2 = 48$

43. $4x^2 + 9y^2 - 16x + 18y - 11 = 0$

44. $x^2 + 2y^2 - 10x + 8y + 29 = 0$

45. $4x^2 + y^2 - 8x - 2y + 1 = 0$

46. $9x^2 + 4y^2 + 54x - 8y + 49 = 0$

*The **eccentricity** of an ellipse is defined as $e = c/a$. For an ellipse, $0 < c < a$, so $0 < e < 1$. When e is close to 0, an ellipse appears to be nearly circular. When e is close to 1, an ellipse is very flat.*

47. Note the shapes of the ellipses in Examples 2 and 4. Which ellipse has the smaller eccentricity? Confirm your answer by computing the eccentricity of each ellipse.

48. Which ellipse has the smaller eccentricity? (Assume that the coordinate systems have the same scale.)

a)   **b)**

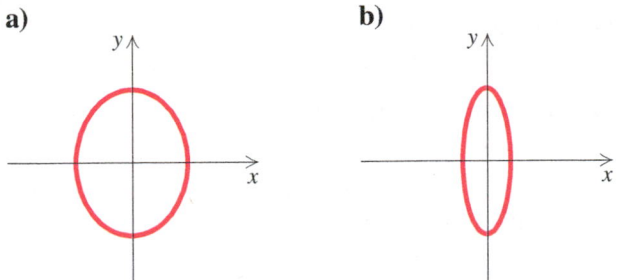

49. Find an equation of an ellipse with vertices $(0, -4)$ and $(0, 4)$ and $e = \frac{1}{4}$.

50. Find an equation of an ellipse with vertices $(-3, 0)$ and $(3, 0)$ and $e = \frac{7}{10}$.

51. *Bridge Supports.* The bridge support shown in the following figure is the top half of an ellipse. Assuming that a coordinate system is superimposed on the drawing in such a way that point Q, the center of the ellipse, is at the origin, find an equation of the ellipse.

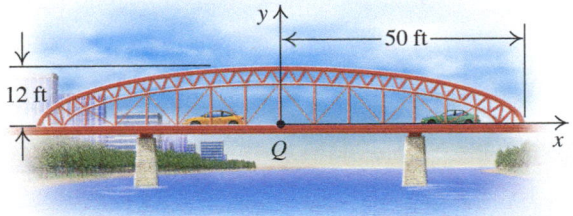

52. *Whispering Gallery.* An art museum is adding a new exhibit room in the shape of an ellipse. The director wants to mark the foci so that a tour guide can stand at one focus and without speaking loudly can be clearly heard by a group touring the museum. If the room is 64 ft long and each focus is 5 ft from the outside wall along the major axis, how high is the ceiling? Round to the nearest tenth of a foot.

53. *Whispering Gallery.* A whispering gallery, often elliptical in shape, has acoustic properties such that a whisper made at one point can be heard at other

distant points. A science museum is designing a new exhibit hall that will illustrate a whispering gallery. The hall will be 90 ft in length with the ceiling 30 ft high at the center. How far are the foci from the center of the ellipse? Round to the nearest tenth of a foot.

54. *The Ellipse.* The lighting of the National Christmas Tree located on the Ellipse, a large grassy area south of the White House, marks the beginning of the holiday season in Washington, D.C. This area of the lawn is actually an ellipse with major axis of length 1048 ft and minor axis of length 898 ft. Assuming that a co-ordinate system is superimposed on the area in such a way that the center is at the origin and the major and minor axes are on the *x*- and *y*-axes of the coordinate system, respectively, find an equation of the ellipse.

55. *The Earth's Orbit.* The maximum distance of the earth from the sun is 9.3×10^7 mi. The minimum distance is 9.1×10^7 mi. The sun is at one focus of the elliptical orbit. Find the distance from the sun to the other focus.

56. *Carpentry.* A carpenter is cutting a 3-ft by 4-ft elliptical sign from a 3-ft by 4-ft piece of plywood. The ellipse will be drawn using a string attached to the board at the foci of the ellipse.

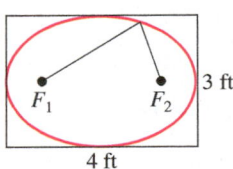

a) How far from the ends of the board should the string be attached?

b) How long should the string be?

▶ Skill Maintenance

Vocabulary Reinforcement

In each of Exercises 57–63, fill in the blank with the correct term. Some of the given choices will not be used.

> piecewise function
> linear equation
> factor
> remainder
> solution
> zero
> x-intercept
> y-intercept
> parabola
> circle
> ellipse
> one real-number solution
> two different real-number solutions
> two different imaginary-number solutions

57. An input c of a function f is a(n) _____ of the function if $f(c) = 0$. **[7.4]**

58. A(n) _____ of the graph of an equation is a point $(0, b)$. **[2.5]**

59. For a quadratic equation $ax^2 + bx + c = 0$, if $b^2 - 4ac > 0$, the equation has _____. **[7.4]**

60. Given a polynomial $f(x)$, then $f(c)$ is the _____ that would be obtained by dividing $f(x)$ by $x - c$. **[8.3]**

61. A(n) _____ is the set of all points in a plane the sum of whose distances from two fixed points is constant. **[11.2]**

62. A(n) _____ is the set of all points in a plane equidistant from a fixed line and a fixed point not on the line. **[11.1]**

63. A(n) _____ is the set of all points in a plane that are at a fixed distance from a fixed point in the plane. **[11.2]**

▶ Synthesis

Find an equation of an ellipse satisfying the given conditions.

64. Vertices: $(-1, -1)$, $(-1, 5)$; endpoints of minor axis: $(-3, 2)$, $(1, 2)$

65. Vertices: $(3, -4)$, $(3, 6)$; endpoints of minor axis: $(1, 1)$, $(5, 1)$

66. Center: $(-2, 3)$; major axis vertical; length of major axis: 4; length of minor axis: 1

67. Vertices: $(-3, 0)$ and $(3, 0)$; passing through $\left(2, \frac{22}{3}\right)$

68. *Bridge Arch.* A bridge with a semielliptical arch spans a river as shown here. What is the clearance 6 ft from the riverbank?

14 ft

50 ft

Mid-Chapter Review

Determine whether the statement is true or false.

1. The graph of $(x + 3)^2 = 8(y - 2)$ is a parabola with vertex $(-3, 2)$. **[11.1]**

2. A parabola must open up or down. **[11.1]**

3. The graph of $(x - 4)^2 + (y + 1)^2 = 9$ is a circle with radius 9. **[11.2]**

4. The major axis of the ellipse $\dfrac{x^2}{4} + \dfrac{y^2}{16} = 1$ is vertical. **[11.2]**

In Exercises 5–12, match the equation with one of the graphs (a)–(h) that follow. **[11.1]**, **[11.2]**

a)
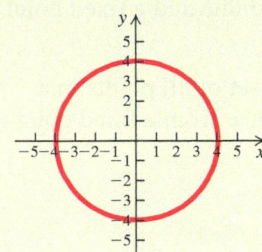

b)
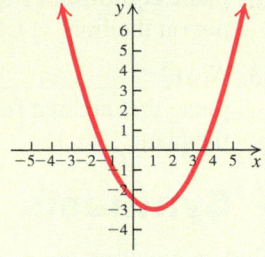

c)
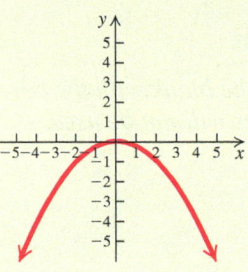

d)
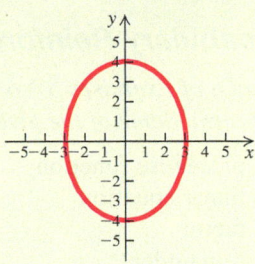

e)
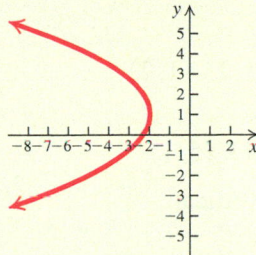

f)
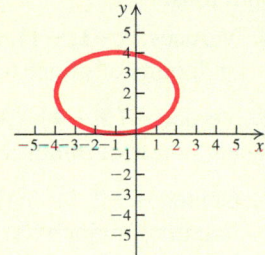

g)
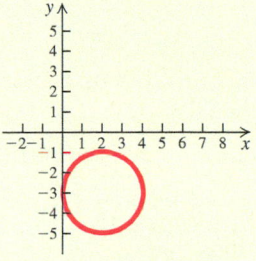

h)
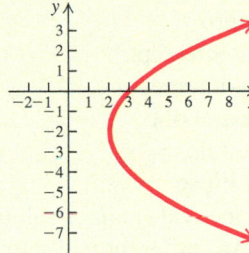

5. $x^2 = -4y$

6. $(y + 2)^2 = 4(x - 2)$

7. $16x^2 + 9y^2 = 144$

8. $x^2 + y^2 = 16$

9. $(x - 1)^2 = 2(y + 3)$

10. $4(x + 1)^2 + 9(y - 2)^2 = 36$

11. $(x - 2)^2 + (y + 3)^2 = 4$

12. $y^2 - 2y + 3x + 7 = 0$

Find the vertex, the focus, and the directrix of the parabola. Then draw the graph. **[11.1]**

13. $y^2 = 12x$

14. $x^2 - 6x - 4y = -17$

Find the equation of a parabola satisfying the given conditions. **[11.1]**

15. Focus: $(0, 3)$; directrix: $y = 1$

16. Focus: $(-4, 6)$; directrix: $x = 2$

Find the center and the radius of the circle. Then draw the graph. **[11.2]**

17. $x^2 + y^2 + 4x - 8y = 5$

18. $x^2 + y^2 - 6x + 2y - 6 = 0$

Find the vertices and the foci of the ellipse. Then draw the graph. **[11.2]**

19. $\dfrac{x^2}{1} + \dfrac{y^2}{9} = 1$

20. $2x^2 + 3y^2 = 12$

21. $\dfrac{(x - 2)^2}{16} + \dfrac{(y + 1)^2}{4} = 1$

22. $25x^2 + 4y^2 - 50x + 8y = 71$

Write an equation of the ellipse satisfying the given conditions. **[11.2]**

23. Vertices: $(-5, 0)$, $(5, 0)$;
foci: $(-2, 0)$, $(2, 0)$

24. Vertices: $(0, -3)$, $(0, 3)$; length of minor axis: 4

25. Foci: $(-3, 0)$, $(3, 0)$;
length of major axis: 8

Collaborative Discussion and Writing

26. Is a parabola always the graph of a function? Why or why not? **[11.1]**

27. Explain how the distance formula is used to find the standard equation of a parabola. **[11.1]**

28. Explain why function notation is not used in Section 11.2. **[11.2]**

29. Is the center of an ellipse part of the graph of the ellipse? Why or why not? **[11.2]**

11.3   The Hyperbola

▶ Given an equation of a hyperbola, complete the square, if necessary, and then find the center, the vertices, and the foci and graph the hyperbola.

The last type of conic section that we will study is the *hyperbola*.

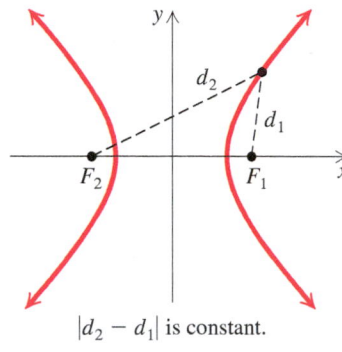

$|d_2 - d_1|$ is constant.

> ### HYPERBOLA
>
> A **hyperbola** is the set of all points in a plane for which the absolute value of the difference of the distances from two fixed points (the **foci**) is constant. The midpoint of the segment between the foci is the **center** of the hyperbola.

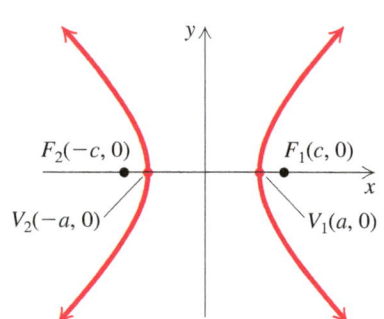

▶ Standard Equations of Hyperbolas

We first consider the equation of a hyperbola with center at the origin. In the figure at left, F_1 and F_2 are the foci. The segment $\overline{V_2V_1}$ is the **transverse axis** and the points V_2 and V_1 are the **vertices**.

> ### STANDARD EQUATION OF A HYPERBOLA WITH CENTER AT THE ORIGIN
>
> ***Transverse Axis Horizontal***
>
> $$\frac{x^2}{a^2} - \frac{y^2}{b^2} = 1$$
>
> Vertices: $(-a, 0), (a, 0)$
> Foci: $(-c, 0), (c, 0)$,
>   where $c^2 = a^2 + b^2$
>
>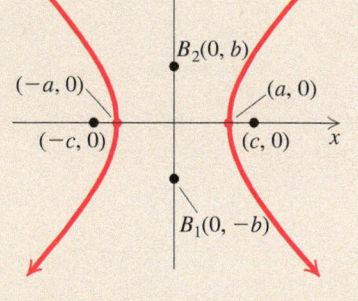
>
> ***Transverse Axis Vertical***
>
> $$\frac{y^2}{a^2} - \frac{x^2}{b^2} = 1$$
>
> Vertices: $(0, -a), (0, a)$
> Foci: $(0, -c), (0, c)$,
>   where $c^2 = a^2 + b^2$
>
>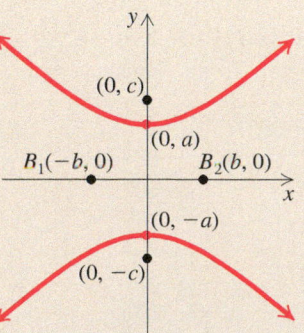

The segment $\overline{B_1B_2}$ is the **conjugate axis** of the hyperbola.

To graph a hyperbola with a horizontal transverse axis, it is helpful to begin by graphing the lines $y = -(b/a)x$ and $y = (b/a)x$. These are the **asymptotes** of the hyperbola. For a hyperbola with a vertical transverse axis, the asymptotes are $y = -(a/b)x$ and $y = (a/b)x$. As $|x|$ gets larger and larger, the graph of the hyperbola gets closer and closer to the asymptotes.

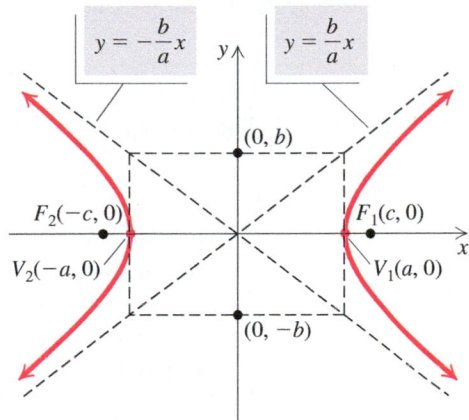

EXAMPLE 1   Find an equation of the hyperbola with vertices $(0, -4)$ and $(0, 4)$ and foci $(0, -6)$ and $(0, 6)$.

Solution   We know that $a = 4$ and $c = 6$. We find b^2:

$$c^2 = a^2 + b^2$$
$$6^2 = 4^2 + b^2$$
$$36 = 16 + b^2$$
$$20 = b^2.$$

Since the vertices and the foci are on the y-axis, we know that the transverse axis is vertical. We can now write the equation of the hyperbola:

$$\frac{y^2}{a^2} - \frac{x^2}{b^2} = 1$$
$$\frac{y^2}{16} - \frac{x^2}{20} = 1.$$

> **Now Try Exercise 7.**

EXAMPLE 2   For the hyperbola given by

$$9x^2 - 16y^2 = 144,$$

find the vertices, the foci, and the asymptotes. Then graph the hyperbola.

Solution   First, we find standard form:

$$9x^2 - 16y^2 = 144$$
$$\frac{1}{144}(9x^2 - 16y^2) = \frac{1}{144} \cdot 144 \qquad \text{Multiplying by } \tfrac{1}{144} \text{ to get 1 on the right side}$$
$$\frac{x^2}{16} - \frac{y^2}{9} = 1$$
$$\frac{x^2}{4^2} - \frac{y^2}{3^2} = 1. \qquad \text{Writing standard form}$$

The hyperbola has a horizontal transverse axis, so the vertices are $(-a, 0)$ and $(a, 0)$, or $(-4, 0)$ and $(4, 0)$. From the standard form of the equation, we know that $a^2 = 4^2$, or 16, and $b^2 = 3^2$, or 9. We find the foci:

$$c^2 = a^2 + b^2$$
$$c^2 = 16 + 9$$
$$c^2 = 25$$
$$c = 5.$$

Thus the foci are $(-5, 0)$ and $(5, 0)$.

Next, we find the asymptotes:

$$y = -\frac{b}{a}x = -\frac{3}{4}x \quad \text{and} \quad y = \frac{b}{a}x = \frac{3}{4}x.$$

To draw the graph, we sketch the asymptotes first. This is easily done by drawing the rectangle with horizontal sides passing through $(0, 3)$ and $(0, -3)$ and vertical sides through $(4, 0)$ and $(-4, 0)$. Then we draw and extend the diagonals of this rectangle. The two extended diagonals are the asymptotes of the hyperbola. Next, we plot the vertices and draw the branches of the hyperbola outward from the vertices toward the asymptotes.

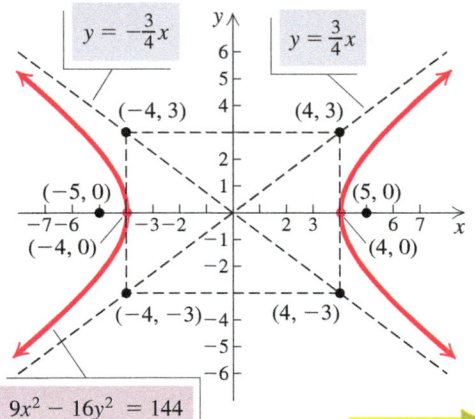

Now Try Exercise 17.

Technology Connection

When the equation of a hyperbola is written in standard form, we can use the Conics HYPERBOLA APP to graph it. The standard equation of the hyperbola in Example 2 is

$$\frac{x^2}{4^2} - \frac{y^2}{3^2} = 1.$$

Note that the center is $(0, 0)$. We enter 4 for A, 3 for B, 0 for H, and 0 for K.

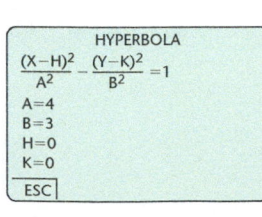

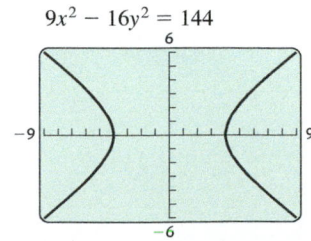

If a hyperbola with center at the origin is translated horizontally $|h|$ units and vertically $|k|$ units, the center is at the point (h, k).

STANDARD EQUATION OF A HYPERBOLA WITH CENTER AT (h, k)

Transverse Axis Horizontal

$$\frac{(x - h)^2}{a^2} - \frac{(y - k)^2}{b^2} = 1$$

Vertices: $(h - a, k), (h + a, k)$

Asymptotes: $y - k = \dfrac{b}{a}(x - h), y - k = -\dfrac{b}{a}(x - h)$

Foci: $(h - c, k), (h + c, k)$, where $c^2 = a^2 + b^2$

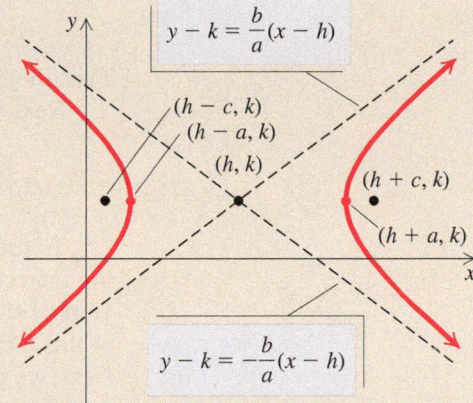

Transverse Axis Vertical

$$\frac{(y - k)^2}{a^2} - \frac{(x - h)^2}{b^2} = 1$$

Vertices: $(h, k - a), (h, k + a)$

Asymptotes: $y - k = \dfrac{a}{b}(x - h), y - k = -\dfrac{a}{b}(x - h)$

Foci: $(h, k - c), (h, k + c)$, where $c^2 = a^2 + b^2$

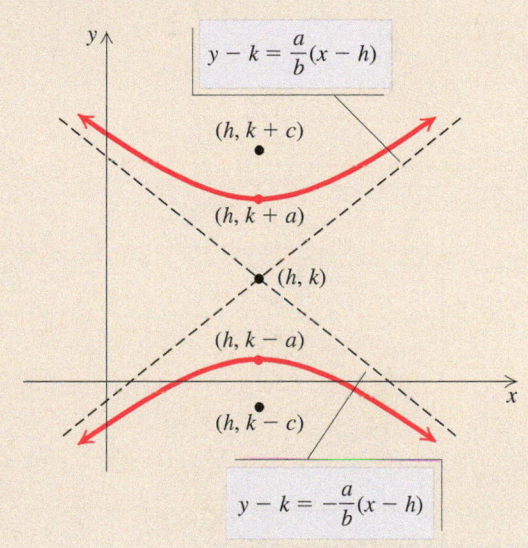

EXAMPLE 3   For the hyperbola given by

$$4y^2 - x^2 + 24y + 4x + 28 = 0,$$

find the center, the vertices, the foci, and the asymptotes. Then draw the graph.

Solution   First, we complete the square to get standard form:

$$4y^2 - x^2 + 24y + 4x + 28 = 0$$
$$4(y^2 + 6y \qquad) - (x^2 - 4x \qquad) = -28$$
$$4(y^2 + 6y + 9 - 9) - (x^2 - 4x + 4 - 4) = -28$$
$$4(y^2 + 6y + 9) + 4(-9) - (x^2 - 4x + 4) - (-4) = -28$$
$$4(y^2 + 6y + 9) - 36 - (x^2 - 4x + 4) + 4 = -28$$
$$4(y^2 + 6y + 9) - (x^2 - 4x + 4) = -28 + 36 - 4$$
$$4(y + 3)^2 - (x - 2)^2 = 4$$
$$\frac{(y + 3)^2}{1} - \frac{(x - 2)^2}{4} = 1 \qquad \text{\color{red}{Dividing by 4}}$$
$$\frac{[y - (-3)]^2}{1^2} - \frac{(x - 2)^2}{2^2} = 1. \qquad \text{\color{red}{Standard form}}$$

The center is $(2, -3)$. Note that $a = 1$ and $b = 2$. The transverse axis is vertical, so the vertices are 1 unit below and above the center:

$$(2, -3 - 1) \text{ and } (2, -3 + 1), \quad \text{or} \quad (2, -4) \text{ and } (2, -2).$$

We know that $c^2 = a^2 + b^2$, so $c^2 = 1^2 + 2^2 = 1 + 4 = 5$ and $c = \sqrt{5}$. Thus the foci are $\sqrt{5}$ units below and above the center:

$$\left(2, -3 - \sqrt{5}\right) \quad \text{and} \quad \left(2, -3 + \sqrt{5}\right).$$

The asymptotes are

$$y - (-3) = \frac{1}{2}(x - 2) \quad \text{and} \quad y - (-3) = -\frac{1}{2}(x - 2),$$

or

$$y + 3 = \frac{1}{2}(x - 2) \quad \text{and} \quad y + 3 = -\frac{1}{2}(x - 2).$$

We sketch the asymptotes, plot the vertices, and draw the graph.

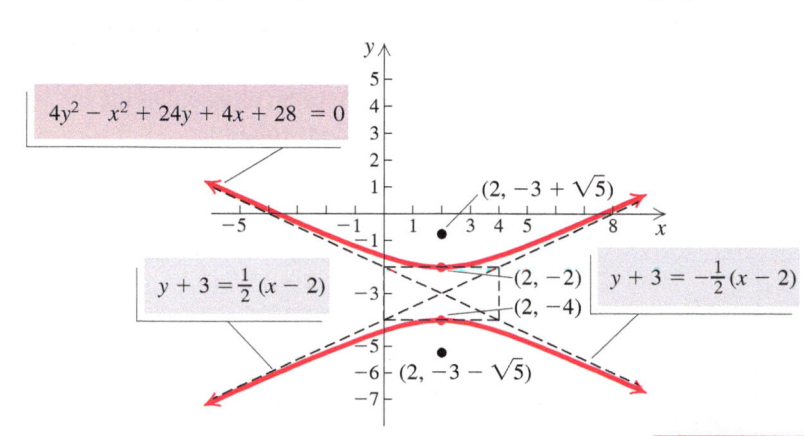

Now Try Exercise 29.

Technology Connection

When the equation of a hyperbola is written in standard form, we can use the Conics HYPERBOLA APP to graph it. The hyperbola in Example 3 is shown below.

$$\frac{[y - (-3)]^2}{1^2} - \frac{(x - 2)^2}{2^2} = 1$$

HYPERBOLA
$$\frac{(Y-K)^2}{A^2} - \frac{(X-H)^2}{B^2} = 1$$

A=1
B=2
H=2
K=-3
ESC

$$4y^2 - x^2 + 24y + 4x + 28 = 0$$

CONNECTING THE CONCEPTS

Classifying Equations of Conic Sections

EQUATION	TYPE OF CONIC SECTION	GRAPH
$x - 4 + 4y = y^2$	Only one variable is squared, so this cannot be a circle, an ellipse, or a hyperbola. Find an equivalent equation: $$x = (y - 2)^2.$$ This is an equation of a parabola.	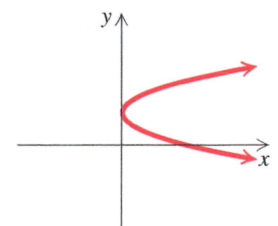
$3x^2 + 3y^2 = 75$	Both variables are squared, so this cannot be a parabola. The squared terms are added, so this cannot be a hyperbola. Divide by 3 on both sides to find an equivalent equation: $$x^2 + y^2 = 25.$$ This is an equation of a circle.	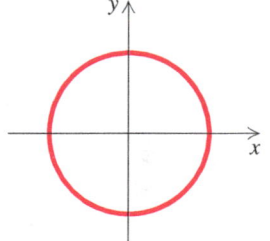
$y^2 = 16 - 4x^2$	Both variables are squared, so this cannot be a parabola. Add $4x^2$ on both sides to find an equivalent equation: $4x^2 + y^2 = 16$. The squared terms are added, so this cannot be a hyperbola. The coefficients of x^2 and y^2 are not the same, so this is not a circle. Divide by 16 on both sides to find an equivalent equation: $$\frac{x^2}{4} + \frac{y^2}{16} = 1.$$ This is an equation of an ellipse.	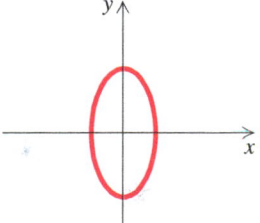
$x^2 = 4y^2 + 36$	Both variables are squared, so this cannot be a parabola. Subtract $4y^2$ on both sides to find an equivalent equation: $x^2 - 4y^2 = 36$. The squared terms are not added, so this cannot be a circle or an ellipse. Divide by 36 on both sides to find an equivalent equation: $$\frac{x^2}{36} - \frac{y^2}{9} = 1.$$ This is an equation of a hyperbola.	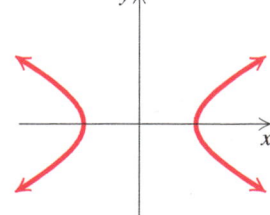

▶ **Applications**

Some comets travel in hyperbolic paths with the sun at one focus. Such comets pass by the sun only one time, unlike those with elliptical orbits, which reappear at intervals. We also see hyperbolas in architecture, such as in a cross section of a planetarium, an amphitheater, or a cooling tower for a steam or nuclear power plant.

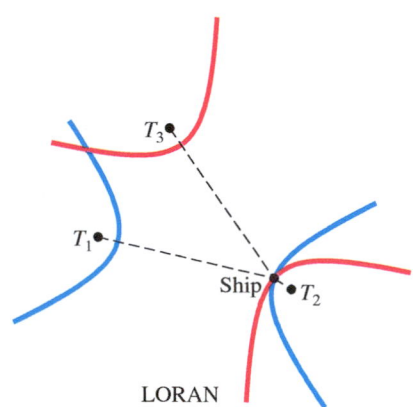

LORAN

Another application of hyperbolas is in the long-range navigation system LORAN. This system uses transmitting stations in three locations to send out simultaneous signals to a ship or an aircraft. The difference in the arrival times of the signals from one pair of transmitters is recorded on the ship or aircraft. This difference is also recorded for signals from another pair of transmitters. For each pair, a computation is performed to determine the difference in the distances from each member of the pair to the ship or aircraft. If each pair of differences is kept constant, two hyperbolas can be drawn. Each has one of the pairs of transmitters as foci, and the ship or aircraft lies on the intersection of two of their branches.

11.3 Exercise Set

In Exercises 1–6, match the equation with one of the graphs (a)–(f) that follow.

a)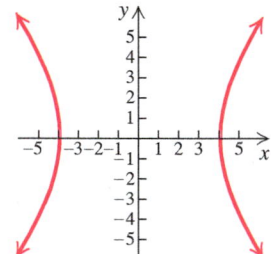

b)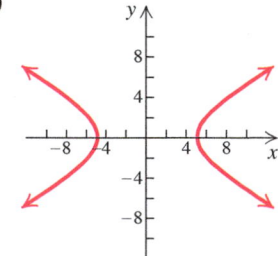

c)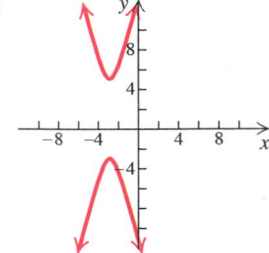

d)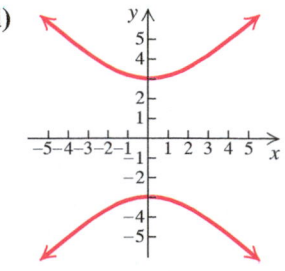

e)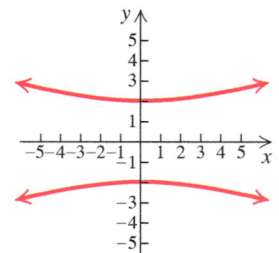

f)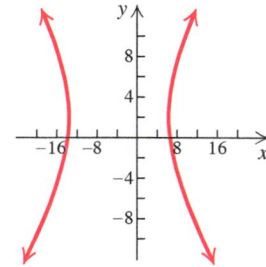

1. $\dfrac{x^2}{25} - \dfrac{y^2}{9} = 1$

2. $\dfrac{y^2}{4} - \dfrac{x^2}{36} = 1$

3. $\dfrac{(y-1)^2}{16} - \dfrac{(x+3)^2}{1} = 1$

4. $\dfrac{(x+4)^2}{100} - \dfrac{(y-2)^2}{81} = 1$

5. $25x^2 - 16y^2 = 400$

6. $y^2 - x^2 = 9$

Find an equation of a hyperbola satisfying the given conditions.

7. Vertices at $(0, 3)$ and $(0, -3)$; foci at $(0, 5)$ and $(0, -5)$

8. Vertices at $(1, 0)$ and $(-1, 0)$; foci at $(2, 0)$ and $(-2, 0)$

9. Asymptotes $y = \frac{3}{2}x$, $y = -\frac{3}{2}x$; one vertex $(2, 0)$

10. Asymptotes $y = \frac{5}{4}x$, $y = -\frac{5}{4}x$; one vertex $(0, 3)$

Find the center, the vertices, the foci, and the asymptotes of the hyperbola. Then draw the graph.

11. $\dfrac{x^2}{4} - \dfrac{y^2}{4} = 1$

12. $\dfrac{x^2}{1} - \dfrac{y^2}{9} = 1$

13. $\dfrac{(x - 2)^2}{9} - \dfrac{(y + 5)^2}{1} = 1$

14. $\dfrac{(x - 5)^2}{16} - \dfrac{(y + 2)^2}{9} = 1$

15. $\dfrac{(y + 3)^2}{4} - \dfrac{(x + 1)^2}{16} = 1$

16. $\dfrac{(y + 4)^2}{25} - \dfrac{(x + 2)^2}{16} = 1$

17. $x^2 - 4y^2 = 4$ **18.** $4x^2 - y^2 = 16$

19. $9y^2 - x^2 = 81$ **20.** $y^2 - 4x^2 = 4$

21. $x^2 - y^2 = 2$ **22.** $x^2 - y^2 = 3$

23. $y^2 - x^2 = \frac{1}{4}$ **24.** $y^2 - x^2 = \frac{1}{9}$

Find the center, the vertices, the foci, and the asymptotes of the hyperbola. Then draw the graph.

25. $x^2 - y^2 - 2x - 4y - 4 = 0$

26. $4x^2 - y^2 + 8x - 4y - 4 = 0$

27. $36x^2 - y^2 - 24x + 6y - 41 = 0$

28. $9x^2 - 4y^2 + 54x + 8y + 41 = 0$

29. $9y^2 - 4x^2 - 18y + 24x - 63 = 0$

30. $x^2 - 25y^2 + 6x - 50y = 41$

31. $x^2 - y^2 - 2x - 4y = 4$

32. $9y^2 - 4x^2 - 54y - 8x + 41 = 0$

33. $y^2 - x^2 - 6x - 8y - 29 = 0$

34. $x^2 - y^2 = 8x - 2y - 13$

The **eccentricity** *of a hyperbola is defined as* $e = c/a$. *For a hyperbola,* $c > a > 0$, *so* $e > 1$. *When e is close to 1, a hyperbola appears to be very narrow. As the eccentricity increases, the hyperbola becomes "wider."*

35. Note the shapes of the hyperbolas in Examples 2 and 3. Which hyperbola has the larger eccentricity? Confirm your answer by computing the eccentricity of each hyperbola.

36. Which hyperbola has the larger eccentricity? (Assume that the coordinate systems have the same scale.)

a)

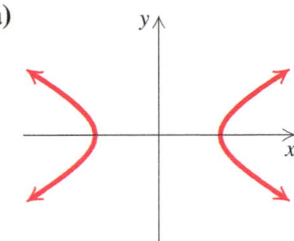

b)

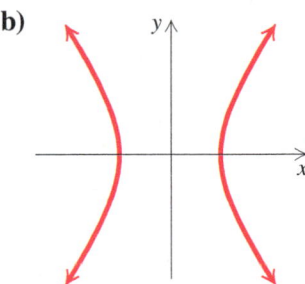

37. Find an equation of a hyperbola with vertices $(3, 7)$ and $(-3, 7)$ and $e = \frac{5}{3}$.

38. Find an equation of a hyperbola with vertices $(-1, 3)$ and $(-1, 7)$ and $e = 4$.

39. *Hyperbolic Mirror.* Certain telescopes contain both a parabolic mirror and a hyperbolic mirror. In the telescope shown in the figure, the parabola and the hyperbola share focus F_1, which is 14 m above the vertex of the parabola. The hyperbola's second focus F_2 is 2 m above the parabola's vertex. The vertex of the hyperbolic mirror is 1 m below F_1. Position a coordinate system with the origin at the center of the hyperbola and with the foci on the y-axis. Then find the equation of the hyperbola.

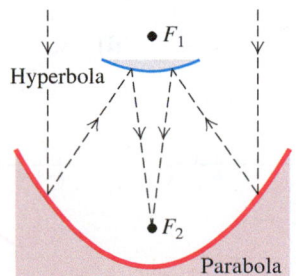

40. *Nuclear Cooling Tower.* A cross section of a nuclear cooling tower is a hyperbola with equation

$$\frac{x^2}{90^2} - \frac{y^2}{130^2} = 1.$$

The tower is 450 ft tall and the distance from the top of the tower to the center of the hyperbola is half the distance from the base of the tower to the center of the hyperbola. Find the diameter of the top and the base of the tower.

▶ **Skill Maintenance**

In Exercises 41–44, given the function:

a) *Determine whether it is one-to-one.* [9.2]
b) *If it is one-to-one, find a formula for the inverse.* [9.2]

41. $f(x) = 2x - 3$ **42.** $f(x) = x^3 + 2$

43. $f(x) = \dfrac{5}{x - 1}$ **44.** $f(x) = \sqrt{x + 4}$

Solve using matrices. [3.2], [3.3], [10.1], [10.3], [10.4]

45. $x + y = 5,$
  $x - y = 7$

46. $3x - 2y = 5,$
  $5x + 2y = 3$

47. $2x - 3y = 7,$
  $3x + 5y = 1$

48. $3x + 2y = -1,$
  $2x + 3y = 6$

▶ **Synthesis**

Find an equation of a hyperbola satisfying the given conditions.

49. Vertices at $(3, -8)$ and $(3, -2)$;
 asymptotes $y = 3x - 14$, $y = -3x + 4$

50. Vertices at $(-9, 4)$ and $(-5, 4)$;
 asymptotes $y = 3x + 25$, $y = -3x - 17$

51. *Navigation.* Two radio transmitters positioned 300 mi apart along the shore send simultaneous signals to a ship that is 200 mi offshore and sailing parallel to the shoreline. The signal from transmitter S reaches the ship 200 microseconds later than the signal from transmitter T. The signals travel at a speed of 186,000 miles per second, or 0.186 mile per microsecond. Find the equation of the hyperbola with foci S and T on which the ship is located. (*Hint*: For any point on the hyperbola, the absolute value of the difference of its distances from the foci is $2a$.)

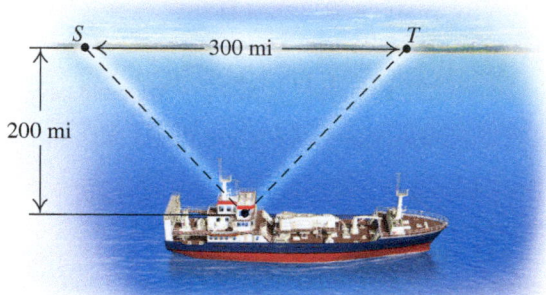

11.4   **Nonlinear Systems of Equations and Inequalities**

▶ Solve a nonlinear system of equations.

▶ Use nonlinear systems of equations to solve applied problems.

▶ Graph nonlinear systems of inequalities.

The systems of equations that we have studied so far have been composed of linear equations. Now we consider systems of two equations in two variables in which at least one equation is not linear.

▶ ## Nonlinear Systems of Equations

The graphs of the equations in a nonlinear system of equations can have no point of intersection or one or more points of intersection. The coordinates of each point of intersection represent a solution of the system of equations. When no point of intersection exists, the system of equations has no real-number solution.

Solutions of nonlinear systems of equations can be found using the substitution method or the elimination method. The substitution method is preferable for a system consisting of one linear equation and one nonlinear equation. The elimination method is preferable in most, but not all, cases when both equations are nonlinear.

EXAMPLE 1　Solve the following system of equations:

$$x^2 + y^2 = 25, \qquad \textbf{(1)} \qquad \text{\color{red}The graph is a circle.}$$
$$3x - 4y = 0. \qquad \textbf{(2)} \qquad \text{\color{red}The graph is a line.}$$

Algebraic Solution

We use the substitution method. First, we solve equation (2) for x:

$$3x - 4y = 0 \qquad \textbf{(2)}$$
$$3x = 4y$$
$$x = \tfrac{4}{3}y. \qquad \textbf{(3)} \qquad \text{\color{red}We could have solved for } y \text{ instead.}$$

Next, we substitute $\tfrac{4}{3}y$ for x in equation (1) and solve for y:

$$\left(\tfrac{4}{3}y\right)^2 + y^2 = 25$$
$$\tfrac{16}{9}y^2 + y^2 = 25$$
$$\tfrac{25}{9}y^2 = 25$$
$$y^2 = 9 \qquad \text{\color{red}Multiplying by } \tfrac{9}{25}$$
$$y = \pm 3.$$

Now we substitute these numbers for y in equation (3) and solve for x:

$$x = \tfrac{4}{3}(3) = 4, \qquad \text{\color{red}The pair } (4, 3) \text{ appears to be a solution.}$$

$$x = \tfrac{4}{3}(-3) = -4. \qquad \text{\color{red}The pair } (-4, -3) \text{ appears to be a solution.}$$

Check:　For $(4, 3)$:

$$\begin{array}{c|c}
x^2 + y^2 = 25 & 3x - 4y = 0 \\
\hline
4^2 + 3^2 \;?\; 25 & 3(4) - 4(3) \;?\; 0 \\
16 + 9 & 12 - 12 \\
25 \;\big|\; 25 \;\; \text{TRUE} & 0 \;\big|\; 0 \;\; \text{TRUE}
\end{array}$$

For $(-4, -3)$:

$$\begin{array}{c|c}
x^2 + y^2 = 25 & 3x - 4y = 0 \\
\hline
(-4)^2 + (-3)^2 \;?\; 25 & 3(-4) - 4(-3) \;?\; 0 \\
16 + 9 & -12 + 12 \\
25 \;\big|\; 25 \;\; \text{TRUE} & 0 \;\big|\; 0 \;\; \text{TRUE}
\end{array}$$

The pairs $(4, 3)$ and $(-4, -3)$ check, so they are the solutions.

Visualizing the Solution

The ordered pairs corresponding to the points of intersection of the graphs of the equations are the solutions of the system of equations.

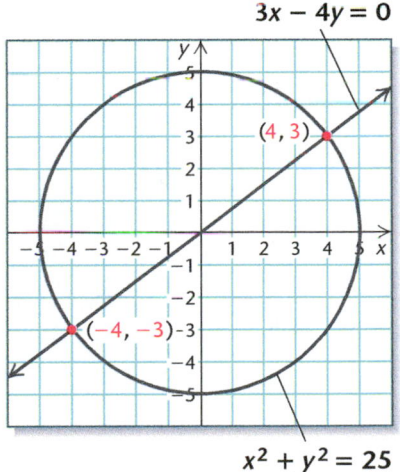

We see that the solutions are $(4, 3)$ and $(-4, -3)$.

Now Try Exercise 7.

Technology Connection

To solve the system of equations in Example 1 on a graphing calculator, we graph both equations in the same viewing window. Note that there are two points of intersection. We can find their coordinates using the INTERSECT feature.

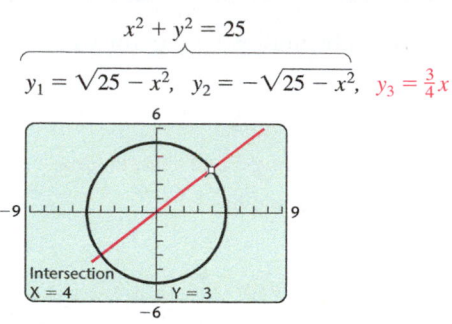

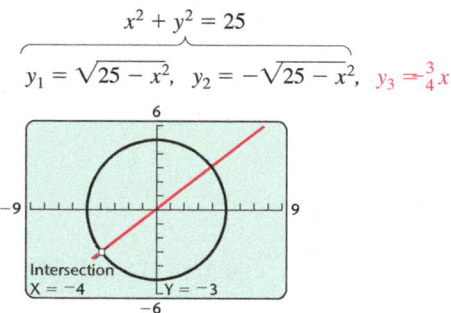

The solutions are $(4, 3)$ and $(-4, -3)$.

In the solution in Example 1, suppose that to find x we had substituted 3 and -3 in equation (1) rather than equation (3). If $y = 3$, $y^2 = 9$, and if $y = -3$, $y^2 = 9$, so both substitutions can be performed at the same time:

$$x^2 + y^2 = 25 \quad \textbf{(1)}$$
$$x^2 + (\pm 3)^2 = 25$$
$$x^2 + 9 = 25$$
$$x^2 = 16$$
$$x = \pm 4.$$

Each y-value produces two values for x. Thus, if $y = 3$, $x = 4$ or $x = -4$, and if $y = -3$, $x = 4$ or $x = -4$. The possible solutions are $(4, 3)$, $(-4, 3)$, $(4, -3)$, and $(-4, -3)$. A check reveals that $(4, -3)$ and $(-4, 3)$ are not solutions of equation (2). Since a circle and a line can intersect in at most two points, it is clear that there can be at most two real-number solutions.

EXAMPLE 2   Solve the following system of equations:

$$x + y = 5, \qquad \textbf{(1)} \qquad \text{\color{red}The graph is a line.}$$
$$y = 3 - x^2. \qquad \textbf{(2)} \qquad \text{\color{red}The graph is a parabola.}$$

Algebraic Solution

We use the substitution method, substituting $3 - x^2$ for y in equation (1):

$$x + 3 - x^2 = 5$$
$$-x^2 + x - 2 = 0 \qquad \text{\color{red}Subtracting 5 and rearranging}$$
$$x^2 - x + 2 = 0. \qquad \text{\color{red}Multiplying by } -1$$

Next, we use the quadratic formula:

$$x = \frac{-b \pm \sqrt{b^2 - 4ac}}{2a}$$
$$= \frac{-(-1) \pm \sqrt{(-1)^2 - 4(1)(2)}}{2(1)}$$
$$= \frac{1 \pm \sqrt{1 - 8}}{2} = \frac{1 \pm \sqrt{-7}}{2}$$
$$= \frac{1 \pm i\sqrt{7}}{2} = \frac{1}{2} \pm \frac{\sqrt{7}}{2}i.$$

Now, we substitute these values for x in equation (1) and solve for y:

$$\frac{1}{2} + \frac{\sqrt{7}}{2}i + y = 5$$
$$y = 5 - \frac{1}{2} - \frac{\sqrt{7}}{2}i$$
$$= \frac{9}{2} - \frac{\sqrt{7}}{2}i$$

and

$$\frac{1}{2} - \frac{\sqrt{7}}{2}i + y = 5$$
$$y = 5 - \frac{1}{2} + \frac{\sqrt{7}}{2}i$$
$$= \frac{9}{2} + \frac{\sqrt{7}}{2}i.$$

The solutions are

$$\left(\frac{1}{2} + \frac{\sqrt{7}}{2}i, \frac{9}{2} - \frac{\sqrt{7}}{2}i \right) \quad \text{and} \quad \left(\frac{1}{2} - \frac{\sqrt{7}}{2}i, \frac{9}{2} + \frac{\sqrt{7}}{2}i \right).$$

There are no real-number solutions.

Visualizing the Solution

We graph the equations, as shown below.

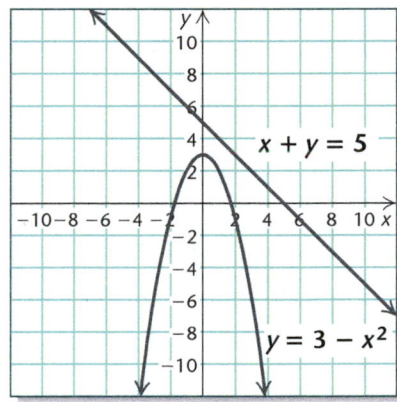

Note that there are no points of intersection. This indicates that there are no real-number solutions.

Now Try Exercise 17.

EXAMPLE 3 Solve the following system of equations:

$$2x^2 + 5y^2 = 39, \quad \textbf{(1)} \qquad \text{The graph is an ellipse.}$$
$$3x^2 - y^2 = -1. \quad \textbf{(2)} \qquad \text{The graph is a hyperbola.}$$

Algebraic Solution

We use the elimination method. First, we multiply equation (2) by 5 and add to eliminate the y^2-term:

$$
\begin{array}{ll}
2x^2 + 5y^2 = 39 & \textbf{(1)} \\
\underline{15x^2 - 5y^2 = -5} & \text{Multiplying (2) by 5} \\
17x^2 \qquad\quad = 34 & \text{Adding} \\
\quad\; x^2 = 2 & \\
\quad\;\; x = \pm\sqrt{2}. &
\end{array}
$$

If $x = \sqrt{2}$, $x^2 = 2$, and if $x = -\sqrt{2}$, $x^2 = 2$. Thus substituting $\sqrt{2}$ or $-\sqrt{2}$ for x in equation (2) gives us

$$
\begin{aligned}
3\left(\pm\sqrt{2}\right)^2 - y^2 &= -1 \\
3 \cdot 2 - y^2 &= -1 \\
6 - y^2 &= -1 \\
-y^2 &= -7 \\
y^2 &= 7 \\
y &= \pm\sqrt{7}.
\end{aligned}
$$

Thus, for $x = \sqrt{2}$, we have $y = \sqrt{7}$ or $y = -\sqrt{7}$, and for $x = -\sqrt{2}$, we have $y = \sqrt{7}$ or $y = -\sqrt{7}$. The possible solutions are $\left(\sqrt{2}, \sqrt{7}\right)$, $\left(\sqrt{2}, -\sqrt{7}\right)$, $\left(-\sqrt{2}, \sqrt{7}\right)$, $\left(-\sqrt{2}, -\sqrt{7}\right)$. All four pairs check, so they are the solutions.

Visualizing the Solution

We graph the equations and note that there are four points of intersection.

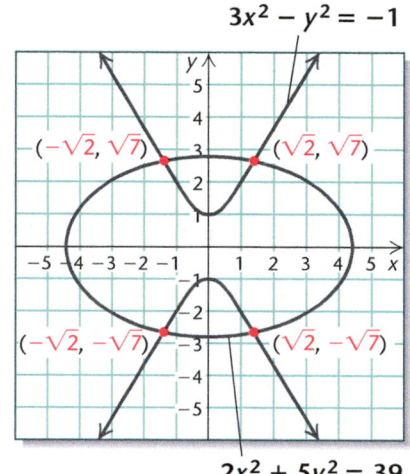

The coordinates of the points of intersection are $\left(\sqrt{2}, \sqrt{7}\right)$, $\left(\sqrt{2}, -\sqrt{7}\right)$, $\left(-\sqrt{2}, \sqrt{7}\right)$, and $\left(-\sqrt{2}, -\sqrt{7}\right)$. These are the solutions of the system of equations.

Now Try Exercise 27.

Technology Connection

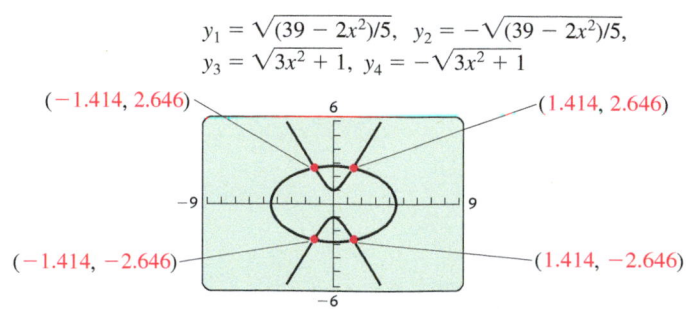

$$y_1 = \sqrt{(39 - 2x^2)/5}, \quad y_2 = -\sqrt{(39 - 2x^2)/5},$$
$$y_3 = \sqrt{3x^2 + 1}, \quad y_4 = -\sqrt{3x^2 + 1}$$

$(-1.414, 2.646)$ $(1.414, 2.646)$

$(-1.414, -2.646)$ $(1.414, -2.646)$

To solve the system of equations in Example 3 on a graphing calculator, we graph the equations in the same viewing window. We can use the INTERSECT feature to find the coordinates of the four points of intersection.

Note that the algebraic solution yields exact solutions, whereas the graphical solution yields decimal approximations of the solutions on most graphing calculators.

The solutions are approximately $(1.414, 2.646)$, $(1.414, -2.646)$, $(-1.414, 2.646)$, and $(-1.414, -2.646)$.

EXAMPLE 4 Solve the following system of equations:

$$x^2 - 3y^2 = 6, \qquad \textbf{(1)}$$
$$xy = 3. \qquad \textbf{(2)}$$

Algebraic Solution

We use the substitution method. First, we solve equation (2) for y:

$$xy = 3 \qquad \textbf{(2)}$$

$$y = \frac{3}{x}. \qquad \textbf{(3)} \qquad \textcolor{red}{\textbf{Dividing by } x}$$

Next, we substitute $3/x$ for y in equation (1) and solve for x:

$$x^2 - 3\left(\frac{3}{x}\right)^2 = 6$$

$$x^2 - 3 \cdot \frac{9}{x^2} = 6$$

$$x^2 - \frac{27}{x^2} = 6$$

$$x^4 - 27 = 6x^2 \qquad \textcolor{red}{\textbf{Multiplying by } x^2}$$

$$x^4 - 6x^2 - 27 = 0$$

$$u^2 - 6u - 27 = 0 \qquad \textcolor{red}{\textbf{Letting } u = x^2}$$

$$(u - 9)(u + 3) = 0 \qquad \textcolor{red}{\textbf{Factoring}}$$

$$u - 9 = 0 \quad or \quad u + 3 = 0 \qquad \textcolor{red}{\textbf{Principle of zero products}}$$

$$u = 9 \quad or \quad u = -3$$

$$x^2 = 9 \quad or \quad x^2 = -3 \qquad \textcolor{red}{\textbf{Substituting } x^2 \textbf{ for } u}$$

$$x = \pm 3 \quad or \quad x = \pm i\sqrt{3}.$$

Since $y = 3/x$,

when $x = 3$, $\qquad y = \dfrac{3}{3} = 1;$

when $x = -3$, $\qquad y = \dfrac{3}{-3} = -1;$

when $x = i\sqrt{3}$, $\qquad y = \dfrac{3}{i\sqrt{3}} = \dfrac{3}{i\sqrt{3}} \cdot \dfrac{-i\sqrt{3}}{-i\sqrt{3}} = -i\sqrt{3};$

when $x = -i\sqrt{3}$, $\quad y = \dfrac{3}{-i\sqrt{3}} = \dfrac{3}{-i\sqrt{3}} \cdot \dfrac{i\sqrt{3}}{i\sqrt{3}} = i\sqrt{3}.$

The pairs $(3, 1)$, $(-3, -1)$, $\left(i\sqrt{3}, -i\sqrt{3}\right)$, and $\left(-i\sqrt{3}, i\sqrt{3}\right)$ check, so they are the solutions.

Visualizing the Solution

The coordinates of the points of intersection of the graphs of the equations give us the real-number solutions of the system of equations. These graphs do not show us the imaginary-number solutions.

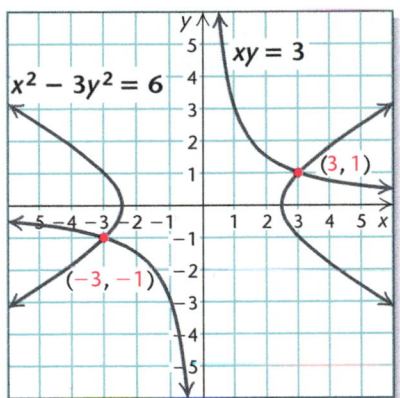

Now Try Exercise 19.

▶ Modeling and Problem Solving

EXAMPLE 5 *Dimensions of a Piece of Land.* For a student recreation building at Southport Community College, an architect wants to lay out a rectangular piece of land that has a perimeter of 204 m and an area of 2565 m². Find the dimensions of the piece of land.

Solution

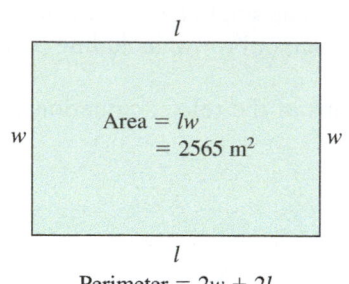

Area = lw
= 2565 m²

Perimeter = $2w + 2l$
= 204 m

1. **Familiarize.** We make a drawing and label it, letting l = the length of the piece of land, in meters, and w = the width, in meters.

2. **Translate.** We now have the following:

   Perimeter: $2w + 2l = 204,$ **(1)**
   Area: $lw = 2565.$ **(2)**

3. **Carry out.** We solve the system of equations

   $$2w + 2l = 204,$$
   $$lw = 2565.$$

   Solving the second equation for l gives us $l = 2565/w$. We then substitute $2565/w$ for l in equation (1) and solve for w:

   $$2w + 2\left(\frac{2565}{w}\right) = 204$$
   $$2w^2 + 2(2565) = 204w \qquad \text{Multiplying by } w$$
   $$2w^2 - 204w + 2(2565) = 0$$
   $$w^2 - 102w + 2565 = 0 \qquad \text{Multiplying by } \tfrac{1}{2}$$
   $$(w - 57)(w - 45) = 0$$
   $$w - 57 = 0 \quad or \quad w - 45 = 0 \qquad \text{Principle of zero products}$$
   $$w = 57 \quad or \qquad w = 45.$$

   If $w = 57$, then $l = 2565/w = 2565/57 = 45$. If $w = 45$, then $l = 2565/w = 2565/45 = 57$. Since length is generally considered to be longer than width, we have the solution $l = 57$ and $w = 45$, or $(57, 45)$.

4. **Check.** If $l = 57$ and $w = 45$, the perimeter is $2 \cdot 45 + 2 \cdot 57$, or 204. The area is $57 \cdot 45$, or 2565. The numbers check.

5. **State.** The length of the piece of land is 57 m and the width is 45 m.

Now Try Exercise 61.

▶ Nonlinear Systems of Inequalities

SYSTEMS OF INEQUALITIES

REVIEW SECTION 3.7

Recall that a solution of a system of inequalities is an ordered pair that is a solution of each inequality in the system. We graphed systems of linear inequalities in Section 3.7. Now we graph a system of nonlinear inequalities.

EXAMPLE 6 Graph the solution set of the system

$$x^2 + y^2 \le 25,$$
$$3x - 4y > 0.$$

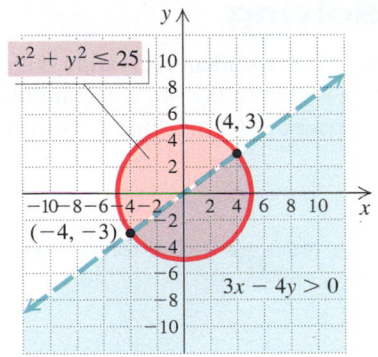

Solution We graph $x^2 + y^2 \leq 25$ by first graphing the related equation of the circle $x^2 + y^2 = 25$. We use a solid line since the inequality symbol is $\leq$. Next, we choose $(0, 0)$ as a test point and find that it is a solution of $x^2 + y^2 \leq 25$, so we shade the region that contains $(0, 0)$ using red. This is the region inside the circle. Now we graph the line $3x - 4y = 0$ using a dashed line since the inequality symbol is $>$. The point $(0, 0)$ is on the line, so we choose another test point, say, $(0, 2)$. We find that this point is not a solution of $3x - 4y > 0$, so we shade the half-plane that does not contain $(0, 2)$ using green. The solution set of the system of inequalities is the region shaded both red and green, or brown, including part of the circle $x^2 + y^2 = 25$.

To find the points of intersection of the graphs of the related equations, we solve the system composed of those equations:

$$x^2 + y^2 = 25,$$
$$3x - 4y = 0.$$

In Example 1, we found that these points are $(4, 3)$ and $(-4, -3)$.

> **Now Try Exercise 75.**

Technology Connection

To use a graphing calculator to graph the system of inequalities in Example 7, we first graph $y_1 = 4 - x^2$ and $y_2 = 2 - x$. Using the test point $(0, 0)$ for each inequality, we find that we should shade below y_1 and above y_2. We can find the points of intersection of the graphs of the related equations, $(-1, 3)$ and $(2, 0)$, using the INTERSECT feature.

$y_1 = 4 - x^2, \quad y_2 = 2 - x$

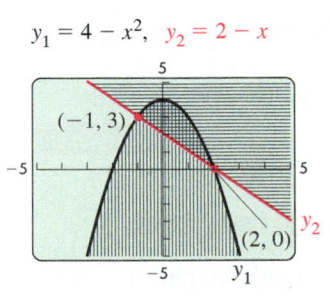

EXAMPLE 7 Graph the solution set of the system

$$y \leq 4 - x^2,$$
$$x + y \geq 2.$$

Solution We graph $y \leq 4 - x^2$ by first graphing the equation of the parabola $y = 4 - x^2$. We use a solid line since the inequality symbol is $\leq$. Next, we choose $(0, 0)$ as a test point and find that it is a solution of $y \leq 4 - x^2$, so we shade the region that contains $(0, 0)$ using red. Now we graph the line $x + y = 2$, again using a solid line since the inequality symbol is $\geq$. We test the point $(0, 0)$ and find that it is not a solution of $x + y \geq 2$, so we shade the half-plane that does not contain $(0, 0)$ using green. The solution set of the system of inequalities is the region shaded both red and green, or brown, including part of the parabola $y = 4 - x^2$ and part of the line $x + y = 2$.

Solving the system of equations

$$y = 4 - x^2,$$
$$x + y = 2,$$

we find that the points of intersection of the graphs of the related equations are $(-1, 3)$ and $(2, 0)$.

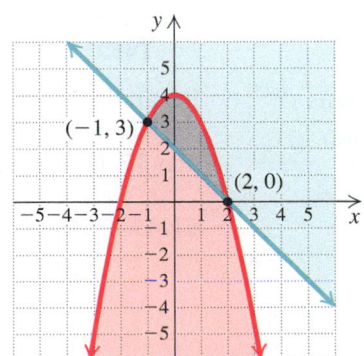

> **Now Try Exercise 81.**

A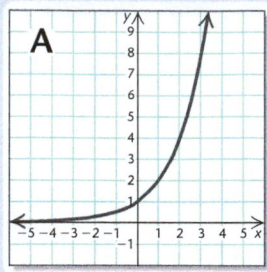

B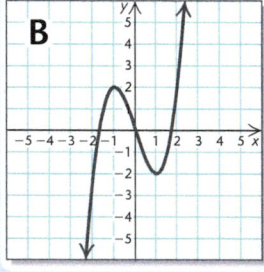

C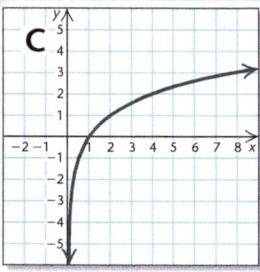

D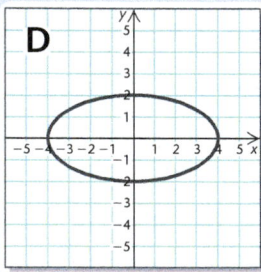

E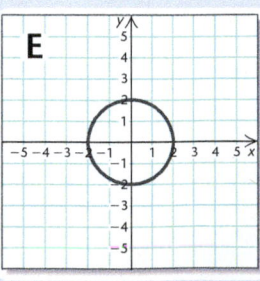

Visualizing the Graph

Match the equation or system of equations with its graph.

1. $y = x^3 - 3x$

2. $y = x^2 + 2x - 3$

3. $y = \dfrac{x - 1}{x^2 - x - 2}$

4. $y = -3x + 2$

5. $x + y = 3,$
   $2x + 5y = 3$

6. $9x^2 - 4y^2 = 36,$
   $x^2 + y^2 = 9$

7. $5x^2 + 5y^2 = 20$

8. $4x^2 + 16y^2 = 64$

9. $y = \log_2 x$

10. $y = 2^x$

F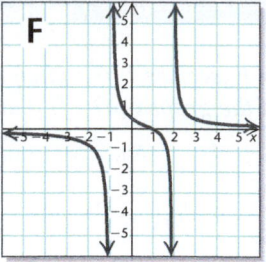

G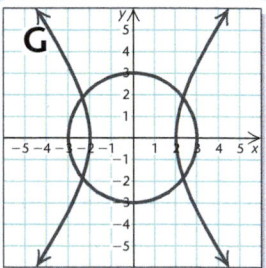

H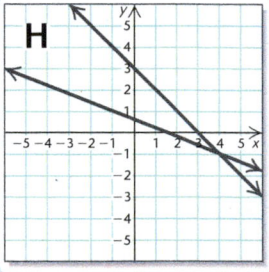

I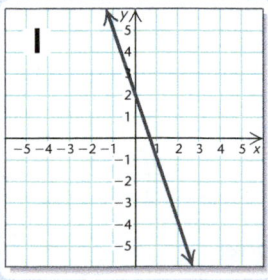

J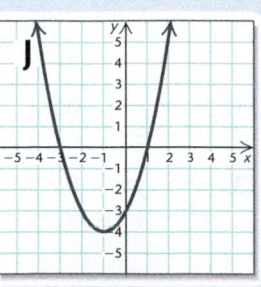

11.4   Exercise Set

In Exercises 1–6, match the system of equations with one of the graphs (a)–(f) that follow.

a)

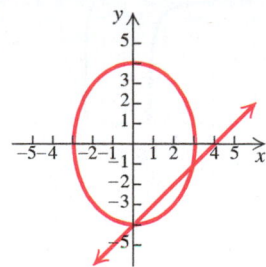

b)

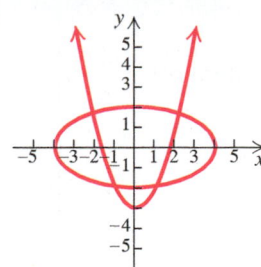

c)

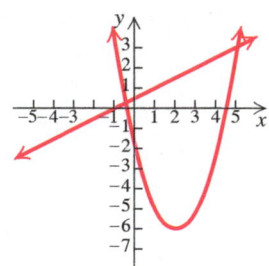

d)

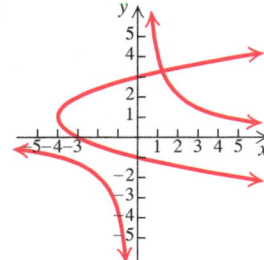

e)

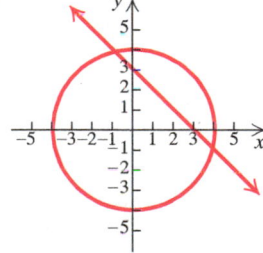

f)

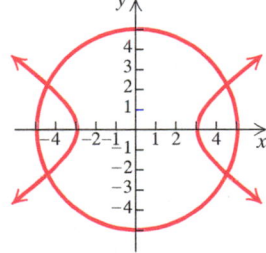

1. $x^2 + y^2 = 16,$
$x + y = 3$

2. $16x^2 + 9y^2 = 144,$
$x - y = 4$

3. $y = x^2 - 4x - 2,$
$2y - x = 1$

4. $4x^2 - 9y^2 = 36,$
$x^2 + y^2 = 25$

5. $y = x^2 - 3,$
$x^2 + 4y^2 = 16$

6. $y^2 - 2y = x + 3,$
$xy = 4$

Solve.

7. $x^2 + y^2 = 25,$
$y - x = 1$

8. $x^2 + y^2 = 100,$
$y - x = 2$

9. $4x^2 + 9y^2 = 36,$
$3y + 2x = 6$

10. $9x^2 + 4y^2 = 36,$
$3x + 2y = 6$

11. $x^2 + y^2 = 25,$
$y^2 = x + 5$

12. $y = x^2,$
$x = y^2$

13. $x^2 + y^2 = 9,$
$x^2 - y^2 = 9$

14. $y^2 - 4x^2 = 4,$
$4x^2 + y^2 = 4$

15. $y^2 - x^2 = 9,$
$2x - 3 = y$

16. $x + y = -6,$
$xy = -7$

17. $y^2 = x + 3,$
$2y = x + 4$

18. $y = x^2,$
$3x = y + 2$

19. $x^2 + y^2 = 25,$
$xy = 12$

20. $x^2 - y^2 = 16,$
$x + y^2 = 4$

21. $x^2 + y^2 = 4,$
$16x^2 + 9y^2 = 144$

22. $x^2 + y^2 = 25,$
$25x^2 + 16y^2 = 400$

23. $x^2 + 4y^2 = 25,$
$x + 2y = 7$

24. $y^2 - x^2 = 16,$
$2x - y = 1$

25. $x^2 - xy + 3y^2 = 27,$
$x - y = 2$

26. $2y^2 + xy + x^2 = 7,$
$x - 2y = 5$

27. $x^2 + y^2 = 16,$
$y^2 - 2x^2 = 10$

28. $x^2 + y^2 = 14,$
$x^2 - y^2 = 4$

29. $x^2 + y^2 = 5,$
$xy = 2$

30. $x^2 + y^2 = 20,$
$xy = 8$

31. $3x + y = 7,$
$4x^2 + 5y = 56$

32. $2y^2 + xy = 5,$
$4y + x = 7$

33. $a + b = 7,$
$ab = 4$

34. $p + q = -4,$
$pq = -5$

35. $x^2 + y^2 = 13,$
$xy = 6$

36. $x^2 + 4y^2 = 20,$
$xy = 4$

37. $x^2 + y^2 + 6y + 5 = 0,$
$x^2 + y^2 - 2x - 8 = 0$

38. $2xy + 3y^2 = 7,$
$3xy - 2y^2 = 4$

39. $2a + b = 1,$
$b = 4 - a^2$

40. $4x^2 + 9y^2 = 36,$
$x + 3y = 3$

41. $a^2 + b^2 = 89,$
$a - b = 3$

42. $xy = 4,$
$x + y = 5$

43. $xy - y^2 = 2,$
$2xy - 3y^2 = 0$

44. $4a^2 - 25b^2 = 0,$
$2a^2 - 10b^2 = 3b + 4$

45. $m^2 - 3mn + n^2 + 1 = 0,$
$3m^2 - mn + 3n^2 = 13$

46. $ab - b^2 = -4,$
$ab - 2b^2 = -6$

47. $x^2 + y^2 = 5,$
$x - y = 8$

48. $4x^2 + 9y^2 = 36,$
$y - x = 8$

49. $a^2 + b^2 = 14,$
$ab = 3\sqrt{5}$

50. $x^2 + xy = 5,$
$2x^2 + xy = 2$

51. $x^2 + y^2 = 25,$
$9x^2 + 4y^2 = 36$

52. $x^2 + y^2 = 1,$
$9x^2 - 16y^2 = 144$

53. $5y^2 - x^2 = 1,$
$xy = 2$

54. $x^2 - 7y^2 = 6,$
$xy = 1$

In Exercises 55–58, determine whether the statement is true or false.

55. A nonlinear system of equations can have both real-number solutions and imaginary-number solutions.

56. If the graph of a nonlinear system of equations consists of a line and a parabola, then the system has two real-number solutions.

57. If the graph of a nonlinear system of equations consists of a line and a circle, then the system has at most two real-number solutions.

58. If the graph of a nonlinear system of equations consists of a line and an ellipse, then it is possible for the system to have exactly one real-number solution.

59. *Photo Dimensions.* Hailey's Frame Shop has been commissioned to frame 5 black-and-white photos for an island resort. Each photo has a perimeter of 68 in. and a diagonal of 26 in. Find the dimensions of the photos.

60. *Sign Dimensions.* Alison's Advertising is building a rectangular sign with an area of 2 yd^2 and a perimeter of 6 yd. Find the dimensions of the sign.

61. *Graphic Design.* Marcia Graham, owner of Graham's Graphics, is designing an advertising brochure for the Art League's spring show. Each page of the brochure is rectangular with an area of 20 in^2 and a perimeter of 18 in. Find the dimensions of the brochure.

62. *Landscaping.* Green Leaf Landscaping is planting a rectangular wildflower garden with a perimeter of

6 m and a diagonal of $\sqrt{5}$ m. Find the dimensions of the garden.

63. *Fencing.* Clark's Country Pet Resort is fencing a new play area for dogs. The manager has purchased 210 yd of fence to enclose a rectangular pen. The area of the pen must be 2250 yd^2. What are the dimensions of the pen?

64. *Carpentry.* Ted Hansen of Hansen Woodworking Designs has been commissioned to make a rectangular tabletop with an area of $\sqrt{2}$ m^2 and a diagonal of $\sqrt{3}$ m for the Decorators' Show House. Find the dimensions of the tabletop.

65. *Banner Design.* A rectangular banner with an area of $\sqrt{3}$ m^2 is being designed to advertise an exhibit at the Davis Gallery. The length of a diagonal is 2 m. Find the dimensions of the banner.

66. *Investment.* Luke made an investment for 1 year that earned $72 simple interest. If the principal had been $240 more and the interest rate 1% less, the interest would have been the same. Find the principal and the interest rate.

67. *Seed Test Plots.* The Burton Seed Company has two square test plots. The sum of their areas is 832 ft^2 and the difference of their areas is 320 ft^2. Find the length of a side of each plot.

68. *Office Dimensions.*   The diagonal of the floor of a rectangular office cubicle is 1 ft longer than the length of the cubicle and 3 ft longer than twice the width. Find the dimensions of the cubicle.

In Exercises 69–74, match the system of inequalities with one of the graphs (a)–(f) that follow.

a)
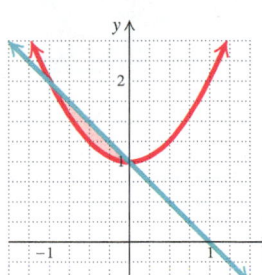

b)
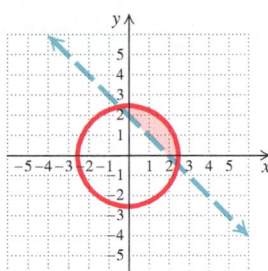

c)
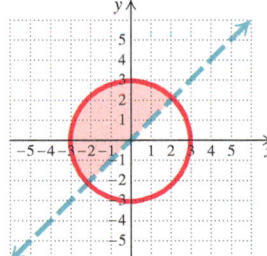

d)
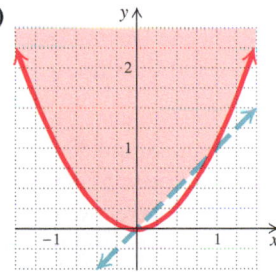

e)
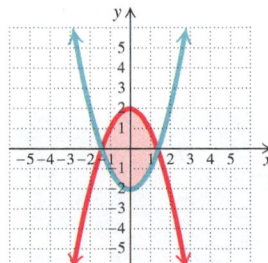

f)
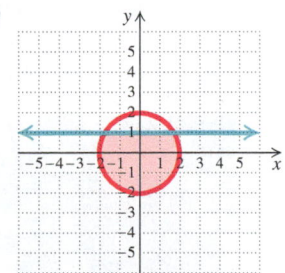

69. $x^2 + y^2 \le 5$,
      $x + y > 2$

70. $y \le 2 - x^2$,
      $y \ge x^2 - 2$

71. $y \ge x^2$,
      $y > x$

72. $x^2 + y^2 \le 4$,
      $y \le 1$

73. $y \ge x^2 + 1$,
      $x + y \le 1$

74. $x^2 + y^2 \le 9$,
      $y > x$

Graph the system of inequalities. Then find the coordinates of the points of intersection of the graphs of the related equations.

75. $x^2 + y^2 \le 16$,
      $y < x$

76. $x^2 + y^2 \le 10$,
      $y > x$

77. $x^2 \le y$,
      $x + y \ge 2$

78. $x \ge y^2$,
      $x - y \le 2$

79. $x^2 + y^2 \le 25$,
      $x - y > 5$

80. $x^2 + y^2 \ge 9$,
      $x - y > 3$

81. $y \ge x^2 - 3$,
      $y \le 2x$

82. $y \le 3 - x^2$,
      $y \ge x + 1$

83. $y \ge x^2$,
      $y < x + 2$

84. $y \le 1 - x^2$,
      $y > x - 1$

▶ **Skill Maintenance**

Solve. **[9.6]**

85. $2^{3x} = 64$

86. $5^x = 27$

87. $\log_3 x = 4$

88. $\log(x - 3) + \log x = 1$

▶ **Synthesis**

89. Find an equation of the circle that passes through the points $(2, 4)$ and $(3, 3)$ and whose center is on the line $3x - y = 3$.

90. Find an equation of the circle that passes through the points $(2, 3)$, $(4, 5)$, and $(0, -3)$.

91. Find an equation of an ellipse centered at the origin that passes through the points $\left(1, \sqrt{3}/2\right)$ and $\left(\sqrt{3}, 1/2\right)$.

92. Find an equation of a hyperbola of the type
$$\frac{x^2}{b^2} - \frac{y^2}{a^2} = 1$$
that passes through the points $\left(-3, -3\sqrt{5}/2\right)$ and $(-3/2, 0)$.

93. Show that a hyperbola does not intersect its asymptotes. That is, solve the system of equations
$$\frac{x^2}{a^2} - \frac{y^2}{b^2} = 1,$$
$$y = \frac{b}{a}x \quad \left(\text{or } y = -\frac{b}{a}x\right).$$

94. *Numerical Relationship.*   Find two numbers whose product is 2 and the sum of whose reciprocals is $\frac{33}{8}$.

95. *Numerical Relationship.* The sum of two numbers is 1, and their product is 1. Find the sum of their cubes. There is a method to solve this problem that is easier than solving a nonlinear system of equations. Can you discover it?

96. *Box Dimensions.* Four squares with sides 5 in. long are cut from the corners of a rectangular metal sheet that has an area of 340 in². The edges are bent up to form an open box with a volume of 350 in³. Find the dimensions of the box.

Solve.

97. $x^3 + y^3 = 72,$
$x + y = 6$

98. $a + b = \dfrac{5}{6},$
$\dfrac{a}{b} + \dfrac{b}{a} = \dfrac{13}{6}$

99. $p^2 + q^2 = 13,$
$\dfrac{1}{pq} = -\dfrac{1}{6}$

100. $e^x - e^{x+y} = 0,$
$e^y - e^{x-y} = 0$

CHAPTER 11 Summary and Review

STUDY GUIDE

KEY TERMS AND CONCEPTS	**EXAMPLES**

SECTION 11.1: THE PARABOLA

Standard Equation of a Parabola with Vertex $(0, 0)$ and Vertical Axis of Symmetry

The standard equation of a parabola with vertex $(0, 0)$ and directrix $y = -p$ is

$$x^2 = 4py.$$

The focus is $(0, p)$ and the y-axis is the axis of symmetry.

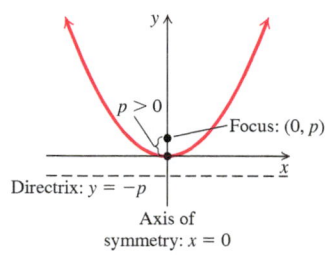

(When $p < 0$, the parabola opens down.)

See also Example 1 on p. 724.

Standard Equation of a Parabola with Vertex $(0, 0)$ and Horizontal Axis of Symmetry

The standard equation of a parabola with vertex $(0, 0)$ and directrix $x = -p$ is

$$y^2 = 4px.$$

The focus is $(p, 0)$ and the x-axis is the axis of symmetry.

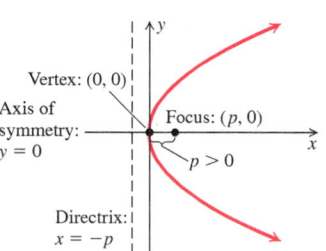

(When $p < 0$, the parabola opens to the left.)

See also Example 2 on p. 724.

Standard Equation of a Parabola with Vertex (h, k) and Vertical Axis of Symmetry

The standard equation of a parabola with vertex (h, k) and vertical axis of symmetry is

$$(x - h)^2 = 4p(y - k),$$

where the vertex is (h, k), the focus is $(h, k + p)$, and the directrix is $y = k - p$.

The parabola opens up if $p > 0$. It opens down if $p < 0$.

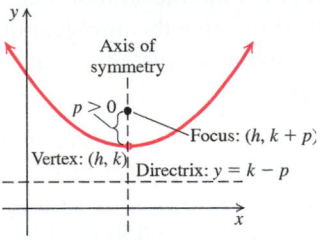

(When $p < 0$, the parabola opens down.)

See also Example 3 on p. 726.

Standard Equation of a Parabola with Vertex (h, k) and Horizontal Axis of Symmetry

The standard equation of a parabola with vertex (h, k) and horizontal axis of symmetry is

$$(y - k)^2 = 4p(x - h),$$

where the vertex is (h, k), the focus is $(h + p, k)$, and the directrix is $x = h - p$.

The parabola opens to the right if $p > 0$. It opens to the left if $p < 0$.

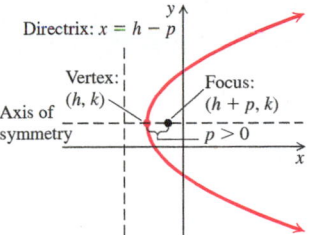

(When $p < 0$, the parabola opens to the left.)

See also Example 4 on pp. 726–727.

SECTION 11.2: THE CIRCLE AND THE ELLIPSE

Standard Equation of a Circle

The standard equation of a circle with center (h, k) and radius r is

$$(x - h)^2 + (y - k)^2 = r^2.$$

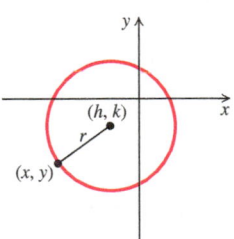

See also Example 1 on pp. 730–731.

Standard Equation of an Ellipse with Center at the Origin

Major Axis Horizontal

$$\frac{x^2}{a^2} + \frac{y^2}{b^2} = 1, \quad a > b > 0$$

Vertices: $(-a, 0), (a, 0)$

y-intercepts: $(0, -b), (0, b)$

Foci: $(-c, 0), (c, 0)$, where
$$c^2 = a^2 - b^2$$

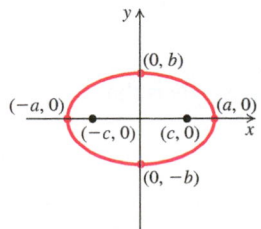

Major Axis Vertical

$$\frac{x^2}{b^2} + \frac{y^2}{a^2} = 1, \quad a > b > 0$$

Vertices: $(0, -a), (0, a)$

x-intercepts: $(-b, 0), (b, 0)$

Foci: $(0, -c), (0, c)$, where
$c^2 = a^2 - b^2$

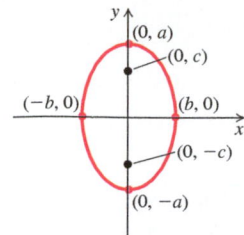

See also Examples 2 and 3 on pp. 733 and 734.

Standard Equation of an Ellipse with Center at (h, k)

Major Axis Horizontal

$$\frac{(x - h)^2}{a^2} + \frac{(y - k)^2}{b^2} = 1, \quad a > b > 0$$

Vertices: $(h - a, k), (h + a, k)$

Length of minor axis: $2b$

Foci: $(h - c, k), (h + c, k)$, where
$c^2 = a^2 - b^2$

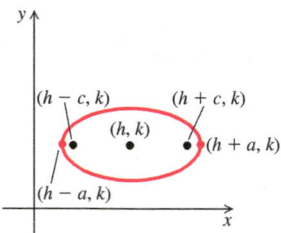

Major Axis Vertical

$$\frac{(x - h)^2}{b^2} + \frac{(y - k)^2}{a^2} = 1, \quad a > b > 0$$

Vertices: $(h, k - a), (h, k + a)$

Length of minor axis: $2b$

Foci: $(h, k - c), (h, k + c)$, where
$c^2 = a^2 - b^2$

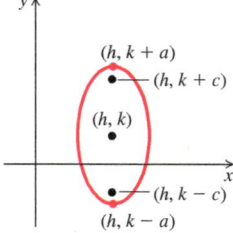

See also Example 4 on pp. 734–735.

SECTION 11.3: THE HYPERBOLA

Standard Equation of a Hyperbola with Center at the Origin

Transverse Axis Horizontal

$$\frac{x^2}{a^2} - \frac{y^2}{b^2} = 1$$

Vertices: $(-a, 0), (a, 0)$

Asymptotes: $y = -\frac{b}{a}x, \; y = \frac{b}{a}x$

Foci: $(-c, 0), (c, 0)$, where $c^2 = a^2 + b^2$

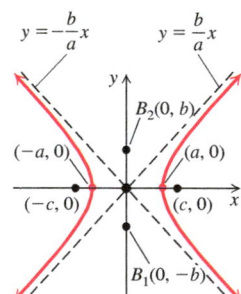

Transverse Axis Vertical

$$\frac{y^2}{a^2} - \frac{x^2}{b^2} = 1$$

Vertices: $(0, -a)$, $(0, a)$

Asymptotes: $y = -\frac{a}{b}x$, $y = \frac{a}{b}x$

Foci: $(0, -c)$, $(0, c)$, where $c^2 = a^2 + b^2$

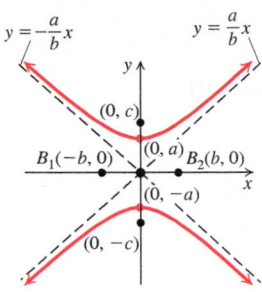

See also Examples 1 and 2 on pp. 742–743.

Standard Equation of a Hyperbola with Center at (h, k)

Transverse Axis Horizontal

$$\frac{(x - h)^2}{a^2} - \frac{(y - k)^2}{b^2} = 1$$

Vertices: $(h - a, k)$, $(h + a, k)$

Asymptotes: $y - k = \frac{b}{a}(x - h)$,

$$y - k = -\frac{b}{a}(x - h)$$

Foci: $(h - c, k)$, $(h + c, k)$,
   where $c^2 = a^2 + b^2$

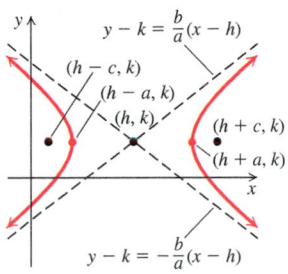

Transverse Axis Vertical

$$\frac{(y - k)^2}{a^2} - \frac{(x - h)^2}{b^2} = 1$$

Vertices: $(h, k - a)$, $(h, k + a)$

Asymptotes: $y - k = \frac{a}{b}(x - h)$,

$$y - k = -\frac{a}{b}(x - h)$$

Foci: $(h, k - c)$, $(h, k + c)$,
   where $c^2 = a^2 + b^2$

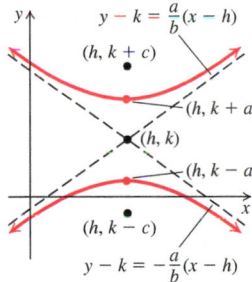

See also Example 3 on p. 745.

SECTION 11.4: NONLINEAR SYSTEMS OF EQUATIONS AND INEQUALITIES

Substitution or elimination can be used to solve **systems of equations containing at least one nonlinear equation**.	Solve: $x^2 - y = 2$, **(1)** **The graph is a parabola.**
	      $x - y = -4$. **(2)** **The graph is a line.**
	
	      $x = y - 4$ **Solving equation (2) for x**
	      $(y - 4)^2 - y = 2$ **Substituting for x in equation (1)**
	      $y^2 - 8y + 16 - y = 2$
	      $y^2 - 9y + 14 = 0$
	      $(y - 2)(y - 7) = 0$
	      $y - 2 = 0 \quad or \quad y - 7 = 0$
	      $y = 2 \quad or \quad y = 7$
	    If $y = 2$, then $x = 2 - 4 = -2$.
	    If $y = 7$, then $x = 7 - 4 = 3$.
	 The pairs $(-2, 2)$ and $(3, 7)$ check, so they are the solutions.

Some applied problems translate to a nonlinear system of equations.	See Example 5 on p. 755.
To graph a **nonlinear system of inequalities**, graph each inequality in the system and then shade the region where their solution sets overlap. To find the point(s) of intersection of the graphs of the related equations, solve the system of equations composed of those equations.	Graph: $x^2 - y \le 2$, $x - y > -4$. 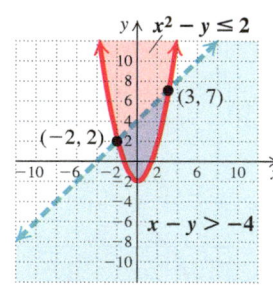 To find the points of intersection of the graphs of the related equations, solve the system of equations $$x^2 - y = 2,$$ $$x - y = -4.$$ We saw in the example above that these points are $(-2, 2)$ and $(3, 7)$.

REVIEW EXERCISES

Determine whether the statement is true or false.

1. The graph of $x + y^2 = 1$ is a parabola that opens to the left. [11.1]

2. The graph of $\dfrac{(x - 2)^2}{4} + \dfrac{(y + 3)^2}{9} = 1$ is an ellipse with center $(-2, 3)$. [11.2]

3. The hyperbola $\dfrac{x^2}{5} - \dfrac{y^2}{10} = 1$ has a horizontal transverse axis. [11.3]

4. Every nonlinear system of equations has at least one real-number solution. [11.4]

5. The graph of a nonlinear system of equations shows all the solutions of the system of equations. [11.4]

In Exercises 6–13, match the equation with one of the graphs (a)–(h) that follow.

a)
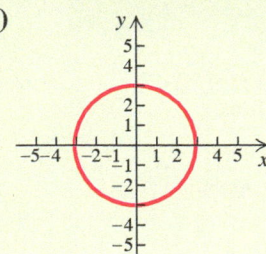

b)
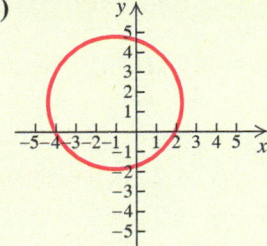

c)
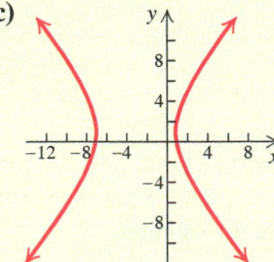

d)
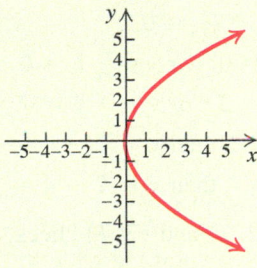

e)
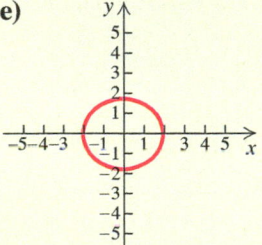

f)
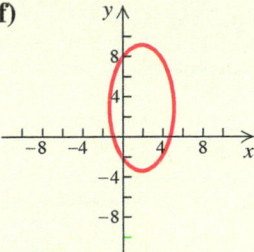

g)
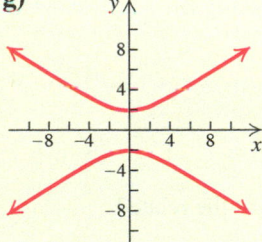

h)
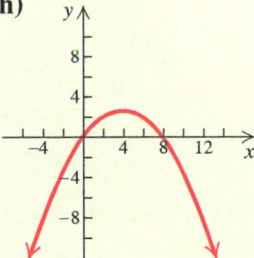

6. $y^2 = 5x$ [11.1]

7. $y^2 = 9 - x^2$ [11.2]

8. $3x^2 + 4y^2 = 12$ [11.2]

9. $9y^2 - 4x^2 = 36$ [11.3]

10. $x^2 + y^2 + 2x - 3y = 8$ [11.2]

11. $4x^2 + y^2 - 16x - 6y = 15$ [11.2]

12. $x^2 - 8x + 6y = 0$ [11.1]

13. $\dfrac{(x + 3)^2}{16} - \dfrac{(y - 1)^2}{25} = 1$ [11.3]

14. Find an equation of the parabola with directrix $y = \frac{3}{2}$ and focus $\left(0, -\frac{3}{2}\right)$. [11.1]

15. Find the focus, the vertex, and the directrix of the parabola given by

$$y^2 = -12x.\ \text{[11.1]}$$

16. Find the vertex, the focus, and the directrix of the parabola given by

$$x^2 + 10x + 2y + 9 = 0.\ \text{[11.1]}$$

17. Find the center, the vertices, and the foci of the ellipse given by

$$16x^2 + 25y^2 - 64x + 50y - 311 = 0.$$

Then draw the graph. [11.2]

18. Find an equation of the ellipse having vertices $(0, -4)$ and $(0, 4)$ with minor axis of length 6. [11.2]

19. Find the center, the vertices, the foci, and the asymptotes of the hyperbola given by

$$x^2 - 2y^2 + 4x + y - \tfrac{1}{8} = 0.\ \text{[11.3]}$$

20. *Spotlight.*   A spotlight has a parabolic cross section that is 2 ft wide at the opening and 1.5 ft deep at the vertex. How far from the vertex is the focus? [11.1]

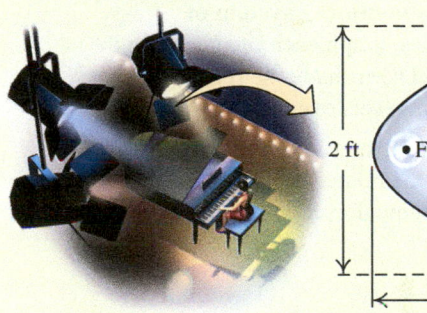

Solve. [11.4]

21. $x^2 - 16y = 0$,
    $x^2 - y^2 = 64$

22. $4x^2 + 4y^2 = 65$,
    $6x^2 - 4y^2 = 25$

23. $x^2 - y^2 = 33$,
    $x + y = 11$

24. $x^2 - 2x + 2y^2 = 8$,
    $2x + y = 6$

25. $x^2 - y = 3$,
    $2x - y = 3$

26. $x^2 + y^2 = 25$,
    $x^2 - y^2 = 7$

27. $x^2 - y^2 = 3$,
    $y = x^2 - 3$

28. $x^2 + y^2 = 18$,
    $2x + y = 3$

29. $x^2 + y^2 = 100$,
    $2x^2 - 3y^2 = -120$

30. $x^2 + 2y^2 = 12$,
    $xy = 4$

31. *Numerical Relationship.*   The sum of two numbers is 11, and the sum of their squares is 65. Find the numbers. [11.4]

32. *Dimensions of a Rectangle.*   A rectangle has a perimeter of 38 m and an area of 84 m². What are the dimensions of the rectangle? [11.4]

33. *Numerical Relationship.* Find two positive integers whose sum is 12 and the sum of whose reciprocals is $\frac{3}{8}$. **[11.4]**

34. *Perimeter.* The perimeter of a square is 12 cm more than the perimeter of another square. The area of the first square exceeds the area of the other by 39 cm^2. Find the perimeter of each square. **[11.4]**

35. *Radius of a Circle.* The sum of the areas of two circles is 130π ft^2. The difference of the areas is 112π ft^2. Find the radius of each circle. **[11.4]**

Graph the system of inequalities. Then find the coordinates of the points of intersection of the graphs of the related equations. **[11.4]**

36. $y \le 4 - x^2$,
$\quad x - y \le 2$

37. $x^2 + y^2 \le 16$,
$\quad x + y < 4$

38. $y \ge x^2 - 1$,
$\quad y < 1$

39. $x^2 + y^2 \le 9$,
$\quad x \le -1$

40. The vertex of the parabola $y^2 - 4y - 12x - 8 = 0$ is which of the following? **[11.1]**

  A. $(1, -2)$ **B.** $(-1, 2)$
  C. $(2, -1)$ **D.** $(-2, 1)$

41. Which of the following cannot be a number of solutions possible for a system of equations representing an ellipse and a straight line? **[11.4]**

  A. 0   **B.** 1
  C. 2   **D.** 4

42. The graph of $x^2 + 4y^2 = 4$ is which of the following? **[11.2], [11.3]**

  A.

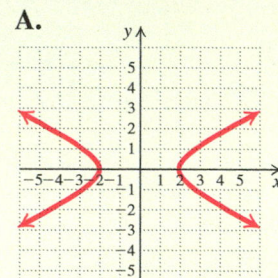

  B.

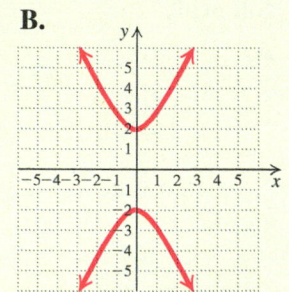

  C.

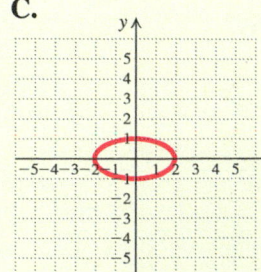

  D.

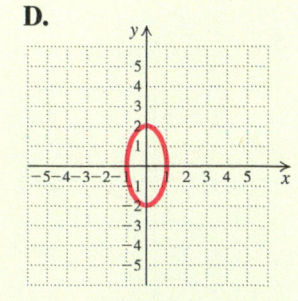

▶ Synthesis

43. Find two numbers whose product is 4 and the sum of whose reciprocals is $\frac{65}{56}$. **[11.4]**

44. Find an equation of the circle that passes through the points $(10, 7)$, $(-6, 7)$, and $(-8, 1)$. **[11.2], [11.4]**

45. Find an equation of the ellipse containing the point $\left(-1/2, 3\sqrt{3}/2\right)$ and with vertices $(0, -3)$ and $(0, 3)$. **[11.2]**

46. *Navigation.* Two radio transmitters positioned 400 mi apart along the shore send simultaneous signals to a ship that is 250 mi offshore and sailing parallel to the shoreline. The signal from transmitter A reaches the ship 300 microseconds before the signal from transmitter B. The signals travel at a speed of 186,000 miles per second, or 0.186 mile per microsecond. Find the equation of the hyperbola with foci A and B on which the ship is located. (*Hint:* For any point on the hyperbola, the absolute value of the difference of its distances from the foci is $2a$.) **[11.3]**

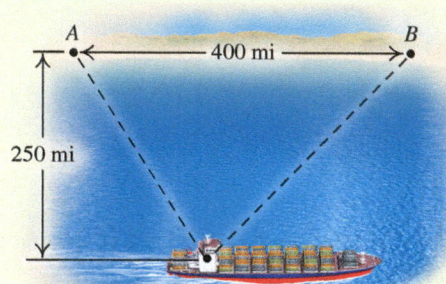

▶ Collaborative Discussion and Writing

47. Is a circle a special type of ellipse? Why or why not? **[11.2]**

48. Are the asymptotes of a hyperbola part of the graph of the hyperbola? Why or why not? **[11.3]**

49. What would you say to a classmate who tells you that it is always possible to visualize all of the solutions of a nonlinear system of equations? **[11.4]**

11   Chapter Test

In Exercises 1–4, match the equation with one of the graphs (a)–(d) that follow.

a)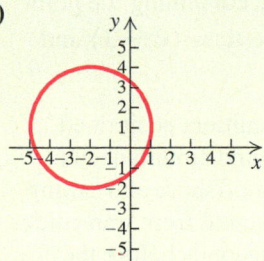

b)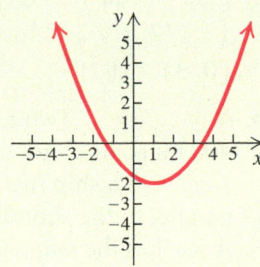

c)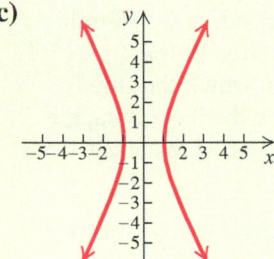

d)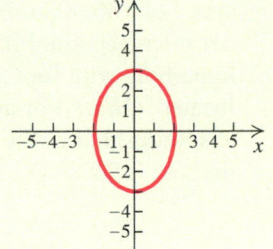

1. $4x^2 - y^2 = 4$ 　　　　**2.** $x^2 - 2x - 3y = 5$

3. $x^2 + 4x + y^2 - 2y - 4 = 0$

4. $9x^2 + 4y^2 = 36$

Find the vertex, the focus, and the directrix of the parabola. Then draw the graph.

5. $x^2 = 12y$

6. $y^2 + 2y - 8x - 7 = 0$

7. Find an equation of the parabola with focus $(0, 2)$ and directrix $y = -2$.

8. Find the center and the radius of the circle given by $x^2 + y^2 + 2x - 6y - 15 = 0$. Then draw the graph.

Find the center, the vertices, and the foci of the ellipse. Then draw the graph.

9. $9x^2 + 16y^2 = 144$

10. $\dfrac{(x + 1)^2}{4} + \dfrac{(y - 2)^2}{9} = 1$

11. Find an equation of the ellipse having vertices $(0, -5)$ and $(0, 5)$ and with minor axis of length 4.

Find the center, the vertices, the foci, and the asymptotes of the hyperbola. Then draw the graph.

12. $4x^2 - y^2 = 4$

13. $\dfrac{(y - 2)^2}{4} - \dfrac{(x + 1)^2}{9} = 1$

14. Find the asymptotes of the hyperbola given by $2y^2 - x^2 = 18$.

15. *Satellite Dish.* A satellite dish has a parabolic cross section that is 18 in. wide at the opening and 6 in. deep at the vertex. How far from the vertex is the focus?

Solve.

16. $2x^2 - 3y^2 = -10,$　　　　**17.** $x^2 + y^2 = 13,$
$\quad x^2 + 2y^2 = 9$ 　　　　　　　　$\quad x + y = 1$

18. $x + y = 5,$
$\quad xy = 6$

19. *Landscaping.* Leisurescape is planting a rectangular flower garden with a perimeter of 18 ft and a diagonal of $\sqrt{41}$ ft. Find the dimensions of the garden.

20. *Fencing.* It will take 210 ft of fencing to enclose a rectangular playground with an area of 2700 ft². Find the dimensions of the playground.

21. Graph the system of inequalities. Then find the coordinates of the points of intersection of the graphs of the related equations.
$$y \ge x^2 - 4,$$
$$y < 2x - 1.$$

22. The graph of $(y - 1)^2 = 4(x + 1)$ is which of the following?

A.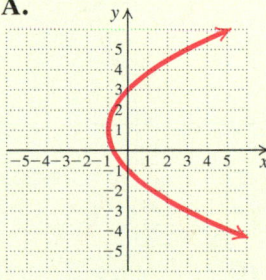

B.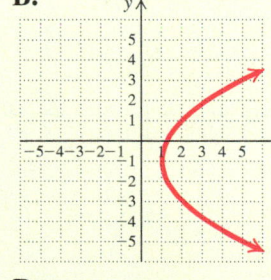

C.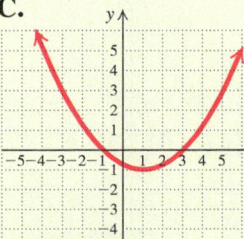

D.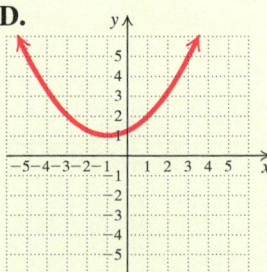

▶ Synthesis

23. Find an equation of the circle for which the endpoints of a diameter are $(1, 1)$ and $(5, -3)$.

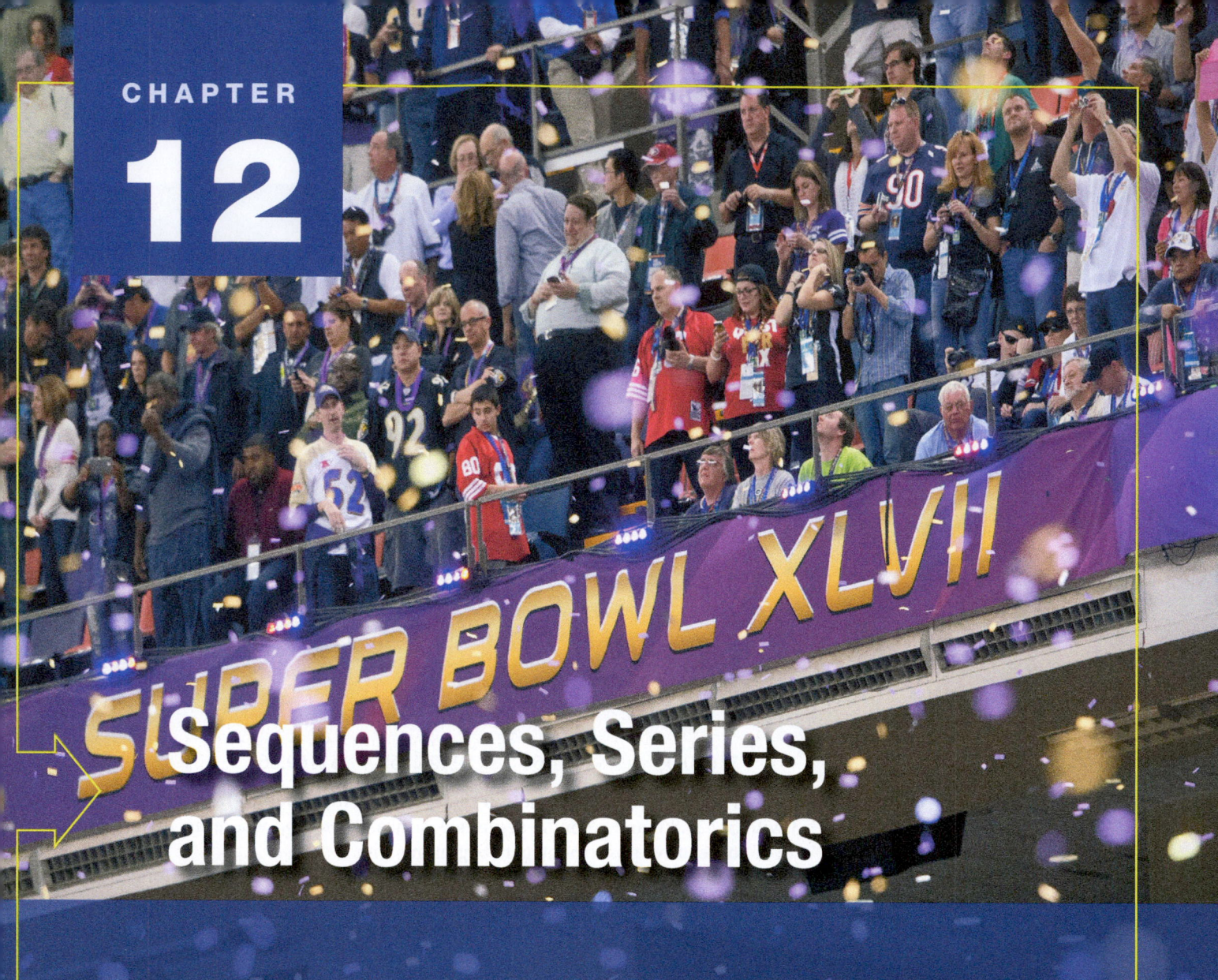

CHAPTER 12

Sequences, Series, and Combinatorics

APPLICATION   **This problem appears as Example 10 in Section 12.3.**

Large sporting events have a significant impact on the economy of the host city. Super Bowl XLVII, hosted by New Orleans, generated a $480-million net impact for the region (*Source*: NewOrleansSaints.com, posted April 18, 2013, Marius M. Mihai, Research Analyst of the Division of Business and Economic Research at the University of New Orleans (DBER)). Assume that 60% of that amount is spent again in the area, and then 60% of that amount is spent again, and so on. This is known as the *economic multiplier effect*. Find the total effect on the economy.

<table>
<tr><td>**12.1**</td><td>**Sequences and Series**</td></tr>
</table>

▶ Find terms of sequences given the *n*th term.

▶ Look for a pattern in a sequence and try to determine a general term.

▶ Convert between sigma notation and other notation for a series.

▶ Construct the terms of a recursively defined sequence.

In this section, we discuss sets or lists of numbers, considered in order, and their sums.

▶ ## Sequences

Suppose that $1000 is invested at 4%, compounded annually. The amounts to which the account will grow after 1 year, 2 years, 3 years, 4 years, and so on, form the following sequence of numbers:

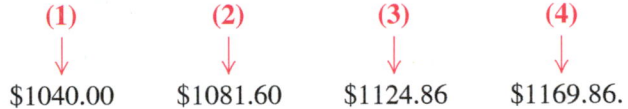

$1040.00 $1081.60 $1124.86 $1169.86.

We can think of this as a function that pairs 1 with $1040.00, 2 with $1081.60, 3 with $1124.86, and so on. A **sequence** is thus a *function*, where the domain is a set of consecutive positive integers beginning with 1.

If we continue to compute the amounts of money in the account forever, we obtain an **infinite sequence** with function values

$1040.00, $1081.60, $1124.86, $1169.86, $1216.65, $1265.32, ...

The dots "..." at the end indicate that the sequence goes on without stopping. If we stop after a certain number of years, we obtain a **finite sequence**:

$1040.00, $1081.60, $1124.86, $1169.86.

SEQUENCES

An **infinite sequence** is a function having for its domain the set of positive integers, $\{1, 2, 3, 4, 5, \ldots\}$.

A **finite sequence** is a function having for its domain a set of positive integers, $\{1, 2, 3, 4, 5, \ldots, n\}$, for some positive integer n.

Consider the sequence given by the formula

$$a(n) = 2^n, \quad \text{or} \quad a_n = 2^n.$$

Some of the function values, also known as the **terms** of the sequence, follow:

$$a_1 = 2^1 = 2,$$
$$a_2 = 2^2 = 4,$$
$$a_3 = 2^3 = 8,$$
$$a_4 = 2^4 = 16,$$
$$a_5 = 2^5 = 32.$$

The first term of the sequence is denoted as a_1, the fifth term as a_5, and the *n*th term, or **general term**, as a_n. This sequence can also be denoted as

$$2, 4, 8, \ldots, \quad \text{or as} \quad 2, 4, 8, \ldots, 2^n, \ldots.$$

EXAMPLE 1 Find the first 4 terms and the 23rd term of the sequence whose general term is given by $a_n = (-1)^n n^2$.

Solution We have $a_n = (-1)^n n^2$, so

$$a_1 = (-1)^1 \cdot 1^2 = -1,$$
$$a_2 = (-1)^2 \cdot 2^2 = 4,$$
$$a_3 = (-1)^3 \cdot 3^2 = -9,$$
$$a_4 = (-1)^4 \cdot 4^2 = 16,$$
$$a_{23} = (-1)^{23} \cdot 23^2 = -529.$$

> **Now Try Exercise 1.**

Note in Example 1 that the power $(-1)^n$ causes the signs of the terms to alternate between positive and negative, depending on whether *n* is even or odd. This kind of sequence is called an **alternating sequence**.

Technology Connection

X	Y₁
1	−1
2	4
3	−9
4	16
23	−529

X =

seq(X/(X+1),X,1,5) ▸ Frac
$\{\frac{1}{2} \quad \frac{2}{3} \quad \frac{3}{4} \quad \frac{4}{5} \quad \frac{5}{6}\}$

We can use a graphing calculator to find the desired terms of the sequence in Example 1. We enter $y_1 = (-1)^x x^2$. We then set up a table in ASK mode and enter 1, 2, 3, 4, and 23 as values for *x*.

We can also use the SEQ feature to find the terms of a sequence. Suppose, for example, that we want to find the first 5 terms of the sequence whose general term is given by $a_n = n/(n+1)$. We select SEQ from the LIST OPS menu and enter the general term, the variable, and the numbers of the first and last terms desired. The calculator will write the terms horizontally as a list. The list can also be written in fraction notation. The first 5 terms of the sequence are $1/2, 2/3, 3/4, 4/5,$ and $5/6$.

We can graph a sequence just as we graph other functions. Consider the function given by $f(x) = x + 1$ and the sequence whose general term is given by $a_n = n + 1$. The graph of $f(x) = x + 1$ is shown on the left below. Since the domain of a sequence is a set of positive integers, the graph of a sequence is a set of points that are not connected. Thus if we use only positive integers for inputs of $f(x) = x + 1$, we have the graph of the sequence $a_n = n + 1$, as shown on the right below.

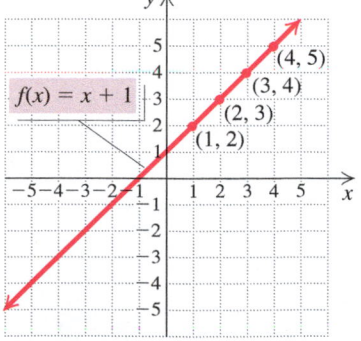

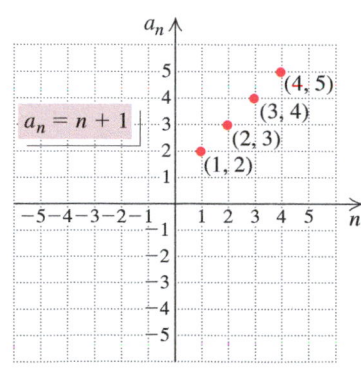

▶ **Finding the General Term**

When only the first few terms of a sequence are known, we do not know for sure what the general term is, but we might be able to make a prediction by looking for a pattern.

EXAMPLE 2   For each of the following sequences, predict the general term.

a) $1, \sqrt{2}, \sqrt{3}, 2, \ldots$ **b)** $-1, 3, -9, 27, -81, \ldots$

c) $2, 4, 8, \ldots$

Solution

a) These are square roots of consecutive integers, so the general term might be $\sqrt{n}$.

b) These are powers of 3 with alternating signs, so the general term might be $(-1)^n 3^{n-1}$.

c) If we see the pattern of powers of 2, we will see 16 as the next term and guess 2^n for the general term. Then the sequence could be written with more terms as

$$2, 4, 8, 16, 32, 64, 128, \ldots.$$

If we see that we can get the second term by adding 2, the third term by adding 4, and the next term by adding 6, and so on, we will see 14 as the next term. A general term for the sequence is $n^2 - n + 2$, and the sequence can be written with more terms as

$$2, 4, 8, 14, 22, 32, 44, 58, \ldots.$$

⟶ **Now Try Exercise 19.**

Example 2(c) illustrates that, in fact, you can never be certain about the general term when only a few terms are given. The fewer the number of given terms, the greater the uncertainty.

▶ **Sums and Series**

SERIES

Given the infinite sequence

$$a_1, a_2, a_3, a_4, \ldots, a_n, \ldots,$$

the sum of the terms

$$a_1 + a_2 + a_3 + \cdots + a_n + \cdots$$

is called an **infinite series**. A **partial sum** is the sum of the first n terms:

$$a_1 + a_2 + a_3 + \cdots + a_n.$$

A partial sum is also called a **finite series**, or **nth partial sum**, and is denoted S_n.

EXAMPLE 3   For the sequence $-2, 4, -6, 8, -10, 12, -14, \ldots$, find each of the following.

a) S_1 **b)** S_4 **c)** S_5

Solution

a) $S_1 = -2$

b) $S_4 = -2 + 4 + (-6) + 8 = 4$

c) $S_5 = -2 + 4 + (-6) + 8 + (-10) = -6$

⟶ **Now Try Exercise 29.**

cumSum(seq($n^2 - 3,n,1,4$))
{-2 -1 5 18}

We can use a graphing calculator to find partial sums of a sequence when a formula for the general term is known. Suppose, for example, that we want to find S_1, S_2, S_3, and S_4 for the sequence whose general term is given by $a_n = n^2 - 3$. We can use the CUMSUM feature from the LIST OPS menu. The calculator will write the partial sums as a list. (Note that the calculator can be set in either FUNCTION mode or SEQUENCE mode. Here we show SEQUENCE mode.)

We have $S_1 = -2$, $S_2 = -1$, $S_3 = 5$, and $S_4 = 18$.

▶ Sigma Notation

The Greek letter Σ (sigma) can be used to denote a sum when the general term of a sequence is a formula. For example, the sum of the first four terms of the sequence $3, 5, 7, 9, \ldots, 2k + 1, \ldots$ can be named as follows, using what is called **sigma notation**, or **summation notation**:

$$\sum_{k=1}^{4} (2k + 1).$$

This is read "the sum as k goes from 1 to 4 of $2k + 1$." The letter k is called the **index of summation**. The index of summation might start at a number other than 1, and letters other than k can be used.

EXAMPLE 4 Find and evaluate each of the following sums.

a) $\displaystyle\sum_{k=1}^{5} k^3$ **b)** $\displaystyle\sum_{k=0}^{4} (-1)^k 5^k$ **c)** $\displaystyle\sum_{i=8}^{11} \left(2 + \frac{1}{i}\right)$

We can combine the SUM and SEQ features on a graphing calculator to add the terms of a sequence. We find the sum in Example 4(a) as shown below.

sum(seq($n^3,n,1,5$))
225

Solution

a) We replace k with 1, 2, 3, 4, and 5. Then we add the results.

$$\sum_{k=1}^{5} k^3 = 1^3 + 2^3 + 3^3 + 4^3 + 5^3$$
$$= 1 + 8 + 27 + 64 + 125$$
$$= 225$$

b) $\displaystyle\sum_{k=0}^{4} (-1)^k 5^k = (-1)^0 5^0 + (-1)^1 5^1 + (-1)^2 5^2 + (-1)^3 5^3 + (-1)^4 5^4$
$$= 1 - 5 + 25 - 125 + 625 = 521$$

c) $\displaystyle\sum_{i=8}^{11} \left(2 + \frac{1}{i}\right) = \left(2 + \frac{1}{8}\right) + \left(2 + \frac{1}{9}\right) + \left(2 + \frac{1}{10}\right) + \left(2 + \frac{1}{11}\right)$
$$= 8\frac{1691}{3960}$$

▶ **Now Try Exercise 33.**

EXAMPLE 5 Write sigma notation for each sum.

a) $1 + 2 + 4 + 8 + 16 + 32 + 64$

b) $-2 + 4 - 6 + 8 - 10$

c) $x + \dfrac{x^2}{2} + \dfrac{x^3}{3} + \dfrac{x^4}{4} + \cdots$

Solution

a) $1 + 2 + 4 + 8 + 16 + 32 + 64$

This is the sum of powers of 2, beginning with 2^0, or 1, and ending with 2^6, or 64. Sigma notation is $\sum_{k=0}^{6} 2^k$.

b) $-2 + 4 - 6 + 8 - 10$

Disregarding the alternating signs, we see that this is the sum of the first 5 even integers. Note that $2k$ is a formula for the kth positive even integer, and $(-1)^k = -1$ when k is odd and $(-1)^k = 1$ when k is even. Thus the general term is $(-1)^k(2k)$. The sum begins with $k = 1$ and ends with $k = 5$, so sigma notation is $\sum_{k=1}^{5}(-1)^k(2k)$.

c) $x + \dfrac{x^2}{2} + \dfrac{x^3}{3} + \dfrac{x^4}{4} + \cdots$

The general term is x^k/k, beginning with $k = 1$. This is also an infinite series. We use the symbol ∞ for infinity and write the series using sigma notation: $\sum_{k=1}^{\infty}(x^k/k)$.

> **Now Try Exercise 51.**

► Recursive Definitions

A sequence can be defined **recursively** or by using a **recursion formula**. Such a definition lists the first term, or the first few terms, and then describes how to determine the remaining terms from the given terms.

EXAMPLE 6   Find the first 5 terms of the sequence defined by

$$a_1 = 5, \qquad a_{n+1} = 2a_n - 3, \quad \text{for } n \geq 1.$$

Solution   We have

$$a_1 = 5,$$
$$a_2 = 2a_1 - 3 = 2 \cdot 5 - 3 = 7,$$
$$a_3 = 2a_2 - 3 = 2 \cdot 7 - 3 = 11,$$
$$a_4 = 2a_3 - 3 = 2 \cdot 11 - 3 = 19,$$
$$a_5 = 2a_4 - 3 = 2 \cdot 19 - 3 = 35.$$

> **Now Try Exercise 61.**

Technology Connection

Many graphing calculators have the capability to work with recursively defined sequences when they are set in SEQUENCE mode. In Example 6, for instance, the function could be entered as $u(n) = 2 * u(n - 1) - 3$ with $u(n\text{Min}) = 5$. We can read the terms of the sequence from a table.

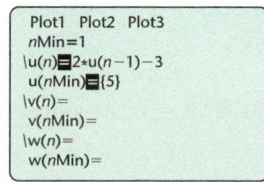

12.1 Exercise Set

In each of the following, the nth term of a sequence is given. Find the first 4 terms, a_{10}, and a_{15}.

1. $a_n = 4n - 1$

2. $a_n = (n - 1)(n - 2)(n - 3)$

3. $a_n = \dfrac{n}{n - 1}, \ n \geq 2$

4. $a_n = n^2 - 1, \ n \geq 3$

5. $a_n = \dfrac{n^2 - 1}{n^2 + 1}$

6. $a_n = \left(-\dfrac{1}{2}\right)^{n-1}$

7. $a_n = (-1)^n n^2$

8. $a_n = (-1)^{n-1}(3n - 5)$

9. $a_n = 5 + \dfrac{(-2)^{n+1}}{2^n}$

10. $a_n = \dfrac{2n - 1}{n^2 + 2n}$

Find the indicated term of the given sequence.

11. $a_n = 5n - 6; \ a_8$

12. $a_n = (3n - 4)(2n + 5); \ a_7$

13. $a_n = (2n + 3)^2; \ a_6$

14. $a_n = (-1)^{n-1}(4.6n - 18.3); \ a_{12}$

15. $a_n = 5n^2(4n - 100); \ a_{11}$

16. $a_n = \left(1 + \dfrac{1}{n}\right)^2; \ a_{80}$

17. $a_n = \ln e^n; \ a_{67}$

18. $a_n = 2 - \dfrac{1000}{n}; \ a_{100}$

Predict the general term, or nth term, a_n, of the sequence. Answers may vary.

19. $2, 4, 6, 8, 10, \ldots$

20. $3, 9, 27, 81, 243, \ldots$

21. $-2, 6, -18, 54, \ldots$

22. $-2, 3, 8, 13, 18, \ldots$

23. $\frac{2}{3}, \frac{3}{4}, \frac{4}{5}, \frac{5}{6}, \frac{6}{7}, \ldots$

24. $\sqrt{2}, 2, \sqrt{6}, 2\sqrt{2}, \sqrt{10}, \ldots$

25. $1 \cdot 2, 2 \cdot 3, 3 \cdot 4, 4 \cdot 5, \ldots$

26. $-1, -4, -7, -10, -13, \ldots$

27. $0, \log 10, \log 100, \log 1000, \ldots$

28. $\ln e^2, \ln e^3, \ln e^4, \ln e^5, \ldots$

Find the indicated partial sums for the sequence.

29. $1, 2, 3, 4, 5, 6, 7, \ldots; \ S_3$ and S_7

30. $1, -3, 5, -7, 9, -11, \ldots; \ S_2$ and S_5

31. $2, 4, 6, 8, \ldots; \ S_4$ and S_5

32. $1, \frac{1}{4}, \frac{1}{9}, \frac{1}{16}, \frac{1}{25}, \ldots; \ S_1$ and S_5

Find and evaluate the sum.

33. $\displaystyle\sum_{k=1}^{5} \dfrac{1}{2k}$

34. $\displaystyle\sum_{i=1}^{6} \dfrac{1}{2i + 1}$

35. $\displaystyle\sum_{i=0}^{6} 2^i$

36. $\displaystyle\sum_{k=4}^{7} \sqrt{2k - 1}$

37. $\displaystyle\sum_{k=7}^{10} \ln k$

38. $\displaystyle\sum_{k=1}^{4} \pi k$

39. $\displaystyle\sum_{k=1}^{8} \dfrac{k}{k + 1}$

40. $\displaystyle\sum_{i=1}^{5} \dfrac{i - 1}{i + 3}$

41. $\displaystyle\sum_{i=1}^{5} (-1)^i$

42. $\displaystyle\sum_{k=0}^{5} (-1)^{k+1}$

43. $\displaystyle\sum_{k=1}^{8} (-1)^{k+1} 3k$

44. $\displaystyle\sum_{k=0}^{7} (-1)^k 4^{k+1}$

45. $\displaystyle\sum_{k=0}^{6} \dfrac{2}{k^2 + 1}$

46. $\displaystyle\sum_{i=1}^{10} i(i + 1)$

47. $\displaystyle\sum_{k=0}^{5} (k^2 - 2k + 3)$

48. $\displaystyle\sum_{k=1}^{10} \dfrac{1}{k(k + 1)}$

49. $\displaystyle\sum_{i=0}^{10} \dfrac{2^i}{2^i + 1}$

50. $\displaystyle\sum_{k=0}^{3} (-2)^{2k}$

Write sigma notation. Answers may vary.

51. $5 + 10 + 15 + 20 + 25 + \cdots$

52. $7 + 14 + 21 + 28 + 35 + \cdots$

53. $2 - 4 + 8 - 16 + 32 - 64$

54. $3 + 6 + 9 + 12 + 15$

55. $-\dfrac{1}{2} + \dfrac{2}{3} - \dfrac{3}{4} + \dfrac{4}{5} - \dfrac{5}{6} + \dfrac{6}{7}$

56. $\dfrac{1}{1^2} + \dfrac{1}{2^2} + \dfrac{1}{3^2} + \dfrac{1}{4^2} + \dfrac{1}{5^2}$

57. $4 - 9 + 16 - 25 + \cdots + (-1)^n n^2$

58. $9 - 16 + 25 + \cdots + (-1)^{n+1} n^2$

59. $\dfrac{1}{1\cdot 2} + \dfrac{1}{2\cdot 3} + \dfrac{1}{3\cdot 4} + \dfrac{1}{4\cdot 5} + \cdots$

60. $\dfrac{1}{1\cdot 2^2} + \dfrac{1}{2\cdot 3^2} + \dfrac{1}{3\cdot 4^2} + \dfrac{1}{4\cdot 5^2} + \cdots$

Find the first 4 terms of the recursively defined sequence.

61. $a_1 = 4, \ a_{n+1} = 1 + \dfrac{1}{a_n}$

62. $a_1 = 256, \ a_{n+1} = \sqrt{a_n}$

63. $a_1 = 6561, \ a_{n+1} = (-1)^n \sqrt{a_n}$

64. $a_1 = e^Q, \ a_{n+1} = \ln a_n$

65. $a_1 = 2, a_2 = 3, \ a_{n+1} = a_n + a_{n-1}$

66. $a_1 = -10, a_2 = 8, \ a_{n+1} = a_n - a_{n-1}$

67. *Compound Interest.* Suppose that $1000 is invested at 6.2%, compounded annually. The value of the investment after n years is given by the sequence model
$$a_n = \$1000(1.062)^n, \quad n = 1, 2, 3, \ldots.$$
a) Find the first 10 terms of the sequence.
b) Find the value of the investment after 20 years.

68. *Salvage Value.* The value of a post-hole digger is $5200. Its salvage value each year is 75% of its value the year before. Give a sequence that lists the salvage value of the machine for each year of a 10-year period.

69. *Wage Sequence.* Adahy is paid $9.80 per hour for working at Red Freight Limited. Each year he receives a $1.10 hourly raise. Give a sequence that lists Adahy's hourly wage over a 10-year period.

70. *Bacteria Growth.* Suppose that a single cell of bacteria divides into two every 15 min. Suppose that the same rate of division is maintained for 4 hr. Give a sequence that lists the number of cells after successive 15-min periods.

71. *Fibonacci Sequence: Rabbit Population Growth.* One of the most famous recursively defined sequences is the **Fibonacci sequence**. In 1202, the Italian mathematician Leonardo da Pisa, also called Fibonacci, proposed the following model for rabbit population growth. Suppose that every month each mature pair of rabbits in the population produces a new pair that

begins reproducing after two months, and also suppose that no rabbits die. Beginning with one pair of newborn rabbits, the population can be modeled by the following recursively defined sequence:
$$a_1 = 1, \quad a_2 = 1, \quad a_n = a_{n-1} + a_{n-2}, \text{ for } n \geq 3,$$
where a_n is the total number of pairs of rabbits in month n. Find the first 7 terms of the Fibonacci sequence.

▶ **Skill Maintenance**

Solve.

72. $3x - 2y = 3,$
$2x + 3y = -11$ **[3.1], [3.2], [3.3], [10.1], [10.3], [10.4]**

73. *Harvesting Pumpkins.* A total of 23,400 acres of pumpkins were harvested in Illinois and Ohio in 2012. The number of acres of pumpkins harvested in Ohio was 9000 fewer than the number of acres of pumpkins harvested in Illinois. (*Source:* U.S. Department of Agriculture) Find the number of acres of pumpkins harvested in Illinois and in Ohio in 2012.   **[3.4]**

Find the center and the radius of the circle with the given equation. **[11.2]**

74. $x^2 + y^2 - 6x + 4y = 3$

75. $x^2 + y^2 + 5x - 8y = 2$

▶ **Synthesis**

Find the first 5 terms of the sequence, and then find S_5.

76. $a_n = \dfrac{1}{2^n} \log 1000^n$

77. $a_n = i^n, i = \sqrt{-1}$

78. $a_n = \ln(1 \cdot 2 \cdot 3 \cdots n)$

For each sequence, find a formula for S_n.

79. $a_n = \ln n$

80. $a_n = \dfrac{1}{n} - \dfrac{1}{n+1}$

12.2 Arithmetic Sequences and Series

▶ For any arithmetic sequence, find the *n*th term when *n* is given and *n* when the *n*th term is given; and given two terms, find the common difference and construct the sequence.

▶ Find the sum of the first *n* terms of an arithmetic sequence.

A sequence in which each term after the first is found by adding the same number to the preceding term is an **arithmetic sequence**.

▶ Arithmetic Sequences

The sequence 2, 5, 8, 11, 14, 17, . . . is arithmetic because adding 3 to any term produces the next term. In other words, the difference between any term and the preceding one is 3. Arithmetic sequences are also called *arithmetic progressions*.

> **ARITHMETIC SEQUENCE**
>
> A sequence is **arithmetic** if there exists a number d, called the **common difference**, such that $a_{n+1} = a_n + d$ for any integer $n \geq 1$.

EXAMPLE 1 For each of the following arithmetic sequences, identify the first term, a_1, and the common difference, d.

a) 4, 9, 14, 19, 24, . . .

b) 34, 27, 20, 13, 6, -1, -8, . . .

c) 2, $2\frac{1}{2}$, 3, $3\frac{1}{2}$, 4, $4\frac{1}{2}$, . . .

Solution The first term, a_1, is the first term listed. To find the common difference, d, we choose any term beyond the first and subtract the preceding term from it.

SEQUENCE	FIRST TERM, a_1	COMMON DIFFERENCE, d
a) 4, 9, 14, 19, 24, . . .	4	5 $(9 - 4 = 5)$
b) 34, 27, 20, 13, 6, -1, -8, . . .	34	-7 $(27 - 34 = -7)$
c) 2, $2\frac{1}{2}$, 3, $3\frac{1}{2}$, 4, $4\frac{1}{2}$, . . .	2	$\frac{1}{2}$ $\left(2\frac{1}{2} - 2 = \frac{1}{2}\right)$

Note that we obtained the common difference by subtracting a_1 from a_2. Had we subtracted a_2 from a_3 or a_3 from a_4, we would have obtained the same values for d. Thus we can check by adding d to each term in a sequence to see if we progress correctly to the next term.

Check:

a) $4 + 5 = 9$, $\quad 9 + 5 = 14$, $\quad 14 + 5 = 19$, $\quad 19 + 5 = 24$

b) $34 + (-7) = 27$, $\quad 27 + (-7) = 20$, $\quad 20 + (-7) = 13$, $13 + (-7) = 6$, $\quad 6 + (-7) = -1$, $\quad -1 + (-7) = -8$

c) $2 + \frac{1}{2} = 2\frac{1}{2}$, $\quad 2\frac{1}{2} + \frac{1}{2} = 3$, $\quad 3 + \frac{1}{2} = 3\frac{1}{2}$, $\quad 3\frac{1}{2} + \frac{1}{2} = 4$, $4 + \frac{1}{2} = 4\frac{1}{2}$

Now Try Exercise 1.

To find a formula for the general, or *n*th, term of any arithmetic sequence, we denote the common difference by *d*, write out the first few terms, and look for a pattern:

$$a_1,$$
$$a_2 = a_1 + d,$$
$$a_3 = a_2 + d = (a_1 + d) + d = a_1 + 2d, \qquad \text{Substituting for } a_2$$
$$a_4 = a_3 + d = (a_1 + 2d) + d = a_1 + 3d. \qquad \text{Substituting for } a_3$$

Note that the coefficient of *d* in each case is 1 less than the subscript.

Generalizing, we obtain the following formula.

---

*n*TH TERM OF AN ARITHMETIC SEQUENCE

The ***n*th term** of an arithmetic sequence is given by

$$a_n = a_1 + (n - 1)d, \quad \text{for any integer } n \geq 1.$$

EXAMPLE 2 Find the 14th term of the arithmetic sequence 4, 7, 10, 13, ...

Solution We first note that $a_1 = 4$, $d = 7 - 4$, or 3, and $n = 14$. Then using the formula for the *n*th term, we obtain

$$a_n = a_1 + (n - 1)d$$
$$a_{14} = 4 + (14 - 1) \cdot 3 \qquad \text{Substituting}$$
$$= 4 + 13 \cdot 3 = 4 + 39$$
$$= 43.$$

The 14th term is 43. **Now Try Exercise 9.**

EXAMPLE 3 In the sequence of Example 2, which term is 301? That is, find *n* if $a_n = 301$.

Solution We substitute 301 for a_n, 4 for a_1, and 3 for *d* in the formula for the *n*th term and solve for *n*:

$$a_n = a_1 + (n - 1)d$$
$$301 = 4 + (n - 1) \cdot 3 \qquad \text{Substituting}$$
$$301 = 4 + 3n - 3$$
$$301 = 3n + 1$$
$$300 = 3n \qquad \text{Solving for } n$$
$$100 = n.$$

The term 301 is the 100th term of the sequence. **Now Try Exercise 15.**

Given two terms and their places in an arithmetic sequence, we can construct the sequence.

EXAMPLE 4 The 3rd term of an arithmetic sequence is 8, and the 16th term is 47. Find a_1 and *d* and construct the sequence.

Solution   We know that $a_3 = 8$ and $a_{16} = 47$. Thus we would have to add d 13 times to get from 8 to 47. That is,

$$8 + 13d = 47. \qquad \textcolor{red}{\textbf{a_3 and a_{16} are } 16 - 3 \textbf{, or 13, terms apart.}}$$

Solving $8 + 13d = 47$, we obtain

$$13d = 39$$
$$d = 3.$$

Since $a_3 = 8$, we subtract d twice to get a_1. Thus,

$$a_1 = 8 - 2 \cdot 3 = 2. \qquad \textcolor{red}{\textbf{a_1 and a_3 are } 3 - 1 \textbf{, or 2, terms apart.}}$$

The sequence is 2, 5, 8, 11, ... Note that we could also subtract d 15 times from a_{16} in order to find a_1.

Now Try Exercise 23.

In general, d should be subtracted $n - 1$ times from a_n in order to find a_1.

▶ Sum of the First *n* Terms of an Arithmetic Sequence

Consider the arithmetic sequence

$$3, 5, 7, 9, \ldots.$$

When we add the first 4 terms of the sequence, we get S_4, which is

$$3 + 5 + 7 + 9, \quad \text{or} \quad 24.$$

This sum is called an **arithmetic series**. To find a formula for the sum of the first n terms, S_n, of an arithmetic sequence, we first denote an arithmetic sequence, as follows:

> **This term is two terms back from the last. If you add d to this term, the result is the next-to-last term, $a_n - d$.**

$$a_1, \quad (a_1 + d), \quad (a_1 + 2d), \quad \ldots, \quad \overbrace{(a_n - 2d)}, \quad \underbrace{(a_n - d)}, \quad a_n.$$

> **This is the next-to-last term. If you add d to this term, the result is a_n.**

Then S_n is given by

$$S_n = a_1 + (a_1 + d) + (a_1 + 2d) + \cdots + (a_n - 2d)$$
$$\qquad + (a_n - d) + a_n. \tag{1}$$

Reversing the order of the addition gives us

$$S_n = a_n + (a_n - d) + (a_n - 2d) + \cdots + (a_1 + 2d)$$
$$\qquad + (a_1 + d) + a_1. \tag{2}$$

If we add corresponding terms of each side of equations (1) and (2), we get

$$2S_n = [a_1 + a_n] + [(a_1 + d) + (a_n - d)] + [(a_1 + 2d) + (a_n - 2d)]$$
$$\qquad + \cdots + [(a_n - 2d) + (a_1 + 2d)]$$
$$\qquad + [(a_n - d) + (a_1 + d)] + [a_n + a_1].$$

In the expression for $2S_n$, there are n expressions in square brackets. Each of these expressions is equivalent to $a_1 + a_n$. Thus the expression for $2S_n$ can be written in simplified form as

$$2S_n = [a_1 + a_n] + [a_1 + a_n] + [a_1 + a_n] + \cdots + [a_n + a_1]$$
$$+ [a_n + a_1] + [a_n + a_1].$$

Since $a_1 + a_n$ is being added n times, it follows that

$$2S_n = n(a_1 + a_n),$$

from which we get the following formula.

SUM OF THE FIRST n TERMS

The sum of the first n terms of an arithmetic sequence is given by

$$S_n = \frac{n}{2}(a_1 + a_n).$$

EXAMPLE 5   Find the sum of the first 100 natural numbers.

Solution   The sum is

$$1 + 2 + 3 + \cdots + 99 + 100.$$

This is the sum of the first 100 terms of the arithmetic sequence for which

$$a_1 = 1, \quad a_n = 100, \quad \text{and} \quad n = 100.$$

Thus substituting into the formula

$$S_n = \frac{n}{2}(a_1 + a_n),$$

we get

$$S_{100} = \frac{100}{2}(1 + 100) = 50(101) = 5050.$$

The sum of the first 100 natural numbers is 5050.

> **Now Try Exercise 27.**

EXAMPLE 6   Find the sum of the first 15 terms of the arithmetic sequence 4, 7, 10, 13, ...

Solution   Note that $a_1 = 4$, $d = 3$, and $n = 15$. Before using the formula

$$S_n = \frac{n}{2}(a_1 + a_n),$$

we find the last term, a_{15}:

$$a_{15} = 4 + (15 - 1)3 \quad \textcolor{red}{\text{Substituting into the formula } a_n = a_1 + (n - 1)d}$$
$$= 4 + 14 \cdot 3 = 46.$$

Thus,

$$S_{15} = \frac{15}{2}(4 + 46) = \frac{15}{2}(50) = 375.$$

The sum of the first 15 terms is 375.

> **Now Try Exercise 25.**

EXAMPLE 7   Find the sum: $\displaystyle\sum_{k=1}^{130}(4k+5)$.

Solution   It is helpful to first write out a few terms:

$$9 + 13 + 17 + \cdots.$$

It appears that this is an arithmetic series coming from an arithmetic sequence with $a_1 = 9$, $d = 4$, and $n = 130$. Before using the formula

$$S_n = \frac{n}{2}(a_1 + a_n),$$

we find the last term, a_{130}:

$$
\begin{aligned}
a_{130} &= 4 \cdot 130 + 5 \qquad \textbf{The \textit{k}th term is } 4k + 5.\\
&= 520 + 5\\
&= 525.
\end{aligned}
$$

Thus,

$$S_{130} = \frac{130}{2}(9 + 525) \qquad \textbf{Substituting into } S_n = \frac{n}{2}(a_1 + a_n)$$

$$= 34{,}710.$$

Now Try Exercise 33.

▶ **Applications**

The translation of some applications and problem-solving situations may involve arithmetic sequences or series. We consider some examples.

EXAMPLE 8   *Hourly Wages.*   Kendall accepts a job, starting with an hourly wage of $14.25, and is promised a raise of 15¢ per hour every 2 months for 5 years. At the end of 5 years, what will Kendall's hourly wage be?

Solution   It helps to first write down the hourly wage for several 2-month time periods:

> Beginning: $14.25,
> After 2 months: $14.40,
> After 4 months: $14.55,
> and so on.

What appears is a sequence of numbers: 14.25, 14.40, 14.55, ... This sequence is arithmetic, because adding 0.15 each time gives us the next term.

We want to find the last term of an arithmetic sequence, so we use the formula $a_n = a_1 + (n - 1)d$. We know that $a_1 = 14.25$ and $d = 0.15$, but what is n? That is, how many terms are in the sequence? Each year there are $12/2$, or 6 raises, since Kendall gets a raise every 2 months. There are 5 years, so the total number of raises will be $5 \cdot 6$, or 30. Thus there will be 31 terms: the original wage and 30 increased rates.

Substituting in the formula $a_n = a_1 + (n - 1)d$ gives us

$$
\begin{aligned}
a_{31} &= 14.25 + (31 - 1) \cdot 0.15\\
&= 18.75.
\end{aligned}
$$

Thus, at the end of 5 years, Kendall's hourly wage will be $18.75.

Now Try Exercise 43.

Often there is more than one valid approach for solving problems like Example 8. In this chapter, we concentrate on the use of sequences and series and their related formulas.

EXAMPLE 9 *Total in a Stack.* A stack of electric poles has 30 poles in the bottom row. There are 29 poles in the second row, 28 in the next row, and so on. How many poles are in the stack if there are 5 poles in the top row?

Solution A drawing will help in this case. The following figure shows the ends of the poles and the way in which they stack.

5 poles in 26th row

28 poles in 3rd row
29 poles in 2nd row
30 poles in 1st row

Since the number of poles decreases from 30 in a row up to 5 in the top row, there must be 26 rows. We want the sum

$$30 + 29 + 28 + \cdots + 5.$$

Thus we have an arithmetic series. We use the formula

$$S_n = \frac{n}{2}(a_1 + a_n),$$

with $n = 26$, $a_1 = 30$, and $a_{26} = 5$.
Substituting, we get

$$S_{26} = \frac{26}{2}(30 + 5) = 455.$$

There are 455 poles in the stack.

Now Try Exercise 39.

12.2   Exercise Set

Find the first term and the common difference.

1. 3, 8, 13, 18, . . .

2. $1.08, $1.16, $1.24, $1.32, . . .

3. 9, 5, 1, −3, . . .

4. −8, −5, −2, 1, 4, . . .

5. $\frac{3}{2}, \frac{9}{4}, 3, \frac{15}{4}, \ldots$

6. $\frac{3}{5}, \frac{1}{10}, -\frac{2}{5}, \ldots$

7. $316, $313, $310, $307, . . .

8. Find the 11th term of the arithmetic sequence 0.07, 0.12, 0.17, ...

9. Find the 12th term of the arithmetic sequence 2, 6, 10, ...

10. Find the 17th term of the arithmetic sequence 7, 4, 1, ...

11. Find the 14th term of the arithmetic sequence 3, $\frac{7}{3}$, $\frac{5}{3}$, ...

12. Find the 13th term of the arithmetic sequence $1200, $964.32, $728.64, ...

13. Find the 10th term of the arithmetic sequence $2345.78, $2967.54, $3589.30, ...

14. In the sequence of Exercise 8, what term is the number 1.67?

15. In the sequence of Exercise 9, what term is the number 106?

16. In the sequence of Exercise 10, what term is −296?

17. In the sequence of Exercise 11, what term is −27?

18. Find a_{20} when $a_1 = 14$ and $d = -3$.

19. Find a_1 when $d = 4$ and $a_8 = 33$.

20. Find d when $a_1 = 8$ and $a_{11} = 26$.

21. Find n when $a_1 = 25$, $d = -14$, and $a_n = -507$.

22. In an arithmetic sequence, $a_{17} = -40$ and $a_{28} = -73$. Find a_1 and d. Write the first 5 terms of the sequence.

23. In an arithmetic sequence, $a_{17} = \frac{25}{3}$ and $a_{32} = \frac{95}{6}$. Find a_1 and d. Write the first 5 terms of the sequence.

24. Find the sum of the first 14 terms of the series $11 + 7 + 3 + \cdots$.

25. Find the sum of the first 20 terms of the series $5 + 8 + 11 + 14 + \cdots$.

26. Find the sum of the first 300 natural numbers.

27. Find the sum of the first 400 even natural numbers.

28. Find the sum of the odd numbers 1 to 199, inclusive.

29. Find the sum of the multiples of 7 from 7 to 98, inclusive.

30. Find the sum of all multiples of 4 that are between 14 and 523.

31. If an arithmetic series has $a_1 = 2$, $d = 5$, and $n = 20$, what is S_n?

32. If an arithmetic series has $a_1 = 7$, $d = -3$, and $n = 32$, what is S_n?

Find the sum.

33. $\displaystyle\sum_{k=1}^{40} (2k + 3)$

34. $\displaystyle\sum_{k=5}^{20} 8k$

35. $\displaystyle\sum_{k=0}^{19} \frac{k - 3}{4}$

36. $\displaystyle\sum_{k=2}^{50} (2000 - 3k)$

37. $\displaystyle\sum_{k=12}^{57} \frac{7 - 4k}{13}$

38. $\displaystyle\sum_{k=101}^{200} (1.14k - 2.8) - \sum_{k=1}^{5} \left(\frac{k + 4}{10}\right)$

39. *Total Savings.* If 10¢ is saved on October 1, 20¢ is saved on October 2, 30¢ on October 3, and so on, how much is saved altogether during the 31 days of October?

40. *Stacking Poles.* How many poles will be in a stack of telephone poles if there are 50 in the first layer, 49 in the second, and so on, with 6 in the top layer?

41. *Auditorium Seating.* Auditoriums are often built with more seats per row as the rows move toward the back. Suppose that the first balcony of a theater has 28 seats in the first row, 32 in the second, 36 in the third, and so on, for 20 rows. How many seats are in the first balcony altogether?

42. *Investment Return.* Brett sets up an investment situation for a client that will return $5000 the first year, $6125 the second year, $7250 the third year, and so on, for 25 years. How much is received from the investment altogether?

43. *Parachutist Free Fall.* When a parachutist jumps from an airplane, the distances, in feet, that the parachutist falls in each successive second before pulling the ripcord to release the parachute are as follows:

16, 48, 80, 112, 144, ...

Is this sequence arithmetic? What is the common difference? What is the total distance fallen in 10 sec?

44. *Lightning Distance.* The following table lists the distance, in miles, from lightning d_n when thunder is heard n seconds after lightning is seen. Is this sequence arithmetic? What is the common difference?

n (in seconds)	d_n (in miles)
5	1
6	1.2
7	1.4
8	1.6
9	1.8
10	2

45. *Garden Plantings.* A gardener is making a planting in the shape of a trapezoid. It will have 35 plants in the front row, 31 in the second row, 27 in the third row, and so on. If the pattern is consistent, how many plants will there be in the last row? How many plants are there altogether?

46. *Band Formation.* A formation of a marching band has 10 marchers in the front row, 12 in the second row, 14 in the third row, and so on, for 8 rows. How many marchers are in the last row? How many marchers are there altogether?

47. *Raw Material Production.* In a manufacturing process, it took 3 units of raw materials to produce 1 unit of a product. The raw material needs thus formed the sequence

$$3, 6, 9, \ldots, 3n, \ldots.$$

Is this sequence arithmetic? What is the common difference?

▶ **Skill Maintenance**

Solve. **[3.1], [3.2], [3.3], [3.5], [10.1], [10.3], [10.4]**

48. $7x - 2y = 4,$
$\quad x + 3y = 17$

49. $2x + y + 3z = 12,$
$\quad x - 3y + 2z = 11,$
$\quad 5x + 2y - 4z = -4$

50. Find the vertices and the foci of the ellipse with the equation $9x^2 + 16y^2 = 144$. **[11.2]**

51. Find an equation of the ellipse with vertices $(0, -5)$ and $(0, 5)$ and minor axis of length 4. **[11.2]**

▶ **Synthesis**

52. *Straight-Line Depreciation.* A company buys an office machine for $5200 on January 1 of a given year. The machine is expected to last for 8 years, at the end of which time its **trade-in value**, or **salvage value**, will be $1100. If the company's accountant figures the decline in value to be the same each year, then its **book values**, or **salvage values**, after t years, $0 \le t \le 8$, form an arithmetic sequence given by

$$a_t = C - t\left(\frac{C - S}{N}\right),$$

where C is the original cost of the item ($5200), N is the number of years of expected life (8), and S is the salvage value ($1100).

a) Find the formula for a_t for the straight-line depreciation of the office machine.

b) Find the salvage value after 0 year, 1 year, 2 years, 3 years, 4 years, 7 years, and 8 years.

53. Find a formula for the sum of the first n odd natural numbers:

$$1 + 3 + 5 + \cdots + (2n - 1).$$

54. Find three numbers in an arithmetic sequence such that the sum of the first and the third is 10 and the product of the first and the second is 15.

55. Find the first term and the common difference for the arithmetic sequence for which

$$a_2 = 40 - 3q \quad \text{and} \quad a_4 = 10p + q.$$

*If p, m, and q form an arithmetic sequence, it can be shown that $m = (p + q)/2$. The number m is the **arithmetic mean**, or **average**, of p and q. Given two numbers p and q, if we find k other numbers $m_1, m_2, \ldots, m_k$ such that*

$$p, m_1, m_2, \ldots, m_k, q$$

forms an arithmetic sequence, we say that we have "inserted k arithmetic means between p and q."

56. Insert three arithmetic means between -3 and 5.

57. Insert four arithmetic means between 4 and 13.

12.3 Geometric Sequences and Series

▶ Identify the common ratio of a geometric sequence, and find a given term and the sum of the first n terms.

▶ Find the sum of an infinite geometric series, if it exists.

A sequence in which each term after the first is found by multiplying the preceding term by the same number is a **geometric sequence**.

▶ Geometric Sequences

Consider the sequence:

$$2, \quad 6, \quad 18, \quad 54, \quad 162, \ldots.$$

Note that multiplying each term by 3 produces the next term. We call the number 3 the **common ratio** because it can be found by dividing any term by the preceding term. A geometric sequence is also called a *geometric progression*.

GEOMETRIC SEQUENCE

A sequence is **geometric** if there is a number r, called the **common ratio**, such that

$$\frac{a_{n+1}}{a_n} = r, \quad \text{or} \quad a_{n+1} = a_n r, \quad \text{for any integer } n \geq 1.$$

EXAMPLE 1 For each of the following geometric sequences, identify the common ratio.

a) 3, 6, 12, 24, 48, ...

b) $1, \ -\dfrac{1}{2}, \ \dfrac{1}{4}, \ -\dfrac{1}{8}, \ldots$

c) \$5200, \$3900, \$2925, \$2193.75, ...

d) \$1000, \$1060, \$1123.60, ...

Solution

SEQUENCE	COMMON RATIO
a) 3, 6, 12, 24, 48, ...	$2 \ \left(\dfrac{6}{3} = 2, \dfrac{12}{6} = 2, \text{and so on}\right)$
b) $1, \ -\dfrac{1}{2}, \ \dfrac{1}{4}, \ -\dfrac{1}{8}, \ldots$	$-\dfrac{1}{2} \ \left(\dfrac{-\frac{1}{2}}{1} = -\dfrac{1}{2}, \dfrac{\frac{1}{4}}{-\frac{1}{2}} = -\dfrac{1}{2}, \text{and so on}\right)$
c) \$5200, \$3900, \$2925, \$2193.75, ...	$0.75 \ \left(\dfrac{\$3900}{\$5200} = 0.75, \dfrac{\$2925}{\$3900} = 0.75, \text{and so on}\right)$
d) \$1000, \$1060, \$1123.60, ...	$1.06 \ \left(\dfrac{\$1060}{\$1000} = 1.06, \dfrac{\$1123.60}{\$1060} = 1.06, \text{and so on}\right)$

Now Try Exercise 1.

We now find a formula for the general, or *n*th, term of a geometric sequence. Let a_1 be the first term and r the common ratio. The first few terms are as follows:

$a_1,$

$a_2 = a_1 r,$

$a_3 = a_2 r = (a_1 r)r = a_1 r^2,$ **Substituting $a_1 r$ for a_2**

$a_4 = a_3 r = (a_1 r^2)r = a_1 r^3.$ **Substituting $a_1 r^2$ for a_3**

Note that the exponent is 1 less than the subscript.

Generalizing, we obtain the following.

> ### *n*TH TERM OF A GEOMETRIC SEQUENCE
>
> The ***n*th term** of a geometric sequence is given by
>
> $$a_n = a_1 r^{n-1}, \quad \text{for any integer } n \geq 1.$$

EXAMPLE 2   Find the 7th term of the geometric sequence 4, 20, 100, ...

Solution   We first note that

$$a_1 = 4 \quad \text{and} \quad n = 7.$$

To find the common ratio, we can divide any term (other than the first) by the preceding term. Since the second term is 20 and the first is 4, we get

$$r = \frac{20}{4}, \quad \text{or} \quad 5.$$

Then using the formula $a_n = a_1 r^{n-1}$, we have

$$a_7 = 4 \cdot 5^{7-1} = 4 \cdot 5^6 = 4 \cdot 15{,}625 = 62{,}500.$$

Thus the 7th term is 62,500.   **Now Try Exercise 11.**

EXAMPLE 3   Find the 10th term of the geometric sequence 64, −32, 16, −8, ...

Solution   We first note that

$$a_1 = 64, \qquad n = 10, \quad \text{and} \quad r = \frac{-32}{64}, \quad \text{or} \quad -\frac{1}{2}.$$

Then using the formula $a_n = a_1 r^{n-1}$, we have

$$a_{10} = 64 \cdot \left(-\frac{1}{2}\right)^{10-1} = 64 \cdot \left(-\frac{1}{2}\right)^9 = 2^6 \cdot \left(-\frac{1}{2^9}\right) = -\frac{1}{2^3} = -\frac{1}{8}.$$

Thus the 10th term is $-\frac{1}{8}$.   **Now Try Exercise 15.**

▶ Sum of the First *n* Terms of a Geometric Sequence

Next, we develop a formula for the sum S_n of the first *n* terms of a geometric sequence:

$$a_1, a_1 r, a_1 r^2, a_1 r^3, \ldots, a_1 r^{n-1}, \ldots.$$

The associated **geometric series** is given by

$$S_n = a_1 + a_1 r + a_1 r^2 + a_1 r^3 + \cdots + a_1 r^{n-1}. \tag{1}$$

We want to find a formula for this sum. If we multiply by r on both sides of equation (1), we have

$$rS_n = a_1 r + a_1 r^2 + a_1 r^3 + a_1 r^4 + \cdots + a_1 r^n. \tag{2}$$

Subtracting equation (2) from equation (1), we see that the differences of the terms shown in red are 0, leaving

$$S_n - rS_n = a_1 - a_1 r^n,$$

or

$$S_n(1 - r) = a_1(1 - r^n). \qquad \text{Factoring}$$

Dividing by $1 - r$ on both sides gives us the following formula.

SUM OF THE FIRST n TERMS

The sum of the first n terms of a geometric sequence is given by

$$S_n = \frac{a_1(1 - r^n)}{1 - r}, \quad \text{for any } r \neq 1.$$

EXAMPLE 4 Find the sum of the first 7 terms of the geometric sequence $3, 15, 75, 375, \ldots$.

Solution We first note that

$$a_1 = 3, \qquad n = 7, \quad \text{and} \quad r = \frac{15}{3}, \text{ or } 5.$$

Then using the formula

$$S_n = \frac{a_1(1 - r^n)}{1 - r},$$

we have

$$S_7 = \frac{3(1 - 5^7)}{1 - 5}$$

$$= \frac{3(1 - 78{,}125)}{-4}$$

$$= 58{,}593.$$

Thus the sum of the first 7 terms is 58,593.   **▶ Now Try Exercise 23.**

EXAMPLE 5 Find the sum: $\displaystyle\sum_{k=1}^{11} (0.3)^k$.

Solution This is a geometric series with $a_1 = 0.3$, $r = 0.3$, and $n = 11$. Thus,

$$S_{11} = \frac{0.3(1 - 0.3^{11})}{1 - 0.3}$$

$$\approx 0.42857. \qquad \qquad \text{▶ Now Try Exercise 41.}$$

▶ **Infinite Geometric Series**

The sum of the terms of an infinite geometric sequence is an **infinite geometric series**. For some geometric series, S_n gets close to a specific number as n gets large. For example, consider the infinite series

$$\frac{1}{2} + \frac{1}{4} + \frac{1}{8} + \frac{1}{16} + \cdots + \frac{1}{2^n} + \cdots.$$

We can visualize S_n by considering the area of a square. For S_1, we shade half the square. For S_2, we shade half the square plus half the remaining half, or $\frac{1}{4}$. For S_3, we shade the parts shaded in S_2 plus half the remaining part. We see that the values of S_n will get close to 1 (shading the complete square).

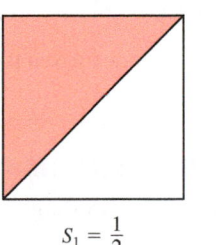 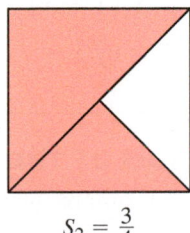  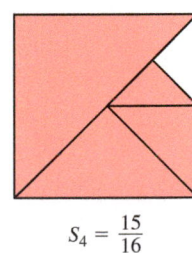

$S_1 = \frac{1}{2}$ $S_2 = \frac{3}{4}$ $S_3 = \frac{7}{8}$ $S_4 = \frac{15}{16}$

TABLE 1

n	S_n
1	0.5
5	0.96875
10	0.9990234375
20	0.9999990463
30	0.9999999991

TABLE 2

n	S_n
1	2
5	62
10	2,046
20	2,097,150
30	2,147,483,646

We examine some partial sums. Note that each of the partial sums in Table 1 is less than 1, but S_n gets very close to 1 as n gets large. We say that 1 is the **limit** of S_n and also that 1 is the **sum of the terms of the infinite geometric sequence**. This sum is denoted S_∞. In this case, $S_\infty = 1$.

Some infinite series do not have limits. Consider the infinite geometric series

$$2 + 4 + 8 + 16 + \cdots + 2^n + \cdots.$$

We again examine some partial sums. Note in Table 2 that as n gets large, S_n gets large without bound. This infinite series does not have a limit.

It can be shown (but we will not do so here) that the limit of an infinite geometric series exists if and only if $|r| < 1$ (that is, the absolute value of the common ratio is less than 1).

To find a formula for the limit of an infinite geometric series, we first consider the sum of the first n terms:

$$S_n = \frac{a_1(1 - r^n)}{1 - r} = \frac{a_1 - a_1 r^n}{1 - r}. \qquad \color{red}{\text{Using the distributive law}}$$

For $|r| < 1$, values of r^n get close to 0 as n gets large. As r^n gets close to 0, so does $a_1 r^n$. Thus, S_n gets close to $a_1/(1 - r)$.

LIMIT OF AN INFINITE GEOMETRIC SERIES

When $|r| < 1$, the limit of an infinite geometric series is given by

$$S_\infty = \frac{a_1}{1 - r}.$$

EXAMPLE 6 Determine whether each of the following infinite geometric series has a limit. If a limit exists, find it.

a) $1 + 3 + 9 + 27 + \cdots$ 　　　　　**b)** $-2 + 1 - \frac{1}{2} + \frac{1}{4} - \frac{1}{8} + \cdots$

Solution

a) Here $r = 3$, so $|r| = |3| = 3$. Since $|r| > 1$, the series *does not* have a limit.

b) Here $r = -\frac{1}{2}$, so $|r| = |-\frac{1}{2}| = \frac{1}{2}$. Since $|r| < 1$, the series *does* have a limit. We find the limit:

$$S_\infty = \frac{a_1}{1 - r} = \frac{-2}{1 - \left(-\frac{1}{2}\right)} = \frac{-2}{\frac{3}{2}} = -\frac{4}{3}.$$

> **Now Try Exercises 33 and 37.**

EXAMPLE 7 Find fraction notation for $0.78787878 \ldots$, or $0.\overline{78}$.

Solution We can express this as

$$0.78 + 0.0078 + 0.000078 + \cdots.$$

Then we see that this is an infinite geometric series, where $a_1 = 0.78$ and $r = 0.01$. Since $|r| < 1$, this series has a limit:

$$S_\infty = \frac{a_1}{1 - r} = \frac{0.78}{1 - 0.01} = \frac{0.78}{0.99} = \frac{78}{99}, \quad \text{or} \quad \frac{26}{33}.$$

Thus fraction notation for $0.78787878\ldots$ is $\frac{26}{33}$. You can check this on your calculator.

> **Now Try Exercise 51.**

▶ Applications

The translation of some applications and problem-solving situations may involve geometric sequences or series. Examples 9 and 10, in particular, show applications in business and economics.

EXAMPLE 8 *A Daily Doubling Salary.* Suppose someone offered you a job for the month of September (30 days) under the following conditions. You will be paid $0.01 for the first day, $0.02 for the second, $0.04 for the third, and so on, doubling your previous day's salary each day. How much would you earn altogether for the month? (Would you take the job? Make a conjecture before reading further.)

Solution You earn $0.01 the first day, $0.01(2)$ the second day, $0.01(2)(2)$ the third day, and so on. The amount earned is the geometric series

$$\$0.01 + \$0.01(2) + \$0.01(2^2) + \$0.01(2^3) + \cdots + \$0.01(2^{29}),$$

where $a_1 = 0.01$, $r = 2$, and $n = 30$. Using the formula

$$S_n = \frac{a_1(1 - r^n)}{1 - r},$$

we have

$$S_{30} = \frac{\$0.01(1 - 2^{30})}{1 - 2} = \$10{,}737{,}418.23.$$

The pay exceeds $10.7 million for the month.

> **Now Try Exercise 57.**

EXAMPLE 9 *The Amount of an Annuity.*   An **annuity** is a sequence of equal payments, made at equal time intervals, that earn interest. Fixed deposits in a savings account are an example of an annuity. Suppose that to save money to buy a car, Jacob deposits $2000 at the *end* of each of 5 years in an account that pays 3% interest, compounded annually. The total amount in the account at the end of 5 years is called the **amount of the annuity**. Find that amount.

Solution   The following time diagram can help visualize the problem. Note that no deposit is made until the end of the first year.

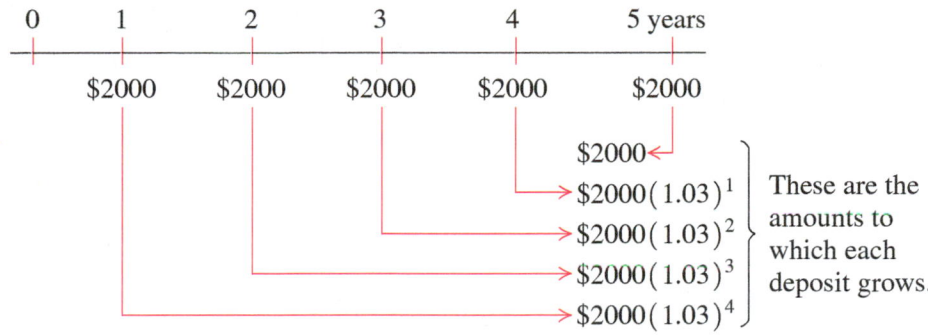

The amount of the annuity is the geometric series

$$\$2000 + \$2000(1.03)^1 + \$2000(1.03)^2 + \$2000(1.03)^3 + \$2000(1.03)^4,$$

where $a_1 = \$2000$, $n = 5$, and $r = 1.03$. Using the formula

$$S_n = \frac{a_1(1 - r^n)}{1 - r},$$

we have

$$S_5 = \frac{\$2000(1 - 1.03^5)}{1 - 1.03} \approx \$10,618.27.$$

The amount of the annuity is $10,618.27. ➤ **Now Try Exercise 61.**

EXAMPLE 10   *The Economic Multiplier.*   Large sporting events have a significant impact on the economy of the host city. Super Bowl XLVII, hosted by New Orleans, generated a $480-million net impact for the region (*Source:* NewOrleansSaints.com, posted April 18, 2013, Marius M. Mihai, Research Analyst of the Division of Business and Economic Research at the University of New Orleans (DBER)). Assume that 60% of that amount is spent again in the area, and then 60% of that amount is spent again, and so on. This is known as the *economic multiplier effect*. Find the total effect on the economy.

Solution   The total economic effect is given by the infinite series

$$\$480,000,000 + \$480,000,000(0.6) + \$480,000,000(0.6)^2 + \cdots.$$

Since $|r| = |0.6| = 0.6 < 1$, the series has a sum. Using the formula for the sum of an infinite geometric series, we have

$$S_\infty = \frac{a_1}{1 - r} = \frac{\$480,000,000}{1 - 0.6} = \$1,200,000,000.$$

The total effect of the spending on the economy is $1,200,000,000.

➤ **Now Try Exercise 67.**

A
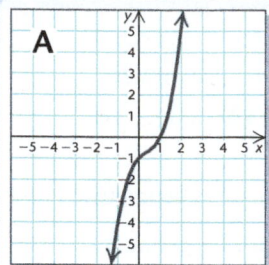

Visualizing the Graph

Match the equation with its graph.

1. $(x - 1)^2 + (y + 2)^2 = 9$

2. $y = x^3 - x^2 + x - 1$

3. $f(x) = 3^x$

4. $f(x) = x$

5. $a_n = n$

6. $y = \log(x + 3)$

7. $f(x) = -(x - 2)^2 + 1$

8. $f(x) = (x - 2)^2 - 1$

9. $y = \dfrac{1}{x - 1}$

10. $y = -3x + 4$

F
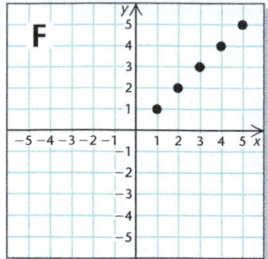

B
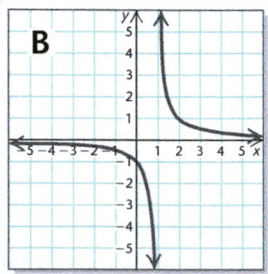

G
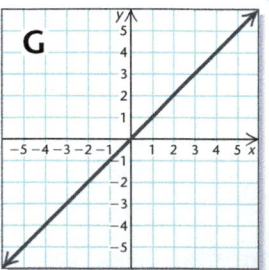

C
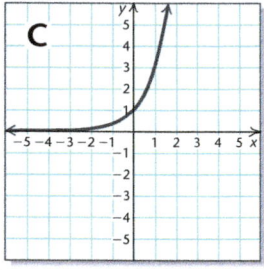

H
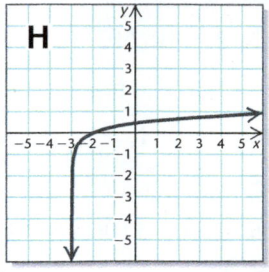

D
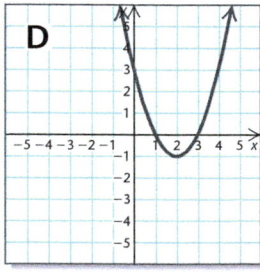

I
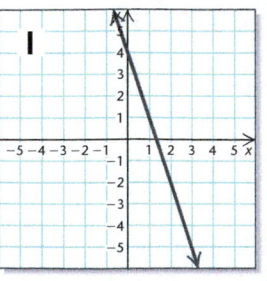

E
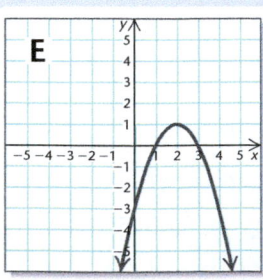

J
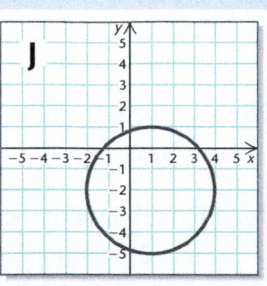

12.3 Exercise Set

Find the common ratio.

1. $2, 4, 8, 16, \ldots$

2. $18, -6, 2, -\frac{2}{3}, \ldots$

3. $-1, 1, -1, 1, \ldots$

4. $-8, -0.8, -0.08, -0.008, \ldots$

5. $\frac{2}{3}, -\frac{4}{3}, \frac{8}{3}, -\frac{16}{3}, \ldots$

6. $75, 15, 3, \frac{3}{5}, \ldots$

7. $6.275, 0.6275, 0.06275, \ldots$

8. $\frac{1}{x}, \frac{1}{x^2}, \frac{1}{x^3}, \ldots$

9. $5, \frac{5a}{2}, \frac{5a^2}{4}, \frac{5a^3}{8}, \ldots$

10. $\$780, \$858, \$943.80, \$1038.18, \ldots$

Find the indicated term.

11. $2, 4, 8, 16, \ldots$; the 7th term

12. $2, -10, 50, -250, \ldots$; the 9th term

13. $2, 2\sqrt{3}, 6, \ldots$; the 9th term

14. $1, -1, 1, -1, \ldots$; the 57th term

15. $\frac{7}{625}, -\frac{7}{25}, \ldots$; the 23rd term

16. $\$1000, \$1060, \$1123.60, \ldots$; the 5th term

Find the nth, or general, term.

17. $1, 3, 9, \ldots$ **18.** $25, 5, 1, \ldots$

19. $1, -1, 1, -1, \ldots$ **20.** $-2, 4, -8, \ldots$

21. $\frac{1}{x}, \frac{1}{x^2}, \frac{1}{x^3}, \ldots$

22. $5, \frac{5a}{2}, \frac{5a^2}{4}, \frac{5a^3}{8}, \ldots$

23. Find the sum of the first 7 terms of the geometric series
$$6 + 12 + 24 + \cdots.$$

24. Find the sum of the first 10 terms of the geometric series
$$16 - 8 + 4 - \cdots.$$

25. Find the sum of the first 9 terms of the geometric series
$$\tfrac{1}{18} - \tfrac{1}{6} + \tfrac{1}{2} - \cdots.$$

26. Find the sum of the geometric series
$$-8 + 4 + (-2) + \cdots + \left(-\tfrac{1}{32}\right).$$

Determine whether the statement is true or false.

27. The sequence $2, -2\sqrt{2}, 4, -4\sqrt{2}, 8, \ldots$ is geometric.

28. The sequence with general term $3n$ is geometric.

29. The sequence with general term 2^n is geometric.

30. Multiplying a term of a geometric sequence by the common ratio produces the next term of the sequence.

31. An infinite geometric series with common ratio -0.75 has a sum.

32. Every infinite geometric series has a limit.

Find the sum, if it exists.

33. $4 + 2 + 1 + \cdots$ **34.** $7 + 3 + \frac{9}{7} + \cdots$

35. $25 + 20 + 16 + \cdots$

36. $100 - 10 + 1 - \frac{1}{10} + \cdots$

37. $8 + 40 + 200 + \cdots$

38. $-6 + 3 - \frac{3}{2} + \frac{3}{4} - \cdots$

39. $0.6 + 0.06 + 0.006 + \cdots$

40. $\displaystyle\sum_{k=0}^{10} 3^k$ **41.** $\displaystyle\sum_{k=1}^{11} 15\left(\frac{2}{3}\right)^k$

42. $\displaystyle\sum_{k=0}^{50} 200(1.08)^k$ **43.** $\displaystyle\sum_{k=1}^{\infty} \left(\frac{1}{2}\right)^{k-1}$

44. $\displaystyle\sum_{k=1}^{\infty} 2^k$ **45.** $\displaystyle\sum_{k=1}^{\infty} 12.5^k$

46. $\displaystyle\sum_{k=1}^{\infty} 400(1.0625)^k$ **47.** $\displaystyle\sum_{k=1}^{\infty} \$500(1.11)^{-k}$

48. $\displaystyle\sum_{k=1}^{\infty} \$1000(1.06)^{-k}$ **49.** $\displaystyle\sum_{k=1}^{\infty} 16(0.1)^{k-1}$

50. $\displaystyle\sum_{k=1}^{\infty} \frac{8}{3}\left(\frac{1}{2}\right)^{k-1}$

Find fraction notation.

51. $0.131313\ldots$, or $0.\overline{13}$ **52.** $0.2222\ldots$, or $0.\overline{2}$

53. $8.9999\overline{9}$ **54.** $6.161616\overline{16}$

55. $3.4125\overline{125}$ **56.** $12.7809\overline{809}$

57. *Daily Doubling Salary.* Suppose that someone offered you a job for the month of February (28 days) under the following conditions. You will be paid $0.01 the 1st day, $0.02 the 2nd, $0.04 the 3rd, and so on, doubling your previous day's salary each day. How much would you earn altogether for the month?

58. *Bouncing Ping-Pong Ball.* A ping-pong ball is dropped from a height of 16 ft and always rebounds $\frac{1}{4}$ of the distance fallen.

  a) How high does it rebound the 6th time?

  b) Find the total sum of the rebound heights of the ball.

59. *Bungee Jumping.* A bungee jumper always rebounds 60% of the distance fallen. A bungee jump is made using a cord that stretches to 200 ft.

  a) After jumping and then rebounding 9 times, how far has a bungee jumper traveled upward (the total rebound distance)?

  b) About how far will a jumper have traveled upward (bounced) before coming to rest?

60. *Population Growth.* A coastal town has a present population of 32,100, and the population is increasing by 3% each year.

  a) What will the population be in 15 years?

  b) How long will it take for the population to double?

61. *Amount of an Annuity.* To save for the down payment on a house, the Clines make a sequence of 10 yearly deposits of $3200 each in a savings account on which interest is compounded annually at 4.6%. Find the amount of the annuity.

62. *Amount of an Annuity.* To create a college fund, a parent makes a sequence of 18 yearly deposits of $1000 each in a savings account on which interest is compounded annually at 3.2%. Find the amount of the annuity.

63. *Doubling the Thickness of Paper.* A piece of paper is 0.01 in. thick. It is cut and stacked repeatedly in such a way that its thickness is doubled each time for 20 times. How thick is the result?

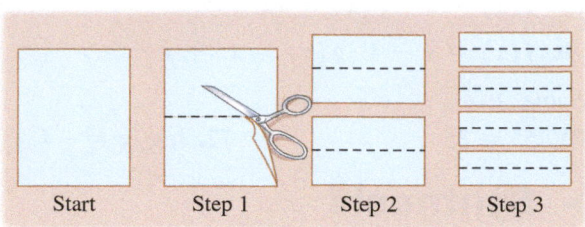

Start   Step 1   Step 2   Step 3

64. *Amount of an Annuity.* A sequence of yearly payments of P dollars is invested at the end of each of N years at interest rate i, compounded annually. The total amount in the account, or the amount of the annuity, is V.

  a) Show that

$$V = \frac{P\left[(1 + i)^N - 1\right]}{i}.$$

  b) Suppose that interest is compounded n times per year and deposits are made every compounding period. Show that the formula for V is then given by

$$V = \frac{P\left[\left(1 + \dfrac{i}{n}\right)^{nN} - 1\right]}{i/n}.$$

65. *Amount of an Annuity.* A sequence of payments of $300 is invested over 12 years at the end of each quarter at 5.1%, compounded quarterly. Find the amount of the annuity. Use the formula in Exercise 64(b).

66. *Amount of an Annuity.* A sequence of yearly payments of $750 is invested at the end of each of 10 years at 4.75%, compounded annually. Find the amount of the annuity. Use the formula in Exercise 64(a).

67. *The Economic Multiplier.* Suppose that the government is making a $13,000,000,000 expenditure to stimulate the economy. If 85% of this is spent again, and so on, what is the total effect on the economy?

68. *Advertising Effect.* Gigi's Cupcake Truck is about to open for business in a city of 3,000,000 people, traveling to several curbside locations in the city each day to sell cupcakes. The owners plan an advertising campaign that they think will induce 30% of the people to buy their cupcakes. They estimate that if those people like the product, they will induce $30\% \cdot 30\% \cdot 3,000,000$ more to buy the product, and those will induce $30\% \cdot 30\% \cdot 30\% \cdot 3,000,000$ and so on. In all, how many people will buy Gigi's cupcakes as a result of the advertising campaign? What percentage of the population is this?

▶ ## Skill Maintenance

For each pair of functions, find $(f \circ g)(x)$ and $(g \circ f)(x)$. **[9.1]**

69. $f(x) = x^2$, $g(x) = 4x + 5$

70. $f(x) = x - 1$, $g(x) = x^2 + x + 3$

Solve. **[9.6]**

71. $5^x = 35$ **72.** $\log_2 x = -4$

▶ ## Synthesis

73. Prove that $\sqrt{3} - \sqrt{2}, 4 - \sqrt{6}$, and $6\sqrt{3} - 2\sqrt{2}$ form a geometric sequence.

74. Assume that $a_1, a_2, a_3, \ldots$ is a geometric sequence. Prove that $\ln a_1, \ln a_2, \ln a_3, \ldots$ is an arithmetic sequence.

75. Consider the sequence $x + 3, x + 7, 4x - 2, \ldots$.
  a) If the sequence is arithmetic, find x and then determine each of the 3 terms and the 4th term.
  b) If the sequence is geometric, find x and then determine each of the 3 terms and the 4th term.

76. Find the sum of the first n terms of
$$1 + x + x^2 + \cdots.$$

77. Find the sum of the first n terms of
$$x^2 - x^3 + x^4 - x^5 + \cdots.$$

78. The sides of a square are 16 cm long. A second square is inscribed by joining the midpoints of the sides, successively. In the second square, we repeat the process, inscribing a third square. If this process is continued indefinitely, what is the sum of all the areas of all the squares? (*Hint*: Use an infinite geometric series.)

12.4

Mathematical Induction

▶ Prove infinite sequences of statements using mathematical induction.

In this section, we learn to prove a sequence of mathematical statements using a procedure called *mathematical induction*.

▶ ### Proving Infinite Sequences of Statements

Infinite sequences of statements occur often in mathematics. In an infinite sequence of statements, there is a statement for each natural number. For example, consider the sequence of statements represented by the following:

"The sum of the first n positive odd integers is n^2,"   or
$$1 + 3 + 5 + \cdots + (2n - 1) = n^2.$$

Let's think of this as $S(n)$, or S_n. Substituting natural numbers for n gives a sequence of statements. We list the first four:

S_1: $1 = 1^2$;
S_2: $1 + 3 = 4 = 2^2$;
S_3: $1 + 3 + 5 = 9 = 3^2$;
S_4: $1 + 3 + 5 + 7 = 16 = 4^2$.

The fact that the statement is true for $n = 1, 2, 3$, and 4 might tempt us to conclude that the statement is true for any natural number n, but we cannot be sure that this is the case. We can, however, use the principle of mathematical induction to prove that the statement is true for all natural numbers.

THE PRINCIPLE OF MATHEMATICAL INDUCTION

We can prove an infinite sequence of statements S_n by showing the following.

(1) *Basis step.* S_1 is true.
(2) *Induction step.* For all natural numbers k, $S_k \rightarrow S_{k+1}$.

Mathematical induction is analogous to lining up a sequence of dominoes. The induction step tells us that if any one domino is knocked over, then the one next to it will be hit and knocked over. The basis step tells us that the first domino can indeed be knocked over. Note that in order for all dominoes to fall, *both* conditions must be satisfied.

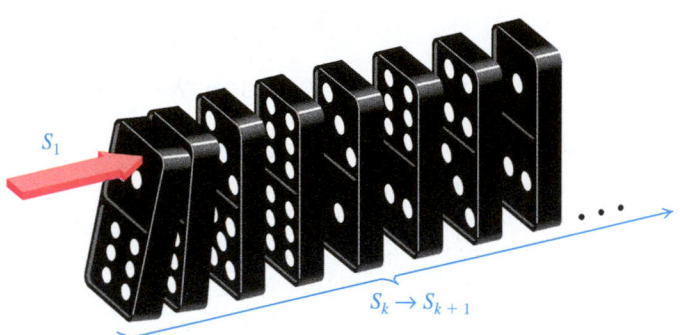

S_1

$S_k \rightarrow S_{k+1}$

When you are learning to do proofs by mathematical induction, it is helpful to first write out S_n, S_1, S_k, and S_{k+1}. This helps to identify what is to be assumed and what is to be deduced.

EXAMPLE 1   Prove: For every natural number n,
$$1 + 3 + 5 + \cdots + (2n - 1) = n^2.$$

Proof.   We first write out S_n, S_1, S_k, and S_{k+1}.

S_n: $1 + 3 + 5 + \cdots + (2n - 1) = n^2$
S_1: $1 = 1^2$
S_k: $1 + 3 + 5 + \cdots + (2k - 1) = k^2$
S_{k+1}: $1 + 3 + 5 + \cdots + (2k - 1) + [2(k + 1) - 1] = (k + 1)^2$

(1) *Basis step.* S_1, as listed, is true since $1 = 1^2$, or $1 = 1$.

(2) *Induction step.* We let k be any natural number. We assume S_k to be true and try to show that it implies that S_{k+1} is true. Now S_k is

$$1 + 3 + 5 + \cdots + (2k - 1) = k^2.$$

Starting with the left side of S_{k+1} and substituting k^2 for $1 + 3 + 5 + \cdots + (2k - 1)$, we have

$$\underbrace{1 + 3 + \cdots + (2k - 1)} + [2(k + 1) - 1]$$

$$\begin{aligned} &= k^2 + [2(k + 1) - 1] \qquad \text{We assume } S_k \text{ is true.}\\ &= k^2 + 2k + 2 - 1 \\ &= k^2 + 2k + 1 \\ &= (k + 1)^2. \end{aligned}$$

We have shown that for all natural numbers k, $S_k \rightarrow S_{k+1}$. This completes the induction step. It and the basis step tell us that the proof is complete.

Now Try Exercise 5.

EXAMPLE 2 Prove: For every natural number n,

$$\frac{1}{2} + \frac{1}{4} + \frac{1}{8} + \cdots + \frac{1}{2^n} = \frac{2^n - 1}{2^n}.$$

Proof. We first list S_n, S_1, S_k, and S_{k+1}.

S_n: $\quad \dfrac{1}{2} + \dfrac{1}{4} + \dfrac{1}{8} + \cdots + \dfrac{1}{2^n} = \dfrac{2^n - 1}{2^n}$

S_1: $\quad \dfrac{1}{2^1} = \dfrac{2^1 - 1}{2^1}$

S_k: $\quad \dfrac{1}{2} + \dfrac{1}{4} + \dfrac{1}{8} + \cdots + \dfrac{1}{2^k} = \dfrac{2^k - 1}{2^k}$

S_{k+1}: $\quad \dfrac{1}{2} + \dfrac{1}{4} + \dfrac{1}{8} + \cdots + \dfrac{1}{2^k} + \dfrac{1}{2^{k+1}} = \dfrac{2^{k+1} - 1}{2^{k+1}}$

(1) *Basis step.* We show S_1 to be true as follows:

$$\frac{2^1 - 1}{2^1} = \frac{2 - 1}{2} = \frac{1}{2}.$$

(2) *Induction step.* We let k be any natural number. We assume S_k to be true and try to show that it implies that S_{k+1} is true. Now S_k is

$$\frac{1}{2} + \frac{1}{4} + \frac{1}{8} + \cdots + \frac{1}{2^k} = \frac{2^k - 1}{2^k}.$$

We start with the left side of S_{k+1}. Since we assume that S_k is true, we can substitute

$$\frac{2^k - 1}{2^k} \quad \text{for} \quad \frac{1}{2} + \frac{1}{4} + \cdots + \frac{1}{2^k}.$$

We have

$$\underbrace{\frac{1}{2} + \frac{1}{4} + \frac{1}{8} + \cdots + \frac{1}{2^k}} + \frac{1}{2^{k+1}}$$

$$= \frac{2^k - 1}{2^k} + \frac{1}{2^{k+1}} = \frac{2^k - 1}{2^k} \cdot \frac{2}{2} + \frac{1}{2^{k+1}}$$

$$= \frac{(2^k - 1) \cdot 2 + 1}{2^{k+1}}$$

$$= \frac{2^{k+1} - 2 + 1}{2^{k+1}}$$

$$= \frac{2^{k+1} - 1}{2^{k+1}}.$$

We have shown that for all natural numbers k, $S_k \to S_{k+1}$. This completes the induction step. It and the basis step tell us that the proof is complete.

Now Try Exercise 15.

EXAMPLE 3 Prove: For every natural number n, $n < 2^n$.

Proof. We first list S_n, S_1, S_k, and S_{k+1}.

S_n: $n < 2^n$

S_1: $1 < 2^1$

S_k: $k < 2^k$

S_{k+1}: $k + 1 < 2^{k+1}$

(1) *Basis step.* S_1, as listed, is true since $2^1 = 2$ and $1 < 2$.

(2) *Induction step.* We let k be any natural number. We assume S_k to be true and try to show that it implies that S_{k+1} is true. Now

$k < 2^k$	This is S_k.
$2k < 2 \cdot 2^k$	Multiplying by 2 on both sides
$2k < 2^{k+1}$	Adding exponents on the right
$k + k < 2^{k+1}.$	Rewriting $2k$ as $k + k$

Since k is any natural number, we know that $1 \leq k$. Thus,

$k + 1 \leq k + k$.   Adding k on both sides of $1 \leq k$

Putting the results $k + 1 \leq k + k$ and $k + k < 2^{k+1}$ together gives us

$k + 1 < 2^{k+1}$.   This is S_{k+1}.

We have shown that for all natural numbers k, $S_k \rightarrow S_{k+1}$. This completes the induction step. It and the basis step tell us that the proof is complete.

Now Try Exercise 11.

12.4   Exercise Set

List the first five statements in the sequence that can be obtained from each of the following. Determine whether each of the five statements is true or false.

1. $n^2 < n^3$

2. $n^2 - n + 41$ is prime. Find a value for n for which the statement is false.

3. A polygon of n sides has $[n(n - 3)]/2$ diagonals.

4. The sum of the angles of a polygon of n sides is $(n - 2) \cdot 180°$.

Use mathematical induction to prove each of the following.

5. $2 + 4 + 6 + \cdots + 2n = n(n + 1)$

6. $4 + 8 + 12 + \cdots + 4n = 2n(n + 1)$

7. $1 + 5 + 9 + \cdots + (4n - 3) = n(2n - 1)$

8. $3 + 6 + 9 + \cdots + 3n = \dfrac{3n(n + 1)}{2}$

9. $2 + 4 + 8 + \cdots + 2^n = 2(2^n - 1)$

10. $2 \leq 2^n$

11. $n < n + 1$

12. $3^n < 3^{n+1}$

13. $2n \leq 2^n$

14. $\dfrac{1}{1 \cdot 2} + \dfrac{1}{2 \cdot 3} + \cdots + \dfrac{1}{n(n + 1)} = \dfrac{n}{n + 1}$

15. $\dfrac{1}{1 \cdot 2 \cdot 3} + \dfrac{1}{2 \cdot 3 \cdot 4} + \dfrac{1}{3 \cdot 4 \cdot 5} + \cdots$

$+ \dfrac{1}{n(n + 1)(n + 2)} = \dfrac{n(n + 3)}{4(n + 1)(n + 2)}$

16. If x is any real number greater than 1, then for any natural number n, $x \leq x^n$.

The following formulas can be used to find sums of powers of natural numbers. Use mathematical induction to prove each formula.

17. $1 + 2 + 3 + \cdots + n = \dfrac{n(n + 1)}{2}$

18. $1^2 + 2^2 + 3^2 + \cdots + n^2 = \dfrac{n(n + 1)(2n + 1)}{6}$

19. $1^3 + 2^3 + 3^3 + \cdots + n^3 = \dfrac{n^2(n + 1)^2}{4}$

20. $1^4 + 2^4 + 3^4 + \cdots + n^4$
$$= \frac{n(n + 1)(2n + 1)(3n^2 + 3n - 1)}{30}$$

Use mathematical induction to prove each of the following.

21. $\displaystyle\sum_{i=1}^{n} i(i + 1) = \frac{n(n + 1)(n + 2)}{3}$

22. $\left(1 + \dfrac{1}{1}\right)\left(1 + \dfrac{1}{2}\right)\left(1 + \dfrac{1}{3}\right) \cdots \left(1 + \dfrac{1}{n}\right)$
$$= n + 1$$

23. The sum of n terms of an arithmetic sequence:

$a_1 + (a_1 + d) + (a_1 + 2d) + \cdots + [a_1 + (n - 1)d]$
$$= \frac{n}{2}[2a_1 + (n - 1)d]$$

▶ **Skill Maintenance**

Solve.

24. $2x - 3y = 1,$
$3x - 4y = 3$ [3.1], [3.2], [3.3], [10.1], [10.3], [10.4]

25. *Investment.*   Clarise received \$104 in simple interest one year from three investments. Part is invested at 1.5%, part at 2%, and part at 3%. The amount invested at 2% is twice the amount invested at 1.5%. There is \$400 more invested at 3% than at 2%. Find the amount invested at each rate. [3.6]

▶ **Synthesis**

Use mathematical induction to prove each of the following.

26. The sum of n terms of a geometric sequence:

$$a_1 + a_1 r + a_1 r^2 + \cdots + a_1 r^{n-1} = \frac{a_1 - a_1 r^n}{1 - r}$$

27. $x + y$ is a factor of $x^{2n} - y^{2n}$.

Prove each of the following using mathematical induction. Do the basis step for $n = 2$.

28. For every natural number $n \geq 2$,
$$2n + 1 < 3^n.$$

29. For every natural number $n \geq 2$,
$$\log_a (b_1 b_2 \cdots b_n)$$
$$= \log_a b_1 + \log_a b_2 + \cdots + \log_a b_n.$$

Prove each of the following for any complex numbers z_1, $z_2, \ldots, z_n$, where $i^2 = -1$ and $\overline{z}$ is the conjugate of z.

30. $\overline{z^n} = \overline{z}^n$

31. $\overline{z_1 + z_2 + \cdots + z_n} = \overline{z_1} + \overline{z_2} + \cdots + \overline{z_n}$

32. *The Tower of Hanoi Problem.*   There are three pegs on a board. On one peg are n disks, each smaller than the one on which it rests. The problem is to move this pile of disks to another peg. The final order must be the same, but you can move only one disk at a time and can never place a larger disk on a smaller one.

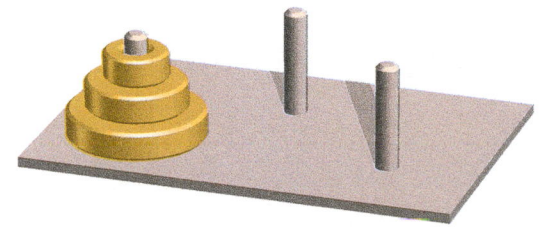

a) What is the *fewest* number of moves needed to move 3 disks? 4 disks? 2 disks? 1 disk?
b) Conjecture a formula for the *fewest* number of moves needed to move n disks. Prove it by mathematical induction.

Mid-Chapter Review

Determine whether the statement is true or false.

1. The general term of the sequence $1, -2, 3, -4, \ldots$ can be expressed as $a_n = n$. [12.1]

2. To find the common difference of an arithmetic sequence, choose any term except the first and then subtract the preceding term from it. [12.2]

3. The sequence $7, 3, -1, -5, \ldots$ is geometric. [12.2], [12.3]

4. If we can show that $S_k \rightarrow S_{k+1}$ for some natural number k, then we know that S_n is true for all natural numbers n. [12.4]

In each of the following, the nth term of a sequence is given. Find the first 4 terms, a_9, and a_{14}. **[12.1]**

5. $a_n = 3n + 5$

6. $a_n = (-1)^{n+1}(n - 1)$

Predict the general term, or nth term, a_n, of the sequence. Answers may vary. **[12.1]**

7. $3, 6, 9, 12, 15, \ldots$

8. $-1, 4, -9, 16, -25, \ldots$

9. Find the partial sum S_4 for the sequence $1, \frac{1}{2}, \frac{1}{4}, \frac{1}{8}, \frac{1}{16}, \ldots$ **[12.1]**

10. Find and evaluate the sum $\sum_{k=1}^{5} k(k + 1)$. **[12.1]**

11. Write sigma notation for the sum $-4 + 8 - 12 + 16 - 20 + \cdots$. **[12.1]**

12. Find the first 4 terms of the sequence defined by $a_1 = 2, a_{n+1} = 4a_n - 2$. **[12.1]**

13. Find the common difference of the arithmetic sequence $12, 7, 2, -3, \ldots$ **[12.2]**

14. Find the 10th term of the arithmetic sequence $4, 6, 8, 10, \ldots$ **[12.2]**

15. In the sequence in Exercise 14, what term is the number 44? **[12.2]**

16. Find the sum of the first 16 terms of the arithmetic series $6 + 11 + 16 + 21 + \cdots$. **[12.2]**

17. Find the common ratio of the geometric sequence $16, -8, 4, -2, 1, \ldots$ **[12.3]**

18. Find **(a)** the 8th term and **(b)** the sum of the first 10 terms of the geometric sequence $\frac{1}{16}, \frac{1}{8}, \frac{1}{4}, \frac{1}{2}, 1, \ldots$ **[12.3]**

Find the sum, if it exists. **[12.3]**

19. $-8 + 4 - 2 + 1 - \cdots$

20. $\sum_{k=0}^{\infty} 5^k$

21. *Landscaping.* A landscaper is planting a triangular flower bed with 36 plants in the first row, 30 plants in the second row, 24 in the third row, and so on, for a total of 6 rows. How many plants will be planted in all? **[12.2]**

22. *Amount of an Annuity.* To save money for adding a bedroom to their home, at the end of each of 4 years the Davidsons deposit $1500 in an account that pays 4% interest, compounded annually. Find the total amount of the annuity. **[12.3]**

23. Prove: For every natural number n,
$1 + 4 + 7 + \cdots + (3n - 2) = \frac{1}{2}n(3n - 1)$. **[12.4]**

Collaborative Discussion and Writing

24. The sum of the first n terms of an arithmetic sequence can be given by

$$S_n = \frac{n}{2}[2a_1 + (n - 1)d].$$

Compare this formula to

$$S_n = \frac{n}{2}(a_1 + a_n).$$

Discuss the reasons for the use of one formula over the other. **[12.2]**

25. It is said that as a young child, the mathematician Karl F. Gauss (1777–1855) was able to compute the sum $1 + 2 + 3 + \cdots + 100$ very quickly in his head to the amazement of a teacher. Explain how Gauss might have done this had he possessed some knowledge of arithmetic sequences and series. Then give a formula for the sum of the first n natural numbers. **[12.2]**

26. Write a problem for a classmate to solve. Devise the problem so that a geometric series is involved and the solution is "The total amount in the bank is $900(1.08)^{40}$, or about $19,552." **[12.3]**

27. Write an explanation of the idea behind mathematical induction for a fellow student. **[12.4]**

12.5 Combinatorics: Permutations

▶ Evaluate factorial notation and permutation notation and solve related applied problems.

In order to study probability, it is first necessary that we learn about **combinatorics**, the theory of counting.

▶ Permutations

In this section, we will consider the part of combinatorics called *permutations*.

> The study of permutations involves *order* and *arrangements*.

EXAMPLE 1 How many 3-letter code symbols can be formed with the letters A, B, C *without* repetition (that is, using each letter only once)?

Solution Consider placing the letters in these boxes.

We can select any of the 3 letters for the first letter in the symbol. Once this letter has been selected, the second must be selected from the 2 remaining letters. After this, the third letter is already determined, since only 1 possibility is left. That is, we can place any of the 3 letters in the first box, either of the remaining 2 letters in the second box, and the only remaining letter in the third box. The possibilities can be determined using a **tree diagram**, as shown below.

TREE DIAGRAM	OUTCOMES	
A < B — C	ABC	
C — B	ACB	
B < A — C	BAC	Each outcome
C — A	BCA	represents one permutation of
C < A — B	CAB	the letters A, B, C.
B — A	CBA	

We see that there are 6 possibilities. The set of all the possibilities is

{ABC, ACB, BAC, BCA, CAB, CBA}.

This is the set of all *permutations* of the letters A, B, C.

Suppose that we perform an experiment such as selecting letters (as in the preceding example), flipping a coin, or drawing a card. The results are called **outcomes**. An **event** is a set of outcomes. The following principle enables us to count actions that are combined to form an event.

> ### THE FUNDAMENTAL COUNTING PRINCIPLE
> Given a combined action, or *event*, in which the first action can be performed in n_1 ways, the second action can be performed in n_2 ways, and so on, the total number of ways in which the combined action can be performed is the product
> $$n_1 \cdot n_2 \cdot n_3 \cdots \cdots n_k.$$

Thus, in Example 1, there are 3 choices for the first letter, 2 for the second letter, and 1 for the third letter, making a total of $3 \cdot 2 \cdot 1$, or 6 possibilities.

EXAMPLE 2 How many 3-letter code symbols can be formed with the letters A, B, C, D, E *with* repetition (that is, allowing letters to be repeated)?

Solution Since repetition is allowed, there are 5 choices for the first letter, 5 choices for the second, and 5 for the third. Thus, by the fundamental counting principle, there are $5 \cdot 5 \cdot 5$, or 125 code symbols.

> ### PERMUTATION
> A **permutation** of a set of n objects is an ordered arrangement of all n objects.

We can use the fundamental counting principle to count the number of permutations of the objects in a set. Consider, for example, a set of 4 objects

$$\{A, B, C, D\}.$$

To find the number of ordered arrangements of the set, we select a first letter: There are 4 choices. Then we select a second letter: There are 3 choices. Then we select a third letter: There are 2 choices. Finally, there is 1 choice for the last selection. Thus, by the fundamental counting principle, there are $4 \cdot 3 \cdot 2 \cdot 1$, or 24, permutations of a set of 4 objects.

We can find a formula for the total number of permutations of all objects in a set of n objects. We have n choices for the first selection, $n - 1$ choices for the second, $n - 2$ for the third, and so on. For the nth selection, there is only 1 choice.

> ### THE TOTAL NUMBER OF PERMUTATIONS OF n OBJECTS
> The total number of permutations of n objects, denoted $_nP_n$, is given by
> $$_nP_n = n(n - 1)(n - 2) \cdots 3 \cdot 2 \cdot 1.$$

Technology Connection

We can find the total number of permutations of n objects, as in Example 3, using the $_nP_r$ operation from the MATH PRB (probability) menu on a graphing calculator.

```
4 nPr 4
                    24
7 nPr 7
                  5040
```

EXAMPLE 3 Find each of the following.

a) $_4P_4$ **b)** $_7P_7$

Solution

Start with 4.

a) $_4P_4 = 4 \cdot 3 \cdot 2 \cdot 1 = 24$

4 factors

b) $_7P_7 = 7 \cdot 6 \cdot 5 \cdot 4 \cdot 3 \cdot 2 \cdot 1 = 5040$

Now Try Exercise 1.

EXAMPLE 4   In how many ways can 9 packages be placed in 9 mailboxes, one package in a box?

Solution   We have

$$_9P_9 = 9 \cdot 8 \cdot 7 \cdot 6 \cdot 5 \cdot 4 \cdot 3 \cdot 2 \cdot 1 = 362{,}880.$$

▶ **Factorial Notation**

We will use products such as $7 \cdot 6 \cdot 5 \cdot 4 \cdot 3 \cdot 2 \cdot 1$ so often that it is convenient to adopt a notation for them. For the product

$$7 \cdot 6 \cdot 5 \cdot 4 \cdot 3 \cdot 2 \cdot 1,$$

we write 7!, read "7 factorial."

We now define factorial notation for natural numbers and for 0.

FACTORIAL NOTATION

For any natural number n,

$$n! = n(n - 1)(n - 2) \cdots 3 \cdot 2 \cdot 1.$$

For the number 0,

$$0! = 1.$$

Now Try Exercise 23.

We define 0! as 1 so that certain formulas can be stated concisely and with a consistent pattern.

Here are some examples of factorial notation.

$$7! = 7 \cdot 6 \cdot 5 \cdot 4 \cdot 3 \cdot 2 \cdot 1 = 5040$$
$$6! = 6 \cdot 5 \cdot 4 \cdot 3 \cdot 2 \cdot 1 = 720$$
$$5! = 5 \cdot 4 \cdot 3 \cdot 2 \cdot 1 = 120$$
$$4! = 4 \cdot 3 \cdot 2 \cdot 1 = 24$$
$$3! = 3 \cdot 2 \cdot 1 = 6$$
$$2! = 2 \cdot 1 = 2$$
$$1! = 1 = 1$$
$$0! = 1 = 1$$

We now see that the following statement is true.

$$_nP_n = n!$$

We will often need to manipulate factorial notation. For example, note that

$$8! = 8 \cdot 7 \cdot 6 \cdot 5 \cdot 4 \cdot 3 \cdot 2 \cdot 1$$
$$= 8 \cdot (7 \cdot 6 \cdot 5 \cdot 4 \cdot 3 \cdot 2 \cdot 1) = 8 \cdot 7!.$$

Generalizing, we get the following.

For any natural number n, $n! = n(n - 1)!$.

By using this result repeatedly, we can further manipulate factorial notation.

EXAMPLE 5 Rewrite 7! with a factor of 5!.

Solution We have $7! = 7 \cdot 6! = 7 \cdot 6 \cdot 5!$.

In general, we have the following.

> For any natural numbers k and n, with $k < n$,
>
> $$n! = \underbrace{n(n-1)(n-2)\cdots[n-(k-1)]}_{k \text{ factors}} \cdot \underbrace{(n-k)!}_{n-k \text{ factors}}.$$

▶ Permutations of *n* Objects Taken *k* at a Time

Consider a set of 5 objects

$$\{A, B, C, D, E\}.$$

How many ordered arrangements can be formed using 3 objects from the set without repetition? Examples of such an arrangement are EBA, CAB, and BCD. There are 5 choices for the first object, 4 choices for the second, and 3 choices for the third. By the fundamental counting principle, there are

$$5 \cdot 4 \cdot 3, \quad \text{or} \quad 60 \text{ } permutations \text{ of a set of 5 objects taken 3 at a time.}$$

Note that

$$5 \cdot 4 \cdot 3 = \frac{5 \cdot 4 \cdot 3 \cdot 2 \cdot 1}{2 \cdot 1}, \quad \text{or} \quad \frac{5!}{2!}.$$

> **PERMUTATION OF *n* OBJECTS TAKEN *k* AT A TIME**
>
> A **permutation** of a set of n objects taken k at a time is an ordered arrangement of k objects taken from the set.

Consider a set of n objects and the selection of an ordered arrangement of k of them. There would be n choices for the first object. Then there would remain $n - 1$ choices for the second, $n - 2$ choices for the third, and so on. We make k choices in all, so there are k factors in the product. By the fundamental counting principle, the total number of permutations is

$$\underbrace{n(n-1)(n-2)\cdots[n-(k-1)]}_{k \text{ factors}}.$$

We can express this in another way by multiplying by 1, as follows:

$$n(n-1)(n-2)\cdots[n-(k-1)] \cdot \frac{(n-k)!}{(n-k)!}$$

$$= \frac{n(n-1)(n-2)\cdots[n-(k-1)](n-k)!}{(n-k)!}$$

$$= \frac{n!}{(n-k)!}.$$

This gives us the following.

THE NUMBER OF PERMUTATIONS OF n OBJECTS TAKEN k AT A TIME

The number of permutations of a set of n objects taken k at a time, denoted $_nP_k$, is given by

$$_nP_k = \underbrace{n(n-1)(n-2)\cdots[n-(k-1)]}_{k \text{ factors}} \quad \textbf{(1)}$$

$$= \frac{n!}{(n-k)!}. \quad \textbf{(2)}$$

Technology Connection

We can evaluate computations like the one in Example 6 using the $_nP_r$ operation from the MATH PRB menu on a graphing calculator.

```
8 nPr 4
           1680
```

EXAMPLE 6   Compute $_8P_4$ using both forms of the formula.

Solution   Using form (1), we have

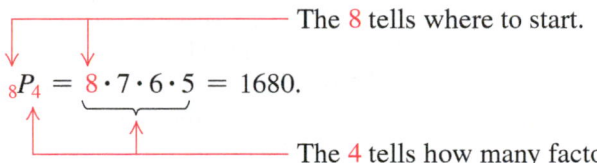

The 8 tells where to start.

$_8P_4 = 8 \cdot 7 \cdot 6 \cdot 5 = 1680.$

The 4 tells how many factors.

Using form (2), we have

$$_8P_4 = \frac{8!}{(8-4)!} = \frac{8!}{4!}$$

$$= \frac{8 \cdot 7 \cdot 6 \cdot 5 \cdot 4!}{4!} = \frac{8 \cdot 7 \cdot 6 \cdot 5 \cdot 4!}{4!}$$

$$= 8 \cdot 7 \cdot 6 \cdot 5 = 1680.$$

> **Now Try Exercise 3.**

EXAMPLE 7   *Flags of Nations.*   The flags of many nations consist of three horizontal stripes. For example, the flag of the Netherlands, shown here, has its first stripe red, its second white, and its third blue.

Suppose that the following 7 colors are available:

{black, yellow, red, white, blue, orange, green}.

How many different flags of three horizontal stripes can be made without repetition of colors in a flag? (This assumes that the order in which the stripes appear is considered.)

Solution   We are determining the number of permutations of 7 objects taken 3 at a time. There is no repetition of colors. Using form (1), we get

$$_7P_3 = 7 \cdot 6 \cdot 5 = 210.$$

> **Now Try Exercise 37(a).**

EXAMPLE 8   *Batting Orders.*   A baseball manager arranges the batting order as follows: The 4 infielders will bat first. Then the 3 outfielders, the catcher, and the pitcher will follow, not necessarily in that order. How many different batting orders are possible?

Solution   The infielders can bat in $_4P_4$ different ways, the rest in $_5P_5$ different ways. Then by the fundamental counting principle, we have

$$_4P_4 \cdot _5P_5 = 4! \cdot 5!, \quad \text{or} \quad 2880 \text{ possible batting orders.}$$

> **Now Try Exercise 31.**

If we allow repetition, a situation like the following can occur.

EXAMPLE 9   How many 5-letter code symbols can be formed with the letters A, B, C, and D if we allow a letter to occur more than once?

Solution   We can select each of the 5 letters in 4 ways. That is, we can select the first letter in 4 ways, the second in 4 ways, and so on. Thus there are 4^5, or 1024 arrangements.

> **Now Try Exercise 37(b).**

> The number of distinct arrangements of n objects taken k at a time, allowing repetition, is n^k.

▶ Permutations of Sets with Nondistinguishable Objects

Consider a set of 7 marbles, 4 of which are blue and 3 of which are red. When they are lined up, one red marble will look just like any other red marble. In this sense, we say that the red marbles are nondistinguishable and, similarly, the blue marbles are nondistinguishable.

We know that there are 7! permutations of this set. Many of them will look alike, however. We develop a formula for finding the number of distinguishable permutations.

Consider a set of n objects in which n_1 are of one kind, n_2 are of a second kind, . . . , and n_k are of a kth kind. The total number of permutations of the set is $n!$, but this includes many that are nondistinguishable. Let N be the total number of distinguishable permutations. For each of these N permutations, there are $n_1!$ actual, nondistinguishable permutations, obtained by permuting the objects of the first kind. For each of these $N \cdot n_1!$ permutations, there are $n_2!$ nondistinguishable permutations, obtained by permuting the objects of the second kind, and so on. By the fundamental counting principle, the total number of permutations, including those that are nondistinguishable, is

$$N \cdot n_1! \cdot n_2! \cdot \cdots \cdot n_k!.$$

Then we have $N \cdot n_1! \cdot n_2! \cdot \cdots \cdot n_k! = n!$. Solving for N, we obtain

$$N = \frac{n!}{n_1! \cdot n_2! \cdot \cdots \cdot n_k!}.$$

Now, to finish our problem with the marbles, we have

$$N = \frac{7!}{4!3!}$$

$$= \frac{7 \cdot 6 \cdot 5 \cdot 4!}{4! \cdot 3 \cdot 2 \cdot 1} = \frac{7 \cdot 3 \cdot 2 \cdot 5 \cdot 4!}{4! \cdot 3 \cdot 2 \cdot 1} = \frac{7 \cdot 5}{1}, \quad \text{or} \quad 35$$

distinguishable permutations of the marbles.

In general, we have the following.

> For a set of n objects in which n_1 are of one kind, n_2 are of another kind, . . . , and n_k are of a kth kind, the number of distinguishable permutations is
>
> $$\frac{n!}{n_1! \cdot n_2! \cdot \cdots \cdot n_k!}.$$

EXAMPLE 10 In how many distinguishable ways can the letters of the word CINCINNATI be arranged?

Solution There are 2 C's, 3 I's, 3 N's, 1 A, and 1 T for a total of 10 letters. Thus,

$$N = \frac{10!}{2! \cdot 3! \cdot 3! \cdot 1! \cdot 1!}, \quad \text{or} \quad 50{,}400.$$

The letters of the word CINCINNATI can be arranged in 50,400 distinguishable ways.

Now Try Exercise 35.

12.5 Exercise Set

Evaluate.

1. $_6P_6$

2. $_4P_3$

3. $_{10}P_7$

4. $_{10}P_3$

5. $5!$

6. $7!$

7. $0!$

8. $1!$

9. $\dfrac{9!}{5!}$

10. $\dfrac{9!}{4!}$

11. $(8-3)!$

12. $(8-5)!$

13. $\dfrac{10!}{7!\,3!}$

14. $\dfrac{7!}{(7-2)!}$

15. $_8P_0$

16. $_{13}P_1$

17. $_{52}P_4$

18. $_{52}P_5$

19. $_nP_3$

20. $_nP_2$

21. $_nP_1$

22. $_nP_0$

In each of Exercises 23–41, give your answer using permutation notation, factorial notation, or other operations. Then evaluate.

How many permutations are there of the letters in each of the following words, if all the letters are used without repetition?

23. CREDIT

24. FRUIT

25. EDUCATION

26. TOURISM

27. How many permutations are there of the letters of the word EDUCATION if the letters are taken 4 at a time?

28. How many permutations are there of the letters of the word TOURISM if the letters are taken 5 at a time?

29. How many 5-digit numbers can be formed using the digits 2, 4, 6, 8, and 9 without repetition? with repetition?

30. In how many ways can 7 athletes be arranged in a straight line?

31. *Program Planning.* A program is planned to have 5 musical numbers and 4 speeches. In how many ways can this be done if a musical number and a speech are to alternate and a musical number is to come first?

32. A professor is going to grade her 24 students on a curve. She will give 3 A's, 5 B's, 9 C's, 4 D's, and 3 F's. In how many ways can she do this?

33. *Phone Numbers.* How many 7-digit phone numbers can be formed with the digits 0, 1, 2, 3, 4, 5, 6, 7, 8, and 9, assuming that the first number cannot be 0 or 1? Accordingly, how many telephone numbers can there be within a given area code, before the area needs to be split with a new area code?

34. How many distinguishable code symbols can be formed from the letters of the word BUSINESS? BIOLOGY? MATHEMATICS?

35. Suppose the expression $a^2b^3c^4$ is rewritten without exponents. In how many distinguishable ways can this be done?

36. *Coin Arrangements.* A penny, a nickel, a dime, and a quarter are arranged in a straight line.

a) Considering just the coins, in how many ways can they be lined up?
b) Considering the coins and heads and tails, in how many ways can they be lined up?

37. How many code symbols can be formed using 5 out of 6 letters of A, B, C, D, E, F if the letters:

a) are not repeated?
b) can be repeated?
c) are not repeated but must begin with D?
d) are not repeated but must begin with DE?

38. *License Plates.* A state forms its license plates by first listing a number that corresponds to the county in which the owner of the car resides. (The names of the counties are alphabetized and the number is its location in that order.) Then the plate lists a letter of the alphabet, and this is followed by a number from 1 to 9999. How many such plates are possible if there are 80 counties?

39. *Zip Codes.* A U.S. postal zip code is a five-digit number.

a) How many zip codes are possible if any of the digits 0 to 9 can be used?
b) If each post office has its own zip code, how many possible post offices can there be?

40. *Zip-Plus-4 Codes.* A zip-plus-4 postal code uses a 9-digit number like 75247-5456. How many 9-digit zip-plus-4 postal codes are possible?

41. *Social Security Numbers.* A social security number is a 9-digit number like 243-47-0825.

a) How many different social security numbers can there be?
b) There are about 324 million people in the United States. Can each person have a unique social security number?

▶ Skill Maintenance

Find the zero(s) of the function.

42. $f(x) = 4x - 9$ **[7.4]**

43. $f(x) = x^2 + x - 6$ **[7.4]**

44. $f(x) = 2x^2 - 3x - 1$ **[7.4]**

45. $f(x) = x^3 - 4x^2 - 7x + 10$ **[8.4]**

▶ Synthesis

Solve for n.

46. $_nP_5 = 7 \cdot {_nP_4}$

47. $_nP_4 = 8 \cdot {_{n-1}P_3}$

48. $_nP_5 = 9 \cdot {_{n-1}P_4}$

49. $_nP_4 = 8 \cdot {_nP_3}$

50. Show that $n! = n(n-1)(n-2)(n-3)!$.

51. *Single-Elimination Tournaments.* In a single-elimination sports tournament consisting of n teams, a team is eliminated when it loses one game. How many games are required to complete the tournament?

52. *Double-Elimination Tournaments.* In a double-elimination softball tournament consisting of n teams, a team is eliminated when it loses two games. At most, how many games are required to complete the tournament?

12.6   Combinatorics: Combinations

▶ Evaluate combination notation and solve related applied problems.

We now consider counting techniques in which order is not considered.

▶ Combinations

We sometimes make a selection from a set *without regard to order.* Such a selection is called a *combination.* If you play cards, for example, you know that in most situations the *order* in which you hold cards is not important. That is,

The hand

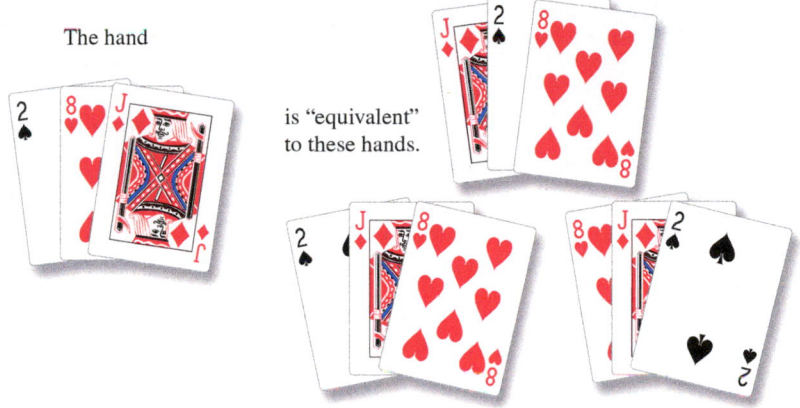

is "equivalent" to these hands.

Each hand contains the same combination of three cards.

EXAMPLE 1   Find all the combinations of 3 letters taken from the set of 5 letters $\{A, B, C, D, E\}$.

Solution   The combinations are

$$\{A, B, C\}, \quad \{A, B, D\},$$
$$\{A, B, E\}, \quad \{A, C, D\},$$
$$\{A, C, E\}, \quad \{A, D, E\},$$
$$\{B, C, D\}, \quad \{B, C, E\},$$
$$\{B, D, E\}, \quad \{C, D, E\}.$$

There are 10 combinations of the 5 letters taken 3 at a time.

When we find all the combinations from a set of 5 objects taken 3 at a time, we are finding all the 3-element subsets. When a set is named, the order of the elements is *not* considered. Thus,

$$\{A, C, B\} \quad \text{names the same set as} \quad \{A, B, C\}.$$

COMBINATION; COMBINATION NOTATION

A **combination** containing k objects chosen from a set of n objects, $k \leq n$, is denoted using **combination notation** $_nC_k$.

We want to derive a general formula for $_nC_k$ for any $k \leq n$. First, it is true that $_nC_n = 1$, because a set with n objects has only 1 subset with n objects, the set itself. Second, $_nC_1 = n$, because a set with n objects has n subsets with 1 element each. Finally, $_nC_0 = 1$, because a set with n objects has only one subset with 0 elements, namely, the empty set $\varnothing$. To consider other possibilities, let's return to Example 1 and compare the number of combinations with the number of permutations.

COMBINATIONS		PERMUTATIONS			
$\{A, B, C\} \rightarrow$ ABC	BCA	CAB	CBA	BAC	ACB
$\{A, B, D\} \rightarrow$ ABD	BDA	DAB	DBA	BAD	ADB
$\{A, B, E\} \rightarrow$ ABE	BEA	EAB	EBA	BAE	AEB
$\{A, C, D\} \rightarrow$ ACD	CDA	DAC	DCA	CAD	ADC
$\{A, C, E\} \rightarrow$ ACE	CEA	EAC	ECA	CAE	AEC
$\{A, D, E\} \rightarrow$ ADE	DEA	EAD	EDA	DAE	AED
$\{B, C, D\} \rightarrow$ BCD	CDB	DBC	DCB	CBD	BDC
$\{B, C, E\} \rightarrow$ BCE	CEB	EBC	ECB	CBE	BEC
$\{B, D, E\} \rightarrow$ BDE	DEB	EBD	EDB	DBE	BED
$\{C, D, E\} \rightarrow$ CDE	DEC	ECD	EDC	DCE	CED

$_5C_3$ of these (left bracket) $3! \cdot {}_5C_3$ of these (right bracket)

Note that each combination of 3 objects yields 6, or 3!, permutations:

$$3! \cdot {}_5C_3 = 60 = {}_5P_3 = 5 \cdot 4 \cdot 3,$$

so

$$_5C_3 = \frac{{}_5P_3}{3!} = \frac{5 \cdot 4 \cdot 3}{3 \cdot 2 \cdot 1} = 10.$$

In general, the number of combinations of n objects taken k at a time, $_nC_k$, times the number of permutations of these objects, $k!$, must equal the number of permutations of n objects taken k at a time:

$$k! \cdot {}_nC_k = {}_nP_k$$

$$_nC_k = \frac{{}_nP_k}{k!}$$

$$= \frac{1}{k!} \cdot {}_nP_k$$

$$= \frac{1}{k!} \cdot \frac{n!}{(n-k)!} = \frac{n!}{k!(n-k)!}.$$

COMBINATIONS OF n OBJECTS TAKEN k AT A TIME

The total number of combinations of n objects taken k at a time, denoted $_nC_k$, is given by

$$_nC_k = \frac{n!}{k!(n-k)!},$$ **(1)**

or

$$_nC_k = \frac{{}_nP_k}{k!} = \frac{n(n-1)(n-2) \cdots [n-(k-1)]}{k!}.$$ **(2)**

Another kind of notation for $_nC_k$ is **binomial coefficient notation**. The reason for such terminology will be seen later.

BINOMIAL COEFFICIENT NOTATION

$$\binom{n}{k} = {}_nC_k$$

You should be able to use either notation and either form of the formula.

EXAMPLE 2 Evaluate $\binom{7}{5}$, using forms (1) and (2).

Solution

a) By form (1),

$$\binom{7}{5} = \frac{7!}{5!\,(7-5)!} = \frac{7!}{5!2!}$$

$$= \frac{7\cdot 6\cdot 5!}{5!\cdot 2!} = \frac{7\cdot 6\cdot 5!}{5!\cdot 2!} = \frac{7\cdot 6}{2\cdot 1} = 21.$$

b) By form (2),

The 7 tells where to start.

$$\binom{7}{5} = \frac{7\cdot 6\cdot 5\cdot 4\cdot 3}{5\cdot 4\cdot 3\cdot 2\cdot 1} = \frac{7\cdot 6}{2\cdot 1} = 21.$$

The 5 tells how many factors there are in both the numerator and the denominator and where to start the denominator.

Now Try Exercise 11.

Be sure to keep in mind that $\binom{n}{k}$ does not mean $n \div k$, or n/k.

EXAMPLE 3 Evaluate $\binom{n}{0}$ and $\binom{n}{2}$.

Solution We use form (1) for the first expression and form (2) for the second. Then

$$\binom{n}{0} = \frac{n!}{0!\,(n-0)!} = \frac{n!}{1\cdot n!} = 1,$$

using form (1), and

$$\binom{n}{2} = \frac{n(n-1)}{2!} = \frac{n(n-1)}{2}, \quad \text{or} \quad \frac{n^2-n}{2},$$

using form (2).

Now Try Exercise 19.

Note that

$$\binom{7}{2} = \frac{7 \cdot 6}{2 \cdot 1} = 21.$$

Using the result of Example 2 gives us

$$\binom{7}{5} = \binom{7}{2}.$$

This says that the number of 5-element subsets of a set of 7 objects is the same as the number of 2-element subsets of a set of 7 objects. When 5 elements are chosen from a set, one also chooses *not* to include 2 elements. To see this, consider the set $\{A, B, C, D, E, F, G\}$:

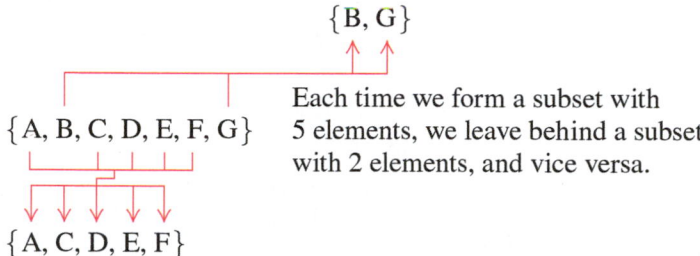

Each time we form a subset with 5 elements, we leave behind a subset with 2 elements, and vice versa.

In general, we have the following. This result provides an alternative way to compute combinations.

SUBSETS OF SIZE k AND OF SIZE $n - k$

$$\binom{n}{k} = \binom{n}{n-k} \quad \text{and} \quad {}_nC_k = {}_nC_{n-k}$$

The number of subsets of size k of a set with n objects is the same as the number of subsets of size $n - k$. The number of combinations of n objects taken k at a time is the same as the number of combinations of n objects taken $n - k$ at a time.

We now solve problems involving combinations.

EXAMPLE 4 *Indiana Lottery.* Run by the state of Indiana, Hoosier Lotto is a twice-weekly lottery game with jackpots starting at $1 million. For a wager of $1, a player can choose 6 numbers from 1 through 48. If the numbers match those drawn by the state, the player wins the jackpot. (*Source*: www.hoosierlottery.com)

a) How many 6-number combinations are there?

b) Suppose it takes you 10 min to pick your numbers and buy a game ticket. How many tickets can you buy in 4 days?

c) How many people would you have to hire for 4 days to buy tickets with all the possible combinations and ensure that you win?

Solution

a) No order is implied here. You pick any 6 different numbers from 1 through 48. Thus the number of combinations is

$$
\begin{aligned}
{}_{48}C_6 = \binom{48}{6} &= \frac{48!}{6!(48-6)!} = \frac{48!}{6!\,42!} \\
&= \frac{48 \cdot 47 \cdot 46 \cdot 45 \cdot 44 \cdot 43 \cdot 42!}{6 \cdot 5 \cdot 4 \cdot 3 \cdot 2 \cdot 1 \cdot 42!} \\
&= \frac{48 \cdot 47 \cdot 46 \cdot 45 \cdot 44 \cdot 43}{6 \cdot 5 \cdot 4 \cdot 3 \cdot 2 \cdot 1} = 12{,}271{,}512.
\end{aligned}
$$

b) First, we find the number of minutes in 4 days:

$$
4 \text{ days} = 4 \text{ days} \cdot \frac{24 \text{ hr}}{1 \text{ day}} \cdot \frac{60 \text{ min}}{1 \text{ hr}} = 5760 \text{ min.}
$$

Thus you could buy $5760/10$, or 576 tickets in 4 days.

c) You would need to hire $12{,}271{,}512/576$, or about 21,305 people, to buy tickets with all the possible combinations and ensure a win. (This presumes lottery tickets can be bought 24 hours a day.) ▶ **Now Try Exercise 23.**

EXAMPLE 5 How many committees can be formed from a group of 5 governors and 7 senators if each committee consists of 3 governors and 4 senators?

Solution The 3 governors can be selected in ${}_5C_3$ ways and the 4 senators can be selected in ${}_7C_4$ ways. If we use the fundamental counting principle, it follows that the number of possible committees is

$$
\begin{aligned}
{}_5C_3 \cdot {}_7C_4 &= \frac{5!}{3!\,2!} \cdot \frac{7!}{4!\,3!} \\
&= \frac{5 \cdot 4 \cdot 3!}{3! \cdot 2 \cdot 1} \cdot \frac{7 \cdot 6 \cdot 5 \cdot 4!}{4! \cdot 3 \cdot 2 \cdot 1} \\
&= \frac{5 \cdot 2 \cdot 2 \cdot 3!}{3! \cdot 2 \cdot 1} \cdot \frac{7 \cdot 3 \cdot 2 \cdot 5 \cdot 4!}{4! \cdot 3 \cdot 2 \cdot 1} \\
&= 10 \cdot 35 \\
&= 350.
\end{aligned}
$$
▶ **Now Try Exercise 27.**

CONNECTING THE CONCEPTS

Permutations and Combinations

PERMUTATIONS	**COMBINATIONS**
Permutations involve order and arrangements of objects.	Combinations do not involve order or arrangements of objects.
Given 5 books, we can arrange 3 of them on a shelf in ${}_5P_3$, or 60 ways.	Given 5 books, we can select 3 of them in ${}_5C_3$, or 10 ways.
Placing the books in different orders produces different arrangements.	The order in which the books are chosen does not matter.

12.6 Exercise Set

Evaluate.

1. $_{13}C_2$

2. $_9C_6$

3. $\begin{pmatrix} 13 \\ 11 \end{pmatrix}$

4. $\begin{pmatrix} 9 \\ 3 \end{pmatrix}$

5. $\begin{pmatrix} 7 \\ 1 \end{pmatrix}$

6. $\begin{pmatrix} 8 \\ 8 \end{pmatrix}$

7. $\dfrac{_5P_3}{3!}$

8. $\dfrac{_{10}P_5}{5!}$

9. $\begin{pmatrix} 6 \\ 0 \end{pmatrix}$

10. $\begin{pmatrix} 6 \\ 1 \end{pmatrix}$

11. $\begin{pmatrix} 6 \\ 2 \end{pmatrix}$

12. $\begin{pmatrix} 6 \\ 3 \end{pmatrix}$

13. $\begin{pmatrix} 7 \\ 0 \end{pmatrix} + \begin{pmatrix} 7 \\ 1 \end{pmatrix} + \begin{pmatrix} 7 \\ 2 \end{pmatrix} + \begin{pmatrix} 7 \\ 3 \end{pmatrix} + \begin{pmatrix} 7 \\ 4 \end{pmatrix} + \begin{pmatrix} 7 \\ 5 \end{pmatrix}$
$+ \begin{pmatrix} 7 \\ 6 \end{pmatrix} + \begin{pmatrix} 7 \\ 7 \end{pmatrix}$

14. $\begin{pmatrix} 6 \\ 0 \end{pmatrix} + \begin{pmatrix} 6 \\ 1 \end{pmatrix} + \begin{pmatrix} 6 \\ 2 \end{pmatrix} + \begin{pmatrix} 6 \\ 3 \end{pmatrix} + \begin{pmatrix} 6 \\ 4 \end{pmatrix}$
$+ \begin{pmatrix} 6 \\ 5 \end{pmatrix} + \begin{pmatrix} 6 \\ 6 \end{pmatrix}$

15. $_{52}C_4$

16. $_{52}C_5$

17. $\begin{pmatrix} 27 \\ 11 \end{pmatrix}$

18. $\begin{pmatrix} 37 \\ 8 \end{pmatrix}$

19. $\begin{pmatrix} n \\ 1 \end{pmatrix}$

20. $\begin{pmatrix} n \\ 3 \end{pmatrix}$

21. $\begin{pmatrix} m \\ m \end{pmatrix}$

22. $\begin{pmatrix} t \\ 4 \end{pmatrix}$

In each of the following exercises, give an expression for the answer using permutation notation, combination notation, factorial notation, or other operations. Then evaluate.

23. *Key Club Officers.* There are 36 students in a high school Key Club, a service organization for teens. How many sets of 4 officers can be selected?

24. *League Games.* How many games can be played in a 9-team sports league if each team plays all other teams once? twice?

25. *Test Options.* On a test, a student is to select 10 out of 13 questions. In how many ways can this be done?

26. *Senate Committees.* Suppose that the Senate of the United States consists of 58 Democrats and 42 Republicans. How many committees can be formed consisting of 6 Democrats and 4 Republicans?

27. *Test Options.* Of the first 10 questions on a test, a student must answer 7. Of the second 5 questions, the student must answer 3. In how many ways can this be done?

28. *Lines and Triangles from Points.* How many lines are determined by 8 points, no 3 of which are collinear? How many triangles are determined by the same points?

29. *Poker Hands.* How many 5-card poker hands are possible with a 52-card deck?

30. *Bridge Hands.* How many 13-card bridge hands are possible with a 52-card deck?

31. *Baskin-Robbins Ice Cream.* Burt Baskin and Irv Robbins began making ice cream in 1945. Initially they developed 31 flavors—one for each day of the month. (*Source:* Baskin-Robbins)

a) How many 2-dip cones are possible using the 31 original flavors if order of flavors is to be considered and no flavor is repeated?

b) How many 2-dip cones are possible if order is to be considered and a flavor can be repeated?

c) How many 2-dip cones are possible if order is not considered and no flavor is repeated?

32. *Powerball®.* Powerball® is a biweekly lottery game in which 5 white balls are drawn from a drum of 59 balls numbered 1–59 and 1 red ball is drawn from a drum of 35 balls numbered 1–35. To win the jackpot, a player must select numbers to match in any order the 5 white balls and the 1 red ball. (*Source:* www.powerball.com) How many 6-number combinations are there?

▶ Skill Maintenance

Solve.

33. $3x - 7 = 5x + 10$ **[1.1]**

34. $2x^2 - x = 3$ **[4.8]**

35. $x^2 + 5x + 1 = 0$ **[7.4]**

36. $x^3 + 3x^2 - 10x = 24$ **[8.4]**

▶ Synthesis

37. *Flush.* A flush in poker consists of a 5-card hand with all cards of the same suit. How many 5-card hands (flushes) are there that consist of all diamonds?

38. *Full House.* A full house in poker consists of three of a kind and a pair (two of a kind). How many full houses are there that consist of 3 aces and 2 queens? (See Section 12.8 for a description of a 52-card deck.)

39. *League Games.* How many games are played in a league with n teams if each team plays each other team once? twice?

40. There are n points on a circle. How many quadrilaterals can be inscribed with these points as vertices?

Solve for n.

41. $\begin{pmatrix} n \\ n-2 \end{pmatrix} = 6$

42. $\begin{pmatrix} n+1 \\ 3 \end{pmatrix} = 2 \cdot \begin{pmatrix} n \\ 2 \end{pmatrix}$

43. $\begin{pmatrix} n+2 \\ 4 \end{pmatrix} = 6 \cdot \begin{pmatrix} n \\ 2 \end{pmatrix}$

44. $\begin{pmatrix} n \\ 3 \end{pmatrix} = 2 \cdot \begin{pmatrix} n-1 \\ 2 \end{pmatrix}$

45. How many line segments are determined by the n vertices of an n-gon? Of these, how many are diagonals? Use mathematical induction to prove the result for the diagonals.

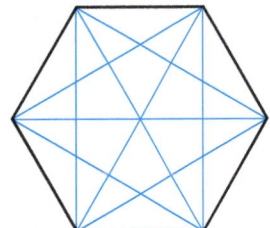

46. Prove that

$$\begin{pmatrix} n \\ k-1 \end{pmatrix} + \begin{pmatrix} n \\ k \end{pmatrix} = \begin{pmatrix} n+1 \\ k \end{pmatrix}$$

for any natural numbers n and k, $k \le n$.

12.7   The Binomial Theorem

▶ Expand a power of a binomial using Pascal's triangle or factorial notation.

▶ Find a specific term of a binomial expansion.

▶ Find the total number of subsets of a set of n objects.

In this section, we consider ways of expanding a binomial $(a + b)^n$.

▶ Binomial Expansion Using Pascal's Triangle

Consider the following expanded powers of $(a + b)^n$, where $a + b$ is any binomial and n is a whole number. Look for patterns.

$$(a + b)^0 = \qquad\qquad\qquad 1$$
$$(a + b)^1 = \qquad\qquad\qquad a + b$$
$$(a + b)^2 = \qquad\qquad a^2 + 2ab + b^2$$
$$(a + b)^3 = \qquad a^3 + 3a^2b + 3ab^2 + b^3$$
$$(a + b)^4 = \quad a^4 + 4a^3b + 6a^2b^2 + 4ab^3 + b^4$$
$$(a + b)^5 = a^5 + 5a^4b + 10a^3b^2 + 10a^2b^3 + 5ab^4 + b^5$$

Each expansion is a polynomial. There are some patterns to be noted.

1. There is one more term than the power of the exponent, n. That is, there are $n + 1$ terms in the expansion of $(a + b)^n$.

2. In each term, the sum of the exponents is n, the power to which the binomial is raised.

3. The exponents of a start with n, the power of the binomial, and decrease to 0. The last term has no factor of a. The first term has no factor of b, so powers of b start with 0 and increase to n.

4. The coefficients start at 1 and increase through certain values about "half"-way and then decrease through these same values back to 1.

Let's explore the coefficients further. Suppose that we want to find an expansion of $(a + b)^6$. The patterns we just noted indicate that there are 7 terms in the expansion:

$$a^6 + c_1a^5b + c_2a^4b^2 + c_3a^3b^3 + c_4a^2b^4 + c_5ab^5 + b^6.$$

How can we determine the value of each coefficient, c_i? We can do so in two ways. The first method involves writing the coefficients in a triangular array, as follows. This is known as **Pascal's triangle**:

$$(a + b)^0: \qquad\qquad\qquad\qquad 1$$
$$(a + b)^1: \qquad\qquad\qquad\quad 1 \quad 1$$
$$(a + b)^2: \qquad\qquad\quad 1 \quad 2 \quad 1$$
$$(a + b)^3: \qquad\quad 1 \quad 3 \quad 3 \quad 1$$
$$(a + b)^4: \quad\quad 1 \quad 4 \quad 6 \quad 4 \quad 1$$
$$(a + b)^5: \quad 1 \quad 5 \quad 10 \quad 10 \quad 5 \quad 1$$

There are many patterns in the triangle. Find as many as you can.

Perhaps you discovered a way to write the next row of numbers, given the numbers in the row above it. There are always 1's on the outside. Each remaining number is the sum of the two numbers above it. Let's try to find an expansion for $(a + b)^6$ by adding another row using the patterns we have discovered:

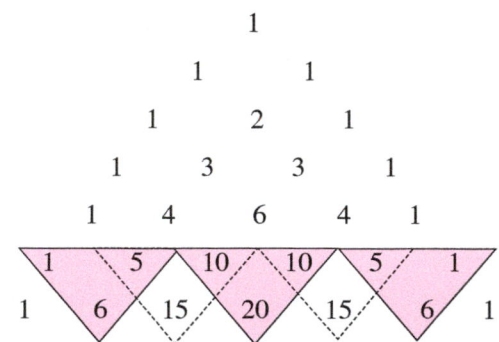

We see that in the last row

the 1st and last numbers are **1**;

the 2nd number is $1 + 5$, or **6**;

the 3rd number is $5 + 10$, or **15**;

the 4th number is $10 + 10$, or **20**;

the 5th number is $10 + 5$, or **15**;   and

the 6th number is $5 + 1$, or **6**.

Thus the expansion for $(a + b)^6$ is

$$(a + b)^6 = 1a^6 + 6a^5b + 15a^4b^2 + 20a^3b^3 + 15a^2b^4 + 6ab^5 + 1b^6.$$

To find an expansion for $(a + b)^8$, we complete two more rows of Pascal's triangle:

$$
\begin{array}{ccccccccccccccccc}
&&&&&&&& 1 \\
&&&&&&& 1 && 1 \\
&&&&&& 1 && 2 && 1 \\
&&&&& 1 && 3 && 3 && 1 \\
&&&& 1 && 4 && 6 && 4 && 1 \\
&&& 1 && 5 && 10 && 10 && 5 && 1 \\
&& 1 && 6 && 15 && 20 && 15 && 6 && 1 \\
& 1 && 7 && 21 && 35 && 35 && 21 && 7 && 1 \\
1 && 8 && 28 && 56 && 70 && 56 && 28 && 8 && 1
\end{array}
$$

Thus the expansion of $(a + b)^8$ is

$$(a + b)^8 = a^8 + 8a^7b + 28a^6b^2 + 56a^5b^3 + 70a^4b^4 + 56a^3b^5 \\ + 28a^2b^6 + 8ab^7 + b^8.$$

We can generalize our results as follows.

THE BINOMIAL THEOREM USING PASCAL'S TRIANGLE

For any binomial $a + b$ and any natural number n,

$$(a + b)^n = c_0 a^n b^0 + c_1 a^{n-1} b^1 + c_2 a^{n-2} b^2 + \cdots \\ + c_{n-1} a^1 b^{n-1} + c_n a^0 b^n,$$

where the numbers $c_0, c_1, c_2, \ldots, c_{n-1}, c_n$ are from the $(n + 1)$st row of Pascal's triangle.

EXAMPLE 1   Expand: $(u - v)^5$.

Solution   We have $(a + b)^n$, where $a = u$, $b = -v$, and $n = 5$. We use the 6th row of Pascal's triangle:

$$1 \quad 5 \quad 10 \quad 10 \quad 5 \quad 1$$

Then we have

$$
\begin{aligned}
(u - v)^5 &= [u + (-v)]^5 \\
&= 1(u)^5 + 5(u)^4(-v)^1 + 10(u)^3(-v)^2 + 10(u)^2(-v)^3 \\
&\quad + 5(u)(-v)^4 + 1(-v)^5 \\
&= u^5 - 5u^4v + 10u^3v^2 - 10u^2v^3 + 5uv^4 - v^5.
\end{aligned}
$$

Note that the signs of the terms alternate between $+$ and $-$. When the power of $-v$ is odd, the sign is $-$.

> **Now Try Exercise 5.**

EXAMPLE 2   Expand: $\left(2t + \dfrac{3}{t}\right)^4$.

Solution   We have $(a + b)^n$, where $a = 2t$, $b = 3/t$, and $n = 4$. We use the 5th row of Pascal's triangle:

$$1 \qquad 4 \qquad 6 \qquad 4 \qquad 1$$

Then we have

$$
\begin{aligned}
\left(2t + \frac{3}{t}\right)^4 &= 1(2t)^4 + 4(2t)^3\left(\frac{3}{t}\right)^1 + 6(2t)^2\left(\frac{3}{t}\right)^2 + 4(2t)^1\left(\frac{3}{t}\right)^3 + 1\left(\frac{3}{t}\right)^4 \\
&= 1(16t^4) + 4(8t^3)\left(\frac{3}{t}\right) + 6(4t^2)\left(\frac{9}{t^2}\right) + 4(2t)\left(\frac{27}{t^3}\right) + 1\left(\frac{81}{t^4}\right) \\
&= 16t^4 + 96t^2 + 216 + 216t^{-2} + 81t^{-4}.
\end{aligned}
$$

> **Now Try Exercise 9.**

▶ **Binomial Expansion Using Factorial Notation**

Suppose that we want to find the expansion of $(a + b)^{11}$. The disadvantage in using Pascal's triangle is that we must compute all the preceding rows of the triangle to obtain the row needed for the expansion. The following method avoids this. It also enables us to find a specific term—say, the 8th term—without computing all the other terms of the expansion. This method is useful in such courses as finite mathematics, calculus, and statistics, and it uses the *binomial coefficient notation* $\dbinom{n}{k}$ developed in Section 12.6.

We can restate the binomial theorem as follows.

THE BINOMIAL THEOREM USING FACTORIAL NOTATION

For any binomial $a + b$ and any natural number n,

$$
(a + b)^n = \binom{n}{0}a^n b^0 + \binom{n}{1}a^{n-1}b^1 + \binom{n}{2}a^{n-2}b^2 + \cdots
$$
$$
+ \binom{n}{n-1}a^1 b^{n-1} + \binom{n}{n}a^0 b^n
$$
$$
= \sum_{k=0}^{n} \binom{n}{k}a^{n-k}b^k.
$$

The binomial theorem can be proved by mathematical induction. (See Exercise 57.)

This form shows why $\dbinom{n}{k}$ is called a *binomial coefficient*.

EXAMPLE 3 Expand: $(x^2 - 2y)^5$.

Solution We have $(a + b)^n$, where $a = x^2$, $b = -2y$, and $n = 5$. Then using the binomial theorem, we have

$$(x^2 - 2y)^5 = \binom{5}{0}(x^2)^5 + \binom{5}{1}(x^2)^4(-2y) + \binom{5}{2}(x^2)^3(-2y)^2$$

$$+ \binom{5}{3}(x^2)^2(-2y)^3 + \binom{5}{4}x^2(-2y)^4 + \binom{5}{5}(-2y)^5$$

$$= \frac{5!}{0!\,5!}x^{10} + \frac{5!}{1!\,4!}x^8(-2y) + \frac{5!}{2!\,3!}x^6(4y^2) + \frac{5!}{3!\,2!}x^4(-8y^3)$$

$$+ \frac{5!}{4!\,1!}x^2(16y^4) + \frac{5!}{5!\,0!}(-32y^5)$$

$$= 1 \cdot x^{10} + 5x^8(-2y) + 10x^6(4y^2) + 10x^4(-8y^3)$$

$$+ 5x^2(16y^4) + 1 \cdot (-32y^5)$$

$$= x^{10} - 10x^8y + 40x^6y^2 - 80x^4y^3 + 80x^2y^4 - 32y^5.$$

> **Now Try Exercise 11.**

EXAMPLE 4 Expand: $\left(\dfrac{2}{x} + 3\sqrt{x}\right)^4$.

Solution We have $(a + b)^n$, where $a = 2/x$, $b = 3\sqrt{x}$, and $n = 4$. Then using the binomial theorem, we have

$$\left(\frac{2}{x} + 3\sqrt{x}\right)^4 = \binom{4}{0}\left(\frac{2}{x}\right)^4 + \binom{4}{1}\left(\frac{2}{x}\right)^3(3\sqrt{x}) + \binom{4}{2}\left(\frac{2}{x}\right)^2(3\sqrt{x})^2$$

$$+ \binom{4}{3}\left(\frac{2}{x}\right)(3\sqrt{x})^3 + \binom{4}{4}(3\sqrt{x})^4$$

$$= \frac{4!}{0!\,4!}\left(\frac{16}{x^4}\right) + \frac{4!}{1!\,3!}\left(\frac{8}{x^3}\right)(3x^{1/2})$$

$$+ \frac{4!}{2!\,2!}\left(\frac{4}{x^2}\right)(9x) + \frac{4!}{3!\,1!}\left(\frac{2}{x}\right)(27x^{3/2})$$

$$+ \frac{4!}{4!\,0!}(81x^2)$$

$$= \frac{16}{x^4} + \frac{96}{x^{5/2}} + \frac{216}{x} + 216x^{1/2} + 81x^2.$$

> **Now Try Exercise 13.**

▶ **Finding a Specific Term**

Suppose that we want to determine only a particular term of an expansion. The method we have developed will allow us to find such a term without computing all the rows of Pascal's triangle or all the preceding coefficients.

Note that in the binomial theorem, $\binom{n}{0}a^n b^0$ gives us the 1st term, $\binom{n}{1}a^{n-1}b^1$ gives us the 2nd term, $\binom{n}{2}a^{n-2}b^2$ gives us the 3rd term, and so on. This can be generalized as follows.

FINDING THE (k + 1)ST TERM

The $(k + 1)$st term of $(a + b)^n$ is $\binom{n}{k}a^{n-k}b^k$.

EXAMPLE 5 Find the 5th term in the expansion of $(2x - 5y)^6$.

Solution First, we note that $5 = 4 + 1$. Thus, $k = 4, a = 2x, b = -5y$, and $n = 6$. Then the 5th term of the expansion is

$$\binom{6}{4}(2x)^{6-4}(-5y)^4, \quad \text{or} \quad \frac{6!}{4!\,2!}(2x)^2(-5y)^4, \quad \text{or} \quad 37{,}500x^2y^4.$$

> **Now Try Exercise 21.**

EXAMPLE 6 Find the 8th term in the expansion of $(3x - 2)^{10}$.

Solution First, we note that $8 = 7 + 1$. Thus, $k = 7, a = 3x, b = -2$, and $n = 10$. Then the 8th term of the expansion is

$$\binom{10}{7}(3x)^{10-7}(-2)^7, \quad \text{or} \quad \frac{10!}{7!\,3!}(3x)^3(-2)^7, \quad \text{or} \quad -414{,}720x^3.$$

> **Now Try Exercise 25.**

▶ Total Number of Subsets

Suppose that a set has n objects. The number of subsets containing k elements is $\binom{n}{k}$ by a result of Section 12.6. The total number of subsets of a set is the number of subsets with 0 elements, plus the number of subsets with 1 element, plus the number of subsets with 2 elements, and so on. The total number of subsets of a set with n elements is

$$\binom{n}{0} + \binom{n}{1} + \binom{n}{2} + \cdots + \binom{n}{n}.$$

Now consider the expansion of $(1 + 1)^n$:

$$(1 + 1)^n = \binom{n}{0} \cdot 1^n + \binom{n}{1} \cdot 1^{n-1} \cdot 1^1 + \binom{n}{2} \cdot 1^{n-2} \cdot 1^2$$

$$+ \cdots + \binom{n}{n} \cdot 1^n$$

$$= \binom{n}{0} + \binom{n}{1} + \binom{n}{2} + \cdots + \binom{n}{n}.$$

Thus the total number of subsets is $(1 + 1)^n$, or 2^n. We have proved the following.

TOTAL NUMBER OF SUBSETS

The total number of subsets of a set with n elements is 2^n.

EXAMPLE 7 The set $\{A, B, C, D, E\}$ has how many subsets?

Solution The set has 5 elements, so the number of subsets is 2^5, or 32.

> **Now Try Exercise 31.**

EXAMPLE 8 Wendy's, a national restaurant chain, offers the following toppings for its hamburgers:

{catsup, mustard, mayonnaise, tomato, lettuce, onions, pickle}.

In how many different ways can Wendy's serve hamburgers, excluding size of hamburger or number of patties?

Solution The toppings on each hamburger are the elements of a subset of the set of all possible toppings, the empty set being a plain hamburger. The total number of possible hamburgers is

$$\binom{7}{0} + \binom{7}{1} + \binom{7}{2} + \cdots + \binom{7}{7} = 2^7 = 128.$$

Thus Wendy's serves hamburgers in 128 different ways.

Now Try Exercise 33.

12.7 Exercise Set

Expand.

1. $(x + 5)^4$

2. $(x - 1)^4$

3. $(x - 3)^5$

4. $(x + 2)^9$

5. $(x - y)^5$

6. $(x + y)^8$

7. $(5x + 4y)^6$

8. $(2x - 3y)^5$

9. $\left(2t + \dfrac{1}{t}\right)^7$

10. $\left(3y - \dfrac{1}{y}\right)^4$

11. $(x^2 - 1)^5$

12. $(1 + 2q^3)^8$

13. $(\sqrt{5} + t)^6$

14. $(x - \sqrt{2})^6$

15. $\left(a - \dfrac{2}{a}\right)^9$

16. $(1 + 3)^n$

17. $(\sqrt{2} + 1)^6 - (\sqrt{2} - 1)^6$

18. $(1 - \sqrt{2})^4 + (1 + \sqrt{2})^4$

19. $(x^{-2} + x^2)^4$

20. $\left(\dfrac{1}{\sqrt{x}} - \sqrt{x}\right)^6$

Find the indicated term of the binomial expansion.

21. 3rd; $(a + b)^7$

22. 6th; $(x + y)^8$

23. 6th; $(x - y)^{10}$

24. 5th; $(p - 2q)^9$

25. 12th; $(a - 2)^{14}$

26. 11th; $(x - 3)^{12}$

27. 5th; $(2x^3 - \sqrt{y})^8$

28. 4th; $\left(\dfrac{1}{b^2} + \dfrac{b}{3}\right)^7$

29. Middle; $(2u - 3v^2)^{10}$

30. Middle two; $(\sqrt{x} + \sqrt{3})^5$

Determine the number of subsets of each of the following.

31. A set of 7 elements

32. A set of 6 members

33. The set of letters of the Greek alphabet, which contains 24 letters

34. The set of letters of the English alphabet, which contains 26 letters

35. What is the degree of $(x^5 + 3)^4$?

36. What is the degree of $(2 - 5x^3)^7$?

Expand each of the following, where $i^2 = -1$.

37. $(3 + i)^5$

38. $(1 + i)^6$

39. $(\sqrt{2} - i)^4$

40. $\left(\dfrac{\sqrt{3}}{2} - \dfrac{1}{2}i\right)^{11}$

41. Find a formula for $(a - b)^n$. Use sigma notation.

42. Expand and simplify:

$$\frac{(x+h)^{13} - x^{13}}{h}.$$

43. Expand and simplify:

$$\frac{(x+h)^n - x^n}{h}.$$

Use sigma notation.

▶ Skill Maintenance

Given that $f(x) = x^2 + 1$ and $g(x) = 2x - 3$, find each of the following.

44. $(f+g)(x)$ [2.4]

45. $(f-g)(x)$ [2.4]

46. $(f \circ g)(x)$ [9.1]

47. $(g \circ f)(x)$ [9.1]

▶ Synthesis

Solve for x.

48. $\displaystyle\sum_{k=0}^{8} \binom{8}{k} x^{8-k} 3^k = 0$

49. $\displaystyle\sum_{k=0}^{4} \binom{4}{k} (-1)^k x^{4-k} 6^k = 81$

50. Find the ratio of the 4th term of

$$\left(p^2 - \frac{1}{2}p\sqrt[3]{q}\right)^5$$

to the 3rd term.

51. Find the term of

$$\left(\sqrt[3]{x} - \frac{1}{\sqrt{x}}\right)^7$$

containing $1/x^{1/6}$.

52. *Money Combinations.*   A money clip contains one each of the following bills: \$1, \$2, \$5, \$10, \$20, \$50, and \$100. How many different sums of money can be formed using the bills?

Find the sum.

53. $_{100}C_0 + {}_{100}C_1 + \cdots + {}_{100}C_{100}$

54. $_nC_0 + {}_nC_1 + \cdots + {}_nC_n$

Simplify.

55. $\displaystyle\sum_{k=0}^{23} \binom{23}{k} (\log_a x)^{23-k} (\log_a t)^k$

56. $\displaystyle\sum_{k=0}^{15} \binom{15}{k} i^{30-2k}$

57. Use mathematical induction and the property

$$\binom{n}{r-1} + \binom{n}{r} = \binom{n+1}{r}$$

to prove the binomial theorem.

12.8   Probability

▶ Compute the probability of a simple event.

When a coin is tossed, we can reason that the chance, or the likelihood, that it will fall heads is 1 out of 2—that is, the **probability** that it will fall heads is $\frac{1}{2}$. Of course, this does not mean that if a coin is tossed 10 times it will necessarily fall heads 5 times. If the coin is a "fair coin" and it is tossed a great many times, however, it will fall heads very nearly half of the time. Here we give an introduction to two kinds of probability, **experimental** and **theoretical**.

▶ Experimental Probability and Theoretical Probability

If we toss a coin a great number of times—say, 1000—and count the number of times it falls heads, we can determine the probability that it will fall heads. If it falls heads 503 times, we would calculate the probability of its falling heads to be

$$\frac{503}{1000}, \quad \text{or} \quad 0.503.$$

This is an **experimental** determination of probability. Such a determination of probability is discovered by the observation and study of data and is quite common and very useful. Here, for example, are some probabilities that have been determined *experimentally*:

1. 60% of all college freshmen entering four-year colleges graduate in 6 years (*Source:* www.satprepct.com, College Planning Partnership's Blog, February 24, 2011, Sam Rosensohn).

2. The probability that a woman will be diagnosed with breast cancer in her lifetime is $\frac{1}{8}$ (*Source*: National Cancer Institute).

3. Anyone who reaches the age of 65 has a 0.4 probability of entering a nursing home during the remaining years of life (*Source:* "Facing the Future," Russ Banham, *Wall Street Journal*).

If we consider a coin and reason that it is just as likely to fall heads as to fall tails, we would calculate the probability that it will fall heads to be $\frac{1}{2}$. This is a **theoretical** determination of probability. Here are some other probabilities that have been determined *theoretically*, using mathematics:

1. If there are 30 people in a room, the probability that two of them have the same birthday (excluding year) is 0.706.

2. While on a trip, you meet someone and, after a period of conversation, discover that you have a common acquaintance. The typical reaction, "It's a small world!", is actually not appropriate, because the probability of such an occurrence is quite high—just over 22%.

In summary, experimental probabilities are determined by making observations and gathering data. Theoretical probabilities are determined by reasoning mathematically. Examples of experimental and theoretical probability like those above, especially those we do not expect, lead us to see the value of a study of probability. You might ask, "What is the *true* probability?" In fact, there is none. Experimentally, we can determine probabilities within certain limits. These may or may not agree with the probabilities that we obtain theoretically. There are situations in which it is much easier to determine one of these types of probabilities than the other. For example, it would be quite difficult to arrive at the probability of college freshmen graduating in 4 years using theoretical probability.

▶ Computing Experimental Probabilities

We first consider experimental determination of probability. The basic principle we use in computing such probabilities is as follows.

> **PRINCIPLE P (EXPERIMENTAL)**
>
> Given an experiment in which *n* observations are made, if a situation, or event, *E* occurs *m* times out of *n* observations, then we say that the **experimental probability** of the event, $P(E)$, is given by
>
> $$P(E) = \frac{m}{n}.$$

EXAMPLE 1 *Television Ratings.* There are an estimated 114,200,000 households in the United States that have at least one television. Each week, viewing information is collected and reported. One week, 28,510,000 households tuned in to the 2013 Grammy Awards ceremony on CBS, and 14,204,000 households tuned in to the action series "NCIS" on CBS (*Source*: Nielsen Media Research). What is the probability that a television household tuned in to the Grammy Awards ceremony during the given week? to "NCIS"?

Solution The probability that a television household was tuned in to the Grammy Awards ceremony is *P*, where

$$P = \frac{28,510,000}{114,200,000} \approx 0.2496 \approx 24.96\%.$$

The probability that a television household was tuned in to "NCIS" is *P*, where

$$P = \frac{14,204,000}{114,200,000} \approx 0.1244 \approx 12.44\%.$$

Now Try Exercise 1.

EXAMPLE 2 *Sociological Survey.* The authors of this text conducted an experimental survey to determine the number of people who are left-handed, right-handed, or both. The results are shown in the graph at left.

a) Determine the probability that a person is right-handed.

b) Determine the probability that a person is left-handed.

c) Determine the probability that a person is ambidextrous (uses both hands with equal ability).

d) There are 120 bowlers in most tournaments held by the Professional Bowlers Association. On the basis of the data in this experiment, how many of the bowlers would you expect to be left-handed?

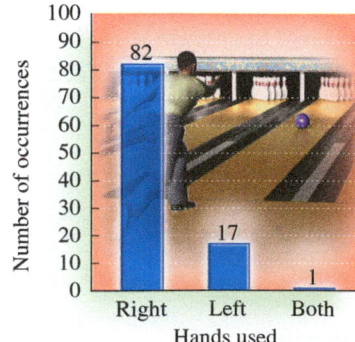

Solution

a) The number of people who are right-handed is 82, the number who are left-handed is 17, and the number who are ambidextrous is 1. The total number of observations is $82 + 17 + 1$, or 100. Thus the probability that a person is right-handed is P, where

$$P = \frac{82}{100}, \quad \text{or} \quad 0.82, \quad \text{or} \quad 82\%.$$

b) The probability that a person is left-handed is P, where

$$P = \frac{17}{100}, \quad \text{or} \quad 0.17, \quad \text{or} \quad 17\%.$$

c) The probability that a person is ambidextrous is P, where

$$P = \frac{1}{100}, \quad \text{or} \quad 0.01, \quad \text{or} \quad 1\%.$$

d) There are 120 bowlers, and from part (b) we can expect 17% to be left-handed. Since

$$17\% \text{ of } 120 = 0.17 \cdot 120 = 20.4,$$

we can expect that about 20 of the bowlers will be left-handed.

 **Now Try Exercise 3.**

▶ **Theoretical Probability**

Suppose that we perform an experiment such as flipping a coin, throwing a dart, drawing a card from a deck, or checking an item off an assembly line for quality. Each possible result of such an experiment is called an **outcome**. The set of all possible outcomes is called the **sample space**. An **event** is a set of outcomes, that is, a subset of the sample space.

EXAMPLE 3   *Dart Throwing.*   Consider the dartboard at left. Assume that the experiment is "throwing a dart" and that the dart hits the board. Find each of the following.

a) The outcomes   b) The sample space

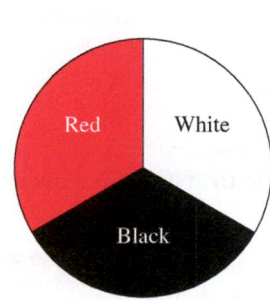

Solution

a) The outcomes are *hitting black* (B), *hitting red* (R), and *hitting white* (W).

b) The sample space is {*hitting black, hitting red, hitting white*}, which can be stated simply as {B, R, W}.

EXAMPLE 4   *Die Rolling.*   A die (pl., dice) is a cube. Each of the six faces has a different number of dots from 1 through 6.

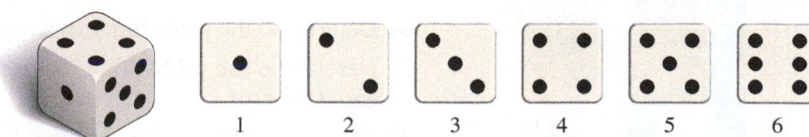

Suppose that a die is rolled. Find each of the following.

a) The outcomes   b) The sample space

Solution

a) The outcomes are 1, 2, 3, 4, 5, 6.

b) The sample space is $\{1, 2, 3, 4, 5, 6\}$.

We denote the probability that an event E occurs as $P(E)$. For example, "a coin falling heads" may be denoted H. Then $P(H)$ represents the probability of the coin falling heads. When all the outcomes of an experiment have the same probability of occurring, we say that they are *equally likely*. To see the distinction between events that are equally likely and those that are not, consider the dart-boards shown below.

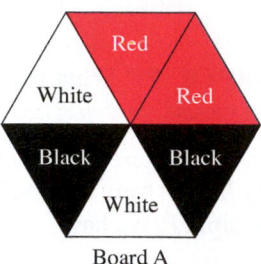
Board A

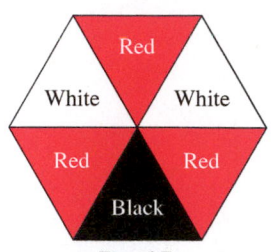
Board B

For board A, the events *hitting black*, *hitting red*, and *hitting white* are equally likely, because the black, red, and white areas are the same. For board B, however, the areas are not the same so these events are not equally likely.

PRINCIPLE P (THEORETICAL)

If an event E can occur m ways out of n possible equally likely outcomes of a sample space S, then the **theoretical probability** of the event, $P(E)$, is given by

$$P(E) = \frac{m}{n}.$$

EXAMPLE 5 Suppose that we select, without looking, one marble from a bag containing 3 red marbles and 4 green marbles. What is the probability of selecting a red marble?

Solution There are 7 equally likely ways of selecting any marble, and since the number of ways of getting a red marble is 3, we have

$$P(\text{selecting a red marble}) = \frac{3}{7}.$$

Now Try Exercise 5(a).

EXAMPLE 6 What is the probability of rolling an even number on a die?

Solution The event is rolling an *even* number. It can occur 3 ways (rolling 2, 4, or 6). The number of equally likely outcomes is 6. By Principle P,

$$P(\text{even}) = \frac{3}{6}, \quad \text{or} \quad \frac{1}{2}.$$

Now Try Exercise 7.

We will use a number of examples related to a standard bridge deck of 52 cards. Such a deck is made up as shown in the following figure.

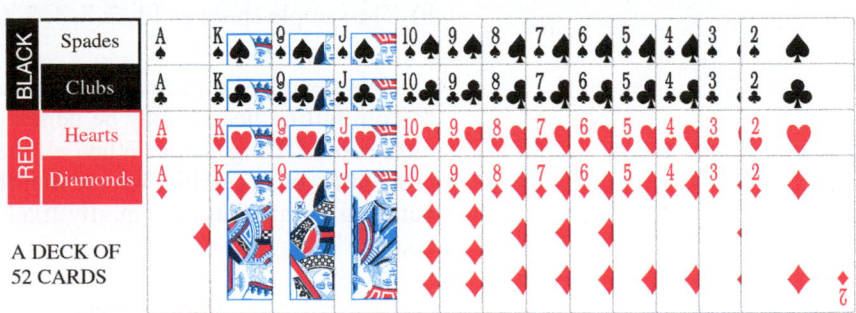

A DECK OF
52 CARDS

EXAMPLE 7 What is the probability of drawing an ace from a well-shuffled deck of cards?

Solution There are 52 outcomes (the number of cards in the deck), they are equally likely (from a well-shuffled deck), and there are 4 ways to obtain an ace, so by Principle *P*, we have

$$P(\text{drawing an ace}) = \frac{4}{52}, \quad \text{or} \quad \frac{1}{13}.$$

> **Now Try Exercise 9(a).**

The following are some results that follow from Principle *P*.

PROBABILITY PROPERTIES

a) If an event E cannot occur, then $P(E) = 0$.
b) If an event E is certain to occur, then $P(E) = 1$.
c) The probability that an event E will occur is a number from 0 to 1: $0 \leq P(E) \leq 1$.

For example, in coin tossing, the event that a coin will land on its edge has probability 0. The event that a coin falls either heads or tails has probability 1.

In the following examples, we use the combinatorics that we studied in Sections 12.5 and 12.6 to calculate theoretical probabilities.

EXAMPLE 8 Suppose that 2 cards are drawn from a well-shuffled deck of 52 cards. What is the probability that both of them are spades?

Solution The number of ways n of drawing 2 cards from a well-shuffled deck of 52 cards is $_{52}C_2$. Since 13 of the 52 cards are spades, the number of ways m of drawing 2 spades is $_{13}C_2$. Thus,

$$P(\text{drawing 2 spades}) = \frac{m}{n} = \frac{_{13}C_2}{_{52}C_2} = \frac{78}{1326} = \frac{1}{17}.$$

> **Now Try Exercise 11.**

EXAMPLE 9 Suppose that 3 people are selected at random from a group that consists of 6 men and 4 women. What is the probability that 1 man and 2 women are selected?

Solution   The number of ways of selecting 3 people from a group of 10 is $_{10}C_3$. One man can be selected in $_6C_1$ ways, and 2 women can be selected in $_4C_2$ ways. By the fundamental counting principle, the number of ways of selecting 1 man and 2 women is $_6C_1 \cdot _4C_2$. Thus the probability that 1 man and 2 women are selected is

$$P = \frac{_6C_1 \cdot _4C_2}{_{10}C_3} = \frac{3}{10}.$$

> **Now Try Exercise 13.**

EXAMPLE 10   *Rolling Two Dice.*   What is the probability of getting a total of 8 on a roll of a pair of dice?

Solution   On each die, there are 6 possible outcomes. The outcomes are paired so there are $6 \cdot 6$, or 36, possible ways in which the two can fall. (Assuming that the dice are different colors—say, one red and one blue—can help in visualizing this.)

6	(1, 6)	(2, 6)	(3, 6)	(4, 6)	(5, 6)	(6, 6)
5	(1, 5)	(2, 5)	(3, 5)	(4, 5)	(5, 5)	(6, 5)
4	(1, 4)	(2, 4)	(3, 4)	(4, 4)	(5, 4)	(6, 4)
3	(1, 3)	(2, 3)	(3, 3)	(4, 3)	(5, 3)	(6, 3)
2	(1, 2)	(2, 2)	(3, 2)	(4, 2)	(5, 2)	(6, 2)
1	(1, 1)	(2, 1)	(3, 1)	(4, 1)	(5, 1)	(6, 1)
	1	2	3	4	5	6

The pairs that total 8 are as shown in the figure above. There are 5 possible ways of getting a total of 8, so the probability is $\frac{5}{36}$.

> **Now Try Exercise 19.**

12.8   Exercise Set

1. *Select a Number.*   In a survey conducted by the authors, 100 people were polled and asked to select a number from 1 to 5. The results are shown in the following table.

Number Chosen	1	2	3	4	5
Number Who Chose That Number	18	24	23	23	12

a) What is the probability that the number chosen is 1? 2? 3? 4? 5?

b) What general conclusion might be made from the results of the experiment?

2. *Mason Dots®.*   Made by the Tootsie Industries of Chicago, Illinois, Mason Dots® is a gumdrop candy. A box was opened by the authors and was found to contain the following number of gumdrops:

Orange	9
Lemon	8
Strawberry	7
Grape	6
Lime	5
Cherry	4

If we take one gumdrop out of the box, what is the probability of getting lemon? lime? orange? grape? strawberry? licorice?

3. *Marketing via E-mail.* In the second quarter of 2013, the probability that a marketing e-mail would be opened was 28.5% (*Source:* Q2 2013 Email Trends and Benchmarks, Epsilon). A business sent a marketing e-mail to 18,200 subscribers. How many of these e-mails can the business expect will be opened?

4. *Linguistics.* An experiment was conducted by the authors to determine the relative occurrence of various letters of the English alphabet. The front page of a newspaper was considered. In all, there were 9136 letters. The number of occurrences of each letter of the alphabet is listed in the following table.

Letter	Number of Occurrences	Probability
A	853	$853/9136 \approx 9.3\%$
B	136	
C	273	
D	286	
E	1229	
F	173	
G	190	
H	399	
I	539	
J	21	
K	57	
L	417	
M	231	
N	597	
O	705	
P	238	
Q	4	
R	609	
S	745	
T	789	
U	240	
V	113	
W	127	
X	20	
Y	124	
Z	21	$21/9136 \approx 0.2\%$

a) Complete the table of probabilities with the percentage, to the nearest tenth of a percent, of the occurrence of each letter.

b) What is the probability of a vowel occurring?

c) What is the probability of a consonant occurring?

5. *Marbles.* Suppose that we select, without looking, one marble from a bag containing 4 red marbles and 10 green marbles. What is the probability of selecting each of the following?

a) A red marble   **b)** A green marble

c) A purple marble

d) A red marble or a green marble

6. *Selecting Coins.* Suppose that we select, without looking, one coin from a bag containing 5 pennies, 3 dimes, and 7 quarters. What is the probability of selecting each of the following?

a) A dime

b) A quarter

c) A nickel

d) A penny, a dime, or a quarter

7. *Rolling a Die.* What is the probability of rolling a number less than 4 on a die?

8. *Rolling a Die.* What is the probability of rolling either a 1 or a 6 on a die?

9. *Drawing a Card.* Suppose that a card is drawn from a well-shuffled deck of 52 cards. What is the probability of drawing each of the following?

a) A queen

b) An ace or a 10

c) A heart

d) A black 6

10. *Drawing a Card.* Suppose that a card is drawn from a well-shuffled deck of 52 cards. What is the probability of drawing each of the following?

a) A 7

b) A jack or a king

c) A black ace

d) A red card

11. *Drawing Cards.* Suppose that 3 cards are drawn from a well-shuffled deck of 52 cards. What is the probability that they are all aces?

12. *Drawing Cards.* Suppose that 4 cards are drawn from a well-shuffled deck of 52 cards. What is the probability that they are all red?

13. *Production Unit.* The sales force of a business consists of 10 men and 10 women. A production unit of 4 people is set up at random. What is the probability that 2 men and 2 women are chosen?

14. *Coin Drawing.* A sack contains 7 dimes, 5 nickels, and 10 quarters, and 8 coins are drawn at random. What is the probability of getting 4 dimes, 3 nickels, and 1 quarter?

Five-Card Poker Hands. *Suppose that 5 cards are drawn from a deck of 52 cards. What is the probability of drawing each of the following?*

15. 3 sevens and 2 kings

16. 5 aces

17. 5 spades

18. 4 aces and 1 five

19. *Tossing Three Coins.* Three coins are flipped. An outcome might be HTH.

    a) Find the sample space.

What is the probability of getting each of the following?

    b) Exactly one head   **c)** At most two tails
    d) At least one head   **e)** Exactly two tails

Roulette. An American roulette wheel contains 38 slots numbered 00, 0, 1, 2, 3, . . . , 35, 36. Eighteen of the slots numbered 1–36 are colored red and 18 are colored black. The 00 and 0 slots are considered to be uncolored. The wheel is spun, and a ball is rolled around the rim until it falls into a slot. What is the probability that the ball falls in each of the following?

20. A red slot   **21.** A black slot

22. The 00 slot   **23.** The 0 slot

24. Either the 00 or the 0 slot

25. A red slot or a black slot

26. The number 24

27. An odd-numbered slot

28. *Dartboard.* The following figure shows a dartboard. A dart is thrown and hits the board. Find the following probabilities.

    a) P (red)
    b) P (green)
    c) P (blue)
    d) P (yellow)

▶ Skill Maintenance

In each of Exercises 29–36, fill in the blank with the correct term. Some of the given choices will be used more than once. Others will not be used.

    range
    domain
    function
    inverse function
    composite function
    direct variation
    inverse variation
    factor
    solution
    zero
    y-intercept
    one-to-one
    rational
    permutation
    combination
    arithmetic sequence
    geometric sequence

29. A(n) _____ of a function is an input for which the output is 0. **[7.4]**

30. A function is _____ if different inputs have different outputs. **[9.2]**

31. A(n) _____ is a correspondence between a first set, called the _____, and a second set, called the _____, such that each member of the _____ corresponds to exactly one member of the _____. **[2.2]**

32. The first coordinate of an *x*-intercept of a function is a(n) _____ of the function. **[7.4]**

33. A selection made from a set without regard to order is a(n) _____. **[12.6]**

34. If we have a function $f(x) = k/x$, where k is a positive constant, we have _____. **[5.8]**

35. For a polynomial function $f(x)$, if $f(c) = 0$, then $x - c$ is a(n) _____ of the polynomial. **[8.3]**

36. We have $\dfrac{a_{n+1}}{a_n} = r$, for any integer $n \geq 1$, in a(n) _____. **[12.3]**

▶ **Synthesis**

Five-Card Poker Hands. *Suppose that 5 cards are drawn from a deck of 52 cards. For the following exercises, give both a reasoned expression and an answer.*

37. *Two Pairs.* A hand with *two pairs* is a hand like Q-Q-3-3-A.

 a) How many are there?
 b) What is the probability of getting two pairs?

38. *Full House.* A *full house* consists of 3 of a kind and a pair such as Q-Q-Q-4-4.

 a) How many full houses are there?
 b) What is the probability of getting a full house?

39. *Three of a Kind.* A *three-of-a-kind* is a 5-card hand in which exactly 3 of the cards are of the same denomination and the other 2 are not a pair, such as Q-Q-Q-10-7.

 a) How many three-of-a-kind hands are there?
 b) What is the probability of getting three of a kind?

40. *Four of a Kind.* A *four-of-a-kind* is a 5-card hand in which exactly 4 of the cards are of the same denomination, such as J-J-J-J-6, 7-7-7-7-A, or 2-2-2-2-5.

 a) How many four-of-a-kind hands are there?
 b) What is the probability of getting four of a kind?

Chapter 12 Summary and Review

STUDY GUIDE

KEY TERMS AND CONCEPTS	EXAMPLES

SECTION 12.1: SEQUENCES AND SERIES

An **infinite sequence** is a function having for its domain the set of positive integers $\{1, 2, 3, 4, 5, \dots\}$.

A **finite sequence** is a function having for its domain a set of positive integers $\{1, 2, 3, 4, 5, \dots, n\}$ for some positive integer n.

The first four terms of the sequence whose general term is given by $a_n = 3n + 2$ are

$$a_1 = 3 \cdot 1 + 2 = 5,$$
$$a_2 = 3 \cdot 2 + 2 = 8,$$
$$a_3 = 3 \cdot 3 + 2 = 11, \quad \text{and}$$
$$a_4 = 3 \cdot 4 + 2 = 14.$$

The sum of the terms of an infinite sequence is an **infinite series**. A **partial sum** is the sum of the first n terms. It is also called a **finite series** or the **nth partial sum** and is denoted S_n.

For the sequence above, $S_4 = 5 + 8 + 11 + 14$, or 38. We can denote this sum using **sigma notation** as

$$\sum_{k=1}^{4} (3k + 2).$$

A sequence can be defined **recursively** by listing the first term, or the first few terms, and then using a **recursion formula** to determine the remaining terms from the given term.

The first four terms of the recursively defined sequence

$$a_1 = 3, \ a_{n+1} = (a_n - 1)^2$$

are

$$a_1 = 3,$$
$$a_2 = (a_1 - 1)^2 = (3 - 1)^2 = 4,$$
$$a_3 = (a_2 - 1)^2 = (4 - 1)^2 = 9, \quad \text{and}$$
$$a_4 = (a_3 - 1)^2 = (9 - 1)^2 = 64.$$

SECTION 12.2: ARITHMETIC SEQUENCES AND SERIES

For an arithmetic sequence:

$a_{n+1} = a_n + d;$ **d is the common difference.**

$a_n = a_1 + (n-1)d;$ **The nth term**

$S_n = \dfrac{n}{2}(a_1 + a_n).$ **The sum of the first n terms**

For the arithmetic sequence 5, 8, 11, 14, . . . :

$a_1 = 5;$

$d = 3$ $(8 - 5 = 3, 11 - 8 = 3,$ and so on$);$

$a_6 = 5 + (6-1)3 = 5 + 15 = 20;$

$S_6 = \dfrac{6}{2}(5 + 20) = 3(25) = 75.$

SECTION 12.3: GEOMETRIC SEQUENCES AND SERIES

For a geometric sequence:

$a_{n+1} = a_n r$ **r is the common ratio.**

$a_n = a_1 r^{n-1};$ **The nth term**

$S_n = \dfrac{a_1(1 - r^n)}{1 - r};$ **The sum of the first n terms**

$S_\infty = \dfrac{a_1}{1 - r}, \quad |r| < 1.$ **The limit of an infinite geometric series**

For the geometric sequence $12, -6, 3, -\frac{3}{2}, \dots$:

$a_1 = 12;$

$r = -\dfrac{1}{2}$ $\left(\dfrac{-6}{12} = -\dfrac{1}{2}, \dfrac{3}{-6} = -\dfrac{1}{2},$ and so on$\right);$

$a_6 = 12\left(-\dfrac{1}{2}\right)^{6-1} = 12\left(-\dfrac{1}{2^5}\right) = -\dfrac{3}{8};$

$S_6 = \dfrac{12\left[1 - \left(-\frac{1}{2}\right)^6\right]}{1 - \left(-\frac{1}{2}\right)} = \dfrac{12\left(1 - \frac{1}{64}\right)}{\frac{3}{2}} = \dfrac{63}{8};$

$|r| = \left|-\frac{1}{2}\right| = \frac{1}{2} < 1,$ so we have

$S_\infty = \dfrac{12}{1 - \left(-\frac{1}{2}\right)} = \dfrac{12}{\frac{3}{2}} = 8.$

SECTION 12.4: MATHEMATICAL INDUCTION

The Principle of Mathematical Induction

We can prove an infinite sequence of statements S_n by showing the following.

(1) *Basis step.* S_1 is true.

(2) *Induction step.* For all natural numbers k, $S_k \rightarrow S_{k+1}.$

See Examples 1–3 on pp. 795–797.

SECTION 12.5: COMBINATORICS: PERMUTATIONS

The Fundamental Counting Principle

Given a combined action, or *event*, in which the first action can be performed in n_1 ways, the second action can be performed in n_2 ways, and so on, the total number of ways in which the combined action can be performed is the product

$$n_1 \cdot n_2 \cdot n_3 \cdots \cdots n_k.$$

The product $n(n-1)(n-2) \cdots 3 \cdot 2 \cdot 1,$ for any natural number $n,$ can also be written in **factorial notation** as $n!.$ For the number 0, $0! = 1.$

(continued)

The total number of permutations, or ordered arrangements, of n objects, denoted $_nP_n$, is given by

$$_nP_n = n(n-1)(n-2)\cdots 3\cdot 2\cdot 1, \quad \text{or} \quad n!.$$

In how many ways can 7 books be arranged in a straight line?

We have

$$_7P_7 = 7! = 7\cdot 6\cdot 5\cdot 4\cdot 3\cdot 2\cdot 1 = 5040.$$

The Number of Permutations of n Objects Taken k at a Time

$$_nP_k = n(n-1)(n-2)\cdots [n-(k-1)] \quad \textbf{(1)}$$

$$= \frac{n!}{(n-k)!} \quad \textbf{(2)}$$

Compute $_7P_4$.

Using form (1), we have

$$_7P_4 = 7\cdot 6\cdot 5\cdot 4 = 840.$$

Using form (2), we have

$$_7P_4 = \frac{7!}{(7-4)!} = \frac{7\cdot 6\cdot 5\cdot 4\cdot 3!}{3!}$$

$$= \frac{7\cdot 6\cdot 5\cdot 4\cdot 3!}{3!} = 840.$$

The number of distinct arrangements of n objects taken k at a time, allowing repetition, is n^k.

The number of 4-number code symbols that can be formed with the numbers 5, 6, 7, 8, and 9, if we allow a number to occur more than once, is 5^4, or 625.

For a set of n objects in which n_1 are of one kind, n_2 are of another kind, . . . , and n_k are of a kth kind, the number of distinguishable permutations is

$$\frac{n!}{n_1!\cdot n_2!\cdot \cdots \cdot n_k!}.$$

Find the number of distinguishable code symbols that can be formed using the letters in the word MISSISSIPPI.

There are 1 M, 4 I's, 4 S's, and 2 P's, for a total of 11 letters, so we have

$$\frac{11!}{1!\,4!\,4!\,2!}, \quad \text{or} \quad 34{,}650.$$

SECTION 12.6: COMBINATORICS: COMBINATIONS

The Number of Combinations of n Objects Taken k at a Time

$$_nC_k = \frac{n!}{k!(n-k)!} \quad \textbf{(1)}$$

$$= \frac{_nP_k}{k!}$$

$$= \frac{n(n-1)(n-2)\cdots [n-(k-1)]}{k!}. \quad \textbf{(2)}$$

We can also use **binomial coefficient notation**:

$$\binom{n}{k} = {_nC_k}.$$

Compute: $_6C_4$, or $\binom{6}{4}$.

Using form (1), we have

$$\binom{6}{4} = \frac{6!}{4!(6-4)!} = \frac{6!}{4!\,2!}$$

$$= \frac{6\cdot 5\cdot 4!}{4!\,2!} = \frac{6\cdot 5\cdot 4!}{4!\cdot 2\cdot 1} = 15.$$

Using form (2), we have

$$\binom{6}{4} = \frac{_6P_4}{4!} = \frac{6\cdot 5\cdot 4\cdot 3}{4\cdot 3\cdot 2\cdot 1} = 15.$$

SECTION 12.7: THE BINOMIAL THEOREM

The Binomial Theorem Using Pascal's Triangle

For any binomial $a + b$ and any natural number n,

$$(a + b)^n = c_0 a^n b^0 + c_1 a^{n-1} b^1 + c_2 a^{n-2} b^2$$
$$+ \cdots + c_{n-1} a^1 b^{n-1} + c_n a^0 b^n,$$

where the numbers $c_0, c_1, c_2, \ldots, c_{n-1}, c_n$ are from the $(n + 1)$st row of Pascal's triangle. (See Pascal's triangle on p. 815.)

Expand: $(x - 2)^4$.

We have $a = x$, $b = -2$, and $n = 4$. We use the fourth row of Pascal's triangle.

$$(x - 2)^4 = 1 \cdot x^4 + 4 \cdot x^3 (-2)^1$$
$$+ 6 \cdot x^2 (-2)^2 + 4 \cdot x^1 (-2)^3 + 1(-2)^4$$
$$= x^4 + 4x^3 (-2) + 6x^2 \cdot 4 + 4x(-8) + 16$$
$$= x^4 - 8x^3 + 24x^2 - 32x + 16$$

The Binomial Theorem Using Factorial Notation

For any binomial $a + b$ and any natural number n,

$$(a + b)^n = \binom{n}{0} a^n b^0 + \binom{n}{1} a^{n-1} b^1$$
$$+ \binom{n}{2} a^{n-2} b^2 + \cdots$$
$$+ \binom{n}{n-1} a^1 b^{n-1} + \binom{n}{n} a^0 b^n$$
$$= \sum_{k=0}^{n} \binom{n}{k} a^{n-k} b^k.$$

Expand: $(x^2 + 3)^3$.

We have $a = x^2$, $b = 3$, and $n = 3$.

$$(x^2 + 3)^3 = \binom{3}{0} (x^2)^3 + \binom{3}{1} (x^2)^2 (3)$$
$$+ \binom{3}{2} (x^2) 3^2 + \binom{3}{3} 3^3$$
$$= \frac{3!}{0! \, 3!} x^6 + \frac{3!}{1! \, 2!} (x^4)(3) + \frac{3!}{2! \, 1!} (x^2)(9)$$
$$+ \frac{3!}{3! \, 0!} (27)$$
$$= 1 \cdot x^6 + 3 \cdot 3x^4 + 3 \cdot 9x^2 + 1 \cdot 27$$
$$= x^6 + 9x^4 + 27x^2 + 27$$

The $(k + 1)$st term of $(a + b)^n$ is
$$\binom{n}{k} a^{n-k} b^k.$$

The third term of $(x^2 + 3)^3$ is

$$\binom{3}{2} (x^2)^{3-2} \cdot 3^2 = 3 \cdot x^2 \cdot 9 = 27x^2. \quad (k = 2)$$

The **total number of subsets** of a set with n elements is 2^n.

How many subsets does the set $\{W, X, Y, Z\}$ have?

The set has 4 elements, so we have

$$2^4, \quad \text{or} \quad 16.$$

SECTION 12.8: PROBABILITY

Principle P (Experimental)

Given an experiment in which n observations are made, if a situation, or event, E occurs m times out of n observations, then we say that the **experimental probability** of the event, $P(E)$, is given by

$$P(E) = \frac{m}{n}.$$

From a batch of 1000 gears, 35 were found to be defective. The probability that a defective gear is produced is

$$\frac{35}{1000} = 0.035, \quad \text{or} \quad 3.5\%.$$

Principle P (Theoretical)

If an event E can occur m ways out of n possible equally likely outcomes of a sample space S, then the **theoretical probability** of the event, $P(E)$, is given by

$$P(E) = \frac{m}{n}.$$

What is the probability of drawing 2 red marbles and 1 green marble from a bag containing 5 red marbles, 6 green marbles, and 4 white marbles?

Number of ways of drawing 3 marbles from a bag of 15: $_{15}C_3$

Number of ways of drawing 2 red marbles from 5 red marbles: $_5C_2$

Number of ways of drawing 1 green marble from 6 green marbles: $_6C_1$

Probability that 2 red marbles and 1 green marble are drawn:

$$\frac{_5C_2 \cdot {_6}C_1}{_{15}C_3} = \frac{10 \cdot 6}{455} = \frac{12}{91}$$

REVIEW EXERCISES

Determine whether the statement is true or false.

1. A sequence is a function. **[12.1]**

2. An infinite geometric series with $r = -1$ has a limit. **[12.3]**

3. Permutations involve order and arrangements of objects. **[12.5]**

4. The total number of subsets of a set with n elements is n^2. **[12.7]**

5. Find the first 4 terms, a_{11}, and a_{23}:

$$a_n = (-1)^n \left(\frac{n^2}{n^4 + 1} \right). \text{ [12.1]}$$

6. Predict the general, or nth, term. Answers may vary.
   $2, -5, 10, -17, 26, \ldots$ **[12.1]**

7. Find and evaluate:

$$\sum_{k=1}^{4} \frac{(-1)^{k+1}3^k}{3^k - 1}. \text{ [12.1]}$$

8. Write sigma notation. Answers may vary.
   $0 + 3 + 8 + 15 + 24 + 35 + 48$ **[12.1]**

9. Find the 10th term of the arithmetic sequence
   $\frac{3}{4}, \frac{13}{12}, \frac{17}{12}, \ldots.$ **[12.2]**

10. Find the 6th term of the arithmetic sequence
    $a - b, a, a + b, \ldots.$ **[12.2]**

11. Find the sum of the first 18 terms of the arithmetic sequence
    $4, 7, 10, \ldots.$ **[12.2]**

12. Find the sum of the first 200 natural numbers. **[12.2]**

13. The 1st term in an arithmetic sequence is 5, and the 17th term is 53. Find the 3rd term. **[12.2]**

14. The common difference in an arithmetic sequence is 3. The 10th term is 23. Find the first term. **[12.2]**

15. For a geometric sequence, $a_1 = -2, r = 2$, and $a_n = -64$. Find n and S_n. **[12.3]**

16. For a geometric sequence, $r = \frac{1}{2}$ and $S_5 = \frac{31}{2}$. Find a_1 and a_5. **[12.3]**

Find the sum, if it exists, of each infinite geometric series. **[12.3]**

17. $25 + 27.5 + 30.25 + 33.275 + \cdots$

18. $0.27 + 0.0027 + 0.000027 + \cdots$

19. $\frac{1}{2} - \frac{1}{6} + \frac{1}{18} - \cdots$

20. Find fraction notation for $2.\overline{43}$. **[12.3]**

21. Insert four arithmetic means between 5 and 9. **[12.2]**

22. *Bouncing Golfball.* A golfball is dropped to the pavement from a height of 30 ft. It always rebounds three-fourths of the distance that it drops. How far (up and down) will the ball have traveled when it hits the pavement for the 6th time? **[12.3]**

23. *The Amount of an Annuity.* To create a college fund, a parent makes a sequence of 18 yearly deposits of $2000 each in a savings account on which interest is compounded annually at 2.8%. Find the amount of the annuity. **[12.3]**

24. *Total Gift.* Suppose you receive 10¢ on the first day of the year, 12¢ on the 2nd day, 14¢ on the 3rd day, and so on.

   a) How much will you receive on the 365th day? **[12.2]**

   b) What is the sum of these 365 gifts? **[12.2]**

25. *The Economic Multiplier.* Suppose that the government is making a $24,000,000,000 expenditure for travel to Mars. If 73% of this amount is spent again, and so on, what is the total effect on the economy? **[12.3]**

Use mathematical induction to prove each of the following. **[12.4]**

26. For every natural number n,

$$1 + 4 + 7 + \cdots + (3n - 2) = \frac{n(3n - 1)}{2}.$$

27. For every natural number n,

$$1 + 3 + 3^2 + \cdots + 3^{n-1} = \frac{3^n - 1}{2}.$$

28. For every natural number $n \geq 2$,

$$\left(1 - \frac{1}{2}\right)\left(1 - \frac{1}{3}\right)\cdots\left(1 - \frac{1}{n}\right) = \frac{1}{n}.$$

29. *Book Arrangements.* In how many ways can 6 books be arranged on a shelf? **[12.5]**

30. *Flag Displays.* If 9 different signal flags are available, how many different displays are possible using 4 flags in a row? **[12.5]**

31. *Prize Choices.* The winner of a contest can choose any 8 of 15 prizes. How many different sets of prizes can be chosen? **[12.6]**

32. *Fraternity–Sorority Names.* The Greek alphabet contains 24 letters. How many fraternity or sorority names can be formed using 3 different letters? **[12.5]**

33. *Letter Arrangements.* In how many distinguishable ways can the letters of the word TENNESSEE be arranged? **[12.5]**

34. *Floor Plans.* A manufacturer of houses has 1 floor plan but achieves variety by having 3 different roofs, 4 different ways of attaching the garage, and 3 different types of entrances. Find the number of different houses that can be produced. **[12.5]**

35. *Code Symbols.* How many code symbols can be formed using 5 out of 6 of the letters of G, H, I, J, K, L if the letters:

   a) cannot be repeated? **[12.5]**

   b) can be repeated? **[12.5]**

   c) cannot be repeated but must begin with K? **[12.5]**

   d) cannot be repeated but must end with IGH? **[12.5]**

36. Determine the number of subsets of a set containing 8 members. **[12.7]**

Expand. **[12.7]**

37. $(m + n)^7$

38. $\left(x - \sqrt{2}\right)^5$

39. $(x^2 - 3y)^4$

40. $\left(a + \dfrac{1}{a}\right)^8$

41. $(1 + 5i)^6$, where $i^2 = -1$

42. Find the 4th term of $(a + x)^{12}$. **[12.7]**

43. Find the 12th term of $(2a - b)^{18}$. Do not multiply out the factorials. **[12.7]**

44. *Rolling Dice.* What is the probability of getting a 10 on a roll of a pair of dice? on a roll of 1 die? **[12.8]**

45. *Drawing a Card.* From a deck of 52 cards, 1 card is drawn at random. What is the probability that it is a club? **[12.8]**

46. *Drawing Three Cards.* From a deck of 52 cards, 3 are drawn at random without replacement. What is the probability that 2 are aces and 1 is a king? **[12.8]**

47. *Election Poll.* Three people were running for mayor in an election campaign. A poll was conducted to see which candidate was favored. During the polling, 86 favored candidate A, 97 favored B, and 23 favored C. Assuming that the poll is a valid indicator of the election results, what is the probability that the election will be won by A? B? C? **[12.8]**

48. Which of the following is the 25th term of the arithmetic sequence 12, 10, 8, 6, . . . ? **[12.2]**

   A. −38   **B.** −36

   C. 32   **D.** 60

49. What is the probability of getting a total of 4 on a roll of a pair of dice? **[12.8]**

   A. $\frac{1}{12}$ **B.** $\frac{1}{9}$

   C. $\frac{1}{6}$ **D.** $\frac{5}{36}$

50. The graph of the sequence whose general term is $a_n = n - 1$ is which of the following? **[12.1]**

   A. 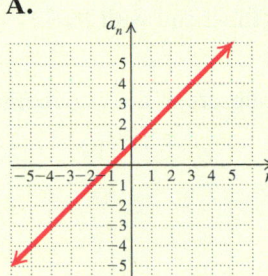   **B.**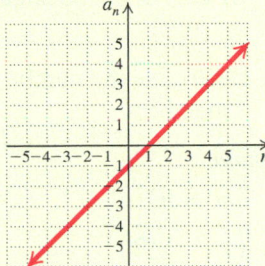

   C. 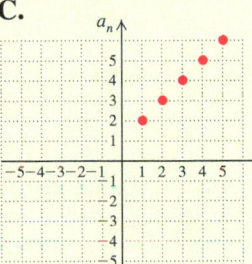   **D.**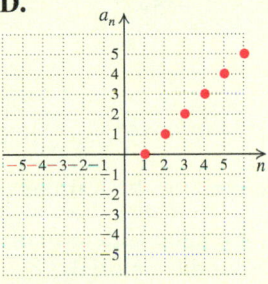

► **Synthesis**

51. Suppose that $a_1, a_2, \ldots, a_n$ is an arithmetic sequence. Is $b_1, b_2, \ldots, b_n$ an arithmetic sequence if:

   a) $b_n = |a_n|$? **[12.2]**

   b) $b_n = a_n + 8$? **[12.2]**

   c) $b_n = 7a_n$? **[12.2]**

   d) $b_n = \dfrac{1}{a_n}$? **[12.2]**

   e) $b_n = \log a_n$? **[12.2]**

   f) $b_n = a_n^3$? **[12.2]**

52. Suppose that $a_1, a_2, \ldots, a_n$ and $b_1, b_2, \ldots, b_n$ are geometric sequences. Prove that $c_1, c_2, \ldots, c_n$ is a geometric sequence, where $c_n = a_n b_n$. **[12.3]**

53. Write the first 3 terms of the infinite geometric series with $r = -\frac{1}{3}$ and $S_\infty = \frac{3}{8}$. **[12.3]**

54. The zeros of this polynomial function form an arithmetic sequence. Find them. **[12.2]**

$$f(x) = x^4 - 4x^3 - 4x^2 + 16x$$

55. Simplify:

$$\sum_{k=0}^{10} (-1)^k \binom{10}{k} (\log x)^{10-k} (\log y)^k. \quad \textbf{[12.6]}$$

Solve for n. **[12.6]**

56. $\dbinom{n}{6} = 3 \cdot \dbinom{n-1}{5}$

57. $\dbinom{n}{n-1} = 36$

58. Solve for a:

$$\sum_{k=0}^{5} \binom{5}{k} 9^{5-k} a^k = 0. \quad \textbf{[12.7]}$$

► **Collaborative Discussion and Writing**

59. *Circular Arrangements.* In how many ways can the numbers on a clock face be arranged? See if you can derive a formula for the number of distinct circular arrangements of *n* objects. Explain your reasoning. **[12.5]**

60. How "long" is 15!? Suppose you own 15 books and decide to make up all the possible arrangements of the books on a shelf. About how long, in years, would it take you if you were to make one arrangement per second? Write out the reasoning you used for this problem in the form of a paragraph. **[12.5]**

61. Explain why a "combination" lock should really be called a "permutation" lock. **[12.6]**

62. Give the reasoning that you might use with a fellow student to explain that

$$\binom{n}{k} = \binom{n}{n-k}. \quad \textbf{[12.6]}$$

1. For the sequence whose nth term is $a_n = (-1)^n(2n + 1)$, find a_{21}.

2. Find the first 5 terms of the sequence with general term
$$a_n = \frac{n + 1}{n + 2}.$$

3. Find and evaluate:
$$\sum_{k=1}^{4} (k^2 + 1).$$

Write sigma notation. Answers may vary.

4. $4 + 8 + 12 + 16 + 20 + 24$

5. $2 + 4 + 8 + 16 + 32 + \cdots$

6. Find the first 4 terms of the recursively defined sequence
$$a_1 = 3, \quad a_{n+1} = 2 + \frac{1}{a_n}.$$

7. Find the 15th term of the arithmetic sequence $2, 5, 8, \ldots$.

8. The 1st term of an arithmetic sequence is 8 and the 21st term is 108. Find the 7th term.

9. Find the sum of the first 20 terms of the series $17 + 13 + 9 + \cdots$.

10. Find the sum: $\sum_{k=1}^{25} (2k + 1)$.

11. Find the 11th term of the geometric sequence $10, -5, \frac{5}{2}, -\frac{5}{4}, \ldots$.

12. For a geometric sequence, $r = 0.2$ and $S_4 = 1248$. Find a_1.

Find the sum, if it exists.

13. $\sum_{k=1}^{8} 2^k$

14. $18 + 6 + 2 + \cdots$

15. Find fraction notation for $0.\overline{56}$.

16. *Salvage Value.* The value of an office machine is $10,000. Its salvage value each year is 80% of its value the year before. Give a sequence that lists the salvage value of the machine for each year of a 6-year period.

17. *Hourly Wage.* William accepts a job, starting with an hourly wage of $12.25, and is promised a raise of 30¢ per hour every three months for 4 years. What will William's hourly wage be at the end of the 4-year period?

18. *Amount of an Annuity.* To create a college fund, a parent makes a sequence of 18 equal yearly deposits of $2500 in a savings account on which interest is compounded annually at 5.6%. Find the amount of the annuity.

19. Use mathematical induction to prove that, for every natural number n,
$$2 + 5 + 8 + \cdots + (3n - 1) = \frac{n(3n + 1)}{2}.$$

Evaluate.

20. $_{15}P_6$

21. $_{21}C_{10}$

22. $\binom{n}{4}$

23. How many 4-digit numbers can be formed using the digits 1, 3, 5, 6, 7, and 9 without repetition?

24. How many code symbols can be formed using 4 of the 6 letters A, B, C, X, Y, Z if the letters:
   a) can be repeated?
   b) are not repeated and must begin with Z?

25. *Scuba Club Officers.* The Bay Woods Scuba Club has 28 members. How many sets of 4 officers can be selected from this group?

26. *Test Options.* On a test with 20 questions, a student must answer 8 of the first 12 questions and 4 of the last 8. In how many ways can this be done?

27. Expand: $(x + 1)^5$.

28. Find the 5th term of the binomial expansion $(x - y)^7$.

29. Determine the number of subsets of a set containing 9 members.

30. *Marbles.* Suppose that we select, without looking, one marble from a bag containing 6 red marbles and 8 blue marbles. What is the probability of selecting a blue marble?

31. *Drawing Coins.* Ethan has 6 pennies, 5 dimes, and 4 quarters in his pocket. Six coins are drawn at random. What is the probability of getting 1 penny, 2 dimes, and 3 quarters?

32. The graph of the sequence whose general term is $a_n = 2n - 2$ is which of the following?

A.

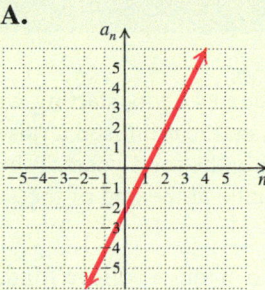

B.

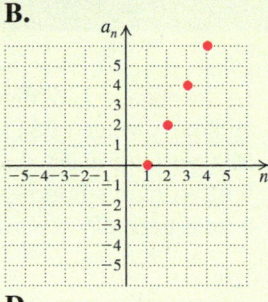

C.

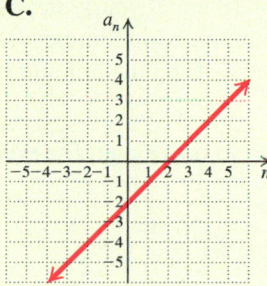

D.

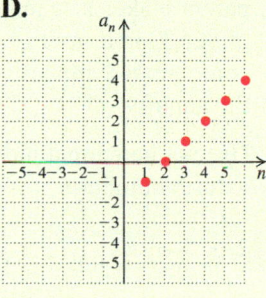

▶ **Synthesis**

33. Solve for n: $_nP_7 = 9 \cdot {_nP_6}$.

Appendix: Partial Fractions

▶ Decompose rational expressions into partial fractions.

There are situations in calculus in which it is useful to write a rational expression as a sum of two or more simpler rational expressions. In the equation

$$\frac{4x - 13}{2x^2 + x - 6} = \frac{3}{x + 2} + \frac{-2}{2x - 3},$$

for example, each fraction on the right side is called a **partial fraction**. The expression on the right side is the **partial fraction decomposition** of the rational expression on the left side. In this section, we learn how such decompositions are created.

▶ Partial Fraction Decompositions

The procedure for finding the partial fraction decomposition of a rational expression involves factoring its denominator into linear factors and quadratic factors.

Procedure for Decomposing a Rational Expression into Partial Fractions

Consider any rational expression $P(x)/Q(x)$ such that $P(x)$ and $Q(x)$ have no common factor other than 1 or -1.

1. If the degree of $P(x)$ is greater than or equal to the degree of $Q(x)$, divide to express $P(x)/Q(x)$ as a quotient + remainder/$Q(x)$ and follow steps (2)–(5) to decompose the resulting rational expression.
2. If the degree of $P(x)$ is less than the degree of $Q(x)$, factor $Q(x)$ into linear factors of the form $(px + q)^n$ and/or quadratic factors of the form $(ax^2 + bx + c)^m$. Any quadratic factor $ax^2 + bx + c$ must be *irreducible*, meaning that it cannot be factored into linear factors with rational coefficients.
3. Assign to each linear factor $(px + q)^n$ the sum of n partial fractions:

$$\frac{A_1}{px + q} + \frac{A_2}{(px + q)^2} + \cdots + \frac{A_n}{(px + q)^n}.$$

4. Assign to each quadratic factor $(ax^2 + bx + c)^m$ the sum of m partial fractions:

$$\frac{B_1 x + C_1}{ax^2 + bx + c} + \frac{B_2 x + C_2}{(ax^2 + bx + c)^2} + \cdots + \frac{B_m x + C_m}{(ax^2 + bx + c)^m}.$$

5. Apply algebraic methods, as illustrated in the following examples, to find the constants in the numerators of the partial fractions.

EXAMPLE 1 Decompose into partial fractions:

$$\frac{4x - 13}{2x^2 + x - 6}.$$

Solution The degree of the numerator is less than the degree of the denominator. We begin by factoring the denominator: $(x + 2)(2x - 3)$. We find constants A and B such that

$$\frac{4x - 13}{(x + 2)(2x - 3)} = \frac{A}{x + 2} + \frac{B}{2x - 3}.$$

To determine A and B, we add the expressions on the right:

$$\frac{4x - 13}{(x + 2)(2x - 3)} = \frac{A(2x - 3) + B(x + 2)}{(x + 2)(2x - 3)}.$$

Next, we equate the numerators:

$$4x - 13 = A(2x - 3) + B(x + 2).$$

Since the last equation containing A and B is true for all x, we can substitute any value of x and still have a true equation. If we choose $x = \frac{3}{2}$, then $2x - 3 = 0$ and A will be eliminated when we make the substitution. This gives us

$$4\left(\tfrac{3}{2}\right) - 13 = A\left(2 \cdot \tfrac{3}{2} - 3\right) + B\left(\tfrac{3}{2} + 2\right)$$
$$-7 = 0 + \tfrac{7}{2}B.$$

Solving, we obtain $B = -2$.

If we choose $x = -2$, then $x + 2 = 0$ and B will be eliminated when we make the substitution. This gives us

$$4(-2) - 13 = A[2(-2) - 3] + B(-2 + 2)$$
$$-21 = -7A + 0.$$

Solving, we obtain $A = 3$.

The decomposition is as follows:

$$\frac{4x - 13}{2x^2 + x - 6} = \frac{3}{x + 2} + \frac{-2}{2x - 3}, \quad \text{or} \quad \frac{3}{x + 2} - \frac{2}{2x - 3}.$$

To check, we can add to see if we get the expression on the left.

Now Try Exercise 3.

Technology Connection

We can use the TABLE feature on a graphing calculator to check a partial fraction decomposition. To check the decomposition in Example 1, we compare values of

$$y_1 = \frac{4x - 13}{2x^2 + x - 6}$$

and

$$y_2 = \frac{3}{x + 2} - \frac{2}{2x - 3}$$

for the same values of x. Since $y_1 = y_2$ for the given values of x as we scroll through the table, the decomposition appears to be correct.

X	Y₁	Y₂
−1	3.4	3.4
0	2.1667	2.1667
1	3	3
2	−1.25	−1.25
3	−.0667	−.0667
4	.1	.1
5	.14286	.14286

X = −1

EXAMPLE 2 Decompose into partial fractions:

$$\frac{7x^2 - 29x + 24}{(2x - 1)(x - 2)^2}.$$

Solution The degree of the numerator is 2 and the degree of the denominator is 3, so the degree of the numerator is less than the degree of the denominator. The denominator is given in factored form. The decomposition has the following form:

$$\frac{7x^2 - 29x + 24}{(2x - 1)(x - 2)^2} = \frac{A}{2x - 1} + \frac{B}{x - 2} + \frac{C}{(x - 2)^2}.$$

As in Example 1, we add the expressions on the right:

$$\frac{7x^2 - 29x + 24}{(2x - 1)(x - 2)^2} = \frac{A(x - 2)^2 + B(2x - 1)(x - 2) + C(2x - 1)}{(2x - 1)(x - 2)^2}.$$

Then we equate the numerators. This gives us

$$7x^2 - 29x + 24 = A(x - 2)^2 + B(2x - 1)(x - 2) + C(2x - 1).$$

Since the equation containing A, B, and C is true for all x, we can substitute any value of x and still have a true equation. In order to have $2x - 1 = 0$, we let $x = \frac{1}{2}$. This gives us

$$7\left(\tfrac{1}{2}\right)^2 - 29 \cdot \tfrac{1}{2} + 24 = A\left(\tfrac{1}{2} - 2\right)^2 + 0 + 0$$
$$\tfrac{45}{4} = \tfrac{9}{4}A.$$

Solving, we obtain $A = 5$.

In order to have $x - 2 = 0$, we let $x = 2$. Substituting gives us

$$7(2)^2 - 29(2) + 24 = 0 + 0 + C(2 \cdot 2 - 1)$$
$$-6 = 3C.$$

Solving, we obtain $C = -2$.

To find B, we choose any value for x except $\frac{1}{2}$ or 2 and replace A with 5 and C with -2. We let $x = 1$:

$$7 \cdot 1^2 - 29 \cdot 1 + 24 = 5(1 - 2)^2 + B(2 \cdot 1 - 1)(1 - 2)$$
$$+ (-2)(2 \cdot 1 - 1)$$
$$2 = 5 - B - 2$$
$$B = 1.$$

The decomposition is as follows:

$$\frac{7x^2 - 29x + 24}{(2x - 1)(x - 2)^2} = \frac{5}{2x - 1} + \frac{1}{x - 2} - \frac{2}{(x - 2)^2}.$$

> **Now Try Exercise 7.**

POLYNOMIAL DIVISION

REVIEW SECTION 8.3

EXAMPLE 3   Decompose into partial fractions:

$$\frac{6x^3 + 5x^2 - 7}{3x^2 - 2x - 1}.$$

Solution   The degree of the numerator is greater than that of the denominator. Therefore, we divide and find an equivalent expression:

$$
\begin{array}{r}
2x + 3 \\
3x^2 - 2x - 1\overline{)6x^3 + 5x^2 - 7} \\
\underline{6x^3 - 4x^2 - 2x} \\
9x^2 + 2x - 7 \\
\underline{9x^2 - 6x - 3} \\
8x - 4
\end{array}
$$

The original expression is thus equivalent to

$$2x + 3 + \frac{8x - 4}{3x^2 - 2x - 1}.$$

We decompose the fraction to get

$$\frac{8x - 4}{(3x + 1)(x - 1)} = \frac{5}{3x + 1} + \frac{1}{x - 1}.$$

The final result is

$$2x + 3 + \frac{5}{3x + 1} + \frac{1}{x - 1}.$$

> **Now Try Exercise 17.**

Systems of equations can also be used to decompose rational expressions. Let's reconsider Example 2.

EXAMPLE 4 Decompose into partial fractions:

$$\frac{7x^2 - 29x + 24}{(2x - 1)(x - 2)^2}.$$

Solution The decomposition has the following form:

$$\frac{A}{2x - 1} + \frac{B}{x - 2} + \frac{C}{(x - 2)^2}.$$

We first add as in Example 2:

$$\frac{7x^2 - 29x + 24}{(2x - 1)(x - 2)^2} = \frac{A}{2x - 1} + \frac{B}{x - 2} + \frac{C}{(x - 2)^2}$$

$$= \frac{A(x - 2)^2 + B(2x - 1)(x - 2) + C(2x - 1)}{(2x - 1)(x - 2)^2}.$$

Then we equate numerators:

$$7x^2 - 29x + 24$$
$$= A(x - 2)^2 + B(2x - 1)(x - 2) + C(2x - 1)$$
$$= A(x^2 - 4x + 4) + B(2x^2 - 5x + 2) + C(2x - 1)$$
$$= Ax^2 - 4Ax + 4A + 2Bx^2 - 5Bx + 2B + 2Cx - C,$$

or, combining like terms,

$$7x^2 - 29x + 24$$
$$= (A + 2B)x^2 + (-4A - 5B + 2C)x + (4A + 2B - C).$$

Next, we equate corresponding coefficients:

$$7 = A + 2B,$$ **The coefficients of the x^2-terms must be the same.**
$$-29 = -4A - 5B + 2C,$$ **The coefficients of the x-terms must be the same.**
$$24 = 4A + 2B - C.$$ **The constant terms must be the same.**

> **SYSTEMS OF EQUATIONS IN THREE VARIABLES**
>
> REVIEW SECTION 3.5, 10.3, OR 10.4

We now have a system of three equations. You should confirm that the solution of the system is

$$A = 5, \quad B = 1, \quad \text{and} \quad C = -2.$$

The decomposition is as follows:

$$\frac{7x^2 - 29x + 24}{(2x - 1)(x - 2)^2} = \frac{5}{2x - 1} + \frac{1}{x - 2} - \frac{2}{(x - 2)^2}.$$

> **Now Try Exercise 15.**

EXAMPLE 5 Decompose into partial fractions:

$$\frac{11x^2 - 8x - 7}{(2x^2 - 1)(x - 3)}.$$

Solution The decomposition has the following form:

$$\frac{11x^2 - 8x - 7}{(2x^2 - 1)(x - 3)} = \frac{Ax + B}{2x^2 - 1} + \frac{C}{x - 3}.$$

Adding and equating numerators, we get

$$11x^2 - 8x - 7 = (Ax + B)(x - 3) + C(2x^2 - 1)$$
$$= Ax^2 - 3Ax + Bx - 3B + 2Cx^2 - C,$$

or $\quad 11x^2 - 8x - 7 = (A + 2C)x^2 + (-3A + B)x + (-3B - C).$

We then equate corresponding coefficients:

$11 = A + 2C,$ **The coefficients of the x^2-terms**

$-8 = -3A + B,$ **The coefficients of the x-terms**

$-7 = -3B - C.$ **The constant terms**

We solve this system of three equations and obtain

$A = 3,$ $B = 1,$ and $C = 4.$

The decomposition is as follows:

$$\frac{11x^2 - 8x - 7}{(2x^2 - 1)(x - 3)} = \frac{3x + 1}{2x^2 - 1} + \frac{4}{x - 3}.$$

Now Try Exercise 13.

Appendix   Exercise Set

Decompose into partial fractions.

1. $\dfrac{x + 7}{(x - 3)(x + 2)}$

2. $\dfrac{2x}{(x + 1)(x - 1)}$

3. $\dfrac{7x - 1}{6x^2 - 5x + 1}$

4. $\dfrac{13x + 46}{12x^2 - 11x - 15}$

5. $\dfrac{3x^2 - 11x - 26}{(x^2 - 4)(x + 1)}$

6. $\dfrac{5x^2 + 9x - 56}{(x - 4)(x - 2)(x + 1)}$

7. $\dfrac{9}{(x + 2)^2(x - 1)}$

8. $\dfrac{x^2 - x - 4}{(x - 2)^3}$

9. $\dfrac{2x^2 + 3x + 1}{(x^2 - 1)(2x - 1)}$

10. $\dfrac{x^2 - 10x + 13}{(x^2 - 5x + 6)(x - 1)}$

11. $\dfrac{x^4 - 3x^3 - 3x^2 + 10}{(x + 1)^2(x - 3)}$

12. $\dfrac{10x^3 - 15x^2 - 35x}{x^2 - x - 6}$

13. $\dfrac{-x^2 + 2x - 13}{(x^2 + 2)(x - 1)}$

14. $\dfrac{26x^2 + 208x}{(x^2 + 1)(x + 5)}$

15. $\dfrac{6 + 26x - x^2}{(2x - 1)(x + 2)^2}$

16. $\dfrac{5x^3 + 6x^2 + 5x}{(x^2 - 1)(x + 1)^3}$

17. $\dfrac{6x^3 + 5x^2 + 6x - 2}{2x^2 + x - 1}$

18. $\dfrac{2x^3 + 3x^2 - 11x - 10}{x^2 + 2x - 3}$

19. $\dfrac{2x^2 - 11x + 5}{(x - 3)(x^2 + 2x - 5)}$

20. $\dfrac{3x^2 - 3x - 8}{(x - 5)(x^2 + x - 4)}$

21. $\dfrac{-4x^2 - 2x + 10}{(3x + 5)(x + 1)^2}$

22. $\dfrac{26x^2 - 36x + 22}{(x - 4)(2x - 1)^2}$

23. $\dfrac{36x + 1}{12x^2 - 7x - 10}$

24. $\dfrac{-17x + 61}{6x^2 + 39x - 21}$

25. $\dfrac{-4x^2 - 9x + 8}{(3x^2 + 1)(x - 2)}$

26. $\dfrac{11x^2 - 39x + 16}{(x^2 + 4)(x - 8)}$

▶ Synthesis

Decompose into partial fractions.

27. $\dfrac{x}{x^4 - a^4}$

28. $\dfrac{9x^3 - 24x^2 + 48x}{(x - 2)^4(x + 1)}$

[*Hint*: Let the expression equal

$$\frac{A}{x + 1} + \frac{P(x)}{(x - 2)^4}$$

and find $P(x)$].

29. $\dfrac{1 + \ln x^2}{(\ln x + 2)(\ln x - 3)^2}$

30. $\dfrac{1}{e^{-x} + 3 + 2e^x}$

Answers

Chapter R

RC1. (c)   **RC2.** (b)   **RC3.** (a)   **RC4.** (f)
RC5. (e)   **RC6.** (d)
1. $1, 12, \sqrt{25}$ **3.** $-6, 0, 1, -\frac{1}{2}, -4, \frac{7}{9}, 12, -\frac{6}{5}, 3.45, 5\frac{1}{2}, \sqrt{25}, -\frac{12}{3}$
5. $-6, 0, 1, -\frac{1}{2}, -4, \frac{7}{9}, 12, -\frac{6}{5}, 3.45, 5\frac{1}{2}, \sqrt{3}, \sqrt{25}, -\frac{12}{3}$,
$0.131331333133331\ldots$ **7.** $12, 0$ **9.** $-11, 12, 0$
11. $-\sqrt{5}, \pi, -3.565665666566665\ldots$ **13.** $\{m, a, t, h\}$
15. $\{1, 2, 3, 4, 5, 6, 7, 8, 9, 10, 11, 12\}$ **17.** $\{2, 4, 6, 8, \ldots\}$
19. $\{x \mid x$ is a whole number less than or equal to $5\}$, or
$\{x \mid x$ is a whole number less than $6\}$
21. $\left\{\frac{p}{q} \,\middle|\, p$ and q are integers and $q \neq 0\right\}$ **23.** $\{x \mid x > -3\}$
25. $>$ **27.** $<$ **29.** $<$ **31.** $<$ **33.** $>$ **35.** $<$ **37.** $>$
39. $<$ **41.** $x < -8$ **43.** $y \geq -12.7$ **45.** False   **47.** True
49. ⟵————)——————————⟶
            -2
51. ⟵————]——————————⟶
            -2
53. ⟵——(┼————————⟶
       -3.3
          -3
55. ⟵————————[—————⟶   **57.** 6   **59.** 28   **61.** 35
                    2
63. $\frac{2}{3}$ **65.** 42.8   **67.** 986   **69.** 0   **71.** $\leq$ **73.** $\leq$
75. $\frac{1}{8}\%, 0.3\%, 0.009, 1\%, 1.1\%, \frac{9}{100}, \frac{1}{11}, \frac{99}{1000}, 0.11, \frac{1}{8}, \frac{2}{7}, 0.286$

RC1. Negative   **RC2.** Positive   **RC3.** Negative
RC4. Positive   **RC5.** Positive   **RC6.** Negative
RC7. Negative   **RC8.** Negative
1. -28 **3.** 5   **5.** -16 **7.** -4 **9.** -10 **11.** -26
13. 1.2   **15.** -8.86 **17.** $-\frac{1}{3}$ **19.** $-\frac{4}{3}$ **21.** $\frac{1}{10}$ **23.** $\frac{7}{20}$
25. 4   **27.** -3.7 **29.** -10 **31.** 0   **33.** -4 **35.** -14
37. 0   **39.** -46 **41.** 5   **43.** 15   **45.** -11.6
47. -29.25 **49.** $-\frac{7}{2}$ **51.** $-\frac{1}{4}$ **53.** $-\frac{19}{12}$ **55.** $-\frac{7}{15}$
57. -21 **59.** -8 **61.** 24   **63.** -112 **65.** 34.2
67. $-\frac{12}{35}$ **69.** 2   **71.** 60   **73.** 26.46   **75.** 1   **77.** $-\frac{8}{27}$
79. -2 **81.** -7 **83.** 7   **85.** 0.3   **87.** Not defined
89. 0   **91.** Not defined   **93.** $\frac{4}{3}$ **95.** $-\frac{8}{7}$ **97.** $\frac{1}{25}$ **99.** 5
101. $-\dfrac{b}{a}$ **103.** $-\frac{6}{77}$ **105.** 25   **107.** -6 **109.** 5
111. -120 **113.** $-\frac{9}{8}$ **115.** $\frac{5}{3}$ **117.** $\frac{3}{2}$ **119.** $\frac{9}{64}$ **121.** -2
123. $\frac{12}{13}$, or $0.\overline{923076}$ **125.** $-\frac{81}{50}$, or -1.62 **127.** Not defined
129. $-\frac{2}{3}, \frac{3}{2}; \frac{5}{4}, -\frac{4}{5}; 0$, does not exist; $-1, 1; 4.5, -\frac{1}{4.5}; -x, \dfrac{1}{x}$
131. $26, 0$ **132.** 26   **133.** $-13, 26, 0$ **134.** $\sqrt{3}, \pi$,
$4.57557555755557\ldots$ **135.** $-12.47, -13, 26$,
$0, -\frac{23}{32}, \frac{7}{11}$ **136.** $\sqrt{3}, -12.47, -13, 26, \pi, 0, -\frac{23}{32}, \frac{7}{11}$,
$4.57557555755557\ldots$ **137.** $<$ **138.** $>$ **139.** $<$
140. $>$ **141.** $\frac{1}{4}$ **143.** 31,250

1. 56   **2.** 96   **3.** 262.5   **4.** $-2.\overline{4}$, or $-\frac{22}{9}$

RC1. False   **RC2.** True   **RC3.** True   **RC4.** False
RC5. False   **RC6.** False   **RC7.** True   **RC8.** True
1. 4^5 **3.** 5^6 **5.** m^3 **7.** $\left(\frac{7}{12}\right)^4$ **9.** $(123.7)^2$ **11.** 128
13. -32 **15.** $\frac{1}{81}$ **17.** -64 **19.** 31.36   **21.** 5   **23.** 1
25. 1   **27.** $\frac{7}{8}$ **29.** 16   **31.** $\frac{27}{8}$ **33.** $\dfrac{1}{y^5}$ **35.** a^2
37. $-\frac{1}{11}$ **39.** 3^{-4} **41.** b^{-3} **43.** $(-16)^{-2}$
45. -4 **47.** -117 **49.** 2   **51.** 8   **53.** -358
55. $144; 74$ **57.** -576 **59.** 2599   **61.** 36   **63.** 5619.712
65. $-200,167,769$ **67.** 3   **69.** 3   **71.** 16   **73.** -310
75. 2   **77.** 1875   **79.** 7804.48   **81.** 12   **83.** 8   **85.** 16
87. -86 **89.** 37   **91.** -1 **93.** 22   **95.** -39 **97.** 12
99. -549 **101.** -144 **103.** 2   **105.** $-\frac{31}{76}$ **107.** $\frac{61}{13}$
109. $\frac{9}{7}$ **110.** 2.3   **111.** 0   **112.** 900   **113.** -33 **114.** -79
115. 33   **116.** -79 **117.** -23 **118.** 23   **119.** -23
120. $\frac{5}{8}$ **121.** $25\frac{1}{4}$ **123.** $9 \cdot 5 + 2 - (8 \cdot 3 + 1) = 22$
125. 3125   **127.** $(2 + 3)^{-1} = (5)^{-1} = \frac{1}{5}; \ 2^{-1} + 3^{-1} =$
$\frac{1}{2} + \frac{1}{3} = \frac{3}{6} + \frac{2}{6} = \frac{5}{6};$ so $(2 + 3)^{-1} \neq 2^{-1} + 3^{-1}.$

RC1. (c)   **RC2.** (c)   **RC3.** (a)   **RC4.** (e)   **RC5.** (a)
RC6. (g)   **RC7.** (b)   **RC8.** (f)
1. $b + 8$, or $8 + b$ **3.** $c - 13.4$ **5.** $5 + q$, or $q + 5$
7. $a + b$, or $b + a$ **9.** $x \div y$, or $\dfrac{x}{y}$ **11.** $x + w$, or $w + x$
13. $n - m$ **15.** $p + q$, or $q + p$ **17.** $3q$ **19.** $-18m$
21. $17\%s$, or $0.17s$ **23.** $75t$ **25.** $\$40 - x$ **27.** -92
29. 3   **31.** 4   **33.** $\frac{45}{2}$, or 22.5   **35.** 16   **37.** 19
39. 57   **41.** $\$440.70$ **43.** $A = 2289.06$ in^2; $C = 169.56$ in.
45. 243   **46.** -243 **47.** 10,000   **48.** 28.09   **49.** $\frac{9}{25}$
50. 1   **51.** 4.5   **52.** $3x$
53. ⟵————)——————————⟶
              -1 0
54. ⟵——————[—————————⟶
                  0
55. $d = r \cdot t$ **57.** 9

RC1. Commutative   **RC2.** Distributive   **RC3.** Associative
RC4. 1   **RC5.** Factors   **RC6.** Terms
1. $-10, -10, 2; 25, 25, -5; 0, 0, 0; 2x + 3x$ and $5x$ are equivalent.
3. $-12, -16, -12; 38.4, 51.2, 38.4; 0, 0, 0; 4x + 8x$ and
$4(x + 2x)$ are equivalent.   **5.** $\dfrac{7x}{8x}$ **7.** $\dfrac{6a}{8a}$ **9.** $\frac{5}{3}$ **11.** -4
13. $3 + w$ **15.** tr **17.** $cd + 4, dc + 4$, or $4 + dc$

19. $x + yz$, $x + zy$, or $zy + x$ **21.** $(m + n) + 2$
23. $7 \cdot (x \cdot y)$ **25.** $a + (8 + b)$, $(a + 8) + b$, $b + (a + 8)$; others are possible   **27.** $(7 \cdot b) \cdot a$, $b \cdot (a \cdot 7)$, $(b \cdot a) \cdot 7$; others are possible   **29.** $4a + 4$ **31.** $8x - 8y$ **33.** $-10a - 15b$
35. $2ab - 2ac + 2ad$ **37.** $2\pi rh + 2\pi r$ **39.** $\frac{1}{2}ha + \frac{1}{2}hb$
41. $4a, -5b, 6$ **43.** $2x, -3y, -2z$ **45.** $24(x + y)$
47. $7(p - 1)$ **49.** $7(x - 3)$ **51.** $x(y + 1)$
53. $2(x - y + z)$ **55.** $3(x + 2y - 1)$
57. $4(w - 3z + 2)$ **59.** $4(5x - 9y - 3)$
61. $a(b + c - d)$ **63.** $\frac{1}{4}\pi r(r + s)$ **65.** $(x + y)^2$
66. $x^2 + y^2$ **67.** $\frac{1}{x^4}$ **68.** n^5 **69.** -26 **70.** 226
71. No   **73.** Yes

Exercise Set R.6, p. R-43

RC1. True   **RC2.** False   **RC3.** True   **RC4.** True
RC5. True   **RC6.** False
1. $12x$ **3.** $-3b$ **5.** $15y$ **7.** $11a$ **9.** $-8t$ **11.** $10x$
13. $11x - 5y$ **15.** $-4c + 12d$ **17.** $22x + 18$
19. $1.19x + 0.93y$ **21.** $-\frac{2}{15}a - \frac{1}{3}b - 27$ **23.** $P = 2l + 2w$
25. $2c$ **27.** $-b - 4$ **29.** $-b + 3$, or $3 - b$
31. $-t + y$, or $y - t$ **33.** $-x - y - z$ **35.** $-8x + 6y - 13$
37. $2c - 5d + 3e - 4f$ **39.** $1.2x - 56.7y + 34z + \frac{1}{4}$
41. $3a + 5$ **43.** $m + 1$ **45.** $9d - 16$ **47.** $-7x + 14$
49. $-9x + 17$ **51.** $17x + 3y - 18$ **53.** $10x - 19$
55. $22a - 15$ **57.** -190 **59.** $12x + 30$ **61.** $3x + 30$
63. $9x - 18$ **65.** $-4x + 808$ **67.** $-14y - 186$ **69.** 13
70. 75 **71.** 94 **72.** -24 **73.** -16 **74.** 16 **75.** -16
76. $-\frac{1}{6}$ **77.** $8a - 8b$ **78.** $-16a + 24b - 32$
79. $6ax - 6bx + 12cx$ **80.** $16x - 8y + 10$ **81.** $24(a - 1)$
82. $8(3a - 2b)$ **83.** $a(b - c + 1)$ **84.** $5(3p + 9q - 2)$
85. $(3 - 8)^2 + 9 = 34$ **87.** $5 \cdot 2^3 \div (3 - 4)^4 = 40$
89. $23a - 18b + 184$ **91.** $-9z + 5x$ **93.** $-x + 19$

Technology Connection, p. R-52

1. 1.2312×10^{-4} **2.** 2.8×10^5 **3.** 3×10^{-6}
4. 1.2×10^{-14}

Exercise Set R.7, p. R-53

RC1. (c)   **RC2.** (b)   **RC3.** (a)   **RC4.** (e)
RC5. (f)   **RC6.** (d)

1. 3^9 **3.** $\frac{1}{6^4}$ **5.** $\frac{1}{8^6}$ **7.** $\frac{1}{b^3}$ **9.** a^3 **11.** $72x^5$
13. $-28m^5n^5$ **15.** $-\frac{14}{x^{11}}$ **17.** $\frac{105}{x^{2t}}$ **19.** $-\frac{8}{y^{6m}}$ **21.** 8^7
23. 6^5 **25.** $\frac{1}{10^9}$ **27.** 9^2 **29.** $\frac{1}{x^{10n}}$ **31.** $\frac{1}{w^{5q}}$ **33.** a^5
35. $-3x^5z^4$ **37.** $-\frac{4x^9}{3y^2}$ **39.** $\frac{3x^3}{2y^2}$ **41.** 4^6 **43.** $\frac{1}{8^{12}}$
45. 6^{12} **47.** $125a^6b^6$ **49.** $\frac{y^{12}}{9x^6}$ **51.** $\frac{a^4}{36b^6c^2}$ **53.** $\frac{1}{4^9 \cdot 3^{12}}$
55. $\frac{8x^9y^3}{27}$ **57.** $\frac{a^{10}b^5}{5^{10}}$ **59.** $\frac{6^{30}2^{12}y^{36}}{z^{48}}$ **61.** $\frac{64}{x^{24}y^{12}}$
63. $\frac{5^7b^{28}}{3^7a^{35}}$ **65.** 10^a **67.** $3a^{-x-4}$ **69.** $\frac{-5x^{a+1}}{y}$

71. 8^{4xy} **73.** 12^{6b-2ab} **75.** $5^{2c}x^{2ac-2c}y^{2bc+2c}$, or $25^c x^{2ac-2c}y^{2bc+2c}$ **77.** $2x^{a+2}y^{b-2}$ **79.** 4.7×10^{10}
81. 1.6×10^{-8} **83.** 2.6×10^9 **85.** 1×10^{-4}
87. $673,000,000$ **89.** 0.000066 cm   **91.** $1,007,000,000$ users
93. 9.66×10^{-5} **95.** 1.3338×10^{-11} **97.** 2.5×10^3
99. 5.0×10^{-4} **101.** 6.3072×10^{10} sec   **103.** 3.33×10^{-2}
105. 1.422×10^{-2} m³   **107.** 1.2×10^{18} calculations;
7.2×10^{19} calculations   **109.** About 2.2×10^{-3} lb
111. $19x + 4y - 20$ **112.** $-23t + 21$ **113.** -11
114. -231 **115.** -8 **116.** 8 **117.** 2^{21}
119. $\frac{1}{a^{14}b^{27}}$ **121.** $4x^{2a}y^{2b}$

Summary and Review: Chapter R, p. R-56

Vocabulary Reinforcement

1. Inequality   **2.** Base   **3.** Variable   **4.** Factors
5. Scientific   **6.** Reciprocals   **7.** Opposites
8. Commutative

Concept Reinforcement

1. False   **2.** True   **3.** True   **4.** False   **5.** False
6. True   **7.** True   **8.** True

Review Exercises

1. $2, -\frac{2}{3}, 0.45\overline{45}, -23.788$ **2.** $\{x | x \text{ is a real number less than or equal to } 46\}$ **3.** $<$ **4.** $x < 19$ **5.** False   **6.** True
7.
8.
9. 7.23 **10.** 0 **11.** -2 **12.** -7.9 **13.** $-\frac{31}{28}$ **14.** -5
15. -26.7 **16.** $\frac{19}{4}$ **17.** 10.26 **18.** $-\frac{3}{7}$ **19.** 168 **20.** -4
21. 21 **22.** -7 **23.** $-\frac{7}{12}$ **24.** $\frac{8}{3}$ **25.** Not defined
26. -24 **27.** 7 **28.** -2.3 **29.** 0 **30.** a^5 **31.** $\left(-\frac{7}{8}\right)^3$
32. $\frac{1}{a^4}$ **33.** x^{-8} **34.** 59 **35.** -116 **36.** $5x$
37. $28\%y$, or $0.28y$ **38.** $t - 9$ **39.** $\frac{a}{b} - 8$ **40.** -17
41. -8 **42.** 84 ft²   **43.** $-4, 16, 36, 6$; $95, 225, 25, 105$;
$-5, 25, 25, 5$; none are equivalent   **44.** $-16, -9, -16, 12$;
$6, 13, 6, 34$; $-14, -7, -14, 14$; $2x - 14$ and $2(x - 7)$ are
equivalent   **45.** $\frac{21x}{9x}$ **46.** -12 **47.** $a + 11$ **48.** $y \cdot 8$
49. $9 + (a + b)$ **50.** $(8x)y$ **51.** $-6x + 3y$
52. $8abc + 4ab$ **53.** $5(x + 2y - z)$ **54.** $pt(r + s)$
55. $-3x + 5y$ **56.** $12c - 4$ **57.** $9c - 4d + 3$ **58.** $x + 3$
59. $6x + 15$ **60.** $22x - 14$ **61.** $-17m - 12$ **62.** $-\frac{10x^7}{y^5}$
63. $-\frac{3y^3}{2x^4}$ **64.** $\frac{a^8}{9b^2c^6}$ **65.** $\frac{81y^{40}}{16x^{24}}$ **66.** 6.875×10^9
67. 1.312×10^{-1} **68.** About 4.08 light-years
69. $\$6.7 \times 10^4$ **70.** D   **71.** A   **72.** x^{12y} **73.** 32
74. (a), (i); (d), (f); (h), (j)
75. Answers may vary. Five rational numbers that are not integers are $\frac{1}{3}, -\frac{3}{4}, 6\frac{5}{8}, -0.001$, and 1.7. They are not integers because they are not whole numbers or opposites of whole numbers.   **76.** The quotient $7/0$ is defined to be the number that gives a result of 7 when multiplied by 0. There is no such number, so we say that the quotient is not defined.   **77.** No; the area is quadrupled. For a triangle with

base b and height h, $A = \frac{1}{2}bh$. For a triangle with base $2b$ and height $2h$, $A = \frac{1}{2} \cdot 2b \cdot 2h = 2bh = 4\left(\frac{1}{2}bh\right)$.   **78.** No; the area is quadrupled. For a parallelogram with base b and height h, $A = bh$. For a parallelogram with base $2b$ and height $2h$, $A = 2b \cdot 2h = 4(bh)$.   **79.** \$5 million in \$20 bills contains $\frac{5 \times 10^6}{20} = 0.25 \times 10^6 = 2.5 \times 10^5$ bills, and 2.5×10^5 bills would weigh $2.5 \times 10^5 \times 2.2 \times 10^{-3} = 5.5 \times 10^2$, or 550 lb. Thus it is not possible that a criminal is carrying \$5 million in \$20 bills in a briefcase.   **80.** For 5^n, where n is a natural number, the ones digit will be 5. Since this is not the case with the given calculator readout, we know that the readout is an approximation.

Test: Chapter R, p. R-59

1. [R.1a] $\sqrt{7}, \pi$ **2.** [R.1a] $\{x \mid x \text{ is a real number greater than 20}\}$ **3.** [R.1b] >   **4.** [R.1b] $5 \geq a$ **5.** [R.1b] True
6. [R.1b] True   **7.** [R.1c]
8. [R.1d] 0   **9.** [R.1d] $\frac{7}{8}$ **10.** [R.2a] -2 **11.** [R.2a] -13.1
12. [R.2a] -6 **13.** [R.2c] -1 **14.** [R.2c] -29.7
15. [R.2c] $\frac{25}{4}$ **16.** [R.2d] -33.62 **17.** [R.2d] $\frac{3}{4}$
18. [R.2d] -528 **19.** [R.2e] 15   **20.** [R.2e] -5
21. [R.2e] $\frac{8}{3}$ **22.** [R.2e] -82 **23.** [R.2e] Not defined
24. [R.2b] 13   **25.** [R.2b] 0   **26.** [R.3a] q^4 **27.** [R.3b] a^{-9}
28. [R.3c] 0   **29.** [R.3c] $-\frac{16}{7}$ **30.** [R.4a] $t + 9$, or $9 + t$
31. [R.4a] $\frac{x}{y} - 12$ **32.** [R.4b] 18   **33.** [R.4b] 3.75 cm^2
34. [R.5a] Yes   **35.** [R.5a] No   **36.** [R.5b] $\frac{27x}{36x}$
37. [R.5b] $\frac{3}{2}$ **38.** [R.5c] qp **39.** [R.5c] $4 + t$
40. [R.5c] $(3 + t) + w$ **41.** [R.5c] $4(ab)$
42. [R.5d] $-6a + 8b$ **43.** [R.5d] $3\pi rs + 3\pi r$
44. [R.5d] $a(b - c + 2d)$ **45.** [R.5d] $h(2a + 1)$
46. [R.6a] $10y - 5x$ **47.** [R.6a] $21a + 14$
48. [R.6b] $9x - 7y + 22$ **49.** [R.6b] $-7x + 14$
50. [R.6b] $10x - 21$ **51.** [R.7a] $-\frac{3y^2}{2x^4}$
52. [R.7a] $-\frac{6a^9}{b^5}$ **53.** [R.7a] $-50a^{9n}$ **54.** [R.7a] $-\frac{5}{x^{4t}}$
55. [R.7b] $\frac{a^{12}}{81b^8c^4}$ **56.** [R.7b] $\frac{16a^{48}}{b^{48}}$
57. [R.7c] 4.37×10^{-5} **58.** [R.7c] 3.741×10^7
59. [R.7c] 1.875×10^{-6} **60.** [R.7c] C
61. [R.5c, d], [R.7b] (b), (e); (d), (f), (h); (i), (j)

Chapter 1

Technology Connection, p. 10

1. Left to the student   **2.** Left to the student

Exercise Set 1.1, p. 11

RC1. (d)   **RC2.** (f)   **RC3.** (b)   **RC4.** (c)
1. Yes   **3.** No   **5.** No   **7.** No   **9.** Yes   **11.** No   **13.** 7
15. -8 **17.** 27   **19.** -39 **21.** 86.86   **23.** $\frac{1}{6}$ **25.** 6
27. -4 **29.** -147 **31.** 32   **33.** -6 **35.** $-\frac{1}{6}$ **37.** 10
39. 11   **41.** -12 **43.** 8   **45.** 2   **47.** 21   **49.** -12

51. No solution   **53.** -1 **55.** $\frac{18}{5}$ **57.** 0   **59.** 1
61. All real numbers   **63.** No solution   **65.** 7   **67.** 2
69. 7   **71.** 5   **73.** $-\frac{3}{2}$ **75.** All real numbers   **77.** 5
79. $\frac{23}{66}$ **81.** $\frac{5}{32}$ **83.** $\frac{79}{32}$ **85.** a^{14} **86.** $\frac{1}{a^{32}}$ **87.** $-\frac{18x^2}{y^{11}}$
88. $-2x^8y^3$ **89.** $12 - 20x$ **90.** $-5 + 6x$
91. $-12x + 8y - 4z$ **92.** $-10x + 35y - 20$ **93.** $2(x - 3y)$
94. $-4(x + 6y)$ **95.** $2(2x - 5y + 1)$
96. $-5(2x - 7y + 4)$ **97.** $\{1, 2, 3, 4, 5, 6, 7, 8, 9\}$; $\{x \mid x \text{ is a positive integer less than 10}\}$
98. $\{-8, -7, -6, -5, -4, -3, -2, -1\}$; $\{x \mid x \text{ is a negative integer greater than } -9\}$ **99.** Approximately -4.176
101. $\frac{3}{2}$ **103.** 8

Exercise Set 1.2, p. 17

RC1. (d)   **RC2.** (b)   **RC3.** (f)   **RC4.** (a)
RC5. (c)   **RC6.** (e)
1. $r = \dfrac{d}{t}$ **3.** $h = \dfrac{A}{b}$ **5.** $w = \dfrac{P - 2l}{2}$, or $\dfrac{P}{2} - l$
7. $b = \dfrac{2A}{h}$ **9.** $a = 2A - b$ **11.** $m = \dfrac{F}{a}$ **13.** $t = \dfrac{I}{Pr}$
15. $c^2 = \dfrac{E}{m}$ **17.** $p = 2Q + q$ **19.** $y = \dfrac{c - Ax}{B}$
21. $N = \dfrac{1.08T}{I}$ **23.** $m = \dfrac{4}{3}C - 5$, or $\dfrac{4C - 15}{3}$
25. $b = 3n - a + c$ **27.** $R = \dfrac{d}{1 - st}$ **29.** $B = \dfrac{T}{1 + qt}$
31. (a) About 1930 calories;   **(b)** $w = \dfrac{R - 66 - 12.7h + 6.8a}{6.23}$
33. (a) About 2340 calories;
(b) $a = \dfrac{1015.25 + 6.74w + 7.29h - K}{7.29}$ **35. (a)** 1614 g;
(b) $a = \dfrac{P + 299}{9.337d}$ **37. (a)** 50 mg; **(b)** $d = \dfrac{c(a + 12)}{a}$, or $c + \dfrac{12c}{a}$ **39.** -5 **40.** 250   **41.** -2 **42.** -25 **43.** $\frac{4}{5}$
44. 6   **45.** -6 **46.** -5 **47.** $s = \dfrac{A - \pi r^2}{\pi r}$, or $\dfrac{A}{\pi r} - r$
49. $V_1 = \dfrac{P_2 V_2 T_1}{P_1 T_2}$; $P_2 = \dfrac{P_1 V_1 T_2}{T_1 V_2}$ **51.** 0.8 year
53. (a) Approximately 120.5 horsepower;
(b) approximately 94.1 horsepower

Exercise Set 1.3, p. 28

RC1. Familiarize   **RC2.** Translate   **RC3.** Solve
RC4. Check   **RC5.** State
1. About 27.5 mi   **3.** $45°, 52°, 83°$ **5.** 26 climbers
7. \$252   **9.** Length: 94 ft; width: 50 ft   **11.** 82 ft
13. \$265,000   **15.** 9, 11, 13   **17.** 229 and 230
19. 90 photos   **21.** \$38,950   **23.** About 4.2 million cases
25. (a) \$54.6 billion; **(b)** about 5 years after 2012, or in 2017
27. 6 min   **29.** Downstream: 1.25 hr; upstream: 1.875 hr
31. Upstream: $\frac{2}{3}$ hr; downstream: $\frac{18}{73}$ hr   **33.** 49   **34.** 208
35. $\frac{78}{1649}$ **36.** $\frac{17}{10}$ **37.** \$115,243   **39.** 25% increase
41. $m\angle 2 = 120°$; $m\angle 1 = 60°$

Mid-Chapter Review: Chapter 1, p. 32

1. True   **2.** True   **3.** False   **4.** False   **5.** Yes   **6.** No
7. No   **8.** Yes   **9.** -3 **10.** -8 **11.** 4   **12.** 2
13. All real numbers   **14.** 2   **15.** $-\frac{5}{2}$ **16.** No solution
17. $-\frac{4}{3}$ **18.** 5   **19.** $\frac{3}{2}$ **20.** $-\frac{3}{16}$ **21.** $n = \dfrac{P}{m}$

22. $t = \dfrac{z - 3w}{3}$, or $\dfrac{z}{3} - w$ **23.** $s = 4N - r$

24. $B = 1.5\dfrac{A}{T}$ **25.** $t = \dfrac{3H + 10}{2}$, or $\dfrac{3H}{2} + 5$

26. $g = \dfrac{f}{1 + hm}$ **27.** 6922 female graduates

28. 140 calories   **29.** Length: 7 ft; width: 5 ft   **30.** 1.5 hr; 3 hr
31. Equivalent expressions have the same value for all possible
replacements. Any replacement that does not make any of
the expressions undefined can be substituted for the variable.
Equivalent equations have the same solution(s).
32. Answers may vary. A walker who knows how far and how long
she walks each day wants to know her average speed each day.
33. Answers may vary. A decorator wants to have a carpet cut
for a bedroom. The perimeter of the room is 54 ft and its length
is 15 ft. How wide should the carpet be?   **34.** We can subtract
by adding an opposite, so we can use the addition principle to
subtract the same number on both sides of an equation. Similarly,
we can divide by multiplying by a reciprocal, so we can use the
multiplication principle to divide both sides of an equation by the
same number.   **35.** The manner in which a guess or an estimate
is manipulated can give insight into the form of the equation to
which the problem will be translated.   **36.** Labeling the variable
clearly makes the *Translate* step more accurate. It also allows us
to determine whether the solution of the equation we translated to
provides the information asked for in the original problem.

Translating for Success, p. 43

1. F   **2.** I   **3.** C   **4.** E   **5.** D   **6.** J   **7.** O   **8.** M
9. B   **10.** L

Exercise Set 1.4, p. 44

RC1. (b)   **RC2.** (h)   **RC3.** (c)   **RC4.** (a)
RC5. (g)   **RC6.** (d)   **1.** No, no, no, yes
3. No, yes, yes, no, no   **5.** $(-\infty, 5)$ **7.** $[-3, 3]$
9. $(-8, -4)$ **11.** $(-2, 5)$ **13.** $(-\sqrt{2}, \infty)$
15. $\{x \mid x > -1\}$, or $(-1, \infty)$ **17.** $\{y \mid y < 6\}$, or $(-\infty, 6)$

19. $\{a \mid a \leq -22\}$, or $(-\infty, -22]$

21. $\{t \mid t \geq -4\}$, or $[-4, \infty)$ **23.** $\{y \mid y > -6\}$, or $(-6, \infty)$

25. $\{x \mid x \leq 9\}$, or $(-\infty, 9]$ **27.** $\{x \mid x \geq 3\}$, or $[3, \infty)$

29. $\{x \mid x < -60\}$, or $(-\infty, -60)$ **31.** $\{x \mid x > 3\}$, or $(3, \infty)$

33. $\{x \mid x \leq 0.9\}$, or $(-\infty, 0.9]$ **35.** $\{x \mid x \leq \frac{5}{6}\}$, or $\left(-\infty, \frac{5}{6}\right]$

37. $\{x \mid x < 6\}$, or $(-\infty, 6)$ **39.** $\{y \mid y \leq -3\}$, or $(-\infty, -3]$
41. $\{y \mid y > \frac{2}{3}\}$, or $\left(\frac{2}{3}, \infty\right)$ **43.** $\{x \mid x \geq 11.25\}$, or
$[11.25, \infty)$ **45.** $\{x \mid x \leq \frac{1}{2}\}$, or $\left(-\infty, \frac{1}{2}\right]$
47. $\{y \mid y \leq -\frac{75}{2}\}$, or $\left(-\infty, -\frac{75}{2}\right]$ **49.** $\{x \mid x > -\frac{2}{17}\}$, or
$\left(-\frac{2}{17}, \infty\right)$ **51.** $\{m \mid m > \frac{7}{3}\}$, or $\left(\frac{7}{3}, \infty\right)$
53. $\{r \mid r < -3\}$, or $(-\infty, -3)$ **55.** $\{x \mid x \geq 2\}$, or $[2, \infty)$
57. $\{y \mid y < 5\}$, or $(-\infty, 5)$ **59.** $\{x \mid x \leq \frac{4}{7}\}$, or $\left(-\infty, \frac{4}{7}\right]$
61. $\{x \mid x < 8\}$, or $(-\infty, 8)$ **63.** $\{x \mid x \geq \frac{13}{2}\}$, or $\left[\frac{13}{2}, \infty\right)$
65. $\{x \mid x < \frac{11}{18}\}$, or $\left(-\infty, \frac{11}{18}\right)$ **67.** $\{x \mid x \geq -\frac{51}{31}\}$, or
$\left[-\frac{51}{31}, \infty\right)$ **69.** $\{a \mid a \leq 2\}$, or $(-\infty, 2]$
71. $\{W \mid W < \text{(approximately) } 136.7 \text{ lb}\}$ **73.** $\{S \mid S \geq 84\}$
75. $\{B \mid B \geq \$11{,}500\}$ **77.** $\{S \mid S > \$7000\}$
79. $\{c \mid c > \$735\}$ **81.** $\{p \mid p > 80\}$ **83.** $\{s \mid s < 980 \text{ ft}^2\}$
85. (a) 2010: \$2333; 2014: \$2661;   (b) more than 13.13 years
since 2005, or $\{t \mid t > 13.13\}$ **87.** $-9a + 30b$
88. $32x - 72y$ **89.** $-8a + 22b$ **90.** $-6a + 17b$
91. $10(3x - 7y - 4)$ **92.** $-6a(2 - 5b)$
93. $-4(2x - 6y + 1)$ **94.** $5(2n - 9mn + 20m)$ **95.** -11.2
96. 6.6   **97.** -11.2 **98.** 6.6   **99.** (a) $\{p \mid p > 10\}$;
(b) $\{p \mid p < 10\}$ **101.** True   **103.** All real numbers
105. All real numbers

Exercise Set 1.5, p. 55

RC1. True   **RC2.** False   **RC3.** True   **RC4.** True
1. $\{9, 11\}$ **3.** $\{b\}$ **5.** $\{9, 10, 11, 13\}$ **7.** $\{a, b, c, d, f, g\}$
9. $\varnothing$ **11.** $\{3, 5, 7\}$ **13.**

$(-4, 1]$ **15.** ; $(1, 6)$

17. $\{x \mid -4 \leq x < 5\}$, or $[-4, 5)$;

19. $\{x \mid x \geq 2\}$, or $[2, \infty)$;

21. $\varnothing$ **23.** $\{x \mid -8 < x < 6\}$, or $(-8, 6)$
25. $\{x \mid -6 < x \leq 2\}$, or $(-6, 2]$ **27.** $\{x \mid -1 < x \leq 6\}$, or
$(-1, 6]$ **29.** $\{y \mid -1 < y \leq 5\}$, or $(-1, 5]$
31. $\{x \mid -\frac{5}{3} \leq x \leq \frac{4}{3}\}$, or $\left[-\frac{5}{3}, \frac{4}{3}\right]$ **33.** $\{x \mid -\frac{7}{2} < x \leq \frac{11}{2}\}$, or
$\left(-\frac{7}{2}, \frac{11}{2}\right]$ **35.** $\{x \mid 10 < x \leq 14\}$, or $(10, 14]$
37. $\{x \mid -\frac{13}{3} \leq x \leq 9\}$, or $\left[-\frac{13}{3}, 9\right]$
39. ; $(-\infty, -2) \cup (1, \infty)$
41. ; $(-\infty, -3] \cup (1, \infty)$
43. $\{x \mid x < -5 \text{ or } x > -1\}$, or $(-\infty, -5) \cup (-1, \infty)$;

45. $\{x \mid x \leq \frac{5}{2} \text{ or } x \geq 4\}$, or $\left(-\infty, \frac{5}{2}\right] \cup [4, \infty)$;

47. $\{x \mid x \geq -3\}$, or $[-3, \infty)$;
49. $\{x \mid x \leq -\frac{5}{4} \text{ or } x > -\frac{1}{2}\}$, or $\left(-\infty, -\frac{5}{4}\right] \cup \left(-\frac{1}{2}, \infty\right)$
51. All real numbers, or $(-\infty, \infty)$
53. $\{x \mid x < -4 \text{ or } x > 2\}$, or $(-\infty, -4) \cup (2, \infty)$
55. $\{x \mid x < \frac{79}{4} \text{ or } x > \frac{89}{4}\}$, or $\left(-\infty, \frac{79}{4}\right) \cup \left(\frac{89}{4}, \infty\right)$
57. $\{x \mid x \leq -\frac{13}{2} \text{ or } x \geq \frac{29}{2}\}$, or $\left(-\infty, -\frac{13}{2}\right] \cup \left[\frac{29}{2}, \infty\right)$
59. $\{d \mid 0 \text{ ft} \leq d \leq 198 \text{ ft}\}$ **61.** Between 23 beats and 27 beats
63. $\{W \mid 101.2 \text{ lb} \leq W \leq 136.2 \text{ lb}\}$
65. $\{d \mid 250 \text{ mg} < d < 500 \text{ mg}\}$ **67.** No solution   **68.** $-\frac{5}{3}$
69. 4   **70.** 9   **71.** All real numbers   **72.** 9

73. $\{x|-4 < x \le 1\}$, or $(-4, 1]$ **75.** $\{x|\frac{2}{5} \le x \le \frac{4}{5}\}$, or $[\frac{2}{5}, \frac{4}{5}]$ **77.** $\{x|-\frac{1}{8} < x < \frac{1}{2}\}$, or $(-\frac{1}{8}, \frac{1}{2})$
79. $\{x|10 < x \le 18\}$, or $(10, 18]$ **81.** True   **83.** False
85. All real numbers; $\varnothing$

Exercise Set 1.6, p. 64

RC1. (f)   **RC2.** (b)   **RC3.** (e)   **RC4.** (c)   **RC5.** (a)

RC6. (d)   **1.** $9|x|$ **3.** $2x^2$ **5.** $2x^2$ **7.** $6|y|$ **9.** $\dfrac{2}{|x|}$

11. $\dfrac{x^2}{|y|}$ **13.** $4|x|$ **15.** $\dfrac{y^2}{3}$ **17.** 38   **19.** 19   **21.** 6.3

23. 5   **25.** $\{-3, 3\}$ **27.** $\varnothing$ **29.** $\{0\}$ **31.** $\{-9, 15\}$
33. $\{-\frac{1}{2}, \frac{7}{2}\}$ **35.** $\{-\frac{5}{4}, \frac{23}{4}\}$ **37.** $\{-11, 11\}$
39. $\{-291, 291\}$ **41.** $\{-8, 8\}$ **43.** $\{-7, 7\}$ **45.** $\{-2, 2\}$
47. $\{-7, 8\}$ **49.** $\{-12, 2\}$ **51.** $\{-\frac{5}{2}, \frac{7}{2}\}$ **53.** $\varnothing$
55. $\{-\frac{13}{54}, -\frac{7}{54}\}$ **57.** $\{-\frac{11}{2}, \frac{3}{4}\}$ **59.** $\{\frac{3}{2}\}$ **61.** $\{5, -\frac{3}{5}\}$
63. All real numbers   **65.** $\{-\frac{3}{2}\}$ **67.** $\{\frac{24}{23}, 0\}$
69. $\{32, \frac{8}{3}\}$ **71.** $\{x|-3 < x < 3\}$, or $(-3, 3)$
73. $\{x|x \le -2 \text{ or } x \ge 2\}$, or $(-\infty, -2] \cup [2, \infty)$
75. $\{x|0 < x < 2\}$, or $(0, 2)$ **77.** $\{x|-6 \le x \le -2\}$, or
$[-6, -2]$ **79.** $\{x|-\frac{1}{2} \le x \le \frac{7}{2}\}$, or $[-\frac{1}{2}, \frac{7}{2}]$
81. $\{y|y < -\frac{3}{2} \text{ or } y > \frac{17}{2}\}$, or $(-\infty, -\frac{3}{2}) \cup (\frac{17}{2}, \infty)$
83. $\{x|x \le -\frac{5}{4} \text{ or } x \ge \frac{23}{4}\}$, or $(-\infty, -\frac{5}{4}] \cup [\frac{23}{4}, \infty)$
85. $\{y|-9 < y < 15\}$, or $(-9, 15)$ **87.** $\{x|-\frac{7}{2} \le x \le \frac{1}{2}\}$,
or $[-\frac{7}{2}, \frac{1}{2}]$ **89.** $\{y|y < -\frac{4}{3} \text{ or } y > 4\}$, or
$(-\infty, -\frac{4}{3}) \cup (4, \infty)$ **91.** $\{x|x \le -\frac{5}{4} \text{ or } x \ge \frac{23}{4}\}$, or
$(-\infty, -\frac{5}{4}] \cup [\frac{23}{4}, \infty)$ **93.** $\{x|-\frac{9}{2} < x < 6\}$, or $(-\frac{9}{2}, 6)$
95. $\{x|x \le -\frac{25}{6} \text{ or } x \ge \frac{23}{6}\}$, or $(-\infty, -\frac{25}{6}] \cup [\frac{23}{6}, \infty)$
97. $\{x|-5 < x < 19\}$, or $(-5, 19)$
99. $\{x|x \le -\frac{2}{15} \text{ or } x \ge \frac{14}{15}\}$, or $(-\infty, -\frac{2}{15}] \cup [\frac{14}{15}, \infty)$
101. $\{m|-12 \le m \le 2\}$, or $[-12, 2]$
103. $\{x|\frac{1}{2} \le x \le \frac{5}{2}\}$, or $[\frac{1}{2}, \frac{5}{2}]$ **105.** $\{x|-1 \le x \le 2\}$, or
$[-1, 2]$ **107.** $\{x|x \le 4\}$, or $(-\infty, -4]$
108. $\{y|y < -\frac{9}{10}\}$, or $(-\infty, -\frac{9}{10})$
109. $\{r|r > 16\}$, or $(16, \infty)$
110. $\{x|-8 < x \le 4\}$, or $(-8, 4]$ **111.** All real numbers,
or $(-\infty, \infty)$ **112.** $\{x|\frac{1}{3} \le x < 4\}$, or $[\frac{1}{3}, 4)$
113. $\{d|5\frac{1}{2} \text{ ft} \le d \le 6\frac{1}{2} \text{ ft}\}$ **115.** All real numbers
117. $\{1, -\frac{1}{4}\}$ **119.** $\varnothing$ **121.** $|x| < 3$ **123.** $|x| \ge 6$
125. $|x + 3| > 5$

Summary and Review: Chapter 1, p. 66

Vocabulary Reinforcement

1. Inequality   **2.** Set-builder   **3.** Interval   **4.** Intersection
5. Conjunction   **6.** Empty set   **7.** Disjoint sets   **8.** Union
9. Disjunction   **10.** Addition principle   **11.** Multiplication
principle   **12.** Distance

Concept Reinforcement

1. True   **2.** False   **3.** False   **4.** False   **5.** True
6. False   **7.** True

Study Guide

1. No   **2.** 8   **3.** $h = \dfrac{4F}{g}$ **4.** -2 is not a solution; 5 is a
solution.   **5. (a)** $(-\infty, -8)$; **(b)** $[-7, 10)$; **(c)** $[3, \infty)$
6. $\{y|y < -2\}$, or $(-\infty, -2)$;
7. $\{z|-2 \le z < 1\}$, or $[-2, 1)$;
8. $\{z|z < -1 \text{ or } z \ge 1\}$, or $(-\infty, -1) \cup [1, \infty)$;

9. $8y^2$ **10.** 28   **11.** $\{-\frac{8}{5}, 2\}$ **12.** $\{3, -\frac{1}{2}\}$
13. (a) $\{x|-4 < x < 1\}$, or $(-4, 1)$;
(b) $\{x|x \le -\frac{10}{3} \text{ or } x \ge 2\}$, or $(-\infty, -\frac{10}{3}] \cup [2, \infty)$

Review Exercises

1. 8   **2.** $\frac{3}{7}$ **3.** $\frac{22}{5}$ **4.** $-\frac{1}{13}$ **5.** -0.2 **6.** 5
7. $d = \frac{11}{4}(C - 3)$ **8.** $b = \dfrac{A - 2a}{-3}$, or $\dfrac{2a - A}{3}$
9. 185 and 186   **10.** 15 m, 12 m   **11.** 160,000
12. 40 sec   **13.** $[-8, 9)$ **14.** $(-\infty, 40]$
15. ; $(-\infty, -2]$
16. ; $(1, \infty)$
17. $\{a|a \le -21\}$, or $(-\infty, -21]$ **18.** $\{y|y \ge -7\}$, or
$[-7, \infty)$ **19.** $\{y|y > -4\}$, or $(-4, \infty)$
20. $\{y|y > -30\}$, or $(-30, \infty)$ **21.** $\{x|x > -3\}$, or
$(-3, \infty)$ **22.** $\{y|y \le -\frac{6}{5}\}$, or $(-\infty, -\frac{6}{5}]$
23. $\{x|x < -3\}$, or $(-\infty, -3)$ **24.** $\{y|y > -10\}$, or
$(-10, \infty)$ **25.** $\{x|x \le -\frac{5}{2}\}$, or $(-\infty, -\frac{5}{2}]$
26. $\{t|t > 4\frac{1}{4} \text{ hr}\}$ **27.** \$10,000
28. ; $[-2, 5)$
29. ; $(-\infty, -2] \cup (5, \infty)$
30. $\{1, 5, 9\}$ **31.** $\{1, 2, 3, 5, 6, 9\}$ **32.** $\varnothing$
33. $\{x|-7 < x \le 2\}$, or $(-7, 2]$ **34.** $\{x|-\frac{5}{4} < x < \frac{5}{2}\}$, or
$(-\frac{5}{4}, \frac{5}{2})$ **35.** $\{x|x < -3 \text{ or } x > 1\}$, or $(-\infty, -3) \cup (1, \infty)$
36. $\{x|x < -11 \text{ or } x \ge -6\}$, or $(-\infty, -11) \cup [-6, \infty)$
37. $\{x|x \le -6 \text{ or } x \ge 8\}$, or $(-\infty, -6] \cup [8, \infty)$
38. $\dfrac{3}{|x|}$ **39.** $\dfrac{2|x|}{y^2}$ **40.** $\dfrac{4}{|y|}$ **41.** 62   **42.** $\{-6, 6\}$
43. $\{-5, 9\}$ **44.** $\{-14, \frac{4}{3}\}$ **45.** $\varnothing$ **46.** $\{x|-\frac{17}{2} < x < \frac{7}{2}\}$,
or $(-\frac{17}{2}, \frac{7}{2})$ **47.** $\{x|x \le -3.5 \text{ or } x \ge 3.5\}$, or
$(-\infty, -3.5] \cup [3.5, \infty)$ **48.** $\{x|x \le -\frac{11}{3} \text{ or } x \ge \frac{19}{3}\}$, or
$(-\infty, -\frac{11}{3}] \cup [\frac{19}{3}, \infty)$ **49.** $\varnothing$ **50.** B   **51.** A
52. $\{x|-\frac{8}{3} \le x \le -2\}$, or $[-\frac{8}{3}, -2]$ **53.** When the signs of
the quantities on either side of the inequality symbol are changed,
their relative positions on the number line are reversed.
54. The distance between x and -5 is $|x - (-5)|$, or $|x + 5|$.
Then the solutions of the inequality $|x + 5| \le 2$ can be
interpreted as "all those numbers x whose distance from -5 is at
most 2 units."   **55.** When $b \ge c$, then the intervals overlap and
$[a, b] \cup [c, d] = [a, d]$.   **56.** The solutions of $|x| \ge 6$ are
those numbers whose distance from 0 is greater than or equal to 6.
In addition to the numbers in $[6, \infty)$, the distance of the numbers
in $(-\infty, -6]$ from 0 is also greater than or equal to 6. Thus,
$[6, \infty)$ is only part of the solution of the inequality.

57. **(1)** $-9(x + 2) = -9x - 18$, not $-9x + 2$. **(2)** This would be correct if (1) were correct except that the inequality symbol should not have been reversed. **(3)** If (2) were correct, the right-hand side would be -5, not 8. **(4)** The inequality symbol should be reversed. The correct solution is

$$7 - 9x + 6x < -9(x + 2) + 10x$$
$$7 - 9x + 6x < -9x - 18 + 10x$$
$$7 - 3x < x - 18$$
$$-4x < -25$$
$$x > \tfrac{25}{4}.$$

58. By definition, the notation $3 < x < 5$ indicates that $3 < x$ and $x < 5$. A solution of the disjunction $3 < x \text{ or } x < 5$ must be in at least one of these sets but not necessarily in both, so the disjunction cannot be written as $3 < x < 5$.

Test: Chapter 1, p. 71

1. [1.1b] -2 **2.** [1.1c] $\tfrac{2}{3}$ **3.** [1.1b] $\tfrac{19}{15}$ **4.** [1.1d] 4

5. [1.1d] 1.1   **6.** [1.1d] -2 **7.** [1.2a] $B = \dfrac{A + C}{3}$

8. [1.2a] $n = \dfrac{m}{1 - t}$ **9.** [1.3a] Length: $14\tfrac{2}{5}$ ft; width: $9\tfrac{3}{5}$ ft

10. [1.3a] 52,000 copies   **11.** [1.3a] 180,000

12. [1.3a] 59°, 60°, 61°   **13.** [1.3b] $2\tfrac{2}{5}$ hr; 4 hr

14. [1.4b] $(-3, 2]$ **15.** [1.4b] $(-4, \infty)$

16. [1.4c] ⟵┼┼┼┼┼┼┼┼┼┼┼┼┨➤; $(-\infty, 6]$

17. [1.4c] ⟵┼┼┼┨┼┼┼┼┼┼┼┼➤; $(-\infty, -2]$

18. [1.4c] $\{x \mid x \geq 10\}$, or $[10, \infty)$ **19.** [1.4c] $\{y \mid y > -50\}$, or $(-50, \infty)$ **20.** [1.4c] $\{a \mid a \leq \tfrac{11}{5}\}$, or $\left(-\infty, \tfrac{11}{5}\right]$ **21.** [1.4c] $\{y \mid y > 1\}$, or $(1, \infty)$

22. [1.4c] $\{x \mid x > \tfrac{5}{2}\}$, or $\left(\tfrac{5}{2}, \infty\right)$ **23.** [1.4c] $\{x \mid x \leq \tfrac{7}{4}\}$, or $\left(-\infty, \tfrac{7}{4}\right]$ **24.** [1.4d] $\{h \mid h > 2\tfrac{1}{10} \text{ hr}\}$

25. [1.5c] $\{d \mid 33 \text{ ft} \leq d \leq 231 \text{ ft}\}$

26. [1.5a] ⟵┼┼┼┠┼┼┼┨┼┼┼➤; $[-3, 4]$

27. [1.5b] ⟵┼┼┼┨┼┼┼┼┼┼┠┼┼┼➤; $(-\infty, -3) \cup (4, \infty)$

28. [1.5a] $\{x \mid x \geq 4\}$, or $[4, \infty)$

29. [1.5a] $\{x \mid -1 < x < 6\}$, or $(-1, 6)$

30. [1.5a] $\{x \mid -\tfrac{2}{5} < x \leq \tfrac{9}{5}\}$, or $\left(-\tfrac{2}{5}, \tfrac{9}{5}\right]$

31. [1.5b] $\{x \mid x < -4 \text{ or } x > -\tfrac{5}{2}\}$, or $(-\infty, -4) \cup \left(-\tfrac{5}{2}, \infty\right)$

32. [1.5b] All real numbers, or $(-\infty, \infty)$

33. [1.5b] $\{x \mid x < 3 \text{ or } x > 6\}$, or $(-\infty, 3) \cup (6, \infty)$

34. [1.6a] $\dfrac{7}{|x|}$ **35.** [1.6a] $2|x|$ **36.** [1.6b] 8.4

37. [1.5a] $\{3, 5\}$ **38.** [1.5b] $\{1, 3, 5, 7, 9, 11, 13\}$

39. [1.6c] $\{-9, 9\}$ **40.** [1.6c] $\{-6, 12\}$ **41.** [1.6d] $\{1\}$

42. [1.6c] $\varnothing$ **43.** [1.6e] $\{x \mid -0.875 < x < 1.375\}$, or $(-0.875, 1.375)$ **44.** [1.6e] $\{x \mid x < -3 \text{ or } x > 3\}$, or $(-\infty, -3) \cup (3, \infty)$ **45.** [1.6e] $\{x \mid -99 \leq x \leq 111\}$, or $[-99, 111]$ **46.** [1.6e] $\{x \mid x \leq -\tfrac{13}{5} \text{ or } x \geq \tfrac{7}{5}\}$, or $\left(-\infty, -\tfrac{13}{5}\right] \cup \left[\tfrac{7}{5}, \infty\right)$ **47.** [1.1d] C   **48.** [1.6e] $\varnothing$

49. [1.5a] $\{x \mid \tfrac{1}{5} < x < \tfrac{4}{5}\}$, or $\left(\tfrac{1}{5}, \tfrac{4}{5}\right)$

Chapter 2

Technology Connection, p. 77

1.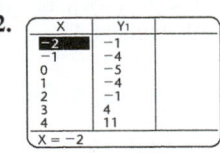

2.

Technology Connection, p. 79

1. $y = 2x - 1$

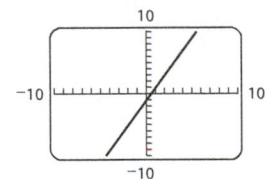

2. $y = -3x + 2$

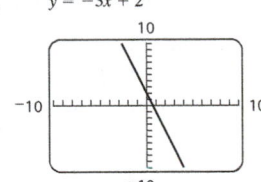

3. $y = 5x - 3$

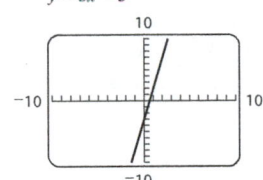

4. $y = -4x + 5$

5. $y = \tfrac{2}{3}x - 3$

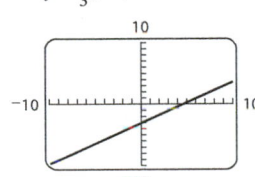

6. $y = -\tfrac{3}{4}x + 4$

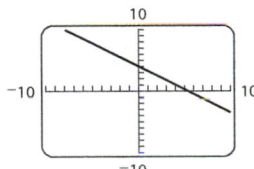

7. $y = 3.104x - 6.21$

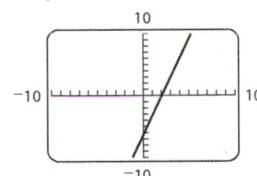

8. $y = -2.98x - 1.75$

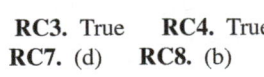

Exercise Set 2.1, p. 82

RC1. False   **RC2.** False   **RC3.** True   **RC4.** True
RC5. True   **RC6.** False   **RC7.** (d)   **RC8.** (b)
RC9. (a)   **RC10.** (c)

1.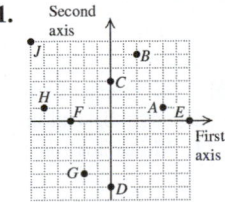

3.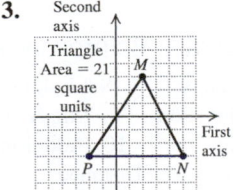

5. Yes   **7.** Yes   **9.** No

11. $y = 4 - x$

$\dfrac{}{5 \; ? \; 4 - (-1)}$
$\quad | \quad 4 + 1$
$\quad | \quad 5 \qquad$ TRUE

$y = 4 - x$
$\dfrac{}{1 \; ? \; 4 - 3}$
$\quad | \quad 1 \qquad$ TRUE

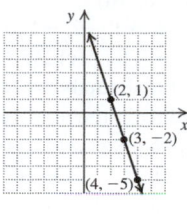

13. $3x + y = 7$

$\dfrac{}{3 \cdot 2 + 1 \; ? \; 7}$
$\quad 6 + 1 \quad |$
$\quad 7 \quad | \qquad$ TRUE

$3x + y = 7$
$\dfrac{}{3 \cdot 4 + (-5) \; ? \; 7}$
$\quad 12 - 5 \quad |$
$\quad 7 \quad | \qquad$ TRUE

15. $6x - 3y = 3$

$\dfrac{}{6 \cdot 1 - 3 \cdot 1 \; ? \; 3}$
$\quad 6 - 3 \quad |$
$\quad 3 \quad | \qquad$ TRUE

$6x - 3y = 3$
$\dfrac{}{6(-1) - 3(-3) \; ? \; 3}$
$\quad -6 + 9 \quad |$
$\quad 3 \quad | \qquad$ TRUE

17.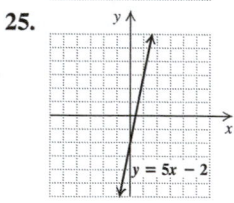

19.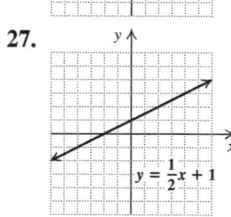

21.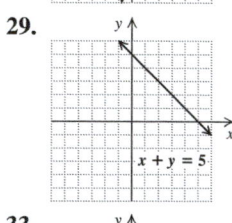

23.

25.

27.

29.

31.

33.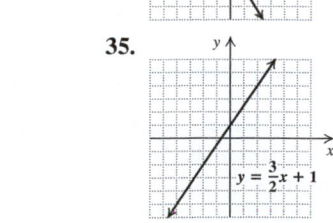

35.

37.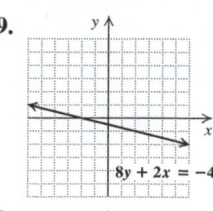

39.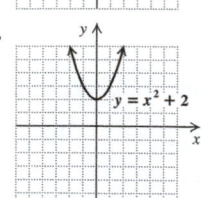

41.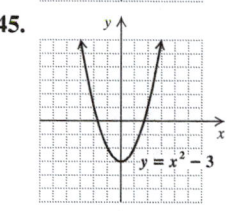

43.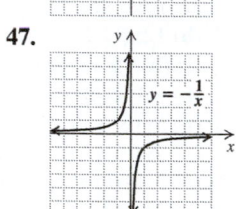

45.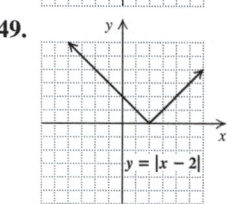

47.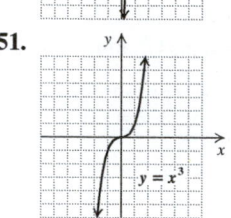

49.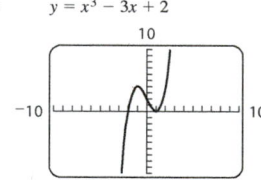

51.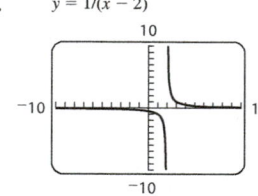

53. $\left\{x \,\middle|\, 1 < x \le \frac{15}{2}\right\}$, or $\left(1, \frac{15}{2}\right]$ **54.** $\{x \,|\, x > -3\}$, or $(-3, \infty)$

55. $\left\{x \,\middle|\, x \le -\frac{7}{3} \text{ or } x \ge \frac{17}{3}\right\}$, or $\left(-\infty, -\frac{7}{3}\right] \cup \left[\frac{17}{3}, \infty\right)$

56. $\{x \,|\, -6 < x < 6\}$, or $(-6, 6)$

57. Kidney: 94,345 people; liver: 15,817 people   **58.** 25 ft

59. $4\frac{3}{4}$ mi   **60.** \$330,000

61. $y = x^3 - 3x + 2$

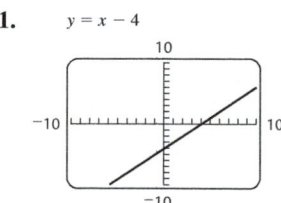

63. $y = 1/(x - 2)$

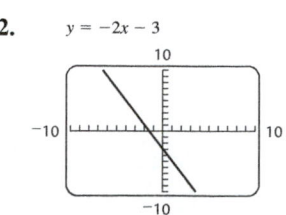

65. $y = -x + 4$ **67.** $y = |x| - 3$

Technology Connection, p. 88

1. 17.3   **2.** 34

Technology Connection, p. 89

1. $y = x - 4$

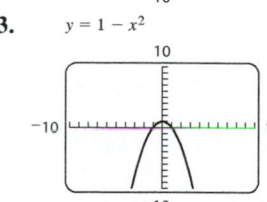

2. $y = -2x - 3$

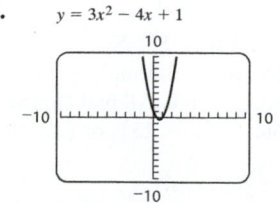

3. $y = 1 - x^2$

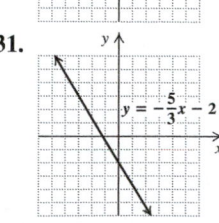

4. $y = 3x^2 - 4x + 1$

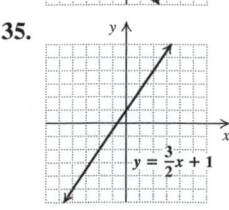

5. 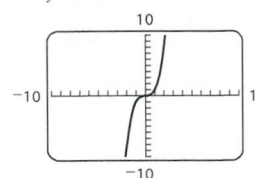 $y = x^3$

6. 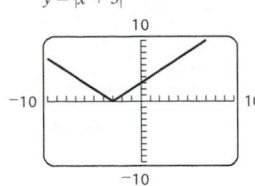 $y = |x + 3|$

Exercise Set 2.2, p. 92

RC1. $f(2) = 3$ **RC2.** $f(0) = 3$ **RC3.** $f(-2) = -5$
RC4. $f(3) = 0$
1. Yes   **3.** Yes   **5.** No   **7.** No   **9.** No   **11.** Yes
13. (a) 9; (b) 12; (c) 2; (d) 5; (e) 7.4; (f) $5\frac{2}{3}$
15. (a) -21; (b) 15; (c) 2; (d) 0; (e) $18a$; (f) $3a + 3$
17. (a) 7; (b) -17; (c) 6; (d) 4; (e) $3a - 2$; (f) $3a + 3h + 4$
19. (a) 0; (b) 5; (c) 2; (d) 170; (e) 65; (f) $32a^2 - 12a$
21. (a) 1; (b) 3; (c) 3; (d) 11; (e) $|a - 1| + 1$; (f) $|a + h| + 1$
23. (a) 0; (b) -1; (c) 8; (d) 1000; (e) -125; (f) $-27a^3$
25. 1980: about 60.5 years; 2013: about 62.0 years
27. $1\frac{20}{33}$ atm; $1\frac{10}{11}$ atm; $4\frac{1}{33}$ atm   **29.** 1.792 cm; 2.8 cm; 11.2 cm
31. 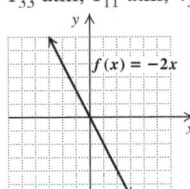 $f(x) = -2x$

33. 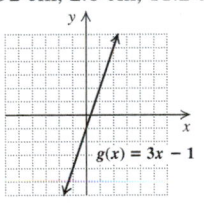 $g(x) = 3x - 1$

35. 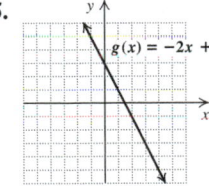 $g(x) = -2x + 3$

37. 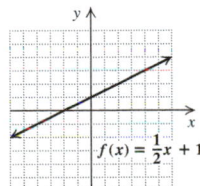 $f(x) = \frac{1}{2}x + 1$

39. 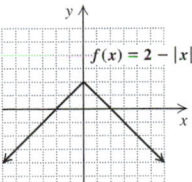 $f(x) = 2 - |x|$

41. 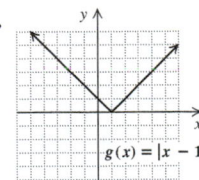 $g(x) = |x - 1|$

43. 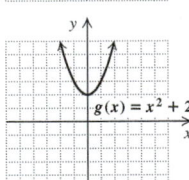 $g(x) = x^2 + 2$

45. 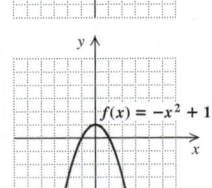 $f(x) = x^2 - 2x - 3$

47. 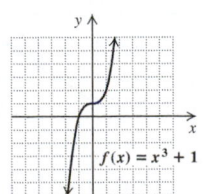 $f(x) = -x^2 + 1$

49. $f(x) = x^3 + 1$

51. Yes   **53.** No   **55.** No   **57.** Yes
59. About 1.8 million children   **61.** About 230,000 pharmacists
63. $\frac{3}{4}$ **64.** All real numbers   **65.** $\{y \mid y < 4\}$
66. $\{x \mid x > 23\}$, or $(23, \infty)$ **67.** No solution   **68.** 26

69. $\{w \mid w \le 11.1\}$, or $(-\infty, 11.1]$ **70.** $\{x \mid x > 2\}$, or
$(2, \infty)$ **71.** $\frac{3}{2}$ **72.** $\frac{80}{9}$ **73.** $g(-2) = 39$ **75.** 26; 99
77. $g(x) = \frac{15}{4}x - \frac{13}{4}$

Exercise Set 2.3, p. 99

RC1. (a)   **RC2.** (b)   **RC3.** (a)   **RC4.** (b)
RC5. (a)   **RC6.** (d)
1. (a) 3; (b) $\{-4, -3, -2, -1, 0, 1, 2\}$; (c) $-2, 0$; (d) $\{1, 2, 3, 4\}$
3. (a) $2\frac{1}{2}$; (b) $[-3, 5]$; (c) $2\frac{1}{4}$; (d) $[1, 4]$
5. (a) 1; (b) all real numbers; (c) 3; (d) all real numbers
7. (a) 1; (b) all real numbers; (c) $-2, 2$; (d) $[0, \infty)$
9. $\{x \mid x$ is a real number $and \ x \ne -3\}$, or $(-\infty, -3) \cup (-3, \infty)$
11. All real numbers   **13.** All real numbers
15. $\left\{x \mid x \text{ is a real number } and \ x \ne \frac{14}{5}\right\}$, or $\left(-\infty, \frac{14}{5}\right) \cup \left(\frac{14}{5}, \infty\right)$
17. All real numbers   **19.** $\left\{x \mid x \text{ is a real number } and \ x \ne \frac{7}{4}\right\}$,
or $\left(-\infty, \frac{7}{4}\right) \cup \left(\frac{7}{4}, \infty\right)$ **21.** $\{x \mid x$ is a real number $and \ x \ne 1\}$,
or $(-\infty, 1) \cup (1, \infty)$ **23.** All real numbers
25. All real numbers   **27.** $\left\{x \mid x \text{ is a real number } and \ x \ne \frac{5}{2}\right\}$,
or $\left(-\infty, \frac{5}{2}\right) \cup \left(\frac{5}{2}, \infty\right)$ **29.** All real numbers
31. $\left\{x \mid x \text{ is a real number } and \ x \ne -\frac{5}{4}\right\}$, or $\left(-\infty, -\frac{5}{4}\right) \cup \left(-\frac{5}{4}, \infty\right)$
33. $-8; 0; -2$ **35.** $\{-8, 8\}$ **36.** $\{\ \}$, or $\varnothing$ **37.** $\{-4, 18\}$
38. $\{-8, 5\}$ **39.** $\left\{\frac{1}{2}, 3\right\}$ **40.** $\left\{-1, \frac{9}{13}\right\}$ **41.** $\{\ \}$, or $\varnothing$
42. $\left\{\frac{8}{3}\right\}$ **43.** $(-\infty, 0) \cup (0, \infty)$; $[2, \infty)$; $[-4, \infty)$; $[0, \infty)$
45. All real numbers

Exercise Set 2.4, p. 104

RC1. (d)   **RC2.** (c)   **RC3.** (b)   **RC4.** (e)
RC5. (a)
1. 1   **3.** 5   **5.** -7 **7.** 1   **9.** -5 **11.** $x^2 - 2x - 2$
13. $x^2 + 2x - 8$ **15.** $x^2 - x + 3$ **17.** 5
19. $-a^3 + 5a^2 + 2a - 10$ **21.** $\dfrac{x^2 - 2}{5 - x}, x \ne 5$ **23.** $\frac{7}{2}$
25. -2 **27.** $\dfrac{15}{2x^3}$ **29.** $\dfrac{10 - 3x}{2x^2}$ **31.** $\dfrac{3x}{10}$
33. $1.2 + 2.9 = 4.1$ million   **35.** $45 - 28 = 17$
37. $\{x \mid x$ is a real number$\}$
39. $\{x \mid x$ is a real number $and \ x \ne -5\}$
41. $\{x \mid x$ is a real number $and \ x \ne 0\}$
43. $\{x \mid x$ is a real number $and \ x \ne 1\}$
45. $\left\{x \mid x \text{ is a real number } and \ x \ne -\frac{9}{2} \text{ and } x \ne 1\right\}$
47. $\{x \mid x$ is a real number $and \ x \ne 3\}$
49. $\{x \mid x$ is a real number $and \ x \ne -4\}$
51. $\{x \mid x$ is a real number $and \ x \ne 4 \text{ and } x \ne 5\}$
53. $\left\{x \mid x \text{ is a real number } and \ x \ne -1 \text{ and } x \ne -\frac{5}{2}\right\}$
55. $4; 3$ **57.** $5; -1$ **59.** $\{x \mid 0 \le x \le 9\}$;
$\{x \mid 3 \le x \le 10\}$; $\{x \mid 3 \le x \le 9\}$; $\{x \mid 3 \le x \le 9\}$
61. $60°, 30°, 90°$ **62.** 11 points   **63.** -18 **64.** 0
65. $\{x \mid x$ is a real number $and \ x \ne 4 \text{ and } x \ne 3 \text{ and }$
$x \ne 2 \text{ and } x \ne -2\}$
67. $\left\{x \mid x \text{ is a real number } and -1 < x < 5 \text{ and } x \ne \frac{3}{2}\right\}$

Mid-Chapter Review: Chapter 2, p. 106

1. True   **2.** False   **3.** True   **4.** True   **5.** False
6. No   **7.** Yes   **8.** Yes   **9.** No
10. Domain: $\{x \mid -3 \le x \le 3\}$, or $[-3, 3]$;
range: $\{y \mid -2 \le y \le 1\}$ **11.** -3 **12.** -7 **13.** 8
14. 9   **15.** 9000   **16.** 0   **17.** Yes   **18.** No   **19.** Yes

20. $\{x\,|\,x$ is a real number *and* $x \neq 4\}$, or $(-\infty, 4) \cup (4, \infty)$
21. All real numbers   **22.** $\{x\,|\,x$ is a real number *and* $x \neq -2\}$, or $(-\infty, -2) \cup (-2, \infty)$ **23.** All real numbers

24.
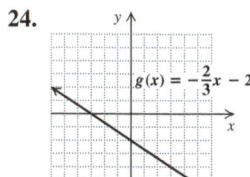
$g(x) = -\frac{2}{3}x - 2$

25.
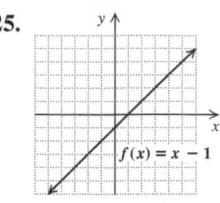
$f(x) = x - 1$

26.
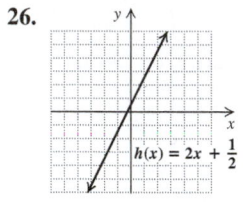
$h(x) = 2x + \frac{1}{2}$

27.
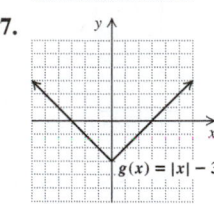
$g(x) = |x| - 3$

28.
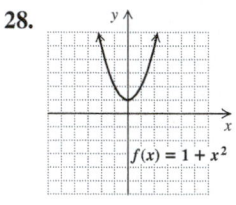
$f(x) = 1 + x^2$

29.
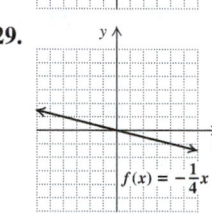
$f(x) = -\frac{1}{4}x$

30. $-x^2 + 3x - 3$ **31.** -42 **32.** $\dfrac{a^2 + 2}{3a - 1}$ **33.** Domain of $f + g$, $f - g$, and $f \cdot g$ is $\{x\,|\,x$ is a real number$\}$; domain of f/g is $\{x\,|\,x$ is a real number *and* $x \neq -4\}$.   **34.** Domain of $f + g, f - g$, and $f \cdot g$ is $\{x\,|\,x$ is a real number *and* $x \neq 9\}$; domain of f/g is $\{x\,|\,x$ is a real number *and* $x \neq 9$ and $x \neq 6\}$.
35. No; since each input has exactly one output, the number of outputs cannot exceed the number of inputs.   **36.** When $x < 0$, then $y < 0$, and the graph contains points in quadrant III. When $0 < x < 30$, then $y < 0$, and the graph contains points in quadrant IV. When $x > 30$, then $y > 0$, and the graph contains points in quadrant I. Thus the graph passes through three quadrants.   **37.** The output -3 corresponds to the input 2. The number -3 in the range is paired with the number 2 in the domain. The point $(2, -3)$ is on the graph of the function.
38. The domain of a function is the set of all inputs, and the range is the set of all outputs.

Technology Connection, p. 109

1. The graph of $y_2 = x + 4$ is the graph of $y_1 = x$ shifted up 4 units.   **2.** The graph of $y_3 = x - 3$ is the graph of $y_1 = x$ shifted down 3 units.

Technology Connection, p. 112

1. The graph of $y = 10x$ will slant up from left to right. It will be steeper than the other graphs.   **2.** The graph of $y = 0.005x$ will slant up from left to right. It will be less steep than the other graphs.   **3.** The graph of $y = -10x$ will slant down from left to right. It will be steeper than the other graphs.   **4.** The graph of $y = -0.005x$ will slant down from left to right. It will be less steep than the other graphs.

Exercise Set 2.5, p. 116

RC1. (f)   **RC2.** (b)   **RC3.** (d)   **RC4.** (c)   **RC5.** (e)
RC6. (a)   **1.** $m = 4$; *y*-intercept: $(0, 5)$ **3.** $m = -2$;

y-intercept: $(0, -6)$ **5.** $m = -\frac{3}{8}$; *y*-intercept: $\left(0, -\frac{1}{5}\right)$
7. $m = 0.5$; *y*-intercept: $(0, -9)$ **9.** $m = \frac{2}{3}$;
y-intercept: $\left(0, -\frac{8}{3}\right)$ **11.** $m = 3$; *y*-intercept: $(0, -2)$
13. $m = -8$; *y*-intercept: $(0, 12)$ **15.** $m = 0$; *y*-intercept: $\left(0, \frac{4}{17}\right)$ **17.** $m = -\frac{1}{2}$ **19.** $m = \frac{1}{3}$ **21.** $m = 2$
23. $m = \frac{2}{3}$ **25.** $m = -\frac{1}{3}$ **27.** $\frac{2}{25}$, or 8%   **29.** $\frac{13}{41}$, or about 31.7%
31. The rate of change is about \$2.74 billion per year.   **33.** The rate of change is $-\$900$ per year.   **35.** The rate of change is about \$116.14 per year.   **37.** -1323 **38.** $45x + 54$
39. $350x - 60y + 120$ **40.** 25   **41.** Square: 15 yd; triangle: 20 yd   **42.** $\left\{x\,\middle|\,x \leq -\frac{24}{5} \text{ or } x \geq 8\right\}$, or $\left(-\infty, -\frac{24}{5}\right] \cup [8, \infty)$
43. $\left\{x\,\middle|\,-\frac{24}{5} < x < 8\right\}$, or $\left(-\frac{24}{5}, 8\right)$ **44.** $\left\{-\frac{24}{5}, 8\right\}$
45. $\{\ \}$, or $\varnothing$

Technology Connection, p. 120

1. $y = -3.2x - 16$
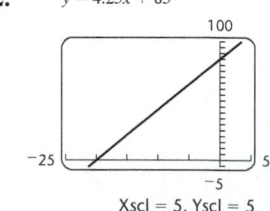
Xscl = 1, Yscl = 2

2. $y = 4.25x + 85$
Xscl = 5, Yscl = 5

3. $y = (-6x + 90)/5$
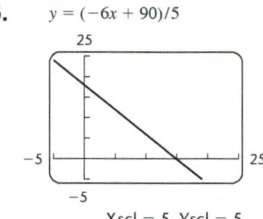
Xscl = 5, Yscl = 5

4. $y = (5x - 30)/6$
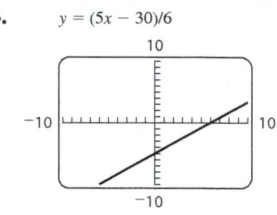

5. $y = (-8x + 9)/3$
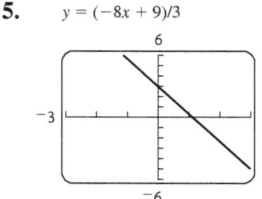

6. $y = 0.4x - 5$
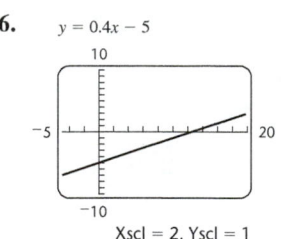
Xscl = 2, Yscl = 1

7. $y = 1.2x - 12$
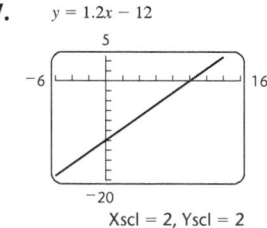
Xscl = 2, Yscl = 2

8. $y = (4x - 2)/5$
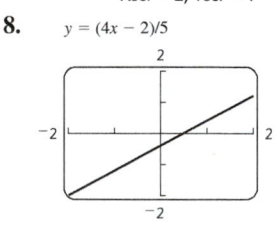

Visualizing the Graph, p. 126

1. D   **2.** I   **3.** H   **4.** C   **5.** F   **6.** A   **7.** G
8. B   **9.** E   **10.** J

Exercise Set 2.6, p. 127

RC1. True   **RC2.** False   **RC3.** False   **RC4.** True
RC5. True   **RC6.** False

1.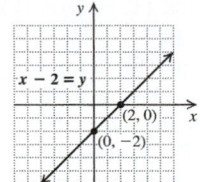

3.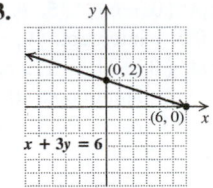

5.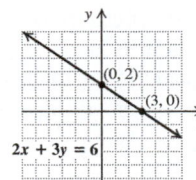

7.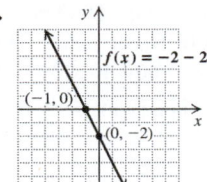

9.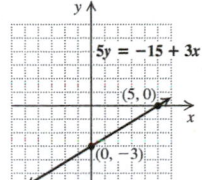

11.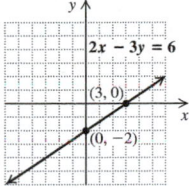

13.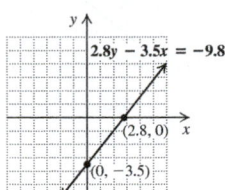

15.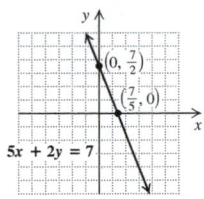

17.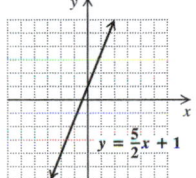

19.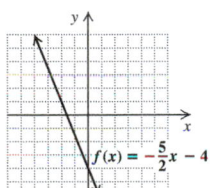

21.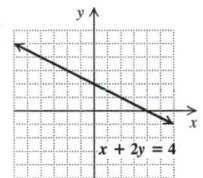

23.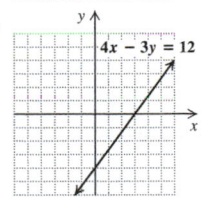

25.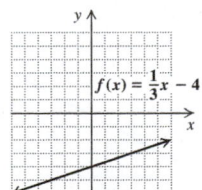

27.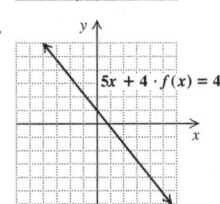

29. Not defined

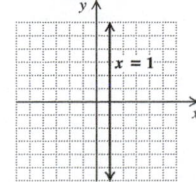

31. $m = 0$

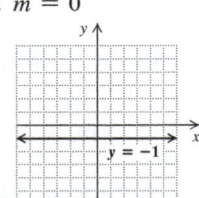

33. $m = 0$

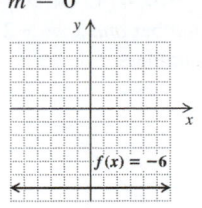

35. $m = 0$

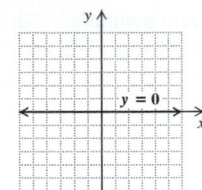

37. $m = 0$

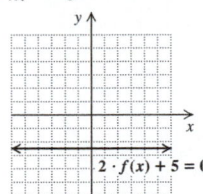

39. Not defined

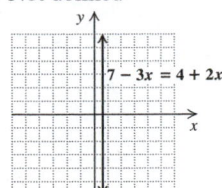

41. Yes **43.** No **45.** Yes **47.** Yes **49.** Yes
51. No **53.** No **55.** Yes **57.** 5.3×10^{10}
58. 4.7×10^{-5} **59.** 1.8×10^{-2} **60.** 9.9902×10^{7}
61. 0.0000213 **62.** $901{,}000{,}000$ **63.** $20{,}000$
64. 0.085677 **65.** $3(3x - 5y)$ **66.** $3a(4 + 7b)$
67. $7p(3 - q + 2)$ **68.** $64(x - 2y + 4)$ **69.** 2009 lb
70. 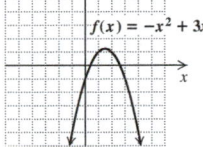   **71.** $a = 2$ **73.** $y = \frac{2}{15}x + \frac{2}{5}$

75. $y = 0$; yes **77.** $m = -\frac{3}{4}$

Exercise Set 2.7, p. 135

RC1. (a) $\frac{4}{11}$; (b) $-\frac{11}{4}$ **RC2.** (a) 0; (b) not defined
RC3. (a) 2; (b) $-\frac{1}{2}$ **RC4.** (a) $-\frac{5}{6}$; (b) $\frac{6}{5}$
RC5. (a) Not defined; (b) 0 **RC6.** (a) -2; (b) $\frac{1}{2}$
1. $y = -8x + 4$ **3.** $y = 2.3x - 1$ **5.** $f(x) = -\frac{7}{3}x - 5$
7. $f(x) = \frac{2}{3}x + \frac{5}{8}$ **9.** $y = 5x - 17$ **11.** $y = -3x + 33$
13. $y = x - 6$ **15.** $y = -2x + 16$ **17.** $y = -7$
19. $y = \frac{2}{3}x - \frac{8}{3}$ **21.** $y = \frac{1}{2}x + \frac{7}{2}$ **23.** $y = x$
25. $y = \frac{7}{4}x + 7$ **27.** $y = \frac{3}{2}x$ **29.** $y = \frac{1}{6}x$
31. $y = 13x - \frac{15}{4}$ **33.** $y = -\frac{1}{2}x + \frac{17}{2}$ **35.** $y = \frac{5}{7}x - \frac{17}{7}$
37. $y = \frac{1}{3}x + 4$ **39.** $y = \frac{1}{2}x + 4$ **41.** $y = \frac{4}{3}x - 6$
43. $y = \frac{5}{2}x + 9$
45. (a) $C(x) = 5x + 10$;

(b)   ; **(c)** $30

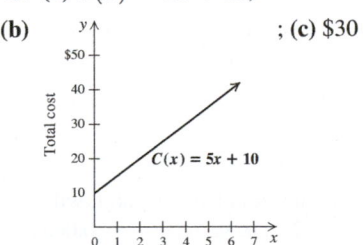

47. (a) $V(t) = 9400 - 85t$;

(b)   ; **(c)** $7870

49. (a) $S(x) = 2x + 11$; (b) \$19 billion; \$37 billion
51. (a) $D(x) = -309.41x + 22,800$; (b) 21,253 dealerships;
(c) about 24 years after 1995, or in 2019
53. (a) $E(t) = 0.346t + 46.56$; (b) about 51.06 years
55. $\{x|x > 24\}$, or $(24, \infty)$ **56.** $\{-27, 24\}$
57. $\{x|x \le 24\}$, or $(-\infty, 24]$ **58.** $\{x|x \ge \frac{7}{3}\}$, or $[\frac{7}{3}, \infty)$
59. $\{x|-8 \le x \le 5\}$, or $[-8, 5]$ **60.** $\{-7, \frac{1}{3}\}$
61. $\{\ \}$, or $\varnothing$ **62.** $\{x|-\frac{15}{2} \le x < 24\}$, or $[-\frac{15}{2}, 24)$
63. -7.75

Summary and Review: Chapter 2, p. 138

Vocabulary Reinforcement

1. Vertical **2.** Point–slope **3.** Function, domain, range,
domain, exactly one, range **4.** Slope **5.** Perpendicular
6. Slope–intercept **7.** Parallel

Concept Reinforcement

1. False **2.** True **3.** False

Study Guide

1. No **2.** $g(0) = -2$; $g(-2) = -3$; $g(6) = 1$
3.
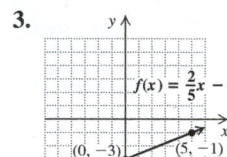
4. Yes **5.** Domain: $[-4, 5]$; range:
$[-2, 4]$ **6.** $\{x|x$ is a real number *and*
$x \ne -3\}$, or $(-\infty, -3) \cup (-3, \infty)$
7. (a) $2x^2 - x + 4$; (b) 30; (c) -15;
(d) 0

8. The domain of $f + g$ = the domain of $f - g$ = the domain of
$f \cdot g = \{x|x$ is a real number *and* $x \ne -\frac{3}{4}\}$; the domain of f/g is
$\{x|x$ is a real number *and* $x \ne -\frac{3}{4}$ *and* $x \ne 2\}$. **9.** -2
10. Slope: $-\frac{1}{2}$; y-intercept: $(0, 2)$
11.
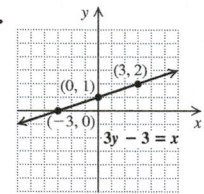
12.
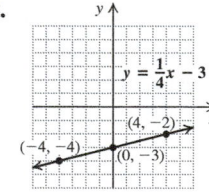
13.
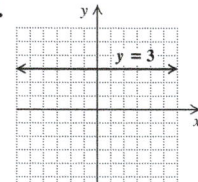
14.
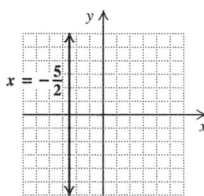
15. Parallel **16.** Perpendicular **17.** $y = -8x + 0.3$
18. $y = -4x - 1$ **19.** $y = -\frac{5}{3}x + \frac{11}{3}$ **20.** $y = \frac{4}{3}x - \frac{23}{3}$
21. $y = -\frac{3}{4}x - \frac{7}{2}$

Review Exercises

1. No **2.** Yes **3.** $g(0) = 5$; $g(-1) = 7$
4. $f(0) = 7$; $f(-1) = 12$ **5.** About \$6810
6.
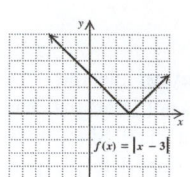
7.
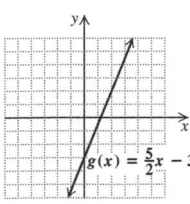

8.
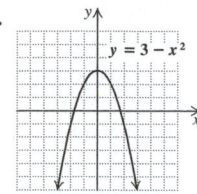
9.
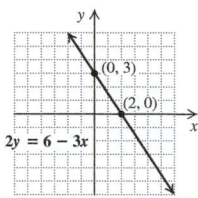
10. Yes **11.** No **12.** (a) $f(2) = 3$; (b) $\{x|-2 \le x \le 4\}$,
or $[-2, 4]$; (c) -1; (d) $\{y|1 \le y \le 5\}$, or $[1, 5]$
13. $\{x|x$ is a real number *and* $x \ne 4\}$, or $(-\infty, 4) \cup (4, \infty)$
14. All real numbers **15.** $\dfrac{x^2 - 3x}{x + 10}$ **16.** 2 **17.** Domain
of $f + g, f - g$, and $f \cdot g$ is $\{x|x$ is a real number$\}$; domain of
f/g is $\{x|x$ is a real number *and* $x \ne -6\}$. **18.** Domain of
$f + g$, $f - g$, and $f \cdot g$ is $\{x|x$ is a real number *and* $x \ne 7\}$;
domain of f/g is $\{x|x$ is a real number *and* $x \ne 7$ *and* $x \ne 5\}$.
19. Slope: -3; y-intercept: $(0, 2)$ **20.** Slope: $-\frac{1}{2}$;
y-intercept: $(0, 2)$ **21.** $\frac{11}{3}$
22.
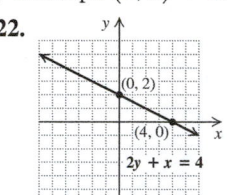
23.
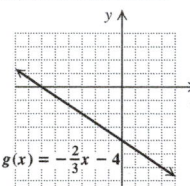
24.
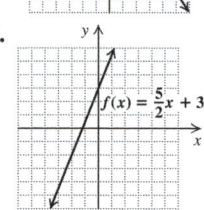
25.
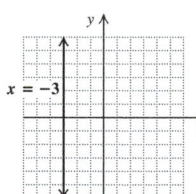
26.
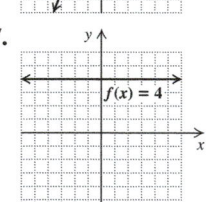
27.

28. Perpendicular **29.** Parallel **30.** Parallel
31. Perpendicular **32.** $f(x) = 4.7x - 23$
33. $y = -3x + 4$ **34.** $y = -\frac{3}{2}x$ **35.** $y = -\frac{5}{7}x + 9$
36. $y = \frac{1}{3}x + \frac{1}{3}$ **37.** (a) $R(x) = -0.018x + 44.66$;
(b) about 44.16 sec; about 43.98 sec **38.** C **39.** A
40. $f(x) = 3.09x + 3.75$ **41.** A line's x- and y-intercepts are
the same only when the line passes through the origin. The equation
for such a line is of the form $y = mx$. **42.** The concept of
slope is useful in describing how a line slants. A line with positive
slope slants up from left to right. A line with negative slope slants
down from left to right. The larger the absolute value of the slope,
the steeper the slant. **43.** Find the slope–intercept form of the
equation:

$$4x + 5y = 12$$
$$5y = -4x + 12$$
$$y = -\frac{4}{5}x + \frac{12}{5}.$$

This form of the equation indicates that the line has a negative
slope and thus should slant down from left to right. The student
may have graphed $y = \frac{4}{5}x + \frac{12}{5}$. **44.** For $R(t) = 50t + 35$,
$m = 50$ and $b = 35$; 50 signifies that the cost per hour of a repair
is \$50; 35 signifies that the minimum cost of a repair job is \$35.

45. $m = \dfrac{\text{change in } y}{\text{change in } x}$

As we move from one point to another on a vertical line, the y-coordinate changes but the x-coordinate does not. Thus the change in y is a nonzero number whereas the change in x is 0. Since division by 0 is not defined, the slope of a vertical line is not defined. As we move from one point to another on a horizontal line, the y-coordinate does not change but the x-coordinate does. Thus the change in y is 0 whereas the change in x is a nonzero number, so the slope is 0.
46. Using algebra, we find that the slope–intercept form of the equation is $y = \frac{5}{2}x - \frac{3}{2}$. This indicates that the y-intercept is $\left(0, -\frac{3}{2}\right)$, so a mistake has been made. It appears that the student graphed $y = \frac{5}{2}x + \frac{3}{2}$.

Test: Chapter 2, p. 148

1. [2.2a] Yes   **2.** [2.2a] No   **3.** [2.2b] -4; 2
4. [2.2b] 7; 8   **5.** [2.2b] -6; -6 **6.** [2.2b] 3; 0
7. [2.2c]   **8.** [2.2c]

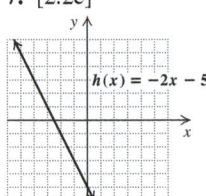   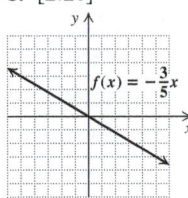

9. [2.2c]   **10.** [2.2c]

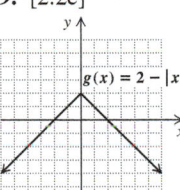   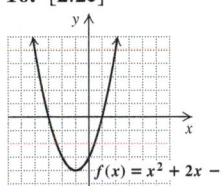

11. [2.6c]   **12.** [2.6c]

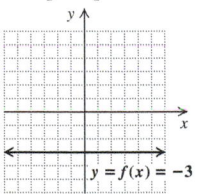   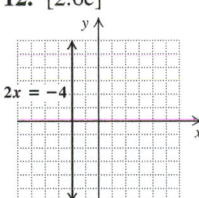

13. [2.2e] **(a)** About 9.4 years; **(b)** 1998   **14.** [2.2d] Yes
15. [2.2d] No   **16.** [2.3a] $\left\{x \mid x \text{ is a real number } and \ x \neq -\frac{3}{2}\right\}$, or $\left(-\infty, -\frac{3}{2}\right) \cup \left(-\frac{3}{2}, \infty\right)$ **17.** [2.3a] $\{x \mid x \text{ is a real number}\}$
18. [2.3a] **(a)** 1; **(b)** $[-3, 4]$; **(c)** -3; **(d)** $[-1, 2]$
19. [2.4a] 8; $\dfrac{-4x + 3}{x^2 - 1}$ **20.** [2.4b] Domain of $f + g$, $f - g$, and $f \cdot g$ is $\{x \mid x \text{ is a real number } and \ x \neq 3\}$; domain of f/g is $\left\{x \mid x \text{ is a real number } and \ x \neq 3 \ and \ x \neq -\frac{1}{2}\right\}$.
21. [2.5b] Slope: $-\frac{3}{5}$; y-intercept: $(0, 12)$
22. [2.5b] Slope: $-\frac{2}{5}$; y-intercept: $\left(0, -\frac{7}{5}\right)$
23. [2.5b] $\frac{5}{8}$ **24.** [2.5b] 0   **25.** [2.5c] $\frac{4}{5}$ km/min
26. [2.6a]   **27.** [2.6b]

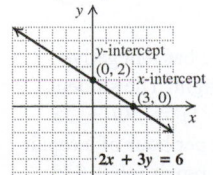

28. [2.6d] Parallel   **29.** [2.6d] Perpendicular
30. [2.7a] $y = -3x + 4.8$ **31.** [2.7a] $f(x) = 5.2x - \frac{5}{8}$
32. [2.7b] $y = -4x + 2$ **33.** [2.7c] $y = -\frac{3}{2}x$
34. [2.7d] $y = \frac{1}{2}x - 3$ **35.** [2.7d] $y = 3x - 1$
36. [2.7e] **(a)** $A(x) = 0.125x + 23.2$; **(b)** 27.95 years; 28.825 years
37. [2.7b] B   **38.** [2.6d] $\frac{24}{5}$ **39.** [2.2b] $f(x) = 3$; answers may vary

Chapter 3

Technology Connection, p. 157

1. $(2, 3)$ **2.** $(-4, -1)$ **3.** $(-1, 5)$ **4.** $(3, -1)$

Exercise Set 3.1, p. 157

RC1. False   **RC2.** True   **RC3.** True   **RC4.** True
1. $(3, 1)$; consistent; independent   **3.** $(1, -2)$; consistent; independent   **5.** $(4, -2)$; consistent; independent
7. $(2, 1)$; consistent; independent   **9.** $\left(\frac{5}{2}, -2\right)$; consistent; independent   **11.** $(3, -2)$; consistent; independent
13. No solution; inconsistent; independent
15. Infinitely many solutions; consistent; dependent
17. $(4, -5)$; consistent; independent   **19.** $(2, -3)$; consistent; independent   **21.** Consistent; independent; F
23. Consistent; dependent; B   **25.** Inconsistent; independent; D
27. -3 **28.** -20 **29.** $\frac{9}{20}$ **30.** -38 **31.** $(2.23, 1.14)$
33. $(3, 3)$, $(-5, 5)$

Exercise Set 3.2, p. 163

RC1. True   **RC2.** True   **RC3.** False   **RC4.** False
1. $(2, -3)$ **3.** $\left(\frac{21}{5}, \frac{12}{5}\right)$ **5.** $(2, -2)$ **7.** $(-2, -6)$
9. $(-2, 1)$ **11.** No solution   **13.** $\left(\frac{19}{8}, \frac{1}{8}\right)$
15. Infinitely many solutions   **17.** $\left(\frac{1}{2}, \frac{1}{2}\right)$
19. Length: 25 m; width: 5 m   **21.** $48°$ and $132°$
23. Wins: 23; ties: 14   **25.** 1.3   **26.** $-15y - 39$
27. $p = \dfrac{7A}{q}$ **28.** $\frac{7}{3}$ **29.** -23 **30.** $\frac{29}{22}$
31. $m = -\frac{1}{2}$; $b = \frac{5}{2}$ **33.** Length: 57.6 in.; width: 20.4 in.

Exercise Set 3.3, p. 170

RC1. Consistent   **RC2.** Inconsistent   **RC3.** Consistent
RC4. Dependent   **RC5.** Inconsistent   **RC6.** Independent
1. $(1, 2)$ **3.** $(-1, 3)$ **5.** $(-1, -2)$ **7.** $(5, 2)$
9. Infinitely many solutions   **11.** $\left(\frac{1}{2}, -\frac{1}{2}\right)$ **13.** $(4, 6)$
15. No solution   **17.** $(10, -8)$ **19.** $(12, 15)$ **21.** $(10, 8)$
23. $(-4, 6)$ **25.** $(10, -5)$ **27.** $(140, 60)$ **29.** 36 and 27
31. 18 and -15 **33.** $48°$ and $42°$ **35.** Two-point shots: 21; three-point shots: 6   **37.** 3-credit courses: 25; 4-credit courses: 8
39. 1   **40.** 5   **41.** 15   **42.** $12a^2 - 2a + 1$
43. $\{x \mid x \text{ is a real number } and \ x \neq -7\}$ **44.** Domain: all real numbers; range: $\{y \mid y \leq 5\}$ **45.** $y = -\frac{3}{5}x - 7$
46. $y = x + 12$ **47.** $(23.12, -12.04)$ **49.** $A = 2, B = 4$
51. $p = 2, q = -\frac{1}{3}$

Translating for Success, p. 180

1. G　**2.** E　**3.** D　**4.** A　**5.** J　**6.** B　**7.** C
8. I　**9.** F　**10.** H

Exercise Set 3.4, p. 181

RC1. 10　**RC2.** 15　**RC3.** $0.15y$　**RC4.** 2
1. Books: 45; games: 23　**3.** Olive oil: $22\frac{1}{2}$ oz; vinegar: $7\frac{1}{2}$ oz
5. Foil balloons; 2; latex balloons: 7　**7.** 5 lb of each
9. 25%-acid: 4 L; 50%-acid: 6 L　**11.** Sweet-pepper packets: 11;
hot-pepper packets: 5　**13.** Whole milk: $169\frac{3}{13}$ lb; cream: $30\frac{10}{13}$ lb
15. $7500 at 6%; $4500 at 3%　**17.** $1800 at 5.5%; $1400 at 4%
19. $5 bills: 7; $1 bills: 15　**21.** 375 mi　**23.** 14 km/h
25. Headwind: 30 mph; plane: 120 mph　**27.** $1\frac{1}{3}$ hr
29. About 1489 mi　**31.** $\{6, 8, 10\}$　**32.** $\{2, 4, 6, 7, 8, 9, 10\}$
33. $3|a|$　**34.** $7x^2$　**35.** $\dfrac{3}{|y|}$　**36.** $\dfrac{a^4}{|c|}$　**37.** $4\frac{4}{7}$ L
39. City: 261 mi; highway: 204 mi

Mid-Chapter Review: Chapter 3, p. 184

1. False　**2.** False　**3.** True　**4.** True
5. $(5, -1)$; consistent; independent　**6.** $(0, 3)$; consistent;
independent　**7.** Infinitely many solutions; consistent; dependent
8. No solution; inconsistent; independent　**9.** $(8, 6)$
10. $(2, -3)$　**11.** $(-3, 5)$　**12.** $(-1, -2)$　**13.** $(2, -2)$
14. $(5, -4)$　**15.** $(-1, -2)$　**16.** $(3, 1)$　**17.** No solution
18. Infinitely many solutions　**19.** $(10, -12)$　**20.** $(-9, 8)$
21. Length: 12 ft; width: 10 ft　**22.** $2100 at 2%; $2900 at 3%
23. 20% acid: 56 L; 50% acid: 28 L　**24.** 26 mph
25. *Graphically*: **1.** Graph $y = \frac{3}{4}x + 2$ and $y = \frac{2}{5}x - 5$ and
find the point of intersection. The first coordinate of this point is
the solution of the original equation. **2.** Rewrite the equation as
$\frac{7}{20}x + 7 = 0$. Then graph $y = \frac{7}{20}x + 7$ and find the x-intercept.
The first coordinate of this point is the solution of the original
equation.　*Algebraically*: **1.** Use the addition and multiplication
principles for equations. **2.** Multiply by 20 to clear the fractions
and then use the addition and multiplication principles for
equations.
26. (a) Answers may vary.
$$x + y = 1,$$
$$x - y = 7$$
　(b) Answers may vary.
$$x + 2y = 5,$$
$$3x + 6y = 10$$
　(c) Answers may vary.
$$x - 2y = 3,$$
$$3x - 6y = 9$$
27. Answers may vary. Form a linear expression in two variables
and set it equal to two different constants. See Exercises 8 and
17 in this review for examples.　**28.** Answers may vary. Let
any linear equation be one equation in the system. Multiply by a
constant on both sides of that equation to get the second equation
in the system. See Exercises 7 and 18 in this review for examples.

Exercise Set 3.5, p. 189

RC1. (b)　**RC2.** (c)　**RC3.** (a)　**RC4.** (a)
1. $(1, 2, -1)$　**3.** $(2, 0, 1)$　**5.** $(3, 1, 2)$
7. $(-3, -4, 2)$　**9.** $(2, 4, 1)$　**11.** $(-3, 0, 4)$　**13.** $(2, 2, 4)$
15. $\left(\frac{1}{2}, 4, -6\right)$　**17.** $(-2, 3, -1)$　**19.** $\left(\frac{1}{2}, \frac{1}{3}, \frac{1}{6}\right)$
21. $(3, -5, 8)$　**23.** $(15, 33, 9)$　**25.** $(4, 1, -2)$
27. $(17, 9, 79)$　**28.** $a = \dfrac{F}{3b}$　**29.** $a = \dfrac{Q - 4b}{4}$, or $\dfrac{Q}{4} - b$
30. $d = \dfrac{tc - 2F}{t}$, or $c - \dfrac{2F}{t}$　**31.** $c = \dfrac{2F + td}{t}$, or $\dfrac{2F}{t} + d$
32. $y = \dfrac{c - Ax}{B}$　**33.** $y = \dfrac{Ax - c}{B}$　**34.** Slope: $-\frac{2}{3}$;
y-intercept: $\left(0, -\frac{5}{4}\right)$　**35.** Slope: -4; y-intercept: $(0, 5)$
36. Slope: $\frac{2}{5}$; y-intercept: $(0, -2)$　**37.** Slope: 1.09375;
y-intercept: $(0, -3.125)$　**39.** $(1, -2, 4, -1)$

Exercise Set 3.6, p. 193

RC1. (c)　**RC2.** (d)　**RC3.** (a)　**RC4.** (b)
1. Reading: 496; math: 514; writing: 488　**3.** $32°, 96°, 52°$
5. $-7, 20, 42$　**7.** Sixteen: 10; Original: 16; Power: 8
9. Egg: 274 mg; cupcake: 19 mg; pizza: 9 mg
11. Automatic transmission: $865; power door locks: $520; air
conditioning: $375　**13.** Dog: $200; cat: $81; bird: $9
15. Roast beef: 2; baked potato: 1; broccoli: 2　**17.** First fund:
$45,000; second fund: $10,000; third fund: $25,000
19. Par-3: 6 holes; par-4: 8 holes; par-5: 4 holes
21. A: 1500 lenses; B: 1900 lenses; C: 2300 lenses
23.

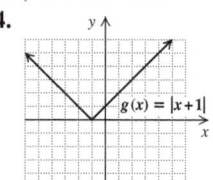

24.

25.

26. No　**27.** Yes　**28.** Yes
29. $180°$　**31.** 464

Visualizing for Success, p. 203

1. D　**2.** B　**3.** E　**4.** C　**5.** I　**6.** G　**7.** F
8. H　**9.** A　**10.** J

Exercise Set 3.7, p. 204

1. (f)　**3.** (h)　**5.** (g)　**7.** (b)
9.

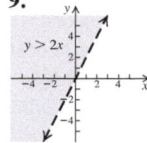

11.

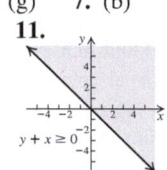

13.

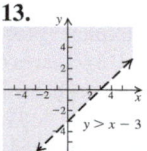

15.
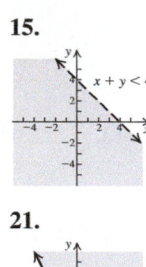
$x + y < 4$

17.
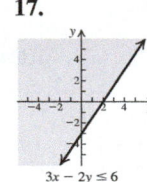
$3x - 2y \le 6$

19.
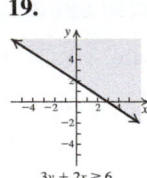
$3y + 2x \ge 6$

21.
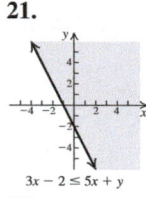
$3x - 2 \le 5x + y$

23.
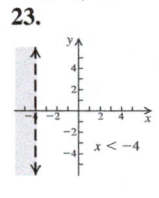
$x < -4$

25.
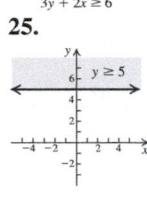
$y \ge 5$

27.
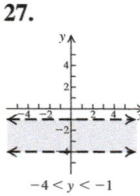
$-4 < y < -1$

29.
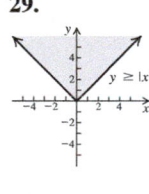
$y \ge |x|$

31. (f)

33. (a)

35. (b)

37. $y \le -x + 4$, $y \le 3x$

39. $x < 2$, $y > -1$

41. $y \le -x + 3$, $y \le x + 1$, $x \ge 0$, $y \ge 0$

43.
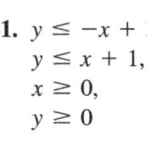
$\left(\frac{3}{2}, \frac{3}{2}\right)$

45.
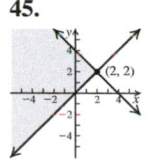
$(2, 2)$

47.
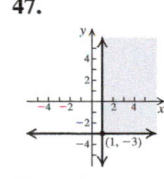
$(1, -3)$

49.
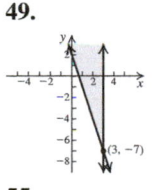
$(3, -7)$

51.
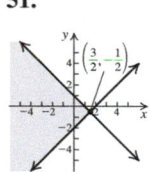
$\left(\frac{3}{2}, \frac{1}{2}\right)$

53.
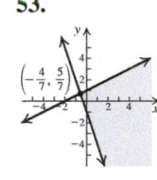
$\left(-\frac{4}{7}, \frac{5}{7}\right)$

55.
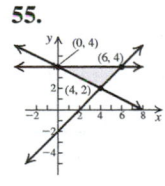
$(0, 4)$, $(6, 4)$, $(4, 2)$

57.
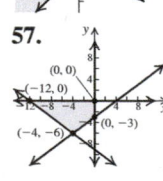
$(0, 0)$, $(-12, 0)$, $(-4, -6)$, $(0, -3)$

59.
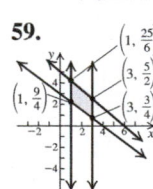
$\left(1, \frac{25}{6}\right)$, $\left(3, \frac{5}{2}\right)$, $\left(1, \frac{9}{4}\right)$, $\left(3, \frac{3}{4}\right)$

61. Maximum: 179 when $x = 7$ and $y = 0$; minimum: 48 when $x = 0$ and $y = 4$ **63.** Maximum: 216 when $x = 0$ and $y = 6$; minimum: 0 when $x = 0$ and $y = 0$ **65.** Maximum number of miles is 480 when the truck uses 9 gal and the moped uses 3 gal. **67.** Maximum profit of $54,800 is achieved when 80 acres of corn and 160 acres of soybeans are planted. **69.** Minimum cost of $51\frac{9}{13}$ is achieved by using $1\frac{11}{13}$ sacks of soybean meal and $1\frac{11}{13}$ sacks of oats. **71.** Maximum income of $1575 is achieved when $10,000 is invested in corporate bonds and $30,000 is invested in municipal bonds. **73.** Minimum cost of $460 thousand is achieved using 30 P_1's and 10 P_2's. **75.** Maximum profit per week of $2210 is achieved when 5 silk organza dresses and 2 lace

dresses are made. **77.** Minimum weekly cost of $19.05 is achieved when 1.5 lb of meat and 3 lb of cheese are used. **79.** Maximum total number of 800 is achieved when there are 550 of A and 250 of B. **81.** $\{x \mid -7 \le x < 2\}$, or $[-7, 2)$ **82.** $\{x \mid x \le 1 \text{ or } x \ge 5\}$, or $(-\infty, 1] \cup [5, \infty)$ **83.** $\{x \mid -1 \le x \le 3\}$, or $[-1, 3]$ **84.** $\{x \mid -3 < x < -2\}$, or $(-3, -2)$

85.
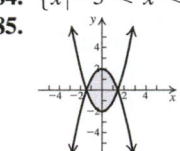

87.
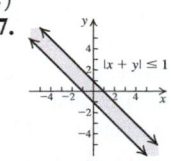
$|x + y| \le 1$

89.
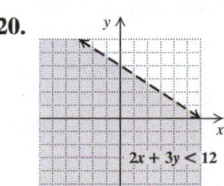
$|x| > |y|$

91. Maximum income of $19,000 is achieved by making 95 chairs and 0 sofas.

Summary and Review: Chapter 3, p. 208

Vocabulary Reinforcement

1. Pair **2.** Consistent **3.** Triple **4.** Independent **5.** Half-plane

Concept Reinforcement

1. False **2.** True **3.** True **4.** False

Study Guide

1. $(4, -1)$; consistent; independent **2.** $(-1, 4)$ **3.** $(-2, 3)$ **4.** $8700 at 6%; $14,300 at 5% **5.** $(3, -5, 1)$

6.
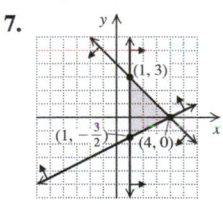
$3x - 2y > 6$

7.
$(1, 3)$, $\left(1, -\frac{3}{2}\right)$, $(4, 0)$

8. Maximum of 6 when $x = 4$ and $y = 2$.

Review Exercises

1. $(-2, 1)$; consistent; independent **2.** Infinitely many solutions; consistent; dependent **3.** No solution; inconsistent; independent **4.** $(1, -1)$ **5.** No solution **6.** $\left(\frac{2}{5}, -\frac{4}{5}\right)$ **7.** $(6, -3)$ **8.** $(2, 2)$ **9.** $(5, -3)$ **10.** Infinitely many solutions **11.** 32 brushes at $8.50; 13 brushes at $9.75 **12.** 5 L of each **13.** $5\frac{1}{2}$ hr **14.** $(10, 4, -8)$ **15.** $(-1, 3, -2)$ **16.** $(2, 0, 4)$ **17.** $\left(2, \frac{1}{3}, -\frac{2}{3}\right)$ **18.** $90°, 67\frac{1}{2}°, 22\frac{1}{2}°$ **19.** Caramel nut crunch: $30; plain $5; mocha choco latte: $14

20.
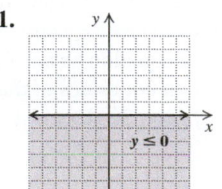
$2x + 3y < 12$

21.
$y \le 0$

22.

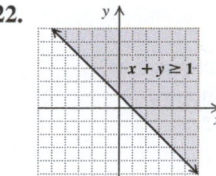

23.

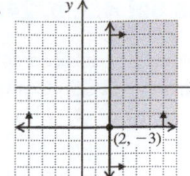

24.

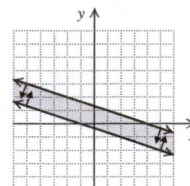

25.

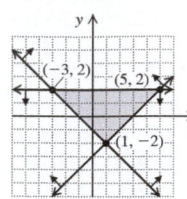

26. Maximum score of 96 is achieved when 0 group A questions and 8 group B questions are answered. **27.** C **28.** A
29. $(0, 2)$ and $(1, 3)$ **30.** Answers may vary. One day, a florist sold a total of 23 hanging baskets and flats of petunias. Hanging baskets cost $10.95 each and flats of petunias cost $12.95 each. The sales totaled $269.85. How many of each were sold?
31. We know that machines A, B, and C can polish 5700 lenses in one week when working together. We also know that A and B together can polish 3400 lenses in one week, so C can polish $5700 - 3400$, or 2300, lenses in one week alone. We also know that B and C together can polish 4200 lenses in one week, so A can polish $5700 - 4200$, or 1500, lenses in one week alone. Also, B can polish $4200 - 2300$, or 1900, lenses in one week alone.
32. Let $x =$ the number of adults in the audience, $y =$ the number of senior citizens, and $z =$ the number of children. The total attendance is 100, so we have equation (1), $x + y + z = 100$. The amount taken in was $100, so equation (2) is $10x + 3y + 0.5z = 100$. There is no other information that can be translated to an equation. Clearing decimals in equation (2) and then eliminating z gives us equation (3), $95x + 25y = 500$. Dividing by 5 on both sides, we have equation (4), $19x + 5y = 100$. Since we have only two equations, it is not possible to eliminate z from another pair of equations. However, in $19x + 5y = 100$, note that 5 is a factor of both $5y$ and 100. Therefore, 5 must also be a factor of $19x$, and hence of x, since 5 is not a factor of 19. Then for some positive integer n, $x = 5n$. (We require n to be positive, since the number of adults clearly cannot be negative and must also be nonzero since the exercise states that the audience consists of adults, senior citizens, and children.) We have

$$19 \cdot 5n + 5y = 100$$
$$19n + y = 20. \quad \text{Dividing by 5}$$

Since n and y must both be positive, $n = 1$. (If $n > 1$, then $19n + y > 20$.) Then $x = 5 \cdot 1$, or 5.

$$19 \cdot 5 + 5y = 100 \quad \text{Substituting in (4)}$$
$$y = 1$$
$$5 + 1 + z = 100 \quad \text{Substituting in (1)}$$
$$z = 94$$

There were 5 adults, 1 senior citizen, and 94 children in the audience. **33.** No; the symbol $\geq$ does not always yield a graph in which the half-plane above the line is shaded. For the inequality $-y \geq 3$, for example, the half-plane below the line $y = -3$ is shaded.

Test: Chapter 3, p. 215

1. [3.1a] $(-2, 1)$; consistent; independent **2.** [3.1a] No solution; inconsistent; independent **3.** [3.1a] Infinitely many solutions; consistent; dependent **4.** [3.2a] $(2, -3)$

5. [3.2a] Infinitely many solutions **6.** [3.2a] $(-4, 5)$
7. [3.3a] $(-1, 1)$ **8.** [3.3a] $\left(-\frac{3}{2}, -\frac{1}{2}\right)$ **9.** [3.3a] No solution
10. [3.2b] Length: 93 ft; width: 51 ft **11.** [3.4b] 120 km/h
12. [3.3b], [3.4a] Buckets: 17; dinners: 11 **13.** [3.4a] 20% solution: 12 L; 45% solution: 8 L **14.** [3.5a] $\left(2, -\frac{1}{2}, -1\right)$
15. [3.6a] 3.5 hr
16. [3.7a]

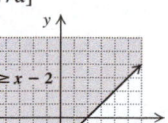

17. [3.7a]

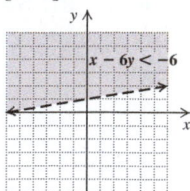

18. [3.7b]

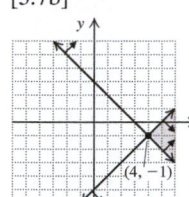

19. [3.7b]

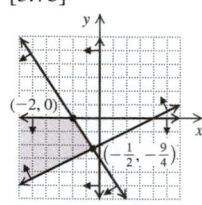

20. [3.7c] Maximum profit of $750 occurs when 25 pound cakes and 75 carrot cakes are prepared. **21.** [3.6a] B
22. [3.3a] $m = 7; b = 10$

Chapter 4

Exercise Set 4.1, p. 224

RC1. (c) **RC2.** (d) **RC3.** (a) **RC4.** (b) **RC5.** (h)
RC6. (f) **RC7.** (g) **RC8.** (e)
1. $-9x^4, -x^3, 7x^2, 6x, -8$; 4, 3, 2, 1, 0; 4; $-9x^4$; -9; -8
3. $t^3, 4t^7, s^2t^4, -2$; 3, 7, 6, 0; 7; $4t^7$; 4; -2
5. $u^7, 8u^2v^6, 3uv, 4u, -1$; 7, 8, 2, 1, 0; 8; $8u^2v^6$; 8; -1
7. $-4y^3 - 6y^2 + 7y + 23$ **9.** $-xy^3 + x^2y^2 + x^3y + 1$
11. $-9b^5y^5 - 8b^2y^3 + 2by$ **13.** $5 + 12x - 4x^3 + 8x^5$
15. $3xy^3 + x^2y^2 - 9x^3y + 2x^4$
17. $-7ab + 4ax - 7ax^2 + 4x^6$ **19.** 45; 21; 5
21. -168; -9; 4; $-7\frac{7}{8}$ **23.** About 288 watts
25. **(a)** About 340 mg; **(b)** about 190 mg; **(c)** $M(5) \approx 65$;
(d) $M(3) \approx 300$ **27.** **(a)** $10,750; **(b)** $18,287.50
29. $P(x) = -x^2 + 280x - 7000$ **31.** 17 **33.** 8 **35.** $2x^2$
37. $3x + 4y$ **39.** $7a + 14$ **41.** $-6a^2b - 2b^2$
43. $9x^2 + 2xy + 15y^2$ **45.** $-x^2y + 4y + 9xy^2$
47. $5x^2 + 2y^2 + 5$ **49.** $6a + b + c$ **51.** $-4a^2 - b^2 + 3c^2$
53. $-3x^2 + 2x + xy - 1$ **55.** $5x^2y - 4xy^2 + 5xy$
57. $9r^2 + 9r - 9$ **59.** $-\frac{2}{15}xy + \frac{19}{12}xy^2 + 1.7x^2y$
61. $-(5x^3 - 7x^2 + 3x - 6); -5x^3 + 7x^2 - 3x + 6$
63. $-(-13y^2 + 6ay^4 - 5by^2); 13y^2 - 6ay^4 + 5by^2$
65. $11x - 7$ **67.** $-4x^2 - 3x + 13$ **69.** $2a + 3c - 4b$
71. $-2x^2 + 6x$ **73.** $-4a^2 + 8ab - 5b^2$
75. $16ab + 8a^2b + 3ab^2$
77. $0.06y^4 + 0.032y^3 - 0.94y^2 + 0.93$
79. $x^4 - x^2 - 1$
81.

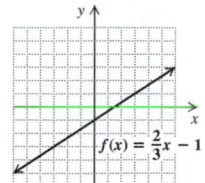

82.

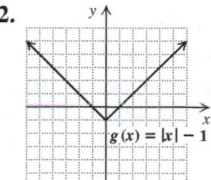

83. 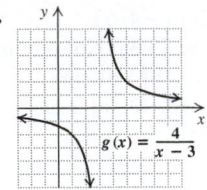   **84.**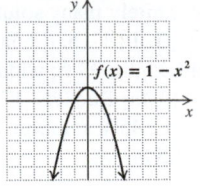

85. $-\frac{1}{2}$ **86.** 3   **87.** No solution   **88.** $\{y \mid y \geq 2\}$, or $[2, \infty)$

89. 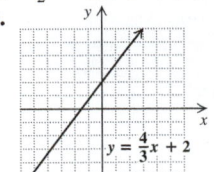   **90.**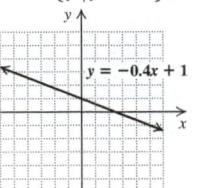

91. 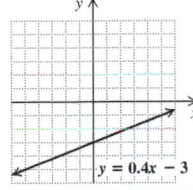   **92.**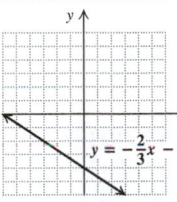

93. 494.55 cm^3 **95.** $f(0) = 5$, but the graph shows that $f(0) = -5$ **97.** $47x^{4a} + 40x^{3a} + 30x^{2a} + x^a + 4$

Technology Connection, p. 233

1. Correct   **2.** Incorrect   **3.** Incorrect   **4.** Correct
5. Incorrect   **6.** Correct

Exercise Set 4.2, p. 235

RC1. True   **RC2.** False   **RC3.** True   **RC4.** True
1. $24y^3$ **3.** $-20x^3y$ **5.** $-10x^6y^7$ **7.** $14z - 2zx$
9. $6a^2b + 6ab^2$ **11.** $15c^3d^2 - 25c^2d^3$ **13.** $15x^2 + x - 2$
15. $s^2 - 9t^2$ **17.** $x^2 - 2xy + y^2$ **19.** $x^6 + 3x^3 - 40$
21. $a^4 - 5a^2b^2 + 6b^4$ **23.** $x^3 - 64$ **25.** $x^3 + y^3$
27. $a^4 + 5a^3 - 2a^2 - 9a + 5$
29. $4a^3b^2 + 4a^3b - 10a^2b^2 - 2a^2b + 3ab^3 + 7ab^2 - 6b^3$
31. $x^2 + \frac{1}{2}x + \frac{1}{16}$ **33.** $\frac{1}{8}x^2 - \frac{2}{9}$ **35.** $3.25x^2 - 0.9xy - 28y^2$
37. $a^2 + 13a + 40$ **39.** $y^2 + 3y - 28$ **41.** $9a^2 + 3a + \frac{1}{4}$
43. $x^2 - 4xy + 4y^2$ **45.** $b^2 - \frac{5}{6}b + \frac{1}{6}$ **47.** $2x^2 + 13x + 18$
49. $400a^2 - 6.4ab + 0.0256b^2$ **51.** $4x^2 - 4xy - 3y^2$
53. $x^6 + 4x^3 + 4$ **55.** $4x^4 - 12x^2y^2 + 9y^4$
57. $a^6b^4 + 2a^3b^2 + 1$ **59.** $0.01a^4 - a^2b + 25b^2$
61. $A = P + 2Pi + Pi^2$ **63.** $d^2 - 64$ **65.** $4c^2 - 9$
67. $36m^2 - 25n^2$ **69.** $x^4 - y^2z^2$ **71.** $m^4 - m^2n^2$
73. $16p^4 - 9p^2t^2$ **75.** $\frac{1}{4}p^2 - \frac{4}{9}n^2$ **77.** $x^4 - 1$
79. $a^4 - 2a^2b^2 + b^4$ **81.** $a^2 + 2ab + b^2 - 1$
83. $4x^2 + 12xy + 9y^2 - 16$
85. $t^2 + 3t - 4, p^2 + 7p + 6, h^2 + 2ah + 5h,$
$t^2 + t - 6 + c, a^2 + 5a + 5$ **87.** $3t^2 - 13t + 18,$
$3p^2 - p + 4, 3h^2 + 6ah - 7h, 3t^2 - 19t + 34 + c,$
$3a^2 - 7a + 13$ **89.** $-t^2 + 7t - 6, -p^2 + 3p + 4,$
$-h^2 - 2ah + 5h, -t^2 + 9t - 14 + c, -a^2 + 5a + 5$
91. $-t^2 + 5t, -p^2 + p + 6, -h^2 - 2ah + 3h,$
$-t^2 + 7t - 6 + c, -a^2 + 3a + 9$
93. 5.5 hr   **94.** 180 mph   **95.** $\left(\frac{4}{3}, -\frac{14}{27}\right)$ **96.** $(1, 3)$
97. Infinitely many solutions   **98.** $\left(\frac{10}{21}, \frac{11}{14}\right)$
99. Left to the student   **101.** z^{5n^5}
103. $r^8 - 2r^4s^4 + s^8$ **105.** $9x^{10} - \frac{30}{11}x^5 + \frac{25}{121}$
107. $x^{4a} - y^{4b}$ **109.** $x^6 - 1$

Exercise Set 4.3, p. 240

RC1. Product   **RC2.** Factors   **RC3.** Factorization
RC4. Prime   **RC5.** Common   **RC6.** Binomial
1. $3a(2a + 1)$ **3.** $x^2(x + 9)$ **5.** $4x^2(2 - x^2)$
7. $4xy(x - 3y)$ **9.** $3(y^2 - y - 3)$ **11.** $2a(2b - 3c + 6d)$
13. $5(2a^4 + 3a^2 - 5a - 6)$ **15.** $3x^2y^4z^3(5y - 4x^2z^4)$
17. $7a^3b^3c^3(2ac^2 + 3b^2c - 5ab)$ **19.** $-5(x + 9)$
21. $-6(a + 14)$ **23.** $-2(x^2 - x + 12)$ **25.** $-3y(y - 8)$
27. $-(a^4 - 2a^3 + 13a^2 + 1)$ **29.** $-3(y^3 - 4y^2 + 5y - 8)$
31. $\pi r^2\left(h + \frac{4}{3}r\right)$, or $\frac{1}{3}\pi r^2(3h + 4r)$
33. (a) $h(t) = -8t(2t - 9)$; (b) $h(2) = 80$ in each
35. $R(x) = 0.4x(700 + x)$ **37.** $(b - 2)(a + c)$
39. $(x - 2)(2x + 13)$ **41.** $(y - 7)(y^7 + 1)$
43. $(c + d)(a + b)$ **45.** $(b - 1)(b^2 + 2)$
47. $(y + 8)(y^2 - 5)$ **49.** $12(x^2 + 3)(2x - 3)$
51. $a(a^3 - a^2 + a + 1)$ **53.** $(y^2 + 3)(2y^2 - 5)$
55. $-7, 13$ **56.** $-8, -\frac{2}{5}$
57. $\{x \mid -10 \leq x \leq 14\}$, or $[-10, 14]$
58. $\left\{y \mid y < \frac{1}{3} \text{ or } y > \frac{13}{3}\right\}$, or $\left(-\infty, \frac{1}{3}\right) \cup \left(\frac{13}{3}, \infty\right)$
59. $\{x \mid 15 \leq x \leq 17\}$, or $[15, 17]$
60. $\left\{x \mid \frac{1}{3} < x < 1\right\}$, or $\left(\frac{1}{3}, 1\right)$
61. $\left\{x \mid x < \frac{1}{3} \text{ or } x > \frac{13}{2}\right\}$, or $\left(-\infty, \frac{1}{3}\right) \cup \left(\frac{13}{2}, \infty\right)$
62. All real numbers, or $(-\infty, \infty)$
63. $x^5y^4 + x^4y^6 = x^3y(x^2y^3 + xy^5)$
65. $(x^2 - x + 5)(r + s)$ **67.** $(x^4 + x^2 + 5)(a^4 + a^2 + 5)$
69. $x^{1/3}(1 - 7x)$ **71.** $x^{1/3}(1 - 5x^{1/6} + 3x^{5/12})$
73. $3a^n(a + 2 - 5a^2)$ **75.** $y^{a+b}(7y^a - 5 + 3y^b)$

Exercise Set 4.4, p. 245

RC1. (c)   **RC2.** (d)   **RC3.** (a)   **RC4.** (b)
1. $(x + 4)(x + 9)$ **3.** $(t - 5)(t - 3)$ **5.** $(x - 11)(x + 3)$
7. $2(y - 4)(y - 4)$ **9.** $(p + 9)(p - 6)$
11. $(x + 3)(x + 9)$ **13.** $\left(y - \frac{1}{3}\right)\left(y - \frac{1}{3}\right)$
15. $(t - 3)(t - 1)$ **17.** $(x + 7)(x - 2)$
19. $(x + 2)(x + 3)$ **21.** $-1(x - 8)(x + 7)$, or
$(-x + 8)(x + 7)$, or $(x - 8)(-x - 7)$
23. $-y(y - 8)(y + 4)$, or $y(-y + 8)(y + 4)$, or
$y(y - 8)(-y - 4)$ **25.** $(x^2 + 16)(x^2 - 5)$
27. Not factorable   **29.** $(x + 9y)(x + 3y)$
31. $2(x - 9)(x + 5)$ **33.** $-1(z + 12)(z - 3)$, or
$(-z - 12)(z - 3)$, or $(z + 12)(-z + 3)$
35. $(x^2 + 49)(x^2 + 1)$ **37.** $(x^3 + 9)(x^3 + 2)$
39. $(x^4 - 3)(x^4 - 8)$ **41.** $(y - 0.4)(y - 0.4)$
43. $(4 + b^{10})(3 - b^{10})$, or $-1(b^{10} + 4)(b^{10} - 3)$
45. Countryside: $9\frac{3}{8}$ lb; Mystic: $15\frac{5}{8}$ lb   **46.** 8 weekdays
47. Yes   **48.** No   **49.** No   **50.** Yes
51. All real numbers   **52.** All real numbers
53. $\left\{x \mid x \text{ is a real number } and \ x \neq \frac{7}{4}\right\}$, or $\left(-\infty, \frac{7}{4}\right) \cup \left(\frac{7}{4}, \infty\right)$
54. All real numbers   **55.** $76, -76, 28, -28, 20, -20$
57. $x - 365$

Mid-Chapter Review: Chapter 4, p. 246

1. True   **2.** False   **3.** True   **4.** False   **5.** True
6. Terms: $-a^7, a^4, -a, 8$; degree of each term: 7, 4, 1, 0; degree of
polynomial: 7; leading term: $-a^7$; leading coefficient: -1; constant
term: 8   **7.** Terms: $3x^4, 2x^3w^5, -12x^2w, 4x^2, -1$; degree of each
term: 4, 8, 3, 2, 0; degree of polynomial: 8; leading term: $2x^3w^5$;
leading coefficient: 2; constant term: -1

8. $5 - 2y - y^3 - 2y^4 + y^9$ **9.** $2x^5 - 4qx^2 + 2qx - 9qr$
10. $h(0) = 5; h(-2) = 21; h\left(\frac{1}{2}\right) = 2\frac{7}{8}$, or $\frac{23}{8}$
11. $f(-1) = 1\frac{1}{2}$, or $\frac{3}{2}; f(1) = -\frac{1}{2}; f(0) = 0$
12. $f(a - 2) = a^2 - 2a - 9; f(a + h) - f(a) = 2ah + h^2 + 2h$ **13.** $-2a^2 - 3b - 4ab - 1$ **14.** $11x^2 + 7x - 8$
15. $b^2 - 11b - 12$ **16.** $3c^4 - c^5$ **17.** $y^8 - 3y^4 - 18$
18. $4y^3 + 6y^2 - 2y$ **19.** $9x - 12$ **20.** $16x^2 - 40x + 25$
21. $4x^2 + 20x + 25$ **22.** $0.11x - 3y$ **23.** $-130x^3y$
24. $x^3 - x^2y + xy^2 + 3y^3$ **25.** $10x^2 + 31x - 63$
26. $81x^2 - 16$ **27.** $h(5h + 7)$ **28.** $(x + 10)(x - 2)$
29. $-(b + 7)(b - 3)$, or $(7 + b)(3 - b)$ **30.** $\left(m + \frac{1}{7}\right)^2$
31. $(2 - x)(xy + 5)$ **32.** $3(w - 1)^2$ **33.** $(t + 3)(t^2 + 1)$
34. $8xy^3z(3y^3z^3 - 2x^3)$ **35.** Not factorable
36. One explanation is as follows. The expression $-(a - b)$ is the opposite of $a - b$. Since $(a - b) + (b - a) = 0$, then $-(a - b) = b - a$.   **37.** No; if the coefficients of at least one pair of like terms are opposites, then the sum is a monomial. For example, $(2x + 3) + (-2x + 1) = 4$, a monomial.
38. No; consider the polynomial $3x^{11} + 5x^7$. All the coefficients and exponents are prime numbers, yet the polynomial can be factored so it is not prime.   **39.** When coefficients and/or exponents are large, a polynomial is more easily evaluated after it has been factored.
40. (a) The middle term, $2 \cdot a \cdot 3$, is missing from the right-hand side.
$$(a + 3)^2 = a^2 + 6a + 9$$
(b) The middle term, $-2ab$, is missing from the right-hand side and the sign preceding b^2 is incorrect.
$$(a - b)(a - b) = a^2 - 2ab + b^2$$
(c) The product of the outside terms and the product of the inside terms are missing from the right-hand side.
$$(x + 3)(x - 4) = x^2 - x - 12$$
(d) There should be a minus sign between the terms of the product.
$$(p + 7)(p - 7) = p^2 - 49$$
(e) The middle term, $-2 \cdot t \cdot 3$, is missing from the right-hand side and the sign preceding 9 is incorrect.
$$(t - 3)^2 = t^2 - 6t + 9$$
41. Answers may vary. For the polynomial $4a^3 - 12a$, an incorrect factorization is $4a(a - 3)$. Evaluating both the polynomial and the factorization for $a = 0$, we get 0 in each case. Thus the evaluation does not catch the mistake.

Exercise Set 4.5, p. 253

RC1. $2x^2$ **RC2.** 1   **RC3.** $-3x$ **RC4.** Negative
RC5. Leading; constant   **RC6.** Product; sum
RC7. $21x$ **RC8.** Factor
1. $(3x + 1)(x - 5)$ **3.** $y(5y - 7)(2y + 3)$
5. $(3c - 8)(c - 4)$ **7.** $(5y + 2)(7y + 4)$
9. $2(5t - 3)(t + 1)$ **11.** $4(2x + 1)(x - 4)$
13. $3(3a - 1)(2a - 5)$ **15.** $5(3t + 1)(2t + 5)$
17. $x(3x - 4)(4x - 5)$ **19.** $x^2(7x + 1)(2x - 3)$
21. $(3a - 4)(a + 1)$ **23.** $(3x + 1)(3x + 4)$
25. $-1(z - 3)(12z + 1)$, or $(-z + 3)(12z + 1)$, or $(z - 3)(-12z - 1)$ **27.** $-1(2t - 3)(2t + 5)$, or $(-2t + 3)(2t + 5)$, or $(2t - 3)(-2t - 5)$
29. $x(3x + 1)(x - 2)$ **31.** $(24x + 1)(x - 2)$
33. $-2t(2t + 5)(2t - 3)$ **35.** $-x(24x + 1)(x - 2)$
37. $(7x + 3)(3x + 4)$ **39.** $4(10x^4 + 4x^2 - 3)$
41. $(4a - 3b)(3a - 2b)$ **43.** $(2x - 3y)(x + 2y)$

45. $2(3x - 4y)(2x - 7y)$ **47.** $(3x - 5y)(3x - 5y)$
49. $(3x^3 - 2)(x^3 + 2)$ **51. (a)** 224 ft; 288 ft; 320 ft; 288 ft; 128 ft;
(b) $h(t) = -16(t - 7)(t + 2)$ **53.** $(2, -1, 0)$
54. $\left(\frac{3}{2}, -4, 3\right)$ **55.** $(1, -1, 2)$ **56.** $(2, 4, 1)$ **57.** Parallel
58. Parallel   **59.** Neither   **60.** Perpendicular
61. $y = -\frac{1}{7}x - \frac{23}{7}$ **62.** $y = -\frac{1}{3}x - \frac{7}{3}$ **63.** $y = -\frac{7}{17}x - \frac{19}{17}$
64. $y = -\frac{5}{2}x - \frac{2}{3}$ **65.** Left to the student
67. $(7a + 6)(ab + 1)$ **69.** $(9xy - 4)(xy + 1)$
71. $(x^a + 8)(x^a - 3)$

Visualizing for Success, p. 261

1. A, E   **2.** F, J   **3.** G, K   **4.** L, S   **5.** P, Q   **6.** C, I
7. D, H   **8.** M, O   **9.** N, T   **10.** B, R

Exercise Set 4.6, p. 262

RC1. Difference of squares   **RC2.** Trinomial square
RC3. Difference of cubes   **RC4.** None of these
RC5. Trinomial square   **RC6.** None of these
RC7. Sum of cubes   **RC8.** Difference of squares
1. $(x - 2)^2$ **3.** $(y + 9)^2$ **5.** $(x + 1)^2$ **7.** $(3y + 2)^2$
9. $y(y - 9)^2$ **11.** $3(2a + 3)^2$ **13.** $2(x - 10)^2$
15. $(1 - 4d)^2$, or $(4d - 1)^2$ **17.** $3a(a - 1)^2$
19. $(0.5x + 0.3)^2$ **21.** $(p - q)^2$ **23.** $(a + 2b)^2$
25. $(5a - 3b)^2$ **27.** $(y^3 + 13)^2$ **29.** $(4x^5 - 1)^2$
31. $(x^2 + y^2)^2$ **33.** $(p + 7)(p - 7)$ **35.** $(y + 2)^2(y - 2)^2$
37. $(pq + 5)(pq - 5)$ **39.** $6(x + y)(x - y)$
41. $4x(y^2 + z^2)(y + z)(y - z)$ **43.** $a(2a + 7)(2a - 7)$
45. $3(x^4 + y^4)(x^2 + y^2)(x + y)(x - y)$
47. $a^2(3a + 5b^2)(3a - 5b^2)$ **49.** $\left(\frac{1}{6} + z\right)\left(\frac{1}{6} - z\right)$
51. $(0.2x + 0.3y)(0.2x - 0.3y)$
53. $(m - 7)(m + 2)(m - 2)$
55. $(a - 2)(a + b)(a - b)$
57. $(a + b + 10)(a + b - 10)$
59. $(12 - p + 8)(12 + p - 8)$, or $(20 - p)(4 + p)$
61. $(a + b + 3)(a + b - 3)$
63. $(r - 1 + 2s)(r - 1 - 2s)$
65. $2(m + n + 5b)(m + n - 5b)$
67. $[3 + (a + b)][3 - (a + b)]$, or $(3 + a + b)(3 - a - b)$
69. $(z + 3)(z^2 - 3z + 9)$ **71.** $(x - 1)(x^2 + x + 1)$
73. $(2 - 3b)(4 + 6b + 9b^2)$ **75.** $(2a + 1)(4a^2 - 2a + 1)$
77. $(2x + 3)(4x^2 - 6x + 9)$ **79.** $(a - b)(a^2 + ab + b^2)$
81. $\left(a + \frac{1}{2}\right)\left(a^2 - \frac{1}{2}a + \frac{1}{4}\right)$ **83.** $(x + 0.1)(x^2 - 0.1x + 0.01)$
85. $2(y - 4)(y^2 + 4y + 16)$ **87.** $3(2a + 1)(4a^2 - 2a + 1)$
89. $r(s + 4)(s^2 - 4s + 16)$
91. $5x^2(x - 2z)(x^2 + 2xz + 4z^2)$
93. $8(2x^2 - t^2)(4x^4 + 2x^2t^2 + t^4)$
95. $(z - 1)(z^2 + z + 1)(z + 1)(z^2 - z + 1)$
97. $(t^2 + 4y^2)(t^4 - 4t^2y^2 + 16y^4)$
99. $(2w^3 - z^3)(4w^6 + 2w^3z^3 + z^6)$
101. $\left(\frac{1}{2}c + d\right)\left(\frac{1}{4}c^2 - \frac{1}{2}cd + d^2\right)$
103. $(0.1x - 0.2y)(0.01x^2 + 0.02xy + 0.04y^2)$ **105.** $\left(-\frac{41}{53}, \frac{148}{53}\right)$
106. $\left(-\frac{26}{7}, -\frac{134}{7}\right)$ **107.** $(1, 13)$ **108.** No solution
109. 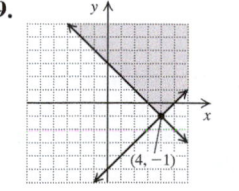   **110.**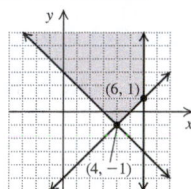

111.   **112.**

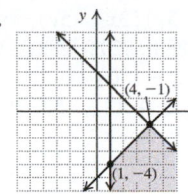

113. $y = x - 2; y = -x - 6$ **114.** $y = \frac{2}{3}x - \frac{23}{3}; y = -\frac{3}{2}x - \frac{11}{2}$
115. $y = -\frac{1}{2}x + 7; y = 2x - 3$
116. $y = \frac{1}{4}x - \frac{3}{2}; y = -4x + 24$
117. $h(3a^2 + 3ah + h^2)$ **119.** (a) $\pi h(R + r)(R - r)$;
(b) $3{,}014{,}400\ \text{cm}^3$ **121.** $5(c^{50} + 4d^{50})(c^{25} + 2d^{25})(c^{25} - 2d^{25})$
123. $(x^{2a} + y^b)(x^{4a} - x^{2a}y^b + y^{2b})$
125. $3(x^a + 2y^b)(x^{2a} - 2x^a y^b + 4y^{2b})$
127. $\frac{1}{3}\left(\frac{1}{2}xy + z\right)\left(\frac{1}{4}x^2y^2 - \frac{1}{2}xyz + z^2\right)$
129. $y(3x^2 + 3xy + y^2)$ **131.** $4(3a^2 + 4)$

Exercise Set 4.7, p. 266

RC1. Common   **RC2.** Difference   **RC3.** Square
RC4. Grouping   **RC5.** Completely   **RC6.** Check
1. $(y + 15)(y - 15)$ **3.** $(2x + 3)(x + 4)$
5. $5(x^2 + 2)(x^2 - 2)$ **7.** $(p + 6)^2$ **9.** $2(x - 11)(x + 6)$
11. $(3x + 5y)(3x - 5y)$ **13.** $4(m^2 + 5)(m^2 - 5)$
15. $6(w - 1)(w + 3)$ **17.** $2x(y + 5)(y - 5)$
19. $(18 - a)(12 + a)$
21. $(m + 1)(m^2 - m + 1)(m - 1)(m^2 + m + 1)$
23. $(x + 3 + y)(x + 3 - y)$
25. $2(5x - 4y)(25x^2 + 20xy + 16y^2)$
27. $(m^3 + 10)(m^3 - 2)$ **29.** $(a + d)(c - b)$
31. $(5b - a)(10b + a)$ **33.** $(2x - 7)(x^2 + 2)$
35. $2(x + 3)(x + 2)(x - 2)$
37. $2(2x + 3y)(4x^2 - 6xy + 9y^2)$ **39.** $-3y(5x + 2)(4x - 1)$,
or $3y(-5x - 2)(4x - 1)$, or $3y(5x + 2)(-4x + 1)$
41. $(a^4 + b^4)(a^2 + b^2)(a + b)(a - b)$.
43. $ab(a + 4b)(a - 4b)$ **45.** $\left(\frac{1}{4}x - \frac{1}{3}y^2\right)^2$
47. $5(x - y)^2(x + y)$ **49.** $(9ab + 2)(3ab + 4)$
51. $y(2y - 5)(4y^2 + 10y + 25)$ **53.** $(a - b - 3)(a + b + 3)$
55. $(q - 5 + r)(q - 5 - r)$ **57.** Correct answers: 55;
incorrect answers: 20   **58.** $\frac{80}{7}$ **59.** $(6y^2 - 5x)(5y^2 - 12x)$
61. $5\left(x - \frac{1}{3}\right)\left(x^2 + \frac{1}{3}x + \frac{1}{9}\right)$ **63.** $x(x - 2p)$
65. $y(y - 1)^2(y - 2)$
67. $(2x + y - r + 3s)(2x + y + r - 3s)$ **69.** $c(c^w + 1)^2$
71. $3x(x + 5)$ **73.** $(x - 1)^3(x^2 + 1)(x + 1)$
75. $y(y^4 + 1)(y^2 + 1)(y + 1)(y - 1)$

Technology Connection, p. 272

1. Left to the student

Translating for Success, p. 275

1. Q   **2.** F   **3.** B   **4.** A   **5.** P   **6.** D   **7.** O   **8.** H
9. I   **10.** N

Exercise Set 4.8, p. 276

RC1. False   **RC2.** False   **RC3.** False   **RC4.** False
1. $-7, 4$ **3.** 3   **5.** -10 **7.** $-5, -4$ **9.** $0, -8$ **11.** $-5, 5$
13. $-12, 12$ **15.** $7, -9$ **17.** $-4, 8$ **19.** $-2, -\frac{2}{3}$ **21.** $\frac{1}{2}, \frac{3}{4}$
23. $0, 6$ **25.** $\frac{2}{3}, -\frac{3}{4}$ **27.** $-1, 1$ **29.** $\frac{2}{3}, -\frac{5}{7}$ **31.** $0, \frac{1}{5}$

33. $7, -2$ **35.** $0, -2, 3$ **37.** $0, -8, 8$ **39.** $5, -5, 1, -1$
41. $-6, 6$ **43.** $-\frac{7}{4}, \frac{4}{3}$ **45.** $-8, -4$ **47.** $-4, \frac{3}{2}$ **49.** $-9, -3$
51. $\{x \mid x \text{ is a real number } and\ x \neq -1\ and\ x \neq 5\}$
53. $\{x \mid x \text{ is a real number } and\ x \neq -3\ and\ x \neq 3\}$
55. $\{x \mid x \text{ is a real number } and\ x \neq \frac{1}{5}\}$
57. $\{x \mid x \text{ is a real number } and\ x \neq 0\ and\ x \neq 2\ and\ x \neq 5\}$
59. x-intercepts: $(-5, 0)$ and $(9, 0)$; solutions: $-5, 9$
61. x-intercepts: $(-4, 0)$ and $(8, 0)$; solutions: $-4, 8$
63. Length: 12 cm; width: 7 cm   **65.** Height: 6 ft; base: 4 ft
67. 6 cm   **69.** 16, 18, 20   **71.** 3 cm
73. Length: 12 ft; width: 8 ft   **75.** $d = 12$ ft; $h = 16$ ft
77. 41 ft   **79.** 6, 8, 10   **81.** 11 sec   **83.** 1   **84.** 1.3
85. $\frac{19}{15}$ **86.** 1023   **87.** $y = \frac{11}{6}x + \frac{32}{3}$ **88.** $y = -\frac{11}{10}x + \frac{24}{5}$
89. $y = -\frac{3}{10}x + \frac{32}{5}$ **90.** $y = \frac{26}{31}x + \frac{934}{31}$
91. $\{-3, 1\}$; $\{x \mid -4 \leq x \leq 2\}$, or $[-4, 2]$
93. (a) 1.2522305, 3.1578935; (b) $-0.3027756, 0, 3.3027756$;
(c) 2.1387475, 2.7238657; (d) $-0.7462555, 3.3276509$

Summary and Review: Chapter 4, p. 280

Vocabulary Reinforcement

1. Ascending   **2.** Factor   **3.** Factor   **4.** Factorization
5. Grouping   **6.** Binomial   **7.** Zero   **8.** Difference

Concept Reinforcement

1. False   **2.** True   **3.** False

Study Guide

1. Terms: $-6x^4, 5x^3, -x^2, 10x, -1$; degree of each term: 4, 3, 2, 1, 0;
degree of polynomial: 4; leading term: $-6x^4$; leading coefficient: -6;
constant term: -1 **2.** $2y^3 + 2y^2 + 17y - 8$
3. $3x^2 + xy - 10y^2$ **4.** $4y^2 + 28y + 49$ **5.** $25d^2 - 100$
6. $f(x + 1) = 3x^2 + 5x + 4; f(a + h) - f(a) =$
$3h^2 + 6ah - h$ **7.** $(y + 3)(y^2 - 8)$ **8.** $(3x - 8)(x + 9)$
9. $(2x - 7)(5x + 1)$ **10.** $(9x - 4)^2$ **11.** $(10t + 1)(10t - 1)$
12. $(6x + 1)(36x^2 - 6x + 1)$
13. $(10y - 3)(100y^2 + 30y + 9)$ **14.** $-2, \frac{7}{3}$

Review Exercises

1. (a) 7, 11, 3, 2; 11; (b) $-7x^8y^3$; -7;
(c) $-3y^2 + 2x^3 + 3x^6y - 7x^8y^3$;
(d) $-7x^8y^3 - 3y^2 + 3x^6y + 2x^3$ **2.** 0; -6 **3.** 4; -31
4. $-7x + 23y$ **5.** $ab + 12ab^2 + 4$ **6.** About 4.9 million
children   **7.** $-x^3 + 2x^2 + 5x + 2$ **8.** $x^3 + 6x^2 - x - 4$
9. $13x^2y - 8xy^2 + 4xy$ **10.** $9x - 7$ **11.** $-2a + 6b + 7c$
12. $16p^2 - 8p$ **13.** $4x^2 - 7xy + 3y^2$ **14.** $-18x^3y^4$
15. $x^8 - x^6 + 5x^2 - 3$ **16.** $8a^2b^2 + 2abc - 3c^2$
17. $4x^2 - 25y^2$ **18.** $4x^2 - 20xy + 25y^2$
19. $20x^4 - 18x^3 - 47x^2 + 69x - 27$ **20.** $x^4 + 8x^2y^3 + 16y^6$
21. $x^3 - 125$ **22.** $x^2 - \frac{1}{2}x + \frac{1}{18}$
23. $a^2 - 4a - 4; 2ah + h^2 - 2h$ **24.** $3y^2(3y^2 - 1)$
25. $3x(5x^3 - 6x^2 + 7x - 3)$ **26.** $(a - 9)(a - 3)$
27. $(3m + 2)(m + 4)$ **28.** $(5x + 2)^2$
29. $4(y + 2)(y - 2)$ **30.** $(a + 2b)(x - y)$
31. $4(x^4 + x^2 + 5)$ **32.** $(3x - 2)(9x^2 + 6x + 4)$
33. $(0.4b - 0.5c)(0.16b^2 + 0.2bc + 0.25c^2)$
34. $y(y^2 + 1)(y + 1)(y - 1)$
35. $2z^6(z^2 - 8)$ **36.** $2y(3x^2 - 1)(9x^4 + 3x^2 + 1)$
37. $(1 + a)(1 - a + a^2)$ **38.** $4(3x - 5)^2$
39. $(3t + p)(2t + 5p)$ **40.** $(x + 2)(x + 3)(x - 3)$

41. $(a - b + 2t)(a - b - 2t)$ **42.** 10   **43.** $\frac{2}{3}, \frac{3}{2}$
44. $0, \frac{7}{4}$ **45.** $-4, 4$ **46.** $-4, 11$
47. $\{x \mid x \text{ is a real number } and\ x \neq \frac{2}{3} \text{ and } x \neq -7\}$
48. Length: 8 in.; width: 5 in.   **49.** $-7, -5, -3; 3, 5, 7$
50. 7   **51.** A   **52.** C
53. $2(2x + y)(4x^2 - 2xy + y^2)(2x - y)(4x^2 + 2xy + y^2)$
54. $2(3x^2 + 1)$ **55.** $a^3 - (b - 1)^3$ **56.** $0, \frac{1}{8}, -\frac{1}{8}$
57. A sum of two squares can be factored when there is a common factor that is a perfect square. For example, consider $4 + 4x^2$:
$$4 + 4x^2 = 4(1 + x^2).$$
58. See the procedure on p. 251 of the text.   **59.** Add the opposite of the polynomial being subtracted.   **60.** To solve $P(x) = 0$, find the first coordinate(s) of the x-intercept(s) of $y = P(x)$. To solve $P(x) = 4$, find the first coordinate(s) of the points of intersection of the graphs of $y_1 = P(x)$ and $y_2 = 4$.
61. To use factoring, write $x^3 - 8 = (x - 2)(x^2 + 2x + 4)$ and $(x - 2)^3 = (x - 2)(x - 2)(x - 2)$. Since $(x - 2)(x^2 + 2x + 4) \neq (x - 2)(x - 2)(x - 2)$, then $x^3 - 8 \neq (x - 2)^3$. To use graphing, graph $y_1 = x^3 - 8$ and $y_2 = (x - 2)^3$, and show that the graphs are different.
62. Both are correct. The factorizations are equivalent:
$$(a - b)(x - y) = -1(b - a)(-1)(y - x)$$
$$= (-1)(-1)(b - a)(y - x)$$
$$= (b - a)(y - x)$$
63.
$$x = 5 \quad or \quad x = -3$$
$$x - 5 = 0 \quad or \quad x + 3 = 0$$
$$(x - 5)(x + 3) = 0$$
$$x^2 - 2x - 15 = 0;$$
No; there cannot be more than two solutions of a quadratic equation. This is because a quadratic equation is factorable into at most two different linear factors. Each of these has one solution when set equal to zero as required by the principle of zero products.   **64.** The discussion could include the following points: **(a)** We can now solve certain polynomial equations. **(b)** Whereas most linear equations have exactly one solution, nonlinear polynomial equations can have more than one solution. **(c)** We used factoring and the principle of zero products to solve polynomial equations.

Test: Chapter 4, p. 286

1. [4.1a] **(a)** 4, 3, 9, 5; 9; **(b)** $5x^5y^4$; 5;
(c) $3xy^3 - 4x^2y - 2x^4y + 5x^5y^4$;
(d) $5x^5y^4 + 3xy^3 - 4x^2y - 2x^4y$, or $5x^5y^4 + 3xy^3 - 2x^4y - 4x^2y$ **2.** [4.1b] 4; 2   **3.** [4.1b] About 250 million tons
4. [4.1c] $3xy + 3xy^2$ **5.** [4.1c] $-3x^3 + 3x^2 - 6y - 7y^2$
6. [4.1c] $7a^3 - 6a^2 + 3a - 3$
7. [4.1c] $7m^3 + 2m^2n + 3mn^2 - 7n^3$ **8.** [4.1d] $6a - 8b$
9. [4.1d] $7x^2 - 7x + 13$ **10.** [4.1d] $2y^2 + 5y + y^3$
11. [4.2a] $64x^3y^3$ **12.** [4.2a] $12a^2 - 4ab - 5b^2$
13. [4.2a] $x^3 - 2x^2y + y^3$ **14.** [4.2a] $-3m^4 - 13m^3 + 5m^2 + 26m - 10$ **15.** [4.2c] $16y^2 - 72y + 81$
16. [4.2d] $x^2 - 4y^2$
17. [4.2e] $a^2 + 15a + 50$; $2ah + h^2 - 5h$
18. [4.3a] $x(9x + 7)$ **19.** [4.3a] $8y^2(3y + 2)$
20. [4.6c] $(y + 5)(y + 2)(y - 2)$
21. [4.4a] $(p - 14)(p + 2)$
22. [4.5a, b] $(6m + 1)(2m + 3)$
23. [4.6b] $(3y + 5)(3y - 5)$
24. [4.6d] $3(r - 1)(r^2 + r + 1)$

25. [4.6a] $(3x - 5)^2$ **26.** [4.6b] $(z + 1 + b)(z + 1 - b)$
27. [4.6b] $(x^4 + y^4)(x^2 + y^2)(x + y)(x - y)$
28. [4.6c] $(y + 4 + 10t)(y + 4 - 10t)$
29. [4.6b] $5(2a + b)(2a - b)$
30. [4.5a, b] $2(4x - 1)(3x - 5)$
31. [4.6d] $2ab(2a^2 + 3b^2)(4a^4 - 6a^2b^2 + 9b^4)$
32. [4.8a] $-3, 6$ **33.** [4.8a] $-5, 5$ **34.** [4.8a] $-\frac{3}{2}, -7$
35. [4.8a] $0, 5$ **36.** [4.8a] $\{x \mid x \text{ is a real number } and\ x \neq -1\}$, or $(-\infty, -1) \cup (-1, \infty)$ **37.** [4.8b] Length: 8 cm; width: 5 cm
38. [4.8b] 24 ft   **39.** [4.3a] $f(n) = \frac{1}{2}n(n - 1)$ **40.** [4.6d] C
41. [4.7a] $(3x^n + 4)(2x^n - 5)$ **42.** [4.2c] 19

Chapter 5

Technology Connection, p. 294

1. Correct   **2.** Correct   **3.** Incorrect
4. Incorrect   **5.** Correct

Exercise Set 5.1, p. 297

RC1. (c)   **RC2.** (g)   **RC3.** (e)   **RC4.** (a)
RC5. (b)   **RC6.** (h)   **RC7.** (f)   **RC8.** (d)
1. $-\frac{17}{3}$ **3.** $-7, -5$ **5.** $\{x \mid x \text{ is a real number } and\ x \neq -7\}$, or $(-\infty, -7) \cup (-7, \infty)$ **7.** $\{x \mid x \text{ is a real number } and\ x \neq 0 \text{ and } x \neq 3\}$, or $(-\infty, 0) \cup (0, 3) \cup (3, \infty)$
9. $\{x \mid x \text{ is a real number } and\ x \neq -\frac{17}{3}\}$, or $\left(-\infty, -\frac{17}{3}\right) \cup \left(-\frac{17}{3}, \infty\right)$ **11.** $\{x \mid x \text{ is a real number } and\ x \neq -7 \text{ and } x \neq -5\}$, or $(-\infty, -7) \cup (-7, -5) \cup (-5, \infty)$
13. $\dfrac{7x(x + 2)}{7x(x + 8)}$ **15.** $\dfrac{(q - 5)(q + 5)}{(q + 3)(q + 5)}$ **17.** $3y$ **19.** $\dfrac{2}{3p^4}$
21. $a - 3$ **23.** $\dfrac{4x - 5}{7}$ **25.** $\dfrac{y - 3}{y + 3}$ **27.** $\dfrac{t + 4}{t - 4}$
29. $\dfrac{x - 8}{x + 4}$ **31.** $\dfrac{w^2 + wz + z^2}{w + z}$ **33.** -1 **35.** $-\dfrac{1}{4}$ **37.** $\dfrac{1}{3x^3}$
39. $\dfrac{(x - 4)(x + 4)}{x(x + 3)}$ **41.** $\dfrac{y + 4}{2}$ **43.** $\dfrac{(2x + 3)(x + 5)}{7x}$
45. $c - 2$ **47.** $\dfrac{1}{x + y}$ **49.** $\dfrac{3x^5}{2y^3}$ **51.** 3
53. $\dfrac{(y - 3)(y + 2)}{y}$ **55.** $\dfrac{2a + 1}{a + 2}$ **57.** $\dfrac{(x + 4)(x + 2)}{3(x - 5)}$
59. $\dfrac{y(y^2 + 3)}{(y + 3)(y - 2)}$ **61.** $\dfrac{x^2 + 4x + 16}{(x + 4)(x + 4)}$
63. $\dfrac{4y^2 - 6y + 9}{(4y - 1)(2y - 3)}$ **65.** $\dfrac{2s}{r + 2s}$ **67.** $\dfrac{y^5}{(y + 2)^3(y + 4)}$
69. Domain $= \{-4, -2, 0, 2, 4, 6\}$; range $= \{-3, -2, 0, 1, 3, 4\}$
70. Domain $= [-4, 5]$; range $= [-3, 2]$
71. Domain $= [-5, 5]$; range $= [-4, 4]$
72. Domain $= [-4, 5]$; range $= [0, 2]$
73. $(7p + 5)(3p - 2)$ **74.** $2(3m + 1)(2m - 5)$
75. $2x(x - 11)(x + 3)$ **76.** $10(y + 13)(y - 5)$
77. $y = -\frac{2}{3}x - 5$ **78.** $y = -\frac{2}{7}x + \frac{48}{7}$
79. $\dfrac{x - 3}{(x + 1)(x + 3)}$ **81.** $\frac{13}{19}$; -3; not defined; $\dfrac{2a + 2h + 3}{4a + 4h - 1}$

Technology Connection, p. 305

Left to the student

Exercise Set 5.2, p. 305

RC1. (a) $3x + 5$; (b) $10x - 3x - 5$; (c) $7x - 5$
RC2. (a) $4 - 9a$; (b) $7 - 4 + 9a$; (c) $3 + 9a$
RC3. (a) $y + 1$; (b) $9y - 2 - y - 1$; (c) $8y - 3$
1. 120   **3.** 144   **5.** 210   **7.** 45   **9.** $\frac{11}{10}$ **11.** $\frac{17}{72}$
13. $\frac{251}{240}$ **15.** $21x^2y$ **17.** $10(y - 10)(y + 10)$ **19.** $30a^3b^2$
21. $5(y - 3)^2$ **23.** $(y + 5)(y - 5)$, or $(y + 5)(5 - y)$
25. $(2r + 3)(r - 4)(3r - 1)(r + 4)$
27. $x^3(x - 2)^2(x^2 + 4)$ **29.** $10x^3(x - 1)^2(x + 1)(x^2 + 1)$
31. $\dfrac{2x + 7y}{x + y}$ **33.** $\dfrac{3y + 5}{y - 2}$ **35.** $a + b$ **37.** $\dfrac{13}{y}$ **39.** $\dfrac{1}{a + 7}$
41. $a^2 + ab + b^2$ **43.** $\dfrac{2(y^2 + 11)}{(y + 4)(y - 5)}$ **45.** $\dfrac{x + y}{x - y}$
47. $\dfrac{3x - 4}{(x - 2)(x - 1)}$ **49.** $\dfrac{8x + 1}{(x + 1)(x - 1)}$ **51.** $\dfrac{2(x - 7)}{15(x + 5)}$
53. $\dfrac{-a^2 + 7ab - b^2}{(a + b)(a - b)}$ **55.** $\dfrac{y}{(y - 2)(y - 3)}$
57. $\dfrac{3y - 10}{(y - 5)(y + 4)}$ **59.** $\dfrac{3y^2 - 3y - 29}{(y + 8)(y - 3)(y - 4)}$
61. $\dfrac{2x^2 - 13x + 7}{(x + 3)(x - 1)(x - 3)}$ **63.** 0   **65.** $\dfrac{4y - 11}{(y + 4)(y - 4)}$
67. $\dfrac{-2y - 3}{(y + 4)(y - 4)}$ **69.** $\dfrac{-3x^2 - 3x - 4}{(x + 1)(x - 1)}$ **71.** $\dfrac{-2}{x - y}$, or
$\dfrac{2}{y - x}$ **73.** $\dfrac{1}{x - 3}$
75. 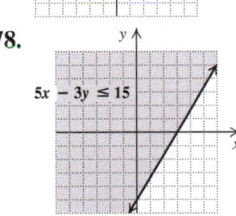   **76.**

77. 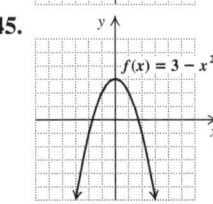   **78.**

79. $(t - 2)(t^2 + 2t + 4)$ **80.** $(q + 5)(q^2 - 5q + 25)$
81. $23x(x + 1)(x^2 - x + 1)$
82. $(4a - 3b)(16a^2 + 12ab + 9b^2)$ **83.** $y = -\frac{8}{3}x + \frac{14}{3}$
84. $y = \frac{5}{4}x + \frac{11}{2}$ **85.** Domain $= (-\infty, 2) \cup (2, \infty)$;
range $= (-\infty, 0) \cup (0, \infty)$
87. $x^4(x + 1)(x - 1)(x^2 + 1)(x^2 + x + 1)(x^2 - x + 1)$
89. -1 **91.** $\dfrac{-x^3 + x^2y + x^2 - xy^2 + xy + y^3}{(x + y)(x + y)(x - y)(x^2 + y^2)}$

Exercise Set 5.3, p. 314

RC1. 4   **RC2.** -6 **RC3.** 7   **RC4.** 6   **RC5.** -4
RC6. -7 **1.** $4x^4 + 3x^3 - 6$ **3.** $9y^5 - 4y^2 + 3$
5. $16a^2b^2 + 7ab - 11$ **7.** $x + 7$ **9.** $a - 12$, R 32, or
$a - 12 + \dfrac{32}{a + 4}$ **11.** $x + 2$, R 4, or $x + 2 + \dfrac{4}{x + 5}$

13. $2y^2 - y + 2$, R 6, or $2y^2 - y + 2 + \dfrac{6}{2y + 4}$
15. $2y^2 + 2y - 1$, R 8, or $2y^2 + 2y - 1 + \dfrac{8}{5y - 2}$
17. $2x^2 - x - 9$, R $(3x + 12)$, or $2x^2 - x - 9 + \dfrac{3x + 12}{x^2 + 2}$
19. $2x^3 + 5x^2 + 17x + 51$, R $152x$, or $2x^3 + 5x^2 + 17x +$
$51 + \dfrac{152x}{x^2 - 3x}$ **21.** $x^2 - x + 1$, R -4, or $x^2 - x + 1 + \dfrac{-4}{x - 1}$
23. $a + 7$, R -47, or $a + 7 + \dfrac{-47}{a + 4}$
25. $x^2 - 5x - 23$, R -43, or $x^2 - 5x - 23 + \dfrac{-43}{x - 2}$
27. $3x^2 - 2x + 2$, R -3, or $3x^2 - 2x + 2 + \dfrac{-3}{x + 3}$
29. $y^2 + 2y + 1$, R 12, or $y^2 + 2y + 1 + \dfrac{12}{y - 2}$
31. $3x^3 + 9x^2 + 2x + 6$ **33.** $x^2 + 2x + 4$
35. $y^3 + 2y^2 + 4y + 8$
37. $y^7 - y^6 + y^5 - y^4 + y^3 - y^2 + y - 1$
39. 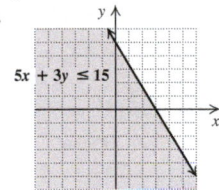   **40.**

41.   **42.**

43. 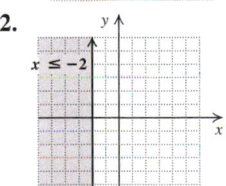   **44.**

45.   **46.**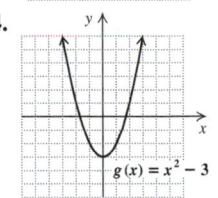

47. 0, 5   **48.** $-\frac{8}{5}, \frac{8}{5}$ **49.** $-\frac{1}{4}, \frac{5}{3}$ **50.** $-\frac{1}{4}, -\frac{2}{3}$
51. $0; -3, -\frac{5}{2}, \frac{3}{2}$ **53.** $-\frac{3}{2}$ **55.** $a^2 + ab$

Exercise Set 5.4, p. 322

RC1. Complex   **RC2.** Numerator   **RC3.** Least common
denominator   **RC4.** Reciprocal
1. $\frac{26}{35}$ **3.** $\frac{88}{15}$ **5.** $\dfrac{x^3}{y^5}$ **7.** $\dfrac{3x + y}{x}$ **9.** $\dfrac{1 + 2a}{1 - a}$ **11.** $\dfrac{x^2 - 1}{x^2 + 1}$
13. $\dfrac{3y + 4x}{4y - 3x}$ **15.** $\dfrac{a^2(b - 3)}{b^2(a - 1)}$ **17.** $\dfrac{1}{a - b}$ **19.** $\dfrac{-1}{x(x + h)}$
21. $\dfrac{(x - 4)(x - 7)}{(x - 5)(x + 6)}$ **23.** $\dfrac{x + 1}{5 - x}$ **25.** $\dfrac{5x - 16}{4x + 1}$

27. $\dfrac{zw(w-z)}{w^2 - wz + z^2}$ **29.** $\dfrac{2x^2 - 11x - 27}{2x^2 + 21x + 13}$

31. 3   **33.** 6   **35.** $\frac{1}{3}$ **37.** $\dfrac{-1}{3x(x+h)}$, or $-\dfrac{1}{3x(x+h)}$

39. $\dfrac{1}{4x(x+h)}$ **41.** $2x + h$ **43.** $-2x - h$

45. $6x + 3h - 2$ **47.** $\dfrac{5|x+h| - 5|x|}{h}$

49. $3x^2 + 3xh + h^2$ **51.** $\dfrac{7}{(x+h+3)(x+3)}$

53. 73,608 pages   **54.** 69%   **55.** $2x(2x^2 + 10x + 3)$
56. $(y+2)(y^2 - 2y + 4)$ **57.** $(y-2)(y^2 + 2y + 4)$
58. $2x(x-9)(x-7)$ **59.** $(10x+1)(100x^2 - 10x + 1)$
60. $(1-10a)(1 + 10a + 100a^2)$
61. $(y-4x)(y^2 + 4xy + 16x^2)$
62. $\left(\frac{1}{2}a - 7\right)\left(\frac{1}{4}a^2 + \frac{7}{2}a + 49\right)$ **63.** $s = 3T - r$
64.

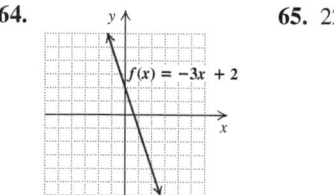

65. 22   **66.** $-1, 6$ **67.** $\frac{5}{6}$

69. $\dfrac{x}{x^3 - 1}$ **71.** $\dfrac{1}{a^2 - ab + b^2}$

Mid-Chapter Review: Chapter 5, p. 324

1. True   **2.** False   **3.** False
4. $\{x\,|\,x$ is a real number $and\ x \neq -10\ and\ x \neq 10\}$, or
$(-\infty, -10) \cup (-10, 10) \cup (10, \infty)$ **5.** $\{x\,|\,x$ is a real number
$and\ x \neq 7\}$, or $(-\infty, 7) \cup (7, \infty)$ **6.** $\{x\,|\,x$ is a real number
$and\ x \neq -9\ and\ x \neq 1\}$, or $(-\infty, -9) \cup (-9, 1) \cup (1, \infty)$
7. $\dfrac{2}{3p^7}$ **8.** $\dfrac{14y - 1}{11}$ **9.** $\dfrac{x - y}{x^2 - xy + y^2}$ **10.** $\dfrac{x + 5}{x + 2}$
11. $\dfrac{a - 2}{a + 2}$ **12.** $\dfrac{-1}{t + 2}$ **13.** $70x^4y^5$
14. $(x-5)^2(x+5)(x+8)$ **15.** $\dfrac{45}{(x+1)^2}$
16. $\dfrac{4x^2 - x + 2}{(x+6)(x-2)}$ **17.** $\dfrac{-3q - 2}{q(q+2)}$
18. $\dfrac{-y^2 - 6y - 3}{(y-1)(y+3)(y+2)}$ **19.** $\dfrac{b}{1+b}$ **20.** $\dfrac{w - z}{5}$
21. $(t-1)(t^2 + 2t + 4)$ **22.** $\dfrac{25c^2 + 6a}{15c}$ **23.** $\dfrac{x + 2}{x - 4}$
24. $3x + 2$, R 17; or $3x + 2 + \dfrac{17}{2x - 3}$ **25.** $x^3 - x^2 + x - 1$
26. $2x^2 - 5x + 15$, R -34; or $2x^2 - 5x + 15 + \dfrac{-34}{x + 2}$
27. $x + 2$
28. $x^3 - 3x^2 + 6x - 18$, R 56; or $x^3 - 3x^2 + 6x - 18 + \dfrac{56}{x + 3}$
29. $3x - 1$, R 7; or $3x - 1 + \dfrac{7}{5x + 1}$ **30.** 4   **31.** $-2x - h$
32. A remainder of 0 indicates that $x - a$ is a factor of the
polynomial. The quotient is a polynomial of one degree less
than the original polynomial and can be factored further, if

possible, using synthetic division again or another factoring
method.   **33.** Addition, subtraction, and multiplication of
polynomials always result in a polynomial, because each is
defined in terms of addition, subtraction, or multiplication of
monomials, and the sum, difference, and product of monomials is
a monomial. Division of polynomials does not always result in a
polynomial, because the quotient is not always a monomial or a
sum of monomials. Example 1 in Section 5.3 in the text illustrates
this.   **34.** No; when we simplify a rational expression by
removing a factor of 1, we are actually reversing the multiplication
process.   **35.** Janine's answer was correct. It is equivalent to the
answer at the back of the book:
$$\dfrac{3 - x}{x - 5} = \dfrac{-x + 3}{x - 5} = \dfrac{-1(-x + 3)}{-1(x - 5)} = \dfrac{x - 3}{-x + 5} = \dfrac{x - 3}{5 - x}.$$
36. Nancy's misconception is that x is a factor of the numerator.
$\left(\dfrac{x + 2}{x} = 3 \text{ only for } x = 1.\right)$ **37.** Most would agree that
it is easier to find the LCM of all the denominators, bd, and
then to multiply by $bd/(bd)$ than it is to add in the numerator,
subtract in the denominator, and then divide the numerator by the
denominator.

Technology Connection, p. 329

1. Left to the student

Exercise Set 5.5, p. 331

RC1. Rational expression   **RC2.** Solutions
RC3. Rational expression   **RC4.** Rational expression
RC5. Solutions   **RC6.** Solutions
RC7. Rational expression   **RC8.** Solutions
1. $\frac{31}{4}$ **3.** $-\frac{12}{7}$ **5.** 144   **7.** $-1, -8$ **9.** 2   **11.** 11
13. 11   **15.** No solution   **17.** 2   **19.** 5   **21.** -145
23. $-\frac{10}{3}$ **25.** -3 **27.** $\frac{31}{5}$ **29.** $\frac{85}{12}$ **31.** $-6, 5$
33. No solution   **35.** 2   **37.** No solution   **39.** $-1, 0$
41. $-\frac{3}{2}, 2$ **43.** $\frac{17}{4}$ **45.** $\frac{3}{5}$ **46.** $4(t+5)(t^2 - 5t + 25)$
47. $(1 - t)(1 + t + t^2)(1 + t)(1 - t + t^2)$
48. $(a + 2b)(a^2 - 2ab + 4b^2)$
49. $(a - 2b)(a^2 + 2ab + 4b^2)$ **50.** 3   **51.** $-4, 3$
52. $-7, 7$ **53.** $\frac{1}{4}, \frac{2}{3}$ **54.** About \$4306 per year
55. A decrease of about 996 permits per year
57. (a) $(-3.5, 1.3)$; **(b), (c)** left to the student

Translating for Success, p. 341

1. N   **2.** B   **3.** A   **4.** C   **5.** E   **6.** G   **7.** I
8. K   **9.** M   **10.** O

Exercise Set 5.6, p. 342

RC1. Corresponding, same, proportional
RC2. Quotient   **RC3.** Proportion
RC4. Rate   **RC5.** Distance
RC6. Distance   **RC7.** Proportional
1. $15\frac{3}{4}$ hr   **3.** $13\frac{1}{3}$ hr   **5.** Machine A: 2 hr; machine B: 6 hr
7. 4.375 hr, or $4\frac{3}{8}$ hr   **9.** Cole: 8 min; Jim: 24 min
11. About 49 three-point field goals   **13.** 7.5 in.   **15.** 28.8 lb
17. 1160 trees   **19.** About 14.5 tons of grapes   **21.** $10\frac{1}{2}$ ft; $17\frac{1}{2}$ ft
23. About 20,658 kg   **25.** 287 trout   **27.** 35 mph   **29.** 7 mph
31. Bus: 60 mph; trolley: 45 mph   **33.** 5.2 ft/sec

35. Domain: $[-5, 5]$; range: $[-4, 3]$
36. Domain: $\{-4, -2, 0, 1, 2, 4\}$; range: $\{-2, 0, 2, 4, 5\}$
37. Domain: $[-5, 5]$; range: $[-5, 3]$
38. Domain: $[-5, 5]$; range: $[-5, 0]$
39.   **40.**

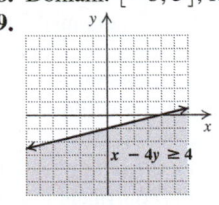   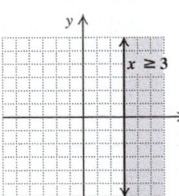

41.   **42.**

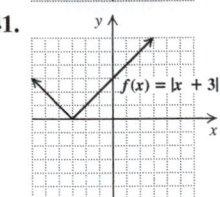   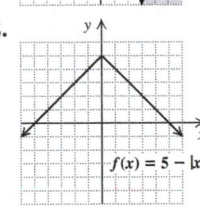

43. $t = \frac{2}{3}$ hr   **45.** City: 261 mi; highway: 204 mi

Exercise Set 5.7, p. 348

RC1. (c)   **RC2.** (a)   **RC3.** (e)   **RC4.** (d)

1. $W_2 = \dfrac{d_2 W_1}{d_1}$ **3.** $r_2 = \dfrac{Rr_1}{r_1 - R}$ **5.** $t = \dfrac{2s}{v_1 + v_2}$

7. $s = \dfrac{Rg}{g - R}$ **9.** $p = \dfrac{qf}{q - f}$ **11.** $a = \dfrac{bt}{b - t}$

13. $E = \dfrac{Inr}{n - I}$ **15.** $H^2 = \dfrac{704.5W}{I}$ **17.** $r = \dfrac{eR}{E - e}$

19. $R = \dfrac{3V + \pi h^3}{3\pi h^2}$ **21.** $h = \dfrac{S - 2\pi r^2}{2\pi r}$ **23.** $t_2 = \dfrac{d_2 - d_1 + t_1 v}{v}$

25. Dimes: 2 rolls; nickels: 5 rolls; quarters: 5 rolls   **26.** -6
27. 6   **28.** 0   **29.** $8a^3 - 2a$ **30.** $-\frac{4}{5}$ **31.** $y = -\frac{4}{5}x + \frac{17}{5}$

Exercise Set 5.8, p. 356

RC1. (f)   **RC2.** (d)   **RC3.** (h)   **RC4.** (i)
RC5. (c)   **RC6.** (a)   **RC7.** (g)   **RC8.** (b)
1. $5; y = 5x$ **3.** $\frac{2}{15}; y = \frac{2}{15}x$ **5.** $\frac{9}{4}; y = \frac{9}{4}x$
7. 175 semi trucks   **9.** 90 g   **11.** 76,361,280 cans

13. 40 kg   **15.** $98; y = \dfrac{98}{x}$ **17.** $36; y = \dfrac{36}{x}$

19. $0.05; y = \dfrac{0.05}{x}$ **21.** 3.5 hr   **23.** $\frac{2}{9}$ ampere   **25.** 960 lb

27. $5\frac{5}{7}$ hr   **29.** $y = 15x^2$ **31.** $y = \dfrac{0.0015}{x^2}$ **33.** $y = xz$

35. $y = \frac{3}{10}xz^2$ **37.** $y = \dfrac{xz}{5wp}$ **39.** 2.5 m   **41.** 199.4 lb

43. 98 earned runs   **45.** 729 gal   **47.** $[-8, \infty)$
48. $(-\infty, \infty)$ **49.** $[-5, 15)$ **50.** $(-\infty, -4) \cup [0, \infty)$
51. $\left(-\infty, -\frac{1}{2}\right)$ **52.** $\left(-\infty, -\frac{5}{8}\right] \cup (2, \infty)$ **53.** $(2, -1)$
54. $(4, 6)$ **55.** (a) Inversely; (b) neither; (c) directly;
(d) directly

Summary and Review: Chapter 5, p. 359

Vocabulary Reinforcement

1. Proportion   **2.** Rational   **3.** Rational
4. Positive, inverse   **5.** Complex   **6.** Positive, direct, constant

Concept Reinforcement

1. True   **2.** True

Study Guide

1. $\{x | x$ is a real number *and* $x \neq -9$ *and* $x \neq 6\}$, or

$(-\infty, -9) \cup (-9, 6) \cup (6, \infty)$ **2.** $\dfrac{b - 3}{b - 8}$

3. $\dfrac{(w - 5)(w^2 + 5w + 25)}{w + 3}$

4. $x^4(x - 3)(x + 3)(2x + 5)$ **5.** $\dfrac{r^2 - 3rs - 9s^2}{(r + 2s)(r - s)(r + s)}$

6. $y - 4$, R 5; or $y - 4 + \dfrac{5}{y - 1}$

7. $x^2 - 8x + 24$, R -73; or $x^2 - 8x + 24 + \dfrac{-73}{x + 3}$

8. $\dfrac{b + 4a}{4b - a}$ **9.** 6   **10.** $-\frac{33}{2}$ **11.** $k = 93; y = 93x$

12. $k = \dfrac{9}{2}; y = \dfrac{9}{2x}$

Review Exercises

1. $-3, 3$ **2.** $\{x | x$ is a real number *and* $x \neq -3$ and $x \neq 3\}$, or

$(-\infty, -3) \cup (-3, 3) \cup (3, \infty)$ **3.** $\dfrac{x - 2}{3x + 2}$ **4.** $\dfrac{1}{a - 2}$

5. $48x^3$ **6.** $(x - 7)(x + 7)(3x + 1)$

7. $(x + 5)(x - 4)(x - 2)$ **8.** $\dfrac{y - 8}{2}$

9. $\dfrac{(x - 2)(x + 5)}{x - 5}$ **10.** $\dfrac{3a - 1}{a - 3}$

11. $\dfrac{(x^2 + 4x + 16)(x - 6)}{(x + 4)(x + 2)}$ **12.** $\dfrac{x - 3}{(x + 1)(x + 3)}$

13. $\dfrac{2x^3 + 2x^2y + 2xy^2 - 2y^3}{(x - y)(x + y)}$ **14.** $\dfrac{-y}{(y + 4)(y - 1)}$

15. $4b^2c - \frac{5}{2}bc^2 + 3abc$ **16.** $y - 14$, R -20; or

$y - 14 + \dfrac{-20}{y - 6}$ **17.** $6x^2 - 9$, R $(5x + 22)$; or

$6x^2 - 9 + \dfrac{5x + 22}{x^2 + 2}$ **18.** $x^2 + 9x + 40$, R 153; or

$x^2 + 9x + 40 + \dfrac{153}{x - 4}$ **19.** $3x^3 - 8x^2 + 8x - 6$, R -1; or

$3x^3 - 8x^2 + 8x - 6 + \dfrac{-1}{x + 1}$ **20.** $\frac{3}{4}$ **21.** $\dfrac{a^2 b^2}{2(a^2 - ab + b^2)}$

22. $\dfrac{(x - 9)(x - 6)}{(x - 3)(x + 6)}$ **23.** $\dfrac{2(2x^2 - 7x + 1)}{3x^2 + 7x - 11}$ **24.** 2

25. $-2x - h$ **26.** $\dfrac{-4}{x(x + h)}$, or $-\dfrac{4}{x(x + h)}$ **27.** $\frac{28}{11}$ **28.** 6

29. No solution   **30.** 3   **31.** $-\frac{11}{3}$ **32.** 2   **33.** $5\frac{1}{7}$ hr
34.

	Distance	Speed	Time
Downstream	50 mi	$x + 6$	t
Upstream	30 mi	$x - 6$	t

24 mph

35. 4000 mi   **36.** $d = \dfrac{Wc}{c - W}; c = \dfrac{Wd}{d - W}$

37. $b = \dfrac{ta}{Sa - p}; t = \dfrac{Sab - pb}{a}$ **38.** $y = 4x$

39. $y = \dfrac{2500}{x}$ **40.** 20 min   **41.** 75   **42.** 500 watts

43. B   **44.** C   **45.** $a^2 + ab + b^2$ **46.** All real numbers except 0 and 13   **47.** When adding or subtracting rational expressions, we use the LCM of the denominators (the LCD). When solving a rational equation or when solving a formula for a given letter, we multiply by the LCM of all the denominators to clear fractions. When simplifying a complex rational expression, we can use the LCM in either of two ways. We can multiply by a/a, where a is the LCM of all the denominators occurring in the expression. Or we can use the LCM to add or subtract as necessary in the numerator and in the denominator.   **48.** Rational equations differ from those previously studied because they contain variables in denominators. Because of this, possible solutions must be checked in the original equation to avoid division by 0.   **49.** Assuming all algebraic procedures have been performed correctly, a possible solution of a rational equation would fail to be an actual solution only if it were not in the domain of one of the rational expressions in the equation. This occurs when the number in question makes a denominator 0.   **50.** Let $y = k_1 x$ and $x = \dfrac{k_2}{z}$. Then $y = k_1 \cdot \dfrac{k_2}{z}$, or $y = \dfrac{k_1 k_2}{z}$, so y varies inversely as z.   **51.** Answers may vary. From Example 4 of Section 5.5, we see that one form of such an equation is $\dfrac{x^2}{x - a} = \dfrac{a^2}{x - a}$.   **52.** Answers may vary. Many would probably argue that it is easier to solve $\dfrac{1}{a} + \dfrac{1}{b} = \dfrac{1}{x}$ since it is easier to multiply a and b than 38 and 47. Others might argue that it is easier to solve $\dfrac{1}{38} + \dfrac{1}{47} = \dfrac{1}{x}$ since it is easier to work with constants than with variables.

Test: Chapter 5, p. 365

1. [5.1a] 1, 2   **2.** [5.1a] $\{x \mid x$ is a real number $and\ x \neq 1$ $and\ x \neq 2\}$, or $(-\infty, 1) \cup (1, 2) \cup (2, \infty)$

3. [5.1c] $\dfrac{3x + 2}{x - 2}$ **4.** [5.1c] $\dfrac{p^2 - p + 1}{p - 2}$

5. [5.2a] $(x + 3)(x - 2)(x + 5)$ **6.** [5.1d] $\dfrac{2(x + 5)}{x - 2}$

7. [5.2b] $\dfrac{x - 6}{(x + 4)(x + 6)}$ **8.** [5.1e] $\dfrac{y + 4}{2}$

9. [5.2b] $x + y$ **10.** [5.2c] $\dfrac{3x}{(x - 1)(x + 1)}$

11. [5.2c] $\dfrac{a^3 + a^2 b + ab^2 + ab - b^2 - 2}{(a - b)(a^2 + ab + b^2)}$

12. [5.3a] $4s^2 + 3s - 2rs^2$ **13.** [5.3b] $y^2 - 5y + 25$

14. [5.3b] $4x^2 + 3x - 4$, R $(-8x + 2)$; or

$4x^2 + 3x - 4 + \dfrac{-8x + 2}{x^2 + 1}$ **15.** [5.3c] $x^2 + 6x + 20$, R 54; or

$x^2 + 6x + 20 + \dfrac{54}{x - 3}$ **16.** [5.3c] $3x^2 + 10x - 40$

17. [5.4a] $\dfrac{x + 1}{x}$ **18.** [5.4a] $\dfrac{b^2 - ab + a^2}{a^2 b^2}$ **19.** [5.4b] $\dfrac{1}{2}$

20. [5.5a] $-1, 4$ **21.** [5.5a] 9   **22.** [5.5a] No solution   **23.** [5.5a] $-\tfrac{7}{3}, 5$ **24.** [5.5a] $\tfrac{17}{8}$ **25.** [5.6a] 2 hr   **26.** [5.6c] $3\tfrac{3}{11}$ mph   **27.** [5.6b] $14\tfrac{2}{17}$ gal

28. [5.7a] $a = \dfrac{Tb}{T - b}; b = \dfrac{Ta}{a + T}$ **29.** [5.7a] $a = \dfrac{2b}{Qb + t}$

30. [5.8e] $Q = \tfrac{5}{2}xy$ **31.** [5.8c] $y = \dfrac{250}{x}$ **32.** [5.8b] $990

33. [5.8d] $7\tfrac{1}{2}$ hr   **34.** [5.8f] 615.44 cm^2 **35.** [5.2a] D   **36.** [5.5a] All real numbers except 0 and 15   **37.** [5.4a], [5.5a] x-intercept: $(11, 0)$; y-intercept: $\left(0, -\tfrac{33}{5}\right)$

Chapter 6

Exercise Set 6.1, p. 374

RC1. (j)   **RC2.** (b)   **RC3.** (h)   **RC4.** (i)   **RC5.** (i)   **RC6.** (g)   **1.** $4, -4$ **3.** $12, -12$ **5.** $20, -20$ **7.** $-\tfrac{7}{6}$ **9.** 14   **11.** 0.06   **13.** Does not exist as a real number

15. 18.628   **17.** 1.962   **19.** $y^2 + 16$ **21.** $\dfrac{x}{y - 1}$

23. $\sqrt{20} \approx 4.472$; 0; does not exist as a real number; does not exist as a real number   **25.** $\sqrt{11} \approx 3.317$; does not exist as a real number; $\sqrt{11} \approx 3.317$; 12   **27.** Domain $= \{x \mid x \geq 2\} = [2, \infty)$ **29.** 21 spaces; 25 spaces

31.   **33.**   **35.**

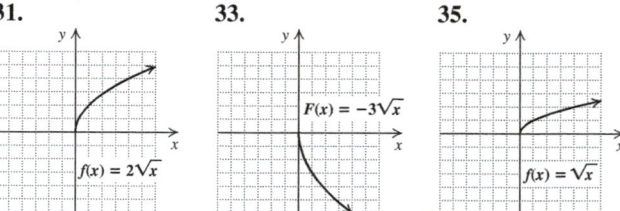

37.   **39.**   **41.**

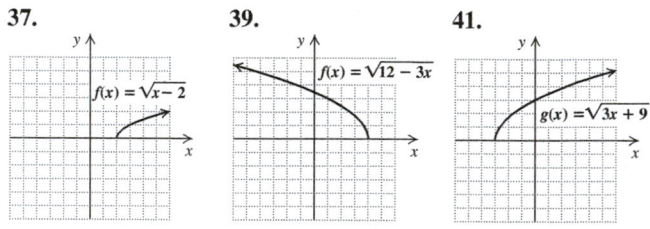

43. $4|x|$ **45.** $12|c|$ **47.** $|p + 3|$ **49.** $|x - 2|$ **51.** 3   **53.** $-4x$ **55.** -6 **57.** $0.7(x + 1)$ **59.** 2; 3; -2; -4 **61.** -1; $-\sqrt[3]{-20}$, or $-\sqrt[3]{-20} \approx 2.714$; -4; -10 **63.** -5 **65.** -1 **67.** $-\tfrac{2}{3}$ **69.** $|x|$ **71.** $5|a|$ **73.** 6   **75.** $|a + b|$ **77.** y **79.** $x - 2$ **81.** $-2, 1$ **82.** $-1, 0$ **83.** $-\tfrac{7}{2}, \tfrac{7}{2}$ **84.** 4, 9   **85.** $-2, \tfrac{5}{3}$ **86.** $\tfrac{5}{2}$ **87.** $0, \tfrac{5}{2}$ **88.** 0, 1   **89.** $a^9 b^6 c^{15}$ **90.** $10a^{10} b^9$ **91.** Domain $= \{x \mid -3 \leq x < 2\} = [-3, 2)$ **93.** 1.7; 2.2; 3.2   **95. (a)** Domain: $(-\infty, \infty)$; range: $(-\infty, \infty)$; **(b)** domain: $(-\infty, \infty)$; range: $(-\infty, \infty)$; **(c)** domain: $[-3, \infty)$; range: $(-\infty, 2]$; **(d)** domain: $[0, \infty)$; range: $[0, \infty)$; **(e)** domain: $[3, \infty)$; range: $[0, \infty)$

Technology Connection, p. 378

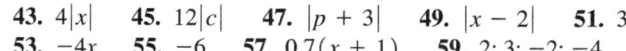

1. 3.344   **2.** 3.281   **3.** 0.283   **4.** 11.053   **5.** 5.527×10^{-5} **6.** 2

Exercise Set 6.2, p. 381

RC1. (h)   **RC2.** (b)   **RC3.** (c)   **RC4.** (g)   **RC5.** (e)
RC6. (d)   **RC7.** (f)   **RC8.** (a)

1. $\sqrt[7]{y}$ **3.** 2   **5.** $\sqrt[5]{a^3b^3}$ **7.** 8   **9.** 343   **11.** $17^{1/2}$
13. $18^{1/3}$ **15.** $(xy^2z)^{1/5}$ **17.** $(3mn)^{3/2}$ **19.** $(8x^2y)^{5/7}$
21. $\frac{1}{3}$ **23.** $\frac{1}{1000}$ **25.** $\frac{3}{x^{1/4}}$ **27.** $\frac{1}{(2rs)^{3/4}}$ **29.** $\frac{2a^{3/4}c^{2/3}}{b^{1/2}}$
31. $\left(\frac{8yz}{7x}\right)^{3/5}$ **33.** $x^{2/3}$ **35.** $\frac{x^4}{2^{1/3}y^{2/7}}$ **37.** $\frac{7x}{z^{1/3}}$ **39.** $\frac{5ac^{1/2}}{3}$
41. $5^{7/8}$ **43.** $7^{1/4}$ **45.** $4.9^{1/2}$ **47.** $6^{3/28}$ **49.** $a^{23/12}$
51. $a^{8/3}b^{5/2}$ **53.** $\frac{1}{x^{2/7}}$ **55.** $\frac{y^{1/3}}{x^{1/2}}$ **57.** $m^{3/5}n^2$ **59.** $\sqrt[3]{a}$
61. x^5 **63.** $\frac{1}{x^3}$ **65.** a^5b^5 **67.** $\sqrt{2}$ **69.** $\sqrt[3]{2x}$ **71.** x^2y^3
73. $2c^2d^3$ **75.** $\sqrt[12]{7^4 \cdot 5^3}$ **77.** $\sqrt[20]{5^5 \cdot 7^4}$ **79.** $\sqrt[6]{4x^5}$
81. a^6b^{12} **83.** $\sqrt[18]{m}$ **85.** $\sqrt[12]{x^4y^3z^2}$ **87.** $\sqrt[30]{\frac{d^{35}}{c^{99}}}$
89. $\{-\frac{4}{7}, 2\}$ **90.** $\{-40, 40\}$ **91.** $\{-\frac{15}{2}, \frac{5}{2}\}$ **92.** $\{-\frac{11}{8}, \frac{3}{8}\}$
93. Left to the student

Exercise Set 6.3, p. 387

RC1. True   **RC2.** False   **RC3.** True   **RC4.** True
1. $2\sqrt{6}$ **3.** $3\sqrt{10}$ **5.** $5\sqrt[3]{2}$ **7.** $6x^2\sqrt{5}$ **9.** $3x^2\sqrt[3]{2x^2}$
11. $2t^2\sqrt[3]{10t^2}$ **13.** $2\sqrt[4]{5}$ **15.** $4a\sqrt{2b}$ **17.** $3x^2y^2\sqrt[4]{3y^2}$
19. $2xy^3\sqrt[5]{3x^2}$ **21.** $5\sqrt{2}$ **23.** $3\sqrt{10}$ **25.** 2   **27.** $30\sqrt{3}$
29. $3x^4\sqrt{2}$ **31.** $5bc^2\sqrt{2b}$ **33.** $a\sqrt[3]{10}$ **35.** $2y^3\sqrt[3]{2}$
37. $4\sqrt[4]{4}$ **39.** $4a^3b\sqrt{6ab}$ **41.** $\sqrt[6]{200}$ **43.** $\sqrt[4]{12}$ **45.** $a\sqrt[4]{a}$
47. $b\sqrt[10]{b^9}$ **49.** $xy\sqrt[6]{xy^5}$ **51.** $2ab\sqrt[4]{2a^3}$ **53.** $3\sqrt{2}$
55. $\sqrt{5}$ **57.** 3   **59.** $y\sqrt{7y}$ **61.** $2\sqrt[3]{a^2b}$ **63.** $4\sqrt{xy}$
65. $2x^2y^2$ **67.** $\frac{1}{\sqrt[6]{a}}$ **69.** $\sqrt[12]{a^5}$ **71.** $\sqrt[12]{x^2y^5}$ **73.** $\frac{5}{6}$
75. $\frac{4}{7}$ **77.** $\frac{5}{3}$ **79.** $\frac{7}{y}$ **81.** $\frac{5y\sqrt{y}}{x^2}$ **83.** $\frac{3y\sqrt[3]{3y^2}}{4}$ **85.** $\frac{3a\sqrt[3]{a}}{2b}$
87. $\frac{3x}{2}$ **89.** $\frac{2a^3}{bc^4}$ **91.** $\frac{2x\sqrt[5]{x^3}}{y^2}$ **93.** $\frac{w\sqrt[5]{w^2}}{z^2}$ **95.** $\frac{x^2\sqrt[6]{x}}{yz^2}$
97. $-10, 9$ **98.** Height: 4 in.; base: 6 in.   **99.** 8
100. No solution   **101.** (a) 1.62 sec; (b) 1.99 sec; (c) 2.20 sec
103. $2yz\sqrt{2z}$

Exercise Set 6.4, p. 391

RC1. Yes   **RC2.** No   **RC3.** No   **RC4.** Yes   **RC5.** Yes
RC6. No   **RC7.** No   **RC8.** Yes
1. $11\sqrt{5}$ **3.** $\sqrt[3]{7}$ **5.** $13\sqrt[3]{y}$ **7.** $-8\sqrt{6}$ **9.** $6\sqrt[3]{3}$
11. $21\sqrt{3}$ **13.** $38\sqrt{5}$ **15.** $122\sqrt{2}$ **17.** $9\sqrt[3]{2}$ **19.** $29\sqrt{2}$
21. $(1 + 6a)\sqrt{5a}$ **23.** $(2 - x)\sqrt[3]{3x}$ **25.** $(21x + 1)\sqrt{3x}$
27. $2 + 3\sqrt{2}$ **29.** $15\sqrt[3]{4}$ **31.** $(x + 1)\sqrt[3]{6x}$ **33.** $3\sqrt{a} - 1$
35. $(x + 3)\sqrt{x - 1}$ **37.** $4\sqrt{5} - 10$ **39.** $\sqrt{6} - \sqrt{21}$
41. $-12 + 6\sqrt{3}$ **43.** $2\sqrt{15} - 6\sqrt{3}$ **45.** -6
47. $6y - 12\sqrt[3]{y^2}$ **49.** $3a\sqrt[3]{2}$ **51.** 1   **53.** -12 **55.** 44
57. 1   **59.** 3   **61.** -19 **63.** $c - d$ **65.** $1 + \sqrt{5}$
67. $7 + 3\sqrt{3}$ **69.** -6 **71.** $a + \sqrt{3a} + \sqrt{2a} + \sqrt{6}$
73. $2\sqrt[3]{9} - 3\sqrt[3]{6} - 2\sqrt[3]{4}$ **75.** $7 + 4\sqrt{3}$
77. $\sqrt[5]{72} + 3 - \sqrt[5]{24} - \sqrt[5]{81}$ **79.** $\frac{x(x^2 + 4)}{(x + 4)(x + 3)}$

80. $\frac{(a + 2)(a + 4)}{a}$ **81.** $a - 2$ **82.** $\frac{(y - 3)(y - 3)}{y + 3}$
83. $\frac{4(3x - 1)}{3(4x + 1)}$ **84.** $\frac{x}{x + 1}$ **85.** $\frac{pq}{q + p}$ **86.** $\frac{a^2b^2}{b^2 - ab + a^2}$
87. $-\frac{29}{3}, 5$ **88.** $\{x \mid -\frac{29}{3} < x < 5\}$, or $\left(-\frac{29}{3}, 5\right)$
89. $\{x \mid x \le -\frac{29}{3} \text{ or } x \ge 5\}$, or $\left(-\infty, -\frac{29}{3}\right] \cup [5, \infty)$
90. $-12, -\frac{2}{5}$ **91.** Domain $= (-\infty, \infty)$ **93.** 6
95. $14 + 2\sqrt{15} - 6\sqrt{2} - 2\sqrt{30}$
97. $3\sqrt[3]{3} + 2\sqrt[3]{9} - 8$

Mid-Chapter Review: Chapter 6, p. 393

1. False   **2.** True   **3.** False   **4.** True   **5.** 9
6. -12 **7.** $\frac{4}{5}$ **8.** Does not exist as a real number
9. 3; does not exist as a real number
10. Domain $= \{x \mid x \le 4\} = (-\infty, 4]$
11.   **12.**

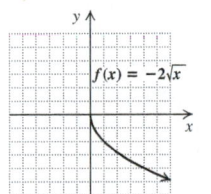

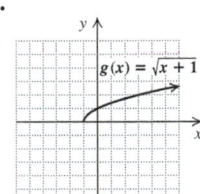

13. $6|z|$ **14.** $|x - 4|$ **15.** -4 **16.** $-3a$ **17.** 2   **18.** $|y|$
19. 5   **20.** $\sqrt[4]{a^3b}$ **21.** $16^{1/5}$ **22.** $(6m^2n)^{1/3}$ **23.** $\frac{1}{3^{3/8}}$
24. $7^{4/5}$ **25.** $\frac{x^{3/2}}{y^{4/3}}$ **26.** $\frac{1}{n^{3/4}}$ **27.** $\sqrt[3]{4}$ **28.** $\sqrt{ab}$
29. $\sqrt[6]{y^5}$ **30.** $\sqrt[15]{a^{10}b^9}$ **31.** $5\sqrt{3}$ **32.** $2xy\sqrt[3]{3y^2}$
33. $2\sqrt[3]{5}$ **34.** $\frac{7a^2\sqrt{a}}{b^4}$ **35.** $11\sqrt{7}$ **36.** $(9x - 24)\sqrt{2x}$
37. $2\sqrt{3} - 15$ **38.** $3 - 4\sqrt{x} + x$ **39.** $m - n$
40. $11 + 4\sqrt{7}$ **41.** $-42 + \sqrt{15}$
42. Yes; since x^2 is nonnegative for any value of x, the nth root of x^2 exists regardless of whether n is even or odd. Thus the nth root of x^2 always exists.   **43.** Formulate an expression containing a radical term with an even index and a radicand R such that the solution of the inequality $R \ge 0$ is $\{x \mid x \le 5\}$. One expression is $\sqrt{5 - x}$. Other expressions could be formulated as $a\sqrt[k]{b(5 - x)} + c$, where $a \ne 0$, $b > 0$, and k is an even integer.   **44.** Since $x^6 \ge 0$ and $x^2 \ge 0$ for any value of x, then $\sqrt[3]{x^6} = x^2$. However, $x^3 \ge 0$ only for $x \ge 0$, so $\sqrt{x^6} = x^3$ only when $x \ge 0$.
45. No; for example, $\frac{\sqrt{8}}{\sqrt{2}} = \sqrt{\frac{8}{2}} = \sqrt{4} = 2$.

Exercise Set 6.5, p. 398

RC1. (g)   **RC2.** (c)   **RC3.** (e)   **RC4.** (h)   **RC5.** (c)
RC6. (d)
1. $\frac{\sqrt{15}}{3}$ **3.** $\frac{\sqrt{22}}{2}$ **5.** $\frac{2\sqrt{15}}{35}$ **7.** $\frac{2\sqrt[3]{6}}{3}$ **9.** $\frac{\sqrt[3]{75ac^2}}{5c}$
11. $\frac{y\sqrt[3]{9yx^2}}{3x^2}$ **13.** $\frac{\sqrt[4]{s^3t^3}}{st}$ **15.** $\frac{\sqrt{15x}}{10}$ **17.** $\frac{\sqrt[3]{100xy}}{5x^2y}$
19. $\frac{\sqrt[4]{2xy}}{2x^2y}$ **21.** $\frac{54 + 9\sqrt{10}}{26}$ **23.** $-2\sqrt{35} + 2\sqrt{21}$
25. $\frac{18\sqrt{6} + 6\sqrt{15}}{13}$ **27.** $\frac{3\sqrt{2} - 3\sqrt{5} + \sqrt{10} - 5}{-3}$

29. $\dfrac{3 + \sqrt{21} - \sqrt{6} - \sqrt{14}}{-4}$ **31.** $\dfrac{\sqrt{15} + 20 - 6\sqrt{2} - 8\sqrt{30}}{-77}$

33. $\dfrac{6 - 5\sqrt{a} + a}{9 - a}$ **35.** $\dfrac{6 + 5\sqrt{x} - 6x}{9 - 4x}$ **37.** $\dfrac{3\sqrt{6} + 4}{2}$

39. $\dfrac{x - 2\sqrt{xy} + y}{x - y}$ **41.** 30 **42.** $-\frac{19}{5}$ **43.** 1 **44.** $\dfrac{x - 2}{x + 3}$

45. Left to the student **47.** $-\dfrac{3\sqrt{a^2 - 3}}{a^2 - 3}$

Technology Connection, p. 401

1. Left to the student

Exercise Set 6.6, p. 405

RC1. Radical **RC2.** Powers **RC3.** Isolate
RC4. Radicands **RC5.** Even
1. $\frac{19}{2}$ **3.** $\frac{49}{6}$ **5.** 57 **7.** $\frac{92}{5}$ **9.** −1 **11.** No solution
13. 3 **15.** 19 **17.** −6 **19.** $\frac{1}{64}$ **21.** 9 **23.** 15
25. 2, 5 **27.** 6 **29.** 5 **31.** 9 **33.** 7 **35.** $\frac{80}{9}$ **37.** 2, 6
39. −1 **41.** No solution **43.** 3 **45.** About 44.1 mi
47. About 680 ft **49.** About 642 ft **51.** 151.25 ft; 281.25 ft
53. About 85°F **55.** About 0.81 ft **57.** About 3.9 ft
59. $4\frac{4}{9}$ hr **60.** Jeff: $1\frac{1}{3}$ hr; Grace: 4 hr **61.** 2808 mi
62. 84 hr **63.** 0, −2.8 **64.** 0, $\frac{5}{3}$ **65.** −8, 8 **66.** −3, $\frac{7}{2}$
67. $2ah + h^2$ **68.** $2ah + h^2 - h$ **69.** $4ah + 2h^2 - 3h$
70. $4ah + 2h^2 + 3h$ **71.** Left to the student **73.** 0
75. −6, −3 **77.** 2 **79.** 0, $\frac{125}{4}$ **81.** 2 **83.** 3

Translating for Success, p. 411

1. J **2.** B **3.** O **4.** M **5.** K **6.** I **7.** G **8.** E
9. F **10.** A

Exercise Set 6.7, p. 412

RC1. (e) **RC2.** (c) **RC3.** (g) **RC4.** (d)
1. $\sqrt{34}$; 5.831 **3.** $\sqrt{450}$; 21.213 **5.** 5 **7.** $\sqrt{43}$; 6.557
9. $\sqrt{12}$; 3.464 **11.** $\sqrt{n - 1}$ **13.** 7.1 ft
15. $\sqrt{116}$ ft; 10.770 ft **17.** $\sqrt{4959}$ ft; about 70.4 ft
19. $\sqrt{10{,}561}$ ft; 102.767 ft **21.** $(3, 0), (−3, 0)$
23. $\left(\sqrt{340} + 8\right)$ ft; 26.439 ft **25.** $\sqrt{420.125}$ in.; 20.497 in.
27. $\sqrt{181}$ cm; 13.454 cm **29.** $s + s\sqrt{2}$
31. Flash: $67\frac{2}{3}$ mph; Crawler: $53\frac{2}{3}$ mph **32.** $3\frac{3}{4}$ mph
33. −7, $\frac{3}{2}$ **34.** 3, 8 **35.** 1 **36.** −2, 2 **37.** 13 **38.** 7
39. 26 packets

Exercise Set 6.8, p. 423

RC1. (a) **RC2.** (c) **RC3.** (b) **RC4.** (b) **RC5.** (c)
1. (a) $(−5, 1)$; (b) $(3, 5)$; (c) $(1, 3)$
3. (a) $(−3, −1), (3, 5)$; (b) $(1, 3)$; (c) $(−5, −3)$
5. (a) $(−\infty, −8), (−3, −2)$; (b) $(−8, −6)$;
(c) $(−6, −3), (−2, \infty)$ **7.** Domain: $[−5, 5]$; range: $[−3, 3]$
9. Domain: $[−5, −1] \cup [1, 5]$; range: $[−4, 6]$
11. Domain: $(−\infty, \infty)$; range: $(−\infty, 3]$
13. Relative maximum: 3.25 at $x = 2.5$; increasing: $(−\infty, 2.5)$;
decreasing: $(2.5, \infty)$ **15.** Relative maximum: 2.370 at
$x = −0.667$; relative minimum: 0 at $x = 2$; increasing:
$(−\infty, −0.667), (2, \infty)$; decreasing: $(−0.667, 2)$

17. $A(x) = x(240 − x)$, or $240x − x^2$
19. $h(d) = \sqrt{d^2 − 3500^2}$ **21.** $A(w) = 10w − \dfrac{w^2}{2}$
23. $d(s) = \dfrac{14}{s}$ **25.** (a) $A(x) = x(240 − 4x)$, or $240x − 4x^2$;
(b) $\{x \mid 0 < x < 60\}$; **(c)** 120 ft by 30 ft
27. (a) $V(x) = x(12 − 2x)(12 − 2x)$, or $4x(6 − x)^2$;
(b) $\{x \mid 0 < x < 6\}$; **(c)** 8 cm by 8 cm by 2 cm
29. $g(−4) = 0; g(0) = 4; g(1) = 5; g(3) = 5$
31. $h(−5) = 1; h(0) = 1; h(1) = 3; h(4) = 6$

33. 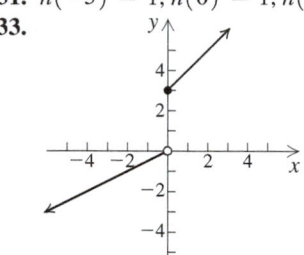 **35.**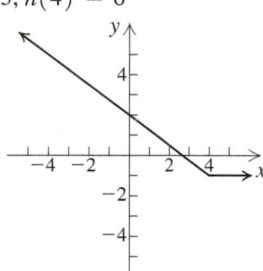

37. 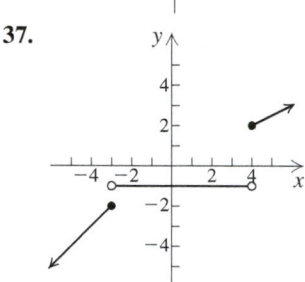 **39.**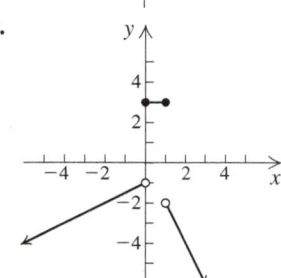

41. 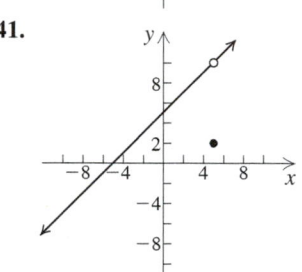 **43.**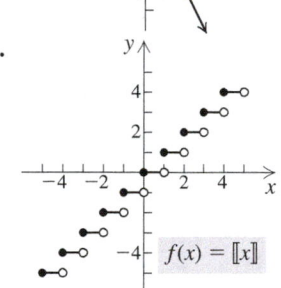

45.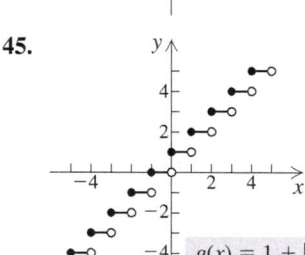

47. Domain:
$(−\infty, \infty)$; range:
$(−\infty, 0) \cup [3, \infty)$
49. Domain: $(−\infty, \infty)$;
range: $[−1, \infty)$

51. Domain: $(−\infty, \infty)$; range: $\{y \mid y \le −2 \text{ or } y = −1 \text{ or } y \ge 2\}$
53. Domain: $(−\infty, \infty)$; range: $\{−5, −2, 4\}$;
$$f(x) = \begin{cases} −2, & \text{for } x < 2, \\ −5, & \text{for } x = 2, \\ 4, & \text{for } x > 2 \end{cases}$$
55. Domain: $(−\infty, \infty)$; range: $(−\infty, −1] \cup [2, \infty)$;
$$g(x) = \begin{cases} x, & \text{for } x \le −1, \\ 2, & \text{for } −1 < x < 2, \\ x, & \text{for } x \ge 2 \end{cases}$$
or
$$g(x) = \begin{cases} x, & \text{for } x \le −1, \\ 2, & \text{for } −1 < x \le 2, \\ x, & \text{for } x > 2 \end{cases}$$

57. Domain: $[-5, 3]$; range: $(-3, 5)$;

$$h(x) = \begin{cases} x + 8, & \text{for } -5 \le x < -3, \\ 3, & \text{for } -3 \le x \le 1, \\ 3x - 6, & \text{for } 1 < x \le 3 \end{cases}$$

59. (a) 38; **(b)** 38; **(c)** $5a^2 - 7$; **(d)** $5a^2 - 7$ **60. (a)** 22;
(b) -22; **(c)** $4a^3 - 5a$; **(d)** $-4a^3 + 5a$ **61.** $y = -\frac{1}{8}x + \frac{7}{8}$
62. Slope; $\frac{2}{9}$; y-intercept; $\left(0, \frac{1}{9}\right)$

63. (a)   **(b)** $C(t) = 3([t] + 1), t > 0$

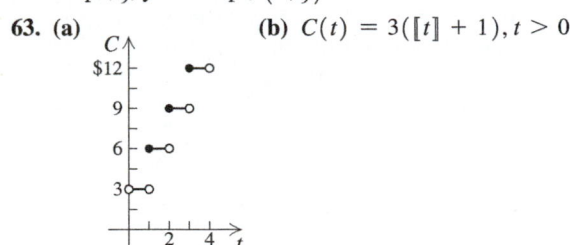

65. $\{x \mid -5 \le x < -4 \text{ or } 5 \le x < 6\}$
67. (a) $h(r) = \dfrac{30 - 5r}{3}$; **(b)** $V(r) = \pi r^2 \left(\dfrac{30 - 5r}{3}\right)$;

(c) $V(h) = \pi h \left(\dfrac{30 - 3h}{5}\right)^2$

Summary and Review: Chapter 6, p. 428

Vocabulary Reinforcement

1. Cube   **2.** Increasing   **3.** Principal   **4.** Rationalizing
5. Radical   **6.** Open   **7.** Square   **8.** Radicand
9. Conjugate   **10.** Index

Concept Reinforcement

1. True   **2.** False   **3.** False   **4.** False   **5.** True
6. True

Study Guide

1. $6|y|$ **2.** $|a + 2|$ **3.** $\sqrt[5]{z^3}$ **4.** $(6ab)^{5/2}$
5. $\dfrac{1}{9^{3/2}} = \dfrac{1}{27}$ **6.** $\sqrt[4]{a^3 b}$ **7.** $5y\sqrt{6}$ **8.** $2\sqrt{a}$ **9.** $2\sqrt{3}$
10. $25 - 10\sqrt{x} + x$ **11.** 5   **12.** 6
13. (a) $(-4, -2)$; **(b)** $(2, 5)$; **(c)** $(-2, 2)$
14. Relative maxima: 4.218 at $x = -1.940$ and 0.575 at
$x = 1.685$; relative minimum: -1.126 at $x = 0.255$
15. $A(h) = h^2 + 5h$

16.

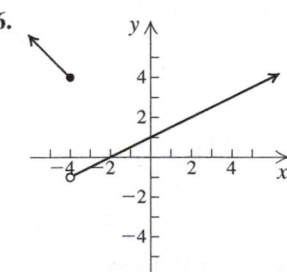

Review Exercises

1. 27.893   **2.** 6.378   **3.** $f(0), f(-1)$, and $f(1)$ do not exist as
real numbers; $f\left(\frac{41}{3}\right) = 5$ **4.** Domain $= \left\{x \mid x \ge \frac{16}{3}\right\}$, or $\left[\frac{16}{3}, \infty\right)$
5. $9|a|$ **6.** $7|z|$ **7.** $|6 - b|$ **8.** $|x + 3|$ **9.** -10

10. $-\frac{1}{3}$ **11.** $2; -2; 3$ **12.** $|x|$ **13.** 3   **14.** $\sqrt[5]{a}$ **15.** 512
16. $31^{1/2}$ **17.** $(a^2 b^3)^{1/5}$ **18.** $\frac{1}{7}$ **19.** $\dfrac{1}{4x^{2/3} y^{2/3}}$ **20.** $\dfrac{5b^{1/2}}{a^{3/4} c^{2/3}}$

21. $\dfrac{3a}{t^{1/4}}$ **22.** $\dfrac{1}{x^{2/5}}$ **23.** $7^{1/6}$ **24.** x^7 **25.** $3x^2$
26. $\sqrt[12]{x^4 y^3}$ **27.** $\sqrt[12]{x^7}$ **28.** $7\sqrt{5}$ **29.** $-3\sqrt[3]{4}$
30. $5b^2 \sqrt[3]{2a^2}$ **31.** $\frac{7}{6}$ **32.** $\dfrac{4x^2}{3}$ **33.** $\dfrac{2x^2}{3y^3}$ **34.** $\sqrt{15xy}$
35. $3a\sqrt[3]{a^2 b^2}$ **36.** $\sqrt[15]{a^5 b^9}$ **37.** $y\sqrt[3]{6}$ **38.** $\frac{5}{2}\sqrt{x}$
39. $\sqrt[12]{x^5}$ **40.** $7\sqrt[3]{x}$ **41.** $3\sqrt{3}$ **42.** $15\sqrt{2}$
43. $(2x + y^2)\sqrt[3]{x}$ **44.** $-43 - 2\sqrt{10}$ **45.** $8 - 2\sqrt{7}$
46. $9 - \sqrt[3]{4}$ **47.** $\dfrac{2\sqrt{6}}{3}$ **48.** $\dfrac{2\sqrt{a} - 2\sqrt{b}}{a - b}$ **49.** 4
50. 13   **51.** 1   **52.** About 4166 rpm   **53.** 4480 rpm
54. 9 cm   **55.** $\sqrt{24}$ ft; 4.899 ft   **56.** 25
57. $\sqrt{46}$; 6.782
58. (a) $(-1, 0), (2, \infty)$; **(b)** $(0, 2)$; **(c)** $(-\infty, -1)$
59. Relative maximum: 4 at $x = 0$; relative minimum: 0 at $x = 2$
60. $A(x) = x(48 - 2x)$, or $48x - 2x^2$
61. $A(x) = 2x\sqrt{4 - x^2}$
62. (a) $A(x) = x^2 + \dfrac{432}{x}$; **(b)** $(0, \infty)$; **(c)** $x = 6$ in.;

height $= 3$ in.

63.

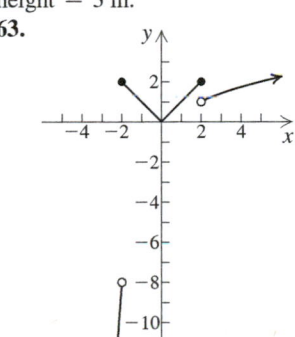

64.

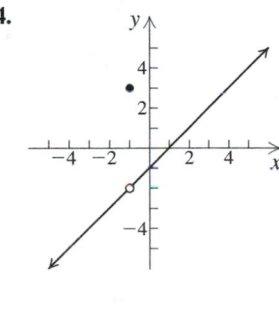

65.

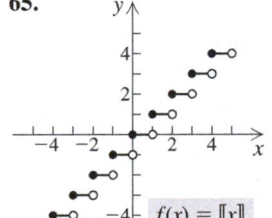

66.

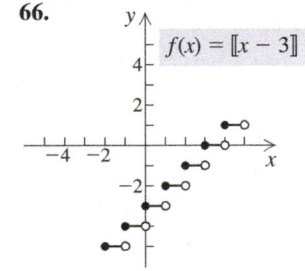

67. $f(-1) = 1; f(5) = 2; f(-2) = 2; f(-3) = -27$
68. $f(-2) = -3; f(-1) = 3; f(0) = -1; f(4) = 3$
69.

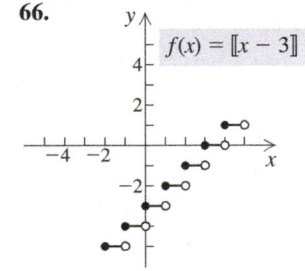

70. B   **71.** 3   **72.** Increasing: $(-3, \infty)$; decreasing:
$(-\infty, -3)$; relative minimum: -5 at $x = -3$ **73.** Increasing:
$(3, \infty)$; decreasing: $(-\infty, 3)$; relative minimum: 1 at $x = 3$

74. $f(x) = (x + 5)^{1/2}(x + 7)^{-1/2}$. Consider $(x + 5)^{1/2}$. Since the exponent is $\frac{1}{2}$, $x + 5$ must be nonnegative. Then $x + 5 \geq 0$, or $x \geq -5$. Consider $(x + 7)^{-1/2}$. Since the exponent is $-\frac{1}{2}$, $x + 7$ must be positive. Then $x + 7 > 0$, or $x > -7$. Then the domain of $f = \{x \mid x \geq -5 \; and \; x > -7\}$, or $\{x \mid x \geq -5\}$. **75.** Since $\sqrt{x}$ exists only for $\{x \mid x \geq 0\}$, this is the domain of $y = \sqrt{x} \cdot \sqrt{x}$. **76.** The distributive law is used to collect radical expressions with the same indices and radicands just as it is used to collect monomials with the same variables and exponents. **77.** No; when n is odd, it is true that if $a^n = b^n$, then $a = b$. **78.** Use a calculator to show that $\dfrac{5 + \sqrt{2}}{\sqrt{18}} \neq 2$. Explain that we multiply by 1 to rationalize a denominator. In this case, we would write 1 as $\sqrt{2}/\sqrt{2}$.

Test: Chapter 6, p. 435

1. [6.1a] 12.166 **2.** [6.1a] 2; does not exist as a real number **3.** [6.1a] Domain $= \{x \mid x \leq 2\}$, or $(-\infty, 2]$ **4.** [6.1b] $3|q|$ **5.** [6.1b] $|x + 5|$ **6.** [6.1c] $-\frac{1}{10}$ **7.** [6.1d] x **8.** [6.1d] 4 **9.** [6.2a] $\sqrt[3]{a^2}$ **10.** [6.2a] 8 **11.** [6.2a] $37^{1/2}$ **12.** [6.2a] $(5xy^2)^{5/2}$ **13.** [6.2b] $\frac{1}{10}$ **14.** [6.2b] $\dfrac{8a^{3/4}}{b^{3/2}c^{2/5}}$ **15.** [6.2c] $\dfrac{x^{8/5}}{y^{9/5}}$ **16.** [6.2c] $\dfrac{1}{2.9^{31/24}}$ **17.** [6.2d] $\sqrt[4]{x}$ **18.** [6.2d] $2x\sqrt{x}$ **19.** [6.2d] $\sqrt[15]{a^6 b^5}$ **20.** [6.2d] $\sqrt[12]{8y^7}$ **21.** [6.3a] $2\sqrt{37}$ **22.** [6.3a] $2\sqrt[4]{5}$ **23.** [6.3a] $2a^3 b^4 \sqrt[3]{3a^2 b}$ **24.** [6.3b] $\dfrac{2x\sqrt[3]{2x^2}}{y^2}$ **25.** [6.3b] $\dfrac{5x}{6y^2}$ **26.** [6.3a] $\sqrt[3]{10xy^2}$ **27.** [6.3a] $xy\sqrt[4]{x}$ **28.** [6.3b] $\sqrt[5]{x^2 y^2}$ **29.** [6.3b] $2\sqrt{a}$ **30.** [6.4a] $38\sqrt{2}$ **31.** [6.4b] -20 **32.** [6.4b] $9 + 6\sqrt{x} + x$ **33.** [6.5b] $\dfrac{13 + 8\sqrt{2}}{-41}$ **34.** [6.6a] 35 **35.** [6.6b] 7 **36.** [6.6a] 5 **37.** [6.7a] 7 ft **38.** [6.6c] 3600 ft **39.** [6.7a] $\sqrt{98}$; 9.899 **40.** [6.7a] 2 **41.** [6.8a] **(a)** $(-5, -2)$; **(b)** $(2, 5)$; **(c)** $(-2, 2)$ **42.** [6.8b] Relative maximum: 3.163 at $x = -1.519$; relative minimum: -1.015 at $x = 2.852$ **43.** [6.8c] $A(b) = \frac{1}{2}b(4b - 6)$, or $2b^2 - 3b$ **44.** [6.8d]

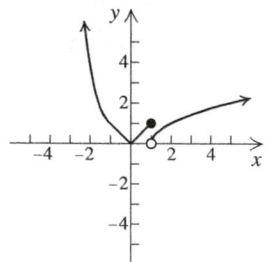

45. [6.8d] $f\left(-\frac{7}{8}\right) = \frac{7}{8}; f(5) = 2; f(-4) = 16$ **46.** [6.6a] A **47.** [6.1a] Domain $= \{x \mid -4 < x \leq 1\}$, or $(-4, 1]$ **48.** [6.6b] 3

Chapter 7

Exercise Set 7.1, p. 443

1. x-axis: no; y-axis: yes; origin: no **3.** x-axis: yes; y-axis: no; origin: no **5.** x-axis: no; y-axis: no; origin: yes **7.** x-axis: no; y-axis: yes; origin: no **9.** x-axis: no; y-axis: no; origin: no **11.** x-axis: no; y-axis: yes; origin: no **13.** x-axis: no; y-axis: no; origin: yes **15.** x-axis: no; y-axis: no; origin: yes **17.** x-axis: yes; y-axis: yes; origin: yes **19.** x-axis: no; y-axis: yes; origin: no **21.** x-axis: yes; y-axis: yes; origin: yes **23.** x-axis: no; y-axis: no; origin: no **25.** x-axis: no; y-axis: no; origin: yes **27.** x-axis: $(-5, -6)$; y-axis: $(5, 6)$; origin: $(5, -6)$ **29.** x-axis: $(-10, 7)$; y-axis: $(10, -7)$; origin: $(10, 7)$ **31.** x-axis: $(0, 4)$; y-axis: $(0, -4)$; origin: $(0, 4)$ **33.** Even **35.** Odd **37.** Neither **39.** Odd **41.** Even **43.** Odd **45.** Neither **47.** Even **49.**

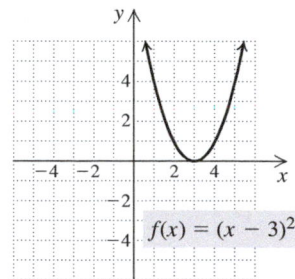

50. University of California–Berkeley: 3576 volunteers; University of Wisconsin–Madison: 3112 volunteers

51. Odd **53.** x-axis: yes; y-axis: no; origin: no **55.** $E(-x) = \dfrac{f(-x) + f(-(-x))}{2} = \dfrac{f(-x) + f(x)}{2} = E(x)$ **57.** **(a)** $E(x) + O(x) = \dfrac{f(x) + f(-x)}{2} + \dfrac{f(x) - f(-x)}{2} = \dfrac{2f(x)}{2} = f(x)$; **(b)** $f(x) = \dfrac{-22x^2 + \sqrt{x} + \sqrt{-x} - 20}{2} + \dfrac{8x^3 + \sqrt{x} - \sqrt{-x}}{2}$ **59.** True

Visualizing the Graph, p. 454

1. C **2.** B **3.** A **4.** E **5.** G **6.** D **7.** H **8.** I **9.** F

Exercise Set 7.2, p. 455

1. Start with the graph of $f(x) = x^2$. Shift it right 3 units. **3.** Start with the graph of $g(x) = x$. Shift it down 3 units.

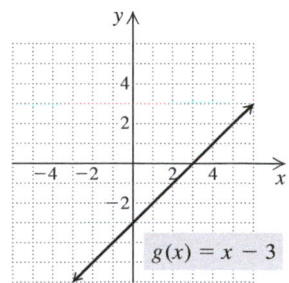

5. Start with the graph of $h(x) = \sqrt{x}$. Reflect it across the *x*-axis.

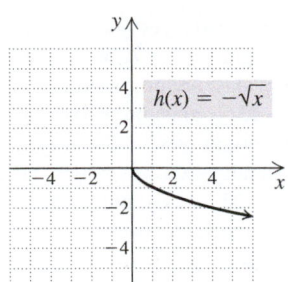

7. Start with the graph of $h(x) = \dfrac{1}{x}$. Shift it up 4 units.

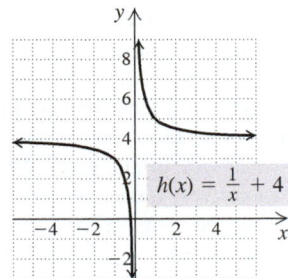

9. Start with the graph of $h(x) = x$. Stretch it vertically by multiplying each *y*-coordinate by 3. Then reflect it across the *x*-axis and shift it up 3 units.

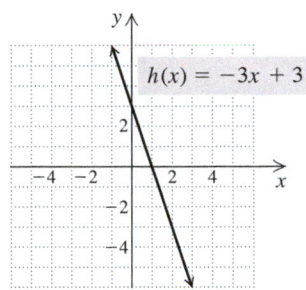

11. Start with the graph of $h(x) = |x|$. Shrink it vertically by multiplying each *y*-coordinate by $\frac{1}{2}$. Then shift it down 2 units.

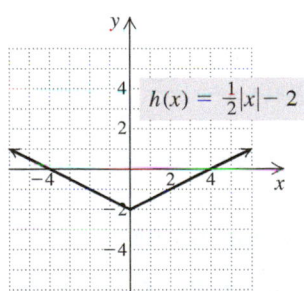

13. Start with the graph of $g(x) = x^3$. Shift it right 2 units. Then reflect it across the *x*-axis.

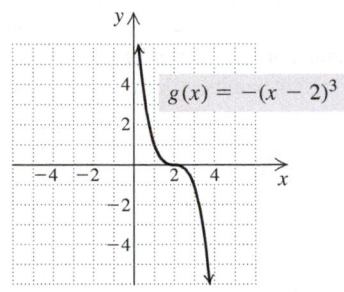

15. Start with the graph of $g(x) = x^2$. Shift it left 1 unit. Then shift it down 1 unit.

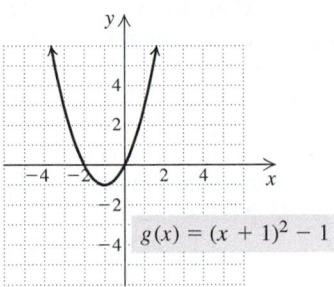

17. Start with the graph of $g(x) = x^3$. Shrink it vertically by multiplying each *y*-coordinate by $\frac{1}{3}$. Then shift it up 2 units.

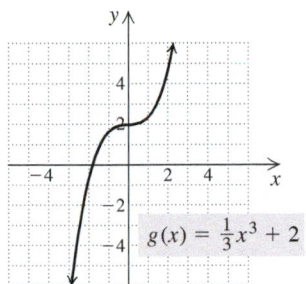

19. Start with the graph of $f(x) = \sqrt{x}$. Shift it left 2 units.

21. Start with the graph of $f(x) = \sqrt[3]{x}$. Shift it down 2 units.

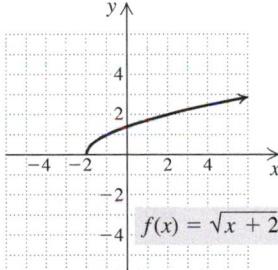 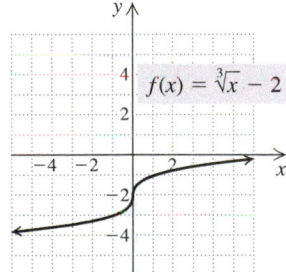

23. Start with the graph of $g(x) = |x|$. Shrink it horizontally by multiplying each *x*-coordinate by $\frac{1}{3}$ (or dividing each *x*-coordinate by 3).   **25.** Start with the graph of $h(x) = \dfrac{1}{x}$. Stretch it vertically by multiplying each *y*-coordinate by 2.   **27.** Start with the graph of $f(x) = \sqrt{x}$. Stretch it vertically by multiplying each *y*-coordinate by 3. Then shift it down 5 units.   **29.** Start with the graph of $g(x) = |x|$. Stretch it horizontally by multiplying each *x*-coordinate by 3. Then shift it down 4 units.   **31.** Start with the graph of $f(x) = x^2$. Shift it right 5 units, shrink it vertically by multiplying each *y*-coordinate by $\frac{1}{4}$, and then reflect it across the *x*-axis.   **33.** Start with the graph of $f(x) = \dfrac{1}{x}$. Shift it left 3 units, then up 2 units.   **35.** Start with the graph of $h(x) = x^2$. Shift it right 3 units. Then reflect it across the *x*-axis and shift it up 5 units.
37. $(-12, 2)$ **39.** $(12, 4)$ **41.** $(-12, 2)$ **43.** $(-12, 16)$
45. B   **47.** A   **49.** $f(x) = -(x - 8)^2$
51. $f(x) = |x + 7| + 2$ **53.** $f(x) = \dfrac{1}{2x} - 3$
55. $f(x) = -(x - 3)^2 + 4$ **57.** $f(x) = \sqrt{-(x + 2)} - 1$

59.

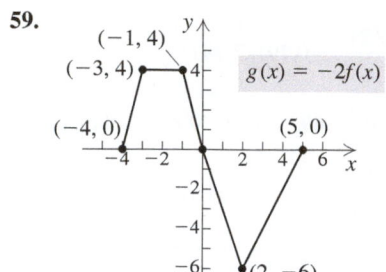

61.

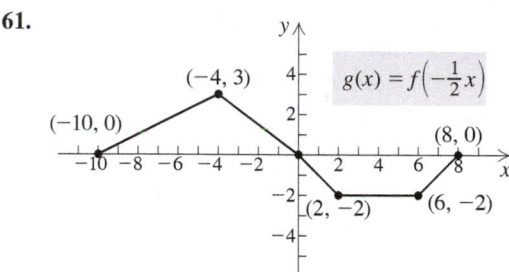

63.

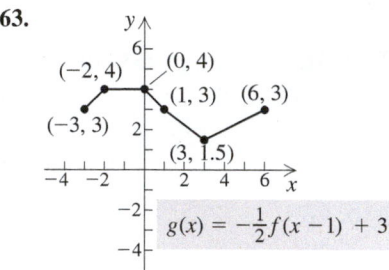

65.

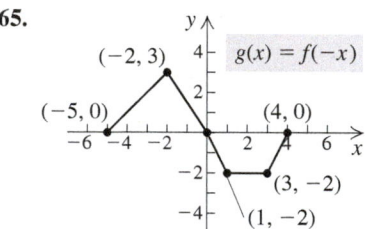

67.

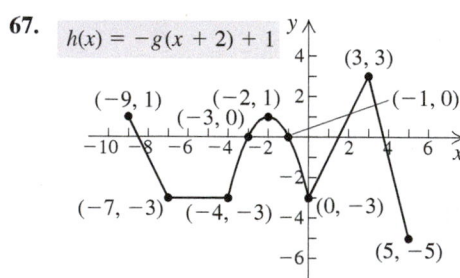

69.

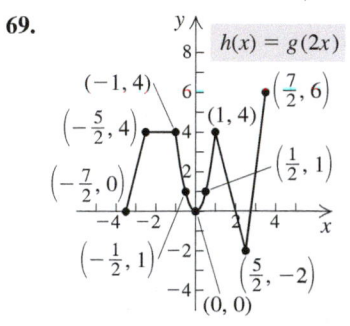

71. (f) **73.** (f) **75.** (d) **77.** (c)
79. $f(-x) = 2(-x)^4 - 35(-x)^3 + 3(-x) - 5 =$
$2x^4 + 35x^3 - 3x - 5 = g(x)$ **81.** $g(x) = x^3 - 3x^2 + 2$
83. $k(x) = (x + 1)^3 - 3(x + 1)^2$ **85.** x-axis, no; y-axis,
yes; origin, no **86.** x-axis, yes; y-axis, no; origin, no
87. x-axis, no; y-axis, no; origin, yes **88.** 40,504 pages
89. 1123 guns **90.** About 29,700 acres
91.   **93.**

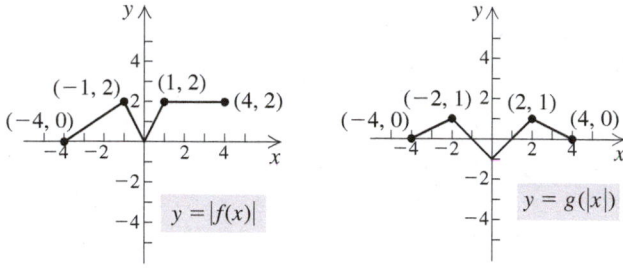

95. 5

Exercise Set 7.3, p. 462

1. $\sqrt{3}i$ **3.** $5i$ **5.** $-\sqrt{33}i$ **7.** $-9i$ **9.** $7\sqrt{2}i$
11. $2 + 11i$ **13.** $5 - 12i$ **15.** $4 + 8i$ **17.** $-4 - 2i$
19. $5 + 9i$ **21.** $5 + 4i$ **23.** $5 + 7i$ **25.** $11 - 5i$
27. $-1 + 5i$ **29.** $2 - 12i$ **31.** -12 **33.** -45
35. $35 + 14i$ **37.** $6 + 16i$ **39.** $13 - i$ **41.** $-11 + 16i$
43. $-10 + 11i$ **45.** $-31 - 34i$ **47.** $-14 + 23i$
49. 41 **51.** 13 **53.** 74 **55.** $12 + 16i$ **57.** $-45 - 28i$
59. $-8 - 6i$ **61.** $2i$ **63.** $-7 + 24i$ **65.** $\frac{15}{146} + \frac{33}{146}i$
67. $\frac{10}{13} - \frac{15}{13}i$ **69.** $-\frac{14}{13} + \frac{5}{13}i$ **71.** $\frac{11}{25} - \frac{27}{25}i$
73. $\dfrac{-4\sqrt{3} + 10}{41} + \dfrac{5\sqrt{3} + 8}{41}i$ **75.** $-\frac{1}{2} + \frac{1}{2}i$
77. $-\frac{1}{2} - \frac{13}{2}i$ **79.** $-i$ **81.** $-i$ **83.** 1 **85.** i
87. 625 **89.** y-intercept **90.** x-intercept **91.** Relation
92. Function **93.** Horizontal lines **94.** Parallel
95. Decreasing **96.** Symmetric with respect to the y-axis
97. True **99.** True **101.** $a^2 + b^2$ **103.** $x^2 - 6x + 25$

Mid-Chapter Review: Chapter 7, p. 464

1. True **2.** False **3.** False **4.** True
5. x-axis: no; y-axis: yes; origin: no **6.** x-axis: yes; y-axis: yes;
origin: yes **7.** x-axis: no; y-axis: no; origin: yes **8.** x-axis:
no; y-axis: no; origin: no **9.** x-axis: $(-7, -2)$; y-axis: $(7, 2)$;
origin: $(7, -2)$ **10.** x-axis: $(0, \frac{1}{3})$; y-axis: $(0, -\frac{1}{3})$; origin:
$(0, \frac{1}{3})$ **11.** Neither **12.** Odd **13.** Even **14.** Even
15. $(9, -2)$ **16.** $(12, -6)$ **17.** $(-9, -6)$ **18.** $(4.5, -3)$
19. $f(x) = (x + 3)^2 - 4$ **20.** $f(x) = -\sqrt{x + 5}$
21.

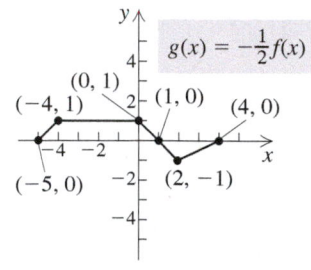

22.

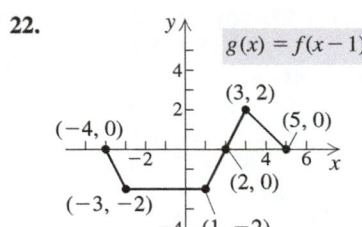

23.

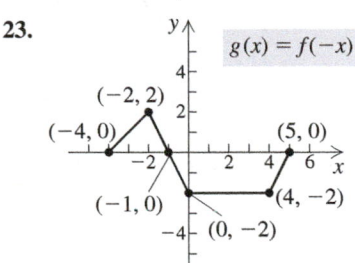

24.

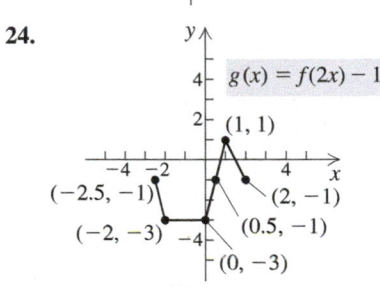

25. $6i$ **26.** $\sqrt{5}i$ **27.** $-4i$ **28.** $4\sqrt{2}i$ **29.** $-1 + i$
30. $-7 + 5i$ **31.** $23 + 2i$ **32.** $-\frac{5}{29} - \frac{17}{29}i$ **33.** i
34. 1 **35.** $-i$ **36.** -64 **37.** The sum of two imaginary
numbers is not always an imaginary number. For example,
$(2 + i) + (3 - i) = 5$, a real number.   **38.** In the graph
of $y = f(cx)$, the constant c stretches or shrinks the graph of
$y = f(x)$ horizontally. The constant c in $y = cf(x)$ stretches
or shrinks the graph of $y = f(x)$ vertically. For $y = f(cx)$, the
x-coordinates of $y = f(x)$ are divided by c; for $y = cf(x)$, the
y-coordinates of $y = f(x)$ are multiplied by c.   **39.** The graph
of $f(x) = 0$ is symmetric with respect to the x-axis, the y-axis,
and the origin. This function is both even and odd.   **40.** If all the
exponents are even numbers, then $f(x)$ is an even function. If $a_0 = 0$
and all the exponents are odd numbers, then $f(x)$ is an odd function.

Exercise Set 7.4, p. 477

1. $\frac{2}{3}, \frac{3}{2}$ **3.** $-2, 10$ **5.** $-1, \frac{2}{3}$ **7.** $-\sqrt{3}, \sqrt{3}$
9. $-\sqrt{7}, \sqrt{7}$ **11.** $-\sqrt{2}i, \sqrt{2}i$ **13.** $-4i, 4i$ **15.** $0, 3$
17. $-\frac{1}{3}, 0, 2$ **19.** $-1, -\frac{1}{7}, 1$ **21. (a)** $(-4, 0), (2, 0)$;
(b) $-4, 2$ **23. (a)** $(-1, 0), (3, 0)$; **(b)** $-1, 3$
25. (a) $(-2, 0), (2, 0)$; **(b)** $-2, 2$ **27. (a)** $(1, 0)$; **(b)** 1
29. $-7, 1$ **31.** $4 \pm \sqrt{7}$ **33.** $-4 \pm 3i$ **35.** $-2, \frac{1}{3}$
37. $-3, 5$ **39.** $-1, \frac{2}{5}$ **41.** $\dfrac{5 \pm \sqrt{7}}{3}$ **43.** $-\dfrac{1}{2} \pm \dfrac{\sqrt{7}}{2}i$
45. $\dfrac{4 \pm \sqrt{31}}{5}$ **47.** $\dfrac{5}{6} \pm \dfrac{\sqrt{23}}{6}i$ **49.** $4 \pm \sqrt{11}$
51. $\dfrac{-1 \pm \sqrt{61}}{6}$ **53.** $\dfrac{5 \pm \sqrt{17}}{4}$ **55.** $-\dfrac{1}{5} \pm \dfrac{3}{5}i$
57. 144; two real   **59.** -7; two imaginary   **61.** 49; two real

63. $-5, -1$ **65.** $\dfrac{3 \pm \sqrt{21}}{2}$; $-0.791, 3.791$
67. $\dfrac{5 \pm \sqrt{21}}{2}$; $0.209, 4.791$ **69.** $-1 \pm \sqrt{6}$; $-3.449, 1.449$
71. $\dfrac{1}{4} \pm \dfrac{\sqrt{31}}{4}i$ **73.** $\dfrac{1 \pm \sqrt{13}}{6}$; $-0.434, 0.768$
75. $\dfrac{1 \pm \sqrt{6}}{5}$; $-0.290, 0.690$ **77.** $\dfrac{-3 \pm \sqrt{57}}{8}$; $-1.319, 0.569$
79. $\pm 1, \pm \sqrt{2}$ **81.** $\pm \sqrt{2}, \pm \sqrt{5}i$ **83.** $\pm 1, \pm \sqrt{5}i$
85. 16 **87.** $-8, 64$ **89.** $1, 16$ **91.** $\frac{5}{2}, 3$ **93.** $-\frac{3}{2}, -1, \frac{1}{2}, 1$
95. 2011 **97.** 1995 **99.** About 10.216 sec
101. Length: 4 ft; width: 3 ft   **103.** 4 and 9; -9 and -4
105. 2 cm   **107.** Length: 8 ft; width: 6 ft   **109.** Linear
111. Quadratic   **113.** Linear   **115.** About $3.95 million
116. About 16 years after 2004, or in 2020   **117.** x-axis: yes;
y-axis: yes; origin: yes   **118.** x-axis: no; y-axis: yes; origin: no
119. Odd   **120.** Neither   **121. (a)** 2; **(b)** $\frac{11}{2}$
123. (a) 2; **(b)** $1 - i$ **125.** 1   **127.** $-\sqrt{7}, -\frac{3}{2}, 0, \frac{1}{3}, \sqrt{7}$
129. $\dfrac{-1 \pm \sqrt{1 + 4\sqrt{2}}}{2}$ **131.** $3 \pm \sqrt{5}$ **133.** 19
135. $-2 \pm \sqrt{2}, \dfrac{1}{2} \pm \dfrac{\sqrt{7}}{2}i$

Visualizing the Graph, p. 489

1. C   **2.** B   **3.** A   **4.** J   **5.** F   **6.** D   **7.** I
8. G   **9.** H   **10.** E

Exercise Set 7.5, p. 490

1. (a) $\left(-\frac{1}{2}, -\frac{9}{4}\right)$; **(b)** $x = -\frac{1}{2}$; **(c)** minimum: $-\frac{9}{4}$
3. (a) $(4, -4)$; **(b)** $x = 4$; **(c)** minimum: -4;
(d)

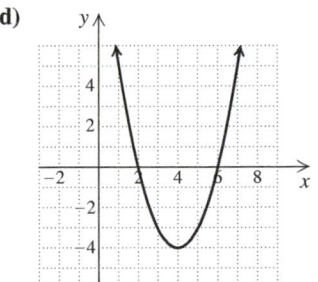

$f(x) = x^2 - 8x + 12$

5. (a) $\left(\frac{7}{2}, -\frac{1}{4}\right)$; **(b)** $x = \frac{7}{2}$; **(c)** minimum: $-\frac{1}{4}$;
(d)

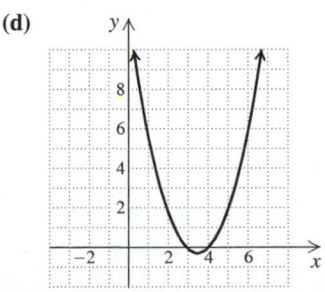

$f(x) = x^2 - 7x + 12$

7. (a) $(-2, 1)$; **(b)** $x = -2$; **(c)** minimum: 1;
(d)

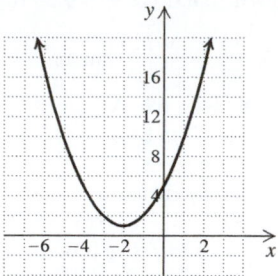

$$f(x) = x^2 + 4x + 5$$

9. (a) $(-4, -2)$; **(b)** $x = -4$; **(c)** minimum: -2;
(d)

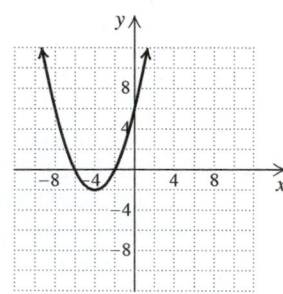

$$g(x) = \frac{x^2}{2} + 4x + 6$$

11. (a) $\left(-\frac{3}{2}, \frac{7}{2}\right)$; **(b)** $x = -\frac{3}{2}$; **(c)** minimum: $\frac{7}{2}$;
(d)

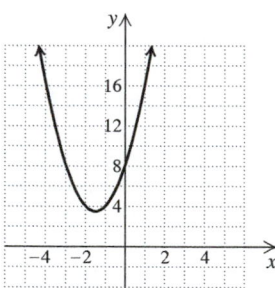

$$g(x) = 2x^2 + 6x + 8$$

13. (a) $(-3, 12)$; **(b)** $x = -3$; **(c)** maximum: 12;
(d)

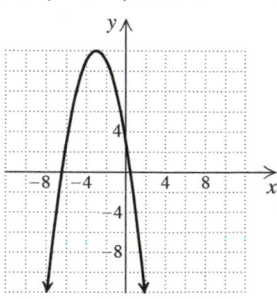

$$f(x) = -x^2 - 6x + 3$$

15. (a) $\left(\frac{1}{2}, \frac{3}{2}\right)$; **(b)** $x = \frac{1}{2}$; **(c)** maximum: $\frac{3}{2}$;

(d)

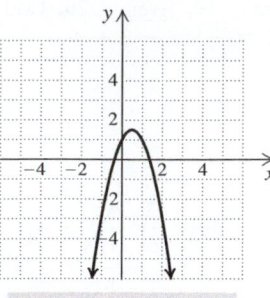

$$g(x) = -2x^2 + 2x + 1$$

17. (f) **19.** (b) **21.** (h) **23.** (c) **25.** True
27. False **29.** True **31. (a)** $(3, -4)$; **(b)** minimum:
-4; **(c)** $[-4, \infty)$; **(d)** increasing: $(3, \infty)$; decreasing:
$(-\infty, 3)$ **33. (a)** $(-1, -18)$; **(b)** minimum: -18;
(c) $[-18, \infty)$; **(d)** increasing: $(-1, \infty)$; decreasing:
$(-\infty, -1)$ **35. (a)** $\left(5, \frac{9}{2}\right)$; **(b)** maximum: $\frac{9}{2}$; **(c)** $\left(-\infty, \frac{9}{2}\right]$;
(d) increasing: $(-\infty, 5)$; decreasing: $(5, \infty)$
37. (a) $(-1, 2)$; **(b)** minimum: 2; **(c)** $[2, \infty)$; **(d)** increasing:
$(-1, \infty)$; decreasing: $(-\infty, -1)$ **39. (a)** $\left(-\frac{3}{2}, 18\right)$;
(b) maximum: 18; **(c)** $(-\infty, 18]$; **(d)** increasing: $\left(-\infty, -\frac{3}{2}\right)$;
decreasing: $\left(-\frac{3}{2}, \infty\right)$ **41.** 0.625 sec; 12.25 ft
43. 3.75 sec; 305 ft **45.** 4.5 in. **47.** Base: 10 cm; height: 10 cm
49. 350 doghouses **51.** \$797; 40 units **53.** 4800 yd^2
55. About 60.5 ft **57.** 3 **58.** $4x + 2h - 1$
59.

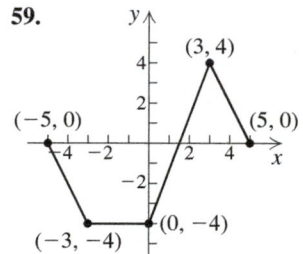

$$g(x) = -2f(x)$$

60.

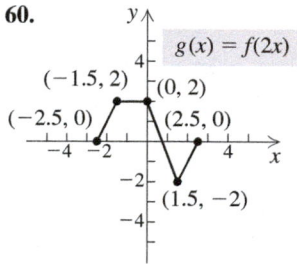

61. -236.25

63.

$$f(x) = (|x| - 5)^2 - 3$$

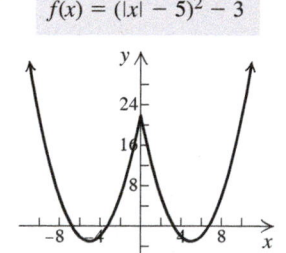

65. Pieces should be $\dfrac{24\pi}{4 + \pi}$ in.
and $\dfrac{96}{4 + \pi}$ in.

Summary and Review: Chapter 7

Review Exercises, p. 500

1. True **2.** True **3.** True **4.** True
5. *x*-axis: yes; *y*-axis: yes; origin: yes **6.** *x*-axis: yes; *y*-axis: yes;
origin: yes **7.** *x*-axis: no; *y*-axis: no; origin: no
8. *x*-axis: no; *y*-axis: yes; origin: no **9.** *x*-axis: no; *y*-axis: no;
origin: yes **10.** *x*-axis: no; *y*-axis: yes; origin: no
11. Even **12.** Even **13.** Odd **14.** Even **15.** Even

16. Neither   **17.** Odd   **18.** Even   **19.** Even   **20.** Odd
21. $f(x) = (x + 3)^2$ **22.** $f(x) = -\sqrt{x - 3} + 4$
23. $f(x) = 2|x - 3|$

24.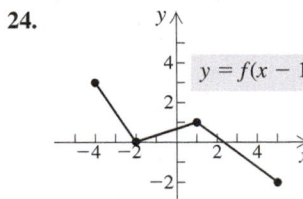

25.

26.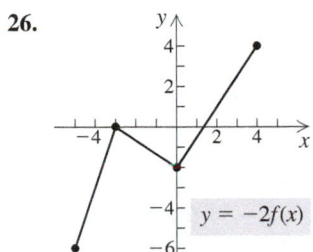

27.

28. $-2\sqrt{10}\,i$ **29.** $-4\sqrt{15}$ **30.** $-\frac{7}{8}$ **31.** $2 - i$ **32.** $1 - 4i$
33. $-18 - 26i$ **34.** $\frac{11}{10} + \frac{3}{10}i$ **35.** $-i$ **36.** $-\frac{5}{2}, \frac{1}{3}$
37. $-5, 1$ **38.** $-2, \frac{4}{3}$ **39.** $-\sqrt{3}, \sqrt{3}$ **40.** $-\sqrt{10}\,i, \sqrt{10}\,i$
41. 1 **42.** $-5, 3$ **43.** $\dfrac{1 \pm \sqrt{41}}{4}$
44. $-\dfrac{1}{3} \pm \dfrac{2\sqrt{2}}{3}i$
45. $x^2 - 3x + \frac{9}{4} = 18 + \frac{9}{4}; \left(x - \frac{3}{2}\right)^2 = \frac{81}{4}; x = \frac{3}{2} \pm \frac{9}{2}; -3, 6$
46. $x^2 - 4x = 2; x^2 - 4x + 4 = 2 + 4; (x - 2)^2 = 6;$
$x = 2 \pm \sqrt{6}; 2 - \sqrt{6}, 2 + \sqrt{6}$ **47.** $-4, \frac{2}{3}$
48. $1 - 3i, 1 + 3i$ **49.** $-2, 5$ **50.** 1 **51.** $\pm\sqrt{\dfrac{3 \pm \sqrt{5}}{2}}$
52. $-\sqrt{3}, 0, \sqrt{3}$ **53.** $-2, -\frac{2}{3}, 3$ **54.** $-5, -2, 2$
55. **(a)** $\left(\frac{3}{8}, -\frac{7}{16}\right)$; **(b)** $x = \frac{3}{8}$; **(c)** maximum: $-\frac{7}{16}$; **(d)** $\left(-\infty, -\frac{7}{16}\right]$;
(e)

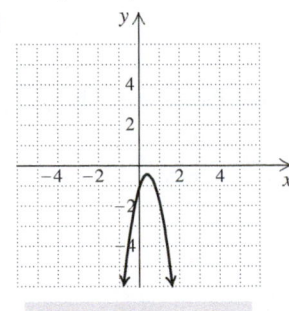

$f(x) = -4x^2 + 3x - 1$

56. **(a)** $(1, -2)$; **(b)** $x = 1$; **(c)** minimum: -2; **(d)** $[-2, \infty)$;
(e)

$f(x) = 5x^2 - 10x + 3$

57. (d)   **58.** (c)   **59.** (b)   **60.** (a)   **61.** 30 ft, 40 ft
62. Rebecca: 15 km/h; Harry: 8 km/h   **63.** $35 - 5\sqrt{33}$ ft, or
about 6.3 ft   **64.** 6 ft by 6 ft   **65.** $\dfrac{15 - \sqrt{115}}{2}$ cm, or
about 2.1 cm   **66.** C   **67.** B   **68.** B   **69.** A
70. Let $f(x)$ and $g(x)$ be odd functions. Then by
definition, $f(-x) = -f(x)$, or $f(x) = -f(-x)$,
and $g(-x) = -g(x)$, or $g(x) = -g(-x)$. Thus,
$(f + g)(x) = f(x) + g(x) = -f(-x) + [-g(-x)] = -[f(-x) + g(-x)] = -(f + g)(-x)$ and $f + g$ is odd.
71. Reflect the graph of $y = f(x)$ across the x-axis and then
across the y-axis.   **72.** $-\frac{1}{4}, 2$ **73.** -1 **74.** 9%   **75.** ± 6
76. **(a)** $4x^3 - 2x + 9$; **(b)** $4x^3 + 24x^2 + 46x + 35$;
(c) $4x^3 - 2x + 42$. **(a)** adds 2 to each function value;
(b) adds 2 to each input before finding a function value;
(c) adds the output for 2 to the output for x **77.** Use the
discriminant. If $b^2 - 4ac < 0$, there are no x-intercepts. If
$b^2 - 4ac = 0$, there is one x-intercept. If $b^2 - 4ac > 0$,
there are two x-intercepts.   **78.** Completing the square was
used in Section 7.4 to solve quadratic equations. It was used
again in Section 7.5 to write quadratic functions in the form
$f(x) = a(x - h)^2 + k$.   **79.** The x-intercepts of $g(x)$ are
also $(x_1, 0)$ and $(x_2, 0)$. This is true because $f(x)$ and $g(x)$ have
the same zeros. Consider $g(x) = 0$, or $-ax^2 - bx - c = 0$.
Multiplying by -1 on both sides, we get an equivalent equation
$ax^2 + bx + c = 0$, or $f(x) = 0$.   **80.** The product of two
imaginary numbers is not always an imaginary number. For
example, $i \cdot i = i^2 = -1$, a real number.

81. No; consider the quadratic formula $x = \dfrac{-b \pm \sqrt{b^2 - 4ac}}{2a}$.
If $b^2 - 4ac = 0$, then $x = \dfrac{-b}{2a}$, so there is one real zero. If
$b^2 - 4ac > 0$, then $\sqrt{b^2 - 4ac}$ is a real number and there are
two real zeros. If $b^2 - 4ac < 0$, then $\sqrt{b^2 - 4ac}$ is an imaginary
number and there are two imaginary zeros. Thus a quadratic
function cannot have one real zero and one imaginary zero.
82. You can conclude that $|a_1| = |a_2|$ since these constants
determine how wide the parabolas are. Nothing can be concluded
about the h's and the k's.

Test: Chapter 7, p. 503

1. [7.1] x-axis: no; y-axis: yes; origin: no   **2.** [7.1] Odd
3. [7.2] $f(x) = (x - 2)^2 - 1$
4. [7.2] $f(x) = (x + 2)^2 - 3$
5. [7.2]

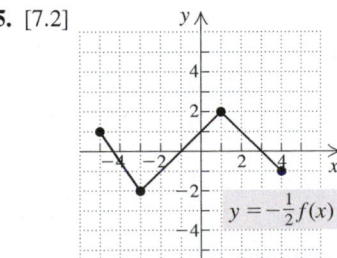

$y = -\frac{1}{2}f(x)$

6. [7.3] $\sqrt{43}\,i$ **7.** [7.3] $-5i$ **8.** [7.3] $3 - 5i$
9. [7.3] $10 + 5i$ **10.** [7.3] $\frac{1}{10} - \frac{1}{5}i$ **11.** [7.3] i
12. [7.4] $\frac{1}{2}, -5$ **13.** [7.4] $-\sqrt{6}, \sqrt{6}$ **14.** [7.4] $-2i, 2i$
15. [7.4] $-1, 3$ **16.** [7.4] $\dfrac{5 \pm \sqrt{13}}{2}$ **17.** [7.4] $\dfrac{3}{4} \pm \dfrac{\sqrt{23}}{4}i$

18. [7.4] 16 **19.** [7.4] $x^2 + 4x = 1$; $x^2 + 4x + 4 = 1 + 4$; $(x + 2)^2 = 5$; $x = -2 \pm \sqrt{5}$; $-2 - \sqrt{5}, -2 + \sqrt{5}$

20. [7.4] $-\frac{1}{4}, 3$ **21.** [7.4] $\dfrac{1 \pm \sqrt{57}}{4}$ **22.** [7.4] About 11.4 sec

23. [7.5] **(a)** $(1, 9)$; **(b)** $x = 1$; **(c)** maximum: 9; **(d)** $(-\infty, 9]$;
(e)

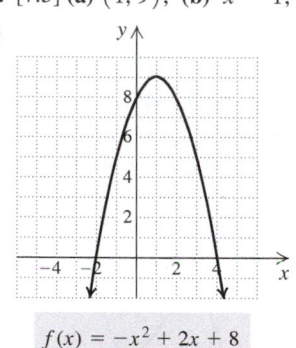

$$f(x) = -x^2 + 2x + 8$$

24. [7.5] 20 ft by 40 ft **25.** [7.2] C **26.** [7.5] C
27. [7.2] $(-1, 1)$

Chapter 8

Exercise Set 8.1, p. 515

1. $\frac{1}{2}x^3$; $\frac{1}{2}$; 3; cubic **3.** $0.9x$; 0.9; 1; linear **5.** $305x^4$; 305; 4; quartic **7.** x^4; 1; 4; quartic **9.** $4x^3$; 4; 3; cubic **11.** (d)
13. (b) **15.** (c) **17.** (a) **19.** (c) **21.** (d)
23. Yes; no; no **25.** No; yes; yes **27.** -3, multiplicity 2; 1, multiplicity 1 **29.** 4, multiplicity 3; -6, multiplicity 1
31. ± 3, each has multiplicity 3 **33.** 0, multiplicity 3; 1, multiplicity 2; -4, multiplicity 1 **35.** 3, multiplicity 2; -4, multiplicity 3; 0, multiplicity 4 **37.** $\pm \sqrt{3}, \pm 1$, each has multiplicity 1 **39.** $-3, -1, 1$, each has multiplicity 1
41. $\pm 2, \frac{1}{2}$, each has multiplicity 1 **43.** False **45.** True
47. 2008: 1.7 million albums; 2012: 4.9 million albums; 2016: 7.9 million albums **49.** 26, 64, and 80 **51.** 5 sec
53. 2003: 684,025 admissions; 2006: 739,119 admissions; 2011: 665,806 admissions **55.** 6.3% **57.** $\{y \mid y \geq 3\}$, or $[3, \infty)$
58. $\{x \mid -2 \leq x < 2\}$, or $[-2, 2)$ **59.** $\{x \mid x < -2 \ or \ x > 2\}$, or $(-\infty, -2) \cup (2, \infty)$ **60.** $\{x \mid -13 \leq x \leq 1\}$, or $[-13, 1]$ **61.** $\{x \mid x \text{ is a real number } and \ x \neq -\frac{1}{3}\}$, or $\left(-\infty, -\frac{1}{3}\right) \cup \left(-\frac{1}{3}, \infty\right)$ **62.** $-\frac{4}{3}$ **63.** 1, 4 **64.** $\frac{3}{2}, -5$
65. $16; x^{16}$

Visualizing the Graph, p. 526

1. H **2.** D **3.** J **4.** B **5.** A **6.** C **7.** I
8. E **9.** G **10.** F

Exercise Set 8.2, p. 527

1. **(a)** 5; **(b)** 5; **(c)** 4 **3.** **(a)** 10; **(b)** 10; **(c)** 9
5. **(a)** 3; **(b)** 3; **(c)** 2 **7.** (d) **9.** (f) **11.** (b)

13.

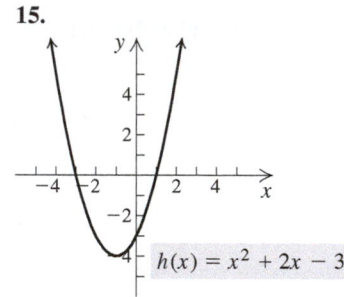

$f(x) = -x^3 - 2x^2$

15.

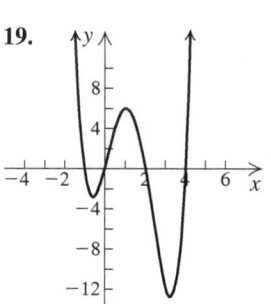

$h(x) = x^2 + 2x - 3$

17.

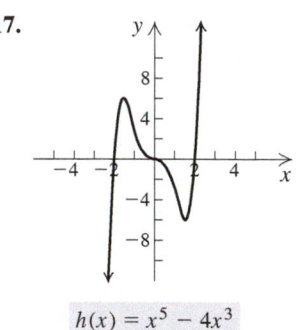

$h(x) = x^5 - 4x^3$

19.

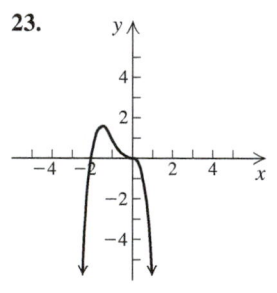

$h(x) = x(x - 4)(x + 1)(x - 2)$

21.

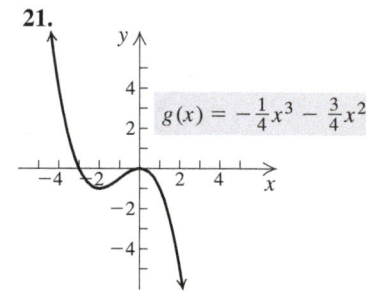

$g(x) = -\frac{1}{4}x^3 - \frac{3}{4}x^2$

23.

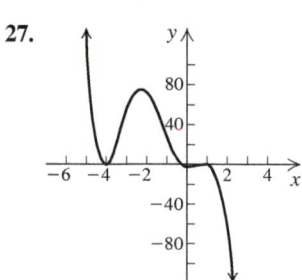

$g(x) = -x^4 - 2x^3$

25.

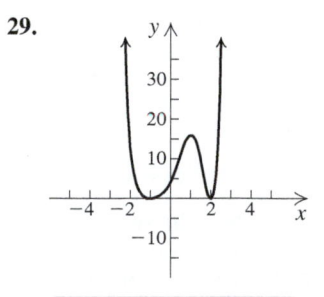

$f(x) = -\frac{1}{2}(x - 2)(x + 1)^2(x - 1)$

27.

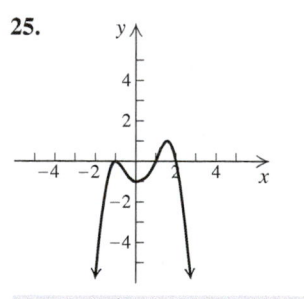

$g(x) = -x(x - 1)^2(x + 4)^2$

29.

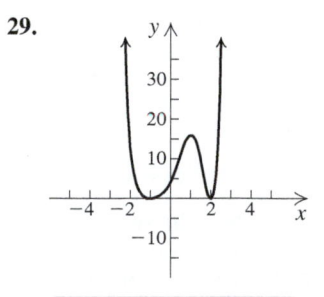

$f(x) = (x - 2)^2(x + 1)^4$

31.

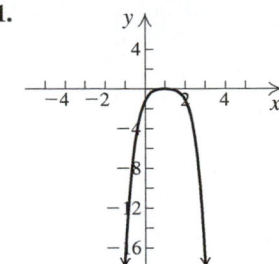

$g(x) = -(x - 1)^4$

33.

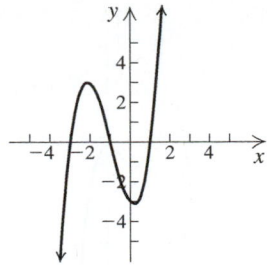

$h(x) = x^3 + 3x^2 - x - 3$

35.

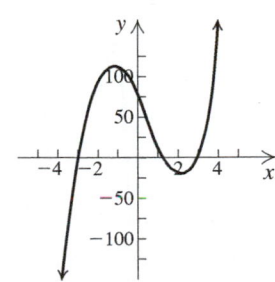

$f(x) = 6x^3 - 8x^2 - 54x + 72$

37.

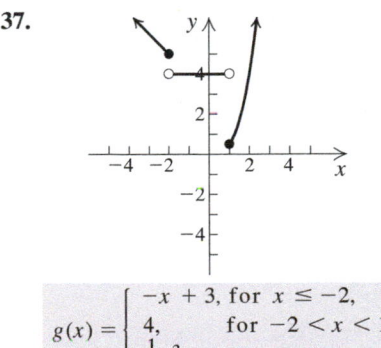

$$g(x) = \begin{cases} -x + 3, & \text{for } x \le -2, \\ 4, & \text{for } -2 < x < 1, \\ \frac{1}{2}x^3, & \text{for } x \ge 1 \end{cases}$$

39. $f(-5) = -18$ and $f(-4) = 7$. By the intermediate value theorem, since $f(-5)$ and $f(-4)$ have opposite signs, then $f(x)$ has a zero between -5 and -4.   **41.** $f(-3) = 22$ and $f(-2) = 5$. Both $f(-3)$ and $f(-2)$ are positive. We cannot use the intermediate value theorem to determine if there is a zero between -3 and -2.   **43.** $f(2) = 2$ and $f(3) = 57$. Both $f(2)$ and $f(3)$ are positive. We cannot use the intermediate value theorem to determine if there is a zero between 2 and 3. **45.** $f(4) = -12$ and $f(5) = 4$. By the intermediate value theorem, since $f(4)$ and $f(5)$ have opposite signs, then $f(x)$ has a zero between 4 and 5.   **47.** (d)   **48.** (f)   **49.** (e)
50. (a)   **51.** (b)   **52.** (c)   **53.** $\frac{9}{10}$ **54.** $-3, 0, 4$
55. $-\frac{5}{3}, \frac{11}{2}$ **56.** $\frac{196}{25}$

Exercise Set 8.3, p. 534

1. (a) No; (b) yes; (c) no   **3.** (a) Yes; (b) no; (c) yes
5. $P(x) = (x + 2)(x^2 - 2x + 4) - 16$
7. $P(x) = (x + 9)(x^2 - 3x + 2) + 0$
9. $P(x) = (x + 2)(x^3 - 2x^2 + 2x - 4) + 11$
11. $Q(x) = 2x^3 + x^2 - 3x + 10, R(x) = -42$
13. $Q(x) = x^2 - 4x + 8, R(x) = -24$
15. $Q(x) = 3x^2 - 4x + 8, R(x) = -18$

17. $Q(x) = x^4 + 3x^3 + 10x^2 + 30x + 89, R(x) = 267$
19. $Q(x) = x^3 + x^2 + x + 1, R(x) = 0$
21. $Q(x) = 2x^3 + x^2 + \frac{7}{2}x + \frac{7}{4}, R(x) = -\frac{1}{8}$
23. $0; -60; 0$ **25.** $10; 80; 998$ **27.** $5,935,988; -772$
29. $0; 0; 65; 1 - 12\sqrt{2}$ **31.** Yes; no   **33.** Yes; yes
35. No; yes   **37.** No; no
39. $f(x) = (x - 1)(x + 2)(x + 3); 1, -2, -3$
41. $f(x) = (x - 2)(x - 5)(x + 1); 2, 5, -1$
43. $f(x) = (x - 2)(x - 3)(x + 4); 2, 3, -4$
45. $f(x) = (x - 3)^3(x + 2); 3, -2$
47. $f(x) = (x - 1)(x - 2)(x - 3)(x + 5); 1, 2, 3, -5$
49.

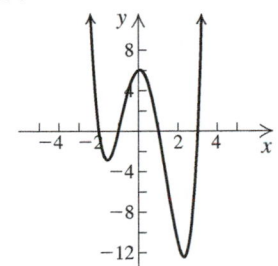

$f(x) = x^4 - x^3 - 7x^2 + x + 6$

51.

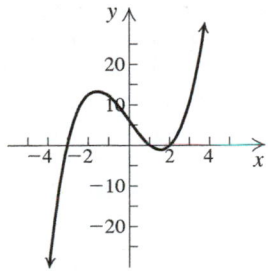

$f(x) = x^3 - 7x + 6$

53.

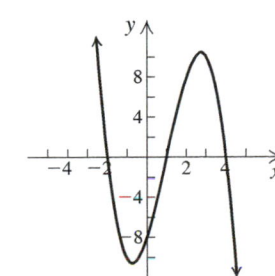

$f(x) = -x^3 + 3x^2 + 6x - 8$

55. $\frac{5}{4} \pm \frac{\sqrt{71}}{4}i$
56. $-1, \frac{3}{7}$ **57.** $-5, 0$
58. 10 **59.** $-3, -2$
60. $f(x) = 0.172x + 2.69$; 1995: \$5.27; 2018: \$9.23
61. $b = 15$ in., $h = 15$ in.

63. (a) $x + 4, x + 3, x - 2, x - 5$;
(b) $P(x) = (x + 4)(x + 3)(x - 2)(x - 5)$;
(c) yes; two examples are $f(x) = c \cdot P(x)$ for any nonzero constant c; and $g(x) = (x - a)P(x)$; (d) no   **65.** $\frac{14}{3}$
67. $0, -6$ **69.** Answers may vary. One possibility is $P(x) = x^{15} - x^{14}$.   **71.** $x^2 + 2ix + (2 - 4i), R -6 - 2i$
73. $x - 3 + i, R\ 6 - 3i$

Mid-Chapter Review: Chapter 8, p. 536

1. False   **2.** True   **3.** True   **4.** False
5. 5; multiplicity 6   **6.** $-5, -\frac{1}{2}, 5$; each has multiplicity 1
7. $\pm 1, \pm \sqrt{2}$; each has multiplicity 1   **8.** 3, multiplicity 2; -4, multiplicity 1   **9.** (d)   **10.** (a)   **11.** (b)   **12.** (c)
13. $f(-2) = -13$ and $f(0) = 3$. By the intermediate value theorem, since $f(-2)$ and $f(0)$ have opposite signs, then $f(x)$ has at least one zero between -2 and 0.   **14.** $f\left(-\frac{1}{2}\right) = 2\frac{3}{8}$ and $f(1) = 2$. Both $f\left(-\frac{1}{2}\right)$ and $f(1)$ are positive. We cannot use the intermediate value theorem to determine if there is a zero between $-\frac{1}{2}$ and 1.   **15.** $P(x) = (x - 1)(x^3 - 5x^2 - 5x - 4) - 6$
16. $Q(x) = 3x^3 + 5x^2 + 12x + 18, R(x) = 42$
17. $Q(x) = x^4 - x^3 + x^2 - x + 1, R(x) = -6$
18. $g(-5) = -380$ **19.** $f\left(\frac{1}{2}\right) = -15$
20. $f\left(-\sqrt{2}\right) = 20 - \sqrt{2}$ **21.** Yes; no   **22.** Yes; yes

23. $h(x) = (x - 1)(x - 8)(x + 7); -7, 1, 8$
24. $g(x) = (x + 1)(x - 2)(x - 4)(x + 3); -3, -1, 2, 4$
25. The range of a polynomial function with an odd degree is $(-\infty, \infty)$. The range of a polynomial function with an even degree is $[s, \infty)$ for some real number s if $a_n > 0$ and is $(-\infty, s]$ for some real number s if $a_n < 0$. **26.** Since we can find $f(0)$ for any polynomial function $f(x)$, it is not possible for the graph of a polynomial function to have no y-intercept. It is possible for a polynomial function to have no x-intercepts. For instance, a function of the form $f(x) = x^2 + a, a > 0$, has no x-intercepts. There are other examples as well. **27.** The zeros of a polynomial function are the first coordinates of the points at which the graph of the function crosses or is tangent to the x-axis. **28.** For a polynomial $P(x)$ of degree n, when we have $P(x) = d(x) \cdot Q(x) + R(x)$, where the degree of $d(x)$ is 1, then the degree of $Q(x)$ must be $n - 1$.

Exercise Set 8.4, p. 544

1. $f(x) = x^3 - 6x^2 - x + 30$ **3.** $f(x) = x^3 + 3x^2 + 4x + 12$
5. $f(x) = x^3 - 3x^2 - 2x + 6$ **7.** $f(x) = x^3 - 6x - 4$
9. $f(x) = x^3 + 2x^2 + 29x + 148$ **11.** $f(x) = x^3 - \frac{5}{3}x^2 - \frac{2}{3}x$
13. $f(x) = x^5 + 2x^4 - 2x^2 - x$
15. $f(x) = x^4 + 3x^3 + 3x^2 + x$ **17.** $-\sqrt{3}$ **19.** $i, 2 + \sqrt{5}$
21. $-3i$ **23.** $-4 + 3i, 2 + \sqrt{3}$ **25.** $-\sqrt{5}, 4i$ **27.** $2 + i$
29. $-3 - 4i, 4 + \sqrt{5}$ **31.** $4 + i$
33. $f(x) = x^3 - 4x^2 + 6x - 4$ **35.** $f(x) = x^2 + 16$
37. $f(x) = x^3 - 5x^2 + 16x - 80$
39. $f(x) = x^4 - 2x^3 - 3x^2 + 10x - 10$
41. $f(x) = x^4 + 4x^2 - 45$ **43.** $-\sqrt{2}, \sqrt{2}$ **45.** $i, 2, 3$
47. $1 + 2i, 1 - 2i$ **49.** ± 1 **51.** $\pm 1, \pm\frac{1}{2}, \pm 2, \pm 4, \pm 8$
53. $\pm 1, \pm 2, \pm\frac{1}{3}, \pm\frac{1}{5}, \pm\frac{2}{3}, \pm\frac{2}{5}, \pm\frac{1}{15}, \pm\frac{2}{15}$
55. (a) Rational: -3; other: $\pm\sqrt{2}$;
(b) $f(x) = (x + 3)(x + \sqrt{2})(x - \sqrt{2})$

57. (a) Rational: $\frac{1}{3}$; other: $\pm\sqrt{5}$;
(b) $f(x) = 3(x - \frac{1}{3})(x + \sqrt{5})(x - \sqrt{5})$
59. (a) Rational: $-2, 1$; other: none;
(b) $f(x) = (x + 2)(x - 1)^2$
61. (a) Rational: $-\frac{3}{2}$; other: $\pm 3i$;
(b) $f(x) = 2(x + \frac{3}{2})(x + 3i)(x - 3i)$
63. (a) Rational: $-\frac{1}{5}, 1$; other: $\pm 2i$;
(b) $f(x) = 5(x + \frac{1}{5})(x - 1)(x + 2i)(x - 2i)$,
or $(5x + 1)(x - 1)(x + 2i)(x - 2i)$
65. (a) Rational: $-2, -1$; other: $3 \pm \sqrt{13}$;
(b) $f(x) = (x + 2)(x + 1)(x - 3 - \sqrt{13})(x - 3 + \sqrt{13})$
67. (a) Rational: 2; other: $1 \pm \sqrt{3}$;
(b) $f(x) = (x - 2)(x - 1 - \sqrt{3})(x - 1 + \sqrt{3})$
69. (a) Rational: -2; other: $1 \pm \sqrt{3}i$;
(b) $f(x) = (x + 2)(x - 1 - \sqrt{3}i)(x - 1 + \sqrt{3}i)$
71. (a) Rational: $\frac{1}{2}$; other: $\frac{1 \pm \sqrt{5}}{2}$;
(b) $f(x) = \frac{1}{3}(x - \frac{1}{2})(x - \frac{1 + \sqrt{5}}{2})(x - \frac{1 - \sqrt{5}}{2})$

73. $1, -3$ **75.** No rational zeros **77.** No rational zeros
79. $-2, 1, 2$ **81.** 3 or 1; 0 **83.** 0; 3 or 1 **85.** 2 or 0; 2 or 0
87. 1; 1 **89.** 1; 0 **91.** 2 or 0; 2 or 0 **93.** 3 or 1; 1
95. 1; 1

97.

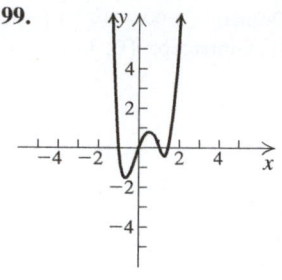

$f(x) = 4x^3 + x^2 - 8x - 2$

99.

$f(x) = 2x^4 - 3x^3 - 2x^2 + 3x$

101. (a) $(4, -6)$; **(b)** $x = 4$; **(c)** minimum: -6 at $x = 4$
102. (a) $(1, -4)$; **(b)** $x = 1$; **(c)** minimum: -4 at $x = 1$
103. $0, -4$ **104.** $-3, 11$ **105.** Cubic; $-x^3$; -1; 3; as $x \to \infty, g(x) \to -\infty$, and as $x \to -\infty, g(x) \to \infty$
106. Quadratic; $-x^2$; -1; 2; as $x \to \infty, f(x) \to -\infty$, and as $x \to -\infty, f(x) \to -\infty$ **107.** Constant; $-\frac{4}{9}$; $-\frac{4}{9}$; zero degree; for all x, $f(x) = -\frac{4}{9}$ **108.** Linear; x; 1; 1; as $x \to \infty, h(x) \to \infty$, and as $x \to -\infty, h(x) \to -\infty$
109. Quartic; x^4; 1; 4; as $x \to \infty, g(x) \to \infty$, and as $x \to -\infty, g(x) \to \infty$ **110.** Cubic; x^3; 1; 3; as $x \to \infty, h(x) \to \infty$, and as $x \to -\infty, h(x) \to -\infty$
111. (a) $-1, \frac{1}{2}, 3$; **(b)** $0, \frac{3}{2}, 4$; **(c)** $-3, -\frac{3}{2}, 1$; **(d)** $-\frac{1}{2}, \frac{1}{4}, \frac{3}{2}$
113. $-8, -\frac{3}{2}, 4, 7, 15$

Visualizing the Graph, p. 560

1. A **2.** C **3.** D **4.** H **5.** G **6.** F **7.** B
8. I **9.** J **10.** E

Exercise Set 8.5, p. 561

1. $\{x | x \neq 2\}$, or $(-\infty, 2) \cup (2, \infty)$
3. $\{x | x \neq 1 \text{ and } x \neq 5\}$, or $(-\infty, 1) \cup (1, 5) \cup (5, \infty)$
5. $\{x | x \neq -5\}$, or $(-\infty, -5) \cup (-5, \infty)$
7. (d); $x = 2, x = -2, y = 0$ **9.** (e); $x = 2, x = -2, y = 0$
11. (c); $x = 2, x = -2, y = 8x$ **13.** $x = 0$ **15.** $x = 2$
17. $x = 4, x = -6$ **19.** $x = \frac{3}{2}, x = -1$ **21.** $y = \frac{3}{4}$
23. $y = 0$ **25.** No horizontal asymptote **27.** $y = x + 1$
29. $y = x$ **31.** $y = x - 3$ **33.** Domain: $(-\infty, 0) \cup (0, \infty)$; no x-intercepts, no y-intercept;

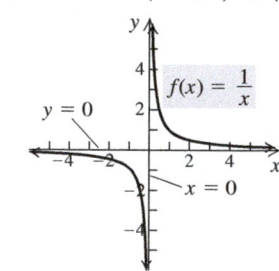

35. Domain: $(-\infty, 0) \cup (0, \infty)$; no x-intercepts, no y-intercept;

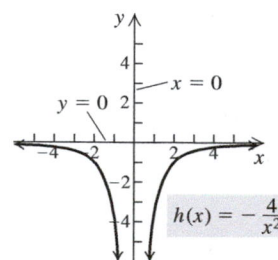

37. Domain: $(-\infty, -1) \cup (-1, \infty)$; x-intercepts: $(1, 0)$ and $(3, 0)$; y-intercept: $(0, 3)$;

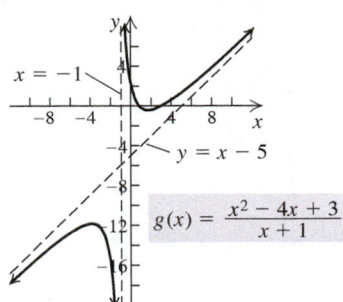

39. Domain: $(-\infty, 5) \cup (5, \infty)$; no x-intercepts, y-intercept: $\left(0, \frac{2}{5}\right)$;

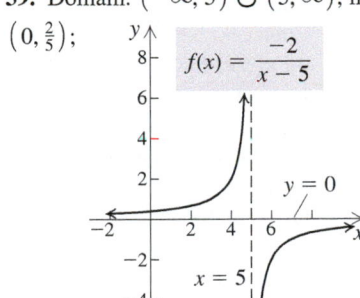

41. Domain: $(-\infty, 0) \cup (0, \infty)$; x-intercept: $\left(-\frac{1}{2}, 0\right)$, no y-intercept;

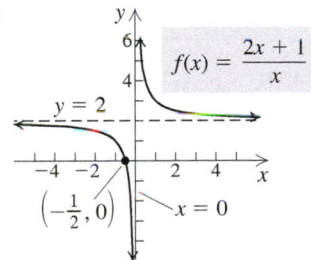

43. Domain: $(-\infty, -3) \cup (-3, 3) \cup (3, \infty)$; no x-intercepts, y-intercept: $\left(0, -\frac{1}{3}\right)$;

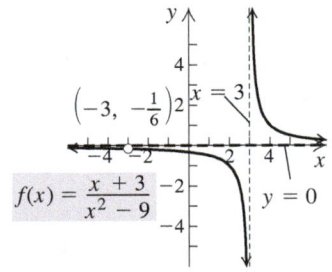

45. Domain: $(-\infty, -3) \cup (-3, 0) \cup (0, \infty)$; no x-intercepts, no y-intercept;

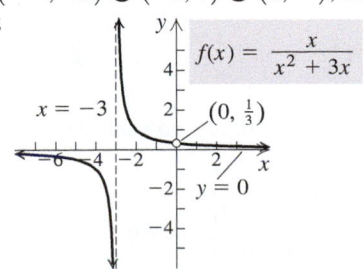

47. Domain: $(-\infty, 2) \cup (2, \infty)$; no x-intercepts, y-intercept: $\left(0, \frac{1}{4}\right)$;

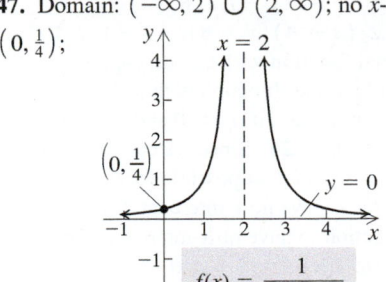

49. Domain: $(-\infty, -3) \cup (-3, -1) \cup (-1, \infty)$; x-intercept: $(1, 0)$, y-intercept: $(0, -1)$;

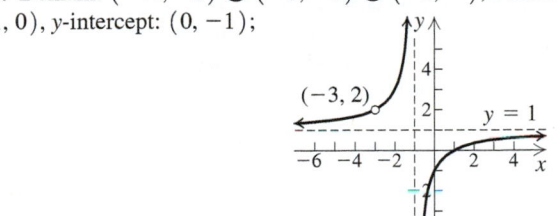

51. Domain: $(-\infty, \infty)$; no x-intercepts, y-intercept: $\left(0, \frac{1}{3}\right)$;

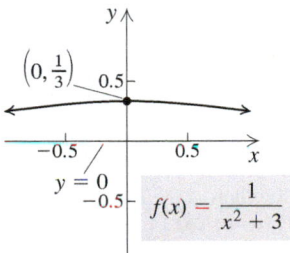

53. Domain: $(-\infty, 2) \cup (2, \infty)$; x-intercept: $(-2, 0)$, y-intercept: $(0, 2)$;

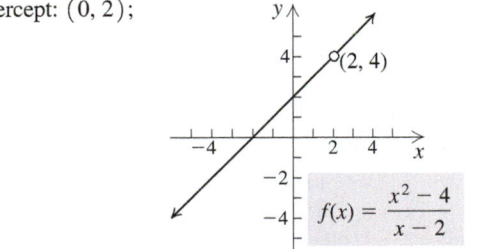

55. Domain: $(-\infty, -2) \cup (-2, \infty)$; x-intercept: $(1, 0)$, y-intercept: $\left(0, -\frac{1}{2}\right)$;

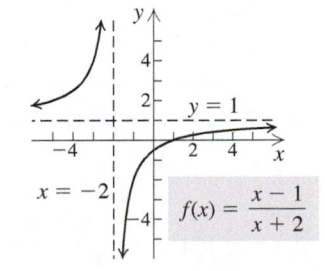

57. Domain: $\left(-\infty, -\frac{1}{2}\right) \cup \left(-\frac{1}{2}, 0\right) \cup (0, 3) \cup (3, \infty)$; x-intercept: $(-3, 0)$, no y-intercept;

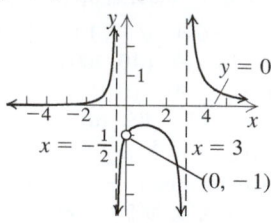

$$f(x) = \frac{x^2 + 3x}{2x^3 - 5x^2 - 3x}$$

59. Domain: $(-\infty, -1) \cup (-1, \infty)$; x-intercepts: $(-3, 0)$ and $(3, 0)$, y-intercept: $(0, -9)$;

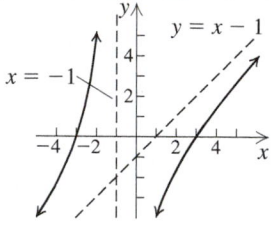

$$f(x) = \frac{x^2 - 9}{x + 1}$$

61. Domain: $(-\infty, \infty)$; x-intercepts: $(-2, 0)$ and $(1, 0)$, y-intercept: $(0, -2)$;

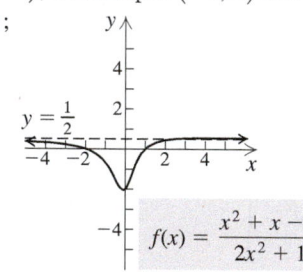

$$f(x) = \frac{x^2 + x - 2}{2x^2 + 1}$$

63. Domain: $(-\infty, 1) \cup (1, \infty)$; x-intercept: $\left(-\frac{2}{3}, 0\right)$, y-intercept: $(0, 2)$;

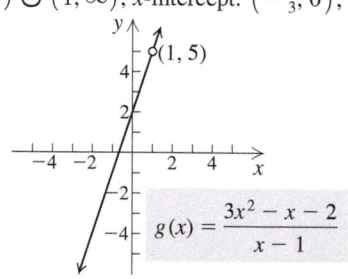

$$g(x) = \frac{3x^2 - x - 2}{x - 1}$$

65. Domain: $(-\infty, -1) \cup (-1, 3) \cup (3, \infty)$; x-intercept: $(1, 0)$, y-intercept: $\left(0, \frac{1}{3}\right)$;

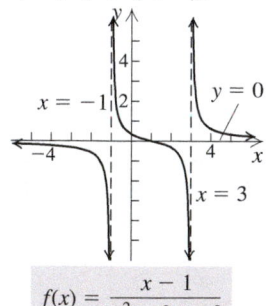

$$f(x) = \frac{x - 1}{x^2 - 2x - 3}$$

67. Domain: $(-\infty, -4) \cup (-4, 2) \cup (2, \infty)$; x-intercept: $\left(\frac{1}{3}, 0\right)$, y-intercept: $\left(0, \frac{1}{2}\right)$;

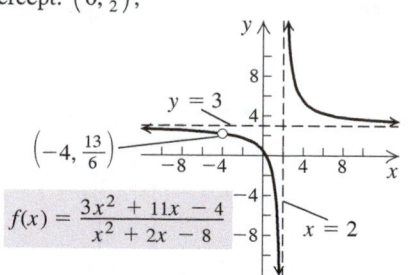

$$f(x) = \frac{3x^2 + 11x - 4}{x^2 + 2x - 8}$$

69. Domain: $(-\infty, -1) \cup (-1, \infty)$; x-intercept: $(3, 0)$, y-intercept: $(0, -3)$;

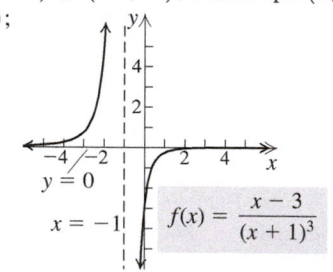

$$f(x) = \frac{x - 3}{(x + 1)^3}$$

71. Domain: $(-\infty, 0) \cup (0, \infty)$; x-intercept: $(-1, 0)$, no y-intercept;

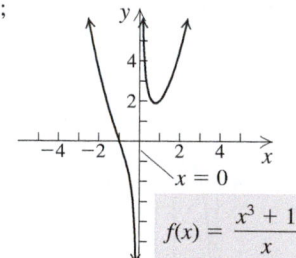

$$f(x) = \frac{x^3 + 1}{x}$$

73. Domain: $(-\infty, -2) \cup (-2, 7) \cup (7, \infty)$; x-intercepts: $(-5, 0)$, $(0, 0)$, and $(3, 0)$, y-intercept: $(0, 0)$;

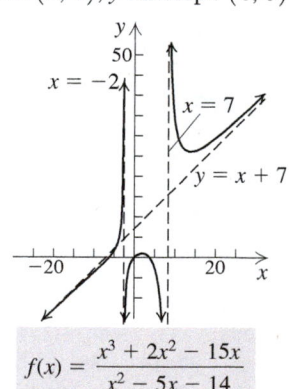

$$f(x) = \frac{x^3 + 2x^2 - 15x}{x^2 - 5x - 14}$$

75. Domain: $(-\infty, \infty)$; x-intercept: $(0, 0)$, y-intercept: $(0, 0)$;

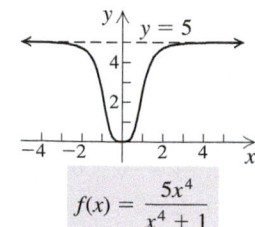

$$f(x) = \frac{5x^4}{x^4 + 1}$$

77. Domain: $(-\infty, -1) \cup (-1, 2) \cup (2, \infty)$; x-intercept: $(0, 0)$, y-intercept: $(0, 0)$;

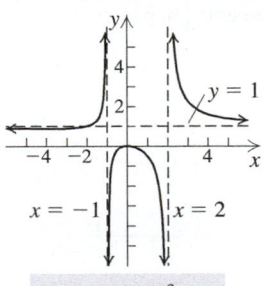

$$f(x) = \frac{x^2}{x^2 - x - 2}$$

79. $f(x) = \dfrac{1}{x^2 - x - 20}$ **81.** $f(x) = \dfrac{3x^2 + 12x + 12}{2x^2 - 2x - 40}$

83. (a) $N(t) \to 0.16$ as $t \to \infty$; **(b)** The medication never completely disappears from the body; a trace amount remains.
85. (a) $P(0) = 0$; $P(1) = 45{,}455$; $P(3) = 55{,}556$; $P(8) = 29{,}197$; **(b)** $P(t) \to 0$ as $t \to \infty$; **(c)** In time, no one lives in this community.   **86.** Domain, range, domain, range
87. Slope   **88.** Slope–intercept equation   **89.** Point–slope equation   **90.** x-intercept   **91.** $f(-x) = -f(x)$
92. Vertical lines   **93.** Difference quotient   **94.** y-intercept
95. $y = x^3 + 4$ **97.** $x = -3$

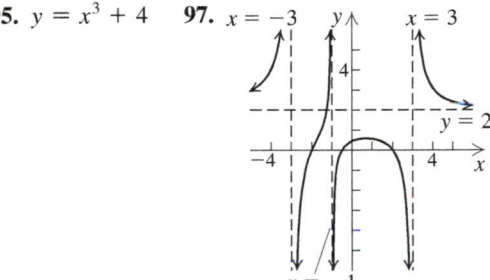

$$f(x) = \frac{2x^3 + x^2 - 8x - 4}{x^3 + x^2 - 9x - 9}$$

Exercise Set 8.6, p. 572

1. $\{-5, 3\}$ **3.** $[-5, 3]$ **5.** $(-\infty, -5] \cup [3, \infty)$
7. $(-\infty, -4) \cup (2, \infty)$ **9.** $(-\infty, -4) \cup [2, \infty)$ **11.** $\{0\}$
13. $(-5, 0] \cup (1, \infty)$ **15.** $(-\infty, -5) \cup (0, 1)$
17. $(-\infty, -3) \cup (0, 3)$ **19.** $(-3, 0) \cup (3, \infty)$
21. $(-\infty, -5) \cup (-3, 2)$ **23.** $(-2, 0] \cup (2, \infty)$ **25.** $(-4, 1)$
27. $(-\infty, -2) \cup (1, \infty)$ **29.** $(-\infty, -1] \cup [3, \infty)$
31. $(-\infty, -5) \cup (5, \infty)$ **33.** $(-\infty, -2] \cup [2, \infty)$
35. $(-\infty, 3) \cup (3, \infty)$ **37.** $\varnothing$ **39.** $\left(-\infty, -\frac{5}{4}\right] \cup [0, 3]$
41. $[-3, -1] \cup [1, \infty)$ **43.** $(-\infty, -2) \cup (1, 3)$
45. $\left[-\sqrt{2}, -1\right] \cup \left[\sqrt{2}, \infty\right)$ **47.** $(-\infty, -1] \cup \left[\frac{3}{2}, 2\right]$
49. $(-\infty, 5]$ **51.** $(-\infty, -1.680) \cup (2.154, 5.526)$
53. -4; $(-4, \infty)$ **55.** $-\frac{5}{2}$; $\left(-\frac{5}{2}, \infty\right)$
57. $0, 4$; $(-\infty, 0] \cup (4, \infty)$ **59.** $2, \frac{7}{2}$; $\left(2, \frac{7}{2}\right]$
61. $-3, -\frac{1}{5}, 1$; $\left(-3, -\frac{1}{5}\right] \cup (1, \infty)$
63. $2, \frac{46}{11}, 5$; $\left(2, \frac{46}{11}\right) \cup (5, \infty)$
65. $1 - \sqrt{2}, 0, 1 + \sqrt{2}$; $\left(1 - \sqrt{2}, 0\right) \cup \left(1 + \sqrt{2}, \infty\right)$
67. $-3, 1, 3, \frac{11}{3}$; $(-\infty, -3) \cup (1, 3) \cup \left[\frac{11}{3}, \infty\right)$
69. 0; $(-\infty, \infty)$ **71.** $-3, \dfrac{1 - \sqrt{61}}{6}, -\dfrac{1}{2}, 0, \dfrac{1 + \sqrt{61}}{6}$;
$\left(-3, \dfrac{1 - \sqrt{61}}{6}\right) \cup \left(-\dfrac{1}{2}, 0\right) \cup \left(\dfrac{1 + \sqrt{61}}{6}, \infty\right)$

73. $-1, 0, \frac{2}{7}, \frac{7}{2}$; $(-1, 0) \cup \left(\frac{2}{7}, \frac{7}{2}\right)$
75. $-6 - \sqrt{33}, -5, -6 + \sqrt{33}, 1, 5$;
$\left[-6 - \sqrt{33}, -5\right) \cup \left[-6 + \sqrt{33}, 1\right) \cup (5, \infty)$
77. $(0.408, 2.449)$ **79. (a)** $(10, 200)$;
(b) $(0, 10) \cup (200, \infty)$ **81.** $\{n \mid 9 \leq n \leq 23\}$
83. $(3, -4)$ **84.** $(-1, 2)$ **85. (a)** $\left(\frac{3}{4}, -\frac{55}{8}\right)$;
(b) maximum: $-\frac{55}{8}$ when $x = \frac{3}{4}$; **(c)** $\left(-\infty, -\frac{55}{8}\right]$
86. (a) $(5, -23)$; **(b)** minimum: -23 when $x = 5$;
(c) $[-23, \infty)$ **87.** $\left[-\sqrt{5}, \sqrt{5}\right]$ **89.** $\left[-\frac{3}{2}, \frac{3}{2}\right]$
91. $\left(-\infty, -\frac{1}{4}\right) \cup \left(\frac{1}{2}, \infty\right)$ **93.** $x^2 + x - 12 < 0$; answers may vary   **95.** $(-\infty, -3) \cup (7, \infty)$

Summary and Review: Chapter 8

Review Exercises, p. 585

1. True   **2.** True   **3.** False   **4.** False   **5.** False
6. $0.45x^4, 0.45, 4$, quartic   **7.** $-25, -25, 0$, constant
8. $-0.5x, -0.5, 1$, linear   **9.** $\frac{1}{3}x^3, \frac{1}{3}, 3$, cubic
10. As $x \to \infty, f(x) \to -\infty$, and as $x \to -\infty, f(x) \to -\infty$.
11. As $x \to \infty, f(x) \to \infty$, and as $x \to -\infty, f(x) \to -\infty$.
12. $\frac{2}{3}$, multiplicity 1; -2, multiplicity 3; 5, multiplicity 2
13. $\pm 1, \pm 5$, each has multiplicity 1   **14.** $\pm 3, -4$, each has multiplicity 1   **15. (a)** 4%; **(b)** 5%

16.

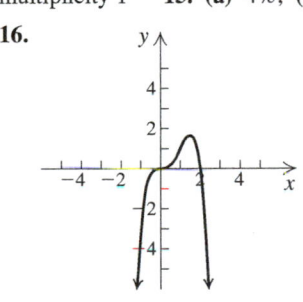

$$f(x) = -x^4 + 2x^3$$

17.

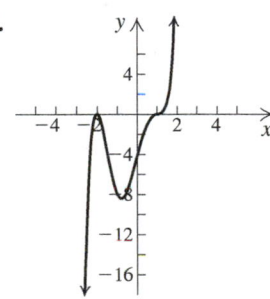

$$g(x) = (x - 1)^3(x + 2)^2$$

18.

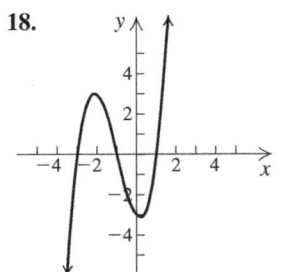

$$h(x) = x^3 + 3x^2 - x - 3$$

19.

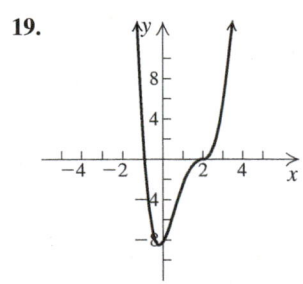

$$f(x) = x^4 - 5x^3 + 6x^2 + 4x - 8$$

20.

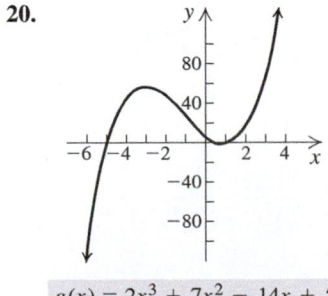

$$g(x) = 2x^3 + 7x^2 - 14x + 5$$

21. $f(1) = -4$ and $f(2) = 3$. Since $f(1)$ and $f(2)$ have opposite signs, $f(x)$ has a zero between 1 and 2.   **22.** $f(-1) = -3.5$ and $f(1) = -0.5$. Since $f(-1)$ and $f(1)$ have the same sign, the intermediate value theorem does not allow us to determine whether there is a zero between -1 and 1.

23. $Q(x) = 6x^2 + 16x + 52, R(x) = 155$;
$P(x) = (x - 3)(6x^2 + 16x + 52) + 155$
24. $Q(x) = x^3 - 3x^2 + 3x - 2, R(x) = 7$;
$P(x) = (x + 1)(x^3 - 3x^2 + 3x - 2) + 7$
25. $x^2 + 7x + 22$, R 120   **26.** $x^3 + x^2 + x + 1$, R 0
27. $x^4 - x^3 + x^2 - x - 1$, R 1   **28.** 36   **29.** 0
30. −141,220   **31.** Yes, no   **32.** No, yes   **33.** Yes, no
34. No, yes   **35.** $f(x) = (x - 1)^2(x + 4)$; −4, 1
36. $f(x) = (x - 2)(x + 3)^2$; −3, 2
37. $f(x) = (x - 2)^2(x - 5)(x + 5)$; −5, 2, 5
38. $f(x) = (x - 1)(x + 1)(x - \sqrt{2})(x + \sqrt{2})$;
$-\sqrt{2}, -1, 1, \sqrt{2}$ **39.** $f(x) = x^3 + 3x^2 - 6x - 8$
40. $f(x) = x^3 + x^2 - 4x + 6$
41. $f(x) = x^3 - \frac{5}{2}x^2 + \frac{1}{2}$, or $2x^3 - 5x^2 + 1$
42. $f(x) = x^4 + \frac{29}{2}x^3 + \frac{135}{2}x^2 + \frac{175}{2}x - \frac{125}{2}$, or
$2x^4 + 29x^3 + 135x^2 + 175x - 125$
43. $f(x) = x^5 + 4x^4 - 3x^3 - 18x^2$ **44.** $-\sqrt{5}, -i$
45. $1 - \sqrt{3}, \sqrt{3}$ **46.** $\sqrt{2}$ **47.** $f(x) = x^2 - 11$
48. $f(x) = x^3 - 6x^2 + x - 6$
49. $f(x) = x^4 - 5x^3 + 4x^2 + 2x - 8$
50. $f(x) = x^4 - x^2 - 20$ **51.** $f(x) = x^3 + \frac{8}{3}x^2 - x$
52. $\pm\frac{1}{4}, \pm\frac{1}{2}, \pm\frac{3}{4}, \pm 1, \pm\frac{3}{2}, \pm 2, \pm 3, \pm 4, \pm 6, \pm 12$
53. $\pm\frac{1}{3}, \pm 1$ **54.** $\pm 1, \pm 2, \pm 3, \pm 4, \pm 6, \pm 8, \pm 12, \pm 24$
55. (a) Rational: 0, −2, $\frac{1}{3}$, 3; other: none;
(b) $f(x) = 3x(x - \frac{1}{3})(x + 2)^2(x - 3)$
56. (a) Rational: 2; other: $\pm\sqrt{3}$;
(b) $f(x) = (x - 2)(x + \sqrt{3})(x - \sqrt{3})$
57. (a) Rational: −1, 1; other: $3 \pm i$;
(b) $f(x) = (x + 1)(x - 1)(x - 3 - i)(x - 3 + i)$
58. (a) Rational: −5; other: $1 \pm \sqrt{2}$;
(b) $f(x) = (x + 5)(x - 1 - \sqrt{2})(x - 1 + \sqrt{2})$
59. (a) Rational: $\frac{2}{3}$, 1; other: none;
(b) $f(x) = 3(x - \frac{2}{3})(x - 1)^2$
60. (a) Rational: 2; other: $1 \pm \sqrt{5}$;
(b) $f(x) = (x - 2)^3(x - 1 + \sqrt{5})(x - 1 - \sqrt{5})$
61. (a) Rational: −4, 0, 3, 4; other: none;
(b) $f(x) = x^2(x + 4)^2(x - 3)(x - 4)$
62. (a) Rational: $\frac{5}{2}$, 1; other: none;
(b) $f(x) = 2(x - \frac{5}{2})(x - 1)^4$, or $(2x - 5)(x - 1)^4$
63. 3 or 1; 0   **64.** 4 or 2 or 0; 2 or 0   **65.** 3 or 1; 0
66. Domain: $(-\infty, -2) \cup (-2, \infty)$; x-intercepts: $(-\sqrt{5}, 0)$
and $(\sqrt{5}, 0)$, y-intercept: $(0, -\frac{5}{2})$

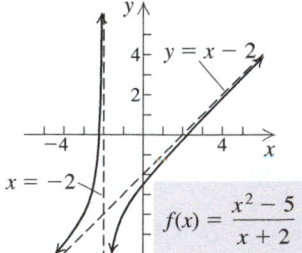
$$f(x) = \frac{x^2 - 5}{x + 2}$$

67. Domain: $(-\infty, 2) \cup (2, \infty)$;
x-intercepts: none,
y-intercept: $(0, \frac{5}{4})$

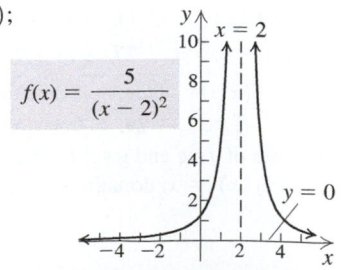
$$f(x) = \frac{5}{(x - 2)^2}$$

68. Domain: $(-\infty, -4) \cup (-4, 5) \cup (5, \infty)$; x-intercepts:
$(-3, 0)$ and $(2, 0)$, y-intercept: $(0, \frac{3}{10})$

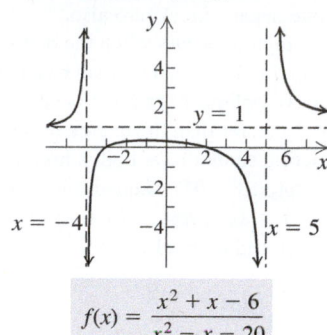
$$f(x) = \frac{x^2 + x - 6}{x^2 - x - 20}$$

69. Domain: $(-\infty, -3) \cup (-3, 5) \cup (5, \infty)$; x-intercept:
$(2, 0)$, y-intercept: $(0, \frac{2}{15})$

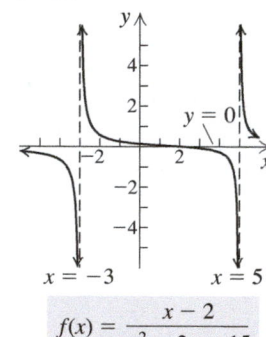
$$f(x) = \frac{x - 2}{x^2 - 2x - 15}$$

70. $f(x) = \dfrac{1}{x^2 - x - 6}$ **71.** $f(x) = \dfrac{4x^2 + 12x}{x^2 - x - 6}$
72. (a) $N(t) \to 0.0875$ as $t \to \infty$; (b) The medication never
completely disappears from the body; a trace amount remains.
73. $(-3, 3)$ **74.** $(-\infty, -\frac{1}{2}) \cup (2, \infty)$
75. $[-4, 1] \cup [2, \infty)$ **76.** $(-\infty, -\frac{14}{3}) \cup (-3, \infty)$

77. (a) $t = 7$; (b) $(2, 3)$ **78.** $\left[\dfrac{5 - \sqrt{15}}{2}, \dfrac{5 + \sqrt{15}}{2}\right]$

79. A   **80.** C   **81.** B
82. $(-\infty, -1 - \sqrt{6}] \cup [-1 + \sqrt{6}, \infty)$
83. $(-\infty, -\frac{1}{2}) \cup (\frac{1}{2}, \infty)$ **84.** $\{1 + i, 1 - i, i, -i\}$

85. $(-\infty, 2)$ **86.** $(x - 1)\left(x + \dfrac{1}{2} - \dfrac{\sqrt{3}}{2}i\right)\left(x + \dfrac{1}{2} + \dfrac{\sqrt{3}}{2}i\right)$

87. 7   **88.** −4   **89.** $(-\infty, -5] \cup [2, \infty)$
90. $(-\infty, 1.1] \cup [2, \infty)$ **91.** $(-1, \frac{3}{7})$
92. A polynomial function is a function that can be defined by
a polynomial expression. A rational function is a function that
can be defined as a quotient of two polynomials.   **93.** No;
since imaginary zeros of polynomials with rational coefficients
occur in conjugate pairs, a third-degree polynomial with rational
coefficients can have at most two imaginary zeros. Thus there
must be at least one real zero.   **94.** If the numerator and the
denominator of a rational function do not share a common
factor, vertical asymptotes occur at any x-values that make the
denominator zero. The graph of a rational function does not cross
any vertical asymptotes. Horizontal asymptotes occur when the
degree of the numerator is less than or equal to the degree of the
denominator. Oblique asymptotes occur when the degree of the
numerator is 1 greater than the degree of the denominator. Graphs
of rational functions may cross horizontal or oblique asymptotes.

95. If $P(x)$ is an even function, then $P(-x) = P(x)$ and thus $P(-x)$ has the same number of sign changes as $P(x)$. Hence, $P(x)$ has one negative real zero also.
96. A horizontal asymptote occurs when the degree of the numerator of a rational function is less than or equal to the degree of the denominator. An oblique asymptote occurs when the degree of the numerator is 1 greater than the degree of the denominator. Thus a rational function cannot have both a horizontal asymptote and an oblique asymptote.   **97.** A quadratic inequality $ax^2 + bx + c \le 0, a > 0$, or $ax^2 + bx + c \ge 0, a < 0$, has a solution set that is a closed interval.

Test: Chapter 8, p. 589

1. [8.1] $-x^4, -1, 4$; quartic   **2.** [8.1] $-4.7x, -4.7, 1$; linear
3. [8.1] $0, \frac{5}{3}$, each has multiplicity 1; 3, multiplicity 2; -1, multiplicity 3   **4.** [8.1] 2008: 329,277 hybrid automobiles; 2011: 275,779 hybrid automobiles
5. [8.2]

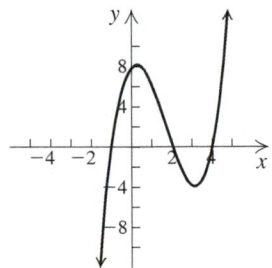

$$f(x) = x^3 - 5x^2 + 2x + 8$$

6. [8.2]

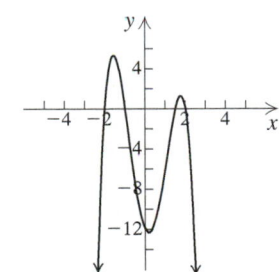

$$f(x) = -2x^4 + x^3 + 11x^2 - 4x - 12$$

7. [8.2] $f(0) = 3$ and $f(2) = -17$. Since $f(0)$ and $f(2)$ have opposite signs, $f(x)$ has a zero between 0 and 2.
8. [8.2] $g(-2) = 5$ and $g(-1) = 1$. Both $g(-2)$ and $g(-1)$ are positive. We cannot use the intermediate value theorem to determine if there is a zero between -2 and -1.
9. [8.3] $Q(x) = x^3 + 4x^2 + 4x + 6, R(x) = 1$;
$P(x) = (x - 1)(x^3 + 4x^2 + 4x + 6) + 1$
10. [8.3] $3x^2 + 15x + 63$, R 322   **11.** [8.3] -115
12. [8.3] Yes   **13.** [8.4] $f(x) = x^4 - 27x^2 - 54x$
14. [8.4] $-\sqrt{3}, 2 + i$
15. [8.4] $f(x) = x^3 + 10x^2 + 9x + 90$
16. [8.4] $f(x) = x^5 - 2x^4 - x^3 + 6x^2 - 6x$
17. [8.4] $\pm 1, \pm 2, \pm 3, \pm 4, \pm 6, \pm 12, \pm\frac{1}{2}, \pm\frac{3}{2}$
18. [8.4] $\pm\frac{1}{10}, \pm\frac{1}{5}, \pm\frac{1}{2}, \pm 1, \pm\frac{5}{2}, \pm 5$
19. [8.4] **(a)** Rational: -1; other: $\pm\sqrt{5}$;
(b) $f(x) = (x + 1)(x - \sqrt{5})(x + \sqrt{5})$
20. [8.4] **(a)** Rational: $-\frac{1}{2}, 1, 2, 3$; other: none;
(b) $f(x) = 2(x + \frac{1}{2})(x - 1)(x - 2)(x - 3)$
21. [8.4] **(a)** Rational: -4; other: $\pm 2i$;
(b) $f(x) = (x - 2i)(x + 2i)(x + 4)$

22. [8.4] **(a)** Rational: $\frac{2}{3}$, 1; other: none;
(b) $f(x) = 3(x - \frac{2}{3})(x - 1)^3$ **23.** [8.4] 2 or 0; 2 or 0
24. [8.5] Domain: $(-\infty, 3) \cup (3, \infty)$; x-intercepts: none, y-intercept: $(0, \frac{2}{9})$;

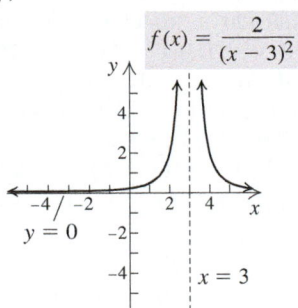

25. [8.5] Domain: $(-\infty, -1) \cup (-1, 4) \cup (4, \infty)$; x-intercept: $(-3, 0)$, y-intercept: $(0, -\frac{3}{4})$;

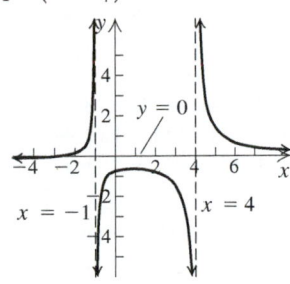

$$f(x) = \frac{x + 3}{x^2 - 3x - 4}$$

26. [8.5] Answers may vary; $f(x) = \dfrac{x + 4}{x^2 - x - 2}$

27. [8.6] $(-\infty, -\frac{1}{2}) \cup (3, \infty)$
28. [8.6] $(-\infty, 4) \cup [\frac{13}{2}, \infty)$
29. **(a)** [8.1] 6 sec; **(b)** [8.1], [8.6] $(1, 3)$ **30.** [8.2] D
31. [8.1], [8.6] $(-\infty, -4] \cup [3, \infty)$

Chapter 9

Exercise Set 9.1, p. 597

1. -8 **3.** 64   **5.** 218   **7.** -80 **9.** -6 **11.** 512
13. -32 **15.** x^9 **17.** $(f \circ g)(x) = (g \circ f)(x) = x$;
domain of $f \circ g$ and $g \circ f$: $(-\infty, \infty)$
19. $(f \circ g)(x) = 3x^2 - 2x$; $(g \circ f)(x) = 3x^2 + 4x$;
domain of $f \circ g$ and $g \circ f$: $(-\infty, \infty)$
21. $(f \circ g)(x) = 16x^2 - 24x + 6$; $(g \circ f)(x) = 4x^2 - 15$;
domain of $f \circ g$ and $g \circ f$: $(-\infty, \infty)$
23. $(f \circ g)(x) = \dfrac{4x}{x - 5}$; $(g \circ f)(x) = \dfrac{1 - 5x}{4}$;
domain of $f \circ g$: $(-\infty, 0) \cup (0, 5) \cup (5, \infty)$;
domain of $g \circ f$: $(-\infty, \frac{1}{5}) \cup (\frac{1}{5}, \infty)$
25. $(f \circ g)(x) = (g \circ f)(x) = x$; domain of $f \circ g$ and $g \circ f$: $(-\infty, \infty)$ **27.** $(f \circ g)(x) = 2\sqrt{x} + 1$;
$(g \circ f)(x) = \sqrt{2x + 1}$; domain of $f \circ g$: $[0, \infty)$; domain of $g \circ f$: $[-\frac{1}{2}, \infty)$ **29.** $(f \circ g)(x) = 20$; $(g \circ f)(x) = 0.05$;
domain of $f \circ g$ and $g \circ f$: $(-\infty, \infty)$ **31.** $(f \circ g)(x) = |x|$;
$(g \circ f)(x) = x$; domain of $f \circ g$: $(-\infty, \infty)$; domain

of $g \circ f$: $[-5, \infty)$ **33.** $(f \circ g)(x) = 5 - x$;
$(g \circ f)(x) = \sqrt{1 - x^2}$; domain of $f \circ g$: $(-\infty, 3]$; domain of
$g \circ f$: $[-1, 1]$ **35.** $(f \circ g)(x) = (g \circ f)(x) = x$; domain of
$f \circ g$: $(-\infty, -1) \cup (-1, \infty)$; domain of $g \circ f$: $(-\infty, 0) \cup (0, \infty)$
37. $(f \circ g)(x) = x^3 - 2x^2 - 4x + 6$;
$(g \circ f)(x) = x^3 - 5x^2 + 3x + 8$; domain of $f \circ g$ and
$g \circ f$: $(-\infty, \infty)$ **39.** $f(x) = x^5$; $g(x) = 4 + 3x$

41. $f(x) = \dfrac{1}{x}$; $g(x) = (x - 2)^4$

43. $f(x) = \dfrac{x - 1}{x + 1}$; $g(x) = x^3$

45. $f(x) = x^6$; $g(x) = \dfrac{2 + x^3}{2 - x^3}$

47. $f(x) = \sqrt{x}$; $g(x) = \dfrac{x - 5}{x + 2}$

49. $f(x) = x^3 - 5x^2 + 3x - 1$; $g(x) = x + 2$
51. **(a)** $r(t) = 3t$; **(b)** $A(r) = \pi r^2$; **(c)** $(A \circ r)(t) = 9\pi t^2$;
the function gives the area of the ripple in terms of time t.
53. $f(x) = x + 1$ **55.** (c)   **56.** None   **57.** (b), (d), (f),
and (h)   **58.** (b)   **59.** (a)   **60.** (c) and (g)   **61.** (c) and (g)
62. (a) and (f)   **63.** Only $(c \circ p)(a)$ makes sense. It represents
the cost of the grass seed required to seed a lawn with area a.

Exercise Set 9.2, p.607

1. $\{(8, 7), (8, -2), (-4, 3), (-8, 8)\}$
3. $\{(-1, -1), (4, -3)\}$ **5.** $x = 4y - 5$
7. $y^3 x = -5$ **9.** $y = x^2 - 2x$

11.

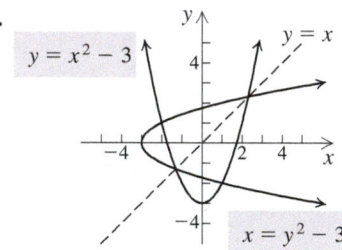

13.

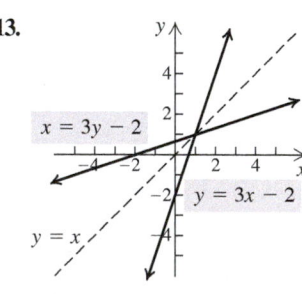

15.

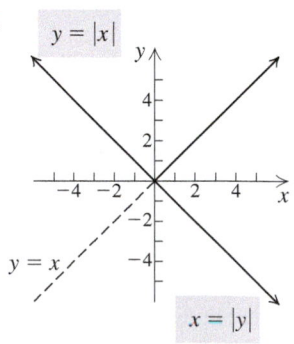

17. Assume $f(a) = f(b)$ for any numbers a and b in the domain
of f. Since $f(a) = \frac{1}{3}a - 6$ and $f(b) = \frac{1}{3}b - 6$, we have
$\frac{1}{3}a - 6 = \frac{1}{3}b - 6$
$\quad \frac{1}{3}a = \frac{1}{3}b$ **Adding 6**
$\quad\quad a = b$.   **Multiplying by 3**
Thus, if $f(a) = f(b)$, then $a = b$ and f is one-to-one.

19. Assume $f(a) = f(b)$ for any numbers a and b in the domain
of f. Since $f(a) = a^3 + \frac{1}{2}$ and $f(b) = b^3 + \frac{1}{2}$, we have
$a^3 + \frac{1}{2} = b^3 + \frac{1}{2}$
$\quad a^3 = b^3$ **Subtracting $\frac{1}{2}$**
$\quad a = b$.   **Taking the cube root**
Thus, if $f(a) = f(b)$, then $a = b$ and f is one-to-one.
21. Find two numbers a and b for which $a \neq b$ and $g(a) = g(b)$.
Two such numbers are -2 and 2, because $g(-2) = g(2) = -3$.
Thus, g is not one-to-one.   **23.** Find two numbers a and b for
which $a \neq b$ and $g(a) = g(b)$. Two such numbers are -1
and 1, because $g(-1) = g(1) = 0$. Thus, g is not
one-to-one.   **25.** Yes   **27.** No   **29.** No   **31.** Yes
33. Yes   **35.** No   **37.** No   **39.** Yes   **41.** No   **43.** No
45. **(a)** One-to-one; **(b)** $f^{-1}(x) = x - 4$

47. **(a)** One-to-one; **(b)** $f^{-1}(x) = \dfrac{x + 1}{2}$

49. **(a)** One-to-one; **(b)** $f^{-1}(x) = \dfrac{4}{x} - 7$

51. **(a)** One-to-one; **(b)** $f^{-1}(x) = \dfrac{3x + 4}{x - 1}$

53. **(a)** One-to-one; **(b)** $f^{-1}(x) = \sqrt[3]{x + 1}$
55. **(a)** Not one-to-one; **(b)** does not have an inverse that is a

function   **57.** **(a)** One-to-one; **(b)** $f^{-1}(x) = \sqrt{\dfrac{x + 2}{5}}$

59. **(a)** One-to-one; **(b)** $f^{-1}(x) = x^2 - 1, x \geq 0$
61. $\frac{1}{3}x$ **63.** $-x$ **65.** $x^3 + 5$

67.

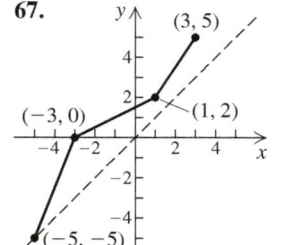

69.

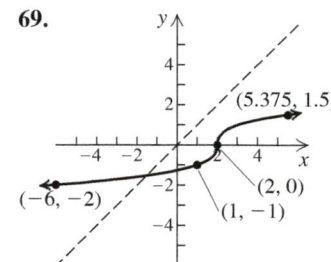

71.

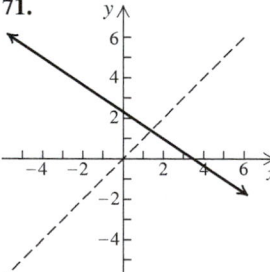

73. $f^{-1}(f(x)) = f^{-1}\left(\frac{7}{8}x\right) = \frac{8}{7} \cdot \frac{7}{8}x = x$;
$f(f^{-1}(x)) = f\left(\frac{8}{7}x\right) = \frac{7}{8} \cdot \frac{8}{7}x = x$

75. $f^{-1}(f(x)) = f^{-1}\left(\dfrac{1 - x}{x}\right) = \dfrac{1}{\dfrac{1 - x}{x} + 1} =$

$\dfrac{1}{\dfrac{1 - x + x}{x}} = \dfrac{1}{\dfrac{1}{x}} = 1 \cdot \dfrac{x}{1} = x$; $f(f^{-1}(x)) = f\left(\dfrac{1}{x + 1}\right) =$

$\dfrac{1 - \dfrac{1}{x + 1}}{\dfrac{1}{x + 1}} = \dfrac{\dfrac{x + 1 - 1}{x + 1}}{\dfrac{1}{x + 1}} = \dfrac{x}{x + 1} \cdot \dfrac{x + 1}{1} = x$

77. $f^{-1}(f(x)) = f^{-1}\left(\dfrac{2}{5}x + 1\right) = \dfrac{5\left(\dfrac{2}{5}x + 1\right) - 5}{2} =$

$\dfrac{2x + 5 - 5}{2} = \dfrac{2x}{2} = x; f(f^{-1}(x)) = f\left(\dfrac{5x - 5}{2}\right) =$

$\dfrac{2}{5}\left(\dfrac{5x - 5}{2}\right) + 1 = x - 1 + 1 = x$

79. $f^{-1}(x) = \frac{1}{5}x + \frac{3}{5}$; domain of f and f^{-1}: $(-\infty, \infty)$; range of f and f^{-1}: $(-\infty, \infty)$;

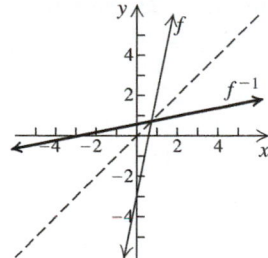

81. $f^{-1}(x) = \dfrac{2}{x}$; domain of f and f^{-1}: $(-\infty, 0)\cup(0, \infty)$;

range of f and f^{-1}: $(-\infty, 0)\cup(0, \infty)$;

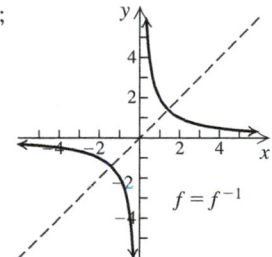

83. $f^{-1}(x) = \sqrt[3]{3x + 6}$; domain of f and f^{-1}: $(-\infty, \infty)$;

range of f and f^{-1}: $(-\infty, \infty)$;

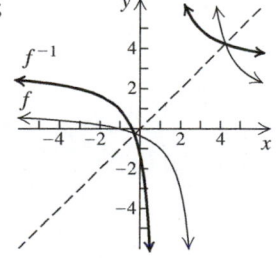

85. $f^{-1}(x) = \dfrac{3x + 1}{x - 1}$; domain of f: $(-\infty, 3) \cup (3, \infty)$;

range of f: $(-\infty, 1) \cup (1, \infty)$; domain of f^{-1}: $(-\infty, 1) \cup (1, \infty)$;

range of f^{-1}: $(-\infty, 3)\cup(3, \infty)$;

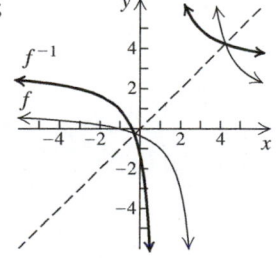

87. $5; a$ **89. (a)** \$38, \$16.40, \$11;   **(b)** $C^{-1}(x) = \dfrac{72}{x - 2}$;

$C^{-1}(x)$ represents the number of players in the group lesson, where x is the cost per player, in dollars;   **(c)** 1 player, 4 players, 8 players   **91. (a)** 2010: \$40.86 billion; 2013: \$60.6 billion;

(b) $H^{-1}(x) = \dfrac{x - 27.7}{6.58}$; $H^{-1}(x)$ represents the number of years

after 2008, where x is the e-commerce holiday season sales, in billions of dollars.   **93. (b)**, (d), (f), (h)   **94. (a)**, (c), (e), (g)   **95. (a)**   **96. (d)**   **97. (f)**   **98. (a)**, (b), (c), (d)   **99.** $f(x) = x^2 - 3$, for inputs $x \geq 0; f^{-1}(x) = \sqrt{x + 3}$, for inputs $x \geq -3$ **101.** Answers may vary; $f(x) = 3/x, f(x) = 1 - x, f(x) = x$

Exercise Set 9.3, p. 618

1. 54.5982   **3.** 0.0856   **5. (f)**   **7. (e)**   **9. (a)**

11.

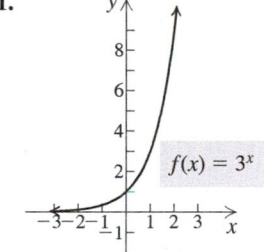

13.

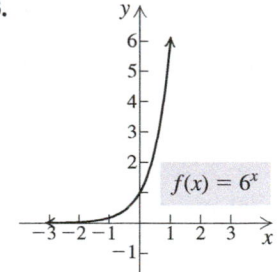

15.

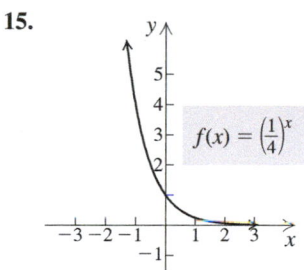

17.

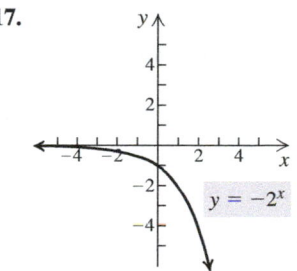

19.

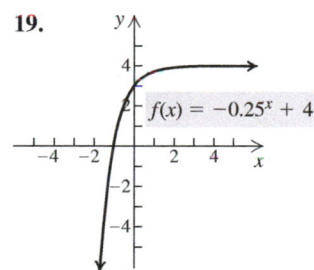

21.

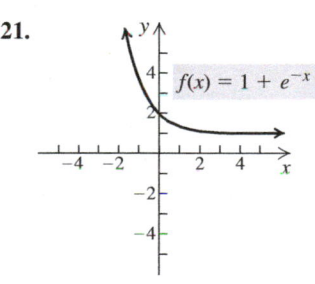

23.

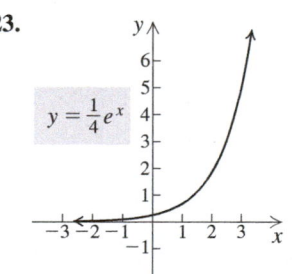

25.

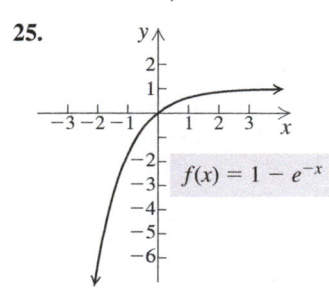

27. Shift the graph of $y = 2^x$ left 1 unit.

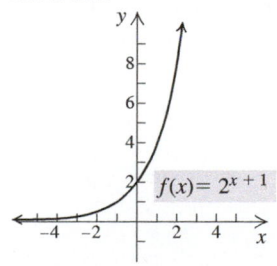

29. Shift the graph of $y = 2^x$ down 3 units.

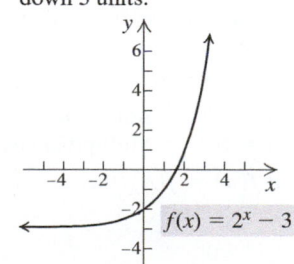

31. Shift the graph of $y = 2^x$ left 1 unit, reflect it across the y-axis, and shift it up 2 units.

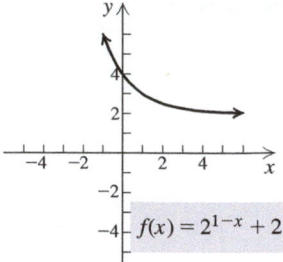

$f(x) = 2^{1-x} + 2$

33. Reflect the graph of $y = 3^x$ across the y-axis and then across the x-axis and then shift it up 4 units.

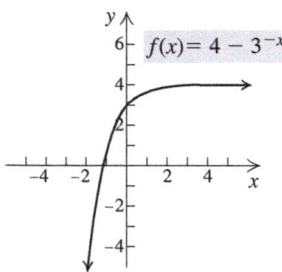

$f(x) = 4 - 3^{-x}$

35. Shift the graph of $y = \left(\frac{3}{2}\right)^x$ right 1 unit.

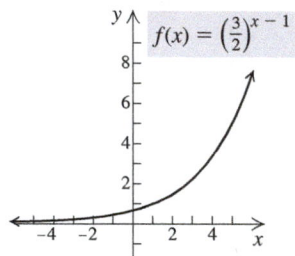

$f(x) = \left(\frac{3}{2}\right)^{x-1}$

37. Shift the graph of $y = 2^x$ left 3 units and then down 5 units.

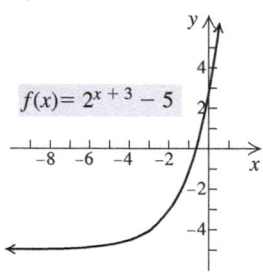

$f(x) = 2^{x+3} - 5$

39. Shift the graph of $y = 2^x$ right 1 unit, stretch it vertically, and shift it up 1 unit.

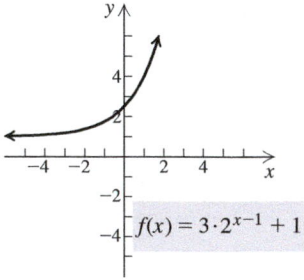

$f(x) = 3 \cdot 2^{x-1} + 1$

41. Shrink the graph of $y = e^x$ horizontally.

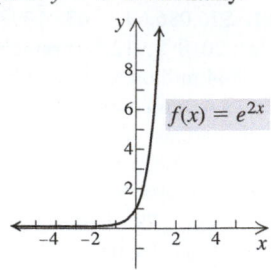

$f(x) = e^{2x}$

43. Reflect the graph of $y = e^x$ across the x-axis, shift it up 1 unit, and shrink it vertically.

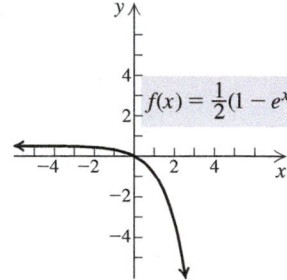

$f(x) = \frac{1}{2}(1 - e^x)$

45. Shift the graph of $y = e^x$ left 1 unit and then reflect it across the y-axis.

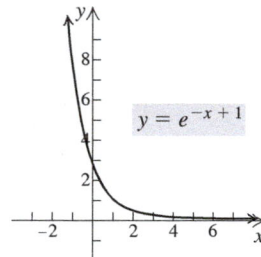

$y = e^{-x+1}$

47. Reflect the graph of $y = e^x$ across the y-axis, then across the x-axis, then shift it up 1 unit, and then stretch it vertically.

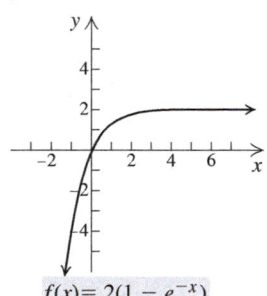

$f(x) = 2(1 - e^{-x})$

49.

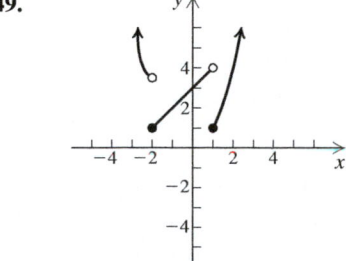

$$f(x) = \begin{cases} e^{-x} - 4, & \text{for } x < -2, \\ x + 3, & \text{for } -2 \le x < 1, \\ x^2, & \text{for } x \ge 1 \end{cases}$$

51. (a) $A(t) = 82{,}000(1.01125)^{4t}$;
(b) \$82,000; \$89,677.22; \$102,561.54; \$128,278.90

53. $4930.86   **55.** $3247.30   **57.** $153,610.15
59. $76,305.59   **61.** $26,086.69   **63.** 1998: 322,420 vehicles;
2010: 938,297 vehicles; 2018: 1,912,580 vehicles   **65.** 2011:
$234 million; 2015: $5844 million, or $5.844 billion   **67.** 2005:
3 million users; 2009: 17 million users; 2012: 54 million users
69. 2020: 101,234 centenarians; 2050: 414,387 centenarians
71. 1982: $48 billion; 1995: $109 billion; 2010: $284 billion
73. $6982; $5935; $5044; $3098; $1903   **75.** About 63%
77. $31 - 22i$ **78.** $\frac{1}{2} - \frac{1}{2}i$ **79.** $\left(-\frac{1}{2}, 0\right), (7, 0); -\frac{1}{2}, 7$
80. $(1, 0); 1$ **81.** $(-1, 0), (0, 0), (1, 0); -1, 0, 1$
82. $(-4, 0), (0, 0), (3, 0); -4, 0, 3$ **83.** $-8, 0, 2$
84. $\dfrac{5 \pm \sqrt{97}}{6}$ **85.** $\pi^7; 70^{80}$

Visualizing the Graph, p. 632

1. J   **2.** F   **3.** H   **4.** B   **5.** E   **6.** A   **7.** C
8. I   **9.** D   **10.** G

Exercise Set 9.4, p. 633

1. 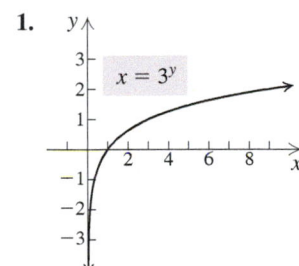   **3.**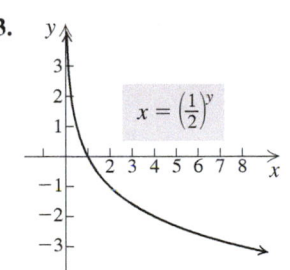

5. 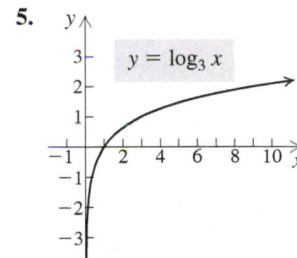   **7.**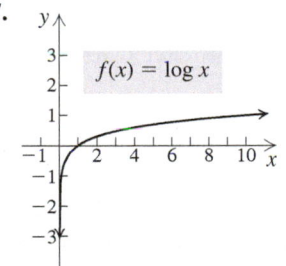

9. 4   **11.** 3   **13.** -3 **15.** -2 **17.** 0   **19.** 1   **21.** 4
23. $\frac{1}{4}$ **25.** -7 **27.** $\frac{1}{2}$ **29.** $\frac{3}{4}$ **31.** 0   **33.** $\frac{1}{2}$
35. $\log_{10} 1000 = 3$, or $\log 1000 = 3$ **37.** $\log_8 2 = \frac{1}{3}$
39. $\log_e t = 3$, or $\ln t = 3$ **41.** $\log_e 7.3891 = 2$, or
$\ln 7.3891 = 2$ **43.** $\log_p 3 = k$ **45.** $5^1 = 5$
47. $10^{-2} = 0.01$ **49.** $e^{3.4012} = 30$ **51.** $a^{-x} = M$
53. $a^x = T^3$ **55.** 0.4771   **57.** 2.7259   **59.** -0.2441
61. Does not exist   **63.** 0.6931   **65.** 6.6962
67. Does not exist   **69.** 3.3219   **71.** -0.2614 **73.** 0.7384
75. 2.2619   **77.** 0.5880

79. 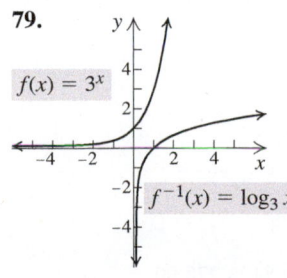   **81.**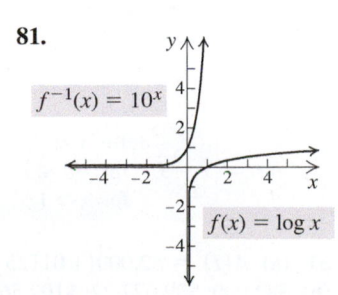

83. Shift the graph of $y = \log_2 x$ left 3 units. Domain: $(-3, \infty)$;
vertical asymptote: $x = -3$;

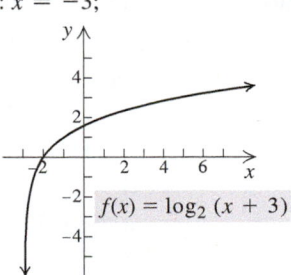

85. Shift the graph of $y = \log_3 x$ down 1 unit. Domain: $(0, \infty)$;
vertical asymptote: $x = 0$;

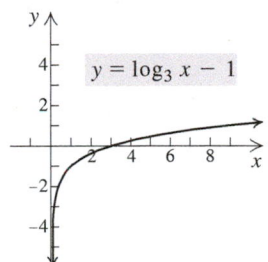

87. Stretch the graph of $y = \ln x$ vertically. Domain: $(0, \infty)$;
vertical asymptote: $x = 0$;

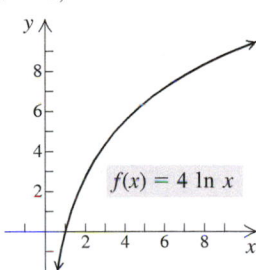

89. Reflect the graph of $y = \ln x$ across the x-axis and shift it up 2
units. Domain: $(0, \infty)$; vertical asymptote: $x = 0$;

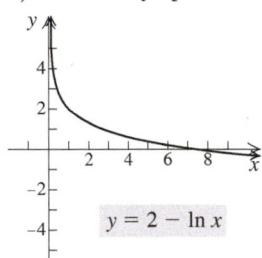

91. Shift the graph of $\log x$ right 1 unit, shrink it vertically, and
shift it down 2 units.

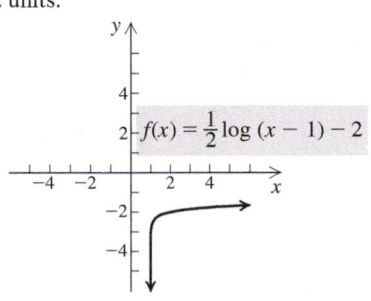

93.

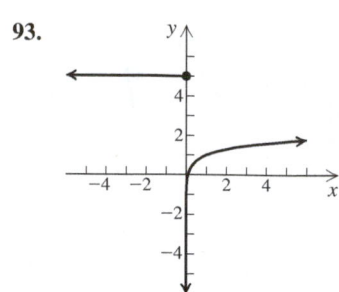

$$g(x) = \begin{cases} 5, & \text{for } x \leq 0, \\ \log x + 1, & \text{for } x > 0 \end{cases}$$

95. (a) 2.5 ft/sec;
(b) 2.8 ft/sec; **(c)** 2.0 ft/sec;
(d) 2.4 ft/sec; **(e)** 2.2 ft/sec;
(f) 2.5 ft/sec; **(g)** 2.3 ft/sec;
(h) 3.1 ft/sec   **97. (a)** 7.7;
(b) 9.5; **(c)** 6.6; **(d)** 7.6;
(e) 8.0; **(f)** 7.9; **(g)** 5.1;
(h) 9.3   **99. (a)** 10^{-7};
(b) 4.0×10^{-6};
(c) 6.3×10^{-4};
(d) 1.6×10^{-5}

101. (a) 140 decibels; **(b)** 115 decibels; **(c)** 40 decibels;
(d) 65 decibels; **(e)** 120 decibels; **(f)** 194 decibels
102. $m = \frac{3}{10}$; y-intercept: $\left(0, -\frac{7}{5}\right)$ **103.** $m = 0$;
y-intercept: $(0, 6)$ **104.** Slope is not defined; no
y-intercept   **105.** -280 **106.** -4 **107.** $f(x) = x^3 - 7x$
108. $f(x) = x^3 - x^2 + 16x - 16$ **109.** 3   **111.** $(0, \infty)$
113. $(-\infty, 0) \cup (0, \infty)$ **115.** $\left(-\frac{5}{2}, -2\right)$ **117.** (d)
119. (b)

Mid-Chapter Review: Chapter 9, p. 635

1. True   **2.** False   **3.** True   **4.** False
5. 6   **6.** 28   **7.** -24 **8.** 102
9. $(f \circ g)(x) = 3x + 2$; $(g \circ f)(x) = 3x + 4$;
domain of $f \circ g$ and $g \circ f$: $(-\infty, \infty)$
10. $(f \circ g)(x) = 3\sqrt{x} + 2$; $(g \circ f)(x) = \sqrt{3x + 2}$;
domain of $f \circ g$: $[0, \infty)$; domain of $g \circ f$: $\left[-\frac{2}{3}, \infty\right)$

11. Yes; $f^{-1}(x) = -\dfrac{2}{x}$ **12.** No   **13.** Yes; $f^{-1}(x) = \dfrac{5}{x} + 2$

14. $(f^{-1} \circ f)(x) = f^{-1}(\sqrt{x - 5}) = (\sqrt{x - 5})^2 + 5 = x - 5 + 5 = x$; $(f \circ f^{-1})(x) = f(x^2 + 5) = \sqrt{(x^2 + 5) - 5} = \sqrt{x^2} = x$ **15.** $f^{-1}(x) = \sqrt[3]{x - 2}$;
domain of f: $(-\infty, \infty)$, range of f^{-1}: $(-\infty, \infty)$; domain of
f^{-1}: $(-\infty, \infty)$, range of f^{-1}: $(-\infty, \infty)$ **16.** (d)   **17.** (h)
18. (c)   **19.** (g)   **20.** (b)   **21.** (f)   **22.** (e)   **23.** (a)
24. \$4185.57   **25.** 0   **26.** $-\frac{4}{5}$ **27.** -2 **28.** 2   **29.** 0
30. -4 **31.** 0   **32.** 3   **33.** $\frac{1}{4}$ **34.** 1   **35.** $\ln 0.0025 = -6$
36. $10^r = T$ **37.** 2.7268   **38.** 2.0115   **39.** This approach
is not valid. Consider Exercise 23 in Section 9.1, for example.
Since $(f \circ g)(x) = \dfrac{4x}{x - 5}$, an examination of only this
composed function would lead to the incorrect conclusion
that the domain of $f \circ g$ is $(-\infty, 5) \cup (5, \infty)$. However, we
must also exclude from the domain of $f \circ g$ those values of
x that are not in the domain of g. Thus the domain of $f \circ g$ is
$(-\infty, 0) \cup (0, 5) \cup (5, \infty)$.   **40.** For an even function f,
$f(x) = f(-x)$, so we have $f(x) = f(-x)$ but $x \neq -x$ (for
$x \neq 0$). Thus, f is not one-to-one and hence it does not have an
inverse.   **41.** In $f(x) = x^3$, the variable x is the base. The range
of f is $(-\infty, \infty)$. In $g(x) = 3^x$, the variable x is the exponent. The
range of g is $(0, \infty)$. The graph of f does not have an asymptote.
The graph of g has an asymptote $y = 0$.
42. If $\log b < 0$, then $0 < b < 1$.

Exercise Set 9.5, p. 643

1. $\log_3 81 + \log_3 27 = 4 + 3 = 7$
3. $\log_5 5 + \log_5 125 = 1 + 3 = 4$

5. $\log_t 8 + \log_t Y$ **7.** $\ln x + \ln y$ **9.** $3 \log_b t$
11. $8 \log y$ **13.** $-6 \log_c K$ **15.** $\frac{1}{3} \ln 4$ **17.** $\log_t M - \log_t 8$
19. $\log x - \log y$ **21.** $\ln r - \ln s$
23. $\log_a 6 + \log_a x + 5 \log_a y + 4 \log_a z$
25. $2 \log_b p + 5 \log_b q - 4 \log_b m - 9$
27. $\ln 2 - \ln 3 - 3 \ln x - \ln y$ **29.** $\frac{3}{2} \log r + \frac{1}{2} \log t$
31. $3 \log_a x - \frac{5}{2} \log_a p - 4 \log_a q$
33. $2 \log_a m + 3 \log_a n - \frac{3}{4} - \frac{5}{4} \log_a b$
35. $\log_a 150$ **37.** $\log 100 = 2$ **39.** $\log m^3 \sqrt{n}$
41. $\log_a x^{-5/2} y^4$, or $\log_a \dfrac{y^4}{x^{5/2}}$ **43.** $\ln x$ **45.** $\ln (x - 2)$
47. $\log \dfrac{x - 7}{x - 2}$ **49.** $\ln \dfrac{x}{(x^2 - 25)^3}$ **51.** $\ln \dfrac{2^{11/5} x^9}{y^8}$
53. -0.74 **55.** 1.991   **57.** 0.356   **59.** 4.827   **61.** -1.792
63. 0.099   **65.** 3   **67.** $|x - 4|$ **69.** $4x$ **71.** w **73.** $8t$
75. $\frac{1}{2}$ **77.** Quartic   **78.** Exponential   **79.** Linear (constant)
80. Exponential   **81.** Rational   **82.** Logarithmic
83. Cubic   **84.** Rational   **85.** Linear   **86.** Quadratic
87. 4   **89.** $\log_a(x^3 - y^3)$ **91.** $\frac{1}{2} \log_a(x - y) - \frac{1}{2} \log_a(x + y)$ **93.** 7   **95.** True   **97.** True   **99.** True
101. -2 **103.** 3   **105.** $e^{-xy} = \dfrac{a}{b}$
107. $\log_a \left(\dfrac{x + \sqrt{x^2 - 5}}{5} \cdot \dfrac{x - \sqrt{x^2 - 5}}{x - \sqrt{x^2 - 5}} \right)$
$= \log_a \dfrac{5}{5(x - \sqrt{x^2 - 5})} = -\log_a (x - \sqrt{x^2 - 5})$

Exercise Set 9.6, p. 653

1. 4   **3.** $\frac{3}{2}$ **5.** 5.044   **7.** $\frac{5}{2}$ **9.** $-3, \frac{1}{2}$ **11.** 0.959
13. 0   **15.** 0   **17.** 6.908   **19.** 84.191   **21.** -1.710
23. 2.844   **25.** $-1.567, 1.567$ **27.** 1.869   **29.** $-1.518, 0.825$
31. 625   **33.** 0.0001   **35.** e **37.** $-\frac{1}{3}$ **39.** $\frac{22}{3}$ **41.** 10
43. 4   **45.** $\frac{1}{63}$ **47.** 2   **49.** $\frac{2}{5}$ **51.** 5   **53.** $\frac{21}{8}$ **55.** $\frac{8}{7}$
57. 6   **59.** 6.192   **61.** 0   **63. (a)** $(0, -6)$; **(b)** $x = 0$;
(c) minimum: -6 when $x = 0$ **64. (a)** $(3, 1)$; **(b)** $x = 3$;
(c) maximum: 1 when $x = 3$ **65. (a)** $(-1, -5)$; **(b)** $x = -1$;
(c) maximum: -5 when $x = -1$ **66. (a)** $(2, 4)$; **(b)** $x = 2$;
(c) minimum: 4 when $x = 2$ **67.** $\dfrac{\ln 2}{2}$, or 0.347   **69.** $1, e^4$ or
$1, 54.598$ **71.** $\frac{1}{3}, 27$ **73.** $1, e^2$ or $1, 7.389$ **75.** $0, \dfrac{\ln 2}{\ln 5}$ or
$0, 0.431$ **77.** e^{-2}, e^2 or $0.135, 7.389$ **79.** $\frac{7}{4}$ **81.** $a = \frac{2}{3}b$

Exercise Set 9.7, p. 664

1. (a) $P(t) = 6.18e^{0.0214t}$, where t is the number of years after
2012 and P is in millions; **(b)** 7.0 million; **(c)** about 12.1 years
after 2012; **(d)** about 32.4 years   **3. (a)** 0.90%; **(b)** 1.63%;
(c) 20.9 years; **(d)** 62.4 years; **(e)** 0.18%; **(f)** 29.9 years;
(g) 54.2 years; **(h)** 0.46%; **(i)** 2.64%; **(j)** 177.7 years
5. About 819 years after 2013   **7. (a)** $P(t) = 10,000e^{0.054t}$;
(b) \$10,554.85; \$11,140.48; \$13,099.64; \$17,160.07;
(c) about 12.8 years   **9.** About 12,320 years
11. (a) 22.4% per minute; **(b)** 3.1% per year; **(c)** 60.3 days;
(d) 10.7 years; **(e)** 2.4% per year; **(f)** 1.0% per year;
(g) 0.0029% per year   **13. (a)** $k \approx 0.0069$, $M(t) = 72.2e^{-0.0069t}$;
(b) 2015: 49.4%; 2018: 48.4%; **(c)** in 2046

15. (a) $k \approx 0.0536$, $C(t) = 1.85e^{0.0536t}$; (b) 7.07 million barrels of oil per day; (c) 12.9 years; (d) 36.4 years after 1980
17. (a) 167; (b) 500; 1758; 3007; 3449; 3495; (c) as $t \to \infty$, $N(t) \to 3500$; the number approaches 3500 but never actually reaches it.   **19.** 46.7°F   **21.** 59.6°F
23. Multiplication principle for inequalities   **24.** Product rule
25. Principle of zero products   **26.** Principle of square roots
27. Power rule   **28.** Multiplication principle for equations
29. $166.16   **31.** $19,609.67   **33.** $t = -\dfrac{L}{R}\left[\ln\left(1 - \dfrac{iR}{V}\right)\right]$

35. Linear

Summary and Review: Chapter 9

Review Exercises, p. 676

1. False   **2.** True   **3.** False   **4.** False   **5.** True
6. False   **7.** True   **8.** 9   **9.** 5   **10.** 128   **11.** 580
12. 7   **13.** −509   **14.** $4x - 3$ **15.** $-24 + 27x^3 - 9x^6 + x^9$
16. (a) $(f \circ g)(x) = \dfrac{4}{(3-2x)^2}$; $(g \circ f)(x) = 3 - \dfrac{8}{x^2}$;
(b) domain of $f \circ g$: $\left(-\infty, \frac{3}{2}\right) \cup \left(\frac{3}{2}, \infty\right)$; domain of $g \circ f$: $(-\infty, 0) \cup (0, \infty)$
17. (a) $(f \circ g)(x) = 12x^2 - 4x - 1$; $(g \circ f)(x) = 6x^2 + 8x - 1$;
(b) domain of $f \circ g$ and $g \circ f$: $(-\infty, \infty)$
18. $f(x) = \sqrt{x}$, $g(x) = 5x + 2$; answers may vary.
19. $f(x) = 4x^2 + 9$, $g(x) = 5x - 1$; answers may vary.
20. $\{(-2.7, 1.3), (-3, 8), (3, -5), (-3, 6), (-5, 7)\}$
21. (a) $x = -2y + 3$; (b) $x = 3y^2 + 2y - 1$;
(c) $0.8y^3 - 5.4x^2 = 3y$ **22.** No   **23.** No   **24.** Yes

25. Yes   **26.** (a) Yes; (b) $f^{-1}(x) = \dfrac{-x + 2}{3}$ **27.** (a) Yes;
(b) $f^{-1}(x) = \dfrac{x + 2}{x - 1}$ **28.** (a) Yes; (b) $f^{-1}(x) = x^2 + 6$, $x \geq 0$

29. (a) Yes; (b) $f^{-1}(x) = \sqrt[3]{x + 8}$ **30.** (a) No
31. (a) Yes; (b) $f^{-1}(x) = \ln x$
32. $f^{-1}(f(x)) = f^{-1}(6x - 5) = \dfrac{6x - 5 + 5}{6} = \dfrac{6x}{6} = x$;
$f(f^{-1}(x)) = f\left(\dfrac{x + 5}{6}\right) = 6\left(\dfrac{x + 5}{6}\right) - 5 = x + 5 - 5 = x$

33. $f^{-1}(f(x)) = f^{-1}\left(\dfrac{x + 1}{x}\right) = \dfrac{1}{\dfrac{x + 1}{x} - 1} =$

$\dfrac{1}{\dfrac{x + 1 - x}{x}} = \dfrac{1}{\dfrac{1}{x}} = x$; $f(f^{-1}(x)) = f\left(\dfrac{1}{x - 1}\right) =$

$\dfrac{\dfrac{1}{x - 1} + 1}{\dfrac{1}{x - 1}} = \dfrac{\dfrac{1 + x - 1}{x - 1}}{\dfrac{1}{x - 1}} = \dfrac{x}{x - 1} \cdot \dfrac{x - 1}{1} = x$

34. $f^{-1}(x) = \dfrac{2 - x}{5}$; domain

of f and f^{-1}: $(-\infty, \infty)$;
range of f and f^{-1}: $(-\infty, \infty)$;

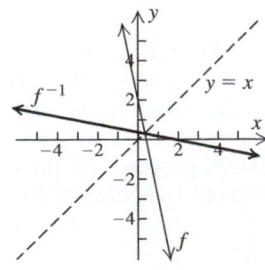

35. $f^{-1}(x) = \dfrac{-2x - 3}{x - 1}$; domain of f: $(-\infty, -2) \cup (-2, \infty)$; range of f: $(-\infty, 1) \cup (1, \infty)$; domain of f^{-1}: $(-\infty, 1) \cup (1, \infty)$; range of f^{-1}: $(-\infty, -2) \cup (-2, \infty)$;

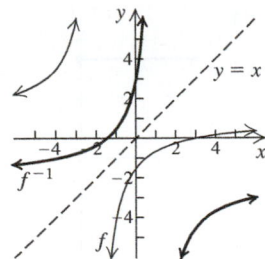

36. 657   **37.** a
38.

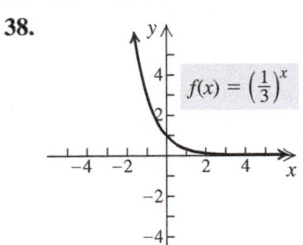

39.

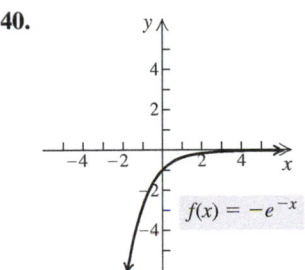

40.

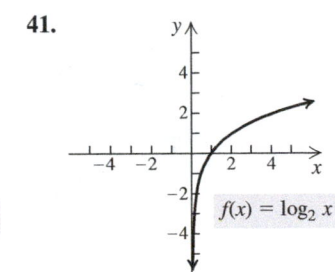

41.

42.

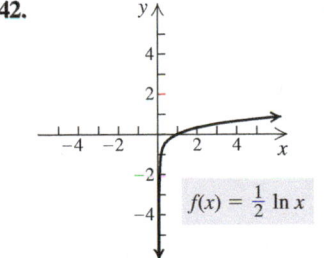

43.

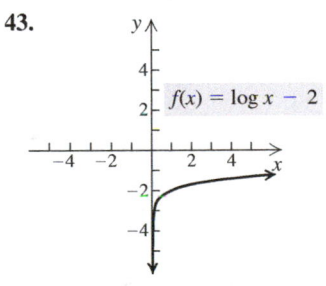

44. (c)   **45.** (a)   **46.** (b)   **47.** (f)   **48.** (e)   **49.** (d)
50. 3   **51.** 5   **52.** 1   **53.** 0   **54.** $\frac{1}{4}$ **55.** $\frac{1}{2}$ **56.** 0
57. 1   **58.** $\frac{1}{3}$ **59.** −2   **60.** $4^2 = x$ **61.** $a^k = Q$
62. $\log_4 \frac{1}{64} = -3$ **63.** $\ln 80 = x$, or $\log_e 80 = x$ **64.** 1.0414
65. −0.6308   **66.** 1.0986   **67.** −3.6119   **68.** Does not exist
69. Does not exist   **70.** 1.9746   **71.** 0.5283   **72.** $\log_b \dfrac{x^3 \sqrt{z}}{y^4}$
73. $\ln(x^2 - 4)$ **74.** $\frac{1}{4}\ln w + \frac{1}{2}\ln r$ **75.** $\frac{2}{3}\log M - \frac{1}{3}\log N$
76. 0.477   **77.** 1.699   **78.** −0.699   **79.** 0.233   **80.** −5k
81. −6t   **82.** 16   **83.** $\frac{1}{5}$ **84.** 4.382   **85.** 2   **86.** $\frac{1}{2}$
87. 5   **88.** 4   **89.** 9   **90.** 1   **91.** 3.912
92. (a) $A(t) = 30,000(1.0105)^{4t}$; (b) $30,000; $38,547.20; $49,529.56; $63,640.87   **93.** 2005: 59.8 GW; 2010: 189.9 GW; 2016: 760.1 GW   **94.** 15.4 years   **95.** 2.7%   **96.** About 2623 years   **97.** 5.6   **98.** 6.3   **99.** 30 decibels
100. (a) 2.2 ft/sec; (b) 8,553,143   **101.** (a) $k \approx 0.1392$; (b) $S(t) = 0.035e^{0.1392t}$, where t is the number of years after 1940 and S is in billions of dollars; (c) 1970: $2.279 billion; 2000: $148.353 billion; 2015: $1197.023 billion, or about $1.197 trillion; (d) in 2019   **102.** (a) $P(t) = 15.2e^{0.0167t}$, where t is the number of years after 2013 and P is in millions; (b) 2017:

16.3 million; 2020: 17.1 million; **(c)** about 10 years after 2013; **(d)** 41.5 years   **103.** D   **104.** A   **105.** D   **106.** B   **107.** $\frac{1}{64}$, 64   **108.** 1   **109.** 16   **110.** Measure the atmospheric pressure P at the top of the building. Substitute that value in the equation $P = 14.7e^{-0.00005a}$, and solve for the height, or altitude, a of the top of the building. Also measure the atmospheric pressure at the base of the building and solve for the altitude of the base. Then subtract to find the height of the building.   **111.** Reflect the graph of $f(x) = \ln x$ across the line $y = x$ to obtain the graph of $h(x) = e^x$. Then shift this graph right 2 units to obtain the graph of $g(x) = e^{x-2}$.   **112.** The inverse of a function $f(x)$ is written $f^{-1}(x)$, whereas $[f(x)]^{-1}$ means $\dfrac{1}{f(x)}$.   **113.** $\log_a ab^3 \neq (\log_a a)(\log_a b^3)$.

If the first step had been correct, then the second step would be as well. The correct procedure follows:

$\log_a ab^3 = \log_a a + \log_a b^3 = 1 + 3\log_a b$.

Test: Chapter 9, p. 679

1. [9.1] 83   **2.** [9.1] 0   **3.** [9.1] 4   **4.** [9.1] $16x + 15$
5. [9.1] $(f \circ g)(x) = \sqrt{x^2 - 4}$; $(g \circ f)(x) = x - 4$
6. [9.1] Domain of $(f \circ g)(x)$: $(-\infty, -2] \cup [2, \infty)$; domain of $(g \circ f)(x)$: $[5, \infty)$ **7.** [9.1] $f(x) = x^4$; $g(x) = 2x - 7$; answers may vary   **8.** [9.2] $\{(5, -2), (3, 4), (-1, 0), (-3, -6)\}$
9. [9.2] No   **10.** [9.2] Yes   **11.** [9.2] **(a)** Yes;
(b) $f^{-1}(x) = \sqrt[3]{x - 1}$ **12.** [9.2] **(a)** Yes;
(b) $f^{-1}(x) = 1 - x$ **13.** [9.2] **(a)** Yes;
(b) $f^{-1}(x) = \dfrac{2x}{1 + x}$ **14.** [9.2] **(a)** No

15. [9.2] $f^{-1}(f(x)) = f^{-1}(-4x + 3) = \dfrac{3 - (-4x + 3)}{4} =$

$\dfrac{4x}{4} = x$; $f(f^{-1}(x)) = f\left(\dfrac{3 - x}{4}\right) = -4\left(\dfrac{3 - x}{4}\right) + 3 =$

$-3 + x + 3 = x$ **16.** [9.2]

$f^{-1}(x) = \dfrac{4x + 1}{x}$; domain of
f: $(-\infty, 4) \cup (4, \infty)$; range
of f: $(-\infty, 0) \cup (0, \infty)$;
domain of f^{-1}:
$(-\infty, 0) \cup (0, \infty)$; range of
f^{-1}: $(-\infty, 4) \cup (4, \infty)$;

17. [9.3]

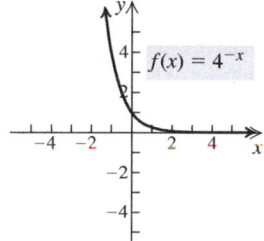

18. [9.4]

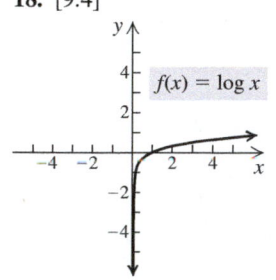

19. [9.3]

20. [9.4]

21. [9.4] -5 **22.** [9.4] 1   **23.** [9.4] 0   **24.** [9.4] $\frac{1}{5}$
25. [9.4] $x = e^4$ **26.** [9.4] $x = \log_3 5.4$ **27.** [9.4] 2.7726
28. [9.4] -0.5331 **29.** [9.4] 1.2851   **30.** [9.5] $\log_a \dfrac{x^2\sqrt{z}}{y}$
31. [9.5] $\frac{2}{5}\ln x + \frac{1}{5}\ln y$ **32.** [9.5] 0.656   **33.** [9.5] $-4t$
34. [9.6] $\frac{1}{2}$ **35.** [9.6] 1   **36.** [9.6] 1   **37.** [9.6] 4.174
38. [9.4] 6.6   **39.** [9.7] 0.0154 or 1.54%   **40.** [9.7] **(a)** 4.5%;
(b) $P(t) = 1000e^{0.045t}$; **(c)** \$1433.33; **(d)** 15.4 years
41. [9.3] C   **42.** [9.6] $\frac{27}{8}$

Chapter 10

Visualizing the Graph, p. 687

1. C   **2.** G   **3.** D   **4.** J   **5.** A   **6.** F
7. I   **8.** B   **9.** H   **10.** E

Exercise Set 10.1, p. 688

1. 3×2 **3.** 1×4 **5.** 3×3 **7.** $\begin{bmatrix} 2 & -1 & | & 7 \\ 1 & 4 & | & -5 \end{bmatrix}$

9. $\begin{bmatrix} 1 & -2 & 3 & | & 12 \\ 2 & 0 & -4 & | & 8 \\ 0 & 3 & 1 & | & 7 \end{bmatrix}$ **11.** $3x - 5y = 1$,
$x + 4y = -2$

13. $2x + y - 4z = 12$,
$3x + 5z = -1$,
$x - y + z = 2$
15. $\left(\frac{3}{2}, \frac{5}{2}\right)$ **17.** $\left(-\frac{63}{29}, -\frac{114}{29}\right)$ **19.** $\left(-1, \frac{5}{2}\right)$ **21.** $(0, 3)$
23. No solution   **25.** Infinitely many solutions; $(3y - 2, y)$
27. $(-1, 2, -2)$ **29.** $\left(\frac{3}{2}, -4, 3\right)$ **31.** $(-1, 6, 3)$
33. Infinitely many solutions; $\left(\frac{1}{2}z + \frac{1}{2}, -\frac{1}{2}z - \frac{1}{2}, z\right)$
35. Infinitely many solutions; $(r - 2, -2r + 3, r)$
37. No solution   **39.** $(1, -3, -2, -1)$ **41.** 8%: \$8000; 10%:
\$12,000; 12%: \$10,000   **43.** 49¢: 160 stamps; 21¢: 40 stamps
45. Perpendicular   **46.** The leading-term test
47. Vertical line   **48.** Horizontal line   **49.** Rational function
50. Inverse variation   **51.** Vertical asymptote
52. Horizontal asymptote   **53.** $y = 3x^2 + \frac{5}{2}x - \frac{15}{2}$
55. $\begin{bmatrix} 1 & 5 \\ 0 & 1 \end{bmatrix}, \begin{bmatrix} 1 & 0 \\ 0 & 1 \end{bmatrix}$ **57.** $\left(-\frac{4}{3}, -\frac{1}{3}, 1\right)$
59. Infinitely many solutions; $\left(-\frac{14}{13}z - 1, \frac{3}{13}z - 2, z\right)$
61. $(-3, 3)$

Exercise Set 10.2, p. 696

1. $x = -3, y = 5$ **3.** $x = -1, y = 1$ **5.** $\begin{bmatrix} -2 & 7 \\ 6 & 2 \end{bmatrix}$

7. $\begin{bmatrix} 1 & 3 \\ 2 & 6 \end{bmatrix}$ **9.** $\begin{bmatrix} 9 & 9 \\ -3 & -3 \end{bmatrix}$ **11.** $\begin{bmatrix} 11 & 13 \\ 5 & 3 \end{bmatrix}$

13. $\begin{bmatrix} -4 & 3 \\ -2 & -4 \end{bmatrix}$ **15.** $\begin{bmatrix} 17 & 9 \\ -2 & 1 \end{bmatrix}$ **17.** $\begin{bmatrix} 0 & 0 \\ 0 & 0 \end{bmatrix}$ **19.** $\begin{bmatrix} 1 & 2 \\ 4 & 3 \end{bmatrix}$

21. $\begin{bmatrix} 1 \\ 40 \end{bmatrix}$ **23.** $\begin{bmatrix} -10 & 28 \\ 14 & -26 \\ 0 & -6 \end{bmatrix}$ **25.** Not defined

27. $\begin{bmatrix} 3 & 16 & 3 \\ 0 & -32 & 0 \\ -6 & 4 & 5 \end{bmatrix}$ **29. (a)** $\begin{bmatrix} 40 & 20 & 30 \end{bmatrix}$;

(b) $\begin{bmatrix} 44 & 22 & 33 \end{bmatrix}$; **(c)** $\begin{bmatrix} 84 & 42 & 63 \end{bmatrix}$; the total amount of each type of produce ordered for both weeks
31. (a) $C = \begin{bmatrix} 140 & 27 & 3 & 13 & 64 \end{bmatrix}$,
$P = \begin{bmatrix} 180 & 4 & 11 & 24 & 662 \end{bmatrix}, B = \begin{bmatrix} 50 & 5 & 1 & 82 & 20 \end{bmatrix}$;
(b) $\begin{bmatrix} 650 & 50 & 28 & 307 & 1448 \end{bmatrix}$; the total nutritional value of a meal of 1 serving of chicken, 1 cup of potato salad, and 3 broccoli spears

33. (a) $\begin{bmatrix} 1.50 & 0.30 & 0.36 & 0.45 & 0.64 \\ 1.55 & 0.28 & 0.48 & 0.57 & 0.75 \\ 1.62 & 0.52 & 0.65 & 0.38 & 0.53 \\ 1.70 & 0.43 & 0.40 & 0.42 & 0.68 \end{bmatrix}$;

(b) $\begin{bmatrix} 65 & 48 & 93 & 57 \end{bmatrix}$;
(c) $\begin{bmatrix} 419.46 & 105.81 & 129.69 & 115.89 & 165.65 \end{bmatrix}$;
(d) the total cost, in dollars, for each item for the day's meals

35. (a) $\begin{bmatrix} 8 & 15 \\ 6 & 10 \\ 4 & 3 \end{bmatrix}$; **(b)** $\begin{bmatrix} 4 & 2.50 & 3 \end{bmatrix}$; **(c)** $\begin{bmatrix} 59 & 94 \end{bmatrix}$;

(d) the total cost, in dollars, of ingredients for each coffee shop
37. (a) $\begin{bmatrix} 7.50 & 4.80 & 6.25 \end{bmatrix}$; **(b)** $PS = \begin{bmatrix} 113.80 & 179.25 \end{bmatrix}$

39. $\begin{bmatrix} 2 & -3 \\ 1 & 5 \end{bmatrix} \begin{bmatrix} x \\ y \end{bmatrix} = \begin{bmatrix} 7 \\ -6 \end{bmatrix}$

41. $\begin{bmatrix} 1 & 1 & -2 \\ 3 & -1 & 1 \\ 2 & 5 & -3 \end{bmatrix} \begin{bmatrix} x \\ y \\ z \end{bmatrix} = \begin{bmatrix} 6 \\ 7 \\ 8 \end{bmatrix}$

43. $\begin{bmatrix} 3 & -2 & 4 \\ 2 & 1 & -5 \end{bmatrix} \begin{bmatrix} x \\ y \\ z \end{bmatrix} = \begin{bmatrix} 17 \\ 13 \end{bmatrix}$

45. $\begin{bmatrix} -4 & 1 & -1 & 2 \\ 1 & 2 & -1 & -1 \\ -1 & 1 & 4 & -3 \\ 2 & 3 & 5 & -7 \end{bmatrix} \begin{bmatrix} w \\ x \\ y \\ z \end{bmatrix} = \begin{bmatrix} 12 \\ 0 \\ 1 \\ 9 \end{bmatrix}$

47. (a) $\left(\frac{1}{2}, -\frac{25}{4} \right)$;   **48. (a)** $\left(\frac{5}{4}, -\frac{49}{8} \right)$;

(b) $x = \frac{1}{2}$;   **(b)** $x = \frac{5}{4}$;
(c) minimum: $-\frac{25}{4}$;   **(c)** minimum: $-\frac{49}{8}$;
(d)   **(d)**

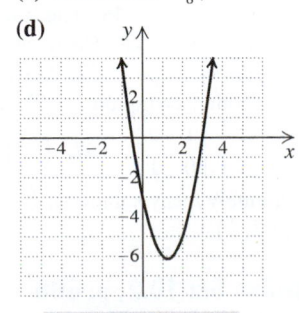

$f(x) = x^2 - x - 6$ $f(x) = 2x^2 - 5x - 3$

49. (a) $\left(-\frac{3}{2}, \frac{17}{4} \right)$;   **50. (a)** $\left(\frac{2}{3}, \frac{16}{3} \right)$;

(b) $x = -\frac{3}{2}$;   **(b)** $x = \frac{2}{3}$;
(c) maximum: $\frac{17}{4}$;   **(c)** maximum: $\frac{16}{3}$;
(d)   **(d)**

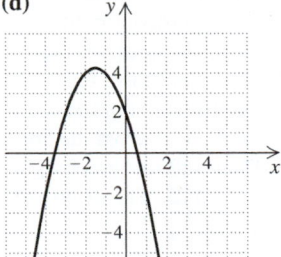 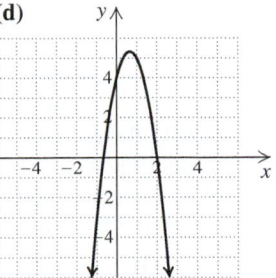

$f(x) = -x^2 - 3x + 2$ $f(x) = -3x^2 + 4x + 4$

51. $(A + B)(A - B) = \begin{bmatrix} -2 & 1 \\ 2 & -1 \end{bmatrix}; A^2 - B^2 = \begin{bmatrix} 0 & 3 \\ 0 & -3 \end{bmatrix}$

53. $(A + B)(A - B) = \begin{bmatrix} -2 & 1 \\ 2 & -1 \end{bmatrix}$
$= A^2 + BA - AB - B^2$

Mid-Chapter Review: Chapter 10, p. 700

1. True   **2.** False   **3.** True   **4.** False
5. $(4, -3)$ **6.** $(1, -2)$ **7.** No solution
8. $\left(x, \dfrac{x-1}{3} \right)$ or $(3y + 1, y)$ **9.** $(-3, 2, -1)$
10. $\left(\frac{1}{3}, -\frac{1}{6}, \frac{4}{3} \right)$ **11.** Under 10 lb: 60 packages; 10 lb up to 15 lb: 70 packages; 15 lb or more: 20 packages

12. $\begin{bmatrix} 1 & 5 \\ 6 & 1 \end{bmatrix}$ **13.** $\begin{bmatrix} -5 & 7 \\ -4 & -7 \end{bmatrix}$ **14.** $\begin{bmatrix} -8 & 12 & 0 \\ 4 & -4 & 8 \\ -12 & 16 & 4 \end{bmatrix}$

15. $\begin{bmatrix} 0 & 16 \\ 13 & -1 \end{bmatrix}$ **16.** $\begin{bmatrix} -7 & 21 \\ -6 & 18 \end{bmatrix}$ **17.** $\begin{bmatrix} 24 & 26 \\ -12 & -13 \end{bmatrix}$

18. $\begin{bmatrix} 20 & 16 & -10 \\ -10 & -8 & 5 \end{bmatrix}$ **19.** Not defined

20. $\begin{bmatrix} 2 & -1 & 3 \\ 1 & 2 & -1 \\ 3 & -4 & 2 \end{bmatrix} \begin{bmatrix} x \\ y \\ z \end{bmatrix} = \begin{bmatrix} 7 \\ 3 \\ 5 \end{bmatrix}$

21. In a matrix in reduced row-echelon form, each column that contains a leading 1 has 0's everywhere else. This property is not required for row-echelon form.   **22.** The last row of the matrix corresponds to the equation $0 = 0$, which is true for all values of x, y, and z. Therefore, the equations are dependent.   **23.** Two matrices can be multiplied only when the number of columns in the first matrix is equal to the number of rows in the second matrix.
24. No; for example, let $A = \begin{bmatrix} 1 & -1 \\ -1 & 1 \end{bmatrix}$ and $B = \begin{bmatrix} 1 & 1 \\ 1 & 1 \end{bmatrix}$;
then $AB = \begin{bmatrix} 0 & 0 \\ 0 & 0 \end{bmatrix}$ and neither A nor B is $\begin{bmatrix} 0 & 0 \\ 0 & 0 \end{bmatrix}$.

Exercise Set 10.3, p. 705

1. Yes **3.** No **5.** $\begin{bmatrix} -3 & 2 \\ 5 & -3 \end{bmatrix}$ **7.** Does not exist

9. $\begin{bmatrix} \frac{2}{5} & -\frac{3}{5} \\ \frac{1}{5} & -\frac{4}{5} \end{bmatrix}$ **11.** $\begin{bmatrix} \frac{3}{8} & -\frac{1}{4} & \frac{1}{8} \\ -\frac{1}{8} & \frac{3}{4} & -\frac{3}{8} \\ -\frac{1}{4} & \frac{1}{2} & \frac{1}{4} \end{bmatrix}$ **13.** Does not exist

15. $\begin{bmatrix} -1 & -1 & -6 \\ 1 & 0 & 2 \\ 0 & 1 & 3 \end{bmatrix}$ **17.** $\begin{bmatrix} 1 & 1 & 2 \\ 1 & 1 & 1 \\ 2 & 3 & 4 \end{bmatrix}$

19. Does not exist **21.** $\begin{bmatrix} 1 & -2 & 3 & 8 \\ 0 & 1 & -3 & 1 \\ 0 & 0 & 1 & -2 \\ 0 & 0 & 0 & -1 \end{bmatrix}$

23. $\begin{bmatrix} 0.25 & 0.25 & 1.25 & -0.25 \\ 0.5 & 1.25 & 1.75 & -1 \\ -0.25 & -0.25 & -0.75 & 0.75 \\ 0.25 & 0.5 & 0.75 & -0.5 \end{bmatrix}$ **25.** $(-23, 83)$

27. $(-1, 5, 1)$ **29.** $(2, -2)$ **31.** $(0, 2)$
33. $(3, -3, -2)$ **35.** $(-1, 0, 1)$ **37.** $(1, -1, 0, 1)$
39. Russian Federation: 80 medals; Ukraine: 25 medals; United
States: 18 medals **41.** Topsoil: \$239; mulch: \$179; pea gravel: \$222

43. -48 **44.** 194 **45.** $\dfrac{-1 \pm \sqrt{57}}{4}$ **46.** $-3, -2$

47. 4 **48.** 9 **49.** $(x + 2)(x - 1)(x - 4)$
50. $(x + 5)(x + 1)(x - 1)(x - 3)$

51. $\mathbf{A}^{-1}$ exists if and only if $x \neq 0$. $\mathbf{A}^{-1} = \begin{bmatrix} \frac{1}{x} \end{bmatrix}$

53. $\mathbf{A}^{-1}$ exists if and only if $xyz \neq 0$. $\mathbf{A}^{-1} = \begin{bmatrix} 0 & 0 & \frac{1}{z} \\ 0 & \frac{1}{y} & 0 \\ \frac{1}{x} & 0 & 0 \end{bmatrix}$

Exercise Set 10.4, p. 713

1. -14 **3.** -2 **5.** -11 **7.** $x^3 - 4x$
9. $M_{11} = 6, M_{32} = -9, M_{22} = -29$
11. $A_{11} = 6, A_{32} = 9, A_{22} = -29$
13. -10 **15.** -10 **17.** $M_{12} = 32, M_{44} = 7$
19. $A_{22} = -10, A_{34} = 1$ **21.** 110 **23.** -109
25. $-x^4 + x^2 - 5x$ **27.** $\left(-\frac{25}{2}, -\frac{11}{2} \right)$ **29.** $(3, 1)$
31. $\left(\frac{1}{2}, -\frac{1}{3} \right)$ **33.** $(1, 1)$ **35.** $\left(\frac{3}{2}, \frac{13}{14}, \frac{33}{14} \right)$
37. $(3, -2, 1)$ **39.** $(1, 3, -2)$ **41.** $\left(\frac{1}{2}, \frac{2}{3}, -\frac{5}{6} \right)$

43. $f^{-1}(x) = \dfrac{x - 2}{3}$ **44.** Not one-to-one

45. Not one-to-one **46.** $f^{-1}(x) = (x - 1)^3$ **47.** $5 - 3i$
48. $6 - 2i$ **49.** $10 - 10i$ **50.** $\frac{9}{25} + \frac{13}{25}i$ **51.** $3, -2$ **53.** 4
55. Answers may vary. **57.** Answers may vary.
$\begin{vmatrix} a & b \\ -b & a \end{vmatrix}$ $\begin{vmatrix} 2\pi r & 2\pi r \\ -h & r \end{vmatrix}$

Summary and Review: Chapter 10

Review Exercises, p. 717

1. True **2.** False **3.** False **4.** True **5.** $(1, 2)$
6. $(-3, 4, -2)$ **7.** $\left(\dfrac{z}{2}, -\dfrac{z}{2}, z \right)$ **8.** $(-4, 1, -2, 3)$
9. Nickels: 31; dimes: 44 **10.** 75, 69, 82
11. $\begin{bmatrix} 0 & -1 & 6 \\ 3 & 1 & -2 \\ -2 & 1 & -2 \end{bmatrix}$ **12.** $\begin{bmatrix} -3 & 3 & 0 \\ -6 & -9 & 6 \\ 6 & 0 & -3 \end{bmatrix}$
13. $\begin{bmatrix} -1 & 1 & 0 \\ -2 & -3 & 2 \\ 2 & 0 & -1 \end{bmatrix}$ **14.** $\begin{bmatrix} -2 & 2 & 6 \\ 1 & -8 & 18 \\ 2 & 1 & -15 \end{bmatrix}$

15. Not defined **16.** $\begin{bmatrix} 2 & -1 & -6 \\ 1 & 5 & -2 \\ -2 & -1 & 4 \end{bmatrix}$

17. $\begin{bmatrix} -13 & 1 & 6 \\ -3 & -7 & 4 \\ 8 & 3 & -5 \end{bmatrix}$ **18.** $\begin{bmatrix} -2 & -1 & 18 \\ 5 & -3 & -2 \\ -2 & 3 & -8 \end{bmatrix}$

19. (a) $\begin{bmatrix} 2.25 & 0.38 & 0.55 & 0.33 & 0.85 \\ 3.09 & 0.42 & 0.46 & 0.48 & 0.51 \\ 2.40 & 0.31 & 0.59 & 0.36 & 0.64 \\ 1.80 & 0.29 & 0.34 & 0.55 & 0.52 \end{bmatrix};$

(b) $\begin{bmatrix} 41 & 18 & 39 & 36 \end{bmatrix}$;
(c) $\begin{bmatrix} 306.27 & 45.67 & 66.08 & 56.01 & 87.71 \end{bmatrix}$;
(d) the total cost, in dollars, for each item for the day's meals

20. $\begin{bmatrix} -\frac{1}{2} & 0 \\ \frac{1}{6} & \frac{1}{3} \end{bmatrix}$ **21.** $\begin{bmatrix} 0 & 0 & \frac{1}{4} \\ 0 & -\frac{1}{2} & 0 \\ \frac{1}{3} & 0 & 0 \end{bmatrix}$

22. $\begin{bmatrix} 1 & 0 & 0 & 0 \\ 0 & \frac{1}{9} & \frac{5}{18} & 0 \\ 0 & -\frac{1}{9} & \frac{2}{9} & 0 \\ 0 & 0 & 0 & 1 \end{bmatrix}$ **23.** $\begin{bmatrix} 3 & -2 & 4 \\ 1 & 5 & -3 \\ 2 & -3 & 7 \end{bmatrix} \begin{bmatrix} x \\ y \\ z \end{bmatrix} = \begin{bmatrix} 13 \\ 7 \\ -8 \end{bmatrix}$

24. $(-8, 7)$ **25.** $(1, -2, 5)$ **26.** $(2, -1, 1, -3)$
27. 10 **28.** -18 **29.** -6 **30.** -1 **31.** $(3, -2)$
32. $(-1, 5)$ **33.** $\left(\frac{3}{2}, \frac{13}{14}, \frac{33}{14} \right)$ **34.** $(2, -1, 3)$ **35.** C
36. A **37.** 4%: \$10,000; 5%: \$12,000; $5\frac{1}{2}$%: \$18,000
38. $\left(\frac{5}{18}, \frac{1}{7} \right)$ **39.** $\left(1, \frac{1}{2}, \frac{1}{3} \right)$ **40.** In general,
$(\mathbf{AB})^2 \neq \mathbf{A}^2 \mathbf{B}^2$. $(\mathbf{AB})^2 = \mathbf{ABAB}$ and $\mathbf{A}^2 \mathbf{B}^2 = \mathbf{AABB}$.
Since matrix multiplication is not commutative,
$\mathbf{BA} \neq \mathbf{AB}$, so $(\mathbf{AB})^2 \neq \mathbf{A}^2 \mathbf{B}^2$.

41. If $\begin{vmatrix} a_1 & b_1 \\ a_2 & b_2 \end{vmatrix} = 0$, then $a_1 = ka_2$ and $b_1 = kb_2$ for some
number k. This means that the equations $a_1 x + b_1 y = c_1$
and $a_2 x + b_2 y = c_2$ are dependent if $c_1 = kc_2$, or the system
is inconsistent if $c_1 \neq kc_2$. **42.** If $a_1 x + b_1 y = c_1$ and
$a_2 x + b_2 y = c_2$ are parallel lines, then $a_1 = ka_2, b_1 = kb_2$,
and $c_1 \neq kc_2$, for some number k.

Then $\begin{vmatrix} a_1 & b_1 \\ a_2 & b_2 \end{vmatrix} = 0, \begin{vmatrix} c_1 & b_1 \\ c_2 & b_2 \end{vmatrix} \neq 0$, and $\begin{vmatrix} a_1 & c_1 \\ a_2 & c_2 \end{vmatrix} \neq 0$.

Test: Chapter 10, p. 718

1. [10.1] $(-3, 5)$ **2.** [10.1] $(x, 2x - 3)$ or $\left(\dfrac{y + 3}{2}, y\right)$

3. [10.1] No solution   **4.** [10.1] $(1, -2)$
5. [10.1] $(-1, 3, 2)$ **6.** [10.1] Student: 462 tickets; nonstudent:
158 tickets   **7.** [10.1] Hui: 120 orders; Ashlyn: 104 orders;

Sheriann: 128 orders   **8.** [10.2] $\begin{bmatrix} -2 & -3 \\ -3 & 4 \end{bmatrix}$

9. [10.2] Not defined   **10.** [10.2] $\begin{bmatrix} -7 & -13 \\ 5 & -1 \end{bmatrix}$

11. [10.2] Not defined   **12.** [10.2] $\begin{bmatrix} 2 & -2 & 6 \\ -4 & 10 & 4 \end{bmatrix}$

13. [10.3] $\begin{bmatrix} 0 & -1 \\ -\frac{1}{4} & -\frac{3}{4} \end{bmatrix}$ **14.** [10.2] **(a)** $\begin{bmatrix} 1.55 & 1.00 & 0.99 \\ 1.70 & 0.95 & 1.01 \\ 1.65 & 0.99 & 0.96 \end{bmatrix}$;

(b) $\begin{bmatrix} 26 & 18 & 23 \end{bmatrix}$; **(c)** $\begin{bmatrix} 108.85 & 65.87 & 66.00 \end{bmatrix}$; **(d)** the total
cost, in dollars, for each type of menu item served on the given day

15. [10.2] $\begin{bmatrix} 3 & -4 & 2 \\ 2 & 3 & 1 \\ 1 & -5 & -3 \end{bmatrix}\begin{bmatrix} x \\ y \\ z \end{bmatrix} = \begin{bmatrix} -8 \\ 7 \\ 3 \end{bmatrix}$

16. [10.3] $(-2, 1, 1)$ **17.** [10.4] 61   **18.** [10.4] -33
19. [10.4] $\left(-\frac{1}{2}, \frac{3}{4}\right)$ **20.** [10.1] $A = 1, B = -3, C = 2$

Chapter 11

Exercise Set 11.1, p. 728

1. (f)   **3.** (b)   **5.** (d)
7. $V: (0, 0); F: (0, 5);$
$D: y = -5$

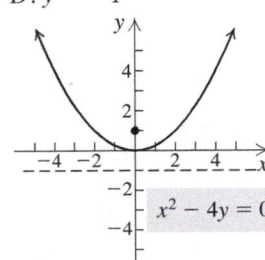
$x^2 = 20y$

9. $V: (0, 0); F: \left(-\frac{3}{2}, 0\right);$
$D: x = \frac{3}{2}$

$y^2 = -6x$

11. $V: (0, 0); F: (0, 1);$
$D: y = -1$

$x^2 - 4y = 0$

13. $V: (0, 0); F: \left(\frac{1}{8}, 0\right);$
$D: x = -\frac{1}{8}$

$x = 2y^2$

15. $y^2 = -12x$ **17.** $y^2 = 28x$ **19.** $x^2 = -4\pi y$
21. $(y - 2)^2 = 14\left(x + \frac{1}{2}\right)$

23. $V: (-2, 1); F: \left(-2, -\frac{1}{2}\right);$
$D: y = \frac{5}{2}$

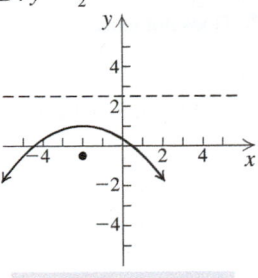
$(x + 2)^2 = -6(y - 1)$

25. $V: (-1, -3);$
$F: \left(-1, -\frac{7}{2}\right); D: y = -\frac{5}{2}$

$x^2 + 2x + 2y + 7 = 0$

27. $V: (0, -2); F: \left(0, -1\frac{3}{4}\right);$
$D: y = -2\frac{1}{4}$

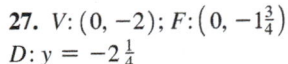
$x^2 - y - 2 = 0$

29. $V: (-2, -1);$
$F: \left(-2, -\frac{3}{4}\right); D: y = -1\frac{1}{4}$

$y = x^2 + 4x + 3$

31. $V: \left(5\frac{3}{4}, \frac{1}{2}\right); F: \left(6, \frac{1}{2}\right); D: x = 5\frac{1}{2}$

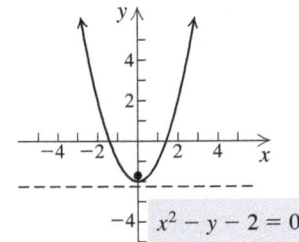
$y^2 - y - x + 6 = 0$

33. **(a)** $y^2 = 16x$; **(b)** $3\frac{33}{64}$ ft
35. $\frac{2}{3}$ ft, or 8 in.   **37.** (h)
38. (d)   **39.** (a), (b), (f), (g)
40. (b)   **41.** (b)   **42.** (f)
43. (a) and (g)   **44.** (a) and (h);
(g) and (h); (b) and (c)

45. $(x + 1)^2 = -4(y - 2)$
47. 10 ft, 11.6 ft, 16.4 ft, 24.4 ft, 35.6 ft, 50 ft

Exercise Set 11.2, p. 736

1. (b)   **3.** (d)   **5.** (a)
7. $(7, -2); 8$

$x^2 + y^2 - 14x + 4y = 11$

9. $(-3, 1); 4$

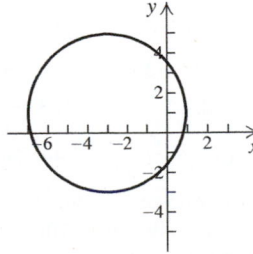
$x^2 + y^2 + 6x - 2y = 6$

11. $(-2, 3)$; 5

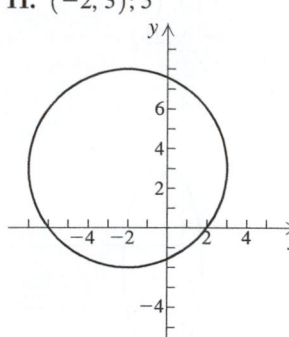

$$x^2 + y^2 + 4x - 6y - 12 = 0$$

13. $(3, 4)$; 3

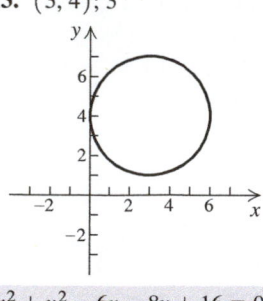

$$x^2 + y^2 - 6x - 8y + 16 = 0$$

15. $(-3, 5)$; $\sqrt{34}$

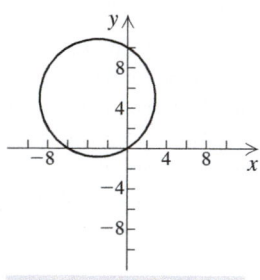

$$x^2 + y^2 + 6x - 10y = 0$$

17. $\left(\frac{9}{2}, -2\right)$; $\frac{5\sqrt{5}}{2}$

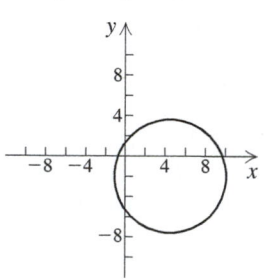

$$x^2 + y^2 - 9x = 7 - 4y$$

19. (c)   **21.** (d)

23. V: $(2, 0)$, $(-2, 0)$; F: $(\sqrt{3}, 0)$, $(-\sqrt{3}, 0)$

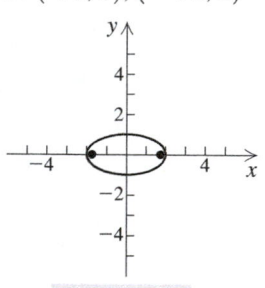

$$\frac{x^2}{4} + \frac{y^2}{1} = 1$$

25. V: $(0, 4)$, $(0, -4)$; F: $(0, \sqrt{7})$, $(0, -\sqrt{7})$

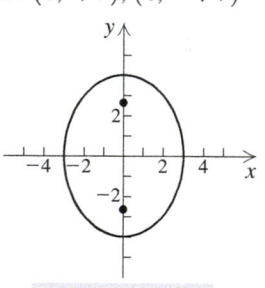

$$16x^2 + 9y^2 = 144$$

27. V: $(-\sqrt{3}, 0)$, $(\sqrt{3}, 0)$; F: $(-1, 0)$, $(1, 0)$

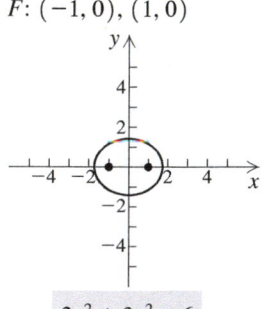

$$2x^2 + 3y^2 = 6$$

29. V: $\left(-\frac{1}{2}, 0\right)$, $\left(\frac{1}{2}, 0\right)$;

F: $\left(-\frac{\sqrt{5}}{6}, 0\right)$, $\left(\frac{\sqrt{5}}{6}, 0\right)$

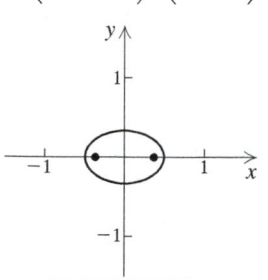

$$4x^2 + 9y^2 = 1$$

31. $\frac{x^2}{49} + \frac{y^2}{40} = 1$ **33.** $\frac{x^2}{25} + \frac{y^2}{64} = 1$ **35.** $\frac{x^2}{9} + \frac{y^2}{5} = 1$

37. C: $(1, 2)$; V: $(4, 2)$, $(-2, 2)$; F: $\left(1 + \sqrt{5}, 2\right)$, $\left(1 - \sqrt{5}, 2\right)$

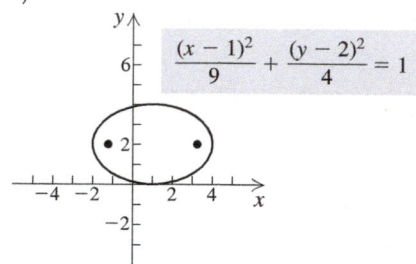

$$\frac{(x - 1)^2}{9} + \frac{(y - 2)^2}{4} = 1$$

39. C: $(-3, 5)$; V: $(-3, 11)$, $(-3, -1)$; F: $\left(-3, 5 + \sqrt{11}\right)$, $\left(-3, 5 - \sqrt{11}\right)$

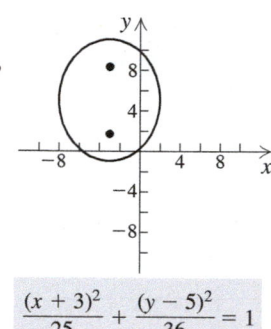

$$\frac{(x + 3)^2}{25} + \frac{(y - 5)^2}{36} = 1$$

41. C: $(-2, 1)$; V: $(-10, 1)$, $(6, 1)$; F: $(-6, 1)$, $(2, 1)$

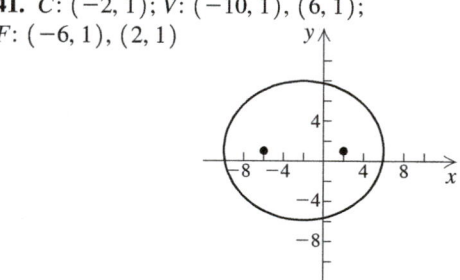

$$3(x + 2)^2 + 4(y - 1)^2 = 192$$

43. C: $(2, -1)$; V: $(-1, -1)$, $(5, -1)$; F: $\left(2 + \sqrt{5}, -1\right)$, $\left(2 - \sqrt{5}, -1\right)$

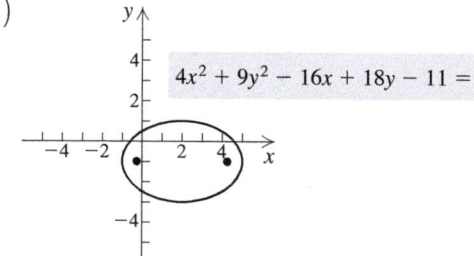

$$4x^2 + 9y^2 - 16x + 18y - 11 = 0$$

45. C: $(1, 1)$; V: $(1, 3)$, $(1, -1)$; F: $\left(1, 1 + \sqrt{3}\right)$, $\left(1, 1 - \sqrt{3}\right)$

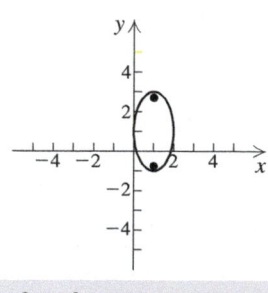

$$4x^2 + y^2 - 8x - 2y + 1 = 0$$

47. Example 2; $\dfrac{3}{5} < \dfrac{\sqrt{12}}{4}$ **49.** $\dfrac{x^2}{15} + \dfrac{y^2}{16} = 1$

51. $\dfrac{x^2}{50^2} + \dfrac{y^2}{12^2} = 1$ **53.** 33.5 ft   **55.** 2×10^6 mi

57. Zero   **58.** *y*-intercept   **59.** Two different real-number solutions   **60.** Remainder   **61.** Ellipse   **62.** Parabola

63. Circle   **65.** $\dfrac{(x-3)^2}{4} + \dfrac{(y-1)^2}{25} = 1$

67. $\dfrac{x^2}{9} + \dfrac{y^2}{484/5} = 1$

Mid-Chapter Review: Chapter 11, p. 739

1. True   **2.** False   **3.** False   **4.** True   **5.** (c)   **6.** (h)
7. (d)   **8.** (a)   **9.** (b)   **10.** (f)   **11.** (g)   **12.** (e)
13. V: $(0,0)$; F: $(3,0)$;   **14.** V: $(3,2)$; F: $(3,3)$;
D: $x = -3$ D: $y = 1$

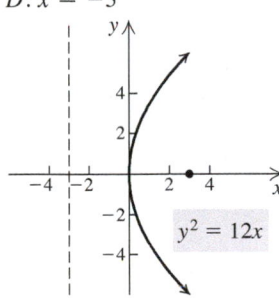

$y^2 = 12x$

$x^2 - 6x - 4y = -17$

15. $x^2 = 4(y-2)$ **16.** $(y-6)^2 = -12(x+1)$
17. $(-2,4)$; 5   **18.** $(3,-1)$; 4

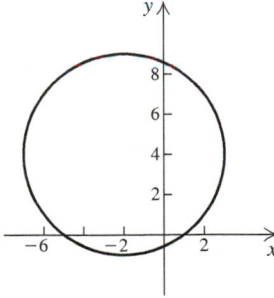

$x^2 + y^2 + 4x - 8y = 5$

$x^2 + y^2 - 6x + 2y - 6 = 0$

19. V: $(0,-3)$, $(0,3)$;   **20.** V: $\left(-\sqrt{6},0\right)$, $\left(\sqrt{6},0\right)$;
F: $\left(0,-2\sqrt{2}\right)$, $\left(0,2\sqrt{2}\right)$ F: $\left(-\sqrt{2},0\right)$, $\left(\sqrt{2},0\right)$

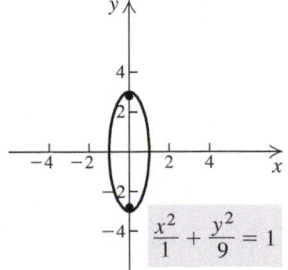

$\dfrac{x^2}{1} + \dfrac{y^2}{9} = 1$

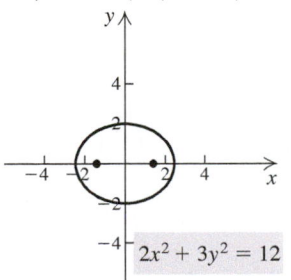

$2x^2 + 3y^2 = 12$

21. V: $(-2,-1)$, $(6,-1)$;
F: $\left(2 - 2\sqrt{3}, -1\right)$,
$\left(2 + 2\sqrt{3}, -1\right)$

22. V: $(1,-6)$, $(1,4)$;
F: $\left(1, -1 - \sqrt{21}\right)$,
$\left(1, -1 + \sqrt{21}\right)$

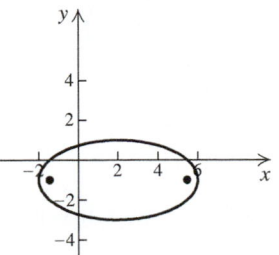

$\dfrac{(x-2)^2}{16} + \dfrac{(y+1)^2}{4} = 1$

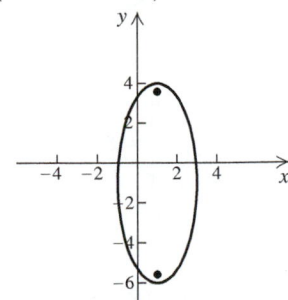

$25x^2 + 4y^2 - 50x + 8y = 71$

23. $\dfrac{x^2}{25} + \dfrac{y^2}{21} = 1$ **24.** $\dfrac{x^2}{4} + \dfrac{y^2}{9} = 1$ **25.** $\dfrac{x^2}{16} + \dfrac{y^2}{7} = 1$

26. No; parabolas with a horizontal axis of symmetry fail the vertical-line test.   **27.** See the development of the formula for the standard form of a parabola that follows Figure 1 at the beginning of Section 11.1.   **28.** Circles and ellipses are not functions.   **29.** No; the center of an ellipse is not part of the graph of the ellipse. Its coordinates do not satisfy the equation of the ellipse.

Exercise Set 11.3, p. 747

1. (b)   **3.** (c)   **5.** (a)   **7.** $\dfrac{y^2}{9} - \dfrac{x^2}{16} = 1$

9. $\dfrac{x^2}{4} - \dfrac{y^2}{9} = 1$

11. C: $(0,0)$; V: $(2,0)$, $(-2,0)$;
F: $(2\sqrt{2},0)$, $(-2\sqrt{2},0)$;
A: $y = x$, $y = -x$

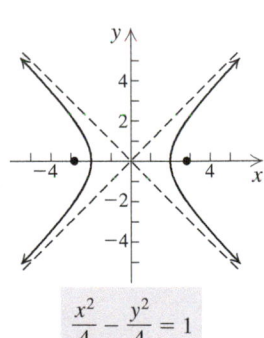

$\dfrac{x^2}{4} - \dfrac{y^2}{4} = 1$

13. C: $(2,-5)$; V: $(-1,-5)$, $(5,-5)$;
F: $\left(2 - \sqrt{10}, -5\right)$, $\left(2 + \sqrt{10}, -5\right)$;
A: $y = -\dfrac{x}{3} - \dfrac{13}{3}$, $y = \dfrac{x}{3} - \dfrac{17}{3}$

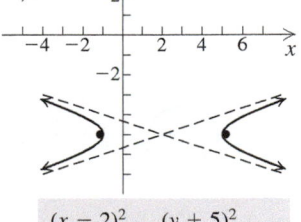

$\dfrac{(x-2)^2}{9} - \dfrac{(y+5)^2}{1} = 1$

15. $C: (-1, -3)$; $V: (-1, -1), (-1, -5)$;
$F: \left(-1, -3 + 2\sqrt{5}\right), \left(-1, -3 - 2\sqrt{5}\right)$;
$A: y = \frac{1}{2}x - \frac{5}{2}, y = -\frac{1}{2}x - \frac{7}{2}$

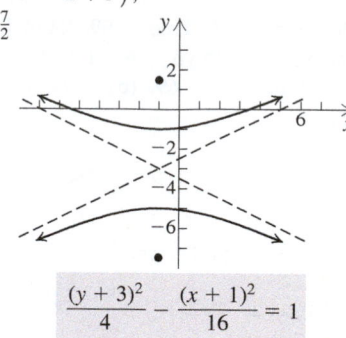

$$\frac{(y+3)^2}{4} - \frac{(x+1)^2}{16} = 1$$

17. $C: (0, 0)$; $V: (-2, 0), (2, 0)$;
$F: \left(-\sqrt{5}, 0\right), \left(\sqrt{5}, 0\right)$;
$A: y = -\frac{1}{2}x, y = \frac{1}{2}x$

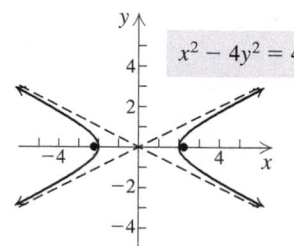

$x^2 - 4y^2 = 4$

19. $C: (0, 0)$; $V: (0, -3), (0, 3)$;
$F: \left(0, -3\sqrt{10}\right), \left(0, 3\sqrt{10}\right)$;
$A: y = \frac{1}{3}x, y = -\frac{1}{3}x$

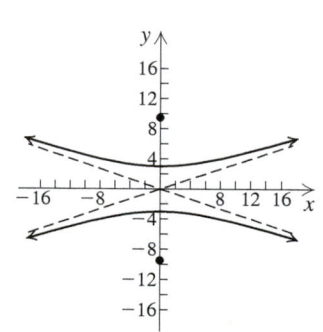

$9y^2 - x^2 = 81$

21. $C: (0, 0)$; $V: \left(-\sqrt{2}, 0\right), \left(\sqrt{2}, 0\right)$; $F: (-2, 0), (2, 0)$;
$A: y = x, y = -x$

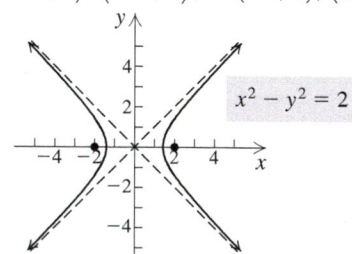

$x^2 - y^2 = 2$

23. $C: (0, 0)$; $V: \left(0, -\frac{1}{2}\right), \left(0, \frac{1}{2}\right)$; $F: \left(0, -\frac{\sqrt{2}}{2}\right), \left(0, \frac{\sqrt{2}}{2}\right)$;
$A: y = x, y = -x$

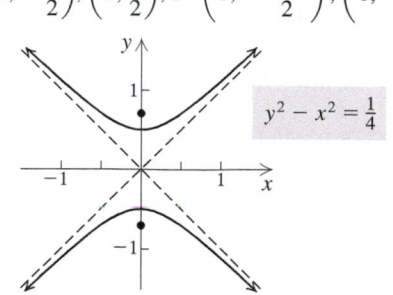

$y^2 - x^2 = \frac{1}{4}$

25. $C: (1, -2)$; $V: (0, -2), (2, -2)$; $F: \left(1 - \sqrt{2}, -2\right)$,
$\left(1 + \sqrt{2}, -2\right)$; $A: y = -x - 1, y = x - 3$

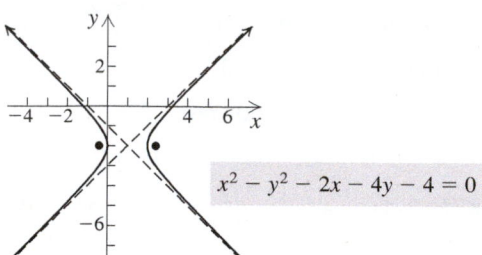

$x^2 - y^2 - 2x - 4y - 4 = 0$

27. $C: \left(\frac{1}{3}, 3\right)$; $V: \left(-\frac{2}{3}, 3\right), \left(\frac{4}{3}, 3\right)$; $F: \left(\frac{1}{3} - \sqrt{37}, 3\right)$,
$\left(\frac{1}{3} + \sqrt{37}, 3\right)$; $A: y = 6x + 1, y = -6x + 5$

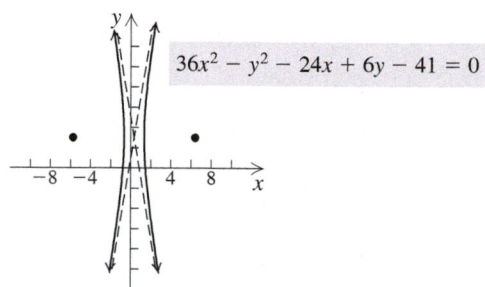

$36x^2 - y^2 - 24x + 6y - 41 = 0$

29. $C: (3, 1)$; $V: (3, 3), (3, -1)$;
$F: \left(3, 1 + \sqrt{13}\right), \left(3, 1 - \sqrt{13}\right)$;
$A: y = \frac{2}{3}x - 1, y = -\frac{2}{3}x + 3$

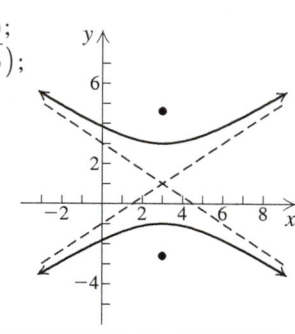

$9y^2 - 4x^2 - 18y + 24x - 63 = 0$

31. $C: (1, -2)$; $V: (2, -2), (0, -2)$;
$F: \left(1 + \sqrt{2}, -2\right), \left(1 - \sqrt{2}, -2\right)$;
$A: y = x - 3, y = -x - 1$

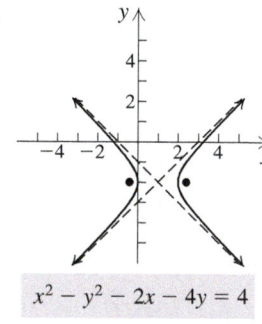

$x^2 - y^2 - 2x - 4y = 4$

33. $C: (-3, 4)$; $V: (-3, 10), (-3, -2)$;
$F: \left(-3, 4 + 6\sqrt{2}\right), \left(-3, 4 - 6\sqrt{2}\right)$;
$A: y = x + 7, y = -x + 1$

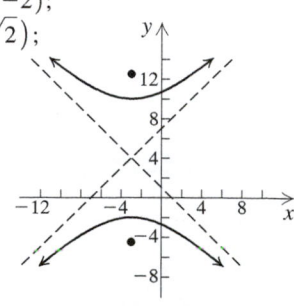

$y^2 - x^2 - 6x - 8y - 29 = 0$

35. Example 3; $\dfrac{\sqrt{5}}{1} > \dfrac{5}{4}$ **37.** $\dfrac{x^2}{9} - \dfrac{(y-7)^2}{16} = 1$

39. $\dfrac{y^2}{25} - \dfrac{x^2}{11} = 1$ **41. (a)** Yes; **(b)** $f^{-1}(x) = \dfrac{x+3}{2}$

42. (a) Yes; **(b)** $f^{-1}(x) = \sqrt[3]{x-2}$ **43. (a)** Yes;

(b) $f^{-1}(x) = \dfrac{5}{x} + 1$, or $\dfrac{5+x}{x}$ **44. (a)** Yes;

(b) $f^{-1}(x) = x^2 - 4, x \geq 0$ **45.** $(6, -1)$ **46.** $(1, -1)$

47. $(2, -1)$ **48.** $(-3, 4)$ **49.** $\dfrac{(y+5)^2}{9} - (x-3)^2 = 1$

51. $\dfrac{x^2}{345.96} - \dfrac{y^2}{22{,}154.04} = 1$

Visualizing the Graph, p. 757

1. B   **2.** J   **3.** F   **4.** I   **5.** H   **6.** G   **7.** E
8. D   **9.** C   **10.** A

Exercise Set 11.4, p. 758

1. (e)   **3.** (c)   **5.** (b)   **7.** $(-4, -3), (3, 4)$
9. $(0, 2), (3, 0)$ **11.** $(-5, 0), (4, 3), (4, -3)$
13. $(3, 0), (-3, 0)$ **15.** $(0, -3), (4, 5)$
17. $(-2, 1)$ **19.** $(3, 4), (-3, -4), (4, 3), (-4, -3)$

21. $\left(\dfrac{6\sqrt{21}}{7}, \dfrac{4i\sqrt{35}}{7}\right), \left(\dfrac{6\sqrt{21}}{7}, -\dfrac{4i\sqrt{35}}{7}\right),$
$\left(-\dfrac{6\sqrt{21}}{7}, \dfrac{4i\sqrt{35}}{7}\right), \left(-\dfrac{6\sqrt{21}}{7}, -\dfrac{4i\sqrt{35}}{7}\right)$

23. $(3, 2), \left(4, \dfrac{3}{2}\right)$

25. $\left(\dfrac{5+\sqrt{70}}{3}, \dfrac{-1+\sqrt{70}}{3}\right), \left(\dfrac{5-\sqrt{70}}{3}, \dfrac{-1-\sqrt{70}}{3}\right)$

27. $(\sqrt{2}, \sqrt{14}), (-\sqrt{2}, \sqrt{14}), (\sqrt{2}, -\sqrt{14}),$
$(-\sqrt{2}, -\sqrt{14})$ **29.** $(1, 2), (-1, -2), (2, 1), (-2, -1)$

31. $\left(\dfrac{15+\sqrt{561}}{8}, \dfrac{11-3\sqrt{561}}{8}\right), \left(\dfrac{15-\sqrt{561}}{8}, \dfrac{11+3\sqrt{561}}{8}\right)$

33. $\left(\dfrac{7-\sqrt{33}}{2}, \dfrac{7+\sqrt{33}}{2}\right), \left(\dfrac{7+\sqrt{33}}{2}, \dfrac{7-\sqrt{33}}{2}\right)$

35. $(3, 2), (-3, -2), (2, 3), (-2, -3)$

37. $\left(\dfrac{5-9\sqrt{15}}{20}, \dfrac{-45+3\sqrt{15}}{20}\right), \left(\dfrac{5+9\sqrt{15}}{20}, \dfrac{-45-3\sqrt{15}}{20}\right)$

39. $(3, -5), (-1, 3)$ **41.** $(8, 5), (-5, -8)$
43. $(3, 2), (-3, -2)$ **45.** $(2, 1), (-2, -1), (1, 2), (-1, -2)$

47. $\left(4 + \dfrac{3i\sqrt{6}}{2}, -4 + \dfrac{3i\sqrt{6}}{2}\right), \left(4 - \dfrac{3i\sqrt{6}}{2}, -4 - \dfrac{3i\sqrt{6}}{2}\right)$

49. $(3, \sqrt{5}), (-3, -\sqrt{5}), (\sqrt{5}, 3), (-\sqrt{5}, -3)$

51. $\left(\dfrac{8\sqrt{5}}{5}i, \dfrac{3\sqrt{105}}{5}\right), \left(\dfrac{8\sqrt{5}}{5}i, -\dfrac{3\sqrt{105}}{5}\right),$
$\left(-\dfrac{8\sqrt{5}}{5}i, \dfrac{3\sqrt{105}}{5}\right), \left(-\dfrac{8\sqrt{5}}{5}i, -\dfrac{3\sqrt{105}}{5}\right)$

53. $(2, 1), (-2, -1), \left(-i\sqrt{5}, \dfrac{2i\sqrt{5}}{5}\right), \left(i\sqrt{5}, -\dfrac{2i\sqrt{5}}{5}\right)$

55. True   **57.** True   **59.** 24 in. by 10 in.   **61.** 4 in. by 5 in.
63. 30 yd by 75 yd   **65.** Length: $\sqrt{3}$ m; width: 1 m
67. 16 ft, 24 ft   **69.** (b)   **71.** (d)   **73.** (a)

75.

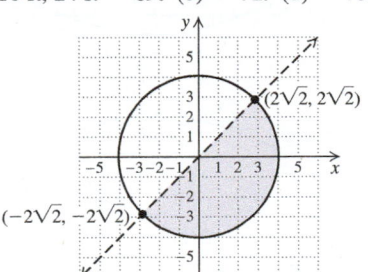

77.

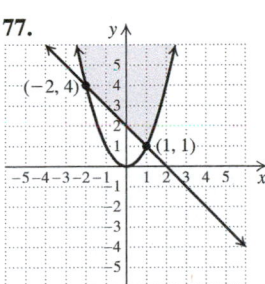

79.

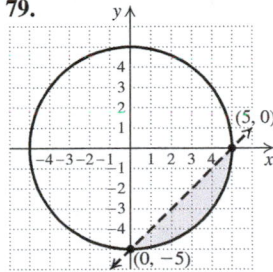

81.

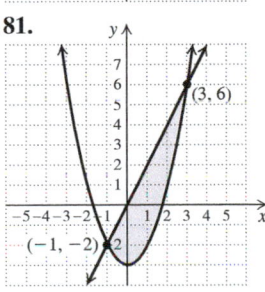

83.

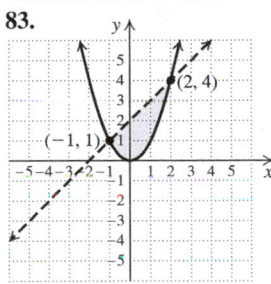

85. 2   **86.** 2.048   **87.** 81   **88.** 5

89. $(x - 2)^2 + (y - 3)^2 = 1$ **91.** $\dfrac{x^2}{4} + y^2 = 1$

93. There is no number x such that $\dfrac{x^2}{a^2} - \dfrac{\left(\dfrac{b}{a}x\right)^2}{b^2} = 1$, because

the left side simplifies to $\dfrac{x^2}{a^2} - \dfrac{x^2}{a^2}$, which is 0.

95. Factor: $x^3 + y^3 = (x + y)(x^2 - xy + y^2)$. We know
that $x + y = 1$, so $(x + y)^2 = x^2 + 2xy + y^2 = 1$,
or $x^2 + y^2 = 1 - 2xy$. We also know that
$xy = 1$, so $x^2 + y^2 = 1 - 2 \cdot 1 = -1$. Then
$x^3 + y^3 = 1 \cdot (-1 - 1) = -2$.   **97.** $(2, 4), (4, 2)$
99. $(3, -2), (-3, 2), (2, -3), (-2, 3)$

Summary and Review: Chapter 11

Review Exercises, p. 765

1. True   **2.** False   **3.** True   **4.** False   **5.** False
6. (d)   **7.** (a)   **8.** (e)   **9.** (g)   **10.** (b)   **11.** (f)
12. (h)   **13.** (c)   **14.** $x^2 = -6y$
15. F: $(-3, 0)$; V: $(0, 0)$; D: $x = 3$
16. V: $(-5, 8)$; F: $\left(-5, \dfrac{15}{2}\right)$; D: $y = \dfrac{17}{2}$

17. C: $(2, -1)$; V: $(-3, -1)$, $(7, -1)$; F: $(-1, -1)$, $(5, -1)$

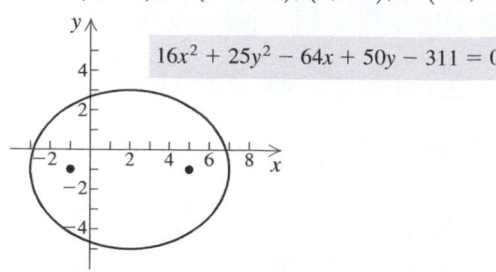
$$16x^2 + 25y^2 - 64x + 50y - 311 = 0$$

18. $\dfrac{x^2}{9} + \dfrac{y^2}{16} = 1$

19. C: $\left(-2, \dfrac{1}{4}\right)$; V: $\left(0, \dfrac{1}{4}\right)$, $\left(-4, \dfrac{1}{4}\right)$;

F: $\left(-2 + \sqrt{6}, \dfrac{1}{4}\right)$, $\left(-2 - \sqrt{6}, \dfrac{1}{4}\right)$;

A: $y - \dfrac{1}{4} = \dfrac{\sqrt{2}}{2}(x + 2)$, $y - \dfrac{1}{4} = -\dfrac{\sqrt{2}}{2}(x + 2)$

20. 0.167 ft   **21.** $\left(-8\sqrt{2}, 8\right)$, $\left(8\sqrt{2}, 8\right)$

22. $\left(3, \dfrac{\sqrt{29}}{2}\right)$, $\left(-3, \dfrac{\sqrt{29}}{2}\right)$, $\left(3, -\dfrac{\sqrt{29}}{2}\right)$, $\left(-3, -\dfrac{\sqrt{29}}{2}\right)$

23. $(7, 4)$ **24.** $(2, 2)$, $\left(\dfrac{32}{9}, -\dfrac{10}{9}\right)$ **25.** $(0, -3)$, $(2, 1)$

26. $(4, 3)$, $(4, -3)$, $(-4, 3)$, $(-4, -3)$

27. $\left(-\sqrt{3}, 0\right)$, $\left(\sqrt{3}, 0\right)$, $(-2, 1)$, $(2, 1)$

28. $\left(-\dfrac{3}{5}, \dfrac{21}{5}\right)$, $(3, -3)$ **29.** $(6, 8)$, $(6, -8)$, $(-6, 8)$, $(-6, -8)$

30. $(2, 2)$, $(-2, -2)$, $(2\sqrt{2}, \sqrt{2})$, $(-2\sqrt{2}, -\sqrt{2})$

31. 7, 4   **32.** 7 m by 12 m   **33.** 4, 8   **34.** 32 cm, 20 cm

35. 11 ft, 3 ft

36.
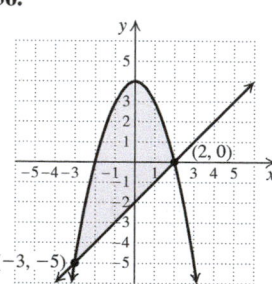

37.
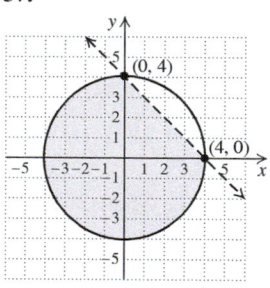

38.
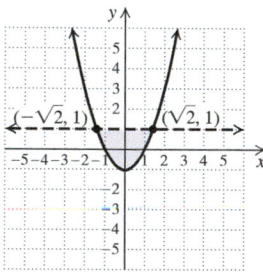

39.
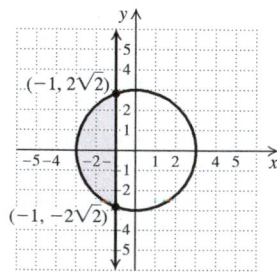

40. B   **41.** D   **42.** C   **43.** $\dfrac{8}{7}, \dfrac{7}{2}$

44. $(x - 2)^2 + (y - 1)^2 = 100$

45. $x^2 + \dfrac{y^2}{9} = 1$ **46.** $\dfrac{x^2}{778.41} - \dfrac{y^2}{39{,}221.59} = 1$

47. The equation of a circle can be written as

$$\frac{(x - h)^2}{a^2} + \frac{(y - k)^2}{b^2} = 1,$$

where $a = b = r$, the radius of the circle. In an ellipse, $a > b$, so a circle is not a special type of ellipse.   **48.** No; the asymptotes of a hyperbola are not part of the graph of the hyperbola. The coordinates of points on the asymptotes do not satisfy the equation of the hyperbola.   **49.** Although we can always visualize the real-number solutions, we cannot visualize the imaginary-number solutions.

Test: Chapter 11, p. 768

1. [11.3] (c)   **2.** [11.1] (b)   **3.** [11.2] (a)   **4.** [11.2] (d)

5. [11.1] V: $(0, 0)$;   **6.** [11.1] V: $(-1, -1)$;
F: $(0, 3)$; D: $y = -3$ F: $(1, -1)$; D: $x = -3$

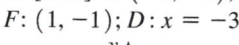
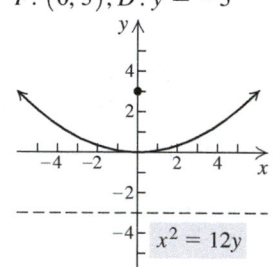

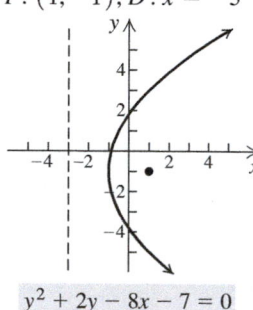

7. [11.1] $x^2 = 8y$

8. [11.2] Center: $(-1, 3)$;   **9.** [11.2] C: $(0, 0)$;
radius: 5 V: $(-4, 0)$, $(4, 0)$;
F: $\left(-\sqrt{7}, 0\right)$, $\left(\sqrt{7}, 0\right)$

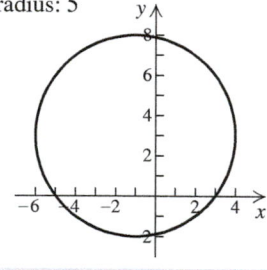

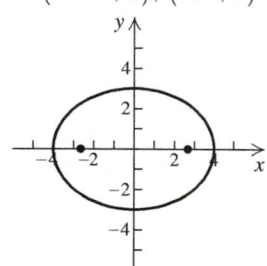

10. [11.2] C: $(-1, 2)$; V: $(-1, -1)$, $(-1, 5)$;
F: $\left(-1, 2 - \sqrt{5}\right)$, $\left(-1, 2 + \sqrt{5}\right)$

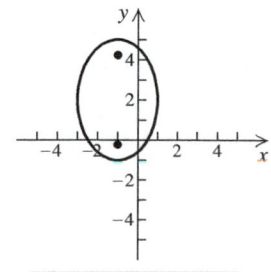

11. [11.2] $\dfrac{x^2}{4} + \dfrac{y^2}{25} = 1$

12. [11.3] C: $(0, 0)$; V: $(-1, 0)$, $(1, 0)$; F: $(-\sqrt{5}, 0)$, $(\sqrt{5}, 0)$; A: $y = -2x, y = 2x$

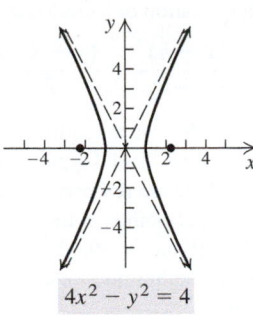

$4x^2 - y^2 = 4$

13. [11.3] C: $(-1, 2)$; V: $(-1, 0)$, $(-1, 4)$; F: $\left(-1, 2 - \sqrt{13}\right)$, $\left(-1, 2 + \sqrt{13}\right)$; A: $y = -\frac{2}{3}x + \frac{4}{3}, y = \frac{2}{3}x + \frac{8}{3}$

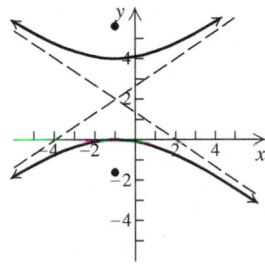

$\dfrac{(y - 2)^2}{4} - \dfrac{(x + 1)^2}{9} = 1$

14. [11.3] $y = \dfrac{\sqrt{2}}{2}x, y = -\dfrac{\sqrt{2}}{2}x$ **15.** [11.1] $\frac{27}{8}$ in.

16. [11.4] $(1, 2)$, $(1, -2)$, $(-1, 2)$, $(-1, -2)$

17. [11.4] $(3, -2)$, $(-2, 3)$ **18.** [11.4] $(2, 3)$, $(3, 2)$

19. [11.4] 5 ft by 4 ft **20.** [11.4] 60 ft by 45 ft

21. [11.4]

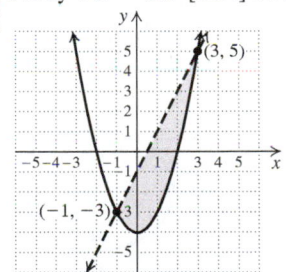

22. [11.1] A **23.** [11.2] $(x - 3)^2 + (y + 1)^2 = 8$

Chapter 12

Exercise Set 12.1, p. 775

1. 3, 7, 11, 15; 39; 59 **3.** 2, $\frac{3}{2}, \frac{4}{3}, \frac{5}{4}; \frac{10}{9}; \frac{15}{14}$ **5.** 0, $\frac{3}{5}, \frac{4}{5}; \frac{15}{17}; \frac{99}{101}, \frac{112}{113}$

7. $-1, 4, -9, 16; 100; -225$ **9.** 7, 3, 7, 3; 3; 7 **11.** 34

13. 225 **15.** $-33,880$ **17.** 67 **19.** $2n$

21. $(-1)^n \cdot 2 \cdot 3^{n-1}$ **23.** $\dfrac{n + 1}{n + 2}$ **25.** $n(n + 1)$

27. $\log 10^{n-1}$, or $n - 1$ **29.** 6; 28 **31.** 20; 30

33. $\frac{1}{2} + \frac{1}{4} + \frac{1}{6} + \frac{1}{8} + \frac{1}{10} = \frac{137}{120}$ **35.** $1 + 2 + 4 + 8 + 16 + 32 + 64 = 127$ **37.** $\ln 7 + \ln 8 + \ln 9 + \ln 10 =$ $\ln(7 \cdot 8 \cdot 9 \cdot 10) = \ln 5040 \approx 8.5252$ **39.** $\frac{1}{2} + \frac{2}{3} + \frac{3}{4} + \frac{4}{5} + \frac{5}{6} + \frac{6}{7} + \frac{7}{8} + \frac{8}{9} = \frac{15,551}{2520}$ **41.** $-1 + 1 - 1 + 1 - 1 = -1$

43. $3 - 6 + 9 - 12 + 15 - 18 + 21 - 24 = -12$

45. $2 + 1 + \frac{2}{5} + \frac{1}{5} + \frac{2}{17} + \frac{1}{13} + \frac{2}{37} = \frac{157,351}{40,885}$

47. $3 + 2 + 3 + 6 + 11 + 18 = 43$ **49.** $\frac{1}{2} + \frac{2}{3} + \frac{4}{5} + \frac{8}{9} + \frac{16}{17} + \frac{32}{33} + \frac{64}{65} + \frac{128}{129} + \frac{256}{257} + \frac{512}{513} + \frac{1024}{1025} \approx 9.736$

51. $\displaystyle\sum_{k=1}^{\infty} 5k$ **53.** $\displaystyle\sum_{k=1}^{6} (-1)^{k+1}2^k$ **55.** $\displaystyle\sum_{k=1}^{6} (-1)^k \frac{k}{k + 1}$

57. $\displaystyle\sum_{k=2}^{n} (-1)^k k^2$ **59.** $\displaystyle\sum_{k=1}^{\infty} \frac{1}{k(k + 1)}$ **61.** 4, $1\frac{1}{4}, 1\frac{4}{5}, 1\frac{5}{9}$

63. 6561, $-81, 9i, -3\sqrt{i}$ **65.** 2, 3, 5, 8 **67.** (a) 1062, 1127.84, 1197.77, 1272.03, 1350.90, 1434.65, 1523.60, 1618.07, 1718.39, 1824.93; (b) $3330.35 **69.** $9.80, $10.90, $12.00, $13.10, $14.20, $15.30, $16.40, $17.50, $18.60, $19.70

71. 1, 1, 2, 3, 5, 8, 13 **72.** $(-1, -3)$

73. Illinois: 16,200 acres; Ohio: 7200 acres **74.** $(3, -2)$; 4

75. $\left(-\dfrac{5}{2}, 4\right)$; $\dfrac{\sqrt{97}}{2}$ **77.** $i, -1, -i, 1, i; i$

79. $\ln(1 \cdot 2 \cdot 3 \cdots n)$

Exercise Set 12.2, p. 782

1. $a_1 = 3, d = 5$ **3.** $a_1 = 9, d = -4$ **5.** $a_1 = \frac{3}{2}, d = \frac{3}{4}$

7. $a_1 = \$316, d = -\$3$ **9.** $a_{12} = 46$ **11.** $a_{14} = -\frac{17}{3}$

13. $a_{10} = \$7941.62$ **15.** 27th term **17.** 46th term

19. $a_1 = 5$ **21.** $n = 39$ **23.** $a_1 = \frac{1}{3}; d = \frac{1}{2}; \frac{1}{3}, \frac{5}{6}, \frac{4}{3}, \frac{11}{6}, \frac{7}{3}$

25. 670 **27.** 160,400 **29.** 735 **31.** 990 **33.** 1760

35. $\frac{65}{2}$ **37.** $-\frac{6026}{13}$ **39.** 4960¢, or $49.60 **41.** 1320 seats

43. Yes; 32; 1600 ft **45.** 3 plants; 171 plants **47.** Yes; 3

48. $(2, 5)$ **49.** $(2, -1, 3)$ **50.** $(-4, 0), (4, 0); (-\sqrt{7}, 0),$ $(\sqrt{7}, 0)$ **51.** $\dfrac{x^2}{4} + \dfrac{y^2}{25} = 1$ **53.** n^2

55. $a_1 = 60 - 5p - 5q; d = 5p + 2q - 20$ **57.** $5\frac{4}{5}, 7\frac{3}{5}, 9\frac{2}{5}, 11\frac{1}{5}$

Visualizing the Graph, p. 791

1. J **2.** A **3.** C **4.** G **5.** F **6.** H **7.** E **8.** D **9.** B **10.** I

Exercise Set 12.3, p. 792

1. 2 **3.** -1 **5.** -2 **7.** 0.1 **9.** $\dfrac{a}{2}$ **11.** 128

13. 162 **15.** $7(5)^{40}$ **17.** 3^{n-1} **19.** $(-1)^{n-1}$ **21.** $\dfrac{1}{x^n}$

23. 762 **25.** $\frac{4921}{18}$ **27.** True **29.** True **31.** True

33. 8 **35.** 125 **37.** Does not exist **39.** $\frac{2}{3}$

41. $S_{11} \approx 29.65317$ **43.** 2 **45.** Does not exist

47. $4545.\overline{45}$ **49.** $\frac{160}{9}$ **51.** $\frac{13}{99}$ **53.** 9 **55.** $\frac{34,091}{9990}$

57. $2,684,354.55 **59.** (a) About 297 ft; (b) 300 ft **61.** $39,505.71 **63.** 10,485.76 in.

65. $19,694.01 **67.** $86,666,666,667 **69.** $(f \circ g)(x) = 16x^2 + 40x + 25; (g \circ f)(x) = 4x^2 + 5$ **70.** $(f \circ g)(x) = x^2 + x + 2; (g \circ f)(x) = x^2 - x + 3$ **71.** 2.209

72. $\frac{1}{16}$ **73.** $(4 - \sqrt{6})/(\sqrt{3} - \sqrt{2}) = 2\sqrt{3} + \sqrt{2}$, $(6\sqrt{3} - 2\sqrt{2})/(4 - \sqrt{6}) = 2\sqrt{3} + \sqrt{2}$; there exists a common ratio, $2\sqrt{3} + \sqrt{2}$; thus the sequence is geometric.

75. (a) $\frac{13}{3}, \frac{22}{3}, \frac{34}{3}, \frac{46}{3}, \frac{58}{3}$; (b) $-\frac{11}{3}; -\frac{2}{3}, \frac{10}{3}, -\frac{50}{3}, \frac{250}{3}$ or 5; 8, 12, 18, 27

77. $S_n = \dfrac{x^2(1 - (-x)^n)}{x + 1}$

Exercise Set 12.4, p. 797

1. $1^2 < 1^3$, false; $2^2 < 2^3$, true; $3^2 < 3^3$, true; $4^2 < 4^3$, true;

$5^2 < 5^3$, true   **3.** A polygon of 3 sides has $\dfrac{3(3-3)}{2}$ diagonals.

True; A polygon of 4 sides has $\dfrac{4(4-3)}{2}$ diagonals. True;

A polygon of 5 sides has $\dfrac{5(5-3)}{2}$ diagonals. True; A polygon of

6 sides has $\dfrac{6(6-3)}{2}$ diagonals. True; A polygon of 7 sides has

$\dfrac{7(7-3)}{2}$ diagonals. True.

5. S_n: $2 + 4 + 6 + \cdots + 2n = n(n+1)$
S_1: $2 = 1(1+1)$
S_k: $2 + 4 + 6 + \cdots + 2k = k(k+1)$
S_{k+1}: $2 + 4 + 6 + \cdots + 2k + 2(k+1)$
      $= (k+1)(k+2)$
(1) *Basis step:* S_1 true by substitution.
(2) *Induction step:* Assume S_k. Deduce S_{k+1}.
   Starting with the left side of S_{k+1}, we have
   $2 + 4 + 6 + \cdots + 2k + 2(k+1)$
   $= k(k+1) + 2(k+1)$ **By S_k**
   $= (k+1)(k+2)$.   **Distributive law**

7. S_n: $1 + 5 + 9 + \cdots + (4n-3) = n(2n-1)$
S_1: $1 = 1(2 \cdot 1 - 1)$
S_k: $1 + 5 + 9 + \cdots + (4k-3) = k(2k-1)$
S_{k+1}: $1 + 5 + 9 + \cdots + (4k-3) + [4(k+1)-3]$
      $= (k+1)[2(k+1)-1]$
      $= (k+1)(2k+1)$
(1) *Basis step:* S_1 true by substitution.
(2) *Induction step:* Assume S_k. Deduce S_{k+1}.
   Starting with the left side of S_{k+1}, we have
   $1 + 5 + 9 + \cdots + (4k-3) + [4(k+1)-3]$
   $= k(2k-1) + [4(k+1)-3]$ **By S_k**
   $= 2k^2 - k + 4k + 4 - 3$
   $= 2k^2 + 3k + 1$
   $= (k+1)(2k+1)$.

9. S_n: $2 + 4 + 8 + \cdots + 2^n = 2(2^n - 1)$
S_1: $2 = 2(2-1)$
S_k: $2 + 4 + 8 + \cdots + 2^k = 2(2^k - 1)$
S_{k+1}: $2 + 4 + 8 + \cdots + 2^k + 2^{k+1} = 2(2^{k+1} - 1)$
(1) *Basis step:* S_1 true by substitution.
(2) *Induction step:* Assume S_k. Deduce S_{k+1}.
   Starting with the left side of S_{k+1}, we have
   $\underbrace{2 + 4 + 8 + \cdots + 2^k} + 2^{k+1}$
   $= \quad 2(2^k - 1) \quad + 2^{k+1}$ **By S_k**
   $= 2^{k+1} - 2 + 2^{k+1}$
   $= 2 \cdot 2^{k+1} - 2$
   $= 2(2^{k+1} - 1)$.

11. S_n: $n < n + 1$
S_1: $1 < 1 + 1$
S_k: $k < k + 1$
S_{k+1}: $k + 1 < (k+1) + 1$
(1) *Basis step:* Since $1 < 1 + 1$, S_1 is true.
(2) *Induction step:* Assume S_k. Deduce S_{k+1}. Now
   $k < k + 1$ **By S_k**
   $k + 1 < k + 1 + 1$ **Adding 1**
   $k + 1 < k + 2$.   **Simplifying**

13. S_n: $2n \leq 2^n$
S_1: $2 \cdot 1 \leq 2^1$
S_k: $2k \leq 2^k$
S_{k+1}: $2(k+1) \leq 2^{k+1}$
(1) *Basis step:* Since $2 = 2$, S_1 is true.
(2) *Induction step:* Let k be any natural number.
   Assume S_k. Deduce S_{k+1}.
   $2k \leq 2^k$ **By S_k**
   $2 \cdot 2k \leq 2 \cdot 2^k$ **Multiplying by 2**
   $4k \leq 2^{k+1}$
   Since $1 \leq k$, $k + 1 \leq k + k$, or $k + 1 \leq 2k$.
   Then $2(k+1) \leq 4k$.   **Multiplying by 2**
   Thus, $2(k+1) \leq 4k \leq 2^{k+1}$, so $2(k+1) \leq 2^{k+1}$.

15. S_n: $\dfrac{1}{1 \cdot 2 \cdot 3} + \dfrac{1}{2 \cdot 3 \cdot 4} + \dfrac{1}{3 \cdot 4 \cdot 5} + \cdots$

      $+ \dfrac{1}{n(n+1)(n+2)} = \dfrac{n(n+3)}{4(n+1)(n+2)}$

S_1: $\dfrac{1}{1 \cdot 2 \cdot 3} = \dfrac{1(1+3)}{4(1+1)(1+2)}$

S_k: $\dfrac{1}{1 \cdot 2 \cdot 3} + \dfrac{1}{2 \cdot 3 \cdot 4} + \cdots + \dfrac{1}{k(k+1)(k+2)}$

      $= \dfrac{k(k+3)}{4(k+1)(k+2)}$

S_{k+1}: $\dfrac{1}{1 \cdot 2 \cdot 3} + \dfrac{1}{2 \cdot 3 \cdot 4} + \cdots + \dfrac{1}{k(k+1)(k+2)}$

      $+ \dfrac{1}{(k+1)(k+2)(k+3)}$

      $= \dfrac{(k+1)(k+1+3)}{4(k+1+1)(k+1+2)} = \dfrac{(k+1)(k+4)}{4(k+2)(k+3)}$

(1) *Basis step:* Since $\dfrac{1}{1 \cdot 2 \cdot 3} = \dfrac{1}{6}$ and

   $\dfrac{1(1+3)}{4(1+1)(1+2)} = \dfrac{1 \cdot 4}{4 \cdot 2 \cdot 3} = \dfrac{1}{6}$, S_1 is true.

(2) *Induction step:* Assume S_k. Deduce S_{k+1}.

   Add $\dfrac{1}{(k+1)(k+2)(k+3)}$ on both sides of S_k and

   simplify the right side. Only the right side is shown here.

   $\dfrac{k(k+3)}{4(k+1)(k+2)} + \dfrac{1}{(k+1)(k+2)(k+3)}$

   $= \dfrac{k(k+3)(k+3) + 4}{4(k+1)(k+2)(k+3)}$

   $= \dfrac{k^3 + 6k^2 + 9k + 4}{4(k+1)(k+2)(k+3)}$

   $= \dfrac{(k+1)^2(k+4)}{4(k+1)(k+2)(k+3)}$

   $= \dfrac{(k+1)(k+4)}{4(k+2)(k+3)}$

17. S_n: $1 + 2 + 3 + \cdots + n = \dfrac{n(n+1)}{2}$

S_1: $1 = \dfrac{1(1+1)}{2}$

S_k: $1 + 2 + 3 + \cdots + k = \dfrac{k(k+1)}{2}$

S_{k+1}: $1 + 2 + 3 + \cdots + k + (k+1) = \dfrac{(k+1)(k+2)}{2}$

(1) *Basis step:* S_1 true by substitution.

(2) *Induction step:* Assume S_k. Deduce S_{k+1}. Starting with the left side of S_{k+1}, we have

$$\underbrace{1 + 2 + 3 + \cdots + k} + (k + 1)$$

$$= \frac{k(k + 1)}{2} + (k + 1) \qquad \textbf{By } S_k$$

$$= \frac{k(k + 1) + 2(k + 1)}{2} \qquad \textbf{Adding}$$

$$= \frac{(k + 1)(k + 2)}{2}. \qquad \textbf{Distributive law}$$

19. S_n: $\quad 1^3 + 2^3 + 3^3 + \cdots + n^3 = \dfrac{n^2(n + 1)^2}{4}$

S_1: $\quad 1^3 = \dfrac{1^2(1 + 1)^2}{4}$

S_k: $\quad 1^3 + 2^3 + 3^3 + \cdots + k^3 = \dfrac{k^2(k + 1)^2}{4}$

S_{k+1}: $\quad 1^3 + 2^3 + 3^3 + \cdots + k^3 + (k + 1)^3$

$$= \frac{(k + 1)^2[(k + 1) + 1]^2}{4}$$

(1) *Basis step:* S_1: $1^3 = \dfrac{1^2(1 + 1)^2}{4} = 1$. True.

(2) *Induction step:* Assume S_k. Deduce S_{k+1}.

$$1^3 + 2^3 + \cdots + k^3 = \frac{k^2(k + 1)^2}{4} \qquad \textbf{By } S_k$$

$$1^3 + 2^3 + \cdots + k^3 + (k + 1)^3 = \frac{k^2(k + 1)^2}{4} + (k + 1)^3$$

$$\textbf{Adding } (k + 1)^3$$

$$= \frac{k^2(k + 1)^2 + 4(k + 1)^3}{4}$$

$$= \frac{(k + 1)^2}{4}[k^2 + 4(k + 1)]$$

$$= \frac{(k + 1)^2}{4}(k^2 + 4k + 4)$$

$$= \frac{(k + 1)^2(k + 2)^2}{4}$$

21. S_n: $\quad 2 + 6 + 12 + \cdots + n(n + 1) = \dfrac{n(n + 1)(n + 2)}{3}$

S_1: $\quad 1(1 + 1) = \dfrac{1(1 + 1)(1 + 2)}{3}$

S_k: $\quad 2 + 6 + 12 + \cdots + k(k + 1) = \dfrac{k(k + 1)(k + 2)}{3}$

S_{k+1}:

$$2 + 6 + 12 + \cdots + k(k + 1) + (k + 1)[(k + 1) + 1]$$

$$= \frac{(k + 1)[(k + 1) + 1][(k + 1) + 2]}{3}$$

(1) *Basis step:* S_1: $1(1 + 1) = \dfrac{1(1 + 1)(1 + 2)}{3}$. True.

(2) *Induction step:* Assume S_k:

$$2 + 6 + 12 + \cdots + k(k + 1) = \frac{k(k + 1)(k + 2)}{3}.$$

Then $2 + 6 + 12 + \cdots + k(k + 1) + (k + 1)(k + 1 + 1)$

$$= \frac{k(k + 1)(k + 2)}{3} + (k + 1)(k + 2)$$

$$= \frac{k(k + 1)(k + 2) + 3(k + 1)(k + 2)}{3}$$

$$= \frac{(k + 1)(k + 2)(k + 3)}{3}$$

$$= \frac{(k + 1)(k + 1 + 1)(k + 1 + 2)}{3}.$$

23. S_n: $\quad a_1 + (a_1 + d) + (a_1 + 2d) + \cdots$
$$+ [a_1 + (n - 1)d] = \frac{n}{2}[2a_1 + (n - 1)d]$$

S_1: $\quad a_1 = \dfrac{1}{2}[2a_1 + (1 - 1)d]$

S_k: $\quad a_1 + (a_1 + d) + (a_1 + 2d) + \cdots$
$$+ [a_1 + (k - 1)d] = \frac{k}{2}[2a_1 + (k - 1)d]$$

S_{k+1}: $\quad a_1 + (a_1 + d) + (a_1 + 2d) + \cdots$
$$+ [a_1 + (k - 1)d] + [a_1 + ((k + 1) - 1)d]$$
$$= \frac{k + 1}{2}[2a_1 + ((k + 1) - 1)d]$$

(1) *Basis step:* Since $\frac{1}{2}[2a_1 + (1 - 1)d] = \frac{1}{2} \cdot 2a_1 = a_1$, S_1 is true.

(2) *Induction step:* Assume S_k. Deduce S_{k+1}. Starting with the left side of S_{k+1}, we have

$$\underbrace{a_1 + (a_1 + d) + \cdots + [a_1 + (k - 1)d]} + [a_1 + kd]$$

$$= \frac{k}{2}[2a_1 + (k - 1)d] \qquad + [a_1 + kd]$$

$$\textbf{By } S_k$$

$$= \frac{k[2a_1 + (k - 1)d]}{2} + \frac{2[a_1 + kd]}{2}$$

$$= \frac{2ka_1 + k(k - 1)d + 2a_1 + 2kd}{2}$$

$$= \frac{2a_1(k + 1) + k(k - 1)d + 2kd}{2}$$

$$= \frac{2a_1(k + 1) + (k - 1 + 2)kd}{2}$$

$$= \frac{2a_1(k + 1) + (k + 1)kd}{2}$$

$$= \frac{k + 1}{2}[2a_1 + kd].$$

24. $(5, 3)$ **25.** 1.5%: \$800; 2%: \$1600; 3%: \$2000

27. S_n: $\quad x + y$ is a factor of $x^{2n} - y^{2n}$.

S_1: $\quad x + y$ is a factor of $x^2 - y^2$.

S_k: $\quad x + y$ is a factor of $x^{2k} - y^{2k}$.

S_{k+1}: $\quad x + y$ is a factor of $x^{2(k+1)} - y^{2(k+1)}$.

(1) *Basis step:* S_1: $x + y$ is a factor of $x^2 - y^2$. True.
$\qquad\qquad\quad$ S_2: $x + y$ is a factor of $x^4 - y^4$. True.

(2) *Induction step:* Assume S_{k-1}: $x + y$ is a factor of $x^{2(k-1)} - y^{2(k-1)}$. Then $x^{2(k-1)} - y^{2(k-1)} = (x + y)Q(x)$ for some polynomial $Q(x)$. Assume S_k: $x + y$ is a factor of $x^{2k} - y^{2k}$. Then $x^{2k} - y^{2k} = (x + y)P(x)$ for some polynomial $P(x)$.

$x^{2(k+1)} - y^{2(k+1)}$
$$= (x^{2k} - y^{2k})(x^2 + y^2) - (x^{2(k-1)} - y^{2(k-1)})(x^2 y^2)$$
$$= (x + y)P(x)(x^2 + y^2) - (x + y)Q(x)(x^2 y^2)$$
$$= (x + y)[P(x)(x^2 + y^2) - Q(x)(x^2 y^2)]$$

so $x + y$ is a factor of $x^{2(k+1)} - y^{2(k+1)}$.

29. S_2: $\quad \log_a (b_1 b_2) = \log_a b_1 + \log_a b_2$

S_k: $\quad \log_a (b_1 b_2 \cdots b_k) = \log_a b_1 + \log_a b_2 + \cdots$
$\quad\quad + \log_a b_k$

S_{k+1}: $\quad \log_a (b_1 b_2 \cdots b_{k+1}) = \log_a b_1 + \log_a b_2 + \cdots$
$\quad\quad + \log_a b_{k+1}$

(1) *Basis step:* S_2 is true by the properties of logarithms.
(2) *Induction step:* Let k be a natural number $k \geq 2$.
Assume S_k. Deduce S_{k+1}.

$\log_a (b_1 b_2 \cdots b_{k+1})$ $\quad$ **Left side of S_{k+1}**
$\quad = \log_a (b_1 b_2 \cdots b_k) + \log_a b_{k+1}$ $\quad$ **By S_2**
$\quad = \log_a b_1 + \log_a b_2 + \cdots + \log_a b_k + \log_a b_{k+1}$

31. S_2: $\quad \overline{z_1 + z_2} = \bar{z}_1 + \bar{z}_2$:
$\overline{(a + bi) + (c + di)} = \overline{(a + c) + (b + d)i}$
$\quad\quad\quad\quad\quad\quad\quad = (a + c) - (b + d)i$
$\overline{(a + bi)} + \overline{(c + di)} = a - bi + c - di$
$\quad\quad\quad\quad\quad\quad\quad = (a + c) - (b + d)i.$

S_k: $\quad \overline{z_1 + z_2 + \cdots + z_k} = \bar{z}_1 + \bar{z}_2 + \cdots + \bar{z}_k.$
$\overline{(z_1 + z_2 + \cdots + z_k) + z_{k+1}}$
$\quad = \overline{(z_1 + z_2 + \cdots + z_k)} + \overline{z_{k+1}}$ $\quad$ **By S_2**
$\quad = \bar{z}_1 + \bar{z}_2 + \cdots + \bar{z}_k + \bar{z}_{k+1}$ $\quad$ **By S_k**

Mid-Chapter Review: Chapter 12, p. 798

1. False **2.** True **3.** False **4.** False **5.** 8, 11, 14, 17;
32; 47 **6.** 0, -1, 2, -3; 8; -13 **7.** $a_n = 3n$
8. $a_n = (-1)^n n^2$ **9.** $1\frac{7}{8}$, or $\frac{15}{8}$ **10.** $2 + 6 + 12 +$
$20 + 30 = 70$ **11.** $\sum_{k=1}^{\infty} (-1)^k 4k$ **12.** 2, 6, 22, 86
13. -5 **14.** 22 **15.** 21 **16.** 696 **17.** $-\frac{1}{2}$
18. (a) 8; **(b)** $\frac{1023}{16}$, or 63.9375 **19.** $-\frac{16}{3}$ **20.** Does not exist
21. 126 plants **22.** $6369.70
23. S_n: $\quad 1 + 4 + 7 + \cdots + (3n - 2) = \frac{1}{2}n(3n - 1)$
S_1: $\quad 3 \cdot 1 - 2 = \frac{1}{2} \cdot 1(3 \cdot 1 - 1)$
S_k: $\quad 1 + 4 + 7 + \cdots + (3k - 2) = \frac{1}{2}k(3k - 1)$
S_{k+1}: $\quad 1 + 4 + 7 + \cdots + (3k - 2) + [3(k + 1) - 2]$
$\quad\quad = \frac{1}{2}(k + 1)[3(k + 1) - 1]$
$\quad\quad = \frac{1}{2}(k + 1)(3k + 2)$
(1) *Basis step:* S_1: $3 \cdot 1 - 2 = \frac{1}{2} \cdot 1(3 \cdot 1 - 1)$. True
(2) *Induction step:* Assume S_k:
$\quad 1 + 4 + 7 + \cdots + (3k - 2) = \frac{1}{2}k(3k - 1).$
Then $1 + 4 + 7 + \cdots + (3k - 2) + [3(k + 1) - 2]$
$\quad = \frac{1}{2}k(3k - 1) + [3(k + 1) - 2]$
$\quad = \frac{3}{2}k^2 - \frac{1}{2}k + 3k + 1$
$\quad = \frac{3}{2}k^2 + \frac{5}{2}k + 1$
$\quad = \frac{1}{2}(3k^2 + 5k + 2)$
$\quad = \frac{1}{2}(k + 1)(3k + 2).$
24. The first formula can be derived from the second by
substituting $a_1 + (n - 1)d$ for a_n. When the first and last terms of
the sum are known, the second formula is the better one to use. If
the last term is not known, the first formula allows us to compute
the sum in one step without first finding a_n.
25. $1 + 2 + 3 + \cdots + 100$
$\quad = (1 + 100) + (2 + 99) + (3 + 98) + \cdots +$
$\quad\quad (50 + 51)$
$\quad = \underbrace{101 + 101 + 101 + \cdots + 101}_{\text{50 addends of 101}}$
$\quad = 50 \cdot 101$
$\quad = 5050$

A formula for the first n natural numbers is $\dfrac{n}{2}(1 + n)$.

26. Answers may vary. One possibility is given. Casey invests
$900 at 8% interest, compounded annually. How much will be in
the account at the end of 40 years? **27.** We can prove an infinite
sequence of statements S_n by showing that a basis statement S_1 is
true and then that for all natural numbers k, if S_k is true, then S_{k+1}
is true.

Exercise Set 12.5, p. 806

1. 720 **3.** 604,800 **5.** 120 **7.** 1 **9.** 3024 **11.** 120
13. 120 **15.** 1 **17.** 6,497,400 **19.** $n(n - 1)(n - 2)$
21. n **23.** $6! = 720$ **25.** $9! = 362,880$ **27.** $_9P_4 = 3024$
29. $_5P_5 = 120$; $5^5 = 3125$ **31.** $_5P_5 \cdot _4P_4 = 2880$
33. $8 \cdot 10^6 = 8,000,000$; 8 million **35.** $\dfrac{9!}{2! \, 3! \, 4!} = 1260$
37. (a) $_6P_5 = 720$; **(b)** $6^5 = 7776$; **(c)** $1 \cdot _5P_4 = 120$;
(d) $1 \cdot 1 \cdot _4P_3 = 24$ **39. (a)** 10^5, or 100,000; **(b)** 100,000
41. (a) $10^9 = 1,000,000,000$; **(b)** yes **42.** $\frac{9}{4}$, or 2.25
43. $-3, 2$ **44.** $\dfrac{3 \pm \sqrt{17}}{4}$ **45.** $-2, 1, 5$ **47.** 8
49. 11 **51.** $n - 1$

Exercise Set 12.6, p. 813

1. 78 **3.** 78 **5.** 7 **7.** 10 **9.** 1 **11.** 15 **13.** 128
15. 270,725 **17.** 13,037,895 **19.** n **21.** 1
23. $_{36}C_4 = 58,905$ **25.** $_{13}C_{10} = 286$
27. $\dbinom{10}{7} \cdot \dbinom{5}{3} = 1200$ **29.** $\dbinom{52}{5} = 2,598,960$
31. (a) $_{31}P_2 = 930$; **(b)** $31^2 = 961$; **(c)** $_{31}C_2 = 465$
33. $-\frac{17}{2}$ **34.** $-1, \frac{3}{2}$ **35.** $\dfrac{-5 \pm \sqrt{21}}{2}$ **36.** $-4, -2, 3$
37. $\dbinom{13}{5} = 1287$ **39.** $\dbinom{n}{2}$; $2\dbinom{n}{2}$ **41.** 4 **43.** 7
45. Line segments:
$$_nC_2 = \frac{n!}{2!(n - 2)!} = \frac{n(n - 1)(n - 2)!}{2 \cdot 1 \cdot (n - 2)!} = \frac{n(n - 1)}{2}$$
Diagonals: The n line segments that form the sides of the n-gon are
not diagonals. Thus the number of diagonals is
$$_nC_2 - n = \frac{n(n - 1)}{2} - n$$
$$= \frac{n^2 - n - 2n}{2} = \frac{n^2 - 3n}{2}$$
$$= \frac{n(n - 3)}{2}, n \geq 4.$$
Let D_n be the number of diagonals of an n-gon. Prove the result
above for diagonals using mathematical induction.

S_n: $\quad D_n = \dfrac{n(n - 3)}{2},$ $\quad$ for $n = 4, 5, 6, \ldots$

S_4: $\quad D_4 = \dfrac{4 \cdot 1}{2}$

S_k: $\quad D_k = \dfrac{k(k - 3)}{2}$

S_{k+1}: $\quad D_{k+1} = \dfrac{(k + 1)(k - 2)}{2}$

(1) *Basis step*: S_4 is true (a quadrilateral has 2 diagonals).

(2) *Induction step*: Assume S_k. Note that when an additional vertex V_{k+1} is added to the k-gon, we gain k segments, 2 of which are sides of the $(k + 1)$-gon, and a former side $\overline{V_1 V_k}$ becomes a diagonal. Thus the additional number of diagonals is $k - 2 + 1$, or $k - 1$. Then the new total of diagonals is $D_k + (k - 1)$, or

$$
\begin{aligned}
D_{k+1} &= D_k + (k - 1) \\
&= \frac{k(k - 3)}{2} + (k - 1) \qquad \textbf{By } S_k \\
&= \frac{(k + 1)(k - 2)}{2}
\end{aligned}
$$

Exercise Set 12.7, p. 820

1. $x^4 + 20x^3 + 150x^2 + 500x + 625$
3. $x^5 - 15x^4 + 90x^3 - 270x^2 + 405x - 243$
5. $x^5 - 5x^4y + 10x^3y^2 - 10x^2y^3 + 5xy^4 - y^5$
7. $15{,}625x^6 + 75{,}000x^5y + 150{,}000x^4y^2 +$
$160{,}000x^3y^3 + 96{,}000x^2y^4 + 30{,}720xy^5 + 4096y^6$
9. $128t^7 + 448t^5 + 672t^3 + 560t + 280t^{-1} + 84t^{-3} +$
$14t^{-5} + t^{-7}$ **11.** $x^{10} - 5x^8 + 10x^6 - 10x^4 + 5x^2 - 1$
13. $125 + 150\sqrt{5}\,t + 375t^2 + 100\sqrt{5}\,t^3 + 75t^4 +$
$6\sqrt{5}\,t^5 + t^6$ **15.** $a^9 - 18a^7 + 144a^5 - 672a^3 + 2016a -$
$4032a^{-1} + 5376a^{-3} - 4608a^{-5} + 2304a^{-7} - 512a^{-9}$
17. $140\sqrt{2}$ **19.** $x^{-8} + 4x^{-4} + 6 + 4x^4 + x^8$ **21.** $21a^5b^2$
23. $-252x^5y^5$ **25.** $-745{,}472a^3$ **27.** $1120x^{12}y^2$
29. $-1{,}959{,}552u^5v^{10}$ **31.** 2^7, or 128 **33.** 2^{24}, or $16{,}777{,}216$
35. 20 **37.** $-12 + 316i$ **39.** $-7 - 4\sqrt{2}i$
41. $\displaystyle\sum_{k=0}^{n}\binom{n}{k}(-1)^k a^{n-k}b^k$ **43.** $\displaystyle\sum_{k=1}^{n}\binom{n}{k}x^{n-k}h^{k-1}$
44. $x^2 + 2x - 2$ **45.** $x^2 - 2x + 4$ **46.** $4x^2 - 12x + 10$
47. $2x^2 - 1$ **49.** $3, 9, 6 \pm 3i$ **51.** $-\dfrac{35}{x^{1/6}}$ **53.** 2^{100}
55. $[\log_a (xt)]^{23}$
57. **(1)** *Basis step*: Since $a + b = (a + b)^1$, S_1 is true.
(2) *Induction step*: Let S_k be the statement of the binomial theorem with n replaced by k. Multiply both sides of S_k by $(a + b)$ to obtain

$$
\begin{aligned}
(a + b)^{k+1} &= \left[a^k + \cdots + \binom{k}{r-1}a^{k-(r-1)}b^{r-1} \right. \\
&\quad \left. + \binom{k}{r}a^{k-r}b^r + \cdots + b^k \right](a + b) \\
&= a^{k+1} + \cdots + \left[\binom{k}{r-1} + \binom{k}{r} \right]a^{(k+1)-r}b^r \\
&\quad + \cdots + b^{k+1} \\
&= a^{k+1} + \cdots + \binom{k+1}{r}a^{(k+1)-r}b^r + \cdots + b^{k+1}.
\end{aligned}
$$

This proves S_{k+1}, assuming S_k. Hence S_n is true for $n = 1, 2, 3, \ldots$.

Exercise Set 12.8, p. 827

1. **(a)** $0.18, 0.24, 0.23, 0.23, 0.12$; **(b)** Opinions may vary, but it seems that people tend not to pick the first or last numbers.
3. 5187 e-mails   **5.** **(a)** $\frac{2}{7}$; **(b)** $\frac{5}{7}$; **(c)** 0; **(d)** 1 **7.** $\frac{1}{2}$
9. **(a)** $\frac{1}{13}$; **(b)** $\frac{2}{13}$; **(c)** $\frac{1}{4}$; **(d)** $\frac{1}{26}$ **11.** $\frac{1}{5525}$ **13.** $\frac{135}{323}$
15. $\frac{1}{108{,}290}$ **17.** $\frac{33}{66{,}640}$ **19.** **(a)** HHH, HHT, HTH, HTT, THH, THT, TTH, TTT; **(b)** $\frac{3}{8}$; **(c)** $\frac{7}{8}$; **(d)** $\frac{7}{8}$; **(e)** $\frac{3}{8}$ **21.** $\frac{9}{19}$
23. $\frac{1}{38}$ **25.** $\frac{18}{19}$ **27.** $\frac{9}{19}$ **29.** Zero   **30.** One-to-one
31. Function; domain; range; domain; range   **32.** Zero

33. Combination   **34.** Inverse variation   **35.** Factor
36. Geometric sequence
37. **(a)** $\dbinom{13}{2} \cdot \dbinom{4}{2} \cdot \dbinom{4}{2} \cdot \dbinom{44}{1} = 123{,}552$;
(b) 0.0475 **39.** **(a)** $13 \cdot \dbinom{4}{3} \cdot \dbinom{48}{2} - 3744$, or $54{,}912$;
(b) $\dfrac{54{,}912}{\dbinom{52}{5}} \approx 0.0211$

Summary and Review: Chapter 12

Review Exercises, p. 834

1. True   **2.** False   **3.** True   **4.** False
5. $-\frac{1}{2}, \frac{4}{17}, -\frac{9}{82}, \frac{16}{257}, -\frac{121}{14{,}642}, -\frac{529}{279{,}842}$ **6.** $(-1)^{n+1}(n^2 + 1)$
7. $\frac{3}{2} - \frac{9}{8} + \frac{27}{26} - \frac{81}{80} = \frac{417}{1040}$ **8.** $\displaystyle\sum_{k=1}^{7}(k^2 - 1)$ **9.** $\frac{15}{4}$
10. $a + 4b$ **11.** 531 **12.** $20{,}100$ **13.** 11 **14.** -4
15. $n = 6, S_n = -126$ **16.** $a_1 = 8, a_5 = \frac{1}{2}$ **17.** Does not exist   **18.** $\frac{3}{11}$ **19.** $\frac{3}{8}$ **20.** $\frac{241}{99}$ **21.** $5\frac{4}{5}, 6\frac{3}{5}, 7\frac{2}{5}, 8\frac{1}{5}$
22. 167.3 ft   **23.** $\$45{,}993.04$ **24.** **(a)** $\$7.38$; **(b)** $\$1365.10$
25. $\$88{,}888{,}888{,}889$

26. S_n: $1 + 4 + 7 + \cdots + (3n - 2) = \dfrac{n(3n - 1)}{2}$

S_1: $1 = \dfrac{1(3 - 1)}{2}$

S_k: $1 + 4 + 7 + \cdots + (3k - 2) = \dfrac{k(3k - 1)}{2}$

S_{k+1}: $1 + 4 + 7 + \cdots + (3k - 2) + \lfloor 3(k + 1) - 2 \rfloor$
$= 1 + 4 + 7 + \cdots + (3k - 2) + (3k + 1)$
$= \dfrac{(k + 1)(3k + 2)}{2}$

(1) *Basis step*: $\dfrac{1(3 - 1)}{2} = \dfrac{2}{2} = 1$ is true.

(2) *Induction step*: Assume S_k. Add $(3k + 1)$ on both sides.
$1 + 4 + 7 + \cdots + (3k - 2) + (3k + 1)$
$= \dfrac{k(3k - 1)}{2} + (3k + 1) = \dfrac{k(3k - 1)}{2} + \dfrac{2(3k + 1)}{2}$
$= \dfrac{3k^2 - k + 6k + 2}{2} = \dfrac{3k^2 + 5k + 2}{2}$
$= \dfrac{(k + 1)(3k + 2)}{2}$

27. S_n: $1 + 3 + 3^2 + \cdots + 3^{n-1} = \dfrac{3^n - 1}{2}$

S_1: $1 = \dfrac{3^1 - 1}{2}$

S_k: $1 + 3 + 3^2 + \cdots + 3^{k-1} = \dfrac{3^k - 1}{2}$

S_{k+1}: $1 + 3 + 3^2 + \cdots + 3^{(k+1)-1} = \dfrac{3^{k+1} - 1}{2}$

(1) *Basis step*: $\dfrac{3^1 - 1}{2} = \dfrac{2}{2} = 1$ is true.

(2) *Induction step*: Assume S_k. Add 3^k on both sides.
$1 + 3 + \cdots + 3^{k-1} + 3^k$
$= \dfrac{3^k - 1}{2} + 3^k = \dfrac{3^k - 1}{2} + 3^k \cdot \dfrac{2}{2}$
$= \dfrac{3 \cdot 3^k - 1}{2} = \dfrac{3^{k+1} - 1}{2}$

28.

$$S_n: \quad \left(1 - \frac{1}{2}\right)\left(1 - \frac{1}{3}\right) \cdots \left(1 - \frac{1}{n}\right) = \frac{1}{n}$$

$$S_2: \quad \left(1 - \frac{1}{2}\right) = \frac{1}{2}$$

$$S_k: \quad \left(1 - \frac{1}{2}\right)\left(1 - \frac{1}{3}\right) \cdots \left(1 - \frac{1}{k}\right) = \frac{1}{k}$$

$$S_{k+1}: \quad \left(1 - \frac{1}{2}\right)\left(1 - \frac{1}{3}\right) \cdots \left(1 - \frac{1}{k}\right)\left(1 - \frac{1}{k+1}\right) = \frac{1}{k+1}.$$

(1) *Basis step*: S_2 is true by substitution.
(2) *Induction step*: Assume S_k. Deduce S_{k+1}. Starting with the left side of S_{k+1}, we have

$$\underbrace{\left(1 - \frac{1}{2}\right)\left(1 - \frac{1}{3}\right) \cdots \left(1 - \frac{1}{k}\right)}\left(1 - \frac{1}{k+1}\right)$$

$$= \frac{1}{k} \cdot \left(1 - \frac{1}{k+1}\right) \qquad \text{By } S_k$$

$$= \frac{1}{k} \cdot \left(\frac{k+1-1}{k+1}\right)$$

$$= \frac{1}{k} \cdot \frac{k}{k+1}$$

$$= \frac{1}{k+1}. \qquad \textbf{Simplifying}$$

29. $6! = 720$ **30.** $9 \cdot 8 \cdot 7 \cdot 6 = 3024$ **31.** $\binom{15}{8} = 6435$

32. $24 \cdot 23 \cdot 22 = 12{,}144$ **33.** $\dfrac{9!}{1! \, 4! \, 2! \, 2!} = 3780$

34. $3 \cdot 4 \cdot 3 = 36$ **35. (a)** $_6P_5 = 720$; **(b)** $6^5 = 7776$; **(c)** $_5P_4 = 120$; **(d)** $_3P_2 = 6$ **36.** 2^8, or 256
37. $m^7 + 7m^6n + 21m^5n^2 + 35m^4n^3 + 35m^3n^4 + 21m^2n^5 + 7mn^6 + n^7$ **38.** $x^5 - 5\sqrt{2}\,x^4 + 20x^3 - 20\sqrt{2}\,x^2 + 20x - 4\sqrt{2}$ **39.** $x^8 - 12x^6y + 54x^4y^2 - 108x^2y^3 + 81y^4$ **40.** $a^8 + 8a^6 + 28a^4 + 56a^2 + 70 + 56a^{-2} + 28a^{-4} + 8a^{-6} + a^{-8}$ **41.** $-6624 + 16{,}280i$

42. $220a^9x^3$ **43.** $-\binom{18}{11}128a^7b^{11}$ **44.** $\frac{1}{12}; 0$ **45.** $\frac{1}{4}$

46. $\frac{6}{5525}$ **47.** $\frac{86}{206} \approx 0.42, \frac{97}{206} \approx 0.47, \frac{23}{206} \approx 0.11$ **48.** B
49. A **50.** D **51. (a)** No (unless a_n is all positive or all negative); **(b)** yes; **(c)** yes; **(d)** no (unless a_n is constant); **(e)** no (unless a_n is constant); **(f)** no (unless a_n is constant)

52. $\dfrac{a_{k+1}}{a_k} = r_1, \dfrac{b_{k+1}}{b_k} = r_2$, so $\dfrac{a_{k+1}b_{k+1}}{a_k b_k} = r_1 r_2$, a constant.

53. $\frac{1}{2}, -\frac{1}{6}, \frac{1}{18}$ **54.** $-2, 0, 2, 4$ **55.** $\left(\log \dfrac{x}{y}\right)^{10}$ **56.** 18

57. 36 **58.** -9 **59.** For each circular arrangement of the numbers on a clock face, there are 12 distinguishable ordered arrangements on a line. The number of arrangements of 12 objects on a line is $_{12}P_{12}$, or $12!$. Thus the number of circular permutations is $\dfrac{_{12}P_{12}}{12} = \dfrac{12!}{12} = 11! = 39{,}916{,}800$. In general, for each circular arrangement of n objects, there are n distinguishable ordered arrangements on a line. The total number of arrangements of n objects on a line is $_nP_n$, or $n!$. Thus the number of circular

permutations is $\dfrac{n!}{n} = \dfrac{n(n-1)!}{n} = (n-1)!$. **60.** Put the following in the form of a paragraph. First find the number of seconds in a year (365 days): $365 \text{ days} \cdot \dfrac{24 \text{ hr}}{1 \text{ day}} \cdot \dfrac{60 \text{ min}}{1 \text{ hr}} \cdot \dfrac{60 \text{ sec}}{1 \text{ min}} = 31{,}536{,}000$ sec. The number of arrangements possible is $15!$. The time is $\dfrac{15!}{31{,}536{,}000} \approx 41{,}466$ years. **61.** Order is considered in a combination lock. **62.** Choosing k objects from a set of n objects is equivalent to not choosing the other $n - k$ objects.

Test: Chapter 12, p. 837

1. [12.1] -43 **2.** [12.1] $\frac{2}{3}, \frac{3}{4}, \frac{4}{5}, \frac{5}{6}, \frac{6}{7}$
3. [12.1] $2 + 5 + 10 + 17 = 34$ **4.** [12.1] $\displaystyle\sum_{k=1}^{6} 4k$
5. [12.1] $\displaystyle\sum_{k=1}^{\infty} 2^k$ **6.** [12.1] $3, 2\frac{1}{3}, 2\frac{3}{7}, 2\frac{7}{17}$ **7.** [12.2] 44
8. [12.2] 38 **9.** [12.2] -420 **10.** [12.2] 675 **11.** [12.3] $\frac{5}{512}$
12. [12.3] 1000 **13.** [12.3] 510 **14.** [12.3] 27 **15.** [12.3] $\frac{56}{99}$
16. [12.1] \$10,000, \$8000, \$6400, \$5120, \$4096, \$3276.80
17. [12.2] \$17.05 **18.** [12.3] \$74,399.77
19. [12.4]

$$S_n: \quad 2 + 5 + 8 + \cdots + (3n - 1) = \frac{n(3n+1)}{2}$$

$$S_1: \quad 2 = \frac{1(3 \cdot 1 + 1)}{2}$$

$$S_k: \quad 2 + 5 + 8 + \cdots + (3k - 1) = \frac{k(3k+1)}{2}$$

$$S_{k+1}: \quad 2 + 5 + 8 + \cdots + (3k - 1) + [3(k+1) - 1]$$
$$= \frac{(k+1)[3(k+1)+1]}{2}$$

(1) *Basis step*: $\dfrac{1(3 \cdot 1 + 1)}{2} = \dfrac{1 \cdot 4}{2} = 2$, so S_1 is true.
(2) *Induction step*:
$$2 + 5 + 8 + \cdots + (3k - 1) + [3(k+1) - 1]$$
$$= \frac{k(3k+1)}{2} + [3k + 3 - 1] \qquad \textbf{By } S_k$$
$$= \frac{3k^2}{2} + \frac{k}{2} + 3k + 2$$
$$= \frac{3k^2}{2} + \frac{7k}{2} + 2 = \frac{3k^2 + 7k + 4}{2}$$
$$= \frac{(k+1)(3k+4)}{2} = \frac{(k+1)[3(k+1)+1]}{2}$$

20. [12.5] 3,603,600 **21.** [12.6] 352,716
22. [12.6] $\dfrac{n(n-1)(n-2)(n-3)}{24}$
23. [12.5] $_6P_4 = 360$ **24.** [12.5] **(a)** $6^4 = 1296$; **(b)** $_5P_3 = 60$
25. [12.6] $_{28}C_4 = 20{,}475$ **26.** [12.6] $_{12}C_8 \cdot _8C_4 = 34{,}650$
27. [12.7] $x^5 + 5x^4 + 10x^3 + 10x^2 + 5x + 1$
28. [12.7] $35x^3y^4$ **29.** [12.7] $2^9 = 512$ **30.** [12.8] $\frac{4}{7}$
31. [12.8] $\frac{48}{1001}$ **32.** [12.1] B **33.** [12.5] 15

Appendix

1. $\dfrac{2}{x-3} - \dfrac{1}{x+2}$ **3.** $\dfrac{5}{2x-1} - \dfrac{4}{3x-1}$

5. $\dfrac{2}{x+2} - \dfrac{3}{x-2} + \dfrac{4}{x+1}$

7. $-\dfrac{3}{(x+2)^2} - \dfrac{1}{x+2} + \dfrac{1}{x-1}$ **9.** $\dfrac{3}{x-1} - \dfrac{4}{2x-1}$

11. $x - 2 + \dfrac{\frac{17}{16}}{x+1} - \dfrac{\frac{11}{4}}{(x+1)^2} - \dfrac{\frac{17}{16}}{x-3}$

13. $\dfrac{3x+5}{x^2+2} - \dfrac{4}{x-1}$ **15.** $\dfrac{3}{2x-1} - \dfrac{2}{x+2} + \dfrac{10}{(x+2)^2}$

17. $3x + 1 + \dfrac{2}{2x-1} + \dfrac{3}{x+1}$

19. $-\dfrac{1}{x-3} + \dfrac{3x}{x^2+2x-5}$

21. $\dfrac{5}{3x+5} - \dfrac{3}{x+1} + \dfrac{4}{(x+1)^2}$ **23.** $\dfrac{8}{4x-5} + \dfrac{3}{3x+2}$

25. $\dfrac{2x-5}{3x^2+1} - \dfrac{2}{x-2}$ **27.** $-\dfrac{\frac{1}{2a^2}x}{x^2+a^2} + \dfrac{\frac{1}{4a^2}}{x-a} + \dfrac{\frac{1}{4a^2}}{x+a}$

29. $-\dfrac{3}{25(\ln x + 2)} + \dfrac{3}{25(\ln x - 3)} + \dfrac{7}{5(\ln x - 3)^2}$

Index of Applications

Index

10 Analytic Geometry

In this chapter, we study a group of curves known as *conic sections*. One conic section, the *ellipse*, has a special reflecting property responsible for "whispering galleries." In a whispering gallery, a person whispering at a certain point in the room, though barely audible close by, can be heard clearly at another point across the room.

The Old House Chamber of the U.S. Capitol, called Statuary Hall, is a whispering gallery. History has it that John Quincy Adams, whose desk was positioned at exactly the right point beneath the ellipsoidal ceiling, often pretended to sleep at his desk as he listened to political opponents whispering strategies in an area across the room. Today, an engraved plate set into the floor marks this point. (*Source: We, the People, The Story of the United States Capitol,* 1991.)

In Section 10.2, we investigate this reflective property of ellipses.

10.1 Parabolas

Conic Sections ▪ **Horizontal Parabolas** ▪ **Geometric Definition and Equations of Parabolas** ▪ **An Application of Parabolas**

Conic Sections *Parabolas, circles, ellipses,* and *hyperbolas* form a group of curves known as **conic sections,** because they are the results of intersecting a cone with a plane. Figure 1 illustrates these curves. We studied circles and some parabolas (those that open up or down, that is, vertical parabolas) in **Chapters 2 and 3.**

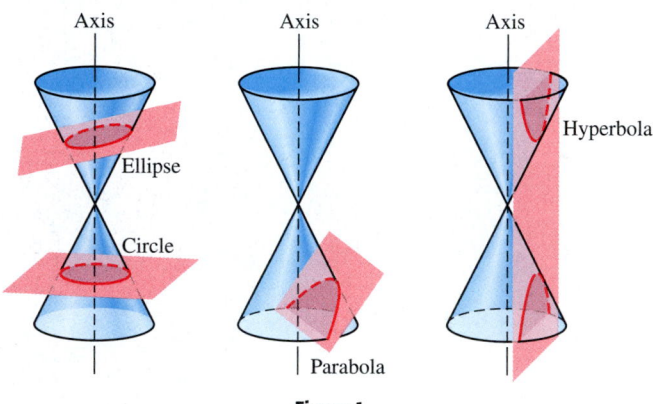

Figure 1

Horizontal Parabolas From **Chapter 3,** we know that the graph of the equation

$$y = a(x - h)^2 + k \quad \text{(Section 3.1)}$$

is a parabola with vertex (h, k) and the vertical line $x = h$ as axis. If we subtract k from each side, this equation becomes

$$y - k = a(x - h)^2. \quad \text{Subtract } k. \quad (1)$$

The equation

$$x - h = a(y - k)^2 \quad \text{Interchange the roles of } x - h \text{ and } y - k. \quad (2)$$

also has a parabola as its graph. While the graph of $y - k = a(x - h)^2$ has a *vertical* axis, the graph of $x - h = a(y - k)^2$ has a *horizontal* axis. The graph of the first equation is the graph of a function (specifically a quadratic function), while the graph of the second equation is not. (Why?)

PARABOLA WITH HORIZONTAL AXIS

The parabola with vertex (h, k) and the horizontal line $y = k$ as axis has an equation of the form

$$x - h = a(y - k)^2.$$

The parabola opens to the right if $a > 0$ and to the left if $a < 0$.

▶ **Note** When the vertex (h, k) is $(0, 0)$ and $a = 1$ in

$$y - k = a(x - h)^2 \quad (1)$$

and $$x - h = a(y - k)^2, \quad (2)$$

the equations $y = x^2$ and $x = y^2$ result. The graphs of these equations are shown in Figure 2. Notice that the graphs are mirror images of each other with respect to the line $y = x$.

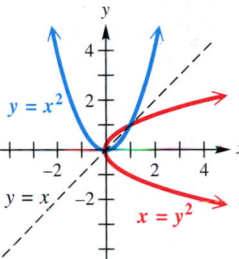

Figure 2

▶ **EXAMPLE 1**   **GRAPHING A PARABOLA WITH HORIZONTAL AXIS**

Graph $x + 3 = (y - 2)^2$. Give the domain and range.

Solution   The graph of $x + 3 = (y - 2)^2$ or $x - (-3) = (y - 2)^2$ has vertex $(-3, 2)$ and opens to the right because $a = 1$, and $1 > 0$. Plotting a few additional points gives the graph shown in Figure 3. Note that the graph is symmetric about its axis, $y = 2$. The domain is $[-3, \infty)$, and the range is $(-\infty, \infty)$.

x	y
-3	2
-2	3
-2	1
1	4
1	0

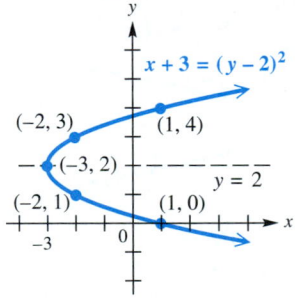

NOW TRY EXERCISE 7. ◀

When an equation of a horizontal parabola is given in the form

$$x = ay^2 + by + c,$$

completing the square on y allows us to write the equation in the form

$$x - h = a(y - k)^2$$

and more easily find the vertex, as shown in Example 2 on the next page.

▶ **EXAMPLE 2** **GRAPHING A PARABOLA WITH HORIZONTAL AXIS**

Graph $x = 2y^2 + 6y + 5$. Give the domain and range.

Algebraic Solution

$$x = 2y^2 + 6y + 5$$

$$x = 2(y^2 + 3y \qquad) + 5 \quad \text{Factor out 2.}$$

$$x = 2\left(y^2 + 3y + \frac{9}{4} - \frac{9}{4}\right) + 5$$

Complete the square; $\left[\frac{1}{2}(3)\right]^2 = \frac{9}{4}$. **(Section 1.4)**

$$x = 2\left(y^2 + 3y + \frac{9}{4}\right) + 2\left(-\frac{9}{4}\right) + 5$$

Distributive property **(Section R.2)**

$$x = 2\left(y + \frac{3}{2}\right)^2 + \frac{1}{2}$$

Factor **(Section R.4)**; simplify.

$$x - \frac{1}{2} = 2\left(y + \frac{3}{2}\right)^2 \quad \text{Subtract } \frac{1}{2}. \text{ (*)}$$

The vertex of the parabola is $\left(\frac{1}{2}, -\frac{3}{2}\right)$. The axis is the horizontal line $y = k$, or $y = -\frac{3}{2}$. Using the vertex and the axis and plotting a few additional points gives the graph in Figure 4. The domain is $\left[\frac{1}{2}, \infty\right)$, and the range is $(-\infty, \infty)$.

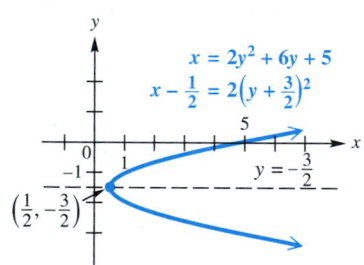

Figure 4

Graphing Calculator Solution

Since a horizontal parabola is *not* the graph of a function, to graph it using a graphing calculator we must write two equations by solving for y.

$$x - \frac{1}{2} = 2\left(y + \frac{3}{2}\right)^2 \quad \text{(*) from algebraic solution}$$

$$x - .5 = 2(y + 1.5)^2 \quad \text{Write with decimals.}$$

$$\frac{x - .5}{2} = (y + 1.5)^2 \quad \text{Divide by 2.}$$

$$\pm\sqrt{\frac{x - .5}{2}} = y + 1.5 \quad \begin{array}{l}\text{Take square roots}\\\text{on both sides.}\\\text{(Section 1.4)}\end{array}$$

$$y = -1.5 \pm \sqrt{\frac{x - .5}{2}} \quad \begin{array}{l}\text{Subtract 1.5;}\\\text{rewrite.}\end{array}$$

Figure 5 shows the graphs of the two functions defined in the final equation. Their union is the graph of $x = 2y^2 + 6y + 5$.

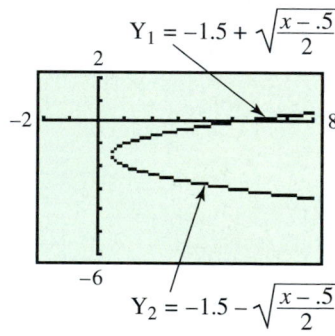

Figure 5

NOW TRY EXERCISE 15. ◀

Geometric Definition and Equations of Parabolas
We can also develop the equation of a parabola from the geometric definition of a parabola as a set of points.

> **PARABOLA**
>
> A **parabola** is the set of points in a plane equidistant from a fixed point and a fixed line. The fixed point is called the **focus,** and the fixed line is called the **directrix** of the parabola.

As shown in Figure 6, the axis of symmetry of a parabola passes through the focus and is perpendicular to the directrix. The vertex is the midpoint of the line segment joining the focus and directrix on the axis.

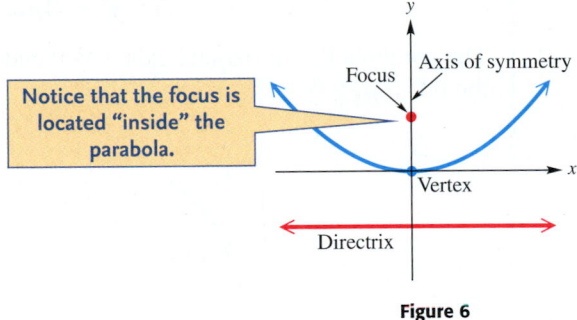

Figure 6

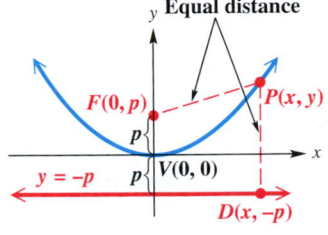

$d(P, F) = d(P, D)$
for all P on the parabola.

Figure 7

We can find an equation of a parabola from the preceding definition. Let p represent the directed distance from the vertex to the focus. Then the directrix is the line $y = -p$ and the focus is the point $F(0, p)$, as shown in Figure 7. To find the equation of the set of points that are the same distance from the line $y = -p$ and the point $(0, p)$, choose one such point P and give it coordinates (x, y). Since $d(P, F)$ and $d(P, D)$ must be equal, using the distance formula gives

$$d(P, F) = d(P, D)$$
$$\sqrt{(x - 0)^2 + (y - p)^2} = \sqrt{(x - x)^2 + (y - (-p))^2}$$

Distance formula **(Section 2.1)**

$$\sqrt{x^2 + (y - p)^2} = \sqrt{(y + p)^2}$$
$$x^2 + y^2 - 2yp + p^2 = y^2 + 2yp + p^2$$

Square both sides **(Section 1.6)**; multiply. **(Section R.3)**

Remember the middle terms.

$$x^2 = 4py.$$

Simplify.

This discussion is summarized as follows.

PARABOLA WITH VERTICAL AXIS AND VERTEX (0, 0)

The parabola with focus $(0, p)$ and directrix $y = -p$ has equation

$$x^2 = 4py.$$

The parabola has vertical axis $x = 0$ and opens up if $p > 0$ or down if $p < 0$.

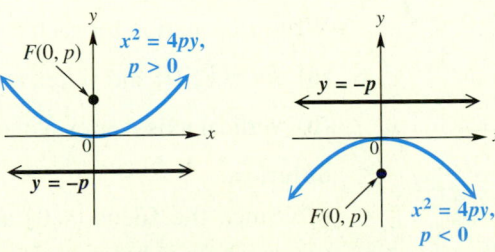

If the directrix is the line $x = -p$ and the focus is $(p, 0)$, a similar procedure leads to the equation of a parabola with a horizontal axis.

PARABOLA WITH HORIZONTAL AXIS AND VERTEX (0,0)

The parabola with focus $(p,0)$ and directrix $x = -p$ has equation

$$y^2 = 4px.$$

The parabola has horizontal axis $y = 0$ and opens to the right if $p > 0$ or to the left if $p < 0$.

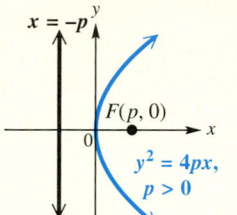 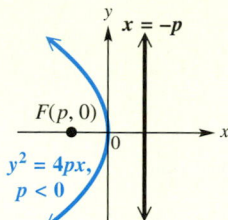

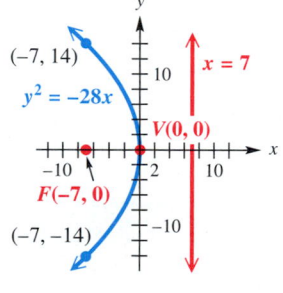

Figure 8

▶ **EXAMPLE 3** **DETERMINING INFORMATION ABOUT PARABOLAS FROM THEIR EQUATIONS**

Find the focus, directrix, vertex, and axis of each parabola. Then use this information to graph the parabola.

(a) $x^2 = 8y$ **(b)** $y^2 = -28x$

Solution

(a) The equation $x^2 = 8y$ has the form $x^2 = 4py$, so $4p = 8$, from which $p = 2$. Since the x-term is squared, the parabola is vertical, with focus $(0, p) = (0, 2)$ and directrix $y = -2$. The vertex is $(0, 0)$, and the axis of the parabola is the y-axis. See Figure 8.

(b) The equation $y^2 = -28x$ has the form $y^2 = 4px$, with $4p = -28$, so $p = -7$. The parabola is horizontal, with focus $(-7, 0)$, directrix $x = 7$, vertex $(0, 0)$, and x-axis as axis of the parabola. Since p is negative, the graph opens to the left, as shown in Figure 9.

Figure 9

NOW TRY EXERCISES 19 AND 23. ◀

▶ **EXAMPLE 4** **WRITING EQUATIONS OF PARABOLAS**

Write an equation for each parabola.

(a) focus $\left(\frac{2}{3}, 0\right)$ and vertex at the origin

(b) vertical axis, vertex at the origin, through the point $(-2, 12)$

Solution

(a) Since the focus $\left(\frac{2}{3}, 0\right)$ and the vertex $(0, 0)$ are both on the x-axis, the parabola is horizontal. It opens to the right because $p = \frac{2}{3}$ is positive. See Figure 10. The equation, which will have the form $y^2 = 4px$, is

$$y^2 = 4\left(\frac{2}{3}\right)x, \quad \text{or} \quad y^2 = \frac{8}{3}x.$$

Figure 10

(b) The parabola will have an equation of the form $x^2 = 4py$ because the axis is vertical and the vertex is $(0, 0)$. Since the point $(-2, 12)$ is on the graph, it must satisfy the equation.

$$x^2 = 4py$$
$$(-2)^2 = 4p(12) \qquad \text{Let } x = -2 \text{ and } y = 12.$$
$$4 = 48p \qquad \text{Multiply.}$$
$$p = \frac{1}{12} \qquad \text{Solve for } p. \text{ (Section 1.1)}$$

Thus,

$$x^2 = 4py$$
$$x^2 = 4\left(\frac{1}{12}\right)y, \qquad \text{Let } p = \frac{1}{12}.$$

which gives the equation

$$x^2 = \frac{1}{3}y, \qquad \text{or} \qquad y = 3x^2.$$

NOW TRY EXERCISES 31 AND 35. ◀

The equations $x^2 = 4py$ and $y^2 = 4px$ can be extended to parabolas having vertex (h, k) by replacing x and y with $x - h$ and $y - k$, respectively.

EQUATION FORMS FOR TRANSLATED PARABOLAS

A parabola with vertex (h, k) has an equation of the form

$$(x - h)^2 = 4p(y - k) \qquad \text{Vertical axis}$$

or

$$(y - k)^2 = 4p(x - h), \qquad \text{Horizontal axis}$$

where the focus is distance $|p|$ from the vertex.

▶ EXAMPLE 5   WRITING AN EQUATION OF A PARABOLA

Write an equation for the parabola with vertex $(1, 3)$ and focus $(-1, 3)$, and graph it. Give the domain and range.

Solution   Since the focus is to the left of the vertex, the axis is horizontal and the parabola opens to the left. See Figure 11. The distance between the vertex and the focus is $-1 - 1$, or -2, so $p = -2$ (since the parabola opens to the left). The equation of the parabola is

$$(y - k)^2 = 4p(x - h) \qquad \text{Parabola with horizontal axis}$$
$$(y - 3)^2 = 4(-2)(x - 1) \qquad \text{Substitute for } p, h, \text{ and } k.$$
$$(y - 3)^2 = -8(x - 1). \qquad \text{Multiply.}$$

The domain is $(-\infty, 1]$, and the range is $(-\infty, \infty)$.

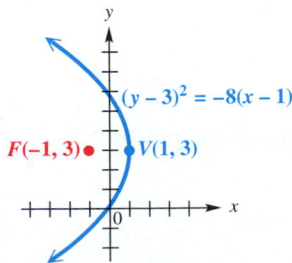

Figure 11

NOW TRY EXERCISE 39. ◀

CONNECTIONS Parabolas have a special reflecting property that makes them useful in the design of telescopes, radar equipment, auto headlights, and solar furnaces. When a ray of light or a sound wave traveling parallel to the axis of a parabolic shape bounces off the parabola, it passes through the focus.

For example, in the solar furnace shown in the figure, a parabolic mirror collects light at the focus and thereby generates intense heat at that point. The reflecting property can be used in reverse. If a light source is placed at the focus, then the reflected light rays will be directed straight ahead. This is why the reflector in a car headlight is parabolic.

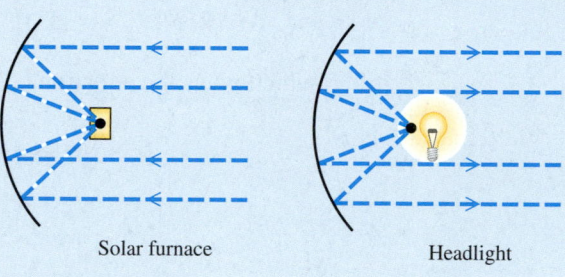

Solar furnace   Headlight

An Application of Parabolas

▶ **EXAMPLE 6** **MODELING THE REFLECTIVE PROPERTY OF PARABOLAS**

The Parkes radio telescope has a parabolic dish shape with diameter 210 ft and depth 32 ft. Because of this parabolic shape, distant rays hitting the dish will be reflected directly toward the focus. A cross section of the dish is shown in Figure 12. (*Source:* Mar, J. and H. Liebowitz, *Structure Technology for Large Radio and Radar Telescope Systems,* The MIT Press, Massachusetts Institute of Technology, 1969.)

(a) Determine an equation that models this cross section by placing the vertex at the origin with the parabola opening up.

(b) The receiver must be placed at the focus of the parabola. How far from the vertex of the parabolic dish should the receiver be located?

Solution

(a) Locate the vertex at the origin as shown in Figure 13. The form of the parabola is $x^2 = 4py$. The parabola must pass through the point $\left(\frac{210}{2}, 32\right) = (105, 32)$. Thus,

$$x^2 = 4py \qquad \text{Vertical parabola}$$
$$(105)^2 = 4p(32) \qquad \text{Let } x = 105 \text{ and } y = 32.$$
$$11{,}025 = 128p \qquad \text{Multiply.}$$
$$p = \frac{11{,}025}{128}. \qquad \text{Solve for } p.$$

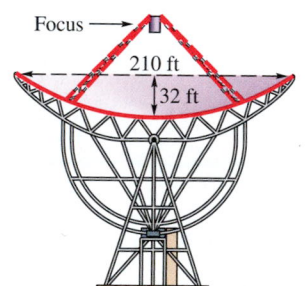

Focus →

210 ft

32 ft

Figure 12

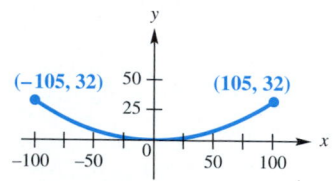

Figure 13

The cross section can be modeled by the equation

$$x^2 = 4py$$

$$x^2 = 4\left(\frac{11{,}025}{128}\right)y \quad \text{Substitute for } p.$$

$$x^2 = \frac{11{,}025}{32}y.$$

(b) The distance between the vertex and the focus is p. In part (a), we found $p = \frac{11{,}025}{128} \approx 86.1$, so the receiver should be located at $(0, 86.1)$ or 86.1 ft above the vertex.

NOW TRY EXERCISE 51. ◀

10.1 Exercises

1. *Concept Check*   Match each equation of a parabola in Column I with its description in Column II.

I

(a) $y - 2 = (x + 4)^2$
(b) $y - 4 = (x + 2)^2$
(c) $y - 2 = -(x + 4)^2$
(d) $y - 4 = -(x + 2)^2$
(e) $x - 2 = (y + 4)^2$
(f) $x - 4 = (y + 2)^2$
(g) $x - 2 = -(y + 4)^2$
(h) $x - 4 = -(y + 2)^2$

II

A. vertex $(-2, 4)$; opens down
B. vertex $(-2, 4)$; opens up
C. vertex $(-4, 2)$; opens down
D. vertex $(-4, 2)$; opens up
E. vertex $(2, -4)$; opens left
F. vertex $(2, -4)$; opens right
G. vertex $(4, -2)$; opens left
H. vertex $(4, -2)$; opens right

2. *Concept Check*   Match each equation of a parabola in Column I with the appropriate description in Column II.

I

(a) $y = 2x^2 + 3x + 9$
(b) $y = -3x^2 + 4x - 2$
(c) $x = 2y^2 - 3y + 9$
(d) $x = -3y^2 - 4y + 2$

II

A. opens right
B. opens up
C. opens left
D. opens down

Graph each horizontal parabola, and give the domain and range. See Examples 1 and 2.

3. $-x = y^2$
4. $x - 2 = y^2$
5. $x = (y - 3)^2$

6. $x = (y + 1)^2$
7. $x - 2 = (y - 4)^2$
8. $x + 1 = (y + 2)^2$

9. $x - 2 = -3(y - 1)^2$
10. $-\frac{1}{2}x = (y + 3)^2$
11. $x - 4 = \frac{1}{2}(y - 1)^2$

12. $x = -\frac{1}{3}(y - 3)^2 + 3$
13. $x = y^2 + 4y + 2$
14. $x = 2y^2 - 4y + 6$

15. $x = -4y^2 - 4y + 3$
16. $-x = 2y^2 - 2y + 3$
17. $x = \frac{1}{2}y^2 - 2y + 3$

18. $x + 3y^2 + 18y + 22 = 0$

Give the focus, directrix, and axis for each parabola. See Example 3.

19. $x^2 = 24y$
20. $x^2 = \frac{1}{8}y$
21. $y = -4x^2$

22. $-9y = x^2$
23. $y^2 = -4x$
24. $y^2 = 16x$

25. $x = -32y^2$ **26.** $x = -16y^2$ **27.** $(y - 3)^2 = 12(x - 1)$

28. $(x + 2)^2 = 20y$ **29.** $(x - 7)^2 = 16(y + 5)$ **30.** $(y - 2)^2 = 24(x - 3)$

Write an equation for each parabola with vertex at the origin. See Example 4.

31. focus $(5, 0)$ **32.** focus $\left(-\frac{1}{2}, 0\right)$

33. directrix $y = -\frac{1}{4}$ **34.** directrix $y = \frac{1}{3}$

35. through $\left(\sqrt{3}, 3\right)$, opening up   **36.** through $\left(-2, -2\sqrt{2}\right)$, opening left

37. through $(3, 2)$, symmetric with respect   **38.** through $(2, -4)$, symmetric with re-
to the x-axis   spect to the y-axis

Write an equation for each parabola. See Example 5.

39. vertex $(4, 3)$, focus $(4, 5)$ **40.** vertex $(-2, 1)$, focus $(-2, -3)$

41. vertex $(-5, 6)$, directrix $x = -12$ **42.** vertex $(1, 2)$, directrix $x = 4$

Determine the two equations necessary to graph each horizontal parabola using a graphing calculator, and graph it in the viewing window specified. See Example 2.

43. $x = 3y^2 + 6y - 4$; $[-10, 2]$ by $[-4, 4]$

44. $x = -2y^2 + 4y + 3$; $[-10, 6]$ by $[-4, 4]$

45. $x + 2 = -(y + 1)^2$; $[-10, 2]$ by $[-4, 4]$

46. $x - 5 = 2(y - 2)^2$; $[-2, 12]$ by $[-2, 6]$

RELATING CONCEPTS

For individual or collaborative investigation
(Exercises 47–50)

Curve Fitting   Given three noncollinear points, an equation of the form $x = ay^2 + by + c$ of the horizontal parabola joining them can be found by solving a system of equations. **Work Exercises 47–50 in order,** to find the equation of the horizontal parabola containing $(-5, 1)$, $(-14, -2)$, and $(-10, 2)$.

47. Write three equations in a, b, and c, by substituting the given values of x and y into the equation $x = ay^2 + by + c$.

48. Solve the system of three equations determined in Exercise 47.

49. Does the horizontal parabola open to the left or to the right? Why?

50. Write the equation of the horizontal parabola.

Solve each problem. See Example 6.

51. *(Modeling) Radio Telescope Design*
The U.S. Naval Research Laboratory designed a giant radio telescope that had diameter 300 ft and maximum depth 44 ft. (*Source:* Mar, J., and H. Liebowitz, *Structure Technology for Large Radio and Radar Telescope Systems,* The MIT Press, 1969.)

(a) Find the equation of a parabola that models the cross section of the dish if the vertex is placed at the origin and the parabola opens up.

(b) The receiver must be placed at the focus of the parabola. How far from the vertex should the receiver be located?

52. *(Modeling) Radio Telescope Design*   Suppose the telescope in Exercise 51 had diameter 400 ft and maximum depth 50 ft.

(a) Find the equation of this parabola.
(b) The receiver must be placed at the focus of the parabola. How far from the vertex should the receiver be located?

53. *Parabolic Arch*   An arch in the shape of a parabola has the dimensions shown in the figure. How wide is the arch 9 ft up?

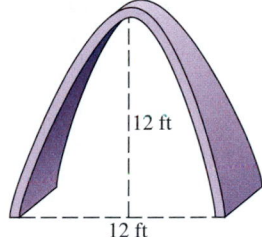

54. *Height of Bridge Cable Supports*   The cable in the center portion of a bridge is supported as shown in the figure to form a parabola. The center vertical cable is 10 ft high, the supports are 210 ft high, and the distance between the two supports is 400 ft. Find the height of the remaining vertical cables, if the vertical cables are evenly spaced. (Ignore the width of the supports and cables.)

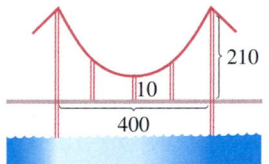

55. *(Modeling) Path of a Cannon Shell*   About 400 yr ago, the physicist Galileo observed that certain projectiles follow a parabolic path. For instance, if a cannon fires a shell at a 45° angle with a speed of v feet per second, then the path of the shell (see the figure on the top) is modeled by the equation

$$y = x - \frac{32}{v^2}x^2.$$

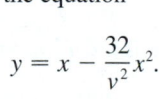

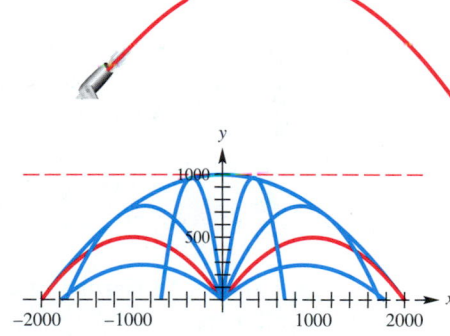

The figure on the bottom shows the paths of shells all fired at the same speed but at different angles. The greatest distance is achieved with a 45° angle. The outline, called the **envelope,** of this family of curves is another parabola with the cannon as focus. The horizontal line through the vertex of the envelope parabola is a directrix for all the other parabolas. Suppose all the shells are fired at a speed of 252.982 ft per sec.

(a) What is the greatest distance that a shell can be fired?
(b) What is the equation of the envelope parabola?
(c) Can a shell reach a helicopter 1500 ft due east of the cannon flying at a height of 450 ft?

56. *(Modeling) Path of an Object*   When an object moves under the influence of a constant force (without air resistance), its path is parabolic. This would occur if a ball is thrown near the surface of a planet or other celestial object. Suppose two balls are simultaneously thrown upward at a 45° angle on two different planets. If their initial velocities are both 30 mph, then their xy-coordinates in feet at time x in seconds can be modeled by the equation at the top of the next page.

$$y = x - \frac{g}{1922}x^2$$

Here g is the acceleration due to gravity. The value of g will vary depending on the mass and size of the planet. (*Source:* Zeilik, M. and S. Gregory, *Introductory Astronomy and Astrophysics,* Fourth Edition, Brooks/Cole, 1998.)

(a) For Earth $g = 32.2$, while on Mars $g = 12.6$. Find the two equations, and graph on the same screen of a graphing calculator the paths of the two balls thrown on Earth and Mars. Use the window $[0, 180]$ by $[0, 120]$. (*Hint:* If possible, set the mode on your graphing calculator to simultaneous.)

(b) Determine the difference in the horizontal distances traveled by the two balls.

57. *(Modeling) Path of an Object* (Refer to Exercise 56.) Suppose the two balls are now thrown upward at a 60° angle on Mars and the moon. If their initial velocity is 60 mph, then their xy-coordinates in feet at time x in seconds can be modeled by the equation

$$y = \frac{19}{11}x - \frac{g}{3872}x^2.$$

(*Source:* Zeilik, M. and S. Gregory, *Introductory Astronomy and Astrophysics,* Fourth Edition, Brooks/Cole, 1998.)

(a) Graph on the same coordinate axes the paths of the balls if $g = 5.2$ for the moon. Use the window $[0, 1500]$ by $[0, 1000]$.

(b) Determine the maximum height of each ball to the nearest foot.

58. Explain how you can tell, just by looking at the equation of a parabola, whether it has a horizontal or a vertical axis.

59. Prove that the parabola with focus $(p, 0)$ and directrix $x = -p$ has the equation $y^2 = 4px$.

60. Write a short paragraph on the appearances of parabolic shapes in your surroundings.

10.2 Ellipses

Equations and Graphs of Ellipses ▪ Translated Ellipses ▪ Eccentricity ▪ Applications of Ellipses

Equations and Graphs of Ellipses Like the parabola, the ellipse is defined as a set of points.

ELLIPSE

An **ellipse** is the set of all points in a plane the sum of whose distances from two fixed points is constant. Each fixed point is called a **focus** (plural, **foci**) of the ellipse.

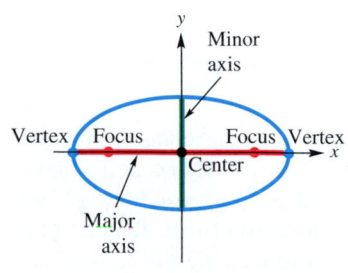

Figure 14

As shown in Figure 14, an ellipse has two axes of symmetry, the **major axis** (the longer one) and the **minor axis** (the shorter one). The foci are always located on the major axis. The midpoint of the major axis is the **center** of the ellipse, and the endpoints of the major axis are the **vertices** of the ellipse. *The graph of an ellipse is not the graph of a function.* (Why?)

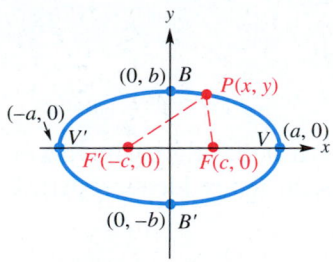

Figure 15

The ellipse in Figure 15 has its center at the origin, foci $F(c, 0)$ and $F'(-c, 0)$, and vertices $V(a, 0)$ and $V'(-a, 0)$. From Figure 15, the distance from V to F is $a - c$ and the distance from V to F' is $a + c$. The sum of these distances is $2a$. Since V is on the ellipse, this sum is the constant referred to in the definition of an ellipse. Thus, for any point $P(x, y)$ on the ellipse,

$$d(P, F) + d(P, F') = 2a.$$

By the distance formula,

$$d(P, F) = \sqrt{(x - c)^2 + y^2},$$

and $\qquad d(P, F') = \sqrt{[x - (-c)]^2 + y^2} = \sqrt{(x + c)^2 + y^2}.$

Thus,

$$\sqrt{(x - c)^2 + y^2} + \sqrt{(x + c)^2 + y^2} = 2a$$

$$\sqrt{(x - c)^2 + y^2} = 2a - \sqrt{(x + c)^2 + y^2}$$

$\qquad\qquad$ Isolate $\sqrt{(x - c)^2 + y^2}$.

$$(x - c)^2 + y^2 = 4a^2 - 4a\sqrt{(x + c)^2 + y^2} + (x + c)^2 + y^2$$

Be careful when squaring.

$\qquad\qquad$ Square both sides. (Sections R.3, 1.6)

$$x^2 - 2cx + c^2 + y^2 = 4a^2 - 4a\sqrt{(x + c)^2 + y^2} + x^2 + 2cx + c^2 + y^2$$

$\qquad\qquad$ Square $x - c$; square $x + c$.

$$4a\sqrt{(x + c)^2 + y^2} = 4a^2 + 4cx \qquad \text{Isolate } 4a\sqrt{(x + c)^2 + y^2}.$$

Divide each term by 4.

$$a\sqrt{(x + c)^2 + y^2} = a^2 + cx \qquad \text{Divide by 4.}$$

$$a^2(x^2 + 2cx + c^2 + y^2) = a^4 + 2ca^2x + c^2x^2$$

$\qquad\qquad$ Square both sides.

$$a^2x^2 + 2ca^2x + a^2c^2 + a^2y^2 = a^4 + 2ca^2x + c^2x^2$$

$\qquad\qquad$ Distributive property (Section R.2)

$$a^2x^2 + a^2c^2 + a^2y^2 = a^4 + c^2x^2 \qquad \text{Subtract } 2ca^2x.$$

$$a^2x^2 - c^2x^2 + a^2y^2 = a^4 - a^2c^2 \qquad \text{Rearrange terms.}$$

$$(a^2 - c^2)x^2 + a^2y^2 = a^2(a^2 - c^2) \qquad \text{Factor. (Section R.4)}$$

$$\frac{x^2}{a^2} + \frac{y^2}{a^2 - c^2} = 1. \qquad (*) \qquad \text{Divide by } a^2(a^2 - c^2).$$

Since $B(0, b)$ is on the ellipse in Figure 15, we have

$$d(B, F) + d(B, F') = 2a$$

$$\sqrt{(-c)^2 + b^2} + \sqrt{c^2 + b^2} = 2a$$

$$2\sqrt{c^2 + b^2} = 2a \qquad \text{Combine terms.}$$

$$\sqrt{c^2 + b^2} = a \qquad \text{Divide by 2.}$$

$$c^2 + b^2 = a^2 \qquad \text{Square both sides.}$$

$$b^2 = a^2 - c^2. \qquad \text{Subtract } c^2.$$

▼ **LOOKING AHEAD TO CALCULUS**
Methods of calculus can be used to solve problems involving ellipses. For example, differentiation is used to find the slope of the tangent line at a point on the ellipse, and integration is used to find the length of any arc of the ellipse.

Replacing $a^2 - c^2$ with b^2 in equation (*) gives

$$\frac{x^2}{a^2} + \frac{y^2}{b^2} = 1,$$

the standard form of the equation of an ellipse centered at the origin with foci on the x-axis. If the vertices and foci were on the y-axis, an almost identical derivation could be used to get the standard form

$$\frac{x^2}{b^2} + \frac{y^2}{a^2} = 1.$$

STANDARD FORMS OF EQUATIONS FOR ELLIPSES

The ellipse with center at the origin and equation

$$\frac{x^2}{a^2} + \frac{y^2}{b^2} = 1 \quad (a > b)$$

has vertices $(\pm a, 0)$, endpoints of the minor axis $(0, \pm b)$, and foci $(\pm c, 0)$, where $c^2 = a^2 - b^2$.

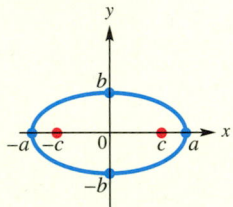

Major axis on x-axis

The ellipse with center at the origin and equation

$$\frac{x^2}{b^2} + \frac{y^2}{a^2} = 1 \quad (a > b)$$

has vertices $(0, \pm a)$, endpoints of the minor axis $(\pm b, 0)$, and foci $(0, \pm c)$, where $c^2 = a^2 - b^2$.

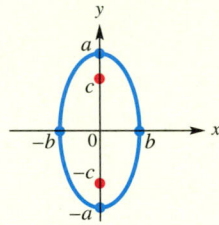

Major axis on y-axis

Do not be confused by the two standard forms—in one case a^2 is associated with x^2; in the other case a^2 is associated with y^2. In practice it is necessary only to find the intercepts of the graph—if the positive x-intercept is larger than the positive y-intercept, then the major axis is horizontal; otherwise, it is vertical. When using the relationship $a^2 - c^2 = b^2$, or $a^2 - b^2 = c^2$, choose a^2 and b^2 so that $a^2 > b^2$.

▶ **EXAMPLE 1** **GRAPHING ELLIPSES CENTERED AT THE ORIGIN**

Graph each ellipse, and find the coordinates of the foci. Give the domain and range.

(a) $4x^2 + 9y^2 = 36$ **(b)** $4x^2 + y^2 = 64$

Solution

(a) Divide each side of $4x^2 + 9y^2 = 36$ by 36 to get

> Divide each term by 36. $\dfrac{x^2}{9} + \dfrac{y^2}{4} = 1.$ Standard form of an ellipse

Thus, the x-intercepts are ± 3, and the y-intercepts are ± 2. The graph of the ellipse is shown in Figure 16.

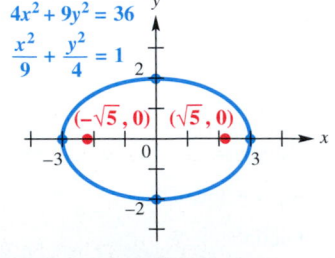

Figure 16

Since $9 > 4$, we find the foci of the ellipse by letting $a^2 = 9$ and $b^2 = 4$ in $c^2 = a^2 - b^2$.

$$c^2 = 9 - 4 = 5, \quad \text{so} \quad c = \sqrt{5}.$$

(By definition, $c > 0$. See Figure 15 on page 961.) The major axis is along the x-axis, so the foci have coordinates $\left(-\sqrt{5}, 0\right)$ and $\left(\sqrt{5}, 0\right)$. The domain of this relation is $[-3, 3]$, and the range is $[-2, 2]$.

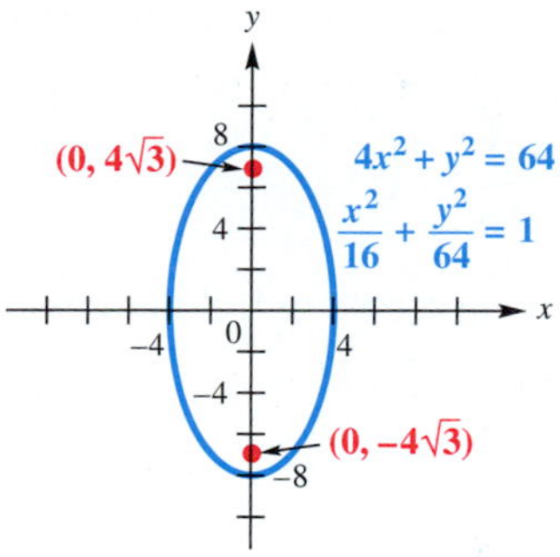

(b) Write the equation $4x^2 + y^2 = 64$ in standard form.

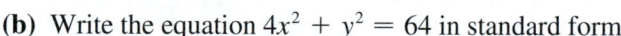

Divide *each* term by 64. $\dfrac{x^2}{16} + \dfrac{y^2}{64} = 1$ Standard form of an ellipse

The x-intercepts are ± 4; the y-intercepts are ± 8. See the graph in Figure 17. Here $64 > 16$, so $a^2 = 64$ and $b^2 = 16$. Thus,

$$c^2 = 64 - 16 = 48, \quad \text{so} \quad c = \sqrt{48} = 4\sqrt{3}. \quad \text{(Section R.7)}$$

The major axis is on the y-axis, so the coordinates of the foci are $\left(0, -4\sqrt{3}\right)$ and $\left(0, 4\sqrt{3}\right)$. The domain of the relation is $[-4, 4]$; the range is $[-8, 8]$.

NOW TRY EXERCISES 7 AND 9. ◀

The graph of an ellipse is **not** the graph of a function. To graph the ellipse in Example 1(a) with a graphing calculator, solve for y in $4x^2 + 9y^2 = 36$ to get equations of the two functions

$$y = 2\sqrt{1 - \dfrac{x^2}{9}} \quad \text{and} \quad y = -2\sqrt{1 - \dfrac{x^2}{9}}.$$

See Figure 18. ■

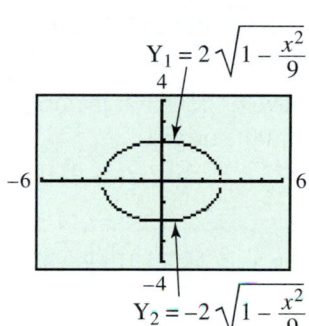

Figure 17

Figure 18

▶ **EXAMPLE 2**   WRITING THE EQUATION OF AN ELLIPSE

Write the equation of the ellipse having center at the origin, foci at $(0, 3)$ and $(0, -3)$, and major axis of length 8 units.

Solution   Since the major axis is 8 units long, $2a = 8$ and $a = 4$. To find b^2, use the relationship $a^2 - b^2 = c^2$, with $a = 4$ and $c = 3$.

$$a^2 - b^2 = c^2$$
$$4^2 - b^2 = 3^2 \quad \text{Substitute for } a \text{ and } c.$$
$$16 - b^2 = 9$$
$$b^2 = 7 \quad \text{Solve for } b^2. \text{ (Section 1.1)}$$

Since the foci are on the y-axis, we use the larger intercept, a, to find the denominator for y^2, giving the equation in standard form as

$$\dfrac{x^2}{7} + \dfrac{y^2}{16} = 1.$$

A graph of this ellipse is shown in Figure 19. The domain of this relation is $\left[-\sqrt{7}, \sqrt{7}\right]$, and the range is $[-4, 4]$.

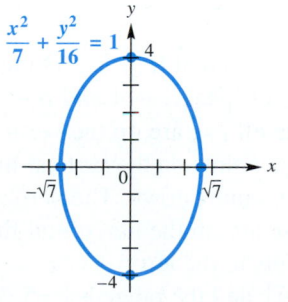

Figure 19

NOW TRY EXERCISE 17. ◀

▶ **EXAMPLE 3** **GRAPHING A HALF-ELLIPSE**

Graph $\dfrac{y}{4} = \sqrt{1 - \dfrac{x^2}{25}}$. Give the domain and range.

Solution Square both sides to get

$$\frac{y^2}{16} = 1 - \frac{x^2}{25}, \quad \text{or} \quad \frac{x^2}{25} + \frac{y^2}{16} = 1,$$

the equation of an ellipse with x-intercepts ± 5 and y-intercepts ± 4. Since

$$\sqrt{1 - \frac{x^2}{25}} \geq 0,$$

the only possible values of y are those making $\frac{y}{4} \geq 0$, giving the half-ellipse shown in Figure 20. The half-ellipse in Figure 20 is the graph of a function. The domain is the interval $[-5, 5]$, and the range is $[0, 4]$.

NOW TRY EXERCISE 29. ◀

Translated Ellipses   Just as a circle need not have its center at the origin, an ellipse may also have its center translated away from the origin.

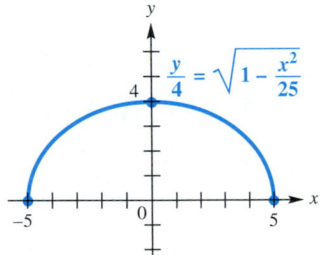

Figure 20

ELLIPSE CENTERED AT (h, k)

An ellipse centered at (h, k) with horizontal major axis of length $2a$ has equation

$$\frac{(x - h)^2}{a^2} + \frac{(y - k)^2}{b^2} = 1.$$

There is a similar result for ellipses having a vertical major axis.

This result can be proved from the definition of an ellipse.

▶ **EXAMPLE 4** **GRAPHING AN ELLIPSE TRANSLATED AWAY FROM THE ORIGIN**

Graph $\dfrac{(x - 2)^2}{9} + \dfrac{(y + 1)^2}{16} = 1$. Give the domain and range.

Solution The graph of this equation is an ellipse centered at $(2, -1)$. As mentioned earlier, ellipses always have $a > b$. For this ellipse, $a = 4$ and $b = 3$. Since $a = 4$ is associated with y^2, the vertices of the ellipse are on the vertical line through $(2, -1)$. Find the vertices by locating two points on the vertical line through $(2, -1)$, one 4 units up from $(2, -1)$ and one 4 units down. The vertices are $(2, 3)$ and $(2, -5)$. Two other points on the ellipse are on the horizontal line through $(2, -1)$, one 3 units to the right and one 3 units to the left.

See the graph in Figure 21. The domain is $[-1, 5]$, and the range is $[-5, 3]$.

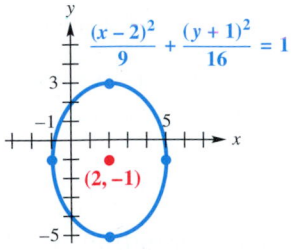

Figure 21

NOW TRY EXERCISE 13. ◀

▶ **Note** As suggested by the graphs in this section, an ellipse is symmetric with respect to its major axis, its minor axis, and its center. *If a = b in the equation of an ellipse, then the graph is a circle.*

Eccentricity The ellipse is the third conic section (or *conic*) we have studied. (The circle and the parabola were the first two.) The fourth conic section, the hyperbola, will be introduced in the next section. All conics can be characterized by one general definition.

CONIC

A **conic** is the set of all points $P(x, y)$ in a plane such that the ratio of the distance from P to a fixed point and the distance from P to a fixed line is constant.

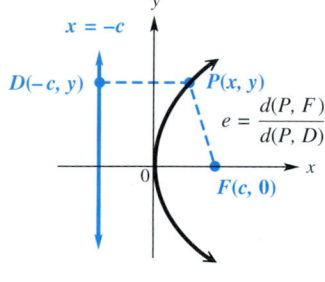

Figure 22

For a parabola, the fixed line is the directrix, and the fixed point is a focus. In Figure 22, the focus is $F(c, 0)$, and the directrix is the line $x = -c$. The constant ratio is called the **eccentricity** of the conic, written *e*. (***This is not the same e as the base of natural logarithms.***)

If the conic is a parabola, then by definition, the distances $d(P, F)$ and $d(P, D)$ in Figure 22 are equal. Thus, every parabola has eccentricity 1.

For an ellipse, eccentricity is a measure of its "roundness." The constant ratio in the definition is $e = \frac{c}{a}$, where (as before) c is the distance from the center of the figure to a focus, and a is the distance from the center to a vertex. By the definition of an ellipse, $a^2 > b^2$ and $c = \sqrt{a^2 - b^2}$. Thus, for the ellipse,

$$0 < c < a$$

$$0 < \frac{c}{a} < 1 \quad \text{Divide by } a. \quad \text{(Section 1.7)}$$

$$0 < e < 1. \quad e = \frac{c}{a}$$

If a is constant, letting c approach 0 would force the ratio $\frac{c}{a}$ to approach 0, which also forces b to approach a (so that $\sqrt{a^2 - b^2} = c$ would approach 0). Since b determines the endpoints of the minor axis, this means that the lengths of the major and minor axes are almost the same, producing an ellipse very close in shape to a circle when e is very close to 0. In a similar manner, if e approaches 1, then b will approach 0.

The path of Earth around the sun is an ellipse that is very nearly circular. In fact, for this ellipse, $e \approx .017$. On the other hand, the path of Halley's comet is a very flat ellipse, with $e \approx .97$. Figure 23 compares ellipses with different eccentricities. The locations of the foci are shown in each case.

The equation of a circle can be written

$$(x - h)^2 + (y - k)^2 = r^2 \quad \text{(Section 2.2)}$$

$$\frac{(x - h)^2}{r^2} + \frac{(y - k)^2}{r^2} = 1. \quad \text{Divide by } r^2.$$

In a circle, the foci coincide with the center, so $a = b$, $c = \sqrt{a^2 - b^2} = 0$, and thus $e = 0$.

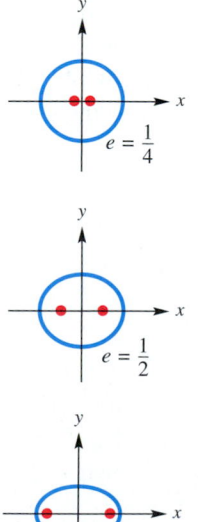

Figure 23

▶ **EXAMPLE 5** **FINDING ECCENTRICITY FROM EQUATIONS OF ELLIPSES**

Find the eccentricity of each ellipse.

(a) $\dfrac{x^2}{9} + \dfrac{y^2}{16} = 1$ **(b)** $5x^2 + 10y^2 = 50$

Solution

(a) Since $16 > 9$, $a^2 = 16$, which gives $a = 4$. Also,

$$c = \sqrt{a^2 - b^2} = \sqrt{16 - 9} = \sqrt{7}.$$

Finally, $\quad e = \dfrac{c}{a} = \dfrac{\sqrt{7}}{4} \approx .66.$

(b) Divide by 50 to obtain $\dfrac{x^2}{10} + \dfrac{y^2}{5} = 1$. Here, $a^2 = 10$, with $a = \sqrt{10}$. Now, find c.

$$c = \sqrt{10 - 5} = \sqrt{5} \quad \text{and} \quad e = \dfrac{\sqrt{5}}{\sqrt{10}} = \sqrt{\dfrac{5}{10}} = \sqrt{\dfrac{1}{2}} \approx .71$$

NOW TRY EXERCISES 37 AND 39. ◀

Applications of Ellipses

▶ **EXAMPLE 6** **APPLYING THE EQUATION OF AN ELLIPSE TO THE ORBIT OF A PLANET**

The orbit of the planet Mars is an ellipse with the sun at one focus. The eccentricity of the ellipse is .0935, and the closest distance that Mars comes to the sun is 128.5 million mi. (*Source: The World Almanac and Book of Facts.*) Find the maximum distance of Mars from the sun.

Solution Figure 24 shows the orbit of Mars with the origin at the center of the ellipse and the sun at one focus. Mars is closest to the sun when Mars is at the right endpoint of the major axis and farthest from the sun when Mars is at the left endpoint. Therefore, the smallest distance is $a - c$, and the greatest distance is $a + c$.

Since $a - c = 128.5$, $c = a - 128.5$. Using $e = \dfrac{c}{a}$,

$$\dfrac{a - 128.5}{a} = .0935$$

$a - 128.5 = .0935a$ Multiply by a.

$.9065a = 128.5$ Subtract $.0935a$; add 128.5.

$a \approx 141.8.$ Divide by .9065.

Then $\qquad\qquad\qquad c = 141.8 - 128.5 = 13.3$

and $\qquad\qquad\qquad a + c = 141.8 + 13.3 = 155.1.$

The maximum distance of Mars from the sun is about 155.1 million mi.

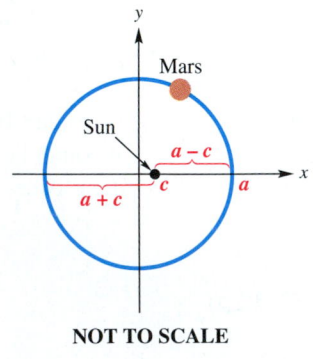

NOT TO SCALE

Figure 24

NOW TRY EXERCISE 45. ◀

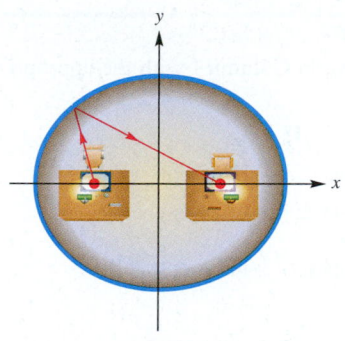

Aerial view of Old House Chamber

Figure 25

When a ray of light or sound emanating from one focus of an ellipse bounces off the ellipse, it passes through the other focus. See Figure 25. As mentioned in the chapter introduction, this reflecting property is responsible for whispering galleries. John Quincy Adams was able to listen in on his opponents' conversations because his desk was positioned at one of the foci beneath the ellipsoidal ceiling and his opponents were located across the room at the other focus.

A lithotripter is a machine used to crush kidney stones using shock waves. The patient is placed in an elliptical tub with the kidney stone at one focus of the ellipse. A beam is projected from the other focus to the tub so that it reflects to hit the kidney stone. See Figures 26 and 27.

▶ **EXAMPLE 7**   MODELING THE REFLECTIVE PROPERTY OF ELLIPSES

If a lithotripter is based on the ellipse

$$\frac{x^2}{36} + \frac{y^2}{27} = 1,$$

determine how many units both the kidney stone and the source of the beam must be placed from the center of the ellipse.

Solution   The kidney stone and the source of the beam must be placed at the foci, $(c, 0)$ and $(-c, 0)$. Here $a^2 = 36$ and $b^2 = 27$, so

$$c = \sqrt{a^2 - b^2} = \sqrt{36 - 27} = \sqrt{9} = 3.$$

Thus, the foci are $(3, 0)$ and $(-3, 0)$, so the kidney stone and the source both must be placed on a line 3 units from the center. See Figure 27.

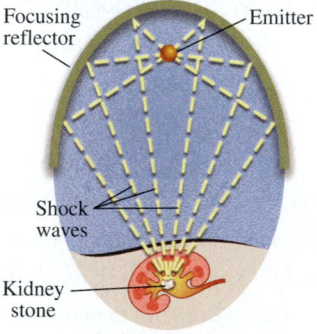

The top of an ellipse is illustrated in this depiction of how a lithotripter crushes a kidney stone.

Figure 26

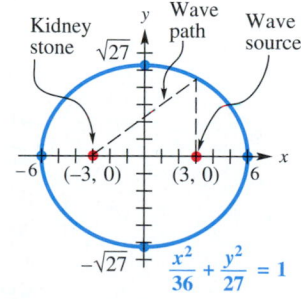

Figure 27

NOW TRY EXERCISE 49. ◀

10.2 Exercises

1. *Concept Check* Match each equation of an ellipse in Column I with the appropriate description in Column II.

I	**II**
(a) $36x^2 + 9y^2 = 324$	**A.** x-intercepts ± 3; y-intercepts ± 6
(b) $9x^2 + 36y^2 = 324$	**B.** x-intercepts ± 4; y-intercepts ± 5
(c) $\dfrac{x^2}{25} = 1 - \dfrac{y^2}{16}$	**C.** x-intercepts ± 6; y-intercepts ± 3
(d) $\dfrac{x^2}{16} = 1 - \dfrac{y^2}{25}$	**D.** x-intercepts ± 5; y-intercepts ± 4

2. *Concept Check* Determine whether or not each equation is that of an ellipse. If it is not, state the kind of graph the equation has.

(a) $x^2 + 4y^2 = 4$ **(b)** $x^2 + y^2 = 4$ **(c)** $x^2 + y = 4$ **(d)** $\dfrac{x}{4} + \dfrac{y}{25} = 1$

Graph each ellipse. Identify the domain, range, center, vertices, endpoints of the minor axis, and the foci in each figure. See Examples 1 and 4.

3. $\dfrac{x^2}{25} + \dfrac{y^2}{9} = 1$ **4.** $\dfrac{x^2}{16} + \dfrac{y^2}{25} = 1$ **5.** $\dfrac{x^2}{9} + y^2 = 1$

6. $\dfrac{x^2}{36} + \dfrac{y^2}{16} = 1$ **7.** $y^2 = 81 - 9x^2$ **8.** $16y^2 = 64 - 4x^2$

9. $4x^2 = 100 - 25y^2$ **10.** $4x^2 = 16 - y^2$

11. $\dfrac{(x-2)^2}{25} + \dfrac{(y-1)^2}{4} = 1$ **12.** $\dfrac{(x+2)^2}{16} + \dfrac{(y+1)^2}{9} = 1$

13. $\dfrac{(x+3)^2}{16} + \dfrac{(y-2)^2}{36} = 1$ **14.** $\dfrac{(x-1)^2}{9} + \dfrac{(y+3)^2}{25} = 1$

Write an equation for each ellipse. See Example 2.

15. x-intercepts ± 5; y-intercepts ± 4

16. x-intercepts $\pm\sqrt{15}$; y-intercepts ± 4

17. major axis with length 6; foci at $(0, 2)$, $(0, -2)$

18. minor axis with length 4; foci at $(-5, 0)$, $(5, 0)$

19. center at $(3, 1)$; minor axis vertical, with length 8; $c = 3$

20. center at $(-2, 7)$; major axis vertical, with length 10; $c = 2$

21. vertices at $(4, 9)$, $(4, 1)$; minor axis with length 6

22. foci at $(-3, -3)$, $(7, -3)$; the point $(2, -7)$ on ellipse

23. foci at $(0, -3)$, $(0, 3)$; the point $(8, 3)$ on ellipse

24. foci at $(-4, 0)$, $(4, 0)$; sum of distances from foci to point on ellipse is 9 (*Hint:* Consider one of the vertices.)

25. foci at $(0, 4)$, $(0, -4)$; sum of distances from foci to point on ellipse is 10

26. eccentricity $\frac{1}{2}$; vertices at $(-4, 0)$, $(4, 0)$

27. eccentricity $\frac{3}{4}$; foci at $(0, -2)$, $(0, 2)$

28. eccentricity $\frac{2}{3}$; foci at $(0, -9)$, $(0, 9)$

Graph each equation. Give the domain and range. Identify any that are graphs of functions. See Example 3.

29. $\dfrac{y}{2} = \sqrt{1 - \dfrac{x^2}{25}}$

30. $\dfrac{x}{4} = \sqrt{1 - \dfrac{y^2}{9}}$

31. $x = -\sqrt{1 - \dfrac{y^2}{64}}$

32. $y = -\sqrt{1 - \dfrac{x^2}{100}}$

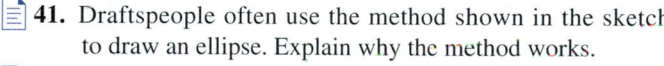 *Determine the two equations necessary to graph each ellipse with a graphing calculator, and graph it in the viewing window indicated. See Figure 18.*

33. $\dfrac{x^2}{16} + \dfrac{y^2}{4} = 1;$
$[-4.7, 4.7]$ by $[-3.1, 3.1]$

34. $\dfrac{x^2}{4} + \dfrac{y^2}{25} = 1;$
$[-9.4, 9.4]$ by $[-6.2, 6.2]$

35. $\dfrac{(x-3)^2}{25} + \dfrac{y^2}{9} = 1;$
$[-9.4, 9.4]$ by $[-6.2, 6.2]$

36. $\dfrac{x^2}{36} + \dfrac{(y+4)^2}{4} = 1;$
$[-9.4, 9.4]$ by $[-6.2, 6.2]$

Find the eccentricity of each ellipse. If necessary, round to the nearest hundredth. See Example 5.

37. $\dfrac{x^2}{3} + \dfrac{y^2}{4} = 1$

38. $\dfrac{x^2}{8} + \dfrac{y^2}{4} = 1$

39. $4x^2 + 7y^2 = 28$

40. $x^2 + 25y^2 = 25$

41. Draftspeople often use the method shown in the sketch to draw an ellipse. Explain why the method works.

42. Explain how the method of Exercise 41 can be modified to draw a circle.

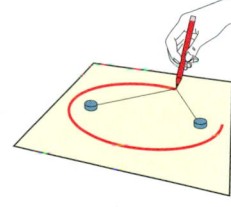

Solve each problem. See Examples 6 and 7.

43. *Height of an Overpass*   A one-way road passes under an overpass in the shape of half an ellipse, 15 ft high at the center and 20 ft wide. Assuming a truck is 12 ft wide, what is the tallest truck that can pass under the overpass?

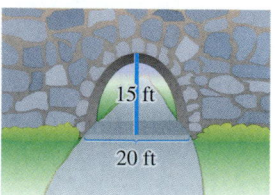

NOT TO SCALE

44. *Height and Width of an Overpass*   An arch has the shape of half an ellipse. The equation of the ellipse is $100x^2 + 324y^2 = 32{,}400$, where x and y are in meters.

 (a) How high is the center of the arch?
 (b) How wide is the arch across the bottom?

NOT TO SCALE

45. *Orbit of Halley's Comet*   The famous Halley's comet last passed by Earth in February 1986 and will next return in 2062. It has an elliptical orbit of eccentricity .9673 with the sun at one focus. The greatest distance of the comet from the sun is 3281 million mi. (*Source: The World Almanac and Book of Facts.*) Find the least distance between Halley's comet and the sun.

46. *(Modeling) Orbit of a Satellite*   The coordinates in miles for the orbit of the artificial satellite Explorer VII can be modeled by the equation

$$\frac{x^2}{a^2} + \frac{y^2}{b^2} = 1,$$

where $a = 4465$ and $b = 4462$. Earth's center is located at one focus of the elliptical orbit. (*Source:* Loh, W., *Dynamics and Thermodynamics of Planetary Entry,* Prentice-Hall, 1963; Thomson, W., *Introduction to Space Dynamics,* John Wiley and Sons, 1961.)

(a) Graph both the orbit of Explorer VII and of Earth on the same coordinate axes if the average radius of Earth is 3960 mi. Use the window $[-6750, 6750]$ by $[-4500, 4500]$.

(b) Determine the maximum and minimum heights of the satellite above Earth's surface.

47. *(Modeling) Orbits of Satellites*   Neptune and Pluto both have elliptical orbits with the sun at one focus. Neptune's orbit has $a = 30.1$ astronomical units (AU) with an eccentricity of $e = .009$, whereas Pluto's orbit has $a = 39.4$ and $e = .249$. (*Source:* Zeilik, M., S. Gregory, and E. Smith, *Introductory Astronomy and Astrophysics,* Fourth Edition, Saunders College Publishers, 1998.)

(a) Position the sun at the origin and determine equations that model each orbit.

(b) Graph both equations on the same coordinate axes. Use the window $[-60, 60]$ by $[-40, 40]$.

48. *(Modeling) The Roman Colosseum*

(a) The Roman Colosseum is an ellipse with major axis 620 ft and minor axis 513 ft. Find the distance between the foci of this ellipse.

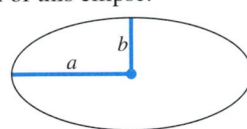

(b) A formula for the approximate circumference of an ellipse is

$$C \approx 2\pi \sqrt{\frac{a^2 + b^2}{2}},$$

where a and b are the lengths shown in the figure. Use this formula to find the approximate circumference of the Roman Colosseum.

49. *Design of a Lithotripter*   Suppose a lithotripter is based on the ellipse with equation

$$\frac{x^2}{36} + \frac{y^2}{9} = 1.$$

How far from the center of the ellipse must the kidney stone and the source of the beam be placed? Give the exact answer.

50. *Design of a Lithotripter*   Rework Exercise 49 if the equation of the ellipse is $9x^2 + 4y^2 = 36$.

Quiz (Sections 10.1–10.2)

1. *Concept Check*   Match each equation of a conic section in Column I with the appropriate description or descriptions in Column II.

I	II
(a) $x + 3 = 4(y - 1)^2$	**A.** circle; center $(-3, 1)$
(b) $(x + 3)^2 + (y - 1)^2 = 81$	**B.** parabola; opens right
(c) $25(x - 2)^2 + (y - 1)^2 = 100$	**C.** ellipse; major axis horizontal
(d) $\dfrac{(x - 2)^2}{16} + \dfrac{(y - 1)^2}{9} = 1$	**D.** parabola; vertex $(-3, 1)$
(e) $-2(x + 3)^2 + 1 = y$	**E.** ellipse; center $(2, 1)$

Write an equation for each conic section.

2. parabola with vertex $(-1, 2)$ and focus $(2, 2)$
3. parabola with vertex at the origin; through $\left(\sqrt{10}, -5\right)$; opening downward
4. ellipse with center $(3, -2)$; $a = 5$; $c = 3$; major axis vertical
5. ellipse with foci at $(-3, 3)$ and $(-3, 11)$; major axis of length 10

Graph each conic section. If it is a parabola, give the vertex, focus, directrix, and axis. If it is an ellipse, give the center, vertices, and foci.

6. $y = (x + 3)^2 - 4$ 7. $4x^2 + 9y^2 = 36$ 8. $8(x + 1) = (y + 3)^2$

9. $\dfrac{(x + 3)^2}{25} + \dfrac{(y + 2)^2}{36} = 1$ 10. $x = -4y^2 - 4y - 3$

10.3 Hyperbolas

Equations and Graphs of Hyperbolas ▪ **Translated Hyperbolas** ▪ **Eccentricity**

Equations and Graphs of Hyperbolas   An ellipse was defined as the set of all points in a plane the sum of whose distances from two fixed points is a constant. A *hyperbola* is defined similarly.

> ### HYPERBOLA
>
> A **hyperbola** is the set of all points in a plane such that the absolute value of the difference of the distances from two fixed points is constant. The two fixed points are called the **foci** of the hyperbola.

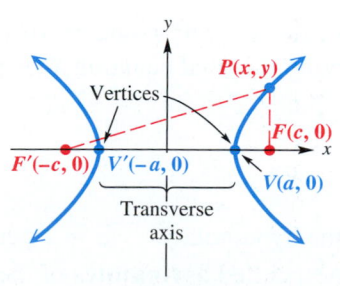

Figure 28

Suppose a hyperbola has center at the origin and foci at $F'(-c, 0)$ and $F(c, 0)$. See Figure 28. The midpoint of the segment $F'F$ is the center of the hyperbola and the points $V'(-a, 0)$ and $V(a, 0)$ are the **vertices** of the hyperbola. The line segment $V'V$ is the **transverse axis** of the hyperbola.

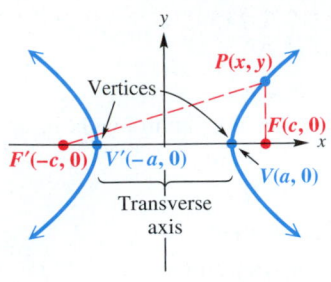

Figure 28 (repeated)

As with the ellipse,

$$d(V, F') - d(V, F) = (c + a) - (c - a) = 2a,$$

so the constant in the definition is $2a$, and

$$\left| d(P, F') - d(P, F) \right| = 2a$$

for any point $P(x, y)$ on the hyperbola. The distance formula and algebraic manipulation similar to that used for finding an equation for an ellipse (see Exercise 60) produce the result

$$\frac{x^2}{a^2} - \frac{y^2}{c^2 - a^2} = 1.$$

Letting $b^2 = c^2 - a^2$ gives

$$\frac{x^2}{a^2} - \frac{y^2}{b^2} = 1$$

as an equation of the hyperbola in Figure 28. Letting $y = 0$ shows that the x-intercepts are $\pm a$. If $x = 0$, the equation becomes

$$\frac{0^2}{a^2} - \frac{y^2}{b^2} = 1 \qquad \text{Let } x = 0.$$

$$-\frac{y^2}{b^2} = 1$$

$$y^2 = -b^2, \quad \text{Multiply by } -b^2.$$

which has no real number solutions, showing that this hyperbola has no y-intercepts. Again, see Figure 28.

Starting with the equation for a hyperbola and solving for y, we get

$$\frac{x^2}{a^2} - \frac{y^2}{b^2} = 1$$

$$\frac{x^2}{a^2} - 1 = \frac{y^2}{b^2} \qquad \text{Subtract 1; add } \frac{y^2}{b^2}.$$

$$\frac{x^2 - a^2}{a^2} = \frac{y^2}{b^2} \qquad \text{Write the left side as a single fraction.}$$

> **Remember both the positive and negative square roots.**

$$y = \pm \frac{b}{a} \sqrt{x^2 - a^2}. \qquad \text{Take square roots on both sides (Section 1.6); multiply by } b.$$

If x^2 is very large in comparison to a^2, the difference $x^2 - a^2$ would be very close to x^2. If this happens, then the points satisfying the final equation above would be very close to one of the lines

$$y = \pm \frac{b}{a} x.$$

Thus, as $|x|$ gets larger and larger, the points of the hyperbola $\frac{x^2}{a^2} - \frac{y^2}{b^2} = 1$ get closer and closer to the lines $y = \pm \frac{b}{a} x$. These lines, called **asymptotes** of the hyperbola, are useful when sketching the graph.

▶ **EXAMPLE 1**   **USING ASYMPTOTES TO GRAPH A HYPERBOLA**

Graph $\dfrac{x^2}{25} - \dfrac{y^2}{49} = 1$. Sketch the asymptotes, and find the coordinates of the vertices and foci. Give the domain and range.

Algebraic Solution

For this hyperbola, $a = 5$ and $b = 7$. With these values,

$$y = \pm \frac{b}{a}x \quad \text{becomes} \quad y = \pm \frac{7}{5}x. \quad \text{Asymptotes}$$

If we choose $x = 5$, then $y = \pm 7$. Choosing $x = -5$ also gives $y = \pm 7$. These four points—$(5, 7)$, $(5, -7)$, $(-5, 7)$, and $(-5, -7)$—are the corners of the rectangle shown in Figure 29.

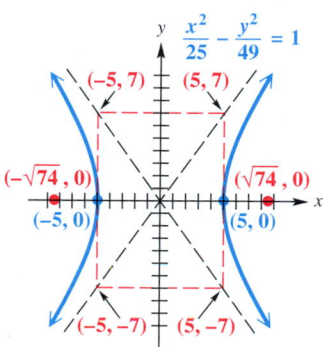

Figure 29

The extended diagonals of this rectangle, called the **fundamental rectangle,** are the asymptotes of the hyperbola. Since $a = 5$, the vertices of the hyperbola are $(5, 0)$ and $(-5, 0)$, as shown in Figure 29. We find the foci by letting

$$c^2 = a^2 + b^2 = 25 + 49 = 74, \quad \text{so} \quad c = \sqrt{74}.$$

Therefore, the foci are $\left(\sqrt{74}, 0\right)$ and $\left(-\sqrt{74}, 0\right)$. The domain is $(-\infty, -5] \cup [5, \infty)$, and the range is $(-\infty, \infty)$.

Graphing Calculator Solution

The graph of a hyperbola is not the graph of a function. We must solve for y in $\frac{x^2}{25} - \frac{y^2}{49} = 1$ to get equations of the **two** functions

$$y = \pm \frac{7}{5} \sqrt{x^2 - 25}.$$

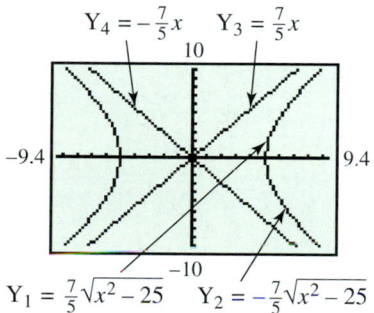

Figure 30

The graph of Y_1 is the upper portion of each branch of the hyperbola shown in Figure 30, and the graph of Y_2 is the lower portion of each branch. We could enter $Y_2 = -Y_1$ to get the part of the graph below the x-axis.

The asymptotes are also shown. We can use tracing to observe how the branches of the hyperbola approach the asymptotes.

NOW TRY EXERCISE 5. ◀

▶ **Note** When graphing hyperbolas, remember that the fundamental rectangle and the asymptotes are not actually parts of the graph. They are simply aids in sketching the graph.

While $a > b$ for an ellipse, examples would show that for hyperbolas, it is possible that $a > b$, $a < b$, or $a = b$. If the foci of a hyperbola are on the y-axis, the equation of the hyperbola has the form

$$\frac{y^2}{a^2} - \frac{x^2}{b^2} = 1, \quad \text{with asymptotes} \quad y = \pm \frac{a}{b}x.$$

> ▶ **Caution** If the foci of a hyperbola are on the x-axis, the asymptotes have equations $y = \pm\frac{b}{a}x$, while foci on the y-axis lead to asymptotes $y = \pm\frac{a}{b}x$. To avoid errors, write the equation of the hyperbola in either the form
>
> $$\frac{x^2}{a^2} - \frac{y^2}{b^2} = 1 \qquad \text{or} \qquad \frac{y^2}{a^2} - \frac{x^2}{b^2} = 1,$$
>
> and replace 1 with 0. Solving the resulting equation for y produces the proper equations for the asymptotes.

The basic information on hyperbolas is summarized as follows.

STANDARD FORMS OF EQUATIONS FOR HYPERBOLAS

The hyperbola with center at the origin and equation

$$\frac{x^2}{a^2} - \frac{y^2}{b^2} = 1$$

has vertices $(\pm a, 0)$, asymptotes $y = \pm\frac{b}{a}x$, and foci $(\pm c, 0)$, where $c^2 = a^2 + b^2$.

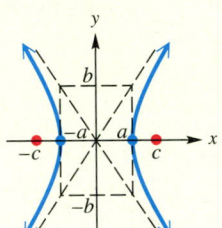

Transverse axis on x-axis

The hyperbola with center at the origin and equation

$$\frac{y^2}{a^2} - \frac{x^2}{b^2} = 1$$

has vertices $(0, \pm a)$, asymptotes $y = \pm\frac{a}{b}x$, and foci $(0, \pm c)$, where $c^2 = a^2 + b^2$.

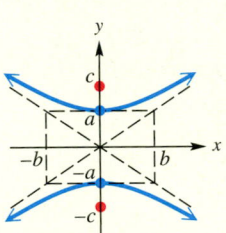

Transverse axis on y-axis

▶ **EXAMPLE 2**   **GRAPHING A HYPERBOLA**

Graph $25y^2 - 4x^2 = 100$. Give the domain and range.

Solution $\dfrac{y^2}{4} - \dfrac{x^2}{25} = 1$ Divide by 100; standard form.

This hyperbola is centered at the origin, has foci on the y-axis, and has vertices $(0, 2)$ and $(0, -2)$. To find the equations of the asymptotes, replace 1 with 0.

$$\frac{y^2}{4} - \frac{x^2}{25} = 0$$

$$\frac{y^2}{4} = \frac{x^2}{25} \qquad \text{Add } \tfrac{x^2}{25}.$$

$$y^2 = \frac{4x^2}{25} \qquad \text{Multiply by 4.}$$

Remember both the positive and negative square roots.

$$y = \pm\frac{2}{5}x \qquad \text{Square root property (Section 1.4)}$$

To graph the asymptotes, use the points $(5, 2)$, $(5, -2)$, $(-5, 2)$, and $(-5, -2)$ to determine the fundamental rectangle. The diagonals of this rectangle are the asymptotes for the graph, as shown in Figure 31. The domain of the relation is $(-\infty, \infty)$, and the range is $(-\infty, -2] \cup [2, \infty)$.

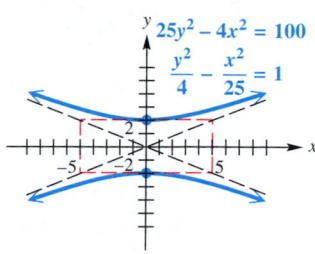

Figure 31

NOW TRY EXERCISE 13. ◄

Translated Hyperbolas

In the graph of each hyperbola considered so far, the center is the origin and the asymptotes pass through the origin. This feature holds in general; *the asymptotes of any hyperbola pass through the center of the hyperbola.* Like an ellipse, a hyperbola can have its center translated away from the origin.

► **EXAMPLE 3**   **GRAPHING A HYPERBOLA TRANSLATED AWAY FROM THE ORIGIN**

Graph $\dfrac{(y + 2)^2}{9} - \dfrac{(x + 3)^2}{4} = 1$. Give the domain and range.

Solution   This equation represents a hyperbola centered at $(-3, -2)$. For this vertical hyperbola, $a = 3$ and $b = 2$. The x-values of the vertices are -3. Locate the y-values of the vertices by taking the y-value of the center, -2, and adding and subtracting 3. Thus, the vertices are $(-3, 1)$ and $(-3, -5)$. The asymptotes have slopes $\pm\frac{3}{2}$ and pass through the center $(-3, -2)$. The equations of the asymptotes can be found either by using the point-slope form of the equation of a line or by replacing 1 with 0 in the equation of the hyperbola as was done in Example 2. Using the point-slope form, we get

$$[y - (-2)] = \pm\frac{3}{2}[x - (-3)] \quad \text{Point-slope form (Section 2.5)}$$

$$y = -2 \pm \frac{3}{2}(x + 3). \quad \text{Solve for } y.$$

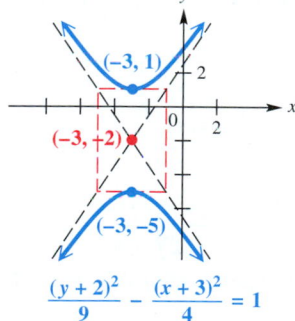

$\dfrac{(y + 2)^2}{9} - \dfrac{(x + 3)^2}{4} = 1$

Figure 32

The graph is shown in Figure 32. The domain of the relation is $(-\infty, \infty)$, and the range is $(-\infty, -5] \cup [1, \infty)$.

NOW TRY EXERCISE 17. ◄

Eccentricity

If we apply the definition of eccentricity from the previous section to the hyperbola, we get

$$e = \frac{\sqrt{a^2 + b^2}}{a} = \frac{c}{a}.$$

Since $c > a$, we have $e > 1$. Narrow hyperbolas have e near 1, and wide hyperbolas have large e. See Figure 33 on the next page.

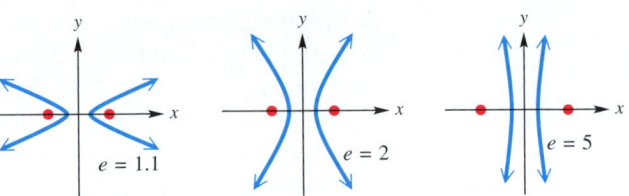

Figure 33

▶ **EXAMPLE 4**   **FINDING ECCENTRICITY FROM THE EQUATION OF A HYPERBOLA**

Find the eccentricity of the hyperbola $\dfrac{x^2}{9} - \dfrac{y^2}{4} = 1$.

Solution   Here $a^2 = 9$; thus, $a = 3$, $c = \sqrt{9 + 4} = \sqrt{13}$, and

$$e = \frac{c}{a} = \frac{\sqrt{13}}{3} \approx 1.2.$$

NOW TRY EXERCISE 27. ◀

▶ **EXAMPLE 5**   **FINDING THE EQUATION OF A HYPERBOLA**

Find the equation of the hyperbola with eccentricity 2 and foci at $(-9, 5)$ and $(-3, 5)$.

Solution   Since the foci have the same y-coordinate, the line through them, and therefore the hyperbola, is horizontal. The center of the hyperbola is halfway between the two foci at $(-6, 5)$. The distance from each focus to the center is $c = 3$. Since $e = \frac{c}{a}$, $a = \frac{c}{e} = \frac{3}{2}$ and $a^2 = \frac{9}{4}$. Thus,

$$b^2 = c^2 - a^2 = 9 - \frac{9}{4} = \frac{27}{4}.$$

The equation of the hyperbola is

$$\frac{(x + 6)^2}{\frac{9}{4}} - \frac{(y - 5)^2}{\frac{27}{4}} = 1, \qquad \text{or} \qquad \frac{4(x + 6)^2}{9} - \frac{4(y - 5)^2}{27} = 1.$$

Simplify complex fractions. **(Section R.5)**

NOW TRY EXERCISE 45. ◀

The following chart summarizes our discussion of eccentricity in this chapter.

Conic	Eccentricity
Parabola	$e = 1$
Circle	$e = 0$
Ellipse	$e = \dfrac{c}{a}$ and $0 < e < 1$
Hyperbola	$e = \dfrac{c}{a}$ and $e > 1$

CONNECTIONS Ships and planes often use a location-finding system called LORAN. With this system, a radio transmitter at *M* in the figure sends out a series of pulses. When each pulse is received at transmitter *S*, it then sends out a pulse. A ship at *P* receives pulses from both *M* and *S*. A receiver on the ship measures the difference in the arrival times of the pulses. The navigator then consults a special map showing hyperbolas that correspond to the differences in arrival times (which give the distances d_1 and d_2 in the figure). In this way the ship can be located as lying on a branch of a particular hyperbola.

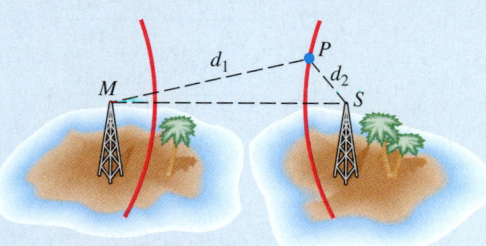

FOR DISCUSSION OR WRITING
Suppose in the figure $d_1 = 80$ mi, $d_2 = 30$ mi, and the distance between the transmitters is 100 mi. Use the definition of a hyperbola to find an equation of the hyperbola on which the ship is located.

10.3 **Exercises** *Concept Check* *Match each equation with the correct graph.*

1. $\dfrac{x^2}{25} + \dfrac{y^2}{9} = 1$

2. $\dfrac{x^2}{9} + \dfrac{y^2}{25} = 1$

3. $\dfrac{x^2}{9} - \dfrac{y^2}{25} = 1$

4. $\dfrac{x^2}{25} - \dfrac{y^2}{9} = 1$

A.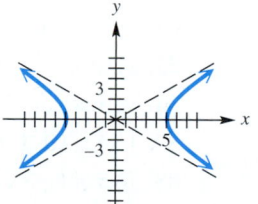

B.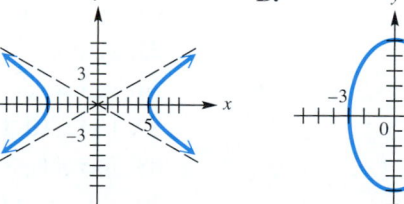

C.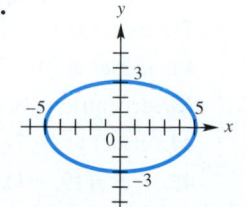

D.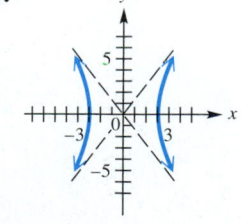

Graph each hyperbola. Give the domain, range, center, vertices, foci, and equations of the asymptotes for each figure. See Examples 1–3.

5. $\dfrac{x^2}{16} - \dfrac{y^2}{9} = 1$

6. $\dfrac{x^2}{25} - \dfrac{y^2}{144} = 1$

7. $\dfrac{y^2}{25} - \dfrac{x^2}{49} = 1$

8. $\dfrac{y^2}{64} - \dfrac{x^2}{4} = 1$

9. $x^2 - y^2 = 9$

10. $x^2 - 4y^2 = 64$

11. $9x^2 - 25y^2 = 225$

12. $y^2 - 4x^2 = 16$

13. $4y^2 - 25x^2 = 100$

14. $\dfrac{x^2}{4} - y^2 = 4$

15. $9x^2 - 4y^2 = 1$

16. $25y^2 - 9x^2 = 1$

17. $\dfrac{(y-7)^2}{36} - \dfrac{(x-4)^2}{64} = 1$

18. $\dfrac{(x+6)^2}{144} - \dfrac{(y+4)^2}{81} = 1$

19. $\dfrac{(x+3)^2}{16} - \dfrac{(y-2)^2}{9} = 1$

20. $\dfrac{(y+5)^2}{4} - \dfrac{(x-1)^2}{16} = 1$

21. $16(x+5)^2 - (y-3)^2 = 1$

22. $4(x+9)^2 - 25(y+6)^2 = 100$

Graph each equation. Give the domain and range. Identify any that are graphs of functions. See Example 3 in the previous section.

23. $\dfrac{y}{3} = \sqrt{1 + \dfrac{x^2}{16}}$

24. $\dfrac{x}{3} = -\sqrt{1 + \dfrac{y^2}{25}}$

25. $5x = -\sqrt{1 + 4y^2}$

26. $3y = \sqrt{4x^2 - 16}$

Find the eccentricity to the nearest tenth of each hyperbola. See Example 4.

27. $\dfrac{x^2}{8} - \dfrac{y^2}{8} = 1$

28. $\dfrac{x^2}{2} - \dfrac{y^2}{18} = 1$

29. $16y^2 - 8x^2 = 16$

30. $8y^2 - 2x^2 = 16$

Write an equation for each hyperbola. See Examples 4 and 5.

31. x-intercepts ± 3; foci at $(-5, 0)$, $(5, 0)$

32. y-intercepts ± 12; foci at $(0, -15)$, $(0, 15)$

33. vertices at $(0, 6)$, $(0, -6)$; asymptotes $y = \pm\frac{1}{2}x$

34. vertices at $(-10, 0)$, $(10, 0)$; asymptotes $y = \pm 5x$

35. vertices at $(-3, 0)$, $(3, 0)$; passing through $(-6, -1)$

36. vertices at $(0, 5)$, $(0, -5)$; passing through $(-3, 10)$

37. foci at $(0, \sqrt{13})$, $(0, -\sqrt{13})$; asymptotes $y = \pm 5x$

38. foci at $(-3\sqrt{5}, 0)$, $(3\sqrt{5}, 0)$; asymptotes $y = \pm 2x$

39. vertices at $(4, 5)$, $(4, 1)$; asymptotes $y = 3 \pm 7(x - 4)$

40. vertices at $(5, -2)$, $(1, -2)$; asymptotes $y = -2 \pm \frac{3}{2}(x - 3)$

41. center at $(1, -2)$; focus at $(-2, -2)$; vertex at $(-1, -2)$

42. center at $(9, -7)$; focus at $(9, -17)$; vertex at $(9, -13)$

43. eccentricity 3; center at $(0, 0)$; vertex at $(0, 7)$

44. center at $(8, 7)$; focus at $(3, 7)$; eccentricity $\frac{5}{3}$

45. foci at $(9, -1)$, $(-11, -1)$; eccentricity $\frac{25}{9}$

46. vertices at $(2, 10)$, $(2, 2)$; eccentricity $\frac{5}{4}$

Determine the two equations necessary to graph each hyperbola with a graphing calculator, and graph it in the viewing window indicated. See Example 1.

47. $\dfrac{x^2}{4} - \dfrac{y^2}{16} = 1$; $[-9.4, 9.4]$ by $[-10, 10]$

48. $\dfrac{x^2}{25} - \dfrac{y^2}{49} = 1$; $[-9.4, 9.4]$ by $[-10, 10]$

49. $4y^2 - 36x^2 = 144$; $[-10, 10]$ by $[-15, 15]$

50. $y^2 - 9x^2 = 9$; $[-10, 10]$ by $[-10, 10]$

RELATING CONCEPTS

For individual or collaborative investigation
(Exercises 51–56)

The graph of $\frac{x^2}{4} - y^2 = 1$ is a hyperbola. We know that the graph of this hyperbola approaches its asymptotes as $|x|$ gets larger and larger. **Work Exercises 51–56 in order,** *to see the relationship between the hyperbola and one of its asymptotes.*

51. Solve $\frac{x^2}{4} - y^2 = 1$ for y, and choose the positive square root.

52. Find the equation of the asymptote with positive slope.

53. Use a calculator to evaluate the y-coordinate of the point where $x = 50$ on the graph of the portion of the hyperbola represented by the equation obtained in Exercise 51. Round your answer to the nearest hundredth.

54. Find the y-coordinate of the point where $x = 50$ on the graph of the asymptote found in Exercise 52.

55. Compare your results in Exercises 53 and 54. How do they support the following statement?

> When $x = 50$, the graph of the function defined by the equation found in Exercise 51 lies *below* the graph of the asymptote found in Exercise 52.

56. What do you think will happen if we choose x-values larger than 50?

Solve each problem.

57. *(Modeling) Atomic Structure* In 1911 Ernest Rutherford discovered the basic structure of the atom by "shooting" positively charged alpha particles with a speed of 10^7 m per sec at a piece of gold foil 6×10^{-7} m thick. Only a small percentage of the alpha particles struck a gold nucleus head-on and were deflected directly back toward their source. The rest of the particles often followed a hyperbolic trajectory because they were repelled by positively charged gold nuclei. As a result of this famous experiment, Rutherford proposed that the atom was composed of mostly empty space with a small and dense nucleus. The figure shows an alpha particle A initially approaching a gold nucleus N and being deflected at an angle $\theta = 90°$. N is located at a focus of the hyperbola, and the trajectory of A passes through a vertex of the hyperbola. (*Source:* Semat, H. and J. Albright, *Introduction to Atomic and Nuclear Physics,* Fifth Edition, International Thomson Publishing, 1972.)

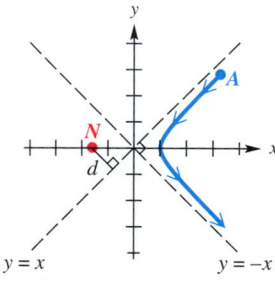

(a) Determine the equation of the trajectory of the alpha particle if $d = 5 \times 10^{-14}$ m.

(b) What was the minimum distance between the centers of the alpha particle and the gold nucleus?

58. *(Modeling) Design of a Sports Complex*   Two buildings in a sports complex are shaped and positioned like a portion of the branches of the hyperbola

$$400x^2 - 625y^2 = 250{,}000,$$

where x and y are in meters.

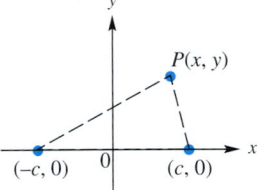

NOT TO SCALE

(a) How far apart are the buildings at their closest point?
(b) Find the distance d in the figure.

59. *Sound Detection*   Microphones are placed at points $(-c, 0)$ and $(c, 0)$. An explosion occurs at point $P(x, y)$ having positive x-coordinate. See the figure. The sound is detected at the closer microphone t seconds before being detected at the farther microphone. Assume that sound travels at a speed of 330 m per sec, and show that P must be on the hyperbola

$$\frac{x^2}{330^2 t^2} - \frac{y^2}{4c^2 - 330^2 t^2} = \frac{1}{4}.$$

60. Suppose a hyperbola has center at the origin, foci at $F'(-c, 0)$ and $F(c, 0)$, and $|d(P, F') - d(P, F)| = 2a$. Let $b^2 = c^2 - a^2$, and show that an equation of the hyperbola is

$$\frac{x^2}{a^2} - \frac{y^2}{b^2} = 1.$$

10.4   Summary of the Conic Sections

Characteristics ▪ Identifying Conic Sections ▪ Geometric Definition of Conic Sections

Characteristics   The graphs of parabolas, circles, ellipses, and hyperbolas are called conic sections since each graph can be obtained by cutting a cone with a plane, as suggested by Figure 1 at the beginning of the chapter. All conic sections of the types presented in this chapter have equations of the general form

$$Ax^2 + Cy^2 + Dx + Ey + F = 0,$$

where either A or C must be nonzero.

The special characteristics of the general equation of each conic section presented earlier are summarized below.

Conic Section	Characteristic	Example
Parabola	Either $A = 0$ or $C = 0$, but not both.	$x^2 - y - 4 = 0$ $-x + y^2 - 4y = 0$
Circle	$A = C \neq 0$	$x^2 + y^2 - 16 = 0$
Ellipse	$A \neq C, AC > 0$	$25x^2 + 16y^2 - 400 = 0$
Hyperbola	$AC < 0$	$x^2 - y^2 - 1 = 0$

The following chart summarizes our work with conic sections.

Equation	Graph	Description	Identification
$(x - h)^2 = 4p(y - k)$ **or** $y - k = a(x - h)^2$	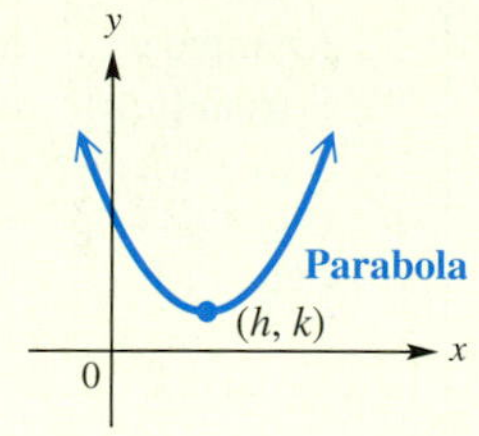	Opens up if $p > 0$ (or $a > 0$), down if $p < 0$ (or $a < 0$). Vertex is (h, k). Axis of symmetry is $x = h$.	x^2-term y is not squared.
$(y - k)^2 = 4p(x - h)$ **or** $x - h = a(y - k)^2$	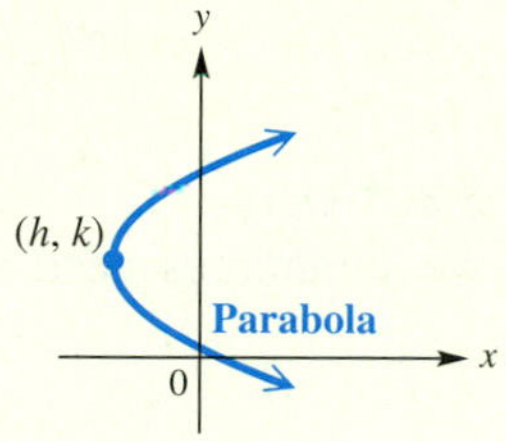	Opens to the right if $p > 0$ (or $a > 0$), to the left if $p < 0$ (or $a < 0$). Vertex is (h, k). Axis of symmetry is $y = k$.	y^2-term x is not squared.
$(x - h)^2 + (y - k)^2 = r^2$	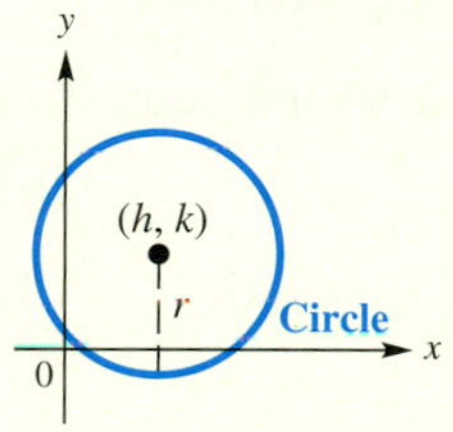	Center is (h, k), radius is r.	x^2- and y^2-terms have the same positive coefficient.
$\dfrac{x^2}{a^2} + \dfrac{y^2}{b^2} = 1 \quad (a > b)$	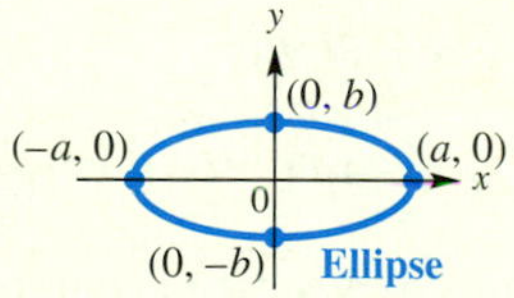	x-intercepts are a and $-a$. y-intercepts are b and $-b$. Horizontal major axis, length $= 2a$.	x^2- and y^2-terms have different positive coefficients.
$\dfrac{x^2}{b^2} + \dfrac{y^2}{a^2} = 1 \quad (a > b)$	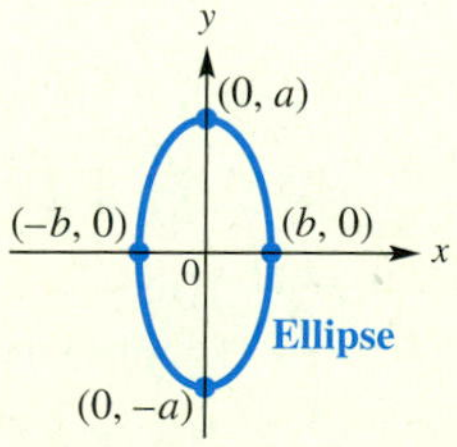	x-intercepts are b and $-b$. y-intercepts are a and $-a$. Vertical major axis, length $= 2a$.	x^2- and y^2-terms have different positive coefficients.
$\dfrac{x^2}{a^2} - \dfrac{y^2}{b^2} = 1$	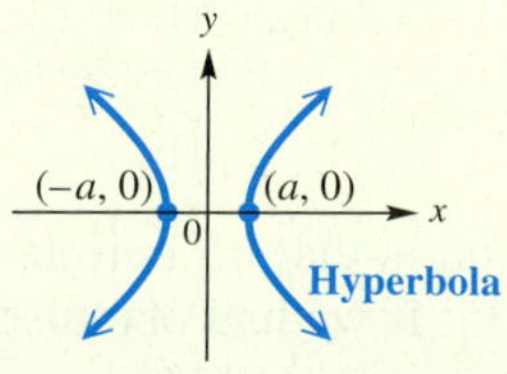	x-intercepts are a and $-a$. Asymptotes are found from (a, b), $(a, -b)$, $(-a, -b)$, and $(-a, b)$.	x^2-term has a positive coefficient. y^2-term has a negative coefficient.

(continued)

Equation	Graph	Description	Identification
$\dfrac{y^2}{a^2} - \dfrac{x^2}{b^2} = 1$	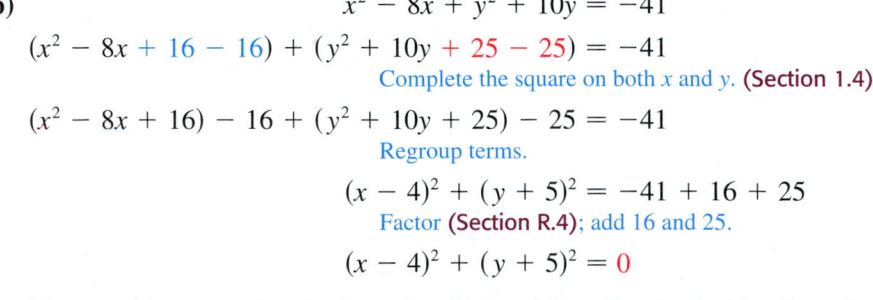 Hyperbola	y-intercepts are a and $-a$. Asymptotes are found from (b, a), $(b, -a)$, $(-b, -a)$, and $(-b, a)$.	y^2-term has a positive coefficient. x^2-term has a negative coefficient.

NOW TRY EXERCISES 1, 5, 7, AND 11. ◀

Identifying Conic Sections To recognize the type of graph that a given conic section has, we sometimes need to transform the equation into a more familiar form.

▶ **EXAMPLE 1** DETERMINING TYPES OF CONIC SECTIONS FROM EQUATIONS

Determine the type of conic section represented by each equation, and graph it.

(a) $x^2 = 25 + 5y^2$ **(b)** $x^2 - 8x + y^2 + 10y = -41$

(c) $4x^2 - 16x + 9y^2 + 54y = -61$ **(d)** $x^2 - 6x + 8y - 7 = 0$

Solution

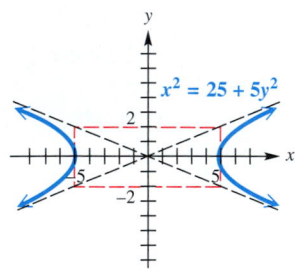

Figure 34

(a)
$$x^2 = 25 + 5y^2$$
$$x^2 - 5y^2 = 25 \qquad \text{Subtract } 5y^2.$$

Divide each term by 25.

$$\frac{x^2}{25} - \frac{y^2}{5} = 1 \qquad \text{Divide by 25.}$$

The equation represents a hyperbola centered at the origin, with asymptotes

$$\frac{x^2}{25} - \frac{y^2}{5} = 0, \qquad \text{or} \qquad y = \pm \frac{\sqrt{5}}{5}x.$$

Remember both the positive and negative square roots.

The x-intercepts are ± 5; the graph is shown in Figure 34.

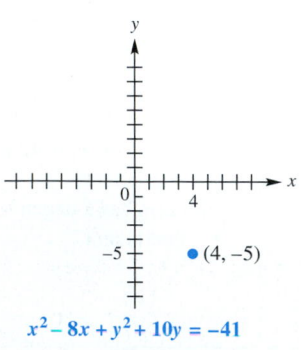

$x^2 - 8x + y^2 + 10y = -41$

Figure 35

(b)
$$x^2 - 8x + y^2 + 10y = -41$$
$$(x^2 - 8x + 16 - 16) + (y^2 + 10y + 25 - 25) = -41$$

Complete the square on both x and y. **(Section 1.4)**

$$(x^2 - 8x + 16) - 16 + (y^2 + 10y + 25) - 25 = -41$$

Regroup terms.

$$(x - 4)^2 + (y + 5)^2 = -41 + 16 + 25$$

Factor **(Section R.4)**; add 16 and 25.

$$(x - 4)^2 + (y + 5)^2 = 0$$

The resulting equation is that of a circle with radius 0; that is, the point $(4, -5)$. See Figure 35. If we had obtained a negative number on the right (instead of 0), the equation would have no solution at all, and there would be no graph.

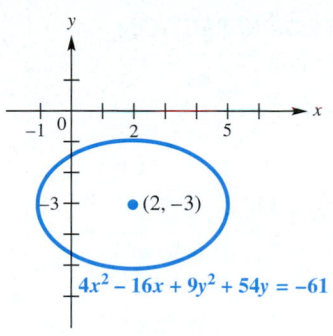

Figure 36

(c) In $4x^2 - 16x + 9y^2 + 54y = -61$, the coefficients of the x^2- and y^2-terms are unequal and both positive, so the equation might represent an ellipse but not a circle. (It might also represent a single point or no points at all.)

$$4x^2 - 16x + 9y^2 + 54y = -61$$

$$4(x^2 - 4x \qquad) + 9(y^2 + 6y \qquad) = -61 \qquad \text{Factor out 4;}$$
$$\text{factor out 9.}$$

$$4(x^2 - 4x + 4 - 4) + 9(y^2 + 6y + 9 - 9) = -61 \qquad \text{Complete the square.}$$

$$4(x^2 - 4x + 4) - 16 + 9(y^2 + 6y + 9) - 81 = -61 \qquad \text{Distributive property}$$
$$\text{(Section R.2)}$$

> **Multiply**
> $4(-4) = -16.$

$$4(x - 2)^2 + 9(y + 3)^2 = 36 \qquad \text{Factor; add 97.}$$

$$\frac{(x - 2)^2}{9} + \frac{(y + 3)^2}{4} = 1 \qquad \text{Divide by 36.}$$

This equation represents an ellipse having center $(2, -3)$ and graph as shown in Figure 36.

(d) Since only one variable in $x^2 - 6x + 8y - 7 = 0$ is squared (x, and not y), the equation represents a parabola. Get the term with y (the variable that is not squared) alone on one side.

$$x^2 - 6x + 8y - 7 = 0$$

$$8y = -x^2 + 6x + 7 \qquad \text{Isolate the } y\text{-term.}$$

$$8y = -(x^2 - 6x \qquad) + 7 \qquad \text{Regroup terms;}$$
$$\text{factor out } -1.$$

$$8y = -(x^2 - 6x + 9 - 9) + 7 \qquad \text{Complete the square.}$$

$$8y = -(x^2 - 6x + 9) + 9 + 7 \qquad \text{Distributive property;}$$
$$-(-9) = +9$$

$$8y = -(x - 3)^2 + 16 \qquad \text{Factor; add.}$$

$$y = -\frac{1}{8}(x - 3)^2 + 2 \qquad \text{Multiply by } \tfrac{1}{8}.$$

$$y - 2 = -\frac{1}{8}(x - 3)^2 \qquad \text{Subtract 2.}$$

The parabola has vertex $(3, 2)$ and opens down, as shown in the graph in Figure 37. An equivalent form for this parabola is

$$(x - 3)^2 = -8(y - 2),$$

as seen in **Section 10.1.**

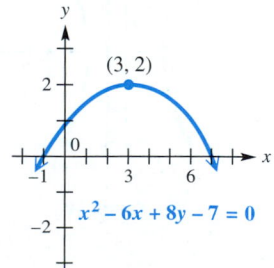

Figure 37

NOW TRY EXERCISES 25, 27, 29, AND 33. ◀

The next example is designed to serve as a warning about a common error.

▶ **EXAMPLE 2** DETERMINING THE TYPE OF CONIC SECTION FROM ITS EQUATION

Identify the graph of $4y^2 - 16y - 9x^2 + 18x = -43$.

Solution

$$4y^2 - 16y - 9x^2 + 18x = -43$$

$$4(y^2 - 4y \qquad) - 9(x^2 - 2x \qquad) = -43 \qquad \text{Factor out 4;}$$
$$\text{factor out } -9.$$

$$4(y^2 - 4y + 4 - 4) - 9(x^2 - 2x + 1 - 1) = -43 \qquad \text{Complete the square.}$$
$$4(y^2 - 4y + 4) - 16 - 9(x^2 - 2x + 1) + 9 = -43 \qquad \text{Distributive property}$$
$$4(y - 2)^2 - 9(x - 1)^2 = -36 \qquad \text{Factor; add 16}$$
$$\text{and subtract 9.}$$

Because of the -36, we might think that this equation does not have a graph. However, the minus sign in the middle on the left shows that the graph is that of a hyperbola.

> **Be careful here!**

$$\frac{(x - 1)^2}{4} - \frac{(y - 2)^2}{9} = 1 \qquad \text{Divide by } -36; \text{ rearrange terms.}$$

This hyperbola has center $(1, 2)$. The graph is shown in Figure 38.

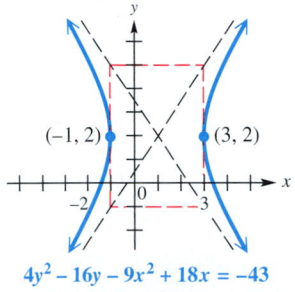

$4y^2 - 16y - 9x^2 + 18x = -43$

Figure 38

> **NOW TRY EXERCISE 35.** ◀

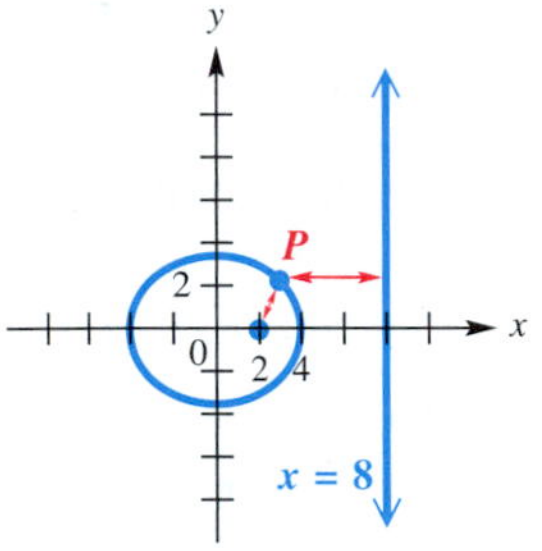

Geometric Definition of Conic Sections In **Section 10.1,** a parabola was defined as the set of points in a plane equidistant from a fixed point (focus) and a fixed line (directrix). A parabola has eccentricity 1. This definition can be generalized to apply to the ellipse and the hyperbola. Figure 39 shows an ellipse with $a = 4$, $c = 2$, and $e = \frac{1}{2}$. The line $x = 8$ is shown also. For any point P on the ellipse,

$$[\text{distance of } P \text{ from the focus}] = \frac{1}{2}[\text{distance of } P \text{ from the line}].$$

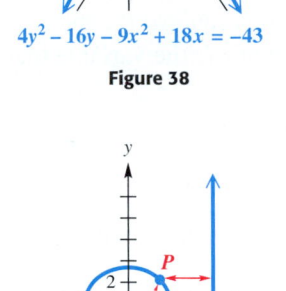

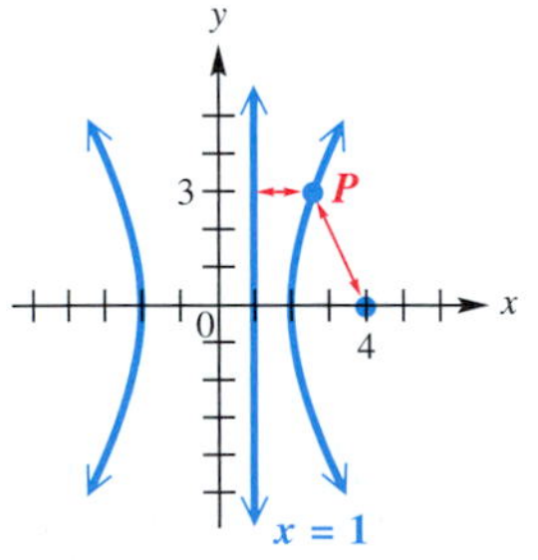

$x = 8$

Figure 39

Figure 40 shows a hyperbola with $a = 2$, $c = 4$, and $e = 2$, along with the line $x = 1$. For any point P on the hyperbola,

$$[\text{distance of } P \text{ from the focus}] = 2[\text{distance of } P \text{ from the line}].$$

The following geometric definition applies to all conic sections except circles, which have $e = 0$.

$x = 1$

Figure 40

GEOMETRIC DEFINITION OF A CONIC SECTION

Given a fixed point F (focus), a fixed line L (directrix), and a positive number e, the set of all points P in the plane such that

$$[\text{distance of } P \text{ from } F] = e \cdot [\text{distance of } P \text{ from } L]$$

is a conic section of eccentricity e. The conic section is a parabola when $e = 1$, an ellipse when $0 < e < 1$, and a hyperbola when $e > 1$.

> **NOW TRY EXERCISE 51.** ◀

10.4 Exercises

The equation of a conic section is given in a familiar form. Identify the type of graph that each equation has, without actually graphing. See the summary chart in this section.

1. $x^2 + y^2 = 144$

2. $(x - 2)^2 + (y + 3)^2 = 25$

3. $y = 2x^2 + 3x - 4$

4. $x = 3y^2 + 5y - 6$

5. $x - 1 = -3(y - 4)^2$

6. $\dfrac{x^2}{25} + \dfrac{y^2}{36} = 1$

7. $\dfrac{x^2}{49} + \dfrac{y^2}{100} = 1$

8. $x^2 - y^2 = 1$

9. $\dfrac{x^2}{4} - \dfrac{y^2}{16} = 1$

10. $\dfrac{(x + 2)^2}{9} + \dfrac{(y - 4)^2}{16} = 1$

11. $\dfrac{x^2}{25} - \dfrac{y^2}{25} = 1$

12. $y + 7 = 4(x + 3)^2$

13. $\dfrac{x^2}{4} = 1 - \dfrac{y^2}{9}$

14. $\dfrac{x^2}{4} = 1 + \dfrac{y^2}{9}$

15. $\dfrac{(x + 3)^2}{16} + \dfrac{(y - 2)^2}{16} = 1$

16. $x^2 = 25 - y^2$

17. $x^2 - 6x + y = 0$

18. $11 - 3x = 2y^2 - 8y$

19. $4(x - 3)^2 + 3(y + 4)^2 = 0$

20. $2x^2 - 8x + 2y^2 + 20y = 12$

21. $x - 4y^2 - 8y = 0$

22. $x^2 + 2x = -4y$

23. $4x^2 - 24x + 5y^2 + 10y + 41 = 0$

24. $6x^2 - 12x + 6y^2 - 18y + 25 = 0$

Determine the type of conic section represented by each equation, and graph it. See Examples 1 and 2.

25. $\dfrac{x^2}{4} + \dfrac{y^2}{4} = -1$

26. $\dfrac{x^2}{4} + \dfrac{y^2}{4} = 1$

27. $x^2 = 25 + y^2$

28. $9x^2 + 36y^2 = 36$

29. $x^2 = 4y - 8$

30. $\dfrac{(x - 4)^2}{8} + \dfrac{(y + 1)^2}{2} = 0$

31. $y^2 - 4y = x + 4$

32. $(x + 7)^2 + (y - 5)^2 + 4 = 0$

33. $3x^2 + 6x + 3y^2 - 12y = 12$

34. $-4x^2 + 8x + y^2 + 6y = -6$

35. $4x^2 - 8x + 9y^2 - 36y = -4$

36. $3x^2 + 12x + 3y^2 = 0$

Solve each problem.

37. Identify the type of conic section consisting of the set of all points in the plane for which the sum of the distances from the points $(5, 0)$ and $(-5, 0)$ is 14.

38. Identify the type of conic section consisting of the set of all points in the plane for which the absolute value of the difference of the distances from the points $(3, 0)$ and $(-3, 0)$ is 2.

39. Identify the type of conic section consisting of the set of all points in the plane for which the distance from the point $(3, 0)$ is one and one-half times the distance from the line $x = \frac{4}{3}$.

40. Identify the type of conic section consisting of the set of all points in the plane for which the distance from the point $(2, 0)$ is one-third of the distance from the line $x = 10$.

Find the eccentricity of each conic section. The point shown on the x-axis is a focus and the line shown is a directrix.

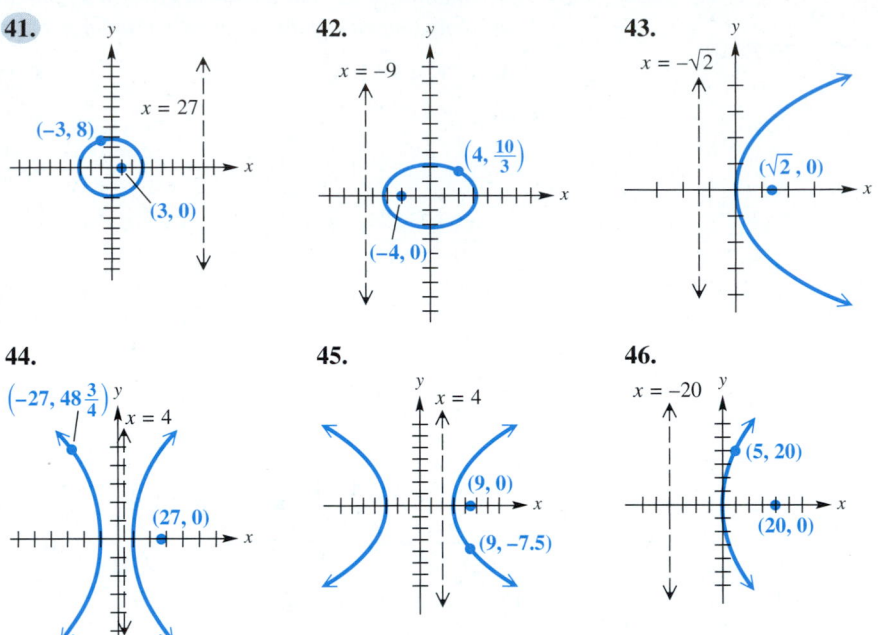

41.

$x = 27$
$(-3, 8)$
$(3, 0)$

42.

$x = -9$
$\left(4, \frac{10}{3}\right)$
$(-4, 0)$

43.

$x = -\sqrt{2}$
$(\sqrt{2}, 0)$

44.

$\left(-27, 48\frac{3}{4}\right)$
$x = 4$
$(27, 0)$

45.

$x = 4$
$(9, 0)$
$(9, -7.5)$

46.

$x = -20$
$(5, 20)$
$(20, 0)$

Satellite Trajectory When a satellite is near Earth, its orbital trajectory may trace out a hyperbola, a parabola, or an ellipse. The type of trajectory depends on the satellite's velocity V in meters per second. It will be hyperbolic if $V > \frac{k}{\sqrt{D}}$, parabolic if $V = \frac{k}{\sqrt{D}}$, and elliptical if $V < \frac{k}{\sqrt{D}}$, where $k = 2.82 \times 10^7$ is a constant and D is the distance in meters from the satellite to the center of Earth. Use this information in Exercises 47–49. (*Source:* Loh, W., *Dynamics and Thermodynamics of Planetary Entry,* Prentice-Hall, 1963; Thomson, W., *Introduction to Space Dynamics,* John Wiley and Sons, 1961.)

47. When the artificial satellite Explorer IV was at a maximum distance D of 42.5×10^6 m from Earth's center, it had a velocity V of 2090 m per sec. Determine the shape of its trajectory.

48. If a satellite is scheduled to leave Earth's gravitational influence, its velocity must be increased so that its trajectory changes from elliptical to hyperbolic. Determine the minimum increase in velocity necessary for Explorer IV to escape Earth's gravitational influence when $D = 42.5 \times 10^6$ m.

49. Explain why it is easier to change a satellite's trajectory from an ellipse to a hyperbola when D is maximum rather than minimum.

50. If $Ax^2 + Cy^2 + Dx + Ey + F = 0$ is the general equation of an ellipse, find its center point by completing the square.

51. Graph the ellipse $\frac{x^2}{16} + \frac{y^2}{12} = 1$ with a graphing calculator. Trace to find the coordinates of several points on the ellipse. For each of these points P, verify that

$$[\text{distance of } P \text{ from } (2, 0)] = \frac{1}{2}[\text{distance of } P \text{ from the line } x = 8].$$

52. Graph the hyperbola $\frac{x^2}{4} - \frac{y^2}{12} = 1$ with a graphing calculator. Trace to find the coordinates of several points on the hyperbola. For each of these points P, verify that

$$[\text{distance of } P \text{ from } (4, 0)] = 2[\text{distance of } P \text{ from the line } x = 1].$$

Chapter 10 Summary

10.1	conic sections	10.2	ellipse		center	10.3	hyperbola
	parabola		foci		vertices		transverse axis
	focus		major axis		conic		asymptotes
	directrix		minor axis		eccentricity		fundamental rectangle

NEW SYMBOLS

e eccentricity

QUICK REVIEW

CONCEPTS	EXAMPLES

10.1 Parabolas

Parabola with Vertical Axis and Vertex (0, 0)
The parabola with focus $(0, p)$ and directrix $y = -p$ has equation $x^2 = 4py$. The parabola has vertical axis $x = 0$ and opens up if $p > 0$ or down if $p < 0$.

Parabola with Horizontal Axis and Vertex (0, 0)
The parabola with focus $(p, 0)$ and directrix $x = -p$ has equation $y^2 = 4px$. The parabola has horizontal axis $y = 0$ and opens to the right if $p > 0$ or to the left if $p < 0$.

Equation Forms for Translated Parabolas
A parabola with vertex (h, k) has an equation of the form

$$(x - h)^2 = 4p(y - k) \quad \text{Vertical axis}$$

or $\quad (y - k)^2 = 4p(x - h), \quad \text{Horizontal axis}$

where the focus is distance p or $-p$ from the vertex.

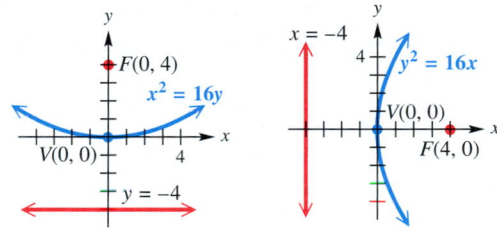

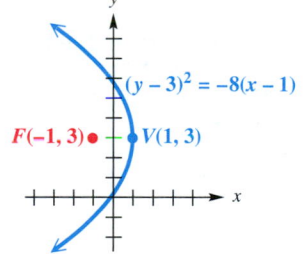

10.2 Ellipses

Standard Forms of Equations for Ellipses
The ellipse with center at the origin and equation

$$\frac{x^2}{a^2} + \frac{y^2}{b^2} = 1 \quad (a > b)$$

has vertices $(\pm a, 0)$, endpoints of the minor axis $(0, \pm b)$, and foci $(\pm c, 0)$, where $c^2 = a^2 - b^2$.
   The ellipse with center at the origin and equation

$$\frac{x^2}{b^2} + \frac{y^2}{a^2} = 1 \quad (a > b)$$

has vertices $(0, \pm a)$, endpoints of the minor axis $(\pm b, 0)$, and foci $(0, \pm c)$, where $c^2 = a^2 - b^2$.

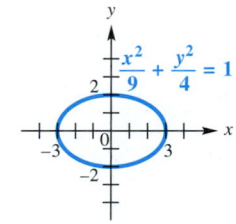

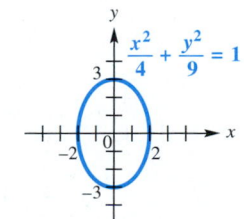

(continued)

CONCEPTS	EXAMPLES

Translated Ellipses

The preceding equations can be extended to ellipses having center (h, k) by replacing x and y with $x - h$ and $y - k$, respectively.

10.3 Hyperbolas

Standard Forms of Equations for Hyperbolas

The hyperbola with center at the origin and equation

$$\frac{x^2}{a^2} - \frac{y^2}{b^2} = 1$$

has vertices $(\pm a, 0)$, asymptotes $y = \pm \frac{b}{a}x$, and foci $(\pm c, 0)$, where $c^2 = a^2 + b^2$.

The hyperbola with center at the origin and equation

$$\frac{y^2}{a^2} - \frac{x^2}{b^2} = 1$$

has vertices $(0, \pm a)$, asymptotes $y = \pm \frac{a}{b}x$, and foci $(0, \pm c)$, where $c^2 = a^2 + b^2$.

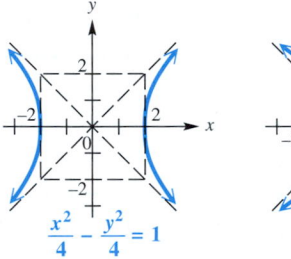
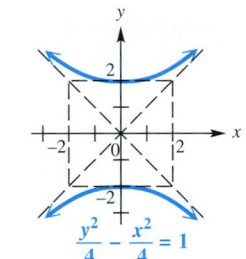

Translated Hyperbolas

The preceding equations can be extended to hyperbolas having center (h, k) by replacing x and y with $x - h$ and $y - k$, respectively.

10.4 Summary of the Conic Sections

Conic sections in this chapter have equations that can be written in the form

$$Ax^2 + Cy^2 + Dx + Ey + F = 0.$$

Conic Section	Characteristic	Example
Parabola	Either $A = 0$ or $C = 0$, but not both.	$x^2 - y - 4 = 0$ $-x + y^2 - 4y = 0$
Circle	$A = C \neq 0$	$x^2 + y^2 - 16 = 0$
Ellipse	$A \neq C, AC > 0$	$25x^2 + 16y^2 - 400 = 0$
Hyperbola	$AC < 0$	$x^2 - y^2 - 1 = 0$

See the summary chart on pages 981 and 982.

CHAPTER 10 ▶ Review Exercises

Graph each parabola. In Exercises 1–4, give the domain, range, vertex, and axis. In Exercises 5–8, give the domain, range, focus, directrix, and axis.

1. $x = 4(y - 5)^2 + 2$ **2.** $x = -(y + 1)^2 - 7$ **3.** $x = 5y^2 - 5y + 3$

4. $x = 2y^2 - 4y + 1$ **5.** $y^2 = -\dfrac{2}{3}x$ **6.** $y^2 = 2x$

7. $3x^2 = y$ **8.** $x^2 + 2y = 0$

Write an equation for each parabola with vertex at the origin.

9. focus $(4, 0)$

10. focus $(0, -3)$

11. through $(-3, 4)$, opening up

12. through $(2, 5)$, opening right

An equation of a conic section is given. Identify the type of conic section. It may be necessary to transform the equation into a more familiar form.

13. $y^2 + 9x^2 = 9$ **14.** $9x^2 - 16y^2 = 144$ **15.** $3y^2 - 5x^2 = 30$

16. $y^2 + x = 4$ **17.** $4x^2 - y = 0$ **18.** $x^2 + y^2 = 25$

19. $4x^2 - 8x + 9y^2 + 36y = -4$ **20.** $9x^2 - 18x - 4y^2 - 16y - 43 = 0$

Concept Check Match each equation with its calculator graph. In all cases except choice B, Xscl = Yscl = 1.

21. $4x^2 + y^2 = 36$

22. $x = 2y^2 + 3$

23. $(x - 2)^2 + (y + 3)^2 = 36$

24. $\dfrac{x^2}{36} + \dfrac{y^2}{9} = 1$

25. $(y - 1)^2 - (x - 2)^2 = 36$

26. $y^2 = 36 + 4x^2$

A.

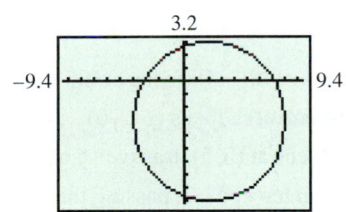

B.

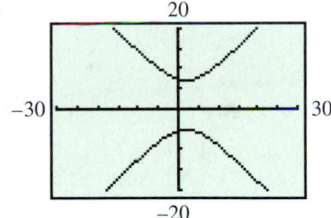

In this screen, Xscl = Yscl = 5.

C.

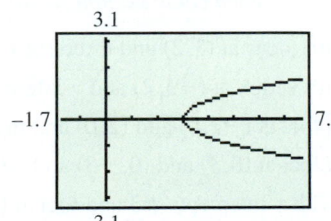

D.

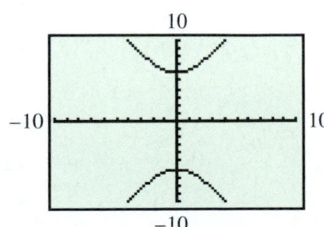

E.

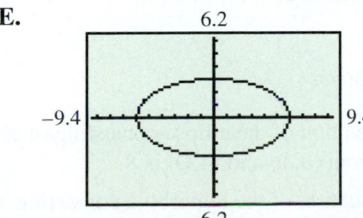

F.

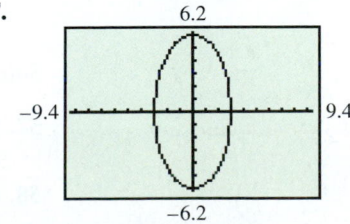

Graph each relation and identify the graph. Give the domain, range, coordinates of the vertices for each ellipse or hyperbola, and equations of the asymptotes for each hyperbola. Give the domain and range for each circle.

27. $\dfrac{x^2}{4} + \dfrac{y^2}{9} = 1$

28. $\dfrac{x^2}{16} + \dfrac{y^2}{4} = 1$

29. $\dfrac{x^2}{64} - \dfrac{y^2}{36} = 1$

30. $\dfrac{y^2}{25} - \dfrac{x^2}{9} = 1$

31. $\dfrac{(x+1)^2}{16} + \dfrac{(y-1)^2}{16} = 1$

32. $(x-3)^2 + (y+2)^2 = 9$

33. $4x^2 + 9y^2 = 36$

34. $x^2 = 16 + y^2$

35. $\dfrac{(x-3)^2}{4} + (y+1)^2 = 1$

36. $\dfrac{(x-2)^2}{9} + \dfrac{(y+3)^2}{4} = 1$

37. $\dfrac{(y+2)^2}{4} - \dfrac{(x+3)^2}{9} = 1$

38. $\dfrac{(x+1)^2}{16} - \dfrac{(y-2)^2}{4} = 1$

39. $x^2 - 4x + y^2 + 6y = -12$

40. $4x^2 + 8x + 25y^2 - 250y = -529$

41. $5x^2 + 20x + 2y^2 - 8y = -18$

42. $-4x^2 + 8x + 4y^2 + 8y = 16$

Graph each relation. Give the domain and range, and identify any that are graphs of functions.

43. $\dfrac{x}{3} = -\sqrt{1 - \dfrac{y^2}{16}}$

44. $x = -\sqrt{1 - \dfrac{y^2}{36}}$

45. $y = -\sqrt{1 + x^2}$

46. $y = -\sqrt{1 - \dfrac{x^2}{25}}$

Write an equation for each conic section with center at the origin.

47. ellipse; vertex at $(0, -4)$, focus at $(0, -2)$

48. ellipse; x-intercept 6, focus at $(2, 0)$

49. hyperbola; focus at $(0, 5)$, transverse axis of length 8

50. hyperbola; y-intercept -2, passing through $(2, 3)$

Write an equation for each conic section satisfying the given conditions.

51. parabola with focus at $(3, 2)$ and directrix $x = -3$

52. parabola with vertex at $(-3, 2)$ and y-intercepts 5 and -1

53. ellipse with foci at $(-2, 0)$ and $(2, 0)$ and major axis of length 10

54. ellipse with foci at $(0, 3)$ and $(0, -3)$ and vertex at $(0, -7)$

55. hyperbola with x-intercepts ± 3 and foci at $(-5, 0)$, $(5, 0)$

56. hyperbola with foci at $(0, 12)$, $(0, -12)$ and asymptotes $y = \pm x$

Solve each problem.

57. Find the equation of the ellipse consisting of all points in the plane the sum of whose distances from $(0, 0)$ and $(4, 0)$ is 8.

58. Find the equation of the hyperbola consisting of all points in the plane for which the absolute value of the difference of the distances from $(0, 0)$ and $(0, 4)$ is 2.

59. Calculator graphs are shown in Figures A–D. Arrange the figures in order so that the first in the list has the smallest eccentricity and the rest have eccentricities in increasing order.

A.

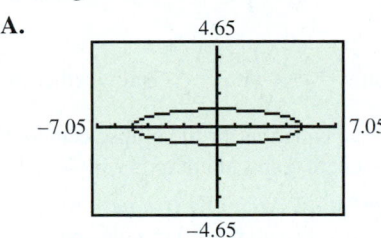

B.

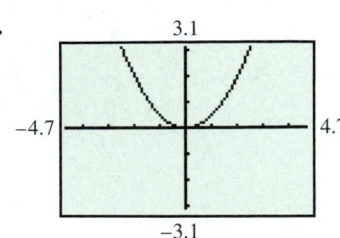

C.

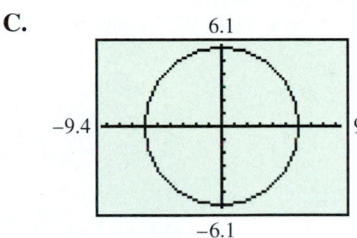

D.

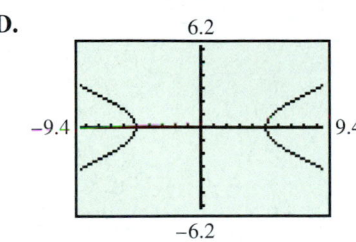

60. *Orbit of Venus* The orbit of Venus is an ellipse with the sun at one of the foci. The eccentricity of the orbit is $e = .006775$, and the major axis has length 134.5 million mi. (*Source: The World Almanac and Book of Facts.*) Find the least and greatest distances of Venus from the sun.

61. *Orbit of a Comet* The comet Swift-Tuttle has an elliptical orbit of eccentricity $e = .964$, with the sun at one of the foci. Find the equation of the comet given that the closest it comes to the sun is 89 million mi.

62. Find the equation of the hyperbola consisting of all points P in the plane for which the absolute value of the difference of the distances of P from $(-5, 0)$ and $(5, 0)$ is 8. Then graph the hyperbola with a graphing calculator and trace to find the coordinates of several points on the graph of the hyperbola. For each of these points, verify that the absolute value of the differences of the distances is indeed 8.

CHAPTER 10 ▶ Test

Graph each parabola. Give the domain, range, vertex, and axis.

1. $y = -x^2 + 6x$

2. $x = 4y^2 + 8y$

3. Give the coordinates of the focus and the equation of the directrix for the parabola with equation $x = 8y^2$.

4. Write an equation for the parabola with vertex $(2, 3)$, passing through the point $(-18, 1)$, and opening to the left.

5. Explain how to determine just by looking at the equation whether a parabola has a vertical or a horizontal axis, and whether it opens up, down, to the left, or to the right.

Graph each ellipse. Give the domain and range.

6. $\dfrac{(x-8)^2}{100} + \dfrac{(y-5)^2}{49} = 1$

7. $16x^2 + 4y^2 = 64$

8. Graph $y = -\sqrt{1 - \dfrac{x^2}{36}}$. Tell whether the graph is that of a function.

9. Write an equation for the ellipse centered at the origin having horizontal major axis with length 6 and minor axis with length 4.

10. *Height of the Arch of a Bridge*   An arch of a bridge has the shape of the top half of an ellipse. The arch is 40 ft wide and 12 ft high at the center. Find the equation of the complete ellipse. Find the height of the arch 10 ft from the center of the bottom.

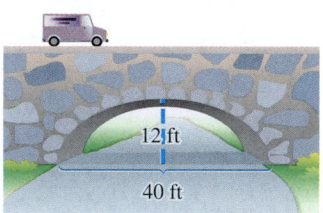

Graph each hyperbola. Give the domain, range, and equations of the asymptotes.

11. $\dfrac{x^2}{4} - \dfrac{y^2}{4} = 1$

12. $9x^2 - 4y^2 = 36$

13. Find the equation of the hyperbola with y-intercepts ± 5 and foci at $(0, -6)$ and $(0, 6)$.

Identify the type of graph, if any, defined by each equation.

14. $x^2 + 8x + y^2 - 4y + 2 = 0$

15. $5x^2 + 10x - 2y^2 - 12y - 23 = 0$

16. $3x^2 + 10y^2 - 30 = 0$

17. $x^2 - 4y = 0$

18. $(x + 9)^2 + (y - 3)^2 = 0$

19. $x^2 + 4x + y^2 - 6y + 30 = 0$

20. The screen shown here gives the graph of

$$\dfrac{x^2}{25} - \dfrac{y^2}{49} = 1$$

as generated by a graphing calculator. What two functions Y_1 and Y_2 were used to obtain the graph?

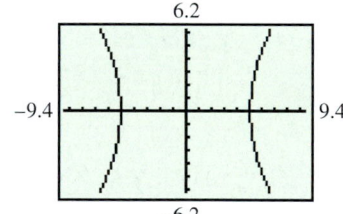

Quantitative Reasoning

Does a rugby player need to know algebra?

A rugby field is similar to a modern football field with the exception that the goalpost, which is 18.5 ft wide, is located on the goal line instead of at the back of the endzone. The rugby equivalent of a touchdown, called a *try,* is scored by touching the ball down beyond the goal line. After a try is scored, the scoring team can earn extra points by kicking the ball through the goalposts. The ball must be placed somewhere on the line perpendicular to the goal line and passing through the point where the try was scored. See the figure below on the left. If that line passes through the goalposts, then the kicker should place the ball at whatever distance he is most comfortable. If the line passes outside the goalposts, then the player might choose the point on the line where angle θ in the figure on the left is as large as possible. The problem of determining this optimal point is similar to a problem posed in 1471 by the astronomer Regiomontanus. (*Source:* Maor, E., *Trigonometric Delights.* Princeton, NJ: Princeton University Press, 1998.)

   The figure on the right below shows a vertical line segment *AB*, where *A* and *B* are *a* and *b* units above the horizontal axis, respectively. If point *P* is located on the axis at a distance of *x* units from point *Q*, then angle θ is greatest when $x = \sqrt{ab}$.

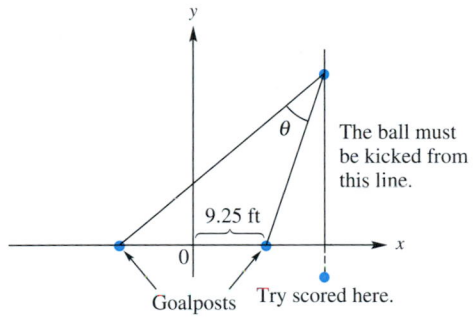

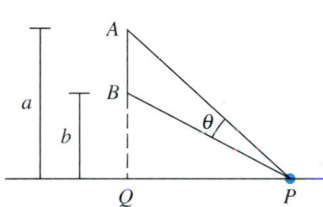

1. Use the result from Regiomontanus' problem to show that when the line is outside the goalposts the optimal location to kick the rugby ball lies on the hyperbola $x^2 - y^2 = 9.25^2$.

2. If the line on which the ball must be kicked is 10 ft to the right of the goalpost, how far from the goal line should the ball be placed to maximize angle θ?

3. Rugby players find it easier to kick the ball from the hyperbola's asymptote. When the line on which the ball must be kicked is 10 ft to the right of the goalpost, how far will this point differ from the exact optimal location?

11 Further Topics in Algebra

Amazing as it may seem, the male honeybee hatches from an unfertilized egg, while the female hatches from a fertilized one. As a result, the "family tree" of a male honeybee exhibits an interesting pattern. If we start with a male honeybee and count the number of bees in successive generations, we obtain the sequence of numbers

$$1, 1, 2, 3, 5, 8, 13, 21, 34, 55, \ldots,$$

known as the *Fibonacci sequence*. This fascinating sequence has countless interesting properties and appears in many places in nature.

In Section 11.1, we further investigate the Fibonacci sequence as we study *sequences* and sums of terms of sequences, known as *series*.

11.1 Sequences and Series

Sequences ▪ Series and Summation Notation ▪ Summation Properties

Sequences A *sequence* is a function that computes an ordered list. For example, the average person in the United States uses 100 gallons of water each day. The function defined by $f(n) = 100n$ generates the terms of the sequence

$$100, 200, 300, 400, 500, 600, 700, \ldots,$$

when $n = 1, 2, 3, 4, 5, 6, 7, \ldots$. This function represents the gallons of water used by the average person after n days.

A second example of a sequence involves investing money. If \$100 is deposited into a savings account paying 5% interest compounded annually, then the function defined by $g(n) = 100(1.05)^n$ calculates the account balance after n years. The terms of the sequence are

$$g(1), g(2), g(3), g(4), g(5), g(6), g(7), \ldots,$$

and can be approximated as

$$105, 110.25, 115.76, 121.55, 127.63, 134.01, 140.71, \ldots.$$

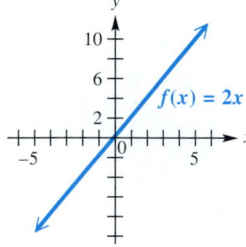

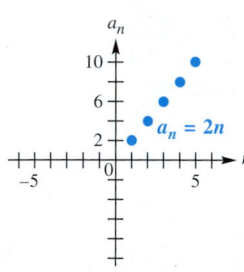

Figure 1

SEQUENCE

A **finite sequence** is a function that has a set of natural numbers of the form $\{1, 2, 3, \ldots, n\}$ as its domain. An **infinite sequence** has the set of natural numbers as its domain.

For example, the sequence of natural-number multiples of 2,

$$2, 4, 6, 8, 10, 12, 14, \ldots, \qquad \text{is infinite,}$$

but the sequence of days in June,

$$1, 2, 3, 4, \ldots, 29, 30, \qquad \text{is finite.}$$

Instead of using $f(x)$ notation to indicate a sequence, it is customary to use a_n, where $a_n = f(n)$. ***The letter n is used instead of x as a reminder that n represents a natural number.*** The elements in the range of a sequence, called the **terms** of the sequence, are $a_1, a_2, a_3, \ldots$. The elements of both the domain and the range of a sequence are *ordered*. The first term is found by letting $n = 1$, the second term is found by letting $n = 2$, and so on. The **general term**, or ***n*th term,** of the sequence is a_n.

Figure 1 shows graphs of $f(x) = 2x$ and $a_n = 2n$. Notice that $f(x)$ is a continuous function, and a_n is discontinuous. To graph a_n, we plot points of the form $(n, 2n)$ for $n = 1, 2, 3, \ldots$.

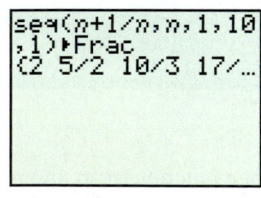

(a)

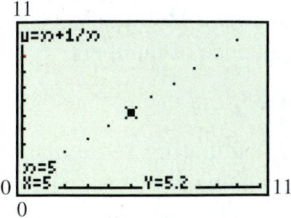

The fifth term is 5.2.

(b)

Figure 2

A graphing calculator can list the terms in a sequence. Using sequence mode to list the first 10 terms of the sequence with general term $a_n = n + \frac{1}{n}$ produces the result shown in Figure 2(a). Additional terms of the sequence can be seen by scrolling to the right. Sequences can also be graphed in sequence mode. Figure 2(b) shows a calculator screen with the graph of $a_n = n + \frac{1}{n}$. Notice that for $n = 5$, the term is $5 + \frac{1}{5} = 5.2$. ▪

▶ **EXAMPLE 1**   **FINDING TERMS OF SEQUENCES**

Write the first five terms for each sequence.

(a) $a_n = \dfrac{n + 1}{n + 2}$ **(b)** $a_n = (-1)^n \cdot n$ **(c)** $a_n = \dfrac{2n + 1}{n^2 + 1}$

Solution

(a) Replacing n in $a_n = \frac{n + 1}{n + 2}$, with 1, 2, 3, 4, and 5 gives

$$n = 1: \quad a_1 = \frac{1 + 1}{1 + 2} = \frac{2}{3}$$

$$n = 2: \quad a_2 = \frac{2 + 1}{2 + 2} = \frac{3}{4}$$

$$n = 3: \quad a_3 = \frac{3 + 1}{3 + 2} = \frac{4}{5}$$

$$n = 4: \quad a_4 = \frac{4 + 1}{4 + 2} = \frac{5}{6}$$

$$n = 5: \quad a_5 = \frac{5 + 1}{5 + 2} = \frac{6}{7}.$$

(b) Replace n in $a_n = (-1)^n \cdot n$ with 1, 2, 3, 4, and 5 to obtain

$$n = 1: \quad a_1 = (-1)^1 \cdot 1 = -1$$

$$n = 2: \quad a_2 = (-1)^2 \cdot 2 = 2$$

$$n = 3: \quad a_3 = (-1)^3 \cdot 3 = -3$$

$$n = 4: \quad a_4 = (-1)^4 \cdot 4 = 4$$

$$n = 5: \quad a_5 = (-1)^5 \cdot 5 = -5.$$

(c) For $a_n = \frac{2n + 1}{n^2 + 1}$, we have

$$a_1 = \frac{3}{2}, \quad a_2 = 1, \quad a_3 = \frac{7}{10}, \quad a_4 = \frac{9}{17}, \quad \text{and} \quad a_5 = \frac{11}{26}.$$

NOW TRY EXERCISES 3, 7, AND 9. ◀

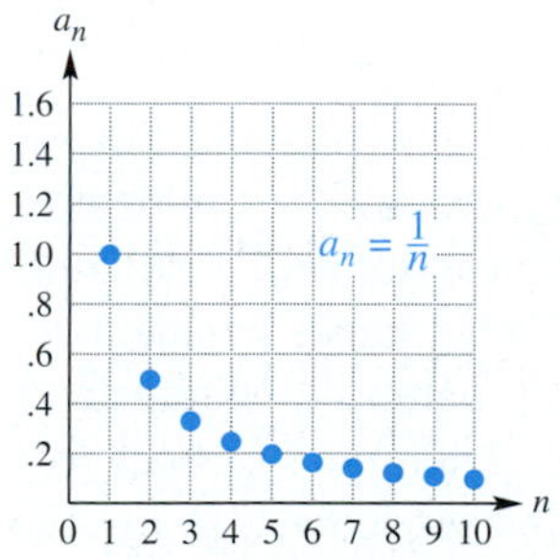

Figure 3

If the terms of an infinite sequence get closer and closer to some real number, the sequence is said to be **convergent** and to **converge** to that real number. For example, the sequence defined by $a_n = \frac{1}{n}$ approaches 0 as n becomes large. Thus, a_n is a convergent sequence that converges to 0. A graph of this sequence for $n = 1, 2, 3, \ldots, 10$ is shown in Figure 3. The terms of a_n approach the horizontal axis.

A sequence that does not converge to any number is **divergent**. The terms of the sequence $a_n = n^2$ are

$$1, 4, 9, 16, 25, 36, 49, 64, 81, \ldots.$$

This sequence is divergent because as n becomes large, the values of a_n do not approach a fixed number; rather, they increase without bound.

Some sequences are defined by a **recursive definition,** one in which each term after the first term or first few terms is defined as an expression involving the previous term or terms. On the other hand, the sequences in Example 1 were defined *explicitly,* with a formula for a_n that does not depend on a previous term.

▶ **EXAMPLE 2** **USING A RECURSION FORMULA**

Find the first four terms of each sequence.

(a) $a_1 = 4$
$a_n = 2 \cdot a_{n-1} + 1$, if $n > 1$

(b) $a_1 = 2$
$a_n = a_{n-1} + n - 1$, if $n > 1$

Solution

(a) This is a recursive definition. We know $a_1 = 4$. Since $a_n = 2 \cdot a_{n-1} + 1$,

$$a_1 = 4$$
$$a_2 = 2 \cdot a_1 + 1 = 2 \cdot 4 + 1 = 9$$
$$a_3 = 2 \cdot a_2 + 1 = 2 \cdot 9 + 1 = 19$$
$$a_4 = 2 \cdot a_3 + 1 = 2 \cdot 19 + 1 = 39.$$

(b) In this recursive definition, $a_1 = 2$ and $a_n = a_{n-1} + n - 1$.

$$a_1 = 2$$
$$a_2 = a_1 + 2 - 1 = 2 + 1 = 3$$
$$a_3 = a_2 + 3 - 1 = 3 + 2 = 5$$
$$a_4 = a_3 + 4 - 1 = 5 + 3 = 8$$

NOW TRY EXERCISES 23 AND 27. ◀

CONNECTIONS One of the most famous sequences in mathematics is the **Fibonacci sequence,**

$$1, 1, 2, 3, 5, 8, 13, 21, 34, 55, \dots ,$$

named for the Italian mathematician Leonardo of Pisa, who was also known as Fibonacci. The Fibonacci sequence is found in numerous places in nature. For example, male honeybees hatch from eggs that have not been fertilized, so a male bee has only one parent, a female. On the other hand, female honeybees hatch from fertilized eggs, so a female has two parents, one male and one female. The number of ancestors in consecutive generations of bees follows the Fibonacci sequence. Successive terms in the sequence also appear in plants, such as in the spirals of the daisy head, pineapple, and pine cone.

FOR DISCUSSION OR WRITING
1. Try to discover the pattern in the Fibonacci sequence.
2. Using the description given above, write a recursive definition that calculates the number of ancestors of a male bee in each generation.

Leonardo of Pisa (Fibonacci)
(1170–1250)

UGA2652087

▶ **EXAMPLE 3**   **MODELING INSECT POPULATION GROWTH**

Frequently the population of a particular insect does not continue to grow indefinitely. Instead, its population grows rapidly at first, and then levels off because of competition for limited resources. In one study, the behavior of the winter moth was modeled with a sequence similar to the following, where a_n represents the population density in thousands per acre during year n. (*Source:* Varley, G. and G. Gradwell, "Population models for the winter moth," Symposium of the Royal Entomological Society of London, 4.)

$$a_1 = 1$$
$$a_n = 2.85a_{n-1} - .19a_{n-1}^2, \quad \text{for } n \geq 2$$

(a) Give a table of values for $n = 1, 2, 3, \ldots, 10$.

(b) Graph the sequence. Describe what happens to the population density.

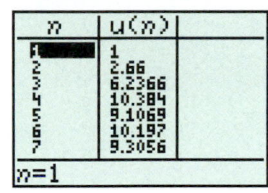

Figure 4

Solution

(a) Evaluate $a_1, a_2, a_3, \ldots, a_{10}$ recursively. Since $a_1 = 1$,

$$a_2 = 2.85a_1 - .19a_1^2 = 2.85(1) - .19(1)^2 = 2.66,$$

and $a_3 = 2.85a_2 - .19a_2^2 = 2.85(2.66) - .19(2.66)^2 \approx 6.24.$

Approximate values for $n = 1, 2, 3, \ldots, 10$ are shown in the table. Figure 4 shows the computation of the sequence, denoted by $u(n)$ rather than a_n, using a calculator.

n	1	2	3	4	5	6	7	8	9	10
a_n	1	2.66	6.24	10.4	9.11	10.2	9.31	10.1	9.43	9.98

(b) The graph of a sequence is a set of discrete points. Plot the points $(1, 1)$, $(2, 2.66)$, $(3, 6.24)$, ..., $(10, 9.98)$, as shown in Figure 5(a). At first, the insect population increases rapidly, and then oscillates about the line $y = 9.7$. (See the Note on the next page.) The oscillations become smaller as n increases, indicating that the population density may stabilize near 9.7 thousand per acre. In Figure 5(b), the first 20 terms have been plotted with a calculator.

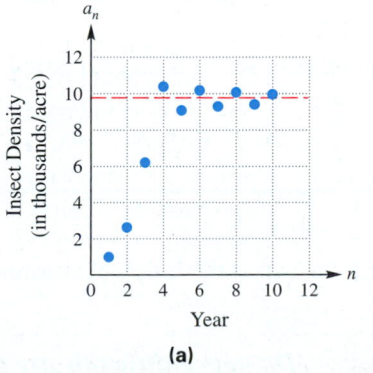

(a)

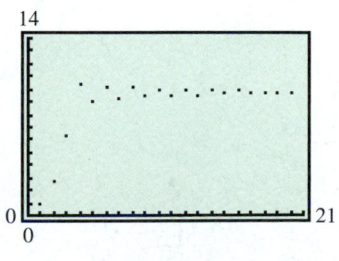

(b)

Figure 5

NOW TRY EXERCISE 87. ◀

> ▶ **Note** In Example 3, the insect population stabilizes near the value $k = 9.7$ thousand. This value of k can be found by solving the quadratic equation $k = 2.85k - .19k^2$. Why is this true?

Series and Summation Notation

Suppose a person has a starting salary of $30,000 and receives a $2000 raise each year. Then,

$$30{,}000, \quad 32{,}000, \quad 34{,}000, \quad 36{,}000, \quad 38{,}000$$

are terms of the sequence that describe this person's salaries over a 5-year period. The total earned is given by the *finite series*

$$30{,}000 + 32{,}000 + 34{,}000 + 36{,}000 + 38{,}000,$$

whose sum is $170,000. Any sequence can be used to define a *series*. For example, the infinite sequence

$$1, \frac{1}{3}, \frac{1}{9}, \frac{1}{27}, \frac{1}{81}, \frac{1}{243}, \dots$$

defines the terms of the *infinite series*

$$1 + \frac{1}{3} + \frac{1}{9} + \frac{1}{27} + \frac{1}{81} + \frac{1}{243} + \cdots.$$

If a sequence has terms $a_1, a_2, a_3, \dots$, then S_n is defined as the sum of the first n terms. That is,

$$S_n = a_1 + a_2 + a_3 + \cdots + a_n.$$

The sum of the terms of a sequence, called a **series,** is written using **summation notation.** The symbol Σ, the Greek capital letter **sigma,** is used to indicate a sum.

▼ **LOOKING AHEAD TO CALCULUS**

An infinite series converges if the sequence of partial sums $S_1, S_2, S_3, \dots$ converges. For example, it can be shown that

$$1 + \frac{1}{2} + \frac{1}{3} + \frac{1}{4} + \cdots \text{ diverges,}$$

while

$$1 - \frac{1}{2} + \frac{1}{3} - \frac{1}{4} + \cdots \text{ converges.}$$

SERIES

A **finite series** is an expression of the form

$$S_n = a_1 + a_2 + a_3 + \cdots + a_n = \sum_{i=1}^{n} a_i,$$

and an **infinite series** is an expression of the form

$$S_\infty = a_1 + a_2 + a_3 + \cdots + a_n + \cdots = \sum_{i=1}^{\infty} a_i.$$

The letter i is called the **index of summation.**

> ▶ **Caution** *Do not confuse this use of i with the use of i to represent the imaginary unit.* Other letters, such as k and j, may be used for the index of summation.

▶ **EXAMPLE 4** **USING SUMMATION NOTATION**

Evaluate the series $\displaystyle\sum_{k=1}^{6} (2^k + 1)$.

Algebraic Solution

Write each of the six terms, then evaluate the sum.

$$\sum_{k=1}^{6} (2^k + 1) = (2^1 + 1) + (2^2 + 1) + (2^3 + 1)$$
$$+ (2^4 + 1) + (2^5 + 1) + (2^6 + 1)$$
$$= (2 + 1) + (4 + 1) + (8 + 1)$$
$$+ (16 + 1) + (32 + 1) + (64 + 1)$$
$$= 3 + 5 + 9 + 17 + 33 + 65$$
$$= 132$$

Graphing Calculator Solution

A graphing calculator can store the sequence into a list, L_1, and then compute the sum of the six terms in the list. See Figure 6.

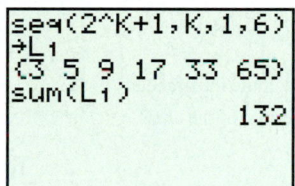

Figure 6

NOW TRY EXERCISE 43. ◀

▶ **EXAMPLE 5** **USING SUMMATION NOTATION WITH SUBSCRIPTS**

Write the terms for each series. Evaluate each sum, if possible.

(a) $\displaystyle\sum_{j=3}^{6} a_j$

(b) $\displaystyle\sum_{i=1}^{3} (6x_i - 2)$, if $x_1 = 2$, $x_2 = 4$, and $x_3 = 6$

(c) $\displaystyle\sum_{i=1}^{4} f(x_i)\Delta x$, if $f(x) = x^2$, $x_1 = 0$, $x_2 = 2$, $x_3 = 4$, $x_4 = 6$, and $\Delta x = 2$

Solution

(a) $\displaystyle\sum_{j=3}^{6} a_j = a_3 + a_4 + a_5 + a_6$

(b) Let $i = 1, 2$, and 3, respectively, to obtain

$$\sum_{i=1}^{3} (6x_i - 2) = (6x_1 - 2) + (6x_2 - 2) + (6x_3 - 2)$$
$$= (6 \cdot 2 - 2) + (6 \cdot 4 - 2) + (6 \cdot 6 - 2) \qquad \text{Substitute the given values for } x_1, x_2, \text{ and } x_3.$$

> **Use the order of operations.**

$$= 10 + 22 + 34$$
$$= 66.$$

(c) $\displaystyle\sum_{i=1}^{4} f(x_i)\Delta x = f(x_1)\Delta x + f(x_2)\Delta x + f(x_3)\Delta x + f(x_4)\Delta x$
$$= x_1^2\Delta x + x_2^2\Delta x + x_3^2\Delta x + x_4^2\Delta x$$
$$= 0^2(2) + 2^2(2) + 4^2(2) + 6^2(2) \qquad f(x) = x^2; \Delta x = 2$$
$$= 0 + 8 + 32 + 72 \qquad\qquad \text{Simplify.}$$
$$= 112 \qquad\qquad\qquad \text{Add.}$$

NOW TRY EXERCISES 51, 53, AND 63. ◀

▼ **LOOKING AHEAD TO CALCULUS**

Summation notation is used in calculus to describe the area under a curve, the volume of a figure rotated about an axis, and in many other applications, as well as in the definition of integral. In the definition of the definite integral, Σ is replaced with an elongated S:

$$\int_a^b f(x)\, dx = \lim_{n \to \infty} \sum_{i=1}^{n} f(x_i)\, \Delta x_i.$$

In some cases, the definite integral can be interpreted as the sum of the areas of rectangles.

Summation Properties   Properties of summation provide useful short-cuts for evaluating series.

SUMMATION PROPERTIES

If $a_1, a_2, a_3, \ldots, a_n$ and $b_1, b_2, b_3, \ldots, b_n$ are two sequences, and c is a constant, then for every positive integer n,

(a) $\displaystyle\sum_{i=1}^{n} c = nc$

(b) $\displaystyle\sum_{i=1}^{n} ca_i = c \sum_{i=1}^{n} a_i$

(c) $\displaystyle\sum_{i=1}^{n} (a_i + b_i) = \sum_{i=1}^{n} a_i + \sum_{i=1}^{n} b_i$

(d) $\displaystyle\sum_{i=1}^{n} (a_i - b_i) = \sum_{i=1}^{n} a_i - \sum_{i=1}^{n} b_i.$

To prove Property (a), expand the series to obtain

$$c + c + c + c + \cdots + c,$$

where there are n terms of c, so the sum is nc.

Property (c) also can be proved by first expanding the series:

$$\sum_{i=1}^{n} (a_i + b_i) = (a_1 + b_1) + (a_2 + b_2) + \cdots + (a_n + b_n)$$

$$= (a_1 + a_2 + \cdots + a_n) + (b_1 + b_2 + \cdots + b_n)$$

Commutative and associative properties (Section R.2)

$$= \sum_{i=1}^{n} a_i + \sum_{i=1}^{n} b_i.$$

Proofs of the other two properties are similar. The following results can be proved by mathematical induction. (See **Section 11.5.**)

SUMMATION RULES

$$\sum_{i=1}^{n} i = 1 + 2 + \cdots + n = \frac{n(n+1)}{2}$$

$$\sum_{i=1}^{n} i^2 = 1^2 + 2^2 + \cdots + n^2 = \frac{n(n+1)(2n+1)}{6}$$

$$\sum_{i=1}^{n} i^3 = 1^3 + 2^3 + \cdots + n^3 = \frac{n^2(n+1)^2}{4}$$

▶ **EXAMPLE 6**   **USING THE SUMMATION PROPERTIES**

Use the summation properties to find each sum.

(a) $\displaystyle\sum_{i=1}^{40} 5$

(b) $\displaystyle\sum_{i=1}^{22} 2i$

(c) $\displaystyle\sum_{i=1}^{14} (2i^2 - 3)$

Solution

(a) $\displaystyle\sum_{i=1}^{40} 5 = 40(5) = 200$ Property (a) with $n = 40$ and $c = 5$

(b) $\displaystyle\sum_{i=1}^{22} 2i = 2\sum_{i=1}^{22} i$ Property (b) with $c = 2$ and $a_i = i$

$$= 2 \cdot \frac{22(22 + 1)}{2}$$ Summation rules

$$= 506$$ Simplify.

(c) $\displaystyle\sum_{i=1}^{14}(2i^2 - 3) = \sum_{i=1}^{14} 2i^2 - \sum_{i=1}^{14} 3$ Property (d) with $a_i = 2i^2$ and $b_i = 3$

$$= 2\sum_{i=1}^{14} i^2 - \sum_{i=1}^{14} 3$$ Property (b) with $c = 2$ and $a_i = i^2$

$$= 2 \cdot \frac{14(14 + 1)(2 \cdot 14 + 1)}{6} - 14(3)$$ Summation rules; Property (a)

$$= 1988$$ Simplify.

> **NOW TRY EXERCISES 67, 69, AND 71.** ◄

▶ **EXAMPLE 7**   **USING THE SUMMATION PROPERTIES**

Evaluate $\displaystyle\sum_{i=1}^{6}(i^2 + 3i + 5)$.

Solution

$$\sum_{i=1}^{6}(i^2 + 3i + 5) = \sum_{i=1}^{6} i^2 + \sum_{i=1}^{6} 3i + \sum_{i=1}^{6} 5$$ Property (c)

$$= \sum_{i=1}^{6} i^2 + 3\sum_{i=1}^{6} i + \sum_{i=1}^{6} 5$$ Property (b)

$$= \sum_{i=1}^{6} i^2 + 3\sum_{i=1}^{6} i + 6(5)$$ Property (a)

$$= \frac{6(6 + 1)(2 \cdot 6 + 1)}{6} + 3\left[\frac{6(6 + 1)}{2}\right] + 6(5)$$ Summation rules

$$= 91 + 3(21) + 6(5) = 184$$ Simplify.

> **NOW TRY EXERCISE 73.** ◄

11.1 Exercises

Write the first five terms of each sequence. See Example 1.

1. $a_n = 4n + 10$

2. $a_n = 6n - 3$

3. $a_n = \dfrac{n + 5}{n + 4}$

4. $a_n = \dfrac{n - 7}{n - 6}$

5. $a_n = \left(\dfrac{1}{3}\right)^n (n - 1)$

6. $a_n = (-2)^n(n)$

7. $a_n = (-1)^n(2n)$

8. $a_n = (-1)^{n-1}(n + 1)$

9. $a_n = \dfrac{4n - 1}{n^2 + 2}$

10. $a_n = \dfrac{n^2 - 1}{n^2 + 1}$

11. $a_n = \dfrac{n^3 + 8}{n + 2}$

12. $a_n = \dfrac{n^3 + 27}{n + 3}$

13. Your friend does not understand what is meant by the nth term or general term of a sequence. How would you explain this idea?

14. *Concept Check* How are sequences related to functions?

Concept Check *Decide whether each sequence is* finite *or* infinite.

15. The sequence of days of the week

16. The sequence of dates in the month of July

17. 1, 2, 3, 4, 5

18. $-1, -2, -3, -4, -5$

19. 1, 2, 3, 4, 5, . . .

20. $-1, -2, -3, -4, -5, . . .$

21. $a_1 = 4$
$a_n = 4 \cdot a_{n-1}$, if $2 \le n \le 10$

22. $a_1 = 2$
$a_2 = 5$
$a_n = a_{n-1} + a_{n-2}$, if $n \ge 3$

Find the first four terms of each sequence. See Example 2.

23. $a_1 = -2$
$a_n = a_{n-1} + 3$, if $n > 1$

24. $a_1 = -1$
$a_n = a_{n-1} - 4$, if $n > 1$

25. $a_1 = 1$
$a_2 = 1$
$a_n = a_{n-1} + a_{n-2}$, if $n \ge 3$
(This is the Fibonacci sequence.)

26. $a_1 = 2$
$a_2 = 5$
$a_n = a_{n-1} + a_{n-2}$, if $n \ge 3$

27. $a_1 = 2$
$a_n = n \cdot a_{n-1}$, if $n > 1$

28. $a_1 = -3$
$a_n = 2n \cdot a_{n-1}$, if $n > 1$

Evaluate each series. See Example 4.

29. $\displaystyle\sum_{i=1}^{5} (2i + 1)$

30. $\displaystyle\sum_{i=1}^{6} (3i - 2)$

31. $\displaystyle\sum_{j=1}^{4} \frac{1}{j}$

32. $\displaystyle\sum_{i=1}^{5} (i + 1)^{-1}$

33. $\displaystyle\sum_{i=1}^{4} i^i$

34. $\displaystyle\sum_{k=1}^{4} (k + 1)^2$

35. $\displaystyle\sum_{k=1}^{6} (-1)^k \cdot k$

36. $\displaystyle\sum_{i=1}^{7} (-1)^{i+1} \cdot i^2$

37. $\displaystyle\sum_{i=2}^{5} (6 - 3i)$

38. $\displaystyle\sum_{i=3}^{7} (5i + 2)$

39. $\displaystyle\sum_{i=-2}^{3} 2(3)^i$

40. $\displaystyle\sum_{i=-1}^{2} 5(2)^i$

41. $\displaystyle\sum_{i=-1}^{5} (i^2 - 2i)$

42. $\displaystyle\sum_{i=3}^{6} (2i^2 + 1)$

43. $\displaystyle\sum_{i=1}^{5} (3^i - 4)$

44. $\displaystyle\sum_{i=1}^{4} [(-2)^i - 3]$

45. $\displaystyle\sum_{i=1}^{3} (i^3 - i)$

46. $\displaystyle\sum_{i=1}^{4} (i^4 - i^3)$

Use a graphing calculator to evaluate each series. See Example 4.

47. $\displaystyle\sum_{i=1}^{10} (4i^2 - 5)$

48. $\displaystyle\sum_{i=1}^{10} (i^3 - 6)$

49. $\displaystyle\sum_{j=3}^{9} (3j - j^2)$

50. $\displaystyle\sum_{k=5}^{10} (k^2 - 4k + 7)$

Write the terms for each series. Evaluate the sum, given that $x_1 = -2$, $x_2 = -1$, $x_3 = 0$, $x_4 = 1$, *and* $x_5 = 2$. *See Examples 5(a) and 5(b).*

51. $\displaystyle\sum_{i=1}^{5} x_i$

52. $\displaystyle\sum_{i=1}^{5} -x_i$

53. $\displaystyle\sum_{i=1}^{5} (2x_i + 3)$

54. $\displaystyle\sum_{i=1}^{4} x_i^2$

55. $\displaystyle\sum_{i=1}^{3} (3x_i - x_i^2)$

56. $\displaystyle\sum_{i=1}^{3} (x_i^2 + 1)$

57. $\displaystyle\sum_{i=2}^{5} \frac{x_i + 1}{x_i + 2}$

58. $\displaystyle\sum_{i=1}^{5} \frac{x_i}{x_i + 3}$

59. $\displaystyle\sum_{i=1}^{4} \frac{x_i^3 + 1000}{x_i + 10}$

 60. Explain how factoring can make the work in Exercises 11, 12, and 59 easier.

Write the terms of $\displaystyle\sum_{i=1}^{4} f(x_i)\Delta x$, with $x_1 = 0$, $x_2 = 2$, $x_3 = 4$, $x_4 = 6$, and $\Delta x = .5$, for each function. Evaluate the sum. See Example 5(c).

61. $f(x) = 4x - 7$

62. $f(x) = 6 + 2x$

63. $f(x) = 2x^2$

64. $f(x) = x^2 - 1$

65. $f(x) = \dfrac{-2}{x + 1}$

66. $f(x) = \dfrac{5}{2x - 1}$

Use the summation properties and rules to evaluate each series. See Examples 6 and 7.

67. $\displaystyle\sum_{i=1}^{100} 6$

68. $\displaystyle\sum_{i=1}^{20} 5$

69. $\displaystyle\sum_{i=1}^{15} i^2$

70. $\displaystyle\sum_{i=1}^{50} 2i^3$

71. $\displaystyle\sum_{i=1}^{5} (5i + 3)$

72. $\displaystyle\sum_{i=1}^{5} (8i - 1)$

73. $\displaystyle\sum_{i=1}^{5} (4i^2 - 2i + 6)$

74. $\displaystyle\sum_{i=1}^{6} (2 + i - i^2)$

75. $\displaystyle\sum_{i=1}^{4} (3i^3 + 2i - 4)$

76. $\displaystyle\sum_{i=1}^{6} (i^2 + 2i^3)$

Concept Check *Use summation notation to write each series.* *

77. $\dfrac{1}{3(1)} + \dfrac{1}{3(2)} + \dfrac{1}{3(3)} + \cdots + \dfrac{1}{3(9)}$

78. $\dfrac{5}{1 + 1} + \dfrac{5}{1 + 2} + \dfrac{5}{1 + 3} + \cdots + \dfrac{5}{1 + 15}$

79. $1 - \dfrac{1}{2} + \dfrac{1}{4} - \dfrac{1}{8} + \cdots - \dfrac{1}{128}$

80. $1 - \dfrac{1}{4} + \dfrac{1}{9} - \dfrac{1}{16} + \cdots - \dfrac{1}{400}$

Use the sequence feature of a graphing calculator to graph the first ten terms of each sequence as defined. Use the graph to make a conjecture as to whether the sequence converges or diverges. If you think it converges, determine the number to which it converges.

81. $a_n = \dfrac{n + 4}{2n}$

82. $a_n = \dfrac{1 + 4n}{2n}$

83. $a_n = 2e^n$

84. $a_n = n(n + 2)$

85. $a_n = \left(1 + \dfrac{1}{n}\right)^n$

86. $a_n = (1 + n)^{1/n}$

Solve each problem involving sequences and series. See Example 3.

87. **(Modeling) Insect Population** Suppose an insect population density in thousands per acre during year n can be modeled by the recursively defined sequence

$$a_1 = 8$$
$$a_n = 2.9a_{n-1} - .2a_{n-1}^2, \quad \text{for } n > 1.$$

  (a) Find the population for $n = 1, 2, 3$.

  (b) Graph the sequence for $n = 1, 2, 3, \ldots, 20$. Use the window $[0, 21]$ by $[0, 14]$. Interpret the graph.

---

*These exercises were suggested by Joe Lloyd Harris, Gulf Coast Community College.

88. *Male Bee Ancestors* As mentioned in the chapter introduction and in the Connections box in this section, one of the most famous sequences in mathematics is the Fibonacci sequence,

1, 1, 2, 3, 5, 8, 13, 21, 34, 55, ...

(Also see Exercise 25.) Recall that male honeybees hatch from eggs that have not been fertilized, so a male bee has only one parent, a female. On the other hand, female honeybees hatch from fertilized eggs, so a female has two parents, one male and one female. The number of ancestors in consecutive generations of bees follows the Fibonacci sequence. Draw a tree showing the number of ancestors of a male bee in each generation following the description given above.

89. *(Modeling) Bacteria Growth* If certain bacteria are cultured in a medium with sufficient nutrients, they will double in size and then divide every 40 min. Let N_1 be the initial number of bacteria cells, N_2 the number after 40 min, N_3 the number after 80 min, and N_j the number after $40(j - 1)$ min. (*Source:* Hoppensteadt, F. and C. Peskin, *Mathematics in Medicine and the Life Sciences,* Springer-Verlag, 1992.)

(a) Write N_{j+1} in terms of N_j for $j \geq 1$.
(b) Determine the number of bacteria after 2 hr if $N_1 = 230$.
(c) Graph the sequence N_j for $j = 1, 2, 3, \ldots, 7$ where $N_1 = 230$. Use the window $[0, 10]$ by $[0, 15{,}000]$.
(d) Describe the growth of these bacteria when there are unlimited nutrients.

90. *(Modeling) Verhulst's Model for Bacteria Growth* Refer to Exercise 89. If the bacteria are not cultured in a medium with sufficient nutrients, competition will ensue and growth will slow. According to Verhulst's model, the number of bacteria N_j at time $40(j - 1)$ in minutes can be determined by the sequence

$$N_{j+1} = \left[\frac{2}{1 + \frac{N_j}{K}} \right] N_j,$$

where K is a constant and $j \geq 1$. (*Source:* Hoppensteadt, F. and C. Peskin, *Mathematics in Medicine and the Life Sciences,* Springer-Verlag, 1992.)

(a) If $N_1 = 230$ and $K = 5000$, make a table of N_j for $j = 1, 2, 3, \ldots, 20$. Round values in the table to the nearest integer.
(b) Graph the sequence N_j for $j = 1, 2, 3, \ldots, 20$. Use the window $[0, 20]$ by $[0, 6000]$.
(c) Describe the growth of these bacteria when there are limited nutrients.
(d) Make a conjecture as to why K is called the **saturation constant.** Test your conjecture by changing the value of K in the given formula.

91. *Approximating ln(1 + x)* The series

$$x - \frac{x^2}{2} + \frac{x^3}{3} - \frac{x^4}{4} + \cdots$$

can be used to approximate the value of $\ln(1 + x)$ for values of x in $(-1, 1]$. Use the first six terms of this series to approximate each expression. Compare this approximation with the value obtained on a calculator.

(a) ln 1.02 $(x = .02)$ **(b)** ln .97 $(x = -.03)$

92. *Approximating* π Find the sum of the first six terms of the series

$$\frac{\pi^4}{90} = \frac{1}{1^4} + \frac{1}{2^4} + \frac{1}{3^4} + \frac{1}{4^4} + \frac{1}{5^4} + \cdots + \frac{1}{n^4} + \cdots .$$

Then multiply the result by 90, and take the fourth root. This will provide an approximation of π. Compare your answer to the actual decimal approximation of π.

93. *Approximating Powers of e*   The series

$$e^a \approx 1 + a + \frac{a^2}{2!} + \frac{a^3}{3!} + \cdots + \frac{a^n}{n!},$$

where $n! = 1 \cdot 2 \cdot 3 \cdot 4 \cdots \cdot n$, can be used to approximate the value of e^a for any real number a. Use the first eight terms of this series to approximate each expression. Compare this approximation with the value obtained on a calculator.

(a) e **(b)** e^{-1}

94. *Approximating Square Roots*   The recursively defined sequence

$$a_1 = k$$

$$a_n = \frac{1}{2}\left(a_{n-1} + \frac{k}{a_{n-1}}\right), \quad \text{if } n > 1$$

can be used to compute $\sqrt{k}$ for any positive number k. This sequence was known to Sumerian mathematicians 4000 yr ago, and it is still used today. Use this sequence to approximate the given square root by finding a_6. Compare your result with the actual value. (*Source:* Heinz-Otto, P., *Chaos and Fractals,* Springer-Verlag, 1993.)

(a) $\sqrt{2}$ **(b)** $\sqrt{11}$

11.2 Arithmetic Sequences and Series

Arithmetic Sequences ▪ Arithmetic Series

Arithmetic Sequences   A sequence in which each term after the first is obtained by adding a fixed number to the previous term is an **arithmetic sequence** (or **arithmetic progression**). The fixed number that is added is the **common difference.** The sequence

$$5, 9, 13, 17, 21, \ldots$$

is an arithmetic sequence since each term after the first is obtained by adding 4 to the previous term. That is,

$$9 = 5 + 4$$
$$13 = 9 + 4$$
$$17 = 13 + 4$$
$$21 = 17 + 4,$$

and so on. The common difference is 4.

If the common difference of an arithmetic sequence is d, then by the definition of an arithmetic sequence,

$$\boldsymbol{d = a_{n+1} - a_n,} \quad \text{Common difference } d$$

for every positive integer n in the domain of the sequence.

▶ **EXAMPLE 1** **FINDING THE COMMON DIFFERENCE**

Find the common difference, d, for the arithmetic sequence

$$-9, -7, -5, -3, -1, \ldots.$$

Solution We find d by choosing any two adjacent terms and subtracting the first from the second. Choosing -7 and -5 gives

$$d = -5 - (-7) = 2.$$

> Be careful when subtracting a negative number.

Choosing -9 and -7 would give $d = -7 - (-9) = 2$, the same result.

NOW TRY EXERCISE 1. ◀

If a_1 and d are known, then all the terms of an arithmetic sequence can be found.

▶ **EXAMPLE 2** **FINDING TERMS GIVEN a_1 AND d**

Find the first five terms for each arithmetic sequence.

(a) The first term is 7, and the common difference is -3.

(b) $a_1 = -12, d = 5$

Solution

(a) $a_1 = 7$ Start with $a_1 = 7$.

$a_2 = 7 + (-3) = 4$ Add $d = -3$.

$a_3 = 4 + (-3) = 1$ Add -3.

$a_4 = 1 + (-3) = -2$ Add -3.

$a_5 = -2 + (-3) = -5.$ Add -3.

(b) $a_1 = -12$ Start with a_1.

$a_2 = -12 + 5 = -7$ Add $d = 5$.

$a_3 = -7 + 5 = -2$

$a_4 = -2 + 5 = 3$

$a_5 = 3 + 5 = 8$

NOW TRY EXERCISES 7 AND 9. ◀

If a_1 is the first term of an arithmetic sequence and d is the common difference, then the terms of the sequence are given by

$$a_1 = a_1$$
$$a_2 = a_1 + d$$
$$a_3 = a_2 + d = a_1 + d + d = a_1 + 2d$$
$$a_4 = a_3 + d = a_1 + 2d + d = a_1 + 3d$$
$$a_5 = a_1 + 4d$$
$$a_6 = a_1 + 5d,$$

and, by this pattern, $a_n = a_1 + (n - 1)d$.

This result can be proved by mathematical induction. (See **Section 11.5.**)

> ### nth TERM OF AN ARITHMETIC SEQUENCE
>
> In an arithmetic sequence with first term a_1 and common difference d, the nth term, a_n, is given by
>
> $$a_n = a_1 + (n - 1)d.$$

▶ **EXAMPLE 3**　**FINDING TERMS OF AN ARITHMETIC SEQUENCE**

Find a_{13} and a_n for the arithmetic sequence $-3, 1, 5, 9, \ldots$.

Solution　Here $a_1 = -3$ and $d = 1 - (-3) = 4$. To find a_{13}, substitute 13 for n in the formula for the nth term.

$$a_n = a_1 + (n - 1)d \qquad \text{Work inside the parentheses first.}$$
$$a_{13} = a_1 + (13 - 1)d \qquad n = 13$$
$$a_{13} = -3 + (12)4 \qquad \text{Let } a_1 = -3, d = 4.$$
$$a_{13} = -3 + 48 \qquad \text{Simplify.}$$
$$a_{13} = 45$$

Find a_n by substituting values for a_1 and d in the formula for a_n.

$$a_n = -3 + (n - 1) \cdot 4 \qquad \text{Let } a_1 = -3, d = 4.$$
$$a_n = -3 + 4n - 4 \qquad \text{Distributive property (Section R.2)}$$
$$a_n = 4n - 7 \qquad \text{Simplify.}$$

NOW TRY EXERCISE 13. ◀

▶ **EXAMPLE 4**　**FINDING TERMS OF AN ARITHMETIC SEQUENCE**

Find a_{18} and a_n for the arithmetic sequence having $a_2 = 9$ and $a_3 = 15$.

Solution　Find d first; $d = a_3 - a_2 = 15 - 9 = 6$.

Since
$$a_2 = a_1 + d,$$
$$9 = a_1 + 6 \qquad \text{Let } a_2 = 9, d = 6.$$
$$a_1 = 3.$$

Then,
$$a_{18} = 3 + (18 - 1)6 \qquad \text{Formula for } a_n; a_1 = 3, n = 18, d = 6$$
$$a_{18} = 105,$$

and
$$a_n = 3 + (n - 1)6$$
$$a_n = 3 + 6n - 6 \qquad \text{Distributive property}$$
$$a_n = 6n - 3.$$

NOW TRY EXERCISE 17. ◀

▶ **EXAMPLE 5** **FINDING THE FIRST TERM OF AN ARITHMETIC SEQUENCE**

Suppose that an arithmetic sequence has $a_8 = -16$ and $a_{16} = -40$. Find a_1.

Solution Since $a_{16} = a_8 + 8d$, it follows that

$$8d = a_{16} - a_8 = -40 - (-16) = -24,$$

and so $d = -3$. To find a_1, use the equation $a_8 = a_1 + 7d$.

$$-16 = a_1 + 7d \qquad \text{Let } a_8 = -16.$$
$$-16 = a_1 + 7(-3) \qquad \text{Let } d = -3.$$
$$a_1 = 5$$

NOW TRY EXERCISE 23. ◀

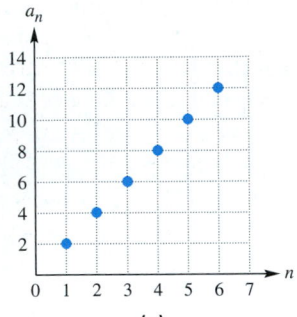

(a)

The graph of any sequence is a scatter diagram. To determine the characteristics of the graph of an arithmetic sequence, start by rewriting the formula for the nth term.

$$a_n = a_1 + (n - 1)d \qquad \text{Formula for the } n\text{th term}$$
$$= a_1 + nd - d \qquad \text{Distributive property}$$
$$= dn + (a_1 - d) \qquad \text{Commutative and associative properties (Section R.2)}$$
$$= dn + c \qquad \text{Let } c = a_1 - d.$$

The points in the graph of an arithmetic sequence are determined by $f(n) = dn + c$, where n is a natural number. Thus, the points in the graph of f must lie on the *line*

$$y = dx + c. \quad \text{(Section 2.5)}$$
$$\uparrow \qquad \uparrow$$
$$\text{slope} \quad y\text{-intercept}$$

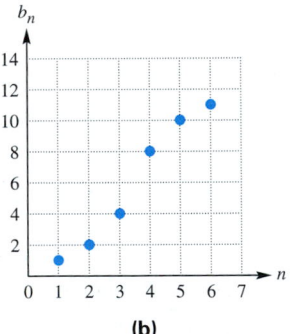

(b)

Figure 7

For example, the sequence a_n shown in Figure 7(a) is an arithmetic sequence because the points that comprise its graph are collinear (lie on a line). The slope determined by these points is 2, so the common difference d equals 2. On the other hand, the sequence b_n shown in Figure 7(b) is not an arithmetic sequence because the points are not collinear.

▶ **EXAMPLE 6** **FINDING THE nth TERM FROM A GRAPH**

Find a formula for the nth term of the sequence a_n shown in Figure 8. What are the domain and range of this sequence?

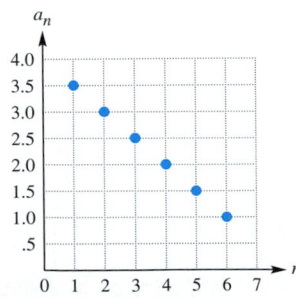

Figure 8

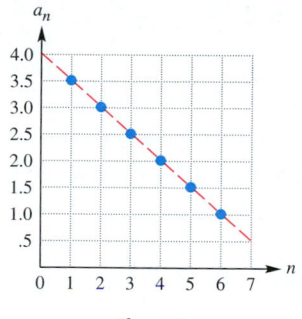

Figure 9

Solution The points in Figure 8 lie on a line, so the sequence is arithmetic. The equation of the dashed line shown in Figure 9 is $y = -.5x + 4$, so the nth term of this sequence is determined by

$$a_n = -.5n + 4.$$

The sequence is comprised of the points

$$(1, 3.5), (2, 3), (3, 2.5), (4, 2), (5, 1.5), (6, 1).$$

Thus, the domain of the sequence is given by $\{1, 2, 3, 4, 5, 6\}$, and the range is given by $\{3.5, 3, 2.5, 2, 1.5, 1\}$.

NOW TRY EXERCISE 27. ◀

Arithmetic Series

The sum of the terms of an arithmetic sequence is an **arithmetic series.** To illustrate, suppose that a person borrows $3000 and agrees to pay $100 per month plus interest of 1% per month on the unpaid balance until the loan is paid off. The first month, $100 is paid to reduce the loan, plus interest of $(.01)3000 = 30$ dollars. The second month, another $100 is paid toward the loan, and $(.01)2900 = 29$ dollars is paid for interest. Since the loan is reduced by $100 each month, interest payments decrease by $(.01)100 = 1$ dollar each month, forming the arithmetic sequence

$$30, 29, 28, \ldots, 3, 2, 1.$$

The total amount of interest paid is given by the sum of the terms of this sequence. Now we develop a formula to find this sum without adding all 30 numbers directly. Since the sequence is arithmetic, we can write the sum of the first n terms as

$$S_n = a_1 + [a_1 + d] + [a_1 + 2d] + \cdots + [a_1 + (n-1)d].$$

We used the formula for the general term in the last expression. Now we write the same sum in reverse order, beginning with a_n and *subtracting d.*

$$S_n = a_n + [a_n - d] + [a_n - 2d] + \cdots + [a_n - (n-1)d]$$

Adding respective sides of these two equations term by term, we obtain

$$S_n + S_n = (a_1 + a_n) + (a_1 + a_n) + \cdots + (a_1 + a_n),$$

or

$$2S_n = n(a_1 + a_n),$$

since there are n terms of $a_1 + a_n$ on the right. Now solve for S_n to get

$$S_n = \frac{n}{2}(a_1 + a_n).$$

Using the formula $a_n = a_1 + (n-1)d$, we can also write this result for S_n as

$$S_n = \frac{n}{2}[a_1 + a_1 + (n-1)d],$$

or

$$S_n = \frac{n}{2}[2a_1 + (n-1)d],$$

which is an alternative formula for the sum of the first n terms of an arithmetic sequence.

A summary of this work follows.

> ### SUM OF THE FIRST n TERMS OF AN ARITHMETIC SEQUENCE
>
> If an arithmetic sequence has first term a_1 and common difference d, then the sum of the first n terms is given by
>
> $$S_n = \frac{n}{2}(a_1 + a_n) \qquad \text{or} \qquad S_n = \frac{n}{2}[2a_1 + (n-1)d].$$

The first formula is used when the first and last terms are known; otherwise the second formula is used.

For example, in the sequence of interest payments discussed earlier, $n = 30$, $a_1 = 30$, and $a_n = 1$. Choosing the first formula,

$$S_n = \frac{n}{2}(a_1 + a_n),$$

gives

$$S_{30} = \frac{30}{2}(30 + 1) = 15(31) = 465,$$

so a total of \$465 interest will be paid over the 30 months.

▶ **EXAMPLE 7** **USING THE SUM FORMULAS**

(a) Evaluate S_{12} for the arithmetic sequence $-9, -5, -1, 3, 7, \ldots$.

(b) Use a formula for S_n to evaluate the sum of the first 60 positive integers.

Solution

(a) We want the sum of the first 12 terms. Using $a_1 = -9$, $n = 12$, and $d = 4$ in the second formula,

$$S_n = \frac{n}{2}[2a_1 + (n-1)d],$$

gives

$$S_{12} = \frac{12}{2}[2(-9) + 11(4)] = 156.$$

(b) The first 60 positive integers form the arithmetic sequence $1, 2, 3, 4, \ldots, 60$. Thus, $n = 60$, $a_1 = 1$, and $a_{60} = 60$, so we use the first formula in the preceding box to find the sum.

$$S_n = \frac{n}{2}(a_1 + a_n)$$

$$S_{60} = \frac{60}{2}(1 + 60) = 1830$$

NOW TRY EXERCISES 33 AND 43. ◀

▶ **EXAMPLE 8**   **USING THE SUM FORMULAS**

The sum of the first 17 terms of an arithmetic sequence is 187. If $a_{17} = -13$, find a_1 and d.

Solution

$$S_{17} = \frac{17}{2}(a_1 + a_{17}) \quad \text{Use the first formula for } S_n, \text{ with } n = 17.$$

$$187 = \frac{17}{2}(a_1 - 13) \quad \text{Let } S_{17} = 187, a_{17} = -13.$$

$$22 = a_1 - 13 \quad \text{Multiply by } \frac{2}{17}.$$

$$a_1 = 35 \quad \text{Add 13; rewrite.}$$

Since $a_{17} = a_1 + (17 - 1)d$,

$$-13 = 35 + 16d \quad \text{Let } a_{17} = -13, a_1 = 35.$$

$$-48 = 16d \quad \text{Subtract 35.}$$

$$d = -3. \quad \text{Divide by 16; rewrite.}$$

NOW TRY EXERCISE 49. ◀

Any sum of the form $\sum_{i=1}^{n}(di + c)$, where d and c are real numbers, represents the sum of the terms of an arithmetic sequence having first term $a_1 = d(1) + c = d + c$ and common difference d. These sums can be evaluated using the formulas in this section.

▶ **EXAMPLE 9**   **USING SUMMATION NOTATION**

Evaluate each sum.

(a) $\sum_{i=1}^{10}(4i + 8)$

(b) $\sum_{k=3}^{9}(4 - 3k)$

Solution

(a) This sum contains the first 10 terms of the arithmetic sequence having

$$a_1 = 4 \cdot 1 + 8 = 12, \quad \text{First term}$$

and

$$a_{10} = 4 \cdot 10 + 8 = 48. \quad \text{Last term}$$

Thus, $\sum_{i=1}^{10}(4i + 8) = S_{10} = \frac{10}{2}(12 + 48) = 5(60) = 300.$

(b) The first few terms are

$$[4 - 3(3)] + [4 - 3(4)] + [4 - 3(5)] + \cdots = -5 + (-8) + (-11) + \cdots.$$

Thus, $a_1 = -5$ and $d = -3$. If the sequence started with $k = 1$, there would be nine terms. Since it starts at 3, two of those terms are missing, so there are seven terms and $n = 7$. Use the second formula for S_n.

$$\sum_{k=3}^{9}(4 - 3k) = \frac{7}{2}[2(-5) + 6(-3)] = -98$$

NOW TRY EXERCISES 55 AND 57. ◀

```
sum(seq(4I+8,I,1
,10,1))
                300
sum(seq(4-3K,K,3
,9,1))
                -98
```

As shown in the previous section, the TI-83/84 Plus will give the sum of a sequence without having to first store the sequence. The screen here illustrates this method for the sequences in Example 9.

11.2 Exercises

Find the common difference d for each arithmetic sequence. See Example 1.

1. $2, 5, 8, 11, \ldots$

2. $4, 10, 16, 22, \ldots$

3. $3, -2, -7, -12, \ldots$

4. $-8, -12, -16, -20, \ldots$

5. $x + 3y, 2x + 5y, 3x + 7y, \ldots$

6. $t^2 + q, -4t^2 + 2q, -9t^2 + 3q, \ldots$

Write the first five terms of each arithmetic sequence. See Example 2.

7. The first term is 8, and the common difference is 6.

8. The first term is -2, and the common difference is 12.

9. $a_1 = 5, d = -2$

10. $a_1 = 4, d = 3$

11. $a_3 = 10, d = -2$

12. $a_1 = 3 - \sqrt{2}, a_2 = 3$

Find a_8 and a_n for each arithmetic sequence. See Examples 3 and 4.

13. $5, 7, 9, \ldots$

14. $-3, -7, -11, \ldots$

15. $a_1 = 5, a_4 = 15$

16. $a_1 = -4, a_5 = 16$

17. $a_{10} = 6, a_{12} = 15$

18. $a_{15} = 8, a_{17} = 2$

19. $a_1 = x, a_2 = x + 3$

20. $a_2 = y + 1, d = -3$

21. $a_4 = s + 6p, d = 2p$

22. *Concept Check* If a_1, a_2, a_3 represents an arithmetic sequence, express a_2 in terms of a_1 and a_3.

Find a_1 for each arithmetic sequence. See Examples 5 and 8.

23. $a_5 = 27, a_{15} = 87$

24. $a_{12} = 60, a_{20} = 84$

25. $S_{16} = -160, a_{16} = -25$

26. $S_{28} = 2926, a_{28} = 199$

Find a formula for the nth term of the finite arithmetic sequence shown in each graph. Then state the domain and range of the sequence. See Example 6.

27.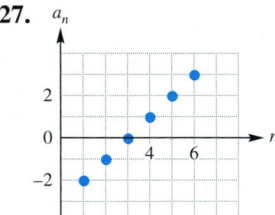

28.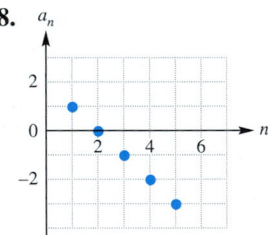

29.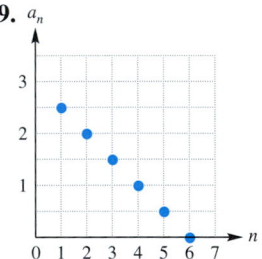

30.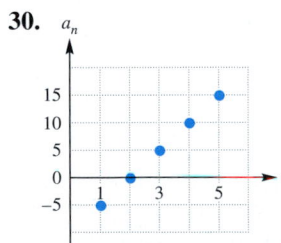

31.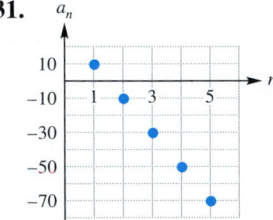

32.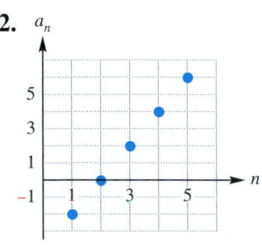

Evaluate S_{10}, the sum of the first ten terms, of each arithmetic sequence. See Example 7(a).

33. $8, 11, 14, \ldots$

34. $-9, -5, -1, \ldots$

35. $5, 9, 13, \ldots$

36. $8, 6, 4, \ldots$

37. $a_2 = 9, a_4 = 13$

38. $a_3 = 5, a_4 = 8$

39. $a_1 = 10, a_{10} = 5.5$

40. $a_1 = -8, a_{10} = -1.25$

41. $a_1 = \pi, a_{10} = 10\pi$

42. *Concept Check* Is this statement accurate? *To find the sum of the first n positive integers, find half the product of n and $n + 1$.*

Find each sum as described. See Example 7(b).

43. the sum of the first 80 positive integers

44. the sum of the first 120 positive integers

45. the sum of the first 50 positive odd integers

46. the sum of the first 90 positive odd integers

47. the sum of the first 60 positive even integers

48. the sum of the first 70 positive even integers

Find a_1 and d for each arithmetic series. See Example 8.

49. $S_{20} = 1090, a_{20} = 102$ **50.** $S_{31} = 5580, a_{31} = 360$

51. $S_{12} = -108, a_{12} = -19$ **52.** $S_{25} = 650, a_{25} = 62$

Evaluate each sum. See Example 9.

53. $\displaystyle\sum_{i=1}^{3} (i + 4)$ **54.** $\displaystyle\sum_{i=1}^{5} (i - 8)$ **55.** $\displaystyle\sum_{j=1}^{10} (2j + 3)$

56. $\displaystyle\sum_{j=1}^{15} (5j - 9)$ **57.** $\displaystyle\sum_{i=4}^{12} (-5 - 8i)$ **58.** $\displaystyle\sum_{k=5}^{19} (-3 - 4k)$

59. $\displaystyle\sum_{i=1}^{1000} i$ **60.** $\displaystyle\sum_{k=1}^{2000} k$ **61.** $\displaystyle\sum_{k=1}^{100} -k$

RELATING CONCEPTS

For individual or collaborative investigation
(Exercises 62–64)

Let $f(x) = mx + b$. **Work Exercises 62–64 in order.**

62. Find $f(1), f(2)$, and $f(3)$.

63. Consider the sequence $f(1), f(2), f(3), \ldots$. Is it an arithmetic sequence? If the sequence is arithmetic, what is the common difference?

64. What is a_n for the sequence described in Exercise 63?

Use the sum and sequence features of a graphing calculator to evaluate the sum of the first ten terms of each arithmetic series with a_n defined as shown. In Exercises 67 and 68, round to the nearest thousandth.

65. $a_n = 4.2n + 9.73$ **66.** $a_n = 8.42n + 36.18$

67. $a_n = \sqrt{8}\,n + \sqrt{3}$ **68.** $a_n = -\sqrt[3]{4}\,n + \sqrt{7}$

Solve each problem.

69. *Integer Sum* Find the sum of all the integers from 51 to 71.

70. *Integer Sum* Find the sum of all the integers from -8 to 30.

71. *Clock Chimes* If a clock strikes the proper number of chimes each hour on the hour, how many times will it chime in a month of 30 days?

72. *Telephone Pole Stack* A stack of telephone poles has 30 in the bottom row, 29 in the next, and so on, with one pole in the top row. How many poles are in the stack?

73. *Population Growth* Five years ago, the population of a city was 49,000. Each year the zoning commission permits an increase of 580 in the population. What will the maximum population be 5 yr from now?

74. *Slide Supports* A super slide of uniform slope is to be built on a level piece of land. There are to be 20 equally spaced vertical supports, with the longest support 15 m long and the shortest 2 m long. Find the total length of all the supports.

75. *Rungs of a Ladder* How much material would be needed for the rungs of a ladder of 31 rungs, if the rungs taper uniformly from 18 in. to 28 in.?

76. *(Modeling) Children's Growth Pattern* The normal growth pattern for children age 3–11 follows that of an arithmetic sequence. An increase in height of about 6 cm per year is expected. Thus, 6 would be the common difference of the sequence. For example, a child who measures 96 cm at age 3 would have his expected height in subsequent years represented by the sequence 102, 108, 114, 120, 126, 132, 138, 144. Each term differs from the adjacent terms by the common difference, 6.

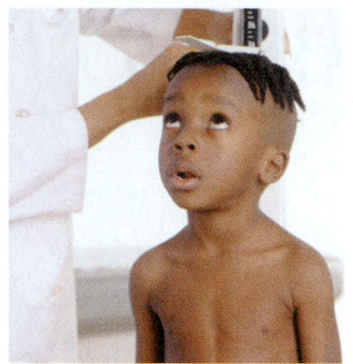

(a) If a child measures 98.2 cm at age 3 and 109.8 cm at age 5, what would be the common difference of the arithmetic sequence describing her yearly height?

(b) What would we expect her height to be at age 8?

77. *Concept Check* Suppose that $a_1, a_2, a_3, a_4, a_5, \ldots$ is an arithmetic sequence. Is $a_1, a_3, a_5, \ldots$ an arithmetic sequence?

78. Explain why the sequence log 2, log 4, log 8, log 16, ... is arithmetic.

11.3 Geometric Sequences and Series

Geometric Sequences ■ Geometric Series ■ Infinite Geometric Series ■ Annuities

Geometric Sequences Suppose you agreed to work for 1¢ the first day, 2¢ the second day, 4¢ the third day, 8¢ the fourth day, and so on, with your wages doubling each day. How much will you earn on day 20, after working 5 days a week for a month? How much will you have earned altogether in 20 days? These questions will be answered in this section.

A **geometric sequence** (or **geometric progression**) is a sequence in which each term after the first is obtained by multiplying the preceding term by a fixed nonzero real number, called the **common ratio.** The sequence discussed above,

$$1, 2, 4, 8, 16, \ldots,$$

is an example of a geometric sequence in which the first term is 1 and the common ratio is 2.

Notice that if we divide any term after the first term by the preceding term, we obtain the common ratio $r = 2$.

$$\frac{a_2}{a_1} = \frac{2}{1} = 2; \quad \frac{a_3}{a_2} = \frac{4}{2} = 2; \quad \frac{a_4}{a_3} = \frac{8}{4} = 2; \quad \frac{a_5}{a_4} = \frac{16}{8} = 2$$

If the common ratio of a geometric sequence is r, then by the definition of a geometric sequence,

$$r = \frac{a_{n+1}}{a_n}, \quad \text{Common ratio } r$$

for every positive integer n. ***Therefore, we find the common ratio by choosing any term except the first and dividing it by the preceding term.***

In the geometric sequence 2, 8, 32, 128, $\ldots$, $r = 4$. Notice that

$$8 = 2 \cdot 4$$
$$32 = 8 \cdot 4 = (2 \cdot 4) \cdot 4 = 2 \cdot 4^2$$
$$128 = 32 \cdot 4 = (2 \cdot 4^2) \cdot 4 = 2 \cdot 4^3.$$

To generalize this, assume that a geometric sequence has first term a_1 and common ratio r. The second term is $a_2 = a_1 r$, the third is $a_3 = a_2 r = (a_1 r)r = a_1 r^2$, and so on. Following this pattern, the nth term is $a_n = a_1 r^{n-1}$. Again, this result can be proved by mathematical induction. (See **Section 11.5.**)

---

nth TERM OF A GEOMETRIC SEQUENCE

In a geometric sequence with first term a_1 and common ratio r, the nth term, a_n, is given by

$$a_n = a_1 r^{n-1}.$$

▶ **EXAMPLE 1** **FINDING THE nth TERM OF A GEOMETRIC SEQUENCE**

Use the formula for the nth term of a geometric sequence to answer the first question posed at the beginning of this section. How much will be earned on day 20 if daily wages follow the sequence 1, 2, 4, 8, 16, $\ldots$ cents?

Solution   To answer the question, let $a_1 = 1$ and $r = 2$, and find a_{20}.

$$a_{20} = a_1 r^{19} = 1(2)^{19} = 524{,}288 \text{ cents,} \quad \text{or} \quad \$5242.88$$

NOW TRY EXERCISE 1(a). ◀

▶ **EXAMPLE 2** **FINDING TERMS OF A GEOMETRIC SEQUENCE**

Find a_5 and a_n for the geometric sequence 4, 12, 36, 108, $\ldots$.

Solution   The first term, a_1, is 4. Find r by choosing any term after the first and dividing it by the preceding term. For example, $r = \frac{36}{12} = 3$. Since $a_4 = 108$,

$$a_5 = 3 \cdot 108 = 324.$$

We could also find the fifth term by using the formula for a_n, $a_n = a_1 r^{n-1}$, and replacing n with 5, r with 3, and a_1 with 4.

$$a_5 = 4 \cdot (3)^{5-1} = 4 \cdot 3^4 = 324$$

By the formula, $a_n = 4 \cdot 3^{n-1}.$

NOW TRY EXERCISE 11. ◀

▶ **EXAMPLE 3** **FINDING TERMS OF A GEOMETRIC SEQUENCE**

Find r and a_1 for the geometric sequence with third term 20 and sixth term 160.

Solution Use the formula for the nth term of a geometric sequence.

$$\text{For } n = 3, \qquad a_3 = a_1 r^2 = 20.$$
$$\text{For } n = 6, \qquad a_6 = a_1 r^5 = 160.$$

Since $a_1 r^2 = 20$, $a_1 = \frac{20}{r^2}$. Substitute this value for a_1 in the second equation.

$$a_1 r^5 = 160$$

$$\left(\frac{20}{r^2}\right) r^5 = 160 \qquad \text{Substitute.}$$

$$20 r^3 = 160 \qquad \text{Quotient rule for exponents (Section R.6)}$$

$$r^3 = 8 \qquad \text{Divide by 20. (Section 1.1)}$$

$$r = 2 \qquad \text{Take cube roots.}$$

Since $a_1 r^2 = 20$ and $r = 2$,

$$a_1(2)^2 = 20 \qquad \text{Substitute.}$$

$$4a_1 = 20$$

$$a_1 = 5. \qquad \text{Divide by 4.}$$

NOW TRY EXERCISE 17. ◀

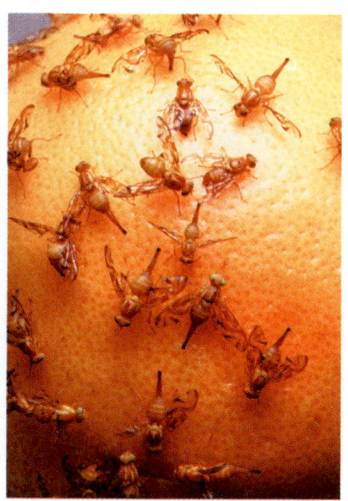

▶ **EXAMPLE 4** **MODELING A POPULATION OF FRUIT FLIES**

A population of fruit flies is growing in such a way that each generation is 1.5 times as large as the last generation. Suppose there were 100 insects in the first generation. How many would there be in the fourth generation?

Solution Write the population of each generation as a geometric sequence with a_1 as the first-generation population, a_2 the second-generation population, and so on. Then the fourth-generation population is a_4. Using the formula for a_n, with $n = 4$, $r = 1.5$, and $a_1 = 100$, gives

$$a_4 = a_1 r^3 = 100(1.5)^3 = 100(3.375) = 337.5.$$

In the fourth generation, the population will number about 338 insects.

NOW TRY EXERCISE 61. ◀

Geometric Series A **geometric series** is the sum of the terms of a geometric sequence. In applications, it may be necessary to find the sum of the terms of such a sequence. For example, a scientist might want to know the total number of insects in four generations of the population discussed in Example 4. This population would equal $a_1 + a_2 + a_3 + a_4$, or

$$100 + 100(1.5) + 100(1.5)^2 + 100(1.5)^3 = 812.5 \approx 813 \text{ insects.}$$

To find a formula for the sum of the first n terms of a geometric sequence, S_n, first write the sum as

$$S_n = a_1 + a_2 + a_3 + \cdots + a_n$$

or $\qquad S_n = a_1 + a_1 r + a_1 r^2 + \cdots + a_1 r^{n-1}. \quad (1)$

If $r = 1$, then $S_n = na_1$, which is a correct formula for this case. If $r \neq 1$, then multiply both sides of equation (1) by r to obtain

$$rS_n = a_1 r + a_1 r^2 + a_1 r^3 + \cdots + a_1 r^n. \quad (2)$$

If equation (2) is subtracted from equation (1),

$$
\begin{aligned}
S_n &= a_1 + a_1 r + a_1 r^2 + \cdots + a_1 r^{n-1} && (1) \\
rS_n &= a_1 r + a_1 r^2 + \cdots + a_1 r^{n-1} + a_1 r^n && (2) \\
\hline
S_n - rS_n &= a_1 \phantom{+ a_1 r + a_1 r^2 + \cdots + a_1 r^{n-1}} - a_1 r^n && \text{Subtract.} \\
S_n(1 - r) &= a_1(1 - r^n) && \text{Factor. (Section R.4)} \\
S_n &= \frac{a_1(1 - r^n)}{1 - r}, \qquad \text{where } r \neq 1. && \text{Divide by } 1 - r. \\
& && \text{(Section 1.1)}
\end{aligned}
$$

> ### SUM OF THE FIRST n TERMS OF A GEOMETRIC SEQUENCE
>
> If a geometric sequence has first term a_1 and common ratio r, then the sum of the first n terms is given by
>
> $$S_n = \frac{a_1(1 - r^n)}{1 - r}, \qquad \text{where } r \neq 1.$$

We can use a geometric series to find the total fruit fly population in Example 4 over the four-generation period. With $n = 4$, $a_1 = 100$, and $r = 1.5$,

$$S_4 = \frac{100(1 - 1.5^4)}{1 - 1.5} = \frac{100(1 - 5.0625)}{-.5} = 812.5 \approx 813 \text{ insects},$$

which agrees with our previous result.

▶ **EXAMPLE 5**   **FINDING THE SUM OF THE FIRST n TERMS**

At the beginning of this section on page 1016, we posed the following question: How much will you have earned altogether after 20 days? Answer this question.

Solution   We must find the total amount earned in 20 days with daily wages of $1, 2, 4, 8, \ldots$ cents. Since $a_1 = 1$ and $r = 2$,

$$S_{20} = \frac{1(1 - 2^{20})}{1 - 2} = \frac{1 - 1{,}048{,}576}{-1} = 1{,}048{,}575 \text{ cents}, \quad \text{or} \quad \$10{,}485.75.$$

Not bad for 20 days of work!

NOW TRY EXERCISE 1(b). ◀

▶ **EXAMPLE 6** **FINDING THE SUM OF THE FIRST n TERMS**

Find $\displaystyle\sum_{i=1}^{6} 2 \cdot 3^i$.

Solution This series is the sum of the first six terms of a geometric sequence having $a_1 = 2 \cdot 3^1 = 6$ and $r = 3$. Using the formula for S_n,

$$S_6 = \frac{6(1 - 3^6)}{1 - 3} = \frac{6(1 - 729)}{-2} = \frac{6(-728)}{-2} = 2184.$$

NOW TRY EXERCISE 29. ◀

Infinite Geometric Series
We extend our discussion of sums of sequences to include infinite geometric sequences such as

$$2, 1, \frac{1}{2}, \frac{1}{4}, \frac{1}{8}, \frac{1}{16}, \cdots,$$

with first term 2 and common ratio $\frac{1}{2}$. Using the formula for S_n gives the following sequence of sums.

$$S_1 = 2, \quad S_2 = 3, \quad S_3 = \frac{7}{2}, \quad S_4 = \frac{15}{4}, \quad S_5 = \frac{31}{8}, \quad S_6 = \frac{63}{16}, \cdots$$

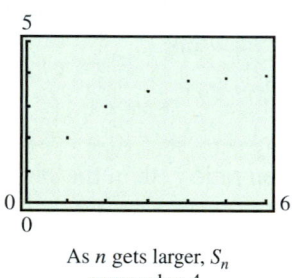

As n gets larger, S_n approaches 4.

Figure 10

The formula for S_n can also be written as $S_n = 4 - \left(\frac{1}{2}\right)^{n-2}$. As this formula and Figure 10 suggest, these sums seem to be getting closer and closer to the number 4. For no value of n is $S_n = 4$. However, if n is large enough, then S_n is as close to 4 as desired. As mentioned earlier, we say the sequence converges to 4. This is expressed as

$$\lim_{n \to \infty} S_n = 4.$$

(Read: "the limit of S_n as n increases without bound is 4.") Since $\lim_{n \to \infty} S_n = 4$, the number 4 is called the *sum of the terms* of the infinite geometric sequence

$$2, 1, \frac{1}{2}, \frac{1}{4}, \cdots$$

and

$$2 + 1 + \frac{1}{2} + \frac{1}{4} + \cdots = 4.$$

▼ **LOOKING AHEAD TO CALCULUS**

In the discussion of

$$\lim_{n \to \infty} S_n = 4,$$

we used the phrases "large enough" and "as close as desired." This description is made more precise in a standard calculus course.

▶ **EXAMPLE 7** **SUMMING THE TERMS OF AN INFINITE GEOMETRIC SERIES**

Evaluate $1 + \dfrac{1}{3} + \dfrac{1}{9} + \dfrac{1}{27} + \cdots$.

Solution Use the formula for the sum of the first n terms of a geometric sequence to obtain

$$S_1 = 1, \quad S_2 = \frac{4}{3}, \quad S_3 = \frac{13}{9}, \quad S_4 = \frac{40}{27},$$

and, in general, $\qquad S_n = \dfrac{1\left[1 - \left(\frac{1}{3}\right)^n\right]}{1 - \frac{1}{3}}.$ 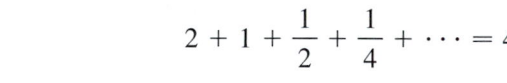 Let $a_1 = 1, r = \frac{1}{3}$.

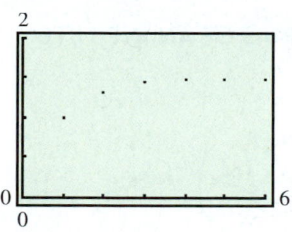

This graph of the first six values of S_n in Example 7 suggests that its value is approaching $\frac{3}{2}$. (The y-scale here is $\frac{1}{2}$.)

The table shows the value of $\left(\frac{1}{3}\right)^n$ for larger and larger values of n.

n	1	10	100	200
$\left(\dfrac{1}{3}\right)^n$	$\dfrac{1}{3}$	1.69×10^{-5}	1.94×10^{-48}	3.76×10^{-96}

As n gets larger and larger, $\left(\frac{1}{3}\right)^n$ gets closer and closer to 0. That is,

$$\lim_{n \to \infty} \left(\frac{1}{3}\right)^n = 0,$$

making it reasonable that

$$\lim_{n \to \infty} S_n = \lim_{n \to \infty} \frac{1\left[1 - \left(\frac{1}{3}\right)^n\right]}{1 - \frac{1}{3}} = \frac{1(1 - 0)}{1 - \frac{1}{3}} = \frac{1}{\frac{2}{3}} = \frac{3}{2}. \qquad \text{Simplify the complex fraction. (Section R.5)}$$

Hence,

$$1 + \frac{1}{3} + \frac{1}{9} + \frac{1}{27} + \cdots = \frac{3}{2}.$$

NOW TRY EXERCISE 39. ◀

▼ LOOKING AHEAD TO CALCULUS

In calculus, functions are sometimes defined in terms of infinite series. Here are three functions we studied earlier in the text defined that way.

$$e^x = \frac{x^0}{0!} + \frac{x^1}{1!} + \frac{x^2}{2!} + \frac{x^3}{3!} + \cdots$$

$$\ln(1 + x) = x - \frac{x^2}{2} + \frac{x^3}{3} - \cdots$$

for x in $(-1, 1)$

$$\frac{1}{1 + x} = 1 - x + x^2 - x^3 + \cdots$$

for x in $(-1, 1)$

If a geometric series has first term a_1 and common ratio r, then

$$S_n = \frac{a_1(1 - r^n)}{1 - r} \qquad (r \neq 1)$$

for every positive integer n. If $-1 < r < 1$, then $\lim\limits_{n \to \infty} r^n = 0$, and

$$\lim_{n \to \infty} S_n = \frac{a_1(1 - 0)}{1 - r} = \frac{a_1}{1 - r}.$$

This quotient, $\dfrac{a_1}{1 - r}$, is called the **sum of the terms of an infinite geometric sequence.** The limit $\lim\limits_{n \to \infty} S_n$ is often expressed as S_∞ or $\sum\limits_{i=1}^{\infty} a_i$.

SUM OF THE TERMS OF AN INFINITE GEOMETRIC SEQUENCE

The sum of the terms of an infinite geometric sequence with first term a_1 and common ratio r, where $-1 < r < 1$, is given by

$$S_\infty = \frac{a_1}{1 - r}.$$

If $|r| > 1$, then the terms get larger and larger in absolute value, so there is no limit as $n \to \infty$. Hence the terms of the sequence will not have a sum.

▶ **EXAMPLE 8** **FINDING THE SUM OF THE TERMS OF AN INFINITE GEOMETRIC SEQUENCE**

Find each sum.

(a) $\displaystyle\sum_{i=1}^{\infty} \left(-\frac{3}{4}\right)\left(-\frac{1}{2}\right)^{i-1}$

(b) $\displaystyle\sum_{i=1}^{\infty} \left(\frac{3}{5}\right)^{i}$

Solution

(a) Here, $a_1 = -\frac{3}{4}$ and $r = -\frac{1}{2}$. Since $-1 < r < 1$, the preceding formula applies.

$$S_{\infty} = \frac{a_1}{1-r} = \frac{-\frac{3}{4}}{1-\left(-\frac{1}{2}\right)} = \frac{-\frac{3}{4}}{\frac{3}{2}} = -\frac{3}{4} \div \frac{3}{2} = -\frac{3}{4} \cdot \frac{2}{3} = -\frac{1}{2}$$

(b) $\displaystyle\sum_{i=1}^{\infty} \left(\frac{3}{5}\right)^{i} = \frac{\frac{3}{5}}{1-\frac{3}{5}} = \frac{\frac{3}{5}}{\frac{2}{5}} = \frac{3}{5} \div \frac{2}{5} = \frac{3}{5} \cdot \frac{5}{2} = \frac{3}{2}$ $\qquad a_1 = \frac{3}{5}, r = \frac{3}{5}$

NOW TRY EXERCISES 45 AND 47. ◀

Annuities A sequence of equal payments made after equal periods of time, such as car payments or house payments, is called an **annuity.** If the payments are accumulated in an account (with no withdrawals), the sum of the payments and interest on the payments is called the **future value** of the annuity.

▶ **EXAMPLE 9** **FINDING THE FUTURE VALUE OF AN ANNUITY**

To save money for a trip, Taylor Wells deposited $1000 at the *end* of each year for 4 yr in an account paying 3% interest, compounded annually. Find the future value of this annuity.

Solution To find the future value, we use the formula for interest compounded annually,

$$A = P(1+r)^{t}. \quad \text{(Section 4.2)}$$

The first payment earns interest for 3 yr, the second payment for 2 yr, and the third payment for 1 yr. The last payment earns no interest. The total amount is

$$1000(1.03)^3 + 1000(1.03)^2 + 1000(1.03) + 1000.$$

This is the sum of the terms of a geometric sequence with first term (starting at the end of the sum as written above) $a_1 = 1000$ and common ratio $r = 1.03$. Using the formula for S_4, the sum of four terms, gives

$$S_4 = \frac{1000[1-(1.03)^4]}{1-1.03} \approx 4183.63.$$

The future value of the annuity is $4183.63.

NOW TRY EXERCISE 71. ◀

The general formula for the future value of an annuity can be stated as follows. (See Exercise 79.)

FUTURE VALUE OF AN ANNUITY

The formula for the future value of an annuity is given by

$$S = R\left[\frac{(1 + i)^n - 1}{i}\right],$$

where S is future value, R is payment at the end of each period, i is interest rate per period, and n is number of periods.

11.3 Exercises

*Refer to the first sentence of this section, which describes a method of payment for a job. Determine (**a**) how much you will earn on the day indicated and (**b**) how much you will have earned altogether after your wages are paid on the day indicated. See Examples 1 and 5.*

1. day 10   **2.** day 12   **3.** day 15   **4.** day 18

Find a_5 and a_n for each geometric sequence. See Example 2.

5. $a_1 = 5, r = -2$ **6.** $a_1 = 8, r = -5$ **7.** $a_2 = -4, r = 3$

8. $a_3 = -2, r = 4$ **9.** $a_4 = 243, r = -3$ **10.** $a_4 = 18, r = 2$

11. $-4, -12, -36, -108, \ldots$ **12.** $-2, 6, -18, 54, \ldots$

13. $\dfrac{4}{5}, 2, 5, \dfrac{25}{2}, \ldots$ **14.** $\dfrac{1}{2}, \dfrac{2}{3}, \dfrac{8}{9}, \dfrac{32}{27}, \ldots$

15. $10, -5, \dfrac{5}{2}, -\dfrac{5}{4}, \ldots$ **16.** $3, -\dfrac{9}{4}, \dfrac{27}{16}, -\dfrac{81}{64}, \ldots$

Find a_1 and r for each geometric sequence. See Example 3.

17. $a_2 = -6, a_7 = -192$ **18.** $a_3 = 5, a_8 = \dfrac{1}{625}$

19. $a_3 = 50, a_7 = .005$ **20.** $a_4 = -\dfrac{1}{4}, a_9 = -\dfrac{1}{128}$

Use the formula for S_n to find the sum of the first five terms of each geometric sequence. In Exercises 25 and 26, round to the nearest hundredth. See Example 5.

21. $2, 8, 32, 128, \ldots$ **22.** $4, 16, 64, 256, \ldots$

23. $18, -9, \dfrac{9}{2}, -\dfrac{9}{4}, \ldots$ **24.** $12, -4, \dfrac{4}{3}, -\dfrac{4}{9}, \ldots$

25. $a_1 = 8.423, r = 2.859$ **26.** $a_1 = -3.772, r = -1.553$

Find each sum. See Example 6.

27. $\displaystyle\sum_{i=1}^{5} 3^i$ **28.** $\displaystyle\sum_{i=1}^{4} (-2)^i$ **29.** $\displaystyle\sum_{j=1}^{6} 48\left(\dfrac{1}{2}\right)^j$

30. $\displaystyle\sum_{j=1}^{5} 243\left(\dfrac{2}{3}\right)^j$ **31.** $\displaystyle\sum_{k=4}^{10} 2^k$ **32.** $\displaystyle\sum_{k=3}^{9} (-3)^k$

33. *Concept Check*   Under what conditions does the sum of an infinite geometric series exist?

34. The number .999. . . can be written as the sum of the terms of an infinite geometric sequence: $.9 + .09 + .009 + \cdots$. Here we have $a_1 = .9$ and $r = .1$. Use the formula for S_∞ to find this sum. Does your intuition indicate that your answer is correct?

Find r for each infinite geometric sequence. Identify any whose sum does not converge.

35. $12, 24, 48, 96, \ldots$

36. $625, 125, 25, 5, \ldots$

37. $-48, -24, -12, -6, \ldots$

38. $2, -10, 50, -250, \ldots$

39. Use $\lim\limits_{n \to \infty} S_n$ to show that $2 + 1 + \frac{1}{2} + \frac{1}{4} + \cdots$ converges to 4. See Example 7.

40. In Example 7, we determined that $1 + \frac{1}{3} + \frac{1}{9} + \frac{1}{27} + \cdots$ converges to $\frac{3}{2}$ using an argument involving limits. Use the formula for the sum of the terms of an infinite geometric sequence to obtain the same result.

Find each sum that converges. See Example 8.

41. $18 + 6 + 2 + \dfrac{2}{3} + \cdots$

42. $100 + 10 + 1 + \cdots$

43. $\dfrac{1}{4} - \dfrac{1}{6} + \dfrac{1}{9} - \dfrac{2}{27} + \cdots$

44. $\dfrac{4}{3} + \dfrac{2}{3} + \dfrac{1}{3} + \cdots$

45. $\displaystyle\sum_{i=1}^{\infty} 3\left(\dfrac{1}{4}\right)^{i-1}$

46. $\displaystyle\sum_{i=1}^{\infty} 5\left(-\dfrac{1}{4}\right)^{i-1}$

47. $\displaystyle\sum_{k=1}^{\infty} (.3)^k$

48. $\displaystyle\sum_{k=1}^{\infty} 10^{-k}$

RELATING CONCEPTS

For individual or collaborative investigation
(Exercises 49–52)

Let $g(x) = ab^x$. **Work Exercises 49–52 in order.**

49. Find $g(1)$, $g(2)$, and $g(3)$.

50. Consider the sequence $g(1), g(2), g(3), \ldots$. Is it a geometric sequence? If so, what is the common ratio?

51. What is the general term of the sequence in Exercise 50?

 52. Explain how geometric sequences are related to exponential functions.

Use the sum and sequence features of a graphing calculator to evaluate each sum. Round to the nearest thousandth.

53. $\displaystyle\sum_{i=1}^{10} (1.4)^i$

54. $\displaystyle\sum_{j=1}^{6} -(3.6)^j$

55. $\displaystyle\sum_{j=3}^{8} 2(.4)^j$

56. $\displaystyle\sum_{i=4}^{9} 3(.25)^i$

Solve each problem. See Examples 1–8.

57. *(Modeling) Investment for Retirement* According to T. Rowe Price Associates, a person with a moderate investment strategy and n years to retirement should have accumulated savings of a_n percent of his or her annual salary. The geometric sequence defined by

$$a_n = 1276(.916)^n$$

gives the appropriate percent for each year n.

(a) Find a_1 and r. Round a_1 to the nearest whole number.

(b) Find and interpret the terms a_{10} and a_{20}. Round to the nearest whole number.

58. *(Modeling) Investment for Retirement* Refer to Exercise 57. For someone who has a conservative investment strategy with n years to retirement, the geometric sequence is $a_n = 1278(.935)^n$. (*Source:* T. Rowe Price Associates.)

(a) Repeat part (a) of Exercise 57. (b) Repeat part (b) of Exercise 57.
(c) Why are the answers in parts (a) and (b) larger than those in Exercise 57?

59. *(Modeling) Bacterial Growth* The strain of bacteria described in Exercise 89 in **Section 11.1** will double in size and then divide every 40 min. Let a_1 be the initial number of bacteria cells, a_2 the number after 40 min, and a_n the number after $40(n-1)$ min. (*Source:* Hoppensteadt, F. and C. Peskin, *Mathematics in Medicine and the Life Sciences,* Springer-Verlag, 1992.)

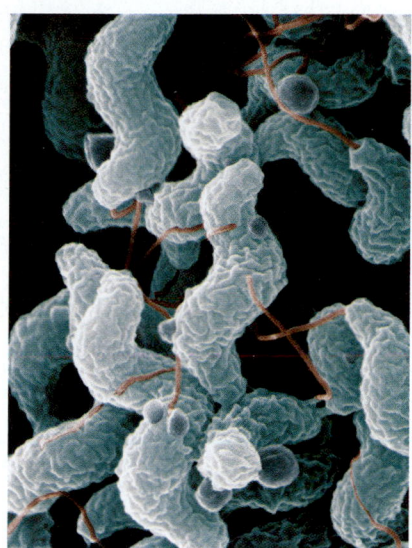

(a) Write a formula for the nth term a_n of the geometric sequence

$$a_1, a_2, a_3, \ldots, a_n, \ldots.$$

(b) Determine the first value for n where $a_n > 1{,}000{,}000$ if $a_1 = 100$.
(c) How long does it take for the number of bacteria to exceed one million?

60. *Photo Processing* The final step in processing a black-and-white photographic print is to immerse the print in a chemical fixer. The print is then washed in running water. Under certain conditions, 98% of the fixer in a print will be removed with 15 min of washing. How much of the original fixer would be left after 1 hr of washing?

61. *(Modeling) Fruit Flies Population* A population of fruit flies is growing in such a way that each generation is 1.25 times as large as the last generation. Suppose there were 200 insects in the first generation. How many would there be in the fifth generation?

62. *Depreciation* Each year a machine loses 20% of the value it had at the beginning of the year. Find the value of the machine at the end of 6 yr if it cost $100,000 new.

63. *Sugar Processing* A sugar factory receives an order for 1000 units of sugar. The production manager thus orders production of 1000 units of sugar. He forgets, however, that the production of sugar requires some sugar (to prime the machines, for example), and so he ends up with only 900 units of sugar. He then orders an additional 100 units, and receives only 90 units. A further order for 10 units produces 9 units. Finally seeing he is wrong, the manager decides to try mathematics. He views the production process as an infinite geometric progression with $a_1 = 1000$ and $r = .1$. Using this, find the number of units of sugar that he should have ordered originally.

64. *Height of a Dropped Ball* Hortense drops a ball from a height of 10 m and notices that on each bounce the ball returns to about $\frac{3}{4}$ of its previous height. About how far will the ball travel before it comes to rest? (*Hint:* Consider the sum of two sequences.)

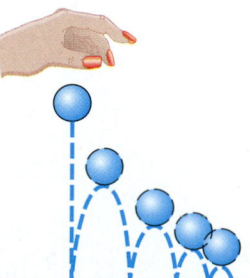

65. *Number of Ancestors* Each person has two parents, four grandparents, eight great-grandparents, and so on. What is the total number of ancestors a person has, going back five generations? ten generations?

66. *Drug Dosage* Certain medical conditions are treated with a fixed dose of a drug administered at regular intervals. Suppose a person is given 2 mg of a drug each day and that during each 24-hr period, the body utilizes 40% of the amount of drug that was present at the beginning of the period.

(a) Show that the amount of the drug present in the body at the end of n days is

$$\sum_{i=1}^{n} 2(.6)^i.$$

(b) What will be the approximate quantity of the drug in the body at the end of each day after the treatment has been administered for a long period of time?

67. *Side Length of a Triangle* A sequence of equilateral triangles is constructed. The first triangle has sides 2 m in length. To get the second triangle, midpoints of the sides of the original triangle are connected. What is the length of each side of the eighth such triangle? See the figure.

   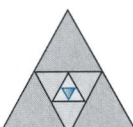

68. *Perimeter and Area of Triangles* In Exercise 67, if the process could be continued indefinitely, what would be the total perimeter of all the triangles? What would be the total area of all the triangles, disregarding the overlapping?

69. *Salaries* You are offered a 6-week summer job and are asked to select one of the following salary options.

Option 1: $5000 for the first day with a $10,000 raise each day for the remaining 29 days (that is, $15,000 for day 2, $25,000 for day 3, and so on)

Option 2: 1¢ for the first day with the pay doubled each day (that is, 2¢ for day 2, 4¢ for day 3, and so on)

Which option would you choose?

70. *Number of Ancestors* Suppose a genealogical Web site allows you to identify all your ancestors that lived during the last 300 yr. Assuming that each generation spans about 25 yr, guess the number of ancestors that would be found during the 12 generations. Then use the formula for a geometric series to find the correct value.

Future Value of an Annuity   *Find the future value of each annuity. See Example 9.*

71. There are payments of $1000 at the end of each year for 9 yr at 3% interest compounded annually.

72. There are payments of $800 at the end of each year for 12 yr at 2% interest compounded annually.

73. There are payments of $2430 at the end of each year for 10 yr at 1% interest compounded annually.

74. There are payments of $1500 at the end of each year for 6 yr at .5% interest compounded annually.

75. Refer to Exercise 73. Use the answer and recursion to find the balance after 11 yr.

76. Refer to Exercise 74. Use the answer and recursion to find the balance after 7 yr.

77. *Individual Retirement Account*   Starting on his fortieth birthday, Michael Branson deposits $2000 per year in an Individual Retirement Account until age 65 (last payment at age 64). Find the total amount in the account if he had a guaranteed interest rate of 3% compounded annually.

78. *Retirement Savings*   To save for retirement, Mort put $3000 at the end of each year into an ordinary annuity for 20 yr at 2.5% annual interest. At the end of the 20 yr, what was the amount of the annuity?

79. Show that the formula for future value of an annuity gives the correct answer when compared to the solution in Example 9.

80. The screen here shows how the TI-83/84 Plus calculator computes the future value of the annuity described in Example 9. Use a calculator with this capability to support your answers in Exercises 71–78.

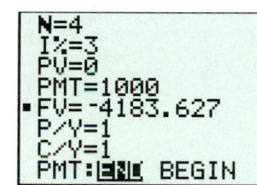

81. *Concept Check*   Suppose that $a_1, a_2, a_3, a_4, a_5, \ldots$ is a geometric sequence. Is $a_1, a_3, a_5, \ldots$ a geometric sequence?

82. Explain why the sequence log 6, log 36, log 1296, log 1,679,616, ... is geometric.

Summary Exercises on Sequences and Series

Use the following guidelines in Exercises 1–16.

> Given a sequence $a_1, a_2, a_3, a_4, a_5, \ldots,$
>
> - If the differences $a_2 - a_1, a_3 - a_2, a_4 - a_3, a_5 - a_4, \ldots$ are all equal to the same number d, then the sequence is *arithmetic,* and d is the *common difference.*
>
> - If the ratios $\dfrac{a_2}{a_1}, \dfrac{a_3}{a_2}, \dfrac{a_4}{a_3}, \dfrac{a_5}{a_4}, \ldots$ are all equal to the same number r, then the sequence is *geometric,* and r is the *common ratio.*

In Exercises 1–9, determine whether the sequence is arithmetic, geometric, *or* neither. *If the sequence is arithmetic, give its common difference d. If the sequence is geometric, give its common ratio r.*

1. 2, 4, 8, 16, 32, ...   **2.** 1, 4, 7, 10, 13, ...   **3.** $3, \dfrac{1}{2}, -2, -\dfrac{9}{2}, -7, \ldots$

4. $1, -2, 3, -4, 5, \ldots$ **5.** $\dfrac{3}{4}, 1, \dfrac{4}{3}, \dfrac{16}{9}, \dfrac{64}{27}, \ldots$ **6.** $4, -12, 36, -108, 324, \ldots$

7. $\dfrac{1}{2}, \dfrac{1}{3}, \dfrac{1}{4}, \dfrac{1}{5}, \dfrac{1}{6}, \ldots$ **8.** $5, 2, -1, -4, -7, \ldots$ **9.** $1, 9, 10, 19, 29, \ldots$

10. *Concept Check*   For the sequence $5, 5, 5, \ldots$, find d and r.

In Exercises 11–16, determine whether the given sequence is either arithmetic *or* geometric. *Then find a_n and $\displaystyle\sum_{i=1}^{10} a_i$.*

11. $3, 6, 12, 24, 48, \ldots$ **12.** $2, 6, 10, 14, 18, \ldots$ **13.** $4, \dfrac{5}{2}, 1, -\dfrac{1}{2}, -2, \ldots$

14. $\dfrac{3}{2}, 1, \dfrac{2}{3}, \dfrac{4}{9}, \dfrac{8}{27}, \ldots$ **15.** $3, -6, 12, -24, 48, \ldots$ **16.** $-5, -8, -11, -14, -17, \ldots$

Evaluate each sum, where possible. Identify any that diverge.

17. $\displaystyle\sum_{i=1}^{\infty} \dfrac{1}{3}(-2)^{i-1}$ **18.** $\displaystyle\sum_{j=1}^{4} 2\left(\dfrac{1}{10}\right)^{j-1}$ **19.** $\displaystyle\sum_{i=1}^{25} (4 - 6i)$

20. $\displaystyle\sum_{i=1}^{6} 3^i$ **21.** $\displaystyle\sum_{i=1}^{\infty} 4\left(-\dfrac{1}{2}\right)^i$ **22.** $\displaystyle\sum_{i=1}^{\infty} (3i - 2)$

23. $\displaystyle\sum_{j=1}^{12} (2j - 1)$ **24.** $\displaystyle\sum_{k=1}^{\infty} 3^{-k}$ **25.** $\displaystyle\sum_{i=1}^{\infty} 1.0001^i$

26. Write $.999 \ldots$ as an infinite geometric series. Find the sum. (This result bothers many people.)

11.4 The Binomial Theorem

A Binomial Expansion Pattern ▪ **Pascal's Triangle** ▪ **n-Factorial** ▪ **Binomial Coefficients** ▪ **The Binomial Theorem** ▪ **kth Term of a Binomial Expansion**

A Binomial Expansion Pattern   In this section, we introduce a method for writing the expansion of expressions of the form $(x + y)^n$, where n is a natural number. Some expansions for various nonnegative integer values of n are given below.

$$(x + y)^0 = 1$$
$$(x + y)^1 = x + y$$
$$(x + y)^2 = x^2 + 2xy + y^2$$
$$(x + y)^3 = x^3 + 3x^2y + 3xy^2 + y^3$$
$$(x + y)^4 = x^4 + 4x^3y + 6x^2y^2 + 4xy^3 + y^4$$
$$(x + y)^5 = x^5 + 5x^4y + 10x^3y^2 + 10x^2y^3 + 5xy^4 + y^5$$

▼ **LOOKING AHEAD TO CALCULUS**

Students taking calculus study the binomial series, which follows from Isaac Newton's extension to the case where the exponent is no longer a positive integer. His result led to a series for $(1 + x)^k$, where k is a real number and $|x| < 1$.

Notice that after the special case $(x + y)^0 = 1$, each expansion begins with x raised to the same power as the binomial itself. That is, the expansion of $(x + y)^1$ has a first term of x^1, $(x + y)^2$ has a first term of x^2, $(x + y)^3$ has a first term of x^3, and so on. Also, the last term in each expansion is y to the same power as the binomial. Thus, the expansion of $(x + y)^n$ should begin with the term x^n and end with the term y^n.

Notice that the exponent on x decreases by 1 in each term after the first, while the exponent on y, beginning with y in the second term, increases by 1 in each succeeding term. That is, the *variables* in the terms of the expansion of $(x + y)^n$ have the following pattern.

$$x^n, x^{n-1}y, x^{n-2}y^2, x^{n-3}y^3, \ldots, xy^{n-1}, y^n$$

This pattern suggests that the sum of the exponents on x and y in each term is n. For example, the third term in the list above is $x^{n-2}y^2$, and the sum of the exponents is $n - 2 + 2 = n$.

Pascal's Triangle   Now, examine the *coefficients* in the terms of the expansion of $(x + y)^n$. Writing the coefficients alone gives the following pattern.

PASCAL'S TRIANGLE

							Row
			1				0
		1		1			1
	1		2		1		2
1		3		3		1	3
1	4		6		4	1	4
1	5	10		10	5	1	5

Blaise Pascal (1623–1662)

With the coefficients arranged in this way, each number in the triangle is the sum of the two numbers directly above it (one to the right and one to the left). For example, in row four, 1 is the sum of 1 (the only number above it), 4 is the sum of 1 and 3, 6 is the sum of 3 and 3, and so on. This triangular array of numbers is called **Pascal's triangle,** in honor of the seventeenth-century mathematician Blaise Pascal. It was, however, known long before his time.

To find the coefficients for $(x + y)^6$, we need to include row six in Pascal's triangle. Adding adjacent numbers, we find that row six is

$$1 \quad 6 \quad 15 \quad 20 \quad 15 \quad 6 \quad 1.$$

Using these coefficients, we obtain the expansion of $(x + y)^6$:

$$(x + y)^6 = x^6 + 6x^5y + 15x^4y^2 + 20x^3y^3 + 15x^2y^4 + 6xy^5 + y^6.$$

n-Factorial   Although it is possible to use Pascal's triangle to find the coefficients of $(x + y)^n$ for any positive integer n, this calculation becomes impractical for large values of n because of the need to write all the preceding rows. A more efficient way of finding these coefficients uses **factorial notation.** The number **$n!$** (read "*n-factorial*") is defined as follows.

---

n-FACTORIAL

For any positive integer n,

$$n! = n(n - 1)(n - 2) \cdots (3)(2)(1) \qquad \text{and} \qquad 0! = 1.$$

For example,

$$5! = 5 \cdot 4 \cdot 3 \cdot 2 \cdot 1 = 120,$$

$$7! = 7 \cdot 6 \cdot 5 \cdot 4 \cdot 3 \cdot 2 \cdot 1 = 5040,$$

and $$2! = 2 \cdot 1 = 2.$$

Binomial Coefficients

Now look at the coefficients of the expansion

$$(x + y)^5 = x^5 + 5x^4y + 10x^3y^2 + 10x^2y^3 + 5xy^4 + y^5.$$

The coefficient of the second term, $5x^4y$, is 5, and the exponents on the variables are 4 and 1. Note that

$$5 = \frac{5!}{4!\,1!}.$$

The coefficient of the third term is 10, with exponents of 3 and 2, and

$$10 = \frac{5!}{3!\,2!}.$$

The last term (the sixth term) can be written as $y^5 = 1x^0y^5$, with coefficient 1, and exponents of 0 and 5. Since $0! = 1$,

$$1 = \frac{5!}{0!\,5!}.$$

Generalizing from these examples, the coefficient for the term of the expansion of $(x + y)^n$ in which the variable part is x^ry^{n-r} (where $r \leq n$) is

$$\frac{n!}{r!(n-r)!}.$$

This number, called a **binomial coefficient,** is often symbolized $\binom{n}{r}$ or $_nC_r$ (read **"n choose r"**).

BINOMIAL COEFFICIENT

For nonnegative integers n and r, with $r \leq n$,

$$_nC_r = \binom{n}{r} = \frac{n!}{r!(n-r)!}.$$

The binomial coefficients are numbers from Pascal's triangle. For example, $\binom{3}{0}$ is the first number in row three, and $\binom{7}{4}$ is the fifth number in row seven.

Graphing calculators are capable of finding binomial coefficients. A calculator with a 10-digit display will give exact values for $n!$ for $n \leq 13$ and approximate values of $n!$ for $14 \leq n \leq 69$. ■

▶ EXAMPLE 1   **EVALUATING BINOMIAL COEFFICIENTS**

Evaluate each binomial coefficient.

(a) $\dbinom{6}{2}$ (b) $\dbinom{8}{0}$ (c) $\dbinom{10}{10}$ (d) $_{12}C_{10}$

Algebraic Solution

(a) $\dbinom{6}{2} = \dfrac{6!}{2!(6-2)!} = \dfrac{6!}{2!4!}$

$= \dfrac{6 \cdot 5 \cdot 4 \cdot 3 \cdot 2 \cdot 1}{2 \cdot 1 \cdot 4 \cdot 3 \cdot 2 \cdot 1} = 15$

(b) $\dbinom{8}{0} = \dfrac{8!}{0!(8-0)!} = \dfrac{8!}{0!8!} = \dfrac{8!}{1 \cdot 8!} = 1$ $0! = 1$

(c) $\dbinom{10}{10} = \dfrac{10!}{10!(10-10)!} = \dfrac{10!}{10!0!} = 1$ $0! = 1$

(d) $_{12}C_{10} = \dfrac{12!}{10!(12-10)!} = \dfrac{12!}{10!\,2!} = 66$

Graphing Calculator Solution

Graphing calculators calculate binomial coefficients using the notation $_nC_r$. For the TI-83/84 Plus, this function is found in the MATH menu. Figure 11 shows the values of the binomial coefficients found in parts (a), (b), and (c).

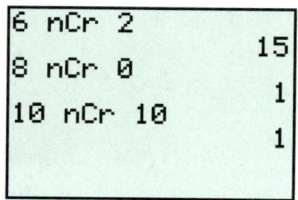

Figure 11

NOW TRY EXERCISES 5, 9, AND 17. ◀

Refer again to Pascal's triangle. Notice the symmetry in each row. This suggests that binomial coefficients should have the same property. That is,

$$\binom{n}{r} = \binom{n}{n-r}.$$

This is true, since

$$\binom{n}{r} = \frac{n!}{r!(n-r)!} \quad \text{and} \quad \binom{n}{n-r} = \frac{n!}{(n-r)!r!}.$$

The Binomial Theorem   Our observations about the expansion of $(x + y)^n$ are summarized as follows.

1. There are $n + 1$ terms in the expansion.

2. The first term is x^n, and the last term is y^n.

3. In each succeeding term, the exponent on x decreases by 1 and the exponent on y increases by 1.

4. The sum of the exponents on x and y in any term is n.

5. The coefficient of the term with $x^r y^{n-r}$ or $x^{n-r} y^r$ is $\binom{n}{r}$.

These observations about the expansion of $(x + y)^n$ for any positive integer value of n suggest the **binomial theorem.** A proof of the binomial theorem using mathematical induction is given in **Section 11.5.**

▼ **LOOKING AHEAD TO CALCULUS**
The binomial theorem is used to show that the derivative of $f(x) = x^n$ is given by the function $f'(x) = nx^{n-1}$. This fact is used extensively in calculus.

BINOMIAL THEOREM

For any positive integer n and any complex numbers x and y,

$$(x + y)^n = x^n + \binom{n}{1}x^{n-1}y + \binom{n}{2}x^{n-2}y^2 + \binom{n}{3}x^{n-3}y^3 + \cdots$$

$$+ \binom{n}{r}x^{n-r}y^r + \cdots + \binom{n}{n-1}xy^{n-1} + y^n.$$

▶ **Note** The binomial theorem looks much more manageable written as a series. The theorem can be summarized as

$$(x + y)^n = \sum_{r=0}^{n} \binom{n}{r}x^{n-r}y^r.$$

▶ **EXAMPLE 2**   **APPLYING THE BINOMIAL THEOREM**

Write the binomial expansion of $(x + y)^9$.

Solution   Using the binomial theorem,

$$(x + y)^9 = x^9 + \binom{9}{1}x^8y + \binom{9}{2}x^7y^2 + \binom{9}{3}x^6y^3 + \binom{9}{4}x^5y^4 + \binom{9}{5}x^4y^5$$

$$+ \binom{9}{6}x^3y^6 + \binom{9}{7}x^2y^7 + \binom{9}{8}xy^8 + y^9.$$

Now evaluate each of the binomial coefficients.

$$(x + y)^9 = x^9 + \frac{9!}{1!8!}x^8y + \frac{9!}{2!7!}x^7y^2 + \frac{9!}{3!6!}x^6y^3 + \frac{9!}{4!5!}x^5y^4 + \frac{9!}{5!4!}x^4y^5$$

$$+ \frac{9!}{6!3!}x^3y^6 + \frac{9!}{7!2!}x^2y^7 + \frac{9!}{8!1!}xy^8 + y^9$$

$$= x^9 + 9x^8y + 36x^7y^2 + 84x^6y^3 + 126x^5y^4 + 126x^4y^5$$

$$+ 84x^3y^6 + 36x^2y^7 + 9xy^8 + y^9$$

NOW TRY EXERCISE 23. ◀

▶ **EXAMPLE 3**   **APPLYING THE BINOMIAL THEOREM**

Expand $\left(a - \dfrac{b}{2} \right)^5$.

Solution   Write the binomial as follows.

$$\left(a - \frac{b}{2} \right)^5 = \left(a + \left(-\frac{b}{2} \right) \right)^5$$

Now use the binomial theorem with $x = a$, $y = -\frac{b}{2}$, and $n = 5$ to obtain

$$\left(a - \frac{b}{2}\right)^5 = a^5 + \binom{5}{1}a^4\left(-\frac{b}{2}\right) + \binom{5}{2}a^3\left(-\frac{b}{2}\right)^2 + \binom{5}{3}a^2\left(-\frac{b}{2}\right)^3 + \binom{5}{4}a\left(-\frac{b}{2}\right)^4 + \left(-\frac{b}{2}\right)^5$$

$$= a^5 + 5a^4\left(-\frac{b}{2}\right) + 10a^3\left(-\frac{b}{2}\right)^2 + 10a^2\left(-\frac{b}{2}\right)^3 + 5a\left(-\frac{b}{2}\right)^4 + \left(-\frac{b}{2}\right)^5$$

$$= a^5 - \frac{5}{2}a^4b + \frac{5}{2}a^3b^2 - \frac{5}{4}a^2b^3 + \frac{5}{16}ab^4 - \frac{1}{32}b^5.$$

NOW TRY EXERCISE 35. ◀

▶ **Note** As Example 3 illustrates, *an expansion of the difference of two terms (that is, x − y) has alternating signs.*

▶ **EXAMPLE 4**   **APPLYING THE BINOMIAL THEOREM**

Expand $\left(\dfrac{3}{m^2} - 2\sqrt{m}\right)^4$. (Assume $m > 0$.)

Solution   By the binomial theorem,

$$\left(\frac{3}{m^2} - 2\sqrt{m}\right)^4 = \left(\frac{3}{m^2}\right)^4 + \binom{4}{1}\left(\frac{3}{m^2}\right)^3(-2\sqrt{m})^1 + \binom{4}{2}\left(\frac{3}{m^2}\right)^2(-2\sqrt{m})^2$$

$$+ \binom{4}{3}\left(\frac{3}{m^2}\right)^1(-2\sqrt{m})^3 + (-2\sqrt{m})^4$$

$$= \frac{81}{m^8} + 4\left(\frac{27}{m^6}\right)(-2m^{1/2}) + 6\left(\frac{9}{m^4}\right)(4m)$$

$$+ 4\left(\frac{3}{m^2}\right)(-8m^{3/2}) + 16m^2 \quad \color{teal}{\sqrt{m} = m^{1/2} \text{ (Section R.7)}}$$

$$= \frac{81}{m^8} - \frac{216}{m^{11/2}} + \frac{216}{m^3} - \frac{96}{m^{1/2}} + 16m^2.$$

NOW TRY EXERCISE 37. ◀

*k*th Term of a Binomial Expansion

Earlier in this section, we wrote the binomial theorem in summation notation as $\sum\limits_{r=0}^{n}\binom{n}{r}x^{n-r}y^r$, which gives the form of each term. We can use this form to write any particular term of a binomial expansion without writing out the entire expansion.

For example, to find the tenth term of $(x + y)^n$, where $n \geq 9$, first notice that in the tenth term y is raised to the ninth power (since y has the power 1 in the second term, the power 2 in the third term, and so on). Because the exponents on x and y in any term must have a sum of n, the exponent on x in the tenth term is $n - 9$. Thus, the tenth term of the expansion is

$$\binom{n}{9}x^{n-9}y^9 = \frac{n!}{9!(n-9)!}x^{n-9}y^9.$$

We give this result in the following theorem.

> ### *k*th TERM OF THE BINOMIAL EXPANSION
>
> The *k*th term of the binomial expansion of $(x + y)^n$, where $n \geq k - 1$, is
>
> $$\binom{n}{k-1} x^{n-(k-1)} y^{k-1}.$$

To find the *k*th term of the binomial expansion, use the following steps.

Step 1 Find $k - 1$. This is the exponent on the second part of the binomial.

Step 2 Subtract the exponent found in Step 1 from n to get the exponent on the first part of the binomial.

Step 3 Determine the coefficient by using the exponents found in the first two steps and n.

▶ **EXAMPLE 5** **FINDING A PARTICULAR TERM OF A BINOMIAL EXPANSION**

Find the seventh term of $(a + 2b)^{10}$.

Solution In the seventh term, $2b$ has an exponent of 6 while a has an exponent of $10 - 6$, or 4. The seventh term is

$$\binom{10}{6} a^4 (2b)^6 = 210 a^4 (64b^6) = 13{,}440 a^4 b^6.$$

NOW TRY EXERCISE 41. ◀

11.4 Exercises

Evaluate each expression, if possible. See Example 1.

1. $\dfrac{6!}{3!\,3!}$ **2.** $\dfrac{5!}{2!\,3!}$ **3.** $\dfrac{7!}{3!\,4!}$ **4.** $\dfrac{8!}{5!\,3!}$

5. $\dbinom{8}{5}$ **6.** $\dbinom{7}{3}$ **7.** $\dbinom{10}{2}$ **8.** $\dbinom{9}{3}$

9. $\dbinom{14}{14}$ **10.** $\dbinom{15}{15}$ **11.** $\dbinom{n}{n-1}$ **12.** $\dbinom{n}{n-2}$

13. $_8C_3$ **14.** $_9C_7$ **15.** $_{100}C_{98}$ **16.** $_{20}C_5$

17. $_9C_0$ **18.** $_5C_1$ **19.** $_{12}C_1$ **20.** $_4C_0$

21. *Concept Check* What are the first and last terms in the expansion of $(2x + 3y)^4$?

22. *Concept Check* Determine the binomial coefficient for the fifth term in the expansion of $(x + y)^9$.

Write the binomial expansion for each expression. See Examples 2–4.

23. $(x + y)^6$ **24.** $(m + n)^4$ **25.** $(p - q)^5$

26. $(a - b)^7$ **27.** $(r^2 + s)^5$ **28.** $(m + n^2)^4$

29. $(p + 2q)^4$ **30.** $(3r - s)^6$ **31.** $(7p + 2q)^4$

32. $(4a - 5b)^5$ **33.** $(3x - 2y)^6$ **34.** $(7k - 9j)^4$

35. $\left(\dfrac{m}{2} - 1\right)^6$ **36.** $\left(3 + \dfrac{y}{3}\right)^5$ **37.** $\left(\sqrt{2}r + \dfrac{1}{m}\right)^4$

38. $\left(\dfrac{1}{k} - \sqrt{3}p\right)^3$ **39.** $\left(\dfrac{1}{x^4} + x^4\right)^4$ **40.** $\left(\dfrac{1}{y^5} - y^5\right)^5$

Write the indicated term of each binomial expansion. See Example 5.

41. Sixth term of $(4h - j)^8$ **42.** Eighth term of $(2c - 3d)^{14}$

43. Fifteenth term of $(a^2 + b)^{22}$ **44.** Twelfth term of $(2x + y^2)^{16}$

45. Fifteenth term of $(x - y^3)^{20}$ **46.** Tenth term of $(a^3 + 3b)^{11}$

Concept Check   *Work Exercises 47–50.*

47. Find the middle term of $(3x^7 + 2y^3)^8$.

48. Find the two middle terms of $(-2m^{-1} + 3n^{-2})^{11}$.

49. Find the value of n for which the coefficients of the fifth and eighth terms in the expansion of $(x + y)^n$ are the same.

50. Find the term(s) in the expansion of $\left(3 + \sqrt{x}\right)^{11}$ that contains x^4.

RELATING CONCEPTS

For individual or collaborative investigation
(Exercises 51–54)

In this section, we saw how the factorial of a positive integer n can be computed as a product: $n! = 1 \cdot 2 \cdot 3 \cdot \cdots \cdot n$. Calculators and computers can evaluate factorials very quickly. Before the days of modern technology, mathematicians developed a formula, called **Stirling's formula,** *for approximating large factorials. Interestingly enough, the formula involves the irrational numbers π and e.*

$$n! \approx \sqrt{2\pi n} \cdot n^n \cdot e^{-n}$$

As an example, the exact value of 5! is 120, and Stirling's formula gives the approximation as 118.019168 with a graphing calculator. This is "off" by less than 2, an error of only 1.65%. **Work Exercises 51–54 in order.**

51. Use a calculator to find the exact value of 10! and its approximation, using Stirling's formula.

52. Subtract the smaller value from the larger value in Exercise 51. Divide it by 10! and convert to a percent. What is the percent error?

53. Repeat Exercises 51 and 52 for $n = 12$.

54. Repeat Exercises 51 and 52 for $n = 13$. What seems to happen as n gets larger?

It can be shown that

$$(1 + x)^n = 1 + nx + \frac{n(n - 1)}{2!}x^2 + \frac{n(n - 1)(n - 2)}{3!}x^3 + \cdots$$

is true for any real number n (not just positive integer values) and any real number x, where $|x| < 1$. Use this series to approximate the given number to the nearest thousandth.

55. $(1.02)^{-3}$ **56.** $(1.04)^{-5}$ **57.** $(1.01)^{1.5}$ **58.** $(1.03)^2$

11.5 Mathematical Induction

Proof by Mathematical Induction ▪ **Proving Statements** ▪ **Generalized Principle of Mathematical Induction** ▪ **Proof of the Binomial Theorem**

Proof by Mathematical Induction Many results in mathematics are claimed true for every positive integer. Any of these results could be checked for $n = 1$, $n = 2$, $n = 3$, and so on, but since the set of positive integers is infinite it would be impossible to check every possible case. For example, let S_n represent the statement that the sum of the first n positive integers is $\frac{n(n+1)}{2}$.

$$S_n: \quad 1 + 2 + 3 + \cdots + n = \frac{n(n+1)}{2}$$

The truth of this statement is easily verified for the first few values of n:

If $n = 1$, then S_1 is $\qquad 1 = \dfrac{1(1+1)}{2},$ $\qquad$ which is true.

If $n = 2$, then S_2 is $\qquad 1 + 2 = \dfrac{2(2+1)}{2},$ $\qquad$ which is true.

If $n = 3$, then S_3 is $\qquad 1 + 2 + 3 = \dfrac{3(3+1)}{2},$ $\qquad$ which is true.

If $n = 4$, then S_4 is $\qquad 1 + 2 + 3 + 4 = \dfrac{4(4+1)}{2},$ $\qquad$ which is true.

Continuing in this way for any amount of time would still not prove that S_n is true for *every* positive integer value of n. To prove that such statements are true for every positive integer value of n, the following principle is often used.

PRINCIPLE OF MATHEMATICAL INDUCTION

Let S_n be a statement concerning the positive integer n. Suppose that

1. S_1 is true;

2. for any positive integer k, $k \leq n$, if S_k is true, then S_{k+1} is also true.

Then S_n is true for every positive integer value of n.

A proof by mathematical induction can be explained as follows. By assumption (1) above, the statement is true when $n = 1$. By assumption (2) above, the fact that the statement is true for $n = 1$ implies that it is true for $n = 1 + 1 = 2$. Using (2) again, the statement is thus true for $2 + 1 = 3$, for $3 + 1 = 4$, for $4 + 1 = 5$, and so on. Continuing in this way shows that the statement must be true for *every* positive integer.

The situation is similar to that of a number of dominoes lined up as shown in Figure 12. If the first domino is pushed over, it pushes the next, which pushes the next, and so on until all are down.

Another example of the principle of mathematical induction might be an infinite ladder. Suppose the rungs are spaced so that whenever you are on a rung, you know you can move to the next rung. Then *if* you can get to the first rung, you can go as high up the ladder as you wish.

Figure 12

Two separate steps are required for a proof by mathematical induction.

> ## PROOF BY MATHEMATICAL INDUCTION
>
> **Step 1** Prove that the statement is true for $n = 1$.
>
> **Step 2** Show that, for any positive integer k, $k \leq n$, if S_k is true, then S_{k+1} is also true.

Proving Statements Mathematical induction is used in the next example to prove the statement S_n mentioned at the beginning of this section.

▶ **EXAMPLE 1** **PROVING AN EQUALITY STATEMENT**

Let S_n represent the statement

$$1 + 2 + 3 + \cdots + n = \frac{n(n + 1)}{2}.$$

Prove that S_n is true for every positive integer n.

Solution The proof by mathematical induction is as follows.

Step 1 Show that the statement is true when $n = 1$. If $n = 1$, S_1 becomes

$$1 = \frac{1(1 + 1)}{2},$$

which is true.

Step 2 Show that S_k implies S_{k+1}, where S_k is the statement

$$1 + 2 + 3 + \cdots + k = \frac{k(k + 1)}{2},$$

and S_{k+1} is the statement

$$1 + 2 + 3 + \cdots + k + (k + 1) = \frac{(k + 1)[(k + 1) + 1]}{2}.$$

Start with S_k and assume it is a true statement.

$$1 + 2 + 3 + \cdots + k = \frac{k(k + 1)}{2}$$

Add $k + 1$ to both sides of this equation to obtain S_{k+1}.

$$1 + 2 + 3 + \cdots + k + (k + 1) = \frac{k(k + 1)}{2} + (k + 1)$$

$$= (k + 1)\left(\frac{k}{2} + 1\right) \qquad \text{Factor out } k + 1. \text{ (Section R.4)}$$

$$= (k + 1)\left(\frac{k + 2}{2}\right) \qquad \text{Add inside the parentheses. (Section R.5)}$$

$$1 + 2 + 3 + \cdots + k + (k + 1) = \frac{(k + 1)[(k + 1) + 1]}{2} \qquad \text{Multiply; } k + 2 = (k + 1) + 1.$$

This final result is the statement for $n = k + 1$; it has been shown that if S_k is true, then S_{k+1} is also true.

The two steps required for a proof by mathematical induction have been completed, so the statement S_n is true for every positive integer value of n.

NOW TRY EXERCISE 1. ◄

> ▶ **Caution** Notice that the left side of the statement S_n always includes *all* the terms up to the *n*th term, as well as the *n*th term.

▶ **EXAMPLE 2** **PROVING AN INEQUALITY STATEMENT**

Prove: If x is a real number between 0 and 1, then for every positive integer n,

$$0 < x^n < 1.$$

Solution

Step 1 Here S_1 is the statement

$$\text{if } 0 < x < 1, \text{ then } 0 < x^1 < 1,$$

which is true.

Step 2 S_k is the statement

$$\text{if } 0 < x < 1, \text{ then } 0 < x^k < 1.$$

To show that this implies that S_{k+1} is true, multiply all three parts of $0 < x^k < 1$ by x to get

$$x \cdot 0 < x \cdot x^k < x \cdot 1. \quad \text{(Section 1.7)}$$

(Here the fact that $0 < x$ is used.) Simplify to obtain

$$0 < x^{k+1} < x.$$

Since $x < 1$,

$$0 < x^{k+1} < x < 1$$

and thus

$$0 < x^{k+1} < 1.$$

This work shows that if S_k is true, then S_{k+1} is true.

Since both steps for a proof by mathematical induction have been completed, the given statement is true for every positive integer n.

NOW TRY EXERCISE 23. ◄

Generalized Principle of Mathematical Induction

Some statements S_n are not true for the first few values of n, but are true for all values of n that are greater than or equal to some fixed integer j. The following slightly generalized form of the principle of mathematical induction takes care of these cases.

GENERALIZED PRINCIPLE OF MATHEMATICAL INDUCTION

Let S_n be a statement concerning the positive integer n. Let j be a fixed positive integer. Suppose that

Step 1 S_j is true;

Step 2 for any positive integer k, $k \geq j$, S_k implies S_{k+1}.

Then S_n is true for all positive integers n, where $n \geq j$.

▶ **EXAMPLE 3** **USING THE GENERALIZED PRINCIPLE**

Let S_n represent the statement $2^n > 2n + 1$. Show that S_n is true for all values of n such that $n \geq 3$.

Solution (Check that S_n is false for $n = 1$ and $n = 2$.)

Step 1 Show that S_n is true for $n = 3$. If $n = 3$, then S_n is

$$2^3 > 2 \cdot 3 + 1,$$

or $$8 > 7.$$

Thus, S_3 is true.

Step 2 Now show that S_k implies S_{k+1}, where $k \geq 3$, and where

$$S_k \quad \text{is} \quad 2^k > 2k + 1,$$

and $$S_{k+1} \text{ is } 2^{k+1} > 2(k + 1) + 1.$$

Multiply both sides of $2^k > 2k + 1$ by 2, obtaining

$$2 \cdot 2^k > 2(2k + 1)$$

$$2^{k+1} > 4k + 2.$$

Rewrite $4k + 2$ as $2k + 2 + 2k = 2(k + 1) + 2k$.

$$2^{k+1} > 2(k + 1) + 2k \quad (1)$$

Since k is a positive integer greater than 3,

$$2k > 1. \quad (2)$$

Adding $2(k + 1)$ to both sides of inequality (2) gives

$$2(k + 1) + 2k > 2(k + 1) + 1. \quad (3)$$

From inequalities (1) and (3),

$$2^{k+1} > 2(k + 1) + 2k > 2(k + 1) + 1,$$

or $$2^{k+1} > 2(k + 1) + 1, \quad \text{as required.}$$

Thus, S_k implies S_{k+1}, and this, together with the fact that S_3 is true, shows that S_n is true for every positive integer value of n greater than or equal to 3.

NOW TRY EXERCISE 21. ◀

Proof of the Binomial Theorem The binomial theorem can be proved by mathematical induction. That is, for any positive integer n and any complex numbers x and y,

$$(x + y)^n = x^n + \binom{n}{1}x^{n-1}y + \binom{n}{2}x^{n-2}y^2 + \binom{n}{3}x^{n-3}y^3$$

$$+ \cdots + \binom{n}{r}x^{n-r}y^r + \cdots + \binom{n}{n-1}xy^{n-1} + y^n. \quad \text{(Section 11.4)} \ (1)$$

Proof Let S_n be statement (1). Begin by verifying S_n for $n = 1$,

$$S_1: \quad (x + y)^1 = x^1 + y^1,$$

which is true.

Now assume that S_n is true for the positive integer k. Statement S_k becomes

$$S_k: \quad (x + y)^k = x^k + \frac{k!}{1!(k-1)!}x^{k-1}y + \frac{k!}{2!(k-2)!}x^{k-2}y^2 \quad \begin{array}{l}\text{Definition of the bi-}\\\text{nomial coefficient}\\\text{(Section 11.4)}\end{array}$$

$$+ \cdots + \frac{k!}{(k-1)!1!}xy^{k-1} + y^k. \qquad (2)$$

Multiply both sides of equation (2) by $x + y$.

$$(x + y)^k \cdot (x + y)$$

$$= x(x + y)^k + y(x + y)^k \quad \text{Distributive property (Section R.2)}$$

$$= \left[x \cdot x^k + \frac{k!}{1!(k-1)!}x^k y + \frac{k!}{2!(k-2)!}x^{k-1}y^2 + \cdots + \frac{k!}{(k-1)!1!}x^2 y^{k-1} + xy^k \right]$$

$$+ \left[x^k \cdot y + \frac{k!}{1!(k-1)!}x^{k-1}y^2 + \cdots + \frac{k!}{(k-1)!1!}xy^k + y \cdot y^k \right]$$

Rearrange terms to get

$$(x + y)^{k+1} = x^{k+1} + \left[\frac{k!}{1!(k-1)!} + 1 \right]x^k y + \left[\frac{k!}{2!(k-2)!} + \frac{k!}{1!(k-1)!} \right]x^{k-1}y^2$$

$$+ \cdots + \left[1 + \frac{k!}{(k-1)!1!} \right]xy^k + y^{k+1}. \quad (3)$$

The first expression in brackets in equation (3) simplifies to $\binom{k+1}{1}$. To see this, note that

$$\binom{k+1}{1} = \frac{(k+1)(k)(k-1)(k-2)\cdots 1}{1 \cdot (k)(k-1)(k-2)\cdots 1} = k + 1.$$

Also, $\quad \dfrac{k!}{1!(k-1)!} + 1 = \dfrac{k(k-1)!}{1(k-1)!} + 1 = k + 1.$

The second expression becomes $\binom{k+1}{2}$, the last $\binom{k+1}{k}$, and so on. The result of equation (3) is just equation (2) with every k replaced by $k + 1$. Thus, the truth of S_n when $n = k$ implies the truth of S_n for $n = k + 1$, which completes the proof of the theorem by mathematical induction.

11.5   Exercises

Write out in full and verify the statements S_1, S_2, S_3, S_4, and S_5 for the following. Then use mathematical induction to prove that each statement is true for every positive integer n. See Example 1.

1. $1 + 3 + 5 + \cdots + (2n - 1) = n^2$ **2.** $2 + 4 + 6 + \cdots + 2n = n(n + 1)$

Assume that n is a positive integer. Use mathematical induction to prove each statement S by following these steps. See Example 1.

(a) *Verify the statement for $n = 1$.*

(b) *Write the statement for $n = k$.*

(c) *Write the statement for $n = k + 1$.*

(d) *Assume the statement is true for $n = k$. Use algebra to change the statement in part (b) to the statement in part (c).*

(e) *Write a conclusion based on Steps (a)–(d).*

3. $3 + 6 + 9 + \cdots + 3n = \dfrac{3n(n + 1)}{2}$

4. $5 + 10 + 15 + \cdots + 5n = \dfrac{5n(n + 1)}{2}$

5. $2 + 4 + 8 + \cdots + 2^n = 2^{n+1} - 2$

6. $3 + 3^2 + 3^3 + \cdots + 3^n = \dfrac{3(3^n - 1)}{2}$

7. $1^2 + 2^2 + 3^2 + \cdots + n^2 = \dfrac{n(n + 1)(2n + 1)}{6}$

8. $1^3 + 2^3 + 3^3 + \cdots + n^3 = \dfrac{n^2(n + 1)^2}{4}$

9. $5 \cdot 6 + 5 \cdot 6^2 + 5 \cdot 6^3 + \cdots + 5 \cdot 6^n = 6(6^n - 1)$

10. $7 \cdot 8 + 7 \cdot 8^2 + 7 \cdot 8^3 + \cdots + 7 \cdot 8^n = 8(8^n - 1)$

11. $\dfrac{1}{1 \cdot 2} + \dfrac{1}{2 \cdot 3} + \dfrac{1}{3 \cdot 4} + \cdots + \dfrac{1}{n(n + 1)} = \dfrac{n}{n + 1}$

12. $\dfrac{1}{1 \cdot 4} + \dfrac{1}{4 \cdot 7} + \dfrac{1}{7 \cdot 10} + \cdots + \dfrac{1}{(3n - 2)(3n + 1)} = \dfrac{n}{3n + 1}$

13. $\dfrac{1}{2} + \dfrac{1}{2^2} + \dfrac{1}{2^3} + \cdots + \dfrac{1}{2^n} = 1 - \dfrac{1}{2^n}$

14. $\dfrac{4}{5} + \dfrac{4}{5^2} + \dfrac{4}{5^3} + \cdots + \dfrac{4}{5^n} = 1 - \dfrac{1}{5^n}$

Find all natural number values for n for which the given statement is false.

15. $2^n > 2n$ **16.** $3^n > 2n + 1$ **17.** $2^n > n^2$ **18.** $n! > 2n$

Prove each statement by mathematical induction. See Examples 2 and 3.

19. $(a^m)^n = a^{mn}$ (Assume a and m are constant.)

20. $(ab)^n = a^n b^n$ (Assume a and b are constant.)

21. $2^n > 2n$, if $n \geq 3$ **22.** $3^n > 2n + 1$, if $n \geq 2$

23. If $a > 1$, then $a^n > 1$.   **24.** If $a > 1$, then $a^n > a^{n-1}$.

25. If $0 < a < 1$, then $a^n < a^{n-1}$.   **26.** $2^n > n^2$, for $n \geq 5$

27. If $n \geq 4$, then $n! > 2^n$, where $n! = n(n - 1)(n - 2) \cdots (3)(2)(1)$.

28. $4^n > n^4$, for $n \geq 5$

Solve each problem.

29. *Number of Handshakes*   Suppose that each of the n ($n \geq 2$) people in a room shakes hands with everyone else, but not with himself. Show that the number of handshakes is $\frac{n^2 - n}{2}$.

30. *Sides of a Polygon*   The series of sketches below starts with an equilateral triangle having sides of length 1. In the following steps, equilateral triangles are constructed on each side of the preceding figure. The length of the sides of each new triangle is $\frac{1}{3}$ the length of the sides of the preceding triangles. Develop a formula for the number of sides of the nth figure. Use mathematical induction to prove your answer.

31. *Perimeter*   Find the perimeter of the nth figure in Exercise 30.

32. *Area*   Show that the area of the nth figure in Exercise 30 is

$$\sqrt{3}\left[\frac{2}{5} - \frac{3}{20}\left(\frac{4}{9}\right)^{n-1}\right].$$

33. *Tower of Hanoi*   A pile of n rings, each ring smaller than the one below it, is on a peg. Two other pegs are attached to a board with this peg. In the game called the *Tower of Hanoi* puzzle, all the rings must be moved to a different peg, with only one ring moved at a time, and with no ring ever placed on top of a smaller ring. Find the least number of moves that would be required. Prove your result with mathematical induction.

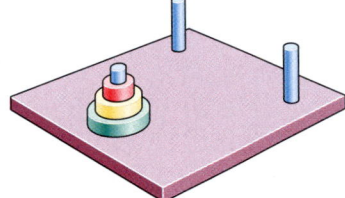

CHAPTER 11 ▶   Quiz (Sections 11.1–11.5)

Write the first five terms of each sequence. State whether the sequence is arithmetic, geometric, *or* neither.

1. $a_n = -4n + 2$

2. $a_n = -2\left(-\frac{1}{2}\right)^n$

3. $a_1 = 5, a_2 = 3, a_n = a_{n-1} + 3a_{n-2},$ for $n \geq 3$

4. A certain arithmetic sequence has $a_1 = -6$ and $a_9 = 18$. Find a_7.

5. Find the sum of the first ten terms of each series.

  (a) arithmetic, $a_1 = -20, d = 14$ **(b)** geometric, $a_1 = -20, r = -\frac{1}{2}$

6. Evaluate each sum that exists.

  (a) $\sum_{i=1}^{30}(-3i + 6)$ **(b)** $\sum_{i=1}^{\infty} 2^i$ **(c)** $\sum_{i=1}^{\infty}\left(\frac{3}{4}\right)^i$

7. Use the binomial theorem to expand $(x - 3y)^5$.

8. Find the fifth term of the expansion of $\left(4x - \frac{1}{2}y\right)^5$.

9. Evaluate each expression.

(a) $9!$ (b) $C(10, 4)$

10. Use mathematical induction to prove that for all positive integers n,

$$6 + 12 + 18 + \cdots + 6n = 3n(n + 1).$$

11.6 Counting Theory

Fundamental Principle of Counting ▪ **Permutations** ▪ **Combinations** ▪ **Distinguishing Between Permutations and Combinations**

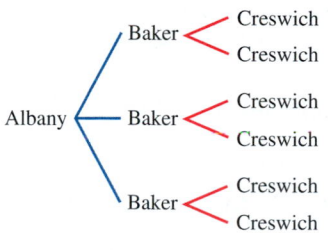

Figure 13

Fundamental Principle of Counting   If there are 3 roads from Albany to Baker and 2 roads from Baker to Creswich, in how many ways can one travel from Albany to Creswich by way of Baker? For each of the 3 roads from Albany to Baker, there are 2 different roads from Baker to Creswich. Hence, there are $3 \cdot 2 = 6$ different ways to make the trip, as shown in the **tree diagram** in Figure 13.

In this situation, each choice of road is an example of an *event.* Two events are **independent events** if neither influences the outcome of the other. The opening example illustrates the fundamental principle of counting with independent events.

FUNDAMENTAL PRINCIPLE OF COUNTING

If n independent events occur, with

m_1 ways for event 1 to occur,

m_2 ways for event 2 to occur,

.

.

.

and m_n ways for event n to occur,

then there are

$$m_1 \cdot m_2 \cdot \cdots \cdot m_n$$

different ways for all n events to occur.

▶ **EXAMPLE 1**   USING THE FUNDAMENTAL PRINCIPLE OF COUNTING

A restaurant offers a choice of 3 salads, 5 main dishes, and 2 desserts. Use the fundamental principle of counting to find the number of different 3-course meals that can be selected.

Solution   Three events are involved: selecting a salad, selecting a main dish, and selecting a dessert. The first event can occur in 3 ways, the second event can occur in 5 ways, and the third event can occur in 2 ways. Thus, there are

$$3 \cdot 5 \cdot 2 = 30 \text{ possible meals.}$$

NOW TRY EXERCISE 23. ◀

> ▶ **EXAMPLE 2**   **USING THE FUNDAMENTAL PRINCIPLE OF COUNTING**

A teacher has 5 different books that he wishes to arrange in a row. How many different arrangements are possible?

Solution   Five events are involved: selecting a book for the first spot, selecting a book for the second spot, and so on. For the first spot the teacher has 5 choices. After a choice has been made, the teacher has 4 choices for the second spot. Continuing in this manner, there are 3 choices for the third spot, 2 for the fourth spot, and 1 for the fifth spot. By the fundamental principle of counting, there are

$$5 \cdot 4 \cdot 3 \cdot 2 \cdot 1 = 120 \text{ different arrangements.}$$

NOW TRY EXERCISE 27. ◀

In using the fundamental principle of counting, products such as

$$5 \cdot 4 \cdot 3 \cdot 2 \cdot 1$$

occur often. We use the symbol **$n!$** (read **"n-factorial"**), for any counting number n, as follows.

$$n! = n(n-1)(n-2) \cdots (3)(2)(1) \quad \text{(Section 11.4)}$$

Thus, $5 \cdot 4 \cdot 3 \cdot 2 \cdot 1$ is written 5! and $3 \cdot 2 \cdot 1$ is written 3!. By the definition of $n!$, $n[(n-1)!] = n!$ for all natural numbers $n \geq 2$. It is convenient to have this relation hold also for $n = 1$, so, by definition,

$$0! = 1. \quad \text{(Section 11.4)}$$

> ▶ **EXAMPLE 3**   **ARRANGING r OF n ITEMS ($r < n$)**

Suppose the teacher in Example 2 wishes to place only 3 of the 5 books in a row. How many arrangements of 3 books are possible?

Solution   The teacher still has 5 ways to fill the first spot, 4 ways to fill the second spot, and 3 ways to fill the third. Since only 3 books will be used, there are only 3 spots to be filled (3 events) instead of 5, with

$$5 \cdot 4 \cdot 3 = 60 \text{ arrangements.}$$

NOW TRY EXERCISE 33. ◀

Permutations   Since each ordering of three books is considered a different *arrangement*, the number 60 in the preceding example is called the number of *permutations* of 5 things taken 3 at a time, written $P(5,3) = 60$. The number of ways of arranging 5 elements from a set of 5 elements, written $P(5,5) = 120$, was found in Example 2.

A **permutation** of n elements taken r at a time is one of the *arrangements* of r elements from a set of n elements. Generalizing from the examples above, the number of permutations of n elements taken r at a time, denoted by $P(n,r)$, is

$$
\begin{aligned}
P(n,r) &= n(n-1)(n-2) \cdots (n-r+1) \\
&= \frac{n(n-1)(n-2) \cdots (n-r+1)(n-r)(n-r-1) \cdots (2)(1)}{(n-r)(n-r-1) \cdots (2)(1)} \\
&= \frac{n!}{(n-r)!}.
\end{aligned}
$$

PERMUTATIONS OF n ELEMENTS TAKEN r AT A TIME

If $P(n, r)$ denotes the number of permutations of n elements taken r at a time, with $r \leq n$, then

$$P(n, r) = \frac{n!}{(n - r)!}.$$

Alternative notations for $P(n, r)$ are P_r^n and $_nP_r$.

▶ **EXAMPLE 4**   USING THE PERMUTATIONS FORMULA

Find each value.

(a) The number of permutations of the letters L, M, and N

(b) The number of permutations of 2 of the letters L, M, and N

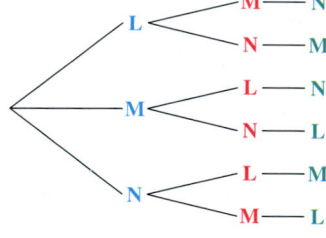

Figure 14

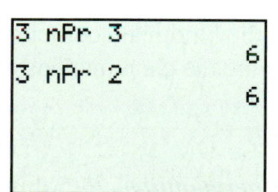

This screen shows how the TI-83/84 Plus calculates $P(3, 3)$ and $P(3, 2)$. See Example 4.

Solution

(a) By the formula for $P(n, r)$, with $n = 3$ and $r = 3$,

$$P(3, 3) = \frac{3!}{(3 - 3)!} = \frac{3!}{0!} = \frac{3!}{1} = 3 \cdot 2 \cdot 1 = 6.$$

As shown in the tree diagram in Figure 14, the 6 permutations are

$$LMN, \ LNM, \ MLN, \ MNL, \ NLM, \ NML.$$

(b) Find $P(3, 2)$.

$$P(3, 2) = \frac{3!}{(3 - 2)!} = \frac{3!}{1!} = \frac{3!}{1} = 6$$

This result is the same as the answer in part (a). After the first two choices are made, the third is already determined since only one letter is left.

NOW TRY EXERCISE 37. ◀

▶ **EXAMPLE 5**   USING THE PERMUTATIONS FORMULA

Suppose 8 people enter an event in a swim meet. In how many ways could the gold, silver, and bronze medals be awarded?

Solution   Using the fundamental principle of counting, there are 3 events, giving $8 \cdot 7 \cdot 6 = 336$ choices. We can also use the formula for $P(n, r)$ to get the same result. There are

$$P(8, 3) = \frac{8!}{5!} = \frac{8 \cdot 7 \cdot 6 \cdot 5 \cdot 4 \cdot 3 \cdot 2 \cdot 1}{5 \cdot 4 \cdot 3 \cdot 2 \cdot 1}$$

$$= 8 \cdot 7 \cdot 6$$

$$= 336 \text{ ways.}$$

NOW TRY EXERCISE 35. ◀

▶ **EXAMPLE 6**   **USING THE PERMUTATIONS FORMULA**

In how many ways can 6 students be seated in a row of 6 desks?

Solution   Use $P(n, r)$ with $n = 6$ and $r = 6$ to get

$$P(6, 6) = 6! = 6 \cdot 5 \cdot 4 \cdot 3 \cdot 2 \cdot 1 = 720 \text{ ways.}$$

NOW TRY EXERCISE 31. ◀

Combinations   In Example 3 we saw that there are 60 ways that a teacher can arrange 3 of 5 different books in a row. That is, there are 60 permutations of 5 things taken 3 at a time. Suppose now that the teacher does not wish to arrange the books in a row but rather wishes to choose, without regard to order, any 3 of the 5 books to donate to a book sale. In how many ways can the teacher do this?

The number 60 counts all possible *arrangements* of 3 books chosen from 5. The following 6 arrangements, however, would all lead to the same set of 3 books being given to the book sale.

mystery-biography-textbook	biography-textbook-mystery
mystery-textbook-biography	textbook-biography-mystery
biography-mystery-textbook	textbook-mystery-biography

The list shows 6 different *arrangements* of 3 books but only one *set* of 3 books. A subset of items selected *without regard to order* is called a **combination.** The number of combinations of 5 things taken 3 at a time is written $\binom{5}{3}$, $_5C_3$, or $C(5, 3)$.

> ▶ **Note**   This combinations notation also represents the binomial coefficient defined in **Section 11.4.** That is, binomial coefficients are the combinations of n elements chosen r at a time.

To evaluate $\binom{5}{3}$ or $C(5, 3)$, start with the $5 \cdot 4 \cdot 3$ *permutations* of 5 things taken 3 at a time. Since order does not matter, and each subset of 3 items from the set of 5 items can have its elements rearranged in $3 \cdot 2 \cdot 1 = 3!$ ways, we find $\binom{5}{3}$ by dividing the number of permutations by $3!$, or

$$\binom{5}{3} = \frac{5 \cdot 4 \cdot 3}{3!} = \frac{5 \cdot 4 \cdot 3}{3 \cdot 2 \cdot 1} = 10.$$

The teacher can choose 3 books for the book sale in 10 ways.

Generalizing this discussion gives the following formula for the number of combinations of n elements taken r at a time:

$$C(n, r) = \binom{n}{r} = \frac{P(n, r)}{r!}.$$

An alternative version of this formula is found as follows.

$$C(n, r) = \binom{n}{r} = \frac{P(n, r)}{r!} = \frac{n!}{(n - r)!} \cdot \frac{1}{r!} = \frac{n!}{(n - r)! \, r!}$$

This version is most useful for calculation and is the one we used earlier to calculate binomial coefficients.

COMBINATIONS OF *n* ELEMENTS TAKEN *r* AT A TIME

If $C(n, r)$ or $\binom{n}{r}$ represents the number of combinations of n elements taken r at a time, with $r \leq n$, then

$$C(n,r) = \binom{n}{r} = \frac{n!}{(n-r)!\,r!}.$$

▶ **Note** The formula for $C(n, r)$ given above is equivalent to the binomial co-efficient formula given in **Section 11.4,** with denominator factors rearranged.

▶ **EXAMPLE 7** USING THE COMBINATIONS FORMULA

How many different committees of 3 people can be chosen from a group of 8 people?

Solution Since a committee is an unordered set, use combinations. There are

$$C(8, 3) = \binom{8}{3} = \frac{8!}{5!\,3!} = \frac{8 \cdot 7 \cdot 6 \cdot 5 \cdot 4 \cdot 3 \cdot 2 \cdot 1}{5 \cdot 4 \cdot 3 \cdot 2 \cdot 1 \cdot 3 \cdot 2 \cdot 1} = 56 \text{ committees.}$$

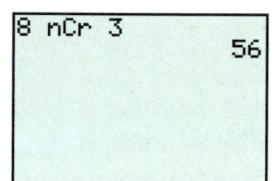

This screen shows how the TI-83/84 Plus calculates $C(8, 3)$. See Example 7.

NOW TRY EXERCISE 39. ◀

▶ **EXAMPLE 8** USING THE COMBINATIONS FORMULA

Three stockbrokers are to be selected from a group of 30 to work on a special project.

(a) In how many different ways can the stockbrokers be selected?

(b) In how many ways can the group of 3 be selected if a particular stockbroker must work on the project?

Solution

(a) Here we wish to know the number of 3-element combinations that can be formed from a set of 30 elements. (We want combinations, not permutations, since order within the group does not matter.)

$$C(30, 3) = \binom{30}{3} = \frac{30!}{27!\,3!} = 4060$$

There are 4060 ways to select the project group.

(b) Since 1 broker has already been selected for the project, the problem is reduced to selecting 2 more from the remaining 29 brokers.

$$C(29, 2) = \binom{29}{2} = \frac{29!}{27!\,2!} = 406$$

In this case, the project group can be selected in 406 ways.

NOW TRY EXERCISE 41. ◀

Distinguishing Between Permutations and Combinations

Students often have difficulty determining whether to use permutations or combinations in solving problems. The following chart lists some of the similarities and differences between these two concepts.

These are combinations. The order of the cards in the hands is *not* important.

Permutations	Combinations
Number of ways of selecting *r* items out of *n* items	
Repetitions are not allowed.	
Order is important.	Order is not important.
Arrangements of *r* items from a set of *n* items	Subsets of *r* items from a set of *n* items
$P(n, r) = \dfrac{n!}{(n - r)!}$	$C(n, r) = \dbinom{n}{r} = \dfrac{n!}{(n - r)!\, r!}$
Clue words: arrangement, schedule, order	Clue words: group, committee, sample, selection

▶ EXAMPLE 9   **DISTINGUISHING BETWEEN PERMUTATIONS AND COMBINATIONS**

Should permutations or combinations be used to solve each problem?

(a) How many 4-digit codes are possible if no digits are repeated?

(b) A sample of 3 light bulbs is randomly selected from a batch of 15 bulbs. How many different samples are possible?

(c) In a basketball tournament with 8 teams, how many games must be played so that each team plays every other team exactly once?

(d) In how many ways can 4 stockbrokers be assigned to 6 offices so that each broker has a private office?

Solution

(a) Since changing the order of the 4 digits results in a different code, permutations should be used.

(b) The order in which the 3 light bulbs are selected is not important. The sample is unchanged if the items are rearranged, so combinations should be used.

(c) Selection of 2 teams for a game is an *unordered* subset of 2 from the set of 8 teams. Use combinations.

(d) The office assignments are an *ordered* selection of 4 offices from the 6 offices. Exchanging the offices of any 2 brokers within a selection of 4 offices gives a different assignment, so permutations should be used.

NOW TRY EXERCISE 21. ◀

To illustrate the distinctions between permutations and combinations in another way, suppose we want to select 2 cans of soup from 4 cans on a shelf: noodle (N), bean (B), mushroom (M), and tomato (T). As shown in Figure 15(a), there are 12 ways to select 2 cans from the 4 cans if order matters (if noodle first and bean second is considered different from bean, then noodle, for example). On the other hand, if order is unimportant, then there are 6 ways to choose 2 cans of soup from the 4, as illustrated in Figure 15(b).

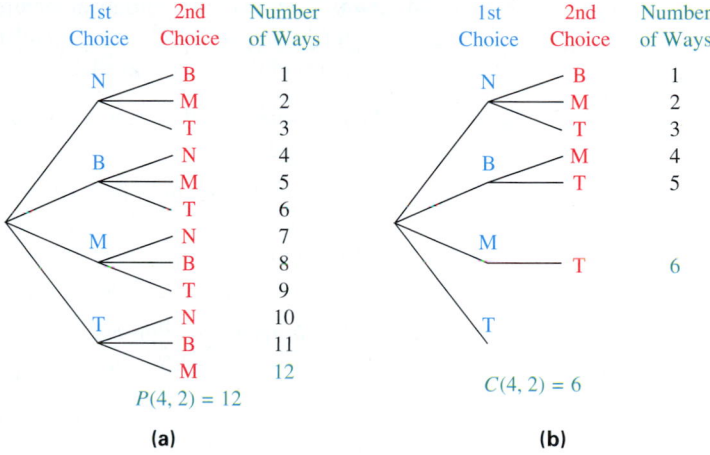

Figure 15

▶ **Caution** Not all counting problems lend themselves to either permutations or combinations. Whenever a tree diagram or the fundamental principle of counting can be used directly, as in the soup example, use it.

11.6 Exercises

Evaluate each expression. See Examples 4–9.

1. $P(12, 8)$ **2.** $P(5, 5)$ **3.** $P(9, 2)$ **4.** $P(10, 9)$

5. $P(5, 1)$ **6.** $P(6, 0)$ **7.** $C(4, 2)$ **8.** $C(9, 3)$

9. $C(6, 0)$ **10.** $C(8, 1)$ **11.** $\binom{12}{4}$ **12.** $\binom{16}{3}$

Use a calculator to evaluate each expression. See Examples 4 and 7.

13. $_{20}P_5$ **14.** $_{100}P_5$ **15.** $_{15}P_8$ **16.** $_{32}P_4$

17. $_{20}C_5$ **18.** $_{100}C_5$ **19.** $\binom{15}{8}$ **20.** $\binom{32}{4}$

21. Decide whether the situation described involves a permutation or a combination of objects. See Example 9.

    (a) a telephone number   **(b)** a Social Security number
    (c) a hand of cards in poker   **(d)** a committee of politicians
    (e) the "combination" on a combination lock
    (f) a lottery choice of six numbers where order does not matter
    (g) an automobile license plate

22. Explain the difference between a permutation and a combination. What should you look for in a problem to decide which is an appropriate method of solution?

Use the fundamental principle of counting or permutations to solve each problem. See Examples 1–6.

23. *Home Plan Choices* How many different types of homes are available if a builder offers a choice of 5 basic plans, 4 roof styles, and 2 exterior finishes?

24. *Auto Varieties* An auto manufacturer produces 7 models, each available in 6 different colors, with 4 different upholstery fabrics, and 5 interior colors. How many varieties of the auto are available?

25. *Radio-Station Call Letters* How many different 4-letter radio-station call letters can be made

(a) if the first letter must be K or W and no letter may be repeated?

(b) if repeats are allowed (but the first letter is K or W)?

(c) How many of the 4-letter call letters (starting with K or W) with no repeats end in R?

26. *Meal Choices* A menu offers a choice of 3 salads, 8 main dishes, and 5 desserts. How many different 3-course meals (salad, main dish, dessert) are possible?

27. *Arranging Blocks* Baby Finley wants to arrange 7 blocks in a row. How many different arrangements can he make?

28. *Names for a Baby* A couple has narrowed down the choice of a name for their new baby to 5 first names and 3 middle names. How many different first- and middle-name combinations are possible?

29. *License Plates* For many years, the state of California used 3 letters followed by 3 digits on its automobile license plates.

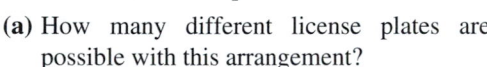

(a) How many different license plates are possible with this arrangement?

(b) When the state ran out of new plates, the order was reversed to 3 digits followed by 3 letters. How many additional plates were then possible?

(c) When the plates described in part (b) were also used up, the state then issued plates with 1 letter followed by 3 digits and then 3 letters. How many plates does this scheme provide?

30. *Telephone Numbers* How many 7-digit telephone numbers are possible if the first digit cannot be 0 and

(a) only odd digits may be used?

(b) the telephone number must be a multiple of 10 (that is, it must end in 0)?

(c) the telephone number must be a multiple of 100?

(d) the first 3 digits are 481?

(e) no repetitions are allowed?

31. *Seating People in a Row*   In an experiment on social interaction, 9 people will sit in 9 seats in a row. In how many ways can this be done?

32. *Genetics Experiment*   In how many ways can 7 of 10 monkeys be arranged in a row for a genetics experiment?

33. *Course Schedule Arrangement*   A business school offers courses in keyboarding, spreadsheets, transcription, business English, technical writing, and accounting. In how many ways can a student arrange a schedule if 3 courses are taken?

34. *Course Schedule Arrangement*   If your college offers 400 courses, 20 of which are in mathematics, and your counselor arranges your schedule of 4 courses by random selection, how many schedules are possible that do not include a math course?

35. *Club Officer Choices*   In a club with 15 members, how many ways can a slate of 3 officers consisting of president, vice-president, and secretary/treasurer be chosen?

36. *Batting Orders*   A baseball team has 20 players. How many 9-player batting orders are possible?

37. *Letter Arrangement*   Consider the word BRUCE.

    (a) In how many ways can all the letters of the word BRUCE be arranged?
    (b) In how many ways can the first 3 letters of the word BRUCE be arranged?

38. *Basketball Positions*   In how many ways can 5 players be assigned to the 5 positions on a basketball team, assuming that any player can play any position? In how many ways can 10 players be assigned to the 5 positions?

Solve each problem involving combinations. See Examples 7 and 8.

39. *Seminar Presenters*   A banker's association has 40 members. If 6 members are selected at random to present a seminar, how many different groups of 6 are possible?

40. *Apple Samples*   How many different samples of 4 apples can be drawn from a crate of 25 apples?

41. *Hamburger Choices*   Howard's Hamburger Heaven sells hamburgers with cheese, relish, lettuce, tomato, mustard, or ketchup.

    (a) How many different hamburgers can be made that use any 4 of the extras?
    (b) How many different hamburgers can be made if one of the 4 extras must be cheese?

42. *Financial Planners*   Four financial planners are to be selected from a group of 12 to participate in a special program. In how many ways can this be done? In how many ways can the group that will not participate be selected?

43. *Card Combinations*   Five cards are marked with the numbers 1, 2, 3, 4, or 5, shuffled, and 2 cards are then drawn. How many different 2-card hands are possible?

44. *Marble Samples*   If a bag contains 15 marbles, how many samples of 2 marbles can be drawn from it? How many samples of 4 marbles can be drawn?

45. *Marble Samples*   In Exercise 44, if the bag contains 3 yellow, 4 white, and 8 blue marbles, how many samples of 2 can be drawn in which both marbles are blue?

46. *Apple Samples*   In Exercise 40, if it is known that there are 5 rotten apples in the crate,

    (a) how many samples of 3 could be drawn in which all 3 are rotten?
    (b) how many samples of 3 could be drawn in which there are 1 rotten apple and 2 good apples?

47. *Convention Delegation Choices*   A city council is composed of 5 liberals and 4 conservatives. Three members are to be selected randomly as delegates to a convention.

   (a) How many delegations are possible?
   (b) How many delegations could have all liberals?
   (c) How many delegations could have 2 liberals and 1 conservative?
   (d) If 1 member of the council serves as mayor, how many delegations are possible that include the mayor?

48. *Delegation Choices*   Seven workers decide to send a delegation of 2 to their supervisor to discuss their grievances.

   (a) How many different delegations are possible?
   (b) If it is decided that a certain employee must be in the delegation, how many different delegations are possible?
   (c) If there are 2 women and 5 men in the group, how many delegations would include at least 1 woman?

Use any or all of the methods described in this section to solve each problem. See Examples 1–9.

49. *Course Schedule*   If Dwight Johnston has 8 courses to choose from, how many ways can he arrange his schedule if he must pick 4 of them?

50. *Pineapple Samples*   How many samples of 9 pineapples can be drawn from a crate of 12?

51. *Soup Ingredients*   Velma specializes in making different vegetable soups with carrots, celery, beans, peas, mushrooms, and potatoes. How many different soups can she make with any 4 ingredients?

52. *Secretary/Manager Assignments*   From a pool of 7 secretaries, 3 are selected to be assigned to 3 managers, 1 secretary to each manager. In how many ways can this be done?

53. *Musical Chairs Seatings*   In a game of musical chairs, 13 children will sit in 12 chairs. (1 will be left out.) How many seatings are possible?

54. *Plant Samples*   In an experiment on plant hardiness, a researcher gathers 6 wheat plants, 3 barley plants, and 2 rye plants. She wishes to select 4 plants at random.

   (a) In how many ways can this be done?
   (b) In how many ways can this be done if exactly 2 wheat plants must be included?

55. *Committee Choices*   In a club with 8 women and 11 men members, how many 5-member committees can be chosen that have the following?

   (a) all women   **(b)** all men
   (c) 3 women and 2 men   **(d)** no more than 3 men

56. *Committee Choices*   From 10 names on a ballot, 4 will be elected to a political party committee. In how many ways can the committee of 4 be formed if each person will have a different responsibility?

57. *Combination Lock*   A briefcase has 2 locks. The combination to each lock consists of a 3-digit number, where digits may be repeated. How many combinations are possible? (*Hint:* The word *combination* is a misnomer. Lock combinations are permutations where the arrangement of the numbers is important.)

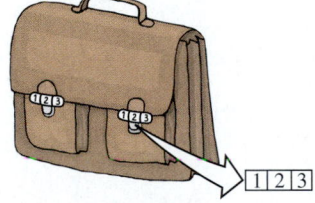

58. *Combination Lock*   A typical "combination" for a padlock consists of 3 numbers from 0 to 39. Find the number of "combinations" that are possible with this type of lock, if a number may be repeated.

59. *Garage Door Openers*   The code for some garage door openers consists of 12 electrical switches that can be set to either 0 or 1 by the owner. With this type of opener, how many codes are possible? (*Source:* Promax.)

60. *Lottery*   To win the jackpot in a lottery game, a person must pick 4 numbers from 0 to 9 in the correct order. If a number can be repeated, how many ways are there to play the game?

61. *Keys*   How many distinguishable ways can 4 keys be put on a circular key ring?

62. *Sitting at a Round Table*   How many ways can 8 people sit at a round table? Assume that a different way means that at least 1 person is sitting next to someone different.

Prove each statement for positive integers n and r, with r ≤ n. (Hint: Use the definitions of permutations and combinations.)

63. $P(n, n - 1) = P(n, n)$ **64.** $P(n, 1) = n$ **65.** $P(n, 0) = 1$

66. $\dbinom{n}{n} = 1$ **67.** $\dbinom{n}{0} = 1$ **68.** $\dbinom{n}{n - 1} = n$

69. $\dbinom{n}{n - r} = \dbinom{n}{r}$

70. Explain why the restriction $r \leq n$ is needed in the formula for $P(n, r)$.

RELATING CONCEPTS

For individual or collaborative investigation
(Exercises 71 and 72)

The value of n! can quickly become too large for most calculators to evaluate. To estimate n! for large values of n, we can use the property of logarithms that

$$\log(n!) = \log(1 \times 2 \times 3 \times \cdots \times n)$$
$$= \log 1 + \log 2 + \log 3 + \cdots + \log n.$$

Using a sum and sequence utility on a calculator, we can then determine a value of r such that $n! \approx 10^r$ since $r = \log n!$. For example, the screen illustrates that a calculator gives the same approximation of 30! using the factorial function and the formula just discussed. Use this technique to approximate each quantity in Exercises 71 and 72. Then, try to compute each value directly on your calculator.

```
30!
       2.652528598E32
10^(sum(seq(log(N
),N,1,30,1)))
       2.652528598E32
```

71. **(a)** 50!   **(b)** 60!   **(c)** 65!

72. **(a)** $P(47, 13)$ **(b)** $P(50, 4)$ **(c)** $P(29, 21)$

11.7 Basics of Probability

Basic Concepts ▪ Complements and Venn Diagrams ▪ Odds ▪ Union of Two Events ▪ Binomial Probability

Basic Concepts Consider an experiment that has one or more possible **outcomes,** each of which is equally likely to occur. For example, the experiment of tossing a fair coin has 2 equally likely possible outcomes: landing heads up (H) or landing tails up (T). Also, the experiment of rolling a fair die has 6 equally likely outcomes: landing so the face that is up shows 1, 2, 3, 4, 5, or 6 dots.

The set S of all possible outcomes of a given experiment is called the **sample space** for the experiment. (In this text, all sample spaces are finite.) One sample space for the experiment of tossing a coin could consist of the outcomes H and T. This sample space can be written as

$$S = \{H, T\}. \quad \text{Use set notation. (Section R.1)}$$

Similarly, a sample space for the experiment of rolling a single die once is

$$S = \{1, 2, 3, 4, 5, 6\}.$$

> **NOW TRY EXERCISES 1 AND 5.** ◀

Any subset of the sample space is called an **event.** In the experiment with the die, for example, "the number showing is a 3" is an event, say E_1, such that $E_1 = \{3\}$. "The number showing is greater than 3" is also an event, say E_2, such that $E_2 = \{4, 5, 6\}$. To represent the number of outcomes that belong to event E, the notation $n(E)$ is used. Then $n(E_1) = 1$ and $n(E_2) = 3$.

The notation $P(E)$ is used for the *probability* of an event E. If the outcomes in the sample space for an experiment are equally likely, then the probability of event E occurring is found as follows.

PROBABILITY OF EVENT E

In a sample space with equally likely outcomes, the **probability** of an event E, written $P(E)$, is the ratio of the number of outcomes in sample space S that belong to event E, $n(E)$, to the total number of outcomes in sample space S, $n(S)$. That is,

$$P(E) = \frac{n(E)}{n(S)}.$$

To use this definition to find the probability of the event E_1 in the die experiment, start with the sample space, $S = \{1, 2, 3, 4, 5, 6\}$, and the desired event, $E_1 = \{3\}$. Since $n(E_1) = 1$ and $n(S) = 6$,

$$P(E_1) = \frac{n(E_1)}{n(S)} = \frac{1}{6}.$$

▶ **EXAMPLE 1** FINDING PROBABILITIES OF EVENTS

A single die is rolled. Write each event in set notation and give the probability of the event.

(a) E_3: the number showing is even

(b) E_4: the number showing is greater than 4

(c) E_5: the number showing is less than 7

(d) E_6: the number showing is 7

Solution

(a) Since $E_3 = \{2, 4, 6\}$, $n(E_3) = 3$. As given earlier, $n(S) = 6$, so

$$P(E_3) = \frac{3}{6} = \frac{1}{2}.$$

(b) Again $n(S) = 6$. Event $E_4 = \{5, 6\}$, with $n(E_4) = 2$.

$$P(E_4) = \frac{2}{6} = \frac{1}{3}$$

(c) $E_5 = \{1, 2, 3, 4, 5, 6\}$ and $P(E_5) = \frac{6}{6} = 1$.

(d) $E_6 = \emptyset$ and $P(E_6) = \frac{0}{6} = 0$.

NOW TRY EXERCISES 7 AND 9. ◀

In Example 1(c), $E_5 = S$. Therefore, the event E_5 is certain to occur every time the experiment is performed. An event that is certain to occur always has probability 1. In Example 1(d), $E_6 = \emptyset$ and $P(E_6) = 0$. The probability of an impossible event, such as E_6, is always 0, since none of the outcomes in the sample space satisfy the event. *For any event E, P(E) is between **0** and **1** inclusive.*

Complements and Venn Diagrams
The set of all outcomes in the sample space that do *not* belong to event E is called the **complement** of E, written **E'**. For example, in the experiment of drawing a single card from a standard deck of 52 cards, let E be the event "the card is an ace." Then E' is the event "the card is not an ace." From the definition of E', for an event E,

$$E \cup E' = S \quad \text{and} \quad E \cap E' = \emptyset.*$$

▶ **Note** A standard deck of 52 cards has four suits: hearts ♥, diamonds ♦, spades ♠, and clubs ♣, with thirteen cards of each suit. Each suit has a jack, a queen, and a king (sometimes called the "face cards"), an ace, and cards numbered from 2 to 10. The hearts and diamonds are red and the spades and clubs are black. We will refer to this standard deck of cards in this section.

---

*The **union** of two sets A and B is the set $A \cup B$ of all elements from either A or B, or both. The **intersection** of sets A and B, written $A \cap B$, includes all elements that belong to both sets. (Section R.1)

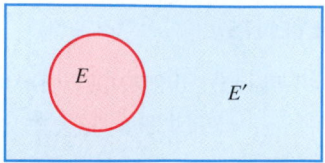

Figure 16

Probability concepts can be illustrated using **Venn diagrams,** as shown in Figure 16. The rectangle in Figure 16 represents the sample space in an experiment. The area inside the circle represents event E, while the area inside the rectangle, but outside the circle, represents event E'.

▶ **EXAMPLE 2**   **USING THE COMPLEMENT**

In the experiment of drawing a card from a well-shuffled deck, find the probabilities of event E, the card is an ace, and of event E'.

Solution   Since there are 4 aces in the deck of 52 cards, $n(E) = 4$ and $n(S) = 52$. Therefore,

$$P(E) = \frac{n(E)}{n(S)} = \frac{4}{52} = \frac{1}{13}.$$

Of the 52 cards, 48 are not aces, so

$$P(E') = \frac{n(E')}{n(S)} = \frac{48}{52} = \frac{12}{13}.$$

NOW TRY EXERCISES 15(a)–(c). ◀

In Example 2, $P(E) + P(E') = \frac{1}{13} + \frac{12}{13} = 1$. This is always true for any event E and its complement E'. That is,

$$P(E) + P(E') = 1.$$

This can be restated as

$$P(E) = 1 - P(E') \quad \text{or} \quad P(E') = 1 - P(E).$$

These two equations suggest an alternative way to compute the probability of an event. For example, if it is known that $P(E) = \frac{1}{10}$, then

$$P(E') = 1 - \frac{1}{10} = \frac{9}{10}.$$

Odds   Sometimes probability statements are expressed in terms of odds, a comparison of $P(E)$ with $P(E')$. The **odds** in favor of an event E are expressed as the ratio of $P(E)$ to $P(E')$ or as the quotient $\frac{P(E)}{P(E')}$. For example, if the probability of rain can be established as $\frac{1}{3}$, the odds that it will rain are

$$P(\text{rain}) \text{ to } P(\text{no rain}) = \frac{1}{3} \text{ to } \frac{2}{3} = \frac{\frac{1}{3}}{\frac{2}{3}} = \frac{1}{2}, \quad \text{or} \quad 1 \text{ to } 2.$$

On the other hand, the odds that it will not rain are 2 to 1 $\left(\text{or } \frac{2}{3} \text{ to } \frac{1}{3}\right)$. If the odds in favor of an event are, say, 3 to 5, then the probability of the event is $\frac{3}{8}$, and the probability of the complement of the event is $\frac{5}{8}$. If the odds favoring event E are m to n, then

$$P(E) = \frac{m}{m+n} \quad \text{and} \quad P(E') = \frac{n}{m+n}.$$

▶ **EXAMPLE 3**   **FINDING ODDS IN FAVOR OF AN EVENT**

A shirt is selected at random from a dark closet containing 6 blue shirts and 4 shirts that are not blue. Find the odds in favor of a blue shirt being selected.

Solution   Let E represent "a blue shirt is selected." Then,

$$P(E) = \frac{6}{10} \quad \text{or} \quad \frac{3}{5} \quad \text{and} \quad P(E') = 1 - \frac{3}{5} = \frac{2}{5}.$$

Therefore, the odds in favor of a blue shirt being selected are

$$P(E) \text{ to } P(E') = \frac{3}{5} \text{ to } \frac{2}{5} = \frac{\frac{3}{5}}{\frac{2}{5}} = \frac{3}{5} \div \frac{2}{5} = \frac{3}{5} \cdot \frac{5}{2} = \frac{3}{2}, \quad \text{or} \quad 3 \text{ to } 2.$$

(Section R.5)

NOW TRY EXERCISES 15(d) AND (e). ◀

Union of Two Events

Since events are sets, we can use set operations to find the union of two events. Suppose a fair die is rolled. Let H be the event "the result is a 3," and K the event "the result is an even number." From the results earlier in this section,

$$H = \{3\} \qquad K = \{2, 4, 6\} \qquad H \cup K = \{2, 3, 4, 6\}$$

$$P(H) = \frac{1}{6} \qquad P(K) = \frac{3}{6} = \frac{1}{2} \qquad P(H \cup K) = \frac{4}{6} = \frac{2}{3}.$$

Notice that $P(H) + P(K) = P(H \cup K)$.

Before assuming that this relationship is true in general, consider another event G for this experiment, "the result is a 2."

$$G = \{2\} \qquad K = \{2, 4, 6\} \qquad G \cup K = \{2, 4, 6\}$$

$$P(G) = \frac{1}{6} \qquad P(K) = \frac{3}{6} = \frac{1}{2} \qquad P(G \cup K) = \frac{3}{6} = \frac{1}{2}$$

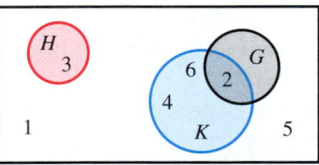

Figure 17

In this case, $P(G) + P(K) \neq P(G \cup K)$. As Figure 17 suggests, the difference in the two examples above comes from the fact that events H and K cannot occur simultaneously. Such events are called **mutually exclusive events.** In fact, $H \cap K = \emptyset$, which is always true for mutually exclusive events. Events G and K, however, can occur simultaneously. Both are satisfied if the result of the roll is a 2, the element in their intersection ($G \cap K = \{2\}$). This example suggests the following property.

PROBABILITY OF THE UNION OF TWO EVENTS

For any events E and F,

$$P(E \text{ or } F) = P(E \cup F) = P(E) + P(F) - P(E \cap F).$$

▶ **EXAMPLE 4**   **FINDING PROBABILITIES OF UNIONS**

One card is drawn from a well-shuffled deck of 52 cards. What is the probability of the following outcomes?

(a) The card is an ace or a spade.   **(b)** The card is a 3 or a king.

Solution

(a) The events "drawing an ace" and "drawing a spade" are not mutually exclusive since it is possible to draw the ace of spades, an outcome satisfying both events. The probability is

$$P(\text{ace or spade}) = P(\text{ace}) + P(\text{spade}) - P(\text{ace and spade})$$
$$= \frac{4}{52} + \frac{13}{52} - \frac{1}{52} = \frac{16}{52} = \frac{4}{13}.$$

(b) "Drawing a 3" and "drawing a king" are mutually exclusive events because it is impossible to draw one card that is both a 3 and a king.

$$P(3 \text{ or } K) = P(3) + P(K) - P(3 \text{ and } K)$$
$$= \frac{4}{52} + \frac{4}{52} - 0 = \frac{8}{52} = \frac{2}{13}$$

NOW TRY EXERCISES 17(a) AND (b). ◀

▶ **EXAMPLE 5**   **FINDING PROBABILITIES OF UNIONS**

Suppose two fair dice are rolled. Find each probability.

(a) The first die shows a 2, or the sum of the two dice is 6 or 7.

(b) The sum of the dots showing is at most 4.

Solution

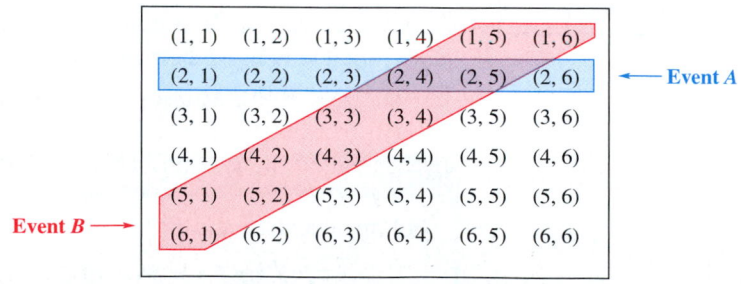

(a) Think of the two dice as being distinguishable, one red and one green for example. (Actually, the sample space is the same even if they are not apparently distinguishable.) A sample space with equally likely outcomes is shown in Figure 18, where $(1, 1)$ represents the event "the first die (red) shows a 1 and the second die (green) shows a 1," $(1, 2)$ represents "the first die shows a 1 and the second die shows a 2," and so on.

(1, 1)	(1, 2)	(1, 3)	(1, 4)	(1, 5)	(1, 6)
(2, 1)	(2, 2)	(2, 3)	(2, 4)	(2, 5)	(2, 6)	⟵ Event *A*
(3, 1)	(3, 2)	(3, 3)	(3, 4)	(3, 5)	(3, 6)
(4, 1)	(4, 2)	(4, 3)	(4, 4)	(4, 5)	(4, 6)
(5, 1)	(5, 2)	(5, 3)	(5, 4)	(5, 5)	(5, 6)
(6, 1)	(6, 2)	(6, 3)	(6, 4)	(6, 5)	(6, 6)

Event *B* ⟶

Figure 18

Let A represent the event "the first die shows a 2," and B represent the event "the sum of the two dice is 6 or 7." See Figure 18. Event A has 6 elements, event B has 11 elements, and the sample space has 36 elements. Thus,

$$P(A) = \frac{6}{36}, \qquad P(B) = \frac{11}{36}, \qquad \text{and} \qquad P(A \cap B) = \frac{2}{36},$$

so

$$P(A \cup B) = P(A) + P(B) - P(A \cap B)$$

$$= \frac{6}{36} + \frac{11}{36} - \frac{2}{36} = \frac{15}{36} = \frac{5}{12}.$$

(b) "At most 4" can be written as "2 or 3 or 4." (A sum of 1 is meaningless here.) Since the events represented by "2," "3," or "4" are mutually exclusive,

$$P(\text{at most 4}) = P(2 \text{ or } 3 \text{ or } 4) = P(2) + P(3) + P(4). \quad \text{(*)}$$

The sample space for this experiment includes the 36 possible pairs of numbers shown in Figure 18. The pair $(1, 1)$ is the only one with a sum of 2, so $P(2) = \frac{1}{36}$. Also $P(3) = \frac{2}{36}$ since both $(1, 2)$ and $(2, 1)$ give a sum of 3. The pairs $(1, 3)$, $(2, 2)$, and $(3, 1)$ have a sum of 4, so $P(4) = \frac{3}{36}$. Substituting into equation (*) above gives

$$P(\text{at most 4}) = \frac{1}{36} + \frac{2}{36} + \frac{3}{36} = \frac{6}{36} = \frac{1}{6}.$$

NOW TRY EXERCISE 17(c). ◀

The properties of probability are summarized as follows.

PROPERTIES OF PROBABILITY

For any events E and F:

1. $0 \leq P(E) \leq 1$ **2.** $P(\text{a certain event}) = 1$

3. $P(\text{an impossible event}) = 0$ **4.** $P(E') = 1 - P(E)$

5. $P(E \text{ or } F) = P(E \cup F) = P(E) + P(F) - P(E \cap F)$.

▶ **Caution** *When finding the probability of a union, remember to subtract the probability of the intersection from the sum of the probabilities of the individual events.*

Binomial Probability A **binomial experiment** is an experiment that consists of repeated independent trials with only two outcomes in each trial, success or failure. Let the probability of success in one trial be p. Then the probability of failure is $1 - p$, and the probability of exactly r successes in n trials is given by

$$\binom{n}{r} p^r (1 - p)^{n-r}.$$

This expression is equivalent to the general term of the binomial expansion given earlier. Thus the terms of the binomial expansion give the probabilities of exactly r successes in n trials, for $0 \le r \le n$, in a binomial experiment.

▶ **EXAMPLE 6** **FINDING PROBABILITIES IN A BINOMIAL EXPERIMENT**

An experiment consists of rolling a die 10 times. Find each probability.

(a) The probability that in exactly 4 of the rolls, the result is a 3

(b) The probability that in exactly 9 of the rolls, the result is not a 3

Algebraic Solution

(a) The probability p of a 3 on one roll is $\frac{1}{6}$. Here $n = 10$ and $r = 4$, so the required probability is

$$\binom{10}{4}\left(\frac{1}{6}\right)^4\left(1 - \frac{1}{6}\right)^{10-4} = 210\left(\frac{1}{6}\right)^4\left(\frac{5}{6}\right)^6$$

$$\approx .054.$$

(b) The probability is

$$\binom{10}{9}\left(\frac{5}{6}\right)^9\left(\frac{1}{6}\right)^1 \approx .323. \quad \text{Use } n = 10, r = 9, \text{ and } p = 1 - \frac{1}{6} = \frac{5}{6}.$$

Graphing Calculator Solution

Graphing calculators, such as the TI-83/84 Plus, that have statistical distribution functions give binomial probabilities. Figure 19 shows the results for parts (a) and (b). The numbers in parentheses separated by commas represent n, p, and r, respectively.

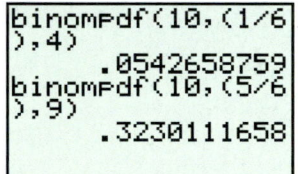

Figure 19

NOW TRY EXERCISE 35. ◀

Write a sample space with equally likely outcomes for each experiment.

1. A two-headed coin is tossed once.

2. Two ordinary coins are tossed.

3. Three ordinary coins are tossed.

4. Slips of paper marked with the numbers 1, 2, 3, 4, and 5 are placed in a box. After mixing well, two slips are drawn.

5. The spinner shown here is spun twice.

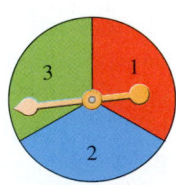

6. A die is rolled and then a coin is tossed.

Write each event in set notation and give the probability of the event. See Example 1.

7. In Exercise 1:

   (a) the result of the toss is heads;   **(b)** the result of the toss is tails.

8. In Exercise 2:

   (a) both coins show the same face;   **(b)** at least one coin turns up heads.

9. In Exercise 5:

   (a) the result is a repeated number;   **(b)** the second number is 1 or 3;

   (c) the first number is even and the second number is odd.

10. In Exercise 4:

   (a) both slips are marked with even numbers;

   (b) both slips are marked with odd numbers;

   (c) both slips are marked with the same number;

   (d) one slip is marked with an odd number, the other with an even number.

11. A student gives the probability of an event in a problem as $\frac{6}{5}$. Explain why this answer must be incorrect.

12. *Concept Check*   If the probability of an event is .857, what is the probability that the event will not occur?

13. *Concept Check*   Associate each probability in parts (a)–(g) with one of the statements in A–F. Choices may be used more than once.

   (a) $P(E) = -.1$ **(b)** $P(E) = .01$ **(c)** $P(E) = 1$ **(d)** $P(E) = 2$

   (e) $P(E) = .99$ **(f)** $P(E) = 0$ **(g)** $P(E) = .5$

   A. The event is certain to occur.   **B.** The event cannot occur.

   C. The event is very likely to occur.   **D.** The event is very unlikely to occur.

   E. The event is just as likely to occur as not occur.

   F. The probability value is impossible.

Work each problem. See Examples 1–6.

14. *Batting Average*   A baseball player with a batting average of .300 comes to bat. What are the odds in favor of the ball player getting a hit?

15. *Drawing a Marble*   A marble is drawn at random from a box containing 3 yellow, 4 white, and 8 blue marbles. Find the probabilities in parts (a)–(c).

   (a) A yellow marble is drawn.   **(b)** A black marble is drawn.

   (c) The marble is yellow or white.

   (d) What are the odds in favor of drawing a yellow marble?

   (e) What are the odds against drawing a blue marble?

16. *Small Business Loan*   The probability that a bank with assets greater than or equal to $30 billion will make a loan to a small business is .002. What are the odds against such a bank making a small business loan? (*Source: The Wall Street Journal* analysis of *CA1 Reports* filed with federal banking authorities.)

17. *Dice Rolls*   Two dice are rolled. Find the probability of each event.

   (a) The sum of the dots is at least 10.

   (b) The sum of the dots is either 7 or at least 10.

   (c) The sum of the dots is 2, or the dice both show the same number.

18. *Languages Spoken in Hispanic Households*   In a recent survey of Hispanic households, 20.4% of the respondents said that only Spanish was spoken at home, 11.9% said that only English was spoken, and the remainder said that both Spanish and English were spoken. What are the odds that English is spoken in a randomly selected Hispanic household? (*Source:* American Demographics.)

19. *U.S. Population Origins* Projected Hispanic and non-Hispanic U.S. populations (in thousands) for the year 2025 are given in the table. (Other populations are all non-Hispanic.) Assume these projections are accurate. Find the probability that a U.S. resident selected at random in 2025 is the following.

(a) of Hispanic origin
(b) not White
(c) Indian (Native American) or Black
(d) What are the odds that a randomly selected U.S. resident is Asian?

Type	Number
Hispanic origin	58,930
White	209,117
Black	43,511
Indian (Native American)	2,744
Asian	20,748

Source: U.S. Census Bureau.

20. *U.S. Population by Region* The U.S. resident population by region (in millions) for selected years is given in the table. Find the probability that a U.S. resident selected at random satisfies the following.

(a) lived in the West in 1997
(b) lived in the Midwest in 1995
(c) lived in the Northeast or Midwest in 1997
(d) lived in the South or West in 1997
(e) What are the odds that a randomly selected U.S. resident in 2000 was not from the South?

Region	1995	1997	2000
Northeast	51.4	51.6	53.6
Midwest	61.8	62.5	64.4
South	91.8	94.2	100.2
West	57.7	59.4	63.2

Source: U.S. Census Bureau.

21. *State Lottery* One game in a state lottery requires you to pick 1 heart, 1 club, 1 diamond, and 1 spade, in that order, from the 13 cards in each suit. What is the probability of getting all four picks correct and winning $5000?

22. *State Lottery* If three of the four selections in Exercise 21 are correct, the player wins $200. Find the probability of this outcome.

23. *Male Life Table* The table is an abbreviated version of a **life table** used by the Office of the Chief Actuary of the Social Security Administration. (The actual table includes every age, not just every tenth age.) Theoretically, this table follows a group of 100,000 males at birth and gives the number still alive at each age.

Exact Age	Number of Lives	Exact Age	Number of Lives
0	100,000	60	82,963
10	98,924	70	66,172
20	98,233	80	39,291
30	96,735	90	10,537
40	94,558	100	468
50	90,757	110	1

Source: Office of the Actuary, Social Security Administration.

(a) What is the probability that a 40-year-old man will live 30 more years?
(b) What is the probability that a 40-year-old man will not live 30 more years?
(c) Consider a group of five 40-year-old men. What is the probability that exactly three of them survive to age 70? (*Hint:* The longevities of the individual men can be considered as independent trials.)

(d) Consider two 40-year-old men. What is the probability that at least one of them survives to age 70? (*Hint:* The probability that both survive is the product of the probabilities that each survives.)

24. *Opinion Survey*   The management of a firm wishes to survey the opinions of its workers, classified as follows for the purpose of an interview:

    30% have worked for the company 5 or more years,
    28% are female,
    65% contribute to a voluntary retirement plan, and 50% of the female workers contribute to the retirement plan.

    Find each probability if a worker is selected at random.

    (a) A male worker is selected.
    (b) A worker is selected who has worked for the company less than 5 yr.
    (c) A worker is selected who contributes to the retirement plan or is female.

25. *Growth in Stock Value*   A financial analyst has determined the possibilities (and their probabilities) for the growth in value of a certain stock during the next year. (Assume these are the only possibilities.) See the table. For instance, the probability of a 5% growth is .15. If you invest $10,000 in the stock, what is the probability that the stock will be worth at least $11,400 by the end of the year?

Percent Growth	Probability
5	.15
8	.20
10	.35
14	.20
18	.10

26. *Growth in Stock Value*   Refer to Exercise 25. Suppose the percents and probabilities in the table are estimates of annual growth during the next 3 yr. What is the probability that an investment of $10,000 will grow in value to *at least* $15,000 during the next 3 yr? (*Hint:* Use the formula for (annual) compound interest discussed in **Section 4.2.**)

College Student Smokers   *The table gives the results of a survey of 14,000 college students who were cigarette smokers in a recent year.*

Number of Cigarettes Per Day	Less than 1	1 to 9	10 to 19	A pack of 20 or more
Percent (as a decimal)	.45	.24	.20	.11

Source: Harvard School of Public Health Study in the *Journal of AMA*.

Using the percents as probabilities, find the probability (to six decimal places) that, out of 10 of these student smokers selected at random, the following were true.

27. Four smoked less than 10 cigarettes per day.

28. Five smoked a pack or more per day.

29. Fewer than 2 smoked between 1 and 19 cigarettes per day.

30. No more than 3 smoked less than 1 cigarette per day.

College Applications   *The table gives the results of a survey of 282,549 freshmen from the class of 2006 at 437 of the nation's baccalaureate colleges and universities.*

Number of Colleges Applied to	1	2 or 3	4–6	7 or more
Percent (as a decimal)	.20	.29	.37	.14

Source: Higher Education Research Institute, UCLA, 2002.

Using the percents as probabilities, find the probability of each event for a randomly selected student.

31. The student applied to fewer than 4 colleges.

32. The student applied to at least 2 colleges.

33. The student applied to more than 3 colleges.

34. The student applied to no colleges.

35. *Color-Blind Males*   The probability that a male will be color-blind is .042. Find the probabilities (to six decimal places) that in a group of 53 men, the following are true.

 (a) Exactly 5 are color-blind.
 (b) No more than 5 are color-blind.
 (c) At least 1 is color-blind.

36. The screens illustrate how the TABLE feature of a graphing calculator can be used to find the probabilities of having 0, 1, 2, 3, or 4 girls in a family of 4 children. (Note that 0 appears for X = 5 and X = 6. Why is this so?)

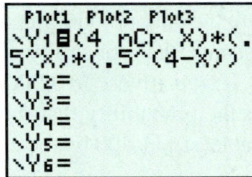

Use this approach to determine the following.

 (a) Find the probabilities of having 0, 1, 2, or 3 boys in a family of 3 children.
 (b) Find the probabilities of having 0, 1, 2, 3, 4, 5, or 6 girls in a family of 6 children.

37. *(Modeling) Spread of Disease*   What will happen when an infectious disease is introduced into a family? Suppose a family has I infected members and S members who are not infected but are susceptible to contracting the disease. The probability P of exactly k people not contracting the disease during a 1-week period can be calculated by the formula

$$P = \binom{S}{k} q^k (1 - q)^{S-k},$$

where $q = (1 - p)^I$, and p is the probability that a susceptible person contracts the disease from an infected person. For example, if $p = .5$, then there is a 50% chance that a susceptible person exposed to 1 infected person for 1 week will contract the disease. (*Source:* Hoppensteadt, F. and C. Peskin, *Mathematics in Medicine and the Life Sciences,* Springer-Verlag, 1992.)

 (a) Compute the probability P of 3 family members not becoming infected within 1 week if there are currently 2 infected and 4 susceptible members. Assume that $p = .1$. (*Hint:* To use the formula, first determine the values of k, I, S, and q.)
 (b) A highly infectious disease can have $p = .5$. Repeat part (a) with this value of p.
 (c) Determine the probability that everyone would become sick in a large family if initially, $I = 1$, $S = 9$, and $p = .5$. Discuss the results.

38. *(Modeling) Spread of Disease*   (Refer to Exercise 37.) Suppose that in a family $I = 2$ and $S = 4$. If the probability P is .25 of there being $k = 2$ uninfected members after 1 week, estimate graphically the possible values of p. (*Hint:* Write P as a function of p.)

Chapter 11 Summary

KEY TERMS

11.1 finite sequence
infinite sequence
terms of a sequence
general term (*n*th term)
convergent sequence
divergent sequence
recursive definition
Fibonacci sequence
series
summation notation
finite series

infinite series
index of summation
11.2 arithmetic sequence (arithmetic progression)
common difference
arithmetic series
11.3 geometric sequence (geometric progression)
common ratio
geometric series

annuity
future value of an annuity
11.4 Pascal's triangle
factorial notation
binomial coefficient
binomial theorem (general binomial expansion)
11.6 tree diagram
independent events
permutation

combination
11.7 outcome
sample space
event
probability
complement
Venn diagram
odds
mutually exclusive events
binomial experiment

NEW SYMBOLS

a_n	*n*th term of a sequence
$\displaystyle\sum_{i=1}^{n} a_i$	summation notation; sum of *n* terms
i	index of summation
S_n	sum of first *n* terms of a sequence
$\displaystyle\sum_{i=1}^{\infty} a_i$	sum of an infinite number of terms
$\displaystyle\lim_{n\to\infty} S_n$	limit of S_n as *n* increases without bound
$n!$	*n*-factorial

$_nC_r$ or $\binom{n}{r}$	binomial coefficient (combinations of *n* elements taken *r* at a time)
$P(n,r)$	permutations of *n* elements taken *r* at a time
$C(n,r)$ or $\binom{n}{r}$	combinations of *n* elements taken *r* at a time
$n(E)$	number of outcomes that belong to event *E*
$P(E)$	probability of event *E*
E'	complement of event *E*

QUICK REVIEW

CONCEPTS	EXAMPLES

11.1 Sequences and Series

Sequence
General Term a_n
Series

The sequence $1, \frac{1}{2}, \frac{1}{3}, \frac{1}{4}, \ldots, \frac{1}{n}$ has general term $a_n = \frac{1}{n}$.

The corresponding series is the *sum*

$$1 + \frac{1}{2} + \frac{1}{3} + \frac{1}{4} + \cdots + \frac{1}{n}.$$

Summation Properties
If $a_1, a_2, a_3, \ldots, a_n$ and $b_1, b_2, b_3, \ldots, b_n$ are sequences and *c* is a constant, then for every positive integer *n*,

(a) $\displaystyle\sum_{i=1}^{n} c = nc$

$$\sum_{i=1}^{6} 5 = 6 \cdot 5 = 30$$

(continued)

CONCEPTS	EXAMPLES

(b) $\displaystyle\sum_{i=1}^{n} ca_i = c \sum_{i=1}^{n} a_i$

$$\sum_{i=1}^{4} 3(2i + 1) = 3 \sum_{i=1}^{4} (2i + 1)$$
$$= 3(3 + 5 + 7 + 9)$$
$$= 72$$

(c) $\displaystyle\sum_{i=1}^{n} (a_i \pm b_i) = \sum_{i=1}^{n} a_i \pm \sum_{i=1}^{n} b_i.$

$$\sum_{i=1}^{3} (5i + 6i^2) = \sum_{i=1}^{3} 5i + \sum_{i=1}^{3} 6i^2$$
$$= (5 + 10 + 15) + (6 + 24 + 54)$$
$$= 30 + 84 = 114$$

11.2 Arithmetic Sequences and Series

Assume a_1 is the first term, a_n is the nth term, and d is the common difference.

The arithmetic sequence 2, 5, 8, 11, . . . has $a_1 = 2$.

Common Difference $d = a_{n+1} - a_n$

$$d = 5 - 2 = 3$$

(Any two successive terms could have been used.)
Suppose that $n = 10$. Then the 10th term is

nth Term $a_n = a_1 + (n - 1)d$

$$a_{10} = 2 + (10 - 1)3$$
$$= 2 + 9 \cdot 3 = 29.$$

Sum of the First n Terms

$$S_n = \frac{n}{2}(a_1 + a_n)$$

or

$$S_n = \frac{n}{2}[2a_1 + (n - 1)d]$$

The sum of the first 10 terms is

$$S_{10} = \frac{10}{2}(a_1 + a_{10})$$
$$= 5(2 + 29) = 5(31) = 155$$

or

$$S_{10} = \frac{10}{2}[2(2) + (10 - 1)3]$$
$$= 5(4 + 9 \cdot 3)$$
$$= 5(4 + 27) = 5(31) = 155.$$

11.3 Geometric Sequences and Series

Assume a_1 is the first term, a_n is the nth term, and r is the common ratio.

The geometric sequence 1, 2, 4, 8, . . . has $a_1 = 1$.

Common Ratio $r = \dfrac{a_{n+1}}{a_n}$

$$r = \frac{8}{4} = 2$$

(Any two successive terms could have been used.)
Suppose that $n = 6$. Then the sixth term is

nth Term $a_n = a_1 r^{n-1}$

$$a_6 = (1)(2)^{6-1} = 1(2)^5 = 32.$$

Sum of the First n Terms

$$S_n = \frac{a_1(1 - r^n)}{1 - r} \qquad (r \neq 1)$$

The sum of the first six terms is

$$S_6 = \frac{1(1 - 2^6)}{1 - 2} = \frac{1 - 64}{-1} = 63.$$

CONCEPTS	EXAMPLES

Sum of the Terms of an Infinite Geometric Sequence with $|r| < 1$

$$S_\infty = \frac{a_1}{1 - r}$$

The sum of the terms of the infinite geometric sequence

$$\sum_{k=0}^{\infty} \left(\frac{1}{2}\right)^k = 1 + \frac{1}{2} + \frac{1}{4} + \cdots$$

is

$$S_\infty = \frac{1}{1 - \frac{1}{2}} = \frac{1}{\frac{1}{2}} = 2.$$

11.4 The Binomial Theorem

For any positive integer n,

$$n! = n(n - 1)(n - 2) \cdots (3)(2)(1).$$

By definition, $\quad 0! = 1.$

$$4! = 4 \cdot 3 \cdot 2 \cdot 1 = 24$$

Binomial Coefficient

For nonnegative integers n and r, with $r \le n$,

$$_nC_r = \binom{n}{r} = \frac{n!}{r!(n - r)!}.$$

$$_5C_3 = \frac{5!}{3!(5 - 3)!} = \frac{5!}{3!\,2!} = \frac{5 \cdot 4 \cdot 3 \cdot 2 \cdot 1}{3 \cdot 2 \cdot 1 \cdot 2 \cdot 1} = 10$$

Binomial Theorem

For any positive integer n and any complex numbers x and y,

$$(x + y)^n = x^n + \binom{n}{1}x^{n-1}y + \binom{n}{2}x^{n-2}y^2 + \binom{n}{3}x^{n-3}y^3 + \cdots$$
$$+ \binom{n}{r}x^{n-r}y^r + \cdots + \binom{n}{n-1}xy^{n-1} + y^n.$$

$$(2m + 3)^4 = (2m)^4 + \frac{4!}{3!\,1!}(2m)^3(3) + \frac{4!}{2!\,2!}(2m)^2(3)^2$$
$$+ \frac{4!}{1!\,3!}(2m)(3)^3 + 3^4$$
$$= 2^4m^4 + 4(2)^3m^3(3) + 6(2)^2m^2(9)$$
$$+ 4(2m)(27) + 81$$
$$= 16m^4 + 12(8)m^3 + 54(4)m^2 + 216m + 81$$
$$= 16m^4 + 96m^3 + 216m^2 + 216m + 81$$

kth Term of the Binomial Expansion of $(x + y)^n$

$$\binom{n}{k - 1}x^{n-(k-1)}y^{k-1} \qquad (n \ge k - 1)$$

The eighth term of $(a - 2b)^{10}$ is

$$\binom{10}{7}a^3(-2b)^7 = \frac{10!}{7!\,3!}a^3(-2)^7b^7$$
$$= 120(-128)a^3b^7$$
$$= -15{,}360a^3b^7.$$

11.5 Mathematical Induction

Principle of Mathematical Induction

Let S_n be a statement concerning the positive integer n. Suppose that

1. S_1 is true;
2. for any positive integer k, $k \le n$, if S_k is true, then S_{k+1} is also true.

Then S_n is true for every positive integer value of n.

See Examples 1 and 2 in **Section 11.5**.
Example 3 in **Section 11.5** illustrates the Generalized Principle of Mathematical Induction.

(continued)

CONCEPTS	EXAMPLES

11.6 Counting Theory

Fundamental Principle of Counting
If n independent events occur, with

$$m_1 \text{ ways for event 1 to occur,}$$

$$m_2 \text{ ways for event 2 to occur,}$$

$$\vdots$$

and $\qquad m_n$ ways for event n to occur,

then there are $m_1 \cdot m_2 \cdots \cdot m_n$ different ways for all n events to occur.

If there are 2 ways to choose a pair of socks and 5 ways to choose a pair of shoes, then there are $2 \cdot 5 = 10$ ways to choose socks and shoes.

Permutations Formula
If $P(n, r)$ denotes the number of permutations of n elements taken r at a time, with $r \leq n$, then

$$P(n, r) = \frac{n!}{(n-r)!}.$$

How many ways are there to arrange the letters of the word *triangle* using 5 letters at a time?
  Here, $n = 8$ and $r = 5$, so the number of ways is

$$P(8, 5) = \frac{8!}{(8-5)!} = \frac{8!}{3!} = 6720.$$

Combinations Formula
The number of combinations of n elements taken r at a time, with $r \leq n$, is

$$C(n, r) = \binom{n}{r} = \frac{n!}{(n-r)!\,r!}.$$

How many committees of 4 senators can be formed from a group of 9 senators?
  Since the arrangement of senators does not matter, this is a combinations problem. The number of committees is

$$C(9, 4) = \binom{9}{4} = \frac{9!}{5!\,4!} = 126.$$

11.7 Basics of Probability

Probability of an Event E
In a sample space S with equally likely outcomes, the probability of an event E is

$$P(E) = \frac{n(E)}{n(S)}.$$

A number is chosen at random from $S = \{1, 2, 3, 4, 5, 6\}$. What is the probability that the number is less than 3?
  The event is $E = \{1, 2\}$; $n(S) = 6$ and $n(E) = 2$, so

$$P(E) = \frac{2}{6} = \frac{1}{3}.$$

Properties of Probability
For any events E and F:
1. $0 \leq P(E) \leq 1$ 　　2. $P(\text{a certain event}) = 1$
3. $P(\text{an impossible event}) = 0$ 　4. $P(E') = 1 - P(E)$
5. $P(E \text{ or } F) = P(E \cup F)$
　　$= P(E) + P(F) - P(E \cap F)$.

What is the probability that the number is 3 or more?
  This event is E'.

$$P(E') = 1 - \frac{1}{3} = \frac{2}{3}$$

Binomial Probability
If the probability of success in a binomial experiment is p, then the probability of r successes in n trials is

$$\binom{n}{r} p^r (1-p)^{n-r}.$$

An experiment consists of rolling a die 8 times. Find the probability that exactly 5 rolls result in a 2.
  Here, we have $n = 8$, $r = 5$, and $p = \frac{1}{6}$.

$$\binom{8}{5}\left(\frac{1}{6}\right)^5\left(1 - \frac{1}{6}\right)^{8-5} = 56\left(\frac{1}{6}\right)^5\left(\frac{5}{6}\right)^3 \approx .00417$$

Review Exercises

Write the first five terms of each sequence. State whether the sequence is arithmetic, geo-metric, *or* neither.

1. $a_n = \dfrac{n}{n+1}$

2. $a_n = (-2)^n$

3. $a_n = 2(n+3)$

4. $a_n = n(n+1)$

5. $a_1 = 5$,
$a_n = a_{n-1} - 3$, if $n \geq 2$

6. $a_1 = 1, a_2 = 3$,
$a_n = a_{n-2} + a_{n-1}$, if $n \geq 3$

7. *Concept Check*   Write an arithmetic sequence that consists of five terms, with first term 4, having sum of the five terms equal to 25.

In Exercises 8–11, write the first five terms of the sequence described.

8. arithmetic, $a_2 = 10, d = -2$

9. arithmetic, $a_3 = \pi, a_4 = 1$

10. geometric, $a_1 = 6, r = 2$

11. geometric, $a_1 = -5, a_2 = -1$

12. An arithmetic sequence has $a_5 = -3$ and $a_{15} = 17$. Find a_1 and a_n.

13. A geometric sequence has $a_1 = -8$ and $a_7 = -\frac{1}{8}$. Find a_4 and a_n.

Find a_8 for each arithmetic sequence.

14. $a_1 = 6, d = 2$

15. $a_1 = 6x - 9, a_2 = 5x + 1$

Find S_{12} for each arithmetic sequence.

16. $a_1 = 2, d = 3$

17. $a_2 = 6, d = 10$

Find a_5 for each geometric sequence.

18. $a_1 = -2, r = 3$

19. $a_3 = 4, r = \dfrac{1}{5}$

Find S_4 for each geometric sequence.

20. $a_1 = 3, r = 2$

21. $a_1 = -1, r = 3$

22. $\dfrac{3}{4}, -\dfrac{1}{2}, \dfrac{1}{3}, \ldots$

Evaluate each sum that exists.

23. $\displaystyle\sum_{i=1}^{7} (-1)^{i-1}$

24. $\displaystyle\sum_{i=1}^{5} (i^2 + i)$

25. $\displaystyle\sum_{i=1}^{4} \dfrac{i+1}{i}$

26. $\displaystyle\sum_{j=1}^{10} (3j - 4)$

27. $\displaystyle\sum_{j=1}^{2500} j$

28. $\displaystyle\sum_{i=1}^{5} 4 \cdot 2^i$

29. $\displaystyle\sum_{i=1}^{\infty} \left(\dfrac{4}{7}\right)^i$

30. $\displaystyle\sum_{i=1}^{\infty} -2\left(\dfrac{6}{5}\right)^i$

31. $\displaystyle\sum_{i=1}^{\infty} 2\left(-\dfrac{2}{3}\right)^i$

32. *Concept Check*   Find an infinite geometric series having common ratio $\frac{3}{4}$ and sum 6.

Evaluate each series that converges. If the series diverges, say so.

33. $24 + 8 + \dfrac{8}{3} + \dfrac{8}{9} + \cdots$

34. $-\dfrac{3}{4} + \dfrac{1}{2} - \dfrac{1}{3} + \dfrac{2}{9} - \cdots$

35. $\dfrac{1}{12} + \dfrac{1}{6} + \dfrac{1}{3} + \dfrac{2}{3} + \cdots$

36. $.9 + .09 + .009 + .0009 + \cdots$

Evaluate each sum where $x_1 = 0$, $x_2 = 1$, $x_3 = 2$, $x_4 = 3$, $x_5 = 4$, and $x_6 = 5$.

37. $\displaystyle\sum_{i=1}^{4} (x_i^2 - 6)$

38. $\displaystyle\sum_{i=1}^{6} f(x_i)\,\Delta x;\ f(x) = (x - 2)^3,\ \Delta x = .1$

Write each sum using summation notation.

39. $4 - 1 - 6 - \cdots - 66$

40. $10 + 14 + 18 + \cdots + 86$

41. $4 + 12 + 36 + \cdots + 972$

42. $\dfrac{5}{6} + \dfrac{6}{7} + \dfrac{7}{8} + \cdots + \dfrac{12}{13}$

Use the binomial theorem to expand each expression.

43. $(x + 2y)^4$

44. $(3z - 5w)^3$

45. $\left(3\sqrt{x} - \dfrac{1}{\sqrt{x}}\right)^5$

46. $(m^3 - m^{-2})^4$

Find the indicated term or terms for each expansion.

47. sixth term of $(4x - y)^8$

48. seventh term of $(m - 3n)^{14}$

49. first four terms of $(x + 2)^{12}$

50. last three terms of $(2a + 5b)^{16}$

Use mathematical induction to prove that each statement is true for every positive integer n.

51. $1 + 3 + 5 + 7 + \cdots + (2n - 1) = n^2$

52. $2 + 6 + 10 + 14 + \cdots + (4n - 2) = 2n^2$

53. $2 + 2^2 + 2^3 + \cdots + 2^n = 2(2^n - 1)$

54. $1^3 + 3^3 + 5^3 + \cdots + (2n - 1)^3 = n^2(2n^2 - 1)$

Find the value of each expression.

55. $P(9, 2)$

56. $P(6, 0)$

57. $\dbinom{8}{3}$

58. $9!$

59. $C(10, 5)$

60. $10 \cdot 9!$

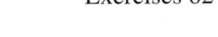 **61.** Explain how you can determine whether to use *permutations* or *combinations* in Exercises 62–68.

Solve each problem.

62. *Wedding Plans*   Two people are planning their wedding. They can select from 2 different chapels, 4 soloists, 3 organists, and 2 ministers. How many different wedding arrangements are possible?

63. *Couch Styles*   Bob Schiffer, who is furnishing his apartment, wants to buy a new couch. He can select from 5 different styles, each available in 3 different fabrics, with 6 color choices. How many different couches are available?

64. *Summer Job Assignments*   Four students are to be assigned to 4 different summer jobs. Each student is qualified for all 4 jobs. In how many ways can the jobs be assigned?

65. *Conference Delegations*   A student body council consists of a president, vice-president, secretary/treasurer, and 3 representatives at large. Three members are to be selected to attend a conference.

(a) How many different such delegations are possible?
(b) How many are possible if the president must attend?

66. *Tournament Outcomes*   Nine football teams are competing for first-, second-, and third-place titles in a statewide tournament. In how many ways can the winners be determined?

67. *License Plates*   How many different license plates can be formed with a letter followed by 3 digits and then 3 letters? How many such license plates have no repeats?

68. *Racetrack Bets*   Most racetracks have "compound" bets on 2 or more horses. An *exacta* is a bet in which the first and second finishers in a race are specified in order. A *quinella* is a bet on the first 2 finishers in a race, with order not specified.

(a) In a field of 9 horses, how many different exacta bets can be placed?

(b) How many different quinella bets can be placed in a field of 9 horses?

69. *Drawing a Marble*   A marble is drawn at random from a box containing 4 green, 5 black, and 6 white marbles. Find the following probabilities.

(a) A green marble is drawn.   (b) A marble that is not black is drawn.

(c) A blue marble is drawn.

(d) What are the odds in favor of drawing a marble that is not white?

70. *Drawing a Card*   A card is drawn from a standard deck of 52 cards. Find the probability of each of the following events.

(a) a black king   (b) a face card or an ace

(c) an ace or a diamond   (d) a card that is not a diamond

(e) What are the odds in favor of drawing an ace?

71. *Political Orientation*   The table describes the political orientation of college freshmen in the class of 2006, as determined from a survey of 282,200 freshmen.

Political Orientation	Number of Freshmen (in thousands)
Far left	7.06
Liberal	71.48
Middle of the road	143.5
Conservative	56.51
Far right	3.673
Total	282.2

Source: Higher Education Research Institute, UCLA, 2002.

(a) What is the probability that a randomly selected student from the class is in the conservative group?

(b) What is the probability that a randomly selected student from the class is on the far left or the far right politically?

(c) What is the probability of a randomly selected student from the class not being politically middle of the road?

72. *Defective Toaster Ovens*   A sample shipment of 5 toaster ovens is chosen. The probability of exactly 0, 1, 2, 3, 4, or 5 toaster ovens being defective is given in the table.

Number Defective	0	1	2	3	4	5
Probability	.31	.25	.18	.12	.08	.06

Find the probability that the given number of toaster ovens are defective.

(a) no more than 3   (b) at least 2   (c) more than 5

73. *Rolling a Die*   A die is rolled 12 times. Find the probability (to three decimal places) that exactly 2 of the rolls result in a 5.

74. *Tossing a Coin*   A coin is tossed 10 times. Find the probability (to three decimal places) that exactly 4 of the tosses result in a tail.

CHAPTER 11 ▶	**Test**

Write the first five terms of each sequence. State whether the sequence is arithmetic, geo-metric, *or neither.*

1. $a_n = (-1)^n(n^2 + 2)$ $\qquad\qquad\qquad$ **2.** $a_n = -3\left(\dfrac{1}{2}\right)^n$

3. $a_1 = 2, a_2 = 3, a_n = a_{n-1} + 2a_{n-2}, \quad$ for $n \geq 3$

4. A certain arithmetic sequence has $a_1 = 1$ and $a_3 = 25$. Find a_5.

5. A certain geometric sequence has $a_1 = 81$ and $r = -\frac{2}{3}$. Find a_6.

Find the sum of the first ten terms of each series.

6. arithmetic, $a_1 = -43, d = 12$ $\qquad\qquad$ **7.** geometric, $a_1 = 5, r = -2$

Evaluate each sum that exists.

8. $\displaystyle\sum_{i=1}^{30}(5i + 2)$ $\qquad$ **9.** $\displaystyle\sum_{i=1}^{5}(-3 \cdot 2^i)$ $\qquad$ **10.** $\displaystyle\sum_{i=1}^{\infty}(2^i) \cdot 4$ $\qquad$ **11.** $\displaystyle\sum_{i=1}^{\infty}54\left(\dfrac{2}{9}\right)^i$

Use the binomial theorem to expand each expression.

12. $(x + y)^6$ $\qquad\qquad\qquad\qquad\qquad$ **13.** $(2x - 3y)^4$

14. Find the third term in the expansion of $(w - 2y)^6$.

Evaluate each expression.

15. $8!$ $\qquad$ **16.** $C(10, 2)$ $\qquad$ **17.** $\displaystyle\binom{7}{3}$ $\qquad$ **18.** $P(11, 3)$

19. Use mathematical induction to prove that for all positive integers n,

$$1 + 7 + 13 + \cdots + (6n - 5) = n(3n - 2).$$

Solve each problem.

20. *Athletic Shoe Styles*   A sports-shoe manufacturer makes athletic shoes in 4 different styles. Each style comes in 3 different colors, and each color comes in 2 different shades. How many different types of shoes can be made?

21. *Seminar Attendees*   A mortgage company has 10 loan officers: 1 black, 2 Asian, and the rest white. In how many ways can 3 of these officers be selected to attend a seminar? How many ways are there if the black officer and exactly 1 Asian officer must be included?

22. *Project Workers*   Refer to Exercise 21. If 4 of the loan officers are women and 6 are men, in how many ways can 2 women and 2 men be selected to work on a special project?

23. *Drawing Cards*   A card is drawn from a standard deck of 52 cards. Find the probability that each of the following is drawn.

(a) a red three   **(b)** a card that is not a face card

(c) a king or a spade

(d) What are the odds in favor of drawing a face card?

24. *Defective Transistors*   A sample of 4 light bulbs is chosen. The probability of exactly 0, 1, 2, 3, or 4 light bulbs being defective is given in the table. Find the probability that at most 2 are defective.

Number Defective	0	1	2	3	4
Probability	.19	.43	.30	.07	.01

25. *Rolling a Die*   Find the probability (to three decimal places) of obtaining 5 on exactly two of six rolls of a single die.

CHAPTER 11 ▶	Quantitative Reasoning

What is the value of a college education?

A high school graduate must decide whether the cost of investing in a college education will be worth it in the long run versus jumping immediately into the job market. Suppose you have estimated the cost of a 4-yr college education, including tuition and living expenses, at $130,000. You have also located the following statistics of annual earnings.

In 2000, the median annual earnings of a person with 4 yr of college was $43,368, and the median annual earnings of a high school graduate with no college attendance was $26,364. The annual median earnings of the two groups have risen at rates of about $994 and $534 per year, respectively. (*Source:* U.S. Bureau of Labor Statistics.)

Now, do the math. Assume the average 18-yr-old high school graduate in 2000 will work until age 65, earning the median amount throughout those years. (Of course, such a person will receive less than the median earnings in the beginning and more than the median earnings in later years. These differences should balance out to produce a reasonable approximation of lifetime earnings.)

1. How much will a person earning the median amount earn until retirement if he or she joins the work force immediately after high school graduation without going to college?

2. How much will a person earning the median amount earn until retirement if he or she attends college for 4 yr and then joins the work force?

3. How much more will a person earning the median amount who attends 4 yr of college earn over his or her lifetime? Is the $130,000 cost worth it?

Appendices

A Polar Form of Conic Sections

Up to this point, we have worked with equations of conic sections in rectangular form. If the focus of a conic section is at the pole, the polar form of its equation is

$$r = \frac{ep}{1 \pm e \cdot f(\theta)},$$

where f is either the sine or cosine function.

POLAR FORMS OF CONIC SECTIONS

A polar equation of the form

$$r = \frac{ep}{1 \pm e \cos \theta} \qquad \text{or} \qquad r = \frac{ep}{1 \pm e \sin \theta}$$

has a conic section as its graph. The eccentricity is e (where $e > 0$), and $|p|$ is the distance between the pole (focus) and the directrix.

We can verify that $r = \frac{ep}{1 + e \cos \theta}$ does indeed satisfy the definition of a conic section. Consider Figure 1, where the directrix is vertical and $p > 0$ units to the right of the focus $F(0, 0°)$. Let $P(r, \theta)$ be a point on the graph. Then the distance between P and the directrix is

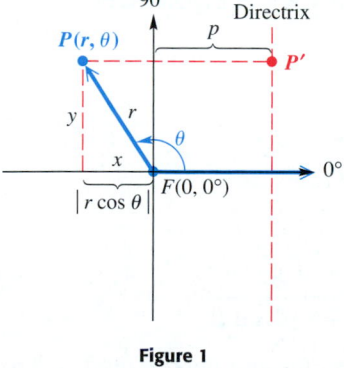

Figure 1

$$PP' = |p - x|$$

$$= |p - r \cos \theta| \qquad \textcolor{blue}{x = r \cos \theta \text{ (Section 8.5)}}$$

$$= \left| p - \left(\frac{ep}{1 + e \cos \theta} \right) \cos \theta \right| \qquad \textcolor{blue}{\text{Use the equation for } r.}$$

$$= \left| \frac{p(1 + e \cos \theta) - ep \cos \theta}{1 + e \cos \theta} \right| \qquad \textcolor{blue}{\text{Write with a common denominator.} \\ \text{(Section R.5)}}$$

$$= \left| \frac{p + ep \cos \theta - ep \cos \theta}{1 + e \cos \theta} \right| \qquad \textcolor{blue}{\text{Distributive property (Section R.2)}}$$

$$PP' = \left| \frac{p}{1 + e \cos \theta} \right|.$$

1075

Since
$$r = \frac{ep}{1 + e \cos \theta},$$

we can multiply each side by $\frac{1}{e}$ to obtain

$$\frac{r}{e} = \frac{p}{1 + e \cos \theta}.$$

We substitute $\frac{r}{e}$ for the expression in the absolute value bars for PP'.

$$PP' = \left| \frac{p}{1 + e \cos \theta} \right| = \left| \frac{r}{e} \right| = \frac{|r|}{|e|} = \frac{|r|}{e}$$

The distance between the pole and P is $PF = |r|$, so the ratio of PF to PP' is

$$\frac{PF}{PP'} = \frac{|r|}{\frac{|r|}{e}} = e. \qquad \text{Simplify the complex fraction. (Section R.5)}$$

Thus, by the definition, the graph has eccentricity e and must be a conic.

In the preceding discussion, we assumed a vertical directrix to the right of the pole. There are three other possible situations, and all four are summarized in the table.

If the equation is:	then the directrix is:
$r = \dfrac{ep}{1 + e \cos \theta}$	*vertical*, p units to the *right* of the pole.
$r = \dfrac{ep}{1 - e \cos \theta}$	*vertical*, p units to the *left* of the pole.
$r = \dfrac{ep}{1 + e \sin \theta}$	*horizontal*, p units *above* the pole.
$r = \dfrac{ep}{1 - e \sin \theta}$	*horizontal*, p units *below* the pole.

▶ **EXAMPLE 1** **GRAPHING A CONIC SECTION WITH EQUATION IN POLAR FORM**

Graph $r = \dfrac{8}{4 + 4 \sin \theta}$.

Algebraic Solution

Divide both numerator and denominator by 4 to get

$$r = \frac{2}{1 + \sin \theta}.$$

Based on the preceding table, this is the equation of a conic with $ep = 2$ and $e = 1$. Thus, $p = 2$. Since $e = 1$, the graph is a parabola. The focus is at the pole, and the directrix is horizontal, 2 units *above* the pole. The vertex must have polar coordinates $(1, 90°)$. Letting $\theta = 0°$ and $\theta = 180°$ gives the additional points $(2, 0°)$ and $(2, 180°)$. See Figure 2 on the next page.

Graphing Calculator Solution

Enter

$$r_1 = \frac{8}{4 + 4 \sin \theta},$$

with the calculator in polar and degree modes. The first two screens in Figure 3 on the next page show the window settings, and the third screen shows the graph. Notice that the point $(1, 90°)$ is indicated at the bottom of the third screen.

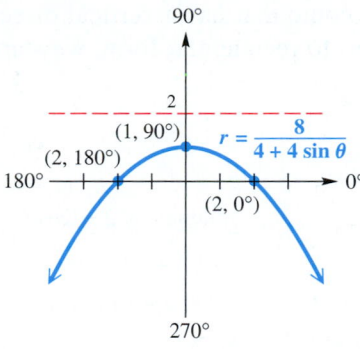

Figure 2

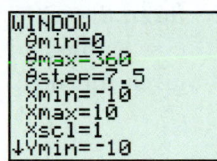

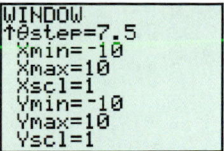

This is a continuation of the screen to the left.

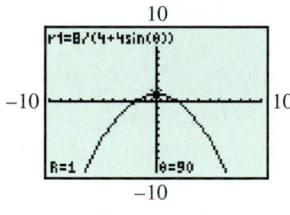

Degree mode

Figure 3

NOW TRY EXERCISE 1. ◄

▶ **EXAMPLE 2**   **FINDING A POLAR EQUATION**

Find the polar equation of a parabola with focus at the pole and vertical directrix 3 units to the left of the pole.

Solution   The eccentricity e must be 1, p must equal 3, and the equation must be of the form

$$r = \frac{ep}{1 - e \cos \theta}.$$

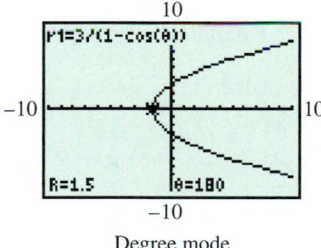

Degree mode

Figure 4

Thus, we have

$$r = \frac{1 \cdot 3}{1 - 1 \cos \theta} = \frac{3}{1 - \cos \theta}.$$

The calculator graph in Figure 4 supports our result. When $\theta = 180°$, $r = 1.5$. The distance from $F(0, 0°)$ to the directrix is $2r = 2(1.5) = 3$ units, as required.

NOW TRY EXERCISE 11. ◄

▶ **EXAMPLE 3**   **IDENTIFYING AND CONVERTING FROM POLAR FORM TO RECTANGULAR FORM**

Identify the type of conic represented by $r = \dfrac{8}{2 - \cos \theta}$. Then convert the equation to rectangular form.

Solution   To identify the type of conic, we divide both the numerator and the denominator on the right side by 2 to obtain

$$r = \frac{4}{1 - \frac{1}{2} \cos \theta}.$$

From the table, we see that this is a conic that has a vertical directrix, with $e = \frac{1}{2}$; thus, it is an ellipse. To convert to rectangular form, we start with the given equation.

$$r = \frac{8}{2 - \cos \theta}$$

$r(2 - \cos \theta) = 8$	Multiply by $2 - \cos \theta$.
$2r - r \cos \theta = 8$	Distributive property
$2r = r \cos \theta + 8$	Add $r \cos \theta$ to each side.
$(2r)^2 = (r \cos \theta + 8)^2$	Square each side. **(Section 1.6)**
$(2r)^2 = (x + 8)^2$	$r \cos \theta = x$
$4r^2 = x^2 + 16x + 64$	Multiply. **(Section R.3)**
$4(x^2 + y^2) = x^2 + 16x + 64$	$r^2 = x^2 + y^2$ **(Section 8.5)**
$4x^2 + 4y^2 = x^2 + 16x + 64$	Distributive property
$3x^2 + 4y^2 - 16x - 64 = 0$	Standard form **(Section 10.2)**

The coefficients of x^2 and y^2 are both positive and are not equal, further supporting our assertion that the graph is an ellipse.

NOW TRY EXERCISE 19. ◀

Appendix A   Exercises

Graph each conic whose equation is given in polar form. See Example 1.

1. $r = \dfrac{6}{3 + 3 \sin \theta}$

2. $r = \dfrac{10}{5 + 5 \sin \theta}$

3. $r = \dfrac{-4}{6 + 2 \cos \theta}$

4. $r = \dfrac{-8}{4 + 2 \cos \theta}$

5. $r = \dfrac{2}{2 - 4 \sin \theta}$

6. $r = \dfrac{6}{2 - 4 \sin \theta}$

7. $r = \dfrac{-1}{1 + 2 \sin \theta}$

8. $r = \dfrac{-1}{1 - 2 \sin \theta}$

9. $r = \dfrac{-1}{2 + \cos \theta}$

10. $r = \dfrac{-1}{2 - \cos \theta}$

Find a polar equation of the parabola with focus at the pole, satisfying the given conditions. See Example 2.

11. Vertical directrix 3 units to the right of the pole

12. Vertical directrix 4 units to the left of the pole

13. Horizontal directrix 5 units below the pole

14. Horizontal directrix 6 units above the pole

Find a polar equation for the conic with focus at the pole, satisfying the given conditions. Also identify the type of conic represented.

15. $e = \frac{4}{5}$; vertical directrix 5 units to the right of the pole

16. $e = \frac{2}{3}$; vertical directrix 6 units to the left of the pole

17. $e = \frac{5}{4}$; horizontal directrix 8 units below the pole

18. $e = \frac{3}{2}$; horizontal directrix 4 units above the pole

Identify the type of conic represented, and convert the equation to rectangular form. See Example 3.

19. $r = \dfrac{6}{3 - \cos\theta}$

20. $r = \dfrac{8}{4 - \cos\theta}$

21. $r = \dfrac{-2}{1 + 2\cos\theta}$

22. $r = \dfrac{-3}{1 + 3\cos\theta}$

23. $r = \dfrac{-6}{4 + 2\sin\theta}$

24. $r = \dfrac{-12}{6 + 3\sin\theta}$

25. $r = \dfrac{10}{2 - 2\sin\theta}$

26. $r = \dfrac{12}{4 - 4\sin\theta}$

B   Rotation of Axes

Derivation of Rotation Equations ▪ Applying a Rotation Equation

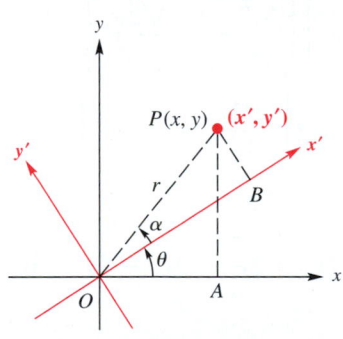

Figure 1

Derivation of Rotation Equations   If we begin with an xy-coordinate system having origin O and rotate the axes about O through an angle θ, the new coordinate system is called a **rotation** of the xy-system. Trigonometric identities can be used to obtain equations for converting the coordinates of a point from the xy-system to the rotated $x'y'$-system. Let P be any point other than the origin, with coordinates (x, y) in the xy-system and (x', y') in the $x'y'$-system. See Figure 1. Let $OP = r$, and let α represent the angle made by OP and the x'-axis. As shown in Figure 1,

$$\cos(\theta + \alpha) = \frac{OA}{r} = \frac{x}{r}, \qquad \sin(\theta + \alpha) = \frac{AP}{r} = \frac{y}{r},$$

$$\cos\alpha = \frac{OB}{r} = \frac{x'}{r}, \qquad \sin\alpha = \frac{PB}{r} = \frac{y'}{r}.$$

(Section 5.2)

These four statements can be written as

$$x = r\cos(\theta + \alpha), \qquad y = r\sin(\theta + \alpha), \qquad x' = r\cos\alpha, \qquad y' = r\sin\alpha.$$

▼ **LOOKING AHEAD TO CALCULUS**

Rotation of axes is a topic traditionally covered in calculus texts, in conjunction with parametric equations and polar coordinates. The coverage in calculus is typically the same as that seen in this section.

Using the trigonometric identity for the cosine of the sum of two angles gives

$$
\begin{aligned}
x &= r\cos(\theta + \alpha) \\
&= r(\cos\theta\cos\alpha - \sin\theta\sin\alpha) \quad \text{(Section 7.3)} \\
&= (r\cos\alpha)\cos\theta - (r\sin\alpha)\sin\theta \quad \text{Distributive property (Section R.2)} \\
&= x'\cos\theta - y'\sin\theta. \quad \text{Substitute.}
\end{aligned}
$$

In the same way, by using the identity for the sine of the sum of two angles, $y = x'\sin\theta + y'\cos\theta$. This proves the following result.

ROTATION EQUATIONS

If the rectangular coordinate axes are rotated about the origin through an angle θ, and if the coordinates of a point P are (x, y) and (x', y') with respect to the xy-system and the $x'y'$-system, respectively, then the **rotation equations** are

$$
x = x'\cos\theta - y'\sin\theta \qquad \text{and} \qquad y = x'\sin\theta + y'\cos\theta.
$$

Applying a Rotation Equation

▶ **EXAMPLE 1** FINDING AN EQUATION AFTER A ROTATION

The equation of a curve is $x^2 + y^2 + 2xy + 2\sqrt{2}x - 2\sqrt{2}y = 0$. Find the resulting equation if the axes are rotated 45°. Graph the equation.

Solution If $\theta = 45°$, then $\sin\theta = \frac{\sqrt{2}}{2}$ and $\cos\theta = \frac{\sqrt{2}}{2}$, and the rotation equations become

$$
x = \frac{\sqrt{2}}{2}x' - \frac{\sqrt{2}}{2}y' \qquad \text{and} \qquad y = \frac{\sqrt{2}}{2}x' + \frac{\sqrt{2}}{2}y'.
$$

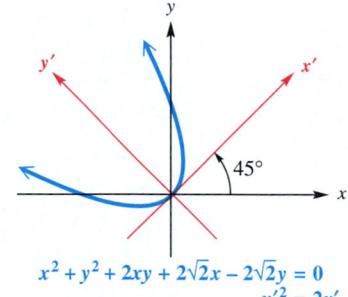

$x^2 + y^2 + 2xy + 2\sqrt{2}x - 2\sqrt{2}y = 0$
$x'^2 = 2y'$

Figure 2

Substituting these values into the given equation yields

$$
x^2 + y^2 + 2xy + 2\sqrt{2}x - 2\sqrt{2}y = 0
$$

$$
\left[\frac{\sqrt{2}}{2}x' - \frac{\sqrt{2}}{2}y'\right]^2 + \left[\frac{\sqrt{2}}{2}x' + \frac{\sqrt{2}}{2}y'\right]^2 + 2\left[\frac{\sqrt{2}}{2}x' - \frac{\sqrt{2}}{2}y'\right]\left[\frac{\sqrt{2}}{2}x' + \frac{\sqrt{2}}{2}y'\right]
$$

$$
+ 2\sqrt{2}\left[\frac{\sqrt{2}}{2}x' - \frac{\sqrt{2}}{2}y'\right] - 2\sqrt{2}\left[\frac{\sqrt{2}}{2}x' + \frac{\sqrt{2}}{2}y'\right] = 0.
$$

Expanding these terms yields

$$
\frac{1}{2}x'^2 - x'y' + \frac{1}{2}y'^2 + \frac{1}{2}x'^2 + x'y' + \frac{1}{2}y'^2 + x'^2 - y'^2 + 2x' - 2y' - 2x' - 2y' = 0.
$$

Collecting terms gives

$$
2x'^2 - 4y' = 0
$$

$$
x'^2 - 2y' = 0 \qquad \text{Divide by 2.}
$$

or, finally

$$
x'^2 = 2y', \qquad \text{(Section 10.1)}
$$

the equation of a parabola. The graph is shown in Figure 2.

NOW TRY EXERCISE 13. ◀

We have graphed equations written in the general form

$$Ax^2 + Cy^2 + Dx + Ey + F = 0. \quad \text{(Section 10.4)}$$

To graph an equation that has an xy-term by hand, it is necessary to find an appropriate **angle of rotation** to eliminate the xy-term. The necessary angle of rotation can be determined by using the following result. The proof is quite lengthy and is not presented here.

ANGLE OF ROTATION

The xy-term is removed from the general equation

$$Ax^2 + Bxy + Cy^2 + Dx + Ey + F = 0$$

by a rotation of the axes through an angle θ, $0° < \theta < 90°$, where

$$\cot 2\theta = \frac{A - C}{B}.$$

NOW TRY EXERCISE 7. ◄

To find the rotation equations, first find $\sin \theta$ and $\cos \theta$. Example 2 illustrates a way to obtain $\sin \theta$ and $\cos \theta$ from $\cot 2\theta$ without first identifying angle θ.

► **EXAMPLE 2** **ROTATING AND GRAPHING**

Remove the xy-term from $52x^2 - 72xy + 73y^2 = 200$ by performing a suitable rotation, and graph the equation.

Solution Here $A = 52$, $B = -72$, and $C = 73$. By substitution,

$$\cot 2\theta = \frac{52 - 73}{-72} = \frac{-21}{-72} = \frac{7}{24}.$$

To find $\sin \theta$ and $\cos \theta$, use the trigonometric identities

$$\sin \theta = \sqrt{\frac{1 - \cos 2\theta}{2}} \quad \text{and} \quad \cos \theta = \sqrt{\frac{1 + \cos 2\theta}{2}}. \quad \text{(Section 7.4)}$$

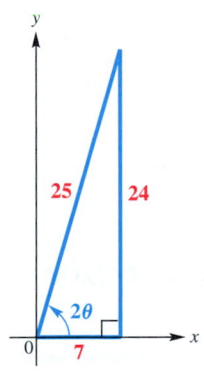

Figure 3

Sketch a right triangle and label it as in Figure 3, to see that $\cos 2\theta = \frac{7}{25}$. (Recall from **Section 5.2** that in the two quadrants for which we are concerned, cosine and cotangent have the same sign.) Then

$$\sin \theta = \sqrt{\frac{1 - \frac{7}{25}}{2}} = \sqrt{\frac{9}{25}} = \frac{3}{5} \quad \text{and} \quad \cos \theta = \sqrt{\frac{1 + \frac{7}{25}}{2}} = \sqrt{\frac{16}{25}} = \frac{4}{5}.$$

Use these values for $\sin \theta$ and $\cos \theta$ to obtain

$$x = \frac{4}{5}x' - \frac{3}{5}y' \quad \text{and} \quad y = \frac{3}{5}x' + \frac{4}{5}y'.$$

Substituting these expressions for x and y into the original equation yields

$$52\left[\frac{4}{5}x' - \frac{3}{5}y'\right]^2 - 72\left[\frac{4}{5}x' - \frac{3}{5}y'\right]\left[\frac{3}{5}x' + \frac{4}{5}y'\right] + 73\left[\frac{3}{5}x' + \frac{4}{5}y'\right]^2 = 200$$

$$52\left[\frac{16}{25}x'^2 - \frac{24}{25}x'y' + \frac{9}{25}y'^2\right] - 72\left[\frac{12}{25}x'^2 + \frac{7}{25}x'y' - \frac{12}{25}y'^2\right] + 73\left[\frac{9}{25}x'^2 + \frac{24}{25}x'y' + \frac{16}{25}y'^2\right] = 200$$

$$25x'^2 + 100y'^2 = 200 \quad \text{Combine terms.}$$

$$\frac{x'^2}{8} + \frac{y'^2}{2} = 1. \quad \text{Divide by 200.}$$

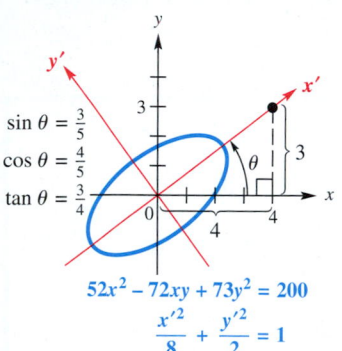

$\sin\theta = \frac{3}{5}$

$\cos\theta = \frac{4}{5}$

$\tan\theta = \frac{3}{4}$

$52x^2 - 72xy + 73y^2 = 200$

$\dfrac{x'^2}{8} + \dfrac{y'^2}{2} = 1$

Figure 4

This is an equation of an ellipse having x'-intercepts $\pm 2\sqrt{2}$ and y'-intercepts $\pm\sqrt{2}$. The graph is shown in Figure 4. To find θ, use the fact that

$$\frac{\sin\theta}{\cos\theta} = \frac{\frac{3}{5}}{\frac{4}{5}} = \frac{3}{4} = \tan\theta, \quad \text{(Section 7.1)}$$

from which $\theta \approx 37°$.

NOW TRY EXERCISE 19. ◀

The following summary enables us to use the general equation to decide on the type of graph to expect.

> ### EQUATION OF A CONIC WITH AN xy-TERM
>
> If the general second-degree equation
>
> $$Ax^2 + Bxy + Cy^2 + Dx + Ey + F = 0$$
>
> has a graph, it will be one of the following:
> **(a)** a circle or an ellipse (or a point) if $B^2 - 4AC < 0$;
> **(b)** a parabola (or one line or two parallel lines) if $B^2 - 4AC = 0$;
> **(c)** a hyperbola (or two intersecting lines) if $B^2 - 4AC > 0$;
> **(d)** a straight line if $A = B = C = 0$, and $D \neq 0$ or $E \neq 0$.

NOW TRY EXERCISE 1. ◀

Appendix B Exercises

Concept Check *Use the summary in this section to predict the type of graph of each second-degree equation.*

1. $4x^2 + 3y^2 + 2xy - 5x = 8$

2. $x^2 + 2xy - 3y^2 + 2y = 12$

3. $2x^2 + 3xy - 4y^2 = 0$

4. $x^2 - 2xy + y^2 + 4x - 8y = 0$

5. $4x^2 + 4xy + y^2 + 15 = 0$

6. $-x^2 + 2xy - y^2 + 16 = 0$

Find the angle of rotation θ that will remove the xy-term in each equation.

7. $2x^2 + \sqrt{3}xy + y^2 + x = 5$

8. $4\sqrt{3}x^2 + xy + 3\sqrt{3}y^2 = 10$

9. $3x^2 + \sqrt{3}xy + 4y^2 + 2x - 3y = 12$

10. $4x^2 + 2xy + 2y^2 + x - 7 = 0$

11. $x^2 - 4xy + 5y^2 = 18$

12. $3\sqrt{3}x^2 - 2xy + \sqrt{3}y^2 = 25$

Use the given angle of rotation to remove the xy-term and graph each equation. See Example 1.

13. $x^2 - xy + y^2 = 6$; $\theta = 45°$

14. $5y^2 + 12xy = 10$; $\sin\theta = \dfrac{3}{\sqrt{13}}$

15. $8x^2 - 4xy + 5y^2 = 36$; $\sin\theta = \dfrac{2}{\sqrt{5}}$

Remove the xy-term from each equation by performing a suitable rotation. Graph each equation. See Example 2.

16. $3x^2 - 2xy + 3y^2 = 8$

17. $x^2 - 4xy + y^2 = -5$

18. $x^2 + 2xy + y^2 + 4\sqrt{2}x - 4\sqrt{2}y = 0$

19. $7x^2 + 6\sqrt{3}xy + 13y^2 = 64$

20. $7x^2 + 2\sqrt{3}xy + 5y^2 = 24$

21. $3x^2 - 2\sqrt{3}xy + y^2 - 2x - 2\sqrt{3}y = 0$

22. $2x^2 + 2\sqrt{3}xy + 4y^2 = 5$

In each equation, remove the xy-term by rotation. Then translate the axes and sketch the graph.

23. $x^2 + 3xy + y^2 - 5\sqrt{2}y = 15$

24. $x^2 - \sqrt{3}xy + 2\sqrt{3}x - 3y - 3 = 0$

25. $4x^2 + 4xy + y^2 - 24x + 38y - 19 = 0$

26. $12x^2 + 24xy + 19y^2 - 12x - 40y + 31 = 0$

27. $16x^2 + 24xy + 9y^2 - 130x + 90y = 0$

28. $9x^2 - 6xy + y^2 - 12\sqrt{10}x - 36\sqrt{10}y = 0$

29. Look at the box titled "Angle of Rotation." Why is no rotation applicable if the value of B is 0?

30. Look at the equation involving $\cot 2\theta$ in the box titled "Angle of Rotation." Why must the angle of rotation be 45° if the coefficients of x^2 and y^2 are equal, and $B \neq 0$?

C   Geometry Formulas

Square

Perimeter: $P = 4s$

Area: $A = s^2$

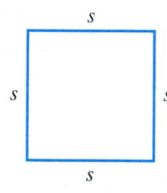

Rectangle

Perimeter: $P = 2L + 2W$

Area: $A = LW$

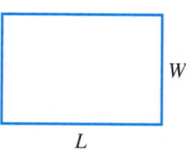

Triangle

Perimeter: $P = a + b + c$

Area: $A = \dfrac{1}{2}bh$

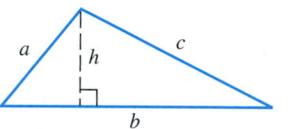

Parallelogram

Perimeter: $P = 2a + 2b$

Area: $A = bh$

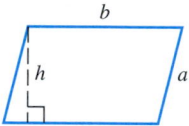

Trapezoid

Perimeter: $P = a + b + c + B$

Area: $A = \dfrac{1}{2}h(B + b)$

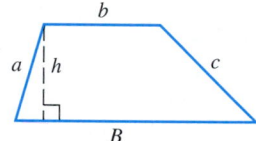

Circle

Diameter: $d = 2r$

Circumference: $C = 2\pi r = \pi d$

Area: $A = \pi r^2$

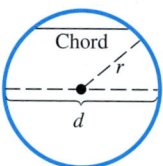

Cube

Volume: $V = e^3$

Surface area: $S = 6e^2$

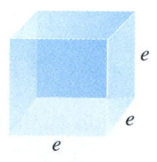

Rectangular Solid

Volume: $V = LWH$

Surface area: $S = 2HW + 2LW + 2LH$

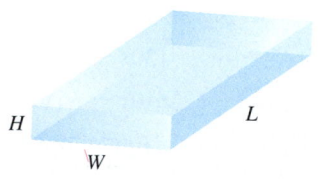

Sphere

Volume: $V = \dfrac{4}{3}\pi r^3$

Surface area: $S = 4\pi r^2$

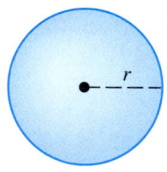

Cone

Volume: $V = \dfrac{1}{3}\pi r^2 h$

Surface area: $S = \pi r\sqrt{r^2 + h^2}$

(excludes the base)

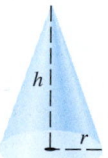

Right Circular Cylinder

Volume: $V = \pi r^2 h$

Surface area: $S = 2\pi rh + 2\pi r^2$

(includes top and bottom)

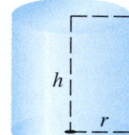

Right Pyramid

Volume: $V = \dfrac{1}{3}Bh$

B = area of the base

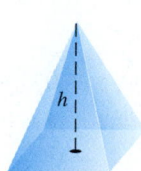

Glossary

A

absolute value The distance on the number line from a number to 0 is called the absolute value of that number. (Sections R.2, 1.8)

absolute value (modulus) of a complex number When a complex number $x + yi$ is written in trigonometric (or polar) form as $r(\cos \theta + i \sin \theta)$, the number r is called the absolute value (or modulus) of the complex number. (Section 8.5)

acute angle An acute angle is an angle measuring between 0° and 90°. (Section 5.1)

addition of ordinates Addition of ordinates is a method for graphing a function that is the sum of two other functions by adding the y-values of the two functions at selected x-values. (Section 6.5)

additive inverse (negative) of a matrix When two matrices are added and a zero matrix results, the matrices are additive inverses (negatives) of each other. (Section 9.7)

airspeed In air navigation, the airspeed of a plane is its speed relative to the air. (Section 8.4)

algebraic expression An algebraic expression is the result of adding, subtracting, multiplying, dividing (except by 0), or raising to powers or taking roots on any combination of variables or constants. (Section R.3)

ambiguous case The situation in which the lengths of two sides of a triangle and the measure of the angle opposite one of them are given (SSA) is called the ambiguous case of the law of sines. Depending on the given measurements, this combination of given parts may result in 0, 1, or 2 possible triangles. (Section 8.1)

amplitude The amplitude of a periodic function is half the difference between the maximum and minimum values of the function. (Section 6.3)

angle An angle is formed by rotating a ray around its endpoint. (Section 5.1)

angle of depression The angle of depression from point X to point Y (below X) is the acute angle formed by ray XY and a horizontal ray with endpoint at X. (Section 5.4)

angle of elevation The angle of elevation from point X to point Y (above X) is the acute angle formed by ray XY and a horizontal ray with endpoint at X. (Section 5.4)

angle of rotation The angle through which the axes of an xy-coordinate system are rotated to obtain a new coordinate system is called the angle of rotation. (Appendix B)

Angle-Side-Angle (ASA) The Angle-Side-Angle (ASA) congruence axiom states that if two angles and the included side of one triangle are equal, respectively, to two angles and the included side of a second triangle, then the triangles are congruent. (Section 8.1)

angle in standard position An angle is in standard position if its vertex is at the origin and its initial side is along the positive x-axis. (Section 5.1)

angle between two vectors The angle between two vectors is defined to be the angle θ, for $0° \le \theta \le 180°$, having the two vectors as its sides. (Section 8.3)

angular speed ω Angular speed ω (omega) measures the speed of rotation and is defined by $\omega = \frac{\theta}{t}$, where θ is the angle of rotation in radians and t is time. (Section 6.2)

annuity An annuity is a sequence of equal payments made at equal periods of time. (Section 11.3)

argument of a complex number When a complex number $x + yi$ is written in trigonometric (or polar) form as $r(\cos \theta + i \sin \theta)$, the angle θ is called the argument of the complex number. (Section 8.5)

argument of a function The argument of a function is the expression containing the independent variable of the function. For example, in the function $y = f(x - d)$, the expression $x - d$ is the argument. (Section 6.4)

argument of a logarithm In the expression $\log_a x$, x is called the argument. (Section 4.3)

arithmetic sequence (arithmetic progression) An arithmetic sequence is a sequence in which each term after the first is obtained by adding a fixed number to the previous term. (Section 11.2)

arithmetic series An arithmetic series is the sum of the terms of an arithmetic sequence. (Section 11.2)

asymptotes of a hyperbola The asymptotes of a hyperbola are two lines that the hyperbola approaches but never touches or intersects. (Section 10.3)

augmented matrix An augmented matrix is a matrix whose elements are the coefficients of the variables and the constants of a system of equations. An augmented matrix is often written with a vertical bar that separates the coefficients of the variables from the constants. (Section 9.2)

average rate of change The average rate of change is an interpretation of slope that applies to many linear models. The slope of a line gives the average rate of change in y per unit of change in x, where the value of y depends on the value of x. (Section 2.4)

axis The line of symmetry for a parabola is called the axis of the parabola. (Sections 3.1, 10.1)

B

base of an exponential The base is the number that is a repeated factor in exponential notation. In the expression a^n, a is the base. (Section R.2)

base of a logarithm In the expression $\log_a x$, a is the base. (Section 4.3)

bearing Bearing is used to identify angles in navigation. One method for expressing bearing uses a single angle, with bearing measured in a clockwise direction from due north. A second method for expressing bearing starts with a north-south line and uses an acute angle to show the direction, either east or west, from this line. (Section 5.4)

binomial A binomial is a polynomial containing exactly two terms. (Section R.3)

binomial coefficient For nonnegative integers n and r, with $r \le n$, the binomial coefficient is the value of $\frac{n!}{r!(n-r)!}$. Binomial coefficients are used in calculating the terms of a binomial expansion. (Section 11.4)

binomial experiment In probability, an experiment that consists of repeated independent trials with only two outcomes in each trial, success or failure, is called a binomial experiment. (Section 11.7)

binomial theorem (general binomial expansion) The binomial theorem is used to expand a binomial raised to a power. (Section 11.4)

boundary A line that separates a plane into two half-planes is called the boundary of each half-plane. (Section 9.6)

break-even point The break-even point is the point where the revenue from selling a product is equal to the cost of producing it. (Sections 1.7, 2.4)

cardioid A cardioid is a heart-shaped curve that is the graph of a polar equation of the form $r = a \pm b \sin \theta$ or $r = a \pm b \cos \theta$, where $\left| \frac{a}{b} \right| = 1$. (Section 8.7)

center of a circle The center of a circle is the given point that is a given distance from all points on the circle. (Section 2.2)

center of an ellipse The center of an ellipse is the midpoint of the major axis. (Section 10.2)

center of a hyperbola The center of a hyperbola is the midpoint of the transverse axis. (Section 10.3)

change in x The change in x is the horizontal difference (the difference in x-coordinates) between two points on a line. (Section 2.4)

change in y The change in y is the vertical difference (the difference in y-coordinates) between two points on a line. (Section 2.4)

circle A circle is the set of all points in a plane that lie a given distance from a given point. (Section 2.2)

circular functions The trigonometric functions of arc lengths, or real numbers, are called circular functions. (Section 6.2)

closed interval A closed interval is an interval that includes both of its endpoints. (Section 1.7)

coefficient (numerical coefficient) The real number in a term of an algebraic expression is called a coefficient. (Section R.3)

cofactor The product of a minor of an element of a square matrix and the number $+1$ (if the sum of the row number and column number is even) or -1 (if the sum of the row number and column number is odd) is called a cofactor. (Section 9.3)

cofunctions The function pairs sine and cosine, tangent and cotangent, and secant and cosecant are called cofunctions. (Section 5.3)

collinear Points are collinear if they lie on the same line. (Section 2.1)

column matrix A matrix with just one column is called a column matrix. (Section 9.7)

combination A subset of items selected *without regard to order* is called a combination. (Section 11.6)

combined variation Variation in which one variable depends on more than one other variable is called combined variation. (Section 3.6)

common difference In an arithmetic sequence, the fixed number that is added to each term to get the next term is called the common difference. (Section 11.2)

common logarithm A base 10 logarithm is called a common logarithm. (Section 4.4)

common ratio In a geometric sequence, the fixed number by which each term is multiplied to get the next term is called the common ratio. (Section 11.3)

complement of an event In probability, the set of all outcomes in a sample space that do *not* belong to an event E is called the complement of E, written E'. (Section 11.7)

complement of a set The set of all elements in the universal set that do not belong to set A is the complement of A, written A'. (Sections R.1, 11.7)

complementary angles (complements) Two positive angles are complementary angles (or complements) if the sum of their measures is 90°. (Section 5.1)

completing the square The process of adding to a binomial the number that makes it a perfect square trinomial is called completing the square. (Sections 1.4, 3.1)

complex conjugates Two complex numbers that differ only in the sign of their imaginary parts are called complex conjugates. (Section 1.3)

complex fraction A complex fraction is a quotient of two rational expressions. (Section R.5)

complex number A complex number is a number of the form $a + bi$, where a and b are real numbers and $i = \sqrt{-1}$. (Section 1.3)

complex plane The complex plane is a two-dimensional representation of the complex numbers in which the horizontal axis is the real axis and the vertical axis is the imaginary axis. (Section 8.5)

composite function (composition) If f and g are functions, then the composite function (or composition) of g and f is defined by $(g \circ f)(x) = g(f(x))$. (Section 2.8)

compound amount In an investment paying compound interest, the compound amount is the balance *after* interest has been earned. (The compound amount is sometimes called the *future value*.) (Section 4.2)

compound interest In compound interest, interest is paid both on the principal and previously earned interest. (Section 4.2)

conditional equation An equation that is satisfied by some numbers but not by others is called a conditional equation. (Section 1.1)

conic (conic section) (geometric definition) A conic is the set of all points $P(x, y)$ in a plane such that the ratio of the distance from P to a fixed point and the distance from P to a fixed line is constant. (Sections 10.2, 10.4)

conic sections (conics) When a plane intersects a double cone at different angles, the figures formed by the intersections are called conic sections. (Section 10.1)

conjugates The expressions $a - b$ and $a + b$ are called conjugates. (Section R.7)

consistent system A consistent system is a system of equations with at least one solution. (Section 9.1)

constant function A function f is constant on an interval I if, for every x_1 and x_2 in I, $f(x_1) = f(x_2)$. (Section 2.3)

constant of variation In a variation equation, such as $y = kx$, $y = \frac{k}{x}$, or $y = kx^n z^m$, the real number k is called the constant of variation. (Section 3.6)

constraints In linear programming, the inequalities that represent restrictions on a particular situation are called the constraints. (Section 9.6)

continuous compounding As the frequency of compounding increases, compound interest approaches a limit, called continuous compounding. (Section 4.2)

continuous function (informal definition) A function is continuous over an interval of its domain if its hand-drawn graph over that interval can be sketched without lifting the pencil from the paper. (Section 2.6)

contradiction An equation that has no solution is called a contradiction. (Section 1.1)

convergent sequence An infinite sequence is convergent if its terms get closer and closer to some real number. (Section 11.1)

coordinate (on a number line) A number that corresponds to a particular point on a number line is called the coordinate of the point. (Section R.2)

coordinate plane (*xy*-plane) The plane into which the rectangular coordinate system is introduced is called the coordinate plane (or *xy*-plane). (Section 2.1)

coordinates (in the *xy*-plane) The coordinates of a point in the *xy*-plane are the numbers in the ordered pair that correspond to that point. (Section 2.1)

coordinate system (on a number line) The correspondence between points on a line and the real numbers is called a coordinate system. (Section R.2)

cosecant Let $P(x, y)$ be a point other than the origin on the terminal side of an angle θ in standard position. Let $r = \sqrt{x^2 + y^2}$ represent the distance from the origin to P. Then the cosecant function is defined by $\csc \theta = \frac{r}{y} \, (y \neq 0)$. (Section 5.2)

cosine Let $P(x, y)$ be a point other than the origin on the terminal side of an angle θ in standard position. Let $r = \sqrt{x^2 + y^2}$ represent the distance from the origin to P. Then the cosine function is defined by $\cos \theta = \frac{x}{r}$. (Section 5.2)

cotangent Let $P(x, y)$ be a point other than the origin on the terminal side of an angle θ in standard position. Let $r = \sqrt{x^2 + y^2}$ represent the distance from the origin to P. Then the cotangent function is defined by $\cot \theta = \frac{x}{y} \, (y \neq 0)$. (Section 5.2)

coterminal angles Two angles that have the same initial side and the same terminal side, but different amounts of rotation, are called coterminal angles. The measures of coterminal angles differ by a multiple of 360°. (Section 5.1)

Cramer's rule Cramer's rule uses determinants to solve systems of linear equations. (Section 9.3)

cycloid A cycloid is a curve that represents the path traced by a fixed point on the circumference of a circle rolling along a line. (Section 8.8)

damped oscillatory motion Damped oscillatory motion is oscillatory motion that has been slowed down (damped) by the force of friction. Friction causes the amplitude of the motion to diminish gradually until the weight comes to rest. (Section 6.6)

decreasing function A function f is decreasing on an interval I if, whenever $x_1 < x_2$ in I, $f(x_1) > f(x_2)$. (Section 2.3)

degree The degree is the most common unit of measure for angles. One degree, written 1°, represents $\frac{1}{360}$ of a rotation. (Section 5.1)

degree of a polynomial The greatest degree of any term in a polynomial is called the degree of the polynomial. (Section R.3)

degree of a term The degree of a term is the sum of the exponents on the variables in the term. (Section R.3)

dependent equations Two equations are dependent if any solution of one equation is also a solution of the other. (Section 9.1)

dependent variable If the value of the variable y depends on the value of the variable x, then y is called the dependent variable. (Section 2.3)

determinant Every $n \times n$ matrix A is associated with a single real number called the determinant of A, written $|A|$. (Section 9.3)

difference quotient If f is a function and h is a positive number, then the expression $\frac{f(x + h) - f(x)}{h}$ is called the difference quotient. (Section 2.8)

direction angle The positive angle between the x-axis and a position vector is the direction angle for the vector. (Section 8.3)

directrix A directrix is a fixed line that, together with a focus, is used to determine the points that form a conic section. (Sections 10.1, 10.4)

discriminant The quantity under the radical in the quadratic formula, $b^2 - 4ac$, is called the discriminant. (Section 1.4)

disjoint sets Two sets that have no elements in common are called disjoint sets. (Section R.1)

divergent sequence If an infinite sequence does not converge to some number, then it is called a divergent sequence. (Section 11.1)

division algorithm The division algorithm is a method for dividing polynomials that is similar to that used for long division of whole numbers. The division algorithm can be stated as follows: Let $f(x)$ and $g(x)$ be polynomials with $g(x)$ of lesser degree than $f(x)$ and $g(x)$ of degree one or more. There exist unique polynomials $q(x)$ and $r(x)$ such that $f(x) = g(x) \cdot q(x) + r(x)$, where either $r(x) = 0$ or the degree of $r(x)$ is less than the degree of $g(x)$. (Section 3.2)

domain of a rational expression The domain of a rational expression is the set of real numbers for which the expression is defined. (Section R.5)

domain of a relation In a relation, the set of all values of the independent variable (x) is called the domain. (Section 2.3)

dominating term In a polynomial function, the dominating term is the term of greatest degree. (Section 3.4)

dot product The dot product of two vectors is the sum of the product of their first components and the product of their second components. The dot product of the two vectors $\mathbf{u} = \langle a, b \rangle$ and $\mathbf{v} = \langle c, d \rangle$ is denoted $\mathbf{u} \cdot \mathbf{v}$ and given by $\mathbf{u} \cdot \mathbf{v} = ac + bd$. (Section 8.3)

doubling time The amount of time that it takes for a quantity that grows exponentially to become twice its initial amount is called its doubling time. (Section 4.6)

eccentricity The eccentricity of a parabola, ellipse, or hyperbola is the fixed ratio of the distance from a point P to a focus compared to the distance from the same point to the directrix. The eccentricity of a circle is 0. (Section 10.2)

element of a matrix Each number in a matrix is called an element of the matrix. (Section 9.2)

elements (members) The objects that belong to a set are called the elements (members) of the set. (Section R.1)

ellipse An ellipse is the set of all points in a plane the sum of whose distances from two fixed points is constant. (Section 10.2)

empty set (null set) The empty set (or null set), written $\emptyset$ or { }, is the set containing no elements. (Sections R.1, 1.1)

end behavior The end behavior of a graph of a polynomial function describes what happens to the values of y as x gets larger and larger in absolute value. (Section 3.4)

endpoint of a ray In a given ray AB, point A is the endpoint of the ray. (Section 5.1)

equation An equation is a statement that two expressions are equal. (Section 1.1)

equilibrant The opposite vector of the resultant of two vectors is called the equilibrant. (Section 8.4)

equivalent equations Equations with the same solution set are called equivalent equations. (Section 1.1)

equivalent systems Equivalent systems are systems of equations that have the same solution set. (Section 9.1)

even function A function f is called an even function if $f(-x) = f(x)$ for all x in the domain of f. (Section 2.7)

event In probability, any subset of the sample space is called an event. (Section 11.7)

exact number A number that represents the result of counting, or a number that results from theoretical work and is not the result of a measurement, is an exact number. (Section 5.4)

expansion by a row or column Expansion by a row or column is a method for evaluating a determinant of a 3×3 or larger matrix. This process involves multiplying each element of any row or column of the matrix by its cofactor and then adding these products. (Section 9.3)

exponent An exponent is a number that indicates how many times a factor is repeated. In the expression a^n, n is the exponent. (Section R.2)

exponential equation An exponential equation is an equation with a variable in an exponent. (Section 4.2)

exponential function If $a > 0$ and $a \neq 1$, then $f(x) = a^x$ defines the exponential function with base a. (Section 4.2)

exponential growth or decay function An exponential growth or decay function models a situation in which a quantity changes at a rate proportional to the amount present. Such a function is defined by an equation of the form $y = y_0 e^{kt}$, where y_0 is the amount or number present at time $t = 0$ and k is a constant. (Section 4.6)

factored completely A polynomial is factored completely when it is written as a product of prime polynomials. (Section R.4)

factored form A polynomial is written in factored form when it is written as a product of prime polynomials. (Section R.4)

factorial notation Factorial notation is a compact way of writing a product of consecutive natural numbers. The symbol $n!$, read "n-factorial," is defined as follows: For any positive integer n, $n! = n(n-1)(n-2)\cdots(3)(2)(1)$, and $0! = 1$. (Section 11.4)

factoring The process of finding polynomials whose product equals a given polynomial is called factoring. (Section R.4)

factoring by grouping Factoring by grouping is a method of grouping the terms of a polynomial in such a way that the polynomial can be factored even though the greatest common factor of its terms is 1. (Section R.4)

Fibonacci sequence The Fibonacci sequence is the sequence 1, 1, 2, 3, 5, 8, 13, ... In this sequence, each term starting with the third term is the sum of the previous two terms. (Section 11.1)

finite sequence A sequence is a finite sequence if its domain is the set $\{1, 2, 3, \ldots, n\}$, where n is a natural number. (Section 11.1)

finite series A finite series is an expression of the form $S_n = a_1 + a_2 + a_3 + \cdots + a_n = \sum_{i=1}^{n} a_i$. (Section 11.1)

finite set A finite set is a set that has a limited number of elements. (Section R.1)

foci (singular, **focus**) Foci are fixed points used to determine the points that form a parabola, an ellipse, or a hyperbola. A parabola has one focus, while an ellipse or a hyperbola has two foci. (Sections 10.1, 10.2, 10.3)

four-leaved rose A four-leaved rose is a curve that is the graph of a polar equation of the form $r = a \sin 2\theta$ or $r = a \cos 2\theta$. (Section 8.7)

frequency In simple harmonic motion, the frequency is the number of cycles per unit of time, or the reciprocal of the period. (Section 6.6)

function A function is a relation (set of ordered pairs) in which, for each value of the first component of the ordered pairs, there is *exactly one* value of the second component. (Section 2.3)

function notation Function notation $f(x)$ (read "f of x") represents the y-value of the function f for the indicated x-value. (Section 2.3)

fundamental rectangle The fundamental rectangle is used as a guide in sketching the graph of a hyperbola. The extended diagonals of this rectangle are the asymptotes of the hyperbola. (Section 10.3)

future value In an investment paying compound interest, the future value is the balance *after* interest has been earned. (The future value is sometimes called the *compound amount*.) (Section 4.2)

future value of an annuity If the payments on an annuity are accumulated in an account (with no withdrawals), then the sum of the payments and interest on the payments is called the future value of the annuity. (Section 11.3)

general term (nth term) In the sequence $a_1, a_2, a_3, \ldots$, the general term (or nth term) is a_n. (Section 11.1)

geometric sequence (geometric progression) A geometric sequence is a sequence in which each term after the first is obtained by multiplying the preceding term by a constant nonzero real number. (Section 11.3)

geometric series A geometric series is the sum of the terms of a geometric sequence. (Section 11.3)

graph of an equation The graph of an equation is the set of all points that correspond to all of the ordered pairs that satisfy the equation. (Section 2.1)

groundspeed In air navigation, the groundspeed of a plane is its speed relative to the ground. (Section 8.4)

half-life The amount of time that it takes for a quantity that decays exponentially to become half its initial amount is called its half-life. (Section 4.6)

half-plane A line separates a plane into two regions, each of which is called a half-plane. (Section 9.6)

horizontal asymptote A horizontal line that a graph approaches as $|x|$ gets larger and larger without bound is called a horizontal asymptote. The line $y = b$ is a horizontal asymptote if $y \to b$ as $|x| \to \infty$. (Section 3.5)

horizontal component When a vector **u** is expressed as an ordered pair in the form $\mathbf{u} = \langle a, b \rangle$, the number a is the horizontal component of the vector. (Section 8.3)

hyperbola A hyperbola is the set of all points in a plane such that the absolute value of the difference of the distances from two fixed points is constant. (Section 10.3)

identity An equation satisfied by every number that is a meaningful replacement for the variable is called an identity. (Section 1.1)

identity matrix (multiplicative identity matrix) An identity matrix is an $n \times n$ matrix with 1s on the main diagonal and 0s everywhere else. (Section 9.8)

imaginary axis In the complex plane, the vertical axis is called the imaginary axis. (Section 8.5)

imaginary part In the complex number $a + bi$, b is called the imaginary part. (Section 1.3)

imaginary unit The number i, defined by $i^2 = -1$ $\left(\text{so } i = \sqrt{-1}\right)$, is called the imaginary unit. (Section 1.3)

inconsistent system An inconsistent system is a system of equations with no solution. (Section 9.1)

increasing function A function f is increasing on an interval I if, whenever $x_1 < x_2$ in I, $f(x_1) < f(x_2)$. (Section 2.3)

independent events Two events are independent events if neither influences the outcomes of the other. (Section 11.6)

independent variable If the value of the variable y depends on the value of the variable x, then x is called the independent variable. (Section 2.3)

index of a radical In a radical of the form $\sqrt[n]{a}$, n is called the index. (Section R.7)

index of summation When using summation notation, such as $\sum\limits_{i=1}^{n} a_i$, the letter i is called the index of summation. Other letters may also be used, such as j and k. (Section 11.1)

inequality An inequality says that one expression is greater than, greater than or equal to, less than, or less than or equal to, another. (Sections R.2, 1.7)

infinite sequence A sequence is an infinite sequence if its domain is the set of *all* natural numbers. (Section 11.1)

infinite series An infinite series is an expression of the form $S_\infty = a_1 + a_2 + a_3 + \cdots + a_n + \cdots = \sum\limits_{i=1}^{\infty} a_i$. (Section 11.1)

infinite set An infinite set is a set that has an unending list of distinct elements. (Section R.1)

initial point When two letters are used to name a vector, the first letter indicates the initial (starting) point of the vector. (Section 8.3)

initial side When a ray is rotated around its endpoint to form an angle, the ray in its initial position is called the initial side of the angle. (Section 5.1)

integers The set of integers is $\{\ldots -3, -2, -1, 0, 1, 2, 3, \ldots\}$. (Section R.2)

intersection The intersection of sets A and B, written $A \cap B$, is the set of elements that belong to both A and B. (Sections R.1, 11.7)

interval An interval is a portion of the real number line, which may or may not include its endpoint(s). (Section 1.7)

interval notation Interval notation is a simplified notation for writing intervals. It uses parentheses and brackets to show whether the endpoints are included. (Section 1.7)

inverse function Let f be a one-to-one function. Then g is the inverse function of f if $(f \circ g)(x) = x$ for every x in the domain of g, and $(g \circ f)(x) = x$ for every x in the domain of f. (Section 4.1)

irrational numbers Real numbers that cannot be represented as quotients of integers are called irrational numbers. (Section R.2)

leading coefficient In a polynomial function of degree n, the leading coefficient is a_n, that is, the coefficient of the term of greatest degree. (Section 3.1)

lemniscate A lemniscate is a figure-eight-shaped curve that is the graph of a polar equation of the form $r^2 = a^2 \sin 2\theta$ or $r^2 = a^2 \cos 2\theta$. (Section 8.7)

like radicals Radicals with the same radicand and the same index are called like radicals. (Section R.7)

like terms Terms with the same variables each raised to the same powers are called like terms. (Section R.3)

limaçon A limaçon is the graph of a polar equation of the form $r = a \pm b \sin \theta$ or $r = a \pm b \cos \theta$. If $\left|\dfrac{a}{b}\right| = 1$, the limaçon is a cardioid. (Section 8.7)

line Two distinct points A and B determine a line called line AB. (Section 5.1)

line segment (segment) Line segment AB (or segment AB) is the portion of line AB between A and B, including A and B themselves. (Section 5.1)

linear cost function A linear cost function is a linear function that has the form $C(x) = mx + b$, where x represents the number of items produced, m represents the variable cost per item, and b represents the fixed cost. (Section 2.4)

linear equation (first-degree equation) in n unknowns Any equation of the form $a_1 x_1 + a_2 x_2 + \cdots + a_n x_n = b$, for real numbers $a_1, a_2, \ldots, a_n$ (not all of which are 0) and b, is a linear equation (or a first-degree equation) in n unknowns. (Section 9.1)

linear equation (first-degree equation) in one variable A linear equation in one variable is an equation that can be written in the form $ax + b = 0$, where a and b are real numbers with $a \neq 0$. (Section 1.1)

linear function A function f is a linear function if $f(x) = ax + b$ for real numbers a and b. (Section 2.4)

linear inequality in one variable A linear inequality in one variable is an inequality that can be written in the form $ax + b > 0$, where a and b are real numbers with $a \neq 0$. (Any of the symbols $<$, $\geq$, or $\leq$ may also be used.) (Section 1.7)

linear inequality in two variables A linear inequality in two variables is an inequality of the form $Ax + By \leq C$, where A, B, and C are real numbers with A and B not both equal to 0. (Any of the symbols $\geq$, $<$, or $>$ may also be used.) (Section 9.6)

linear programming Linear programming, an application of mathematics to business and social science, is a method for finding an optimum value, such as minimum cost or maximum profit. (Section 9.6)

linear speed v The linear speed v measures the distance traveled per unit of time. (Section 6.2)

literal equation A literal equation is an equation that relates two or more variables (letters). (Section 1.1)

logarithm A logarithm is an exponent. The expression $\log_a x$ is the power to which the base a must be raised to obtain x. (Section 4.3)

logarithmic equation A logarithmic equation is an equation with a logarithm in at least one term. (Section 4.3)

logarithmic function If $a > 0$, $a \neq 1$, and $x > 0$, then $f(x) = \log_a x$ defines the logarithmic function with base a. (Section 4.3)

lowest terms A rational expression is in lowest terms when the greatest common factor of its numerator and its denominator is 1. (Section R.5)

M

magnitude The length of a vector represents the magnitude of the vector quantity. (Section 8.3)

major axis The major axis of an ellipse is its longer axis of symmetry. (Section 10.2)

mathematical induction Mathematical induction is a method for proving that a statement S_n is true for every positive integer n. (Section 11.5)

mathematical model A mathematical model is an equation (or inequality) that describes the relationship between two or more quantities. (Section 1.2)

matrix (plural, **matrices**) A matrix is a rectangular array of numbers enclosed in brackets. (Section 9.2)

minor In an $n \times n$ matrix with $n \geq 3$, the minor of a particular element is the determinant of the $(n - 1) \times (n - 1)$ matrix that results when the row and column that contain the chosen element are eliminated. (Section 9.3)

minor axis The minor axis of an ellipse is its shorter axis of symmetry. (Section 10.2)

minute One minute, written $1'$, is $\frac{1}{60}$ of a degree. (Section 5.1)

monomial A monomial is a polynomial containing only one term. (Section R.3).

multiplicative inverse of a matrix If A is an $n \times n$ matrix, then its multiplicative inverse, written A^{-1}, must satisfy both $AA^{-1} = I_n$ and $A^{-1}A = I_n$. (Section 9.8)

mutually exclusive events In probability, two events that cannot occur simultaneously are called mutually exclusive events. (Section 11.7)

N

natural logarithm A base e logarithm is called a natural logarithm. (Section 4.4)

natural numbers (counting numbers) The natural numbers or counting numbers form the set of numbers $\{1, 2, 3, 4, \ldots\}$. (Sections R.1, R.2)

negative angle A negative angle is an angle that is formed by clockwise rotation around its endpoint. (Section 5.1)

nonlinear system A system of equations in which at least one equation is *not* linear is called a nonlinear system. (Section 9.5)

nonreal complex number A complex number $a + bi$ in which $b \neq 0$ is called a nonreal complex number. (Section 1.3)

nonstrict inequality An inequality in which the symbol is either $\leq$ or $\geq$ is called a nonstrict inequality. (Section 1.7)

nth root of a complex number For a positive integer n, the complex number $a + bi$ is an nth root of the complex number $x + yi$ if $(a + bi)^n = x + yi$. (Section 8.6)

O

objective function In linear programming, the function to be maximized or minimized is called the objective function. (Section 9.6)

oblique asymptote A nonvertical, nonhorizontal line that a graph approaches as $|x|$ gets larger and larger without bound is called an oblique asymptote. (Section 3.5)

oblique triangle A triangle that is not a right triangle is called an oblique triangle. (Section 8.1)

obtuse angle An obtuse angle is an angle measuring more than 90° but less than 180°. (Section 5.1)

odd function A function f is called an odd function if $f(-x) = -f(x)$ for all x in the domain of f. (Section 2.7)

odds The odds in favor of an event are expressed as the ratio of the probability of the event to the probability of the complement of the event. (Section 11.7)

one-to-one function In a one-to-one function, each x-value corresponds to only one y-value, and each y-value corresponds to only one x-value. (Section 4.1)

open interval An open interval is an interval that does not include its endpoint(s). (Section 1.7)

opposite of a vector The opposite of a vector $\mathbf{v}$ is a vector $-\mathbf{v}$ that has the same magnitude as $\mathbf{v}$ but opposite direction. (Section 8.3)

ordered pair An ordered pair consists of two components, written inside parentheses, in which the order of the components is important. Ordered pairs are used to identify points in the coordinate plane. (Section 2.1)

ordered triple An ordered triple consists of three components, written inside parentheses, in which the order of the components is important. Ordered triples are used to identify points in space. (Section 9.1)

origin The point of intersection of the x-axis and the y-axis of a rectangular coordinate system is called the origin. (Section 2.1)

orthogonal vectors Orthogonal vectors are vectors that are perpendicular, that is, the angle between the two vectors is 90°. (Section 8.3)

outcome In probability, a possible result of each trial in an experiment is called an outcome of the experiment. (Section 11.7)

P

parabola A parabola is a curve that is the graph of a quadratic function. (Sections 2.5, 3.1)

parabola (geometric definition) A parabola is the set of all points in a plane equidistant from a fixed point (called the focus) and a fixed line (called the directrix). (Section 10.1)

parallelogram rule The parallelogram rule is a way to find the sum of two vectors. If the two vectors are placed so that their initial points coincide and a parallelogram is completed that has these two vectors as two of its sides, then the diagonal vector of the parallelogram that has the same initial point as the two vectors is their sum. (Section 8.3)

parameter A parameter is a variable in terms of which two or more other variables are expressed. In a pair of parametric equations $x = f(t)$ and $y = g(t)$, the variable t is the parameter. (Section 8.8)

parametric equations of a plane curve A pair of equations $x = f(t)$ and $y = g(t)$ are parametric equations of a plane curve. (Section 8.8)

partial fraction Each term in the decomposition of a rational expression is called a partial fraction. (Section 9.4)

partial fraction decomposition When one rational expression is expressed as the sum of two or more rational expressions, the sum is called the partial fraction decomposition. (Section 9.4)

Pascal's triangle Pascal's triangle is a triangular array of numbers that is helpful in expanding binomials. The numbers in the triangle are the binomial coefficients. (Section 11.4)

period For a periodic function such that $f(x) = f(x + np)$, the smallest possible positive value of p is the period of the function. (Section 6.3)

periodic function A periodic function is a function f such that $f(x) = f(x + np)$, for every real number x in the domain of f, every integer n, and some positive real number p. (Section 6.3)

permutation A permutation of n elements taken r at a time is one of the *arrangements* of r elements from a set of n elements. (Section 11.6)

phase shift With trigonometric functions, a horizontal translation is called a phase shift. (Section 6.4)

piecewise-defined function A piecewise-defined function is a function that is defined by different rules over different parts of its domain. (Section 2.6)

plane curve A plane curve is a set of points (x, y) such that $x = f(t)$ and $y = g(t)$, and f and g are both defined on an interval I. (Section 8.8)

point-slope form The point-slope form of the equation of the line with slope m through the point (x_1, y_1) is $y - y_1 = m(x - x_1)$. (Section 2.5)

polar axis The polar axis is a specific ray in the polar coordinate system that has the pole as its endpoint. The polar axis is usually drawn in the direction of the positive x-axis. (Section 8.7)

polar coordinates In the polar coordinate system, the ordered pair (r, θ) gives polar coordinates of point P, where r is the directed distance from the pole to P and θ is the directed angle from the positive x-axis to ray OP. (Section 8.7)

polar coordinate system The polar coordinate system is a coordinate system based on a point (the pole) and a ray (the polar axis). (Section 8.7)

polar equation A polar equation is an equation that uses polar coordinates. The variables are r and θ. (Section 8.7)

pole The pole is the single fixed point in the polar coordinate system that is the endpoint of the polar axis. The pole is usually placed at the origin of a rectangular coordinate system. (Section 8.7)

polynomial A polynomial is a term or a finite sum of terms, with only positive or zero integer exponents permitted on the variables. (Section R.3)

polynomial function of degree n A polynomial function of degree n, where n is a nonnegative integer, is a function defined by an expression of the form $f(x) = a_n x^n + a_{n-1} x^{n-1} + \cdots + a_1 x + a_0$, where $a_n, a_{n-1}, \ldots, a_1$, and a_0 are real numbers, with $a_n \neq 0$. (Section 3.1)

position vector A vector with its initial point at the origin is called a position vector. (Section 8.3)

positive angle A positive angle is an angle that is formed by counterclockwise rotation around its endpoint. (Section 5.1)

power (exponential expression, exponential) An expression of the form a^n is called a power, an exponential expression, or an exponential. (Section R.2)

present value In an investment paying compound interest, the principal is sometimes called the present value. (Section 4.2)

prime polynomial A polynomial with variable terms that cannot be written as a product of two polynomials of lesser degree is a prime polynomial. (Section R.4)

principal nth root For even values of n (square roots, fourth roots, and so on), when a is positive, there are two real nth roots, one positive and one negative. In such cases, the notation $\sqrt[n]{a}$ represents the positive root, or principal nth root. (Section R.7)

probability of an event In a sample space with equally likely outcomes, the probability of an event is the ratio of the number of outcomes in the sample space that belong to the event to the number of outcomes in the sample space. (Section 11.7)

pure imaginary number A complex number $a + bi$ in which $a = 0$ and $b \neq 0$ is called a pure imaginary number. (Section 1.3)

Pythagorean theorem The Pythagorean theorem states that in a right triangle, the sum of the squares of the lengths of the legs is equal to the square of the length of the hypotenuse. (Section 1.5)

Q

quadrantal angle A quadrantal angle is an angle that, when placed in standard position, has its terminal side along the x-axis or the y-axis. (Section 5.1)

quadrants The quadrants are the four regions into which the x-axis and y-axis divide the coordinate plane. (Section 2.1)

quadratic equation (second-degree equation) An equation that can be written in the form $ax^2 + bx + c = 0$, where a, b, and c are real numbers with $a \neq 0$, is a quadratic equation. (Section 1.4)

quadratic in form An equation is said to be quadratic in form if it can be written as $au^2 + bu + c = 0$, where $a \neq 0$ and u is some algebraic expression. (Section 1.6)

quadratic formula The quadratic formula $x = \dfrac{-b \pm \sqrt{b^2 - 4ac}}{2a}$ is a general formula that can be used to solve any quadratic equation. (Section 1.4)

quadratic function A function f is a quadratic function if $f(x) = ax^2 + bx + c$, where a, b, and c are real numbers, with $a \neq 0$. (Section 3.1)

quadratic inequality A quadratic inequality is an inequality that can be written in the form $ax^2 + bx + c < 0$, for real numbers a, b, and c with $a \neq 0$. (The symbol $<$ can be replaced with $>$, $\leq$, or $\geq$.) (Section 1.7)

R

radian A radian is a unit of measure for angles. An angle with its vertex at the center of a circle that intercepts an arc on the circle equal in length to the radius of the circle has a measure of 1 radian. (Section 6.1)

radicand The number or expression under a radical sign is called the radicand. (Section R.7)

radius The radius of a circle is the given distance between the center and any point on the circle. (Section 2.2)

range In a relation, the set of all values of the dependent variable (y) is called the range. (Section 2.3)

rational equation A rational equation is an equation that has a rational expression for one or more of its terms. (Section 1.6)

rational expression The quotient of two polynomials P and Q, with $Q \neq 0$, is called a rational expression. (Section R.5)

rational function A function f of the form $f(x) = \frac{p(x)}{q(x)}$, where $p(x)$ and $q(x)$ are polynomials, with $q(x) \neq 0$, is called a rational function. (Section 3.5)

rational inequality A rational inequality is an inequality in which one or both sides contain rational expressions. (Section 1.7)

rationalizing the denominator Rationalizing a denominator is a way of simplifying a radical expression so that there are no radicals in the denominator. (Section R.7)

rational numbers The rational numbers are the set of numbers $\frac{p}{q}$, where p and q are integers and $q \neq 0$. (Section R.2)

ray The portion of line AB that starts at A and continues through B, and on past B, is called ray AB. (Section 5.1)

real axis In the complex plane, the horizontal axis is called the real axis. (Section 8.5)

real numbers The set of all numbers that correspond to points on a number line is called the real numbers. (Section R.2)

real part In the complex number $a + bi$, a is called the real part. (Section 1.3)

reciprocal function The function defined by $f(x) = \frac{1}{x}$ is called the reciprocal function. (Section 3.5)

rectangular (Cartesian) coordinate system The x-axis and y-axis together make up a rectangular coordinate system. (Section 2.1)

rectangular (Cartesian) equation A rectangular (or Cartesian) equation is an equation that uses rectangular coordinates. If it is an equation in two variables, the variables are x and y. (Section 8.7)

rectangular form (standard form) of a complex number The rectangular form (or standard form) of a complex number is $a + bi$, where a and b are real numbers. (Section 8.5)

recursive definition A sequence is defined by a recursive definition if each term after the first term or first few terms is defined as an expression involving the previous term or terms. (Section 11.1)

reference angle The reference angle for an angle θ, written θ', is the positive acute angle made by the terminal side of angle θ and the x-axis. (Section 5.3)

reference arc The reference arc for a point on the unit circle is the shortest arc from the point itself to the nearest point on the x-axis. (Section 6.2)

region of feasible solutions In linear programming, the region of feasible solutions is the region of the graph that satisfies all of the constraints. (Section 9.6)

relation A relation is a set of ordered pairs. (Section 2.3)

resultant If $\mathbf{A}$ and $\mathbf{B}$ are vectors, the vector sum $\mathbf{A} + \mathbf{B}$ is called the resultant of vectors $\mathbf{A}$ and $\mathbf{B}$. (Section 8.3)

right angle A right angle is an angle measuring exactly 90°. (Section 5.1)

root (solution) of an equation A zero of $f(x)$ is called a root (or solution) of the equation $f(x) = 0$. (Section 3.2)

rose curve A rose curve is a member of a family of curves that resemble flowers. It is the graph of a polar equation of the form $r = a \sin n\theta$ or $r = a \cos n\theta$. (Section 8.7)

rotation of axes If the axes in an xy-coordinate system having origin O are rotated about O through an angle θ, the new coordinate system is called a rotation of the xy-system. (Appendix B)

row matrix A matrix with just one row is called a row matrix. (Section 9.7)

S

sample space In probability, the set of all possible outcomes of a given experiment is called the sample space. (Section 11.7)

scalar In work with matrices or vectors, a real number is called a scalar to distinguish it from a matrix or vector. (Sections 8.3, 9.7)

scalar product The scalar product of a real number (or scalar) k and a vector $\mathbf{u}$ is the vector $k \cdot \mathbf{u}$, which has magnitude $|k|$ times the magnitude of $\mathbf{u}$. (Section 8.3)

scatter diagram A scatter diagram is a graph of specific ordered pairs of data. (Section 2.5)

secant Let $P(x, y)$ be a point other than the origin on the terminal side of an angle θ in standard position. Let $r = \sqrt{x^2 + y^2}$ represent the distance from the origin to P. Then the secant function is defined by $\sec \theta = \frac{r}{x}$ ($x \neq 0$). (Section 5.2)

second One second, written $1''$, is $\frac{1}{60}$ of a minute. (Section 5.1)

sector of a circle A sector of a circle is the portion of the interior of a circle intercepted by a central angle. (Section 6.1)

semiperimeter The semiperimeter is half the sum of the lengths of the three sides of a triangle. (Section 8.2)

sequence A sequence is a function that has a set of natural numbers of the form $\{1, 2, 3, \ldots, n\}$ or $\{1, 2, 3, \ldots, n, \ldots\}$ as its domain. (Section 11.1)

series A series is the sum of the terms of a sequence. (Section 11.1)

set A set is a collection of objects. (Section R.1)

set-builder notation Set-builder notation uses the form $\{x \mid x$ has a certain property$\}$ to describe a set without having to list all of it elements. (Section R.1)

set operations The processes of finding the complement of a set, the intersection of two sets, and the union of two sets are called set operations. (Section R.1)

side of an angle One of the two rays (or line segments) with a common endpoint that form an angle is called a side of the angle. (Section 5.1)

Side-Angle-Side (SAS) The Side-Angle-Side (SAS) congruence axiom states that if two sides and the included angle of one triangle are equal, respectively, to two sides and the included angle of a second triangle, then the triangles are congruent. (Section 8.1)

Side-Side-Side (SSS) The Side-Side-Side (SSS) congruence axiom states that if three sides of one triangle are equal, respectively, to three sides of a second triangle, then the triangles are congruent. (Section 8.1)

significant digit A significant digit is a digit obtained by actual measurement. (Section 5.4)

simple harmonic motion Simple harmonic motion is oscillatory motion about an equilibrium position. If friction is neglected, this motion can be described by a sinusoid. (Section 6.5)

simple interest In simple interest, interest is paid only on the principal, not on previously earned interest. (Section 1.1)

sine Let $P(x, y)$ be a point other than the origin on the terminal side of an angle θ in standard position. Let $r = \sqrt{x^2 + y^2}$ represent the distance from the origin to P. Then the sine function is defined by $\sin \theta = \frac{y}{r}$. (Section 5.2)

sine wave (sinusoid) The graph of a sine function is called a sine wave (or sinusoid). (Section 6.3)

size (order, dimension) of a matrix The size of a matrix indicates the number of rows and columns that the matrix has, with the number of rows given first. (Section 9.2)

slope The slope of a nonvertical line is the ratio of the change in y to the change in x. (Section 2.4)

slope-intercept form The slope-intercept form of the equation of the line with slope m and y-intercept b is $y = mx + b$. (Section 2.5)

solution (root) A solution (or root) of an equation is a number that makes the equation a true statement. (Section 1.1)

solution set The solution set of an equation is the set of all numbers that satisfy the equation. (Section 1.1)

solutions of a system of equations The solutions of a system of equations must satisfy every equation in the system. (Section 9.1).

spiral of Archimedes A spiral of Archimedes is an infinite curve that is the graph of a polar equation of the form $r = n\theta$. (Section 8.7)

square matrix An $n \times n$ matrix, that is, a matrix with the same number of columns as rows, is called a square matrix. (Section 9.7)

standard form of a complex number A complex number written in the form $a + bi$ (or $a + ib$) is in standard form. (Section 1.3)

standard form of a linear equation The form $Ax + By = C$ is called the standard form of a linear equation. (Section 2.4)

step function A step function is a function whose graph looks like a series of steps. (Section 2.6)

straight angle A straight angle is an angle measuring exactly 180°. (Section 5.1)

strict inequality An inequality in which the symbol is either $<$ or $>$ is called a strict inequality. (Section 1.7)

subset If every element of set A is also an element of set B, then A is a subset of B, written $A \subseteq B$. (Section R.1)

summation notation Summation notation is a compact way of writing a series using the general term of the corresponding sequence and the symbol Σ, the Greek capital letter sigma. (Section 11.1)

supplementary angles (supplements) Two positive angles are supplementary angles (or supplements) if the sum of their measures is 180°. (Section 5.1)

synthetic division Synthetic division is a shortcut method of dividing a polynomial by a binomial of the form $x - k$. (Section 3.2)

system of equations A set of equations that are considered at the same time is called a system of equations. (Section 9.1)

system of inequalities A set of inequalities that are considered at the same time is called a system of inequalities. (Section 9.6)

system of linear equations (linear system) If all the equations in a system are linear, then the system is a system of linear equations, or a linear system. (Section 9.1)

tangent Let $P(x, y)$ be a point other than the origin on the terminal side of an angle θ in standard position. Let $r = \sqrt{x^2 + y^2}$ represent the distance from the origin to P. Then the tangent function is defined by $\tan \theta = \frac{y}{x}$ $(x \neq 0)$. (Section 5.2)

term The product of a real number and one or more variables raised to powers is called a term. (Section R.3)

terminal point When two letters are used to name a vector, the second letter indicates the terminal (ending) point of the vector. (Section 8.3)

terminal side When a ray is rotated around its endpoint to form an angle, the ray in its location after rotation is called the terminal side of the angle. (Section 5.1)

terms of a sequence The elements in the range of a sequence are called the terms of the sequence. (Section 11.1)

translation A translation is a horizontal or vertical shift of a graph. (Section 2.7)

transverse axis The line segment that has the vertices of a hyperbola as endpoints is called the transverse axis of the hyperbola. (Section 10.3)

tree diagram A tree diagram is a diagram with branches that is used to systematically list all the outcomes of a counting situation or probability experiment. (Section 11.6)

trigonometric (polar) form of a complex number The expression $r(\cos \theta + i \sin \theta)$ is called the trigonometric form (or polar form) of the complex number $x + yi$. The expression $\cos \theta + i \sin \theta$ is sometimes abbreviated as cis θ. (Section 8.5)

trinomial A trinomial is a polynomial containing exactly three terms. (Section R.3)

turning points The points on the graph of a function where the function changes from increasing to decreasing or from decreasing to increasing are called turning points. (Section 3.4)

union The union of sets A and B, written $A \cup B$, is the set of all elements that belong to set A or set B (or both). (Sections R.1, 1.7, 11.7)

unit circle The unit circle is the circle with center at the origin and radius 1. (Section 6.2)

unit vector A unit vector is a vector that has magnitude 1. Two useful unit vectors are $\mathbf{i} = \langle 1, 0 \rangle$ and $\mathbf{j} = \langle 0, 1 \rangle$. (Section 8.3)

universal set The universal set, written U, contains all the elements appearing in any set used in a given problem. (Section R.1)

V

varies directly (directly proportional to) y varies directly as x, or y is directly proportional to x, if there exists a nonzero real number k such that $y = kx$. (Section 3.6)

varies inversely (inversely proportional to) y varies inversely as x, or y is inversely proportional to x, if there exists a nonzero real number k such that $y = \frac{k}{x}$. (Section 3.6)

varies jointly In joint variation, a variable depends on the product of two or more other variables. If m and n are real numbers, then y varies jointly as the nth power of x and the mth power of z if there exists a nonzero real number k such that $y = kx^nz^m$. (Section 3.6)

vector A vector is a directed line segment that represents a vector quantity. (Section 8.3)

vector quantities Quantities that involve both magnitude and direction are called vector quantities. (Section 8.3)

Venn diagram A Venn diagram is a diagram used to illustrate relationships among sets or probability concepts. (Sections R.1, 11.7)

vertex (corner point) In linear programming, a vertex (or corner) point is one of the vertices of the region of feasible solutions. (Section 9.6)

vertex of an angle The vertex of an angle is the endpoint of the ray that is rotated to form the angle. (Section 5.1)

vertex of a parabola The vertex of a parabola is the point where the axis of symmetry intersects the parabola. This is the turning point of the parabola. (Sections 2.6, 3.1)

vertical asymptote A vertical line that a graph approaches, but never touches or intersects, is called a vertical asymptote. The line $x = a$ is a vertical asymptote for a function f if $|f(x)| \to \infty$ as $x \to a$. (Section 3.5)

vertical component When a vector **u** is expressed as an ordered pair in the form $\mathbf{u} = \langle a, b \rangle$, the number b is the vertical component of the vector. (Section 8.3)

vertices of an ellipse The vertices of an ellipse are the endpoints of the major axis. (Section 10.2)

vertices of a hyperbola The vertices of a hyperbola are the two points on the hyperbola that are closest to the center. (Section 10.3)

whole numbers The set of whole numbers $\{0, 1, 2, 3, 4, \ldots\}$ is formed by combining the set of natural numbers and the number 0. (Section R.2)

x-axis The horizontal number line in a rectangular coordinate system is called the x-axis. (Section 2.1)

x-intercept An x-intercept is an x-value of a point where the graph of an equation intersects the x-axis. (Section 2.1)

y-axis The vertical number line in a rectangular coordinate system is called the y-axis. (Section 2.1)

y-intercept A y-intercept is a y-value of a point where the graph of an equation intersects the y-axis. (Section 2.1)

Z

zero-factor property The zero-factor property states that if the product of two (or more) complex numbers is 0, then at least one of the numbers must be 0. (Section 1.4)

zero matrix A matrix containing only zero elements is called a zero matrix. (Section 9.7)

zero of multiplicity n A polynomial function has a zero k of multiplicity n if the zero occurs n times, that is, the polynomial has n factors of $x - k$. (Section 3.3)

zero polynomial The function defined by $f(x) = 0$ is called the zero polynomial. (Section 3.1)

zero of a polynomial function A zero of a polynomial function f is a number k such that $f(k) = 0$. (Section 3.2)

zero vector The zero vector is the vector with magnitude 0. (Section 8.3)

Solutions to Selected Exercises

CHAPTER R   REVIEW OF BASIC CONCEPTS

R.2 Exercises *(pages 17–21)*

55. No; in general $a - b \neq b - a$. *Examples:*

$a = 15, b = 0: a - b = 15 - 0 = 15$, but
$b - a = 0 - 15 = -15$.

$a = 12, b = 7: a - b = 12 - 7 = 5$, but
$b - a = 7 - 12 = -5$.

$a = -6, b = 4: a - b = -6 - 4 = -10$, but
$b - a = 4 - (-6) = 10$.

$a = -18, b = -3: a - b = -18 - (-3) = -15$, but
$b - a = -3 - (-18) = 15$.

67. The process in your head should be like the following:

$$72 \cdot 17 + 28 \cdot 17 = 17(72 + 28)$$
$$= 17(100)$$
$$= 1700.$$

111. $\dfrac{x^3}{y} > 0$

The quotient of two numbers is positive if they have the same sign (both positive or both negative). The sign of x^3 is the same as the sign of x. Therefore, $\dfrac{x^3}{y} > 0$ if x and y have the same sign.

R.3 Exercises *(pages 30–33)*

23. $-(4m^3n^0)^2 = -[4^2(m^3)^2(n^0)^2]$ Power Rule 2

$= -(4^2)m^6n^0$ Power Rule 1

$= -(4^2)m^6 \cdot 1$ Zero exponent

$= -4^2m^6$,   or $-16m^6$

47. $(6m^4 - 3m^2 + m) - (2m^3 + 5m^2 + 4m) + (m^2 - m)$

$= (6m^4 - 3m^2 + m) + (-2m^3 - 5m^2 - 4m)$
$\quad + (m^2 - m)$

$= 6m^4 - 2m^3 + (-3 - 5 + 1)m^2 + (1 - 4 - 1)m$

$= 6m^4 - 2m^3 - 7m^2 - 4m$

67. $[(2p - 3) + q]^2$

$= (2p - 3)^2 + 2(2p - 3)(q) + q^2$
   Square of a binomial, treating $(2p - 3)$ as one term

$= (2p)^2 - 2(2p)(3) + 3^2 + 2(2p - 3)(q) + q^2$
   Square the binomial $(2p - 3)$.

$= 4p^2 - 12p + 9 + 4pq - 6q + q^2$

87. $\dfrac{-4x^7 - 14x^6 + 10x^4 - 14x^2}{-2x^2}$

$= \dfrac{-4x^7}{-2x^2} + \dfrac{-14x^6}{-2x^2} + \dfrac{10x^4}{-2x^2} + \dfrac{-14x^2}{-2x^2}$

$= 2x^5 + 7x^4 - 5x^2 + 7$

R.4 Exercises *(pages 40–43)*

39. $24a^4 + 10a^3b - 4a^2b^2$

$= 2a^2(12a^2 + 5ab - 2b^2)$ Factor out the GCF, $2a^2$.

$= 2a^2(4a - b)(3a + 2b)$ Factor the trinomial.

47. $(a - 3b)^2 - 6(a - 3b) + 9$

$= [(a - 3b) - 3]^2$ Factor the perfect square trinomial.

$= (a - 3b - 3)^2$

69. $27 - (m + 2n)^3$

$= 3^3 - (m + 2n)^3$ Write as a difference of cubes.

$= [3 - (m + 2n)][3^2 + 3(m + 2n) + (m + 2n)^2]$
   Factor the difference of cubes.

$= (3 - m - 2n)(9 + 3m + 6n + m^2 + 4mn + 4n^2)$
   Distributive property; square the binomial $(m + 2n)$.

83. $9(a - 4)^2 + 30(a - 4) + 25$

$= 9u^2 + 30u + 25$ Replace $a - 4$ with u.

$= (3u)^2 + 2(3u)(5) + 5^2$

$= (3u + 5)^2$ Factor the perfect square trinomial.

$= [3(a - 4) + 5]^2$ Replace u with $a - 4$.

$= (3a - 12 + 5)^2$

$= (3a - 7)^2$

R.5 Exercises *(pages 50–52)*

19. $\dfrac{8m^2 + 6m - 9}{16m^2 - 9} = \dfrac{(2m + 3)(4m - 3)}{(4m + 3)(4m - 3)}$ Factor.

$= \dfrac{2m + 3}{4m + 3}$ Lowest terms

35. $\dfrac{x^3 + y^3}{x^3 - y^3} \cdot \dfrac{x^2 - y^2}{x^2 + 2xy + y^2}$

$= \dfrac{(x + y)(x^2 - xy + y^2)}{(x - y)(x^2 + xy + y^2)} \cdot \dfrac{(x + y)(x - y)}{(x + y)(x + y)}$ Factor.

$= \dfrac{x^2 - xy + y^2}{x^2 + xy + y^2}$ Lowest terms

55. $\dfrac{4}{x + 1} + \dfrac{1}{x^2 - x + 1} - \dfrac{12}{x^3 + 1}$

$= \dfrac{4}{x + 1} + \dfrac{1}{x^2 - x + 1} - \dfrac{12}{(x + 1)(x^2 - x + 1)}$

   Factor the sum of cubes.

$= \dfrac{4(x^2 - x + 1)}{(x + 1)(x^2 - x + 1)} + \dfrac{1(x + 1)}{(x + 1)(x^2 - x + 1)}$

$- \dfrac{12}{(x + 1)(x^2 - x + 1)}$

   Write each fraction with the common denominator.

$= \dfrac{4(x^2 - x + 1) + 1(x + 1) - 12}{(x + 1)(x^2 - x + 1)}$

   Add and subtract numerators.

$= \dfrac{4x^2 - 4x + 4 + x + 1 - 12}{(x + 1)(x^2 - x + 1)}$

   Distributive property

$= \dfrac{4x^2 - 3x - 7}{(x + 1)(x^2 - x + 1)}$ Combine like terms.

$= \dfrac{(4x - 7)(x + 1)}{(x + 1)(x^2 - x + 1)}$ Factor the numerator.

$= \dfrac{4x - 7}{x^2 - x + 1}$ Lowest terms

67. $\dfrac{\dfrac{1}{x + h} - \dfrac{1}{x}}{h}$

Multiply both numerator and denominator by the LCD of all the fractions, $x(x + h)$.

$\dfrac{\dfrac{1}{x + h} - \dfrac{1}{x}}{h} = \dfrac{x(x + h)\left(\dfrac{1}{x + h} - \dfrac{1}{x}\right)}{x(x + h)(h)}$

$= \dfrac{x(x + h)\left(\dfrac{1}{x + h}\right) - x(x + h)\left(\dfrac{1}{x}\right)}{x(x + h)(h)}$

   Distributive property

$= \dfrac{x - (x + h)}{x(x + h)(h)}$

$= \dfrac{-h}{x(x + h)(h)}$

$= \dfrac{-1}{x(x + h)}$ Lowest terms

R.6 Exercises *(pages 59–62)*

41. $\left(-\dfrac{64}{27}\right)^{1/3} = -\dfrac{4}{3}$ because $\left(-\dfrac{4}{3}\right)^3 = -\dfrac{64}{27}$.

69. $\dfrac{p^{1/5}p^{7/10}p^{1/2}}{(p^3)^{-1/5}} = \dfrac{p^{1/5 + 7/10 + 1/2}}{p^{-3/5}}$ Product rule; power rule 1

$= \dfrac{p^{2/10 + 7/10 + 5/10}}{p^{-6/10}}$ Write all fractions with the LCD, 10.

$= \dfrac{p^{14/10}}{p^{-6/10}}$

$= p^{(14/10) - (-6/10)}$ Quotient rule

$= p^{20/10}$

$= p^2$

79. $(r^{1/2} - r^{-1/2})^2 = (r^{1/2})^2 - 2(r^{1/2})(r^{-1/2}) + (r^{-1/2})^2$

   Square of a binomial

$= r - 2r^0 + r^{-1}$ Power rule 1; product rule

$= r - 2 \cdot 1 + r^{-1}$ Zero exponent

$= r - 2 + r^{-1}$, or $r - 2 + \dfrac{1}{r}$

   Negative exponent

101. $\dfrac{x - 9y^{-1}}{(x - 3y^{-1})(x + 3y^{-1})}$

$= \dfrac{x - \dfrac{9}{y}}{\left(x - \dfrac{3}{y}\right)\left(x + \dfrac{3}{y}\right)}$ Definition of negative exponent

$= \dfrac{x - \dfrac{9}{y}}{x^2 - \dfrac{9}{y^2}}$ Multiply in the denominator.

$= \dfrac{y^2\left(x - \dfrac{9}{y}\right)}{y^2\left(x^2 - \dfrac{9}{y^2}\right)}$ Multiply numerator and denominator by the LCD, y^2.

$= \dfrac{y^2x - 9y}{y^2x^2 - 9}$ Distributive property

$= \dfrac{y(xy - 9)}{x^2y^2 - 9}$ Factor numerator.

R.7 Exercises *(pages 70–72)*

53. $\sqrt[4]{\dfrac{g^3h^5}{9r^6}} = \dfrac{\sqrt[4]{g^3h^5}}{\sqrt[4]{9r^6}}$

  Quotient rule

$= \dfrac{\sqrt[4]{h^4(g^3h)}}{\sqrt[4]{r^4(9r^2)}}$

  Factor out perfect fourth powers.

$= \dfrac{h\sqrt[4]{g^3h}}{r\sqrt[4]{9r^2}}$

  Remove all perfect fourth powers from the radicals.

$= \dfrac{h\sqrt[4]{g^3h}}{r\sqrt[4]{9r^2}} \cdot \dfrac{\sqrt[4]{9r^2}}{\sqrt[4]{9r^2}}$

  Rationalize the denominator.

$= \dfrac{h\sqrt[4]{9g^3hr^2}}{r\sqrt[4]{81r^4}}$

$= \dfrac{h\sqrt[4]{9g^3hr^2}}{3r^2}$

71. $\left(\sqrt[3]{11} - 1\right)\left(\sqrt[3]{11^2} + \sqrt[3]{11} + 1\right)$

This product has the pattern

$$(x - y)(x^2 + xy + y^2) = x^3 - y^3,$$

the difference of cubes. Thus,

$\left(\sqrt[3]{11} - 1\right)\left(\sqrt[3]{11^2} + \sqrt[3]{11} + 1\right) = \left(\sqrt[3]{11}\right)^3 - 1^3$

$= 11 - 1$

$= 10.$

73. $\left(\sqrt{3} + \sqrt{8}\right)^2 = \left(\sqrt{3}\right)^2 + 2\left(\sqrt{3}\right)\left(\sqrt{8}\right) + \left(\sqrt{8}\right)^2$

  Square of a binomial

$= 3 + 2\sqrt{24} + 8$

$= 3 + 2\sqrt{4 \cdot 6} + 8$

$= 3 + 2\left(2\sqrt{6}\right) + 8$ Product rule

$= 11 + 4\sqrt{6}$

83. $\dfrac{-4}{\sqrt[3]{3}} + \dfrac{1}{\sqrt[3]{24}} - \dfrac{2}{\sqrt[3]{81}} = \dfrac{-4}{\sqrt[3]{3}} + \dfrac{1}{\sqrt[3]{8 \cdot 3}} - \dfrac{2}{\sqrt[3]{27 \cdot 3}}$

$= \dfrac{-4}{\sqrt[3]{3}} + \dfrac{1}{2\sqrt[3]{3}} - \dfrac{2}{3\sqrt[3]{3}}$

  Simplify radicals.

$= \dfrac{-4 \cdot 6}{\sqrt[3]{3} \cdot 6} + \dfrac{1 \cdot 3}{2\sqrt[3]{3} \cdot 3} - \dfrac{2 \cdot 2}{3\sqrt[3]{3} \cdot 2}$

  Write fractions with common denominator, $6\sqrt[3]{3}$.

$= \dfrac{-24}{6\sqrt[3]{3}} + \dfrac{3}{6\sqrt[3]{3}} - \dfrac{4}{6\sqrt[3]{3}}$

$= \dfrac{-25}{6\sqrt[3]{3}}$

$= \dfrac{-25}{6\sqrt[3]{3}} \cdot \dfrac{\sqrt[3]{3^2}}{\sqrt[3]{3^2}}$

  Rationalize the denominator.

$= \dfrac{-25\sqrt[3]{9}}{6\sqrt[3]{27}}$

$= \dfrac{-25\sqrt[3]{9}}{6 \cdot 3}$

$= \dfrac{-25\sqrt[3]{9}}{18}$

87. $\dfrac{\sqrt{7} - 1}{2\sqrt{7} + 4\sqrt{2}}$

$= \dfrac{\sqrt{7} - 1}{2\sqrt{7} + 4\sqrt{2}} \cdot \dfrac{2\sqrt{7} - 4\sqrt{2}}{2\sqrt{7} - 4\sqrt{2}}$

  Multiply numerator and denominator by the conjugate of the denominator.

$= \dfrac{\left(\sqrt{7} - 1\right)\left(2\sqrt{7} - 4\sqrt{2}\right)}{\left(2\sqrt{7} + 4\sqrt{2}\right)\left(2\sqrt{7} - 4\sqrt{2}\right)}$

  Multiply numerators; multiply denominators.

$= \dfrac{\sqrt{7} \cdot 2\sqrt{7} - \sqrt{7} \cdot 4\sqrt{2} - 1 \cdot 2\sqrt{7} + 1 \cdot 4\sqrt{2}}{\left(2\sqrt{7}\right)^2 - \left(4\sqrt{2}\right)^2}$

  Use FOIL in the numerator; product of the sum and difference of two terms in the denominator.

$= \dfrac{2 \cdot 7 - 4\sqrt{14} - 2\sqrt{7} + 4\sqrt{2}}{4 \cdot 7 - 16 \cdot 2}$

$= \dfrac{14 - 4\sqrt{14} - 2\sqrt{7} + 4\sqrt{2}}{-4}$

$= \dfrac{-2\left(-7 + 2\sqrt{14} + \sqrt{7} - 2\sqrt{2}\right)}{-2 \cdot 2}$

  Factor the numerator and denominator.

$= \dfrac{-7 + 2\sqrt{14} + \sqrt{7} - 2\sqrt{2}}{2}$

CHAPTER 1   EQUATIONS AND INEQUALITIES

1.1 Exercises *(pages 88–90)*

25. $.5x + \dfrac{4}{3}x = x + 10$

$\dfrac{1}{2}x + \dfrac{4}{3}x = x + 10$ Change decimal to fraction.

$6\left(\dfrac{1}{2}x + \dfrac{4}{3}x\right) = 6(x + 10)$ Multiply by the LCD, 6.

$3x + 8x = 6x + 60$ Distributive property

$11x = 6x + 60$ Combine terms.

$5x = 60$ Subtract $6x$.

$x = 12$ Divide by 5.

Solution set: $\{12\}$

33. $.3(x + 2) - .5(x + 2) = -.2x - .4$

$10[.3(x + 2) - .5(x + 2)] = 10(-.2x - .4)$

Multiply by 10 to clear decimals.

$3(x + 2) - 5(x + 2) = -2x - 4$

Distributive property

$3x + 6 - 5x - 10 = -2x - 4$

Distributive property

$-2x - 4 = -2x - 4$ Combine terms.

$0 = 0$ Add $2x$; add 4.

$0 = 0$ is a true statement, so the equation is an identity.
Solution set: {all real numbers}

57. $3x = (2x - 1)(m + 4)$,   for x

$3x = 2xm + 8x - m - 4$ FOIL

$m + 4 = 2xm + 5x$ Add $m + 4$; subtract $3x$.

$m + 4 = x(2m + 5)$ Factor out x.

$\dfrac{m + 4}{2m + 5} = x$,   or $x = \dfrac{m + 4}{2m + 5}$ Divide by $2m + 5$.

1.2 Exercises *(pages 97–103)*

17. Let h = the height of the box. Use the formula for the surface area of a rectangular box.

$S = 2lw + 2wh + 2hl$

$496 = 2 \cdot 18 \cdot 8 + 2 \cdot 8 \cdot h + 2 \cdot h \cdot 18$

Let $S = 496$, $l = 18$, $w = 8$.

$496 = 288 + 16h + 36h$

$496 = 288 + 52h$

$208 = 52h$

$4 = h$

The height of the box is 4 ft.

21. Let x = David's biking speed.

Then $x + 4.5$ = David's driving speed.

Summarize the given information in a table, using the equation $d = rt$. Because the speeds are given in miles per hour, the times must be changed from minutes to hours.

	r	t	d
Driving	$x + 4.5$	$\frac{1}{3}$	$\frac{1}{3}(x + 4.5)$
Biking	x	$\frac{3}{4}$	$\frac{3}{4}x$

Distance driving = Distance biking

$\dfrac{1}{3}(x + 4.5) = \dfrac{3}{4}x$

$12\left(\dfrac{1}{3}(x + 4.5)\right) = 12\left(\dfrac{3}{4}x\right)$

Multiply by the LCD, 12.

$4(x + 4.5) = 9x$

$4x + 18 = 9x$ Distributive property

$18 = 5x$ Subtract $4x$.

$\dfrac{18}{5} = x$ Divide by 5.

Now find the distance.

$d = rt = \dfrac{3}{4}x = \dfrac{3}{4}\left(\dfrac{18}{5}\right) = \dfrac{27}{10} = 2.7$

David travels 2.7 mi to work.

33. Let x = the number of milliliters of water to be added.

Strength	Milliliters of Solution	Milliliters of Salt
6%	8	.06(8)
0%	x	$0(x)$
4%	$8 + x$	$.04(8 + x)$

The number of milliliters of salt in the 6% solution plus the number of milliliters of salt in the water (0% solution) must equal the number of milliliters of salt in the 4% solution, so

$.06(8) + 0(x) = .04(8 + x).$

$.48 = .32 + .04x$

$.16 = .04x$

$4 = x$

To reduce the saline concentration to 4%, 4 mL of water should be added.

43. **(a)** $V = lwh = (10 \text{ ft})(10 \text{ ft})(8 \text{ ft}) = 800 \text{ ft}^3$

(b)   Area of panel = $(8 \text{ ft})(4 \text{ ft}) = 32 \text{ ft}^2$

$32 \text{ ft}^2\left(\dfrac{3365 \, \mu g}{\text{ft}^2}\right) = 107{,}680 \, \mu g$

(c) $F = 107{,}680x$

(d) $107{,}680x = 33(800)$

$x = \dfrac{33(800)}{107{,}680} \approx .25$

It will take approximately .25 day, or 6 hr.

1.3 Exercises *(pages 109–110)*

49. $-i - 2 - (6 - 4i) - (5 - 2i)$

$= (-2 - 6 - 5) + [-1 - (-4) - (-2)]i$

$= -13 + 5i$

67. $(2 + i)(2 - i)(4 + 3i)$

$= [(2 + i)(2 - i)](4 + 3i)$ Associative property

$= (2^2 - i^2)(4 + 3i)$ Product of the sum and difference of two terms

$= [4 - (-1)](4 + 3i)$ $i^2 = -1$

$= 5(4 + 3i)$

$= 20 + 15i$ Distributive property

79. $\dfrac{1}{i^{-11}} = i^{11}$

$= i^8 \cdot i^3$

$= (i^4)^2 \cdot i^3$

$= 1(-i)$

$= -i$

95. $\left(\dfrac{\sqrt{2}}{2} + \dfrac{\sqrt{2}}{2}i\right)^2 = \left(\dfrac{\sqrt{2}}{2}\right)^2 + 2 \cdot \dfrac{\sqrt{2}}{2} \cdot \dfrac{\sqrt{2}}{2}i + \left(\dfrac{\sqrt{2}}{2}i\right)^2$

$\qquad\qquad\qquad\qquad$ Square of a binomial

$= \dfrac{2}{4} + 2 \cdot \dfrac{2}{4}i + \dfrac{2}{4}i^2$

$= \dfrac{1}{2} + i + \dfrac{1}{2}i^2$

$= \dfrac{1}{2} + i + \dfrac{1}{2}(-1) \quad i^2 = -1$

$= \dfrac{1}{2} + i - \dfrac{1}{2}$

$= i$

Thus, $\dfrac{\sqrt{2}}{2} + \dfrac{\sqrt{2}}{2}i$ is a square root of i.

1.4 Exercises *(pages 119–121)*

17. $-4x^2 + x = -3$

$0 = 4x^2 - x - 3 \qquad$ Standard form

$0 = (4x + 3)(x - 1) \qquad$ Factor.

$4x + 3 = 0 \qquad$ or $\qquad x - 1 = 0$

$\qquad\qquad\qquad\qquad$ Zero-factor property

$x = -\dfrac{3}{4} \qquad$ or $\qquad x = 1$

Solution set: $\left\{-\dfrac{3}{4}, 1\right\}$

53. $\dfrac{1}{2}x^2 + \dfrac{1}{4}x - 3 = 0$

$4\left(\dfrac{1}{2}x^2 + \dfrac{1}{4}x - 3\right) = 4 \cdot 0 \quad$ Multiply by the LCD, 4.

$2x^2 + x - 12 = 0 \qquad$ Distributive property;
$\qquad\qquad\qquad\qquad\qquad$ standard form

$x = \dfrac{-b \pm \sqrt{b^2 - 4ac}}{2a}$

$\qquad\qquad\qquad$ Quadratic formula

$x = \dfrac{-1 \pm \sqrt{1^2 - 4(2)(-12)}}{2(2)}$

$\qquad\qquad\qquad a = 2, b = 1, c = -12$

$x = \dfrac{-1 \pm \sqrt{97}}{4}$

Solution set: $\left\{\dfrac{-1 \pm \sqrt{97}}{4}\right\}$

71. $4x^2 - 2xy + 3y^2 = 2$

(a) Solve for x in terms of y.

$4x^2 - 2yx + 3y^2 - 2 = 0 \qquad$ Standard form

$4x^2 - (2y)x + (3y^2 - 2) = 0$

$\qquad\qquad\qquad a = 4, b = -2y, c = 3y^2 - 2$

$x = \dfrac{-b \pm \sqrt{b^2 - 4ac}}{2a}$

$= \dfrac{-(-2y) \pm \sqrt{(-2y)^2 - 4(4)(3y^2 - 2)}}{2(4)}$

$= \dfrac{2y \pm \sqrt{4y^2 - 16(3y^2 - 2)}}{8}$

$= \dfrac{2y \pm \sqrt{4y^2 - 48y^2 + 32}}{8}$

$= \dfrac{2y \pm \sqrt{32 - 44y^2}}{8}$

$= \dfrac{2y \pm \sqrt{4(8 - 11y^2)}}{8}$

$= \dfrac{2y \pm 2\sqrt{8 - 11y^2}}{8}$

$= \dfrac{2(y \pm \sqrt{8 - 11y^2})}{2(4)}$

$x = \dfrac{y \pm \sqrt{8 - 11y^2}}{4}$

(b) Solve for y in terms of x.

$3y^2 - 2xy + 4x^2 - 2 = 0 \qquad$ Standard form

$3y^2 - (2x)y + (4x^2 - 2) = 0$

$\qquad\qquad\qquad a = 3, b = -2x, c = 4x^2 - 2$

$y = \dfrac{-b \pm \sqrt{b^2 - 4ac}}{2a}$

$= \dfrac{-(-2x) \pm \sqrt{(-2x)^2 - 4(3)(4x^2 - 2)}}{2(3)}$

$= \dfrac{2x \pm \sqrt{4x^2 - 12(4x^2 - 2)}}{6}$

$= \dfrac{2x \pm \sqrt{4x^2 - 48x^2 + 24}}{6}$

$= \dfrac{2x \pm \sqrt{24 - 44x^2}}{6}$

$= \dfrac{2x \pm \sqrt{4(6 - 11x^2)}}{6}$

$= \dfrac{2x \pm 2\sqrt{6 - 11x^2}}{6}$

$= \dfrac{2(x \pm \sqrt{6 - 11x^2})}{2(3)}$

$y = \dfrac{x \pm \sqrt{6 - 11x^2}}{3}$

85. $x = 4$ or $x = 5$

$x - 4 = 0$ or $x - 5 = 0$ Zero-factor property

$(x - 4)(x - 5) = 0$

$x^2 - 9x + 20 = 0$

$a = 1, b = -9, c = 20$

(Any nonzero constant multiple of these numbers will also work.)

1.5 Exercises *(pages 126–132)*

27. $S = 2\pi rh + 2\pi r^2$ Surface area of a cylinder

$8\pi = 2\pi r \cdot 3 + 2\pi r^2$ Let $S = 8\pi, h = 3$.

$8\pi = 6\pi r + 2\pi r^2$

$0 = 2\pi r^2 + 6\pi r - 8\pi$

$0 = 2\pi(r^2 + 3r - 4)$

$0 = 2\pi(r + 4)(r - 1)$

$r + 4 = 0$ or $r - 1 = 0$

$r = -4$ or $r = 1$

Because r represents the radius of a cylinder, -4 is not reasonable. The radius is 1 ft.

33. Let r = radius of circle.

Let x = length of side of square.

From the figure, the radius is $\frac{1}{2}$ the length of the diagonal of the square.

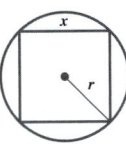

$a^2 + b^2 = c^2$ Pythagorean theorem

$x^2 + x^2 = (2r)^2$ Let $a = x, b = x, c = 2r$.

$2x^2 = 4r^2$

$x^2 = 2r^2$

$800 = 2r^2$ Area of square = $x^2 = 800$

$400 = r^2$

$r = \pm\sqrt{400} = \pm 20$

Because r represents the radius of a circle, -20 is not reasonable. The radius is 20 ft.

57. Let x = number of passengers in excess of 75. Then $225 - 5x$ = the cost per passenger (in dollars) and $75 + x$ = the number of passengers.

(Cost per passenger) (Number of passengers)

= Revenue

$(225 - 5x)(75 + x) = 16,000$

$16,875 - 150x - 5x^2 = 16,000$

$0 = 5x^2 + 150x - 875$

$0 = x^2 + 30x - 175$

$0 = (x + 35)(x - 5)$

$x + 35 = 0$ or $x - 5 = 0$

$x = -35$ or $x = 5$

The negative solution is not meaningful. Since there are 5 passengers in excess of 75, the total number of passengers is 80.

1.6 Exercises *(pages 142–145)*

13. $\dfrac{4}{x^2 + x - 6} - \dfrac{1}{x^2 - 4} = \dfrac{2}{x^2 + 5x + 6}$

$\dfrac{4}{(x + 3)(x - 2)} - \dfrac{1}{(x + 2)(x - 2)} = \dfrac{2}{(x + 3)(x + 2)}$

Factor denominators.

$(x + 3)(x - 2)(x + 2)\left(\dfrac{4}{(x + 3)(x - 2)} - \dfrac{1}{(x + 2)(x - 2)}\right)$

$= (x + 3)(x - 2)(x + 2)\left(\dfrac{2}{(x + 3)(x + 2)}\right)$

Multiply by the LCD, $(x + 3)(x - 2)(x + 2)$, where $x \neq -3, x \neq 2, x \neq -2$.

$4(x + 2) - 1(x + 3) = 2(x - 2)$

Simplify on both sides.

$4x + 8 - x - 3 = 2x - 4$

Distributive property

$3x + 5 = 2x - 4$

Combine terms.

$x = -9$

Subtract $2x$; subtract 5.

The restrictions $x \neq -3, x \neq 2, x \neq -2$ do not affect the result.

Solution set: $\{-9\}$

31. Let x = the number of hours to fill the pool with both pipes open.

	Rate	Time	Part of the Job Accomplished
Inlet Pipe	$\frac{1}{5}$	x	$\frac{1}{5}x$
Outlet Pipe	$\frac{1}{8}$	x	$\frac{1}{8}x$

Filling the pool is 1 whole job, but because the outlet pipe empties the pool, its contribution should be subtracted from the contribution of the inlet pipe.

$\dfrac{1}{5}x - \dfrac{1}{8}x = 1$

$40\left(\dfrac{1}{5}x - \dfrac{1}{8}x\right) = 40 \cdot 1$ Multiply by the LCD, 40.

$8x - 5x = 40$ Distributive property

$3x = 40$ Combine terms.

$x = \dfrac{40}{3} = 13\dfrac{1}{3}$ Divide by 3.

It took $13\frac{1}{3}$ hr to fill the pool.

53. $\sqrt{2\sqrt{7x + 2}} = \sqrt{3x + 2}$

$\left(\sqrt{2\sqrt{7x + 2}}\right)^2 = \left(\sqrt{3x + 2}\right)^2$ Square both sides.

$2\sqrt{7x + 2} = 3x + 2$

$\left(2\sqrt{7x + 2}\right)^2 = (3x + 2)^2$ Square both sides again.

$$4(7x + 2) = 9x^2 + 12x + 4$$

         Square the binomial on the right.

$$28x + 8 = 9x^2 + 12x + 4$$

$$0 = 9x^2 - 16x - 4$$

$$0 = (9x + 2)(x - 2)$$

$$9x + 2 = 0 \quad \text{or} \quad x - 2 = 0$$

$$x = -\frac{2}{9} \quad \text{or} \quad x = 2$$

Check both proposed solutions by substituting first $-\frac{2}{9}$ and then 2 in the *original* equation. These checks will verify that both of these numbers are solutions.

Solution set: $\left\{-\dfrac{2}{9}, 2\right\}$

85. $x^{-2/3} + x^{-1/3} - 6 = 0$

Since $(x^{-1/3})^2 = x^{-2/3}$, let $u = x^{-1/3}$.

$$u^2 + u - 6 = 0 \quad \text{Substitute.}$$

$$(u + 3)(u - 2) = 0 \quad \text{Factor.}$$

$$u + 3 = 0 \quad \text{or} \quad u - 2 = 0$$

$$u = -3 \quad \text{or} \quad u = 2$$

Now replace u with $x^{-1/3}$.

$$x^{-1/3} = -3 \quad \text{or} \quad x^{-1/3} = 2$$

$$(x^{-1/3})^{-3} = (-3)^{-3} \quad \text{or} \quad (x^{-1/3})^{-3} = 2^{-3}$$

     Raise both sides of each equation to the -3 power.

$$x = \frac{1}{(-3)^3} \quad \text{or} \quad x = \frac{1}{2^3}$$

$$x = -\frac{1}{27} \quad \text{or} \quad x = \frac{1}{8}$$

A check will show that both $-\frac{1}{27}$ and $\frac{1}{8}$ satisfy the original equation.

Solution set: $\left\{-\dfrac{1}{27}, \dfrac{1}{8}\right\}$

95. $m^{3/4} + n^{3/4} = 1$, for m

$$m^{3/4} = 1 - n^{3/4}$$

$$(m^{3/4})^{4/3} = (1 - n^{3/4})^{4/3} \quad \text{Raise both sides to the } \tfrac{4}{3}$$

                                       power.

$$m = (1 - n^{3/4})^{4/3}$$

1.7 Exercises *(pages 154–158)*

51. $x^2 - 2x \le 1$

Step 1   Solve the corresponding quadratic equation.

$$x^2 - 2x = 1$$

$$x^2 - 2x - 1 = 0$$

The trinomial $x^2 - 2x - 1$ cannot be factored, so solve this equation with the quadratic formula.

$$x = \frac{-b \pm \sqrt{b^2 - 4ac}}{2a}$$

$$x = \frac{-(-2) \pm \sqrt{(-2)^2 - 4(1)(-1)}}{2(1)}$$

                        $a = 1, b = -2, c = -1$

Simplify to obtain $x = 1 \pm \sqrt{2}$, that is, $x = 1 - \sqrt{2}$ or $x = 1 + \sqrt{2}$.

Step 2   Identify the intervals determined by the solutions of the equation. The values $1 - \sqrt{2}$ and $1 + \sqrt{2}$ divide a number line into three intervals: $(-\infty, 1 - \sqrt{2})$, $(1 - \sqrt{2}, 1 + \sqrt{2})$, and $(1 + \sqrt{2}, \infty)$. Use solid circles on $1 - \sqrt{2}$ and $1 + \sqrt{2}$ because the inequality symbol includes equality. (Note that $1 - \sqrt{2} \approx -.4$ and $1 + \sqrt{2} \approx 2.4$.)

Step 3   Use a test value from each interval to determine which intervals form the solution set.

Interval	Test Value	Is $x^2 - 2x \le 1$ True or False?
A: $(-\infty, 1 - \sqrt{2})$	-1	$(-1)^2 - 2(-1) \le 1$? $3 \le 1$ False
B: $(1 - \sqrt{2}, 1 + \sqrt{2})$	0	$0^2 - 2(0) \le 1$? $0 \le 1$ True
C: $(1 + \sqrt{2}, \infty)$	3	$3^2 - 2(3) \le 1$? $3 \le 1$ False

Only Interval B makes the inequality true. Both endpoints are included because the given inequality is a nonstrict inequality.

Solution set: $\left[1 - \sqrt{2}, 1 + \sqrt{2}\right]$

61. $4x - x^3 \ge 0$

Step 1 $4x - x^3 = 0$ Corresponding equation

                 $x(4 - x^2) = 0$ Factor out the GCF, x.

$$x(2 - x)(2 + x) = 0$$

             Factor the difference of squares.

$$x = 0 \quad \text{or} \quad 2 - x = 0 \quad \text{or} \quad 2 + x = 0$$

               Zero-factor property

$$x = 0 \quad \text{or} \quad x = 2 \quad \text{or} \quad x = -2$$

(continued)

Step 2 The values -2, 0, and 2 divide the number line into four intervals.

Interval A: $(-\infty, -2)$; Interval B: $(-2, 0)$;

Interval C: $(0, 2)$; Interval D: $(2, \infty)$

Step 3 Use a test value from each interval to determine which intervals form the solution set.

Interval	Test Value	Is $4x - x^3 \geq 0$ True or False?
A: $(-\infty, -2)$	-3	$4(-3) - (-3)^3 \geq 0$? $15 \geq 0$ True
B: $(-2, 0)$	-1	$4(-1) - (-1)^3 \geq 0$? $-3 \geq 0$ False
C: $(0, 2)$	1	$4(1) - 1^3 \geq 0$? $3 \geq 0$ True
D: $(2, \infty)$	3	$4(3) - 3^3 \geq 0$? $-15 \geq 0$ False

Intervals A and C make the inequality true. The endpoints -2, 0, and 2 are all included because the given inequality is a nonstrict inequality.

Solution set: $(-\infty, -2] \cup [0, 2]$

79. $\dfrac{7}{x + 2} \geq \dfrac{1}{x + 2}$

Step 1 Rewrite the inequality so that 0 is on one side and there is a single fraction on the other side.

$$\frac{7}{x + 2} - \frac{1}{x + 2} \geq 0$$

$$\frac{6}{x + 2} \geq 0$$

Step 2 Determine the values that will cause either the numerator or the denominator to equal 0.

Since $6 \neq 0$, the numerator is never equal to 0.

The denominator is equal to 0 when $x + 2 = 0$, or $x = -2$.

-2 divides a number line into two intervals, $(-\infty, -2)$ and $(-2, \infty)$.

Use an open circle on -2 because it makes the denominator 0.

| Interval A | Interval B |
| $(-\infty, -2)$ | $(-2, \infty)$ |

(number line with open circle at -2 and 0 marked)

Step 3 Use a test value from each interval to determine which intervals form the solution set.

Interval	Test Value	Is $\dfrac{7}{x+2} \geq \dfrac{1}{x+2}$ True or False?
A: $(-\infty, -2)$	-3	$\dfrac{7}{-3 + 2} \geq \dfrac{1}{-3 + 2}$? $-7 \geq -1$ False
B: $(-2, \infty)$	0	$\dfrac{7}{0 + 2} \geq \dfrac{1}{0 + 2}$? $\dfrac{7}{2} \geq \dfrac{1}{2}$ True

Interval B satisfies the original inequality. The endpoint -2 is not included because it makes the denominator 0.

Solution set: $(-2, \infty)$

87. $\dfrac{2x - 3}{x^2 + 1} \geq 0$

The inequality already has 0 on one side, so set the numerator and denominator equal to 0. If $2x - 3 = 0$, then $x = \frac{3}{2}$. $x^2 + 1 = 0$ has no real solutions. $\frac{3}{2}$ divides a number line into two intervals, $\left(-\infty, \frac{3}{2}\right)$ and $\left(\frac{3}{2}, \infty\right)$.

Interval	Test Value	Is $\dfrac{2x - 3}{x^2 + 1} \geq 0$ True or False?
A: $\left(-\infty, \frac{3}{2}\right)$	0	$\dfrac{2(0) - 3}{0^2 + 1} \geq 0$? $-3 \geq 0$ False
B: $\left(\frac{3}{2}, \infty\right)$	2	$\dfrac{2(2) - 3}{2^2 + 1} \geq 0$? $\dfrac{1}{5} \geq 0$ True

Interval B satisfies the original inequality, along with the endpoint $\frac{3}{2}$.

Solution set: $\left[\dfrac{3}{2}, \infty\right)$

1.8 Exercises *(pages 163–165)*

33. $4|x - 3| > 12$

$\qquad |x - 3| > 3$ Divide by 4.

$\qquad x - 3 < -3 \qquad$ or $\qquad x - 3 > 3 \quad$ Property 4

$\qquad\quad x < 0 \qquad$ or $\qquad\quad x > 6 \quad$ Add 3.

Solution set: $(-\infty, 0) \cup (6, \infty)$

49. $\left|5x + \dfrac{1}{2}\right| - 2 < 5$

$\qquad \left|5x + \dfrac{1}{2}\right| < 7$ Add 2.

$\qquad -7 < 5x + \dfrac{1}{2} < 7 \quad$ Property 3

$\qquad 2(-7) < 2\left(5x + \dfrac{1}{2}\right) < 2(7)$

          Multiply each part by 2.

$\qquad -14 < 10x + 1 < 14$

          Distributive property

$$-15 < 10x < 13$$ Subtract 1 from each part.

$$\frac{-15}{10} < x < \frac{13}{10}$$ Divide each part by 10.

$$-\frac{3}{2} < x < \frac{13}{10}$$ Lowest terms

Solution set: $\left(-\dfrac{3}{2}, \dfrac{13}{10}\right)$

63. $|2x + 1| \le 0$

Since the absolute value of a number is always nonnegative, $|2x + 1| < 0$ is never true, so $|2x + 1| \le 0$ is only true when $|2x + 1| = 0$. Solve this equation.

$$|2x + 1| = 0$$
$$2x + 1 = 0$$
$$x = -\frac{1}{2}$$

Solution set: $\left\{-\dfrac{1}{2}\right\}$

73.
$$|x^2 + 1| - |2x| = 0$$
$$|x^2 + 1| = |2x|$$
$$x^2 + 1 = 2x \quad \text{or} \quad x^2 + 1 = -2x$$
    Property 2
$$x^2 - 2x + 1 = 0 \quad \text{or} \quad x^2 + 2x + 1 = 0$$
$$(x - 1)^2 = 0 \quad \text{or} \quad (x + 1)^2 = 0$$
$$x - 1 = 0 \quad \text{or} \quad x + 1 = 0$$
$$x = 1 \quad \text{or} \quad x = -1$$

Solution set: $\{-1, 1\}$

CHAPTER 2   GRAPHS AND FUNCTIONS

2.1 Exercises *(pages 190–193)*

17. $P(3\sqrt{2}, 4\sqrt{5}), Q(\sqrt{2}, -\sqrt{5})$

(a) $d(P, Q) = \sqrt{(\sqrt{2} - 3\sqrt{2})^2 + (-\sqrt{5} - 4\sqrt{5})^2}$

Let $x_1 = 3\sqrt{2}, y_1 = 4\sqrt{5}, x_2 = \sqrt{2}, y_2 = -\sqrt{5}.$
$$= \sqrt{(-2\sqrt{2})^2 + (-5\sqrt{5})^2}$$
$$= \sqrt{8 + 125}$$
$$= \sqrt{133}$$

(b) The midpoint of the segment PQ has coordinates
$$\left(\frac{3\sqrt{2} + \sqrt{2}}{2}, \frac{4\sqrt{5} + (-\sqrt{5})}{2}\right) = \left(2\sqrt{2}, \frac{3\sqrt{5}}{2}\right).$$

2.2 Exercises *(pages 198–200)*

45. Let $P(x, y)$ be a point whose distance from $A(1, 0)$ is $\sqrt{10}$ and whose distance from $B(5, 4)$ is also $\sqrt{10}$.

$d(P, A) = \sqrt{10}$, so
$$\sqrt{(1 - x)^2 + (0 - y)^2} = \sqrt{10}$$ Distance formula
$$(1 - x)^2 + y^2 = 10.$$ Square both sides.

$d(P, B) = \sqrt{10}$, so
$$\sqrt{(5 - x)^2 + (4 - y)^2} = \sqrt{10}$$
$$(5 - x)^2 + (4 - y)^2 = 10.$$

Thus,
$$(1 - x)^2 + y^2 = (5 - x)^2 + (4 - y)^2$$
$$1 - 2x + x^2 + y^2 = 25 - 10x + x^2 + 16 - 8y + y^2$$
$$1 - 2x = 41 - 10x - 8y$$
$$8y = 40 - 8x$$
$$y = 5 - x.$$

Substitute $5 - x$ for y in the equation $(1 - x)^2 + y^2 = 10$ and solve for x.
$$(1 - x)^2 + (5 - x)^2 = 10$$
$$1 - 2x + x^2 + 25 - 10x + x^2 = 10$$
$$2x^2 - 12x + 26 = 10$$
$$2x^2 - 12x + 16 = 0$$
$$x^2 - 6x + 8 = 0$$
$$(x - 2)(x - 4) = 0$$
$$x - 2 = 0 \quad \text{or} \quad x - 4 = 0$$
$$x = 2 \quad \text{or} \quad x = 4$$

To find the corresponding values of y, substitute in the equation $y = 5 - x$.

If $x = 2$, then $y = 5 - 2 = 3$.

If $x = 4$, then $y = 5 - 4 = 1$.

The points satisfying the given conditions are $(2, 3)$ and $(4, 1)$.

2.4 Exercises *(pages 225–231)*

49. $5x - 2y = 10$

Find two ordered pairs that are solutions of the equation. If $x = 0$, then $5(0) - 2y = 10$, or $-2y = 10$, so $y = -5$. If $y = 0$, then $5x - 2(0) = 10$, or $5x = 10$, so $x = 2$. Thus, the two ordered pairs are $(0, -5)$ and $(2, 0)$. The slope is

$$m = \frac{\text{rise}}{\text{run}} = \frac{0 - (-5)}{2 - 0} = \frac{5}{2}.$$

Plot the points $(0, -5)$ and $(2, 0)$ and draw a line through them.

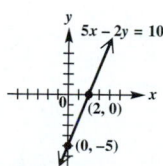

91. fixed cost = \$1650; variable cost = \$400; price of item = \$305

(a) $C(x) = 400x + 1650$ $m = 400, b = 1650$

(b) $R(x) = 305x$

(continued)

(c) $P(x) = R(x) - C(x)$

$$= 305x - (400x + 1650)$$

$$P(x) = -95x - 1650$$

(d) $\qquad C(x) = R(x)$

$$400x + 1650 = 305x$$

$$95x = -1650$$

$$x \approx -17.4$$

This result indicates a negative "break-even point," but the number of units produced must be a positive number. A calculator graph of the lines $y = 400x + 1650$ and $y = 305x$ on the same screen or solving the inequality $305x < 400x + 1650$ will show that $R(x) < C(x)$ for all positive values of x (in fact whenever x is greater than about -17.4). Do not produce the product since it is impossible to make a profit.

2.5 Exercises (pages 242–247)

15. Since the x-intercept is 3 and the y-intercept is -2, the line passes through the points $(3, 0)$ and $(0, -2)$. Use these points to find the slope.

$$m = \frac{-2 - 0}{0 - 3} = \frac{-2}{-3} = \frac{2}{3}$$

The slope is $\frac{2}{3}$ and the y-intercept is -2, so the equation of the line in slope-intercept form is

$$y = \frac{2}{3}x - 2.$$

55. (a) Find the slope of the line $3y + 2x = 6$.

$$3y + 2x = 6$$

$$3y = -2x + 6$$

$$y = -\frac{2}{3}x + 2$$

Thus, $m = -\frac{2}{3}$. A line parallel to $3y + 2x = 6$ will also have slope $-\frac{2}{3}$. Using the points $(4, -1)$ and $(k, 2)$ and the definition of slope,

$$\frac{2 - (-1)}{k - 4} = -\frac{2}{3}.$$

Solve this equation for k.

$$\frac{3}{k - 4} = -\frac{2}{3}$$

$$3(k - 4)\left(\frac{3}{k - 4}\right) = 3(k - 4)\left(-\frac{2}{3}\right)$$

$$9 = -2(k - 4)$$

$$9 = -2k + 8$$

$$2k = -1$$

$$k = -\frac{1}{2}$$

(b) Find the slope of the line $2y - 5x = 1$.

$$2y - 5x = 1$$

$$2y = 5x + 1$$

$$y = \frac{5}{2}x + \frac{1}{2}$$

Thus, $m = \frac{5}{2}$. A line perpendicular to $2y - 5x = 1$ will have slope $-\frac{2}{5}$ since $\frac{5}{2}\left(-\frac{2}{5}\right) = 1$. Using the points $(4, -1)$ and $(k, 2)$ and the definition of slope,

$$\frac{2 - (-1)}{k - 4} = -\frac{2}{5}.$$

Solve this equation for k.

$$\frac{3}{k - 4} = -\frac{2}{5}$$

$$5(k - 4)\left(\frac{3}{k - 4}\right) = 5(k - 4)\left(-\frac{2}{5}\right)$$

$$15 = -2(k - 4)$$

$$15 = -2k + 8$$

$$2k = -7$$

$$k = -\frac{7}{2}$$

81. $A(-1, 4)$, $B(-2, -1)$, $C(1, 14)$

For A and B, $m = \dfrac{-1 - 4}{-2 - (-1)} = \dfrac{-5}{-1} = 5.$

For B and C, $m = \dfrac{14 - (-1)}{1 - (-2)} = \dfrac{15}{3} = 5.$

For A and C, $m = \dfrac{14 - 4}{1 - (-1)} = \dfrac{10}{2} = 5.$

All three slopes are the same, so by Exercise 79, the three points are collinear.

2.6 Exercises (pages 255–259)

35. First, notice that for all x-values less than or equal to 0, the function value is -1. Next, notice for all x-values greater than 0, the function value is 1. So, we have a function f that is defined in two pieces:

$$f(x) = \begin{cases} -1 & \text{if } x \leq 0 \\ 1 & \text{if } x > 0 \end{cases}.$$

2.7 Exercises (pages 270–273)

73. $f(x) = 2x + 5$

Translate the graph of $f(x)$ up 2 units to obtain the graph of

$$t(x) = (2x + 5) + 2 = 2x + 7.$$

Now translate the graph of $t(x)$ left 3 units to obtain the graph of

$$g(x) = 2(x + 3) + 7$$

$$= 2x + 6 + 7$$

$$= 2x + 13.$$

(Note that if the original graph is first translated left 3 units and then up 2 units, the final result will be the same.)

2.8 Exercises *(pages 282–287)*

25. (a) From the graph, $f(-1) = 0$ and $g(-1) = 3$, so
$$(f + g)(-1) = f(-1) + g(-1)$$
$$= 0 + 3 = 3.$$

(b) From the graph, $f(-2) = -1$ and $g(-2) = 4$, so
$$(f - g)(-2) = f(-2) - g(-2)$$
$$= -1 - 4 = -5.$$

(c) From the graph, $f(0) = 1$ and $g(0) = 2$, so
$$(fg)(0) = f(0) \cdot g(0)$$
$$= 1 \cdot 2 = 2.$$

(d) From the graph, $f(2) = 3$ and $g(2) = 0$, so
$$\left(\frac{f}{g}\right)(2) = \frac{f(2)}{g(2)} = \frac{3}{0},$$

which is undefined.

39. $f(x) = \dfrac{1}{x}$

(a) $f(x + h) = \dfrac{1}{x + h}$

(b) $f(x + h) - f(x) = \dfrac{1}{x + h} - \dfrac{1}{x}$
$$= \frac{x - (x + h)}{x(x + h)}$$
$$= \frac{-h}{x(x + h)}$$

(c) $\dfrac{f(x + h) - f(x)}{h} = \dfrac{-h}{x(x + h)} \cdot \dfrac{1}{h}$
$$= \frac{-1}{x(x + h)}$$

73. $g(f(2)) = g(1) = 2$; $g(f(3)) = g(2) = 5$
Since $g(f(1)) = 7$ and $f(1) = 3$, $g(3) = 7$.
Completed table:

x	$f(x)$	$g(x)$	$g(f(x))$
1	3	2	7
2	1	5	2
3	2	7	5

CHAPTER 3   POLYNOMIAL AND RATIONAL FUNCTIONS

3.1 Exercises *(pages 311–320)*

3. $f(x) = -2(x + 3)^2 + 2$

(a) domain: $(-\infty, \infty)$
range: $(-\infty, 2]$

(b) vertex: $(h, k) = (-3, 2)$

(c) axis: $x = -3$

(d) $y = -2(0 + 3)^2 + 2$ Let $x = 0$.
$$= -16$$
y-intercept: -16

(e) $0 = -2(x + 3)^2 + 2$ Let $f(x) = 0$.
$$2(x + 3)^2 = 2$$
$$(x + 3)^2 = 1$$
$$x + 3 = \pm\sqrt{1}$$
$$x + 3 = 1 \quad \text{or} \quad x + 3 = -1$$
$$x = -2 \quad \text{or} \quad x = -4$$
x-intercepts: $-4, -2$

61. $y = \dfrac{-16x^2}{.434v^2} + 1.15x + 8$

(a) $10 = \dfrac{-16(15)^2}{.434v^2} + 1.15(15) + 8$
    Let $y = 10$, $x = 15$.
$$10 = \frac{-3600}{.434v^2} + 17.25 + 8$$
$$\frac{3600}{.434v^2} = 15.25$$
$$3600 = 6.6185v^2$$
$$v^2 = \frac{3600}{6.6185}$$
$$v = \pm\sqrt{\frac{3600}{6.6185}} \approx \pm 23.32$$

Because v represents velocity, only the positive square root is meaningful. The basketball should have an initial velocity of approximately 23.32 ft per sec.

(b) $y = \dfrac{-16x^2}{.434(23.32)^2} + 1.15x + 8$

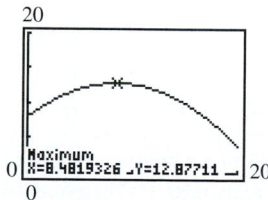

Graph this function in an appropriate window, such as $[0, 20]$ by $[0, 20]$, with X scale $= 5$, Y scale $= 5$. Use the calculator to find the vertex of the parabola, which is the maximum point. The y-coordinate of this point is approximately 12.88, so the maximum height of the basketball is approximately 12.88 ft.

73. $y = x^2 - 10x + c$

An x-intercept occurs where $y = 0$, or
$$0 = x^2 - 10x + c.$$

(continued)

There will be exactly one x-intercept if this equation has exactly one solution, or the discriminant is 0.

$$b^2 - 4ac = 0$$
$$(-10)^2 - 4(1)c = 0$$
$$100 = 4c$$
$$c = 25$$

3.2 Exercises *(pages 326–328)*

7. $\dfrac{x^5 + 3x^4 + 2x^3 + 2x^2 + 3x + 1}{x + 2}$

$x + 2 = x - (-2)$

$$
\begin{array}{r|rrrrrr}
-2) & 1 & 3 & 2 & 2 & 3 & 1 \\
   &   & -2 & -2 & 0 & -4 & 2 \\
\hline
   & 1 & 1 & 0 & 2 & -1 & 3
\end{array}
$$

In the last row of the synthetic division, all numbers except the last one give the coefficients of the quotient, and the last number gives the remainder. The quotient is $1x^4 + 1x^3 + 0x^2 + 2x - 1$ or $x^4 + x^3 + 2x - 1$, and the remainder is 3. Thus,

$$\frac{x^5 + 3x^4 + 2x^3 + 2x^2 + 3x + 1}{x + 2} = x^4 + x^3 + 2x - 1 + \frac{3}{x + 2}.$$

19. $f(x) = 2x^3 + x^2 + x - 8; \quad k = -1$

$$
\begin{array}{r|rrrr}
-1) & 2 & 1 & 1 & -8 \\
    &   & -2 & 1 & -2 \\
\hline
    & 2 & -1 & 2 & -10
\end{array}
$$

$$f(x) = (x + 1)(2x^2 - x + 2) + (-10)$$
$$= (x + 1)(2x^2 - x + 2) - 10$$

27. $f(x) = x^2 + 5x + 6; \quad k = -2$

$$
\begin{array}{r|rrr}
-2) & 1 & 5 & 6 \\
    &   & -2 & -6 \\
\hline
    & 1 & 3 & 0
\end{array}
$$

The last number in the bottom row of the synthetic division gives the remainder, 0. Therefore, by the remainder theorem, $f(-2) = 0$.

3.3 Exercises *(pages 337–339)*

23. $f(x) = x^3 + (7 - 3i)x^2 + (12 - 21i)x - 36i; \quad k = 3i$

$$
\begin{array}{r|rrrr}
3i) & 1 & 7 - 3i & 12 - 21i & -36i \\
    &   & 3i & 21i & 36i \\
\hline
    & 1 & 7 & 12 & 0
\end{array}
$$

The quotient is $x^2 + 7x + 12$, so

$$f(x) = (x - 3i)(x^2 + 7x + 12)$$
$$= (x - 3i)(x + 4)(x + 3).$$

45. $f(x) = 3(x - 2)(x + 3)(x^2 - 1)$

$$0 = 3(x - 2)(x + 3)(x^2 - 1) \qquad \text{Let } f(x) = 0.$$
$$0 = 3(x - 2)(x + 3)(x - 1)(x + 1)$$
$$\qquad\qquad\qquad \text{Factor the difference of squares.}$$

$x - 2 = 0$ or $x + 3 = 0$ or $x - 1 = 0$ or $x + 1 = 0$
$\qquad\qquad\qquad\qquad$ Zero-factor property

$x = 2$ or $x = -3$ or $x = 1$ or $x = -1$

The zeros are 2, -3, 1, and -1, all of multiplicity 1.

51. Zeros of -2, 1, and 0; $f(-1) = -1$

The factors of $f(x)$ are $x - (-2) = x + 2$, $x - 1$, and $x - 0 = x$.

$$f(x) = a(x + 2)(x - 1)(x)$$
$$f(-1) = a(-1 + 2)(-1 - 1)(-1) = -1$$
$$a(1)(-2)(-1) = -1$$
$$2a = -1$$
$$a = -\frac{1}{2}$$

Therefore,

$$f(x) = -\frac{1}{2}(x + 2)(x - 1)(x)$$
$$= -\frac{1}{2}(x^2 + x - 2)(x)$$
$$= -\frac{1}{2}(x^3 + x^2 - 2x)$$
$$f(x) = -\frac{1}{2}x^3 - \frac{1}{2}x^2 + x.$$

3.4 Exercises *(pages 351–358)*

37. $f(x) = x^3 + 5x^2 - x - 5$

$$= x^2(x + 5) - 1(x + 5)$$
$$= (x + 5)(x^2 - 1) \qquad \text{Factor by grouping.}$$
$$= (x + 5)(x + 1)(x - 1) \qquad \text{Factor the difference of squares.}$$

Find the real zeros of f. (All three zeros of this function are real.)

$x + 5 = 0$ or $x + 1 = 0$ or $x - 1 = 0$
$\qquad\qquad\qquad\qquad\qquad$ Zero-factor property

$x = -5$ or $\qquad$ $x = -1$ or $\qquad$ $x = 1$

The zeros of f are -5, -1, and 1, so the x-intercepts of the graph are also -5, -1, and 1. Plot the points

$$(-5, 0), (-1, 0), \text{ and } (1, 0).$$

Now find the y-intercept.

$$f(0) = 0^3 + 5 \cdot 0^2 - 0 - 5 = -5,$$

so the y-intercept is -5. Plot the point $(0, -5)$.

The x-intercepts divide the x-axis into four intervals:

$$(-\infty, -5), (-5, -1), (-1, 1), (1, \infty).$$

Test a point in each interval to find the sign of $f(x)$ in each interval. See the table.

Interval	Test Point	Value of $f(x)$	Sign of $f(x)$	Graph Above or Below x-axis
$(-\infty, -5)$	-6	-35	Negative	Below
$(-5, -1)$	-2	9	Positive	Above
$(-1, 1)$	0	-5	Negative	Below
$(1, \infty)$	2	21	Positive	Above

Plot the points $(-6, -35)$, $(-2, 9)$, and $(2, 21)$. Note that $(0, -5)$ was already plotted when the y-intercept was found. Connect these test points, the zeros (or x-intercepts), and the y-intercept with a smooth curve to obtain the graph.

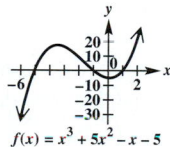

$$f(x) = x^3 + 5x^2 - x - 5$$

83. $f(x) = x^3 + 4x^2 - 8x - 8; \quad [-3.8, -3]$

Graph this function in a window that will produce a comprehensive graph, such as $[-10, 10]$ by $[-50, 50]$, with X scale = 1, Y scale = 10.

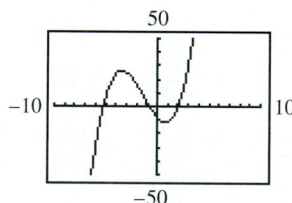

From this graph, we can see that the turning point in the interval $[-3.8, -3]$ is a maximum. Use "maximum" from the CALC menu to approximate the coordinates of this turning point.

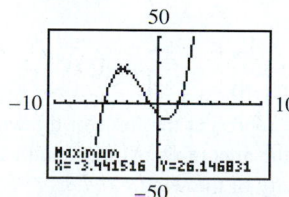

To the nearest hundredth, the turning point in the interval $[-3.8, -3]$ is $(-3.44, 26.15)$.

101. Use the following volume formulas:

$$V_{\text{cylinder}} = \pi r^2 h$$

$$V_{\text{hemisphere}} = \frac{1}{2} V_{\text{sphere}} = \frac{1}{2} \left(\frac{4}{3} \pi r^3 \right) = \frac{2}{3} \pi r^3$$

$$\pi r^2 h + 2 \left(\frac{2}{3} \pi r^3 \right) = \text{Total volume of tank}$$

$$\pi x^2 (12) + \frac{4}{3} \pi x^3 = 144\pi \quad \text{Let } V = 144\pi, h = 12, \text{ and } r = x.$$

$$\frac{4}{3} \pi x^3 + 12\pi x^2 - 144\pi = 0$$

$$\frac{4}{3} x^3 + 12 x^2 - 144 = 0 \quad \text{Divide by } \pi.$$

$$4 x^3 + 36 x^2 - 432 = 0 \quad \text{Multiply by 3.}$$

$$x^3 + 9 x^2 - 108 = 0 \quad \text{Divide by 4.}$$

Synthetic division or graphing $f(x) = x^3 + 9x^2 - 108$ with a graphing calculator will show that this function has zeros of -6 (multiplicity 2) and 3. Because x represents the radius of the hemispheres, a negative solution for the equation $x^3 + 9x^2 - 108 = 0$ is not meaningful. In order to get a volume of 144π ft^3, a radius of 3 ft should be used.

3.5 Exercises (pages 371–379)

81. $f(x) = \dfrac{1}{x^2 + 1}$

Because $x^2 + 1 = 0$ has no real solutions, the denominator is never 0, so there are no vertical asymptotes.

As $|x| \to \infty$, $y \to 0$, so the line $y = 0$ (the x-axis) is the horizontal asymptote.

$$f(0) = \frac{1}{0^2 + 1} = 1, \text{ so the } y\text{-intercept is } 1.$$

The equation $f(x) = 0$ has no solutions (notice that the numerator has no zeros), so there are no x-intercepts.

$$f(-x) = \frac{1}{(-x)^2 + 1} = \frac{1}{x^2 + 1} = f(x), \text{ so the graph is}$$

symmetric with respect to the y-axis.

Use a table of values to obtain several additional points on the graph. Use the asymptote, y-intercept, and these additional points (which reflect the symmetry of the graph) to sketch the graph of the function.

x	y
$\pm .5$	$.8$
± 1	$.5$
± 2	$.2$

89. $f(x) = \dfrac{x^2 + 2x}{2x - 1}$

Step 1 Find any vertical asymptotes.

$$2x - 1 = 0$$

$$x = \frac{1}{2}$$

(continued)

Step 2 Find any horizontal or oblique asymptotes. Because the numerator has degree exactly one more than the denominator, there is an oblique asymptote. Because $2x - 1$ is not of the form $x - a$, use polynomial long division rather than synthetic division to find the equation of this asymptote.

$$
\begin{array}{r}
\frac{1}{2}x + \frac{5}{4} \\
2x - 1 \overline{) x^2 + 2x + 0} \\
\underline{x^2 - \frac{1}{2}x} \\
\frac{5}{2}x + 0 \\
\underline{\frac{5}{2}x - \frac{5}{4}} \\
\frac{5}{4}
\end{array}
$$

Disregard the remainder. The equation of the oblique asymptote is $y = \frac{1}{2}x + \frac{5}{4}$.

Step 3 Find the y-intercept.

$$f(0) = \frac{0^2 + 2(0)}{2(0) - 1} = \frac{0}{-1} = 0,$$

so the y-intercept is 0.

Step 4 Find the x-intercepts, if any, by finding the zeros of the numerator.

$$x^2 + 2x = 0$$
$$x(x + 2) = 0$$
$$x = 0 \quad \text{or} \quad x = -2$$

There are two x-intercepts, -2 and 0.

Step 5 The graph does not intersect its oblique asymptote because the equation

$$\frac{x^2 + 2x}{2x - 1} = \frac{1}{2}x + \frac{5}{4},$$

which is equivalent to

$$x^2 + 2x = x^2 + 2x - \frac{5}{4},$$

has no solution.

Step 6

x	y
-3	$-\frac{3}{7} \approx -.43$
-1	$\frac{1}{3} \approx .33$
$\frac{1}{4} = .25$	$-\frac{9}{8} = -1.125$
1	3
3	3

Use the asymptotes, the intercepts, and these additional points to sketch the graph.

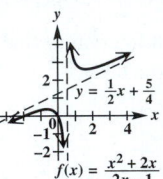

$$f(x) = \frac{x^2 + 2x}{2x - 1}$$

101. The graph has one vertical asymptote, $x = 2$, so $x - 2$ is a factor of the denominator of the rational expression. There is a point of discontinuity ("hole") in the graph at $x = -2$, so there is a factor of $x + 2$ in both numerator and denominator. There is one x-intercept, 3, so 3 is a zero of the numerator, which means that $x - 3$ is a factor of the numerator. Putting all of this information together, we have a possible function for the graph:

$$f(x) = \frac{(x - 3)(x + 2)}{(x - 2)(x + 2)}, \quad \text{or} \quad f(x) = \frac{x^2 - x - 6}{x^2 - 4}.$$

Note: From the second form of the function, we can see that the graph of this function has a horizontal asymptote of $y = 1$, which is consistent with the given graph.

3.6 Exercises *(pages 383–387)*

19. Step 1 Write the general relationship among the variables as an equation. Use the constant k.

$$a = \frac{kmn^2}{y^3}$$

Step 2 Substitute $a = 9$, $m = 4$, $n = 9$, and $y = 3$ to find k.

$$9 = \frac{k \cdot 4 \cdot 9^2}{3^3}$$
$$9 = 12k$$
$$k = \frac{3}{4}$$

Step 3 Substitute this value of k into the equation from Step 1, obtaining a specific formula.

$$a = \frac{3}{4} \cdot \frac{mn^2}{y^3}$$
$$a = \frac{3mn^2}{4y^3}$$

Step 4 Substitute $m = 6$, $n = 2$, and $y = 5$ and solve for a.

$$a = \frac{3mn^2}{4y^3} = \frac{3 \cdot 6 \cdot 2^2}{4 \cdot 5^3} = \frac{18}{125}$$

35. Step 1 Let F represent the force of the wind, A represent the area of the surface, and v represent the velocity of the wind.

$$F = kAv^2$$

Step 2 $50 = k\left(\dfrac{1}{2}\right)(40)^2$ Let $F = 50$, $A = \frac{1}{2}$, $v = 40$.

$$50 = 800k$$

$$k = \frac{50}{800} = \frac{1}{16}$$

Step 3 $F = \dfrac{1}{16}Av^2$

Step 4 $F = \dfrac{1}{16} \cdot 2 \cdot 80^2 = 800$

The force of the wind would be 800 lb.

43. Let t represent the Kelvin temperature.

$$R = kt^4$$

$$213.73 = k \cdot 293^4 \quad \text{Let } R = 213.73,\ t = 293.$$

$$k = \frac{213.73}{293^4} \approx 2.9 \times 10^{-8}$$

Thus, $R = (2.9 \times 10^{-8})t^4$.

If $t = 335$, then

$$R = (2.9 \times 10^{-8})(335^4) \approx 365.24.$$

CHAPTER 4   INVERSE, EXPONENTIAL, AND LOGARITHMIC FUNCTIONS

4.1 Exercises *(pages 411–415)*

47. $f(x) = \dfrac{2}{x + 6}$, $g(x) = \dfrac{6x + 2}{x}$

$$(f \circ g) = f(g(x))$$

$$= f\left(\frac{6x + 2}{x}\right)$$

$$= \frac{2}{\dfrac{6x + 2}{x} + 6}$$

$$= \frac{2}{\dfrac{6x + 2 + 6x}{x}}$$

$$= \frac{2}{1} \cdot \frac{x}{12x + 2}$$

$$= \frac{2x}{12x + 2}$$

$$= \frac{2x}{2(6x + 1)}$$

$$= \frac{x}{6x + 1} \neq x$$

Since $(f \circ g)(x) \neq x$, the functions are not inverses. It is not necessary to check $(g \circ f)(x)$.

65. $f(x) = \dfrac{1}{x - 3}$

(a) $y = \dfrac{1}{x - 3}$ $y = f(x)$

$$x = \frac{1}{y - 3} \qquad \text{Interchange } x \text{ and } y.$$

$$x(y - 3) = 1 \qquad \text{Solve for } y.$$

$$xy - 3x = 1$$

$$xy = 1 + 3x$$

$$y = \frac{1 + 3x}{x}$$

$$f^{-1}(x) = \frac{1 + 3x}{x} \qquad \text{Replace } y \text{ with } f^{-1}(x).$$

(b)

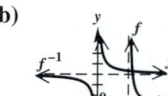

(c)   For both f and f^{-1}, the domain contains all real numbers except those for which the denominator equals 0.

Domain of f = range of $f^{-1} = (-\infty, 3) \cup (3, \infty)$

Domain of f^{-1} = range of $f = (-\infty, 0) \cup (0, \infty)$

4.2 Exercises *(pages 427–431)*

7. $g(x) = \left(\dfrac{1}{4}\right)^x$

$$g(-2) = \left(\frac{1}{4}\right)^{-2} = 4^2 = 16$$

61. $\left(\dfrac{1}{e}\right)^{-x} = \left(\dfrac{1}{e^2}\right)^{x+1}$

$$(e^{-1})^{-x} = (e^{-2})^{x+1} \qquad \text{Definition of negative exponent}$$

$$e^x = e^{-2(x+1)} \qquad (a^m)^n = a^{mn}$$

$$e^x = e^{-2x-2} \qquad \text{Distributive property}$$

$$x = -2x - 2 \qquad \text{Property (b)}$$

$$3x = -2 \qquad \text{Add } 2x.$$

$$x = -\frac{2}{3} \qquad \text{Divide by 3.}$$

Solution set: $\left\{-\dfrac{2}{3}\right\}$

63. $\left(\sqrt{2}\right)^{x+4} = 4^x$

$\left(2^{1/2}\right)^{x+4} = \left(2^2\right)^x$ Definition of $a^{1/n}$; write both sides as powers of a common base.

$2^{(1/2)(x+4)} = 2^{2x}$ $(a^m)^n = a^{mn}$

$2^{(1/2)x+2} = 2^{2x}$ Distributive property

$\dfrac{1}{2}x + 2 = 2x$ Property (b)

$2 = \dfrac{3}{2}x$ Subtract $\frac{1}{2}x$.

$\dfrac{4}{3} = x$ Multiply by $\frac{2}{3}$.

Solution set: $\left\{\dfrac{4}{3}\right\}$

65. $\dfrac{1}{27} = b^{-3}$

$3^{-3} = b^{-3}$

$b = 3$

Solution set: $\{3\}$

Alternative solution:

$\dfrac{1}{27} = b^{-3}$

$\dfrac{1}{27} = \dfrac{1}{b^3}$

$27 = b^3$

$b = \sqrt[3]{27}$

$b = 3$

Solution set: $\{3\}$

4.3 Exercises *(pages 441–445)*

23. $\log_x 25 = -2$

$x^{-2} = 25$ Write in exponential form.

$(x^{-2})^{-1/2} = 25^{-1/2}$ Raise both sides to the same power.

$x = \dfrac{1}{25^{1/2}}$

$x = \dfrac{1}{5}$

Solution set: $\left\{\dfrac{1}{5}\right\}$

79. $-\dfrac{2}{3}\log_5 5m^2 + \dfrac{1}{2}\log_5 25m^2$

$= \log_5(5m^2)^{-2/3} + \log_5(25m^2)^{1/2}$ Power property

$= \log_5[(5m^2)^{-2/3} \cdot (25m^2)^{1/2}]$ Product property

$= \log_5(5^{-2/3}m^{-4/3} \cdot 5m)$ Properties of exponents

$= \log_5(5^{-2/3} \cdot 5^1 \cdot m^{-4/3} \cdot m^1)$

$= \log_5(5^{1/3} \cdot m^{-1/3})$ $a^m \cdot a^n = a^{m+n}$

$= \log_5\dfrac{5^{1/3}}{m^{1/3}}$, or $\log_5\sqrt[3]{\dfrac{5}{m}}$

87. $\log_{10}\sqrt{30} = \log_{10} 30^{1/2}$ Definition of $a^{1/n}$

$= \dfrac{1}{2}\log_{10} 30$ Power property

$= \dfrac{1}{2}\log_{10}(10 \cdot 3)$

$= \dfrac{1}{2}(\log_{10} 10 + \log_{10} 3)$ Product property

$= \dfrac{1}{2}(1 + .4771)$ $\log_a a = 1$; Substitute.

$= \dfrac{1}{2}(1.4771) \approx .7386$

4.4 Exercises *(pages 453–457)*

69. $\log_{\sqrt{13}} 12 = \dfrac{\ln 12}{\ln\sqrt{13}}$ Change-of-base theorem

$= \dfrac{\ln 12}{\ln 13^{1/2}}$ Definition of $a^{1/n}$

$= \dfrac{\ln 12}{.5\ln 13}$ Power property

≈ 1.9376 Use a calculator; round answer to four decimal places.

The required logarithm can also be found by entering $\ln\sqrt{13}$ into the calculator directly:

$\log_{\sqrt{13}} 12 = \dfrac{\ln 12}{\ln\sqrt{13}}$ Change-of-base theorem

≈ 1.9376 Use a calculator; round answer to four decimal places.

73. $\ln\left(b^4\sqrt{a}\right) = \ln(b^4 a^{1/2})$ Definition of $a^{1/n}$

$= \ln b^4 + \ln a^{1/2}$ Product property

$= 4\ln b + \dfrac{1}{2}\ln a$ Power property

$= 4v + \dfrac{1}{2}u$ Substitute v for $\ln b$ and u for $\ln a$.

4.5 Exercises *(pages 464–468)*

23. $3(2)^{x-2} + 1 = 100$

$3(2)^{x-2} = 99$ Subtract 1.

$2^{x-2} = 33$ Divide by 3.

$\log 2^{x-2} = \log 33$ Take logarithms on both sides.

$(x - 2)\log 2 = \log 33$ Power property

$x\log 2 - 2\log 2 = \log 33$ Distributive property

$x\log 2 = \log 33 + 2\log 2$ Add 2 log 2.

$x = \dfrac{\log 33 + 2\log 2}{\log 2}$ Divide by log 2.

$x \approx 7.044$

Solution set: $\{7.044\}$

53. $\log_2(\log_2 x) = 1$

$\quad\quad \log_2 x = 2^1$ Write in exponential form.

$\quad\quad\quad\quad x = 2^2$ Write in exponential form.

$\quad\quad\quad\quad x = 4$ Evaluate.

Solution set: {4}

55. $\quad\quad \log x^2 = (\log x)^2$

$\quad\quad 2\log x = (\log x)^2$ Power property

$(\log x)^2 - 2\log x = 0$ Subtract $2\log x$; rewrite.

$(\log x)(\log x - 2) = 0$ Factor.

$\log x = 0$ or $\log x - 2 = 0$

$\quad\quad\quad\quad\quad\quad\quad$ Zero-factor property

$\quad\quad\quad\quad\quad \log x = 2$

$x = 10^0$ or $\quad\quad x = 10^2$

$\quad\quad\quad\quad\quad$ Write in exponential form.

$x = 1$ or $\quad\quad x = 100$

Solution set: {1, 100}

59. $\quad\quad p = a + \dfrac{k}{\ln x}, \quad$ for x

$\quad\quad p - a = \dfrac{k}{\ln x}$ Subtract a.

$(\ln x)\,(p - a) = k$ Multiply by $\ln x$.

$\quad\quad \ln x = \dfrac{k}{p - a}$ Divide by $p - a$.

$\quad\quad\quad x = e^{k/(p-a)}$ Change to exponential form.

4.6 Exercises *(pages 475–481)*

23. $\quad A = Pe^{rt}$ Continuous compounding formula

$\quad 3P = Pe^{.05t}$ Let $A = 3P$ and $r = .05$.

$\quad\quad 3 = e^{.05t}$ Divide by P.

$\ln 3 = \ln e^{.05t}$ Take logarithms on both sides.

$\ln 3 = .05t$ $\ln e^x = x$

$\dfrac{\ln 3}{.05} = t$ Divide by .05.

$\quad t \approx 21.97$ Use a calculator.

It will take about 21.97 yr for the investment to triple.

37. $f(t) = 200(.90)^{t-1}$

Find t when $f(t) = 50$.

$\quad\quad 50 = 200(.90)^{t-1}$ Let $f(t) = 50$.

$\quad\quad .25 = (.90)^{t-1}$ Divide by 200.

$\ln .25 = \ln(.90)^{t-1}$ Take logarithms on both sides.

$\ln .25 = (t - 1)\ln .90$ Power property

$\dfrac{\ln .25}{\ln .90} = t - 1$ Divide by $\ln .90$.

$\dfrac{\ln .25}{\ln .90} + 1 = t$ Add 1.

$\quad t \approx 14.2$ Use a calculator.

The initial dose will reach a level of 50 mg in about 14.2 hr.

CHAPTER 5   TRIGONOMETRIC FUNCTIONS

5.1 Exercises *(pages 499–502)*

43. $90° - 72° 58' 11''$

$\quad\quad 89° 59' 60''$ Write $90°$ as $89° 59' 60''$.

$\quad \underline{-72° 58' 11''}$

$\quad\quad 17° 01' 49''$

Thus, $90° - 72° 58' 11'' = 17° 1' 49''$.

129. 600 rotations per min

$\quad = \dfrac{600}{60}$ rotations per sec

$\quad = 10$ rotations per sec

$\quad = 5$ rotations per $\frac{1}{2}$ sec

$\quad = 5(360°)$ per $\frac{1}{2}$ sec

$\quad = 1800°$ per $\frac{1}{2}$ sec

A point on the edge of the tire will move $1800°$ in $\frac{1}{2}$ sec.

5.2 Exercises *(pages 512–515)*

45. Evaluate $\tan 360° + 4\sin 180° + 5\cos^2 180°$.

$\quad \tan 360° = \tan 0° = \dfrac{y}{x} = \dfrac{0}{1} = 0$

$\quad \sin 180° = \dfrac{y}{r} = \dfrac{0}{1} = 0$

$\quad \cos 180° = \dfrac{x}{r} = \dfrac{-1}{1} = -1$

$\tan 360° + 4\sin 180° + 5\cos^2 180° = 0 + 4(0) + 5(-1)^2$

$\quad\quad\quad\quad\quad\quad\quad$ Substitute; $\cos^2 x = (\cos x)^2$.

$\quad\quad\quad\quad\quad\quad\quad\quad\quad = 5$

101. Given $\tan\theta = -\dfrac{15}{8}$, with θ in quadrant II

Draw θ in standard position in quadrant II. Because $\tan\theta = \frac{y}{x}$ and θ is in quadrant II, we can use the values $y = 15$ and $x = -8$ for a point on its terminal side.

$r = \sqrt{x^2 + y^2} = \sqrt{(-8)^2 + 15^2} = \sqrt{64 + 225}$

$\quad = \sqrt{289} = 17$

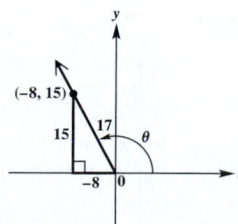

Use the values of x, y, and r and the definitions of the trigonometric functions to find the six trigonometric function values for θ.

(continued)

$$\sin \theta = \frac{y}{r} = \frac{15}{17} \qquad \csc \theta = \frac{r}{y} = \frac{17}{15}$$

$$\cos \theta = \frac{x}{r} = \frac{-8}{17} = -\frac{8}{17} \qquad \sec \theta = \frac{r}{x} = \frac{17}{-8} = -\frac{17}{8}$$

$$\tan \theta = \frac{y}{x} = \frac{15}{-8} = -\frac{15}{8} \qquad \cot \theta = \frac{x}{y} = \frac{-8}{15} = -\frac{8}{15}$$

117. Multiply the compound inequality $90° < \theta < 180°$ by 2 to find that $180° < 2\theta < 360°$. Thus, 2θ must lie in either quadrant III or IV. In both of these quadrants, the sine function is negative, and so $\sin 2\theta$ must be negative.

5.3 Exercises *(pages 524–529)*

49. One point on the line $y = \sqrt{3}x$ is the origin, $(0, 0)$. Let (x, y) be any other point on this line. Then, by the definition of slope, $m = \frac{y - 0}{x - 0} = \frac{y}{x} = \sqrt{3}$, but also, by the definition of tangent, $\tan \theta = \frac{y}{x}$. Thus, $\tan \theta = \sqrt{3}$. Because $\tan 60° = \sqrt{3}$, the line $y = \sqrt{3}x$ makes a $60°$ angle with the positive x-axis. (See Exercise 46.)

51. Apply the relationships between the lengths of the sides of a $30°$–$60°$ right triangle first to the triangle on the left to find the values of x and y, and then to the triangle on the right to find the values of z and w. In a $30°$–$60°$ right triangle, the side opposite the $30°$ angle is $\frac{1}{2}$ the length of the hypotenuse. The longer leg is $\sqrt{3}$ times the shorter leg.

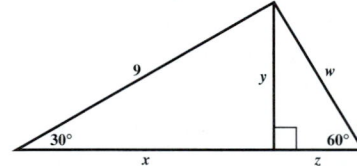

Thus,

$$y = \frac{1}{2}(9) = \frac{9}{2} \quad \text{and} \quad x = y\sqrt{3} = \frac{9\sqrt{3}}{2}.$$

Since $y = z\sqrt{3}$,

$$z = \frac{y}{\sqrt{3}} = \frac{\frac{9}{2}}{\sqrt{3}} = \frac{9}{2\sqrt{3}} \cdot \frac{\sqrt{3}}{\sqrt{3}} = \frac{9\sqrt{3}}{6} = \frac{3\sqrt{3}}{2},$$

and $\quad w = 2z = 2\left(\frac{3\sqrt{3}}{2}\right) = 3\sqrt{3}.$

75. To find the reference angle for $-300°$, sketch this angle in standard position.

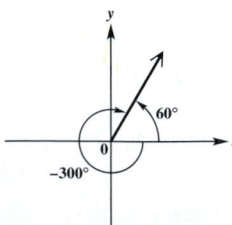

The reference angle for $-300°$ is $-300° + 360° = 60°$. Because $-300°$ is in quadrant I, the values of all its trigonometric functions will be positive, so these values will be identical to the trigonometric function values for

$60°$. (See the Function Values of Special Angles table that follows Example 3 in **Section 5.3**.)

$$\sin(-300°) = \frac{\sqrt{3}}{2} \qquad \csc(-300°) = \frac{2\sqrt{3}}{3}$$

$$\cos(-300°) = \frac{1}{2} \qquad \sec(-300°) = 2$$

$$\tan(-300°) = \sqrt{3} \qquad \cot(-300°) = \frac{\sqrt{3}}{3}$$

141. For parts (a) and (b), $\theta = 3°$, $g = 32.2$, and $f = .14$.

(a) Since 45 mph = 66 ft per sec,

$$R = \frac{V^2}{g(f + \tan \theta)}$$

$$= \frac{66^2}{32.2(.14 + \tan 3°)}$$

$$\approx 703 \text{ ft.}$$

(b) Since 70 mph = $\frac{70(5280)}{3600}$ ft per sec = 102.67 ft per sec,

$$R = \frac{V^2}{g(f + \tan \theta)}$$

$$= \frac{102.67^2}{32.2(.14 + \tan 3°)}$$

$$\approx 1701 \text{ ft.}$$

(c) Intuitively, increasing θ would make it easier to negotiate the curve at a higher speed much like is done at a race track. Mathematically, a larger value of θ (acute) will lead to a larger value for $\tan \theta$. If $\tan \theta$ increases, then the ratio determining R will *decrease*. Thus, the radius can be smaller and the curve sharper if θ is increased.

$$R = \frac{V^2}{g(f + \tan \theta)}$$

$$= \frac{66^2}{32.2(.14 + \tan 4°)}$$

$$\approx 644 \text{ ft}$$

$$R = \frac{V^2}{g(f + \tan \theta)}$$

$$= \frac{102.67^2}{32.2(.14 + \tan 4°)}$$

$$\approx 1559 \text{ ft}$$

As predicted, both values are smaller.

5.4 Exercises *(pages 538–545)*

15. Solve the right triangle with $B = 73.0°$, $b = 128$ in., and $C = 90°$.

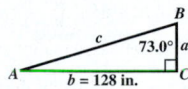

$$A = 90° - 73.0° = 17.0°$$

$$\tan 73.0° = \frac{128}{a}$$

$$a = \frac{128}{\tan 73.0°} \approx 39.1 \text{ in.}$$ Three significant digits

$$\sin 73.0° = \frac{128}{c}$$

$$c = \frac{128}{\sin 73.0°} \approx 134 \text{ in.}$$ Three significant digits

33. Let x represent the horizontal distance between the two buildings and y represent the height of the portion of the building across the street that is higher than the window.

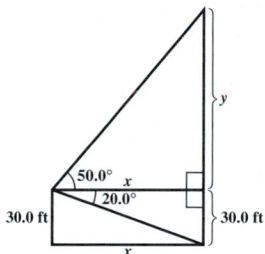

$$\tan 20.0° = \frac{30.0}{x}$$

$$x = \frac{30.0}{\tan 20.0°} \approx 82.4$$

$$\tan 50.0° = \frac{y}{x}$$

$$y = x \tan 50.0° = \left(\frac{30.0}{\tan 20.0°}\right) \tan 50.0° \approx 98.2$$

$$\text{height} = y + 30.0 = \left(\frac{30.0}{\tan 20.0°}\right) \tan 50.0° + 30.0 \approx 128$$

Three significant digits

The height of the building across the street is about 128 ft.

39. Let h represent the height of the tower.

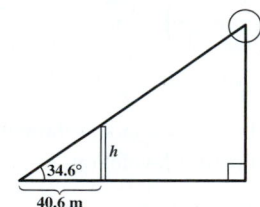

$$\tan 34.6° = \frac{h}{40.6}$$

$$h = 40.6 \tan 34.6° \approx 28.0$$

Three significant digits

The height of the tower is about 28.0 m.

53. Let $x = $ the distance between the two ships. The angle between the bearings of the ships is

$$180° - (28° \, 10' + 61° \, 50') = 90°.$$

The triangle formed is a right triangle.

Distance traveled at 24.0 mph:

$$(4 \text{ hr})(24.0 \text{ mph}) = 96 \text{ mi}$$

Distance traveled at 28.0 mph:

$$(4 \text{ hr})(28.0 \text{ mph}) = 112 \text{ mi}$$

Applying the Pythagorean theorem gives

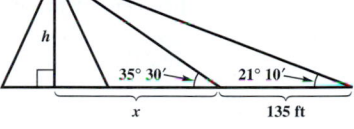

$$x^2 = 96^2 + 112^2$$

$$x^2 = 21{,}760$$

$$x \approx 148.$$

The ships are 148 mi apart.

57. Let $x = $ the distance from the closer point on the ground to the base of height h of the pyramid.

In the larger right triangle:

$$\tan 21° \, 10' = \frac{h}{135 + x}$$

$$h = (135 + x) \tan 21° \, 10'$$

In the smaller right triangle:

$$\tan 35° \, 30' = \frac{h}{x}$$

$$h = x \tan 35° \, 30'$$

Substitute for h in this equation, and solve for x.

$$(135 + x) \tan 21° \, 10' = x \tan 35° \, 30'$$

$$135 \tan 21° \, 10' + x \tan 21° \, 10' = x \tan 35° \, 30'$$

$$135 \tan 21° \, 10' = x \tan 35° \, 30' - x \tan 21° \, 10'$$

$$135 \tan 21° \, 10' = x(\tan 35° \, 30' - \tan 21° \, 10')$$

$$\frac{135 \tan 21° \, 10'}{\tan 35° \, 30' - \tan 21° \, 10'} = x$$

Then substitute for x in the equation for the smaller triangle.

$$h = \frac{135 \tan 21° \, 10'(\tan 35° \, 30')}{\tan 35° \, 30' - \tan 21° \, 10'}$$

$$h \approx 114$$

The height of the pyramid is 114 ft.

CHAPTER 6   THE CIRCULAR FUNCTIONS AND THEIR GRAPHS

6.1 Exercises *(pages 564–571)*

97. For the large gear and pedal,

$$s = r\theta = 4.72\pi. \quad 180° = \pi \text{ radians}$$

Thus, the chain moves 4.72π in. Find the angle through which the small gear rotates.

$$\theta = \frac{s}{r} = \frac{4.72\pi}{1.38} \approx 3.42\pi$$

(continued)

The angle θ for the wheel and for the small gear are the same, so for the wheel,

$$s = r\theta = 13.6(3.42\pi) \approx 146 \text{ in.}$$

The bicycle will move about 146 in.

121. (a)

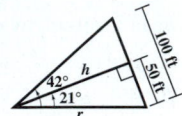

The triangle formed by the sides of the central angle and the chord is isosceles. Therefore, the bisector of the central angle is also the perpendicular bisector of the chord and divides the larger triangle into two congruent right triangles.

$$\sin 21° = \frac{50}{r}$$

$$r = \frac{50}{\sin 21°} \approx 140 \text{ ft}$$

The radius of the curve is about 140 ft.

(b) $r = \dfrac{50}{\sin 21°}$; $\theta = 42°$

$$42° = 42\left(\frac{\pi}{180} \text{ radian}\right) = \frac{7\pi}{30} \text{ radian}$$

$$s = r\theta = \frac{50}{\sin 21°} \cdot \frac{7\pi}{30} = \frac{35\pi}{3 \sin 21°} \approx 102 \text{ ft}$$

The length of the arc determined by the 100-ft chord is about 102 ft.

(c) The portion of the circle bounded by the arc and the 100-ft chord is the shaded region in the figure below.

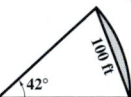

The area of the portion of the circle can be found by subtracting the area of the triangle from the area of the sector. From the figure in part (a),

$$\tan 21° = \frac{50}{h}, \quad \text{so} \quad h = \frac{50}{\tan 21°}.$$

$$A_{\text{sector}} = \frac{1}{2}r^2\theta$$

$$= \frac{1}{2}\left(\frac{50}{\sin 21°}\right)^2\left(\frac{7\pi}{30}\right) \quad \begin{array}{l}\text{From part (b),}\\ 42° = \frac{7\pi}{30}.\end{array}$$

$$\approx 7135 \text{ ft}^2$$

$$A_{\text{triangle}} = \frac{1}{2}bh = \frac{1}{2}(100)\left(\frac{50}{\tan 21°}\right)$$

$$\approx 6513 \text{ ft}^2$$

$$A_{\text{portion}} = A_{\text{sector}} - A_{\text{triangle}} \approx 7135 \text{ ft}^2 - 6513 \text{ ft}^2$$

$$= 622 \text{ ft}^2$$

The area of the portion is about 622 ft².

123. Use the Pythagorean theorem to find the hypotenuse of the right triangle, which is also the radius of the sector of the circle.

$$r^2 = 30^2 + 40^2 = 900 + 1600 = 2500$$

$$r = \sqrt{2500} = 50$$

$$A_{\text{triangle}} = \frac{1}{2}bh = \frac{1}{2}(30)(40)$$

$$= 600 \text{ yd}^2$$

$$A_{\text{sector}} = \frac{1}{2}r^2\theta$$

$$= \frac{1}{2}(50)^2 \cdot \frac{\pi}{3} \quad 60° = \frac{\pi}{3}$$

$$= \frac{1250\pi}{3} \text{ yd}^2$$

$$\text{Total area} = A_{\text{triangle}} + A_{\text{sector}} = 600 \text{ yd}^2 + \frac{1250\pi}{3} \text{ yd}^2$$

$$\approx 1900 \text{ yd}^2$$

6.2 Exercises *(pages 580–584)*

45. cos 2

$\frac{\pi}{2} \approx 1.57$ and $\pi \approx 3.14$, so $\frac{\pi}{2} < 2 < \pi$. Thus, an angle of 2 radians is in quadrant II. (The figure for Exercises 35–44 also shows that 2 radians is in quadrant II.) Because values of the cosine function are negative in quadrant II, cos 2 is negative.

63. $\left[\pi, \dfrac{3\pi}{2}\right]$; $\quad \tan s = \sqrt{3}$

Recall that $\tan \frac{\pi}{3} = \sqrt{3}$ and in quadrant III tan s is positive. Therefore,

$$\tan\left(\pi + \frac{\pi}{3}\right) = \tan \frac{4\pi}{3} = \sqrt{3},$$

and thus, $s = \frac{4\pi}{3}$.

91. The hour hand of a clock moves through an angle of 2π radians (one complete revolution) in 12 hr, so

$$\omega = \frac{\theta}{t} = \frac{2\pi}{12} = \frac{\pi}{6} \text{ radian per hr.}$$

101. At 215 revolutions per min, the bicycle tire is moving $215(2\pi) = 430\pi$ radians per min. This is the angular velocity ω. The linear velocity of the bicycle is

$$v = r\omega = 13(430\pi) = 5590\pi \text{ in. per min.}$$

Convert this velocity to miles per hour.

$$v = \frac{5590\pi \text{ in.}}{\text{min}} \cdot \frac{60 \text{ min}}{\text{hr}} \cdot \frac{1 \text{ ft}}{12 \text{ in.}} \cdot \frac{1 \text{ mi}}{5280 \text{ ft}} \approx 16.6 \text{ mph}$$

6.3 Exercises *(pages 593–598)*

51. $E = 5 \cos 120\pi t$

(a) The amplitude is $|5| = 5$, and the period is
$$\frac{2\pi}{120\pi} = \frac{1}{60}.$$

(b) Since the period is $\frac{1}{60}$, one cycle is completed in $\frac{1}{60}$ sec. Therefore, in 1 sec, 60 cycles are completed.

(c) For $t = 0$, $E = 5 \cos 120\pi(0) = 5 \cos 0 = 5$.

For $t = .03$, $E = 5 \cos 120\pi(.03) \approx 1.545$.

For $t = .06$, $E = 5 \cos 120\pi(.06) \approx -4.045$.

For $t = .09$, $E \approx -4.045$.

For $t = .12$, $E \approx 1.545$.

(d)

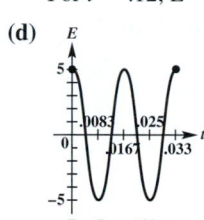

$E = 5 \cos 120\pi t$

6.4 Exercises *(pages 604–607)*

53. $y = \dfrac{1}{2} + \sin 2\left(x + \dfrac{\pi}{4}\right)$

This equation has the form $y = c + a \sin b(x - d)$ with $c = \frac{1}{2}$, $a = 1$, $b = 2$, and $d = -\frac{\pi}{4}$. Start with the graph of $y = \sin x$ and modify it to take into account the amplitude, period, and translations required to obtain the desired graph.

Amplitude: $|a| = 1$;   Period: $\dfrac{2\pi}{b} = \dfrac{2\pi}{2} = \pi$

Vertical translation: $\dfrac{1}{2}$ unit up

Phase shift (horizontal translation): $\dfrac{\pi}{4}$ units to the left

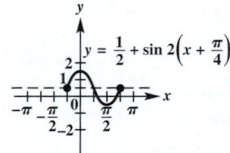

$y = \frac{1}{2} + \sin 2\left(x + \frac{\pi}{4}\right)$

6.5 Exercises *(pages 620–623)*

29. $y = -1 + \dfrac{1}{2} \cot(2x - 3\pi)$

$$y = -1 + \frac{1}{2} \cot 2\left(x - \frac{3\pi}{2}\right) \qquad \begin{array}{l} \text{Rewrite } 2x - 3\pi \\ \text{as } 2\left(x - \frac{3\pi}{2}\right). \end{array}$$

Period: $\dfrac{\pi}{b} = \dfrac{\pi}{2}$

Vertical translation: 1 unit down

Phase shift (horizontal translation): $\dfrac{3\pi}{2}$ units to the right

Because the function is to be graphed over a two-period interval, locate three adjacent vertical asymptotes. Because asymptotes of the graph of $y = \cot x$ occur at multiples of π, the following equations can be solved to locate asymptotes:

$$2\left(x - \frac{3\pi}{2}\right) = -2\pi, \quad 2\left(x - \frac{3\pi}{2}\right) = -\pi, \quad \text{and}$$

$$2\left(x - \frac{3\pi}{2}\right) = 0.$$

Solve each of these equations.

$$2\left(x - \frac{3\pi}{2}\right) = -2\pi$$

$$x - \frac{3\pi}{2} = -\pi$$

$$x = -\pi + \frac{3\pi}{2} = \frac{\pi}{2}$$

$$2\left(x - \frac{3\pi}{2}\right) = -\pi$$

$$x - \frac{3\pi}{2} = -\frac{\pi}{2}$$

$$x = -\frac{\pi}{2} + \frac{3\pi}{2} = \frac{2\pi}{2} = \pi$$

$$2\left(x - \frac{3\pi}{2}\right) = 0$$

$$x - \frac{3\pi}{2} = 0$$

$$x = \frac{3\pi}{2}$$

Divide the interval $\left(\frac{\pi}{2}, \pi\right)$ into four equal parts to obtain the following key x-values:

first-quarter value: $\dfrac{5\pi}{8}$;   middle value: $\dfrac{3\pi}{4}$;

third-quarter value: $\dfrac{7\pi}{8}$.

Evaluating the given function at these three key x-values gives the following points:

$$\left(\frac{5\pi}{8}, -\frac{1}{2}\right), \quad \left(\frac{3\pi}{4}, -1\right), \quad \left(\frac{7\pi}{8}, -\frac{3}{2}\right).$$

(continued)

Connect these points with a smooth curve and continue the graph to approach the asymptotes $x = \frac{\pi}{2}$ and $x = \pi$ to complete one period of the graph. Sketch an identical curve between the asymptotes $x = \pi$ and $x = \frac{3\pi}{2}$ to complete a second period of the graph.

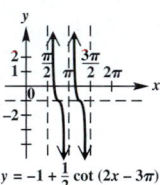

$$y = -1 + \frac{1}{2} \cot (2x - 3\pi)$$

41. $\tan (-x) = \dfrac{\sin (-x)}{\cos (-x)}$

$= \dfrac{-\sin x}{\cos x}$

$= -\dfrac{\sin x}{\cos x}$

$= -\tan x$

71. $\sec (-x) = \dfrac{1}{\cos (-x)}$

$= \dfrac{1}{\cos x}$

$= \sec x$

6.6 Exercises *(pages 626–628)*

19. (a) We will use a model of the form $s(t) = a \cos \omega t$ with $a = -3$. Since

$$s(0) = -3 \cos(0\omega) = -3 \cos 0 = -3 \cdot 1 = -3,$$

using a cosine function rather than a sine function will avoid the need for a phase shift.

Since the frequency $= \frac{6}{\pi}$ cycles per sec, by definition,

$$\frac{6}{\pi} = \frac{\omega}{2\pi}$$

$$6 \cdot 2\pi = \pi\omega$$

$$12\pi = \pi\omega$$

$$\omega = 12. \quad \text{Divide by } \pi; \text{ rewrite.}$$

Therefore, a model for the position of the weight at time t seconds is

$$s(t) = -3 \cos 12t.$$

(b) Period $= \dfrac{1}{\frac{6}{\pi}} = \dfrac{\pi}{6}$ sec

CHAPTER 7   TRIGONOMETRIC IDENTITIES AND EQUATIONS

7.1 Exercises *(pages 645–648)*

29. $\cot \theta = \dfrac{4}{3}$, $\sin \theta > 0$

Since $\cot \theta > 0$ and $\sin \theta > 0$, θ is in quadrant I, so all the function values are positive.

$\tan \theta = \dfrac{1}{\cot \theta} = \dfrac{1}{\frac{4}{3}} = \dfrac{3}{4}$

$\sec^2 \theta = \tan^2 \theta + 1 \qquad$ Pythagorean identity

$= \left(\dfrac{3}{4}\right)^2 + 1 = \dfrac{9}{16} + \dfrac{16}{16} = \dfrac{25}{16}$

$\sec \theta = \sqrt{\dfrac{25}{16}} = \dfrac{5}{4} \qquad \sec \theta > 0$

$\cos \theta = \dfrac{1}{\sec \theta} = \dfrac{1}{\frac{5}{4}} = \dfrac{4}{5}$

$\sin^2 \theta = 1 - \cos^2 \theta \qquad$ Alternative form of Pythagorean identity

$= 1 - \left(\dfrac{4}{5}\right)^2 = \dfrac{9}{25}$

$\sin \theta = \sqrt{\dfrac{9}{25}} = \dfrac{3}{5} \qquad \sin \theta > 0$

$\csc \theta = \dfrac{1}{\sin \theta} = \dfrac{1}{\frac{3}{5}} = \dfrac{5}{3}$

Thus, $\sin \theta = \frac{3}{5}$, $\cos \theta = \frac{4}{5}$, $\tan \theta = \frac{3}{4}$, $\sec \theta = \frac{5}{4}$, and $\csc \theta = \frac{5}{3}$.

51. $\csc x = \dfrac{1}{\sin x} \qquad$ Reciprocal identity

$= \dfrac{1}{\pm\sqrt{1 - \cos^2 x}} \qquad$ Alternative form of Pythagorean identity

$= \dfrac{\pm 1}{\sqrt{1 - \cos^2 x}}$

$= \dfrac{\pm 1}{\sqrt{1 - \cos^2 x}} \cdot \dfrac{\sqrt{1 - \cos^2 x}}{\sqrt{1 - \cos^2 x}} \qquad$ Rationalize the denominator.

$= \dfrac{\pm\sqrt{1 - \cos^2 x}}{1 - \cos^2 x}$

63. $\sec \theta - \cos \theta = \dfrac{1}{\cos \theta} - \cos \theta$

$= \dfrac{1}{\cos \theta} - \dfrac{\cos^2 \theta}{\cos \theta} \qquad$ Get a common denominator.

$= \dfrac{1 - \cos^2 \theta}{\cos \theta} \qquad$ Subtract fractions.

$$= \frac{\sin^2 \theta}{\cos \theta} \qquad\qquad 1 - \cos^2 \theta = \sin^2 \theta$$

$$= \frac{\sin \theta}{\cos \theta} \cdot \sin \theta$$

$$= \tan \theta \sin \theta$$

69. Since $\cos x = \frac{1}{5} > 0$, x is in quadrant I or quadrant IV.

$$\sin x = \pm\sqrt{1 - \cos^2 x} = \pm\sqrt{1 - \left(\frac{1}{5}\right)^2}$$

$$= \pm\sqrt{\frac{24}{25}} = \pm\frac{2\sqrt{6}}{5}$$

$$\tan x = \frac{\sin x}{\cos x} = \frac{\pm\frac{2\sqrt{6}}{5}}{\frac{1}{5}} = \pm 2\sqrt{6}$$

$$\sec x = \frac{1}{\cos x} = \frac{1}{\frac{1}{5}} = 5$$

Quadrant I:

$$\frac{\sec x - \tan x}{\sin x} = \frac{5 - 2\sqrt{6}}{\frac{2\sqrt{6}}{5}} = \frac{5(5 - 2\sqrt{6})}{2\sqrt{6}}$$

$$= \frac{25 - 10\sqrt{6}}{2\sqrt{6}} \cdot \frac{\sqrt{6}}{\sqrt{6}} = \frac{25\sqrt{6} - 60}{12}$$

Quadrant IV:

$$\frac{\sec x - \tan x}{\sin x} = \frac{5 - (-2\sqrt{6})}{-\frac{2\sqrt{6}}{5}} = \frac{5(5 + 2\sqrt{6})}{-2\sqrt{6}}$$

$$= \frac{25 + 10\sqrt{6}}{-2\sqrt{6}} \cdot \frac{-\sqrt{6}}{-\sqrt{6}} = \frac{-25\sqrt{6} - 60}{12}$$

7.2 Exercises *(pages 654–657)*

11. $\dfrac{1}{1 + \cos x} - \dfrac{1}{1 - \cos x} = \dfrac{1(1 - \cos x) - 1(1 + \cos x)}{(1 + \cos x)(1 - \cos x)}$

$$= \frac{1 - \cos x - 1 - \cos x}{1 - \cos^2 x}$$

$$= \frac{-2 \cos x}{\sin^2 x} = -\frac{2 \cos x}{\sin^2 x}$$

or $\qquad -\dfrac{2 \cos x}{\sin^2 x} = -2\left(\dfrac{\cos x}{\sin x}\right)\left(\dfrac{1}{\sin x}\right)$

$$= -2 \cot x \csc x$$

15. $(\sin x + 1)^2 - (\sin x - 1)^2$

$$= (\sin^2 x + 2 \sin x + 1) - (\sin^2 x - 2 \sin x + 1)$$
$$\qquad\qquad\qquad\qquad\qquad\text{Square the binomials.}$$

$$= \sin^2 x + 2 \sin x + 1 - \sin^2 x + 2 \sin x - 1$$

$$= 4 \sin x$$

59. Verify that $\dfrac{\tan^2 t - 1}{\sec^2 t} = \dfrac{\tan t - \cot t}{\tan t + \cot t}$ is an identity.

Work with the right side.

$$\frac{\tan t - \cot t}{\tan t + \cot t} = \frac{\tan t - \dfrac{1}{\tan t}}{\tan t + \dfrac{1}{\tan t}} \qquad \cot t = \frac{1}{\tan t}$$

$$= \frac{\tan t}{\tan t}\left(\frac{\tan t - \dfrac{1}{\tan t}}{\tan t + \dfrac{1}{\tan t}}\right)$$

Multiply numerator and denominator of the complex fraction by the LCD, tan t.

$$= \frac{\tan^2 t - 1}{\tan^2 t + 1} \qquad \text{Distributive property}$$

$$= \frac{\tan^2 t - 1}{\sec^2 t} \qquad \tan^2 t + 1 = \sec^2 t$$

87. Show that $\sin(\csc s) = 1$ is not an identity.

We need find only one value for which the statement is false. Let $s = 2$. Use a calculator to find that $\sin(\csc 2) \approx .891094$, which is not equal to 1. Thus, $\sin(\csc s) = 1$ is not true for *all* real numbers s, so it is not an identity.

7.3 Exercises *(pages 666–671)*

23. $\sec \theta = \csc\left(\dfrac{\theta}{2} + 20°\right)$

By a cofunction identity, $\sec \theta = \csc(90° - \theta)$. Thus,

$$\csc\left(\frac{\theta}{2} + 20°\right) = \csc(90° - \theta) \qquad \text{Substitute.}$$

$$\frac{\theta}{2} + 20° = 90° - \theta$$

$$\frac{3\theta}{2} = 70° \qquad \text{Add } \theta; \text{ subtract } 20°.$$

$$\theta = \frac{2}{3}(70°) = \frac{140°}{3}. \quad \text{Multiply by } \tfrac{2}{3}.$$

87. $\cos s = -\dfrac{8}{17}$ and $\cos t = -\dfrac{3}{5}$, s and t in quadrant III

In order to substitute into sum and difference identities, we need to find the values of sin s and sin t, and also the values of tan s and tan t. Because s and t are both in quadrant III, the values of sin s and sin t will be negative, while tan s and tan t will be positive.

$$\sin s = -\sqrt{1 - \cos^2 s} = -\sqrt{1 - \left(-\frac{8}{17}\right)^2}$$

$$= -\sqrt{\frac{225}{289}} = -\frac{15}{17}$$

$$\sin t = -\sqrt{1 - \cos^2 t} = -\sqrt{1 - \left(-\frac{3}{5}\right)^2}$$

(continued)

$$= -\sqrt{\frac{16}{25}} = -\frac{4}{5}$$

$$\tan s = \frac{\sin s}{\cos s} = \frac{-\frac{15}{17}}{-\frac{8}{17}} = \frac{15}{8}$$

$$\tan t = \frac{\sin t}{\cos t} = \frac{-\frac{4}{5}}{-\frac{3}{5}} = \frac{4}{3}$$

(a) $\sin(s + t) = \sin s \cos t + \cos s \sin t$

$$= \left(-\frac{15}{17}\right)\left(-\frac{3}{5}\right) + \left(-\frac{8}{17}\right)\left(-\frac{4}{5}\right)$$

$$= \frac{45}{85} + \frac{32}{85} = \frac{77}{85}$$

(b) $\tan(s + t) = \dfrac{\tan s + \tan t}{1 - \tan s \tan t} = \dfrac{\frac{15}{8} + \frac{4}{3}}{1 - \left(\frac{15}{8}\right)\left(\frac{4}{3}\right)}$

$$= \frac{\frac{45}{24} + \frac{32}{24}}{1 - \frac{60}{24}} = \frac{\frac{77}{24}}{-\frac{36}{24}} = -\frac{77}{36}$$

(c) From parts (a) and (b), $\sin(s + t) > 0$ and $\tan(s + t) < 0$. The only quadrant in which values of sine are positive and values of tangent are negative is quadrant II.

93. $\tan \dfrac{11\pi}{12} = \tan\left(\dfrac{3\pi}{4} + \dfrac{\pi}{6}\right)$ $\dfrac{3\pi}{4} = \dfrac{9\pi}{12}; \dfrac{\pi}{6} = \dfrac{2\pi}{12}$

$$= \frac{\tan \frac{3\pi}{4} + \tan \frac{\pi}{6}}{1 - \tan \frac{3\pi}{4} \tan \frac{\pi}{6}}$$ Tangent sum identity

$$= \frac{-1 + \frac{\sqrt{3}}{3}}{1 - (-1)\left(\frac{\sqrt{3}}{3}\right)}$$ $\tan \frac{3\pi}{4} = -1;$

$\tan \frac{\pi}{6} = \frac{\sqrt{3}}{3}$

$$= \frac{-1 + \frac{\sqrt{3}}{3}}{1 + \frac{\sqrt{3}}{3}}$$ Simplify.

$$= \frac{-1 + \frac{\sqrt{3}}{3}}{1 + \frac{\sqrt{3}}{3}} \cdot \frac{3}{3}$$ Multiply numerator and denominator by 3.

$$= \frac{-3 + \sqrt{3}}{3 + \sqrt{3}}$$

$$= \frac{-3 + \sqrt{3}}{3 + \sqrt{3}} \cdot \frac{3 - \sqrt{3}}{3 - \sqrt{3}}$$ Rationalize the denominator.

$$= \frac{-9 + 6\sqrt{3} - 3}{9 - 3}$$ FOIL

$$= \frac{-12 + 6\sqrt{3}}{6}$$ Combine terms.

$$= \frac{6(-2 + \sqrt{3})}{6}$$ Factor the numerator.

$$= -2 + \sqrt{3}$$ Lowest terms

107. Verify that $\dfrac{\sin(x - y)}{\sin(x + y)} = \dfrac{\tan x - \tan y}{\tan x + \tan y}$ is an identity.

Work with the left side.

$$\frac{\sin(x - y)}{\sin(x + y)} = \frac{\sin x \cos y - \cos x \sin y}{\sin x \cos y + \cos x \sin y}$$

Sine sum and difference identities

$$= \frac{\dfrac{\sin x \cos y}{\cos x \cos y} - \dfrac{\cos x \sin y}{\cos x \cos y}}{\dfrac{\sin x \cos y}{\cos x \cos y} + \dfrac{\cos x \sin y}{\cos x \cos y}}$$

Divide numerator and denominator by cos x cos y.

$$= \frac{\dfrac{\sin x}{\cos x} \cdot 1 - 1 \cdot \dfrac{\sin y}{\cos y}}{\dfrac{\sin x}{\cos x} \cdot 1 + 1 \cdot \dfrac{\sin y}{\cos y}}$$

$$= \frac{\tan x - \tan y}{\tan x + \tan y}$$ Tangent quotient identity

7.4 Exercises *(pages 681–685)*

25. $\dfrac{1}{4} - \dfrac{1}{2} \sin^2 47.1° = \dfrac{1}{4}(1 - 2 \sin^2 47.1°)$ Factor out $\frac{1}{4}$.

$$= \frac{1}{4} \cos 2(47.1°)$$

$$\cos 2A = 1 - 2 \sin^2 A$$

$$= \frac{1}{4} \cos 94.2°$$

31. $\tan 3x = \tan(2x + x)$

$$= \frac{\tan 2x + \tan x}{1 - \tan 2x \tan x}$$ Tangent sum identity

$$= \frac{\dfrac{2 \tan x}{1 - \tan^2 x} + \tan x}{1 - \dfrac{2 \tan x}{1 - \tan^2 x} \cdot \tan x}$$ Tangent double-angle identity

$$= \frac{\dfrac{2 \tan x + (1 - \tan^2 x)\tan x}{1 - \tan^2 x}}{\dfrac{1 - \tan^2 x - 2 \tan^2 x}{1 - \tan^2 x}}$$ Add and subtract using the common denominator.

$$= \frac{2 \tan x + \tan x - \tan^3 x}{1 - \tan^2 x - 2 \tan^2 x}$$

Multiply numerator and denominator by $1 - \tan^2 x$.

$$= \frac{3 \tan x - \tan^3 x}{1 - 3 \tan^2 x}$$ Combine terms.

43. Verify that $\sin 4x = 4 \sin x \cos x \cos 2x$ is an identity.

Work with the left side.

$\sin 4x = \sin 2(2x)$

$$= 2 \sin 2x \cos 2x$$ Sine double-angle identity

$$= 2(2 \sin x \cos x)\cos 2x$$ Sine double-angle identity

$$= 4 \sin x \cos x \cos 2x$$

77. Find $\tan \dfrac{\theta}{2}$, given $\sin \theta = \dfrac{3}{5}$, with $90° < \theta < 180°$.

To find $\tan \frac{\theta}{2}$, we need the values of $\sin \theta$ and $\cos \theta$. We know $\sin \theta = \frac{3}{5}$.

$$\cos \theta = \pm\sqrt{1 - \sin^2 \theta} \qquad \text{Fundamental identity}$$

$$= \pm\sqrt{1 - \left(\frac{3}{5}\right)^2} \qquad \text{Substitute.}$$

$$= \pm\sqrt{\frac{16}{25}}$$

$$\cos \theta = -\frac{4}{5} \qquad \theta \text{ is in quadrant II.}$$

Thus,

$$\tan \frac{\theta}{2} = \frac{\sin \theta}{1 + \cos \theta} \qquad \text{Half-angle identity}$$

$$= \frac{\frac{3}{5}}{1 - \frac{4}{5}} = 3. \quad \text{Simplify.}$$

103. Verify that $\sec^2 \dfrac{x}{2} = \dfrac{2}{1 + \cos x}$ is an identity.

Work with the left side.

$$\sec^2 \frac{x}{2} = \frac{1}{\cos^2 \frac{x}{2}} \qquad \text{Reciprocal identity}$$

$$= \frac{1}{\left(\pm\sqrt{\dfrac{1 + \cos x}{2}}\right)^2} \qquad \begin{array}{l}\text{Cosine half-angle} \\ \text{identity}\end{array}$$

$$= \frac{1}{\dfrac{1 + \cos x}{2}}$$

$$= \frac{2}{1 + \cos x}$$

7.5 Exercises *(pages 696–700)*

85. $\sin\left(2 \cos^{-1} \dfrac{1}{5}\right)$

Let $\theta = \cos^{-1} \frac{1}{5}$, so $\cos \theta = \frac{1}{5}$. The inverse cosine function yields values only in quadrants I and II, and since $\frac{1}{5}$ is positive, θ is in quadrant I. Sketch θ in quadrant I, and label the sides of a right triangle. By the Pythagorean theorem, the length of the side opposite θ will be $\sqrt{5^2 - 1^2} = \sqrt{24} = 2\sqrt{6}$.

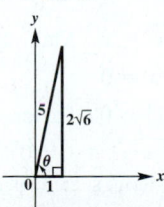

From the figure, $\sin \theta = \dfrac{2\sqrt{6}}{5}$. Then,

$$\sin\left(2 \cos^{-1} \frac{1}{5}\right) = \sin 2\theta$$

$$= 2 \sin \theta \cos \theta$$
$$\text{Sine double-angle identity}$$

$$= 2\left(\frac{2\sqrt{6}}{5}\right)\left(\frac{1}{5}\right) = \frac{4\sqrt{6}}{25}.$$

91. $\sin\left(\sin^{-1} \dfrac{1}{2} + \tan^{-1}(-3)\right)$

Let $\sin^{-1} \frac{1}{2} = A$ and $\tan^{-1}(-3) = B$; then $\sin A = \frac{1}{2}$ and $\tan B = -3$. Sketch angle A in quadrant I and angle B in quadrant IV.

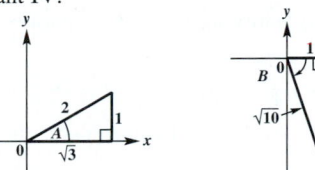

$$\sin\left(\sin^{-1} \frac{1}{2} + \tan^{-1}(-3)\right) = \sin(A + B)$$

$$= \sin A \cos B + \cos A \sin B$$
$$\text{Sine sum identity}$$

$$= \frac{1}{2} \cdot \frac{1}{\sqrt{10}} + \frac{\sqrt{3}}{2} \cdot \frac{-3}{\sqrt{10}}$$

$$= \frac{1 - 3\sqrt{3}}{2\sqrt{10}} = \frac{\sqrt{10} - 3\sqrt{30}}{20}$$

7.6 Exercises *(pages 708–712)*

15. $\tan^2 x + 3 = 0$, so $\tan^2 x = -3$.

The square of a real number cannot be negative, so this equation has no solution. Solution set: $\emptyset$

25.
$$2 \sin \theta - 1 = \csc \theta$$

$$2 \sin \theta - 1 = \frac{1}{\sin \theta} \qquad \text{Reciprocal identity}$$

$$2 \sin^2 \theta - \sin \theta = 1 \qquad \text{Multiply by } \sin \theta.$$

$$2 \sin^2 \theta - \sin \theta - 1 = 0 \qquad \text{Subtract 1.}$$

$$(2 \sin \theta + 1)(\sin \theta - 1) = 0 \qquad \text{Factor.}$$

$$2 \sin \theta + 1 = 0 \quad \text{or} \quad \sin \theta - 1 = 0$$
$$\text{Zero-factor property}$$

$$\sin \theta = -\frac{1}{2} \quad \text{or} \quad \sin \theta = 1$$

Over the interval $[0°, 360°)$, the equation $\sin \theta = -\frac{1}{2}$ has two solutions, the angles in quadrants III and IV that have reference angle $30°$. These are $210°$ and $330°$. In the same interval, the only angle θ for which $\sin \theta = 1$ is $90°$.

Solution set: $\{90°, 210°, 330°\}$

49.
$$\frac{2 \tan \theta}{3 - \tan^2 \theta} = 1$$
$$2 \tan \theta = 3 - \tan^2 \theta$$
$$\tan^2 \theta + 2 \tan \theta - 3 = 0$$
$$(\tan \theta - 1)(\tan \theta + 3) = 0$$
$$\tan \theta = 1 \quad \text{or} \quad \tan \theta = -3$$

Over the interval $[0°, 360°)$, the equation $\tan \theta = 1$ has two solutions, $45°$ and $225°$. Over the same interval, the equation $\tan \theta = -3$ has two solutions that are approximately $-71.6° + 180° = 108.4°$ and $-71.6° + 360° = 288.4°$.

The period of the tangent function is $180°$, so the solution set is $\{45° + 180°n, 108.4° + 180°n, \text{where } n \text{ is any integer}\}$.

73. $\cos 2x + \cos x = 0$

We choose the identity for $\cos 2x$ that involves only the cosine function.

$$\cos 2x + \cos x = 0$$
$$2 \cos^2 x - 1 + \cos x = 0 \quad \text{Cosine double-angle identity}$$
$$2 \cos^2 x + \cos x - 1 = 0 \quad \text{Standard quadratic form}$$
$$(2 \cos x - 1)(\cos x + 1) = 0 \quad \text{Factor.}$$
$$2 \cos x - 1 = 0 \quad \text{or} \quad \cos x + 1 = 0 \quad \text{Zero-factor property}$$
$$2 \cos x = 1 \quad \text{or} \quad \cos x = -1$$
$$\cos x = \frac{1}{2} \quad \text{or} \quad x = \pi$$
$$x = \frac{\pi}{3} \text{ or } \frac{5\pi}{3}$$

Solution set: $\left\{\dfrac{\pi}{3}, \pi, \dfrac{5\pi}{3}\right\}$

81.
$$2 \sin \theta = 2 \cos 2\theta$$
$$\sin \theta = \cos 2\theta \quad \text{Divide by 2.}$$
$$\sin \theta = 1 - 2 \sin^2 \theta \quad \text{Cosine double-angle identity}$$
$$2 \sin^2 \theta + \sin \theta - 1 = 0$$
$$(2 \sin \theta - 1)(\sin \theta + 1) = 0$$
$$2 \sin \theta - 1 = 0 \quad \text{or} \quad \sin \theta + 1 = 0$$
$$\sin \theta = \frac{1}{2} \quad \text{or} \quad \sin \theta = -1$$

Over the interval $[0°, 360°)$, the equation $\sin \theta = \frac{1}{2}$ has two solutions, $30°$ and $150°$. Over the same interval, the equation $\sin \theta = -1$ has one solution, $270°$.

Solution set: $\{30°, 150°, 270°\}$

7.7 Exercises *(pages 715–718)*

15.
$$y = \cos(x + 3)$$
$$x + 3 = \arccos y$$
$$x = -3 + \arccos y$$

37. $\arccos x + 2 \arcsin \dfrac{\sqrt{3}}{2} = \pi$

$$\arccos x = \pi - 2 \arcsin \frac{\sqrt{3}}{2}$$
$$\arccos x = \pi - 2\left(\frac{\pi}{3}\right) \qquad \arcsin \tfrac{\sqrt{3}}{2} = \tfrac{\pi}{3}$$
$$\arccos x = \pi - \frac{2\pi}{3}$$
$$\arccos x = \frac{\pi}{3}$$
$$x = \cos \frac{\pi}{3} = \frac{1}{2}$$

Solution set: $\left\{\dfrac{1}{2}\right\}$

41. $\cos^{-1} x + \tan^{-1} x = \dfrac{\pi}{2}$

$$\cos^{-1} x = \frac{\pi}{2} - \tan^{-1} x$$
$$x = \cos\left(\frac{\pi}{2} - \tan^{-1} x\right)$$
Definition of $\cos^{-1} x$
$$x = \cos \frac{\pi}{2} \cdot \cos(\tan^{-1} x)$$
$$+ \sin \frac{\pi}{2} \cdot \sin(\tan^{-1} x)$$
Cosine difference identity
$$x = 0 \cdot \cos(\tan^{-1} x) + 1 \cdot \sin(\tan^{-1} x)$$
$\cos \tfrac{\pi}{2} = 0; \sin \tfrac{\pi}{2} = 1$
$$x = \sin(\tan^{-1} x)$$

Let $u = \tan^{-1} x$, so $\tan u = x$.

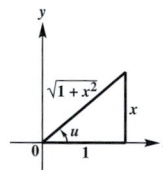

From the triangle, we find that $\sin u = \dfrac{x}{\sqrt{1 + x^2}}$, so the equation $x = \sin(\tan^{-1} x)$ becomes $x = \dfrac{x}{\sqrt{1 + x^2}}$. Solve this equation.

$$x = \frac{x}{\sqrt{1 + x^2}}$$
$$x\sqrt{1 + x^2} = x \quad \text{Multiply by } \sqrt{1 + x^2}.$$
$$x\sqrt{1 + x^2} - x = 0$$
$$x\left(\sqrt{1 + x^2} - 1\right) = 0 \quad \text{Factor.}$$
$$x = 0 \quad \text{or} \quad \sqrt{1 + x^2} - 1 = 0 \quad \text{Zero-factor property}$$
$$\sqrt{1 + x^2} = 1 \quad \text{Isolate the radical.}$$

$$1 + x^2 = 1 \quad \text{Square both sides.}$$
$$x^2 = 0$$
$$x = 0$$

Solution set: $\{0\}$

CHAPTER 8 APPLICATIONS OF TRIGONOMETRY

8.1 Exercises (pages 741–748)

31.
$$\frac{\sin B}{b} = \frac{\sin A}{a} \quad \text{Alternative form of the law of sines}$$

$$\frac{\sin B}{2} = \frac{\sin 60°}{\sqrt{6}} \quad \text{Substitute values from the figure.}$$

$$\sin B = \frac{2 \sin 60°}{\sqrt{6}} \quad \text{Multiply by 2.}$$

$$\sin B = \frac{2 \cdot \frac{\sqrt{3}}{2}}{\sqrt{6}} \quad \sin 60° = \frac{\sqrt{3}}{2}$$

$$\sin B = \frac{\sqrt{3}}{\sqrt{6}} = \sqrt{\frac{1}{2}} = \frac{\sqrt{2}}{2}$$

$$B = 45° \quad \sin 45° = \frac{\sqrt{2}}{2}$$

There is another angle between $0°$ and $180°$ whose sine is $\frac{\sqrt{2}}{2}$: $180° - 45° = 135°$. However, this is too large because $A = 60°$ and $60° + 135° = 195°$. Since $195° > 180°$, there is only one solution, $B = 45°$.

39. $A = 142.13°$, $b = 5.432$ ft, $a = 7.297$ ft

$$\frac{\sin B}{b} = \frac{\sin A}{a} \quad \text{Alternative form of the law of sines}$$

$$\sin B = \frac{b \sin A}{a}$$

$$\sin B = \frac{5.432 \sin 142.13°}{7.297} \quad \text{Substitute given values.}$$

$$\sin B \approx .45697580$$

$$B \approx 27.19° \quad \text{Use the inverse sine function.}$$

Because angle A is obtuse, angle B must be acute, so this is the only possible value for B and there is one triangle with the given measurements.

$$C = 180° - A - B \quad \text{Sum of the angles of any triangle is } 180°.$$

$$C \approx 180° - 142.13° - 27.19°$$

$$C \approx 10.68°$$

Thus, $B \approx 27.19°$ and $C \approx 10.68°$.

61. We cannot find θ directly because the length of the side opposite angle θ is not given. Redraw the triangle shown in the figure and label the third angle as α.

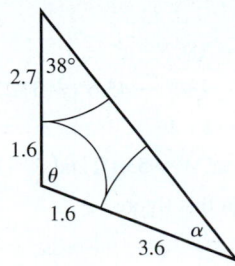

$$\frac{\sin \alpha}{1.6 + 2.7} = \frac{\sin 38°}{1.6 + 3.6} \quad \begin{array}{l}\text{Alternative form of the}\\ \text{law of sines}\end{array}$$

$$\frac{\sin \alpha}{4.3} = \frac{\sin 38°}{5.2}$$

$$\sin \alpha = \frac{4.3 \sin 38°}{5.2} \approx .50910468$$

$$\alpha \approx 31°$$

Then
$$\theta \approx 180° - 38° - 31°$$
$$\theta \approx 111°.$$

73. To find the area of the triangle, use $\mathcal{A} = \frac{1}{2}bh$, with $b = 1$ and $h = \sqrt{2}$.

$$\mathcal{A} = \frac{1}{2}(1)\left(\sqrt{2}\right) = \frac{\sqrt{2}}{2}$$

Now use $\mathcal{A} = \frac{1}{2}ab \sin C$, with $a = 2$, $b = 1$, and $C = 45°$.

$$\mathcal{A} = \frac{1}{2}(2)(1) \sin 45° = \sin 45° = \frac{\sqrt{2}}{2}$$

Both formulas show that the area is $\frac{\sqrt{2}}{2}$ sq unit.

8.2 Exercises (pages 755–762)

21. $C = 45.6°$, $b = 8.94$ m, $a = 7.23$ m

First find c.
$$c^2 = a^2 + b^2 - 2ab \cos C \quad \text{Law of cosines}$$
$$c^2 = 7.23^2 + 8.94^2 - 2(7.23)(8.94) \cos 45.6°$$
$$\qquad\qquad\qquad\qquad\qquad\qquad \text{Substitute given values.}$$
$$c^2 \approx 41.7493$$
$$c \approx 6.46$$

Find A next since angle A is smaller than angle B (because $a < b$), and thus angle A must be acute.

$$\frac{\sin A}{a} = \frac{\sin C}{c} \quad \begin{array}{l}\text{Alternative form of the}\\ \text{law of sines}\end{array}$$

$$\sin A = \frac{a \sin C}{c}$$

$$\sin A = \frac{7.23 \sin 45.6°}{6.46}$$

$$\sin A \approx .79963428$$

$$A \approx 53.1°$$

(continued)

Finally, find B.

$$B = 180° - C - A$$
$$B \approx 180° - 45.6° - 53.1°$$
$$B \approx 81.3°$$

Thus, $c \approx 6.46$ m, $A \approx 53.1°$, and $B \approx 81.3°$.

41. Find AC, or b, in this figure.

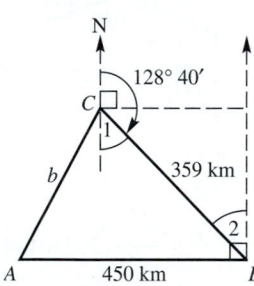

Angle $1 = 180° - 128° 40' = 51° 20'$

Angles 1 and 2 are alternate interior angles formed when two parallel lines (the north lines) are cut by a transversal, line BC, so angle $2 =$ angle $1 = 51° 20'$.

angle $ABC = 90° -$ angle $2 = 90° - 51° 20' = 38° 40'$
Complementary angles

$$b^2 = a^2 + c^2 - 2ac \cos B$$
Law of cosines

$$b^2 = 359^2 + 450^2 - 2(359)(450) \cos 38° 40'$$
Substitute values from the figure.

$$b^2 \approx 79,106$$
$$b \approx 281$$

C is about 281 km from A.

8.3 Exercises *(pages 770–773)*

19. Use the figure to find the components of **a** and **b**: $\mathbf{a} = \langle -8, 8 \rangle$ and $\mathbf{b} = \langle 4, 8 \rangle$.

(a) $\mathbf{a} + \mathbf{b} = \langle -8, 8 \rangle + \langle 4, 8 \rangle = \langle -8 + 4, 8 + 8 \rangle = \langle -4, 16 \rangle$

(b) $\mathbf{a} - \mathbf{b} = \langle -8, 8 \rangle - \langle 4, 8 \rangle = \langle -8 - 4, 8 - 8 \rangle$
$$= \langle -12, 0 \rangle$$

(c) $-\mathbf{a} = -\langle -8, 8 \rangle = \langle 8, -8 \rangle$

47. $\mathbf{v} = \langle a, b \rangle = \langle 5 \cos(-35°), 5 \sin(-35°) \rangle$
$$= \langle 4.0958, -2.8679 \rangle$$

81. First write the given vectors in component form.

$$3\mathbf{i} + 4\mathbf{j} = \langle 3, 4 \rangle; \quad \mathbf{j} = \langle 0, 1 \rangle$$

$$\cos \theta = \frac{\langle 3, 4 \rangle \cdot \langle 0, 1 \rangle}{|\langle 3, 4 \rangle| \, |\langle 0, 1 \rangle|}$$

$$\cos \theta = \frac{3(0) + 4(1)}{\sqrt{9 + 16} \cdot \sqrt{0 + 1}} = \frac{4}{5} = .8$$

$$\theta = \cos^{-1} .8 \approx 36.87°$$

8.4 Exercises *(pages 776–779)*

5. Use the parallelogram rule. In the figure, **x** represents the second force and **v** is the resultant.

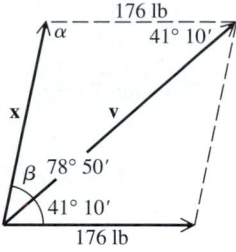

$$\alpha = 180° - 78° 50'$$
$$= 101° 10'$$

$$\beta = 78° 50' - 41° 10'$$
$$= 37° 40'$$

$$\frac{|\mathbf{x}|}{\sin 41° 10'} = \frac{176}{\sin 37° 40'}$$
Law of sines

$$|\mathbf{x}| = \frac{176 \sin 41° 10'}{\sin 37° 40'} \approx 190$$

$$\frac{|\mathbf{v}|}{\sin \alpha} = \frac{176}{\sin 37° 40'}$$
Law of sines

$$|\mathbf{v}| = \frac{176 \sin 101° 10'}{\sin 37° 40'} \approx 283$$

Thus, the magnitude of the second force is about 190 lb, and the magnitude of the resultant is about 283 lb.

27. Let **v** represent the airspeed vector.

The groundspeed is $\dfrac{400 \text{ mi}}{2.5 \text{ hr}} = 160$ mph.

angle $BAC = 328° - 180° = 148°$

$$|\mathbf{v}|^2 = 11^2 + 160^2 - 2(11)(160) \cos 148°$$
Law of cosines

$$|\mathbf{v}|^2 \approx 28,706$$
$$|\mathbf{v}| \approx 169.4$$

The airspeed must be approximately 170 mph.

$$\frac{\sin B}{11} = \frac{\sin 148°}{169.4}$$
Law of sines

$$\sin B = \frac{11 \sin 148°}{169.4} \approx .03441034$$

$$B \approx 2°$$

The bearing must be approximately $360° - 2° = 358°$.

8.5 Exercises *(pages 788–791)*

29. $3 \operatorname{cis} 150° = 3(\cos 150° + i \sin 150°)$

$$= 3\left(-\frac{\sqrt{3}}{2} + i \cdot \frac{1}{2} \right)$$

$$= -\frac{3\sqrt{3}}{2} + \frac{3}{2} i$$
Rectangular form

41. $-5 - 5i$

Sketch the graph of $-5 - 5i$ in the complex plane.

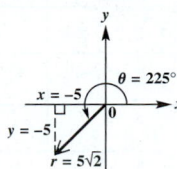

Since $x = -5$ and $y = -5$,

$$r = \sqrt{x^2 + y^2} = \sqrt{(-5)^2 + (-5)^2} = \sqrt{50} = 5\sqrt{2}$$

and

$$\tan \theta = \frac{y}{x} = \frac{-5}{-5} = 1.$$

Since $\tan \theta = 1$, the reference angle for θ is $45°$. The graph shows that θ is in quadrant III, so $\theta = 180° + 45° = 225°$. Therefore,

$$-5 - 5i = 5\sqrt{2}(\cos 225° + i \sin 225°).$$

61. $[4(\cos 60° + i \sin 60°)][6(\cos 330° + i \sin 330°)]$

$$= 4 \cdot 6[\cos(60° + 330°) + i \sin(60° + 330°)]$$
Product theorem

$$= 24(\cos 390° + i \sin 390°) \quad \text{Multiply and add.}$$

$$= 24(\cos 30° + i \sin 30°) \quad \begin{array}{l}390° \text{ and } 30° \text{ are}\\ \text{coterminal angles.}\end{array}$$

$$= 24\left(\frac{\sqrt{3}}{2} + i \cdot \frac{1}{2}\right) \quad \cos 30° = \frac{\sqrt{3}}{2}; \sin 30° = \frac{1}{2}$$

$$= 12\sqrt{3} + 12i$$

77. $\dfrac{-i}{1 + i}$

Numerator: $-i = 0 - 1i$

$$r = \sqrt{0^2 + (-1)^2} = 1$$

$\theta = 270°$ since $\cos 270° = 0$ and $\sin 270° = -1$, so $-i = 1 \operatorname{cis} 270°$.

Denominator: $1 + i = 1 + 1i$

$$r = \sqrt{1^2 + 1^2} = \sqrt{2}$$

$$\tan \theta = \frac{y}{x} = \frac{1}{1} = 1$$

Since x and y are both positive, θ is in quadrant I, so $\theta = \tan^{-1} 1 = 45°$. Thus, $1 + i = \sqrt{2} \operatorname{cis} 45°$.

$$\frac{-i}{1 + i} = \frac{1 \operatorname{cis} 270°}{\sqrt{2} \operatorname{cis} 45°}$$

$$= \frac{1}{\sqrt{2}} \operatorname{cis}(270° - 45°) \quad \text{Quotient theorem}$$

$$= \frac{\sqrt{2}}{2} \operatorname{cis} 225°$$

$$= \frac{\sqrt{2}}{2}(\cos 225° + i \sin 225°)$$

$$= \frac{\sqrt{2}}{2}\left(-\frac{\sqrt{2}}{2} - i \cdot \frac{\sqrt{2}}{2}\right)$$

$$= -\frac{1}{2} - \frac{1}{2}i \qquad \text{Rectangular form}$$

8.6 Exercises (*pages 796–798*)

11. $(-2 - 2i)^5$

First write $-2 - 2i$ in trigonometric form.

$$r = \sqrt{(-2)^2 + (-2)^2} = \sqrt{8} = 2\sqrt{2}$$

$$\tan \theta = \frac{y}{x} = \frac{-2}{-2} = 1$$

Because x and y are both negative, θ is in quadrant III, so $\theta = 225°$.

$$-2 - 2i = 2\sqrt{2}(\cos 225° + i \sin 225°)$$

$$(-2 - 2i)^5 = \left[2\sqrt{2}(\cos 225° + i \sin 225°)\right]^5$$

$$= (2\sqrt{2})^5[\cos(5 \cdot 225°) + i \sin(5 \cdot 225°)]$$
De Moivre's theorem

$$= 32 \cdot 4\sqrt{2}(\cos 1125° + i \sin 1125°)$$

$$= 128\sqrt{2}(\cos 1125° + i \sin 1125°)$$

$$= 128\sqrt{2}(\cos 45° + i \sin 45°)$$
1125° and 45° are coterminal.

$$= 128\sqrt{2}\left(\frac{\sqrt{2}}{2} + i \cdot \frac{\sqrt{2}}{2}\right)$$
$\cos 45° = \frac{\sqrt{2}}{2}; \sin 45° = \frac{\sqrt{2}}{2}$

$$= 128 + 128i \quad \text{Rectangular form}$$

41. $x^3 - \left(4 + 4i\sqrt{3}\right) = 0$

$$x^3 = 4 + 4i\sqrt{3}$$

$$r = \sqrt{4^2 + (4\sqrt{3})^2} = \sqrt{16 + 48} = \sqrt{64} = 8$$

$$\tan \theta = \frac{4\sqrt{3}}{4} = \sqrt{3}$$

θ is in quadrant I, so $\theta = 60°$.

$$x^3 = 4 + 4i\sqrt{3}$$

$$x^3 = 8\left(\frac{1}{2} + i\frac{\sqrt{3}}{2}\right)$$

$$r^3(\cos 3\alpha + i \sin 3\alpha) = 8(\cos 60° + i \sin 60°)$$

$r^3 = 8$, so $r = 2$.

$$\alpha = \frac{60°}{3} + \frac{360° \cdot k}{3}, \text{ } k \text{ any integer} \quad n\text{th root theorem}$$

$$\alpha = 20° + 120° \cdot k, \text{ } k \text{ any integer}$$

If $k = 0$, then $\alpha = 20° + 0° = 20°$.

If $k = 1$, then $\alpha = 20° + 120° = 140°$.

If $k = 2$, then $\alpha = 20° + 240° = 260°$.

Solution set: $\{2(\cos 20° + i \sin 20°),$
$2(\cos 140° + i \sin 140°), 2(\cos 260° + i \sin 260°)\}$

8.7 Exercises *(pages 808–812)*

53.

$$r = 2 \sin \theta$$

$r^2 = 2r \sin \theta$	Multiply by r.
$x^2 + y^2 = 2y$	$r^2 = x^2 + y^2,\ r \sin \theta = y$
$x^2 + y^2 - 2y = 0$	Subtract $2y$.
$x^2 + y^2 - 2y + 1 = 1$	Add 1 to complete the square on y.
$x^2 + (y - 1)^2 = 1$	Factor the perfect square trinomial.

The graph is a circle with center $(0, 1)$ and radius 1.

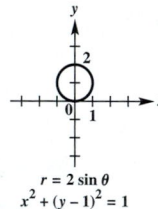

$r = 2 \sin \theta$
$x^2 + (y - 1)^2 = 1$

59.

$$r = 2 \sec \theta$$

$r = \dfrac{2}{\cos \theta}$	Reciprocal identity
$r \cos \theta = 2$	Multiply by $\cos \theta$.
$x = 2$	$r \cos \theta = x$

The graph is the vertical line through $(2, 0)$.

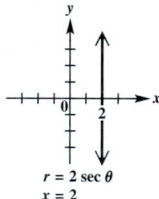

$r = 2 \sec \theta$
$x = 2$

8.8 Exercises *(pages 818–821)*

9. $x = t^3 + 1,\ y = t^3 - 1,$ for t in $(-\infty, \infty)$

(a)

t	x	y
-2	-7	-9
-1	0	-2
0	1	-1
1	2	0
2	9	7
3	28	26

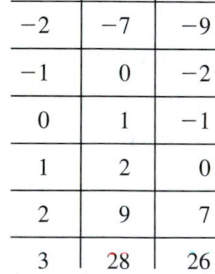

$x = t^3 + 1$
$y = t^3 - 1$
for t in $(-\infty, \infty)$

(b)

$x = t^3 + 1$

$\underline{y = t^3 - 1}$

$x - y = 2$	Subtract equations to eliminate t.
$y = x - 2$	Solve for y.

The rectangular equation is $y = x - 2$, for x in $(-\infty, \infty)$.
The graph is a line with slope 1 and y-intercept -2.

13. $x = 3 \tan t,\ y = 2 \sec t,$ for t in $\left(-\frac{\pi}{2}, \frac{\pi}{2}\right)$

(a)

t	x	y
$-\frac{\pi}{3}$	$-3\sqrt{3} \approx -5.2$	4
$-\frac{\pi}{6}$	$-\sqrt{3} \approx -1.7$	$\frac{4\sqrt{3}}{3} \approx 2.3$
0	0	2
$\frac{\pi}{6}$	$\sqrt{3} \approx 1.7$	$\frac{4\sqrt{3}}{3} \approx 2.3$
$\frac{\pi}{3}$	$3\sqrt{3} \approx 5.2$	4

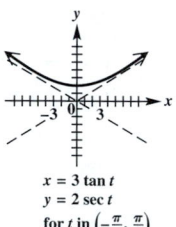

$x = 3 \tan t$
$y = 2 \sec t$
for t in $\left(-\frac{\pi}{2}, \frac{\pi}{2}\right)$

(b) $x = 3 \tan t$, so $\dfrac{x}{3} = \tan t$.

$y = 2 \sec t$, so $\dfrac{y}{2} = \sec t$.

$1 + \tan^2 t = \sec^2 t$	Pythagorean identity
$1 + \left(\dfrac{x}{3}\right)^2 = \left(\dfrac{y}{2}\right)^2$	Substitute expressions for $\tan t$ and $\sec t$.
$1 + \dfrac{x^2}{9} = \dfrac{y^2}{4}$	
$y^2 = 4\left(1 + \dfrac{x^2}{9}\right)$	Multiply by 4.
$y = 2\sqrt{1 + \dfrac{x^2}{9}}$	Use the positive square root because $y > 0$ in the given interval for t.

The rectangular equation is $y = 2\sqrt{1 + \dfrac{x^2}{9}}$, for x in $(-\infty, \infty)$. The graph is the upper half of a hyperbola. (See Chapter 10.)

CHAPTER 9   SYSTEMS AND MATRICES

9.1 Exercises *(pages 848–854)*

27. $\dfrac{x}{2} + \dfrac{y}{3} = 4$ (1)

$\dfrac{3x}{2} + \dfrac{3y}{2} = 15$ (2)

It is easier to work with equations that do not contain fractions. To clear fractions, multiply each equation by the LCD of all the fractions in the equation.

$6\left(\dfrac{x}{2} + \dfrac{y}{3}\right) = 6(4)$ Multiply by the LCD, 6.

$2\left(\dfrac{3x}{2} + \dfrac{3y}{2}\right) = 2(15)$ Multiply by the LCD, 2.

This gives the system

$$3x + 2y = 24 \quad (3)$$
$$3x + 3y = 30. \quad (4)$$

To solve this system by elimination, multiply equation (3) by -1 and add the result to equation (4), eliminating x.

$$\begin{array}{r} -3x - 2y = -24 \\ \underline{3x + 3y = 30} \\ y = 6 \end{array}$$

Substitute 6 for y in equation (1).

$$\frac{x}{2} + \frac{6}{3} = 4 \quad \text{Let } y = 6.$$

$$\frac{x}{2} + 2 = 4$$

$$\frac{x}{2} = 2$$

$$x = 4$$

Solution set: $\{(4, 6)\}$

69. $\dfrac{2}{x} + \dfrac{1}{y} = \dfrac{3}{2} \quad (1)$

$\dfrac{3}{x} - \dfrac{1}{y} = 1 \quad (2)$

Let $\frac{1}{x} = t$ and $\frac{1}{y} = u$. With these substitutions, the system becomes

$$2t + u = \frac{3}{2} \quad (3)$$

$$3t - u = 1. \quad (4)$$

Add these equations, eliminating u, and solve for t.

$$5t = \frac{5}{2}$$

$$t = \frac{1}{2} \quad \text{Multiply by } \tfrac{1}{5}.$$

Substitute $\frac{1}{2}$ for t in equation (3) and solve for u.

$$2\left(\frac{1}{2}\right) + u = \frac{3}{2} \quad \text{Let } t = \tfrac{1}{2}.$$

$$1 + u = \frac{3}{2}$$

$$u = \frac{1}{2}$$

Now find the values of x and y, the variables in the original system. Since $t = \frac{1}{x}$, $tx = 1$, and $x = \frac{1}{t}$. Likewise, $y = \frac{1}{u}$.

$$x = \frac{1}{t} = \frac{1}{\frac{1}{2}} = 2 \quad \text{Let } t = \tfrac{1}{2}.$$

$$y = \frac{1}{u} = \frac{1}{\frac{1}{2}} = 2 \quad \text{Let } u = \tfrac{1}{2}.$$

Solution set: $\{(2, 2)\}$

9.2 Exercises *(pages 862–866)*

47. $\dfrac{1}{(x - 1)(x + 1)} = \dfrac{A}{x - 1} + \dfrac{B}{x + 1}$

$\dfrac{1}{(x - 1)(x + 1)}$

$= \dfrac{A(x + 1)}{(x - 1)(x + 1)} + \dfrac{B(x - 1)}{(x - 1)(x + 1)}$

Write terms on the right with the LCD, $(x - 1)(x + 1)$.

$= \dfrac{A(x + 1) + B(x - 1)}{(x - 1)(x + 1)}$

Combine terms on the right.

$1 = A(x + 1) + B(x - 1)$ Denominators are equal, so numerators must be equal.

$1 = Ax + A + Bx - B$ Distributive property

$1 = (Ax + Bx) + (A - B)$ Group like terms.

$1 = (A + B)x + (A - B)$ Factor $Ax + Bx$.

Since $1 = 0x + 1$, we can equate the coefficients of like powers of x to obtain the system

$$A + B = 0$$
$$A - B = 1.$$

Solve this system by the Gauss-Jordan method.

$$\begin{bmatrix} 1 & 1 & | & 0 \\ 1 & -1 & | & 1 \end{bmatrix}$$

$$\begin{bmatrix} 1 & 1 & | & 0 \\ 0 & -2 & | & 1 \end{bmatrix} \quad -R1 + R2$$

$$\begin{bmatrix} 1 & 1 & | & 0 \\ 0 & 1 & | & -\frac{1}{2} \end{bmatrix} \quad -\tfrac{1}{2}R2$$

$$\begin{bmatrix} 1 & 0 & | & \frac{1}{2} \\ 0 & 1 & | & -\frac{1}{2} \end{bmatrix} \quad -R2 + R1$$

From the final matrix, $A = \frac{1}{2}$ and $B = -\frac{1}{2}$.

55. Let x = number of cubic centimeters of 2% solution and y = number of cubic centimeters of 7% solution.

From the given information, we can write the system

$$x + y = 40$$
$$.02x + .07y = .032(40),$$

or

$$x + y = 40$$
$$.02x + .07y = 1.28.$$

(continued)

Solve this system by the Gauss-Jordan method.

$$\begin{bmatrix} 1 & 1 & | & 40 \\ .02 & .07 & | & 1.28 \end{bmatrix}$$

$$\begin{bmatrix} 1 & 1 & | & 40 \\ 0 & .05 & | & .48 \end{bmatrix} \quad -.02R1 + R2$$

$$\begin{bmatrix} 1 & 1 & | & 40 \\ 0 & 1 & | & 9.6 \end{bmatrix} \quad \frac{1}{.05}R2 \text{ (or } 20R2)$$

$$\begin{bmatrix} 1 & 0 & | & 30.4 \\ 0 & 1 & | & 9.6 \end{bmatrix} \quad -R2 + R1$$

From the final matrix, $x = 30.4$ and $y = 9.6$, so the chemist should mix 30.4 cm^3 of the 2% solution with 9.6 cm^3 of the 7% solution.

9.3 Exercises *(pages 874–878)*

53.
$$\begin{vmatrix} -4 & 1 & 4 \\ 2 & 0 & 1 \\ 0 & 2 & 4 \end{vmatrix} = \begin{vmatrix} 0 & 1 & 6 \\ 2 & 0 & 1 \\ 0 & 2 & 4 \end{vmatrix} \quad \begin{array}{l} \text{Use determinant} \\ \text{theorem 6; } 2R2 + R1. \end{array}$$

$$= 0\begin{vmatrix} 0 & 1 \\ 2 & 4 \end{vmatrix} - 2\begin{vmatrix} 1 & 6 \\ 2 & 4 \end{vmatrix} + 0\begin{vmatrix} 1 & 6 \\ 0 & 1 \end{vmatrix}$$

Expand about column 1.

$$= 0 - 2(4 - 12) + 0$$

$$= -2(-8) = 16$$

67. $1.5x + 3y = 5$ (1)

   $2x + 4y = 3$ (2)

$$D = \begin{vmatrix} 1.5 & 3 \\ 2 & 4 \end{vmatrix} = 1.5(4) - 2(3) = 6 - 6 = 0$$

Because $D = 0$, Cramer's rule does not apply. To determine whether the system is inconsistent or has infinitely many solutions, use the elimination method.

   $6x + 12y = 20$ Multiply equation (1) by 4.

  $\underline{-6x - 12y = -9}$ Multiply equation (2) by -3.

        $0 = 11$ False

The system is inconsistent.

Solution set: $\emptyset$

9.4 Exercises *(pages 884–885)*

13. $\dfrac{x^2}{x^2 + 2x + 1}$

This is not a proper fraction; the numerator has degree greater than or equal to that of the denominator. Divide the numerator by the denominator.

$$\begin{array}{r} 1 \\ x^2 + 2x + 1 \overline{) x^2 } \\ \underline{x^2 + 2x + 1} \\ -2x - 1 \end{array}$$

Find the partial fraction decomposition for

$$\frac{-2x - 1}{x^2 + 2x + 1} = \frac{-2x - 1}{(x + 1)^2}.$$

$$\frac{-2x - 1}{(x + 1)^2} = \frac{A}{x + 1} + \frac{B}{(x + 1)^2} \quad \begin{array}{l} \text{Factor the denominator} \\ \text{of the given fraction.} \end{array}$$

$$-2x - 1 = A(x + 1) + B \quad \begin{array}{l} \text{Multiply by the LCD,} \\ (x + 1)^2. \end{array}$$

Use this equation to find the value of B.

$$-2(-1) - 1 = A(-1 + 1) + B \quad \text{Let } x = -1.$$

$$1 = B$$

Now use the same equation and the value of B to find the value of A.

$$-2(2) - 1 = A(2 + 1) + 1 \quad \text{Let } x = 2 \text{ and } B = 1.$$

$$-5 = 3A + 1$$

$$-6 = 3A$$

$$A = -2$$

Thus,

$$\frac{x^2}{x^2 + 2x + 1} = 1 + \frac{-2}{x + 1} + \frac{1}{(x + 1)^2}.$$

23. $\dfrac{1}{x(2x + 1)(3x^2 + 4)}$

The denominator contains two linear factors and one quadratic factor. All factors are distinct, and $3x^2 + 4$ cannot be factored, so it is irreducible. The partial fraction decomposition is of the form

$$\frac{1}{x(2x + 1)(3x^2 + 4)} = \frac{A}{x} + \frac{B}{2x + 1} + \frac{Cx + D}{3x^2 + 4}.$$

We need to find the values of A, B, C, and D.

$$1 = A(2x + 1)(3x^2 + 4) + Bx(3x^2 + 4)$$

$$+ (Cx + D)(x)(2x + 1) \quad \begin{array}{l} \text{Multiply by the LCD,} \\ x(2x + 1)(3x^2 + 4). \end{array}$$

$$1 = A(6x^3 + 3x^2 + 8x + 4) + B(3x^3 + 4x)$$

$$+ C(2x^3 + x^2) + D(2x^2 + x) \quad \text{Multiply on the right.}$$

$$1 = 6Ax^3 + 3Ax^2 + 8Ax + 4A + 3Bx^3 + 4Bx$$

$$+ 2Cx^3 + Cx^2 + 2Dx^2 + Dx \quad \text{Distributive property}$$

$$1 = (6A + 3B + 2C)x^3 + (3A + C + 2D)x^2$$

$$+ (8A + 4B + D)x + 4A \quad \text{Collect like terms.}$$

Since $1 = 0x^3 + 0x^2 + 0x + 1$, equating the coefficients of like powers of x produces the following system of equations:

$$6A + 3B + 2C = 0$$

$$3A + C + 2D = 0$$

$$8A + 4B + D = 0$$

$$4A = 1.$$

Any of the methods from this chapter can be used to solve this system. If the Gauss-Jordan method is used, begin by representing the system with the augmented matrix

$$\begin{bmatrix} 6 & 3 & 2 & 0 & 0 \\ 3 & 0 & 1 & 2 & 0 \\ 8 & 4 & 0 & 1 & 0 \\ 4 & 0 & 0 & 0 & 1 \end{bmatrix}.$$

After performing a series of row operations, we obtain the following final matrix:

$$\begin{bmatrix} 1 & 0 & 0 & 0 & \frac{1}{4} \\ 0 & 1 & 0 & 0 & -\frac{8}{19} \\ 0 & 0 & 1 & 0 & -\frac{9}{76} \\ 0 & 0 & 0 & 1 & -\frac{6}{19} \end{bmatrix},$$

from which we read the values of the four variables:
$A = \frac{1}{4},\, B = -\frac{8}{19},\, C = -\frac{9}{76},\, D = -\frac{6}{19}.$

Substitute these values into the form given at the beginning of this solution for the partial fraction decomposition.

$$\frac{1}{x(2x+1)(3x^2+4)} = \frac{A}{x} + \frac{B}{2x+1} + \frac{Cx+D}{3x^2+4}$$

$$\frac{1}{x(2x+1)(3x^2+4)} = \frac{\frac{1}{4}}{x} + \frac{-\frac{8}{19}}{2x+1} + \frac{-\frac{9}{76}x - \frac{6}{19}}{3x^2+4}$$

$$= \frac{\frac{1}{4}}{x} + \frac{-\frac{8}{19}}{2x+1} + \frac{-\frac{9}{76}x - \frac{24}{76}}{3x^2+4}$$

Get a common denominator for the numerator of the last term.

$$= \frac{1}{4x} + \frac{-8}{19(2x+1)} + \frac{-9x-24}{76(3x^2+4)}$$

Simplify the complex fractions.

9.5 Exercises *(pages 893–896)*

51. Let x and y represent the two numbers.

$$\frac{x}{y} = \frac{9}{2} \quad \text{(1)}$$

$$xy = 162 \quad \text{(2)}$$

Solve equation (1) for x.

$$y\left(\frac{x}{y}\right) = y\left(\frac{9}{2}\right) \quad \text{Multiply by } y.$$

$$x = \frac{9}{2}y$$

Substitute $\frac{9}{2}y$ for x in equation (2).

$$\left(\frac{9}{2}y\right)y = 162$$

$$\frac{9}{2}y^2 = 162$$

$$\frac{2}{9}\left(\frac{9}{2}y^2\right) = \frac{2}{9}(162) \quad \text{Multiply by } \frac{2}{9}.$$

$$y^2 = 36$$

$$y = \pm 6$$

If $y = 6$, then $x = \frac{9}{2}(6) = 27$. If $y = -6$, then $x = \frac{9}{2}(-6) = -27$. The two numbers are either 27 and 6, or -27 and -6.

63. supply: $p = \sqrt{.1q + 9} - 2$

demand: $p = \sqrt{25 - .1q}$

(a) Equilibrium occurs when supply = demand, so solve the system formed by the supply and demand equations. This system can be solved by substitution. Substitute $\sqrt{.1q + 9} - 2$ for p in the demand equation and solve the resulting equation for q.

$$\sqrt{.1q + 9} - 2 = \sqrt{25 - .1q}$$

$$\left(\sqrt{.1q + 9} - 2\right)^2 = \left(\sqrt{25 - .1q}\right)^2$$

Square both sides.

$$.1q + 9 - 4\sqrt{.1q + 9} + 4 = 25 - .1q$$

$(x - y)^2 = x^2 - 2xy + y^2$

$$.2q - 12 = 4\sqrt{.1q + 9}$$

Combine like terms; isolate radical.

$$(.2q - 12)^2 = \left(4\sqrt{.1q + 9}\right)^2$$

Square both sides again.

$$.04q^2 - 4.8q + 144 = 16(.1q + 9)$$

$$.04q^2 - 4.8q + 144 = 1.6q + 144$$

Distributive property

$$.04q^2 - 6.4q = 0 \quad \text{Combine like terms.}$$

$$.04q(q - 160) = 0 \quad \text{Factor.}$$

$$q = 0 \quad \text{or} \quad q = 160 \quad \text{Zero-factor property}$$

Disregard an equilibrium demand of 0. The equilibrium demand is 160 units.

(b) Substitute 160 for q in either equation and solve for p. Using the supply equation, we obtain

$$p = \sqrt{.1q + 9} - 2$$

$$= \sqrt{.1(160) + 9} - 2 \quad \text{Let } q = 160.$$

$$= \sqrt{16 + 9} - 2$$

$$= \sqrt{25} - 2$$

$$= 5 - 2 = 3.$$

The equilibrium price is $3.

9.6 Exercises *(pages 904–908)*

55. $y \le \log x$

$y \ge |x - 2|$

Graph $y = \log x$ as a solid curve because $y \le \log x$ is a nonstrict inequality.

(Recall that "$\log x$" means $\log_{10} x$.) This graph contains the points $(.1, -1)$ and $(1, 0)$ because $10^{-1} = .1$ and $10^0 = 1$. Use a calculator to approximate other points on the graph, such as $(2, .30)$ and $(4, .60)$. Because the symbol is $\le$, shade the region *below* the curve.

(continued)

Now graph $y = |x - 2|$. Make this boundary solid because $y \geq |x - 2|$ is also a nonstrict inequality. This graph can be obtained by translating the graph of $y = |x|$ to the right 2 units. It contains the points $(0, 2)$, $(2, 0)$, and $(4, 2)$. Because the symbol is $\geq$, shade the region *above* the absolute value graph.

The solution set of the system is the intersection (or overlap) of the two shaded regions, which is shown in the final graph.

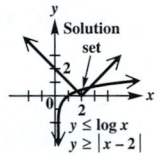

79. Let x = number of cabinet A

and y = number of cabinet B.

The cost constraint is

$$10x + 20y \leq 140.$$

The space constraint is

$$6x + 8y \leq 72.$$

Since the numbers of cabinets cannot be negative, we also have the constraints $x \geq 0$ and $y \geq 0$. We want to maximize the objective function,

$$\text{storage capacity} = 8x + 12y.$$

Find the region of feasible solutions by graphing the system of inequalities that is made up of the constraints.

To graph $10x + 20y \leq 140$, draw the line with x-intercept 14 and y-intercept 7 as a solid line. Because the symbol is $\leq$, shade the region *below* the line.

To graph $6x + 8y \leq 72$, draw the line with x-intercept 12 and y-intercept 9 as a solid line. Because the symbol is $\leq$, shade the region *below* the line.

The constraints $x \geq 0$ and $y \geq 0$ restrict the graph to the first quadrant. The graph of the feasible region is the intersection of the regions that are the graphs of the individual constraints.

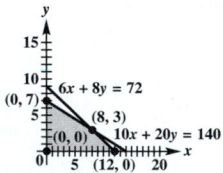

From the graph, observe that three of the vertices are $(0, 0)$, $(0, 7)$, and $(12, 0)$. The fourth vertex is the intersection point of the lines $10x + 20y = 140$ and $6x + 8y = 72$. To find this point, solve the system

$$10x + 20y = 140$$

$$6x + 8y = 72$$

to obtain the ordered pair $(8, 3)$. Evaluate the objective function at each vertex.

Point	Storage Capacity = $8x + 12y$
$(0, 0)$	$8(0) + 12(0) = 0$
$(0, 7)$	$8(0) + 12(7) = 84$
$(8, 3)$	$8(8) + 12(3) = 100$
$(12, 0)$	$8(12) + 12(0) = 96$

The maximum value of $8x + 12y$ occurs at the vertex $(8, 3)$, so the office manager should buy 8 of cabinet A and 3 of cabinet B for a total storage capacity of 100 ft^3.

9.8 Exercises *(pages 931–935)*

23. $A = \begin{bmatrix} 1 & 1 & 0 & 2 \\ 2 & -1 & 1 & -1 \\ 3 & 3 & 2 & -2 \\ 1 & 2 & 1 & 0 \end{bmatrix}$

$\left[\begin{array}{cccc|cccc} 1 & 1 & 0 & 2 & 1 & 0 & 0 & 0 \\ 2 & -1 & 1 & -1 & 0 & 1 & 0 & 0 \\ 3 & 3 & 2 & -2 & 0 & 0 & 1 & 0 \\ 1 & 2 & 1 & 0 & 0 & 0 & 0 & 1 \end{array}\right]$ Write the augmented matrix $[A \,|\, I_4]$.

$\left[\begin{array}{cccc|cccc} 1 & 1 & 0 & 2 & 1 & 0 & 0 & 0 \\ 0 & -3 & 1 & -5 & -2 & 1 & 0 & 0 \\ 0 & 0 & 2 & -8 & -3 & 0 & 1 & 0 \\ 0 & 1 & 1 & -2 & -1 & 0 & 0 & 1 \end{array}\right]$ $\begin{array}{l} -2R1 + R2 \\ -3R1 + R3 \\ -1R1 + R4 \end{array}$

$\left[\begin{array}{cccc|cccc} 1 & 1 & 0 & 2 & 1 & 0 & 0 & 0 \\ 0 & 1 & -\frac{1}{3} & \frac{5}{3} & \frac{2}{3} & -\frac{1}{3} & 0 & 0 \\ 0 & 0 & 2 & -8 & -3 & 0 & 1 & 0 \\ 0 & 1 & 1 & -2 & -1 & 0 & 0 & 1 \end{array}\right]$ $-\frac{1}{3}R2$

$\left[\begin{array}{cccc|cccc} 1 & 0 & \frac{1}{3} & \frac{1}{3} & \frac{1}{3} & \frac{1}{3} & 0 & 0 \\ 0 & 1 & -\frac{1}{3} & \frac{5}{3} & \frac{2}{3} & -\frac{1}{3} & 0 & 0 \\ 0 & 0 & 2 & -8 & -3 & 0 & 1 & 0 \\ 0 & 0 & \frac{4}{3} & -\frac{11}{3} & -\frac{5}{3} & \frac{1}{3} & 0 & 1 \end{array}\right]$ $\begin{array}{l} -1R2 + R1 \\ \\ \\ -1R2 + R4 \end{array}$

$\left[\begin{array}{cccc|cccc} 1 & 0 & \frac{1}{3} & \frac{1}{3} & \frac{1}{3} & \frac{1}{3} & 0 & 0 \\ 0 & 1 & -\frac{1}{3} & \frac{5}{3} & \frac{2}{3} & -\frac{1}{3} & 0 & 0 \\ 0 & 0 & 1 & -4 & -\frac{3}{2} & 0 & \frac{1}{2} & 0 \\ 0 & 0 & \frac{4}{3} & -\frac{11}{3} & -\frac{5}{3} & \frac{1}{3} & 0 & 1 \end{array}\right]$ $\frac{1}{2}R3$

$\left[\begin{array}{cccc|cccc} 1 & 0 & 0 & \frac{5}{3} & \frac{5}{6} & \frac{1}{3} & -\frac{1}{6} & 0 \\ 0 & 1 & 0 & \frac{1}{3} & \frac{1}{6} & -\frac{1}{3} & \frac{1}{6} & 0 \\ 0 & 0 & 1 & -4 & -\frac{3}{2} & 0 & \frac{1}{2} & 0 \\ 0 & 0 & 0 & \frac{5}{3} & \frac{1}{3} & \frac{1}{3} & -\frac{2}{3} & 1 \end{array}\right]$ $\begin{array}{l} -\frac{1}{3}R3 + R1 \\ \frac{1}{3}R3 + R2 \\ \\ -\frac{4}{3}R3 + R4 \end{array}$

$$\begin{bmatrix} 1 & 0 & 0 & \frac{5}{3} & \frac{5}{6} & \frac{1}{3} & -\frac{1}{6} & 0 \\ 0 & 1 & 0 & \frac{1}{3} & \frac{1}{6} & -\frac{1}{3} & \frac{1}{6} & 0 \\ 0 & 0 & 1 & -4 & -\frac{3}{2} & 0 & \frac{1}{2} & 0 \\ 0 & 0 & 0 & 1 & \frac{1}{5} & \frac{1}{5} & -\frac{2}{5} & \frac{3}{5} \end{bmatrix} \quad \frac{3}{5}\text{R4}$$

$$\begin{bmatrix} 1 & 0 & 0 & 0 & \frac{1}{2} & 0 & \frac{1}{2} & -1 \\ 0 & 1 & 0 & 0 & \frac{1}{10} & -\frac{2}{5} & \frac{3}{10} & -\frac{1}{5} \\ 0 & 0 & 1 & 0 & -\frac{7}{10} & \frac{4}{5} & -\frac{11}{10} & \frac{12}{5} \\ 0 & 0 & 0 & 1 & \frac{1}{5} & \frac{1}{5} & -\frac{2}{5} & \frac{3}{5} \end{bmatrix} \quad \begin{array}{l} -\frac{5}{3}\text{R4} + \text{R1} \\ -\frac{1}{3}\text{R4} + \text{R2} \\ 4\text{R4} + \text{R3} \end{array}$$

$$A^{-1} = \begin{bmatrix} \frac{1}{2} & 0 & \frac{1}{2} & -1 \\ \frac{1}{10} & -\frac{2}{5} & \frac{3}{10} & -\frac{1}{5} \\ -\frac{7}{10} & \frac{4}{5} & -\frac{11}{10} & \frac{12}{5} \\ \frac{1}{5} & \frac{1}{5} & -\frac{2}{5} & \frac{3}{5} \end{bmatrix}$$

41. $.2x + .3y = -1.9$

$.7x - .2y = 4.6$

$$A = \begin{bmatrix} .2 & .3 \\ .7 & -.2 \end{bmatrix}, \quad X = \begin{bmatrix} x \\ y \end{bmatrix}, \quad B = \begin{bmatrix} -1.9 \\ 4.6 \end{bmatrix}$$

Find A^{-1}.

$$\begin{bmatrix} .2 & .3 & 1 & 0 \\ .7 & -.2 & 0 & 1 \end{bmatrix} \qquad \text{Write the augmented}$$
$$\text{matrix } [A \,|\, I_2].$$

$$\begin{bmatrix} 1 & 1.5 & 5 & 0 \\ .7 & -.2 & 0 & 1 \end{bmatrix} \qquad 5\text{R1}$$

$$\begin{bmatrix} 1 & 1.5 & 5 & 0 \\ 0 & -1.25 & -3.5 & 1 \end{bmatrix} \quad -.7\text{R1} + \text{R2}$$

$$\begin{bmatrix} 1 & 1.5 & 5 & 0 \\ 0 & 1 & 2.8 & -.8 \end{bmatrix} \quad \frac{1}{-1.25}\text{R2, or } -.8\text{R2}$$

$$\begin{bmatrix} 1 & 0 & .8 & 1.2 \\ 0 & 1 & 2.8 & -.8 \end{bmatrix} \quad -1.5\text{R2} + \text{R1}$$

$$A^{-1} = \begin{bmatrix} .8 & 1.2 \\ 2.8 & -.8 \end{bmatrix}$$

$$X = A^{-1}B = \begin{bmatrix} .8 & 1.2 \\ 2.8 & -.8 \end{bmatrix}\begin{bmatrix} -1.9 \\ 4.6 \end{bmatrix} = \begin{bmatrix} 4 \\ -9 \end{bmatrix}$$

Solution set: $\{(4, -9)\}$

CHAPTER 10   ANALYTIC GEOMETRY

10.1 Exercises *(pages 957–960)*

29. $(x - 7)^2 = 16(y + 5)$

This equation can be rewritten as

$$(x - 7)^2 = 16[y - (-5)].$$

Thus, it has the form

$$(x - h)^2 = 4p(y - k),$$

with $h = 7$, $k = -5$, and $4p = 16$, so $p = 4$. The graph of the given equation is a parabola with vertical axis. The vertex (h, k) is $(7, -5)$. Because this parabola has a vertical axis and $p > 0$, the parabola opens up, so the focus is distance $p = 4$ units above the vertex. Thus, the focus is the point $(7, -1)$.

The directrix is a horizontal line $p = 4$ units below the vertex, so the directrix is the line $y = -9$. The axis is the vertical line through the vertex, so the equation of the axis is $x = 7$.

37. Through $(3, 2)$, symmetric with respect to the x-axis

This parabola has a horizontal axis (the x-axis) because of the symmetry and vertex $(0, 0)$, so its equation can be written in the form $y^2 = 4px$. Use this equation with the coordinates of the point $(3, 2)$ to find the value of p.

$$y^2 = 4px$$
$$2^2 = 4p \cdot 3 \quad \text{Let } x = 3 \text{ and } y = 2.$$
$$4 = 12p$$
$$p = \frac{1}{3}$$

Thus, the equation of the parabola is

$$y^2 = 4px = 4\left(\frac{1}{3}\right)x = \frac{4}{3}x.$$

53. Place the parabola that represents the arch on a coordinate system with the center of the bottom of the arch at the origin. Then the vertex will be at $(0, 12)$ and the points $(-6, 0)$ and $(6, 0)$ will also be on the parabola.

Because the axis of the parabola is the y-axis and the vertex is $(0, 12)$, the equation will have the form $x^2 = 4p(y - 12)$. Use the coordinates of the point $(6, 0)$ to find the value of p.

$$x^2 = 4p(y - 12)$$
$$6^2 = 4p(0 - 12) \quad \text{Let } x = 6 \text{ and } y = 0.$$
$$36 = -48p$$
$$p = -\frac{3}{4}$$

Thus, the equation of the parabola is

$$x^2 = 4\left(-\frac{3}{4}\right)(y - 12), \quad \text{or} \quad x^2 = -3(y - 12).$$

Now find the x-coordinate of a point whose y-coordinate is 9 and whose x-coordinate is positive.

$$x^2 = -3(y - 12)$$
$$x^2 = -3(9 - 12) \quad \text{Let } y = 9.$$
$$x^2 = 9$$
$$x = \sqrt{9} = 3 \qquad x > 0$$

Using symmetry, the width of the arch 9 ft up is $2(3 \text{ ft}) = 6 \text{ ft}$.

10.2 Exercises *(pages 968–970)*

25. foci at $(0, 4)$, $(0, -4)$; sum of distances from foci to point on ellipse is 10

Center: $(0, 0)$

Distance from center to either focus is $4 = c$.

Sum of distances from foci to point on ellipse is $2a = 10$, so $a = 5$.

Form of equation of ellipse with center at origin and vertical major axis:

$$\frac{x^2}{b^2} + \frac{y^2}{a^2} = 1$$

$c^2 = a^2 - b^2$, so $b^2 = a^2 - c^2$.

$$b^2 = 5^2 - 4^2 = 25 - 16 = 9$$

Equation of ellipse: $\dfrac{x^2}{9} + \dfrac{y^2}{25} = 1$

43. Place the half-ellipse that represents the overpass on a coordinate system with the center of the bottom of the overpass at the origin. If the complete ellipse were drawn, the center of the ellipse would also be at $(0, 0)$. Then the half-ellipse will include the points $(0, 15)$, $(-10, 0)$, and $(10, 0)$. Thus, for the complete ellipse, $a = 15$ and $b = 10$, and the equation would be

$$\frac{x^2}{b^2} + \frac{y^2}{a^2} = 1$$

$$\frac{x^2}{10^2} + \frac{y^2}{15^2} = 1 \quad \text{Let } a = 15 \text{ and } b = 10.$$

$$\frac{x^2}{100} + \frac{y^2}{225} = 1.$$

To find the equation of the half-ellipse, solve this equation for y and use the positive square root since the overpass is represented by the upper half of the ellipse.

$$\frac{y^2}{225} = 1 - \frac{x^2}{100}$$

$$y^2 = 225\left(1 - \frac{x^2}{100}\right)$$

$$y = \sqrt{225\left(1 - \frac{x^2}{100}\right)}$$

$$y = 15\sqrt{1 - \frac{x^2}{100}}$$

Find the y-coordinate of the point whose x-intercept is $\frac{1}{2}(12) = 6$.

$$y = 15\sqrt{1 - \frac{x^2}{100}} \quad \text{Equation of half-ellipse}$$

$$y = 15\sqrt{1 - \frac{6^2}{100}} \quad \text{Let } x = 6.$$

$$y = 15\sqrt{1 - \frac{36}{100}} = 15\sqrt{\frac{64}{100}} = 15(.8) = 12$$

The tallest truck that can pass under the overpass is 12 ft tall.

10.3 Exercises *(pages 977–980)*

23. $\dfrac{y}{3} = \sqrt{1 + \dfrac{x^2}{16}}$

Square both sides to get

$$\frac{y^2}{9} = 1 + \frac{x^2}{16}, \quad \text{or} \quad \frac{y^2}{9} - \frac{x^2}{16} = 1.$$

This is the equation of a hyperbola with center $(0, 0)$, vertices $(0, 3)$ and $(0, -3)$, and asymptotes $y = \pm\frac{3}{4}x$.

The original equation represents the top half of the hyperbola. The domain is $(-\infty, \infty)$, and the range is $[3, \infty)$. The vertical line test shows that this is the graph of a function.

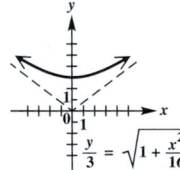

33. vertices at $(0, 6)$ and $(0, -6)$; asymptotes $y = \pm\frac{1}{2}x$

From the given vertices, the equation is of the form

$$\frac{y^2}{a^2} - \frac{x^2}{b^2} = 1$$

$$\frac{y^2}{6^2} - \frac{x^2}{b^2} = 1$$

$$\frac{y^2}{36} - \frac{x^2}{b^2} = 1.$$

The slopes of the asymptotes are $\pm\frac{1}{2}$. Use the positive slope to find the value of b.

$$\frac{a}{b} = \frac{1}{2}$$

$$\frac{6}{b} = \frac{1}{2} \quad \text{Let } a = 6.$$

$$b = 12$$

Use the values of a and b to write the equation of the hyperbola.

$$\frac{y^2}{6^2} - \frac{x^2}{12^2} = 1 \quad \text{Let } a = 6 \text{ and } b = 12.$$

$$\frac{y^2}{36} - \frac{x^2}{144} = 1$$

43. eccentricity 3; center at $(0, 0)$; vertex at $(0, 7)$

Since the center and the given vertex lie on the y-axis, the equation is of the form

$$\frac{y^2}{a^2} - \frac{x^2}{b^2} = 1.$$

The distance between the center and a vertex is 7 units, so $a = 7$. Use the given eccentricity to find the value of c.

$$e = \frac{c}{a}$$

$$3 = \frac{c}{7} \quad \text{Let } e = 3 \text{ and } a = 7.$$

$$c = 21$$

Now find the value of b^2. Since $c^2 = a^2 + b^2$,

$$b^2 = c^2 - a^2 = 21^2 - 7^2 = 441 - 49 = 392.$$

The equation of the hyperbola is $\dfrac{y^2}{49} - \dfrac{x^2}{392} = 1$.

10.4 Exercises *(pages 985–986)*

31. $y^2 - 4y = x + 4$

$y^2 - 4y + 4 = x + 4 + 4$ Add 4 to complete the square.

$(y - 2)^2 = x + 8$ Factor on the left; add on the right.

The equation is a parabola of the form $(y - k)^2 = 4p(x - h)$ with $p = \frac{1}{4}$, $h = -8$, and $k = 2$. Thus, the vertex is $(-8, 2)$ and the parabola opens to the right.

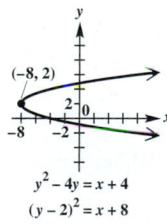

$y^2 - 4y = x + 4$
$(y - 2)^2 = x + 8$

41. From the graph, the coordinates of P (a point on the graph) are $(-3, 8)$, the coordinates of F (a focus) are $(3, 0)$, and the equation of L (the directrix) is $x = 27$.

By the distance formula, the distance from P to F is

$$\sqrt{(x_2 - x_1)^2 + (y_2 - y_1)^2} = \sqrt{[3 - (-3)]^2 + (0 - 8)^2}$$
$$= \sqrt{6^2 + (-8)^2}$$
$$= \sqrt{36 + 64} = \sqrt{100} = 10.$$

The distance between a point and a line is defined as the perpendicular distance, so the distance from P to L is $|27 - (-3)| = 30$.

$$e = \frac{\text{distance from } P \text{ to } F}{\text{distance from } P \text{ to } L} = \frac{10}{30} = \frac{1}{3}$$

CHAPTER 11   FURTHER TOPICS IN ALGEBRA

11.2 Exercises *(pages 1014–1016)*

25. $S_{16} = -160$, $a_{16} = -25$

$$S_n = \frac{n}{2}(a_1 + a_n)$$

$$S_{16} = \frac{16}{2}(a_1 + a_{16}) \quad \text{Let } n = 16.$$

$$-160 = 8(a_1 - 25) \quad \text{Let } S_{16} = -160, a_{16} = -25.$$

$$-20 = a_1 - 25 \quad \text{Divide by 8.}$$

$$a_1 = 5 \quad \text{Add 25.}$$

47. The positive even integers form the arithmetic sequence $2, 4, 6, 8, \ldots$, with $a_1 = 2$ and $d = 2$. Find the sum of the first 60 terms of this sequence.

$$S_n = \frac{n}{2}[2a_1 + (n - 1)d]$$

$$S_{60} = \frac{60}{2}(2a_1 + 59d) \quad \text{Let } n = 60.$$

$$= \frac{60}{2}(2 \cdot 2 + 59 \cdot 2) \quad \text{Let } a_1 = 2, d = 2.$$

$$= 30(4 + 118) \quad \text{Multiply inside parentheses.}$$

$$= 30(122) \quad \text{Add inside parentheses.}$$

$$S_{60} = 3660 \quad \text{Multiply.}$$

11.3 Exercises *(pages 1023–1027)*

31. $\displaystyle\sum_{k=4}^{10} 2^k$

This series is the sum of the fourth through tenth terms of a geometric sequence with $a_1 = 2^1 = 2$ and $r = 2$. To find this sum, find the difference between the sum of the first ten terms and the sum of the first three terms.

$$\sum_{k=4}^{10} 2^k = \sum_{k=1}^{10} 2^k - \sum_{k=1}^{3} 2^k$$

$$= \frac{2(1 - 2^{10})}{1 - 2} - \frac{2(1 - 2^3)}{1 - 2}$$

$$= \frac{2(1 - 1024)}{-1} - \frac{2(1 - 8)}{-1}$$

$$= \frac{2(-1023)}{-1} - \frac{2(-7)}{-1}$$

$$= 2046 - 14$$

$$\sum_{k=4}^{10} 2^k = 2032$$

43. $\dfrac{1}{4} - \dfrac{1}{6} + \dfrac{1}{9} - \dfrac{2}{27} + \cdots$

For this infinite geometric series, $a_1 = \frac{1}{4}$ and

$$r = \frac{-\frac{1}{6}}{\frac{1}{4}} = -\frac{1}{6} \cdot \frac{4}{1} = -\frac{2}{3}.$$

Because $-1 < r < 1$, this series converges.

$$S_\infty = \frac{a_1}{1 - r} = \frac{\frac{1}{4}}{1 - \left(-\frac{2}{3}\right)} = \frac{\frac{1}{4}}{\frac{5}{3}} = \frac{1}{4} \cdot \frac{3}{5} = \frac{3}{20}$$

$$\text{Let } a_1 = \tfrac{1}{4}, r = -\tfrac{2}{3}.$$

69. Option 1 is modeled by the arithmetic sequence
$a_n = 5000 + 10{,}000(n - 1)$ with sum

$$S_n = \frac{n}{2}[2a_1 + (n - 1)d]$$

$$S_{30} = \frac{30}{2}[2a_1 + 29d] \qquad \text{Let } n = 30.$$

$$= \frac{30}{2}(2 \cdot 5000 + 29 \cdot 10{,}000) \quad \text{Let } a_1 = 5000 \text{ and } d = 10{,}000.$$

$$= 15(10{,}000 + 290{,}000)$$

$$S_{30} = 15(300{,}000) = 4{,}500{,}000.$$

Thus, Option 1 pays you a total of \$4,500,000.00.

Option 2 is modeled by the geometric sequence
$a_n = .01(2)^{n-1}$ with sum

$$S_n = \frac{a_1(1 - r^n)}{1 - r}$$

$$S_{30} = \frac{.01(1 - 2^{30})}{1 - 2} = 10{,}737{,}418.23.$$

$$\text{Let } n = 30, \, a_1 = .01, \, r = 2.$$

Thus, Option 2 pays you a total of \$10,737,418.23.

You should choose Option 2.

11.4 Exercises *(pages 1034–1035)*

11. $\dbinom{n}{n - 1} = \dfrac{n!}{(n - 1)![n - (n - 1)]!}$

$$= \frac{n!}{(n - 1)!1!} = \frac{n!}{(n - 1)!}$$

$$= \frac{n(n - 1)!}{(n - 1)!} = n$$

45. Fifteenth term of $(x - y^3)^{20}$

Here, $n = 20$ and $k = 15$, so $k - 1 = 14$ and
$n - (k - 1) = 6$. The fifteenth term of the expansion is

$$\binom{20}{14}x^6(-y^3)^{14} = 38{,}760x^6y^{42}.$$

47. Middle term of $(3x^7 + 2y^3)^8$

This expansion has nine terms, so the middle term is the
fifth term. Here, $n = 8$ and $k = 5$, so $k - 1 = 4$ and
$n - (k - 1) = 4$. The fifth term of the expansion is

$$\binom{8}{4}(3x^7)^4(2y^3)^4 = 70(81x^{28})(16y^{12}) = 90{,}720x^{28}y^{12}.$$

11.5 Exercises *(pages 1041–1042)*

29. Let S_n represent the statement that the number of handshakes for n people is $\dfrac{n^2 - n}{2}$.

We need to prove that this statement is true for every
positive integer $n \geq 2$, so we will use the generalized
principle of mathematical induction.

Step 1 Show that the statement is true when $n = 2$.
S_2 is the statement that for 2 people, the number
of handshakes is

$$\frac{2^2 - 2}{1} = \frac{4 - 2}{2} = \frac{2}{2} = 1,$$

which is true.

Step 2 Show that S_k implies S_{k+1}, where S_k is the statement that the number of handshakes for k people
is $\dfrac{k^2 - k}{2}$ and S_{k+1} is the statement that the number of handshakes for $(k + 1)$ people is
$\dfrac{(k + 1)^2 - (k + 1)}{2}$.

Start with S_k and assume it is a true statement:

For k people, there are $\dfrac{k^2 - k}{2}$ handshakes.

If one more person enters the room that already
contains k people, this person will shake hands
once with each of the k people who were
already in the room, so there will be k additional
handshakes. Thus, the number of handshakes for
$(k + 1)$ people is

$$\frac{k^2 - k}{2} + k = \frac{k^2 - k}{2} + \frac{2k}{2}$$

Get a common denominator.

$$= \frac{k^2 - k + 2k}{2}$$

Add rational expressions.

$$= \frac{k^2 + k}{2}$$

Combine like terms in the numerator.

$$= \frac{(k^2 + 2k + 1) - k - 1}{2}$$

*Write k as $2k - k$; add 1 to
complete the square; subtract 1.*

$$= \frac{(k + 1)^2 - (k + 1)}{2}.$$

Factor out -1 in the numerator.

This work shows that if S_k is true, then S_{k+1}
is true.

Since both steps for a proof by the generalized principle
of mathematical induction have been completed, the
given statement is true for every positive integer $n \geq 2$.

11.6 Exercises *(pages 1049–1053)*

25. (a) There are two choices for the first letter, K and W.

The second letter can be any of the 26 letters of the
alphabet except for the one chosen for the first
letter, so there are 25 choices. The third letter can
be any of the remaining 24 letters of the alphabet,

so there are 24 choices. The fourth letter can be any of the remaining 23 letters of the alphabet, so there are 23 choices.

Therefore, by the fundamental principle of counting, the number of possible 4-letter radio-station call letters is

$$2 \cdot 25 \cdot 24 \cdot 23 = 27{,}600.$$

(b) There are two choices for the first letter.

Because repeats are allowed, there are 26 choices for each of the remaining three letters.

Therefore, by the fundamental principle of counting, the number of possible 4-letter radio-station call letters is

$$2 \cdot 26 \cdot 26 \cdot 26 = 35{,}152.$$

(c) There are two choices for the first letter.

There are 24 choices for the second letter since it cannot repeat the first letter and cannot be R. There are 23 choices for the third letter since it cannot repeat either of the first two letters and cannot be R. There is only one choice for the last letter since it must be R.

Therefore, by the fundamental principle of counting, the number of possible 4-letter radio-station call letters is

$$2 \cdot 24 \cdot 23 \cdot 1 = 1104.$$

55. (c) Choosing 3 women and 2 men involves two independent events, each of which involves combinations.

First, select the women. There are 8 women in the club, so the number of ways to do this is

$$C(8, 3) = \binom{8}{3} = \frac{8!}{5!\,3!} = 56.$$

Now select the men. There are 11 men in the club, so the number of ways to do this is

$$C(11, 2) = \binom{11}{2} = \frac{11!}{9!\,2!} = 55.$$

To find the number of committees, use the fundamental principle of counting. The number of delegations with 3 women and 2 men is $56 \cdot 55 = 3080$.

61. Because the keys are arranged in a circle, there is no "first" key. The number of distinguishable arrangements is the number of ways to arrange the other three keys in relation to any one of the keys, which is

$$P(3, 3) = 3! = 6.$$

11.7 Exercises (pages 1060–1064)

23. (a) A 40-yr-old man who lives 30 more yr would be 70 yr old.

Let E be the event "selected man will live to be 70"; then $n(E) = 66{,}172$. For this situation, the sample

space S is the set of all 40-yr-old men, so $n(S) = 94{,}558$. Thus, the probability that a 40-yr-old man will live 30 more yr is

$$P(E) = \frac{n(E)}{n(S)} = \frac{66{,}172}{94{,}558} \approx .6998.$$

(b) Using the notation and result from part (a), the probability that a 40-yr-old man will not live 30 more yr is

$$P(E') = 1 - P(E) = 1 - .6998 = .3002.$$

(c) Use the notation and results from parts (a) and (b). In this binomial experiment, we call "a 40-yr-old man survives to age 70" a success. Then,

$$p = P(E) = .6998$$

and

$$1 - p = P(E') = .3002.$$

There are 5 independent trials and we need the probability of 3 successes, so $n = 5$ and $r = 3$. The probability that exactly 3 of the 40-yr-old men survive to age 70 is

$$\binom{n}{r} p^r (1 - p)^{n-r} = \binom{5}{3} (.6998)^3 (.3002)^2$$

$$= 10(.6998)^3 (.3002)^2$$

$$\approx .3088.$$

(d) Let F be the event "at least one man survives to age 70." The easiest way to find $P(F)$ is to first find the probability of the complementary event F': "neither man survives to age 70."

$$P(F') = P(E') \cdot P(E') = (.3002)^2 = .0901$$

Then,

$$P(F) = 1 - P(F') \approx 1 - .0901 = .9099.$$

29. $P(\text{smoked between 1 and 19})$

$$= P(\text{smoked 1 to 9, or smoked 10 to 19})$$

$$= P(\text{smoked 1 to 9}) + P(\text{smoked 10 to 19})$$

$$= .24 + .20$$

$$= .44$$

In this binomial experiment, call "smoked between 1 and 19" a success. Then $p = .44$ and $1 - p = .56$. "Fewer than 2" means 0 or 1, so

$P(0 \text{ smoked between 1 and 19}) + P(1 \text{ smoked between 1 and 19})$

$$= \binom{10}{0} (.44)^0 (.56)^{10} + \binom{10}{1} (.44)^1 (.56)^9$$

$$\approx .003033 + .023831$$

$$= .026864.$$

Answers to Selected Exercises

CHAPTER R   REVIEW OF BASIC CONCEPTS

R.1 Exercises *(pages 6–7)*

1. $\{12, 13, 14, 15, 16, 17, 18, 19, 20\}$ **3.** $\left\{1, \dfrac{1}{2}, \dfrac{1}{4}, \dfrac{1}{8}, \dfrac{1}{16}, \dfrac{1}{32}\right\}$ **5.** $\{17, 22, 27, 32, 37, 42, 47\}$ **7.** $\{8, 9, 10, 11, 12, 13, 14\}$

9. finite   **11.** infinite   **13.** infinite   **15.** infinite   **17.** $\in$ **19.** $\notin$ **21.** $\in$ **23.** $\notin$ **25.** $\notin$ **27.** $\notin$

29. false   **31.** true   **33.** true   **35.** true   **37.** false   **39.** true   **41.** true   **43.** false   **45.** true   **47.** false

49. true   **51.** true   **53.** true   **55.** true   **57.** false   **59.** true   **61.** true   **63.** false   **65.** $\subseteq$ **67.** $\nsubseteq$

69. $\subseteq$ **71.** $\{0, 2, 4\}$ **73.** $\{0, 1, 2, 3, 4, 5, 6, 7, 8, 9, 11, 13\}$ **75.** $\emptyset$; M and N are disjoint sets.

77. $\{0, 1, 2, 3, 4, 5, 7, 9, 11, 13\}$ **79.** Q, or $\{0, 2, 4, 6, 8, 10, 12\}$ **81.** $\{10, 12\}$ **83.** $\emptyset$; $\emptyset$ and R are disjoint.

85. N, or $\{1, 3, 5, 7, 9, 11, 13\}$; N and $\emptyset$ are disjoint.   **87.** R, or $\{0, 1, 2, 3, 4\}$ **89.** $\{0, 1, 2, 3, 4, 6, 8\}$ **91.** R, or $\{0, 1, 2, 3, 4\}$

93. $\emptyset$; Q' and $(N' \cap U)$ are disjoint.   **95.** all students in this school who are not taking this course   **97.** all students in this school who are taking calculus and history   **99.** all students in this school who are taking this course or history (or both)

R.2 Exercises *(pages 17–21)*

1. (a) B, C, D, F **(b)** A, B, C, D, F **(c)** D, F **(d)** A, B, C, D, F **(e)** E, F **(f)** D, F   **3.** false; Some are whole numbers, but negative integers are not.   **5.** false; No irrational number is an integer.   **7.** true   **9.** true   **11.** 1, 3

13. $-6, -\dfrac{12}{4}$ (or -3), 0, 1, 3   **15.** -16 **17.** 16   **19.** -243 **21.** -162 **23.** -6 **25.** -60 **27.** -12

29. $-\dfrac{25}{36}$ **31.** $-\dfrac{6}{7}$ **33.** 36   **35.** $-\dfrac{1}{2}$ **37.** $-\dfrac{23}{20}$ **39.** $-\dfrac{13}{3}$ **41.** 92.9   **43.** 86.0   **45.** .031

47. .024; .023; Increased weight results in lower BACs.   **49.** distributive   **51.** inverse   **53.** identity

57. $(8 - 14)p = -6p$ **59.** $-4z + 4y$ **61.** $20z$ **63.** $m + 11$ **65.** $\dfrac{2}{3}y + \dfrac{4}{9}z - \dfrac{5}{3}$ **67.** 1700   **69.** 150

71. false; $|6 - 8| = |8| - |6|$ **73.** true   **75.** false; $|a - b| = |b| - |a|$ **77.** 10   **79.** $-\dfrac{4}{7}$ **81.** 6   **83.** 4

85. -1 **87.** -5 **89.** property 2   **91.** property 3   **93.** property 1   **95.** 8; This represents the number of strokes between their scores.   **97.** 9   **99.** 47°F   **101.** 22°F   **103.** 3   **105.** 9   **107.** x and y have the same sign.

109. x and y have different signs.   **111.** x and y have the same sign.

R.3 Exercises *(pages 30–33)*

1. incorrect; $(mn)^2 = m^2n^2$ **3.** incorrect; $\left(\dfrac{k}{5}\right)^3 = \dfrac{k^3}{5^3}$ **5.** incorrect; $4^5 \cdot 4^2 = 4^{5+2} = 4^7$ **7.** incorrect; $cd^0 = c \cdot 1 = c$

9. correct **11.** 9^8 **13.** $-16x^7$ **15.** n^{11} **17.** $72m^{11}$ **19.** 2^{10} **21.** $(-6)^3x^6$, or $-216x^6$ **23.** -4^2m^6, or $-16m^6$

25. $\dfrac{r^{24}}{s^6}$ **27.** $\dfrac{(-4)^4m^8}{t^4}$, or $\dfrac{256m^8}{t^4}$ **29.** **(a)** B **(b)** C **(c)** B **(d)** C **33.** polynomial; degree 11; monomial

35. polynomial; degree 6; binomial **37.** polynomial; degree 6; binomial **39.** polynomial; degree 6; trinomial

41. not a polynomial **43.** $x^2 - x + 2$ **45.** $12y^2 + 4$ **47.** $6m^4 - 2m^3 - 7m^2 - 4m$ **49.** $28r^2 + r - 2$

51. $15x^4 - \dfrac{7}{3}x^3 - \dfrac{2}{9}x^2$ **53.** $12x^5 + 8x^4 - 20x^3 + 4x^2$ **55.** $-2z^3 + 7z^2 - 11z + 4$

57. $m^2 + mn - 2n^2 - 2km + 5kn - 3k^2$ **59.** $4m^2 - 9$ **61.** $16x^4 - 25y^2$ **63.** $16m^2 + 16mn + 4n^2$

65. $25r^2 - 30rt^2 + 9t^4$ **67.** $4p^2 - 12p + 9 + 4pq - 6q + q^2$ **69.** $9q^2 + 30q + 25 - p^2$

71. $9a^2 + 6ab + b^2 - 6a - 2b + 1$ **73.** $y^3 + 6y^2 + 12y + 8$ **75.** $q^4 - 8q^3 + 24q^2 - 32q + 16$

77. $p^3 - 7p^2 - p - 7$ **79.** $49m^2 - 4n^2$ **81.** $-14q^2 + 11q - 14$ **83.** $4p^2 - 16$ **85.** $11y^3 - 18y^2 + 4y$

87. $2x^5 + 7x^4 - 5x^2 + 7$ **89.** $-5x^2 + 8 + \dfrac{2}{x^2}$ **91.** $2m^2 + m - 2 + \dfrac{6}{3m + 2}$ **93.** $x^3 - x^2 - x + 4 + \dfrac{-17}{3x + 3}$

95. 9999 **96.** 3591 **97.** 10,404 **98.** 5041 **99.** **(a)** $(x + y)^2$ **(b)** $x^2 + 2xy + y^2$ **(d)** the special product for squaring

a binomial **101.** **(a)** approximately 60,501,000 ft^3 **(b)** The shape becomes a rectangular box with a square base, with volume

$V = b^2h$. **(c)** If we let $a = b$, then $V = \dfrac{1}{3}h(a^2 + ab + b^2)$ becomes $V = \dfrac{1}{3}h(b^2 + bb + b^2)$, which simplifies to $V = hb^2$. Yes,

the Egyptian formula gives the same result. **103.** 6.2; .1 high **105.** 2.3; 0, exact **107.** 1,000,000 **109.** 32

R.4 Exercises *(pages 40–43)*

1. $12(m + 5)$ **3.** $8k(k^2 + 3)$ **5.** $xy(1 - 5y)$ **7.** $-2p^2q^4(2p + q)$ **9.** $4k^2m^3(1 + 2k^2 - 3m)$

11. $2(a + b)(1 + 2m)$ **13.** $(r + 3)(3r - 5)$ **15.** $(m - 1)(2m^2 - 7m + 7)$ **17.** The *completely* factored form is

$4xy^3(xy^2 - 2)$. **19.** $(2s + 3)(3t - 5)$ **21.** $(m^4 + 3)(2 - a)$ **23.** $(p^2 - 2)(q^2 + 5)$ **25.** $(2a - 1)(3a - 4)$

27. $(3m + 2)(m + 4)$ **29.** prime **31.** $2a(3a + 7)(2a - 3)$ **33.** $(3k - 2p)(2k + 3p)$ **35.** $(5a + 3b)(a - 2b)$

37. $(4x + y)(3x - y)$ **39.** $2a^2(4a - b)(3a + 2b)$ **41.** $(3m - 2)^2$ **43.** $2(4a + 3b)^2$ **45.** $(2xy + 7)^2$

47. $(a - 3b - 3)^2$ **49.** **(a)** B **(b)** C **(c)** A **(d)** D **51.** $(3a + 4)(3a - 4)$ **53.** $\left(6x + \dfrac{4}{5}\right)\left(6x - \dfrac{4}{5}\right)$

55. $(5s^2 + 3t)(5s^2 - 3t)$ **57.** $(a + b + 4)(a + b - 4)$ **59.** $(p^2 + 25)(p + 5)(p - 5)$ **61.** $(2 - a)(4 + 2a + a^2)$

63. $(5x - 3)(25x^2 + 15x + 9)$ **65.** $(3y^3 + 5z^2)(9y^6 - 15y^3z^2 + 25z^4)$ **67.** $r(r^2 + 18r + 108)$

69. $(3 - m - 2n)(9 + 3m + 6n + m^2 + 4mn + 4n^2)$ **71.** B **73.** $(x - 1)(x^2 + x + 1)(x + 1)(x^2 - x + 1)$

74. $(x - 1)(x + 1)(x^4 + x^2 + 1)$ **75.** $(x^2 - x + 1)(x^2 + x + 1)$ **76.** additive inverse property (0 in the form $x^2 - x^2$

was added on the right.); associative property of addition; factoring a perfect square trinomial; factoring a difference of squares;

commutative property of addition **77.** They are the same. **78.** $(x^4 - x^2 + 1)(x^2 + x + 1)(x^2 - x + 1)$

79. $(m^2 - 5)(m^2 + 2)$ **81.** $9(7k - 3)(k + 1)$ **83.** $(3a - 7)^2$ **85.** $(2b + c + 4)(2b + c - 4)$ **87.** $(x + y)(x - 5)$

89. $(m - 2n)(p^4 + q)$ **91.** $(2z + 7)^2$ **93.** $(10x + 7y)(100x^2 - 70xy + 49y^2)$ **95.** $(5m^2 - 6)(25m^4 + 30m^2 + 36)$

97. $9(x + 2)(3x^2 + 4)$ **99.** $\left(\dfrac{2}{5}x + 7y\right)\left(\dfrac{2}{5}x - 7y\right)$ **101.** prime **103.** $4xy$ **107.** ± 36 **109.** 9

R.5 Exercises *(pages 50–52)*

1. $\{x \mid x \neq 6\}$ **3.** $\left\{x \mid x \neq -\dfrac{1}{2}, 1\right\}$ **5.** $\{x \mid x \neq -2, -3\}$ **7.** (a) $\dfrac{3}{4}$ (b) $\dfrac{1}{6}$ **8.** No; $\dfrac{1}{x} + \dfrac{1}{y} \neq \dfrac{1}{x + y}$.   **9.** (a) $\dfrac{2}{15}$

(b) $-\dfrac{1}{2}$ **10.** No; $\dfrac{1}{x} - \dfrac{1}{y} \neq \dfrac{1}{x - y}$.   **11.** $\dfrac{8}{9}$ **13.** $\dfrac{-3}{t + 5}$ **15.** $\dfrac{2x + 4}{x}$ **17.** $\dfrac{m - 2}{m + 3}$ **19.** $\dfrac{2m + 3}{4m + 3}$ **21.** $\dfrac{25p^2}{9}$

23. $\dfrac{2}{9}$ **25.** $\dfrac{5x}{y}$ **27.** $\dfrac{2a + 8}{a - 3}$, or $\dfrac{2(a + 4)}{a - 3}$ **29.** 1   **31.** $\dfrac{m + 6}{m + 3}$ **33.** $\dfrac{x + 2y}{4 - x}$ **35.** $\dfrac{x^2 - xy + y^2}{x^2 + xy + y^2}$ **37.** B, C

39. $\dfrac{19}{6k}$ **41.** $\dfrac{137}{30m}$ **43.** $\dfrac{a - b}{a^2}$ **45.** $\dfrac{5 - 22x}{12x^2y}$ **47.** 3   **49.** $\dfrac{2x}{(x + z)(x - z)}$ **51.** $\dfrac{4}{a - 2}$, or $\dfrac{-4}{2 - a}$

53. $\dfrac{3x + y}{2x - y}$, or $\dfrac{-3x - y}{y - 2x}$ **55.** $\dfrac{4x - 7}{x^2 - x + 1}$ **57.** $\dfrac{2x^2 - 9x}{(x - 3)(x + 4)(x - 4)}$ **59.** $\dfrac{x + 1}{x - 1}$ **61.** $\dfrac{-1}{x + 1}$

63. $\dfrac{(2 - b)(1 + b)}{b(1 - b)}$ **65.** $\dfrac{m^3 - 4m - 1}{m - 2}$ **67.** $\dfrac{-1}{x(x + h)}$ **69.** $\dfrac{y^2 - 2y - 3}{y^2 + y - 1}$ **71.** about 2305 mi

R.6 Exercises *(pages 59–62)*

1. (a) B (b) D (c) B (d) D   **3.** $\dfrac{1}{(-4)^3}$, or $-\dfrac{1}{64}$ **5.** $-\dfrac{1}{5^4}$, or $-\dfrac{1}{625}$ **7.** 3^2, or 9   **9.** $\dfrac{1}{16x^2}$ **11.** $\dfrac{4}{x^2}$ **13.** $-\dfrac{1}{a^3}$

15. 4^2, or 16   **17.** x^4 **19.** $\dfrac{1}{r^3}$ **21.** 6^6 **23.** $\dfrac{2r^3}{3}$ **25.** $\dfrac{4n^7}{3m^7}$ **27.** $-4r^6$ **29.** $\dfrac{5^4}{a^{10}}$ **31.** $\dfrac{p^4}{5}$ **33.** $\dfrac{1}{2pq}$

35. $\dfrac{4}{a^2}$ **37.** 13   **39.** 2   **41.** $-\dfrac{4}{3}$ **43.** This expression is not a real number.   **45.** (a) E (b) G (c) F (d) F

47. 4   **49.** 1000   **51.** -27 **53.** $\dfrac{256}{81}$ **55.** 9   **57.** 4   **59.** y **61.** $k^{2/3}$ **63.** x^3y^8 **65.** $\dfrac{1}{x^{10/3}}$

67. $\dfrac{6}{m^{1/4}n^{3/4}}$ **69.** p^2 **71.** (a) approximately 250 sec (b) $\dfrac{1}{2^{1.5}} \approx .3536$ **73.** $y - 10y^2$ **75.** $-4k^{10/3} + 24k^{4/3}$

77. $x^2 - x$ **79.** $r - 2 + r^{-1}$, or $r - 2 + \dfrac{1}{r}$ **81.** $k^{-2}(4k + 1)$ **83.** $4t^{-4}(t^2 + 2)$ **85.** $z^{-1/2}(9 + 2z)$

87. $p^{-7/4}(p - 2)$ **89.** $4a^{-7/5}(-a + 4)$ **91.** $(p + 4)^{-3/2}(p^2 + 9p + 21)$ **93.** $6(3x + 1)^{-3/2}(9x^2 + 8x + 2)$

95. $2x(2x + 3)^{-5/9}(-16x^4 - 48x^3 - 30x^2 + 9x + 2)$ **97.** $b + a$ **99.** -1 **101.** $\dfrac{y(xy - 9)}{x^2y^2 - 9}$

103. $\dfrac{2x(1 - 3x^2)}{(x^2 + 1)^5}$ **105.** $\dfrac{1 + 2x^3 - 2x}{4}$ **107.** $\dfrac{3x - 5}{(2x - 3)^{4/3}}$ **109.** 27,000   **111.** 27   **113.** 4   **115.** $\dfrac{1}{100}$

R.7 Exercises *(pages 70–72)*

1. (a) F (b) H (c) G (d) C   **3.** $\sqrt[3]{m^2}$, or $\left(\sqrt[3]{m}\right)^2$ **5.** $\sqrt[3]{(2m + p)^2}$, or $\left(\sqrt[3]{2m + p}\right)^2$ **7.** $k^{2/5}$ **9.** $-3 \cdot 5^{1/2}p^{3/2}$ **11.** A

13. $x \geq 0$ **15.** 5   **17.** $5k^2|m|$ **19.** $|4x - y|$ **21.** 5   **23.** This expression is not a real number.   **25.** $3\sqrt[3]{3}$

27. $-2\sqrt[4]{2}$ **29.** $\sqrt{42pqr}$ **31.** $\sqrt[3]{14xy}$ **33.** $-\dfrac{3}{5}$ **35.** $-\dfrac{\sqrt[3]{5}}{2}$ **37.** $\dfrac{\sqrt[4]{m}}{n}$ **39.** -15 **41.** $32\sqrt[3]{2}$

43. $2x^2z^4\sqrt{2x}$ **45.** This expression cannot be simplified further.   **47.** $\dfrac{\sqrt{6x}}{3x}$ **49.** $\dfrac{x^2y\sqrt{xy}}{z}$ **51.** $\dfrac{2\sqrt[3]{x}}{x}$

53. $\dfrac{h\sqrt[4]{9g^3hr^2}}{3r^2}$ **55.** $\sqrt{3}$ **57.** $\sqrt[3]{2}$ **59.** $\sqrt[12]{2}$ **61.** This expression cannot be simplified further.   **63.** $12\sqrt{2x}$

65. $7\sqrt[3]{3}$ **67.** $3x\sqrt[4]{x^2y^3} - 2x^2\sqrt[4]{x^2y^3}$ **69.** -7 **71.** 10   **73.** $11 + 4\sqrt{6}$ **75.** $5\sqrt{6}$ **77.** $\dfrac{m\sqrt[3]{n^2}}{n}$

79. $\dfrac{x\sqrt[3]{2} - \sqrt[3]{5}}{x^3}$ **81.** $\dfrac{11\sqrt{2}}{8}$ **83.** $-\dfrac{25\sqrt[3]{9}}{18}$ **85.** $\dfrac{\sqrt{15} - 3}{2}$ **87.** $\dfrac{-7 + 2\sqrt{14} + \sqrt{7} - 2\sqrt{2}}{2}$ **89.** $\dfrac{p(\sqrt{p} - 2)}{p - 4}$

91. $\dfrac{5\sqrt{x}\left(2\sqrt{x} - \sqrt{y}\right)}{4x - y}$ **93.** $\dfrac{3m\left(2 - \sqrt{m + n}\right)}{4 - m - n}$ **95.** 19.1 ft per sec **97.** $-12.3°$; The table gives $-12°$.

99. 2 **101.** 2 **103.** 3 **105.** It gives six decimal places of accuracy.

Chapter R Review Exercises *(pages 77–81)*

1. $\{6, 8, 10, 12, 14, 16, 18, 20\}$ **3.** true **5.** true **7.** false **9.** true **11.** true **13.** $\{2, 6, 9, 10\}$ **15.** $\emptyset$

17. $\emptyset$ **19.** $\{1, 2, 3, 4, 6, 8\}$ **21.** $-12, -6, -\sqrt{4}$ (or -2), $0, 6$ **23.** whole number, integer, rational number, real number

25. irrational number, real number **31.** commutative **33.** associative **35.** identity **37.** 3750 **39.** 32

41. $-\dfrac{37}{18}$ **43.** $-\dfrac{12}{5}$ **45.** -32 **47.** -13 **49.** $7q^3 - 9q^2 - 8q + 9$ **51.** $16y^2 + 42y - 49$

53. $9k^2 - 30km + 25m^2$ **55.** (a) 51 million (b) 52 million (c) The approximation is 1 million high. **57.** (a) 183 million

(b) 183 million (c) They are the same. **59.** $6m^2 - 3m + 5$ **61.** $3b - 8 + \dfrac{2}{b^2 + 4}$ **63.** $3(z - 4)^2(3z - 11)$

65. $(z - 8k)(z + 2k)$ **67.** $6a^6(4a + 5b)(2a - 3b)$ **69.** $(7m^4 + 3n)(7m^4 - 3n)$ **71.** $3(9r - 10)(2r + 1)$

73. $(3x - 4)(9x - 34)$ **75.** $\dfrac{1}{2k^2(k - 1)}$ **77.** $\dfrac{x + 1}{x + 4}$ **79.** $\dfrac{(p + q)(p + 6q)^2}{5p}$ **81.** $\dfrac{2m}{m - 4}$, or $\dfrac{-2m}{4 - m}$ **83.** $\dfrac{q + p}{pq - 1}$

85. $\dfrac{1}{64}$ **87.** $\dfrac{16}{25}$ **89.** $-10z^8$ **91.** 1 **93.** $-8y^{11}p$ **95.** $\dfrac{1}{(p + q)^5}$ **97.** $-14r^{17/12}$ **99.** $y^{1/2}$

101. $10z^{7/3} - 4z^{1/3}$ **103.** $10\sqrt{2}$ **105.** $5\sqrt[4]{2}$ **107.** $-\dfrac{\sqrt[3]{50p}}{5p}$ **109.** $\sqrt[12]{m}$ **111.** 66 **113.** $-9m\sqrt{2m} + 5m\sqrt{m}$,

or $m\left(-9\sqrt{2m} + 5\sqrt{m}\right)$ **115.** $\dfrac{6(3 + \sqrt{2})}{7}$

In Exercises 117–127, we give only the corrected right-hand sides of the equations.

117. $x^3 + 5x$ **119.** m^6 **121.** $\dfrac{a}{2b}$ **123.** One possible answer is $\dfrac{\sqrt{a} - \sqrt{b}}{a - b}$. **125.** $4 - t - 1$, or $3 - t$ **127.** 5^2

Chapter R Test *(pages 81–82)*

[R.1] 1. false **2.** true **3.** false **4.** false **5.** true **[R.2] 6.** (a) $-13, -\dfrac{12}{4}$ (or -3), $0, \sqrt{49}$ (or 7) (b) $-13, -\dfrac{12}{4}$

(or -3), $0, \dfrac{3}{5}, 5.9, \sqrt{49}$ (or 7) (c) All are real numbers. **7.** 4 **8.** (a) associative (b) commutative (c) distributive

(d) inverse **9.** 87.9 **[R.3] 10.** $11x^2 - x + 2$ **11.** $36r^2 - 60r + 25$ **12.** $3t^3 + 5t^2 + 2t + 8$

13. $2x^2 - x - 5 + \dfrac{3}{x - 5}$ **14.** \$8401 **15.** \$8797 **[R.4] 16.** $(3x - 7)(2x - 1)$ **17.** $(x^2 + 4)(x + 2)(x - 2)$

18. $2m(4m + 3)(3m - 4)$ **19.** $(x - 2)(x^2 + 2x + 4)(y + 3)(y - 3)$ **[R.5] 20.** $\dfrac{x^4(x + 1)}{3(x^2 + 1)}$

21. $\dfrac{x(4x + 1)}{(x + 2)(x + 1)(2x - 3)}$ **22.** $\dfrac{2a}{2a - 3}$, or $\dfrac{-2a}{3 - 2a}$ **23.** $\dfrac{y}{y + 2}$ **[R.6] 24.** $\dfrac{y}{x}$ **25.** $\dfrac{9}{16}$ **[R.7] 26.** $3x^2y^4\sqrt{2x}$

27. $2\sqrt{2x}$ **28.** $x - y$ **29.** $\dfrac{7\left(\sqrt{11} + \sqrt{7}\right)}{2}$ **30.** approximately 2.1 sec

CHAPTER 1 EQUATIONS AND INEQUALITIES

1.1 Exercises (pages 88–90)

1. true   **3.** false   **7.** B   **9.** $\{-4\}$ **11.** $\{1\}$ **13.** $\left\{-\dfrac{2}{7}\right\}$ **15.** $\left\{-\dfrac{7}{8}\right\}$ **17.** $\{-1\}$ **19.** $\{10\}$ **21.** $\{75\}$

23. $\{0\}$ **25.** $\{12\}$ **27.** $\{50\}$ **29.** identity; {all real numbers}   **31.** conditional equation; $\{0\}$

33. identity; {all real numbers}   **35.** contradiction; $\emptyset$ **39.** $l = \dfrac{V}{wh}$ **41.** $c = P - a - b$

43. $B = \dfrac{2A - hb}{h}$, or $B = \dfrac{2A}{h} - b$ **45.** $h = \dfrac{S - 2\pi r^2}{2\pi r}$, or $h = \dfrac{S}{2\pi r} - r$ **47.** $h = \dfrac{S - 2lw}{2w + 2l}$

Answers in Exercises 49–57 exist in equivalent forms as well.

49. $x = -3a + b$ **51.** $x = \dfrac{3a + b}{3 - a}$ **53.** $x = \dfrac{3 - 3a}{a^2 - a - 1}$ **55.** $x = \dfrac{2a^2}{a^2 + 3}$ **57.** $x = \dfrac{m + 4}{2m + 5}$

59. (a) \$126 **(b)** \$3276   **61.** $104°F$ **63.** $15°C$ **65.** $37.8°C$ **67.** $463.9°C$ **69.** $-14°C$

1.2 Exercises (pages 97–103)

1. 25 mi   **3.** \$40   **5.** A   **7.** D   **9.** 90 cm   **11.** 6 cm   **13.** 600 ft, 800 ft, 1000 ft

15. width: 1.28 cm; length: 1.7 cm   **17.** 4 ft   **19.** 50 mi   **21.** 2.7 mi   **23.** 45 min

25. 1 hr, 8 min, 12 sec; It is about $\dfrac{1}{2}$ the world record time.   **27.** 35 km per hr   **29.** $7\dfrac{1}{2}$ gal   **31.** 2 L   **33.** 4 mL

35. short-term note: \$100,000; long-term note: \$140,000   **37.** \$10,000 at 2.5%; \$20,000 at 3%

39. \$50,000 at 1.5%; \$90,000 at 4%   **41. (a)** .0352 **(b)** approximately .015, or 1.5% **(c)** approximately 1 case

43. (a) 800 ft^3 **(b)** 107,680 μg **(c)** $F = 107,680x$ **(d)** approximately .25 day, or 6 hr

45. (a) 16.8 million **(b)** 2009 **(c)** They are quite close. **(d)** 13.5 million

1.3 Exercises (pages 109–110)

1. true   **3.** true   **5.** false; *Every* real number is a complex number.   **7.** real, complex   **9.** complex, pure imaginary, nonreal complex   **11.** complex, nonreal complex   **13.** real, complex   **15.** complex, pure imaginary, nonreal complex

17. $5i$ **19.** $i\sqrt{10}$ **21.** $12i\sqrt{2}$ **23.** $-3i\sqrt{2}$ **25.** -13 **27.** $-2\sqrt{6}$ **29.** $\sqrt{3}$ **31.** $i\sqrt{3}$ **33.** $\dfrac{1}{2}$

35. -2 **37.** $-3 - i\sqrt{6}$ **39.** $2 + 2i\sqrt{2}$ **41.** $-\dfrac{1}{8} + \dfrac{\sqrt{2}}{8}i$ **43.** $12 - i$ **45.** 2   **47.** $1 - 10i$

49. $-13 + 5i$ **51.** $8 - i$ **53.** $-14 + 2i$ **55.** $5 - 12i$ **57.** 10   **59.** 13   **61.** 7   **63.** $25i$ **65.** $12 + 9i$

67. $20 + 15i$ **69.** i **71.** -1 **73.** $-i$ **75.** 1   **77.** $-i$ **79.** $-i$ **83.** $2 - 2i$ **85.** $\dfrac{3}{5} - \dfrac{4}{5}i$

87. $-1 - 2i$ **89.** $5i$ **91.** $8i$ **93.** $-\dfrac{2}{3}i$ **97.** $4 + 6i$

1.4 Exercises (pages 119–121)

1. G   **3.** C   **5.** H   **7.** D   **9.** D; $\left\{\dfrac{1}{3}, 7\right\}$ **11.** C; $\{-4, 3\}$ **13.** $\{2, 3\}$ **15.** $\left\{-\dfrac{2}{5}, 1\right\}$ **17.** $\left\{-\dfrac{3}{4}, 1\right\}$

19. $\{\pm 4\}$ **21.** $\{\pm 3\sqrt{3}\}$ **23.** $\{\pm 9i\}$ **25.** $\left\{\dfrac{1 \pm 2\sqrt{3}}{3}\right\}$ **27.** $\{-5 \pm i\sqrt{3}\}$ **29.** $\left\{\dfrac{3}{5} \pm \dfrac{\sqrt{3}}{5}i\right\}$ **31.** $\{1, 3\}$

33. $\left\{-\dfrac{7}{2}, 4\right\}$ **35.** $\{1 \pm \sqrt{3}\}$ **37.** $\left\{-\dfrac{5}{2}, 2\right\}$ **39.** $\left\{\dfrac{2 \pm \sqrt{10}}{2}\right\}$ **41.** $\left\{1 \pm \dfrac{\sqrt{3}}{2}i\right\}$

43. He is incorrect because $c = 0$.   **45.** $\left\{\dfrac{1 \pm \sqrt{5}}{2}\right\}$ **47.** $\{3 \pm \sqrt{2}\}$ **49.** $\{1 \pm 2i\}$ **51.** $\left\{\dfrac{3}{2} \pm \dfrac{\sqrt{2}}{2}i\right\}$

53. $\left\{\dfrac{-1 \pm \sqrt{97}}{4}\right\}$ **55.** $\left\{\dfrac{-2 \pm \sqrt{10}}{2}\right\}$ **57.** $\left\{\dfrac{-3 \pm \sqrt{41}}{8}\right\}$ **59.** $\{5\}$ **61.** $\left\{2, -1 \pm i\sqrt{3}\right\}$

63. $\left\{-3, \dfrac{3}{2} \pm \dfrac{3\sqrt{3}}{2}i\right\}$ **65.** $t = \dfrac{\pm\sqrt{2sg}}{g}$ **67.** $v = \dfrac{\pm\sqrt{FrkM}}{kM}$ **69.** $t = \dfrac{v_0 \pm \sqrt{v_0^2 - 64h + 64s_0}}{32}$

71. (a) $x = \dfrac{y \pm \sqrt{8 - 11y^2}}{4}$ **(b)** $y = \dfrac{x \pm \sqrt{6 - 11x^2}}{3}$ **73.** 0; one rational solution (a double solution)

75. 1; two distinct rational solutions   **77.** 84; two distinct irrational solutions

79. -23; two distinct nonreal complex solutions   **81.** 2304; two distinct rational solutions

In Exercises 85 and 87, there are other possible answers.

85. $a = 1, b = -9, c = 20$ **87.** $a = 1, b = -2, c = -1$

Chapter 1 Quiz *(page 121)*

[1.1] 1. $\{2\}$ **2. (a)** contradiction; $\emptyset$ **(b)** identity; $\{$all real numbers$\}$ **(c)** conditional equation; $\left\{\dfrac{11}{4}\right\}$ **3.** $y = \dfrac{3x}{a - 1}$

[1.2] 4. \$10,000 at 2.5\%; \$20,000 at 3\%   **5.** \$5.46; The model predicts a wage that is \$.31 greater than the actual wage.

[1.3] 6. $-\dfrac{1}{2} + \dfrac{\sqrt{6}}{4}i$ **7.** $\dfrac{3}{10} - \dfrac{8}{5}i$ **[1.4] 8.** $\left\{\dfrac{1}{6} \pm \dfrac{\sqrt{11}}{6}i\right\}$ **9.** $\left\{\pm\sqrt{29}\right\}$ **10.** $r = \dfrac{\pm\sqrt{A\pi}}{\pi}$

1.5 Exercises *(pages 126–132)*

1. A   **3.** D   **5.** 7, 8 or $-8, -7$ **7.** 12, 14 or $-14, -12$ **9.** 7, 9 or $-9, -7$ **11.** 5, 6 or $-6, -5$

13. 9, 11 or $-11, -9$ **15.** 6, 8, 10   **17.** 87 in., 10 in.   **19.** 100 yd by 400 yd   **21.** 9 ft by 12 ft   **23.** 20 in. by 30 in.

25. 3.75 cm   **27.** 1 ft   **29.** 4   **31.** 5 ft   **33.** $10\sqrt{2}$ ft   **35.** 16.4 ft   **37.** 3000 yd   **39. (a)** 1 sec, 5 sec

(b) 6 sec   **41. (a)** It will not reach 80 ft. **(b)** 2 sec   **43. (a)** .19 sec, 10.92 sec **(b)** 11.32 sec

45. (a) approximately 19.2 hr **(b)** 84.3 ppm (109.8 is not in the interval $[50, 100]$.)   **47. (a)** 10.3 hr **(b)** 722.2 ppm (857.8 is

not in the interval $[500, 800]$.)   **49.** 23.93 million   **51.** $80 - x$ **52.** $300 + 20x$ **53.** $(80 - x)(300 + 20x) =$

$24{,}000 + 1300x - 20x^2$ **54.** $20x^2 - 1300x + 11{,}000 = 0$ **55.** $\{10, 55\}$; Because of the restriction, only $x = 10$ is valid.

The number of apartments rented is 70.   **57.** 80

1.6 Exercises *(pages 142–145)*

1. $-\dfrac{3}{2}, 6$ **3.** $2, -1$ **5.** 0   **7.** $\{-10\}$ **9.** $\emptyset$ **11.** $\emptyset$ **13.** $\{-9\}$ **15.** $\{-2\}$ **17.** $\emptyset$ **19.** $\left\{-\dfrac{5}{2}, \dfrac{1}{9}\right\}$

21. $\left\{\dfrac{3}{4}, 1\right\}$ **23.** $\{3, 5\}$ **25.** $\left\{-2, \dfrac{5}{4}\right\}$ **27.** $1\dfrac{7}{8}$ hr   **29.** 78 hr   **31.** $13\dfrac{1}{3}$ hr   **33.** 10 min   **35.** $\{3\}$

37. $\{-1\}$ **39.** $\{5\}$ **41.** $\{9\}$ **43.** $\{9\}$ **45.** $\emptyset$ **47.** $\{\pm 2\}$ **49.** $\{0, 3\}$ **51.** $\{-2\}$ **53.** $\left\{-\dfrac{2}{9}, 2\right\}$ **55.** $\{4\}$

57. $\{-2\}$ **59.** $\left\{\dfrac{2}{5}, 1\right\}$ **61.** $\left\{\dfrac{3}{2}\right\}$ **63.** $\{31\}$ **65.** $\{-3, 1\}$ **67.** $\{-27, 3\}$ **69.** $\left\{\pm 1, \pm \dfrac{\sqrt{10}}{2}\right\}$

71. $\left\{\pm\sqrt{3}, \pm i\sqrt{5}\right\}$ **73.** $\left\{\dfrac{1}{4}, 1\right\}$ **75.** $\{0, 8\}$ **77.** $\{-63, 28\}$ **79.** $\{0, 31\}$ **81.** $\left\{\dfrac{-6 \pm 2\sqrt{3}}{3}, \dfrac{-4 \pm \sqrt{2}}{2}\right\}$

83. $\left\{-\dfrac{2}{7}, 5\right\}$ **85.** $\left\{-\dfrac{1}{27}, \dfrac{1}{8}\right\}$ **87.** $\left\{\pm\dfrac{1}{2}, \pm 4\right\}$ **89.** $\{16\}$; $u = -3$ does not lead to a solution of the equation.

90. $\{16\}$; 9 does not satisfy the equation.   **92.** $\{4\}$ **93.** $h = \dfrac{d^2}{k^2}$ **95.** $m = (1 - n^{3/4})^{4/3}$ **97.** $e = \dfrac{Er}{R + r}$

Summary Exercises on Solving Equations *(page 145)*

1. $\{3\}$ **2.** $\{-1\}$ **3.** $\left\{-3 \pm 3\sqrt{2}\right\}$ **4.** $\{2, 6\}$ **5.** $\emptyset$ **6.** $\{-31\}$ **7.** $\{-6\}$ **8.** $\{6\}$ **9.** $\left\{\dfrac{1}{5} \pm \dfrac{2}{5}i\right\}$

10. $\{-2, 1\}$ **11.** $\left\{-\dfrac{1}{243}, \dfrac{1}{3125}\right\}$ **12.** $\{-1\}$ **13.** $\{\pm i, \pm 2\}$ **14.** $\{-2.4\}$ **15.** $\{4\}$ **16.** $\left\{\dfrac{1}{3} \pm \dfrac{\sqrt{2}}{3}i\right\}$

17. $\left\{\dfrac{15}{7}\right\}$ **18.** $\{4\}$ **19.** $\{3, 11\}$ **20.** $\{1\}$ **21.** $\{x \mid x \neq 3\}$ **22.** $a = \sqrt[3]{c^3 - b^3}$

1.7 Exercises *(pages 154–158)*

1. F   **3.** A   **5.** I   **7.** B   **9.** E   **13.** $(-\infty, 4]$;   **15.** $[-1, \infty)$;

17. $(-\infty, 6]$;   **19.** $(-\infty, 4)$;   **21.** $\left[-\dfrac{11}{5}, \infty\right)$;

23. $\left(-\infty, \dfrac{48}{7}\right]$;   **25.** $(-5, 3)$;   **27.** $[3, 6]$;

29. $(4, 6)$;   **31.** $[-9, 9]$;   **33.** $(-16, 19]$;   **35.** $[500, \infty)$

37. $[45, \infty)$ **39.** $(-\infty, -2) \cup (3, \infty)$ **41.** $\left[-\dfrac{3}{2}, 6\right]$ **43.** $(-\infty, -3] \cup [-1, \infty)$ **45.** $[-2, 3]$ **47.** $[-3, 3]$

49. $\left(\dfrac{-5 - \sqrt{33}}{2}, \dfrac{-5 + \sqrt{33}}{2}\right)$ **51.** $\left[1 - \sqrt{2}, 1 + \sqrt{2}\right]$ **53.** A   **55.** $\left\{\dfrac{4}{3}, -2, -6\right\}$ **56.**

57. In the interval $(-\infty, -6)$, choose $x = -10$, for example. It satisfies the original inequality. In the interval $(-6, -2)$, choose $x = -4$, for example. It does not satisfy the inequality. In the interval $\left(-2, \dfrac{4}{3}\right)$, choose $x = 0$, for example. It satisfies the original inequality. In the interval $\left(\dfrac{4}{3}, \infty\right)$, choose $x = 4$, for example. It does not satisfy the original inequality.

58. $(-\infty, -6] \cup \left[-2, \dfrac{4}{3}\right]$ **59.** $\left[-2, \dfrac{3}{2}\right] \cup [3, \infty)$ **61.** $(-\infty, -2] \cup [0, 2]$

63. $(-\infty, -1) \cup (-1, 3)$ **65.** $[-4, -3] \cup [3, \infty)$ **67.** $(-\infty, \infty)$ **69.** $(-5, 3]$ **71.** $(-\infty, -2)$

73. $(-\infty, 6) \cup \left[\dfrac{15}{2}, \infty\right)$ **75.** $(-\infty, 1) \cup \left(\dfrac{9}{5}, \infty\right)$ **77.** $\left(-\infty, -\dfrac{3}{2}\right) \cup \left[-\dfrac{1}{2}, \infty\right)$ **79.** $(-2, \infty)$

81. $\left(0, \dfrac{4}{11}\right) \cup \left(\dfrac{1}{2}, \infty\right)$ **83.** $(-\infty, -2] \cup (1, 2)$ **85.** $(-\infty, 5)$ **87.** $\left[\dfrac{3}{2}, \infty\right)$ **89.** $\left(\dfrac{5}{2}, \infty\right)$

91. $\left[-\dfrac{8}{3}, \dfrac{3}{2}\right] \cup (6, \infty)$ **93.** **(a)** 1992   **(b)** 1998   **95.** between (and inclusive of) 4 sec and 9.75 sec

97. **(a)** $2.08 \times 10^{-5} \leq \dfrac{R}{72} \leq 8.33 \times 10^{-5}$ **(b)** between 6400 and 25,800

1.8 Exercises *(pages 163–165)*

1. F   **3.** D   **5.** G   **7.** C   **9.** $\left\{-\dfrac{1}{3}, 1\right\}$ **11.** $\left\{\dfrac{2}{3}, \dfrac{8}{3}\right\}$ **13.** $\{-6, 14\}$ **15.** $\left\{\dfrac{5}{2}, \dfrac{7}{2}\right\}$ **17.** $\left\{-\dfrac{4}{3}, \dfrac{2}{9}\right\}$

19. $\left\{-\dfrac{7}{3}, -\dfrac{1}{7}\right\}$ **21.** $\{1\}$ **23.** $(-\infty, \infty)$ **27.** $(-4, -1)$ **29.** $(-\infty, -4] \cup [-1, \infty)$ **31.** $\left(-\dfrac{3}{2}, \dfrac{5}{2}\right)$

33. $(-\infty, 0) \cup (6, \infty)$ **35.** $\left(-\infty, -\dfrac{2}{3}\right) \cup (4, \infty)$ **37.** $\left[-\dfrac{2}{3}, 4\right]$ **39.** $\left[-1, -\dfrac{1}{2}\right]$ **41.** $(-101, -99)$

43. $\left\{-1, -\dfrac{1}{2}\right\}$ **45.** $\{2, 4\}$ **47.** $\left(-\dfrac{4}{3}, \dfrac{2}{3}\right)$ **49.** $\left(-\dfrac{3}{2}, \dfrac{13}{10}\right)$ **51.** $\left(-\infty, \dfrac{3}{2}\right] \cup \left[\dfrac{7}{2}, \infty\right)$ **53.** $\varnothing$

55. $(-\infty, \infty)$ **57.** $\varnothing$ **59.** $\left\{-\dfrac{5}{8}\right\}$ **61.** $\varnothing$ **63.** $\left\{-\dfrac{1}{2}\right\}$ **65.** $\left(-\infty, -\dfrac{2}{3}\right) \cup \left(-\dfrac{2}{3}, \infty\right)$ **67.** -6 or 6

68. $x^2 - x = 6; \{-2, 3\}$ **69.** $x^2 - x = -6; \left\{\dfrac{1}{2} \pm \dfrac{\sqrt{23}}{2}i\right\}$ **70.** $\left\{-2, 3, \dfrac{1}{2} \pm \dfrac{\sqrt{23}}{2}i\right\}$ **71.** $\left\{-\dfrac{1}{4}, 6\right\}$

73. $\{-1, 1\}$ **75.** $\varnothing$ **77.** $\left(-\infty, -\dfrac{1}{3}\right) \cup \left(-\dfrac{1}{3}, \infty\right)$

In Exercises 79–85, the expression in absolute value bars may be replaced by its additive inverse. For example, in

Exercise 79, $p - q$ may be written $q - p$.   **79.** $|p - q| = 2$ **81.** $|m - 7| \le 2$ **83.** $|p - 6| < .0001$

85. $|r - 29| \ge 1$ **87.** $(.9996, 1.0004)$ **89.** $[6.7, 9.7]$ **91.** $|F - 730| \le 50$

93. $25.33 \le R_L \le 28.17; \ 36.58 \le R_E \le 40.92$

Chapter 1 Review Exercises *(pages 170–177)*

1. $\{6\}$ **3.** $\left\{-\dfrac{11}{3}\right\}$ **5.** $f = \dfrac{AB(p + 1)}{24}$ **7.** A, B   **9.** 13 in. on each side   **11.** $3\dfrac{3}{7}$ L   **13.** 15 mph

15. (a) $A = 36.525x$ **(b)** 2629.8 mg   **17. (a)** \$3.60; The model gives a figure that is \$.25 more than the actual figure of

\$3.35.   **(b)** 35.5 yr after 1955, which is mid-1990; This is consistent with the minimum wage changing to \$4.25 in 1991.

19. $13 - 3i$ **21.** $-14 + 13i$ **23.** $19 + 17i$ **25.** 146   **27.** $-30 - 40i$ **29.** $1 - 2i$ **31.** $-i$ **33.** i

35. i **37.** $\{-7 \pm \sqrt{5}\}$ **39.** $\left\{-3, \dfrac{5}{2}\right\}$ **41.** $\left\{-\dfrac{3}{2}, 7\right\}$ **43.** $\{2 \pm \sqrt{6}\}$ **45.** $\left\{\dfrac{\sqrt{5} \pm 3}{2}\right\}$ **47.** D   **49.** A

51. 76; two distinct irrational solutions   **53.** -124; two distinct nonreal complex solutions   **55.** 0; one rational solution

(a double solution)   **57.** 6.25 sec and 7.5 sec   **59.** $\dfrac{1}{2}$ ft   **61.** 15,056   **63.** $\left\{\pm i, \pm \dfrac{1}{2}\right\}$ **65.** $\left\{-\dfrac{7}{24}\right\}$ **67.** $\varnothing$

69. $\left\{-\dfrac{7}{4}\right\}$ **71.** $\left\{-15, \dfrac{5}{2}\right\}$ **73.** $\{3\}$ **75.** $\{-2, -1\}$ **77.** $\varnothing$ **79.** $\{-4, 1\}$ **81.** $\{-1\}$ **83.** $\left(-\dfrac{7}{13}, \infty\right)$

85. $(-\infty, 1]$ **87.** $[4, 5]$ **89.** $[-4, 1]$ **91.** $\left(-\dfrac{2}{3}, \dfrac{5}{2}\right)$ **93.** $(-\infty, -4] \cup [0, 4]$ **95.** $(-\infty, -2) \cup (5, \infty)$

97. $(-2, 0)$ **99.** $(-3, 1) \cup [7, \infty)$ **101. (b)** 87.7 ppb   **103. (a)** 20 sec **(b)** between 2 sec and 18 sec   **107.** $W \ge 65$

109. $a \le 100{,}000$ **111.** $\{-11, 3\}$ **113.** $\left\{\dfrac{11}{27}, \dfrac{25}{27}\right\}$ **15.** $\left\{-\dfrac{2}{7}, \dfrac{4}{3}\right\}$ **117.** $[-6, -3]$ **119.** $\left(-\infty, -\dfrac{1}{7}\right) \cup (1, \infty)$

121. $\left\{-4, -\dfrac{2}{3}\right\}$ **123.** $(-\infty, \infty)$ **125.** $\{0, -4\}$ **127.** $|k - 6| = 12$ (or $|6 - k| = 12$)

129. $|t - 5| \ge .01$ (or $|5 - t| \ge .01$)

Chapter 1 Test *(pages 177–178)*

[1.1] 1. $\{0\}$ **2.** $\{-12\}$ **[1.4] 3.** $\left\{-\dfrac{1}{2}, \dfrac{7}{3}\right\}$ **4.** $\left\{\dfrac{-1 \pm 2\sqrt{2}}{3}\right\}$ **5.** $\left\{-\dfrac{1}{3} \pm \dfrac{\sqrt{5}}{3}i\right\}$ **[1.6] 6.** $\varnothing$ **7.** $\left\{-\dfrac{3}{4}\right\}$

8. $\{4\}$ **9.** $\{-3, 1\}$ **10.** $\{-2\}$ **11.** $\{\pm 1, \pm 4\}$ **12.** $\{-30, 5\}$ **[1.8] 13.** $\left\{-\dfrac{5}{2}, 1\right\}$ **14.** $\left\{-6, \dfrac{4}{3}\right\}$

[1.1] 15. $W = \dfrac{S - 2LH}{2H + 2L}$ **[1.3] 16. (a)** $5 - 8i$ **(b)** $-29 - 3i$ **(c)** $55 + 48i$ **(d)** $6 + i$ **17. (a)** -1 **(b)** i **(c)** i

[1.2] 18. (a) $A = 806{,}400x$ **(b)** 24,192,000 gal **(c)** $P = 40.32x$; approximately 40 pools **(d)** approximately 24.8 days

19. length: 200 m; width: 110 m **20.** cashews: $23\dfrac{1}{3}$ lb; walnuts: $11\dfrac{2}{3}$ lb **21.** 560 km per hr

[1.5] 22. (a) 1 sec and 5 sec **(b)** 6 sec **[1.2, 1.5] 23.** B **[1.7] 24.** $(-3, \infty)$ **25.** $[-10, 2]$ **26.** $(-\infty, -1] \cup \left[\dfrac{3}{2}, \infty\right)$

27. $(-\infty, 3) \cup (4, \infty)$ **[1.8] 28.** $(-2, 7)$ **29.** $(-\infty, -6] \cup [5, \infty)$ **30.** $\left\{-\dfrac{7}{3}\right\}$

CHAPTER 2 GRAPHS AND FUNCTIONS

2.1 Exercises *(pages 190–193)*

1. false; $(-1, 3)$ lies in quadrant II. **3.** true **5.** true **7.** any three of the following: $(2, -5), (-1, 7), (3, -9), (5, -17),$
$(6, -21)$ **9.** any three of the following: $(1993, 31), (1995, 35), (1997, 37), (1999, 35), (2001, 28), (2003, 25)$ **11. (a)** $8\sqrt{2}$
(b) $(-9, -3)$ **13. (a)** $\sqrt{34}$ **(b)** $\left(\dfrac{11}{2}, \dfrac{7}{2}\right)$ **15. (a)** $3\sqrt{41}$ **(b)** $\left(0, \dfrac{5}{2}\right)$ **17. (a)** $\sqrt{133}$ **(b)** $\left(2\sqrt{2}, \dfrac{3\sqrt{5}}{2}\right)$ **19.** yes
21. no **23.** yes **25.** yes **27.** no **29.** no **31.** $(-3, 6)$ **33.** $(5, -4)$ **35.** $(2a - p, 2b - q)$
37. 24.15%; This is close to the actual figure of 24.4%. **39.** \$11,563
Other ordered pairs are possible in Exercises 43–53.

43. (a)

x	y
0	-2
4	0
2	-1

(b)

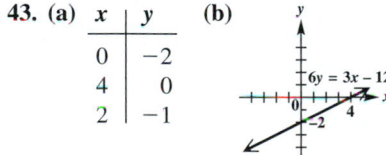

45. (a)

x	y
0	$\dfrac{5}{3}$
$\dfrac{5}{2}$	0
4	-1

(b)

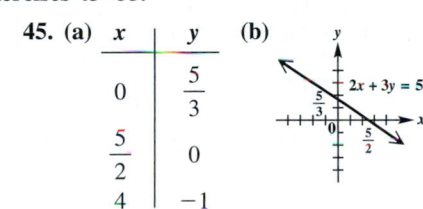

47. (a)

x	y
0	0
1	1
-2	4

(b)

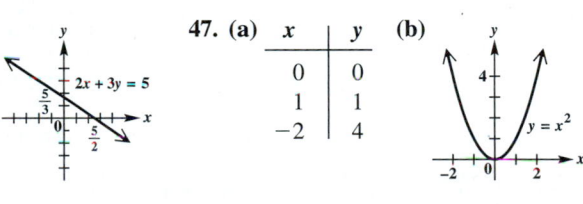

49. (a)

x	y
3	0
4	1
7	2

(b)

51. (a)

x	y
4	2
-2	4
0	2

(b)

53. (a)

x	y
0	0
-1	-1
2	8

(b)

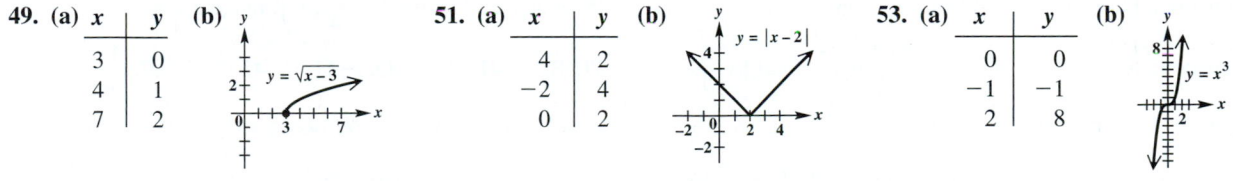

55. $(4, 0)$ **57.** III; I; IV; IV **59.** yes; no

2.2 Exercises *(pages 198–200)*

1. (a) $x^2 + y^2 = 36$ **(b)**   **3. (a)** $(x - 2)^2 + y^2 = 36$ **(b)**

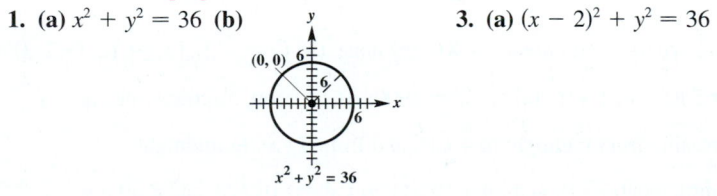

5. (a) $(x + 2)^2 + (y - 5)^2 = 16$ **(b)**   **7. (a)** $(x - 5)^2 + (y + 4)^2 = 49$ **(b)**

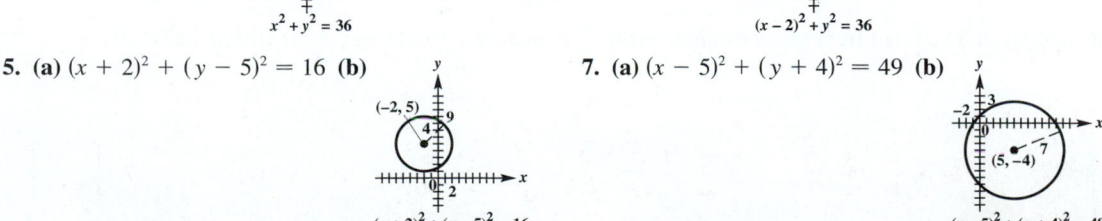

9. (a) $x^2 + (y - 4)^2 = 16$ **(b)** 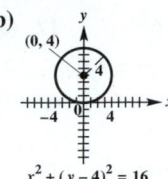 **11. (a)** $(x - \sqrt{2})^2 + (y - \sqrt{2})^2 = 2$ **(b)**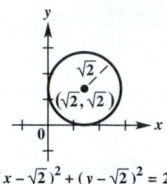

13. (a) $(x - 3)^2 + (y - 1)^2 = 4$ **(b)** $x^2 + y^2 - 6x - 2y + 6 = 0$ **15. (a)** $(x + 2)^2 + (y - 2)^2 = 4$

(b) $x^2 + y^2 + 4x - 4y + 4 = 0$ **17.** B **19.** yes; center: $(-3, -4)$; radius: 4 **21.** yes; center: $(2, -6)$; radius: 6

23. yes; center: $\left(-\dfrac{1}{2}, 2\right)$; radius: 3 **25.** no; The graph is nonexistent. **27.** no; The graph is the point $(3, 3)$.

29. yes; center: $(-2, 0)$; radius: $\dfrac{2}{3}$ **31.** $(2, -3)$ **32.** $3\sqrt{5}$ **33.** $3\sqrt{5}$ **34.** $3\sqrt{5}$ **35.** $(x - 2)^2 + (y + 3)^2 = 45$

36. $(x + 2)^2 + (y + 1)^2 = 41$ **37.** at $(3, 1)$ **39.** at $(-2, -2)$ **41.** $(x - 3)^2 + (y - 2)^2 = 4$

43. $(2 + \sqrt{7}, 2 + \sqrt{7}), (2 - \sqrt{7}, 2 - \sqrt{7})$ **45.** $(2, 3)$ and $(4, 1)$ **47.** $9 + \sqrt{119}, 9 - \sqrt{119}$ **49.** $\sqrt{113} - 5$

2.3 Exercises *(pages 213–217)*

1. function **3.** not a function **5.** function **7.** function **9.** not a function; domain: $\{0, 1, 2\}$;

range: $\{-4, -1, 0, 1, 4\}$ **11.** function; domain: $\{2, 3, 5, 11, 17\}$; range: $\{1, 7, 20\}$ **13.** function; domain: $\{0, -1, -2\}$;

range: $\{0, 1, 2\}$ **15.** function; domain: $\{2001, 2002, 2003, 2004\}$; range: $\{4,400,823, \ 4,339,139, \ 4,464,400, \ 4,672,911\}$

17. function; domain: $(-\infty, \infty)$; range: $(-\infty, \infty)$ **19.** not a function; domain: $[3, \infty)$; range: $(-\infty, \infty)$

21. not a function; domain: $[-4, 4]$; range: $[-3, 3]$ **23.** function; domain: $(-\infty, \infty)$; range: $[0, \infty)$

25. not a function; domain: $[0, \infty)$; range: $(-\infty, \infty)$ **27.** function; domain: $(-\infty, \infty)$; range: $(-\infty, \infty)$

29. not a function; domain: $(-\infty, \infty)$; range: $(-\infty, \infty)$ **31.** function; domain: $[0, \infty)$; range: $[0, \infty)$

33. function; domain: $(-\infty, 0) \cup (0, \infty)$; range: $(-\infty, 0) \cup (0, \infty)$ **35.** function; domain: $\left[-\dfrac{1}{4}, \infty\right)$; range: $[0, \infty)$

37. function; domain: $(-\infty, 3) \cup (3, \infty)$; range: $(-\infty, 0) \cup (0, \infty)$ **39.** B **41.** 4 **43.** -11 **45.** 3 **47.** $\dfrac{11}{4}$

49. $-3p + 4$ **51.** $3x + 4$ **53.** $-3x - 2$ **55.** $-6m + 13$ **57. (a)** 2 **(b)** 3 **59. (a)** 15 **(b)** 10

61. (a) 3 **(b)** -3 **63. (a)** $f(x) = -\dfrac{1}{3}x + 4$ **(b)** 3 **65. (a)** $f(x) = -2x^2 - x + 3$ **(b)** -18

67. (a) $f(x) = \dfrac{4}{3}x - \dfrac{8}{3}$ **(b)** $\dfrac{4}{3}$ **69.** $f(3) = 4$ **71.** -4 **73. (a)** 0 **(b)** 4 **(c)** 2 **(d)** 4 **75. (a)** -3 **(b)** -2 **(c)** 0 **(d)** 2

77. (a) $[4, \infty)$ **(b)** $(-\infty, -1]$ **(c)** $[-1, 4]$ **79. (a)** $(-\infty, 4]$ **(b)** $[4, \infty)$ **(c)** none **81. (a)** none **(b)** $(-\infty, -2]; [3, \infty)$ **(c)** $(-2, 3)$

83. (a) yes **(b)** $[0, 24]$ **(c)** 1200 megawatts **(d)** at 17 hr or 5 P.M.; at 4 A.M. **(e)** $f(12) = 1900$; At 12 noon, electricity use is

1900 megawatts. **(f)** increasing from 4 A.M. to 5 P.M.; decreasing from midnight to 4 A.M. and from 5 P.M. to midnight

85. (a) about 12 noon to about 8 P.M. **(b)** from midnight until about 6 A.M. and after 10 P.M. **(c)** about 10 A.M. and 8:30 P.M.

2.4 Exercises *(pages 225–231)*

1. B **3.** C **5.** A

In Exercises 7–23, we give the domain first and then the range.

7. $(-\infty, \infty); (-\infty, \infty)$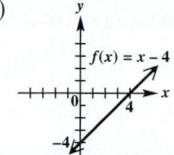

9. $(-\infty, \infty); (-\infty, \infty)$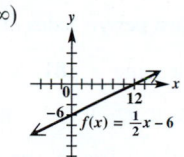

11. $(-\infty, \infty); (-\infty, \infty)$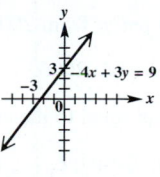

13. $(-\infty, \infty); (-\infty, \infty)$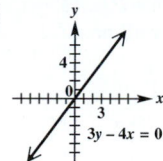

15. $(-\infty, \infty); (-\infty, \infty)$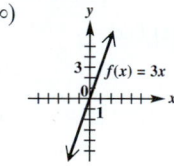

17. $(-\infty, \infty); \{-4\}$; constant function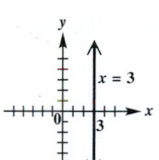

19. $\{3\}; (-\infty, \infty)$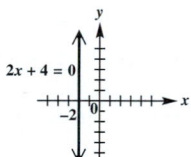

21. $\{-2\}; (-\infty, \infty)$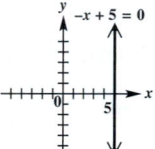

23. $\{5\}; (-\infty, \infty)$

25. A   **27.** D

29.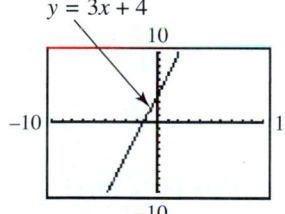

31.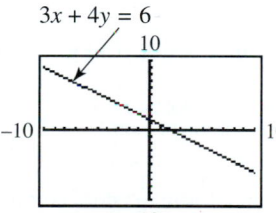

33. A, C, D, E   **35.** $\dfrac{2}{5}$ **37.** 0

39. 0   **41.** undefined   **45.** $m = 3$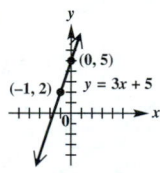

47. $m = -\dfrac{3}{2}$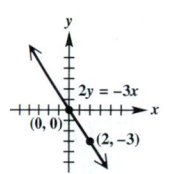

49. $m = \dfrac{5}{2}$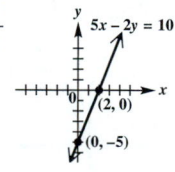

53. 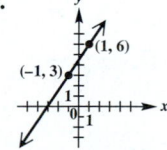   **55.** 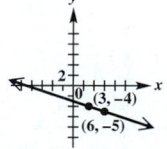   **57.** 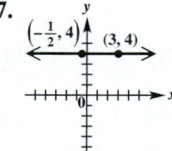   **59.** D   **61.** A   **63.** E

65. $-\$4000$ per yr; The value of the machine is decreasing \$4000 each year during these years.

67. 0% per yr (or no change); The percent of pay raise is not changing—it is 3% each year during these years.   **69.** zero

71. **(b)** 189.5; This means that the average rate of change in the number of radio stations per year is an increase of 189.5.

73. **(a)** $-.57$ million recipients per yr **(b)** The negative slope means the numbers of recipients *decreased* by .57 million each year.

75. $-.575\%$ per yr; The percent of freshmen listing computer science as their probable field of study decreased an average of .575% per yr from 2000 to 2004.   **77.** 2.16 million per yr; Sales of DVD players increased an average of 2.16 million per year from 1997 to 2006.   **78.** 3   **79.** 3   **80.** the same   **81.** $\sqrt{10}$ **82.** $2\sqrt{10}$ **83.** $3\sqrt{10}$ **84.** The sum is $3\sqrt{10}$, which is equal to the answer in Exercise 83.   **85.** B; C; A; C **86.** The midpoint is $(3, 3)$, which is the same as the middle entry in the table.   **87.** 7.5   **89. (a)** $C(x) = 11x + 180$ **(b)** $R(x) = 20x$ **(c)** $P(x) = 9x - 180$ **(d)** 20 units; produce   **91. (a)** $C(x) = 400x + 1650$ **(b)** $R(x) = 305x$ **(c)** $P(x) = -95x - 1650$ **(d)** $R(x) < C(x)$ for all positive x; don't produce, impossible to make a profit

Chapter 2 Quiz *(page 231)*

[2.1] 1. $\sqrt{41}$ **2.** 1985: 4.85 million; 1995: 5.50 million   **3.**   **[2.2] 4.**

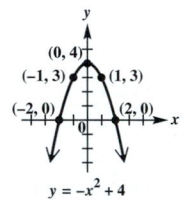

5. radius: $\sqrt{17}$; center: $(2, -4)$ **[2.3] 6.** 2   **7.** domain: $(-\infty, \infty)$; range: $[0, \infty)$ **8. (a)** $(-\infty, -3]$ **(b)** $[-3, \infty)$ **(c)** none   **[2.4] 9. (a)** $\dfrac{3}{2}$ **(b)** 0 **(c)** undefined   **10.** $-.49\%$ per yr; The number of college freshmen age 18 or younger on Dec. 31 decreased an average of .49% per yr from 1982 to 2002.

2.5 Exercises *(pages 242–247)*

1. D   **3.** C   **5.** $2x + y = 5$ **7.** $3x + 2y = -7$ **9.** $x = -8$ **11.** $y = -8$ **13.** $y = \dfrac{1}{4}x + \dfrac{13}{4}$

15. $y = \dfrac{2}{3}x - 2$ **17.** $x = -6$ (cannot be written in slope-intercept form)   **19.** $y = 4$ **21.** $y = 5x + 15$

23. $y = -\dfrac{2}{3}x - \dfrac{4}{5}$ **25.** $y = \dfrac{3}{2}$ **27.** -2; does not; undefined; $\dfrac{1}{2}$; does not; 0   **29. (a)** B **(b)** D **(c)** A **(d)** C

31. slope: 3; y-intercept: -1 **33.** slope: 4; y-intercept: -7 **35.** slope: $-\dfrac{3}{4}$; y-intercept: 0

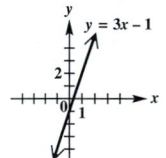   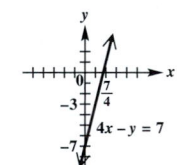   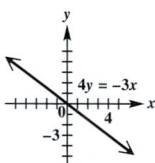

37. slope: $-\dfrac{1}{2}$; y-intercept: -2 **39.** slope: $\dfrac{3}{2}$; y-intercept: 1

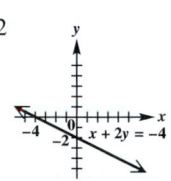   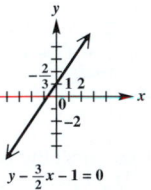

41. (a) -2; 1; $\dfrac{1}{2}$ **(b)** $f(x) = -2x + 1$ **43. (a)** $-\dfrac{1}{3}$; 2; 6 **(b)** $f(x) = -\dfrac{1}{3}x + 2$ **45. (a)** -200; 300; $\dfrac{3}{2}$

(b) $f(x) = -200x + 300$ **47. (a)** $x + 3y = 11$ **(b)** $y = -\dfrac{1}{3}x + \dfrac{11}{3}$ **49. (a)** $5x - 3y = -13$ **(b)** $y = \dfrac{5}{3}x + \dfrac{13}{3}$

51. (a) $y = 1$ **(b)** $y = 1$ **53. (a)** $y = 6$ **(b)** $y = 6$ **55. (a)** $-\dfrac{1}{2}$ **(b)** $-\dfrac{7}{2}$

57. $y = .457x - 856.99$; 59.8%; This figure is very close to the actual figure.

59. **(a)** $f(x) = 874.9x + 11,719$

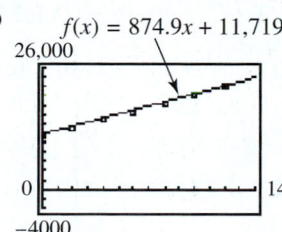

The average tuition increase is about \$875 per year for the period, because this is the slope of the line. **(b)** $f(11) = \$21,343$; This is a fairly good approximation. **(c)** $f(x) = 877.1x + 11,322$

61. **(a)** $F = \dfrac{9}{5}C + 32$ **(b)** $C = \dfrac{5}{9}(F - 32)$ **(c)** $-40°$ **63.** **(a)** $C = .952I - 1634$ **(b)** $.952$ **65.** $\{1\}$ **67.** $\{4\}$

69. **(a)** $\{12\}$ **70.** the Pythagorean theorem and its converse **71.** $\sqrt{x_1^2 + m_1^2 x_1^2}$ **72.** $\sqrt{x_2^2 + m_2^2 x_2^2}$

73. $\sqrt{(x_2 - x_1)^2 + (m_2 x_2 - m_1 x_1)^2}$ **75.** $-2x_1 x_2(m_1 m_2 + 1) = 0$ **76.** Since $x_1 \neq 0$, $x_2 \neq 0$, we have $m_1 m_2 + 1 = 0$,

implying that $m_1 m_2 = -1$. **77.** If two nonvertical lines are perpendicular, then the product of the slopes of these lines is -1.

81. yes **83.** no

Summary Exercises on Graphs, Circles, Functions, and Equations *(pages 247–248)*

1. **(a)** $\sqrt{65}$ **(b)** $\left(\dfrac{5}{2}, 1\right)$ **(c)** $y = 8x - 19$ **2.** **(a)** $\sqrt{29}$ **(b)** $\left(\dfrac{3}{2}, -1\right)$ **(c)** $y = -\dfrac{2}{5}x - \dfrac{2}{5}$ **3.** **(a)** 5 **(b)** $\left(\dfrac{1}{2}, 2\right)$

(c) $y = 2$ **4.** **(a)** $\sqrt{10}$ **(b)** $\left(\dfrac{3\sqrt{2}}{2}, 2\sqrt{2}\right)$ **(c)** $y = -2x + 5\sqrt{2}$ **5.** **(a)** 2 **(b)** $(5, 0)$ **(c)** $x = 5$ **6.** **(a)** $4\sqrt{2}$

(b) $(-1, -1)$ **(c)** $y = x$ **7.** **(a)** $4\sqrt{3}$ **(b)** $\left(4\sqrt{3}, 3\sqrt{5}\right)$ **(c)** $y = 3\sqrt{5}$ **8.** **(a)** $\sqrt{34}$ **(b)** $\left(\dfrac{3}{2}, -\dfrac{3}{2}\right)$ **(c)** $y = \dfrac{5}{3}x - 4$

9. $y = -\dfrac{1}{3}x + \dfrac{1}{3}$ 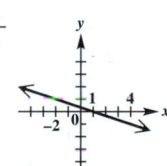 **10.** $y = 3$ 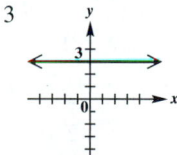 **11.** $(x - 2)^2 + (y + 1)^2 = 9$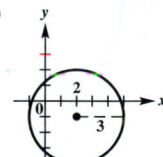

12. $x^2 + (y - 2)^2 = 4$ 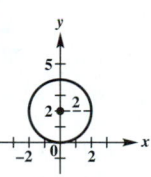 **13.** $y = -\dfrac{5}{6}x - \dfrac{5}{2}$ 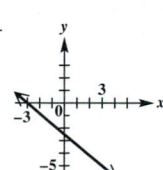 **14.** $y = -\dfrac{4}{3}x$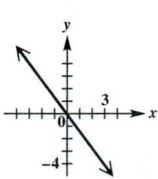

15. $y = -\dfrac{2}{3}x$ 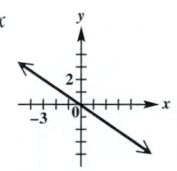 **16.** $x = -4$ 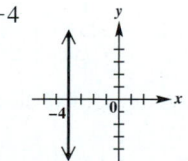 **17.** yes; center: $(2, -1)$; radius: 3 **18.** no

19. yes; center: $(6, 0)$; radius: 4 **20.** yes; center: $(-1, -8)$; radius: 2 **21.** no **22.** yes; center: $(0, 4)$; radius: 5

23. $(4 - \sqrt{7}, 2)$, $(4 + \sqrt{7}, 2)$ **24.** 8 **25.** **(a)** domain: $(-\infty, \infty)$; range: $(-\infty, \infty)$ **(b)** $f(x) = \dfrac{1}{4}x + \dfrac{3}{2}$; 1

26. **(a)** domain: $[-5, \infty)$; range: $(-\infty, \infty)$ **(b)** y is not a function of x. **27.** **(a)** domain: $[-7, 3]$; range: $[-5, 5]$

(b) y is not a function of x. **28.** **(a)** domain: $(-\infty, \infty)$; range: $\left[-\dfrac{3}{2}, \infty\right)$ **(b)** $f(x) = \dfrac{1}{2}x^2 - \dfrac{3}{2}$; $\dfrac{1}{2}$

2.6 Exercises *(pages 255–259)*

1. E; $(-\infty, \infty)$ **3.** A; $(-\infty, \infty)$ **5.** F; $y = x$ **7.** H; no   **9.** B; $\{\ldots, -3, -2, -1, 0, 1, 2, 3, \ldots\}$ **11.** $(-\infty, \infty)$

13. $[0, \infty)$ **15.** $(-\infty, 1)$; $[1, \infty)$ **17. (a)** -10 **(b)** -2 **(c)** -1 **(d)** 2 **19. (a)** -3 **(b)** 1 **(c)** 0 **(d)** 9

21.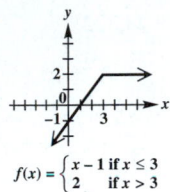

$f(x) = \begin{cases} x - 1 & \text{if } x \le 3 \\ 2 & \text{if } x > 3 \end{cases}$

23.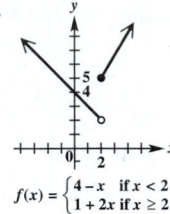

$f(x) = \begin{cases} 4 - x & \text{if } x < 2 \\ 1 + 2x & \text{if } x \ge 2 \end{cases}$

25.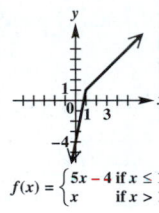

$f(x) = \begin{cases} 5x - 4 & \text{if } x \le 1 \\ x & \text{if } x > 1 \end{cases}$

27.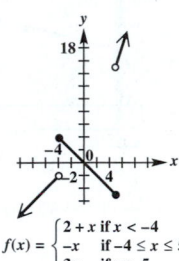

$f(x) = \begin{cases} 2 + x & \text{if } x < -4 \\ -x & \text{if } -4 \le x \le 5 \\ 3x & \text{if } x > 5 \end{cases}$

29.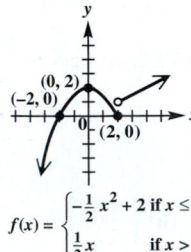

$f(x) = \begin{cases} -\frac{1}{2}x^2 + 2 & \text{if } x \le 2 \\ \frac{1}{2}x & \text{if } x > 2 \end{cases}$

31.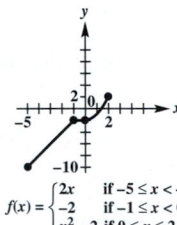

$f(x) = \begin{cases} 2x & \text{if } -5 \le x < -1 \\ -2 & \text{if } -1 \le x < 0 \\ x^2 - 2 & \text{if } 0 \le x \le 2 \end{cases}$

33.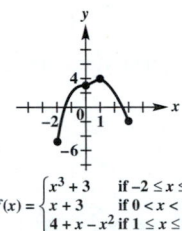

$f(x) = \begin{cases} x^3 + 3 & \text{if } -2 \le x \le 0 \\ x + 3 & \text{if } 0 < x < 1 \\ 4 + x - x^2 & \text{if } 1 \le x \le 3 \end{cases}$

In Exercises 35–41 we give one of the possible rules, then the domain, and then the range.

35. $f(x) = \begin{cases} -1 & \text{if } x \le 0 \\ 1 & \text{if } x > 0 \end{cases}$; $(-\infty, \infty)$; $\{-1, 1\}$ **37.** $f(x) = \begin{cases} 2 & \text{if } x \le 0 \\ -1 & \text{if } x > 1 \end{cases}$; $(-\infty, 0] \cup (1, \infty)$; $\{-1, 2\}$

39. $f(x) = \begin{cases} x & \text{if } x \le 0 \\ 2 & \text{if } x > 0 \end{cases}$; $(-\infty, \infty)$; $(-\infty, 0] \cup \{2\}$ **41.** $f(x) = \begin{cases} \sqrt[3]{x} & \text{if } x < 1 \\ x + 1 & \text{if } x \ge 1 \end{cases}$; $(-\infty, \infty)$; $(-\infty, 1) \cup [2, \infty)$

43. $(-\infty, \infty)$; $\{\ldots, -2, -1, 0, 1, 2, \ldots\}$

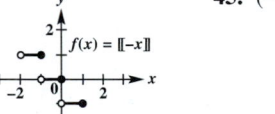

45. $(-\infty, \infty)$; $\{\ldots, -2, -1, 0, 1, 2, \ldots\}$

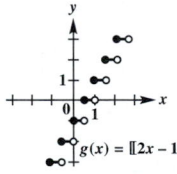

47.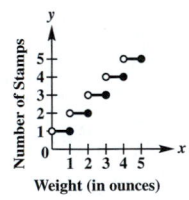

49.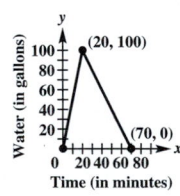

51. (a) for $[0, 4]$: $y = -.9x + 42.8$; for $(4, 8]$: $y = -1.625x + 45.7$

(b) $f(x) = \begin{cases} -.9x + 42.8 & \text{if } 0 \le x \le 4 \\ -1.625x + 45.7 & \text{if } 4 < x \le 8 \end{cases}$

53. (a) 50,000 gal; 30,000 gal **(b)** during the first and fourth days **(c)** 45,000; 40,000 **(d)** 5000 gal per day

55. (a) $f(x) = .8 \left[\!\left[\dfrac{x}{2} \right]\!\right]$ if $6 \le x \le 18$ **(b)** $3.20; $5.60

2.7 Exercises *(pages 270–273)*

1. (a) B **(b)** D **(c)** E **(d)** A **(e)** C   **3. (a)** B **(b)** A **(c)** G **(d)** C **(e)** F **(f)** D **(g)** H **(h)** E **(i)** I

5.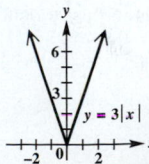

7.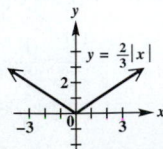

9.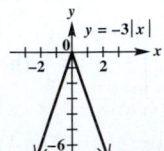

11.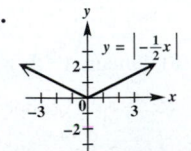

13.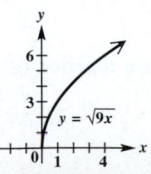

15. **(a)** $(4, 12)$ **(b)** $(8, 16)$ **17.** **(a)** $(2, 12)$ **(b)** $(32, 12)$ **19.** 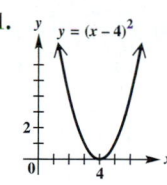   **21.**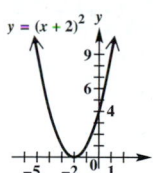

23. y-axis   **25.** x-axis, y-axis, origin   **27.** origin   **29.** none of these   **31.** odd   **33.** even   **35.** neither

37. 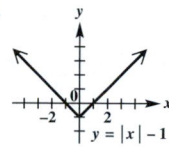   **39.** 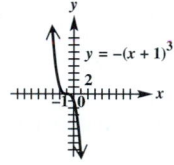   **41.** 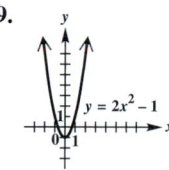   **43.**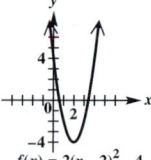

45. 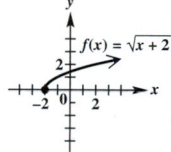   **47.** 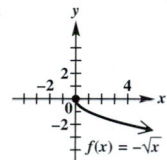   **49.** 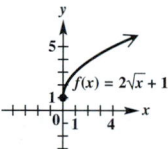   **51.**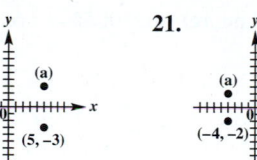

53. 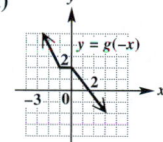   **55.** 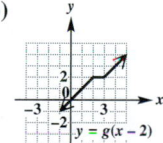   **57.**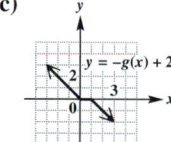

59. **(a)**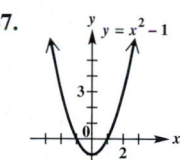

The graph of $g(x)$ is reflected across the y-axis.

(b)

The graph of $g(x)$ is translated to the right 2 units.

(c)

The graph of $g(x)$ is reflected across the x-axis and translated 2 units up.

61. It is the graph of $f(x) = |x|$ translated 1 unit to the left, reflected across the x-axis, and translated 3 units up. The equation is $y = -|x + 1| + 3$.   **63.** It is the graph of $g(x) = \sqrt{x}$ translated 1 unit to the right and translated 3 units down. The equation is $y = \sqrt{x - 1} - 3$.   **65.** It is the graph of $g(x) = \sqrt{x}$ translated 4 units to the left, stretched vertically by a factor of 2, and translated 4 units down. The equation is $y = 2\sqrt{x + 4} - 4$.   **67.** $f(-3) = -6$ **69.** $f(9) = 6$ **71.** $f(-3) = -6$

73. $g(x) = 2x + 13$ **75.** **(a)** 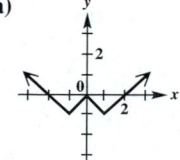   **(b)**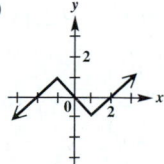

Chapter 2 Quiz *(pages 273–274)*

[2.5] 1. **(a)** $y = 2x + 11$ **(b)** $-\dfrac{11}{2}$ **2.** $y = -\dfrac{2}{3}x$ **3.** **(a)** $x = -8$ **(b)** $y = 5$ **[2.6] 4.** **(a)** cubing function; domain: $(-\infty, \infty)$; range: $(-\infty, \infty)$; increasing over $(-\infty, \infty)$ **(b)** absolute value function; domain: $(-\infty, \infty)$; range: $[0, \infty)$; decreasing over $(-\infty, 0]$; increasing over $[0, \infty)$ **(c)** cube root function; domain: $(-\infty, \infty)$; range: $(-\infty, \infty)$; increasing over $(-\infty, \infty)$

5. (a) 55 mph; 30 mph **(b)** 12 mi **(c)** 40; 30; 55   **6.**   **[2.7] 7.**

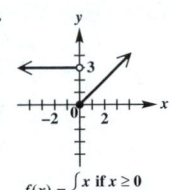

$$f(x) = \begin{cases} x \text{ if } x \geq 0 \\ 3 \text{ if } x < 0 \end{cases}$$

8.

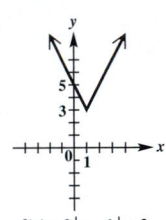

$f(x) = 2|x - 1| + 3$

9. $y = -\sqrt{x + 4} - 2$ **10. (a)** y-axis **(b)** x-axis **(c)** x-axis, y-axis, origin

2.8 Exercises *(pages 282–287)*

1. 12   **3.** -4 **5.** -38 **7.** $\dfrac{1}{2}$ **9.** $5x - 1; x + 9; 6x^2 - 7x - 20; \dfrac{3x + 4}{2x - 5}$; all domains are $(-\infty, \infty)$

except for that of $\dfrac{f}{g}$, which is $\left(-\infty, \dfrac{5}{2}\right) \cup \left(\dfrac{5}{2}, \infty\right)$.   **11.** $3x^2 - 4x + 3; x^2 - 2x - 3; 2x^4 - 5x^3 + 9x^2 - 9x; \dfrac{2x^2 - 3x}{x^2 - x + 3}$;

all domains are $(-\infty, \infty)$.   **13.** $\sqrt{4x - 1} + \dfrac{1}{x}; \sqrt{4x - 1} - \dfrac{1}{x}; \dfrac{\sqrt{4x - 1}}{x}; x\sqrt{4x - 1}$; all domains are $\left[\dfrac{1}{4}, \infty\right)$.

15. 7.7; 11.8; 19.5   **17.** 1996–2006   **19.** 6; It represents the dollars in billions spent for general science in 2000.

21. space and other technologies; 1995–2000   **23. (a)** 2 **(b)** 4 **(c)** 0 **(d)** $-\dfrac{1}{3}$ **25. (a)** 3 **(b)** -5 **(c)** 2 **(d)** undefined

27. (a) 5 **(b)** 5 **(c)** 0 **(d)** undefined   **29.**

x	$(f + g)(x)$	$(f - g)(x)$	$(fg)(x)$	$\left(\dfrac{f}{g}\right)(x)$
-2	6	-6	0	0
0	5	5	0	undefined
2	5	9	-14	-3.5
4	15	5	50	2

33. (a) $2 - x - h$ **(b)** $-h$ **(c)** -1 **35. (a)** $6x + 6h + 2$ **(b)** $6h$ **(c)** 6   **37. (a)** $-2x - 2h + 5$ **(b)** $-2h$ **(c)** -2

39. (a) $\dfrac{1}{x + h}$ **(b)** $\dfrac{-h}{x(x + h)}$ **(c)** $\dfrac{-1}{x(x + h)}$ **41.** -5 **43.** 7   **45.** 6   **47.** -1 **49.** 1   **51.** 9

53. 1   **55.** $g(1) = 9$, and $f(9)$ cannot be determined from the table given.   **57. (a)** $-30x - 33; (-\infty, \infty)$

(b) $-30x + 52; (-\infty, \infty)$ **59. (a)** $\sqrt{x + 3}; [-3, \infty)$ **(b)** $\sqrt{x} + 3; [0, \infty)$ **61. (a)** $(x^2 + 3x - 1)^3; (-\infty, \infty)$

(b) $x^6 + 3x^3 - 1; (-\infty, \infty)$ **63. (a)** $\sqrt{3x - 1}; \left[\dfrac{1}{3}, \infty\right)$ **(b)** $3\sqrt{x} - 1; [1, \infty)$ **65. (a)** $\dfrac{2}{x + 1}; (-\infty, -1) \cup (-1, \infty)$

(b) $\dfrac{2}{x} + 1; (-\infty, 0) \cup (0, \infty)$ **67. (a)** $\sqrt{-\dfrac{1}{x} + 2}; (-\infty, 0) \cup \left[\dfrac{1}{2}, \infty\right)$ **(b)** $-\dfrac{1}{\sqrt{x + 2}}; (-2, \infty)$ **69. (a)** $\sqrt{\dfrac{1}{x + 5}}; (-5, \infty)$

(b) $\dfrac{1}{\sqrt{x} + 5}; [0, \infty)$ **71. (a)** $\dfrac{x}{1 - 2x}; (-\infty, 0) \cup \left(0, \dfrac{1}{2}\right) \cup \left(\dfrac{1}{2}, \infty\right)$ **(b)** $x - 2; (-\infty, 2) \cup (2, \infty)$

73.

x	$f(x)$	$g(x)$	$g(f(x))$
1	3	2	7
2	1	5	2
3	2	7	5

In Exercises 81–85, we give only one of the many possible ways.

81. $g(x) = 6x - 2, f(x) = x^2$ **83.** $g(x) = x^2 - 1, f(x) = \sqrt{x}$ **85.** $g(x) = 6x, f(x) = \sqrt{x} + 12$

87. $(f \circ g)(x) = 63{,}360x$ computes the number of inches in x miles.   **89. (a)** $A(2x) = \sqrt{3}x^2$ **(b)** $64\sqrt{3}$ square units

91. (a) $(A \circ r)(t) = 16\pi t^2$ **(b)** It defines the area of the leak in terms of the time t, in minutes. **(c)** 144π ft^2

93. (a) $N(x) = 100 - x$ **(b)** $G(x) = 20 + 5x$ **(c)** $C(x) = (100 - x)(20 + 5x)$ **(d)** $\$9600$

Chapter 2 Review Exercises *(pages 292–297)*

1. $\sqrt{85}; \left(-\dfrac{1}{2}, 2\right)$ **3.** $5; \left(-6, \dfrac{11}{2}\right)$ **5.** $-7; -1; 8; 23$ **7.** $(x + 2)^2 + (y - 3)^2 = 225$

9. $(x + 8)^2 + (y - 1)^2 = 289$ **11.** $x^2 + y^2 = 34$ **13.** $x^2 + (y - 3)^2 = 13$ **15.** $(2, -3); 1$ **17.** $\left(-\dfrac{7}{2}, -\dfrac{3}{2}\right); \dfrac{3\sqrt{6}}{2}$

19. $3 + 2\sqrt{5}; 3 - 2\sqrt{5}$ **21.** no; $[-6, 6]; [-6, 6]$ **23.** no; $(-\infty, \infty); (-\infty, -1] \cup [1, \infty)$ **25.** no; $[0, \infty); (-\infty, \infty)$

27. function of x **29.** not function of x **31.** $(-\infty, 8) \cup (8, \infty)$ **33. (a)** $[2, \infty)$ **(b)** $(-\infty, -2]$

35. -15 **37.** $-2k^2 + 3k - 6$ **39.** 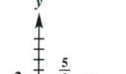   **41.** 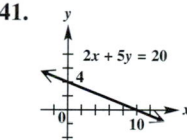   **43.**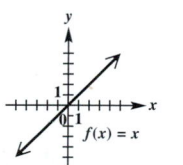

45. 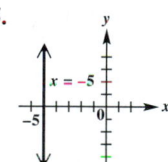   **47.** 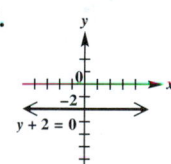   **49.** 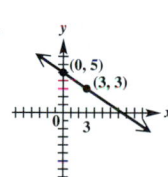   **51.** -2 **53.** 0 **55.** $-\dfrac{11}{2}$

57. undefined   **59.** Initially, the car is at home. After traveling 30 mph for 1 hr, the car is 30 mi away from home. During the second hour the car travels 20 mph until it is 50 mi away. During the third hour the car travels toward home at 30 mph until it is 20 mi away. During the fourth hour the car travels away from home at 40 mph until it is 60 mi away from home. During the last hour, the car travels 60 mi at 60 mph until it arrives home.   **61. (a)** $y = 8x + 30.7$; The slope, 8, indicates that the percent of returns filed electronically rose 8% per year during this period. **(b)** 62.7%   **63. (a)** $y = -2x + 1$ **(b)** $2x + y = 1$

65. (a) $y = 3x - 7$ **(b)** $3x - y = 7$ **67. (a)** $y = -10$ **(b)** $y = -10$ **69. (a)** not possible **(b)** $x = -7$

71. 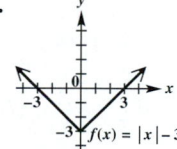   **73.** 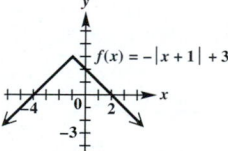   **75.** 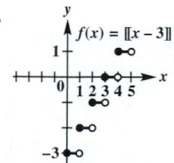   **77.**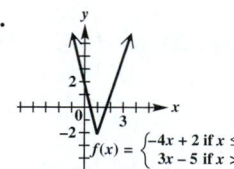

79. 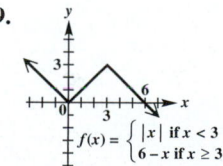   **81.** true   **83.** false; For example, $f(x) = x^2$ is even, and $(2, 4)$ is on the graph but $(2, -4)$ is not.

85. true   **87.** x-axis   **89.** y-axis   **91.** none of these   **93.** Reflect the graph of $f(x) = |x|$ across the x-axis.   **95.** Translate the graph of $f(x) = |x|$ to the right 4 units and stretch vertically by a factor of 2.   **97.** $y = -3x - 4$

99. (a)

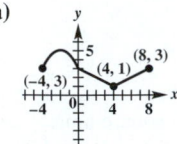

(b)

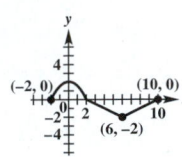

(c)

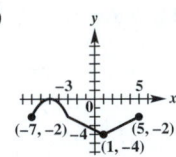

(d)

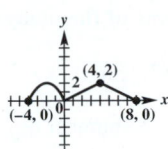

101. $3x^4 - 9x^3 - 16x^2 + 12x + 16$ **103.** 68   **105.** $-\dfrac{23}{4}$ **107.** $(-\infty, \infty)$ **109.** C and D   **111.** $2x + h - 5$

113. $x - 2$ **115.** 1   **117.** 8   **119.** -6 **121.** 2   **123.** 1   **125.** $f(x) = 36x$; $g(x) = 1760x$;

$(g \circ f)(x) = g(f(x)) = 1760(36x) = 63{,}360x$ **127.** $V(r) = \dfrac{4}{3}\pi(r + 3)^3 - \dfrac{4}{3}\pi r^3$

Chapter 2 Test *(pages 297–299)*

[2.3] 1. (a) D **(b)** D **(c)** C **(d)** B **(e)** C **(f)** C **(g)** C **(h)** D **(i)** D **(j)** C   **[2.4] 2.** $\dfrac{3}{5}$ **[2.1] 3.** $\sqrt{34}$ **4.** $\left(\dfrac{1}{2}, \dfrac{5}{2}\right)$

[2.5] 5. $3x - 5y = -11$ **6.** $f(x) = \dfrac{3}{5}x + \dfrac{11}{5}$ **[2.2] 7. (a)** $x^2 + y^2 = 4$ **(b)** $(x - 1)^2 + (y - 4)^2 = 1$

8.

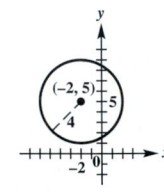

$x^2 + y^2 + 4x - 10y + 13 = 0$

[2.3] 9. (a) not a function; domain: $[0, 4]$; range: $[-4, 4]$ **(b)** function; domain: $(-\infty, -1) \cup (-1, \infty)$;

range: $(-\infty, 0) \cup (0, \infty)$; decreasing on $(-\infty, -1)$ and on $(-1, \infty)$

[2.5] 10. (a) $x = 5$ **(b)** $y = -3$ **11. (a)** $y = -3x + 9$ **(b)** $y = \dfrac{1}{3}x + \dfrac{7}{3}$

[2.3] 12. (a) $(-\infty, -3)$ **(b)** $(4, \infty)$ **(c)** $[-3, 4]$ **(d)** $(-\infty, -3)$; $[-3, 4]$; $(4, \infty)$ **(e)** $(-\infty, \infty)$ **(f)** $(-\infty, 2)$

[2.6, 2.7] 13.

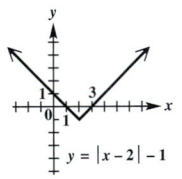

$y = |x - 2| - 1$

14.

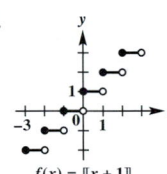

$f(x) = [\![x + 1]\!]$

15.

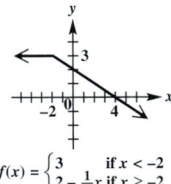

$f(x) = \begin{cases} 3 & \text{if } x < -2 \\ 2 - \frac{1}{2}x & \text{if } x \geq -2 \end{cases}$

16. (a)

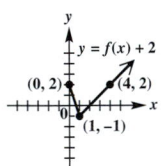

(b)

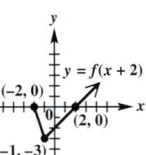

(c)

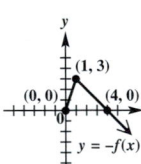

(d)

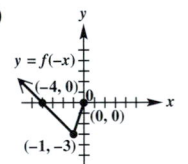

(e)

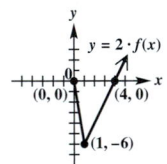

17. It is translated 2 units to the left, stretched vertically by a factor of 2, reflected across the *x*-axis, and translated 3 units down.

18. (a) yes **(b)** yes **(c)** yes   **[2.8] 19. (a)** $2x^2 - x + 1$ **(b)** $\dfrac{2x^2 - 3x + 2}{-2x + 1}$ **(c)** $\left(-\infty, \dfrac{1}{2}\right) \cup \left(\dfrac{1}{2}, \infty\right)$

(d) $4x + 2h - 3$ **20. (a)** 0 **(b)** -12 **(c)** 1   **21.** $\sqrt{2x - 6}$; $[3, \infty)$ **22.** $2\sqrt{x + 1} - 7$; $[-1, \infty)$

[2.6] 23. \$2.75   **[2.4] 24. (a)** $C(x) = 3300 + 4.50x$ **(b)** $R(x) = 10.50x$ **(c)** $R(x) - C(x) = 6.00x - 3300$ **(d)** 551

CHAPTER 3   POLYNOMIAL AND RATIONAL FUNCTIONS

3.1 Exercises *(pages 311–320)*

1. (a) domain: $(-\infty, \infty)$; range: $[-4, \infty)$ **(b)** $(-3, -4)$ **(c)** $x = -3$ **(d)** 5 **(e)** $-5, -1$ **3. (a)** domain: $(-\infty, \infty)$; range: $(-\infty, 2]$
(b) $(-3, 2)$ **(c)** $x = -3$ **(d)** -16 **(e)** $-4, -2$ **5.** B   **7.** D   **9.** 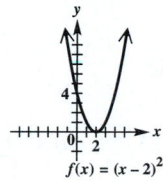   **(e)** If the absolute value of the coefficient is greater than 1, it causes the graph to be stretched vertically, so it is narrower. If the absolute value of the coefficient is between 0 and 1, it causes the graph to shrink vertically, so it is broader.

11. 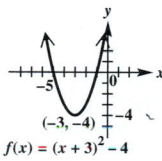   **(e)** The graph of $(x - h)^2$ is translated h units to the right if h is positive and $|h|$ units to the left if h is negative.   **13.** vertex: $(2, 0)$; axis: $x = 2$; domain: $(-\infty, \infty)$; range: $[0, \infty)$

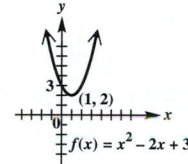

$f(x) = (x - 2)^2$

15. vertex: $(-3, -4)$; axis: $x = -3$; domain: $(-\infty, \infty)$; range: $[-4, \infty)$

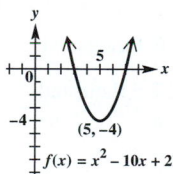

$f(x) = (x + 3)^2 - 4$

17. vertex: $(-1, -3)$; axis: $x = -1$; domain: $(-\infty, \infty)$; range: $(-\infty, -3]$

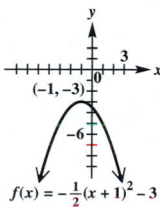

$f(x) = -\frac{1}{2}(x + 1)^2 - 3$

19. vertex: $(1, 2)$; axis: $x = 1$; domain: $(-\infty, \infty)$; range: $[2, \infty)$

$f(x) = x^2 - 2x + 3$

21. vertex: $(5, -4)$; axis: $x = 5$; domain: $(-\infty, \infty)$; range: $[-4, \infty)$

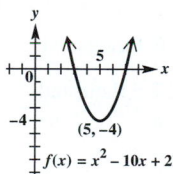

$f(x) = x^2 - 10x + 21$

23. vertex: $(-3, 2)$; axis: $x = -3$; domain: $(-\infty, \infty)$; range: $(-\infty, 2]$

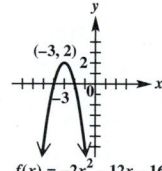

$f(x) = -2x^2 - 12x - 16$

25. vertex: $(-3, 4)$; axis: $x = -3$; domain: $(-\infty, \infty)$; range: $(-\infty, 4]$

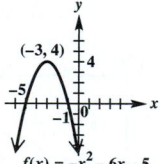

$f(x) = -x^2 - 6x - 5$

27. 3   **29.** none   **31.** E   **33.** D   **35.** C   **37.** $f(x) = \frac{1}{4}(x - 2)^2 - 1$, or $f(x) = \frac{1}{4}x^2 - x$

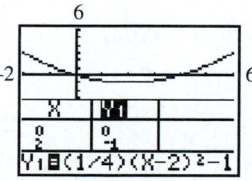

39. $f(x) = -2(x - 1)^2 + 4$, or $f(x) = -2x^2 + 4x + 2$ **41.** quadratic; negative

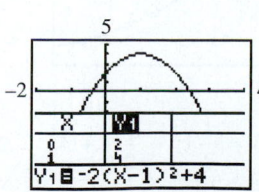

43. quadratic; positive   **45.** linear; positive   **47.** 6 and 6   **49.** 20 units; $210   **51.** 5 in.; 25 million mosquitos

53. **(a)** $f(t) = -16t^2 + 200t + 50$ **(b)** 6.25 sec; 675 ft **(c)** between approximately 1.4 and 11.1 sec **(d)** approximately 12.75 sec

55. **(a)** $640 - 2x$ **(b)** $0 < x < 320$ **(c)** $A(x) = -2x^2 + 640x$ **(d)** between approximately 57.04 ft and 85.17 ft or 234.83 ft and

262.96 ft **(e)** 160 ft by 320 ft; The maximum area is 51,200 ft².   **57.** **(a)** $2x$ **(b)** length: $2x - 4$; width: $x - 4$; $x > 4$

(c) $V(x) = 4x^2 - 24x + 32$ **(d)** 8 in. by 20 in. **(e)** 13.0 in. to 14.2 in.   **59.** **(a)** 3.5 ft **(b)** approximately .2 ft and 2.3 ft

(c) 1.25 ft **(d)** approximately 3.78 ft   **61.** **(a)** approximately 23.32 ft per sec **(b)** approximately 12.88 ft   **63.** **(a)** 38.6 (percent)

(b) No. Based on the model, the number of births to unmarried mothers is expected to rise after 2004.   **65.** 2000

67. **(a)** 1,000,000   **(b)** quadratic; The data increases at a different rate each year.

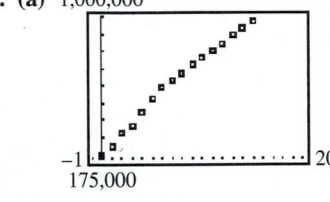

(c) $f(x) = -1239x^2 + 70,792x + 183,081$ **(d)** 1,000,000   The graph of f models the data extremely well.

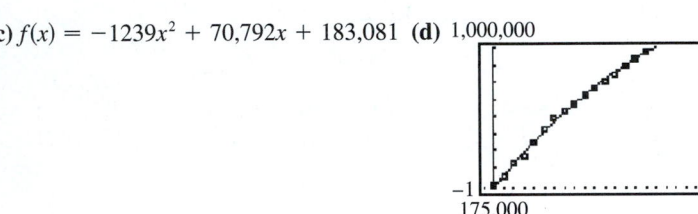

(e) 2009: 1,080,850; 2010: 1,103,321 **(f)** 22,471   **69.** **(a)** 120.2   **(b)** $f(x) = .6(x - 4)^2 + 50$

(c) There is a good fit. $f(x) = .6(x - 4)^2 + 50$ **(d)** $g(x) = .402x^2 - 1.175x + 48.343$ **(e)** $f(16) = 136.4$ thousand;

$g(16) \approx 132.5$ thousand

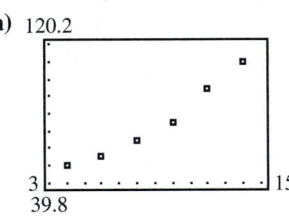

71. **(a)** 35   **(b)** $g(x) = .0074x^2 - 1.185x + 59.02$ models the data very well.   **(c)** $g(70) \approx 12.33$ sec

$g(x) = .0074x^2 - 1.185x + 59.02$

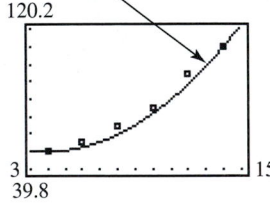

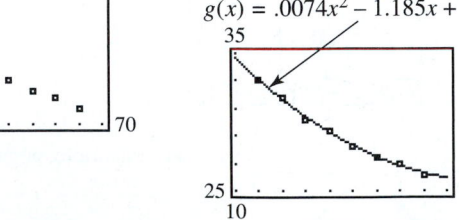

(d) about 39.1 mph   **73.** $c = 25$ **75.** $f(x) = \frac{1}{2}x^2 - \frac{7}{2}x + 5$ **77.** $(3, 6)$

79. The x-intercepts are -4 and 2.   **80.** the open interval $(-4, 2)$ **81.**

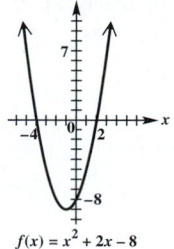

$f(x) = x^2 + 2x - 8$

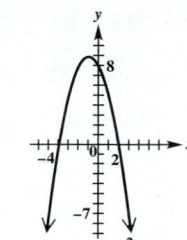

$g(x) = -f(x) = -x^2 - 2x + 8$

The graph of g is obtained by reflecting the graph of f across the x-axis.

82. the open interval $(-4, 2)$ **83.** They are the same.

3.2 Exercises *(pages 326–328)*

1. $x^2 + 2x + 9$ **3.** $5x^3 + 2x - 3$ **5.** $x^3 + 2x + 1$ **7.** $x^4 + x^3 + 2x - 1 + \dfrac{3}{x + 2}$

9. $-9x^2 - 10x - 27 + \dfrac{-52}{x - 2}$ **11.** $\dfrac{1}{3}x^2 - \dfrac{1}{9}x + \dfrac{1}{x - \dfrac{1}{3}}$ **13.** $x^3 - x^2 - 6x$ **15.** $x^2 + x + 1$

17. $x^4 - x^3 + x^2 - x + 1$ **19.** $f(x) = (x + 1)(2x^2 - x + 2) - 10$ **21.** $f(x) = (x + 2)(x^2 + 2x + 1) + 0$

23. $f(x) = (x - 3)(4x^3 + 9x^2 + 7x + 20) + 60$ **25.** $f(x) = (x + 1)(3x^3 + x^2 - 11x + 11) + 4$ **27.** 0   **29.** -1

31. -6 **33.** -5 **35.** 7   **37.** $-6 - i$ **39.** 0   **41.** yes   **43.** yes   **45.** no; -9 **47.** yes   **49.** yes

51. no; $\dfrac{357}{125}$ **53.** yes   **55.** no; $13 + 7i$ **57.** no; $-2 + 7i$ **59.** $-12; (-2, -12)$ **60.** $0; (-1, 0)$ **61.** $2; (0, 2)$

62. $0; (1, 0)$ **63.** $-\dfrac{5}{8}; \left(\dfrac{3}{2}, -\dfrac{5}{8}\right)$ **64.** $0; (2, 0)$ **65.** $8; (3, 8)$ **66.**

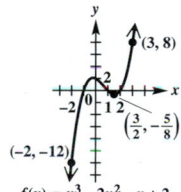

$f(x) = x^3 - 2x^2 - x + 2$

3.3 Exercises *(pages 337–339)*

1. true   **3.** false; -2 is a zero of multiplicity 4.   **5.** yes   **7.** no   **9.** yes   **11.** no   **13.** yes   **15.** yes

17. $f(x) = (x - 2)(2x - 5)(x + 3)$ **19.** $f(x) = (x + 3)(3x - 1)(2x - 1)$ **21.** $f(x) = (x + 4)(3x - 1)(2x + 1)$

23. $f(x) = (x - 3i)(x + 4)(x + 3)$ **25.** $f(x) = [x - (1 + i)](2x - 1)(x + 3)$ **27.** $f(x) = (x + 2)^2(x + 1)(x - 3)$

29. $-1 \pm i$ **31.** $3, 2 + i$ **33.** $i, \pm 2i$ **35.** **(a)** $\pm 1, \pm 2, \pm 5, \pm 10$ **(b)** $-1, -2, 5$ **(c)** $f(x) = (x + 1)(x + 2)(x - 5)$

37. **(a)** $\pm 1, \pm 2, \pm 3, \pm 5, \pm 6, \pm 10, \pm 15, \pm 30$ **(b)** $-5, -3, 2$ **(c)** $f(x) = (x + 5)(x + 3)(x - 2)$

39. **(a)** $\pm 1, \pm 2, \pm 3, \pm 4, \pm 6, \pm 12, \pm\dfrac{1}{2}, \pm\dfrac{3}{2}, \pm\dfrac{1}{3}, \pm\dfrac{2}{3}, \pm\dfrac{4}{3}, \pm\dfrac{1}{6}$ **(b)** $-4, -\dfrac{1}{3}, \dfrac{3}{2}$ **(c)** $f(x) = (x + 4)(3x + 1)(2x - 3)$

41. **(a)** $\pm 1, \pm 2, \pm 3, \pm 4, \pm 6, \pm 12, \pm\dfrac{1}{2}, \pm\dfrac{3}{2}, \pm\dfrac{1}{3}, \pm\dfrac{2}{3}, \pm\dfrac{4}{3}, \pm\dfrac{1}{4}, \pm\dfrac{3}{4}, \pm\dfrac{1}{6}, \pm\dfrac{1}{8}, \pm\dfrac{3}{8}, \pm\dfrac{1}{12}, \pm\dfrac{1}{24}$ **(b)** $-\dfrac{3}{2}, -\dfrac{2}{3}, \dfrac{1}{2}$

(c) $f(x) = 2(2x + 3)(3x + 2)(2x - 1)$ **43.** $0, \pm\dfrac{\sqrt{7}}{7}i$ **45.** $2, -3, 1, -1$ **47.** -2 (multiplicity 5), 1 (multiplicity 5),

$1 - \sqrt{3}$ (multiplicity 2)   **49.** $f(x) = -3x^3 + 6x^2 + 33x - 36$ **51.** $f(x) = -\dfrac{1}{2}x^3 - \dfrac{1}{2}x^2 + x$

53. $f(x) = \dfrac{1}{6}x^3 + \dfrac{3}{2}x^2 + \dfrac{9}{2}x + \dfrac{9}{2}$

In Exercises 55–71, we give only one possible answer.

55. $f(x) = x^2 - 10x + 26$ **57.** $f(x) = x^3 - 4x^2 + 6x - 4$ **59.** $f(x) = x^3 - 3x^2 + x + 1$

61. $f(x) = x^4 - 6x^3 + 10x^2 + 2x - 15$ **63.** $f(x) = x^3 - 8x^2 + 22x - 20$ **65.** $f(x) = x^4 - 4x^3 + 5x^2 - 2x - 2$

67. $f(x) = x^4 - 16x^3 + 98x^2 - 240x + 225$ **69.** $f(x) = x^5 - 12x^4 + 74x^3 - 248x^2 + 445x - 500$

71. $f(x) = x^4 - 6x^3 + 17x^2 - 28x + 20$ **73.** 2 or 0 positive; 1 negative   **75.** 1 positive; 1 negative

77. 2 or 0 positive; 3 or 1 negative   **79.** $-5, 3, \pm 2i\sqrt{3}$ **81.** $1, 1, 1, -4$ **83.** $-3, -3, 0, \dfrac{1 \pm i\sqrt{31}}{4}$

85. $2, 2, 2, \pm i\sqrt{2}$ **87.** $-\dfrac{1}{2}, 1, \pm 2i$ **89.** $-\dfrac{1}{5}, 1 \pm i\sqrt{5}$ **91.** $\pm 2i, \pm 5i$ **93.** $\pm i, \pm i$ **95.** $0, 0, 3 \pm \sqrt{2}$

97. $3, 3, 1 \pm i\sqrt{7}$ **99.** $\pm 2, \pm 3, \pm 2i$

3.4 Exercises *(pages 351–358)*

1. A   **3.** one   **5.** B and D   **7.** $f(x) = x(x + 5)^2(x - 3)$

9. 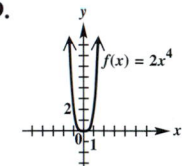   **11.** 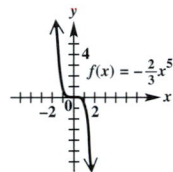   **13.** 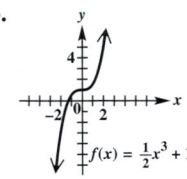   **15.** 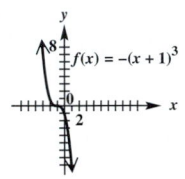   **17.**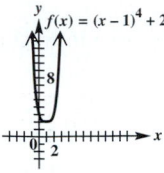

19. 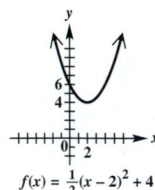   **21.** 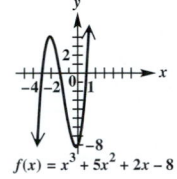   **23.** 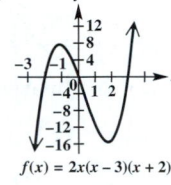   **25.** 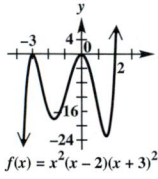   **27.** 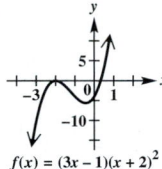   **29.** 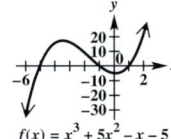   **31.**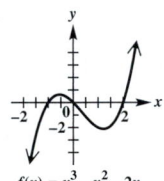

33. 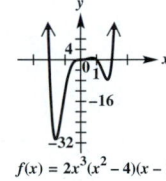   **35.** 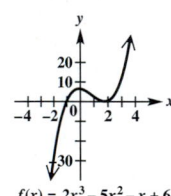   **37.** 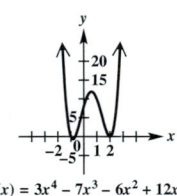   **39.**   **41.**

43. 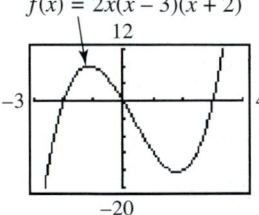   **45.** 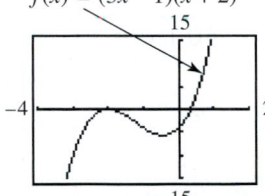   **47.** $f(2) = -2 < 0; f(3) = 1 > 0$ **49.** $f(0) = 7 > 0; f(1) = -1 < 0$

51. $f(1) = -6 < 0; f(2) = 16 > 0$ **53.** $f(3.2) = -3.8144 < 0; f(3.3) = 7.1891 > 0$

55. $f(-1) = -35 < 0; f(0) = 12 > 0$ **65.** $f(x) = \dfrac{1}{2}(x + 6)(x - 2)(x - 5)$, or $f(x) = \dfrac{1}{2}x^3 - \dfrac{1}{2}x^2 - 16x + 30$

67. $f(x) = (x - 1)^3(x + 1)^3$, or $f(x) = x^6 - 3x^4 + 3x^2 - 1$ **69.** $f(x) = (x - 3)^2(x + 3)^2$, or $f(x) = x^4 - 18x^2 + 81$

71. $f(1.25) = -14.21875$ **73.** $f(1.25) = 29.046875$ **75.** 2.7807764   **77.** 1.543689   **79.** $-3.0, -1.4, 1.4$

$f(x) = 2x(x - 3)(x + 2)$

$f(x) = (3x - 1)(x + 2)^2$

81. $-1.1, 1.2$ **83.** $(-3.44, 26.15)$ **85.** $(-.09, 1.05)$ **87.** $(-.20, -28.62)$ **89.** Answers will vary.

91.

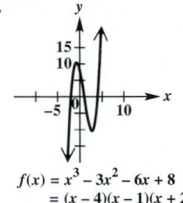

$f(x) = x^3 - 3x^2 - 6x + 8$
$= (x - 4)(x - 1)(x + 2)$

(a) $\{-2, 1, 4\}$

(b) $(-\infty, -2) \cup (1, 4)$

(c) $(-2, 1) \cup (4, \infty)$

92.

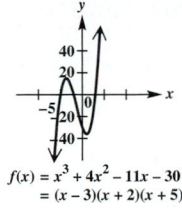

$f(x) = x^3 + 4x^2 - 11x - 30$
$= (x - 3)(x + 2)(x + 5)$

(a) $\{-5, -2, 3\}$

(b) $(-\infty, -5) \cup (-2, 3)$

(c) $(-5, -2) \cup (3, \infty)$

93.

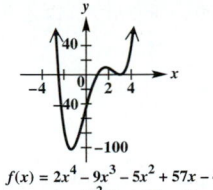

$f(x) = 2x^4 - 9x^3 - 5x^2 + 57x - 45$
$= (x - 3)^2(2x + 5)(x - 1)$

(a) $\{-2.5, 1, 3 \text{ (multiplicity 2)}\}$

(b) $(-2.5, 1)$

(c) $(-\infty, -2.5) \cup (1, 3) \cup (3, \infty)$

94.

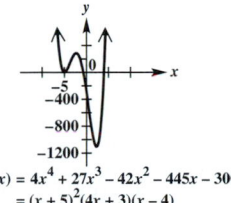

$f(x) = 4x^4 + 27x^3 - 42x^2 - 445x - 300$
$= (x + 5)^2(4x + 3)(x - 4)$

(a) $\{-5 \text{ (multiplicity 2)}, -.75, 4\}$

(b) $(-.75, 4)$

(c) $(-\infty, -5) \cup (-5, -.75) \cup (4, \infty)$

95.

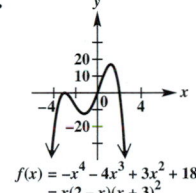

$f(x) = -x^4 - 4x^3 + 3x^2 + 18x$
$= x(2 - x)(x + 3)^2$

(a) $\{-3 \text{ (multiplicity 2)}, 0, 2\}$

(b) $\{-3\} \cup [0, 2]$

(c) $(-\infty, 0] \cup [2, \infty)$

96.

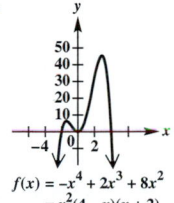

$f(x) = -x^4 + 2x^3 + 8x^2$
$= x^2(4 - x)(x + 2)$

(a) $\{-2, 0 \text{ (multiplicity 2)}, 4\}$

(b) $[-2, 4]$

(c) $(-\infty, -2] \cup \{0\} \cup [4, \infty)$

97. (a) $0 < x < 6$ **(b)** $V(x) = x(18 - 2x)(12 - 2x)$, or $V(x) = 4x^3 - 60x^2 + 216x$ **(c)** $x \approx 2.35$; about 228.16 in.^{3}

(d) $.42 < x < 5$ **99. (a)** $x - 1$; $(1, \infty)$ **(b)** $\sqrt{x^2 - (x - 1)^2}$ **(c)** $2x^3 - 5x^2 + 4x - 28,225 = 0$ **(d)** hypotenuse: 25 in.;

legs: 24 in. and 7 in. **101.** 3 ft **103. (a)** about 7.13 cm; The ball floats partly above the surface. **(b)** The sphere is more

dense than water and sinks below the surface. **(c)** 10 cm; The balloon is submerged with its top even with the surface.

105. (a)

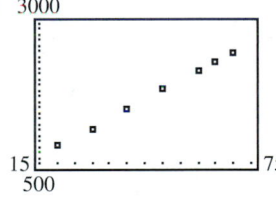

(b)

$y = 33.93x + 113.4$

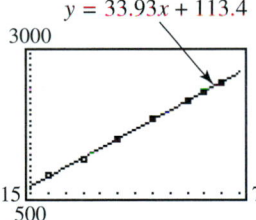

(c)

$y = -.0032x^3 + .4245x^2 + 16.64x$
$+ 323.1$

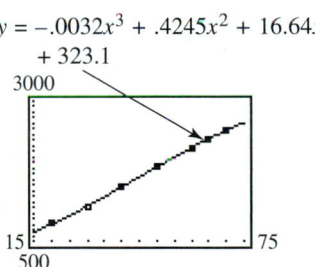

(d) linear: 1572 ft; cubic: 1569 ft **(e)** The cubic function appears slightly better because only one data point is not on the curve.

107. B

Summary Exercises on Polynomial Functions, Zeros, and Graphs *(page 358)*

1. (a) positive zeros: 1; negative zeros: 3 or 1 **(b)** $\pm 1, \pm 2, \pm 3, \pm 6$ **(c)** $-3, -1$ (multiplicity 2), 2 **(d)** no other real zeros

(e) no other complex zeros **(f)** $-3, -1, 2$ **(g)** -6 **(h)** $f(4) = 350$; $(4, 350)$ **(i)** ↳ ↱ **(j)**

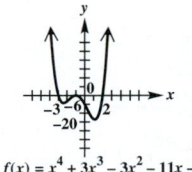

$f(x) = x^4 + 3x^3 - 3x^2 - 11x - 6$

2. (a) positive zeros: 3 or 1; negative zeros: 2 or 0 **(b)** $\pm 1, \pm 3, \pm 5, \pm 9, \pm 15, \pm 45, \pm \dfrac{1}{2}, \pm \dfrac{3}{2}, \pm \dfrac{5}{2}, \pm \dfrac{9}{2}, \pm \dfrac{15}{2}, \pm \dfrac{45}{2}$

(c) $-3, \dfrac{1}{2}, 5$ **(d)** $-\sqrt{3}, \sqrt{3}$ **(e)** no other complex zeros **(f)** $-3, \dfrac{1}{2}, 5, -\sqrt{3}, \sqrt{3}$ **(g)** 45 **(h)** $f(4) = 637$; $(4, 637)$ **(i)** ↘ ↘

(j)

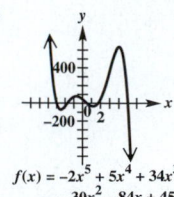

$f(x) = -2x^5 + 5x^4 + 34x^3$
$- 30x^2 - 84x + 45$

3. (a) positive zeros: 4, 2, or 0; negative zeros: 1 **(b)** $\pm 1, \pm 5, \pm \dfrac{1}{2}, \pm \dfrac{5}{2}$ **(c)** 5 **(d)** $-\dfrac{\sqrt{2}}{2}, \dfrac{\sqrt{2}}{2}$ **(e)** $-i, i$ **(f)** $-\dfrac{\sqrt{2}}{2}, \dfrac{\sqrt{2}}{2}, 5$

(g) 5 **(h)** $f(4) = -527$; $(4, -527)$ **(i)** 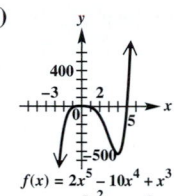 **(j)**

$f(x) = 2x^5 - 10x^4 + x^3$
$- 5x^2 - x + 5$

4. (a) positive zeros: 2 or 0; negative zeros: 2 or 0 **(b)** $\pm 1, \pm 2, \pm 3, \pm 6, \pm 9, \pm 18, \pm \dfrac{1}{3}, \pm \dfrac{2}{3}$ **(c)** $-\dfrac{2}{3}, 3$ **(d)** $\dfrac{-1 + \sqrt{13}}{2}$,

$\dfrac{-1 - \sqrt{13}}{2}$ **(e)** no other complex zeros **(f)** $-\dfrac{2}{3}, 3, \dfrac{-1 \pm \sqrt{13}}{2}$ **(g)** 18 **(h)** $f(4) = 238$; $(4, 238)$ **(i)** 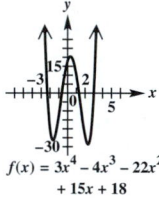 **(j)**

$f(x) = 3x^4 - 4x^3 - 22x^2$
$+ 15x + 18$

5. (a) positive zeros: 1; negative zeros: 3 or 1 **(b)** $\pm 1, \pm 2, \pm \dfrac{1}{2}$ **(c)** $-1, 1$ **(d)** no other real zeros

(e) $-\dfrac{1}{4} + \dfrac{\sqrt{15}}{4}i, -\dfrac{1}{4} - \dfrac{\sqrt{15}}{4}i$ **(f)** $-1, 1$ **(g)** 2 **(h)** $f(4) = -570$; $(4, -570)$ **(i)** 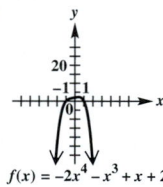 **(j)**

$f(x) = -2x^4 - x^3 + x + 2$

6. (a) positive zeros: 0; negative zeros: 4, 2, or 0 **(b)** $0, \pm 1, \pm 3, \pm 9, \pm 27, \pm \dfrac{1}{2}, \pm \dfrac{3}{2}, \pm \dfrac{9}{2}, \pm \dfrac{27}{2}, \pm \dfrac{1}{4}, \pm \dfrac{3}{4}, \pm \dfrac{9}{4}, \pm \dfrac{27}{4}$

(c) $0, -\dfrac{3}{2}$ (multiplicity 2) **(d)** no other real zeros **(e)** $\dfrac{1}{2} + \dfrac{\sqrt{11}}{2}i, \dfrac{1}{2} - \dfrac{\sqrt{11}}{2}i$ **(f)** $0, -\dfrac{3}{2}$ **(g)** 0 **(h)** $f(4) = 7260$; $(4, 7260)$

(i) 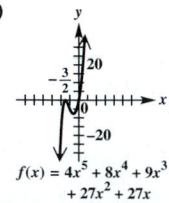 **(j)**

$f(x) = 4x^5 + 8x^4 + 9x^3$
$+ 27x^2 + 27x$

7. (a) positive zeros: 1; negative zeros: 1 **(b)** $\pm 1, \pm 5, \pm \dfrac{1}{3}, \pm \dfrac{5}{3}$ **(c)** no rational zeros **(d)** $-\sqrt{5}, \sqrt{5}$ **(e)** $-\dfrac{\sqrt{3}}{3}i, \dfrac{\sqrt{3}}{3}i$

(f) $-\sqrt{5}, \sqrt{5}$ **(g)** -5 **(h)** $f(4) = 539; (4, 539)$ **(i)** 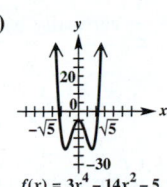 **(j)**

$f(x) = 3x^4 - 14x^2 - 5$

8. (a) positive zeros: 2 or 0; negative zeros: 3 or 1 **(b)** $\pm 1, \pm 3, \pm 9$ **(c)** $-3, -1$ (multiplicity 2), 1, 3 **(d)** no other real zeros **(e)** no other complex zeros **(f)** $-3, -1, 1, 3$ **(g)** -9 **(h)** $f(4) = -525; (4, -525)$ **(i)** 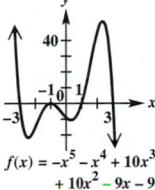 **(j)**

$f(x) = -x^5 - x^4 + 10x^3 + 10x^2 - 9x - 9$

9. (a) positive zeros: 4, 2, or 0; negative zeros: 0 **(b)** $\pm 1, \pm 2, \pm 3, \pm 4, \pm 6, \pm 12, \pm \dfrac{1}{3}, \pm \dfrac{2}{3}, \pm \dfrac{4}{3}$ **(c)** $\dfrac{1}{3}, 2$ (multiplicity 2), 3

(d) no other real zeros **(e)** no other complex zeros **(f)** $\dfrac{1}{3}, 2, 3$ **(g)** -12 **(h)** $f(4) = -44; (4, -44)$ **(i)**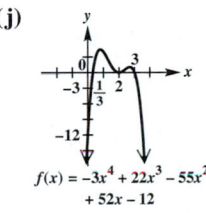
(j)

$f(x) = -3x^4 + 22x^3 - 55x^2 + 52x - 12$

10. For the function in Exercise 2: ± 1.732; for the function in Exercise 3: $\pm.707$; for the function in Exercise 4: $-2.303, 1.303$; for the function in Exercise 7: ± 2.236

3.5 Exercises *(pages 371–379)*

1. $(-\infty, 0) \cup (0, \infty); (-\infty, 0) \cup (0, \infty)$ **3.** none; $(-\infty, 0) \cup (0, \infty)$; none   **5.** $x = 3; y = 2$

7. even; symmetry with respect to the y-axis   **9.** A, B, C   **11.** A   **13.** A   **15.** A, C, D

17. To obtain the graph of f, stretch the graph of $y = \dfrac{1}{x}$ vertically by a factor of 2.

     domain: $(-\infty, 0) \cup (0, \infty)$; range: $(-\infty, 0) \cup (0, \infty)$

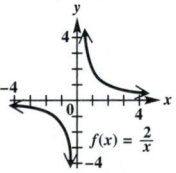

$f(x) = \dfrac{2}{x}$

19. To obtain the graph of f, shift the graph of $y = \dfrac{1}{x}$ to the left 2 units.

     domain: $(-\infty, -2) \cup (-2, \infty)$; range: $(-\infty, 0) \cup (0, \infty)$

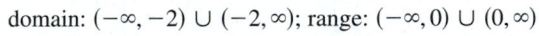
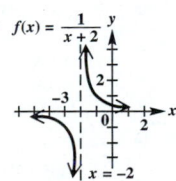

$f(x) = \dfrac{1}{x + 2}$

21. To obtain the graph of f, shift the graph of $y = \dfrac{1}{x}$ up 1 unit.

     domain: $(-\infty, 0) \cup (0, \infty)$; range: $(-\infty, 1) \cup (1, \infty)$

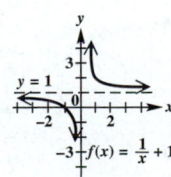

$f(x) = \dfrac{1}{x} + 1$

23. To obtain the graph of f, stretch the graph of $y = \dfrac{1}{x^2}$ vertically by a factor of 2, and reflect across the x-axis.

domain: $(-\infty, 0) \cup (0, \infty)$; range: $(-\infty, 0)$

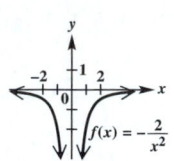

25. To obtain the graph of f, shift the graph of $y = \dfrac{1}{x^2}$ to the right 3 units.

domain: $(-\infty, 3) \cup (3, \infty)$; range: $(0, \infty)$

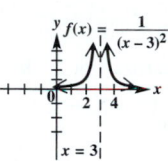

27. To obtain the graph of f, shift the graph of $y = \dfrac{1}{x^2}$ to the left 2 units,

reflect across the x-axis, and shift 3 units down.

domain: $(-\infty, -2) \cup (-2, \infty)$; range: $(-\infty, -3)$

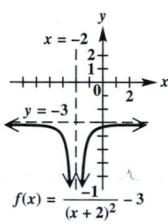

29. D   **31.** G   **33.** E   **35.** F

In Exercises 37–45 and 53–59, V.A. represents vertical asymptote, H.A. represents horizontal asymptote, and O.A. represents oblique asymptote.

37. V.A.: $x = 5$; H.A.: $y = 0$ **39.** V.A.: $x = -\dfrac{1}{2}$; H.A.: $y = -\dfrac{3}{2}$ **41.** V.A.: $x = -3$; O.A.: $y = x - 3$

43. V.A.: $x = -2$, $x = \dfrac{5}{2}$; H.A.: $y = \dfrac{1}{2}$ **45.** V.A.: none; H.A.: $y = 1$ **47. (a)** $f(x) = \dfrac{2x - 5}{x - 3}$ **(b)** $\dfrac{5}{2}$

(c) horizontal asymptote: $y = 2$; vertical asymptote: $x = 3$ **49. (a)** $y = x + 1$ **(b)** at $x = 0$ and $x = 1$ **(c)** above

51. A   **53.** V.A.: $x = 2$; H.A.: $y = 4$; $(-\infty, 2) \cup (2, \infty)$ **55.** V.A.: $x = \pm 2$; H.A.: $y = -4$; $(-\infty, -2) \cup (-2, 2) \cup (2, \infty)$

57. V.A.: none; H.A.: $y = 0$; $(-\infty, \infty)$ **59.** V.A.: $x = -1$; O.A.: $y = x - 1$; $(-\infty, -1) \cup (-1, \infty)$

61.

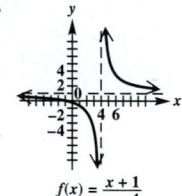

$f(x) = \dfrac{x + 1}{x - 4}$

63.

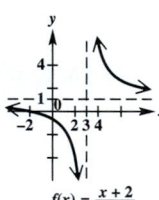

$f(x) = \dfrac{x + 2}{x - 3}$

65.

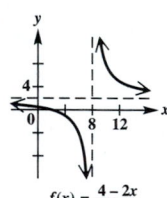

$f(x) = \dfrac{4 - 2x}{8 - x}$

67.

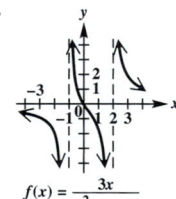

$f(x) = \dfrac{3x}{x^2 - x - 2}$

69.

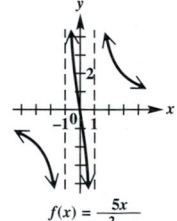

$f(x) = \dfrac{5x}{x^2 - 1}$

71.

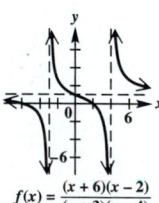

$f(x) = \dfrac{(x + 6)(x - 2)}{(x + 3)(x - 4)}$

73.

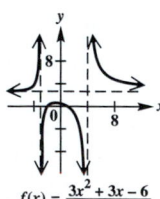

$f(x) = \dfrac{3x^2 + 3x - 6}{x^2 - x - 12}$

75.

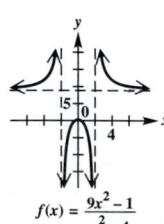

$f(x) = \dfrac{9x^2 - 1}{x^2 - 4}$

77.

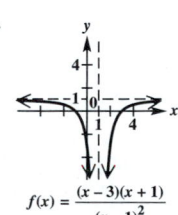

$f(x) = \dfrac{(x - 3)(x + 1)}{(x - 1)^2}$

79.

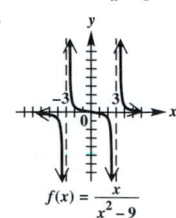

$f(x) = \dfrac{x}{x^2 - 9}$

81.

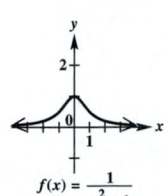

$f(x) = \dfrac{1}{x^2 + 1}$

83.

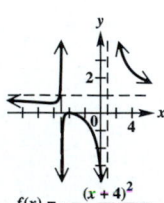

$f(x) = \dfrac{(x + 4)^2}{(x - 1)(x + 5)}$

85.

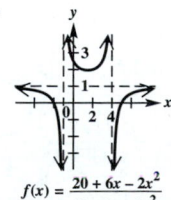

$f(x) = \dfrac{20 + 6x - 2x^2}{8 + 6x - 2x^2}$

87.

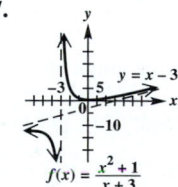

$f(x) = \dfrac{x^2 + 1}{x + 3}$

89.

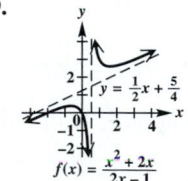

$f(x) = \dfrac{x^2 + 2x}{2x - 1}$

91.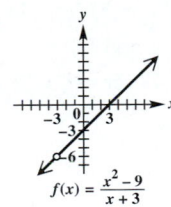

$f(x) = \dfrac{x^2 - 9}{x + 3}$

93.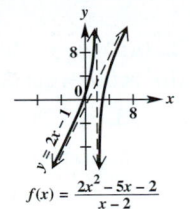

$f(x) = \dfrac{2x^2 - 5x - 2}{x - 2}$

95.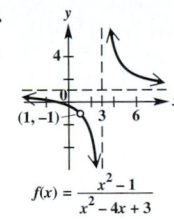

$f(x) = \dfrac{x^2 - 1}{x^2 - 4x + 3}$

97.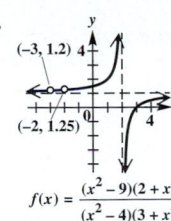

$f(x) = \dfrac{(x^2 - 9)(2 + x)}{(x^2 - 4)(3 + x)}$

99.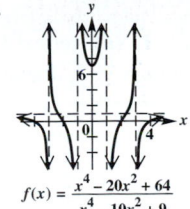

$f(x) = \dfrac{x^4 - 20x^2 + 64}{x^4 - 10x^2 + 9}$

101. $f(x) = \dfrac{(x - 3)(x + 2)}{(x - 2)(x + 2)}$, or $f(x) = \dfrac{x^2 - x - 6}{x^2 - 4}$

103. $f(x) = \dfrac{x - 2}{x(x - 4)}$, or $f(x) = \dfrac{x - 2}{x^2 - 4x}$

105. Several answers are possible. One answer is $f(x) = \dfrac{(x - 3)(x + 1)}{(x - 1)^2}$.

107. $f(1.25) = -.\overline{81}$

109. $f(1.25) = 2.708\overline{3}$

111. (a) 26 per min **(b)** 5 park attendants

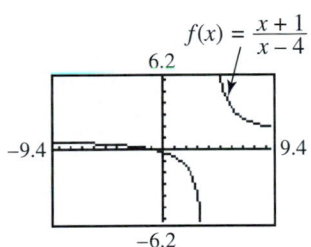

$f(x) = \dfrac{x + 1}{x - 4}$

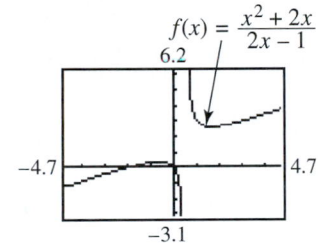

$f(x) = \dfrac{x^2 + 2x}{2x - 1}$

For $r = x$,

$y = T(r) = \dfrac{2x - 25}{2x^2 - 50x}$ $y = .5$

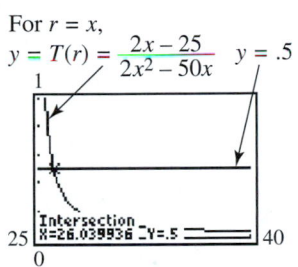

113. (a) approximately 52.1 mph

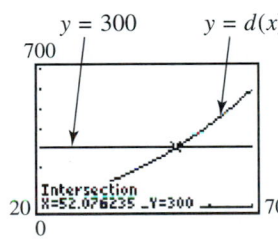

$y = 300$ $\quad y = d(x)$

x	$d(x)$	x	$d(x)$
20	**34**	50	**273**
25	**56**	55	**340**
30	**85**	60	**415**
35	**121**	65	**499**
40	**164**	70	**591**
45	**215**		

115. All answers are given in tens of millions. **(a)** \$65.5 **(b)** \$64 **(c)** \$60 **(d)** \$40 **(e)** \$0

(f)

$R(x) = \dfrac{80x - 8000}{x - 110}$

117. $y = 1$ **118.** $(x + 4)(x + 1)(x - 3)(x - 5)$

119. (a) $(x - 1)(x - 2)(x + 2)(x - 5)$ **(b)** $f(x) = \dfrac{(x + 4)(x + 1)(x - 3)(x - 5)}{(x - 1)(x - 2)(x + 2)(x - 5)}$

120. (a) $x - 5$ **(b)** 5 **121.** $-4, -1, 3$ **122.** -3 **123.** $x = 1, x = 2, x = -2$

124. $\left(\dfrac{7 + \sqrt{241}}{6}, 1\right), \left(\dfrac{7 - \sqrt{241}}{6}, 1\right)$

125.

$f(x) = \dfrac{x^4 - 3x^3 - 21x^2 + 43x + 60}{x^4 - 6x^3 + x^2 + 24x - 20}$

126. (a) $(-4, -2) \cup (-1, 1) \cup (2, 3)$ **(b)** $(-\infty, -4) \cup (-2, -1) \cup (1, 2) \cup (3, 5) \cup (5, \infty)$

Chapter 3 Quiz *(page 379)*

[3.1] 1. (a) vertex: $(-3, -1)$; axis: $x = -3$; domain: $(-\infty, \infty)$; range: $(-\infty, -1]$

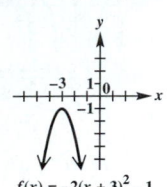

$f(x) = -2(x + 3)^2 - 1$

(b) vertex: $(2, -5)$; axis: $x = 2$; domain: $(-\infty, \infty)$; range: $[-5, \infty)$

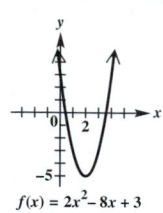

$f(x) = 2x^2 - 8x + 3$

2. (a) $s(t) = -16t^2 + 64t + 200$

(b) between approximately .78 sec and 3.22 sec

[3.2] 3. no; 38 **4.** yes

[3.3] 5. $f(x) = x^4 - 7x^3 + 10x^2 + 26x - 60$

[3.4] 6.

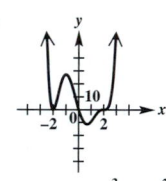

$f(x) = x(x - 2)^3(x + 2)^2$

7.

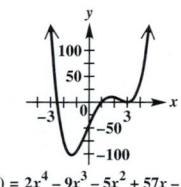

$f(x) = 2x^4 - 9x^3 - 5x^2 + 57x - 45$
$= (x - 3)^2(2x + 5)(x - 1)$

8.

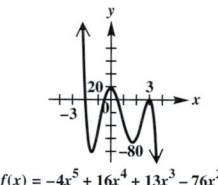

$f(x) = -4x^5 + 16x^4 + 13x^3 - 76x^2 - 3x + 18$
$= -(x + 2)(2x + 1)(2x - 1)(x - 3)^2$

[3.5] 9.

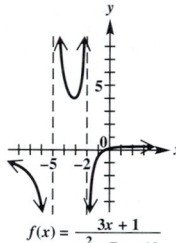

$f(x) = \dfrac{3x + 1}{x^2 + 7x + 10}$

10.

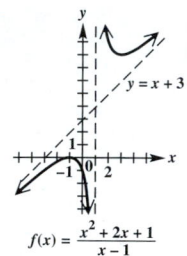

$f(x) = \dfrac{x^2 + 2x + 1}{x - 1}$

3.6 Exercises *(pages 383–387)*

1. The circumference of a circle varies directly as (or is proportional to) its radius. **3.** The speed varies directly as (or is proportional to) the distance traveled and inversely as the time. **5.** The strength of a muscle varies directly as (or is proportional to) the cube of its length. **7.** C **9.** A **11.** -30 **13.** $\dfrac{220}{7}$ **15.** $\dfrac{3}{2}$ **17.** $\dfrac{12}{5}$ **19.** $\dfrac{18}{125}$

21. 69.08 in. **23.** 850 ohms **25.** 8 lb **27.** 16 in. **29.** 90 revolutions per minute **31.** .0444 ohm

33. $1375 **35.** 800 lb **37.** $\dfrac{8}{9}$ metric ton **39.** $\dfrac{66\pi}{17}$ sec **41.** 21 **43.** 365.24 **45.** 92; undernourished

47. increases; decreases **49.** y is half as large as before. **51.** y is one-third as large as before.

53. p is $\dfrac{1}{32}$ as large as before.

Chapter 3 Review Exercises *(pages 393–398)*

1. vertex: $(-4, -5)$; axis: $x = -4$; x-intercepts: $\dfrac{-12 \pm \sqrt{15}}{3}$; y-intercept: 43; domain: $(-\infty, \infty)$; range: $[-5, \infty)$

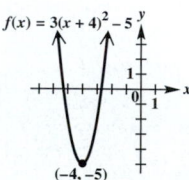

$f(x) = 3(x + 4)^2 - 5$

3. vertex: $(-2, 11)$; axis $x = -2$; x-intercepts: $\dfrac{-6 \pm \sqrt{33}}{3}$; y-intercept: -1; domain: $(-\infty, \infty)$; range: $(-\infty, 11]$

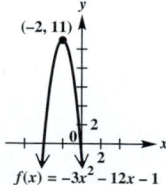

$f(x) = -3x^2 - 12x - 1$

5. (h, k) **7.** $k \le 0; h \pm \sqrt{\dfrac{-k}{a}}$ **9.** 90 m by 45 m   **11. (a)** 120 **(b)** 40 **(c)** 22 **(d)** 84 **(e)** 146

(f) minimum at $x = 8$ (August)   **13.** Because the discriminant is 67.3033, a positive number, there are two x-intercepts.

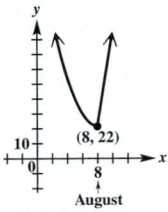

15. (a) the open interval $(-.52, 2.59)$ **(b)** $(-\infty, -.52) \cup (2.59, \infty)$ **17.** $x^2 + 4x + 1 + \dfrac{-7}{x - 3}$

19. $2x^2 - 8x + 31 + \dfrac{-118}{x + 4}$ **21.** $(x - 2)(5x^2 + 7x + 16) + 26$ **23.** -1 **25.** 28   **27.** yes   **29.** $7 - 2i$

In Exercises 31 and 33, other answers are possible.

31. $f(x) = x^3 - 10x^2 + 17x + 28$ **33.** $f(x) = x^4 - 5x^3 + 3x^2 + 15x - 18$ **35.** $\dfrac{1}{2}, -1, 5$

37. (a) $f(-1) = -10 < 0; f(0) = 2 > 0$ **(b)** $f(2) = -4 < 0; f(3) = 14 > 0$ **41.** yes

43. $f(x) = -2x^3 + 6x^2 + 12x - 16$ **45.** $1, -\dfrac{1}{2}, \pm 2i$ **47.** $\dfrac{13}{2}$

49. Any polynomial that can be factored into $a(x - b)^3$ works. One example is $f(x) = 2(x - 1)^3$.

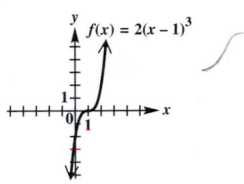

51. (a) $(-\infty, \infty)$ **(b)** $(-\infty, \infty)$ **(c)** $f(x) \to \infty$ as $x \to \infty$, $f(x) \to -\infty$ as $x \to -\infty$: **(d)** at most 7 **(e)** at most 6

53. C   **55.** E   **57.** B   **59.**   **61.**   **63.**

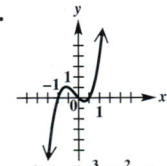
$f(x) = (x - 2)^2(x + 3)$

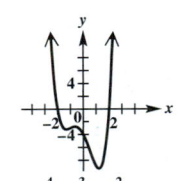
$f(x) = 2x^3 + x^2 - x$

$f(x) = x^4 + x^3 - 3x^2 - 4x - 4$

65. 7.6533119, 1, $-.6533119$

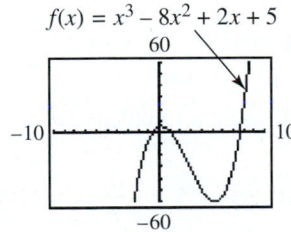
$f(x) = x^3 - 8x^2 + 2x + 5$

67. (a)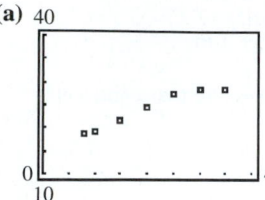

(b) $f(x) = -.011x^2 + .869x + 11.9$

(c) $f(x) = -.00087x^3 + .0456x^2 - .219x + 17.8$

(d) $f(x) = -.011x^2 + .869x + 11.9$ $f(x) = -.00087x^3 + .0456x^2 - .219x + 17.8$

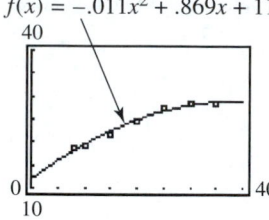   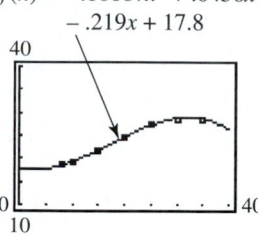

(e) Both functions approximate the data well. The quadratic function is probably better for prediction because it is unlikely that the percent of out-of-pocket spending would decrease after 2025 (as the cubic function shows) unless changes were made in Medicare law.   **69.** 12 in. × 4 in. × 15 in.

71. 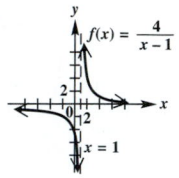   **73.** 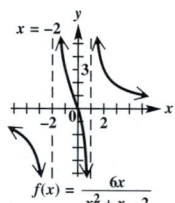   **75.** 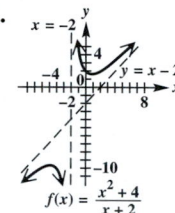   **77.**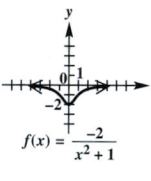

79. (a) 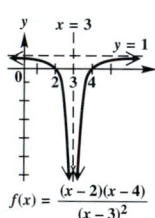   **(b)** One possibility is $f(x) = \dfrac{(x-2)(x-4)}{(x-3)^2}$.   **81.** $f(x) = \dfrac{-3x + 6}{x - 1}$

83. (a) 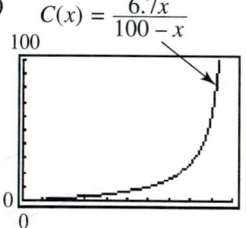   **(b)** approximately \$127.3 thousand   **85.** 35   **87.** .75   **89.** 27   **91.** 7500 lb

Chapter 3 Test *(pages 398–400)*

[3.1] 1. 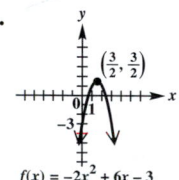 x-intercepts: $\dfrac{-3 \pm \sqrt{3}}{-2}$ $\left(\text{or } \dfrac{3 \pm \sqrt{3}}{2}\right)$; y-intercept: -3; vertex: $\left(\dfrac{3}{2}, \dfrac{3}{2}\right)$; axis: $x = \dfrac{3}{2}$;

domain: $(-\infty, \infty)$; range: $\left(-\infty, \dfrac{3}{2}\right]$ **2.** about 8.1 million

[3.2] 3. $3x^2 - 2x - 5 + \dfrac{16}{x + 2}$ **4.** $2x^2 - x - 5 + \dfrac{3}{x - 5}$ **5.** 53

[3.3] 6. It is a factor. The other factor is $6x^3 + 7x^2 - 14x - 8$.   **7.** $-2, -3 - 2i, -3 + 2i$

8. $f(x) = 2x^4 - 2x^3 - 2x^2 - 2x - 4$ **9.** Because $f(x) > 0$ for all x, the graph never intersects or touches the x-axis, so $f(x)$ has no real zeros.   **[3.3, 3.4] 10. (a)** $f(1) = 5 > 0; f(2) = -1 < 0$ **(b)** 2 or 0 positive zeros; 1 negative zero
(c) 4.0937635, 1.8370381, $-.9308016$

[3.4] 11. 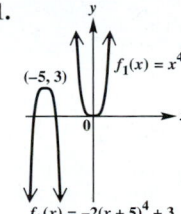 To obtain the graph of f_2, translate the graph of f_1 5 units to the left, stretch by a factor of 2, reflect across the x-axis, and translate 3 units up.

12. C **13.** 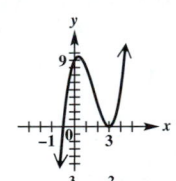 **14.** 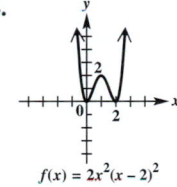 **15.**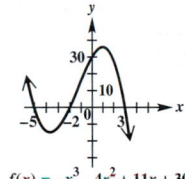

16. $f(x) = 2(x - 2)^2(x + 3)$, or $f(x) = 2x^3 - 2x^2 - 16x + 24$ **17. (a)** 270.08 **(b)** increasing from $t = 0$ to $t = 5.9$ and $t = 9.5$ to $t = 15$; decreasing from $t = 5.9$ to $t = 9.5$

[3.5] 18. 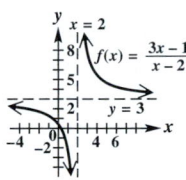 **19.** 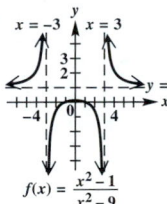 **20. (a)** $y = 2x + 3$ **(b)** $-2, \dfrac{3}{2}$ **(c)** 6 **(d)** $x = 1$

(e) 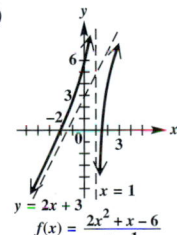 **[3.6] 21.** 60 **22.** $\dfrac{640}{9}$ kg

CHAPTER 4 INVERSE, EXPONENTIAL, AND LOGARITHMIC FUNCTIONS

4.1 Exercises *(pages 411–415)*

1. Yes, it is one-to-one, because every number in the list of registered passenger cars is used only once. **3.** one-to-one **5.** one-to-one **7.** not one-to-one **9.** one-to-one **11.** not one-to-one **13.** one-to-one **15.** one-to-one **17.** not one-to-one **19.** one-to-one **21.** range; domain **23.** false **25.** -3 **29.** untying your shoelaces **31.** leaving a room **33.** unscrewing a light bulb **35.** inverses **37.** not inverses **39.** inverses **41.** inverses **43.** not inverses **45.** inverses **47.** not inverses **49.** inverses **51.** $\{(6, -3), (1, 2), (8, 5)\}$ **53.** not one-to-one

55. (a) $f^{-1}(x) = \dfrac{1}{3}x + \dfrac{4}{3}$ **(b)** **(c)** Domains and ranges of both f and f^{-1} are $(-\infty, \infty)$.

57. (a) $f^{-1}(x) = -\dfrac{1}{4}x + \dfrac{3}{4}$ **(b)** 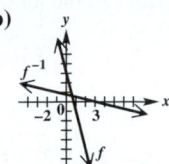   **(c)** Domains and ranges of both f and f^{-1} are $(-\infty, \infty)$.

59. (a) $f^{-1}(x) = \sqrt[3]{x - 1}$ **(b)** 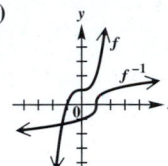   **(c)** Domains and ranges of both f and f^{-1} are $(-\infty, \infty)$.

61. not one-to-one   **63. (a)** $f^{-1}(x) = \dfrac{1}{x}$ **(b)** 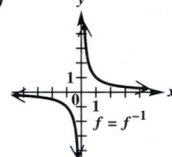   **(c)** Domains and ranges of both f and f^{-1} are $(-\infty, 0) \cup (0, \infty)$.

65. (a) $f^{-1}(x) = \dfrac{1 + 3x}{x}$ **(b)** 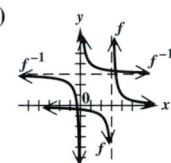   **(c)** Domain of f = range of f^{-1} = $(-\infty, 3) \cup (3, \infty)$; Domain of f^{-1} = range of f = $(-\infty, 0) \cup (0, \infty)$.

67. (a) $f^{-1}(x) = \dfrac{3x + 1}{x - 1}$ **(b)** 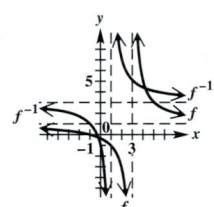   **(c)** Domain of f = range of f^{-1} = $(-\infty, 3) \cup (3, \infty)$; Domain of f^{-1} = range of f = $(-\infty, 1) \cup (1, \infty)$.

69. (a) $f^{-1}(x) = x^2 - 6, x \geq 0$ **(b)** 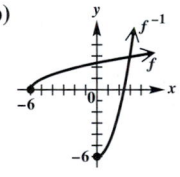   **(c)** Domain of f = range of f^{-1} = $[-6, \infty)$; Domain of f^{-1} = range of f = $[0, \infty)$.

71. 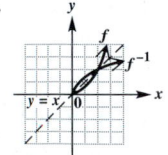   **73.** 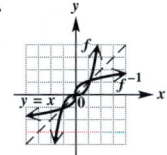   **75.** 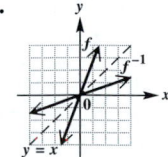   **77.** 4   **79.** 2   **81.** -2

83. It represents the cost, in dollars, of building 1000 cars.   **85.** $\dfrac{1}{a}$ **87.** not one-to-one

89. one-to-one; $f^{-1}(x) = \dfrac{-5 - 3x}{x - 1}$ **91.** $f^{-1}(x) = \dfrac{1}{3}x + \dfrac{2}{3}$; MIGUEL HAS ARRIVED

93. 6858   124   2743   63   511   124   1727   4095; $f^{-1}(x) = \sqrt[3]{x + 1}$

4.2 Exercises *(pages 427–431)*

1. 9   **3.** $\frac{1}{9}$ **5.** $\frac{1}{16}$ **7.** 16   **9.** 5.196   **11.** .039   **13.** 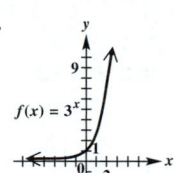   **15.**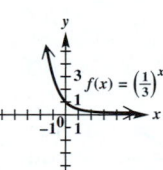

17. 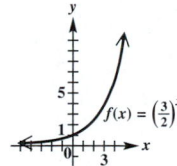   **19.** 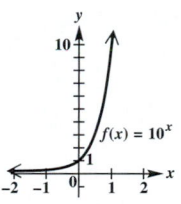   **21.** 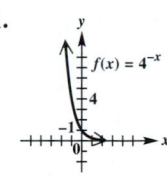   **23.** 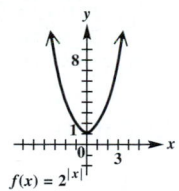   **25.**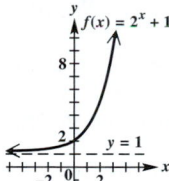

27. 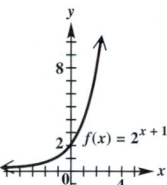   **29.** 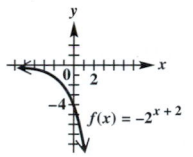   **31.** 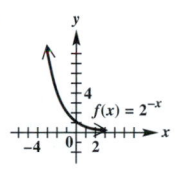   **33.**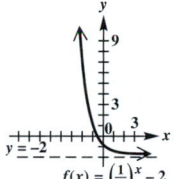

35. 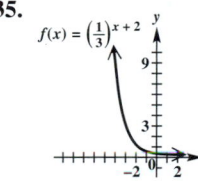   **37.** 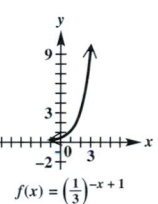   **39.** 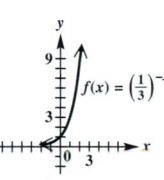   **41.** $\frac{1}{a}$ **43.** $f(x) = 2^{x+3} - 1$

45. $f(x) = -2^{x+2} + 3$ **47.** $f(x) = 3^{-x} + 1$ **49.** $\left\{\frac{1}{2}\right\}$ **51.** $\{-2\}$ **53.** $\{0\}$ **55.** $\left\{\frac{1}{2}\right\}$

57. $\left\{\frac{1}{5}\right\}$ **59.** $\{-7\}$ **61.** $\left\{-\frac{2}{3}\right\}$ **63.** $\left\{\frac{4}{3}\right\}$ **65.** $\{3\}$ **67.** $\{-8, 8\}$ **69.** $\{-27\}$

71. (a) \$13,891.16; \$4984.62 **(b)** \$13,968.24; \$5061.70   **73.** \$21,223.33   **75.** \$3528.81   **77.** 4.5%

79. Bank A (even though it has the highest stated rate)   **81. (a)** 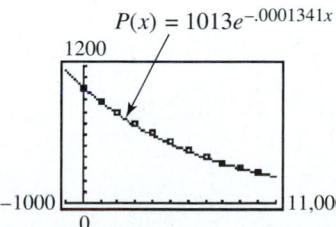   **(b)** exponential

(c) 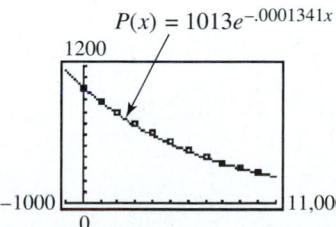 $P(x) = 1013e^{-.0001341x}$ **(d)** $P(1500) \approx 828$ mb; $P(11,000) \approx 232$ mb

83. (a) about 63,000 **(b)** about 42,000 **(c)** about 21,000   **85.** $\{.9\}$ **87.** $\{-.5, 1.3\}$ **91.** $f(x) = 2^x$

93. $f(x) = \left(\dfrac{1}{4}\right)^x$ **95.** $f(t) = 27 \cdot 9^t$ **97.** $f(t) = \left(\dfrac{1}{3}\right)9^t$ **101.** yes; an inverse function

102.
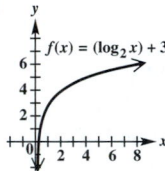
  103. $x = a^y$ **104.** $x = 10^y$ **105.** $x = e^y$ **106.** (q, p)

4.3 Exercises *(pages 441–445)*

1. (a) C **(b)** A **(c)** E **(d)** B **(e)** F **(f)** D   **3.** $\log_3 81 = 4$ **5.** $\log_{2/3} \dfrac{27}{8} = -3$ **7.** $6^2 = 36$ **9.** $\left(\sqrt{3}\right)^8 = 81$

13. $\{-4\}$ **15.** $\left\{\dfrac{1}{2}\right\}$ **17.** $\left\{\dfrac{1}{4}\right\}$ **19.** $\{8\}$ **21.** $\{9\}$ **23.** $\left\{\dfrac{1}{5}\right\}$ **25.** $\{64\}$ **27.** $\left\{\dfrac{2}{3}\right\}$ **29.** $\{243\}$

33. $(0, \infty); (-\infty, \infty)$
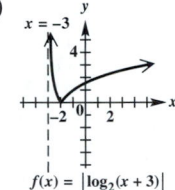
   35. $(-3, \infty); [0, \infty)$ **37.** $(2, \infty); (-\infty, \infty)$
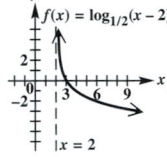

39. E   **41.** B   **43.** F   **45.**
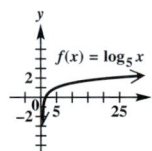
   47.
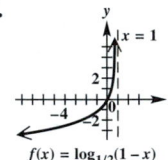
   49.
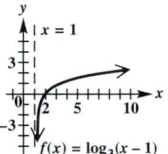

51. $f(x) = \log_2(x + 1) - 3$ **53.** $f(x) = \log_2(-x + 3) - 2$ **55.** $f(x) = -\log_3(x - 1)$

57.
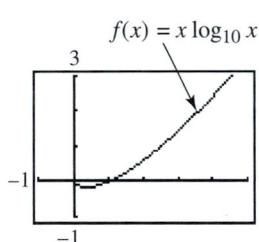
   59. $\log_2 6 + \log_2 x - \log_2 y$ **61.** $1 + \dfrac{1}{2}\log_5 7 - \log_5 3$

63. This cannot be simplified.   **65.** $\dfrac{1}{2}(\log_m 5 + 3\log_m r - 5\log_m z)$ **67.** $\log_2 a + \log_2 b - \log_2 c - \log_2 d$

69. $\dfrac{1}{2}\log_3 x + \dfrac{1}{3}\log_3 y - 2\log_3 w - \dfrac{1}{2}\log_3 z$ **71.** $\log_a \dfrac{xy}{m}$ **73.** $\log_a \dfrac{m}{nt}$ **75.** $\log_m \dfrac{a^2}{b^6}$ **77.** $\log_a [(z + 1)^2(3z + 2)]$

79. $\log_5 \dfrac{5^{1/3}}{m^{1/3}}$, or $\log_5 \sqrt[3]{\dfrac{5}{m}}$ **81.** .7781   **83.** .1761   **85.** .3522   **87.** .7386

89. (a)
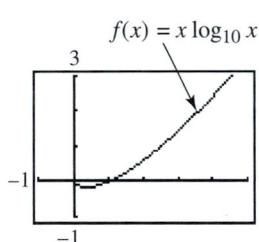
   (b) logarithmic   **91. (a)** -4 **(b)** 6 **(c)** 4 **(d)** -1 **95.** $\{.01, 2.38\}$

Summary Exercises on Inverse, Exponential, and Logarithmic Functions *(pages 445–446)*

1. They are inverses.   **2.** They are not inverses.   **3.** They are inverses.   **4.** They are inverses.   **5.**

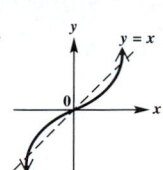

6. 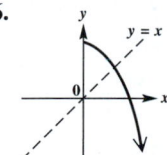   **7.** They are not one-to-one.   **8.** 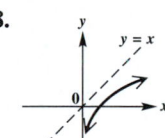   **9.** B   **10.** D   **11.** C   **12.** A

13. The functions in Exercises 9 and 12 are inverses of one another. The functions in Exercises 10 and 11 are inverses of one

another. 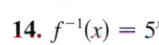   **14.** $f^{-1}(x) = 5^x$ **15.** $f^{-1}(x) = \dfrac{1}{3}x + 2$; domains and ranges of both f and f^{-1} are $(-\infty, \infty)$.

16. $f^{-1}(x) = \sqrt[3]{\dfrac{x}{2}} - 1$; domains and ranges of both f and f^{-1} are $(-\infty, \infty)$.   **17.** f is not one-to-one.

18. $f^{-1}(x) = \dfrac{5x + 1}{2 + 3x}$; domain of f = range of $f^{-1} = \left(-\infty, \dfrac{5}{3}\right) \cup \left(\dfrac{5}{3}, \infty\right)$; domain of f^{-1} = range of $f =$

$\left(-\infty, -\dfrac{2}{3}\right) \cup \left(-\dfrac{2}{3}, \infty\right)$ **19.** f is not one-to-one.   **20.** $f^{-1}(x) = \sqrt{x^2 + 9}$, $x \geq 0$; domain of f = range of $f^{-1} = [3, \infty)$;

domain of f^{-1} = range of $f = [0, \infty)$.   **21.** $\log_{1/10} 1000 = -3$ **22.** $\log_a c = b$ **23.** $\log_{\sqrt{3}} 9 = 4$

24. $\log_4 \dfrac{1}{8} = -\dfrac{3}{2}$ **25.** $\log_2 32 = x$ **26.** $\log_{27} 81 = \dfrac{4}{3}$ **27.** $\{2\}$ **28.** $\{-3\}$ **29.** $\{-3\}$ **30.** $\{25\}$

31. $\{-2\}$ **32.** $\left\{\dfrac{1}{3}\right\}$ **33.** $(0, 1) \cup (1, \infty)$ **34.** $\left\{\dfrac{3}{2}\right\}$ **35.** $\{5\}$ **36.** $\{243\}$ **37.** $\{1\}$ **38.** $\{-2\}$ **39.** $\{1\}$

40. $\{2\}$ **41.** $\{2\}$ **42.** $\left\{\dfrac{1}{9}\right\}$ **43.** $\left\{-\dfrac{1}{3}\right\}$ **44.** $(-\infty, \infty)$

4.4 Exercises *(pages 453–457)*

1. increasing   **3.** $f^{-1}(x) = \log_5 x$ **5.** natural; common   **7.** There is no power of 2 that yields a result of 0.

9. $\log 8 = .90308999$ **11.** 1.7243   **13.** -2.8861 **15.** 3.9703   **17.** -6.6454 **19.** 4.4914   **21.** -5.3010

23. 5.7918   **25.** 3.9494   **29.** 3.2   **31.** 8.4   **33.** 2.0×10^{-3} **35.** 1.6×10^{-5} **37.** poor fen   **39.** bog

41. rich fen   **43.** **(a)** 2.60031933 **(b)** 1.60031933 **(c)** .6003193298 **(d)** The whole number parts will vary, but the decimal

parts will be the same.   **45.** **(a)** 20 **(b)** 30 **(c)** 50 **(d)** 60 **(e)** about 3 decibels   **47.** **(a)** 3 **(b)** 6 **(c)** 8

49. $398,107,171 I_0$ **51.** 66.6 million; We must assume that the model continues to be logarithmic.

53. **(a)** 2 **(b)** 2 **(c)** 2 **(d)** 1   **55.** 1   **57.** between 7°F and 11°F   **59.** 1.126 billion yr   **61.** 2.3219   **63.** $-.2537$

65. -1.5850 **67.** .8736   **69.** 1.9376   **71.** -1.4125 **73.** $4v + \dfrac{1}{2}u$ **75.** $\dfrac{3}{2}u - \dfrac{5}{2}v$

77. **(a)** 4 **(b)** 5^2, or 25 **(c)** $\dfrac{1}{e}$ **79.** **(a)** 6 **(b)** $\ln 3$ **(c)** $2 \ln 3$, or $\ln 9$ **81.** D   **83.** domain: $(-\infty, 0) \cup (0, \infty)$; range:

$(-\infty, \infty)$; symmetric with respect to the y-axis   **85.** $f(x) = 2 + \ln x$, so it is the graph of $g(x) = \ln x$ translated 2 units up.

87. $f(x) = \ln x - 2$, so it is the graph of $g(x) = \ln x$ translated 2 units down.

Chapter 4 Quiz *(page 458)*

[4.1] 1. $f^{-1}(x) = \dfrac{x^3 + 6}{3}$ **[4.2] 2.** $\{4\}$ **3.** 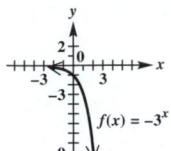 domain: $(-\infty, \infty)$; range: $(-\infty, 0)$

$f(x) = -3^x$

[4.3] 4. 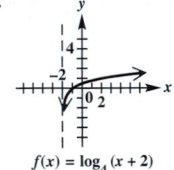 domain: $(-2, \infty)$; range: $(-\infty, \infty)$ **[4.2] 5.** $2148.62 **[4.4] 6. (a)** 1.5386 **(b)** 3.5427

$f(x) = \log_4 (x + 2)$

[4.3] 7. The expression $\log_6 25$ represents the exponent to which 6 must be raised in order to obtain 25. **8. (a)** $\{4\}$ **(b)** $\{5\}$

(c) $\left\{\dfrac{1}{16}\right\}$ **[4.4] 9.** $\dfrac{1}{2}\log_3 x + \log_3 y - \log_3 p - 4\log_3 q$ **10.** 7.8137 **11.** 3.3578 **12.** 12

4.5 Exercises *(pages 464–468)*

1. $\log_7 19;\ \dfrac{\log 19}{\log 7};\ \dfrac{\ln 19}{\ln 7}$ **3.** $\log_{1/2} 12;\ \dfrac{\log 12}{\log\left(\dfrac{1}{2}\right)};\ \dfrac{\ln 12}{\ln\left(\dfrac{1}{2}\right)}$ **5.** $\{1.771\}$ **7.** $\{-2.322\}$ **9.** $\{-6.213\}$

11. $\{-1.710\}$ **13.** $\{3.240\}$ **15.** $\{\pm 2.146\}$ **17.** $\{9.386\}$ **19.** $\varnothing$ **21.** $\{32.950\}$ **23.** $\{7.044\}$

25. $\{25.677\}$ **27.** $\{2011.568\}$ **29.** $\{e^2\}$ **31.** $\left\{\dfrac{e^{1.5}}{4}\right\}$ **33.** $\{2 - \sqrt{10}\}$ **35.** $\{16\}$ **37.** $\{3\}$ **39.** $\{e\}$

41. $\{25\}$ **43.** $\{5\}$ **45.** $\{2.5\}$ **47.** $\{3\}$ **49.** $\left\{\dfrac{1 + \sqrt{41}}{4}\right\}$ **51.** $\{6\}$ **53.** $\{4\}$ **55.** $\{1, 100\}$

57. Proposed solutions that cause *any argument of a logarithm* to be negative or zero must be rejected. The statement is not correct. For example, the solution set of $\log(-x + 99) = 2$ is $\{-1\}$. **59.** $x = e^{k/(p-a)}$ **61.** $t = -\dfrac{1}{k}\log\left(\dfrac{T - T_0}{T_1 - T_0}\right)$

63. $t = -\dfrac{2}{R}\ln\left(1 - \dfrac{RI}{E}\right)$ **65.** $x = \dfrac{\ln\left(\dfrac{A + B - y}{B}\right)}{-C}$ **67.** $A = \dfrac{B}{x^C}$ **69.** $t = \dfrac{\log\left(\dfrac{A}{P}\right)}{n\log\left(1 + \dfrac{r}{n}\right)}$

71. $11,611.84 **73.** 2.6 yr **75.** 6.48% **77. (a)** 10.9% **(b)** 35.8% **(c)** 84.1%

79. during 2011 **81. (a)** about 24% **(b)** 1963 **83. (a)** $P(T) = 1 - e^{-.0034 - .0053T}$

(b) 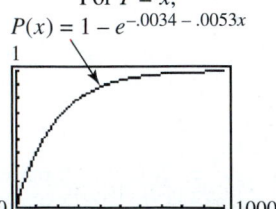 For $T = x$, $P(x) = 1 - e^{-.0034 - .0053x}$ **(c)** $P(60) \approx .275$ or 27.5%; The reduction in carbon emissions from a tax of $60 per ton of carbon is 27.5%. **(d)** $T = $130.14

85. $f^{-1}(x) = \ln(x + 4) - 1$; domain: $(-4, \infty)$; range: $(-\infty, \infty)$ **87.** $\{1.52\}$ **89.** $\{0\}$ **91.** $\{2.45, 5.66\}$

93. By the power rule for exponents, $(a^m)^n = a^{mn}$.   **94.** $(e^x - 1)(e^x - 3) = 0$ **95.** $\{0, \ln 3\}$

96. The graph intersects the x-axis at 0 and $\ln 3 \approx 1.099$.

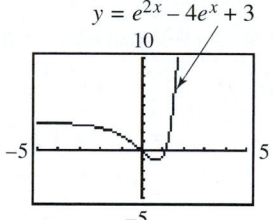

$y = e^{2x} - 4e^x + 3$

4.6 Exercises *(pages 475–481)*

1. B   **3.** C   **5. (a)** 440 g **(b)** 387 g **(c)** 264 g **(d)** 21.66 yr   **7.** 1611.97 yr   **9.** 3.57 g   **11.** about 9000 yr

13. about 15,600 yr   **15. (a)** $f(x) = .05(1.04)^{x-1950}$ **(b)** 4%   **17.** 6.25°C   **19. (a)** 7% compounded quarterly

(b) $800.32   **21.** about 27.73 yr   **23.** about 21.97 yr   **25. (a)** 315 **(b)** 229 **(c)** 142   **27. (a)** $P = 1$; $a \approx 1.01355$

(b) 1.14 billion **(c)** 2030   **29. (a)** 25.7 billion **(b)** 2003   **31. (a)** 13.92 **(b)** 31.05 **(c)** 107.8   **33. (a)** 15,000

(b) 9098 **(c)** 5249   **35. (a)** 611 million **(b)** 746 million **(c)** 1007 million   **37.** about 14.2 hr   **39.** 2013   **41.** 6.9 yr

43. 11.6 yr   **45. (a)** .065; .82; Among people age 25, 6.5% have some CHD, while among people age 65, 82% have some CHD.

(b) about 48 yr

Summary Exercises on Functions: Domains and Defining Equations *(pages 481–483)*

1. $(-\infty, \infty)$ **2.** $\left[\dfrac{7}{2}, \infty\right)$ **3.** $(-\infty, \infty)$ **4.** $(-\infty, 6) \cup (6, \infty)$ **5.** $(-\infty, \infty)$ **6.** $(-\infty, -3] \cup [3, \infty)$

7. $(-\infty, -3) \cup (-3, 3) \cup (3, \infty)$ **8.** $(-\infty, \infty)$ **9.** $(-4, 4)$ **10.** $(-\infty, -7) \cup (3, \infty)$ **11.** $(-\infty, -1] \cup [8, \infty)$

12. $(-\infty, 0) \cup (0, \infty)$ **13.** $(-\infty, \infty)$ **14.** $(-\infty, -5) \cup (-5, \infty)$ **15.** $[1, \infty)$ **16.** $(-\infty, -\sqrt{5}) \cup (-\sqrt{5}, \sqrt{5}) \cup (\sqrt{5}, \infty)$

17. $(-\infty, \infty)$ **18.** $(-\infty, -1) \cup (-1, 1) \cup (1, \infty)$ **19.** $(-\infty, 1)$ **20.** $(-\infty, 2) \cup (2, \infty)$ **21.** $(-\infty, \infty)$

22. $[-2, 3] \cup [4, \infty)$ **23.** $(-\infty, -2) \cup (-2, 3) \cup (3, \infty)$ **24.** $[-3, \infty)$ **25.** $(-\infty, 0) \cup (0, \infty)$

26. $(-\infty, -\sqrt{7}) \cup (-\sqrt{7}, \sqrt{7}) \cup (\sqrt{7}, \infty)$ **27.** $(-\infty, \infty)$ **28.** $\emptyset$ **29.** $[-2, 2]$ **30.** $(-\infty, \infty)$

31. $(-\infty, -7] \cup (-4, 3) \cup [9, \infty)$ **32.** $(-\infty, \infty)$ **33.** $(-\infty, 5]$ **34.** $(-\infty, 3)$ **35.** $(-\infty, 4) \cup (4, \infty)$

36. $(-\infty, \infty)$ **37.** $(-\infty, -5] \cup [5, \infty)$ **38.** $(-\infty, \infty)$ **39.** $(-2, 6)$ **40.** $(0, 1) \cup (1, \infty)$ **41.** A   **42.** B

43. C   **44.** D   **45.** A   **46.** B   **47.** D   **48.** C   **49.** C   **50.** B

Chapter 4 Review Exercises *(pages 487–490)*

1. not one-to-one   **3.** one-to-one   **5.** not one-to-one   **7.** $f^{-1}(x) = \sqrt[3]{x + 3}$ **9.** It represents the number of years after

2004 for the investment to reach $50,000.   **11.** one-to-one   **13.** B   **15.** C   **17.** $\log_2 32 = 5$ **19.** $\log_{3/4} \dfrac{4}{3} = -1$

21. $\log_3 4$ **23.** $10^3 = 1000$ **25.** 3   **27.** $\log_3 m + \log_3 n - \log_3 5 - \log_3 r$ **29.** This cannot be simplified.

31. -1.3862 **33.** 11.8776   **35.** 1.1592   **37.** $\left\{\dfrac{22}{5}\right\}$ **39.** $\{3.667\}$ **41.** $\{-13.257\}$ **43.** $\{-.485\}$

45. $\{2.102\}$ **47.** $\{-2.487\}$ **49.** $\{3\}$ **51.** $\{e^{13/3}\}$ **53.** $\left\{\dfrac{\sqrt[4]{10}-7}{2}\right\}$ **55.** $\left\{1, \dfrac{10}{3}\right\}$ **57.** $\{2\}$

59. $\{-3\}$ **61.** $I_0 = \dfrac{I}{10^{d/10}}$ **63.** $\{1.315\}$ **65. (a)** about $200{,}000{,}000 I_0$ **(b)** about $13{,}000{,}000 I_0$ **(c)** The 1906 earthquake

had a magnitude almost 16 times greater than the 1989 earthquake.   **67.** 5.1% **69.** $\$60{,}602.77$ **71.** 17.3 yr

73. 2007 **75. (a)** $\log_4(2x^2 - x) = \dfrac{\ln(2x^2 - x)}{\ln 4}$ **(b)**   **(c)** $-\dfrac{1}{2}, 1$ **(d)** $x = 0, x = \dfrac{1}{2}$

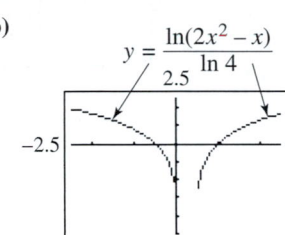

Chapter 4 Test *(pages 490–491)*

[4.1] 1. (a) $(-\infty, \infty); (-\infty, \infty)$ **(b)** The graph is a stretched translation of $y = \sqrt[3]{x}$, which passes the horizontal line test and is thus

a one-to-one function. **(c)** $f^{-1}(x) = \dfrac{x^3 + 7}{2}$ **(d)** $(-\infty, \infty); (-\infty, \infty)$ **(e)**   The graphs are reflections of

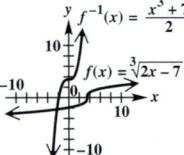

each other across the line $y = x$.   **[4.2, 4.3] 2. (a)** B **(b)** A **(c)** C **(d)** D   **[4.2] 3.** $\left\{\dfrac{1}{2}\right\}$

[4.3] 4. (a) $\log_4 8 = \dfrac{3}{2}$ **(b)** $8^{2/3} = 4$ **[4.1–4.3] 5.** They are inverses.

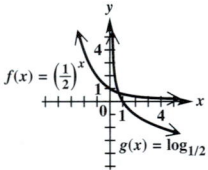

[4.3] 6. $2 \log_7 x + \dfrac{1}{4} \log_7 y - 3 \log_7 z$ **[4.4] 7.** 3.3780 **8.** 7.7782 **9.** 1.1674 **[4.3] 10.** $\left\{\dfrac{3}{4}\right\}$

[4.5] 11. $\{.631\}$ **12.** $\{12.548\}$ **13.** $\{2.811\}$ **14.** $\{2\}$ **15.** $\varnothing$ **16.** $\left\{\dfrac{7}{2}\right\}$ **[4.6] 18.** 10 sec

19. (a) 18.9 yr **(b)** 18.8 yr   **20.** 16.2 yr   **21. (a)** 329.3 g **(b)** 13.9 days   **22.** 2012

CHAPTER 5   TRIGONOMETRIC FUNCTIONS

5.1 Exercises *(pages 499–502)*

1. (a) $60°$ **(b)** $150°$ **3. (a)** $45°$ **(b)** $135°$ **5. (a)** $36°$ **(b)** $126°$ **7. (a)** $89°$ **(b)** $179°$ **9. (a)** $75° \, 40'$ **(b)** $165° \, 40'$

11. (a) $69° \, 49' \, 30''$ **(b)** $159° \, 49' \, 30''$ **13.** $70°; 110°$ **15.** $30°; 60°$ **17.** $40°; 140°$ **19.** $107°; 73°$ **21.** $69°; 21°$

23. $45°$ **25.** $150°$ **27.** $7° \, 30'$ **29.** $(90 - x)°$ **31.** $(x - 360)°$ **33.** $83° \, 59'$ **35.** $23° \, 49'$ **37.** $38° \, 32'$

39. $60° \, 34'$ **41.** $30° \, 27'$ **43.** $17° \, 1' \, 49''$ **45.** $35.5°$ **47.** $112.25°$ **49.** $-60.2°$ **51.** $20.9°$ **53.** $91.598°$

55. $274.316°$ **57.** $39° \, 15' \, 00''$ **59.** $126° \, 45' \, 36''$ **61.** $-18° \, 30' \, 54''$ **63.** $31° \, 25' \, 47''$ **65.** $89° \, 54' \, 1''$

67. $178° \, 35' \, 58''$ **69.** $392°$ **71.** $386° \, 30'$ **73.** $320°$ **75.** $235°$ **77.** $1°$ **79.** $359°$ **81.** $179°$

83. $130°$ **85.** $240°$ **87.** $120°$

In Exercises 89 and 91, answers may vary.

89. $450°, 810°; -270°, -630°$ **91.** $360°, 720°; -360°, -720°$ **93.** $30° + n \cdot 360°$ **95.** $135° + n \cdot 360°$

97. $-90° + n \cdot 360°$ **99.** $0° + n \cdot 360°$, or $n \cdot 360°$

Angles other than those given are possible in Exercises 103–113.

103.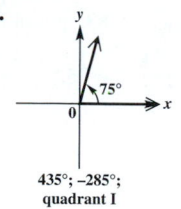

435°; –285°;
quadrant I

105.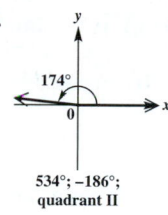

534°; –186°;
quadrant II

107.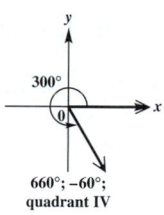

660°; –60°;
quadrant IV

109.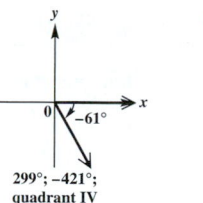

299°; –421°;
quadrant IV

111.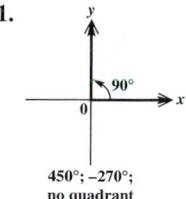

450°; –270°;
no quadrant

113.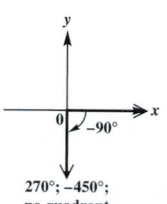

270°; –450°;
no quadrant

115. $3\sqrt{2}$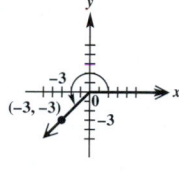

117. $\sqrt{34}$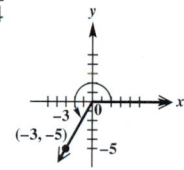

119. 2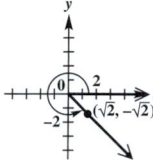

121. 2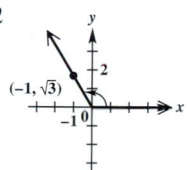

123. 4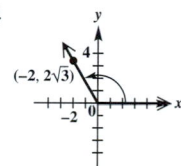

125. 4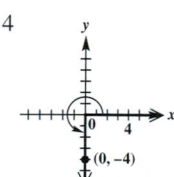

127. $\dfrac{3}{4}$ **129.** $1800°$

131. 12.5 rotations per hr

133. 4 sec

5.2 Exercises *(pages 512–515)*

In Exercises 1–11 and 23–27, we give, in order, sine, cosine, tangent, cotangent, secant, and cosecant.

1.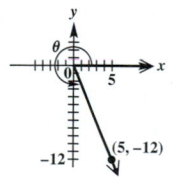

$-\dfrac{12}{13}; \dfrac{5}{13}; -\dfrac{12}{5};$

$-\dfrac{5}{12}; \dfrac{13}{5}; -\dfrac{13}{12}$

3.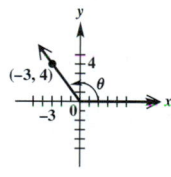

$\dfrac{4}{5}; -\dfrac{3}{5}; -\dfrac{4}{3};$

$-\dfrac{3}{4}; -\dfrac{5}{3}; \dfrac{5}{4}$

5.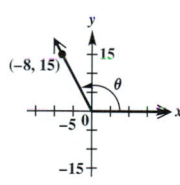

$\dfrac{15}{17}; -\dfrac{8}{17}; -\dfrac{15}{8};$

$-\dfrac{8}{15}; -\dfrac{17}{8}; \dfrac{17}{15}$

7.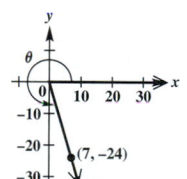

$-\dfrac{24}{25}; \dfrac{7}{25}; -\dfrac{24}{7};$

$-\dfrac{7}{24}; \dfrac{25}{7}; -\dfrac{25}{24}$

9.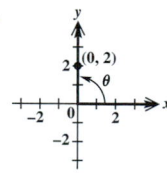

1; 0; undefined;
0; undefined; 1

11.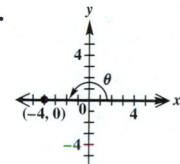

0; −1; 0; undefined;
−1; undefined

15. negative   **17.** negative

19. positive   **21.** positive

23.

$-\dfrac{2\sqrt{5}}{5}; \dfrac{\sqrt{5}}{5}; -2;$

$-\dfrac{1}{2}; \sqrt{5}; -\dfrac{\sqrt{5}}{2}$

25.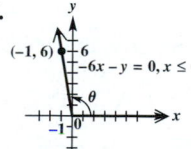

$\dfrac{6\sqrt{37}}{37}; -\dfrac{\sqrt{37}}{37}; -6;$

$-\dfrac{1}{6}; -\sqrt{37}; \dfrac{\sqrt{37}}{6}$

27.

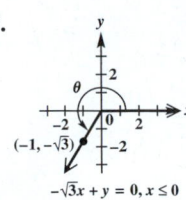

$-\sqrt{3}x + y = 0, x \le 0$

$-\dfrac{\sqrt{3}}{2}; -\dfrac{1}{2}; \sqrt{3};$

$\dfrac{\sqrt{3}}{3}; -2; -\dfrac{2\sqrt{3}}{3}$

29. 0 **31.** 0 **33.** -1 **35.** 1 **37.** undefined **39.** -1 **41.** -3 **43.** -3 **45.** 5

47. 0 **49.** 0 **51.** -1 **53.** $\dfrac{3}{2}$ **55.** $-\dfrac{7}{3}$ **57.** $\dfrac{1}{5}$ **59.** $\dfrac{\sqrt{2}}{2}$ **61.** .70069071

65. All are positive. **67.** Tangent and cotangent are positive; all others are negative.

69. Sine and cosecant are positive; all others are negative. **71.** Cosine and secant are positive; all others

are negative. **73.** I, II **75.** II **77.** I **79.** III, IV **81.** impossible **83.** possible

85. possible **87.** impossible **89.** possible **91.** impossible **93.** $-\dfrac{4}{5}$ **95.** $-\dfrac{\sqrt{5}}{2}$

97. $-\dfrac{\sqrt{3}}{3}$ **99.** 3.44701905

In Exercises 101–111, we give, in order, sine, cosine, tangent, cotangent, secant, and cosecant.

101. $\dfrac{15}{17}; -\dfrac{8}{17}; -\dfrac{15}{8}; -\dfrac{8}{15}; -\dfrac{17}{8}; \dfrac{17}{15}$ **103.** $\dfrac{\sqrt{5}}{7}; \dfrac{2\sqrt{11}}{7}; \dfrac{\sqrt{55}}{22}; \dfrac{2\sqrt{55}}{5}; \dfrac{7\sqrt{11}}{22}; \dfrac{7\sqrt{5}}{5}$ **105.** $\dfrac{8\sqrt{67}}{67}; \dfrac{\sqrt{201}}{67}; \dfrac{8\sqrt{3}}{3}; \dfrac{\sqrt{3}}{8};$

$\dfrac{\sqrt{201}}{3}; \dfrac{\sqrt{67}}{8}$ **107.** $\dfrac{\sqrt{2}}{6}; -\dfrac{\sqrt{34}}{6}; -\dfrac{\sqrt{17}}{17}; -\sqrt{17}; -\dfrac{3\sqrt{34}}{17}; 3\sqrt{2}$ **109.** $\dfrac{\sqrt{15}}{4}; -\dfrac{1}{4}; -\sqrt{15}; -\dfrac{\sqrt{15}}{15}; -4; \dfrac{4\sqrt{15}}{15}$

111. .164215; $-$.986425; $-$.166475; -6.00691; -1.01376; 6.08958 **115.** This statement is false. For example,

$\sin 180° + \cos 180° = 0 + (-1) = -1 \ne 1.$ **117.** negative **119.** negative **121.** positive **123.** positive

5.3 Exercises *(pages 524–529)*

In Exercises 1 and 3, we give, in order, sine, cosine, and tangent.

1. $\dfrac{21}{29}; \dfrac{20}{29}; \dfrac{21}{20}$ **3.** $\dfrac{n}{p}; \dfrac{m}{p}; \dfrac{n}{m}$ **5.** C **7.** B **9.** E

In Exercises 11 and 13, we give, in order, the unknown side, sine, cosine, tangent, cotangent, secant, and cosecant.

11. $c = 13; \dfrac{12}{13}; \dfrac{5}{13}; \dfrac{12}{5}; \dfrac{5}{12}; \dfrac{13}{5}; \dfrac{13}{12}$ **13.** $b = \sqrt{13}; \dfrac{\sqrt{13}}{7}; \dfrac{6}{7}; \dfrac{\sqrt{13}}{6}; \dfrac{6\sqrt{13}}{13}; \dfrac{7}{6}; \dfrac{7\sqrt{13}}{13}$

15. $\sin A = \cos(90° - A); \cos A = \sin(90° - A); \tan A = \cot(90° - A); \cot A = \tan(90° - A); \sec A = \csc(90° - A);$

$\csc A = \sec(90° - A)$ **17.** $\csc 51°$ **19.** $\csc(75° - \theta)$ **21.** $\tan(100° - \theta)$ **23.** $\cos 51.3°$

25. $\dfrac{\sqrt{3}}{3}$ **27.** $\dfrac{1}{2}$ **29.** $\dfrac{2\sqrt{3}}{3}$ **31.** $\sqrt{2}$ **33.** $\dfrac{\sqrt{2}}{2}$ **35.** 1 **37.** $\dfrac{\sqrt{3}}{2}$ **39.** $\sqrt{3}$

41.

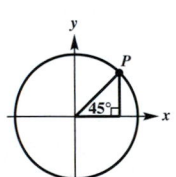

42.

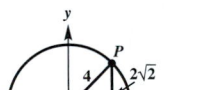

43. the legs; $(2\sqrt{2}, 2\sqrt{2})$ **44.** $(1, \sqrt{3})$ **47.** $y = \dfrac{\sqrt{3}}{3}x$ **49.** 60°

51. $x = \dfrac{9\sqrt{3}}{2}; y = \dfrac{9}{2}; z = \dfrac{3\sqrt{3}}{2}; w = 3\sqrt{3}$

53. $p = 15; r = 15\sqrt{2}; q = 5\sqrt{6}; t = 10\sqrt{6}$ **55.** $A = \dfrac{s^2}{2}$

57. C **59.** A **61.** D **65.** $\dfrac{\sqrt{2}}{2}; \dfrac{\sqrt{2}}{2}; \sqrt{2}; \sqrt{2}$ **67.** $-\dfrac{1}{2}; -\dfrac{\sqrt{3}}{3}; -2$ **69.** $\dfrac{1}{2}; -\sqrt{3}; -\dfrac{2\sqrt{3}}{3}$ **71.** $\sqrt{3}; \dfrac{\sqrt{3}}{3}$

In Exercises 73–89, we give, in order, sine, cosine, tangent, cotangent, secant, and cosecant.

73. $-\dfrac{\sqrt{2}}{2}; \dfrac{\sqrt{2}}{2}; -1, -1; \sqrt{2}; -\sqrt{2}$ **75.** $\dfrac{\sqrt{3}}{2}; \dfrac{1}{2}; \sqrt{3}; \dfrac{\sqrt{3}}{3}; 2; \dfrac{2\sqrt{3}}{3}$ **77.** $\dfrac{\sqrt{3}}{2}; -\dfrac{1}{2}; -\sqrt{3}; -\dfrac{\sqrt{3}}{3}; -2; \dfrac{2\sqrt{3}}{3}$

79. $-\dfrac{1}{2}; -\dfrac{\sqrt{3}}{2}; \dfrac{\sqrt{3}}{3}; \sqrt{3}; -\dfrac{2\sqrt{3}}{3}; -2$ **81.** $-\dfrac{\sqrt{2}}{2}; -\dfrac{\sqrt{2}}{2}; 1; 1; -\sqrt{2}; -\sqrt{2}$ **83.** $\dfrac{1}{2}; -\dfrac{\sqrt{3}}{2}; -\dfrac{\sqrt{3}}{3}; -\sqrt{3}; -\dfrac{2\sqrt{3}}{3}; 2$

85. $-\dfrac{1}{2}; -\dfrac{\sqrt{3}}{2}; \dfrac{\sqrt{3}}{3}; \sqrt{3}; -\dfrac{2\sqrt{3}}{3}; -2$ **87.** $\dfrac{1}{2}; -\dfrac{\sqrt{3}}{2}; -\dfrac{\sqrt{3}}{3}; -\sqrt{3}; -\dfrac{2\sqrt{3}}{3}; 2$ **89.** $-\dfrac{\sqrt{3}}{2}; \dfrac{1}{2}; -\sqrt{3}; -\dfrac{\sqrt{3}}{3};$

$2; -\dfrac{2\sqrt{3}}{3}$ **91.** $-\dfrac{\sqrt{3}}{2}$ **93.** $\dfrac{\sqrt{3}}{2}$ **95.** $-\sqrt{2}$ **97.** -1

In Exercises 99–111, the number of decimal places may vary depending on the calculator used.

99. .6252427   **101.** 1.0273488   **103.** 15.055723   **105.** 1.4887142   **107.** .6743024   **109.** .9999905

111. $\tan 23.4° \approx .4327386$ **113.** 55.845496°   **115.** 16.166641°   **117.** 38.491580°   **119.** 68.673241°

121. 45.526434°   **123.** 30°; 150°   **125.** 120°; 300°   **127.** 45°; 315°   **129.** 210°; 330°   **131.** 30°; 210°

133. 225°; 315°   **135.** A: 68.94 mph; B: 65.78 mph   **137.** 65.96 lb   **139.** −2.87°   **141.** **(a)** 703 ft **(b)** 1701 ft

(c) R would decrease; 644 ft, 1559 ft

Chapter 5 Quiz *(pages 529–530)*

[5.1] 1. (a) 71° **(b)** 161°   **2.** 65°; 115°   **3.** 26°; 64°   **4. (a)** 77.2025° **(b)** 22° 1′ 30″   **5. (a)** 50° **(b)** 300° **(c)** 170°

(d) 417°   **6.** 1800°   **[5.2] 7.** $\sin \theta = \dfrac{7}{25}$; $\cos \theta = -\dfrac{24}{25}$; $\tan \theta = -\dfrac{7}{24}$; $\cot \theta = -\dfrac{24}{7}$; $\sec \theta = -\dfrac{25}{24}$; $\csc \theta = \dfrac{25}{7}$

[5.3] 8. $\sin A = \dfrac{3}{5}$; $\cos A = \dfrac{4}{5}$; $\tan A = \dfrac{3}{4}$ **9.**

θ	$\sin \theta$	$\cos \theta$	$\tan \theta$	$\cot \theta$	$\sec \theta$	$\csc \theta$
30°	$\dfrac{1}{2}$	$\dfrac{\sqrt{3}}{2}$	$\dfrac{\sqrt{3}}{3}$	$\sqrt{3}$	$\dfrac{2\sqrt{3}}{3}$	2
45°	$\dfrac{\sqrt{2}}{2}$	$\dfrac{\sqrt{2}}{2}$	1	1	$\sqrt{2}$	$\sqrt{2}$
60°	$\dfrac{\sqrt{3}}{2}$	$\dfrac{1}{2}$	$\sqrt{3}$	$\dfrac{\sqrt{3}}{3}$	2	$\dfrac{2\sqrt{3}}{3}$

10. $w = 18$; $x = 18\sqrt{3}$; $y = 18$; $z = 18\sqrt{2}$ **11.** $3x^2 \sin \theta$

In Exercises 12–14, we give, in order, sine, cosine, tangent, cotangent, secant, and cosecant.

12. $\dfrac{\sqrt{2}}{2}$; $-\dfrac{\sqrt{2}}{2}$; -1; -1; $-\sqrt{2}$; $\sqrt{2}$ **13.** $-\dfrac{1}{2}$; $-\dfrac{\sqrt{3}}{2}$; $\dfrac{\sqrt{3}}{3}$; $\sqrt{3}$; $-\dfrac{2\sqrt{3}}{3}$; -2 **14.** $-\dfrac{\sqrt{3}}{2}$; $\dfrac{1}{2}$; $-\sqrt{3}$; $-\dfrac{\sqrt{3}}{3}$; 2; $-\dfrac{2\sqrt{3}}{3}$

15. 60°; 120°   **16.** 135°; 225°   **17.** .67301251   **18.** −1.18176327   **19.** 69.497888°   **20.** 24.777233°

5.4 Exercises *(pages 538–545)*

1. 22,894.5 to 22,895.5   **3.** 8958.5 to 8959.5

Note to student: While most of the measures resulting from solving triangles in this chapter are approximations, for convenience we use = rather than ≈.

5. $B = 53° 40'$; $a = 571$ m; $b = 777$ m   **7.** $M = 38.8°$; $n = 154$ m; $p = 198$ m   **9.** $A = 47.9108°$; $c = 84.816$ cm;

$a = 62.942$ cm   **11.** $A = 37° 40'$; $B = 52° 20'$; $c = 20.5$ ft   **13.** $B = 62.0°$; $a = 8.17$ ft; $b = 15.4$ ft   **15.** $A = 17.0°$;

$a = 39.1$ in.; $c = 134$ in.   **17.** $B = 27.5°$; $b = 6.61$ m; $c = 14.3$ m   **19.** $A = 36°$; $B = 54°$; $b = 18$ m

21. $c = 85.9$ yd; $A = 62° 50'$; $B = 27° 10'$ **23.** $b = 42.3$ cm; $A = 24° 10'$; $B = 65° 50'$ **25.** $B = 36° 36'$; $a = 310.8$ ft;

$b = 230.8$ ft   **27.** $A = 50° 51'$; $a = .4832$ m; $b = .3934$ m   **31.** 9.35 m   **33.** 128 ft   **35.** 26.92 in.   **37.** 22°

39. 28.0 m   **41.** 13.3 ft   **43.** 146 m   **45. (a)** 29,000 ft **(b)** shorter   **47.** 220 mi   **49.** 47 nautical mi   **51.** 140 mi

53. 148 mi   **55.** 433 ft   **57.** 114 ft   **59.** 5.18 m   **61. (a)** $d = \dfrac{b}{2}\left(\cot \dfrac{\alpha}{2} + \cot \dfrac{\beta}{2}\right)$ **(b)** 345.4 cm   **63.** 10.8 ft

65. (a) 320 ft **(b)** $R\left(1 - \cos \dfrac{\theta}{2}\right)$

Chapter 5 Review Exercises *(pages 549–552)*

1. complement: $55°$; supplement: $145°$ **3.** $186°$ **5.** $x = 30; y = 30$ **7.** $9360°$ **9.** $119.134°$ **11.** $275° \, 6' \, 2''$

13. $40°; 60°; 80°$

In Exercises 15–27 and 31 and 33, we give, in order, sine, cosine, tangent, cotangent, secant, and cosecant.

15. $-\dfrac{\sqrt{3}}{2}; \dfrac{1}{2}; -\sqrt{3}; -\dfrac{\sqrt{3}}{3}; 2; -\dfrac{2\sqrt{3}}{3}$ **17.** $-\dfrac{4}{5}; \dfrac{3}{5}; -\dfrac{4}{3}; -\dfrac{3}{4}; \dfrac{5}{3}; -\dfrac{5}{4}$ **19.** $-\dfrac{1}{2}; \dfrac{\sqrt{3}}{2}; -\dfrac{\sqrt{3}}{3}; -\sqrt{3}; \dfrac{2\sqrt{3}}{3}; -2$

21. $0; -1; 0;$ undefined; $-1;$ undefined   **23.** $-\dfrac{\sqrt{39}}{8}; -\dfrac{5}{8}; \dfrac{\sqrt{39}}{5}; \dfrac{5\sqrt{39}}{39}; -\dfrac{8}{5}; -\dfrac{8\sqrt{39}}{39}$

25. $\dfrac{2\sqrt{5}}{5}; -\dfrac{\sqrt{5}}{5}; -2; -\dfrac{1}{2}; -\sqrt{5}; \dfrac{\sqrt{5}}{2}$ **27.** $\dfrac{60}{61}; \dfrac{11}{61}; \dfrac{60}{11}; \dfrac{11}{60}; \dfrac{61}{11}; \dfrac{61}{60}$ **31.** $-\dfrac{\sqrt{3}}{2}; \dfrac{1}{2}; -\sqrt{3}; -\dfrac{\sqrt{3}}{3}; 2; -\dfrac{2\sqrt{3}}{3}$

33. $-\dfrac{1}{2}; \dfrac{\sqrt{3}}{2}; -\dfrac{\sqrt{3}}{3}; -\sqrt{3}; \dfrac{2\sqrt{3}}{3}; -2$ **35.** $120°; 240°$ **37.** $150°; 210°$ **39.** $.95371695$ **41.** $-.71592968$

43. $12.733938°$ **45.** $47.1°; 132.9°$ **47.** $B = 31° \, 30'; a = 638; b = 391$ **49.** $B = 50.28°; a = 32.38$ m; $c = 50.66$ m

51. 73.7 ft   **53.** 18.75 cm   **55.** 1200 m   **57.** 140 mi

Chapter 5 Test *(pages 553–554)*

[5.1] 1. (a) $23°$ **(b)** $113°$ **2. (a)** $74.31°$ **(b)** $45° \, 12' \, 9''$ **3. (a)** $30°$ **(b)** $280°$ **(c)** $90°$ **4.** $2700°$

[5.2] 5.

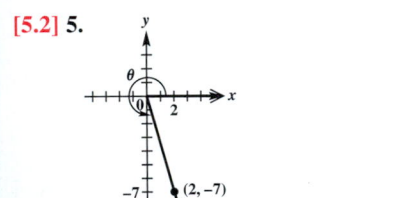

$$\sin \theta = -\frac{7\sqrt{53}}{53}; \cos \theta = \frac{2\sqrt{53}}{53}; \tan \theta = -\frac{7}{2};$$
$$\cot \theta = -\frac{2}{7}; \sec \theta = \frac{\sqrt{53}}{2}; \csc \theta = -\frac{\sqrt{53}}{7}$$

6.

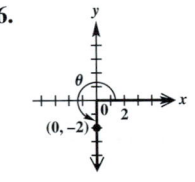

$\sin \theta = -1; \cos \theta = 0; \tan \theta$ is undefined;
$\cot \theta = 0; \sec \theta$ is undefined; $\csc \theta = -1$

7.

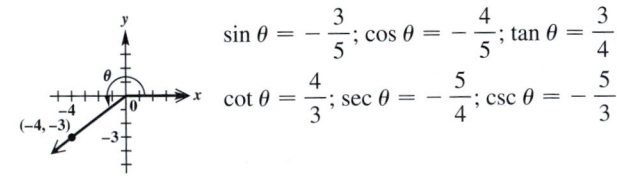

$\sin \theta = -\dfrac{3}{5}; \cos \theta = -\dfrac{4}{5}; \tan \theta = \dfrac{3}{4};$
$\cot \theta = \dfrac{4}{3}; \sec \theta = -\dfrac{5}{4}; \csc \theta = -\dfrac{5}{3}$

8. row 1: $1, 0,$ undefined, $0,$ undefined, 1;
row 2: $0, 1, 0,$ undefined, $1,$ undefined;
row 3: $-1, 0,$ undefined, $0,$ undefined, -1

9. (a) I **(b)** III, IV **(c)** III   **10.** $\cos \theta = -\dfrac{2\sqrt{10}}{7}; \tan \theta = -\dfrac{3\sqrt{10}}{20}; \cot \theta = -\dfrac{2\sqrt{10}}{3}; \sec \theta = -\dfrac{7\sqrt{10}}{20}; \csc \theta = \dfrac{7}{3}$

[5.3] 11. $\sin A = \dfrac{12}{13}; \cos A = \dfrac{5}{13}; \tan A = \dfrac{12}{5}; \cot A = \dfrac{5}{12}; \sec A = \dfrac{13}{5}; \csc A = \dfrac{13}{12}$ **12.** $x = 4; y = 4\sqrt{3}; z = 4\sqrt{2}; w = 8$

In Exercises 13 and 14, we give, in order, sine, cosine, tangent, cotangent, secant, and cosecant.

13. $-\dfrac{\sqrt{3}}{2}; -\dfrac{1}{2}; \sqrt{3}; \dfrac{\sqrt{3}}{3}; -2; -\dfrac{2\sqrt{3}}{3}$ **14.** $-\dfrac{\sqrt{2}}{2}; -\dfrac{\sqrt{2}}{2}; 1; 1; -\sqrt{2}; -\sqrt{2}$ **15.** $135°; 225°$ **16.** $240°; 300°$

17. (a) $.97939940$ **(b)** 1.9362132 **18.** $16.16664145°$ **[5.4] 19.** $B = 31° \, 30'; c = 877; b = 458$ **20.** $67.1°,$ or $67° \, 10'$

21. 15.5 ft   **22.** 72 nautical mi   **23.** 92 km   **24.** 448 m

CHAPTER 6   THE CIRCULAR FUNCTIONS AND THEIR GRAPHS

6.1 Exercises *(pages 564–571)*

1. 1   **3.** 3   **5.** −3   **7.** $\dfrac{\pi}{3}$ **9.** $\dfrac{\pi}{2}$ **11.** $\dfrac{5\pi}{6}$ **13.** $-\dfrac{5\pi}{3}$ **15.** $\dfrac{5\pi}{2}$ **17.** 10π **25.** $60°$ **27.** $315°$

29. $330°$ **31.** $-30°$ **33.** $126°$ **35.** $-48°$ **37.** $153°$ **39.** $-900°$ **41.** .68   **43.** .742   **45.** 2.43

47. 1.122   **49.** .9847   **51.** .832391   **53.** $114°\ 35'$ **55.** $99°\ 42'$ **57.** $19°\ 35'$ **59.** $-287°\ 6'$ **61.** We begin

the answers with the blank next to $30°$, and then proceed counterclockwise from there: $\dfrac{\pi}{6}$; 45; $\dfrac{\pi}{3}$; 120; 135; $\dfrac{5\pi}{6}$; π; $\dfrac{7\pi}{6}$; $\dfrac{5\pi}{4}$; 240;

300; $\dfrac{7\pi}{4}$; $\dfrac{11\pi}{6}$.   **63.** 2π **65.** 20π **67.** 6   **69.** 1   **71.** 2   **73.** 25.8 cm   **75.** 3.61 ft   **77.** 5.05 m

79. 55.3 in.   **81.** The length is doubled.   **83.** 3500 km   **85.** 5900 km   **87.** $44°$ N   **89.** $156°$ **91.** $38.5°$

93. 18.7 cm   **95.** (a) 11.6 in. (b) $37°\ 5'$ **97.** 146 in.   **99.** .20 km   **101.** 6π **103.** 72π **105.** $60°$ **107.** 1.5

109. 1116.1 m^2 **111.** 706.9 ft^2 **113.** 114.0 cm^2 **115.** 1885.0 mi^2 **117.** 3.6   **119.** (a) $13\dfrac{1}{3}^{\circ}$; $\dfrac{2\pi}{27}$ (b) 478 ft

(c) 17.7 ft (d) approximately 672 ft^2 **121.** (a) 140 ft (b) 102 ft (c) 622 ft^2 **123.** 1900 yd^2 **125.** radius: 3950 mi;

circumference: 24,800 mi   **127.** The area is quadrupled.

6.2 Exercises *(pages 580–584)*

1. (a) 1 (b) 0 (c) undefined   **3.** (a) 0 (b) 1 (c) 0   **5.** (a) 0 (b) −1 (c) 0   **7.** $-\dfrac{1}{2}$ **9.** −1   **11.** −2

13. $-\dfrac{1}{2}$ **15.** $\dfrac{\sqrt{2}}{2}$ **17.** $\dfrac{\sqrt{3}}{2}$ **19.** $\dfrac{2\sqrt{3}}{3}$ **21.** $-\dfrac{\sqrt{3}}{3}$ **23.** .5736   **25.** .4068   **27.** 1.2065   **29.** 14.3338

31. −1.0460   **33.** −3.8665   **35.** .7   **37.** .9   **39.** −.6   **41.** 2.3 or 4.0   **43.** .8 or 2.4   **45.** negative

47. negative   **49.** positive   **51.** $\sin\theta = \dfrac{\sqrt{2}}{2}$; $\cos\theta = \dfrac{\sqrt{2}}{2}$; $\tan\theta = 1$; $\cot\theta = 1$; $\sec\theta = \sqrt{2}$; $\csc\theta = \sqrt{2}$

53. $\sin\theta = -\dfrac{12}{13}$; $\cos\theta = \dfrac{5}{13}$; $\tan\theta = -\dfrac{12}{5}$; $\cot\theta = -\dfrac{5}{12}$; $\sec\theta = \dfrac{13}{5}$; $\csc\theta = -\dfrac{13}{12}$ **55.** .2095   **57.** 1.4426   **59.** .3887

61. $\dfrac{5\pi}{6}$ **63.** $\dfrac{4\pi}{3}$ **65.** $\dfrac{7\pi}{4}$ **67.** 2π sec   **69.** (a) $\dfrac{\pi}{2}$ radians (b) 10π cm (c) $\dfrac{5\pi}{3}$ cm per sec   **71.** 2π radians

73. $\dfrac{3\pi}{32}$ radian per sec   **75.** $\dfrac{6}{5}$ min   **77.** .180311 radian per sec   **79.** 8π m per sec   **81.** $\dfrac{9}{5}$ radians per sec

83. 1.83333 radians per sec   **85.** 18π cm   **87.** 12 sec   **89.** $\dfrac{3\pi}{32}$ radian per sec   **91.** $\dfrac{\pi}{6}$ radian per hr

93. $\dfrac{\pi}{30}$ radian per min   **95.** $\dfrac{7\pi}{30}$ cm per min   **97.** 168π m per min   **99.** 1500π m per min   **101.** 16.6 mph

103. (a) $\dfrac{2\pi}{365}$ radian (b) $\dfrac{\pi}{4380}$ radian per hr (c) about 67,000 mph   **105.** (a) 3.1 cm per sec (b) .24 radian per sec

107. 3.73 cm   **109.** 523.6 radians per sec

6.3 Exercises *(pages 593–598)*

1. G   **3.** E   **5.** B   **7.** F   **9.** D   **11.** C   **13.** 2   **15.** $\dfrac{2}{3}$ **17.** 1

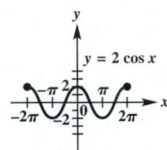 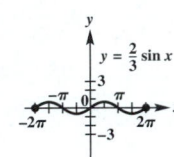 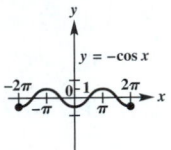

19. 2   **21.** 1   **23.** 4π; 1   **25.** $\dfrac{8\pi}{3}$; 1   **27.** $\dfrac{2\pi}{3}$; 1

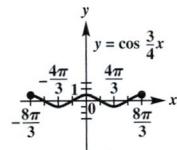

29. 8π; 2   **31.** $\dfrac{2\pi}{3}$; 2   **33.** 2; 1   **35.** 1; 2   **37.** 4; $\dfrac{1}{2}$

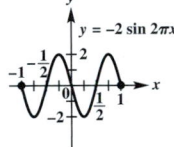

39. 2; π **41.** $y = 2\cos 2x$ **43.** $y = -3\cos\dfrac{1}{2}x$ **45.** $y = 3\sin 4x$ **47.** **(a)** 80°; 50° **(b)** 15

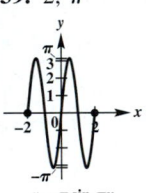

(c) about 35,000 yr **(d)** downward   **49.** **(a)** about 2 hr **(b)** 1 yr

51. **(a)** 5; $\dfrac{1}{60}$ **(b)** 60

(c) 5; 1.545; −4.045; −4.045; 1.545

(d)

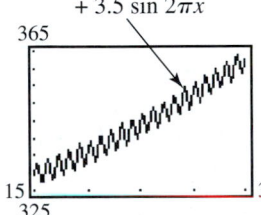

53. **(a)** $L(x) = .022x^2 + .55x + 316$ **(b)** maximums: $x = \dfrac{1}{4}, \dfrac{5}{4}, \dfrac{9}{4}, \ldots$;
$\qquad\qquad + 3.5 \sin 2\pi x$

minimums: $x = \dfrac{3}{4}, \dfrac{7}{4}, \dfrac{11}{4}, \ldots$

55. **(a)** 8° **(b)** 21° **(c)** 62° **(d)** 61° **(e)** 31° **(f)** −11°   **57.** 24 hr   **59.** approximately 6:00 P.M.; approximately .2 ft

61. approximately 2:00 A.M.; approximately 2.6 ft   **63.** 1; 240°, or $\dfrac{4\pi}{3}$

6.4 Exercises *(pages 604–607)*

1. D **3.** H **5.** B **7.** F **9.** C **11.** A **15.** B **17.** C **19.** $y = -1 + \sin x$ **21.** $y = \cos\left(x - \dfrac{\pi}{3}\right)$

23. 2; 2π; none; π to the right **25.** 4; 4π; none; π to the left **27.** 3; 4; none; $\dfrac{1}{2}$ to the right **29.** 1; $\dfrac{2\pi}{3}$; up 2; $\dfrac{\pi}{15}$ to the right

31. 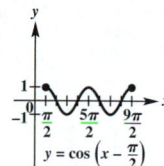 **33.** 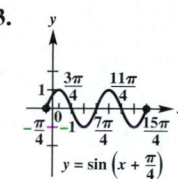 **35.** 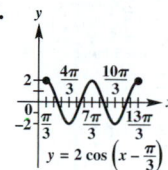 **37.** 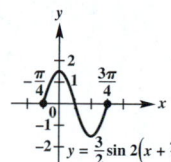 **39.**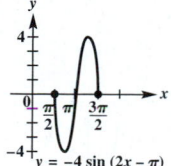

41. 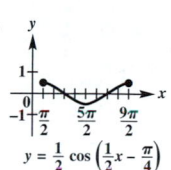 **43.** 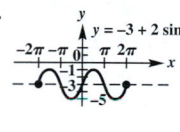 **45.** 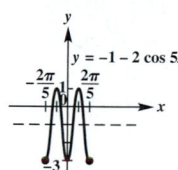 **47.**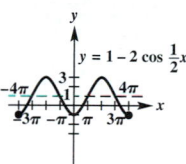

49. 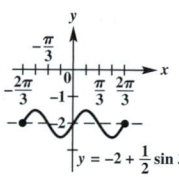 **51.** 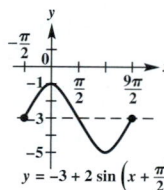 **53.**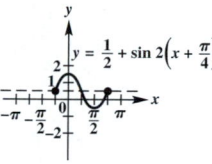

55. (a) yes 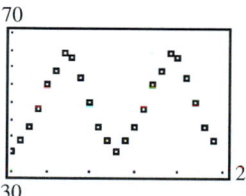 **(b)** It represents the average yearly temperature.

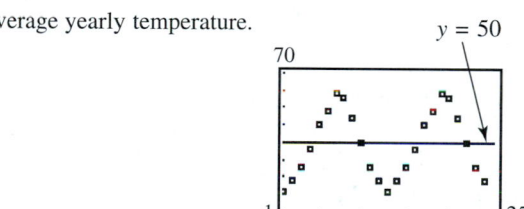

(c) 14; 12; 4.2 **(d)** $f(x) = 14 \sin\left[\dfrac{\pi}{6}(x - 4.2)\right] + 50$

(e) The function gives an excellent model for the given data. 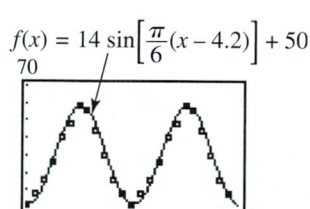 **(f)**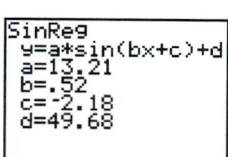

$f(x) = 14 \sin\left[\dfrac{\pi}{6}(x - 4.2)\right] + 50$

TI-83/84 Plus fixed to the nearest hundredth

Chapter 6 Quiz *(pages 607–608)*

[6.1] 1. $\dfrac{5\pi}{4}$ **2.** $-210°$ **3.** 1.5 **4.** 67,500 in.² **[6.2] 5.** $-\dfrac{1}{2}$ **6.** 0 **7.** $\dfrac{2\pi}{3}$

[6.3] 8. 2π; 4 **9.** $\pi; \dfrac{1}{2}$ **[6.4] 10.** 2π; 2 **11.** π; 1 **[6.3] 12.** $y = \cos 2x$

13. $y = -\sin x$

[6.3, 6.4] 14. 73°F

15. 60°F; 84°F

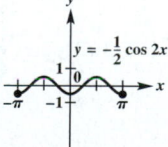 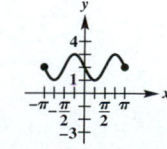

6.5 Exercises *(pages 620–623)*

1. C   **3.** B   **5.** F   **7.** 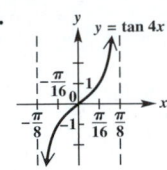   **9.** 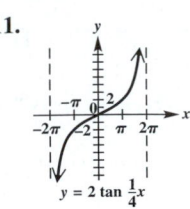   **11.** 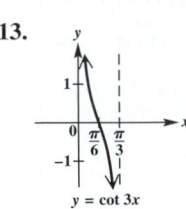   **13.**

15. 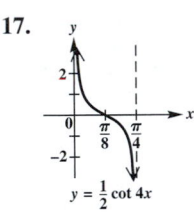   **17.**   **19.** 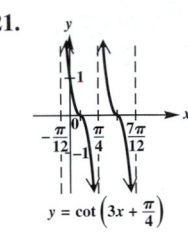   **21.** 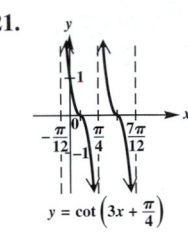   **23.**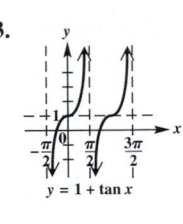

25. 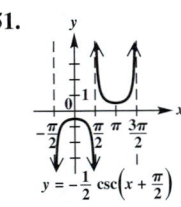   **27.**   **29.** 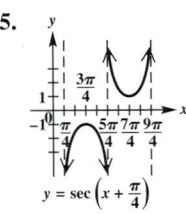   **31.**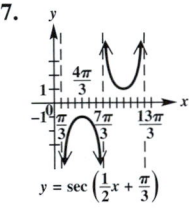

33. $y = -2 \tan x$ **35.** $y = \cot 3x$ **37.** true   **39.** four

43. **(a)** 0 m **(b)** −2.9 m **(c)** −12.3 m **(d)** 12.3 m **(e)** It leads to $\tan \frac{\pi}{2}$, which is undefined.   **45.** B   **47.** D

49. 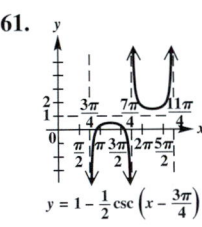   **51.**    **53.**   **55.**   **57.**

59. 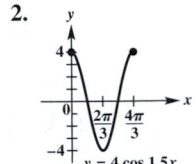   **61.** 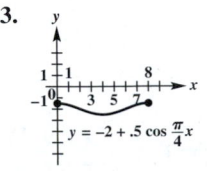   **63.** $y = \sec 4x$ **75.**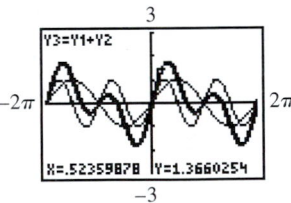

65. $y = -2 + \csc x$ **67.** true

69. none   **73.** **(a)** 4 m

(b) 6.3 m   **(c)** 63.7 m

Summary Exercises on Graphing Circular Functions *(page 623)*

1. 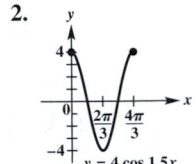   **2.** 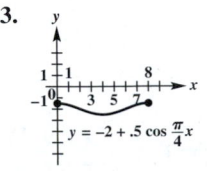   **3.** 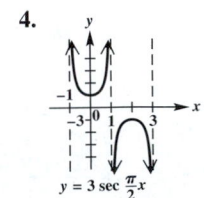   **4.** 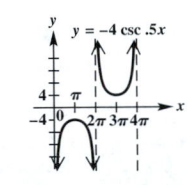   **5.**

6.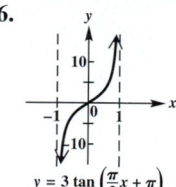
$y = 3 \tan\left(\frac{\pi}{2}x + \pi\right)$

7.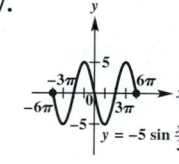
$y = -5 \sin\frac{x}{3}$

8.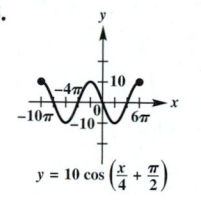
$y = 10 \cos\left(\frac{x}{4} + \frac{\pi}{2}\right)$

9.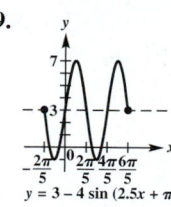
$y = 3 - 4 \sin(2.5x + \pi)$

10.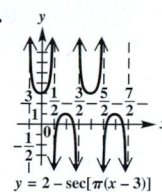
$y = 2 - \sec[\pi(x - 3)]$

6.6 Exercises *(pages 626–628)*

1. (a) $s(t) = 2 \cos 4\pi t$ **(b)** $s(1) = 2$; The weight is neither moving upward nor downward. At $t = 1$, the motion of the weight is changing from up to down.   **3. (a)** $s(t) = -3 \cos 2.5\pi t$ **(b)** $s(1) = 0$; upward

5. $s(t) = .21 \cos 55\pi t$
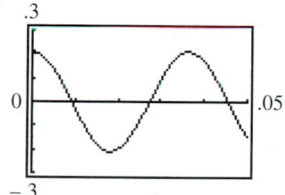

7. $s(t) = .14 \cos 110\pi t$
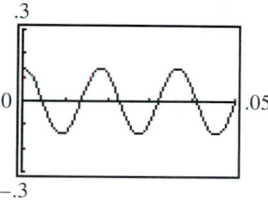

9. (a) $s(t) = -4 \cos\frac{2\pi}{3}t$ **(b)** 3.46 units **(c)** $\frac{1}{3}$ oscillation per sec   **11. (a)** $s(t) = 2 \sin 2t$; amplitude: 2; period: π; frequency: $\frac{1}{\pi}$ rotation per sec **(b)** $s(t) = 2 \sin 4t$; amplitude: 2; period: $\frac{\pi}{2}$; frequency: $\frac{2}{\pi}$ rotation per sec   **13.** $\frac{8}{\pi^2}$ ft

15. (a) 4 in. **(b)** after $\frac{1}{8}$ sec **(c)** 4 cycles per sec; $\frac{1}{4}$ sec   **17. (a)** 5 in. **(b)** 2 cycles per sec; $\frac{1}{2}$ sec **(c)** after $\frac{1}{4}$ sec

(d) approximately 4; After 1.3 sec, the weight is about 4 in. above the equilibrium position.

19. (a) $s(t) = -3 \cos 12t$ **(b)** $\frac{\pi}{6}$ sec   **21.** 0; π; They are the same.

Chapter 6 Review Exercises *(pages 632–638)*

1. An angle of 1 radian is larger.   **3.** Three of many possible answers are $1 + 2\pi$, $1 + 4\pi$, and $1 + 6\pi$.   **5.** $\frac{\pi}{4}$ **7.** $\frac{35\pi}{36}$

9. $\frac{40\pi}{9}$ **11.** 225°   **13.** 480°   **15.** −110°   **17.** π in.   **19.** 12π in.   **21.** 35.8 cm   **23.** 7.683 cm

25. 273 m²   **27.** 4500 km   **29.** $\frac{3}{4}$; 1.5 sq units   **31.** $\sqrt{3}$ **33.** $-\frac{1}{2}$ **35.** 2   **37.** .8660   **39.** .9703

41. 1.9513   **43.** .3898   **45.** .5148   **47.** 1.1054   **49.** $\frac{\pi}{4}$ **51.** $\frac{7\pi}{6}$ **53.** $\frac{15}{32}$ sec   **55.** $\frac{\pi}{20}$ radian per sec

57. 285.3 cm   **59.** $\frac{\pi}{36}$ radian per sec   **61. (b)** $\frac{\pi}{6}$ **(c)** There is less ultraviolet light when $\theta = \frac{\pi}{3}$.   **63.** D   **65.** 2; 2π;

none; none   **67.** $\frac{1}{2}$; $\frac{2\pi}{3}$; none; none   **69.** 2; 8π; 1 up; none   **71.** 3; 2π; none; $\frac{\pi}{2}$ to the left   **73.** not applicable; π; none;

$\frac{\pi}{8}$ to the right   **75.** not applicable; $\frac{\pi}{3}$; none; $\frac{\pi}{9}$ to the right   **77.** tangent   **79.** cosine   **81.** cotangent

83.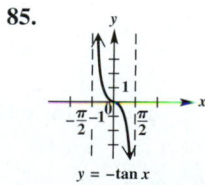
$y = 3 \sin x$

85.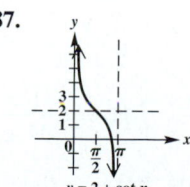
$y = -\tan x$

87.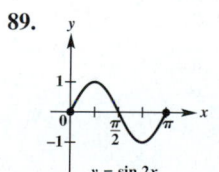
$y = 2 + \cot x$

89.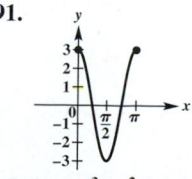
$y = \sin 2x$

91.
$y = 3 \cos 2x$

93.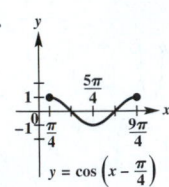

95.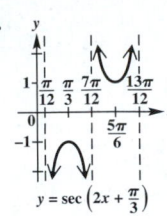

97.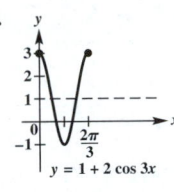

99.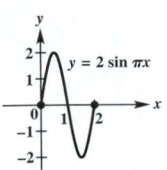

101. (b)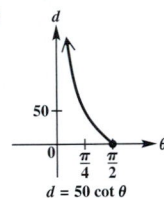

103. (a) $30°$ **(b)** $60°$ **(c)** $75°$ **(d)** $86°$ **(e)** $86°$ **(f)** $60°$ **105. (a)** 100 **(b)** 258 **(c)** 122 **(d)** 296

107. amplitude: 4; period: 2; frequency: $\dfrac{1}{2}$ cycle per sec   **109.** The frequency is the number of cycles in one unit of time; $-4; 0; -2\sqrt{2}$

Chapter 6 Test *(pages 638–640)*

[6.1] 1. $\dfrac{2\pi}{3}$ **2.** $-\dfrac{\pi}{4}$ **3.** $.09$ **4.** $135°$ **5.** $-210°$ **6.** $229.18°$ **7. (a)** $\dfrac{4}{3}$ **(b)** $15{,}000$ cm^2 **8.** 2 radians

[6.2] 9. $\dfrac{\sqrt{2}}{2}$ **10.** $-\dfrac{\sqrt{3}}{2}$ **11.** undefined   **12.** -2 **13.** 0 **14.** 0 **15.** $\sin\dfrac{7\pi}{6} = -\dfrac{1}{2}$; $\cos\dfrac{7\pi}{6} = -\dfrac{\sqrt{3}}{2}$;

$\tan\dfrac{7\pi}{6} = \dfrac{\sqrt{3}}{3}$ **16. (a)** $.9716$ **(b)** $\dfrac{\pi}{3}$ **17. (a)** $\dfrac{2\pi}{3}$ radians **(b)** 40π cm **(c)** 5π cm per sec   **18. (a)** 75 ft

(b) $\dfrac{\pi}{45}$ radian per sec   **[6.3–6.5] 19. (a)** $y = \sec x$ **(b)** $y = \sin x$ **(c)** $y = \cos x$ **(d)** $y = \tan x$ **(e)** $y = \csc x$ **(f)** $y = \cot x$

[6.4] 20. (a) π **(b)** 6 **(c)** $[-3, 9]$ **(d)** -3 **(e)** $\dfrac{\pi}{4}$ to the left $\left(\text{that is, } -\dfrac{\pi}{4}\right)$

[6.3, 6.4] 21.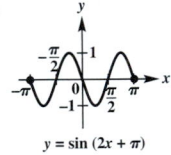

22.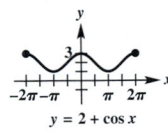

23.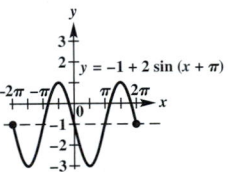

[6.5] 24.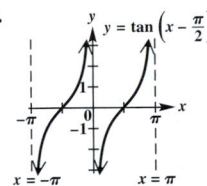

25.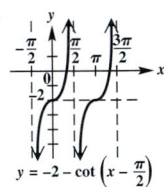

26.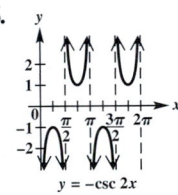

[6.4] 27. (a) $f(x) = 17.5 \sin\left[\dfrac{\pi}{6}(x-4)\right] + 67.5$

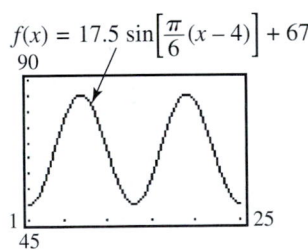

(b) 17.5; 12; 4 to the right; 67.5 up **(c)** approximately $52°$F **(d)** $50°$F in January; $85°$F in July **(e)** approximately $67.5°$; This is the vertical translation.

[6.6] 28. (a) 4 in. **(b)** after $\dfrac{1}{8}$ sec **(c)** 4 cycles per sec; $\dfrac{1}{4}$ sec

CHAPTER 7   TRIGONOMETRIC IDENTITIES AND EQUATIONS

7.1 Exercises *(pages 645–648)*

1. -2.6 **3.** $.625$ **5.** $\dfrac{2}{3}$ **7.** $\dfrac{\sqrt{7}}{4}$ **9.** $-\dfrac{2\sqrt{5}}{5}$ **11.** $-\dfrac{\sqrt{105}}{11}$ **15.** $-\sin x$ **16.** odd

17. $\cos x$ **18.** even   **19.** $-\tan x$ **20.** odd   **21.** $f(-x) = f(x)$ **23.** $f(-x) = -f(x)$

25. $\cos\theta = -\dfrac{\sqrt{5}}{3}$; $\tan\theta = -\dfrac{2\sqrt{5}}{5}$; $\cot\theta = -\dfrac{\sqrt{5}}{2}$; $\sec\theta = -\dfrac{3\sqrt{5}}{5}$; $\csc\theta = \dfrac{3}{2}$ **27.** $\sin\theta = -\dfrac{\sqrt{17}}{17}$; $\cos\theta = \dfrac{4\sqrt{17}}{17}$;

$\cot\theta = -4$; $\sec\theta = \dfrac{\sqrt{17}}{4}$; $\csc\theta = -\sqrt{17}$ **29.** $\sin\theta = \dfrac{3}{5}$; $\cos\theta = \dfrac{4}{5}$; $\tan\theta = \dfrac{3}{4}$; $\sec\theta = \dfrac{5}{4}$; $\csc\theta = \dfrac{5}{3}$

31. $\sin\theta = -\dfrac{\sqrt{7}}{4}$; $\cos\theta = \dfrac{3}{4}$; $\tan\theta = -\dfrac{\sqrt{7}}{3}$; $\cot\theta = -\dfrac{3\sqrt{7}}{7}$; $\csc\theta = -\dfrac{4\sqrt{7}}{7}$ **33.** B   **35.** E   **37.** A

39. A   **41.** D   **45.** $\sin\theta = \dfrac{\pm\sqrt{2x+1}}{x+1}$ **47.** $\sin x = \pm\sqrt{1 - \cos^2 x}$ **49.** $\tan x = \pm\sqrt{\sec^2 x - 1}$

51. $\csc x = \dfrac{\pm\sqrt{1 - \cos^2 x}}{1 - \cos^2 x}$ **53.** $\cos\theta$ **55.** $\cot\theta$ **57.** $\cos^2\theta$ **59.** $\sec\theta - \cos\theta$ **61.** $\cot\theta - \tan\theta$

63. $\tan\theta\sin\theta$ **65.** $\cos^2\theta$ **67.** $\sec^2\theta$ **69.** $\dfrac{25\sqrt{6}-60}{12}$; $\dfrac{-25\sqrt{6}-60}{12}$ **71.** $-\sin(2x)$ **72.** It is the negative of $\sin(2x)$.

73. $\cos(4x)$ **74.** It is the same function.   **75.** (a) $y = -\sin(4x)$ (b) $y = \cos(2x)$ (c) $y = 5\sin(3x)$ **77.** identity

79. not an identity   **81.** not an identity

7.2 Exercises *(pages 654–657)*

1. $\csc\theta\sec\theta$, or $\dfrac{1}{\sin\theta\cos\theta}$ **3.** $1 + \sec s$ **5.** 1   **7.** 1   **9.** $2 + 2\sin t$ **11.** $-\dfrac{2\cos x}{\sin^2 x}$, or $-2\cot x\csc x$

13. $(\sin\theta + 1)(\sin\theta - 1)$ **15.** $4\sin x$ **17.** $(2\sin x + 1)(\sin x + 1)$ **19.** $(\cos^2 x + 1)^2$

21. $(\sin x - \cos x)(1 + \sin x\cos x)$ **23.** $\sin\theta$ **25.** 1   **27.** $\tan^2\beta$ **29.** $\tan^2 x$ **31.** $\sec^2 x$ **33.** $\cos^2 x$

79. $(\sec\theta + \tan\theta)(1 - \sin\theta) = \cos\theta$ **81.** $\dfrac{\cos\theta + 1}{\sin\theta + \tan\theta} = \cot\theta$ **83.** identity   **85.** not an identity

91. It is true when $\sin x \geq 0$.   **93.** (a) $I = k(1 - \sin^2\theta)$ (b) For $\theta = 2\pi n$ and all integers n, $\cos^2\theta = 1$, its maximum value,

and I attains a maximum value of k.   **95.** (a) The sum of L and C equals 3.   (b) Let $Y_1 = L(t)$, $Y_2 = C(t)$, and   (c) $E(t) = 3$
$Y_3 = E(t)$. $Y_3 = 3$ for all inputs.

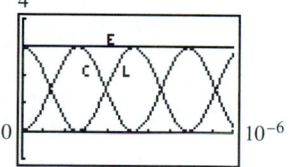

7.3 Exercises *(pages 666–671)*

1. F   **3.** E   **5.** $\dfrac{\sqrt{6}-\sqrt{2}}{4}$ **7.** $\dfrac{\sqrt{2}-\sqrt{6}}{4}$ **9.** $\dfrac{\sqrt{2}-\sqrt{6}}{4}$ **11.** 0   **13.** The calculator gives a value of 0 for the

expression.   **15.** $\tan$ **17.** $\cos$ **19.** $\csc$

For Exercises 21–25, other answers are possible.

21. $15°$ **23.** $\dfrac{140°}{3}$ **25.** $20°$ **27.** $\cos\theta$ **29.** $-\cos\theta$ **31.** $\cos\theta$ **33.** $-\cos\theta$ **35.** $\dfrac{4 - 6\sqrt{6}}{25}$; $\dfrac{4 + 6\sqrt{6}}{25}$

37. $\dfrac{16}{65}$; $-\dfrac{56}{65}$ **39.** $\dfrac{2\sqrt{638}-\sqrt{30}}{56}$; $\dfrac{2\sqrt{638}+\sqrt{30}}{56}$ **42.** $\dfrac{-\sqrt{6}-\sqrt{2}}{4}$ **43.** $\dfrac{-\sqrt{6}-\sqrt{2}}{4}$ **44.** (a) $\dfrac{\sqrt{2}-\sqrt{6}}{4}$

(b) $\dfrac{-\sqrt{6}-\sqrt{2}}{4}$ **45.** C   **47.** E   **49.** B   **51.** $\dfrac{\sqrt{6}+\sqrt{2}}{4}$ **53.** $2 - \sqrt{3}$ **55.** $\dfrac{-\sqrt{6}-\sqrt{2}}{4}$ **57.** $\dfrac{\sqrt{2}}{2}$

59. -1 **61.** 0   **63.** 1   **65.** $\dfrac{\sqrt{3}\cos\theta - \sin\theta}{2}$ **67.** $\dfrac{\cos\theta - \sqrt{3}\sin\theta}{2}$ **69.** $\dfrac{\sqrt{2}\,(\sin x - \cos x)}{2}$ **71.** $\dfrac{\sqrt{3}\tan\theta + 1}{\sqrt{3} - \tan\theta}$

73. $\dfrac{\sqrt{2}\,(\cos x + \sin x)}{2}$ **75.** $-\cos\theta$ **77.** $-\tan\theta$ **79.** $-\tan\theta$ **83.** (a) $\dfrac{63}{65}$ (b) $\dfrac{63}{16}$ (c) I

85. (a) $\dfrac{4\sqrt{2} + \sqrt{5}}{9}$ (b) $\dfrac{-8\sqrt{5} - 5\sqrt{2}}{20 - 2\sqrt{10}}$ (Other forms are possible.) (c) II   **87.** (a) $\dfrac{77}{85}$ (b) $-\dfrac{77}{36}$ (c) II   **89.** $\dfrac{\sqrt{6} - \sqrt{2}}{4}$

91. $\dfrac{-\sqrt{6} - \sqrt{2}}{4}$ **93.** $-2 + \sqrt{3}$ **95.** $\sin\left(\dfrac{\pi}{2} + \theta\right) = \cos\theta$ **97.** $\tan\left(\dfrac{\pi}{2} + \theta\right) = -\cot\theta$ **111.** (a) 425 lb (c) 0°

113. (a) For $x = t$,

$$V = V_1 + V_2$$
$$= 30\sin 120\pi t + 40\cos 120\pi t$$

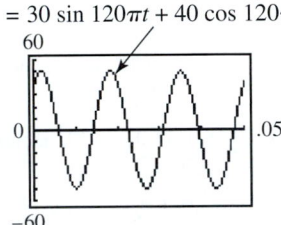

(b) $a = 50; \phi \approx -5.353$ **115.** (a) 3 (b) 163 and -163; no

Chapter 7 Quiz *(page 671)*

[7.1] 1. $\cos\theta = \dfrac{24}{25}; \tan\theta = -\dfrac{7}{24}; \cot\theta = -\dfrac{24}{7}; \sec\theta = \dfrac{25}{24}; \csc\theta = -\dfrac{25}{7}$ **2.** $\dfrac{\cos^2 x + 1}{\sin^2 x}$ **[7.3] 3.** $\dfrac{-\sqrt{6} - \sqrt{2}}{4}$

4. $-\cos\theta$ **5.** (a) $-\dfrac{16}{65}$ (b) $-\dfrac{63}{65}$ (c) III   **6.** $\dfrac{-1 + \tan x}{1 + \tan x}$

7.4 Exercises *(pages 681–685)*

1. C   **3.** B   **5.** C   **7.** $\cos\theta = \dfrac{2\sqrt{5}}{5}; \sin\theta = \dfrac{\sqrt{5}}{5}$ **9.** $\cos\theta = -\dfrac{\sqrt{42}}{12}; \sin\theta = \dfrac{\sqrt{102}}{12}$ **11.** $\cos 2\theta = \dfrac{17}{25};$

$\sin 2\theta = -\dfrac{4\sqrt{21}}{25}$ **13.** $\cos 2x = -\dfrac{3}{5}; \sin 2x = \dfrac{4}{5}$ **15.** $\cos 2\theta = \dfrac{39}{49}; \sin 2\theta = -\dfrac{4\sqrt{55}}{49}$ **17.** $\dfrac{\sqrt{3}}{2}$ **19.** $\dfrac{\sqrt{3}}{2}$

21. $-\dfrac{\sqrt{2}}{2}$ **23.** $\dfrac{1}{2}\tan 102°$ **25.** $\dfrac{1}{4}\cos 94.2°$ **27.** $-\cos\dfrac{4\pi}{5}$ **29.** $\sin 4x = 4\sin x \cos^3 x - 4\sin^3 x \cos x$

31. $\tan 3x = \dfrac{3\tan x - \tan^3 x}{1 - 3\tan^2 x}$ **33.** $\cos^4 x - \sin^4 x = \cos 2x$ **35.** $\dfrac{2\tan x}{2 - \sec^2 x} = \tan 2x$ **51.** $\sin 160° - \sin 44°$

53. $\sin\dfrac{\pi}{2} - \sin\dfrac{\pi}{6}$ **55.** $3\cos x - 3\cos 9x$ **57.** $-2\sin 3x \sin x$ **59.** $-2\sin 11.5° \cos 36.5°$ **61.** $2\cos 6x \cos 2x$

63. C   **65.** D   **67.** F   **69.** $\dfrac{\sqrt{2 + \sqrt{2}}}{2}$ **71.** $-\dfrac{\sqrt{2 + \sqrt{3}}}{2}$ **73.** $-\dfrac{\sqrt{2 + \sqrt{3}}}{2}$ **75.** $\dfrac{\sqrt{10}}{4}$

77. 3   **79.** $\dfrac{\sqrt{50 - 10\sqrt{5}}}{10}$ **81.** $-\sqrt{7}$ **83.** $\dfrac{\sqrt{5}}{5}$ **85.** $-\dfrac{\sqrt{42}}{12}$ **87.** $\sin 20°$ **89.** $\tan 73.5°$

91. $\tan 29.87°$ **93.** $\cos 9x$ **95.** $\tan 4\theta$ **97.** $\cos\dfrac{x}{8}$ **99.** $\dfrac{\sin x}{1 + \cos x} = \tan\dfrac{x}{2}$ **101.** $\dfrac{\tan\frac{x}{2} + \cot\frac{x}{2}}{\cot\frac{x}{2} - \tan\frac{x}{2}} = \sec x$

113. 106°   **115.** 2   **117.** $a = -885.6; c = 885.6; \omega = 240\pi$

7.5 Exercises *(pages 696–700)*

1. one-to-one   **3.** domain   **5.** π **7.** (a) $[-1, 1]$ (b) $\left[-\dfrac{\pi}{2}, \dfrac{\pi}{2}\right]$ (c) increasing (d) -2 is not in the domain.

9. (a) $(-\infty, \infty)$ (b) $\left(-\dfrac{\pi}{2}, \dfrac{\pi}{2}\right)$ (c) increasing (d) no   **11.** $\cos^{-1}\dfrac{1}{a}$ **13.** 0   **15.** π **17.** $-\dfrac{\pi}{2}$ **19.** 0   **21.** $\dfrac{\pi}{2}$

23. $\dfrac{\pi}{4}$ **25.** $\dfrac{5\pi}{6}$ **27.** $\dfrac{3\pi}{4}$ **29.** $-\dfrac{\pi}{6}$ **31.** $\dfrac{\pi}{6}$ **33.** 0   **35.** $-45°$ **37.** $-60°$ **39.** $120°$ **41.** $120°$

43. $60°$ **45.** $\sin^{-1} 2$ does not exist.   **47.** $-7.6713835°$ **49.** $113.500970°$ **51.** $30.987961°$ **53.** $121.267893°$

55. $-82.678329°$ **57.** $.83798122$ **59.** 2.3154725 **61.** 1.1900238 **63.** 1.9033723 **65.** 3.1144804

67. 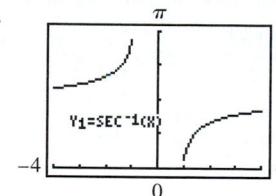   **69.** 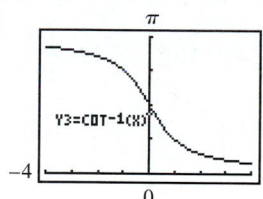   **71.**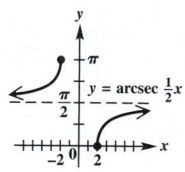

73. 1.003 is not in the domain of $y = \sin^{-1} x$.   **74.** In both cases, the result is x. In each case, the graph is a straight line

bisecting quadrants I and III (i.e., the line $y = x$).   **75.** It is the graph of $y = x$.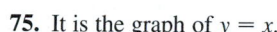

76. It does not agree because the range of the inverse tangent function is $\left(-\dfrac{\pi}{2}, \dfrac{\pi}{2} \right)$, not $(-\infty, \infty)$, as was the case in

Exercise 75.   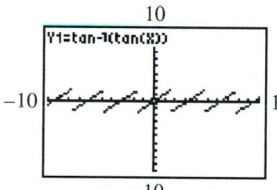   **77.** $\dfrac{\sqrt{7}}{3}$ **79.** $\dfrac{\sqrt{5}}{5}$ **81.** $\dfrac{120}{169}$ **83.** $-\dfrac{7}{25}$ **85.** $\dfrac{4\sqrt{6}}{25}$ **87.** 2

89. $\dfrac{63}{65}$ **91.** $\dfrac{\sqrt{10} - 3\sqrt{30}}{20}$ **93.** $.894427191$ **95.** $.1234399811$

97. $\sqrt{1 - u^2}$ **99.** $\sqrt{1 - u^2}$ **101.** $\dfrac{4\sqrt{u^2 - 4}}{u^2}$ **103.** $\dfrac{u\sqrt{2}}{2}$

105. $\dfrac{2\sqrt{4 - u^2}}{4 - u^2}$ **107.** (a) $45°$ (b) $\theta = 45°$ **109.** (a) $18°$ (b) $18°$ (c) $15°$ (e) 1.4142151 m (Note: Due to the

computational routine, there may be a discrepancy in the last few decimal places.)   (f) $\sqrt{2}$

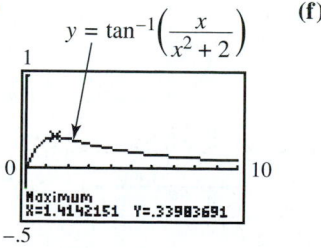

Radian mode

7.6 Exercises *(pages 708–712)*

1. Solve the linear equation for $\cot x$.   **3.** Solve the quadratic equation for $\sec x$ by factoring.   **5.** Solve the quadratic

equation for $\sin x$ using the quadratic formula.   **7.** Use an identity to rewrite as an equation with one trigonometric function.

11. $\left\{ \dfrac{3\pi}{4}, \dfrac{7\pi}{4} \right\}$ **13.** $\left\{ \dfrac{\pi}{6}, \dfrac{5\pi}{6} \right\}$ **15.** $\emptyset$ **17.** $\left\{ \dfrac{\pi}{4}, \dfrac{2\pi}{3}, \dfrac{5\pi}{4}, \dfrac{5\pi}{3} \right\}$ **19.** $\{\pi\}$ **21.** $\left\{ \dfrac{7\pi}{6}, \dfrac{3\pi}{2}, \dfrac{11\pi}{6} \right\}$

23. $\{30°, 210°, 240°, 300°\}$ **25.** $\{90°, 210°, 330°\}$ **27.** $\{45°, 135°, 225°, 315°\}$ **29.** $\{45°, 225°\}$

31. $\{0°, 30°, 150°, 180°\}$ **33.** $\{0°, 45°, 135°, 180°, 225°, 315°\}$ **35.** $\{53.6°, 126.4°, 187.9°, 352.1°\}$

37. $\{149.6°, 329.6°, 106.3°, 286.3°\}$ **39.** $\emptyset$ **41.** $\{57.7°, 159.2°\}$ **43.** $\{.8751 + 2n\pi, 2.2665 + 2n\pi,$

$3.5908 + 2n\pi, 5.8340 + 2n\pi,$ where n is any integer$\}$ **45.** $\left\{\dfrac{\pi}{3} + n\pi, \dfrac{2\pi}{3} + n\pi, \text{ where } n \text{ is any integer}\right\}$

47. $\{33.6° + 360°n, 326.4° + 360°n, \text{ where } n \text{ is any integer}\}$ **49.** $\{45° + 180°n, 108.4° + 180°n, \text{ where } n \text{ is any integer}\}$

51. $\{.6806, 1.4159\}$ **53.** $\left\{\dfrac{\pi}{3}, \pi, \dfrac{4\pi}{3}\right\}$ **55.** $\{60°, 210°, 240°, 310°\}$ **57.** $\left\{\dfrac{\pi}{12}, \dfrac{11\pi}{12}, \dfrac{13\pi}{12}, \dfrac{23\pi}{12}\right\}$ **59.** $\left\{\dfrac{\pi}{2}, \dfrac{7\pi}{6}, \dfrac{11\pi}{6}\right\}$

61. $\left\{\dfrac{\pi}{18}, \dfrac{7\pi}{18}, \dfrac{13\pi}{18}, \dfrac{19\pi}{18}, \dfrac{25\pi}{18}, \dfrac{31\pi}{18}\right\}$ **63.** $\left\{\dfrac{3\pi}{8}, \dfrac{5\pi}{8}, \dfrac{11\pi}{8}, \dfrac{13\pi}{8}\right\}$ **65.** $\left\{\dfrac{\pi}{2}, \dfrac{3\pi}{2}\right\}$ **67.** $\left\{0, \dfrac{\pi}{3}, \pi, \dfrac{5\pi}{3}\right\}$ **69.** $\emptyset$

71. $\left\{\dfrac{\pi}{2}\right\}$ **73.** $\left\{\dfrac{\pi}{3}, \pi, \dfrac{5\pi}{3}\right\}$ **75.** $\{15°, 45°, 135°, 165°, 255°, 285°\}$ **77.** $\{0°\}$ **79.** $\{120°, 240°\}$

81. $\{30°, 150°, 270°\}$ **83.** $\{180°n, 30° + 360°n, 150° + 360°n, \text{ where } n \text{ is any integer}\}$ **85.** $\{60° + 360°n, 300° + 360°n,$

where n is any integer$\}$ **87.** $\{11.8° + 180°n, 78.2° + 180°n, \text{ where } n \text{ is any integer}\}$ **89.** $\{30° + 180°n, 90° + 180°n,$

$150° + 180°n, \text{ where } n \text{ is any integer}\}$ **91.** $\{1.2802\}$ **93. (a)** .00164 and .00355 **(b)** [.00164, .00355] **(c)** outward

95. (a) 3 beats per sec **(b)** 4 beats per sec **(c)** The number of beats is equal to the absolute value

For $x = t$,
$P(t) = .005 \sin 440\pi t +$
$.005 \sin 446\pi t$

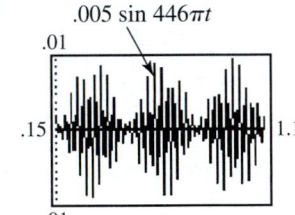

For $x = t$,
$P(t) = .005 \sin 440\pi t +$
$.005 \sin 432\pi t$

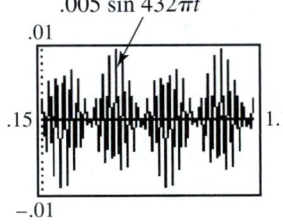

of the difference in the frequencies of the two tones.

97. .001 sec **99.** .004 sec **101.** 14°

103. (a) 2 sec **(b)** $3\dfrac{1}{3}$ sec

Chapter 7 Quiz *(page 712)*

[7.5] 1. $[-1, 1]; [0, \pi]$

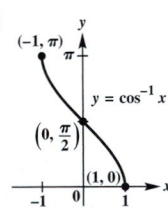

2. (a) $-\dfrac{\pi}{4}$ **(b)** $\dfrac{\pi}{3}$ **(c)** $\dfrac{5\pi}{6}$ **3. (a)** 22.568922° **(b)** 137.431085°

4. (a) $\dfrac{5\sqrt{41}}{41}$ **(b)** $\dfrac{\sqrt{3}}{2}$ **[7.6] 5.** $\{60°, 120°\}$ **6.** $\{60°, 180°, 300°\}$

7. $\{.6089, 1.3424, 3.7505, 4.4840\}$ **8.** $\left\{\dfrac{\pi}{6}, \dfrac{2\pi}{3}, \dfrac{7\pi}{6}, \dfrac{5\pi}{3}\right\}$

9. $\left\{\dfrac{5\pi}{3} + 4n\pi, \dfrac{7\pi}{3} + 4n\pi, \text{ where } n \text{ is any integer}\right\}$ **10. (a)** 0 sec **(b)** .20 sec

7.7 Exercises *(pages 715–718)*

1. C **3.** C **5.** $x = \arccos \dfrac{y}{5}$ **7.** $x = \dfrac{1}{3} \operatorname{arccot} 2y$ **9.** $x = \dfrac{1}{2} \arctan \dfrac{y}{3}$ **11.** $x = 4 \arccos \dfrac{y}{6}$

13. $x = \dfrac{1}{5} \arccos\left(-\dfrac{y}{2}\right)$ **15.** $x = -3 + \arccos y$ **17.** $x = \arcsin(y + 2)$ **19.** $x = \arcsin\left(\dfrac{y + 4}{2}\right)$

21. $x = \dfrac{1}{2} \sec^{-1}\left(\dfrac{y - \sqrt{2}}{3}\right)$ **25.** $\left\{-\dfrac{\sqrt{2}}{2}\right\}$ **27.** $\{-2\sqrt{2}\}$ **29.** $\{\pi - 3\}$ **31.** $\left\{\dfrac{3}{5}\right\}$ **33.** $\left\{\dfrac{4}{5}\right\}$ **35.** $\{0\}$

37. $\left\{\dfrac{1}{2}\right\}$ **39.** $\left\{-\dfrac{1}{2}\right\}$ **41.** $\{0\}$ **43.** $Y = \arcsin X - \arccos X - \dfrac{\pi}{6}$ **45.** $\{4.4622\}$

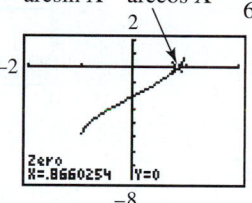

47. (a) $A \approx .00506$, $\phi \approx .484$; $P = .00506 \sin(440\pi t + .484)$

(b) The two graphs are the same.

For $x = t$,
$P(t) = .00506 \sin(440\pi t + .484)$
$P_1(t) + P_2(t) = .0012 \sin(440\pi t + .052) +$
$.004 \sin(440\pi t + .61)$

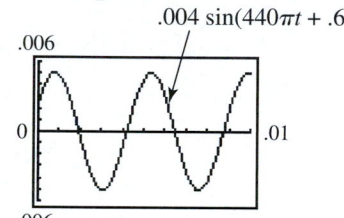

49. (a) $\tan \alpha = \dfrac{x}{z}$; $\tan \beta = \dfrac{x + y}{z}$ **(b)** $\dfrac{x}{\tan \alpha} = \dfrac{x + y}{\tan \beta}$

(c) $\alpha = \arctan\left(\dfrac{x \tan \beta}{x + y}\right)$ **(d)** $\beta = \arctan\left(\dfrac{(x + y) \tan \alpha}{x}\right)$

51. (a) $t = \dfrac{1}{2\pi f} \arcsin \dfrac{E}{E_{\max}}$ **(b)** $.00068$ sec

53. (a) $t = \dfrac{3}{4\pi} \arcsin 3y$ **(b)** $.27$ sec

Chapter 7 Review Exercises *(pages 722–727)*

1. B **3.** C **5.** D **7.** 1 **9.** $\dfrac{1}{\cos^2 \theta}$ **11.** $-\dfrac{\cos \theta}{\sin \theta}$ **13.** $\sin x = -\dfrac{4}{5}$; $\tan x = -\dfrac{4}{3}$; $\cot(-x) = \dfrac{3}{4}$

15. E **17.** J **19.** I **21.** H **23.** G **25.** $\dfrac{117}{125}; \dfrac{4}{5}; -\dfrac{117}{44}$; II

In Exercises 27 and 29, other forms are possible for $\tan(x + y)$.

27. $\dfrac{2 + 3\sqrt{7}}{10}; \dfrac{2\sqrt{3} + \sqrt{21}}{10}; \dfrac{2 + 3\sqrt{7}}{2\sqrt{3} - \sqrt{21}}$; II **29.** $\dfrac{4 - 9\sqrt{11}}{50}; \dfrac{12\sqrt{11} - 3}{50}; \dfrac{4 - 9\sqrt{11}}{12\sqrt{11} + 3}$; IV

31. $\sin \theta = \dfrac{\sqrt{14}}{4}$; $\cos \theta = \dfrac{\sqrt{2}}{4}$ **33.** $\sin 2x = \dfrac{3}{5}$; $\cos 2x = -\dfrac{4}{5}$ **35.** $\dfrac{1}{2}$ **37.** $\dfrac{\sqrt{5} - 1}{2}$ **39.** $.5$

41. $\dfrac{\sin x}{1 - \cos x} = \cot \dfrac{x}{2}$ **43.** $\dfrac{2(\sin x - \sin^3 x)}{\cos x} = \sin 2x$ **67.** $\dfrac{\pi}{4}$ **69.** $-\dfrac{\pi}{3}$ **71.** $\dfrac{3\pi}{4}$ **73.** $\dfrac{2\pi}{3}$ **75.** $\dfrac{3\pi}{4}$

77. $-60°$ **79.** $60.67924514°$ **81.** $36.4895081°$ **83.** $73.26220613°$ **85.** -1 **87.** $\dfrac{3\pi}{4}$ **89.** $\dfrac{\pi}{4}$ **91.** $\dfrac{\sqrt{7}}{4}$

93. $\dfrac{\sqrt{3}}{2}$ **95.** $\dfrac{294 + 125\sqrt{6}}{92}$ **97.** $\dfrac{1}{u}$ **99.** $\{.4636, 3.6052\}$ **101.** $\left\{\dfrac{\pi}{4}, \dfrac{3\pi}{4}, \dfrac{5\pi}{4}, \dfrac{7\pi}{4}\right\}$

103. $\left\{\dfrac{\pi}{8}, \dfrac{3\pi}{8}, \dfrac{5\pi}{8}, \dfrac{7\pi}{8}, \dfrac{9\pi}{8}, \dfrac{11\pi}{8}, \dfrac{13\pi}{8}, \dfrac{15\pi}{8}\right\}$ **105.** $\left\{\dfrac{\pi}{3} + 2n\pi, \pi + 2n\pi, \dfrac{5\pi}{3} + 2n\pi, \text{ where } n \text{ is any integer}\right\}$

107. $\{270°\}$ **109.** $\{45°, 90°, 225°, 270°\}$ **111.** $\{70.5°, 180°, 289.5°\}$ **113.** $x = \arcsin 2y$

115. $x = \left(\dfrac{1}{3} \arctan 2y\right) - \dfrac{2}{3}$ **117.** $\emptyset$ **119.** $\left\{-\dfrac{1}{2}\right\}$

121. (b) 8.6602567 ft; There may be a discrepancy in the final digits. $f(x) = \arctan\left(\frac{15}{x}\right) - \arctan\left(\frac{5}{x}\right)$

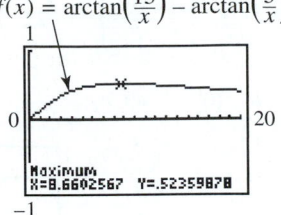

123. The light beam is completely underwater.   **125.**

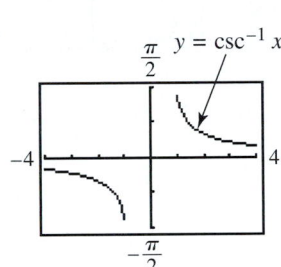

Radian mode

Chapter 7 Test *(pages 727–728)*

[7.1] 1. $\sin\theta = -\frac{7}{25}$; $\tan\theta = -\frac{7}{24}$; $\cot\theta = -\frac{24}{7}$; $\sec\theta = \frac{25}{24}$; $\csc\theta = -\frac{25}{7}$ **2.** $\cos\theta$ **3.** -1 **[7.3] 4.** $\frac{\sqrt{6} - \sqrt{2}}{4}$

5. (a) $-\sin x$ **(b)** $\tan x$ **[7.4] 6.** $-\frac{\sqrt{2 - \sqrt{2}}}{2}$ **[7.3] 7. (a)** $\frac{33}{65}$ **(b)** $-\frac{56}{65}$ **(c)** $\frac{63}{16}$ **(d)** II

[7.4] 8. (a) $-\frac{7}{25}$ **(b)** $-\frac{24}{25}$ **(c)** $\frac{24}{7}$ **(d)** $\frac{\sqrt{5}}{5}$ **(e)** 2 **[7.3] 13. (a)** $V = 163\cos\left(\frac{\pi}{2} - \omega t\right)$ **(b)** 163 volts; $\frac{1}{240}$ sec

[7.5] 14. $[-1, 1]$; $\left[-\frac{\pi}{2}, \frac{\pi}{2}\right]$ **15. (a)** $\frac{2\pi}{3}$ **(b)** $-\frac{\pi}{3}$ **(c)** 0 **(d)** $\frac{2\pi}{3}$ **16. (a)** $30°$ **(b)** $-45°$ **(c)** $135°$

(d) $-60°$ **17. (a)** $42.54°$ **(b)** $22.72°$ **(c)** $125.47°$

18. (a) $\frac{\sqrt{5}}{3}$ **(b)** $\frac{4\sqrt{2}}{9}$ **19.** $\frac{u\sqrt{1 - u^2}}{1 - u^2}$ **[7.6] 20.** $\{30°, 330°\}$ **21.** $\{90°, 270°\}$ **22.** $\{18.4°, 135°, 198.4°, 315°\}$

23. $\left\{0, \frac{2\pi}{3}, \frac{4\pi}{3}\right\}$ **24.** $\left\{\frac{\pi}{12}, \frac{7\pi}{12}, \frac{3\pi}{4}, \frac{5\pi}{4}, \frac{17\pi}{12}, \frac{23\pi}{12}\right\}$ **25.** $\{.3649, 1.2059, 3.5065, 4.3475\}$ **26.** $\{90° + 180°n,$

where n is any integer$\}$ **27.** $\left\{\frac{2\pi}{3} + 4n\pi, \frac{4\pi}{3} + 4n\pi, \text{where } n \text{ is any integer}\right\}$ **[7.7] 28.** $x = \frac{1}{3}\arccos y$ **29.** $\left\{\frac{4}{5}\right\}$

30. $\frac{5}{6}$ sec, $\frac{11}{6}$ sec, $\frac{17}{6}$ sec

CHAPTER 8   APPLICATIONS OF TRIGONOMETRY

Note to student: While most of the measurements resulting from solving triangles in this chapter are approximations, for convenience we use = rather than ≈ in the answers.

8.1 Exercises *(pages 741–748)*

1. C   **3.** $\sqrt{3}$ **5.** $C = 95°, b = 13$ m, $a = 11$ m   **7.** $B = 37.3°, a = 38.5$ ft, $b = 51.0$ ft   **9.** $C = 57.36°,$

$b = 11.13$ ft, $c = 11.55$ ft   **11.** $B = 18.5°, a = 239$ yd, $c = 230$ yd   **13.** $A = 56° \, 00', AB = 361$ ft, $BC = 308$ ft

15. $B = 110.0°, a = 27.01$ m, $c = 21.37$ m   **17.** $A = 34.72°, a = 3326$ ft, $c = 5704$ ft   **19.** $C = 97° \, 34', b = 283.2$ m,

$c = 415.2$ m   **21.** A   **23.** **(a)** $4 < h < 5$ **(b)** $h = 4$ or $h > 5$ **(c)** $h < 4$ **25.** 1   **27.** 2   **29.** 0   **31.** 45°

33. $B_1 = 49.1°, C_1 = 101.2°, B_2 = 130.9°, C_2 = 19.4°$ **35.** $B = 26° \, 30', A = 112° \, 10'$ **37.** no such triangle

39. $B = 27.19°, C = 10.68°$ **41.** $B = 20.6°, C = 116.9°, c = 20.6$ ft   **43.** no such triangle   **45.** $B_1 = 49° \, 20',$

$C_1 = 92° \, 00', c_1 = 15.5$ km; $B_2 = 130° \, 40', C_2 = 10° \, 40', c_2 = 2.88$ km   **47.** $B = 37.77°, C = 45.43°, c = 4.174$ ft

49. $A_1 = 53.23°, C_1 = 87.09°, c_1 = 37.16$ m; $A_2 = 126.77°, C_2 = 13.55°, c_2 = 8.719$ m   **51.** 1; 90°; a right triangle   **55.** 118 m

57. 17.8 km   **59.** 10.4 in.   **61.** 111°   **63.** first location: 5.1 mi; second location: 7.2 mi   **65.** 664 m   **67.** 218 ft

69. about 419,000 km, which compares favorably to the actual value   **71.** $\dfrac{\sqrt{3}}{2}$ sq unit   **73.** $\dfrac{\sqrt{2}}{2}$ sq unit

75. 46.4 m²   **77.** 356 cm²   **79.** 722.9 in.²   **81.** 65.94 cm²   **83.** 100 m²   **85.** $a = \sin A, b = \sin B, c = \sin C$

87. $x = \dfrac{d \sin \alpha \sin \beta}{\sin(\beta - \alpha)}$

8.2 Exercises *(pages 755–762)*

1. **(a)** SAS **(b)** law of cosines   **3.** **(a)** SSA **(b)** law of sines   **5.** **(a)** ASA **(b)** law of sines   **7.** **(a)** ASA **(b)** law of sines

9. 5   **11.** 120°   **13.** $a = 7.0, B = 37.6°, C = 21.4°$ **15.** $A = 73.7°, B = 53.1°, C = 53.1°$ (The angles do not sum to

180° due to rounding.)   **17.** $b = 88.2, A = 56.7°, C = 68.3°$ **19.** $a = 2.60$ yd, $B = 45.1°, C = 93.5°$ **21.** $c = 6.46$ m,

$A = 53.1°, B = 81.3°$ **23.** $A = 82°, B = 37°, C = 61°$ **25.** $C = 102° \, 10', B = 35° \, 50', A = 42° \, 00'$ **27.** $C = 84° \, 30',$

$B = 44° \, 40', A = 50° \, 50'$ **29.** $a = 156$ cm, $B = 64° \, 50', C = 34° \, 30'$ **31.** $b = 9.529$ in., $A = 64.59°, C = 40.61°$

33. $a = 15.7$ m, $B = 21.6°, C = 45.6°$ **35.** $A = 30°, B = 56°, C = 94°$ **37.** The value of $\cos \theta$ will be greater than 1;

your calculator will give you an error message (or a nonreal complex number) when using the inverse cosine function.

39. 257 m   **41.** 281 km   **43.** 10.8 mi   **45.** 40°   **47.** 26° and 36°   **49.** second base: 66.8 ft;

first and third bases: 63.7 ft   **51.** 39.2 km   **53.** 350°   **55.** approximately 47.5 ft   **57.** 163.5°   **59.** 22 ft

61. 16.26°   **63.** $24\sqrt{3}$ sq units   **65.** 78 m²   **67.** 12,600 cm²   **69.** 3650 ft²   **71.** 25.24983 mi

73. Area and perimeter are both 36.   **75.** 390,000 mi²   **77.** **(a)** 87.8° and 92.2° both appear possible. **(b)** 92.2° **(c)** With the

law of cosines we are required to find the inverse cosine of a negative number. Therefore, we know that angle C is greater than 90°.

81.

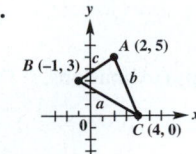

82. $a = \sqrt{34}, b = \sqrt{29}, c = \sqrt{13}$ **83.** 9.5 sq units

84. 9.5 sq units   **85.** 9.5 sq units

Chapter 8 Quiz *(page 762)*

[8.1] 1. 131.3°   **[8.2] 2.** 201 m   **3.** 48.0°   **[8.1] 4.** 15.75 sq units   **[8.2] 5.** 189 km² **[8.1] 6.** 41.6°, 138.4°

7. $a = 648, b = 456, C = 28°$ **8.** 3.6 mi

8.3 Exercises *(pages 770–773)*

1. **m** and **p**; **n** and **r**   **3.** **m** and **p** equal 2**t**, or **t** is one-half **m** or **p**; also **m** = 1**p** and **n** = 1**r**

5.   **7.**   **9.** 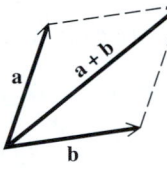   **11.** 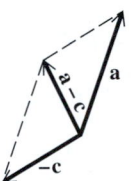   **13.**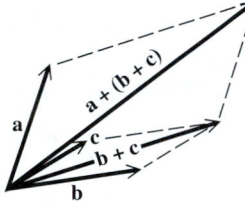

15. 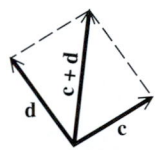   **17.** Yes, it appears that vector addition is associative (and this is true, in general).

19. (a) $\langle -4, 16 \rangle$ (b) $\langle -12, 0 \rangle$ (c) $\langle 8, -8 \rangle$ **21.** (a) $\langle 8, 0 \rangle$ (b) $\langle 0, 16 \rangle$ (c) $\langle -4, -8 \rangle$

23. (a) $\langle 0, 12 \rangle$ (b) $\langle -16, -4 \rangle$ (c) $\langle 8, -4 \rangle$ **25.** (a) 4**i** (b) 7**i** + 3**j** (c) −5**i** + **j**

27. (a) $\langle -2, 4 \rangle$ (b) $\langle 7, 4 \rangle$ (c) $\langle 6, -6 \rangle$

29. 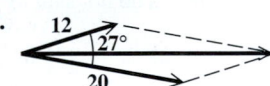   **31.** 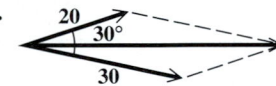   **33.** 17; 331.9°   **35.** 8; 120°   **37.** 47, 17

39. 38.8, 28.0   **41.** 123, 155   **43.** $\left\langle \dfrac{5\sqrt{3}}{2}, \dfrac{5}{2} \right\rangle$ **45.** $\langle 3.0642, 2.5712 \rangle$ **47.** $\langle 4.0958, -2.8679 \rangle$ **49.** 530 newtons

51. 88.2 lb   **53.** 94.2 lb   **55.** 24.4 lb   **57.** $\langle a + c, b + d \rangle$ **59.** $\langle 2, 8 \rangle$ **61.** $\langle 8, -20 \rangle$ **63.** $\langle -30, -3 \rangle$

65. $\langle 8, -7 \rangle$ **67.** −5**i** + 8**j**   **69.** 2**i**, or 2**i** + 0**j**   **71.** 7   **73.** −3   **75.** 20   **77.** 135°   **79.** 90°

81. 36.87°   **83.** −6   **85.** −24   **87.** orthogonal   **89.** not orthogonal   **91.** not orthogonal

In Exercises 93–97, answers may vary due to rounding.

93. magnitude: 9.5208; direction angle: 119.0647°   **94.** $\langle -4.1042, 11.2763 \rangle$ **95.** $\langle -.5209, -2.9544 \rangle$ **96.** $\langle -4.6252, 8.3219 \rangle$

97. magnitude: 9.5208; direction angle: 119.0647°   **98.** They are the same. Preference of method is an individual choice.

8.4 Exercises *(pages 776–779)*

1. 2640 lb at an angle of 167.2° with the 1480-lb force   **3.** 93.9°   **5.** 190 lb and 283 lb, respectively   **7.** 18°   **9.** 2.4 tons

11. 17.5°   **13.** weight: 64.8 lb; tension: 61.9 lb   **15.** 13.5 mi; 50.4°   **17.** 39.2 km   **19.** current: 3.5 mph; motorboat:

19.7 mph   **21.** bearing: 237°; groundspeed: 470 mph   **23.** groundspeed: 161 mph; airspeed: 156 mph   **25.** bearing: 74°;

groundspeed: 202 mph   **27.** bearing: 358°; airspeed: 170 mph   **29.** groundspeed: 230 km per hr; bearing: 167°

31. (a) $|\mathbf{R}| = \sqrt{5} \approx 2.2$, $|\mathbf{A}| = \sqrt{1.25} \approx 1.1$; About 2.2 in. of rain fell. The area of the opening of the rain gauge is about 1.1 in.^{2}. **(b)** $V = 1.5$; The volume of rain was 1.5 in.^{3}. **(c) R** and **A** should be parallel and point in opposite directions.

Summary Exercises on Applications of Trigonometry and Vectors *(pages 779–780)*

1. 29 ft, 38 ft   **2.** 43 ft   **3.** 38.3 cm   **4.** 5856 m   **5.** 15.8 ft per sec; 71.6°   **6.** 42 lb   **7.** 7200 ft

8. (a) 10 mph **(b)** $3\mathbf{v} = 18\mathbf{i} + 24\mathbf{j}$; This represents a 30 mph wind in the direction of **v**. **(c) u** represents a southeast wind of $\sqrt{128} \approx 11.3$ mph.   **9.** It cannot exist.   **10.** Other angles can be 36° 10′, 115° 40′, third side 40.5, or other angles can be 143° 50′, 8° 00′, third side 6.25. (Lengths are in yards.)

8.5 Exercises *(pages 788–791)*

1. 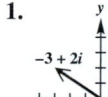   **3.** 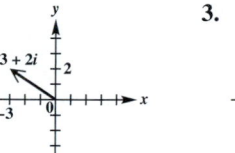   **5.** 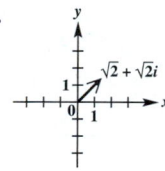   **7.** 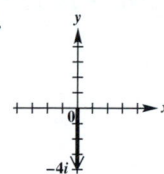   **9.** $1 - 4i$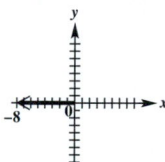

11. $3 - i$ 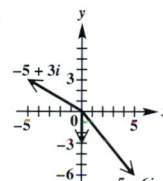   **13.** $-3i$ 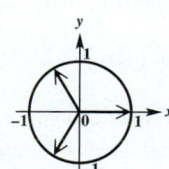   **15.** $-3 + 3i$ **17.** $2 + 4i$ **19.** $7 + 9i$

21. $\frac{7}{6} + \frac{7}{6}i$ **23.** $\sqrt{2} + i\sqrt{2}$ **25.** $10i$ **27.** $-2 - 2i\sqrt{3}$ **29.** $-\frac{3\sqrt{3}}{2} + \frac{3}{2}i$ **31.** $\frac{5}{2} - \frac{5\sqrt{3}}{2}i$ **33.** $-1 - i$

35. $2\sqrt{3} - 2i$ **37.** $6(\cos 240° + i \sin 240°)$ **39.** $2(\cos 330° + i \sin 330°)$ **41.** $5\sqrt{2}(\cos 225° + i \sin 225°)$

43. $2\sqrt{2}(\cos 45° + i \sin 45°)$ **45.** $5(\cos 90° + i \sin 90°)$ **47.** $4(\cos 180° + i \sin 180°)$

49. $\sqrt{13}(\cos 56.31° + i \sin 56.31°)$ **51.** $-1.0261 - 2.8191i$ **53.** $12(\cos 90° + i \sin 90°)$

55. $\sqrt{34}(\cos 59.04° + i \sin 59.04°)$ **57.** yes   **59.** $-3\sqrt{3} + 3i$ **61.** $12\sqrt{3} + 12i$ **63.** $-4i$ **65.** $-3i$

67. $-\frac{15\sqrt{2}}{2} + \frac{15\sqrt{2}}{2}i$ **69.** $\sqrt{3} - i$ **71.** -2 **73.** $-\frac{1}{6} - \frac{\sqrt{3}}{6}i$ **75.** $2\sqrt{3} - 2i$ **77.** $-\frac{1}{2} - \frac{1}{2}i$

79. $\sqrt{3} + i$ **81.** $.6537 + 7.4715i$ **83.** $30.8580 + 18.5414i$ **85.** $.2091 + 1.9890i$ **87.** $-3.7588 - 1.3681i$

89. 2   **90.** $w = \sqrt{2}$ cis 135°; $z = \sqrt{2}$ cis 225°   **91.** 2 cis 0°   **92.** 2; It is the same.   **93.** $-i$ **94.** cis$(-90°)$

95. $-i$; It is the same.   **97.** $1.18 - .14i$ **99.** approximately $27.43 + 11.5i$

8.6 Exercises *(pages 796–798)*

1. $27i$ **3.** 1   **5.** $\frac{27}{2} - \frac{27\sqrt{3}}{2}i$ **7.** $-16\sqrt{3} + 16i$ **9.** $4096i$ **11.** $128 + 128i$

13. (a) $\cos 0° + i \sin 0°$, $\cos 120° + i \sin 120°$, $\cos 240° + i \sin 240°$ **(b)**

15. (a) 2 cis 20°, 2 cis 140°, 2 cis 260° **(b)**

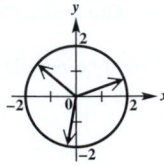

17. (a) 2(cos 90° + i sin 90°), 2(cos 210° + i sin 210°), 2(cos 330° + i sin 330°) **(b)**

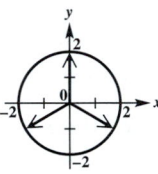

19. (a) 4(cos 60° + i sin 60°), 4(cos 180° + i sin 180°), 4(cos 300° + i sin 300°) **(b)**

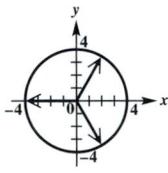

21. (a) $\sqrt[3]{2}$ (cos 20° + i sin 20°), $\sqrt[3]{2}$ (cos 140° + i sin 140°), $\sqrt[3]{2}$ (cos 260° + i sin 260°) **(b)**

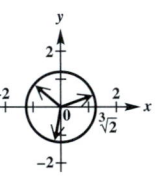

23. (a) $\sqrt[3]{4}$ (cos 50° + i sin 50°), $\sqrt[3]{4}$ (cos 170° + i sin 170°), $\sqrt[3]{4}$ (cos 290° + i sin 290°) **(b)**

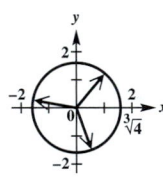

25. cos 0° + i sin 0°,
cos 180° + i sin 180°

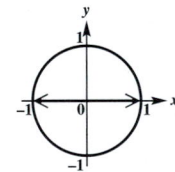

27. cos 0° + i sin 0°, cos 60° + i sin 60°,
cos 120° + i sin 120°, cos 180° + i sin 180°,
cos 240° + i sin 240°, cos 300° + i sin 300°

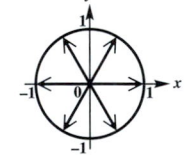

29. cos 30° + i sin 30°,
cos 150° + i sin 150°,
cos 270° + i sin 270°

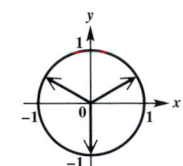

31. {cos 0° + i sin 0°, cos 120° + i sin 120°, cos 240° + i sin 240°}

33. {cos 90° + i sin 90°, cos 210° + i sin 210°, cos 330° + i sin 330°}

35. {2(cos 0° + i sin 0°), 2(cos 120° + i sin 120°), 2(cos 240° + i sin 240°)}

37. {cos 45° + i sin 45°, cos 135° + i sin 135°, cos 225° + i sin 225°, cos 315° + i sin 315°}   **39.** {cos 22.5° + i sin 22.5°,

cos 112.5° + i sin 112.5°, cos 202.5° + i sin 202.5°, cos 292.5° + i sin 292.5°}   **41.** {2(cos 20° + i sin 20°),

2(cos 140° + i sin 140°), 2(cos 260° + i sin 260°)}   **43.** $1, -\dfrac{1}{2} + \dfrac{\sqrt{3}}{2}i, -\dfrac{1}{2} - \dfrac{\sqrt{3}}{2}i$ **45.** cos 2θ + i sin 2θ

46. (cos² θ − sin² θ) + i(2 cos θ sin θ) = cos 2θ + i sin 2θ **47.** cos 2θ = cos² θ − sin² θ **48.** sin 2θ = 2 cos θ sin θ

49. (a) yes **(b)** no **(c)** yes   **51.** $1, .30901699 + .95105652i, -.809017 + .58778525i, -.809017 - .5877853i,$

$.30901699 - .9510565i$ **53.** $-4, 2 - 2i\sqrt{3}$ **55.** $\{-1.8174 + .5503i, 1.8174 - .5503i\}$ **57.** $\{.87708 + .94922i,$

$-.63173 + 1.1275i, -1.2675 - .25240i, -.15164 - 1.28347i, 1.1738 - .54083i\}$ **59.** false

Chapter 8 Quiz *(page 799)*

[8.3] 1. (a) $\langle -3, 12 \rangle$ **(b)** $\langle -14, 12 \rangle$ **(c)** $\sqrt{17}$ **(d)** 3 **(e)** $82.23°$ **[8.4] 2.** 30 lb   **[8.5] 3.** $-1 + 6i$

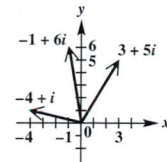

4. (a) $4(\cos 270° + i \sin 270°)$ **(b)** $2(\cos 300° + i \sin 300°)$ **(c)** $\sqrt{10}(\cos 198.4° + i \sin 198.4°)$

5. (a) $2 + 2i\sqrt{3}$ **(b)** $-3.2139 + 3.8302i$ **(c)** $-7i$, or $0 - 7i$ **[8.5, 8.6] 6. (a)** $36(\cos 130° + i \sin 130°)$ **(b)** $2\sqrt{3} + 2i$

(c) $-\dfrac{27\sqrt{3}}{2} + \dfrac{27}{2}i$ **[8.6] 7.** $-2 - 2i$ **8.** $2(\cos 45° + i \sin 45°), 2(\cos 135° + i \sin 135°), 2(\cos 225° + i \sin 225°),$

$2(\cos 315° + i \sin 315°); \sqrt{2} + i\sqrt{2}, -\sqrt{2} + i\sqrt{2}, -\sqrt{2} - i\sqrt{2}, \sqrt{2} - i\sqrt{2}$

8.7 Exercises *(pages 808–812)*

1. (a) II **(b)** I **(c)** IV **(d)** III

Graphs for Exercises 3(a), 5(a), 7(a), 9(a), 11(a)

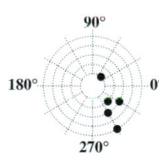

Answers may vary in Exercises 3(b)–11(b).

3. (b) $(1, 405°), (-1, 225°)$ **(c)** $\left(\dfrac{\sqrt{2}}{2}, \dfrac{\sqrt{2}}{2}\right)$ **5. (b)** $(-2, 495°), (2, 315°)$ **(c)** $\left(\sqrt{2}, -\sqrt{2}\right)$ **7. (b)** $(5, 300°), (-5, 120°)$

(c) $\left(\dfrac{5}{2}, -\dfrac{5\sqrt{3}}{2}\right)$ **9. (b)** $(-3, 150°), (3, -30°)$ **(c)** $\left(\dfrac{3\sqrt{3}}{2}, -\dfrac{3}{2}\right)$ **11. (b)** $\left(3, \dfrac{11\pi}{3}\right), \left(-3, \dfrac{2\pi}{3}\right)$ **(c)** $\left(\dfrac{3}{2}, -\dfrac{3\sqrt{3}}{2}\right)$

Graphs for Exercises 13(a), 15(a), 17(a), 19(a), 21(a)

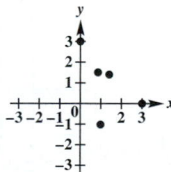

Answers may vary in Exercises 13(b)–21(b).

13. (b) $\left(\sqrt{2}, 315°\right), \left(-\sqrt{2}, 135°\right)$ **15. (b)** $(3, 90°), (-3, 270°)$ **17. (b)** $(2, 45°), (-2, 225°)$

19. (b) $\left(\sqrt{3}, 60°\right), \left(-\sqrt{3}, 240°\right)$ **21. (b)** $(3, 0°), (-3, 180°)$ **23.** $r = \dfrac{4}{\cos \theta - \sin \theta}$

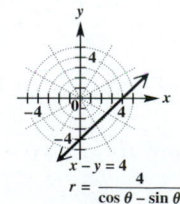

25. $r = 4$ or $r = -4$

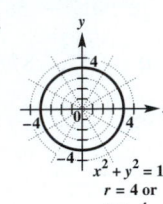

$x^2 + y^2 = 16$
$r = 4$ or
$r = -4$

27. $r = \dfrac{5}{2\cos\theta + \sin\theta}$

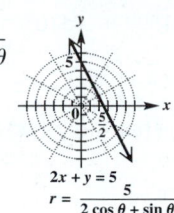

$2x + y = 5$
$r = \dfrac{5}{2\cos\theta + \sin\theta}$

29. $r\sin\theta = k$ **30.** $r = \dfrac{k}{\sin\theta}$

31. $r = k\csc\theta$ **32.**

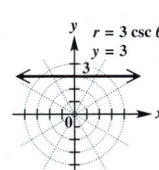

$r = 3\csc\theta$ $y = 3$
$y = 3$

33. $r\cos\theta = k$ **34.** $r = \dfrac{k}{\cos\theta}$ **35.** $r = k\sec\theta$

36.

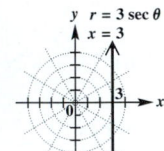

$r = 3\sec\theta$ $x = 3$
$x = 3$

37. C   **39.** A   **41.** cardioid

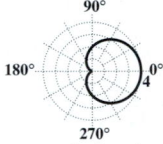

$r = 2 + 2\cos\theta$

43. limaçon

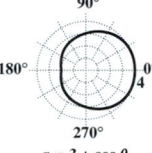

$r = 3 + \cos\theta$

45. four-leaved rose

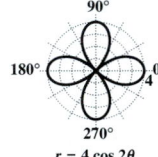

$r = 4\cos 2\theta$

47. lemniscate

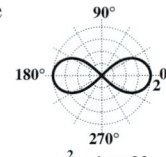

$r^2 = 4\cos 2\theta$

49. cardioid

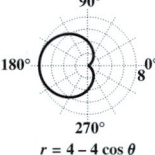

$r = 4 - 4\cos\theta$

51.

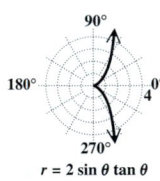

$r = 2\sin\theta\tan\theta$

53. $x^2 + (y - 1)^2 = 1$

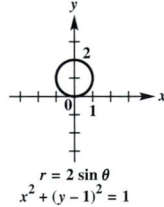

$r = 2\sin\theta$
$x^2 + (y - 1)^2 = 1$

55. $y^2 = 4(x + 1)$

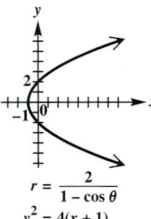

$r = \dfrac{2}{1 - \cos\theta}$
$y^2 = 4(x + 1)$

57. $(x + 1)^2 + (y + 1)^2 = 2$

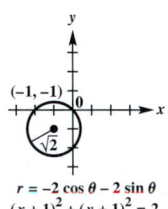

$(-1, -1)$
$\sqrt{2}$
$r = -2\cos\theta - 2\sin\theta$
$(x + 1)^2 + (y + 1)^2 = 2$

59. $x = 2$

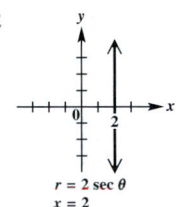

$r = 2\sec\theta$
$x = 2$

61. $x + y = 2$

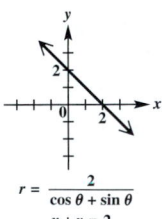

$r = \dfrac{2}{\cos\theta + \sin\theta}$
$x + y = 2$

63.

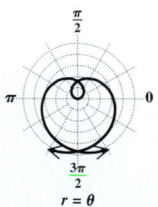

$r = \theta$

65. $r = \dfrac{2}{2\cos\theta + \sin\theta}$

67. (a) $(r, -\theta)$ (b) $(r, \pi - \theta)$ or $(-r, -\theta)$ (c) $(r, \pi + \theta)$ or $(-r, \theta)$

69. $r = \theta, 0 \le \theta \le 4\pi$

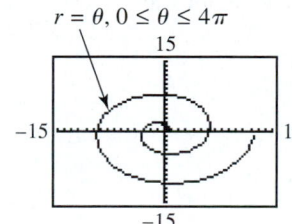

71. $r = 1.5\theta, -4\pi \le \theta \le 4\pi$

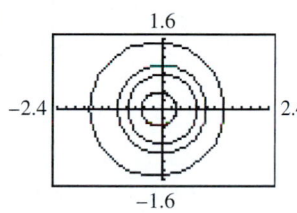

73. $\left(2, \dfrac{\pi}{6}\right), \left(2, \dfrac{5\pi}{6}\right)$

75. $\left(\dfrac{4 + \sqrt{2}}{2}, \dfrac{\pi}{4}\right), \left(\dfrac{4 - \sqrt{2}}{2}, \dfrac{5\pi}{4}\right)$ **77. (a)**

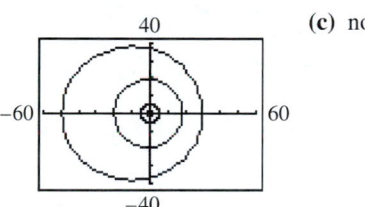

(b)

(c) no

Earth is closest to the sun.

8.8 Exercises *(pages 818–821)*

1. C   **3.** A   **5. (a)**

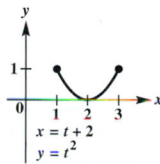

(b) $y = x^2 - 4x + 4$, for x in $[1, 3]$

7. (a)

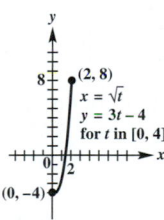

(b) $y = 3x^2 - 4$, for x in $[0, 2]$ **9. (a)**

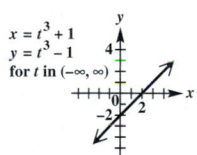

(b) $y = x - 2$, for x in $(-\infty, \infty)$

11. (a)

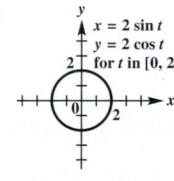

(b) $x^2 + y^2 = 4$, for x in $[-2, 2]$ **13. (a)**

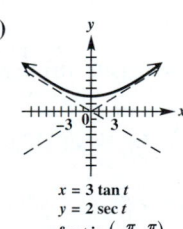

(b) $y = 2\sqrt{1 + \dfrac{x^2}{9}}$, for x in $(-\infty, \infty)$

15. (a)

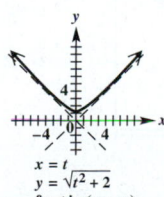

(b) $y = \dfrac{1}{x}$, for x in $(0, 1]$ **17. (a)**

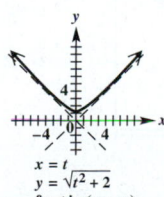

(b) $y = \sqrt{x^2 + 2}$, for x in $(-\infty, \infty)$

19. (a) 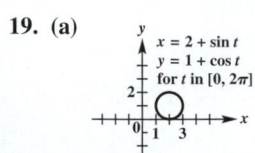 **(b)** $(x - 2)^2 + (y - 1)^2 = 1$, for x in $[1, 3]$

21. (a) 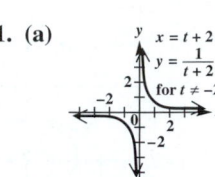 **(b)** $y = \dfrac{1}{x}$, for x in $(-\infty, 0) \cup (0, \infty)$

23. (a) 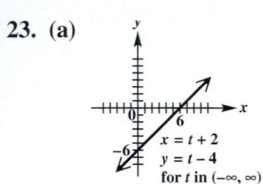 **(b)** $y = x - 6$, for x in $(-\infty, \infty)$

25. 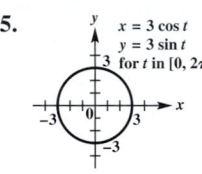 $x^2 + y^2 = 9$

27. 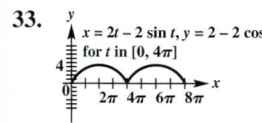 $\dfrac{x^2}{9} + \dfrac{y^2}{4} = 1$

Answers may vary for Exercises 29 and 31.

29. $x = t, y = (t + 3)^2 - 1$, for t in $(-\infty, \infty)$; $x = t - 3, y = t^2 - 1$, for t in $(-\infty, \infty)$

31. $x = t, y = t^2 - 2t + 3$, for t in $(-\infty, \infty)$; $x = t + 1, y = t^2 + 2$, for t in $(-\infty, \infty)$

33.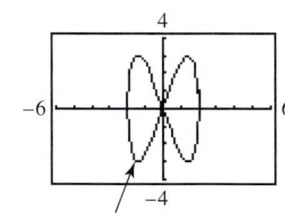

35.

$x = 2 \cos t, y = 3 \sin 2t$, for t in $[0, 6.5]$

37.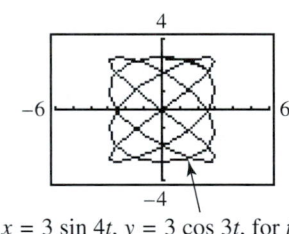

$x = 3 \sin 4t, y = 3 \cos 3t$, for t in $[0, 6.5]$

39. (a) $x = 24t, y = -16t^2 + 24\sqrt{3}t$ **(b)** $y = -\dfrac{1}{36}x^2 + \sqrt{3}x$ **(c)** 2.6 sec; 62 ft

41. (a) $x = (88 \cos 20°)t, y = 2 - 16t^2 + (88 \sin 20°)t$ **(b)** $y = 2 - \dfrac{x^2}{484 \cos^2 20°} + (\tan 20°)x$ **(c)** 1.9 sec; 161 ft

43. (a) $y = -\dfrac{1}{256}x^2 + \sqrt{3}x + 8$; parabolic path **(b)** approximately 7 sec; approximately 448 ft

45. (a) $x = 32t, y = 32\sqrt{3}t - 16t^2 + 3$ **(b)** about 112.6 ft **(c)** 51 ft maximum height; The ball had traveled horizontally about 55.4 ft. **(d)** yes

47. (a) $x = 56.56530965t$
$y = -16t^2 + 67.41191099t$

(b) 50.0° **(c)** $x = (88 \cos 50.0°)t, y = -16t^2 + (88 \sin 50.0°)t$

49. Many answers are possible; for example, $y = a(t - h)^2 + k, x = t$ and $y = at^2 + k, x = t + h$. **51.** Many answers are possible; for example, $x = a \sin t, y = b \cos t$ and $x = t, y^2 = b^2\left(1 - \dfrac{t^2}{a^2}\right)$. **55.** The graph is translated horizontally by c units.

Chapter 8 Review Exercises *(pages 827–832)*

1. 63.7 m   **3.** 41.7°   **5.** 54° 20′ or 125° 40′   **9. (a)** $b = 5, b \geq 10$ **(b)** $5 < b < 10$ **(c)** $b < 5$

11. 19.87°, or 19° 52′   **13.** 55.5 m   **15.** 19 cm   **17.** $B = 17.3°, C = 137.5°, c = 11.0$ yd   **19.** $c = 18.7$ cm,

$A = 91° 40′, B = 45° 50′$ **21.** 153,600 m^2 **23.** .234 km^2 **25.** 58.6 ft   **27.** 13 m   **29.** 53.2 ft   **31.** 115 km

33. 25 sq units   **35.** 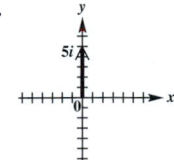   **37. (a)** true **(b)** false   **39.** 207 lb   **41.** 869; 418

43. 15; 126.9°   **45.** −9; 142.1°   **47.** $\left\langle \dfrac{5}{13}, \dfrac{12}{13} \right\rangle$ **49.** 29 lb

51. bearing: 306°; speed: 524 mph   **53.** $AB = 1978.28$ ft; $BC = 975.05$ ft

55. $-3 - 3i\sqrt{3}$ **57.** -2 **59.** $-128 + 128i$

61. 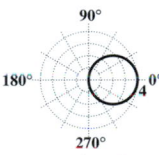   **63.** 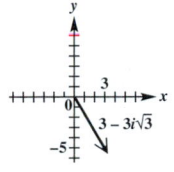   **65.** $2\sqrt{2}(\cos 135° + i \sin 135°)$ **67.** $-\sqrt{2} - i\sqrt{2}$

69. $\sqrt{2}(\cos 315° + i \sin 315°)$ **71.** $4(\cos 270° + i \sin 270°)$

73. It is the line $y = -x$.

75. $\sqrt[3]{2}(\cos 105° + i \sin 105°), \sqrt[3]{2}(\cos 225° + i \sin 225°), \sqrt[3]{2}(\cos 345° + i \sin 345°)$ **77.** none

79. $\{2(\cos 45° + i \sin 45°), 2(\cos 135° + i \sin 135°), 2(\cos 225° + i \sin 225°), 2(\cos 315° + i \sin 315°)\}$ **81.** $(2, 120°)$

83. circle

r = 4 cos θ

85. eight-leaved rose

r = 2 sin 4θ

87. $y^2 = -6\left(x - \dfrac{3}{2}\right)$, or $y^2 + 6x - 9 = 0$

89. $x^2 + y^2 = 4$ **91.** $r = \tan \theta \sec \theta$, or $r = \dfrac{\tan \theta}{\cos \theta}$ **93.** $r = 2 \sec \theta$, or $r = \dfrac{2}{\cos \theta}$ **95.** $r = \dfrac{4}{\cos \theta + 2 \sin \theta}$

97. 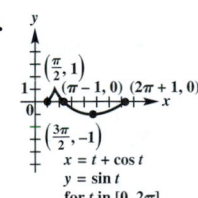   **99.** $y = \sqrt{x^2 + 1}$, for x in $[0, \infty)$ **101.** $y = 3\sqrt{1 + \dfrac{x^2}{25}}$, for x in $(-\infty, \infty)$

103. $x = 3 + 5 \cos t, y = 4 + 5 \sin t$, for t in $[0, 2\pi]$ **105. (a)** $x = (118 \cos 27°)t$,

$y = 3.2 - 16t^2 + (118 \sin 27°)t$ **(b)** $y = 3.2 - \dfrac{4x^2}{3481 \cos^2 27°} + (\tan 27°)x$ **(c)** 3.4 sec; 358 ft

Chapter 8 Test *(pages 833–834)*

[8.1] 1. 137.5°   **[8.2] 2.** 179 km   **3.** 49.0°   **4.** 168 sq units   **[8.1] 5.** 18 sq units   **6. (a)** $b > 10$

(b) none **(c)** $b \leq 10$ **[8.1, 8.2] 7.** $a = 40$ m, $B = 41°, C = 79°$ **8.** $B_1 = 58° 30′, A_1 = 83° 00′, a_1 = 1250$ in.;

$B_2 = 121° 30′, A_2 = 20° 00′, a_2 = 431$ in.   **[8.3] 9.** $|\mathbf{v}| = 10; \theta = 126.9°$ **10. (a)** $\langle 1, -3 \rangle$ **(b)** $\langle -6, 18 \rangle$ **(c)** -20 **(d)** $\sqrt{10}$

[8.1] 11. 2.7 mi   **[8.3] 12.** $\langle -346, 451 \rangle$ **[8.4] 13.** 1.91 mi   **[8.1] 14.** 14 m

[8.5] 15. $7 - 3i$ **16. (a)** $3(\cos 90° + i \sin 90°)$ **(b)** $\sqrt{5}$ cis 63.43° **(c)** $2(\cos 240° + i \sin 240°)$

17. (a) $\dfrac{3\sqrt{3}}{2} + \dfrac{3}{2}i$ (b) $3.06 + 2.57i$ (c) $3i$ **[8.5, 8.6] 18.** (a) $16(\cos 50° + i \sin 50°)$ (b) $2\sqrt{3} + 2i$ (c) $4\sqrt{3} + 4i$

[8.6] 19. 2 cis 67.5°, 2 cis 157.5°, 2 cis 247.5°, 2 cis 337.5°   **[8.7] 20.** Answers may vary. (a) $(5, 90°), (5, -270°)$

(b) $\left(2\sqrt{2}, 225°\right), \left(2\sqrt{2}, -135°\right)$ **21.** (a) $\left(\dfrac{3\sqrt{2}}{2}, -\dfrac{3\sqrt{2}}{2}\right)$ (b) $(0, -4)$ **22.** cardioid

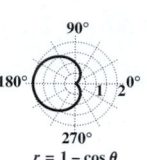

$r = 1 - \cos \theta$

23. three-leaved rose   **24.** (a) $x - 2y = -4$ (b) $x^2 + y^2 = 36$

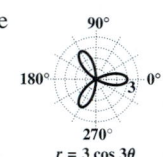

$r = 3 \cos 3\theta$

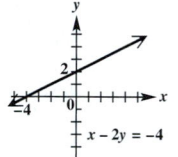

$x - 2y = -4$

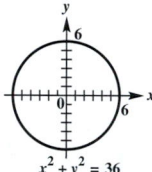

$x^2 + y^2 = 36$

[8.8] 25.   **26.**

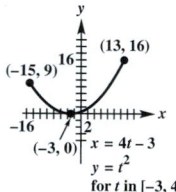

$x = 4t - 3$
$y = t^2$
for t in $[-3, 4]$

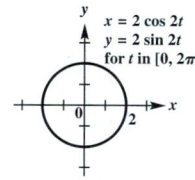

$x = 2 \cos 2t$
$y = 2 \sin 2t$
for t in $[0, 2\pi]$

CHAPTER 9   SYSTEMS AND MATRICES

9.1 Exercises *(pages 848–854)*

1. 2005   **3.** $\{(1998, .060), (2000, .059), (2001, .058), (2003, .059), (2005, .056)\}$ **5.** year; ozone, ppm   **7.** $\{(-4, 1)\}$

9. $\{(48, 8)\}$ **11.** $\{(1, 3)\}$ **13.** $\{(-1, 3)\}$ **15.** $\{(3, -4)\}$ **17.** $\{(0, 2)\}$ **19.** $\{(0, 4)\}$ **21.** $\{(1, 3)\}$ **23.** $\{(4, -2)\}$

25. $\{(2, -2)\}$ **27.** $\{(4, 6)\}$ **29.** $\{(5, 2)\}$ **31.** $\varnothing$; inconsistent system   **33.** $\left\{\left(\dfrac{y + 9}{4}, y\right)\right\}$; infinitely many solutions

35. $\varnothing$; inconsistent system   **37.** $\left\{\left(\dfrac{6 - 2y}{7}, y\right)\right\}$; infinitely many solutions   **39.** A   **41.** $x - 3y = -3$
$3x + 2y = 6$

43. $\{(.820, -2.508)\}$ **45.** $\{(.892, .453)\}$ **47.** $\{(1, 2, -1)\}$ **49.** $\{(2, 0, 3)\}$ **51.** $\{(1, 2, 3)\}$ **53.** $\{(4, 1, 2)\}$

55. $\left\{\left(\dfrac{1}{2}, \dfrac{2}{3}, -1\right)\right\}$ **57.** $\{(-3, 1, 6)\}$ **59.** $\{(x, -4x + 3, -3x + 4)\}$ **61.** $\{(x, -x + 15, -9x + 69)\}$

63. $\left\{\left(x, \dfrac{41 - 7x}{9}, \dfrac{47 - x}{9}\right)\right\}$ **65.** $\varnothing$; inconsistent system   **67.** $\left\{\left(-\dfrac{z}{9}, \dfrac{z}{9}, z\right)\right\}$; infinitely many solutions

69. $\{(2, 2)\}$ **71.** $\left\{\left(\dfrac{1}{5}, 1\right)\right\}$ **73.** $\{(4, 6, 1)\}$ **75.** Other answers are possible. **(a)** One example is $x + 2y + z = 5$,

$2x - y + 3z = 4$. **(b)** One example is $x + y + z = 5$, $2x - y + 3z = 4$. **(c)** One example is $2x + 2y + 2z = 8$,

$2x - y + 3z = 4$.   **77.** $y = \dfrac{3}{4}x^2 + \dfrac{1}{4}x - \dfrac{1}{2}$ **79.** $y = 3x - 1$ **81.** $y = -\dfrac{1}{2}x^2 + x + \dfrac{1}{4}$

83. $x^2 + y^2 - 4x + 2y - 20 = 0$ **85.** $x^2 + y^2 + x - 7y = 0$ **87.** **(a)** $a = \dfrac{7}{800}, b = \dfrac{39}{40}, c = 318$;

$C = \dfrac{7}{800}t^2 + \dfrac{39}{40}t + 318$ **(b)** near the end of 2104   **89.** 23, 24   **91.** baseball: $171.20; football: $329.82

93. 120 gal of $9.00; 60 gal of $3.00; 120 gal of $4.50   **95.** 28 in.; 17 in.; 14 in.   **97.** $100,000 at 3%; $40,000 at 2.5%; $60,000 at 1.5%   **99. (a)** $16 **(b)** $11 **(c)** $6   **100. (a)** 8 **(b)** 4 **(c)** 0   **101.** See the answer to Exercise 103.

102. (a) 0 **(b)** $\dfrac{40}{3}$ **(c)** $\dfrac{80}{3}$ **103.**

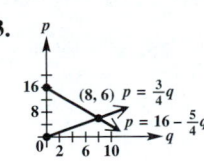

104. price: $6; demand: 8   **105.** $\{(40, 15, 30)\}$

107. 11.92 lb of Arabian Mocha Sanani; 14.23 lb of Organic Shade Grown Mexico; 23.85 lb of Guatemala Antigua (Answers are approximations.)

9.2 Exercises *(pages 862–866)*

1. $\begin{bmatrix} 2 & 4 \\ 0 & -1 \end{bmatrix}$ **3.** $\begin{bmatrix} 1 & 5 & 6 \\ 0 & 13 & 11 \\ 4 & 7 & 0 \end{bmatrix}$ **5.** $\left[\begin{array}{cc|c} 2 & 3 & 11 \\ 1 & 2 & 8 \end{array}\right]$; 2×3 **7.** $\left[\begin{array}{ccc|c} 2 & 1 & 1 & 3 \\ 3 & -4 & 2 & -7 \\ 1 & 1 & 1 & 2 \end{array}\right]$; 3×4

9. $3x + 2y + z = 1$
$2y + 4z = 22$
$-x - 2y + 3z = 15$
    11. $x = 2$
$y = 3$
$z = -2$
    13. $x + y = 3$
$2y + z = -4$
$x - z = 5$
    15. $\{(2, 3)\}$ **17.** $\{(-3, 0)\}$ **19.** $\{(2, -7)\}$ **21.** $\emptyset$

23. $\left\{\left(\dfrac{4}{3}y + \dfrac{7}{3}, y\right)\right\}$ **25.** $\{(-1, -2, 3)\}$ **27.** $\{(-1, 23, 16)\}$ **29.** $\{(2, 4, 5)\}$ **31.** $\left\{\left(\dfrac{1}{2}, 1, -\dfrac{1}{2}\right)\right\}$ **33.** $\{(2, 1, -1)\}$

35. $\emptyset$ **37.** $\left\{\left(-\dfrac{15}{23}z - \dfrac{12}{23}, \dfrac{1}{23}z - \dfrac{13}{23}, z\right)\right\}$ **39.** $\{(1, 1, 2, 0)\}$ **41.** $\{(0, 2, -2, 1)\}$ **43.** $\{(.571, 7.041, 11.442)\}$

45. none   **47.** $A = \dfrac{1}{2}, B = -\dfrac{1}{2}$ **49.** $A = \dfrac{1}{2}, B = \dfrac{1}{2}$ **51.** day laborer: $152; concrete finisher: $160   **53.** 12, 6, 2

55. 9.6 cm³ of 7%; 30.4 cm³ of 2%   **59.** 44.4 g of A; 133.3 g of B; 222.2 g of C   **61. (a)** 65 or older: $y = .00184x + .124$; ages 25–34: $y = -.000178x + .134$ **(b)** $\{(4.9554, .1331)\}$; 2009; 13.3%   **63. (a)** using the first equation, approximately 245 lb; using the second, approximately 253 lb **(b)** for the first, 7.46 lb; for the second, 7.93 lb **(c)** at approximately 66 in. and 118 lb

65. $a + 871b + 11.5c + 3d = 239$
$a + 847b + 12.2c + 2d = 234$
$a + 685b + 10.6c + 5d = 192$
$a + 969b + 14.2c + 1d = 343$
    66. $\left[\begin{array}{cccc|c} 1 & 871 & 11.5 & 3 & 239 \\ 1 & 847 & 12.2 & 2 & 234 \\ 1 & 685 & 10.6 & 5 & 192 \\ 1 & 969 & 14.2 & 1 & 343 \end{array}\right]$; The solution is $a \approx -715.457$, $b \approx .34756$, $c \approx 48.6585$, and $d \approx 30.71951$.

67. $F = -715.457 + .34756A + 48.6585P + 30.71951W$ **68.** approximately 323

9.3 Exercises *(pages 874–878)*

1. -31 **3.** 7   **5.** 0   **7.** -26 **9.** 0   **11.** $2, -6, 4$ **13.** $-6, 0, -6$ **15.** 186   **17.** 17   **19.** 166

21. 0   **23.** 0   **25.** 1   **27.** 2   **29.** -5.5 **32.** 102   **33.** 102; yes   **34.** no   **35.** $\left\{-\dfrac{4}{3}\right\}$ **37.** $\{-1, 4\}$

39. $\{-4\}$ **41.** $\{13\}$ **43.** 1   **45.** 9.5   **47.** approximately 19,328.3 ft²   **49.** 0   **51.** 0   **53.** 16   **55.** 298

57. -88 **59. (a)** D **(b)** A **(c)** C **(d)** B   **61.** $\{(2, 2)\}$ **63.** $\{(2, -5)\}$ **65.** $\{(2, 0)\}$ **67.** Cramer's rule does not apply, since $D = 0$; $\emptyset$ **69.** Cramer's rule does not apply, since $D = 0$; $\left\{\left(\dfrac{4 - 2y}{3}, y\right)\right\}$ **71.** $\{(-4, 12)\}$ **73.** $\{(-3, 4, 2)\}$

75. $\{(0, 0, -1)\}$ **77.** Cramer's rule does not apply, since $D = 0$; $\emptyset$ **79.** Cramer's rule does not apply, since $D = 0$; $\left\{\left(\dfrac{-32 + 19z}{4}, \dfrac{24 - 13z}{4}, z\right)\right\}$ **81.** $\{(0, 4, 2)\}$ **83.** $\left\{\left(\dfrac{31}{5}, \dfrac{19}{10}, -\dfrac{29}{10}\right)\right\}$ **85.** $W_1 = W_2 = \dfrac{100\sqrt{3}}{3} \approx 58$ lb

87. $\{(-a - b, a^2 + ab + b^2)\}$ **89.** $\{(1, 0)\}$

9.4 Exercises *(pages 884–885)*

1. $\dfrac{5}{3x} + \dfrac{-10}{3(2x + 1)}$ **3.** $\dfrac{6}{5(x + 2)} + \dfrac{8}{5(2x - 1)}$ **5.** $\dfrac{5}{6(x + 5)} + \dfrac{1}{6(x - 1)}$ **7.** $\dfrac{-2}{x + 1} + \dfrac{2}{x + 2} + \dfrac{4}{(x + 2)^2}$

9. $\dfrac{4}{x} + \dfrac{4}{1 - x}$ **11.** $\dfrac{2}{(x + 2)^2} + \dfrac{-3}{(x + 2)^3}$ **13.** $1 + \dfrac{-2}{x + 1} + \dfrac{1}{(x + 1)^2}$ **15.** $x^3 - x^2 + \dfrac{-1}{3(2x + 1)} + \dfrac{2}{3(x + 2)}$

17. $\dfrac{1}{9} + \dfrac{-1}{x} + \dfrac{25}{18(3x + 2)} + \dfrac{29}{18(3x - 2)}$ **19.** $\dfrac{-3}{5x^2} + \dfrac{3}{5(x^2 + 5)}$ **21.** $\dfrac{-2}{7(x + 4)} + \dfrac{6x - 3}{7(3x^2 + 1)}$

23. $\dfrac{1}{4x} + \dfrac{-8}{19(2x + 1)} + \dfrac{-9x - 24}{76(3x^2 + 4)}$ **25.** $\dfrac{-1}{x} + \dfrac{2x}{2x^2 + 1} + \dfrac{2x + 3}{(2x^2 + 1)^2}$ **27.** $\dfrac{-1}{x + 2} + \dfrac{3}{(x^2 + 4)^2}$

29. $5x^2 + \dfrac{3}{x} + \dfrac{-1}{x + 3} + \dfrac{2}{x - 1}$ **31.** graphs coincide; correct   **33.** graphs do not coincide; not correct

Chapter 9 Quiz *(pages 885–886)*

[9.1] 1. $\{(-2, 0)\}$ **2.** $\left\{\left(x, \dfrac{2 - x}{2}\right)\right\}$ or $\{(2 - 2y, y)\}$ **3.** $\emptyset$ **4.** $\{(3, -4)\}$ **[9.2] 5.** $\{(-5, 2)\}$ **[9.3] 6.** $\{(-3, 6)\}$

[9.1] 7. $\{(-2, 1, 2)\}$ **[9.2] 8.** $\{(2, 1, -1)\}$ **[9.3] 9.** Cramer's rule does not apply, since $D = 0$; $\left\{\left(\dfrac{2y + 6}{9}, y, \dfrac{23y + 6}{9}\right)\right\}$

[9.1] 10. 11.03 million with stereo sound; 21 million without stereo sound   **11.** $1000 at 8%; $1500 at 11%; $2500 at 14%

[9.3] 12. -3 **13.** 59   **[9.4] 14.** $\dfrac{7}{x - 5} + \dfrac{3}{x + 4}$ **15.** $\dfrac{-1}{x + 2} + \dfrac{6}{x + 4} + \dfrac{-3}{x - 1}$

9.5 Exercises *(pages 893–896)*

7. Consider the graphs; a line and a parabola cannot intersect in more than two points.   **9.** $\{(1, 1), (-2, 4)\}$

11. $\left\{(2, 1), \left(\dfrac{1}{3}, \dfrac{4}{9}\right)\right\}$ **13.** $\{(2, 12), (-4, 0)\}$ **15.** $\left\{\left(-\dfrac{3}{5}, \dfrac{7}{5}\right), (-1, 1)\right\}$ **17.** $\{(2, 2), (2, -2), (-2, 2), (-2, -2)\}$

19. $\{(0, 0)\}$ **21.** $\left\{\left(i, \sqrt{6}\right), \left(-i, \sqrt{6}\right), \left(i, -\sqrt{6}\right), \left(-i, -\sqrt{6}\right)\right\}$ **23.** $\{(1, -1), (-1, 1), (1, 1), (-1, -1)\}$ **25.** $\emptyset$

27. $\{(2, 0), (-2, 0)\}$ **29.** $\left\{(-3, 5), \left(\dfrac{15}{4}, -4\right)\right\}$ **31.** $\left\{\left(4, -\dfrac{1}{8}\right), \left(-2, \dfrac{1}{4}\right)\right\}$

33. $\left\{(3, 2), (-3, -2), \left(4, \dfrac{3}{2}\right), \left(-4, -\dfrac{3}{2}\right)\right\}$ **35.** $\left\{\left(\sqrt{5}, 0\right), \left(-\sqrt{5}, 0\right), \left(\sqrt{5}, \sqrt{5}\right), \left(-\sqrt{5}, -\sqrt{5}\right)\right\}$

37. $\{(3, 5), (-3, -5), (5i, -3i), (-5i, 3i)\}$ **39.** $\{(3, -3), (3, 3)\}$ **41.** $\{(2, 2), (-2, -2), (2, -2), (-2, 2)\}$

43. $\{(-.79, .62), (.88, .77)\}$ **45.** $\{(.06, 2.88)\}$ **47.** 14 and 3   **49.** 8 and 6, 8 and -6, -8 and 6, -8 and -6

51. 27 and 6, -27 and -6 **53.** 5 m and 12 m   **55.** yes   **57.** $y = 9$ **59.** length = width: 6 ft; height: 10 ft

63. (a) 160 units **(b)** $3   **65. (a)** The carbon emissions are increasing with time. The carbon emissions from the former USSR and Eastern Europe have surpassed the emissions of Western Europe. **(b)** They were equal in 1963 when the levels were approximately 400 million metric tons. **(c)** year: 1962; emissions level: approximately 414 million metric tons

66. Translate the graph of $y = |x|$ one unit to the right.   **67.** Translate the graph of $y = x^2$ four units down.

68. $y = \begin{cases} x - 1 & \text{if } x \geq 1 \\ 1 - x & \text{if } x < 1 \end{cases}$ **69.** $x^2 - 4 = x - 1 \ (x \geq 1)$; $x^2 - 4 = 1 - x \ (x < 1)$ **70.** $\dfrac{1 + \sqrt{13}}{2}; \dfrac{-1 - \sqrt{21}}{2}$

71. $\left\{\left(\dfrac{1 + \sqrt{13}}{2}, \dfrac{-1 + \sqrt{13}}{2}\right), \left(\dfrac{-1 - \sqrt{21}}{2}, \dfrac{3 + \sqrt{21}}{2}\right)\right\}$

Summary Exercises on Systems of Equations *(page 897)*

1. $\{(-3, 2)\}$ **2.** $\{(-5, -4)\}$ **3.** $\{(0, \sqrt{5}), (0, -\sqrt{5})\}$ **4.** $\{(-10, -6, 25)\}$ **5.** $\left\{\left(\frac{4}{3}, 3\right), \left(-\frac{1}{2}, -8\right)\right\}$

6. $\{(-1, 2, -4)\}$ **7.** $\{(-13 + 5z, 9 - 3z, z)\}$ **8.** $\left\{(-2, -2), \left(\frac{14}{5}, -\frac{2}{5}\right)\right\}$ **9.** $\emptyset$ **10.** $\{(-3, 3, -3)\}$

11. $\left\{\left(-3 + i\sqrt{7}, 3 + i\sqrt{7}\right), \left(-3 - i\sqrt{7}, 3 - i\sqrt{7}\right)\right\}$ **12.** $\{(2, -5, 3)\}$ **13.** $\{(1, 2), (-5, 14)\}$ **14.** $\emptyset$ **15.** $\{(0, 1, 0)\}$

16. $\left\{\left(\frac{3}{8}, -1\right)\right\}$ **17.** $\{(-1, 0, 0)\}$ **18.** $\{(-56 - 8z, 13 + z, z)\}$ **19.** $\left\{(0, 3), \left(-\frac{36}{17}, -\frac{3}{17}\right)\right\}$ **20.** $\{(3 + 2y, y)\}$

21. $\{(1, 2, 3)\}$ **22.** $\left\{\left(2 + 2i\sqrt{2}, 5 - i\sqrt{2}\right), \left(2 - 2i\sqrt{2}, 5 + i\sqrt{2}\right)\right\}$ **23.** $\{(-3, 5, -6)\}$ **24.** $\emptyset$ **25.** $\{(4, 2), (-1, -3)\}$

26. $\{(2, -3, 1)\}$ **27.** $\left\{\left(\sqrt{13}, i\sqrt{2}\right), \left(-\sqrt{13}, i\sqrt{2}\right), \left(\sqrt{13}, -i\sqrt{2}\right), \left(-\sqrt{13}, -i\sqrt{2}\right)\right\}$ **28.** $\{(-6 - z, 5, z)\}$

29. $\{(1, -3, 4)\}$ **30.** $\left\{\left(-\frac{16}{3}, -\frac{11}{3}\right), (4, 1)\right\}$ **31.** $\{(-1, -2, 5)\}$ **32.** $\emptyset$ **33.** $\{(1, -3), (-3, 1)\}$ **34.** $\{(1, -6, 2)\}$

35. $\{(-6, 9), (-1, 4)\}$ **36.** $\left\{(0, -3), \left(\frac{12}{5}, \frac{9}{5}\right)\right\}$ **37.** $\{(2, 2, 5)\}$ **38.** $\{(-2, -4, 0)\}$ **39.** $\{(2, 3)\}$

9.6 Exercises *(pages 904–908)*

1. 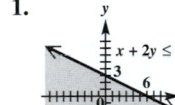   **3.** 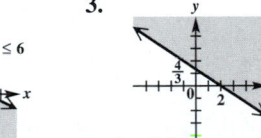   **5.** 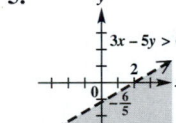   **7.** 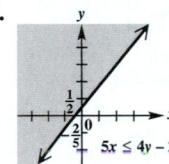   **9.**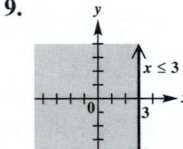

11. 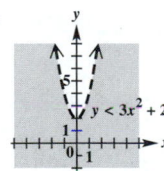   **13.** 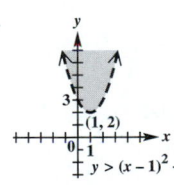   **15.** 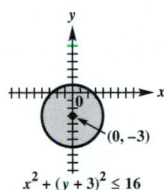   **17.** 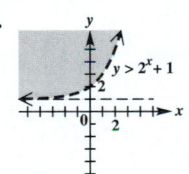   **21.** above   **23.** B

25. C   **27.** A   **29.** 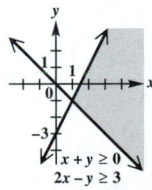   **31.** 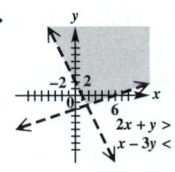   **33.** 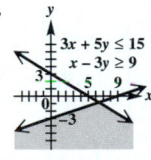   **35.**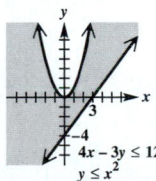

37. 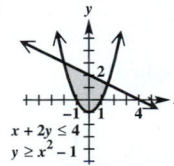   **39.** 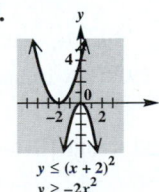   **41.** 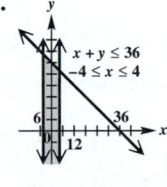   **43.** The solution set is $\emptyset$.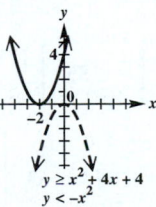

45. 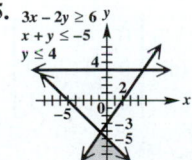   **47.** 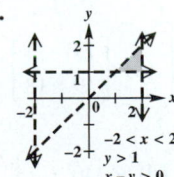   **49.** 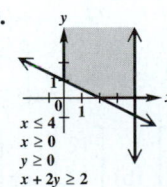   **51.** 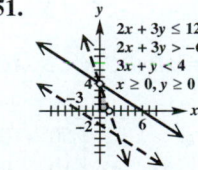   **53.**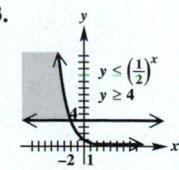

55. 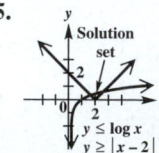   **57.** 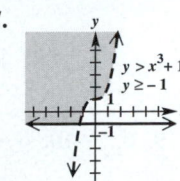   **59.** $x + 2y \le 4$ **61.** $x^2 + y^2 \le 16$ **63.** $x > 0$
$3x - 4y \le 12$ $y \ge 2$ $y > 0$
$x^2 + y^2 \le 4$
$y > \dfrac{3}{2}x - 1$

65. $3x + 2y \ge 6$ **67.** $x + y \ge 2, x + y \le 6$ **69.** maximum of 65 at $(5, 10)$; minimum of 8 at $(1, 1)$

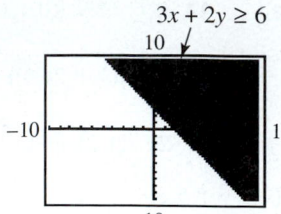   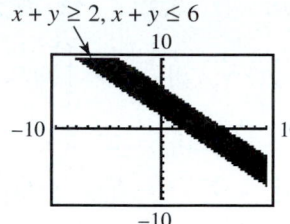

71. maximum of 66 at $(7, 9)$; minimum of 3 at $(1, 0)$ **73.** maximum of 100 at $(1, 10)$; minimum of 0 at $(1, 0)$

75. Let x = number of Brand X pills and y = number of Brand Y pills. Then $3000x + 1000y \ge 6000$, $45x + 50y \ge 195$,

$75x + 200y \ge 600, x \ge 0, y \ge 0$.   **77.** **(a)** 300 cartons of food, 400 cartons of clothes **(b)** 6200 people

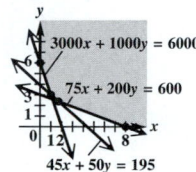

79. 8 of A and 3 of B for 100 ft^3 of storage   **81.** $3\dfrac{3}{4}$ servings of A and $1\dfrac{7}{8}$ servings of B, for a minimum cost of \$1.69

9.7 Exercises *(pages 918–923)*

1. $a = 0, b = 4, c = -3, d = 5$ **3.** $x = -4, y = 14, z = 3, w = -2$ **5.** $w = 2, x = 6, y = -2, z = 8$

7. This cannot be true.   **9.** $z = 18, r = 3, s = 3, p = 3, a = \dfrac{3}{4}$ **11.** size; equal   **13.** 2×2; square   **15.** 3×4

17. 2×1; column   **21.** $\begin{bmatrix} -2 & -5 \\ 17 & 4 \end{bmatrix}$ **23.** $\begin{bmatrix} -2 & -7 & 7 \\ 10 & -2 & 7 \end{bmatrix}$ **25.** They cannot be added.   **27.** $\begin{bmatrix} -6 & 8 \\ 4 & 2 \end{bmatrix}$ **29.** $\begin{bmatrix} 4 \\ -5 \\ 4 \end{bmatrix}$

31. They cannot be subtracted.   **33.** $\begin{bmatrix} -\sqrt{3} & -13 \\ 4 & -2\sqrt{5} \\ -1 & -\sqrt{2} \end{bmatrix}$ **35.** $\begin{bmatrix} 5x + y & x + y & 7x + y \\ 8x + 2y & x + 3y & 3x + y \end{bmatrix}$ **37.** $\begin{bmatrix} -4 & 8 \\ 0 & 6 \end{bmatrix}$

39. $\begin{bmatrix} -9 & 3 \\ 6 & 0 \end{bmatrix}$ **41.** $\begin{bmatrix} 2 & 6 \\ -4 & 6 \end{bmatrix}$ **43.** $\begin{bmatrix} -1 & -3 \\ 2 & -3 \end{bmatrix}$ **45.** yes; 2×5 **47.** no   **49.** yes; 3×2 **51.** $\begin{bmatrix} 13 \\ 25 \end{bmatrix}$

53. $\begin{bmatrix} -17 \\ -1 \end{bmatrix}$ **55.** $\begin{bmatrix} 17\sqrt{2} & -4\sqrt{2} \\ 35\sqrt{3} & 26\sqrt{3} \end{bmatrix}$ **57.** $\begin{bmatrix} 3 + 4\sqrt{3} & -3\sqrt{2} \\ 2\sqrt{15} + 12\sqrt{6} & -2\sqrt{30} \end{bmatrix}$ **59.** They cannot be multiplied.

61. $[2 \quad 7 \quad -4]$ **63.** $\begin{bmatrix} -15 & -16 & 3 \\ -1 & 0 & 9 \\ 7 & 6 & 12 \end{bmatrix}$ **65.** $\begin{bmatrix} 23 & -9 \\ -6 & -2 \\ 33 & 1 \end{bmatrix}$ **67.** $\begin{bmatrix} -25 & 23 & 11 \\ 0 & -6 & -12 \\ -15 & 33 & 45 \end{bmatrix}$ **69.** They cannot be multiplied.

71. $\begin{bmatrix} 10 & -10 \\ 15 & -5 \end{bmatrix}$ **73.** $BA \ne AB, BC \ne CB, AC \ne CA$ **75.** **(a)** $\begin{bmatrix} 38 & -8 \\ -7 & -2 \end{bmatrix}$ **(b)** $\begin{bmatrix} 18 & 24 \\ 19 & 18 \end{bmatrix}$ **77.** **(a)** $\begin{bmatrix} 0 & 1 & -1 \\ 0 & 1 & 0 \\ 0 & 0 & 1 \end{bmatrix}$

(b) $\begin{bmatrix} 0 & 1 & -1 \\ 0 & 1 & 0 \\ 0 & 0 & 1 \end{bmatrix}$ **79.** 1   **81.** **(a)** $\begin{bmatrix} 50 & 100 & 30 \\ 10 & 90 & 50 \\ 60 & 120 & 40 \end{bmatrix}$ **(b)** $\begin{bmatrix} 12 \\ 10 \\ 15 \end{bmatrix}$ **(c)** $\begin{bmatrix} 2050 \\ 1770 \\ 2520 \end{bmatrix}$ **(d)** \$6340

83. Answers will vary a little if intermediate steps are rounded. **(a)** 2940, 2909, 2861, 2814, 2767

(b) The northern spotted owl will become extinct. **(c)** 3023, 3052, 3079, 3107, 3135

9.8 Exercises *(pages 931–935)*

3. yes   **5.** no   **7.** no   **9.** yes   **11.** $\begin{bmatrix} -\frac{1}{5} & -\frac{2}{5} \\ \frac{2}{5} & -\frac{1}{5} \end{bmatrix}$ **13.** $\begin{bmatrix} 2 & 1 \\ -\frac{3}{2} & -\frac{1}{2} \end{bmatrix}$ **15.** The inverse does not exist.

17. $\begin{bmatrix} -1 & 1 & 1 \\ 0 & -1 & 0 \\ 2 & -1 & -1 \end{bmatrix}$ **19.** $\begin{bmatrix} 7 & -3 & -3 \\ -1 & 1 & 0 \\ -1 & 0 & 1 \end{bmatrix}$ **21.** $\begin{bmatrix} -\frac{15}{4} & -\frac{1}{4} & -3 \\ \frac{5}{4} & \frac{1}{4} & 1 \\ -\frac{3}{2} & 0 & -1 \end{bmatrix}$ **23.** $\begin{bmatrix} \frac{1}{2} & 0 & \frac{1}{2} & -1 \\ \frac{1}{10} & -\frac{2}{5} & \frac{3}{10} & -\frac{1}{5} \\ -\frac{7}{10} & \frac{4}{5} & -\frac{11}{10} & \frac{12}{5} \\ \frac{1}{5} & \frac{1}{5} & -\frac{2}{5} & \frac{3}{5} \end{bmatrix}$

25. $\begin{bmatrix} 2 & 9 \\ 1 & 5 \end{bmatrix}$ **27.** $ad - bc$ is the determinant of A.   **28.** $\begin{bmatrix} \frac{d}{|A|} & \frac{-b}{|A|} \\ \frac{-c}{|A|} & \frac{a}{|A|} \end{bmatrix}$ **29.** $A^{-1} = \frac{1}{|A|} \begin{bmatrix} d & -b \\ -c & a \end{bmatrix}$

31. $A^{-1} = \begin{bmatrix} -\frac{3}{2} & 1 \\ \frac{7}{2} & -2 \end{bmatrix}$ **32.** zero   **33.** $\{(2, 3)\}$ **35.** $\{(-2, 4)\}$ **37.** $\{(4, -6)\}$ **39.** $\left\{\left(-\frac{1}{2}, \frac{2}{3}\right)\right\}$

41. $\{(4, -9)\}$ **43.** $\{(3, 1, 2)\}$ **45.** $\{(10, -1, -2)\}$ **47.** $\{(11, -1, 2)\}$ **49.** $\{(1, 0, 2, 1)\}$

51. (a) $602.7 = a + 5.543b + 37.14c$ **(b)** $a \approx -490.547$, $b = -89$, $c = 42.71875$ **(c)** $S = -490.547 - 89A + 42.71875B$
$656.7 = a + 6.933b + 41.30c$
$778.5 = a + 7.638b + 45.62c$

(d) $S \approx 843.5$ **(e)** $S \approx 1547.5$ **53.** Answers will vary.   **55.** $\begin{bmatrix} -.1215875322 & .0491390161 \\ 1.544369078 & -.046799063 \end{bmatrix}$

57. $\begin{bmatrix} 2 & -2 & 0 \\ -4 & 0 & 4 \\ 3 & 3 & -3 \end{bmatrix}$ **59.** $\{(1.68717058, -1.306990242)\}$ **61.** $\{(13.58736702, 3.929011993, -5.342780076)\}$

65. $A = \begin{bmatrix} 1 & 0 \\ 1 & 1 \end{bmatrix}, B = \begin{bmatrix} 1 & 1 \\ 0 & 1 \end{bmatrix}$ (Other answers are possible.)   **67.** $\begin{bmatrix} \frac{1}{a} & 0 & 0 \\ 0 & \frac{1}{b} & 0 \\ 0 & 0 & \frac{1}{c} \end{bmatrix}$ **69.** I_n, $-A^{-1}$, $\frac{1}{k}A^{-1}$

Chapter 9 Review Exercises *(pages 941–946)*

1. $\{(0, 1)\}$ **3.** $\{(9 - 5y, y)\}$; infinitely many solutions   **5.** $\emptyset$; inconsistent system   **7.** $\left\{\left(\frac{1}{3}, \frac{1}{2}\right)\right\}$ **9.** $\{(3, 2, 1)\}$

11. $\{(5, -1, 0)\}$ **13.** One possible answer is $\begin{matrix} x + y = 2 \\ x + y = 3 \end{matrix}$.   **15.** $\frac{1}{3}$ cup of rice, $\frac{1}{5}$ cup of soybeans   **17.** 5 blankets, 3 rugs, 8 skirts   **19.** $x \approx 177.1$, $y \approx 174.9$; If an athlete's maximum heart rate is 180 beats per minute (bpm), then it will be about 177 bpm 5 sec after stopping and 175 bpm 10 sec after stopping.   **21.** $Y_1 = 2.4X^2 - 6.2X + 1.5$

23. $\{(x, 1 - 2x, 6 - 11x)\}$ **25.** $\{(-2, 0)\}$ **27.** $\{(0, 1, 0)\}$ **29.** 10 lb of \$4.60 tea, 8 lb of \$5.75 tea, 2 lb of \$6.50 tea

31. 1979; approximately 10 million   **33.** -25 **35.** -1 **37.** $\left\{\frac{8}{19}\right\}$ **39.** $\{(-4, 2)\}$ **41.** $\emptyset$; inconsistent system

43. $\{(14, -15, 35)\}$ **45.** $\frac{2}{x - 1} - \frac{6}{3x - 2}$ **47.** $\frac{1}{x - 1} - \frac{x + 3}{x^2 + 2}$ **49.** $\{(-3, 4), (1, 12)\}$

51. $\{(-4, 1), (-4, -1), (4, -1), (4, 1)\}$ **53.** $\left\{(5, -2), \left(-4, \frac{5}{2}\right)\right\}$ **55.** $\{(-2, 0), (1, 1)\}$ **57.** $b = \pm 5\sqrt{10}$

59.

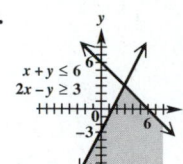

61. maximum of 24 at $(0, 6)$ **63.** 3 units of food A and 4 units of food B; Minimum cost is \$1.02 per serving.   **65.** $a = 5, x = \dfrac{3}{2}, y = 0, z = 9$ **67.** $\begin{bmatrix} -4 \\ 6 \\ 1 \end{bmatrix}$ **69.** They cannot be subtracted.

71. $\begin{bmatrix} 3 & -4 \\ 4 & 48 \end{bmatrix}$ **73.** $\begin{bmatrix} -9 & 3 \\ 10 & 6 \end{bmatrix}$ **75.** $\begin{bmatrix} -2 & 5 & -3 \\ 3 & 4 & -4 \\ 6 & -1 & -2 \end{bmatrix}$ **77.** $\begin{bmatrix} 3 & -1 \\ -5 & 2 \end{bmatrix}$ **79.** $\begin{bmatrix} \frac{2}{3} & 0 & -\frac{1}{3} \\ \frac{1}{3} & 0 & -\frac{2}{3} \\ -\frac{2}{3} & 1 & \frac{1}{3} \end{bmatrix}$ **81.** $\{(-3, 2, 0)\}$

Chapter 9 Test *(pages 946–948)*

[9.1] 1. $\{(4, 3)\}$ **2.** $\left\{\left(\dfrac{-3y - 7}{2}, y\right)\right\}$; infinitely many solutions   **3.** $\{(1, 2)\}$ **4.** $\emptyset$; inconsistent system   **5.** $\{(2, 0, -1)\}$

[9.2] 6. $\{(5, 1)\}$ **7.** $\{(5, 3, 6)\}$ **[9.1] 8.** $y = 2x^2 - 8x + 11$ **9.** 22 units from Toronto, 56 units from Montreal, 22 units from Ottawa   **[9.3] 10.** -58 **11.** -844 **12.** $\{(-6, 7)\}$ **13.** $\{(1, -2, 3)\}$ **[9.4] 14.** $\dfrac{2}{x} + \dfrac{-2}{x + 1} + \dfrac{-1}{(x + 1)^2}$

[9.5] 15. $\begin{array}{l} y = x^2 - 4x + 4 \\ x - 3y = -2 \end{array}$ **16.** $\{(1, 2), (-1, 2), (1, -2), (-1, -2)\}$ **17.** $\{(3, 4), (4, 3)\}$ **18.** 5 and -6

[9.6] 19.

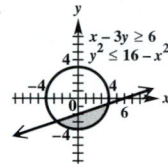

20. maximum of 42 at $(12, 6)$ **21.** 0 VIP rings and 24 SST rings; Maximum profit is \$960.

[9.7] 22. $x = -1; y = 7; w = -3$ **23.** $\begin{bmatrix} 8 & 3 \\ 0 & -11 \\ 15 & 19 \end{bmatrix}$ **24.** The matrices cannot be added.

25. $\begin{bmatrix} -5 & 16 \\ 19 & 2 \end{bmatrix}$ **26.** The matrices cannot be multiplied.   **27.** A   **[9.8] 28.** $\begin{bmatrix} -2 & -5 \\ -3 & -8 \end{bmatrix}$

29. The inverse does not exist.   **30.** $\begin{bmatrix} -9 & 1 & -4 \\ -2 & 1 & 0 \\ 4 & -1 & 1 \end{bmatrix}$ **31.** $\{(-7, 8)\}$ **32.** $\{(0, 5, -9)\}$

CHAPTER 10 ANALYTIC GEOMETRY

10.1 Exercises *(pages 957–960)*

1. (a) D **(b)** B **(c)** C **(d)** A **(e)** F **(f)** H **(g)** E **(h)** G

In Exercises 3–17, we give the domain first, then the range.

3. $(-\infty, 0]; (-\infty, \infty)$ **5.** $[0, \infty); (-\infty, \infty)$ **7.** $[2, \infty); (-\infty, \infty)$ **9.** $(-\infty, 2]; (-\infty, \infty)$ **11.** $[4, \infty); (-\infty, \infty)$

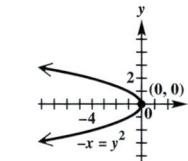 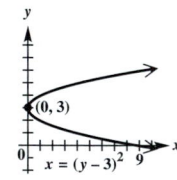 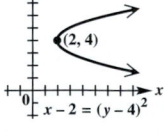 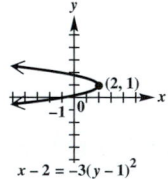 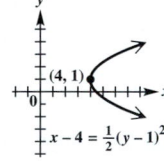

13. $[-2, \infty); (-\infty, \infty)$ **15.** $(-\infty, 4]; (-\infty, \infty)$ **17.** $[1, \infty); (-\infty, \infty)$ **19.** $(0, 6), y = -6, y\text{-axis}$

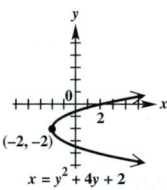 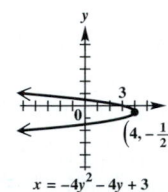 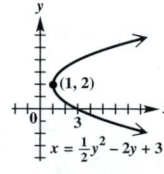

21. $\left(0, -\dfrac{1}{16}\right), y = \dfrac{1}{16}, y\text{-axis}$ **23.** $(-1, 0), x = 1, x\text{-axis}$ **25.** $\left(-\dfrac{1}{128}, 0\right), x = \dfrac{1}{128}, x\text{-axis}$ **27.** $(4, 3), x = -2, y = 3$

29. $(7, -1)$, $y = -9$, $x = 7$ **31.** $y^2 = 20x$ **33.** $x^2 = y$ **35.** $x^2 = y$ **37.** $y^2 = \dfrac{4}{3}x$ **39.** $(x - 4)^2 = 8(y - 3)$

41. $(y - 6)^2 = 28(x + 5)$ **43.**

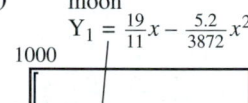
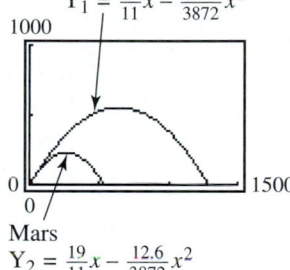

45.

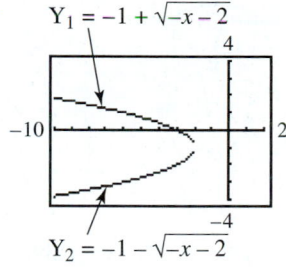

47. $a + b + c = -5$; $4a - 2b + c = -14$; $4a + 2b + c = -10$ **48.** $\{(-2, 1, -4)\}$ **49.** to the left, because $a = -2$,

and $-2 < 0$ **50.** $x = -2y^2 + y - 4$ **51.** **(a)** $y = \dfrac{11}{5625}x^2$ **(b)** 127.8 ft   **53.** 6 ft   **55.** **(a)** 2000 ft

(b) $y = 1000 - .00025x^2$ **(c)** no   **57.** **(a)**

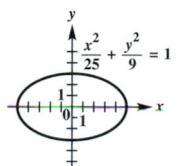

(b) Mars: approximately 229 ft; moon: approximately

555 ft

10.2 Exercises *(pages 968–970)*

1. **(a)** A **(b)** C **(c)** D **(d)** B   **3.** $[-5, 5]$; $[-3, 3]$; $(0, 0)$;   **5.** $[-3, 3]$; $[-1, 1]$; $(0, 0)$,   **7.** $[-3, 3]$; $[-9, 9]$; $(0, 0)$;

$(-5, 0)$, $(5, 0)$; $(0, -3)$, $(-3, 0)$, $(3, 0)$; $(0, -1)$, $(0, -9)$, $(0, 9)$; $(-3, 0)$,

$(0, 3)$; $(-4, 0)$, $(4, 0)$ $(0, 1)$; $\left(-2\sqrt{2}, 0\right)$, $\left(2\sqrt{2}, 0\right)$ $(3, 0)$; $\left(0, -6\sqrt{2}\right)$, $\left(0, 6\sqrt{2}\right)$

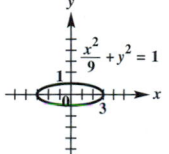
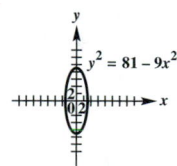

9. $[-5, 5]$; $[-2, 2]$; $(0, 0)$;   **11.** $[-3, 7]$; $[-1, 3]$; $(2, 1)$;   **13.** $[-7, 1]$; $[-4, 8]$; $(-3, 2)$;

$(-5, 0)$, $(5, 0)$; $(0, -2)$, $(-3, 1)$, $(7, 1)$; $(2, -1)$, $(2, 3)$; $(-3, -4)$, $(-3, 8)$; $(-7, 2)$, $(1, 2)$;

$(0, 2)$; $\left(-\sqrt{21}, 0\right)$, $\left(\sqrt{21}, 0\right)$ $\left(2 - \sqrt{21}, 1\right)$, $\left(2 + \sqrt{21}, 1\right)$ $\left(-3, 2 - 2\sqrt{5}\right)$, $\left(-3, 2 + 2\sqrt{5}\right)$

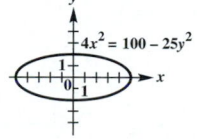
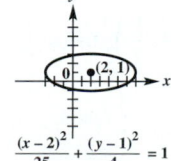
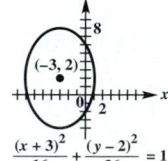

15. $\dfrac{x^2}{25} + \dfrac{y^2}{16} = 1$ **17.** $\dfrac{x^2}{5} + \dfrac{y^2}{9} = 1$ **19.** $\dfrac{(x - 3)^2}{25} + \dfrac{(y - 1)^2}{16} = 1$ **21.** $\dfrac{(x - 4)^2}{9} + \dfrac{(y - 5)^2}{16} = 1$

23. $\dfrac{x^2}{72} + \dfrac{y^2}{81} = 1$ **25.** $\dfrac{x^2}{9} + \dfrac{y^2}{25} = 1$ **27.** $\dfrac{9x^2}{28} + \dfrac{9y^2}{64} = 1$

29. $[-5, 5]$; $[0, 2]$; function

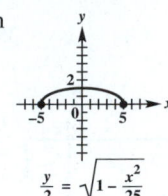

$$\frac{y}{2} = \sqrt{1 - \frac{x^2}{25}}$$

31. $[-1, 0]$; $[-8, 8]$

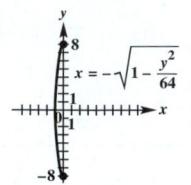

$$x = -\sqrt{1 - \frac{y^2}{64}}$$

33.

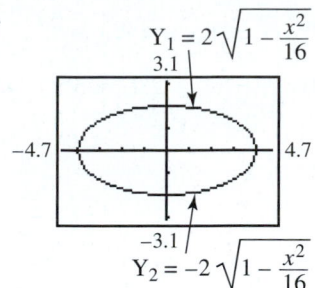

$$Y_1 = 2\sqrt{1 - \frac{x^2}{16}}$$

$$Y_2 = -2\sqrt{1 - \frac{x^2}{16}}$$

35.

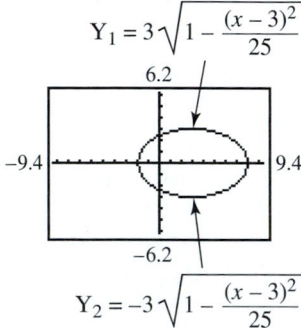

$$Y_1 = 3\sqrt{1 - \frac{(x-3)^2}{25}}$$

$$Y_2 = -3\sqrt{1 - \frac{(x-3)^2}{25}}$$

37. $\dfrac{1}{2}$ **39.** .65   **43.** 12 ft tall   **45.** approximately 55 million mi

47. (a) Neptune: $\dfrac{(x - .2709)^2}{30.1^2} + \dfrac{y^2}{30.1^2} = 1$;   Pluto: $\dfrac{(x - 9.8106)^2}{39.4^2} + \dfrac{y^2}{38.16^2} = 1$

(b)

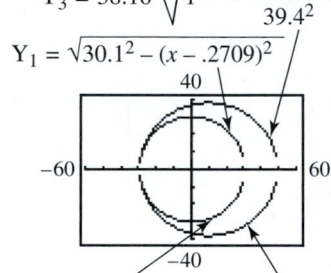

$$Y_3 = 38.16\sqrt{1 - \frac{(x - 9.8106)^2}{39.4^2}}$$

$$Y_1 = \sqrt{30.1^2 - (x - .2709)^2}$$

$$Y_2 = -Y_1 \qquad Y_4 = -Y_3$$

49. $3\sqrt{3}$ units

Chapter 10 Quiz *(page 971)*

[10.1, 10.2] 1. (a) B, D **(b)** A **(c)** E **(d)** C, E **(e)** D   **[10.1] 2.** $(y - 2)^2 = 12(x + 1)$ **3.** $y = -\dfrac{1}{2}x^2$

[10.2] 4. $\dfrac{(x - 3)^2}{16} + \dfrac{(y + 2)^2}{25} = 1$ **5.** $\dfrac{(x + 3)^2}{9} + \dfrac{(y - 7)^2}{25} = 1$

[10.1, 10.2] 6. vertex: $(-3, -4)$; focus: $\left(-3, -\dfrac{15}{4}\right)$;

directrix: $y = -\dfrac{17}{4}$; axis: $x = -3$

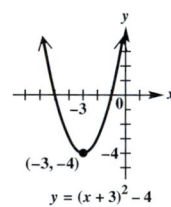

$$y = (x + 3)^2 - 4$$

7. center: $(0, 0)$;

vertices: $(-3, 0)$, $(3, 0)$;

foci: $(-\sqrt{5}, 0)$, $(\sqrt{5}, 0)$

$$4x^2 + 9y^2 = 36$$

8. vertex: $(-1, -3)$; focus: $(1, -3)$;

directrix: $x = -3$; axis: $y = -3$

$$8(x + 1) = (y + 3)^2$$

9. center: $(-3, -2)$; vertices: $(-3, 4)$, $(-3, -8)$;

foci: $(-3, -2 + \sqrt{11})$, $(-3, -2 - \sqrt{11})$

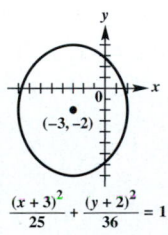

$$\frac{(x + 3)^2}{25} + \frac{(y + 2)^2}{36} = 1$$

10. vertex: $\left(-2, -\dfrac{1}{2}\right)$; focus: $\left(-\dfrac{33}{16}, -\dfrac{1}{2}\right)$;

directrix: $x = -\dfrac{31}{16}$; axis: $y = -\dfrac{1}{2}$

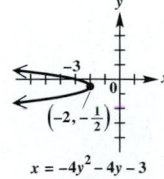

$$x = -4y^2 - 4y - 3$$

10.3 Exercises *(pages 977–980)*

1. C   **3.** D   **5.** $(-\infty, -4] \cup [4, \infty)$; $(-\infty, \infty)$; $(0,0)$; $(-4,0)$, $(4,0)$; $(-5,0)$, $(5,0)$; $y = \pm \dfrac{3}{4}x$

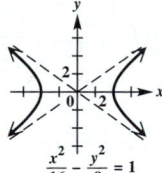

$$\dfrac{x^2}{16} - \dfrac{y^2}{9} = 1$$

7. $(-\infty, \infty)$; $(-\infty, -5] \cup [5, \infty)$; $(0,0)$; $(0,-5)$, $(0,5)$; $\left(0, -\sqrt{74}\right)$, $\left(0, \sqrt{74}\right)$; $y = \pm \dfrac{5}{7}x$

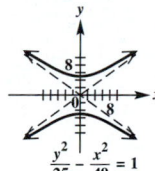

$$\dfrac{y^2}{25} - \dfrac{x^2}{49} = 1$$

9. $(-\infty, -3] \cup [3, \infty)$; $(-\infty, \infty)$; $(0,0)$; $(-3,0)$, $(3,0)$; $\left(-3\sqrt{2}, 0\right)$, $\left(3\sqrt{2}, 0\right)$; $y = \pm x$

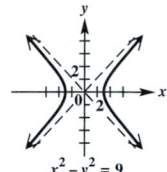

$$x^2 - y^2 = 9$$

11. $(-\infty, -5] \cup [5, \infty)$; $(-\infty, \infty)$; $(0,0)$; $(-5,0)$, $(5,0)$; $\left(-\sqrt{34}, 0\right)$, $\left(\sqrt{34}, 0\right)$; $y = \pm \dfrac{3}{5}x$

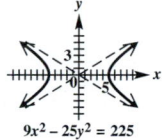

$$9x^2 - 25y^2 = 225$$

13. $(-\infty, \infty)$; $(-\infty, -5] \cup [5, \infty)$; $(0,0)$; $(0,-5)$, $(0,5)$; $\left(0, -\sqrt{29}\right)$, $\left(0, \sqrt{29}\right)$; $y = \pm \dfrac{5}{2}x$

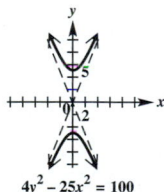

$$4y^2 - 25x^2 = 100$$

15. $\left(-\infty, -\dfrac{1}{3}\right] \cup \left[\dfrac{1}{3}, \infty\right)$; $(-\infty, \infty)$; $(0,0)$; $\left(-\dfrac{1}{3}, 0\right)$, $\left(\dfrac{1}{3}, 0\right)$; $\left(-\dfrac{\sqrt{13}}{6}, 0\right)$, $\left(\dfrac{\sqrt{13}}{6}, 0\right)$; $y = \pm \dfrac{3}{2}x$

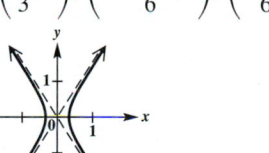

$$9x^2 - 4y^2 = 1$$

17. $(-\infty, \infty)$; $(-\infty, 1] \cup [13, \infty)$; $(4,7)$; $(4,1)$, $(4,13)$; $(4,-3)$, $(4,17)$; $y = 7 \pm \dfrac{3}{4}(x - 4)$

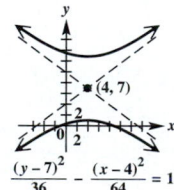

$$\dfrac{(y-7)^2}{36} - \dfrac{(x-4)^2}{64} = 1$$

19. $(-\infty, -7] \cup [1, \infty)$; $(-\infty, \infty)$; $(-3,2)$; $(-7,2)$, $(1,2)$; $(-8,2)$, $(2,2)$; $y = 2 \pm \dfrac{3}{4}(x + 3)$

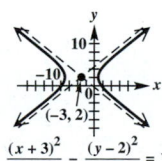

$$\dfrac{(x+3)^2}{16} - \dfrac{(y-2)^2}{9} = 1$$

21. $\left(-\infty, -\dfrac{21}{4}\right] \cup \left[-\dfrac{19}{4}, \infty\right)$; $(-\infty, \infty)$; $(-5,3)$; $\left(-\dfrac{21}{4}, 3\right)$, $\left(-\dfrac{19}{4}, 3\right)$; $\left(-5 - \dfrac{\sqrt{17}}{4}, 3\right)$, $\left(-5 + \dfrac{\sqrt{17}}{4}, 3\right)$; $y = 3 \pm 4(x + 5)$

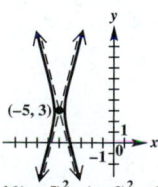

$$16(x+5)^2 - (y-3)^2 = 1$$

23. $(-\infty, \infty)$; $[3, \infty)$; function   **25.** $\left(-\infty, -\dfrac{1}{5}\right]$; $(-\infty, \infty)$ **27.** 1.4   **29.** 1.7   **31.** $\dfrac{x^2}{9} - \dfrac{y^2}{16} = 1$ **33.** $\dfrac{y^2}{36} - \dfrac{x^2}{144} = 1$

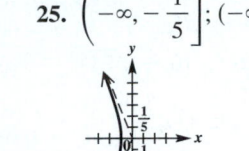

35. $\dfrac{x^2}{9} - 3y^2 = 1$ **37.** $\dfrac{2y^2}{25} - 2x^2 = 1$

39. $\dfrac{(y-3)^2}{4} - \dfrac{49(x-4)^2}{4} = 1$ **41.** $\dfrac{(x-1)^2}{4} - \dfrac{(y+2)^2}{5} = 1$

43. $\dfrac{y^2}{49} - \dfrac{x^2}{392} = 1$ **45.** $\dfrac{25(x+1)^2}{324} - \dfrac{25(y+1)^2}{2176} = 1$

47. 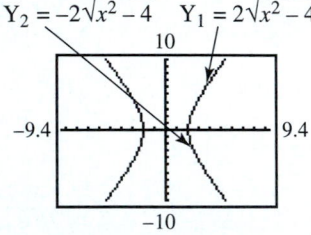   **49.** 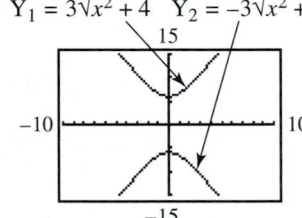   **51.** $y = \dfrac{1}{2}\sqrt{x^2 - 4}$ **52.** $y = \dfrac{1}{2}x$

53. $y \approx 24.98$ **54.** $y = 25$ **55.** Because $24.98 < 25$, the graph of $y = \dfrac{1}{2}\sqrt{x^2 - 4}$ lies below the graph of $y = \dfrac{1}{2}x$ when $x = 50$.   **56.** The y-values on the hyperbola will approach the y-values on the asymptote.   **57. (a)** $x = \sqrt{y^2 + 2.5 \times 10^{-27}}$ **(b)** approximately 1.2×10^{-13} m

10.4 Exercises *(pages 985–986)*

1. circle   **3.** parabola   **5.** parabola   **7.** ellipse   **9.** hyperbola   **11.** hyperbola   **13.** ellipse   **15.** circle

17. parabola   **19.** point   **21.** parabola   **23.** point   **25.** no graph

27. hyperbola   **29.** parabola   **31.** parabola   **33.** circle

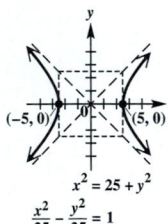   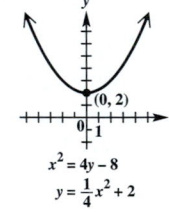   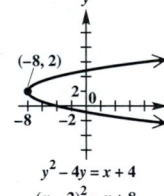   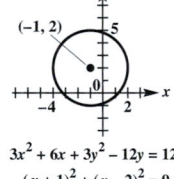

35. ellipse   **37.** ellipse   **39.** hyperbola   **41.** $\dfrac{1}{3}$ **43.** 1   **45.** 1.5   **47.** elliptical

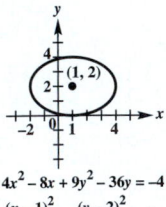

Chapter 10 Review Exercises *(pages 989–991)*

1. $[2, \infty)$; $(-\infty, \infty)$; $(2, 5)$;
$y = 5$

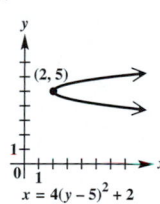

3. $\left[\dfrac{7}{4}, \infty\right)$; $(-\infty, \infty)$; $\left(\dfrac{7}{4}, \dfrac{1}{2}\right)$;
$y = \dfrac{1}{2}$

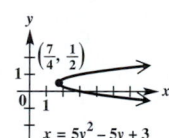

5. $(-\infty, 0]$; $(-\infty, \infty)$; $\left(-\dfrac{1}{6}, 0\right)$;
$x = \dfrac{1}{6}$; x-axis

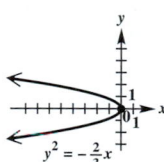

7. $(-\infty, \infty)$; $[0, \infty)$; $\left(0, \dfrac{1}{12}\right)$;
$y = -\dfrac{1}{12}$; y-axis

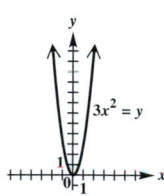

9. $y^2 = 16x$ **11.** $x^2 = \dfrac{9}{4}y$ **13.** ellipse   **15.** hyperbola   **17.** parabola

19. ellipse   **21.** F   **23.** A   **25.** B

27. ellipse; $[-2, 2]$; $[-3, 3]$;
$(0, -3)$, $(0, 3)$

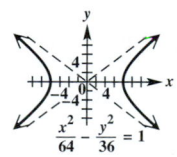

29. hyperbola; $(-\infty, -8] \cup [8, \infty)$;
$(-\infty, \infty)$; $(-8, 0)$, $(8, 0)$; $y = \pm \dfrac{3}{4}x$

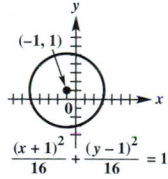

31. circle; $[-5, 3]$; $[-3, 5]$

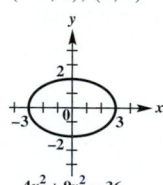

33. ellipse; $[-3, 3]$; $[-2, 2]$;
$(-3, 0)$, $(3, 0)$

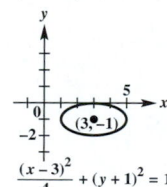

35. ellipse; $[1, 5]$; $[-2, 0]$;
$(1, -1)$, $(5, -1)$

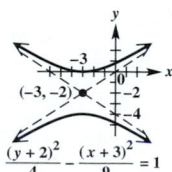

37. hyperbola; $(-\infty, \infty)$;
$(-\infty, -4] \cup [0, \infty)$; $(-3, -4)$,
$(-3, 0)$; $y = -2 \pm \dfrac{2}{3}(x + 3)$

39. circle; $[1, 3]$; $[-4, -2]$

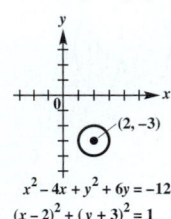

$x^2 - 4x + y^2 + 6y = -12$
$(x - 2)^2 + (y + 3)^2 = 1$

41. ellipse; $[-2 - \sqrt{2}, -2 + \sqrt{2}]$;
$[2 - \sqrt{5}, 2 + \sqrt{5}]$; $(-2, 2 - \sqrt{5})$,
$(-2, 2 + \sqrt{5})$

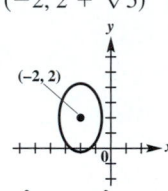

$5x^2 + 20x + 2y^2 - 8y = -18$
$\frac{(x + 2)^2}{2} + \frac{(y - 2)^2}{5} = 1$

43. $[-3, 0]$; $[-4, 4]$

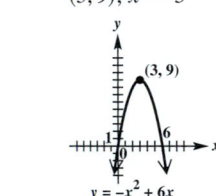

$\frac{x}{3} = -\sqrt{1 - \frac{y^2}{16}}$

45. $(-\infty, \infty)$; $(-\infty, -1]$; function

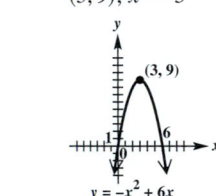

$y = -\sqrt{1 + x^2}$

47. $\frac{x^2}{12} + \frac{y^2}{16} = 1$

49. $\frac{y^2}{16} - \frac{x^2}{9} = 1$

51. $(y - 2)^2 = 12x$

53. $\frac{x^2}{25} + \frac{y^2}{21} = 1$

55. $\frac{x^2}{9} - \frac{y^2}{16} = 1$

57. $\frac{(x - 2)^2}{16} + \frac{y^2}{12} = 1$

59. C, A, B, D

61. $\frac{x^2}{6,111,883} + \frac{y^2}{432,135} = 1$

Chapter 10 Test *(pages 991–992)*

[10.1] 1. $(-\infty, \infty)$; $(-\infty, 9]$;
$(3, 9)$; $x = 3$

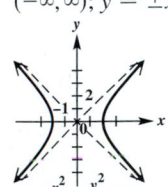

$y = -x^2 + 6x$

2. $[-4, \infty)$; $(-\infty, \infty)$;
$(-4, -1)$; $y = -1$

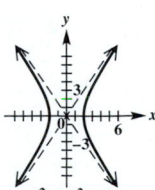

$x = 4y^2 + 8y$

3. $\left(\frac{1}{32}, 0\right)$; $x = -\frac{1}{32}$

4. $x - 2 = -5(y - 3)^2$

[10.2] 6. $[-2, 18]$; $[-2, 12]$

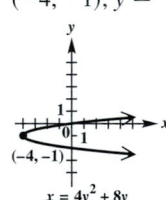

$\frac{(x - 8)^2}{100} + \frac{(y - 5)^2}{49} = 1$

7. $[-2, 2]$; $[-4, 4]$

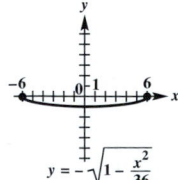

$16x^2 + 4y^2 = 64$

8. It is the graph of a function.

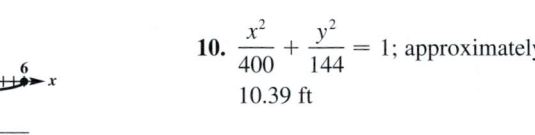

$y = -\sqrt{1 - \frac{x^2}{36}}$

9. $\frac{x^2}{9} + \frac{y^2}{4} = 1$

10. $\frac{x^2}{400} + \frac{y^2}{144} = 1$; approximately
10.39 ft

[10.3] 11. $(-\infty, -2] \cup [2, \infty)$;
$(-\infty, \infty)$; $y = \pm x$

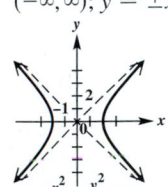

$\frac{x^2}{4} - \frac{y^2}{4} = 1$

12. $(-\infty, -2] \cup [2, \infty)$;
$(-\infty, \infty)$; $y = \pm\frac{3}{2}x$

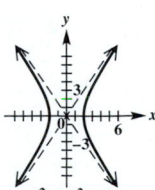

$9x^2 - 4y^2 = 36$

13. $\frac{y^2}{25} - \frac{x^2}{11} = 1$

[10.4] 14. circle

15. hyperbola

16. ellipse

17. parabola

18. point

19. no graph

[10.3] 20. $Y_1 = 7\sqrt{\frac{x^2}{25} - 1}$, $Y_2 = -7\sqrt{\frac{x^2}{25} - 1}$

CHAPTER 11   FURTHER TOPICS IN ALGEBRA

11.1 Exercises *(pages 1003–1007)*

1. 14, 18, 22, 26, 30   **3.** $\dfrac{6}{5}, \dfrac{7}{6}, \dfrac{8}{7}, \dfrac{9}{8}, \dfrac{10}{9}$ **5.** $0, \dfrac{1}{9}, \dfrac{2}{27}, \dfrac{1}{27}, \dfrac{4}{243}$ **7.** $-2, 4, -6, 8, -10$ **9.** $1, \dfrac{7}{6}, 1, \dfrac{5}{6}, \dfrac{19}{27}$

11. 3, 4, 7, 12, 19   **15.** finite   **17.** finite   **19.** infinite   **21.** finite   **23.** $-2, 1, 4, 7$ **25.** 1, 1, 2, 3

27. 2, 4, 12, 48   **29.** 35   **31.** $\dfrac{25}{12}$ **33.** 288   **35.** 3   **37.** -18 **39.** $\dfrac{728}{9}$ **41.** 28   **43.** 343   **45.** 30

47. 1490   **49.** -154 **51.** $-2 + (-1) + 0 + 1 + 2; 0$ **53.** $-1 + 1 + 3 + 5 + 7; 15$ **55.** $-10 - 4 + 0; -14$

57. $0 + \dfrac{1}{2} + \dfrac{2}{3} + \dfrac{3}{4}; \dfrac{23}{12}$ **59.** $124 + 111 + 100 + 91; 426$ **61.** $-3.5 + .5 + 4.5 + 8.5; 10$

63. $0 + 4 + 16 + 36; 56$ **65.** $-1 - \dfrac{1}{3} - \dfrac{1}{5} - \dfrac{1}{7}; -\dfrac{176}{105}$ **67.** 600   **69.** 1240   **71.** 90   **73.** 220   **75.** 304

There are other acceptable forms of the answers in Exercises 77 and 79.

77. $\displaystyle\sum_{i=1}^{9} \dfrac{1}{3i}$ **79.** $\displaystyle\sum_{k=1}^{8} \left(-\dfrac{1}{2}\right)^{k-1}$ **81.** converges to $\dfrac{1}{2}$ **83.** diverges   **85.** converges to $e \approx 2.71828$

87. (a) $a_1 = 8$ thousand per acre, $a_2 = 10.4$ thousand per acre, $a_3 = 8.528$ thousand per acre **(b)** The population density oscillates above and below 9.5 thousand per acre (approximately).

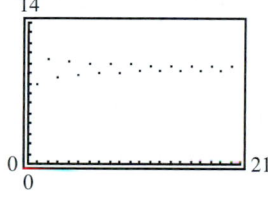

89. (a) $N_{j+1} = 2N_j$ for $j \geq 1$ **(b)** 1840 **(c)**

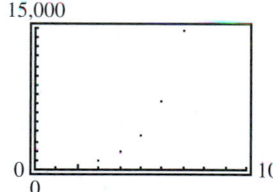

91. (a) .0198026273; $\ln 1.02 \approx .0198026273$ **(b)** $-.0304592075$; $\ln .97 \approx -.0304592075$

93. (a) 2.718254; $e \approx 2.718282$ **(b)** .367857; $e^{-1} \approx .367879$

11.2 Exercises *(pages 1014–1016)*

1. 3   **3.** -5 **5.** $x + 2y$ **7.** 8, 14, 20, 26, 32   **9.** 5, 3, 1, -1, -3 **11.** 14, 12, 10, 8, 6

13. $a_8 = 19; a_n = 3 + 2n$ **15.** $a_8 = \dfrac{85}{3}; a_n = \dfrac{5}{3} + \dfrac{10}{3}n$ **17.** $a_8 = -3; a_n = -39 + \dfrac{9}{2}n$

19. $a_8 = x + 21; a_n = x + 3n - 3$ **21.** $a_8 = s + 14p; a_n = s + 2pn - 2p$ **23.** 3   **25.** 5

In Exercises 27–31, D is the domain and R is the range.

27. $a_n = n - 3; D: \{1, 2, 3, 4, 5, 6\}; R: \{-2, -1, 0, 1, 2, 3\}$ **29.** $a_n = 3 - \dfrac{1}{2}n; D: \{1, 2, 3, 4, 5, 6\}; R: \{0, .5, 1, 1.5, 2, 2.5\}$

31. $a_n = 30 - 20n; D: \{1, 2, 3, 4, 5\}; R: \{-70, -50, -30, -10, 10\}$ **33.** 215   **35.** 230   **37.** 160   **39.** 77.5   **41.** 55π

43. 3240   **45.** 2500   **47.** 3660   **49.** $a_1 = 7, d = 5$ **51.** $a_1 = 1, d = -\dfrac{20}{11}$ **53.** 18   **55.** 140   **57.** -621

59. 500,500   **61.** -5050 **62.** $f(1) = m + b; f(2) = 2m + b; f(3) = 3m + b$ **63.** yes; m **64.** $a_n = mn + b$

65. 328.3   **67.** 172.884   **69.** 1281   **71.** 4680   **73.** 54,800   **75.** 713 in.   **77.** yes

11.3 Exercises *(pages 1023–1027)*

1. (a) $5.12 **(b)** $10.23   **3. (a)** $163.84 **(b)** $327.67

In Exercises 5–15, there may be other ways to express a_n.

5. $a_5 = 80; a_n = 5(-2)^{n-1}$ **7.** $a_5 = -108; a_n = -\dfrac{4}{3}(3)^{n-1}$ **9.** $a_5 = -729; a_n = -9(-3)^{n-1}$

11. $a_5 = -324; a_n = -4(3)^{n-1}$ **13.** $a_5 = \dfrac{125}{4}; a_n = \dfrac{4}{5}\left(\dfrac{5}{2}\right)^{n-1}$ **15.** $a_5 = \dfrac{5}{8}; a_n = 10\left(-\dfrac{1}{2}\right)^{n-1}$ **17.** $-3; 2$

19. $5000; \pm.1$ **21.** 682 **23.** $\dfrac{99}{8}$ **25.** 860.95 **27.** 363 **29.** $\dfrac{189}{4}$ **31.** 2032 **33.** The sum exists if $|r| < 1$.

35. 2; does not converge   **37.** $\dfrac{1}{2}$ **41.** 27 **43.** $\dfrac{3}{20}$ **45.** 4 **47.** $\dfrac{3}{7}$ **49.** $g(1) = ab; g(2) = ab^2; g(3) = ab^3$

50. yes; The common ratio is b.   **51.** $a_n = ab^n$ **53.** 97.739 **55.** $.212$ **57. (a)** $a_1 = 1169; r = .916$

(b) $a_{10} = 531; a_{20} = 221$; This means that a person who is 10 yr from retirement should have savings of 531% of his or her

annual salary; a person 20 yr from retirement should have savings of 221% of his or her annual salary.

59. (a) $a_n = a_1 \cdot 2^{n-1}$ **(b)** 15 (rounded from 14.28) **(c)** 560 min or 9 hr, 20 min   **61.** about 488   **63.** $\dfrac{10{,}000}{9}$ units

65. $62; 2046$ **67.** $\dfrac{1}{64}$ m   **69.** Option 2 pays better.   **71.** $10,159.11   **73.** $25,423.18   **75.** $28,107.41

77. $72,918.53   **81.** yes

Summary Exercises on Sequences and Series *(pages 1027–1028)*

1. geometric; $r = 2$ **2.** arithmetic; $d = 3$ **3.** arithmetic; $d = -\dfrac{5}{2}$ **4.** neither   **5.** geometric; $r = \dfrac{4}{3}$

6. geometric; $r = -3$ **7.** neither   **8.** arithmetic; $d = -3$ **9.** neither   **10.** $d = 0; r = 1$ **11.** geometric; $3(2)^{n-1}; 3069$

12. arithmetic; $-2 + 4n; 200$ **13.** arithmetic; $\dfrac{11}{2} - \dfrac{3}{2}n; -\dfrac{55}{2}$ **14.** geometric; $\dfrac{3}{2}\left(\dfrac{2}{3}\right)^{n-1}$ or $\left(\dfrac{2}{3}\right)^{n-2}; \dfrac{58{,}025}{13{,}122}$

15. geometric; $3(-2)^{n-1}; -1023$ **16.** arithmetic; $-2 - 3n; -185$ **17.** diverges   **18.** $\dfrac{1111}{500}$ **19.** -1850

20. 1092 **21.** $-\dfrac{4}{3}$ **22.** diverges   **23.** 144 **24.** $\dfrac{1}{2}$ **25.** diverges   **26.** $.9 + .09 + .009 + \cdots$; The sum is 1.

11.4 Exercises *(pages 1034–1035)*

1. 20 **3.** 35 **5.** 56 **7.** 45 **9.** 1 **11.** n **13.** 56 **15.** 4950 **17.** 1 **19.** 12 **21.** $16x^4; 81y^4$

23. $x^6 + 6x^5y + 15x^4y^2 + 20x^3y^3 + 15x^2y^4 + 6xy^5 + y^6$ **25.** $p^5 - 5p^4q + 10p^3q^2 - 10p^2q^3 + 5pq^4 - q^5$

27. $r^{10} + 5r^8s + 10r^6s^2 + 10r^4s^3 + 5r^2s^4 + s^5$ **29.** $p^4 + 8p^3q + 24p^2q^2 + 32pq^3 + 16q^4$

31. $2401p^4 + 2744p^3q + 1176p^2q^2 + 224pq^3 + 16q^4$

33. $729x^6 - 2916x^5y + 4860x^4y^2 - 4320x^3y^3 + 2160x^2y^4 - 576xy^5 + 64y^6$

35. $\dfrac{m^6}{64} - \dfrac{3m^5}{16} + \dfrac{15m^4}{16} - \dfrac{5m^3}{2} + \dfrac{15m^2}{4} - 3m + 1$ **37.** $4r^4 + \dfrac{8\sqrt{2}r^3}{m} + \dfrac{12r^2}{m^2} + \dfrac{4\sqrt{2}r}{m^3} + \dfrac{1}{m^4}$ **39.** $\dfrac{1}{x^{16}} + \dfrac{4}{x^8} + 6 + 4x^8 + x^{16}$

41. $-3584h^3j^5$ **43.** $319{,}770a^{16}b^{14}$ **45.** $38{,}760x^6y^{42}$ **47.** $90{,}720x^{28}y^{12}$ **49.** 11 **51.** exact: $3{,}628{,}800$;

approximate: $3{,}598{,}695.619$ **52.** about $.830\%$ **53.** exact: $479{,}001{,}600$; approximate: $475{,}687{,}486.5$; about $.692\%$

54. exact: $6{,}227{,}020{,}800$; approximate: $6{,}187{,}239{,}475$; about $.639\%$; As n gets larger, the percent error decreases.

55. $.942$ **57.** 1.015

11.5 Exercises *(pages 1041–1042)*

1. S_1: $1 = 1^2$; S_2: $1 + 3 = 2^2$; S_3: $1 + 3 + 5 = 3^2$; S_4: $1 + 3 + 5 + 7 = 4^2$; S_5: $1 + 3 + 5 + 7 + 9 = 5^2$

Although we do not usually give proofs, the answers for Exercises 3 and 11 are given here.

3. (a) $3(1) = 3$ and $\dfrac{3(1)(1+1)}{2} = \dfrac{6}{2} = 3$, so S is true for $n = 1$. **(b)** $3 + 6 + 9 + \cdots + 3k = \dfrac{3(k)(k+1)}{2}$

(c) $3 + 6 + 9 + \cdots + 3(k+1) = \dfrac{3(k+1)((k+1)+1)}{2}$ **(d)** Add $3(k+1)$ to both sides of the equation in part (b). Simplify

the expression on the right side to match the right side of the equation in part (c). **(e)** Since S is true for $n = 1$ and S is true for

$n = k + 1$ when it is true for $n = k$, S is true for every positive integer n.

11. (a) $\dfrac{1}{1 \cdot 2} = \dfrac{1}{2}$ and $\dfrac{1}{1+1} = \dfrac{1}{2}$, so S is true for $n = 1$. **(b)** $\dfrac{1}{1 \cdot 2} + \dfrac{1}{2 \cdot 3} + \dfrac{1}{3 \cdot 4} + \cdots + \dfrac{1}{k(k+1)} = \dfrac{k}{k+1}$

(c) $\dfrac{1}{1 \cdot 2} + \dfrac{1}{2 \cdot 3} + \cdots + \dfrac{1}{(k+1)((k+1)+1)} = \dfrac{k+1}{(k+1)+1}$ **(d)** Add the last term on the left of the equation in part (c) to

both sides of the equation in part (b). Simplify the right side until it matches the right side in part (c). **(e)** Since S is true for $n = 1$

and S is true for $n = k + 1$ when it is true for $n = k$, S is true for every positive integer n.   **15.** $n = 1$ or 2 **17.** $n = 2, 3,$ or 4

For Exercises 19–27, we show only the proof for Exercise 19.

19. (a) $(a^m)^1 = a^m$ and $a^{m(1)} = a^m$, so S is true for $n = 1$. **(b)** $(a^m)^k = a^{mk}$ **(c)** $(a^m)^{(k+1)} = a^{m(k+1)}$

(d) $(a^m)^k \cdot (a^m)^1 = a^{mk} \cdot (a^m)^1$

$(a^m)^{(k+1)} = a^{(mk+m)}$ Product rule for exponents

$(a^m)^{(k+1)} = a^{m(k+1)}$ Factor.

(e) Since S is true for $n = 1$ and S is true for $n = k + 1$ when it is true for $n = k$, S is true for every positive integer n.

31. $\dfrac{4^{n-1}}{3^{n-2}}$ or $3\left(\dfrac{4}{3}\right)^{n-1}$ **33.** $2^n - 1$

Chapter 11 Quiz *(pages 1042–1043)*

[11.1–11.3] 1. $-2, -6, -10, -14, -18$; arithmetic   **2.** $1, -\dfrac{1}{2}, \dfrac{1}{4}, -\dfrac{1}{8}, \dfrac{1}{16}$; geometric   **3.** $5, 3, 18, 27, 81$; neither

[11.2] 4. 12 **[11.2, 11.3] 5. (a)** 430 **(b)** $-\dfrac{1705}{128}$ **6. (a)** -1215 **(b)** does not exist **(c)** 3

[11.4] 7. $x^5 - 15x^4y + 90x^3y^2 - 270x^2y^3 + 405xy^4 - 243y^5$ **8.** $\dfrac{5}{4}xy^4$ **9. (a)** $362{,}880$ **(b)** 210

11.6 Exercises *(pages 1049–1053)*

1. $19{,}958{,}400$ **3.** 72 **5.** 5 **7.** 6 **9.** 1 **11.** 495 **13.** $1{,}860{,}480$ **15.** $259{,}459{,}200$ **17.** $15{,}504$

19. 6435 **21. (a)** permutation **(b)** permutation **(c)** combination **(d)** combination **(e)** permutation **(f)** combination

(g) permutation   **23.** 40 **25. (a)** $27{,}600$ **(b)** $35{,}152$ **(c)** 1104 **27.** 5040

29. (a) $17{,}576{,}000$ **(b)** $17{,}576{,}000$ **(c)** $456{,}976{,}000$ **31.** $362{,}880$ **33.** 120 **35.** 2730 **37. (a)** 120 **(b)** 6

39. $3{,}838{,}380$ **41. (a)** 15 **(b)** 10 **43.** 10 **45.** 28 **47. (a)** 84 **(b)** 10 **(c)** 40 **(d)** 28 **49.** 1680 **51.** 15

53. $6{,}227{,}020{,}800$ **55. (a)** 56 **(b)** 462 **(c)** 3080 **(d)** 8526 **57.** $1{,}000{,}000$ **59.** 4096 **61.** 6

71. (a) $3.04140932 \times 10^{64}$ **(b)** $8.320987113 \times 10^{81}$ **(c)** $8.247650592 \times 10^{90}$

72. (a) $8.759976613 \times 10^{20}$ **(b)** $5{,}527{,}200$ **(c)** $2.19289732 \times 10^{26}$

11.7 Exercises *(pages 1060–1064)*

1. $\{H\}$ **3.** $\{(H,H,H),(H,H,T),(H,T,H),(T,H,H),(H,T,T),(T,H,T),(T,T,H),(T,T,T)\}$

5. $\{(1,1),(1,2),(1,3),(2,1),(2,2),(2,3),(3,1),(3,2),(3,3)\}$ **7. (a)** $\{H\}$; 1 **(b)** $\emptyset$; 0

9. (a) $\{(1,1),(2,2),(3,3)\}$; $\dfrac{1}{3}$ **(b)** $\{(1,1),(1,3),(2,1),(2,3),(3,1),(3,3)\}$; $\dfrac{2}{3}$ **(c)** $\{(2,1),(2,3)\}$; $\dfrac{2}{9}$

13. (a) F **(b)** D **(c)** A **(d)** F **(e)** C **(f)** B **(g)** E   **15. (a)** $\dfrac{1}{5}$ **(b)** 0 **(c)** $\dfrac{7}{15}$ **(d)** 1 to 4 **(e)** 7 to 8

17. (a) $\dfrac{1}{6}$ **(b)** $\dfrac{1}{3}$ **(c)** $\dfrac{1}{6}$ **19. (a)** .176 **(b)** .376 **(c)** .138 **(d)** 10,374 to 157,151 or about 1 to 15   **21.** $\dfrac{1}{28{,}561} \approx .000035$

23. (a) about .6998 **(b)** about .3002 **(c)** about .3088 **(d)** about .9099   **25.** .3   **27.** .042246   **29.** .026864

31. .49   **33.** .51   **35. (a)** .047822 **(b)** .976710 **(c)** .897110   **37. (a)** about .404 **(b)** about .047 **(c)** about .002

Chapter 11 Review Exercises *(pages 1069–1072)*

1. $\dfrac{1}{2},\dfrac{2}{3},\dfrac{3}{4},\dfrac{4}{5},\dfrac{5}{6}$; neither   **3.** 8, 10, 12, 14, 16; arithmetic   **5.** 5, 2, -1, -4, -7; arithmetic   **7.** 4, 4.5, 5, 5.5, 6

9. $3\pi - 2, 2\pi - 1, \pi, 1, -\pi + 2$ **11.** $-5, -1, -\dfrac{1}{5}, -\dfrac{1}{25}, -\dfrac{1}{125}$ **13.** ± 1; $-8\left(\dfrac{1}{2}\right)^{n-1} = -\left(\dfrac{1}{2}\right)^{n-4}$ or

$-8\left(-\dfrac{1}{2}\right)^{n-1} = \left(-\dfrac{1}{2}\right)^{n-4}$ **15.** $-x + 61$ **17.** 612   **19.** $\dfrac{4}{25}$ **21.** -40 **23.** 1   **25.** $\dfrac{73}{12}$ **27.** 3,126,250

29. $\dfrac{4}{3}$ **31.** $-\dfrac{4}{5}$ **33.** 36   **35.** diverges   **37.** -10

In Exercises 39 and 41, other answers are possible.

39. $\displaystyle\sum_{i=1}^{15} (-5i + 9)$ **41.** $\displaystyle\sum_{i=1}^{6} 4(3)^{i-1}$ **43.** $x^4 + 8x^3y + 24x^2y^2 + 32xy^3 + 16y^4$

45. $243x^{5/2} - 405x^{3/2} + 270x^{1/2} - 90x^{-1/2} + 15x^{-3/2} - x^{-5/2}$ **47.** $-3584x^3y^5$ **49.** $x^{12} + 24x^{11} + 264x^{10} + 1760x^9$

55. 72   **57.** 56   **59.** 252   **63.** 90   **65. (a)** 20 **(b)** 10   **67.** 456,976,000; 258,336,000

69. (a) $\dfrac{4}{15}$ **(b)** $\dfrac{2}{3}$ **(c)** 0 **(d)** 3 to 2   **71. (a)** .2002 **(b)** .0380 **(c)** .4915   **73.** .296

Chapter 11 Test *(pages 1072–1073)*

[11.1–11.3] 1. $-3, 6, -11, 18, -27$; neither   **2.** $-\dfrac{3}{2}, -\dfrac{3}{4}, -\dfrac{3}{8}, -\dfrac{3}{16}, -\dfrac{3}{32}$; geometric   **3.** 2, 3, 7, 13, 27; neither

[11.2] 4. 49   **[11.3] 5.** $-\dfrac{32}{3}$ **[11.2] 6.** 110   **[11.3] 7.** -1705 **[11.2] 8.** 2385   **[11.3] 9.** -186

10. does not exist   **11.** $\dfrac{108}{7}$ **[11.4] 12.** $x^6 + 6x^5y + 15x^4y^2 + 20x^3y^3 + 15x^2y^4 + 6xy^5 + y^6$

13. $16x^4 - 96x^3y + 216x^2y^2 - 216xy^3 + 81y^4$ **14.** $60w^4y^2$ **[11.4, 11.6] 15.** 40,320   **16.** 45   **17.** 35   **18.** 990

[11.6] 20. 24   **21.** 120; 14   **22.** 90   **[11.7] 23. (a)** $\dfrac{1}{26}$ **(b)** $\dfrac{10}{13}$ **(c)** $\dfrac{4}{13}$ **(d)** 3 to 10   **24.** .92   **25.** .201

Appendix A Exercises *(pags 1078–1079)*

1. 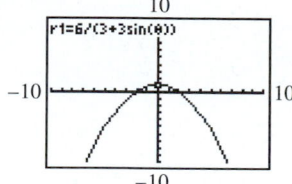   **3.** 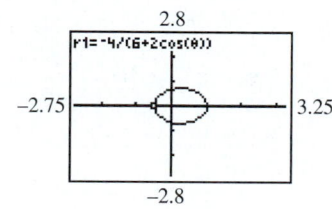   **5.**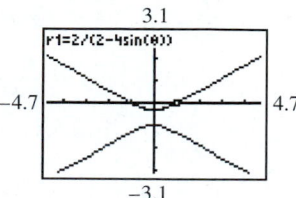

7. 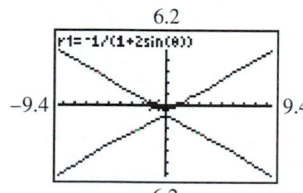   **9.**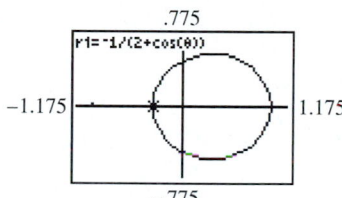

11. $r = \dfrac{3}{1 + \cos\theta}$ **13.** $r = \dfrac{5}{1 - \sin\theta}$ **15.** $r = \dfrac{20}{5 + 4\cos\theta}$; ellipse   **17.** $r = \dfrac{40}{4 - 5\sin\theta}$; hyperbola

19. ellipse; $8x^2 + 9y^2 - 12x - 36 = 0$ **21.** hyperbola; $3x^2 - y^2 + 8x + 4 = 0$ **23.** ellipse; $4x^2 + 3y^2 - 6y - 9 = 0$

25. parabola; $x^2 - 10y - 25 = 0$

Appendix B Exercises *(pages 1082–1083)*

1. circle or ellipse or a point   **3.** hyperbola or two intersecting lines   **5.** parabola or one line or two parallel lines

7. 30°   **9.** 60°   **11.** 22.5°

13.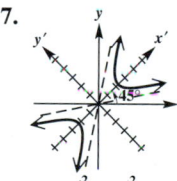
$$\frac{x'^2}{12} + \frac{y'^2}{4} = 1$$

15.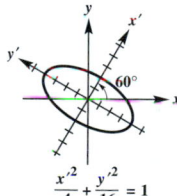
$$\frac{x'^2}{9} + \frac{y'^2}{4} = 1$$

17.
$$\frac{x'^2}{5} - \frac{3y'^2}{5} = 1$$

19.
$$\frac{x'^2}{4} + \frac{y'^2}{16} = 1$$

21.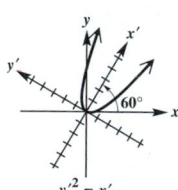
$$y'^2 = x'$$

23.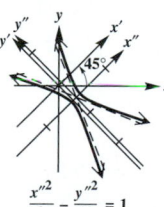
$$\frac{x''^2}{2} - \frac{y''^2}{10} = 1$$

25.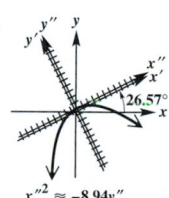
$$x''^2 \approx -8.94y''$$

27.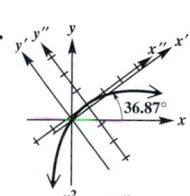
$$x''^2 = -6y''$$

29. If $B = 0$, cot 2θ is undefined. The graph may be translated but is not rotated.

Index of Applications

Index

Photo Credits